AF872573

Audi TT

Service Manual

2000, 2001, 2002, 2003, 2004, 2005, 2006
1.8 liter turbo, 3.2 liter
including roadster and quattro

[B] BentleyPublishers®
.com

BENTLEY PUBLISHERS® | Automotive Reference™

Bentley Publishers, a division of Robert Bentley, Inc.
1734 Massachusetts Avenue
Cambridge, MA 02138 USA
800-423-4595 / 617-547-4170

Information that makes the difference®

BentleyPublishers®.com

Technical contact information

We welcome your feedback. Please submit corrections and additions to our Audi technical discussion forum at:

`http://www.BentleyPublishers.com`

Updates and corrections

We will evaluate submissions and post appropriate editorial changes online as updates or tech discussion. Appropriate updates and corrections will be added to the book in future printings. Check for updates and corrections for this book before beginning work on your vehicle. See the following web address for additional information:

`http://www.BentleyPublishers.com/updates/`

WARNING—IMPORTANT SAFETY NOTICE

Do not use this manual unless you are familiar with basic automotive repair procedures and safe workshop practices. This manual illustrates the workshop procedures required for most service work. It is not a substitute for full and up-to-date information from the vehicle manufacturer or for proper training as an automotive technician. Note that it is not possible for us to anticipate all of the ways or conditions under which vehicles may be serviced or to provide cautions as to all of the possible hazards that may result.

We have endeavored to ensure the accuracy of the information in this manual. Please note, however, that considering the vast quantity and the complexity of the service information involved, we cannot warrant the accuracy or completeness of the information contained in this manual.

FOR THESE REASONS, NEITHER THE PUBLISHER NOR THE AUTHOR MAKES ANY WARRANTIES, EXPRESS OR IMPLIED, THAT THE INFORMATION IN THIS BOOK IS FREE OF ERRORS OR OMISSIONS, AND WE EXPRESSLY DISCLAIM THE IMPLIED WARRANTIES OF MERCHANTABILITY AND OF FITNESS FOR A PARTICULAR PURPOSE, EVEN IF THE PUBLISHER OR AUTHOR HAVE BEEN ADVISED OF A PARTICULAR PURPOSE, AND EVEN IF A PARTICULAR PURPOSE IS INDICATED IN THE MANUAL. THE PUBLISHER AND AUTHOR ALSO DISCLAIM ALL LIABILITY FOR DIRECT, INDIRECT, INCIDENTAL OR CONSEQUENTIAL DAMAGES THAT RESULT FROM ANY USE OF THE EXAMPLES, INSTRUCTIONS OR OTHER INFORMATION IN THIS BOOK. IN NO EVENT SHALL OUR LIABILITY WHETHER IN TORT, CONTRACT OR OTHERWISE EXCEED THE COST OF THIS MANUAL.

Before attempting any work on your Audi, read **00 Warnings and Cautions** and any **WARNING** or **CAUTION** that accompanies a procedure in the service manual. Review the **WARNINGS** and **CAUTIONS** each time you prepare to work on your Audi.

Your common sense and good judgment are crucial to safe and successful service work. Read procedures through before starting them. Think about whether the condition of your car, your level of mechanical skill, or your level of reading comprehension might result in or contribute in some way to an occurrence which might cause you injury, damage your car, or result in an unsafe repair. If you have doubts for these or other reasons about your ability to perform safe repair work on your car, have the work done at an authorized Audi dealer or other qualified shop.

Part numbers listed in this manual are for identification purposes only, not for ordering. Always check with your authorized Audi dealer to verify part numbers and availability before beginning service work that may require new parts.

Special tools required to perform certain service operations are identified in the manual and are recommended for use. Use of tools other than those recommended in this manual may be detrimental to the car's safe operation as well as the safety of the person servicing the car.

The vehicle manufacturer will continue to issue service information updates and parts retrofits after the editorial closing of this manual. Some of these updates and retrofits will apply to procedures and specifications in this manual. We regret that we cannot supply updates to purchasers of this manual.

This manual is prepared, published and distributed by Bentley Publishers, 1734 Massachusetts Avenue, Cambridge, Massachusetts 02138 USA. All information contained in this maual is based on information available to the publisher from Audi at the time of editorial closing. Audi has not reviewed and does not vouch for the accuracy of completeness of the technical specifications and work procedures described and given in this manual.

© 2006 Robert Bentley, Inc. Bentley Publishers is a trademark of Robert Bentley, Inc.

[B]® is a registered trademark of Bentley Publishers and Robert Bentley, Inc.

All rights reserved. The right is reserved to make changes at any time without notice. No part of this publication may be reproduced, stored in a retrieval system, or transmitted in any form or by any means, electronic, mechanical, photocopying, recording, or otherwise, without the prior written consent of the publisher. This includes text, figures, and tables. All rights reserved under Berne and Pan-American Copyright conventions. This manual is simultaneously published in Canada.

ISBN: 978-0-8376-1625-4 **Job Code: AT06-08** Editorial closing 06/2005

Library of Congress Cataloging-in-Publication Data for 2006 edition

Audi TT service manual : 2000, 2001, 2002, 2003, 2004, 2005, 2006, 1.8 liter turbo, 3.2 liter including Roadster and Quattro.
p. cm.
Includes index.
ISBN-13: 978-0-8376-1500-4 (alk. paper)
ISBN-10: 0-8376-1500-3 (alk. paper)
1. Audi TT automobile--Maintenance and repair--Handbooks, manuals, etc.
TL215.A828A835 2006
629.28'722--dc22 2006021565

Cover Design and use of blue band on spine and back cover are trade dress and trademarks of Bentley Publishers™. All Rights reserved.

The paper used in this publication is acid free and meets the requirements of the National Standard for Information Sciences—Permanence of Paper for Printed Library Materials. ∞

Manufactured in the United States of America.

202608R002

Foreword

For the Audi owner with automotive mechanical skills, and for independent service professionals, this manual includes maintenance and repair procedures the Audi TT is likely to need during its service life, including many procedures and specifications that were available in an authorized Audi dealer service department at the time this manual was prepared. The Audi owner with no intention of working on his or her car will find that owning and referring to this manual will make it possible to be better informed and to more knowledgeably discuss repairs with a professional automotive technician.

For those intending to do maintenance and repair on their Audi, it is essential that safety equipment be used and safety precautions observed when working on your vehicle. A minimum safety equipment list includes hand protection, eye protection, and a fire extinguisher. A selection of good quality hand tools is also needed. This includes a torque wrench to ensure that fasteners are tightened in accordance with specifications. In some cases, the text refers to special tools that are recommended or required to accomplish adjustments or repairs. These tools are often identified by their Volkswagen special tool number and illustrated.

Disclaimer

We have endeavored to ensure the accuracy of the information in this manual. When the vast array of data presented in the manual is taken into account, however, no claim to infallibility can be made. We therefore cannot be responsible for the result of any errors that may have crept into the text. Please also read **WARNING–Important Safety Notice** on the copyright page at the beginning of this book.

Prior to starting a repair procedure, read the procedure, **00 WARNINIG and CAUTIONS** and the **WARNINGS** and **CAUTIONS** that accompany the procedure. Reading a procedure before beginning work helps you determine in advance the need for specific skills, identify hazards, prepare for appropriate capture and handling of hazardous materials, and the need for particular tools and replacement parts such as gaskets.

Bentley Publishers encourages comments from the readers of this manual with regard to errors, and suggestions for improvement of our product. These communications have been and will be carefully considered in the preparation of this and other manuals. If you identify inconsistencies in the manual, you may have found an error. Please contact the publisher and we will endeavor to post applicable corrections on our website. Review updates and corrections that we have posted before beginning work. Please see the following web address:

`http://www.BentleyPublishers.com/updates/`

Audi continues to issue service information and parts retrofits after the editorial closing of this manual. Some of this updated information may apply to procedures and specifications in this manual. For the latest information, please see the following web address:

`https://erwin.audi.com`

Audi offers extensive warranties, especially on components of the fuel delivery and emission control systems. Therefore, before deciding to repair a Audi that may be covered wholly or in part by any warranties issued by Audi of America, Inc., consult your authorized Audi dealer. You may find that the dealer can make the repair either free or at minimum cost. Regardless of its age, or whether it is under warranty, your Audi is both an easy car to service and an easy car to get serviced. So if at any time a repair is needed that you feel is too difficult to do yourself, a trained Audi technician is ready to do the job for you.

Bentley Publishers

00 Warnings and Cautions

Please read these warnings and cautions before proceeding with maintenance and repair work.

WARNINGS—
*See also **CAUTIONS** on next page.*

- Read the important safety notice on the copyright page at the beginning of the book.
- Some repairs may be beyond your capability. If you lack the skills, tools and equipment, or a suitable workplace for any procedure described in this manual, we suggest you leave such repairs to an authorized Audi dealer service department or other qualified shop.
- Thoroughly read each procedure and the **WARNINGS** and **CAUTIONS** that accompany the procedure. Also review posted corrections at `www.BentleyPublishers.com/updates/` before beginning work.
- If any procedure, tightening torque, wear limit, specification or data presented in this manual does not appear to be appropriate for a specific application, contact the publisher or the vehicle manufacturer for clarification before using the information in question.
- Audi is constantly improving its cars. Sometimes these changes, both in parts and specifications, are made applicable to earlier models. Therefore, before starting any major jobs or repairs to components on which passenger safety may depend, consult your authorized Audi dealer about technical bulletins that may have been issued.
- Do not reuse any fasteners that are worn or deformed in normal use. Many fasteners are designed to be used only once and become unreliable and may fail when used a second time. This includes, but is not limited to, nuts, bolts, washers, self-locking nuts or bolts, circlips and cotter pins. Replace these fasteners with new parts.
- Do not work under a lifted car unless it is solidly supported on stands designed for the purpose. Do not support a car on cinder blocks, hollow tiles or other props that may crumble under continuous load. Do not work under a car that is supported solely by a jack. Do not work under the car while the engine is running.
- If you are going to work under a car on the ground, make sure that the ground is level. Block the wheels to keep the car from rolling. Disconnect the battery negative (–) terminal (ground strap) to prevent others from starting the car while you are under it.
- Do not run the engine unless the work area is well ventilated. Carbon monoxide kills.
- Remove finger rings, bracelets and other jewelry so that they cannot cause electrical shorts, get caught in running machinery or be crushed by heavy parts.
- Tie long hair behind your head. Do not wear a necktie, a scarf, loose clothing, or a necklace when you work near machine tools or running engines. If your hair, clothing, or jewelry were to get caught in the machinery, severe injury could result.
- Do not attempt to work on your car if you do not feel well. You increase the danger of injury to yourself and others if you are tired, upset or have taken medication or any other substance that may keep you from being fully alert.
- Illuminate your work area adequately but safely. Use a portable safety light for working inside or under the car. Make sure the bulb is enclosed by a wire cage. The hot filament of an accidentally broken bulb can ignite spilled fuel, vapors or oil.
- Catch draining fuel, oil or brake fluid in suitable containers. Do not use food or beverage containers that might mislead someone into drinking from them. Store flammable fluids away from fire hazards. Wipe up spills at once but do not store the oily rags, which can ignite and burn spontaneously.
- Observe good workshop practices. Wear goggles when you operate machine tools or work with battery acid. Wear gloves or other protective clothing whenever the job requires working with harmful substances.
- Greases, lubricants and other automotive chemicals contain toxic substances, many of which are absorbed directly through the skin. Read the manufacturer's instructions and warnings carefully. Use hand and eye protection. Avoid direct skin contact.
- Disconnect the battery negative (–) terminal (ground strap) whenever you work on the fuel system or the electrical system. Do not smoke or work near heaters or other fire hazards. Keep an approved fire extinguisher handy.
- Friction components (such as brake pads or shoes or clutch discs) contain asbestos fibers or other friction materials. Do not create dust by grinding, sanding, or by cleaning with compressed air. Avoid breathing dust. Breathing any friction material dust can lead to serious diseases and may result in death.
- Batteries give off explosive hydrogen gas during charging. Keep sparks, lighted matches and open flame away from the top of the battery. If hydrogen gas escaping from the cap vents is ignited, it will ignite gas trapped in the cells and cause the battery to explode.
- Battery acid (electrolyte) can cause severe burns. Flush contact area with water and seek medical attention.
- Do not quick-charge the battery (for boost starting) for longer than one minute. Wait at least one minute before boosting the battery a second time.
- Connect and disconnect a battery charger only with the battery charger switched OFF.
- Connect and disconnect battery cables, jumper cables or a battery charger only with the ignition switched OFF. Do not disconnect the battery while the engine is running.
- Do not allow the battery charging voltage to exceed 16.5 volts. If the battery begins producing gas or boiling violently, reduce the charging rate. Boosting a sulfated battery at a high rate can cause an explosion.

continued on next page

Please read these warnings and cautions before proceeding with maintenance and repair work.

WARNINGS— (continued)

- The air-conditioning system is filled with a hazardous refrigerant. Make sure the system is serviced only by a trained technician using approved refrigerant recovery/recycling equipment, trained in related safety precautions, and familiar with regulations governing the discharging and disposal of automotive chemical refrigerants.
- Do not expose any part of the air-conditioning system to high temperatures such as open flame. Excessive heat increases system pressure and may cause the system to burst.
- Some aerosol tire inflators are highly flammable. Be extremely cautious when repairing a tire that may have been inflated using an aerosol tire inflator. Keep sparks, open flame or other sources of ignition away from the tire repair area. Inflate and deflate the tire at least four times before breaking the bead from the rim. Completely remove the tire from the rim before attempting any repair.
- Cars covered by this manual are equipped with a supplemental restraint system, that automatically deploys airbags and pyrotechnic seat belt reels in case of a frontal or side impact. These are explosive devices. Handled improperly or without adequate safeguards, they can be accidently activated and cause serious injury.
- The ignition system produces high voltages that can be fatal. Avoid contact with exposed terminals and use extreme care when working on a car with the engine running or the ignition switched ON.
- Place jack stands only at locations specified by the manufacturer. The vehicle lifting jack supplied with the vehicle is intended for tire changes only. Use a heavy duty floor jack to lift the vehicle before installing jack stands. See **03 Maintenance**.
- Aerosol cleaners and solvents may contain hazardous or deadly vapors and are highly flammable. Use only in a well ventilated area. Do not use on hot surfaces (engines, brakes, etc.).
- Due to risk of personal injury, be sure the engine is cold before beginning work on the cooling system.
- Use extreme care when draining and disposing of engine coolant. Coolant is poisonous and lethal to humans and pets. Pets are attracted to coolant because of its sweet smell and taste. Seek medical attention immediately if coolant is ingested.

CAUTIONS—
*See also **WARNINGS** on previous page.*

- If you lack the skills, tools and equipment, or a suitable workshop for any procedure described in this manual, we suggest you leave such repairs to an authorized Audi dealer or other qualified shop.
- Audi is constantly improving its cars and sometimes these changes, both in parts and specifications, are made applicable to earlier models. Therefore, part numbers listed in this manual are for reference only. Always check with your authorized Audi dealer parts department for the latest information.
- Before starting a job, make certain that you have the necessary tools and parts on hand. Read the instructions thoroughly and do not attempt shortcuts. Use tools appropriate to the work and use replacement parts meeting Audi specifications. Makeshift tools, parts and procedures do not make good repairs
- Do not use pneumatic and electric tools to tighten fasteners, especially on light alloy parts. Use a torque wrench to tighten fasteners to the tightening torque specification listed.
- Be mindful of the environment. Before you drain the crankcase, find out the proper way to dispose of the oil. Do not pour oil onto the ground, down a drain, or into a stream, pond or lake. Dispose of in accordance with Federal, State and Local laws.
- Do not expose electronic control modules to temperatures from a paint-drying booth or a heat lamp in excess of 203°F (95°C). Do not expose control modules to temperatures in excess of 185°F (85°C) for more than two hours.
- Before doing any electrical welding on cars equipped with ABS, disconnect the battery negative (–) terminal (ground strap) and the ABS control module connector.
- Make sure ignition is OFF before disconnecting the battery.
- Label battery cables before disconnecting. On some models, battery cables are not color coded.
- Disconnecting the battery may erase fault code(s) stored in control module memory. Use Audi diagnostic equipment or equivalent to check for fault codes prior to disconnecting the battery cables. If the Check Engine light (malfunction indicator light or MIL) is illuminated or any other system faults are detected (indicated by an illuminated warning light), see an authorized Audi dealer.
- If a normal or rapid charger is used to charge battery, disconnect the battery and remove from the vehicle in order to avoid damaging paint and upholstery.
- Do not quick-charge the battery (for boost starting) for longer than one minute. Wait at least one minute before boosting the battery a second time.
- Slow-charge a sealed or "maintenance free" battery at an amperage rate that is approximately 10% of the battery's ampere-hour (Ah) rating.

01 Vehicle Identification and VIN Decoder

Vehicle Identification Number (VIN), decoding

Some of the information in this manual applies only to cars of a particular model year or range of years. For example, 2005 refers to the 2005 model year but does not necessarily match the calendar year in which the car was manufactured or sold. To be sure of the model year of a particular car, check the Vehicle Identification Number (VIN) on the car.

The VIN is a unique sequence of 17 characters assigned by Audi to identify each individual car. When decoded, the VIN tells the country and year of manufacture; make, model and serial number; assembly plant and even some equipment specifications.

The Audi VIN is on a plate mounted on the top of the dashboard, on the driver's side where the number can be seen through the windshield. The 10th character is the model year code. The letters I, O, Q and U are not used for model year designation. Examples: V for 1997, W for 1998, X for 1999, Y for 2000, 1 for 2001, 2 for 2002, etc. The table below explains some of the codes in the VIN for 2000 - 2006 Audi TT models covered by this manual.

Sample VIN: TRUUT28N811005098
position: 1 2 3 4 5 6 7 8 9 10 11 12 - 17

VIN position	Description	Decoding information	
1	Manufacturing Country	T	Hungary
2	Manufacturer	R	Audi Hungary
3	Vehicle Type	U	Passenger vehicle
4	Series	S T U W	TT coupe, front wheel drive TT roadster, front wheel drive TT roadster, all wheel drive TT coupe, all wheel drive
5	Engine code	C F S T X	1.8 liter 4-cylinder, 180 hp (2001 - 2006 m.y.) 3.2 liter 6-cylinder, 250 hp (2003 - 2006 m.y.) 1.8 liter 4-cylinder, 180 hp (2000 m.y.) 1.8 liter 4-cylinder, 225 hp (2000 - 2006 m.y.) 1.8 liter 4-cylinder, 180 hp (2001 - 2002 m.y.)
6	Restraint system	2	Varies with model year
7-8	Model type	8N	TT
9	Check digit		0 - 9 or X, calculated by NHTSA
10	Model year	Y 1 2 3 4 5 6	2000 2001 2002 2003 2004 2005 2006
11	Assembly plant	1	Gyor, Hungary
12-17	Serial number		Sequential production number for specific vehicle

02 Product Familiarization

General

The information included in this section is based on introductory and preliminary information for Audi TT vehicles sold in the USA and Canada. The content provided here is intended to serve as a product familiarization guide. Note that the information presented is subject to change and should be used as a general reference only.

> ***WARNING—***
> - *Always check the Audi factory repair information or the publisher's website at www.bentleypublishers.com for information that may supersede any information included in this section.*

BentleyPublishers.com

©2006 BentleyPublishers.com—All Rights Reserved

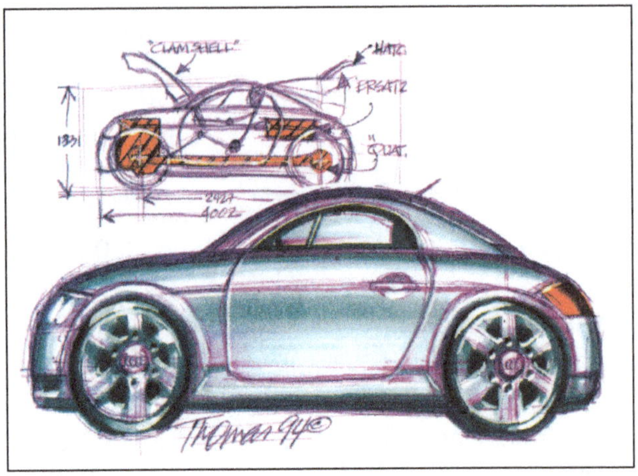

The TT started as a sketch, just a simple idea of how an elemental sports car might look. From the sketch came a concept car. In a matter of months, a full scale design was developed, and shown for the first time in September, 1995 at the German motor show in Frankfurt. Production of the TT started in the fall of 1998. The TT is the first sports coupe from Audi since the coupe quattro of the early 1990s.

Product Overview

The Audi TT was introduced in model year 2000 in coupe form with a 180 hp 4-cylinder, 5-valve, engine, and a choice of front wheel drive or the Audi all-wheel drive quattro system.

General specifications, model year 2000	
Engine type	dual overhead cam, 20-valve turbo-charged inline 4-cylinder
Engine size	1.8 liters / 107 cu. in.
Horsepower	180 @ 5500 rpm
Torque (lb-ft)	173 @ 1950 through 4700 rpm
Transmission	5-speed manual
Wheelbase / Length	95.4 in. / 159.1 in.
Curb weight	2655 lbs.
Fuel capacity	14.5 gal.
Fuel requirement	unleaded premium (91 octane)
Tires	P225/45 YR17
Brakes, front/rear	vented disc / solid disc, antilock standard
Suspension, front	independent MacPherson strut
Suspension, rear	fwd: torsion beam axle with trailing arms and coil springs
	quattro: parallelogram, multi-link, fully independent, separate coil spring and shock absorber, stabilizer bar
Drivetrain	front engine, front-wheel drive or all-wheel drive.

In model year 2001, the TT roadster was introduced.

The safety concept

Audi TT is equipped with advanced safety features. Occupants are surrounded by a sturdy safety cage and further protected by front and side airbags. Four-wheel antilock disc brakes stop it quickly and effectively.

- ABS (Anti-lock Braking System) with traction control
- Front and side impact airbags

BentleyPublishers.com
© 2006 BentleyPublishers.com—All Rights Reserved

- More passive safety. Additional impact absorbing material in A-pillar area to reduce chance of head injury. Crash energy design to move pedals away from driver's feet to reduce foot injury.
- Safety belts with added features. Pyrotechnic belt tensioners with force limiters to help moderate chest forces in serious accidents. 3-point rear seat belts.
- Central locking system revisions. In an accident serious enough to deploy an airbag, the central locking system unlocks the doors and turns on the emergency flashers and interior lights. Fuel supply is also shut off.
- Tailored material thicknesses are employed to add strength where it does most good without the added weight penalty. Strong intrusion beams (shown in yellow) add to the protection of the supplemental restraint airbags.

The interior concept

Many of the functional elements of the interior are crafted from aluminum. The padded aluminum grab handles by the lower console are also structural. They provide additional stiffness to the body.

The front seats are a supportive sport design. The rear seats, while designed for children, can be folded down for a more generous luggage space and a virtually flat load floor.

The useable cargo space is made even more useful with the addition of standard items such as the integrated tie down eyelets, easy to remove hatshelf and a wide opening rear deck lid.

Storage and convenience features include a covered tray in the console for small items, a glove box and a tray underneath to handle larger objects, mesh door pockets and twin aluminum cup holders.

BentleyPublishers
.com
© 2006 BentleyPublishers.com—All Rights Reserved

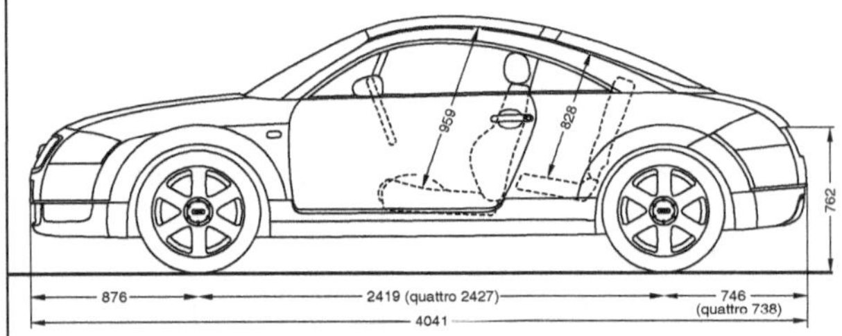

Exterior dimensions and weights (2000 coupe)

The basic dimensions of the Audi TT body are shown at left and detailed in the tables below.

Length	159.1 in (4041 mm)
Width	60.6 in (1525 mm)
Height	53.0 in (1346 mm)
Wheelbase	95.4 in (2422 mm), quattro: 95.6 in (2429 mm)
Front track width	60.0 in (1525 mm)
Rear track width	59.6 in (1513 mm) quattro: 59.3 (1507 mm)
Max. permissible gross weight	3472 lb (1575 kg) quattro: 3891 lb (1765 kg)
Empty weight	2655 lb (1205 kg) quattro: 2910 lb (1320 kg)
Front seat headroom	37.8 in (959 mm)
Rear seat headroom	37.0 in (828 mm)
Cargo volume	fwd: seat up: 13.8 ft^3 (391 $liter^3$) rear seat folded: 24.2 ft^3 (685 $liter^3$)
	quattro: seat up: 10.8 ft^3 (306 $liter^3$) rear seat folded: 18.6 ft^3 (587 $liter^3$)

Technical specifications (2000 coupe)

EPA class	Minicompact
Body type	Fully galvanized unitized construction, with aluminum hood
Front suspension	MacPherson struts with lower control arms, subframe, anti-roll bar
Rear suspension	Torsion beam, coil springs, toe-correcting axle bearing, tubular anti-roll bar
Brakes	Power assisted, dual circuit, vented front discs and solid rear discs
Anti-lock Braking System (ABS)	Standard, all four wheels, with electronic brake force distribution
Wheels	7J x 16 in. alloy wheels, 5-spoke 7.5J x 17 in. alloy wheels, 6-spoke (opt.)
Tires	205/55 R16 high performance summer 225/45 R17 high perform. summer (opt.)
Drag coefficient	0.34
Fuel economy (EPA)	data n. a.
0-60 mph	7.4 seconds
Top speed	130 mph (209 km/h) (Max. speed electronically limited in North America)

BentleyPublishers.com
© 2006 BentleyPublishers.com—All Rights Reserved

TT 3.2 quattro

The Audi TT 3.2 quattro, introduced in model year 2004, was the first-ever TT with a V6 engine. It was als the first Audi vehicle to use the 02E Direct Shift Gearbox (DSG), the revolutionary racing-derived six-speed transmission that combines the best qualities of both manual and automatic transmissions.

Also new or modified for 2004 are:

- Steering wheel gearshift paddles
- Electronic Stabilization Program (ESP) and Anti-lock Braking System (ABS) with Electronic Differential Locks (EDL) and Electronic rear Brake-force Distribution (EBD).
- Larger dual-piston floating calipers with ventilated rotors
- Modified exhaust system
- Sport suspension
- Larger rear spoiler
- Lower front grille
- Rear diffuser with honeycomb pattern
- Xenon headlights
- Aluminum shift cover
- Tire mobility system

Audi TT 3.2 quattro

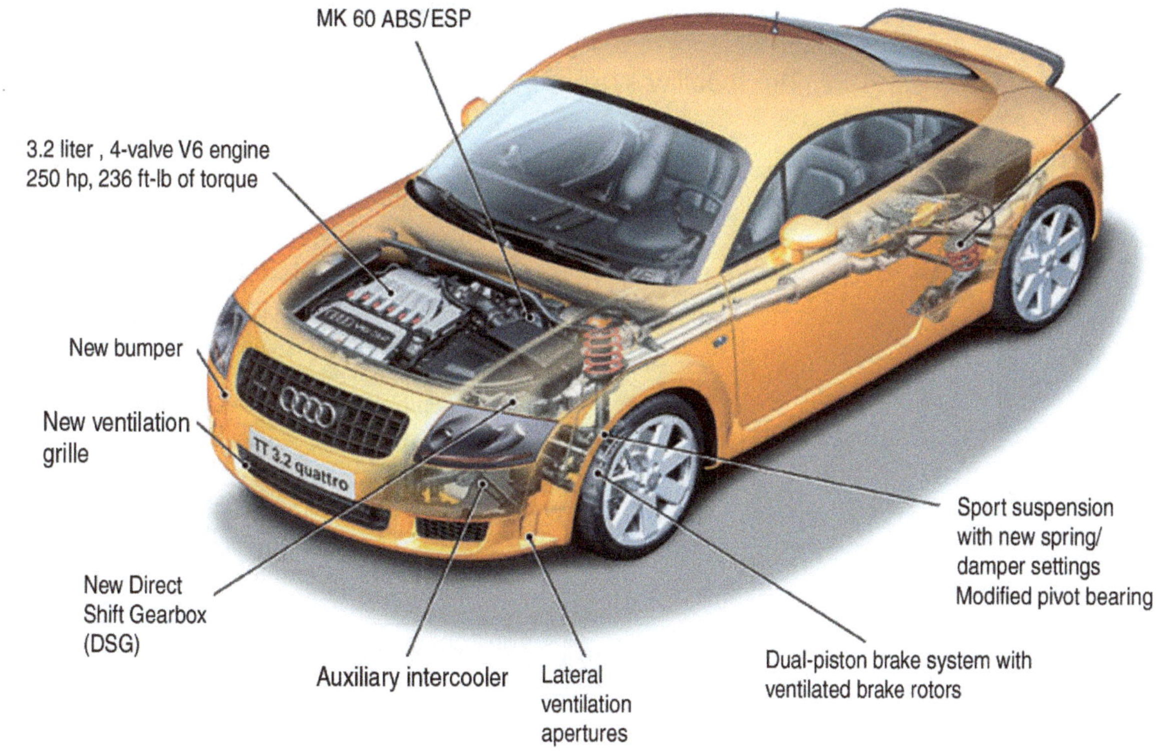

BentleyPublishers.com
©2006 BentleyPublishers.com—All Rights Reserved

With the V6 engine and quattro, the TT 3.2 covers 0 - 60 mph in 6.4 seconds as a coupe, and 6.6 seconds as a roadster. Top speed for both models is electronically governed at 130 mph.

Audi TT 3.2 quattro

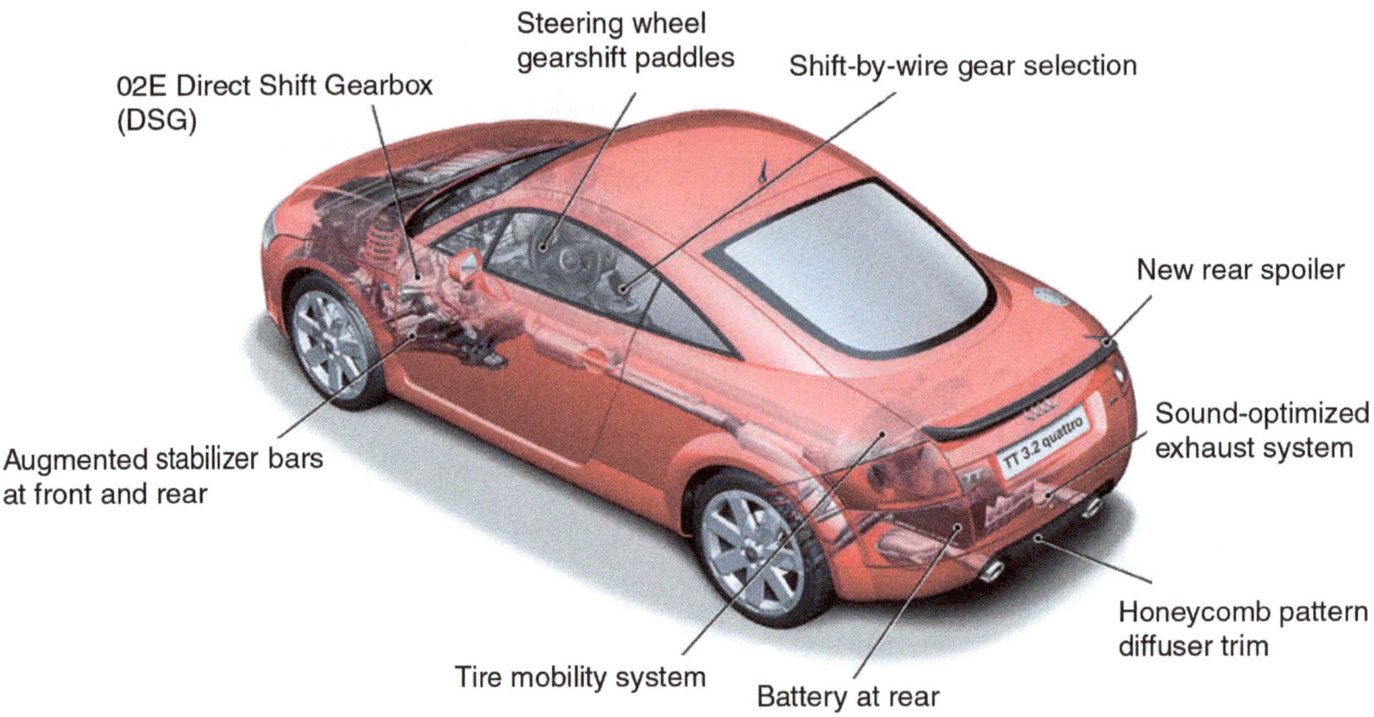

1 Engine
2 Engine Management

Multiple engine and transaxles are available for 2000 through 2006 Audi TT models. For 2000 engine choices were limited to the 180 hp 1.8 liter engine. The 225 hp 1.8 liter and 250 hp 3.2 liter engine became available in subsequent years.

Engine applications	Engine code
1.8 liter 4-cylinder, 180 hp	ATC (2000) AWP (2001 - on)
1.8 liter 4-cylinder, 225 hp	AMU (2000 - 2002) BEA (2003 - on)
3.2 liter 6-cylinder, 250 hp	BHE (2004 - on)

BentleyPublishers.com
©2006 BentleyPublishers.com—All Rights Reserved

1.8 liter turbocharged five-valves-per-cylinder engine

1.8 liter engine

The 1.8 liter turbo engine combines the advantages of Bosch ME 7.5 injection with turbocharging technology. The result is an extremely responsive engine. There are two basic variants of the 1.8 liter engines; a 180 hp version and a 225 hp version for all-wheel drive models.

Technical data, 1.8 liter engine	**Engine Code**	
	ATC (2000) **AWP (2001 - on)**	**AMU (2000 - 2002)** **BEA (2003 - 2006)**
Type	4-cylinder inline, 5 valves/cylinder	4-cylinder inline, 5 valves/cylinder
Displacement	1.8L liter (1781 cc)	1.8 liter (1781 cc)
Bore	3.18 in (81.0 mm)	3.18 in (81.0 mm)
Stroke	3.40 in (88.4 mm)	3.40 in (88.4 mm)
Compression ratio	9.5 : 1	9.0:1
Power	180 HP (134 kW) @ 5500 rpm	225 HP (167 kW) @ 5900 rpm
Torque	173 ft-lb (234 Nm) @1950 to 4700 rpm	207 ft-lb (280 Nm) @ 2200 to 5500 rpm
Engine management	Bosch Motronic ME 7.5	Bosch Motronic ME 7.5

BentleyPublishers.com
©2006 BentleyPublishers.com—All Rights Reserved

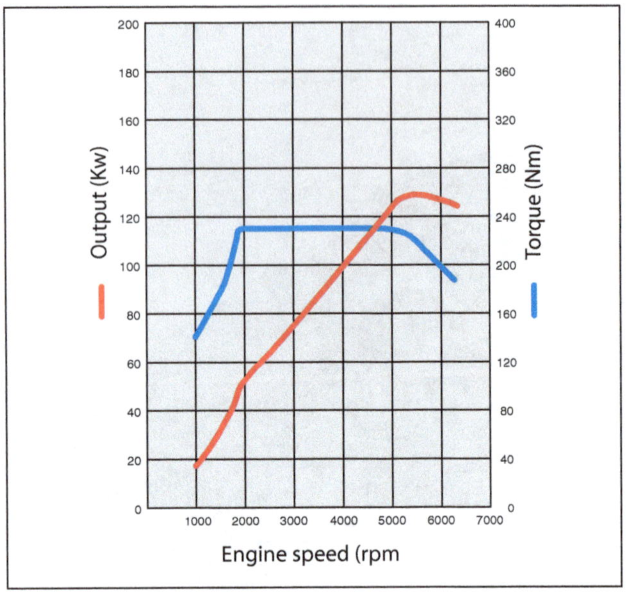

1.8 liter (180 hp) engine performance graph.

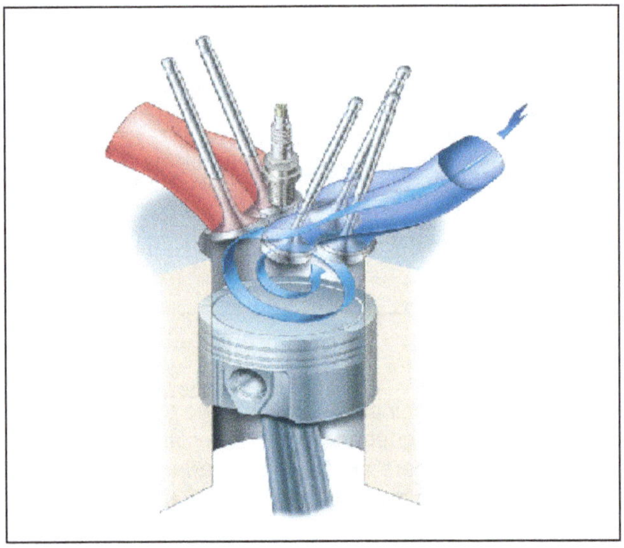

Major design elements of the 1.8 liter engine include tumble duct intake manifold to improve mixture in the combustion chamber.

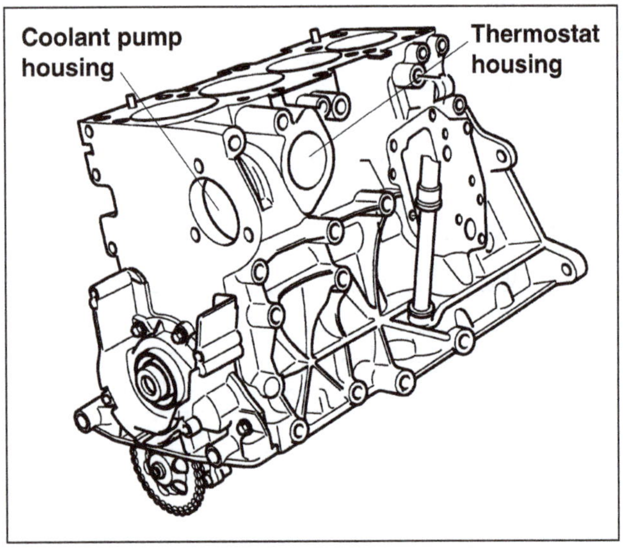

Low block engine design with integral thermostat housing and coolant pump housing reduces the overall height of the engine.

BentleyPublishers.com
© 2006 BentleyPublishers.com—All Rights Reserved

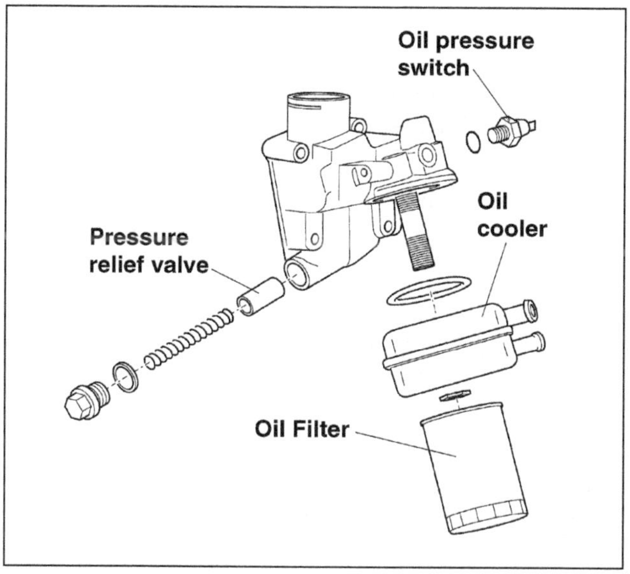

The oil pressure relief valve is located in the oil filter flange housing. A single 1.4 bar (21 psi) oil pressure switch monitors oil pressure.

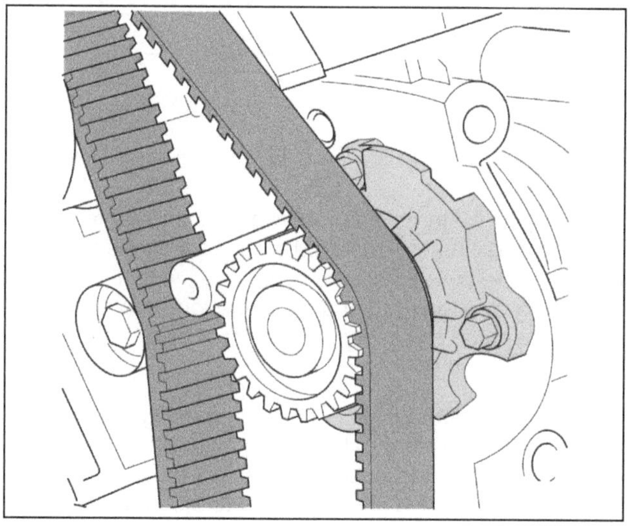

A coolant pump with plastic impeller installed directly in the cylinder block is driven by the toothed camshaft timing belt (toothed belt). This results in fewer components and lighter weight.

BentleyPublishers.com
©2006 BentleyPublishers.com—All Rights Reserved

3.2 liter V6 engine

The Audi 3.2 liter V6 engine is an extremely compact package. A cylinder angle of just 15 degrees makes the four valves per cylinder engine suitable for a transverse-mount application used in the TT 3.2 quattro.

This four-stroke V6 develops an impressive 250 hp at 6300 rpm, and has a broad peak torque range of 236 lb-ft at 2800 to 3200 rpm.

Technical data	Engine Code: BHE
Type	6-cylinder inline, 4 valves/cylinder
Displacement	3.2L, (3195 cc)
Bore	3.31 in (84.0 mm)
Stroke	3.78 in (95.9 mm)
Compression ratio	11.3:1
Power	250 HP (186 kW) @6300 rpm
Torque	236 ft-lb (320 Nm) @2800 to 3200 rpm
Engine management	Bosch Motronic ME 7.1.1

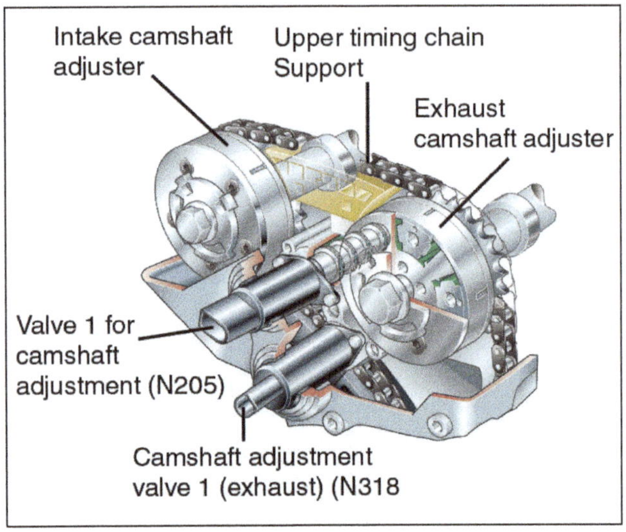

Variable valve timing

The 3.2-liter V6 engine features a variable valve timing system that enables the intake and exhaust cams to be controlled independently. Adjustable valve timing systems overcome the limitations of static valve timing by altering the points in the four-stroke cycle when the valves open and close. This allows for greater torque throughout a wider rpm range.

The camshaft timing adjustment is accomplished using a hydraulic circuit which passes oil pressure to the timing adjusters. This is achieved with the aid of the two solenoids. Valve 1 for Camshaft Adjustment (N205) controls the passage of oil to the intake cam, and Camshaft Adjustment Valve 1 (exhaust) (N318) controls the passage of oil to the exhaust cam. The Motronic Engine Control Module (J220) governs the operation of both solenoids.

Variable path intake manifold

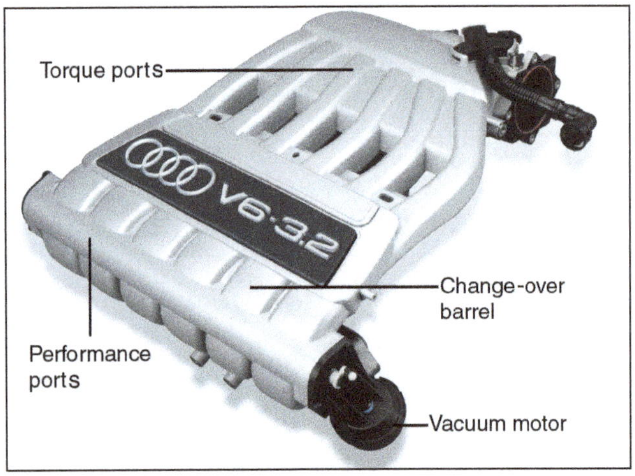

The variable path intake manifold design increases low rpm torque and high rpm power by taking advantage of the self-charging (ram effect) that exists at some engine speeds.

By tuning the intake manifold length, engineers can produce this ram effect for a given rpm range. A manifold that has two different lengths can produce the ram effect over a broader rpm range.

The 3.2-liter V6 engine uses two lengths of air ducts but not in the same way as the dual path manifolds used on other engines.

BentleyPublishers.com
© 2006 BentleyPublishers.com—All Rights Reserved

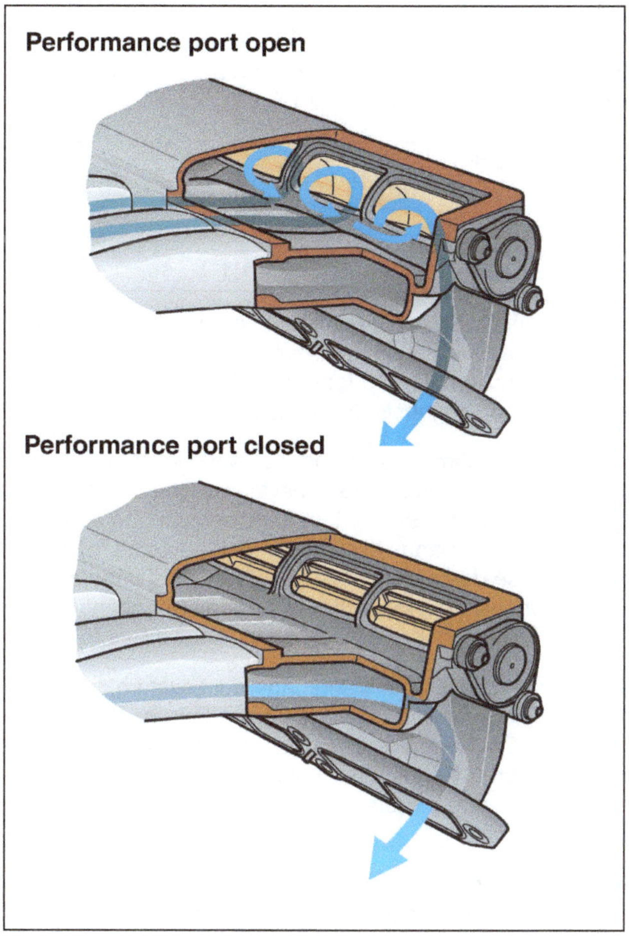

Instead of using high velocity air flow in a long narrow manifold duct to ram more air into an engine at low rpm and then opening a short, large diameter duct for high rpm, the 3.2-liter V6 engine takes advantage of the pressure wave created by the pressure differential that exists between the combustion chamber and the intake manifold. All air enters the intake manifold plenum and torque port, then is drawn down the long intake ducts to the cylinders.

A second plenum called the performance port, which is attached to a set of short manifold ducts, joins the long intake ducts near the cylinder head. A performance port valve, similar in design to a throttle valve, separates the performance port from the short ducts. Note that the performance port does not have any other passages to the intake manifold other than through the performance port valve. It does not have access to the torque port and does not admit any more air into the cylinders than what is already drawn down the long intake ducts.

At engine speeds below 900 rpm, the performance port valve is actuated and the performance port is open for idling. At engine speeds between 900 rpm and 4100 rpm the performance port is closed and the engine produces its maximum low end torque (the performance port valve is not actuated). At engine speeds above 4100 rpm the performance port is open (the performance port valve is actuated).

Cylinder head

The cylinder head is a cross-flow type and has four valves per cylinder. The camshafts are hollow with separate cam lobes for reduced weight. These cam lobes are fitted on the cam during manufacture using hydraulic pressure (hydro-forming).

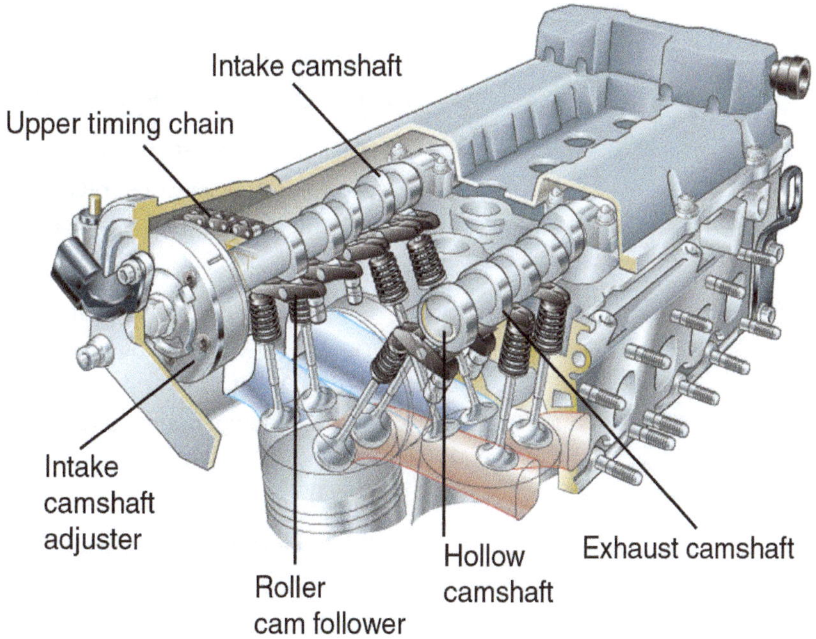

BentleyPublishers.com
© 2006 BentleyPublishers.com—All Rights Reserved

Fuel injection

Two variants of the Bosch Motroinc engine management systems used on the engines used in the Audi TT.

Engine (engine code)	
1.8 liter (ATC, AWP, AMU, BEA)	Motronic ME 7.5
3.2 liter (BHE)	Motronic ME 7.1.1

The Bosch Motronic ME 7 family of engine management systems are referred to as torque-based engine management systems. These systems fully comply with federal and state government mandated On-Board Diagnostic (OBD) II standards.

In older engine management systems, demands for torque and efficiency are handled by separate subsystems. The interaction between all of the different subsystems has become increasingly complicated as new demands are placed on the engine management system. These demands can include control of transmission, traction control and vehicle dynamics in addition to enhanced links with existing subsystems.

The Bosch Motronic ME 7 engine management system uses a new internal circuit design to calculate engine torque. The engine control module (ECM) receives all of the relevant signal data from the input sensors and various other sources and computes the required amount of engine torque needed based on pre-established priorities.

The ECM is able to obtain the calculated engine torque through precise coordination of the following output components:

- Throttle valve control module (throttle angle)
- Turbo wastegate regulator solenoid valve (boost control)
- Fuel injectors (injection time and fuel cut-off)
- Power output stages with ignition coils (ignition timing)

Main sensor inputs to the ME 7 ECM (J220) include:

- Mass airflow sensor (G70)
- Speed / reference sensor (G28)
- Camshaft position sensor (G40)
- Pre-catalyst heated oxygen sensor (G39)
- Post-catalyst heated oxygen sensor (G108)
- Throttle valve control module (J338) with position sensors (G187) and (G188)
- Intake air temperature sensor (G42)
- Coolant temperature sensors (G2) and (G62)
- Knock sensor, cylinders 1 - 2 (G61)
- Knock sensor, cylinders 3 - 4 (G66)
- Accelerator pedal position sensors (G79) and (G185)
- Turbo boost pressure sensor (G31)
- Brake light (F) and pedal switches (F47)
- Clutch pedal switch (F36)
- Leak detection pump (V144)

BentleyPublishers.com
© 2006 BentleyPublishers.com—All Rights Reserved

Additional inputs are received from the cruise control switches, transmission control module, vehicle speed sensor, instrument cluster, anti-lock brake system control module, fan control module, airbag control module and air conditioning system.

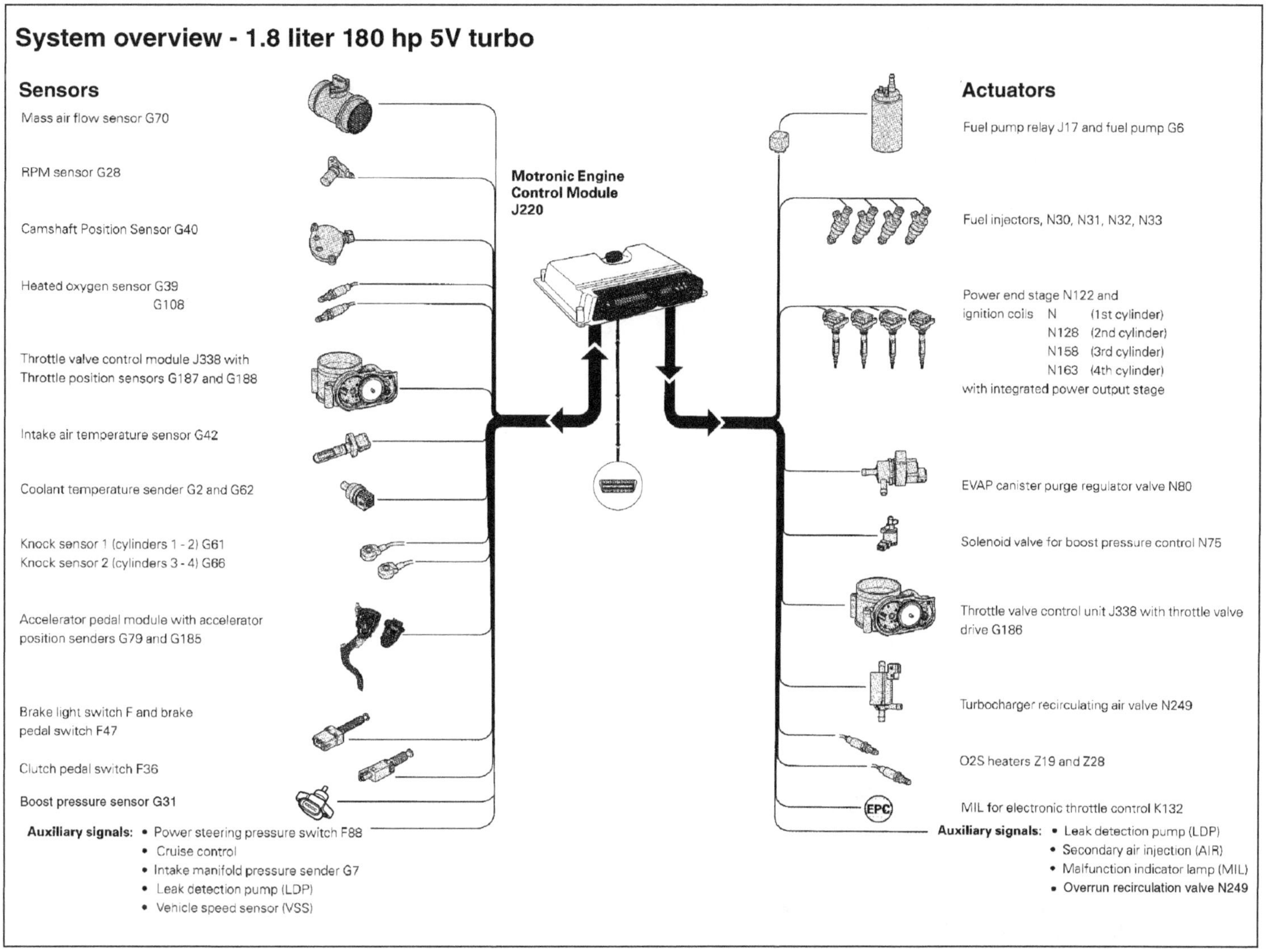

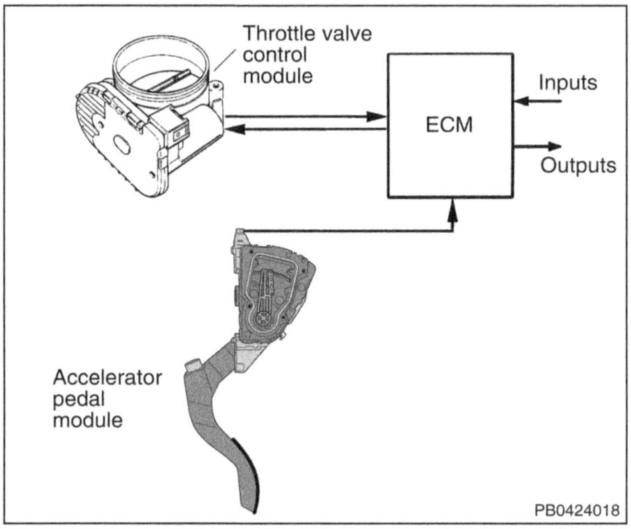

With ME 7, there is no longer a throttle cable between the accelerator pedal and the throttle valve. This has been replaced by an electronic control system.

The system has the following components:

- Throttle valve control module
- Accelerator pedal module

The electronic throttle valve control system is used to reduce and increase torque without adversely affecting the exhaust emissions.

The ideal combination of throttle opening and charge pressure can be controlled for greatest efficiency. Under acceleration, throttle movement is programmed to best increase intake charge velocity. On deceleration, the throttle valve can be held open to reduce emissions.

BentleyPublishers.com
© 2006 BentleyPublishers.com—All Rights Reserved

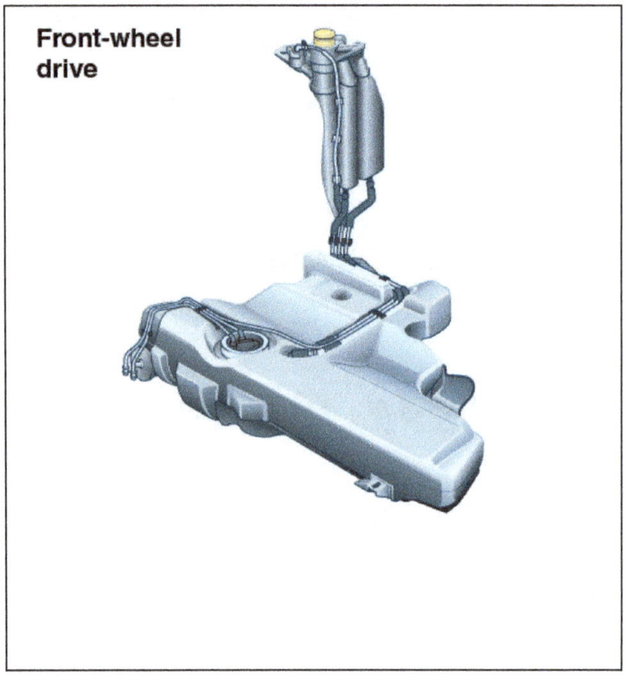

Fuel supply

Different fuel tanks are used in the front-wheel drive and the quattro versions. Both types are constructed of plastic.

◄ The front-drive tank has a capacity of 14.5 gallons (55 liters).

◄ The quattro tank has a capacity of 16.3 gallons (62 liters).

BentleyPublishers.com

© 2006 BentleyPublishers.com—All Rights Reserved

3 Clutch and Transmission

Manual transaxles are available with 5-speeds or 6-speeds. Automatic transaxles are available as a 6-speed Tiptronic or a 6-speed Tiptronic DSG (Direct Shift Gearbox).

Transmission applications	Transmission code
5-speed manual, front wheel drive	02J
5- and 6-speed manual, all wheel drive (quattro)	02M 02Y
6-speed automatic (Tiptronic), front wheel drive	09G
6-speed direct shift automatic (DSG), all wheel drive (quattro)	02E

Manual transmission

Two manual gearbox variants are used for power transmission. Both feature cable operated shift linkage to help reduce the transfer of noise and vibration to the interior of the vehicle. The hydraulically operated clutch is mounted to a dual mass flywheel to dampen crankshaft oscillations.

Manual transmission (cable-operated)

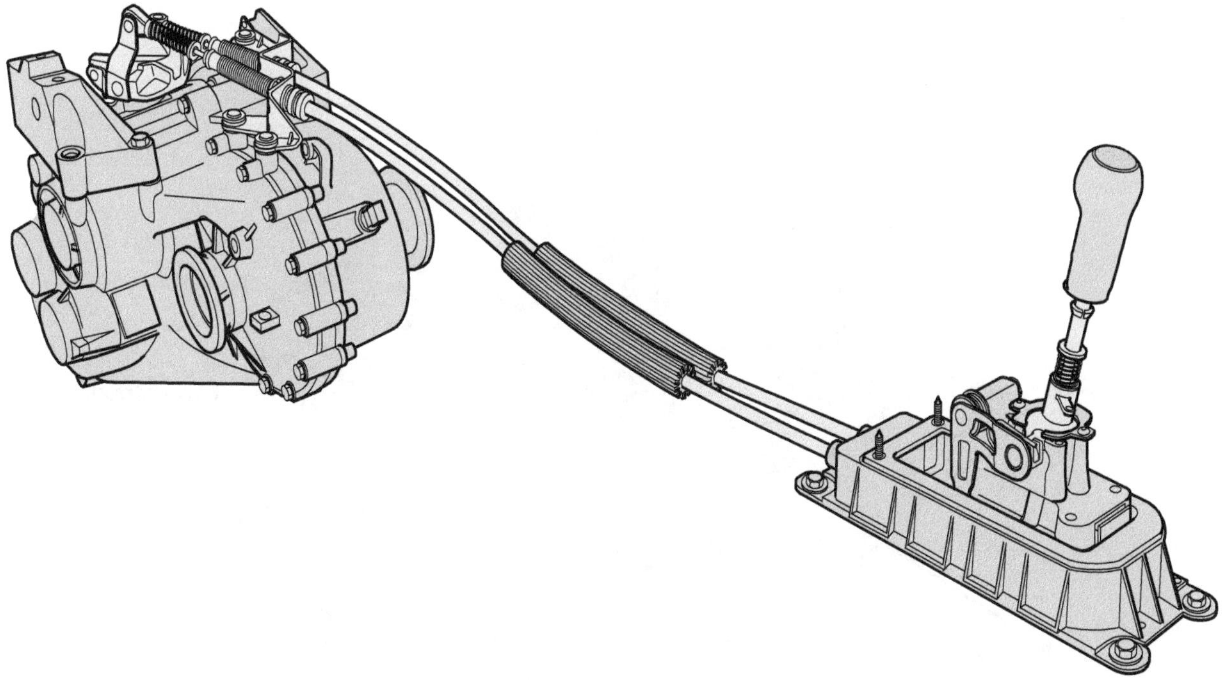

BentleyPublishers.com

© 2006 BentleyPublishers.com—All Rights Reserved

Automatic transmission

S318_024

6-speed automatic transmission 09G

The 6-speed automatic transmission is transverse mounted, compact, lightweight and electronically controlled. Features include:

- Maximum torque 229 ft-lb (310 Nm)
- Weight 185 lb (84 kg)
- Length 13.8 in (350 mm)
- Torque converter with lockup
- Automatic and Tiptronic operation

Forward and reverse gears have simple planetary gear sets with a double Ravigneaux planetary gear set. Modulating valves are activated by the transmission control module to control pressure build-up in the multi-plate clutches and plate brakes. By allowing a delayed pressure build-up, the modulating valves deliver fast response times and smooth shifts.

6-speed Direct Shift Gearbox (DSG) 02E

The 02E Direct Shift Gearbox (DSG) is a compact six-speed transmission that combines the best features of both a manual transmission and an automatic transmission with Tiptronic. The 02E also provides better fuel economy than typical manual transmissions.

DSG Transmission

BentleyPublishers.com

BentleyPublishers.com
© 2006 BentleyPublishers.com—All Rights Reserved

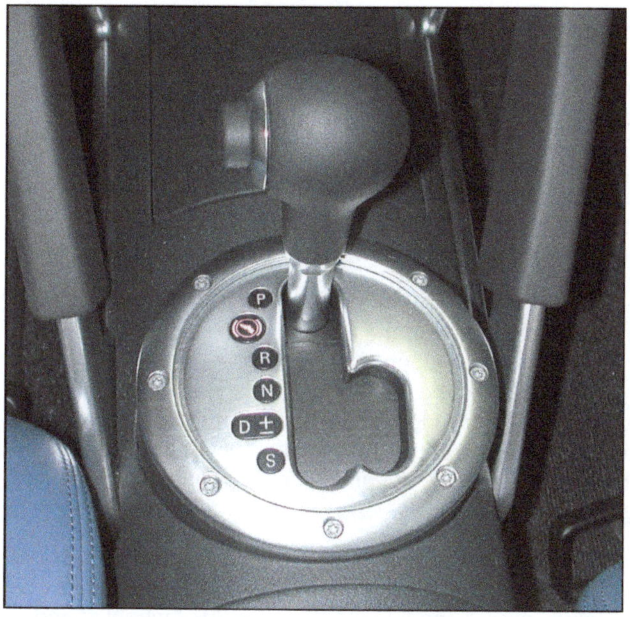

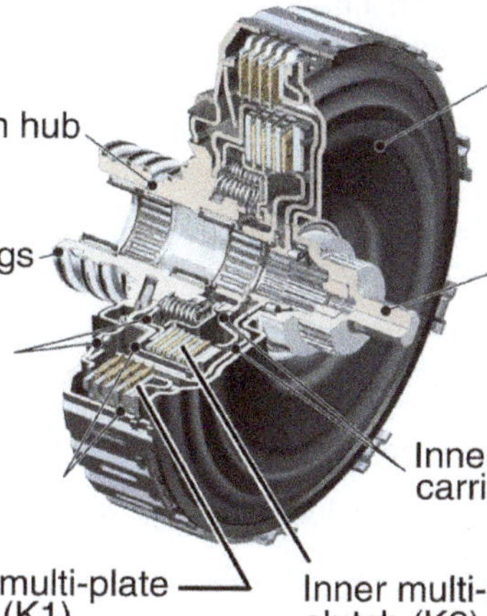

DSG features include:

- 6 forward and one reverse gear
- Normal shift control program (D) and Sport shift control program (S)
- Tiptronic shift control and steering wheel shift paddles
- Mechatronics: electronic and hydraulic control modules in a single unit are integral to the transmission
- Oil cooler mounted on transmission
- Maximum torque 258 ft-lb (349 Nm)

The selector lever position is no longer transmitted mechanically to the transmission by way of the selector lever cable and multi-function switch (driving range sensor). Instead, the DSG uses a shift-by-wire control concept in which signals to initiate upshifts and downshifts are sent to the transmission Mechatronics control module via the drivetrain CAN-bus. No cable is involved in the gearshift operation. Only the parking lock is engaged mechanically by way of the selector lever cable for P.

At first glance, the DSG shift gate looks like a regular Tiptronic transmission with standard P, R, N and D plus an added S for sport mode with a gate to the right for manual shifting. Place the DSG in D and it behaves just like a regular automatic for the most part, except that at a stop the car goes into neutral.

When you start to ease off the brake, the clutch smoothly engages just like a manual transmission. In D mode, the DSG shifts through the gears very quickly in city driving, making 6th gear to 2nd gear changes in 0.9 seconds. In these situations, the DSG downshifts immediately into the power band. Normal upshifts are quick and smooth, with no jerkiness.

The DSG delivers uninterrupted power flow during gear changes. This was accomplished by replacing the conventional dry clutch with two multi-plate hydraulic wet clutches like those that control the shifting in an automatic transmission.

The dual multi-plate wet clutch, comprised of clutch 1 (K1) and clutch 2 (K2), is a complex and innovative component. The two electronically controlled, hydraulically actuated clutches are packed inside each other for maximum space economy.

DSG's dual clutch offers greater thermal capacity and better regulation than dry clutches. Its pressure is regulated hydraulically by solenoid valves that are controlled by the Mechatronics control module.

BentleyPublishers.com
©2006 BentleyPublishers.com—All Rights Reserved

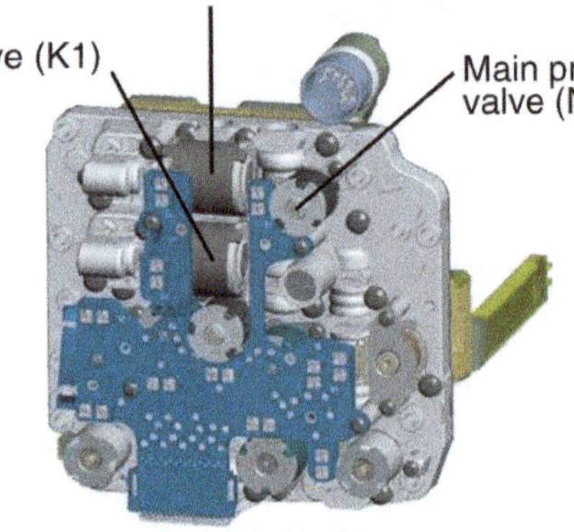

The most complex component of the DSG is the Mechatronics control module, which is located at the front of the transmission. The Mechatronics control module is the hub for the transmission's mechanical, hydraulic and electronic systems. The Mechatronics control module controls 13 sensors and four actuators. It manages data for controlling the clutches, shift points, input and output shafts, cooling, individual gears, hydraulic pressures, and various malfunction programs.

Located inside the transmission, the Mechatronics control module is immersed in oil that can heat up to 285° F. To protect the semi-conductor elements of the control module, a heat exchanger bolted on the outside of the transmission ensures the correct thermal condition.

Power transmission - Haldex coupling

The four wheel drive powertrain used in the TT Coupe quattro uses a slip-dependant force distribution control for both axles called a Haldex coupling.

The Haldex coupling is mounted on the rear axle differential and is driven by the drive shaft. It has its own oil circuit, hydraulic components, an electro-hydraulic control valve and an electrical control module.

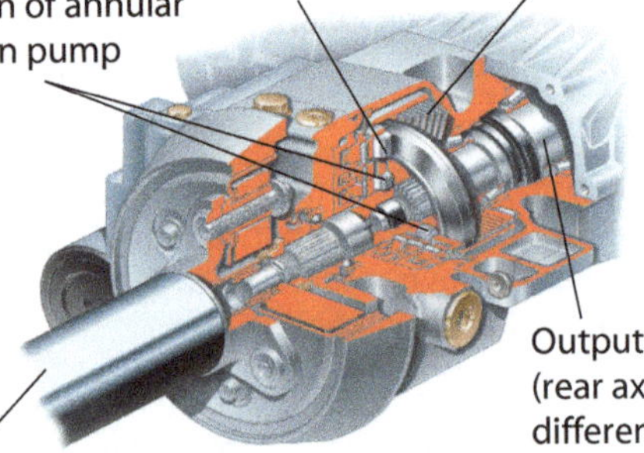

Engine torque is transmitted to the drive shaft through the gearbox, the front axle differential, and the front drive axles. The driveshaft is connected to the input shaft of the Haldex couplling. In the Haldex coupling, the input shaft is separated from the output shaft to the rear axle differential. Torque can only be transmitted to the rear axle differential when the Haldex coupling clutch plates are engaged.

With the Haldex coupling, slip is no longer the decisive factor in the distribution of drive forces. The car's dynamic state is also a factor. The Haldex control module monitors the ABS wheel speed sensors and the accelerator pedal signal (from engine control module) via the CAN-bus. This data provides the engine control module with all the information it needs on road speed, cornering, coasting or traction mode, and can respond optimally to any driving situation.

Characteristics of the Haldex coupling:

- Permanent, fully automatic all-wheel drive with electronically controlled multi-plate clutch
- Torque split can be varied by the control module
- Rear axle torque of up to 3200 Nm
- Acceleration with high directional stability
- Handling is neutral with a slight tendency to understeer
- Communication via CAN-BUS

BentleyPublishers.com
© 2006 BentleyPublishers.com—All Rights Reserved

Haldex coupling (quattro)

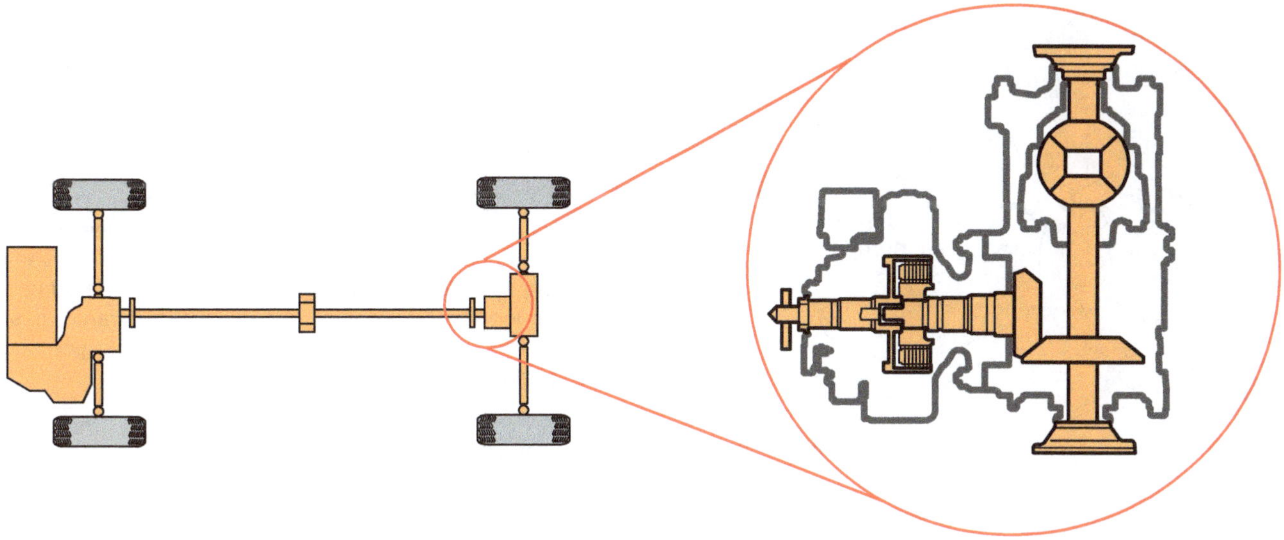

4 Suspension, Steering and Brakes

The suspension system of the Audi TT provides unsurpassed driving comfort and dynamics.

BentleyPublishers
.com
© 2006 BentleyPublishers.com—All Rights Reserved

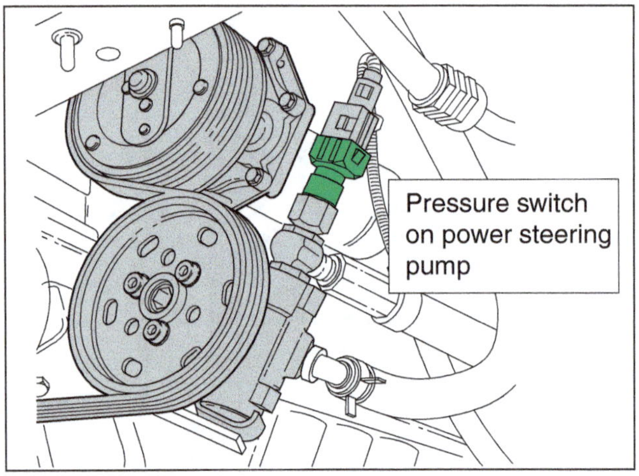

Steering

The safety steering column is adjustable for rake and reach. The rack and pinion steering gear is power assisted. A steering damper reduces vibrations transmitted from the wheels to the steering gear. The vibration-optimized mounting of the steering column to the dash panel crossmember keeps the steering wheel free of unwanted vibrations.

On 4-cylinder engines, the power steering pump is fitted with a pressure switch to signal the engine control unit when the power steering pump is subject to a load. The pump is driven by the engine by a ribbed belt. At full steering lock, the pump places a higher load on the engine, and idling speed can drop sharply. The signal from the pressure switch enables the engine control module to regulate engine torque at idling speed.

Front axle

To ensure track stability, a new forged crossmember as well as modified wheel bearing housings with tie rod attachments were implemented in the front axle.

The features of the front axle are:

- Low friction tie rod ends
- Ball joints reinforced by thicker journals
- Anti-roll bar link attached directly to suspension strut for better response

Front axle

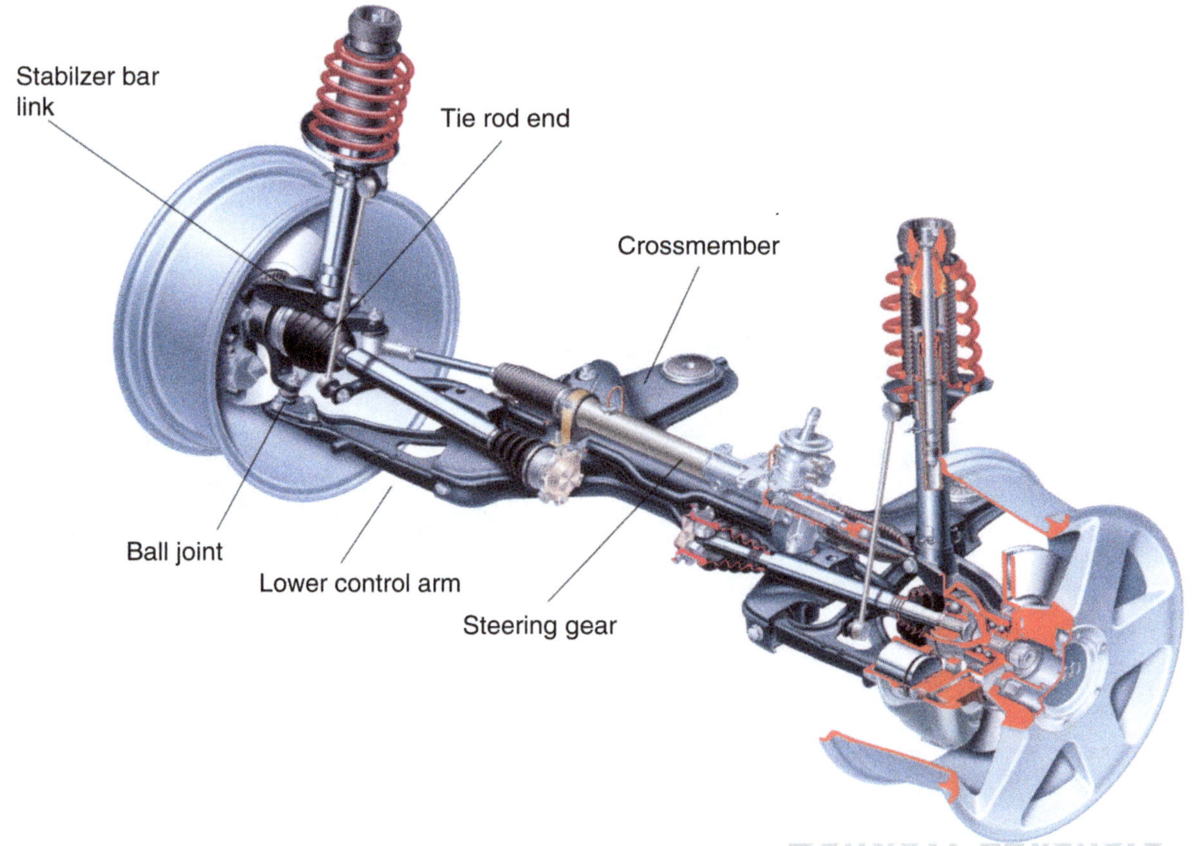

© 2006 BentleyPublishers.com—All Rights Reserved

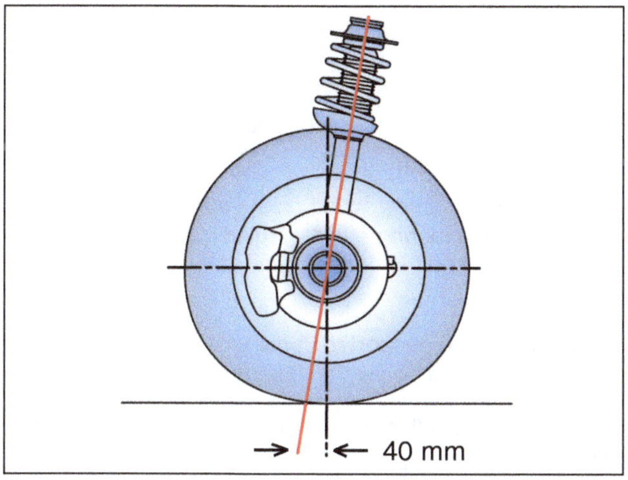

Front wheel alignment

The front strut attachment to the front wheel bearing housing does not provide for camber adjustment. Camber can be equalized, however, by moving the crossmember from side to side.

Caster offset of 40 mm (1.57 in.) helps ensure good steering directional stability. Power assisted steering, which is standard equipment on the TT, compensates for what would otherwise be higher steering effort. The caster is not adjustable. Toe can be adjusted by loosening and turning either tie rod.

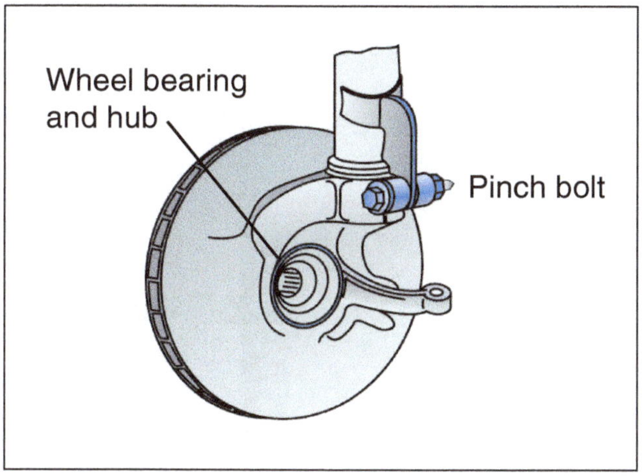

Front Strut

The two-piece strut assembly consists of a separate shock absorber housing and wheel bearing housing.

The shock absorber portion of the the strut is clamped in the wheel bearing housing with a pinch bolt.

BentleyPublishers.com
© 2006 BentleyPublishers.com—All Rights Reserved

Rear axle (front wheel drive)

Front-wheel drive models use a rear torsion beam axle with internal stabilizer bar. The rear axle is attached to the body with mounts angled at 25° to the transverse axis of the vehicle.

- Track width: 1507 mm (59.3 in.)
- Modified axle plates for increased camber and modified toe-in
- Track correcting axle bearing

When cornering, lateral forces acting on the suspension alter the track and cause a self-steering effect. Angled mounts reduce the tendency to self-steer and increase vehicle stability as a turn is completed.

The shock absorbers are supported in the wheel housings and the coil springs are supported below the side member.

The axle is stabilized by a tubular stabilizer bar located within the cross section of the transverse bar. Due to the separate layout of the springs and shock absorbers, the vehicle has a large luggage compartment and driving noise in the passenger compartment is reduced.

Rear axle (front wheel drive)

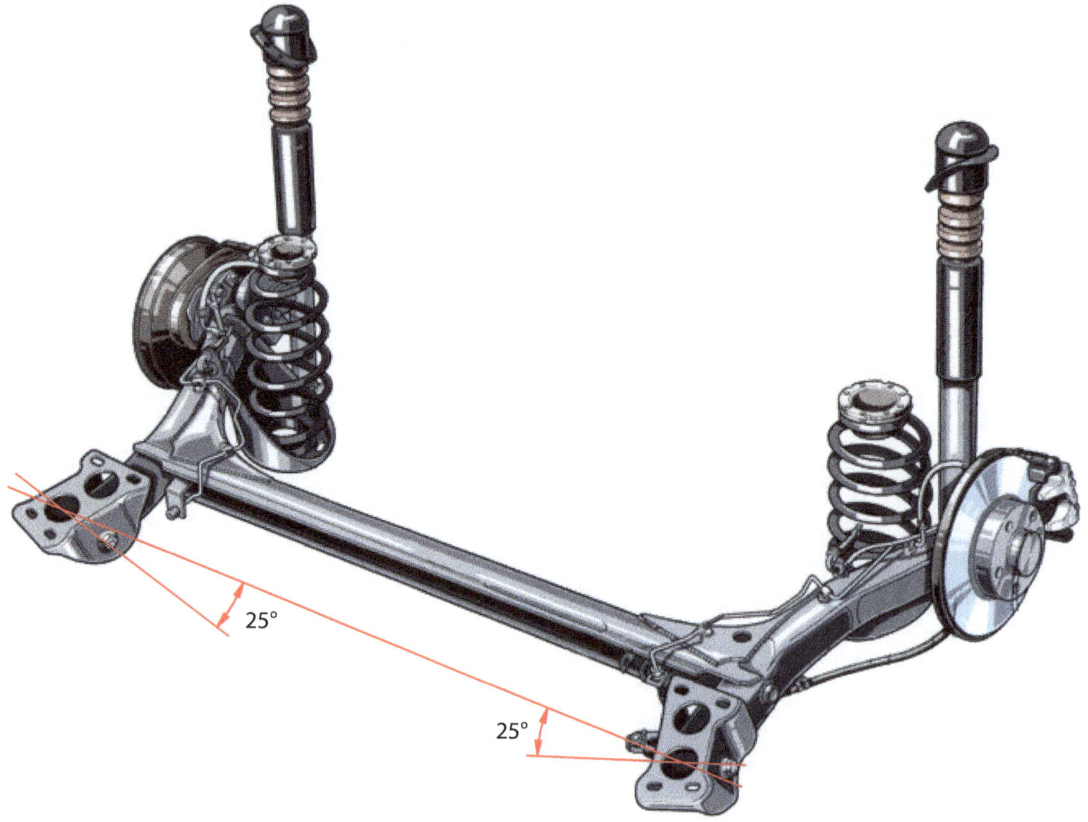

BentleyPublishers.com
© 2006 BentleyPublishers.com—All Rights Reserved

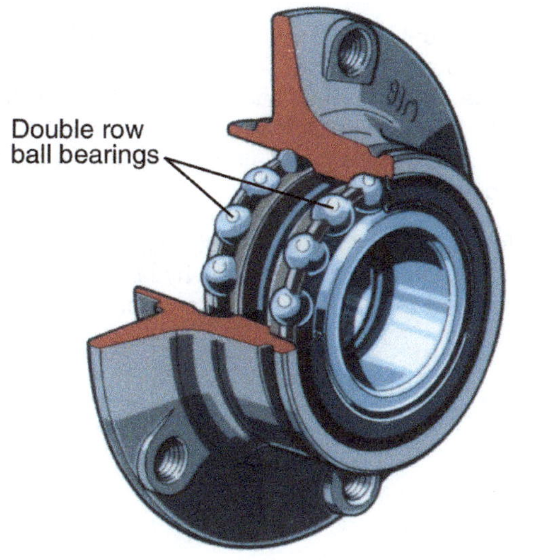

A double-row rear wheel bearing consists of two inner races, two sets of ball bearings, and a hub, which doubles as the bearing outer race. This assembly is pressed onto the stub axle, and the brake disc mounts to the hub.

Rear axle (quattro drive)

The dual link trailing arm axle on the TT quattro is fixed via the subframe (4-point attachment) and each of the track correcting bushings are attached to the trailing arm.

For stabilization purposes, a transverse anti-roll bar is located on the axle sub-frame. The rear shock absorbers are fitted at approximately a 45°. The wheel bearings are the same as those used in the rear axle of front-wheel drive models.

Rear axle (quattro)

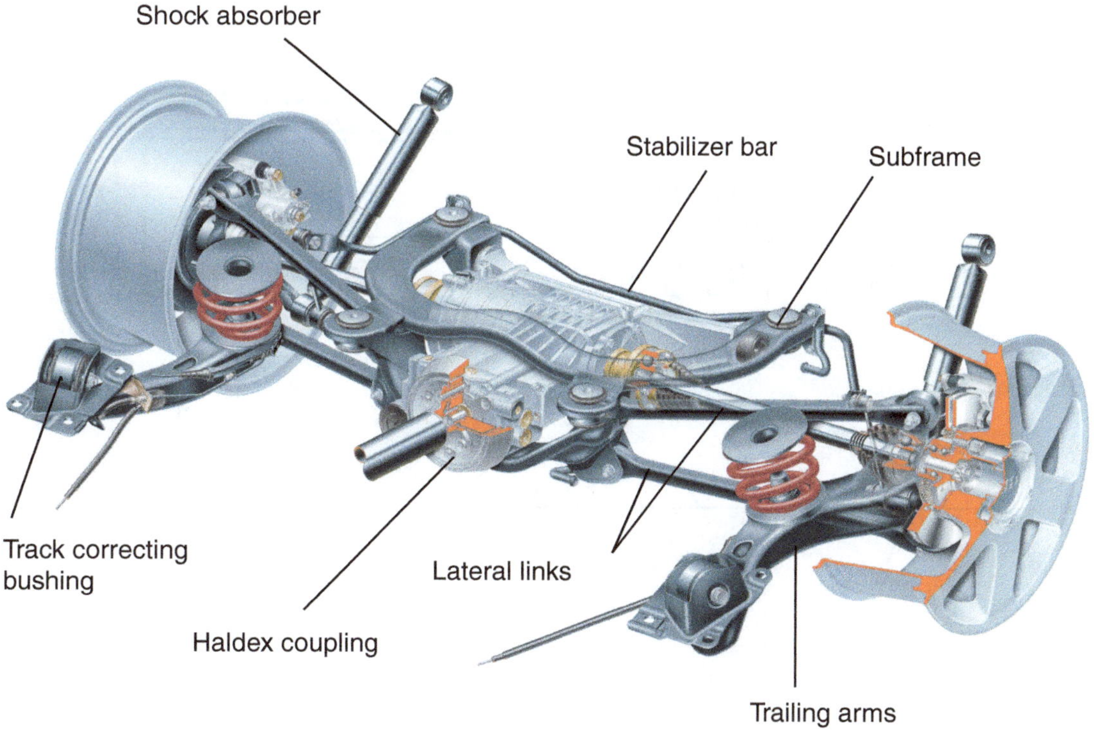

BentleyPublishers.com
©2006 BentleyPublishers.com—All Rights Reserved

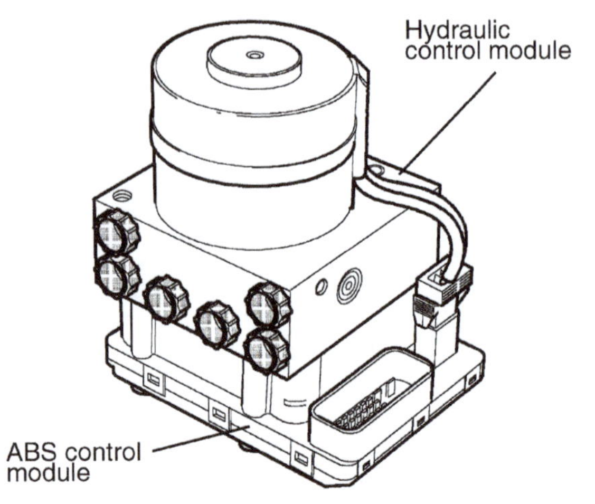

Brakes

The brake system is a dual-diagonal circuit with Anti-lock Braking (ABS) and Electronic Brake Pressure Distribution (EBD). Additional features include Anti Slip Regulation (ASR), ESP (Electronic Stability Program) and Electronic Differential Lock (EDL). All models use discs at all 4 wheels (ventilated front discs) and a 10 in. brake servo (booster).

The TT uses the Teves 20 IE or Teves 60 ABS, depending on model year and model. On both systems, the hydraulic control module is integrated with the ABS electronic control module.

Diagonal-split dual circuit brake system with ABS

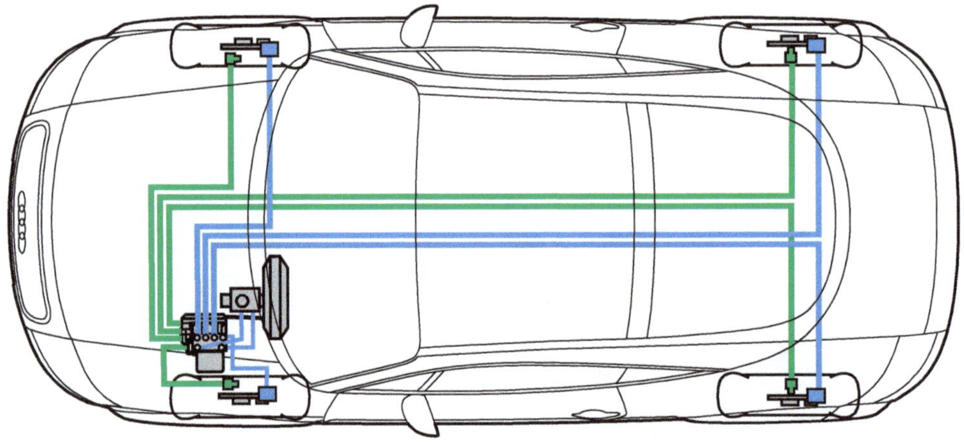

BentleyPublishers.com
© 2006 BentleyPublishers.com—All Rights Reserved

5 Body – Assembly
6 Body – Components and Accessories

Front bumper

The front bumper cover is in two parts: the flexible cover panel and the decorative grille. The bumper reinforcement beam is made of aluminum and bolted to the long members by impact absorbing elements. Guide profiles attached to the left and right fenders ensure an even gap all around. A seamless transition to the fenders (zero joint) is created by attaching the bumper cover to the fender with the threaded bolts and nut/washer combinations.

Rear bumper

The rear bumper assembly is in four parts: flexible cover panel, the lower cover panel, an aluminum cross-member reinforcement beam, and the central locating element. The rear cover is available in two versions, depending on engine variant. A seamless transition to the body side section (zero joint) is produced by means of 2 bolts on each side.

Front Bumper

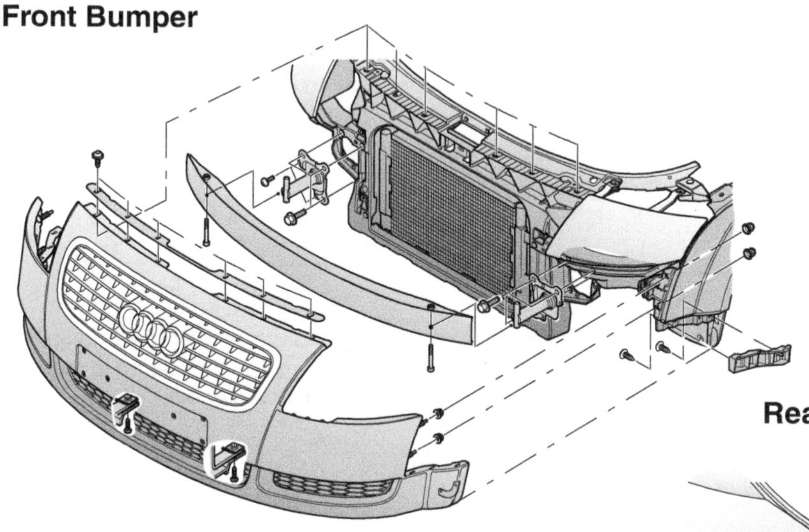

Rear bumper

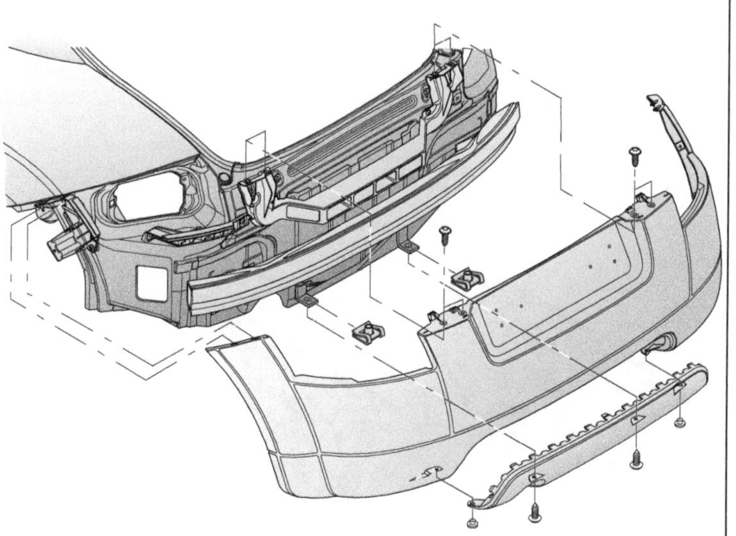

BentleyPublishers.com

©2006 BentleyPublishers.com—All Rights Reserved

Body (coupe)

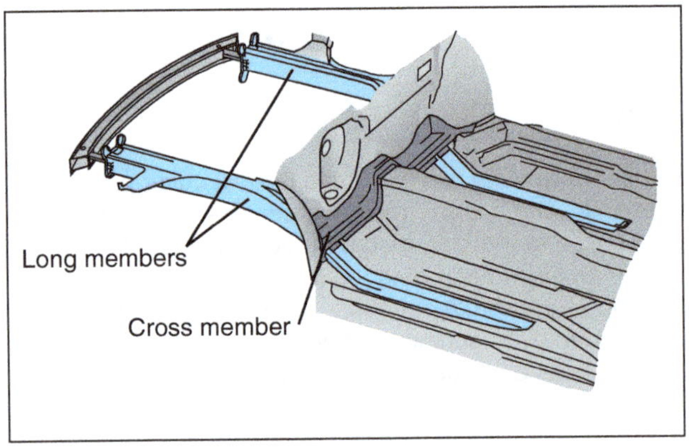

The TT hood is made of aluminum.

The vehicle front end area is designed to deform in a predefined manner, absorbing the impact energy without impairing the stability of the occupant cell. The long members are manufactured from 2-, 3-, and 1.5-mm thick mash welded metal plates. In a side impact, the strong crossmembers will also deform on the side of the body facing away from the impact to absorb some of the impact energy.

The structure of the rear end was designed around two key priorities. The first is to protect the integrity of the fuel system by securing it in a protected area designed to remain largely intact. Second, the crush zones of the structure are designed to absorb and dissipate crash energy, thus keeping the load on the occupants to a minimum.

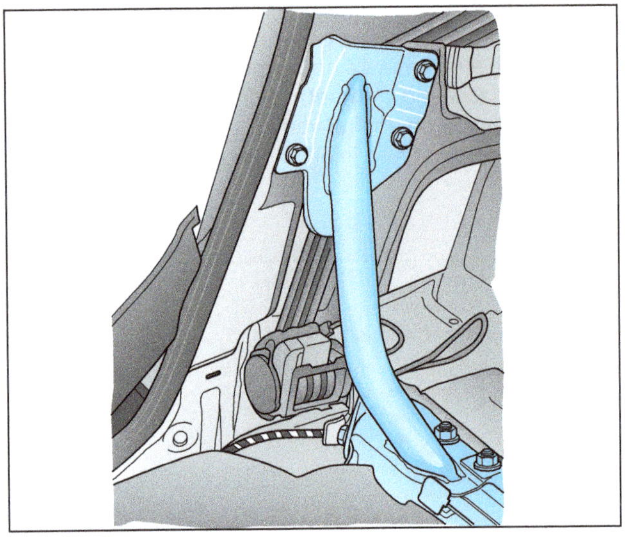

To absorb the load resulting from a side impact, the body structure is extremely rigid, even though the B-pillar does not continue directly into the roof. An additional transverse support extending from the base of the B-pillar to the rear seat crossmember minimizes cell deformation and the rate of intrusion into the side structure.

Since the available crush zone in a side impact is limited, various design measures are necessary to effectively perform this task.

During a side impact, the forces acting on the vehicle are distributed to the sill and the A- and B-pillars via side reinforcements in the door.

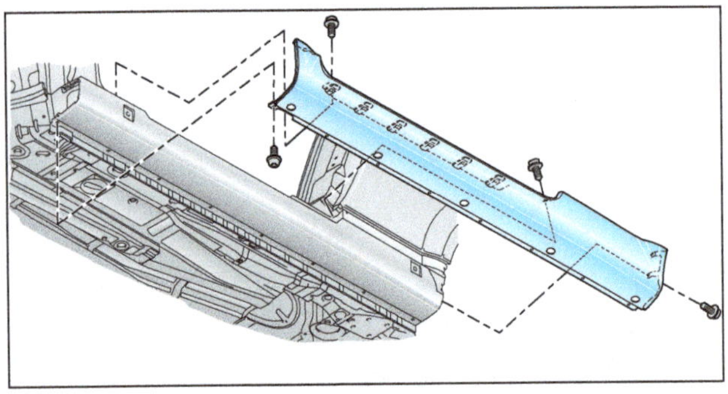

The strong sill also absorbs energy and simultaneously transmits this energy to the stable floor pan assembly. The sill trim is made of steel and is secured to the sill using 17 fasteners.

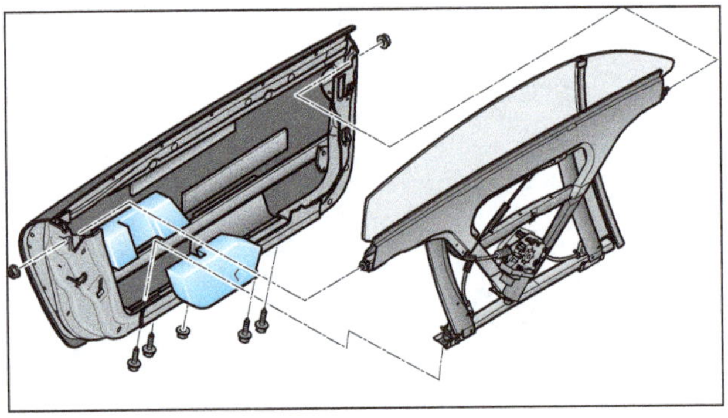

The frameless doors on the TT coupe are a two piece construction. The door shell is made of steel with a bolted-in high strength extruded aluminum side reinforcement. The door component carrier is made of aluminum. It can be adjusted for length, height, and inclination.

BentleyPublishers
.com
© 2006 BentleyPublishers.com—All Rights Reserved

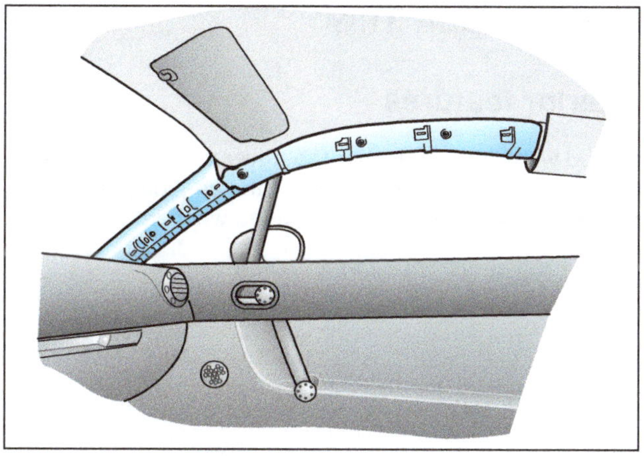

For additional head protection, padding is integrated in the roof area. An additional deformation element is welded into the A-pillar.

Body (roadster)

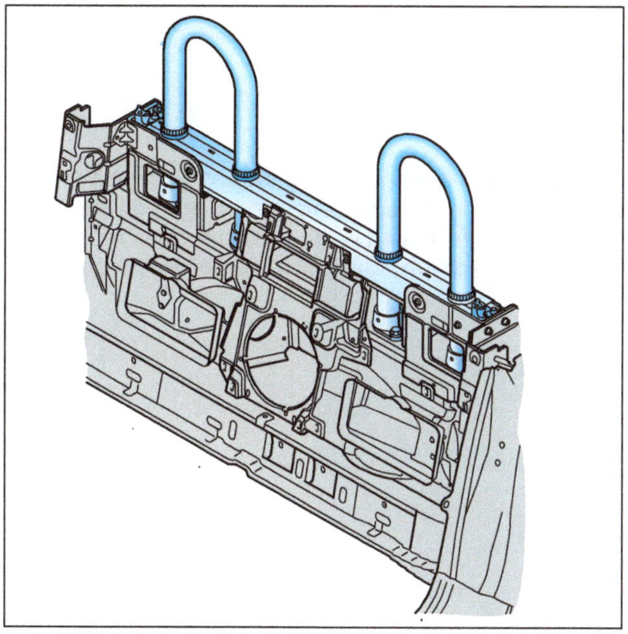

Passengers are provided with protection in the event of a roll over by a safety cage. Two tubular aluminum roll bars anchored to the vehicle body and projecting beyond the two head restraints, in combination with the ultra-rigid A-pillar / windshield cross member, provide the vehicle occupants effective protection.

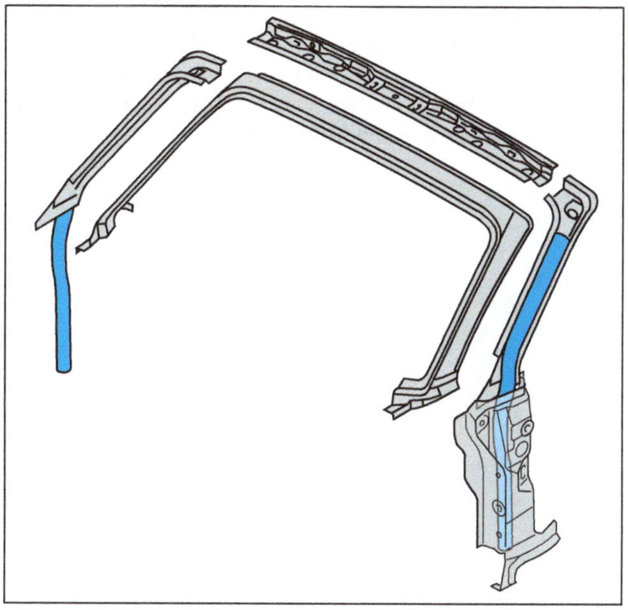

The A-pillar comprises a pipe-in-pipe system, reinforced with high tensile steel.

BentleyPublishers
.com
© 2006 BentleyPublishers.com—All Rights Reserved

7 Body – Interior Trim

Body interior features

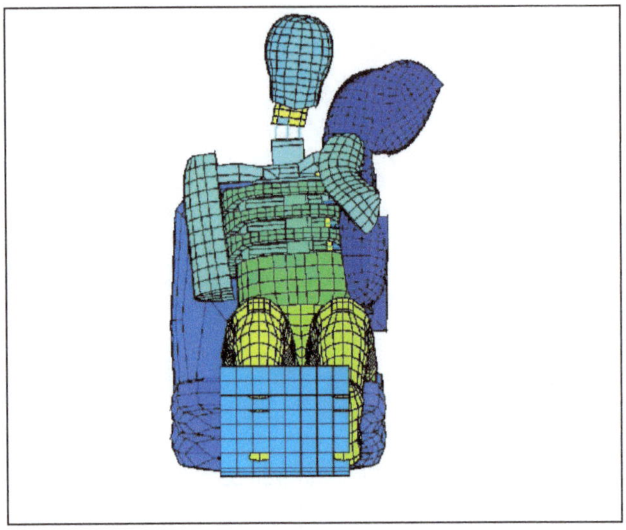

The TT is fitted with side airbags for the driver and the front passenger. The side airbags are integrated in the seat backrests and extend from the rib cage area up to the head when inflated.

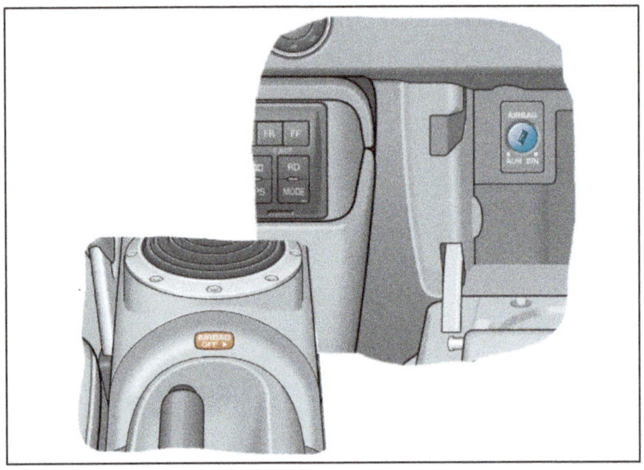

The TT is fitted with a passenger airbag disable function. When using a child seat on the front passenger seat, or when children 12 years old or younger are seated in the front, the driver must disable the front airbag with the vehicle key via the key switch inside the glovebox.

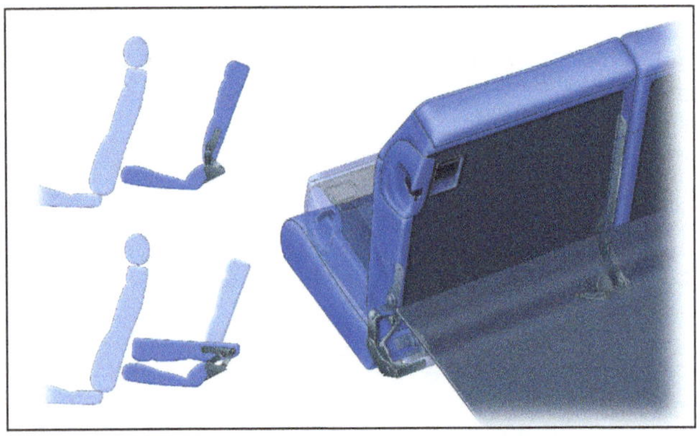

The coupe rear seats were designed to offer occasional seating for children. The rear seat can be folded on a 50/50 basis to provide variability in load and passenger carrying capability. A four-bar link style hinge allows the seat back to move rearward as it is folded. This provides a flat extension to the cargo area without obstructing the full range of movement of the front seats.

© 2006 BentleyPublishers.com—All Rights Reserved

8 Heating and Air-conditioning

The TT's fully automatic climate control system provides automatic temperature and air flow rate control.

Pressing the defroster button automatically disables the air recirculation mode. Only fresh air (no recirculation) is available in defrost mode.

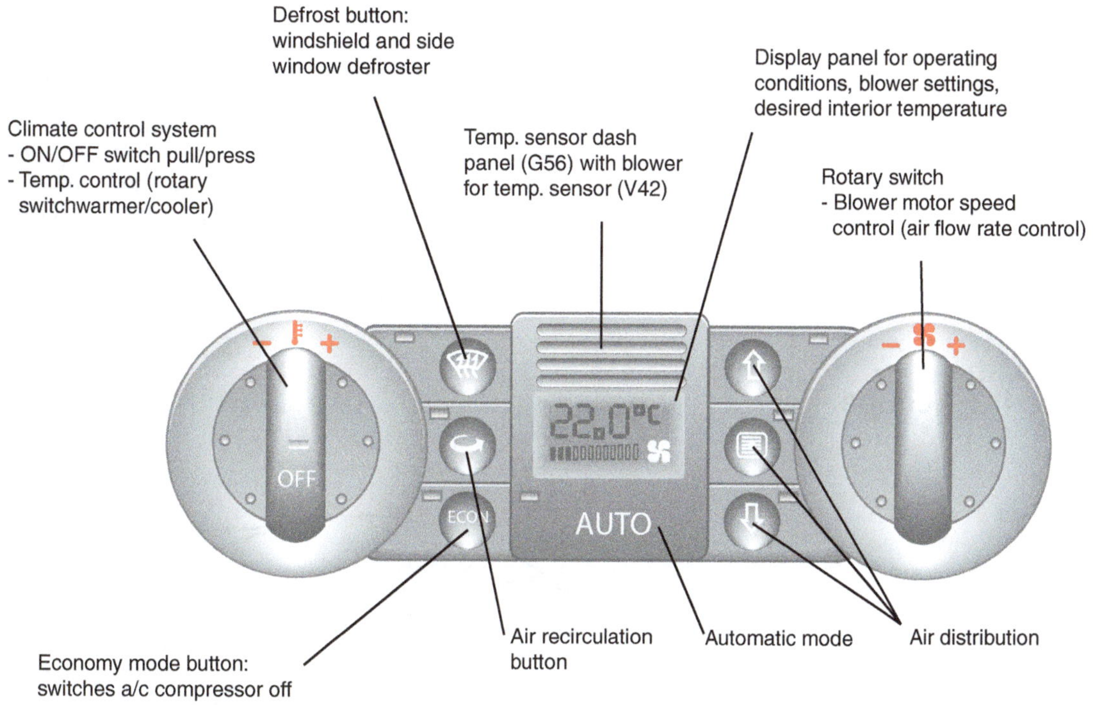

Air distribution components

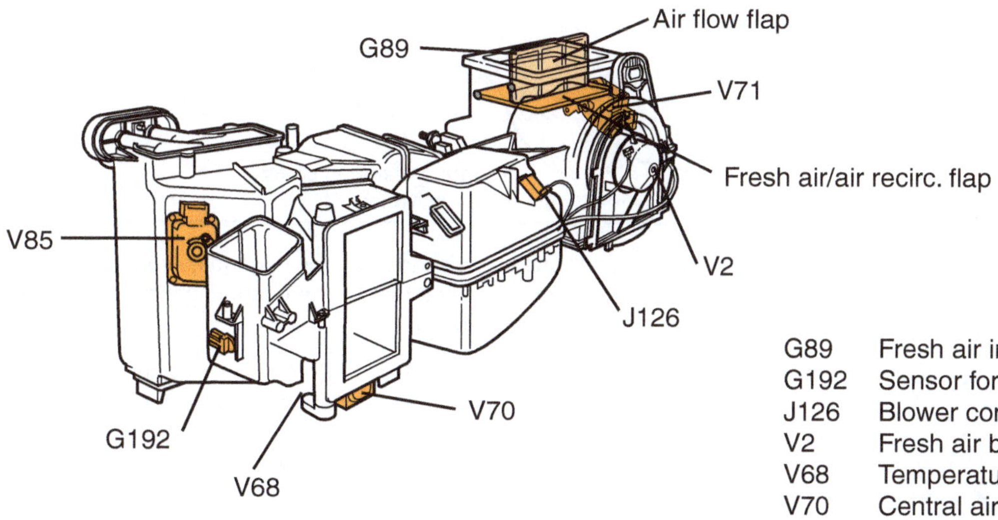

G89	Fresh air intake duct temperature sensor
G192	Sensor for footwell air outlet temperature
J126	Blower control module
V2	Fresh air blower
V68	Temperature regulator flap motor
V70	Central air flap motor
V71	Air flow flap motor
V85	Footwell / defroster flap motor

BentleyPublishers.com

© 2006 BentleyPublishers.com—All Rights Reserved

9 Electrical System

Interior monitoring and alarm system (coupe models)

In addition to the alarm system that sounds if the hood, trunk, or doors are opened, the vehicle interior is monitored by an ultrasonic motion sensor.

If a break-in disturbs the interior, even if the doors are not opened, the anti-theft system alarm sounds and the hazard lights flash.

The system communicates with the anti-theft warning system regarding activation / deactivation as well as triggering of the alarm. Located on the central console is the interior monitoring switch. It switches off the interior monitoring for a single locking operation (it must be switched off each time the vehicle is locked). This can be useful to prevent unwanted activation if, for example, a pet must be left in a locked car.

Alarm system with interior monitoring - coupe

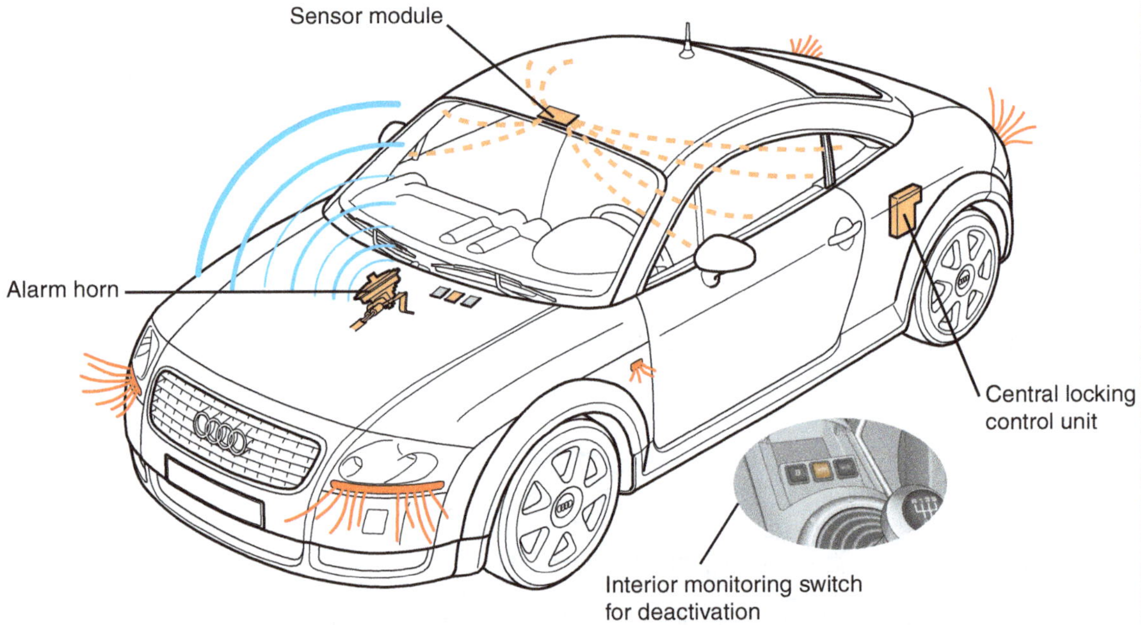

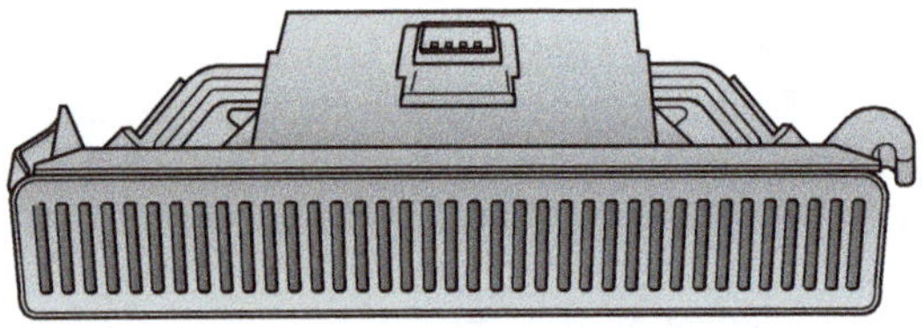

The sensor unit has a sender, a receiver and evaluation electronics. The sensor is located behind the interior lighting in the vehicle headliner (coupe models).

The anti-theft warning system LED (located in the driver's door panel) indicates the status of the system. If after locking the car, the status light blinks quickly for approximately 30 seconds, then slowly, the alarm is armed and the motion sensor is active. If the status light is continuously on after locking the car, the motion sensor is deactivated.

BentleyPublishers.com

© 2006 BentleyPublishers.com—All Rights Reserved

Sound system

The illustration below shows locations for radio, loudspeakers, and antenna, including Bose amplifier, 6-disc CD changer, and car phone, when fitted.

Audi concert radio (standard version) includes features such as:

- RDS displays station format and call letters.
- GALA (graduated audio level adjustment) adjusts sound output according to vehicle speed.
- CD changer operation capability.
- Four speakers in the doors and two 2-way speakers in the rear.
- 80 watts of amplifier power.

The Bose sound system includes all of the above plus:

- Sound output custom engineered to the TT's own acoustic characteristics.
- Upgraded Bose speakers.
- Center speaker in dash panel.
- 175-watt, 4-channel amplifier

Sound system - coupe

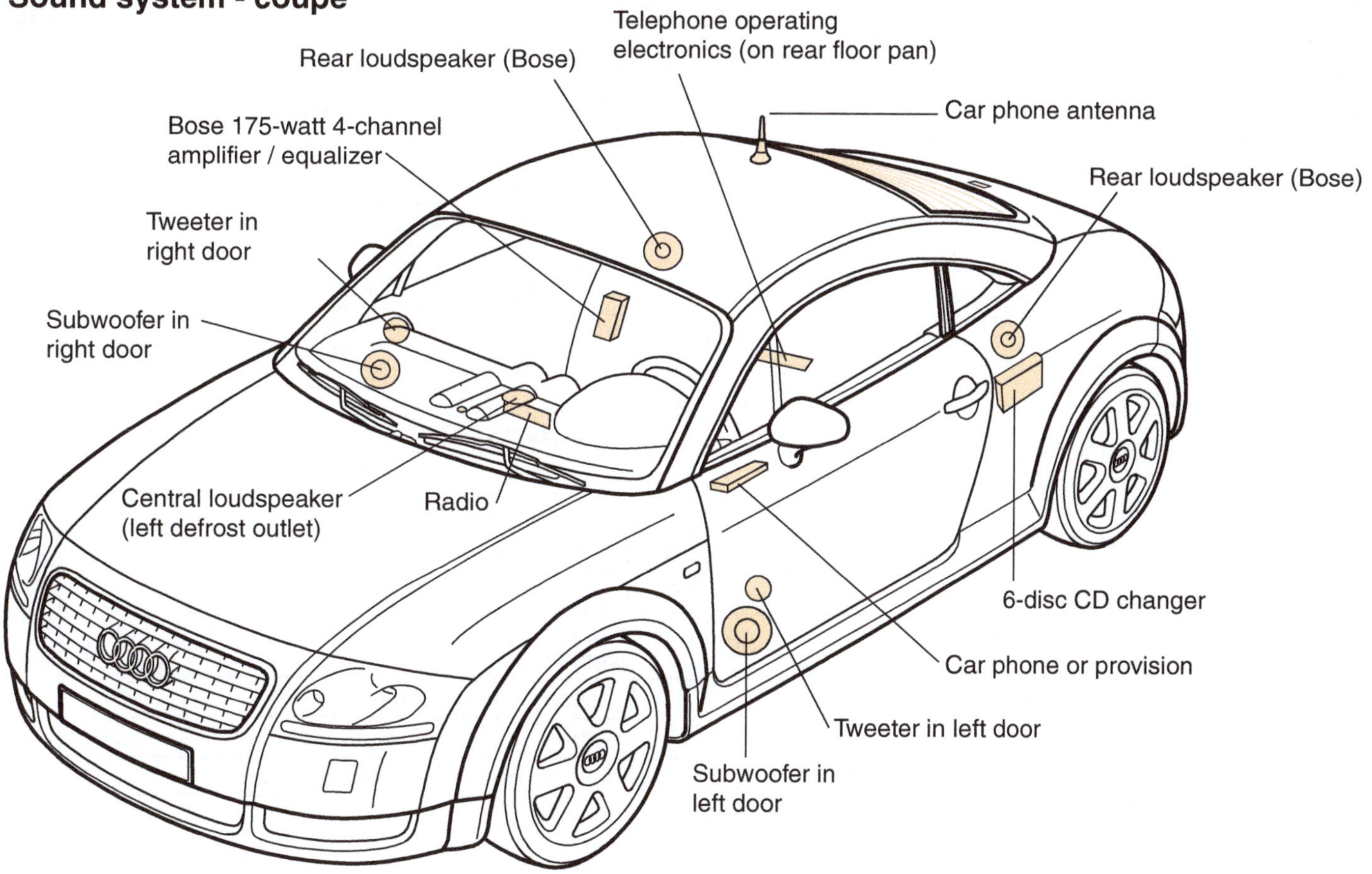

BentleyPublishers.com
©2006 BentleyPublishers.com—All Rights Reserved

Instrument panel overview

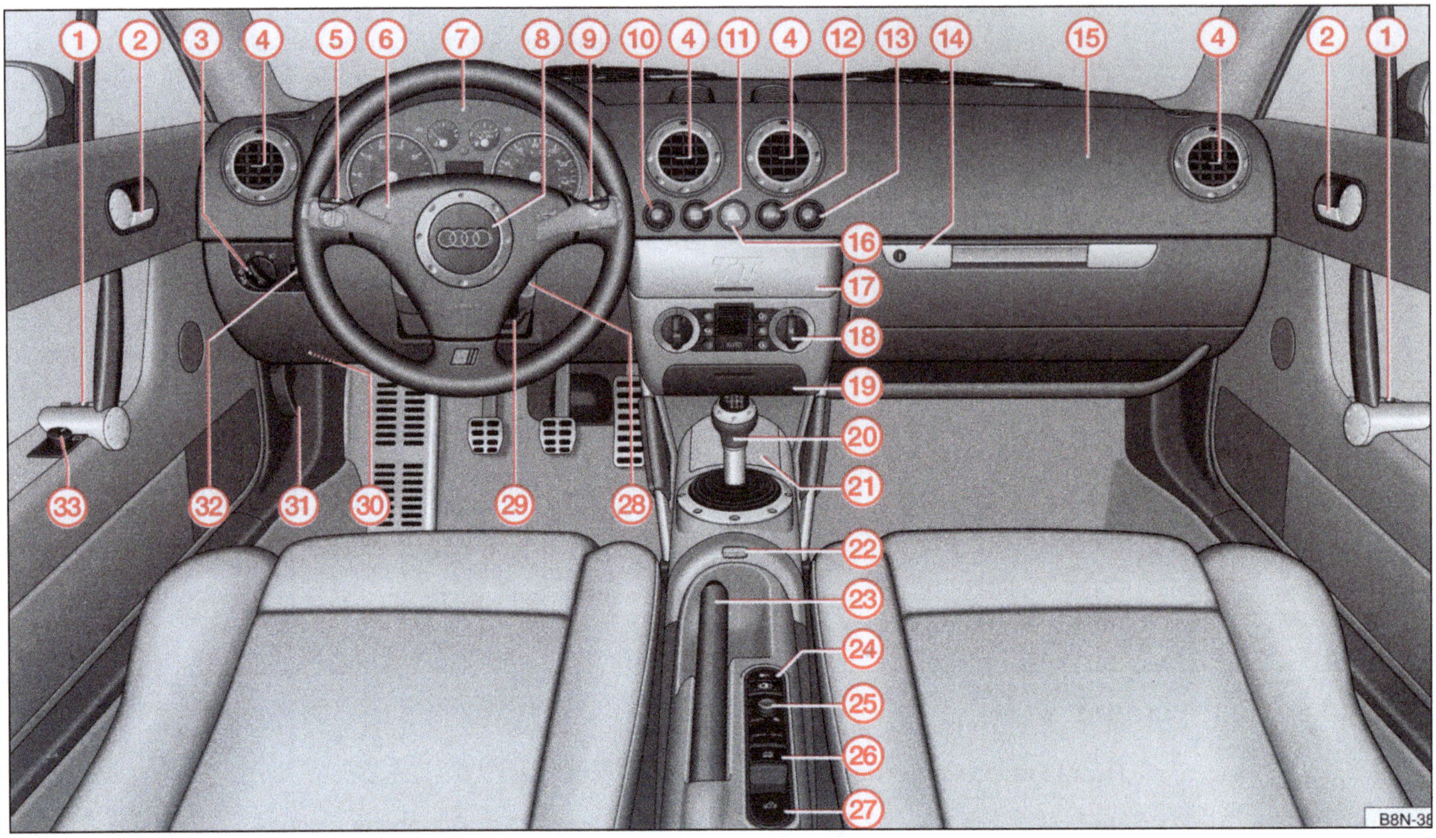

1	Power window switches
2	Door handle
3	Light switch
4	Air vents
5	Turn signals / headlight dimmer switch lever Cruise control
6	Horn
7	Instrument cluster Warning / indicator lights
8	Driver's airbag
9	Windshield wiper / washer lever Trip computer function control switch
10	Seat heater switch (driver's seat)
11	Switch for rear window defogger
12	Electronic Stabilization Program (ESP)
13	Seat heater switch (passenger's seat)
14	Glove compartment (lockable)
15	Passenger's airbag
16	Emergency flasher
17	Radio cover
18	Climate controls
19	Ashtray Cigarette lighter

20	Selector lever or Gearshift lever
21	Fuel filler flap release switch, Rear lid release switch
22	Warning light "PASSENGER AIRBAG OFF"
23	Parking brake lever
24	Power locking switch
25	Storage compartment or Navigation system switch*
26	Switch for wind deflector
27	Switch for automatic folding top
28	Steering lock / ignition / starter switch
29	Adjustable steering wheel
30	Data Link Connector (DLC) for On Board Diagnostics (OBD II)
31	Release lever for engine hood
32	Instrument illumination
33	Adjuster control for outside mirrors

Notes

- Some features mentioned are standard equipment on some models only and options on others.
- For technical reasons, the arrangement of the switches and indicator/warning lights may differ. The governing factor is the symbol on the switch itself.

BentleyPublishers.com
© 2006 BentleyPublishers.com—All Rights Reserved

03 Maintenance

MAINTENANCE TABLES

General

All maintenance work described in this repair group is important and should be performed at the specified time or mileage interval. The Owner's Manual, Maintenance Record, and Warranty Booklet originally supplied with the vehicle contain maintenance schedules that apply to your Audi. Following these schedules will ensure safe and dependable operation. In addition, many maintenance procedures are necessary to maintain warranty protection.

NOTE—
- *Audi continuously updates their recommended maintenance procedures and requirements. The information contained here may not include updates or revisions made by Audi since publication of the Owner's Manual, Maintenance Record, and Warranty Booklet supplied with the vehicle. If there is any doubt about what procedures apply to a specific model or model year, or what intervals should be followed, consult an authorized Audi dealer for the latest maintenance information. Current Maintenance Schedules are included at the end of this section.*

WARNING—
- *Disconnecting the negative (–) battery cable may erase fault codes and basic settings in the engine management and automatic transmission control modules along with settings in other computer-controlled systems. Some driveability problems may be noticed until the system re-adapts to operating conditions. OBD II readiness codes, which may be required for emissions testing, may also be erased. Convenience electronics (alarm system, interior light control, power locks, mirrors, and windows) may need to be re-set using a VAG 1551/1552, VAS 5051/5052 or equivalent scan tool or scan tool program.*

How To Use This Manual

This manual is divided into 12 main sections, or partitions:

0 General, Maintenance
1 Engine
2 Engine Management, Exhaust and Engine Electrical
3 Clutch, and Transmission
4 Suspension, Brakes and Steering
5 Body–Assembly
6 Body–Components and Accessories
7 Body–Interior Trim
8 Heating and Air Conditioning
9 Electrical System
EWD Electrical Wiring Diagrams
OBD On Board Diagnostics

A master listing of the 12 partitions and the corresponding specific repair groups can be found on the inside front cover of this manual.

Thumb tabs are used on the first page of each repair group to help locate the groups quickly. Page numbers throughout the manual are organized according to the repair group system. A comprehensive **Index** can be found on the last pages of the manual.

BentleyPublishers
.com
© 2006 BentleyPublishers.com—All Rights Reserved

Warnings, Cautions and Notes

Throughout this manual, there are numerous paragraphs with the headings **WARNING, CAUTION** or **NOTE**. These very important headings have different meanings.

> ***WARNING—***
> - *Text under this heading warns of unsafe practices that are very likely to cause injury, either by direct threat to the person(s) doing the work or by increased risk of accident or mechanical failure while driving. Warnings are always contained in a box.*

> ***CAUTION—***
> - *Text under this heading also calls attention to important precautions to be observed during the repair work that will help prevent accidentally damaging the car or its parts.*

NOTE—
- *A note contains information, tips, or pointers that will help in doing a better job and completing it more easily.*

Read every **WARNING, CAUTION** and **NOTE** before you begin any maintenance or repair job.

Work safety

Although a modern automobile presents many potential hazards, common sense and good equipment can help ensure safety. Many accidents happen because of carelessness. Pay attention and stick to these few important safety rules.

> ***WARNING—***
> - *Never run the engine in the work area unless it is well-ventilated. The exhaust should be vented to the outside. Carbon monoxide (CO) is an odorless and colorless gas. Carbon monoxide (CO) in the exhaust kills.*
> - *Remove all neckties, scarfs, loose clothing, and jewelry when working near running engines or power tools. Tuck in shirts. Tie back long hair and secure it under a cap. Severe injury can result from these things being caught in rotating parts.*
> - *Remove rings, watches, and bracelets. Aside from the dangers of moving parts, metallic jewelry conducts electricity and may cause shorts, sparks, burns, or damage to the electrical system when accidentally contacting the battery or other electrical terminals.*
> - *Disconnect the battery negative (–) (GND) cable whenever working on or near the fuel system, airbag system or anything that is electrically powered. Accidental electrical contact may damage the electrical system, cause fire, or result in serious personal injury.*
> - *Never work under a lifted car unless it is solidly supported on jack stands that are intended for the purpose. Do not support on cinder blocks, bricks, or objects that may shift or crumble under continuous load. Never work under a car that is supported only by a jack.*

BentleyPublishers
.com
© 2006 BentleyPublishers.com—All Rights Reserved

> ***CAUTION—***
> - *The fuel system retains pressure even when the ignition is off. Loosen the fuel lines very slowly to allow the residual pressure to dissipate gradually. Cover fittings with a suitable shop cloth to avoid spraying fuel.*
> - *Fuel is highly flammable. When working around fuel, do not smoke or work near fire hazards. Keep an approved fire extinguisher handy. Be aware that water and building heaters may have standing pilot lights.*
> - *Illuminate the work area adequately and safely. Use a portable safety light for working inside or under the car. A fluorescent type light is best because it gives off less heat. If using a light with a normal incandescent bulb, use rough service bulbs to avoid breakage. The hot filament of an accidentally broken bulb can ignite spilled fuel or oil.*
> - *Keep sparks, lighted matches, and open flame away from the top of the battery. Hydrogen gas emitted by the battery is highly flammable. Any nearby source of ignition may cause the battery to explode.*
> - *Never lay tools or parts in the engine compartment or on top of the battery. They may fall and be difficult to retrieve, become caught in belts or other rotating parts, or cause electrical shorts and damage to the electrical system.*

IDENTIFICATION PLATES AND LABELS

Vehicle Identification Number (VIN)

Vehicle year, model, engine type, along with other pertinent data can be decoded from the Vehicle Identification Number (VIN). Data plates and stickers containing the VIN are attached to the vehicle at several locations.

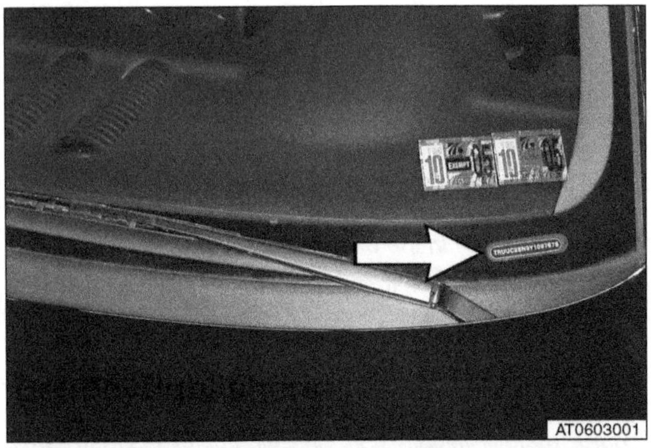

AT0603001

◄ United States Federal law requires VIN information to be visible from outside the vehicle and at a standard location. A plate with the complete 17 digit VIN is attached to the padded dash at the base of the windshield (**arrow**) and is visible through a mask in the glass from the outside. See **01 VIN Identifier** in the front of this manual for detailed VIN decoding information.

AT0603002

◄ VIN is also located in the engine compartment on the bulkhead at center (**arrow**) and is visible through a plastic window. A rubber weather strip and molding retain the window panel and may obscure the VIN. Clean the window or lift up weatherstrip and pull panel aside for access.

BentleyPublishers
.com
© 2006 BentleyPublishers.com—All Rights Reserved

AT0603003

Door “B” pillar certification sticker (**arrow**) lists full VIN, vehicle weight ratings, production date and emission compliance certification.

AT0603004

Vehicle data sticker (**left arrow**) is located in the luggage compartment on left side of spare tire well. The data plate contains the following vehicle information:

- Production control number
- Vehicle identification number
- Model identification number
- Model explanation/engine output
- Engine and transmission code letters
- Paint number/interior equipment identification number
- Optional equipment identification numbers

Note spare tire use and warning label (**right arrow**) attached to stowed tire/rim.

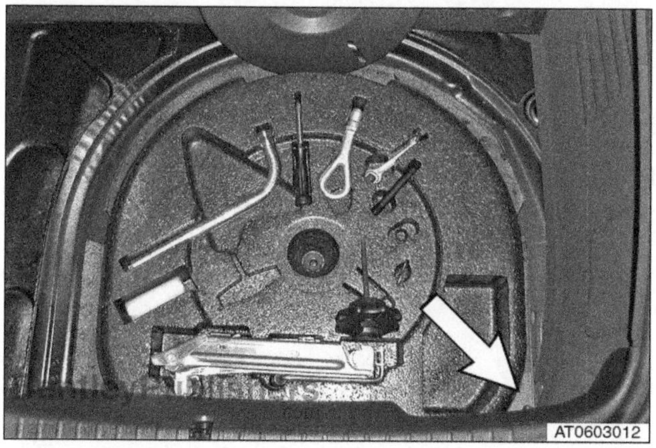

AT0603012

Some vehicles will have the vehicle data sticker on the other side of the spare tire well (**arrow**).

AT0603005

A tire pressure inflation sticker is attached to the fuel filler door on the right rear fender. Recommended cold tire inflation pressures will vary from car to car due to differing equipment levels. Later production vehicles may have tire inflation pressure sticker on the driver’s side “B” pillar.

BentleyPublishers
.com
© 2006 BentleyPublishers.com—All Rights Reserved

Engine identification

The engines used in vehicles covered by this manual are identified by a three letter code followed by a six digit serial number such as: BGP 001 188. Engine code and engine serial number are located on the engine block. A sticker with engine code and serial number is often attached to the upper timing belt cover or cylinder head cover. In addition, engine code may be stamped into the cylinder head to allow for quick identification.

Engine Type	Engine Codes
1.8L 4-cylinder 180-hp turbo gasoline	ATC, AWP
1.8L 4-cylinder 225-hp turbo gasoline	AMU, BEA
3.2L 6-cylinder 250-hp gasoline	BHE

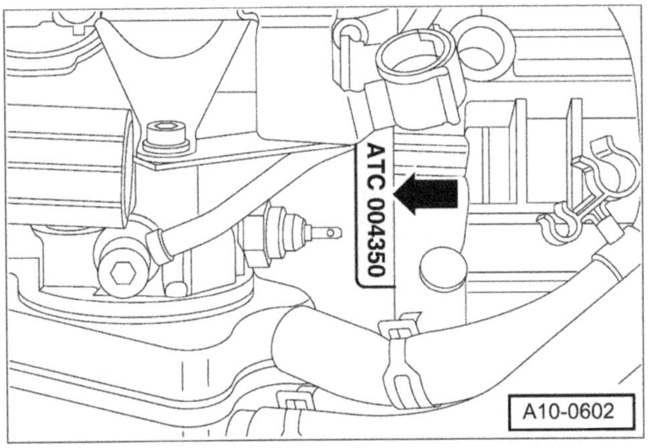

◀ **1.8L 4-cylinder 180-hp** engine code letters and serial numbers (**arrow**) are located at rear of cylinder block. This information is stamped onto a machined area on the cylinder block casting near the oil filter flange close to the joint between engine and transmission.

◀ **1.8L 4-cylinder 180-hp** engine code and serial number can also be found on a sticker (**upper arrow**) on upper section of timing belt guard. 3-digit engine code is stamped on machined surface on cylinder head (**lower arrow**). Sound absorber panels must be removed to completely see this sticker.

NOTE—
- *Information on sticker generally includes engine code, serial number, and build date.*

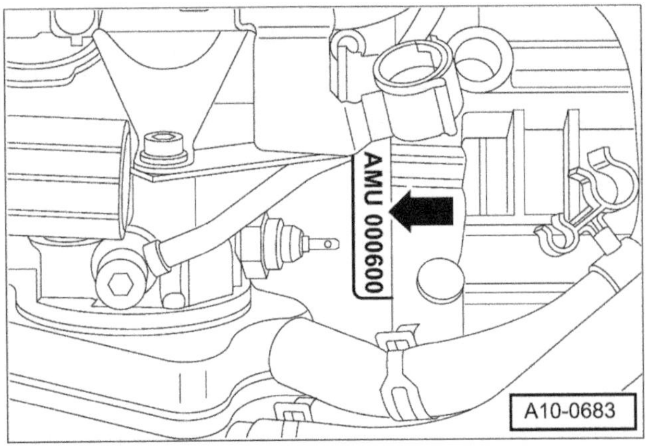

◀ **1.8L 4-cylinder 225-hp** engine code letters and serial numbers (**arrow**) are located at rear of cylinder block. This information is stamped onto a machined area on the cylinder block casting near the oil filter flange close to the joint between engine and transmission.

BentleyPublishers.com
©2006 BentleyPublishers.com—All Rights Reserved

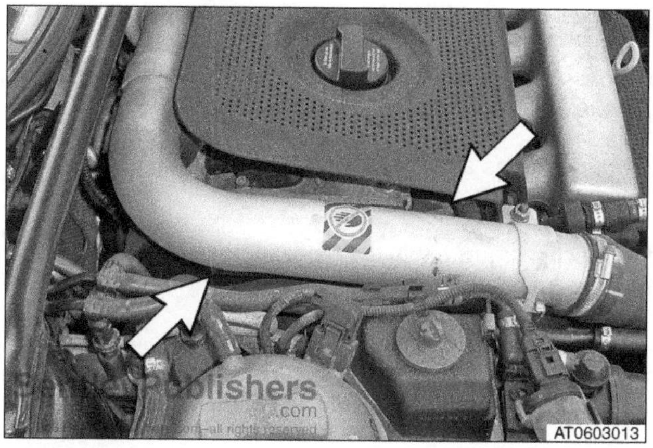

1.8L 4-cylinder 225-hp engine code and serial number can also be found on a sticker (**left arrow**) on upper section of timing belt guard. 3-digit engine code is stamped on machined surface on cylinder head (**right arrow**). Sound absorber panels must be removed to completely see this sticker.

NOTE—

- *Information on sticker generally includes engine code, serial number, and build date.*

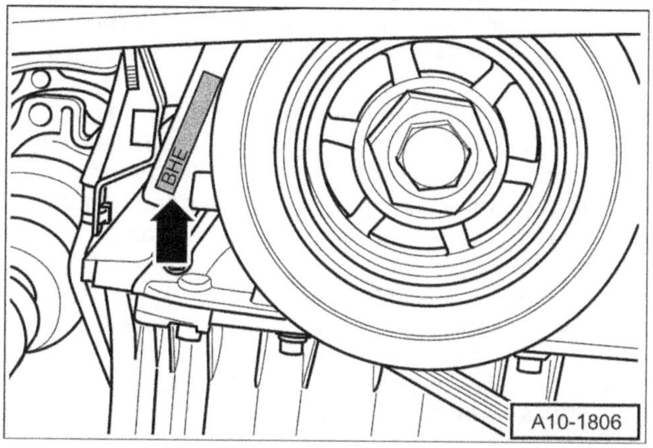

3.2L 6-cylinder 250-hp engine code letters and serial numbers are stamped into a machined area of cylinder block on the back side at the rear near the engine/transmission joint (**arrow**).

3.2L 6-cylinder 250-hp engine code information can be found on a sticker that is intermittently installed on the engine. When installed, the sticker will be attached to the intake manifold (**left arrow**) or the vapor separator near the oil fill cap (**right arrow**).

NOTE—

- *Information on sticker includes engine code and serial number.*

BentleyPublishers
.com
©2006 BentleyPublishers.com—All Rights Reserved

Transmission identification

The transmissions and rear differentials used in vehicles covered by this manual are identified by type, code, and number.

The **type** number denotes a family of transmissions that have generally similar characteristics.

The **code** is a specific group within the type and identifies individual traits such as gear ratios.

The **number** is a five digit string such as 12025. The digits indicate transmission build date and all transmissions built on any given day will have the same number.

The code and number are stamped together into a machined area or onto a data plate riveted to the transmission. Both code and number make up what is generally referred to as the "transmission code".

A typical "transmission code" of HAH 30604 is broken down as follows: the code is HAH and it was manufactured on the thirtieth day of the sixth month of 2004 (30-06-04). The type is 02E as noted in the specific transmission repair groups of this manual and by noting the identification marks cast into the transmission housings.

Five different transmissions types are available as described below. Each transmission type comes in numerous versions, depending on engine type. For a listing of various transmission codes and specifications see **34a Manual Transmission**, **34b DSG Transmission** or **37 Automatic Transmission**.

Transmission	Transmission type
5-speed manual, front wheel drive	02J
5- and 6- speed manual, all wheel drive (quattro)	02M
6-speed direct shift automatic (DSG), all wheel drive (quattro)	02E
6-speed automatic, front wheel drive	09G
Rear final drive, all wheel drive (quattro)	02D

5-speed manual, front wheel drive (02J) code and number is stamped into machined area of case near shift mechanism (**arrow**). Transmission type information is cast into the housing at several locations.

5-and 6-speed manual, all wheel drive (02M) code and number is stamped into machined area on top of case (**arrow**). Transmission type information is cast into housing at several locations including area near left side drive flange. Removal of air filter assembly allows access to top of transmission.

BentleyPublishers
.com
©2006 BentleyPublishers.com—All Rights Reserved

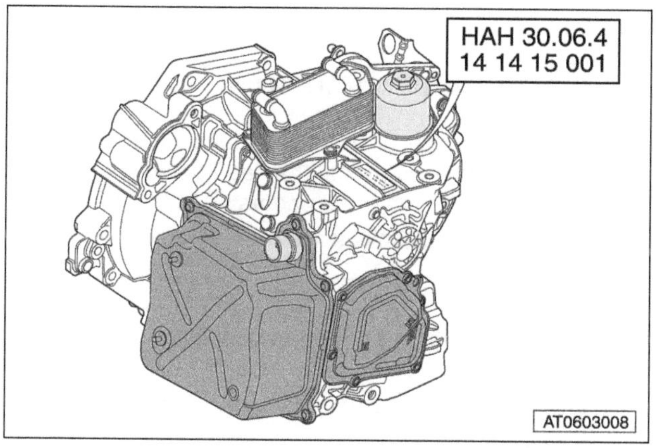

6-speed direct shift automatic, all wheel drive (02E) code and number is stamped into machined area on top of case near oil cooler (**arrow**).

NOTE—

- *Transmission numbers for direct shift gearboxes do not follow the usual conventions. The production date is listed on the same line as the code in the format dd.mm.yy. The second line data includes a plant code, time code, and serial number.*

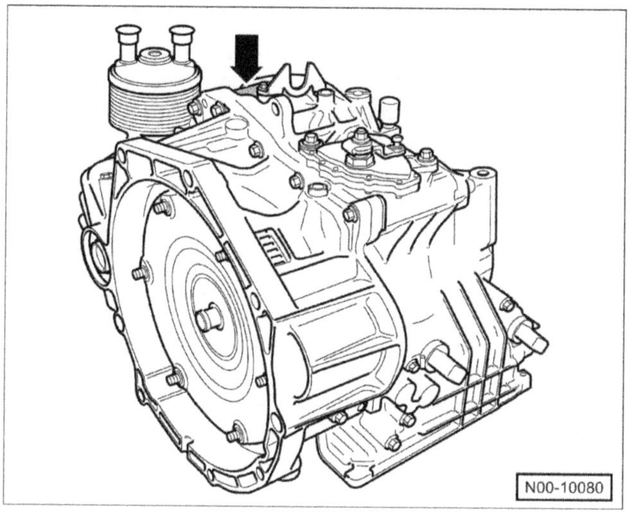

6-speed automatic, front wheel drive (09G) code is stamped into machined area of case near oil cooler. Other numbers may be stamped or cast into the case but are not relevant to overall code.

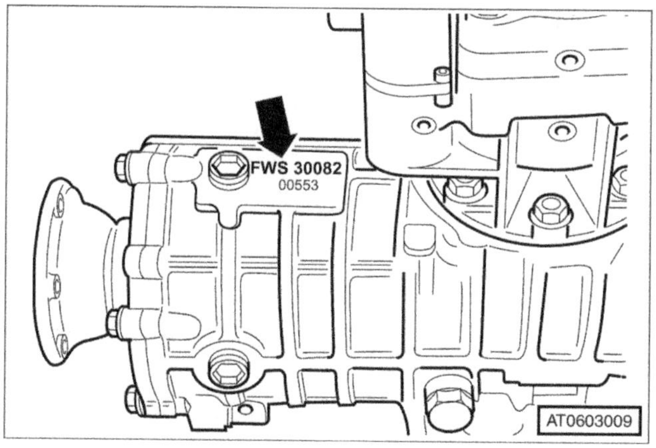

Rear differential, all wheel drive, all quattro (02D) code is stamped into machined area of case near filler plug and may be partially obstructed by mountings. Other numbers may be stamped or cast into the case, but are not relevant to overall code.

BentleyPublishers
.com
©2006 BentleyPublishers.com—All Rights Reserved

Towing

Emergency towing with commercial equipment

AT0603010

The following information is to be used by commercial flatbed tow truck operators who know how to operate their equipment safely. To prevent damage to the vehicle, read and understand all of the following information before proceeding.

> ***WARNING —***
> - *All Audi TTs require loading and transport on flatbed equipment.*
> - *Do not use conventional sling-type equipment or wheel dollies.*
> - *Improper attachment of vehicle to the flatbed carrier may cause damage to underbody components.*
> - *Never allow passengers to ride in a towed vehicle for any reason.*
> - *To reduce the risk of accident and serious personal injury, always stay within manufacturer's rated equipment capacities. Exceeding manufacturer's design specifications is dangerous.*

Towing eyes

Front towing eye mount is located at right front of vehicle. A detachable towing eye is stored in the vehicle tool kit.

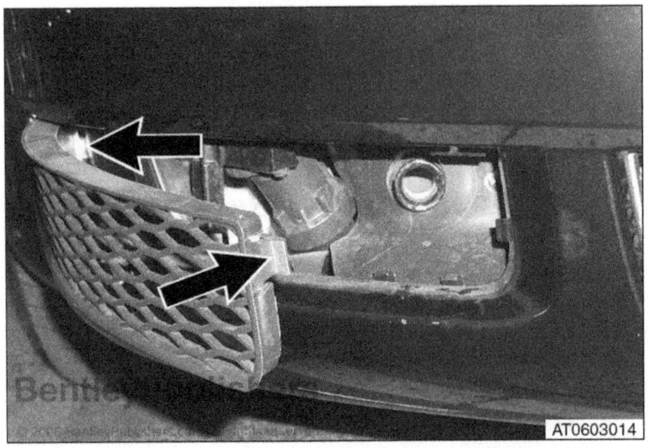
AT0603014

◄ Working at the passenger's side front of the vehicle, remove cover by carefully inserting a screwdriver between grille and bumper cover and releasing clip (**lower arrow**). Gently pry grille out and away from bumper cover.

NOTE —
- *Cover pivots (hinges) from opposite side (**upper arrow**).*

– Locate towing eye in tool kit.

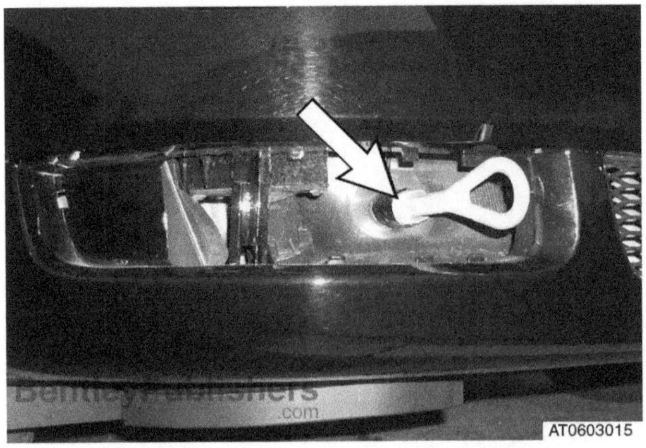
AT0603015

◄ Thread towing eye (**arrow**) fully into mounting hole and tighten with wheel lug wrench (also from tool kit).

– After use, unscrew towing eye, return to vehicle tool kit, and stow tool kit in luggage compartment.

– Reinstall cover.

BentleyPublishers
.com
© 2006 BentleyPublishers.com—All Rights Reserved

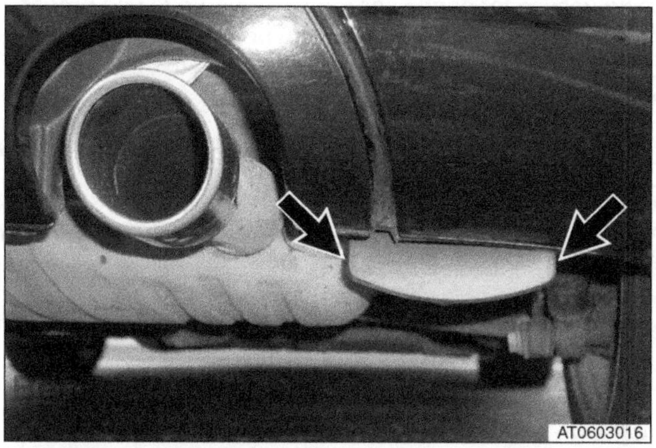

The rear towing eye mount is concealed behind a plastic cover at the passenger's side rear. Grasp both edges of cover (**arrows**), pull securing tabs out, and pull cover straight out to access mounting hole for the towing eye.

Thread towing eye fully into rear mounting hole (**arrow**) and tighten with wheel lug wrench (also from tool kit).

– After use, unscrew towing eye, return to vehicle tool kit, and stow tool kit in luggage compartment.

– Reinstall cover.

LIFTING VEHICLE

Raising with hoist, lift, or floor jack

> ***WARNING —***
> - *Before driving a vehicle onto a lift, jack or hoist, confirm that vehicle weight does not exceed the allowable lifting capacity.*
> - *If work is to be performed under vehicle while it is raised, it must be supported by suitable jack stands.*
> - *Lift vehicle only at points indicated in order to avoid damaging vehicle floor pan and to prevent tipping.*
> - *Never start engine and engage a gear with vehicle lifted if any drive wheel has contact with the floor, this is especially important for vehicles with all wheel drive (quattro). There is danger of an accident due to possible vehicle movement.*
> - *Exercise care while running the engine with the vehicle on a lift, hoist, or jack. Engine vibration and vehicle movement could cause the vehicle to slip off of the lift, hoist, or jack.*
> - *Before driving on to a vehicle hoist or lift, be sure there is enough clearance under vehicle so that vehicle does not drag the lift arms. Pay particular attention to front spoilers and aprons, and all wheel drive system components.*
> - *Observe all lift, hoist, or floor jack manufacturers safety instructions and guidelines before raising vehicle.*

BentleyPublishers
.com
© 2006 BentleyPublishers.com—All Rights Reserved

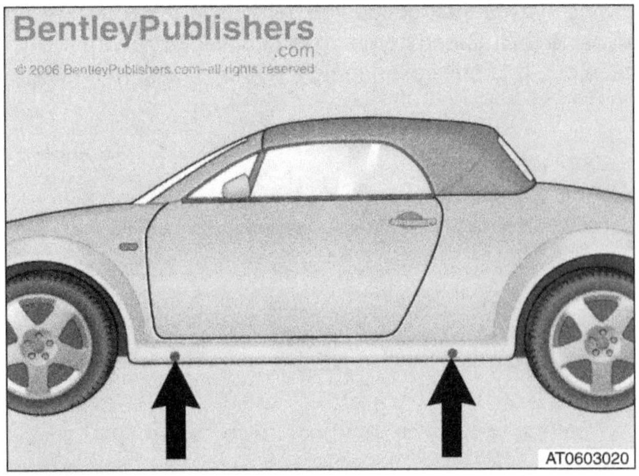

When raising vehicle for service or repairs, be sure to use proper lifting/jacking points. There are two jacking points (**arrows**) on each side. Each is on a reinforced body seam and is marked by an indentation in the outside lower (rocker) panel sheet metal.

NOTE —

- *Lower sheet metal (rocker) panels are sprayed with a polyester coating to protect against damage from road debris and stones. Coating may make indentations in sheet metal panels difficult to see.*
- *Depending upon lift configuration, it may be necessary to "flip up" hoist pads. This may be required to prevent damage to sheet metal panels from hoist lifting pad.*
- *Where available, the use of rubber-coated lift pads is highly recommended.*
- *Plastic aerodynamic panels cover some of the bottom area, especially towards the rear. Use care during lifting/hoisting to avoid damage.*

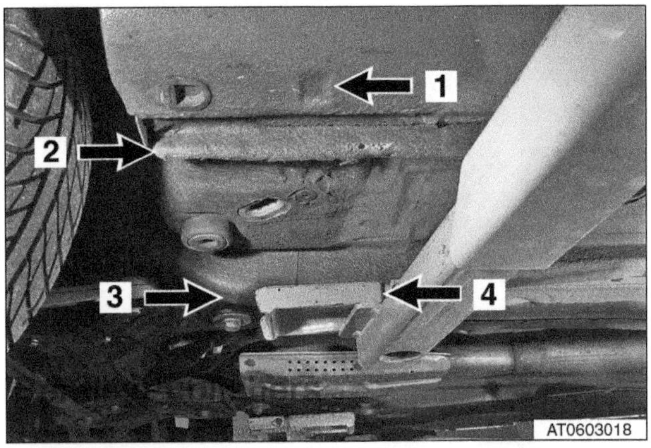

At the front of the vehicle, points for lifting include the above-noted body seam (**arrow 2**) indicated by lift point indentation (**arrow 1**). Body seam is reinforced at this point with additional stiffener plate. Longitudinal chassis member under vehicle (**arrow 3**) is suitable for a floor jack or hoist pad (**arrow 4**). Left front of vehicle shown, right side is similar.

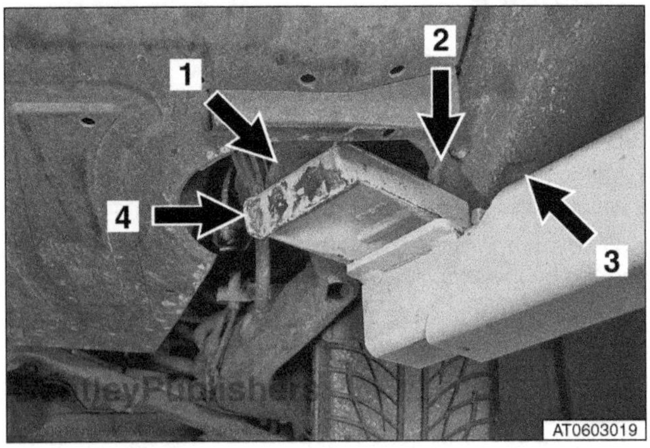

To lift rear of vehicle, place floor jack or hoist at appropriate rear lifting point on seam as shown (**arrow 1**). Body seam is reinforced at this point with additional stiffener plate (**arrow 2**) in the same manner as front. Note indentation in sill denoting lift point (**arrow 3**). Left rear of vehicle shown, right side is similar.

WARNING —

- *Do not lift vehicle at engine oil pan, transmission, fuel tank, rear final drive or axles. Serious damage may result.*

– Place vehicle on a flat level area with a surface capable of supporting weight of vehicle, jack, and jackstands, e.g. concrete.

– Block wheels to prevent vehicle from rolling.

– Place jack at proper lifting points as shown.

– Operate jack or hoist and raise vehicle slowly.

– If working under vehicle, support vehicle with approved jack stands.

WARNING —

- *Vehicle underbody may be coated with a slippery wax anti-rust material. Clean area prior to supporting with jackstands.*
- *Do not use concrete blocks or similar materials to support the vehicle due to the danger from crumbling.*

BentleyPublishers
.com
© 2006 BentleyPublishers.com—All Rights Reserved

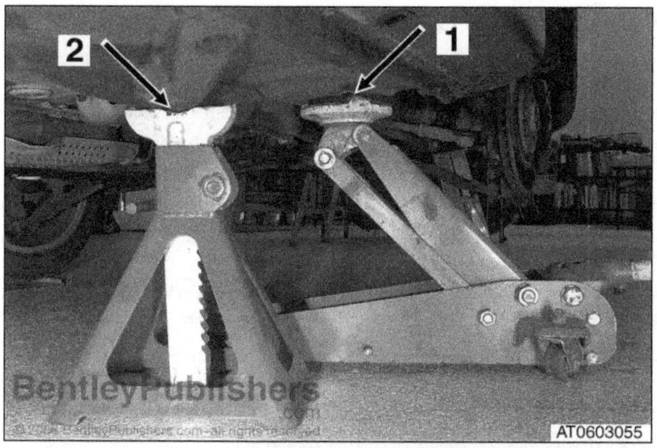

In the front, place floor jack with rubber pad (**1**) under front or along reinforced seam in line with lift point on sill. Place jack stand under longitudinal (fore-aft) reinforcement (**2**). Right front shown, left front is similar.

NOTE —

- *Floor jacks and jack stands with rubber coated lifting pads generally reduce the chance of slipping and damage.*
- *If working under vehicle while on a hoist or lift, place jack stand under solid area of underbody. Do not place under springs, plastic aero panels, or sheet metal panels or shields.*
- *Use care when lifting vehicle on arms attached to vehicle coil springs.*

In the rear, lift vehicle with floor jack under suspension (**1**) and place jack stand along reinforced seam (**2**) in line with lift point on sill (**3**). Use care when lifting vehicle on arms attached to coil springs. Right rear shown, left rear is similar.

Working under the vehicle

When working under any vehicle, the following points should be observed as a matter of safety and good practice:

- Disconnect negative (GND) battery cable so that no one else can start vehicle. Let others know that you will be under vehicle.
- Place at least two jack stands under vehicle. A jack is a temporary lifting device and should not be used alone to support vehicle while you are under it. Use positively locking jack stands that are designed for the purpose of supporting a vehicle.
- If you are using a hoist, be sure that safety locks are engaged and that the weight of the vehicle is resting on the locks, not the hydraulic system.
- Lower vehicle slowly until its weight is fully supported by jack stands or hoist. Watch to make sure that jack stands do not tip or lean as vehicle settles on them, and that jack stands and hoist arms are placed solidly and will not move.
- Check to make sure vehicle is stable before working under car.
- Observe all jacking precautions again when raising vehicle to remove jack stands.

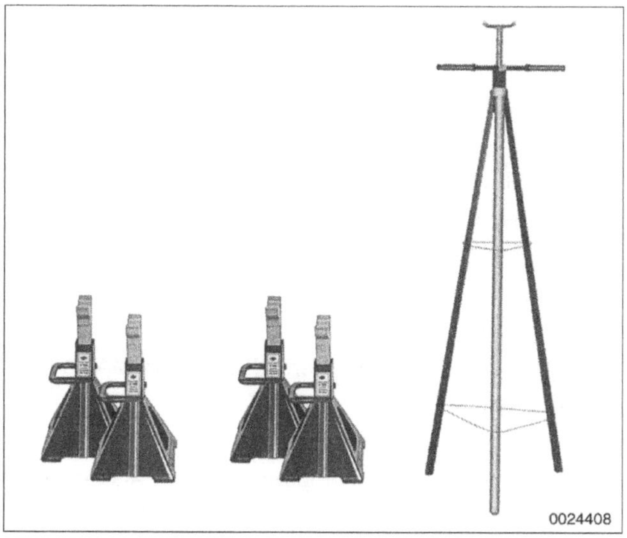

WARNING —

- *Use care when removing major (heavy) components from one end of the vehicle. The sudden change in weight and balance can cause vehicle to tip off of hoist or jack stands.*
- *Do not support vehicle at engine oil pan, transmission, fuel tank, or on front or rear axle as serious damage may result.*

BentleyPublishers
.com
© 2006 BentleyPublishers.com—All Rights Reserved

Engine Compartment

Visually inspect engine and engine compartment for leaks. Inspect lines, hoses, and connections for fuel, brake, cooling, heating, air conditioning systems for leaks, worn areas, porosity, and deterioration.

Fluid levels must also be between specified MIN and MAX marks on appropriate reservoirs and tanks.

The following illustrations show representative engine compartment layouts for the three major engine types. There may be minor differences between model years, such as engine sound absorber configuration, but these slight variations will be readily apparent and understood. Passenger side sound absorber panel has been removed to show power steering fluid reservoir.

1.8L 4-cylinder 180-hp

1. Coolant expansion reservoir
2. Power steering fluid reservoir
3. Engine oil filler cap
4. Engine oil dipstick
5. Brake fluid reservoir (not visible, behind and under MAF sensor)
6. Battery and main fuse box (under plastic cover)
7. Windshield washer fluid reservoir

1.8L 4-cylinder 225-hp

1. Coolant expansion reservoir
2. Power steering fluid reservoir
3. Engine oil filler cap
4. Engine oil dipstick
5. Brake fluid reservoir (not visible, behind and under MAF sensor)
6. Battery and main fuse box (under plastic cover)
7. Windshield washer fluid reservoir

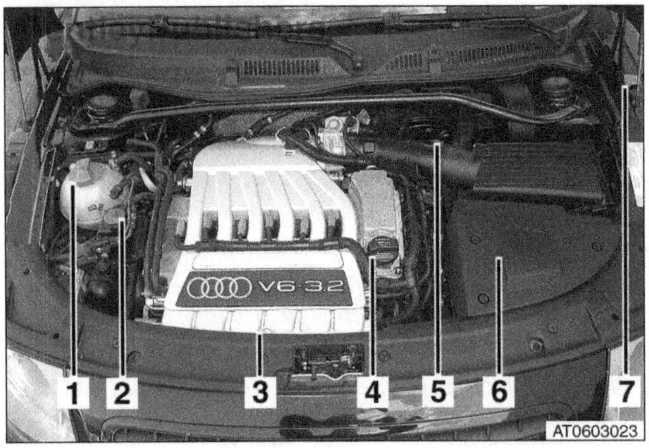

3.2L 6-cylinder 250-hp

1. Coolant expansion reservoir
2. Power steering fluid reservoir
3. Engine oil dipstick
4. Engine oil filler cap
5. Brake fluid reservoir (not visible, behind and under MAF sensor)
6. Main fuse box (under plastic cover)
7. Windshield washer fluid reservoir

* Battery is under rear luggage compartment floor. AGM-type battery is completely sealed and requires no maintenance.

BentleyPublishers.com
© 2006 BentleyPublishers.com—All Rights Reserved

Sound absorber panels, removing and installing

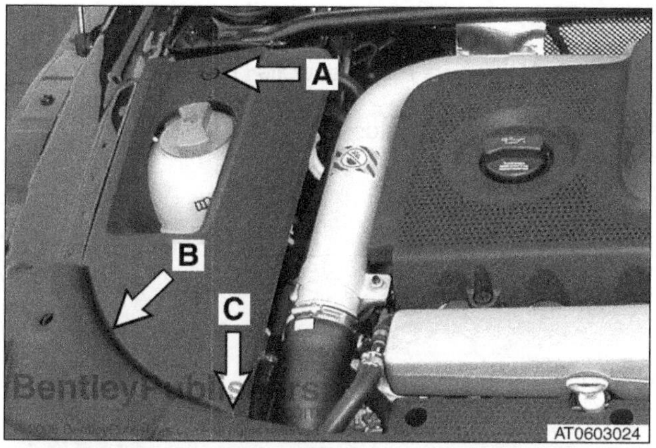

AT0603024

All engines, passenger side sound absorber panel

– All engine versions use a cover on the passenger's side in the engine compartment. Remove plastic head bolt (**arrow A**). Pull panel up from the rear, unhook along side (**arrow B**), and slide tab out of adjacent panel (**arrow C**).

– This panel must be removed for access to power steering fluid reservoir.

AT0603025

1.8L 4-cylinder 180-hp, upper

– Release retainers (**arrows**) by turning approximately ¼-turn counter-clockwise. Lift upper sound absorber panel off.

– Carefully position panel and with retainers onto brackets. Turn retainers approximately ¼-turn clockwise to secure.

NOTE —

• Retainers are clipped into panel and should stay with it.

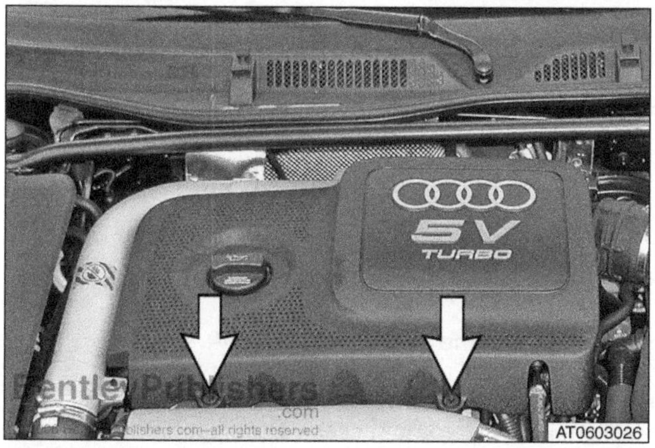

AT0603026

1.8L 4-cylinder 225-hp, upper

– Release retainers (**arrows**) by turning approximately ¼-turn counter-clockwise. Lift upper sound absorber panel off.

– Carefully position panel and with retainers onto brackets. Turn retainers approximately ¼-turn clockwise to secure.

NOTE —

• Retainers are clipped into panel and should stay with it.

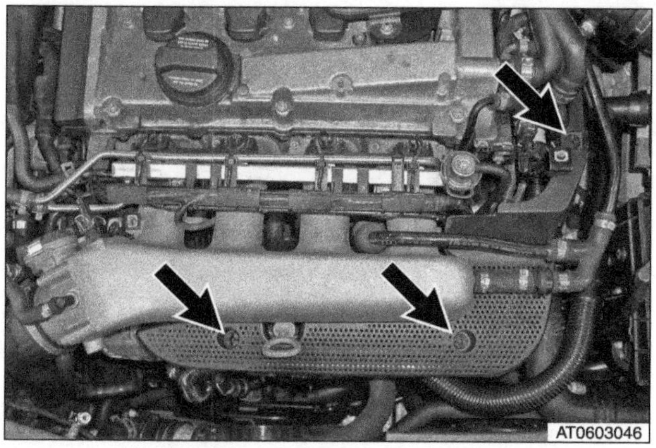
AT0603046

1.8L 4-cylinder 180-hp and 225-hp engines, upper

– After removal of upper sound absorber panel, remove additional panel. Release retainers (**arrows**) by turning approximately ¼-turn counter-clockwise. Lift additional sound absorber panel around hoses/lines and off.

– Carefully position panel and with retainers onto brackets. Turn retainers approximately ¼-turn clockwise to secure

NOTE —

• Retainers are clipped into panel and should stay with it.

BentleyPublishers
.com
©2006 BentleyPublishers.com—All Rights Reserved

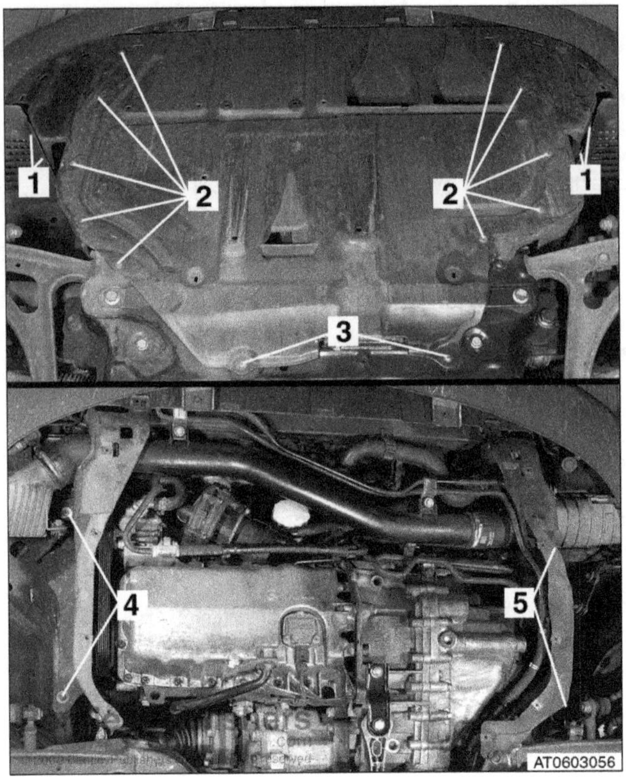

1.8L 4-cylinder 180-hp and 225-hp engines, lower

- Loosen plastic ¼-turn fasteners at wheel wells (**1**).
- Remove screws along both lower sides (**2**).
- Remove screws at the rear (**3**).
- Pull panel down slightly at rear and side rearward to remove.
- Installation is the reverse of removal.
- When removing lower side panels, remove large lower panel first.
- Remove speed nuts along the passenger's side (**4**).
- Remove speed clips along the driver's side (**5**). It may be necessary to loosen headlight level sensor for access if equipped.
- Pull panels down and out.
- Installation is the reverse of removal.

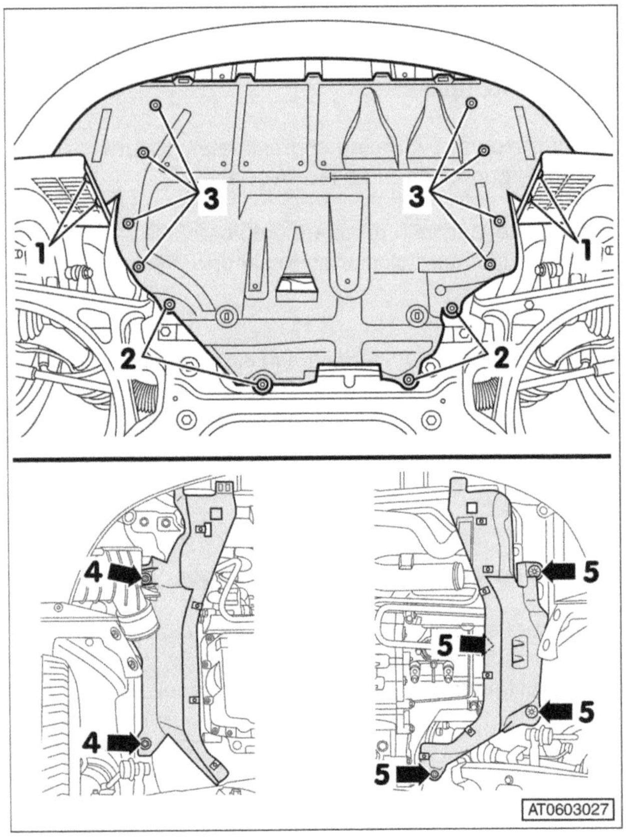

3.2L 6-cylinder 250-hp engine, lower

- Remove screws or loosen plastic ¼-turn fasteners at wheel wells (**1**).
- Remove screws along both lower sides (**3**).
- Remove screws at the rear (**2**).
- Pull panel down slightly at rear and side rearward to remove.
- Installation is the reverse of removal.
- When removing lower side panels, remove large lower panel first.
- Remove speed nuts along the passenger's side (**4**).
- Remove speed clips along the driver's side (**5**). It may be necessary to loosen headlight level sensor for access if equipped.
- Pull panels down and out.
- Installation is the reverse of removal.

BentleyPublishers
.com
©2006 BentleyPublishers.com—All Rights Reserved

All vehicles, rear sound absorber/aero panel(s)

- Remove speed clips (**1**).
- Remove speed nuts (**2**).
- Lower panel down.
- Installation is the reverse of removal.

Fluid leaks, visual inspection

Remove upper and/or lower sound absorber panels as required and check engine compartment for signs of fluid leaks. Fluid leaks attract dust making them easier to spot. Many expensive repairs can be avoided by prompt repair of minor fluid leaks.

Visually inspect for oil and ATF leaks on engine and transmission. Also inspect cooling, fuel, heating and air conditioning systems for leakage. Visually inspect hoses and hose connections for leaks, worn areas, porosity and brittleness. Also see **Fluid and exhaust leaks, visual inspection** given in this repair group.

ENGINE OIL

Lubricating oil for the engines covered by this manual must meet certain quality and performance standards as specified by Audi. These standards are sufficiently high so that generally, only synthetic oils will meet them.

Engine oils meeting specified performance standards are available from Audi and several other sources.

Typical oil container label showing approvals (**A**) and viscosity (**B**). If the container of the oil in question is not marked with appropriate specification(s), the manufacturer advises that it must be assumed that the oil is not suitable for the application. Use of oils other than those specified may cause engine damage. Oil specifications and capacities differ between model year and engine type; consult the following tables.

NOTE —

- *Although some model years do not list a specific Audi oil specification, experience shows that synthetic oils meeting the VW 502 00 specification seem to work best in the long term. This applies to both 4- and 6-cylinder engines.*

BentleyPublishers.com
© 2006 BentleyPublishers.com—All Rights Reserved

Engine oil specifications

1.8L 4-cylinder engines, model years 2000 - 2002	
Audi specification	none listed
Industry specifications	SAE/API SJ or SL
Capacity (approximate)	with filter change 4.5 liters (4.8 qt.)
Viscosity	5W40 (preferred) 5W30

1.8L 4-cylinder engines, model year 2003	
Audi specification	VW 500 00 VW 501 01
Industry specifications	SAE/API SJ or SL
Capacity (approximate)	with filter change 4.5 liters (4.8 qt.)
Viscosity	5W40 (preferred) 5W30

1.8L 4-cylinder engines, model year 2004	
Audi specification	none listed
Industry specifications	SAE/API SJ or SL ACEA A2 or A3
Capacity (approximate)	with filter change 4.5 liters (4.8 qt.)
Viscosity	5W40 (preferred) 5W30

1.8L 4-cylinder engines, model years 2005 - 2006	
Audi specification	VW 502 00
Industry specifications	none listed
Capacity (approximate)	with filter change 4.5 liters (4.8 qt.)
Viscosity	5W40 (preferred) 5W30, 0W40
Oil can be topped-off with industry specification ACEA A3, A5, B5; SAE/API SL; or ILSAC GF-3 to a maximum of 0.5 liter (0.5 quart).	

3.2L 6-cylinder engines, model year 2004	
Audi specification	none listed
Industry specifications	SAE/API SJ or SL ACEA A2 or A3
Capacity (approximate)	with filter change 5.5 liters (5.8 qt.)
Viscosity	5W40 (preferred) 5W30

BentleyPublishers.com
© 2006 BentleyPublishers.com—All Rights Reserved

3.2L 6-cylinder engines, model year 2005 - 2006	
Audi specification	VW 502 00
Industry specifications	none listed
Capacity (approximate)	with filter change 5.5 liters (5.8 qt.)
Viscosity	5W40 (preferred) 5W30, 0W40
Oil can be topped-off with industry specification ACEA A3, A5, B5; SAE/API SL; or ILSAC GF-3 to a maximum of 0.5 liter (0.5 quart).	

Engine oil level, checking

It is normal for any engine to consume a small amount of oil. Oil consumption varies from engine to engine and may change significantly over the life of the engine. The rate of consumption depends on oil quality, viscosity, and operating conditions. Always check oil level at every fuel filling and especially before starting out on a long trip.

– Switch ignition OFF.

– After switching ignition OFF, wait a few minutes to allow oil to flow back into the oil pan.

– Remove dipstick located in front of intake manifold, wipe with a clean cloth and fully re-insert into tube.

– Remove dipstick again and read oil level.

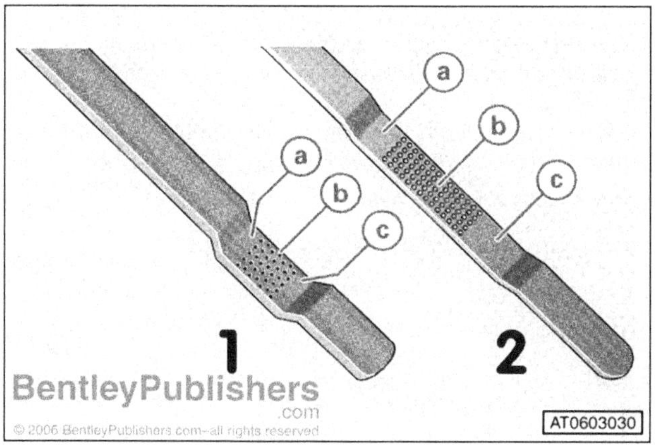

◄ Two basic types of dipsticks are used; both are read in the same way:

- **MAX** mark (**a**). Oil **MUST NOT** be topped off.
- **Normal** operating range (**b**). Oil does not have to be topped off, but can be as long as level does not exceed MAX mark (**a**).
- **MIN** mark (**c**). Oil **MUST** be topped off. After topping off, level is correct when it is within normal area (**b**).

– Engine oil level must not exceed maximum mark (**a**). If oil level is at or below minimum (**c**), fill to maximum mark (**a**).

> ***WARNING—***
> - *Do not overfill. Overfilling can cause misfire Diagnostic Trouble Codes (DTCs) to be stored in Engine Control Module (ECM).*
> - *If engine oil level is above maximum mark, catalytic converter can be damaged.*

– Observe oil specifications given in tables above. Use only synthetic oils. Oils for all engines must meet specific Audi quality standards. Preferred viscosity (weight) for all engines is SAE 5W40. Do not overfill.

BentleyPublishers
.com
© 2006 BentleyPublishers.com—All Rights Reserved

Engine oil and filter, changing (1.8L 4-cylinder engines)

– Warm engine to normal operating temperature.

– Shut off ignition and apply parking brake.

– Raise vehicle and support on jackstands or ramps designed for the purpose or use a lift or hoist.

> ***WARNING —***
> - *Jack stand and ramps should be placed on hard level surface (e.g. concrete).*

– Remove lower sound absorber panel.

◄ Position suitable drain pan under engine, remove drain plug from rear of oil pan (**arrow**) and allow oil to drain into pan.

> ***WARNING —***
> - *Hot oil can scald. Wear protective clothing, gloves and eye protection.*

– When oil has completely drained, install new drain plug and tighten.

Tightening torque	
Drain plug to oil pan (always replace)	30 Nm (22 ft-lb)

NOTE —
- *Oil drain plug is equipped with a permanent seal. As a result, drain plug must always be replaced.*

◄ Engine oil filter (**arrow**) is attached to an oil cooler on the filter flange on the front of the engine. Access is from under the vehicle.

– Using an oil filter wrench such as Hazet 2171-1 or 3417, loosen and remove oil filter and allow oil to drain from cooler and flange assembly.

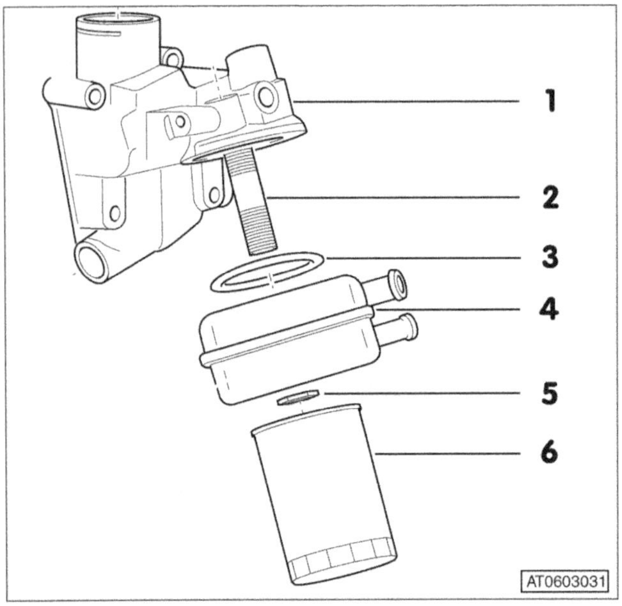

◄ Oil filter arrangement is similar for 180-hp and 225-hp engines.

1. Oil filter flange
2. Mounting stud/tube
3. O-ring seal
4. Oil cooler
5. Nut
   - No service normally required during oil change
   - Tighten to 25 Nm (18 ft-lb), if needed
6. Spin-on oil filter
   - Tighten to 20 Nm (15 ft-lb)

– When oil has stopped draining from flange and cooler, clean sealing surface on oil cooler.

– Apply a light film of clean oil to oil filter gasket and thread onto oil cooler mounting stud and tighten by hand.

BentleyPublishers
.com
© 2006 BentleyPublishers.com—All Rights Reserved

◄ Using a suitable tool, (**3417** shown), tighten oil filter to specification.

Tightening torque	
Oil filter to flange	20 Nm (15 ft-lb)

NOTE—
- *Tool 3417 is designed to work with factory oil filters and has 15 flutes or "flats" to match the factory filter design.*

– Fill engine with oil of proper viscosity and rating as noted.
- To prevent overfilling, check engine oil capacity tables and fill to approximately ½ liter less than listed capacity. After vehicle is lowered and sitting on a level area, start engine, allow to run until temperature is approximately 60°C (140°F) and switch OFF ignition. Wait approximately 3 minutes, check dipstick, then carefully fill to MAX mark.

– Start engine and inspect oil filter and drain plug area for leaks.
- After changing oil and filter, start and run engine at idle only until oil pressure light goes out. DO NOT race or accelerate engine. Insufficient oil supply at this first engine start could cause damage or failure of turbocharger.

– Switch off engine.

– Install lower sound absorber panel under engine.

– Remove vehicle from jackstands or ramps or lower from lift or hoist. Be sure to set handbrake to prevent vehicle from rolling.

– Properly dispose of waste oil, oily rags, and filter.

WARNING—
- *Always dispose of used oil and oil filters properly at a facility equipped for recycling or storage.*
- *Always observe all applicable government regulations regarding storage, recycling, and disposal.*

– Observe oil specifications as noted. Synthetic oil is highly recommended for all TTs. Oils for all engines must meet specific Audi oil quality standards. Preferred viscosity (weight) for all engines is SAE 5W40. If not available, SAE 5W30 is permissible. Do not overfill.

Engine oil and filter, changing (3.2L 6-cylinder engines)

– Warm engine to normal operating temperature.

– Shut off ignition and apply parking brake.

– Raise vehicle and support on jackstands or ramps designed for the purpose or use a lift or hoist.

WARNING—
- *Jack stand and ramps should be placed on hard level surface (e.g. concrete).*

– Remove lower sound absorber panel.

BentleyPublishers.com
©2006 BentleyPublishers.com—All Rights Reserved

◀ Position suitable drain pan under engine, remove drain plug from rear of oil pan (**arrow**) and allow oil to drain into pan.

> ***WARNING —***
> • *Hot oil scalds. Wear protective clothing, gloves and eye protection.*

– When oil has completely drained, install new drain plug and tighten.

Tightening torque	
Drain plug to oil pan (always replace)	30 Nm (22 ft-lb)

NOTE —

• *Oil drain plug is equipped with a permanent seal. As a result, drain plug must always be replaced.*

◀ Cartridge type engine oil filter is contained in a filter flange on front of the engine (**arrow**). Access is from under vehicle.

– Using a 6 mm hex key (Allen wrench), remove drain plug in center of sealing cap (where equipped) and allow oil to drain from housing.

– Using a 36 mm wrench, remove sealing cap from oil filter housing.

NOTE —

• *Oil filter cartridge may come out with sealing cap.*

– Remove filter cartridge from housing if it did not come out with sealing cap.

– Remove O-rings and clean sealing cap, housing, and sealing cap drain plug (if equipped).

– Install new O-rings on sealing cap and sealing cap drain plug (where equipped) and apply a light film of clean oil to them.

– Install new filter cartridge and thread sealing cap into housing by hand.

– Tighten sealing cap and drain plug (where equipped) to specification.

Tightening torque	
Sealing cap to housing	25 Nm (18 ft-lb)
Drain plug in sealing cap	10 Nm (7 ft-lb)

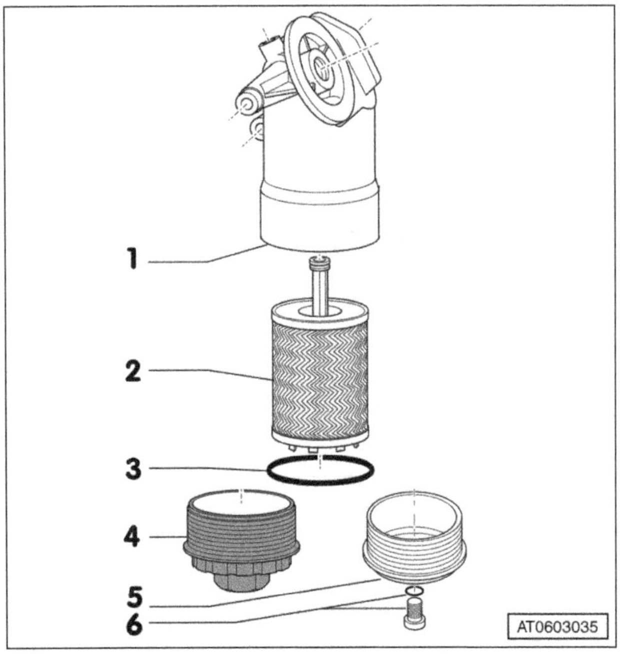

◀ Cartridge oil filter configurations for 6-cylinder engines.

1. Oil filter flange/housing
2. Cartridge oil filter
3. O-ring seal
   • Always replace
4. Sealing cap
   • Tighten to 25 Nm (18 ft-lb)
5. Sealing cap with drain plug
   • Tighten to 25 Nm (18 ft-lb)
6. Drain plug with O-ring
   • Tighten drain plug to 10 Nm (7 ft-lb)
   • Always replace drain plug O-ring

BentleyPublishers.com
© 2006 BentleyPublishers.com—All Rights Reserved

— Fill engine with oil of proper viscosity and rating as noted.

- To prevent overfilling, check engine oil capacity tables and fill to approximately ½ liter less than listed capacity. After vehicle is lowered and sitting on a level area, start engine, allow to run until temperature is approximately 60°C (140°F) and switch OFF ignition. Wait approximately 3 minutes, check dipstick, then carefully fill to MAX mark.

— Start engine and inspect oil filter and drain plug area for leaks.

- After changing oil and filter, start and run engine at idle only until oil pressure light goes out. DO NOT race or accelerate engine.

— Switch off engine.

— Install lower sound absorber panel under engine.

— Remove vehicle from jackstands or ramps or lower from lift or hoist. Be sure to set handbrake to prevent vehicle from rolling.

— Properly dispose of waste oil, oily rags, and filter.

> ***WARNING —***
> - *Always dispose of used oil and oil filters properly at a facility equipped for recycling or storage.*
> - *Always observe all applicable government regulations regarding storage, recycling, and disposal.*

— Observe oil specifications as noted. Synthetic oil is highly recommended for all TTs. Oils for all engines must meet specific Audi oil quality standards. Preferred viscosity (weight) for all engines is SAE 5W40. If not available, SAE 5W30 is permissible. Do not overfill.

Engine oil and filter, changing by extracting

Engine oil can be changed using an extraction system such as VAG 1307A, VAG 1358A, or VAG 1782A. This method involves siphoning oil out of engine through the dipstick tube. In some instances this can save considerable time. When using these devices, always follow manufacturers instructions.

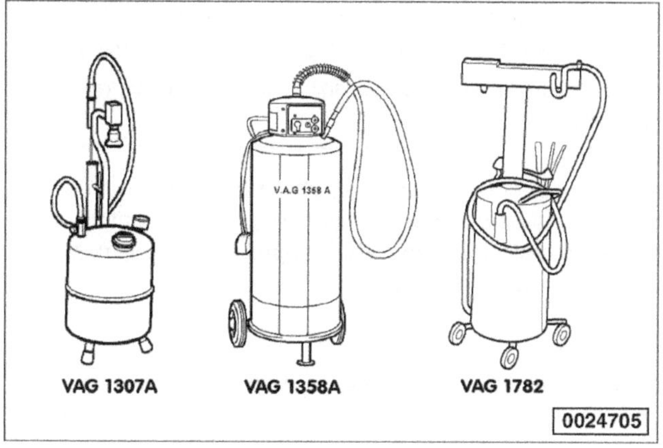

◀ Three different engine oil extraction systems are offered by Audi. Size is determined by service volume.

— Remove and replace engine oil filter as described in the appropriate engine section.

— Extract remaining engine oil according to extraction system manufacturers' instructions.

> ***WARNING —***
> - *Always dispose of used oil and oil filters properly at a facility equipped for recycling or storage.*

— Fill engine with oil of proper viscosity and rating as noted.

- To prevent overfilling, check engine oil capacity tables and fill to approximately ½ liter less than listed capacity. After vehicle is lowered and sitting on a level area, start engine, allow to run until temperature is approximately 60°C (140°F) and switch OFF ignition. Wait approximately 3 minutes, check dipstick, then carefully fill to MAX mark.

BentleyPublishers.com
© 2006 BentleyPublishers.com—All Rights Reserved

– Start engine and inspect oil filter area for leaks.

• After changing oil and filter, start and run engine at idle only until oil pressure light goes out. DO NOT race or accelerate engine.

– Observe correct oil specifications as given in tables above. Use of synthetic oils only is highly recommended for TTs. Oils for all engines must meet specific Audi oil quality standards. Preferred viscosity (weight) for all engines is SAE 5W40. If not available, SAE 5W30 is permissible. Do not overfill.

Cooling System

NOTE —

• Cooling system maintenance consists of maintaining coolant level, checking coolant freezing point, and inspecting hoses. Coolant flushing is not part of Audi's scheduled maintenance due in part, to installation of enhanced G12 coolant which is identified by it's pink or purple color.

• Advantages of G12 over earlier types include improvements in corrosion protection, thermal stability, heat transfer/control, hard water tolerance and environmental protection.

WARNING —

• Hot coolant can scald. Do not work on cooling system until it has fully cooled.

• Use extreme care when draining and disposing of coolant. Coolant is poisonous and lethal. Children and pets are attracted to it because of its color, sweet smell and taste. See a doctor or veterinarian immediately if any amount is ingested.

• Use only Audi original anti-freeze or equivalent when filling cooling system. Use of any other anti-freeze may be harmful to the cooling system. Do not use an anti-freeze containing phosphates.

• Never mix green coolant (G11) with purple coolant (G12). Mixing can cause serious engine damage.

• Contamination of G12 with other colored coolants is identifiable by discoloration (brown, gray, etc.). This mixture causes a foamy deposit to appear in the expansion tank/radiator and MUST be drained immediately. The cooling system must be completely free of this mixture before refilling with the correct type of coolant/antifreeze.

• Do not use tap water in cooling system. Use distilled water only to mix anti-freeze.

0024690g

Audi uses one type of coolant in vehicles covered by this manual. It can be identified by its pink or purple color. This coolant/antifreeze is phosphate free and silicate free. When supplied by Audi, this coolant/antifreeze is known as G12.

NOTE —

• Audi does not require coolant replacement as part of routine maintenance. The cooling system has been filled at the factory with a permanent coolant.

Cooling system capacity	
1.8L 4-cylinder 180-hp and 225-hp engines	5.0 liters (5.3 qt.)
3.2L 6-cylinder 250-hp engine	8.0 liters (8.5 qt.)

BentleyPublishers.com
© 2006 BentleyPublishers.com—All Rights Reserved

Coolant level, checking

A translucent expansion tank, or overflow reservoir, on right side of engine compartment provides easy monitoring of coolant level without opening the system. Coolant level should always be checked when engine is cold because level will change as engine warms up. Coolant level should be within the area above **min** mark (**lower arrow**), but below maximum line just above min mark (**upper arrow**) on expansion tank.

NOTE—

- *Plastic sound absorber panel removed for photographic clarity; coolant level is visible with panel installed.*

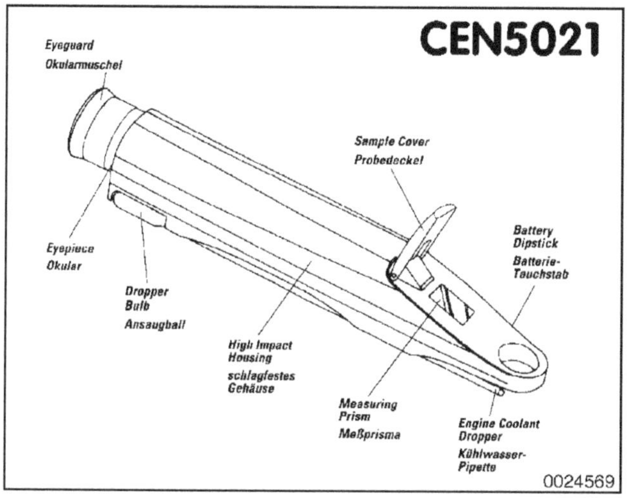

Coolant/antifreeze concentration checking

Using a refractometer (VW tools CEN5021, T10007, or equivalent), take a small amount of coolant using **dropper bulb** and place it onto **measuring prism**. Close **sample cover** and hold tester up to a light source.

– Look into **eyepiece** and read scale for ethylene glycol.

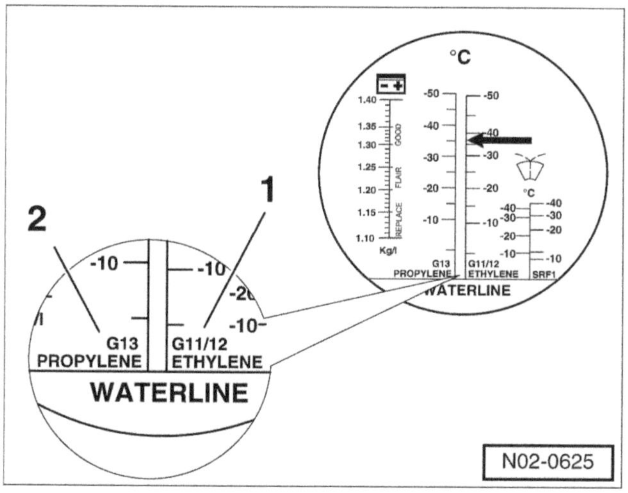

Measure point on refractometer is where bright and dark come together (boundary). For G12, A 50/50 mixture is generally preferred (**scale 1**) and will provide freeze protection to -34°F (-37°C) (**arrow**). Refractometer T10007 is shown, CEN 5021 and others are similar.

NOTE—

- *Refractometers of this type can also generally be used to measure specific gravity of a battery (far left scale), washer fluid strength (far right scale) and other types of coolant/antifreeze not used in vehicles covered by this manual (**scale 2**).*
- *For protection to approximately -40°C (-40°F), percentage of anti-freeze may be increased up to 60%. The percentage of anti-freeze must not exceed 60%, as higher amounts will decrease frost protection and reduce cooling ability.*

– Rinse tester clean.

– Periodically check tester calibration by placing distilled water onto the prism and closing the cover. The measure point should be at 32°F (0°C) which is the freezing point for pure water. If this reading is not obtained, follow instructions that came with tester for calibration and adjustment.

BentleyPublishers
.com
© 2006 BentleyPublishers.com—All Rights Reserved

Cooling System

Antifreeze Protection Levels (Quarts and °F)								
		Quarts of antifreeze required for protection to temperature shown (°F)						
		2	3	4	5	6	7	8
Total cooling system capacity (quarts)	5	-12	-62					
	6	0	-34					
	7	6	-17	-54				
	8		-7	-34	-69			
	9		0	-21	-50			
	10		4	-12	-34	-62		
	11		8	-6	-23	-47		
	12		10	0	-15	-34	-57	
	13			3	-9	-25	-45	
	14			6	-5	-18	-34	-54

Notes:
For best overall protection, antifreeze and water mixture should be within shaded areas. This represents an antifreeze concentration of between 40% and 60% which is preferred. Under no circumstances should engine be run with 100% antifreeze (or other high concentrations). Engine will overheat and heater output will be minimal. Maximum heat transfer only occurs when antifreeze is mixed with pure water.
Under no circumstances should engine be run with 100% water. Freeze, boil over, and corrosion protection will be eliminated and severe engine damage will result.

◄ Freeze protection chart for Audi-supplied G12 antifreeze in quarts and degrees Fahrenheit.

Antifreeze Protection Levels (Liters and °C)								
		Liters of antifreeze required for protection to temperature shown (°C)						
		2	3	4	5	6	7	8
Total cooling system capacity (liters)	5	-24	-52					
	6	-18	-37					
	7	-14	-27	-48				
	8		-22	-37	-58			
	9		-18	-29	-46			
	10		-16	-24	-37	-52		
	11		-13	-21	-31	-44		
	12		-12	-18	-25	-37	-50	
	13			-16	-23	-32	-43	
	14			-14	-21	-28	-37	-48

Notes:
See notes in Quarts/Fahrenheit table given previously.

◄ Freeze protection chart for Audi-supplied G12 antifreeze in liters and degrees Celsius.

BentleyPublishers.com
©2006 BentleyPublishers.com—All Rights Reserved

Cooling system hoses, inspecting

Inspect hoses by first checking that all connections are tight and dry. Coolant seepage indicates that either the hose clamp is loose, hose is damaged, or connection is dirty or corroded. Dried coolant has a chalky appearance.

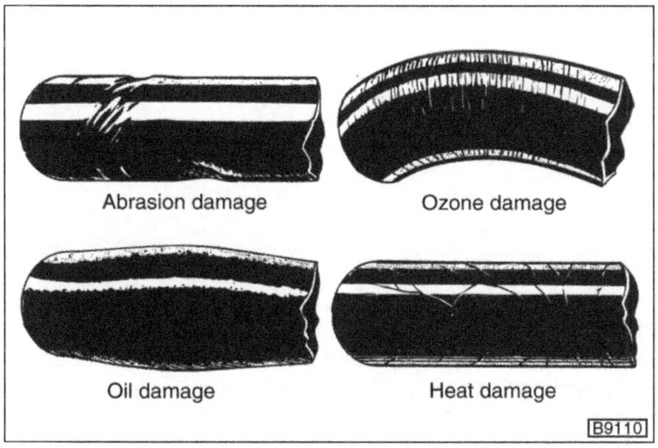

◄ Check hose condition by a visual and tactile inspection as described. Hoses should be firm and springy. Replace hoses that exhibit conditions noted below.

- Leakage - Dripping, moisture, or seepage near clamps or connectors.
- Electromechanical Degradation (ECD) - Difficult to see, but detectable by squeezing hose and feeling for cracks, weak areas, and voids.
- Oil damage - Soft and spongy to touch, visible bulges and swelling.
- Abrasion damage - Wear, abrasion, or scuffing, often due to contact with components in engine compartment.
- Heat damage - Internal and external damage generally due to high underhood temperatures or overheating. Internal heat damage is often indicated by swelling with external damage marked by hardened and cracked areas.
- Ozone damage - Small, parallel cracks in outer layers, but without hardening. Due to exposure to atmospheric conditions.

NOTE —
- *Always use clamps specified by Audi. Some clamp types do not allow for heat expansion and contraction and will often leak when hot or cold.*

BRAKE FLUID

Brake fluid level, checking

Routine maintenance of the brake system includes maintaining an adequate level of brake fluid in reservoir. Additional service items include replacing brake fluid every 2 years, checking brake pads for wear, checking hand brake function, and inspecting system for fluid leaks or other damage. See **Under Car Maintenance** for pad inspection and brake system inspection.

WARNING —
- *Use only new, previously unopened brake fluid conforming to US Standard FMVSS 116 DOT 4.*
- *Brake fluid is poisonous. DO NOT ingest brake fluid. Wash thoroughly with soap and water if brake fluid comes into contact with skin.*
- *DO NOT let brake fluid come in contact with paint. Wash immediately with soap and water.*
- *Brake fluid absorbs moisture from the air and must be stored in an airtight container.*
- *DO NOT allow brake fluid to exceed maximum level in fluid reservoir.*
- *Observe all regulations concerning storage and waste disposal of brake fluid.*

BentleyPublishers.com
© 2006 BentleyPublishers.com—All Rights Reserved

AT0603059

Check fluid level at brake fluid reservoir, located on driver's side of engine compartment, towards rear. Brake fluid level should be between **MIN** and **MAX** marks (inset) and will drop slightly as brake pad material wears.

0024435

Audi specifies use of brake fluid that meets Department of Transportation (DOT) 4 Super specification in addition to Federal Motor Vehicle Safety Standard (FMVSS) 116 and Society of Automotive Engineers (SAE) standard J1703. DOT 4 Super is also known as DOT 4+. Brake fluid to this specification is also available in 30 and 50 liter containers for use in pressure bleeders.

Brake fluid specifications	
Audi specification	DOT 4 Super (DOT 4+)
Audi part numbers	B 000 700 A1 B 000 700 A7 B 000 700 A8
Minimum wet boiling point	329°F (165°C)
Dry boiling point (exceeds)	500°F (260°C)

Brake fluid, replacing

Brake fluid absorbs moisture from the atmosphere and so is said to be hygroscopic. If moisture content of brake fluid is too high, corrosion inside brake system may result. The boiling point of brake fluid will also be considerably lowered, resulting in the possible vaporization of brake fluid and reduction in braking force under some conditions.

Brake fluid is also used as the operating fluid for the hydraulic clutch on vehicles with a manual transmission. Fluid is drawn into clutch master cylinder via a small fitting in upper part of reservoir. In the event of a hydraulic clutch leak, fluid loss will only be to the level of the fitting, which retains an adequate reserve for the brake system. Because the hydraulic clutch and the brake system share a common fluid and reservoir, the clutch hydraulic system must be considered when replacing the brake fluid.

0024434

Audi currently specifies that brake system bleeding and brake fluid replacement be performed with VAS 5234 or VAG 1869 brake system filling/bleeding machine (pressure bleeder). Use of a brake pedal depressor such as VAG 1869/2 is also required.

BentleyPublishers.com
©2006 BentleyPublishers.com—All Rights Reserved

> ***WARNING —***
> - *Do not use DOT 5 (silicone) brake fluid in any Audi brake system under any circumstances. Brake system performance is designed around specific characteristics of DOT 4 Super brake fluid only. Safety and performance may be compromised if DOT 5 brake fluid is used instead.*
> - *Do not mix DOT 5 (silicone) brake fluid with DOT 4 Super brake fluid. They are not compatible and severe component corrosion will result. Such corrosion could lead to brake system failure.*
> - *DOT 3 and standard DOT 4 do not meet previously noted specifications. The wet boiling point for DOT 3 is 284°F (140°C) while DOT 4 is 311°F (155°C). The dry boiling point for DOT 3 is 401°F (205°C) with DOT 4 at 446°F (230°C).*

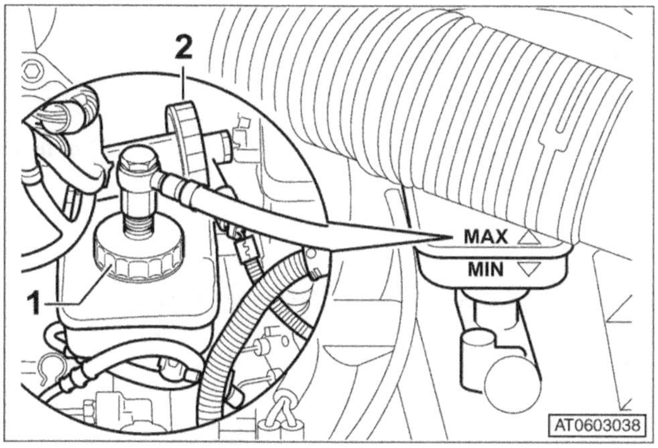

- Remove level sensor harness connector and unscrew brake fluid reservoir cap (**2**).
- Using suction hose from brake filler and bleeder unit VAS 5234 or other suitable device, extract as much brake fluid as possible.
- Screw adapter (**1**) onto brake fluid reservoir.
- Connect filler hose from pressure bleeder to adapter.

NOTE —
- *Follow manufacturers instructions for proper operation of VAS 5234 and VAG 1869.*

- Install brake pedal depressor VAG 1869/2 or equivalent between drivers seat and brake pedal and apply tension so that brake pedal remains depressed.

Vehicles with manual transmission

- Pull cover cap off of manual transmission slave cylinder bleeder valve.
- Attach collecting bottle hose to clutch slave cylinder bleeder valve (**arrow**).

NOTE —
- *Use a suitable bleed hose. It must seat tightly on bleed screw so that no air can enter brake system.*
- *Removal of air cleaner housing allows easier access to components on top of transmission.*

- Open **clutch slave cylinder** bleeder valve and allow approximately 100 ml (3.5 oz.) of fluid to flow out.
- Close clutch slave cylinder bleeder valve, remove collecting bottle hose, and reinstall cover cap.
- Depress clutch pedal several times to ensure proper operation.

Continued for all vehicles

- Pull cover caps off bleeder screws of all brake calipers.

BentleyPublishers
.com
© 2006 BentleyPublishers.com—All Rights Reserved

< Attach collecting bottle hose to **right rear** brake caliper bleeder valve (**arrow**). Right rear shown, left rear is symmetrically opposite.

NOTE—

- *Use a suitable bleed hose. It must seat tightly on bleed screw so that no air can enter brake system.*

– Open bleeder valve and allow approximately 200 ml (7 oz.) of fluid to flow out. Close bleeder valve.

– Repeat procedure for **left rear**.

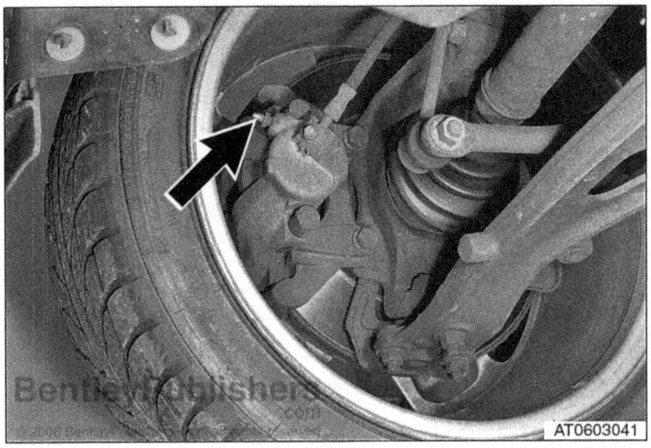

< Attach collecting bottle hose to **right front** brake caliper bleeder valve (**arrow**). Right front shown, left front is symmetrically opposite.

– Open bleeder valve and allow approximately 200 ml (7 oz.) of fluid to flow out. Close bleeder valve.

– Repeat procedure for **left front**.

– Install cover caps on bleeder screws of all brake calipers.

– Remove all pressure bleeder equipment according to manufacturer's instruction.

– Check fluid level in brake fluid reservoir and adjust as needed.

– Install brake fluid reservoir cap and harness connector for level sensor.

– Remove brake pedal depressor and check pedal free play. Maximum pedal free play is approximately ⅓ of total pedal travel.

< Approximate brake fluid quantity for brake and hydraulic clutch.

– If fluid losses or leakage are greater than can reasonably be expected, determine cause and repair.

Brake bleeding sequence and approximate bleed quantity	
Caliper sequence order:	Brake fluid quantity to be bled:
Right rear	200 ml (7 oz.)
Left rear	200 ml (7 oz.)
Right front	200 ml (7 oz.)
Left front	200 ml (7 oz.)
Approximate total quantity to be bled from brake system: (includes fluid siphoned from reservoir)	1.0 liters (1.1 qts.)
Approximate quantity to be bled from clutch hydraulic system (in addition to brake system):	100 ml (3.5 oz.)

BentleyPublishers.com

© 2006 BentleyPublishers.com—All Rights Reserved

POWER STEERING

Power steering hydraulic fluid can be checked when either hot or cold.

If fluid is **cold**, do not start engine, but steer wheels to the straight-ahead position.

If fluid is at **operating temperature (over 50°C)**, start engine and steer wheels to the straight-ahead position.

– Remove passenger side sound absorber panel. See **Sound absorber panels, removing and installing** in this repair group.

AT0603050

◄ Remove filler cap (**arrow**) using a screwdriver. Dipstick is attached to cap.

– Clean dipstick with a clean lint-free shop cloth.

– Screw cap hand tight onto reservoir and remove.

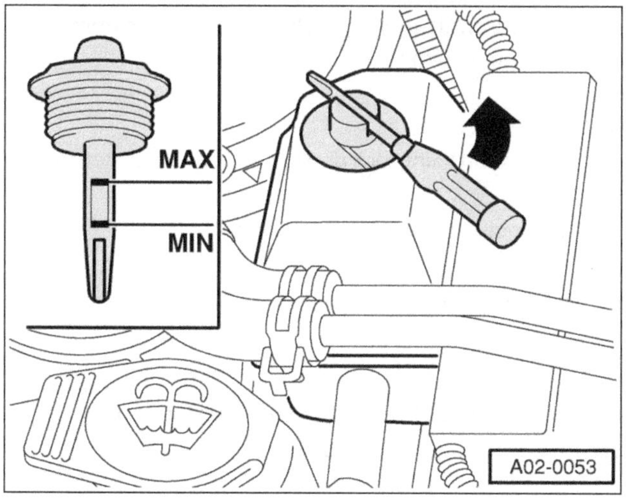

A02-0053

◄ Read level on stick:

- If fluid meets requirements for **COLD**, level should be plus or minus 2 mm of MIN mark
- If fluid meets requirements for **OPERATING TEMPERATURE**, level should be between MIN and MAX marks.

NOTE—

- *If fluid level is above (approx. 2 mm) the MIN mark, some fluid must be extracted.*
- *If fluid level is more than 2 mm below the MIN mark, hydraulic system must be checked for leaks. It is not sufficient to top up with fluid.*
- *Cap must be fully screwed in to obtain accurate readings.*

Hydraulic (power steering) fluid	
Audi part number	G 002 000
Notes: Do not reuse drained power steering fluid. Do not use ATF.	

WARNING—

- *Use only Audi hydraulic oil G 002 000 in the power steering system. Do not use ATF or other non-approved types of power steering fluid. If the wrong fluid is used, power steering components may fail.*
- *Audi part numbers are given for reference only! Always consult your Audi parts department or aftermarket parts specialist for the latest parts information.*

BentleyPublishers.com
©2006 BentleyPublishers.com—All Rights Reserved

FUEL FILTER

Fuel filters do not have a specified replacement interval. They should be replaced as conditions and/or situations dictate. Much of the fuel system is made of non-rusting materials, and screens protect areas such as fuel pump inlets. This, combined with the relatively large size of the filter, generally results in long service life for this vital component.

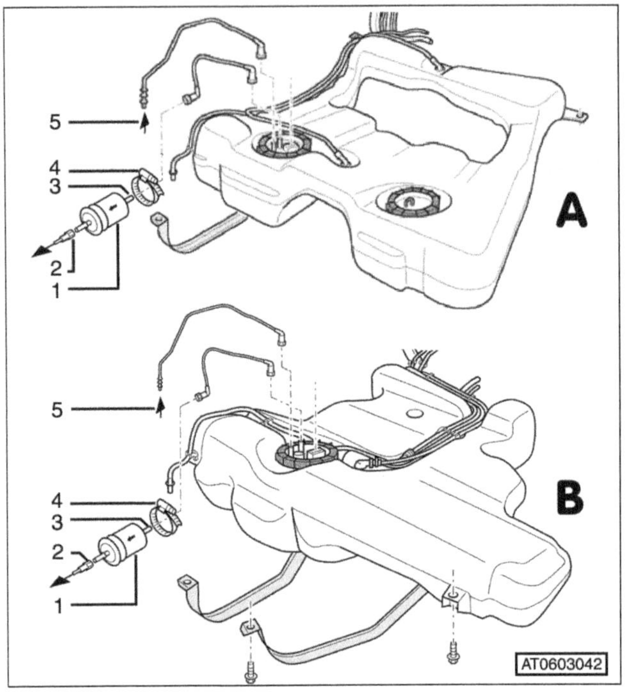

Fuel filter, replacing

NOTE—
- *Filter location and replacement procedures are similar for front wheel drive and quattro versions, and 4- and 6-cylinder engines.*
- *Observe all rules for cleanliness and all safety precautions for working on fuel systems. See* **20 Fuel Storage and Supply**.

Removing

< Fuel filter mounting and location for quattro (**A**) and front wheel drive (**B**) vehicles.

1. Fuel filter
   - Location: under right rear of vehicle, near fuel tank
   - Note fuel flow direction arrows on filter housing, do not interchange fuel connections
   - Note installed position: pin on filter housing engages detent on on filter mounting bracket
2. Fuel filter outlet line
   - Black: from fuel filter to engine fuel rail
   - Press release tabs on fitting lock ring to disconnect line from filter
   - Flow is in direction of **arrow**
3. Fuel filter inlet line
   - Black: from fuel tank
   - Press release tabs on fitting lock ring to disconnect line from filter
4. Retaining clamp
   - Note installed position of screw mechanism
5. Fuel return line
   - Blue: from fuel rail to tank

Removing

– Place suitable drip tray under fuel filter to catch any spillage.

WARNING—
- *Wear eye protection and protective clothing to avoid injuries from contact with fuel.*

NOTE—
- *Before removing a hose connection wrap a cloth around connection.*

< Disconnect fuel lines:
- Press release tabs on lock ring on inlet line fittings (**1**)
- Press release tabs on lock ring on outlet line fittings (**2**)

– Loosen clamp (**3**) on filter.

– Slide fuel filter out of clamp.

BentleyPublishers.com
© 2006 BentleyPublishers.com—All Rights Reserved

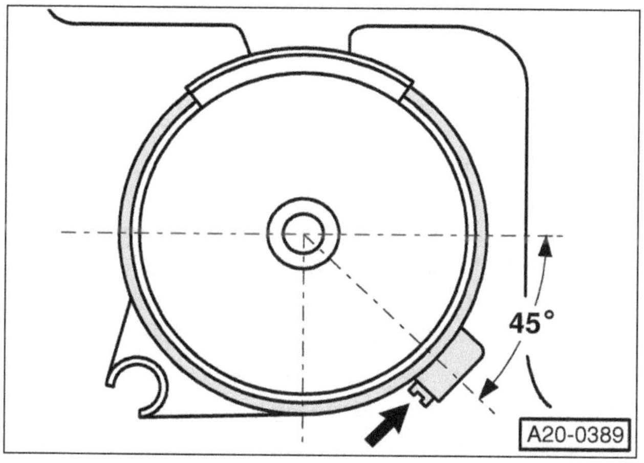

Installing

– Installation is the reverse of removal noting the following:

◄ Install new filter so that clamp screw mechanism (**arrow**) is positioned at 45° as shown. View is looking at front of filter.

NOTE —
- *Arrows on fuel filter show direction of flow.*

– Make sure that all disturbed fuel lines are properly positioned.

– Check for fuel leaks upon completion of repair.

BATTERY SERVICE

Routine battery maintenance depends on battery type. Vehicles with 4-cylinder engines use a traditional lead-acid battery mounted under the hood in the left front of the vehicle. Vehicles with 6-cylinder engines use batteries with Absorbed Glass Mat (AGM) technology mounted in the left rear of the luggage compartment in the spare tire well.

Warnings, cautions and guidelines

Warnings, cautions and guidelines generally apply to all types of vehicle batteries.

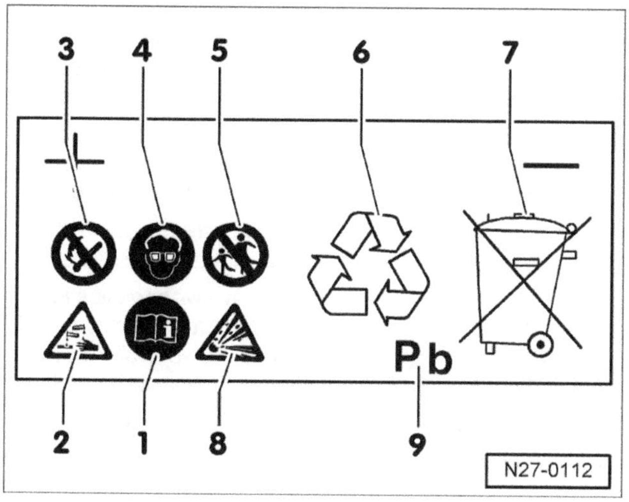

◄ A warning and safety label similar to this is found on most original equipment Audi batteries.

1. Follow recommendations on battery, in vehicle glovebox owners's manual, and in this manual.
2. Danger of acid burns. Follow warnings listed for working with acid.
3. Keep open flames, sparks, unenclosed light bulbs and lit smoking materials away from battery. Avoid electrical short circuits.
4. Wear eye protection.
5. Keep children away.
6. Dispose of old batteries properly. Recycle when possible. Comply with all applicable laws regarding hazardous waste disposal due to lead and acid content.
7. Do not discard in household trash.
8. Explosion hazard.
9. Contains lead (chemical symbol Pb)

When servicing battery, observe the following points and guidelines:

- On batteries with cell caps, all cell caps must be equipped with an O-ring seal.
- Electrolyte level on batteries with visible minimum and maximum markings on outside of translucent battery case is checked by visual inspection.
- Electrolyte level must be above minimum marking or just reach maximum marking.
- On batteries where exterior markings are difficult to see (due to installation location, battery case opaqueness or dirt), remove cell caps and visually inspect individual cell electrolyte levels.
- Electrolyte level must align with internal electrolyte level indicator (lip). This equates to external "maximum" marking on battery case.

BentleyPublishers.com
© 2006 BentleyPublishers.com—All Rights Reserved

- When electrolyte level is too low and cell plates are exposed, a loss of battery capacity (power) will result. If cell plates are not completely covered by electrolyte, (sulfuric acid/water mixture) corrosion will occur, on plates, plate bridges and cell connectors. Optimum battery function is not possible under these conditions!
- When electrolyte level is too high, electrolyte (sulfuric acid/water mixture) may leak out and damage surrounding areas such as battery tray and engine compartment.
- Only use distilled water to top up batteries. This prevents electrolyte impurities which cause self-discharging.
- Do not overfill battery. Overfilled batteries can boil over.
- Too little electrolyte reduces service life of battery.
- Excess electrolyte must be extracted using a hydrometer.
- Dispose of electrolyte (sulfuric acid/water mixture) properly. Waste electrolyte must only be disposed of in appropriate waste disposal sites. Refer to local regulations pertaining to electrolyte disposal.
- Audi recommends installation of new generation batteries only.
- For technical reasons, some batteries are equipped with sealing plugs with plastic foil. These batteries are considered maintenance free.

WARNING —
- *Always wear eye protection. Explosive gasses (hydrogen) produced during normal operation can cause blindness.*
- *Always shield your eyes and avoid leaning over battery whenever possible.*
- *If battery acid or gasses get into eyes, flush eyes immediately with large quantities of cold water for several minutes and get medical help fast. Sulfuric acid can cause blindness and/or severe burns.*
- *Do not let battery acid or lead particles get on skin, clothing, or painted surfaces. If battery acid gets onto these surfaces, immediately rinse and flush with cold water for several minutes.*
- *Do not smoke when working with or around batteries. Do not expose battery to open flames, electric sparks or open lights.*
- *Since battery acid contains sulfuric acid, always wear gloves in addition to eye protection. If battery acid gets onto skin, flush immediately with large quantities of cold water for several minutes and get medical help fast. Sulfuric acid can cause severe burns.*
- *If the slightest amount battery acid is ingested, get medical help fast.*
- *Do not reverse battery polarity.*
- *If battery is removed from vehicle or not in its proper place in the vehicle, always keep battery well out of reach of children.*
- *When a battery is charged, it produces hydrogen gas which is explosive. Exploding hydrogen gas can cause serious personal injury.*
- *When working on or near battery, be sure not to short circuit terminals with tools or other metallic objects. Battery could overheat and cause damage, fire or explosion and serious personal injury.*
- *When replacing battery, be sure to use a type and style approved by vehicle manufacturer.*
- *Before working on electrical system, disconnect negative (ground) battery cable.*
- *Observe all other applicable warnings and cautions when working in vehicle's engine compartment.*

BentleyPublishers
.com
© 2006 BentleyPublishers.com—All Rights Reserved

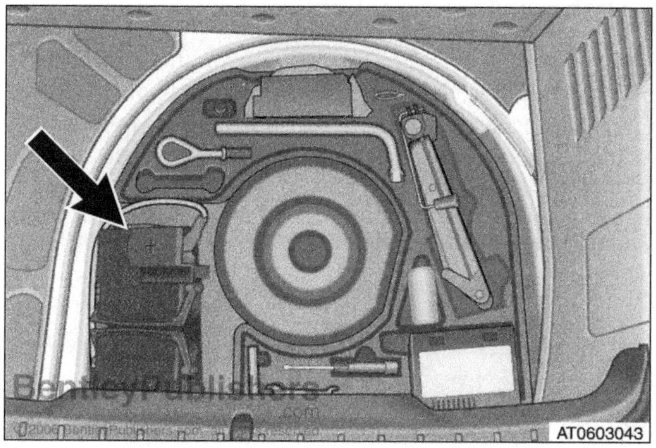

Absorbed glass mat (AGM) battery (6-cylinder vehicles)

Under normal operating conditions, AGM batteries are maintenance free. Unlike conventional lead-acid batteries, AGM batteries are totally sealed and use a fiberglass mat construction which completely absorbs the electrolyte. As a result, no excess electrolyte is present and it is not possible to check the electrolyte level or otherwise service the electrolyte. Routine servicing consists of visual checks of hold-down clamp, cable clamps, and central gas venting hose.

For battery testing, service and installation procedures, see **27 Battery, Alternator, Starter**.

AGM battery (**arrow**) in spare tire well of 6-cylinder TT.

If it is determined by testing that an AGM battery is malfunctioning, replacement is necessary.

> ***WARNING —***
> • *Do not attempt to open an AGM battery.*

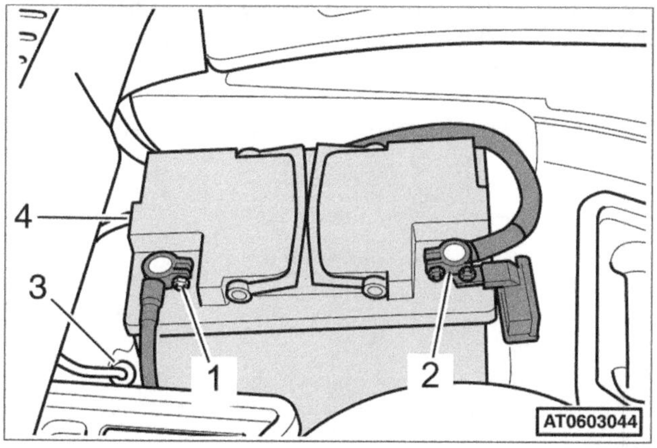

Maintenance, AGM battery

- Fold luggage compartment floor panel forward.

Inspect negative (-) terminal (**1**) and positive (+) terminal (**2**) cable clamps for tightness and corrosion.

- Inspect central gas vent hose for integrity of connection at battery (**3**) and at exit from spare tire well (**4**).

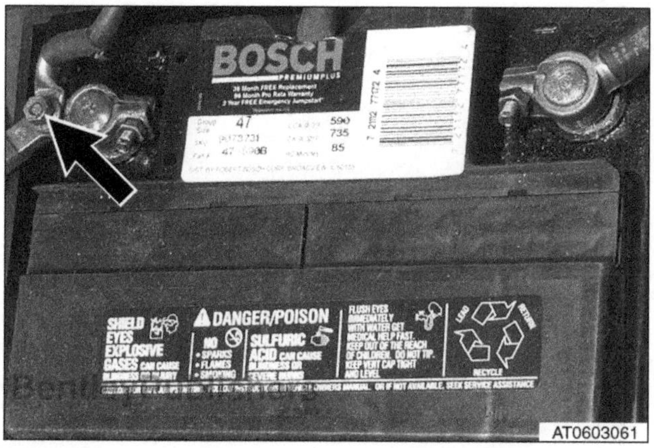

Inspect additional cables (**arrow**) attached to positive (+) and negative (–) cables (where applicable) for tightness and corrosion.

> ***WARNING —***
> • *3.2L vehicles are equipped with an Absorbed Glass Mat (AGM) battery. AGM batteries must not be quick charged under any circumstances.*

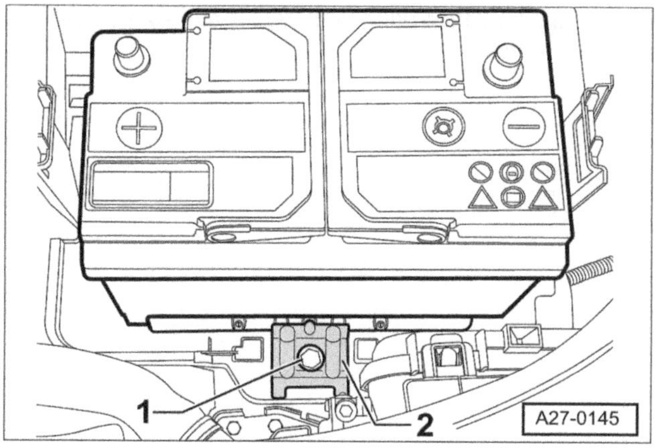

Inspect battery hold-down clamp bolt (**1**) for tightness and proper positioning of hold-down clamp (**2**).

Tightening torques	
Bolts, positive and negative cable clamps and additional cables on clamps	6 Nm (53 in-lb)
Bolt, battery hold-down clamp	20 Nm (15 ft-lb)

BentleyPublishers.com
© 2006 BentleyPublishers.com—All Rights Reserved

Lead-acid battery (4-cylinder vehicles)

Under normal operating conditions, modern lead-acid batteries require little maintenance. Routine servicing consists of visual checks of hold-down clamp, cable clamps, central gas venting hose, electrolyte level. Access to the battery for electrolyte level checking requires removal of cover on driver's side of engine compartment and the fuse box on top of the battery. Two different styles of fuse panel are used on top of the battery.

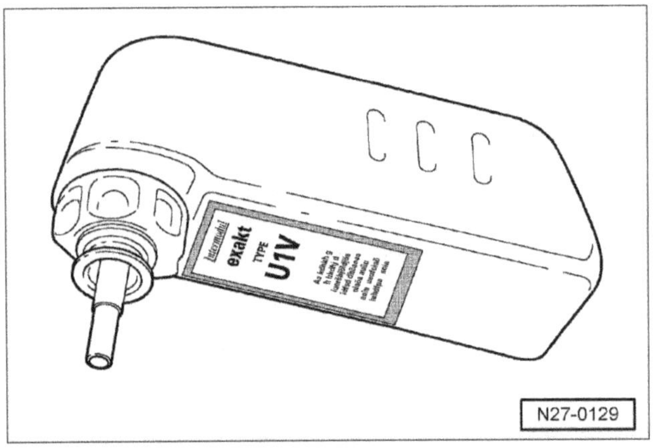

A battery filling bottle such as VAS 5045 is specifically designed for battery service and will prevent overfilling of battery.

For battery testing, service and installation procedures, see **27 Battery, Alternator, Starter**.

After disconnecting and re-connecting battery, some vehicle systems may require initialization (reset of basic settings), see **27 Battery, Alternator, Starter**.

Loosen ¼-turn fasteners (**arrows**), and lift off trim panel.

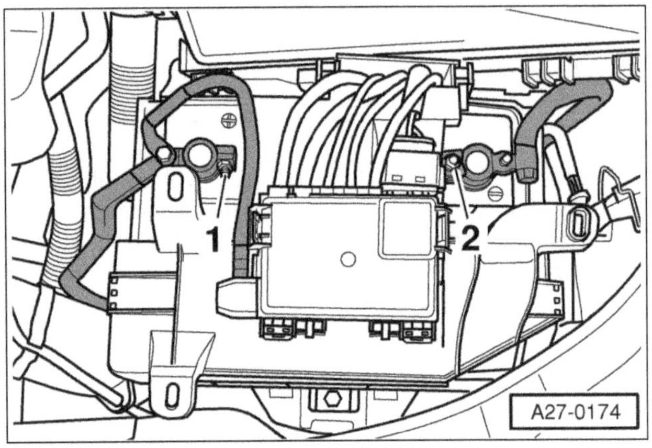

Version I: first, disconnect battery negative cable (**2**), then disconnect battery positive cable (**1**).

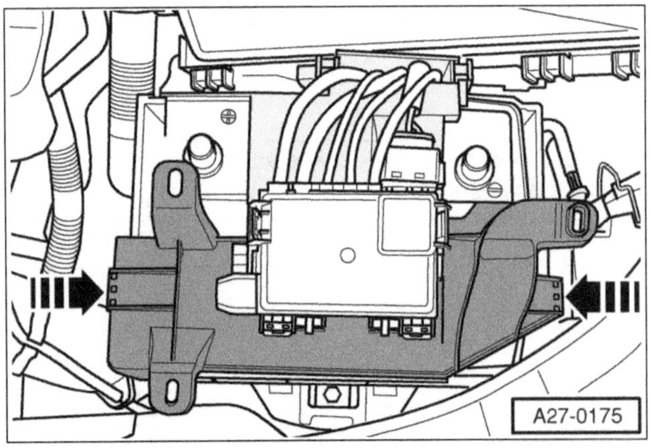

Version I: push and hold fuse panel retainers (**dashed arrows**) and pivot fuse panel out to the rear.

BentleyPublishers
.com
©2006 BentleyPublishers.com—All Rights Reserved

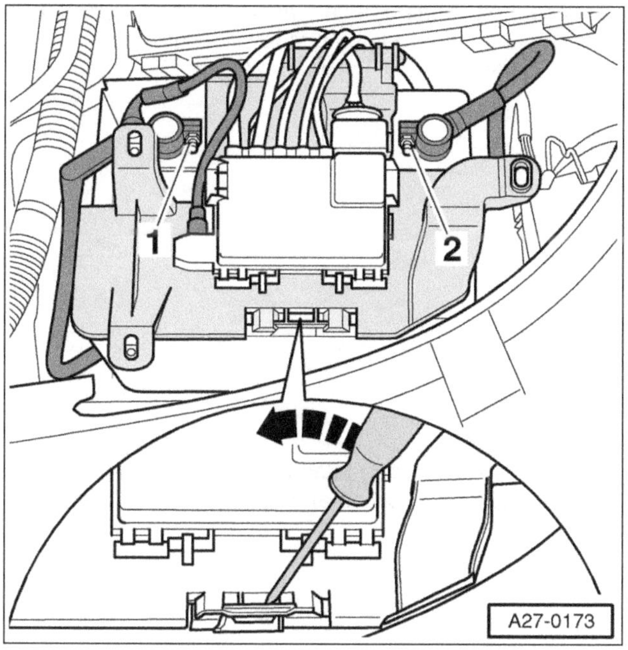

Version II: first, disconnect battery negative cable (**2**), then disconnect battery positive cable (**1**).

– Using a screwdriver or similar, push retaining clip in direction of **dashed arrow** and pivot fuse panel out to the rear.

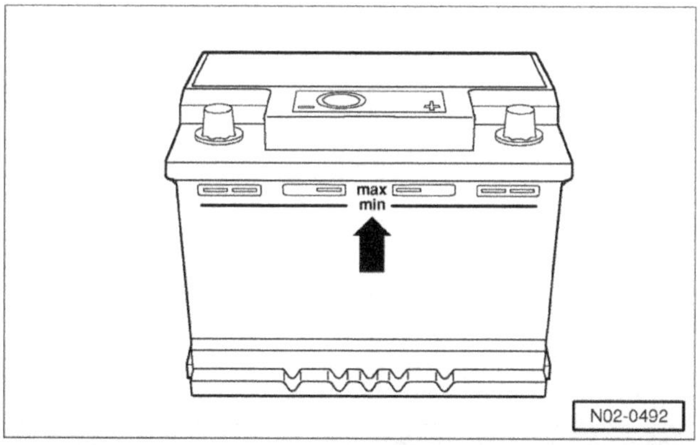

Remove individual cell caps or multiple cell caps and inspect electrolyte level. Level should be between **MIN** and **MAX** marks on translucent case batteries or just above battery plates and their separators on opaque case batteries. Battery plates can be seen once filler caps are removed. If electrolyte level is low, replenish it by adding distilled water only.

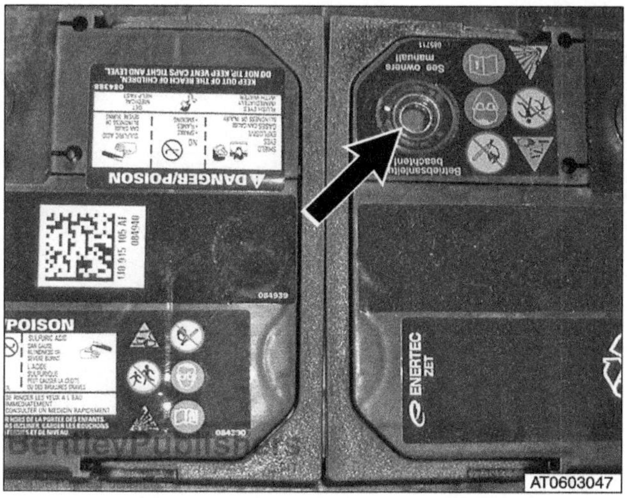

Some batteries have a charge indicator (**arrow**) or magic eye on top that displays electrolyte level and charge condition.

- Green, charge and electrolyte level are OK.
- Black, there is insufficient or no charge.
- Yellow or colorless, electrolyte level is critically low and should be topped off with distilled water immediately.

NOTE—

- *If charge indicator on a battery over 5 years old is colorless, do not attempt to top off or recharge battery. Battery should be replaced.*
- *Air bubbles that occur normally during battery charging (even during vehicle operation) may adversely affect charge indicator reading. To obtain an accurate reading, gently tap charge indicator with a screwdriver handle or rock vehicle to displace any possible air bubbles that have formed.*
- *Since charge indicator is located in a single cell, the indication is valid for that cell only. If battery condition is in doubt, confirm condition with a load test and further testing.*

– When electrolyte level is determined to be satisfactory, reinstall fuse panel in reverse of removal.

BentleyPublishers.com
© 2006 BentleyPublishers.com—All Rights Reserved

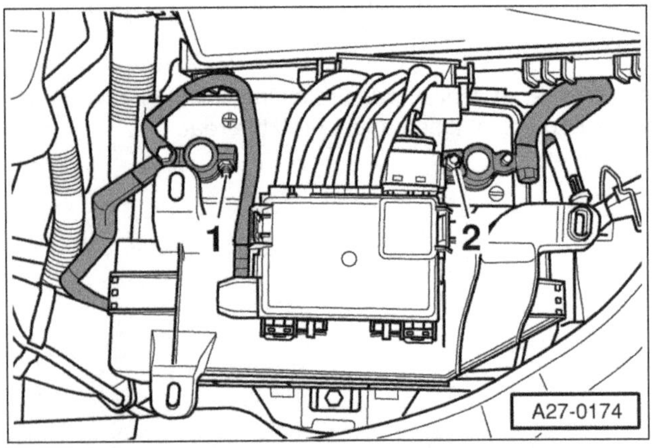

For both versions, first connect battery positive cable (**1**), and then battery negative cable (**2**) and tighten to specification.

Inspect additional cables (**arrow**) attached to positive (+) and negative (-) cables (where applicable) for tightness and corrosion.

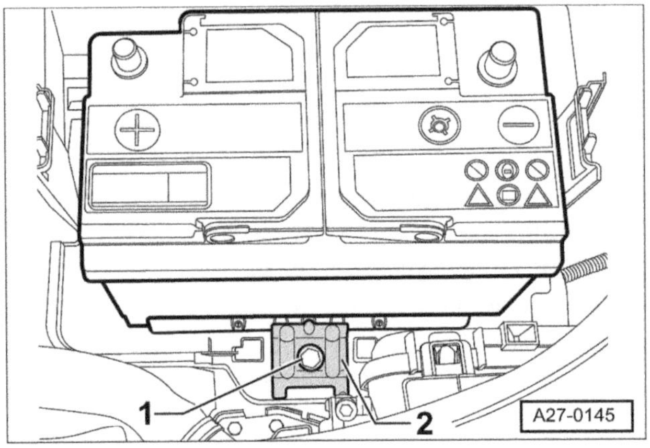

Inspect battery hold-down clamp bolt (**1**) for tightness and proper positioning of hold-down clamp (**2**).

Tightening torques	
Bolts, positive and negative cable clamps and additional cables on clamps	6 Nm (80 in-lb)
Bolt, battery hold-down clamp	20 Nm (15 ft-lb)

– Instal fuse panel in reverse order of removal.

– Install trim panels in reverse order of removal.

BentleyPublishers
.com
©2006 BentleyPublishers.com—All Rights Reserved

Ribbed Belt

Ribbed belt, checking

Engines covered by this manual use doubled-sided ribbed V-belts to drive engine accessories. Although belts of this design are known for their smooth running, high efficiency, and long life; periodic inspection is advisable.

– Remove upper sound absorber panel as needed. See **Sound absorber panels, removing and installing** in this repair group.

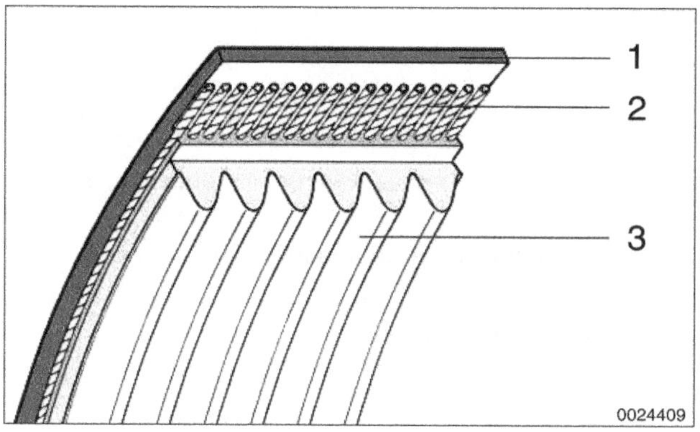

Cross section of typical single-sided ribbed V-belt. Double sided is similar.

1. Top layer, cover
2. Tension cords
3. Base material (substrata)

Although specific mileage intervals for inspection are not indicated, ribbed V-belt(s) should be periodically checked. Manually turn engine at vibration damper or pulley with a suitable wrench. Inspect belt from above or below for the following conditions:

- Splits in base material (substrata) including cracks, core fractures and cross section fractures.
- Separation between layers, especially between top cover and tension cord layers.
- Chunking or breaking of base or lower layer material.
- Fraying of tension cords.
- Flank (side) wear including flaking, fraying, chunking, glazing, hardening, and surface cracking.
- Oil and grease contamination.

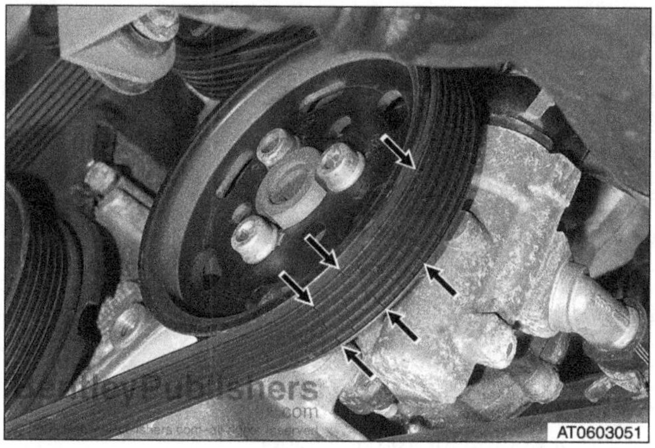

If any of above conditions exist, such as excessive cracking of the base material as shown here (**arrows**), the belt should be replaced. Consult maintenance tables at end of this section for appropriate routine replacement intervals.

Ribbed belt, replacing

1.8L 4-cylinder engines

– Switch off engine.

– Remove upper and lower sound absorber panels. See **Sound absorber panels, removing and installing** in this repair group.

– Remove upper turbocharger air duct where necessary.

NOTE—

- *If belt is to be re-used, such as for an emergency spare, mark running direction before removal.*

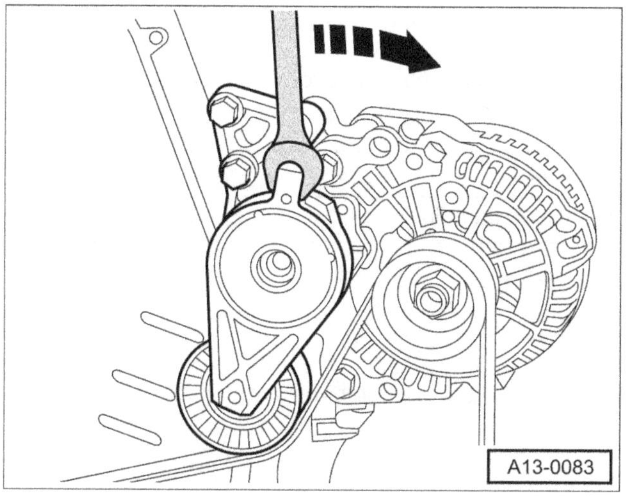

Using a 15 mm open-end wrench, pull tensioner roller away from belt (**dashed arrow**) and hold in that position.

BentleyPublishers.com
©2006 BentleyPublishers.com—All Rights Reserved

Ribbed Belt

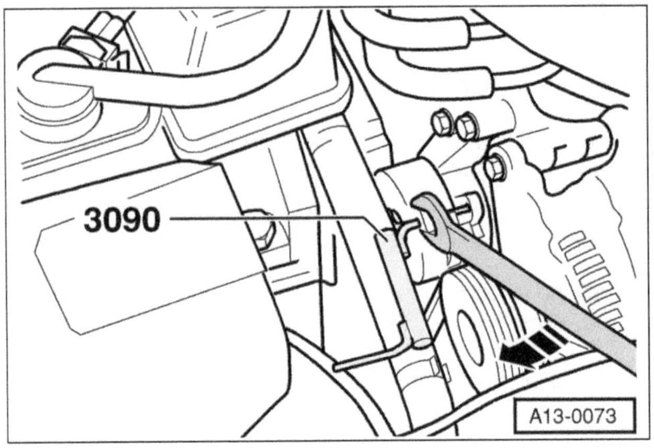

While holding tension on roller with wrench (**dashed arrow**), align hole in tensioner with corresponding hole in tensioner mount and secure with a lock pin.

NOTE —

- *Tool* **3090** *shown, but other tools can be used including the later* **T10069** *and derivatives. Tool is approximately 4.5 mm (3/16 in.) in diameter and approximately 55 mm (2 in.) long. Common tool box items such as drifts, punches, and hex (Allen) keys are often suitable.*

– Remove ribbed belt.

– Installation is the reverse of removal noting the following points:
  - When installing, note if belt is marked with running direction. Install in same direction.
  - Be sure that all drive pulleys run true and are tight.
  - Be sure belt is seated completely in all pulleys.
  - Be sure belt is installed with correct path.

– Remove locking tool.

– Start engine and check belt to ensure proper running.

– Install sound absorber panels.

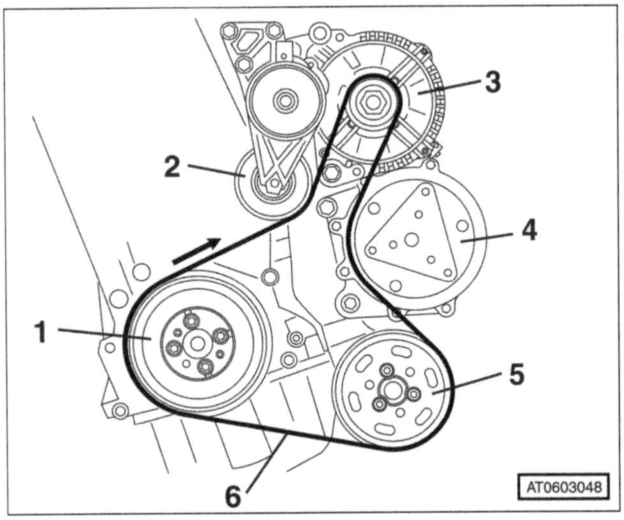

1.8L 4-cylinder engine, ribbed belt path. **Arrow** indicates belt running direction.

1. Crankshaft belt pulley
2. Tensioner roller
3. Alternator (generator)
4. A/C compressor
5. Power steering pump
6. Ribbed belt

– If tensioner replacement is needed, remove belt as described and remove tool 3090 or equivalent from tensioner.

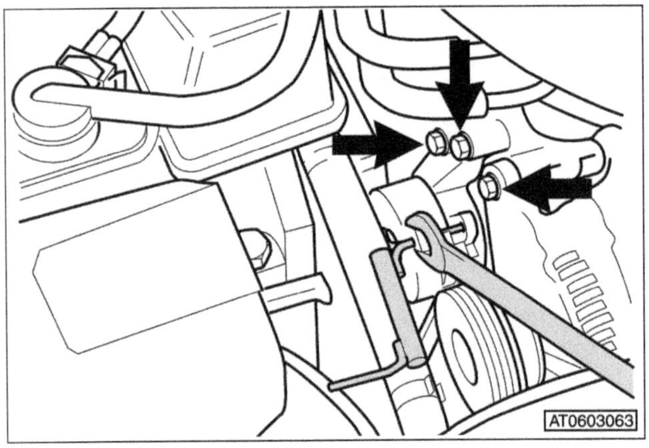

Remove three bolts (**arrows**) and lift tensioner out.

– Install new tensioner, tighten bolts, and lock with tool 3090 or equivalent.

Tightening torque	
Bolt, tensioner to cylinder block	25 Nm (18 ft-lb)

– Install/replace ribbed belt.

– Remove locking tool.

– Balance of installation is the reverse of removal.

BentleyPublishers.com
© 2006 BentleyPublishers.com—All Rights Reserved

3.2L 6-cylinder engine

– Switch off engine.

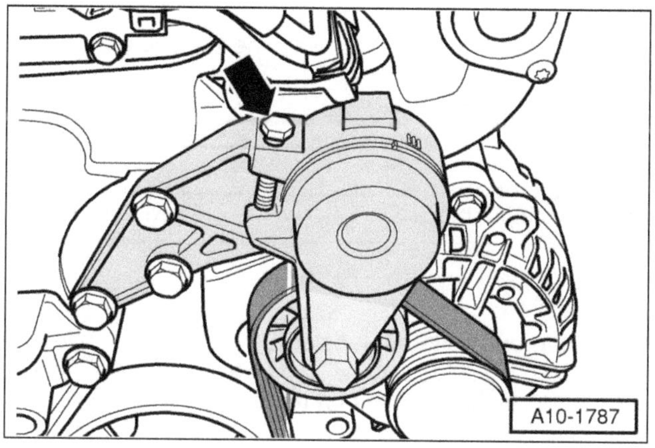

– Remove passenger side sound absorber panel, lower, and right side lower sound absorber panels. See **Sound absorber panels, removing and installing** in this repair group.

NOTE —

- *If belt is to be re-used, such as for an emergency spare, mark running direction prior to removal.*

◄ Release belt tensioner. Thread an M8x40 all-thread bolt (**arrow**) into tensioner as shown.

– Remove belt from alternator pulley.

– Working from under vehicle, and with sound absorber panels removed, slip ribbed belt off of remaining pulleys.

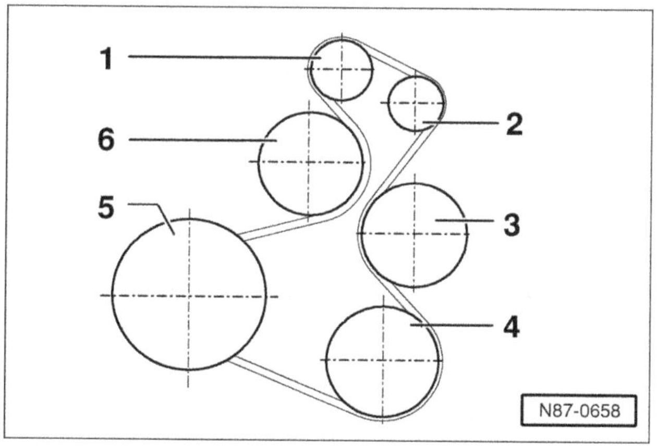

◄ Installation is the reverse of removal noting the following points:

- When installing, note if belt is marked with running direction. Install in same direction.
- Install ribbed belt onto pulleys in the order 1 to 6 with the coolant pump pulley (**6**) being last.
- Be sure that all drive pulleys run true and are tight.
- Be sure belt is seated completely in all pulleys.
- Be sure belt is installed with correct path.

– Remove M8x40 all-thread bolt from tensioner.

– Start engine and check belt to ensure proper running.

– Install upper and lower sound absorber panels.

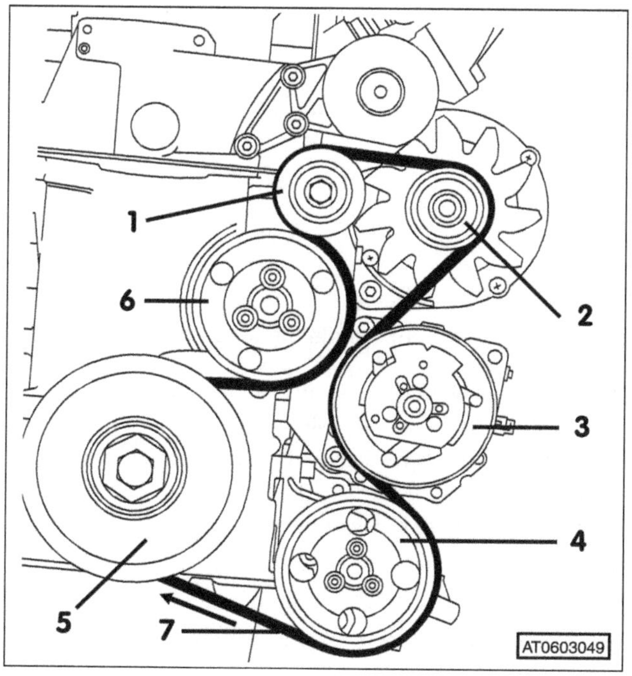

◄ **3.2L 6-cylinder engine**, ribbed belt path. **Arrow** indicates belt running direction.

1. Tensioner roller
2. Alternator (generator)
3. A/C compressor
4. Power steering pump
5. Crankshaft
6. Coolant pump
7. Ribbed belt

BentleyPublishers .com
©2006 BentleyPublishers.com—All Rights Reserved

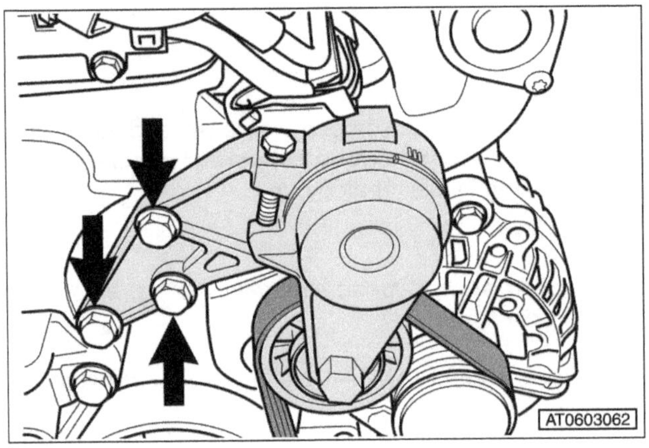

If tensioner replacement is needed, remove belt as described and leave the M8x40 all-thread bolt installed in tensioner.

- Remove three mounting bolts (**arrows**) and lift tensioner out.
- Remove M8x40 all-thread bolt and transfer to new tensioner.
- Install new tensioner, tighten bolts.

Tightening torque	
Bolt, tensioner to cylinder block	23 Nm (17 ft-lb)

- Install/replace ribbed belt.
- Remove M8x40 all-thread bolt from tensioner.
- Balance of installation is the reverse of removal.

TIMING BELT (TOOTHED BELT)

Timing belt (toothed belt), checking

The timing belt, also called the spur, camshaft, or toothed belt, should be inspected periodically. Belt inspection information contained in this section applies to the 1.8L 4-cylinder engines.

NOTE —
- *The 3.2L 6-cylinder engine uses timing chains instead of a timing belt. No maintenance service is required. See* **15b Cylinder Head (3.2L Engine)** *for more information.*

- Remove upper sound absorber panel. See **Sound absorber panels, removing and installing** in this repair group.

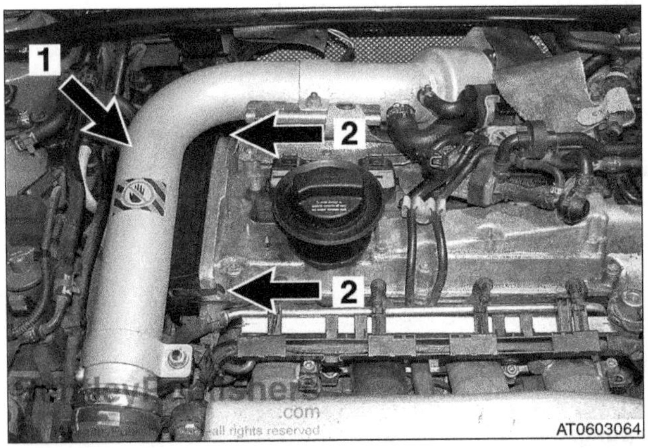

Remove upper turbocharger air duct (**1**) where necessary.

- Release clips (**2**) for upper belt cover.

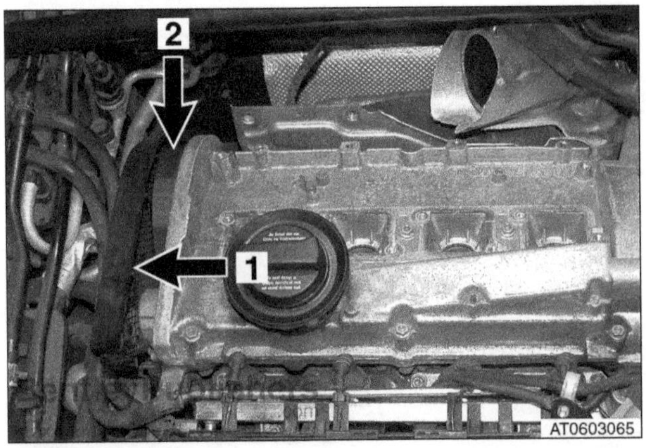

Carefully pull cover (**1**) away from cylinder head to expose belt (**2**).

- Use the following cross section illustration and inspect belt from above for the following general conditions:
  - Separation between layers, especially between top cover and tension cord layers
  - Chunking or breaking of base or lower layer material
  - Oil and grease contamination
  - Missing and/or damaged drive cogs
  - Insufficient tension as evidenced by contact with belt covers

BentleyPublishers.com
© 2006 BentleyPublishers.com—All Rights Reserved

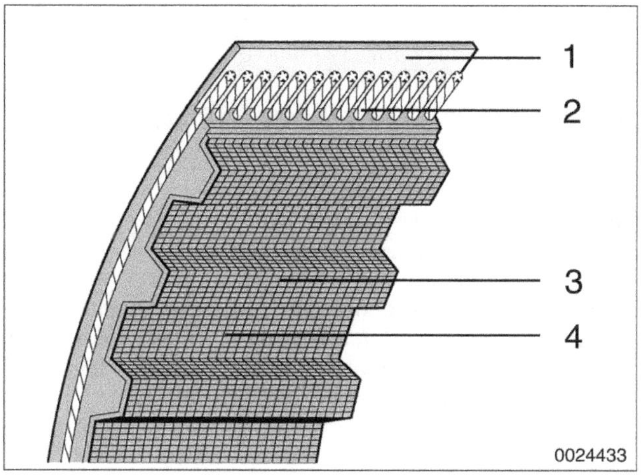

Cross section of typical Audi timing belt.

1. Top layer, backing, cover
2. Tension cords
3. Base material and drive teeth (drive cogs)
4. Base material fabric

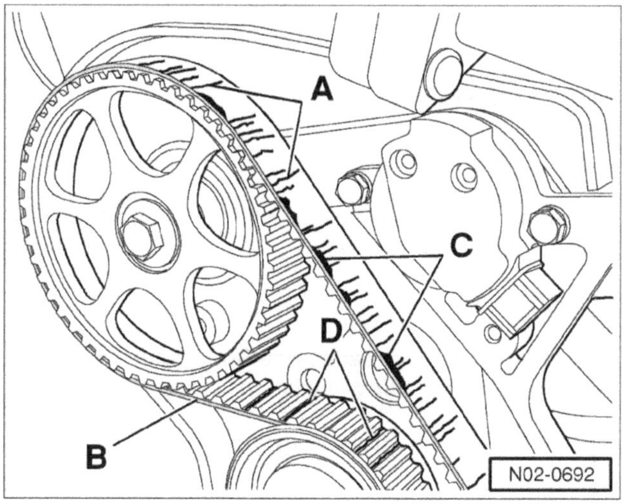

Specific conditions and areas to inspect include:

A. Cracks in cover material, including splits and fractures. Also glazing from belt cover contact.
B. Chafing and wear on edges.
C. Fraying of cord material.
D. Cracks on drive (gear) side.

In addition to overall condition of belt, inspect tensioner for wear and damage. If any of the above conditions exist, belt must be replaced. Consult maintenance tables at the end of this section for appropriate routine replacement intervals. Refer to specific Cylinder Head sections for additional information and belt replacement procedures.

> ***WARNING—***
> - *If damage is found, replace belt. Failure of camshaft belt will result in serious engine damage.*

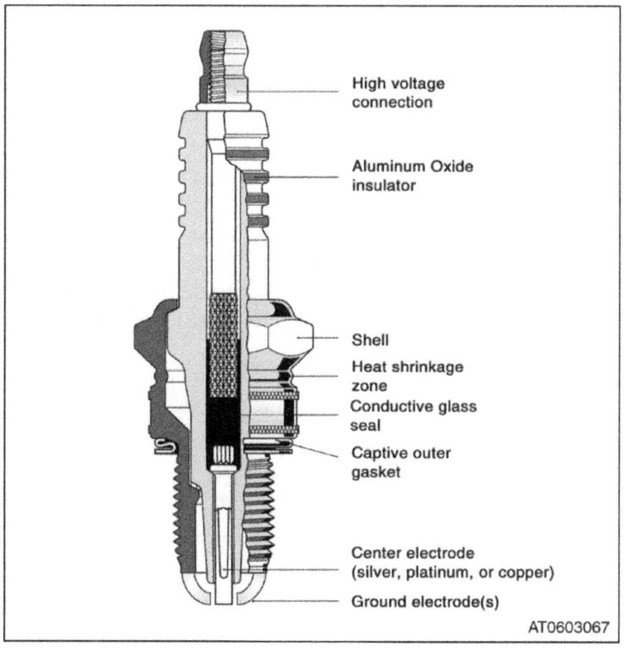

Spark Plugs

Gasoline engines require periodic spark plug replacement. Removal and installation procedures are similar.

Audi specifies spark plugs made by several different manufacturers. Because specifications and part numbers can change, always check with an authorized Audi Dealer's parts department or aftermarket parts specialist for the latest application information.

Engine	Spark plug specifications
1.8L 4-cyl. 180-hp 225-hp	AoA number....101 000 063 AA NGK number....PFR6Q Gap....0.8 mm (0.032 in.) maximum Torque....30 Nm (22 ft-lb) Firing order....1-3-4-2
3.2L 6-cyl. 250-hp	AoA number....101 905 606 A NGK number....IZKR7B Gap....1.1 mm (0.043 in.) maximum Torque....30 Nm (22 ft-lb) Firing order....1-5-3-6-2-4

BentleyPublishers.com
© 2006 BentleyPublishers.com—All Rights Reserved

NOTE —
- *Part number information is current at time of publication, but can change due to supersessions, etc.*
- *Audi part numbers are given for reference only! Always consult with your Audi of America Parts Department or aftermarket parts specialist for the latest parts information.*
- *Spark plugs are generally pre-gapped by the original manufacturer during production but should be checked before installation.*

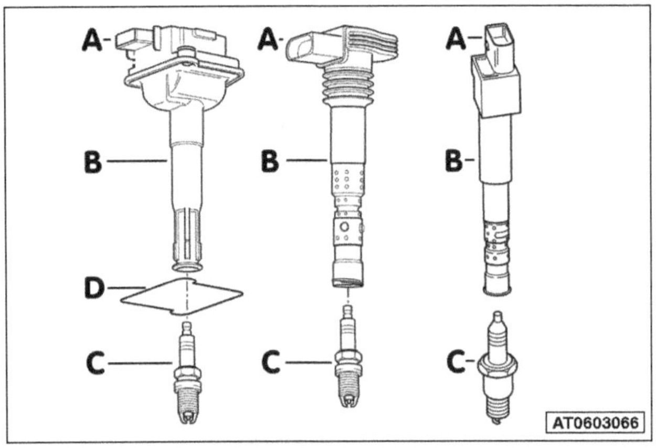

Spark plug access requires disconnecting harness connector from ignition coil and removal of ignition coil. Early 4-cylinder coils (**left**), are bolted in place and use a separate gasket. Later 4-cylinder (**center**) and 6-cylinder versions (**right**) are push-in.

A. Wiring harness connection
B. Ignition coil
C. Spark plug
D. Gasket

Spark plugs, replacing

Spark plugs, replacing (1.8L)

Access to the spark plugs is similar for both the 180-hp and 225-hp 1.8L engines. 225-hp engine shown.

– Switch off engine.

– Remove upper sound absorber panel. See **Sound absorber panels, removing and installing** in this repair group.

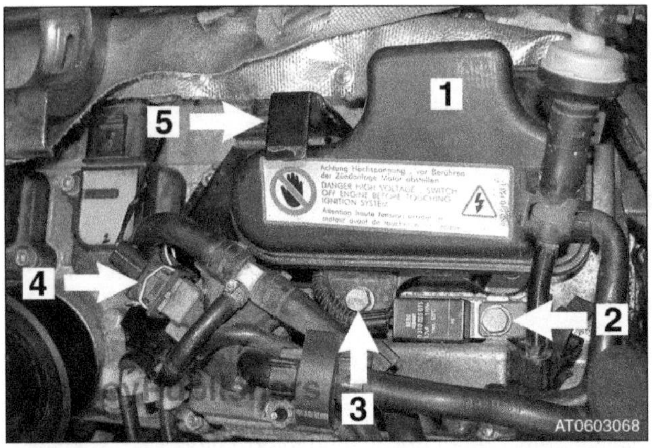

Remove vacuum reservoir (**1**) and surpressor by removing bolts (**2**) and (**3**).

– Disconnect harness connector (**4**) where applicable.

– Pivot vacuum reservoir (**1**) out of rear retaining bracket (**5**) and carefully move aside.

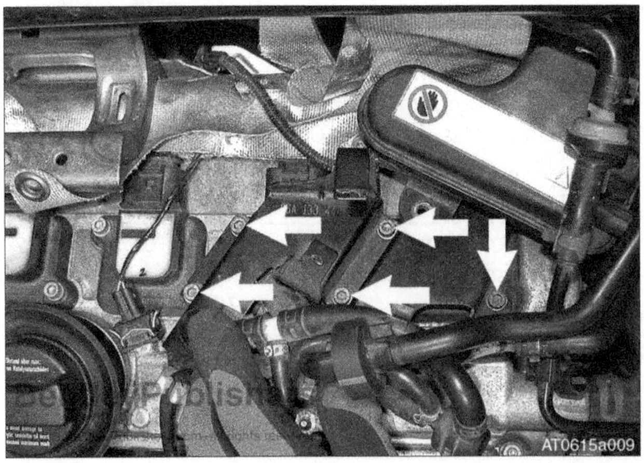

Remove mounting bracket screws (**arrows**) and mounting bracket to allow access to ignition coils for cylinders three (3) and four (4).

BentleyPublishers.com
©2006 BentleyPublishers.com—All Rights Reserved

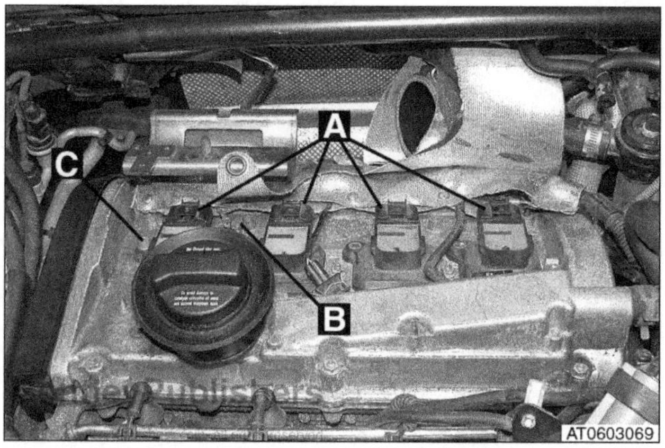

Bolt-in ignition coils

< Disconnect individual ignition coil harness connectors (**A**) and ground connection (**B**) and carefully push complete harness assembly to the rear.

– Remove bolts (**C**) from coils and lift out.

NOTE —
- *Mark coils to ensure installation in original locations.*

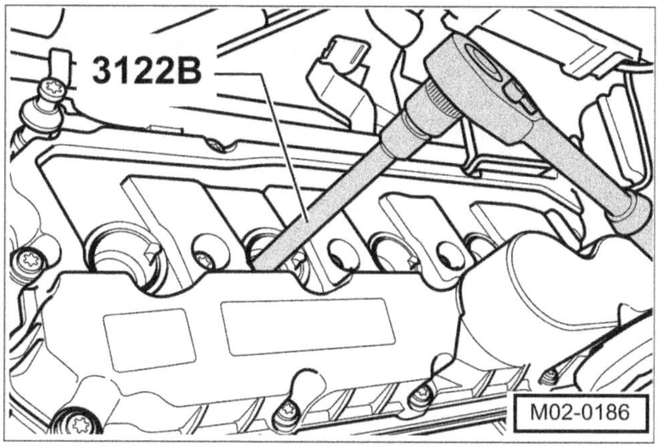

< Remove and install spark plugs for each cylinder with socket **3122B** or equivalent.

NOTE —
- *A 16 mm or 5/8 in socket with appropriate extensions is suitable for spark plug removal and installation.*

Tightening torque	
Spark plugs, in cylinder head	30 Nm (22 ft-lb)

– When all spark plugs have been replaced and torqued, insert all ignition coils into spark plug shafts and press evenly onto spark plugs by hand.

– Tighten ignition coils into cylinder heads.

Tightening torque	
Ignition coils, to cylinder head cover	10 Nm (7 ft-lb)

– Carefully push complete harness into position and install connectors onto ignition coils.

– Install canister mounting bracket.

– Install canister, surpressor, and harness connector where applicable.

– Install upper sound absorber panel. See **Sound absorber panels, removing and installing** in this repair group.

Push-in ignition coils

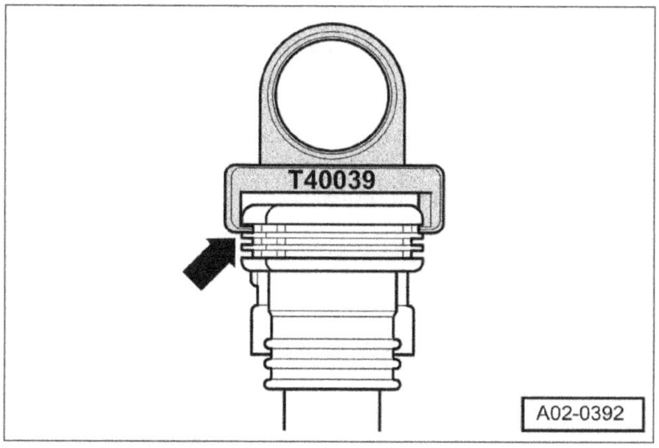

< Attach puller **T40039** or equivalent to uppermost thick rib (**arrow**) of each ignition coil.

NOTE —
- *Lower ribs of coil may be damaged if they are used to attempt removal.*
- *Harness connectors remain connected at this time.*

BentleyPublishers
.com
© 2006 BentleyPublishers.com–All Rights Reserved

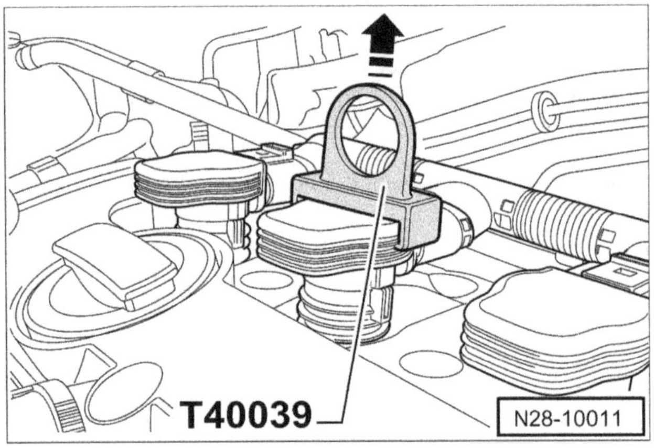

Pull all ignition coils approximately 30 mm (1¼ inch) out (**arrow**) of spark plug shaft using puller **T40039** attached as noted.

NOTE—

- *Two flat-bladed screwdrivers can also be used to carefully remove coils.*

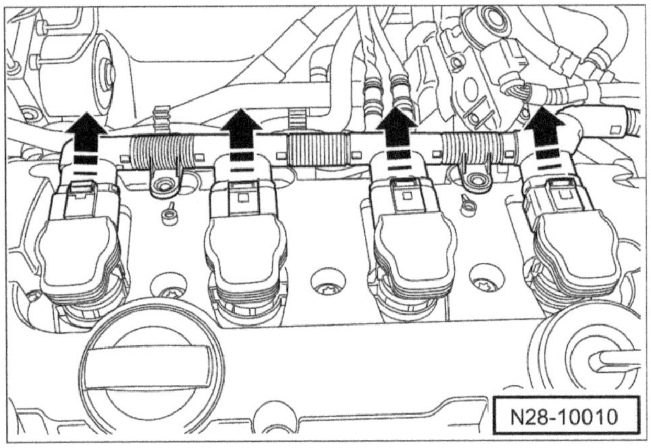

Disconnect harness connectors (**arrows**) and pull all connectors off ignition coils simultaneously.

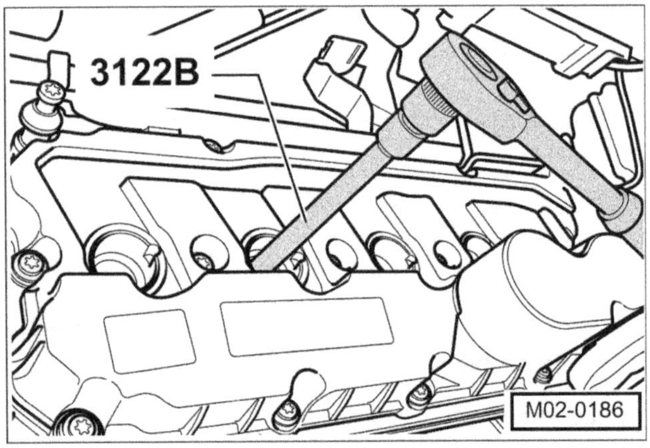

Remove and install spark plugs for each cylinder with socket **3122B** or equivalent.

NOTE—

- *A 16 mm or 5/8 in deep socket with appropriate extensions is suitable for spark plug removal and installation.*

– When all spark plugs have been replaced and torqued, insert all ignition coils into spark plug shafts and press evenly onto spark plugs by hand.

Tightening torque	
Spark plugs, in cylinder head	30 Nm (22 ft-lb)

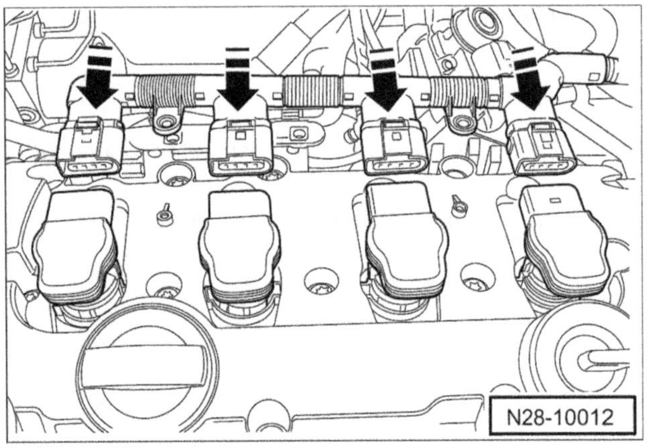

Align ignition coils to connectors (**arrows**) and install all connectors to ignition coils simultaneously.

– Align ignition coils to recesses in cylinder head cover.

– Install vacuum reservoir mounting bracket.

– Install vacuum reservoir, surpressor, and harness connector where applicable.

– Install upper sound absorber panel. See **Sound absorber panels, removing and installing** in this repair group.

BentleyPublishers
.com
© 2006 BentleyPublishers.com—All Rights Reserved

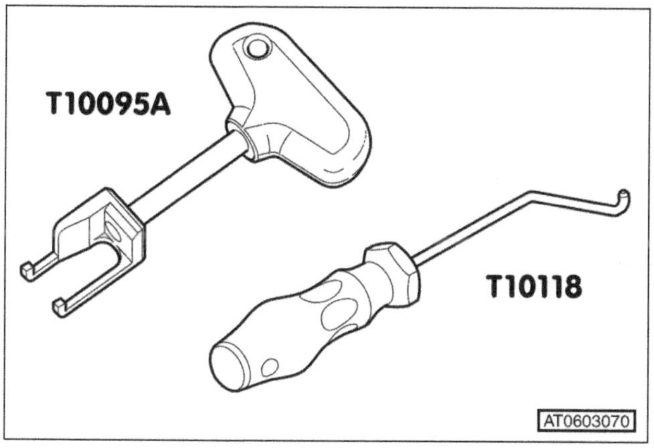

Spark plugs, replacing (3.2L)

Access to spark plugs requires ignition coil removal. Special tools allow access without damage. **T10118** facilitates removal of wiring harness connector. **T10095A** allows ignition coil removal.

– Switch off engine.

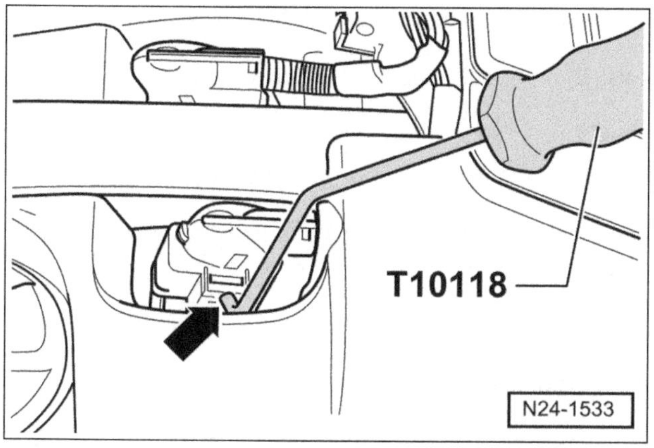

Disconnect harness connectors from ignition coils by releasing tabs (**arrow**) with **T10118** or equivalent.

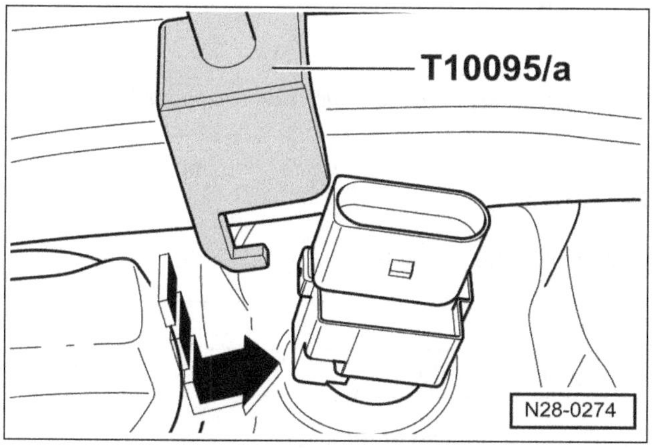

Hook puller **T10095A** under ignition coil (**arrow**) and remove by pulling straight up. Repeat for each coil.

– Remove and install spark plugs for each cylinder with socket **3122B** or equivalent.

Tightening torque	
Spark plugs, in cylinder head	30 Nm (22 ft-lb)

NOTE —

- *A 16 mm or 5/8 in deep socket with appropriate extensions is suitable for spark plug removal and installation.*

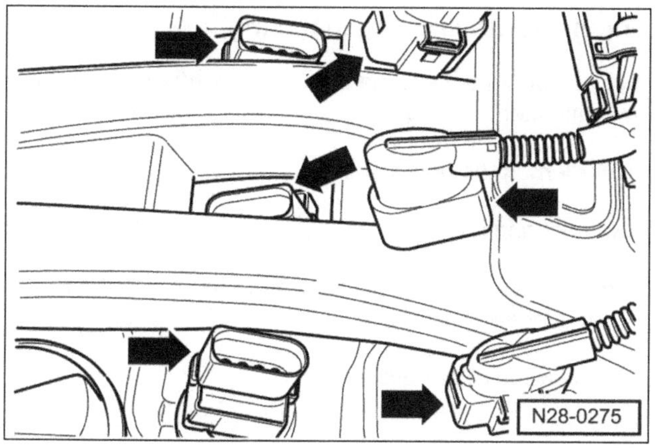

When all spark plugs have been replaced and torqued, insert all ignition coils into spark plug shafts noting alternating position of harness connectors on coils (**arrows**). Press coils evenly onto spark plugs by hand.

– Install harness connectors on coils.

BentleyPublishers.com
©2006 BentleyPublishers.com—All Rights Reserved

Air Filter

Air filter, replacing

The air filter element for 4-cylinder and 6-cylinder engines is contained in a housing on the driver's side of the engine compartment. All versions are similar.

- Switch off engine.

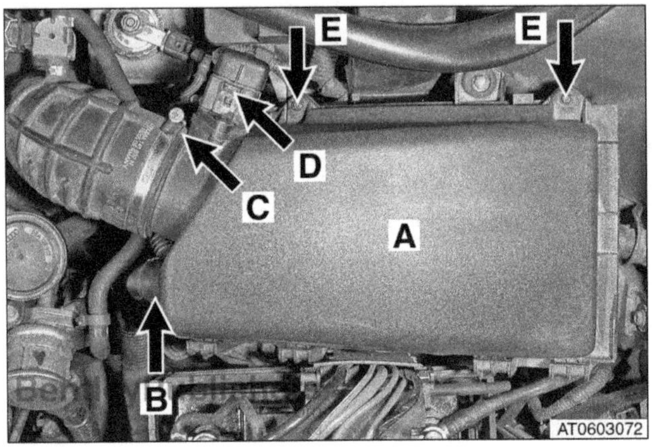

< Depress tabs on lock ring on secondary air injection hose (**B**) and remove hose.

- Loosen clamp on turbocharger inlet hose (**C**) and remove hose from mass air flow sensor.
- Disconnect harness connector for mass air flow sensor (**D**).
- Loosen screws (**E**) on upper section of filter housing.

NOTE—
- *Screws are captive and should not be completely removed.*

- Lift upper section of air filter housing (**A**) in area of screws (**E**) and pivot away from lower section.
- Remove paper air filter element.
- Clean both sections of filter housing.
- Install new air filter element into lower section of air filter housing and install upper part of air filter housing.

NOTE—
- *Observe any/all waste disposal regulations when discarding old air filter element.*
- *Make sure upper and lower parts of air filter housing are correctly positioned and seated.*

- Tighten screws (**E**) in upper section.

Tightening torque	
Screws, upper air filter housing	9 Nm (80 in-lb)

- Balance of installation is reverse of removal.

Air filter housing, removing and installing

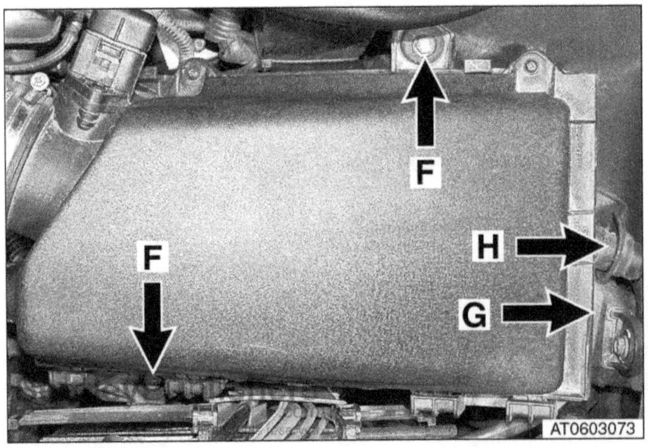

< Access to other components near the air cleaner is made easier by removal of lower air filter housing.

- Remove air filter element as noted above.
- Remove screws securing lower air filter housing to body (**F**).
- Lift lower air filter housing up, out and away from fresh air intake connection (**G**) and locating pin (**H**) in body.
- Installation is the reverse of removal.

BentleyPublishers
.com
©2006 BentleyPublishers.com—All Rights Reserved

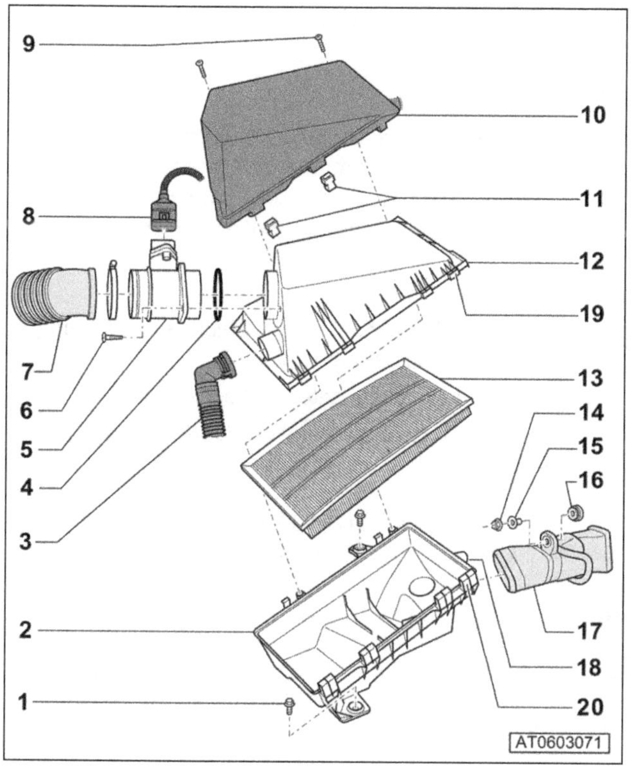

Air filter, assembly overview

1. Bolt
   - Qty. 2
   - Tighten to 10 Nm (89 in-lb)
   - Secures lower section of air filter housing to body
2. Air filter housing, lower section
   - Remove for access to other components in area
   - Secured with screws (item 1) and locating pin
3. Secondary air injection hose
   - To secondary air injection pump
   - Depress tabs on lock ring to remove
4. O-ring
   - Replace if damaged
   - Ensure proper seating before installing
5. Mass air flow (MAF) sensor (G70)
6. Screw
   - Qty. 2
7. Intake hose
   - To turbocharger inlet
8. Harness connector
   - To mass air flow sensor
9. Screws
   - Qty. 2
   - Secure upper and lower air filter housing sections together
   - Tighten to 9 Nm (80 in-lb)
   - Retained by upper air filter housing upper (captive)
10. Trim cover
    - Not installed on all engine versions
11. Retaining clips
    - For trim cover
12. Air filter housing, upper section
13. Air filter element
    - Paper
14. Nut
    - Tighten to 10 Nm (7 ft-lb)
15. Bushing sleeve
16. Rubber bushing
17. Fresh air intake connection
18. Locating pin
    - Engages slot in body
19. Securing tabs
    - Upper
    - Qty. 4
20. Securing tabs
    - Lower
    - Qty. 4

BentleyPublishers.com
© 2006 BentleyPublishers.com—All Rights Reserved

UNDER CAR MAINTENANCE

Brake system, visual inspection

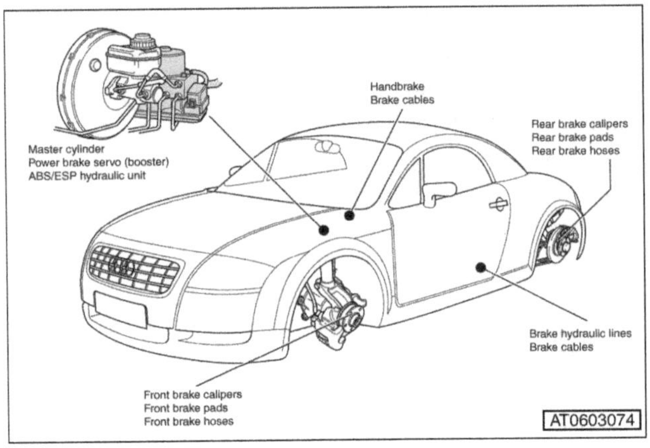

◀ Visually inspect power brake servo (booster) and brake master cylinder, ABS/ESP hydraulic unit, front and rear brake calipers and hoses, and brake lines for leaks and damage. Also inspect brake cables and handbrake for proper operation and damage. Note the following:

- Brake hoses must not be twisted.
- Brake hoses must not touch any part of the vehicle when steering is at full lock and suspension is at the limits.
- Check hoses and lines for porosity, deterioration and chafing.
- Check brake connections and attachments for correct seating, leaks and corrosion.

Correct problems noted during the visual inspection immediately. For parking brake adjustment, see **46 Brakes–Mechanical**.

Brake pads, checking

Periodic maintenance of the brake system includes inspection of front and rear brake pad thickness. The wear area of the brake pad is called lining or friction material.

NOTE—

- *Vehicles covered by this manual may use wheel locks. The key for the locks is usually stored in a pocket in the tool kit pouch in the trunk area with spare tire. See* **Tire and wheel service** *in this section. If using a lift or hoist, remove pouch before raising vehicle.*

Front brake pads

Depending on wheel type, brake pad material can be inspected visually through the wheel spokes or by removing the wheel. If the wheel is left in position, only the outer pad can generally be seen. Accurate inspection of front brake pad thickness requires removal of the front wheels which allows measurement of both outer and inner pads.

Front wheel installed

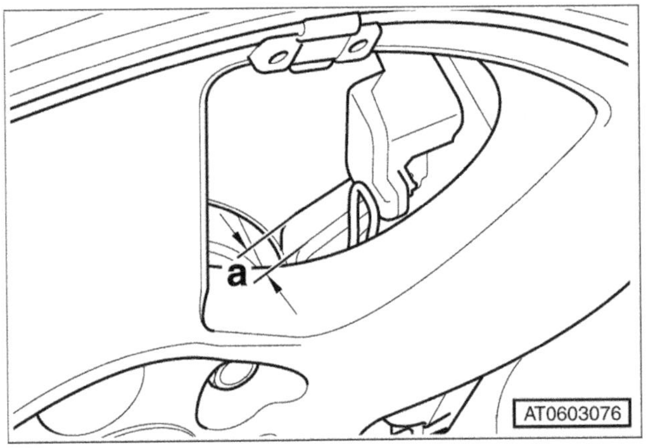

◀ Using a flashlight, illuminate area behind opening in wheel at brake caliper. Determine thickness of outer pad (**a**) by visual inspection. Measurement (**a**) does not include backing plate or any shims. Compare measurement to wear limit.

Front wheel removed

– Remove wheel bolt caps, full wheel covers, or alloy wheel center cap as required. Before removing a wheel, mark it's position in relation to brake rotor so wheel can be reinstalled in same position.

– Using hand tools, loosen wheel bolts slightly before raising vehicle.

– Raise vehicle and properly support and chock.

– Remove wheel bolts and wheel.

BentleyPublishers.com
© 2006 BentleyPublishers.com—All Rights Reserved

Measure inner and outer pad thickness (distance between **arrows**). Measurement does not include backing plate or any shims.

Wear limit	
Front brake pad, minimum lining thickness (without backing plate)	2 mm (0.079 in)

New front brake pads have friction material approximately 14 mm (0.551) thick. When friction material has reached the minimum thickness of 2 mm (0.079 in), brake pads have reached their wear limit and should be replaced. These measurements do not include the backing plate or any shims. When replacing brake pads, be sure to inspect brake discs (rotors) for wear. See **46 Brakes–Mechanical**. Continued use beyond the wear limit can result in damage to the brake disc or rotor.

– Install wheel noting reference mark made on removal. Install wheel bolts and torque to specification in proper sequence. See **Tire and wheel service** in this repair group.

Tightening torque	
Wheel bolts to hub	120 Nm (89 ft-lb)

– Install wheel bolt caps, full wheel covers, or alloy wheel center cap as required.

– Lower vehicle. Stow jack, tools, and locking wheel bolt key in trunk.

Rear brake pads

Depending on wheel type, brake pad material can be inspected visually through the wheel spokes or by removing the wheel. If the wheel is left in position, only the outer pad can generally be seen. Accurate inspection of front brake pad thickness requires removal of the front wheels which allows measurement of both outer and inner pads.

Rear wheel installed

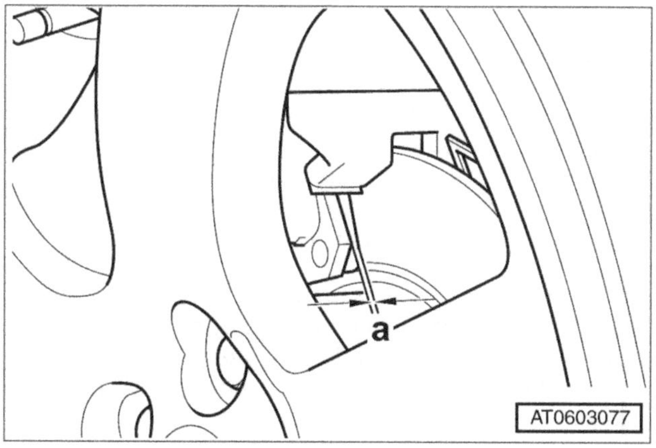

Using a flashlight, illuminate area behind opening in wheel at brake caliper. Determine thickness of outer pad (**a**) by visual inspection. Measurement (**a**) does not include backing plate or any shims.

Rear wheel removed

– Remove wheel bolt caps, full wheel covers, or alloy wheel center cap as required. Before removing a wheel, mark it's position in relation to brake rotor so wheel can be reinstalled in same position.

– Using hand tools, loosen wheel bolts slightly before raising vehicle.

– Raise vehicle and properly support and chock.

– Remove wheel bolts and wheel.

Using a flashlight as necessary, determine thickness of pads by visual inspection. Measurement (distance between **arrows**) does not include backing plate or any shims.

– Repeat process for inner brake pad using a mirror.

BentleyPublishers
.com
© 2006 BentleyPublishers.com—All Rights Reserved

Wear limit	
Rear brake pad, minimum lining thickness (without backing plate)	2 mm (0.079 in)

New rear brake pads have friction material approximately 12 mm (0.472) thick. When friction material has reached the minimum thickness of 2 mm (0.079 in), brake pads have reached their wear limit and should be replaced. These measurements do not include the backing plate or any shims. When replacing brake pads, be sure to inspect brake discs (rotors) for wear. See **46 Brakes–Mechanical**. Continued use beyond the wear limit can result in damage to the brake disc or rotor.

– Install wheel noting reference mark made on removal. Install wheel bolts and torque to specification in proper sequence. See **Tire and wheel service** in this section.

Tightening torque	
Wheel bolts to hub	120 Nm (89 ft-lb)

– Install wheel bolt caps, full wheel covers, or alloy wheel center cap as required.

– Lower vehicle. Stow jack, tools, and locking wheel bolt key in trunk.

Tire and wheel service

Tire maintenance

Tire maintenance should be carried out at least as often as every maintenance service. Pressures should be checked more often and it should be remembered that pressures will change with seasonal temperature changes. A "walk around" visual vehicle inspection including the tires, is advisable before vehicle operation.

Check tire pressures when the tires are cold due to the normal pressure rise associated with heat generated by driving. Tire pressures vary between vehicle with different engine types and between different model years. Refer to data label on fuel filler door or on driver's side "B" pillar for proper inflation pressures. See **Identification Plates and Labels** at the beginning of this section. Be sure to also check spare tire pressure where equipped.

Always replace dust caps on tire valves when checking tire pressures. If valve extensions are used (as on steel wheels and wheel covers), inspect extensions for damage and replace as required.

Tire inspection includes the following:

- Check tire tread surface and side walls for any signs of damage and remove any foreign objects from the tread as required.
- Check that all tires are the same type, size, and tread pattern, including spare.
- Measure tread depth. If tread wear has exceeded minimum specification listed, tire should be replaced.
- Check for scuffing, cupping, feathering, irregular tread wear, sidewall checking, dry rot, cuts and fractures.
- Rims (both steel and alloy) should also be checked for damage.

BentleyPublishers
.com
© 2006 BentleyPublishers.com—All Rights Reserved

All modern tires have tread wear indicators (TWI) spaced around the tire, perpendicular to the tread, at marked intervals. When the TWIs start to become visible, the tire is reaching the wear limit. When the tops of the tread wear indicators are even with the tread, the tire should be replaced.

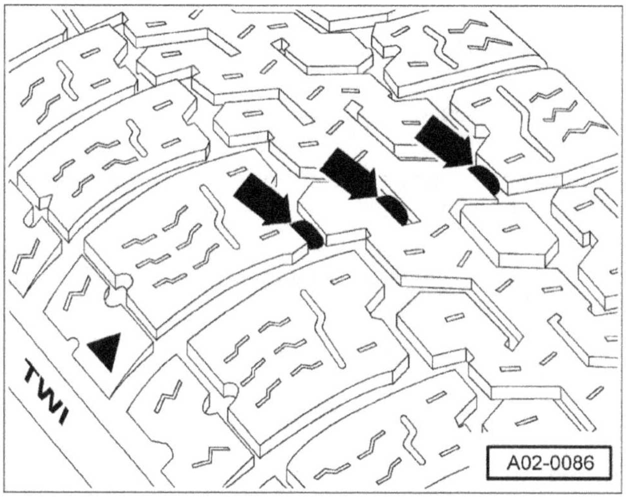

Tread wear indicator band on tire (**arrows**). Note bold **TWI** lettering and location **triangle** on sidewall at location of TWI bands.

NOTE —

- *Some tire manufacturers use symbols in place of TWI lettering on sidewalls, however, TWI bands are still at that location on tire.*
- *New summer and all season tires generally have $^{10}/_{32}$ of an inch tread depth while snow tires generally have $^{14}/_{32}$ of an inch (tread depth is measured in 32nds of an inch). Tread depth can, however, vary considerably between different designs and manufacturers.*
- *Specifications below are minimum as specified by vehicle and tire manufacturers. State, Provincial, and Local laws may specify a different tread depth minimum which should always be observed.*

Wear limit	
Tire tread depth, minimum remaining	
Summer & All-season	1.6 mm (0.063 or $^{2}/_{32}$ in)
Snow	4.0 mm (0.157 or $^{5}/_{32}$ in)

Snow tires (or winter tires) have a deep lug tread pattern and MUST be used on all 4 wheels. They are not generally suitable for warm weather use and have different speed ratings.

The factory recommends that because winter tires improve vehicle handling on snow and ice, they should be used when temperatures below 45°F (7°C) are expected. Tire pressure should generally be increased by 0.2 bar (3 psi). Always consult tire manufacturer and vehicle owner's manual for additional information and applications.

If irregular wear patterns are identified, determine causes by checking wheel alignment (front and rear) and inspecting suspension for damage and worn components.

Vibration felt through the steering wheel can be an indication of an out of balance condition which should be checked and corrected. An out of balance condition can shorten tire life, cause wear on steering and suspension components, and may result in an uncomfortable ride.

Tires should be rotated periodically for maximum service life and quiet running. Audi specifies tire rotation at the first 5000 mile service and then defers to the tire manufacturers recommendations from that point on. These recommendations can vary slightly between manufacturers, but experience has shown that rotation at 5000 to 7500 mile intervals results in optimal wear.

Audi specifies that under most conditions, tires should remain on same side of vehicle when they are rotated (front to back).

Old tires (6 years or more) may not function or perform as originally intended. These tires should only be used with caution. Audi specifies replacement after this time period regardless of tread wear.

BentleyPublishers
.com
© 2006 BentleyPublishers.com—All Rights Reserved

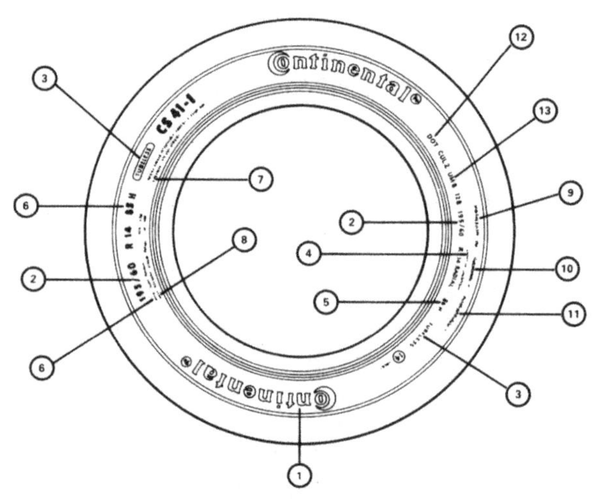

1. Manufacturers brand
2. Tire dimensions (size)
3. Type (tubeless)
4. Country of origin
5. Load index and speed rating
6. Load range (A, B, C)
7. Tread & sidewall ply and construction
8. Maximum cold inflation pressure
9. Treadwear index number
10. Traction index number (AA, A, B, C)
11. Temperature index number (A, B, C)
12. USA DOT compliant
13. Manufacturers production coding

0024416

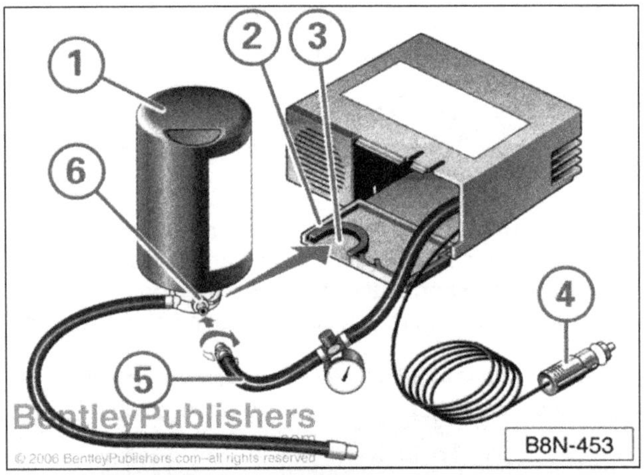

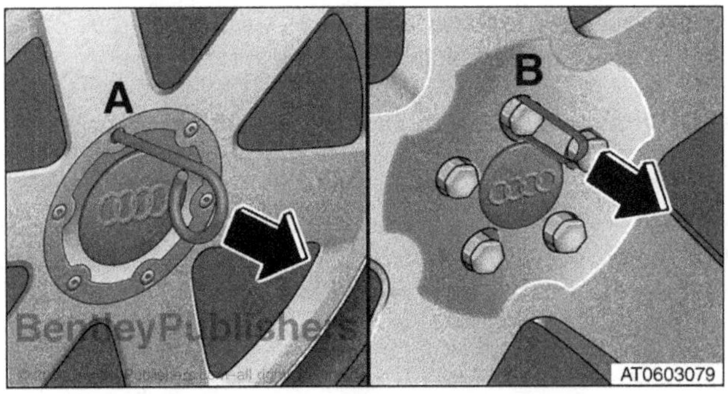

Overview of typical information found on tire sidewall that complies with United States Federal Motor Vehicle Safety Standard (FMVSS) 109. When installing new tires, the following points should be observed:

- Audi specifies that only radial tires be used.
- For safety reasons, replace tires in pairs and on the same axle. However, replacement of all tires at the same time is preferred.
- New tires should be balanced and their rims (wheels) fitted with new valve stems.
- New tires tend to be slippery and require a break-in period of at least 350 miles (560 Km).
- New tires should be the same size as originally installed. If you change sizes, be aware that gear ratios and speedometer readings will change.
- For safety reasons, only tires of the same type and tread pattern should be installed on a vehicle.
- Do not use tires of a lesser speed rating and/or load range than originally installed.
- Wheel and tires may require re-balancing as they accumulate mileage due to settling and wear.
- Make sure that wheel covers are installed correctly so as to not cut valve stems and/or extensions.
- Audi specifies that vehicles with quattro must always have tires of the same size, construction, and tread type.

Tire Mobility System

Vehicles equipped with 3.2L engines do not have a spare tire. As a result, a tire repair set called the Tire Mobility System is installed in the vehicle. A canister containing sealant is a key component and should be checked for "use by" or expiration dates since effectiveness diminishes over time. Replace sealant as indicated to insure proper function if use is needed. See glovebox owner's manual for additional information

Tire mobility system components

1. **1.** Sealant container with date stamp
2. **2.** Air compressor
3. **3.** Mounting cutout in air compressor flap
4. **4.** Power connector plug
5. **5.** Air compressor outlet hose with gauge
6. **6.** Sealant container manifold fitting with air hose to tire valve

Wheel maintenance

Visually inspect steel and alloy wheels for damage such as dents and missing balance weights. Ensure that all wheel bolts are present and tight.

From the factory, alloy road wheels could have either a center cap (version **A**) or individual lug bolt covers (version **B**). Removal is required to access the lug bolts.

– To remove center cap (version **A**), insert hook tool from vehicle tool kit into hole in cap and pull cap off (**arrow**).

– To remove lug covers (version **B**), slip U-shaped plastic puller tool from vehicle tool kit over plastic lug cover and pull straight off (**arrow**).

© 2006 BentleyPublishers.com—All Rights Reserved

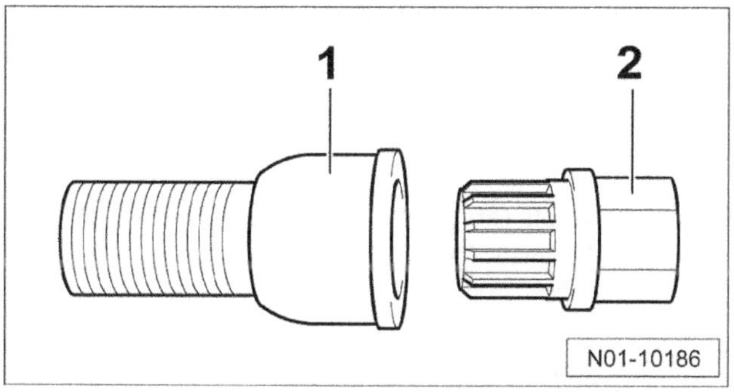

Some vehicles use wheel locks. Locking lug bolts (**1**) require a key (**2**) for removal. Several different wheel lock key types are available, but all function in a similar manner. Key is usually stored with the tool kit in the trunk area with spare tire. If using a lift or hoist, remove pouch before raising vehicle.

When installing wheel bolts, they should be snugged down in the standard diagonal pattern as shown and then tightened to final torque with the same pattern.

Tightening torque	
Wheel bolts to hub	120 Nm (89 ft-lb)

When removing and installing road wheels (alloy and steel), note the following points to be sure that wheels remain secure at all times:

- Always tighten wheel bolts in specified pattern and to specified torque. Excessive torque can distort wheel bolt seat in wheel.
- When installing a wheel, do not torque down first wheel bolt immediately (for instance, with an impact wrench). This would prevent other bolts from centering the wheel properly when they are threaded in. The rim could then come loose, even though bolts have been tightened to specified torque.
- Wheel bolts must be clean and easy to turn by hand. Do not apply grease or oil.
- Before installing wheels, examine seats of wheel bolts and contact surfaces between rims and hubs for rust, corrosion and paint. Clean off these parts if necessary, otherwise this can affect clamping force of bolts after installation, and rims can come loose.
- A puller is available in vehicle tool kit for removing wheel covers where applicable.
- A plastic mounting pin is available in vehicle tool kit that threads into one lug bolt hole to facilitate attaching wheel to rotor.

WARNING —
- *If the points listed above are not observed, wheel mountings can come loose and wheel bolt seats can become distorted.*
- *This distortion may be very slight and not visible to the naked eye. Even this slight distortion of wheel bolt seats can prevent rims from being held firmly and cause the wheels to come loose.*

Tire rotation

Periodic tire rotation should be performed according to recommendations in vehicle owner's manual. Generally, tires should be rotated to equalize wear. This keeps rolling circumference the same which is important to some vehicle systems.

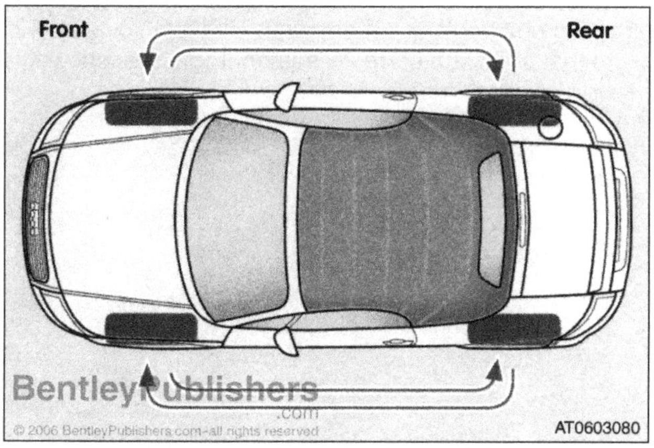

Rotate tires using 4-tire pattern keeping them on same side of vehicle; front-to-rear and rear-to-front.

BentleyPublishers.com
© 2006 BentleyPublishers.com—All Rights Reserved

Front suspension components, checking

When vehicle is safely raised and wheels are off the ground, inspect rubber boots protecting suspension and driveline components as noted.

◄ Inspect lower ball joint boots (**arrow**) for leakage, cuts and damage. Ensure that boot is correctly seated.

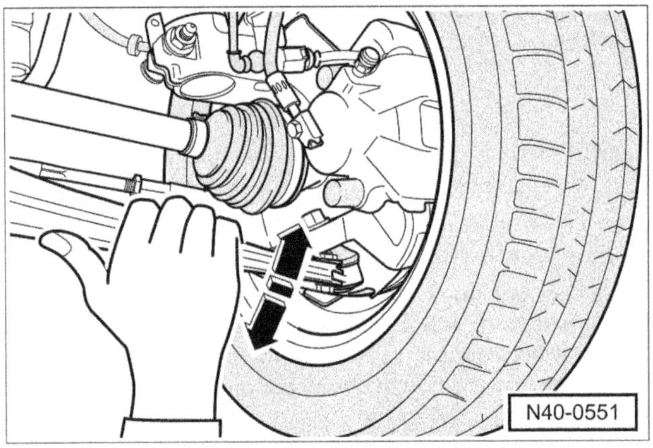

◄ Inspect lower ball joint axial play:

- Pull control arm firmly downward (direction of **arrow**) and push back up again. There should be no noticeable or visible up/down "play" in the joint.

NOTE —

- *Observe lower ball joint while checking.*
- *Allow for any wheel bearing or upper strut mount "play".*

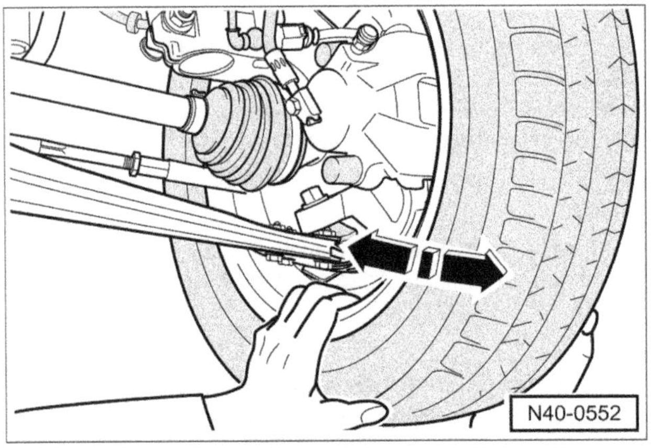

◄ Inspect lower ball joint radial clearance:

- Push wheel firmly in and out at bottom (direction of **arrow**). There should be no noticeable or visible in/out "play" in joint.

NOTE —

- *Observe lower ball joint while checking.*
- *Allow for any wheel bearing or upper strut mount "play".*

◄ Inspect outer tie rod end boots (inset **arrow**) for leakage, cuts and damage. Ensure that boots are correctly seated. Right side shown, left side is similar.

BentleyPublishers
.com
© 2006 BentleyPublishers.com—All Rights Reserved

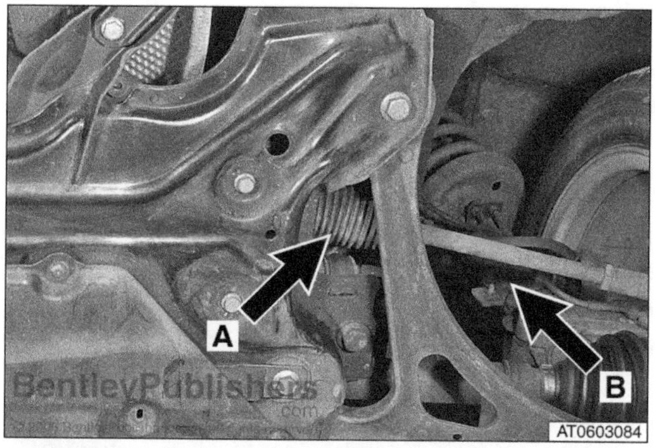

Inspect inner tie rod end boots (**A**) for leakage, cuts and damage. Ensure that boots are correctly seated. Right side shown, left side is similar.Inspect front strut (**B**) for evidence of leakage and damage. Right side shown, left side is similar.

– If inspection of front suspension components reveals wear, leakage or damage as outlined above, replacement is indicated. See **40 Front Suspension and Drive Axles** and **48 Steering** for suspension component replacement.

Drive axles, checking boots

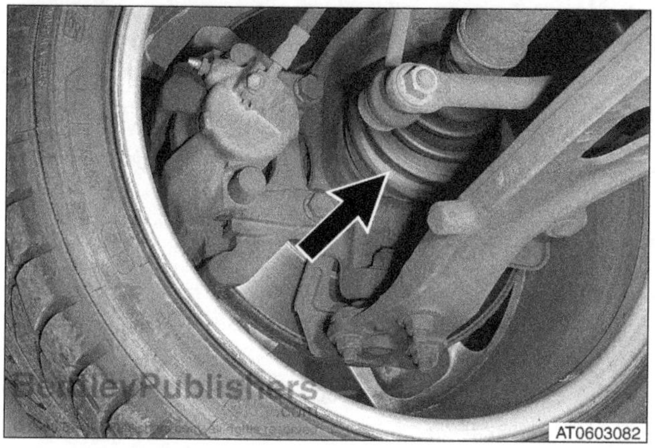

Inspect front outer constant velocity (CV) joint boots (**arrow**) for leakage, cuts and damage. Ensure that boots are correctly seated. Right front outer shown, left front outer is similar.

Inspect front inner constant velocity (CV) joint boots (**arrow**) for leakage, cuts and damage. Ensure that boots are correctly seated. Right side inner is shown, left side inner is similar.

On quattro equipped vehicles, inspect rear outer (**arrow**) and inner constant velocity (CV) joint boots for leakage, cuts and damage. Ensure that boots are correctly seated. Right rear is shown, left rear is similar.

– If inspection of drive axle components reveals wear, leakage or damage as outlined above, replacement is indicated. See **40 Front Suspension and Drive Axles** for suspension component replacement.

BentleyPublishers.com
© 2006 BentleyPublishers.com—All Rights Reserved

Fluid and exhaust leaks, visual inspection

The engine and transmission, along with components in the engine compartment should be visually inspected for leakage and damage. Fluid leaks attract dust and dirt making them somewhat easier to spot.

A visual inspection should include:

- Engine
- Transmission (manual and automatic)
- Fuel system
- Cooling and heating systems
- Brake system
- Exhaust system

Check engine and components in engine compartment from above and below for leaks and damage. Check hoses, pipes and connections for routing, signs of leaks, abrasions, chafing, porosity and brittleness.

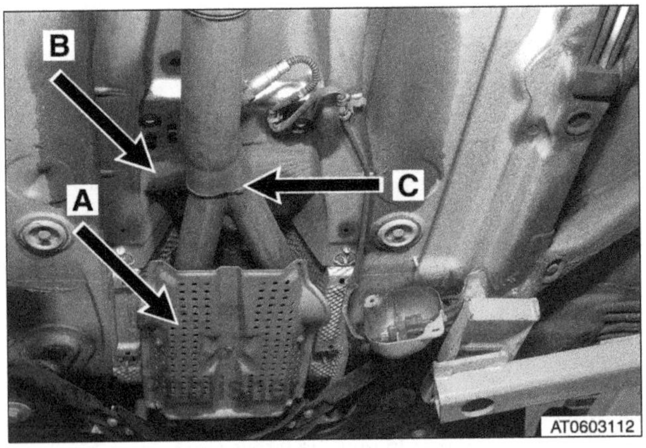

Check exhaust system for damaged and missing lower (**A**) and upper heat shields (**B**). Inspect exhaust pipes (**C**) at welds and clamps for leakage and damage.

NOTE—

- *A small amount of dampness may be considered normal in some cases, especially around axle and pulley seals since the contained fluid helps the seal work properly.*
- *Many expensive repairs can be avoided by prompt repair of minor fluid leaks.*

If fluid losses are greater than can be reasonably expected, determine cause and repair. See appropriate section for additional information.

Underbody, visual inspection

The factory recommends that underside of vehicle be inspected for damage. Damage can be caused by normal wear and tear or by driving over road debris.

Whenever vehicle is raised on a lift or a hoist, it is advisable to visually inspect underbody, wheel wells, and sill/rocker panels for damage to underbody sealants and coatings. Inspection should also be made after major repairs to vehicle systems.

Any damage or defects found must be promptly repaired. Only use wax-based or tar-based anti-corrosion compounds as specified. Do not use oil-based anti-corrosion sprays due to possible incompatibility with factory applied protection. Check with an authorized Audi Dealer's Parts Department for appropriate sealing compounds.

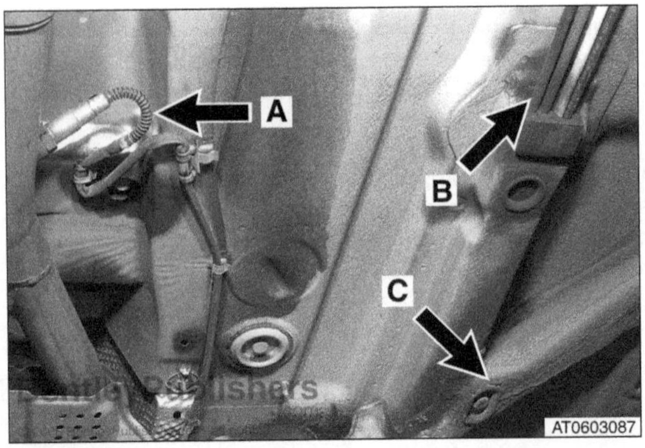

Special attention should be given to underbody components such as electrical components (**A**), fuel and vent lines (**B**). Underbody coatings should be inspected for peeling and damage (**C**).

BentleyPublishers.com
© 2006 BentleyPublishers.com—All Rights Reserved

Transmission Oil/Fluid

Transmission oil (manual transmission)

On front wheel drive vehicles with manual transmissions, the gear box and final drive share a common oil supply. Checking oil level of transmission also checks level of final drive (differential). All wheel drive (quattro) vehicles with manual transmissions are configured the same as front wheel drive vehicles, but add a bevel gear transfer case which shares the same gear oil supply.

Access to the gear oil supply is through fill plug on side of transmission. For manual transmission gear oil specifications and quantities, see **34a Manual Transmission**.

◄ Removal of front wheel drive manual transmission and rear differential oil fill plug requires a 17 mm hex key or hex key socket (**left**). Removal of quattro manual transmission fill plug requires either a 17 mm hex key described above or a triple-square socket driver such as Audi tool 3357. A combination tool (**right**) with both 16 mm anti-tamper triple-square (XZN) and 17 mm ends is commonly available as MN2567.

Front wheel drive manual transmission oil, checking and /or changing

NOTE —

- *Vehicle must sit level during this operation. If front of vehicle is higher than rear, such as when driven up on ramps, oil inside transmission will pool towards rear, and level cannot be accurately determined.*

– Raise vehicle. See **Lifting Vehicle** in this repair group.

– Remove lower sound absorber panel. See **Sound absorber panels, removing and installing** in this repair group.

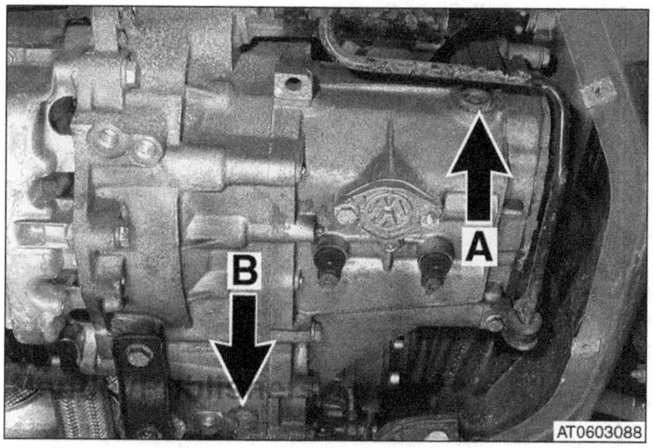

◄ Remove fill plug (**A**) on front of transmission. Oil level is correct when transmission is filled up to lower edge of fill hole. Note drain plug (**B**).

– If checking gear oil level, install fill plug (**A**) and tighten.

Tightening torque	
Front wheel drive transmission fill plug	25 Nm (18 ft-lb)

– If changing gear oil, position suitable drain pan under transmission, remove drain plug (**B**) and allow oil to drain into pan.

– When oil has completely drained, install drain plug (**B**) and tighten.

– When filling with oil, top off oil to lower edge of fill hole. Screw in fill plug, start engine, engage a gear and allow transmission to turn slowly for approximately 2 minutes.

– Switch off engine and remove fill plug. Top off transmission oil again up to lower edge of fill hole. Install fill plug and tighten.

– Install lower sound absorber panel.

BentleyPublishers
.com
© 2006 BentleyPublishers.com—All Rights Reserved

All wheel drive manual transmission oil, checking and/or changing

NOTE—
- *Vehicle must sit level during this operation. If front of vehicle is higher than rear, such as when driven up on ramps, oil inside transmission will pool towards rear, and level cannot be accurately determined.*
- *When checking transmission oil level, level of bevel gear transfer case is also checked.*

– Switch off ignition.

– Raise vehicle. See **Lifting Vehicle** in this repair group.

– Remove lower sound insulation. See **Sound absorber panels, removing and installing** in this repair group.

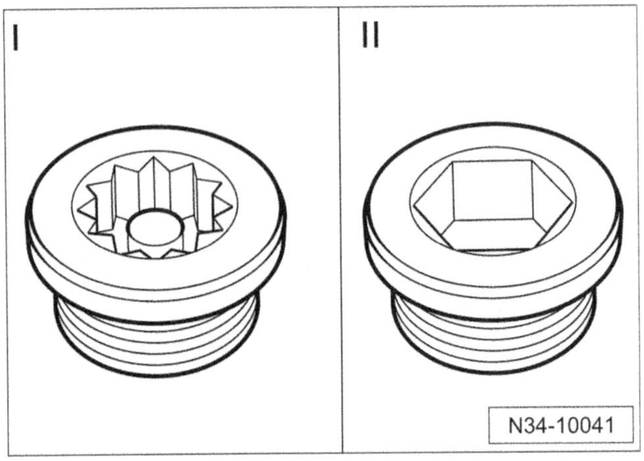

Drain and fill plug will be either of two different versions. Both are equipped with sealing rings.
- **I** - 16 mm triple-square (XZN) with (center) anti-tamper pin.
- **II** - 17 mm internal hex.

NOTE—
- *Replace sealing ring whenever fill plug is removed.*

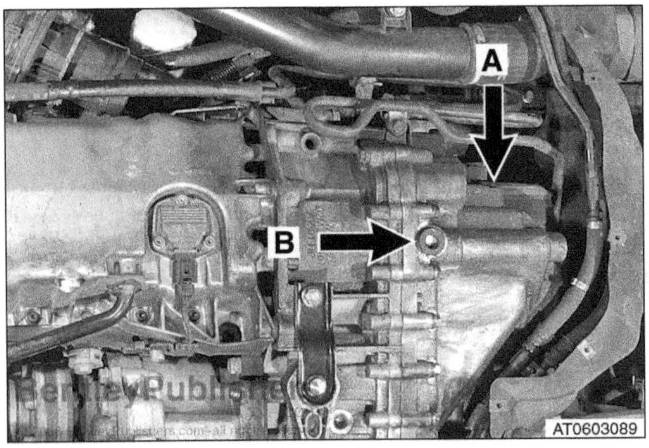

Remove fill plug (**A**) on front of transmission. Oil level is correct when transmission is filled up to lower edge of fill hole.

– If checking gear oil level, install fill plug (**A**) with new sealing ring and tighten.

Tightening torque	
All wheel drive transmission fill plug I (16 mm triple-square)	45 Nm (33 ft-lb)
All wheel drive transmission fill plug II (17 mm internal hex)	30 Nm (22 ft-lb)

– If changing gear oil, position suitable drain pan under transmission, remove drain plug (**B**) and allow oil to drain into pan.

– When oil has completely drained, install drain plug (**B**) with new sealing ring and tighten.

– When filling with oil, top off oil to lower edge of fill hole. Screw in fill plug, start engine, engage a gear and allow transmission to turn slowly for approximately 2 minutes.

– Switch off engine and remove fill plug. Top off transmission oil again up to lower edge of fill hole. Install fill plug with new sealing ring and tighten.

– Install lower sound absorber panel.

BentleyPublishers
.com
© 2006 BentleyPublishers.com—All Rights Reserved

Rear final drive and Haldex (all wheel drive)

Rear differential oil, checking and/or changing

– Raise vehicle. See **Lifting Vehicle** in this repair group.

NOTE —

- *Vehicle must sit level during this operation. If front of vehicle is higher than rear, such as when driven up on ramps, oil inside transmission will pool towards rear, and level cannot be accurately determined.*

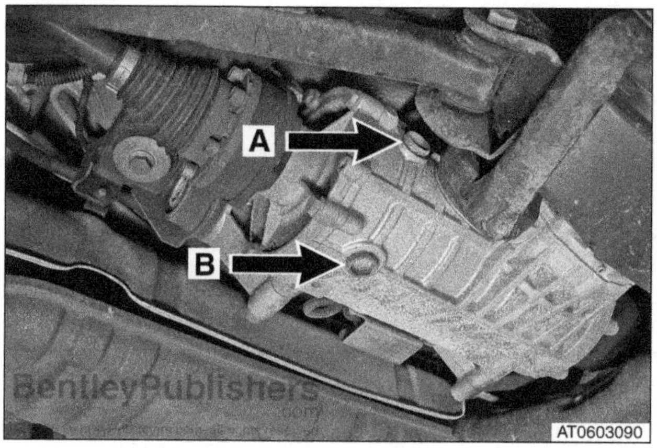

AT0603090

< Remove fill plug (**A**) on front of differential. Oil level is correct when differential is filled up to lower edge of fill hole.

– If changing gear oil, position suitable drain pan under transmission, remove drain plug (**B**) and allow oil to drain into pan.

– When oil has completely drained, install drain plug (**B**) and tighten.

Tightening torque	
Rear differential fill and drain plug	25 Nm (18 ft-lb)

– When filling with oil, top off oil to lower edge of fill hole. Screw in fill plug (**A**) and tighten.

Haldex oil, checking and/or changing

Checking or changing oil in the Haldex unit on the rear final drive assembly requires measurement of oil temperature. Use a scan tool connected to the data link connector and configured as follows:

- Address word 22 (4 wheel drive electronics)
- Function 08 (read measuring value blocks)
- Display group 002
- Display field 2

Read Measuring Value Block			002
XXX	30°C	XXX	XXX

AT0603201

< Proper oil level can only be obtained when temperature displayed in field 2 is between 20°C and 40°C (corresponding to 68°F and 104°F). Vehicle can be driven to bring temperature up. It should be allowed to cool if too hot. Disregard information in other display fields.

NOTE —

- *Vehicle must sit level during this operation. If front of vehicle is higher than rear, such as when driven up on ramps, oil inside transmission will pool towards rear, and level cannot be accurately determined.*
- *If changing Haldex oil and filter, replace oil filter before changing oil.*

– Raise vehicle. See **Lifting Vehicle** in this repair group.

AT0603092

< Using a 5 mm hex key, remove inspection plug (**A**). Oil level is correct when Haldex clutch is filled to the lower edge, or to within approximately 3 mm (1/8 in) of the lower edge of the inspection hole.

– If level is correct, install new seal on inspection plug (**A**) and tighten.

Tightening torque	
Rear differential fill and drain plug	30 Nm (22 ft-lb)

– If level is not correct, change Haldex oil.

BentleyPublishers.com
© 2006 BentleyPublishers.com—All Rights Reserved

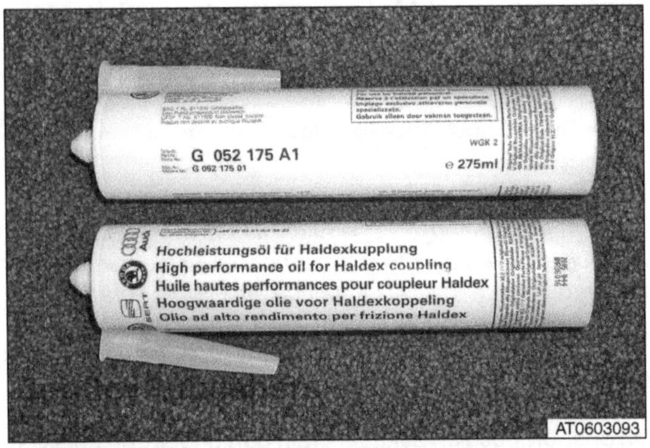

AT0603093

◄ Special high performance oil is used in the Haldex section of the rear differential. Due to the need for filling through the drain hole in the bottom of the Haldex, this special oil is packaged in cartridges with a threaded-on nozzle. A cartridge gun such as the VAG 1628 is specified by the factory, however, cartridges are a standard size and will generally work properly in conventional "caulking" type guns with a manually operated plunger. One cartridge (or tube) of 275 ml will supply enough oil for the Haldex unit (250 ml) and still allow for normal spillage and residual oil remaining in cartridge when the plunger is pushed fully.

– Confirm oil temperature with a scan tool as given above.

– Position suitable drain pan under differential.

AT0603092

◄ Using a 5 mm hex key, remove drain plug (**B**) and allow oil to drain into pan.

> ***CAUTION—***
> - *Do not use inspection hole (**A**) to fill Haldex unit. Unit should only be filled through drain hole (**B**) due to the possibility of air becoming trapped resulting in insufficient lubrication.*

– While oil is draining, clean drain plug and install new seal. Put aside for quick access.

– Insert cartridge into cartridge gun, hold vertically, and snip off end of cartridge.

– Thread nozzle onto cartridge and cut nozzle to appropriate size.

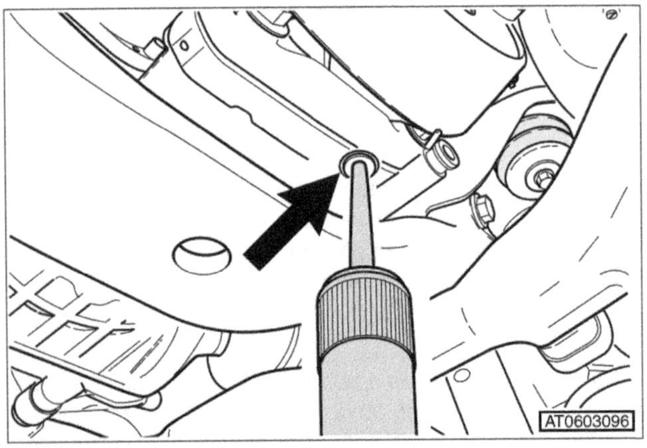
AT0603096

◄ Insert cartridge vertically into drain hole (**arrow**) and quickly pump/push oil into Haldex clutch.

– When oil has been completely emptied from cartridge, immediately remove cartridge from drain and install drain plug with new seal.

Tightening torque	
Rear differential fill and drain plug	30 Nm (22 ft-lb)

– Recheck Haldex oil level.

Haldex oil filter, replacing

NOTE—
- *If changing Haldex oil and filter, replace oil filter before changing oil.*

AT0603091

◄ Haldex oil filter (**arrow**) is located on the passenger's side of the differential above the differential mounts. Access to the filter is severely restricted. Cover is constructed of aluminum or plastic.

– Remove passenger's side rear sound absorber/aero panel. See **Sound absorber panels, removing and installing** in this repair group.

BentleyPublishers
.com
© 2006 BentleyPublishers.com—All Rights Reserved

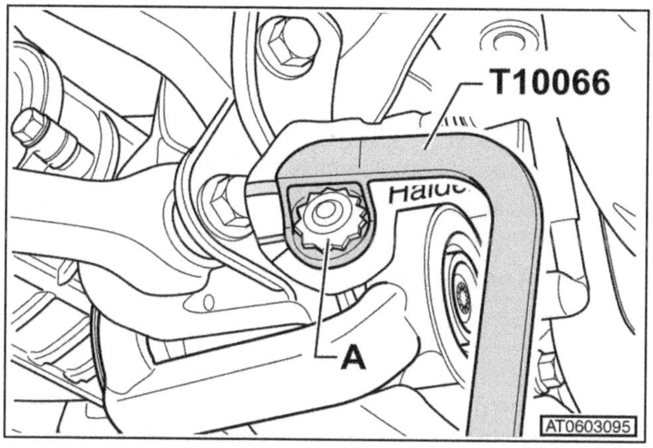

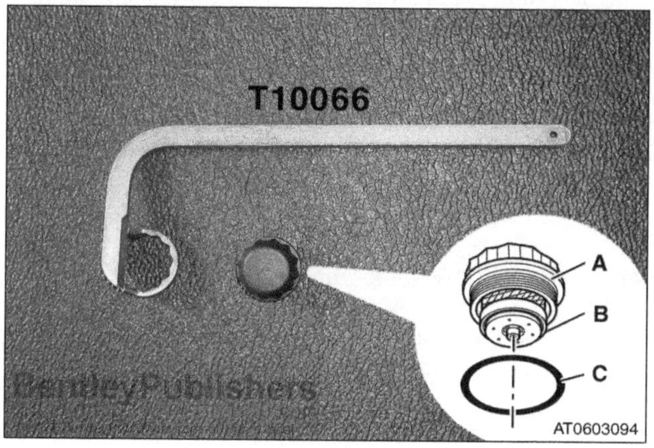

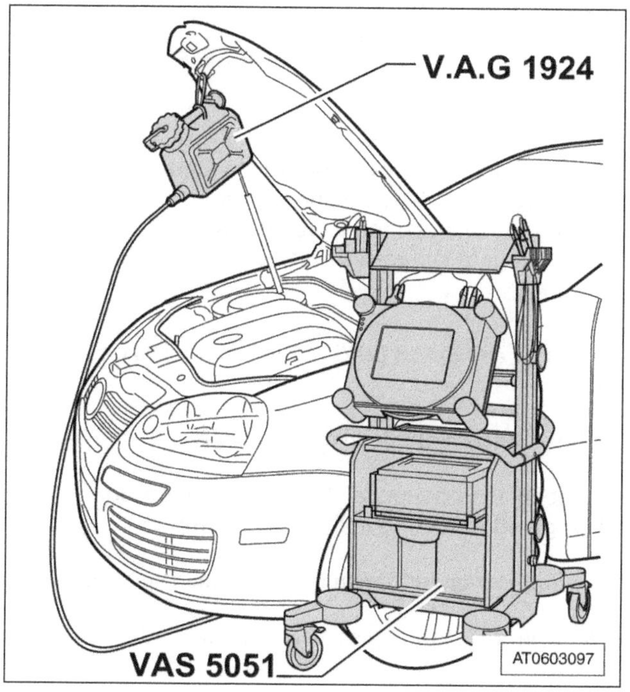

Special wrench T10066 allows removal of Haldex oil filter. Carefully insert wrench into space between housing and mounts from the front taking care not to force it.

NOTE—

- *Wrench placement on the filter cover can be difficult to see. Illustration shows relative position of wrench installation that cannot be seen because of suspension interference.*

– Loosen oil filter cover and remove filter assembly.

NOTE—

- *Wrench movement is limited to only a few degrees of travel. Repositioning wrench on cover must be done several times to remove cover. After "breaking free", it is usually possible to hand loosen filter assembly.*
- *Do not pry against components attached to Haldex housing. Control unit and pump are secured to housing near space where wrench is inserted.*

Special 12-point wrench T10066 is approximately 46 mm (1 13/16 in) and fits on either plastic or aluminum cover. Filter supplied by the factory consists of cover (**A**), filter element (**B**), and O-ring (**C**).

– When installing new filter, lightly coat O-ring with Haldex oil and screw filter assembly into housing and tighten.

Tightening torque	
Haldex oil filter to housing	35 Nm (26 ft-lb)

– Recheck Haldex oil level.

– Install passenger's side rear sound absorber/aero panel. See **Sound absorber panels, removing and installing** in this repair group.

Transmission fluid (automatic transmission)

On vehicles with a 6-speed automatic transmission, the automatic section and the final drive share a common fluid supply. Checking ATF level of transmission also checks fluid level of final drive (differential).

Checking level

ATF level is checked by overflow method which is temperature dependent. ATF level will vary according to temperature as volume increases with a rise in temperature.

Checking ATF level or adding ATF requires the ability to check ATF temperature via transmission control module on-board diagnostics. The VAS 5051 shown here can be used, as can numerous other scan tools and scan tool computer programs. A means of adding fluid is also needed since fluid is added from underneath. A gravity-fed system such as VAG 1924 shown here or an equivalent should be used.

BentleyPublishers.com
©2006 BentleyPublishers.com—All Rights Reserved

Before starting an ATF level check, observe the following:

- All electrical consumers must be switched off.
- A/C system must be switched off.
- Transmission not in emergency running mode.
- ATF temperature not above approximately 30°C (86°F).
- Vehicle in a level, horizontal position, with all 4 supports of hoist or lift at same height.
- Selector lever in Park position, engine at idle.

NOTE —

- *Vehicle must sit level during this operation. If front of vehicle is higher than rear, such as when driven up on ramps, oil inside transmission will pool towards rear, and level cannot be accurately determined.*

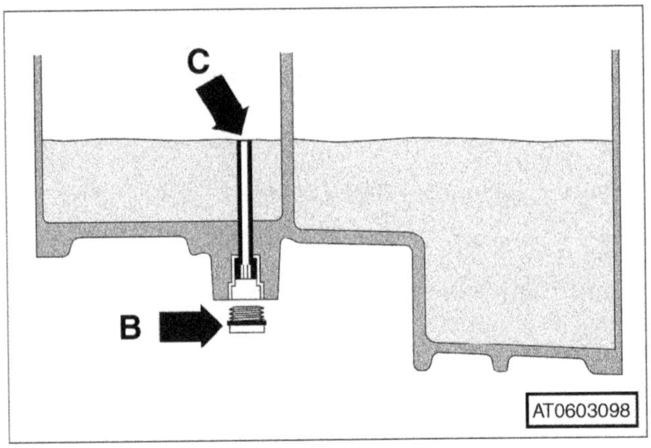

Cross sectional overview of transmission oil pan shows inspection plug (**B**) and overflow pipe (**C**). ATF level is correct when it reaches overflow pipe at given temperatures.

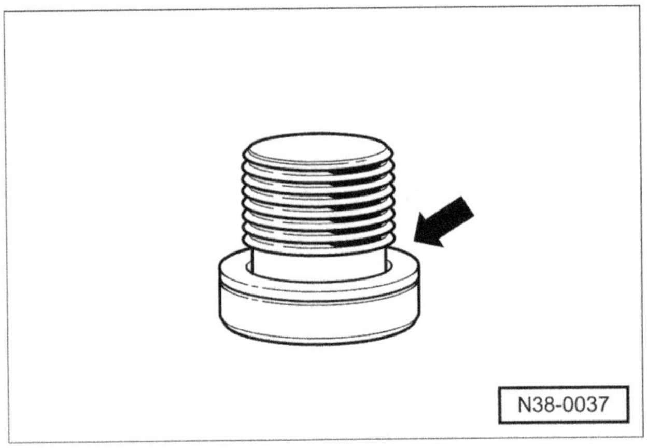

Before starting an ATF level check, be sure to have a new inspection plug seal available. Always replace seal (**arrow**) every time level is checked.

NOTE —

- *Checking ATF level at an ATF temperature that is too low (cold) leads to overfilling. Performing ATF level check when ATF temperature is too high (hot) will cause insufficient filling.*
- *If ATF level is too high or too low it will adversely affect operation of transmission.*

– Secure container of ATF fill tool VAG 1924 to vehicle so that it is higher than transmission.

– Connect diagnostic scan tool to Data Link Connector (DLC) and start program so that ATF temperature can be read. Scan tool can be configured as follows:

- Address word 02 (transmission electronics)
- Function 08 (read measuring value blocks)
- Display group 006
- Display field 1

Read Measuring Value Block			006
40°C	XXX	XXX	XXX

AT0603202

Proper oil level can only be obtained when temperature displayed in field 1 is between 35°C and 45°C (corresponding to 95°F and 113°F). Vehicle can be driven to bring temperature up. It should be allowed to cool if too hot. Disregard information in other display fields.

NOTE —

- *Temperature can be as high as 50°C (122°F) in tropical or hot climate areas.*

– Raise vehicle. See **Lifting Vehicle** in this repair group.

– Remove lower sound absorber panel. See **Sound absorber panels, removing and installing** in this repair group.

– Place a suitable drip tray under transmission to catch spillage.

– With selector lever in "P", start engine and let it run at idle.

BentleyPublishers.com
© 2006 BentleyPublishers.com—All Rights Reserved

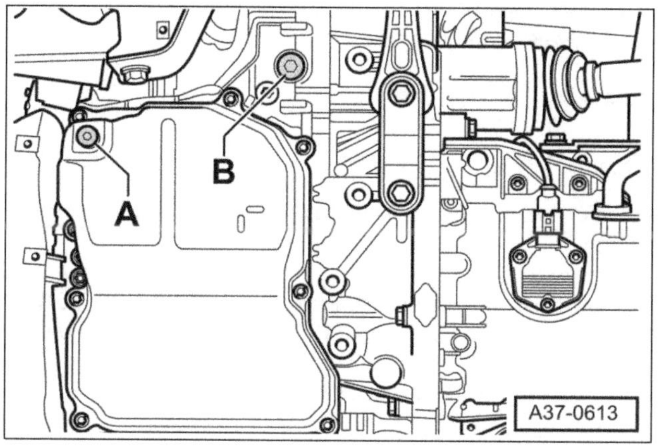

When ATF temperature reads between 35°C and 45°C (corresponding to 95°F and 113°F) on scan tool display, remove inspection plug (**A**) from transmission pan. Do not remove drain plug (**B**).

NOTE—
- *A small amount of ATF contained in the overflow pipe may drip out.*

– If ATF drips from inspection plug hole before ATF temperature reaches 40°C, ATF level is correct. No additional ATF is needed.

– Before temperature rises out of range, reinstall inspection plug with new seal and tighten.

Tightening torque	
Transmission inspection plug to transmission	27 Nm (20 ft-lb)

– If no ATF drips out by the time ATF temperature reaches 45°C, ATF must be topped off. Screw inspection plug with old seal in hand-tight and switch engine off immediately so that ATF temperature does not rise unnecessarily. See **ATF, topping off** below.

ATF, topping off

Continuing from ATF checking procedure above, engine remains switched off.

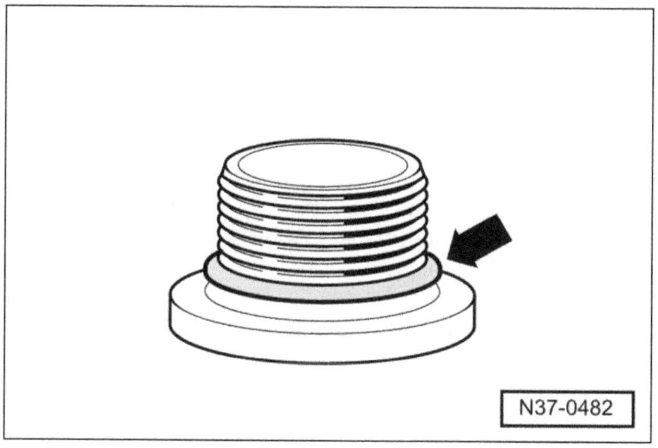

Before topping off ATF, be sure to have a new fill plug seal available. Always replace seal (**arrow**) every time fill plug is removed.

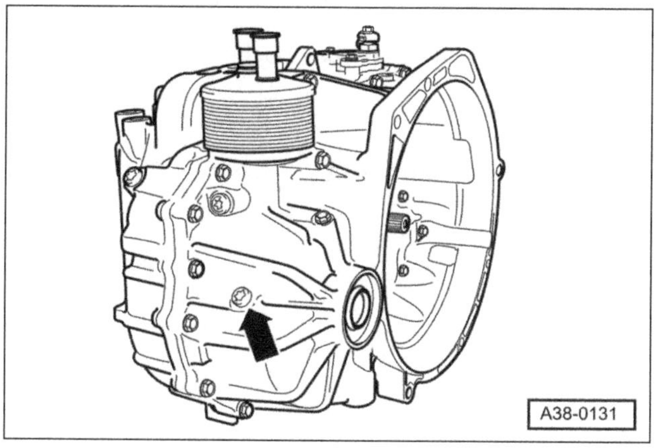

Using a T55 wrench, remove ATF fill plug (**arrow**) from rear of transmission housing.

BentleyPublishers
.com
©2006 BentleyPublishers.com—All Rights Reserved

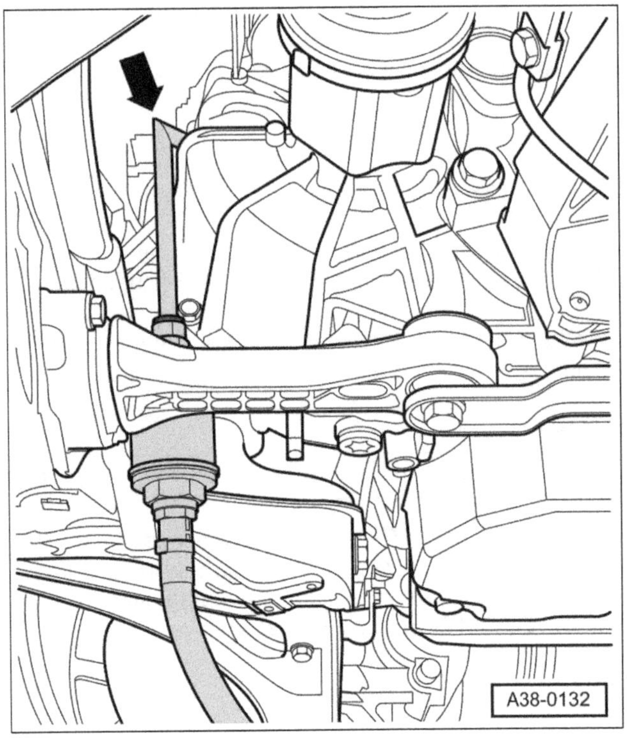

◀ Hang filler nozzle (**arrow**) of gravity-fed fluid filler system such as VAG 1924, or equivalent, in filler hole as shown.

– Re-start engine.

– Re-start scan tool.

– Remove inspection plug (previously installed hand-tight only) from transmission pan.

– Add ATF until fluid drips out when temperature reads between 35°C and 45°C (corresponding to 95°F and 113°F) on scan tool display.

– Reinstall inspection plug with new seal and tighten.

Tightening torque	
Transmission inspection plug to pan	27 Nm (20 ft-lb)

– Remove ATF filler equipment from transmission housing.

– Reinstall fill plug with new seal and tighten.

Tightening torque	
Transmission fill plug to transmission	39.5 Nm (29 ft-lb)

– Install lower sound absorber panel. See **Sound absorber panels, removing and installing** in this repair group.

– Disconnect scan tool.

ATF, changing

Automatic transmission fluid can be changed without removal of transmission oil pan or replacement of filter. ATF does not require changing as part of regular maintenance. Replace ATF after repairs requiring removal of the ATF oil pan.

– Raise vehicle. See **Lifting Vehicle** in this repair group.

– Remove lower sound absorber panel. See **Sound absorber panels, removing and installing** in this repair group.

– Place a suitable drip tray under transmission to catch spillage.

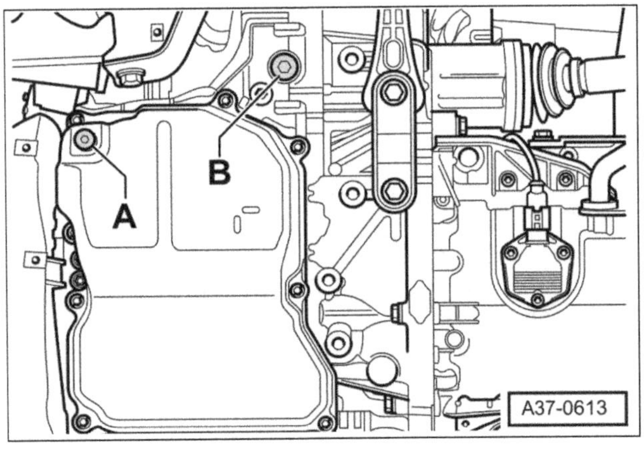

◀ Remove drain plug (**B**) from transmission housing.

– Drain ATF into suitable container.

– Remove inspection plug (**A**) from transmission pan.

BentleyPublishers.com
© 2006 BentleyPublishers.com—All Rights Reserved

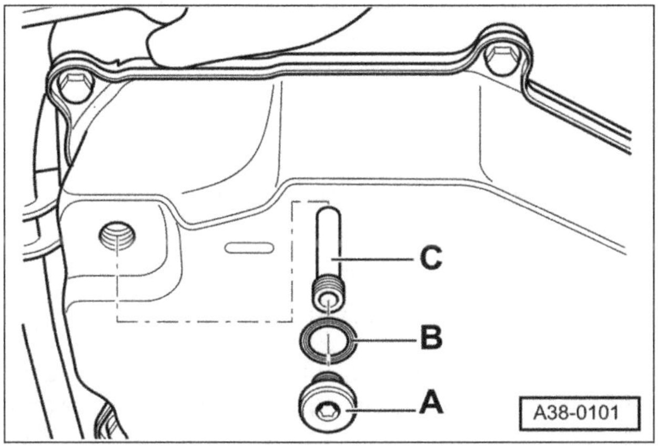

With inspection plug (**A**) and seal (**B**) removed, unscrew overflow tube (**C**) using a 5 mm hex key through inspection hole.

– Drain remaining ATF into suitable container.

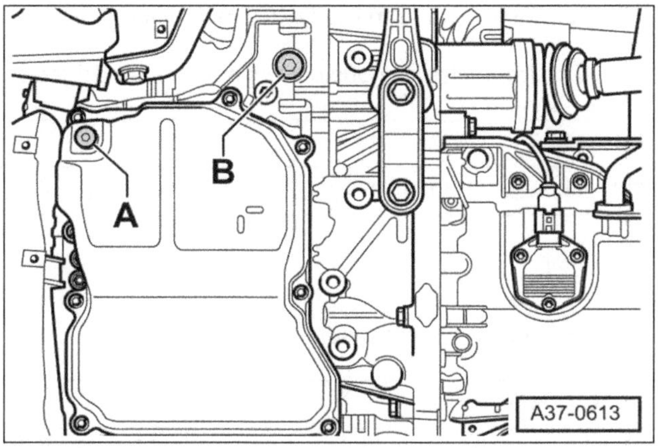

When ATF is completely drained, reinstall drain plug (**B**) with new seal into transmission housing and tighten.

Tightening torque	
Transmission drain plug to transmission	39 Nm (29 ft-lb)

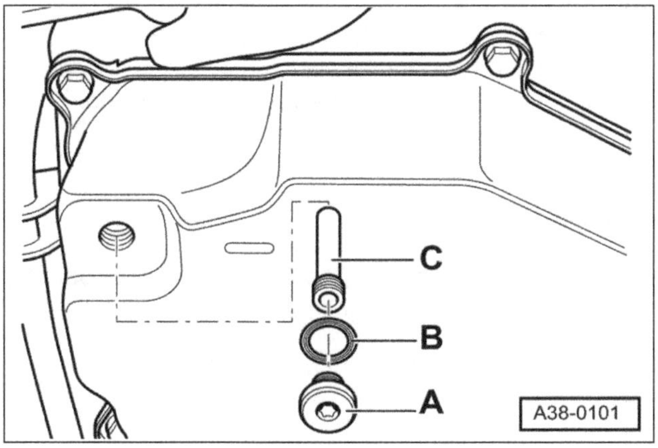

When ATF is completely drained, reinstall overflow tube (**C**) and tighten.

Tightening torque	
Transmission overflow tube to pan	1 Nm (9 in-lb)

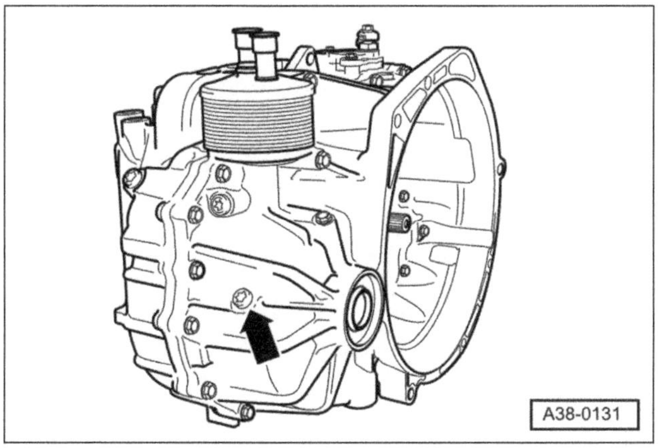

Using a T55 wrench, remove ATF fill plug (**arrow**) from rear of transmission housing.

BentleyPublishers
.com
© 2006 BentleyPublishers.com—All Rights Reserved

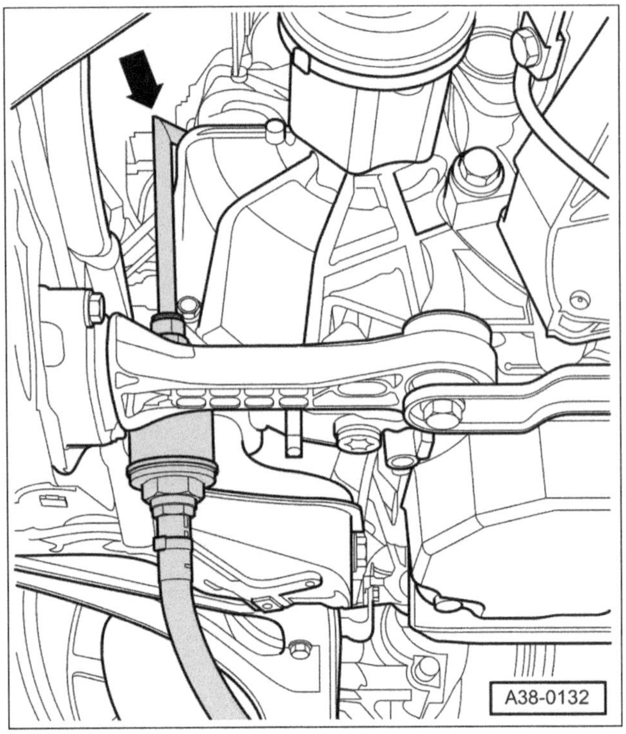

◀ Hang filler nozzle (**arrow**) of gravity-fed fluid filler system such as VAG 1924 or equivalent in filler hole as shown.

– Add ATF until it drips out of overflow pipe.

NOTE—
- *Do not run engine with fluid drained from transmission.*

– Start engine.

– Start scan tool and configure to read ATF temperature.

– Continue adding ATF until fluid drips out when temperature reads between 35°C and 45°C (corresponding to 95°F and 113°F) on scan tool display.

– Remove ATF filler equipment from transmission housing.

– Reinstall fill plug on transmission housing with new seal and tighten.

Tightening torque	
Transmission fill plug to transmission	39.5 Nm (29 ft-lb)

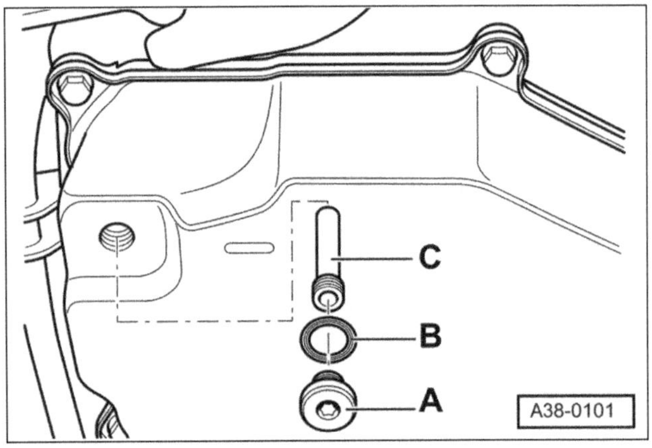

◀ Install inspection plug (**A**) with new seal (**B**) and tighten.

Tightening torque	
Transmission inspection plug to pan	27 Nm (20 ft-lb)

– Start engine and shift once through all selector lever positions, remaining in each shifted position for approximately 3 seconds, with vehicle at a standstill. Return selector lever to Park position.

– Recheck ATF level and top off as required.

– Check for ATF leaks.

– Install lower sound absorber panel. See **Sound absorber panels, removing and installing** in this repair group.

– Lower vehicle, disconnect scan tool or scan tool computer program.

BentleyPublishers .com

© 2006 BentleyPublishers.com—All Rights Reserved

Transmission oil and filter (direct shift gearbox)

The 6-speed direct shift gearbox (DSG) is also known as a double-clutch transmission. Engine torque is transferred to transmission by way of the dual-mass flywheel. The transmission is constructed as a 6-speed manual transmission. Alternating hydraulic actuation of the two wet multi-plate clutches permits operation similar to that of an automatic transmission. The gears are engaged automatically or manually via the Tiptronic mode. There is no clutch pedal.

The selector lever position is not transferred mechanically via selector lever cable as with the 09G automatic transmission. Selector lever positions or gear shifts are transferred by way of a separate control module for the shift mechanism through the CAN-Bus to the transmission control module. While gear shifting occurs without a cable, when the selector lever is in position "P", a parking lock acting on the differential carrier is engaged mechanically by the selector lever cable.

◄ The 6-speed gearbox section and final drive share a common fluid supply which is specifically designed for the DSG transmission. Checking oil level of gearbox also checks oil level of final drive (differential).

> ***WARNING —***
> - *Do not start engine if there is little or no oil in transmission such as immediately following repairs or after severe loss or leakage of transmission oil.*
> - *If there is little or no oil in transmission, add sufficient quantity with engine cold and not running before proceeding with a diagnosis.*
> - *Use only oil specified for direct shift gearboxes (DSG). Always consult your Audi Parts Department or aftermarket parts specialist for the latest information. Use of oils other than those specified will lead to malfunctions or failure of the transmission.*
> - *Observe all possible care when working on the DSG.*
> - *Observe all rules for cleanliness when working on the DSG.*
> - *Special tools are required for some operations. Review all repair procedures before starting.*
> - *Wear eye protection and acid-resistant protection gloves.*

NOTE —
- *Transmission oil level changes with transmission oil temperature.*
- *Checking oil level at an oil temperature that is too low (cold) leads to overfilling. Performing oil level check when oil temperature is too high (hot) will cause insufficient filling.*
- *Oil level that is too high or too low it will adversely affect operation of the transmission.*
- *Always replace sealing rings for drain and check plugs.*

– Vehicle must sit level during these operations. If front of vehicle is higher than rear, such as when driven up on ramps, oil inside transmission will pool towards rear, and level cannot be accurately determined.

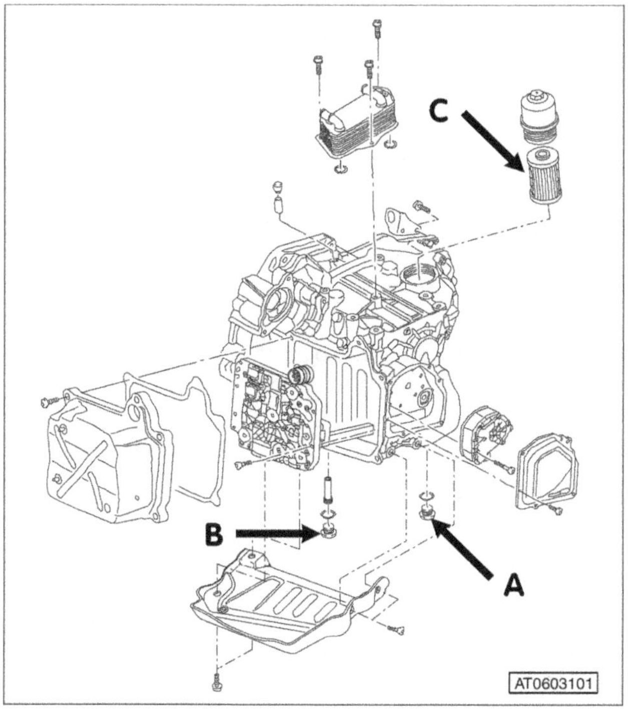

◄ When servicing, note difference between drain plug (**A**) and check plug (**B**). Check plug (**B**) is located near pendulum support. Do not mistake one for the other. Not all transmissions will have drain plug (**A**).

– DSG transmission oil filter (**C**) is located on top under a cover.

BentleyPublishers.com

© 2006 BentleyPublishers.com—All Rights Reserved

Prerequisites

Before beginning work involving fluid in the DSG, observe the following:

- Ignition, A/C, and all electrical consumers switched off.
- Vehicle in a level, horizontal position, with all 4 supports of hoist or lift at same height.
- Scan tool or scan tool computer program connected.
- At the start, ATF temperature not above approximately 30°C (86°F).

DSG transmission oil, checking level

– Be sure that appropriate pre-requisites listed above are met.

– Connect scan tool to Data Link Connector and start program so that ATF temperature can be read. Configured scan tool as follows:
  - Address word 02 (transmission electronics)
  - Function 08 (read measuring value blocks)
  - Display group 019
  - Display field 1

Read Measuring Value Block			019
40°C	XXX	XXX	XXX

AT0603203

◀ Proper oil level readings can only be obtained when temperature displayed in field 1 is between 35°C and 45°C (corresponds to 95°F and 113°F). Vehicle can be driven to bring temperature up. It should be allowed to cool if too hot. Disregard information in other display fields.

– Start engine and shift once through all selector lever positions, remaining in each shifted position for approximately 3 seconds, with vehicle at a standstill. Move shifter lever back into "P" position and leave engine running at idle. Do not switch off engine!

– Raise vehicle. See **Lifting Vehicle** in this repair group.

– Remove lower sound absorber panel. See **Sound absorber panels, removing and installing** in this repair group.

– Place a suitable drip tray under transmission to catch spillage.

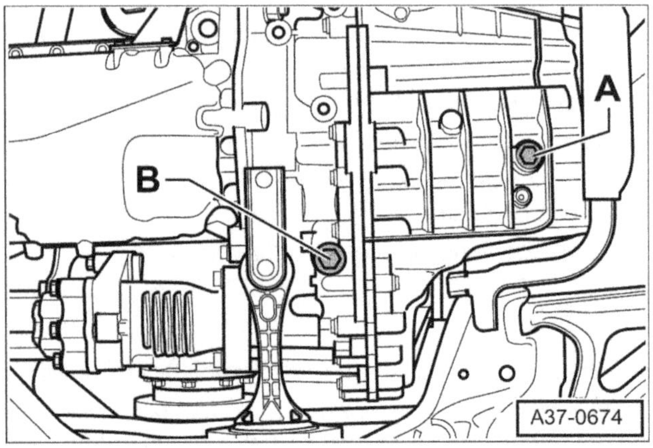

> ***WARNING —***
> - *Electric radiator fan may switch on at any time. Always maintain a safe working distance.*
> - *Wear eye protection and acid-resistant protection gloves.*

◀ When transmission oil temperature reaches 35°C (corresponds to 95°F) on scan tool display, remove check plug (**B**) from transmission.

NOTE —
- *Note difference between the drain plug (**A**) and check plug (**B**). Check plug (**B**) is located near the pendulum support. Do not mistake one for the other.*
- *When oil check plug (**B**) is removed, overflow tube will be accessible.*

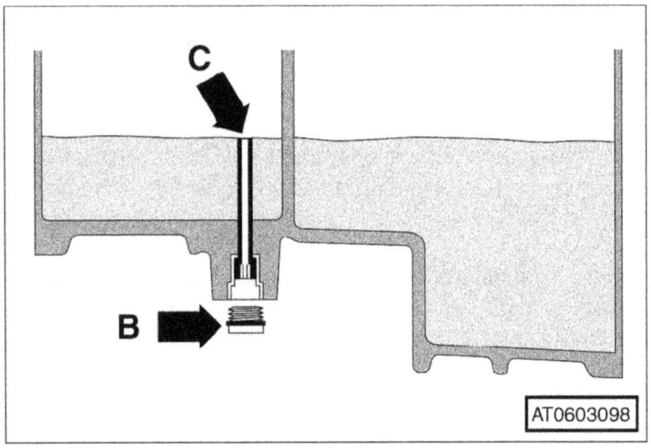

◀ Allow excess oil in overflow pipe (**C**) to drain out.

– If a small amount of gear oil drips out when transmission oil temperature is between 35°C and 45°C (corresponds to 95°F and 113°F), transmission oil level is correct. This occurs because fluid expands and level rises as it heats up. In this case, no transmission fluid needs to be added; fluid level is correct.

BentleyPublishers.com
© 2006 BentleyPublishers.com—All Rights Reserved

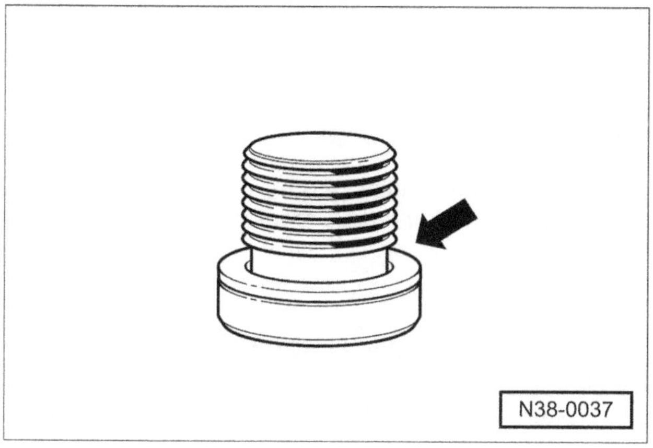

NOTE —

- *Every 30 seconds, a small surge of oil will exit overflow tube, independent of the height of the transmission oil level. This is due to cool oil impulse to clutches. This surge of oil is **not** indicative of excessive transmission oil level and must not be considered when determining transmission oil level.*
- *Transmission oil that drains out must not be reused. Dispose of oil in accordance with all applicable regulations.*

If gear oil level is correct, install check plug with a new seal (**arrow**) and tighten.

Tightening torque	
Transmission check plug	45 Nm (33 ft-lb)

- Check for leaks.
- Install lower sound absorber panels.
- Lower vehicle.
- Remove scan tool.
- If no gear oil has dripped out by the time the oil temperature reaches 45°C (corresponds to 113°F), level is low and needs to be topped up.

DSG transmission oil, topping up

- If no gear oil has dripped out (except for excess oil in overflow pipe) when oil temperature is between 35°C and 45°C (corresponds to 95°F and 113°F), transmission oil level is low, and needs to be topped up.

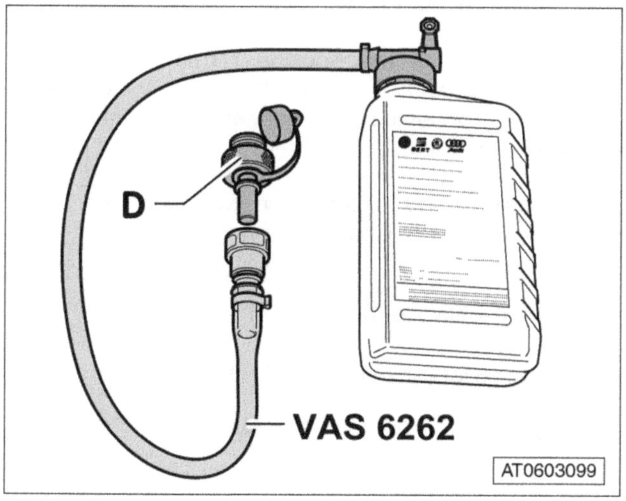

Gear oil is added from below through the oil check plug using special filling apparatus such as the VAS 6262 which attaches to the container of gear oil as shown.

NOTE —

- *Make sure that hose and adapter fitting of VAS 6262 are clean. Do not allow DSG transmission oil to be mixed or contaminated by any other type of oil or fluid.*

- Thread adapter (**D**) into check plug opening finger tight.

NOTE —

- Attach VAS 6262 hose assembly with bottle to adapter (**D**) and hold as high as possible over transmission.

- Add gear oil from below.
- Remove filling apparatus, reinstall check plug.
- Recheck gear oil level.
- When gear oil level is correct, install check plug with a new seal and tighten.

Tightening torque	
Transmission check plug	45 Nm (33 ft-lb)

BentleyPublishers
.com
©2006 BentleyPublishers.com—All Rights Reserved

– Check for leaks.

– Install lower sound absorber panels.

– Lower vehicle.

– Remove scan tool.

DSG transmission oil filter, changing

DSG oil and filter require changing as part of regular maintenance. Also replace transmission oil after repairs requiring removal of the oil pan.

NOTE —
- *DSG oil filter is accessed from above and requires removal of air cleaner assembly.*
- *Be sure that appropriate prerequisites listed previously are met.*
- *If changing gear oil, replace DSG oil filter first. This allows fluid to drain back into transmission.*

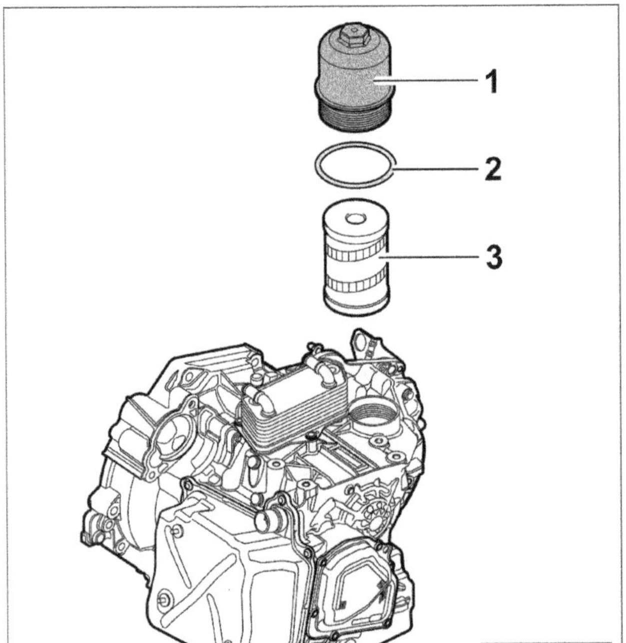

M34-0328

– Switch off ignition.

– Disconnect battery ground strap in rear luggage compartment. See **27 Battery, Alternator, Starter**.

– Remove air filter and air filter housing. See **Air Filter**, given in this repair group.

– Place a suitable drip tray under transmission to catch any spillage.

< Using a 24 mm wrench, unscrew filter housing (**1**) from transmission to the point where it is just ready to take off, but do not remove. Tilt filter housing slightly to allow air in so that gear oil can flow back into transmission, minimizing spillage.

– After filter housing has drained sufficiently, remove filter housing (**1**), clean and install new O-ring (**2**). Coat new O-ring with fresh transmission oil.

– Remove old oil filter (**3**) by pulling out.

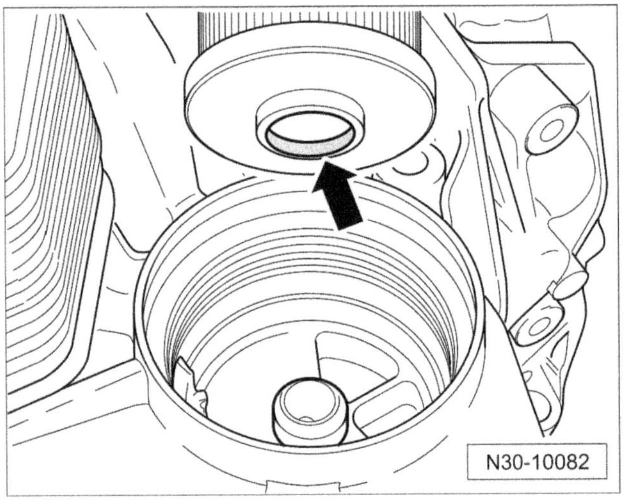
N30-10082

< Coat O-ring in suction collar (**arrow**) of a new filter with gear oil.

– Install new filter with suction collar (**arrow**) down.

– Install and tighten filter housing.

Tightening torque	
DSG filter housing to transmission	20 Nm (15 ft-lb)

– Install air filter and air filter housing. See **Air Filter**, given in this repair group.

– Connect battery ground strap in rear luggage compartment. See **27 Battery, Alternator, Starter**.

– Add gear oil to transmission. See **DSG transmission oil, topping up**, given in this repair group.

BentleyPublishers.com
© 2006 BentleyPublishers.com—All Rights Reserved

DSG transmission gear oil, changing or filling after repairs

> ***WARNING—***
> - *Do not start engine if there is little or no oil in transmission such as immediately following repairs or after severe loss or leakage of transmission oil.*
> - *If there is little or no oil in transmission, add sufficient quantity with engine cold and not running before proceeding with a diagnosis.*
> - *Use only oil specified for direct shift gearboxes (DSG). Always consult your Audi Parts Department or aftermarket parts specialist for the latest information. Use of oils other than those specified will lead to malfunctions or failure of the transmission.*
> - *Observe all possible care when working on the DSG.*
> - *Observe all rules for cleanliness when working on the DSG.*
> - *Special tools are required for some operations. Review all repair procedures before starting.*
> - *Wear eye protection and acid-resistant protection gloves.*

– Vehicle must sit level during this operations. If front of vehicle is higher than rear, such as when driven up on ramps, oil inside transmission will pool towards rear, and level cannot be accurately determined.

– Move shifter lever into “P” position.

– Switch off ignition.

– If changing gear oil, replace DSG oil filter first.

– Raise vehicle. See **Lifting Vehicle** in this repair group.

– Remove lower sound absorber panel. See **Sound absorber panels, removing and installing** in this repair group.

– Position suitable drain pan under transmission.

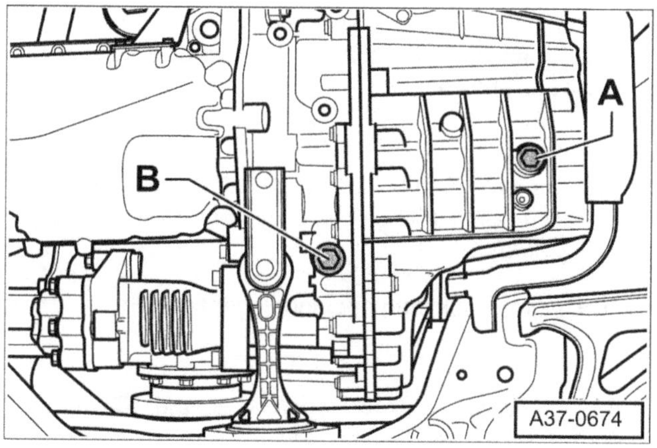

◄ Depending on production date, transmission will have either a drain plug and a check plug (**A** and **B**), or just a check plug (**B**).

– If equipped, remove drain plug (**A**) from transmission housing and allow oil to drain into pan.

> ***WARNING—***
> - *Hot oil can scald. Wear protective clothing, gloves and eye protection.*

– In both cases, remove check plug (**B**).

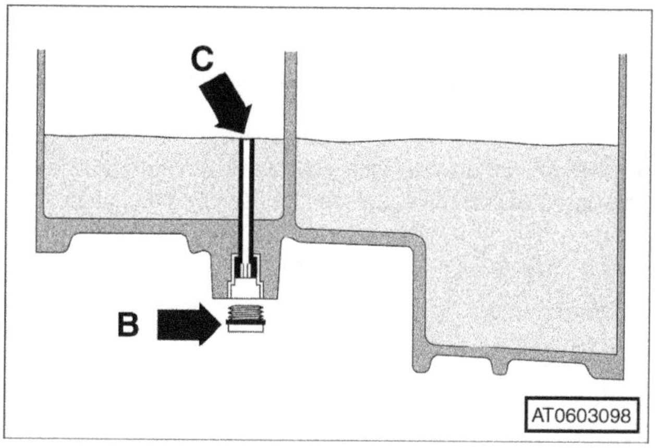

◄ After removal of check plug (**B**), insert an 8 mm hex key into check plug opening and remove overflow pipe (**C**) to drain gear oil from transmission.

– Allow gear oil to completely drain out.

NOTE—
- *Approximately 5 liters (5.3 qts) of transmission oil will drain out if level was at proper specification.*

BentleyPublishers.com
©2006 BentleyPublishers.com—All Rights Reserved

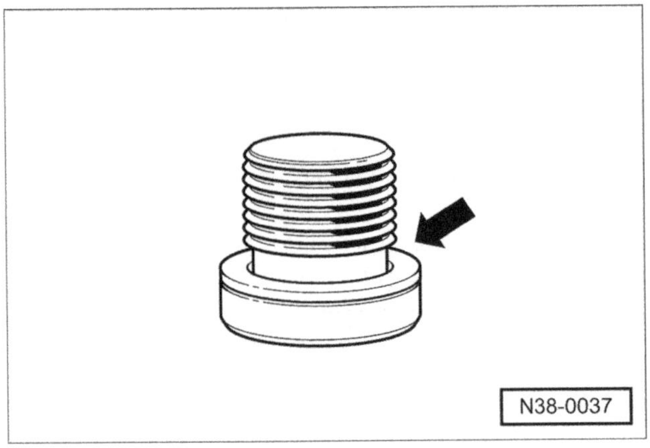

When gear oil has completely drained, install drain plug (where equipped) with a new seal (**arrow**) and tighten.

Tightening torque	
Transmission drain plug	45 Nm (33 ft-lb)

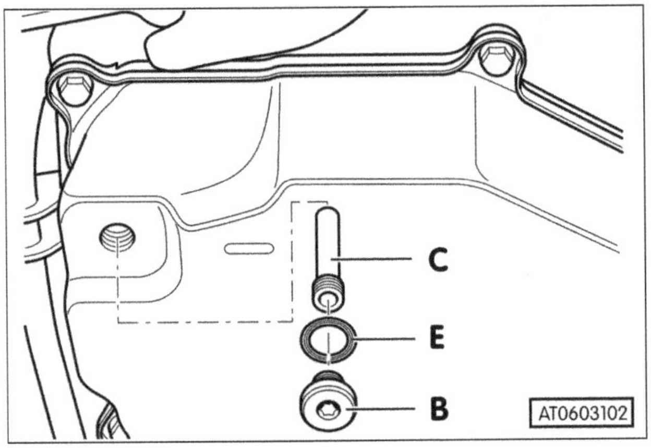

Install overflow tube (**C**) and tighten.

Tightening torque	
Transmission overflow tube	3 Nm (27 in-lb)

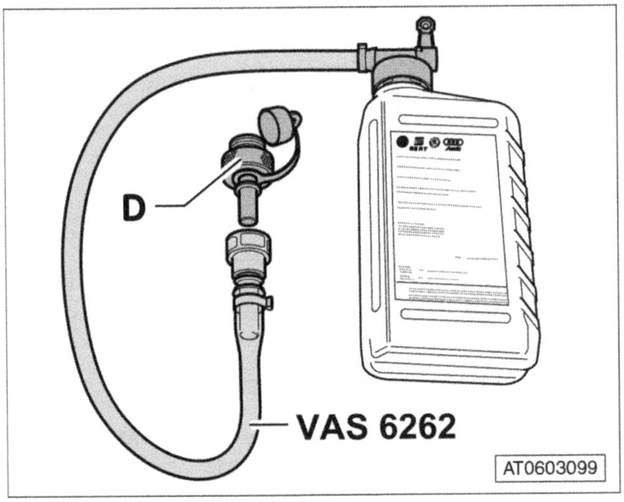

Add gear oil from below through the oil check plug using special filling apparatus such as the VAS 6262 which attaches to the container of gear oil as shown.

NOTE —

- *Make sure that hose and adapter fitting of VAS 6262 are clean. Do not allow DSG transmission oil to be mixed or contaminated by any other type of oil or fluid.*

– Thread adapter (**D**) into check plug opening finger tight.

NOTE —

- *Attach VAS 6262 hose assembly with bottle to adapter (**D**) and hold as high as possible over transmission.*

– Add approximately 5.5 liters of specified oil to transmission, closing valve at oil container end as required to change containers.

– Remove filling apparatus and temporarily install check plug with old seal.

– Connect diagnostic scan tool to data link connector and start program so that ATF temperature can be read. Scan tool can be configured as follows:

- Address word 02 (transmission electronics)
- Function 08 (read measuring value blocks)
- Display group 019
- Display field 1

BentleyPublishers.com
© 2006 BentleyPublishers.com—All Rights Reserved

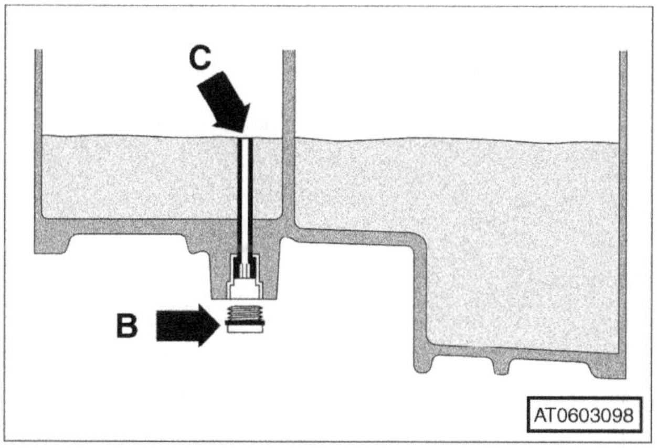

– Start engine and shift once through all selector lever positions, remaining in each shifted position for approximately 3 seconds, with vehicle at a standstill. Move shifter lever back into "P" position and leave engine running at idle. Do not switch off engine!

◄ Remove check plug and allow excess oil in overflow pipe (**C**) to drain out.

– If a small amount of gear oil drips out when transmission oil temperature is between 35°C and 45°C (corresponds to 95°F and 113°F), transmission oil level is correct.

– Repeat as needed so that gear oil level is correct at prescribed temperature.

NOTE —

- *Every 30 seconds, a small surge of oil will exit overflow tube, independent of the height of the transmission oil level. This is due to cool oil impulse to clutches. This surge of oil is* ***not*** *indicative of excessive transmission oil level and must not be considered when determining transmission oil level.*

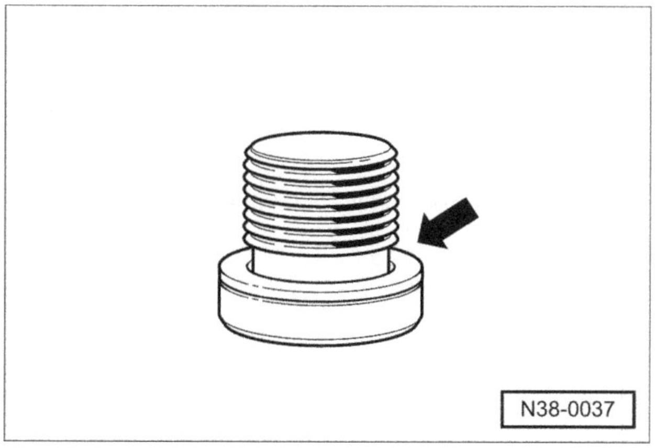

◄ When gear oil level is correct, install check plug with an new seal (**arrow**) and tighten.

Tightening torque	
Transmission check plug	45 Nm (33 ft-lb)

– Check for ATF leaks.

– Install lower sound absorber panel. See **Sound absorber panels, removing and installing** in this repair group.

– Lower vehicle.

– Disconnect scan tool or scan tool computer program.

DSG transmission bevel box gear oil, checking and topping up

Vehicles equipped with the DSG use a bevel gear box to direct power to the rear differential. Unlike manual transmissions, where the transmission, front differential and center differential all share a common transmission gear oil, the bevel gear box does not share gear oil with the transmission.

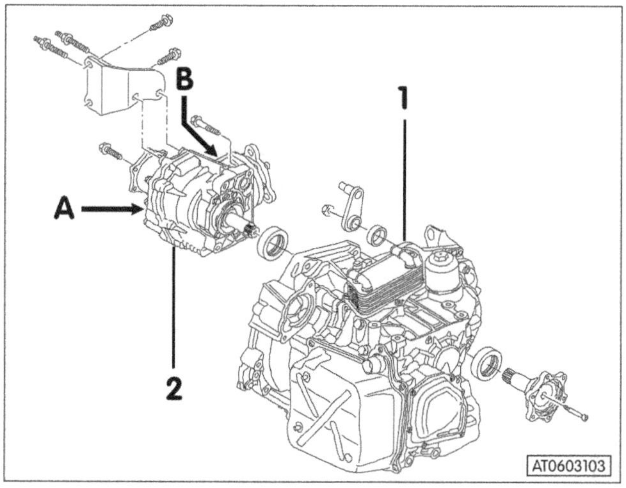

◄ Bevel box (**2**) is attached to right side of the transmission housing (**1**). Fill plug (**B**) is located between drive flange for right side drive axle and rear drive shaft and is used for checking oil level. Drain plug (**A**) is located below right side drive flange.

– Be sure that appropriate pre-requisites listed above are met.

– Before starting, have a new fill plug available.

– Switch off ignition.

– Raise vehicle. See **Lifting Vehicle** in this repair group.

– Remove lower sound absorber panel. See **Sound absorber panels, removing and installing** in this repair group.

BentleyPublishers.com
© 2006 BentleyPublishers.com—All Rights Reserved

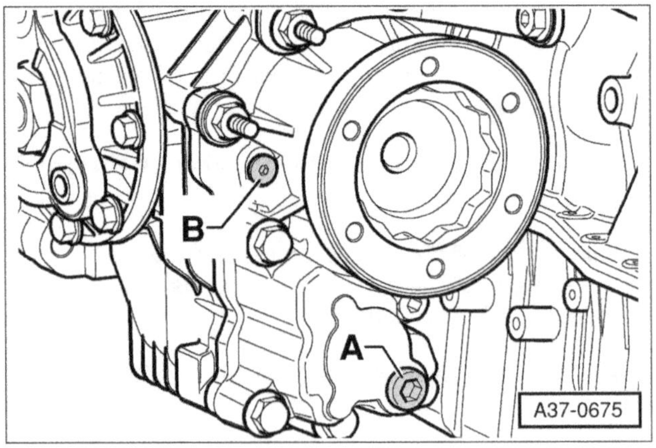

- Place a suitable drip tray under transmission to catch spillage.
- Remove fill plug (**B**) near drive flanges.
- Oil level is correct when bevel box is filled up to lower edge of fill hole.
- Install new fill plug (**B**) and tighten.

Tightening torque	
Bevel box fill plug	15 Nm (11 ft-lb)

- If level is low, top up oil through fill plug hole and install old fill plug.
- Start engine and shift into any gear. With foot on brake pedal, allow engine to run for approximately 2 minutes.
- Switch off ignition.
- Remove fill plug (**B**) near drive flanges and recheck bevel box oil level.
- Oil level is correct when bevel box is filled up to lower edge of fill hole.
- Install new fill plug (**B**) and tighten.

Tightening torque	
Bevel box fill plug	15 Nm (11 ft-lb)

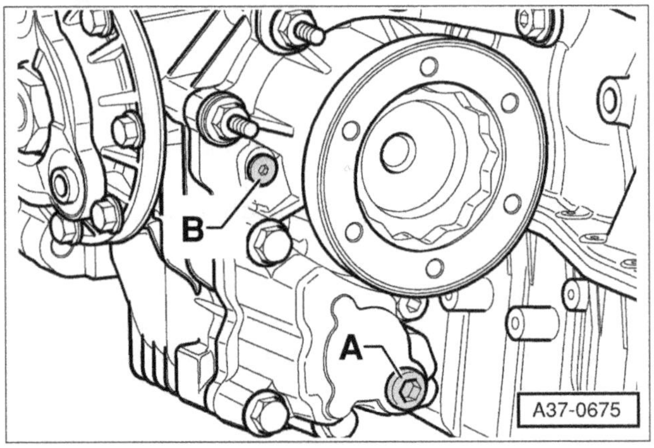

- If changing bevel box gear oil, place a suitable drain pan under transmission.
- Remove drain plug (**A**) near right side drive flange.
- Allow oil to drain completely.
- Install drain plug and tighten.

Tightening torque	
Bevel box drain plug	60 Nm (44 ft-lb)

- Refill bevel box with oil.
- Reheck oil level.
- Check for ATF leaks.
- Install lower sound absorber panel. See **Sound absorber panels, removing and installing** in this repair group.
- Lower vehicle.

BentleyPublishers
.com
©2006 BentleyPublishers.com—All Rights Reserved

Body and Interior maintenance

Airbags, visual inspection

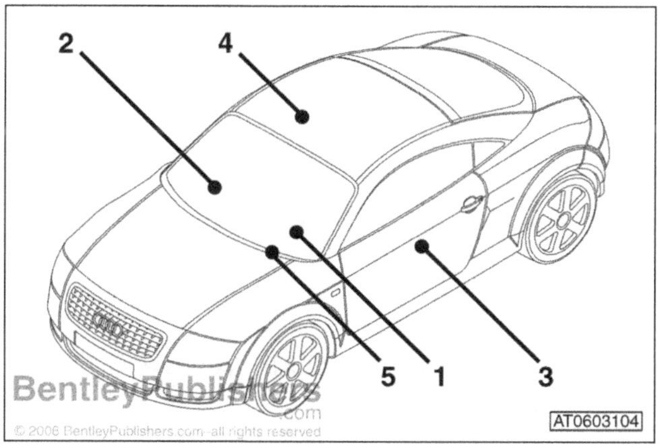

◄ Airbags are installed in several locations in passenger compartment:

- Driver and passenger side front airbags (**1** and **2**).
- Side airbags built into front seat backrests (**3** and **4**).

– Inspect padded airbag covers on steering wheel and instrument panel for signs of external damage. Ensure that they are not covered over or have any objects attached to them such as stickers, etc.

– Inspect front seats for covers that may obstruct or hinder proper deployment of side airbag.

– Do not apply any chemical treatment to airbag unit covers. Clean with a dry or water moistened cloth only.

– Observe airbag warning light in instrument cluster (**5**) when starting vehicle. It should come on for a few seconds when engine is started, as an electronic systems self-test, and immediately go off and stay off. It should only illuminate during this self-test or if there is a malfunction.

For proper operation of airbag warning light in instrument cluster, consult applicable vehicle glovebox owner's manual.

Door hinge and check mechanism, lubricating

Lubricate door hinges and check mechanisms to insure smooth operation and to prevent creaking and groaning noises associated with normal chassis flexing over irregular road surfaces. Lubricate door checks and hinges on both sides of vehicle.

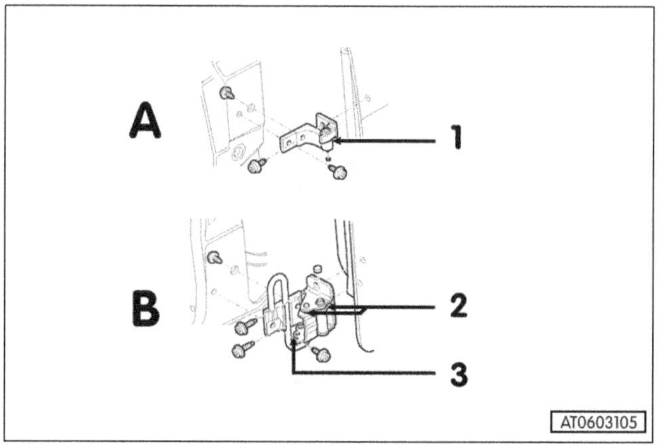

◄ Lubricate upper hinge (**A**) and lower hinge (**B**) on both sides of vehicle. Apply lubricant to hinge areas (**1** and **2**) and check mechanism (**3**).

Door hinge lubricant	
Audi part number	G 052 778 A2

NOTE —

- *Audi part numbers are given for reference only. Always consult with your Audi parts department or aftermarket parts specialist for the latest parts information.*

Dust and pollen filter element, replacing

A paper dust and pollen filter is located at the base of the windshield on the passenger's side. Access is from the engine compartment. Before removing dust and pollen filter, be sure that area around filter opening is clear of dust, dirt, and pollen.

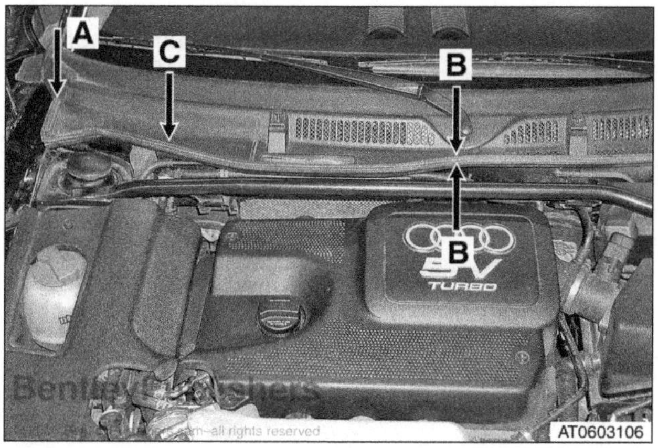

◄ Working from inside engine compartment, pull hood seal off of metal retaining lip on body from the end (**A**) to approximately the middle (**B**) and lay aside.

– Lift up black plastic closeout cover from the front (**C**) and hold up so that filter element can be accessed.

BentleyPublishers.com
©2006 BentleyPublishers.com—All Rights Reserved

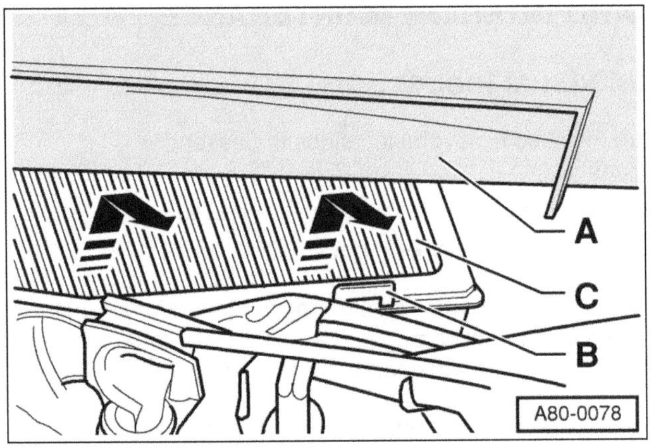

◀ While holding closeout cover (**A**), unsnap both retaining clips (**B**) and lift filter element (**C**) up and out.

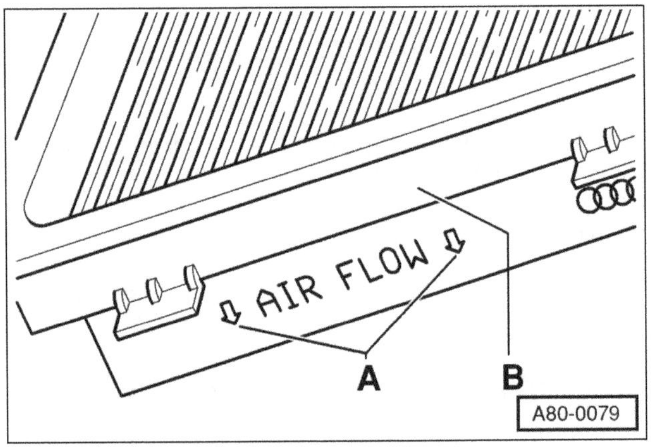

◀ Install new dust and pollen filter (**B**) into filter frame noting air flow direction arrows (**A**) on element.

– Snap element (**A**) into place.

– Install closeout cover into position taking care to ensure proper fit.

– Slide hood seal onto metal retaining lip on body.

Windshield washers and wipers, checking

Windshield washers, adjusting

Operate windshield washers and observe spray pattern. If pattern spray pattern does not contact windshield as noted, adjust jets.

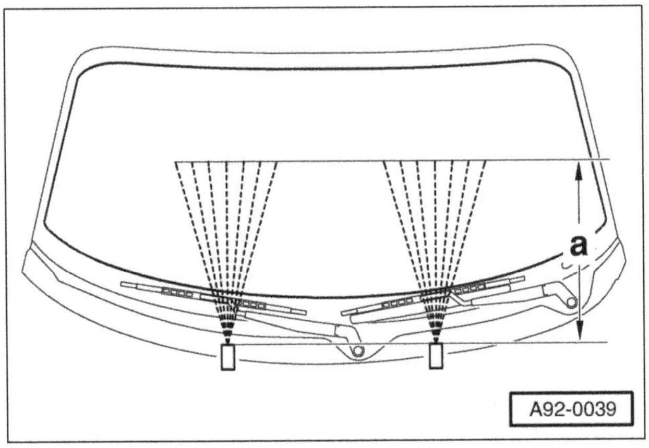

◀ Measure from windshield washer jet to a location on windshield approximately 400 mm away (15¾ in.) (distance **a**).

• Mark windshield with a removable crayon or marker.

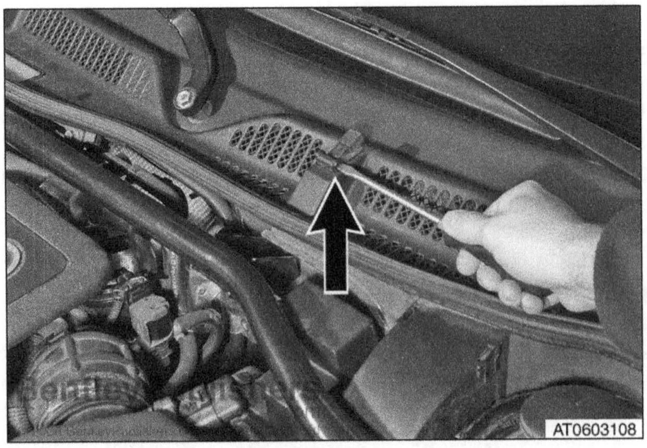

◀ Using a screwdriver, adjust the height setting of each washer jet (**arrow**) so that the top of the main part of the fan spray aligns at the marks on the windshield.

NOTE—

• Avoid using pins or sharp objects in washer jets because of the potential for damage.

BentleyPublishers.com

©2006 BentleyPublishers.com—All Rights Reserved

Windshield washers, filling reservoir

◄ Fill reservoir for windshield washers (**arrow**).

- Fluid level in tank is difficult to see; use a flashlight and look down into container.
- When adding windshield washer fluid, be sure that the level of freeze protection is adequate for the area of vehicle operation.
- Some windshield washer fluid brands are used as-is, straight from the container. Some are in concentrate form and require mixing with water. Refer to product label for proper freeze protection levels.
- Most windshield washer fluids contain additives that help remove insect residue which improves visibility in all seasons.
- Windshield washer reservoir supplies fluid to both windshield washers and headlight washers.

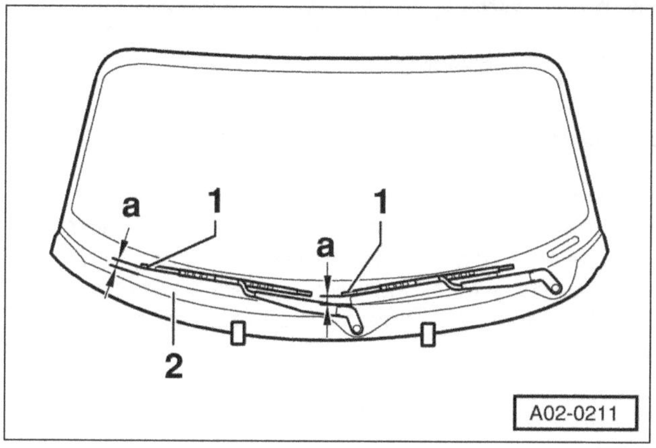

Windshield wipers, checking park position

– Switch windshield wipers on and off, allowing wiper arms to return to park position at base of windshield.

– Switch off ignition.

◄ **Early version** (open frame):

- Check distances (**a**) between ends of wiper blades (**1**) and plastic plenum chamber cover (**2**) at bottom of windshield.
  Distance **a** = 15 - 25 mm (⁹⁄₁₆ - 1 in)
- If distance **a** is not correct, pull wiper arm off of wiper pivot and reposition arm on pivot. Wiper arm puller tool T10130 is available to facilitate removal of arm.

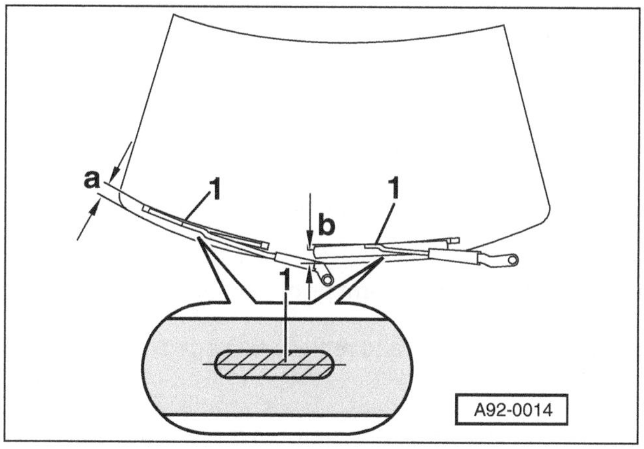

◄ **Later version** (aero):

- Ensure that wiper blades rest against marking (**1**) on windshield.
- Check distances (**a**, **b**) between ends of wiper blades and plastic plenum chamber cover at bottom of windshield.
  Distance **a** = 25 mm ± 5 mm (1 in ± ³⁄₁₆)
  Distance **b** = 35 mm ± 5 mm (1 ³⁄₈ in ± ³⁄₁₆)
- If distance **a** or **b** is not correct, pull wiper arm off of wiper pivot and reposition arm on pivot. Wiper arm puller tool T10130 is available to facilitate removal of arm.

– Operate wipers and recheck distances.

Tightening torque	
Wiper arm nut to pivot	16 Nm (12 ft-lb)

Windshield wiper blades, removing and installing

Two different types of wiper blades are used on vehicle covered by this manual. Procedures are similar for both early and late versions.

Early version (open frame)

– Switch windshield wipers on and off, allowing wiper arms to return to park position at base of windshield.

– Switch off ignition.

– Fold wiper arm away from windshield.

BentleyPublishers.com
© 2006 BentleyPublishers.com—All Rights Reserved

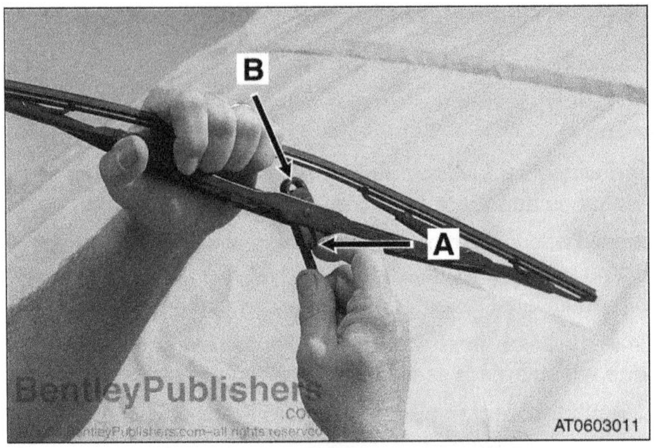

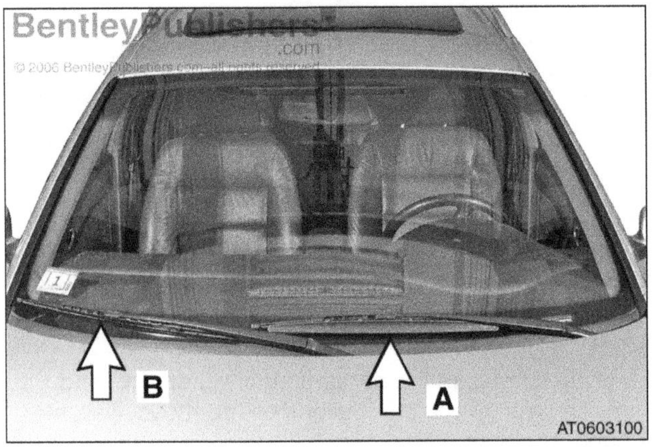

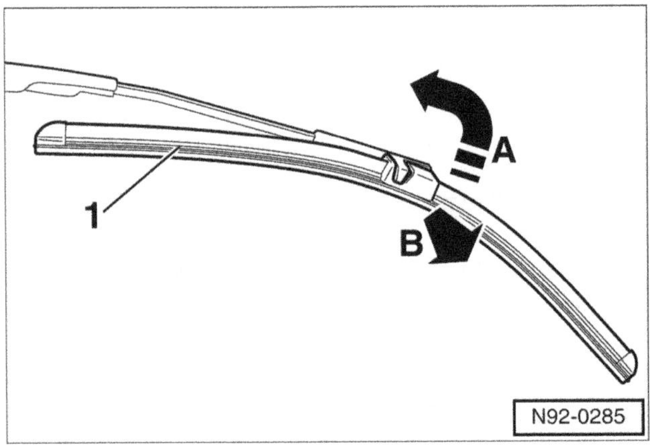

◀ Remove wiper blade by turning approximately ¼-turn on the arm. Depress locking tab (**A**) and slide blade out of loop end of arm (**B**).

– Move loop end of arm out of wiper blade.

NOTE —
- *Use caution when wiper arm is folded away from windshield. If arm snaps back against windshield glass, windshield damage can result.*

– Pull off wiper blade (**1**) from wiper arm pivot (**B**).

– Inspect windshield wiper blades and replace damaged, torn or deteriorated parts.

– Installation is reverse of removal. Use caution when folding wiper arm back onto windshield.

◀ Note blade installation positions
- Air deflector (or spoiler) on driver's side (**A**) is below blade
- Curvature of passenger's side (**B**) faces up ("cup is up")

– Some types of wiper blades have provision for replacing the rubber part only. These rubber parts are commonly called "wiper inserts" or "refills" and can be an economical alternative to replacing the complete blade assembly. Most original equipment blades allow for replacement of the inserts and come packaged with appropriate installation instructions.

Late version (aero)

– Switch windshield wipers on and off, allowing wiper arms to return to park position at base of windshield.

– Switch off ignition.

– Fold wiper arm away from windshield.

◀ Remove wiper blade (**1**) by turning approximately ¼-turn until it reaches stop (**A**).

– Pull off wiper blade (**1**) from wiper arm pivot (**B**).

– Inspect windshield wiper blades and replace damaged, torn or deteriorated parts.

– Installation is reverse of removal; use caution when folding wiper arm back onto windshield.

Headlights, adjusting

Proper adjustment (aim) of headlights (and foglights) requires specialized equipment which varies from service shop to service shop and in some cases will require a scan tool or computer scan tool program. Adjustment parameters can vary between different State and Provincial governments.

Because of these variables, always follow the equipment manufacturers instructions along with all applicable governmental regulations when adjusting headlight aim.

BentleyPublishers
.com
© 2006 BentleyPublishers.com—All Rights Reserved

Regardless of equipment or regulations, the following points are common and preliminary to the proper adjustment of headlights:

- Inflate tires to correct pressures.
- Vehicle must be standing on a level surface.
- Headlight lenses and reflectors clean and undamaged with correct type and wattage bulbs.
- Vehicle must be properly loaded with approximately 75 Kg (165 lb) on driver's seat.
- Vehicle at curb weight with tool kit, jack, spare tire, etc. and 90% fuel load. If fuel tank is not at least 90% full, add weight to compensation.
- Vehicle bounced several times to settle suspension.

Adjusting halogen and xenon HID headlights, and fog lights

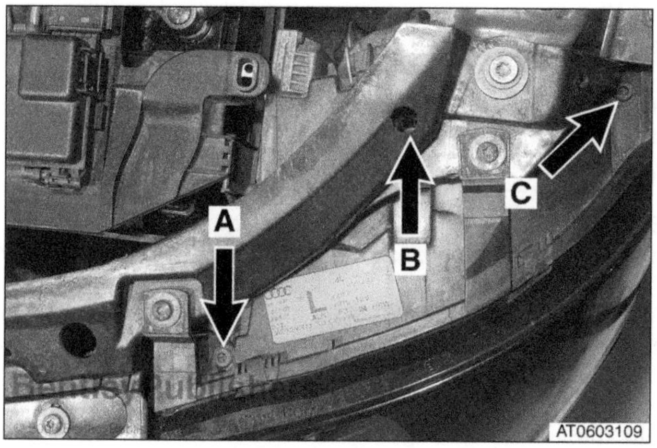

Adjustment points for halogen and xenon HID lights are the same.

◄ Left side headlight:

A. Fog light vertical (height) adjustment screw
B. Headlight vertical (height) adjustment screw
C. Headlight lateral (side-to-side) adjustment screw

– When adjusting headlights, turn adjustment screws (**B** and **C**) the same amount and in the same direction.

NOTE—

- *Adjustment screws for right headlight are symmetrically opposite.*
- *Fog lights are only adjustable for vertical aim, there is no lateral adjustment.*

Service reminder display

Most vehicles covered by this manual are equipped with a service reminder display (also called a service interval display) that will come on starting at approximately 1200 miles (2000 kilometers) before a scheduled service is required when the engine is started. The display will read "SERVICE IN 1,200 MLS" as the interval approaches. The 1,200 MLS will be reduced at each successive engine start until the service is due. At that point, the display will simply read "SERVICE". Kilometers are substituted where applicable. The display will return to normal operation a few seconds after the engine starts and should be reset after the appropriate service has been completed.

– Switch ignition off.

◄ Press and hold trip odometer reset button (**2**) while switching ignition on; "SERVICE" will appear in the display.

– Press the digital clock set button (**1**) until "SERVICE" text is cleared from the display. The next service event should now be showing.

NOTE—

- *Do not reset the display between intervals. This will result in inaccurate displays.*
- *The display information remains in memory even if the battery is disconnected.*
- *Malfunctions in the instrument cluster will result in a continuous display of "dEF" and should be diagnosed as soon as possible.*

BentleyPublishers.com
© 2006 BentleyPublishers.com—All Rights Reserved

QUALITY REVIEW

When you have finished working under the hood or around other areas of the vehicle, take a moment to quality check or review your work. This helps to insure that the operation or repair has been completed properly with all affected systems functioning within normal parameters. A check with a scan tool and/or a road test is usually required.

On Board Diagnostics (OBD)

Audi maintenance specifies checking On Board Diagnostic (OBD) system memory. The ODB system is integrated into many vehicle systems including engine management. These systems monitor emissions related and other electrical components, including automatic transmission, ABS system, and airbag systems. Checking OBD memory for Diagnostic Trouble Codes (DTCs) requires special tools and training. If you lack the tools or expertise, see an authorized Audi dealer or qualified independent repair shop for assistance.

AT0603111

Data Link Connector (DLC) (**arrow**) is located near the hood release handle on driver's side footwell kick panel.

A list of diagnostic trouble codes for scan tools can be found at the back of this manual in the SAE standard J1930 format or in this manufacturer's VAG format. See **OBD On Board Diagnosis.**

More manufacturer supplied diagnostic scan tool information for using the factory supplied VAG 1551/1552, VAS 5051/5052 and derivatives or aftermarket equivalent scan tool or computer scan tool programs, is available in the Official Factory Repair Manual, electronic edition. This can be purchased at www.bentleypublishers.com or by contacting the publisher at the address listed in the front of this manual.

Road test

Upon completion of maintenance work, vehicle should be road tested. Check the following during road test:

Exterior lights

– With the help of an assistant, as required, check operation of all front and rear external lights including high and low beam headlights, side marker and running lights, tail lights, brake and back-up lights, turn signals, emergency flashers, and their corresponding indicator lights.

Automatic transmission shift lock

– Turn ignition on, but do not start engine. Apply parking brake.

– With gearshift selector in park attempt to shift into a drive or reverse. Selector lever should not move.

– Press brake pedal. Selector lever should move into gear.

– Place gear selector lever in neutral and release brake pedal.

BentleyPublishers.com
©2006 BentleyPublishers.com—All Rights Reserved

– Attempt to shift into gear. Selector lever should not move.

– Press brake pedal. Selector lever should move into gear.

Automatic park/neutral position switch

– Apply parking brake and step on brake pedal. Place gear shift selector in park and attempt to start car. Repeat for all gears. Engine should only start in park or neutral. Check for backup light operation in reverse only.

Brake pedal and parking brake

– Check pedal and handbrake travel (free play) and operation.

– Check starter operation with clutch pedal depressed only.

Manual gearshift

– Check for smooth, easy operation of transmission and gearshift.

Steering

– Check steering play with vehicle weight on wheels by turning steering wheel back and forth (wheels in straight ahead position).

Automatic transmission kickdown operation

– Depress accelerator pedal fully to floor. Depending on vehicle speed and engine speed, upshift is either delayed or transmission shifts down into next lower gear.

Air conditioner and heater

– Check full range of heater and air conditioner for proper function. Cold or warm air must flow out of vents.

Additional checks

Perform the following checks in accordance with vehicle equipment and road conditions available during the road test:

- Engine performance, idle speed, acceleration, cold and hot starting.
- Clutch operation, smoothness and pedal pressure.
- Brake operation, noise and function of ABS.
- Cruise control system operation.
- Radio, CD and SAT operation, and reception.
- General vehicle handling and dynamics such as pulling to the side, cornering, vibrations in steering wheel and unusual noises.

BentleyPublishers.com
© 2006 BentleyPublishers.com—All Rights Reserved

MAINTENANCE SCHEDULES

Aside from keeping your Audi in the best possible condition, scheduled maintenance plays a role in maintaining full coverage under Audi's extensive warranties. If in doubt about terms and conditions of your vehicle warranty, consult an authorized Audi dealer.

Maintenance schedules list all routine maintenance specified by Audi, and the time and/or mileage intervals when they should be performed.

In addition to the factory requirements, the publisher has adjusted the time/mileage interval for some operations to reflect current operating conditions based on actual experience.

NOTE —

- *Audi continually updates maintenance schedules to suit changing conditions through the issuance of Maintenance Schedule Service Circulars. If in doubt about any of the requirements for your vehicle, consult an authorized Audi dealer.*
- *Maintenance tables on the following pages are engine specific.*
- *For continued service beyond the mileage range of the individual table, repeat the table for remaining life of vehicle.*
- *Routine maintenance operations within each engine group can vary depending on vehicle equipment level; not all operations apply to all vehicles.*

CAUTION—

- *To avoid costly damage to engine mechanical components as a result of timing belt or tensioner failure, periodically inspect the timing belt for wear, cracks or other damage. Replace the belt and tensioning system every 60,000 - 65,000 miles (96,000 - 104,000 kilometers) on engines so equipped.*

BentleyPublishers
.com
© 2006 BentleyPublishers.com—All Rights Reserved

Table a. 2000 m.y. Maintenance Table

Miles (in thousands)	5	10	15	20	25	30	35	40	45	50	55	60	65	70	75	80	85	90	95	100	105
Kilometers (in thousands)	**8**	**16**	**24**	**32**	**40**	**48**	**56**	**64**	**72**	**80**	**88**	**96**	**104**	**112**	**120**	**128**	**136**	**144**	**152**	**160**	**168**
Engine: Change engine oil and oil filter	•	•	•	•	•	•	•	•	•	•	•	•	•	•	•	•	•	•	•	•	•
Check fluid levels, top off as needed	•	•	•	•	•	•	•	•	•	•	•	•	•	•	•	•	•	•	•	•	•
Service interval display: Reset (*where applicable*)	•	•	•	•	•	•	•	•	•	•	•	•	•	•	•	•	•	•	•	•	•
Engine: Check for leaks		•		•		•		•		•		•		•		•		•		•	
Battery: Check electrolyte level, add if necessary		•		•		•		•		•		•		•		•		•		•	
Door check/hinge mechanism: Lubricate		•		•		•		•		•		•		•		•		•		•	
Windshield washer fluid: Check level; add if necessary		•		•		•		•		•		•		•		•		•		•	
Dust and pollen filter: replace (*where applicable)*		•		•		•		•		•		•				•		•		•	
Automatic shift lock: Check operation including park/neutral position switch	•	•	•	•	•	•	•	•	•	•	•	•	•	•	•	•	•	•	•	•	•
Automatic transmission: Check for leaks		•		•		•		•		•		•		•		•		•		•	
Manual transmission: Check for leaks		•		•		•		•		•		•		•		•		•		•	
Manual transmission: Check gearshift mechanism and clutch for proper interlock operation	•	•	•	•	•	•	•	•	•	•	•	•	•	•	•	•	•	•	•	•	•
Haldex clutch (quattro), Change oil				•				•				•				•				•	
Tires/spare/wheels: Check condition (tread depth) and pressure, adjust if necessary		•		•		•		•		•		•		•		•		•		•	
Wheels: Rotate front to rear	•																				
Drive shafts: Check boots		•		•		•		•		•		•		•		•		•		•	
On-Board Diagnostic system: Check fault memory, erase if necessary		•		•		•		•		•		•		•		•		•		•	
Brake system; Check for damage/leaks, brake pad thickness, brake fluid level		•		•		•		•		•		•		•		•		•		•	
Cooling system: Check coolant level, add if necessary		•		•		•		•		•		•		•		•		•		•	
Front suspension: Check dust seals on ball joints and tie rod ends,				•				•				•				•				•	
Air cleaner: Replace filter				•				•				•				•				•	
Headlights and exterior lights: Check and adjust if necessary		•		•		•		•		•		•		•		•		•		•	
Exhaust system: Check for damage and leaks		•		•		•		•		•		•		•		•		•		•	
Power steering: Check fluid level		•		•		•		•		•		•		•		•		•		•	
Transmission final drives Check oil levels; add if necessary		•		•		•		•		•		•		•		•		•		•	
Timing belt and tensioner system: Replace												•									
Spark plugs; Replace						•						•						•			
Ribbed belt: Replace																•					
Road test: Check kick-down, braking, steering, heating/ventilation, air conditioning, power accessories and electrical systems		•		•		•		•		•		•		•		•		•		•	
Every 2 years, regardless of mileage (kilometers): Change brake fluid																					

AMS-06-01, AMS-01-14, AMC2000

BentleyPublishers.com

©2006 BentleyPublishers.com—All Rights Reserved

Maintenance Schedules

Table b. 2001 m.y. Maintenance Table

Miles (in thousands)	5	10	20	30	40	50	60	70	80	90	100
Kilometers (in thousands)	8	16	32	48	64	80	96	112	128	144	160
Engine: Change engine oil and oil filter	•	•	•	•	•	•	•	•	•	•	•
Check fluid levels, top off as needed	•	•		•		•		•		•	
Service interval display: Reset (where applicable)	•	•	•	•	•	•	•	•	•	•	•
Engine: Check for leaks			•		•		•		•		•
Battery: Check electrolyte level, add if necessary		•	•	•	•	•	•	•	•	•	•
Door check/hinge mechanism: Lubricate		•	•	•	•	•	•	•	•	•	•
Windshield washer fluid: Check level; add if necessary		•	•	•	•	•	•	•	•	•	•
Dust and pollen filter: replace (*where applicable)*			•		•		•		•		•
Manual transmission: Check for leaks			•		•		•		•		•
Manual transmission: Check gearshift mechanism and clutch for proper interlock operation	•	•	•	•	•	•	•	•	•	•	•
Haldex clutch (quattro), Change oil and filter			•		•		•		•		•
Tires/spare/wheels: Check condition (tread depth) and pressure, adjust if necessary		•	•	•	•	•	•	•	•	•	•
Wheels: Rotate front to rear	•										
Drive shafts: Check boots		•	•	•	•	•	•	•	•	•	•
On-Board Diagnostic system: Check fault memory, erase if necessary		•	•	•	•	•	•	•	•	•	•
Brake system; Check for damage/leaks, brake pad thickness, brake fluid level	•	•	•	•	•	•	•	•	•	•	•
Cooling system: Check coolant level, add if necessary		•	•	•	•	•	•	•	•	•	
Front suspension: Check dust seals on ball joints and tie rod ends			•		•		•		•		•
Air cleaner: Replace filter					•				•		
Headlights and exterior lights: Check and adjust if necessary		•	•	•	•	•	•	•	•	•	•
Exhaust system: Check for damage and leaks			•		•		•		•		•
Power steering: Check fluid level					•				•		
Manual transmission final drives Check oil levels; add if necessary			•		•		•		•		•
Timing belt and tensioner system: Replace							•				
Spark plugs; Replace					•				•		
Ribbed belt: Replace									•		
Road test: Check kick-down, braking, steering, heating/ventilation, air conditioning, power accessories and electrical systems		•		•	•	•	•	•	•	•	•
Every 2 years, regardless of mileage (kilometers): Change brake fluid											
At 120,000 miles (192,000 kilometers): Replace timing belt											

AMS-02-02, AMC2001

BentleyPublishers
.com
©2006 BentleyPublishers.com—All Rights Reserved

Table c. 2002 m.y. Maintenance Table

Miles (in thousands)	10	20	30	40	50	60	70	80	90	100	105
Kilometers (in thousands)	15	30	45	60	75	90	105	120	135	150	155
Engine: Change engine oil and oil filter	•	•	•	•	•	•	•	•	•	•	
Check fluid levels, top off as needed	•	•	•	•	•	•	•	•	•	•	
Service interval display: Reset (where applicable)	•	•	•	•	•	•	•	•	•	•	
Engine: Check for leaks		•		•		•		•		•	
Battery: Check electrolyte level, add if necessary	•	•	•	•	•	•	•	•	•	•	
Door check/hinge mechanism: Lubricate	•	•	•	•	•	•	•	•	•	•	
Windshield washer fluid: Check level; add if necessary	•	•	•	•	•	•	•	•	•	•	
Dust and pollen filter: replace (*where applicable*)		•		•		•		•		•	
Manual transmission: Check for leaks		•		•		•		•		•	
Manual transmission: Check gearshift mechanism and clutch for proper interlock operation	•	•	•	•	•	•	•	•	•	•	
Haldex clutch (quattro), Change oil and filter		•		•		•		•		•	
Tires/spare/wheels: Check condition (tread depth) and pressure, adjust if necessary	•	•	•	•	•	•	•	•	•	•	
Drive shafts: Check boots	•	•	•	•	•	•	•	•	•	•	
On-Board Diagnostic system: Check fault memory, erase if necessary	•	•	•	•	•	•	•	•	•	•	
Brake system; Check for damage/leaks, brake pad thickness, brake fluid level	•	•	•	•	•	•	•	•	•	•	
Cooling system: Check coolant level, add if necessary	•	•	•	•	•	•	•	•	•	•	
Front suspension: Check dust seals on ball joints and tie rod ends		•		•		•		•		•	
Air cleaner: Replace filter				•				•			
Headlights and exterior lights: Check and adjust if necessary	•	•	•	•	•	•	•	•	•	•	
Exhaust system: Check for damage and leaks	•	•	•	•	•	•	•	•	•	•	
Power steering: Check fluid level				•				•			
Manual transmission final drives Check oil levels; add if necessary		•		•		•		•		•	
Spark plugs; Replace				•				•			
Road test: Check kick-down, braking, steering, heating/ventilation, air conditioning, power accessories and electrical systems	•	•	•	•	•	•	•	•	•	•	
Ribbed belt: Check											•
Timing belt and tensioner system: Replace						•					
Every 2 years, regardless of mileage (kilometers): Change brake fluid											

AMS-02-01, AMC2002

BentleyPublishers.com
©2006 BentleyPublishers.com—All Rights Reserved

Maintenance Schedules

Table d. 2003 m.y. Maintenance Table (through chassis #021775)

Miles (in thousands)	10	20	30	40	50	60	70	80	90	100	105
Kilometers (in thousands)	**15**	**30**	**45**	**60**	**75**	**90**	**105**	**120**	**135**	**150**	**155**
Engine: Change engine oil and oil filter	•	•	•	•	•	•	•	•	•	•	
Check fluid levels, top off as needed	•	•	•	•	•	•	•	•	•	•	
Service interval display: Reset (where applicable)	•	•	•	•	•	•	•	•	•	•	
Engine: Check for leaks		•		•		•		•		•	
Battery: Check electrolyte level, add if necessary	•	•	•	•	•	•	•	•	•	•	
Door check/hinge mechanism: Lubricate	•	•	•	•	•	•	•	•	•	•	
Windshield washer fluid: Check level; add if necessary	•	•	•	•	•	•	•	•	•	•	
Dust and pollen filter: replace (*where applicable)*		•		•		•		•		•	
Automatic transmission: Check for leaks		•		•		•		•		•	
Automatic shift lock: Check operation including park/neutral position switch		•		•		•		•		•	
Manual transmission: Check for leaks		•		•		•		•		•	
Manual transmission: Check gearshift mechanism and clutch for proper interlock operation	•	•	•	•	•	•	•	•	•	•	
Haldex clutch (quattro), Change oil and filter		•		•		•		•		•	
Tires/spare/wheels: Check condition (tread depth) and pressure, adjust if necessary	•	•	•	•	•	•	•	•	•	•	
Drive shafts: Check boots	•	•	•	•	•	•	•	•	•	•	
On-Board Diagnostic system: Check fault memory, erase if necessary	•	•	•	•	•	•	•	•	•	•	
Brake system; Check for damage/leaks, brake pad thickness, brake fluid level	•	•	•	•	•	•	•	•	•	•	
Cooling system: Check coolant level, add if necessary	•	•	•	•	•	•	•	•	•	•	
Front suspension: Check dust seals on ball joints and tie rod ends		•		•		•		•		•	
Air cleaner: Replace filter				•				•			
Headlights and exterior lights: Check and adjust if necessary	•	•	•	•	•	•	•	•	•	•	
Exhaust system: Check for damage and leaks	•	•	•	•	•	•	•	•	•	•	
Power steering: Check fluid level				•				•			
Manual transmission final drives Check oil levels; add if necessary		•		•		•		•		•	
Spark plugs; Replace				•				•			
Road test: Check kick-down, braking, steering, heating/ventilation, air conditioning, power accessories and electrical systems	•	•	•	•	•	•	•	•	•	•	
Ribbed belt: Check											•
Timing belt and tensioner system: Replace						•					
Every 2 years, regardless of mileage (kilometers): Change brake fluid											

AMS-06-02, AMC2003

BentleyPublishers.com
©2006 BentleyPublishers.com—All Rights Reserved

Table e. 2003 m.y. Maintenance Table (from chassis #021776)

Miles (in thousands)	5	15	25	35	45	55	65	75	85	95
Kilometers (in thousands)	8	25	40	55	70	85	100	115	130	145
Engine: Change engine oil and oil filter	•	•	•	•	•	•	•	•	•	•
Service interval display: Reset (where applicable)	•	•	•	•	•	•	•	•	•	•
Engine: Check for leaks		•		•		•		•		•
Battery: Check electrolyte level, add if necessary		•		•		•		•		•
Door check/hinge mechanism: Lubricate		•		•		•		•		•
Windshield washer fluid: Check level; add if necessary	•	•	•	•	•	•	•	•	•	•
Dust and pollen filter: replace		•		•		•		•		•
Manual transmission: Check for leaks		•		•		•		•		•
Automatic transmission: Check for leaks		•		•		•		•		•
Haldex clutch (quattro), Change oil and filter		•		•		•		•		•
Tires/spare/wheels: Check condition (tread depth) and pressure, adjust if necessary	•	•	•	•	•	•	•	•	•	•
Wheels: Rotate front to rear	•									
Wheels: Check condition and expiration date on tire mobility system sealant	•	•	•	•	•	•	•	•	•	•
Drive shafts: Check boots		•		•		•		•		•
On-Board Diagnostic system: Check fault memory, erase if necessary		•		•		•		•		•
Brake system; Check for damage/leaks, brake pad thickness, brake fluid level	•	•	•	•	•	•	•	•	•	•
Cooling system: Check coolant level, add if necessary		•		•		•		•		•
Front suspension: Check dust seals on ball joints and tie rod ends		•		•		•		•		•
Air cleaner: Replace filter				•				•		
Headlights and exterior lights: Check and adjust if necessary		•		•		•		•		•
Exhaust system: Check for damage and leaks		•		•		•		•		•
Power steering: Check fluid level				•				•		
Manual transmission final drives Check oil levels; add if necessary		•		•		•		•		•
Automatic transmission final drive oil: Check level and for leaks; add if necessary		•		•		•		•		•
Direct shift gearbox: Change oil and filter				•				•		
Spark plugs; Replace				•				•		
Ribbed belt: Replace (1.8L)								•		
Ribbed belt: Check (3.2)								•		
Timing belt and tensioner system: Replace							•			
Road test: Check kick-down, braking, steering, heating/ventilation, air conditioning, power accessories and electrical systems	•	•	•	•	•	•	•	•	•	•
Every 2 years, regardless of mileage (kilometers): Change brake fluid										

AMS-06-03

BentleyPublishers.com
© 2006 BentleyPublishers.com—All Rights Reserved

Maintenance Schedules

Table f. 2004 m.y. Maintenance Table

Miles (in thousands)	5	15	25	35	45	55	65	75	85	95
Kilometers (in thousands)	8	25	40	55	70	85	100	115	130	145
Engine: Change engine oil and oil filter	•	•	•	•	•	•	•	•	•	•
Service interval display: Reset (where applicable)	•	•	•	•	•	•	•	•	•	•
Engine: Check for leaks		•		•		•		•		•
Battery: Check electrolyte level, add if necessary		•		•		•		•		•
Door check/hinge mechanism: Lubricate		•		•		•		•		•
Windshield washer fluid: Check level; add if necessary	•	•	•	•	•	•	•	•	•	•
Dust and pollen filter: replace		•		•		•		•		•
Haldex clutch (quattro), Change oil		•		•		•		•		•
Haldex clutch (quattro), Change filter				•				•		
Tires/spare/wheels: Check condition (tread depth) and pressure, adjust if necessary	•	•	•	•	•	•	•	•	•	•
Wheels: Rotate front to rear	•									
Wheels: Check condition and expiration date on tire mobility system sealant		•		•		•		•		•
Drive shafts: Check boots		•		•		•		•		•
On-Board Diagnostic system: Check fault memory, erase if necessary		•		•		•		•		•
Brake system; Check for damage/leaks, brake pad thickness, brake fluid level	•	•	•	•	•	•	•	•	•	•
Cooling system: Check coolant level, add if necessary		•		•		•		•		•
Front suspension: Check dust seals on ball joints and tie rod ends		•		•		•		•		•
Air cleaner: Replace filter				•				•		
Headlights and exterior lights: Check and adjust if necessary		•		•		•		•		•
Exhaust system: Check for damage and leaks		•		•		•		•		•
Power steering: Check fluid level				•				•		
Manual transmission final drives Check for leaks		•		•		•		•		•
Automatic transmission final drive oil: Check for leaks		•		•		•		•		•
Direct shift gearbox: Change oil and filter				•				•		
Spark plugs; Replace				•				•		
Ribbed belt: Replace (1.8L)								•		
Ribbed belt: Check (3.2L)								•		
Timing belt and tensioner system: Replace							•			
Road test: Check kick-down, braking, steering, heating/ventilation, air conditioning, power accessories and electrical systems		•		•		•		•		•
Every 2 years, regardless of mileage (kilometers): Change brake fluid										

AMS-05-22, AMC2004

BentleyPublishers.com
© 2006 BentleyPublishers.com—All Rights Reserved

Table g. 2005 m.y. Maintenance Table

Miles (in thousands)	5	15	25	35	45	55	65	75	85	95
Kilometers (in thousands)	8	25	40	55	70	85	100	115	130	145
Engine: Change engine oil and oil filter	•	•	•	•	•	•	•	•	•	•
Service interval display: Reset (where applicable)	•	•	•	•	•	•	•	•	•	•
Engine: Check for leaks		•		•		•		•		•
Battery: Check electrolyte level, add if necessary		•		•		•		•		•
Door check/hinge mechanism: Lubricate		•		•		•		•		•
Windshield washer fluid: Check level; add if necessary	•	•	•	•	•	•	•	•	•	•
Windshield wiper arm pivot points: Lubricate	•	•	•	•	•	•	•	•	•	•
Dust and pollen filter: replace		•		•		•		•		•
Haldex clutch (quattro), Change oil		•		•		•		•		•
Haldex clutch (quattro), Change filter				•				•		
Tires/spare/wheels: Check condition (tread depth) and pressure, adjust if necessary	•	•	•	•	•	•	•	•	•	•
Wheels: Rotate front to rear	•									
Wheels: Check condition and expiration date on tire mobility system sealant		•		•		•		•		•
Drive shafts: Check boots		•		•		•		•		•
Drive shafts: Check front and rear for excessive play		•		•		•		•		•
On-Board Diagnostic system: Check fault memory, erase if necessary		•		•		•		•		•
Brake system; Check for damage/leaks, brake pad thickness, brake fluid level	•	•	•	•	•	•	•	•	•	•
Cooling system: Check coolant level, add if necessary		•		•		•		•		•
Front suspension: Check dust seals on ball joints and tie rod ends		•		•		•		•		•
Air cleaner: Replace filter				•				•		
Headlights and exterior lights: Check and adjust if necessary		•		•		•		•		•
Undercarriage: Visual inspection for damage		•		•		•		•		•
Drive shafts: Check boots		•		•		•		•		•
Power steering: Check fluid level				•				•		
Manual transmission final drives Check for leaks		•		•		•		•		•
Automatic transmission final drive oil: Check for leaks		•		•		•		•		•
Direct shift gearbox: Change oil and filter				•				•		
Spark plugs; Replace				•				•		
Ribbed belt: Replace (1.8L)								•		
Ribbed belt: Check (3.2L)								•		
Timing belt and tensioner system: Replace							•			
Road test: Check kick-down, braking, steering, heating/ventilation, air conditioning, power accessories and electrical systems		•		•		•		•		•
Every 2 years, regardless of mileage (kilometers): Change brake fluid										

AMS-05-23

BentleyPublishers.com

© 2006 BentleyPublishers.com—All Rights Reserved

Table h. 2006 m.y. Maintenance Table

Miles (in thousands)	5	15	25	35	45	55	65	75	85	95
Kilometers (in thousands)	8	25	40	55	70	85	100	115	130	145
Engine: Change engine oil and oil filter	•	•	•	•	•	•	•	•	•	•
Service interval display: Reset (where applicable)	•	•	•	•	•	•	•	•	•	•
Engine: Check for leaks		•		•		•		•		•
Battery: Check electrolyte level, add if necessary		•		•		•		•		•
Door check/hinge mechanism: Lubricate		•		•		•		•		•
Windshield washer fluid: Check level; add if necessary	•	•	•	•	•	•	•	•	•	•
Windshield wiper arm pivot points: Lubricate	•	•	•	•	•	•	•	•	•	•
Dust and pollen filter: replace		•		•		•		•		•
Haldex clutch (quattro), Change oil		•		•		•		•		•
Haldex clutch (quattro), Change filter				•				•		
Tires/spare/wheels: Check condition (tread depth) and pressure, adjust if necessary	•	•	•	•	•	•	•	•	•	•
Wheels: Rotate front to rear	•									
Wheels: Check condition and expiration date on tire mobility system sealant		•		•		•		•		•
Drive shafts: Check boots		•		•		•		•		•
Drive shafts: Check front and rear for excessive play		•		•		•		•		•
On-Board Diagnostic system: Check fault memory, erase if necessary		•		•		•		•		•
Brake system; Check for damage/leaks, brake pad thickness, brake fluid level	•	•	•	•	•	•	•	•	•	•
Cooling system: Check coolant level, add if necessary		•		•		•		•		•
Front suspension: Check dust seals on ball joints and tie rod ends		•		•		•		•		•
Air cleaner: Replace filter				•				•		
Headlights and exterior lights: Check and adjust if necessary		•		•		•		•		•
Undercarriage: Visual inspection for damage		•		•		•		•		•
Drive shafts: Check boots		•		•		•		•		•
Power steering: Check fluid level				•				•		
Manual transmission final drives Check for leaks		•		•		•		•		•
Automatic transmission final drive oil: Check for leaks		•		•		•		•		•
Direct shift gearbox: Change oil and filter				•				•		
Spark plugs; Replace				•				•		
Ribbed belt: Replace (1.8L)								•		
Ribbed belt: Check (3.2L)								•		
Timing belt and tensioner system: Replace							•			
Road test: Check kick-down, braking, steering, heating/ventilation, air conditioning, power accessories and electrical systems		•		•		•		•		•
Every 2 years, regardless of mileage (kilometers): Change brake fluid										

AMS-05-20

BentleyPublishers.com

©2006 BentleyPublishers.com—All Rights Reserved

15a Cylinder Head (1.8L Engine)

General

This section covers cylinder head and camshaft drive belt (timing belt) service and repair work for the 1.8 liter engine. Most of the operations described in this repair group require special equipment and experience. If you lack the skills, tools, or a suitable workplace for servicing or repairing the cylinder head, then leave these repairs to an authorized Audi dealer or other qualified shop.

Engine type	Engine code
1.8L 4-cylinder turbo (180 hp) 1.8L 4-cylinder turbo (225 hp)	ATC, AWP AMU, BEA

WARNING—
- *When using an exchanged cylinder head with camshaft installed, lubricate the contact surfaces between the lifters and cam lobes before installing cylinder head cover.*
- *When installing a new factory supplied cylinder head, remove the plastic protectors installed to protect the open valves immediately before installing cylinder head.*
- *After replacing cylinder head, flush and replace all coolant.*
- *Cylinder heads with cracks between the valve seats can continue to be used without reducing service life, as long as the cracks have a width of max. 0.5 mm (0.02 in.).*
- *Do not mill the valve seats in the cylinder head; only lapping of the valves is permissible.*
- *After installing the camshafts, do not start the engine for approx. 30 minutes. This allows the hydraulic lifters to seat themselves. Otherwise the valves will crash into the pistons.*
- *After working on the valvetrain and lifters, slowly rotate the crankshaft by hand at least 2 full revolutions before starting engine to be sure that valves do not strike the pistons.*
- *Always replace gaskets and seals.*

BentleyPublishers.com

© 2006 BentleyPublishers.com—All Rights Reserved

Diagnostic Testing

The tests that follow can be used to help isolate engine problems, to better understand a problem before starting expensive and extensive repairs, or to just periodically check engine condition.

Compression Test

A compression test will tell a lot about the overall condition of the engine without the need for taking it apart. Testing is relatively simple and straightforward.

– Warm-up engine until oil temperature is a minimum of 30°C (86°F).

– Remove upper sound absorber panel. See **03 Maintenance**.

< Disconnect harness connector (**arrow**) from camshaft position sensor (G40).

NOTE —

- *Audi identifies electrical components by a letter and/or a number in the electrical schematics. See* **EWD Electrical Wiring Diagrams**. *These electrical identifiers are listed in parentheses as an aid to electrical troubleshooting.*

– Use compressed air to clean area around ignition coils.

– Remove individual ignition coils. See **03 Maintenance, Spark Plugs, replacing**.

– Using a deep socket and extension remove all spark plugs and lay aside in proper order.

< Install compression tester into spark plug hole for cylinder #1.

– Have a helper:
- Depress clutch pedal fully, as applicable.
- Put transmission in neutral or park, as applicable.
- Crank engine over with starter motor.

NOTE —

- *Cranking engine over with fuel injection system components disconnected may cause Diagnostic Trouble Codes (DTCs) to be stored in engine management system memory.*

– Crank engine a minimum of 4 to 5 revolutions with throttle valve open.

– Record readings, release pressure in gauge and repeat procedure for each cylinder.

– Compare readings to specification.

Compression pressure (1.8L engine, 180 hp or 225 hp)	
New	10.0 to 13.0 bar (145 - 189 psi)
Wear limit	7.0 bar (102 psi)
Max. difference between cylinders	3.0 bar (44 psi)

BentleyPublishers.com
© 2006 BentleyPublishers.com—All Rights Reserved

– When all cylinders have been checked, reinstall spark plugs, ignition coils and harness connector for camshaft position sensor. Install upper engine cover and any other components that were removed.

Tightening torques	
Ignition coils to cylinder head, as applicable	10 Nm (7 ft-lb)
Spark plug to cylinder head	30 Nm (22 ft-lb)
Vacuum reservoir to cylinder head	30 Nm (22 ft-lb)

Cylinder leakdown test

The most conclusive method for diagnosing low compression symptoms is a cylinder leak-down test. Using a special tester and a supply of compressed air, each cylinder is pressurized. The rate at which air leaks out of cylinder, as well as the sound and location of escaping air can more accurately pinpoint the magnitude and source of leakage. Any engine compression diagnosis that will require major disassembly should first be confirmed by a cylinder leak-down test. Because this test requires special equipment and experience, it should be performed by an Audi dealer or other qualified repair shop. Leakdown limit specifications are usually supplied by the equipment manufacturer.

CYLINDER HEAD SERVICE

The following page contains an assembly view of internal and external cylinder head components. Removal and installation of the cylinder head cover (valve cover) and timing belt are also covered in this section. For timing belt replacement interval, see Maintenance Schedules in **03 Maintenance**.

BentleyPublishers
.com
©2006 BentleyPublishers.com—All Rights Reserved

Cylinder head, assembly

1. **Cylinder head**
   - Replace coolant if head or head gasket is replaced
2. **Oil deflector**
   - Note installation position: above intake camshaft between #1 and #2 cylinders
3. **Cylinder head bolt**
   - 40 Nm (30 ft-lb) plus ½ turn (180°)
   - See **Cylinder head bolts, removing and installing** in this repair group
   - Always replace
4. **Cylinder head cover gaskets**
   - Replace if damaged or leaking
   - Before installing, apply sealant (D 454 300 02) to sealing surfaces
   - See **Cylinder head cover, removing and installing** in this repair group
5. **Cylinder head cover**
   - Always clean sealing surfaces of cylinder head with a clean cloth before installing
   - See **Cylinder head cover, removing and installing** in this repair group
6. **Gasket**
   - For ignition coil
   - Replace if damaged
7. **Ignition coil**
8. **Bolt**
   - 10 Nm (7 ft-lb)
9. **Oil fill cap**
10. **Seal**
    - Replace if damaged
11. **Cylinder head cover bolts**
    - 10 Nm (7 ft-lb)
    - See **Cylinder head cover, removing and installing** in this repair group
12. **Intake manifold gasket**
    - Always replace
13. **Cylinder head gasket**
    - Part number on gasket should be facing cylinder head
    - If gasket is replaced, replace all engine coolant

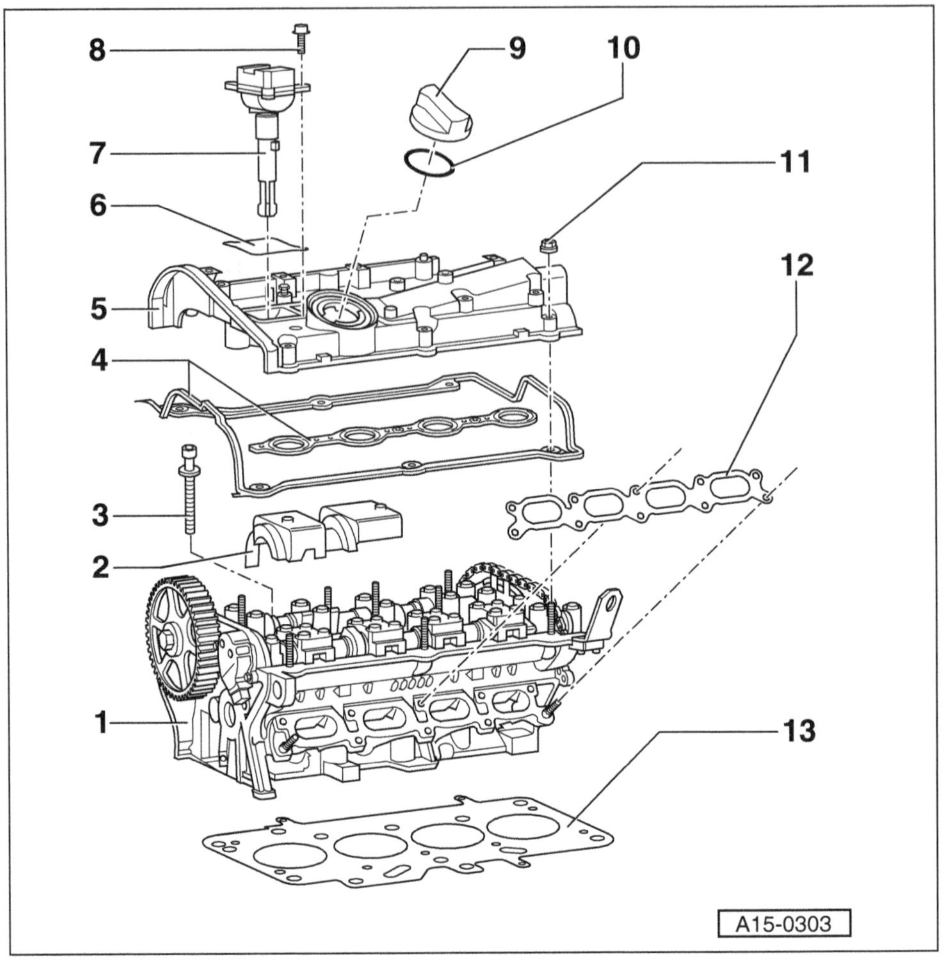

Cylinder head bolts, removing and installing

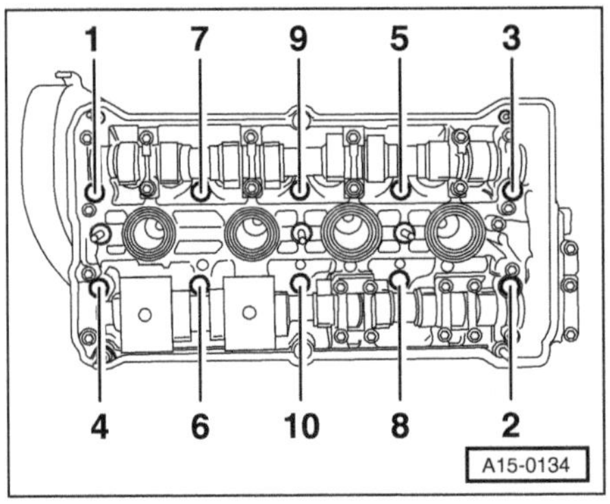

Remove cylinder head bolts using sequence shown, starting with #**1**.

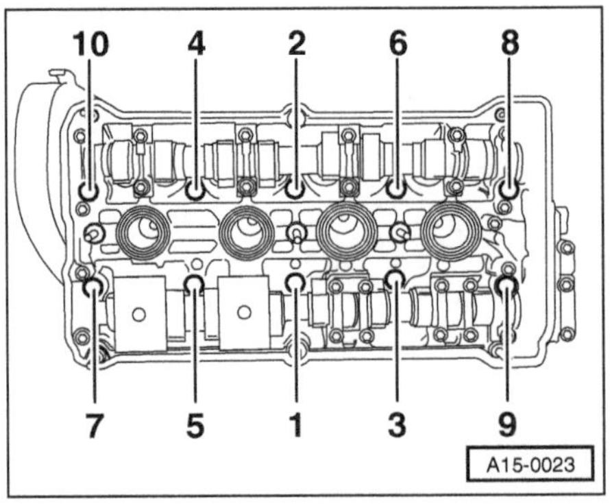

Install cylinder head bolts using sequence shown, starting with #**1**.

BentleyPublishers.com
© 2006 BentleyPublishers.com—All Rights Reserved

Cylinder head cover, removing and installing

– Remove upper sound absorber panel from engine. See **03 Maintenance**.

225 hp engine (code: AMU, BEA)

Remove intake air duct by disconnecting hoses and clamps at **arrows**.

NOTE —

- *Seal off openings of charge air circuit with clean rags to prevent entry of dirt particles.*

Continued for all engines

Remove mounting bolts (**arrows**) and vacuum reservoir from top of cylinder head cover.

- Tilt vacuum reservoir toward rear to remove and place out of way.
- Engine code AMU shown, others are similar.

225 hp engine (code: AMU, BEA)

– Disconnect harness connector (**left arrow**) for turbocharger recirculation valve (N249).

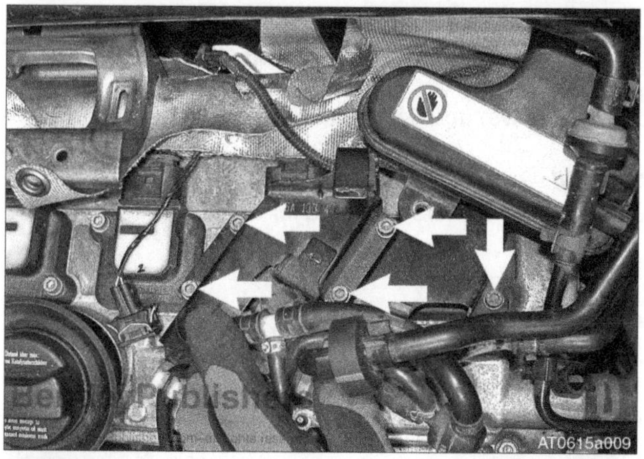

Remove mounting bolts (**arrows**) and position bracket for turbocharger recirculation valve out of way.

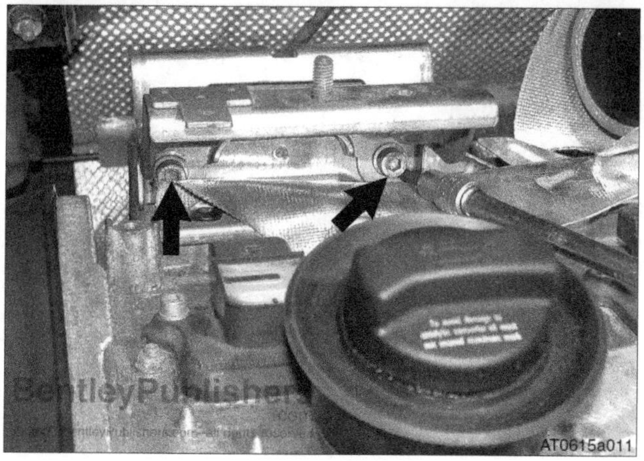

Remove mounting bolts (**arrows**) and rear bracket for intake air duct.

BentleyPublishers.com
© 2006 BentleyPublishers.com—All Rights Reserved

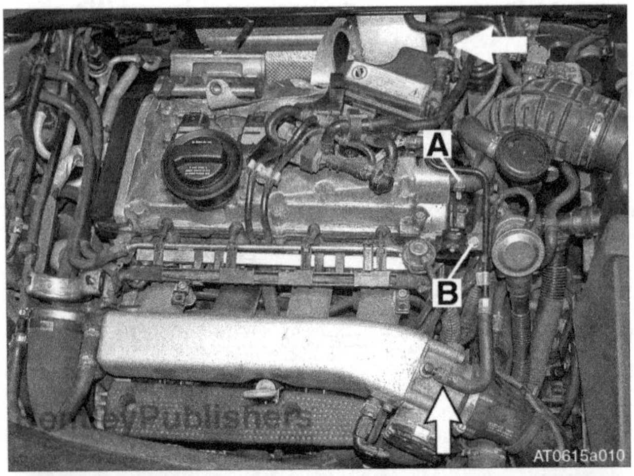

Continued for all engines

Loosen clamp (**A**) and remove breather hose from cylinder head cover.

225 hp engine (code: AMU, BEA)

– Loosen clamps (**arrows**), remove bolt (**B**) and remove vacuum line for throttle valve control module.

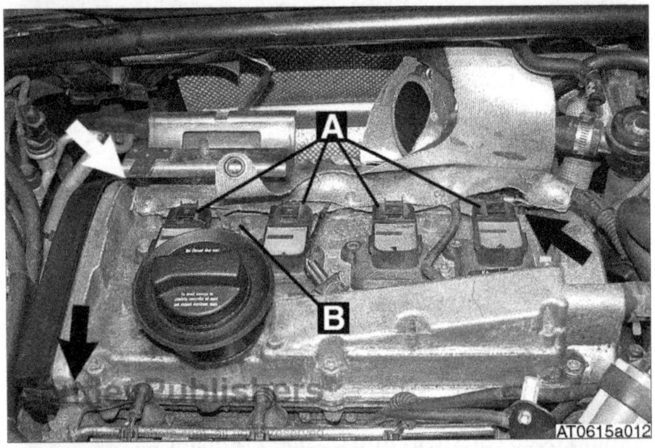

Continued for all engines

Disconnect harness connectors (**A**) and unbolt ground wire (**B**) for ignition coils.

– Pry out retaining tab (**right arrow**) and position ignition coil wiring harness out of way.

– Unclip (**left arrows**) upper timing belt cover.

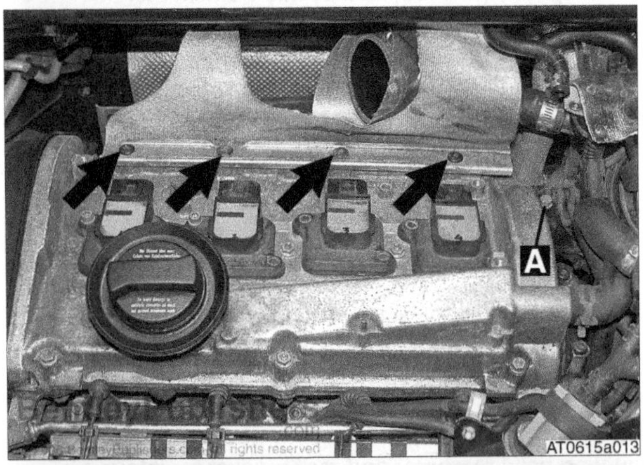

Remove bolt (**A**) and position bracket out of way.

– Remove screws (**arrows**) and heat insulation wrap.

– Remove remaining coil mounting bolts and ignition coils.
  - Coils are shown with numbers indicated for cylinder #.

Unbolt cylinder head cover mounting bolts (**arrows**).

– Remove cylinder head cover.

BentleyPublishers.com

© 2006 BentleyPublishers.com—All Rights Reserved

Cylinder head cover, installing

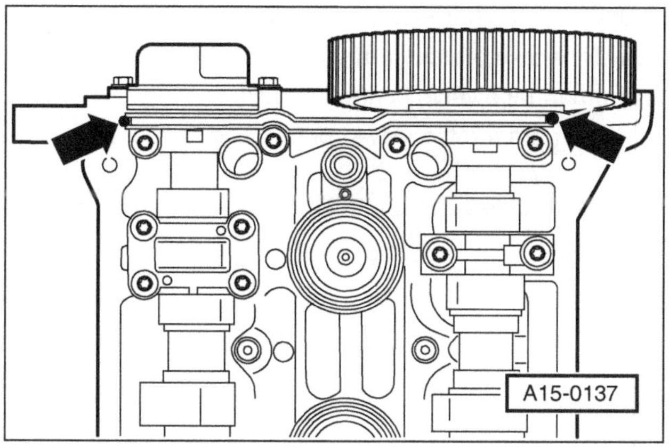

◄ Before installing cylinder head cover, apply drops of sealant (approx. 5 mm (¼ in.) diameter) at both corners (**arrows**) of cylinder head sealing surfaces. Use Audi sealant (part no. AMV 174 004 01).

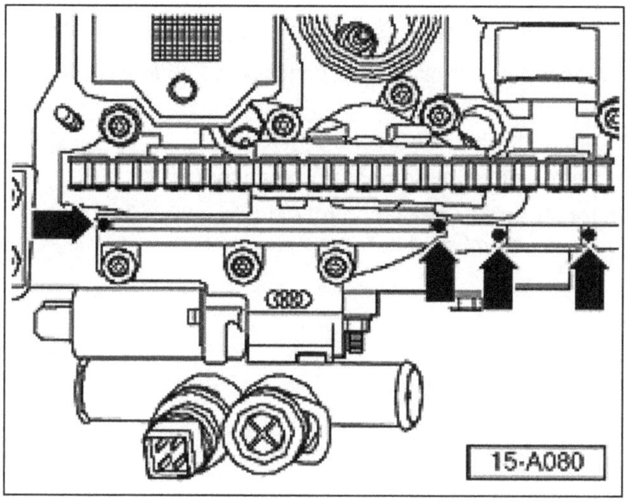

◄ Before installing cylinder head cover, apply drops of sealant (approx. 5 mm (¼ in.) diameter) at sides of joints (**arrows**) of cylinder head sealing surfaces. Use Audi sealant (part no. AMV 174 004 01).

– Replace cylinder head cover gasket if damaged.

– Tighten cylinder head cover bolts in diagonal sequence to specified torque.

Tightening torque	
Cylinder head cover to cylinder head	10 Nm (7 ft-lb)

– Remaining installation is reverse of removal.

Intake manifold, removing and installing

NOTE —

- *Replace and properly reposition any cable ties that were removed in their original locations when installing components.*

– Disconnect battery ground (GND) strap from battery negative (–) terminal. See **27 Battery, Alternator, Starter** before disconnecting battery.

- Be sure to have the anti-theft radio code on hand before disconnecting the battery.

NOTE —

- *Disconnecting the negative (–) battery cable may erase fault codes and basic settings in the engine management and automatic transmission control modules. Some driveability problems may be noticed until the system re-adapts to operating conditions. OBD II readiness codes, which may be required for emissions testing, may also be erased.*

– Remove upper sound absorber panel from engine. See **03 Maintenance**.

BentleyPublishers.com
©2006 BentleyPublishers.com—All Rights Reserved

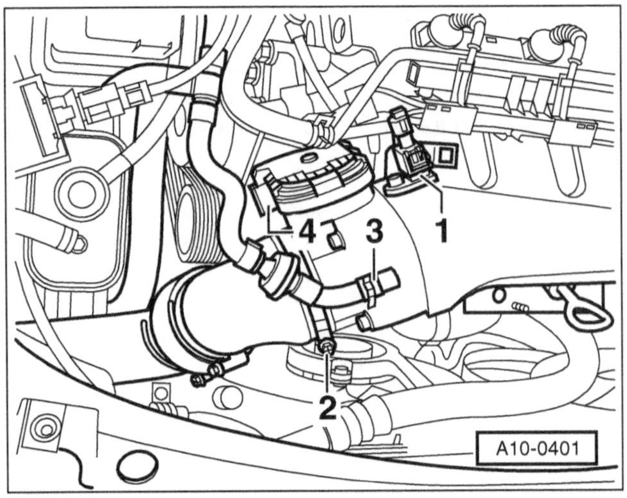

180 hp engine (code: ATC, AWP)

Disconnect the following harness connectors and hoses:

1. Connector for intake air temperature sensor (G42).
2. Air hose from throttle valve control module.
3. Vacuum hose from throttle valve control module to EVAP canister.
4. Connector for throttle valve control module (J338) (below throttle valve control module).

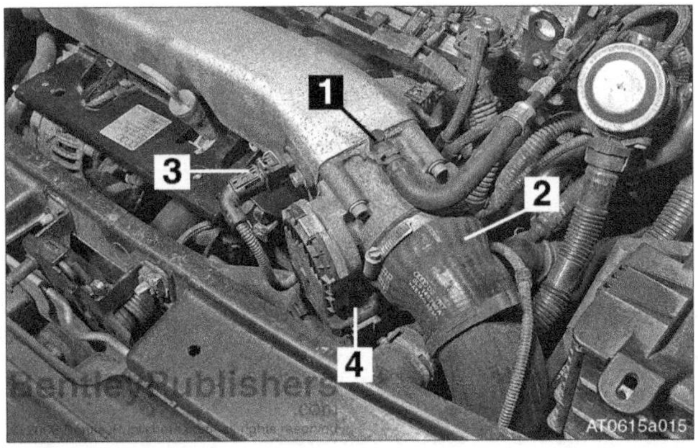

225 hp engine (code: AMU, BEA)

Disconnect the following harness connectors and hoses:

1. Vacuum hose from throttle valve control module to EVAP canister.
2. Air hose from throttle valve control module.
3. Connector for intake air temperature sensor (G42).
4. Connector for throttle valve control module (J338) (below throttle valve control module).

Remove intake air duct by disconnecting hoses and clamps at **arrows**.

NOTE—

- *Seal off openings of charge air circuit with clean rags to prevent entry of dirt particles.*

Continued for all vehicles

Disconnect harness connector (**arrow**) from camshaft position sensor (G40).

BentleyPublishers
.com
©2006 BentleyPublishers.com—All Rights Reserved

◄ Disconnect harness connectors (**1**) for fuel injectors and unclip (**2**) support bar from fuel rail.

– Carefully lay fuel injector harness to side.

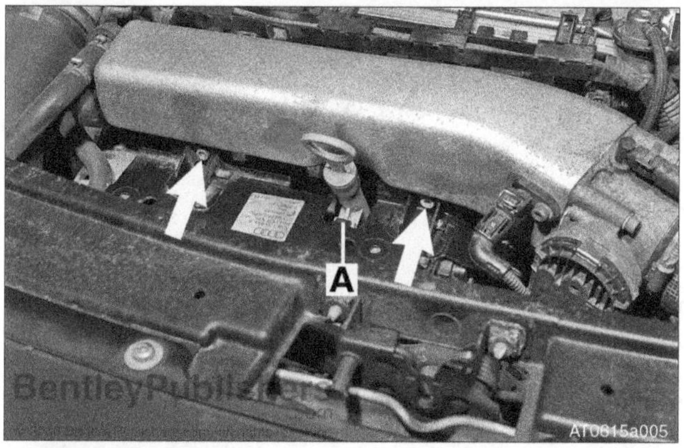

◄ Unbolt (**arrows**) mounting bracket for solenoid valves from front of intake manifold.

– Detach dipstick guide tube (**A**).

◄ Detach harness connector and unscrew mounting bolts (**arrows**) for solenoid valve.

– Unclip vacuum line from retaining clip (**right arrow**).

– Remove solenoid valve mounting bracket.

◄ Press release tabs and pull fuel supply line (**1**) and fuel return line (**2**) off connections.

- Fuel supply line has white markings
- Fuel return line has blue markings

> ***WARNING —***
> - *Fuel system is under pressure! Before opening system place a cloth around the connection. Then release pressure by carefully loosening connection.*

BentleyPublishers
.com
© 2006 BentleyPublishers.com—All Rights Reserved

– Unbolt intake manifold support from intake manifold.

– Pull out oil dipstick.

◄ Unbolt intake manifold mounting bolts (**arrows**) and remove manifold.

– Install intake manifold in reverse order of removal.

– After connecting battery terminals, enter anti-theft code for radio.

– Close power windows all the way to top positions using window switches. Then operate power windows again for at least one second in the "close" direction to activate the automatic one-touch function.

– Set clock to correct time.

– Perform adaptation of throttle valve control module. See **24a Fuel Injection (1.8L)**.

– Check for Diagnostic Trouble Codes (DTCs).

NOTE—
- *Diagnostic Trouble Codes (DTCs) will likely result from harness connectors and components that have been disconnected. Use a diagnostic scan tool to read and erase DTC memory after installation of harness connectors and components. See* **OBD On Board Diagnostics** *for more information.*

BentleyPublishers.com
© 2006 BentleyPublishers.com—All Rights Reserved

TIMING BELT

Timing belt, overview

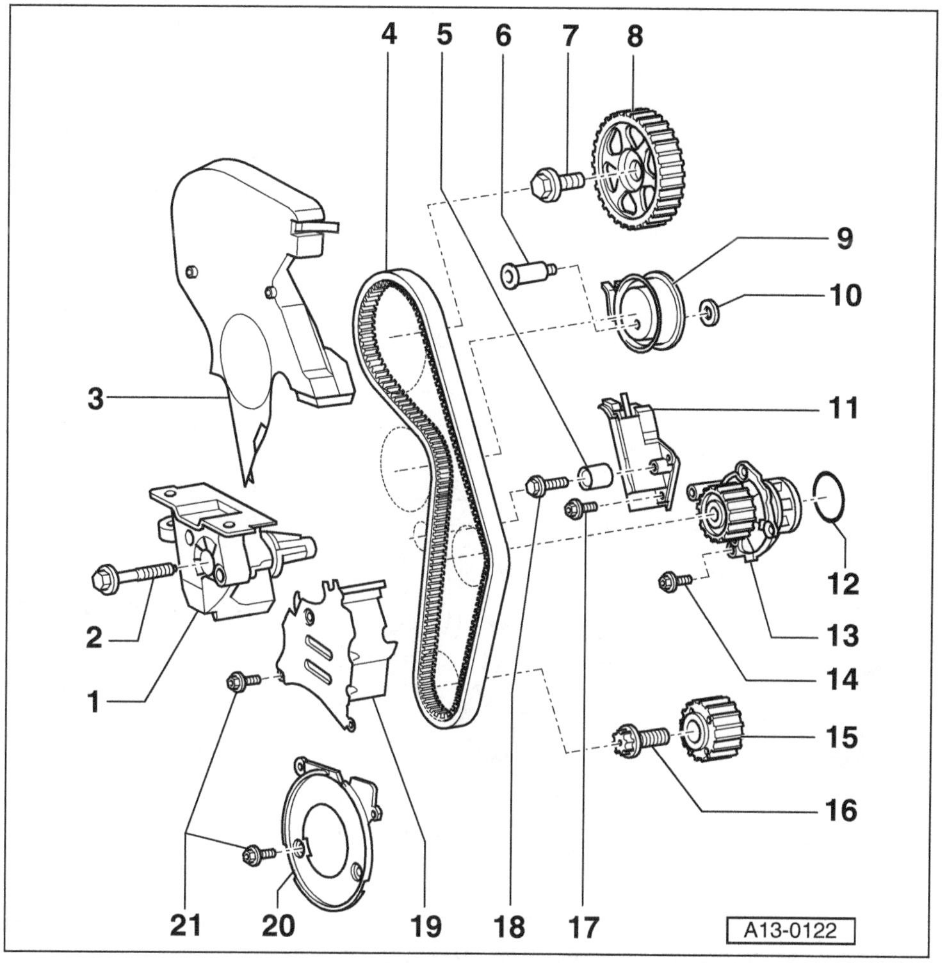

1. **Engine mount**
2. **Bolt**
   - 45 Nm (35 ft-lb)
3. **Toothed belt guard, upper**
   - To remove belt guard, it may be necessary to unbolt retainer for turbocharger coolant return line
4. **Toothed belt**
   - Mark direction of rotation before removing if re-using belt
   - Check for wear
   - Do not kink
5. **Idler roller**
6. **Tensioner mounting bolt**
   - 27 Nm (20 ft-lb)
7. **Bolt**
   - 65 Nm (48 ft-lb)
   - Use special tool 3036 (counter holder) to loosen/tighten
8. **Camshaft sprocket**
   - For exhaust camshaft
   - Note installation position: The sprocket is installed with the smaller side facing out and the TDC mark for cylinder #1 visible
9. **Tensioning roller**
10. **Washer**
11. **Tensioning dampener**
    - With pressure piston to automatically adjust belt tension
12. **O-ring**
    - Always replace
    - Lightly coat with coolant before installing
13. **Coolant pump**
14. **Bolt**
    - 15 Nm (11 ft-lb)
15. **Crankshaft sprocket**
    - Contact surface between sprocket and crankshaft must be free of oil
    - Can only be installed in one position
16. **Bolt**
    - 90 Nm (66 ft-lb) plus an additional 1/4 turn (90 °)
    - Always replace
    - Use special tool 3099 (counter holder) to loosen/ tighten
    - Place 2 washers between sprocket and retainer when installing retainer
17. **Bolt**
    - 15 Nm (11 ft-lb)
18. **Bolt**
    - 20 Nm (15 ft-lb)
19. **Toothed belt guard, center**
20. **Toothed belt guard, lower**
21. **Bolt**
    - 10 Nm (7 ft-lb)

BentleyPublishers.com
© 2006 BentleyPublishers.com—All Rights Reserved

Timing belt, removing

The timing belt, also known as the toothed belt or camshaft drive belt, should be inspected and replaced periodically. Inspection and replacement intervals are covered in **03 Maintenance**. Belt replacement is covered here. Regular belt inspection and replacement will prevent damage to the engine due to belt stretch or breakage.

NOTE—

- *Removal and installation of the timing belt requires the use of a number of special tools. Read the procedure through and have these special tools on hand before starting work.*

– Remove engine cover. See **03 Maintenance**.

– Loosen lug bolts for right front wheel. Raise vehicle with lift or jack and support with jackstands. See **03 Maintenance.** Remove right front wheel and wheel housing liner.

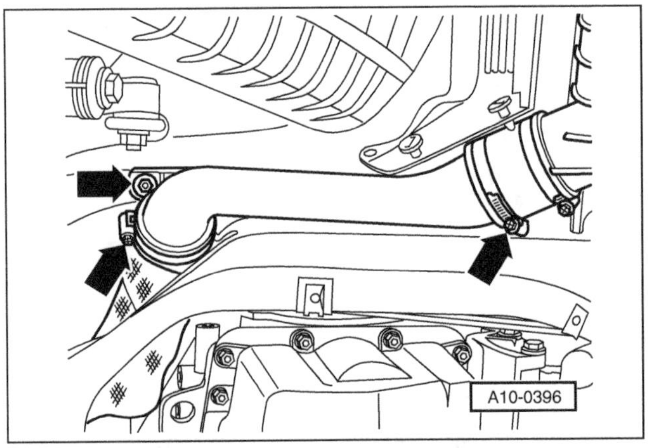

180 hp engine (code: ATC, AWP)

< Remove connecting duct between intercooler and intake manifold. Disconnect duct at **arrows** to remove.

NOTE—

- *Seal off openings of charge air circuit with a clean rag to prevent entry of dirt particles.*

225 hp engine (code: AMU, BEA)

< Remove intake air duct by disconnecting hoses and clamps at **arrows**.

NOTE—

- *Seal off openings of charge air circuit with clean rags to prevent entry of dirt particles.*

Continued for all vehicles

< Pivot ribbed belt tensioner with 17 mm open end wrench (**arrow**) to relieve tension on belt. Remove ribbed belt.

NOTE—

- *Mark running direction of ribbed belt if reusing.*

– Remove lower sound absorber panel from underneath center of engine. Also remove right side lower sound absorber panel. See **03 Maintenance**.

BentleyPublishers.com
© 2006 BentleyPublishers.com—All Rights Reserved

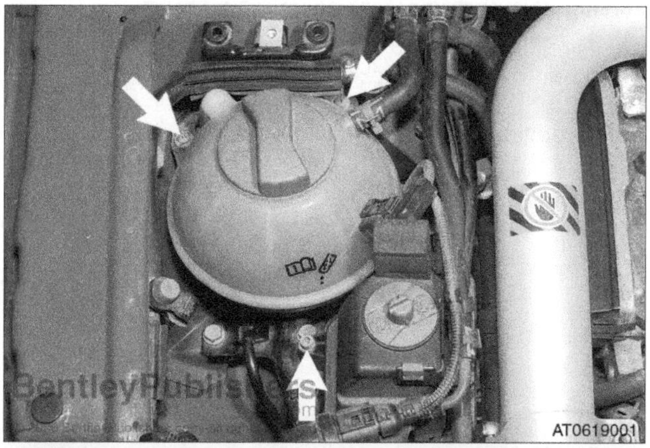

◀ Unbolt (**arrows**) coolant expansion tank and power steering reservoir and move to side with hoses still connected.

– Disconnect vacuum hose from throttle valve control module.

– Unclip and remove upper toothed belt guard.
- To remove belt guard, it may be necessary to unbolt retainer for turbocharger coolant return line.

◀ Install engine support bar with appropriate adapters and support weight of engine.

– Lift engine slightly with spindle of engine support bar.

NOTE —
- *Lift coolant expansion tank and power steering reservoir and tie them in place on the engine support bracket.*

◀ Turn center bolt of camshaft pulley in clockwise direction until TDC marks (**arrows**) for cylinder #1 line up.

– Remove vibration damper (crankshaft pulley), making sure not to change the TDC position.

NOTE —
- *The vibration dampener is also be referred to as the crankshaft ribbed belt pulley. Removal allows access to the timing belt for removal.*

– Remove center and lower sections of toothed belt guard.

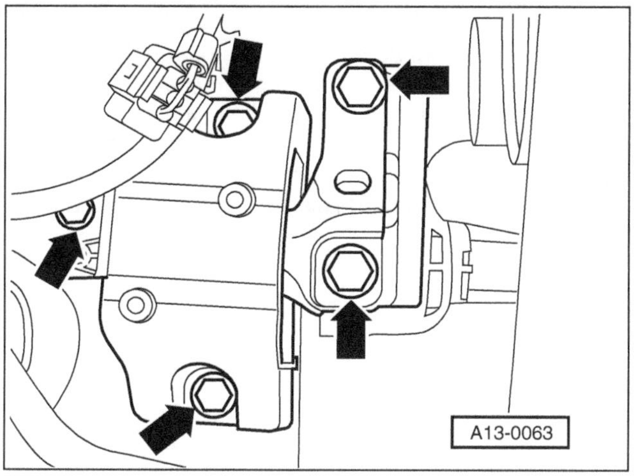

◀ Unbolt (**arrows**) engine mount from engine and body.

– Unbolt engine mount connecting piece between engine and body.

– Remove engine mount.

BentleyPublishers
.com
©2006 BentleyPublishers.com—All Rights Reserved

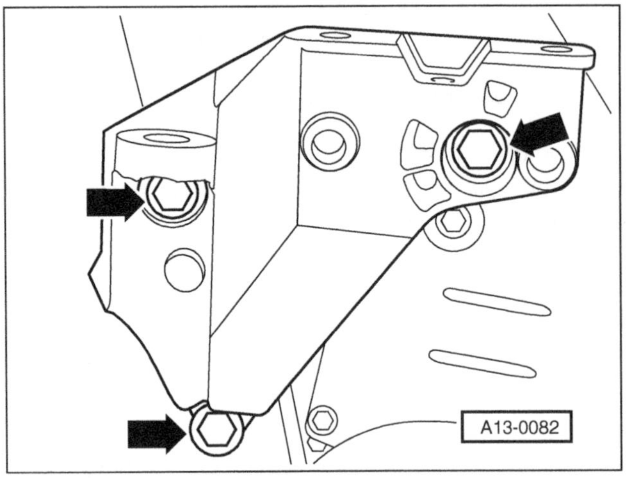

- Remove mounting bolts (**arrows**) for engine mount bracket.
  - Raise or lower engine slightly as necessary using spindles of engine support bar to access bracket mounting bolts.
- Mark direction of rotation of timing belt if re-using belt.

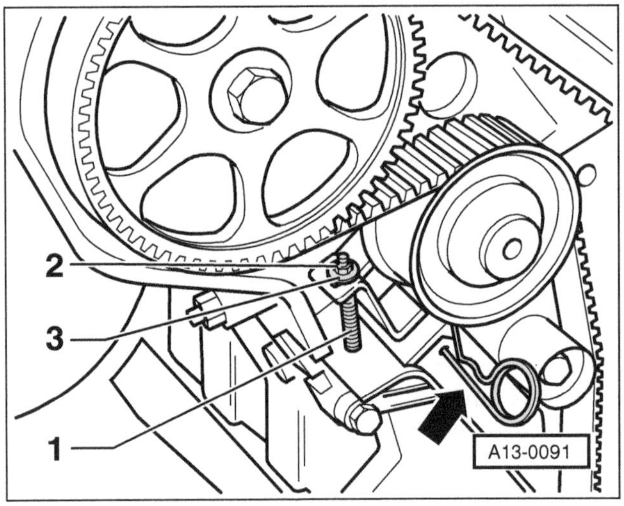

- Relieve tension on timing belt.
  - Screw a 5x55 mm threaded stud (**1**), or special tool T10092, into timing belt tensioning damper housing and install a nut (**2**) and washer (**3**) to the other end.
  - Line up the hole in the tensioning damper housing with the hole in the pressure piston by turning the pressure piston. Use a suitable tool to turn the pressure piston before tightening the nut on the threaded stud.
  - Tighten nut (**2**) on threaded stud until the hole in the pressure piston aligns with the hole in the tension damper housing.
  - Secure pressure piston in position with a pin, or special tool T40011 (**arrow**), inserted through hole in tensioning damper housing.
- Remove timing belt.
- Turn crankshaft back off of TDC slightly.

> ***CAUTION—***
> - *Due to engine design, care must be used when turning the camshaft with the timing belt removed. If valves are allowed to open with the piston at TDC, serious internal damage will result.*

Timing belt, installing

> ***CAUTION—***
> - *Timing belt tension must not be adjusted on a hot engine. Allow to cool sufficiently before proceeding. Engine temperature must no hotter than warm to the touch.*

- Make sure mark on camshaft sprocket (**arrow**) is still aligned with mark on upper rear timing belt guard. This positions the camshaft at TDC for cylinder number #1.

BentleyPublishers
.com
©2006 BentleyPublishers.com—All Rights Reserved

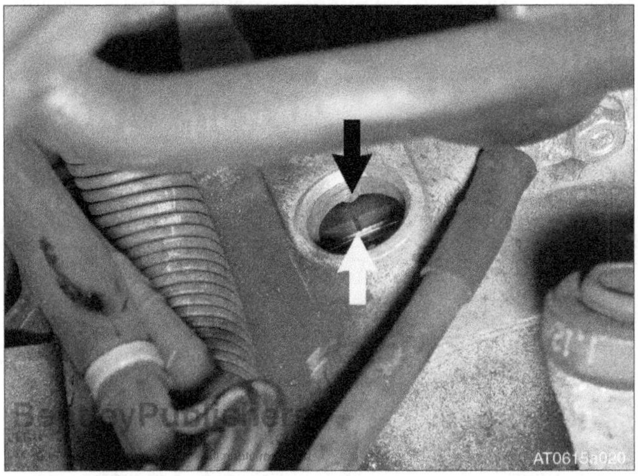

Remove plastic cover on transmission inspection hole and confirm TDC for cylinder #1 by checking alignment of mark on flywheel or torque converter to notch (**arrows**).

– Install timing belt onto the crankshaft (not camshaft) sprocket only.

NOTE —

- *If re-using an old belt, be sure to note running direction marks placed on it before removal.*

– Install engine mount bracket on cylinder block.

Tightening torque	
Engine mount bracket to cylinder block	45 Nm (33 ft-lb)

– Install lower timing belt guard.

– Install crankshaft ribbed belt pulley/vibration damper to crankshaft sprocket.

Tightening torque	
Vibration damper to crankshaft sprocket	25 Nm (18 ft-lb)

– Ensure that camshaft is still at TDC for cylinder #1.

– Install the timing belt onto the tensioner, coolant pump and camshaft drive sprockets. Check for correct placement of belt on all sprockets.

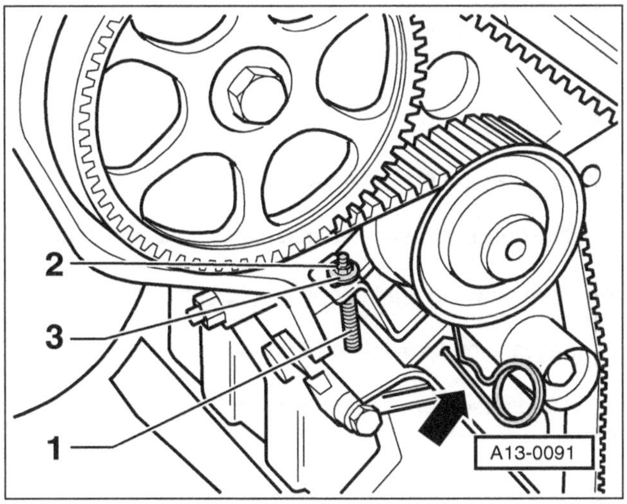

Apply tension to timing belt.

- Remove securing pin (**arrow**) from tensioner damper housing.
- Loosen nut on threaded stud (**1**) and allow belt tension to tighten.
- Remove threaded stud, nut and washer from tensioning damper housing.

– Ensure that camshaft and flywheel/torque converter marks are still at TDC for cylinder #1.

– Turn engine two full turns in the running direction and confirm that TDC reference marks still align.

NOTE —

- *Some movement of the sprockets and their marks is to be expected as belt tension is adjusted. Keep in mind that the smallest possible increment of adjustment is one whole tooth of the belt or sprocket.*

– Install center and upper timing belt guards.

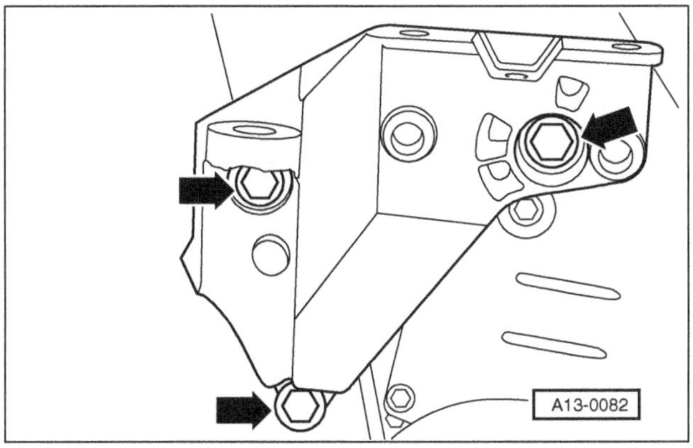

Install mounting bolts (**arrows**) for engine mount bracket.

- Raise or lower engine slightly as necessary using spindles of engine support bar to install bracket mounting bolts.

Tightening torque	
Engine mount bracket to engine block	45 Nm (33 ft-lb)

BentleyPublishers
.com
© 2006 BentleyPublishers.com—All Rights Reserved

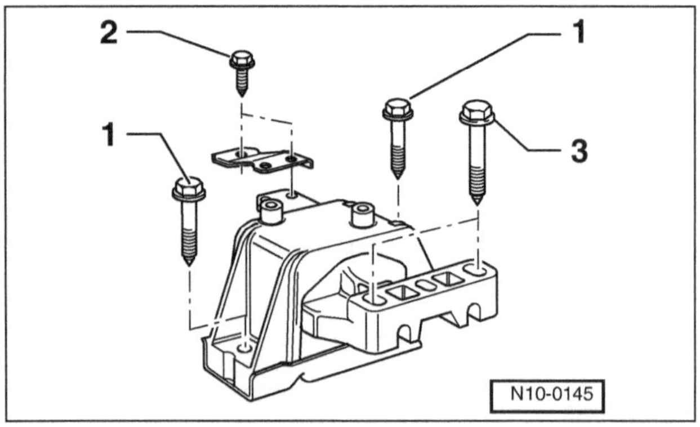

Install engine mount to body with mounting bolts (**1** and **2**). Replace bolts as specified.

Tightening torques	
Engine mount to body (bolts **1** at left) always replace	40 Nm (30 ft-lb) plus 90° (¼ turn)
Mount bracket to body (bolts **2** at left)	25 Nm (18 ft-lb)

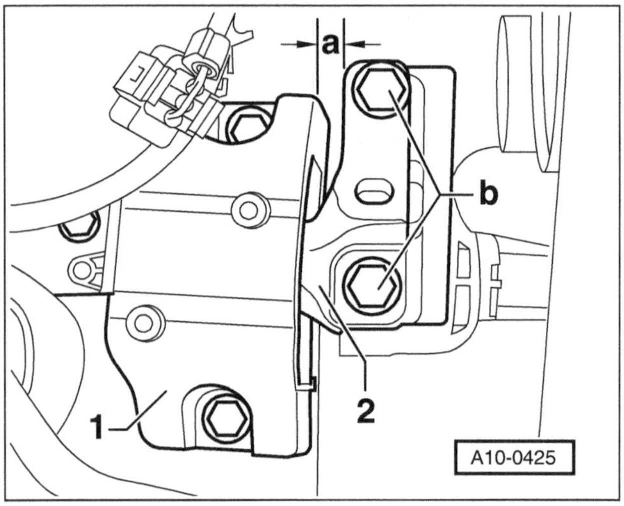

Lower engine to installation position. Bring engine mount and bracket into contact by slowly lowering engine using support bar. Install new mounting bolts (**b**) for engine mount to bracket.

- Before tightening engine mount bolts (**b**), move engine as necessary to obtain proper gap (**a**):
  - **a** = approx. 13 mm (0.51 in.)

NOTE —
- *A 12mm thick flat bar, or Audi special tool 2011, can be used to help obtain the proper engine mount gap.*

- Tighten engine mount to bracket bolts (**b**) to final torque.

Tightening torque	
Engine mount to engine bracket always replace	60 Nm (44 ft-lb) plus 90° (¼ turn)

- Remove engine support bar.
- Install ribbed belt. See **03 Maintenance**.
- Install connecting duct between intercooler and intake manifold (180 hp engine).
  Install intake air duct at top right of engine (225 hp engine).
- Install coolant expansion tank and power steering reservoir.
- Install lower sound absorber panels (belly pan) and upper sound absorber panel.
- Install any remaining parts and quality check your work.

BentleyPublishers
.com
©2006 BentleyPublishers.com—All Rights Reserved

Valvetrain Assembly

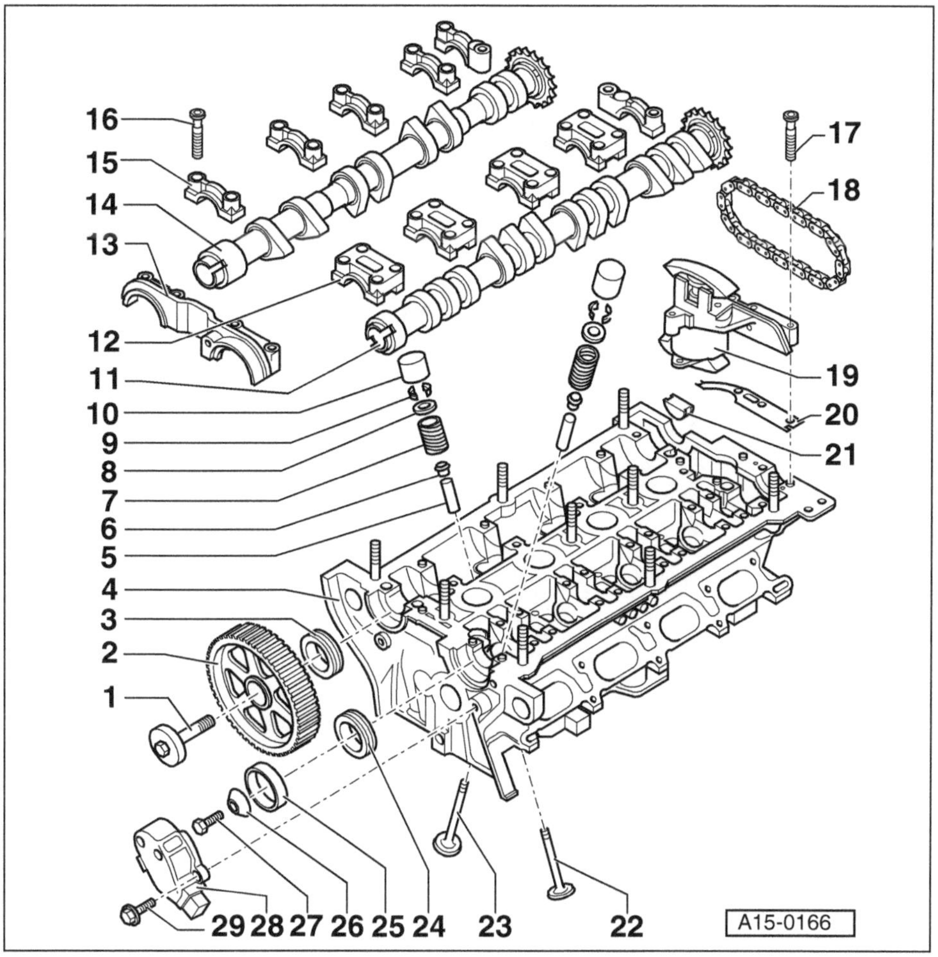

1. **Bolt**
   - 65 Nm (48 ft-lb)
   - Use special tool 3036 (counter holder) to loosen and tighten
2. **Camshaft sprocket**
   - For exhaust camshaft
   - Note installation position: smaller side faces out and TDC marking for cylinder #1 is visible
3. **Oil seal**
   - For exhaust camshaft
4. **Cylinder head**
5. **Valve guide**
6. **Valve stem seal**
7. **Valve springs**
8. **Valve spring plate**
9. **Valve keepers**
10. **Hydraulic valve lifter**
    - Check by pressing down with plastic or wooden wedge, if 0.20 mm (0.0079 in.) feeler gauge can be inserted between camshaft and lifter, replace lifter
    - Do not interchange
    - Oil contact surface
    - Store with contact surface down
11. **Intake camshaft**
    - Axial clearance: 0.20mm (0.008 in.)
    - Check radial clearance with Plastigage (valve lifters not installed)
      Wear limit: 0.1 mm (0.004 in.)
      Run-out max: 0.01 mm (0.0004 in.)
12. **Bearing cap**
    - Must be located on dowel sleeves
13. **Double bearing cap**
    - Must be located on dowel sleeves
    - Seal connections between double bearing cap and cylinder head
    - Coat sealing surface lightly with sealant (D 454 300 02)
14. **Exhaust camshaft**
    - Axial clearance: 0.20mm (0.008 in.)
    - Check radial clearance with Plastigage (valve lifters not installed)
      Wear limit: 0.1 mm (0.004 in.)
      Run-out max: 0.01 mm (0.0004 in.)
15. **Bearing cap**
    - Must be located on dowel sleeves
16. **Bolt**
    - 10 Nm (7 ft-lb)
17. **Bolt**
    - 10 Nm (7 ft-lb)
18. **Drive chain**
    - Check for wear
    - Before removing, mark running direction (installation position)
19. **Hydraulic chain tensioner**
    - Before removing, lock in position with retainer V.A.G 3366
    - Seal connections between hydraulic chain tensioner and cylinder head
20. **Metal/rubber gasket**
    - Always replace
21. **Gasket**
    - Always replace
22. **Exhaust valve**
    - Sodium filled
    - Do not rework; only hand lap
23. **Intake valve**
    - Do not rework; only hand lap
24. **Oil seal**
25. **Hall sensor rotor**
    - Note installation position: locate notch in camshaft
26. **Washer**
    - With conical taper
    - Note installation position
27. **Bolt**
    - 25 Nm (18 ft-lb)
28. **Camshaft position sensor (G40)**
29. **Bolt**
    - 10 Nm (7 ft-lb)

BentleyPublishers.com
© 2006 BentleyPublishers.com—All Rights Reserved

15b Cylinder Head (3.2L Engine)

GENERAL

This section covers cylinder head and related component overviews for the 3.2 liter engine. Removal and installation of these components requires special equipment and experience. If you lack the skills, tools, or a suitable workplace for servicing or repairing the cylinder head, then leave these repairs to an authorized Audi dealer or other qualified shop. More detailed removal and installation information is included in our electronic version (CD-ROM or web subscription) of the Official Factory Repair Manual, available at www.bentleypublishers.com.

Engine type	Engine code
3.2L 6-cylinder (250 hp)	BHE

WARNING —
- *When using an exchanged cylinder head with camshaft installed, lubricate the contact surfaces between the lifters and cam lobes before installing cylinder head cover.*
- *When installing a new factory supplied cylinder head, remove the plastic protectors installed to protect the open valves immediately before installing cylinder head.*
- *After replacing cylinder head, flush and replace all coolant.*
- *Cylinder heads with cracks between the valve seats can continue to be used without reducing service life, as long as the cracks have a width of max. 0.5 mm (0.02 in.).*
- *Do not mill the valve seats in the cylinder head; only lapping of the valves is permissible.*
- *After installing the camshafts, do not start the engine for approx. 30 minutes. This allows the hydraulic lifters to seat themselves. Otherwise the valves will crash into the pistons.*
- *After working on the valvetrain and lifters, slowly rotate the crankshaft by hand at least 2 full revolutions before starting engine to be sure that valves do not strike the pistons.*
- *Always replace gaskets and seals.*

BentleyPublishers
.com
© 2006 BentleyPublishers.com—All Rights Reserved

Diagnostic Testing

Compression Test

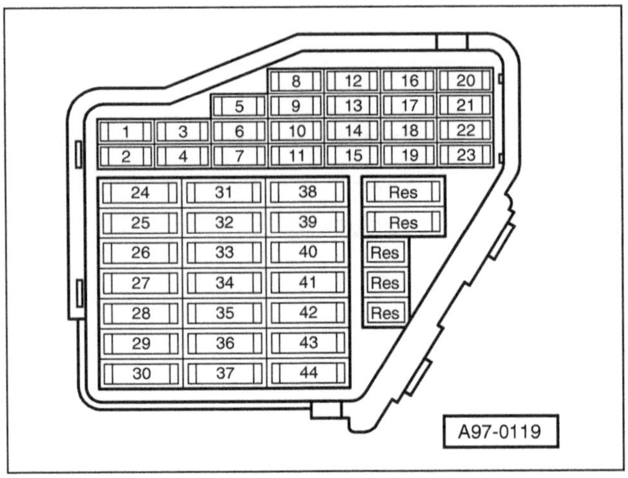

A compression test will tell a lot about the overall condition of the engine without the need for taking it apart. Testing is relatively simple and straightforward.

– Warm-up engine until oil temperature is a minimum of 30°C (86°F).

– Switch ignition off and remove fuse #28 from the fuse panel located at the left side end of the dash board.

– Remove upper sound absorber panel. See **03 Maintenance**.

– Remove individual ignition coils. See **03 Maintenance, Spark Plugs, replacing**.

– Use compressed air to clear area around spark plugs.

– Using a deep socket and extension remove all spark plugs and lay aside in proper order.

– Install compression tester into spark plug hole for cylinder #1.
  - 1.8L engine shown, 3.2L is similar

– Have a helper:
  - Depress clutch pedal fully, as applicable.
  - Put transmission in neutral or park, as applicable.
  - Crank engine over with starter motor.

NOTE —
- *Cranking engine over with fuel injection system components disconnected may cause Diagnostic Trouble Codes (DTCs) to be stored in engine management system memory.*

– Crank engine a minimum of 4 to 5 revolutions with throttle valve open.

– Record readings, release pressure in gauge and repeat procedure for each cylinder.

– Compare readings to specification.

Compression pressures (1.8L engine, 180 hp or 225 hp)	
New	10.0 to 13.0 bar (145 - 189 psi)
Wear limit	7.5 bar (109 psi)
Max. difference between cylinders	3.0 bar (44 psi)

– When all cylinders have been checked, reinstall spark plugs, ignition coils and harness connector for camshaft position sensor. Install upper engine cover and any other components that were removed.

Tightening torque	
Spark plug to cylinder head	30 Nm (22 ft-lb)

BentleyPublishers.com
© 2006 BentleyPublishers.com—All Rights Reserved

CYLINDER HEAD

Cylinder head, assembly

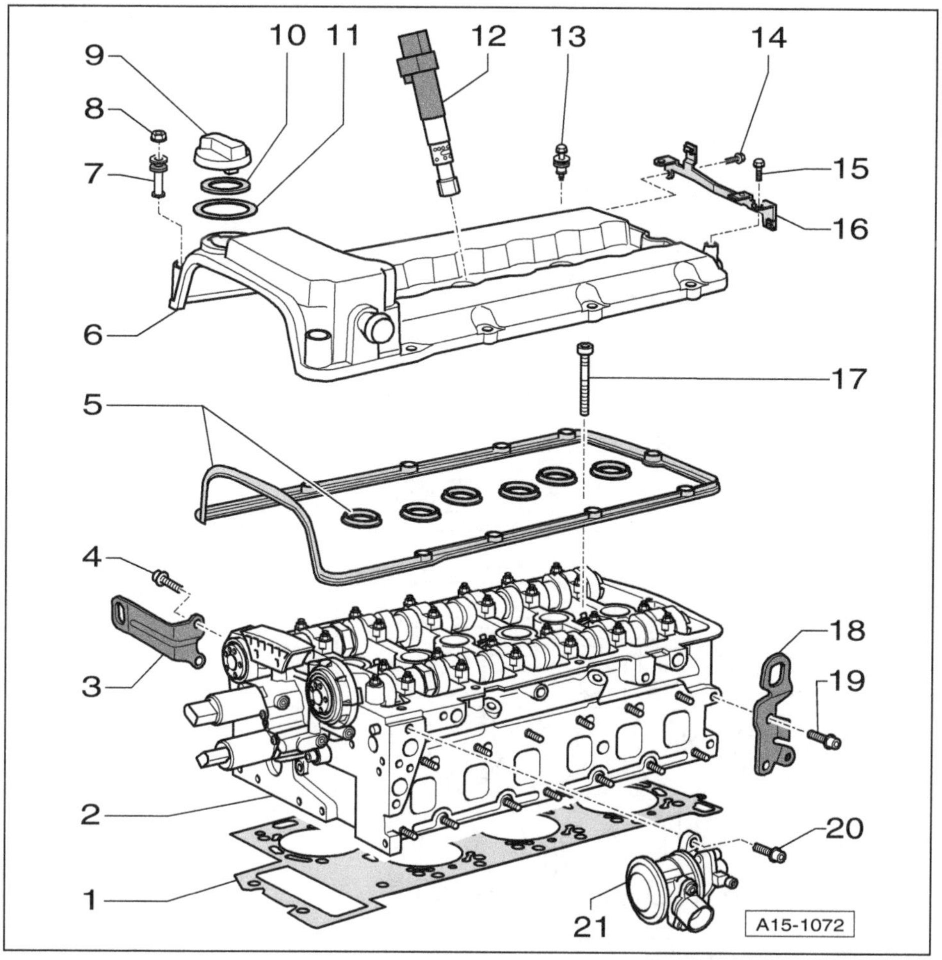

1. **Cylinder head gasket**
   - Metal gasket
   - Always replace
   - If gasket is replaced, replace all engine coolant
2. **Cylinder head**
   - Replace all coolant if head is replaced
   - Check for distortion at lower mounting surface: max. 0.05 mm
3. **Lifting eye**
4. **Bolt**
   - 23 Nm (17 ft-lb)
5. **Cylinder head cover gaskets**
   - Replace if damaged or leaking
   - Part number must be visible from intake side
6. **Cylinder head cover**
   - Always clean sealing surfaces of cylinder head with a clean cloth before installing
   - To remove, the following items must first be removed:
     - Front lock carrier upper trim
     - V-belt
     - Lower sound absorber panels
     - Lower front cross support
     - Radiator fan assembly
     - Power steering pump
     - Secondary air pump and bracket
     - A/C compressor
     - Alternator
     - Dipstick and guide tube
     - Vacuum reservoir at manifold
     - Intake manifold
     - Coolant pump mount bracket
     - Ignition coils
     - Front strut mount brace
     - Intake hose
     - Crankcase breather hose
     - Throttle valve
   - See **Cylinder head cover bolts, removing and installing** in this repair group
7. **Spacing sleeve**
   - Replace seal if damaged
8. **Nut**
   - 10 Nm (7 ft-lb)
9. **Oil filler cap**
10. **Seal**
    - Replace if damaged
11. **Rubber buffer**
    - Replace if damaged
12. **Ignition coil**
    - See **03 Maintenance** for removal and installation
13. **Bolt**
    - With spacer sleeve and seal
    - 10 Nm (7 ft-lb)
14. **Bolt**
    - 10 Nm (7 ft-lb)
15. **Bolt**
    - 10 Nm (7 ft-lb)
16. **Wiring harness retainer**
17. **Cylinder head bolt**
    - See **Cylinder head bolts, removing and installing** in this repair group
    - Always replace
18. **Lifting eye**
19. **Bolt**
    - 23 Nm (17 ft-lb)
20. **Bolt**
    - 10 Nm (7 ft-lb)
21. **Combination valve**
    - For secondary air injection system

BentleyPublishers.com
© 2006 BentleyPublishers.com—All Rights Reserved

Cylinder head cover (valve cover) bolts, removing and installing

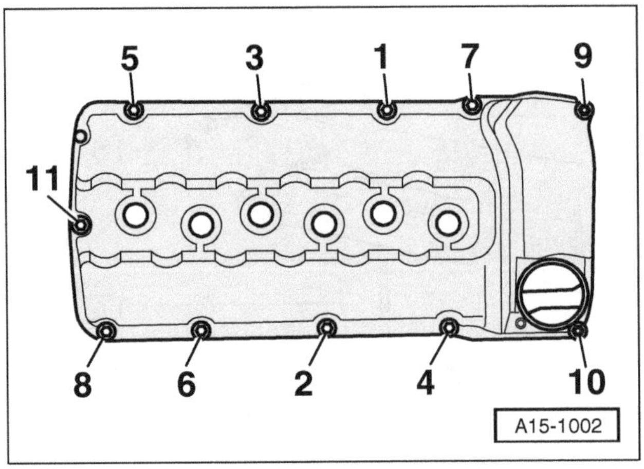

Cylinder head cover bolts loosening and tightening sequence.

- Loosen bolts starting with #**11** and ending with #**1**.
- Tighten bolts starting with #**1** and ending with #**11**.

Tightening torque	
Cylinder head cover to cylinder head	10 Nm (7 ft-lb)

Cylinder head bolts, removing and installing

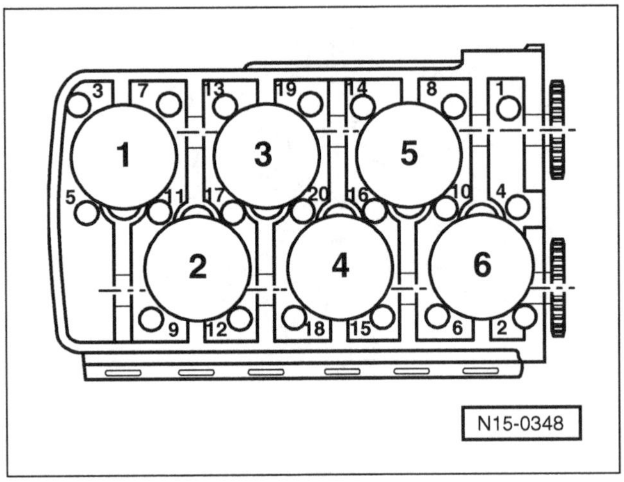

Remove cylinder head bolts using sequence shown, starting with #**1**.

NOTE —

- *Larger numbers (**1-6**) show cylinder numbers in illustration at left.*

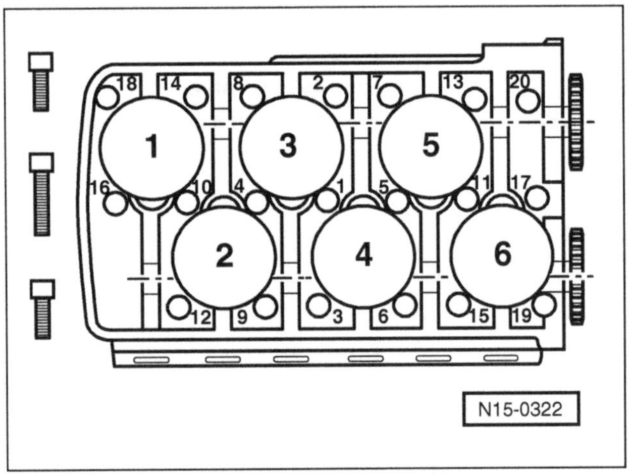

Install cylinder head bolts using sequence shown, starting with #**1**.

NOTE —

- *The long cylinder head bolts are used in the middle bores of the cylinder head.*
- *Always replace cylinder head bolts.*
- *Two old head bolts, with hex heads removed, can be used as alignment dowels during cylinder head installation.*
- *Larger numbers (**1-6**) show cylinder numbers in illustration at left.*

Tightening torque	
Cylinder head to engine block (always replace)	
Stage 1	30 Nm (22 ft-lb)
Stage 2	50 Nm (37 ft-lb)
Stage 3	plus 90° (¼ turn)
Stage 4	plus 90° (¼ turn)

NOTE —

- *Re-tightening of cylinder head bolts is not necessary after installation.*

BentleyPublishers.com

© 2006 BentleyPublishers.com—All Rights Reserved

Timing chain assembly

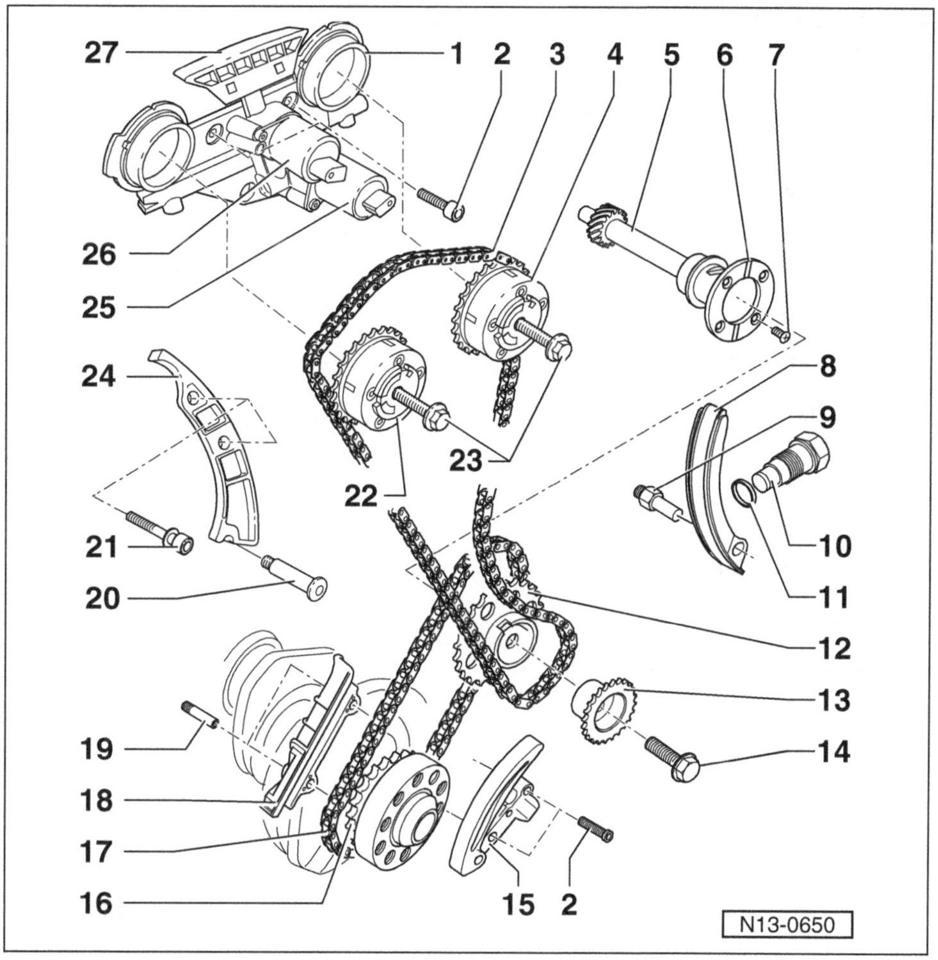

NOTE —
- *Audi identifies electrical components by a letter and/or a number in the electrical schematics. See* **EWD Electrical Wiring Diagrams**. *These electrical identifiers are listed in parentheses as an aid to electrical troubleshooting.*
- *Removal and installation of timing chains requires removal of transmission or engine.*

1. **Control housing**
   - Check small oval shaped filter on back of housing for dirt or grime
   - Lightly coat contact surfaces of oil seals before installing
2. **Bolt**
   - 8 Nm (71 in-lb)
3. **Camshaft timing chain, upper**
   - Mark direction of rotation before removing
4. **Exhaust camshaft sprocket and timing adjuster**
   - Identification: 32A
   - Turn engine over only when timing adjuster is installed
5. **Intermediate shaft**
   - Drives engine oil pump shaft
6. **Thrust washer**
7. **Screw**
   - 8 Nm (71 in-lb)
   - Install with locking compound
8. **Guide rail, upper chain**
   - Upper and lower timing chain covers must be removed to access
9. **Pivot pin**
10. **Chain tensioner, upper**
    - 40 Nm (30 ft-lb)
    - Turn engine over only when tensioner is installed
11. **Seal**
    - Replace if damaged or leaking
12. **Intermediate shaft sprocket, inner**
    - For lower chain
13. **Intermediate shaft sprocket, outer**
    - For upper chain
14. **Bolt**
    - Always replace
    - 60 Nm (44 ft-lb) plus 90° (¼ turn)
    - Counter hold vibration damper with special tool T10069 to remove
15. **Chain tensioner, lower**
    - Turn engine over only when tensioner is installed
    - Before installing, release locking spline in chain tensioner with a small screwdriver and press guide rail against tensioner
16. **Crankshaft with integral chain sprocket**
    - Beveled (ground) tooth used as TDC timing mark when aligned with bearing cap joint
17. **Camshaft timing chain, lower**
    - Mark direction of rotation before removing
18. **Guide rail, lower chain**
19. **Locating pin**
    - For lower chain guide rail
    - 10 Nm (7 ft-lb)
20. **Bolt**
    - 18 Nm (13 ft-lb)
21. **Bolt**
    - 23 Nm (17 ft-lb)
22. **Intake camshaft sprocket and timing adjuster**
    - Identification: 24E
    - Turn engine over only when timing adjuster is installed
23. **Bolts**
    - Always replace
    - 60 Nm (44 ft-lb) plus 90° (¼ turn)
    - Use 32 mm open end wrench on camshaft flats to counter hold when removing or installing
24. **Guide rail, upper chain**
25. **Exhaust camshaft adjustment valve (N318)**
26. **Intake camshaft adjustment valve (N205)**
27. **Guide rail**
    - Clipped onto control housing

BentleyPublishers.com
© 2006 BentleyPublishers.com—All Rights Reserved

Intake Manifold Assembly

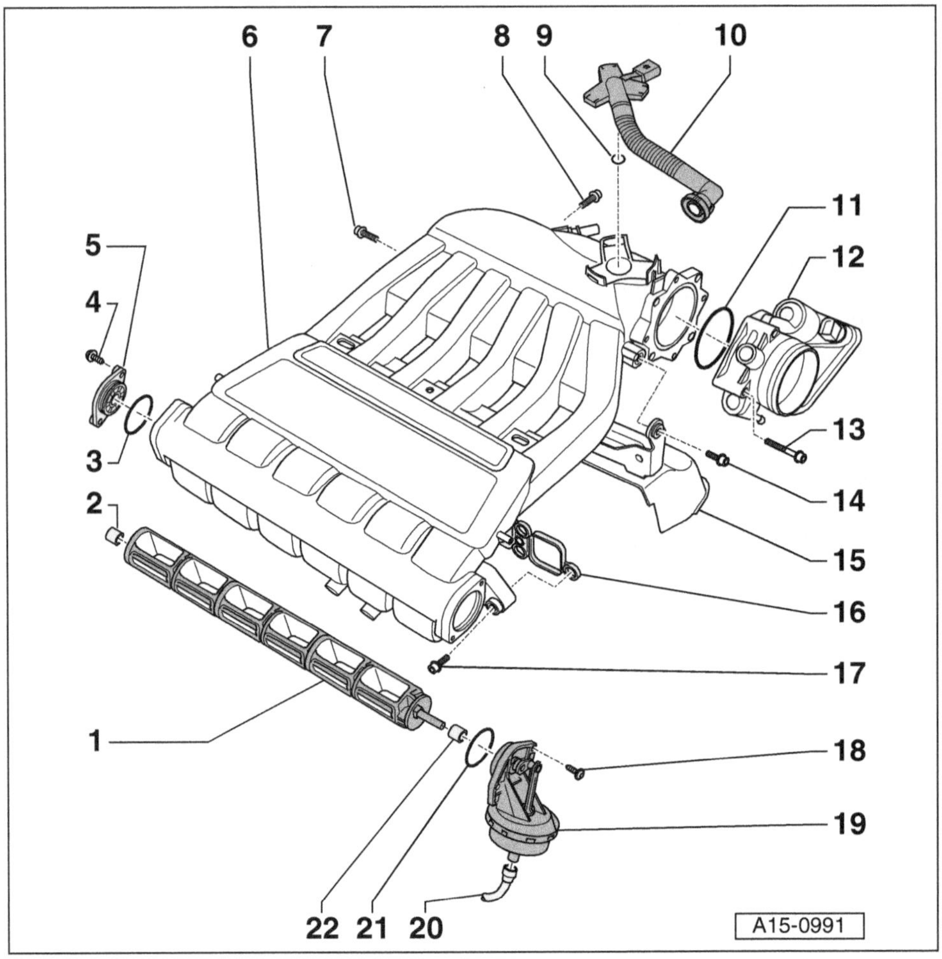

1. **Intake manifold change-over valve (N156)**
2. **Busing**
3. **O-ring**
   - Always replace
4. **Panel bolt**
5. **Bearing cap**
6. **Intake manifold**
7. **Bolt**
   - 20 Nm (15 ft-lb)
8. **Bolt**
   - 20 Nm (15 ft-lb)
9. **Seal**
10. **Crankcase breather connection**
    - With positive crankcase ventilation (PCV) heating element (N79)
11. **O-ring**
12. **Throttle valve control module (J338)**
13. **Bolt**
    - 10 Nm (7 ft-lb)
14. **Bolt**
    - 20 Nm (15 ft-lb)
15. **Heat shield for exhaust manifold**
    - With support bracket for intake manifold
16. **Gasket**
    - Always replace
    - Insert into manifold
17. **Bolt**
    - 13 Nm (10 ft-lb)
18. **Panel bolt**
19. **Vacuum unit**
    - For intake manifold change-over valve
20. **Vacuum connection**
21. **O-ring**
    - Always replace
22. **Bushing**

BentleyPublishers.com

© 2006 BentleyPublishers.com—All Rights Reserved

Valvetrain Assembly

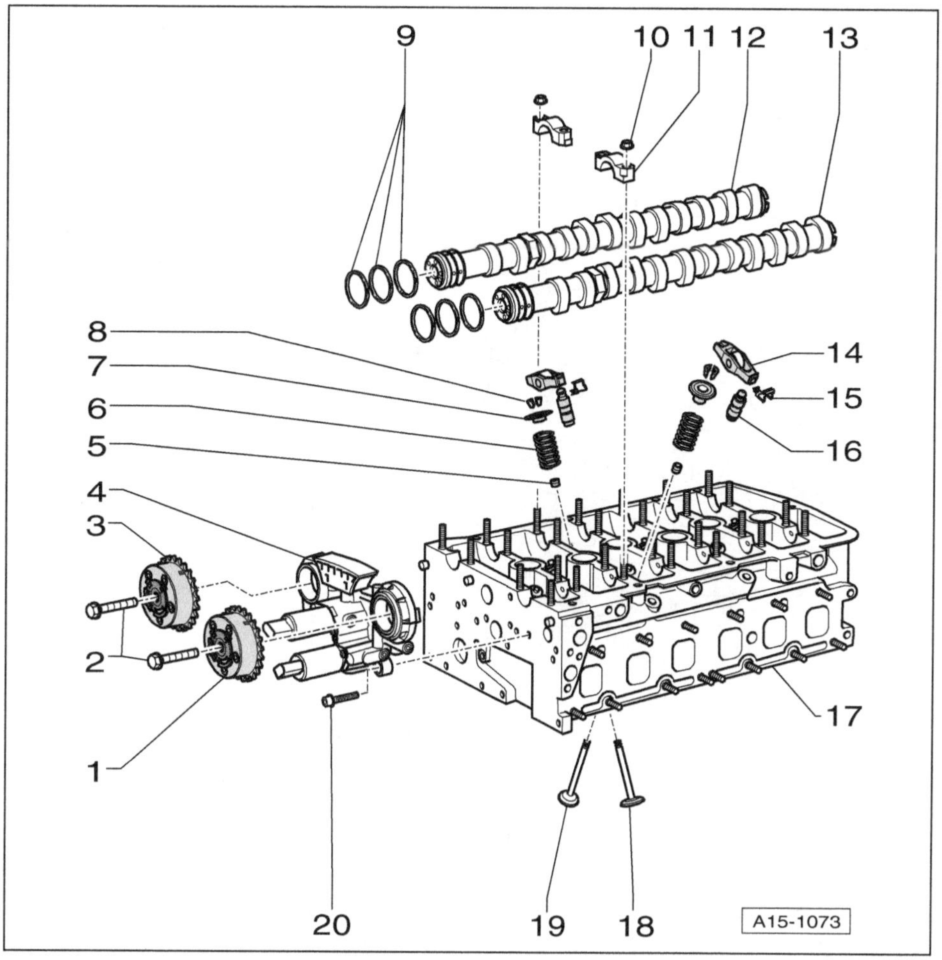

1. **Camshaft adjuster, exhaust**
   - Identification: 32A
2. **Bolt**
   - Always replace
   - 60 Nm (44 ft-lb) plus 90° (¼ turn)
   - Use 32 mm open end wrench on camshaft flats to counter hold when removing or installing
3. **Camshaft adjuster, intake**
   - Identification: 24E
4. **Control housing**
   - Check small oval shaped filter on back of housing for dirt or grime
   - Lightly coat contact surfaces of oil seals before installing
5. **Valve stem oil seal**
6. **Valve spring**
   - Install with larger diameter on bottom
7. **Valve spring plate**
8. **Valve keepers**
9. **Camshaft seals**
   - 3 per camshaft
   - Replace all if leak detected
   - Offset gaps by 120° when installing
   - Do not stretch seals too wide when replacing
10. **Bolt**
    - 5 Nm (44 in-lb) plus 45° (⅛ turn)
11. **Bearing cap**
    - Numbered
    - Pointed tip on bearing cap ends points toward exhaust manifold
12. **Intake camshaft**
    - Identification: 022
      Index: 102
    - Axial clearance: 0.1 mm (0.004 in.)
    - Check radial clearance with Plastigage (valve lifters not installed)
      Wear limit: 0.1 mm (0.004 in.)
      Run-out max: 0.01 mm (0.0004 in.)
13. **Exhaust camshaft**
    - Identification: 022
      Index: 101
    - Axial clearance: 0.1 mm (0.004 in.)
    - Check radial clearance with Plastigage (valve lifters not installed)
      Wear limit: 0.1 mm (0.004 in.)
      Run-out max: 0.01 mm (0.0004 in.)
14. **Roller rocker finger**
    - Do not interchange
    - Check roller bearing
    - Before installing, check camshaft axial clearance
15. **Securing clip**
    - Check for firm attachment
16. **Hydraulic valve adjuster**
    - Do not interchange
    - Before installing, check camshaft axial clearance
    - Lubricate contact surface
17. **Cylinder head**
18. **Intake valve**
    - Must not be machined, only hand lap
19. **Exhaust valve**
    - Must not be machined, only hand lap
20. **Bolt**
    - 20 Nm (15 ft-lb)

BentleyPublishers.com
©2006 BentleyPublishers.com—All Rights Reserved

Engine Mounts

Torque specifications and engine mount alignment specifications are provided here as resource in case the engine mounts need to be removed for any reason.

> **WARNING —**
> - *Before removing any engine mount bolts, the weight of the engine must be supported by an engine support bar.*

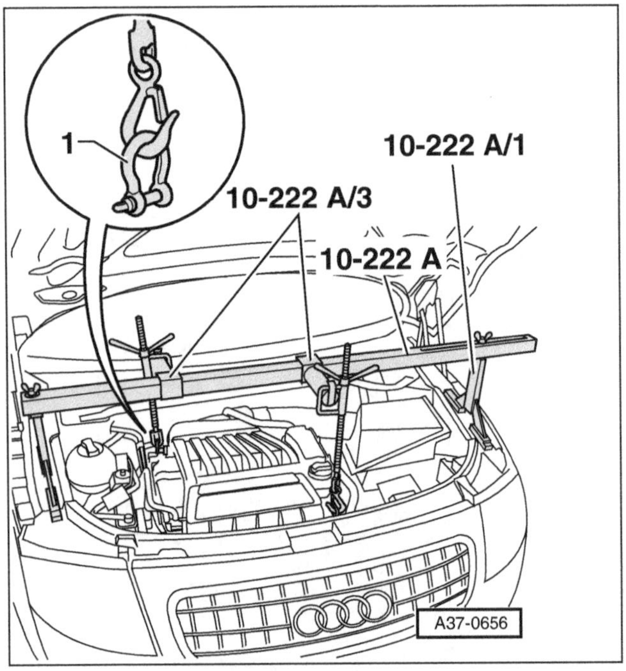

◀ Attach engine support bar (Audi special tool 10-222 A shown) to engine lifting eyes.

– Tighten nuts on lifting spindles so that weight of engine is supported by engine support bar.

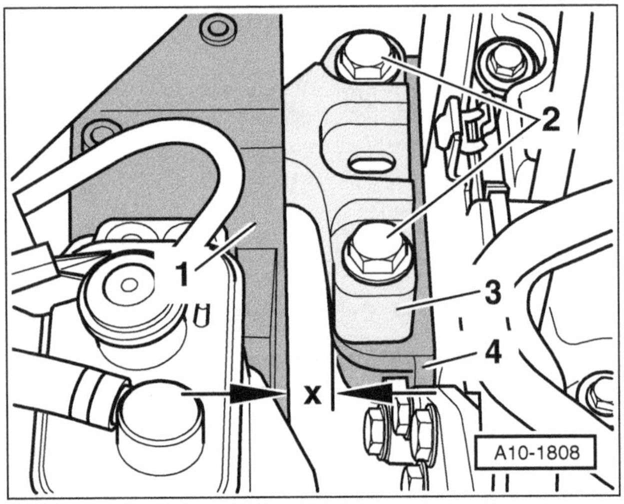

◀ Loosen engine mount bolts (**2**) enough to move engine slightly.

NOTE —
- *It may also be necessary to loosen transmission mount bolts (shown below) in order to provide enough movement to align the engine mount.*

– Move engine slightly using a large lever.
- Both bolt heads (**2**) must be parallel to edge of support arm (**3**).
- Distance (**x**) = 10 -13 mm (0.40 - 0.51 in.)

– Tighten engine mount bolts to specification.

Tighttening torques	
Engine mount to bracket (bolts **2**)	65 Nm (48 ft-lb)
Remaining engine/transmission mount bolts M8 M10 M12	 20 Nm (16 ft-lb) 45 Nm (33 ft-lb) 65 Nm (48 ft-lb)

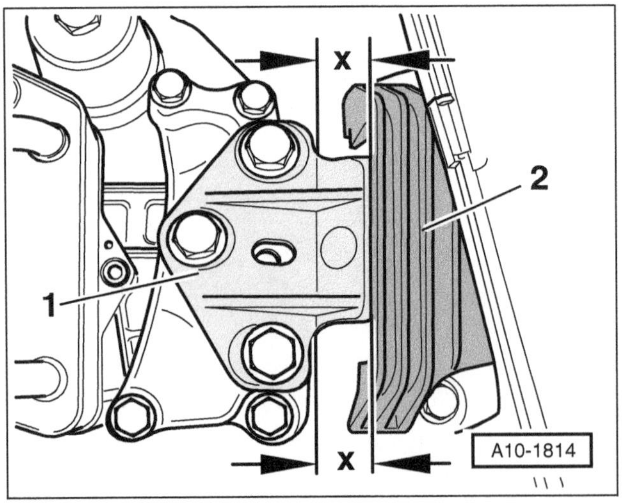

◀ Loosen mounting bolts from transmission mount support arm (**1**).
- Distance (**x**) must be the same (parallel) at top and bottom.

– Tighten transmission mount bolts to specification.

Tightening torques	
Transmission support arm to bracket	65 Nm (48 ft-lb)
Pendulum mount to transmission (not shown) always replace	40 Nm (30 ft-lb) + 90° (¼ turn)
Pendulum mount to subframe (not shown) always replace	20 Nm (15 ft-lb) + 90° (¼ turn)

BentleyPublishers.com
© 2006 BentleyPublishers.com—All Rights Reserved

17 Engine–Lubrication System

GENERAL

Proper engine lubrication relies on a constant supply of oil, fed to the moving parts under pressure. Pressure is supplied by a gear-type oil pump located inside the engine oil pan. Engine oil returns to the oil pan by way of internal passages in the cylinder block and head. Returned oil collects in the oil pan where it is stored for eventual pickup by the oil pump. Engines with turbo-chargers also have external oil supply and return lines.

Engine	Engine Code
1.8L 4-cylinder turbo (180 hp)	ATC, AWP
1.8L 4-cylinder turbo (225 hp)	AMU, BEA
3.2L 6-cylinder gasoline (250 hp)	BHE

Engine	Oil Capacity with filter
1.8L 4-cylinder turbo (180 hp)	4.5 liters (4.8 qt.)
1.8L 4-cylinder turbo (225 hp)	4.5 liters (4.8 qt.)
3.2L 6-cylinder gasoline (250 hp)	5.5 liters (5.8 qt.)

Tightening torque	
Oil drain plug (1.8L, 3.2L) always replace	30 Nm (22 ft-lb)

CAUTION—
- *Oil capacities may vary. Oil level should ALWAYS be checked using the dipstick.*

NOTE —
- *For changing oil/filter and oil specifications, see* **03 Maintenance**.

BentleyPublishers
.com
© 2006 BentleyPublishers.com–All Rights Reserved

0024485

NOTE —
- *If, when repairing an engine, metal shavings or large amounts of small metal particles are found in the engine oil, caused for example by partial seizure of crankshaft or connecting rod bearings, do the following to prevent further damage:*
- *Thoroughly clean all oil passages and insides of cylinder block and head.*
- *Replace oil spray jets.*
- *Replace oil cooler.*
- *Replace oil filter.*
- *Replace oil non-return (check) valve.*

Engine Oil

Some engine codes covered by this manual require oils of a type specifically developed for their special needs. These oils must have the ability to withstand the high loads caused by the pressures generated by the individual pump injectors on the camshaft.

Also note that all 1.8L turbo-charged gasoline engines covered by this manual require engine oils of the highest quality due to the very high operating demands imposed by the turbo-charger. These high standards can only met by synthetic oils.

With all engine codes, failure to use oils meeting the factory requirements can result in shortened engine life and/or engine failure. For this reason, close attention should be paid to the engine oil viscosity requirements and specifications as listed in **03 Maintenance**. If in doubt, always consult an authorized Audi dealer.

Oil Pressure Warning System

To prevent serious engine damage, a dynamic oil pressure warning system warns the driver of insufficient oil pressure.

For gasoline engines, this system includes a single oil pressure switch with a nominal closing pressure of 1.4 bar (20 psi). An electronic control unit with a buzzer and warning light completes the system and is integrated into the instrument cluster.

Oil pressure warning system, checking

With the ignition off, the single oil pressure switch is open (no continuity between the terminal connection and the switch body). When the ignition is switched on, but the engine is not started, the oil pressure warning light should come on for approximately 3 seconds and then go out.

When the engine is started and the oil pressure rises, the oil pressure switch should close at the specified pressure completing a circuit to ground. When the engine reaches approximately 1500 rpm, the electronic control unit checks to see if the ground is complete and if so, the system is functioning normally. If the ground is not complete, the warning light and buzzer are activated. Several small time delays are programmed into the system to prevent accidental warnings due to normal minor fluctuations in pressure.

BentleyPublishers
.com
© 2006 BentleyPublishers.com—All Rights Reserved

CAUTION—
- *If the warning indicators stay on after engine is started, or while driving, always assume that there is insufficient oil pressure. Check oil level and oil pressure before proceeding.*

A quick-check of the system can be accomplished as follows:

– Check instrument cluster warning light.
  - Turn key on, but do not start engine.
  - Warning light should come on for about 3 seconds and then go out.

– Turn key off. Disconnect the wire at the oil pressure switch and ground the wire. A jumper wire with alligator clips works well.

– Turn key on, but do not start engine.

– Warning light should flash and the buzzer should beep 3 times.

– Turn key off. Disconnect the jumper wire but do not re-connect to oil pressure switch (leave unplugged).

– Start engine and let idle.
  - Warning light and buzzer should not come on.

– Increase engine speed to approx. 1500-2000 rpm.
  - Warning light and buzzer should come on after a 2-3 second delay.

If the above tests give results as noted, the oil pressure warning system is functioning within normal parameters. If the above test results deviate from specification, continue by testing the oil pressure and the oil pressure switch.

NOTE—
- *Since oil temperature can have a considerable influence on oil pressure, it may be advisable to conduct the above noted tests on both a cold and a warmed-up engine.*
- *Thin engine oils (oils with lower viscosity numbers) will also have a considerable influence on oil pressure. If in doubt, consult the owner's manual and change the engine oil to the correct viscosity.*

Oil pressure and oil pressure switch, testing

It is possible for the oil pressure warning system to be functioning properly and still give an occasional indication of a malfunction. In this case, it is advisable to check the oil pressure switch. This switch is located on the oil filter bracket. See **Oil filter assembly** for the specific engine application in this repair group.

– Disconnect wire from oil pressure switch and leave it unplugged for this test.

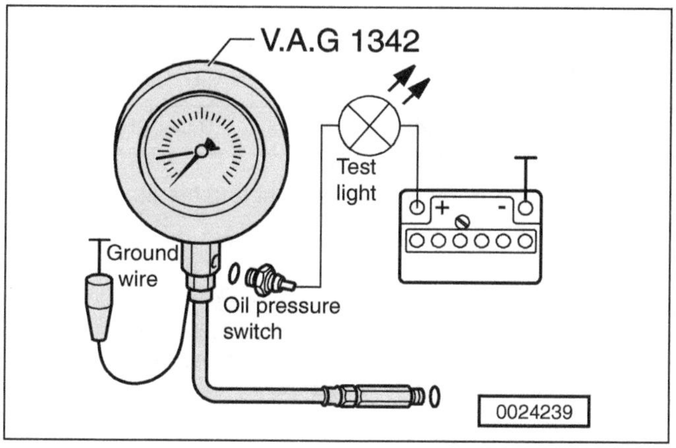

◄ Remove oil pressure switch and thread it into tester (VAG 1342 shown).

– Thread oil pressure tester hose into hole previously occupied by switch.

– Connect ground wire from tester to a good ground.

BentleyPublishers
.com
© 2006 BentleyPublishers.com—All Rights Reserved

– Connect a test light from battery positive to oil pressure switch terminal. Light should not come on at this time; if it does, oil pressure switch is defective and must be replaced.

– Start engine and let it warm up.

– Disregard oil pressure light in instrument cluster for this test.

– When radiator fan has cycled at least once, accelerate engine until test light comes on and observe pressure gauge. The light must come on at the specified value. If light does not illuminate at specified value, switch is defective and should be replaced.

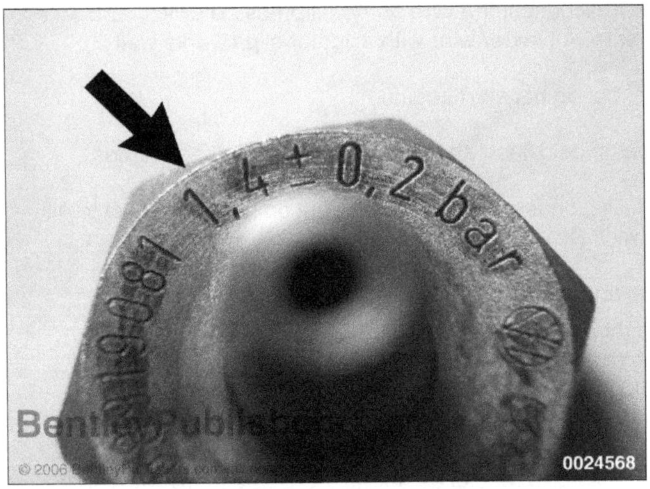

Factory installed and original equipment replacement switches can usually be identified by their color and by their pressure rating (**arrow**) marked on the switch body. 1.4 bar switch shown.

Oil pressure switch closing pressure	
1.8L, 3.2L gasoline (1.4 bar switch)	1.2-1.6 bar (17.5-23 psi)

– Continue system test by checking oil pressure. With tester still attached, run engine at specified speeds and compare readings on gauge to specification.
- If specification is not reached, oil pump or other mechanical deficiencies are indicated.
- If oil pressure exceeds maximum specification, malfunctions with the high pressure relief valve are indicated.

NOTE —
- *Incorrect engine oil viscosity can influence engine oil pressures. Before performing repairs, ensure that engine has a sufficient quantity of fresh oil of correct viscosity.*
- *Excessive oil pressure can cause problems with hydraulic valve lifters along with oil seal and oil filter leakage.*

Oil pressure specifications at 80°C (176°F) and specified engine speed	
1.8L turbo at idle 1.8L turbo at 3000 RPM	min. 1.0 bar (14.5 psi) 3.5 - 4.5 bar (51 -65 psi)
3.2L gasoline at idle 3.2L gasoline at 2000 RPM	min. 2.0 bar (29 psi) min. 3.0 bar (43 psi)
Maximum oil pressure (all engines)	7.0 bar (102 psi)

– When testing is completed, remove the test light and tester. Install the oil pressure switch and the wiring.

NOTE —
If an oil pressure switch seal is leaking, a new sealing washer should be installed. The sealing washer is held captive on the switch and must be cut off. New sealing rings are available from authorized Audi parts dealers or aftermarket parts specialists.

BentleyPublishers.com
© 2006 BentleyPublishers.com—All Rights Reserved

Lubrication System Components (1.8L)

Oil pan and pump assembly

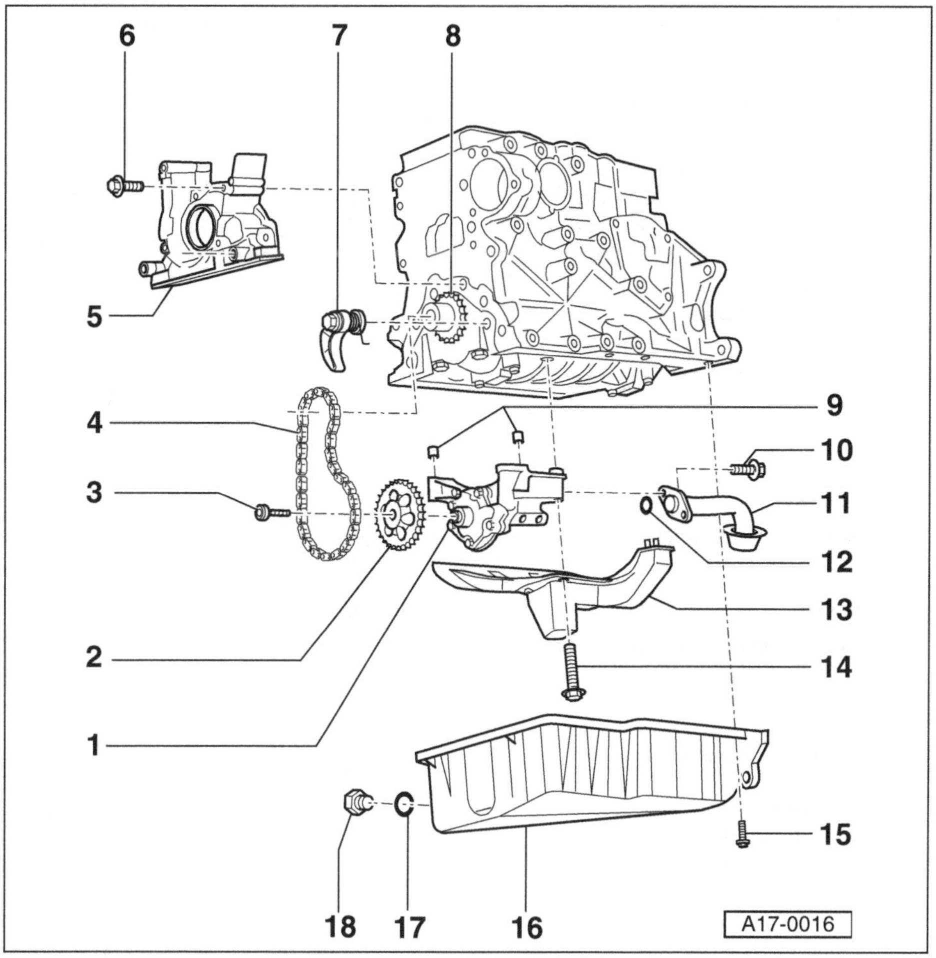

1. **Oil pump**
   - With pressure relief valve (12 bar)
   - Before installing, check to be sure both alignment bushings are present (for centering oil pump to cylinder block)
   - Replace if running surfaces and gears are scored
2. **Chain sprocket for oil pump**
   - Sprocket can only be installed in one position
   - Check for wear
3. **Bolt**
   - 25 Nm (18 ft-lb)
4. **Chain**
   - Mark direction of rotation before removing
   - Check for wear
5. **Sealing flange**
   - With integrated seal
   - Before installing, clean crankshaft journal and sealing surface
   - Install with silicone sealant (Audi part # D 176 404 A2)
   - Must be located on dowel sleeves
   - Do oil or grease sealing lip
6. **Bolt**
   - 15 Nm (11 ft-lb)
7. **Chain tensioner**
   - 15 Nm (11 ft-lb)
   - Note installation position
   - When installing, pretension spring and hook in
   - If spring is broken, replace complete tensioner
8. **Chain sprocket**
   - For oil pump drive
   - Check for wear
9. **Dowel sleeves**
   - For oil pump alignment
10. **Bolt**
    - 15 Nm (11 ft-lb)
11. **Suction line**
    - Clean strainer if soiled
12. **O-ring**
    - Always replace
13. **Baffle plate**
    - Prevents oil starvation during cornering
    - Only install with oil pump installed
14. **Bolt**
    - 15 Nm (11 ft-lb)
15. **Bolt**
    - 15 Nm (11 ft-lb)
16. **Oil pan**
    - Clean sealing surface before installing
    - Install with silicone sealant (Audi part # D 176 404 A2)
17. **Seal**
    - Always replace
18. **Oil drain plug**
    - 30 Nm (22 ft-lb)

BentleyPublishers.com
©2006 BentleyPublishers.com—All Rights Reserved

Oil filter assembly

NOTE—

- *Audi identifies electrical components by a letter and/or a number in the electrical schematics. See* **EWD Electrical Wiring Diagrams**. *These electrical identifiers are listed in parenthesis as an aid to electrical troubleshooting.*

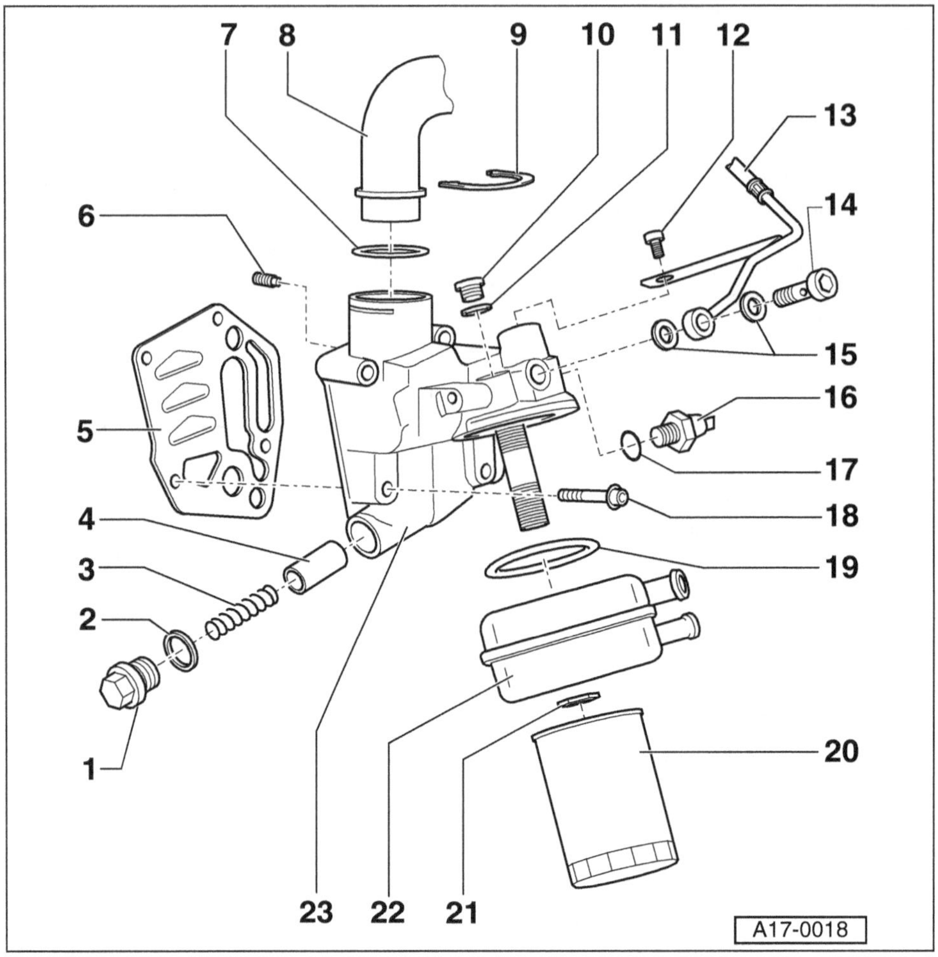

1. **Oil filter bracket**
2. **Seal**
   - Always replace
3. **Spring**
   - For pressure relief valve, approx. 4 bar (58 psi)
4. **Piston**
   - For pressure relief valve, approx. 4 bar (58 psi)
5. **Gasket**
   - Always replace
6. **Oil retention valve**
   - 8 Nm (71 in-lb)
   - Built into oil filter bracket
7. **O-ring**
   - Always replace
   - Slide on until flush with collar on pipe (item 8)
8. **Pipe**
9. **Retaining clip**
10. **Screw plug**
    - 15 Nm (11 ft-lb)
11. **Seal**
    - If seal is leaking, cut open and replace
12. **Bolt**
    - 20 Nm (15 ft-lb)
13. **Oil supply line**
    - To turbocharger
14. **Banjo bolt**
    - 30 Nm (22 ft-lb)
15. **Seals**
    - Always replace
16. **Oil pressure switch (F1)**
    - 1.4 bar switch: black
    - 25 Nm (18 ft-lb)
17. **Seal**
    - If seal is leaking cut off and replace
18. **Bolt**
    - 15 Nm (11 ft-lb) plus an additional 90° (¼ turn)
    - Always replace
19. **Gasket**
    - Always replace
20. **Oil filter**
    - Loosen with oil filter wrench (3417)
    - 20 Nm (15 ft-lb)
    - Observe installation instructions on filter
    - Observe change intervals, see **03 Maintenance**
21. **Nut**
    - 25 Nm (18 ft-lb)
22. **Oil cooler**
    - Make sure there is sufficient clearance for surrounding components
    - Coat contact area to flange, outside the seal with AMV 188 001 02
23. **Oil filter bracket**
    - With pressure relief valve, approx. 4 bar (58 psi)

BentleyPublishers.com
© 2006 BentleyPublishers.com—All Rights Reserved

Oil pan and pump, removing and installing

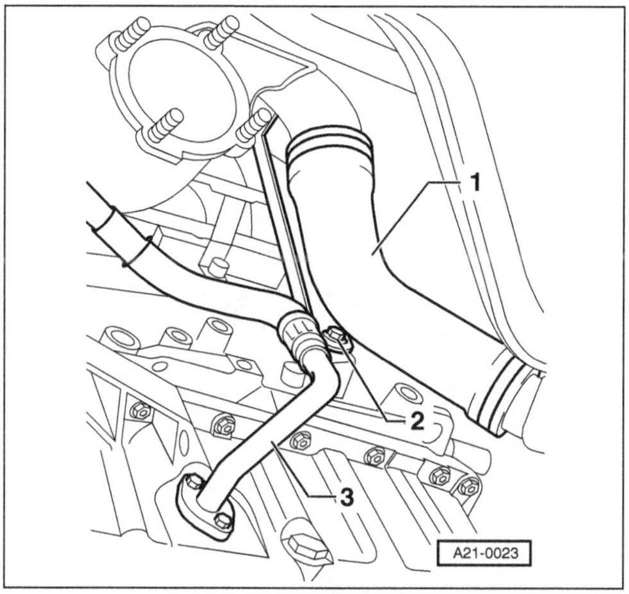

– Remove lower center sound insulation panel and drain engine oil. See **03 Maintenance**.

◄ Remove bracket (**2**) and oil return line (**3**) from oil pan.

– Unbolt and remove oil pan. If necessary loosen oil pan from block by striking it lightly with a rubber mallet.

– Remove old sealant using a flat scraper. If necessary use a rotating plastic brush on a drill to remove old sealant. Clean sealing surfaces so they are free of oil and grease.

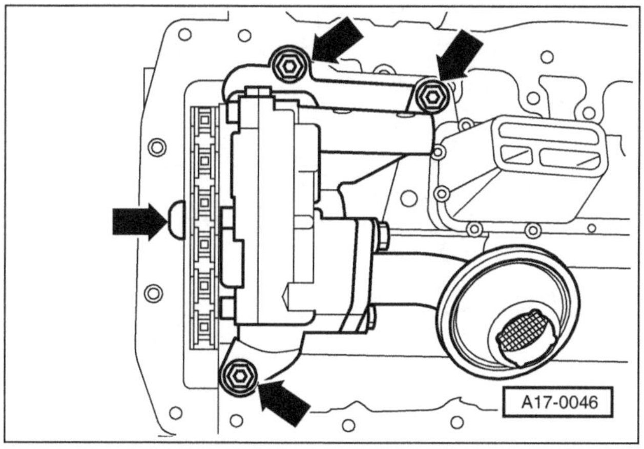

◄ To remove oil pump, loosen mounting bolts (**arrows**). Detach oil pump but leave chain sprocket in place.

– Install pump in reverse order of removal.

- Chain sprocket can only be installed in one position on oil pump shaft.

Tightening torque	
Oil pump to cylinder block	15 Nm (11 ft-lb)

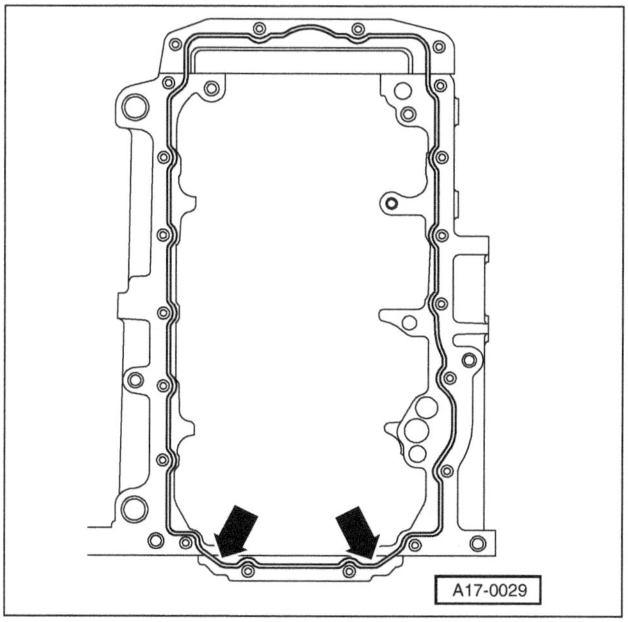

◄ Apply 2 to 3 mm wide bead of silicone sealant (D 176 404 A2) inside of oil pan bolt holes as shown (**arrows**).

NOTE —

- *If sealant bead is too thick, the excess sealant could get in oil pan and block oil suction pipe screen.*

– Install oil pan immediately and bolt in place. Start by tightening all bolts by hand.

– Tighten oil pan to transmission bolts first. Tighten oil pan to cylinder block bolts diagonally. Install oil return line.

Tightening torque	
Oil pan to cylinder block	15 Nm (11 ft-lb)
Oil pan to transmission	45 Nm (33 ft-lb)
Oil return line to oil pan	15 Nm (11 ft-lb)

– Wait 30 minutes for sealant to dry before putting engine oil in.

– Install drain plug, fill engine with oil and replace lower sound absorber panel.

BentleyPublishers.com
© 2006 BentleyPublishers.com—All Rights Reserved

Lubrication System Components (3.2L)

Oil pan and pump assembly

1. **Bolt**
   - 5 Nm (44 in-lb)
   - Secured to intake manifold
2. **Oil dipstick**
   - Oil level must not be above max. mark
3. **Guide tube**
4. **Cylinder block**
5. **Oil return valve**
6. **Bolt**
   - 8 Nm (71 in-lb)
7. **Cover for oil pump drive**
8. **O-ring**
   - Always replace
   - Lubricate before installing
9. **Oil pump drive**
10. **Intermediate shaft**
11. **Thrust washer**
12. **Bolt**
    - 10 Nm (7 ft-lb)
    - Install with locking fluid
13. **Oil spray jet**
    - For crankshaft bearings #2-7
    - Opening pressure 2.0 bar (29 psi)
14. **Input shaft for oil pump drive**
15. **Oil pump**
    - Coat oil pressure pipe at cylinder block and at oil pump housing with sealing paste
16. **Bolt**
    - 23 Nm (17 ft-lb)
17. **Oil drain plug**
    - 30 Nm (22 ft-lb)
    - With permanent sealing ring
    - Always replace
18. **Sump (oil pan)**
19. **Oil level and thermal sensor (G266)**
20. **Bolt**
    - 12 Nm (9 ft-lb)
21. **Oil filter assembly**

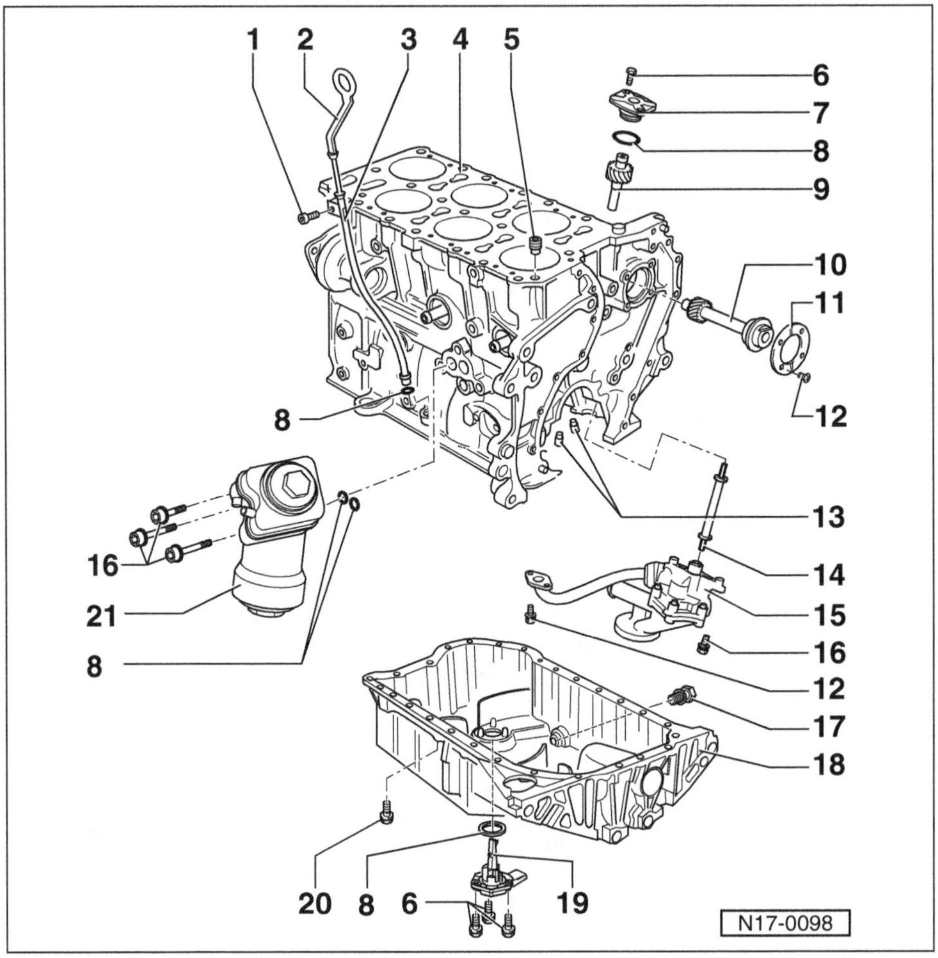

BentleyPublishers.com
© 2006 BentleyPublishers.com—All Rights Reserved

Oil filter assembly

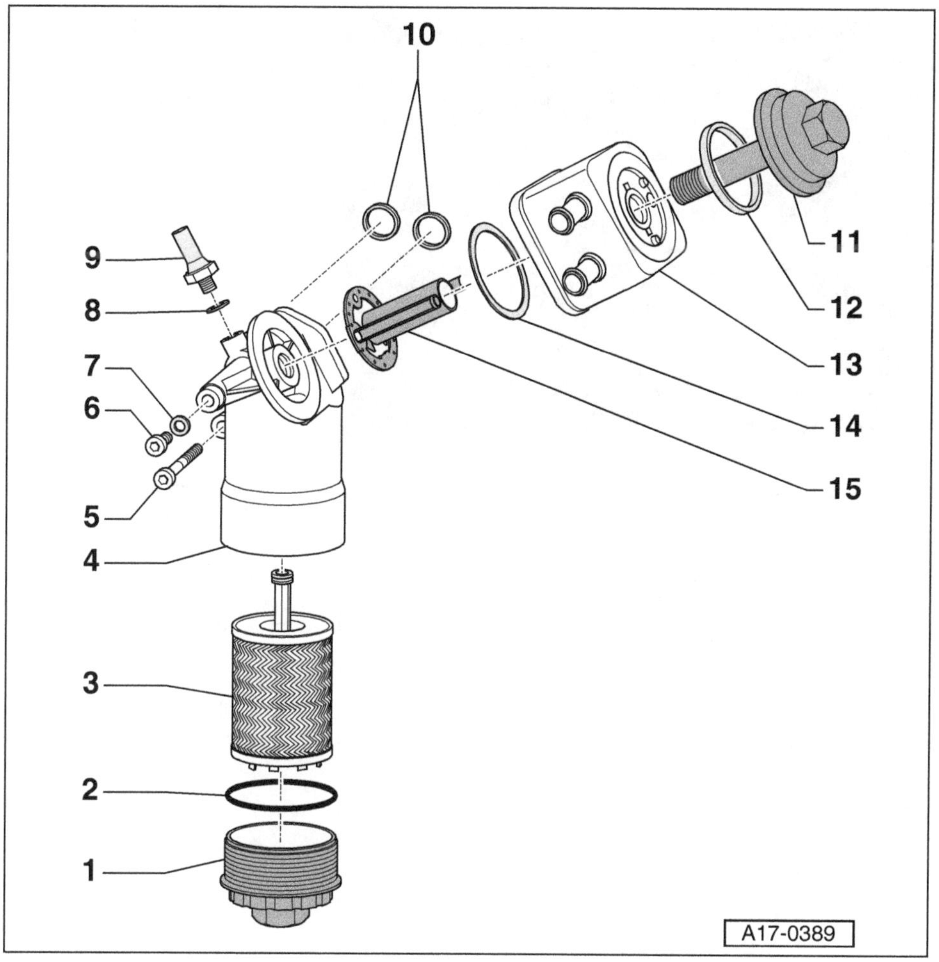

1. **Sealing cap**
   - 25 Nm (18 ft-lb)
2. **O-ring**
   - Always replace
3. **Oil filter element**
   - Observe change intervals, see **03 Maintenance**
4. **Oil filter bracket**
   - With oil retention valve
5. **Bolt**
   - 23 Nm (17 ft-lb)
6. **Screw plugs**
   - 15 Nm (11 ft-lb)
7. **Oil seal**
   - If seal is leaking, cut open and replace
8. **Oil seal**
   - If seal is leaking, cut open and replace
9. **Oil pressure switch (F1)**
   - 1.4 bar, black
   - 25 Nm (18 ft-lb)
10. **Oil seals**
    - Always replace
11. **Sealing cap**
    - 25 Nm (18 ft-lb)
12. **Oil seal**
    - Always replace
    - Lubricate with oil before installing
13. **Oil cooler**
    - Make sure there is sufficient clearance for surrounding components
14. **Oil seal**
    - Lubricate with oil before installing
    - Press onto oil cooler with plastic element (item 15)
15. **Plastic element**
    - Install into oil cooler with oil seal (item 14)

Oil seal and plastic element in oil cooler, installing

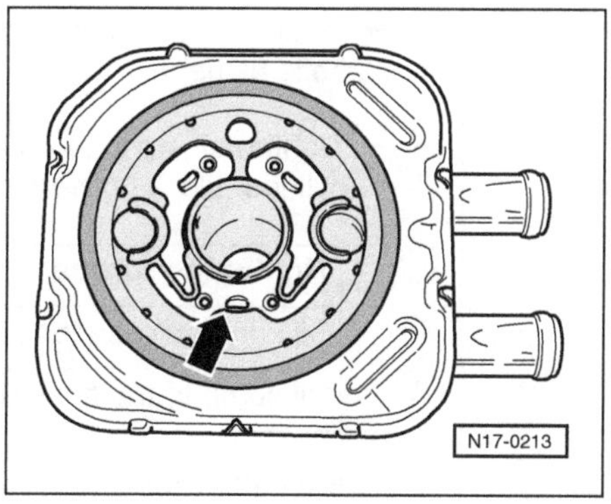

– Coat seal with oil and insert into oil cooler.

– Insert plastic element into oil cooler.

◀ Side of plastic element with support must align with lug (**arrow**) in oil cooler.

Bentley Publishers .com
© 2006 BentleyPublishers.com—All Rights Reserved

Oil sump (pan), removing and installing

– Remove lower sound insulation panels and drain engine oil, see **03 Maintenance**.

◄ Disconnect harness connector (**arrow**) for oil level thermal sensor (G266).

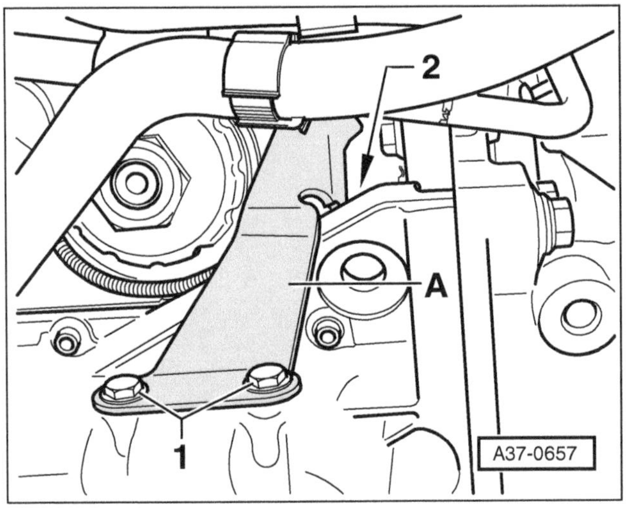

◄ Remove bolts (**1** and **2**) and move secondary air pump bracket (**A**) out of way.

– Remove oil sump to transmission bolts first.

– Loosen oil sump to cylinder block mounting bolts in a diagonal sequence.

– Unbolt and remove oil pan. If necessary loosen oil pan from block by striking it lightly with a rubber mallet.

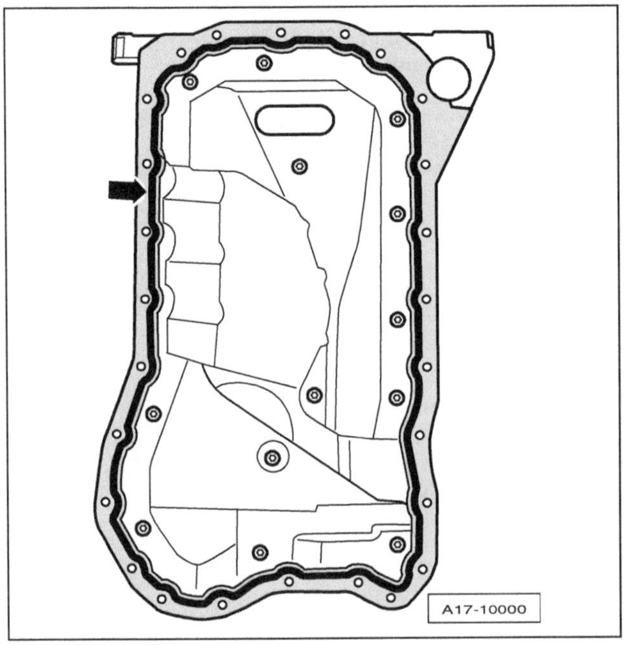

◄ Apply 2 mm wide bead of silicone sealant (D 176 404 A2) inside of oil pan bolt holes as shown. Use extra care applying sealant near timing chain cover area (**arrows**).

NOTE —
- *If sealant bead is too thick, the excess sealant could get in oil pan and block oil suction pipe screen.*

– Install oil pan immediately and bolt in place. Start by tightening all bolts by hand.

– Tighten oil sump to transmission bolts first. Tighten oil sump to cylinder block bolts diagonally.

Tightening torque	
Oil sump to cylinder block	12 Nm (11 ft-lb)
Oil sump to transmission	45 Nm (33 ft-lb)

– Wait 30 minutes for sealant to dry before putting engine oil in.

– Install drain plug, fill engine with oil and replace lower sound absorber panels.

BentleyPublishers.com
©2006 BentleyPublishers.com—All Rights Reserved

19 Engine–Cooling System

General

This section covers repairs and troubleshooting for the engine cooling system. For heater core and related heating and air conditioning components, see **87 Heating and Air Conditioning**. For information on the engine oil cooler, see **17 Engine–Lubrication**.

Engine	Engine Code
1.8L 4-cylinder turbo (180 hp)	ATC, AWP
1.8L 4-cylinder turbo (225 hp)	AMU, BEA
3.2L 6-cylinder gasoline (250 hp)	BHE

Engine	Cooling system capacity
1.8L 4-cylinder turbo (180 hp)	5.0 liters (5.3 qt.)
1.8L 4-cylinder turbo (225 hp)	5.0 liters (5.3 qt.)
3.2L 6-cylinder gasoline (250 hp)	8.0 liters (8.5 qt.)

NOTE—

- *For checking coolant level and concentration, see* **03 Maintenance**.

BentleyPublishers.com

© 2006 BentleyPublishers.com—All Rights Reserved

Warnings and Cautions

The following Warnings and Cautions should be observed when working on the cooling system.

WARNING—
- *Hot coolant can scald and result in serious personal injury. Do not work on the cooling system until it has fully cooled.*
- *At normal operating temperature the cooling system is pressurized. Allow the system to cool as long as possible before opening, a minimum of an hour, then release the cap slowly to allow safe release of the pressure.*
- *Releasing the cooling system pressure lowers the coolant's boiling point, and the coolant may boil suddenly. Use heavy gloves and wear eye and face protection to guard against scalding.*
- *Use extreme care when working at or near the radiator cooling fans when the engine is hot. The fans can come on at any time, even if the ignition key is switched off.*
- *Use extreme care when draining and disposing of engine coolant. Coolant is poisonous and lethal. Children and pets are attracted to coolant because of its sweet smell and taste. See a doctor or veterinarian immediately if any amount of coolant is ingested.*
- *Disposal of used coolant must be done in conformance with all applicable local, state and federal laws. Do not pour into the ground or down drains or sewers. Contamination of ground water will result.*

CAUTION—
- *Avoid adding cold water to the cooling system while the engine is hot or overheated. If it is necessary to add coolant to a hot system, do so only with the engine running and the coolant pump operating.*
- *Use only Audi original coolant or equivalent when filling the cooling system. Use of any other coolant may be harmful to the cooling system. Do not use a coolant containing phosphates.*
- *Never mix green coolant (G11) with purple coolant (G12). Mixing can cause serious engine damage.*
- *Contamination of G12 with other colored coolants is identifiable by discoloration (brown, gray, etc.) of coolant. This mixture causes a foamy deposit to appear in the expansion tank/radiator and MUST be drained immediately. The cooling system must be completely free of this mixture before refilling with correct type of coolant.*
- *Do not use tap water in cooling system. Use distilled water only to mix coolant.*

TROUBLESHOOTING

When investigating the cause of overheating or coolant loss, begin with a visual inspection of the system. Check the coolant level and inspect for evidence of coolant leaks.

The system becomes pressurized at normal operating temperature. Leaks may prevent the system from becoming pressurized and allow coolant to boil at a lower temperature. If visual evidence is

BentleyPublishers.com
© 2006 BentleyPublishers.com—All Rights Reserved

inconclusive, a cooling system pressure test will determine whether the system leaks and may help to indicate the source.

If the cooling system holds pressure, the most probable causes of overheating are an electrical fault with the cooling fans, a faulty thermostat, or poor coolant circulation due to restrictions in the system.

The **Cooling System Troubleshooting** table below lists overheating and underheating symptoms, their probable causes, and suggested corrective actions. The bold type refers to areas outside of this repair group for suggested repairs.

Cooling System Troubleshooting

Symptom	Probable cause	Corrective action
Engine overheats	Low coolant level	Fill expansion tank to the MAX mark on a cold engine. Check cooling system for leaks with pressure tester.
	Incorrect coolant concentration	Adjust coolant concentration to a mix of 50% coolant additive and 50% distilled water.
	Poor air flow through radiator	Check for debris (leaves, bugs, etc.) buildup on front of radiator. Clean radiator exterior using compressed air.
	Radiator fan not switching on	Test thermoswitch and fan. Replace faulty parts.
	Faulty radiator cap	Pressure test radiator cap. Replace faulty cap.
	Faulty thermostat	Remove and test thermostat. Replace faulty thermostat.
	Coolant pump faulty	Remove and inspect for spun impeller or pulley hub.
	Radiator hose restricted (lower hose may collapse at high engine or highway speeds)	Check hoses for soft, spongy areas. Replace faulty hoses.
	Clogged radiator	Clean or replace faulty radiator.
	Internal engine mechanical fault	Check internal engine condition, see **15 Cylinder Head**, for appropriate engine.
Temperature gauge reads low, poor heater output	Incorrect coolant concentration	Adjust coolant concentration to a mix of 50% coolant additive and 50% distilled water.
	Faulty thermostat	Remove and test thermostat. Replace faulty thermostat.
	Radiator fan not switching off	Test thermoswitch. Replace faulty parts.
Temperature gauge reads low, heater output normal	Faulty instrument cluster or coolant temperature sensor	Test instrument cluster and sensor. Replace parts as required.
Temperature gauge reads normal, poor heater output	Installed position of heater hoses reversed	Inspect routing of heater hoses, see **87 Heating and Air Conditioning**.
	Heater hose restriction	Inspect heater hoses for ply separation and clogging. Replace hoses as required.
	Heat core restricted or clogged	Clean or replace faulty heater core.
	Heater or A/C ventilation controls or flaps not operating correctly	Check operation and adjustment of controls, see **87 Heating and Air Conditioning**.

Cooling system, pressure testing

A pressure test uses a special tester in place of the cap on the expansion tank to pressurize the system and simulate normal operating conditions. If the system is unable to hold pressure, the engine will overheat more easily and fluid will be lost.

NOTE —

- *A pressure test also checks for internal leakage. Some of the common sources of internal leakage include a faulty cylinder head gasket, a cracked cylinder head, and a cracked cylinder block.*

BentleyPublishers.com
© 2006 BentleyPublishers.com—All Rights Reserved

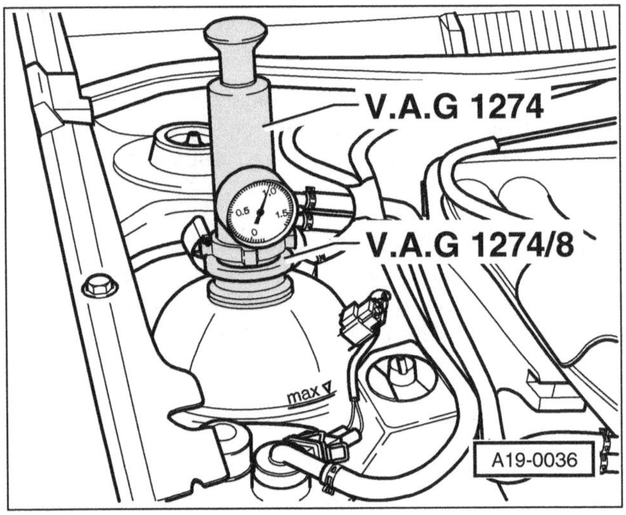

With engine at normal operating temperature, pressure test the system using a cooling system pressure tester with suitable adapters. Generate a positive pressure of 1.0 bar (14.5 psi) using the hand pump on tester. (V.A.G. 1274 pressure tester shown at left).

– Observe the gauge reading. If the pressure drops, there is a leak in the system. A rapid drop may indicate a faulty cylinder head gasket if no external leaks can be found. Remove the spark plugs or glow plugs to inspect for coolant in the cylinders.

– Using the correct tester adapter, test coolant expansion tank cap opening pressure. Compare opening pressure to specification and replace the cap if opening pressure is outside of specification.

NOTE—
- *It is not unusual for cap opening pressure to be slightly above specification when pressure is first applied due to slight sticking of the internal valve. Subsequent pressure applications should open the valve within specification, otherwise, replace the cap.*

Cooling system test pressures	
Testing pressure	1.0 bar (14.5 psi)
Cap opening pressure	1.4 - 1.6 bar (20 - 23 psi)

– Inspect gasket in expansion tank cap for cuts and damage.

Thermostat, testing

A thermostat that is stuck open will cause the engine to warm up slowly and generally run below normal operating temperature at highway speed. A thermostat that is stuck closed will restrict coolant flow to the radiator and cause overheating.

Accurate testing of the thermostat requires removal and observation while being warmed in a container of hot water. See **Thermostat, removing and installing**. An accurate thermometer is used to note the temperatures at which it begins to open and at which it is fully open.

Thermostat specifications	
Opening starts at approx.	87°C (189°F)
Fully open by approx.	102°C (216°F)
Valve travel, min.	7mm (0.28 in.)

To quickly check if thermostat is opening and if coolant is circulating through radiator, allow a cold engine to run at idle and warm up. As the temperature of the coolant rises, carefully feel the heater and expansion tank hoses. They will get hot. Feel the radiator hoses, particularly the lower hose. They will stay relatively cool to the touch. When the thermostat opens, the radiator hoses will quickly get hot and the radiator fan should cycle on shortly thereafter.

Check the radiator hoses. If they are not hot, then the thermostat has not opened or the radiator is clogged. If the hoses are hot, but the radiator fan did not come on, check the radiator fans using the wiring diagrams, see **EWD Electrical Wiring Diagrams**.

BentleyPublishers.com
© 2006 BentleyPublishers.com—All Rights Reserved

> ***WARNING —***
> - *If engine runs long enough for warning light in instrument cluster to indicate an overheat condition, shut down engine immediately.*

Cooling System, Draining and Filling

– See **Warnings and Cautions** at the beginning of this section.

– Remove lower sound absorber panel (belly pan). See **03 Maintenance**.

– With engine fully cold, remove cap from the coolant expansion tank.

> ***WARNING —***
> - *Hot steam may escape when opening coolant expansion tank if engine is warm. Wear protective goggles and protective clothing to prevent damage to eyes and scalding. Cover cap with a rag and open very carefully.*

– Position a drain pan to catch coolant.

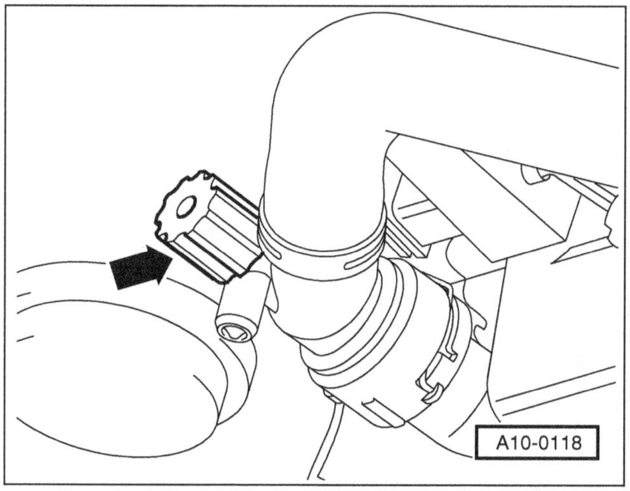

< Unscrew drain plug (**arrow**) at lower radiator hose fitting and allow coolant to drain.
- It may be necessary to run a small hose from coolant drain fitting to prevent coolant from spilling onto lower left sound absorber panel.

< In addition, to drain coolant from 3.2L engine, disconnect coolant hose at after-run coolant pump (**arrow**).

– Dispose of used coolant properly.

– Close drain plug on lower radiator hose fitting.

– Attach coolant hose at after-run coolant pump on 3.2L engine.

BentleyPublishers
.com
©2006 BentleyPublishers.com—All Rights Reserved

– Slowly fill expansion tank with the appropriate premixed concentration of G-12 coolant and water while allowing the air to escape.

Engine	Cooling system capacity
1.8L 4-cylinder turbo (180 hp)	5.0 liters (5.3 qt.)
1.8L 4-cylinder turbo (225 hp)	5.0 liters (5.3 qt.)
3.2L 6-cylinder gasoline (250 hp)	8.0 liters (8.5 qt.)

Freeze protection	Coolant concentration
To -25°C (-13°F)	40%
To -35°C (-31°F)	50%
To -40°C (-40°F)	60%

CAUTION—

- *Under no circumstances should straight coolant/anti-freeze or straight water be used. A 50/50 mix of coolant/anti-freeze to water provides freeze protection to -35°C (-31°F) and boiling protection to 108°C (226°F). The coolant/anti-freeze percentage should always be a minimum of 40% and a maximum of 60%.*
- *Do not use tap water in cooling system. Use distilled water only to mix coolant.*
- *Never mix green coolant (G11) with purple coolant (G12). Mixing can cause serious engine damage.*
- *Contamination of G12 with other colored coolants is identifiable by discoloration (brown, gray, etc.) of coolant. This mixture causes a foamy deposit to appear in the expansion tank/radiator and MUST be drained immediately. The cooling system must be completely free of this mixture before refilling with correct type of coolant.*

NOTE—

- *If a cooling system pressure tester is available, connect it to expansion tank and pressurize the system. This will force coolant past the thermostat and into many of the drained hoses which will speed up filling process.*

– Install expansion tank cap, start engine and allow engine to run until the radiator fans cycle at least once.

NOTE—

- *If expansion tank empties completely during warm-up, shut off engine, carefully remove expansion tank cap and refill to **max** mark. Restart engine and run until fans cycle.*

– Shut off engine.

◀ Recheck coolant level in expansion tank when engine has cooled. Coolant level should be between max. mark (**upper arrow**) and min. mark (**lower arrow**).

BentleyPublishers.com
© 2006 BentleyPublishers.com—All Rights Reserved

Cooling System Components (1.8L)

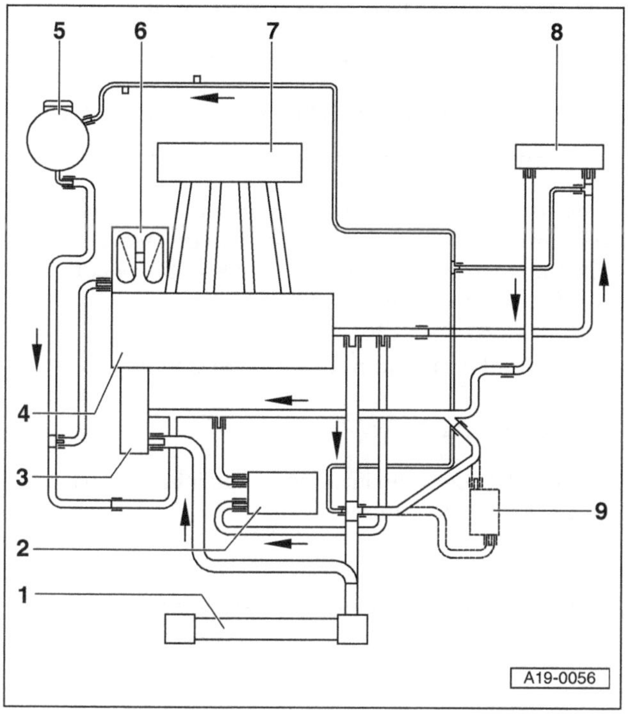

Cooling system overview (180 hp engine)

1. Radiator
   - If replacing, also replace all coolant
2. Oil cooler
   - See **17 Engine–Lubrication System** for more information
3. Coolant pump, and thermostat
4. Cylinder block
5. Expansion tank
6. Turbocharger
7. Intake manifold
8. Heat exchanger
   - For heating system
9. Transmission oil cooler
   - For automatic transmission

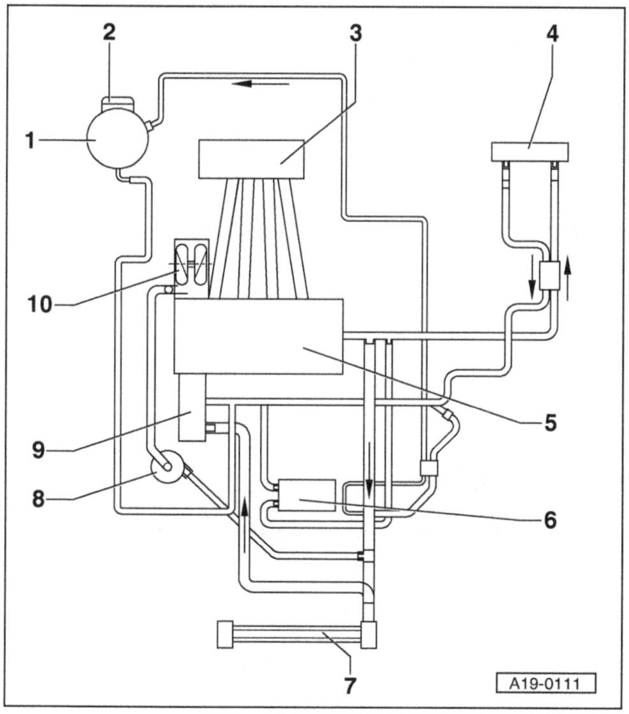

Cooling system overview (225 hp engine)

1. Expansion tank
2. Filler cap for expansion tank
3. Intake manifold
4. Heat exchanger
   - For heating system
5. Cylinder block
6. Oil cooler
   - See **17 Engine–Lubrication System** for more information
7. Radiator
   - If replacing, also replace all coolant
8. After-run coolant pump
9. Coolant pump, and thermostat
10. Turbocharger

BentleyPublishers.com
©2006 BentleyPublishers.com–All Rights Reserved

Coolant pump, removing and installing

– Remove upper engine cover, see **03 Maintenance**.

– With engine fully cold, remove cap from the coolant expansion tank.

> ***WARNING —***
> *• Hot steam may escape when opening coolant expansion tank if engine is warm. Wear protective goggles and protective clothing to prevent damage to eyes and scalding. Cover cap with a rag and open very carefully.*

– Drain coolant. See **Cooling System, Draining and Filling**.

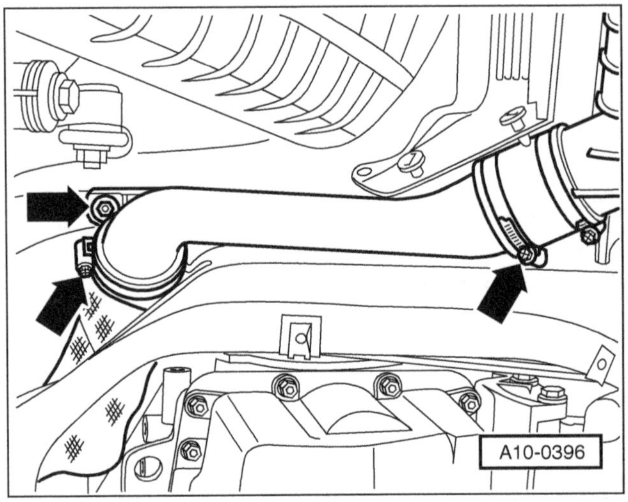

180 hp engine (code: ATC, AWP)

< Remove connecting duct between intercooler and intake manifold. Disconnect duct at **arrows** to remove.

NOTE —
• Seal off openings of charge air circuit with a clean rag to prevent entry of dirt particles.

225 hp engine (code: AMU, BEA)

< Remove intake air duct by disconnecting hoses and clamps at **arrows**.

NOTE —
• Seal off openings of charge air circuit with a clean rag to prevent entry of dirt particles.

Continued for all vehicles

< Pivot ribbed belt tensioner with 17 mm open end wrench (**arrow**) to relieve tension on belt. Remove ribbed belt.

NOTE —
• Mark running direction of ribbed belt if reusing.

BentleyPublishers.com
©2006 BentleyPublishers.com—All Rights Reserved

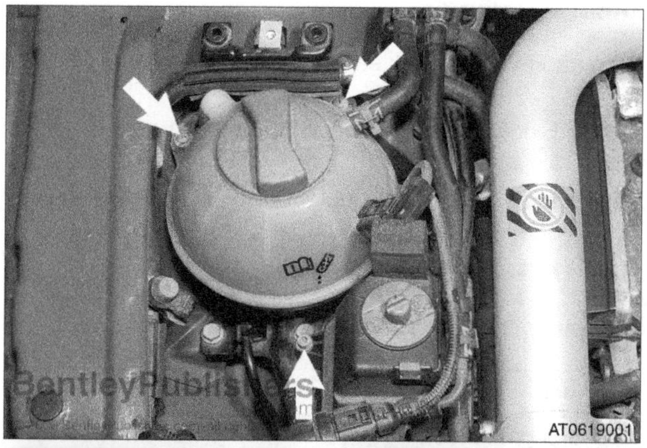

◄ Unbolt (**arrows**) coolant expansion tank and power steering reservoir and move to side with hoses still connected.

– Remove upper and center toothed belt guards.

– Put transmission in neutral (manual shift) or Park (automatic).

◄ turn crankshaft sprocket in a clockwise direction (direction of belt rotation). Align mark on camshaft sprocket (**lower arrow**) with mark on upper rear timing belt guard (**upper arrow**). This brings the camshaft to TDC for cylinder number #1.

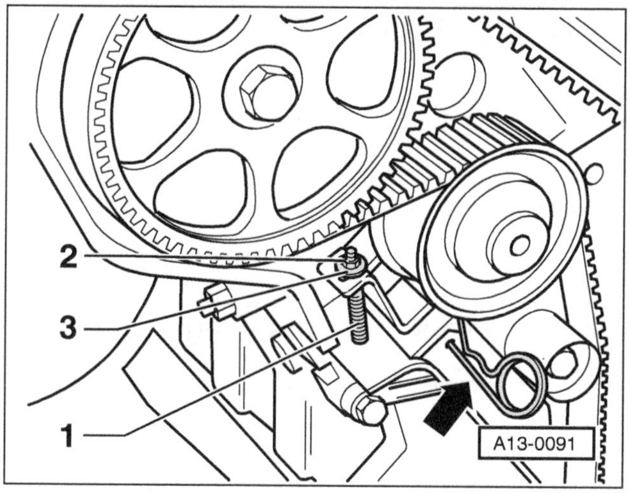

◄ Relieve tension on timing belt.

- Screw a 5x55 mm threaded stud (**1**), or special tool T10092, into timing belt tensioning damper housing and install a nut (**2**) and washer (**3**) to the other end.
- Line up the hole in the tensioning damper housing with the hole in the pressure piston by turning the pressure piston. Use a suitable tool to turn the pressure piston before tightening the nut on the threaded stud.
- Tighten nut (**2**) on threaded stud until the hole in the pressure piston aligns with the hole in the tension damper housing.
- Secure pressure piston in position with a pin, or special tool T40011 (**arrow**), inserted through hole in tensioning damper housing.

– Remove timing belt from camshaft sprocket only.

NOTE—

- *The vibration damper/crankshaft pulley do not need to be removed.*
- *Leave the timing belt in position on the crankshaft sprocket.*
- *Cover toothed belt with a cloth to protect it from coolant when removing coolant pump.*

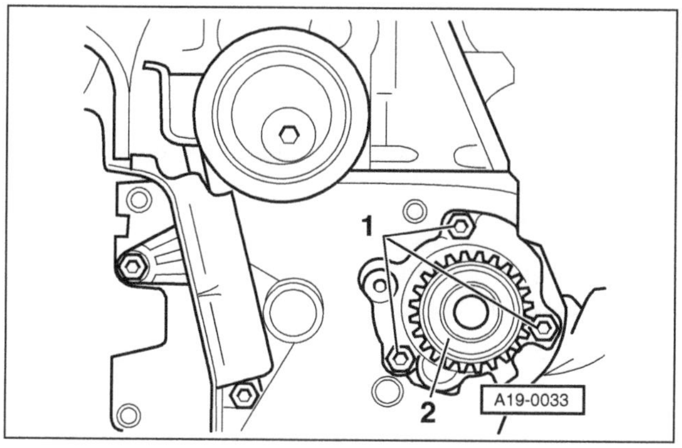

◄ Remove coolant pump mounting bolts (**1**) and coolant pump (**2**).

BentleyPublishers
.com
©2006 BentleyPublishers.com—All Rights Reserved

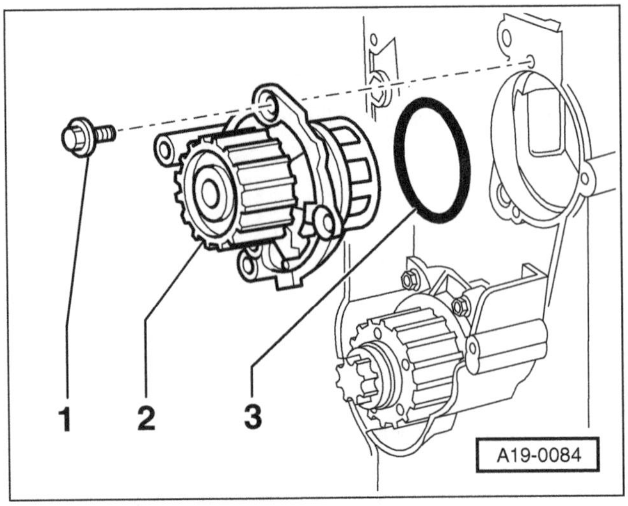

Coolant pump, installing

< Install coolant pump in reverse order of removal, noting the following:

- Clean and smooth down sealing surface for coolant pump O-ring (**3**) as needed.
- Moisten coolant pump O-ring with coolant when installing.
- Install timing belt and adjust valve timing if necessary. See **15a Cylinder Head (1.8L Engine)**.

– Tighten coolant pump mounting bolts to specification.

Tightening torque	
Coolant pump to engine block	15 Nm (11 ft-lb)

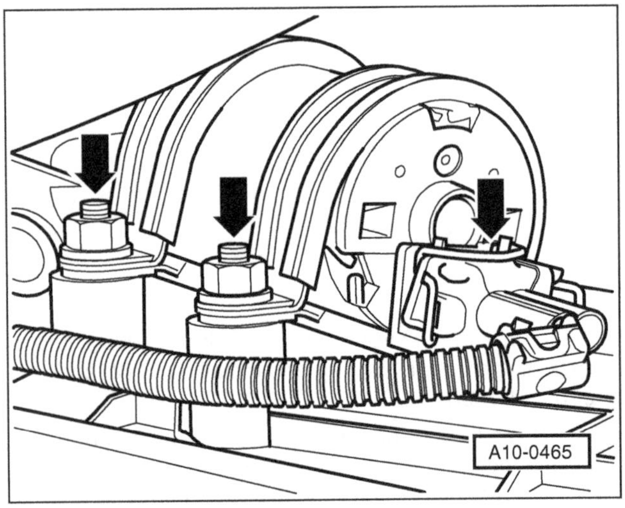

After-run coolant pump, checking (1.8L 225 hp engine)

NOTE —

- *After-run coolant pump (V51) is located on the fan cowl to the right of the radiator fan.*
- *The after-run coolant pump provides additional cooling for higher horsepower engines and turbocharger cooling. It also runs for pre-determined period of time after the engine is shut off (after-run).*

– Switch on ignition.

< After-run coolant pump can be felt or possibly heard running.

– If after-run coolant pump does not run, check fuse and wiring using wiring diagram. See **EWD Electrical Wiring Diagrams**.

– If fuse and wiring is OK and pump still does not run with the key on, replace after-run coolant pump. Mounting bolts are shown at **arrows**.

Thermostat, removing and installing

– With engine fully cold, remove cap from the coolant expansion tank.

> ***WARNING —***
>
> - *Hot steam may escape when opening coolant expansion tank if engine is warm. Wear protective goggles and protective clothing to prevent damage to eyes and scalding. Cover cap with a rag and open very carefully.*

– Drain coolant. See **Cooling System, Draining and Filling**.

– Detach coolant hose at bracket above alternator.

< Unbolt (**arrows**) bracket for solenoid valves from front of intake manifold. Detach dipstick guide tube (**A**).

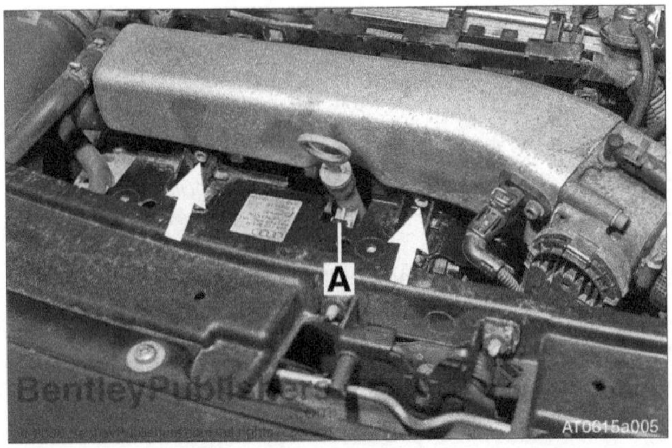

BentleyPublishers.com
©2006 BentleyPublishers.com—All Rights Reserved

◀ Detach harness connector and unscrew mounting bolts (**arrows**) for solenoid valve. Unclip vacuum line from bracket (**right arrow**). Remove bracket.

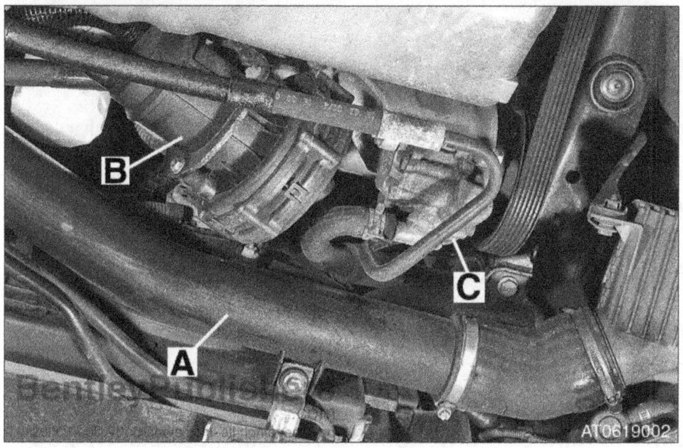

◀ Working below engine, remove connecting pipe (**A**) between intercoolers (225 hp engine only).

– Remove secondary air injection pump (**B**) and bracket.

– Unbolt power steering pump (**C**) and position out of way. Secure pump to vehicle with safety wire leaving hydraulic lines connected.

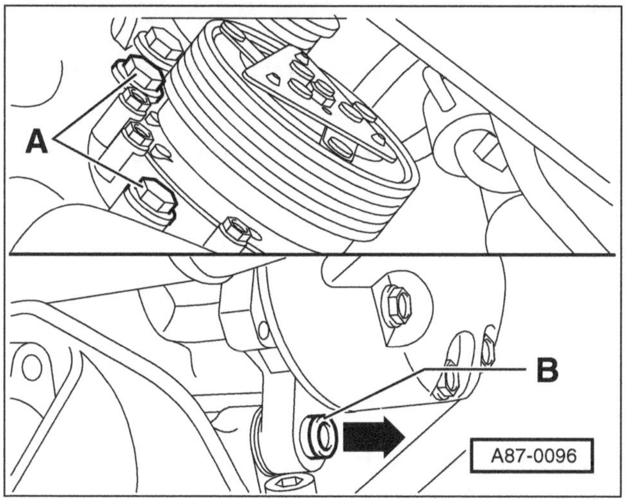

◀ Unbolt A/C compressor mounting bolts (**A**). Lift compressor out and secure to vehicle with safety wire, leaving refrigerant lines connected.

– Before installing compressor, push bushing (**B**) all the way into hole in compressor.

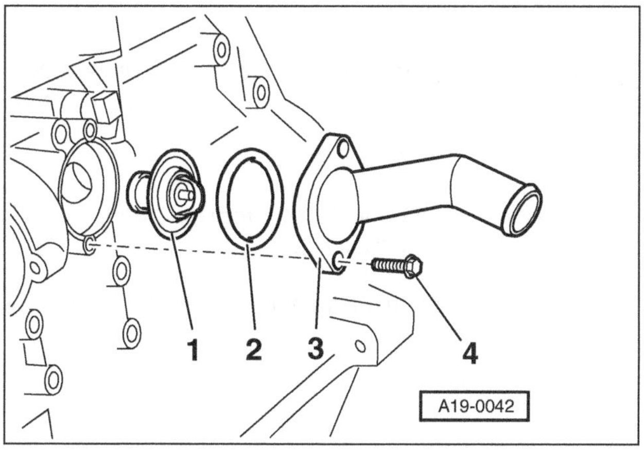

◀ Disconnect coolant hose from thermostat housing (**3**).

– Remove mounting bolts (**4**), thermostat housing (**3**), O-ring (**2**) and thermostat (**1**). Note installed position of thermostat.

BentleyPublishers.com
© 2006 BentleyPublishers.com–All Rights Reserved

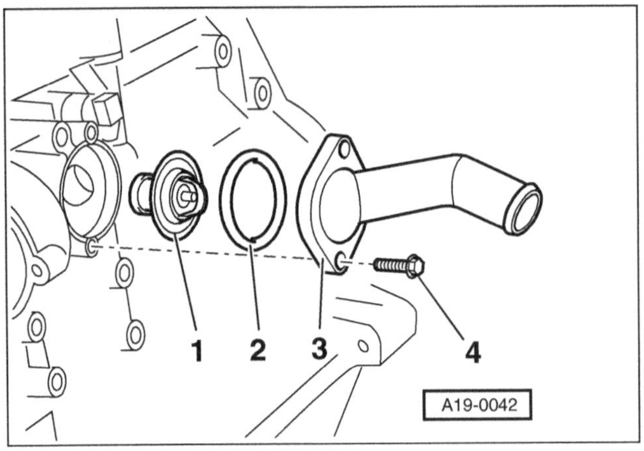

When installing thermostat:

- Coat new O-ring (**2**) with coolant.
- Position thermostat (**1**) so that outer support brace is as close to vertical as possible.

Tightening torque	
Thermostat housing to engine block	15 Nm (11 ft-lb)

– Install remaining components in reverse order of removal and refill system with coolant. See **Cooling System, Draining and Filling**.

Radiator, removing and installing

– With engine fully cold, remove cap from the coolant expansion tank.

> ***WARNING—***
> *• Hot steam may escape when opening coolant expansion tank if engine is warm. Wear protective goggles and protective clothing to prevent damage to eyes and scalding. Cover cap with a rag and open very carefully.*

– Remove upper and lower engine sound insulation panels. Also remove left and right side lower sound insulation panels. See **03 Maintenance**.

– Drain coolant. See **Cooling System, Draining and Filling** in this repair group.

– Release clamps and disconnect coolant hoses from connections on radiator (top and bottom).

– Remove front bumper cover. See **63 Bumpers**.

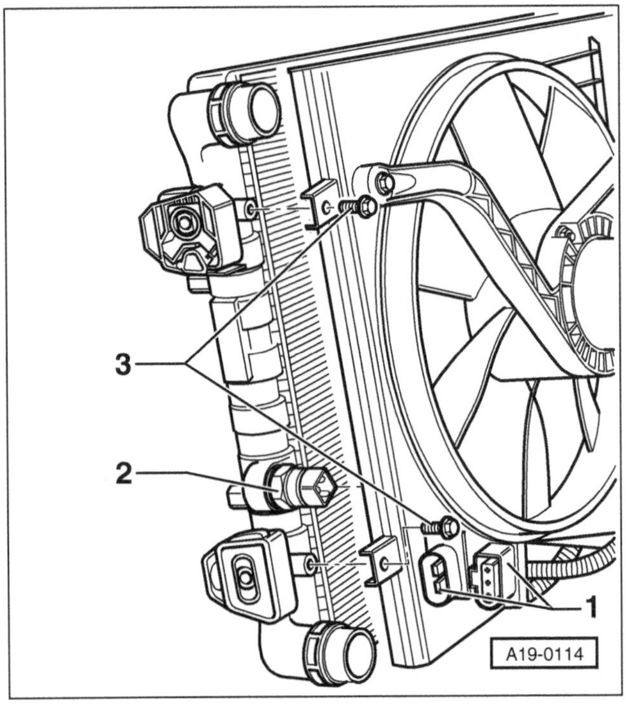

Disconnect harness connectors from electrical switches (**1** and **2**).

– Remove mounting bolts (**3**) and fan housing.

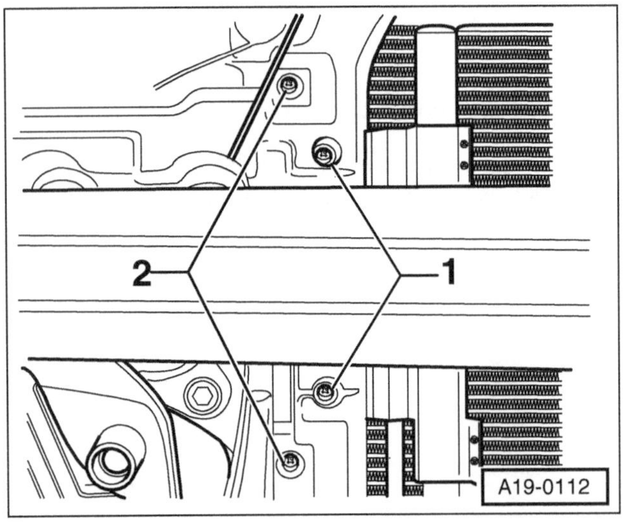

Remove four mounting bolts (**1**) for condenser. Do not remove condenser or disconnect refrigerant lines.

NOTE—
• To prevent damage to the condenser and refrigerant lines/hoses, ensure that the lines and hoses are not stretched, kinked or bent.

– Remove four mounting bolts (**2**) for radiator.

– Remove coolant fan control module (J293) from left-hand longitudinal member (in front of battery).

– Carefully lower radiator between engine and condenser and remove from below.

– Install in reverse of removal.

BentleyPublishers.com
© 2006 BentleyPublishers.com—All Rights Reserved

Cooling System Components (3.2L)

Cooling system overview

NOTE—
- *Volkswagen identifies electrical components by a letter and/or a number in the electrical schematics. See* **EWD Electrical Wiring Diagrams**. *These electrical identifiers are listed in parentheses as an aid to electrical troubleshooting.*

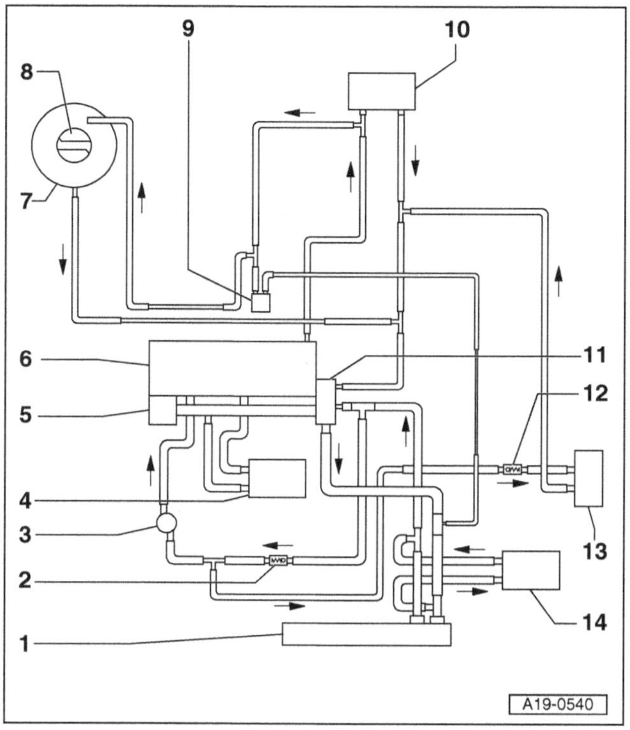

1. Radiator
2. Non-return valve
3. After-run coolant pump (V51)
4. Oil cooler
5. Coolant pump
6. Cylinder head/cylinder block
7. Expansion tank
8. Expansion tank filler cap
   - With pressure relief valve
9. Throttle valve control module (J338)
10. Heat exchanger
11. Thermostat
12. Non-return valve
13. ATF cooler
14. Auxiliary radiator

Coolant pump, removing and installing

– With engine fully cold, remove cap from the coolant expansion tank.

WARNING—
- *Hot steam may escape when opening coolant expansion tank if engine is warm. Wear protective goggles and protective clothing to prevent damage to eyes and scalding. Cover cap with a rag and open very carefully.*

◀ Loosen hose clamp (**arrow**) and disconnect intake air hose at throttle valve control module.

NOTE—
- *Air hose is disconnected to avoid damaging it when lifting engine.*

– Loosen lug bolts for right front wheel. Lift vehicle and support safely on jack stands. See **03 Maintenance**. Remove right front wheel.

– Remove upper and lower engine sound insulation panels. Remove lower right engine sound insulation panel. See **03 Maintenance**.

– Drain coolant. See **Cooling System, Draining and Filling** in this repair group.

– Remove lower right front wheel housing liner. See **66 Body–Exterior Equipment**, if necessary.

BentleyPublishers.com
© 2006 BentleyPublishers.com—All Rights Reserved

– Remove plastic sound insulation panel above coolant reservoir in engine compartment.

◄ Turn fasteners (**arrows**) ¼ turn and remove sound insulation panel at front right side of engine compartment (on lock carrier).

– Remove ribbed belt, see **03 Maintenance**.
  - Mark direction of rotation on belt if re-using.

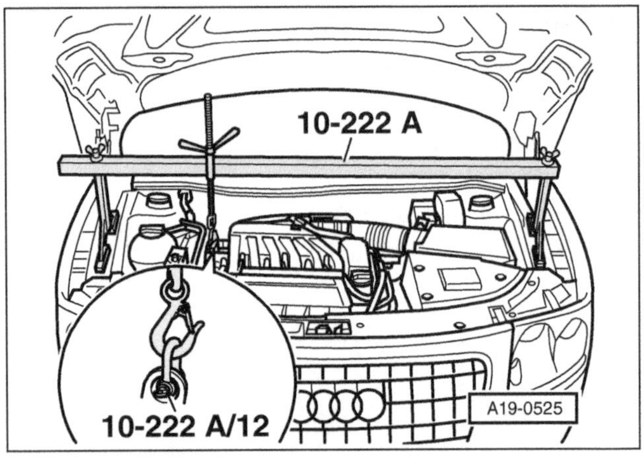

◄ Attach engine support bar and shackle assembly to right rear engine lifting eye.
  - VW/Audi special tool 10-222A support bar shown. An aftermarket engine support bar may be used. Follow manufacturer's instructions for proper use of support bar.

– Pretension engine support bar spindle slightly.

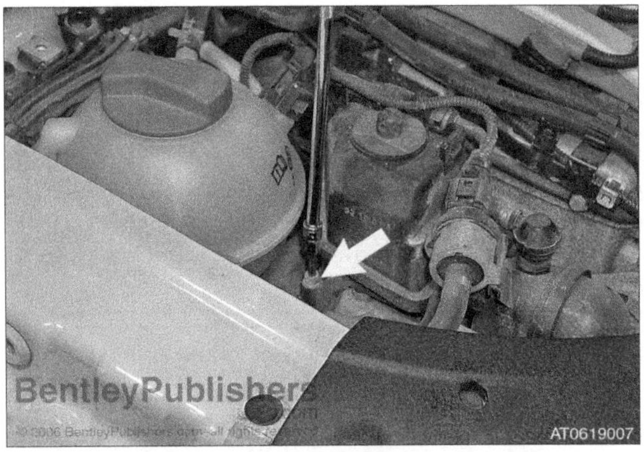

◄ Remove mounting bolt (**arrow**) for power steering reservoir. Leave power steering lines connected and position reservoir out of way.

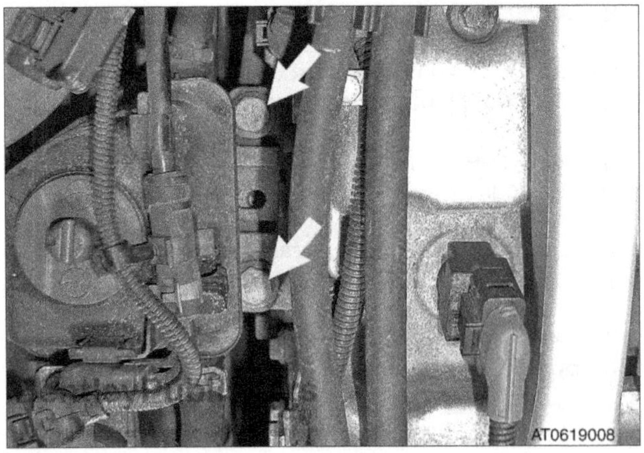

◄ Working below power steering reservoir, remove mounting bolts (**arrows**) for front engine mount.

BentleyPublishers
.com
©2006 BentleyPublishers.com—All Rights Reserved

Working under engine, remove mounting bolts (**arrows**) for pendulum mount.

– Use support bar spindle to lower engine enough to allow access to coolant pump through right front wheel housing.
  - Lower engine approx. 65 mm (2.5 inches).

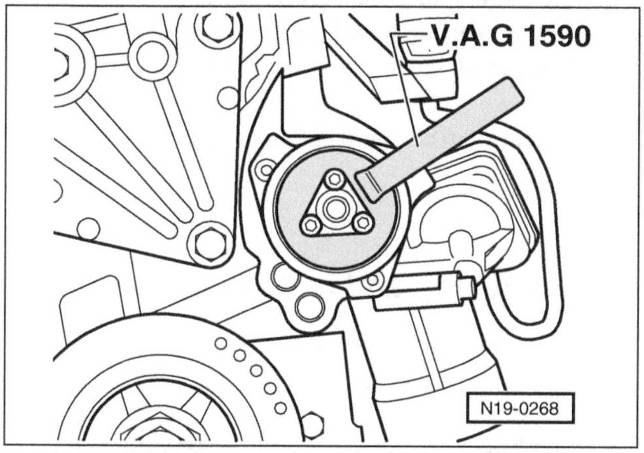

Remove pulley from coolant pump. Counterhold pulley using special coolant pump wrench (VW special tool VAG 1590).

NOTE —

- *Coolant pump mounting bolts are accessible through holes in pulley (as shown in illustration below). However, the coolant pump pulley must be removing in order to enough have enough clearance to remove pump from engine.*

Remove coolant pump mounting bolts (**arrows**) and slide coolant pump out of cylinder block. It may be necessary to use a lever to move the engine toward one side enough to remove coolant pump.

– Install in reverse order of removal, noting the following:
  - Clean coolant pump mating surfaces.
  - Moisten new O-ring with coolant.
  - Align engine mount and torque bolts to proper specification. See **15b Cylinder Head (3.2L Engine)** for engine mount torques.

Tightening torques	
Coolant pump to engine block	8 Nm (71 in-lb)
Coolant pulley to coolant pump	20 Nm (15 ft-lb)

BentleyPublishers.com
©2006 BentleyPublishers.com–All Rights Reserved

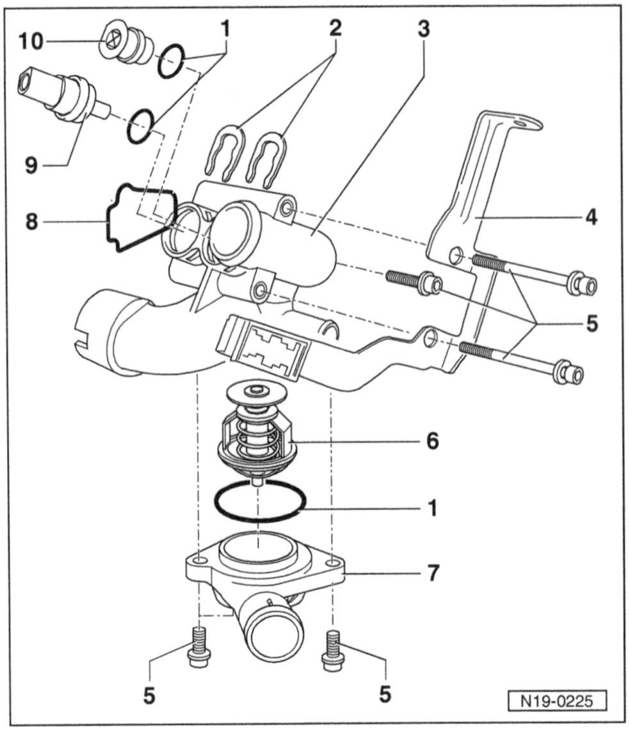

Thermostat housing overview

1. O-ring
   - Always replace
2. Retaining clip
3. Thermostat housing
4. Bracket
5. Bolt
   - 10 Nm (7 ft-lb)
6. Thermostat
7. Connection
8. Gasket
   - Always replace
9. Engine coolant temperature sensor (G62)
   - With integrated coolant temperature gauge sensor (G2)
   - Release cooling system pressure before removing
10. Sealing plug
    - Release cooling system pressure before removing

Thermostat, removing and installing

– With engine fully cold, remove cap from the coolant expansion tank.

> ***WARNING —***
> - *Hot steam may escape when opening coolant expansion tank if engine is warm. Wear protective goggles and protective clothing to prevent damage to eyes and scalding. Cover cap with a rag and open very carefully.*

– Disconnect battery ground (GND) strap from battery negative (–) terminal. See **27 Battery, Alternator, Starter** before disconnecting battery.

– Remove upper and lower engine sound insulation panels. See **03 Maintenance**.

– Drain coolant. See **Cooling System, Draining and Filling** in this repair group.

◄ Turn fasteners (**arrows**) ¼ turn and remove sound insulation panels from left side of engine compartment.

BentleyPublishers
.com
© 2006 BentleyPublishers.com—All Rights Reserved

◀ Disconnect air hose (**1**) at throttle valve control module.

– Disconnect harness connector from mass air flow sensor (**2**).

– Remove air cleaner mounting bolts (**arrows**) and remove air cleaner assembly.

◀ Release retaining tabs (**arrows**) and pivot fuse box upward.
- Electrical wires remain connected.

– Disconnect fuse box from carrier at rear hinge and position out of way.
- Use tie wrap or similar to keep fuse box out of way.

◀ Remove mounting bolts (**arrows**) and fuse box carrier.

BentleyPublishers.com
©2006 BentleyPublishers.com–All Rights Reserved

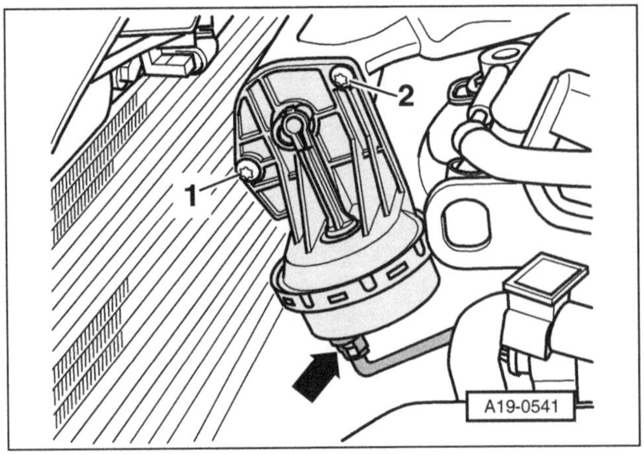

Disconnect vacuum hose (**arrow**) from intake manifold change-over valve.

- Remove mounting bolts (**1** and **2**) and intake manifold change-over assembly.

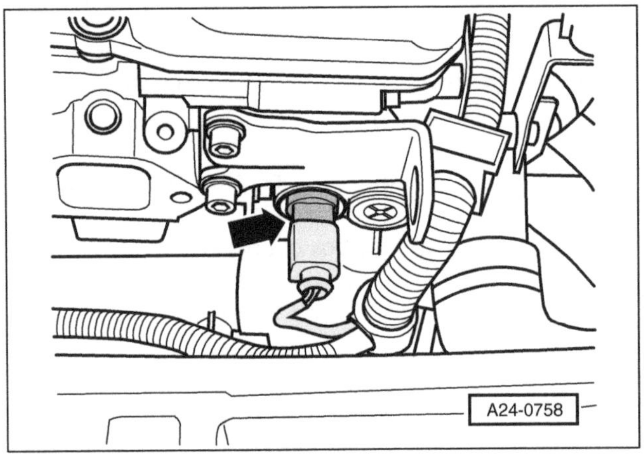

Working at front left side of engine, disconnect engine coolant temperature sensor harness connector (**arrow**).

Working at left side of engine, disconnect secondary air hose (**arrow**) and move out of way.

Disconnect coolant hoses at connections indicated (**arrows**).

BentleyPublishers.com
© 2006 BentleyPublishers.com—All Rights Reserved

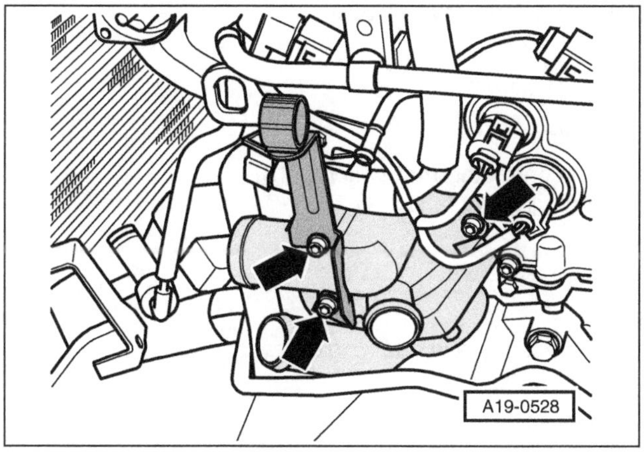

< Remove mounting bolts (**arrows**) and bracket for secondary air hoses.

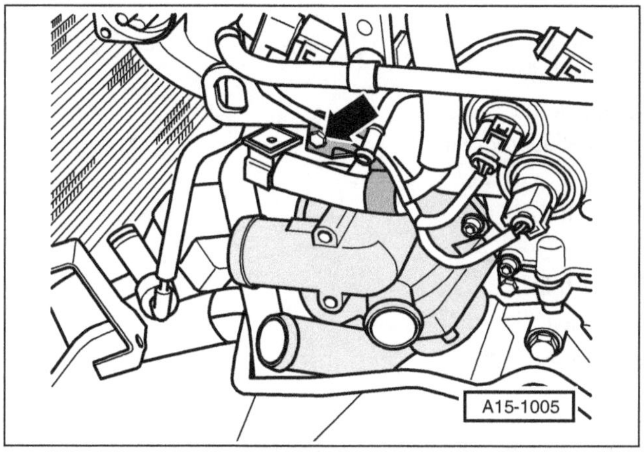

< Remove mounting bolt (**arrow**) on bracket for wiring harness.

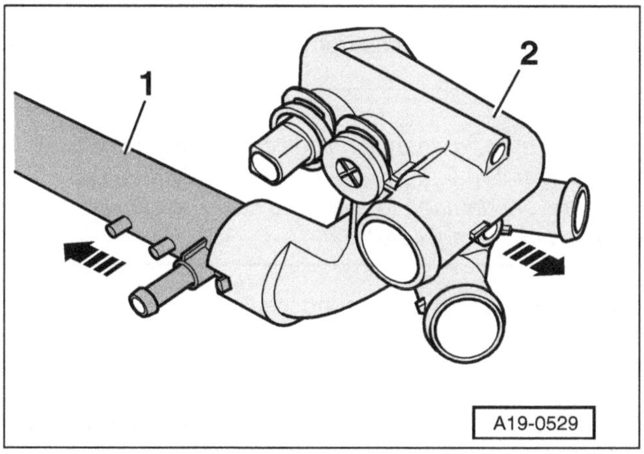

< Remove thermostat housing (**2**).

- Push coolant pipe (**1**) toward coolant pump (direction of **arrow**) to prevent it from also being removed.

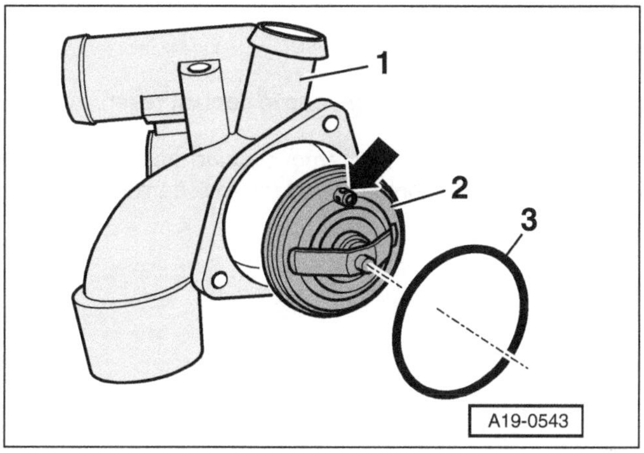

< Remove O-ring (**3**) and thermostat (**2**) from housing.

– Install in reverse order of removal, noting the following:

- Note installation position of thermostat (**arrow**).
- Replace O-ring and gaskets.

Tightening torques	
Thermostat housing to engine	10 Nm (7 ft-lb)
Wiring harness bracket to engine	10 Nm (7 ft-lb)

BentleyPublishers
.com
© 2006 BentleyPublishers.com—All Rights Reserved

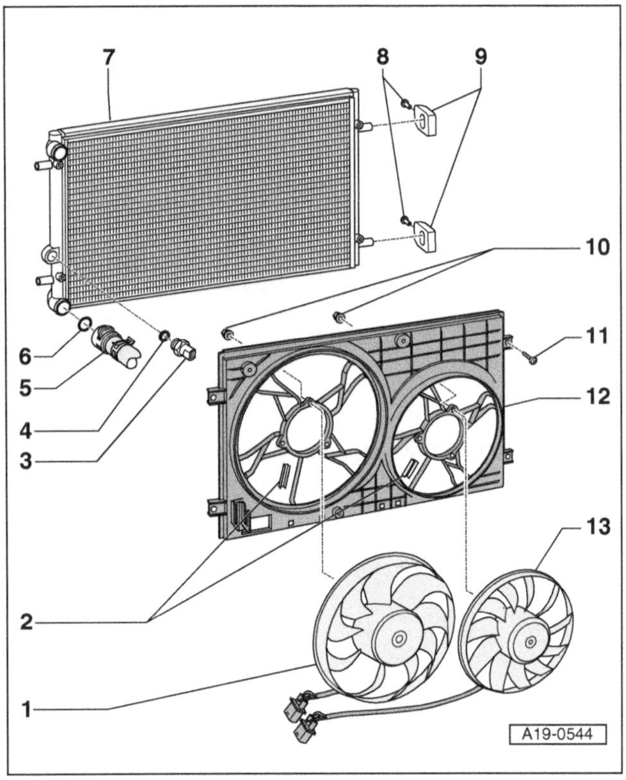

Radiator and fan overview

1. Coolant fan (V7)
2. Retaining clips
   - For fan wiring
3. Coolant fan control thermal switch (F18)
   - 35 Nm (26 ft-lb)
4. Seal
   - Always replace
5. Coolant hose
   - To disconnect, release retaining clip
6. O-ring
   - Always replace
7. Radiator
8. Bolt
   - 15 Nm (11 ft-lb)
9. Mounting bracket
   - Note installation position
10. Nut
    - 10 Nm (7 ft-lb)
11. Bolt
    - 10 Nm (7 ft-lb)
12. Radiator cowl
13. Coolant fan 2 (V177)

Radiator, removing and installing

– With engine fully cold, remove cap from the coolant expansion tank.

> ***WARNING —***
> - *Hot steam may escape when opening coolant expansion tank if engine is warm. Wear protective goggles and protective clothing to prevent damage to eyes and scalding. Cover cap with a rag and open very carefully.*

– Remove upper and lower engine sound insulation panels. Also remove left and right side lower sound insulation panels. See **03 Maintenance**.

– Drain coolant. See **Cooling System, Draining and Filling** in this repair group.

– Remove lower coolant hose from radiator.

– Remove sound insulation panel from above coolant reservoir.

◄ Turn fasteners (**arrows**) ¼ turn and remove sound insulation panels at front right side of engine compartment (on lock carrier).

BentleyPublishers.com
© 2006 BentleyPublishers.com—All Rights Reserved

◄ Turn fasteners (**arrows**) ¼ turn and remove sound insulation panels from left side of engine compartment.

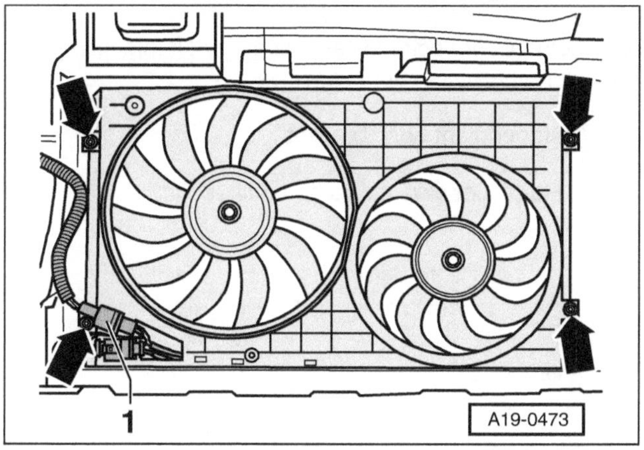

◄ Remove upper securing bolts (**upper arrows**) for radiator cowl.

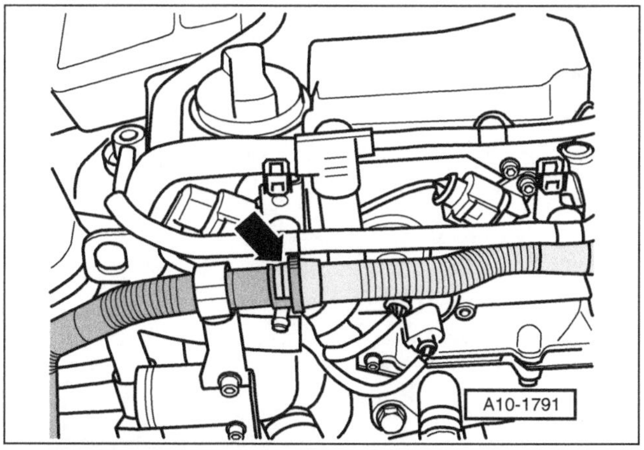

◄ Disconnect secondary air hose (**arrow**) and move out of way.

◄ Remove mounting bolts (**arrows**) and cross piece.

- Mounting bolts shown for left side, right side similar.

BentleyPublishers
.com
©2006 BentleyPublishers.com—All Rights Reserved

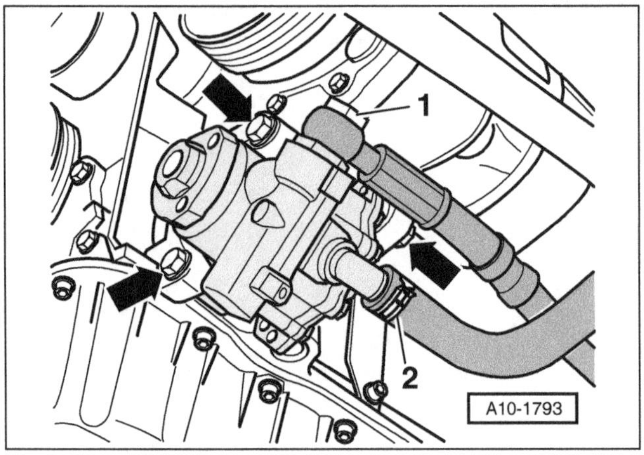

– Remove ribbed belt, see **03 Maintenance**.
  - Mark direction of rotation on belt if re-using.
– Remove power steering pump pulley (counterhold with mandrel).
◄ Remove mounting bolts (**arrows**) and remove power steering pump from bracket.
– Use safety wire to hang power steering pump from vehicle with hydraulic hoses (**1** and **2**) still attached.

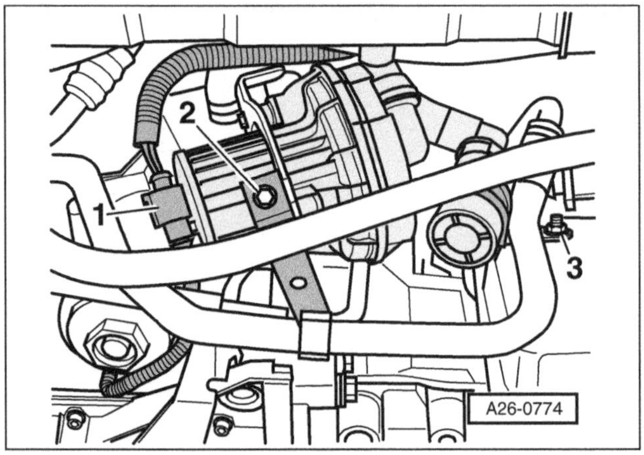

◄ Disconnect electrical connector (**1**) at secondary air injection pump motor. Move wiring harness out of way.
– Remove bolt (**2**) for coolant hose bracket.
– Remove bolt (**3**) for secondary air pump air filter.

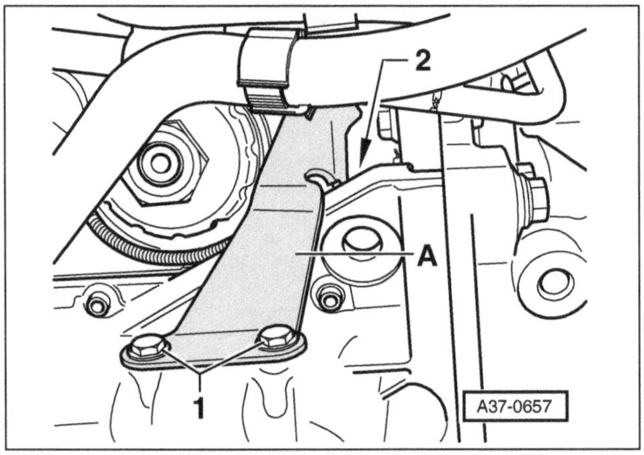

◄ Remove bolts (**1** and **2**) and remove secondary air pump and mounting bracket (**A**).

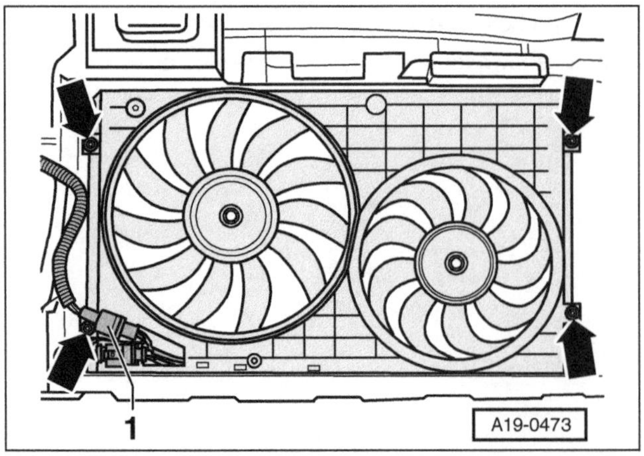

◄ Disconnect harness connectors (**1**) for coolant fans at bottom of radiator cowl.
– Unscrew lower securing bolts (**lower arrows**) for radiator fan cowl.
– Remove fan cowl from below.

BentleyPublishers.com
©2006 BentleyPublishers.com—All Rights Reserved

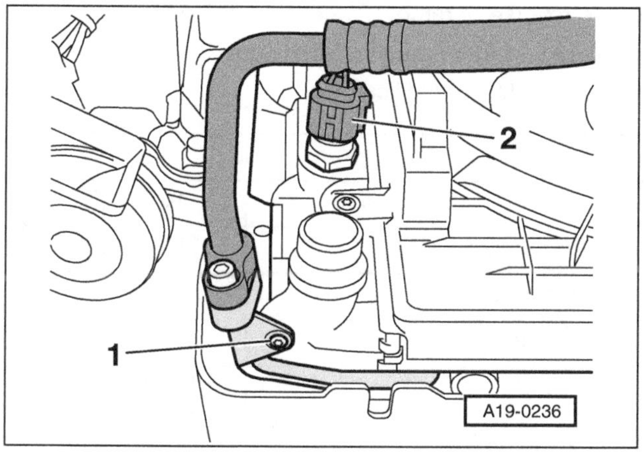

< Working under engine, front left side, disconnect harness connector (**2**) at coolant fan control thermal switch.

– Unbolt (**1**) refrigerant line from radiator.

NOTE —

- *The air conditioning refrigerant circuit must not be opened.*
- *To prevent damage to the condenser and the refrigerant lines/hoses, ensure that A/C lines and hoses are not stretched, kinked or bent.*

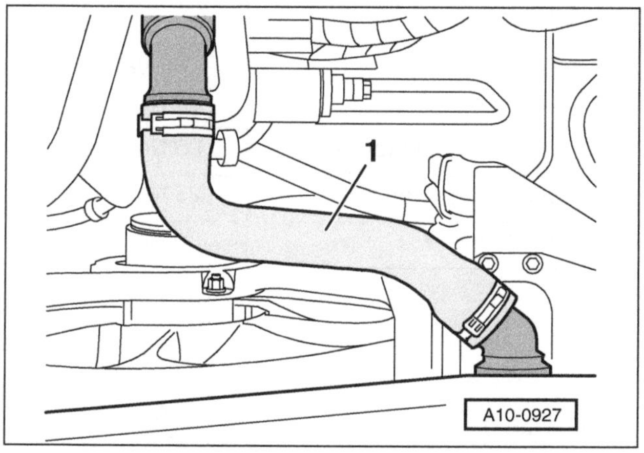

< Remove upper coolant hose (**1**) from radiator.

– Remove front bumper cover. See **63 Bumpers**.

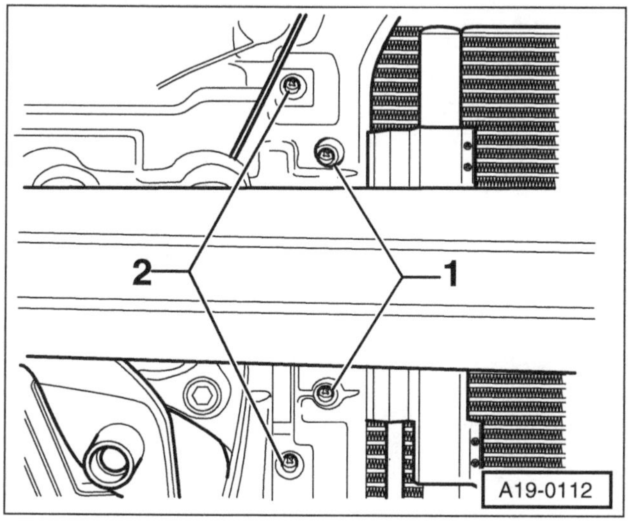

< Remove four mounting bolts (**1**) for condenser. Do not remove condenser or disconnect refrigerant lines.

NOTE —

- *To prevent damage to the condenser and refrigerant lines/hoses, ensure that the lines and hoses are not stretched, kinked or bent.*

– Remove four mounting bolts (**2**) for radiator.

– Remove coolant fan control module from left-hand longitudinal member.

– Remove front mounting bolts for lock carrier. Loosen upper bolts for lock carrier 2 turns. See **50 Body–Front**.

– Carefully lower radiator between engine and condenser and remove from below.

BentleyPublishers.com
©2006 BentleyPublishers.com—All Rights Reserved

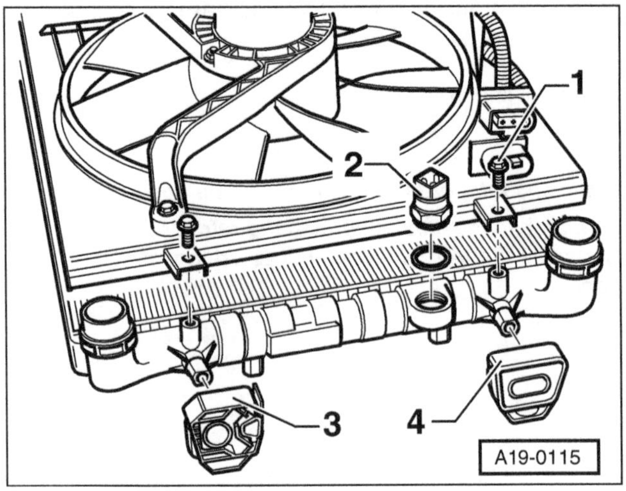

◀ Installation is the reverse of removal, noting the following:

- Install brackets (**3** and **4**) on radiator as shown.
- Tighten mounting bolts to specification.

Tightening torques	
Fan cowl to radiator	10 Nm (7 ft-lb)
Thermal switch to radiator	35 Nm (25 ft-lb)
A/C refrigerant line to radiator	10 Nm (7 ft-lb)
Condenser to radiator	10 Nm (7 ft-lb)
Radiator to lock carrier	15 Nm (11 ft-lb)
Secondary air injection pump to bracket	10 Nm (7 ft-lb)
Secondary air injection pump bracket to engine M6 bolts M8 bolts	 10 Nm (7 ft-lb) 20 Nm (15 ft-lb)
Power steering pump to mount bracket	25 Nm (18 ft-lb)
Power steering pump pulley to pump	25 Nm (18 ft-lb)
Cross member to subframe	30 Nm (22 ft-lb)

BentleyPublishers
.com
© 2006 BentleyPublishers.com—All Rights Reserved

20 Fuel Storage and Supply

General

This repair group covers fuel storage (fuel tank) for front wheel drive and all wheel drive vehicles. Also covered are fuel supply components and troubleshooting.

NOTE—

- *For fuel filter replacement, see* **03 Maintenance**.

Fuel tank capacity	
Vehicles with front wheel drive	14.5 gallons (55 liters)
Vehicles with all wheel drive	16.3 gallons (62 liters)

BentleyPublishers.com

©2006 BentleyPublishers.com—All Rights Reserved

Warnings and Cautions

Read the following warnings and cautions before starting work on your fuel system.

WARNING *—*

- *Disconnect negative (-) battery cable and cover terminal with insulated material when working on fuel related components.*
- *Gasoline is dangerous to your health. Wear suitable hand, skin and eye protection when working on fuel system. Do not breath fuel vapors. Work in a well-ventilated area.*
- *Fuel and fuel vapors may leak during many operations described in this repair group. Do not smoke or create sparks. Be aware of pilot lights in gas operated equipment (heating systems, water heaters, etc.). Have an approved fire extinguisher handy.*
- *Fuel supply systems are designed to maintain pressure in the system after the engine is turned off. Fuel is expelled under pressure as fuel lines are disconnected. This can be a fire hazard, especially if the engine is warm. Wrap a clean shop rag around any fuel line fitting before loosening or disconnecting it.*
- *Exercise extreme caution when using spray-type cleaners on a warm engine. Observe all manufacturer recommendations.*

CAUTION *—*

- *Cleanliness is essential when working on any part of the fuel system. Thoroughly clean fuel line unions and hose fittings before disconnecting them. Use only clean tools.*
- *Use only spring-type clamps on fuel hoses. Do not use screw-type hose clamps. They do not expand and contract with engine heat and will leak.*
- *Keep removed components clean. Seal or cover them with plastic or paper, especially if repair cannot be completed immediately. Seal open fuel supply and return lines to prevent contamination.*
- *When replacing parts, install only new, clean components. Replace seals and O-rings.*

NOTE *—*

- *Audi identifies electrical components by a letter and/or a number in the electrical schematics. See* **EWD Electrical Wiring Diagrams***. These electrical identifiers are listed in parentheses as an aid to electrical troubleshooting.*

BentleyPublishers
.com
© 2006 BentleyPublishers.com—All Rights Reserved

Fuel Storage

Fuel tank assembly (vehicles with front wheel drive)

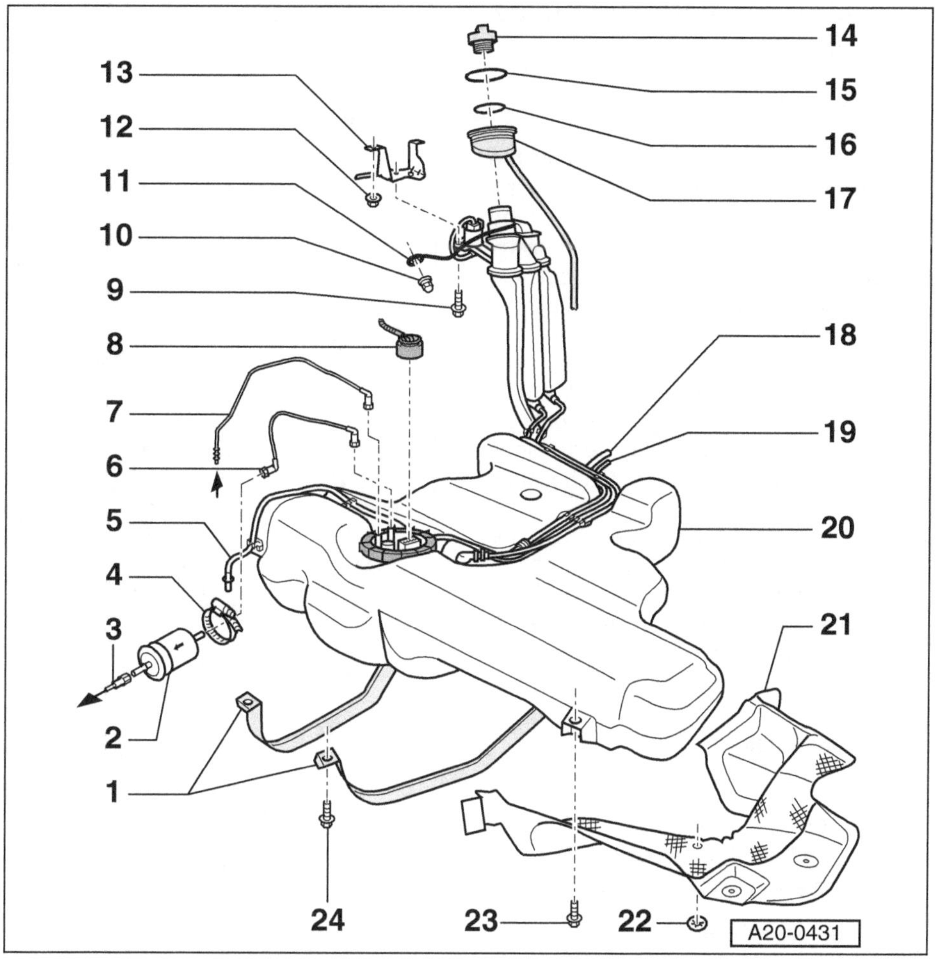

1. **Mounting straps**
   - Note varying lengths
2. **Fuel filter**
   - Arrow points in direction of flow
   - See **Fuel filter mounting position** in this repair group
3. **Fuel supply line**
   - To fuel rail
   - Black
   - To remove, press release buttons on collar
4. **Screw clamp**
   - See **Fuel filter mounting position** in this repair group
5. **Vent line**
   - White
   - From EVAP canister
   - To EVAP canister purge regulator valve
6. **Fuel supply line**
   - To fuel filter
   - Black
   - Attaches to fuel pump/sender assembly at "V" mark
   - To remove, press release buttons on collar
7. **Fuel return line**
   - From fuel rail
   - Blue
   - Attaches to fuel pump/sender assembly at "R" mark
   - To remove, press release buttons on collar
8. **4-pin harness connector**
9. **Bolt**
   - 11 Nm (8 ft-lb)
10. **Cap nut**
    - 10 Nm (7 ft-lb)
11. **Ground connection**
    - Check for tight fit of connector at metal ring of filler tube
    - Check ground connection with multimeter after installing
12. **Nut**
    - 10 Nm (7 ft-lb)
13. **Bracket**
    - For fuel filler tube
14. **Gas cap**
15. **Large tensioning ring**
16. **Small tensioning ring**
17. **Rubber cup**
    - With overflow hose
18. **Vent line**
    - White
    - To EVAP canister
    - Only remove together with tank protection valve and gravity valve
19. **Vent line**
    - White
    - From EVAP canister
    - To EVAP canister purge regulator valve
20. **Fuel tank**
    - Support with engine / transmission jack (VAG 1383A or equivalent) when removing
21. **Heat shield**
22. **Clamping washer**
23. **Bolt**
    - 26 Nm (19 ft-lb)
24. **Bolt**
    - 25 Nm (18 ft-lb)

BentleyPublishers
.com
© 2006 BentleyPublishers.com—All Rights Reserved

Fuel tank assembly (vehicles with all wheel drive)

Vehicles equipped with all wheel drive (quattro) have a uniquely shaped fuel tank to accommodate the rear drive components. There is also a secondary fuel pump located on the left-hand side of the fuel tank which pumps fuel to the right-hand side of the tank, upstream of the baffle housing of the fuel delivery unit.

The secondary fuel pump is driven by the primary fuel pump. It can only be removed after detaching the fuel lines and tank sender from baffle housing.

Tank senders (for the fuel gauge) are located on both left- and right-hand sides of the fuel tank. They are connected in series and fuel tank level signals are evaluated by the instrument cluster microprocessor.

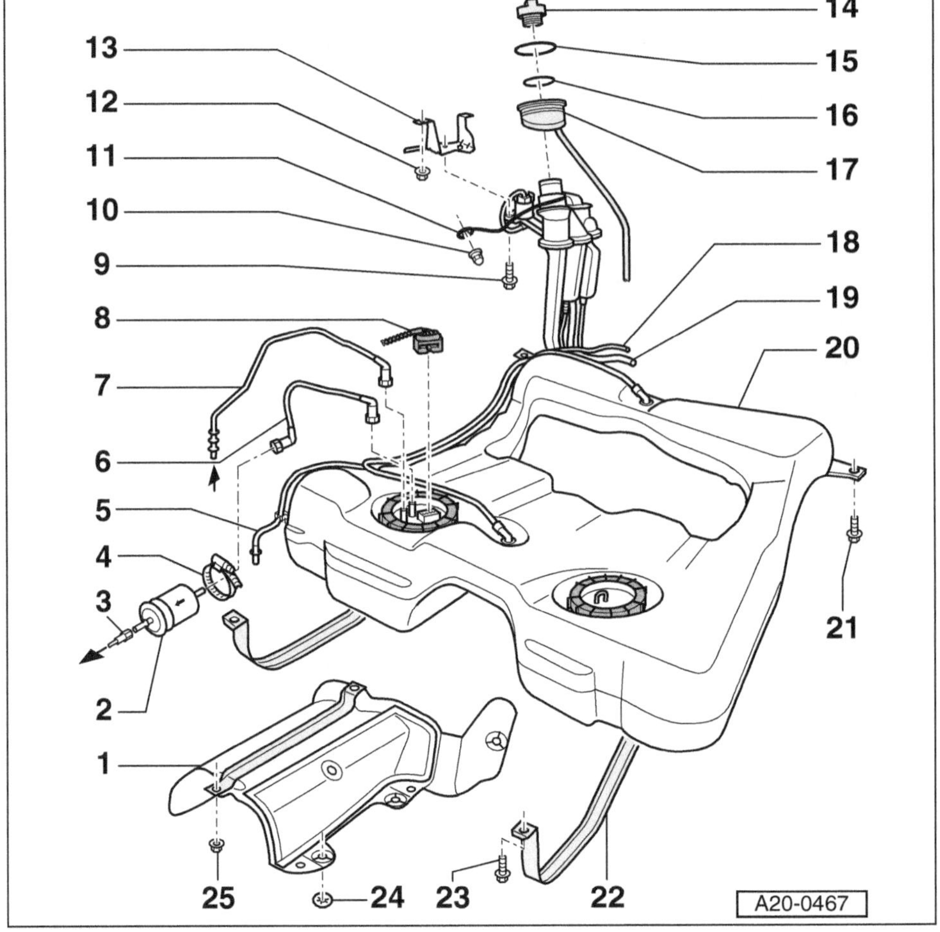

1. **Heat shield**
2. **Fuel filter**
   - Arrow points in direction of flow
   - See **Fuel filter mounting position** in this repair group
3. **Fuel supply line**
   - To fuel rail
   - Black
   - To remove, press release buttons on collar
4. **Screw clamp**
   - See **Fuel filter mounting position** in this repair group
5. **Vent line**
   - White
   - From EVAP canister
   - To EVAP canister purge regulator valve
6. **Fuel supply line**
   - To fuel filter
   - Black
   - Attaches to fuel delivery unit at "V" mark
   - To remove, press release buttons on collar
7. **Fuel return line**
   - From fuel rail
   - Blue
   - Attaches to fuel delivery unit at "R" mark
   - To remove, press release buttons on collar
8. **4-pin harness connector**
9. **Bolt**
   - 11 Nm (8 ft-lb)
10. **Cap nut**
    - 10 Nm (7 ft-lb)
11. **Ground connection**
    - Check for tight fit of connector at metal ring of filler tube
    - Check ground connection with multimeter after installing
12. **Nut**
    - 10 Nm (7 ft-lb)
13. **Fuel filler tube bracket**
14. **Gas cap**
15. **Large tensioning ring**
16. **Small tensioning ring**
17. **Rubber cup**
    - With overflow hose
18. **Vent line**
    - White
    - To EVAP canister
19. **Vent line**
    - White
    - From EVAP canister
    - To EVAP canister purge regulator valve
20. **Fuel tank**
    - Support with engine / transmission jack (VAG 1383A or equivalent) when removing
21. **Bolt**
    - 26 Nm (19 ft-lb)
22. **Mounting straps**
    - Note varying lengths
23. **Bolt**
    - 26 Nm (19 ft-lb)
24. **Clamping washer**
25. **Bolt**
    - 26 Nm (19 ft-lb)

BentleyPublishers.com

© 2006 BentleyPublishers.com—All Rights Reserved

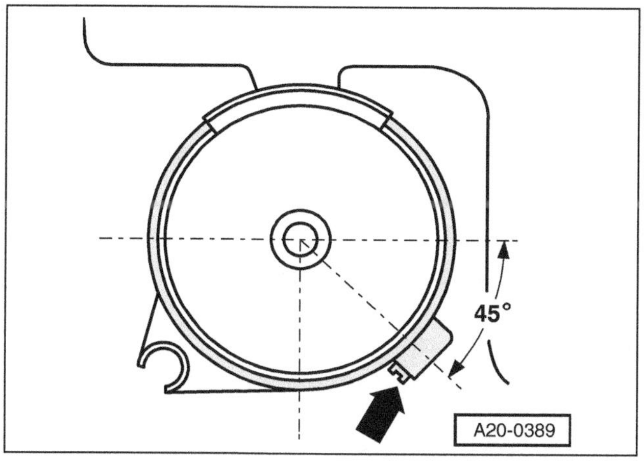

Fuel filter mounting position

◄ Install screw clamp (arrow) as shown.

- Illustration shows fuel filter from front.

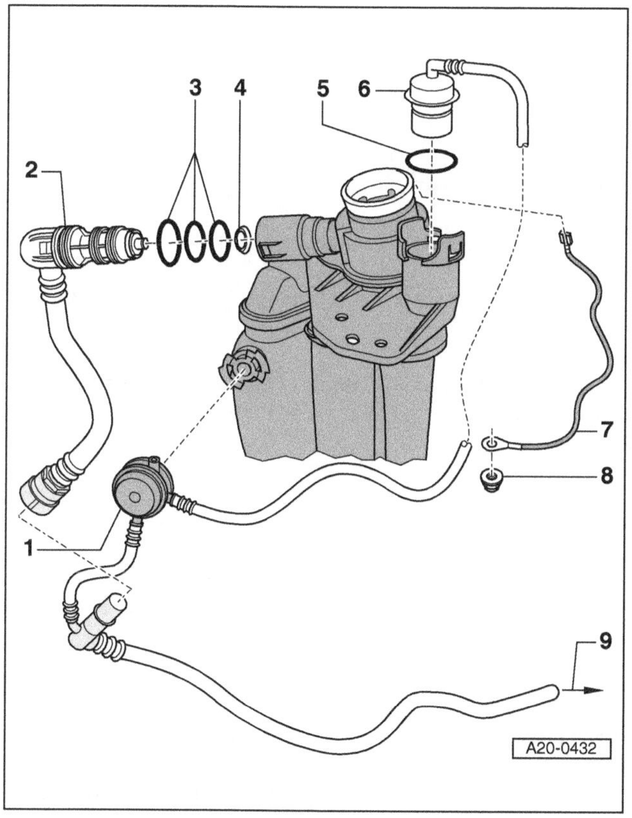

Fuel filler tube overview

1. Pressure retention valve
   - Attached to fuel filler unit
   - Closes vent line when vacuum from intake manifold is present
   - Remove with vent line and gravity valve still attached
2. Vent valve
   - Check by blowing air through valve
     With button on end of valve pressed in, it must be possible to blow air through
     With button extended, it should not be possible to blow air through
3. O-rings
   - Note different sizes
   - Always replace
4. Seal
   - Sealing lip faces fuel filler unit
5. O-ring
   - Always replace
6. Gravity valve
   - Remove with vent line and pressure retention valve still attached
7. Ground wire
   - Make sure wire is securely attached at both ends
8. Cap nut
   - 10 Nm (7 ft-lb)
9. Vent line
   - To EVAP canister

BentleyPublishers
.com
©2006 BentleyPublishers.com—All Rights Reserved

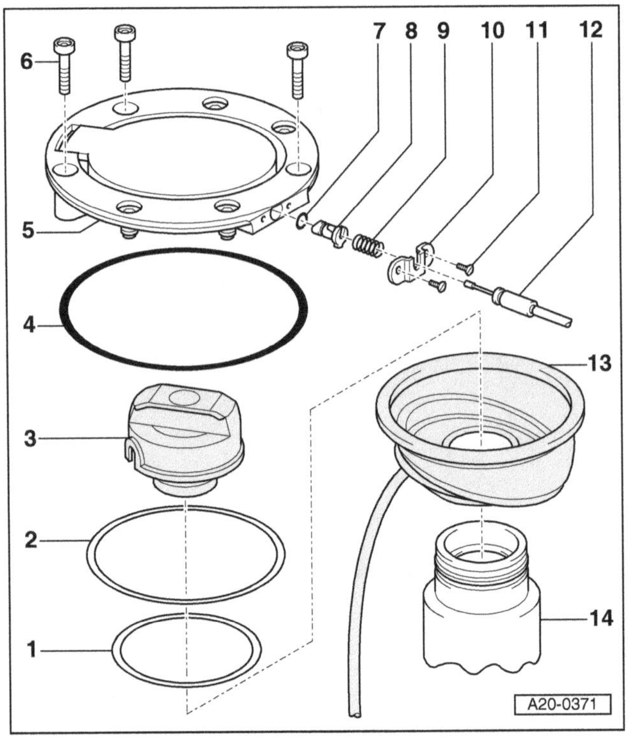

Fuel filler flap overview

1. Tensioning ring, small
2. Tensioning ring, large
3. Gas cap
4. O-ring
5. Fuel filler flap
6. Bolt
   - 4 Nm (35 in-lb)
   - Remove only 3 bolts shown to remove flap
7. O-ring
   - Replace if damaged
8. Striker
9. Compression spring
10. Retaining plate
11. Screw
12. Release cable
13. Rubber cup
    - With overflow hose
14. Fuel filler unit

Fuel filler flap, removing and installing

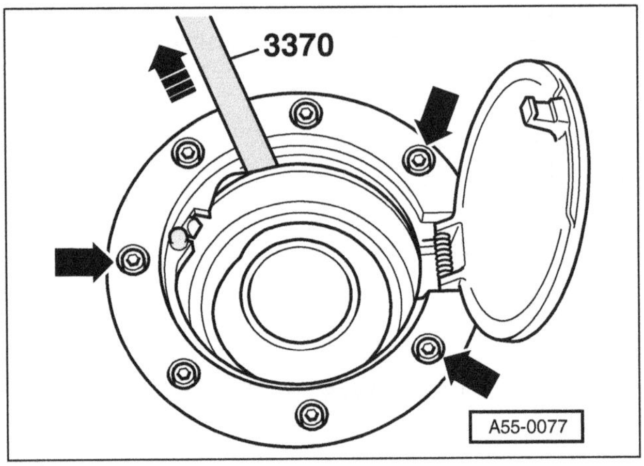

◄ Remove three mounting bolts (**arrows**) for fuel filler flap.

– Remove fuel tank flap upward using hook tool (special tool 3370 shown).

– Unscrew release cable retaining screws. See item #11 under **Fuel filler flap overview**.

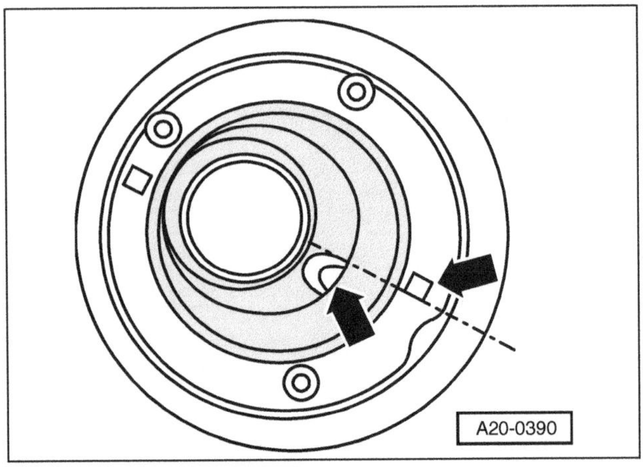

– When installing be sure to route overflow hose without kinks.

◄ Edge of overflow hole must be flush with the edge of the square hole as shown (**arrows**).

– Install mounting bolts and tighten to specification.

Tightening torque	
Fuel filler flap to body	4 Nm (35 in-lb)

BentleyPublishers.com
©2006 BentleyPublishers.com—All Rights Reserved

FUEL SUPPLY

The fuel supply system is an integral part of the operation of the fuel injection system. Problems such as a no-start condition, hesitation, or stalling may be due to poor fuel delivery. The fuel pump itself is not directly monitored as part of the OBD II system, but the fuel pump relay is monitored.

In case of poor fuel delivery, the fuel injection system will attempt to adjust for the resultant lean running condition. This may exceed the system's ability to compensate and will store diagnostic trouble codes (DTCs) in the engine control module (ECM) memory. Investigate suspected fuel pump problems with a diagnostic scan tools. See **OBD On Board Diagnostics** for more information.

When the driver's door is opened, the fuel system activates the fuel pump for a few seconds to build up fuel pressure, depending on various conditions. This improves vehicle start characteristics.

Vehicles covered by this manual are equipped with an emergency fuel shut-off system designed to reduce the danger of a fire in a crash by switching the fuel pump OFF via the fuel pump relay.

Fuel pump, electrical testing

Requirements:

- Battery voltage at least 11.5 Volts
- Fuse #28 for fuel pump OK. See **97 Fuses, Relays, Component Locations** for fuse locations.
- Relay #4 OK, located in relay panel above brake pedal. See **97 Fuses, Relays, Component Locations** for relay locations.

– Switch ignition ON. Have a helper listen carefully for fuel pump to briefly start running.

- Fuel pump runs quietly, it may be necessary to lift up rear seat bottom (coupe) or remove rear trim (roadster) to hear fuel pump.

– Switch ignition off. If fuel pump does not run, remove rear seat bottom (coupe) or rear trim (roadster). See **72 Seats** or **70 Trim–Interior**.

◄ Remove cover of fuel delivery unit and disconnect harness connector (**arrow**).

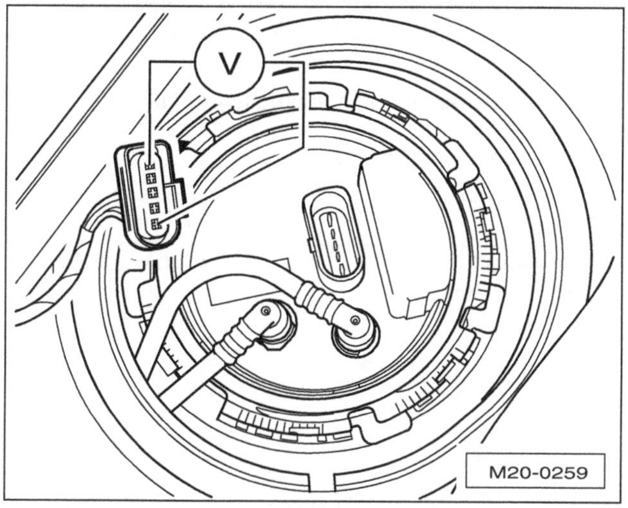

◄ Connect LED voltage tester (V.A.G. 1527 B or equivalent) to outer terminals of fuel delivery unit harness connector.

– Switch ignition ON. LED must light up briefly.

– If LED does not light up briefly, locate and repair open circuit in wiring, see **EWD Electrical Wiring Diagrams**.

– If LED lights up briefly fuel pump voltage supply is OK. Further diagnosis of fuel pump problems can be accomplished checking fuel pressure and delivery rate. See **Fuel pressure, checking** in this repair group.

BentleyPublishers
.com
©2006 BentleyPublishers.com—All Rights Reserved

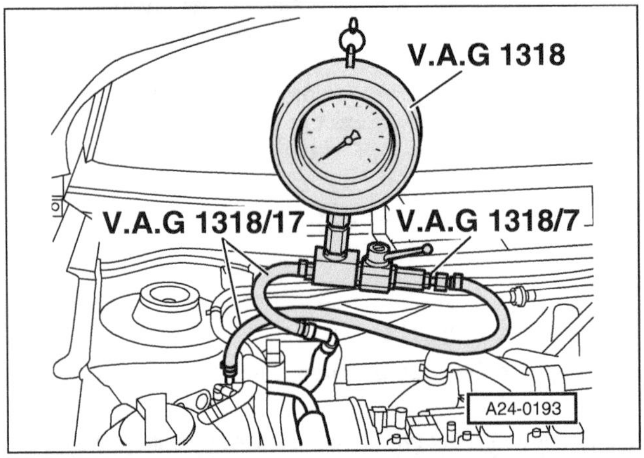

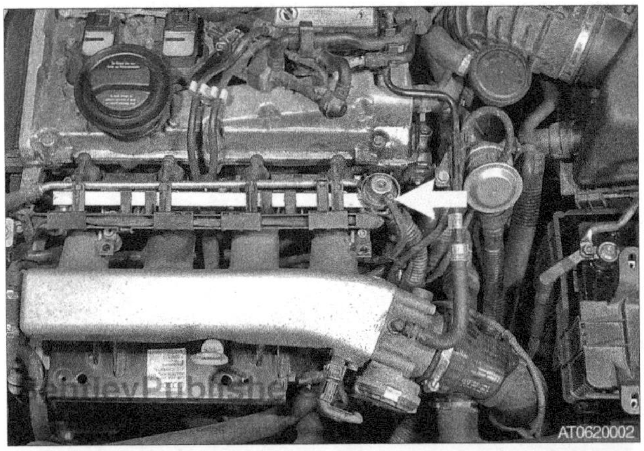

Fuel pressure, checking

NOTE —

- *The fuel pressure regulator regulates fuel pressure according to intake manifold pressure. This ensures that the pressure drop at the fuel injectors remains the same for various speed and load range demands of the engine.*

◀ Disconnect fuel supply line (**1**, white markings) by pressing release buttons on retaining collar.

> ***WARNING —***
> - *The fuel system is designed to maintain pressure after the engine is turned off. Fuel is be expelled under pressure when fuel lines are disconnected. This can be a fire hazard, especially if the engine is warm. Wrap a clean shop rag around any fuel line fitting before disconnecting it.*

◀ Connect pressure gauge (V.A.G. 1318 shown) with appropriate adapters (V.A.G. 1381/7 and 1381/17 shown) to fuel supply line.

– Open shut-off valve of pressure gauge.
  - Shut-off lever will point in direction of fuel flow.

– Start engine and run at idle.

– Measure fuel pressure and compare to specification.

Specifications	
Fuel pressure at idle (1.8L)	about 2.5 bar (36 psi)
Fuel pressure at idle (3.2L)	about 3.5 bar (51 psi)

1.8L engine

◀ Disconnect vacuum hose (**arrow**) at fuel pressure regulator.

– Fuel pressure must rise to specification.

Specification	
Fuel pressure at idle (1.8L)	about 3.0 bar (43 psi)

3.2L engine

◀ Disconnect vacuum hose (**arrow**) at fuel pressure regulator.

– Fuel pressure must rise to specification.

Specification	
Fuel pressure at idle (3.2L)	about 4.0 bar (58 psi)

BentleyPublishers.com
© 2006 BentleyPublishers.com—All Rights Reserved

If fuel pressure specification is not obtained (all engines):

– Switch off ignition.

– See **Fuel delivery rate, checking** in this repair group.

– If fuel delivery rate is OK but fuel pressure is still too low, replace fuel pressure regulator.

If fuel pressure specification is obtained:

– Switch off ignition.

– Check for leaks and residual fuel pressure. Observe pressure drop on fuel pressure gauge after 10 minutes.

Specifications	
Residual fuel pressure (1.8L)	1.5 bar (22 psi)
Residual fuel pressure (3.2L)	3.0 bar (43 psi)

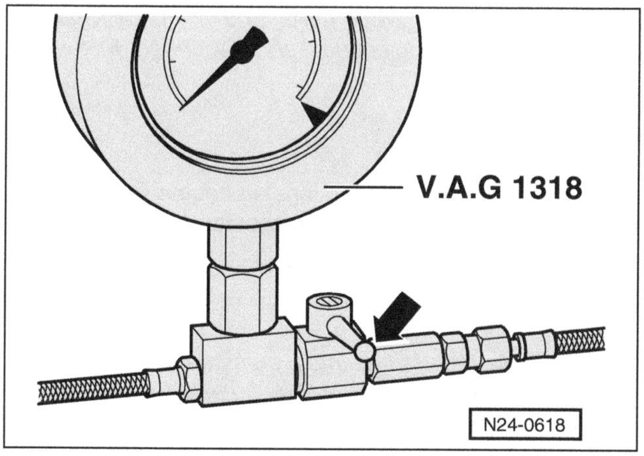

If residual fuel pressure drops below specification:

– Start engine and run at idle speed.

◄ Wait until fuel pressure has risen, and then switch off ignition. At same time, close fuel pressure gauge shut-off valve (**arrow**).

- Lever should be perpendicular to direction of fuel flow.

– Observe pressure drop (engine side) on fuel pressure gauge.

If residual fuel pressure does not drop to specification:

- Check fuel line connections to fuel tank.
- Check fuel delivery unit check valve. Check valve and fuel delivery unit are replaced as one unit.
- If check valve is OK, replace fuel pressure regulator.

If the residual pressure drops below specification again:

– Open shut-off valve of fuel pressure gauge.

- Lever should be parallel to direction of fuel flow.

– Start engine and let run at idle.

– Wait until fuel pressure has built up and then switch off ignition. Immediately clamp fuel return line (blue markings) closed.

If residual fuel pressure does not drop to specification:

– Replace fuel pressure regulator.

If the residual pressure drops below specification again:

– Check harness connectors and O-rings at fuel rail and fuel injectors for leaks.

- Check fuel pressure gauge for leaks.

NOTE —

- *Before removing fuel pressure gauge, put clean rag around fuel line connection to absorb fuel.*

BentleyPublishers.com
©2006 BentleyPublishers.com—All Rights Reserved

Fuel delivery rate, checking

Requirements:

- Battery voltage at least 12.5 Volts.
- Fuel filter OK.
- Ignition off.

– Open fuse box cover at left side of instrument panel.

– Remove fuse #28 for fuel pump. See **97 Fuses, Relays, Component Locations** for fuse locations.

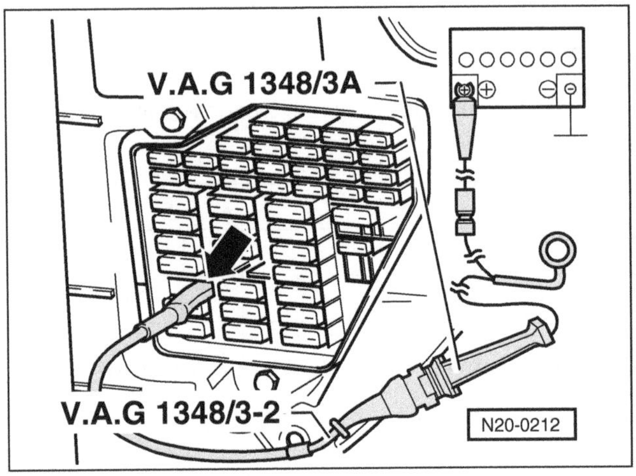

◀ Connect fuel pump remote control (V.A.G 1348/3A) using adapter cable (V.A.G. 1348/3-2) at right side terminal for fuse #28.

– Connect power lead from remote control to Battery positive (+) terminal.

NOTE —
- *A homemade jumper wire with a switch can also be used. The wire should be at least 1.5 mm (14 gauge AWG) and include a 15-amp in-line fuse for safety. Be sure to connect and disconnect wire with switch in off position.*

– Remove gas cap.

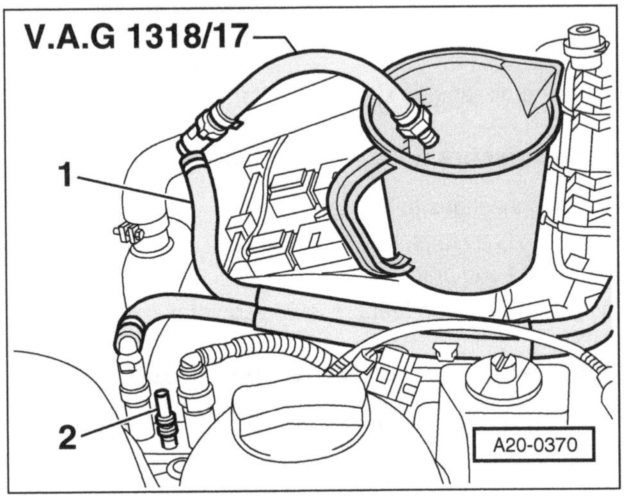

◀ Disconnect fuel return line (**1**; blue markings) in engine compartment and catch fuel in a suitable container.
- Squeeze securing ring to disengage fuel line.

> ***WARNING —***
> - *The fuel system is designed to maintain pressure after the engine is turned off. Fuel is be expelled under pressure when fuel lines are disconnected. This can be a fire hazard, especially if the engine is warm. Wrap a clean shop rag around any fuel line fitting before disconnecting it.*

– Connect adapter (V.A.G 1318/17) to fuel filter/pressure regulator and hold end of adapter in a measuring container.

– Delivery rate of fuel pump is dependent on voltage supply. Therefore, connect multimeter to battery.

– Operate remote control for 15 seconds and measure battery voltage.

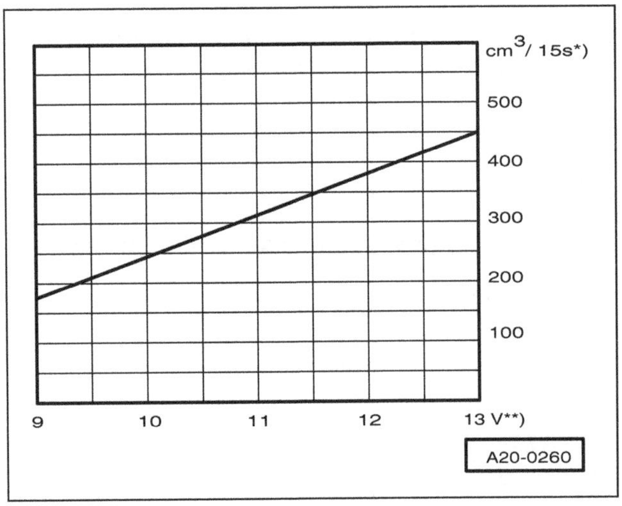

◀ Compare quantity of fuel delivered with specification.
- * Minimum delivery rate in cm^3 per 15 seconds
- ** Voltage at fuel pump with engine off and pump running (approx. 2 Volts less than battery voltage).

Example: During the test, a voltage of 12.5 volts is measured at the battery. As the voltage at the fuel pump is approx. 2 volts less than the battery voltage then this will equate to a minimum delivery of 280 cm^3 /15 seconds.

NOTE —
- *Fuel pump current draw specification is 9 amps maximum for gasoline engines.*

BentleyPublishers.com
© 2006 BentleyPublishers.com—All Rights Reserved

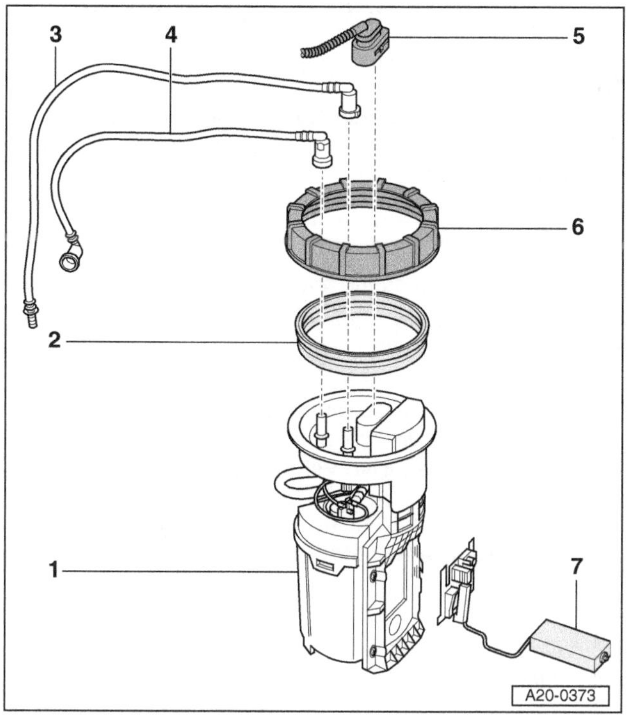

Fuel delivery unit assembly (front wheel drive)

1. Fuel pump
2. Seal
   - Always replace
   - To install, place dry seal in opening of fuel tank
3. Fuel return line
   - Blue
   - From fuel rail
   - Attaches to fuel delivery unit at "R" mark
4. Fuel supply line
   - Black
   - To fuel filter
   - Attaches to fuel delivery unit at "V" mark
5. 4-pin harness connector
   - For fuel pump and fuel level sensor
6. Retaining ring
   - 80 Nm (59 ft-lb)
7. Fuel level sensor (G)

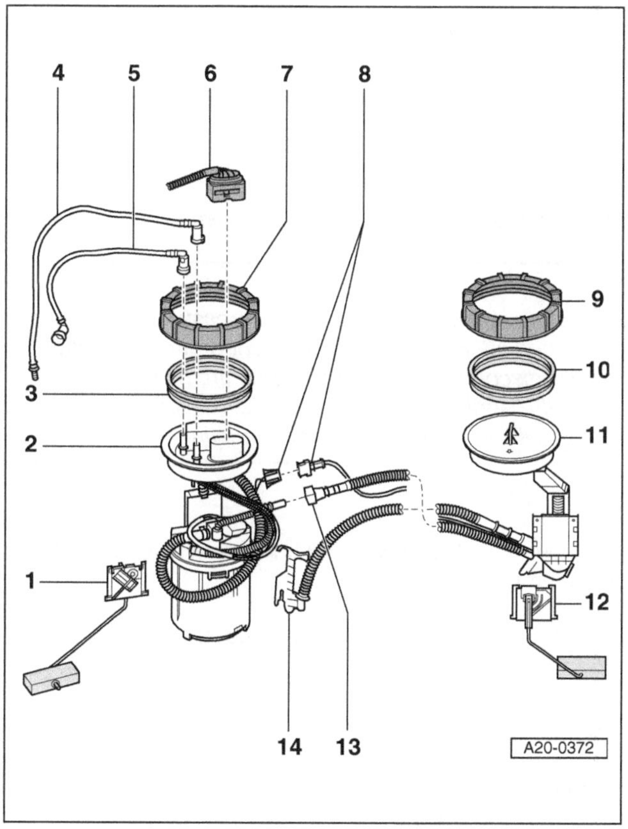

Fuel delivery unit assembly (all wheel drive)

1. Fuel level sensor (G)
2. Primary fuel pump
3. Seal
   - Always replace
   - To install, place dry seal in opening of fuel tank
4. Fuel return line
   - Blue
   - From fuel rail
   - Attaches to fuel delivery unit at "R" mark
5. Fuel supply line
   - Black
   - To fuel filter
   - Attaches to fuel delivery unit at "V" mark
6. 4-pin harness connector
   - For fuel pump and fuel level sensor
7. Retaining ring
   - 80 Nm (59 ft-lb)
8. 2-pin harness connector
   - For fuel level sensor 2 (G169)
9. Retaining ring
   - 80 Nm (59 ft-lb)
10. Seal
    - Always replace
    - To install, place dry seal in opening of fuel tank
11. Secondary fuel pump
12. Fuel level sensor 2 (G169)
13. Hose coupling
14. Retaining strap

BentleyPublishers
.com
© 2006 BentleyPublishers.com—All Rights Reserved

Fuel delivery unit, removing and installing

WARNING—
- *Be sure the fuel tank is less than ½ full before starting this procedure. If necessary siphon out fuel using VAS 5190 fuel extraction unit, or equivalent.*

NOTE—
- *The fuel delivery unit contains the fuel pump and fuel level sensor.*
- *The procedure and illustrations shown below apply to the primary fuel pump. Vehicles with all wheel drive also have a secondary fuel pump located on the left side of the fuel tank. Removal and installation of the secondary fuel pump is similar. For more information on the secondary fuel pump, see* **Secondary fuel pump, function (all wheel drive vehicles)** *in this repair group.*

– Disconnect battery ground (GND) strap from battery negative (–) terminal. See **27 Battery, Alternator, Starter**.

NOTE—
- *Be sure to have the anti-theft radio code on hand before disconnecting the battery.*

– Remove rear seat bottom (coupe) or rear trim (roadster). See **72 Seats** or **70 Trim–Interior**.

– Remove fuel delivery unit cover.

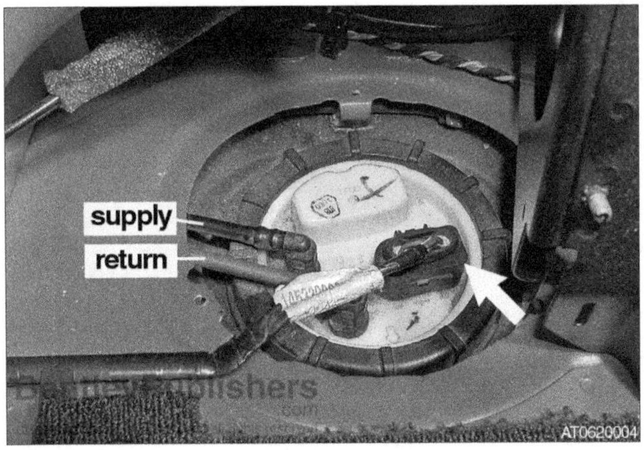

◄ Disconnect fuel delivery unit harness connector (**arrow**), black fuel **supply** line and blue fuel **return** line.

NOTE—
- *Press release buttons on retaining collar to remove fuel lines.*

WARNING—
- *The fuel system is designed to maintain pressure after the engine is turned off. Fuel is be expelled under pressure when fuel lines are disconnected. This can be a fire hazard, especially if the engine is warm. Wrap a clean shop rag around any fuel line fitting before disconnecting it.*

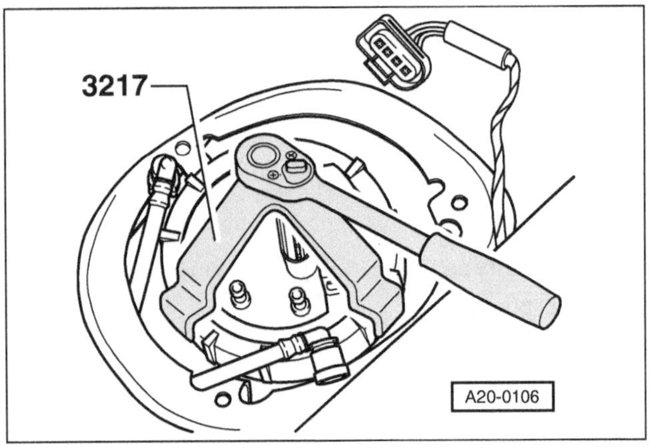

◄ Loosen fuel delivery unit retaining ring using special tool 3217. Remove retaining ring.

– Pull fuel delivery unit and seal out through fuel tank opening.

NOTE—
- *When removing fuel delivery unit, do not bend fuel gauge sensor.*

– Empty fuel from delivery unit into a suitable container.

BentleyPublishers.com
© 2006 BentleyPublishers.com—All Rights Reserved

All wheel drive vehicles

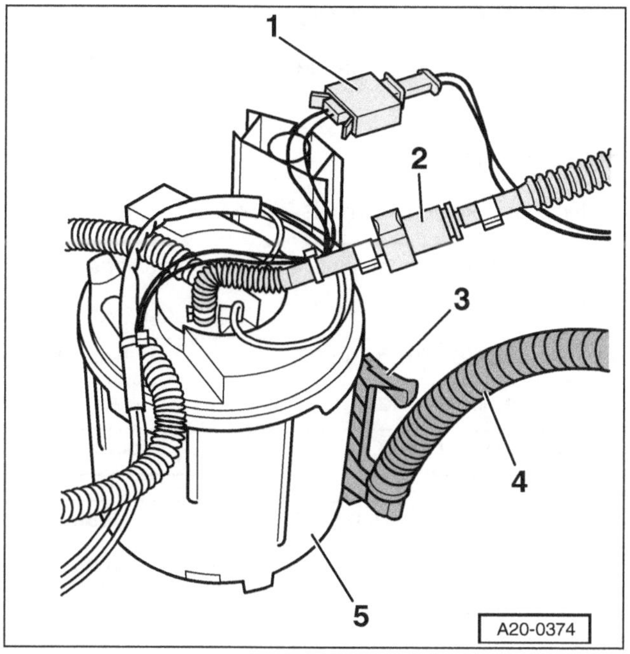

◄ Reach into fuel tank and disconnect harness connector (**1**).

– Press release button of hose coupling and disconnect wire (**2**).

– Pull retaining strap (**3**) upward and disengage strap and wire (**4**) from fuel pump (**5**).

NOTE—
- *Fuel delivery unit is shown removed from fuel tank in illustration.*

– Pull fuel delivery unit and seal out through fuel tank opening.

– Install connectors and retaining strap in reverse order of removal.
- Retaining strap must engage properly to side of fuel pump.

– Install fuel delivery unit in reverse order of removal, noting the following:
- Install new, dry fuel delivery unit seal into opening of fuel tank.
- Be sure not to bend fuel level sensor.

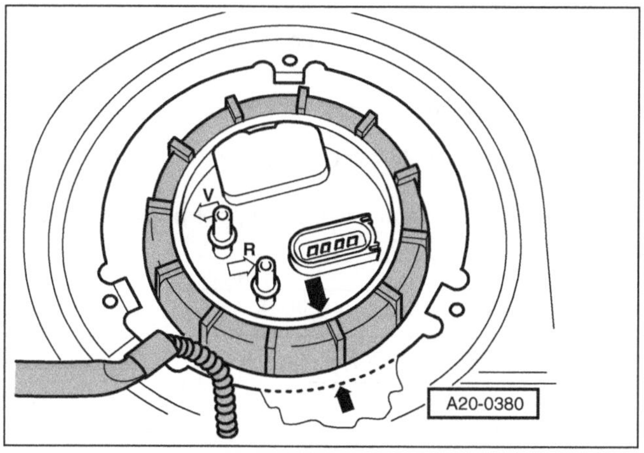

◄ Install fuel delivery unit with **arrow** aligned with **arrow** on fuel tank.

– Tighten retaining ring by hand.

Front wheel drive vehicles

– Install fuel delivery unit in reverse order of removal, noting the following:
- Install new, dry fuel delivery unit seal into opening of fuel tank.
- Be sure not to bend fuel level sensor.

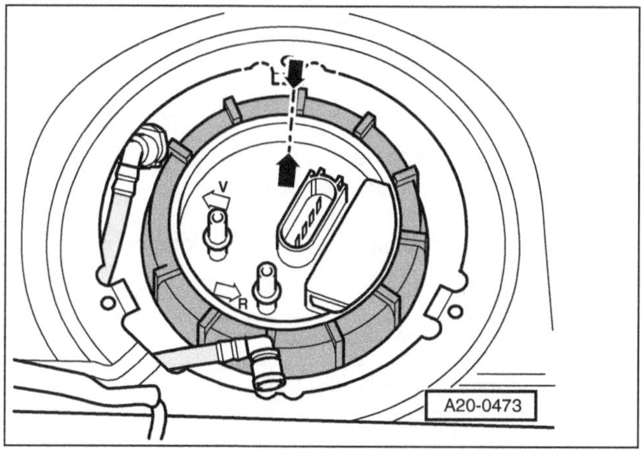

◄ Install fuel delivery unit with **arrow** aligned with **arrow** on fuel tank.

– Tighten retaining ring by hand.

Continued for all vehicles

– Tighten fuel delivery unit retaining ring with special tool 3217.

Tightening Torque	
Fuel delivery unit retaining ring to fuel tank	80 Nm (59 ft-lb)

– Install fuel supply and return lines. Do not interchange lines. Check to make sure fuel lines are fastened securely.

– Install rear seat (coupe) or rear trim (roadster) in reverse order of removal. See **72 Seats** or **70 Trim–Interior**.

– Connect negative (GND) strap to battery negative (–) terminal. See **27 Battery, Alternator, Starter.**

BentleyPublishers.com
© 2006 BentleyPublishers.com—All Rights Reserved

Fuel level sensor, checking

– Remove fuel delivery unit. See **Fuel delivery unit, removing and installing**.

Front wheel drive vehicles

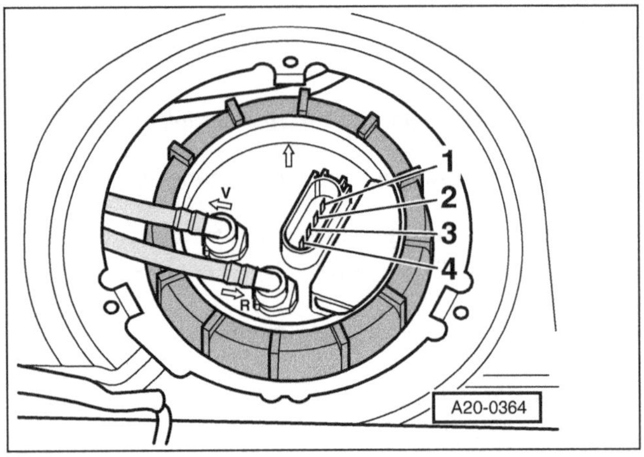

◄ Connect multimeter between terminals **2** and **3**.

– Position fuel level sensor float at lower and upper positions and measure resistance.

Fuel level float position (through 05/02)	Resistance
Lower (simulates empty fuel tank)	approx. 54 Ω
Upper (simulates full fuel tank)	approx. 290 Ω

Fuel level float position (from 05/02)	Resistance
Lower (simulates empty fuel tank)	approx. 290Ω
Upper (simulates full fuel tank)	approx. 54 Ω

A measured value of 0 Ω indicates a short circuit.
A measured value of ∞ Ω indicates an open circuit.

All wheel drive vehicles

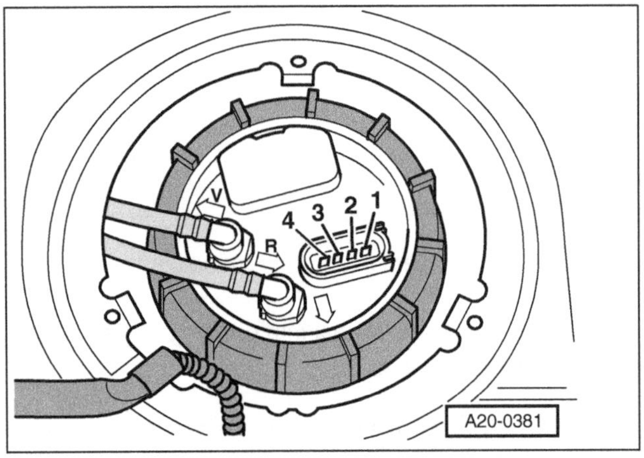

◄ Connect multimeter between terminals **2** and **3**.

– Position fuel level sensor float at lower and upper positions and measure resistance.

Fuel level float position (through 05/02)	Resistance
Lower (simulates empty fuel tank)	approx. 29 Ω
Upper (simulates full fuel tank)	approx. 147 Ω

Fuel level float position (from 05/02)	Resistance
Lower (simulates fuel empty fuel tank)	approx. 147 Ω
Upper (simulates full fuel tank)	approx. 29 Ω

A measured value of 0 Ω indicates a short circuit.
A measured value of ∞ Ω indicates an open circuit.

– Repeat procedure for fuel level sensor 2 (G169), located in left side of fuel tank.

Fuel level float position (through 05/02)	Resistance
Sensor 2, lower (simulates empty fuel tank)	approx. 27 Ω
Sensor 2, upper (simulates full fuel tank)	approx. 140 Ω

Fuel level float position (from 05/02)	Resistance
Sensor 2, lower (simulates empty fuel tank)	approx. 140 Ω
Sensor 2, upper (simulates full fuel tank)	approx. 27 Ω

BentleyPublishers.com
© 2006 BentleyPublishers.com—All Rights Reserved

Continued for all vehicles

– If specifications are not obtained, replace fuel level sensor.

Fuel delivery unit harness connector pin identification is provided below for further testing.

Fuel delivery unit connector pin i.d.	Pin number
Fuel pump voltage supply (B+)	1
Fuel sender ground (–)	2
Sender to fuel gauge signal	3
Fuel pump ground (–)	4

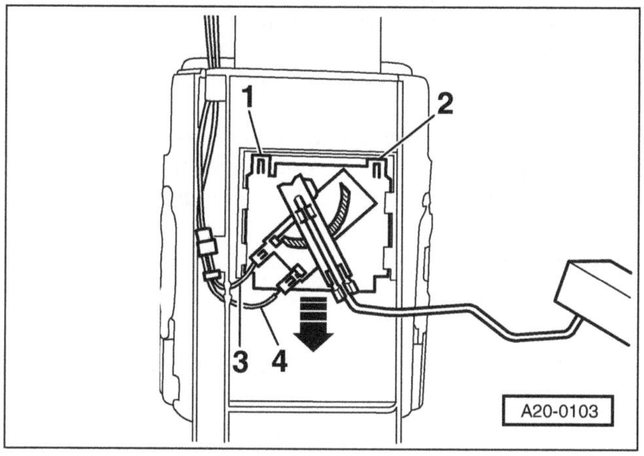

Fuel level sensor, removing and installing

– Remove fuel delivery unit. See **Fuel delivery unit, removing and installing**.

< Disengage electrical connector prongs (**3** and **4**) and disconnect wires.

– Pry off retaining straps (**1** and **2**) using small screwdriver and remove fuel level sensor toward bottom (**arrow**).

– Install in reverse order of removal.

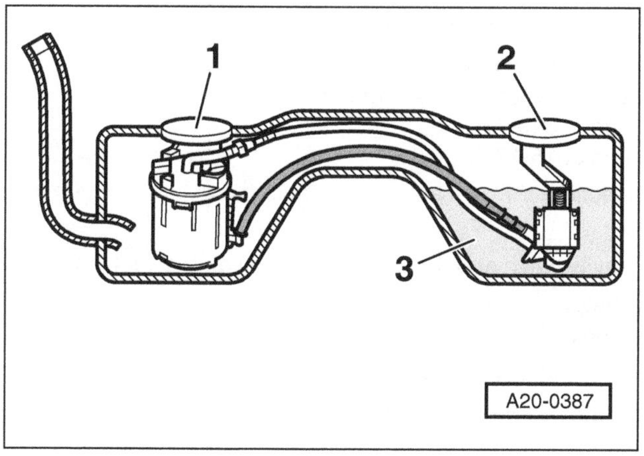

Secondary fuel pump function (vehicles with all wheel drive)

The primary fuel pump creates a pressure stream that is forced into the secondary fuel pump via a jet. This pressure causes fuel in the left side of the tank to be drawn into the right side of the tank.

< If the secondary fuel pump (**2**) is faulty the vehicle may not run when the fuel gauge indicates the tank is about ¼ full. In this case, up to 15 liters (4 gallons) of fuel (**3**) can remain in the left side of the tank that cannot be drawn out by the primary fuel pump (**1**).

If the vehicle will not run due to lack of fuel, but the tank is about ¼ full, proceed as follows:

– Check primary fuel pump. See **Fuel pump, electrical testing**.

– If primary fuel pump is OK, but no fuel is being pumped, add about 5 liters (1 gallon) of fuel to tank.

– Start engine.

– If engine now starts, replace secondary fuel pump. See **Fuel delivery unit, removing and installing**.

BentleyPublishers.com
© 2006 BentleyPublishers.com—All Rights Reserved

Evaporative Emissions Control (EVAP) System

Vehicles covered by this manual are equipped with evaporative emissions system control and monitoring. This system includes on-board refueling vapor recovery (ORVR) and a leak detection pump (LDP). ORVR filters fuel vapors through the evaporative emission control (EVAP) canister to prevent these vapors from escaping directly into the atmosphere.

The engine control module (ECM) determines when the appropriate conditions are met and activates the LDP to pump a small amount of air into fuel tank and lines. The pump runs until a predetermined pressure or time is reached. If the system does not hold pressure, a diagnostic trouble code (DTC) is stored in ECM memory and the malfunction indicator light (MIL) is illuminated.

Often, when the MIL illuminates after refueling, it is due to the filler cap being inadvertently left off or loose. Proper reinstallation of the cap allows the system to maintain pressure and turn the MIL off.

Begin EVAP system basic troubleshooting with a visual inspection of hoses and components with particular attention paid to fuel cap and seal. For additional testing use a diagnostic scan tool.

EVAP system overview

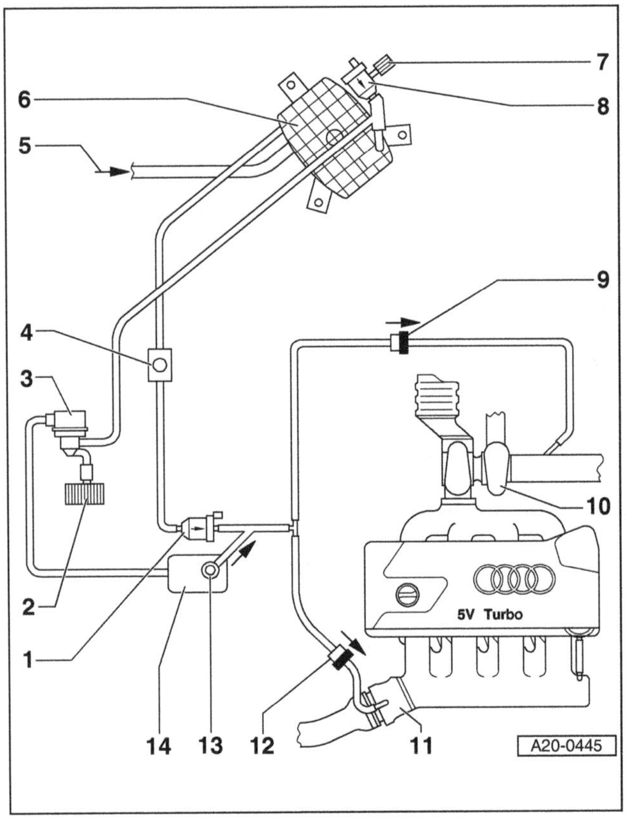

NOTE —
- *1.8L (engine code: ATC, AWP) shown, other engines are similar.*

1. EVAP canister purge regulator valve (N80)
    - Located in engine compartment near power steering reservoir
    - Test resistance at 2-pin connector, specification: 22 - 30 Ω
2. Alr filter for leak detection pump
3. Leak detection pump (V144)
    - Test resistance at harness connector
      Pins 1 and 3: 640-720 Ω
      Pins 2 and 3: 15-17 Ω
4. Test connection
5. Vent line
    - From gravity valve and pressure retention valve at fuel tank
6. EVAP canister
    - Located below spare tire well
7. Air filter for EVAP canister purge regulator valve
8. EVAP canister purge regulator valve (N115)
9. Check valve
    - 1.8L only
10. Turbocharger
    - 1.8L only
11. Throttle valve control module (J338)
12. Check valve
    - 1.8L only
13. Check valve
14. Vacuum reservoir
    - For leak detection pump

BentleyPublishers.com
© 2006 BentleyPublishers.com—All Rights Reserved

EVAP canister assembly

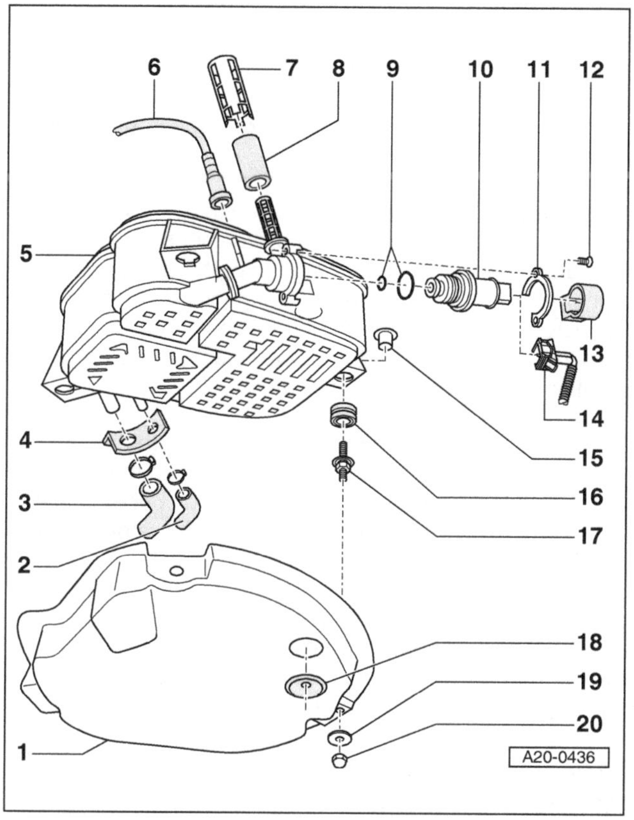

1. Cover
2. Vent line to EVAP canister purge regulator valve (N80)
3. Vent line from gravity valve at fuel tank
4. Sealing piece
5. EVAP canister
6. Vacuum line to leak detection pump (V144)
7. Bracket
8. Air filter for leak detection pump
9. O-rings
10. EVAP canister purge regulator valve (N115)
11. Retaining ring
12. Screw
    - 5 Nm (44 in-lb)
13. Bracket
14. 2-pin harness connector
15. Bushing
16. Grommet
17. Pin bolt
    - 16 Nm (12 ft-lb)
18. Grommet
19. Washer
20. Cap nut
    - 9 Nm (80 in-lb)

EVAP canister purge regulator valve, overview

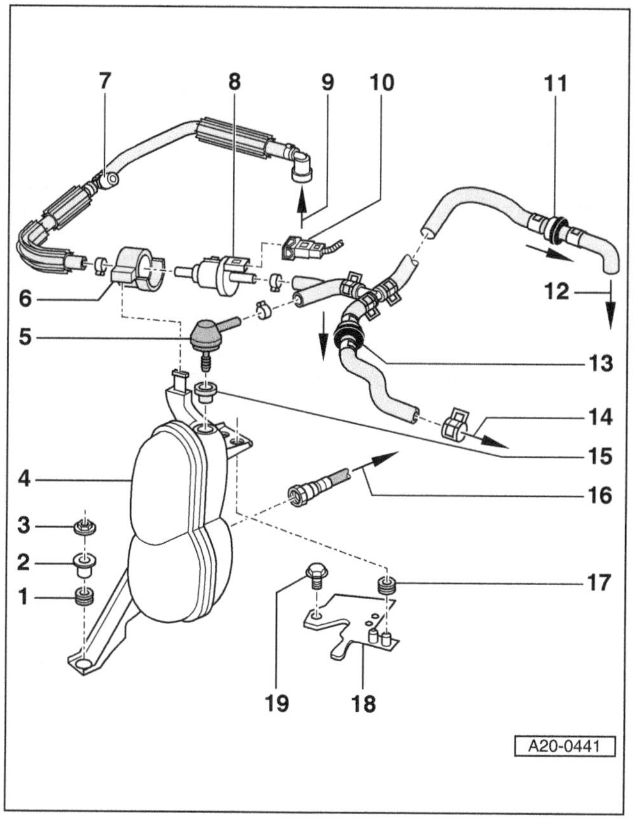

1. Grommet
2. Bushing
3. Nut
    - 8 Nm (17 in-lb)
4. Vacuum reservoir
5. Check valve
6. Bracket
7. Test connection
8. EVAP canister purge regulator valve (N80)
9. From EVAP canister
10. 2-pin harness connector
11. Check valve
12. To intake tubes of turbocharger (1.8L only)
13. Check valve (1.8L only)
    - Arrow points in direction of flow
14. To intake manifold
15. Grommet
16. To EVAP canister
17. Grommet
18. Bracket
19. Bolt
    - 8 Nm (71 in-lb)

BentleyPublishers
.com
© 2006 BentleyPublishers.com—All Rights Reserved

Electronic Power Control (EPC) System

Vehicles covered by this manual use an electronic power control (EPC) system. In the EPC system, commonly known as drive-by-wire, the throttle valve is not operated via a cable from the accelerator pedal, but rather by an electric motor in the throttle valve control module.

The accelerator pedal position is communicated to the engine control module via two accelerator pedal position sensors that are connected to the accelerator pedal. Operation of the throttle valve occurs throughout the entire rpm and load range via an electric motor (throttle valve actuator) in the throttle valve control module. The throttle valve is operated by the throttle valve actuator according to the specifications of the engine control module (ECM).

With the engine at standstill and ignition switched ON, the ECM activates the throttle valve actuator exactly according to the specifications of the accelerator pedal position sensor. For example, if the pedal is depressed halfway, the throttle valve actuator opens the throttle valve to the same degree.

EPC is a system containing all components that contribute to determining, regulating and monitoring the throttle valve position (i.e. accelerator pedal position sensor, throttle valve control module, EPC warning lamp, and engine control module).

Accelerator pedal assembly, removing and installing

- Remove driver's side storage compartment. See **70 Trim-Interior**.

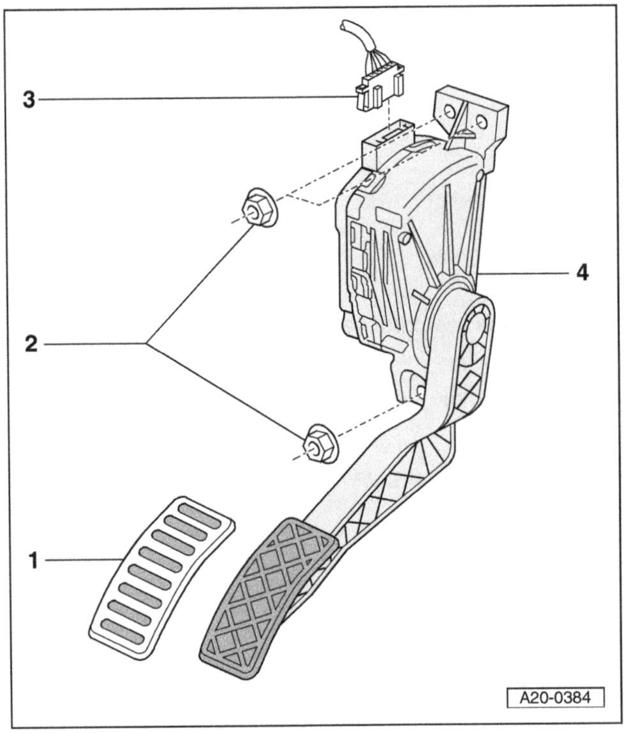

- ◀ Disconnect harness connector (**3**).
- Remove mounting nuts (**2**) and accelerator pedal assembly.
- Install in reverse order of removal.

Tightening torque	
Accelerator pedal to body	10 Nm (7 ft-lb)

- If accelerator pedal assembly was replaced in vehicles with automatic transmission or direct shift transmission, kickdown function must be adapted (learned) using diagnostic scan tool.

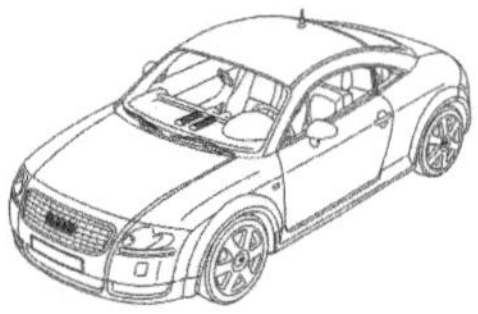

BentleyPublishers
.com
© 2006 BentleyPublishers.com—All Rights Reserved

21 Turbocharger and Intercooler (1.8L)

General

This section covers turbocharger and intercooler service and repair for the 1.8L engines. Turbocharger boost pressure is not adjustable and separate parts for the turbocharger are usually not available from Audi.

Engine	Engine code
1.8L 4-cylinder turbo (180 hp)	ATC, AWP
1.8L 4-cylinder turbo (225 hp)	AMU, BEA

NOTE—

- *The turbocharger and related components operate at very high temperature. Always allow the system to cool or use proper protective clothing to prevent severe burns.*
- *Thoroughly clean all joints, pipe unions and connections, before disconnecting or reconnecting the turbocharger or any of its related components.*
- *Take measures to prevent dust and dirt contamination. Cover all components with dust-free paper or seal them in plastic bags. Do not use cloth material. Avoid nearby use of compressed air. Do not move the car or work in dusty conditions while the turbocharger is open or removed. Install only clean components. Do not remove components from packaging until just before installation. Avoid use of components that have been stored loose.*

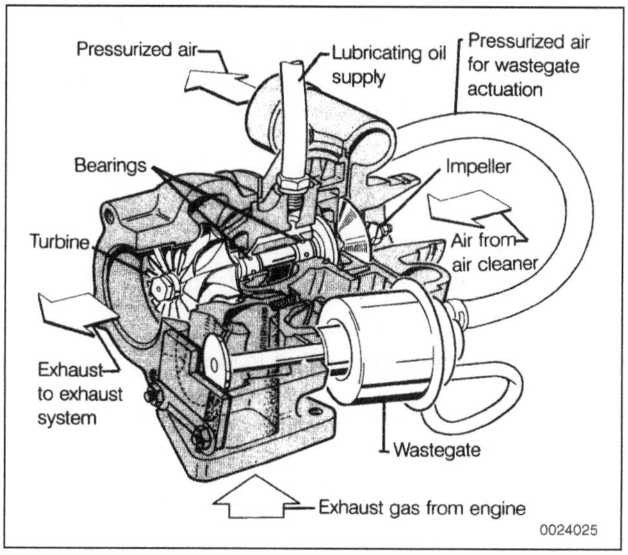

A cutaway view of a typical turbocharger is shown at left.

The turbocharger system consists of the following components:

- Exhaust gas turbocharger
- Charge air cooler (intercooler)
- Charge air pressure control system
- Air divert control system in overrun

BentleyPublishers.com
©2006 BentleyPublishers.com—All Rights Reserved

The exhaust-driven turbocharger vanes spin at very high speeds and are precisely balanced. The turbocharger is cooled and lubricated by engine oil with additional cooling supplied by a connection to the engine cooling system.

The intake air temperature is increased by compression, but reduced again in the charge air intercooler. Cool air has a higher density so the amount of fuel-air mixture entering the engine is greater too. The result is an increase in power output which can be especially beneficial for smaller displacement engines.

Charge pressure increases in proportion to turbocharger speed. To prolong the engine's life, the pressure is limited by the charge pressure control system.

The air divert control system prevents the turbocharger slowing down unnecessarily if the throttle valve closes suddenly. Maintaining high turbocharger charger speed allows it to deliver power quickly without "turbo lag".

The ECM calculates the charge pressure setpoint based on engine torque demand. Boost pressure is controlled by the wastegate bypass regulator valve operated by the Engine Control Module (ECM). When the boost pressure exceeds programmed values, the ECM opens the wastegate to bypass some of the exhaust gases around the turbine. If the ECM control fails, the maximum charge pressure is limited to a basic charge pressure (mechanical charge pressure).

When the wastegate is closed, the charge pressure rises. In the lower engine speed range, the turbocharger supplies the charge pressure required to develop high torque or the required volume of air.

When the throttle valve is closed, it produces a back pressure in the compressor circuit due to the charge pressure still present in the system. This causes the turbine and impeller to decelerate rapidly. When the throttle valve is again opened, the turbocharger speed must again be increased. The time that this takes is called "turbo lag". The air divert control in the overrun prevents turbo lag.

Hose connections

NOTE—
- *Turbocharger air hoses and lines as well as their connections must be free of oil and grease before re-installing.*
- *Do not assemble turbocharger air hoses and lines with lubricants containing oils, only coat with water (without additives) if necessary.*
- *Hose connections for turbocharger air system are secured by screw or spring-type clamps or locking clip style couplings.*
- *Pliers for spring clamps (VAS 5024 A) or hose clamp pliers (V.A.G 1921), or equivalent, are recommended for removing/installing spring clamps.*

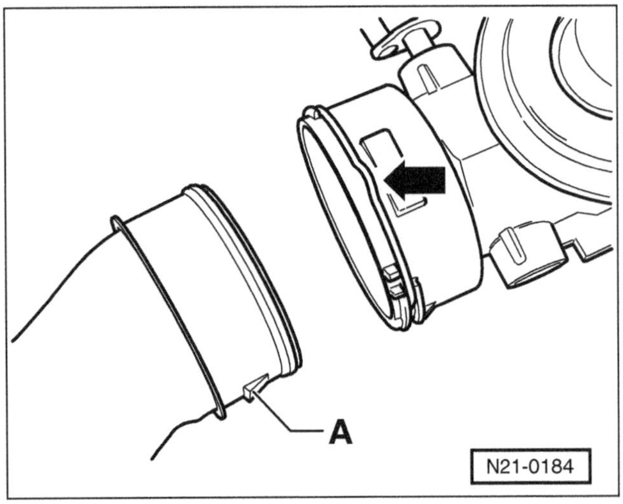

To release locking style couplings pull on locking clip (**arrow**).

– When assembling, make sure retaining tabs (**A**) are engaged securely.

– Pay attention to any installation marks on hoses and components.

BentleyPublishers.com
©2006 BentleyPublishers.com—All Rights Reserved

Turbocharger and Intercooler (Engine code: ATC, AWP)

NOTE—
- *Audi identifies electrical components by a letter and/or a number in the electrical schematics. See* **EWD Electrical Wiring Diagrams**. *These electrical identifiers are listed in parentheses as an aid to electrical troubleshooting.*

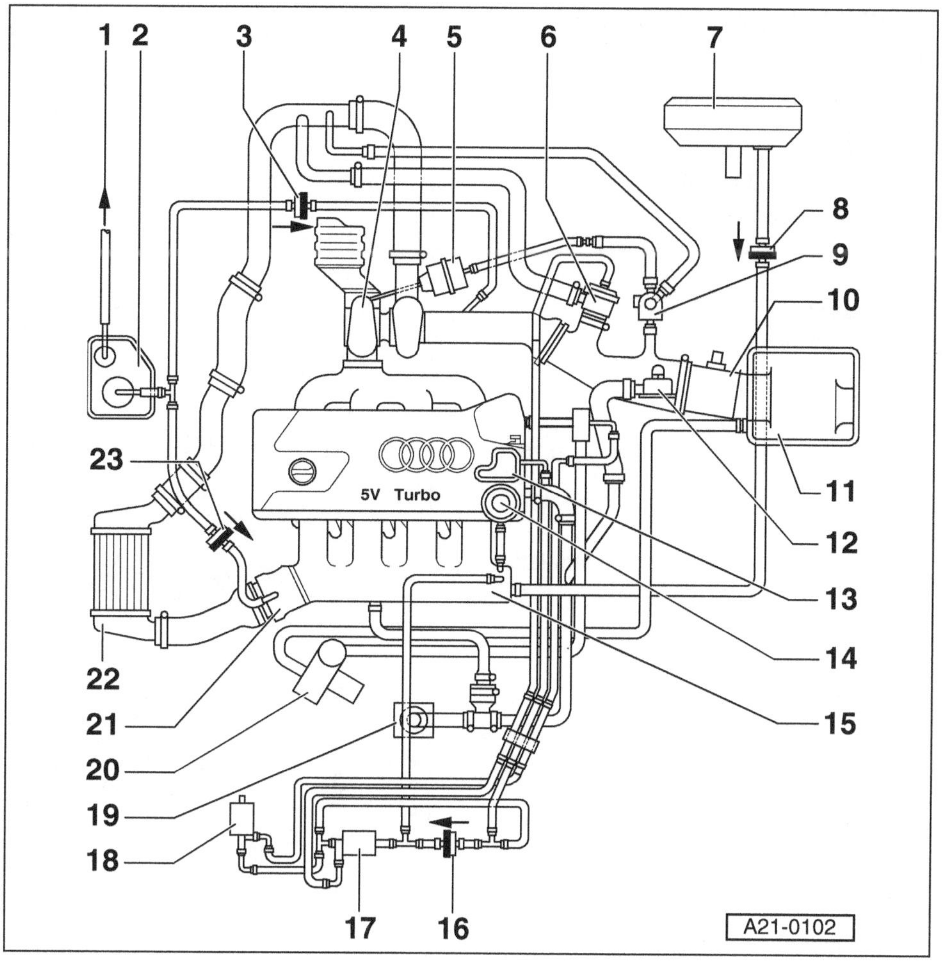

Turbocharger system overview

1. **From fuel tank**
2. **Evaporative emission (EVAP) canister**
   - With EVAP canister purge regulator valve (N80)
3. **Non-return valve for EVAP canister**
   - Between EVAP canister and intake line upstream of turbocharger
   - Install with light and dark sides in positions shown in illustration. **Arrow** points in direction of flow.
4. **Turbocharger**
5. **Pressure unit for charge air pressure control**
6. **Charge air pressure bypass valve**
7. **Brake servo**
8. **Non-return valve**
   - Between brake servo and intake manifold
   - Install with light and dark sides in positions shown in illustration. **Arrow** points in direction of flow.
9. **Wastegate bypass regulator valve (N75)**
   - Check internal resistance at 2-pin harness connector terminal: 25 - 35 Ω
10. **Mass Air Flow (MAF) sensor (G70)**
11. **Air cleaner**
12. **Crankcase breather pressure regulating valve**
13. **Vacuum reservoir**
    - Bolted to cylinder head cover
14. **Fuel pressure regulator**
15. **Intake manifold**
    - With Intake Air Temperature (IAT) sensor (G42)
16. **Non-return valve**
    - Install with light and dark sides in positions shown in illustration. **Arrow** points in direction of flow.
17. **Recirculating valve for turbocharger (N249)**
    - Check internal resistance at 2-pin harness connector terminal: 27 - 30 Ω
18. **Secondary air injection solenoid valve (N112)**
19. **Crankcase breather**
20. **Secondary air injection pump motor (V101)**
21. **Throttle valve control module (J338)**
22. **Charge air cooler**
    - With charge air pressure sensor (G31)
23. **Non-return valve**
    - Between EVAP canister and intake manifold
    - Install with light and dark sides in positions shown in illustration. **Arrow** points in direction of flow.

BentleyPublishers.com
© 2006 BentleyPublishers.com—All Rights Reserved

Turbocharger and Intercooler (Engine code: ATC, AWP)

Turbocharger assembly, part 1

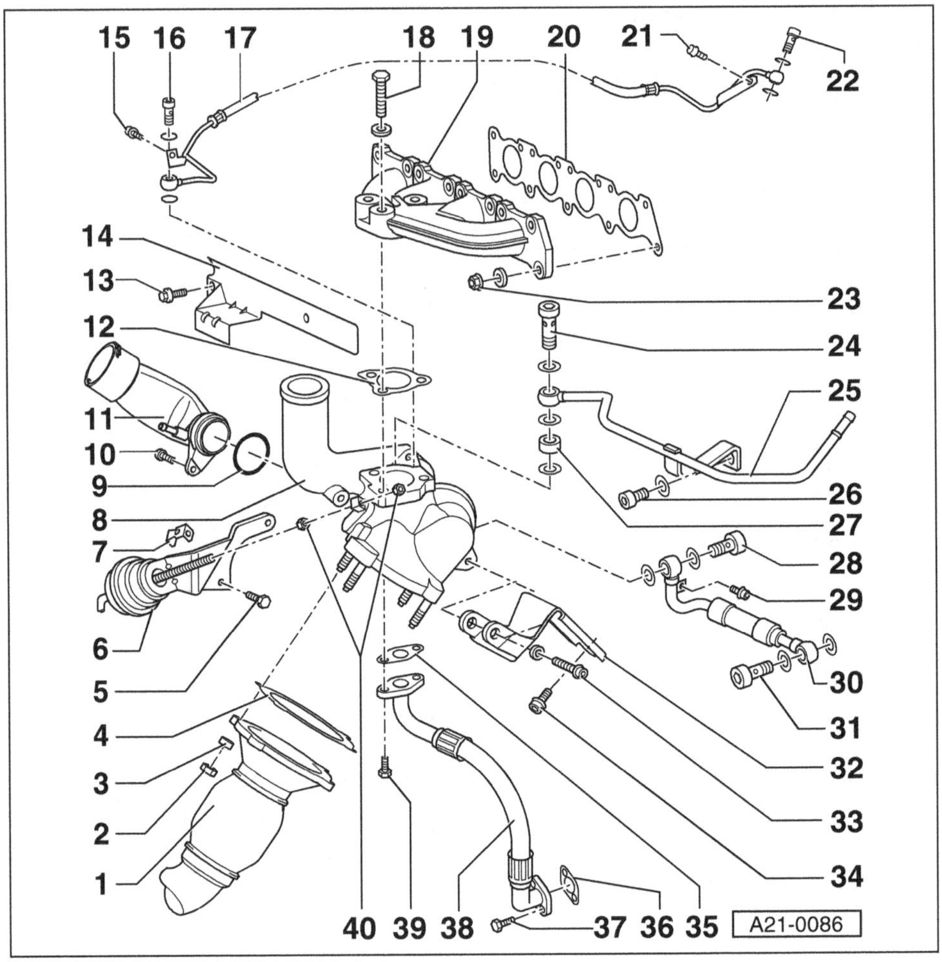

1. **Front exhaust pipe**
2. **Nut**
   - 40 Nm (30 ft-lb)
   - Always replace
3. **Spacer sleeve**
4. **Gasket**
   - Always replace
5. **Bolt**
   - 10 Nm (7 ft-lb)
   - Install with locking fluid D 000 600 A2
6. **Pressure unit**
   - For charge air pressure regulating valve
7. **Securing clip**
8. **Turbocharger**
9. **O-ring**
   - Always replace
10. **Bolt**
    - 20 Nm (15 ft-lb)
11. **Connection**
12. **Gasket**
    - Always replace
    - Note installed position
13. **Bolt**
    - 10 Nm (7 ft-lb)
14. **Heat shield**
15. **Bolt**
    - 10 Nm (7 ft-lb)
16. **Banjo bolt**
    - 30 Nm (22 ft-lb)
17. **Oil supply line**
    - From oil filter bracket
18. **Bolt**
    - 30 Nm (22 ft-lb)
    - Always replace
    - Coat threads and bolt head seating surface with G 000 500 (anti-seize compound)
19. **Exhaust manifold**
20. **Gasket**
    - Always replace
    - Note installed position
21. **Bolt**
    - 20 Nm (15 ft-lb)
22. **Banjo bolt**
    - 30 Nm (22 ft-lb)
23. **Nut**
    - 25 Nm (18 ft-lb)
    - Always replace
24. **Banjo bolt**
    - 35 Nm (26 ft-lb)
25. **Coolant return line**
26. **Bolt**
    - 25 Nm (18 ft-lb)
27. **Spacer sleeve**
28. **Banjo bolt**
    - 35 Nm (26 ft-lb)
29. **Bolt**
    - 10 Nm (7 ft-lb)
30. **Coolant supply line**
31. **Banjo bolt**
    - 35 Nm (26 ft-lb)
32. **Bracket**
33. **Bolt**
    - 30 Nm (22 ft-lb)
    - Use only genuine bolts
34. **Bolt**
    - 25 Nm (18 ft-lb)
35. **Gasket**
    - Always replace
36. **Gasket**
    - Always replace
37. **Bolt**
    - 10 Nm (7 ft-lb)
38. **Oil return line**
    - To oil pan
39. **Bolt**
    - 10 Nm (7 ft-lb)
40. **Nuts**
    - 10 Nm (7 ft-lb)
    - Install with locking fluid D 000 600 A2

BentleyPublishers.com
© 2006 BentleyPublishers.com—All Rights Reserved

Turbocharger assembly, part 2

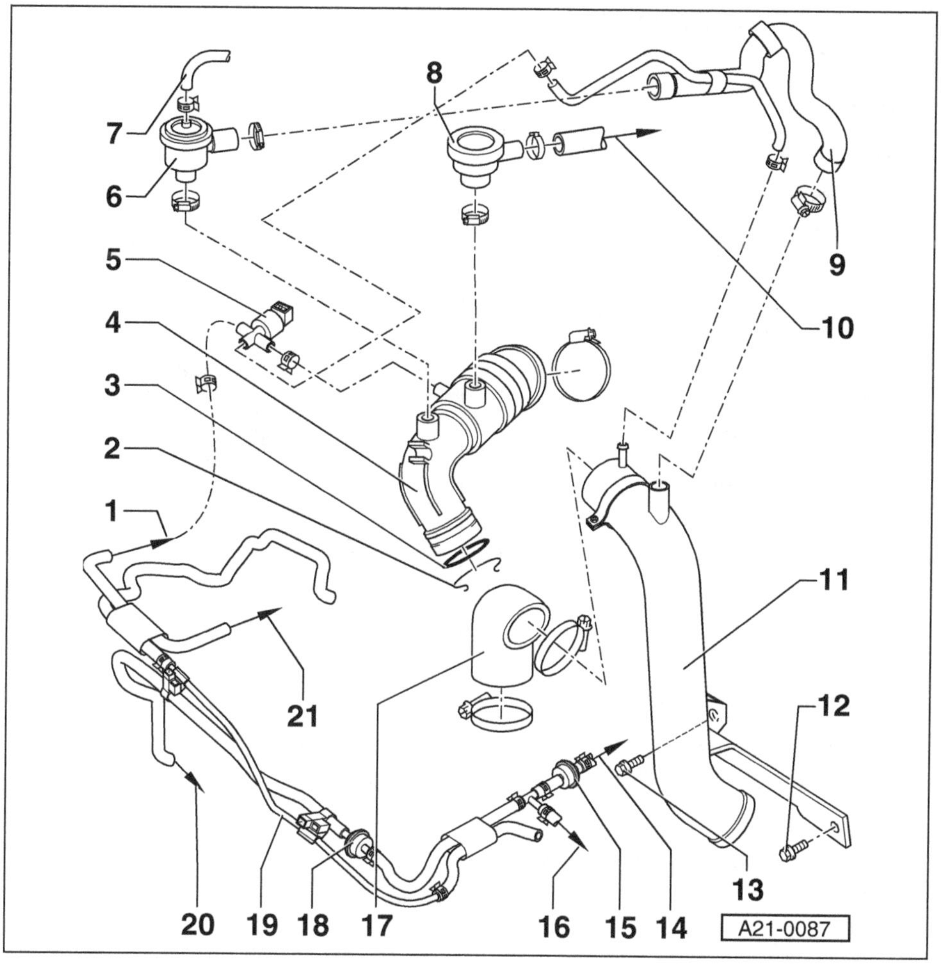

1. **To charge air pressure control valve**
2. **Locking clip**
3. **O-ring**
   - Always replace
4. **Air intake hose**
   - To connection on turbocharger
5. **Wastegate bypass regulator valve (N75)**
6. **Charge air pressure bypass valve**
   - Check the charge air pressure bypass valve if the engine is not producing full power, or jerking when the throttle is opened and closed. See **Charge air pressure valve, checking**
7. **Hose**
   - To recirculating valve for turbocharger (N249)
8. **Pressure regulating valve**
   - For crankcase breather
9. **Hose**
   - Between charge air pressure bypass valve and upper air line
10. **To crankcase breather**
11. **Upper air line**
12. **Bolt**
    - 25 Nm (18 ft-lb)
13. **Bolt**
    - 25 Nm (18 ft-lb)
14. **To throttle valve housing**
15. **Non-return valve**
    - Install with light and dark sides in positions shown in illustration
16. **To EVAP canister purge regulator valve (N80)**
17. **Hose**
18. **Non-return valve**
    - Install with light and dark sides in positions shown in illustration
19. **Coolant line**
20. **To connection on turbocharger**
21. **To pressure unit for charge air pressure control valve**

Charge air pressure bypass valve, checking

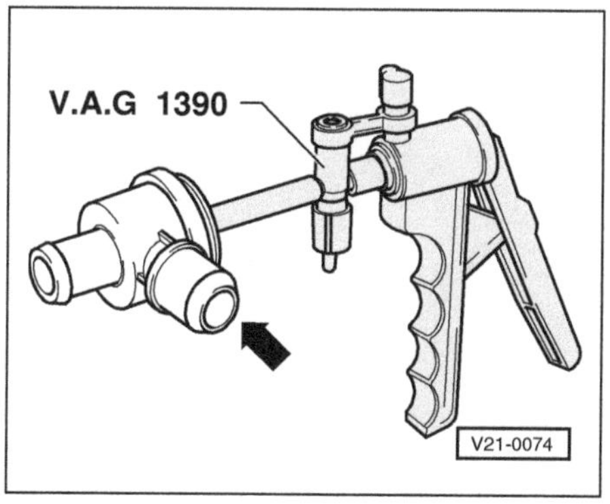

◄ Connect vacuum pump V.A.G. 1390 to charge air pressure bypass valve.

– Operate vacuum pump.
  - Charge air pressure bypass valve (**arrow**) should open.

– Release vacuum after about 30 seconds
  - Charge air pressure bypass valve (**arrow**) should close.

– If the charge air pressure bypass valve does not open and close as specified, or if the valve plate does not seal properly when the valve is closed, replace charge air pressure bypass valve.

– Secure connections on valve with screw-type clips.

BentleyPublishers .com
© 2006 BentleyPublishers.com—All Rights Reserved

Turbocharger and Intercooler (Engine code: ATC, AWP)

Charge air cooling system assembly

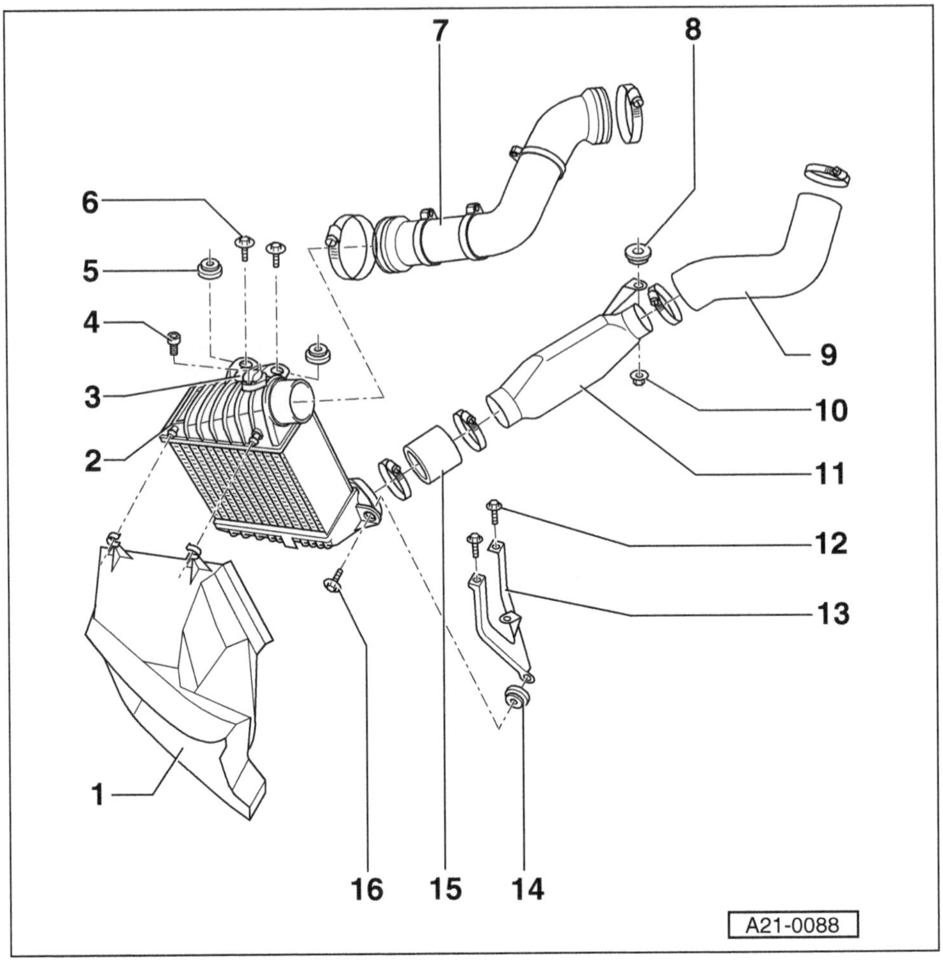

1. **Air duct**
2. **Charge air cooler (intercooler)**
3. **Charge air pressure sensor (G31)**
   - Check for voltage (approx. 5V) across harness connector at pins #1 and #3 with ignition on
   - Check for ground at harness connector pin #4
4. **Bolt**
   - 10 Nm (7 ft-lb)
5. **Rubber grommet**
   - With sleeve
6. **Bolt**
   - 10 Nm (7 ft-lb)
7. **Hose**
   - Between intake manifold and charge air cooler (intercooler)
8. **Rubber grommet**
   - With sleeve
9. **Hose**
   - Between upper air line and lower air line
10. **Bolt**
    - 10 Nm (7 ft-lb)
11. **Lower air line**
12. **Bolt**
    - 10 Nm (7 ft-lb)
13. **Bracket**
14. **Rubber grommet**
    - With sleeve
15. **Hose**
16. **Bolt**
    - 10 Nm (7 ft-lb)

BentleyPublishers.com
© 2006 BentleyPublishers.com—All Rights Reserved

Intercooler, removing and installing

- Remove lower sound insulation panel. See **03 Maintenance**.
- Remove front bumper. See **50 Body–Front**.
- Remove right side headlight assembly. See **94 Exterior Lights**.
- Remove EVAP canister. See **20 Fuel Storage and Supply**.
- Disconnect air hose from top of intercooler.

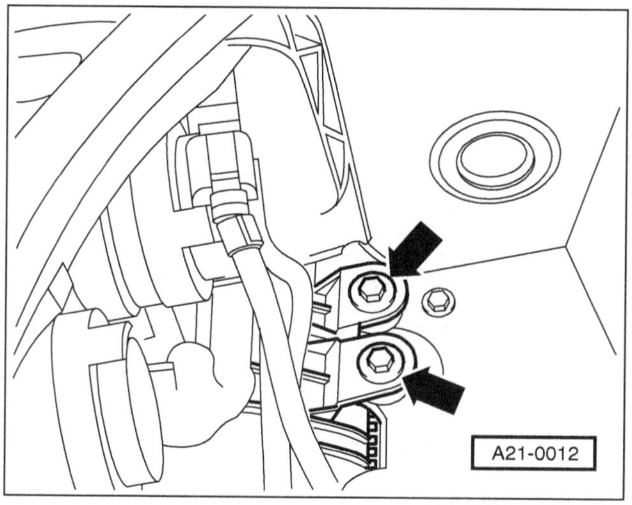

- Remove upper mounting bolts (**arrows**) for intercooler.
- Unclip cowl from intercooler.

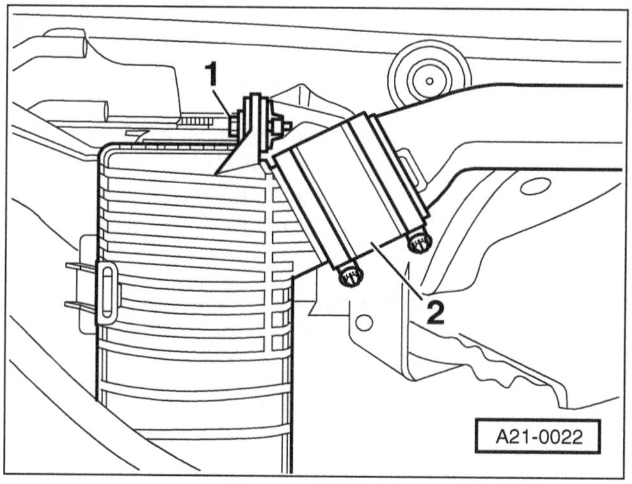

- Disconnect air hose (**2**) from bottom of intercooler.
- Remove lower mounting bolt (**1**) and intercooler from below.
- Installation is performed in reverse order of removal, noting the following:
  - Thoroughly clean all joints, pipe unions and connections, before disconnecting or reconnecting the turbocharger or any of its related components.
  - Take measures to prevent dust and dirt contamination. Cover all components with dust-free paper or seal them in plastic bags. Do not use cloth material. Avoid nearby use of compressed air. Do not move the car or work in dusty conditions while the turbocharger is open or removed. Install only clean components. Do not remove components from packaging until just before installation. Avoid use of components that have been stored loose.
- Adjust headlights as necessary. See **94 Exterior Lights**.

BentleyPublishers
.com
© 2006 BentleyPublishers.com—All Rights Reserved

Turbocharger and Intercooler (Engine code: AMU, BEA)

NOTE—

- *Audi identifies electrical components by a letter and/or a number in the electrical schematics. See* **EWD Electrical Wiring Diagrams**. *These electrical identifiers are listed in parentheses as an aid to electrical troubleshooting.*

Turbocharger system overview

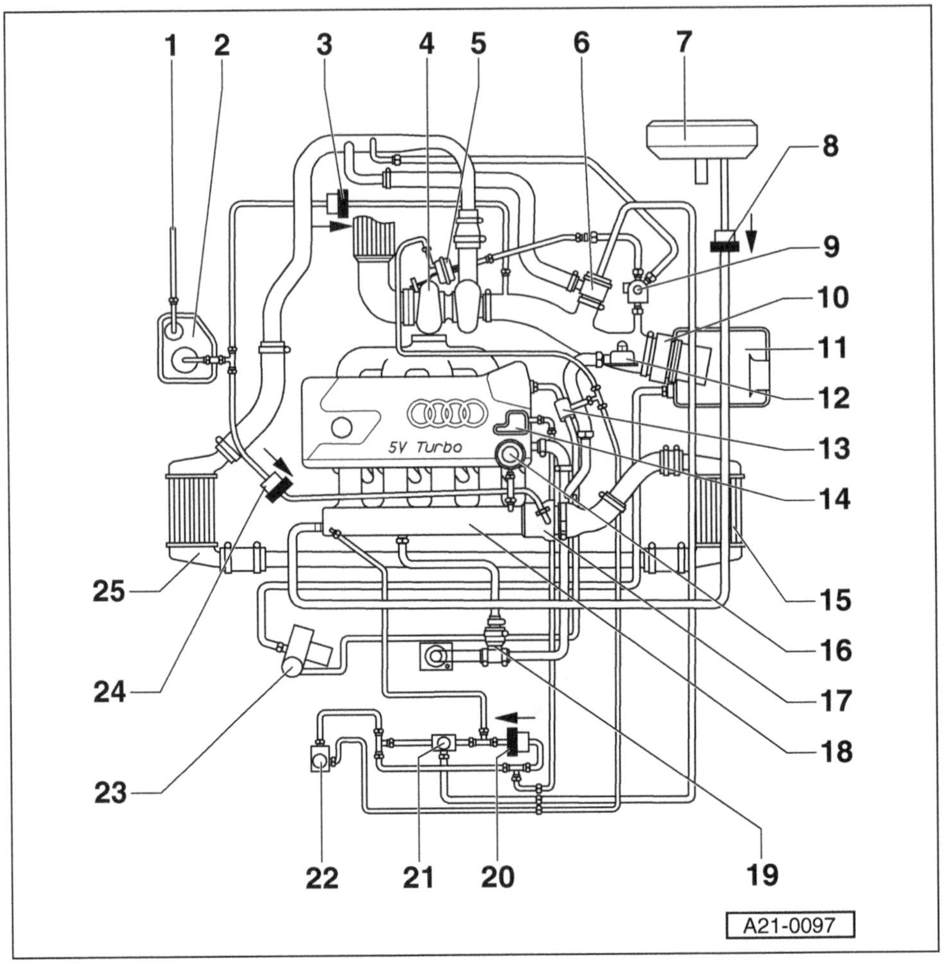

1. **From fuel tank**
2. **EVAP canister**
   - With EVAP canister purge regulator valve (N80)
3. **Non-return valve for EVAP canister**
   - Between EVAP canister and intake line upstream of turbocharger
   - Install with light and dark sides in positions shown in illustration. **Arrow** points in direction of flow.
4. **Turbocharger**
5. **Pressure unit for charge air pressure control**
6. **Charge air pressure bypass valve**
7. **Brake servo**
8. **Non-return valve**
   - Between brake servo and intake manifold
   - Install with light and dark sides in positions shown in illustration. **Arrow** points in direction of flow.
9. **Wastegate bypass regulator valve (N75)**
   - Check internal resistance at 2-pin harness connector terminal: 25 - 35 Ω
10. **Mass Air Flow (MAF) Sensor (G70)**
11. **Air cleaner**
12. **Crankcase breather pressure regulating valve**
13. **Mechanical secondary air valve**
14. **Vacuum reservoir**
    - Bolted to cylinder head cover
15. **Charge air cooler (intercooler)**
    - With charge air pressure sensor (G31)
    - Check G31 for voltage (approx. 5V) across harness connector at pins #1 and #3 with ignition on
    - Check G31 harness connector pin #4 for ground
16. **Fuel pressure regulator**
17. **Throttle valve control module (J338)**
18. **Intake manifold**
    - With Intake Air Temperature (IAT) sensor (G42)
19. **Crankcase breather**
20. **Non-return valve**
    - Install with light and dark sides in positions shown in illustration. **Arrow** points in direction of flow.
21. **Recirculating valve for turbocharger (N249)**
    - Check internal resistance at 2-pin harness connector terminal: 27 - 30 Ω
22. **Secondary air injection solenoid valve (N112)**
23. **Secondary air injection pump motor (V101)**
24. **Non-return valve**
    - Between activated charcoal filter and intake manifold
    - Install with light and dark sides in positions shown in illustration. **Arrow** points in direction of flow.
25. **Charge air cooler (intercooler)**

BentleyPublishers.com

© 2006 BentleyPublishers.com—All Rights Reserved

Turbocharger assembly

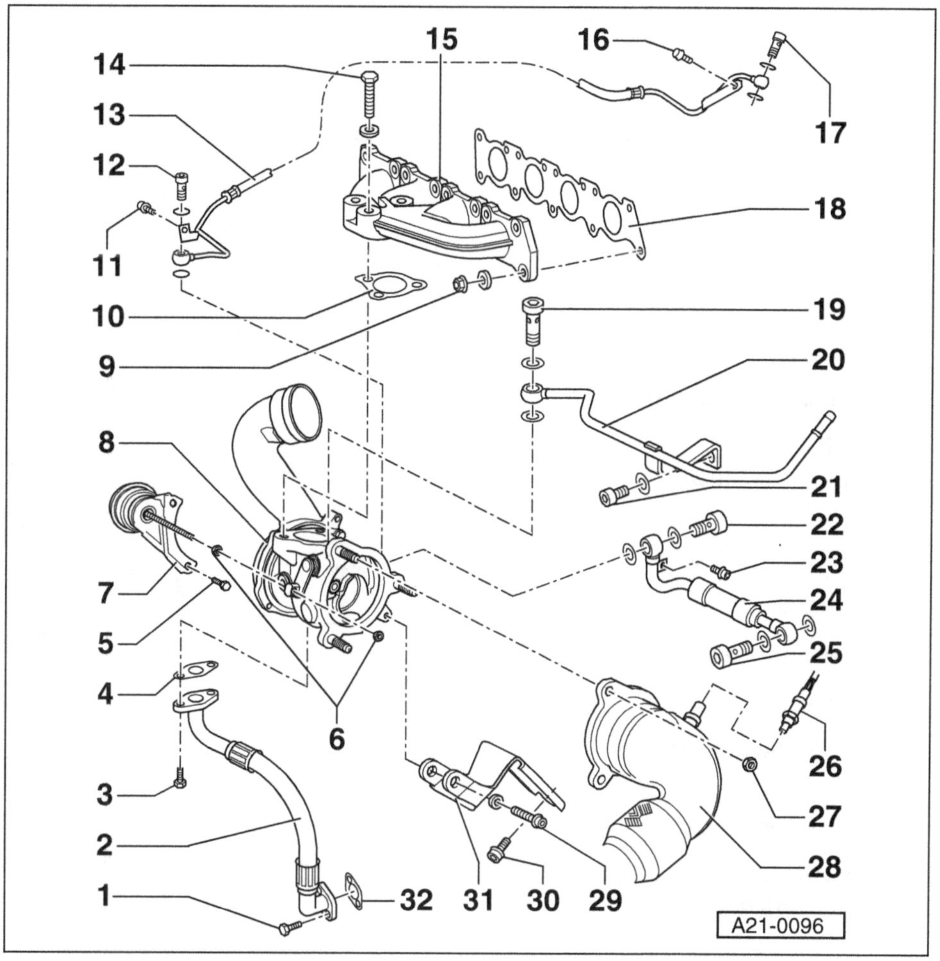

1. **Bolt**
   - 10 Nm (7 ft-lb)
2. **Oil return line**
   - To sump
3. **Bolt**
   - 10 Nm (7 ft-lb)
   - Install with locking fluid D 000 600 A2
4. **Gasket**
   - Always replace
5. **Bolt**
   - 10 Nm (7 ft-lb)
6. **Nut**
   - 10 Nm (7 ft-lb)
7. **Pressure unit**
   - For wastegate bypass regulator valve
8. **Turbocharger**
9. **Nut**
   - 25 Nm (18 ft-lb)
   - Always replace
10. **Gasket**
    - Always replace
    - Note installation position
11. **Bolt**
    - 10 Nm (7 ft-lb)
12. **Banjo bolt**
    - 30 Nm (22 ft-lb)
13. **Oil supply line**
14. **Bolt**
    - 30 Nm (22 ft-lb)
    - Always replace
    - Coat threads and bolt head seating surface with G 000 500 (anti-seize compound)
15. **Exhaust manifold**
16. **Bolt**
    - 10 Nm (7 ft-lb)
17. **Banjo bolt**
    - 35 Nm (26 ft-lb)
18. **Gasket**
    - Always replace
    - Note installation position
19. **Banjo bolt**
    - 35 Nm (26 ft-lb)
20. **Coolant return pipe**
21. **Bolt**
    - 20 Nm (15 ft-lb)
22. **Banjo bolt**
    - 30 Nm (22 ft-lb)
23. **Bolt**
    - 10 Nm (7 ft-lb)
24. **Coolant supply line**
25. **Banjo bolt**
    - 35 Nm (26 ft-lb)
26. **Oxygen sensor**
27. **Nut**
    - 40 Nm (30 ft-lb)
    - Always replace
28. **Front exhaust pipe**
    - Protect from damage by knocks and impact
    - With de-coupling element
    - De-coupling element may only be bent slightly - not more than 10°
29. **Bolt**
    - 30 Nm (22 ft-lb)
    - Use only genuine bolts
30. **Bolt**
    - 20 Nm (15 ft-lb)
31. **Bracket**
    - For turbocharger
32. **Gasket**
    - Always replace
    - Install with locking fluid D 000 600 A2

BentleyPublishers.com
© 2006 BentleyPublishers.com—All Rights Reserved

Turbocharger and Intercooler (Engine code: AMU, BEA)

Charge air cooling system assembly, part 1

1. **Pressure hose**
2. **To connection on charge air cooler**
3. **Bolt**
   - 20 Nm (15 ft-lb)
4. **Bottom bracket for pressure line**
5. **Pressure pipe**
6. **Top bracket for pressure line**
7. **Bolt**
   - 20 Nm (15 ft-lb)
8. **Bolt**
   - 20 Nm (15 ft-lb)
9. **Top bracket for pressure line**
10. **Pressure hose**
11. **Turbocharger**
12. **To crankcase breather**
13. **Pressure regulating valve for crankcase breather**
14. **Pressure unit for charge air pressure control**
15. **Air intake hose**
16. **To air cleaner**
17. **To Recirculating valve for turbocharger (N249)**
18. **Charge air pressure bypass valve**
    - Check the charge air pressure bypass valve if the engine is not producing full power, or jerking when the throttle is opened and closed. See **Charge air pressure valve, checking**
19. **Connecting piece**
20. **Recirculating valve for turbocharger (N249)**
21. **Bottom bracket for pressure line**

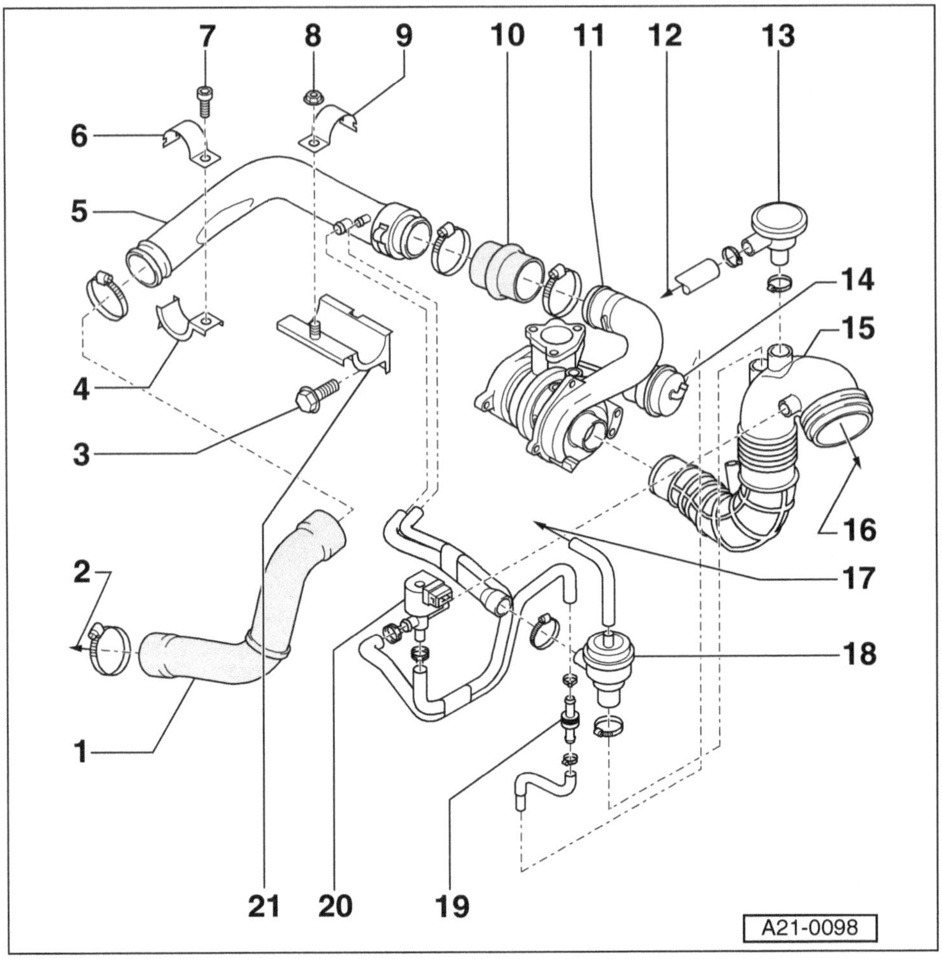

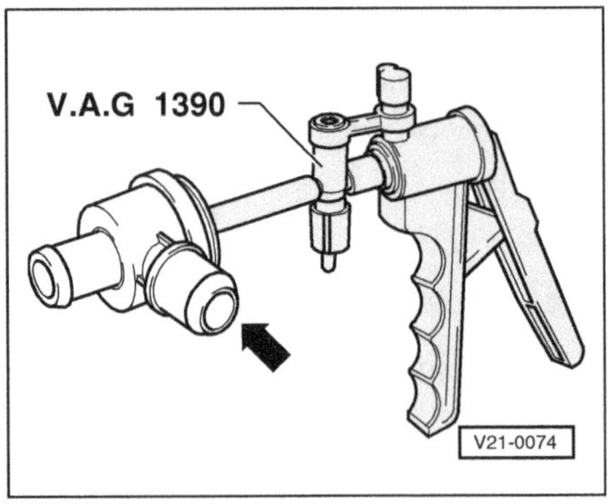

Charge air pressure bypass valve, checking

◀ Connect vacuum pump V.A.G. 1390 to charge air pressure bypass valve.

– Operate vacuum pump.
  - Charge air pressure bypass valve (**arrow**) should open.

– Release vacuum after about 30 seconds
  - Charge air pressure bypass valve (**arrow**) should close.

– If the charge air pressure bypass valve does not open and close as specified, or if the valve plate does not seal properly when the valve is closed, replace charge air pressure bypass valve.

– Secure connections on valve with screw-type clips.

BentleyPublishers
.com
© 2006 BentleyPublishers.com—All Rights Reserved

Charge air cooling system assembly, part 2

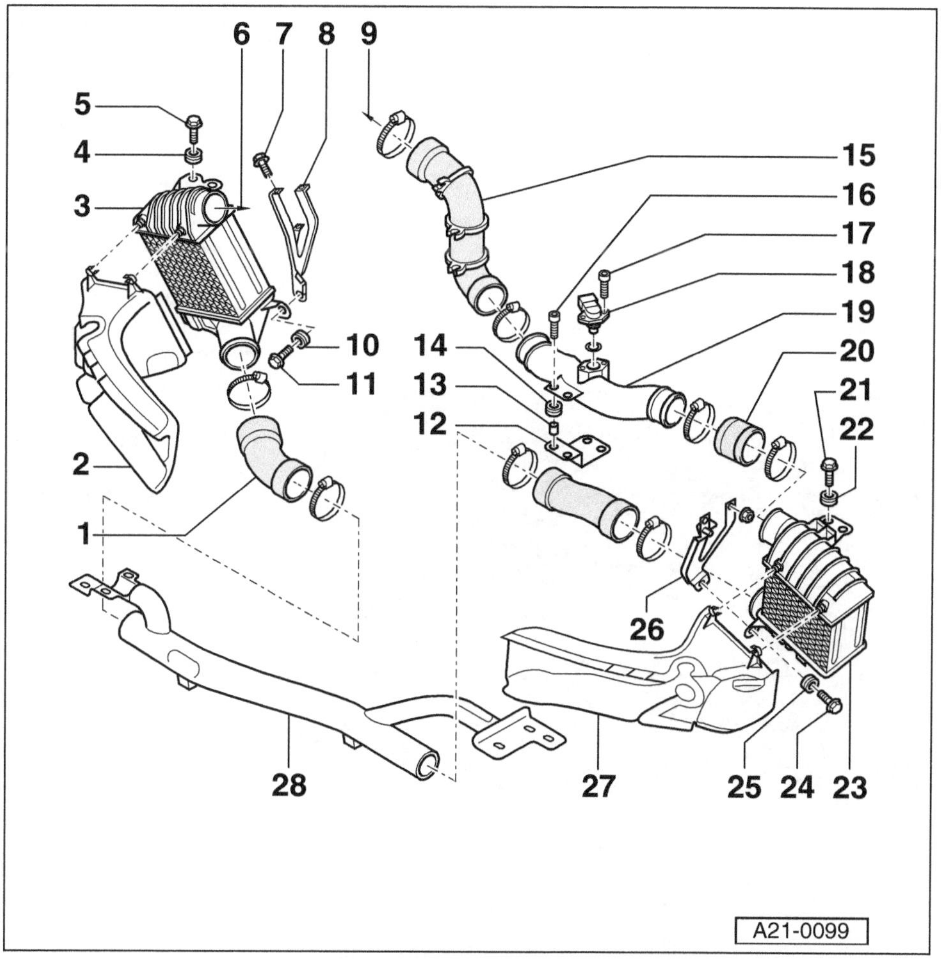

1. **Pressure hose**
2. **Air duct**
3. **Charge air cooler (Intercooler), right**
4. **Rubber grommet**
   - With sleeve
5. **Bolt**
   - 10 Nm (7 ft-lb)
6. **To turbocharger**
7. **Bolt**
   - 10 Nm (7 ft-lb)
8. **Mounting bracket**
9. **Connection on intake manifold**
10. **Rubber grommet**
    - With sleeve
11. **Bolt**
    - 10 Nm (7 ft-lb)
12. **Bracket**
    - Secured on longitudinal member
13. **Sleeve**
14. **Rubber grommet**
15. **Pressure hose**
    - With retaining clips for wire to charge air pressure sensor (G31)
16. **Bolt**
    - 10 Nm (7 ft-lb)
17. **Bolt**
    - 10 Nm (7 ft-lb)
18. **Charge air pressure sensor (G31)**
19. **Pressure line**
20. **Pressure hose**
21. **Bolt**
    - 10 Nm (7 ft-lb)
22. **Rubber grommet**
23. **Charge air cooler (intercooler), left**
24. **Bolt**
    - 10 Nm (7 ft-lb)
25. **Rubber grommet**
26. **Mounting bracket**
27. **Air duct**
28. **Connecting line between left and right intercoolers**

BentleyPublishers.com
© 2006 BentleyPublishers.com—All Rights Reserved

Intercoolers, removing and installing

– Remove lower sound insulation panels. See **03 Maintenance**.

– Remove front bumper. See **50 Body–Front**.

– Remove right and/or left side headlight assemblies. See **94 Exterior Lights**.

– Remove EVAP canister (only if removing right side intercooler). See **20 Fuel Storage and Supply**.

– Remove batter and battery tray (only if removing left side intercooler). See **27 Battery, Alternator, Starter**.

– Disconnect air hose(s) from top of intercooler(s).

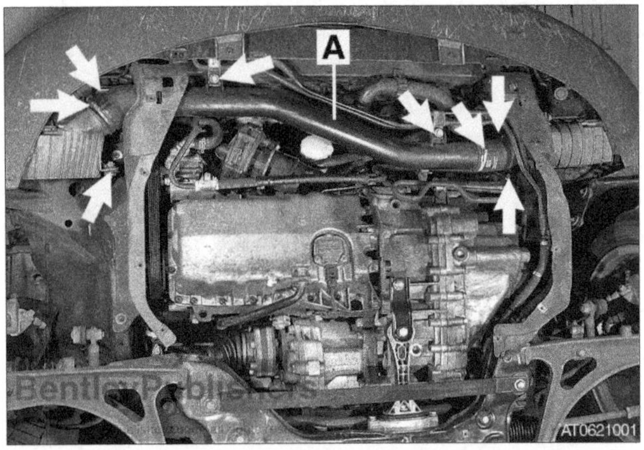

◄ Remove connecting line (**A**) between right and left intercoolers.

• Some mounting bolts (**arrows**) are located above connecting line.

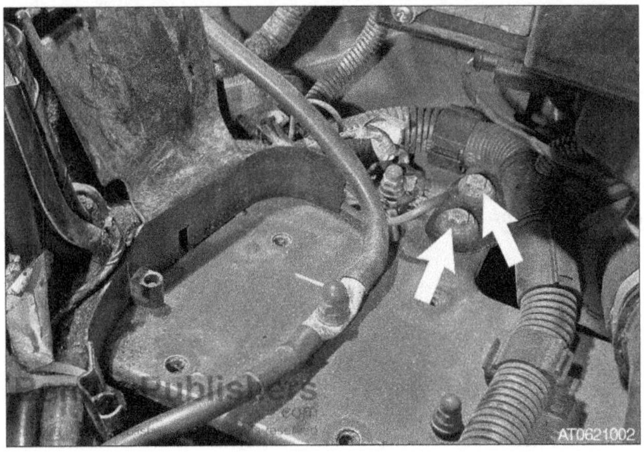

◄ Remove upper mounting bolts (**arrows**) for intercooler.

• Left side shown, right side similar.

– Remove intercooler from below.

– Installation is performed in reverse order of removal, noting the following:

• Thoroughly clean all joints, pipe unions and connections, before disconnecting or reconnecting the turbocharger or any of its related components.

• Take measures to prevent dust and dirt contamination. Cover all components with dust-free paper or seal them in plastic bags. Do not use cloth material. Avoid nearby use of compressed air. Do not move the car or work in dusty conditions while the turbocharger is open or removed. Install only clean components. Do not remove components from packaging until just before installation. Avoid use of components that have been stored loose.

– Adjust headlights as necessary. See **94 Exterior Lights**.

BentleyPublishers.com

© 2006 BentleyPublishers.com—All Rights Reserved

24a Fuel Injection (1.8L Engine)

GENERAL

This repair group covers fuel injection and engine management systems for the 1.8L gasoline engines. Special testing equipment may be necessary for some tests and repair procedures given here.

Engine	Engine code
1.8L 4-cylinder turbo (180 hp)	ATC, AWP
1.8L 4-cylinder turbo (225 hp)	AMU, BEA

Also see the following:

- **20 Fuel Storage and Supply** for fuel pump and fuel tank testing and repair, electronic power control (drive-by-wire or EPC) and evaporative emission control (EVAP) information.
- **26 Exhaust System** for catalytic converter, oxygen sensor and exhaust gas recirculation (EGR) information.
- **28 Ignition System** for ignition system testing and repair.
- **OBD On Board Diagnostics** for information on system fault diagnosis and use of diagnostic scan tool.

Motronic system

The Motronic fuel injection system combines ignition, fuel injection and electronic accelerator pedal functions into one system managed by the engine control module (ECM). There are no basic adjustments or settings that can be made to the systems without specialized equipment.

BentleyPublishers.com

©2006 BentleyPublishers.com—All Rights Reserved

The engine management systems are fully adaptive and feature built-in diagnostics that are capable of detecting and storing coded fault information. To gain access to system fault memory, attach a diagnostic scan too to the data link connector (DLC) (**arrow**) under left lower dashboard. See **OBD On Board Diagnostics** for more information. **Table a** lists some basic information about the fuel injection systems for the 1.8L engines covered by this manual.

Table a. Engine management applications and technical data	
1.8L 180 hp engine (code: ATC, AWP)	
Fuel injection system	Motronic ME7.5
Idle speed	640 - 920 rpm
Max. rpm	approx. 6800
Fuel injector delivery rate for 30 seconds	133 - 157 ml
1.8L 225 hp engine (code: AMU, BEA)	
Fuel injection system	Motronic ME7.5
Idle speed	700 - 820 rpm
Max. rpm	approx. 6800
Fuel injector delivery rate for 30 seconds	165 - 193 ml

Warning and Cautions

WARNING —
- *Fuel may be discharged during fuel system repairs. Do not smoke or work near heaters or other fire hazards. Have a fire extinguisher handy. Work only in a well-ventilated area.*
- *Wear suitable hand and eye protection when working with gasoline. Prolonged contact with fuel can cause illness and skin disorders.*

CAUTION—
- *Secure engine compartment fuel hoses with spring-type clips. The use of clamp or screw-type clips damages fuel lines and cause fuel leaks.*
- *Connect and disconnect wires and test equipment only with ignition switched off.*
- *Before making any electrical tests that require engine to be cranked using the starter, disable ignition system as described in* **28 Ignition System.**
- *Do not use sealants containing silicones. Particles of silicone drawn into engine do not burn and instead damage the oxygen sensors. Use only sealants and engine chemicals marked as safe for oxygen sensors.*
- *During testing, it is possible for the ECM to store a diagnostic trouble code (DTC) due solely to the testing procedure. Therefore, after completing repairs, clear the DTC memory using diagnostic scan tool.*
- *Cleanliness is essential when working on the fuel system. Thoroughly clean fuel line connections and surrounding areas before loosening. Avoid using compressed air. Avoid moving vehicle. Only install clean components.*

BentleyPublishers.com
© 2006 BentleyPublishers.com—All Rights Reserved

FUEL INJECTION COMPONENTS

NOTE—

- *Audi identifies electrical components by a letter and/or a number in the electrical schematics. See* **EWD Electrical Wiring Diagrams***. These electrical identifiers are listed in parentheses as an aid to electrical troubleshooting.*

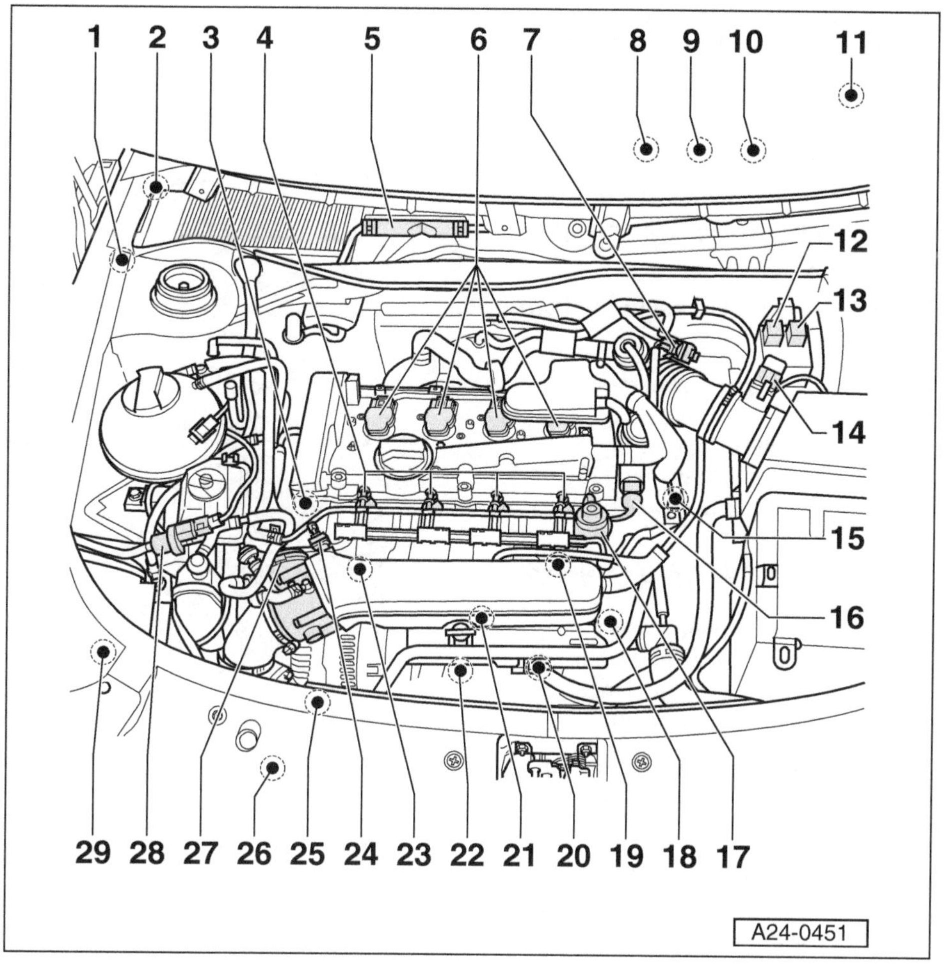

Component overview (engine code: ATC, AWP)

1. **Leak detection pump (V144)**
   - Located under right front wheel housing liner
2. **Oxygen sensor and oxygen sensor heater**
   - Oxygen sensor (G39) and heater (Z19) are located before catalyst
   - Oxygen sensor (G130) and heater (Z29) are located after catalyst
   - Testing and evaluation must be done with a diagnostic scan tool
   - Use special oxygen sensor socket for removal and installation (Audi special tool 3337 or equivalent)
3. **Camshaft position sensor 2 (G163)**
4. **Fuel injectors**
   - Cylinder #1 (N30)
   - Cylinder #2 (N31)
   - Cylinder #3 (N32)
   - Cylinder #4 (N33)
   - See **Fuel Rail and Injectors** in this repair group for removal, installation and testing
5. **Motronic engine control module (J220)**
   - See **Engine control module, removing and installing**
6. **Ignition coils with power output stages**
   - See **03 Maintenance, spark plugs, replacing** for removal and installation
   - Ignition coil #1 (N)
   - Ignition coil #2 (N128)
   - Ignition coil #3 (N158)
   - Ignition coil #4 (N163)
7. **Wastegate bypass regulator valve (N75)**
8. **Throttle position sensor (G79) and accelerator position sender (G185)**
   - Part of accelerator pedal assembly. For removal and installation, see **20 Fuel Storage and Supply, Electronic Power Control (EPC) System.**
9. **Brake light switch (F) and brake pedal position switch (F47)**
   - Located on top of brake pedal
10. **Clutch vacuum vent valve switch (F36)**
    - Located on top of clutch pedal
11. **Instrument cluster**
    - With (MIL) Malfunction Indicator Lamp (K83), aka. check engine light
    - With (EPC) Electronic Power Control lamp (K132)
12. **Power supply relay (J271) for engine control module**
13. **Power supply relay (J299) for secondary air injection pump**
14. **Mass Air Flow (MAF) sensor (G70)**
15. **Engine coolant temperature (ECT) sensor (G62)**
16. **Valve 1 for camshaft adjustment (N205)**
17. **Fuel pressure regulator**
18. **Engine speed rpm sensor (G28)**
19. **Knock sensor 2 (G66)**
20. **Turbocharger recirculating valve (N249)**
    - Installed directly on turbocharger
21. **3-pin harness connectors**
    - For knock sensors 1 and 2, and vehicle speed sensor
22. **Secondary air injection solenoid valve (N112)**
23. **Knock sensor 1 (G61)**

(continued on following page)

BentleyPublishers .com
©2006 BentleyPublishers.com—All Rights Reserved

Component overview (engine code: ATC, AWP)

(continued from previous page)

24. **Intake air temperature sensor (G42)**
25. **Secondary air injection pump motor (V101)**
26. **Power steering pressure switch (F88)**
27. **Throttle valve control module (J338)**
    - With throttle drive (power accelerator actuation) (G186)
    - With throttle drive angle sensor 1 (G187)
    - With throttle drive angle sensor 2 (G188)
    - After replacement, erase adaptation values and adapt ECM to throttle valve control module using diagnostic scan tool
28. **Evaporative emissions control (EVAP) canister purge regulator valve (N80)**
29. **Charge air pressure sensor (G31)**
    - Located on top of charge air cooler (intercooler)

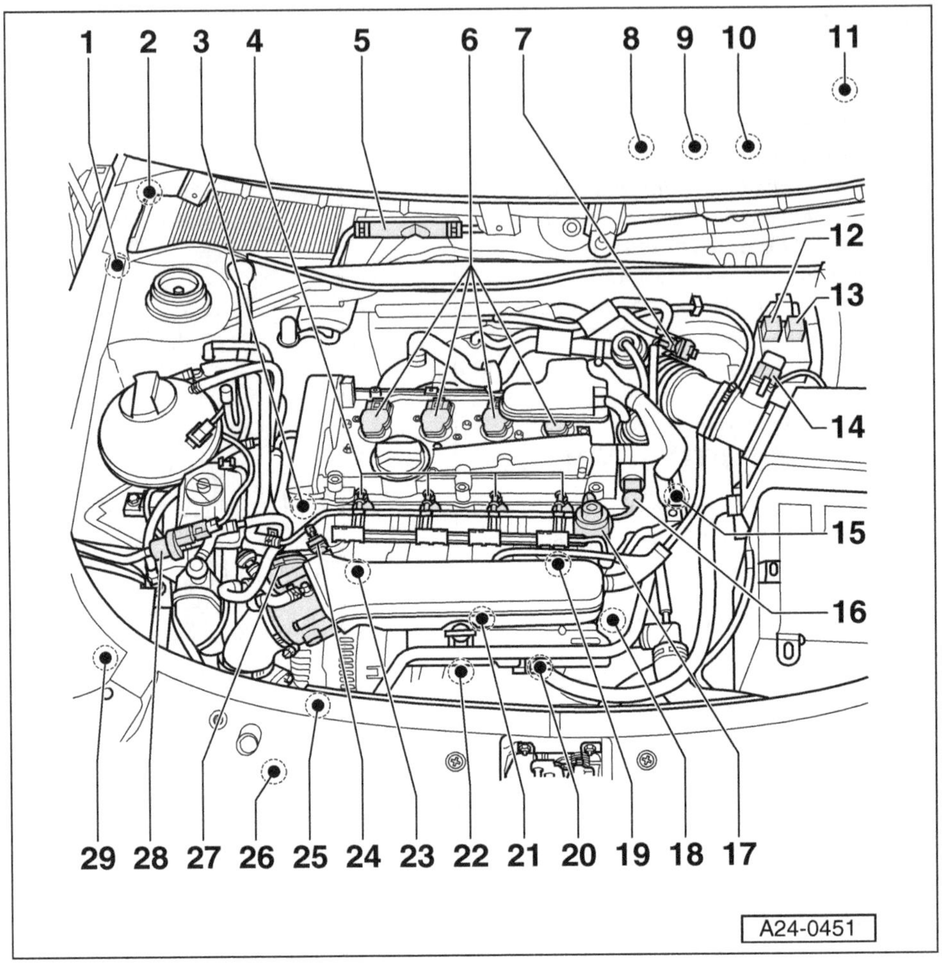

BentleyPublishers.com

©2006 BentleyPublishers.com—All Rights Reserved

Component overview (engine code: AMU, BEA)

NOTE—
- *Audi identifies electrical components by a letter and/or a number in the electrical schematics. See* **EWD Electrical Wiring Diagrams**. *These electrical identifiers are listed in parentheses as an aid to electrical troubleshooting.*

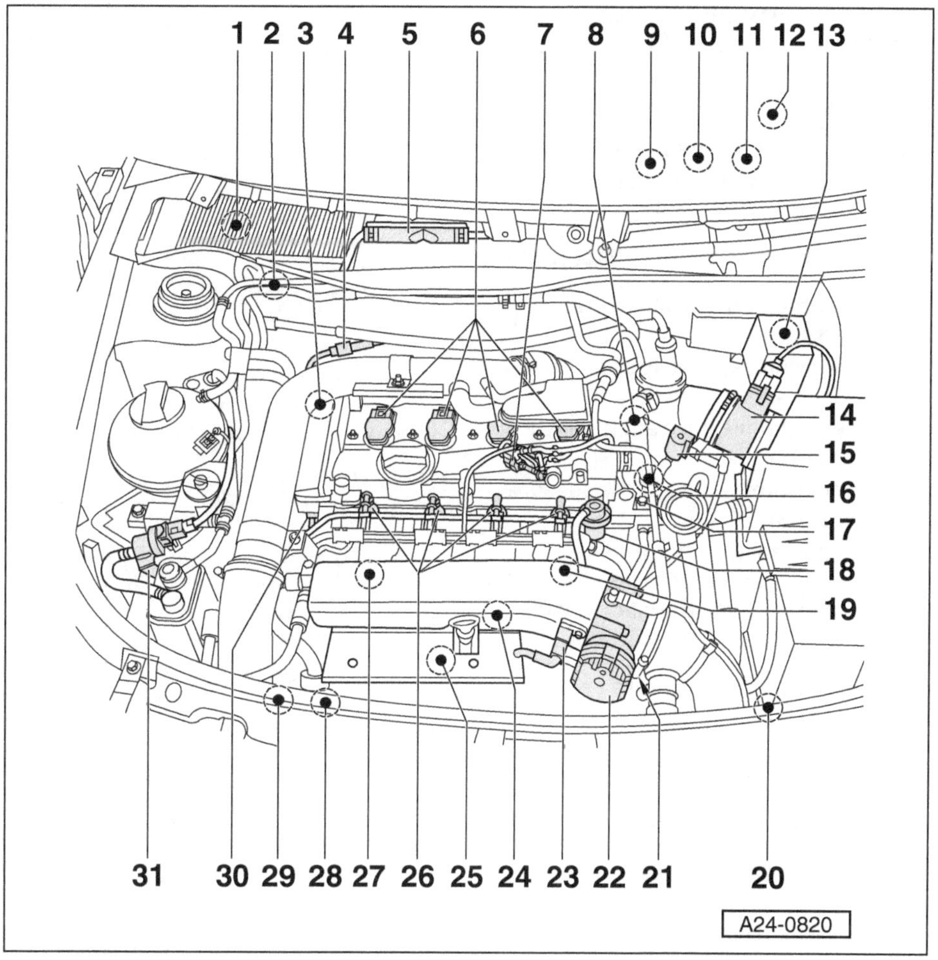

1. **Oxygen sensor (G130) and oxygen sensor heater (Z29)**
   - Located after catalyst
   - Testing and evaluation must be done with a diagnostic scan tool
   - Use special oxygen sensor socket for removal and installation (Audi special tool 3337 or equivalent)
2. **4-pin harness connector**
   - For G130 and Z29
3. **Oxygen sensor (G39) and oxygen sensor heater (Z19)**
   - Located before catalyst
   - Testing and evaluation must be done with a diagnostic scan tool
   - Use special oxygen sensor socket for removal and installation (Audi special tool 3337 or equivalent)
4. **6-pin harness connector**
   - For G39 and Z19
5. **Motronic engine control module (J220)**
   - See **Engine control module, removing and installing**
6. **Ignition coils with power output stages**
   - See **03 Maintenance, spark plugs, replacing** for removal and installation
   - Ignition coil #1 (N)
   - Ignition coil #2 (N128)
   - Ignition coil #3 (N158)
   - Ignition coil #4 (N163)
7. **Turbocharger recirculating valve (N249)**
   - Installed directly on turbocharger
8. **Combination valve for secondary air injection**
9. **Throttle position sensor (G79) and accelerator position sender (G185)**
   - Part of accelerator pedal assembly. For removal and installation, see **20 Fuel Storage and Supply, Electronic Power Control (EPC) System**.
10. **Brake light switch (F) and brake pedal position switch (F47)**
    - Located on top of brake pedal
11. **Clutch vacuum vent valve switch (F36)**
    - Located on top of clutch pedal
12. **Instrument cluster**
    - With (MIL) Malfunction Indicator Lamp (K83), aka. check engine light
    - With (EPC) Electronic Power Control lamp (K132)
13. **Power supply relays**
    - For engine control module (J271)
    - For secondary air injection pump (J299)
14. **Mass Air Flow (MAF) sensor (G70)**
15. **Wastegate bypass regulator valve (N75)**
16. **Engine coolant temperature (ECT) sensor (G62)**
17. **Valve 1 for camshaft adjustment (N205)**
18. **Fuel pressure regulator**
19. **Knock sensor 2 (G66)**

(continued on following page)

BentleyPublishers.com
©2006 BentleyPublishers.com—All Rights Reserved

Component overview (engine code: AMU, BEA)

(continued from previous page)

20. **Charge air pressure sensor (G31)**
    - Located on top of charge air cooler (intercooler)
21. **Engine speed rpm sensor (G28)**
22. **Throttle valve control module (J338)**
    - With throttle drive (power accelerator actuation) (G186)
    - With throttle drive angle sensor 1 (G187)
    - With throttle drive angle sensor 2 (G188)
    - After replacement, erase adaptation values and adapt ECM to throttle valve control module using diagnostic scan tool
23. **Intake air temperature sensor (G42)**
24. **3-pin harness connectors**
    - For knock sensors 1 and 2, and vehicle speed sensor
25. **Sensor 1 for exhaust temperature (G235)**
    - Located in turbocharger
26. **Fuel injectors**
    - Cylinder #1 (N30)
    - Cylinder #2 (N31)
    - Cylinder #3 (N32)
    - Cylinder #4 (N33)
    - See **Fuel Rail and Injectors** in this repair group for removal, installation and testing
27. **Secondary air injection pump motor (V101)**
28. **Knock sensor 1 (G61)**
29. **Power steering pressure switch (F88)**
30. **Camshaft position sensor 2 (G163)**
31. **Evaporative emissions control (EVAP) canister purge regulator valve (N80)**

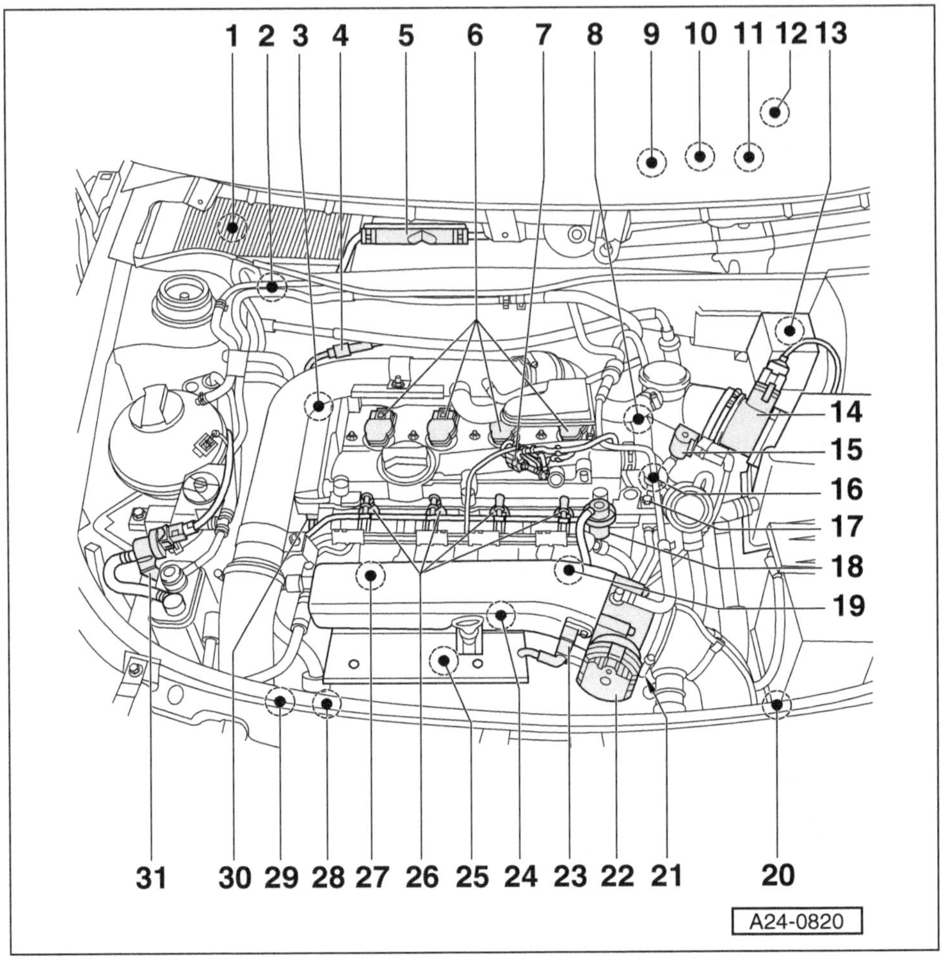

BentleyPublishers.com

© 2006 BentleyPublishers.com—All Rights Reserved

Component details and testing

Various components of the fuel injection system are identified here. For more component location information, including fuses and relays, see **97 Fuses, Relays, Component Locations**.

Applicable testing specifications are also provided. Testing components will typically involve the use of a digital multimeter and/or a diagnostic scan tool. For more information on electrical testing with a multimeter, see **9 Electrical System–General**. For more information on using a diagnostic scan tool, see **OBD On Board Diagnostics**.

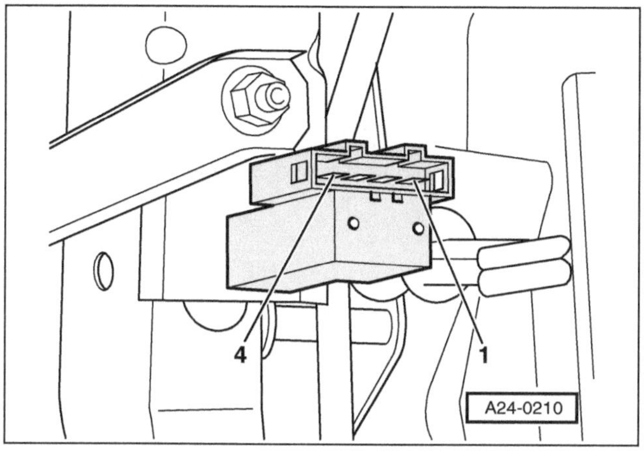

Brake light switch (F) and brake pedal switch (F47)

- Located above brake pedal.

– Disconnect harness connector to test resistance of switch.

– Connect multimeter to switch terminals 1 and 4.

- Specified value: ∞ Ω (no continuity)

– Depress brake pedal.

- Specified value: approx. 0 Ω

– Connect multimeter to switch terminals 2 and 3.

- Specified value: approx. 0 Ω

– Depress brake pedal.

- Specified value: ∞ Ω (no continuity)

– If specified values are not obtained, check switch wiring using wiring diagrams. See **EWD Electrical Wiring Diagrams**.

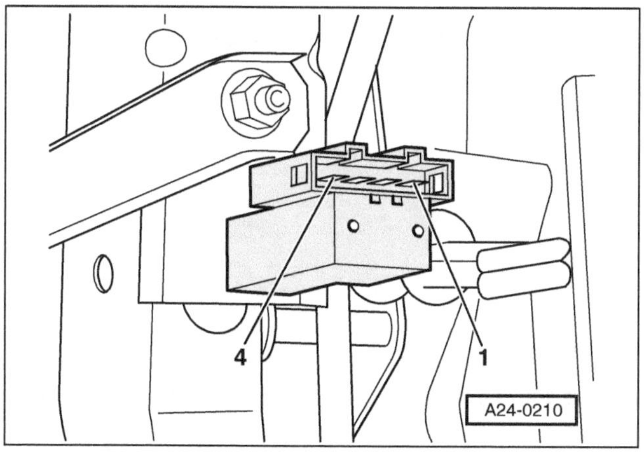

Clutch vacuum vent valve switch (F36)

- Located above clutch pedal (only for vehicles with manual transmission).

– Connect multimeter to switch terminals 1 and 2.

- Specified value: approx. 0 Ω

– Depress clutch pedal.

- Specified value: ∞ Ω (no continuity)

– If specified values are not obtained, check switch wiring using wiring diagrams. See **EWD Electrical Wiring Diagrams**.

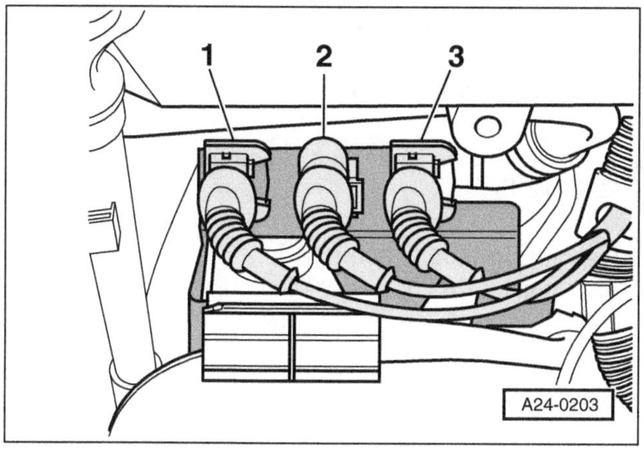

Connectors, 3-pin

- Located at front of intake manifold, under mounting bracket for secondary air injection solenoid valve, to right of engine oil dipstick.

1. For knock sensor 1 (G61)
2. For engine speed RPM sensor (G28)
3. For knock sensor 2 (G66)

BentleyPublishers.com

© 2006 BentleyPublishers.com—All Rights Reserved

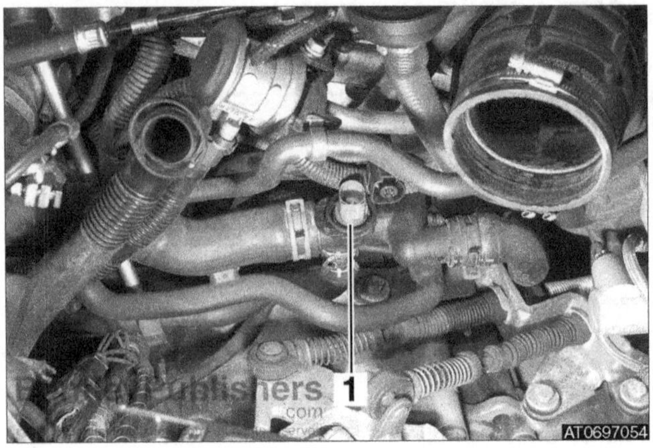

Engine coolant temperature (ECT) sensor (G2/G62)

- ECT sensor (**1**) is located on coolant flange, to left of cylinder head.
- Later style sensor shown on 2000 m.y. vehicle (engine code: AMU).

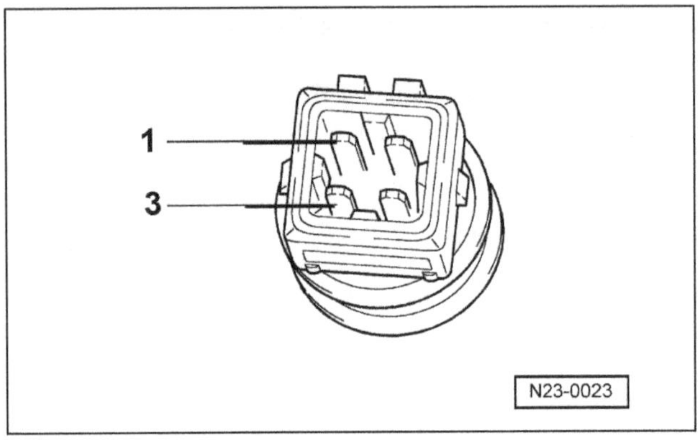

Early style ECT sensor.

- Measure resistance at terminals **1** and **3** of sensor.

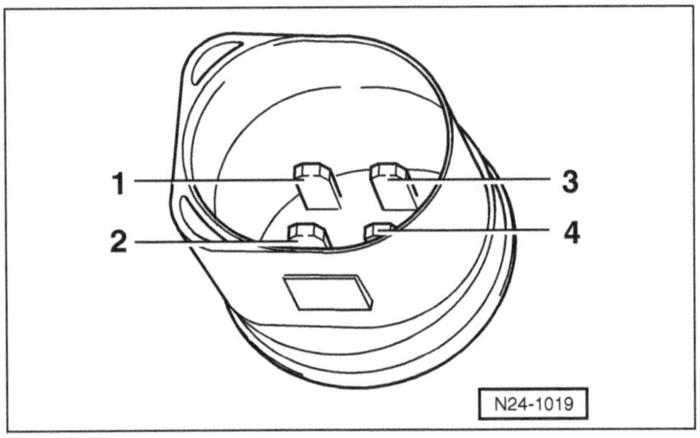

Later style ECT sensor.

- Measure resistance at terminals **3** and **4** of sensor.

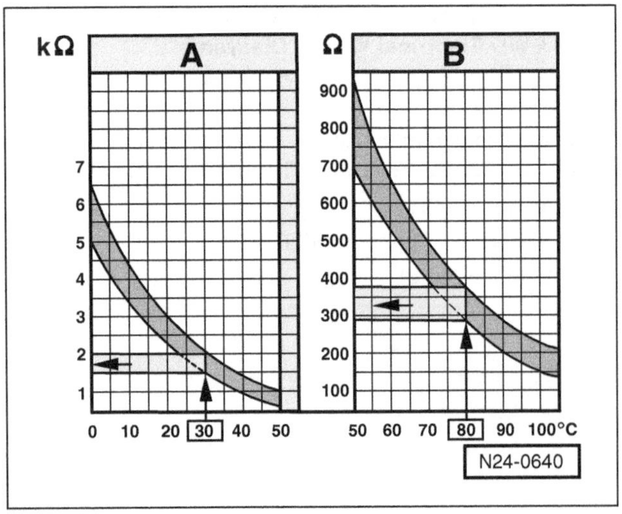

– Measure temperature of engine coolant using diagnostic scan tool or thermometer.

Compare ECT sensor resistance and coolant temperature readings to graph.

- Scale **A** shows resistance for coolant temperature range 0 - 50°C (32 - 122°F).
- Scale **B** shows resistance for coolant temperature range 50 - 100°C (122 - 212°F).
- Example:
  30°C corresponds to 1500 - 2000 Ω
  80°C corresponds to 275 - 375 Ω

BentleyPublishers.com

©2006 BentleyPublishers.com—All Rights Reserved

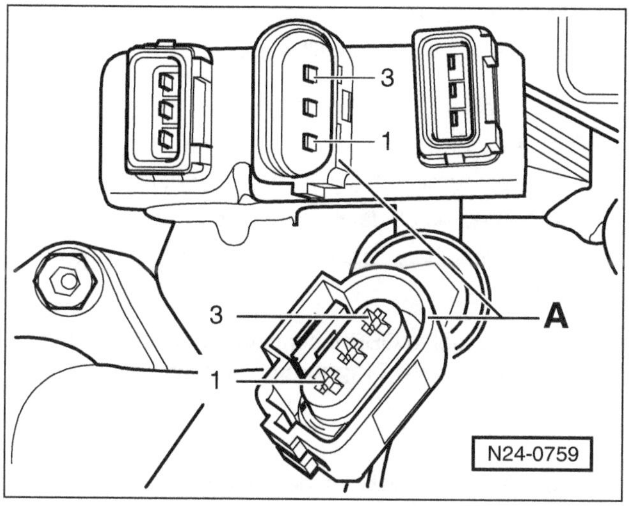

Engine speed (RPM) sensor (G28), measuring resistance

- Connector is located at front of intake manifold, under mounting bracket for secondary air injection solenoid valve, to right of engine oil dipstick.
- Center connector (**A**) is for G28

– Unplug grey 3-pin connector (**A**) and measure resistance between the various terminals, on "male" side of connector.

Engine speed (RPM) sensor resistance	Specification
Resistance between terminals 1 and 3	∞ Ω (no continuity)
Resistance between terminals 1 and 2	∞ Ω (no continuity)
Resistance between terminals 2 and 3	480 to 1000 Ω

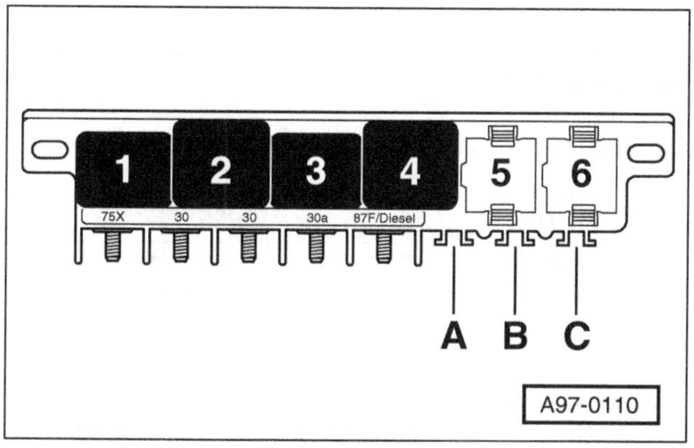

Fuel pump relay (J17)

- Located in micro central electronic panel under left side of dashboard, position #4

Intake air temperature (IAT) sensor (G42), measuring resistance

- IAT sensor (**arrow**) is located at intake manifold near throttle valve control module.
- AMU/BEA engine shown. Sensor for ATC/AWP engines is located at other end of intake manifold, near throttle valve control module.

– Disconnect harness connector from sensor and measure resistance between two terminals of IAT sensor.

– Measure temperature of intake air using diagnostic scan tool or thermometer.

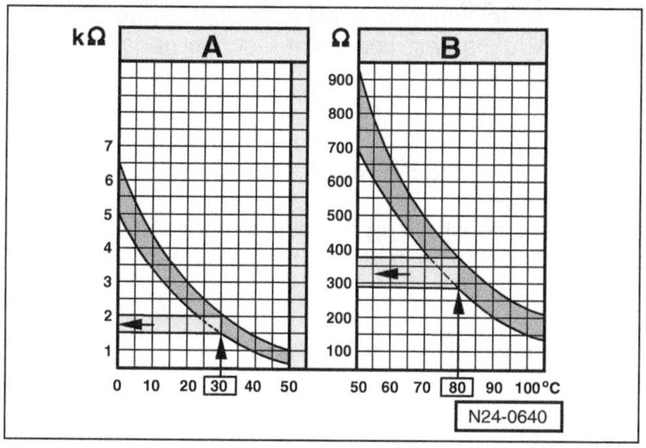

Compare IAT sensor resistance and intake air temperature readings to graph.

- Scale **A** shows resistance for intake air temperature range 0 - 50°C (32 - 122°F).
- Scale **B** shows resistance for intake air temperature range 50 - 100°C (122 - 212°F).
- Example:
  30°C corresponds to 1500 - 2000 Ω
  80°C corresponds to 275 - 375 Ω

BentleyPublishers.com
©2006 BentleyPublishers.com—All Rights Reserved

Mass air flow (MAF) sensor (G70), testing

- MAF sensor (**arrow**) is located on air filter housing.

Requirements for testing:

- Battery voltage at least 11.5 volts.
- All electrical consumers turned off (lights, A/C, rear defogger, etc.).
- Parking brake applied (turns off daytime running lights).
- Coolant temperature must be at least 80°C (176°F), check using diagnostic scan tool.

– Connect diagnostic scan tool to OBD II connector under left side of dashboard. See **OBD On Board Diagnostics** for more information.

– Start engine and let it run at idle.

– Select address word 33, “Diagnostic mode 1: Checking measuring values” on scan tool.

– Select the measuring value, “PID 16: Air flow quantity at Mass Air Flow (MAF) sensor at idle”.

– Check the specified value of air flow quantity displayed on scan tool.

Diagnostic test PID 16	Specification
Air flow quantity at MAF sensor, at idle	2.00 to 4.50 g/sec.

– End diagnosis and switch off ignition.

– If specification was not obtained, replace MAF. If specified value was obtained and a diagnostic trouble code still exists relating to the MAF, then continue checking electrical wiring of MAF as described below.

– Disconnect wiring harness from MAF.

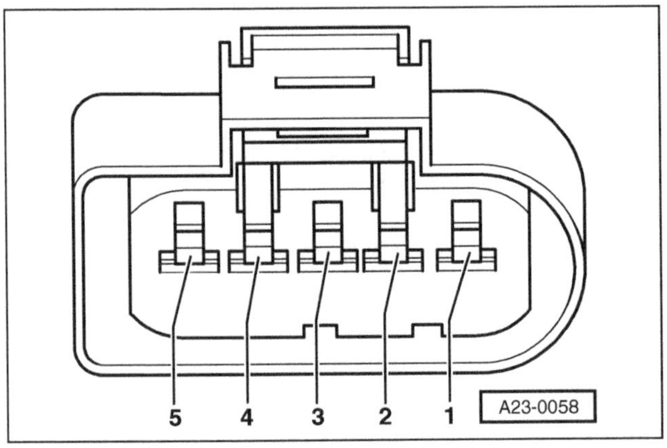

Connect multimeter positive lead to terminal **2** of MAF harness connector. Connect negative lead to engine ground (–).

– Briefly operate starter.

- Multimeter should read approx. battery voltage.

– If no voltage is present, check wiring harness from MAF terminal **2** to fuel pump relay (J17) using wiring diagrams. See **EWD Electrical Wiring Diagrams**.

– If voltage OK, continue testing by connecting multimeter positive lead to terminal **4** of MAF harness connector. Connect negative lead to engine ground (–).

– Switch ignition on.

- Multimeter should read 4.5 to 5.5 volts.

– If no voltage is present, check wiring harness from the various MAF terminals for short circuits to each other, to ground (–), or battery positive (+).

– Repair wiring or replace MAF as necessary.

BentleyPublishers.com

©2006 BentleyPublishers.com—All Rights Reserved

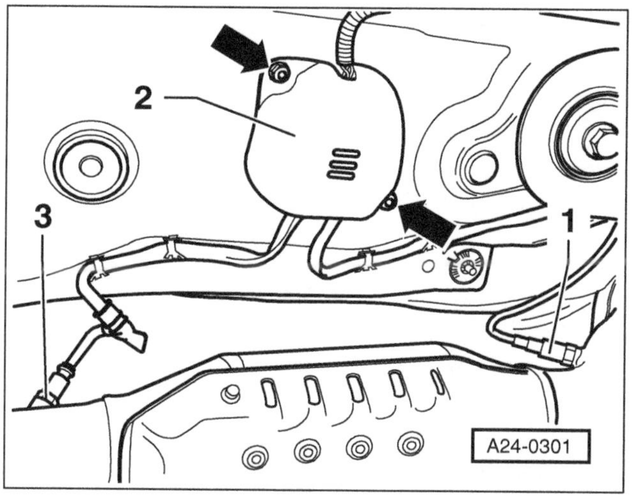

Oxygen sensor harness connections

- Connector cover (**2**) is located under vehicle near oxygen sensors (**1** and **3**).
- Vehicle with ATC/AWP engine shown. AMU/BEA engine is similar.

– Remove mounting screws (**arrows**) and cover (**2**) to access connectors.

- Black connector: For oxygen sensor (G39) and oxygen sensor heater (Z19), before catalytic converter (**1**).
- Brown connector: For oxygen sensor (G130) and oxygen sensor heater (Z29), after catalytic converter (**3**).

Power supply relays in engine compartment

1. Secondary air injection pump power supply relay (J299)
2. Engine control module power supply relay (J271)

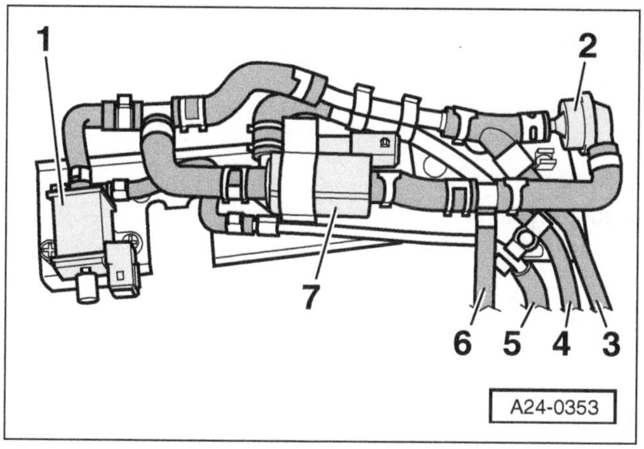

Secondary air injection components

- Mounted to bottom of metal bracket at front of intake manifold.

1. Secondary air injection solenoid valve (N112)
2. Non-return valve
3. To vacuum reservoir
4. To mechanical air recirculation valve
5. To combination valve of secondary air system
6. To connection on intake manifold
7. Recirculating valve for turbocharger (N249)

BentleyPublishers.com
© 2006 BentleyPublishers.com—All Rights Reserved

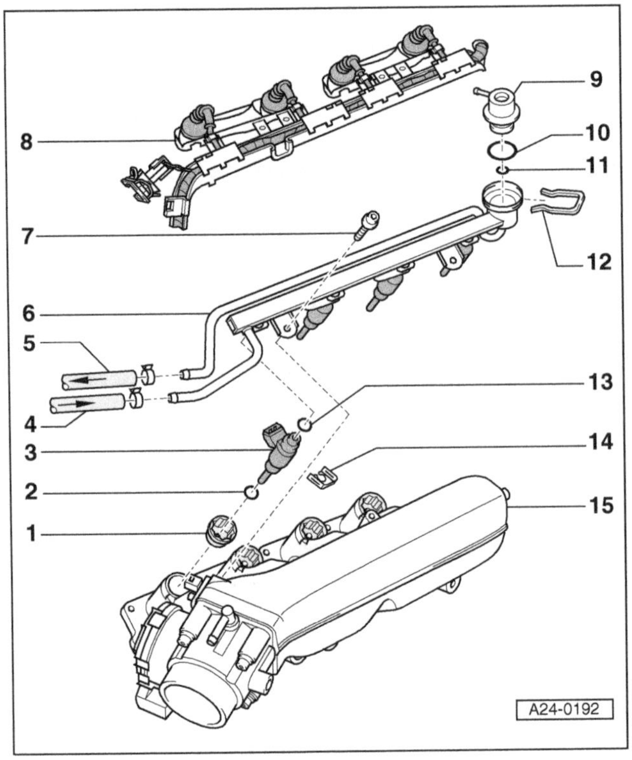

Fuel Rail and Injectors

Fuel rail and intake manifold (engine code: ATC, AWP)

1. Fuel injector insert
   - Use locking fluid (D 000 600 A2) when installing
2. O-ring
3. Fuel injector
4. Fuel supply line
5. Fuel return line
6. Fuel rail
7. Bolt
   - 10 Nm (7 ft-lb)
8. Fuel injector harness
9. Fuel pressure regulator
10. O-ring
11. O-ring
12. Retaining clip
13. O-ring
14. Retaining clip
15. Intake manifold
   - Removing/installing, see **15a Cylinder Head (1.8L Engine)**

Fuel rail and intake manifold (engine code: AMU, BEA)

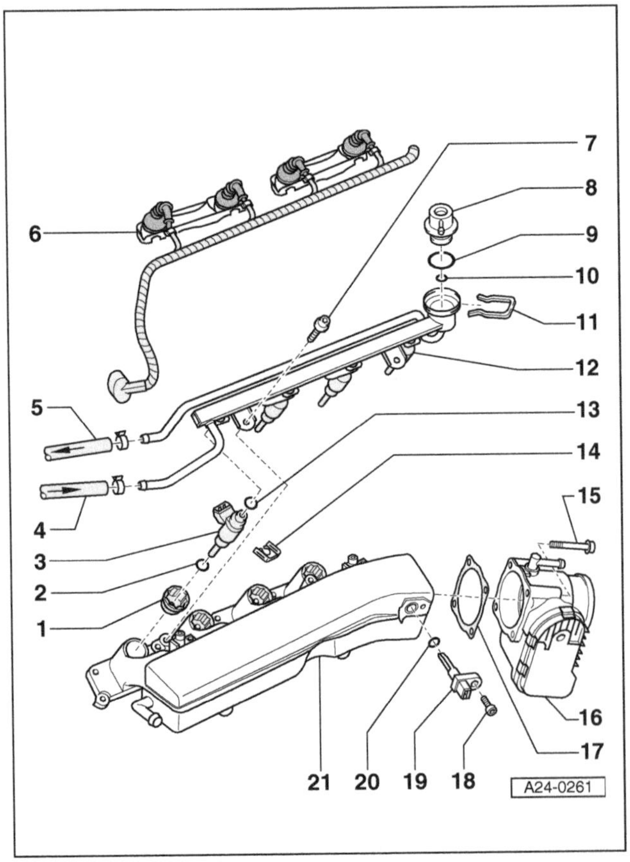

1. Fuel injector insert
   - Use locking fluid (D 000 600 A2) when installing
2. O-ring
3. Fuel injectors
4. Fuel supply line
5. Fuel return line
6. Fuel injector harness
7. Bolt
   - 10 Nm (7 ft-lb)
8. Fuel pressure regulator
9. O-ring
10. O-ring
11. Retaining clip
12. Fuel rail
13. O-ring
14. Retaining clip
15. Bolt
   - 10 Nm (7 ft-lb)
16. Throttle valve control module (J338)
17. Gasket
18. Bolt
   - 10 Nm (7 ft-lb)
19. Intake air temperature sensor (G42)
20. O-ring
21. Intake manifold
   - Removing/installing, see **15a Cylinder Head (1.8L Engine)**

BentleyPublishers.com
©2006 BentleyPublishers.com—All Rights Reserved

Fuel injectors, removing and installing

NOTE—
- *Always replace all fuel system O-rings during installation.*
- *Coat new O-rings with clean engine oil for lubrication.*

– Disconnect battery ground (GND) strap from battery negative (–) terminal.

– Remove engine cover. See **03 Maintenance**.

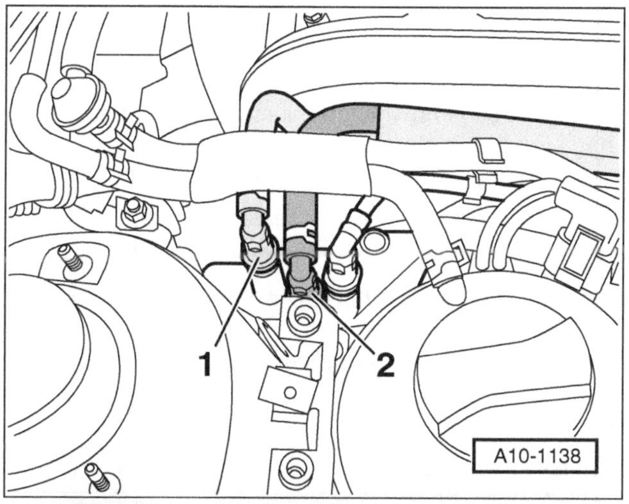

Engine code: ATC, AWP

◄ Disconnect fuel supply line (**1**) and return line (**2**). Press in securing ring to disengage lines.
- Fuel supply line has white markings.
- Fuel return line has blue markings.

Engine code: AMU, BEA

◄ Disconnect fuel supply line (**1**) and return line (**2**). Press in securing ring to disengage lines.
- Fuel supply line has white markings.
- Fuel return line has blue markings.

> ***WARNING—***
> - *Fuel supply line is under pressure. Wear eye protection and protective clothing in order to avoid injuries. Before removing a hose connection wrap a cloth around the connection. Then release pressure by carefully pulling hose off connection.*

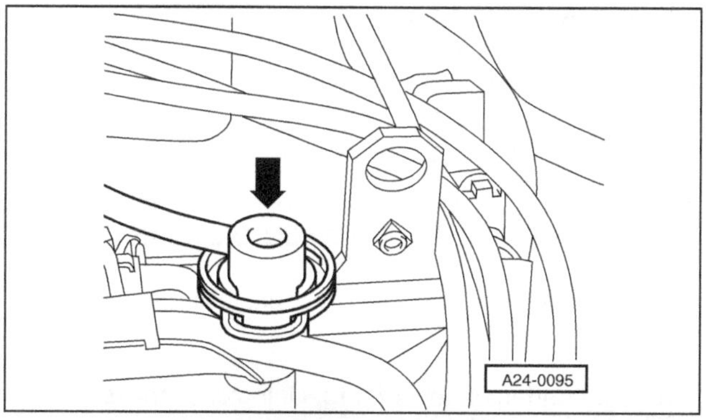

Continued for all engines

◄ Disconnect vacuum hose from fuel pressure regulator (**arrow**).

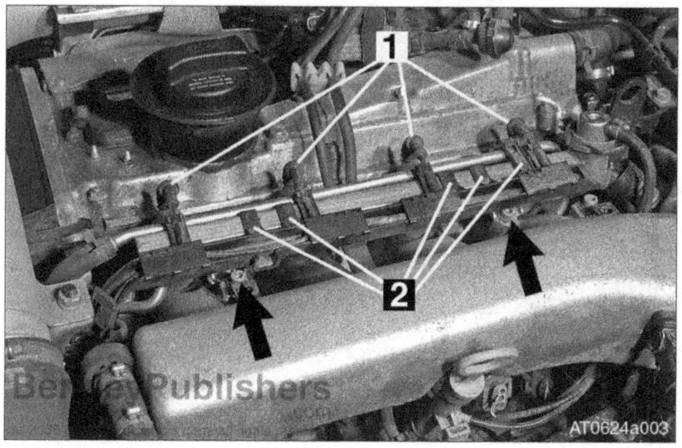

◄ Disconnect fuel injector harness connectors (**1**) and unclip (**2**) support bar from fuel rail.

– Position fuel injector harness out of way.

– Remove mounting bolts (**arrows**) for fuel rail.

– Pull fuel rail and injectors off intake manifold. Pull off evenly from the top and place assembly on a clean cloth.

– Remove retaining clips and injectors from fuel rail.

BentleyPublishers
.com
©2006 BentleyPublishers.com—All Rights Reserved

– Install fuel injectors and fuel rail in reverse order of removal, noting the following:
  - Replace all O-rings
    When replacing the lower fuel injector O-ring, make sure not to remove the plastic cap from the fuel injector head. Carefully pull lower O-ring off over the plastic cap.
  - Lubricate new O-rings with clean engine oil.
  - Make sure injectors are installed and seated correctly.
  - Make sure injector retaining clips are properly seated.
  - Position fuel rail with injectors against intake manifold and press into place evenly.
  - Tighten fuel rail mounting bolts to specification.

Tightening torque	
Fuel rail to intake manifold	10 Nm (7 ft-lb)

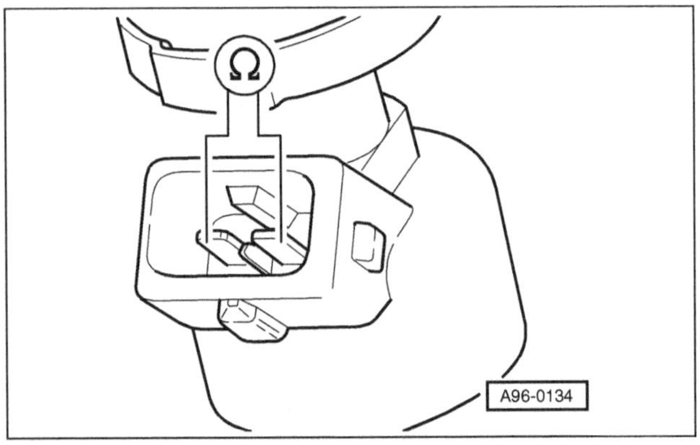

Fuel injectors, checking resistance and activation

– Disconnect fuel injector harness connector.

◄ Connect multimeter and measure resistance across two terminals of fuel injector.

Specification	
Fuel injector resistance, at room temperature	12 -13 Ω

– If specified value is not obtained, replace fuel injector.

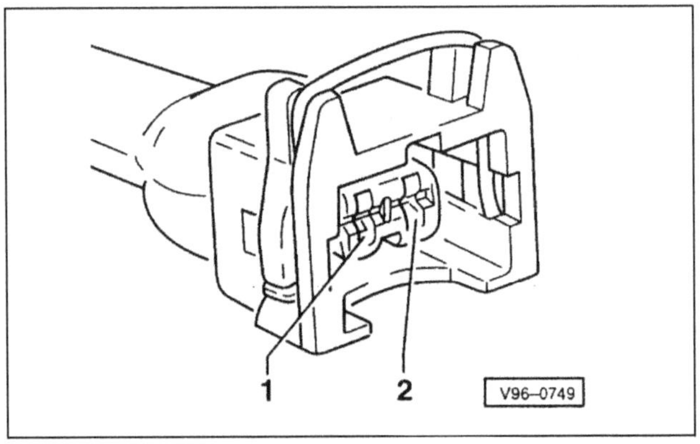

◄ Continue testing for fuel injector voltage supply by connecting LED voltage tester (VAG 1527B or equivalent) between fuel injector harness connector lead #**1** and ground.

– Operate starter briefly (engine can also start).

– LED of voltage tester must blink.

– If LED does not light, check wiring for open or short circuit using wiring diagrams. See **EWD Electrical Wiring Diagrams**.

– Continue testing for injector activation by re-installing fuel injector harness connector.

– Slide back rubber connector cover for fuel injector harness.

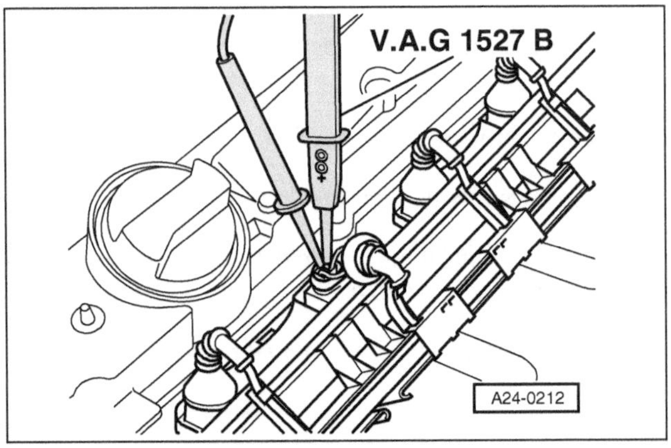

◄ Connect LED voltage tester (VAG 1527 B shown) between two fuel injector harness connector terminals.

– Operate starter briefly (engine can also start).

– LED of voltage tester must blink.
  - LED voltage testers do not go out completely between activations by the ECM, but rather continue to glow and then get slightly brighter during activation.

– If LED does not light, check wiring for open or short circuit using wiring diagrams. See **EWD Electrical Wiring Diagrams**.

BentleyPublishers.com
©2006 BentleyPublishers.com—All Rights Reserved

Engine Control Module (ECM)

ECM, removing and installing

The ECM is in the air plenum near the base of windshield in the engine compartment.

– Before replacing ECM, connect VAG 1551/1552 or VAS 5051/5052 diagnostic scan tool, or equivalent, to diagnostic plug under left side of dashboard and record (print) ECM coding data. See **OBD On Board Diagnostics** for more information on using scan tool.

NOTE—
- *When the connector is detached from the ECM, adaptation values are erased but contents of DTC memory remain intact.*

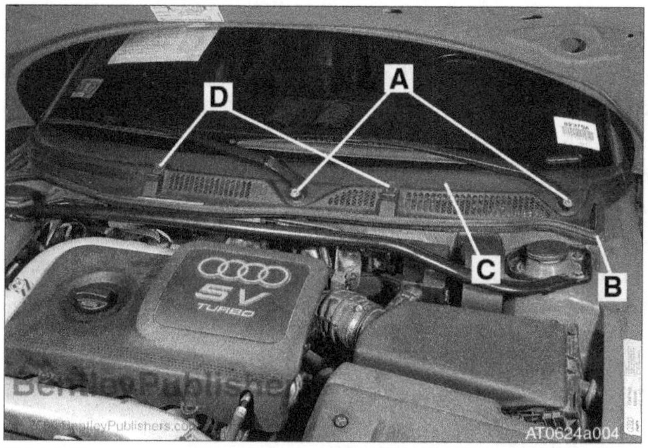

– Switch ignition OFF.

◄ Remove windshield wipers (**A**). See **92 Wipers and Washers**.

– Remove rubber weather-strip (**B**) from plenum base.

– Remove plenum cover (**C**) by carefully pulling away from base of windshield.

> ***CAUTION—***
> - *Do not use screwdriver or pry tool against windshield glass. Glass may crack.*

– Disconnect washer fluid hoses from back of washer jets (**D**) when removing plenum cover (**C**).

ECM without metal anti-theft housing

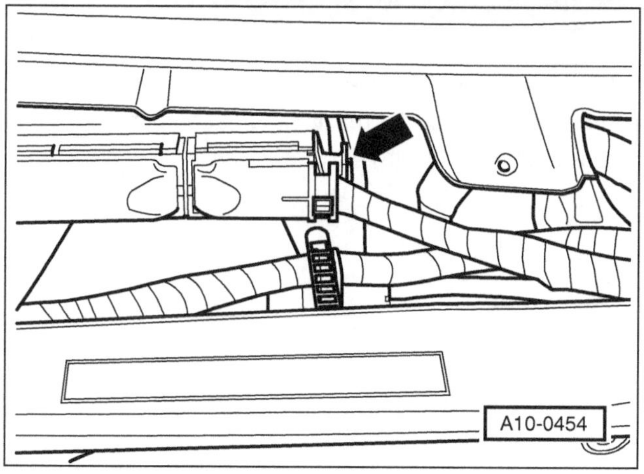

◄ Release ECM harness connectors at **arrow**.

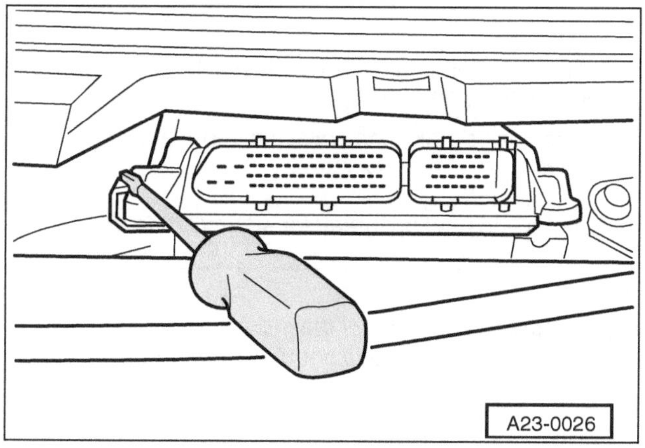

◄ Use screwdriver to press locking lugs of bracket to the side and then remove ECM toward front.

BentleyPublishers
.com
© 2006 BentleyPublishers.com—All Rights Reserved

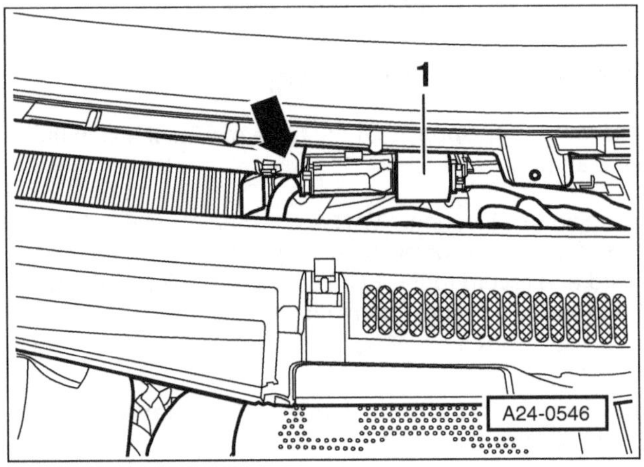

ECM with metal anti-theft housing

NOTE—
- *A heat gun with a semicircular nozzle is necessary to remove the ECM from the anti-theft housing. A heat gun from the VAS 1978 wiring harness repair kit is shown below. An equivalent heat gun may be used instead.*

Use a screwdriver to press locking lug of bracket (**arrow**) to the side and then remove ECM with anti-theft housing toward front.

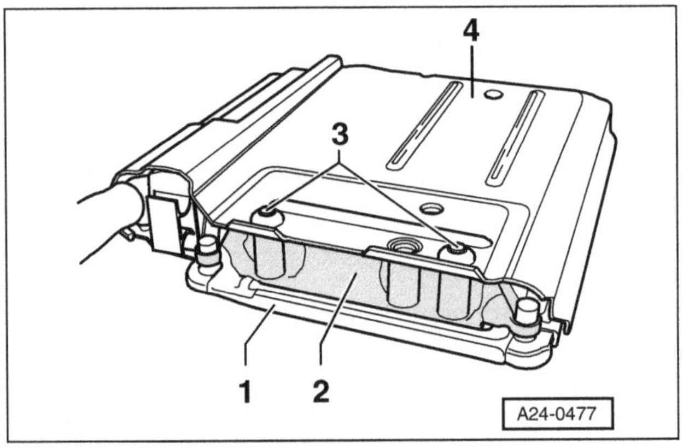

To restrict accessibility (for theft prevention), the ECM (**1**) is screwed to a metal anti-theft housing (**4**) using a locking mechanism (**2**) and snap-off screws (**3**).
- The threads of the snap-off screws are coated with locking compound to make it more difficult to remove the snap-off screws.
- To disconnect the harness connectors from the ECM (e.g. when connecting a test box or when replacing the ECM), the ECM must be removed from the metal anti-theft housing.

– Position the ECM so that the locking mechanism (**2**) is accessible. Place a clean rag below the ECM.

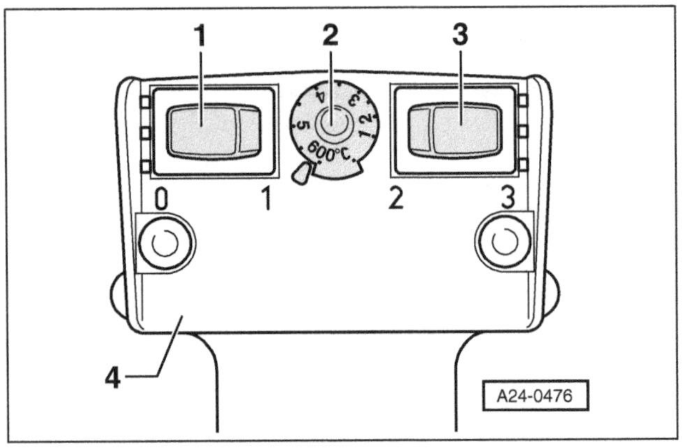

Adjust heat gun as shown in illustration.
- Temperature adjustment (**2**) for maximum heat output
- Two-stage switch for air flow (**3**) set to position 3

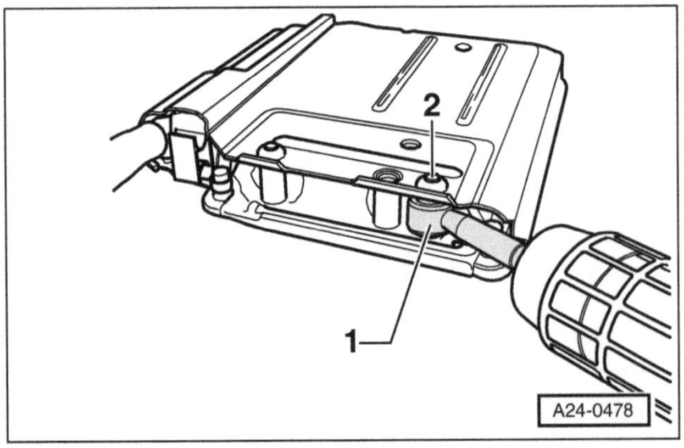

Guide nozzle (**1**) of heat gun onto the locking mechanism threads so that the nozzle "encloses" the threads.
- It is OK to let the nozzle make contact at the upper end of the metal anti-theft housing.

– Switch on heat gun and heat up thread for approx. 20 to 25 seconds.

> ***WARNING—***
> - *The snap-off screws and parts of the metal anti-theft housing can become very hot. Be careful not to get burned!*

BentleyPublishers.com
©2006 BentleyPublishers.com—All Rights Reserved

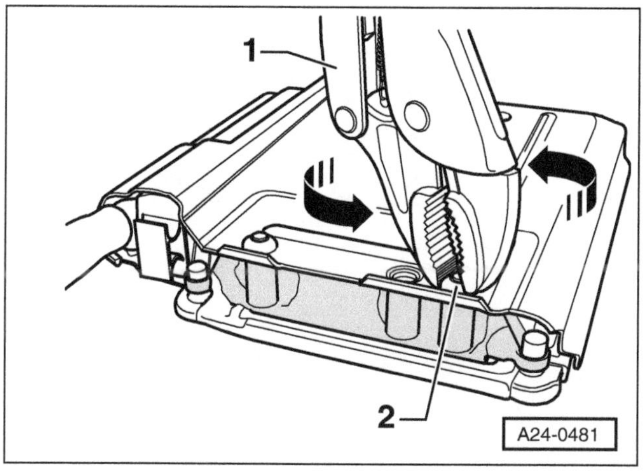

Use vice grips to grab screw head (**2**) and remove snap-off screw by turning in counter clockwise direction (**arrows**).

- Repeat procedure with heat gun and vice grips for other snap-off screw.
- Remove ECM locking mechanism.
- The ECM can now be slid out from the metal anti-theft housing.
- Disconnect harness connectors from ECM.
- Installation is the reverse order of removal, noting the following:
  - After repairs are complete, the metal anti-theft housing must be reinstalled. Use new snap-off screws.
  - Bend back locking lug of bracket.

Continued for all versions

- Install ECM in reverse order of removal.
- If ECM was replaced, use diagnostic scan tool to:
  - Check Diagnostic Trouble Code (DTC) memory and erase if need be
  - Code new ECM
  - Adapt throttle valve control module
  - Adapt kickdown point for vehicles with automatic transmission
  - Generate readiness code

See **OBD On Board Diagnostics** for more information.

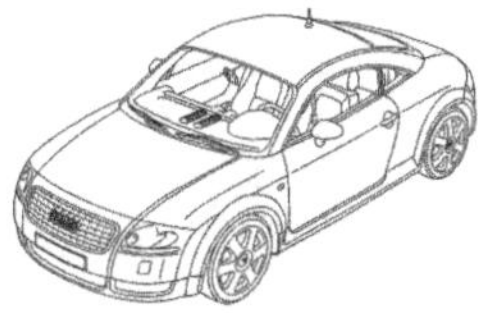

BentleyPublishers
.com
©2006 BentleyPublishers.com—All Rights Reserved

24b Fuel Injection (3.2L Engine)

GENERAL

This repair group covers fuel injection and engine management systems for the 1.8L gasoline engines. Special testing equipment may be necessary for some tests and repair procedures given here.

Engine	Engine code
3.2L 6-cylinder turbo (250 hp)	BHE

Also see the following:

- **20 Fuel Storage and Supply** for fuel pump and fuel tank testing and repair, electronic power control (drive-by-wire or EPC) and evaporative emission control (EVAP) information.
- **26 Exhaust System** for catalytic converter, oxygen sensor and exhaust gas recirculation (EGR) information.
- **28 Ignition System** for ignition system testing and repair.
- **OBD On Board Diagnostics** for information on system fault diagnosis and use of diagnostic scan tool.

Motronic system

The Motronic fuel injection system combines ignition, fuel injection and electronic accelerator pedal functions into one system managed by the engine control module (ECM). There are no basic adjustments or settings that can be made to the systems without specialized equipment.

BentleyPublishers.com
© 2006 BentleyPublishers.com—All Rights Reserved

The engine management systems are fully adaptive and feature built-in diagnostics that are capable of detecting and storing coded fault information. To gain access to system fault memory, attach a diagnostic scan too to the data link connector (DLC) (**arrow**) under left lower dashboard. See **OBD On Board Diagnostics** for more information. **Table a** lists some basic information about the fuel injection systems for the 3.2L engine covered by this manual.

Table a. Engine management applications and technical data	
3.2L 250 hp engine (code: BHE)	
Fuel injection system	Motronic ME7.1.1
Idle speed	640 - 750 rpm
Max. rpm	approx. 6500
Fuel injector delivery rate for 30 seconds	128 - 140 ml

Warning and Cautions

WARNING —
- *Fuel may be discharged during fuel system repairs. Do not smoke or work near heaters or other fire hazards. Have a fire extinguisher handy. Work only in a well-ventilated area.*
- *Wear suitable hand and eye protection when working with gasoline. Prolonged contact with fuel can cause illness and skin disorders.*

CAUTION—
- *Secure engine compartment fuel hoses with spring-type clips. The use of clamp or screw-type clips damages fuel lines and cause fuel leaks.*
- *Connect and disconnect wires and test equipment only with ignition switched off.*
- *Before making any electrical tests that require engine to be cranked using the starter, disable ignition system as described in* **28 Ignition System.**
- *Do not use sealants containing silicones. Particles of silicone drawn into engine do not burn and instead damage the oxygen sensors. Use only sealants and engine chemicals marked as safe for oxygen sensors.*
- *During testing, it is possible for the ECM to store a diagnostic trouble code (DTC) due solely to the testing procedure. Therefore, after completing repairs, clear the DTC memory using diagnostic scan tool.*
- *Cleanliness is essential when working on the fuel system. Thoroughly clean fuel line connections and surrounding areas before loosening. Avoid using compressed air. Avoid moving vehicle. Only install clean components.*

BentleyPublishers
.com
© 2006 BentleyPublishers.com—All Rights Reserved

FUEL INJECTION COMPONENTS

NOTE—

- *Audi identifies electrical components by a letter and/or a number in the electrical schematics. See* **EWD Electrical Wiring Diagrams***. These electrical identifiers are listed in parentheses as an aid to electrical troubleshooting.*

Component overview

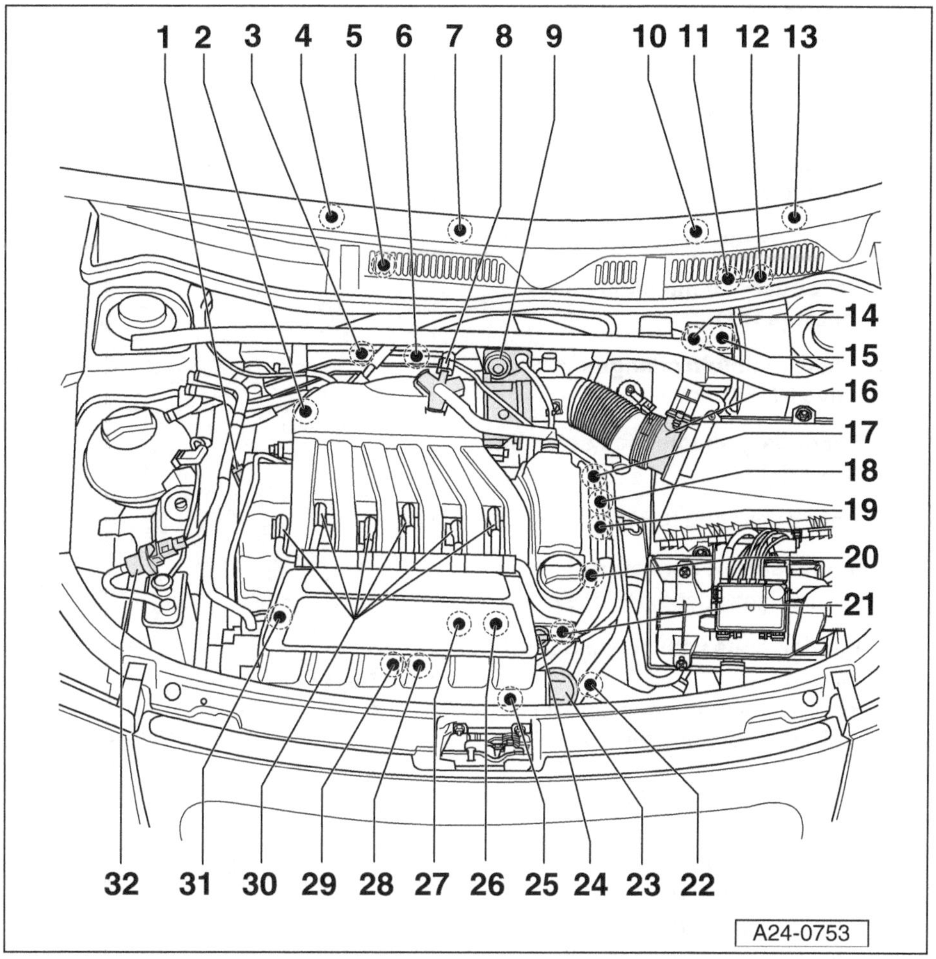

1. **3-pin harness connector**
   - For knock sensor 1
2. **Knock sensor 1 (G61)**
3. **Oxygen sensor (G39) and oxygen sensor heater (Z19)**
   - Located before catalyst, bank 1
   - Test with diagnostic scan tool
4. **Oxygen sensor (G130) and oxygen sensor heater (Z29)**
   - Located after catalyst, bank 1
   - Test with diagnostic scan tool
5. **Motronic engine control module (J220)**
   - See **Engine control module, removing and installing**
6. **Oxygen sensor (G108) and oxygen sensor heater (Z28)**
   - Located before catalyst, bank 2
   - Test with diagnostic scan tool
7. **Oxygen sensor (G108) and oxygen sensor heater (Z28)**
   - Located before catalyst, bank 2
   - Test with diagnostic scan tool
8. **Positive crankcase ventilation (PCV) heating element (N79)**
9. **Throttle valve control module (J338)**
   - With throttle drive (power accelerator actuation) (G186)
   - With throttle drive angle sensor 1 (G187)
   - With throttle drive angle sensor 2 (G188)
   - After replacement, erase adaptation values and adapt ECM to throttle valve control module using diagnostic scan tool
10. **Instrument cluster**
    - With (MIL) Malfunction Indicator Lamp (K83), aka. check engine light
    - With (EPC) Electronic Power Control lamp (K132)
11. **Throttle position sensor (G79) and accelerator position sender (G185)**
    - Part of accelerator pedal assembly. For removal and installation, see **20 Fuel Storage and Supply, Electronic Power Control (EPC) System**.
12. **Brake light switch (F) and brake pedal position switch (F47)**
    - Located on top of brake pedal
13. **Fuel pump relay (J17)**
14. **Power supply relay (J299) for secondary air injection pump**
15. **Power supply relay (J271) for engine control module**
16. **Mass Air Flow (MAF) sensor (G70)**
    - Integral intake air temperature sensor
17. **Camshaft position sensor 2 (G163)**
    - Black connector
    - For exhaust camshaft
18. **Camshaft adjustment valve**
    - Brown connector
    - For exhaust camshaft
19. **Valve 1 for camshaft adjustment (N205)**
    - Black connector
    - For intake camshaft
20. **Camshaft position sensor 1 (G40)**
    - Black connector
    - For intake camshaft
21. **Engine coolant temperature (ECT) sensor (G62)**

(continued on following page)

BentleyPublishers.com
© 2006 BentleyPublishers.com—All Rights Reserved

Component overview

(continued from previous page)

22. **Engine speed rpm sensor (G28)**
23. **Vacuum unit for intake manifold change-over**
24. **Fuel pressure regulator**
25. **Secondary air injection pump motor (V101)**
26. **Intake manifold change-over valve (N156)**
27. **Secondary air injection solenoid valve (N112)**
28. **Knock sensor 2 (G66)**
29. **3-pin harness connector**
    - For engine speed (RPM) sensor
30. **Ignition coils with power output stages**
    - See **03 Maintenance, spark plugs, replacing** for removal and installation
    - Ignition coil #1 (N70)
    - Ignition coil #2 (N127)
    - Ignition coil #3 (N291)
    - Ignition coil #4 (N292)
    - Ignition coil #5 (N323)
    - Ignition coil #6 (N324)
31. **Fuel injectors**
    - Cylinder #1 (N30)
    - Cylinder #2 (N31)
    - Cylinder #3 (N32)
    - Cylinder #4 (N33)
    - Cylinder #5 (N83)
    - Cylinder #6 (N84)
32. **Evaporative emissions control (EVAP) canister purge regulator valve (N80)**

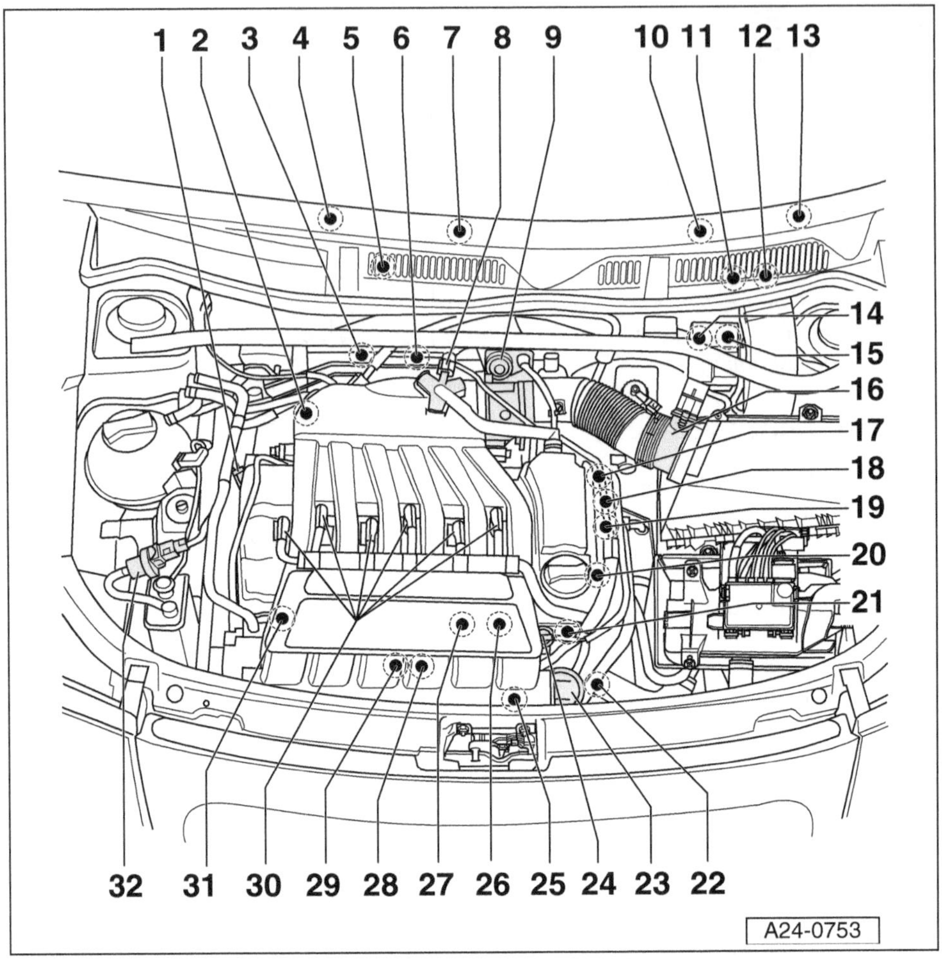

BentleyPublishers.com

©2006 BentleyPublishers.com—All Rights Reserved

Component details and testing

Various components of the fuel injection system are identified here. For more component location information, including fuses and relays, see **97 Fuses, Relays, Component Locations**.

Applicable testing specifications are also provided. Testing components will typically involve the use of a digital multimeter and/or a diagnostic scan tool. For more information on electrical testing with a multimeter, see **9 Electrical System–General**. For more information on using a diagnostic scan tool, see **OBD On Board Diagnostics**.

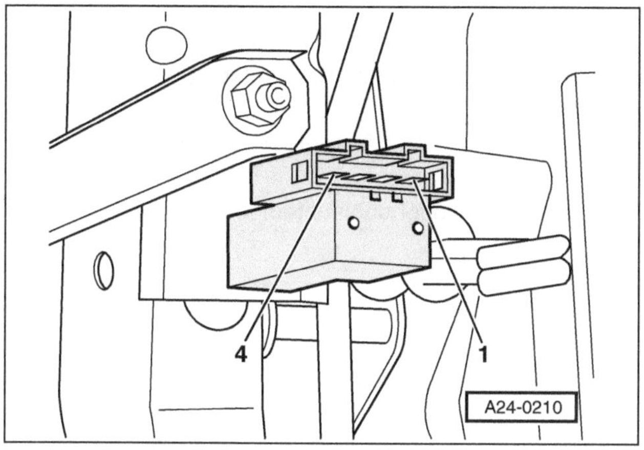

◀ **Brake light switch (F) and brake pedal switch (F47)**

- Located above brake pedal.

– Disconnect harness connector to test resistance of switch.

– Connect multimeter to switch terminals 1 and 4.

- Specified value: ∞ Ω (no continuity)

– Depress brake pedal.

- Specified value: approx. 0 Ω

– Connect multimeter to switch terminals 2 and 3.

- Specified value: approx. 0 Ω

– Depress brake pedal.

- Specified value: ∞ Ω (no continuity)

– If specified values are not obtained, check switch wiring using wiring diagrams. See **EWD Electrical Wiring Diagrams**.

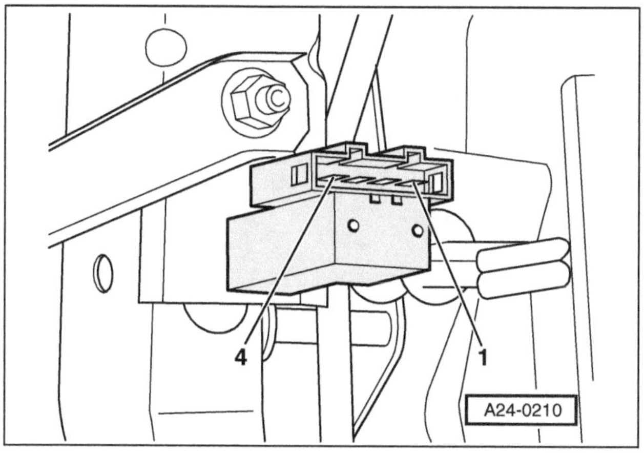

◀ **Clutch vacuum vent valve switch (F36)**

- Located above clutch pedal (only for vehicles with manual transmission).

– Connect multimeter to switch terminals 1 and 2.

- Specified value: approx. 0 Ω

– Depress clutch pedal.

- Specified value: ∞ Ω (no continuity)

– If specified values are not obtained, check switch wiring using wiring diagrams. See **EWD Electrical Wiring Diagrams**.

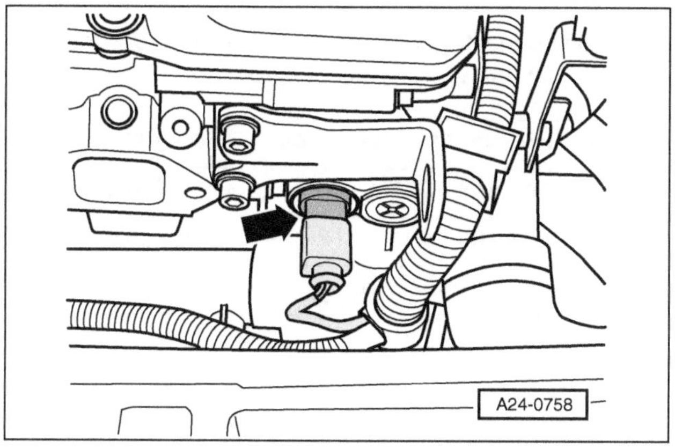

◀ **Engine coolant temperature (ECT) sensor (G62)**

- ECT sensor (**arrow**) is located on end of cylinder head.

BentleyPublishers.com
© 2006 BentleyPublishers.com—All Rights Reserved

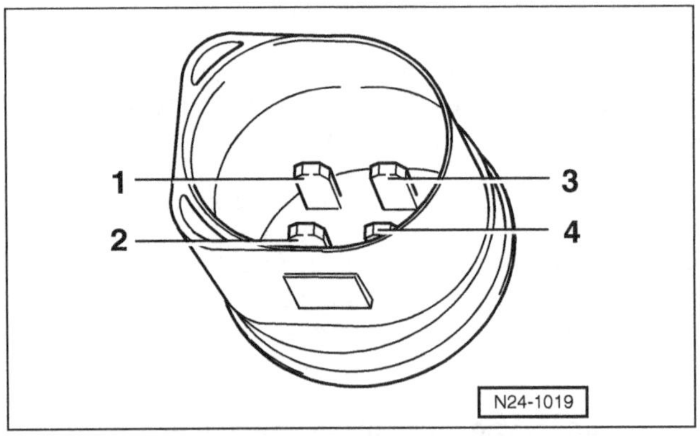

ECT sensor harness connector.

- Measure resistance at terminals **3** and **4** of sensor.

– Measure temperature of engine coolant using diagnostic scan tool or thermometer.

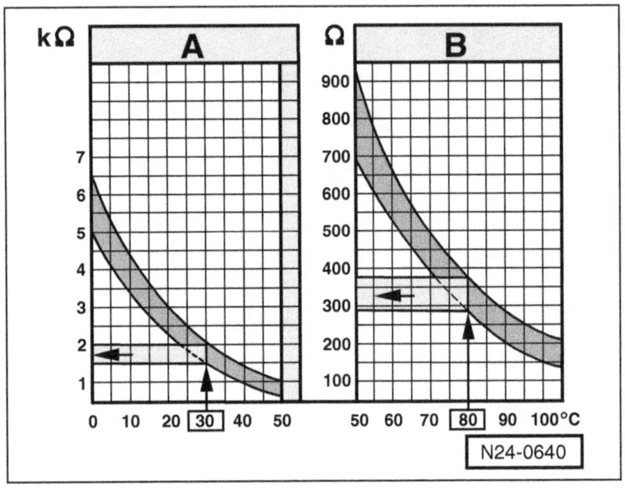

Compare ECT sensor resistance and coolant temperature readings to graph.

- Scale **A** shows resistance for coolant temperature range 0 - 50°C (32 - 122°F).
- Scale **B** shows resistance for coolant temperature range 50 - 100°C (122 - 212°F).
- Example:
  30°C corresponds to 1500 - 2000 Ω
  80°C corresponds to 275 - 375 Ω

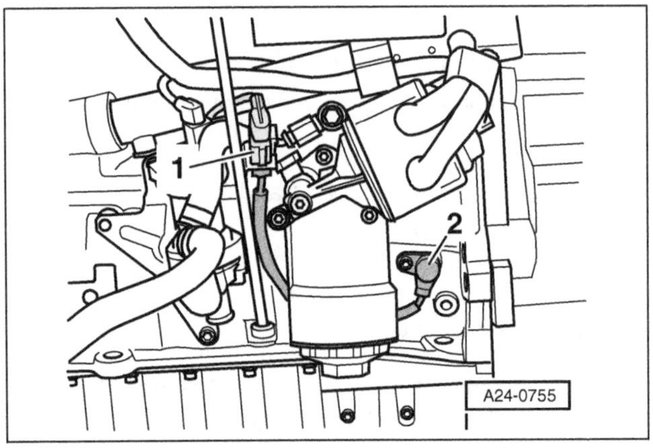

Engine speed (RPM) sensor (G28), measuring resistance

- RPM sensor (**2**) is located on engine block near oil filter.
- Grey 3-pin harness connector (**1**) is for RPM sensor.

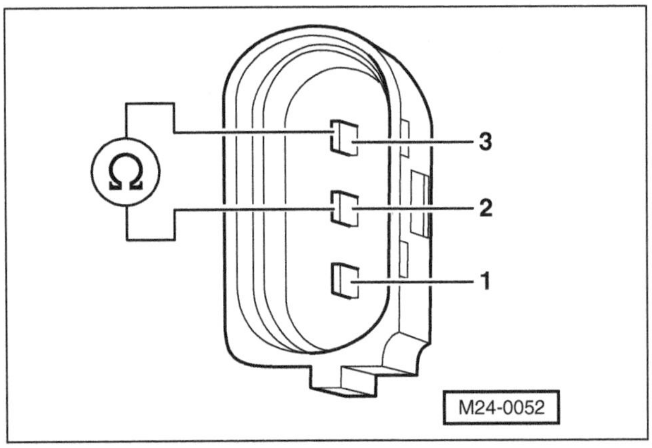

Unplug grey 3-pin connector and measure resistance between the various terminals, on "male" side of connector.

Engine speed (RPM) sensor resistance	Specification
Resistance between terminals 1 and 3	∞ Ω (no continuity)
Resistance between terminals 1 and 2	∞ Ω (no continuity)
Resistance between terminals 2 and 3	730 to 1000 Ω

BentleyPublishers.com
©2006 BentleyPublishers.com—All Rights Reserved

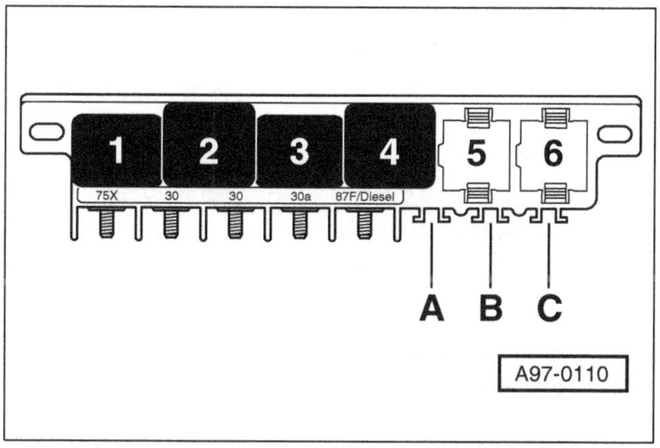

A97-0110

◀ **Fuel pump relay (J17)**

- Located in micro central electronic panel under left side of dashboard, position #4

AT0624a002

◀ **Intake air temperature (IAT) sensor (G42), measuring resistance**

- IAT sensor is integral with the Mass Air Flow (MAF) sensor (**arrow**) and is located in the upper section of air filter housing.

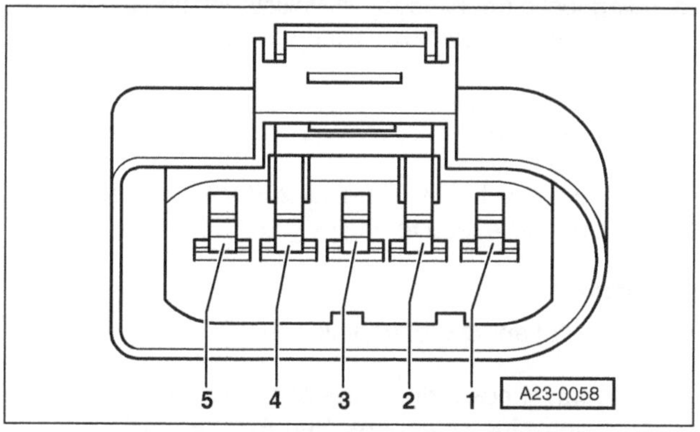

A23-0058

◀ Disconnect harness connector from IAT/MAF sensor and measure resistance between terminals 1 and 3 of sensor.

– Measure temperature of intake air using diagnostic scan tool or thermometer.

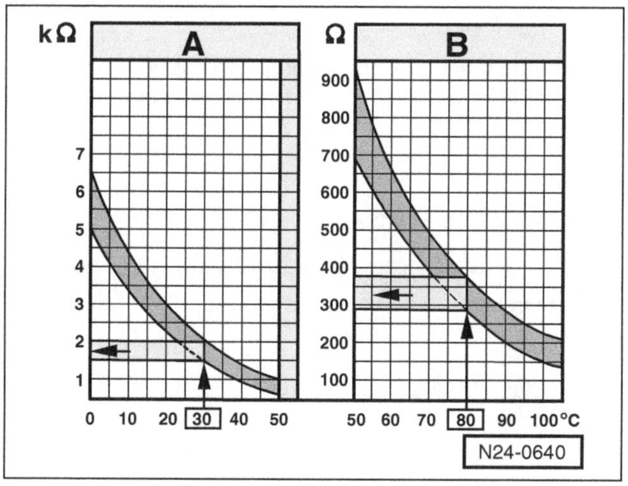

N24-0640

◀ Compare IAT/MAF sensor resistance and intake air temperature readings to graph.

- Scale **A** shows resistance for intake air temperature range 0 - 50°C (32 - 122°F).
- Scale **B** shows resistance for intake air temperature range 50 - 100°C (122 - 212°F).
- Example:
  30°C corresponds to 1500 - 2000 Ω
  80°C corresponds to 275 - 375 Ω

BentleyPublishers.com
©2006 BentleyPublishers.com—All Rights Reserved

Mass air flow (MAF) sensor (G70), testing

- MAF sensor (**arrow**) is located on air filter housing.

Requirements for testing:

- Battery voltage at least 11.5 volts.
- All electrical consumers turned off (lights, A/C, rear defogger, etc.).
- Parking brake applied (turns off daytime running lights).
- Coolant temperature must be at least 80°C (176°F), check using diagnostic scan tool.

– Connect diagnostic scan tool to OBD II connector under left side of dashboard. See **OBD On Board Diagnostics** for more information.

– Start engine and let it run at idle.

– Select address word 33, “Diagnostic mode 1: Checking measuring values” on scan tool.

– Select the measuring value, “PID 16: Air flow quantity at Mass Air Flow (MAF) sensor at idle”.

– Check the specified value of air flow quantity displayed on scan tool.

Diagnostic test PID 16	Specification
Air flow quantity at MAF sensor, at idle	2.00 to 7.00 g/sec.

– End diagnosis and switch off ignition.

– If specification was not obtained, replace MAF. If specified value was obtained and a diagnostic trouble code still exists relating to the MAF, then continue checking electrical wiring of MAF as described below.

– Disconnect wiring harness from MAF.

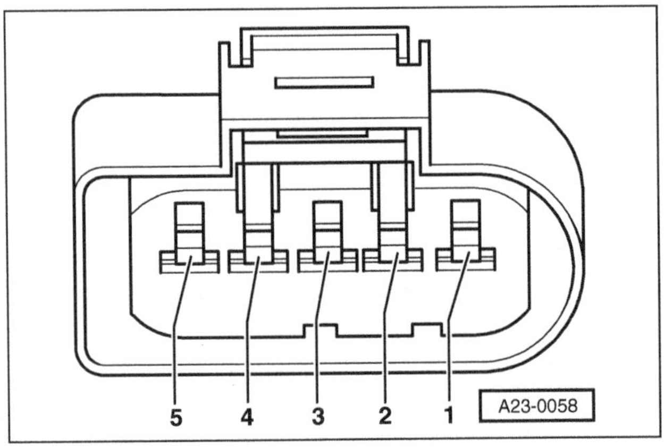

Connect multimeter positive lead to terminal **2** of MAF harness connector. Connect negative lead to engine ground (–).

– Briefly operate starter.

- Multimeter should read approx. battery voltage.

– If no voltage is present, check wiring harness from MAF terminal **2** for voltage supply using wiring diagrams. See **EWD Electrical Wiring Diagrams**.

– If voltage OK, continue testing by connecting multimeter positive lead to terminal **4** of MAF harness connector. Connect negative lead to engine ground (–).

– Switch ignition on.

- Multimeter should read 4.5 to 5.5 volts.

– If no voltage is present, check wiring harness from the various MAF terminals for short circuits to each other, to ground (–), or battery positive (+).

– Repair wiring or replace MAF as necessary.

BentleyPublishers
.com
© 2006 BentleyPublishers.com—All Rights Reserved

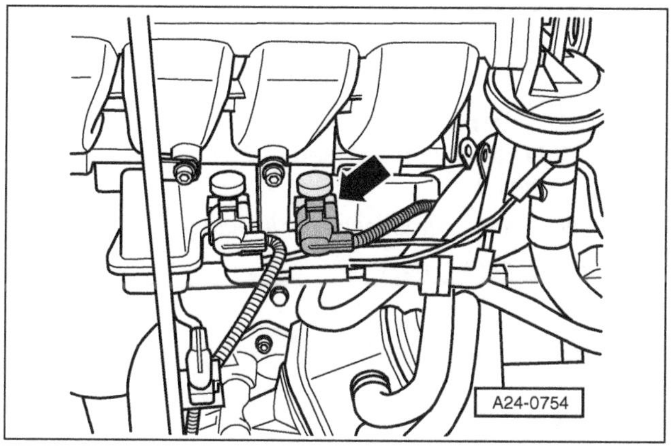

Intake manifold change-over valve (N156)

- Valve (**arrow**) is located in front of intake manifold.

– Remove harness connector from intake manifold change-over valve and measure resistance between two terminals of valve.

Specification	
Intake manifold change-over valve resistance	25 - 35 Ω

– If specified value is not obtained, replace N156.

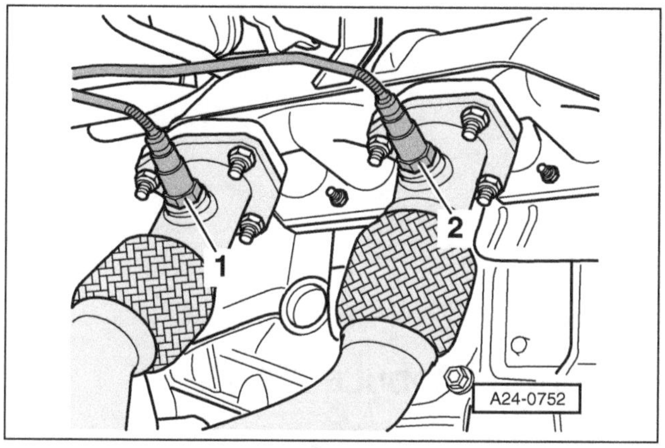

Oxygen sensor locations (pre-cat)

1. Heated oxygen sensor (G108) and oxygen sensor heater (Z28) for cylinder bank #2
2. Heated oxygen sensor (G39) and oxygen sensor heater (Z19) for cylinder bank #1

- Use special tool 3337 to remove/install oxygen sensors

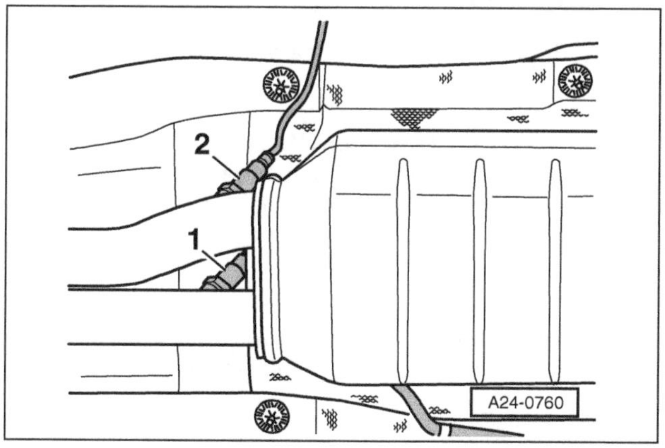

Oxygen sensor 2 locations (post-cat)

1. Heated oxygen sensor 2 (G131) and oxygen sensor heater 2 (Z30) for cylinder bank #2
2. Heated oxygen sensor 2 (G130) and oxygen sensor heater 2 (Z29) for cylinder bank #1

- Use special tool 3337 to remove/install oxygen sensors

Power supply relays in engine compartment

1. Secondary air injection pump power supply relay (J299)
2. Engine control module power supply relay (J271)

BentleyPublishers.com
© 2006 BentleyPublishers.com—All Rights Reserved

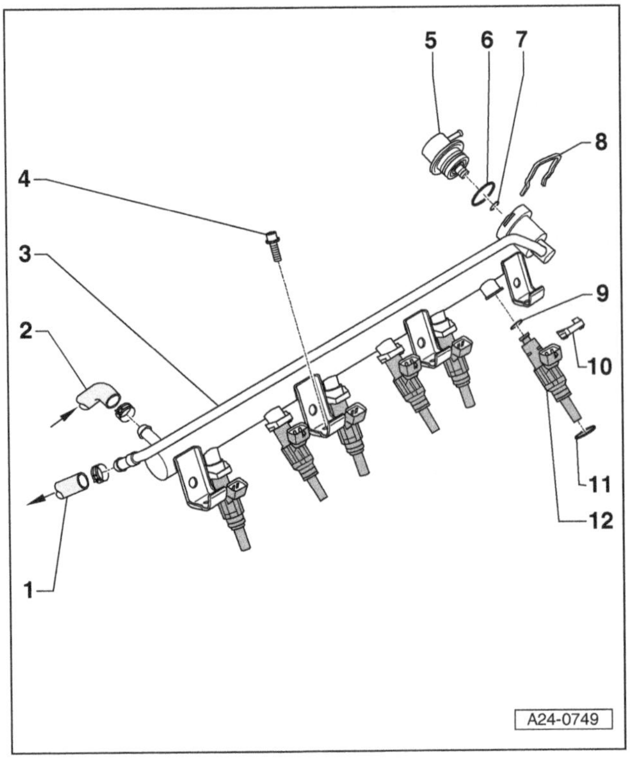

Fuel Rail and Injectors

NOTE—

- *The intake manifold must be removed in order to remove the fuel rail and injectors. See* **15b Cylinder Head (3.2L Engine)** *for overview of intake manifold.*
- *Replace all fuel rail O-rings when installing injectors.*
- *Lubricate O-rings with clean engine oil.*

1. Fuel return line
2. Fuel supply line
3. Fuel rail
4. Socket head bolt
   - 10 Nm (7 ft-lb)
5. Fuel pressure regulator
6. O-ring
7. O-ring
8. Retaining clip
9. O-ring
10. Retaining clip
11. O-ring
12. Fuel injector

Engine Control Module (ECM)

ECM, removing and installing

The ECM is in the air plenum near the base of windshield in the engine compartment.

– Before replacing ECM, connect VAG 1551/1552 or VAS 5051/5052 diagnostic scan tool, or equivalent, to diagnostic plug under left side of dashboard and record (print) ECM coding data. See **OBD On Board Diagnostics** for more information on using scan tool.

NOTE—

- *When the connector is detached from the ECM, adaptation values are erased but contents of DTC memory remain intact.*

– Switch ignition OFF.

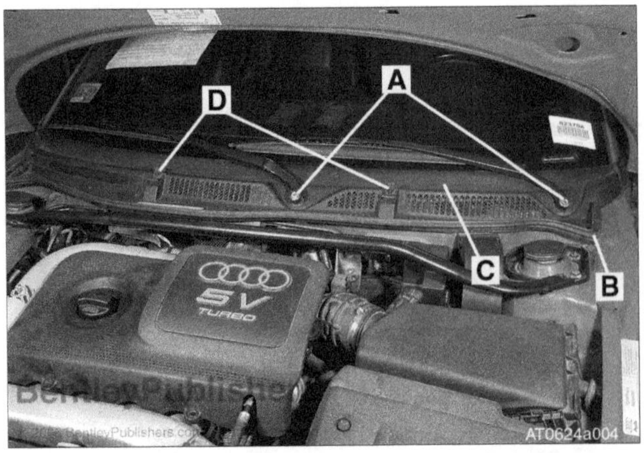

< Remove windshield wipers (**A**). See **92 Wipers and Washers**.

– Remove rubber weather-strip (**B**) from plenum base.

– Remove plenum cover (**C**) by carefully pulling away from base of windshield.

CAUTION—
• *Do not use screwdriver or pry tool against windshield glass. Glass may crack.*

– Disconnect washer fluid hoses from back of washer jets (**D**) when removing plenum cover (**C**).

BentleyPublishers.com
©2006 BentleyPublishers.com—All Rights Reserved

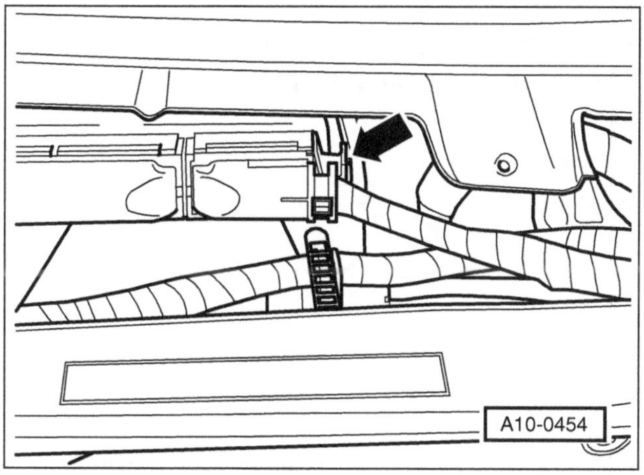

ECM without metal anti-theft housing

◀ Release ECM harness connectors at **arrow**.

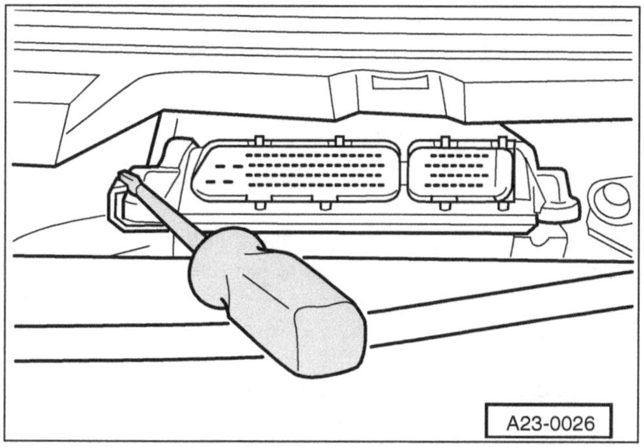

◀ Use screwdriver to press locking lugs of bracket to the side and then remove ECM toward front.

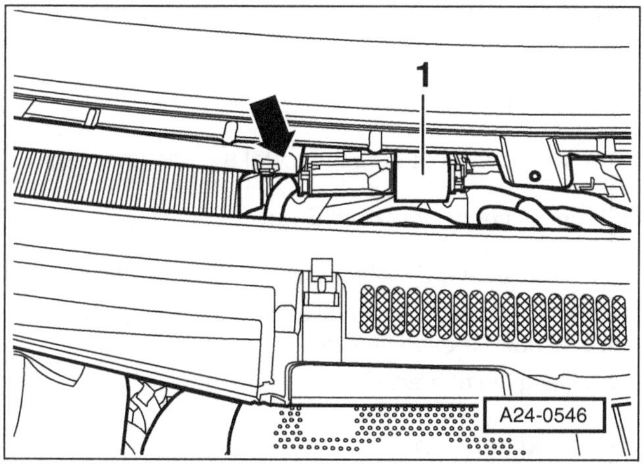

ECM with metal anti-theft housing

NOTE—

- *A heat gun with a semicircular nozzle is necessary to remove the ECM from the anti-theft housing. A heat gun from the VAS 1978 wiring harness repair kit is shown below. An equivalent heat gun may be used instead.*

◀ Use a screwdriver to press locking lug of bracket (**arrow**) to the side and then remove ECM with anti-theft housing toward front.

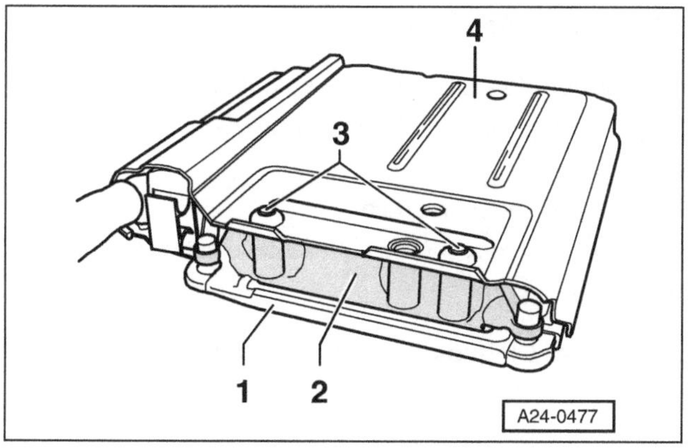

◀ To restrict accessibility (for theft prevention), the ECM (**1**) is screwed to a metal anti-theft housing (**4**) using a locking mechanism (**2**) and snap-off screws (**3**).

- The threads of the snap-off screws are coated with locking compound to make it more difficult to remove the snap-off screws.
- To disconnect the harness connectors from the ECM (e.g. when connecting a test box or when replacing the ECM), the ECM must be removed from the metal anti-theft housing.

– Position the ECM so that the locking mechanism (**2**) is accessible. Place a clean rag below the ECM.

BentleyPublishers
.com
© 2006 BentleyPublishers.com—All Rights Reserved

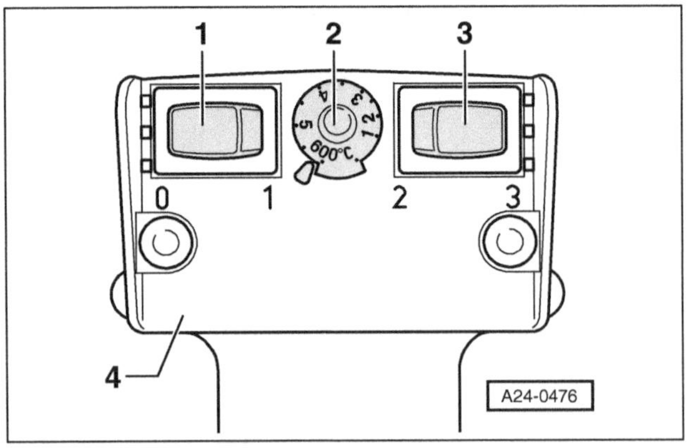

◄ Adjust heat gun as shown in illustration.
- Temperature adjustment (**2**) for maximum heat output
- Two-stage switch for air flow (**3**) set to position 3

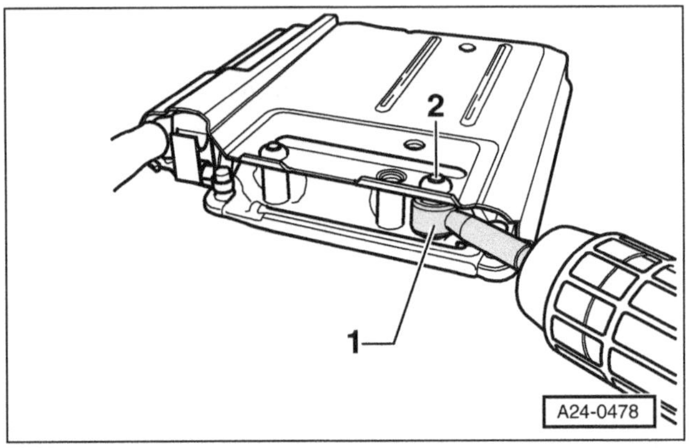

◄ Guide nozzle (**1**) of heat gun onto the locking mechanism threads so that the nozzle "encloses" the threads.
- It is OK to let the nozzle make contact at the upper end of the metal anti-theft housing.

— Switch on heat gun and heat up thread for approx. 20 to 25 seconds.

> ***WARNING —***
> - *The snap-off screws and parts of the metal anti-theft housing can become very hot. Be careful not to get burned!*

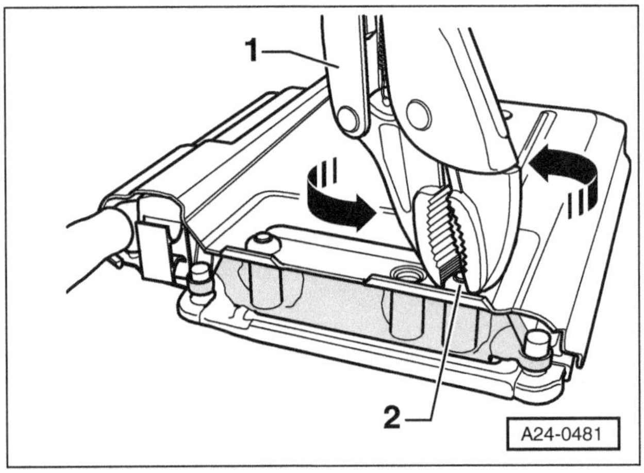

◄ Use vice grips to grab screw head (**2**) and remove snap-off screw by turning in counter clockwise direction (**arrows**).

— Repeat procedure with heat gun and vice grips for other snap-off screw.

— Remove ECM locking mechanism.

— The ECM can now be slid out from the metal anti-theft housing.

— Disconnect harness connectors from ECM.

— Installation is the reverse order of removal, noting the following:
- After repairs are complete, the metal anti-theft housing must be reinstalled. Use new snap-off screws.
- Bend back locking lug of bracket.

Continued for all versions

— Install ECM in reverse order of removal.

— If ECM was replaced, use diagnostic scan tool to:
- Check Diagnostic Trouble Code (DTC) memory and erase if need be
- Code new ECM
- Adapt throttle valve control module
- Adapt kickdown point for vehicles with automatic transmission
- Generate readiness code

See **OBD On Board Diagnostics** for more information.

BentleyPublishers.com
©2006 BentleyPublishers.com—All Rights Reserved

26 Exhaust System

26

General

This repair group covers removal and installation of the exhaust system and secondary air injection system (AIR) components.

Engine	Engine code
1.8L 4-cylinder turbo (180 hp)	ATC, AWP
1.8L 4-cylinder turbo (225 hp)	AMU, BEA
3.2L 6-cylinder gasoline (250 hp)	BHE

NOTE—

- *Vehicles covered by this manual have a self-diagnostic program to detect emissions-related problems and store coded fault information in electronic memory. When diagnosing an emissions-related problem, first check for stored fault codes and troubleshoot using Audi scan tools VAG 1551/1552, diagnostic computer VAS 5051/5052, 5051B or equivalent scan tool or computer program.*
- *Vehicles covered by this manual have a vacuum hose routing diagram (sticker) located in the engine compartment which can be helpful when working with vacuum hoses.*

BentleyPublishers.com
© 2006 BentleyPublishers.com—All Rights Reserved

Warnings and Cautions

To help guard against personal injury or damage to vehicles components, the following warnings apply when servicing exhaust system.

WARNING—
- *Exhaust gases are colorless, odorless, and very toxic. Run the engine only in a well-ventilated area. Immediately repair any leaks in exhaust system or structural damage to car body that might allow exhaust gases to enter passenger compartment.*
- *The exhaust system, catalytic converter and other emission control systems operate at very high temperatures. Allow these components to cool before servicing, or wear protective clothing to prevent burns. Do not use flammable chemicals near a hot catalytic converter.*
- *Old corroded exhaust system components crumble easily and often have exposed sharp edges. To avoid injury, wear eye protection and heavy gloves when working with such parts.*
- *Do not work under a lifted car unless it is solidly supported on jack stands designed for that purpose. Never work under a car that is supported solely by a jack.*
- *Inspect the exhaust system heat shields on the car underbody and repair or replace damaged parts. Heat shields protect the car occupants, undercoating, and various other components from excessive heat. Damaged or missing shields, particularly those above the catalytic converter, increase interior temperatures and create a fire hazard.*

CAUTION—
- *Do not operate the starter for long periods if the engine fails to start. Extended cranking may allow excess fuel to enter the catalytic converter, creating a fire hazard and possibly damaging the converter.*

EXHAUST SYSTEM (ALL ENGINES)

For safe operation, check exhaust components to be sure that all are free of holes and all connections airtight. Check the system immediately if it becomes noisy or an exhaust odor is detected inside the car.

Excessive exhaust system back pressure can cause driveability problems such as a rough idle or stalling. Back pressure problems are caused by external damage such as crushed or collapsed parts, or internal damage such as a plugged catalytic converter.

Exhaust system replacement

The exhaust systems for the various engines available in the cars covered by this manual are illustrated in several exploded views in this repair group. Installation details for the various systems are also included.

A liberal application of penetrating oil to cold exhaust system nuts, bolts and slip joints makes removal easier. Use new fasteners, clamps and rubber mounts when replacing exhaust components.

BentleyPublishers.com

© 2006 BentleyPublishers.com—All Rights Reserved

Replace gaskets whenever flange joints are disconnected. Use high-temperature anti-seize compound on threaded fasteners to extend their service life and make future replacement easier. Some smoke and odor is normal as new exhaust system parts become hot for the first few times.

NOTE —

- *After performing repairs on the exhaust system, ensure that the system is not under stress and that it has sufficient clearance from the body. If necessary, loosen the double clamps and align the muffler and exhaust pipe so sufficient clearance is maintained to the body while the support rings are evenly loaded.*
- *When performing repairs, always replace self-locking nuts.*

Exhaust system, installation details

The figures below show various details relevant to installing exhaust components in cars covered by this manual.

– Raise car and support safely. See **03 Maintenance**.

> ***WARNING —***
> - *Make sure the car is stable and well supported at all times. Use a professional automotive lift or jack stands designed for the purpose. A floor jack is not adequate support.*

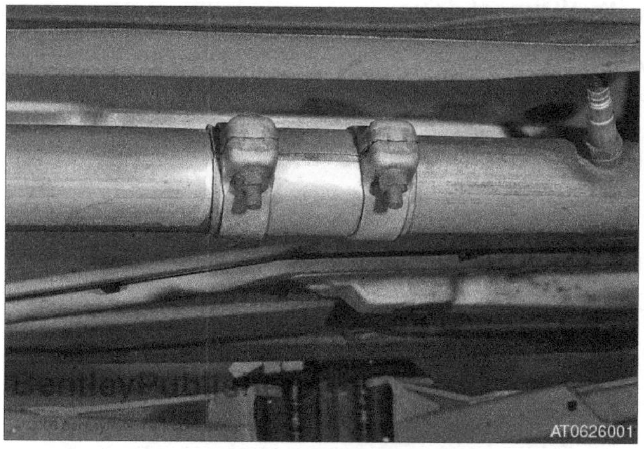

◄ Installation position of double clamp

- Do not allow mounting bolts to project beyond bottom edge of clamps.

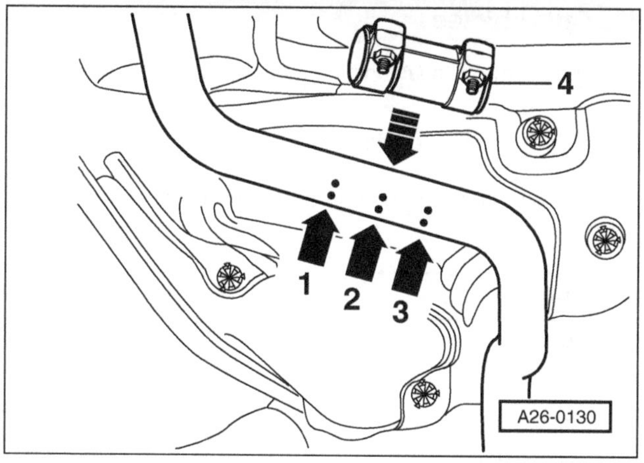

◄ Separation point of front and rear exhaust pipes, used if replacing front or rear muffler separately:

- Cut exhaust pipe at separation point (**2**) using body repair saw at right angle to exhaust pipe.
- Use an assistant to install repair clamp and help hold exhaust components in place.
- Attach double clamp (**4**) joining front exhaust pipe and catalytic converter loosely.
- Align rear muffler horizontally and hold in position.
- Position double clamp (**4**) between **1** and **3** over separation point.

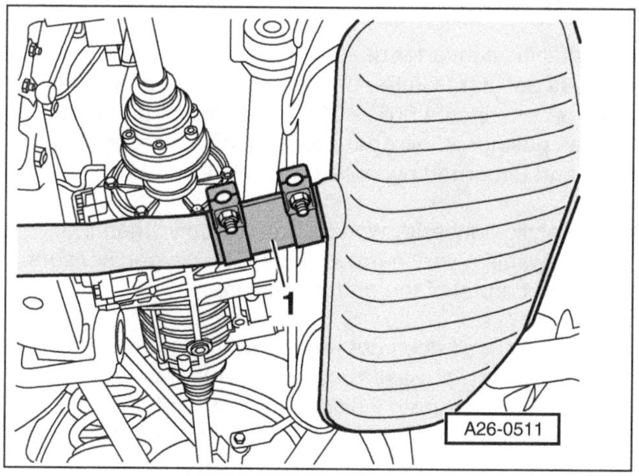

◄ A double clamp (**1**) may also be used on the rear of the exhaust system to separate the center muffler and pipe from the rear muffler.

– When installing double clamp, be sure to align exhaust as shown in **Exhaust system, aligning**.

– Tighten mounting bolts to specification.

Tightening torque	
Exhaust double clamp bolts (1.8L)	25 Nm (18 ft-lb)
Exhaust double clamp bolts (3.2L)	40 Nm (30 ft-lb)

BentleyPublishers.com
©2006 BentleyPublishers.com—All Rights Reserved

Exhaust system, aligning

– Install front double clamp as described above. Tighten mounting bolts hand tight.

Vehicles with front wheel drive

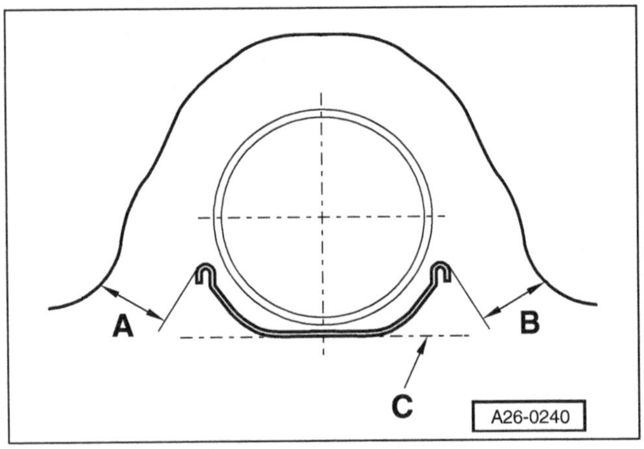

◀ Align catalytic converter horizontally (**C**) and make sure that gaps (**A** and **B**) between catalytic converter and body tunnel are equal.

Vehicles with all wheel drive

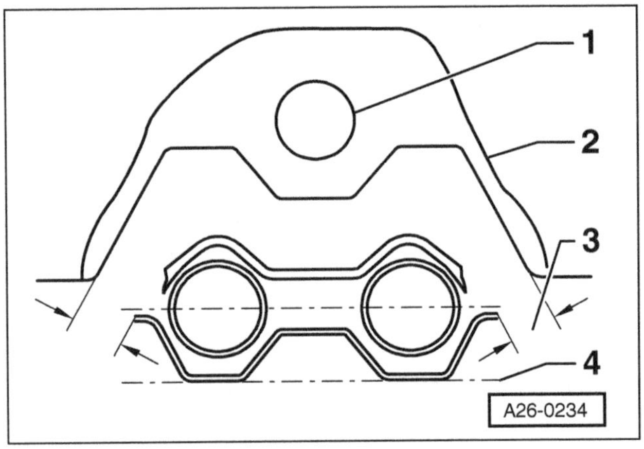

◀ Align catalytic converter horizontally (**4**) and make sure that gaps (**3**) between catalytic converter and heat shield in body tunnel (**2**) are equal.

- Driveshaft shown at **1**.

Continued for all vehicles

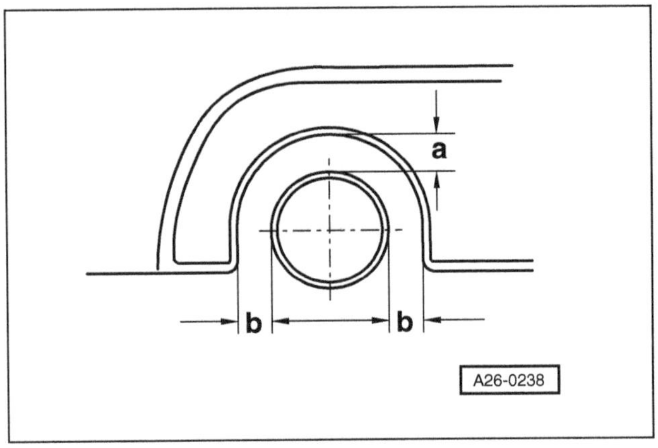

◀ Align rear muffler so that gaps (**A** and **B**) between tail pipe and bumper cut out are equal.

- To align tailpipe, loosen rear muffler mount.

– Hold exhaust system in aligned position and tighten double clamp uniformly to specification.

Tightening torque	
Double clamp mounting bolts	25 Nm (18 ft-lb)

Catalytic converter

The three-way catalytic converter used on gasoline engines chemically reduces pollutants in the engine exhaust. A properly operating converter provides a 90% - 95% reduction of the three major exhaust gas pollutants, nitrogen oxides (NO_x), carbon monoxide (CO), and unburned hydrocarbons (HC).

The three-way catalytic converter works correctly only when the air/fuel ratio is kept within a very narrow range. The oxygen sensors provide feedback for control of the air/fuel mixture.

The catalytic converter itself does not require routine maintenance. However, any problem that increases converter temperature beyond normal operating range (incorrect ignition timing or air/fuel mixture, engine misfire, prolonged idling or extended high engine loads) can

BentleyPublishers.com
© 2006 BentleyPublishers.com—All Rights Reserved

damage the converter. Reduced power, stalling, exhaust system rattles and excessive emissions are symptoms that may be caused by a faulty catalytic converter.

Catalytic converters on vehicles covered by this manual use a second heated oxygen sensor (post-catalyst) to monitor the operation of the catalyst. If the converter fails to properly oxidize the spent combustion gasses, the oxygen sensor detects this malfunction and a Diagnostic Trouble Code (DTC) is stored in the Engine Control Module (ECM).

Exhaust leaks anywhere in the system, but especially between the cylinder head and the catalytic converter, can cause running problems, false DTCs or converter failure.

Accurately testing the function of the catalytic converter requires an infrared exhaust gas analyzer to check the air/fuel mixture adjustment (% CO) at the exhaust system test port upstream of the catalytic converter, and then again at the tailpipe. If the converter is working properly, the tailpipe reading will be lower than the test port reading. If the reading is the same or only slightly less, this indicates a faulty catalytic converter.

Oxygen sensor, removing and installing

Testing the oxygen sensor requires specialized scan tool equipment. Scan tools are available from Audi and aftermarket suppliers. For a list of diagnostic scan tool suppliers, see **OBD On Board Diagnostics**.

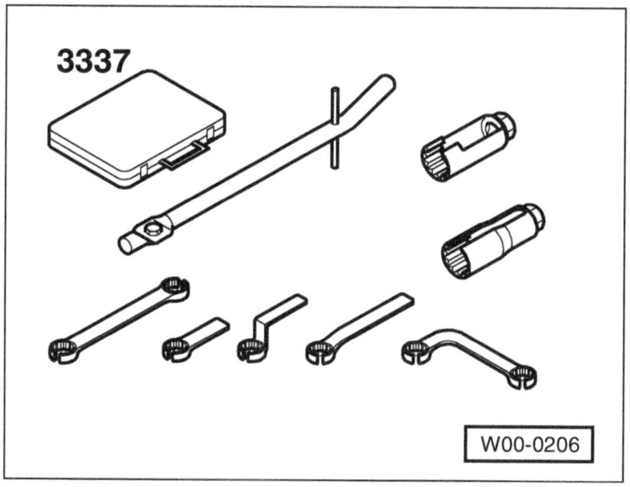

Removing and installing the oxygen sensor requires Audi special tool 3337, or equivalent, to avoid damaging the sensor.

> ***CAUTION—***
> - *When installing oxygen sensor be sure to use high temperature anti-seize compound that contains no silicone and is marked "safe for oxygen sensors" (Audi part # G 052 112 A3).*
> - *Audi part numbers are given for reference only. Always consult with your Audi Parts Department or aftermarket parts specialist for the latest parts information.*
> - *Do not allow anti-seize compound to enter slots on oxygen sensor body.*

Aftermarket oxygen sensor socket (**arrow**) shown in use.

BentleyPublishers.com
©2006 BentleyPublishers.com—All Rights Reserved

Exhaust System (1.8L, Engine Code: ATC, AWP)

NOTE—

- *Audi identifies electrical components by a letter and/or a number in the electrical schematics. See* **EWD Electrical Wiring Diagrams**. *These electrical identifiers are listed in parentheses as an aid to electrical troubleshooting.*

Exhaust system overview (front wheel drive)

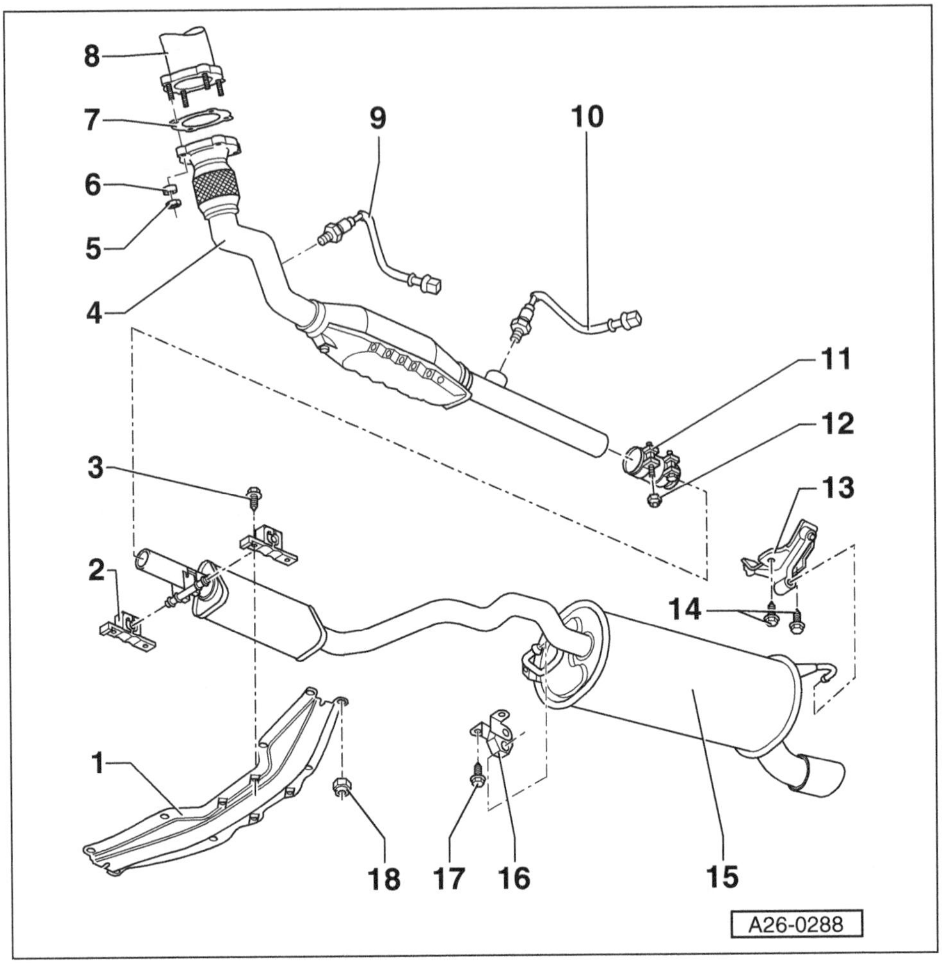

1. **Tunnel cross-piece**
2. **Front mounts**
3. **Bolt**
   - 25 Nm (18 ft-lb)
4. **Front exhaust pipe**
   - With catalytic converter
   - With de-coupling element
   - Do not bend de-coupling element more than 10°
   - Protect from impact damage
5. **Bolt**
   - Quantity: 4
   - 40 Nm (30 ft-lb)
   - Always replace
6. **Spacer sleeve**
   - Quantity: 4
7. **Gasket**
   - Always replace
8. **Turbocharger**
9. **Heated oxygen sensor, pre-cat (G39)**
   - 55 Nm (41 ft-lb)
   - Coat threads with anti-seize paste (Audi part # G 052 112 A3)
   - Do not get anti-seize in slots of sensor body
10. **Heated oxygen sensor, post-cat (G130)**
    - 55 Nm (41 ft-lb)
    - Coat threads with anti-seize paste (Audi part # G 052 112 A3)
    - Do not get anti-seize in slots of sensor body
11. **Double clamp**
    - Note installation position: bolted connection on left side
12. **Nut**
    - 25 Nm (18 ft-lb)
13. **Rear mount**
14. **Bolts**
    - 25 Nm (18 ft-lb)
15. **Rear muffler**
    - Can be replaced separately from front muffler/pipe
16. **Middle mount**
17. **Bolt**
    - 25 Nm (18 ft-lb)
18. **Nut**
    - 25 Nm (18 ft-lb)

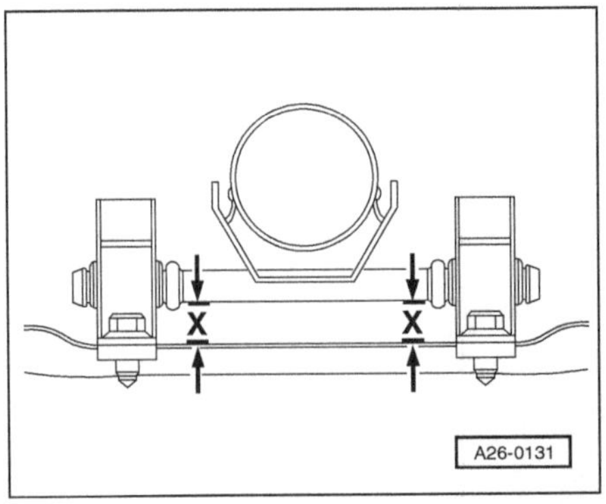

Installation details

◄ Mounting pin must be parallel to tunnel cross piece.

- Dimension **X** must be the same on both sides.

BentleyPublishers.com

© 2006 BentleyPublishers.com—All Rights Reserved

Exhaust system overview (all wheel drive)

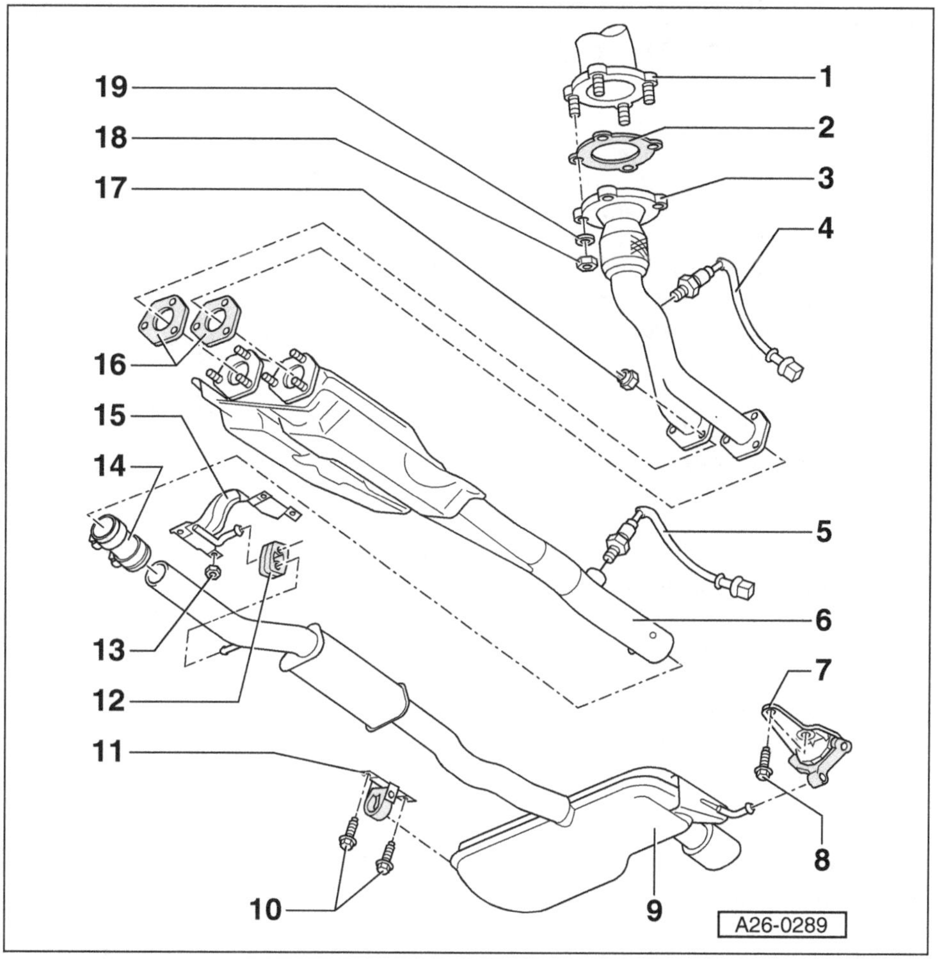

1. **Turbocharger**
2. **Gasket**
   - Always replace
3. **Front exhaust pipe**
   - With catalytic converter
   - With de-coupling element
   - Do not bend de-coupling element more than 10°
   - Protect from impact damage
   - Gap between front pipe and heat shield for steering gear should be at least 24 mm (1 inch)
4. **Heated oxygen sensor, pre-cat (G39)**
   - 55 Nm (41 ft-lb)
   - Coat threads with anti-seize paste (Audi part # G 052 112 A3)
   - Do not get anti-seize in slots of sensor body
5. **Heated oxygen sensor, post-cat (G130)**
   - 55 Nm (41 ft-lb)
   - Coat threads with anti-seize paste (Audi part # G 052 112 A3)
   - Do not get anti-seize in slots of sensor body
6. **Catalytic converter**
   - Protect from impact damage
7. **Rear mount**
   - Loosen mount to center tailpipe in bumper opening
8. **Bolt**
   - 25 Nm (18 ft-lb)
9. **Rear muffler**
   - Can be replaced separately from front muffler/pipe
10. **Bolts**
    - 25 Nm (18 ft-lb)
11. **Mount**
12. **Rubber hanger**
13. **Nut**
    - 25 Nm (18 ft-lb)
14. **Double clamp**
    - Note installation position: bolted connection on right side
15. **Front mount**
16. **Gaskets**
    - Always replace
17. **Nut**
    - Quantity: 6
    - 25 Nm (18 ft-lb)
18. **Nut**
    - Quantity: 4
    - 40 Nm (30 ft-lb)
19. **Spacer sleeve**
    - Quantity: 4

Installation details

◄ Angled end (**arrow**) of front mount faces toward front of vehicle.

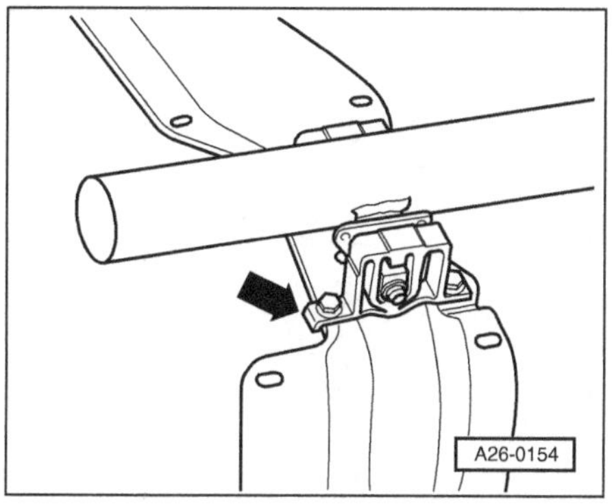

BentleyPublishers.com
©2006 BentleyPublishers.com—All Rights Reserved

Exhaust System (1.8L, Engine Code: AMU, BEA)

NOTE—

- *Audi identifies electrical components by a letter and/or a number in the electrical schematics. See* **EWD Electrical Wiring Diagrams**. *These electrical identifiers are listed in parentheses as an aid to electrical troubleshooting.*

Exhaust system overview (all wheel drive)

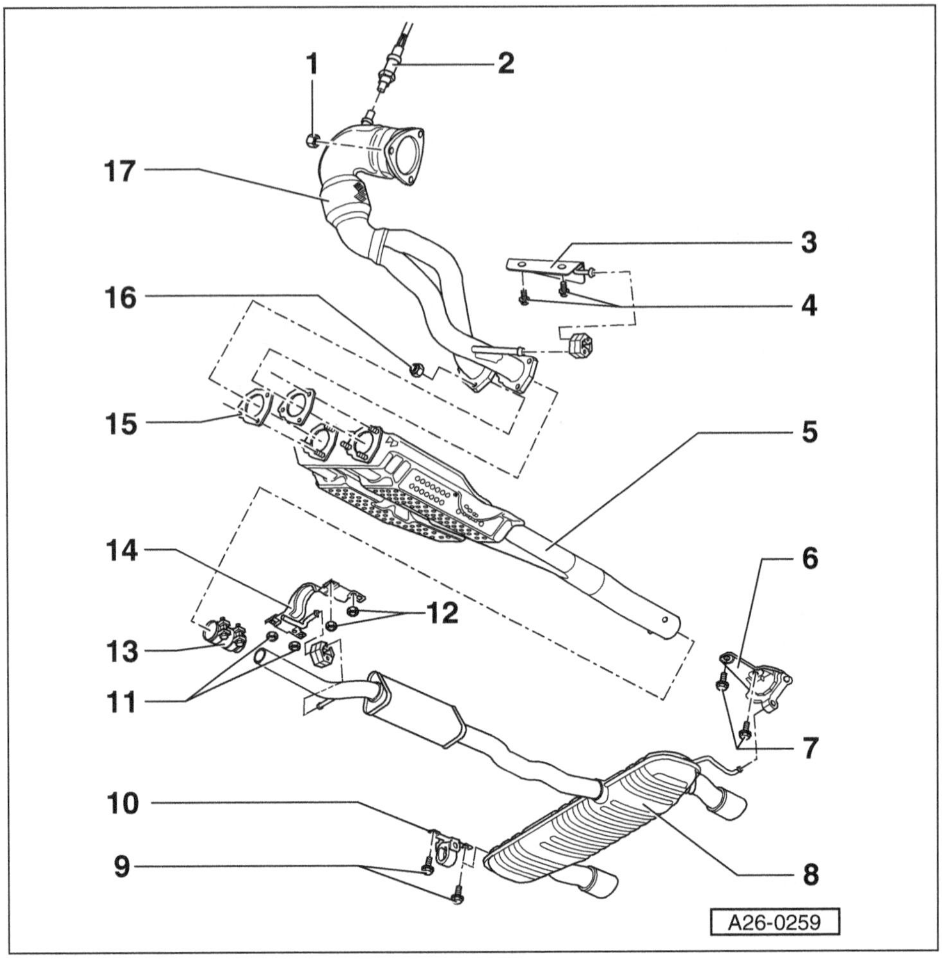

1. **Bolt**
   - 40 Nm (30 ft-lb)
2. **Heated oxygen sensor, pre-cat (G39)**
   - 55 Nm (41 ft-lb)
   - Coat threads with anti-seize paste (Audi part # G 052 112 A3)
   - Do not get anti-seize in slots of sensor body
3. **Mount**
4. **Bolts**
   - 25 Nm (18 ft-lb)
5. **Catalytic converter**
   - Protect from impact damage
6. **Mount**
7. **Bolts**
   - 25 Nm (18 ft-lb)
8. **Rear muffler**
   - Can be replaced separately from front muffler/pipe
9. **Bolts**
   - 25 Nm (18 ft-lb)
10. **Mount**
11. **Nuts**
    - 25 Nm (18 ft-lb)
12. **Nuts**
    - 25 Nm (18 ft-lb)
13. **Double clamp**
    - Note installation position: bolted connection on left side
14. **Front mount**
15. **Gaskets**
    - Always replace
16. **Nut**
    - Quantity: 6
    - 25 Nm (18 ft-lb)
17. **Front exhaust pipe**
    - With catalytic converter
    - With de-coupling element
    - Do not bend de-coupling element more than 10°
    - Protect from impact damage
    - Gap between front pipe and heat shield for steering gear should be approx. 30 mm (1.2 inches)

Installation details

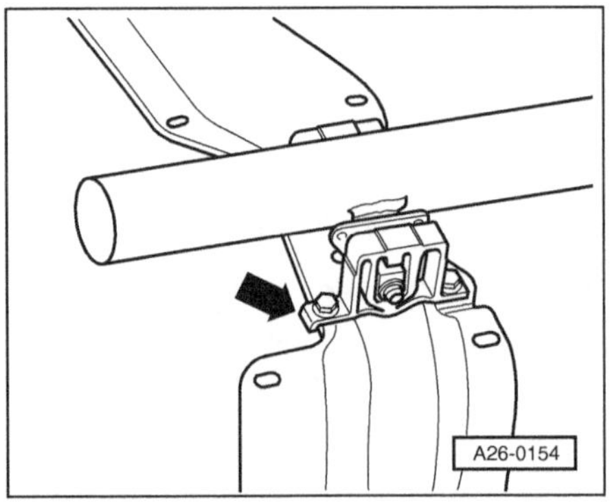

Angled end (**arrow**) of front mount faces toward front of vehicle.

BentleyPublishers.com

© 2006 BentleyPublishers.com—All Rights Reserved

Exhaust System (3.2L Engine)

Exhaust system overview (all wheel drive)

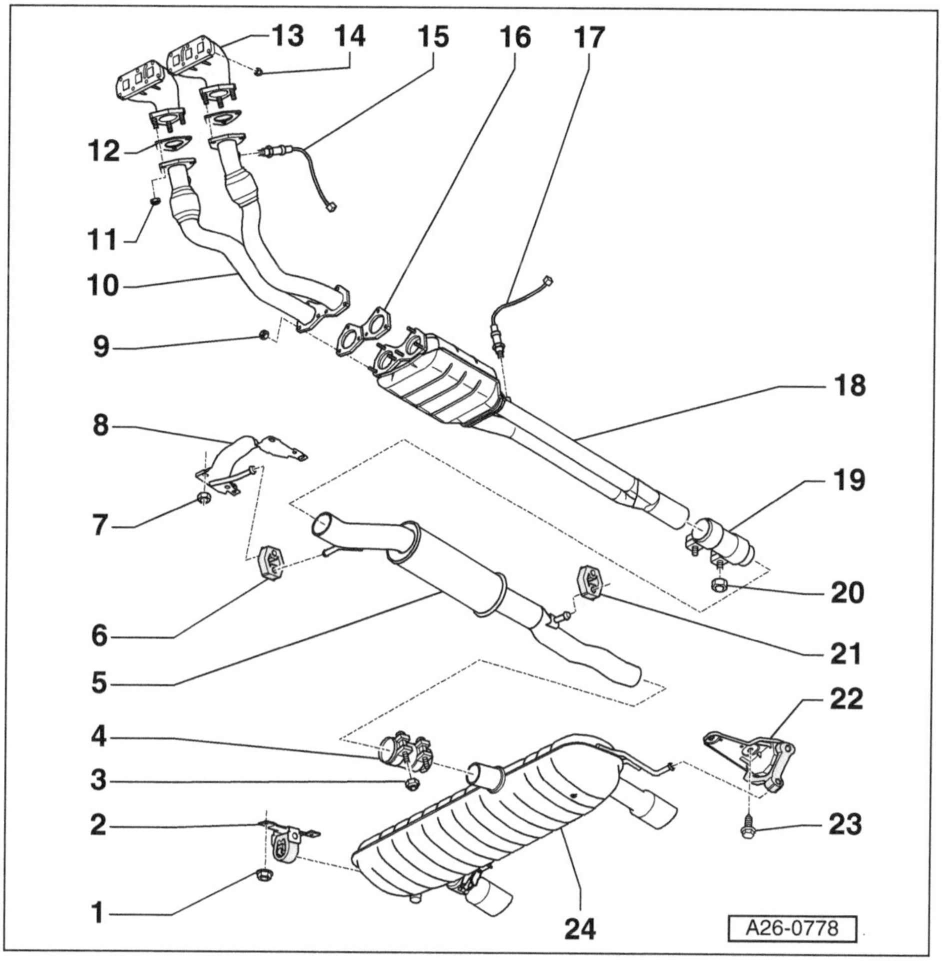

1. **Nut**
   - 23 Nm (17 ft-lb)
2. **Rear mount**
3. **Nut**
   - 40 Nm (30 ft-lb)
4. **Rear double clamp**
   - For separation of center muffler/pipe from rear muffler
   - Note installation position: bolted connection on left side
5. **Center muffler**
6. **Rubber mount**
7. **Nut**
   - 23 Nm (17 ft-lb)
8. **Bracket**
9. **Nut**
   - Quantity: 6
   - 23 Nm (17 ft-lb)
   - Always replace
10. **Front exhaust pipe**
    - With catalytic converter
    - With de-coupling element
    - Do not bend de-coupling element more than 10°
11. **Nut**
    - Quantity: 6
    - 40 Nm (30 ft-lb)
12. **Gaskets**
    - Always replace
13. **Exhaust manifold**
14. **Nut**
    - 23 Nm (17 ft-lb)
15. **Heated oxygen sensor, pre-cat (G39, G108)**
    - Quantity: 2
    - 55 Nm (41 ft-lb)
    - Coat threads with anti-seize paste (Audi part # G 052 112 A3)
    - Do not get anti-seize in slots of sensor body
16. **Gasket**
17. **Heated oxygen sensor, post-cat (G130, G131)**
    - Quantity: 2
    - 55 Nm (41 ft-lb)
    - Coat threads with anti-seize paste (Audi part # G 052 112 A3)
    - Do not get anti-seize in slots of sensor body
18. **Catalytic converter**
    - Protect from impact damage
19. **Double clamp**
    - Note installation position: bolted connection on right side
20. **Bolt**
    - 40 Nm (30 ft-lb)
21. **Rubber mount**
22. **Mount**
23. **Bolt**
    - 23 Nm (17 ft-lb)
24. **Rear muffler**
    - Can be replaced separately from front muffler/pipe

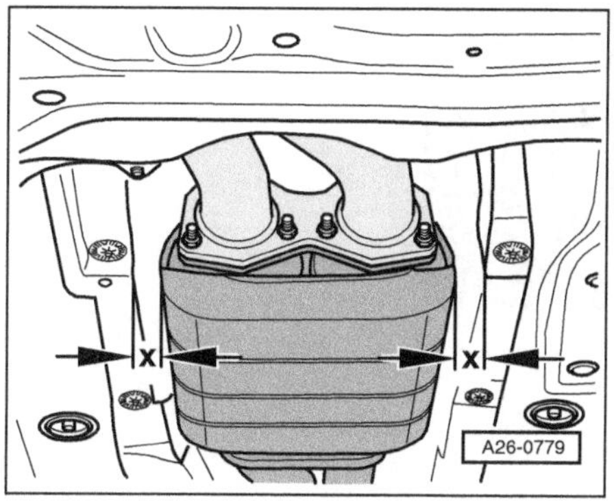

Installation details

◀ Align catalytic converter so that distance (**x**) is the same on both sides.

BentleyPublishers.com
© 2006 BentleyPublishers.com—All Rights Reserved

Secondary Air Injection (AIR) System

The secondary air injection (AIR) system uses an electrically driven air pump to inject fresh air behind the exhaust valves. The purpose of this system is to reduce exhaust emissions during engine warm-up when the Motronic engine management system is in open loop. The AIR system produces an oxygen rich exhaust gas, causes after-burning and reduces duration of the catalyst heat-up phase. Activation initiates from the Motronic ECM via the secondary air injection pump relay and secondary air injection solenoid valve.

System overview (1.8L)

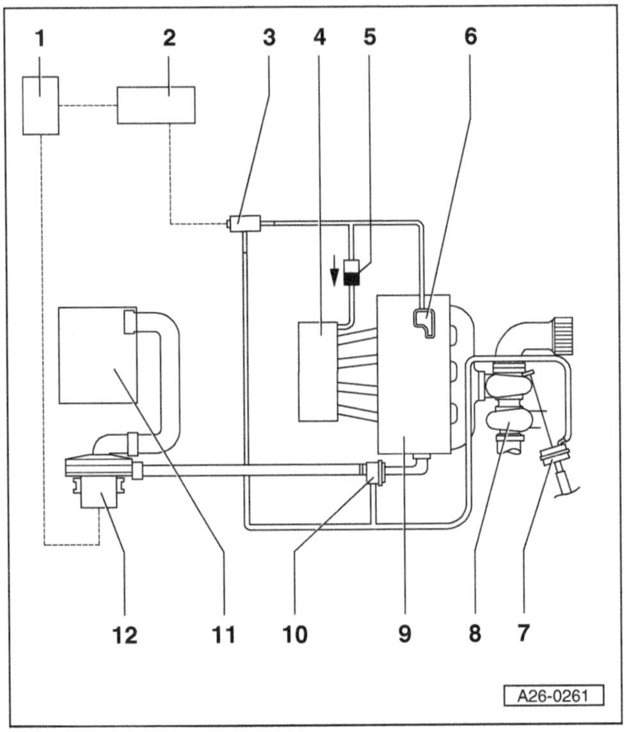

Secondary air injection system diagram

1. Secondary air injection pump relay (J299)
   - See **97 Fuses, Relays, Component Locations** for location
2. Motronic engine control module (J220)
3. Secondary air injection solenoid valve (N112)
4. Intake manifold
5. Check valve
   - Note installation position: light/dark sides installed as shown in diagram **arrow** points in direction of flow
6. Vacuum reservoir
   - Mounted on cylinder head cover
7. Pressure can for charge air pressure regulation
8. Turbocharger
9. Cylinder head
10. Combination valve
11. Air filter
12. Secondary air injection pump motor (V101)

Secondary air injection pump motor, removing and installing (1.8L)

– Remove upper sound absorber panels from top and front of engine. See **03 Maintenance**.

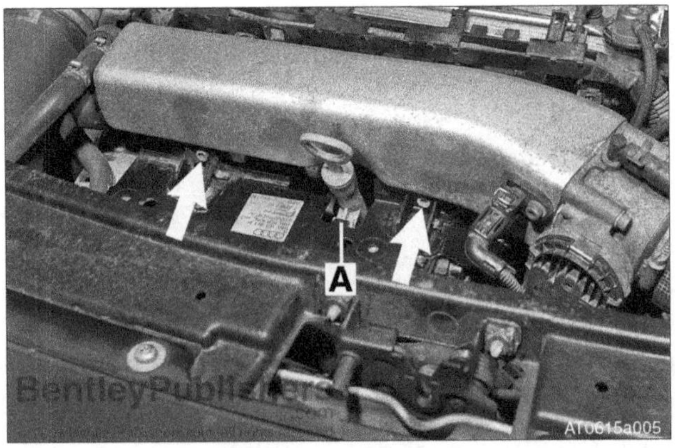

Unbolt (**arrows**) mounting bracket for solenoid valves from front of intake manifold.

– Detach dipstick guide tube (**A**).

BentleyPublishers.com
©2006 BentleyPublishers.com—All Rights Reserved

◀ Detach harness connector and unscrew mounting bolts (**arrows**) for secondary air injection solenoid valve.

– Unclip vacuum line from retaining clip (**right arrow**).

– Remove solenoid valve mounting bracket.

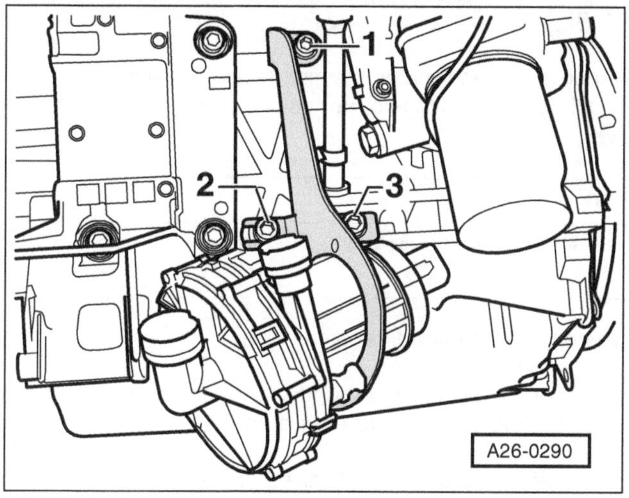

◀ Loosen secondary air pump bracket upper mounting bolt (**1**) approx. two turns.

- Illustration shows secondary air pump motor and mounting bracket with engine out of vehicle for clarity.

– Raise car and support safely. See **03 Maintenance**.

> ***WARNING —***
> - *Make sure the car is stable and well supported at all times. Use a professional automotive lift or jack stands designed for the purpose. A floor jack is not adequate support.*

– Remove sound absorber panel from under engine. See **03 Maintenance**.

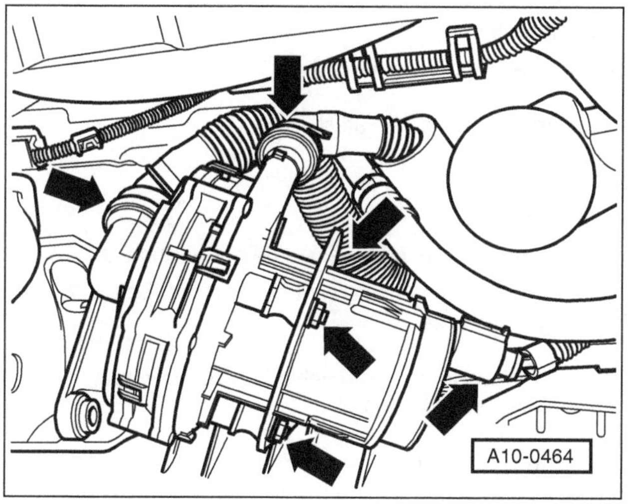

◀ Disconnect secondary air injection motor harness connector and hoses (**arrows**).

– Remove upper and lower mounting bolts for secondary air pump motor mounting bracket.

– Swing bracket to side and remove secondary air pump motor downward.

– Unbolt secondary air pump from mounting bracket.

– Install in reverse order of removal. Make sure pressure and intake hoses are securely engaged to secondary air injection motor.

Tightening torque	
Secondary air injection pump motor bracket to engine block	10 Nm (7 ft-lb)

BentleyPublishers.com
©2006 BentleyPublishers.com—All Rights Reserved

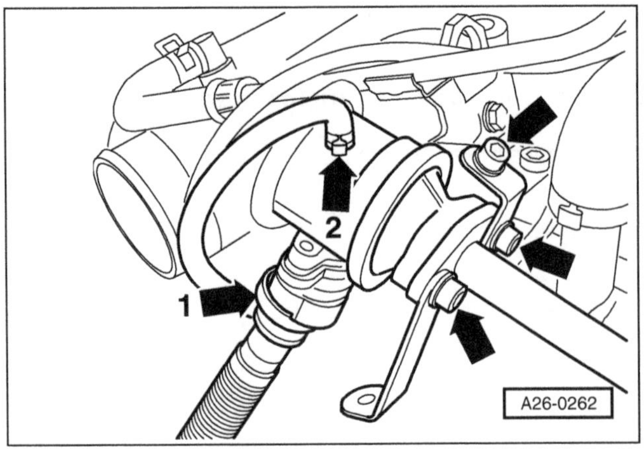

Secondary air injection combination valve, removing and installing (1.8L)

– Remove upper sound absorber panels from top of engine. See **03 Maintenance**.

◄ Remove vacuum line (**arrow 1**) and connecting line (**arrow 2)** at combination valve.

– Unbolt mounting bolts (**arrows**) and remove valve from bracket.

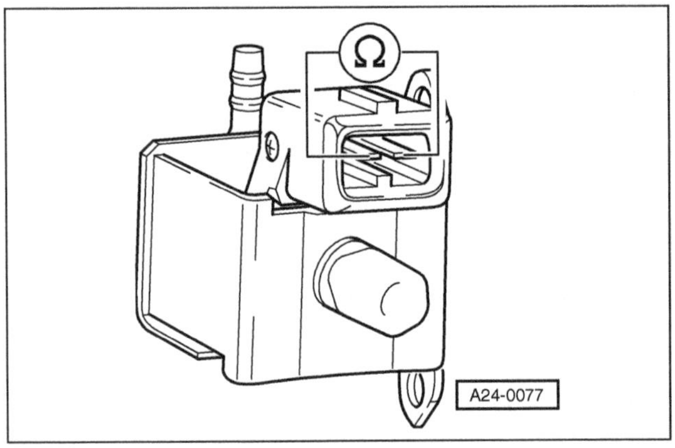

Secondary air injection solenoid valve (N112), testing (1.8L)

◄ Check resistance of secondary air injection solenoid valve by connecting multimeter leads to two harness connector terminals of solenoid valve.

Specification	
Secondary air injection solenoid valve resistance	25 -35 Ω

– If specified value is not obtained, replace valve.

– Continue testing solenoid valve for ground and voltage signals using LED voltage tester (V.A.G. 1527B or equivalent).

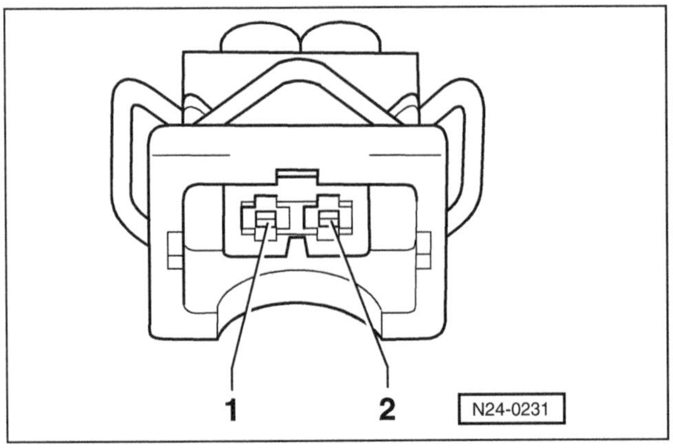

◄ Connect LED tester leads to solenoid valve harness connector terminal **1** and ground.

1. Ground (–) connection
2. Activation (voltage) signal from ECM

– Operate starter briefly (engine can start).
  - LED should light up.

– If LED does not light up:
  - Check fuse #43 in fuse panel at driver's side of dashboard
  - Check wiring for short circuits using wiring diagrams. See **EWD Electrical Wiring Diagrams**.

– Connect LED tester leads to solenoid valve harness connector terminal **2** and battery positive (B+).

– Use diagnostic scan tool to initiate output diagnostic test mode and activate secondary air injection solenoid valve. See **OBD On Board Diagnostics** for diagnostic scan tool information.
  - LED should light up.

– If LED does not light up:
  - Check fuse #43 in fuse panel at driver's side of dashboard
  - Check wiring for short circuits using wiring diagrams. See **EWD Electrical Wiring Diagrams**.

BentleyPublishers.com
© 2006 BentleyPublishers.com—All Rights Reserved

Secondary air injection system components (3.2L)

1. **Intake manifold**
2. **Vacuum unit**
   - For intake manifold change-over
3. **To intake manifold**
4. **To fuel pressure regulator**
5. **Combination valve**
6. **Gasket**
   - Always replace
7. **Air channel**
   - In cylinder head
8. **Bolts/nuts**
   - 10 Nm (7 ft-lb)
9. **Combination valve connection**
10. **Bolt**
    - 5 Nm (44 in-lb)
11. **Pressure hose**
    - Check for secure connection
12. **O-ring**
13. **Connection**
14. **Bracket**
    - For secondary air pump motor
15. **Bolt**
    - 20 Nm (15 ft-lb)
16. **Electrical harness connector**
17. **Rubber bushing**
18. **Secondary air injection pump motor (V101)**
19. **Air cleaner**
    - Clean if dirty
20. **Retaining clamp**
21. **Intake hose**
22. **Non-return valve**
    - Note installation position: light/dark sides installed as shown in diagram arrow points in direction of flow

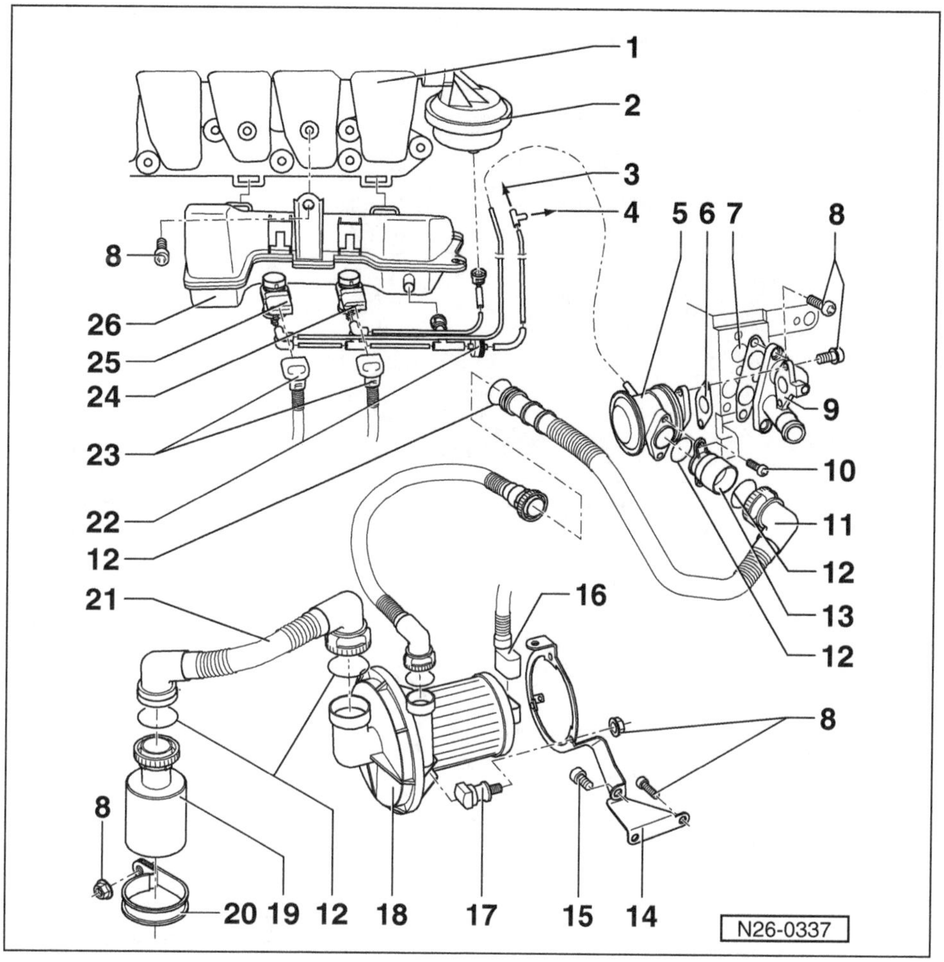

23. **Electrical harness connectors**
24. **Intake manifold change-over valve (N156)**
25. **Secondary air injection solenoid valve (N112)**
26. **Vacuum reservoir**

BentleyPublishers.com
©2006 BentleyPublishers.com—All Rights Reserved

Secondary air combination valve, checking (3.2L)

◀ Loosen hose clamps (**arrows**) and remove intake air hose.

– Disconnect vacuum line at combination valve.

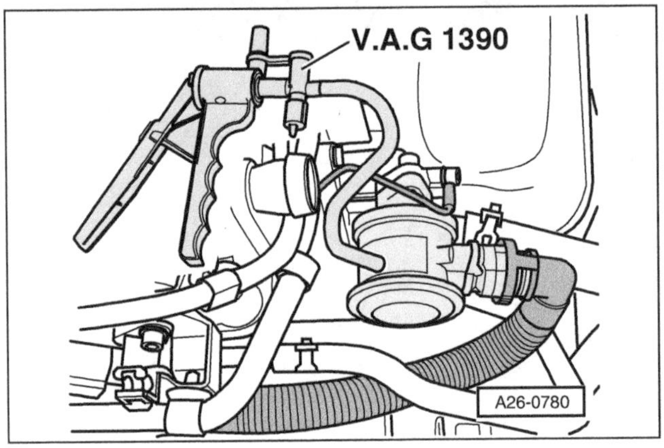

◀ Connect hand operated vacuum pump (V.A.G. 1390 shown) to vacuum line of combination valve.

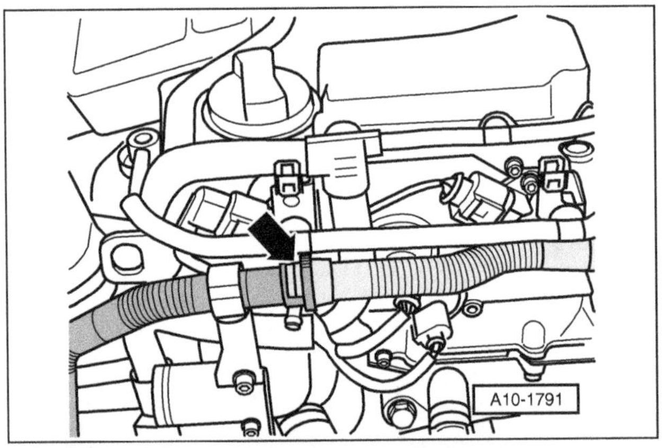

◀ Disconnect secondary air hose at **arrow**.

– Blow gently into hose for combination valve (do not use compressed air).
  - Combination valve must be closed, it should not be possible to blow air through.

– Operate hand vacuum pump.
  - Combination valve should open, it should now be possible to blow air through.

– If the combination valve does not open, replace valve.

Secondary air combination valve, removing and installing (3.2L)

◀ Disconnect air hose (**1**) at throttle valve control module.

– Disconnect harness connector from mass air flow sensor (**2**).

– Remove air cleaner mounting bolts (**arrows**) and remove air cleaner assembly.

BentleyPublishers
.com
© 2006 BentleyPublishers.com—All Rights Reserved

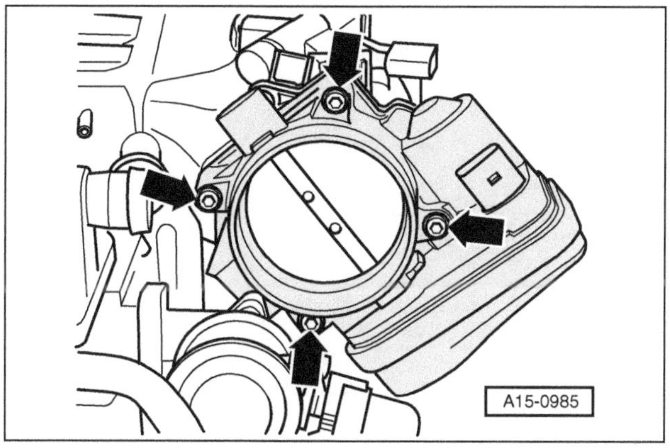

◀ Remove mounting bolts (**arrows**) and throttle valve control module from intake manifold. Set to side.

- Coolant hoses remain connected to throttle valve control module.

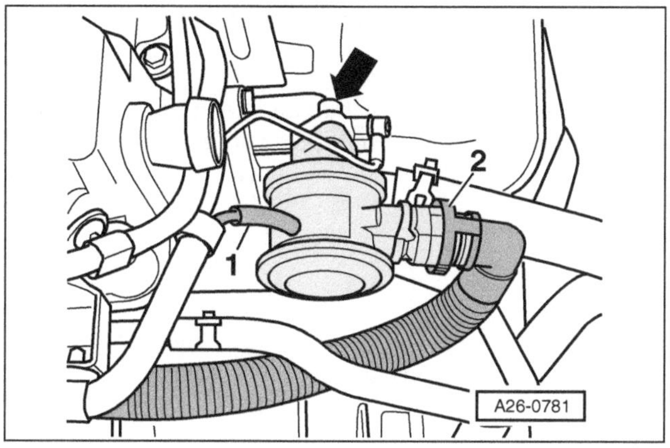

◀ Disconnect air hose (**2**) and vacuum hose (**1**) at combination valve.

– Remove mounting bolts (**arrows**) and combination valve.

– Install in reverse order of removal.

Tightening torques	
Combination valve to mounting bracket	10 Nm (7 ft-lb)
Throttle valve control module to intake manifold	10 Nm (7 ft-lb)

Secondary air injection pump, removing and installing (3.2L)

– Remove upper sound absorber panels from top of engine and top of coolant expansion tank. See **03 Maintenance**.

– Release tension of ribbed belt. See **03 Maintenance**.

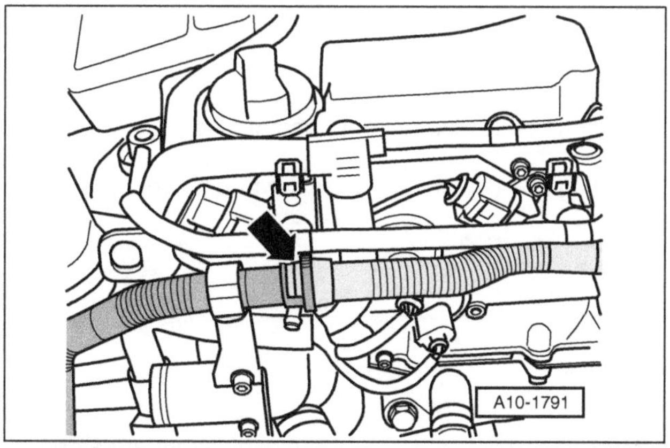

◀ Disconnect secondary air hose at **arrow**.

– Raise car and support safely. See **03 Maintenance**.

> ***WARNING —***
> - *Make sure the car is stable and well supported at all times. Use a professional automotive lift or jack stands designed for the purpose. A floor jack is not adequate support.*

– Remove sound absorber panel from under engine. Remove left and right side lower sound insulation panels. See **03 Maintenance**.

◀ Remove mounting bolts (**arrows**) and cross piece.

- Mounting bolts shown for left side, right side similar.

BentleyPublishers.com
©2006 BentleyPublishers.com—All Rights Reserved

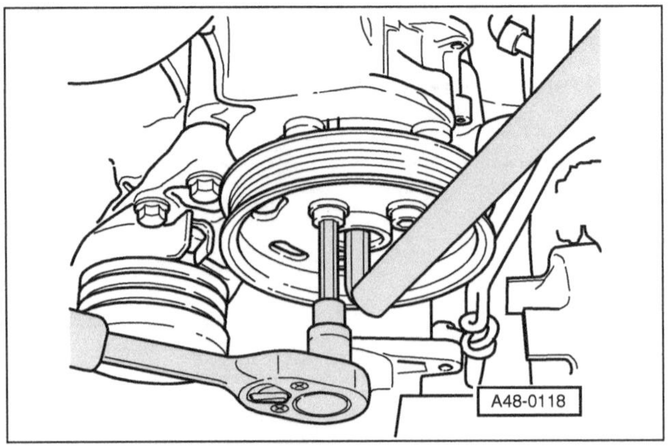

Remove pulley from power steering pump.

- Counterhold pulley with Allen key.

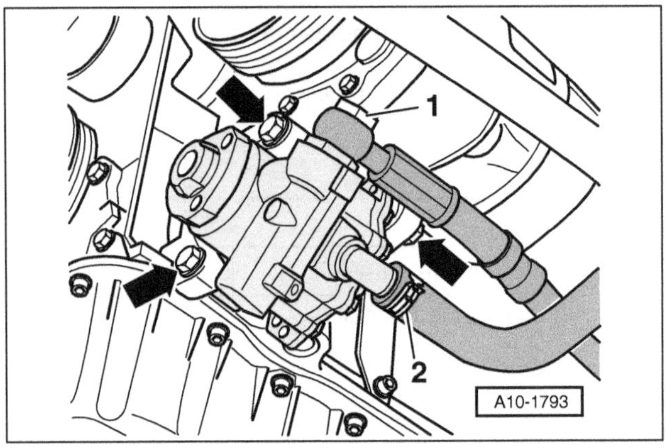

Remove mounting bolts (**arrows**) and power steering pump from bracket.

– Leave hydraulic lines (**1** and **2**) attached and hang power steering pump from vehicle using stiff wire or heavy duty zip ties.

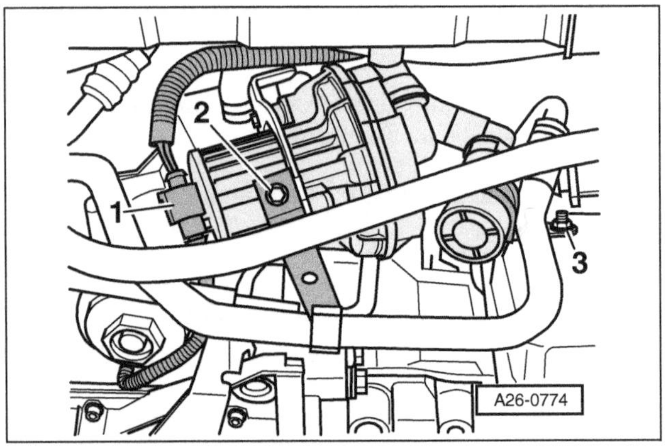

Disconnect electrical harness connector (**1**) from secondary air pump motor.

– Move wiring harness out of way.

– Remove mounting bolt (**2**) for coolant hose bracket.

– Remove mounting bolt (**3**) for secondary air pump filter bracket.

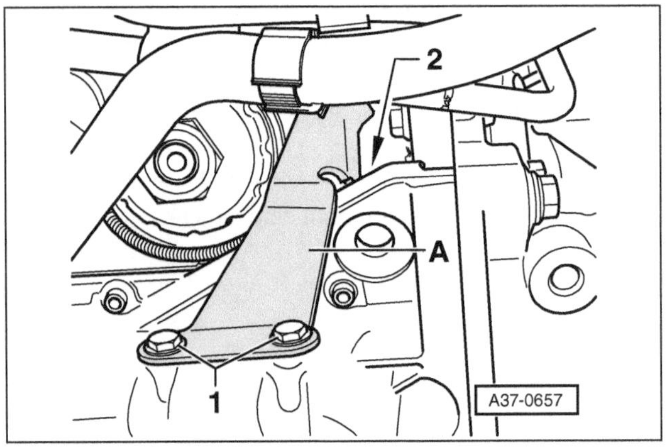

Remove mounting bolts (**1** and **2**) and remove secondary air pump motor with bracket (**A**).

– Remove secondary air pump motor from bracket.

– Install in reverse order of removal.

Tightening torques	
Secondary air pump to bracket	10 Nm (7 ft-lb)
Bracket to engine block	10 Nm (7 ft-lb)

BentleyPublishers.com
©2006 BentleyPublishers.com—All Rights Reserved

27 Battery, Alternator, Starter

TABLES

GENERAL

This repair group covers battery, alternator, and starter service and repair.

Also see the following:

- **03 Maintenance** for battery notes and ribbed belt replacement.
- **9 Electrical System–General** for bus information and electrical diagnostics.
- **EWD Electrical Wiring Diagrams** for battery, charging system and starter motor schematics.

NOTE—

- *The alternator is identified as generator by the vehicle manufacturer.*

Cruise control

Cruise control functions are controlled by the engine control module (ECM). The only serviceable cruise control components are the steering wheel, clutch and brake pedal switches and related wiring. See the following repair groups:

- **30 Clutch** for clutch pedal position switch.
- **46 Brakes–Mechanical** for brake light switch.
- **96 Interior Lights, Switches, Anti-theft** for steering column switches.

BentleyPublishers.com

© 2006 BentleyPublishers.com—All Rights Reserved

Warnings and Cautions

WARNING —

- *Wear goggles, rubber gloves, and a rubber apron when working around the battery and battery acid (electrolyte). Battery acid contains sulfuric acid and can cause skin irritation and burning.*
- *If battery acid is spilled on your skin or clothing, flush the area at once with large quantities of water. If electrolyte gets into your eyes, bathe them with large quantities of clean water for several minutes and call a physician.*
- *Batteries that are being charged or are fully charged give off explosive hydrogen gas. Keep sparks and open flames away. Do not smoke.*
- *Allow a frozen battery to thaw before attempting to recharge it.*
- *The gasses given off by the battery during charging are explosive. Do not smoke. Keep open flames away from top of battery, and prevent electrical sparks by turning off battery charger before connecting or disconnecting it.*
- *If battery begins gassing (boiling) violently when charging, reduce charging rate immediately.*
- *Ensure that battery is charged in a well ventilated area.*
- *When removing battery from vehicle, disconnect negative (-) terminal first.*
- *Before beginning repairs on the electrical system:*
  -Switch electrical consumers OFF.
  -Switch ignition OFF and remove ignition key.

CAUTION—

- *Do not disconnect the battery cables while the engine is running. The alternator will be damaged.*
- *Do not operate the alternator with its output terminal (B+ or 30) disconnected and the other terminals connected. Do not short, bridge, or ground any terminals of the charging system.*
- *Disconnect the negative (–) battery cable when working at or near the alternator. Battery voltage is always present at rear of alternator, even with ignition key OFF.*
- *Disconnect the battery cables during battery charging. This prevents damage to the alternator and other solid-state components.*
- *Loosen battery cell caps before charging battery but leave them on battery to prevent battery acid splatter.*
- *Do not reverse battery terminals. Even a momentary wrong connection can damage the alternator or electrical components.*
- *Replace the battery if case is cracked or leaking. Spilled electrolyte can damage the car. If electrolyte is spilled, clean area with a solution of baking soda and water.*
- *Batteries are damaged by quick charging. Use quick charging as a last alternative when slow charging is not possible.*
- *Do not quick charge a totally discharged battery.*
- *Do not allow battery charging rate to exceed 16.5 volts.*
- *Do not store precision tools in same room where batteries are being charged. Tools may corrode due to chemical reaction.*

BentleyPublishers.com
© 2006 BentleyPublishers.com—All Rights Reserved

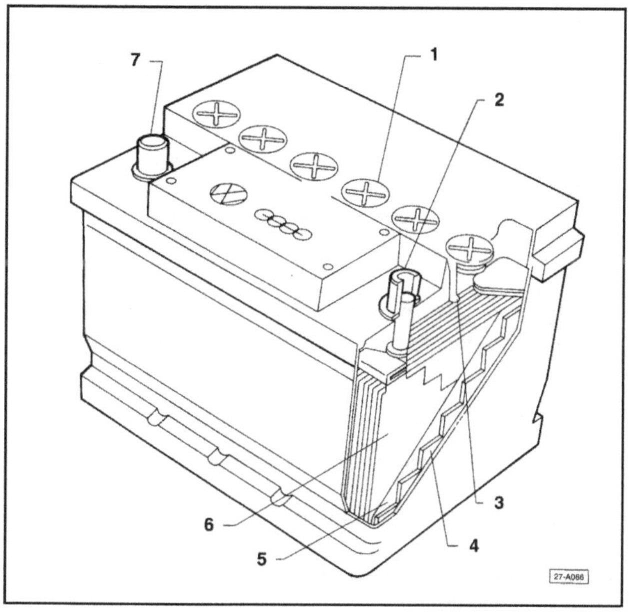

Battery

Battery components:

1. Filler cap
2. Positive (B+) terminal
3. Electrolyte level indicator
4. Negative plate (grey)
5. Separator (insulator)
6. Positive plate (dark brown)
7. Negative (–) terminal

The six-cell, 12-volt lead-acid battery capacity is rated in ampere / hours (Ah) and cold cranking amps (CCA). The Ah rating is determined by the average amount of current the battery can deliver over time without dropping below a specified voltage. The CCA rating is determined by the battery's ability to deliver starting current at 0°F (–18°C).

If the battery discharges when vehicle is not driven, there may be a constant drain or current draw causing it to discharge when ignition is OFF. Depending on the amount of draw and battery condition, a full discharge can happen overnight or it may take a few weeks. Although a small static drain on the battery is normal (example: to operate clock or radio memory), a large drain such as a relay sticking on or a faulty switch will cause the battery to discharge quickly. Perform a static current draw test first when experiencing battery discharge. See **Static current draw, checking** in this repair group.

If current draw on the battery is not excessive but it still discharges, the battery condition should be tested. Battery testing determines the state of battery charge. The most common methods are open-circuit and load voltage testing. See **Open-circuit voltage test** and **Load voltage, testing** in this repair group. Batteries with removable filler caps can also be tested by checking specific gravity of the electrolyte. See **Electrolyte specific gravity, checking (1.8L) (1.8L)** in this repair group. Inexpensive specific gravity testers are available at most auto supply stores.

Some vehicles are equipped with batteries that have a central gas venting system with anti-flash protection. The anti-flash protection consists of a small round fiberglass mat. Its purpose is to vent gasses that form in the battery through a vent opening on the upper side of battery cover. The anti-flash protection is also installed to prevent ignition of flammable gasses in battery.

> ***CAUTION—***
> - *Use genuine Audi battery filler caps fitted with an O-ring seal.*
> - *Before diagnosing electrical problems, visually inspect battery terminals for clean and tight connections.*

Electrolyte level, checking

Visually check battery electrolyte level on batteries with a recognizable minimum and maximum marking (**arrow**).

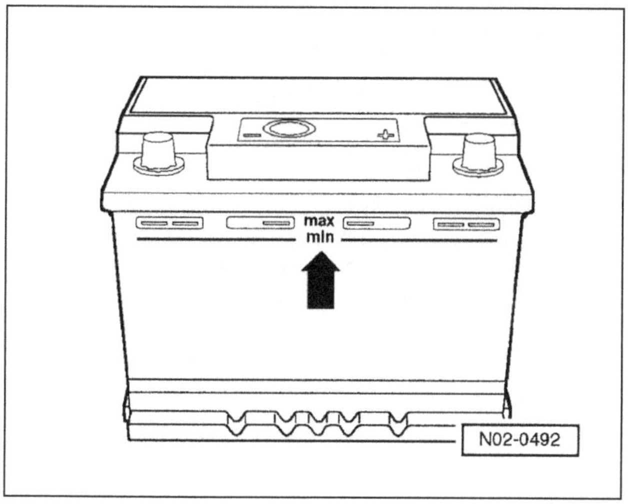

BentleyPublishers
.com
© 2006 BentleyPublishers.com—All Rights Reserved

– If there are no min. or max. markings on battery housing, or if electrolyte level cannot be read, remove battery filler caps to view electrolyte level. Check that level is at same height as visible plastic peg.

– Top off each battery cell as necessary with distilled water using a battery filling bottle. Use a hydrometer to remove any excess electrolyte. Be sure to install original sealing caps with O-rings.

NOTE—
- *Use of distilled water to fill battery cells prevents contamination of battery electrolyte and decreases the likelihood of self-discharge.*

Static current draw, checking

– Make sure ignition and all electrical accessories are switched OFF.

– Disconnect negative (–) cable from battery.

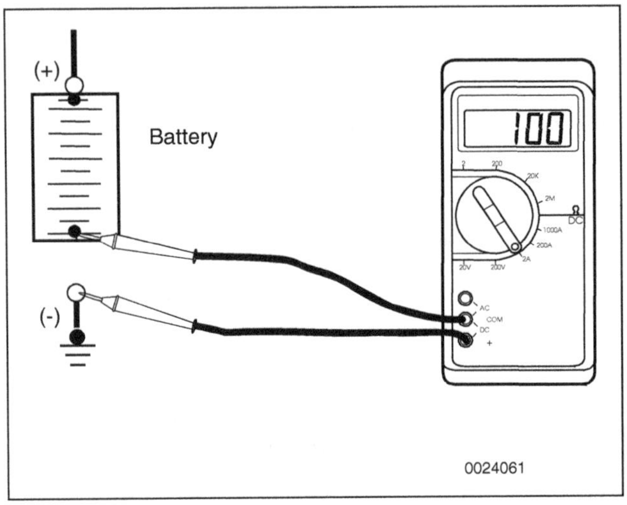

◄ Connect a digital ammeter between battery negative post and negative battery cable and measure current draw.

A range of about 0 to 100 milliamperes is normal, depending on the number of accessories that need constant power. A current of 500 milliampere (0.5 amp) or more indicates a problem. To determine the circuit or component causing the problem, remove one fuse at a time until the current drops to a normal range. Use the wiring diagrams shown in **EWD Electrical Wiring Diagrams** to locate wiring or component faults.

Open-circuit voltage test

An open-circuit voltage test checks battery voltage by connecting an accurate digital voltmeter to the battery posts after disconnecting the battery ground cable. Before making an open-circuit voltage test on a battery, first load the battery with 15 amps for one minute, i.e. by turning on the headlights with engine off. See **Table a** for open-circuit voltage levels and state of charge.

Table a. Open circuit voltage and battery charge

Open circuit voltage	State of charge
12.6 volts or more	Fully charged
12.4 volts	75% charged
12.2 volts	50% charged
12.0 volts	25% charged
11.7 volts or less	Fully discharged

The battery is in satisfactory condition if open-circuit voltage is at least 12.5 volts. If open-circuit voltage is at this level or above, but battery still lacks power for starting, perform a load voltage test to determine battery's service condition. See **Open-circuit voltage test** in this repair group. If open-circuit voltage is below 12.5 volts, recharge battery. If battery cannot be fully recharged, replace it.

BentleyPublishers.com
© 2006 BentleyPublishers.com—All Rights Reserved

Electrolyte specific gravity, checking (1.8L)

Checking the specific gravity of each battery cells can provide accurate information about the battery condition.

Vehicles with the 3.2L engine use batteries with an Absorbed Glass Mat (AGM) that are totally sealed. As a result, no excess electrolyte is present and it is not possible to check the electrolyte level or otherwise service the electrolyte.

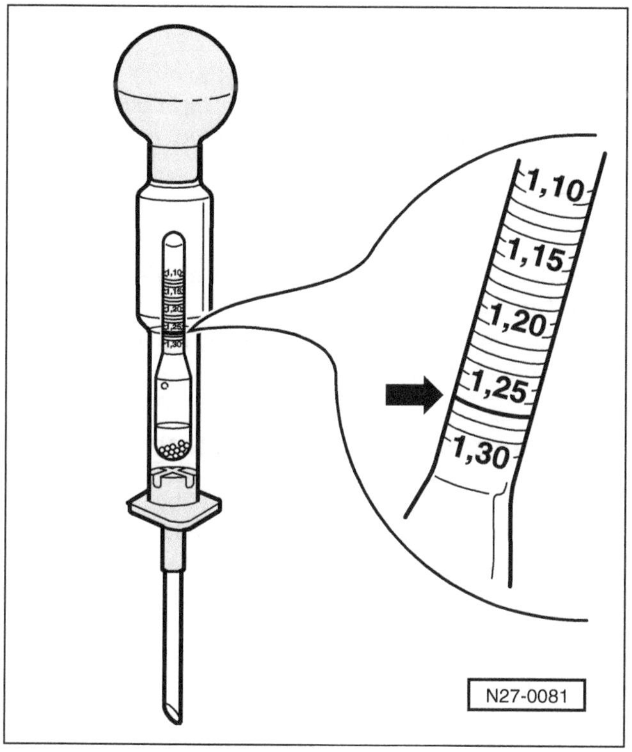

– Turn ignition OFF and remove battery filler caps.

– Check that battery electrolyte temperature is at least 10°C (50°F).

< Immerse hydrometer in a cell and extract sufficient electrolyte so that indicator floats free in electrolyte. Read specific gravity and compare to specification.

Electrolyte density	Charge condition
1.28 kg / liter	100%
1.21 kg / liter	60%
1.18 kg / liter	40%
1.10 kg / liter	0%

– If electrolyte density is below 1.24 kg / liter, charge battery.

– The specific gravity of each battery cell must not differentiate by more than 0.03 kg / liter. Replace battery if not within specification.

– Install filler caps with O-rings and wipe up spilled electrolyte.

Load voltage, testing

A load voltage battery test is made by connecting a specific resistive load to the battery terminals and then measuring battery voltage.

– Make sure battery is fully charged and at room temperature.

– Disconnect negative (–) battery cable.

– Connect battery load tester and apply specified load (**Table b**, column 3) for 15 seconds and read off battery voltage (**Table b**, column 4). If voltage is below specification, or minimum voltage does not stay at a steady reading, replace battery.

Table b. Battery load current and minimum voltage			
Battery capacity (Ah)	**Cold cranking amps (CCA)**	**Load current (amps)**	**Minimum voltage**
36 Ah	340	100	10.0
40 Ah-49 Ah	220	200	9.2
50 Ah-60 Ah	265-280	200	9.4
61 Ah-80 Ah	300-380	300	9.0
81 Ah-110 Ah	380-500	300	9.5

BentleyPublishers
.com
©2006 BentleyPublishers.com—All Rights Reserved

Battery, charging

Recharge a discharged battery using a battery charger. Read and follow the instructions provided by the battery charger's manufacturer. A slow-charging rate (10% of battery capacity) is best to prevent battery damage caused by overheating.

> ***WARNING —***
> - *3.2L vehicles are equipped with an Absorbed Glass Mat (AGM) battery. The AGM battery must not be quick charged under any circumstances.*

> ***CAUTION—***
> - *Prolonged battery charging evaporates the electrolyte to a level that can damage the battery.*
> - *Loosen battery cell caps before charging battery but leave them on battery to prevent battery acid splatter.*

Battery, disconnecting and connecting

Make sure the vehicle electrical system is protected by disconnecting the battery negative (–) terminal prior to working on any electrical system components.

> ***CAUTION—***
> - *Do not loosen or remove ground strap from body. Disconnect terminal from battery only.*

1.8L engine

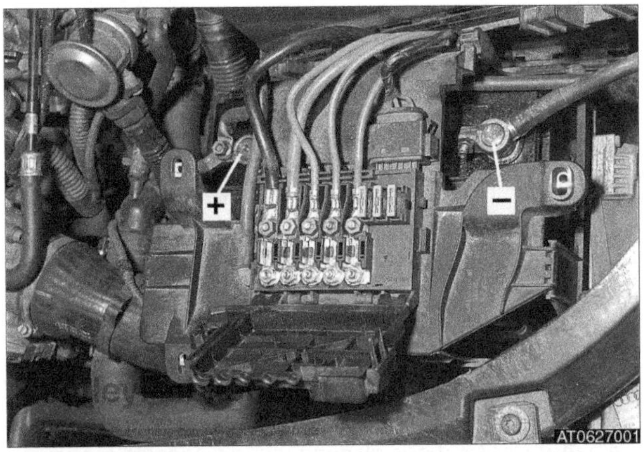

◄ Unbolt and disconnect battery negative (–) cable. Position cable to side where it cannot come in contact with battery negative terminal.

3.2L engine

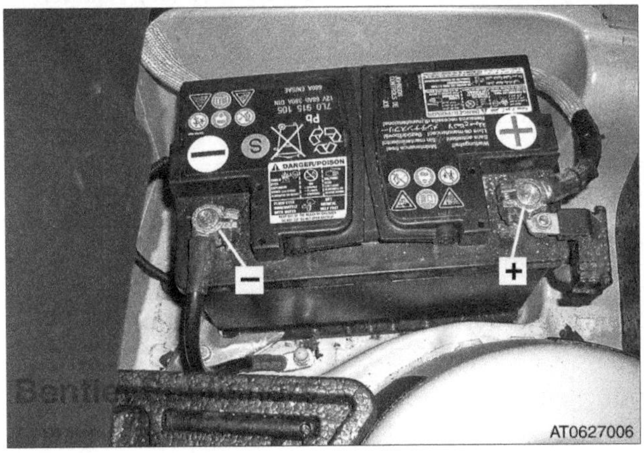

◄ Unbolt and disconnect battery negative (–) cable. Position cable to side where it cannot come in contact with battery negative terminal.

– If necessary, detach battery positive (+) cable and place aside (example: removing or replacing battery).

– When reconnecting battery terminals, connect positive (+) terminal first, then negative (–) terminal.

Tightening torque	
Battery cable to battery (M6)	6 Nm (53 in-lb)

– After connecting battery terminals:
- Switch ignition ON, then OFF.
- Check for diagnostic trouble codes (DTCs) using Audi scan tool or equivalent.
- Reset clock to local time.
- Completely open and close all power windows and sunroof. This re-initializes pinch-protection.
- Check function of all electrical consumers.
- Reset basic settings using diagnostic scan tool.

BentleyPublishers.com
© 2006 BentleyPublishers.com—All Rights Reserved

Battery, removing and installing

1.8L engine

◄ Loosen ¼-turn fasteners (**arrows**) and lift off trim panel above battery.

– Disconnect battery negative (–) and positive (+) cables. See **Battery, disconnecting and connecting** in this repair group. Position battery cables to side where they cannot come in contact with battery terminals.

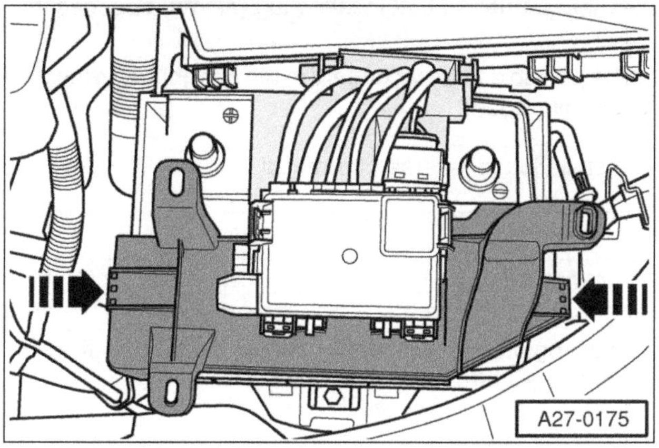

– Position fuse panel on top of battery out of way.

- Two versions of battery fuse panel are used as described below.

◄ **Version I:** push and hold fuse panel retainers (**dashed arrows**) and pivot fuse panel out to the rear.

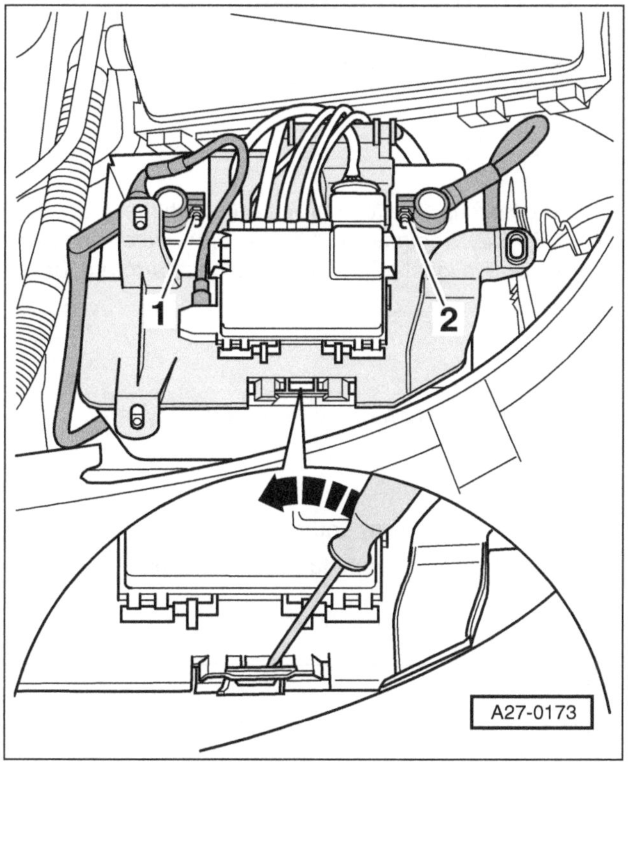

◄ **Version II:** Using a screwdriver or similar, push retaining clip in direction of **dashed arrow** and pivot fuse panel out to the rear.

BentleyPublishers.com
©2006 BentleyPublishers.com—All Rights Reserved

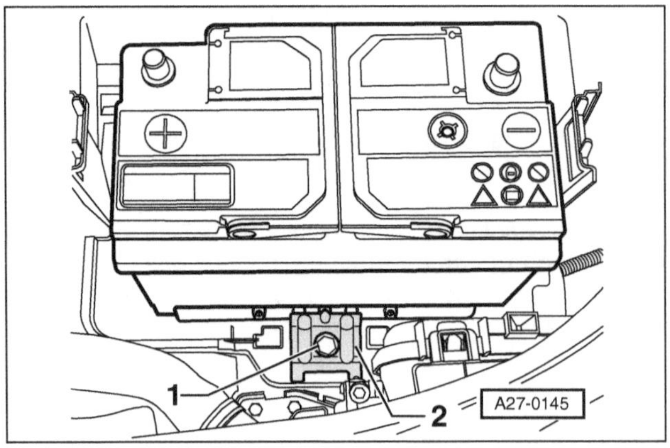

◄ Remove battery mounting bolt (**1**) and retaining bracket (**2**).

– Carefully lift battery out using plastic handles on top of battery, if so equipped.

Battery, installing

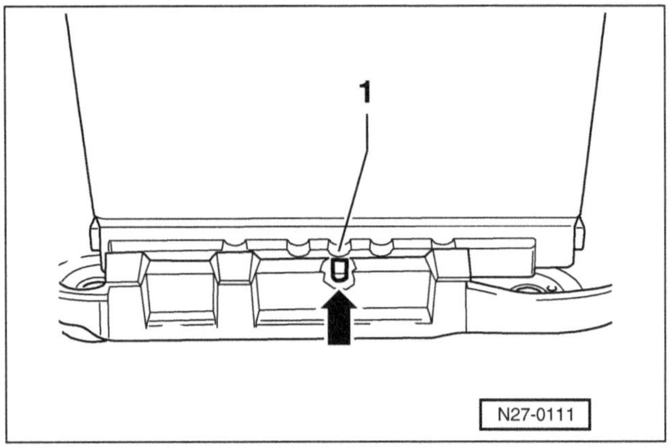

◄ Place battery into carrier so that locating lug (**arrow**) engages in notch (**1**).

- It must not be possible to move battery to left or right when locating lug is properly engaged.

– Remaining installation is reverse of removal.

Tightening torques	
Battery mounting bolt to body	20 Nm (15 ft-lb)

– When reconnecting battery terminals, connect positive (+) terminal first, then negative (–) terminal.

Battery tray, removing and installing

– For 1.8L engines, remove battery as described previously. For 3.2L engines, disconnect the battery negative (–) terminal as described previously.

NOTE —

- *The procedure below applies to vehicles with both the 1.8L and 3.2L engines. For 3.2L engines the "battery tray" is referred to as a fuse box carrier and is located in the engine compartment, below the fuse box.*

Coupe

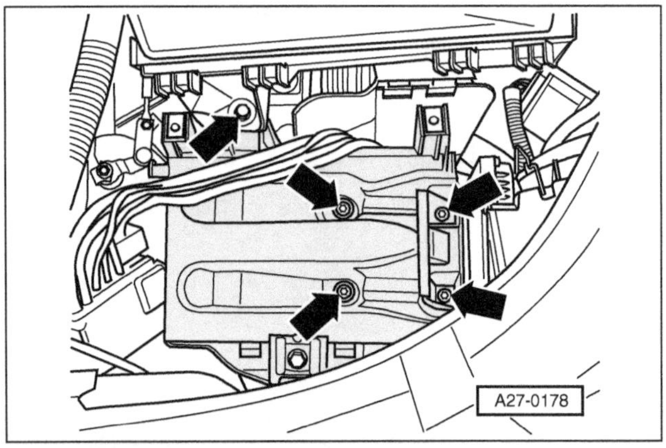

◄ Remove mounting bolts (**arrows**) and remove battery tray.

Roadster

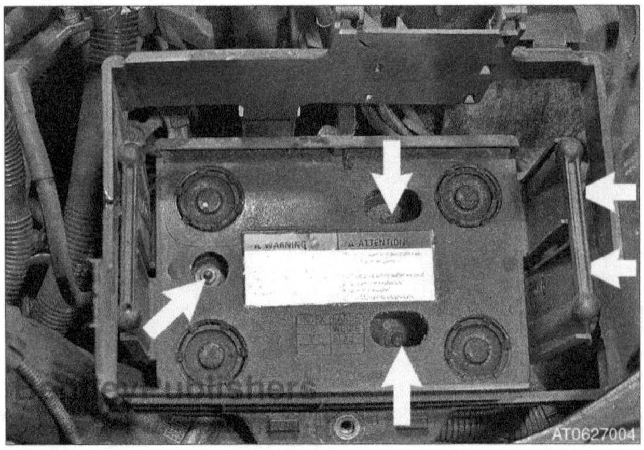

◄ Remove mounting bolts (**arrows**) and remove battery tray.

- Mounting bolts on right side are hidden in illustration.

– Install battery tray in reverse order of removal.

Tightening torque	
Battery tray to body	10 Nm (7 ft-lb)

BentleyPublishers
.com
© 2006 BentleyPublishers.com—All Rights Reserved

ALTERNATOR

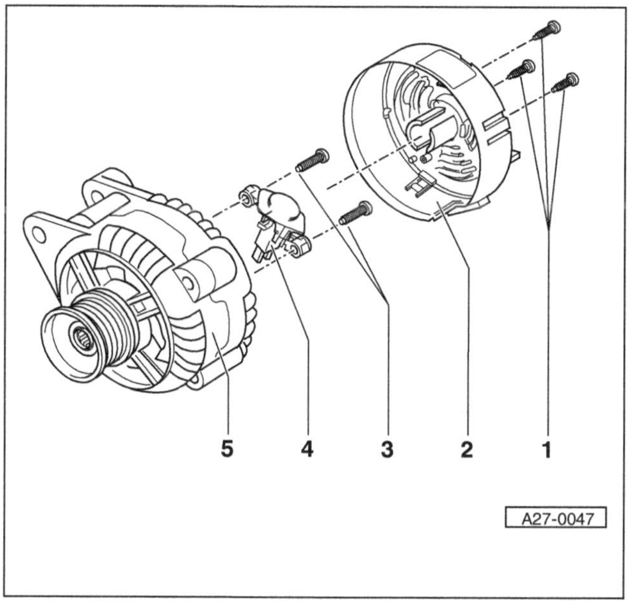

The charging system consists of a belt-driven 14-volt, 120 amp alternator with an internal voltage regulator. The charging system provides the current necessary to keep the battery charged and to operate the vehicle's electrical accessories. The alternator is also referred to as the generator.

Alternator and voltage regulator (Bosch)

1. Screws
   - 1 Nm (9 in-lb)
2. Protective cap
   - With 3 retaining lugs
3. Screws
   - 2 Nm (18 in-lb)
4. Voltage regulator
   - Wear limit of carbon brushes: 5 mm
5. Alternator

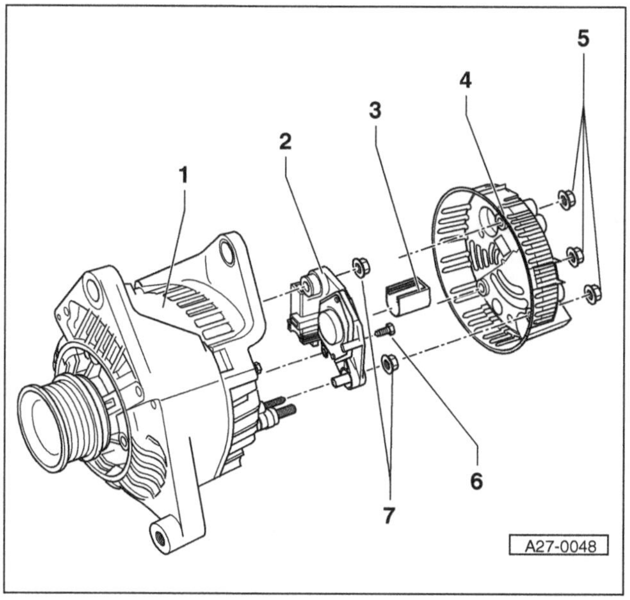

Alternator and voltage regulator (Valeo)

1. Alternator
2. Voltage regulator
   - Wear limit of carbon brushes: 5 mm
3. Cover
   - Press fit to regulator
4. Protective cover
5. Nuts
   - 2 Nm (18 in-lb)
6. Hex bolt
   - 2 Nm (18 in-lb)
7. Nuts
   - 3.5 Nm (31 in-lb)

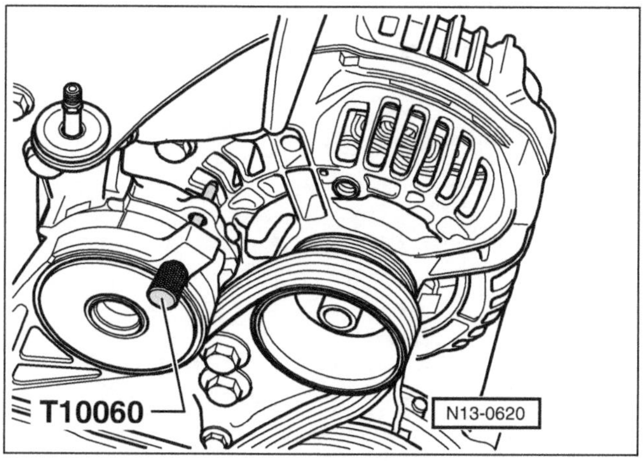

Alternator, removing and installing (1.8L)

– Remove sound absorber panels from engine, battery, top of lock carrier and coolant reservoir. See **03 Maintenance**.

– Disconnect battery negative (–) terminal. See **Battery, disconnecting and connecting** in this repair group.

– Release tension on ribbed belt. **See 03 Maintenance**.

– Mark running direction on ribbed belt if re-using.

◀ Lock ribbed belt tensioner with appropriate sized drift (or special tool T10060 as shown). Remove ribbed belt.

BentleyPublishers
.com
© 2006 BentleyPublishers.com—All Rights Reserved

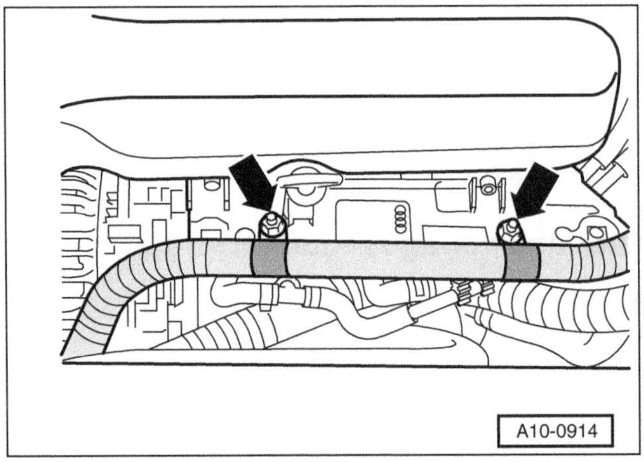

Vehicles with 180 hp engine

Remove mounting bolts (**arrows**) at front of engine. Unclip and remove secondary air injection hose.

Vehicles with 225 hp engine

Remove intake air duct by disconnecting hoses and clamps at **arrows**.

NOTE —

- *Seal off openings of charge air circuit with clean rags to prevent entry of dirt particles.*

Loosen clamp (**arrow**) and remove vacuum hose from intake manifold.

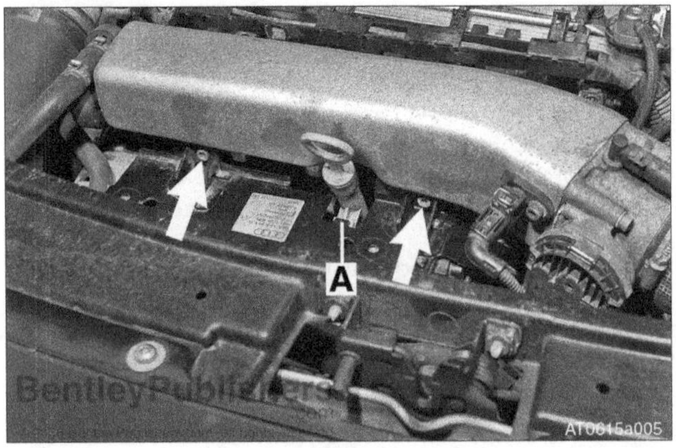

Continued for all vehicles

Unbolt (**arrows**) mounting bracket for solenoid valves from front of intake manifold.

– Detach dipstick guide tube (**A**).

BentleyPublishers
.com
© 2006 BentleyPublishers.com—All Rights Reserved

Detach harness connector and unscrew mounting bolts (**arrows**) for solenoid valve.

– Unclip vacuum line from retaining clip (**right arrow**).

– Remove solenoid valve mounting bracket.

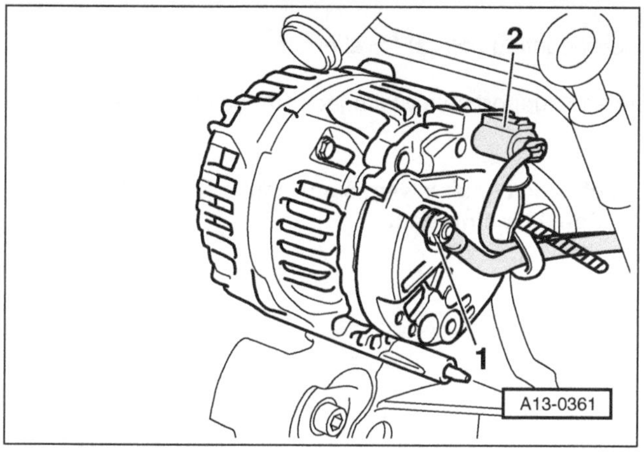

Use a small screwdriver to release locking tab and remove harness connector (**2**).

– Remove nut from positive (B+) alternator terminal (**1**).

– Disconnect wiring harness retainer from back of alternator. Position wiring to side.

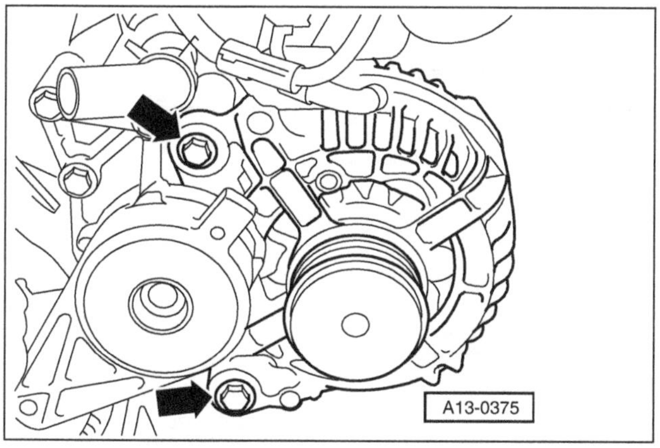

Remove alternator mounting bolts (**arrows**).

– Lift alternator upward from mounting bracket.

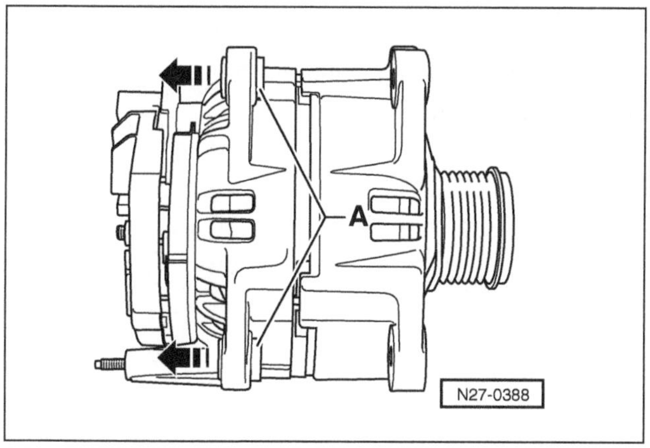

Before installing alternator, drive threaded sleeves (**A**) approximately 4 mm (0.16 in) in direction of **arrows**.

– Install in reverse order of removal, noting the following:
  - If belt was removed, install so running direction is the same as previously marked. See **03 Maintenance** for ribbed belt routing.
  - Make sure ribbed belt is correctly seated in pulleys.

Tightening torque	
Alternator to mounting bracket	23 Nm (17 ft-lb)

– After installing alternator, reconnect battery and run engine briefly. Shut engine off and check that ribbed belt remains correctly seated in belt pulleys.

BentleyPublishers.com
© 2006 BentleyPublishers.com—All Rights Reserved

Alternator, removing and installing (3.2L)

– Remove sound absorber panels from engine, top of lock carrier and coolant reservoir. See **03 Maintenance**.

– Disconnect battery negative (–) terminal. See **Battery, disconnecting and connecting** in this repair group.

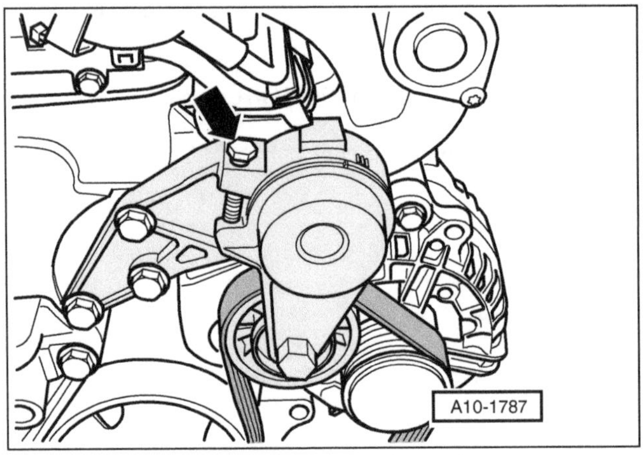

◄ Thread M8 x 40mm bolt (**arrow**) into ribbed belt tensioner to relieve belt tension.

– Mark running direction of ribbed belt if re-using.

– Remove ribbed belt.

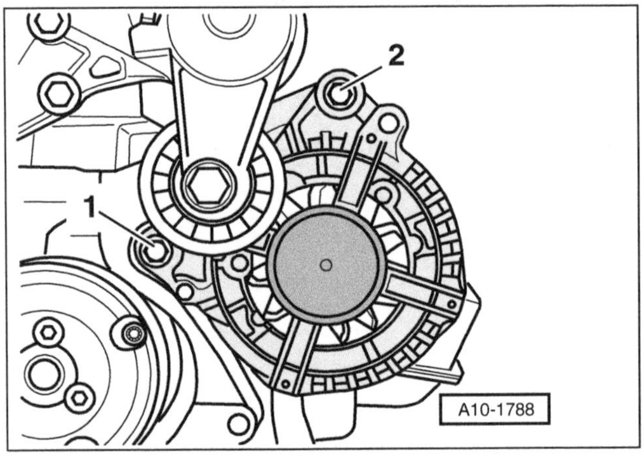

◄ Remove upper mounting bolt (**2**) for alternator.

– Raise vehicle and support safely with jack stands or lift. See **03 Maintenance**.

– Remove left, right and center lower sound insulation panels from under engine. See **03 Maintenance**.

◄ Remove mounting bolts (**arrows**) and cross piece.

- Mounting bolts shown for left side, right side similar.

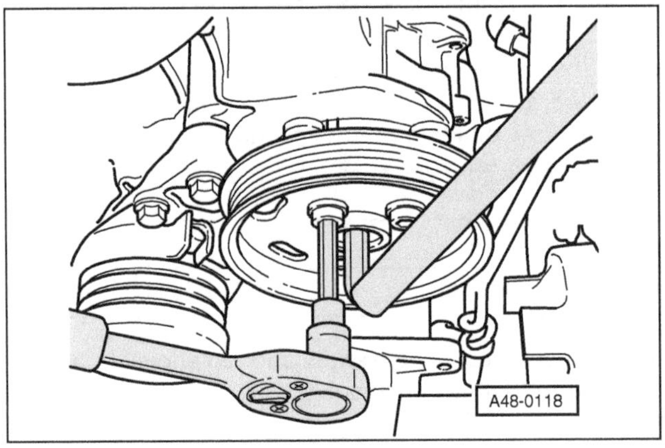

◄ Remove pulley from power steering pump.

- Counterhold pulley with Allen key.

BentleyPublishers.com
© 2006 BentleyPublishers.com—All Rights Reserved

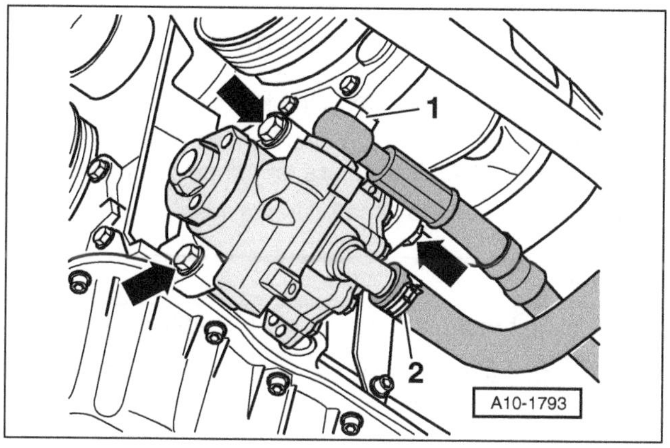

◀ Remove mounting bolts (**arrows**) and power steering pump from bracket.

– Leave hydraulic lines (**1** and **2**) attached and hang power steering pump from vehicle using stiff wire or heavy duty zip ties.

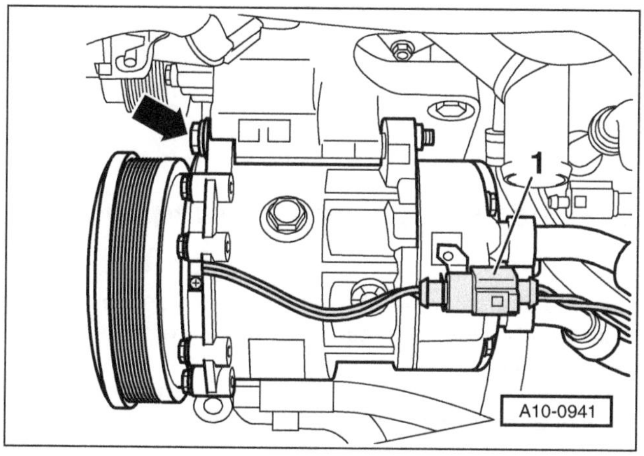

◀ Disconnect electrical connector (**1**) from A/C compressor.

– Remove A/C compressor mounting bolt (**arrow**).

– Leave A/C refrigerant lines connected to compressor. Position compressor out of way and suspend from body with heavy wire.
  - It may be necessary to have a helper use a lever to move the engine slightly to left in order to remove A/C compressor.
  - If necessary loosen engine mount bolts. Tightening torques for engine mounts are provided in **15b Cylinder Head (3.2L engine)**.

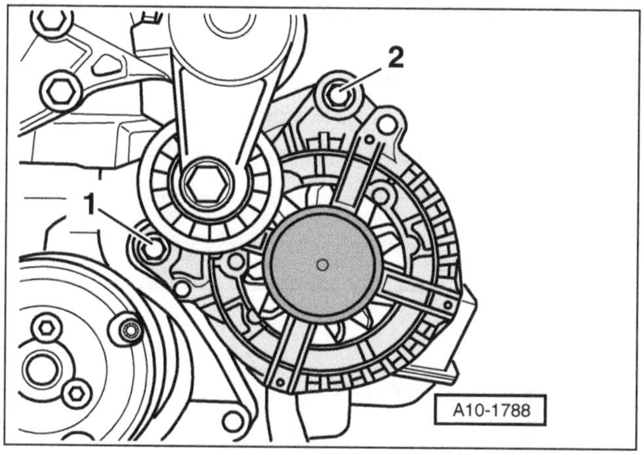

◀ Remove lower mounting bolt (**1**) for alternator. Remove alternator from mounting bracket. Leave wiring harness connected to back of alternator for now.

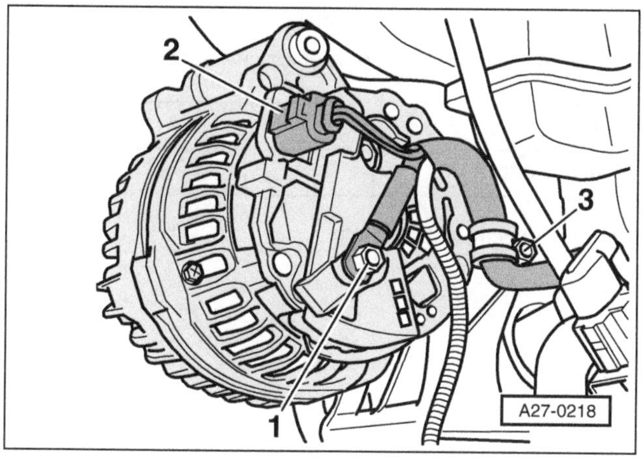

◀ Remove harness retainer (**3**) from back of alternator.

– Remove nut (**1**) and terminal B+ wire.

– Use a small screwdriver to release locking tab and remove harness connector (**2**).

– Remove alternator.

BentleyPublishers.com
©2006 BentleyPublishers.com—All Rights Reserved

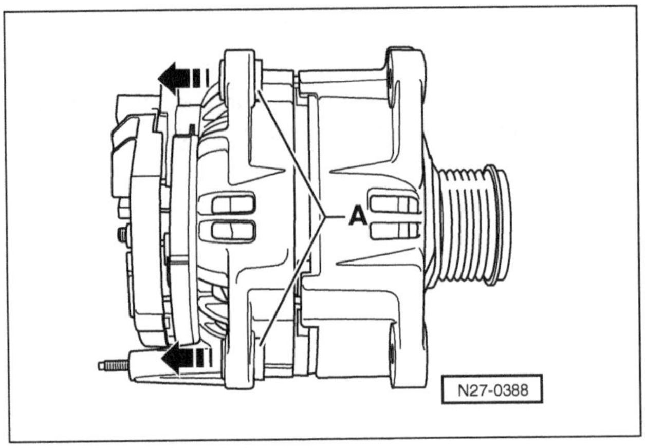

◀ Before installing alternator, drive threaded sleeves (**A**) approximately 4 mm (0.16 in) in direction of **arrows**.

– Install in reverse order of removal, noting the following:
  - If belt was removed, install so running direction is the same as previously marked. See **03 Maintenance** for ribbed belt routing.
  - Make sure ribbed belt is correctly seated in pulleys.

Tightening torques	
Alternator to mounting bracket	23 Nm (17 ft-lb)
Terminal B+ mounting nut to alternator	16 Nm (12 ft-lb)
A/C compressor to bracket	45 Nm (33 ft-lb)
Power steering pump to bracket	23 Nm (17 ft-lb)

– After installing alternator, reconnect battery and run engine briefly. Shut engine off and check that ribbed belt remains correctly seated in belt pulleys.

Alternator ribbed belt pulley, removing and installing

When replacing the alternator the ribbed belt pulley must be removed and installed on the new alternator. There are two types of pulleys as shown below.

– Remove alternator as described previously.

Ribbed belt pulley without freewheel

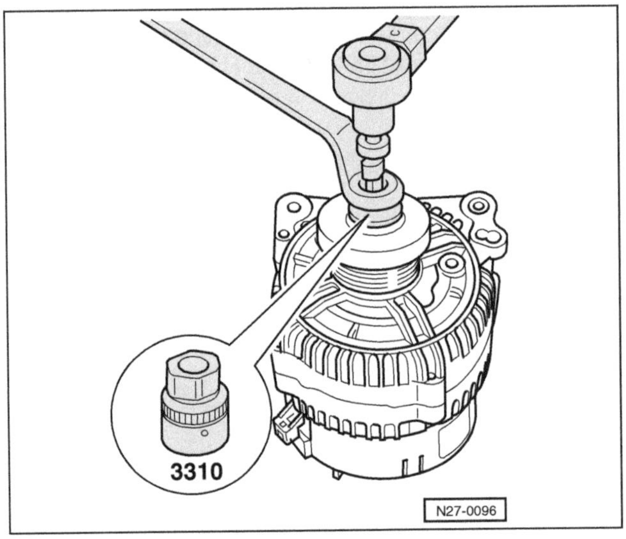

◀ Counter hold securing nut with special tool 3310 and wrench.
  - To remove, turn alternator shaft in clockwise direction.
  - To install, turn alternator shaft in counter-clockwise direction.

Tightening torque	
Alternator pulley to alternator	65 Nm (48 ft-lb)

Ribbed belt pulley with freewheel

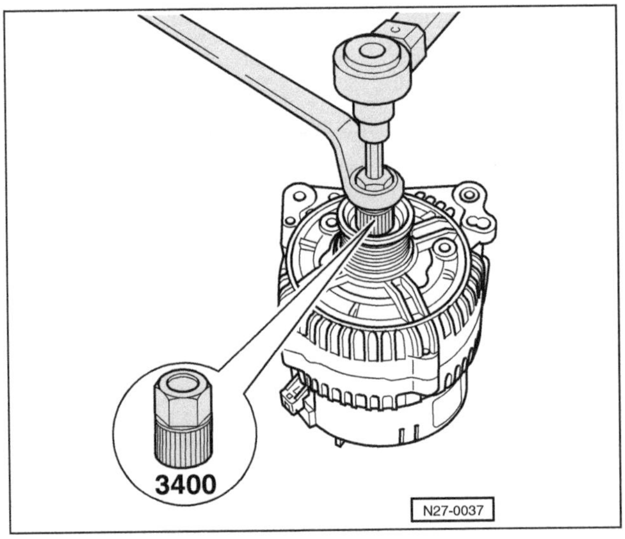

◀ Counter hold securing nut with special tool 3400 and wrench.
  - To remove, turn alternator shaft in clockwise direction.
  - To install, turn alternator shaft in counter-clockwise direction.

Tightening torque	
Alternator pulley to alternator	80 Nm (59 ft-lb)

– Clip protective cap onto freewheel pulley.

BentleyPublishers.com
© 2006 BentleyPublishers.com—All Rights Reserved

Starter

Starter, removing and installing (1.8L)

– Disconnect battery negative (–) terminal. See **Battery, disconnecting and connecting** in this repair group.

– Remove upper and front engine sound absorber panels. See **03 Maintenance**.

– Remove battery and battery tray as described earlier in this repair group.

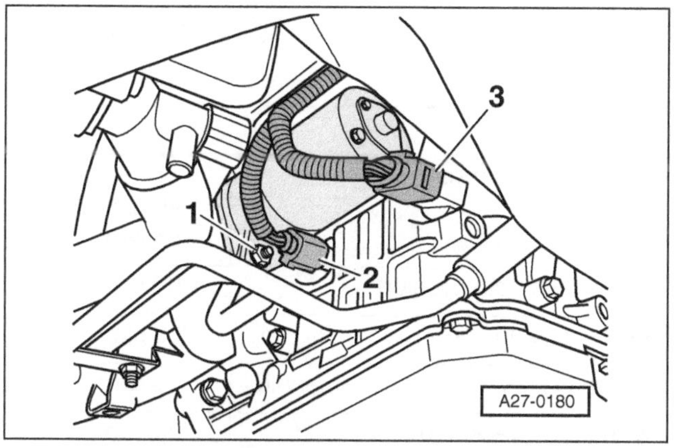

Vehicles with automatic transmission

< Disconnect harness connectors (**2** and **3**) from below starter.

Continued for all vehicles

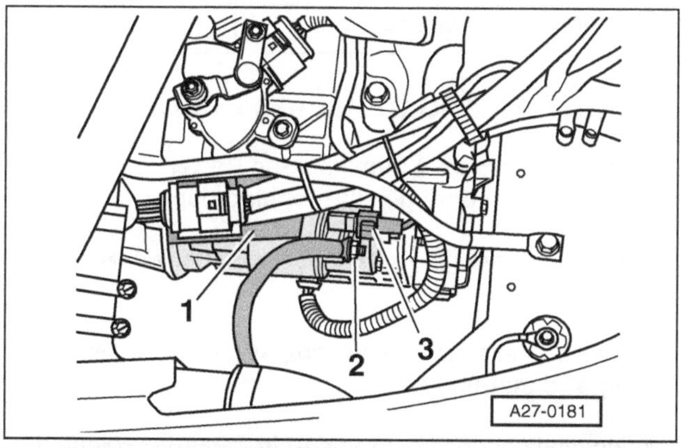

< Disconnect harness connector (**3**) and unbolt positive (B+) cable (**2**) from starter solenoid.

- Wire guide shown at (**1**).

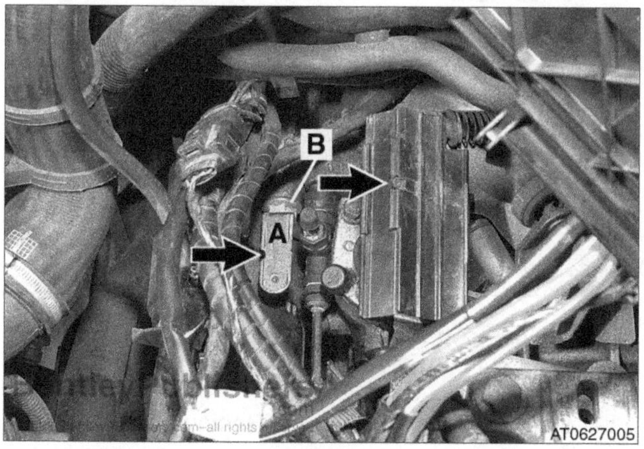

< Unclip wire guide from bracket (**A**) and slide off. Position guide out of way.

- Wire guide clips on at **arrows**.

– Remove mounting bolt and bracket (**A**).

– Remove upper mounting bolt (**B**) for starter.

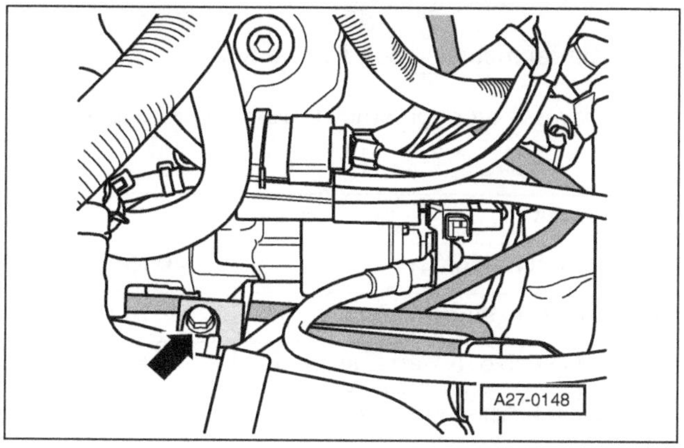

< Where applicable, remove bolt (**arrow**) and move power steering lines out of way.

© 2006 BentleyPublishers.com—All Rights Reserved

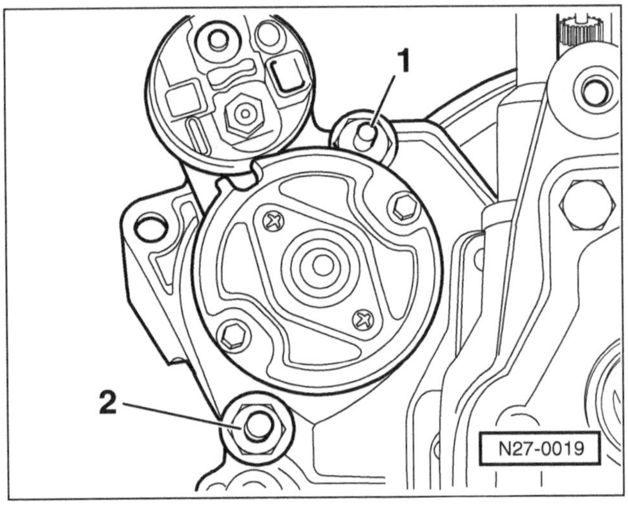

Remove lower starter mounting bolt (**2**) and starter.

– Replace starter bushing in transmission housing, as applicable.

– Install starter in reverse order of removal. Tighten all fasteners as specified below:

Tightening torques	
Starter to manual transmission (M10 bolt) (M12 bolt)	 45 Nm (33 ft-lb) 65 Nm (48 ft-lb)
Starter to automatic transmission	80 Nm (59 ft-lb)
Wiring guide bracket to starter mounting bolt	22 Nm (16 ft-lb)
Battery positive (+) cable to starter solenoid	16 Nm (12 ft-lb)
Power steering line bracket to starter	22 Nm (16 ft-lb)
Ground wire to automatic transmission	15 Nm (11 ft-lb)

Starter, removing and installing (3.2L)

– Disconnect battery negative (–) terminal. See **Battery, disconnecting and connecting** in this repair group.

– Remove upper and front engine sound absorber panels. See **03 Maintenance**.

– Remove fuse box carrier from engine compartment. See **Battery tray, removing and installing**.

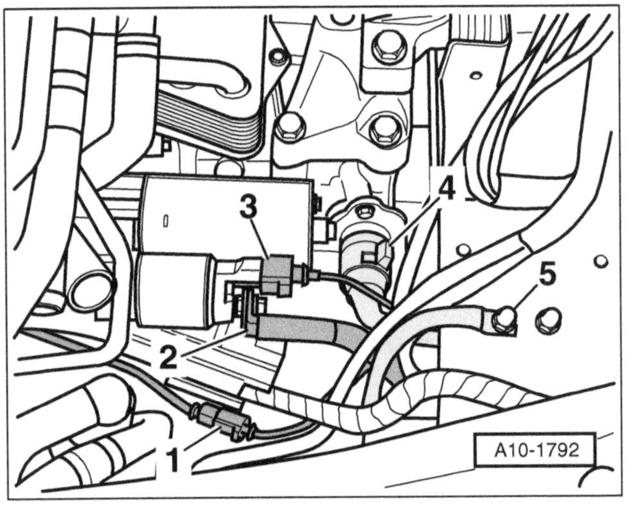

Disconnect electrical connectors (**3** and **4**) from starter solenoid.

– Remove positive (B+) cable (**2**).

– Remove harness connector (**1**) from wiring guide.

– Unclip and remove wiring guide.

> ***CAUTION—***
> *• Do not loosen or remove ground (–) connection (**5**).*

– Remove mounting nut for power steering line bracket and rotate bracket upward and out of way.

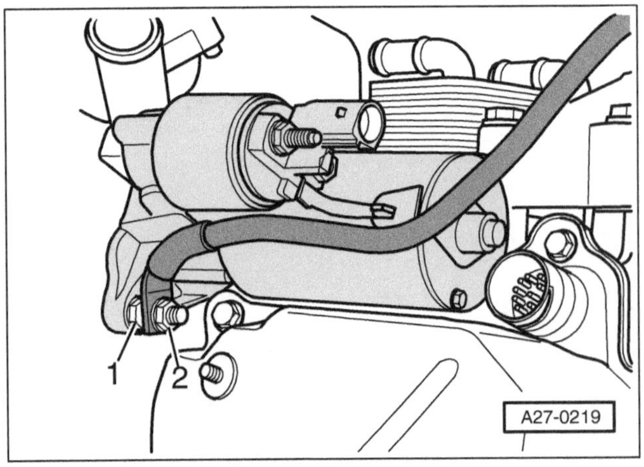

Remove outer nut (**2**) and ground (–) wire. Remove lower starter mounting bolt (**1**).

– Remove upper starter mounting bolt and starter.

– Install starter in reverse order of removal. Tighten all fasteners as specified below:

Tightening torques	
Starter to DSG transmission	40 Nm (30 ft-lb)
Wiring guide bracket to starter mounting bolt	22 Nm (16 ft-lb)
Battery positive (+) cable to starter solenoid	16 Nm (12 ft-lb)
Power steering line bracket to starter	22 Nm (16 ft-lb)

BentleyPublishers.com
©2006 BentleyPublishers.com—All Rights Reserved

28 Ignition System

GENERAL

The ignition systems of the 1.8L and 3.2L gasoline engines are controlled by the Motronic engine management system. The engine control module computes ignition timing based on inputs from various sensors. Crankshaft speed from the crankshaft position sensor and engine load from the mass air flow (MAF) sensor are the two main inputs used to calculate the basic ignition timing point. The other sensors, including two knock sensors, are used by the ECM to modify and correct the ignition timing point based on changing operating conditions.

Ignition firing order	
1.8L turbo gasoline	1-3-4-2
3.2L gasoline	1-5-3-6-2-4

NOTE—
- *The Motronic fuel systems are all OBD II compliant. It is recommended that fault diagnosis and troubleshooting be carried out using Audi scan tools VAG 1551/1552, diagnostic computer VAS 5051/5052, or equivalent scan tool or computer program. See* **OBD On Board Diagnostics** *for a list of diagnostic trouble codes (DTCs) and scan tool suppliers.*
- *For spark plug and ignition coil removal and installation, see* **03 Maintenance***.*

Service precautions

Ignition system service and repair work must be carried out carefully. The ignition system contains sensitive electronic components. To guard against system damage, and for personal and general safety, the following warnings and cautions apply to any ignition system troubleshooting, maintenance or repair work.

BentleyPublishers.com

© 2006 BentleyPublishers.com—All Rights Reserved

> ***WARNING —***
> - *Ignition systems operate in a dangerous voltage range that could prove to be fatal if exposed terminals or live parts are contacted. Use extreme caution when working on a vehicle with the ignition on or the engine running.*
> - *Connect and disconnect ignition coil connectors, terminal connectors and test equipment leads only while the ignition is off unless specifically instructed otherwise.*

Disabling ignition system

If the engine must be turned at starter speed without starting, the ignition system should be disabled to prevent the discharge of dangerously high voltage. This should be done any time repairs or maintenance are performed on the engine with the ignition key on or if the starter needs to be operated without running the engine.

– Remove upper sound absorber panel from top of engine, see **03 Maintenance**.

1.8L engine

◀ Disconnect harness connectors (**A**) from each individual ignition coil to disable ignition system. See **03 Maintenance, Spark Plugs**, for detailed ignition coil removal instructions.

3.2L engine

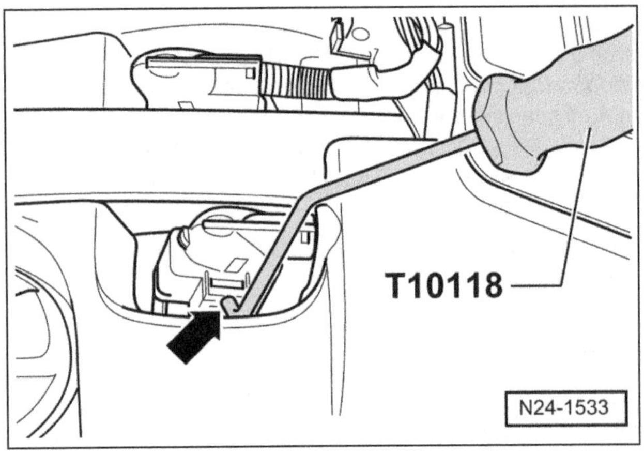

◀ Disconnect harness connectors from ignition coils by releasing tabs (**arrow**) with **T10118** or equivalent.

BentleyPublishers.com
© 2006 BentleyPublishers.com—All Rights Reserved

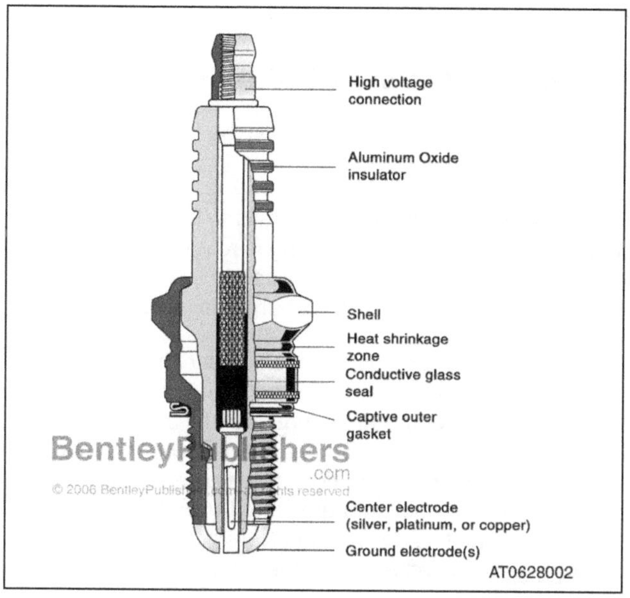

Ignition Coils and Spark Plugs

Audi specifies spark plugs made by several different manufacturers. Because specifications and part numbers can change, always check with an authorized Audi Dealer's parts department or aftermarket parts specialist for the latest application information.

Engine	Spark plug specifications
1.8L 4-cyl. 180-hp 225-hp	AoA number....101 000 063 AA NGK number....PFR6Q Gap....0.8 mm (0.032 in.) maximum Torque....30 Nm (22 ft-lb) Firing order....1-3-4-2
3.2L 6-cyl. 250-hp	AoA number....101 905 606 A NGK number....IZKR7B Gap....1.1 mm (0.043 in.) maximum Torque....30 Nm (22 ft-lb) Firing order....1-5-3-6-2-4

NOTE—

- *Part number information is current at time of publication, but can change due to supersessions, etc.*
- *Audi part numbers are given for reference only! Always consult with your Audi of America Parts Department or aftermarket parts specialist for the latest parts information.*
- *Spark plugs are generally pre-gapped by the original manufacturer during production but should be checked before installation.*

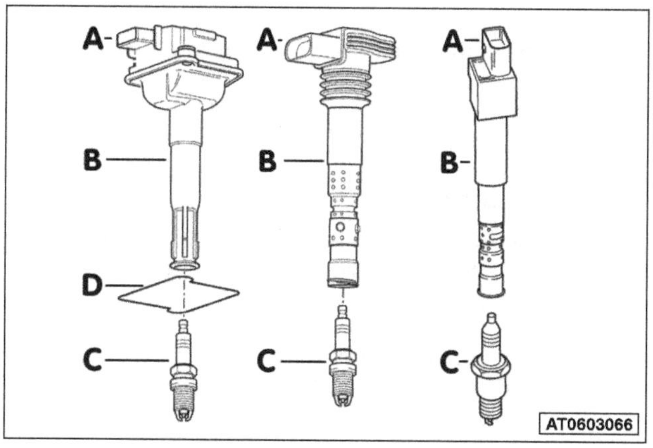

Spark plug access requires disconnecting harness connector from ignition coil and removal of ignition coil. Early 4-cylinder coils (**left**), are bolted in place and use a separate gasket. Later 4-cylinder (**center**) and 6-cylinder versions (**right**) are push-in.

A. Wiring harness connection
B. Ignition coil
C. Spark plug
D. Gasket

BentleyPublishers.com
©2006 BentleyPublishers.com—All Rights Reserved

IGNITION COMPONENTS (1.8L)

Camshaft position sensor (G163), testing

The camshaft position sensor (**1**) is located on the right side of the cylinder head on 1.8L engines.

If the camshaft position (Hall effect) sensor fails to function, knock control is switched off and the ignition timing is retarded slightly because the signals can no longer be assigned to the cylinders. In addition the charge pressure of the turbocharger is reduced.

Even without a signal from the camshaft position sensor the engine will continue to run and can also be re-started. When a malfunction is detected the engine control module produces one spark per crankshaft revolution for each cylinder. The fact that the injection system is one revolution out of phase does not have any noticeable effect on the fuel injection. If this happens, the fuel is injected upstream of the closed inlet valve instead of while the inlet valve is open. This has only a minor influence on the quality of the air/fuel mixture.

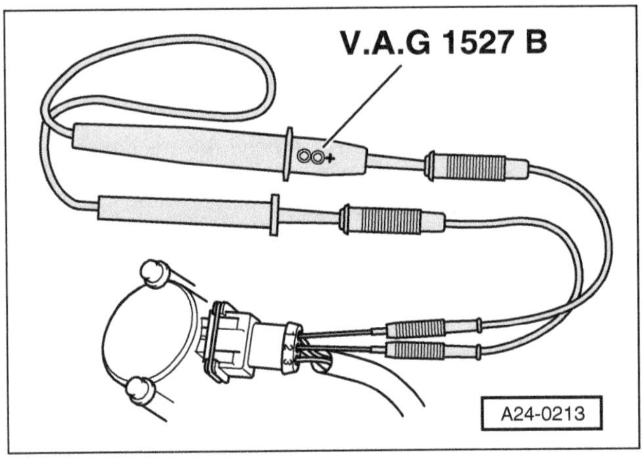

Slide back rubber sleeve (if applicable) from camshaft position sensor harness connector. Connect test leads of LED voltage tester (V.A.G. 1527B shown) to terminals 1 (positive) and terminal 2 (signal).

- The connector terminals on numbered on the back of the connector.

– Operate starter briefly.

– LED of voltage tester must blink.

- LED voltage testers do not go out completely between activations by the ECM, but rather continue to glow and then get slightly brighter during activation.

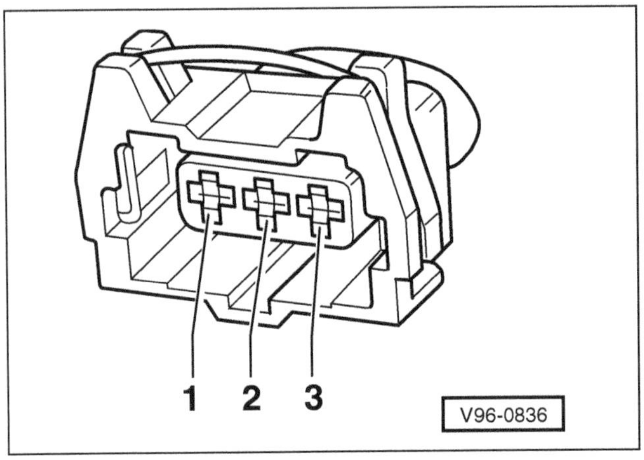

If LED does not light, unplug connector from sensor and connect multimeter to harness connector terminal **1** and engine ground (–).

1. Voltage supply (+)
2. Signal
3. Ground (–)

– Switch on ignition.

- Specification: approximately 5 volts.

– If specification is not obtained, check signal wire by connecting multimeter to harness connector terminal **2** and engine ground (–).

– Switch on ignition.

- Specification: approximately battery voltage (12 volts).

– If specification is not obtained, check ground wire by connecting multimeter to harness connector terminal **3** and engine ground (–).

- Check for continuity to ground.

– If specification is not obtained, check wiring for open or short circuit using wiring diagrams. See **EWD Electrical Wiring Diagrams**.

– If above specifications are not obtained and no wiring faults can be found, replace camshaft position sensor.

BentleyPublishers.com
© 2006 BentleyPublishers.com—All Rights Reserved

Knock sensors (G61 / G66), testing

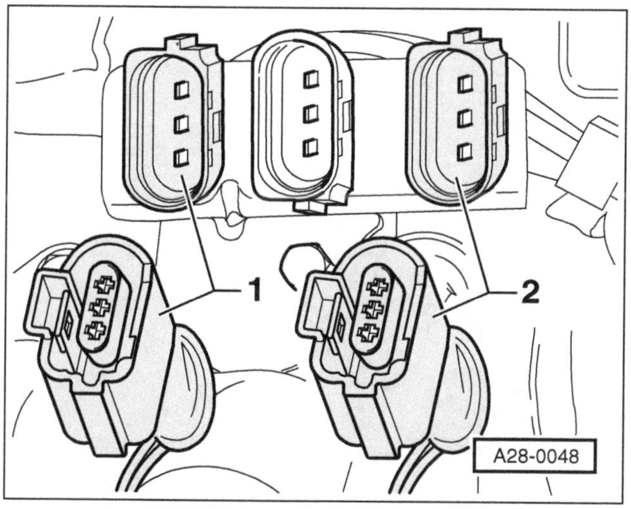

Knock sensor control must be tested using a diagnostic scan tool (VAG 1551/1552 or VAS 5051/5052). The wiring for the knock sensors can be checked for electrical short circuits as described below.

NOTE —

- *Audi identifies electrical components by a letter and/or a number in the electrical schematics. See* **EWD Electrical Wiring Diagrams**. *These electrical identifiers are listed in parentheses as an aid to electrical troubleshooting.*

◄ Unplug harness connector for knock sensor **1** (G61) or knock sensor **2** (G66).

- Knock sensor harness connectors are located in front of intake manifold.

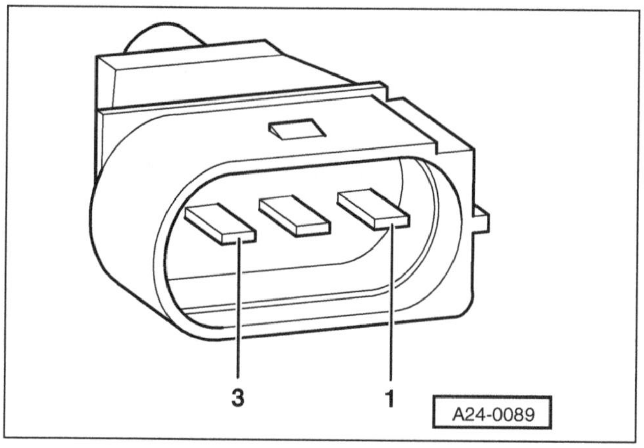

◄ Test for short circuits between all three terminals of knock sensor harness connector (male side).

- Specification: ∞ Ω (no continuity)

– If there is continuity between any of the terminals, replace knock sensor.

– Knock sensors must be tightened to exact specification for proper operation. Use of an accurate torque wrench is necessary when installing.

Tightening torque	
Knock sensor to cylinder block	20 Nm (15 ft-lb)

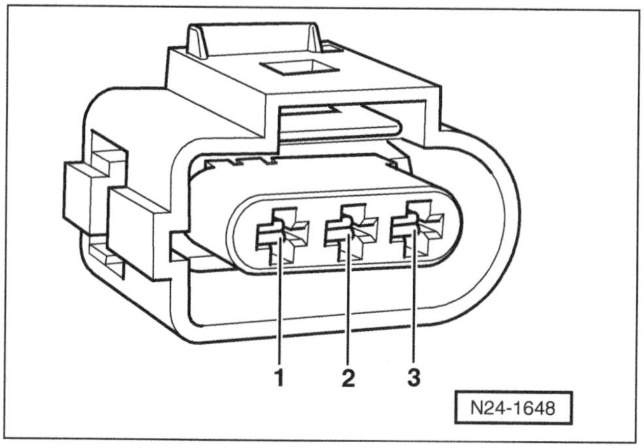

◄ Test for short circuits between all three terminals of knock sensor harness connector (female side).

- Specification: ∞ Ω (no continuity)

1. Signal
2. Ground (–)
3. Screening

– If there is continuity between any of the terminals, repair wiring using wiring diagrams. See **EWD Electrical Wiring Diagrams**.

BentleyPublishers.com
©2006 BentleyPublishers.com—All Rights Reserved

Ignition Components (3.2L)

Ignition system components

> **NOTE—**
> - *Audi identifies electrical components by a letter and/or a number in the electrical schematics. See* **EWD Electrical Wiring Diagrams**. *These electrical identifiers are listed in parentheses as an aid to electrical troubleshooting.*

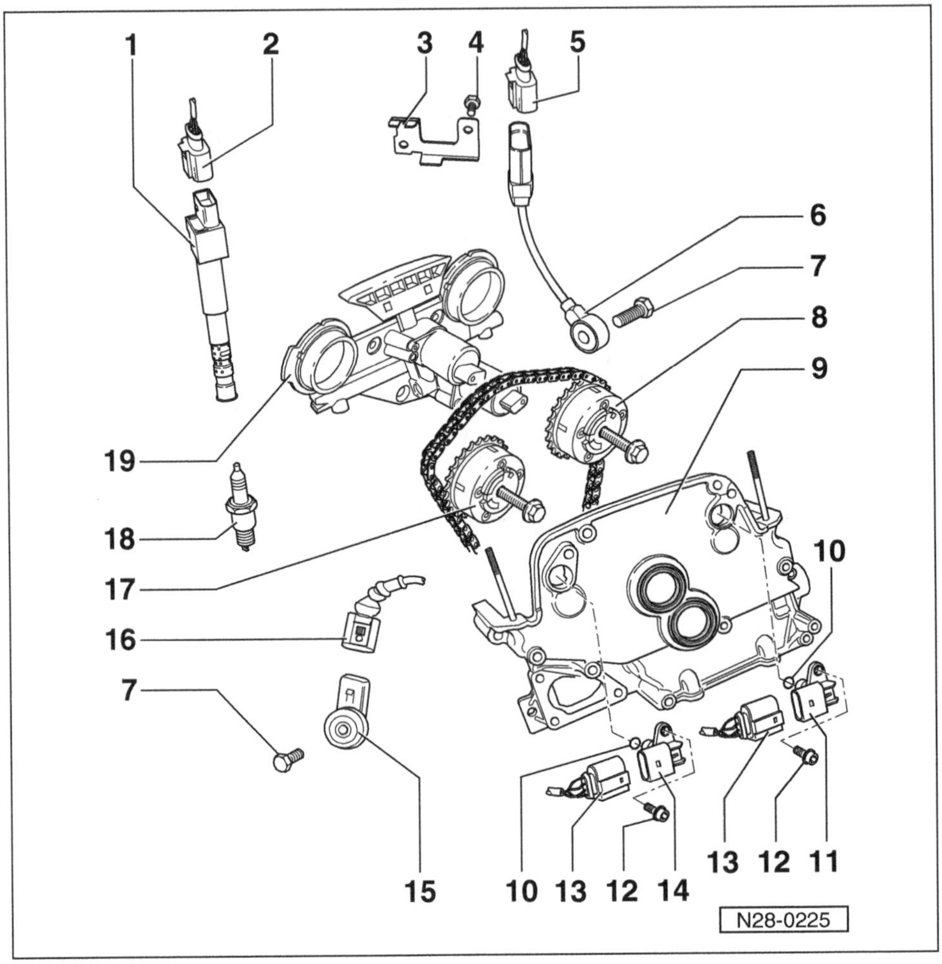

1. **Ignition coils with power output stage**
   - Cylinder #1 (N70)
   - Cylinder #2 (N127)
   - Cylinder #3 (N291)
   - Cylinder #4 (N292)
   - Cylinder #5 (N323)
   - Cylinder #6 (N324)
   - Removing and installing, see **03 Maintenance**
2. **4-pin harness connector**
   - For ignition coil
3. **Bracket**
   - For knock sensor connector
4. **Bolt**
   - 10 Nm (7 ft-lb)
5. **3-pin harness connector**
   - Black
6. **Knock sensor 1 (G61)**
7. **Bolt**
   - 20 Nm (15 ft-lb)
8. **Camshaft adjuster**
   - For exhaust camshaft
   - With sender wheel for camshaft position sensor 2 (G163)
   - After re-installing, adjust valve timing
9. **Cover (upper) for camshaft timing chain**
10. **Seal**
    - Always replace
11. **Camshaft position sensor 2 (G163)**
    - For exhaust camshaft
12. **Bolt**
    - 10 Nm (7 ft-lb)
13. **3-pin harness connector**
    - Black
14. **Camshaft position sensor (G40)**
    - For intake camshaft
15. **Knock sensor 2 (G66)**
16. **2-pin harness connector**
    - Black
17. **Camshaft adjuster**
    - For intake camshaft
    - With sender wheel for camshaft position sensor (G40)
    - After re-installing, adjust valve timing
18. **Spark plug**
    - 30 Nm (22 ft-lb)
    - Remove/install with spark plug socket 3122 B or 16 mm (5/8 inch) spark plug socket and extension
19. **Valve timing housing**
    - For variable valve timing

BentleyPublishers.com
© 2006 BentleyPublishers.com—All Rights Reserved

Camshaft position sensors (G40 / G163), testing

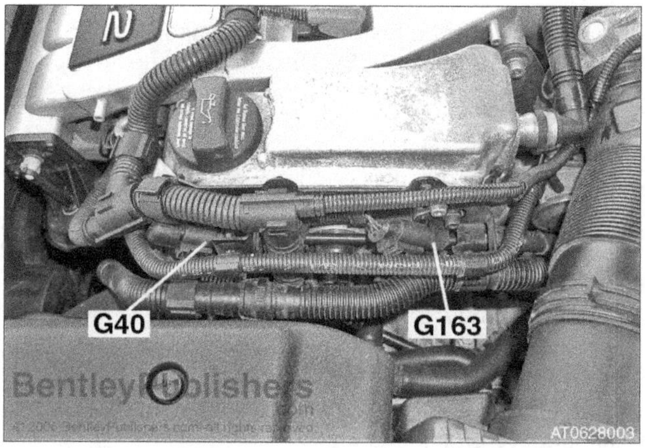

◀ The camshaft position sensors are located on the left side of the 3.2L cylinder head.

- **G40** - camshaft position sensor for intake camshaft
- **G163** - camshaft position sensor for exhaust camshaft

– Disconnect harness connector for camshaft position sensor.

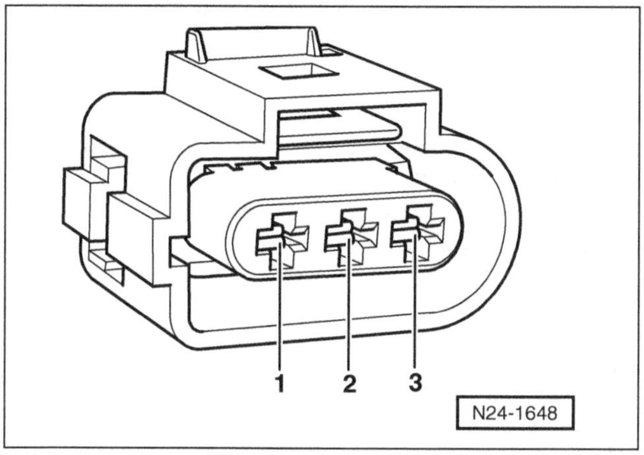

◀ Connect test leads of multimeter between terminal 1 (B+) and terminal 3 (ground).

1. Voltage supply (+)
2. Signal
3. Ground (–)

– Switch ignition on.

- Specified value: minimum 4.5 V

– If specification is not obtained, check wiring for open or short circuit using wiring diagrams. See **EWD Electrical Wiring Diagrams**.

– If above specifications are not obtained and no wiring faults can be found, replace camshaft position sensor.

Camshaft adjustment valves (N205 / N318), testing

– Disconnect 2-pin harness connector from camshaft adjustment valve.

- N205 - valve 1 for camshaft adjuster (intake camshaft)
- N318 - camshaft adjustment valve 1 (exhast camshaft)

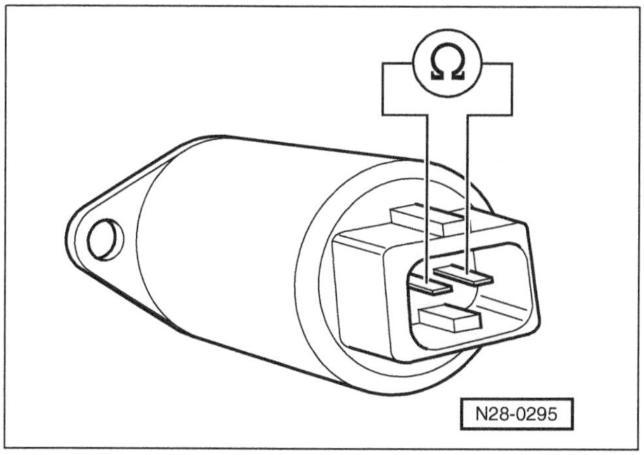

◀ Use a multimeter to measure resistance between two terminals of camshaft adjustment valve.

- Specification: 10 - 18 Ω

– If specification is obtained, check wiring for open or short circuit using wiring diagrams. See **EWD Electrical Wiring Diagrams**.

– If specification is not obtained and no wiring faults can be found, replace control housing for camshaft adjustment with integral camshaft adjustment valves.

BentleyPublishers.com
© 2006 BentleyPublishers.com—All Rights Reserved

Knock sensors (G61 / G66), testing

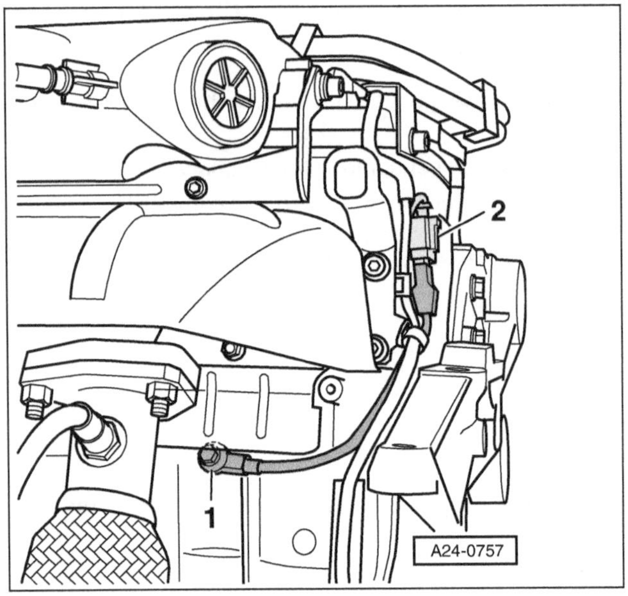

Knock sensor control must be tested using a diagnostic scan tool (VAG 1551/1552 or VAS 5051/5052). The wiring for the knock sensors can be checked for electrical short circuits as described below.

NOTE —
- *Audi identifies electrical components by a letter and/or a number in the electrical schematics. See* **EWD Electrical Wiring Diagrams**. *These electrical identifiers are listed in parentheses as an aid to electrical troubleshooting.*

Unplug 3-pin harness connector (**2**) for knock sensor **1** (G61).

- Knock sensor 1 harness connector is located at right side of cylinder head, above engine mount.

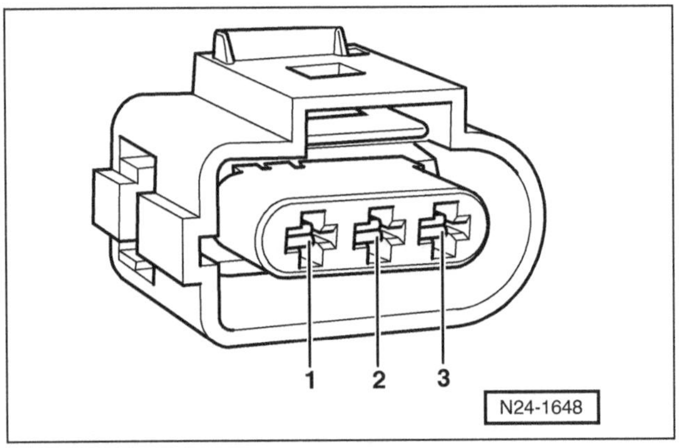

Check wiring for open or short circuit to each other, to ground (–) and to battery positive (+). See **EWD Electrical Wiring Diagrams**.

- Specification: ∞ Ω (no continuity)
- WIre resistance max.: 1.5 Ω

1. Signal
2. Ground (–)
3. Screening

– Test for short circuits between all three terminals of knock sensor harness connector (male side).

- Specification: ∞ Ω (no continuity)

– If there is continuity between any of the terminals, replace knock sensor.

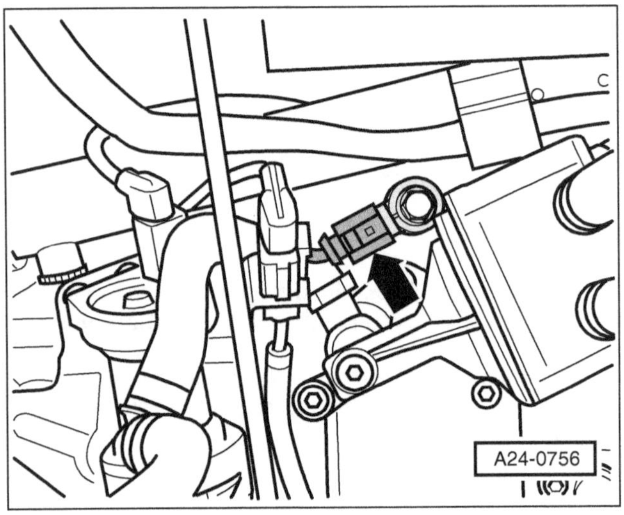

Unplug 2-pin harness connector (**arrow**) for knock sensor 2.

– Check wiring for open or short circuit to each other, to ground (–) and to battery positive (+). See **EWD Electrical Wiring Diagrams**.

- Specification: ∞ Ω (no continuity)
- WIre resistance max.: 1.5 Ω

– Test for short circuits between two terminals of knock sensor harness connector (male side).

- Specification: ∞ Ω (no continuity)

– If there is continuity between the terminals, replace knock sensor.

– Knock sensors must be tightened to exact specification for proper operation. Use of an accurate torque wrench is necessary when installing.

Tightening torque	
Knock sensor to cylinder block	20 Nm (15 ft-lb)

BentleyPublishers
.com
©2006 BentleyPublishers.com—All Rights Reserved

30 Clutch

30

General

This repair group covers clutch hydraulic and mechanical repairs.

Service to the clutch assembly requires removal of the transmission from the vehicle. Special tools and equipment are required to remove the transmission, service the clutch, and bleed the clutch hydraulic system. See **34a Manual Transmission** for transmission removal procedures.

NOTE—

- *The Direct Shift Gearbox (DSG) is a fully automatic transmission. But it is equipped with a wet multi-disc clutch which is used in place of a torque converter. The DSG clutch assembly is illustrated in this repair group.*

Clutch Pedal

The illustration on the following page shows clutch pedal assembly components for vehicles equipped with front wheel and all wheel drive manual transmissions. When servicing, replace self-locking nuts, circlips, and gaskets as indicated.

BentleyPublishers.com

© 2006 BentleyPublishers.com—All Rights Reserved

Clutch pedal assembly

1. **Bulkhead**
   - With support for mounting bracket and clutch master cylinder
2. **Seal**
   - Always replace
3. **Mounting bracket (support)**
   - For clutch pedal
4. **Pivot bolt**
5. **Mounting bracket (support)**
   - For brake and accelerator pedals
6. **Nut, hex self-locking**
   - Always replace
   - Tighten to 25 Nm (18 ft-lb)
7. **Connecting bracket**
8. **Nut, hex self-locking**
   - Always replace
   - Tighten to 25 Nm (18 ft-lb)
9. **Pivot bushings**
10. **Fulcrum pin**
11. **Clutch pedal**
    - See **Removing and installing**
12. **Pushrod retainer**
13. **Nut, hex self-locking**
    - Always replace
    - Tighten to 25 Nm (18 ft-lb)
14. **Clutch master cylinder**
    - See **Removing and installing**
15. **Fluid supply hose**
    - From brake fluid reservoir
16. **Over-center spring**
    - Remove and install with special tools T10111 or 3317
17. **Over-center spring seat**
    - Installs in mounting bracket
    - Remove and install clutch master cylinder to replace
18. **Clutch pedal position switch F194, clutch cruise control switch F36**
    - Different versions installed identified by housing shape, beveled or round
    - Switch is designed to be installed one time only; will not fit tightly into mounting bracket if reused

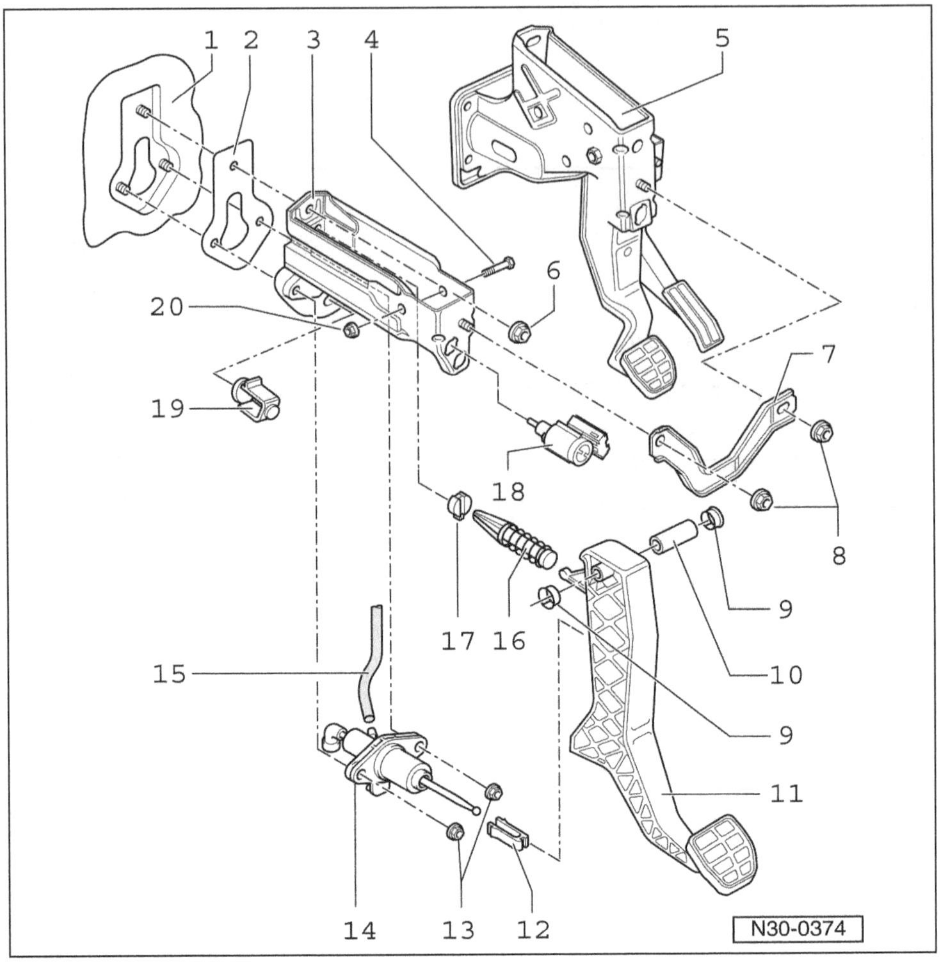

19. **Clutch pedal stop**
20. **Nut, hex self-locking**
    - Always replace
    - Tighten to 25 Nm (18 ft-lb)

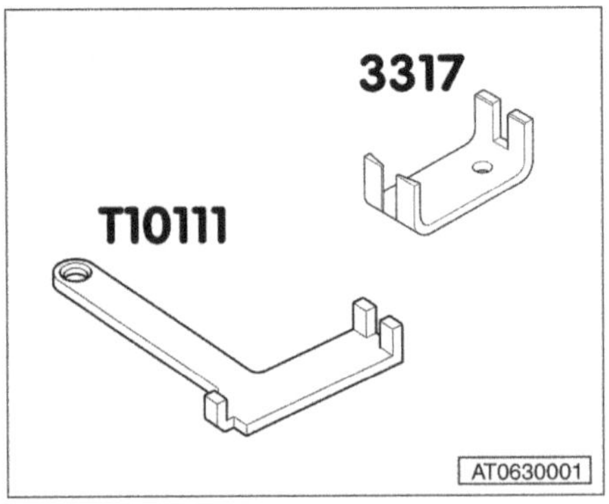

◀ Special tools needed to remove over-center spring (item 16). Either can be used, T10111 is later version.

BentleyPublishers.com
© 2006 BentleyPublishers.com—All Rights Reserved

NOTE—

- *Audi identifies electrical components by a letter and/or a number in the electrical schematics. See* **EWD Electrical Wiring Diagrams**. *These electrical identifiers are listed in parenthesis as an aid to electrical troubleshooting.*

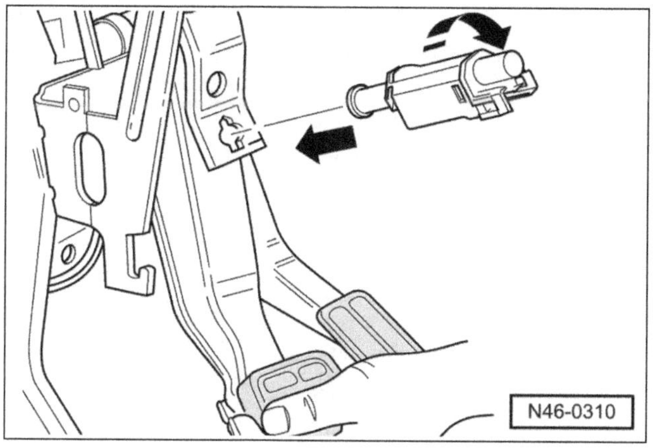

Clutch pedal position switch (F194) and/or clutch cruise control switch (F36), removing and installing (bevel housing)

- To ensure a tight fit, switch is designed to be installed one time only.

– Remove left side lower under-dash (close-out) panels. See **70 Trim-Interior**.

– Disconnect harness connector.

– Remove switch by twisting ¼ turn (90°) counter-clockwise.

– To install new switch, pull plunger out to the stop by hand.

– Depress clutch pedal fully by hand.

– Insert switch into mounting bracket opening (**left arrow**) and rotate ¼ turn (90°) clockwise (**right arrow**) to secure in mounting bracket.

– Release clutch pedal to self adjust switch.

– Install left side lower under-dash (close-out) panels.

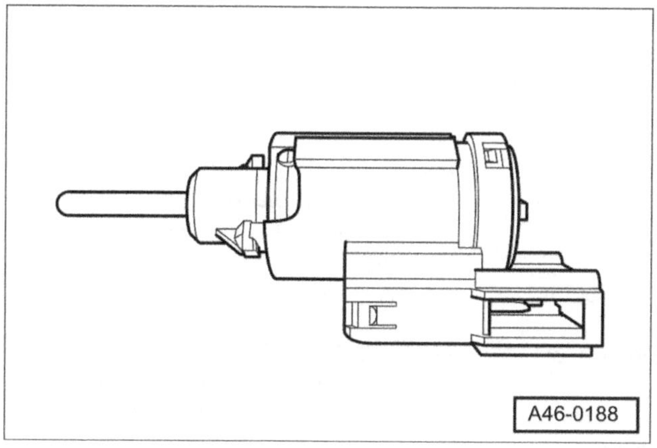

Clutch pedal clutch cruise control switch (F36), removing and installing (round housing)

- To ensure a tight fit, switch is designed to be installed one time only.

– Remove left side lower under-dash (close-out) panels.

– Disconnect harness connector.

– Remove switch by twisting ¼ turn (90°) counter-clockwise.

– To install new switch, pull plunger out to the stop by hand.
  - **Do not** depress clutch pedal.

– With pedal in at-rest position (not operated) position, guide switch through bracket opening while pressing plunger against pedal and rotate ¼ turn (90°) clockwise to secure in mounting bracket.

– Install left side lower under-dash (close-out) panels.

BentleyPublishers.com
© 2006 BentleyPublishers.com—All Rights Reserved

Clutch Hydraulic System

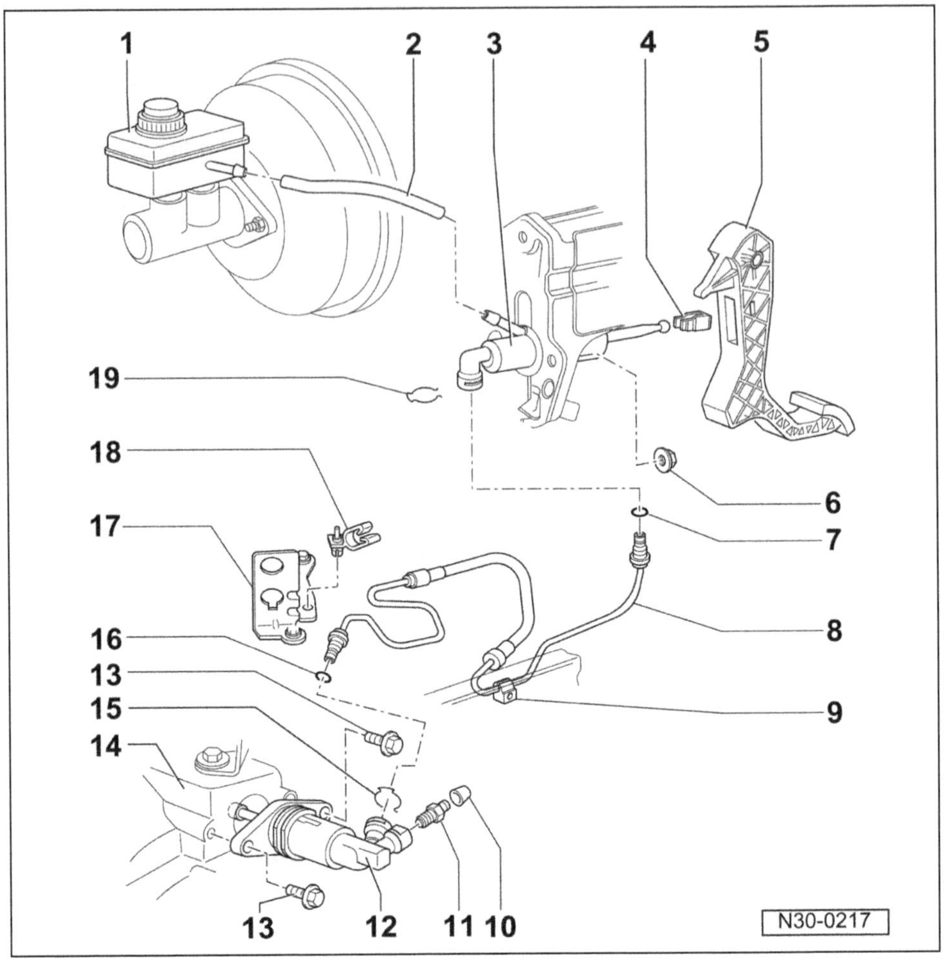

Clutch hydraulic system and release bearing (front wheel drive transmission)

1. **Brake fluid reservoir**
   - Brake system and clutch hydraulics share a common reservoir
   - System design keeps brake fluid at safe level if faults occur in clutch hydraulic system
   - Fluid level should be between **MIN** and **MAX** markings on reservoir
2. **Supply hose**
   - From reservoir to clutch master cylinder
3. **Clutch master cylinder**
   - See **Clutch master cylinder O-rings**
4. **Pushrod retainer**
   - Replace with master cylinder or pedal removed
   - See **Pushrod retainer, removing**
   - See **Pushrod retainer, installing**
5. **Clutch pedal**
6. **Nut, hex self-locking**
   - Always replace
   - Tighten to 25 Nm (18 ft-lb)
7. **O-ring seal**
   - Always replace
   - Moisten with brake fluid before installation
   - Pull onto line/hose connection
8. **Hydraulic line/hose assembly**
9. **Bracket**
   - Attached to body
10. **Bleeder valve dust cap**
    - Always reinstall when finished bleeding
11. **Bleeder valve**
    - See **Clutch hydraulic system, bleeding**
12. **Clutch slave cylinder**
    - Remove the following components before removing clutch slave cylinder:
      Gear selector cable from selector lever
      Relay lever and actuating arm for gate selector cable
    - For selector cable components, see **34a Manual Transmission**
    - Adjust gear selector mechanism after reinstalling components
13. **Bolt, shouldered hex**
    - Tighten to 25 Nm (18 ft-lb)
14. **Transmission housing**
15. **Securing clip**
    - Pull clip out and onto stop to release hydraulic line hose assembly
16. **O-ring seal**
    - Always replace
    - Moisten with brake fluid before installation
    - Pull onto line/hose connection
17. **Support bracket**
    - Size, shape and placement vary slightly with engine type
18. **Hose bracket**
    - Secure to cable support bracket
19. **Securing clip**
    - Pull clip out and onto stop to release hydraulic line hose assembly

BentleyPublishers
.com
© 2006 BentleyPublishers.com—All Rights Reserved

Clutch hydraulic system and release bearing (all wheel drive transmission)

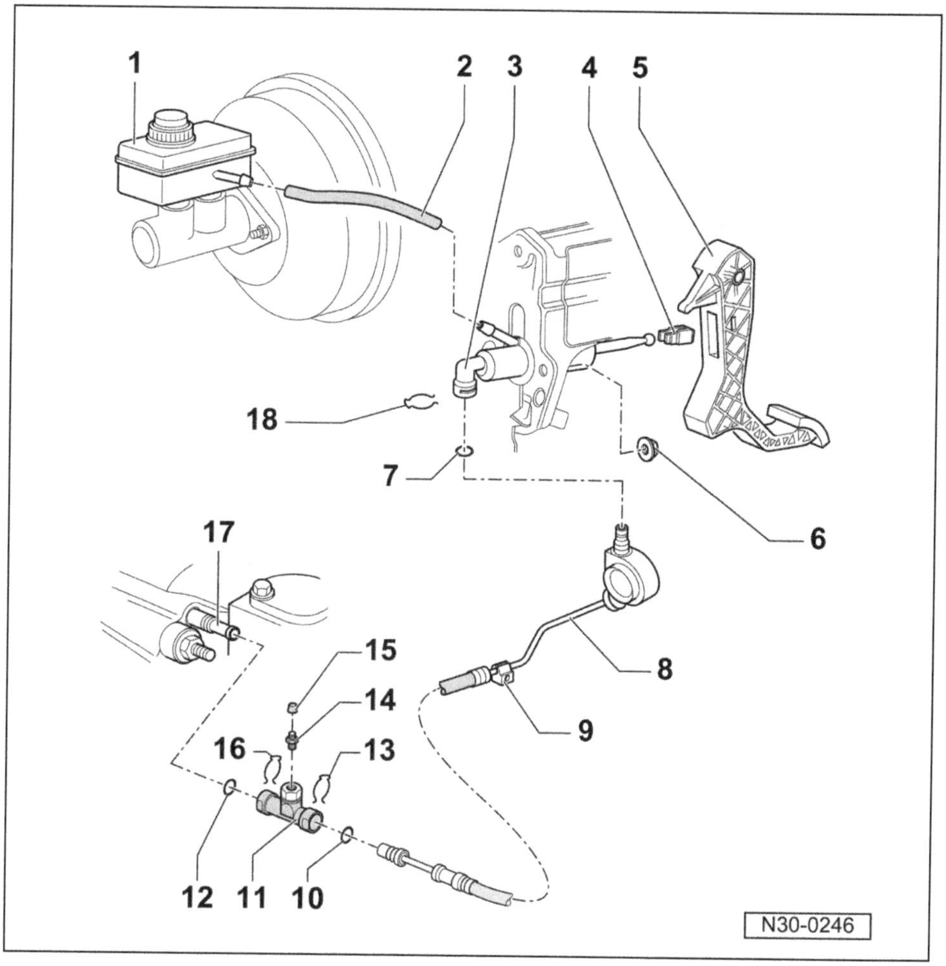

1. **Brake fluid reservoir**
   - Brake system and clutch hydraulics share a common reservoir
   - System design keeps brake fluid at safe level if faults occur in clutch hydraulic system
   - Fluid level should be between MIN and MAX markings on reservoir
2. **Supply hose**
   - From reservoir to clutch master cylinder
3. **Clutch master cylinder**
   - See **Clutch master cylinder O-rings**
4. **Pushrod retainer**
   - Replace with master cylinder or pedal removed
   - See **Pushrod retainer, removing**
   - See **Pushrod retainer, installing**
5. **Clutch pedal**
6. **Nut, hex self-locking**
   - Always replace
   - Tighten to 25 Nm (18 ft-lb)
7. **O-ring seal**
   - Always replace
   - Moisten with brake fluid before installation
   - Pull onto line/hose connection
8. **Hydraulic line/hose assembly**
9. **Bracket**
   - Attached to body
10. **O-ring seal**
    - Always replace
    - Moisten with brake fluid before installation
    - Pull onto line/hose connection
11. **Bleeder housing/coupling**
12. **O-ring seal**
    - Always replace
    - Moisten with brake fluid before installation
    - Pull onto line/hose connection
13. **Securing clip**
    - Pull clip out and onto stop to release hydraulic line hose assembly
14. **Bleeder valve**
    - See **Clutch hydraulic system, bleeding**
15. **Bleeder valve dust cap**
    - Always reinstall when finished bleeding
16. **Securing clip**
    - Pull clip out and onto stop to release hydraulic line hose assembly
17. **Clutch slave cylinder**
    - Located within bellhousing of transmission. See **Clutch release mechanism (all wheel drive)**
    - Requires removal of transmission for access. See **34a Manual Transmission**
18. **Securing clip**
    - Pull clip out and onto stop to release hydraulic line hose assembly

BentleyPublishers
.com
© 2006 BentleyPublishers.com—All Rights Reserved

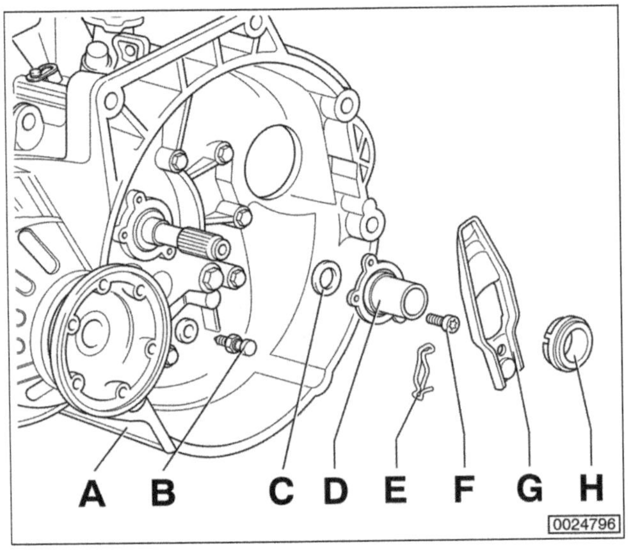

Clutch release mechanism (front wheel drive)

A. Transmission housing

B. Ball stud:
- Pivot for clutch release lever
- Tighten to 25 Nm (18 ft-lb).
- Lubricate with MoS_2 (moly) grease.

C. Input shaft oil seal.

D. Guide sleeve:
- With vulcanized O-ring.
- If damaged, replace guide sleeve and O-ring.
- Lubricate guide sleeve at release bearing with MoS_2 (moly) grease.

E. Retaining spring:
- Secures clutch release lever to ball stud.

F. Bolt, polygon socket head:
- Tighten to 20 Nm (15 ft-lb).

G. Clutch release lever

H. Release (throw-out) bearing:
- Do not wash bearing, wipe clean only.
- If noisy, replace.
- Lubricate with MoS_2 (moly) grease.

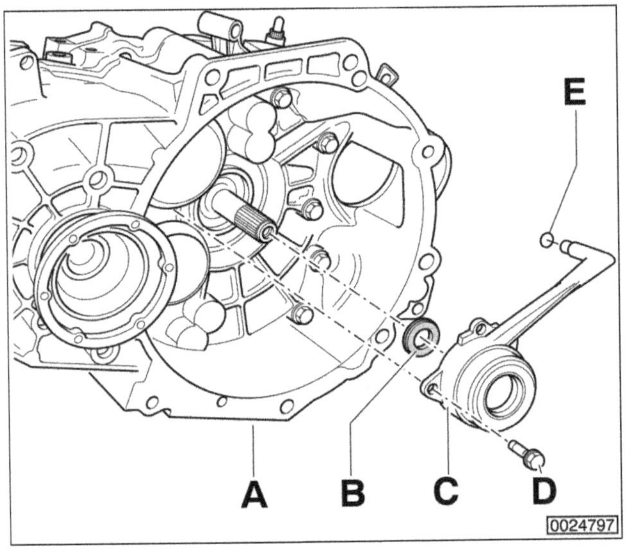

Clutch release mechanism (all wheel drive)

A. Transmission housing

B. Input shaft oil seal

C. One piece slave cylinder and release bearing:
- To gain access, remove transmission.
- If bearing is noisy, replace complete assembly.
- Do not wash bearing, wipe only.

D. Bolt:
- Metal slave cylinder: Tighten to 12 Nm (9 ft-lb).
- Plastic slave cylinder: Tighten to 15 Nm (11 ft-lb).

E. O-ring seal:
- Moisten with brake fluid to install.
- Install on fluid line.

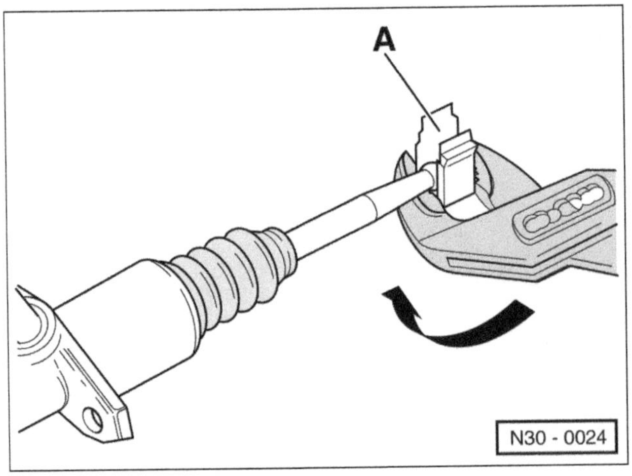

Pushrod retainer, removing
- Pry retainer (**A**) off in direction of **arrow**

BentleyPublishers.com
©2006 BentleyPublishers.com—All Rights Reserved

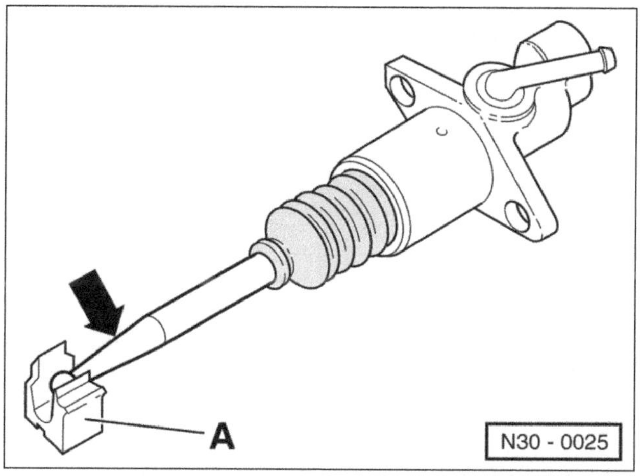

Pushrod retainer, installing

- Press pushrod into retainer (**A**) in direction of **arrow**

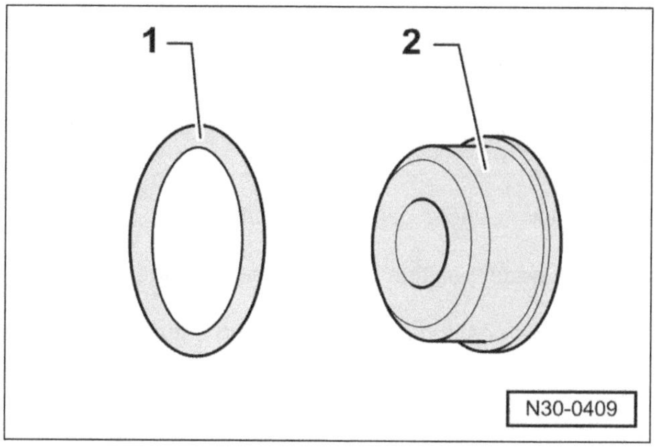

Clutch master cylinder O-rings

1. Use with metal line connections
2. Use with plastic line connections

Clutch master cylinder, removing and installing

Removal of the clutch master cylinder requires removal of the entire pedal cluster for access.

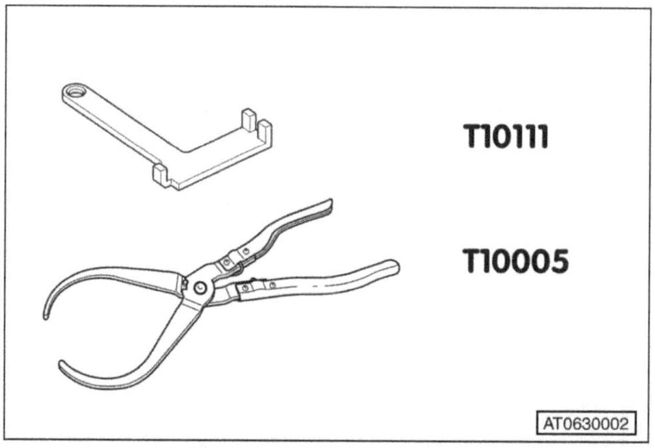

Several special tools are required. Release tools **T10111** or 3317 and **T10005** are used along with line clamps 3094.

Removing

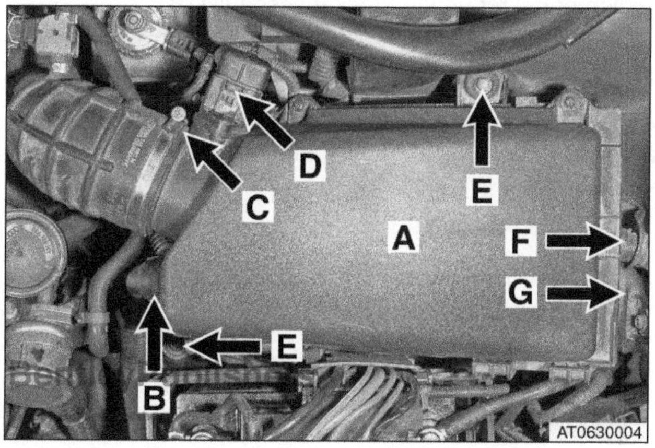

Remove complete air filter assembly (**A**) to allow access.

– Depress tabs on lock ring on secondary air injection hose (**B**) and remove hose.

– Loosen clamp on turbocharger inlet hose (**C**) and remove hose from mass air flow sensor. On 6-cylinder engines, additional access is available by also removing hose at throttle valve control module.

– Disconnect harness connector for mass air flow sensor (**D**).

– Remove screws (**E**) securing air filter housing to body.

BentleyPublishers
.com
© 2006 BentleyPublishers.com—All Rights Reserved

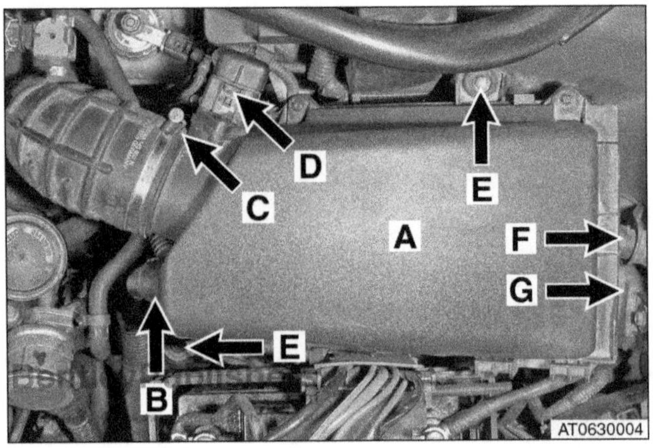

Lift air filter housing up, out, and away from fresh air intake connection (**G**) and locating pin (**F**) in body.

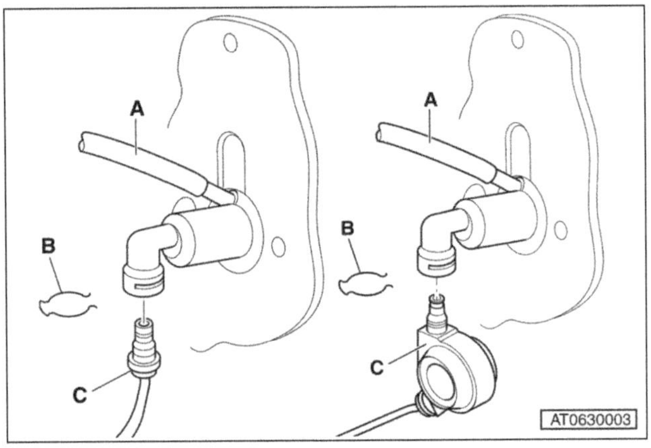

Clamp supply hose (**A**) from reservoir with clamp 3094 or equivalent.

- Remove supply hose (**A**) from clutch master cylinder fitting.
- Remove clip (**B**) and pull pipe hose assembly out of clutch master cylinder (**C**).
- Remove left side lower under-dash (close-out) panels.
- Disconnect clutch pedal position switch harness connector.
- Remove clutch pedal position switch by twisting ¼ turn (90°) counter-clockwise.

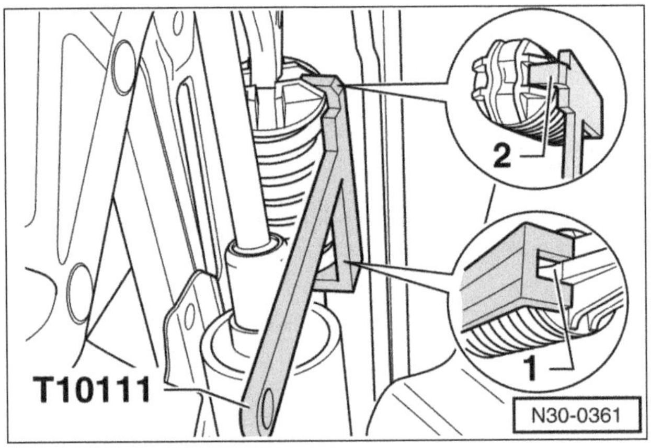

In some instances, access may be easier if clutch pedal over-center spring is removed. Insert assembly clamp **T10011** or 3317 onto over-center spring from right side of clutch pedal mounting bracket.

- Position rear part of **T10111** or 3317 on over-center spring so that claw fits over web (**1**).
- Push clutch pedal slightly towards bulkhead and hook tab or jaw of tool into cut-out(s) of over-center spring.
- Remove spring.

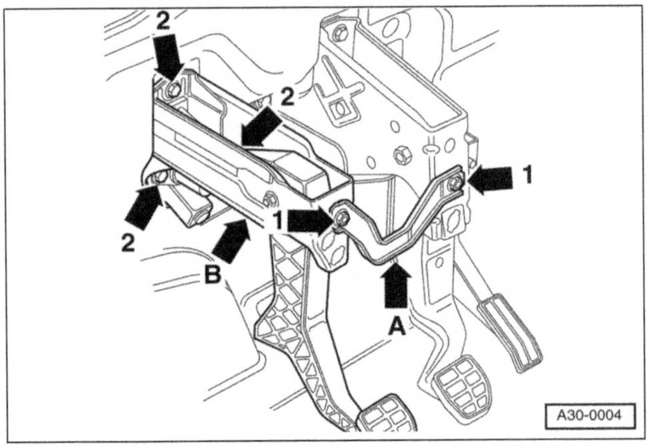

Remove clutch pedal mounting bracket assembly (**B**).

- Remove nuts (**1**) and take off connecting bracket (**A**).
- Remove securing nuts (**2**) and slide mounting bracket complete with clutch master cylinder away from bulkhead and out.

BentleyPublishers.com
©2006 BentleyPublishers.com—All Rights Reserved

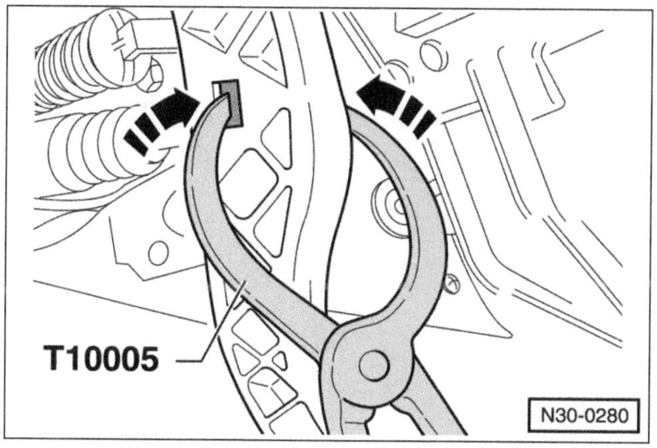

◀ Disconnect clutch master cylinder operating rod from pedal. Pull clutch pedal slightly towards the at-rest position and insert tips of tool **T10005** into retainer recesses (**dashed arrows**). Squeeze pliers while pulling slightly on pedal.

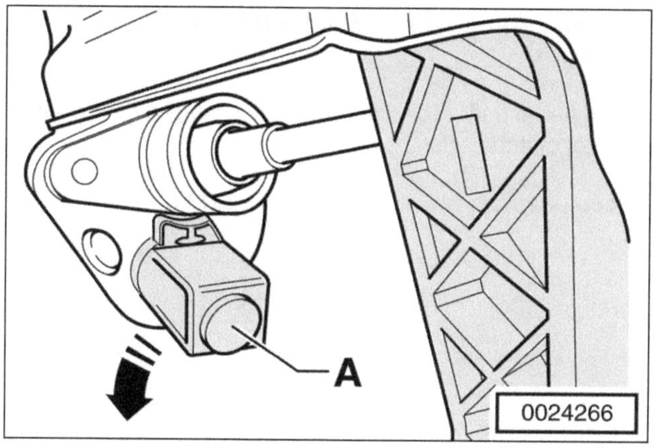

◀ Remove clutch pedal stop (**A**) by turning counter-clockwise (**arrow**) and pulling out.

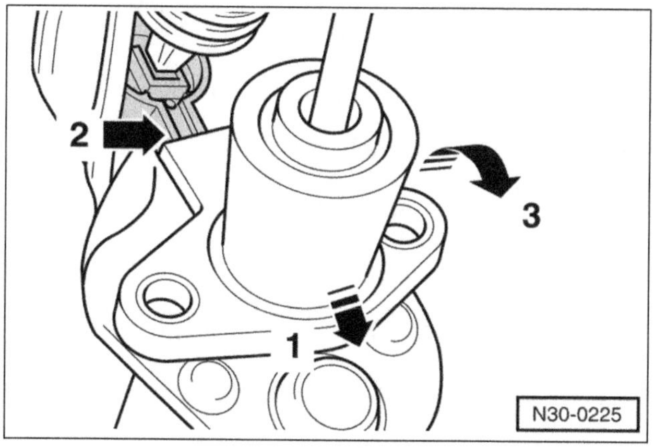

◀ Push clutch master cylinder down (**arrow 1**) onto stop so that top of master cylinder is not covered by the over-center spring mount (**arrow 2**).

– Swing master cylinder down and out (**arrow 3**) of mounting bracket.

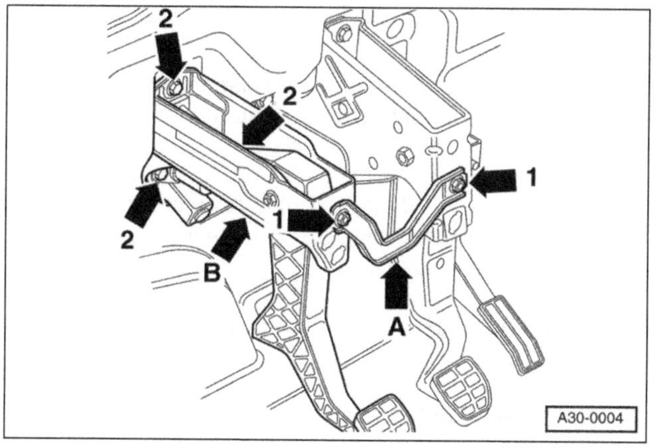

Clutch master cylinder, installing

– Install clutch master cylinder into mounting bracket.

◀ Install clutch pedal mounting bracket assembly (**B**) with new nuts (**2**) tighten.

– Install connecting bracket (**A**) with new nuts and tighten.

Tightening torque	
Nuts, mounting bracket (always replace)	25 Nm (18 ft-lb)

BentleyPublishers.com
© 2006 BentleyPublishers.com—All Rights Reserved

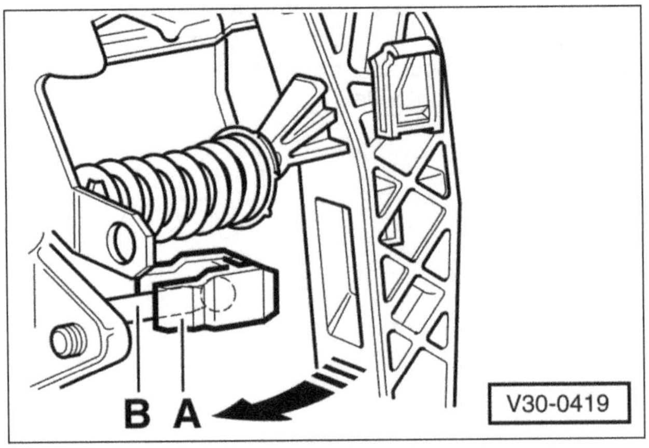

◀ If retainer (**A**) is not already installed, position retainer onto master cylinder pushrod (**B**).

– Depress clutch pedal (**arrow**) to engage and seat retainer in pedal.

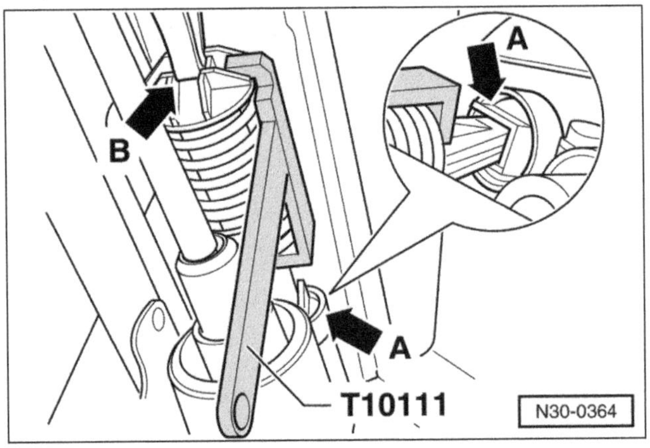

◀ If removed, insert over-center spring with **T10111** installed, into rear mounting (**arrow A**).

– Operate clutch pedal until over-center spring is seated on mounting tab (**arrow B**) of clutch pedal.

– When over-center spring is seated, remove **T10111**.

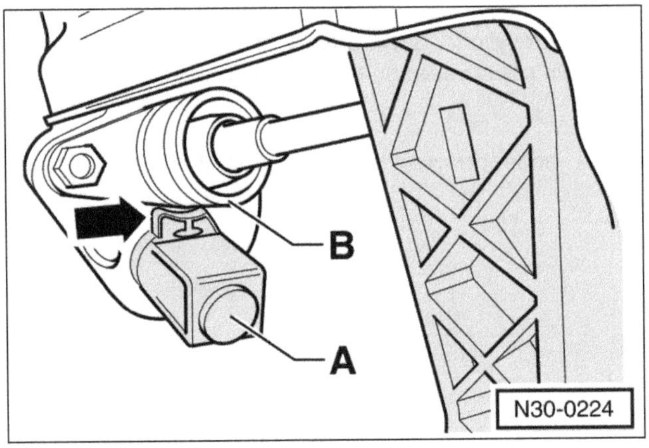

◀ Install clutch pedal stop (**A**) by pushing in and turning clockwise (**arrow**) until tab (**B**) is seated on clutch master cylinder.

– Install new clutch switch by twisting ¼ turn (90°) clockwise.

– Connect clutch switch harness connector.

– Install left side lower under-dash (close-out) panels.

– Bleed clutch hydraulic system.

– Install complete air filter assembly.

Clutch hydraulic system, bleeding

The brake system and hydraulic clutch share brake fluid. Pre-filling the system is not required, but the system must be bled after repair work. For brake fluid specifications, see **03 Maintenance**.

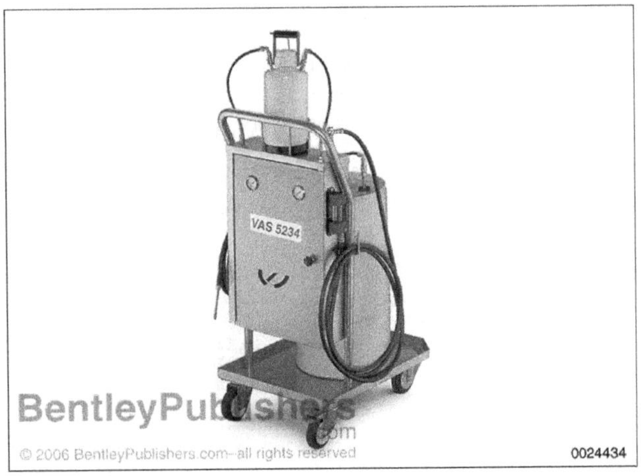

BentleyPublishers.com
© 2006 BentleyPublishers.com–all rights reserved

◀ To bleed clutch hydraulics, use VW pressure bleeder VAS 5234, VAG 1869, or equivalent. Follow manufacturers instructions for operation.

– Catch bled fluid using a bottle. Connect VAG 1238 B3 hose to bleeder valve, or use approximately 670 mm (26 in) of hose sized to fit over bleeder valve on slave cylinder. Place end of hose in clear glass or plastic bottle half-filled with brake fluid.

– If not already removed for other repairs, remove air cleaner assembly. See **03 Maintenance**.

– Connect brake bleeder to brake fluid reservoir.

BentleyPublishers.com
© 2006 BentleyPublishers.com—All Rights Reserved

– Remove cap on bleeder valve

Front wheel drive transmission

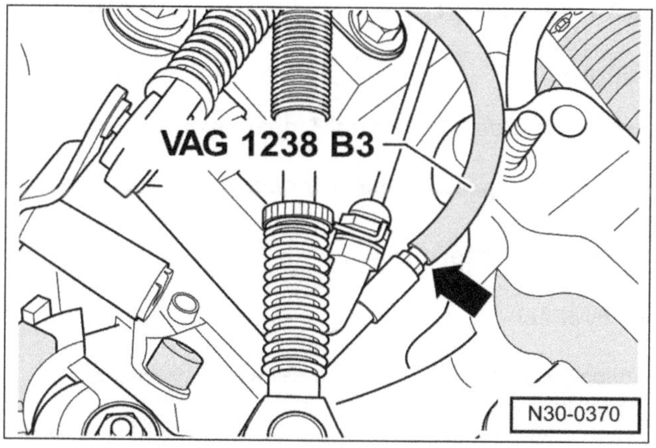

◄ Connect bleeder hose from catch bottle to slave cylinder bleeder valve (**arrow**). Avoid spilling brake fluid onto transmission during bleeding.

All wheel drive transmission

◄ Connect bleeder hose from catch bottle to slave cylinder bleeder valve (**arrow**). Avoid spilling brake fluid onto transmission during bleeding.

Continued for all vehicles

– Pressurize system to 2 bar (29 psi).

– Open bleeder valve.

– Bleed out approximately 100 cc (4 oz) of brake fluid.

– Close bleeder valve.

– Operate clutch pedal quickly 15 times through full pedal travel.

– Open bleeder valve again.

– Bleed again and bleed out approximately 50 cc (2 oz) of brake fluid.

– Close bleeder valve.

– Operate clutch pedal several times through full pedal travel.

– Repeat procedure if needed.

– Remove bleeder and top up brake fluid reservoir as needed.

– Install cap on bleeder valve.

– Install complete air filter assembly.

BentleyPublishers
.com
© 2006 BentleyPublishers.com—All Rights Reserved

Dual-Mass Flywheel

Dual-mass flywheel, removing and installing

4-cylinder engine

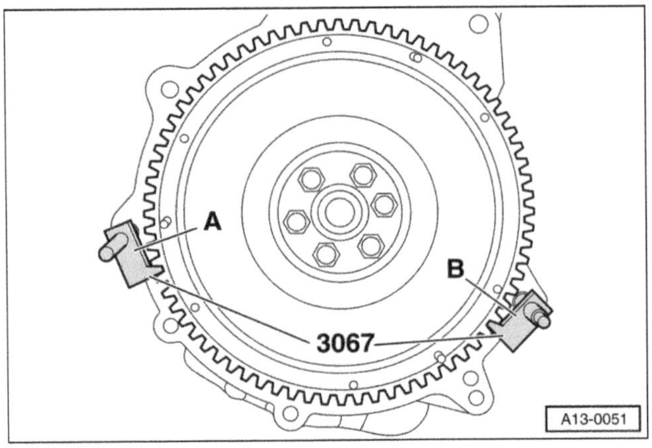

– Remove transmission. See **34a Manual Transmission**.

◄ Insert flywheel retainer, special tool **3067** or equivalent, into hole in cylinder block.

- Installed location of flywheel retainer:
  A - to tighten flywheel
  B - to loosen flywheel

– Mark installed location of dual-mass flywheel with a felt-tipped marker before removing.

NOTE —
- *Bolt pattern may have one offset hole to ensure installation in the same position as removal.*

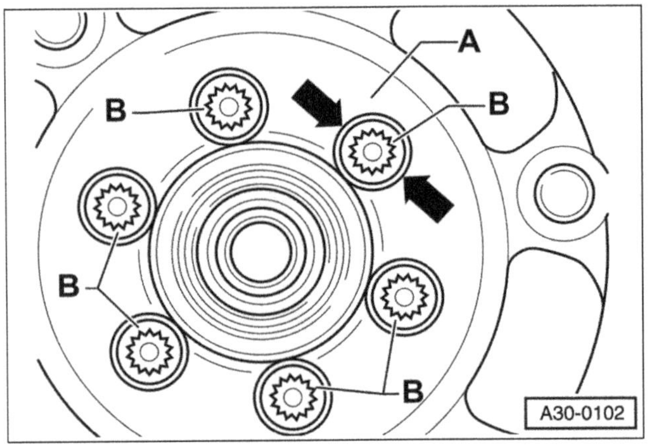

◄ Remove flywheel mounting bolts (**B**).

NOTE —
- *To prevent damage to dual-mass flywheel when removing bolts (**B**), do not use air-powered tools or impact driver. Remove bolts by hand only.*
- *When removing bolts, make sure that bolt heads do not make contact with dual-mass flywheel. Damage to flywheel could occur.*

– Rotate dual-mass flywheel (**A**) so that bolts are centered in holes (**arrows**).

– Installation is reverse of removal. Use new mounting bolts.

Tightening torque	
Dual-mass flywheel to crankshaft (replace bolts)	
• Stage 1	60 Nm (44 ft-lb)
• Stage 2	¼ turn (90°)

6-cylinder engine

– Remove transmission. See **34a Manual Transmission**.

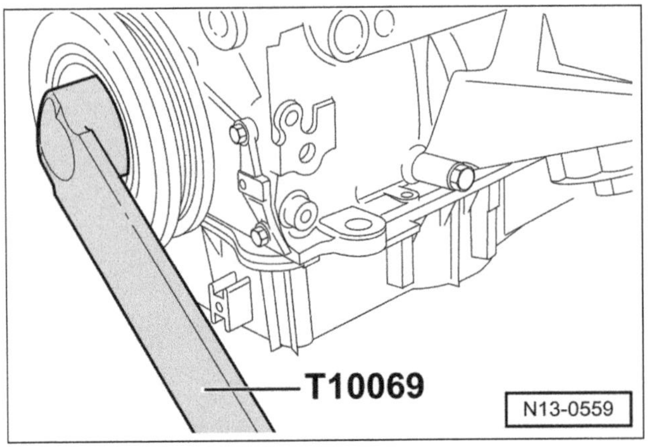

◄ Lock crankshaft from the pulley end with counterholder **T10069** or equivalent.

– Mark installed location of dual-mass flywheel with a felt-tipped marker before removing.

BentleyPublishers.com
© 2006 BentleyPublishers.com—All Rights Reserved

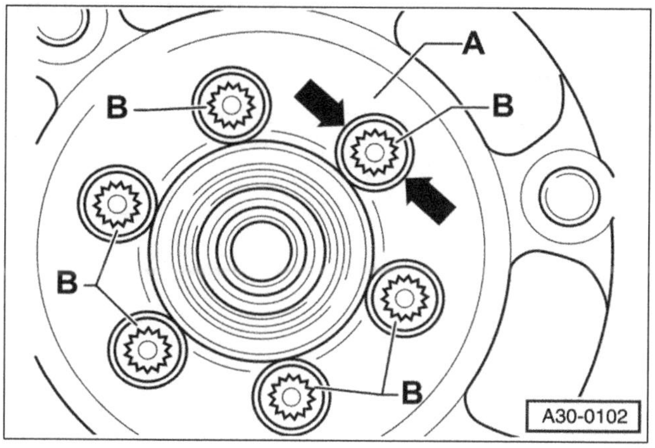

◀ Remove flywheel mounting bolts (**B**).

NOTE —

- *To prevent damage to dual-mass flywheel when removing bolts (**B**), do not use an air-powered tools or impact driver. Remove bolts by hand only.*
- *When removing bolts, make sure that bolt heads do not make contact with dual-mass flywheel. Damage to flywheel could occur.*

– Rotate dual-mass flywheel (**A**) so that bolts are centered in holes (**arrows**).

– Installation is reverse of removal. Use new mounting bolts.

Tightening torque	
Dual-mass flywheel to crankshaft (replace bolts)	
• Stage 1	60 Nm (44 ft-lb)
• Stage 2	¼ turn (90°)

Clutch Assembly (manual transmission)

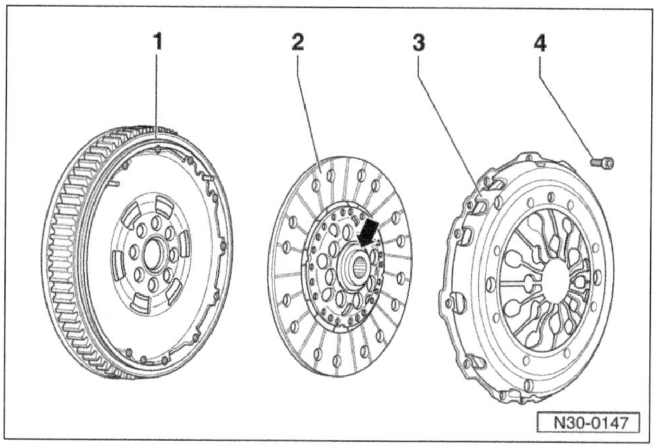

Clutch assembly

1. Dual-mass flywheel
   - Ensure dowel pins fit securely.
   - Disc contact surface must be free of grooves, oil, and grease.
2. Clutch disc
   - Installation position: short end of hub (**arrow**) faces toward pressure plate.
3. Clutch pressure plate (cover)
   - If reusing pressure plate, inspect diaphragm spring.
   - If reusing pressure plate, inspect spring connections and rivets.
4. Bolt, special
   - Loosen bolts in clockwise sequence ¼ turn (90°) at a time continuing until pressure plate is free.
   - Install bolts in counter-clockwise sequence ¼ turn (90°) at a time continuing until pressure plate snug on flywheel. Tighten according to size.
   - M6 - tighten to 13 Nm (10 ft-lb)
   - M7 - tighten to 20 Nm (15 ft-lb)
   - M8 - tighten to 22 Nm (16 ft-lb)

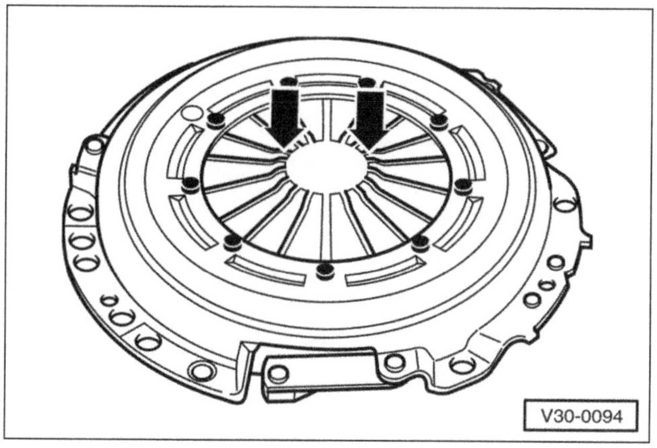

◀ **Clutch pressure plate diaphragm spring, inspecting**

- Check pressure plate diaphragm spring ends (**arrows**) for wear.
- Maximum wear (wear limit): up to one half (½) of the original spring material thickness.

BentleyPublishers
.com
© 2006 BentleyPublishers.com—All Rights Reserved

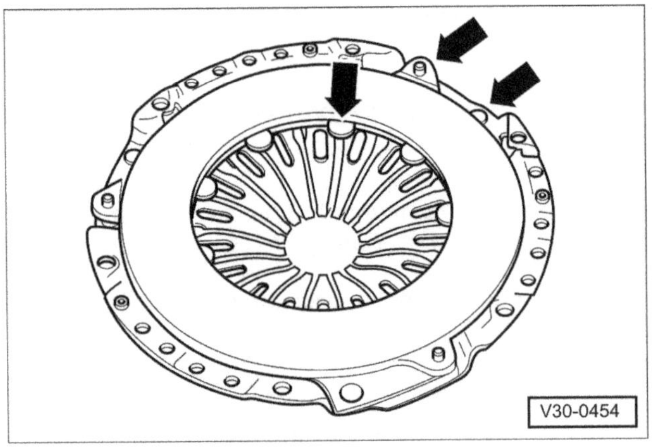
V30-0454

Clutch pressure plate spring connections and rivets, inspecting

- Check pressure plate spring connections and rivets (**arrows**) for looseness and tearing.
- Replace pressure plate if above conditions are found.

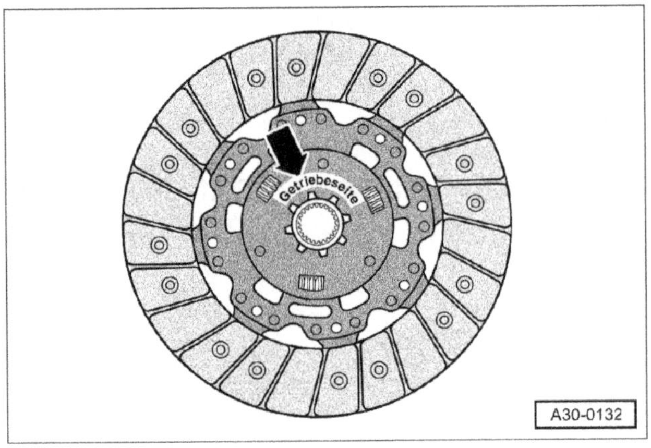

A30-0132

Clutch disc installation position

- Hub area marked **Getriebeseite** (transmission side) (**arrow**) must face transmission.
- Lubricate splines with G 000 100 grease.
- Center disc assembly according to transmission type, as described below.

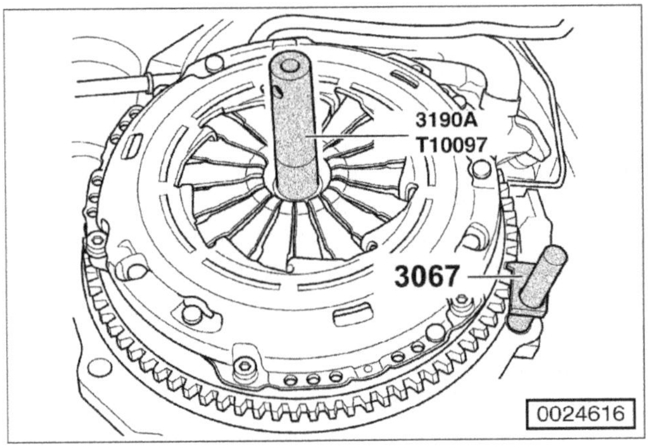

0024616

Clutch disc centering

- Center clutch disc with **3190A** or **T10097** as appropriate for clutch version
- Flywheel lock, special tool **3067**, is shown in position during clutch installation. Reverse position for clutch removal. Use T10069 on pulley end of engine where appropriate.
- If not seated and torqued evenly, chattering and engagement problems can result.

BentleyPublishers.com
©2006 BentleyPublishers.com—All Rights Reserved

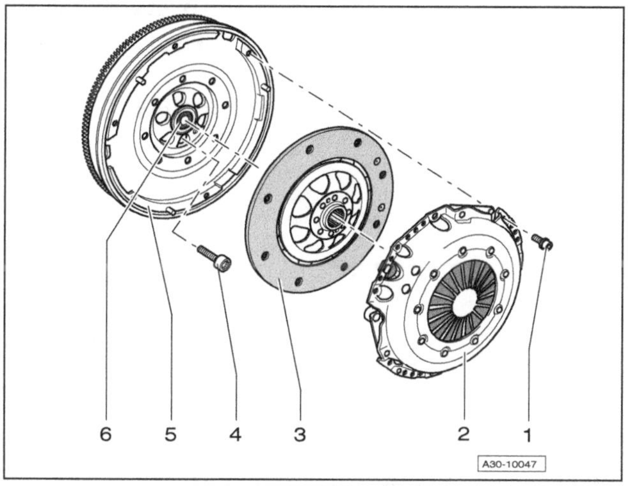

Clutch assembly, Sachs (self-adjusting)

1. Bolt, special:
   - Loosen bolts in clockwise sequence ¼ turn (90°) at a time continuing until pressure plate is free.
   - Install bolts by hand until heads contact surface. Tighten in clockwise sequence ¼ turn (90°) at a time continuing until pressure plate fits snug on flywheel. Impact position sensor must lift from pressure plate.
   - Tighten according to size.
     M6 - tighten to 13 Nm (10 ft-lb)
     M7 - tighten to 20 Nm (15 ft-lb)
     M8 - tighten to 22 Nm (16 ft-lb)
2. Clutch pressure plate (cover):
   - Self-adjusting
   - Only replace as a set with clutch disc.
   - When installing, fit onto locating pins.
3. Clutch disc
   - Only replace as a set with clutch pressure plate.
   - Ensure full contact with flywheel.
4. Bolt
5. Dual-mass flywheel:
   - Ensure dowel pins fit securely.
   - Disc contact surface must be free of grooves, oil, and grease.
6. Bushing
   - Not used in this application.

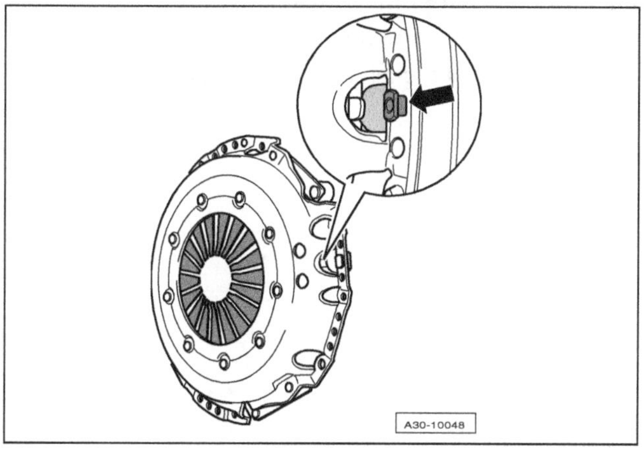

Clutch pressure self-adjusting mechanism

- With impact (position) sensor (**arrow**).

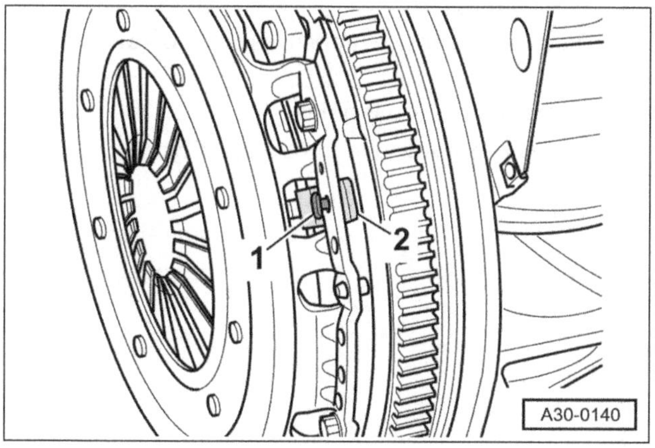

Clutch pressure self-adjusting mechanism, removing

- When removing, impact (position) sensor (**2**) must loosen together with pin (**1**).

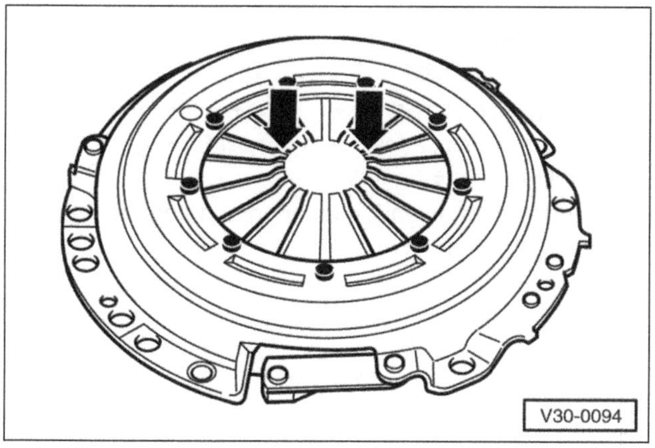

Clutch pressure plate diaphragm spring, inspecting

- Check pressure plate diaphragm spring ends (**arrows**) for wear.
- Maximum wear (wear limit): up to one half (½) of the original spring material thickness.

BentleyPublishers
.com
©2006 BentleyPublishers.com—All Rights Reserved

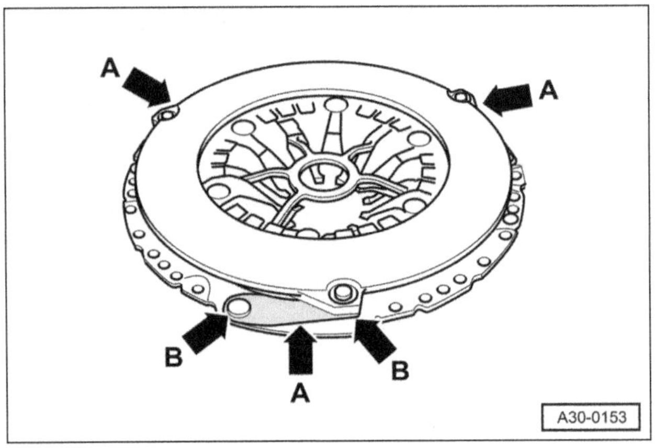

Clutch pressure plate spring connections and rivets, inspecting

- Check pressure plate spring connections (**A**) and rivets (**B**) for looseness and tearing.
- Replace pressure plate if above conditions are found.

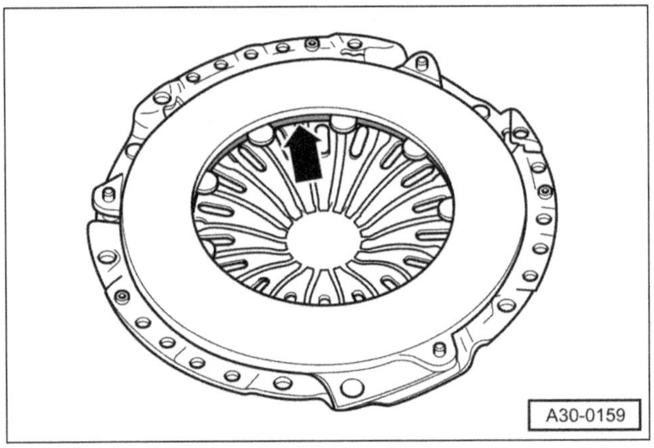

Clutch pressure plate wire ring, inspecting

- Check wire ring (**arrow**) in pressure plate for damage.
- Replace pressure plate if broken wire ring is found.

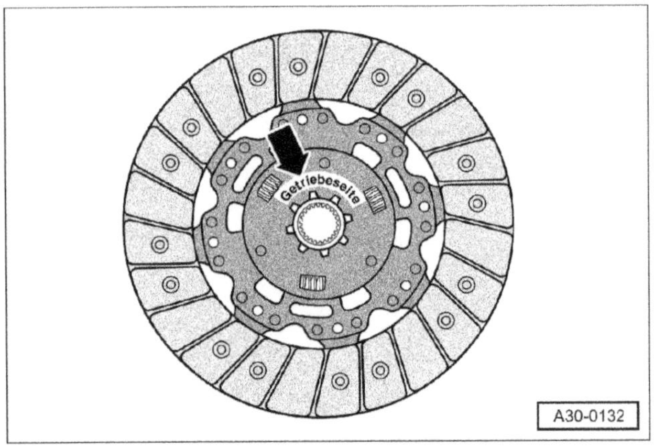

Clutch disc installation position

- Hub area marked **Getriebeseite** (transmission side) (**arrow**) must face transmission.
- Lubricate splines with G 000 100 grease.
- Center disc assembly according to transmission type:

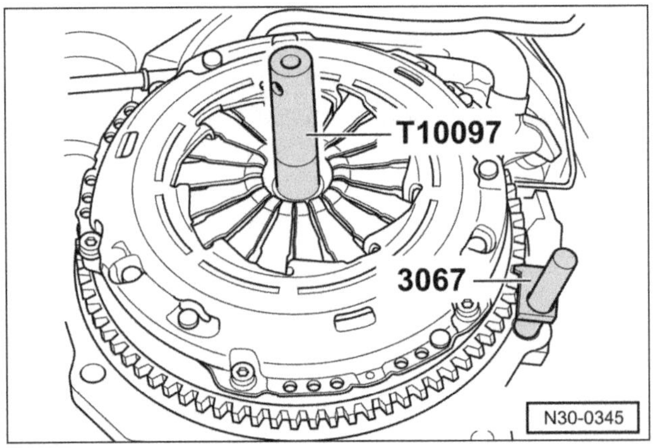

Clutch disc centering

- Center clutch disc with **T10097**.
- Flywheel lock, special tool **3067**, is shown in position during clutch installation. Reverse position for clutch removal. Use T10069 on pulley end of engine where appropriate.
- If not seated and torqued evenly, chattering and engagement problems can result.

BentleyPublishers
.com
©2006 BentleyPublishers.com—All Rights Reserved

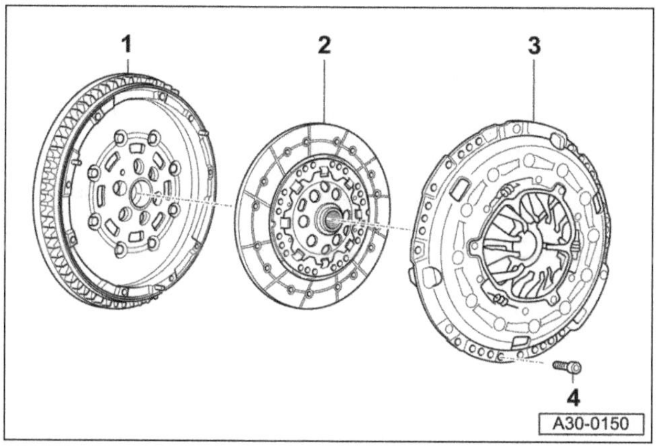

Clutch assembly, LuK (self-adjusting)

1. Dual-mass flywheel
   - Ensure dowel pins fit securely
   - Disc contact surface must be free of grooves, oil and grease
2. Clutch plate (disc)
   - Replace clutch disc and cover together
   - Do not interchange parts from different manufactures
3. Clutch pressure plate (cover), self-adjusting clutch (SAC)
   - Replace clutch disc and cover together
   - Do not interchange parts from different manufactures.
   - When installing, fit onto locating pins.
4. Bolt, special
   - Loosen bolts in clockwise sequence ¼ turn (90°) at a time continuing until pressure plate is free.
   - Install bolts by hand until heads contact surface. Tighten in clockwise sequence ¼ turn (90°) at a time continuing until pressure plate fits snug on flywheel.
   - Tighten according to size.
     M6 - tighten to 13 Nm (10 ft-lb)
     M7 - tighten to 20 Nm (15 ft-lb)
     M8 - tighten to 22 Nm (16 ft-lb)

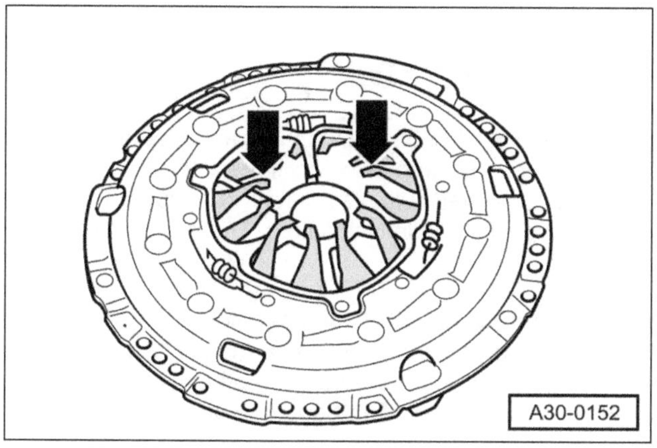

Clutch pressure plate diaphragm spring, inspecting

- Check pressure plate diaphragm spring ends (**arrows**) for wear.
- Maximum wear (wear limit): up to one half (½) of the original spring material thickness.

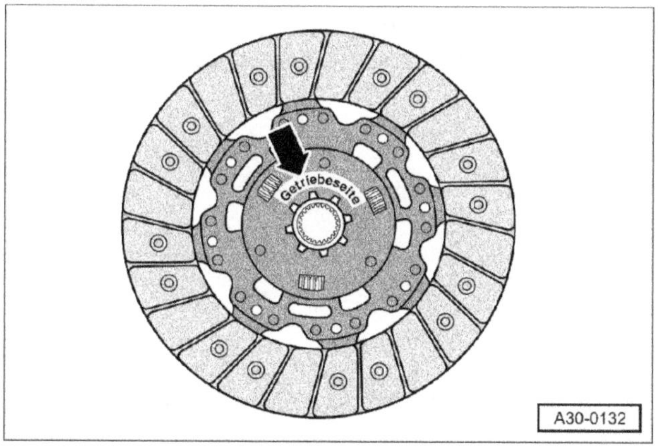

Clutch disc installation position

- Hub area marked **Getriebeseite** (transmission side) (**arrow**) must face transmission.
- Lubricate splines with G 000 100 grease.
- Center disc assembly according to transmission type:

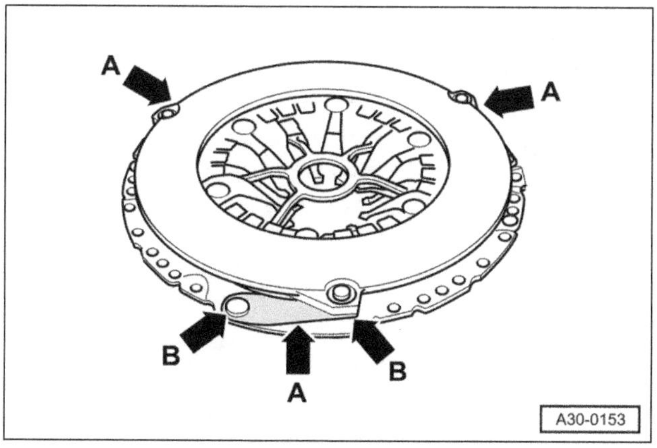

Clutch pressure plate spring connections and rivets, inspecting

- Check pressure plate spring connections (**A**) and rivets (**B**) for looseness and tearing.
- Replace pressure plate if above conditions are found.

BentleyPublishers.com
©2006 BentleyPublishers.com—All Rights Reserved

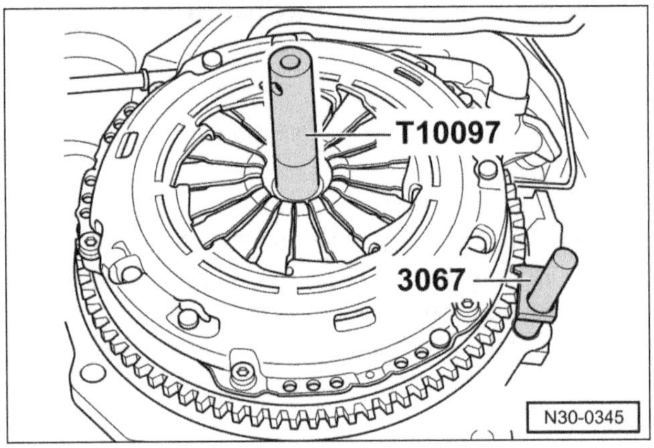

Clutch disc centering

- Center clutch disc with **T10097**.
- Flywheel lock, special tool **3067**, is shown in position during clutch installation. Reverse position for clutch removal. Use T10069 on pulley end of engine where appropriate.
- If not seated and torqued evenly, chattering and engagement problems can result.

CLUTCH ASSEMBLY (DSG)

The Direct Shift Gearbox (DSG) is a manual transmission designed to operate like an automatic transmission. In place of a dry clutch (or a torque converter), two wet multi-disc clutches are used. These wet clutches rotate in special transmission fluid. There is no clutch pedal.

Transmission electronics control clutch operation and gear selection, although the transmission can be manually shifted through the selector lever in the console or optional paddles (**arrows**) in the steering wheel. The transmission is always engaged.

Repair and replacement of the wet multi-disc clutch requires the use of specialized tools and procedures and is outside the scope of this manual.

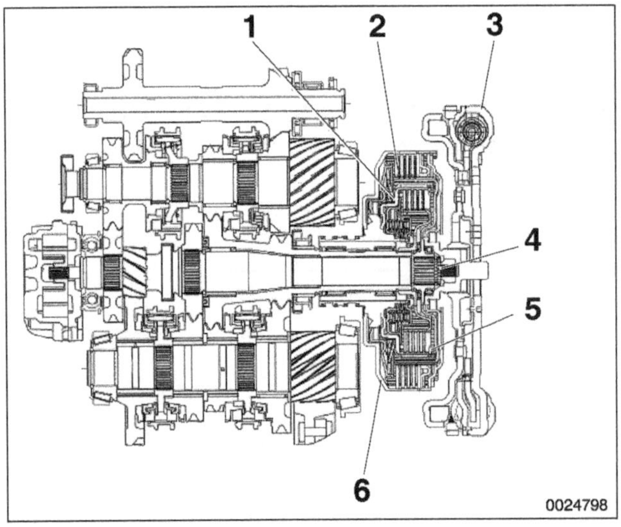

Cutaway overview of DSG transmission showing clutch arrangement:

1. K2 disc carrier
2. K1 disc carrier
3. Dual-mass flywheel
4. Input hub
5. Clutch driving disc
6. Main hub

BentleyPublishers.com
©2006 BentleyPublishers.com—All Rights Reserved

32 Torque Converter

General

Service to the automatic transmission torque converter requires removal of the transmission from the vehicle and separation from the engine. Special tools and equipment are required to remove the automatic transmission and service the torque converter. Removal of the automatic transmission is outside the scope of this manual.

Diagnostic trouble codes (DTC) for some torque converter malfunctions can be stored in the Transmission Control Module (TCM) and the Engine Control Module (ECM),and can be accessed with scan tools VAS 5051/5052 and derivatives or equivalent scan tools or scan tool programs. See **OBD On Board Diagnostics**.

Only one fully automatic transmission type is available in the TT.

Transmission	Transmission Type
6-speed automatic	09G

WARNING —
- *Disconnecting the negative (–) battery cable has ramifications with many vehicle systems. Before proceeding, be familiar with the warnings associated with this procedure in* **03 Maintenance** *and* **69 Seatbelts, Airbags**. *These systems may need to be re-set using a VAG 1551/1552, VAS 5051/5052 or equivalent scan tool or computer scan tool program.*

Torque Converter

The torque converter is a type of fluid coupling connecting the engine and the transmission. It is driven by a steel drive plate attached to the engine crankshaft. In addition to connecting the engine and transmission, the torque converter drives the ATF pump within the transmission. The pump delivers automatic transmission fluid (ATF) under pressure. Pressurized ATF is the hydraulic operating fluid.

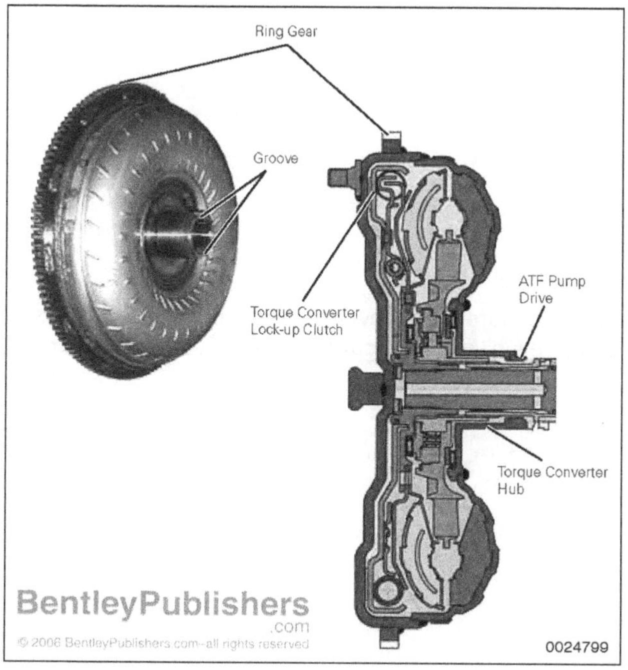

09G torque converter showing internal construction and externally mounted components.

BentleyPublishers.com
©2006 BentleyPublishers.com—All Rights Reserved

Torque converter identification

Different versions of the torque converter are installed depending on engine type. These can be identified by a 3-digit code. For correct applications, check with an authorized Audi dealer parts department or aftermarket parts specialist.

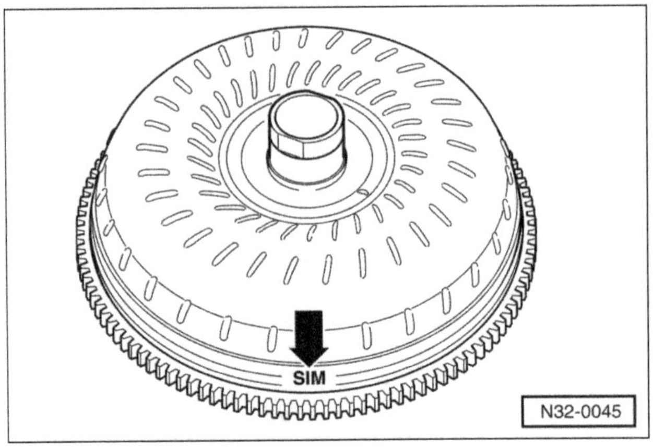

Torque converter identification (**SIM** in example) is stamped into circumference as shown (**arrow**).

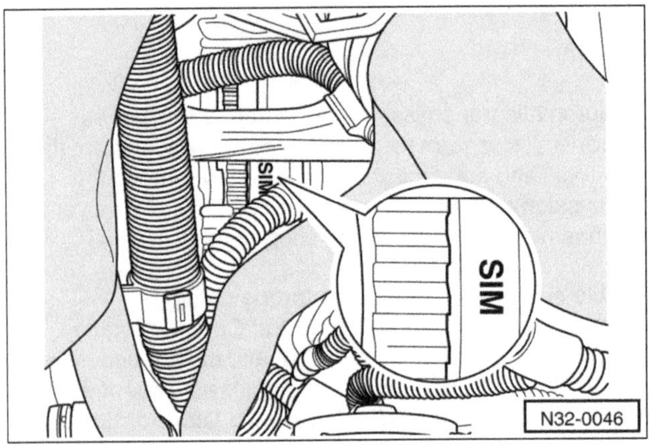

Torque converter identification can be made on 09A transmissions without removal. Turn engine over by hand until the 3-digit code (**SIM** in example) shows through the slot in the transmission housing as shown.

Torque converter, oil seal replacement

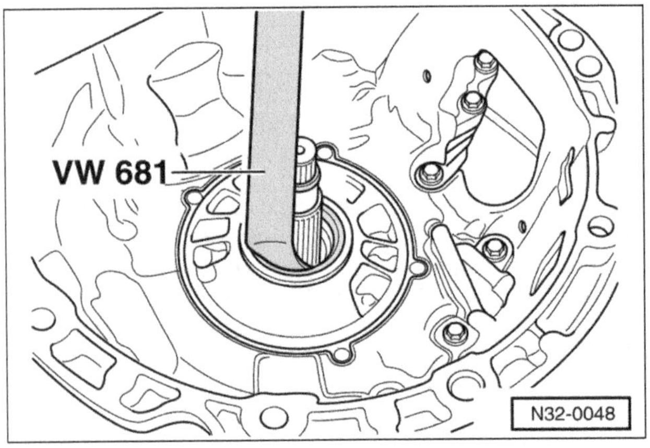

Carefully pry out oil seal with **VW 681** or equivalent. Use care to avoid gouging the housing with the tip of the removal tool. Seal is usually destroyed during removal.

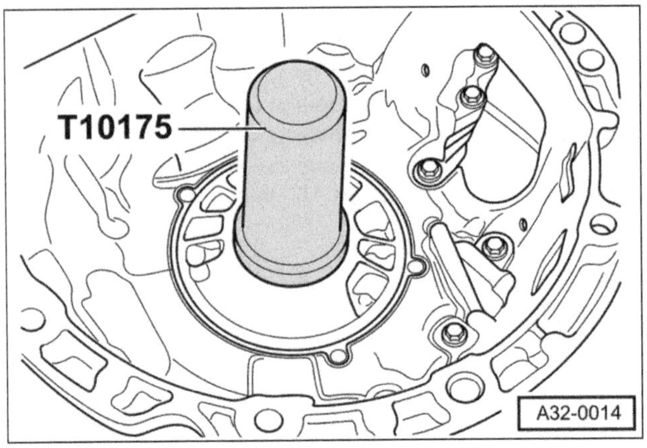

Install new oil seal squarely into cover with seal installer **T10175** or equivalent. Drive new oil seal into cover until it is flush with cover edge.

BentleyPublishers.com
© 2006 BentleyPublishers.com—All Rights Reserved

Torque converter, checking

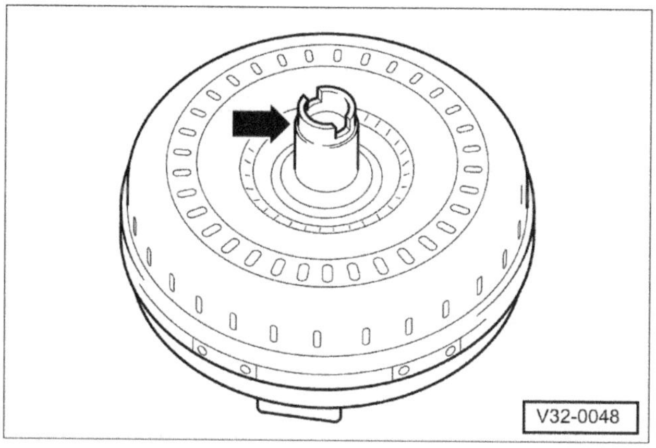

⊲ Inspect torque converter hub for wear grooves and scoring in area of oil seal (**arrow**).

Torque converter, installing

– Lightly lubricate seals with ATF. Other types of lubricant can cause problems with the hydraulic operation of the transmission.

⊲ When installing, carefully push torque converter hub through oil seal as far as possible. This will usually result in the converter being only partially inserted into position. Lightly press converter into transmission while turning it to engage drive lugs on hub (**arrows**) into oil pump in transmission. Torque converter must be felt to slide into position.

- The torque converter is correctly installed when it can be easily turned by hand and is squarely and equally spaced from the transmission all around the circumference.

NOTE —

- *If torque converter is not properly installed, the drive lugs on the torque converter and the ATF pump will be severely damaged when the transmission is attached to the engine.*

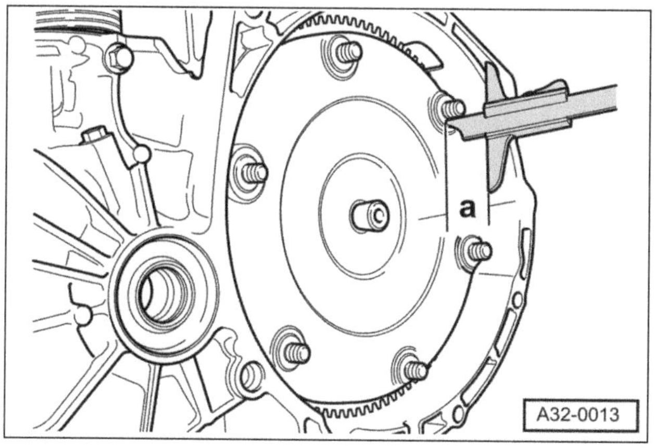

⊲ Measure the distance from the edge of the transmission housing down to the contact surface of the torque converter attachment stud.

- Distance **a** = 22 mm (0.866 in).

NOTE —

- *Once in position, torque converter can slide out of transmission. Ensure that torque converter remains fully seated during movement and installation of transmission.*
- *If the torque converter is not installed correctly, the torque converter drive plate or the ATF pump will be seriously damaged when the transmission is joined to the engine.*
- *Before and during tightening of bolts on engine/transmission drive plate, continually check that the torque converter behind the drive plate can be turned. If the converter cannot be turned, it must be assumed that it has not been installed correctly and the drive plate of the converter or the ATF pump will be damaged during final tightening of bolted connections.*

BentleyPublishers
.com
© 2006 BentleyPublishers.com—All Rights Reserved

Torque converter, draining

– If the torque converter has become contaminated by friction material wear particles or if required during major overhaul procedure, the torque converter must be completely drained. This is best accomplished by use of the VAG 1358, VAG 1782, or equivalent.

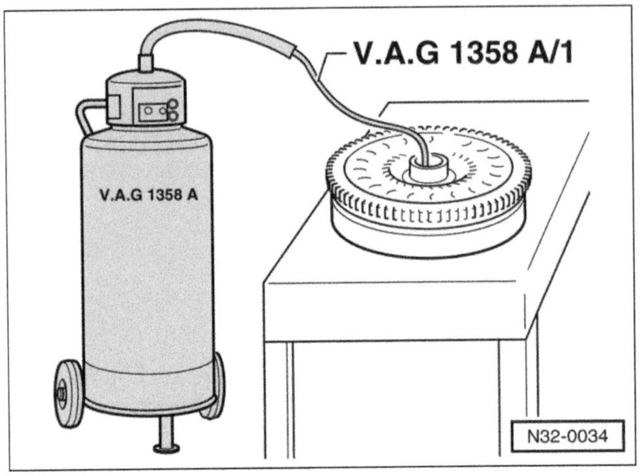

◀ VAG 1358 shown with adapter VAG 1358 A/1. Always follow manufacturers instructions for operation.

NOTE—
- *Be sure to allow sufficient time for fluid to drain from all internal areas of torque converter.*

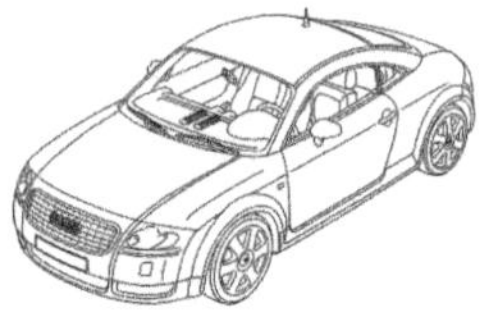

BentleyPublishers.com
©2006 BentleyPublishers.com—All Rights Reserved

34a Manual Transmission

TABLES

GENERAL

This repair group covers repair information for cable-operated gear shift mechanisms and removal and installation of the all wheel drive manual transmission used in combination with the 4-cylinder engine. This repair group does not cover transaxle or transmission tear down, disassembly, or internal repairs.

- For servicing information on hydraulic and mechanical clutch components, see **30 Clutch**.
- For servicing information on drive axles, including drive flange oil seals, see **40 Front Suspension and Drive Axles** and **42 Rear Suspension**.
- For transmission identification information and transmission oil level servicing, see **03 Maintenance**.

> ***WARNING —***
> - *Disconnecting negative (–) battery cable has ramifications with many vehicle systems. Before proceeding, be familiar with warnings associated with this procedure in* ***03 Maintenance*** *and* ***69 Seatbelts, Airbags****. These systems may need to be re-set using a VAG 1551/1552, VAS 5051/5052 or equivalent scan tool or computer scan tool program.*

Transmission	Transmission type
5-speed manual (front wheel drive)	02J
5- and 6-speed manual (all wheel drive)	02M
5- and 6-speed manual (all wheel drive)	02Y

BentleyPublishers.com

© 2006 BentleyPublishers.com—All Rights Reserved

Table a. 02J manual transmission specifications (front wheel drive)

Type (Code)	02J (DZF)	02J (EGZ)	02J (EWV)
Engine application, horsepower	1.8L, 180 hp	1.8L, 180 hp	1.8L, 180 hp
Gear ratio: Z_2:Z_1 Final drive	63:16 = 3.938	63:16 = 3.938	63:16 = 3.938
1st gear	33:10 = 3.300	33:10 = 3.300	33:10 = 3.300
2nd gear	35:18 = 1.944	35:18 = 1.944	35:18 = 1.944
3rd gear	34:26 = 1.308	34:26 = 1.308	34:26 = 1.308
4th gear	30:29 = 1.034	30:29 = 1.034	30:29 = 1.034
5th gear	31:37 = 0.838	31:37 = 0.838	31:37 = 0.838
6th gear	n/a	n/a	n/a
Reverse gear	17:10 x 36:20 = 3.060	17:10 x 36:20 = 3.060	17:10 x 36:20 = 3.060
Speedometer drive	13:22 = 0.591	13:22 = 0.591	13:22 = 0.591
Lubricant Capacity	2.0 liters (2.1 qt.)		
Audi part number	G 052 726 A2		
Clutch control	hydraulic		
Clutch disc diameter[3]	219 mm		
Driveshaft flange diameter	108 mm		
Ratio, overall in top gear	3.299		

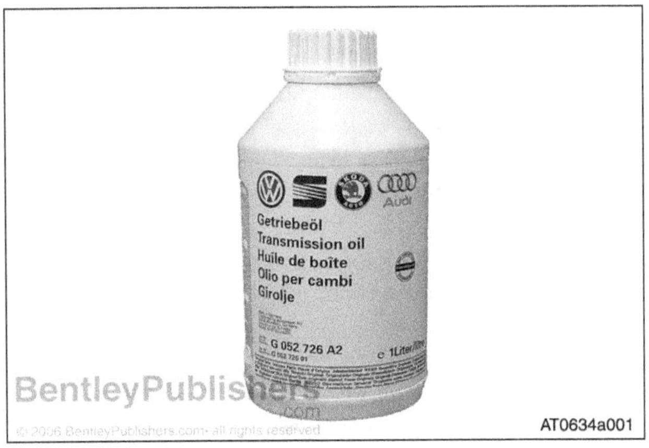

BentleyPublishers.com
© 2006 BentleyPublishers.com - all rights reserved
AT0634a001

Special manual transmission oil G 052 726 A2 as supplied by Audi for use in front wheel drive transmissions. Viscosity and type are not specified in factory service information.

BentleyPublishers.com
©2006 BentleyPublishers.com—All Rights Reserved

Table b. 02M manual transmission specifications (all wheel drive)

Type (Code)	02M, (DQB, FHB)	02M (DXW, FHC)	
Engine application, horsepower	1.8L, 225 hp	1.8L, 180 hp	
Gear ratio: Z_2:Z_1			
Final drive	63:15 = 4.200 (1-4)[1] 63:19 = 3.316 (5,6,R)	63:15 = 4.200 (1-4)[1] 63:19 = 3.316 (5,R)	
1st gear	41:12 = 3.417	42:11 = 3.818	
2nd gear	40:19 = 2.105	40:19 = 2.105	
3rd gear	40:28 = 1.429	39:29 = 1.345	
4th gear	37:34 = 1.088	35:36 = 0.972	
5th gear	34:31 = 1.097	32:33 = 0.970	
6th gear	31:34 = 0.912	n/a	
Reverse gear	30:12 x 23:14 = 4.108	31:11 x 23:14 = 4.630	
Speedometer drive[2]	n/a	n/a	
Lubricant			
Capacity	2.6 liters (2.75 qt.)		
Audi part number	G 052 171 A2		
Clutch control	hydraulic		
Clutch disc diameter	240 mm		
Driveshaft flange diameter	108 mm		
Rear differential type	02D		
Ratio, overall in top gear	3.023	3.215	

1. Ring gear is driven by 2 separate pinion gears on individual output shafts, dependent on gear selected.
   Output shaft 1 is used by gears 1, 2, 3, 4.
   Output shaft 2 is used by gears 5, 6, R.
2. Speedometer does not use separate sender, signal is derived from ABS/ESP wheel speed sensors via CAN-bus.

◄ Special manual transmission oil G 052 171 A2 as supplied by Audi for use in all wheel drive transmissions. Viscosity and type are not specified in factory service information.

BentleyPublishers.com
© 2006 BentleyPublishers.com—All Rights Reserved

Table c. 02Y manual transmission specifications (all wheel drive)

Type (Code)	02Y (FMN, FZL)	02Y (FMR)	02Y (FMT, FZP)
Engine application, horsepower	1.8L, 225 hp	1.8L, 180 hp	1.8L, 180 hp
Gear ratio: Z_2:Z_1 Final drive	72:17 = 4.235 (1-4)[1] 72:22 = 3.273 (5,6,R)	72:17 = 4.235 (1-4)[1] 72:22 = 3.273 (5,R)	72:17 = 4.235 (1-4)[1] 72:22 = 3.273 (5,6,R)
1st gear	47:14 = 3.357	49:13 = 3.769	47:14 = 3.357
2nd gear	48:23 = 2.087	48:23 = 2.087	48:23 = 2.087
3rd gear	47:32 = 1.469	45:34 = 1.324	47:32 = 1.469
4th gear	45:41 = 1.098	42:43 = 0.977	46:40 = 1.150
5th gear	41:37 = 1.108	39:40 = 0.975	43:36 = 1.194
6th gear	38:41 = 0.927	n/a	39:40 = 0.975
Reverse gear	23:14 x 34:14 = 3.990	23:14 x 36:13 = 4.549	23:14 x 34:14 = 3.990
Speedometer drive[2]	n/a	n/a	n/a
Lubricant Capacity	2.6 liters (2.75 qt.)		
Audi part number	G 052 171 A2		
Clutch control	hydraulic		
Clutch disc diameter	240 mm		
Driveshaft flange diameter	108 mm		
Rear differential type	02D		
Ratio, overall in top gear	3.033	3.191	3.191

1. Ring gear is driven by 2 separate pinion gears on individual output shafts, dependent on gear selected.
   Output shaft 1 is used by gears 1, 2, 3, 4.
   Output shaft 2 is used by gears 5, 6, R.
2. Speedometer does not use separate sender, signal is derived from ABS/ESP wheel speed sensors via CAN-bus.

Special manual transmission oil G 052 171 A2 as supplied by Audi for use in all wheel drive transmissions. Viscosity and type are not specified in factory service information.

BentleyPublishers.com

©2006 BentleyPublishers.com—All Rights Reserved

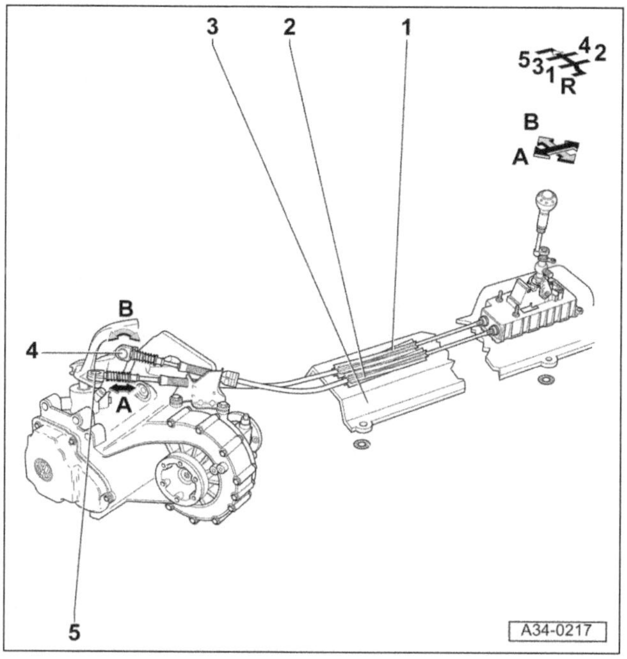

Gear Selector Mechanism

Selector mechanism, operational overview

5-speed manual transmission (02J)

1. Gear change cable
   - **A arrows** (⇔) indicate fore and aft movement of gear shift lever and selector linkage mechanism.
2. Gear selector cable
   - **B arrows** (⇔) indicate side to side movement of gear shift lever and up and down movement of selector mechanism.
3. Heat shield
   - Remove heat shield before removing selector mechanism
4. Relay lever
5. Transmission shift lever
   - With vibration damper weight

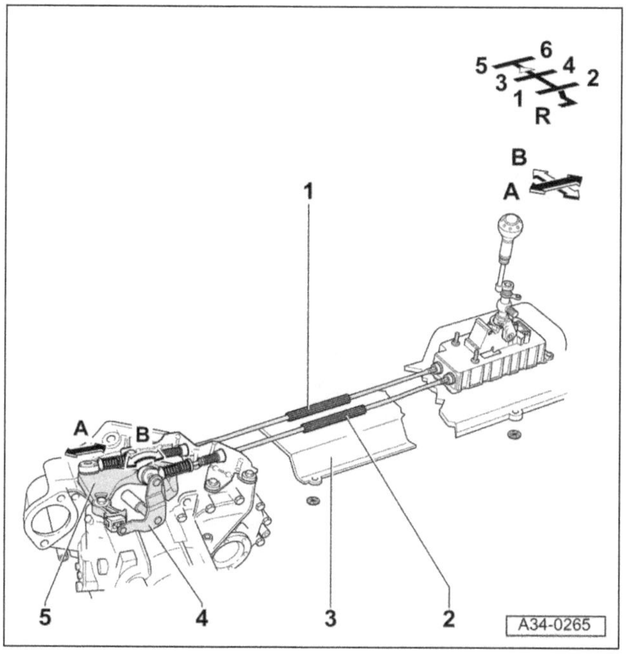

6-speed manual transmission (02M and 02Y)

1. Gear change cable
   - **A arrows** (⇔) indicate fore and aft movement of gear shift lever and selector linkage mechanism.
2. Gear selector cable
   - **B arrows** (⇔) indicate side to side movement of gear shift lever and up and down movement of selector mechanism.
3. Heat shield
   - Remove heat shield before removing selector mechanism
4. Relay lever
5. Transmission shift lever
   - With vibration damper weight

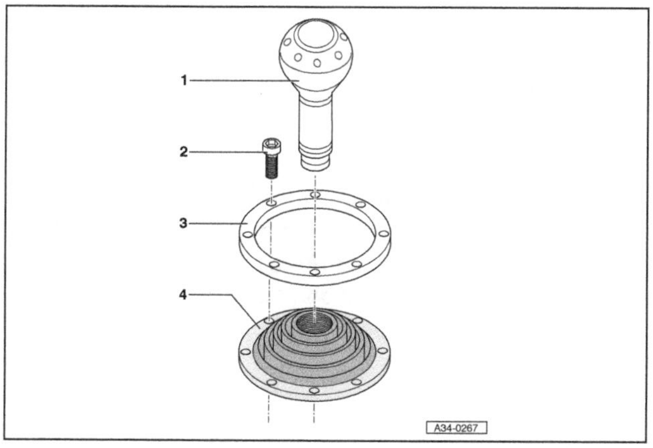

Shift knob/boot assembly

- Gear shift knob/boot assembly is the same for front wheel and all wheel drive transmissions.

1. Gear shift knob with shift pattern
   - Unscrew to remove
2. Bolt
   - Threaded through retaining ring and boot into mount in floor
   - Qty: 8, M4x10
3. Retaining ring
   - Lift off to remove
4. Cover (boot)
   - Lift off to remove

BentleyPublishers.com
© 2006 BentleyPublishers.com—All Rights Reserved

Shift lever and selector housing assembly

1. **Screw, Torx head**
   - Tighten to 6 Nm (53 in-lb)
2. **Cover**
3. **Shift lever**
4. **Insulation damper**
5. **Bearing shell**
   - Snaps into insulation damper
   - Open side faces shift housing
6. **Gasket**
   - Self adhesive between selector housing and floor
   - Adhere to selector housing
7. **Bracket**
   - Secured to body on some vehicles
8. **Bolt**
   - Tighten to 25 Nm (18 ft-lb)
9. **Nut**
   - Tighten to 25 Nm (18 ft-lb)
10. **Insulation**
11. **Shifter housing**
12. **Bushing**
13. **Pivot pin**
14. **Guide bushing**
15. **Compression spring**
    - See **Compression spring, installing**
16. **Selector bracket**
17. **Screw, Torx head**
    - Tighten to 6 Nm (53 in-lb)
18. **Base plate**
    - Bend tabs open to remove
    - Always replace
19. **Lock washer**
20. **Selector cable**
21. **Shift cable**
22. **Nut**
    - Tighten to 25 Nm (18 ft-lb)
23. **Bearing bushing**
    - Installs in one position
24. **Lock washer**
    - See **Gate selector bracket lock washer, removing**

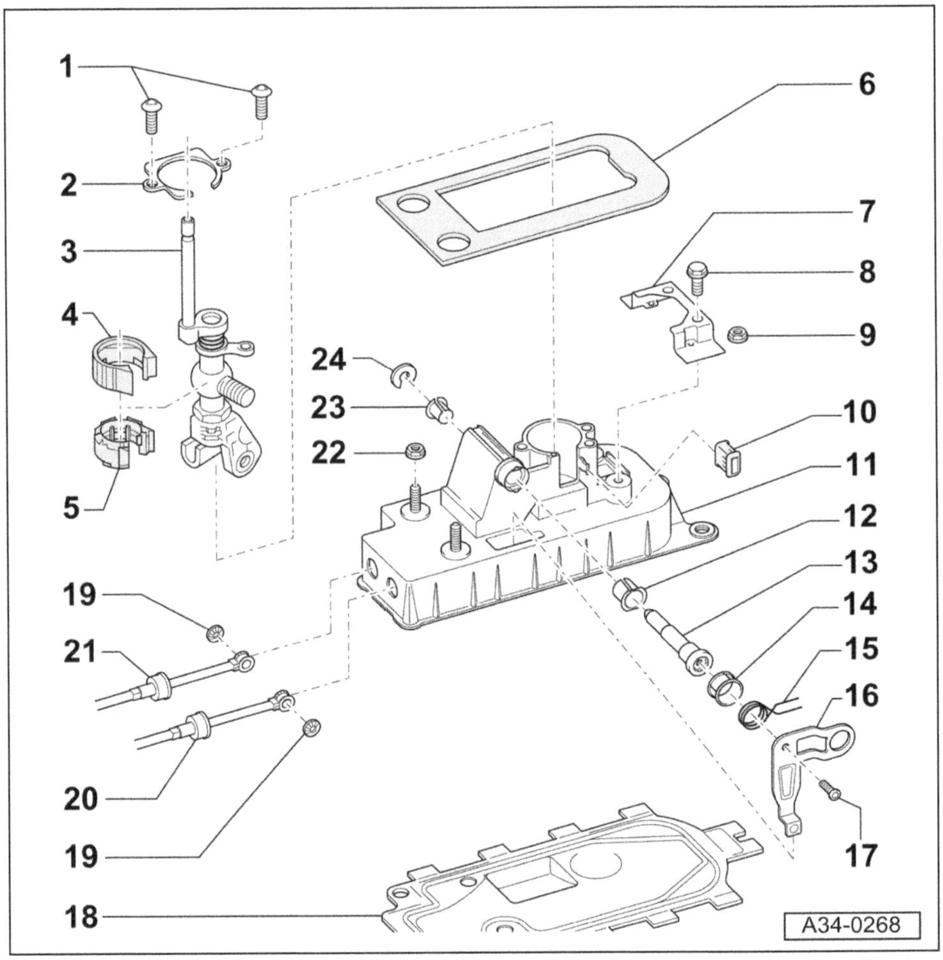

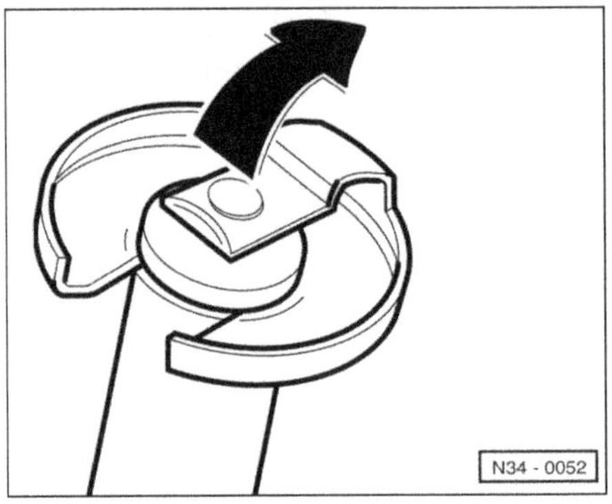

Gate selector bracket lock washer, removing

- Lift center (**arrow**) and slide off of pivot pin.

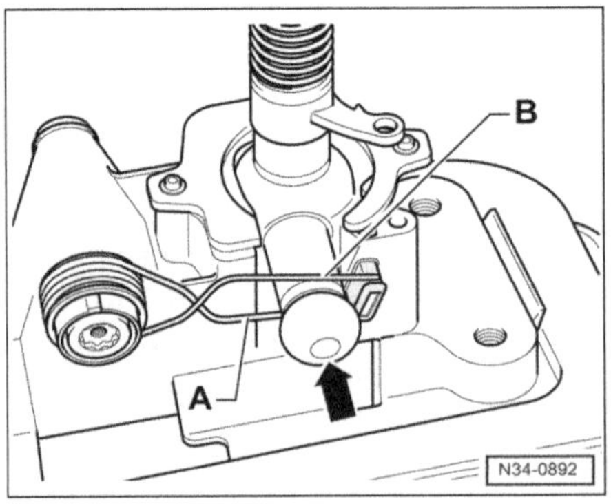

Compression spring, installing

- Insert compression spring so that arm (**A**) contacts lower part of pin (**arrow**).
- Pull up arm (**B**) far enough so that it reaches upper part of pin (**arrow**)
- Ensure arms of spring rest on insulation (item 10)

BentleyPublishers.com
© 2006 BentleyPublishers.com—All Rights Reserved

Selector cable assembly, (front wheel drive)

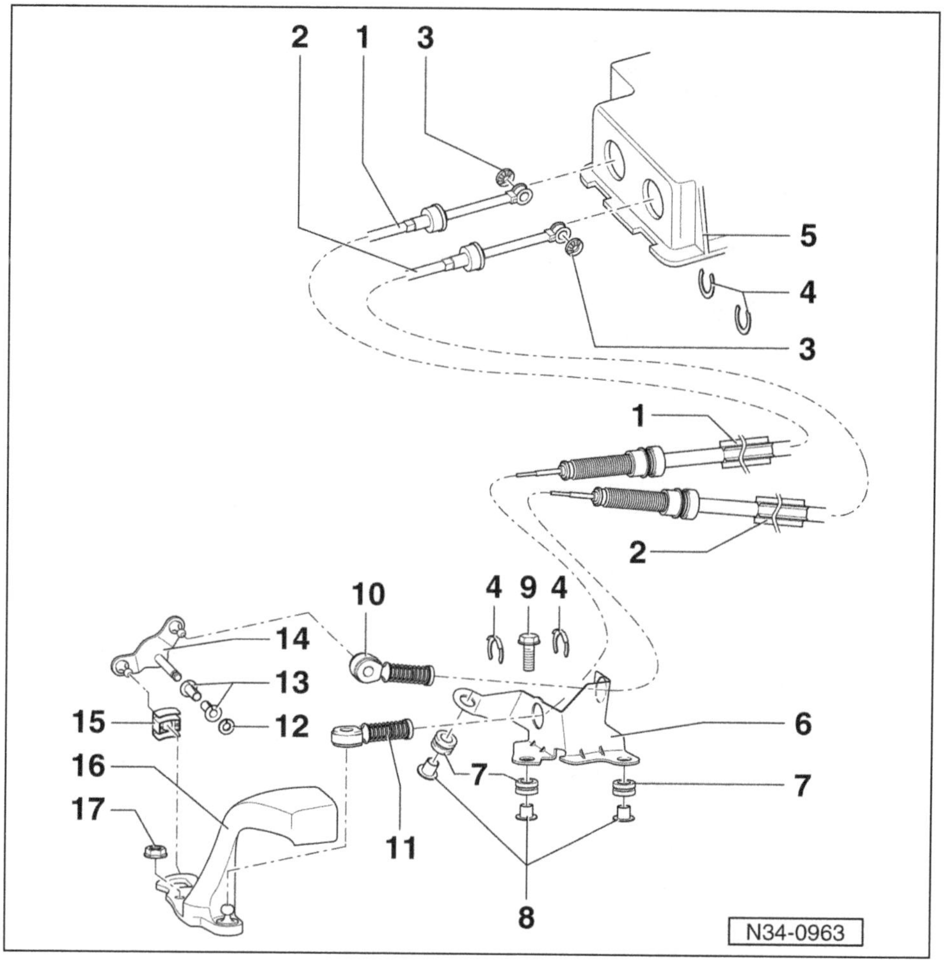

1. **Gear selector cable**
   - Press onto gear shift guide
   - See **Installation position of gear selector mechanism**
2. **Gate selector cable**
   - On gate selector bracket
   - See **Installation position of gear selector mechanism**
3. **Circlip**
4. **Circlip**
   - When removing use caution to avoid damage to cables
5. **Selector housing**
6. **Support bracket**
7. **Grommet**
   - Support bracket mounting to transmission
8. **Spacer**
9. **Bolt**
   - Tighten to 25 Nm (18 ft-lb)
   - For support bracket
10. **Cable end attachment/locking piece**
    - For gate selector cable to relay lever, 2 versions
    - Through 06/01, snapped in place Always replace after removing from relay lever (item **14**)
    - From production 07/01 with through hole, secured with clip. Replacement not required
    - Both versions, see **Installation position of gear selector mechanism**
11. **Cable end attachment/locking piece**
    - For gear selector cable to gear selector lever, 2 versions
    - Through production 06/01, snapped in place Always replace after removing from relay lever (item **16**)
    - From production 07/01 with through hole, secured with clip Replacement not required
    - Both versions, see **Installation position of gear selector mechanism**
12. **Lock washer**
    - See **Gate selector bracket lock washer, removing**
13. **Bushing**
14. **Relay lever**
    - See **Installation position of gear selector mechanism**
15. **Sliding shoe**
16. **Transmission selector lever**
    - With balance weight
    - Install so master spline aligns with matching position on selector shaft. See **Installation position of gear selector mechanism**
    - Adjust selector mechanism after installing, see **Adjusting gear selector mechanism**
    - See **Transmission selector lever, installing**
17. **Self-locking nut**
    - Always replace
    - Tighten to 20 Nm (15 ft lb)

NOTE—

- *Lubricate all bearing areas, mounting and contact surfaces with polycarbamide grease (Audi part no. G 052 142 A2).*

BentleyPublishers.com
© 2006 BentleyPublishers.com—All Rights Reserved

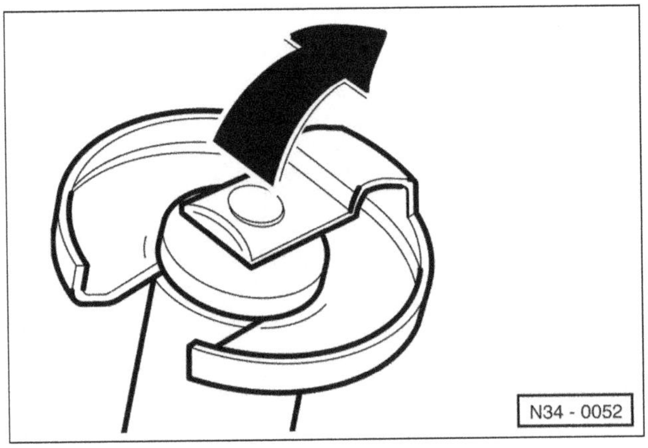

Gate selector bracket lock washer, removing

- Lift center (**arrow**) and slide off of pivot pin.

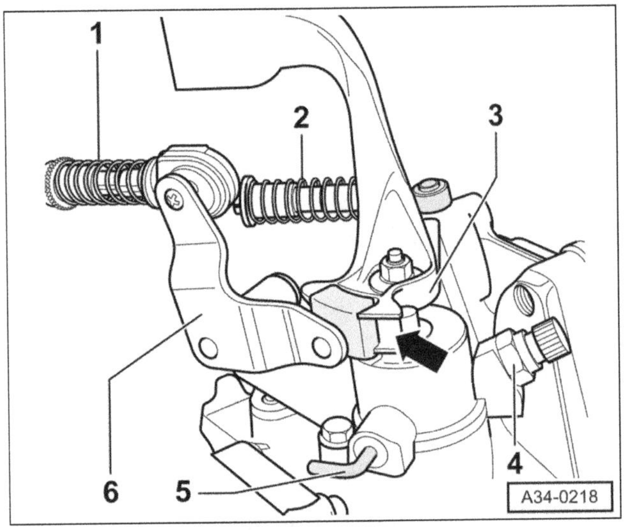

Installation position of gear selector mechanism

1. Cable end attachment/locking piece for gate selector cable
2. Cable end attachment/locking piece for gear selector cable
3. Transmission selector lever with vibration dampener weight
4. Back-up light switch (F4)
5. Adjuster pin
    - Part of mechanism
    - Turn to up position when adjusting shift mechanism
6. Relay lever
    - Locates in guide rail of gear selector lever (**arrow**) with a sliding block
    - Lubricate with MoS_2 (moly) grease

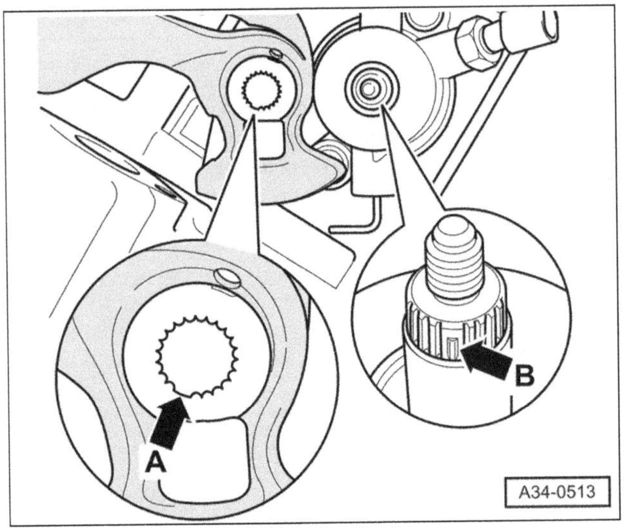

Installing transmission selector lever

- Install selector lever onto selector shaft so that gap of master spline (**arrow A**) on aligns with missing splines on (**arrow B**).
- Selector lever should slide onto selector shaft without being forced.

BentleyPublishers.com
©2006 BentleyPublishers.com—All Rights Reserved

Selector mechanism, adjusting (front wheel drive)

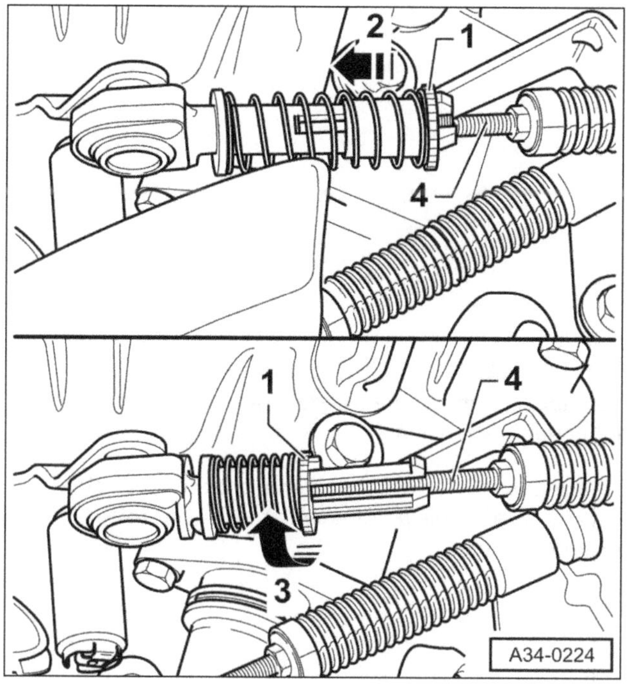

Before adjusting the gear selector mechanism, note the following:

- All moving parts of the selector mechanism and transmission shift mechanism must be in proper working order and lubricated.
- Selector mechanism must move freely.
- Transmission, clutch and clutch mechanism must also be in proper working order.

NOTE—

- *Alignment pin tool (T10027) is required to position shift linkage in passenger compartment. Each leg of this "L" shaped tool is approx. 65 mm (2.5 in.) long with a diameter of approx. 5.0 mm (0.197 in.).*

– Place transmission in neutral. Gear selector mechanism is adjusted with transmission and shift linkage both locked in neutral.

– Remove air filter assembly to allow access to top of transmission. See **03 Maintenance**.

< Secure locking mechanism on gear selector cable and gate selector cable ends by pushing collar (**1**) forward as far as it will go (**arrow 2**)

– Turn collar (**1**) as far as it will go to the left to lock (**arrow 3**).

– Cable (**4**) must be free to move within cable end.

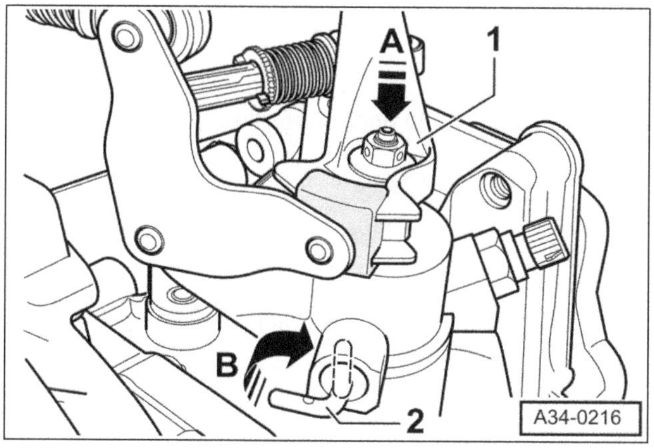

< Lock selector shaft (**1**) by pressing shaft down (**arrow A**) while at the same time turning shaft lock (**2**) approximately 90° (**dashed arrow B**) so that it is in the up position.

– Shaft lock (**2**) is part of, and built into selector housing. Selector shaft is locked in neutral when shaft lock (**2**) arm is pointed up and pushed in.

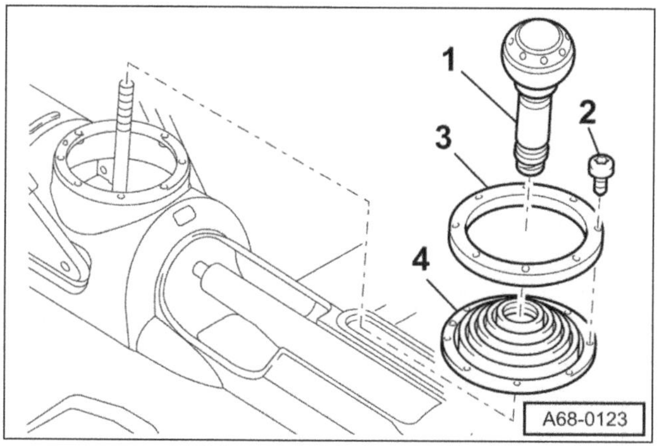

< Unscrew gearshift knob (**1**).

– Remove bolts (**2**).

– Lift off retaining ring (**3**) and boot (**4**).

BentleyPublishers.com
© 2006 BentleyPublishers.com—All Rights Reserved

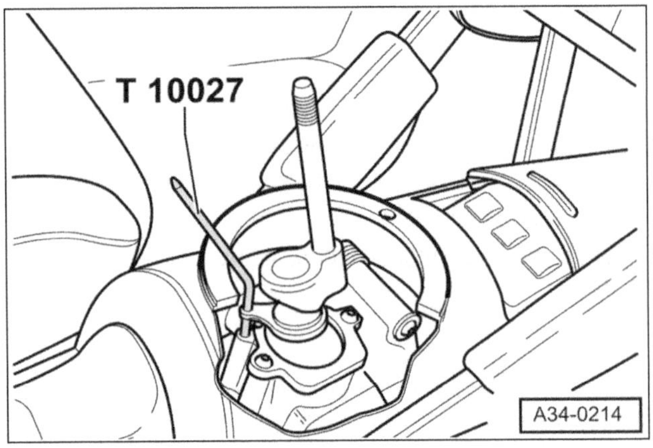

◀ Lock gear lever in passenger compartment by selecting neutral position and sliding **T10027** locking pin, or equivalent, through holes.

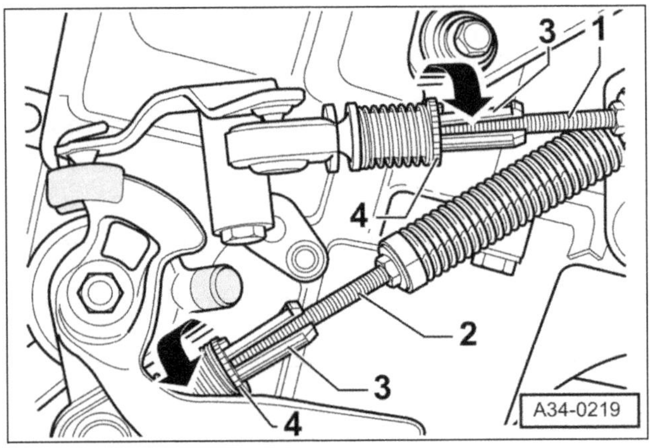

◀ Ensure that cables (**1** and **2**) are free to move within cable ends (**3**) and are not under any tension.

– While mechanism is still locked at shifter and on transmission, turn collars (**4**) as far as they will go (dashed arrows) and release. This locks cable end to cable.

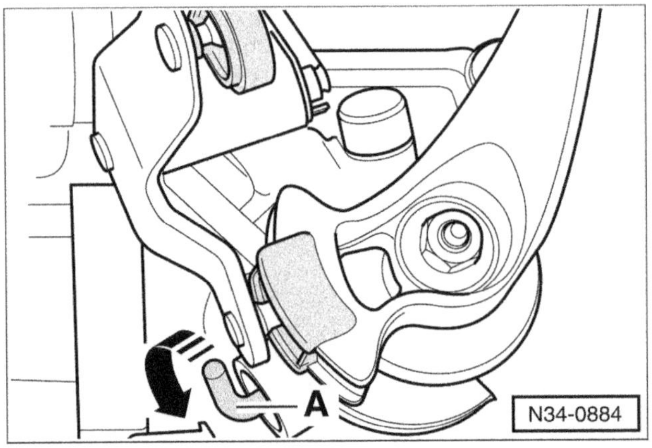

◀ Turn shaft lock arm (**A**) back to original position (**dashed arrow**).

– Shaft lock arm (**A**) must be horizontal.

– Remove retaining pin **T10027**, or equivalent, from gear lever in passenger compartment.

– Install boot, retaining ring and gearshift knob.

– Install air filter assembly.

– Functional check selector mechanism.

BentleyPublishers.com
©2006 BentleyPublishers.com—All Rights Reserved

Selector mechanism, functional check (front wheel drive)

The correct at-rest position for the gear shift lever is in neutral, between 3rd and 4th gears.

– Depress clutch pedal.

– Shift through the shift pattern several times. Pay particular attention to the operation of the reverse gear lock-out, 5th gear and 6th gear operation as applicable.

  • It should only be possible to engage reverse gear after pressing gear shift lever down to overcome reverse gear lock out.

  • If a gear fails to engage smoothly after repeated selection, the selector shaft play (lift, at the transmission), should be checked as described below.

– Shift into first gear.

– Press gear shift lever to the left stop and release. At the same time, a second technician must observe the selector shaft on the transmission.

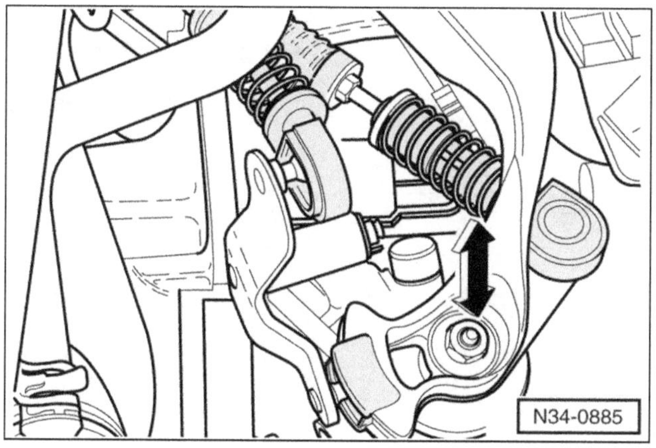

◀ When moving gear lever, selector shaft must move approximately 1 mm in direction of **arrow** (up/down).

– If movement is not present, re-adjust selector mechanism and re-check shift operation as needed.

– Check backup light operation.

– Install air filter assembly.

BentleyPublishers.com
©2006 BentleyPublishers.com—All Rights Reserved

Selector cable assembly (all wheel drive)

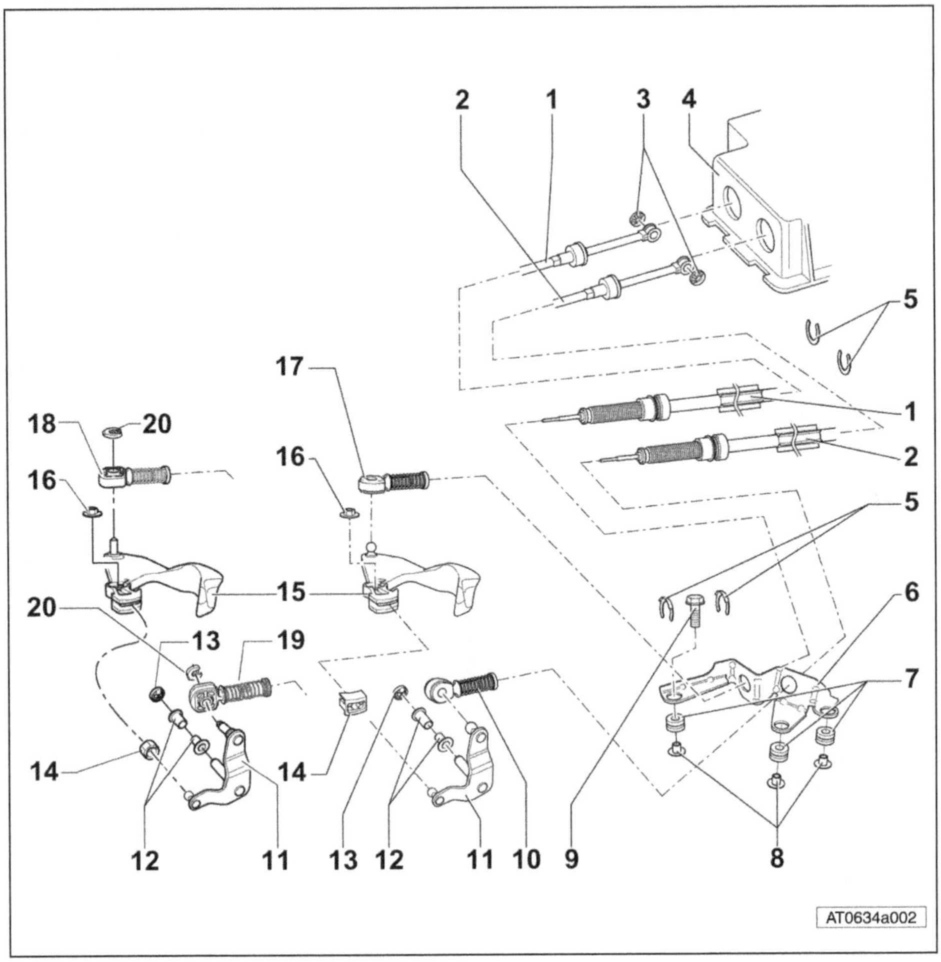

1. **Gear selector cable**
   - See **Installation position of gear selector mechanism**
2. **Gate selector cable**
   - See **Installation position of gear selector mechanism**
3. **Circlip**
4. **Selector housing**
5. **Circlip**
   - When removing use caution to avoid damage to cables
6. **Support bracket**
   - Early version shown, later versions similar
7. **Grommet**
   - Support bracket mounting to transmission
8. **Spacer**
9. **Bolt**
   - Tighten to 25 Nm (18 ft-lb)
   - For support bracket
10. **Cable end attachment/locking piece (through 06/01 production)**
    - Gate selector cable to relay lever
    - Snapped in place, removing destroys. Always replace after removing from relay lever (item **11**)
    - See **Removing cable end attachment/locking piece (through 06/01 production)**
    - See **Installation position of gear selector mechanism (through 06/01 production)**
11. **Relay lever**
12. **Bushings**
13. **Circlip**
14. **Sliding shoe**
15. **Transmission selector lever**
    - With balance weight
    - Install so master spline aligns with matching position on selector shaft. See **Installing transmission selector lever** given earlier
    - Adjust selector mechanism after installing, see **Selector mechanism, adjusting (all wheel drive)**
    - See **Transmission selector lever, installing**
16. **Nut**
    - Through 06/01 production
      Tighten to 25 Nm (18 ft-lb)
    - From 07/01 production
      Always replace
      Tighten to 23 Nm (17 ft-lb)
17. **Cable end attachment/locking piece (through 06/01 production)**
    - Gear selector cable to gear selector lever
    - Snapped in place, removing destroys. Always replace after removing from selector lever (item **15**)
    - See **Removing cable end attachment/locking piece (through 06/01 production)**
    - See **Installation position of gear selector mechanism (through 06/01 production)**
18. **Cable end attachment/locking piece (from 07/01 production)**
    - Gear selector cable to gear selector lever
    - With through hole, secured with clip, replacement not required
    - Mounting hole diameter: 10 mm
    - See **Installation position of gear selector mechanism (from 07/01 production)**
19. **Cable end attachment/locking piece (from 07/01 production)**
    - Gate selector cable to relay lever
    - With through hole, secured with clip, replacement not required
    - Mounting hole diameter: 8 mm
    - See **Installation position of gear selector mechanism (from 07/01 production)**
20. **Lock washers**
    - Always replace

NOTE—

- *Lubricate all bearing areas, mounting and contact surfaces with polycarbamide grease (Audi part no. G 052 142 A2).*

BentleyPublishers.com

© 2006 BentleyPublishers.com—All Rights Reserved

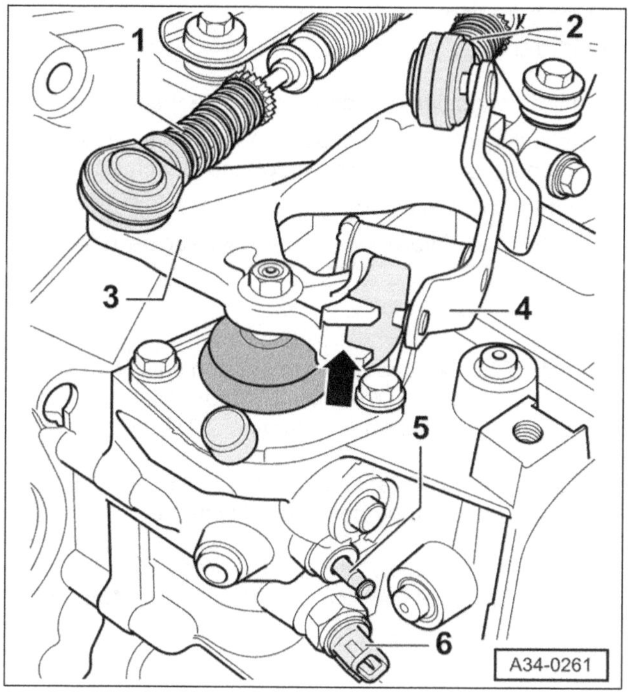

Installation position of gear selector mechanism (through 06/01 production)

1. Cable end attachment/locking piece for gear selector cable
2. Cable end attachment/locking piece for gate selector cable
3. Transmission selector lever with vibration dampener weight
4. Relay lever
   - Locates in guide rail of gear selector lever (**arrow**) with a sliding block
5. Adjuster pin
   - Part of mechanism
   - Used when adjusting shift mechanism
6. Back-up light switch (F4)

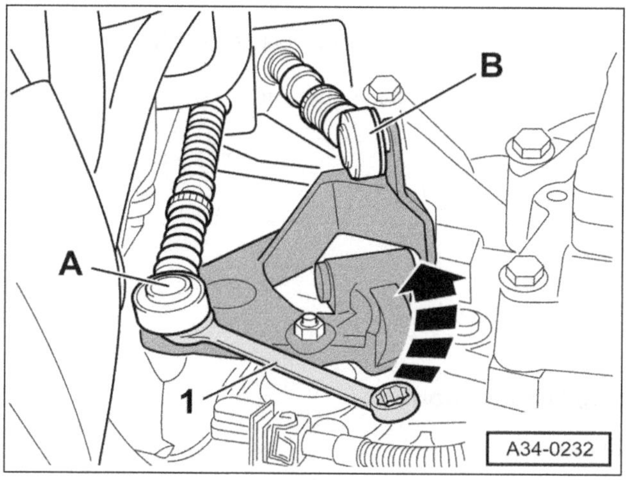

Removing cable end attachment/locking piece (through 06/01 production)

NOTE —

- *Prying cable end attachments off of relay lever and/or selector lever destroys locking mechanism. Replacement will be required.*

– Pry cable end attachment/locking piece for gear selector cable (**A**) off of selector lever with a 13 mm open ended wrench (**1**).

– Repeat procedure for cable end attachment/locking piece for gate selector cable (**B**) on relay lever.

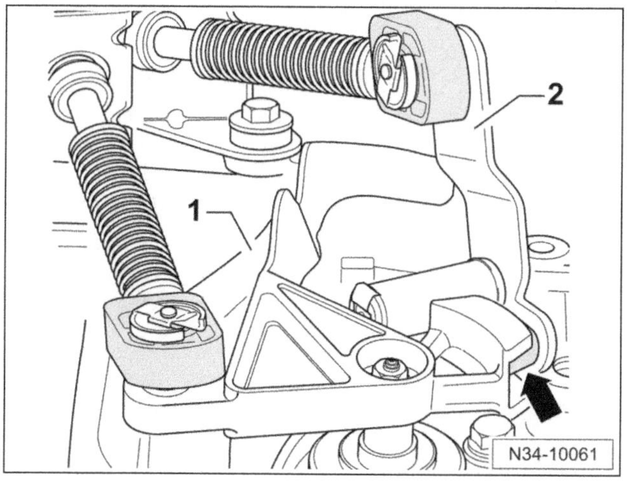

Installation position of gear selector mechanism (from 07/01 production)

1. Cable end attachment/locking piece for gear selector cable on gear selector lever.
2. Cable end attachment/locking piece for gate selector cable on relay lever.
   - Note position of sliding shoe (**arrow**).

BentleyPublishers.com
© 2006 BentleyPublishers.com—All Rights Reserved

Selector mechanism, adjusting (all wheel drive)

Before adjusting the gear selector mechanism, note the following:

- All moving parts of the selector mechanism and transmission shift mechanism must be in proper working order and lubricated.
- Selector mechanism must move freely.
- Transmission, clutch and clutch mechanism must also be in proper working order.

NOTE—

- *Alignment pin tool (T10027) is required to position shift linkage in passenger compartment. Each leg of this "L" shaped tool is approx. 65 mm (2.5 in.) long with a diameter of approx. 5.0 mm (0.197 in.).*

– Place transmission in neutral. Gear selector mechanism is adjusted with transmission and shift linkage both locked in neutral.

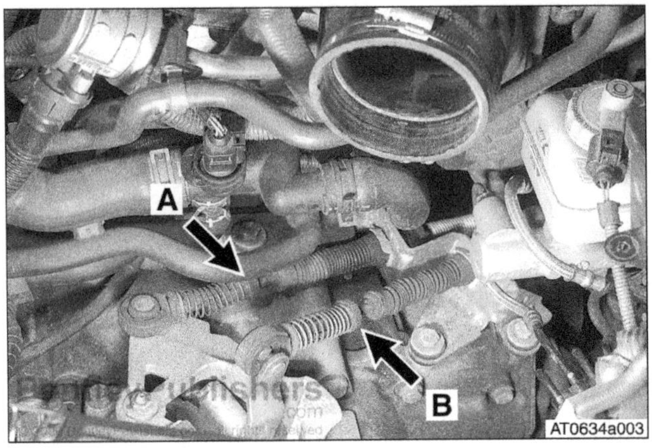

– Remove air filter assembly to allow access to gear (**A**) and gate (**B**) selector cable collars on top of transmission. See **03 Maintenance**.

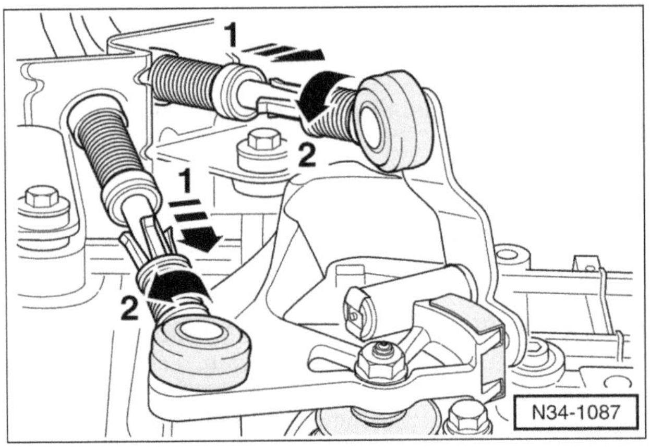

– Secure locking mechanism on gear selector cable and gate selector cable ends by pushing collar forward (**arrow 1**) as far as it will go.

– Turn collar as far as it will go to the left to lock (**arrow 2**).

– Cables must be free to move within cable end.

Through 03/03 production

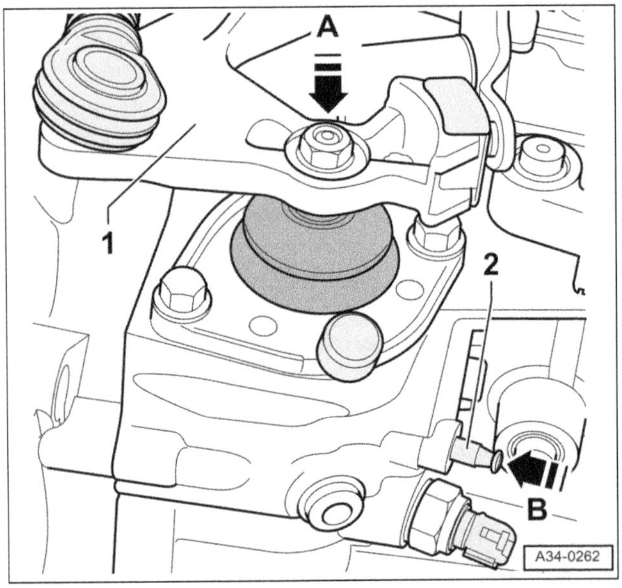

– Secure shift lever

- Press shift rod lever (**1**) down (**arrow A**) into 1st/2nd gear gate.
- At the same time, press lock pin (**2**) in (**arrow B**) until it engages and locks shift linkage.

BentleyPublishers.com
© 2006 BentleyPublishers.com—All Rights Reserved

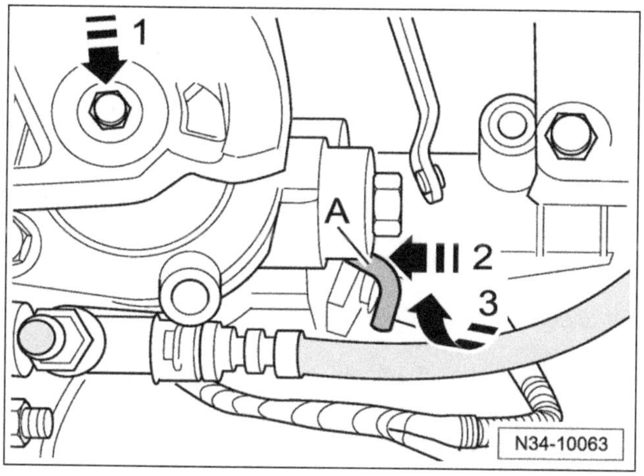

From 04/03 production

Secure shift lever

- Press shift rod lever down (**arrow 1**) into 1st/2nd gear gate.
- At the same time, rotate locking pin (**A**) upward (**arrow 3**) and press lock bolt (**arrow 2**) in until it engages and locks shift linkage.

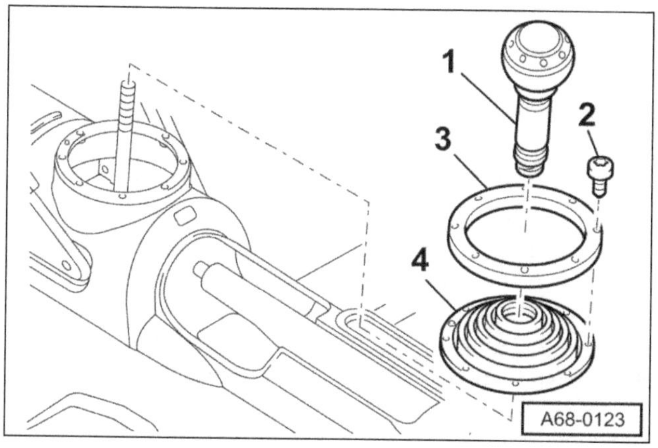

Continued for all versions

For all versions, unscrew gearshift knob (**1**).

– Remove bolts (**2**).

– Lift off retaining ring (**3**) and boot (**4**).

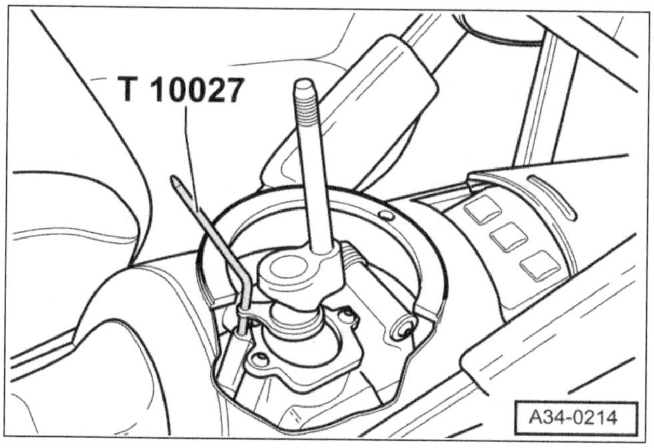

Lock gear lever in passenger compartment by selecting neutral position and sliding **T10027** locking pin or equivalent through holes.

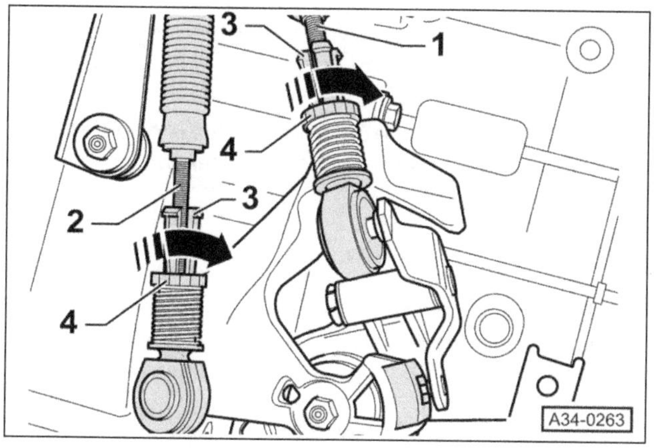

Ensure that cables (**1** and **2**) are free to move within cable ends (**3**) and are not under any tension.

– While mechanism is still locked at shifter and on transmission, turn collars (**4**) as far as they will go (**dashed arrows**) and release. This locks cable end to cable.

BentleyPublishers.com
©2006 BentleyPublishers.com—All Rights Reserved

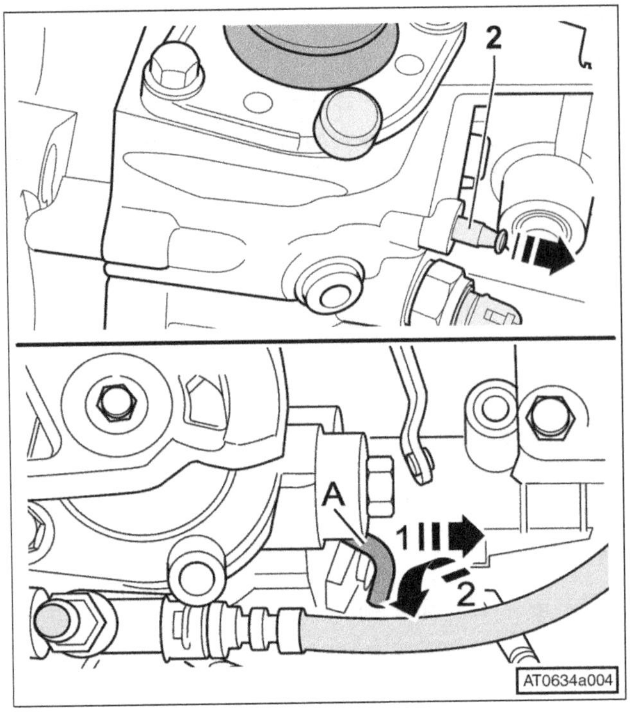

Unlock shift lever at transmission:

- Through 03/03 production (upper) - pull lock pin (**2**) out (**dashed arrow**) and back to original position.
- From 04/03 production (lower) - rotate locking pin (**A**) back to horizontal position (**arrow 2**) while pulling pin out (**arrow B**).

– Remove retaining pin T10027, or equivalent, from gear lever in interior.

– Install boot, retaining ring and gearshift knob.

– Install air filter assembly.

– Functional check selector mechanism.

Selector mechanism, functional check (all wheel drive)

– Depress clutch pedal.

– Shift through the pattern several times. Pay close attention to operation of the reverse gear lock-out and 5th/6th gear operation (as applicable).

- The correct at-rest position for the gear shift lever is in neutral, between 3rd and 4th gears.
- It should only be possible to engage reverse gear after pressing gear shift lever down to overcome reverse gear lock out. Gear shift lever should return to neutral between 3rd and 4th gears.
- If a gear fails to engage smoothly after repeated selection, the selector mechanism should be re-adjusted and re-checked as needed.

– Check backup light operation.

Transmission, Removing and Installing (All Wheel Drive)

This procedure describes removal and installation of the all wheel drive manual transmission installed in combination with the 1.8L engine. It is typical of procedures used for all engine/transmission combinations.

In addition to several special tools, the factory specifies special engine lifting and jacking equipment to support and reposition the engine as the transmission is removed from below.

Transmission, removing

NOTE —

- *All cable ties (tie wraps) opened, cut, or removed during transmission removal must be reinstalled and/or replaced at the same location during transmission installation.*
- *Transmission removal is best done on a vehicle that is cold (i.e. not warmed up). Hot coolant and fluids can cause scalding and will make handling difficult if spilled.*

– Shift transmission into neutral position.

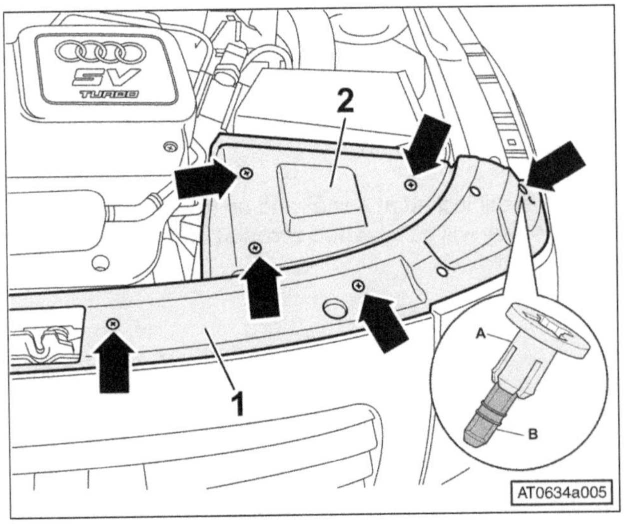

Release expanding pins (**arrows**) and remove covers above lock carrier (**1**) and battery (**2**). Expanding pins (**inset**) are released by pushing inner pin (**B**) down into body (**A**) approximately 4 mm (5/16 in) and pulling complete pin out.

BentleyPublishers.com
© 2006 BentleyPublishers.com—All Rights Reserved

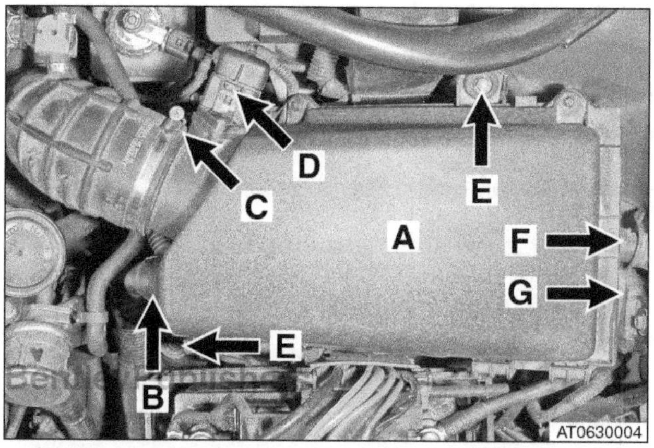

- With ignition switched off, disconnect battery ground strap.
- Open coolant expansion tank cap to relieve pressure.
- Remove upper sound absorber panels, see **03 Maintenance**.
- ◄ Remove complete air filter assembly (**A**) to allow access.
- Depress tabs on lock ring on secondary air injection hose (**B**) and remove hose.
- Loosen clamp on turbocharger inlet hose (**C**) and remove hose from mass air flow sensor. On 6-cylinder engines, additional access is available by also removing hose at throttle valve control module.
- Disconnect harness connector for mass air flow sensor (**D**).
- Remove screws (**E**) securing air filter housing to body.
- Lift air filter housing up, out, and away from fresh air intake connection (**G**) and locating pin (**F**) in body.
- Remove battery, housing, and tray. See **27 Battery, Alternator, Starter**.
- ◄ Remove cable mounting bracket bolts (**arrows**).

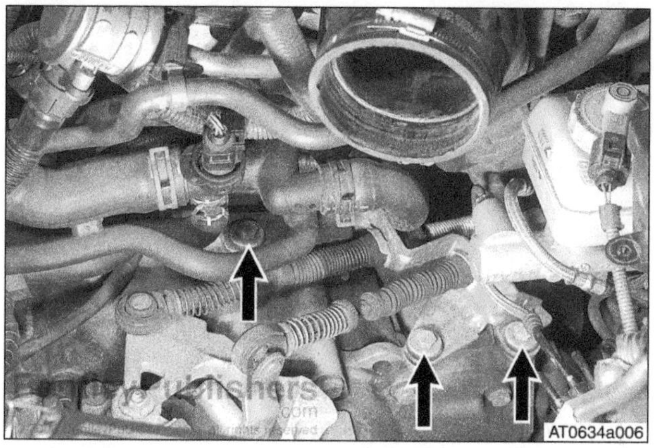

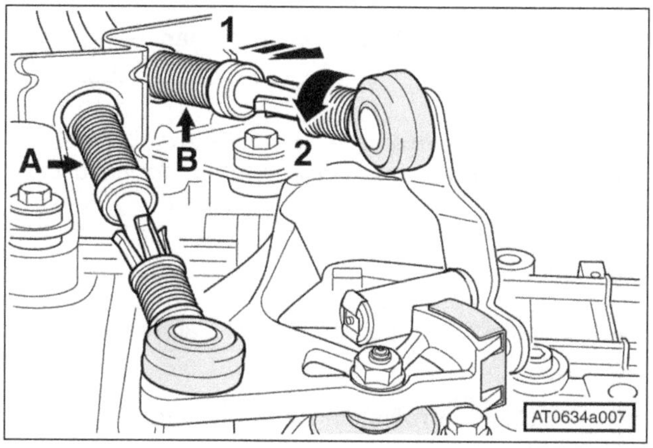

- ◄ Early vehicles (through 06/01 production) - secure locking mechanism on gate selector cable end (**B**) by pushing collar forward (**arrow 1**) as far as it will go.
- Turn collar as far as it will go to the left to lock (**arrow 2**).
- Pull gate selector cable (**B**) out of cable end. Gear shift cable (**A**) remains attached to shift lever.

NOTE —

- *Prying cable end attachments off of relay lever and/or selector lever destroys locking mechanism. Replacement will be required.*

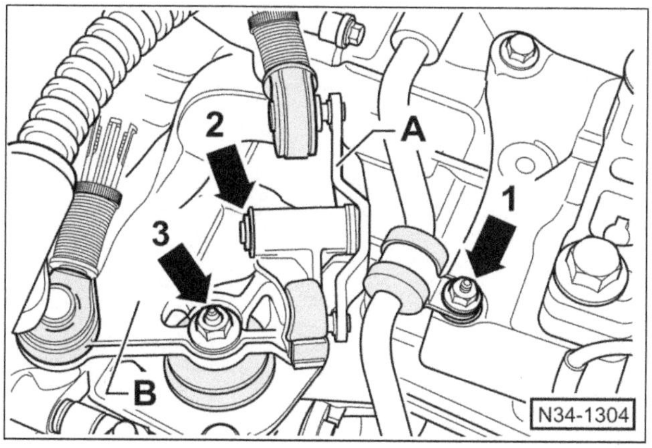

- ◄ Later vehicles (from 07/01 production) - remove locking washer (**arrow 2**) and remove relay lever (**A**) with attached gate selector cable
- Continued, all vehicles - remove nut (**arrow 1**) for power steering line.
- Remove nut (**arrow 3**) for shift lever (**B**) and pull off lever with attached gear selector cable. Use a puller if necessary.
- Tie gear selector cable, gate selector cable, bracket, and lever up and out of the way using wire or tie wraps.
- Disconnect harness connector from vehicle speed sensor at rear of transmission (where applicable).

BentleyPublishers.com
© 2006 BentleyPublishers.com—All Rights Reserved

Transmission, Removing and Installing (All Wheel Drive)

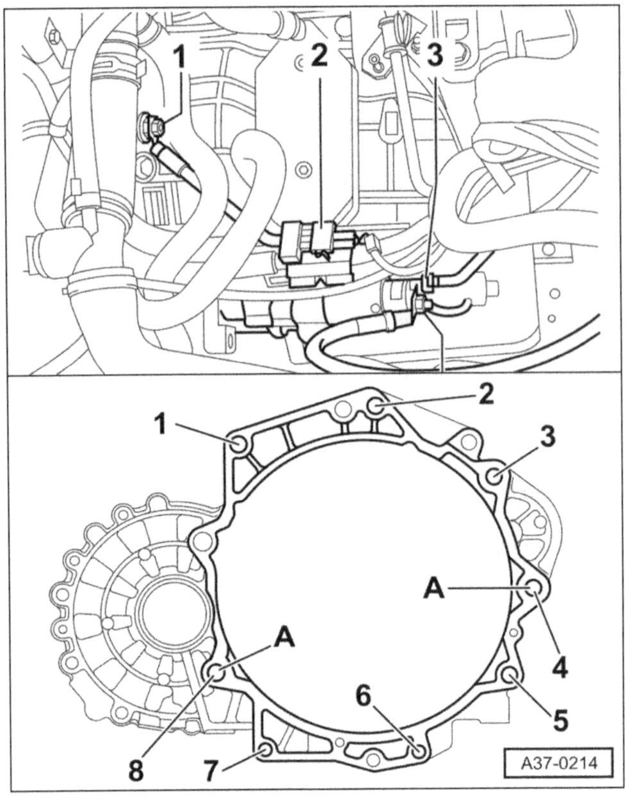

◄ Remove ground strap eyelet (**1**) from engine/transmission mounting bolt.

– Disconnect harness connector (**3**) from starter. Heavy gauge positive wire (**4**) remains connected.

– Disconnect harness connector (**2**) and detach from bracket.

– Detach wires from bracket at starter and lay to the side.

– Remove bracket at starter.

– Remove upper starter mounting bolt.

◄ Remove engine/transmission connecting bolts accessible from above.

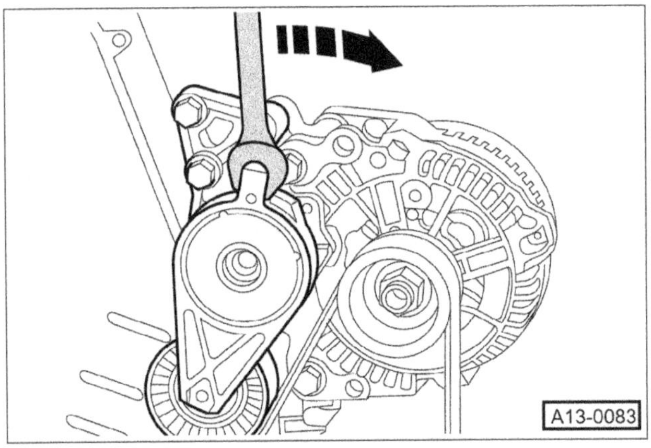

◄ Mark running direction of ribbed belt. Using a 15 mm open end wrench, release tensioner (**dashed arrow**) and remove ribbed belt.

– Raise vehicle and support safely. See **03 Maintenance**.

> ***WARNING —***
> *• Make sure vehicle is stable and well supported at all times. Use a professional automotive lift or jack stands designed for the purpose. A floor jack is not adequate support.*

– Remove both front wheels.

– Remove lower sound absorber panels. See **03 Maintenance**.

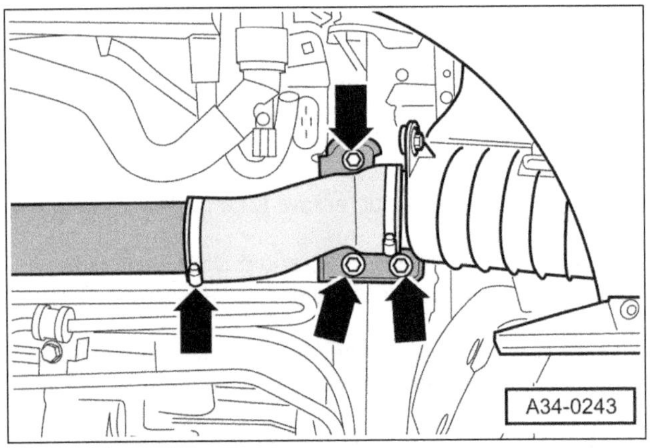

◄ Remove bolts and clamp (**arrows**) and remove intercooler connecting duct.

BentleyPublishers.com
© 2006 BentleyPublishers.com—All Rights Reserved

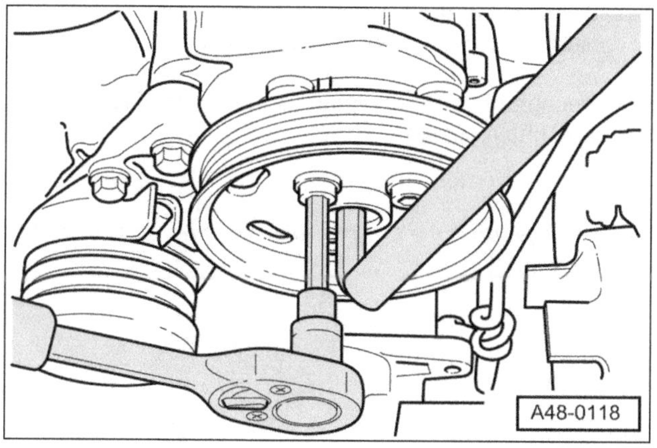

◀ Counterhold power steering pump shaft with a hex key and remove pulley.

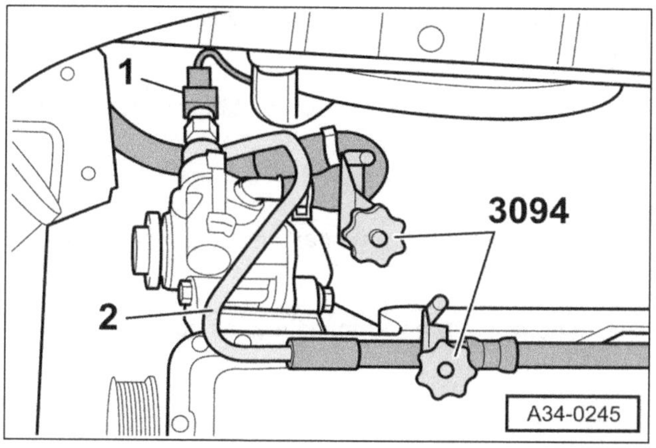

◀ Clamp supply and pressure lines at power steering pump lines with **3094** or equivalent clamps with 25 mm (1 in) diameter line capacity.

- Disconnect harness connector from power steering pressure switch (**1**).
- Disconnect and remove pressure line (**2**) from power steering pump.
- Disconnect and remove supply line from power steering pump.

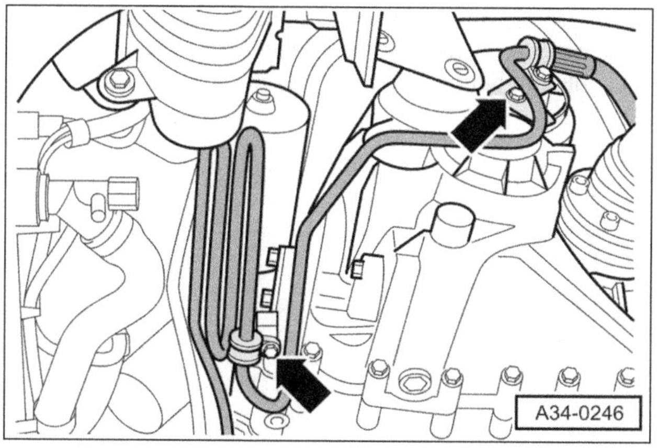

◀ Remove power steering line clamp and bracket bolts (**arrows**). Remove clamps, brackets and additional support brackets (if installed).

- Remove lower starter mounting bolts and starter. See **27 Battery, Alternator, Starter**

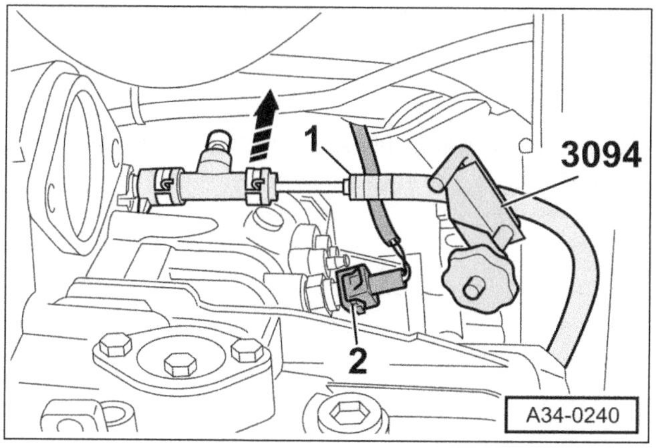

◀ Clamp clutch slave cylinder line with **3094** or equivalent clamp with 25 mm (1 in) diameter line capacity.

- Pull retainer clip upward (**dashed arrow**) and pull pressure line (**1**) out of clutch slave cylinder connection.

NOTE —

- *Do not depress clutch pedal while line is disconnected.*

- Disconnect harness connector for back up light switch (**2**).

BentleyPublishers.com
©2006 BentleyPublishers.com—All Rights Reserved

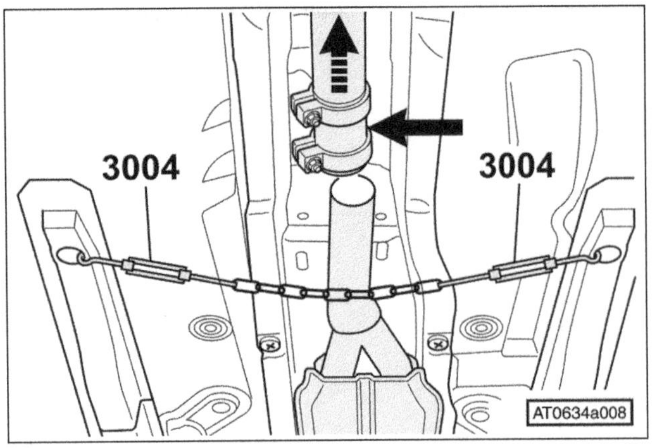

- Loosen double clamp on exhaust system (**arrow**) and slide clamp to the rear of vehicle (**dashed arrow**) to disconnect.

- Support front section of exhaust system using turnbuckles and chains (**3004** and chain shown).

NOTE —

- *To prevent damage, do not allow flex joint for exhaust system to bend more than 10°.*

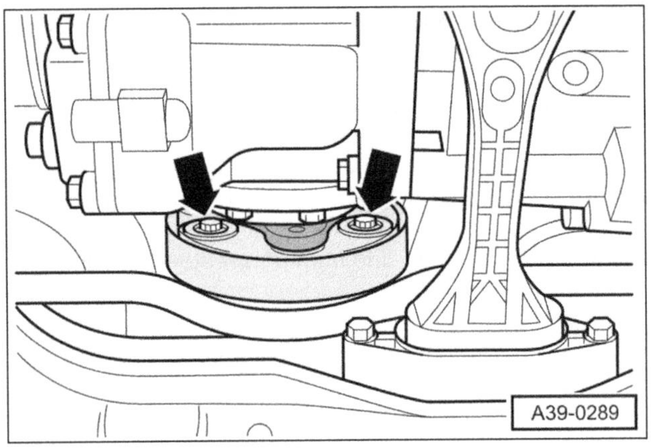

- Mark position of drive shaft at flexible disc coupling at bevel gear transfer case.

- Unbolt driveshaft from flexible disc coupling at bevel gear transfer case (**arrows**).

NOTE —

- *To remove bolts, counterhold driveshaft at triangular drive flange using a lever.*

- Push driveshaft horizontally as far to the rear as possible.

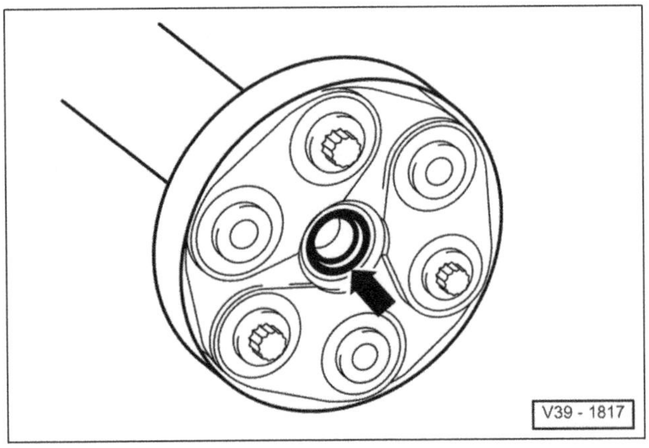

- When working around driveshaft, be sure not to damage seals (**arrow**) in flange of drive axle, especially when removing and installing. If seals are damaged, driveshaft must be replaced.

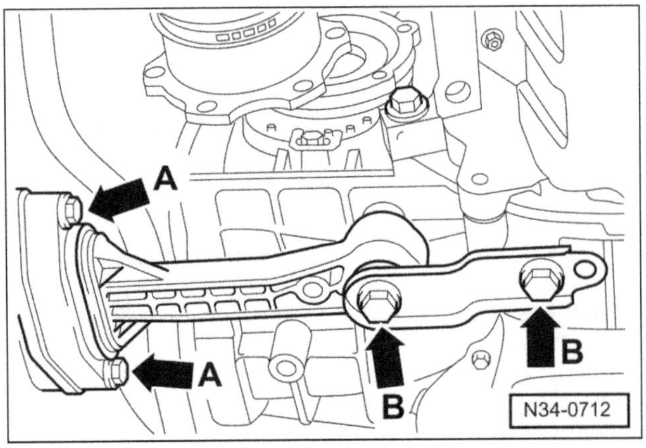

- Remove pendulum support. First remove bolts (**A**) and then bolts (**B**).

NOTE —

- *After removing bolts (**A**), engine/transmission assembly will move slightly forward.*

BentleyPublishers.com
© 2006 BentleyPublishers.com—All Rights Reserved

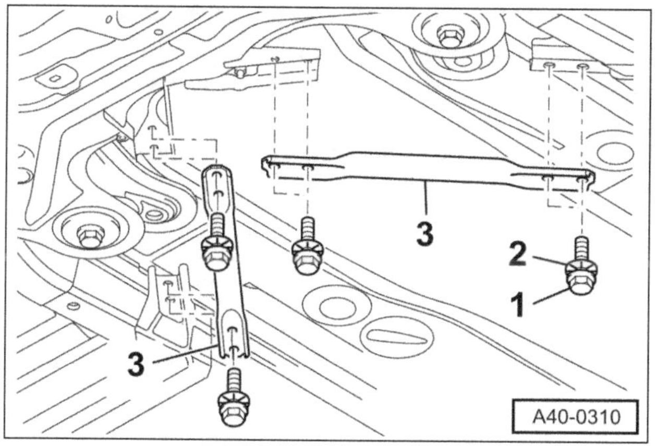

◀ Where applicable, remove 8 bolts (**1**) with locking washers (**2**) and remove subframe diagonal support members (**3**).

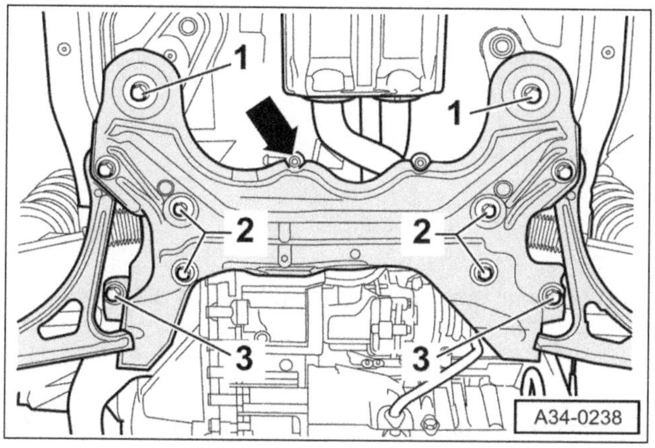

◀ Remove front exhaust system bracket from subframe (**arrow**) where installed.

– Remove steering gear bolts (**2**).

– Carefully pry off steering gear from subframe.

– Place engine/transmission jack under subframe for support.

– Remove subframe mounting bolts (**1** and **3**).

– Carefully lower subframe and let it hang on ball joints and connecting links.

– Tie steering gear up and out of the way using wire or tie wraps.

– Remove drive axles from transmission drive flanges and lay behind on subframe taking care to avoid damage to subframe coatings and constant velocity joint boots.

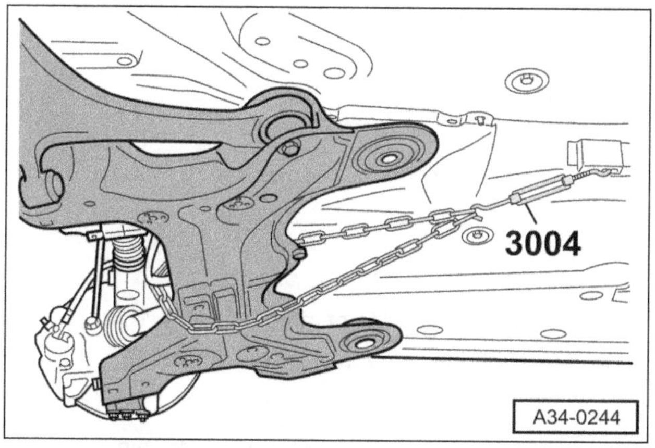

◀ Tie subframe toward the rear using a chain with a turnbuckle (**3004**) arrangement. Remove plug from underbody as needed and hook chain into opening in vehicle floor

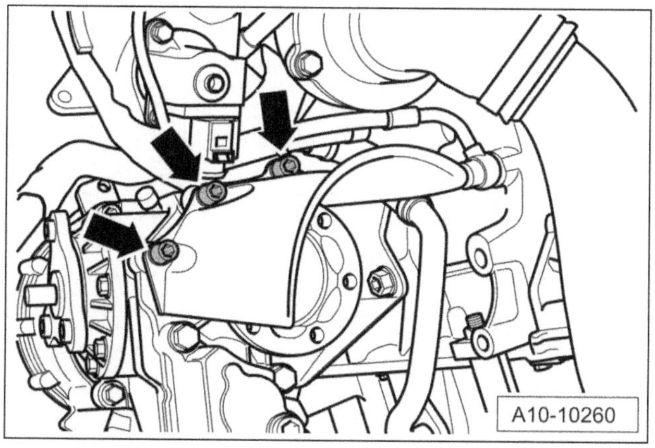

◀ Unbolt (**arrows**) and remove heatshield for right side drive axle from bevel gear transfer case.

BentleyPublishers.com
© 2006 BentleyPublishers.com—All Rights Reserved

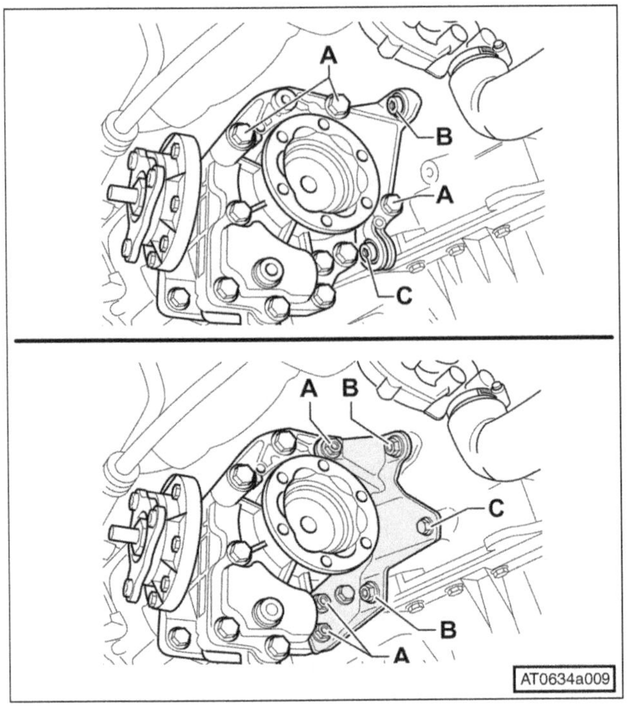

◄ Unbolt carrier mounting bracket for bevel gear transfer case.

- Cast bracket (upper) - remove bolts (**A**, **B**, **C**) at transfer case bracket. Bracket can not be removed at this point, leave in position but unbolted.
- Steel bracket (lower) - remove bolts (**A**,**B**, **C**) and remove carrier mounting bracket.

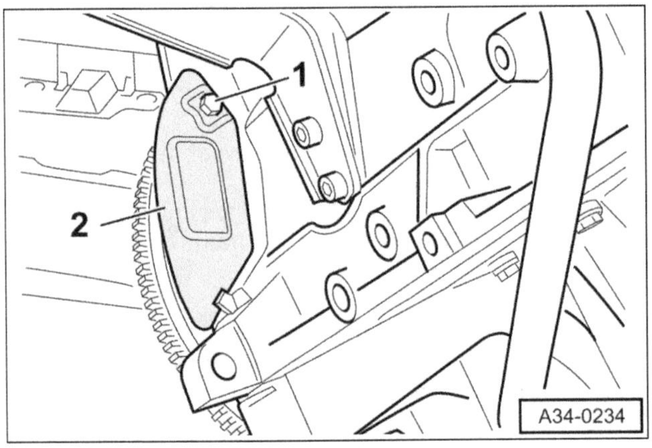

◄ Remove bolt (**1**) securing small cover plate (**2**) behind bevel gear transfer case.

NOTE —

- *Small cover plate (**2**) is located behind bevel gear transfer case and cannot be seen when bevel gear transfer case is installed. Illustration shows small cover plate with bevel gear transfer case removed.*

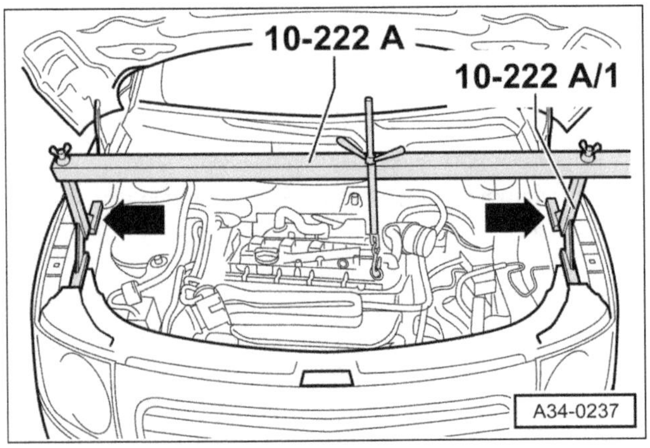

◄ Install engine support bridge on edge of inner fender (**arrows**).

- Audi support bridge **10-222A** with attachments shown for reference.

– Pretension engine/transmission assembly slightly with spindle to relieve tension on transmission support.

BentleyPublishers.com
©2006 BentleyPublishers.com—All Rights Reserved

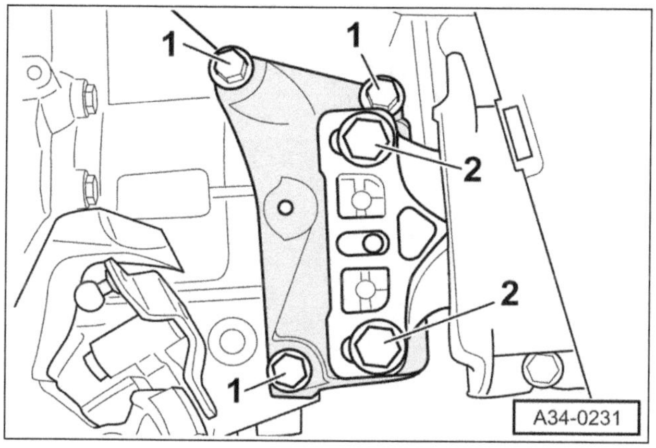

◀ Remove left transmission mount bolts (**1** and **2**) and remove mount.

– Using spindle on support bridge, lower engine/transmission slightly so that gap between left transmission mount and attaching surface on transmission is approximately 100 to 110 mm (4 - 4½ in).

– Position suitable transmission jack under transmission.

– Align transmission jack support arms to secure and support transmission. Bolt transmission to jack where necessary.

◀ Remove bolt (**arrow**) above bevel gear transfer case.

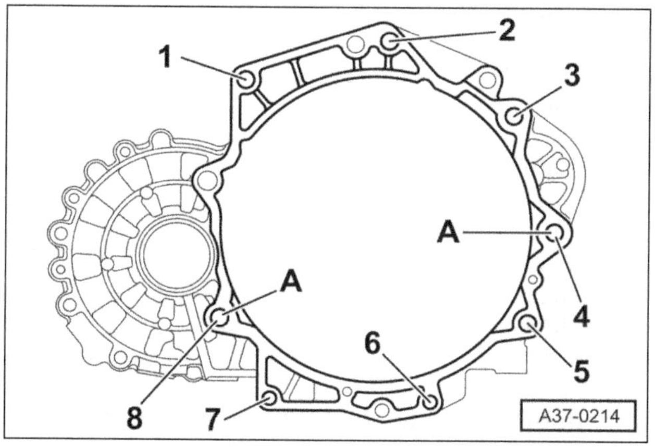

◀ Remove remaining engine/transmission connecting bolts accessible from below.

– Press transmission off of alignment sleeves (dowel pins).

– Where applicable, remove carrier mounting bracket for bevel gear transfer case.

– Remove transmission from engine starting with bevel gear transfer case side and working to the front.

– When completely separated, lower transmission.

Transmission, installing

Install transmission in reverse order of removal noting the following:

- Replace self-locking nuts and bolts during assembly work.
- Always replace bolts, O-rings and gaskets as appropriate.
- Secure all hose connections using appropriate hose clamps.
- All cable ties opened or cut during engine removal must be reinstalled at the same locations during installation.
- Clean input shaft splines and (in case of used clutch disc) clean hub splines, remove corrosion and apply a very thin coating of lubricant G 000 100 on splines. Always remove excessively applied grease.
- Clutch plate must move freely on input shaft.
- Make sure centering sleeves (dowels) for engine to transmission are installed in cylinder block, reinstall where necessary.
- Inspect clutch release bearing for wear and replace if necessary.

BentleyPublishers.com
© 2006 BentleyPublishers.com—All Rights Reserved

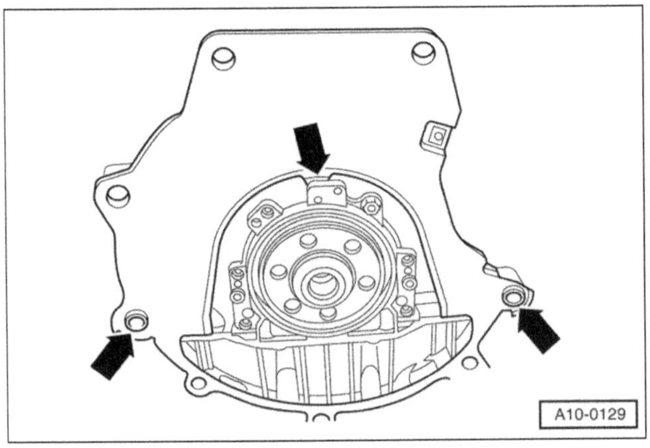

◄ Ensure intermediate plate seats at sealing flange (**center arrow**).

– Ensure that alignment sleeves (**outer arrows**) are installed and that intermediate plate is aligned over them.

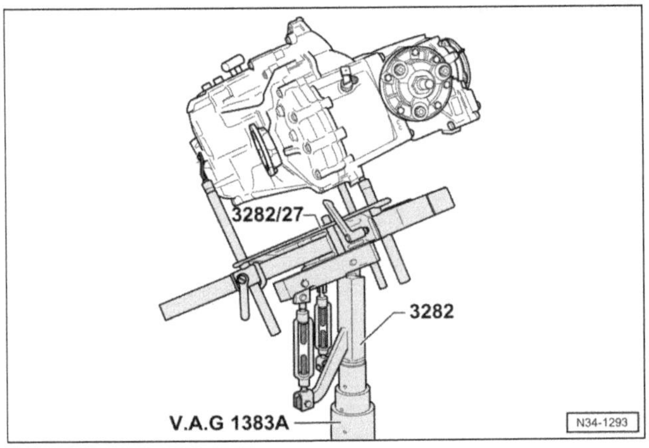

◄ Install transmission onto transmission jack at an angle so that left side is down approximately as shown.

– With vehicle in position, align transmission jack to installation position close to engine.

– Press transmission against engine, starting at the front side and then at the bevel gear transfer case side over the flywheel.

– When in proper position, push transmission against engine.

Cast carrier mounting bracket for bevel gear transfer case

– Insert cast carrier mounting bracket for bevel gear transfer case.
  - Cast bracket must be installed before tightening engine/transmission bolts. It cannot be installed later.

– Install engine/transmission connecting bolts accessible from below and tighten to specification.

– Install bevel gear transfer case.

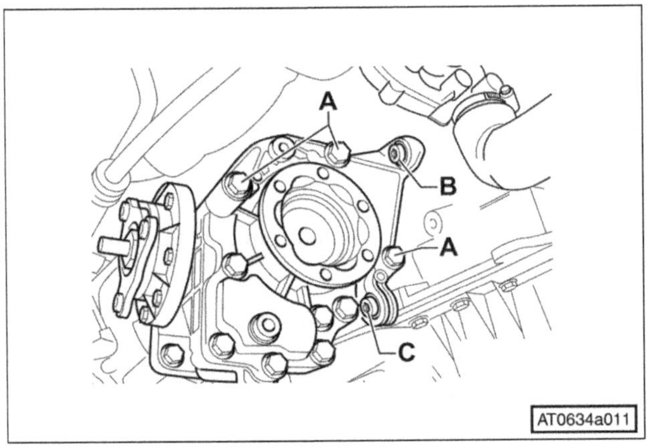

◄ Install bolts for carrier mounting bracket for bevel gear transfer case.
  - Loosely install bolts (**A**, **B**, **C**) and tighten in sequence:
    Tighten **A** to 3 Nm (27 in-lb)
    Tighten **B** and **C** to 35 Nm (26 ft-lb)
    Retighten **A** to 45 Nm (33 ft-lb)

Steel carrier mounting bracket for bevel gear transfer case

– Install engine/transmission connecting bolts accessible from below and tighten to specification.

– Install bevel gear transfer case.

– Insert steel carrier mounting bracket for bevel gear transfer case.

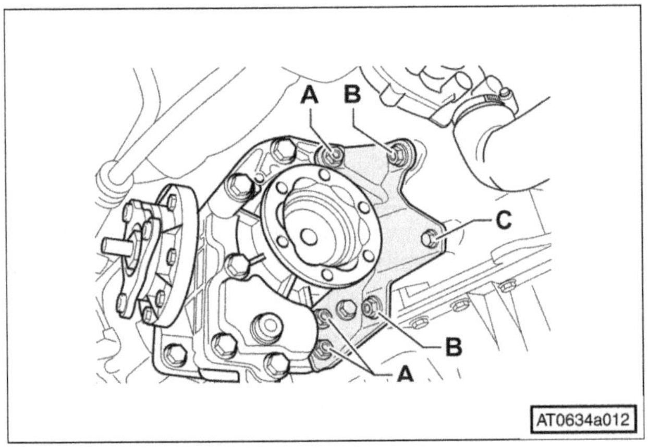

◄ Install bolts for carrier mounting bracket for bevel gear transfer case.
  - Loosely install bolts (**A**, **B**, **C**) and tighten in sequence:
    Tighten **A** to 3 Nm (27 in-lb)
    Tighten **B** to 35 Nm (26 ft-lb)
    Tighten **C** to 22 Nm (16 ft-lb)
    Retighten **A** to 45 Nm (33 ft-lb)

BentleyPublishers.com
© 2006 BentleyPublishers.com—All Rights Reserved

Continued for all transmissions

- Remove transmission jack.

- Install small cover plate for flywheel and tighten bolt.

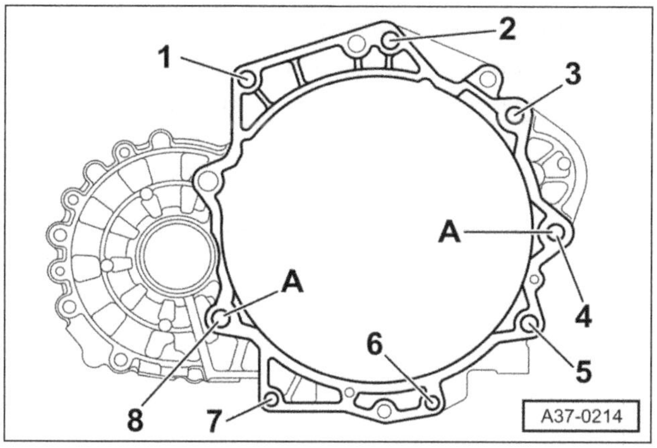

- Install engine/transmission connecting bolts accessible from above and tighten to specification.

Table d. Transmission to engine tightening torques

Fastener	Size	Tightening torque
1	M12 X 55	80 Nm (59 ft-lb)[2]
2[1]	M12 X 55	80 Nm (59 ft-lb)[2]
3[1] + 4[1]	M12 X 165	80 Nm (59 ft-lb)[2]
5	M10 X 105	40 Nm (30 ft-lb)[2]
6 + 7	M10 X 50	40 Nm (30 ft-lb)[2]
8	M12 X 70	80 Nm (59 ft-lb)[2]
A	Locating dowel sleeves	

1. With additional 8 mm threaded pin.
2. Always replace bolts.

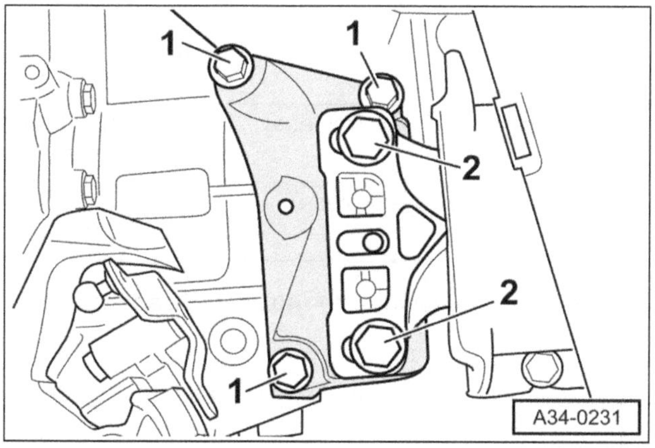

- Install subframe mount with new bolts (**1**).

- Bolt transmission mount to transmission with pendulum support still removed.

- Lift transmission up to transmission mount on body using support bridge spindle.

> ***CAUTION—***
> - *Before installing bolts, be sure that mounting holes in transmission are aligned with holes in mount on body so threads will not be damaged. Lift transmission at rear using floor jack, if necessary*

- Install new transmission mount bolts (**2**) and tighten.

Tightening torque	
Bolts (**1**), transmission subframe (always replace)	40 Nm (30 ft-lb) plus ¼ turn (90°)
Bolts (**2**), transmission mount (always replace)	60 Nm (44 ft-lb) plus ¼ turn (90°)

> ***CAUTION—***
> - *Only remove engine support bridge when all mount bolts are installed and tightened to specification.*

- Remove engine support bridge.

BentleyPublishers.com
© 2006 BentleyPublishers.com—All Rights Reserved

Install subframe:

- Position both drive axles on transmission drive flanges and install bolts hand-tight only.
- Move subframe into installed position.
- Install new subframe bolts and hand-tighten.
- Position steering gear on subframe and install new bolts hand-tight.
- Be sure sleeve (**left inset**, **1**) is positioned in hole in subframe.
- Be sure rubber buffer (**right inset**) is seated properly on subframe.

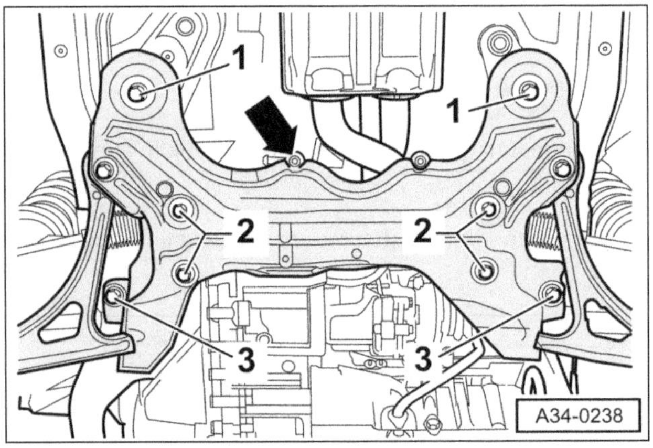

Tighten steering gear bolts (**2**).

– Tighten subframe bolts (**1** and **3**).

Tightening torques	
Bolts (**1** and **3**), subframe (always replace)	100 Nm (74 ft-lb) plus ¼ turn (90°)
Bolts (**2**), steering gear (always replace)	20 Nm (15 ft-lb) plus ¼ turn (90°)

– Tighten drive axles on drive flanges, see **40 Front Suspension and Drive Axles**.

– Where applicable, install diagonal subframe support members.

– Install right side driveshaft heat shield.

– Using new bolts, attach pendulum support to transmission.

– Push engine/transmission assembly forward and carefully install driveshaft onto guide pin on transmission drive flange.

NOTE—
- *Be careful not to damage seals in flange of drive axle when removing and installing. If seals are damaged, driveshaft must be replaced.*

– Using new bolts, attach pendulum support to subframe.

– Tighten driveshaft bolts.

– Connect exhaust system and adjust positioning so that there is no pretension. Attach to subframe where applicable. Remove turnbuckle supports.

– Connect harness connector for back up light switch.

BentleyPublishers.com
©2006 BentleyPublishers.com—All Rights Reserved

– Connect pressure line to clutch slave cylinder and remove line clamp.

NOTE —
• Do not depress clutch pedal while line is disconnected.

– Install starter with new lower mounting bolt.

– Install power steering line clamps, bracket, and additional support brackets (if installed) to transmission.

– Install power steering lines on pump.
  - Install banjo bolt with new sealing washers.
  - Install supply line.
  - Remove line clamps.

– Connect harness connector to power steering pressure switch.

– Counterhold power steering pump shaft with a hex key and install pulley and ribbed belt.

– Install intercooler connecting duct.

– Check gear oil level in bevel gear transfer case, see **03 Maintenance**.

– Install lower sound absorber panels. See **03 Maintenance**.

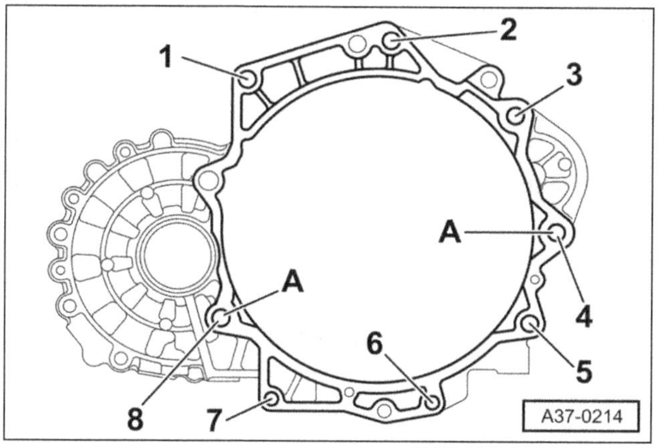

◄ Install new upper starter mounting bolt. Tighten upper and lower starter bolts to specification.

– Install engine/transmission connecting bolts accessible from below and tighten to specification.

Table d. Transmission to engine tightening torque

Fastener	Size	Tightening torque
1	M12 X 55	80 Nm (59 ft-lb)[2]
2[1]	M12 X 55	80 Nm (59 ft-lb)[2]
3[1] + 4[1]	M12 X 165	80 Nm (59 ft-lb)[2]
5	M10 X 105	40 Nm (30 ft-lb)[2]
6 + 7	M10 X 50	40 Nm (30 ft-lb)[2]
8	M12 X 70	80 Nm (59 ft-lb)[2]
A	Locating dowel sleeves	

1. With additional 8 mm threaded pin.
2. Always replace bolts.

– Connect harness connector to starter.

– Install bracket at starter and install wires into bracket.

– Bolt ground strap eyelet to engine/transmission mounting bolt.

– Install harness connector for vehicle speed sensor at rear of transmission (where applicable).

BentleyPublishers.com
© 2006 BentleyPublishers.com—All Rights Reserved

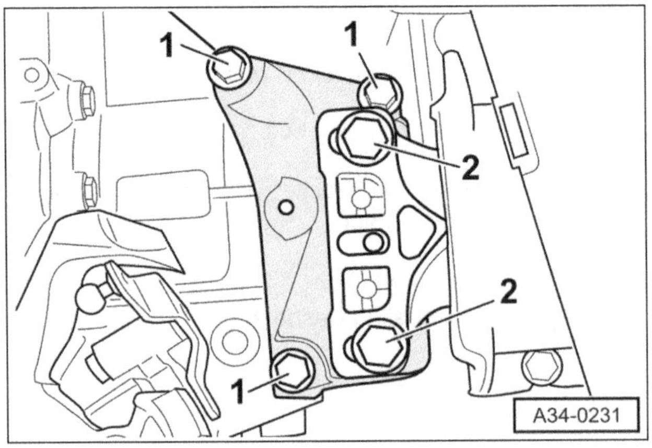

– Install gear selector cable, gate selector cable, bracket, and shift lever.

– Install nut for power steering line.

– Check subframe to be sure that there is no pretension.

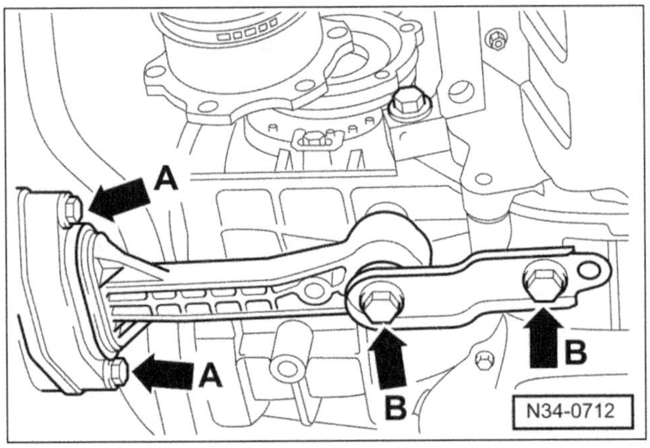

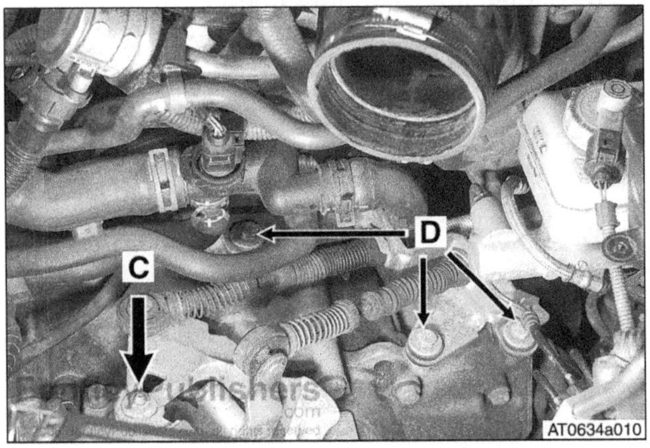

Table e. Transmission removal and installation tightening torques

Fastener	Tightening torque
Small cover plate behind bevel gear transfer case	10 Nm (7 ft-lb)
Bolts (**1**), transmission subframe (always replace)	40 Nm (30 ft-lb) plus ¼ turn (90°)
Bolts (**2**), transmission mount (always replace)	60 Nm (44 ft-lb) plus ¼ turn (90°)
Bolts (**A**), pendulum support to subframe (always replace)	20 Nm (15 ft-lb) plus ¼ turn (90°)
Bolts (**B**), pendulum support to transmission (always replace)	40 Nm (30 ft-lb) plus ¼ turn (90°)
Nut (**C**), shift rod lever to shift rod (always replace)	20 Nm (15 ft-lb)
Bolts (**D**), cable mounting bracket to transmission	23 Nm (17 ft-lb)
Bolts, heatshield right side drive axle	25 Nm (18 ft-lb)
Nuts, intercooler connecting duct.	10 Nm (7 ft-lb)
Banjo bolt, power steering high pressure line on pump (always replace seals	38 Nm (28 ft-lb)

– Bleed clutch hydraulic system. See **30 Clutch**.

– Adjust shift linkage as needed.

– Install battery tray, housing, and battery. Connect battery. See **27 Battery, Alternator, Starter**.

– Install complete air filter assembly.

– Check coolant level and secure reservoir cap.

– Install upper sound absorber panels, see **03 Maintenance**.

– Reset expanding pins by pushing center section back through pin body. Insert pins into covers, install covers above lock carrier and battery and push pin center to lock.

– Road test.

BentleyPublishers.com
© 2006 BentleyPublishers.com—All Rights Reserved

34b DSG Transmission

TABLES

GENERAL

This repair group covers repair information for external components and shift mechanism of the Direct Shift Gearbox (DSG). This transmission is installed in Coupes and Roadsters with 3.2L 6-cylinder engines. Numerous special tools and equipment are required for removal, installation and internal repair of the DSG. These repairs are outside the scope of this manual.

- For servicing information on drive axles, see **40 Front Suspension and Drive Axles** and **42 Rear Suspension**.
- For transmission identification information, see **03 Maintenance.**
- For information on checking and changing transmission oil and filter, and checking and changing bevel gearbox oil, see **03 Maintenance**.
- For information on checking and changing rear final drive oil and Haldex oil and filter, see **03 Maintenance**.

> ***WARNING —***
> - *Disconnecting negative (–) battery cable has ramifications with many vehicle systems. Before proceeding, be familiar with warnings associated with this procedure in **03 Maintenance** and **69 Seatbelts, Airbags**. These systems may need to be re-set using a VAG 1551/1552, VAS 5051/5052 or equivalent scan tool or computer scan tool program.*

Transmission	Transmission type
6-speed direct shift gearbox (all wheel drive)	02E

BentleyPublishers .com
© 2006 BentleyPublishers.com—All Rights Reserved

Table a. 02E direct shift gearbox specifications (all wheel drive)

Type (Code)	02E (GYV)	02E (GYU)	02E (HAH)	02E (HYB)
Engine application, horsepower	3.2L, 250 hp	3.2L, 250 hp	3.2L, 250 hp	3.2L, 250 hp
Installation period	07/03 - 03/04	03/04 - 05/04	06/04 - 10/05	11/05 - 06/06
Gear ratio: Z_2:Z_1				
Final drive	69:17 = 4.059 (1-4)[1] 69:22 = 3.136 (5,6,R)	69:17 = 4.059 (1-4)[1] 69:22 = 3.136 (5,6,R)	69:17 = 4.059 (1-4)[1] 69:22 = 3.136 (5,6,R)	n/a
1st gear	45:13 = 3.46	45:13 = 3.46	n/a	n/a
2nd gear	45:22 = 2.05	45:22 = 2.05	n/a	n/a
3rd gear	45:31 = 1.45	45:31 = 1.45	n/a	n/a
4th gear	41:38 = 1.079	41:38 = 1.079	n/a	n/a
5th gear	39:35 = 1.11	39:35 = 1.11	n/a	n/a
6th gear	35:38 = 0.92	35:38 = 0.92	n/a	n/a
Reverse gear	22:14 x 33:13 = 3.99	22:14 x 33:13 = 3.99	n/a	n/a
Lubricant (gearbox)				
Capacity	Overall 7.2 liters (7.6 qt.), Change 6.0 liters (6.3 qt.)			
Audi part number	G 052 182 A2			
Lubricant (bevel gearbox)				
Capacity	Overall 0.9 liters (1.0 qt.)			
Audi part number	G 052 145 S2			
Driveshaft flange diameter	108 mm			
Rear differential type	02D			
Ratio, overall in top gear	2.89		n/a	

1. Ring gear is driven by 2 separate pinion gears on individual output shafts, dependent on gear selected.
   Output shaft 1 is used by gears 1, 2, 3, 4.
   Output shaft 2 is used by gears 5, 6, R.

Special transmission oil G 052 182 A2 (**left**) as supplied by Audi for use in direct shift gearboxes. Viscosity and type are not specified in factory service information. Bevel gearbox oil G 052 145 S2 (**right**) is identified as synthetic, but with no viscosity specification.

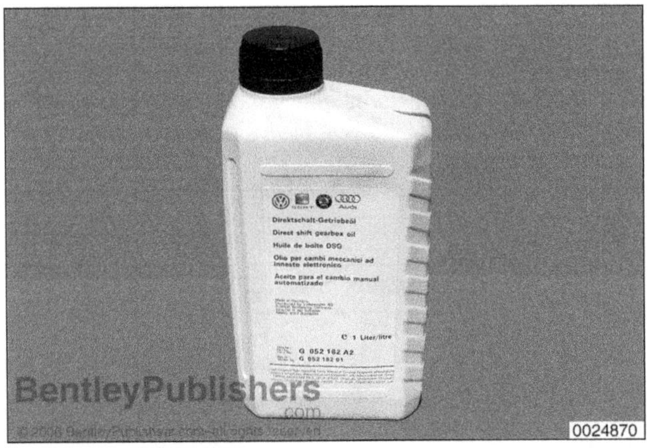

0024870

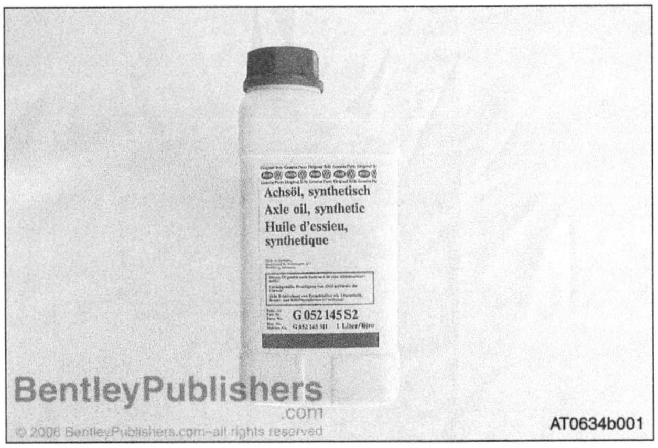

AT0634b001

BentleyPublishers
.com
© 2006 BentleyPublishers.com—All Rights Reserved

DSG Electrical Components

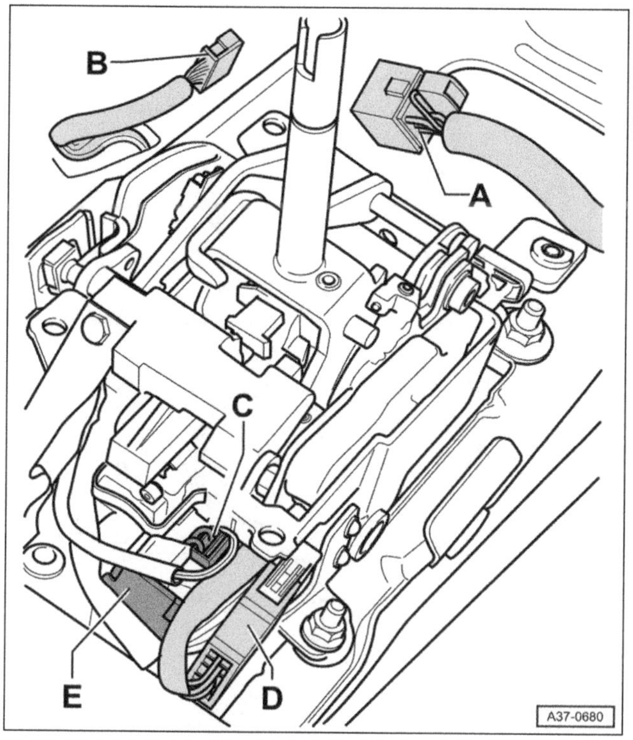

Gear selector mechanism (E313) electrical components and connections

A. Connector from vehicle wiring harness to selector lever (E313)
 - With CAN-bus wiring

B. Connector for selector mechanism cover

C. Connector for shift lock solenoid (N110)

D. Connector for Tiptronic switch (F189)

E. Shift lock solenoid (N110)

Additional components:
- Data link connector (T16)
- Back-up light relay (J219)
- Starter interlock relay (J207)
- Upshift and downshift buttons in steering wheel (E438, E439)
- Transmission range selector lever display (Y6)
- Brake light switch (F)
- Kickdown switch (F8)

NOTE —

- *Audi identifies electrical components by a letter and/or a number in the electrical schematics. See* **EWD Electrical Wiring Diagrams**. *These electrical identifiers are listed in parentheses as an aid to electrical troubleshooting.*

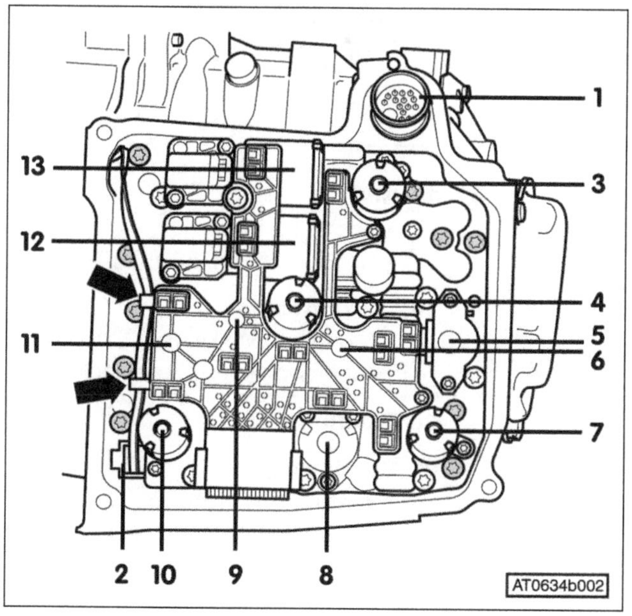

Mechatronic unit (J743) electrical components

1. Connector for vehicle wiring harness
 - With CAN-bus wiring
2. Connector for RPM sensor (G182) and transmission fluid temperature sensor (G93)
 - Wiring harness held in position by retainers (**arrows**)
3. Pressure control valve (N217)
4. Pressure control valve (N218)
5. Solenoid valve (N92)
6. Solenoid valve (N89)
 - Under circuit board
7. Pressure control valve (N233)
8. Solenoid valve (N88)
9. Solenoid valve (N90)
 - Under circuit board
10. Pressure control valve (N371)
11. Solenoid valve (N91)
 - Under circuit board
12. Pressure control valve (N215)
13. Pressure control valve (N216)

Additional internal components:
- Multi-plate clutch oil temperature sensor (G509)
- Transmission output RPM sensors (G195, G196)
- Driveshaft speed sensors (G501, G502)
- Hydraulic pressure sensors (G193, G194)
- TCM temperature sensor (G510)
- Gear position distances sensors (G487, G488, G489, G490)

BentleyPublishers
.com
© 2006 BentleyPublishers.com—All Rights Reserved

Shift Mechanism

Selector lever and selector housing assembly

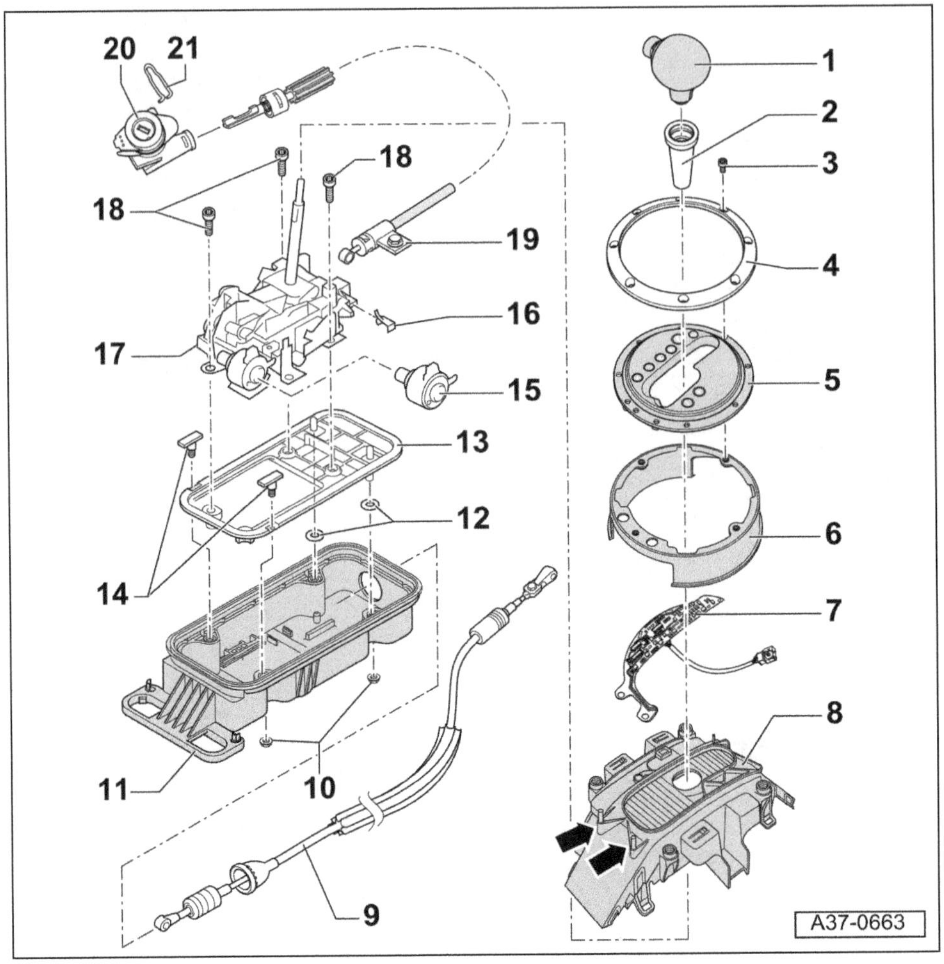

1. **Gear shift knob**
   - See **Gear shift knob, removing and installing**
2. **Trim/locking sleeve**
   - Clips into gear shift knob
   - Locks knob to prevent knob from pulling off
3. **Bolt**
   - Threaded through retaining ring and upper cover into side cover
   - Qty: 8, M4x10
   - Tighten to 2 Nm (18 in-lb)
4. **Retaining ring**
5. **Upper cover**
   - With integral symbol insert
   - Positioned on frame (item **8**) (**arrows**)
6. **Side cover**
   - Positioned on frame (item **8**) (**arrows**)
7. **Printed circuit board**
   - Positioned on frame (item **8**) (**arrows**)
   - See **Printed circuit board, installed position**
8. **Frame**
   - With locating pins (**arrows**)
   - Pull 4 integral hooks outward to remove
   - Remove complete with retaining ring, upper cover, side cover, and printed circuit board (items **4**, **5**, **6**, **7**)
9. **Selector cable**
   - Do not bend or kink
   - Do not lubricate cable eye or ball socket
   - See **Selector cable, adjusting**
10. **Nut, self locking**
    - Qty: 4
    - Always replace
    - Tighten to 10 Nm (7 ft-lb)
11. **Housing**
    - With glued-on gasket
12. **Washer**
13. **Intermediate plate**
14. **Bolt**
    - Inserted into intermediate plate
15. **Shift lock solenoid (N110)**
    - With integral tiptronic switch (F189)
    - See **Shift lock solenoid, removing and installing**
16. **Retainer**
    - Secures cable
17. **Selector mechanism**
18. **Bolt**
    - Qty: 3
    - Tighten to 6 Nm (53 in-lb)
19. **Locking cable**
    - Do not bend or kink
    - For ignition key removal lock
20. **Steering lock**
21. **Locking clip**
    - Install clip from above
    - Angled ends point towards steering lock

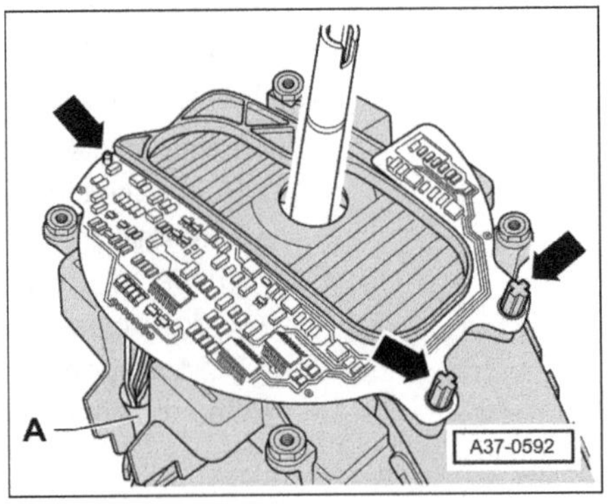

Printed circuit board, installed position

- Position holes on printed circuit board onto front and rear guide pins (**arrows**) on frame.
- Ensure wiring harness is properly positioned and that printed circuit board is not deformed.

BentleyPublishers.com
© 2006 BentleyPublishers.com—All Rights Reserved

Shift lock solenoid, removing and installing

- Disconnect harness connector (**B**) from shift lock solenoid (N110).
- Depress retaining tab (**A**) to rear while simultaneously turning solenoid counter-clockwise (**dashed arrow**).
- Pull shift lock solenoid off.

Gear shift knob, removing and installing

Several slightly different style gear shift knobs are used, but all are removed in the same manner.

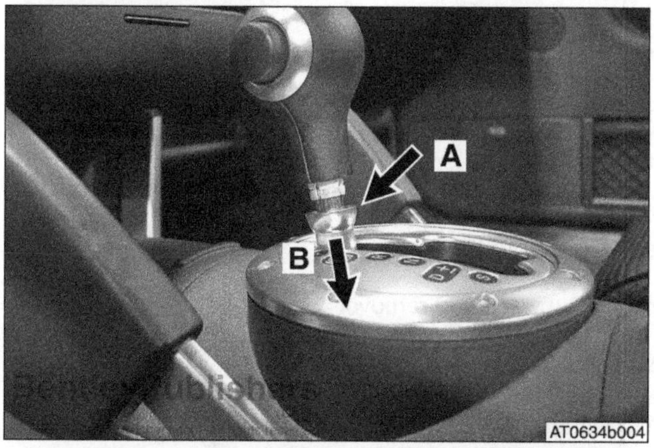

With a firm grip, grasp trim/locking sleeve (**A**) and pull down on shaft (**B**).

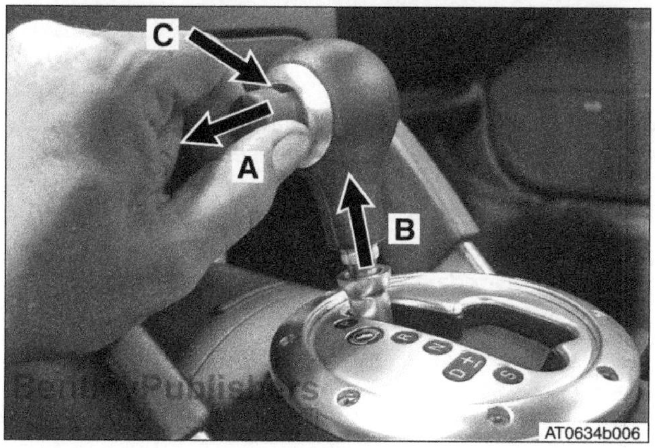

Pull release button out (**A**) of shift knob as far as possible and lock in place with a wire clip in the opening on top (**C**).

– While holding button out, pull up (**B**) on knob to remove from shaft.

– If release button slips back into knob, use a pick tool to pull button out and lock in place with wire clip.

BentleyPublishers.com

© 2006 BentleyPublishers.com—All Rights Reserved

– Installation is the reverse of removal, noting the following:
  - Install trim/locking sleeve onto shaft.
  - Release button must be pulled out to install knob.
  - Installed position: button points to driver.
  - Slide knob onto shaft and move around to engage mechanism.
  - When knob is felt to engage mechanism, push down fully onto shaft.
  - While holding knob down, grasp trim/locking sleeve, slide up and lock onto knob.

Backup light and neutral start function

Operation of the backup lights and the ability to allow starter operation in park and neutral positions only, has typically been a direct function of the selector mechanism. DSG transmissions use a combination of selector mechanism (E313) and Mechatronic unit (J743) control to operate relays which enable these functions.

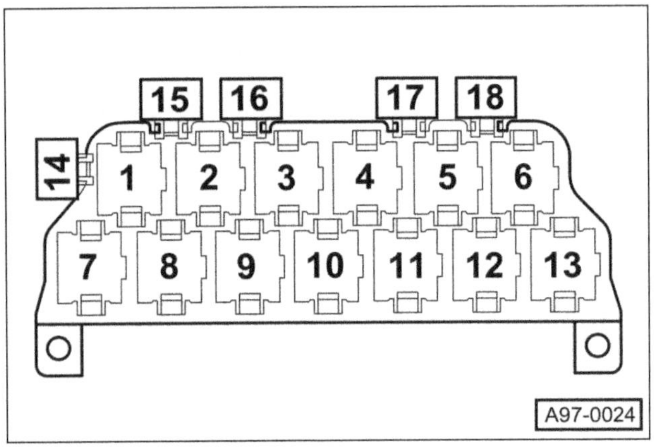

◄ The 13-fold relay panel is located above the micro central electric panel under the driver's side of the padded dash.
  - Position 11 - Starter interlock relay (J207)
  - Position 13 - Backup light relay (J219)

SELECTOR CABLE

Selector cable, checking

– With ignition switched off, disconnect battery ground strap. See **27 Battery, Alternator, Starter**.

– Remove air cleaner assembly. See **03 Maintenance**.

– Shift selector lever into **P** position.

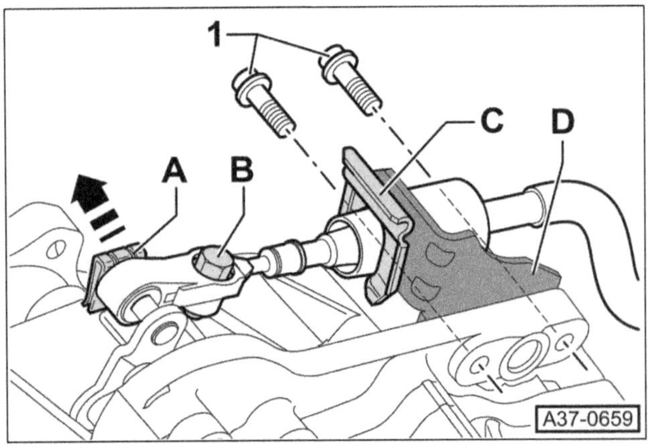

◄ Push circlip (**A**) upward and remove.

– Press selector lever cable off of selector shaft lever (**arrow**) and place on top to allow free movement of cable and selector mechanism without shifting transmission.

NOTE —
- *Do not loosen adjustment bolt (**B**).*

– Lay selector lever cable aside in such a way that end can move freely. Use care to avoid bending or kinking selector lever cable.

– Shift selector lever from **P** to **S**.
  - Check protective sleeve on selector mechanism (on front of selector lever cable) for damage. Replace cable if damaged.

BentleyPublishers.com
©2006 BentleyPublishers.com—All Rights Reserved

– Shift selector lever into **P** position.
  - Selector mechanism and selector lever cable should move smoothly. Replace selector lever cable or service selector mechanism as needed.

– With selector lever in **P**, carefully press selector lever cable back onto selector shaft lever and install circlip (**A**).

– Install air cleaner assembly.

– Connect battery.

– Check backup light operation.

Selector cable, adjusting

– With ignition switched off, disconnect battery ground strap. See **27 Battery, Alternator, Starter**.

– Remove air cleaner assembly. See **03 Maintenance**.

– Shift selector lever into **P** position.

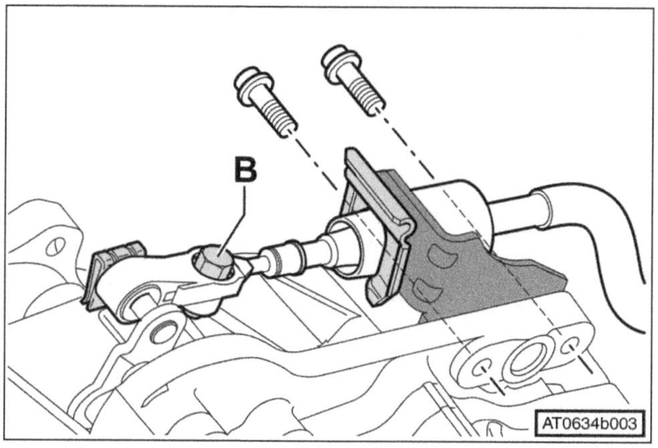

◄ Loosen bolt (**B**).

– Move selector shaft lever on transmission into **P** position.

– Turn (rotate) both front wheels in one direction by hand (if lifted) or by pushing vehicle, until detent lever in transmission engages parking lock gear and wheels lock. This will confirm that internal transmission mechanism is fully in park position.

NOTE—
- *The selector lever must be in* **P** *position when the clamping bolt (***B***) is loosened; if not, adjustment will be incorrect.*

– Carefully move selector lever slightly forward and backward, without shifting into a different selector lever position, to centralize shift lever.

– Tighten bolt (**B**) smoothly and evenly.

Tightening torque	
Bolt, selector cable adjustment	13 Nm (10 ft-lb)

NOTE—
- *When tightening bolt (***B***), make sure the selector lever cable does not move out of position.*

– Install air cleaner assembly.

– Connect battery.

– Functional check selector cable.

– Check backup light operation.

BentleyPublishers
.com
© 2006 BentleyPublishers.com—All Rights Reserved

Selector cable, checking adjustment

– Depress button on gear shift knob while pulling selector lever back approximately 5 mm (3⁄16 in) to the rear and out of **P** position. Do not allow selector lever to be shifted into **R**.
  - Release selector lever. Selector lever must spring back into **P** position automatically.
  - If necessary, re-adjust selector cable and/or inspect selector mechanism for free movement.

– Depress button on gear shift knob while pulling selector lever **back** approximately 5 mm (3⁄16 in) to the rear and out of **N** position. Do not allow selector lever to be shifted into **D**.
  - Release selector lever. Selector lever must spring back into **N** position automatically.
  - If necessary, re-adjust selector cable and/or inspect selector mechanism for free movement.

– Depress button on gear shift knob while pushing selector lever **forward** approximately 5 mm (3⁄16 in) to the front and out of **N** position. Do not allow selector lever to be shifted into **R**.
  - Release selector lever. Selector lever must spring back into **R** position automatically.
  - If necessary, re-adjust selector cable and/or inspect selector mechanism for free movement.

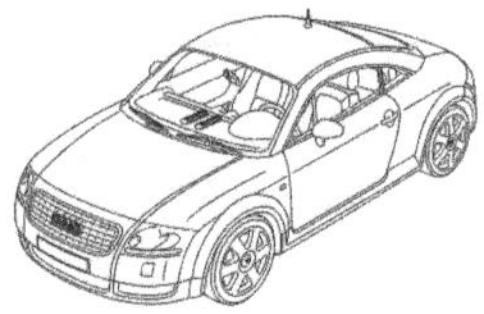

BentleyPublishers.com
© 2006 BentleyPublishers.com—All Rights Reserved

37 Automatic Transmission

TABLE

GENERAL

This repair group covers repair information for external components and shift mechanism of the 09G automatic transmission. This front wheel drive transmission is installed in coupes and roadsters with 180 hp 1.8L 4-cylinder engines. Numerous special tools and equipment are required for removal, installation and internal repair of this transmission. These repairs are outside the scope of this manual.

NOTE—

- *For servicing information on drive axles, see* **40 Front Suspension and Drive Axles**.
- *For transmission identification information, see* **03 Maintenance.**
- *For information on checking and changing automatic transmission fluid, see* **03 Maintenance**.

> ***WARNING—***
> - *Disconnecting negative (–) battery cable has ramifications with many vehicle systems. Before proceeding, be familiar with warnings associated with this procedure in* ***03 Maintenance*** *and* ***69 Seatbelts, Airbags****. These systems may need to be re-set using a VAG 1551/1552, VAS 5051/5052 or equivalent scan tool or computer scan tool program.*

Transmission	Transmission type
6-speed automatic transmission (front wheel drive)	09G

© 2006 BentleyPublishers.com—All Rights Reserved

Table a. 09G automatic transmission specifications (front wheel drive)

Type (Code)	09G (FXA)	09G (HFY)	
Engine application, horsepower	1.8L, 180 hp	1.8L, 180 hp	
Installation period	09/02 - 07/04	08/04 - 06/06	
Gear ratio: Z_2:Z_1 Final drive	61:15 = 4.067	n/a	
1st gear	4.044	n/a	
2nd gear	2.371	n/a	
3rd gear	1.556	n/a	
4th gear	1.159	n/a	
5th gear	0.852	n/a	
6th gear	0.676	n/a	
Reverse gear	3.173	n/a	
Lubricant (gearbox) Capacity	Overall 7.0 liters (7.4 qt.)		
Audi part number	G 055 025 A2		
Torque converter (code)	QBAA		
Driveshaft flange diameter	n/a		
Ratio, overall in top gear	2.749	n/a	

BentleyPublishers.com
© 2006 BentleyPublishers.com—All rights reserved
AT0637005

Special transmission oil G 055 025 A2 as supplied by Audi for use in 6-speed automatic transmissions. This ATF is optimized for use in 09G transmissions.

BentleyPublishers.com
© 2006 BentleyPublishers.com—All Rights Reserved

Internal and external component overview

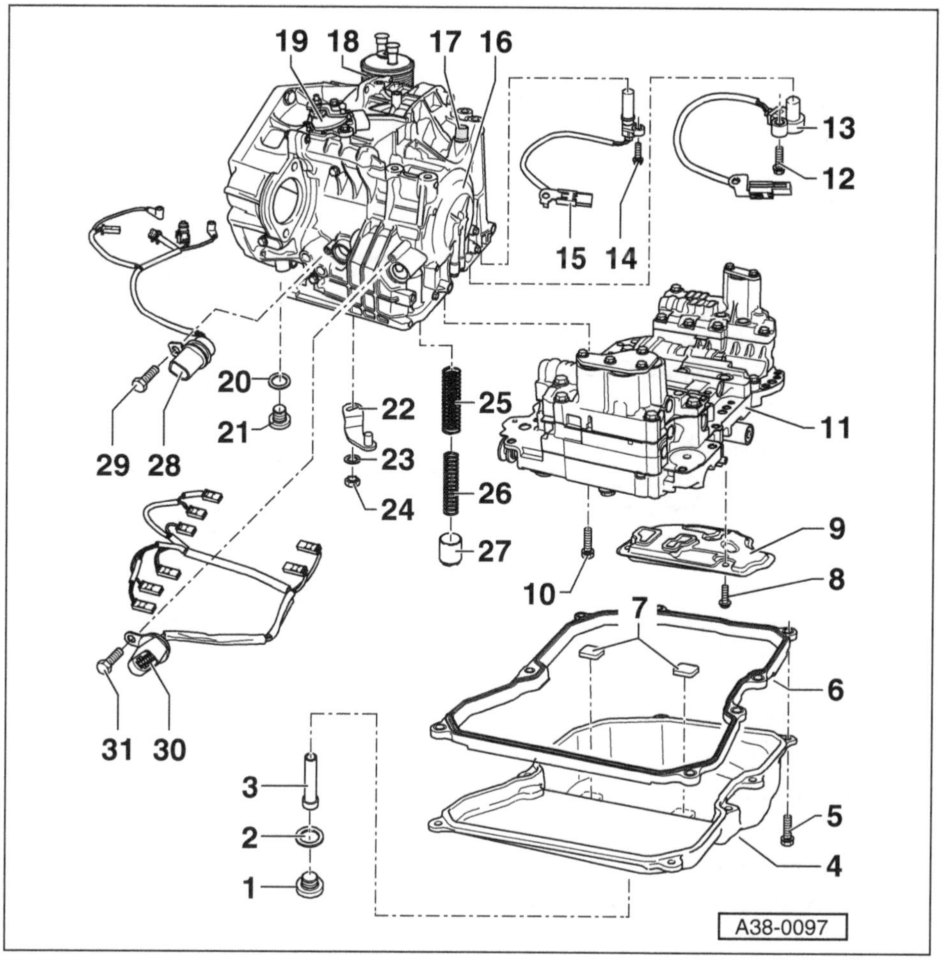

1. **ATF inspection plug**
   - M10
   - Tighten to 27 Nm (20 ft-lb)
2. **ATF inspection plug seal**
   - Always replace
3. **Overflow pipe**
   - Threaded into ATF pan
   - Service with 5 mm hex key
   - Tighten to 1 Nm (9 in-lb)
4. **ATF oil pan**
5. **Bolts, M6**
   - Secure ATF oil pan
   - Qty: 8
   - Tighten to 7 Nm (62 in-lb)
6. **ATF oil pan seal/gasket**
7. **Magnets**
   - Retained by flanges in ATF oil pan
   - Qty: 2
   - Clean before installing
8. **Bolts, M6**
   - Secure ATF screen
   - Qty: 3
   - Tighten to 11 Nm (97 in-lb)
9. **ATF screen**
   - Can be reused if seals are not damaged
   - Coat seals with ATF before installing
10. **Bolts, M6**
    - Secures valve body to transmission housing
    - Qty: 12, different lengths
    - Always replace
    - Tighten to 8 Nm + 90° (71 in-lb + ¼ turn)
    - Remove and install in prescribed sequence, see **Valve body bolts, removing and installing pattern**
11. **Valve body**
12. **Bolt, M6**
    - Tighten to 5.5 Nm (49 in-lb)
13. **Transmission RPM sensor (G182)**
    - See **Transmission RPM sensors**
    - See **Transmission component and connector identification**
14. **Bolt, M5**
    - Tighten to 6.5 Nm (58 in-lb)
15. **Transmission output RPM speed sensor (G195)**
    - See **Transmission RPM sensors**
    - See **Transmission component and connector identification**
16. **Transmission housing**
17. **ATF breather (vent pipe)**
18. **ATF cooler**
19. **Multifunction transmission range switch, TR switch (F125)**
    - Located on transmission housing, see **Transmission range switch**
20. **Sealing ring**
    - Always replace
21. **ATF drain plug**
    - M18
    - Tighten to 39 Nm (29 ft-lb)
22. **Lever**
    - Connects selector lever shaft to valve body
23. **Washer**
24. **Nut**
    - Tighten to 10 Nm (89 in-lb)
25. **Large spring**
26. **Small spring**
27. **Shock absorber piston**
28. **Wiring harness**
    - With integral transmission temperature sensor (G93)
    - With O-ring seal, always replace
    - See **Valve body component and connector identification**
29. **Bolt, M6**
    - Tighten to 5.5 Nm (49 in-lb)
30. **Wiring harness**
    - With O-ring seal, always replace
    - See **Valve body solenoid valve identification,**
31. **Bolt, M6**
    - Tighten to 5.5 Nm (49 in-lb)

Bentley Publishers
.com
© 2006 BentleyPublishers.com—All Rights Reserved

AUTOMATIC TRANSMISSION ELECTRICAL COMPONENTS

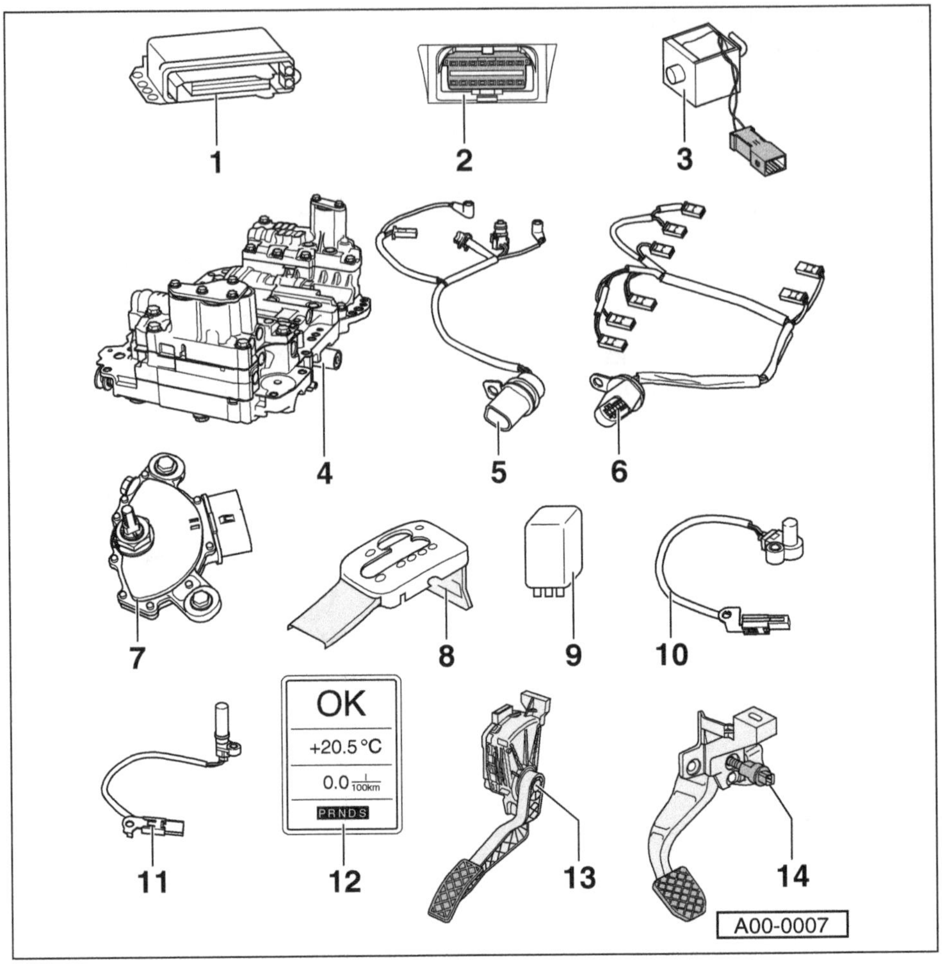

1. **Automatic Transmission Control Module, TCM (J217)**
   - Installed in air plenum at base of windshield near ECM
   - See **Transmission control module, removing and installing**
2. **Data link connector**
   - Located under left side of instrument panel, see **OBD On Board Diagnostics**
3. **Shift lock solenoid (N110)**
   - See **Shift lock solenoid, removing and installing**
4. **Valve body**
   - Installed in transmission housing
   - See **Valve body solenoid valve identification**
5. **Wiring harness**
   - With integral transmission temperature sensor (G93)
   - See **Valve body component and connector identification**
6. **Wiring harness**
   - See **Valve body solenoid valve identification,**
7. **Multifunction transmission range switch, TR switch (F125)**
   - Located on transmission housing, see **Multifunction transmission range switch**
8. **Shift mechanism cover**
   - With Tiptronic switch (F189) in printed circuit board under cover
   - See **Tiptronic switch and printed circuit board**
9. **Starter interlock relay (J207)**
   - Allows starter operation in neutral and park only
   - See **Starter interlock relay (J207), removing and installing**
10. **Transmission RPM sensor (G182)**
    - See **Transmission RPM sensors**
    - See **Transmission component and connector identification**
11. **Transmission output RPM speed sensor (G195)**
    - See **Transmission RPM sensors**
    - See **Transmission component and connector identification**
12. **Transmission range (selector lever) display (Y6)**
    - If display does not illuminate, transmission is in emergency running mode with TCM inactive
    - If display illuminates with all segments on, transmission is in emergency running mode but with TCM active
13. **Kickdown switch (F8)**
    - Integral with accelerator pedal position sensors G79/G185
    - Integral with accelerator pedal
    - Data transfer via CAN-bus
14. **Brake light switch (F)**
    - Located on pedal cluster
    - See **45 Anti-Lock Brakes**

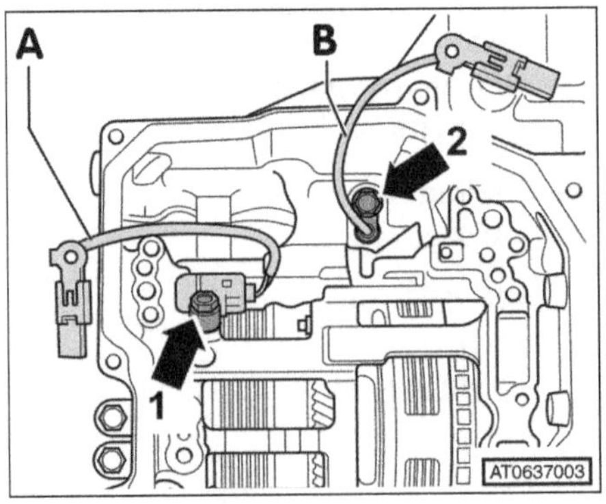

Transmission RPM sensors

Shown with valve body removed.

1. Transmission RPM sensor (G182)
2. Transmission output RPM speed sensor (G195)

A. Harness and connector for G182

B. Harness and connector for G195

BentleyPublishers.com
© 2006 BentleyPublishers.com—All Rights Reserved

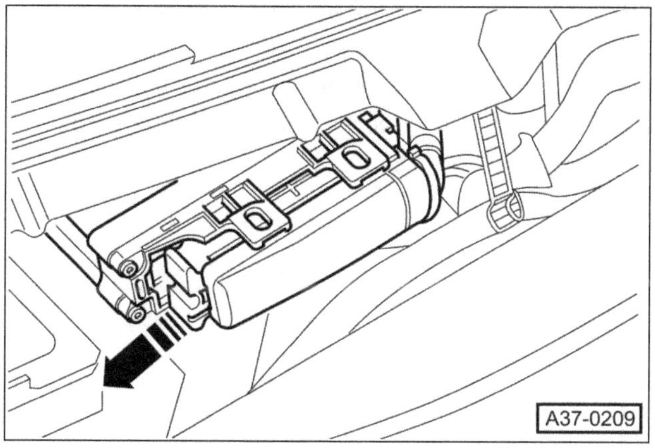

Transmission control module (J217), removing and installing

- Remove wiper arms, see **92 Wipers and Washers**
- Remove rubber hood seal and cover over plenum chamber.
- Disconnect harness connectors to heated washer nozzles.
- Switch ignition on and operate wipers until pivots point to the full right position, switch off ignition as pivots reach that point. This step allows easier removal of TCM.
- Allow at least 30 seconds after switching off ignition, and slide connector off (**arrow**) of TCM.
- Remove 4 securing nuts and turn TCM slightly to remove.
- Install in reverse order of removal.

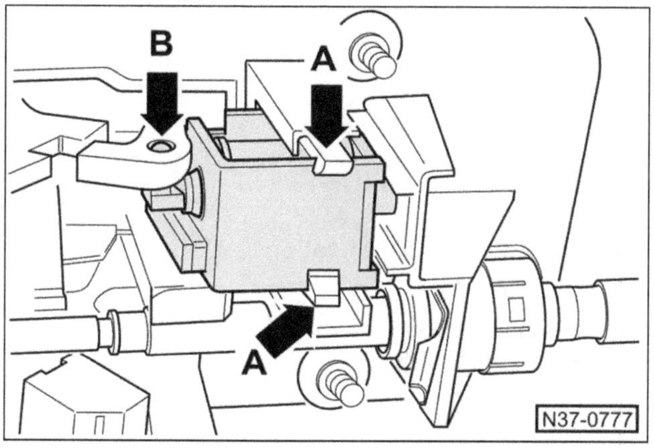

Shift lock solenoid (N110), removing and installing

- Disconnect harness connector from shift lock solenoid.
- Unclip shift lock solenoid from retaining tabs (**A**)
- Disconnect shift lock solenoid from locking pawl (**B**).
- Pull shift lock solenoid out from below.

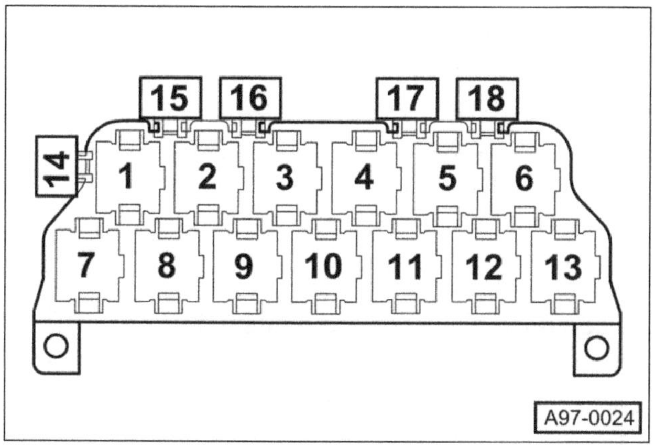

Starter interlock relay (J207), removing and installing

- The 13-fold relay panel is located above the micro central electric panel under the driver's side dashboard.
- Position 11 - Starter interlock relay (J207)

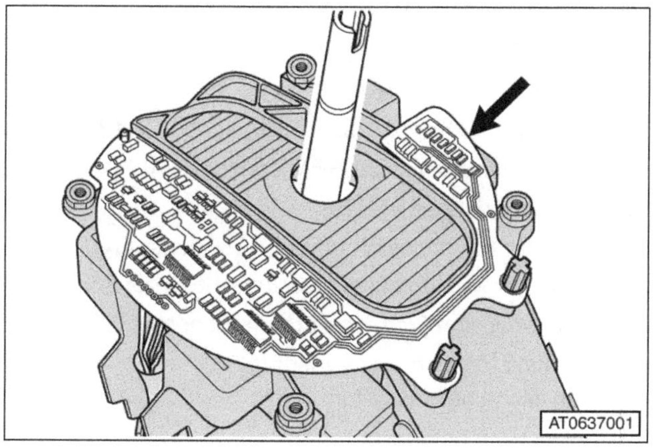

Tiptronic switch (F189) and printed circuit board

- Integral with printed circuit board (**arrow**) under shift cover.

BentleyPublishers
.com
© 2006 BentleyPublishers.com—All Rights Reserved

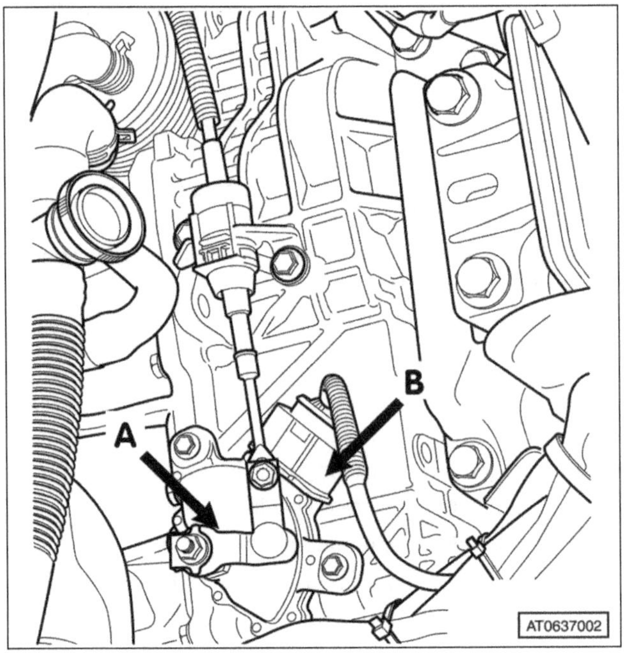

Multifunction transmission range switch, TR switch (F125)

- Switch (**A**) located on top of transmission.
- Harness connector (**B**) plugs directly into switch.
- Remove air filter assembly for access. See **03 Maintenance**.
- Adjust with gauge tool T10173.

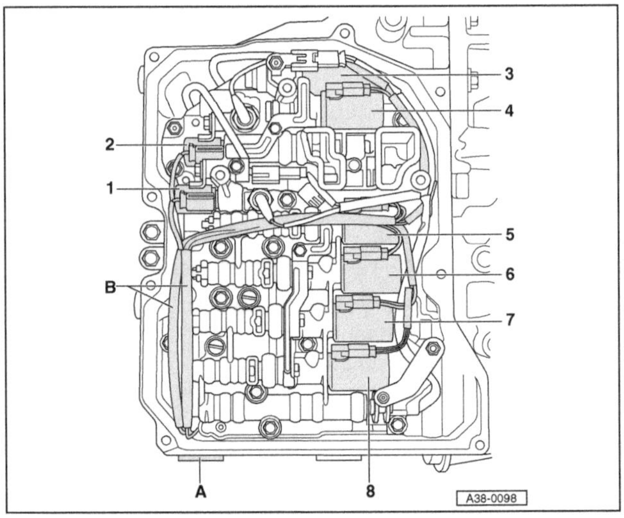

Valve body solenoid valve identification

A. Harness connector, 14-pin
B. Component wiring harness
- Note routing on valve body

1. Solenoid valve 1 (N88)
2. Solenoid valve 2 (N89)
3. Solenoid valve 4 (N91)
4. Solenoid valve 6 (N93)
5. Solenoid valve 5 (N92)
6. Solenoid valve 9 (N282)
7. Solenoid valve 10 (N283)
8. Solenoid valve 3 (N90)

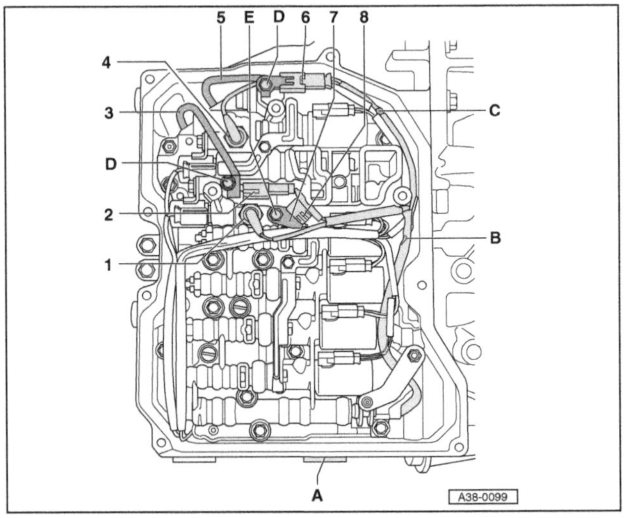

Valve body component and connector identification

A. Harness connector, 14-pin
B. Component wiring harness, note routing on valve body
C. Wiring harness securing bracket
D. Bolt, M5, tighten to 7 Nm (62 in-lb)
E. Bolt, M6, tighten to 11 Nm (97 in-lb)

1. Hydraulic pressure sensor I (G193)
2. Harness connector for transmission RPM sensor (G182)
3. Wiring harness for transmission RPM sensor (G182)
4. Hydraulic pressure sensor II (G194)
5. Wiring harness for transmission output RPM sensor (G195)
6. Harness connector for transmission output RPM sensor (G195)
7. Bracket for transmission fluid temperature sensor (G93)
8. Transmission fluid temperature sensor (G93)
   - Integral with wiring harness (Item **B**).

BentleyPublishers.com
©2006 BentleyPublishers.com—All Rights Reserved

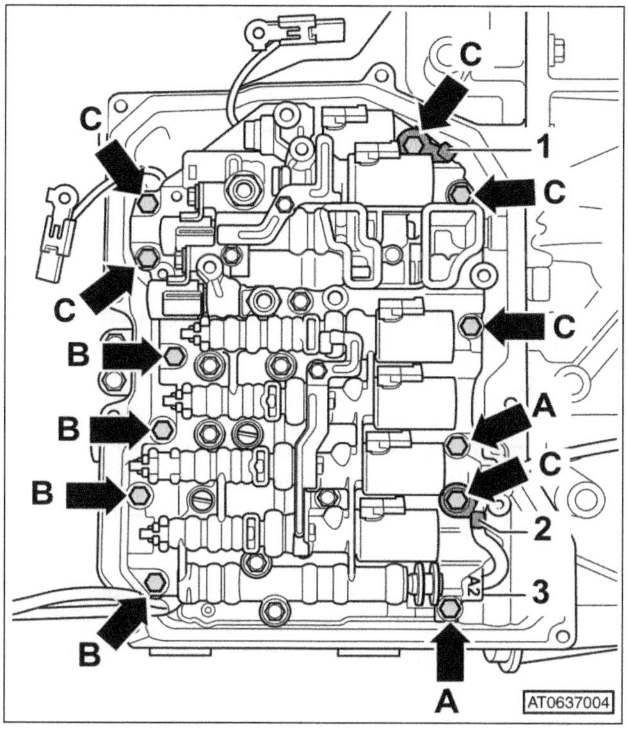

Valve body bolts, removing and installing

A. M6x16 (always replace)
B. M6x28 (always replace)
C. M6x21 (always replace)
1. Wiring harness bracket
2. Wiring harness bracket
3. Valve body designation

– Loosen and remove bolts in a diagonal pattern. Only loosen bolts indicated (A, B, C). Loosening other bolts can affect operation.

– Install new bolts hand tight only.

– When all components are installed, tighten valve body bolts (**A**, **B**, **C**) starting from the outside and working inwards.

Tightening torque	
Bolts M6, all lengths (**A, B, C**), valve body to transmission housing (always replace)	8 Nm (71 in-lb) plus ¼ turn (90°)
Nut, gear lever on selector shaft	10 Nm (7 ft-lb)
Bolts M6, (**1**, **2**) transmission fluid temperature sensor wiring harness brackets	11 Nm (97 in-lb)
Bolts M5, speed sensor connector brackets	7 Nm (62 in-lb)

SHIFT MECHANISM

Selector mechanism, operation

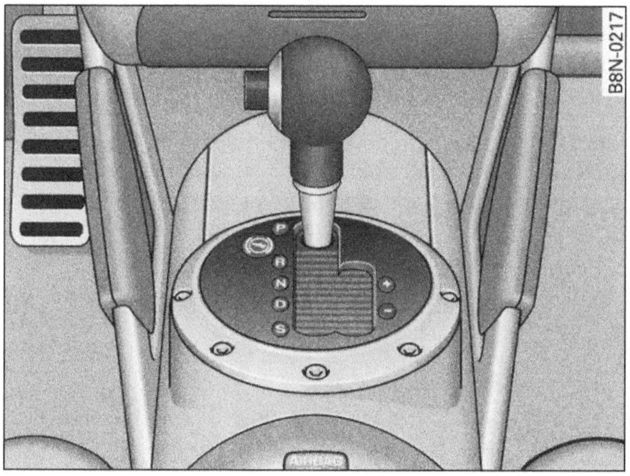

Basic operational characteristics of the 09G 6-speed automatic transmission shifter include the following:

- The starter should not operate if the selector lever is in the **S**, **D** or **R** positions, or in the Tiptronic gate (**+**/**-**). The starter should operate in **P** or **N** only.
- At speeds over 4 mph (5 kph) and with the selector lever in position **N**, the shift lock solenoid should not engage and lock the selector lever. At this low speed, selector lever can be shifted into a driving gear without depressing the brake pedal.
- When travelling at very slow speeds where the vehicle is almost stopped (approximately 4 mph [5 kph]), and shifting into selector lever position **N**, shift lock solenoid engagement should be delayed until after approximately 1 second. It should not be possible to shift the selector lever out of **N** position until brake pedal is depressed.

Selector lever in P position and ignition switched on:

Brake pedal not depressed:

- The selector lever is locked and cannot be shifted from the **P** position. The shift lock solenoid locks the selector lever.

Brake pedal depressed:

- Shift lock solenoid releases selector lever. It is possible to shift into a driving gear. Slowly shift the selector lever from **P** through **R**, **N**, **D**, and **S**, checking whether the display on the instrument cluster corresponds to the selector lever position.

BentleyPublishers
.com
©2006 BentleyPublishers.com—All Rights Reserved

Selector lever in N position and ignition switched on:

Brake pedal not depressed:

- The selector lever is locked and cannot be shifted from the **N** position. The shift lock solenoid locks the selector lever.

Brake pedal depressed:

- Shift lock solenoid releases selector lever. It is possible to shift into a driving gear.

Selector lever in D position, ignition switched on, and lights switched on:

Selector lever shifted into the Tiptronic gate:

The **D** symbol in the selector lever display (**left arrow**) in the instrument cluster should no longer be illuminated and the **+** and **-** symbols in the upper cover of the shifter housing should be illuminated.

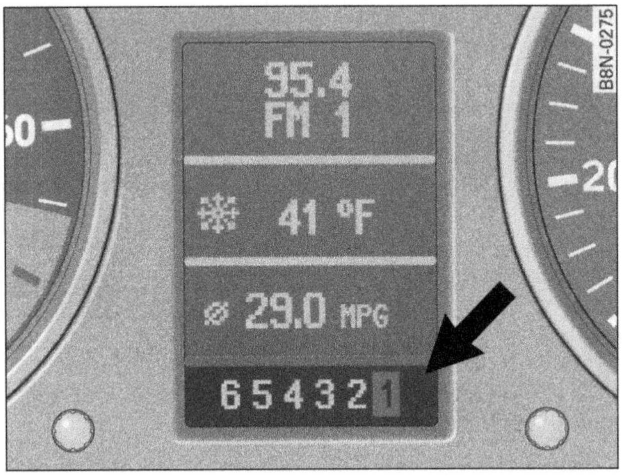

The selector lever position display in the instrument cluster must switch from **PRDNS** to **654321** (**right arrow**) when the selector lever is shifted into the Tiptronic gate and the gear selected is highlighted.

- While driving, momentary movement of the shifter towards **+** or **-** symbols in the upper cover of the shifter housing should result in a gear change. Selector lever display in instrument cluster should reflect actual gear engaged.

While driving, momentary contact of **+** or **-** buttons on either side of he steering wheel crossbar should result in a gear change. Selector lever display should reflect actual gear engaged.

Neutral start and backup light function

Operation of the backup lights and the ability to allow starter operation in park and neutral positions only, has typically been a direct function of the selector mechanism. 09G automatic transmissions use the multi-function transmission range switch (F125) to operate a relay (J207) to enable the neutral start function. See **Starter interlock relay (J207), removing and installing** given earlier.

Backup lights are controlled directly by the multi-function transmission range switch (F125).

BentleyPublishers
.com
© 2006 BentleyPublishers.com—All Rights Reserved

Selector cable, checking and adjusting

Checking

– Remove air cleaner assembly, see **03 Maintenance**.

– Move selector lever to position P.

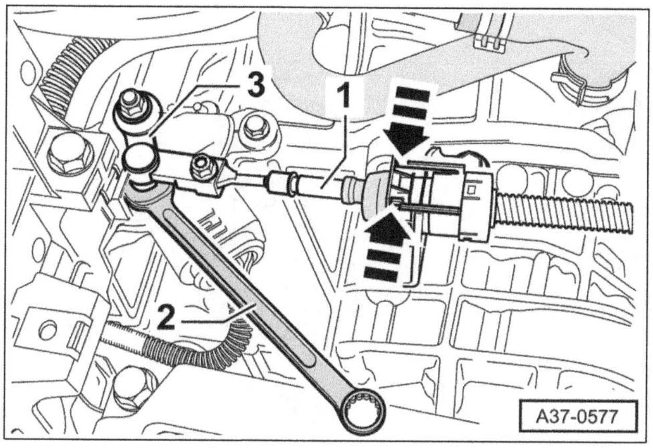

◄ Pry off selector cable (**1**) from lever/selector shaft (**3**) using a 10 mm open end wrench (**2**) from underneath.

– Press catches on support bracket of selector lever cable together (**dashed arrows**) and pull selector cable out of support bracket on transmission.

– Place selector cable to one side so that end can move freely.

NOTE —
• Do not bend or kink selector lever cable.

– Move selector lever from **P** to **S**.

– Check protective sleeve on front of selector cable for damage and replace cable if necessary.

– Move selector lever to position P.

– Shift mechanism and selector cable must move freely. If necessary replace selector cable and/or service shift mechanism.

– Carefully press selector cable onto lever/selector shaft and insert in support bracket on transmission.

Adjusting

– Move selector lever to position P.

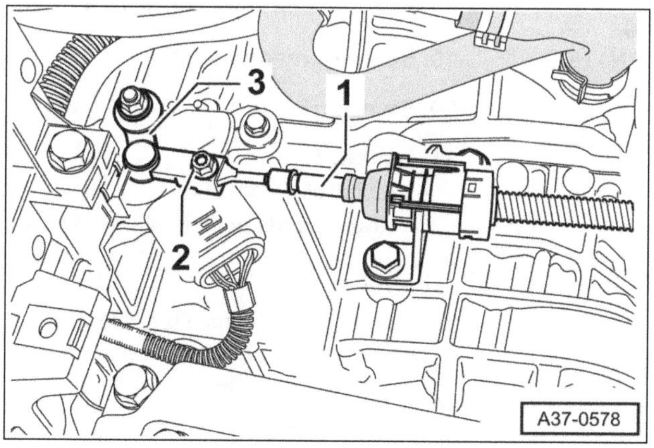

◄ Loosen securing/adjusting bolt (**2**) at front ball socket of selector lever cable (**1**).

– Set lever/selector shaft (**3**) on transmission to P. Lever must engage detent in parking lock gear indicated by both front wheels being locked (cannot be turned together in one direction).

– Tighten securing/adjusting bolt (**2**).

Tightening torque	
Bolt, selector cable adjustment	13 Nm (10 ft-lb)

– Check shift mechanism.

– Install air cleaner assembly.

BentleyPublishers
.com
© 2006 BentleyPublishers.com—All Rights Reserved

Selector lever and housing assembly

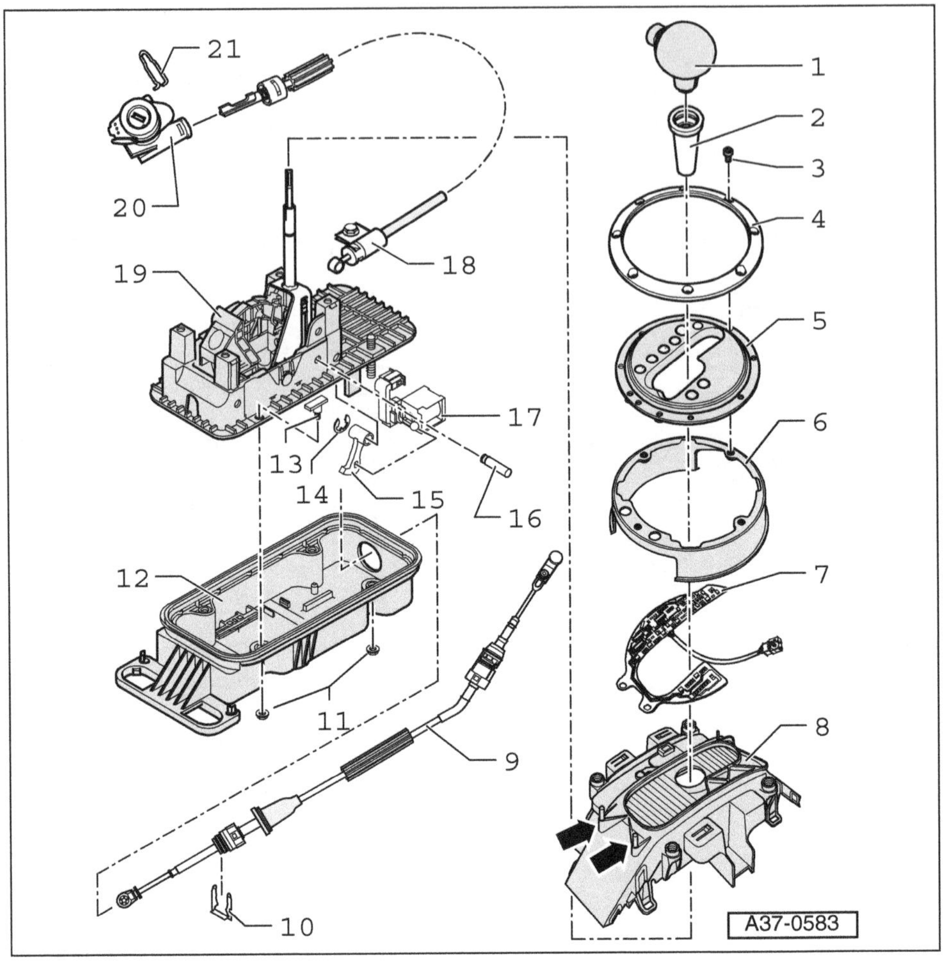

1. **Gear shift knob**
   - Removal same as DSG transmission. See **34b DSG Transmission**
2. **Trim/locking sleeve**
   - Clips into gear shift knob
   - Locks knob to prevent knob from pulling off
3. **Bolt**
   - Threaded through retaining ring and upper cover into side cover
   - Qty: 8, M4x10
   - Tighten to 4 Nm (35 in-lb)
4. **Retaining ring**
5. **Upper cover**
   - With integral symbol insert
   - Positioned on frame (item 8) (**arrows**)
6. **Side cover**
   - Positioned on frame (item 8) (**arrows**)
7. **Printed circuit board**
   - Positioned on frame (item 8) (**arrows**)
   - See **Tiptronic switch (F189) and printed circuit board** given earlier
8. **Frame**
   - With locating pins (**arrows**)
   - See **Frame assembly**
9. **Selector cable**
   - Do not bend or kink
   - Do not lubricate cable eye or ball socket
   - See **Selector cable, checking and adjusting**
10. **Locking clip**
    - Always replace
11. **Nut, self locking**
    - Qty: 4
    - Always replace
    - Tighten to 10 Nm (7 ft-lb)
12. **Housing**
    - With glued-on gasket
13. **Bolt**
    - Inserted into intermediate plate
14. **Locking clip**
15. **Locking pawl**
    - For selector lever lock
16. **Pivot pin**
    - For locking pawl
17. **Shift lock solenoid (N110)**
    - See **Shift lock solenoid (N110), removing and installing**
18. **Locking cable**
    - Do not bend or kink
    - For ignition key removal lock
19. **Selector mechanism**
20. **Steering lock**
21. **Locking clip**
    - Install clip from above
    - Angled ends point to steering lock

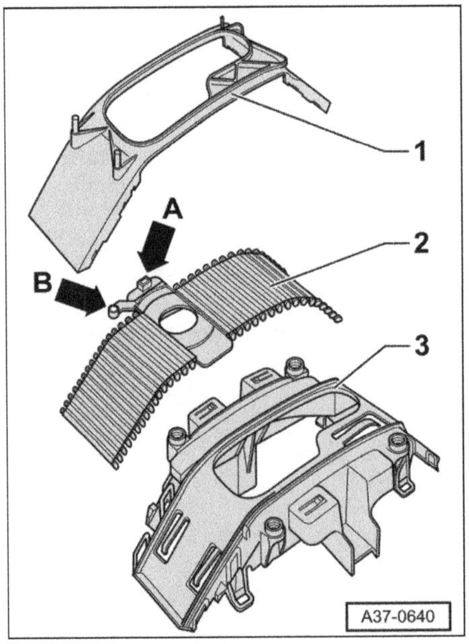

Frame assembly

Sliding screen is installed into lower section and upper and lower sections are clipped together.

1. Upper section
2. Sliding screen
3. Lower section

A. Solenoid, Tiptronic switch

B. Solenoid, Tiptronic switch

Ensure that solenoids **A** and **B** are securely attached to sliding screen. Missing solenoids will render Tiptronic switch inoperative.

BentleyPublishers.com
© 2006 BentleyPublishers.com—All Rights Reserved

40 Front Suspension

General

This section covers repairs to the front suspension and related components. For rear suspension servicing and repair information, see **42 Rear Suspension**. For wheel alignment specifications, see **44 Wheels, Tires, Wheel Alignment**.

Special tools, equipment and procedures are required for most front suspension repair and component replacement. On suspension components with bonded rubber bushings, fasteners must be tightened with vehicle at curb weight position. In addition, front wheel alignment is almost always disturbed when suspension components are removed or replaced. Plan on a wheel alignment after most suspension repairs.

Warnings

WARNING —
- *Do not reuse any fasteners that are worn or deformed in normal use. Most fasteners are designed to be used only once and become unreliable and may fail when used a second time. This includes, but is not limited to, nuts, bolts, washers, circlips, cotter pins, self-locking nuts and bolts. For replacements, always use new parts.*
- *Do not reinstall bolts and nuts coated with undercoating wax, as correct tightening torque cannot be ensured. Always clean the threads of bolts and nuts with a suitable solvent before installing.*
- *Always replace rusted or corroded bolts, nuts and washers even if not specifically indicated.*
- *Do not attempt to straighten or weld suspension struts, wheel bearing housings, control arms or any other wheel locating or load bearing components of the front suspension.*

BentleyPublishers.com

© 2006 BentleyPublishers.com—All Rights Reserved

Front Suspension

NOTE—

- *Vehicles without drive axle installed must not be moved or wheel bearing will be damaged.*
- *If vehicle does have to be moved, install an outer joint in place of drive axle and tighten to specification for drive axle.*
- *Bonded rubber bushings can only be turned to a limited extent. The bolted connections on suspension links should only be tightened when vehicle is standing on the ground.*

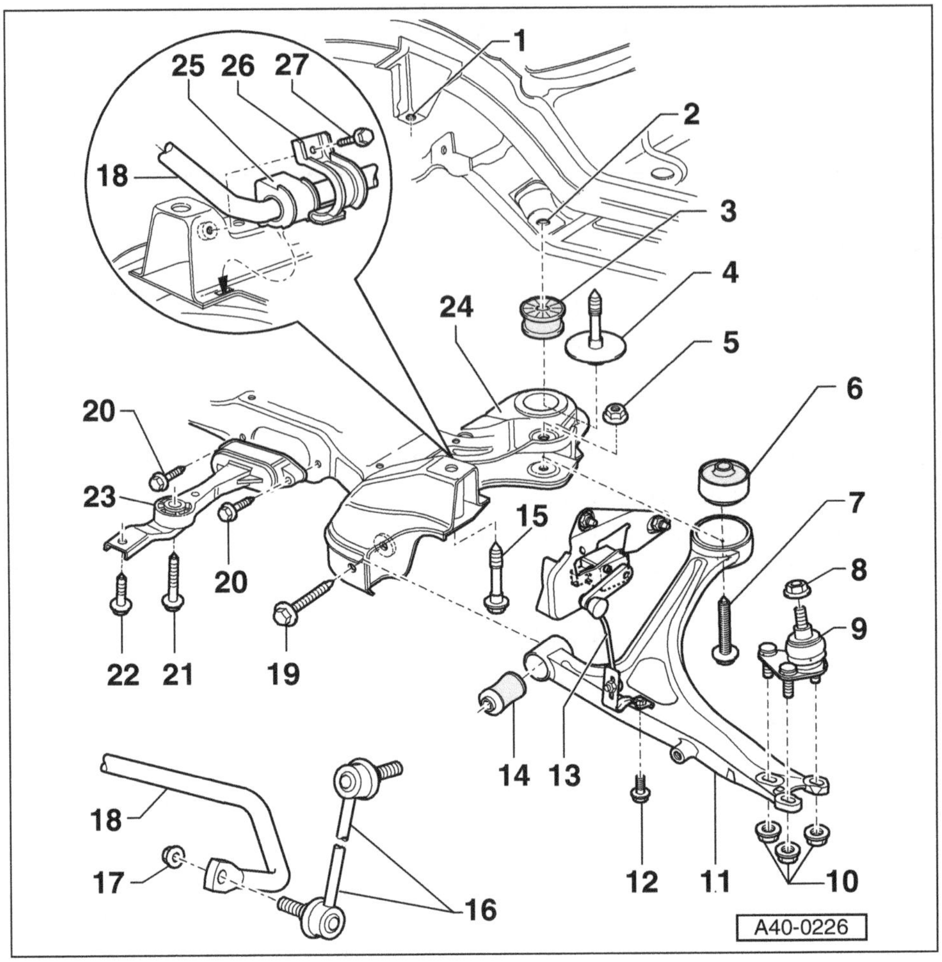

Front suspension components

1. **Subframe mounting point**
2. **Captive nut in body**
   - Can be repaired with Helicoil insert
3. **Bushing**
4. **Bolt**
   - Always replace
   - 100 Nm (74 ft-lb) + 90° (¼ turn)
5. **Self-locking nut**
   - Always replace
6. **Rear bushing for control arm**
7. **Bolt**
   - 70 Nm (52 ft-lb) + 90° (¼ turn)
   - Always replace
8. **Self-locking nut**
   - Tightening torque
     3.2L roadster: 150 Nm (111 ft-lb)
     all other models: 75 Nm (55 ft-lb)
   - Always replace
9. **Ball joint**
   - Mark installation position; if control arm is replaced, set to center of elongated hole and check toe setting and camber
10. **Nut**
    - 75 Nm (55 ft-lb)
    - Always replace
11. **Control arm**
    - Different versions available, the same version must be installed on both sides
12. **Bolt**
    - 25 Nm (16 ft-lb)
13. **Linkage for headlight range control (xenon headlights)**
    - Tighten nuts to 4 Nm (35 in-lb)
14. **Front bushing for control arm**
15. **Bolt**
    - 100 Nm (74 ft-lb) + 90° (¼ turn)
    - Always replace
    - The captive nut cannot be reworked. If the nut is damaged, the complete mounting bracket must be replaced.
16. **Coupling link**
    - Link between stabilizer bar and suspension strut
17. **Nut**
    - 90 Nm (66 ft-lb)
    - Always replace
18. **Stabilizer bar**
19. **Bolt**
    - 70 Nm (52 ft-lb) + 90° (¼ turn)
    - Always replace
20. **Bolts**
    - 25 Nm (16 ft-lb)
21. **Bolt**
    - 50 Nm (37 ft-lb)
22. **Bolt**
    - 50 Nm (37 ft-lb)
23. **Pendulum mount**
24. **Subframe**
25. **Rubber bushing**
26. **Clamp**
27. **Bolt**
    - 25 Nm (16 ft-lb)

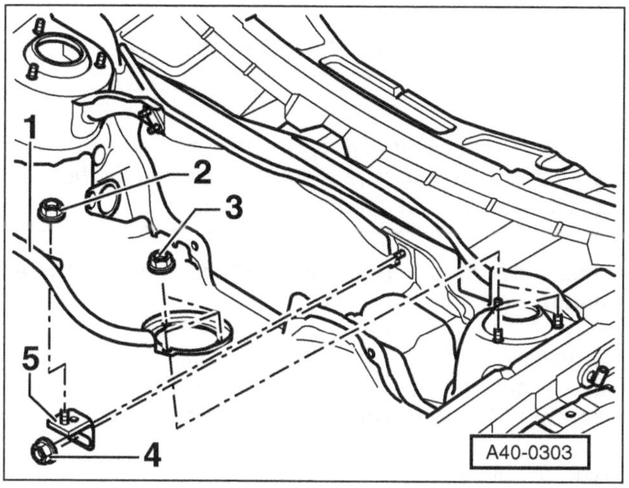

Strut tower bar, installation details

Strut tower bar (**1**) is mounted in engine compartment between tops of suspension strut towers.

1. Strut tower bar
2. Nut
   - 20 Nm (15 ft-lb)
3. Nut
   - 20 Nm (15 ft-lb)
4. Nut
   - 20 Nm (15 ft-lb)
   - Nut can be loosened to adjust bar
5. Mounting bracket

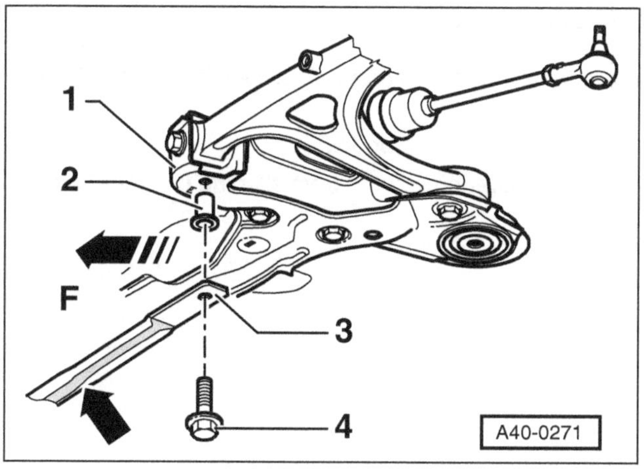

Subframe cross support, installation details

Note orientation of components when removing and installing cross support for subframe.

- **F** = direction of travel

1. Subframe
2. Pop rivet nut (M10)
3. Cross support
   - Install with marking "L" or "links" on left (driver's) side
   - Install with foam strip (**arrow**) on bottom
4. Bolt
   - 30 Nm (22 ft-lb) + 90° (¼ turn)
   - Always replace
   - Use locking fluid when installing

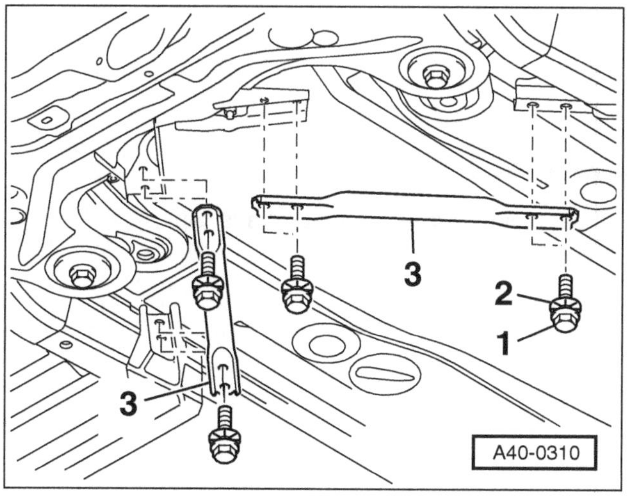

Subframe diagonal supports, installation details

Note orientation of components when removing and installing diagonal supports for subframe.

1. Bolts
   - 35 Nm (26 ft-lb)
2. Lock washer
3. Diagonal supports

BentleyPublishers.com
©2006 BentleyPublishers.com—All Rights Reserved

Stabilizer bar, removing and installing

– Raise car and support safely. See **03 Maintenance.**

> ***WARNING —***
> - *Make sure car is stable and well supported at all times. Use a professional automotive lift or jack stands designed for the purpose. A floor jack is not adequate support.*

– Remove center, left and right side lower sound absorber panels. See **03 Maintenance**.

– If applicable (vehicles with Xenon headlights), detach linkage for vehicle level sensor from left side control arm. See **Vehicle level sensor assembly (Xenon headlights)** in this repair group.

– Unbolt (**arrow**) coupling link from stabilizer bar.

NOTE —
- *Rear stabilizer bar shown. Front is similar.*

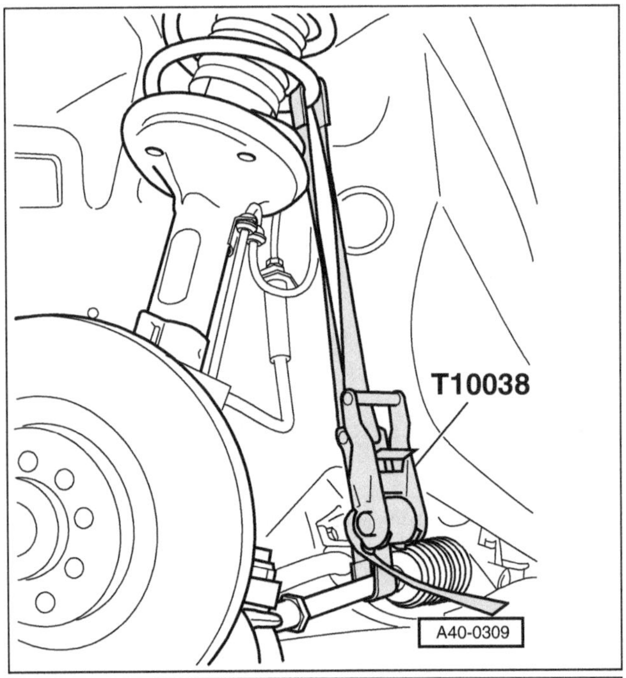

– Secure tie rod to suspension strut using tensioning straps (**T10038** shown).
- Install strap next inner tie rod boots.
- This prevents the steering column from slipping off steering gear splines.

Vehicles with 3.2L engine

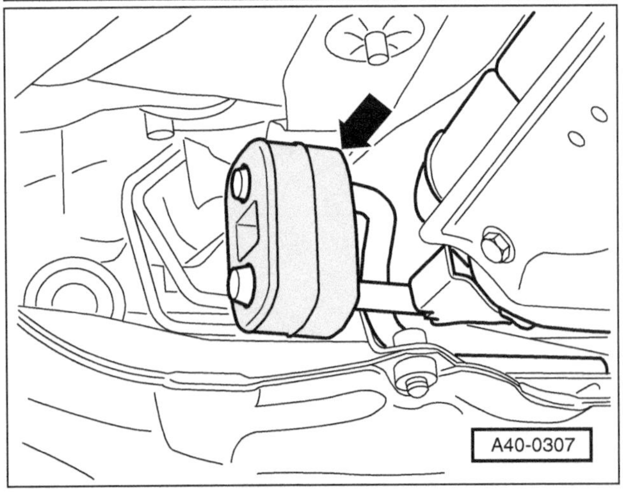

– Remove rubber mount (**arrow**) at front exhaust pipe.
- Note installation position

BentleyPublishers
.com
© 2006 BentleyPublishers.com—All Rights Reserved

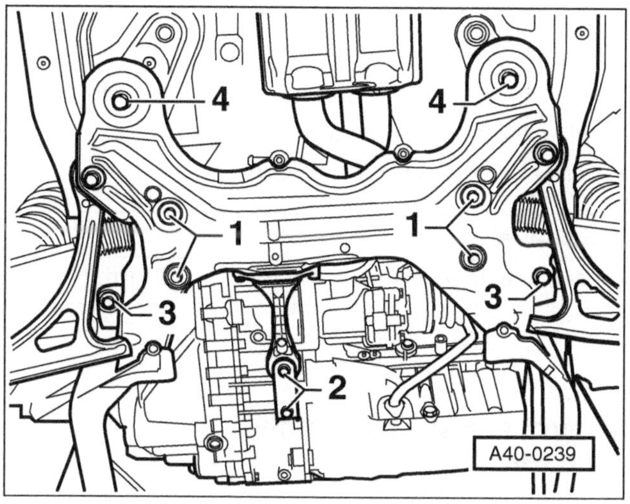

Continued for all vehicles

– Remove cross support and diagonal support bars from subframe. See **Subframe cross support, installation details** and **Subframe diagonal supports, installation details** in this repair group.

< Remove mounting bolts (**1**) for steering gear.

– Position engine/transmission jack (VAG 1383A or equivalent) with wood block under subframe.

– Remove mounting bolts (**2**) for pendulum mount to engine.

– Removing mounting bolts (**3** and **4**) for subframe to body.

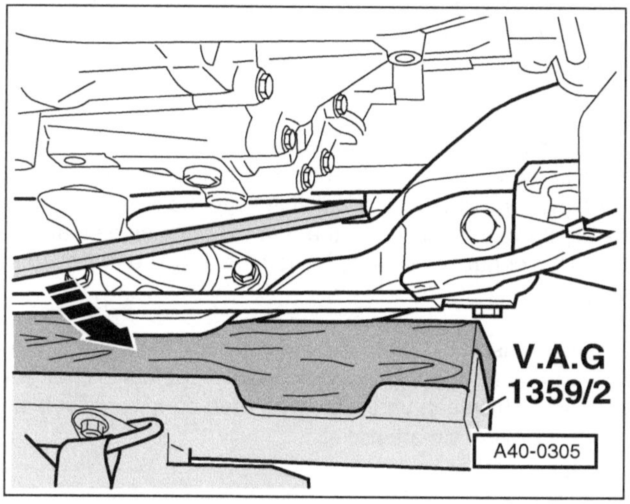

< Carefully pry subframe down from body using a large lever in direction of **arrow**.

– Carefully lower subframe using engine/transmission jack.

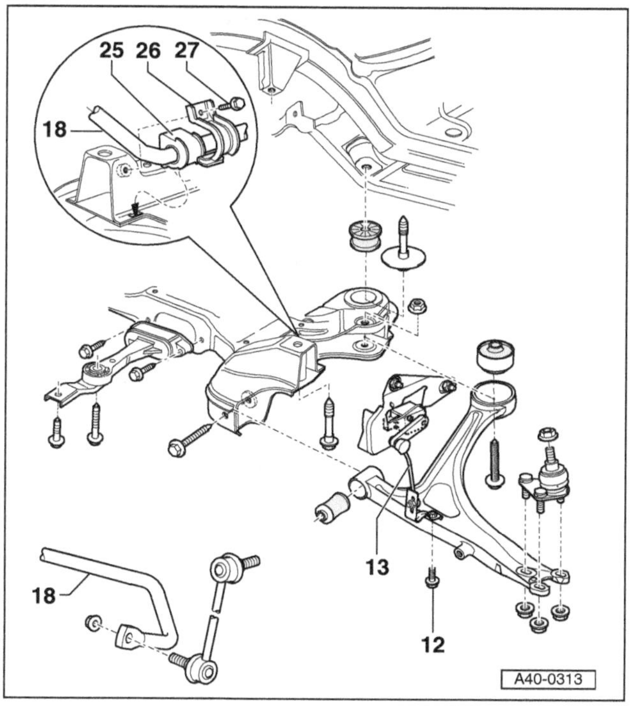

< Remove mounting bolt (**27**) on left and right sides and stabilizer bar clamp (**26**).

– Remove stabilizer bar to rear.

BentleyPublishers
.com
© 2006 BentleyPublishers.com—All Rights Reserved

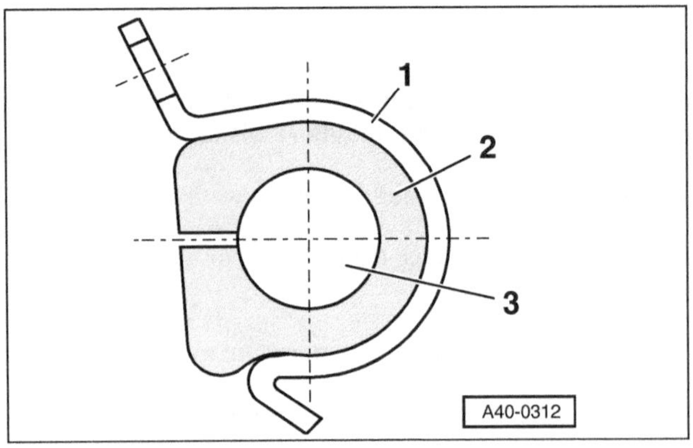

When installing stabilizer bar (**3**), position bushing (**2**) and clamp (**1**) as shown.

– Remaining installation is reverse of removal. Remember to:
  - Mount bushing and stabilizer bar without grease.
  - Match stabilizer bar to suspension version.

Tightening torque	
Stabilizer bar clamp to body (replace nuts)	25 Nm (18 ft-lb)

Vehicle level sensor assembly (Xenon headlights)

Vehicles with gas discharge (Xenon) headlights have automatic headlight range control as standard equipment. This system uses information regarding compression and rebound travel at the front and rear suspension in order to function correctly.

The position of the lower left front control arm, in relation to the body, is transferred to the left front level control system sensor (G78) via a coupling rod. This transmits electrical signals to the control module for the headlight range control. On the rear axle, the signals are transmitted to the control module by the left rear level control system sensor (G76).

These signals are used to determine the current ride height of the vehicle. The automatic headlight range control responds to changes in the ride height which can vary based on road conditions, vehicle load or different driving situations.

The headlight basic setting must be checked using a diagnostic scan tool. It should be checked whenever:
- The suspension control arm has been removed or replaced.
- The vehicle level sensor has been replaced.
- The coupling rod has been removed.

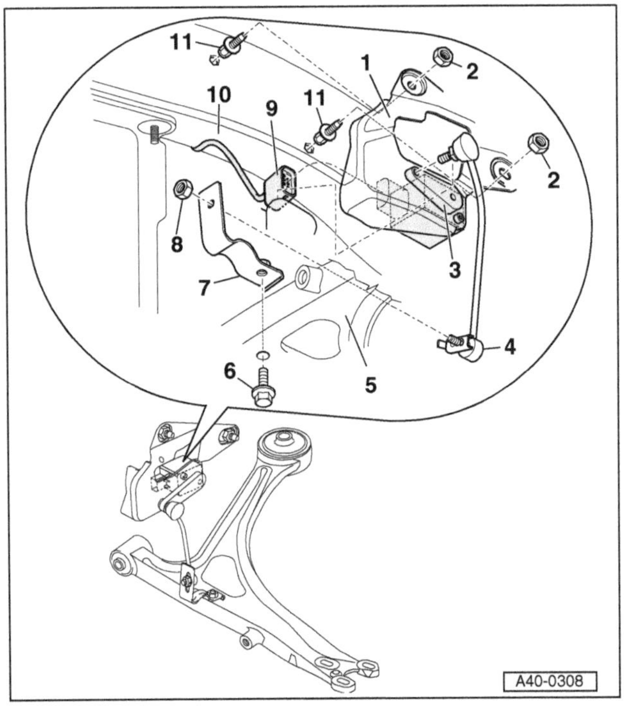

1. Bracket
2. Nut
   - 8 Nm (71 in-lb)
3. Left front headlight control system sensor (G78)
4. Coupling rod
5. Control arm
6. Bolt
   - 25 Nm (21 ft-lb)
7. Coupling bracket
8. Nut
   - 4 Nm (35 in-lb)
9. Connector
10. Longitudinal member (body)
11. Pop-rivet bolt
    - Can be installed with pop rivet tool (VAG 1765B)

BentleyPublishers
.com
© 2006 BentleyPublishers.com—All Rights Reserved

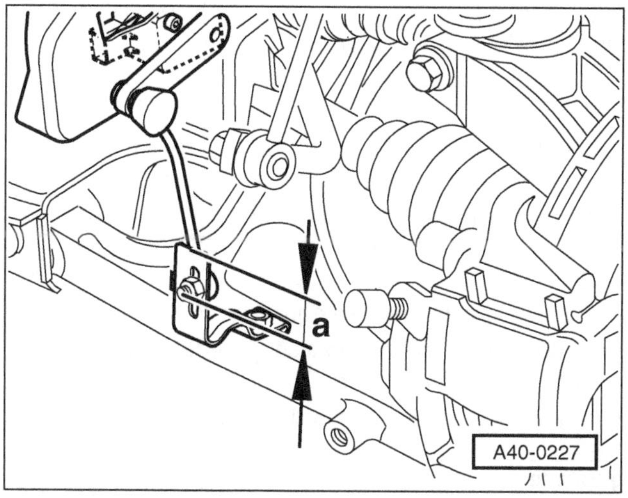

Xenon headlight range control linkage, adjusting

◀ Adjustment for headlight range control coupling bracket at control arm.

- Dimension **a** = 19 mm (¾ in.)

Ball joint, removing and installing

– Raise car and support safely. See **03 Maintenance**.

> ***WARNING —***
> - *Make sure car is stable and well supported at all times. Use a professional automotive lift or jack stands designed for the purpose. A floor jack is not adequate support.*

– Remove front wheel(s).

◀ Mark positions of ball joint lower mounting nuts (**lower arrows**). Remove nuts.

– Loosen upper nut (**upper arrow**) from ball joint. Leave nut on end of threaded shaft for now.

3.2L Roadster only

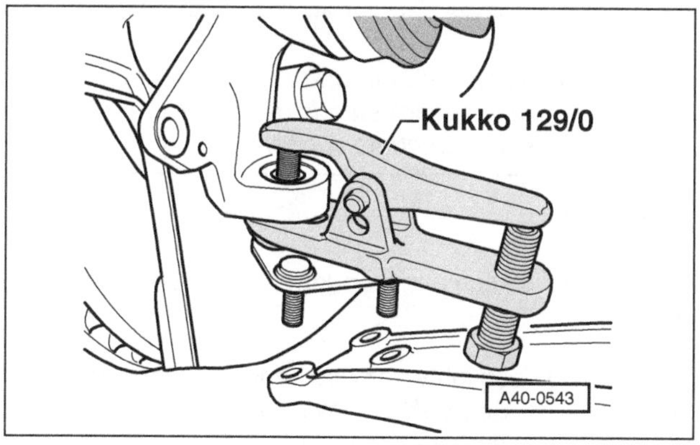

– Detach control arm from ball joint.

◀ Use ball joint puller Kukko 129/0 as shown in the illustration and press out ball joint.

- Leave nut on top of ball joint a few turns to protect threads.

NOTE —
- *Use the ball joint puller specified here and make sure it is seated correctly, otherwise the sleeve in the wheel bearing housing will be pulled out together with the ball joint.*

– Remove upper mounting nut and ball joint.

All other vehicles

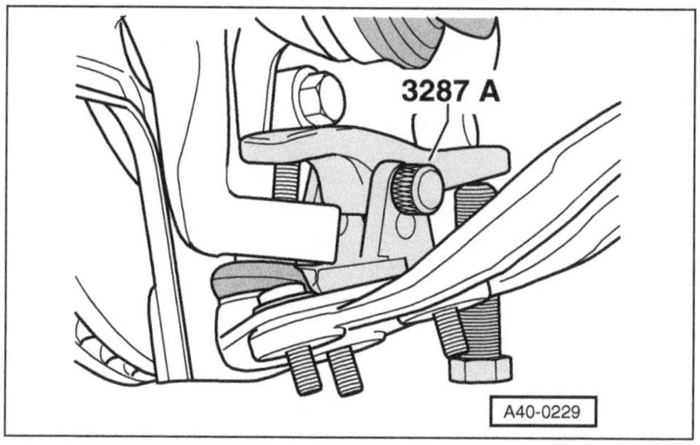

◀ Attach ball joint puller (3287A or equivalent) as shown in the illustration and press out ball joint.

- Leave nut on top of ball joint a few turns to protect threads.

– Remove upper mounting nut and ball joint.

BentleyPublishers
.com
©2006 BentleyPublishers.com—All Rights Reserved

– Begin installation by mounting ball joint to control arm. Position lower mounting nuts at previously marked positions.

Tightening torque	
Ball joint to control arm (always replace nuts)	75 Nm (55 ft-lb)

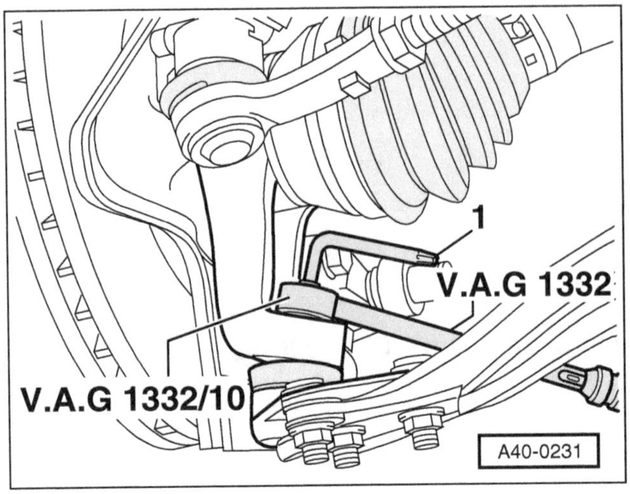

Use new self-locking nut to install ball joint to wheel bearing housing. Counter hold ball joint with T40 Torx wrench (**1**) while tightening upper mounting nut.

- VAG 1332 = torque wrench, may use equivalent
- VAG1332/10 = box wrench head, can also use equivalent 18 mm box wrench head (i.e. Stahlwille 732/10 or Hazet 6630c-18).
- Make sure ball joint boot is not damaged or twisted.

Tightening torque	
Ball joint to wheel bearing housing 3.2L roadster All other vehicles	 150 Nm (111 ft-lb) 75 Nm (55 ft-lb)

– Align vehicle. See **44 Wheels-Tires, Wheel alignment** for wheel alignment specifications.

Control arm, removing and installing

NOTE —
- *There are different versions of the control arm. The same version must always be installed on both sides.*

– Raise car and support safely. See **03 Maintenance**.

> ***WARNING —***
> - *Make sure car is stable and well supported at all times. Use a professional automotive lift or jack stands designed for the purpose. A floor jack is not adequate support.*

– Remove front wheel(s).

– Remove lower sound insulation panel under engine. See **03 Maintenance**.

– For vehicles with Xenon headlights, remove headlight range control linkage at left control arm. See **Vehicle level sensor assembly (Xenon headlights)** given earlier.

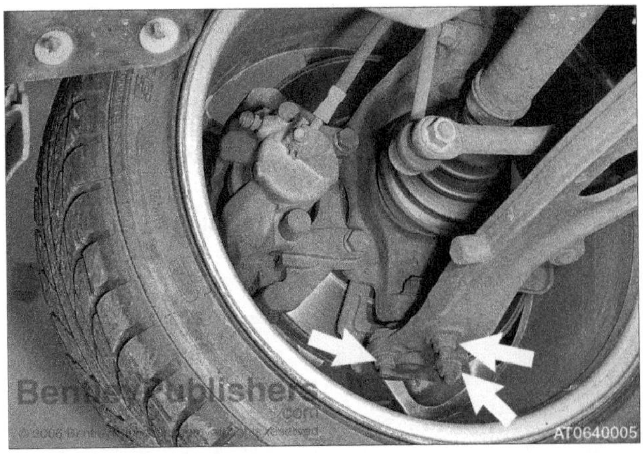

Mark positions of ball joint lower mounting nuts (**arrows**) and remove nuts.

BentleyPublishers.com
© 2006 BentleyPublishers.com—All Rights Reserved

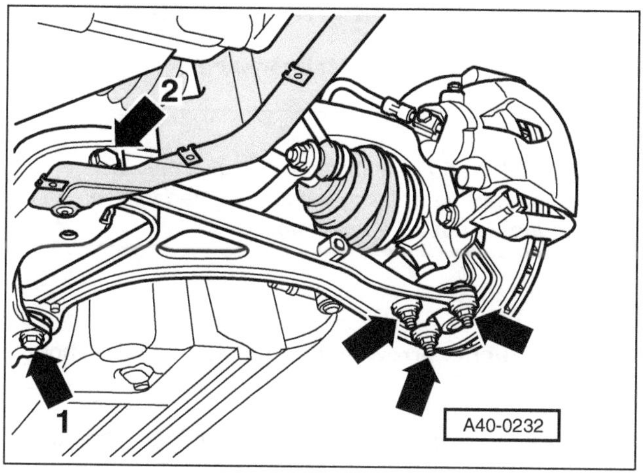

◀ Remove control arm mounting bolts (**1** and **2**) and control arm.

- Bolt (**1**) is secured from above with a nut.

– When installing, note the following points:

- Install control arm into subframe and secure.
- Position lower ball joint mounting nuts at previously marked positions.
- Bonded rubber bushings can only be turned to a limited extent. The bolted connections on suspension links should only be tightened when vehicle is standing on the ground.
- See **Front suspension components** for tightening torques.

– Align vehicle. See **44 Wheels-Tires, Wheel alignment** for wheel alignment specifications.

Suspension strut assembly

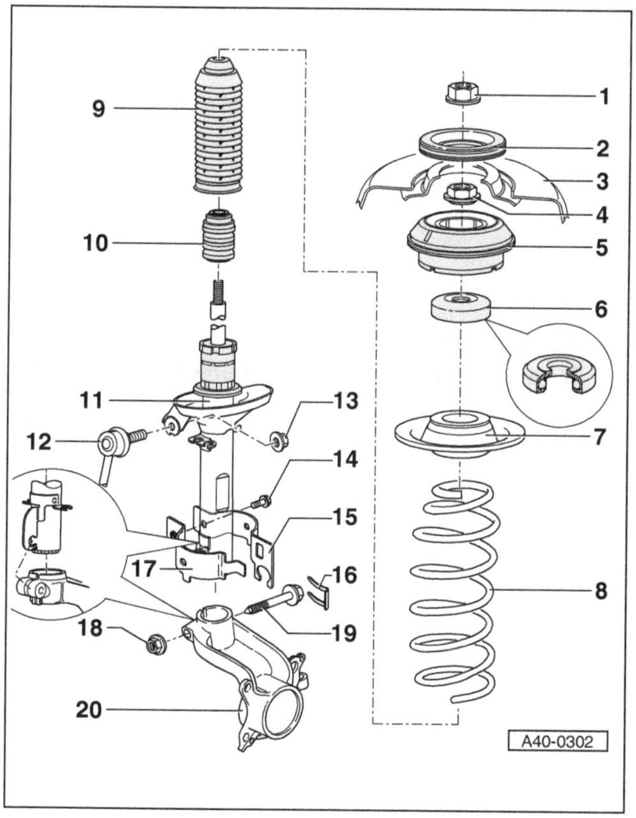

1. Nut, self-locking
   - 60 Nm (44 ft-lb)
2. Stop shell
3. Body (suspension strut upper mounting point)
4. Nut
   - 60 Nm (44 ft-lb)
5. Strut bearing
   - Coat with assembly lubricant (G 294 421 A1)
6. Deep-groove thrust ball bearing
7. Spring plate
8. Coil spring
   - Different versions
   - Vehicle data sticker indicates suspension type, see **44 Wheels-Tires, Wheel Alignment**
9. Protective sleeve
10. Stop buffer
11. Shock absorber
    - Different versions
    - Vehicle data sticker indicates suspension type, see **44 Wheels-Tires, Wheel Alignment**
12. Coupling link
    - Between stabilizer bar and suspension strut
13. Nut, self-locking
    - 90 Nm (86 ft-lb)
14. Bolt
    - 10 Nm (7 ft-lb)
15. Bracket
    - For brake hose
16. Spring clip
17. Clamp
18. Nut, self-locking
    - 60 Nm (44 ft-lb) + 90° (¼ turn)
    - Always replace
19. Bolt
    - Always replace
20. Wheel bearing housing

BentleyPublishers.com
©2006 BentleyPublishers.com—All Rights Reserved

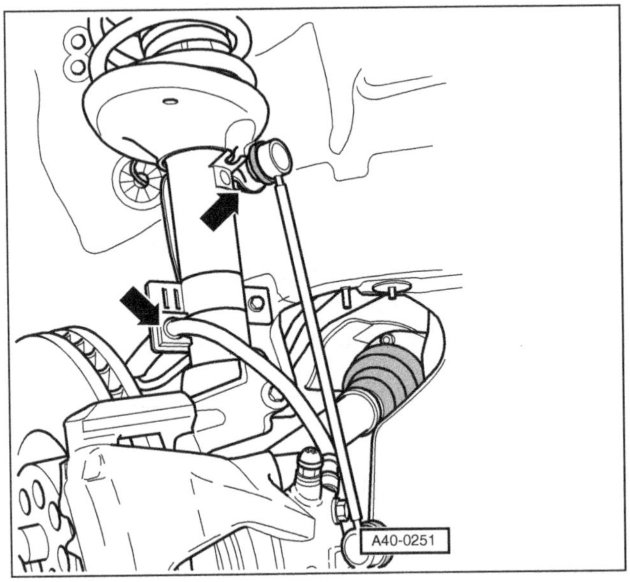

Suspension strut, removing and installing

– Raise car and support safely. See **03 Maintenance**.

> ***WARNING —***
> - *Make sure car is stable and well supported at all times. Use a professional automotive lift or jack stands designed for the purpose. A floor jack is not adequate support.*

– Remove front wheels.

◀ Remove coupling rod (**upper arrow**) from suspension strut on both sides of vehicle.

– Pull spring clip (**lower arrow**) out from bracket and detach brake hose.

– Detach wire for speed sensor from suspension strut.

– Remove lower sound insulation panel from under engine.

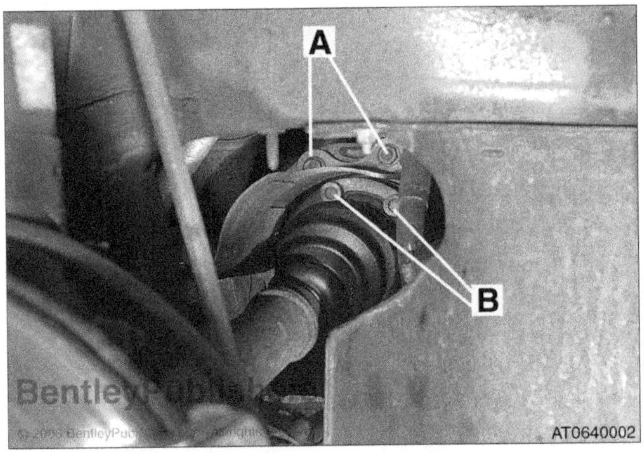

◀ Remove mounting bolts (**A**) and protective cover above axle boot, if applicable.

– Unbolt drive axle from transmission flange.
  - Only two mounting bolts (**B**) are shown for drive axle. Rotate wheel to access remaining mounting bolts.

NOTE —
- *If the drive axle does not have mounting bolts as shown, see* **Front Drive Axles, Drive axles with splines on inner joint,** *given later in this repair group.*

– Secure drive axle to vehicle with wire.

– Remove bolt and nut (**arrow**) securing wheel bearing housing to lower suspension strut.
  - Note installation position of nut/bolt; nut is on front side

– Insert spreader tool **3424** into slot of wheel bearing housing where lower strut is mounted.

– Insert ratchet into 3424 and rotate 90° (¼ turn) to pry slot open.

– Press brake disc (rotor) in direction of suspension strut by hand to prevent lower shock absorber tube from canting in wheel bearing housing opening.

– Pull wheel bearing housing downward from strut tube.

BentleyPublishers.com
© 2006 BentleyPublishers.com—All Rights Reserved

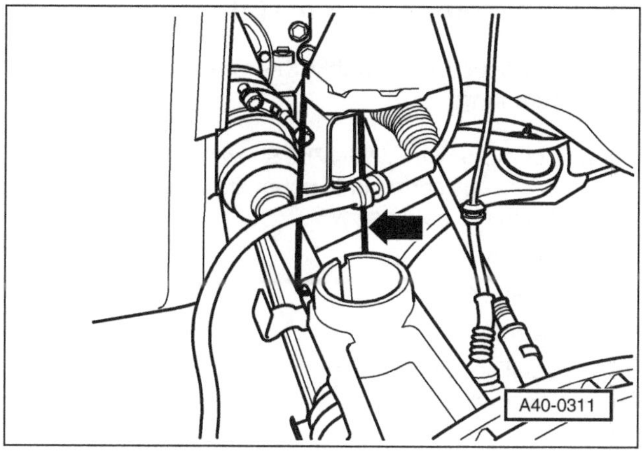

◄ Tie wheel bearing housing to subframe with wire (**arrow**).

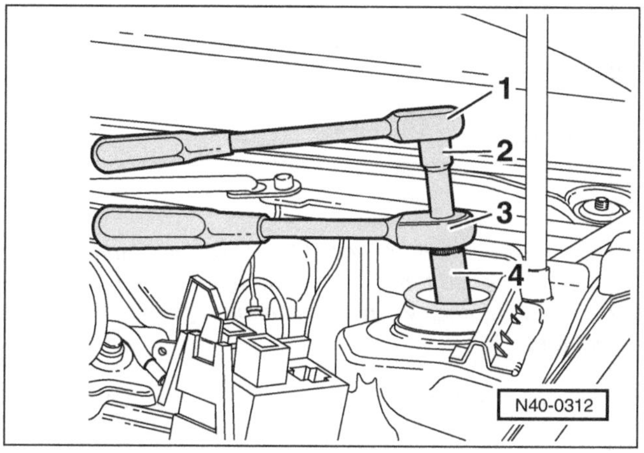

◄ Remove hex nut for upper strut mount using special tools.

1. Ratchet handle/driver
2. T10001/8 hex key (7 mm)
3. T10001/11 driver handle
4. T10001/5 special socket (21 mm) can also use special tool 3186

– Remove strut assembly.

– When installing strut, coat upper strut bearing with assembly lubricant (G 294 421 A1).

– Install suspension strut upper mounting bolt first.

Tightening torque	
Strut to body, upper mount	60 Nm (44 ft-lb)

– Remove wire used to hang wheel bearing housing.

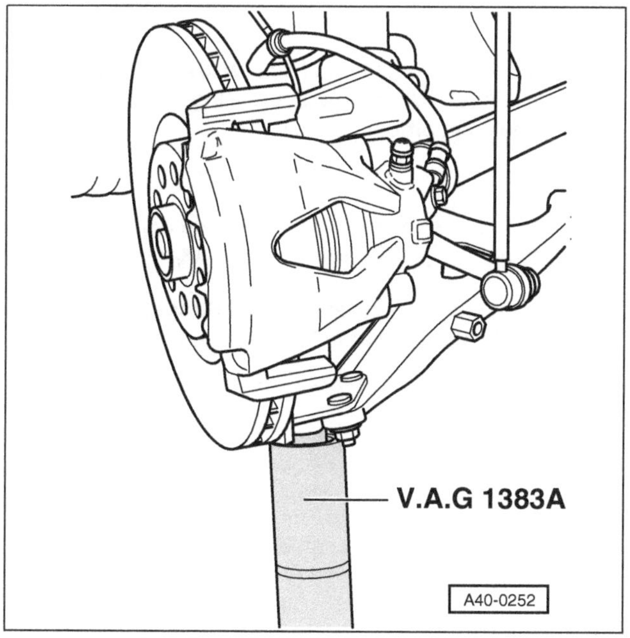

◄ Using engine/transmission jack or threaded support stand, carefully raise wheel bearing housing until lower strut mounting bolt can be installed. Remove spreader tool 3424.

NOTE —

• Do not press against ball joint with the engine/transmission jack.

– Press brake disc (rotor) in direction of suspension strut by hand to help lower shock absorber tube slide into wheel bearing housing opening.

– Tighten lower strut mounting bolt, coupling rod and drive axles to specification.

Tightening torque	
Wheel bearing housing to strut (lower mount) (always replace)	60 Nm (44 ft-lb) + 90° (¼ turn)
Coupling rod to suspension strut (always replace)	90 Nm (66 ft-lb)
Drive axle to transmission flange	70 Nm (52 ft-lb)

BentleyPublishers
.com
© 2006 BentleyPublishers.com—All Rights Reserved

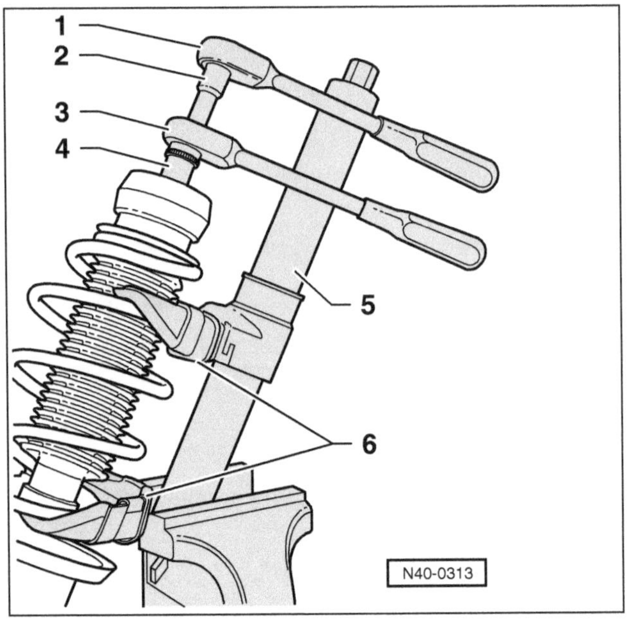

Coil spring, removing and installing

– Remove complete front strut assembly. See **Suspension strut, removing and installing**.

◀ Use a coil spring compressor (VAG 1752 shown) and appropriate adapters to compress coil spring until upper spring seat is free and remove inner nut with the setup shown.

1. Ratchet handle/driver
2. T10001/8 hex key (7 mm)
3. T10001/11 driver handle
4. T10001/5 special socket (21 mm) can also use special tool 3186
5. VAG 1752/1 spring compressor
6. VAG 1752/4 adapter

> ***WARNING —***
> • *Do not remove inner nut from top of strut assembly without a spring compressor.*

NOTE —

• *A special offset box wrench or special 21 mm socket (such as VW tool 3186) can be used to turn the outer strut nut while holding the strut shaft with a 7 mm hex wrench, as an alternative to the above listed special tools.*

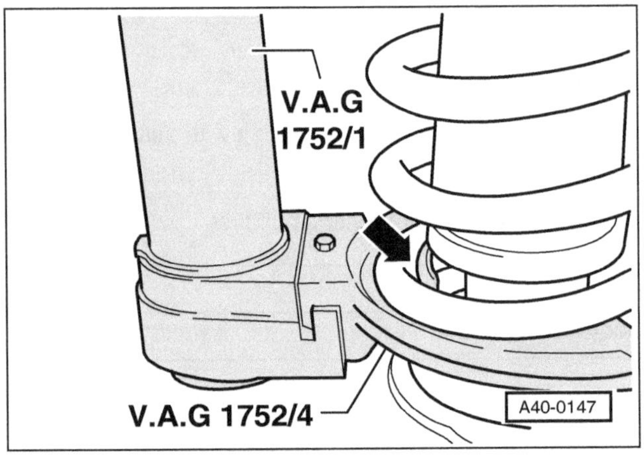

◀ Make sure coil spring is seated correctly (**arrow**) in adapter (VAG 1752/4).

– Remove components of suspension strut keeping parts in order as they are removed.

– Inspect removed components for wear and damage. Pay particular attention to the upper suspension strut mounts, bump stops and protection boots.

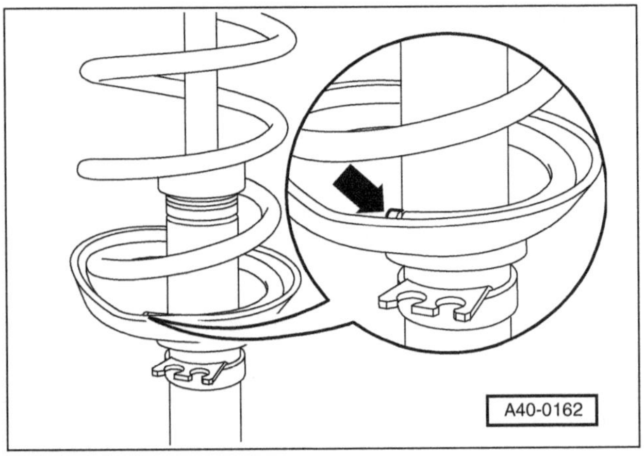

◀ When re-installing coil spring, fit end of coil (**arrow**) against stop in lower spring seat on strut. When de-compressing spring, ensure that end of coil remains in proper position.

NOTE —

• *If coil springs are to be replaced, note identifying paint marks on coils for proper matching, particularly if only one spring is replaced.*

– Installation is the reverse of disassembly. See **Suspension strut assembly** earlier in this repair group for tightening torques.

BentleyPublishers.com
© 2006 BentleyPublishers.com—All Rights Reserved

Front Drive Axles

Vehicles with drive axle secured by 12-point nut

The drive axle/wheel hub splines can either be "bonded" with locking fluid, or on newer-version wheel hubs, greased. The wheel hub version determines whether the wheel hub is bonded or greased.

Old-version wheel hubs (Part No. 8N0 407 613; wheel hub stamped with 8N0 615) can still be used with drive axles that were bonded, but these must then still be bonded using locking fluid (D 185 400 A2).

Newer wheel hubs (Part No. 8N0 407 613 A or B; wheel hub stamped with 8N0 615A or B) must not be bonded. These wheel hubs may be coated only with solid grease (G 052 142 A2).

It is acceptable to have a bonded drive axle/wheel hub on one side of the vehicle and a greased drive axle/wheel hub on the other side. A bonded wheel hub may be replaced with a greased one. However, a greased wheel hub must not be replaced with a bonded one.

Drive axle, removing and installing

– Remove hubcap. For alloy wheels, pull off center cap (use pulling hook in vehicle tool kit). **See 03 Maintenance, Tire and wheel service**.

– Loosen flange bolt or 12-point drive axle nut a maximum of 90° (¼ turn) with vehicle standing on its wheels.

- Flange bolt or 12-point nut are accessible through center of wheel, after center cap is removed.

– Raise car enough to relieve load on front wheels and support safely.

> ***WARNING —***
> - *Make sure car is stable and well supported at all times. Use a professional automotive lift or jack stands designed for the purpose. A floor jack is not adequate support.*

– Have an assistant press brake pedal to keep wheel from turning.

– Remove flange bolt or 12-point nut from drive axle.

– Remove wheel and raise vehicle high enough to allow access underneath. Safely support vehicle.

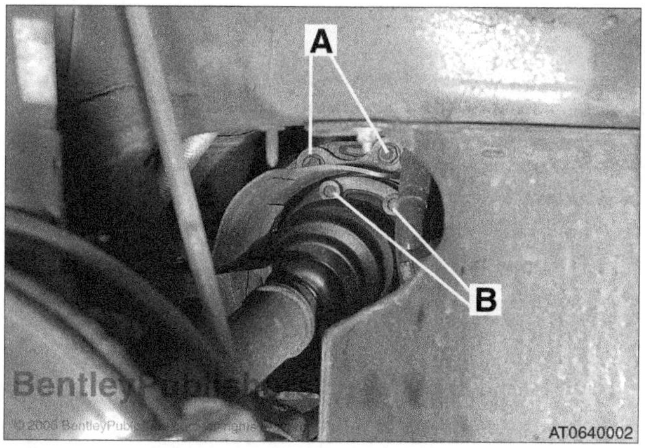

AT0640002

◄ Remove mounting bolts (**A**) and protective cover above axle boot, if applicable.

– Unbolt drive axle from transmission flange.

- Only two mounting bolts (**B**) are shown. Rotate axle to access remaining mounting bolts.

NOTE —
- *If the drive axle does not have mounting bolts as shown, see* **Drive axles with splines on inner joint**, *later in this procedure.*

BentleyPublishers.com
© 2006 BentleyPublishers.com—All Rights Reserved

– Use wire or heavy duty zip ties to temporarily suspend axle from body. Do not let axle hang without support as the inner joint can be damaged.

◄ Mark positions of ball joint lower mounting nuts (**arrows**) and remove nuts.

– Remove control arm from ball joint.

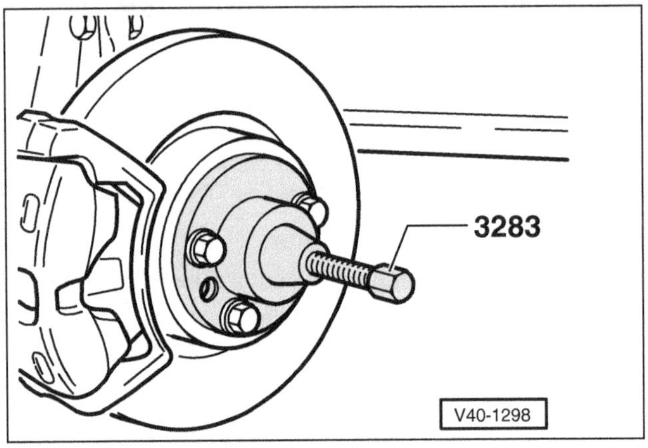

◄ Press drive axle out of wheel bearing housing. If need be, use a press tool (3283 shown) to free axle.

NOTE—
- *Some drive axles use an adhesive to bond them to the wheel bearing housing. The press tool (3283) will be necessary to free bonded axles.*

Drive axles with splines on inner joint

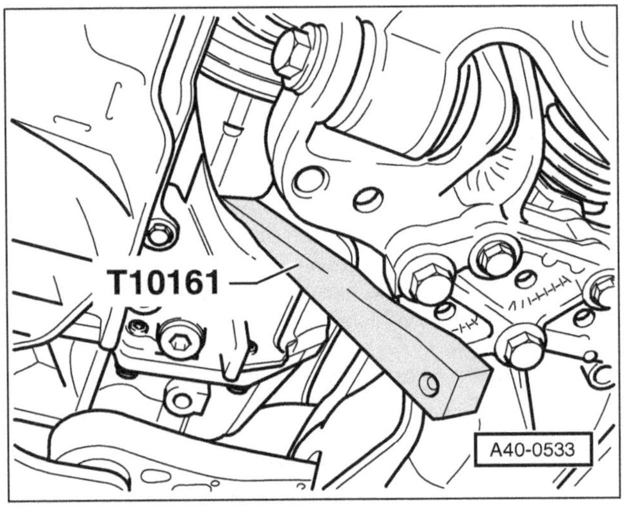

◄ Place aluminum wedge (**T10161**) between transmission housing and inner triple roller joint.

– Force drive axle out of transmission with a firm blow to wedge tool.

– Replace circlip on end of joint body (with splines) if circlip is removed.

Continued for all models

– Remove drive axle from vehicle.

> ***WARNING—***
> - *Do not move vehicle after drive axle has been removed. Doing so will result in wheel bearing damage. If vehicle must be moved, install an outer joint in place of the drive axle and tighten 12-point nut or flange bolt to 190 Nm (140 ft-lb).*

– Before installing, clean splines of drive axle. Check axle (CV) boots for damage or cracks. Replace as necessary.

BentleyPublishers.com
© 2006 BentleyPublishers.com—All Rights Reserved

– Install components in reverse order of removal.

Tightening torques	
Drive axle to transmission: M8 bolt M10 bolt	 40 Nm (30 ft-lb) 70 Nm (52 ft-lb)

– For drive axles with bonded spline apply locking fluid (D 185 400 A2) to outer spline.

– Install new 12-point nut or flange bolt as far as possible by hand.

– Lower vehicle, if necessary, and have an assistant press brake pedal to keep drive axle from turning.

< Tighten 12-point nut or flange bolt for drive axle to initial torque specification. Install wheel (without center cap) and tighten lug bolts to initial torque specification using diagonal pattern (shown at left).

Tightening torques (initial)	
Drive axle to wheel bearing housing (always replace nut/bolt) Bonded wheel hub Greased wheel hub	 240 Nm (177 ft-lb) 190 Nm (140 ft-lb)
Wheel to hub	50 Nm (37 ft-lb)

– Lower vehicle so weight is resting on wheels.

– Tighten wheel lugs to final toque specification using diagonal pattern. Tighten 12-point nut or flange bolt for drive axle to final torque specification.

Tightening torques (final)	
Drive axle to wheel bearing housing Bonded wheel hub Greased wheel hub	 265 Nm (195 ft-lb) 190 Nm (140 ft-lb) + 90° (¼ turn)
Wheel to hub	120 Nm (89 ft-lb)

BentleyPublishers.com
© 2006 BentleyPublishers.com—All Rights Reserved

Drive axle assembly with CV joint

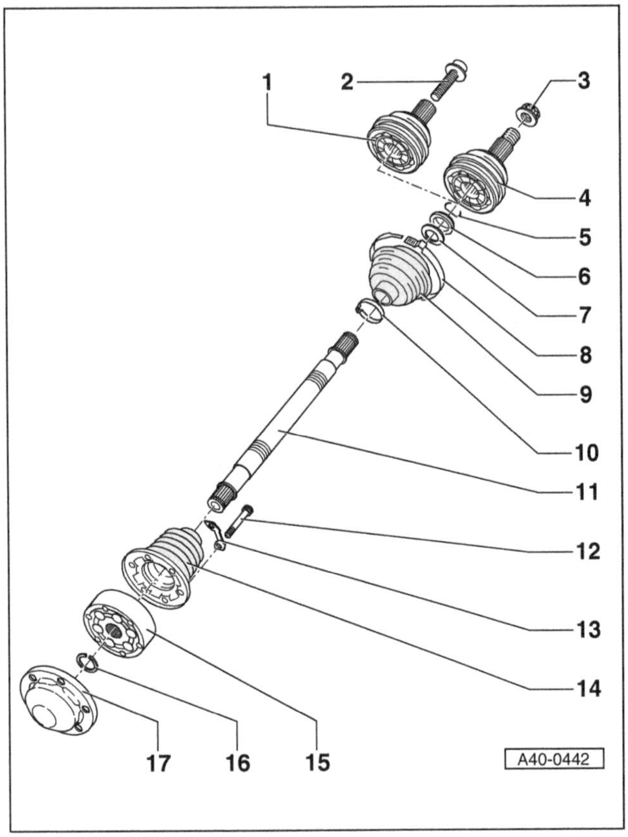

1. Outer CV joint
2. Flange bolt
   - Always replace
3. Nut, 12-point
   - Always replace
4. Outer CV joint
5. Circlip
   - Always replace
6. Spacer ring (plastic)
7. Convex washer
8. Boot clamp
9. Boot for CV joint
10. Boot clamp
11. Drive axle shaft
12. Bolt
    - 70 Nm (52 ft-lb)
13. Lock plate
14. Boot for CV joint
15. Inner CV joint
    - Apply sealant (D 454 300 A2) between CV joint and cover
16. Circlip
    - Always replace
17. Cover
    - Drive off CV joint using drift

Drive axle assembly with triple roller joint

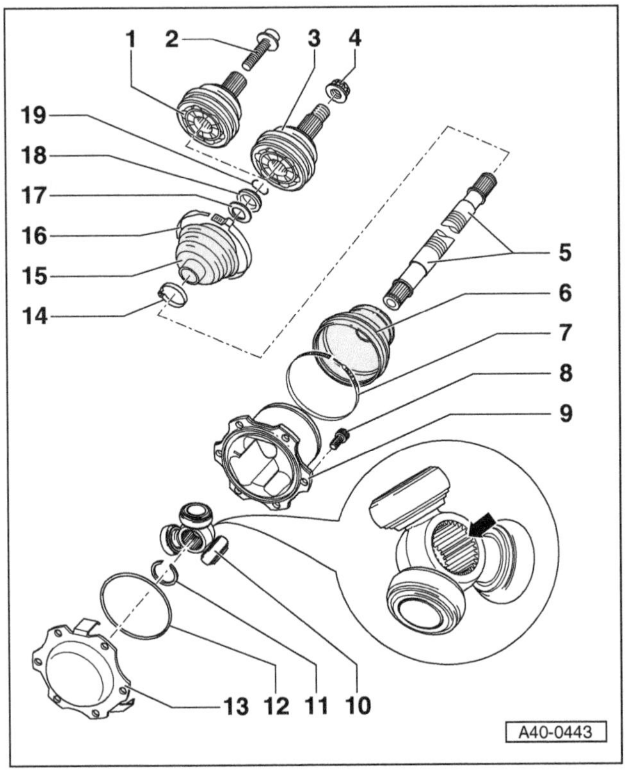

1. Outer CV joint
2. Bolt (always replace)
3. Nut, 12-point (always replace)
4. Outer CV joint
5. Drive axle shaft
6. Boot for triple roller joint
7. Boot clamp
8. Bolt
   - 70 Nm (52 ft-lb)
9. Cover
10. Triple roller joint
    - Chamfer (**arrow**) faces toward drive axle splines
11. Circlip (always replace)
12. O-ring
13. Cover (always replace)
14. Boot clamp
15. Boot for CV joint
16. Boot clamp
17. Convex washer
18. Spacer ring (plastic)
19. Circlip (always replace)

BentleyPublishers.com
© 2006 BentleyPublishers.com—All Rights Reserved

Drive axle assembly with triple roller joint and joint body

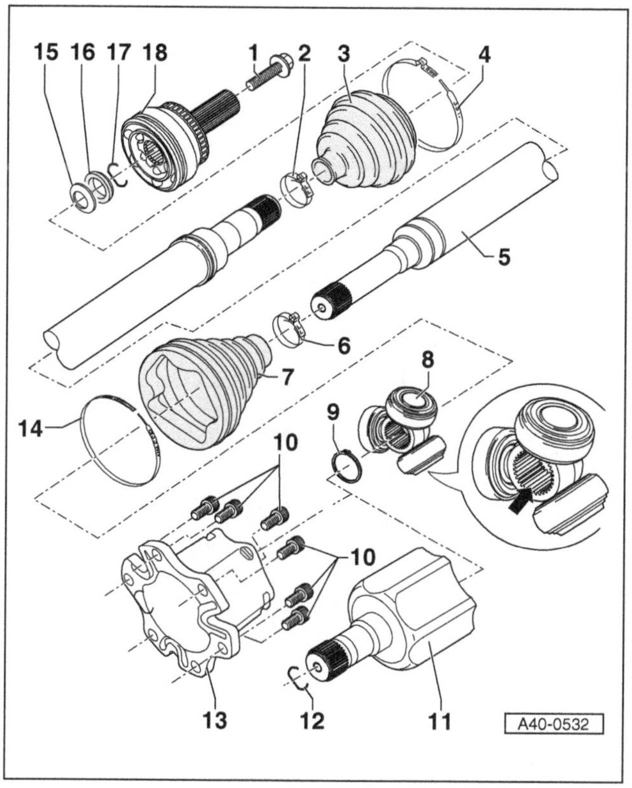

1. Bolt with washer (always replace)
2. Boot clamp
3. Boot for CV joint
4. Boot clamp
5. Drive axle shaft
6. Boot clamp
7. Boot for triple roller joint
8. Triple roller joint (AAR 3300i)
   - Chamfer (**arrow**) faces toward drive axle splines
9. Circlip (always replace)
10. Bolt
    - M8: 40 Nm (30 ft-lb)
    - M10: 70 Nm (52 ft-lb)
11. Joint body (version with splines)
    - Before installing, grease splines on drive axle
12. Circlip (always replace)
13. Joint body (version with bolted flange)
14. Boot clamp
15. Convex washer
16. Spacer ring (plastic)
17. Circlip (always replace)
18. Outer CV joint

BentleyPublishers.com
©2006 BentleyPublishers.com—All Rights Reserved

FRONT WHEEL BEARING HOUSING

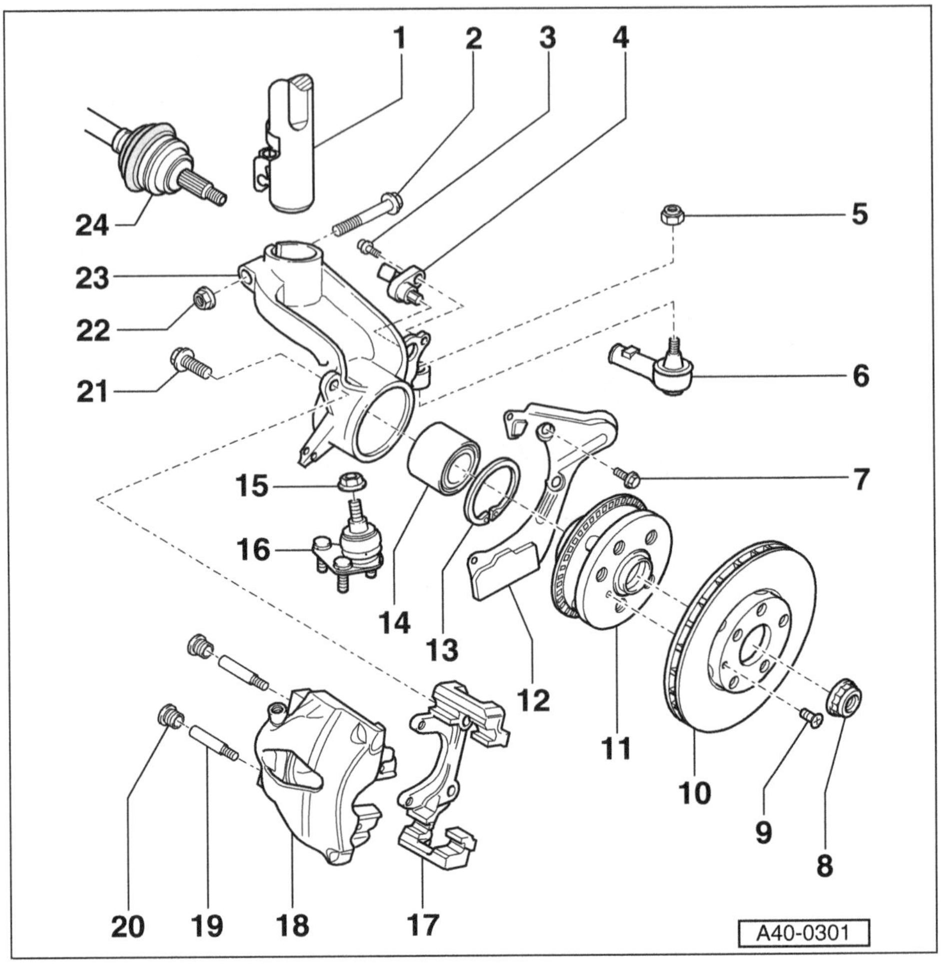

1. **Suspension strut**
2. **Bolt**
3. **Bolt**
   - 8 Nm (71 in-lb)
4. **ABS wheel speed sensor**
5. **Self-locking nut**
   - Always replace
   - Tightening torque:
     3.2L roadster: 75 Nm (55 ft-lb)
     All other models: 45 Nm (33 ft-lb)
6. **Ball joint (tie rod end)**
7. **Bolt**
   - 10 Nm (7 ft-lb)
8. **Self-locking 12-point nut**
   - Always replace
   - See **Drive Axle, removing and installing**, earlier in this repair group
9. **Screw**
   - 4 Nm (35 in-lb)
10. **Brake disc, ventilated**
11. **Wheel hub**
    - With wheel speed sensor rotor welded to hub
12. **Backing plate**
13. **Snap ring**
    - Check for proper seating
14. **Wheel bearing**
    - Before installing wheel bearing, coat entire inner surface of bore in wheel bearing housing with Molypaste (part number AOS 115 000 01)
    - A grease packet containing 3 grams of molypaste is included with the Audi-supplied wheel bearing kit
15. **Self-locking nut**
    - Always replace
    - Tightening torque:
      3.2L roadster: 150 Nm (111 ft-lb)
      All other models: 75 Nm (55 ft-lb)
16. **Ball joint**
17. **Brake carrier**
18. **Brake caliper**
19. **Guide pins**
    - 25 Nm (18 ft-lb)
20. **Protective cap**
21. **Ribbed bolt**
    - 125 Nm (92 ft-lb)
    - Clean ribs on underside of bolt head
    - Do not twist or kink brake line when removing or installing
22. **Self-locking nut**
    - 60 Nm (44 ft-lb) plus 90° (¼ turn)
23. **Wheel bearing housing**
    - Remove housing to remove/install wheel bearing
    - Wheel bearing can also be removed/installed with housing still on vehicle using special tools. See **Wheel Bearing, special tools for removing with housing installed on vehicle.**
24. **Drive axle**

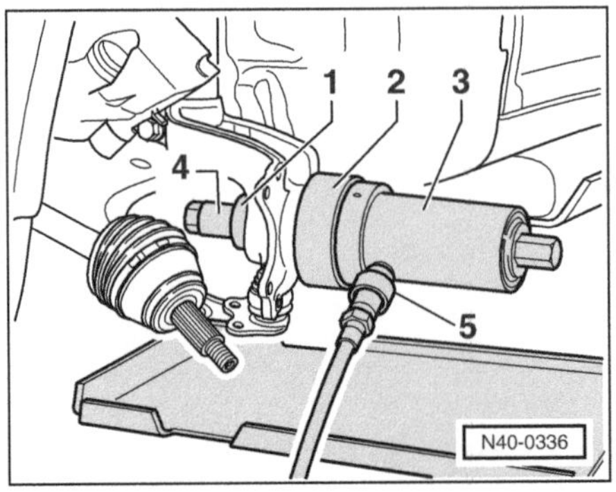

◄ **Wheel bearing, special tools for removing with housing installed on vehicle.**

1. Thrust bolt E-15
2. Support E-40
3. Hollow piston cylinder HZK-15
4. Special nut E-8-214 and pull rod
5. High pressure hose with quick release coupling attached to foot pump (VAG 1389A)

42 Rear Suspension

General

This section covers repairs to the rear suspension and related components. For front suspension servicing and repair information, see **40 Front Suspension and Drive Axles**. For wheel alignment specifications, see **44 Wheels, Tires, Wheel Alignment**.

Special tools, equipment and procedures are required for most rear suspension repair and component replacement. In addition, wheel alignment is often disturbed when rear suspension components are removed or replaced.

WARNING —
- *Do not re-use any fasteners that are worn or deformed. Most fasteners are designed to be used only once and become unreliable and may fail when used a second time. This includes, but is not limited to, nuts, bolts, washers, circlips, cotter pins, self-locking nuts and bolts. For replacements, always use new parts.*
- *Do not reinstall bolts and nuts coated with undercoating wax, as correct tightening torque cannot be ensured. Clean the threads of removed bolts and nuts that are to be re-used with a suitable solvent before installation. Make sure new parts are clean.*
- *Do not use standard nuts and bolts in place of green colored parts. The green color denotes a special anti-corrosion process known as "dacromet" or "delta-tone". Torque specifications are given based on use of these parts only.*
- *Always replace rusted or corroded bolts, nuts and washers even if not specifically indicated.*
- *Do not attempt to straighten or weld suspension struts, wheel bearing housings, control arms or any other wheel locating or load bearing components of the front suspension.*

BentleyPublishers.com

© 2006 BentleyPublishers.com—All Rights Reserved

Rear Suspension (vehicles with front wheel drive)

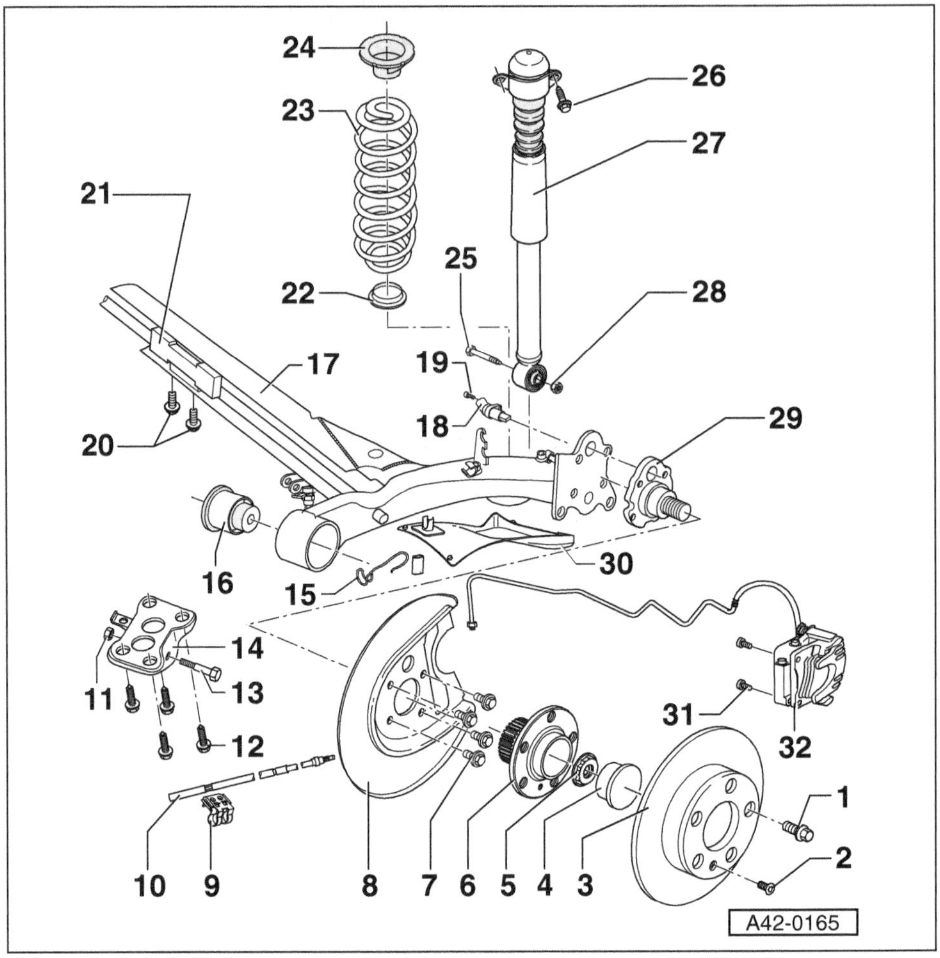

Rear axle assembly

1. **Wheel bolt**
   - 120 Nm (89 ft-lb)
2. **Screw**
3. **Brake disc**
4. **Protective cap**
   - Always replace, only a new cap will seal properly to wheel hub
5. **12-point stub axle nut**
   - 220 Nm (162 ft-lb)
6. **Wheel hub with bearing**
   - Bearing and hub are combined in one housing
   - No maintenance or adjustment is necessary
7. **Bolt**
   - 50 Nm (37 ft-lb) + 90° (¼ turn)
8. **Splash plate**
   - Always replace
9. **Bracket for parking brake cable**
10. **Parking brake cable**
11. **Nut**
    - 45 Nm (33 ft-lb) + 90° (¼ turn)
12. **Bolt**
    - 30 Nm (22 ft-lb) + 90° (¼ turn)
    - Always replace
13. **Bolt**
    - 45 Nm (33 ft-lb) + 90° (¼ turn)
14. **Mounting bracket for rear axle**
15. **Retaining clip for parking brake cable**
16. **Bonded rubber bushing**
17. **Axle beam**
18. **ABS wheel speed sensor**
19. **Bolt**
    - 8 Nm (71 in-lb)
20. **Bolt**
    - 20 Nm (15 ft-lb) + 45° (⅛ turn)
21. **Vibration damper**
22. **Lower spring plate**
    - Material: Zinc
23. **Coil spring**
24. **Upper spring plate**
25. **Bolt**
    - Always replace
    - 40 Nm (30 ft-lb) + 90° (¼ turn)
    - Vehicle weight must be on wheels when tightening
26. **Bolt**
    - 30 Nm (22 ft-lb) + 90° (¼ turn)
27. **Shock absorber**
    - The same version must be installed on both sides
28. **Nut**
    - Always replace if removed
    - 40 Nm (30 ft-lb) + 90° (¼ turn)
    - Vehicle weight must be on wheels when tightening
29. **Stub axle**
30. **Stone deflector**
31. **Bolt**
    - 65 Nm (48 ft-lb)
32. **Brake caliper**

BentleyPublishers.com

© 2006 BentleyPublishers.com—All Rights Reserved

Shock absorber, removing and installing

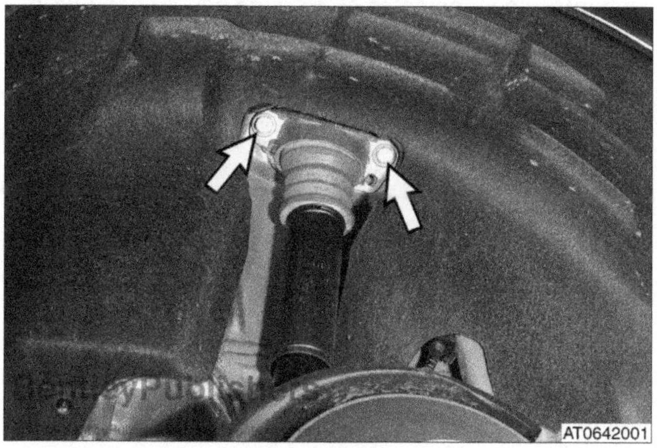

Models covered by this manual are equipped with gas-pressurized telescopic rear shock absorbers bolted inside the wheel wells. Shock absorbers can be removed without removing rear coil springs if suspension is not allowed to extend fully.

– Raise vehicle enough to access upper shock mounting bolts above rear wheel. Vehicle wheels should still be on the ground.

< Remove shock absorber upper mounting bolts (**arrows**).

– Raise vehicle slightly to relieve coil spring pressure, if coil spring has not been removed.

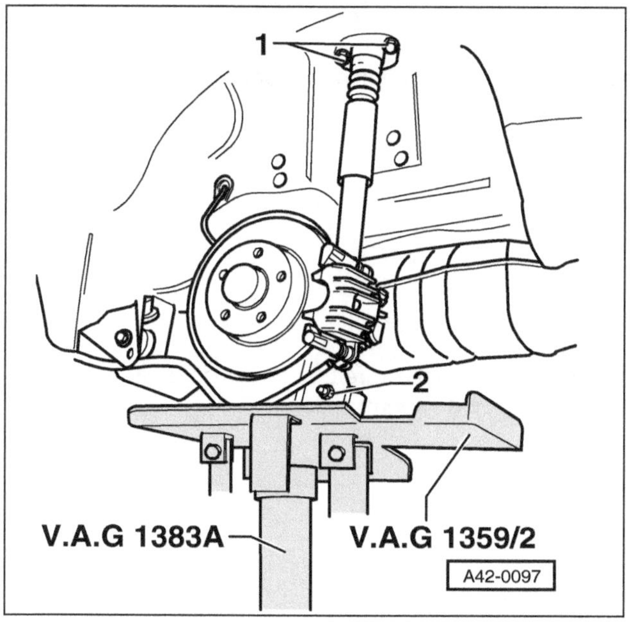

< Remove lower shock absorber mounting bolt (**2**).

- Illustration shown with rear wheel removed for clarity.
- VAG 1383 engine/transmission jack shown to support rear axle when vehicle is positioned higher on a lift.

– Remove shock absorber.

– Installation is performed in the reverse order of removal. Note the following:

- Bring rear axle into curb weight position by lowering vehicle. An engine/transmission jack (VAG 1383 shown) can also be used to raise the rear axle into the curb weight position if vehicle is on a lift.
- Do not bolt shock absorber to wheel bearing housing until curb weight position has been achieved.

> ***CAUTION—***
> - *Failure to return rear suspension to curb weight position before tightening bolts/nuts of bonded rubber bushing suspension parts will result in bushing pre-load which will shorten service life and may result in noise.*

< Insert rear shock absorber and tighten upper shock mounting bolts (**arrows**).

– Lower vehicle, or raise rear axle, and tighten lower shock absorber bolt.

Tightening torques	
Upper shock bolts to body	30 Nm (22 ft-lb) + ¼ turn (90°)
Lower shock bolt to wheel bearing housing (tighten in curb weight position)	40 Nm (30 ft-lb) + ¼ turn (90°)

– If applicable, replace coil spring. See **Coil spring, removing and installing** in this repair group.

BentleyPublishers .com
© 2006 BentleyPublishers.com—All Rights Reserved

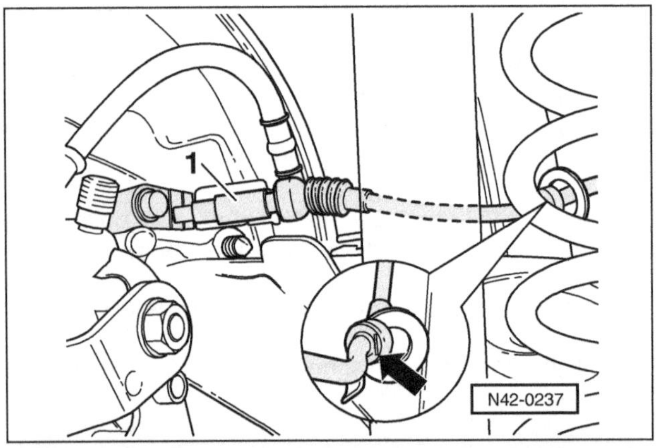

Coil spring, removing and installing

◀ Disconnect harness connector for ABS speed sensor and unclip from mounting bracket.

– Raise vehicle enough to access upper shock mounting bolts above rear wheel. Vehicle wheels should still be on the ground.

◀ Remove shock absorber upper mounting bolts (**arrows**).

– Raise vehicle slightly to relieve coil spring pressure and remove spring.

– Before installing, check spring and lower spring plate for damage. Replace as needed.

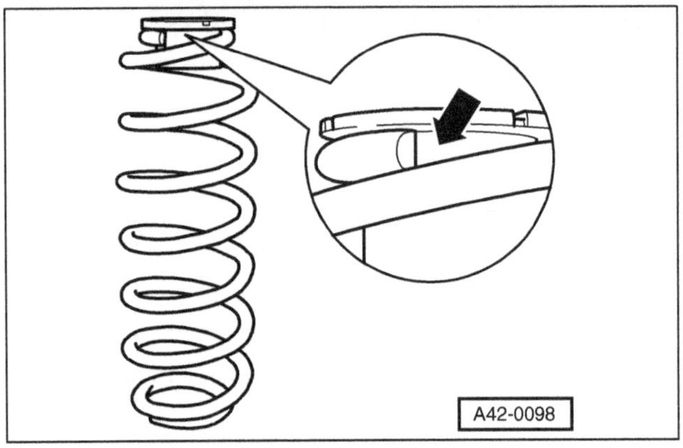

◀ Before installing spring, make sure that end of spring (**arrow**) is positioned against stop on spring plate.

– Install spring.

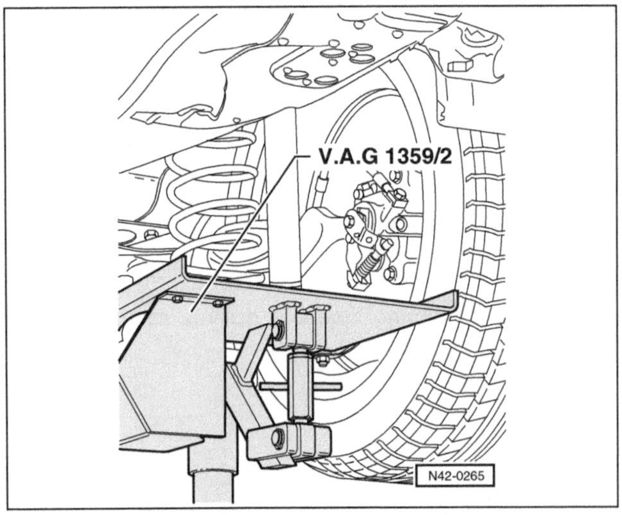

◀ Use transmission jack (VAG 1359/2 shown) to lift rear axle upward.

– Bolt upper mounting bolts for shock absorber to body.

Tightening torques	
Upper shock bolts to body	30 Nm (22 ft-lb) + ¼ turn (90°)
Rear wheel to hub	120 Nm (89 ft-lb)

BentleyPublishers.com
©2006 BentleyPublishers.com—All Rights Reserved

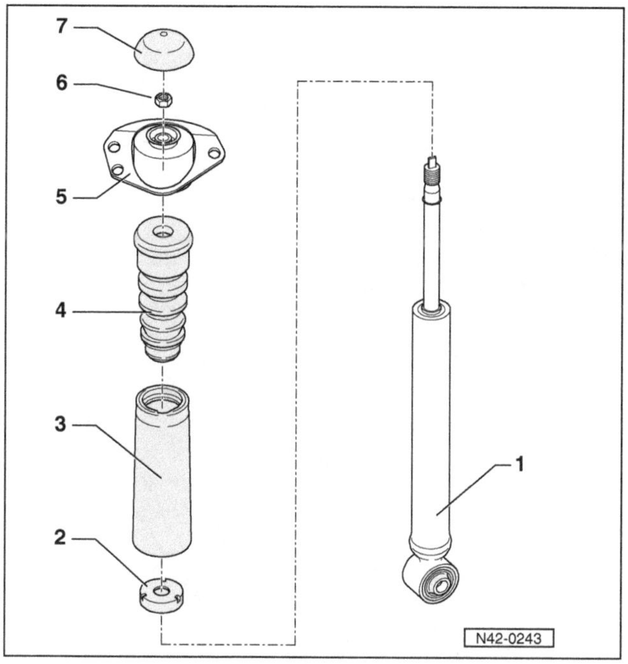

Shock absorber assembly

1. Shock absorber
   - See **Shock absorber, removing and installing**
2. Protective cap
3. Protective tube
4. Stop buffer
5. Shock absorber mount
6. Retaining nut
   - 25 Nm (18 ft-lb)
   - Always replace
7. Cover

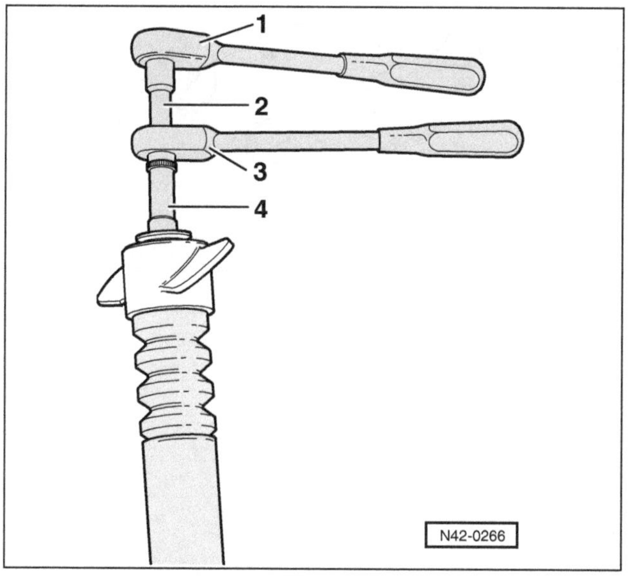

Use tools as shown to remove shock absorber retaining nut.

1. Ratchet handle/driver
2. T10001/9
   special tool 3079 counter holder, can also be used
3. T10001/11 driver handle
4. T10001/1 special socket

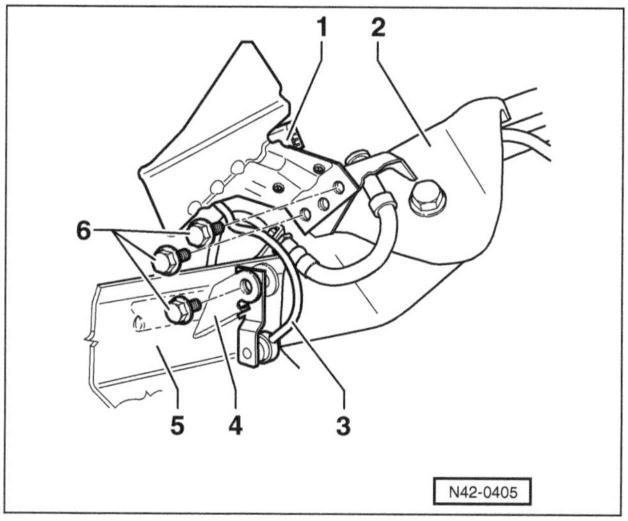

Vehicle level sensor assembly (Xenon headlights)

NOTE—

- *Audi identifies electrical components by a letter and/or a number in the electrical schematics. See* **EWD Electrical Wiring Diagrams**. *These electrical identifiers are listed in parentheses as an aid to electrical troubleshooting.*

1. Left rear level control system sensor (G76)
2. Mounting bracket
3. Coupling rod
4. Bracket
5. Rear axle beam
6. Bolt
   - 20 Nm (15 ft-lb)

BentleyPublishers.com
©2006 BentleyPublishers.com—All Rights Reserved

Rear Suspension (vehicles with all wheel drive)

Rear subframe assembly

1. **Subframe**
2. **Rear final drive**
   - Checking and changing oil, see **03 Maintenance**
3. **Cross member**
4. **Stabilizer bar**
5. **Shock absorber**
6. **Wheel bearing**
7. **Trailing arm**
8. **Transverse links**
9. **Coil spring**
   - Use spring compressor to remove and install
10. **Rear drive axle**
11. **12-point bolts**
    - 60 Nm (44 ft-lb)
    - For drive shaft

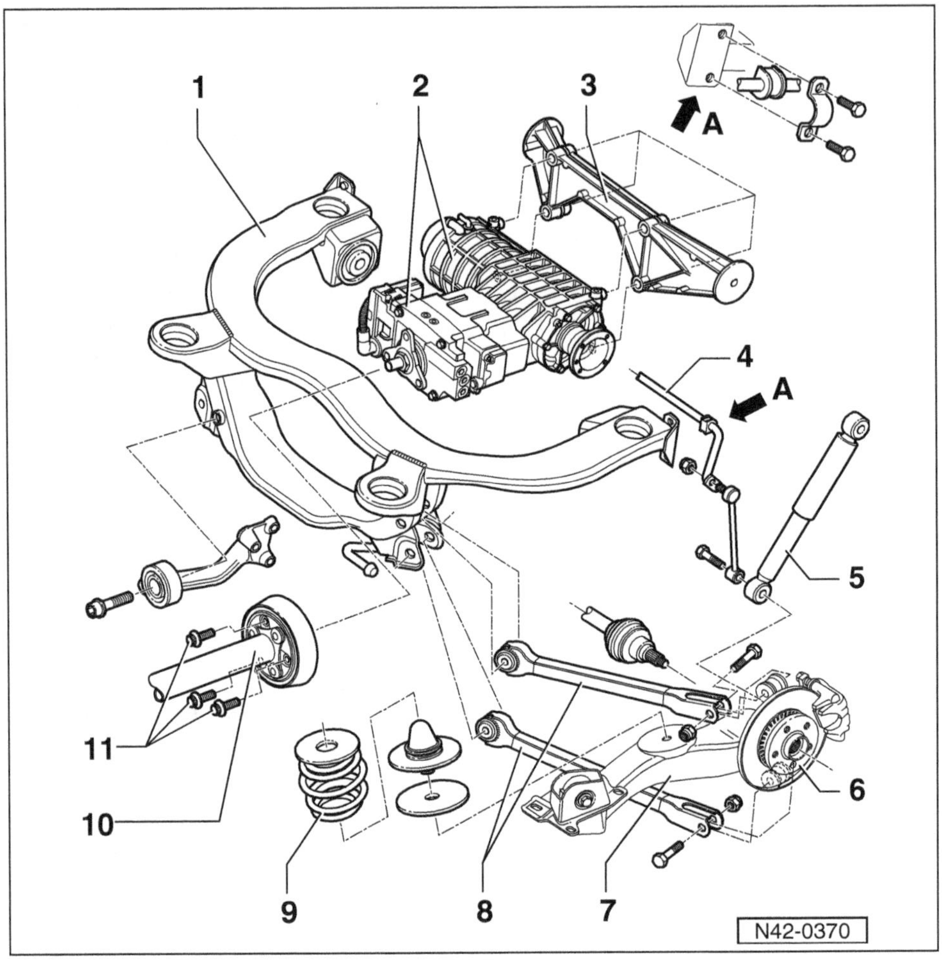

BentleyPublishers
.com
©2006 BentleyPublishers.com–All Rights Reserved

Rear trailing arm and transverse link assembly

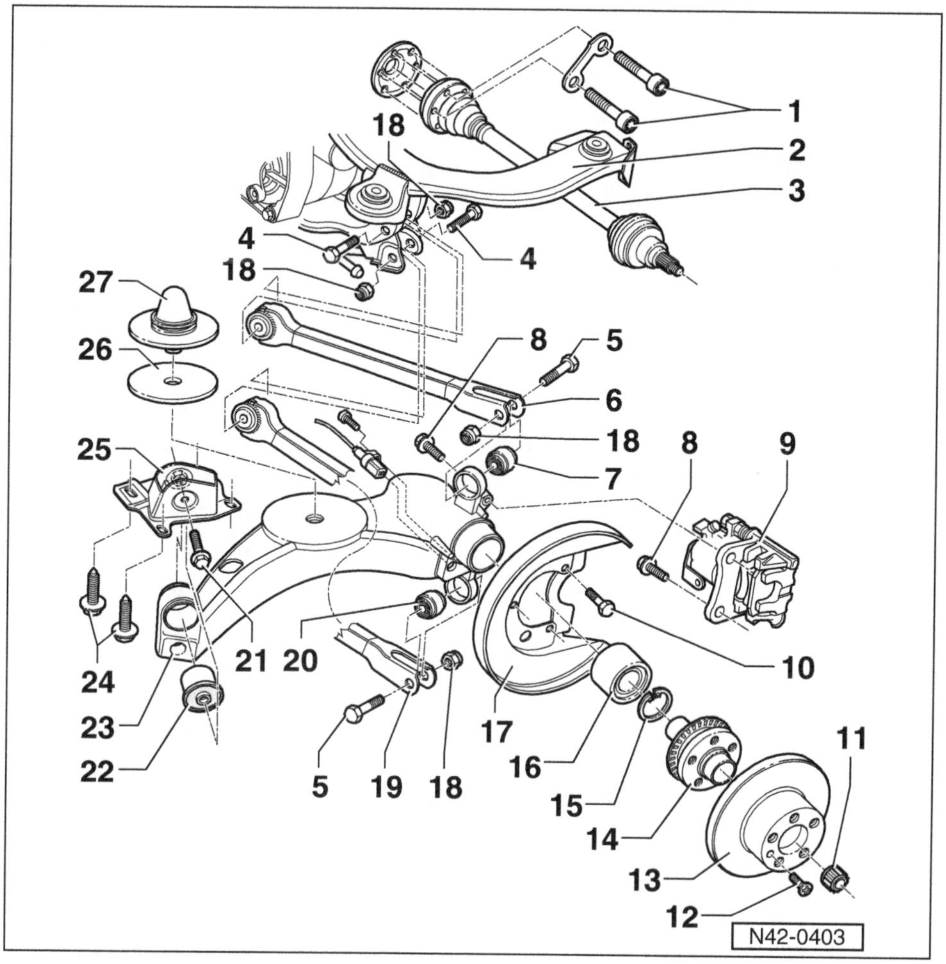

1. **Socket head bolts**
   - 40 Nm (30 ft-lb)
2. **Subframe**
3. **Rear drive axle**
4. **Bolt**
   - M12 x 1.5 x 80
   - Always replace
5. **Bolt**
   - M12 x 1.5 x 75
   - Always replace
6. **Upper transverse link**
7. **Ball joint**
   - Check rubber boot for damage
8. **Socket head bolt**
   - 65 Nm (48 ft-lb)
9. **Brake caliper**
10. **Bolt**
    - 10 Nm (7 ft-lb)
11. **12-point axle nut**
12. **Screw**
    - 4 Nm (35 in-lb)
13. **Brake disc**
14. **Wheel hub**
    - With welded-in rotor for ABS wheel speed sensor
15. **Circlip**
16. **Wheel bearing**
17. **Splash plate**
18. **Nut**
    - 70 Nm (52 ft-lb) + ¼ turn (90°)
    - Always replace
19. **Lower transverse link**
    - Different versions for ball joint or bonded rubber bushing
20. **Ball joint**
    - Check rubber boot for damage
    - Replaced with bonded rubber bushing in 2003 m.y.
21. **Bolt**
    - 90 Nm (66 ft-lb)
22. **Bonded rubber bushing**
23. **Trailing arm**
24. **Bolt**
    - 75 Nm (55 ft-lb)
25. **Mounting bracket for rear axle**
    - Toe setting can be adjusted by moving bracket
26. **Spacer**
27. **Bump stop**
    - 10 Nm (7 ft-lb)

BentleyPublishers.com
© 2006 BentleyPublishers.com—All Rights Reserved

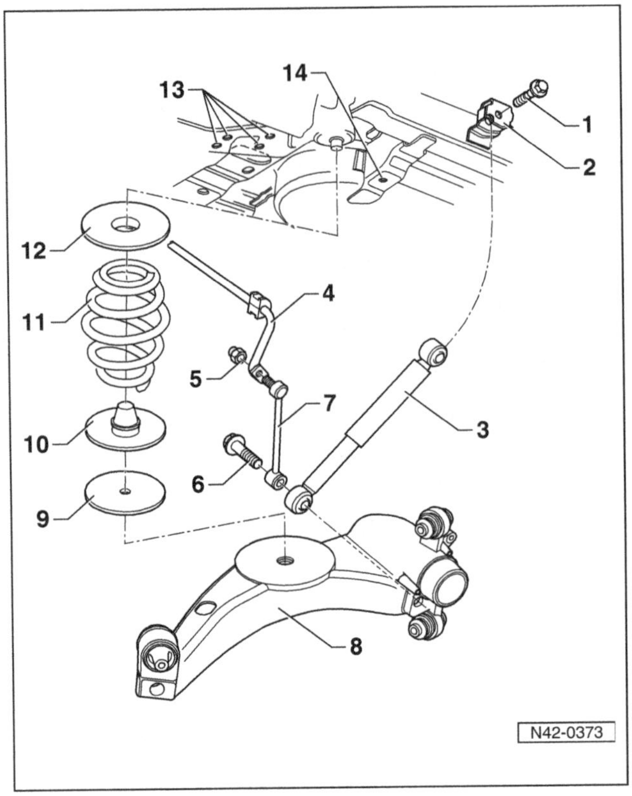

Coil spring and shock overview

1. Bolt
   - 60 Nm (44 ft-lb)
   - Always replace
2. Mounting bracket for shock
3. Shock absorber
   - See **Shock absorber, removing and installing**
4. Stabilizer bar
5. Self-locking nut
   - 25 Nm (18 ft-lb)
   - Always replace
6. Bolt
   - 110 Nm (81 ft-lb)
   - Always replace
7. Coupling rod
8. Trailing arm
9. Spacer
10. Stop buffer
    - 10 Nm (7 ft-lb)
11. Coil spring
12. Spring plate
13. Captive mounting nuts in body
    - Can be repaired with Helicoil insert
14. Captive mounting nut in body
    - Can be repaired with Helicoil insert

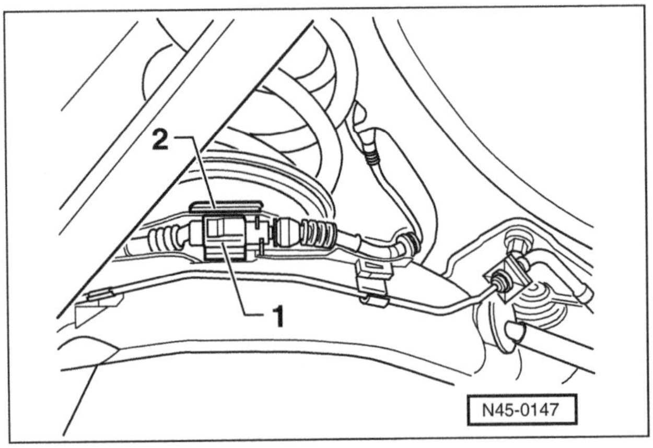

Shock absorber, removing and installing

– Raise vehicle and support safely. See **03 Maintenance**. Remove wheel.

◄ Unclip ABS wheel speed sensor wire from bracket.
- Open cover (**2**) to disconnect wire (**1**).

– Remove rear wheel housing liner.

– Place engine/transmission jack under wheel bearing housing and raise slightly.

> ***WARNING —***
> - *Do not lift wheel bearing housing more than necessary to prevent any risk of vehicle falling off lift or jackstands.*

◄ Remove shock absorber upper mounting bolt (**arrow**).

BentleyPublishers .com

©2006 BentleyPublishers.com—All Rights Reserved

< Remove shock absorber lower mounting bolt (**arrow**).

> ***CAUTION—***
> - *Make sure that headlight level control linkage is not over-stretched (left side only).*

– Remove shock absorber.

– When installing shock absorber note the following:
  - Removal of coupling rod (**A**) may make installation of lower shock mounting bolt easier.
  - Bring rear subframe into curb weight position by lowering vehicle before tightening bonded rubber bushing mounting bolts. An engine/transmission jack can also be used to carefully raise the rear axle into the curb weight position if vehicle is on a lift.

> ***CAUTION—***
> - *Failure to return rear suspension to curb weight position before tightening bolts/nuts of bonded rubber bushing suspension parts will result in bushing pre-load which will shorten service life and may result in noise.*

– Install shock absorber upper mounting bolt.

– Lower vehicle so wheel is on the ground.

– Install shock absorber lower mounting bolt and coupling rod upper mounting bolt. Tighten bolts to specification.

Tightening torques	
Coupling rod to sway bar	25 Nm (17 ft-lb)
Upper shock bolt to body (always replace)	60 Nm (44 ft-lb)
Lower shock bolt to wheel bearing housing (tighten in curb weight position) (always replace)	110 Nm (81 ft-lb)
Rear wheel to hub	120 Nm (89 ft-lb)

Drive axle, removing and installing

– Remove hubcap. For alloy wheels, pull off center cap (use pulling hook in vehicle tool kit). **See 03 Maintenance, Tire and wheel service**.

– Loosen 12-point drive axle nut a maximum of 90° (¼ turn) with vehicle standing on its wheels.
  - 12-point nut is accessible through center of wheel, after center cap is removed.

– Apply parking brake to keep rear drive axle from turning.

– Raise car enough to relieve load on rear wheels and support safely.

> ***WARNING—***
> - *Make sure car is stable and well supported at all times. Use a professional automotive lift or jack stands designed for the purpose. A floor jack is not adequate support.*

BentleyPublishers.com

© 2006 BentleyPublishers.com—All Rights Reserved

Rear Suspension (vehicles with all wheel drive)

– Remove 12-point axle nut from drive axle and release parking brake.

– Remove wheel and raise vehicle high enough to allow access underneath. Safely support vehicle.

◄ Unbolt drive axle from rear final drive flange.

- Only three mounting bolts (**arrows**) are shown for drive axle. Rotate axle to access remaining mounting bolts.

– Lower inner part of drive axle away from rear final drive flange.

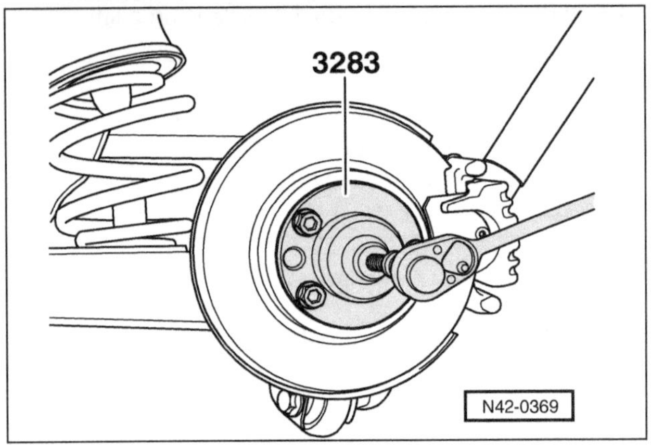

◄ Press drive axle out of wheel hub. If need be, use a press tool (3283 shown) to free axle.

– Remove drive axle.

- It may be necessary to carefully move exhaust pipe using a pry bar to provide enough clearance to remove left-side drive axle.

– Remove any paint residue or corrosion on threads and splines of outer joint.

– Before installing drive axle, use clean oil to lightly lubricate:

- Splines on outer joint and in wheel hub.
- Threads of outer joint.
- Contact surface and threads of 12-point drive axle mounting nut.

– Begin installation by inserting outer joint as far as possible into wheel hub splines.

– Pull outer joint into wheel hub until it makes contact with wheel bearing.

– Position inner joint of drive axle onto rear final drive flange and tighten mounting bolts to specification using a diagonal pattern. Install wheel and tighten lug bolts to specification using a diagonal pattern.

Tightening torques	
Drive axle to rear final drive	40 Nm (30 ft-lb)
Rear wheel to hub	120 Nm (89 ft-lb)

– Lower vehicle until the wheel touches the ground.

– Tighten 12-point axle nut to 200 Nm (148 ft-lb) and then loosen ½ turn (180°).

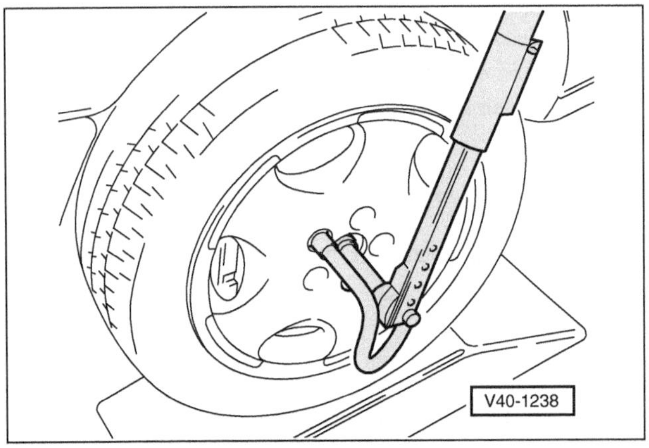

◄ Use a torque angle wrench (i.e. VAG 1756) to tighten 12-point axle nut to specification.

Tightening torque	
12-point axle nut to axle (using torque angle wrench)	50 Nm (37 ft-lb) + ⅙ turn (60°)

BentleyPublishers.com
© 2006 BentleyPublishers.com—All Rights Reserved

If a torque angle wrench is not available, use the following procedure to tighten 12-point axle nut:

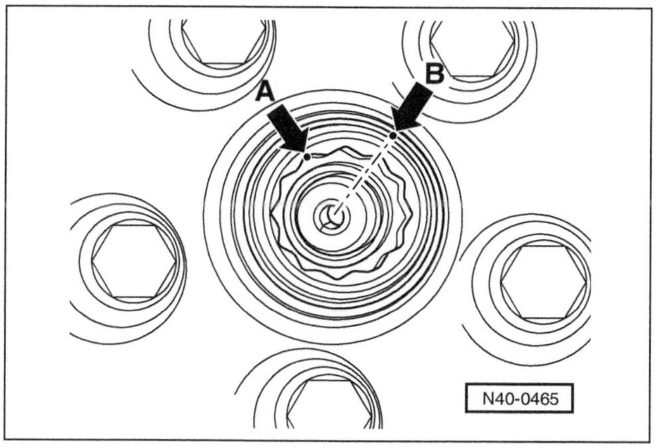

- Tighten 12-point axle nut to 200 Nm (148 ft-lb) and then loosen ½ turn (180°).
- Tighten 12-point axle nut to 50 Nm (37 ft-lb).
- Mark edge of one of the 12-point nut tips with a line (**arrow A**).
- Make a second mark (**arrow B**) on the inside of the wheel hub as illustrated.
  - Second mark (**arrow B**) should be positioned "2 tips" to right of first mark (**arrow A**).
- Tighten 12-point axle nut until the two marks line up.
- Install center cap on wheel.

Drive axle assembly with CV joint

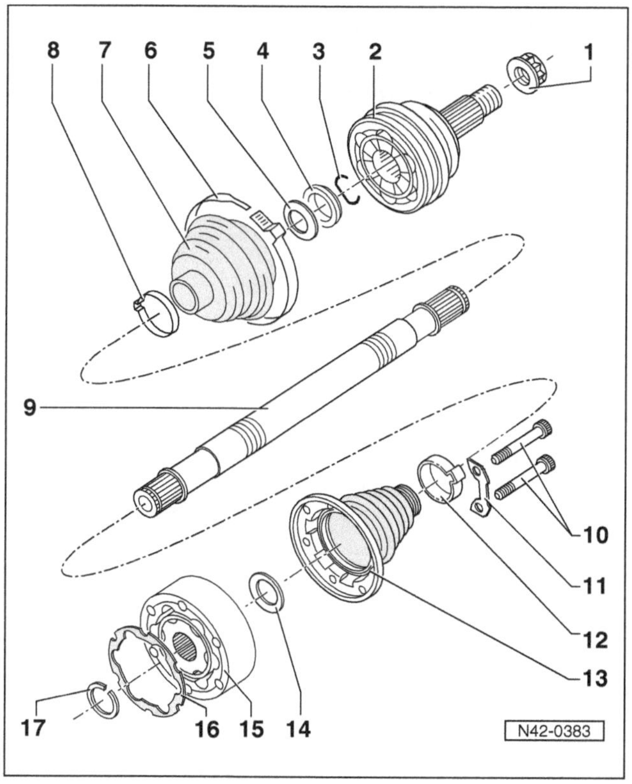

1. 12-point axle nut
   - Always replace
2. Outer CV joint
   - Replace as complete unit
   - Drive off CV joint using drift
3. Circlip
   - Always replace
4. Thrust washer
5. Convex washer
6. Boot clamp
7. Boot for CV joint
8. Boot clamp
9. Drive axle
10. Socket head bolts
    - 40 Nm (30 ft-lb)
11. Lock plate
12. Boot clamp
13. Boot for CV joint
14. Convex washer
15. Inner CV joint
    - Replace as complete unit
    - Drive off CV joint using drift
16. Gasket
    - Pull of protective foil before installing new gasket
17. Snap ring

BentleyPublishers
.com
©2006 BentleyPublishers.com—All Rights Reserved

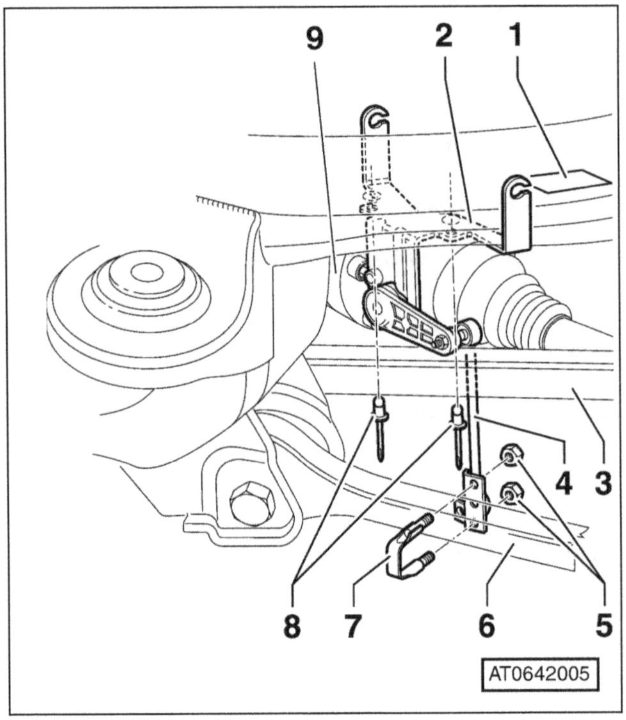

Vehicle level sensor assembly (Xenon headlights)

NOTE—
- *Audi identifies electrical components by a letter and/or a number in the electrical schematics. See* **EWD Electrical Wiring Diagrams**. *These electrical identifiers are listed in parentheses as an aid to electrical troubleshooting.*

1. Subframe
2. Bracket
3. Upper transverse link
4. Coupling rod
5. Self-locking nut
    - 4 Nm (35 in-lb)
    - Always replace
6. Lower transverse link
7. Clamp
    - Mark installation position before removing
8. Rivet
    - Replace with self tapping screw if necessary
9. Left rear level control system sensor (G76)
    - Do not remove coupling rod from ball head

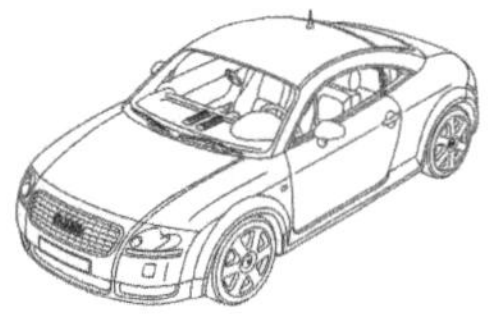

BentleyPublishers
.com
©2006 BentleyPublishers.com—All Rights Reserved

44 Wheels, Tires, Wheel Alignment

TABLES

GENERAL

This repair group covers basic tire, wheel, and wheel alignment information. Also covered here is wheel alignment specifications to be used in conjunction with professional alignment tools and measuring equipment.

WHEELS AND TIRES

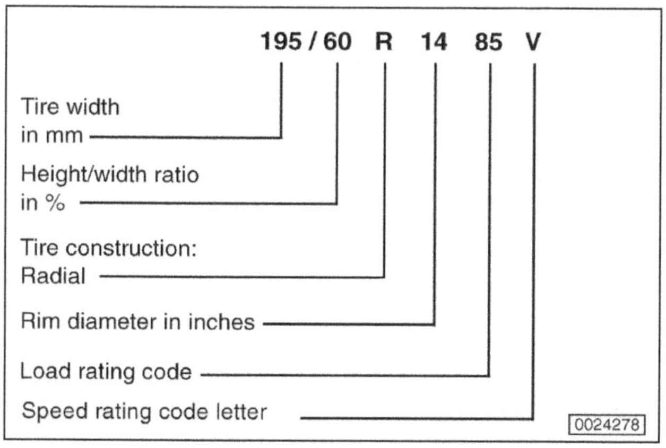

Wheels and tires approved by the manufacturer have been matched to the vehicle and contribute largely to the road handling and driving characteristics. Replace tires with tires having the same specifications with regard to size, design, load carrying capacity, speed rating, tread pattern, tread depth, etc. This information can be found on the tire's sidewall.

Audi recommends that tires be rotated front to rear, with the tires remaining on the same side of the vehicle. For tire rotation schedules and general tire and wheel service information, see **03 Maintenance**.

When installing wheel bolts, they should be snugged down in a diagonal pattern as shown and then tightened to final torque using the same pattern

Tightening torque	
Wheel bolt to wheel hub	120 Nm (89 ft-lb)

BentleyPublishers.com

© 2006 BentleyPublishers.com—All Rights Reserved

Audi TT roadster with 7.5J x 17 inch, 5-spoke alloy wheels.

A number of factory supplied alloy wheels are available for the models covered by this manual. There are also two versions of steel wheels available: a road wheel and a temporary spare. All wheels use a 5 by 100mm (5/100) bolt pattern. Specifications for the various wheels can be found below.

Wheel specifications	Offset
3,5J x 18 inch (temporary spare)	not specified
6J x 16 inch (steel road wheel)	31 mm
7J x 16 inch (7-spoke alloy)	31 mm
7.5J x 17 inch (6-spoke alloy)	32 mm
7.5J x 17 inch (5-spoke alloy)	32 mm
7.5J x 18 inch (7-spoke alloy)	32 mm
8J x 18 inch (9-spoke alloy)	33 mm

Tire mounting

Use tire mounting equipment designed for use with alloy wheels to avoid damage to the wheel and tire.

Wheel balancing

Alloy wheels should have a maximum of 60 grams in wheel balancing weights on each side (front and back).

Wheel bolts

Wheel bolt threads should be cleaned with a wire brush and lightly greased to prevent corrosion.

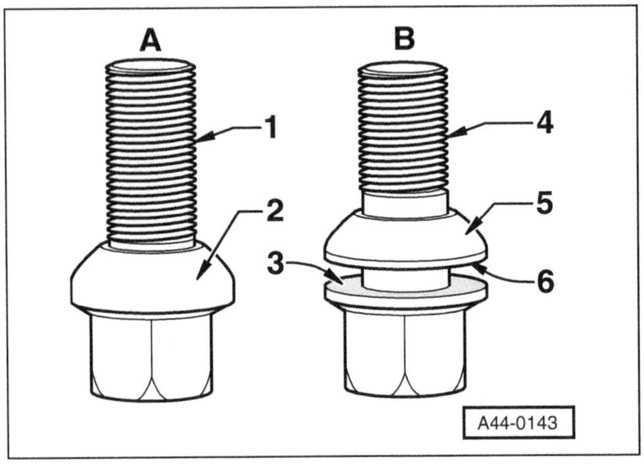

Two types of wheel bolts are used: one-piece (**A**) and two-piece (**B**). Lightly grease the following areas:

- Threads (**1** and **4**)
- Convex shoulder (**2**)
- Area between bolt head contact surface (**3**) and back of convex washer (**6**)
- DO NOT grease convex surface (**5**) on two-piece bolt (**B**).

Wheel seat protection

When changing an alloy or steel wheel, the centering seat (hub centric) should be protected against corrosion. Failure to protect this area will allow rust and/or corrosion to form which will make it difficult to remove the wheel.

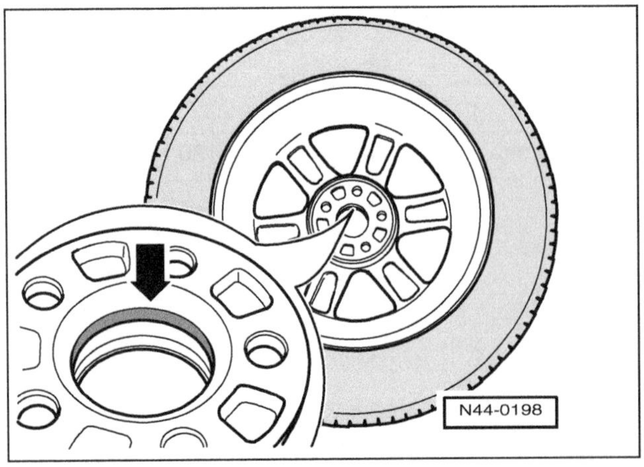

Apply wax to centering area (**arrow**). Factory recommended spray wax **D 322 000 A2** or equivalent should be applied after area has been cleaned with a wire brush.

> ***WARNING —***
> - *Apply wax to center seat area of wheel only. Avoid getting wax on mounting surfaces of wheel and brake rotor, brake pad contact surfaces and threaded surfaces. Wax on brake pad contact surfaces can result in poor brake performance.*

BentleyPublishers.com
©2006 BentleyPublishers.com—All Rights Reserved

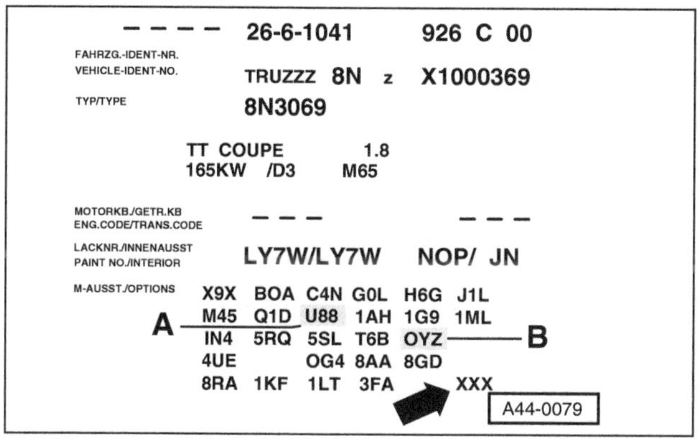

Wheel Alignment

Alignment specifications

Wheel alignment specifications depend on suspension type. Various suspension versions are available depending on model, engine and equipment. The suspension installed is identified by PR numbers (**arrow**) on the vehicle data label. The label is located in the spare wheel recess or in the glovebox owner's manual.

Table a. Wheel Alignment Data—Front Suspension

Suspension (version)	**Standard (FWD and AWD)**	**Sport (FWD and AWD)**
PR number (from vehicle data label)	G17, G78, G79, G91, G97	PQZ, G98
Toe setting at each wheel (wheels not pressed)	4' ± 3.5'	
Total toe (wheels not pressed)	8' ± 7'	
Camber (in straight-ahead position)	–45' ± 30'	–58' ± 30'
Maximum permissible difference between left and right	max. 30'	
Toe angle difference at 20° lock from left to right	1° 31" ± 20'	
Caster (not adjustable)	+7° 58'	+8° 15'
Maximum permissible difference between left and right	max. 30'	

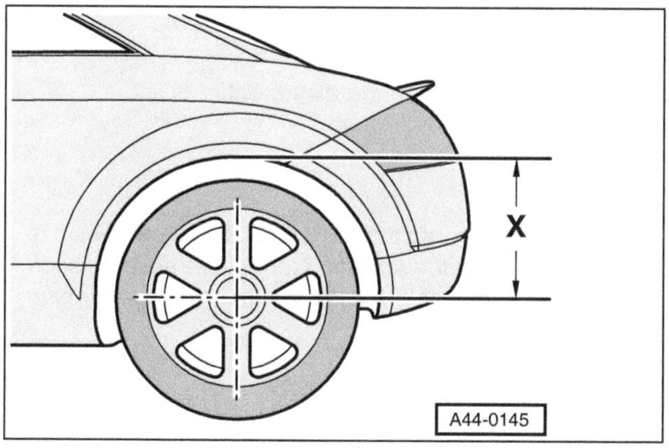

Rear axle camber various according to suspension height of vehicle. Be sure to measure the rear suspension height (**X**) before checking wheel alignment. Rear camber is only adjustable on vehicles with All Wheel Drive (AWD).

Rear suspension height and camber value (AWD)			
Standard suspension		**Sport suspension**	
Height	**Camber**	**Height**	**Camber**
352 mm	-2° 10'	332 mm	-2° 48'
355 mm	-2° 04'	335 mm	-2° 42'
360 mm	-1° 54'	340 mm	-2° 32'
365 mm	-1° 44'	345 mm	-2° 22'
370 mm	-1° 34'	350 mm	-2° 12'
372 mm	-1° 30'	352 mm	-2° 08'
375 mm	-1° 24'	355 mm	-2° 02'
380 mm	-1° 14'	360 mm	-1° 52'
382 mm	-1° 10'	362 mm	-1° 48'

BentleyPublishers
.com
©2006 BentleyPublishers.com—All Rights Reserved

Table b. Wheel Alignment Data—Rear Suspension

Suspension (version)	**Standard (FWD)**	**Sport (FWD)**	**Standard (AWD)**	**Sport (AWD)**
PR number (from vehicle data label)	G17, G78, G79, G91, G97	PQZ, G98	G17, G78, G79, G91, G97	PQZ, G98
Camber (only adjustable on AWD)	-2° ± 20'	-2° ± 20'	see **Rear suspension height and camber values** table	
Maximum permissible difference between left and right	max. 30'	max. 30'	max. 20'	max. 20'
Toe setting at each wheel (only adjustable on AWD)	+14' ± 5'	+19.5' ± 5'	$+7.5'\,^{+7.5'}/_{-5'}$	$+7.5'\,^{+7.5'}/_{-5'}$
Total toe (with specified camber)	+28' ± 10'	+39' ± 10'	$+15'\,^{+15'}/_{-10'}$	$+15'\,^{+15'}/_{-10'}$
Maximum permissible deviation from direction of travel (thrust angle)	max. 15'	max. 15'	max. 20'	max. 20'

Alignment adjustments

Alignment should be checked whenever the vehicle is not tracking properly, if there is uneven tire wear, if suspension damage is suspected, or if a repair may have altered the alignment.

The procedural information provided below details how the various alignment adjustments are acheived with the use of special alignment tools. Accurate wheel alignment is best performed by your Audi dealer or certified alignment shop.

Requirements

- Install brake pedal depressor.
- Make sure no one is in vehicle during alignment.
- Check wheel rim runout.

NOTE —
- *A certain amount of lateral runout at the wheel rims is permissible, but this may already exceed the specified toe-in tolerance. In such cases it is not possible to correctly set the toe-in without performing the wheel runout compensation.*

Order of work

1. Check camber at front wheels and center if necessary.
2. Check camber at rear wheels.
3. Check toe setting at rear wheels and adjust if necessary.
   - The toe setting at the rear wheels is not adjustable on vehicles with front wheel drive.
4. Check toe setting at front wheels and adjust if necessary.
5. If settings have been altered at the front wheels, perform zero position compensation for steering angle sensor (G85) using a diagnostic scan tool.
   - Only for vehicles with Electronic Stability Program (ESP)

BentleyPublishers.com
© 2006 BentleyPublishers.com—All Rights Reserved

Camber at front wheels, adjusting

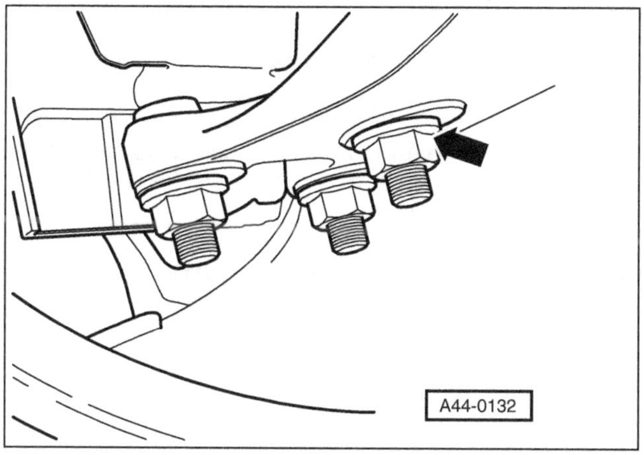

◀ Remove inner nut (**arrow**) for ball joint mount.

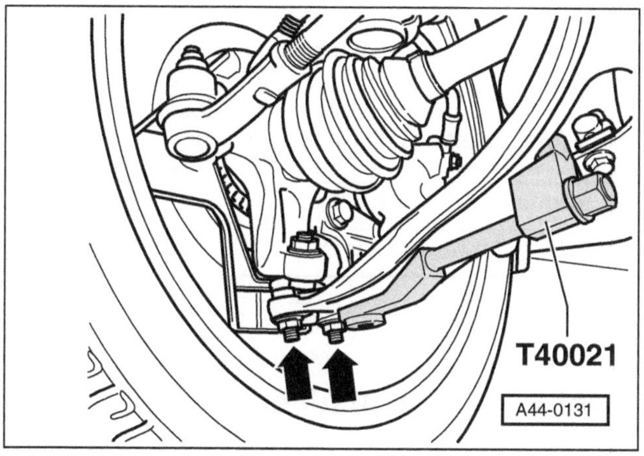

◀ Attach tool special tool **T40021** to control arm as shown.

– Remove two remaining ball joint mounting nuts (**arrows**).

– Camber can be adjusted by turning the adjustment nut on special tool T40021.

– Install new mounting nuts for ball joint and tighten to specification.

Tightening torque	
Ball joint to control arm (always replace nuts)	75 Nm (55 ft-lb)

Camber at rear wheels, adjusting

For vehicles with front wheel drive, camber cannot be adjusted at the rear wheels. If measurements are outside the permitted tolerances, the rear axle beam must be checked for damage and replaced if necessary.

For vehicles with all wheel drive, the camber can be adjusted by means of the hole tolerance at the transverse links.

– Loosen bolts securing transverse links to trailing arm. See **42 Rear Suspension**.

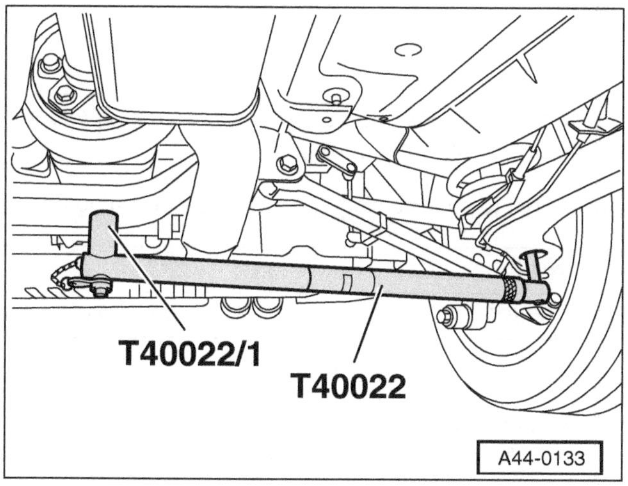

◀ Attach alignment tools as shown.

– Adjust camber to specification using alignment tool (T40022) and tighten transverse link mounting bolts to specification.

Tightening torque	
Transverse links to trailing arm (always replace)	70 Nm (52 ft-lb) plus ¼ turn (90°)

BentleyPublishers.com
©2006 BentleyPublishers.com—All Rights Reserved

Toe setting at rear wheels, adjusting

On vehicles with front wheel drive, the total toe setting at the rear wheels cannot be adjusted. The individual toe values can be centered by moving the rear axle within the elongated mounting holes. See **42 Rear Suspension**. If toe measurements are outside the permitted tolerances, the rear axle beam must be checked for damage and replaced if necessary.

Toe setting at rear wheels, adjusting (AWD)

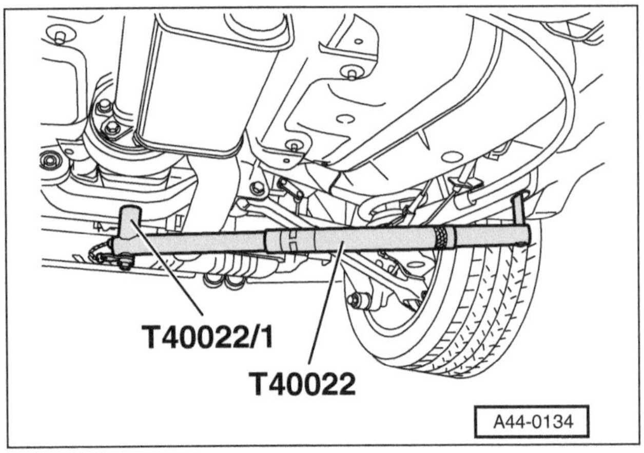

◄ Attach alignment tools as shown.

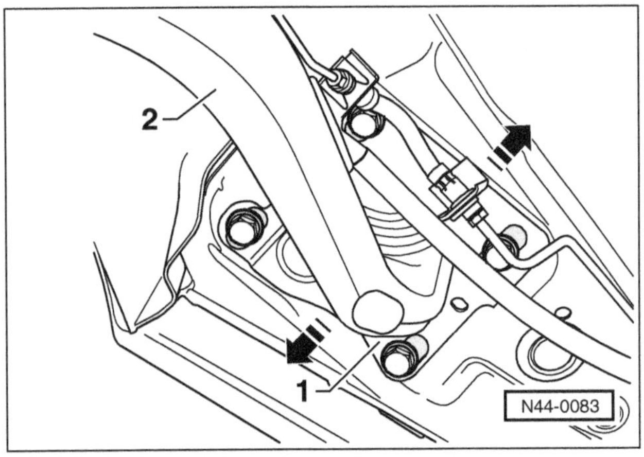

◄ Loosen all bolts for trailing arm mounting bracket (**1**).

– Move trailing arm (**2**) in direction of **arrows** using alignment tool T40022 to adjust toe.

– Tighten trailing arm mounting bolts to specification.

Tightening torque	
Trailing arm to body (always replace)	75 Nm (55 ft-lb)

Toe at front wheels, adjusting

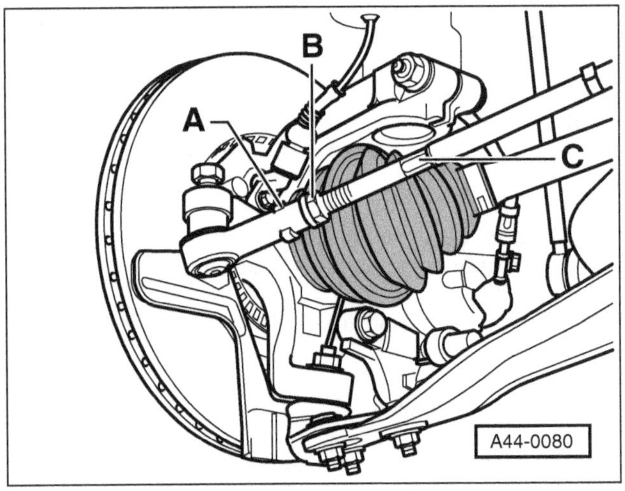

◄ Counterhold tie rod end (**A**) with a 24mm wrench when loosening and tightening lock nut (**B**).

– Loosen lock nut (**B**).

– Adjust toe setting at left and right-hand wheels by turning hex flats (**C**) on tie rod.

NOTE—
- *Make sure that rubber boots are not twisted after turning tie rods. Twisted rubber boots wear out quickly.*

– Tighten lock nut (**B**) to specification and check toe-in setting again.

Tightening torque	
Tie rod end to tie rod	50 Nm (37 ft-lb)

NOTE—
- *The setting may change slightly after tightening lock nut. Adjustment is correct if measured toe value is still inside the tolerance.*

BentleyPublishers.com
© 2006 BentleyPublishers.com—All Rights Reserved

45 Antilock Brakes (ABS)

TABLE

GENERAL

This repair group covers the antilock brake systems. For related information see:

- **46 Brakes–Mechanical**
- **47 Brakes–Hydraulic**

ABS variants

Electronically controlled basic ABS maintains vehicle stability and control during emergency braking by preventing wheel lock-up. Vehicles covered by this manual are equipped with refined versions of the ABS system. The capabilities of these systems go beyond antilock braking and include additional stability aids:

- **Mark 20: ABS/EDL, ABS/EDL/ASR, ABS/EDL/ASR/ESP**
- **Mark 60: ABS/EDL/ASR/ESP**

The following explains commonly used acronyms:

- **ABS (antilock braking system)**, the basic system, prevents the wheels from locking during critical braking maneuvers, allowing the driver improved brake effect, track stability and steering control.
- **ASR (antislip regulation)** inhibits drive wheel spin while accelerating by intervening with the engine management system and applying brake pressure to the front drive wheels. ASR is only used on vehicles with front wheel drive.
- **EDL (electronic differential lock)**. On a road surface where each wheel has a different degree of traction, the wheel that spins is braked and the wheel with the most traction is supplied torque. EDL can be used on front wheel drive or all wheel drive equipped vehicles, but only functions on the front brakes.
- **ESP (electronic stabilization program)** helps control the vehicle in critical maneuvers, such as negotiating curves, by preventing the wheels from losing traction. This is done by intervening with the engine management system and applying different brake pressure to different wheels.

BentleyPublishers.com

© 2006 BentleyPublishers.com—All Rights Reserved

Antislip regulation (ASR)

ASR uses ABS and the engine management system to inhibit drive wheel spin during acceleration. Drive wheel speeds and engine torque are controlled to a level that corresponds with the available traction on the road surface.

ASR slows down a spinning wheel using the braking system, then sends a message from the ABS control module through the CAN-bus to the engine control module (ECM). The ECM then tries to decrease excessive wheel spin by:

- Adjusting throttle valve position
- Retarding ignition timing
- Suppressing individual injector pulses

ASR may be switched ON or OFF by pressing the dash mounted ASR button (**arrow**). However, when the ignition is cycled, ASR automatically switches ON. Under normal conditions, leave ASR switched ON. Switch OFF in certain circumstances where wheel spin can offer advantages, such as:

- Driving using the compact spare tire
- Rocking the vehicle back and forth
- Starting off on an uphill gradient where the amount of traction is severely reduced on one side of the vehicle

When ASR intervenes, the ASR indicator in the instrument cluster flashes. When ASR is switched OFF, the ASR indicator light (K86) remains illuminated.

Electronic stabilization program (ESP)

ESP is active throughout the driving experience, unlike ASR which is only active during acceleration and braking. ESP helps to stabilize the vehicle in cornering and avoidance maneuvers by adjusting throttle, ignition, fuel injection and application of brake pressure to individual wheels.

ESP uses many principles and components of the ABS / ASR system, in addition to the following components:

- Transverse acceleration sensor (G200) and rotation rate (yaw) sensor (G202). See **Transverse acceleration sensor (G200) and rotation rate (yaw) sensor (G202), removing and installing**, in this repair group. Note, G200 and G202 were integrated into the ESP sensor (G419) as of approximately 09/01.
- Steering angle sensor (G85). See **48 Steering** for functional information. See **96 Interior Lights, Switches, Anti-theft** for removal and installation.

The ESP motion sensor unit uses various inputs to determine vehicle instability during braking, cornering, or reduced traction situations. Based upon these inputs the motion sensor sends outputs to the engine control module and the ABS hydraulic unit to activate torque reduction protocols and braking intervention.

ESP may be toggled ON or OFF by pressing the dash mounted ESP switch (**arrow**). Turning ESP off does not disable other ABS functions. However, when the ignition is cycled, ESP automatically switches ON. Under normal conditions, leave ESP switched ON. Switch OFF in certain circumstances where wheel spin can offer advantages.

BentleyPublishers.com

©2006 BentleyPublishers.com—All Rights Reserved

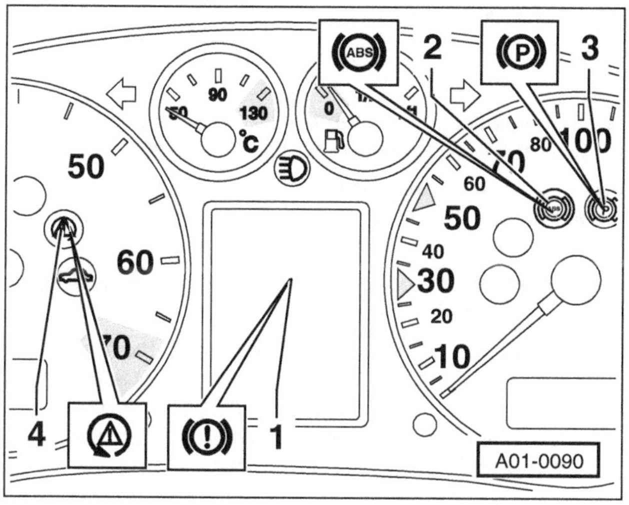

ABS warning lights

Warning lights for ABS system.

Item	Designation
1	Brake system warning light (K118)
2	ABS warning light (K47)
3	Parking brake indicator light (K14)
4	Traction control indicator light (K86 or K155)

Table a. ABS warning light troubleshooting

Warning light	Condition	Possible explanation
Brake system (K118)	Warning light does not go out after ignition switched ON.	Parking brake applied.
		Low brake fluid (warning light blinks and warning tones sound).
		Short in wiring to brake system warning light.
ABS warning (K47)	Warning light does not go out after ignition switched ON.	Voltage supply below 10 volts.
		Malfunction in ABS, conventional brake system remains functional.
		Temporary wheel speed sensor malfunction.
		Open circuit in connection from instrument cluster to ABS control module.
		Instrument cluster faulty.
Parking brake (K14)	Warning light does not go out after ignition switched ON.	Parking brake applied.
		Parking brake light switch malfunction.
Traction control (K86 or K155)	Light flashes while driving.	System in intervention mode and functioning correctly.
	Light does not light when ignition switched ON.	Indicator light malfunction.
	Light does not go out after ignition switched ON.	Malfunction of ASR or ESP. Other systems remain functional.
		Short circuit to B+ in ASR or ESP button.
		Indicator light malfunction.
		ASR or ESP switched off by console button.

WARNING —
- *If both the ABS and BRAKE warning lights come on at the same time, the rear wheels could lock up under hard braking. This could cause loss of vehicle control. Have vehicle repaired immediately by an authorized Audi dealer if both warning lights come on. Drive slowly and avoid sudden, hard braking.*

BentleyPublishers.com
© 2006 BentleyPublishers.com—All Rights Reserved

Warnings and Cautions

WARNING —

- *ABS is a vehicle safety system; appropriate knowledge and special equipment are necessary to work on the system.*
- *If ABS and brake system warning lights illuminate, the ABS system is compromised and wheels may lock up prematurely when braking.*
- *Bleed the ABS system after opening the brake hydraulic system. See* **47 Brakes–Hydraulic System**.
- *If an ABS fault is detected by On-Board Diagnostics, the function of the brake system may be limited and there is a risk of accident. The brake pressure at the rear wheels is no longer controlled by the EBD function. This can result in excessive braking at the rear, and the vehicle may skid unexpectedly under braking.*
- *Brake fluid is poisonous.*
- *The ABS and brake system warning lights are not capable of monitoring all brake system functions; visual inspections are still required for the system.*
- *Do not use silicone-based brake fluids (DOT 5). This fluid is incompatible with the brake system and even the smallest trace can cause severe corrosion.*
- *Absolute cleanliness is required when working on the ABS system. Do not use any products that contain mineral oil such as oil, grease, etc.*
- *Carefully seal or cover opened components if repairs cannot be completed immediately. Sealing plugs are contained in factory supplied repair kit.*
- *Upon completion of repairs, road test the vehicle so that the ABS is felt to engage through the pulsing of the brake pedal.*
- *Do not reuse fasteners that are worn or deformed.*

CAUTION—

- *For ASR to work properly, make sure all four wheels are fitted with identical tires. Any differences in the rolling radius of the tires can cause ASR to reduce engine power.*
- *Keep brake fluid off painted surfaces; it removes paint.*
- *An illuminated ABS warning light indicates a fault in the system. Check the fault memory with an Audi diagnostic scan tool, or equivalent, before disconnecting the battery.*
- *Before working on the ABS, disconnect the battery ground cable.*
- *Brake fluid absorbs moisture from the surrounding air. Replace every two years. Use only new brake fluid that complies with Federal Motor Vehicle Safety Standards 116, DOT 4 Super or DOT 4 Plus.*
- *Before doing any electric welding to the vehicle, disconnect the ABS control module.*
- *Do not expose the ABS control module to high heat (as in a drying booth) for a prolonged period. The module may be exposed to a max. temperature of 95°C (203°F) for a short period and to a max. temperature of 85°C (185°F) for up to 2 hours.*
- *Do not let brake fluid enter wiring harness connectors.*

BentleyPublishers.com

©2006 BentleyPublishers.com—All Rights Reserved

ABS Components

ABS electrical components

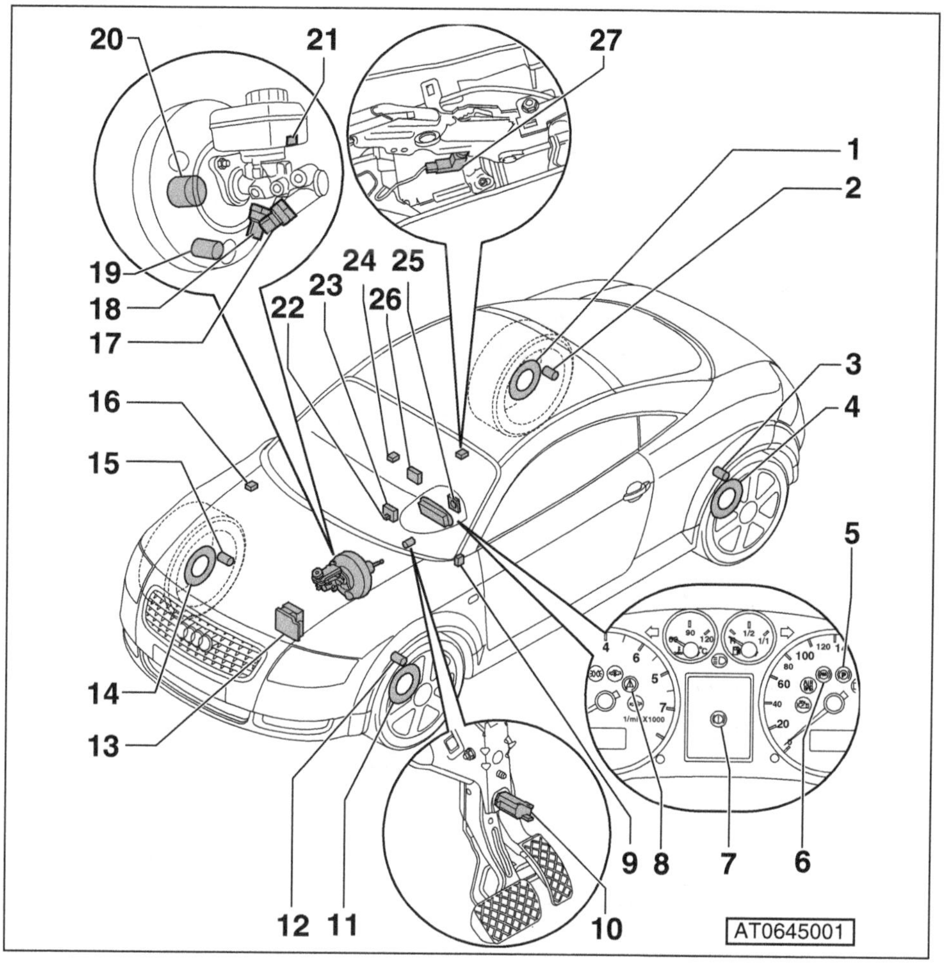

1. **Rotor for ABS wheel speed sensor**
   - Right rear
   - Rotor is welded to wheel hub
2. **ABS wheel speed sensor (G44)**
   - Right rear
   - Check using diagnostic scan tool
   - See **ABS wheel speed sensors, removing and installing** in this repair group
3. **ABS wheel speed sensor (G46)**
   - Left rear
4. **Rotor for ABS wheel speed sensor**
   - Left rear
5. **Parking brake indicator light (K14)**
   - Located in instrument cluster
6. **ABS warning light (K47)**
   - Located in instrument cluster
7. **Brake system warning light (K118)**
   - Located in instrument cluster display
8. **Traction control indicator light (K86)**
   - Located in instrument cluster
9. **Brake light disable relay (J508)**
   - Located at fuse box
   - Only for vehicles with ESP
10. **Brake light switch (F)**
    - See **Brake light switch, removing and installing** in this repair group
11. **Rotor for ABS wheel speed sensor**
    - Left front
12. **ABS wheel speed sensor (G46)**
    - Left front
13. **ABS hydraulic unit (N55) with ABS control module (J104)**
    - Hydraulic unit consists of:
      ABS hydraulic pump (V64)
      ABS hydraulic unit (N55)
    - Do not separate hydraulic pump and valve block
    - Check using diagnostic scan tool
14. **Rotor for ABS wheel speed sensor**
    - Right front
15. **ABS wheel speed sensor (G45)**
    - Right front
16. **Longitudinal acceleration sensor (G251)**
    - Only for vehicles with all wheel drive
17. **Brake pressure sensor 2 (G214)**
18. **Brake pressure sensor 1 (G201)**
19. **Brake pressure release solenoid (F84)**
    - Brake booster assembly must be replaced if solenoid is defective
    - Only for ITT Mark 20 ABS system
20. **Magnetic coil for brake pressure (N247)**
    - Brake booster assembly must be replaced if solenoid is defective
    - Only for ITT Mark 20 ABS system
21. **Brake fluid level warning switch (F34)**
22. **Transverse acceleration sensor (G200)**
    - Integral with ESP sensor unit (G419), from 09/01
23. **Rotation rate sensor (G202)**
    - Integral with ESP sensor unit (G419), from 09/01
24. **Anti-slip control switch (E132)**
    - Located on dash board
    - ASR or ESP, depending on system
25. **Steering angle sensor (G85)**
    - Located on steering column
    - Check using diagnostic scan tool
26. **ESP motion sensor unit (G419)**
    - Installed from 09/01
27. **Parking brake warning light switch (F9)**
    - Located on parking brake

BentleyPublishers
.com
©2006 BentleyPublishers.com—All Rights Reserved

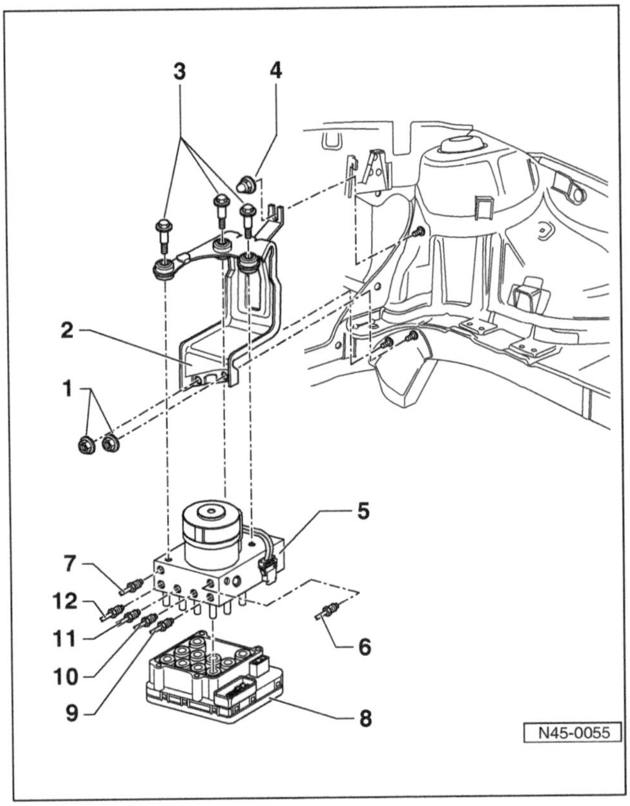

ABS hydraulic components (Mark 20)

1. Nut, self-locking (always replace)
   - 20 Nm (15 ft-lb)
2. Bracket
3. Bolt
   - 8 Nm (70 in-lb)
4. Cap nut
   - 20 Nm (15 ft-lb)
5. ABS hydraulic unit (N55)
   - Mark hydraulic lines before removing
6. Brake line
   - Brake master cylinder/secondary piston circuit to hydraulic unit
7. Brake line
   - Brake master cylinder/primary piston circuit to hydraulic unit
8. ABS control module (J104)
9. Brake line
   - To left front brake caliper
10. Brake line
    - To right rear brake caliper
11. Brake line
    - To left rear brake caliper
12. Brake line
    - To right front brake caliper

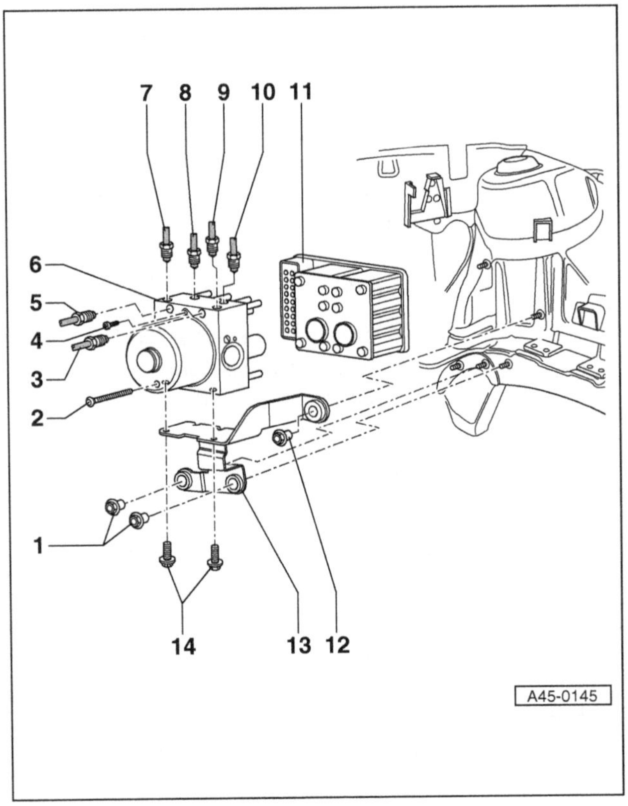

ABS hydraulic components (Mark 60)

1. Cap nut
   - 20 Nm (15 ft-lb)
2. Torx head socket bolt
   - 4 Nm (35 in-lb)
3. Brake line
   - To left rear brake caliper
4. Torx head socket bolt
   - 4 Nm (35 in-lb)
5. Brake line
   - To right rear brake caliper
6. ABS hydraulic unit (N55)
   - Mark hydraulic lines before removing
7. Brake line
   - Brake master cylinder/primary piston circuit to hydraulic unit
8. Brake line
   - To left front brake caliper
9. Brake line
   - To right front brake caliper
10. Brake line
    - Brake master cylinder/secondary piston circuit to hydraulic unit
11. ABS control module (J104)
12. Cap nut
    - 20 Nm (15 ft-lb)
13. Retainer
14. Bolt
    - 8 Nm (70 in-lb)

BentleyPublishers.com
©2006 BentleyPublishers.com—All Rights Reserved

ABS Component Replacement

For information on steering angle sensor (G85), see **48 Steering**. For steering angle sensor replacement, see **96 Interior Lights, Switches, Anti-theft, Steering Column Switches**.

ABS wheel speed sensors, removing and installing

Each wheel has a separate speed sensor to supply data to the ABS control module. Sensors are monitored for malfunctions by the On Board Diagnostic (OBD) system. The system checks the sensors for proper resistance when the ignition is first switched ON and again for correct signal when the vehicle is in motion. Malfunctions will cause a Diagnostic Trouble Code (DTC) to be stored in the OBD system.

A wheel speed sensor rotor is welded to each wheel hub. The rotors are not available separately.

– Raise car and support safely. See **03 Maintenance**.

> ***WARNING —***
> - *Make sure car is stable and well supported at all times. Use a professional automotive lift or jack stands designed for the purpose. A floor jack is not adequate support.*

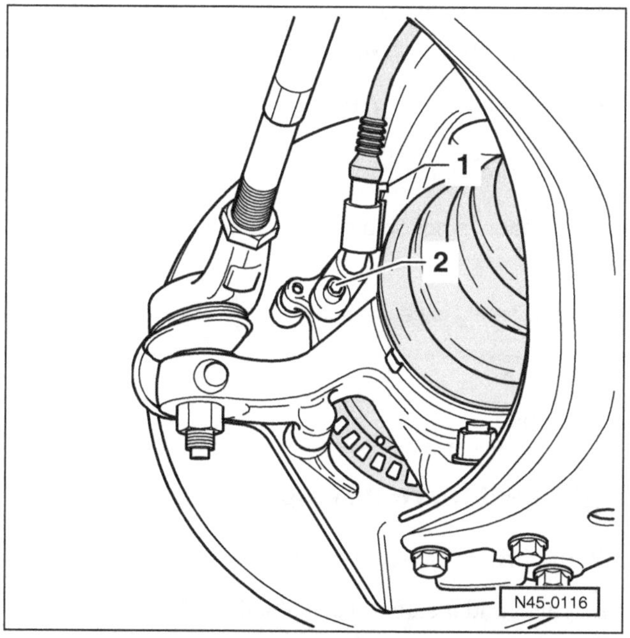

Front sensor:

- Working at front wheel, disconnect wheel speed senor harness connector (**1**).
- Remove bolt (**2**) from wheel bearing housing.

Rear sensor:

- Working at rear wheel, unclip sensor wire from retainers (as applicable) and disconnect harness connector.
- Remove bolt (**arrow**) from wheel bearing housing.
- All wheel drive model shown, front wheel drive model is similar.

– Remove wheel speed sensor from wheel bearing housing.

– Prior to inserting new wheel speed sensor, clean inner surface of bore hole and coat sensor bore with securing grease (Audi G 000 650 or equivalent).

– Insert wheel speed sensor into hole in wheel bearing housing.

Tightening torque	
Wheel speed sensor to wheel bearing housing	8 Nm (70 in-lb)

– Reconnect speed sensor harness connector.

– Turn front wheels fully to left and right to check sensor wire clearance.

BentleyPublishers.com
©2006 BentleyPublishers.com—All Rights Reserved

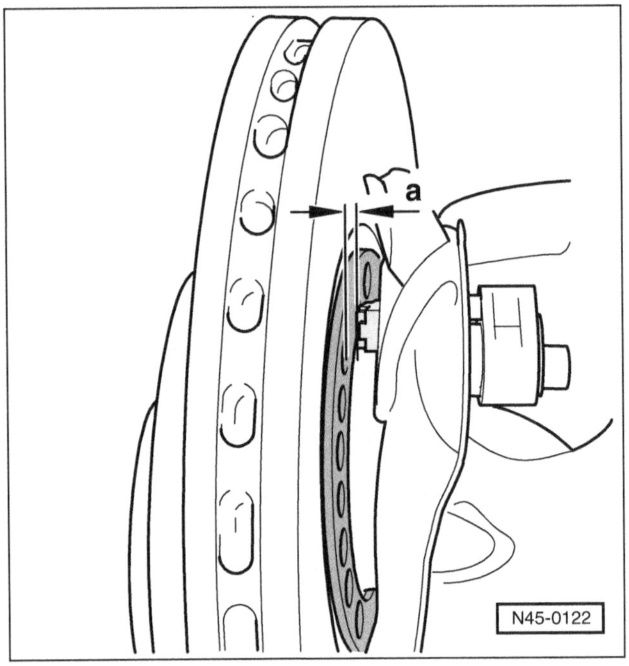

ABS wheel speed sensor rotors, checking

A wheel speed sensor rotor is welded to each wheel hub. The rotors are not available separately.

– Raise car and support safely. See **03 Maintenance**.

> ***WARNING —***
> • *Make sure car is stable and well supported at all times. Use a professional automotive lift or jack stands designed for the purpose. A floor jack is not adequate support.*

– Remove wheel.

– Check rotor for dirt and damage by rotating brake disc.
  - If rotor is damaged, wheel hub with welded rotor must be replaced.

◄ Check rotor for lateral run-out by turning wheel hub and checking if gap (**a**) is uniform.
  - Distance **a** = 0.3 mm (0.012 in)

– If run-out specification is not obtained, replace wheel hub with welded ABS rotor.

– Install wheels and safely lower vehicle.

Tightening torque	
Wheel to wheel hub	120 Nm (89 ft-lb)

Brake light switch, removing and installing

Brake light switch with square housing

– Remove trim panel under steering column. See **70 Trim–Interior**.

– Unplug harness connector from switch.

– Remove brake light switch by turning it counter-clockwise 90° and pulling out from brake pedal.

– Before installing, pull plunger out from brake light switch.

– Press brake pedal down as far as possible by hand.

◄ Position switch in brake pedal and turn clockwise 90° (**arrows**) to install.

– Release brake pedal.

– Check function of brake lights when brake pedal is pressed and released.

> ***NOTE —***
> • *Brake light switch should only be installed once to ensure secure mounting. Replace switch if necessary.*

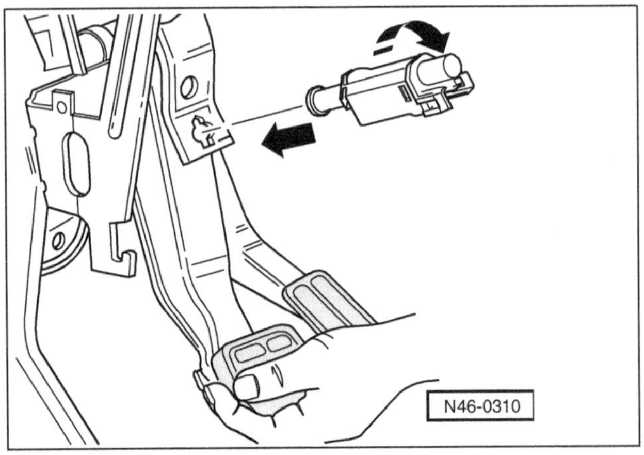

BentleyPublishers.com
©2006 BentleyPublishers.com—All Rights Reserved

Brake light switch with round housing

Some vehicles have a brake light switch with a round housing. Removal and installation of this switch only requires a 45° rotation.

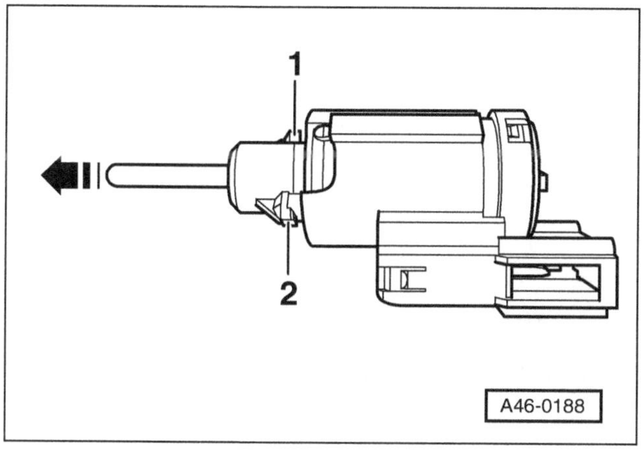

< Before installing, pull plunger (**arrow**) out from brake light switch.

– Position retainer tabs (**1** and **2**) into recesses of installation hole and secure switch by turning clockwise 45°.

– When installing switch with round housing, do not operate brake pedal.

– The plunger is automatically adjusted when the brake light switch is installed into pedal.

– After installation, check function of brake lights when brake pedal is pressed and released.

NOTE —

- *Brake light switch should only be installed once to ensure secure mounting. Replace switch if necessary.*

ESP motion sensors, removing and installing

The ESP system utilizes 2 or 3 motion sensors, depending on drivetrain version:

- Transverse acceleration sensor (G200)
- Rotation rate sensor (G202)
- Longitudinal acceleration sensor (G251) (all wheel drive only)

From approximately 09/01 the motion sensors are integrated into the ESP sensor unit (G419).

> ***CAUTION—***
> - *Excessive shaking (e.g. dropping, impact) may damage the motion sensors. If in doubt, replace sensors or sensor unit.*

Transverse acceleration sensor (G200) and rotation rate sensor (G202), removing and installing (through 08/01)

– Remove trim panel under steering column. See **70 Trim–Interior**.

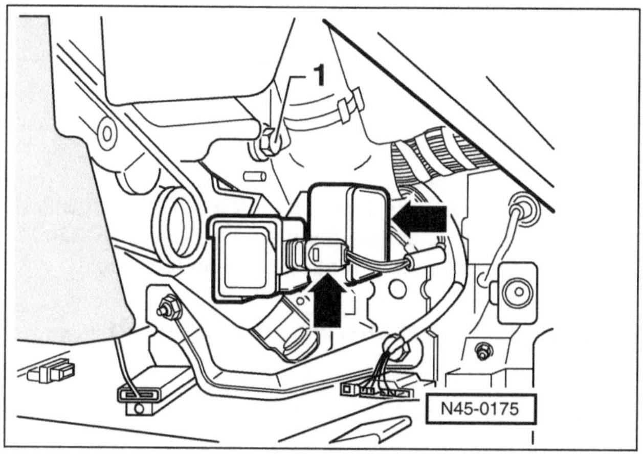

< Unplug harness connectors (**arrows**) from motion sensors (G200 and G202).

– Remove bolt (**1**) and remove mounting bracket with sensors attached.

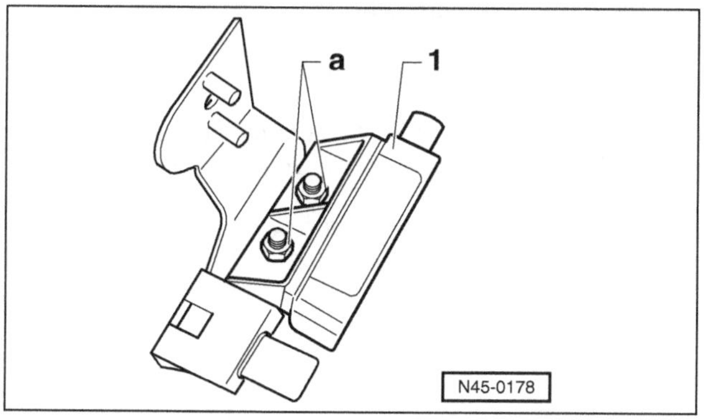

< Remove nuts (**a**) and remove rotation rate sensor (**1**).

– To install, insert rotation rate sensor into mounting bracket and align straight before tightening mounting nuts.

Tightening torque	
Rotation rate sensor to mounting bracket	9 Nm (80 in-lb)

Bentley Publishers .com
© 2006 BentleyPublishers.com—All Rights Reserved

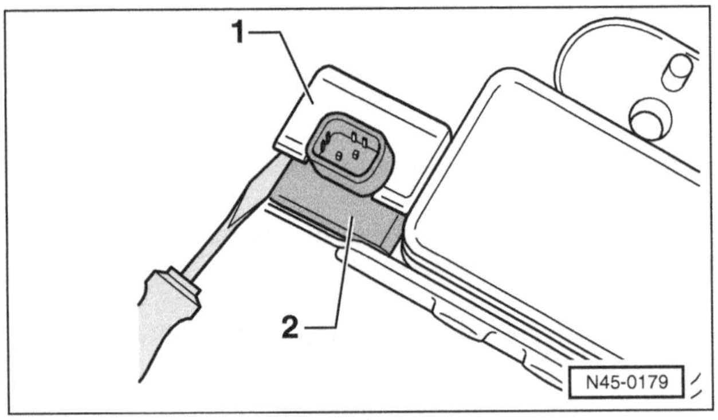

- Use screwdriver to pry mounting bracket (**1**) off transverse acceleration sensor (**2**) and remove sensor.

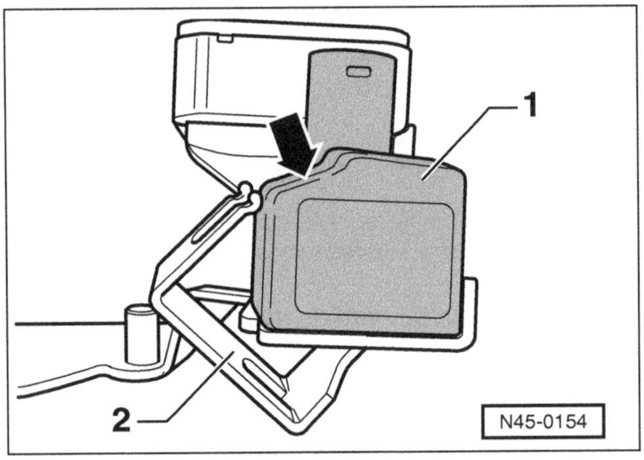

- Install transverse acceleration sensor (**1**) into mounting bracket.
  - Sloping side (**arrow**) of sensor must be positioned as shown.
- Slip mounting bracket (**2**) over sensor until it engages with a click.
- Install mounting bracket with sensors in reverse order of removal.

Tightening torques	
Motion sensor mounting bracket to steering column	20 Nm (15 ft-lb)

- Use a diagnostic scan tool (VAG 1551/5051 or equivalent) to initiate basic setting for motion sensors. See **OBD On Board Diagnostics**.

Longitudinal acceleration sensor (G251), removing and installing (through 08/01)

- Remove glove box. See **70 Trim–Interior**.

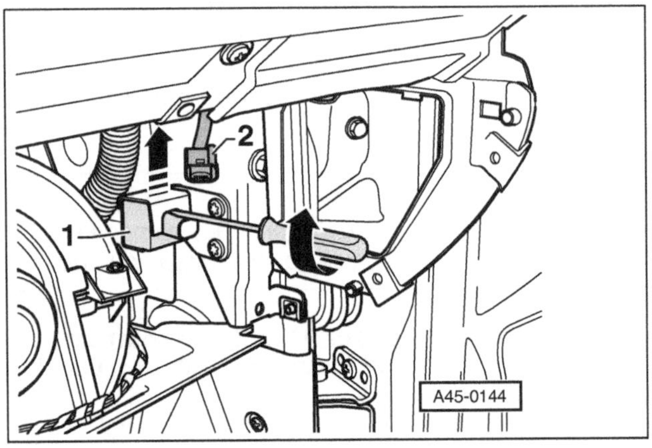

- Unplug harness connector (**2**).
- Use a screwdriver to pry (**arrows**) mount off longitudinal acceleration sensor and remove sensor (**1**).
- Install in reverse order of removal.
- Use a diagnostic scan tool (VAG 1551/5051 or equivalent) to initiate basic setting for longitudinal acceleration sensor. See **OBD On Board Diagnostics** for more information on scan tool use.

ESP sensor unit (G419), removing and installing (from 09/01)

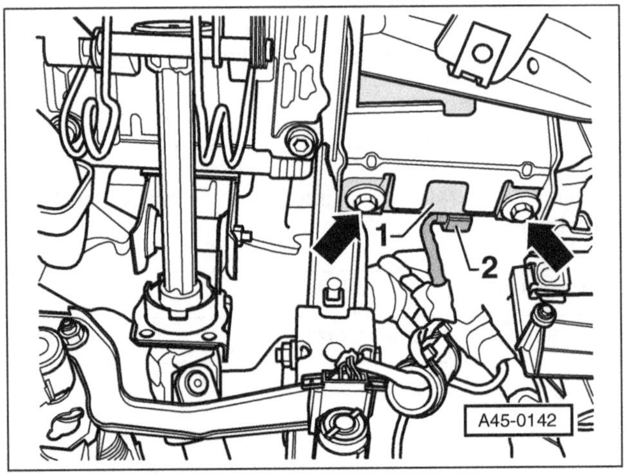

- Disconnect harness connector (**2**) from ESP motion sensor unit (**1**).
- Remove two mounting bolts (**arrows**).
- Remove ESP motion sensor unit (G419).
- Installation is reverse of removal. Make sure sensor unit is properly seated on bracket and is free of tension. Do not force into position.
- Use a diagnostic scan tool (VAG 1551/5051 or equivalent) to initiate basic setting for ESP sensor unit. See **OBD On Board Diagnostics** for more information on scan tool use.

BentleyPublishers.com
© 2006 BentleyPublishers.com—All Rights Reserved

46 Brakes–Mechanical

TABLES

General

This repair group covers mechanical brake components including front and rear brake pads, calipers, brake pedal and parking brake. For additional brake system information, see:

- **03 Maintenance** for brake pad inspection
- **45 Antilock Brakes (ABS)** for ABS related components
- **47 Brakes–Hydraulic** for brake calipers, brake fluid bleeding

Brake discs

Brake discs should be inspected for cracks, scoring, glazing and warpage. Replace discs (in pairs) if either disc is worn below the minimum thickness or if any of the above listed defects are found.

For brake disc application information, see **Table a. Brake caliper and disc applications** in this repair group.

Wear limits	
Front or rear brake pads minimum thickness, without backing plate (all models)	2 mm (0.079 in)
1.8L front brake disc minimum thickness	23 mm (0.90 in)
3.2L front brake disc minimum thickness	30 mm (1.18 in)
1.8L (180 hp) rear disc minimum thickness	7 mm (0.27 in)
1.8L (225 hp) rear disc minimum thickness	20 mm (0.79 in)
3.2L rear brake disc minimum thickness	20 mm (0.79 in)

NOTE—
- *Minimum brake rotor thickness is also stamped on rotor hub.*

© 2006 BentleyPublishers.com—All Rights Reserved

Table a. Brake caliper and disc applications		
Engine	**PR number (vehicle data label)**	**Caliper, disc**
Front		
1.8L (180 hp)	1LT	FN3, 312 x 25 mm ventilated
1.8L (225 hp)	G1A, 1LV	FN3, 312 x 25 mm ventilated
3.2L (250 hp)	1ZT	2FN, 334 x 32 mm ventilated
Rear		
1.8L (180 hp, FWD)	1KK, 2EK	C 38, 232 x 9 mm non-ventilated
1.8L (180 hp, AWD)	1KE	C 41, 239 x 9 mm non-ventilated
1.8L (225 hp, AWD)	1KF	C 38, 256 x 22 mm ventilated
3.2L (250 hp)	1KF	C 38, 256 x 22 mm ventilated

Warnings

WARNING —
- *If the red brake/parking light illuminates while driving, check brake system immediately.*
- *Brake fluid is poisonous. Wear safety glasses when working with brake fluid, and wear rubber gloves to prevent brake fluid from entering the bloodstream through cuts or scratches. Do not siphon brake fluid by mouth.*
- *New brake pads require some breaking in. Allow for slightly longer stopping distances for the first 100 to 150 miles of city driving, and avoid hard stops.*
- *All brake work must be done with cleanliness, careful attention to specifications, and proper working procedures. If you lack the skills, the tools or a clean workplace for servicing the brake system, we suggest you leave these repairs to an authorized VW dealer or other qualified shop.*
- *After replacing brake components, depress the brake pedal firmly several times to seat the brakes in their normal operating position. The pedal should be firm and at its normal height, if not, further work is required before driving vehicle.*
- *Brake fluid damages paint. Immediately wipe up any brake fluid that spills on the vehicle and rinse area with water.*

BentleyPublishers.com
© 2006 BentleyPublishers.com—All Rights Reserved

FRONT BRAKES (1.8L)

Front brake assembly

NOTE—

- *After replacing brake pads, depress brake pedal firmly several times with vehicle stationary so that the brake pads are properly seated in their normal operating position.*
- *Use brake filler/bleeder unit (VAS 5234 or equivalent) to extract brake fluid from brake fluid reservoir.*
- *Before removing brake caliper or disconnecting brake hose, brake pedal depressor V.A.G 1869/2 or equivalent) must be installed (this relieves hydraulic pressure).*

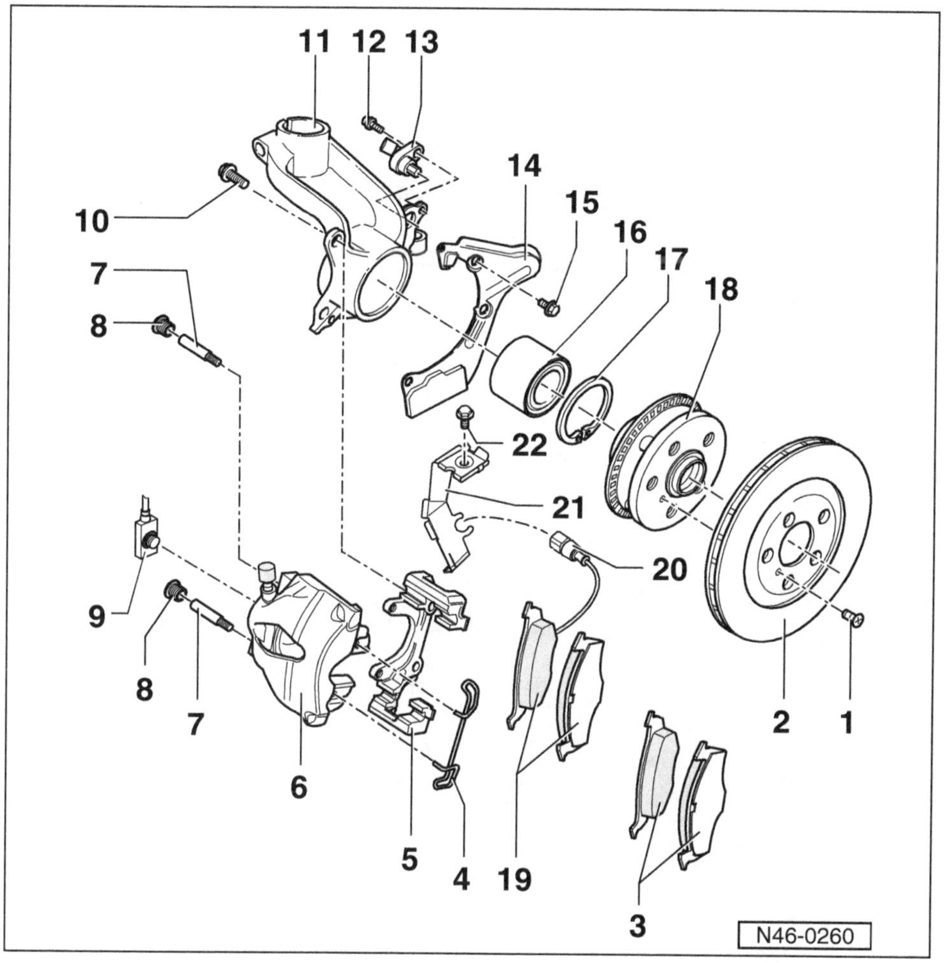

1. **Screw**
   - 4 Nm (35 in-lb)
2. **Brake disc**
   - Diameter: 312 mm (12.3 in)
   - Thickness: 25 mm (0.98 in)
   - Wear limit: 23 mm (0.90 in)
   - Always replace both sides
   - Disc can be damaged if removed forcefully, use penetrating oil
3. **Brake pads**
   - Thickness: 14 mm (0.55 in) (without backing plate)
   - Wear limit: 2 mm (0.079 in) (without backing plate)
   - See **Front brake pads, removing and installing** in this repair group
4. **Retaining spring**
   - Insert in both holes of brake caliper
5. **Brake carrier**
   - Replacement carriers are supplied with sufficient grease on guide pins
6. **Brake caliper FN3**
   - Do not disconnect brake hose when changing brake pads
   - See **47 Brakes–Hydraulic**
7. **Guide pins**
   - 28 Nm (21 ft-lb)
   - Lubricate when installing
8. **Guide pin cap**
9. **Brake hose with banjo fitting and union bolt**
   - 35 Nm (26 ft-lb)
10. **Bolt**
    - 125 Nm (92 ft-lb)
    - Clean if re-using
11. **Wheel bearing housing**
12. **Bolt**
    - 8 Nm (71 in-lb)
13. **ABS wheel speed sensor**
    - Removing and installing, see **45 Anti-Lock Brakes (ABS)**
14. **Backing plate**
15. **Hex head bolt**
    - 10 Nm (7 ft-lb)
16. **Wheel bearing**
17. **Circlip**
18. **Wheel hub**
    - ABS wheel speed rotor is welded to wheel hub
19. **Brake pads**
    - With wear indicator
    - Thickness: 14 mm (0.55 in) (without backing plate)
    - Wear limit: 2 mm (0.079 in) (without backing plate)
    - See **Front brake pads, removing and installing** in this repair group
20. **Harness connector**
21. **Bracket**
22. **Screw**

BentleyPublishers
.com
©2006 BentleyPublishers.com–All Rights Reserved

Front brake pads, removing and installing

Removing

– Loosen wheel lugs with vehicle standing on the ground.

– Raise vehicle and support safely. Remove wheels.

> ***WARNING —***
> • *Make sure vehicle is stable and well supported at all times. Use a professional automotive lift or jack stands designed for the purpose. A floor jack is not adequate support.*

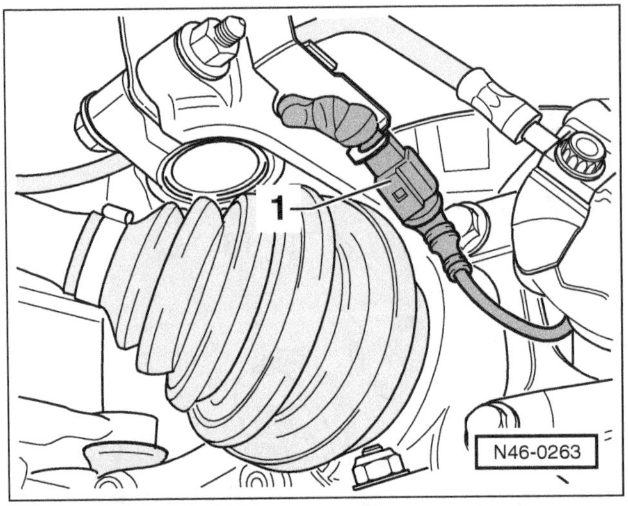

< Disconnect harness connector (**1**) for brake pad wear indicator.

< Using a screwdriver, pry off (at **arrows**) brake pad retaining spring from brake caliper and remove.

< Remove caliper guide pin caps (**arrows**).

– Unbolt and remove guide pins from brake caliper.

– Remove brake caliper and secure with wire so that weight of brake caliper does not damage brake hose.

– Remove brake pads from brake caliper and brake carrier.

NOTE —
• *If reusing brake pads, mark location and install in original position.*

– Clean brake caliper, especially the adhesion surface for the brake pad, it must be free of residual adhesive and grease. Use only appropriate solvents for cleaning brake caliper.

BentleyPublishers .com
© 2006 BentleyPublishers.com—All Rights Reserved

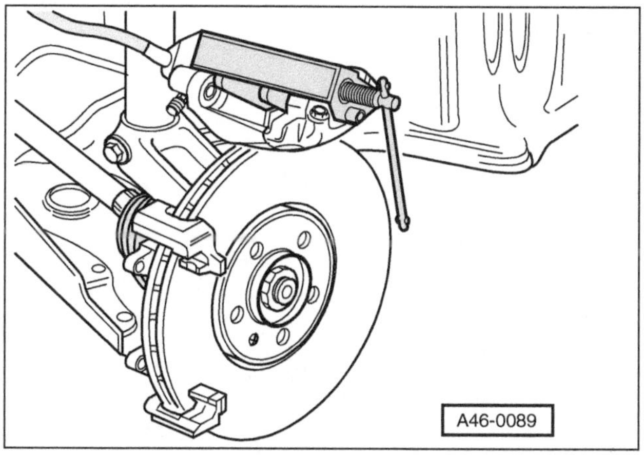

Installing

◄ Press piston back into caliper housing. Be careful not to damage piston seal.

> ***CAUTION—***
> - *Before pressing piston into cylinder, extract fluid from brake fluid reservoir. Otherwise, fluid can overflow and cause damage.*

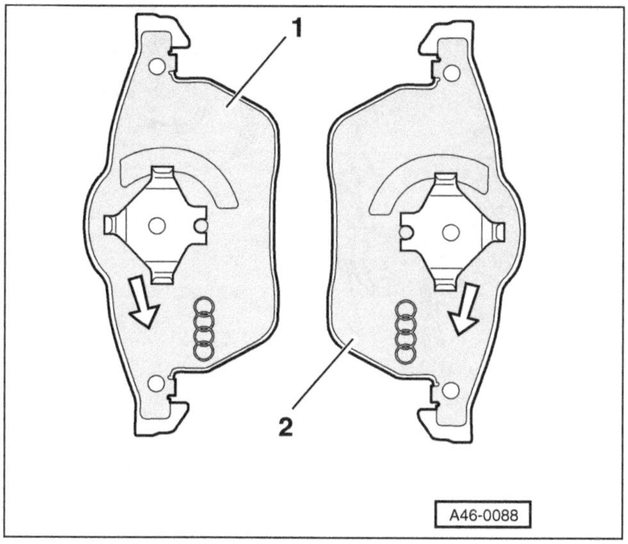

◄ Note installation position of brake pads.

- **Arrow** on backing plate must face downward.

1. Right side pad
2. Left side pad

– Remove protective foil from backing plate of outer brake pad.

– Install brake pads in carrier.

– When installing brake caliper, make sure outer brake pad does not stick to brake caliper and interfere with proper installation.

– Do not damage adhesion surface on outer brake pad.

– Lubricate guide pins with multi-purpose grease.

– Install guide pins and tighten to specification.

Tightening torque	
Caliper guide pins to brake carrier	28 Nm (21 ft-lb)

– Install caps on caliper guide pins.

◄ Install retaining spring in brake caliper.

– Connect harness connector of brake pad wear indicator.

– Install wheels.

Tightening torque	
Wheel bolt to wheel hub	120 Nm (90 ft-lb)

NOTE—
- *After replacing brake pads, depress brake pedal firmly several times with vehicle stationary so that the brake pads are properly seated in their normal operating position.*

– Check brake fluid level and top off if necessary.

BentleyPublishers
.com
© 2006 BentleyPublishers.com—All Rights Reserved

FRONT BRAKES (3.2L)

Front brake assembly

NOTE—
- *After replacing brake pads, depress brake pedal firmly several times with vehicle stationary so that the brake pads are properly seated in their normal operating position.*
- *Use brake filler/bleeder unit (VAS 5234 or equivalent) to extract brake fluid from brake fluid reservoir.*
- *Before removing brake caliper or disconnecting brake hose, brake pedal depressor V.A.G 1869/2 or equivalent) must be installed (this relieves hydraulic pressure).*

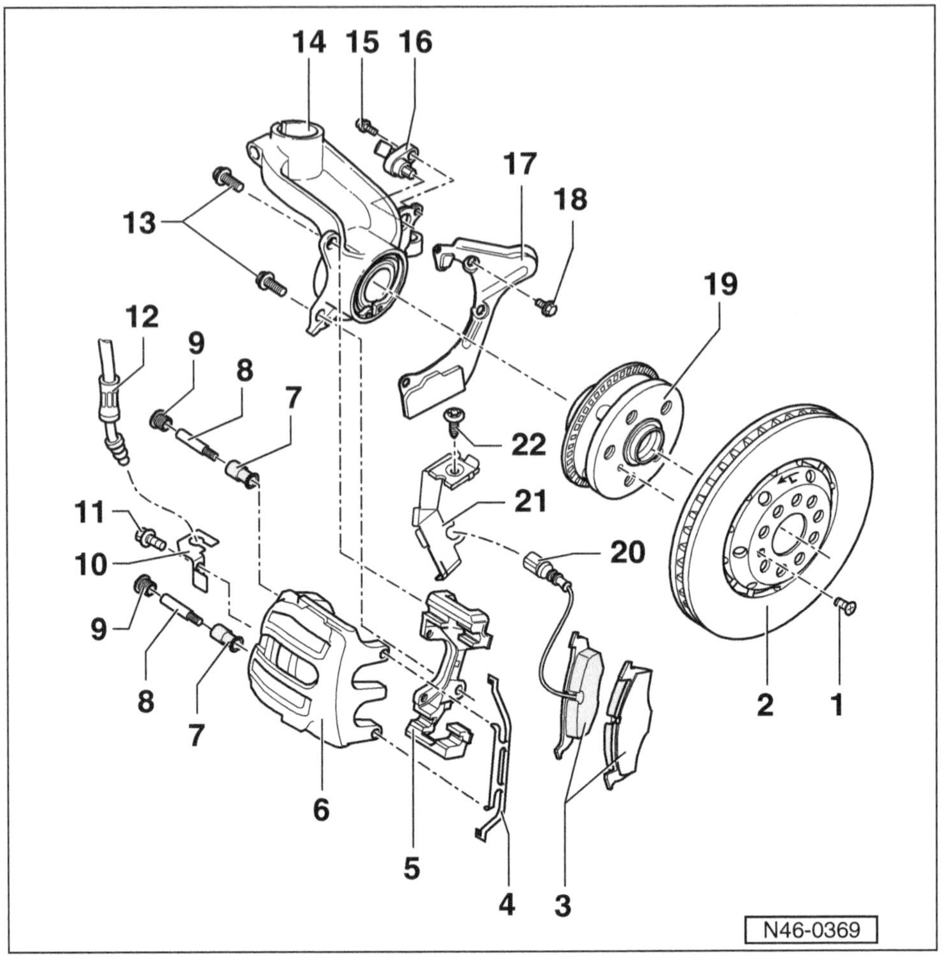

1. **Screw**
   - 4 Nm (35 in-lb)
2. **Brake disc (ventilated)**
   - Diameter: 334 mm (13.1 in)
   - Thickness: 32 mm (1.26 in)
   - Wear limit: 30 mm (1.18 in)
   - Always replace both sides
   - Disc can be damaged if removed forcefully, use penetrating oil
3. **Brake pads**
   - With wear indicator
   - Thickness: 14 mm (0.55 in) (without backing plate)
   - Wear limit: 2 mm (0.079 in) (without backing plate)
   - See **Front brake pads, removing and installing** in this repair group
4. **Retaining spring**
   - Press under brake carrier and insert in both holes of brake caliper
5. **Brake carrier**
   - Bolted to wheel bearing housing
6. **Brake caliper 2FN**
   - Do not disconnect brake hose when changing brake pads
   - See **47 Brakes–Hydraulic**
7. **Bushing**
8. **Guide pins**
   - 25 Nm (18 ft-lb)
   - Lubricate when installing
9. **Guide pin cap**
10. **Bracket**
    - Bolted to brake caliper housing
11. **Bolt**
    - 8 Nm (71 in-lb)
12. **Brake line**
    - 15 Nm (11 ft-lb)
    - Do not disconnect when changing pads
    - Do not twist when installing
13. **Ribbed bolt**
    - 125 Nm (92 ft-lb)
    - Clean if re-using
14. **Wheel bearing housing**
15. **Bolt**
    - 8 Nm (71 in-lb)
16. **ABS wheel speed sensor**
    - Removing and installing, see **45 Anti-Lock Brakes (ABS)**
17. **Backing plate**
18. **Hex head bolt**
    - 10 Nm (7 ft-lb)
19. **Wheel hub**
    - ABS wheel speed rotor is welded to wheel hub
20. **Connector**
21. **Bracket**
22. **Screw**

BentleyPublishers.com
© 2006 BentleyPublishers.com—All Rights Reserved

Front brake pads, removing and installing

Removing

- Loosen wheel lugs with vehicle standing on the ground.
- Raise vehicle and support safely. Remove wheels.

> ***WARNING —***
> - *Make sure vehicle is stable and well supported at all times. Use a professional automotive lift or jack stands designed for the purpose. A floor jack is not adequate support.*

- Secure brake disc with one wheel bolt.

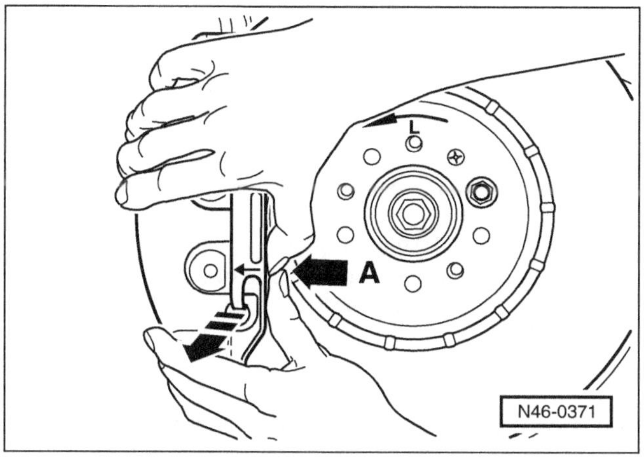

- Push retaining spring in (**A**) until it can be pulled out of hole in brake caliper.

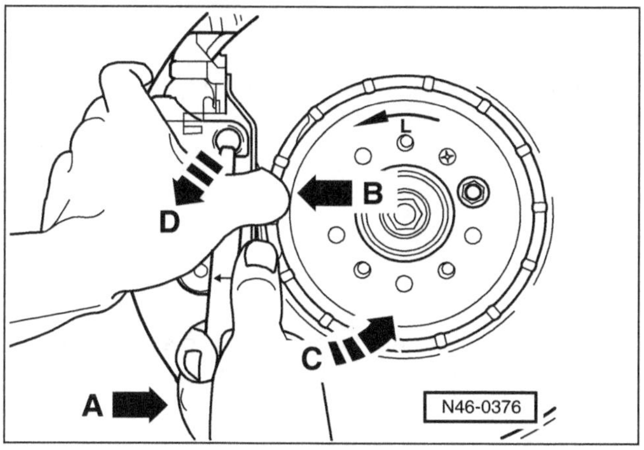

- Pull bottom of retaining spring in direction of **arrow A** and hold in this position. Push top of retaining spring in direction of **arrow B**. At the same time, turn spring slightly counter-clockwise (**arrow C**) and pull it out of hole in caliper (**arrow D**).

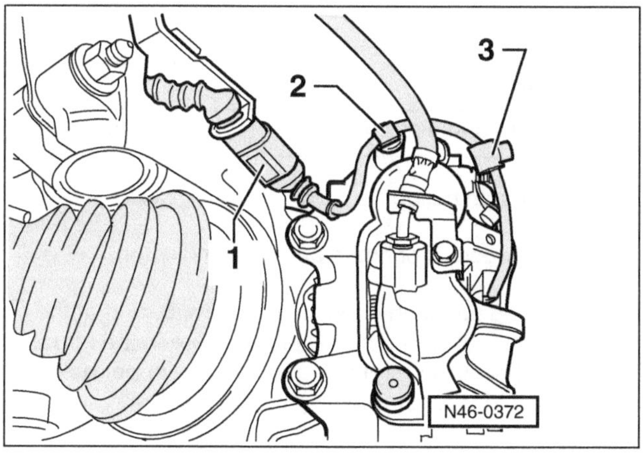

- Disconnect harness connector (**1**) for brake pad wear indicator.
- Detach wire for pad wear indicator from retainer (**2**) on upper guide pin cap and from retainer (**3**) on dust cap.
- Remove caliper guide pin caps.
  - Upper cap is combined with retainer for wire of brake pad wear indicator.

BentleyPublishers
.com
© 2006 BentleyPublishers.com—All Rights Reserved

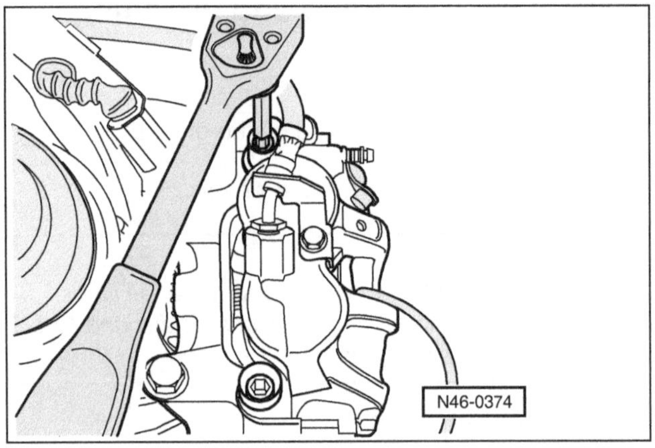

Unbolt and remove guide pins from brake caliper.

– Remove brake caliper and secure with wire so that weight of brake caliper does not damage brake hose.

– Remove brake pads from brake caliper and brake carrier.

NOTE—

- *If reusing brake pads, mark location and install in original position.*

– Clean brake caliper, especially the adhesion surface for the brake pad, it must be free of residual adhesive and grease. Use only appropriate solvents for cleaning brake caliper.

Installing

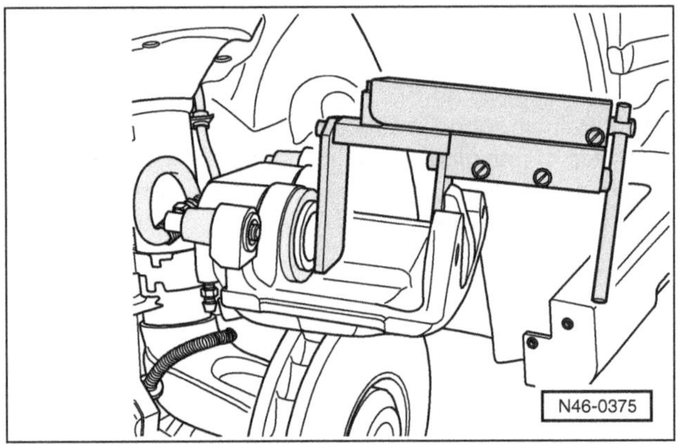

Press piston back into caliper housing. Be careful not to damage piston seal.

> ***CAUTION—***
>
> - *Before pressing piston into cylinder, extract fluid from brake fluid reservoir. Otherwise, fluid can overflow and cause damage.*

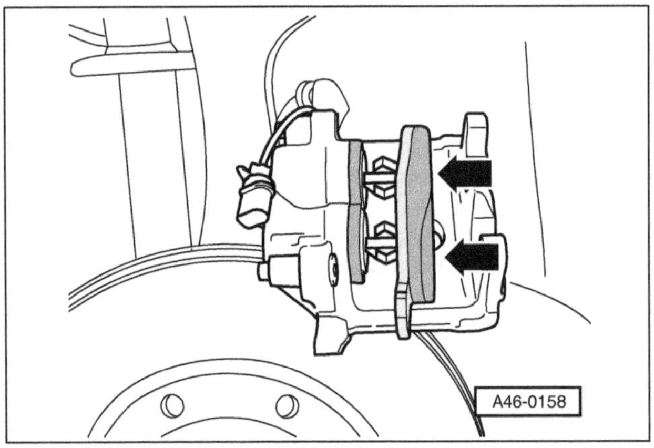

Install inner brake pad in carrier. Position retaining springs into caliper pistons and push in direction of **arrows**. Install outer brake pad.

– Lubricate guide pins with multi-purpose grease. Install guide pins and tighten to specification.

Tightening torque	
Caliper guide pins to brake carrier	25 Nm (21 ft-lb)

– Install guide pin caps on caliper.

– Connect harness connector of brake pad wear indicator.

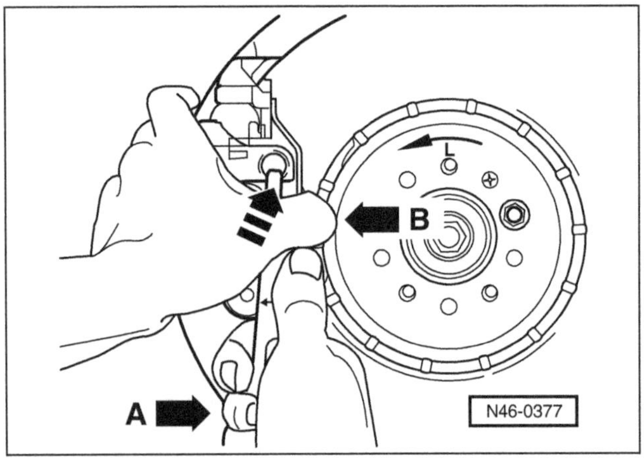

Press top of retaining spring in direction or **arrow B**, and at the same time hold bottom of spring and exert pressure in the opposite direction (**arrow A**). Now push retaining spring into upper hole of brake caliper. Repeat for lower retaining spring hole.

– Install wheels.

Tightening torque	
Wheel bolt to wheel hub	120 Nm (90 ft-lb)

NOTE—

- *After replacing brake pads, depress brake pedal firmly several times with vehicle stationary so that the brake pads are properly seated in their normal operating position.*

– Check brake fluid level and top off if necessary.

BentleyPublishers.com
© 2006 BentleyPublishers.com—All Rights Reserved

Rear Brakes

Rear brake assembly (vehicles with front wheel drive)

NOTE—
- *After replacing brake pads, depress brake pedal firmly several times with vehicle stationary so that the brake pads are properly seated in their normal operating position.*
- *Use brake charger/bleeder unit (VAS 5234 or equivalent) to extract brake fluid from brake fluid reservoir.*
- *Before removing brake caliper or disconnecting brake hose, brake pedal depressor (V.A.G 1869/2 or equivalent) must be installed (this relieves hydraulic pressure).*

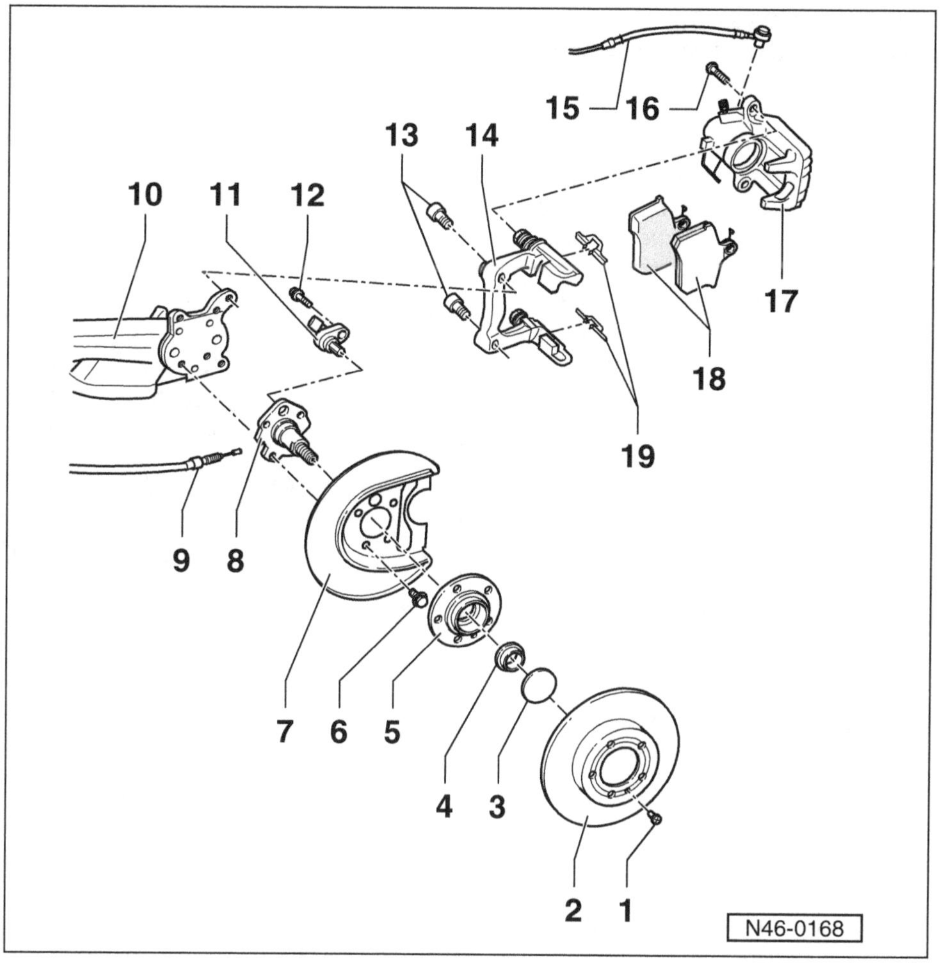

1. **Screw**
   - 4 Nm (35 in-lb)
2. **Brake disc**
   - See **Table a. Brake caliper and disc applications** earlier in this repair group for specifications
   - Replace both sides if worn
   - Disc can be damaged if removed forcefully, use penetrating oil
3. **Dust cap**
4. **12-point stub axle nut**
   - 220 Nm (162 ft-lb)
5. **Wheel hub**
   - With wheel bearing and ABS rotor
6. **Bolt**
   - 60 Nm (44 ft-lb)
7. **Backing plate**
8. **Stub axle**
9. **Parking brake cable**
10. **Axle beam**
11. **ABS wheel speed sensor**
    - Before inserting sensor, clean inner surface of bore hole and coat with hot bolt paste G 052 112 A3
12. **Bolt**
    - 8 Nm (71 in-lb)
13. **Socket head bolts**
    - 65 Nm (48 ft-lb)
14. **Brake carrier**
    - Replacement carriers are supplied with sufficient grease on guide pins
15. **Brake line**
    - Replace if banjo bolt is loosened
    - Tighten banjo bolt to 38 Nm (28 ft-lb)
16. **Caliper mounting bolt**
    - 35 Nm (26 ft-lb)
    - Always replace
    - Counter-hold guide pin when removing/installing
17. **Brake caliper**
    - Do not disconnect brake hose when changing brake pads
18. **Brake pads**
    - Thickness: 12 mm (0.47 in) (without backing plate)
    - Wear limit: 2 mm (0.078 in) (without backing plate)
    - See **Rear brake pads, removing and installing** in this repair group
19. **Anti-rattle springs**
    - Always replace

BentleyPublishers
.com
© 2006 BentleyPublishers.com—All Rights Reserved

Rear brake assembly (vehicles with all wheel drive)

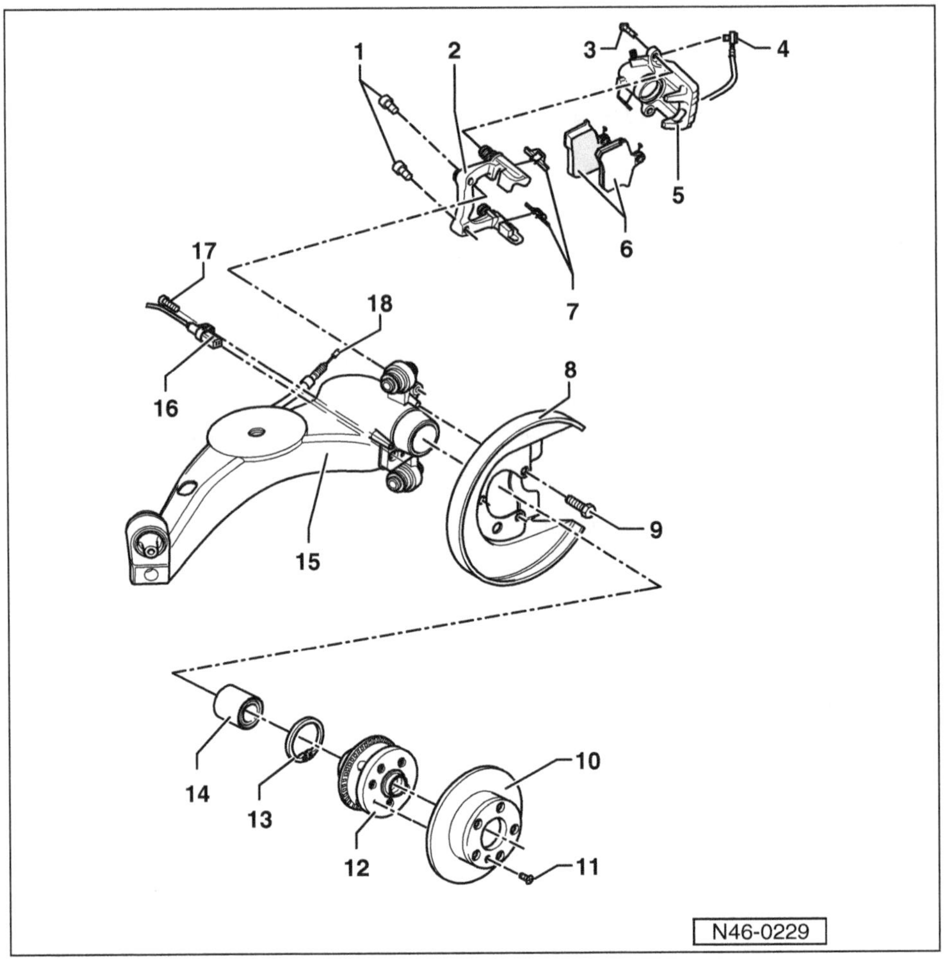

NOTE—

- *After replacing brake pads, depress brake pedal firmly several times with vehicle stationary so that the brake pads are properly seated in their normal operating position.*
- *Use brake charger/bleeder unit (VAS 5234 or equivalent) to extract brake fluid from brake fluid reservoir.*
- *Before removing brake caliper or disconnecting brake hose, brake pedal depressor (V.A.G 1869/2 or equivalent) must be installed (this relieves hydraulic pressure).*

1. **Socket head bolts**
   - 65 Nm (48 ft-lb)
2. **Brake carrier**
   - Replacement carriers are supplied with sufficient grease on guide pins
3. **Caliper mounting bolt**
   - 35 Nm (26 ft-lb)
   - Always replace
   - Counter-hold guide pin when removing/installing
4. **Brake line**
   - Replace if banjo bolt is loosened
   - Tighten banjo bolt to 38 Nm (28 ft-lb)
5. **Brake caliper**
   - Do not disconnect brake hose when changing brake pads
6. **Brake pads**
   - Thickness: 12 mm (0.47 in) (without backing plate)
   - Wear limit: 2 mm (0.078 in) (without backing plate)
   - See **Rear brake pads, removing and installing** in this repair group
7. **Anti-rattle springs**
   - Always replace
8. **Backing plate**
9. **Bolt**
   - 10 Nm (7 ft-lb)
10. **Brake disc**
    - See **Table a. Brake caliper and disc applications** earlier in this repair group for specifications
    - Replace both sides if worn
    - Disc can be damaged if removed forcefully, use penetrating oil
11. **Screw**
    - 4 Nm (35 in-lb)
12. **Wheel hub**
    - With ABS rotor
13. **Circlip**
14. **Wheel bearing**
15. **Trailing arm**
16. **ABS wheel speed sensor**
    - Before inserting sensor, clean inner surface of bore hole and coat with hot bolt paste G 052 112 A3
17. **Bolt**
    - 8 Nm (71 in-lb)
18. **Parking brake cable**

BentleyPublishers.com

© 2006 BentleyPublishers.com—All Rights Reserved

Rear brake pads, removing and installing

Removing

– Loosen wheel lugs with vehicle standing on the ground.

– Raise vehicle and support safely. Remove wheels.

> ***WARNING —***
> - *Make sure vehicle is stable and well supported at all times. Use a professional automotive lift or jack stands designed for the purpose. A floor jack is not adequate support.*

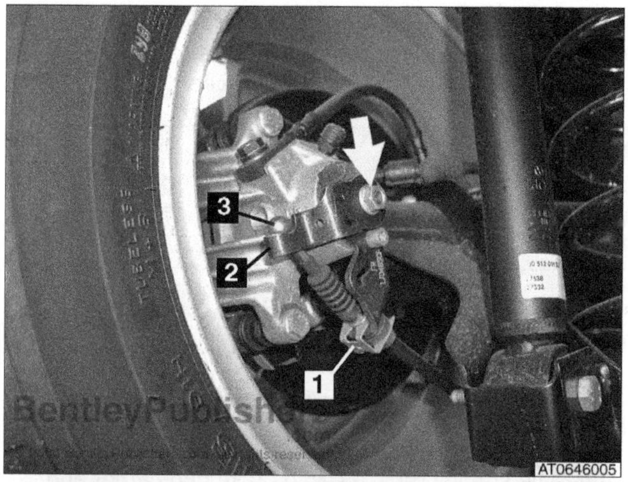

◄ Pry off retaining clip (**1**) for parking brake cable. Use a wrench on retaining nut (**arrow**) to rotate brake lever (**2**) downward and unhook end of parking brake cable (**3**).

◄ Counter-hold guide pins with open-end wrench and remove brake caliper mounting bolts.

– Remove brake caliper and secure with wire so that weight of brake caliper does not damage brake hose.

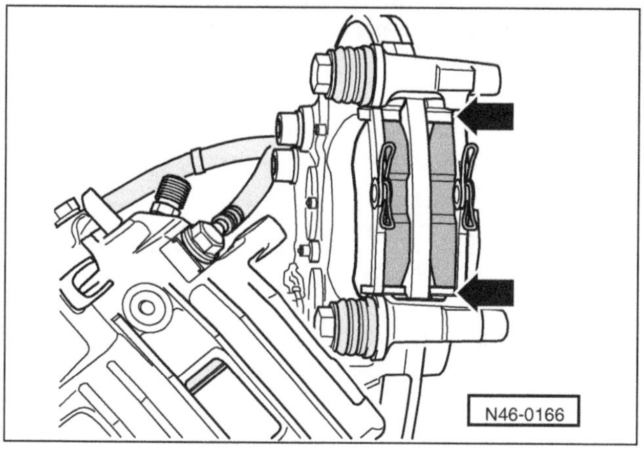

◄ Remove brake pads and anti-rattle springs (**arrows**).

NOTE —
- *If reusing brake pads, mark location and install in original position.*

BentleyPublishers.com
© 2006 BentleyPublishers.com—All Rights Reserved

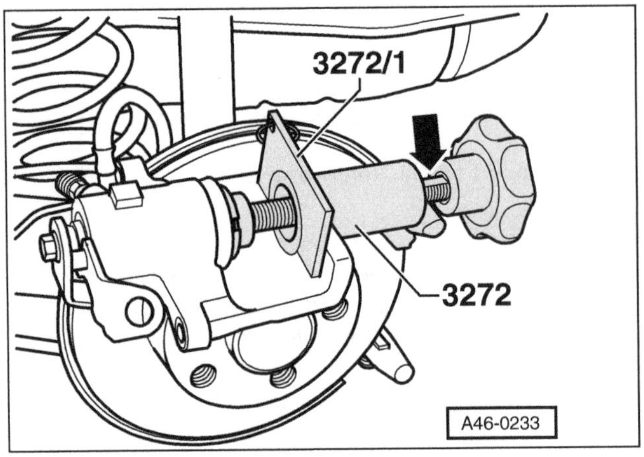

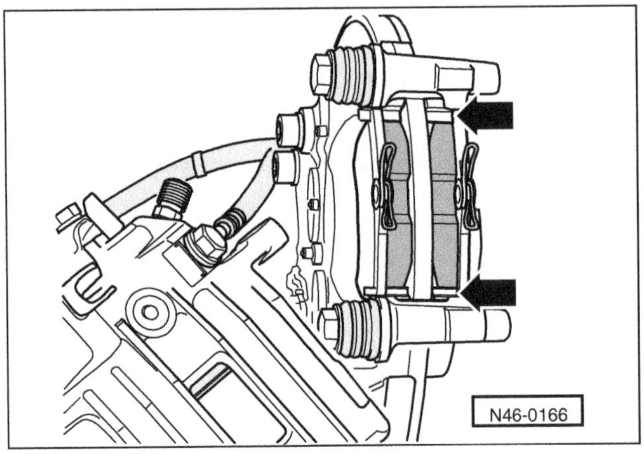

Installing

◀ Using rear brake caliper piston resetting tool (3272 and adapter plate 3272/1), screw piston back into caliper housing by turning knurled wheel clockwise, do not damage piston seal.

> ***CAUTION—***
> - *Before pressing piston into cylinder, extract fluid from brake fluid reservoir. Otherwise, fluid can overflow and cause damage.*

– For pistons that are difficult to move, use a 13 mm open end wrench on the flats provided (**arrow**).

– Before installing new brake pads, remove protective foil from outer brake pad backing plate, if applicable.

◀ Install new anti-rattle springs (**arrows**) into brake pad carrier and install brake pads.

– Lubricate guide pins with multi-purpose grease.

– Install brake caliper using new self-locking bolts.

Tightening torque	
Caliper guide pins to brake carrier (always replace)	35 Nm (26 ft-lb)

NOTE—
- *The factory repair kit includes four self-locking hex bolts which must be installed in all cases.*

– Position parking brake cable in brake lever and install retaining clip in reverse order of removal.

– Install wheels.

Tightening torque	
Wheel bolt to wheel hub	120 Nm (90 ft-lb)

NOTE—
- *After replacing brake pads, depress brake pedal firmly several times with vehicle stationary so that the brake pads are properly seated in their normal operating position.*

– Check brake fluid level and top off.

BentleyPublishers.com
©2006 BentleyPublishers.com—All Rights Reserved

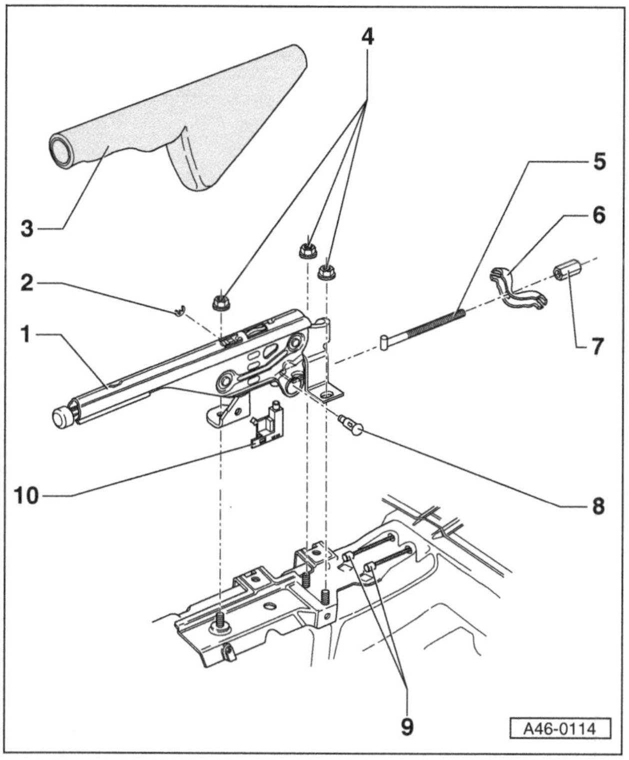

PARKING BRAKE

Parking brake assembly

1. Parking brake lever
2. Lock washer
3. Trim for parking brake handle
4. Hex nut
   - 18 Nm (13 ft-lb)
5. Pull rod
6. Compensator bracket
7. Adjustment nut
   - 8 Nm (71 in-lb)
   - See **Parking brake, adjusting** in this repair group
8. Pivot pin
9. Parking brake cables
10. Parking brake switch

Parking brake, adjusting

Normally, parking brake adjustment is necessary only after replacing brake cables, brake calipers or brake discs.

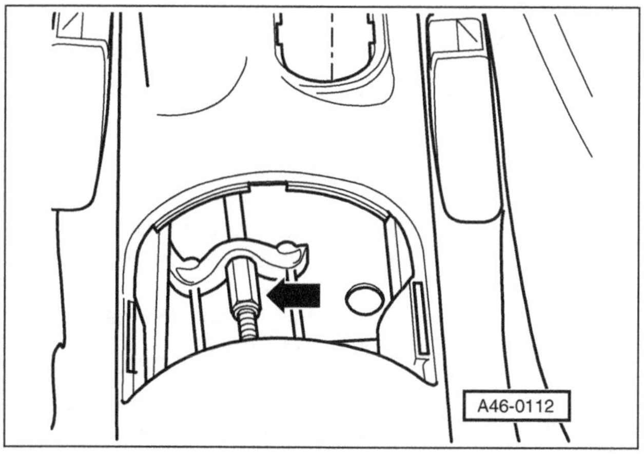

– Working inside vehicle, remove rear center console. See **70 Interior–Trim**.

– Operate foot brake forcefully at least three times.

– Apply parking brake three times and then release.

◄ With parking brake in rest position, tighten adjustment nut (**arrow**) until brake lever lifts off rear caliper stop.
- See illustration below.

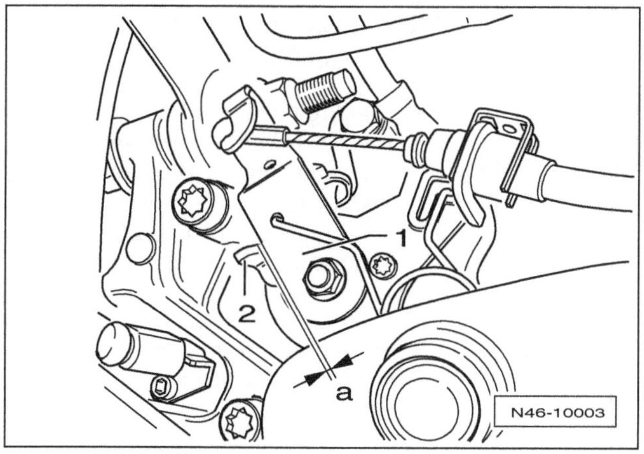

◄ Tighten parking brake adjustment nut until lever (**1**) on brake caliper lifts off from stop (**2**).
- Distance (**a**) = 1 mm (0.039 in) to 3 mm (0.12 in)

– Check that both wheels turn freely.

– Once initial adjustment is performed, parking brake is self-adjusting.

BentleyPublishers.com
©2006 BentleyPublishers.com—All Rights Reserved

Brake Pedal Assembly

1. **Mounting bracket**
2. **Collapsing support**
3. **Nuts**
   - 28 Nm (21 ft-lb)
4. **Brake light switch**
   - Removing and installing, see **45 Anti-Lock Brakes (ABS)**
5. **Accelerator pedal**
6. **Bolt**
7. **Brake pedal**
   - Grease bearings/bushings with polycarbamide grease (G 052 142 A2) when installing
   - Use special tool T10006 to separate pedal from brake booster
8. **Pedal pad**
9. **Nut**
   - 20 Nm (15 ft-lb)
10. **Mounting bushing**
    - Grease bearings/bushings with polycarbamide grease (G 052 142 A2) when installing
11. **Pivot pin**
    - Grease bearings/bushings with polycarbamide grease (G 052 142 A2) when installing

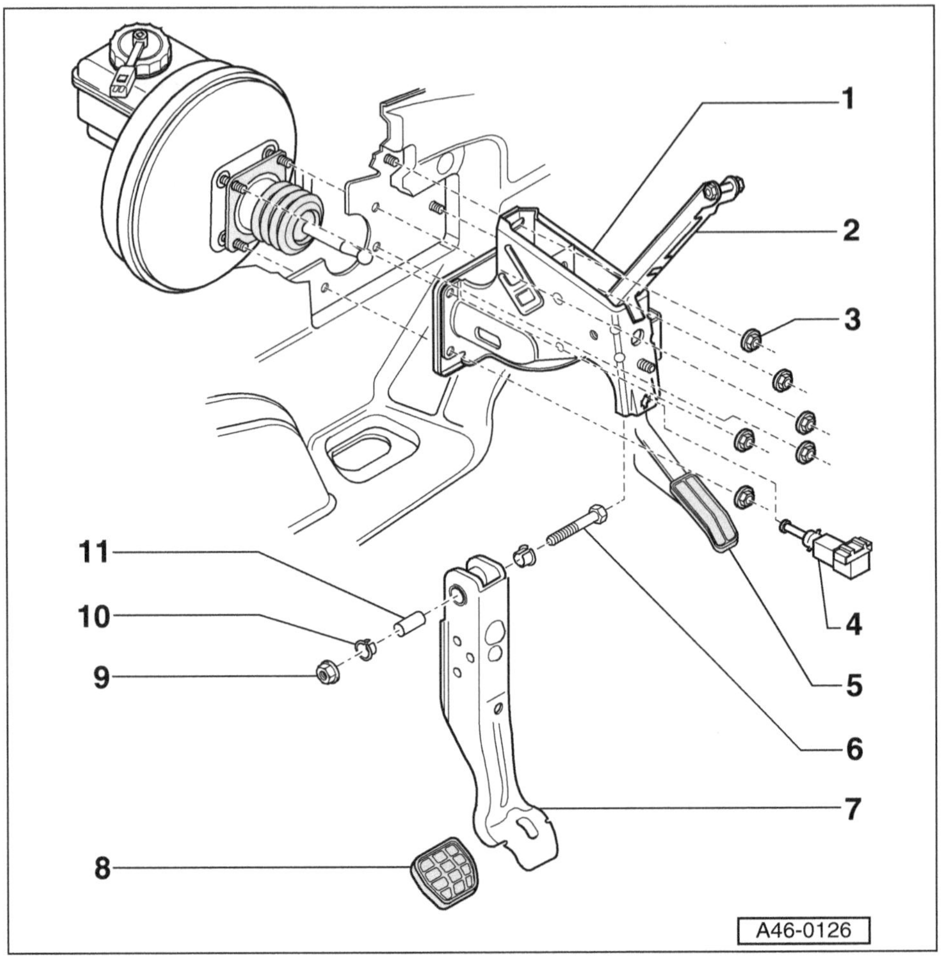

Brake pedal, separating from brake booster

– Remove brake light switch. See **45 Anti-Lock Brakes (ABS)**.

– Disconnect harness connectors from clutch pedal switch.

– Remove connecting plate between clutch and brake pedals.

◄ Press brake pedal (**1**) in direction of brake booster and hold.

– Insert release tool T10006 as shown and pull toward driver's seat while providing support at brake pedal (**1**) (pedal must not be allowed to move to rear at this point). This presses retainer tabs (**3**) off ball head.

– Pull both release tool T10006 and brake pedal toward driver's seat (this pulls brake pedal off push rod ball head.

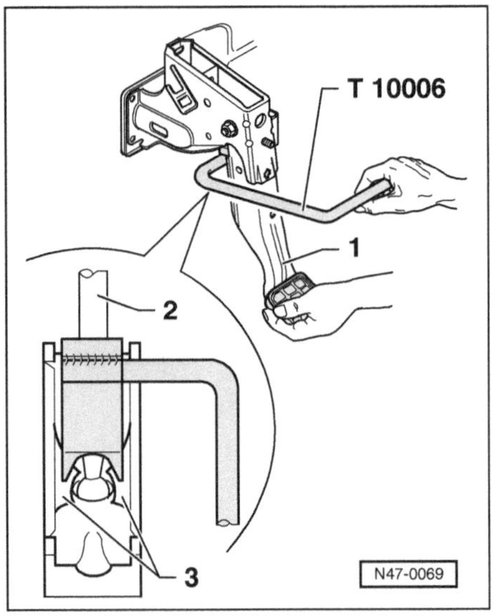

BentleyPublishers.com
© 2006 BentleyPublishers.com–All Rights Reserved

47 Brakes–Hydraulic

GENERAL

This repair group covers brake bleeding and replacement procedures for the vacuum pump and master cylinder. For additional brake system information, see the following repair groups:

- **03 Maintenance** for brake fluid check and brake fluid changing.
- **45 Antilock Brakes (ABS)** for ABS system and traction control.
- **46 Brakes–Mechanical** for brake pads, rotors and parking brake.

System description

Models covered by this manual are equipped with power assisted four-wheel disc brakes. A vacuum brake booster provides power assistance to a dual-circuit master cylinder.

Troubleshooting

Brake performance is mainly affected by three things:

- Level and condition of brake fluid
- Brake system's ability to create and maintain hydraulic pressure
- Condition of friction components

Air in the brake fluid will make the brake pedal feel spongy during braking or will increase the brake pedal force required to stop. Fluid contaminated by moisture or dirt can corrode components. Inspect the brake fluid. If fluid is dirty or murky, or over two years old, fluid should be replaced.

Visually check hydraulic system starting at the master cylinder. To check the function of master cylinder hold brake pedal down hard with car stopped and engine running. The pedal should feel solid and stay solid. If pedal slowly falls to floor, either master cylinder is leaking internally, or fluid is escaping from other points. If no leaks can be found, the master cylinder is faulty and should be replaced.

BentleyPublishers.com
© 2006 BentleyPublishers.com—All Rights Reserved

Check all brake fluid lines and couplings for leaks, kinks, chafing or corrosion.

Check brake booster by pumping brake pedal approximately 10 times with engine off. Hold pedal down and start engine. Pedal should fall slightly. If not, check for any visible faults before suspecting a faulty brake booster, and check the one-way check-valve at the hose connection on the vacuum booster.

Warnings

> ***WARNING—***
> - *Use only new, previously unopened DOT 4 Super brake fluid.*
> - *Brake fluid is poisonous. DO NOT ingest brake fluid. Wash thoroughly with soap and water if brake fluid comes into contact with skin.*
> - *DO NOT let brake fluid come in contact with paint. Wash immediately with soap and water.*
> - *Brake fluid absorbs moisture from the air and must be stored in an airtight container.*
> - *DO NOT allow brake fluid to exceed maximum level in fluid reservoir.*
> - *Observe all regulations concerning storage and waste disposal of brake fluid.*

BLEEDING BRAKES

Brake bleeding is usually done for one of two reasons: either to replace old brake fluid as part of routine maintenance or to expel trapped air in the system that resulted from opening the brake hydraulic system during repairs.

When adding or replacing brake fluid, add new brake fluid from an unopened container. If you are certain no air was introduced into the master cylinder or ABS hydraulic unit, bleed brakes at the calipers using a pressure bleeder. See **03 Maintenance** for more information.

If air was introduced into the master cylinder or ABS hydraulic unit, or you are unsure, bleed brakes using Audi service tester (VAS 5051 or equivalent), or bring the car to an authorized Audi repair facility. See **Brake system, bleeding** in this repair group.

> ***WARNING—***
> - *The ABS system uses electronic controls and a sophisticated hydraulic unit. Once air enters the hydraulic unit, it cannot be removed using traditional bleeding methods. For this reason, pressure bleed the brake using VAS 5234 pressure bleeder and VAS 5051 diagnostic scan tool.*
> - *When flushing brake fluid from the system, use extreme care to not let the brake fluid reservoir run dry. If air enters the hydraulic unit, the Audi service tester MUST be used to bleed the brake system before the vehicle is driven.*

BentleyPublishers
.com
©2006 BentleyPublishers.com—All Rights Reserved

The design of the ABS hydraulic unit requires that brakes be bleed in the following (non-traditional) order:

- left front brake
- right front brake
- left rear brake
- right rear brake

If this order is not followed, proper bleeding cannot be ensured.

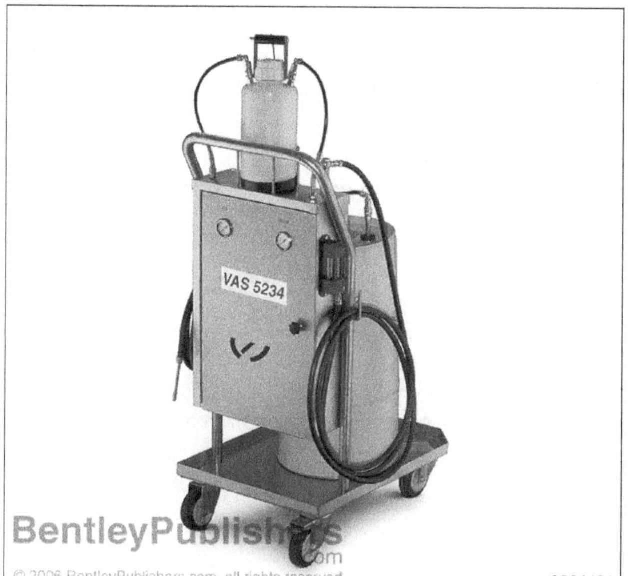

0024434

Volkswagen currently specifies that brake system bleeding and brake fluid replacement be performed with VAS 5234 brake system filling/bleeding machine (pressure bleeder) and adapter VAG1869/4. The use of a brake pedal depressor such as VAG 1869/2 is also required. Models covered by this manual require DOT 4 Super brake fluid.

Brake fluid specifications	
Audi specification	DOT 4 Super (DOT 4+)
Audi part numbers	B 000 700 A1 B 000 700 A7 B 000 700 A8
Minimum wet boiling point	329°F (165°C)
Dry boiling point (exceeds)	500°F (260°C)

Brake caliper, pre-bleeding

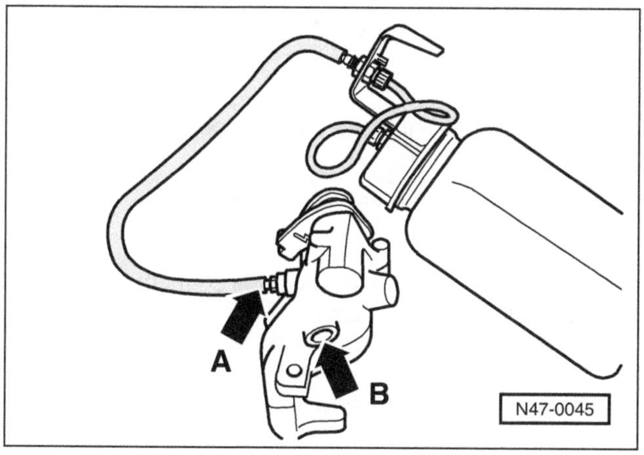

N47-0045

Pre-bleed brake caliper whenever caliper has been drained of brake fluid. Procedure is the same for front or rear caliper.

– Position brake caliper as illustrated.

Open bleeder valve (**A**) and fill caliper with brake fluid using standard bleeder bottle until brake fluid flows out of threaded hole (end of brake hose) (**B**) without bubbles.

– Close bleeder valve and fully bleed brake system. See **Brake system**, bleeding in this repair group.

BentleyPublishers.com
©2006 BentleyPublishers.com—All Rights Reserved

Brake system, bleeding

This procedure requires using a VW service tester (VAS 5051 or equivalent) and brake filler/bleeder tool VAS 5234.

> ***WARNING —***
> - *Only use new brake fluid conforming to DOT 4 Super standards.*
> - *Brake fluid is poisonous. Also due to its corrosive effect brake fluid must not come into contact with paintwork.*
> - *Brake fluid is hygroscopic, which means it absorbs moisture from the ambient air and should always be stored in air tight containers.*
> - *Rinse off spillages, by using plenty of water.*
> - *There must be a maximum filling pressure of 1 bar to bleed the hydraulic unit.*

NOTE —
- *If one chamber of brake fluid reservoir has been run completely empty on vehicles with ABS/ASR or ABS/EDL/ASR/ESP (e.g. leaks in brake system), a pre-bleeding of the brake system must be performed first. See* **Prebleeding** *in this repair group.*

– Raise car and support safely. Remove wheels

> ***WARNING —***
> - *Make sure vehicle is stable and well supported at all times. Use a professional automotive lift or jack stands designed for the purpose. A floor jack is not adequate support.*

Prebleeding

– Connect brake filler/bleeder unit VAS 5234.

Bleeding sequence:

1. Bleed left front and right front brake caliper together simultaneously
2. Bleed left rear and right rear brake caliper together simultaneously

– Leave bleeder screw open, with brake fluid hose connected, until brake fluid discharges free of air bubbles.

– The hydraulic unit must be bled once more via function “Basic setting” using Vehicle Diagnosis, Testing and Information System function of the VAS 5051.

– Initiate basic setting (to bleed brake system) using VAS 5051.

– Then bleed brake system normally. See **Bleeding (normal)** in this repair group.

BentleyPublishers.com
©2006 BentleyPublishers.com—All Rights Reserved

Bleeding (normal)

NOTE —
• Adhere strictly to work sequence when bleeding brake system.

– Connect brake filler and bleeder unit VAS 5234.

– Open bleeder valves in the specified order and bleed brake calipers.
  1. Left front brake caliper
  2. Right front brake caliper
  3. Left rear brake caliper
  4. Right rear brake caliper

– Use suitable bleeder hose. It must fit tightly on bleeder valve so that no air can enter brake system.

– Leave brake caliper bleeder valve open, with bleeder hose connected, until brake fluid discharges free of air bubbles.

Subsequent bleeding

With the help of an assistant:
- Depress brake pedal forcefully and hold.
- Open bleeder valve at brake caliper.
- Press brake pedal down onto stop.
- Close bleeder screw with pedal depressed.
- Release brake pedal slowly.

– Repeat this procedure 5 times per brake caliper.

– Bleeding sequence:
  1. Left front brake caliper
  2. Right front brake caliper
  3. Left rear brake caliper
  4. Right rear brake caliper

– Install wheels.

Tightening torque	
Wheel bolt to wheel hub	120 Nm (89 ft-lb)

– Perform a road test after bleeding. During test, perform at least one stop that requires ABS regulation.

BentleyPublishers.com
© 2006 BentleyPublishers.com—All Rights Reserved

Vacuum Pump

Models with automatic transmission are equipped with an electric pump to provide vacuum to the brake booster.

NOTE—
- *Models with manual transmission supply vacuum to the brake booster via the intake manifold.*

Vacuum pump, removing and installing

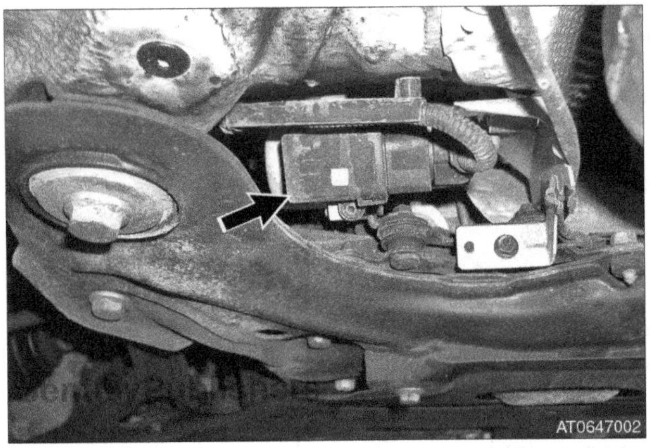

The brake system vacuum pump (V192) is located on left side of front subframe, in front of the steering gear. The vacuum pump is not serviceable. If there is a malfunction, replace vacuum pump (**arrow**).

Check vacuum pump using diagnostic scan tool (VAG 1551/VAS 5051 or equivalent). It is also possible to feel the pump running briefly upon initial start-up of the engine. A helper will be needed to do this.

Removing

- Raise vehicle and support safely. See **03 Maintenance**.

> ***WARNING—***
> - *Make sure vehicle is stable and well supported at all times. Use a professional automotive lift or jack stands designed for the purpose. A floor jack is not adequate support.*

- Remove lower sound absorber panel. See **03 Maintenance**.
- Working under front left side of vehicle, remove guard plate under vacuum pump.

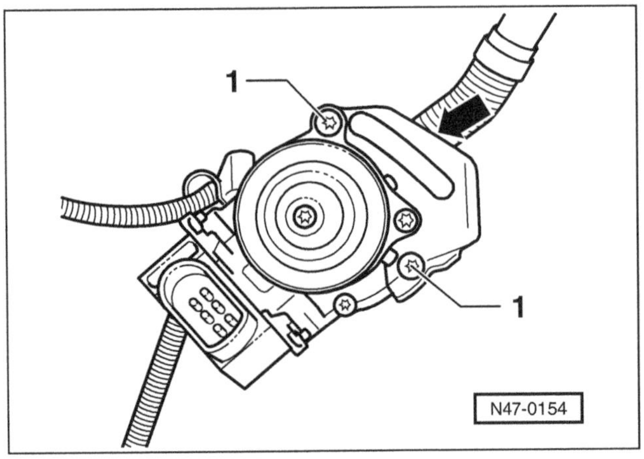

Disconnect vacuum hose (**arrow**) from vacuum pump.

- Disconnect harness connector from pump.
- Remove mounting bolts (**1**) and vacuum pump.

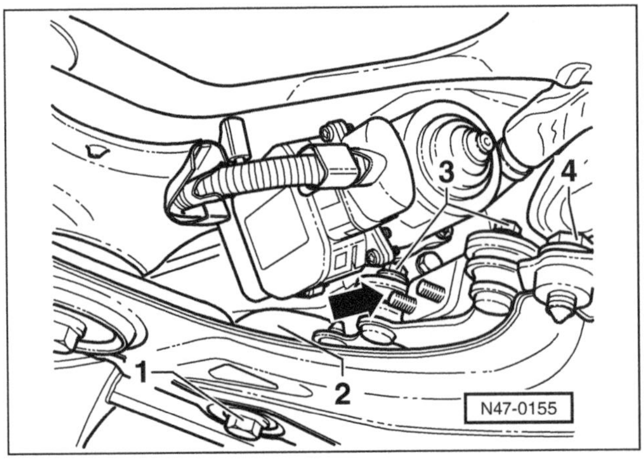

To continue removing vacuum pump mounting bracket, unscrew nut (**2**) from bolt (**1**).

- Remove bolt (**4**) and bracket. Bracket can be further disassembled by removing bolts (**3**) if necessary.
- Install in reverse order of removal.

Tightening torques	
Brake vacuum pump (V192) to bracket	8 Nm (6 ft-lb)
Bracket to subframe	20 Nm (15 ft-lb)

BentleyPublishers
.com
© 2006 BentleyPublishers.com—All Rights Reserved

Master Cylinder

Brake booster and master cylinder assembly

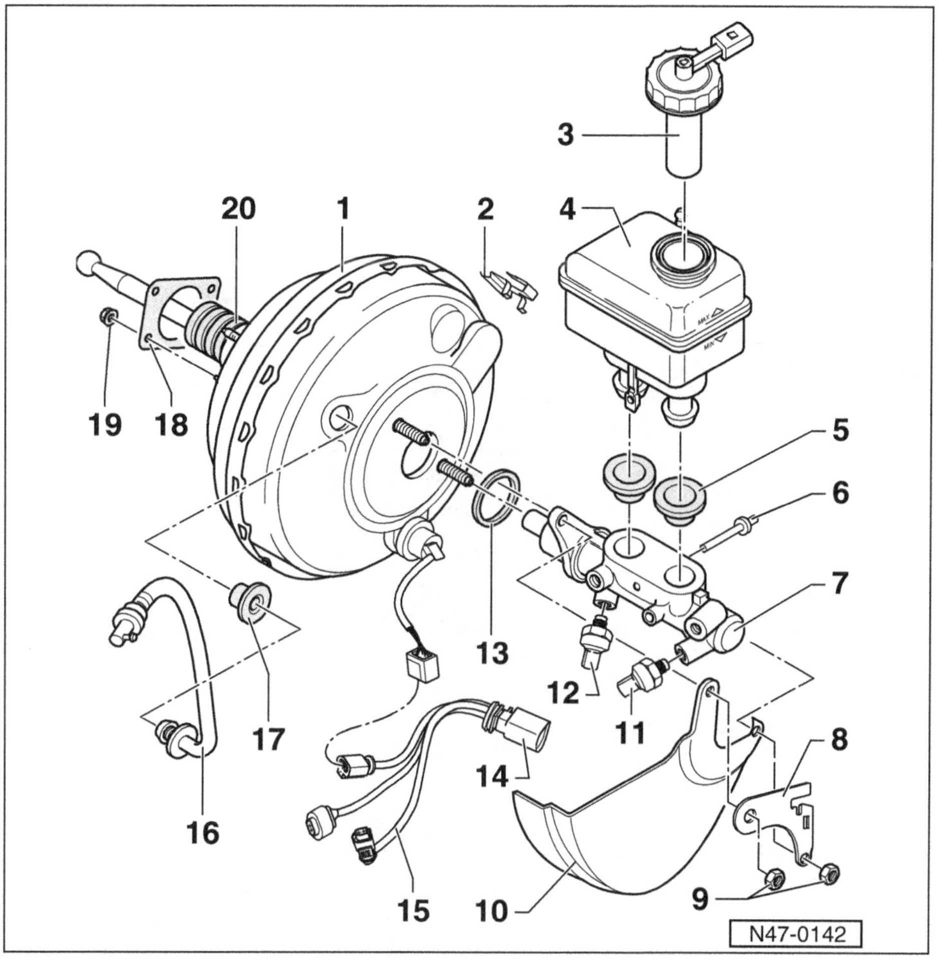

1. **Vacuum brake booster**
   - On manual transmission vehicles vacuum is taken from intake manifold
   - Vehicles with automatic transmission include a brake system vacuum pump (V192)
   - Functional check:
     With engine switched off, depress brake pedal firmly several times (to deplete vacuum in brake booster).
     Now depress brake pedal with average foot pressure, hold and start engine. If the booster unit is working properly, the pedal will be felt to give slightly under foot (boost is functioning).
   - Replace if faulty.
   - See **Brake booster, removing and installing** in this repair group
2. **Bracket**
   - ITT Mark 20 only
   - Retains electrical wire (item 15)
3. **Cap**
   - With low level indicator
4. **Brake fluid reservoir**
5. **Sealing plugs**
   - Coat with brake fluid
6. **Retaining pin**
   - Holds fluid reservoir in place
7. **Brake master cylinder**
   - Cannot be repaired. If faulty, replace completely
   - See **Brake master cylinder, removing and installing** in this repair group
8. **Mounting plate**
   - Different versions available
9. **Nut**
   - 4 Nm (35 in-lb)
10. **Heat shield**
11. **Brake pressure sensor 2 (G214)**
12. **Brake pressure sensor 1 (G201)**
13. **Seal**
    - Always replace
14. **Electrical connector**
    - ITT Mark 20 only
15. **Electrical wire**
    - ITT Mark 20 only
16. **Vacuum hose**
    - With non-return valve
17. **Sealing plug**
    - Coat with brake fluid
18. **Gasket**
    - For brake booster
19. **Self-locking nut**
    - 28 Nm (21 ft-lb)
    - Always replace
20. **Boot**
    - If unusually loud suction noise is heard when braking, check for proper position of boot on booster

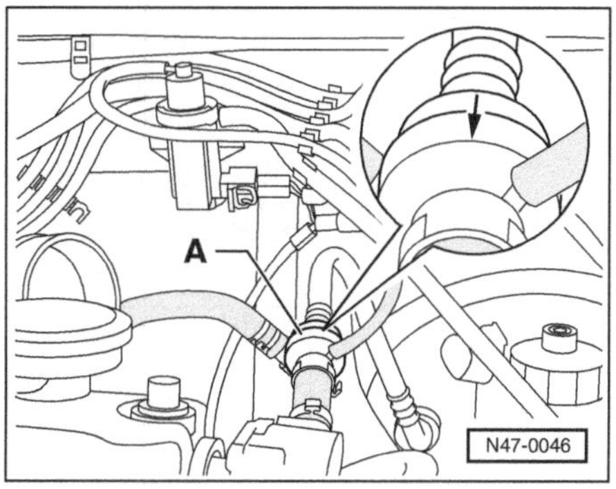

Non-return valve, checking

Non-return valve (**A**) should open and allow passage of air in direction of **arrow** when blowing into valve.

- Valve should remain closed and not allow air to pass when blowing air in direction opposite of arrow.
- Do not used compressed air to test.
- Note installation position.

BentleyPublishers.com
© 2006 BentleyPublishers.com—All Rights Reserved

Brake master cylinder, removing and installing

– Remove engine cover and air filter assembly. See **03 Maintenance**.

– Place lint-free cloths in area under master cylinder to absorb any leaking fluid.

– Extract as much brake fluid as possible from brake fluid reservoir using brake filler and bleeder unit VAS 5234 or equivalent.

For vehicles with manual transmission:

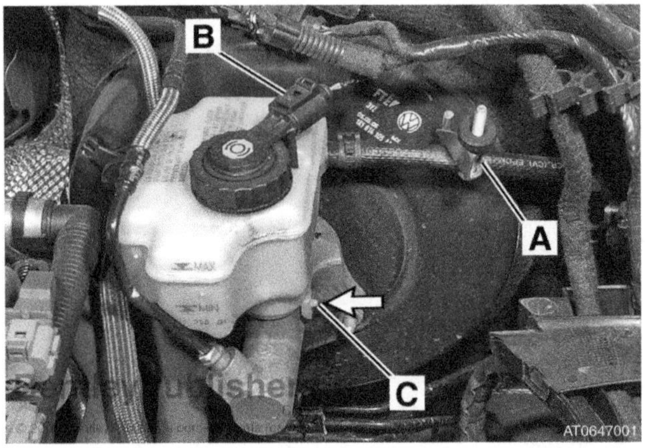

< Clamp supply hose (**A**) of clutch master cylinder shut using hose clamp 3094 or equivalent.

– Disconnect supply hose for clutch master cylinder.

Continued for all vehicles:

– Screw cap with brake fluid level warning switch back on to brake fluid reservoir.

– Disconnect electrical connector (**B**) for brake fluid level warning switch.

– Press out pin for mounting brake fluid reservoir (**C**) in direction of **arrow**.

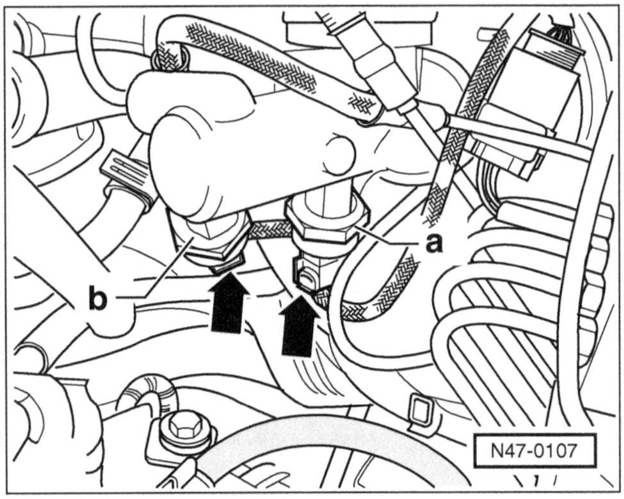

< Disconnect harness connectors (**a** and **b**) from brake pressure sensors.

– Disconnect brake lines on brake master cylinder and seal openings using sealing plugs.

– Remove mounting nuts from brake master cylinder.

– If equipped, remove heat shield from vacuum brake booster.

– Disconnect brake pedal from brake booster. See **46 Brakes–Mechanical**.

– Carefully remove brake master cylinder from brake booster.

– Install in reverse of removal, noting the following:
- Make sure brake pedal push rod is correctly located in brake master cylinder.
- Replace sealing ring between master cylinder and booster.
- Bleed braking system. See **Brake system, bleeding**.
- Bleed hydraulic clutch, if applicable. See **30 Clutch**.
- Use diagnostic scan tool to initiate basic settings for brake pressure sensors.

Tightening torques	
Brake master cylinder to brake booster (use new nuts)	20 Nm (15 ft-lb)
Brake lines to brake master cylinder	14 Nm (10 ft-lb)
Brake pressure sensors to master cylinder	14 Nm (10 ft-lb)

BentleyPublishers.com
© 2006 BentleyPublishers.com—All Rights Reserved

Brake booster, removing and installing

– Operate brake pedal several times with engine off to release brake booster vacuum.

– Remove brake master cylinder. **See Brake master cylinder, removing and installing**.

– Pull vacuum hose of brake booster.

– Disconnect electrical harness connectors from booster, if applicable.

– Remove trim panels above and in front of pedal cluster. See **70 Trim–Interior**.

– Disconnect brake pedal from brake booster. See **46 Brakes–Mechanical**.

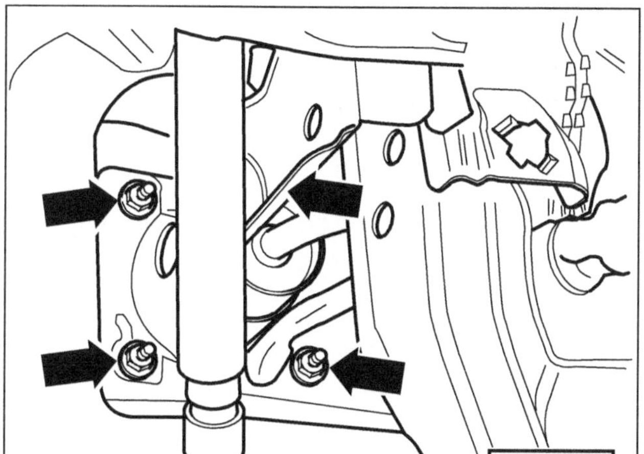

A47-0032

◄ Remove mounting nuts for brake booster.

– Remove brake booster.

– Install in reverse of removal, noting the following:
  - Make sure brake pedal push rod is correctly located in brake master cylinder.
  - Replace sealing ring between master cylinder and booster.
  - Bleed braking system. See **Brake system, bleeding**.
  - Bleed hydraulic clutch, if applicable. See **30 Clutch**.
  - Use diagnostic scan tool to initiate basic settings for brake pressure sensors.

Tightening torques	
Brake booster to bulkhead (always replace nuts)	28 Nm (21 ft-lb)

Brake lines and hoses

Installation details and tightening torques are provided here for various brake lines and hoses.

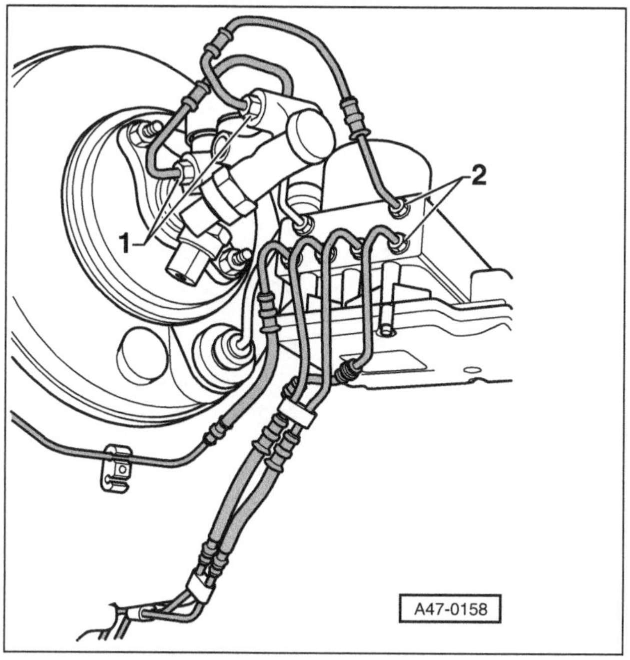

A47-0158

◄ ITT Mark 20 master cylinder lines.

1. Brake line to master cylinder
   14 Nm (10 ft-lb)
2. Brake line to hydraulic unit
   14 Nm (10 ft-lb)

BentleyPublishers.com
©2006 BentleyPublishers.com–All Rights Reserved

Master Cylinder

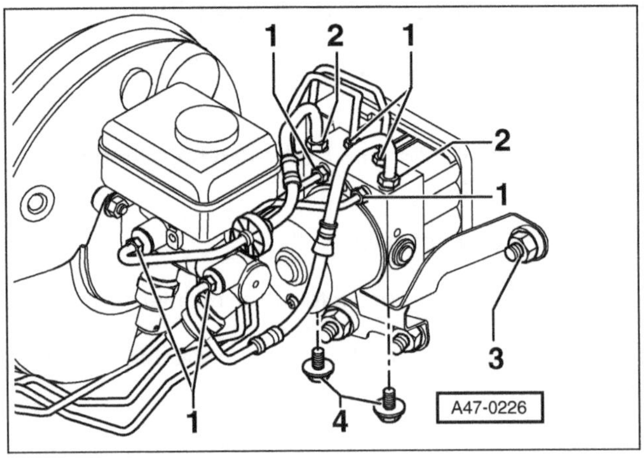

ITT Mark 60 master cylinder lines.

1. Tightening torque: 14 Nm (10 ft-lb)
2. Tightening torque: 24 Nm (18 ft-lb)
3. Tightening torque: 8 Nm (71 in-lb)
4. Tightening torque: 20 Nm (15 ft-lb)

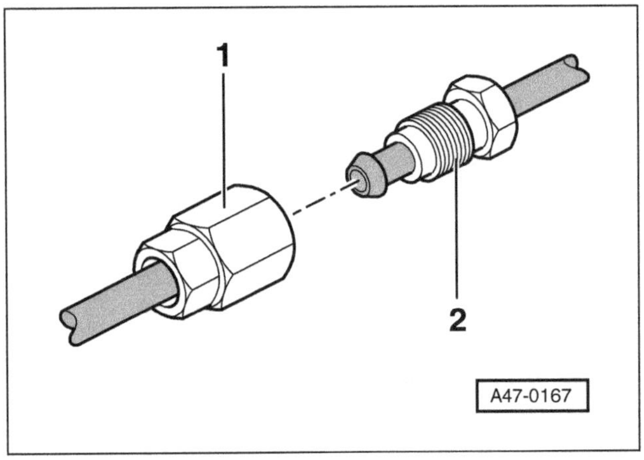

Brake line connection.

- Counter-hold union nut (**1**) when tightening banjo bolt (**2**).
- Tightening torque: 14 Nm (10 ft-lb)

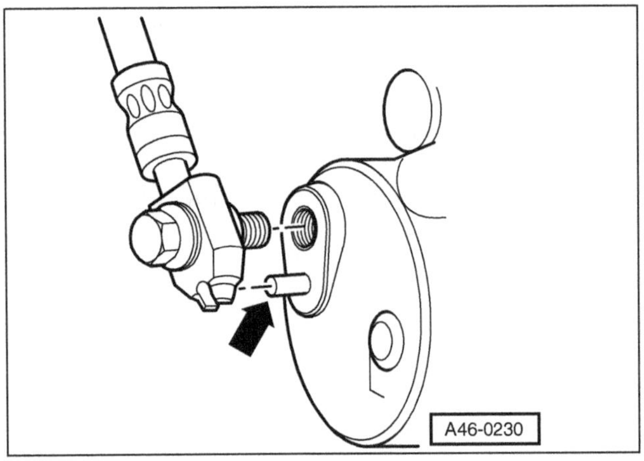

Brake hose with anti-twist connector.

- Make sure that groove in anti-twist connector is installed on locating lug (**arrow**).

BentleyPublishers.com
©2006 BentleyPublishers.com—All Rights Reserved

48 Steering

General

This repair group covers steering wheel removal, steering column assemblies, steering rack, tie rods and power steering pump.

All vehicles covered by this manual have airbag equipped steering wheels. For removal and installation of the driver's side airbag see **69 Seatbelts, Airbags**.

Warnings

WARNING —
- *Welding and straightening work on supporting or wheel carrying components of the suspension is not permitted.*
- *Always replace self-locking nuts.*
- *Always replace corroded bolts/nuts.*
- *Checking, removing, installing and servicing airbags may ONLY be performed by qualified personnel.*
- *Disconnect battery ground (GND) strap before performing any work on the airbag system. No waiting time is required after disconnecting the battery. When connecting airbag system to a voltage source, no one should be in the vehicle interior.*
- *Before handling (touching) the airbag unit, the technician must be electrostatically discharged. This is accomplished by touching grounded metal like e.g. water pipes, heating pipes or metal carriers.*
- *Always install airbag units in the vehicle, as soon as they are removed from the packaging.*
- *If installation work is interrupted, immediately return the airbag unit to its original packaging.*
- *Do not leave an airbag unit unattended.*
- *Once removed, lay the airbag down with impact absorbing pad facing up.*

BentleyPublishers
.com
© 2006 BentleyPublishers.com—All Rights Reserved

Steering Wheel and Column

Specifications	
Steering gear	Power assist, maintenance-free rack-and-pinion
Number of turns, lock to lock	2.75
Maximum steering angle at wheel on inside of curve	37° 54'
Turning circle diameter	approx. 10.5 meters (34.5 feet)

In the event of an accident, the steering column should be checked for damage. Check the following:

- Visual damage of any parts of steering column.
- Smooth operation across entire range of steering.
- Ability to adjust height and reach of steering column.

Replace steering column if any defects are found.

Steering angle sensor (G85)

The steering angle sensor is part of the electronic stability program (ESP). It is located on the steering column between the steering column switches and the steering wheel. The sensor reports the current steering angle and steering speed to the ABS control module (J104) via the CAN-bus.

For steering angle sensor replacement, see **96 Interior Lights, Switches, Anti-theft**.

Steering wheel, removing and installing

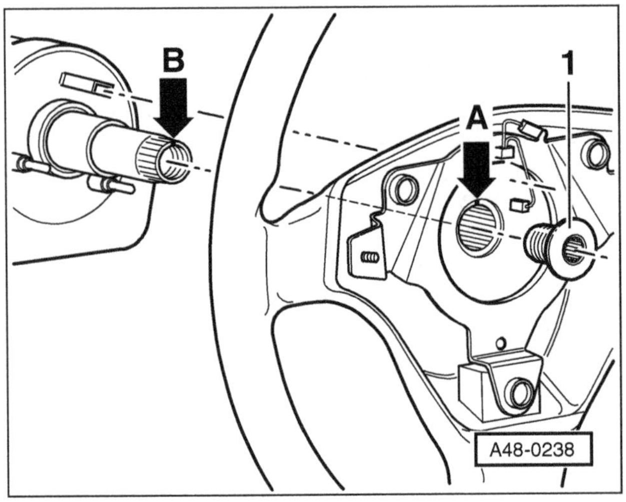

– Remove steering wheel airbag. See **69 Seatbelts, Airbags**.

– Center steering wheel (wheels in straight ahead position).

◄ Remove mounting bolt (**1**) and pull steering wheel off steering column.

– To install, place steering wheel onto steering column.
  - Center markings for steering wheel and steering column (**A** and **B**) must align.

– Secure steering wheel with new bolt (**1**).

Tightening torque	
Steering wheel to steering column (always replace bolt)	50 Nm (37 ft-lb)

– Install steering wheel airbag. See **69 Seatbelts, Airbags**.

BentleyPublishers.com
© 2006 BentleyPublishers.com—All Rights Reserved

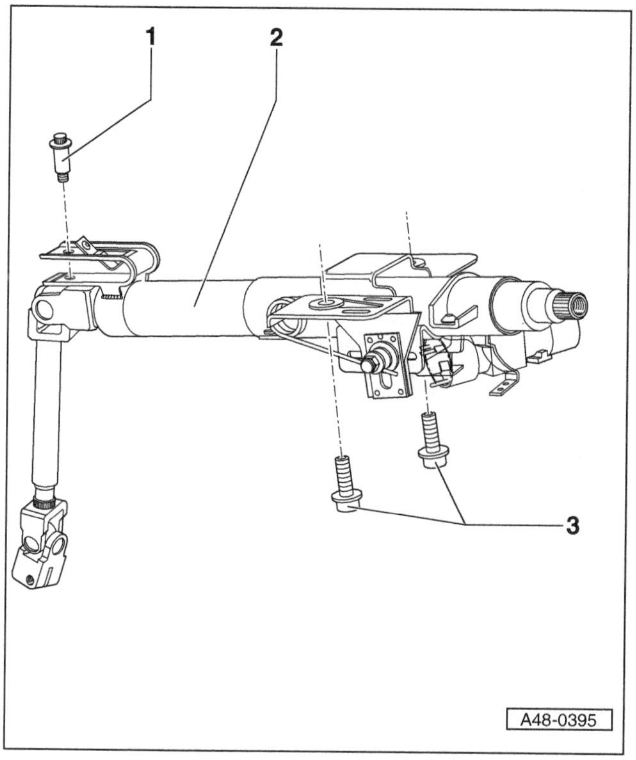

Steering column, version 1

1. Shear bolt
   - Tighten until head shears off, approx. 7-12 Nm
2. Steering column
3. Bolt with washer
   - 20 Nm (15 ft-lb) + 5 Nm (44 in-lb)

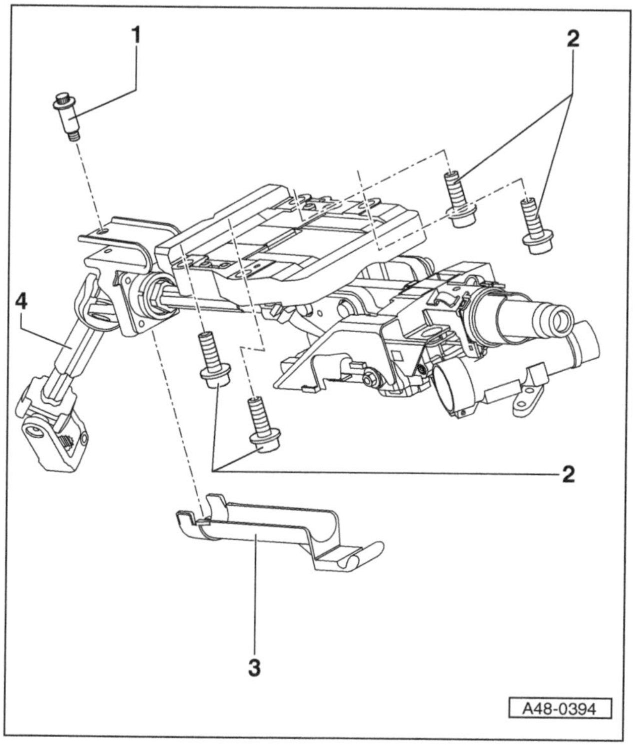

Steering column, version 2

1. Shear bolt
   - Tighten until head shears off, approx. 7-12 Nm
2. Bolt with washer
   - 22 Nm (16 ft-lb)
3. Transport lock
   - Remove during installation of new column
4. Steering column

BentleyPublishers
.com
© 2006 BentleyPublishers.com—All Rights Reserved

Steering Rack

Steering rack assembly

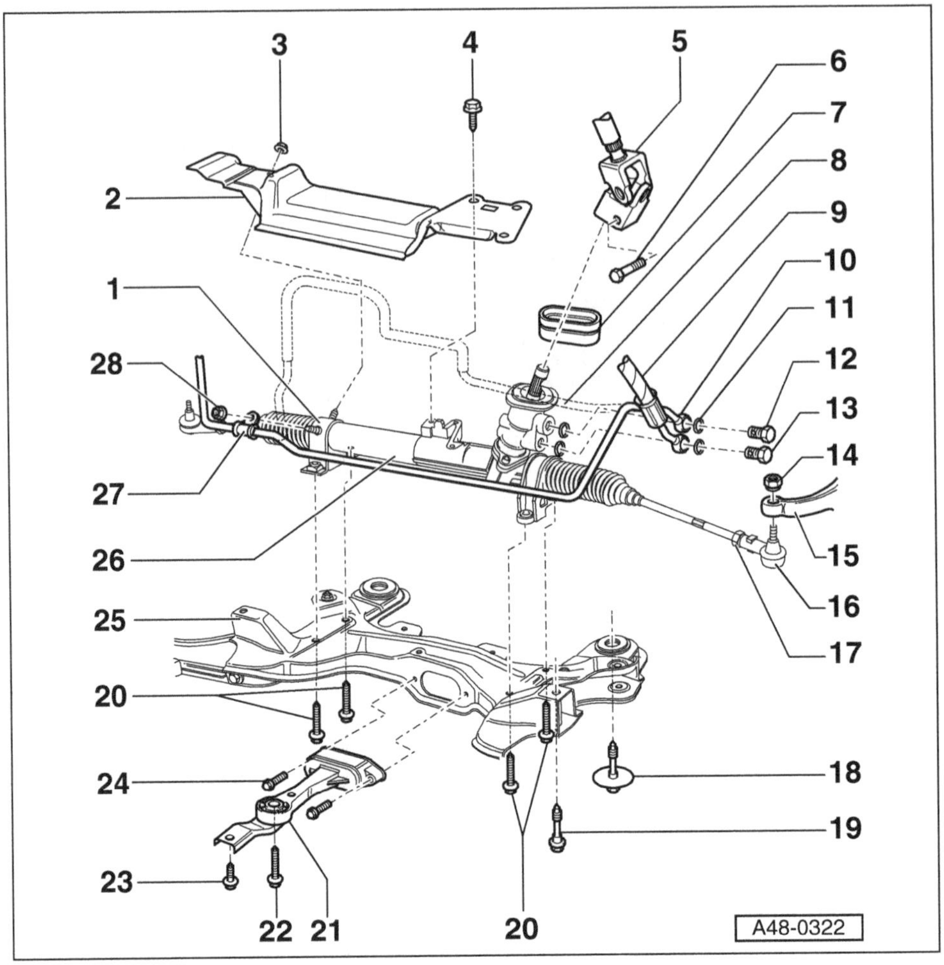

1. **Clamp with rubber insert**
   - Arrow on clamp points in direction of travel
2. **Heat shield**
3. **Nut**
   - 22 Nm (16 ft-lb)
4. **Bolt**
   - 25 Nm (19 ft-lb)
5. **Universal joint**
6. **Bolt**
   - 30 Nm (22 ft-lb)
7. **Sealing collar**
8. **Return line**
   - Only for vehicles with all wheel drive
   - Steering gear to fluid reservoir
9. **Expansion hose**
   - Power steering pump to steering gear
10. **Return line**
    - Only for vehicles with front wheel drive
    - Steering gear to fluid reservoir
11. **Seals**
    - Always replace
12. **Banjo bolt**
    - 48 Nm (35 ft-lb)
13. **Banjo bolt**
    - 38 Nm (28 ft-lb)
14. **Self-locking nut**
    - 3.2L roadster: 75 Nm (55 ft-lb)
    - All other models: 45 Nm (33 ft-lb)
    - Always replace
15. **Steering knuckle**
    - Located on wheel bearing housing
16. **Tie rod end**
    - Check seal for damage
17. **Lock nut**
    - 50 Nm (37 ft-lb)
    - Counterhold at flats on tie rod end when loosening and tightening
18. **Bolt with washer**
    - 100 Nm (74 ft-lb)
    - Always replace
19. **Combination bolt**
    - 100 Nm (74 ft-lb) + ¼ turn (90°)
    - Always replace
20. **Bolt**
    - 20 Nm (15 ft-lb) + ¼ turn (90°)
    - Always replace
21. **Pendulum mount**
22. **Bolt**
    - 50 Nm (37 ft-lb)
23. **Bolt**
    - 50 Nm (37 ft-lb)
24. **Bolt**
    - 25 Nm (18 ft-lb)
25. **Subframe**
26. **Power steering gear**
27. **Clamp**
28. **Nut**
    - 20 Nm (15 ft-lb)

BentleyPublishers.com
© 2006 BentleyPublishers.com—All Rights Reserved

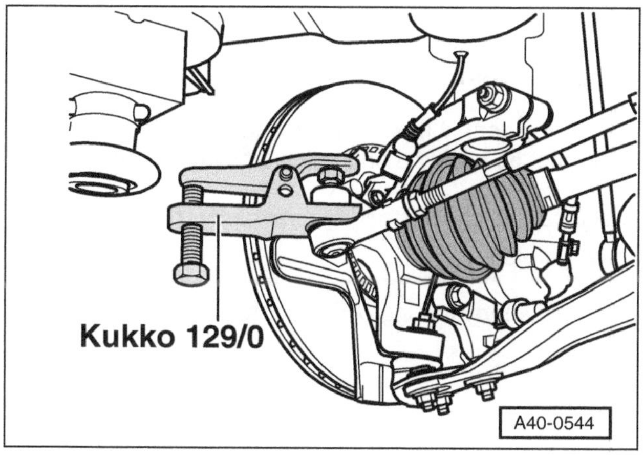

Tie rod ends, removing and installing

– Loosen mounting nut at tie rod end.
  • Leave nut threaded onto tie rod end to protect threads.

◄ Remove tie rods from steering arms using puller (Kukko 129/0 shown).

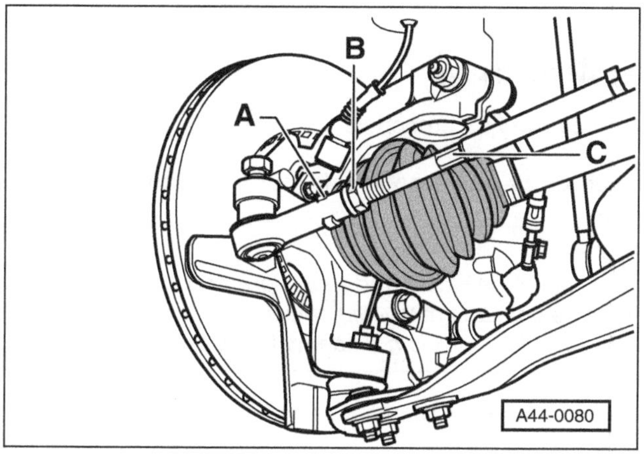

◄ Counterhold tie rod end (**A**) with a 24mm wrench and loosen lock nut (**B**).

– Remove mounting nut from tie rod end.

– Remove tie rod end.

– Install in reverse order of removal. Make sure that tie rod rubber boots are not twisted.

– Tighten tie rod mounting nut to specification.

Tightening torque	
Tie rod to wheel bearing housing	50 Nm (37 ft-lb)

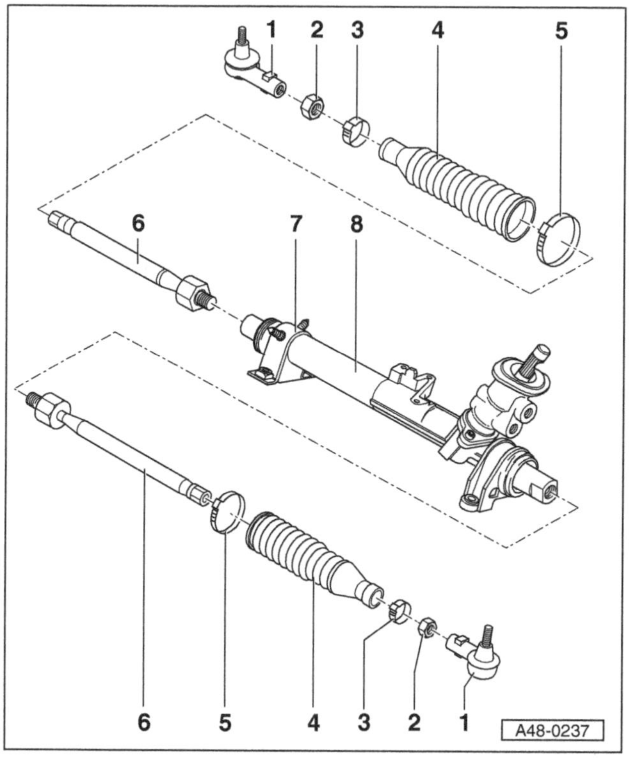

Steering gear assembly

1. Tie rod end
   • Different versions for left and right sides
2. Lock nut
   • 50 Nm (37 ft-lb)
   • Counter-hold at flats on tie rod end when loosening/tightening
3. Hose clamp
   • Install new clamp with hose clamp pliers (VAG 1275 or equivalent)
4. Boot
   • Check for slits and cracks
   • Check sealing surfaces for dirt and damage
5. Hose clamp
   • Install new clamp with hose clamp pliers (VAG 1275 or equivalent)
6. Tie rod
   • Tightening torque at rack: 75 Nm (55 ft-lb)
   • Counter-hold at flats on tie rod end when loosening/tightening
   • If replaced, check center position of steering gear
7. Clamp with rubber insert
   • For mounting steering gear to subframe
   • Arrow on clamp points in direction of travel
8. Power steering gear

BentleyPublishers
.com
© 2006 BentleyPublishers.com—All Rights Reserved

Steering gear, adjusting

Checking center position

If tie rod or steering gear is replaced, the center position of the steering gear must be checked and adjusted if necessary.

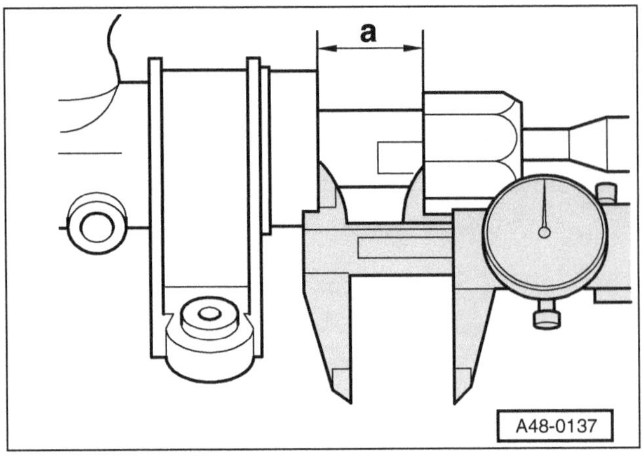

- Measure distance **a** to check steering gear center position.
  - **a** = 33.5 mm (1.32 in.)

– Adjust center position by turning pinion shaft.

Adjusting steering play

A helper is necessary for properly adjusting steering gear play.

– Raise front of vehicle and support safely. See **03 Maintenance**.

> ***WARNING —***
> *• Make sure vehicle is stable and well supported at all times. Use a professional automotive lift or jack stands designed for the purpose. A floor jack is not adequate support.*

– Position wheels in straight ahead position.

– If there is excessive play in the steering, a knocking noise will be heard when steering wheel is turned back and forth (about 30°) across center position.

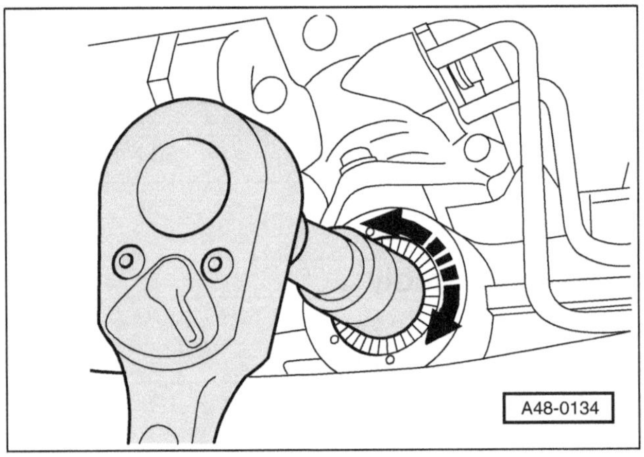

- If knocking noise is heard, screw in adjusting screw (**arrow**) into cover until knocking noise can no longer be heard inside vehicle.

– Test drive vehicle and check that steering centers smoothly by itself back to straight-ahead position after a turn. Correct the adjustment if necessary.

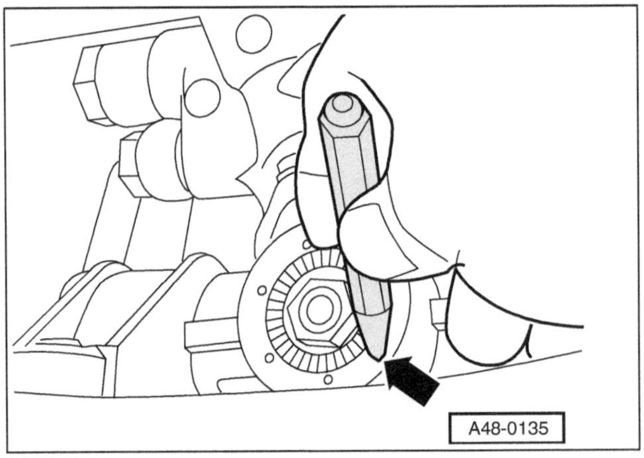

- Secure adjusting nut to prevent it from turning by making a center punch mark (**arrow**) in collar of steering gear housing.

BentleyPublishers.com
© 2006 BentleyPublishers.com—All Rights Reserved

Steering Pump

Power steering pump, removing and installing

For checking power steering fluid level, see **03 Maintenance**.

There is no provision for repairing the power steering pump. In the event of problems, trace the malfunction by performing a leak test and pressure test. If there is a malfunction, replace the power steering pump.

- Power steering pump delivery pressure (engine at idle): 110 – 120 bar (1595 – 1740 psi)

Replacement pumps are not filled with fluid. It is essential to fill the pump with hydraulic fluid (G 002 000) and to rotate the pump by hand before installing, otherwise the pump may be damaged or be noisy during operation.

- Quantity of fluid in system: 0.7 to 0.9 Liters.
- Do not reuse old hydraulic fluid which has been drained off.
- Always replace seals.
- Dispose of used components properly.

– Remove lower center and right (vertical) sound absorber panels. See **03 Maintenance**.

– Release tension on ribbed belt. **See 03 Maintenance**.

– Mark running direction on ribbed belt if re-using.

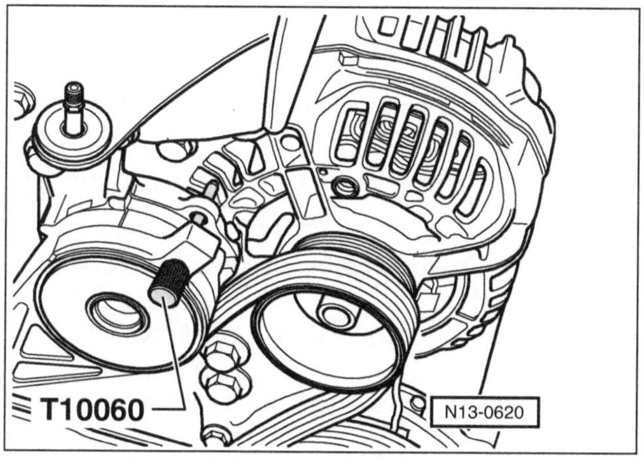

◀ Lock ribbed belt tensioner with appropriate sized drift (or special tool T10060 as shown). Remove ribbed belt.

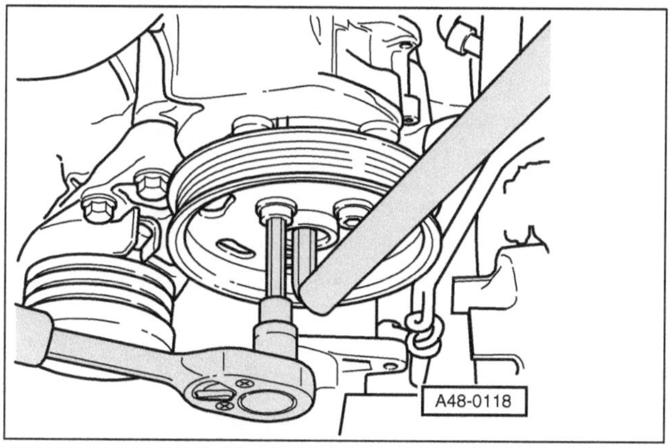

◀ Remove pulley from power steering pump.

- Counterhold pulley with Allen (hex) key.
- Pulley is removed to access pump mounting bolts.

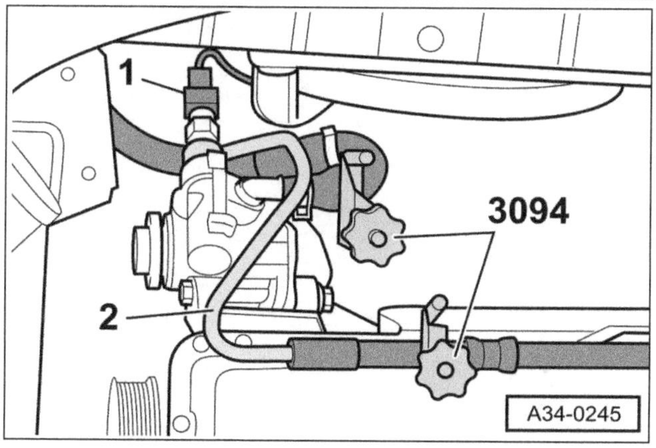

◀ Clamp supply and pressure lines for power steering pump with clamp **3094**, or equivalent clamps with 25 mm (1 in.) diameter line capacity.

– Disconnect harness connector from power steering pressure switch (**1**), if applicable.

NOTE—

- *On vehicles with 1.8L engine, it may be necessary to remove secondary air pump to allow access to power steering lines at pump. See* **26 Exhaust System**.

BentleyPublishers.com
© 2006 BentleyPublishers.com—All Rights Reserved

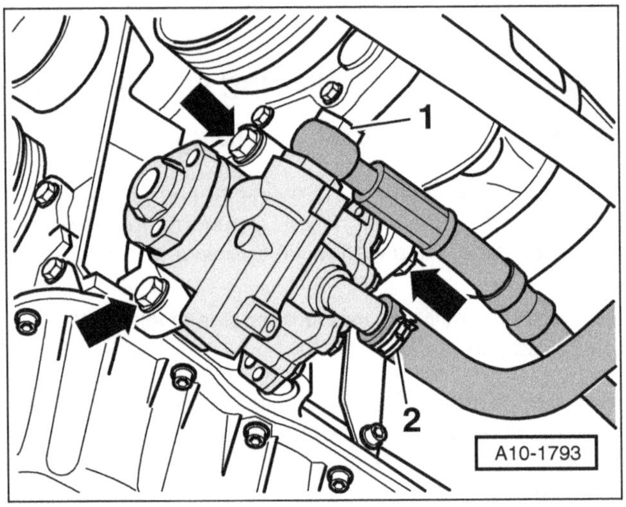

Unbolt banjo bolt (**1**) and remove pressure line from pump.

– Release spring-type hose clamp (**2**) and remove supply line from pump.

– Remove mounting bolts (**arrows**) and power steering pump from bracket.

– If replacing pump, fill new pump with hydraulic fluid (G 002 000) and rotate pump by hand before installing, otherwise the pump may be damaged or be noisy during operation.
  - Quantity of fluid in system: 0.7 to 0.9 Liters.
  - Do not reuse old hydraulic fluid which has been drained off.
  - Always replace seals.

– Install power steering pump in reverse order of removal. Tighten mounting bolts to specification.

Tightening torques	
Power steering pump to mounting bracket	25 Nm (18 ft-lb)
Power steering pump pulley to pump	25 Nm (18 ft-lb)
Banjo bolt (pressure line) to pump	38 Nm (28 ft-lb)

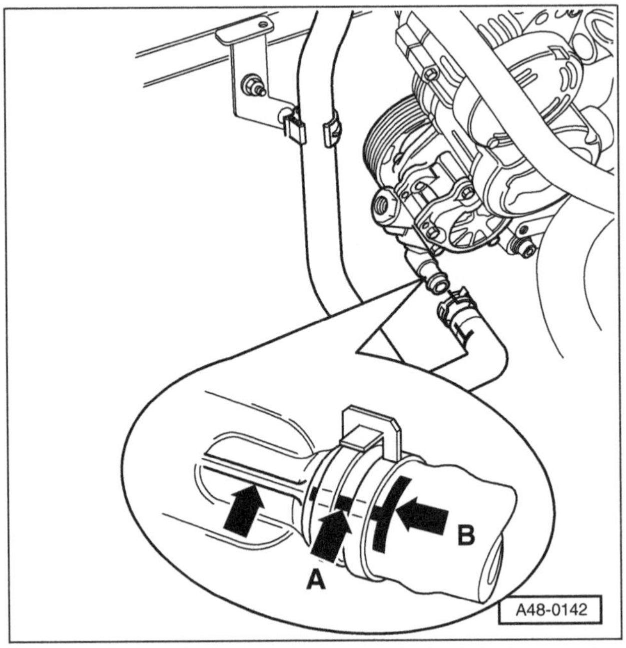

When installing supply line to pump, attach spring-type hose clamp as shown.
- Lengthways mark (**A**) on supply hose must be aligned with seam on pump housing.
- Do not install spring-type hose clamp beyond transverse mark (**B**).

BentleyPublishers.com
© 2006 BentleyPublishers.com—All Rights Reserved

50 Body–Front

General

This repair group covers the front body components, including the lock carrier (front sub-assembly) which can be moved for accessing components located in the front of the engine compartment. Also covered are basic platform and chassis dimensions, and the front fenders.

NOTE—

- *For information on front and rear bumpers, see* **63 Bumpers**.
- *For information on lower sound absorber panel, see* **03 Maintenance**
- *For information on plenum chamber covers, see* **92 Wipers and Washers**

Basic platform dimensions

Basic platform dimensions are used to determine suspension position and body integrity. This can be useful in situations where components do not fit properly during repairs, holes do not line up, or suspension alignment can not be brought into specification. In such cases, it may be necessary to proceed to advanced measurements through the use of specialized equipment intended solely for that purpose. Additional measuring points and precise dimensions will generally be supplied by the equipment manufacturer and apply to that manufacturer's equipment.

All measurements in the following illustrations are from the centerline of the bolt hole or chassis opening.

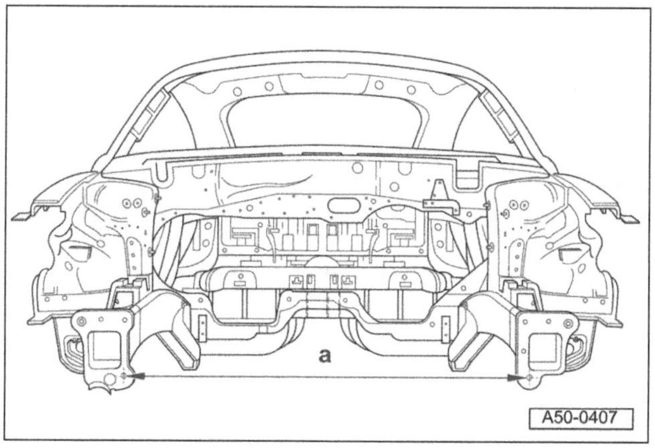

Front platform group

Distance between front longmembers

a = 867 mm ± 2 mm

BentleyPublishers.com

© 2006 BentleyPublishers.com—All Rights Reserved

General

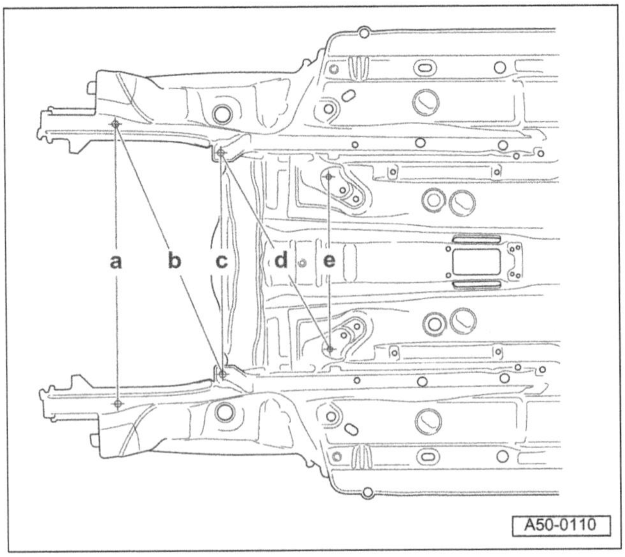

Front platform group

Distance between front attachment points of front suspension

a = 956 mm ± 2 mm

Diagonal distance between front longmember and mounting points of front axle

b = 949 mm ± 2 mm

Distance between front mounting points of front axle

c = 753 mm ± 2 mm

Diagonal distance between front mounting points of front axle

d = 772 mm ± 2 mm

Distance between rear mounting points of front axle

e = 610 mm ± 2 mm

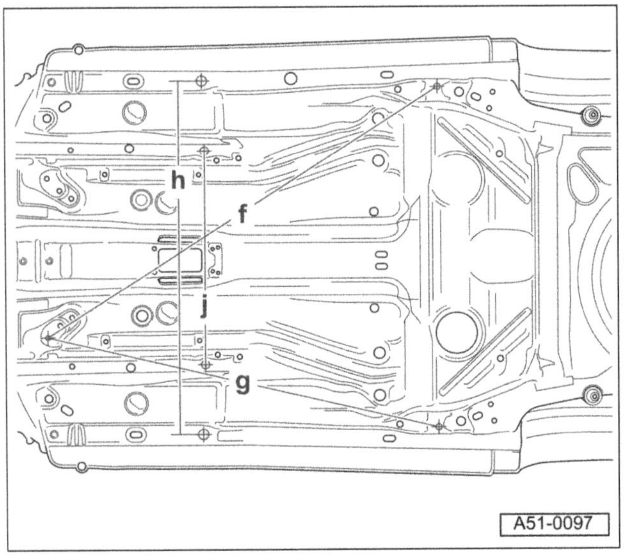

Center platform group

Diagonal distance between rear mounting points of front suspension and rear axle mount

f = 1620 mm ± 2mm

Diagonal distance between front longmember and mounting points of front axle

g = 1360 mm ± 2mm

Distance between front mounting points of front axle

h = 1360 mm ± 2mm

Diagonal distance between front mounting points of front axle

j = 830 mm ± 2mm

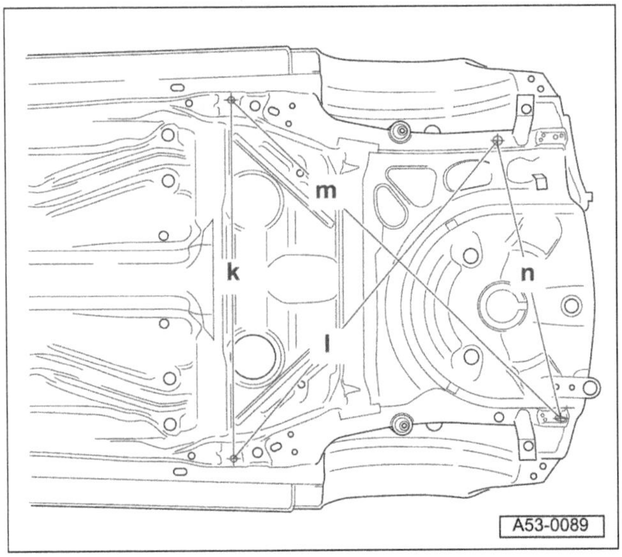

Rear platform group

Distance between rear axle mounts

k = 1270 mm ± 2mm

Diagonal distance between axle mount and longmember

l = 1753 mm ± 2mm

Diagonal distance between rear mounting points of rear axle and rear longmember

m = 1853 mm ± 2mm

Diagonal distance between factory mounting points and longmember

n = 962 mm ± 2mm

BentleyPublishers.com
© 2006 BentleyPublishers.com—All Rights Reserved

Lock Carrier

Lock carrier with attachments

1. **Bolt**
   - Tighten to 23 Nm (17 ft-lb)
2. **Bumper**
   - Does not need to be detached from bumper mounts when removing lock carrier
3. **Bolt**
   - Tighten to 35 Nm (26 ft-lb)
4. **Torx bolt**
   - T30
   - Tighten to 6 Nm (53 in-lb)
5. **Fender**
6. **Bolt**
   - Tighten to 23 Nm (17 ft-lb)
7. **Lock carrier**
   - Remove trim covers for access. See **Trim covers, overview** and **Trim covers, mounting**
   - Disconnect hood cable, see **55 Hoods and Lids**
   - To remove, unscrew bolts (items 3, 4, 6). Do not remove locating bolts (item 8)
   - Remove bumper cover first. See **63 Bumpers**
   - Remove A/C condenser before removing lock carrier. See **87 Heating and Air Conditioning**
   - Lock carrier can be removed with headlights and radiator installed
8. **Bolts**
   - Locate bumper on lock carrier
   - Tighten to 4 Nm (35 in-lb)
9. **Nut**
   - Tighten to 23 Nm (17 ft-lb)

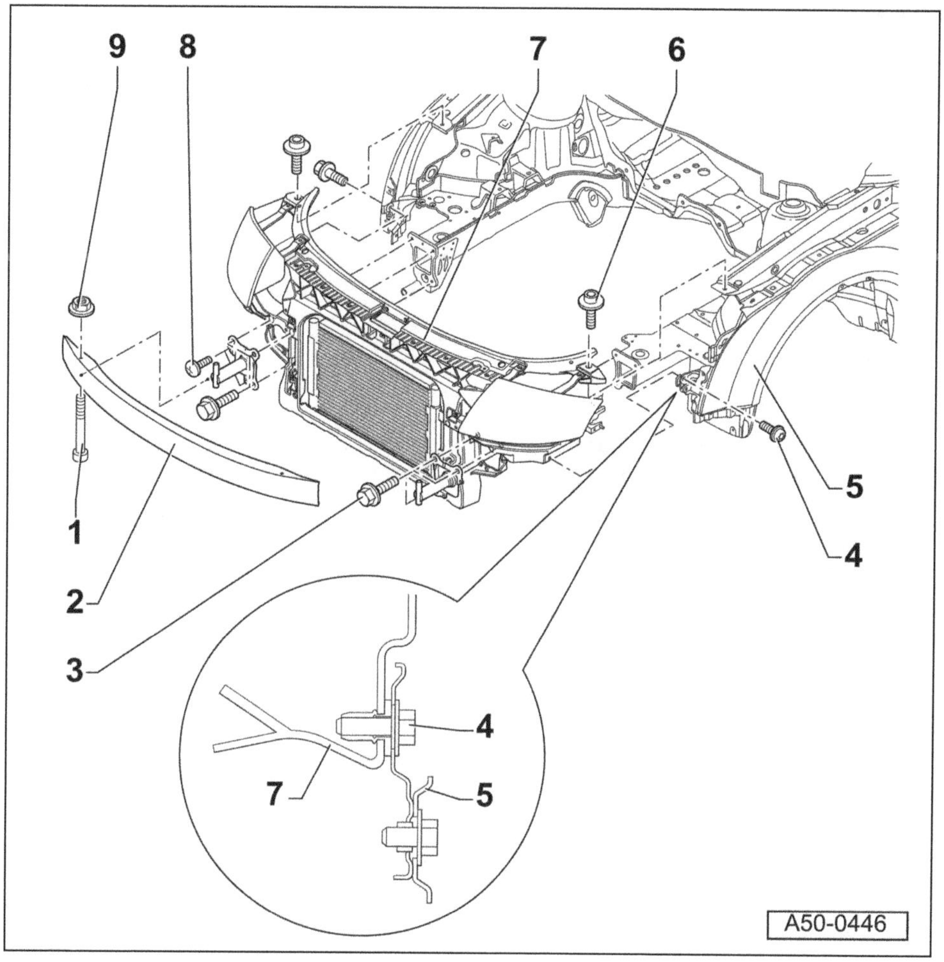

Trim covers, overview

- Remove covers (**A**, **B**, **C**, **D**) to allow access to lock carrier.

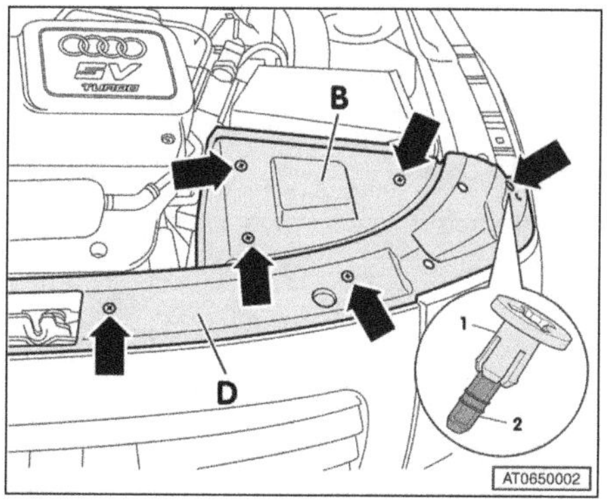

Trim covers, mounting

- Trim covers (**B**, **D**) are secured with either expanding pins or threaded fasteners (**arrows**).
- Expanding pins (**inset**) are released by pushing inner pin (**2**) down into body (**1**) approximately 4 mm (5/16 in) and pulling complete pin out.
- Left side panels shown, right side panels are similar.

BentleyPublishers.com
© 2006 BentleyPublishers.com—All Rights Reserved

Front Fender

1. **Bolt**
   - Tighten to 10 Nm (7 ft-lb)
2. **Support brace**
3. **Bolt**
   - Tighten to 15 Nm (11 ft-lb)
4. **Fender**

   Removing:
   - Remove bumper cover, see **63 Bumpers**
   - Remove wheelhouse lining, see **66 Body-Exterior Equipment**
   - Remove fender end plate, see **Fender end plate**
   - Detach headlights
   - Remove mounting bolts (item 3)
5. **Speed nuts**
   - Inserted into fender mounting flange

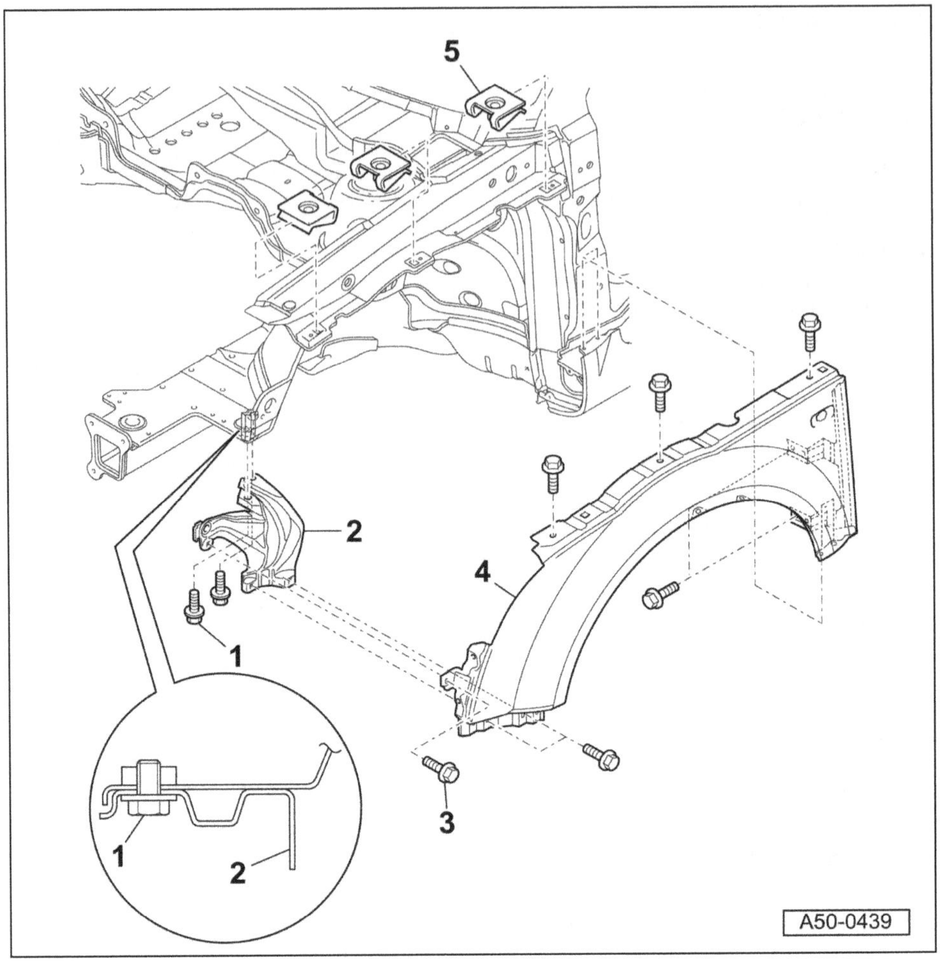

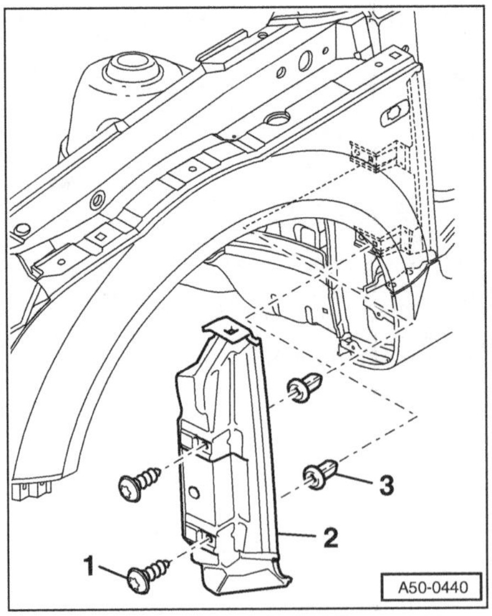

Fender end plate

A. Bolts

B. Fender end plate

C. Expanding clips

BentleyPublishers.com
© 2006 BentleyPublishers.com—All Rights Reserved

55 Hoods and Lids

GENERAL

This repair group covers removal/installation and components associated with the hood, trunk lid, and fuel filler flap.

The TT hood is made of aluminum and requires special handling to avoid damage from contact corrosion. Contact corrosion is caused by contact between dissimilar metals and may result if the wrong fastening components (screws, bolts, nuts, washers, rivets, plugs, grommets, adhesives, etc.) are installed. Because of this, the manufacturer only installs fasteners that have a special surface coating, recognized by its greenish color, and only uses rubber parts, plastic parts and adhesives that are electrically non-conductive. These parts, which are compatible with aluminum, are available as replacement parts.

NOTE—

- *When removing large body components such as the hood or trunk lid, using a helper can help avoid damage to the body and component.*

FUEL FILLER FLAP

Fuel filler flap removing and installing

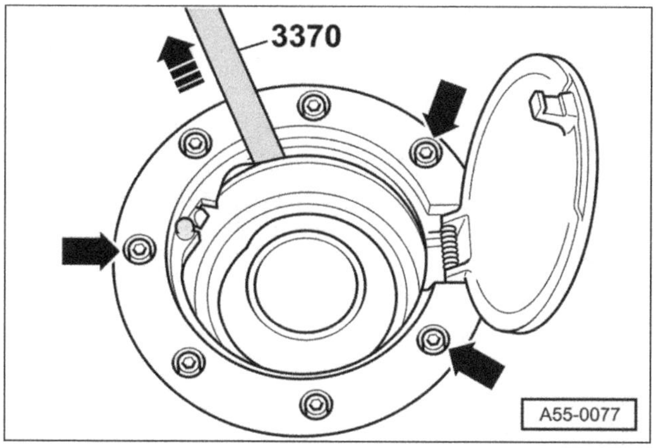

◄ Unscrew and remove the three screws indicated (**arrows**). Remaining screws may be simulated screws.

– Lift tank flap assembly straight up and out using tool **3370** or equivalent.

– Remove two screws securing cable to flap assembly.

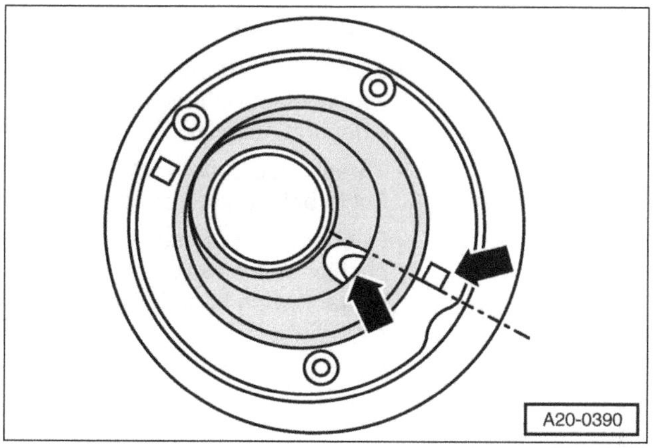

◄ When installing, align rubber cup to avoid kinking overflow hose. Edge of overflow hose must be aligned with edge of square hole as shown (**arrows**).

– Install screws and tighten to 4 Nm (35 in-lb).

BentleyPublishers.com
© 2006 BentleyPublishers.com—All Rights Reserved

Hood

Hood assembly

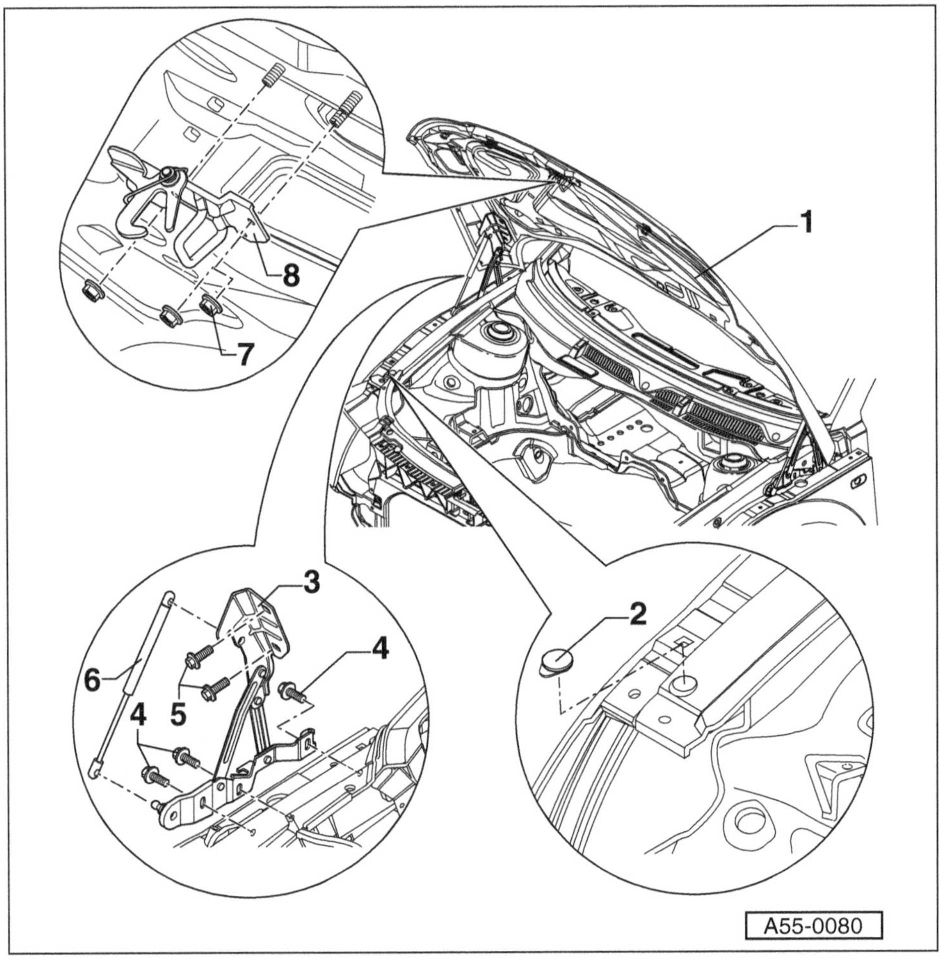

1. **Hood**
   - Removing:
     Use a helper(s)
     Detach gas-charged strut
     Remove bolts (item 5)
   - Installing:
     Install in reverse order
   - Adjusting:
     Align hood between fenders, see **Panel gaps**
     Adjust overall height at hood latch, see **Lower hood latch, removing and installing**
     Adjust height at fenders, see **Hood stop buffers, adjusting**
   - With sound absorber panel. See **Hood sound absorber panel**
2. **Protective cap**
   - Self adhesive
3. **Hood hinge**
   - Remove or properly support hood before servicing
   - Remove gas-charged strut
   - Remove bolts (items 4 and 5)
4. **Bolt**
   - Tighten to 21 Nm (15 ft-lb)
5. **Bolt**
   - Tighten to 21 Nm (15 ft-lb)
6. **Gas-charged strut**
   - Tube end goes toward hood
   - See **Gas-charged strut, removing and Installing**
   - See **Gas-charged strut, releasing charge**
7. **Nut**
   - Tighten to 8 Nm (71 in-lb)
8. **Upper hood latch with secondary hood latch**

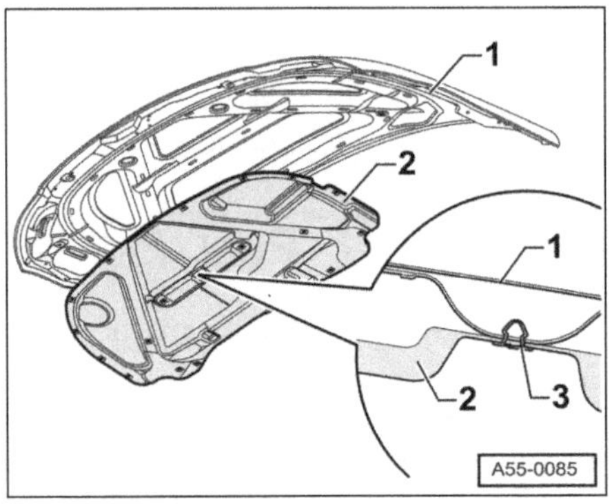

Hood sound absorber panel
- Secured with clips (**3**), Qty: 18
- Squeeze clips (**3**) to remove panel (**2**) from hood (**1**).

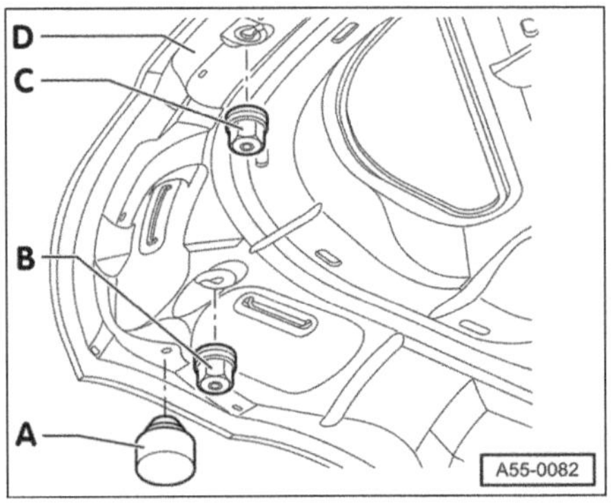

Hood stop buffers, adjusting
- Stop buffers (**B** and **C**) must be in light contact with the latch carrier when the hood (**D**) is closed
- Thread stop buffers (**B** and **C**) in or out to adjust hood height
- Rubber buffer (**A**) serves as impact cushion and should not make contact when hood is closed

BentleyPublishers.com
© 2006 BentleyPublishers.com—All Rights Reserved

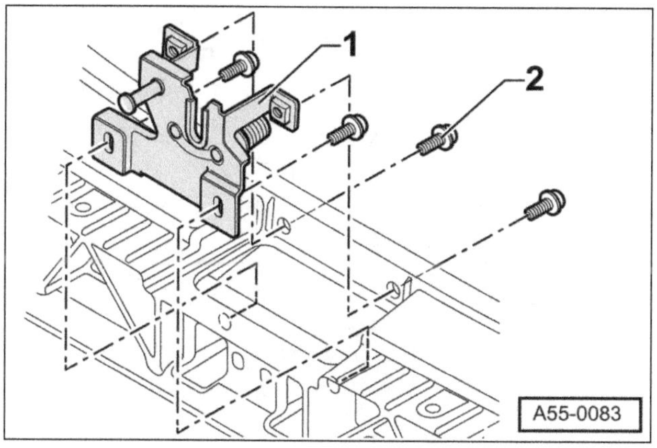

Lower hood latch, removing and installing

- Release hood latch cable
- Unscrew bolts (**2**), Qty: 4
- Remove lower hood latch (**1**)
- To adjust hood, move lower section of latch up or down as required.
- Install hood latch in reverse order
- Tighten bolts (**2**) to 10 Nm (7 ft-lb)

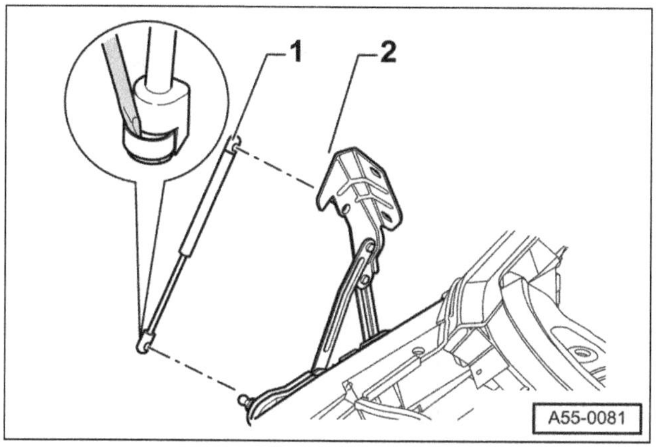

Gas-charged strut, removing and Installing

- Prop up and safely secure hood in open position
- Using a small screwdriver (**inset**), partially pry out retaining clip while pulling gas-charged strut (**1**) off of ball stud at lower mounting
- Repeat for upper connection (**2**)

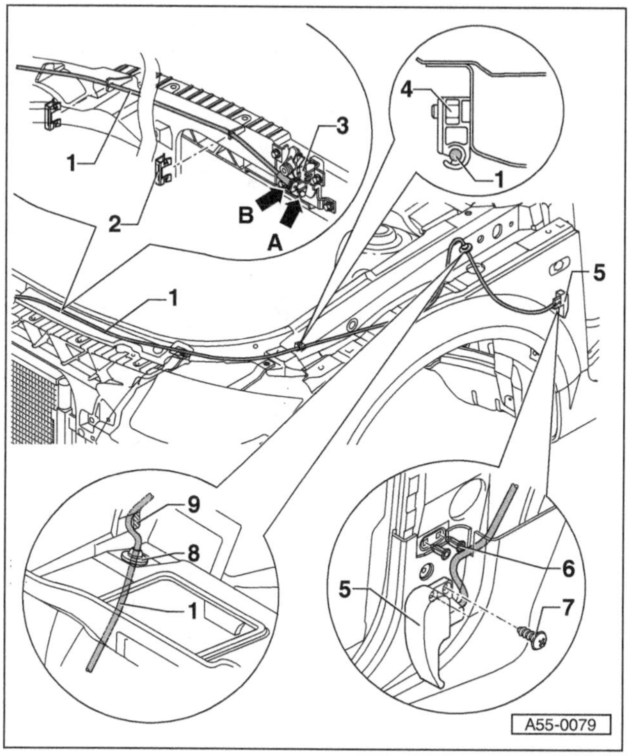

Hood latch cable, removing and installing

1. Hood latch cable
   Removing:
   - Remove left side under dash panels, see **70 Trim–Interior**
   - Remove mounting bolts (**3**)
   - Detach cable casing at support (**B**) and release cable end (**A**) from lower section of hood latch (**3**)
   - Remove cable securing clips (**2**)
   - Remove cable from retaining clip (**4**)
   - Pull cable (**1**) out from passenger compartment
   Installing:
   - Installation is the reverse of removal
2. Securing clip
3. Lower hood latch
4. Cable retaining clip
5. Hood latch cable release handle
6. Expanding nut
7. Screw, self-tapping
   - Tighten to 1.5 Nm (13 in-lb)
8. Grommet
   - Ensure proper fit
9. Corner guide

BentleyPublishers.com
©2006 BentleyPublishers.com—All Rights Reserved

Trunk Lid

Trunk lid assembly (coupe)

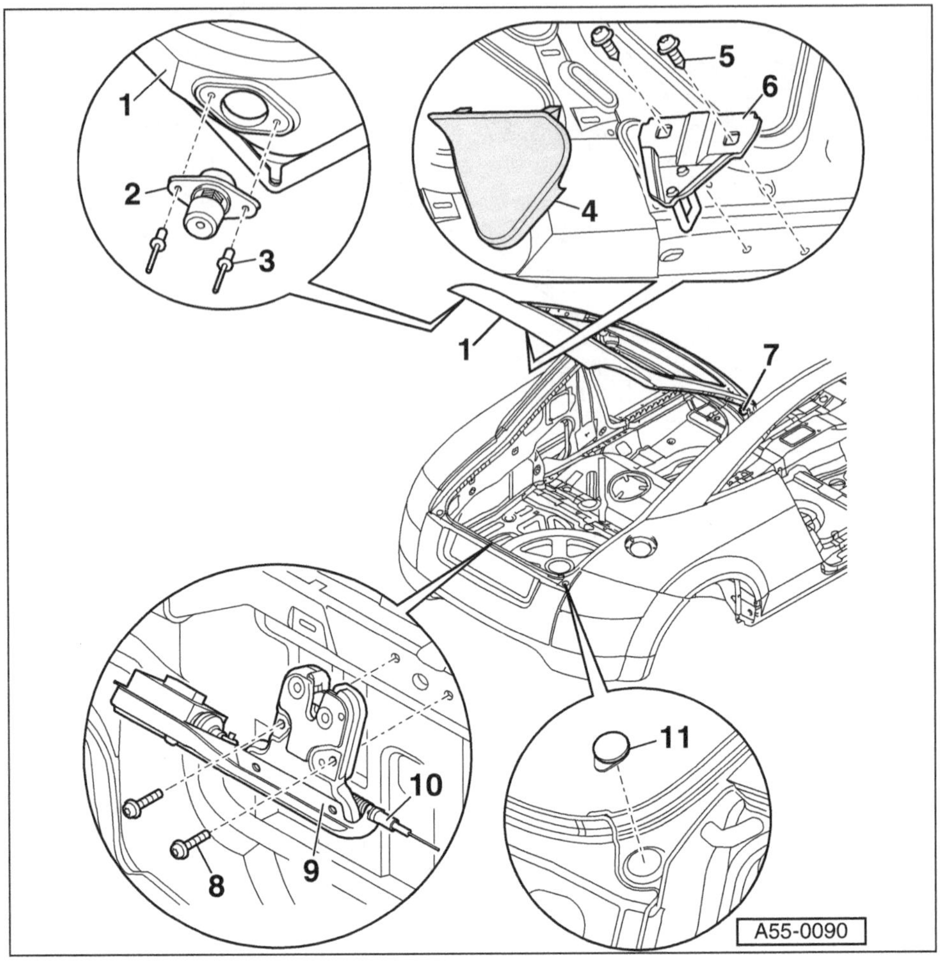

1. **Trunk lid**
   - Removing:
     Remove lining
     Disconnect and remove all electrical wiring from inside lid
     Use a helper(s)
     Remove both gas-charged struts
     Remove hinge attaching bolts bolts, see **Trunk lid hinges, removing and installing**
   - Installing:
     Install in reverse order
   - Adjusting:
     Align trunk vertically with hinges, see **Panel gaps**
     Align trunk height with latch, see **Panel gaps**
     Adjust overall height at hood latch, see **Lower hood latch, removing and installing**
     Adjust height at fenders, see **Trunk stop buffers, adjusting**
2. **Stop buffers**
   - Qty: 2
   - See **Trunk stop buffers, adjusting**
3. **Pop rivet**
   - Drill out to remove
4. **Trim cover**
   - Unclip to remove
5. **Bolt**
   - Tighten to 12 Nm (9 ft-lb)
6. **Striker plate**
   - To adjust, tighten bolt (item 5) finger tight so that striker plate can still be moved
   - Gently close tailgate until positioned with rear side panels
   - Carefully open trunk lid and tighten bolts to specification
   - Repeat as needed until satisfactory adjustment is obtained
7. **Trunk lid hinge**
   - See **Trunk lid hinges, removing and installing**
8. **Bolt**
   - Tighten to 8 Nm (71 in-lb)
9. **Trunk lid latch**
   - To remove, remove bolts (item 8)
   - Release cable for latch
10. **Trunk lid latch cable**
    - See **Trunk cable overview**
11. **Protective cap**
    - Self adhesive

*. **Gas-charged strut** (not shown)
   - Tube end goes toward body
   - See **Gas-charged strut, removing and installing**
   - See **Gas-charged strut, releasing charge**

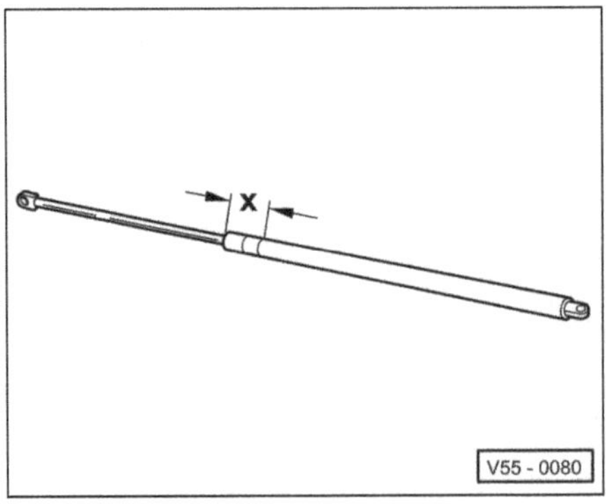

Gas-charged strut, releasing charge

- Wear protective goggles when sawing cylinder open and cover area with a rag to catch escaping fluid.
- Only clamp in area as indicated
- Clamp in a vise at area **X** (**arrows**).
- Make a saw cut in the cylinder tube at a point within approximately ⅓ of the distance up from the piston rod end to release gas charge.

BentleyPublishers.com
© 2006 BentleyPublishers.com—All Rights Reserved

Trunk lid assembly (roadster)

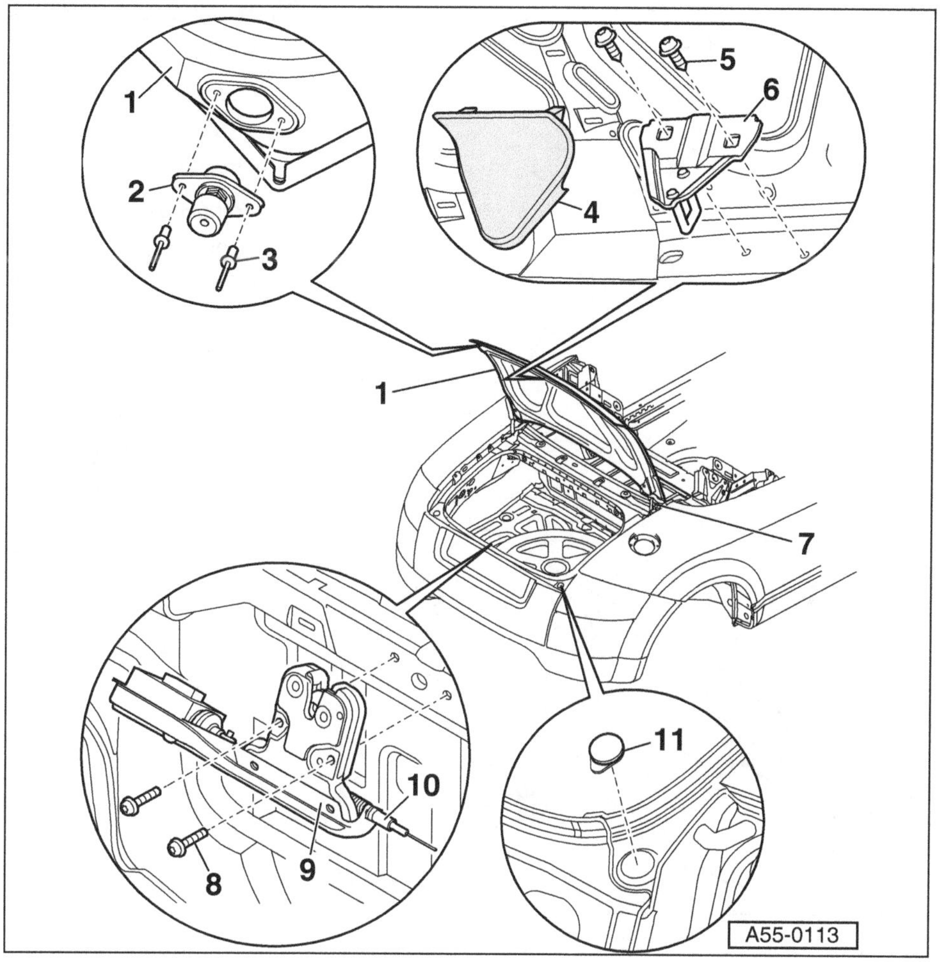

1. **Trunk lid**
   - Removing:
     Remove lining
     Disconnect and remove all electrical wiring from inside lid
     Use a helper(s)
     Remove both gas-charged struts
     Remove hinge attaching bolts bolts, see **Trunk lid hinges, removing and installing**
   - Installing:
     Install in reverse order
   - Adjusting:
     Align trunk vertically with hinges, see **Panel gaps, trunk lid**
     Align trunk height with latch, see **Panel gaps, trunk lid**
     Adjust overall height at hood latch, see **Lower hood latch, removing and installing**
     Adjust height at fenders, see **Trunk stop buffers, adjusting**
2. **Stop buffers**
   - Qty: 2
   - See **Trunk stop buffers, adjusting**
3. **Pop rivet**
   - Drill out to remove
4. **Trim cover**
   - Unclip to remove
5. **Bolt**
   - Tighten to 12 Nm (9 ft-lb)
6. **Striker plate**
   - To adjust, tighten bolt (item 5) finger tight so that striker plate can still be moved
   - Gently close tailgate until positioned with rear side panels
   - Carefully open trunk lid and tighten bolts to specification
   - Repeat as needed until satisfactory adjustment is obtained
7. **Trunk lid hinge**
   - See **Trunk lid hinges, removing and installing**
8. **Bolt**
   - Tighten to 8 Nm (71 in-lb)
9. **Trunk lid latch**
   - To remove, remove bolts (item 8)
   - Release cable for latch
10. **Trunk lid latch cable**
    - See **Trunk cable overview**
11. **Protective cap**
    - Qty: 2
    - Self adhesive

*. **Gas-charged strut** (not shown)
   - Tube end goes toward trunk lid
   - See **Gas-charged strut, removing and installing**
   - See **Gas-charged strut, releasing charge**

BentleyPublishers.com
© 2006 BentleyPublishers.com—All Rights Reserved

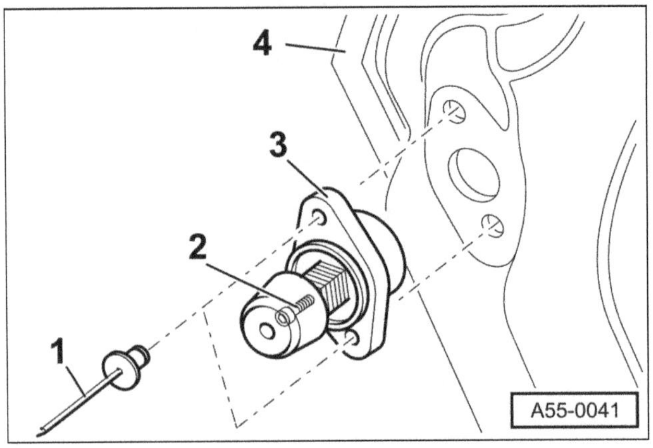

Trunk stop buffers, removing, installing, and adjusting

Removing:

- Drill out pop rivet head (**1**) and tap shank out.
- Pull out adjuster stop (**3**).
- Install in reverse order.

Adjusting:

- Loosen clamping screw (**2**) and pull adjuster stop out slightly.
- Close trunk lid (**4**) to the flush position; tailgate latch should engage in second stop in striker.
- Set adjuster stop by opening trunk lid and pulling adjuster stop out by one notch.
- Tighten clamping screw (**2**) to 18 Nm (13 ft-lb)
- Close trunk lid.
- Check panel gap.

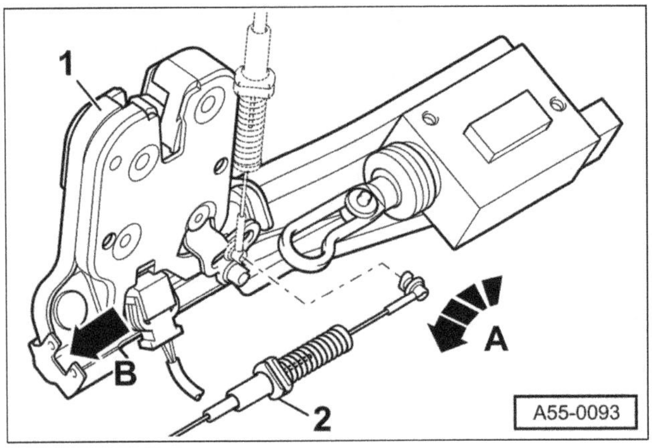

Trunk lid latch, disconnecting cable

- Remove rear lid latch (**1**).
- Turn cable casing (**2**) in support bracket 90° (¼ turn) and remove upward.
- Pivot cable 90° (¼ turn) out of trunk lid latch connecting loop (**A**).

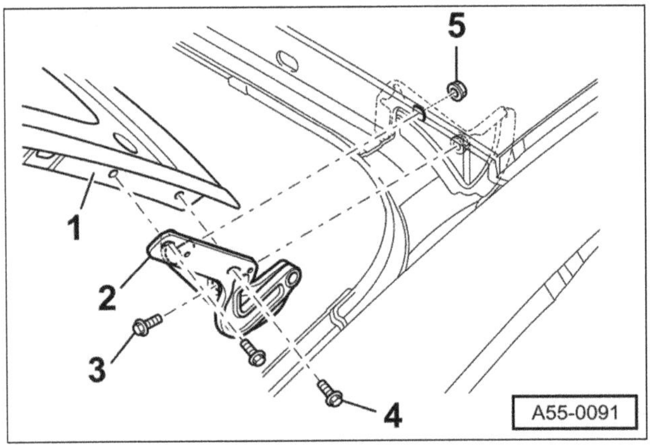

Trunk lid hinge, removing and installing (coupe)

- Pry out trim cover at the bottom and detach from trunk lid.
- Remove roof rail trim.
- Unscrew bolts (**4**) and lift off trunk lid (**1**).
- Remove bolts (**3**) and nuts (**5**).
- Remove hinge (**2**).
- Install in reverse order of removal.
- Tighten nut (**5**) and bolts (**3 and 4**) to 21 Nm (15 ft-lb).

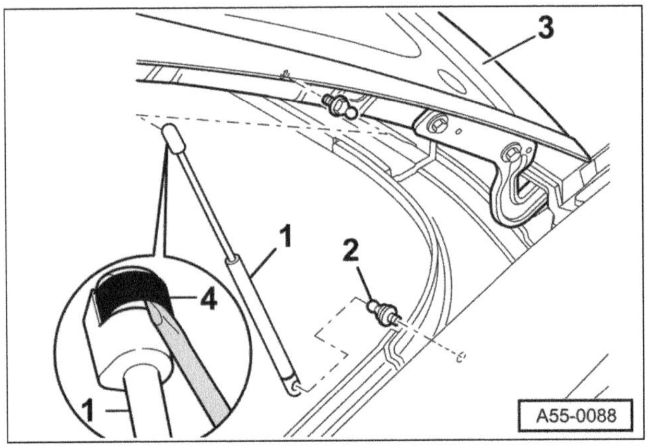

Gas-charged strut, removing and installing (coupe)

- Prop up and safely secure trunk lid (**3**) in open position.
- Slightly push up on trunk lid to relieve tension of ball studs (**2**).
- Using a small screwdriver (**inset**), partially pry out retaining clip (**4**) while pulling gas-charged strut (**1**) off of ball stud mountings (**2**).
- Install in reverse order of removal.
  Tube end goes toward body of vehicle (down).
  Clip onto bottom ball stud first.
  If ball stud is removed, install and tighten to 15 Nm (11 ft-lb).

BentleyPublishers.com
© 2006 BentleyPublishers.com—All Rights Reserved

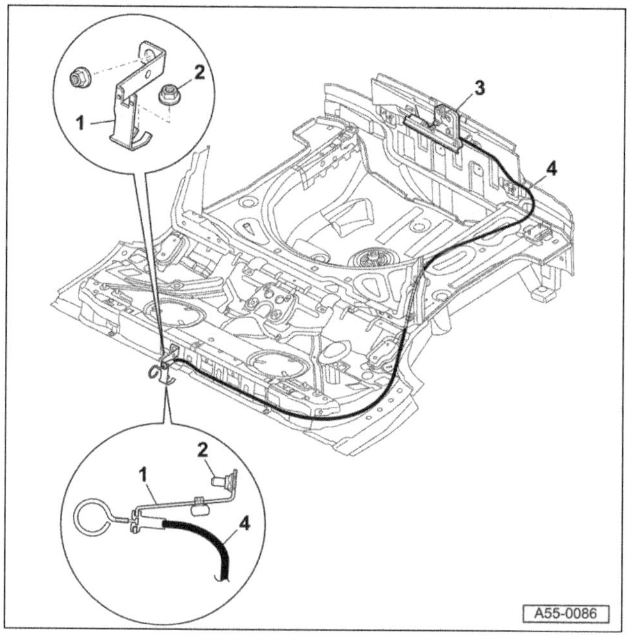

Trunk cable overview (coupe)

1. Bracket for trunk latch cable
2. Nut,
   - Tighten to 5 Nm (44 in-lb)
3. Trunk lid latch
4. Trunk lid latch cable

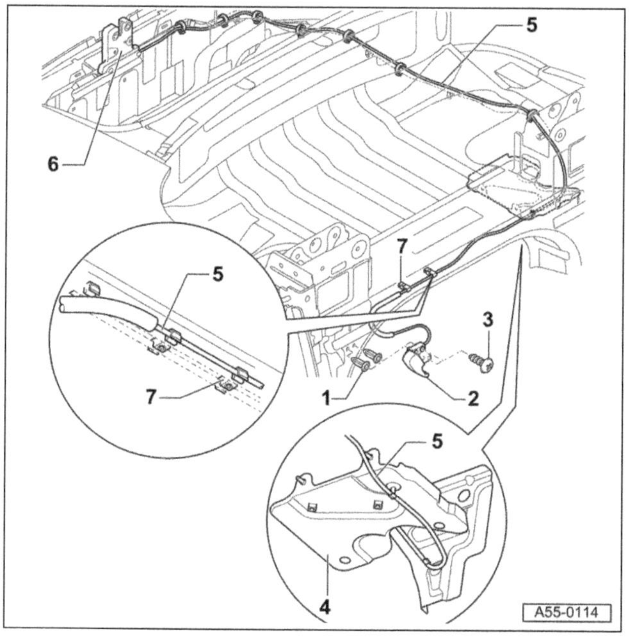

Trunk cable overview (roadster)

1. Expanding inserts
2. Trunk release handle
   - Route through clips as shown
3. Screw
   - Tighten to 1.5 Nm (13 in-lb)
4. Noise chamber panel, lower section
   - Can only be removed after hydraulic pump is removed
5. Trunk lid latch cable
6. Trunk lid latch
7. Clip

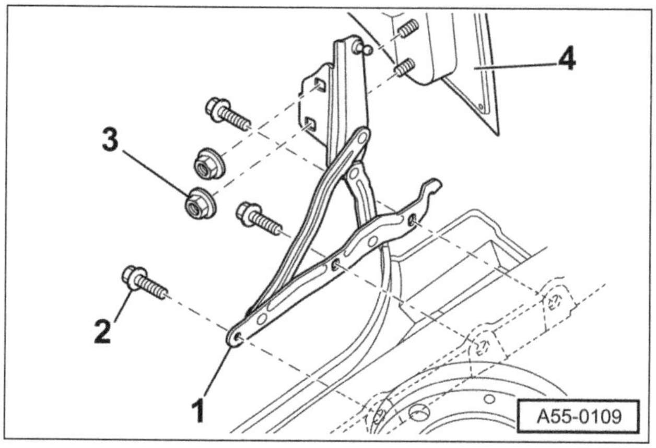

Trunk lid hinge, removing and installing (roadster)

- Remove nuts (**3**) and lift off trunk lid (**4**).
- Remove bolts (**2**) and remove hinge (**1**).
- Install in reverse order of removal.
- Tighten nut (**3**) and bolts (**2**) to 21 Nm (15 ft-lb).

BentleyPublishers
.com
©2006 BentleyPublishers.com—All Rights Reserved

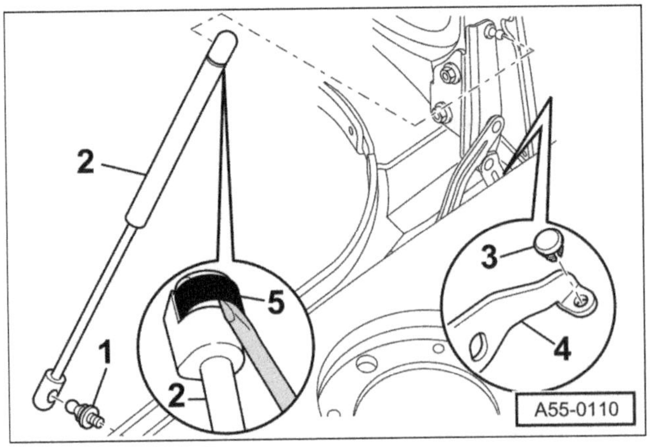

Gas-charged strut, removing and installing (roadster)

- Prop up and safely secure trunk lid in open position.
- Slightly push up on trunk lid to relieve tension of ball studs (**1**).
- Using a small screwdriver (**inset**), partially pry out retaining clip (**5**) while pulling gas-charged strut (**2**) off of ball stud mountings (**1**).
- Install in reverse order.
  Tube end goes toward trunk lid (up).
  Clip onto bottom ball stud first.
  If ball stud is removed, 15 Nm (11 ft-lb).
  Ensure that stop (**3**) is clipped into leg of hinge (**4**).

PANEL GAPS

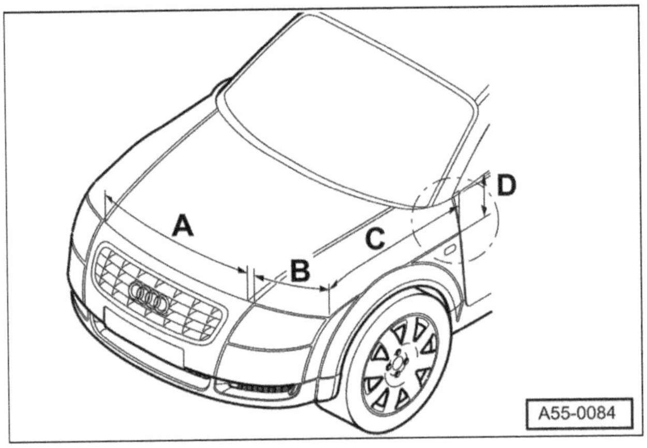

Hood and front panel gaps

A = 4.0 mm ± 1 mm
B = 4.0 mm ± 1 mm
C = 4.0 mm ± 1 mm
D = 4.0 mm ± 1 mm

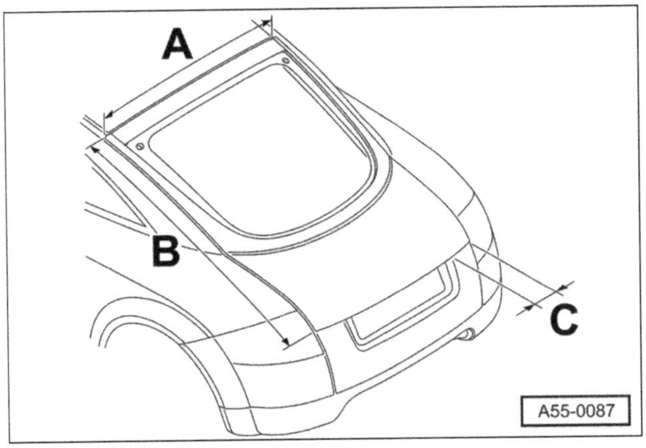

Rear lid panel gaps (coupe)

A = 4.5 mm + 1 mm
B = 3.5 mm + 1 mm
C = 5.0 mm + 1 mm

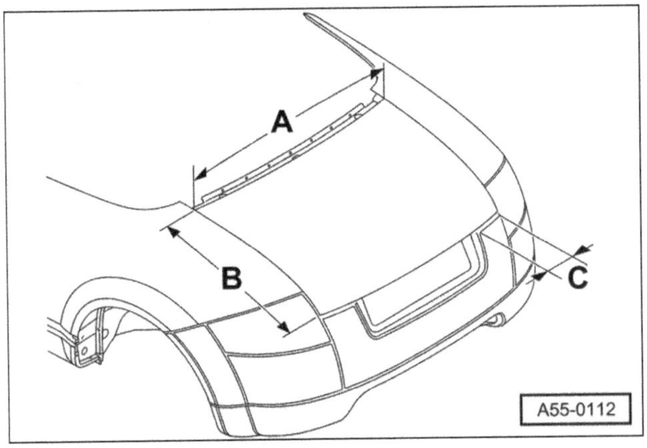

Rear lid panel gaps (roadster)

A = 3.5 mm + 1 mm
B = 3.5 mm + 1 mm
C = 5.0 mm + 1 mm

BentleyPublishers
.com
©2006 BentleyPublishers.com—All Rights Reserved

57 Doors and Central Locking

GENERAL

This section covers door assemblies and central locking components. The doors on the TT consist of an outer panel, an inner subframe for mounting hardware and the interior door trim panel. The window motor is accessed from the inside by removing the interior trim panel.

- For information on door window switches, see **96 Interior Lights, Switches, Anti-theft**.
- For information on door trim panels, see **70 Trim–Interior**.
- For information on door stereo speakers, see **91 Radio and Communication**.

REMOTE CONTROL KEY

Remote control key battery, replacing

Remote control keys use batteries. Battery type varies with model year, but replacement for all model years is similar.

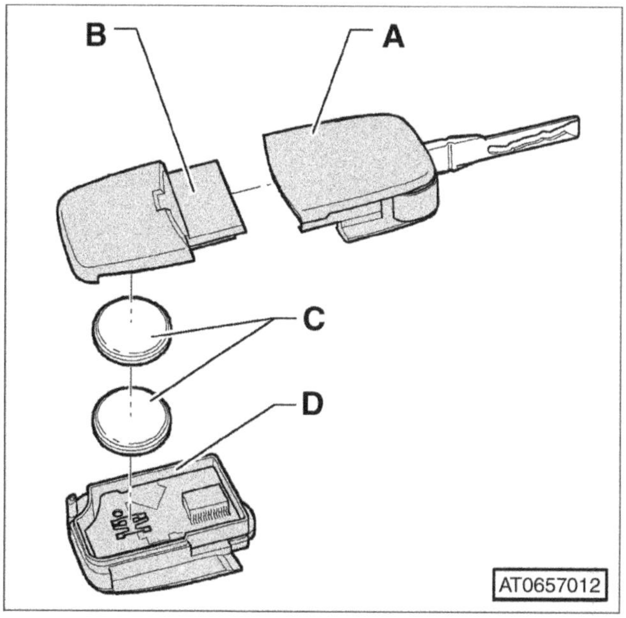

After separating the transmitter unit (**B**) and the main folding key body (**A**), remove the lower section of the transmitter (**D**) by carefully prying apart.

– Batteries (**C**) lift out for replacement. Battery type varies by model year. Remote control keys use two 3 Volt lithium batteries

– On reassembly, observe battery polarity.

NOTE—

- *Dispose of all dead batteries properly, recycle when possible.*
- *Comply with all applicable laws regarding disposal.*

BentleyPublishers.com

©2006 BentleyPublishers.com—All Rights Reserved

Front Doors

Front door and door subframe overview

NOTE—
- *Coupe door shown; coupe and roadster door assemblies are similar.*

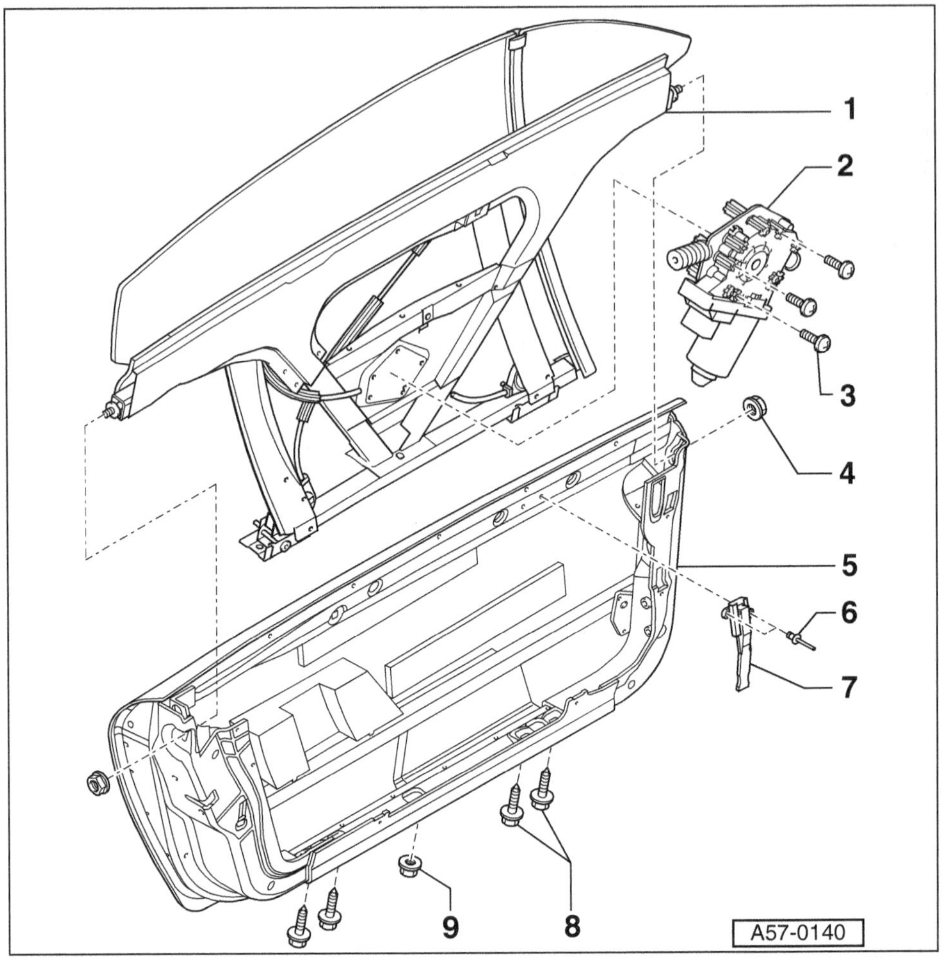

1. **Front door subframe**
   - See **Door subframe, removing and installing**
   - See **Door subframe, adjusting overview**
2. **Window motor**
3. **Torx bolt**
   - Qty: 3
   - Tighten to 3 Nm (27 in-lb)
4. **Nut**
   - Qty: 2
   - Tighten to 30 Nm (22 ft-lb)
   - Pry off trim covers for access
5. **Front door**
   - See **Front door adjusting**
6. **Rivet**
   - Qty: 2
   - To remove, drill out rivet head and tap shank out
7. **Support**
   - At base of window divider bar between front door subframe and inner door structure
   - To remove, drill out rivets
   - See **Adjusting support**
8. **Bolt**
   - Qty: 4
   - Tighten to 25 Nm (18 ft-lb)
   - Pry off covers (not shown) for access
9. **Nut**
   - Qty: 1
   - Tighten to 30 Nm (22 ft-lb)
   - Pry off covers (not shown) for access

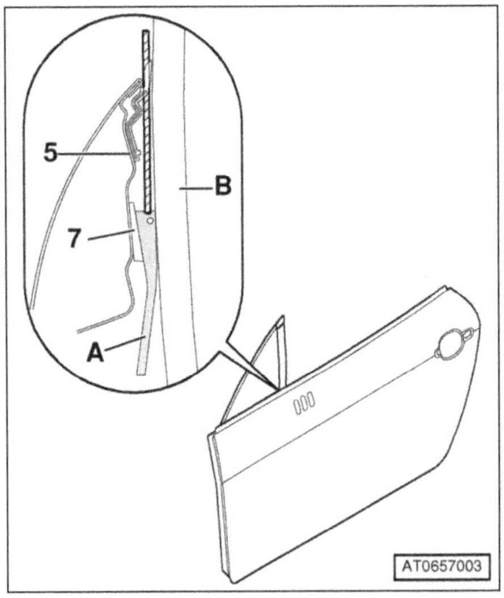

◀ **Adjusting support**
- Adjust door subframe first.
- Push adjuster part (**A**) of support (**7**) up as far as it will go.
- Correct position is between door (**5**) and divider bar of window (**B**).

BentleyPublishers.com
© 2006 BentleyPublishers.com—All Rights Reserved

Door subframe, removing and installing

Removing

NOTE—

- *Callout numbers in illustrations AT0657001 and AT0657002 correspond to the callout numbers in illustration A57-0140.*

— Before removing, mark position of subframe on the door using a fine point marker (Sharpie®) or equivalent. Mark sufficiently to allow re-assembly at the same locations and positions.

— If removing window glass, raise window fully.

— Remove door panel, see **70 Trim–Interior**.

— Disconnect wiring and unclip from door subframe.

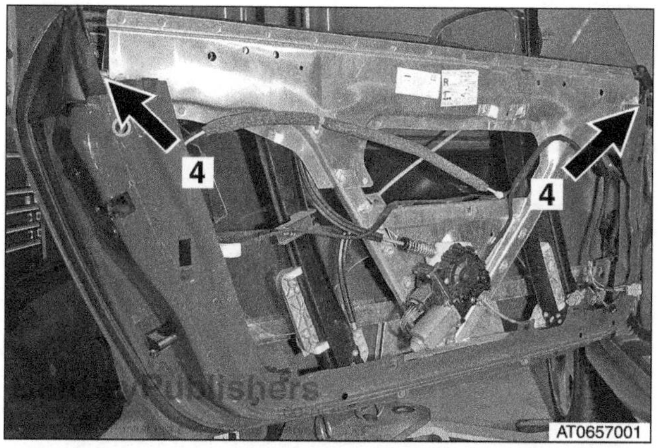

AT0657001

◀ Pry off round trim covers and loosen nuts (**4**).

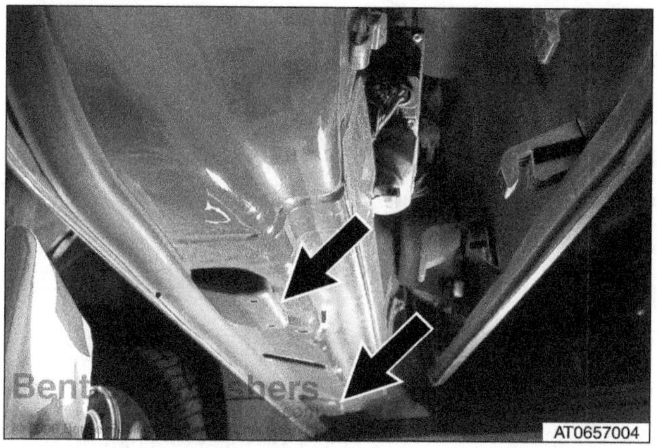
AT0657004

◀ Pry off trim covers (**arrows**) on bottom of door.

AT0657002

◀ Remove bolts (**8**).

— Remove nuts (**9**).

— Lift off subframe.

Installing

— Adjust front door before installing subframe. See **Front door adjusting**.

— Install subframe into door and align with markings made before removal.

— Adjust door subframe, see **Door subframe, adjusting**.

BentleyPublishers.com
©2006 BentleyPublishers.com—All Rights Reserved

Door subframe overview

NOTE—
- *Coupe door shown: coupe and roadster door assemblies are similar.*

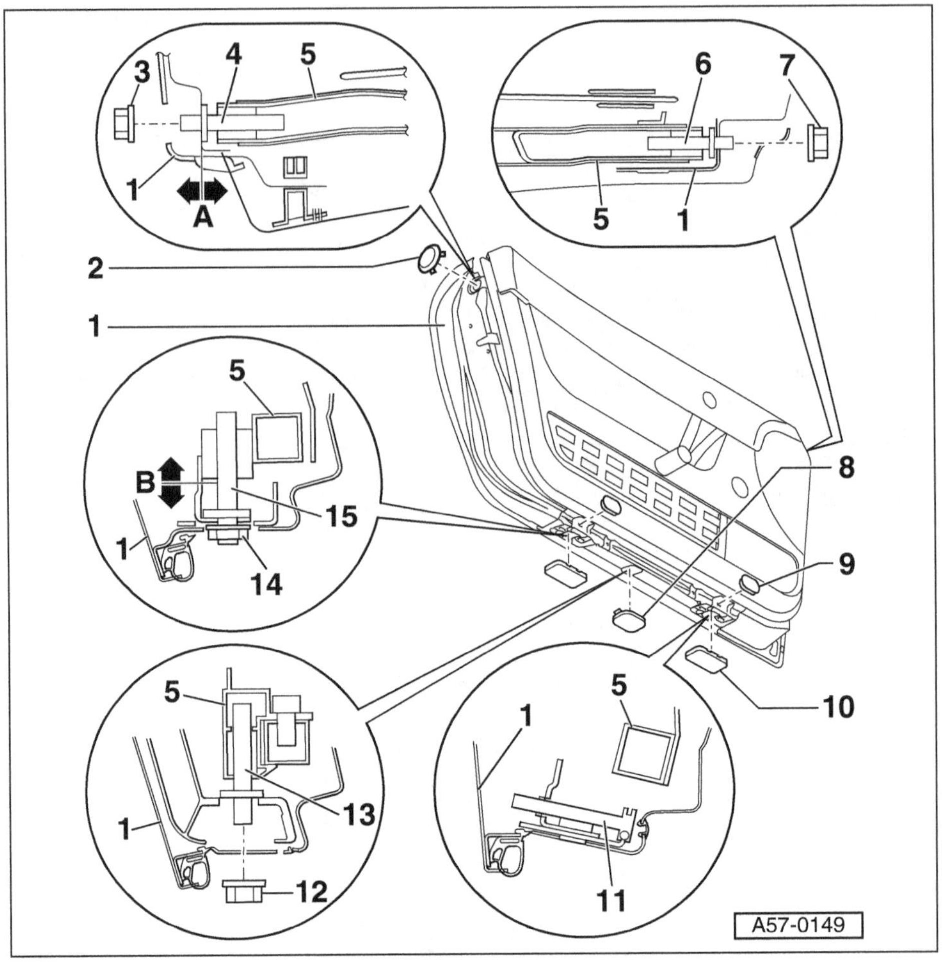

1. **Front door with subframe**
2. **Trim cover**
   - Pry out
3. **Lock nut, rear**
   - Tighten to 30 Nm (22 ft-lb)
4. **Adjuster screw, rear**
   - Allows for longitudinal (front-back) adjustments
   - Adjustment range (**A**) is approximately 5 mm from center position (10 mm overall)
   - Counterhold when tightening lock nut (item **3**) to prevent changes in adjustment
   - Move in conjunction with front adjuster screw (item **6**)
   - Loosen lower subframe mounting bolts before adjusting (reference item 8 in illustration AT0657002)
5. **Front door subframe**
6. **Adjuster screw, front**
   - Allows for longitudinal (front-back) adjustments
   - Adjustment range (**A**) is approximately 5 mm from center position (10 mm overall)
   - Counterhold when tightening lock nut (item **3**) to prevent changes in adjustment
   - Move in conjunction with rear adjuster screw (item **4**)
   - Loosen lower subframe mounting bolts before adjusting (reference item 8 in illustration AT0657002)
7. **Lock nut, front**
   - Tighten to 30 Nm (22 ft-lb)
8. **Trim cover**
   - Qty: 1 per door
   - Pry out
9. **Trim cover**
   - Qty: 2 per door
   - Pry out
10. **Trim cover**
    - Qty: 2 per door
    - Pry out
11. **Adjuster screw**
    - Allows for angle (in-out at the top) adjustments
    - Remove trim cover (item **9**) for access
    - Front adjuster shown, rear adjuster is similar
    - Observe horizontal positioning, do not confuse with front height adjuster screw
12. **Lock nut, rear**
    - Only tighten after stop bolt (item **13**) is back to contact position
    - Tighten to 30 Nm (22 ft-lb)
13. **Stop bolt**
    - After obtaining proper height with adjuster screw (item **15**), thread back down until contact is made with door sheet metal
14. **Lock nut**
    - Tighten to 30 Nm (22 ft-lb)
15. **Adjuster screw**
    - Allows for height adjustments
    - Adjustment range (**B**) is approximately 4 mm from center position (8 mm overall)
    - Remove trim cover (item **10**) for access
    - Rear adjuster shown, front adjuster is similar
    - Observe vertical positioning, do not confuse with rear angle adjuster screw

BentleyPublishers
.com
© 2006 BentleyPublishers.com—All Rights Reserved

Doors, adjusting

Adjust doors using standard and customary body shop procedures to the specified panel gaps. Door striker adjustment is only used for inward/outward alignment. Do not use for vertical adjustment; this may cause damage to the latching mechanism and will cause the door to drag resulting in difficult opening and/or closing. Incomplete latching may also result.

Use of a plastic feeler gauge such as **3371**, which is designed for measuring panel gaps, will minimize the chance of paint damage and insure accurate panel gaps.

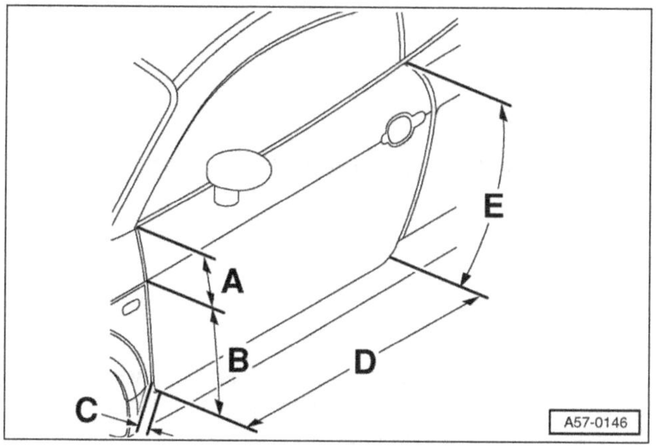

Left front door panel gaps; right front gaps are the same dimensions.

A = 4 mm + 1 mm

B = 4 mm + 1 mm

C = 4 mm ± 0.5 mm

D = 5 mm ± 1.5 mm

E = 3.5 mm + 1mm

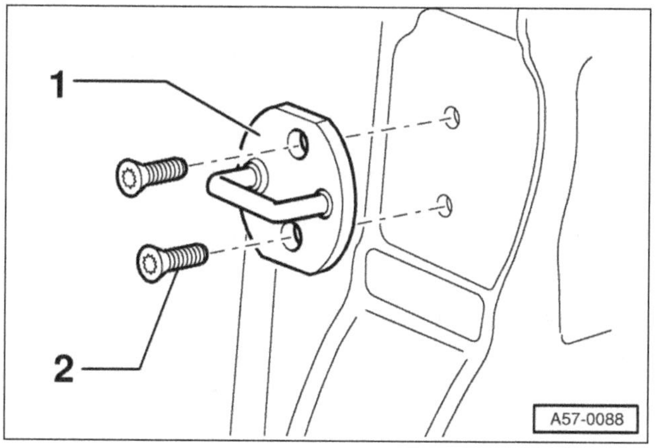

Adjust striker plate (1).

– Loosen bolts (**2**) just enough to allow striker plate to move.

– Move striker plate until door is flush with body contours, but not dragging.

– When door is flush, tighten bolts.
  - Tighten to 20 Nm (15 ft-lb)

NOTE—

- *Experience suggests that door panel gap measurements are best made with the striker plate removed. This allows for accurate adjustments without the striker mechanism bearing improperly against the door structure.*

BentleyPublishers.com
© 2006 BentleyPublishers.com—All Rights Reserved

Door subframe, adjusting

Coupe door shown; coupe and roadster door assemblies are similar. Also see **Front door subframe overview** given earlier in this section. Adjusting subframe also adjusts the door window glass.

– Door must be properly adjusted before adjusting door subframe. See **Doors, adjusting**.

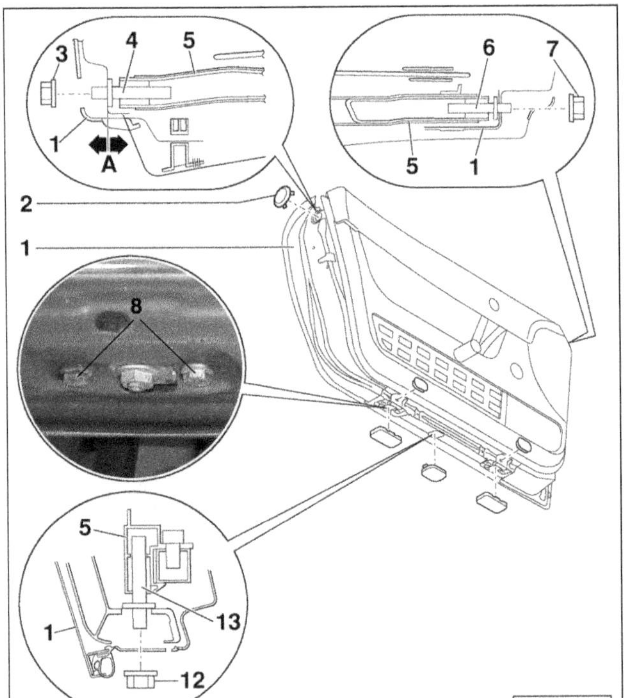

AT0657006

Longitudinal adjustment (A)

◄ Remove trim covers and loosen lock nuts (**3**, **7**, **12**).

– Loosen rear subframe securing bolts (**8**), repeat for front bolts.

– **Adjusting to the rear** - screw in rear adjuster (**4**). One full turn equal approximately 1 mm travel. Screw out front adjuster (**6**) to move subframe to the rear until rear adjuster (**4**) touches door. Check and readjust until desired position is achieved. When adjustment is correct, both adjusters (**4** and **6**) should just contact door.

– **Adjusting to the front** - screw in front adjuster (**6**). One full turn equal approximately 1 mm travel. Screw out rear adjuster (**4**) to move subframe to the front until front adjuster (**6**) touches door. Check and readjust until desired position is achieved. When adjustment is correct, both adjusters (**6** and **4**) should just contact door.

– When tightening lock nuts (**3**, **7**), counterhold adjuster screws (**4**, **6**) with a Torx T30 to prevent movement.

– When tightening lock nuts (**12**), counterhold stop bolt (**13**) to prevent movement.

– Install trim covers.

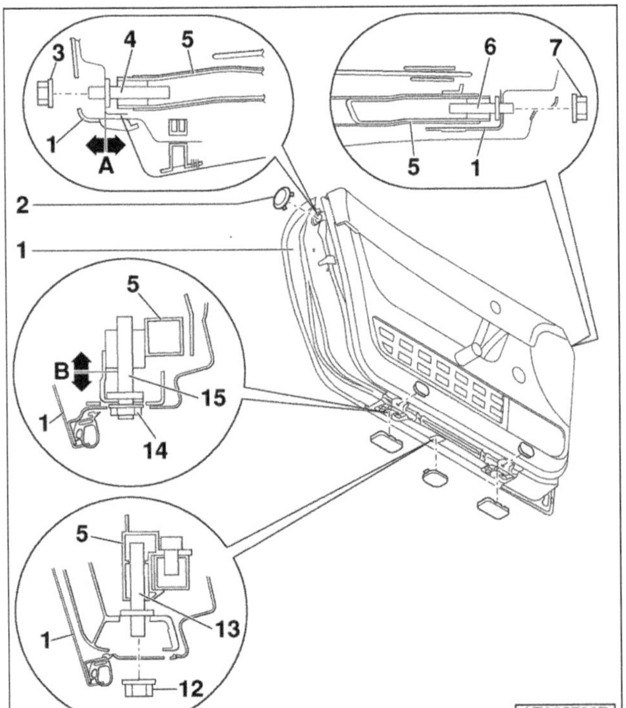

AT0657007

Height adjustment (B)

◄ Remove trim covers and loosen lock nuts (**3**, **7**, **12**). Loosen rear adjuster locknut (**14**) and repeat for front lock nut (**14**). If downward adjustment is needed, also loosen stop bolt (**13**).

– Adjust front and rear height with adjuster screws (**15**).

– After completing height adjustment, back stop bolt (**13**) out as far as possible without forcing it.

– When tightening lock nuts (**3**, **7**), counterhold adjuster screws (**4**, **6**) with a Torx T30 to prevent movement.

– When tightening lock nut (**12**), counterhold stop bolt (**13**) to prevent movement.

– When tightening lock nuts (**14**), counterhold adjuster screws (**15**) to prevent movement.

– Install trim covers.

BentleyPublishers.com
© 2006 BentleyPublishers.com—All Rights Reserved

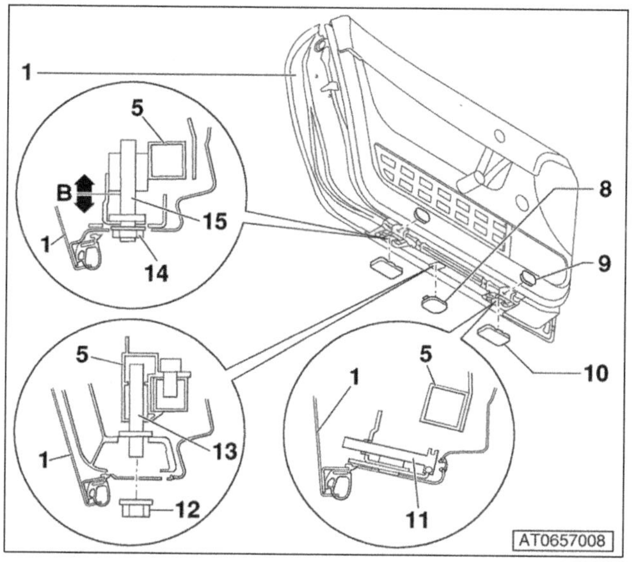

Angle adjustment (in/out at top)

- Remove trim covers (**8**, **9**, **10**).
- Loosen lock nut (**12**).
- Loosen rear adjuster locknut (**14**) and repeat for front lock nut (**14**).
- Using front and/or rear adjuster screws (**11**), adjust angle at the top of glass. Note that correct adjusters are behind trim covers (**9**).
- When tightening lock nut (**12**), counterhold stop bolt (**13**) to prevent movement.
- When tightening lock nuts (**14**), counterhold adjuster screws (**15**) to prevent movement.
- Install trim covers.

Door hinges, removing and installing

When working with the door hinges, do not leave the door unsupported. If removing doors, using a helper will make the job easier. See **03 Maintenance** for additional information.

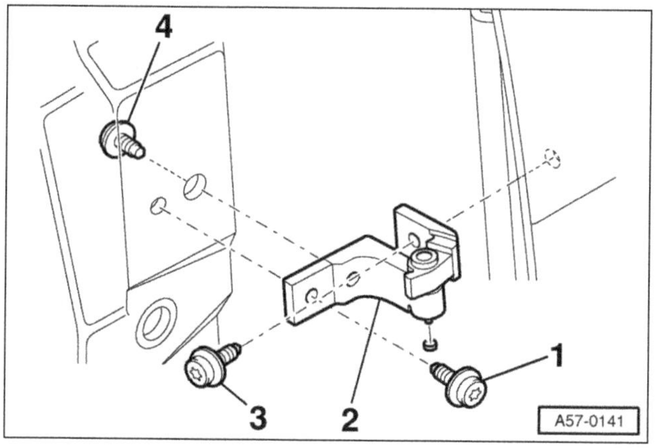

Upper door hinge. Right side shown, left side is similar.

1. Bolt
   - Tighten to 30 Nm (22 ft-lb)
2. Upper hinge assembly
3. Torx bolt (T45)
   - Tighten to 30 Nm (22 ft-lb)
4. Bolt, installed from inside vehicle
   - Tighten to 30 Nm (22 ft-lb)

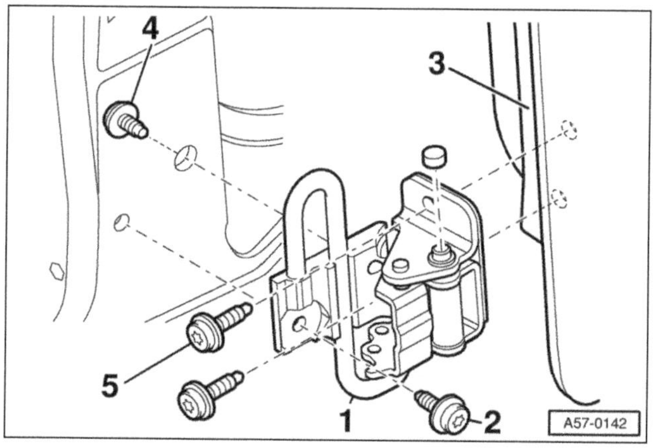

Lower door hinge. Right side shown, left side is similar.

1. Lower door hinge
2. Bolt
   - Tighten to 30 Nm (22 ft-lb)
3. Front door
4. Bolt
   - Tighten to 30 Nm (22 ft-lb)
5. Bolt, Torx
   - Tighten to 30 Nm (22 ft-lb)

BentleyPublishers
.com
©2006 BentleyPublishers.com—All Rights Reserved

Door Handles and Latches

Door handles and latches, overview

1. **Door latch**
   - Remove door trim panel. See **70 Trim-Interior**
   - Latch can be removed without removing subframe assembly
   - See **Door Latch, removing and installing**
2. **Release cable**
   - Connects door handle with door latch
3. **Bearing bracket**
   - Remove door handle (item **5**) and lock cylinder (items **6**, **7**) to access
   - Remove Phillips screw (item **4**) and slide bracket slightly to the rear to remove from door
4. **Screw, Phillips**
   - Tighten to 4.5 Nm (40 in-lb)
5. **Door Handle with base**
   - See **Door handle, removing and installing**
6. **Cover trim**
   - Clipped on
7. **Lock cylinder housing**
   - See **Lock cylinder, removing and installing**
   - Lock cylinder integral with housing
8. **Backing gasket**
9. **Bolt, Torx T20**
   - Use tool **T10011** or equivalent
   - Do not remove bolt. Do not screw bolt fully in without lock cylinder (item **7**) installed, locking ring may fall into door
   - Retains lock cylinder (item **7**) in bearing bracket (item **3**)
10. **Bolt, Torx T20**
    - Tighten to 20 Nm (15 ft-lb)

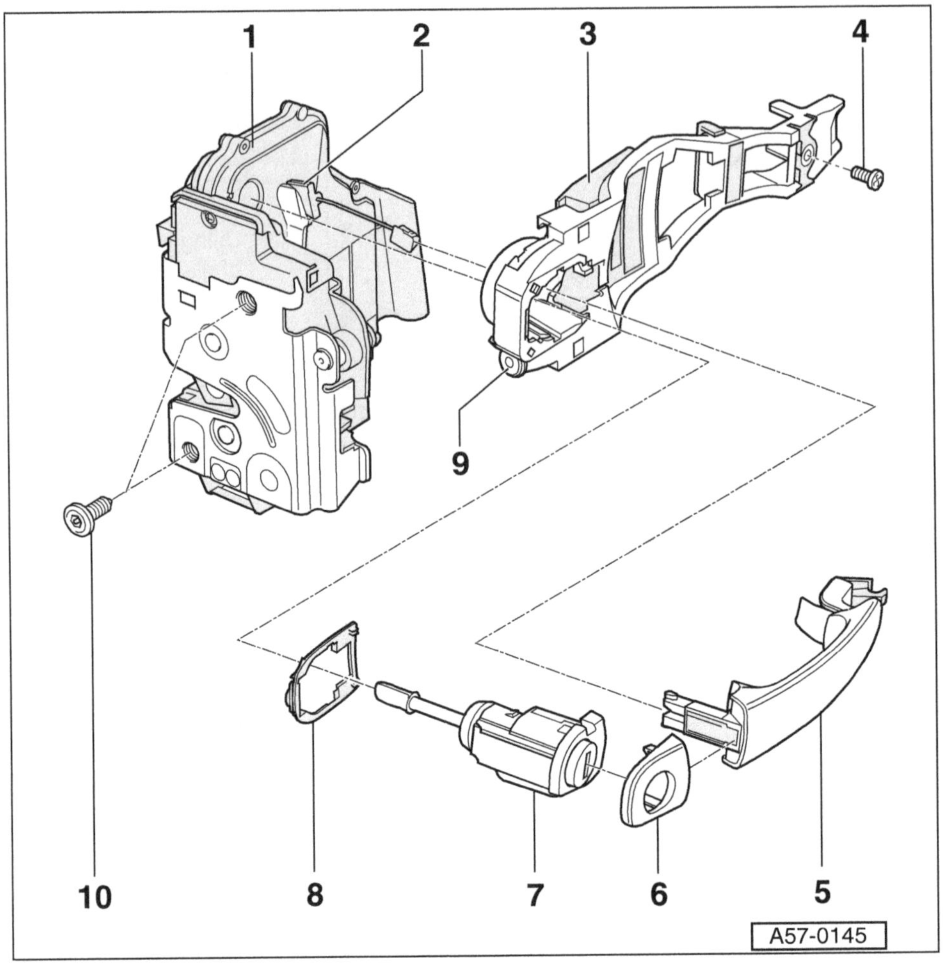

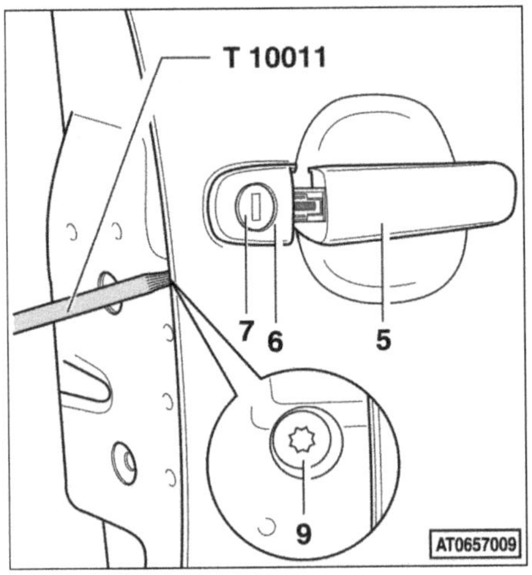

Lock cylinder, removing and installing

- Pull door handle (**5**) out and hold.
- Unscrew bolt (**9**) to the stop using **T10011** or equivalent. This releases lock cylinder housing (**7**). Do not remove bolt. Do not screw bolt fully in without lock cylinder installed, locking ring may fall into door.
- Pull lock cylinder housing (**7**) and cover trim (**6**) straight out.
- Do not let backing gasket fall into door.

BentleyPublishers.com

©2006 BentleyPublishers.com—All Rights Reserved

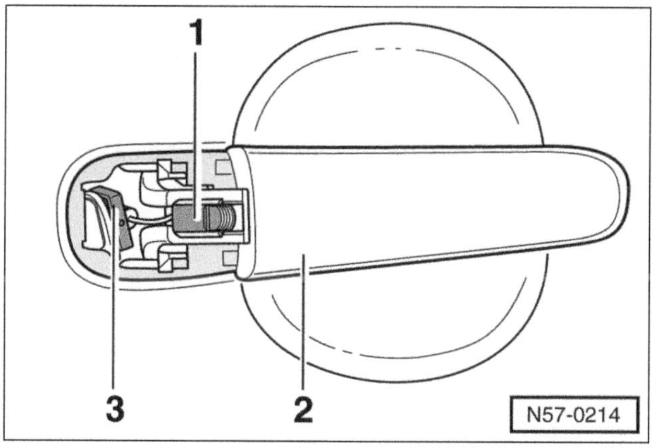

Door handle, removing and installing

Removing

- Remove lock cylinder. See **Lock cylinder, removing and installing**
- Using a pick tool or similar, unclip cable end (**1**) from door handle (**2**).
- Pivot door handle (**2**) out and away from door.

Installing

- Pivot door handle (**2**) into door, reversing removal procedure.
- Clip cable end (**1**) onto door handle (**2**) and snap into place. Do not pull cable tight from operating lever (**3**).
- Install lock cylinder.
- Check door handle for proper operation. An incorrectly installed cable (too tight) will not allow the door to be opened.

Door latch, removing and installing

Removing

– Remove door trim panel. See **70 Trim-Interior**. This will disconnect latch from inner door release mechanism.

– Remove lock cylinder. See **Lock cylinder, removing and installing**.

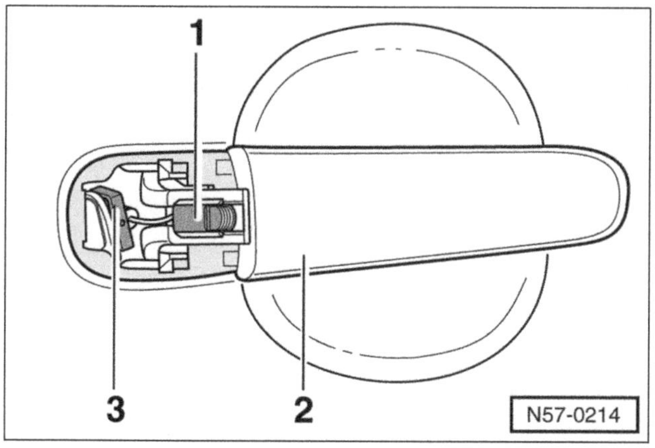

Using a pick tool or similar, unclip cable end (**1**) from door handle (**2**).

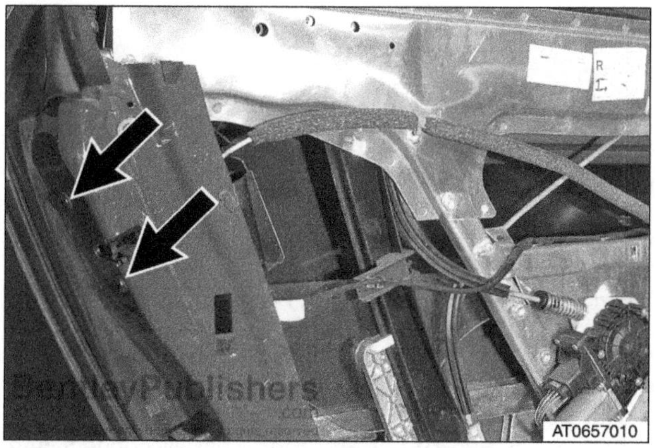

Remove screws (**arrows**) from door.

– Lift latch mechanism, complete with inside release cable, out of door.

BentleyPublishers
.com
© 2006 BentleyPublishers.com—All Rights Reserved

Installing

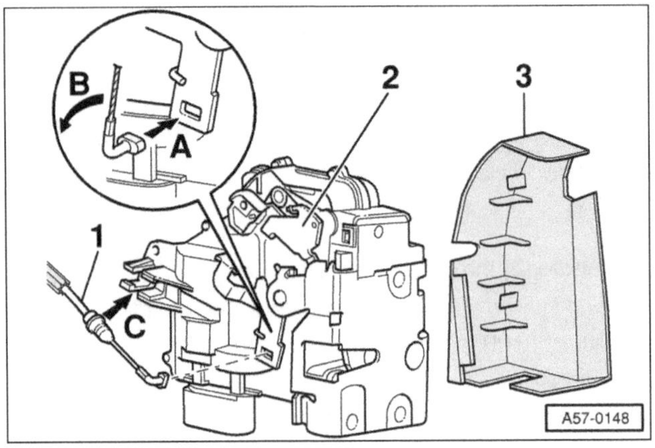

- Before installing door latch assembly (**1**), ensure that inside release cable is correctly attached to inside release lever.

- Insert cable (**1**) into lever in direction of arrow (**A**) and turn 90° (¼ turn) into position (**B**).

- Clip cable (**1**) into slot (**C**) on latch (**2**).

- Snap cover (**3**) into place.

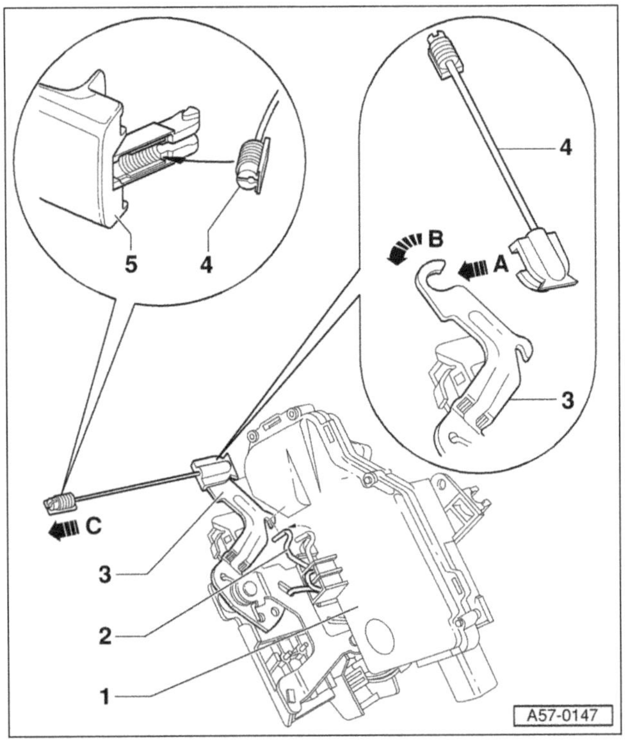

- Confirm that release cable (**4**) is clipped (**A**) into lock lever (**3**) and is free to move (**B**).

- Bring mechanism to basic position for installation. Pull lock lever (**3**) out so that spring (**2**) can engage in slot on lever (**3**). Spring will hold lock lever in basic position.

- Install door latch assembly (**1**) complete with inside release cables into door. Verify that release cable (**4**) can be reached from outside opening in door.

- Install screws (**arrows**) and tighten.
  - Tighten to 20 Nm (15 ft-lb)

- Install door handle. See **Door handle, removing and installing**.

- Install lock cylinder. See **Lock cylinder, removing and installing**.

- Verify proper operation of inside and outside release mechanisms. Use unattached cable loop to check inside operation.

- Install door trim panel. See **70 Trim-Interior**. This will connect inner door release mechanism.

BentleyPublishers
.com
©2006 BentleyPublishers.com—All Rights Reserved

Door Window Glass and Regulator

Door glass, overview

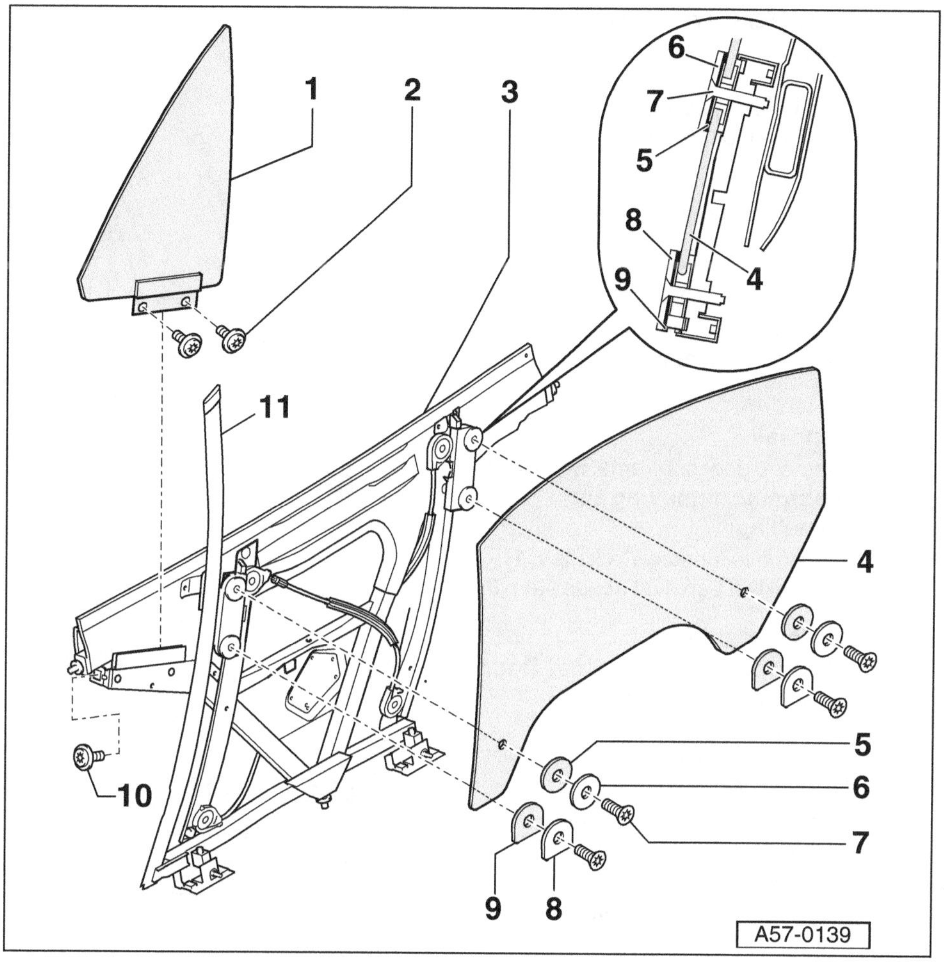

1. **Side window glass**
   - Remove door subframe. See **Door subframe, removing and installing**
2. **Torx bolts**
   - Tighten to 8 Nm (71 in-lb)
3. **Door subframe**
   - See **Door subframe, removing and installing**
   - See **Door subframe, adjusting overview**
4. **Door window glass**
   - Removing:
     Raise window fully.
     Remove door subframe.
     Mark position of window glass at mountings using a fine point marker (Sharpie®) or equivalent.
     Remove Torx bolts (item **7**), pull glass toward the rear and out of guide rail .
   - Installing:
     Push door window glass into guide rail (item **11**) and center on lift pads of lifter mechanism.
     Loosely install washers (item **6**), clamping washers (item **5**), and Torx bolts (item **7**).
     Tighten bolts to 6 Nm (53 in-lb)
   - Installing new glass:
     Position window glass on lift pads of lifter mechanism.
     Slide window into guide rail (item **11**) as far as it will go.
     Keeping window straight, push it upward in guide rail until there is a smooth transition to side window glass.
     Pull door window glass out of guide rail approximately 0.5 to 1.0 mm.
     Loosely install washers (item **6**), clamping washers (item **5**), and Torx bolts (item **7**).
     Tighten bolts to 6 Nm (53 in-lb).
5. **Clamping washer, top**
6. **Washer, top**
7. **Torx bolts**
   - Tighten to 6 Nm (53 in-lb)
8. **Washer, bottom**
9. **Clamping washer, bottom**
10. **Torx bolt**
    - Tighten bolts to 8 Nm (71 in-lb)
    - Can be used to adjust side window glass laterally in relation to windshield (**A**) pillar.

BentleyPublishers.com
© 2006 BentleyPublishers.com—All Rights Reserved

Door window regulator, overview

View is back side of subframe; the side side facing the door sheet metal.

1. **Bolt**
   - Tighten to 8 Nm (71 in-lb)
2. **Seal**
   - Pull out to remove
   - To ease installation, lubricate with soap solution and slip into guide rail
3. **Guide rail**
   - Remove door subframe. See **Door subframe, removing and installing**
   - To remove, unscrew bolt (item **1**) and drill off pop rivet heads (item **8**)
4. **Window regulator**
   - Remove door subframe. See **Door subframe, removing and installing**
   - To remove, drill off pop rivet heads (items **5**, **6**), and lift out
   - Note position of cables, see **Window regulator cable position**
   - Motor removal, see **Window regulator motor, removing and installing**
5. **Rivet**
6. **Rivet**
7. **Door subframe**
   - See **Door subframe, removing and installing**
8. **Pop rivet**

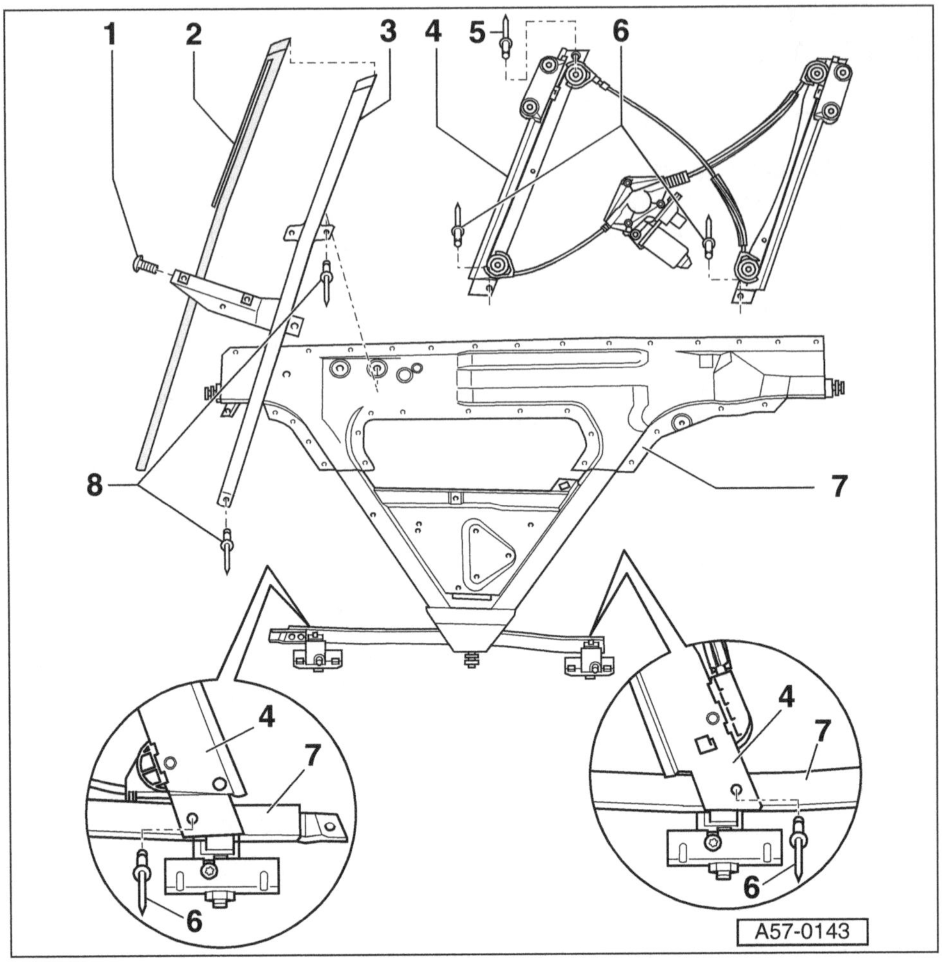

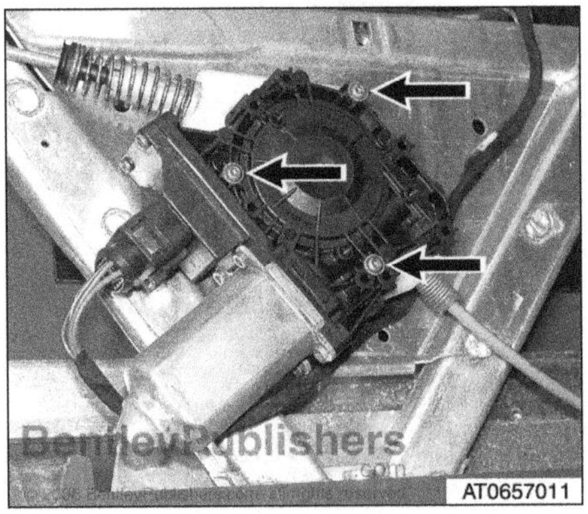

Window regulator motor, removing and installing

- Unplug harness connector, remove 3 Torx bolts (**arrows**) and pull motor out.
- Tighten to 3 Nm (27 in-lb) + 0.5 Nm (4 in-lb)
- Window motor can be removed and installed with glass in any position.
- After repairs, activate automatic functions, see **Activating automatic window lifting and lowering functions**.

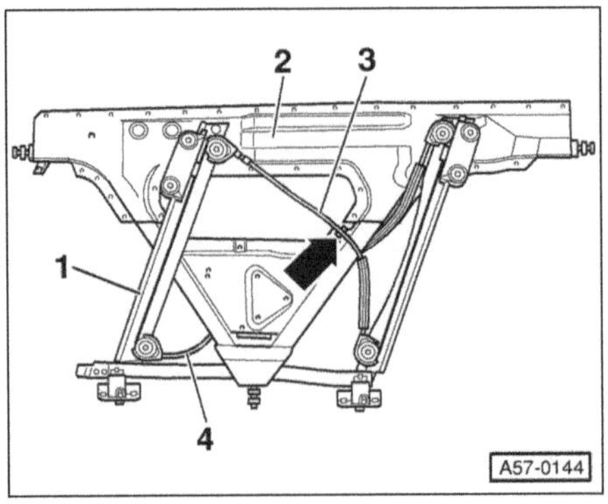

Window regulator cable position

- Install window regulator (**1**) in such a way that cable guide (**3**) is on the inside of subframe (**2**), (faces door sheetmetal).
- Cable guide (**4**) must be on the outside of subframe (connects to motor).
- Clip cable guide (**3**) into holder (**arrow**).

BentleyPublishers.com

© 2006 BentleyPublishers.com—All Rights Reserved

Basic setting activation

Basic window setting

The following repair procedure must be completed after installation of replacement parts.

- All sealing and trim must be fully installed.
- Door and window regulator must be adjusted.
- Door must be fully assembled and closed.

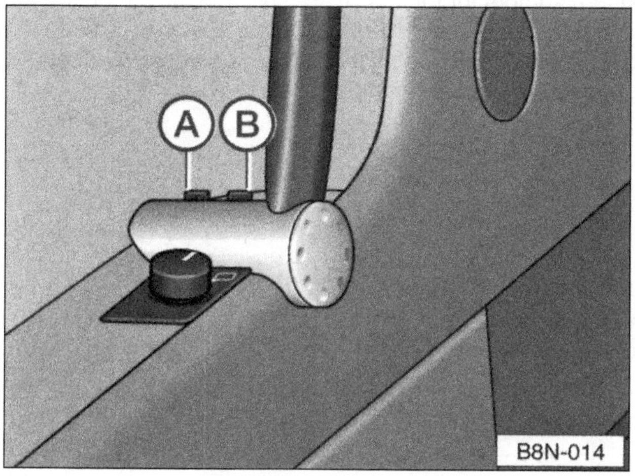

The following repair procedure must be performed carefully and without interruption. Breaks longer than 10 seconds are considered an interruption in "activation mode".

- Close door and switch ignition on.
- Using appropriate window switch (**A** or **B**), run window up to top position (fully closed).
- Release window switch.
- Pull window switch up again, and hold in position longer than 1 second.
- Using window switch, run window down to bottom position (fully open) by pressing window regulator switch down.
- Run window up to top position (fully closed).
- Release window switch.
- Pull window regulator switch up again, and hold in position longer than 1 second.
- Basic setting for power windows is now activated. One-touch up and down functions should be operational.

Basic setting for windows, clearing

This procedure may be needed if basic setting was interrupted. This is indicated by a window that unintentionally reverses itself during closing. This procedure must be completed 3 times within one minute.

- Close door and switch ignition on.
- Open door window, and place a piece of wood approximately 8 -10 cm (3¼ - 4 inch) high between door window glass and upper door molding (at the top of window travel).
- Close door window by pulling window switch up.
- Door window will stop upon contacting the wood, the pinch protection will activate and the window will reverse direction.
- Close door window a second time by pulling window switch up.
- Door window will again stop upon contacting the wood, pinch protection will again be activated, and the window will reverse direction.
- Close door window a third time by pulling window switch up.

Automatic reversing feature is now deactivated. Door window will remain in contact with the wood and will lower approximately ½ second after releasing door window switch.

- Close door window a fourth time by pulling window switch up.
- Door window will again stop upon contacting wood, but will not go down once window switch is released.
- Open door window by pressing window switch.
- Release window switch.
- Press window switch down for at least 1 second.
- Remove wood and close door window to top position (fully closed).

BentleyPublishers.com
© 2006 BentleyPublishers.com—All Rights Reserved

- Release door window switch.
- Immediately pull window switch up again for at least 1 second.
- Repeat this procedure two more times.
- When complete, activate basic setting for window regulator motor.

Central Locking System

Central locking system actuators

Central locking actuators for the doors are built into latch mechanisms. Additional actuators are integral with trunk latches. Roadsters have central locking actuators for a storage compartment and a CD changer compartment behind the seats

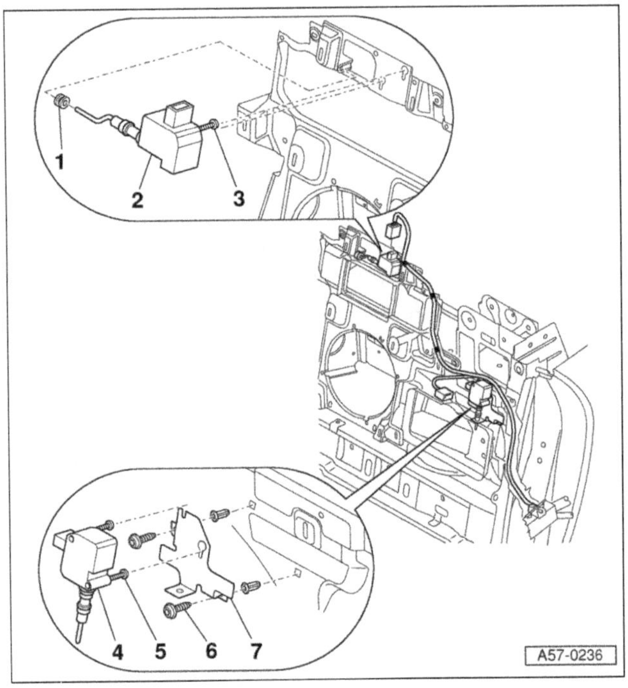

◄ Rear compartment actuators (roadster)

1. Grommet
2. Actuator, for storage compartment, center
   - Remove center trim of rear panel
   - Disconnect harness connector
   - Loosen attaching screws (item **3**) and slide actuator out of slotted holes in mounting plate
3. Torx bolt
   - Tighten to 2 Nm (18 in-lb)
4. Actuator, for CD compartment, left
   - Remove side trim of center compartment
   - Disconnect harness connector
   - Loosen attaching screws (item **5**) and slide actuator out of slotted holes in mounting plate
5. Torx bolt
   - Tighten to 2 Nm (18 in-lb)
6. Torx bolt
   - Tighten to 2 Nm (18 in-lb)
7. Mounting bracket

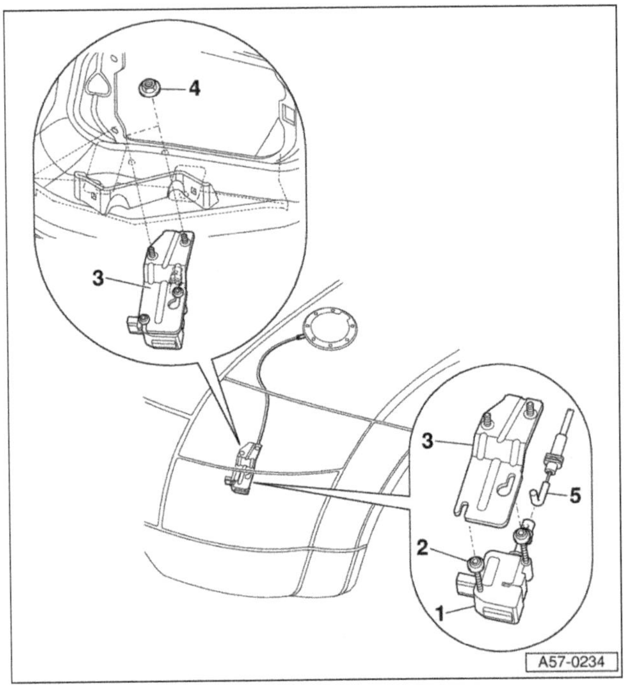

◄ Fuel filler flap actuator (coupe and roadster)

1. Actuator for fuel filler flap
   - Remove side luggage compartment trim
   - Disconnect harness connector
   - Loosen attaching screws (item **2**) and slide actuator out of slotted holes in mounting plate
   - Remove actuator from bracket (item **3**) and unclip cable (item **5**)
2. Torx bolt
   - Tighten to 2 Nm (18 in-lb)
3. Mounting bracket
4. Nut
   - Tighten to 8 Nm (71 in-lb)
5. Release cable for fuel filler flap

BentleyPublishers
.com
© 2006 BentleyPublishers.com—All Rights Reserved

61 Convertible Top

General

This repair group covers removal, installation and adjustment of the convertible top on TT roadsters. It also covers the hydraulic system components on power operated convertible tops.

The convertible top consists of the outer top cover, the mounting frame and an electrically heated rear window. There is also an optional retractable glass wind deflector located behind the roll bars.

On manual top versions, a pair of gas struts slow the top movement during storage and assists when unfolding the top. On power top versions, a hydraulic pump (V118) operates the top mechanism by way of double-acting hydraulic cylinders on each side. The hydraulic pump is controlled by a convertible top operation switch (E137), a relay (J321), and a control module (J429).

NOTE—

- *Audi identifies electrical components by a letter and/or a number in the electrical schematics. See* **EWD Electrical Wiring Diagrams**. *These electrical identifiers are listed in parentheses as an aid to electrical troubleshooting.*
- *Convertible top components can be found in the electrical wiring diagram entitled, "Central Locking System / Anti Theft System / Radar Interior Monitor (Roadster)".*

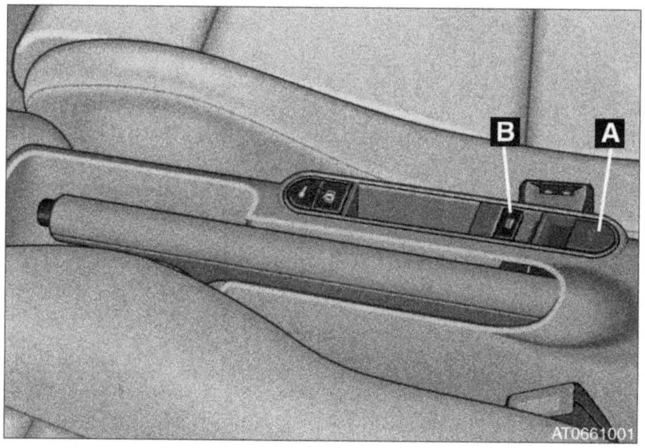

The power convertible top operation switch (**A**) is mounted in the rear section of the center console, next to the parking brake. The optional wind deflector switch (**B**) is located in front of the power convertible top switch. For removal and installation of the switches, see **96 Interior Lights, Switches, Anti-theft**.

Operation of the convertible top is covered in the glovebox owner's manual included with the vehicle.

BentleyPublishers.com
©2006 BentleyPublishers.com—All Rights Reserved

Convertible Top Assembly

Assembly overview

1. **Convertible top**
2. **Bolt**
   - 20 Nm (15 ft-lb)
3. **Main bearing**
   - Has gradations to facilitate adjustment
   - Before removing top, mark position relative to gradations
   - Use marks to adjust main bearing during installation of top
4. **Stop**
   - Convertible top must be removed before removing/installing
5. **Pop rivet**
   - Install using VAG 1753A special high force pop rivet pliers
6. **Tensioning strut**
   - To remove, remove cover (item 8) and combination bolts (item 7)
   - Press bracket toward front and down until tensioning strut is resting loosely in convertible top storage compartment
7. **Combination bolt**
   - 8 Nm (71 in-lb)
8. **Cover for tensioning strut bearing**
9. **Seal mount, left and right**
10. **Seal mount, center**
11. **Bracket**
    - Has marks to facilitate height adjustment
    - Note positions of marks before removal to set same height during installation
12. **Torx bolt**
    - 15 Nm (11 ft-lb)
13. **Torx bolt**
    - 4.5 Nm (40 in-lb)
14. **Tension strap**
15. **Storage compartment trim**

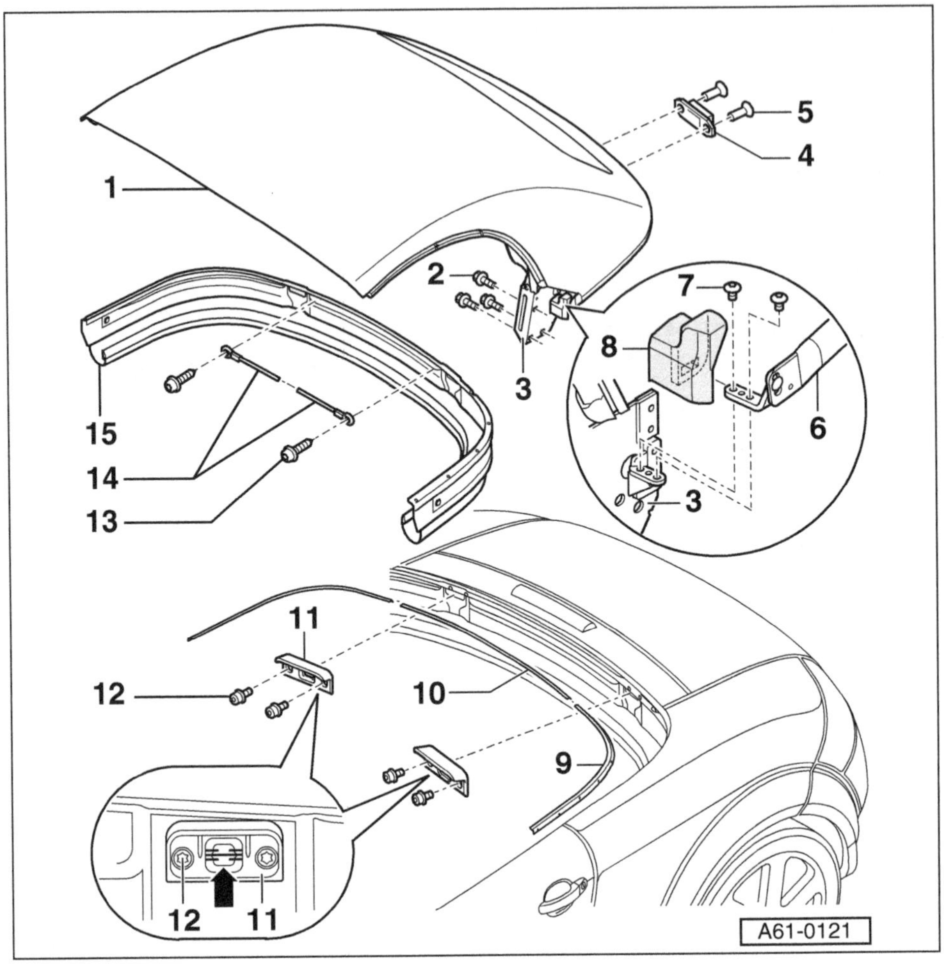

BentleyPublishers.com
© 2006 BentleyPublishers.com—All Rights Reserved

Top and frame components

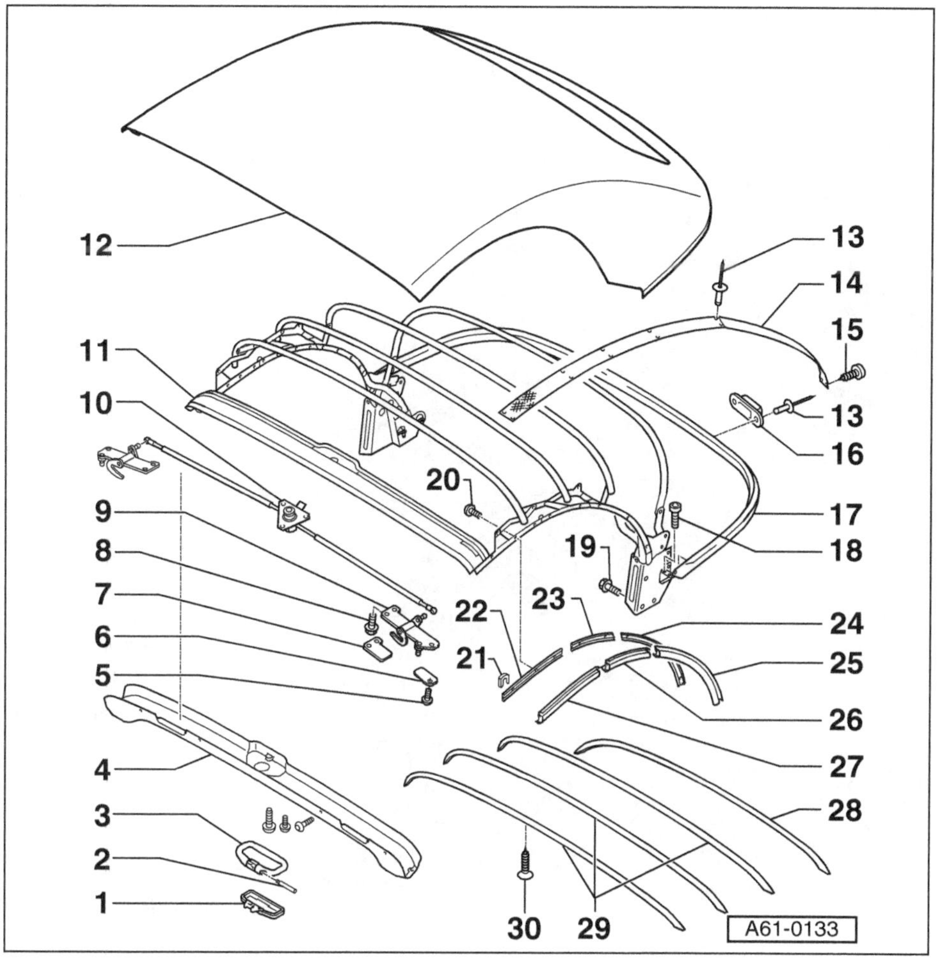

1. **Face plate**
2. **Bolt**
3. **Rotating handle**
4. **Front cover**
5. **Combination bolt**
   - 10 Nm (7 ft-lb)
6. **Base plate**
   - For height adjustment of top lock
7. **Base plate**
   - For height adjustment of top lock
8. **Combination bolt**
   - 10 Nm (7 ft-lb)
9. **Convertible top lock**
10. **Linkage for convertible top lock**
11. **Front roof rail**
12. **Convertible top fabric**
13. **Pop rivet**
    - Replaced by bolts shortly after start of production
14. **Tension strap**
15. **Bolt**
16. **Stop for tensioning strut**
17. **Tensioning strut**
18. **Combination bolt**
    - 8 Nm (71 in-lb)
19. **Bolt**
    - 20 Nm (15 ft-lb)
20. **Torx bolt**
    - 2.5 Nm (22 in-lb)
21. **Spacer**
    - For adjusting retaining tracks and seals from side
22. **Front retaining track**
23. **Center retaining track**
24. **Rear retaining track**
25. **Rear seal**
26. **Center seal**
27. **Front seal**
28. **Bow casing**
    - For bow #4
29. **Bow casings**
    - For bows #1, 2 and 3
30. **Bolt**
    - For securing bow casings

Convertible top lock, adjusting

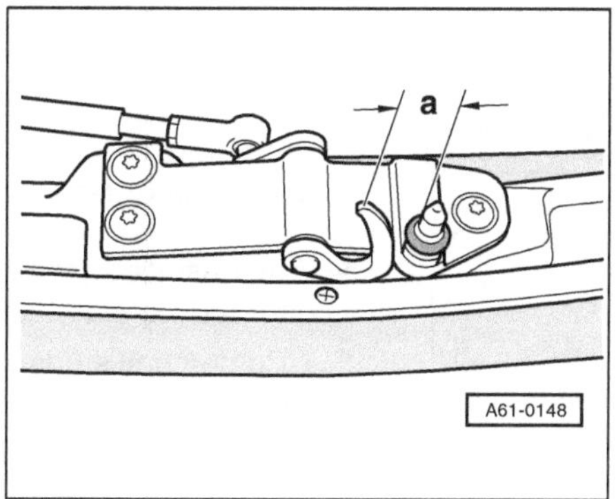

Adjust lock at ball stud so that gap (**a**) between tip of latch hook and inner edge of centering pin = 79 mm (3.1 in)

BentleyPublishers.com
© 2006 BentleyPublishers.com—All Rights Reserved

Convertible Top, Removing and Installing

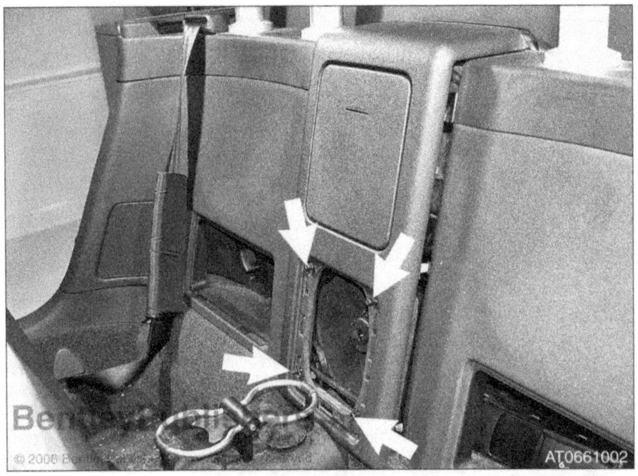

– Pry out grille for rear center speaker from rear center trim panel.

◀ Remove lower mounting screws (**arrows**) for center rear trim panel. Pry trim panel away from body.

- It may be helpful to reach inside center trim panel upper storage compartment and pull trim panel forward.

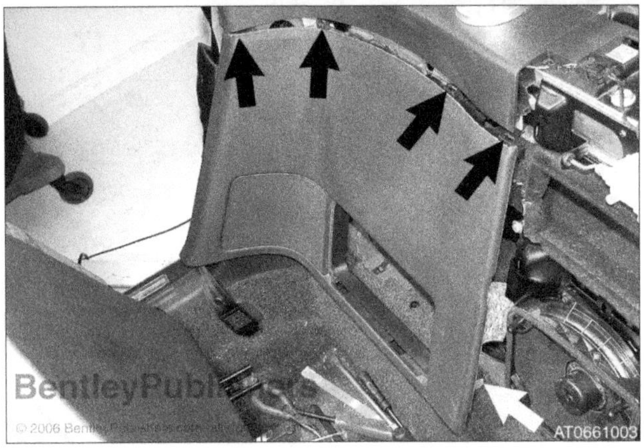

◀ Remove left and right rear trim panels.

- Pull trim panel forward at bottom and then downward from rollbar trim.
- Note location of visible trim clips (**arrows**) in illustration.
- Also see illustration showing back of trim panel given later in this procedure.
- See **70 Trim–Interior** for more trim panel details.

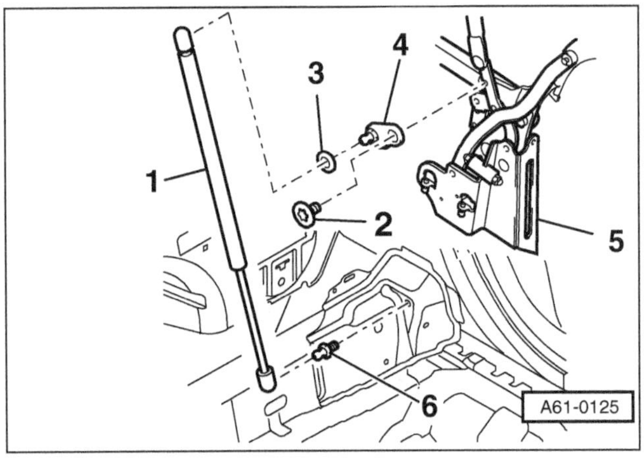

Vehicles with manually operated convertible top

◀ With gas-filled strut (**1**) extended (top closed), pry strut from upper ball stud (**4**) at left and right sides.

- Washer (**3**) must be pushed onto ball stud before attaching gas-filled strut, during installation.
- When installing ball stud (**4**), tighten bolt (**2**) to 6 Nm (53 in-lb).
- Item **5** is the main bearing for convertible top.
- Tighten ball stud (**6**) to 14 Nm (10 ft-lb) when installing.

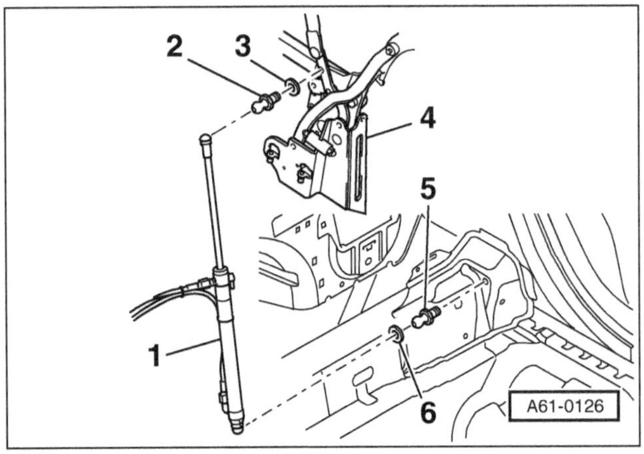

Vehicles with power operated convertible top

◀ With gas-filled strut (**1**) extended (top closed), pry strut from upper ball stud (**2**) at left and right sides.

- Tighten ball studs (**2** and **5**) to 23 Nm (17 ft-lb) when installing.
- Washer (**3**) must be positioned under ball stud during installation.
- Item **4** is the main bearing for convertible top.
- Rubber washer (**6**) must be pushed onto lower ball stud (**5**) before installing gas-filled strut.

– Disconnect harness connector for rear window defroster.

BentleyPublishers.com
© 2006 BentleyPublishers.com—All Rights Reserved

Continued for all models

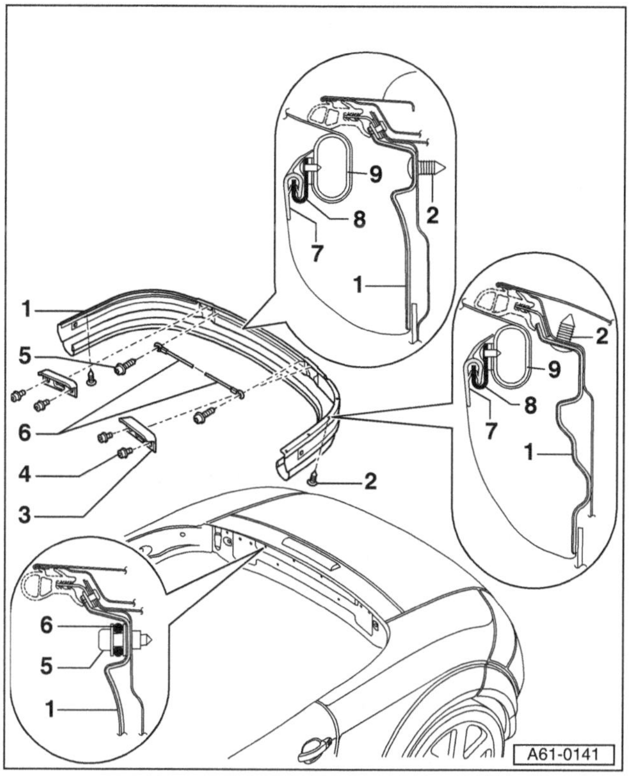

◄ Unclip trim from retaining strip (**8**) at tensioning strut (**9**), on interior side of vehicle.

1. Storage compartment trim
   - Convertible top must be removed before removing trim
   - To remove trim:
     Remove bolts (**4**) and bracket (**3**)
     Unscrew Torx bolts (**5**) and remove tension strap (**6**)
     Remove clips (**2**) from trim
     Remove trim from storage compartment
2. Clip
   - Quantity: 14
3. Bracket
   - Note markings before removal of top for height adjustment
4. Torx bolt
   - 15 Nm (11 ft-lb)
5. Torx bolt
   - 4.5 Nm (40 in-lb)
6. Tension strap
7. Seal
   - Part of trim
8. Retaining strip
   - Two pieces, each bolted to tensioning strut
9. Tensioning strut

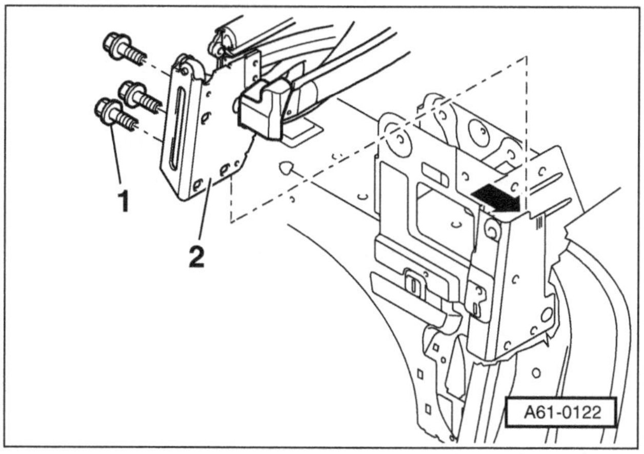

◄ Remove mounting bolts (**1**) for left and right main bearings (**2**).

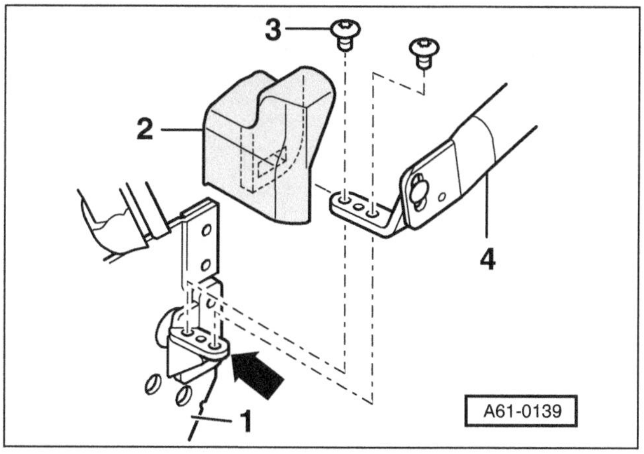

◄ Push cover (**2**) back slightly and remove from main bearing.

– Remove mounting bolts (**3**) for left and right tensioning struts (**4**).

– Press tensioning strut forward under bolting surface (**arrow**) at main bearing until tensioning strut is resting loosely.

BentleyPublishers
.com
© 2006 BentleyPublishers.com—All Rights Reserved

Convertible Top, Removing and Installing

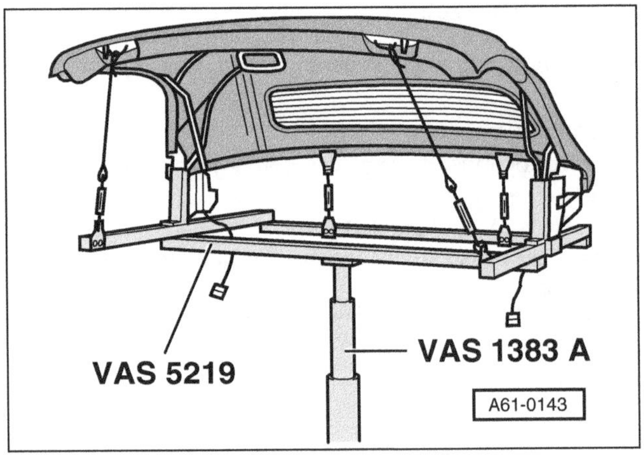

- Lift convertible top sufficiently at rear until tensioning strut can be lifted out.

◀ Remove convertible top and place on VAS 5219, or similar assembly device.

- Secure tensioning strut at main bearing.

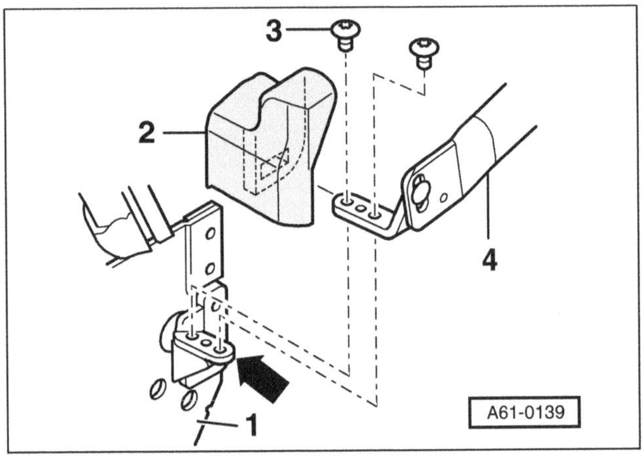

Convertible top, installing

◀ If necessary, remove tensioning strut (**4**) from main bearing (**1**).

- Lift convertible top over vehicle and guide tensioning strut into storage compartment on one side.
- Place convertible top down in storage compartment diagonally toward front.
- Attach tensioning strut to main bearing and tighten bolts to specification.
- Install convertible top and tighten mounting bolts at main bearing to specification.

Tightening torques	
Tensioning strut to main bearing	8 Nm (71 in-lb)
Main bearing to body	20 Nm (15 ft-lb)

- Reattach trim for convertible top storage compartment at rear tensioning strut.
- Reconnect electrical connections.
- Perform function test.

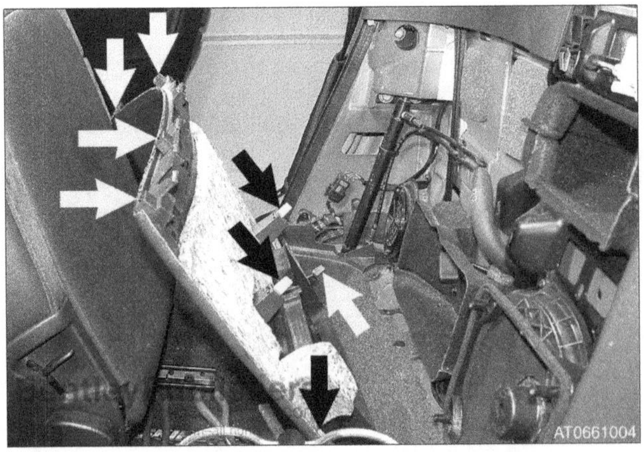

◀ Install various trim pieces. Note location of retaining clips (**arrows**) on left and right rear trim panels. See **70 Trim–Interior** for more trim panel installation information.

©2006 BentleyPublishers.com—All Rights Reserved

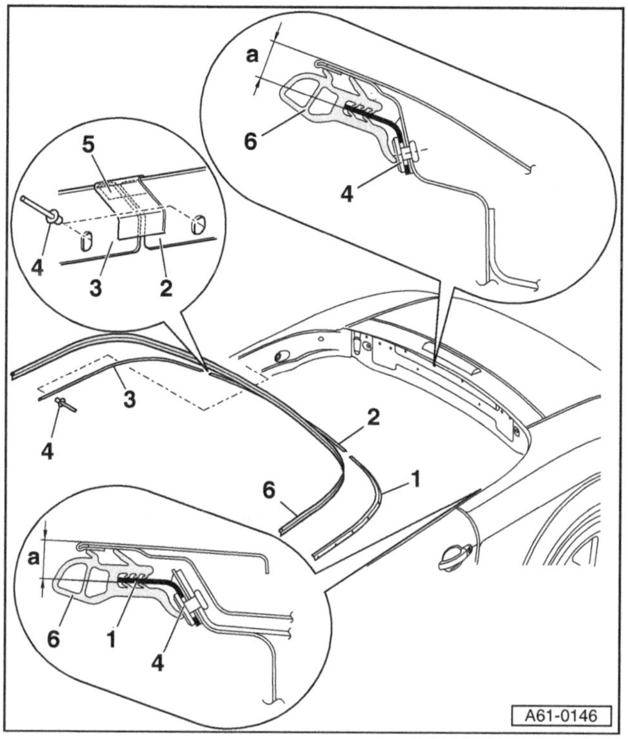

Tensioning strut seal and seal mount, removing and installing

◄ Details of tensioning strut seal and seal mount.

1. Seal mount, left
   - Can only be removed or attached with convertible top removed
   - Install, left, center and right seal mounts
   - Dimension (**a**) = 10 mm ± 0.5 mm, as measured from lower lip of seal mount to upper edge of convertible top storage compartment
2. Seal mount, center
   - Seal abutting edges of seal mounts using sealant (part # AKD 512 000 05)
   - Apply adhesive tape (**5**) to abutting edges from lower side before applying sealant
3. Seal mount, right
4. Pop rivets
   - Quantity: 21
5. Adhesive tape
6. Tensioning strut seal
   - Can be removed or pressed on with convertible top opened halfway

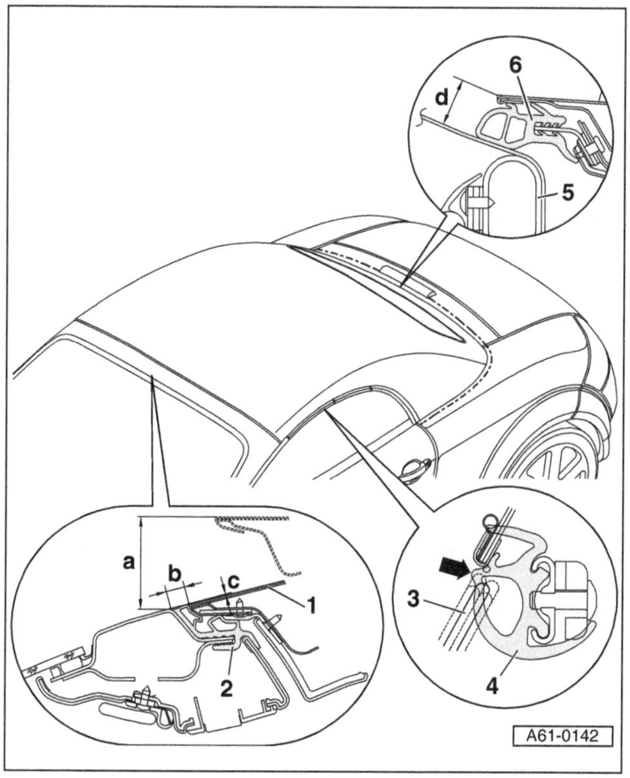

CONVERTIBLE TOP, ADJUSTING

◄ Details for adjusting convertible top gaps and positioning dimensions.

1. Convertible top
   - Dimension (**a**) = 35 mm ± 10 mm
     This is the "jump gap" for the convertible top
   - When the convertible top is opened, it must jump up approx. 35 mm
   - At this dimension, the forces required to close the convertible top are acceptable
   - Roof inset, dimension (**c**) = 1 mm ± 1 mm
2. Seal for front roof frame
   - Dimension (**b**) = 9 mm + 2 mm
3. Window
   - Window must only rise up to bulge (**arrow**) in seal
4. Seal
   - Seals can be adjusted from the side using shims
5. Tensioning strut
6. Seal for tensioning strut
   - Dimension (**d**) = 15 mm ± 2 mm

BentleyPublishers
.com
© 2006 BentleyPublishers.com—All Rights Reserved

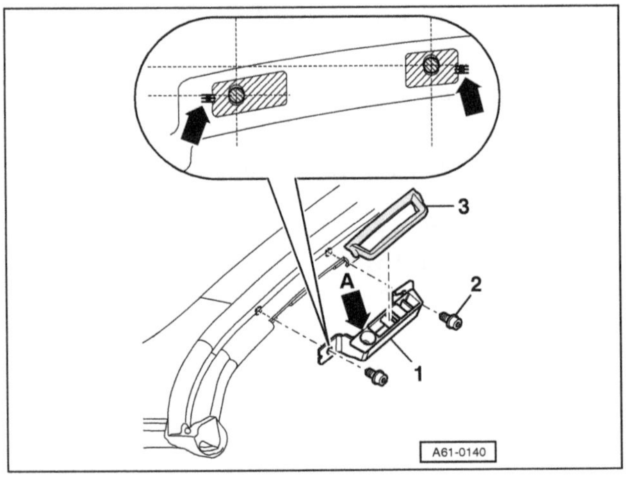

Convertible top, adjusting

Lengthwise adjustment

◀ First determine how far convertible top must be shifted at end plate (**1**).

- Striker pin on convertible top must align with center of hole (**arrow A**).

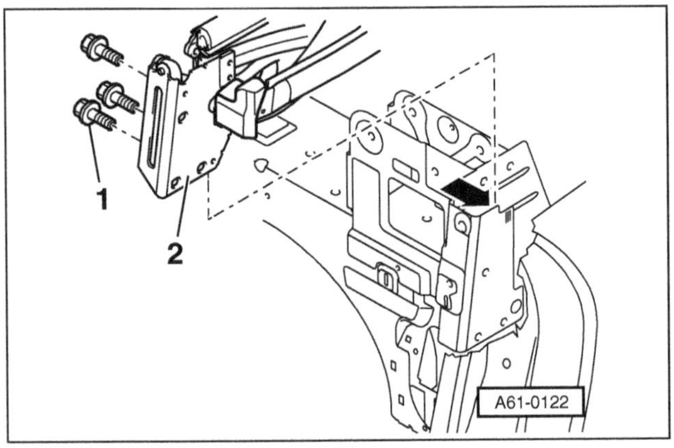

◀ To adjust convertible top lengthwise, loosen mounting bolts (**1**) at left and right main bearing (**2**).

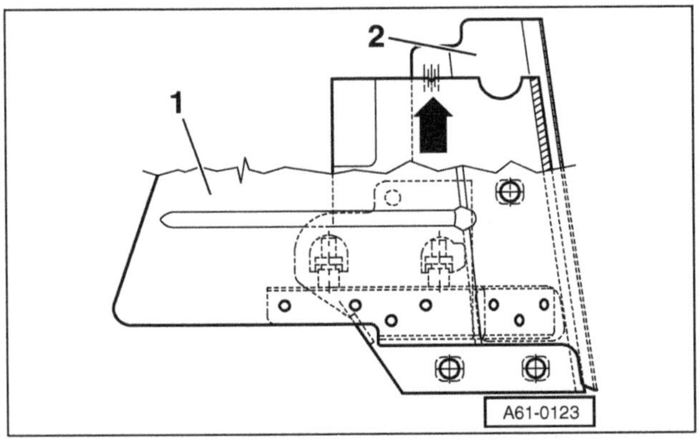

◀ Shift main bearing (**1**) of convertible top by the previously determined amount using marks (**arrow**) on chassis (**2**).

- Striker pin on front of convertible top must align with center of hole on end plate.

Tightening torque	
Main bearing to body	20 Nm (15 ft-lb)

Height adjustment

– Remove roof frame trim. See **70 Trim–Interior**.

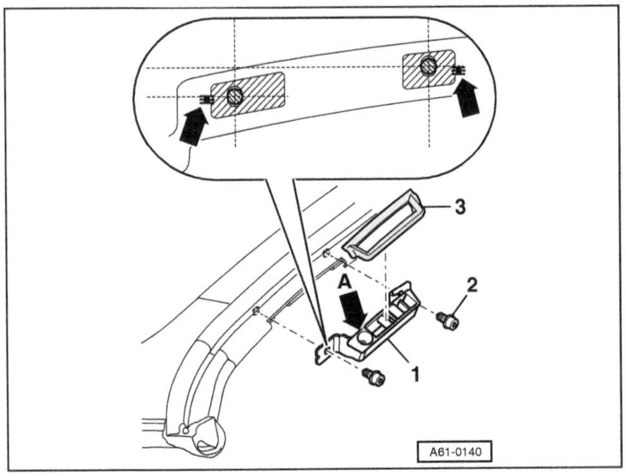

◀ To adjust height of convertible top at front, shift height of end plate (**1**).

- Marks (**arrows**) are present for more precise height adjustments.

BentleyPublishers.com
© 2006 BentleyPublishers.com—All Rights Reserved

Hydraulic System

System overview

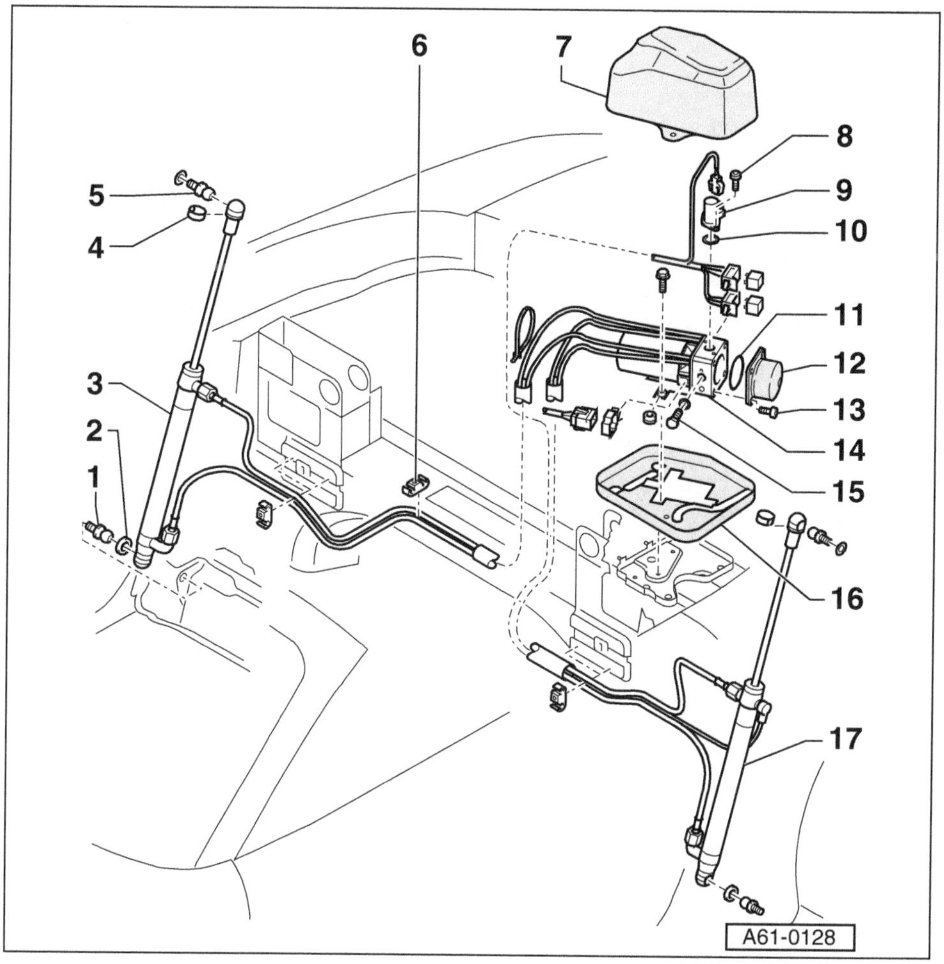

1. **Ball stud**
2. **Rubber washer**
   - Rubber washer must be pushed onto ball stud (**1**) before attaching hydraulic cylinder (**3**)
3. **Right hydraulic cylinder**
4. **Spring clip**
   - To remove hydraulic cylinder, lift spring clip slightly with screwdriver and pull cylinder off ball stud
5. **Ball stud**
6. **Clip**
7. **Pump cover**
   - For noise insulation
8. **Torx bolt**
   - 8 Nm (71 in-lb)
9. **Solenoid valve**
   - Solenoid valves must only be threaded into the hydraulic pump in activated (energized state) using diagnostic scan tool in output diagnostic test mode
10. **Sealing ring for solenoid valve**
    - Always replace
11. **Sealing ring for oil reservoir**
12. **Oil reservoir**
    - Lightly oil sealing ring before assembly
    - To assemble, twist oil reservoir while pushing on
13. **Torx bolt**
    - 2 Nm (18 in-lb)
14. **Hydraulic pump**
    - Remove convertible top before disassembly
15. **Oil filler plug**
    - Oil level must be between "MIN" and "MAX" marks
    - Use hydraulic oil G 002 000
16. **Insulation**
17. **Left hydraulic cylinder**
    - With connection for microswitch

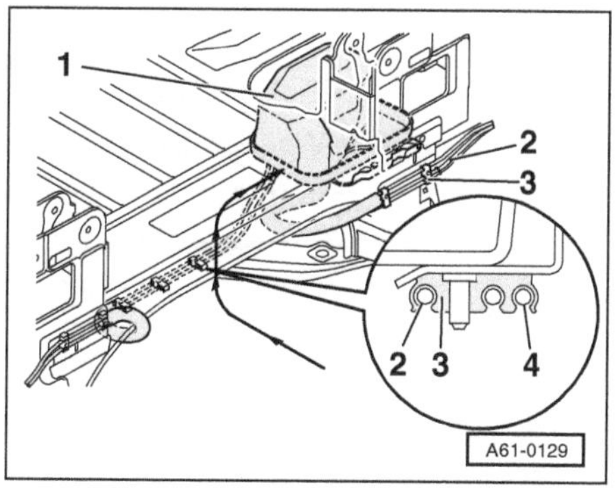

Hydraulic line routing

◄ Hydraulic pump with cover (**1**).

- Hydraulic line for left cylinder has control wire for microswitch
- Wires must be unclipped from retaining clamp (**3**) before removing hydraulic pump

BentleyPublishers.com
© 2006 BentleyPublishers.com—All Rights Reserved

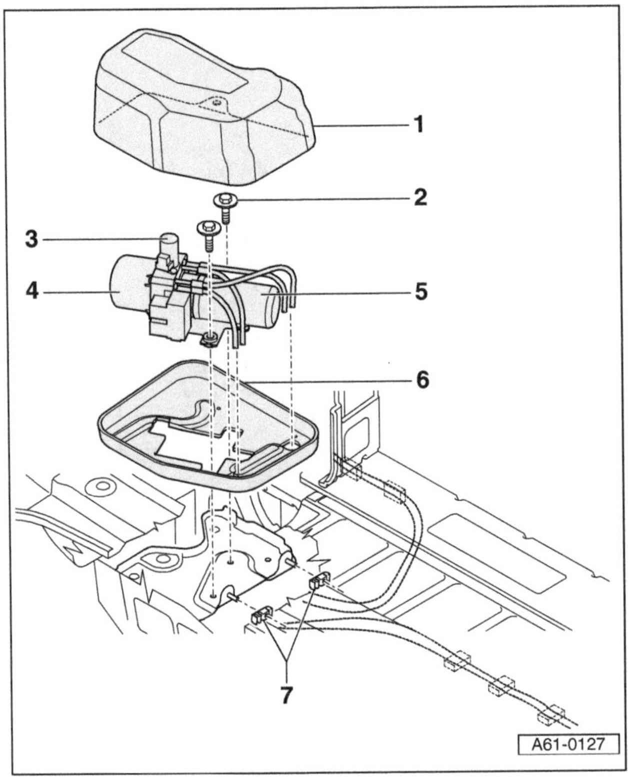

Pump assembly

1. Pump cover
2. Mounting bolts
   - 8 Nm (71 in-lb)
3. Solenoid valve
   - Solenoid valves must only be threaded into the hydraulic pump in activated (energized state) using diagnostic scan tool in output diagnostic test mode
4. Oil reservoir
   - Tighten mounting bolts to 2 Nm (18 in-lb) when installing oil reservoir
5. Hydraulic pump
   - Remove convertible top before disassembly
   - Removing:
     Remove pump cover (**1**) first
     Remove mounting bolts (**2**)
     Lift hydraulic pump up and out through hole in floor of convertible top storage compartment
6. Insulation
7. Clips

Hydraulic strut, removing and installing

– Close convertible top.

– Remove center, left and right rear trim panels. See **Convertible top, removing and installing** earlier in this repair group. Also see **70 Trim-Interior**.

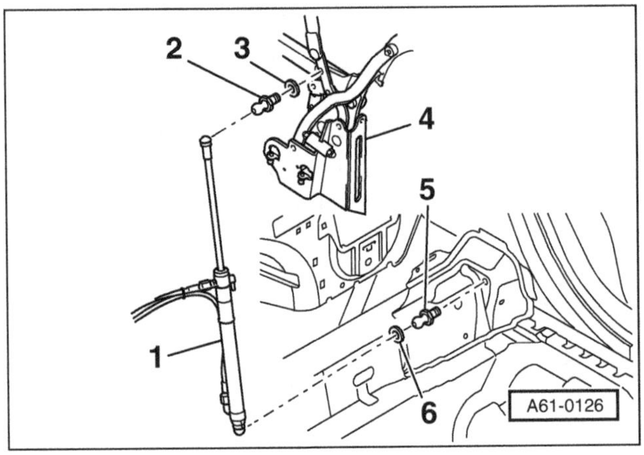

◀ With gas-filled strut (**1**) extended (top closed), pry strut from upper (**2**) and lower (**5**) ball studs.

- Tighten ball studs (**2** and **5**) to 23 Nm (17 ft-lb) when installing.
- Washer (**3**) must be positioned under ball stud during installation.
- Item **4** is the main bearing for convertible top.
- Rubber washer (**6**) must be pushed onto lower ball stud (**5**) before installing gas-filled strut.

BentleyPublishers
.com
©2006 BentleyPublishers.com—All Rights Reserved

63 Bumpers

General

The bumpers consist of a beam that is bolted directly to the body structure. The bumper cover must first be removed to access the bumper beam. This section covers removal and installation of front and rear bumper components.

Front Bumper

Front grilles

NOTE—

- *Front bumper cover must be removed before removing front grilles. See* ***Front bumper cover assembly****.*

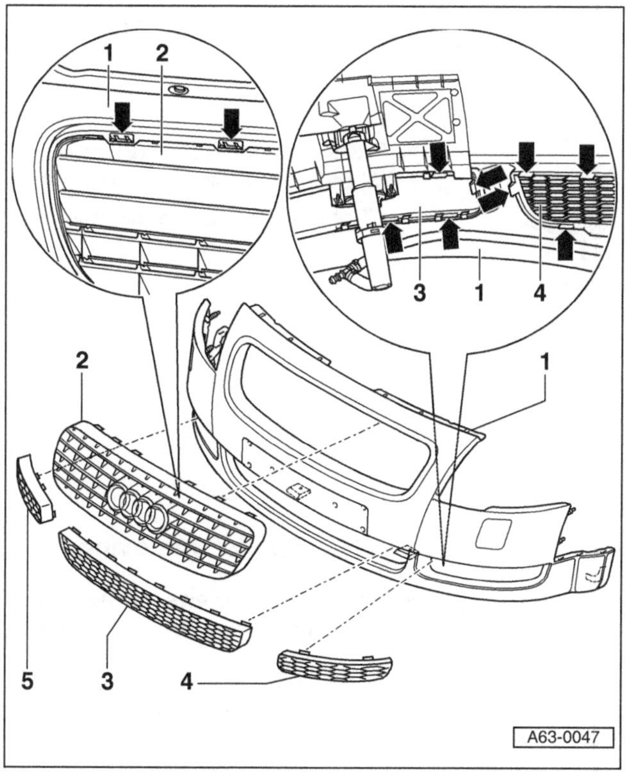

Front grilles

1. Bumper cover
2. Radiator grille
   - To remove, release retaining tabs (**arrows**)
3. Center air intake grille
4. Left air intake grille
5. Right air intake grille

BentleyPublishers.com
© 2006 BentleyPublishers.com—All Rights Reserved

Front bumper cover assembly

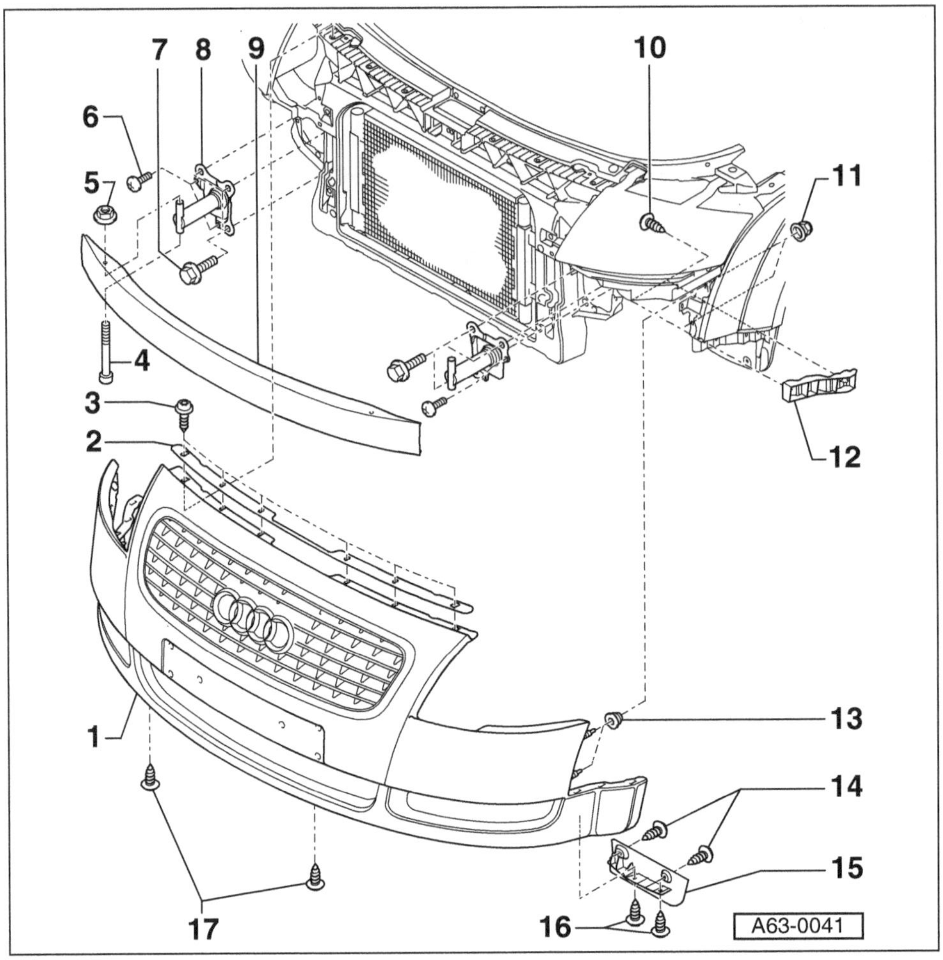

1. **Bumper cover**

   Removing:
   - Detach front part of wheelhousing liners
   - Remove upper screws (**3**) and securing strip (**2**)
   - Remove screws (**14 and 16**) and detach wheel spoiler
   - Release quick-release fasteners from wheel housing liners
   - Remove lower screws (**17**)
   - Remove nuts (**11**) from inner side of wheel housing
   - Loosen screws (**10**) and press guide bracket (**12**) downward slightly to disengage from cover
   - Detach headlight washer hoses at connection, where equipped
   - Remove bumper cover
2. **Securing strip**
3. **Screw**
4. **Bolt**
   - 35 Nm (26 ft-lb)
   - For mounting bumper to bracket
5. **Nut**
6. **Screw**
7. **Bolt**
   - 35 Nm (26 ft-lb)
8. **Bumper bracket**
9. **Bumper**
10. **Screw**
    - Loosen to disengage guide bracket from bumper cover
11. **Nut with collar**
    - Accessible after removal of wheel housing liner
12. **Guide bracket**
    - Press upward and tighten screw (**10**) before installing cover (**1**)
13. **Expanding clip**
    - Press into fender before installing bumper cover
14. **Screw**
15. **Wheel spoiler**
16. **Screw**
17. **Bolt**

Expanding clips

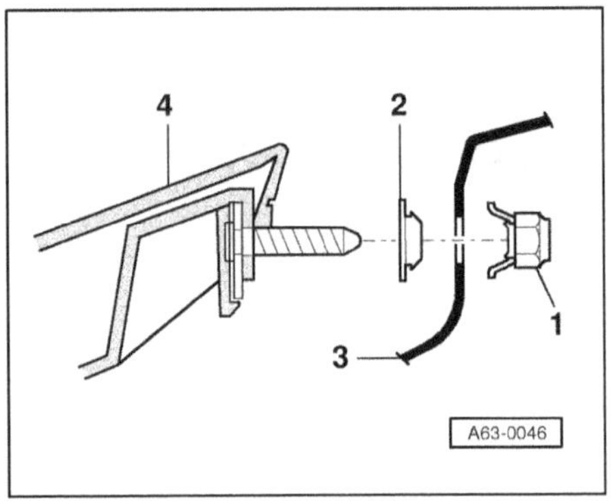

Press expanding clips (**2**) into fender (**3**) before installing bumper cover (**4**).

- Tighten nut to 6 Nm (53 in-lb)

Wheel spoiler, removing

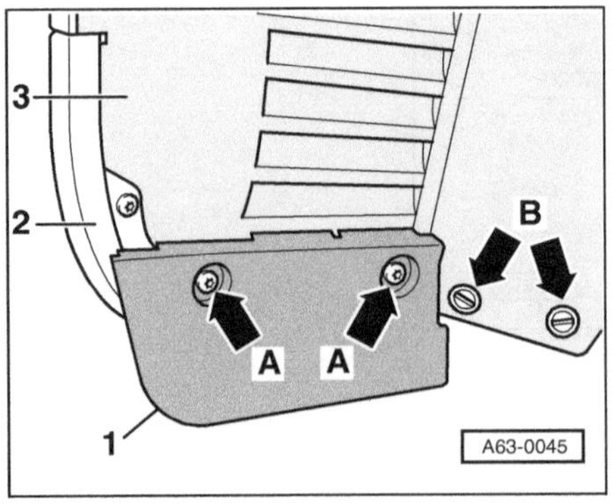

Remove Torx screws (**arrow A**) and 2 mounting screws from underneath.

– Release quick release fasteners (**arrow B**) to detach wheel housing liner.

BentleyPublishers.com

© 2006 BentleyPublishers.com—All Rights Reserved

Rear Bumper

Rear bumper cover assembly

1. **Rear bumper**
2. **Nuts**
   - 23 Nm (17 ft-lb)
   - For mounting rear bumper
3. **Speed nuts**
4. **Left trim strip**
   - To remove, unclip upward
5. **Right trim strip**
   - To remove, unclip upward
6. **Screw**
7. **Bumper cover**

   Removing:
   - Remove rear wheels
   - Remove rear wheel housing liners
   - Unclip left and right trim strips
   - Remove screws (**6** and **8**)
   - Loosen nuts (**10**)
   - Loosen mounting screws for guide brackets (**11**) and press brackets downward to disengage from bumper cover
   - Remove bumper cover
8. **Screw**
9. **Lower trim**
10. **Nut**
11. **Guide bracket**
    - Install nuts (**10**) and bracket mounting screws before installing bumper cover (**7**)

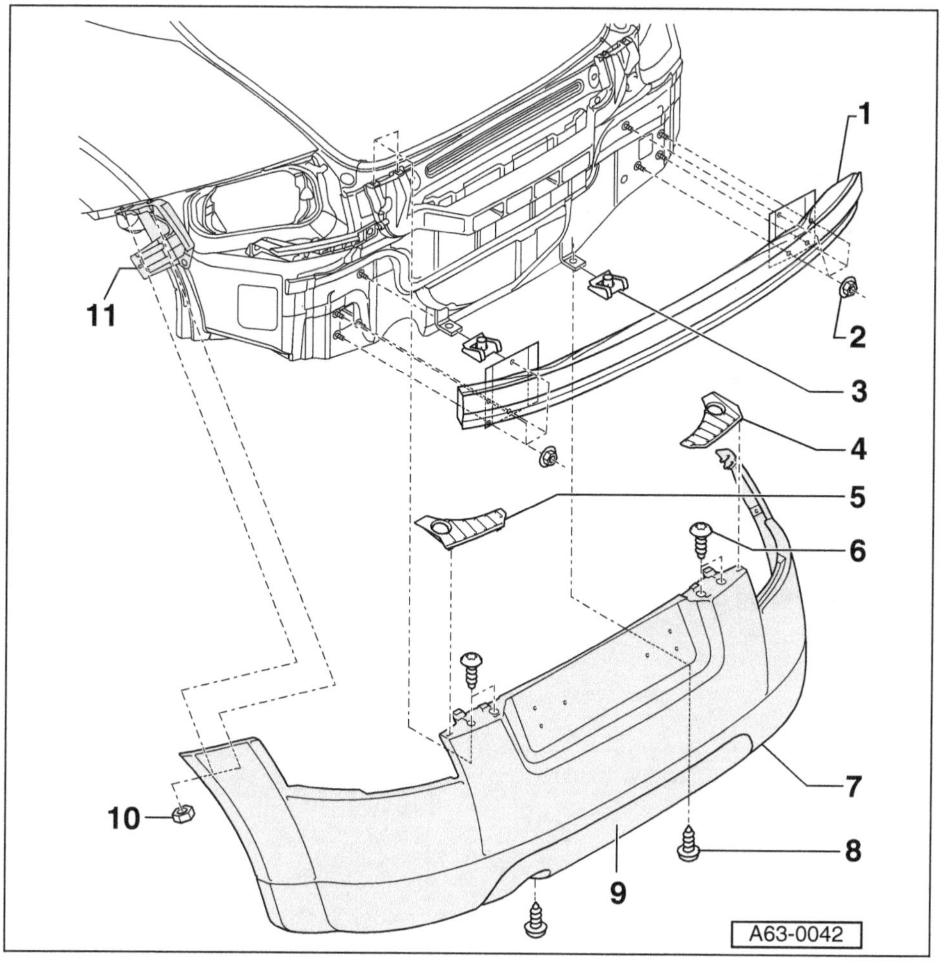

Lower trim, removing

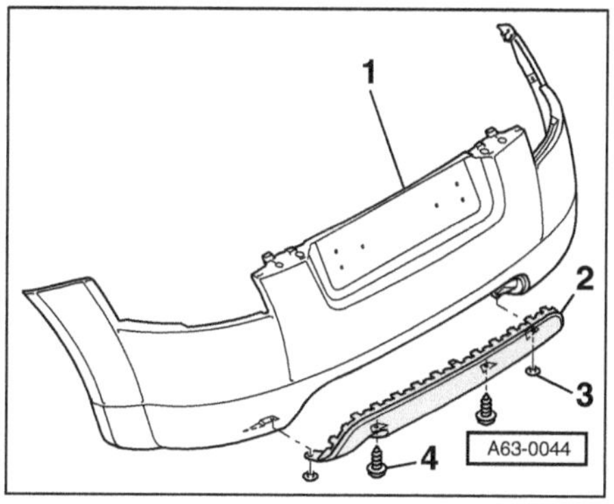

◀ Lower trim (**3**) can be removed without removing bumper cover (**1**).

– Pry out clips (**3**) to remove.

– Remove screws (**4**) and lower trim.

BentleyPublishers.com
©2006 BentleyPublishers.com—All Rights Reserved

Rear bumper cover brackets

1. **Outer guide bracket**
   - To remove, unscrew nuts (**3**) and bolts (**2**)
2. **Bolt**
3. **Nut**
4. **Washer**
5. **Speed nut**
6. **Center guide bracket**
7. **Bolt**
8. **Bolt**
9. **Left guide bracket**
   - To remove, unscrew bolts (**8**)
10. **Expanding nut**

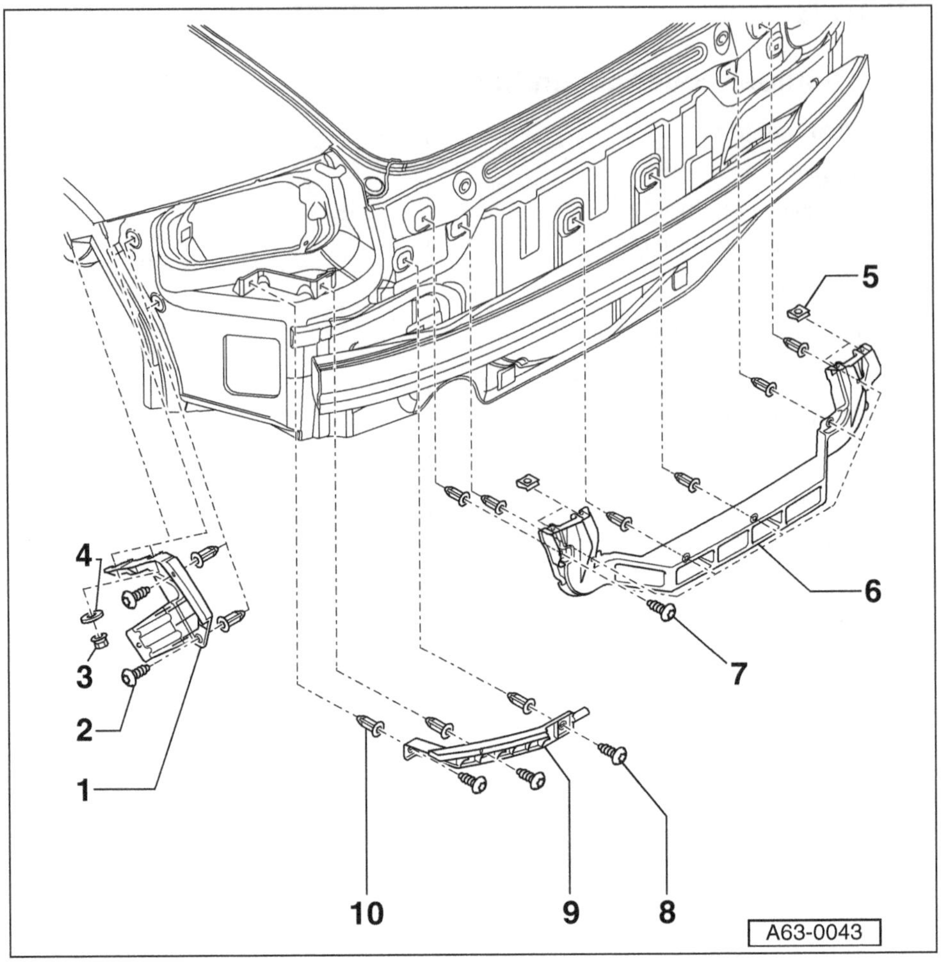

BentleyPublishers
.com
© 2006 BentleyPublishers.com—All Rights Reserved

66 Body–Exterior Equipment

General

This repair group covers exterior components including the outside rear view mirrors, door sill trim and wheel housing liners. For additional information see the following repair groups:

- **63 Bumpers**
- **94 Exterior Lights**

Roof Trim

Convertible top cover molding

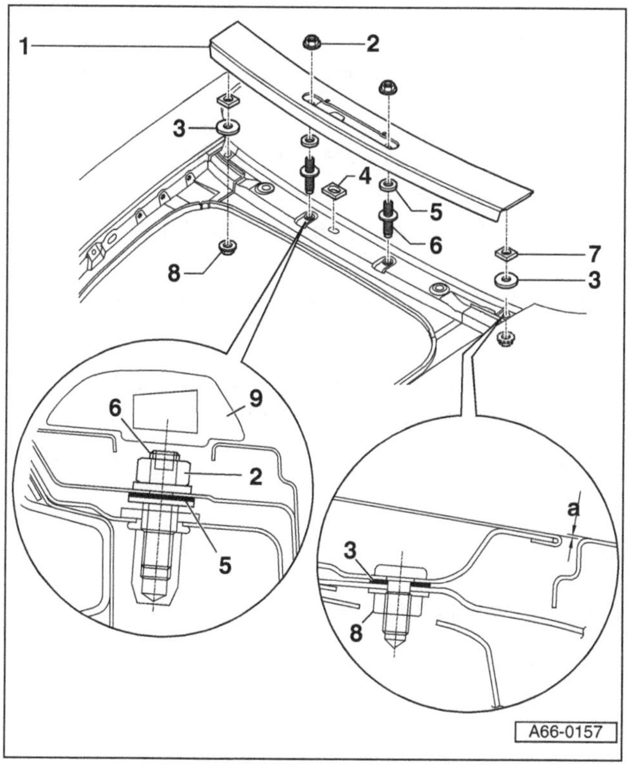

Convertible top cover molding (**1**) houses third brake light.

1. Convertible top cover molding
   - Remove third brake light before removing molding. See **94 Exterior Lights**.
   - Remove nuts (**2**) and loosen nuts (**8**). Press molding toward front and remove upward.
2. Nuts
   - 20 Nm (15 ft-lb)
3. Washer
   - Gap **a** = 0.1 mm ± 0.3 mm
4. Spacer
5. Adhesive disk
   - Glued to cover molding
6. Adjusting screw
   - For adjusting cover molding height
7. Spacer
8. Nut
   - 8 Nm (71 in-lb)
9. Third brake light

BentleyPublishers.com

©2006 BentleyPublishers.com–All Rights Reserved

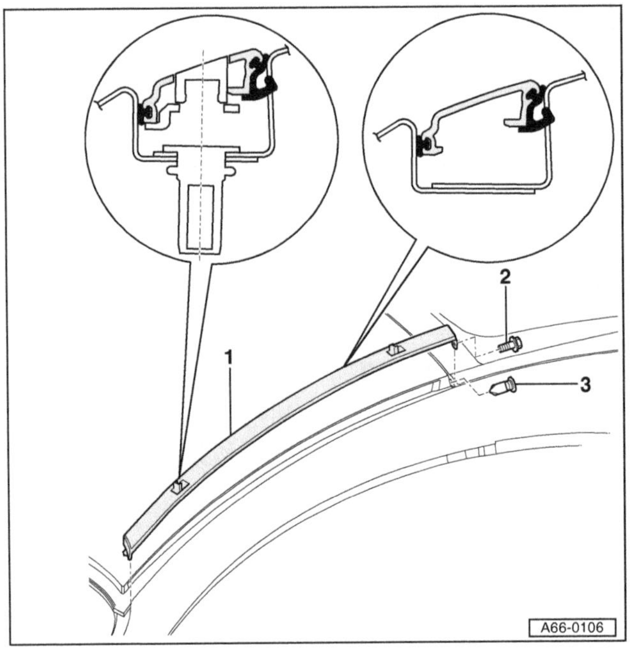

Roof seam molding, removing and installing

◀ Remove mounting bolt (**2**) from expanding nut (**3**).

– Carefully pry molding (**1**) out of roof seam, working from back to front.

– When installing, apply a lubricant such as liquid soap to profile of seal.

– Begin installation by inserting roof seam molding below trim cover at windshield and then press fully into roof seam working from front to back.

– Install mounting bolt.

Roof rack hardware, installing (coupe models through 2001)

NOTE —
- *As of VIN# TRU ZZZ 8N X1 021700 (mid-2001 production) the roof rack adapter pieces are installed during production.*
- *The roof rack is only available for coupe models.*

Roof rack hardware kit (part # 8N8 898 115) includes the following:
- Retainer housing (Qty. 4)
- Adapter (Qty. 4), M8 external thread, M6 core thread
- Pins (Qty. 4) for anchoring flap
- Threaded pins (Qty. 4), plastic M6 x 10

NOTE —
- *Audi part numbers are given for reference only! Always consult your Audi parts department or aftermarket parts specialist for the latest parts information.*

– Check to make sure roof rack hardware is not already installed.
  - Open anchor flap on roof seam molding.
  - If threaded screw with M6 thread can be installed, hardware is installed.
  - If threaded screw with M8 thread can be installed, hardware is not installed.

– Remove roof seam molding from left and right sides of roof. See **Roof seam molding, removing and installing**.

– Remove plastic threaded screw (M8 x 30) from securing nut in edge molding.

– De-grease securing nut.

– If necessary, remove sealant from adapter contact surfaces.

– Apply thread locking compound to external threads of adapter.

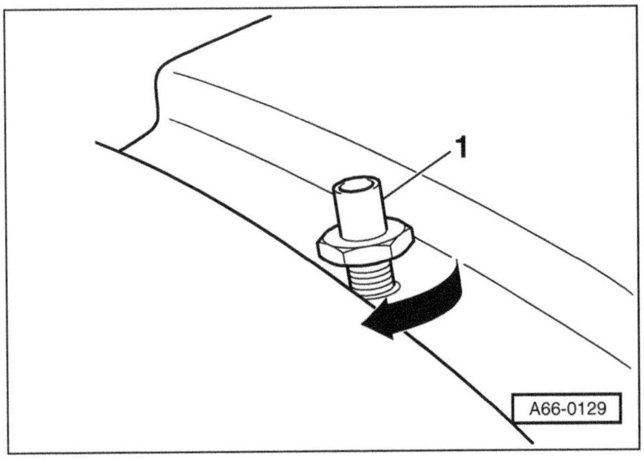

◀ Install adapter (**1**) into securing nut and tighten to 16 Nm (12 ft-lb).

BentleyPublishers.com
© 2006 BentleyPublishers.com—All Rights Reserved

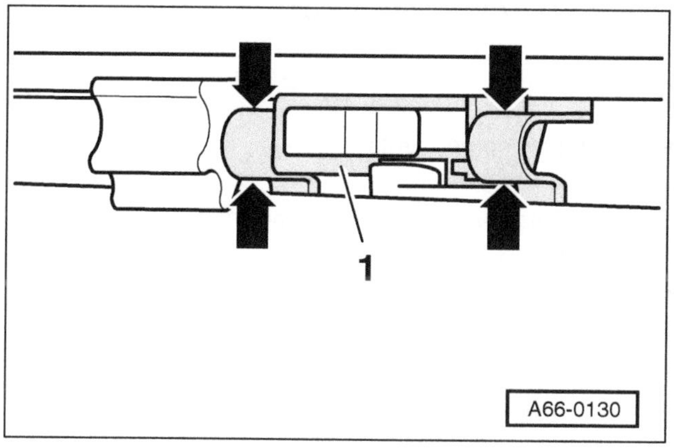

– Turn plastic threaded pin (M6 x 10) into adapter piece until contact is reached.

< Press together inner clips of roof seam molding in retainer housing (**1**) at **arrows** and remove.

– Press securing pin of flap to side and remove from retainer housing.

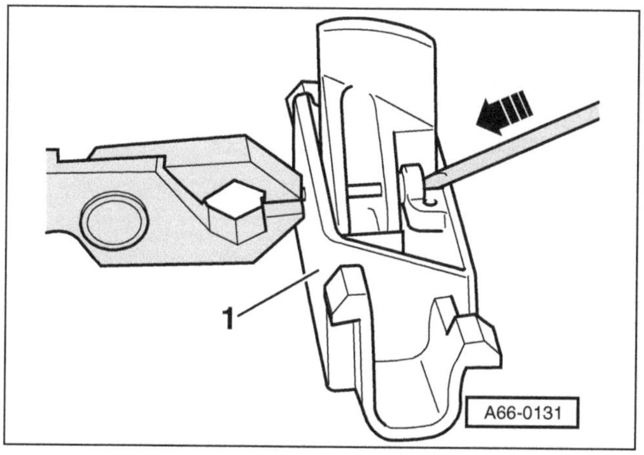

< Using pliers, pull securing pin completely out of retainer housing (**1**).

– Place removed flap into new retainer housing.

– Replace pin and secure flap in new retainer housing.

– Clip retainer housing back into roof seam molding.

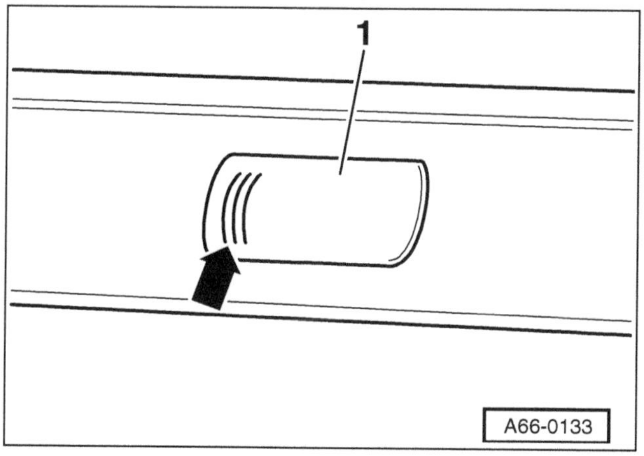

< The textured (ridged) end (**arrow**) of flap (**1**) must point forward when installed.

– Install roof seam molding. See **Roof seam molding, removing and installing**.

BentleyPublishers.com
© 2006 BentleyPublishers.com—All Rights Reserved

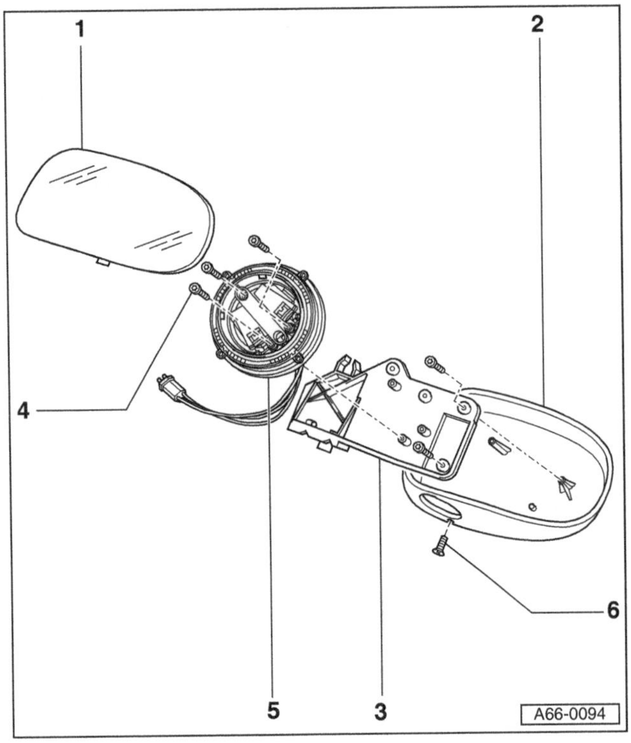

Outside Rear View Mirror

Overview

Outside rear view mirror components.

1. Mirror glass
   - Always use protective gloves when removing/installing
   - Only exert pressure at center of glass
2. Mirror housing
3. Mounting plate
4. Torx bolts
   - Remove to detach mirror adjustor unit
5. Mirror adjustor unit
   - To remove:
     Pry off mirror glass
     Pull off electrical connectors for mirror heater
     Remove Torx mounting bolts (**4**)
     Remove adjuster unit from mirror housing (**2**)
6. Torx bolt

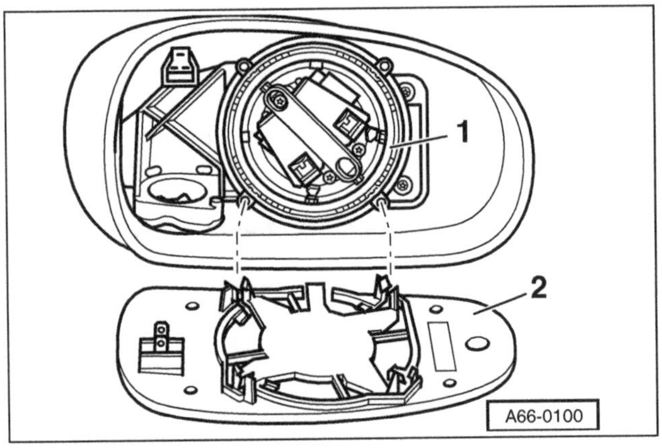

Mirror housing, removing and installing

– Disconnect battery. See **27 Battery, Alternator, Starter**.

Press mirror glass (**2**) out of mirror adjuster unit (**1**) using special tool 80-200 (removal lever).

- Pry mirror starting at bottom and then at top.
- Use fabric adhesive tape to protect mirror housing at top and bottom.

– Pull off electrical connectors.

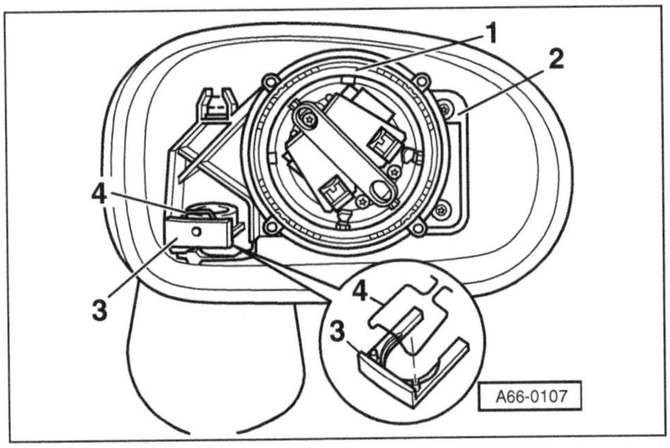

Release spring (**4**) upward and remove.

– Pull clamping wedge (**3**) out of mirror base and take out entire mirror housing with adjuster unit (**1**) and mounting plate (**2**) upward.

– Remove mounting screws (3x) from mirror adjuster unit.

BentleyPublishers.com
©2006 BentleyPublishers.com—All Rights Reserved

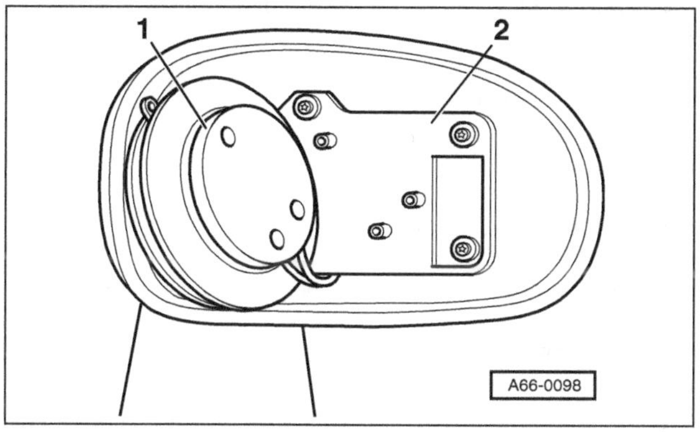

◀ Pivot mirror adjuster unit (**1**) to side and remove screws from mounting plate (**2**).

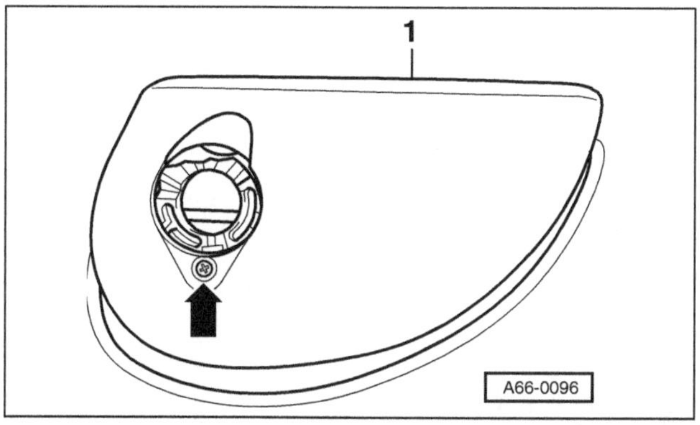

◀ Turn housing around and remove screw (**arrow**).

– Take mirror adjuster unit and mounting plate out of housing.

Mirror base, removing and installing

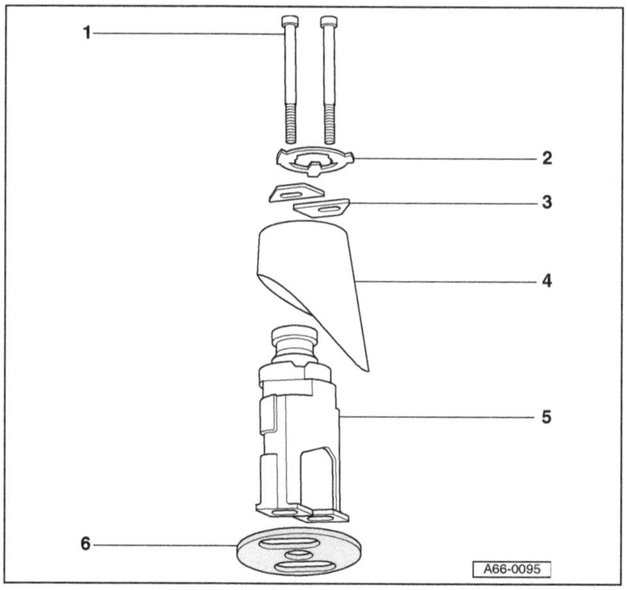

◀ Outside rear view mirror base and mounting hardware.

1. Mounting bolts
   - 6 Nm (53 in-lb)
2. Adapter ring
3. Shims
4. Mirror base cover
5. Mirror base
6. Seal
   - Must make uniform contact between mirror base cover (**4**) and outer door panel

– Remove mirror housing. See **Mirror housing, removing and installing** earlier in this repair group.

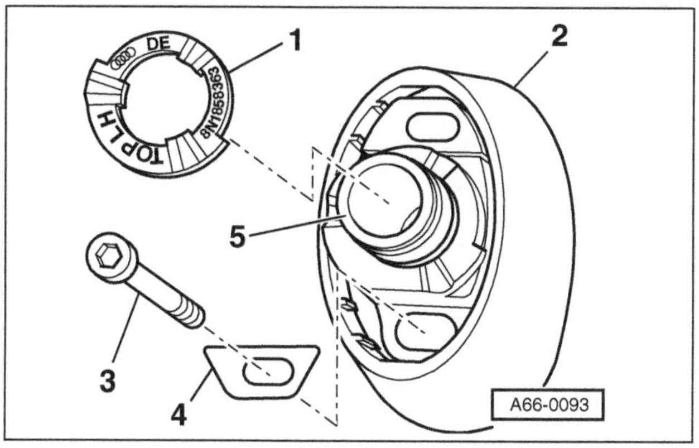

◀ Unscrew mounting bolts (**3**) and pull off mirror base (**5**) and cover (**2**) upward over wiring.

– Remove shims (**4**) from cover (**2**).

– Note position of adapter ring (**1**).
- Different adapter rings are used for left- and right-hand drive vehicles.
- Install left-hand drive adapter ring as shown.

BentleyPublishers
.com
© 2006 BentleyPublishers.com—All Rights Reserved

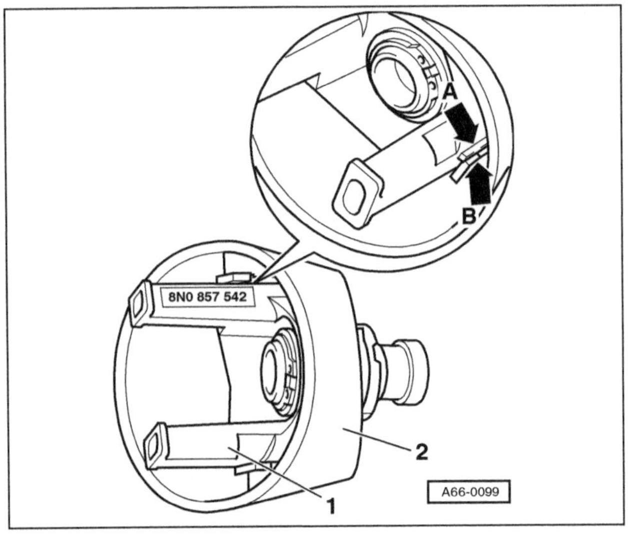

◄ Press mirror base (**1**) sideways at an angle over retainer tabs of cover (**arrow**).

– Pull off cover upward over mirror base.

Installing

– Pull adjustment cables through mirror base.

– Place mirror base on outer door panel and install mounting bolts.

– Make sure seal is uniformly positioned between cover and outer door panel.

Door Sill Trim

Sill molding

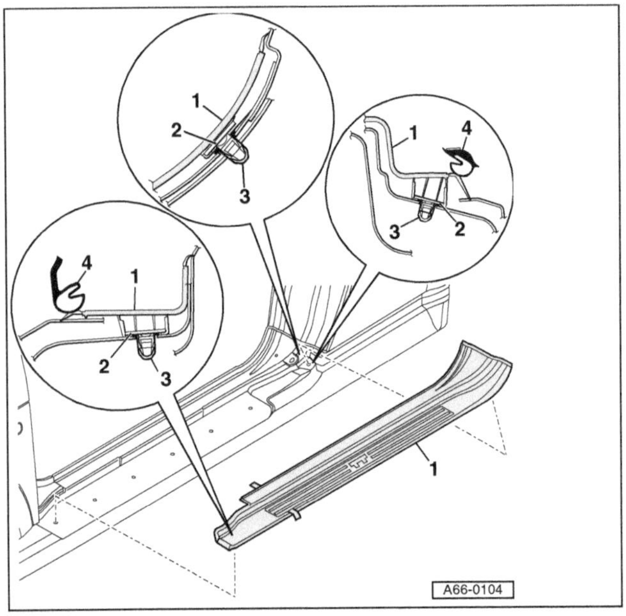

◄ Aluminum sill molding and mounting hardware.

1. Outer sill molding
   - Removing:
     Unclip at rear first and then front
     Heat bonded surfaces with heat gun and pull off sill molding
   - Installing:
     Clean bonding surfaces of dust and grease
     Press new grommets (**3**) into body at front and rear
     Heat sill molding with heat gun to approx. 40°C (104°F)
     Pull off backing paper
     Press sill into place, lining up clips (**2**) with grommets (**3**)
2. Clip
3. Grommet
   - Always replace
4. Door seal

Sill trim

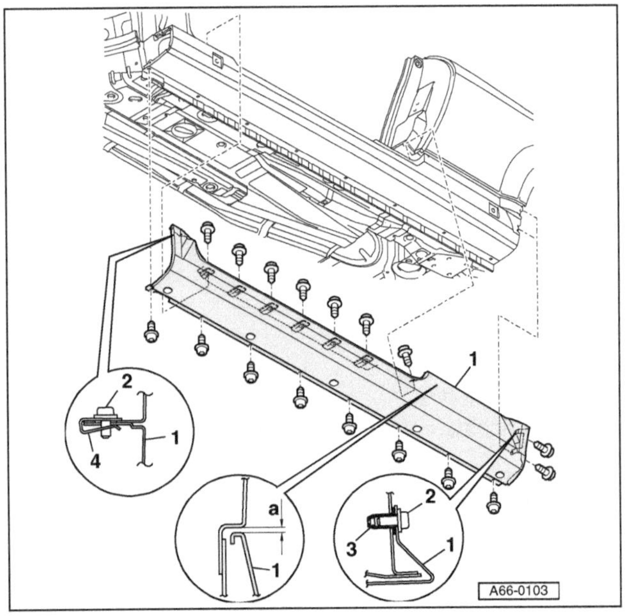

◄ Sill panel outer trim and mounting hardware.

1. Sill panel trim
   - Remove outer sill molding before removing sill panel trim
   - To remove, unscrew mounting bolts (**2**) and detach trim
2. Mounting bolts
   - Qty: 17
   - 4 Nm (35 in-lb)
3. Pop rivet nut
   - Install with pop rivet tool VAG 1618A

BentleyPublishers.com
©2006 BentleyPublishers.com–All Rights Reserved

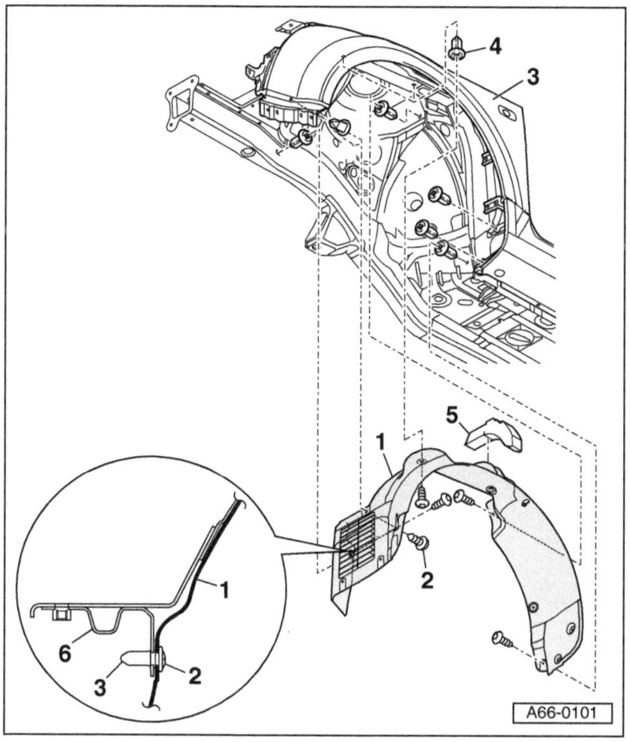

Wheel Housing Liners

Front wheel housing liner assembly

NOTE—
- *Left side wheel housing liner shown. Right side is similar.*

1. Wheel housing liner
2. Torx bolts (T25)
    - Qty: 13
    - 1.5 Nm (10 in-lb)
3. Fender
4. Expanding clips
    - Qty: 7
5. Seal
    - Self adhesive, glued onto wheelhousing liner
6. Fender brace

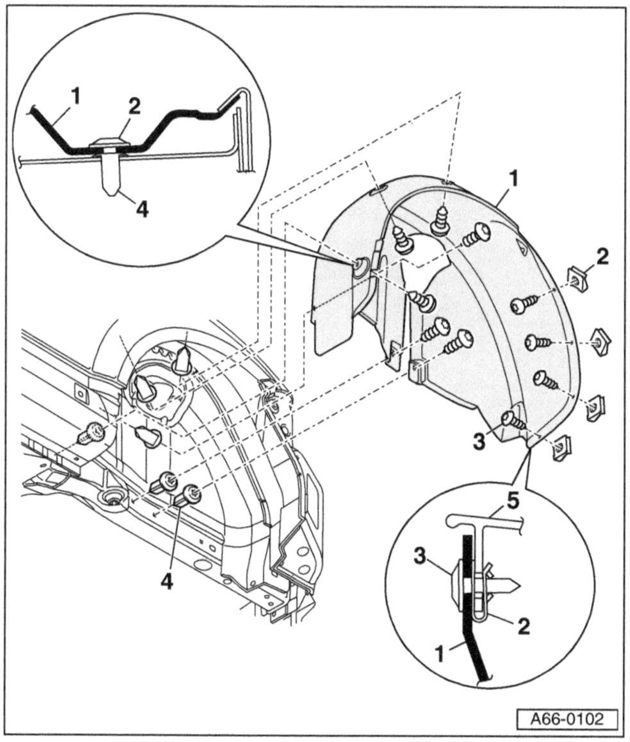

Rear wheel housing liner assembly

NOTE—
- *Left side wheel housing liner shown. Right side is similar.*

1. Wheel housing liner
2. Speed nuts
    - Qty: 4
3. Self tapping screws
    - Qty: 10
    - 1.5 Nm (10 in-lb)
4. Expanding nuts
5. Body

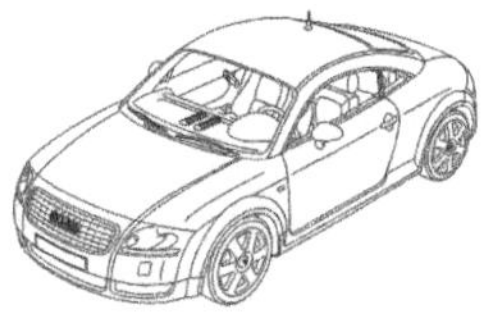

BentleyPublishers.com
©2006 BentleyPublishers.com—All Rights Reserved

69 Seatbelts, Airbags

GENERAL

This repair group covers front and rear seatbelts and front airbag system components. This repair group does not cover airbag system or pyrotechnic seatbelt fault diagnosis or repair. Service and repair to these systems requires special test equipment, knowledge and training and should only be carried out by an authorized Audi dealer. Before starting repairs involving these systems, always read and observe all warnings, cautions and notes.

Passenger compartment safety systems installed in Audi TTs include:

- Front airbags
- Side airbags (located in side bolster of front seats)
- Front seat pyrotechnic seatbelts

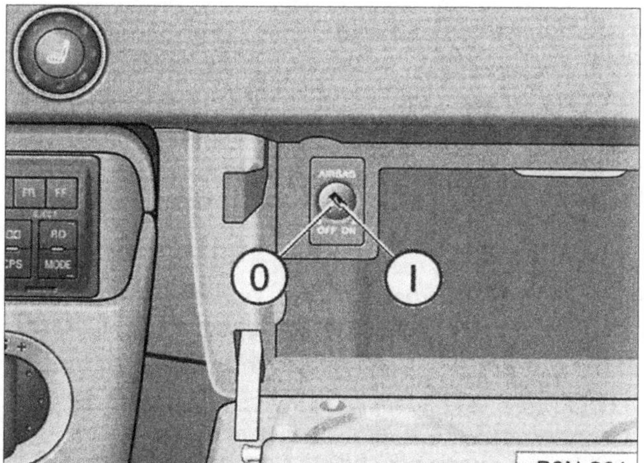

B8N-264

TT models covered by this manual have a front passenger airbag deactivation switch located inside the glove compartment. This is used when it is necessary to put small children in the front seat. Consult the glovebox owner's manual included with the vehicle for more information about airbags and passenger safety.

NOTE —

- *Audi identifies electrical components by a letter and/or a number in the electrical schematics. See* **EWD Electrical Wiring Diagrams**. *These electrical identifiers are listed in parentheses as an aid to electrical troubleshooting.*

BentleyPublishers.com
© 2006 BentleyPublishers.com—All Rights Reserved

Warnings

In addition to the warnings listed below, watch for airbag identification tags at various locations around the vehicle interior. The number and placement of tags varies with vehicle equipment.

> ***WARNING —***
> - *Always disconnect the battery ground (GND) strap before starting to work on airbag or pyrotechnic seatbelt systems. Work can begin immediately after disconnecting battery, no waiting time is needed.*
> - *NEVER let anyone sit in passenger compartment when the airbag or seatbelt system is being worked on, especially when either system is being connected to a power source or the battery.*
> - *Never test, remove, repair or install airbag units or pyrotechnic seatbelts or work on either system unless you have proper training, qualifications and tools to do the work safely and correctly.*
> - *Always discharge static electricity from your body before handling or even touching an airbag unit. Place your hands on a suitable metal object that is grounded such as a water or heating pipe or metal frame.*
> - *NEVER use impact wrenches to remove airbag or pyrotechnic seatbelt tensioners.*
> - *Always install airbag unit immediately after removing it from its shipping container.*
> - *If installation work is interrupted, immediately return airbag unit to its original packaging.*
> - *NEVER leave an airbag unit unattended. If work is interrupted, always return airbag unit to original packaging or shipping container.*
> - *NEVER place airbag unit on a workbench or other surface with the side that points towards inside of passenger compartment facing down. If airbag unit should deploy, it can cause personal injury. Always make sure that the side of the airbag unit that faces passenger compartment faces up.*
> - *NEVER install airbag or pyrotechnic seatbelt units that have dropped onto a hard surface, or which have signs of damage.*
> - *NEVER open or attempt to repair airbag or seatbelt tensioner components; always use new parts (danger of injury).*
> - *Always replace airbag and seatbelt tensioner units which have experienced mechanical damage (dents, cracks).*
> - *Always put airbag units that have not deployed or have not completely deployed in original airbag shipping container and mark outside of container so that others will know that there is still a possibility of airbag deployment.*
> - *Always store and handle airbag units according to all applicable legal requirements.*
> - *Seatbelt pyrotechnic ignition charge does not have an expiration date, i.e. it has an infinite shelf life and is maintenance free.*
> - *NEVER treat a pyrotechnic seatbelt tensioner unit with grease, spray lubricants, cleaning products or similar products.*
> - *DO NOT expose pyrotechnic seatbelt tensioner or airbag units to temperatures above 212°F (100°C) even for a few seconds.*
> - *Airbag and pyrotechnic seatbelt tensioner storage and transportation are covered under several explosive substance laws. Always be aware and observant of applicable regulations.*

BentleyPublishers
.com
©2006 BentleyPublishers.com—All Rights Reserved

Seatbelt inspection

Seatbelts should periodically be checked for proper operation, wear and damage. Check points include visual and operational inspection of the following:

- Belt webbing
- Automatic retraction (stop function)
- Belt latch mechanisms.
- Guide stop and seatbelt locking tongue
- Anchoring points and components

> ***WARNING —***
> - *Inspect seatbelts very carefully after any accident whether in use or not. Pay particular attention to seatbelts that were in use at the time. Any damage found should be reported to the appropriate involved party (insurance, repair shop, owner, etc.).*

Belt webbing

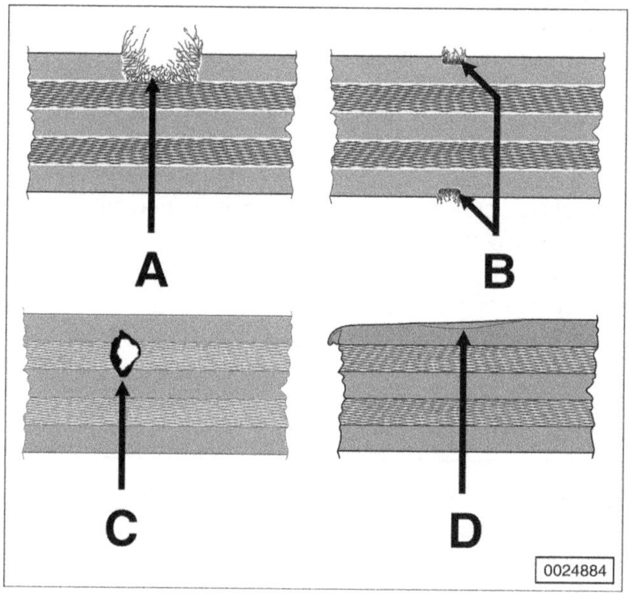

– Pull belt webbing completely off of automatic belt retractor.

– Check belt webbing for soiling and wash with mild soap and water solution as needed. Avoid harsh chemical solutions.

◄ Inspect webbing for damage.

A. Webbing cut, torn or chafed.

B. Webbing loops on belt edges torn.

C. Burned spots from cigarettes or other sources.

D. One side of belt edge deformed, or area of belt edge is wavy or rolled over.

– **A**, **B** - If noted in a post-accident inspection, replace complete belt assembly and height adjuster where applicable.

– **A**, **B**, **C**, **D**, - If noted during routine inspection and vehicle has not been involved in an accident, replace damaged parts as needed.

Automatic retraction (stop function)

Pyrotechnic seatbelts that have been activated do not retract. These units are no longer functional and require replacement.

The automatic belt retractor has two locking functions triggered by either occupant or vehicle movement.

– Function 1 (occupant movement): Locking is triggered by rapid unrolling of belt from automatic belt retractor reel (belt extraction acceleration).

- To check, pull belt out of automatic belt retractor with firm jerk. Seatbelt should lock while being pulled out. If not, replace seatbelt and belt latch.

NOTE —
- *If problems are noted during function test 1, check mounting/installation position of automatic retractor. Incorrect installation position can influence operation.*

BentleyPublishers.com

© 2006 BentleyPublishers.com—All Rights Reserved

Automatic retraction (stop function) continued

– Function 2 (vehicle movement): Locking function triggered by a change in vehicle movement (vehicle-dependent locking function).
  - To check, securely fasten belt and drive vehicle on a clear, level, even surface to a speed of 12 mph (20 kph).
  - Apply brake pedal fully in a panic-style application.
  - Seatbelt should lock as vehicle suddenly stops while braking. If not, replace seatbelt and belt latch.

> ***WARNING —***
> - *For safety reasons, perform road test in an area without traffic, pedestrians, or obstacles such as curbs, light and sign poles, or parked vehicles.*

Belt latch mechanisms

– Inspect seatbelt latch for cracks and fractures.
  - If damaged, replace seatbelt and belt latch.

– Check operation of seatbelt tongue into latch by inserting until an audible engagement is heard. Pull seatbelt webbing forcefully to be sure of proper engagement in lock mechanism.
  - Repeat test at least five (5) times. If seatbelt fails to engage even one (1) time, replace seatbelt and belt latch.

– Check release operation by pressing release button with finger while seatbelt is in a relaxed state. Locking tongue must spring out of seatbelt lock completely and without assistance.
  - Repeat test at least five (5) times. If seatbelt fails to release even one (1) time, replace seatbelt and belt latch.

> ***WARNING —***
> - *Do not lubricate seatbelt tongue, buttons, or release mechanisms to repair noises or stiffness.*

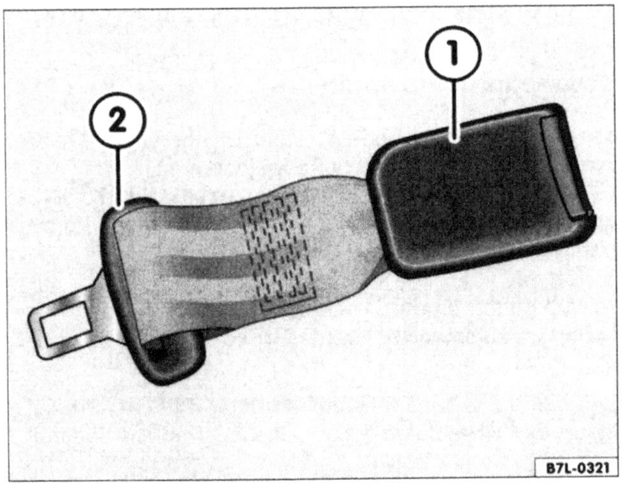

Inspect seatbelt extender (if equipped) lock tongue (**2**) and belt latch mechanisms (**1**) for the same conditions found with seatbelt latch mechanisms and seatbelt lock tongue.

Guides and seatbelt lock tongue

– Inspect all plastic coated seatbelt guides and plastic coating on seatbelt lock tongue. Check for deformation, cracking, tearing and wear.
  - If damaged, replace seatbelt and belt latch.

Anchoring points and components

– Visually inspect and functionally test lock strap, height adjuster and anchor points on seats, pillars, and floor. Check for deformation, secure and correct mounting, and proper operation.
  - If damage is found at components, replace seatbelt and belt latch.
  - If damage is found on anchoring points, replace components as required.
  - If damage is found that results from general wear, and not due to accident, replace only damaged component(s).

BentleyPublishers.com
© 2006 BentleyPublishers.com—All Rights Reserved

Child seat anchors

Vehicle manufactures are required to have an anchorage point installed in vehicles to facilitate installation of a tether strap for child seats. Lower anchor points in combination with upper tether anchor points are known as "**L**ower **A**nchor and **T**ether for **CH**ildren" or simply "**LATCH**" points. For information regarding proper installation of child seats booster seats, infant seats, and restraints, always consult vehicle owner's manual.

– Inspect LATCH points for damage or deformation after an accident.
  - Child seat anchors welded into chassis must not be repaired or adjusted.
  - Child seat anchors bolted to chassis must be replaced if damaged or deformed.

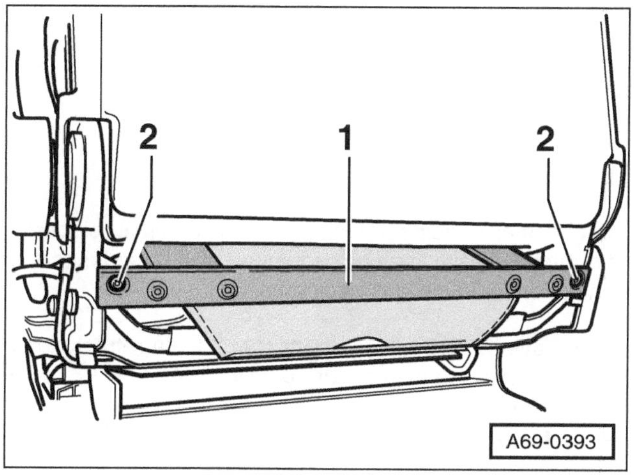

Lower child seat anchor bracket, removing and installing (roadster)

– Remove front passenger seat and spring cover trim. See **72 Seats**.

< Remove mounting bolts (**2**) and LATCH bracket (**1**) from seat bottom.

– Installation is reverse of removal. Always replace bolts.
  - Tighten bolts (**2**) to 10 Nm (7 ft-lb).

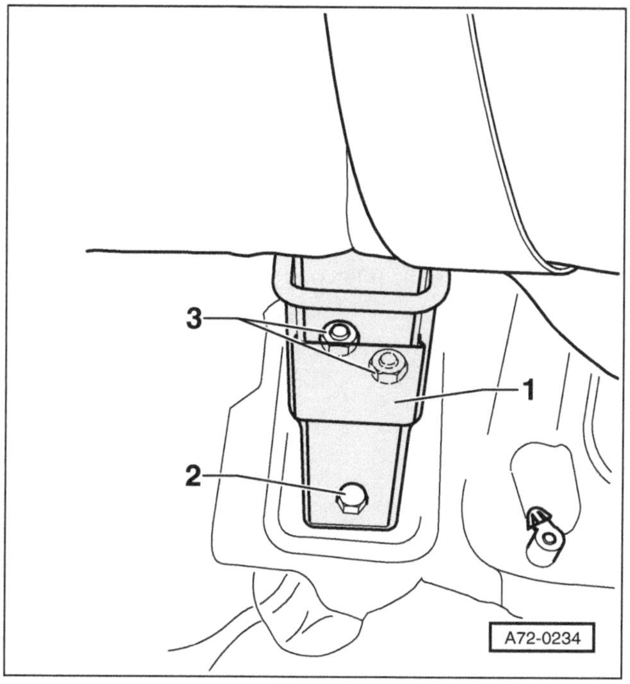

Lower child seat anchor bracket, removing and installing (coupe)

– Remove rear seat bottom and backrest. See **72 Seats**.

< Remove mounting bolt (**2**) and nuts (**3**).

– Remove LATCH bracket (**1**) from below seat bottom.

– Installation is reverse of removal. Always replace bolts.
  - Tighten nuts and bolts (**2 and 3**) to 23 Nm (17 ft-lb).

BentleyPublishers
.com
© 2006 BentleyPublishers.com—All Rights Reserved

Lower child seat anchors, removing and installing (coupe)

- Working at back side of rear seat backrest, pry out cover (**4**) on bottom side in direction of **arrow**.
- Remove bolt (**3**) and anchor (**2**).
- To remove strap guide (**1**), remove rear seat upholstery. See **72 Seats**.

SEATBELTS

The front seatbelts are equipped with pyrotechnic belt tensioners that retract during a crash. They are equipped with force limiters to moderate chest forces in serious accidents.

The rear seats have three point seatbelts that have a retractor with a ratcheting mechanism to help secure a child seat.

Component overview (coupe)

- Right front belt assembly is shown, left front is similar.
- Coupe model shown, roadster is similar with the exception of the rear seatbelts.

1. Front seatbelt
   - With belt tensioner
2. Anchorage point, outer floor panel
3. Anchorage point, latch
4. Anchorage point, for guide fitting
5. Rear seatbelts
6. Anchorage point, center floor latches

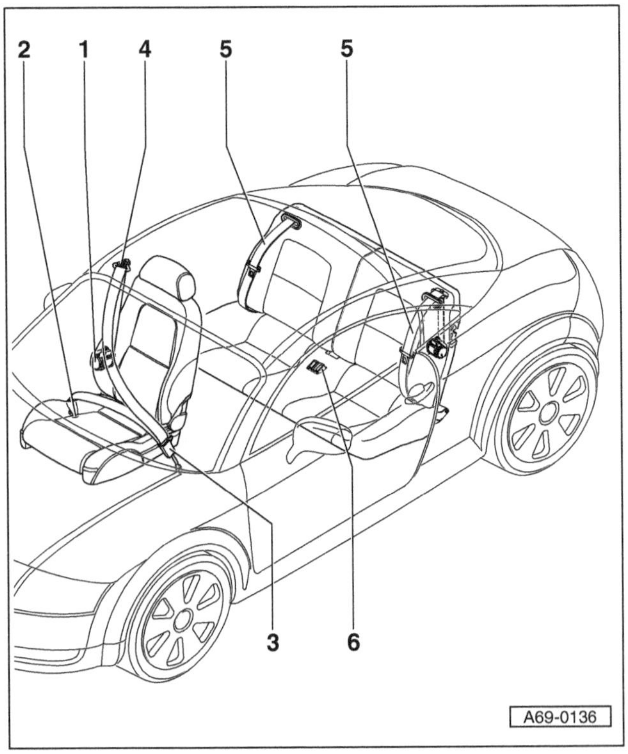

BentleyPublishers.com
©2006 BentleyPublishers.com—All Rights Reserved

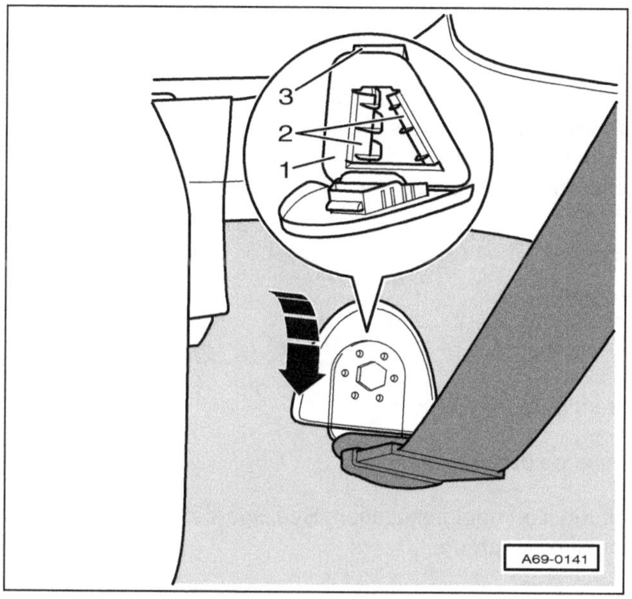

Front seatbelt, removing and installing (coupe)

– Observe all **Warnings** at the beginning of this section.

– Disconnect battery negative (–) cable.

– Remove rear seat bottom and backrest. See **72 Seats**.

< Open flap of cover for anchorage point on outer floor panel.
- Pull cover (**1**) in direction of **arrow** to open.

– If necessary, pull cover (**1**) downward to remove.
- When installing place tab (**3**) under floor covering.

– Unbolt seatbelt from floor.
- When installing, tighten bolt to 55 Nm (41 ft-lb)

– Remove rear side trim. See **70 Trim–Interior**.

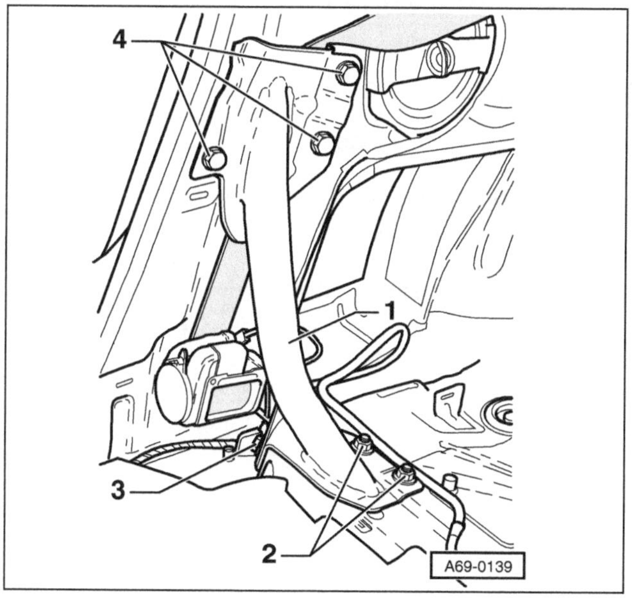

< Remove mounting nuts (**2**), bolts (**3** and **4**) and B-pillar brace (**1**).
- When installing, tighten nuts and bolts to 40 Nm (30 ft-lb).

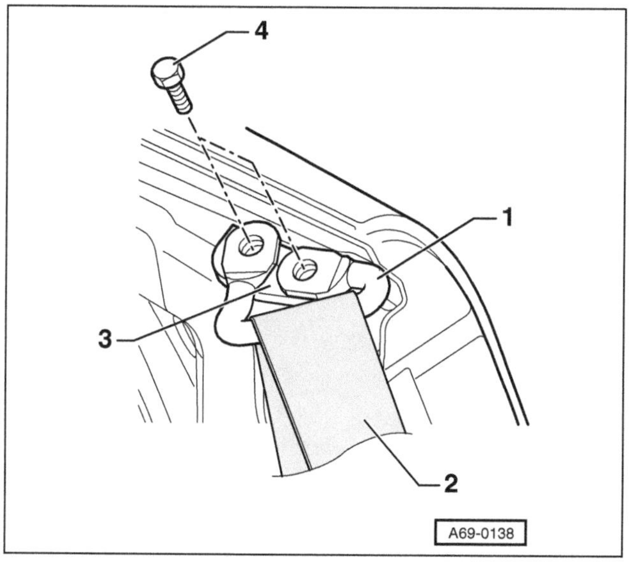

< Remove mounting bolts (**4**) and belt guide (**1**).
- When installing:
  Be sure to install washer (**3**).
  Tighten mounting bolts to 55 Nm (41 ft-lb).

BentleyPublishers.com
©2006 BentleyPublishers.com—All Rights Reserved

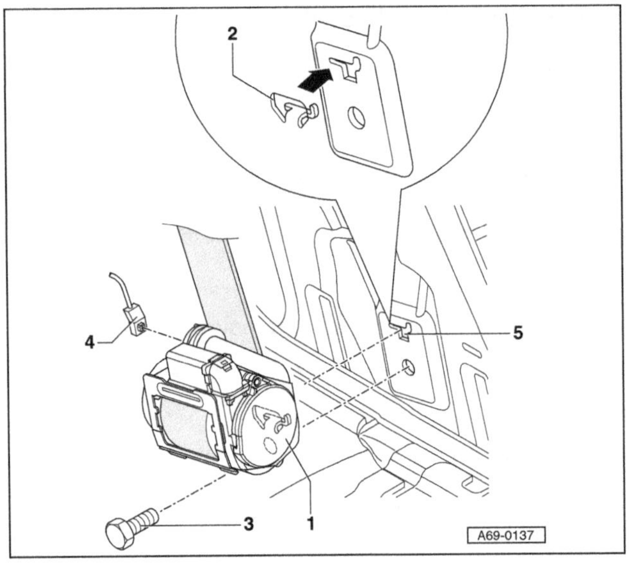

◄ Disconnect harness connector (**4**) from belt tensioner (**1**).

> ***WARNING —***
> • *Static electricity discharge may result in involuntary deployment of seatbelt tensioner. Always discharge static charge buildup by briefly touching chassis or door striker mounting before disconnecting any seatbelt tensioner wiring harness.*

– Remove mounting bolt (**3**) and belt tensioner.
  • When installing:
  Locate mounting tab (**2**) as shown.
  Tighten mounting bolt to 55 Nm (41 ft-lb).

– Remove seat belt from vehicle.

– Install in reverse order of removal.

– Check belt function after installation. See **Seatbelt inspection** earlier in this repair group.

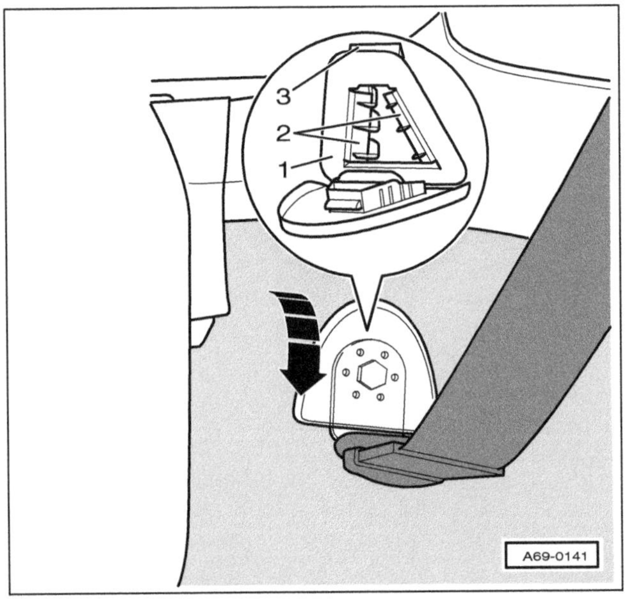

Front seatbelt, removing and installing (roadster)

– Observe all **Warnings** at the beginning of this section.

– Disconnect battery negative (–) cable.

◄ Open flap of cover for anchorage point on outer floor panel.
  • Pull cover (**1**) in direction of **arrow** to open.

– If necessary, pull cover (**1**) downward to remove.
  • When installing place tab (**3**) under floor covering.

– Unbolt seatbelt from floor.
  • When installing, tighten bolt to 55 Nm (41 ft-lb).

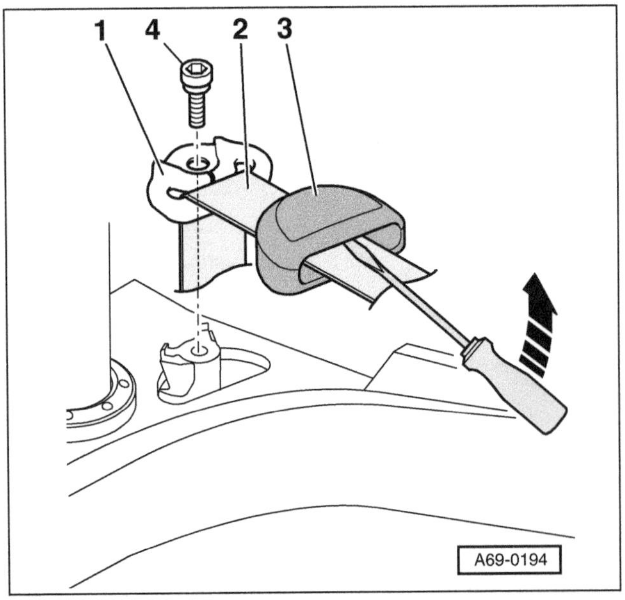

◄ Using a screwdriver, carefully pry seatbelt guide trim (**3**) in direction of **arrow**.

– Remove mounting bolt (**4**).
  • When installing, tighten mounting bolt to 55 Nm (41 ft-lb).

– Remove rear side trim and rear storage compartment. See **70 Trim–Interior**.

BentleyPublishers.com
© 2006 BentleyPublishers.com—All Rights Reserved

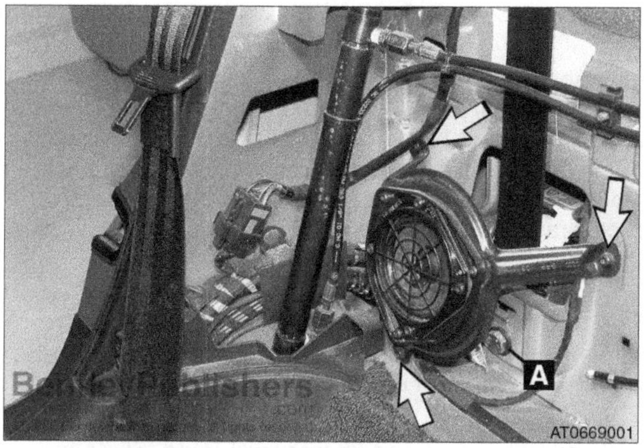

◄ Remove mounting bolts (**arrows**) and speaker.

– Remove mounting bolt (**A**) for belt tensioner.
  - When installing, tighten mounting bolt to 55 Nm (41 ft-lb).

– Remove belt tensioner and disconnect harness connector.

> ***WARNING —***
> - *Static electricity discharge may result in involuntary deployment of seatbelt tensioner. Always discharge static charge buildup by briefly touching chassis or door striker mounting before disconnecting any seatbelt tensioner wiring harness.*

– Remove seatbelt from vehicle.

– Install in reverse order of removal.

– Check belt function after installation. See **Seatbelt inspection** earlier in this repair group.

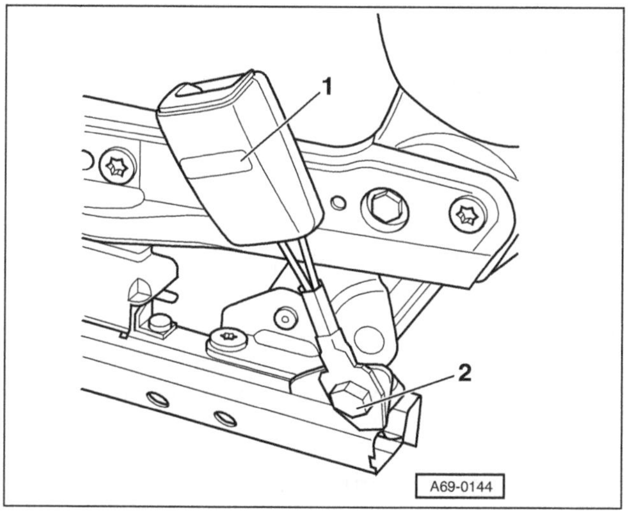

Front seatbelt latch, removing and installing

– Observe all **Warnings** at the beginning of this section.

– Disconnect battery negative (–) cable.

– Remove front seat. See **72 Seats**.

◄ Remove mounting bolt (**2**) for latch (**1**).
  - When installing, tighten mounting bolt to 24 Nm (18 ft-lb)

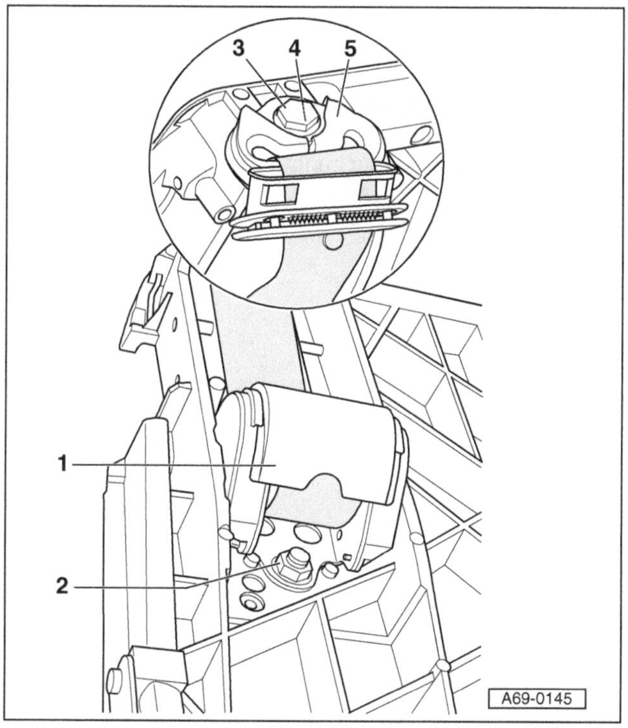

Rear seatbelt assembly (coupe)

◄ The rear seatbelt tensioner reel is located inside the rear seat back. To remove the rear seatbelts, the rear seat back and upholstery must first be removed. See **72 Seats**.

1. Belt reel
2. Mounting nut
   - 55 Nm (41 ft-lb)
3. Mounting bolt
   - 55 Nm (41 ft-lb)
4. Spacer ring
5. Guide fitting

BentleyPublishers
.com
© 2006 BentleyPublishers.com—All Rights Reserved

AIRBAGS

Vehicles covered by this manual are equipped with airbags installed in four locations in the front passenger compartment. The driver and front seat passenger airbags have a reduced initial deployment force to further protect occupants from injury. Side impact airbags are located in the lower sides of the front seats. These airbags extend upward to offer coverage for the head and torso (thorax) area.

Airbag system overview

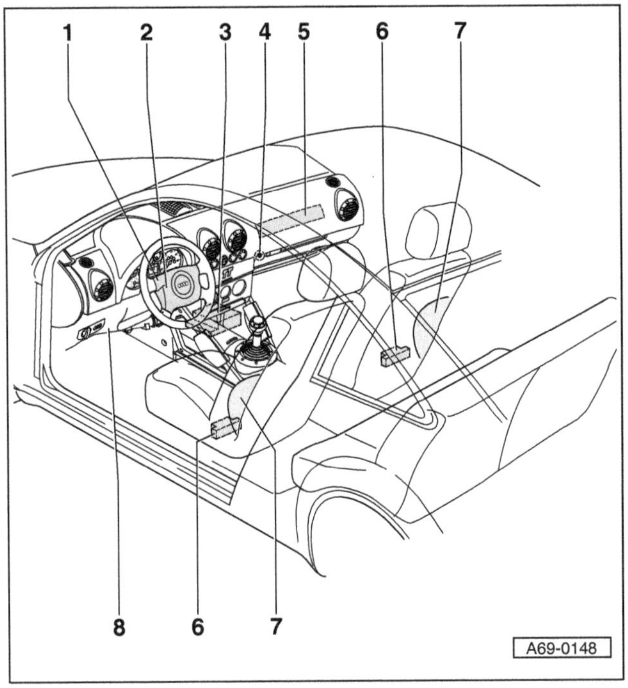

1. Driver's airbag unit
2. Airbag spiral spring and slip ring (F138)
   - In steering column
   - Removal and installation, see **96 Interior Lights, Switches, Anti-theft**
3. Airbag control module (J234)
   - Under dashboard, on tunnel
4. Deactivation switch for passenger side airbag
   - In glove compartment
   - Operate using vehicle key
5. Passenger airbag unit
6. Crash sensor
   - Driver's side, under seat (G179)
   - Passenger side, under seat (G180)
7. Side airbag
8. Data Link Connector (DLC)
   - Under dash
   - See **OBD On Board Diagnostics** for more information

Airbag control module, removing and installing

– Observe all **Warnings** at the beginning of this section.

– Disconnect battery negative (–) cable.

– Remove footwell vent trim and footwell vent. See **70 Trim–Interior**.

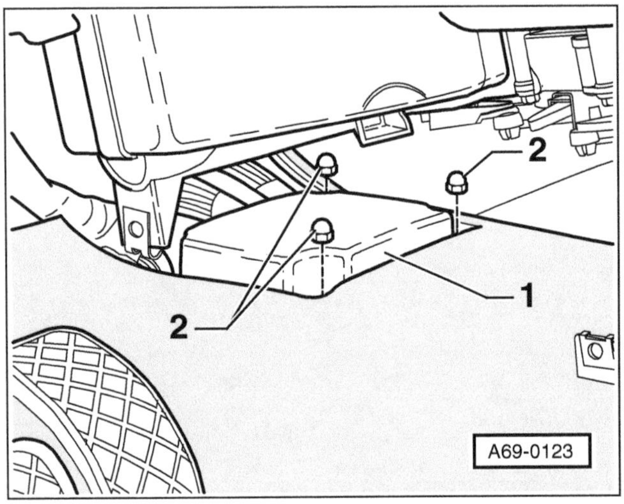

◄ Remove mounting nuts (**2**) for airbag control module (**1**).

BentleyPublishers.com
© 2006 BentleyPublishers.com—All Rights Reserved

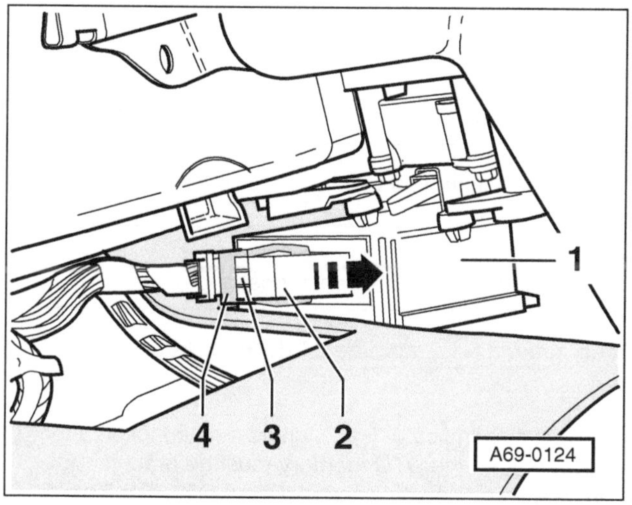

< Lift control module up enough to access harness connector.

- Press harness connector release tab (**3**) and pull retaining lever forward (**arrow**).
- Pull connector (**2**) off of control module.
- Remove control module.
- If control module is replaced, enter control module code using diagnostic scan tool. See **OBD On Board Diagnostics**.
- Tighten control module mounting nuts to specification.

Tightening torque	
Airbag control module to body	6 Nm (53 in-lb)

- Switch ignition on and close doors.

> ***WARNING —***
> *• Make sure no one is in vehicle when connecting battery.*

- Connect battery ground (–) strap. See **27 Battery, Alternator, Starter**.
- If the airbag malfunction indicator light signals a malfunction after installing the airbag control module, then the DTC memory must be erased and checked again using a diagnostic scan tool. See **OBD On Board Diagnostics**.

Driver's airbag, removing and installing

- Observe all **Warnings** at the beginning of this section.
- Disconnect battery negative (–) cable.
- Turn steering wheel as necessary to access airbag mounting bolts on back side of steering wheel.

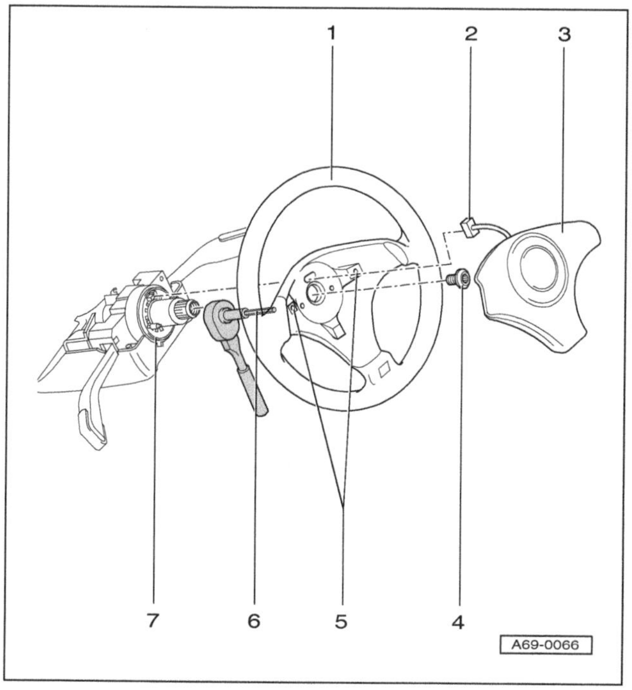

< Remove airbag mounting bolts (**5**) using Torx T30 wrench (**6**).

- Pull airbag (**3**) partially out of steering wheel.

> ***WARNING —***
> *• Static electricity discharge may result in involuntary deployment of airbag. Always discharge static charge buildup by briefly touching chassis or door striker mounting before disconnecting any airbag wiring harness.*

- Disconnect airbag harness connector (**2**) and remove airbag.
- Store airbag with padded horn side facing upward. Place in secure and safe location.
- Install in reverse order of removal. Make sure that airbag harness connector engages securely when installing.

BentleyPublishers
.com
©2006 BentleyPublishers.com—All Rights Reserved

– Tighten airbag mounting bolts to specification.

Tightening torque	
Airbag to steering wheel	7 Nm (62 in-lb)

– Switch ignition on and close doors.

> ***WARNING —***
> • *Make sure no one is in vehicle when connecting battery.*

– Connect battery ground (–) strap. See **27 Battery, Alternator, Starter**.

– If the airbag malfunction indicator light signals a malfunction after installing the airbag, then the DTC memory must be erased and checked again using a diagnostic scan tool. See **OBD On Board Diagnostics**.

Passenger's side front airbag, removing and installing

When working around airbag systems, observe all **Warnings** at the beginning of this section. Always disconnect negative (–) battery cable. See **27 Battery, Alternator, Starter**.

NOTE —
• *Removal of passenger's side airbag requires removal of glovebox. See* ***70 Trim–Interior****.*

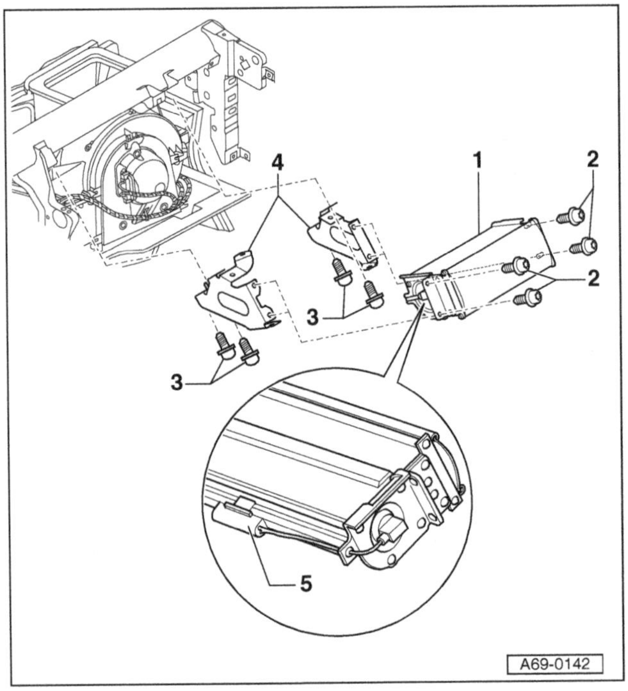

◄ Passenger's side front airbag components

1. Airbag unit
   • Disconnect harness connector (**5**) to remove airbag
   • Remove bolts (**3**) and airbag still attached to brackets (**4**)
   • Never unplug red 2-pin harness connector
2. Bolts
   • Qty: 4
   • Tighten to 12 Nm (10 ft-lb)
3. Bolts
   • Qty: 4
   • Tighten to 12 Nm (10 ft-lb)
4. Mounting brackets
   • Must be replaced if airbag has been triggered
5. Connector
   • See detail below for disconnecting

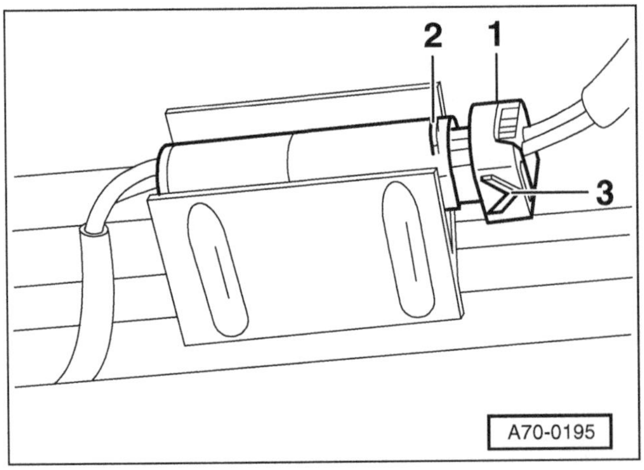

◄ Use screwdriver to lift airbag connector (**1**) over retaining tab (**2**) and pull off connector in direction of arrow (**3**).

> ***WARNING —***
> • *Static electricity discharge may result in involuntary deployment of airbag. Always discharge static charge buildup by briefly touching chassis or door striker mounting before disconnecting any airbag wiring harness.*

BentleyPublishers.com
© 2006 BentleyPublishers.com—All Rights Reserved

– After installing airbag and glove compartment, switch ignition on and close doors.

> ***WARNING —***
> *• Make sure no one is in vehicle when connecting battery.*

– Connect battery ground (–) strap. See **27 Battery, Alternator, Starter**.

– If the airbag malfunction indicator light signals a malfunction after installing airbag, then the DTC memory must be erased and checked again using a diagnostic scan tool. See **OBD On Board Diagnostics**.

Airbag crash sensors, removing and installing

The airbag crash sensors, also known as lateral acceleration sensors, are located under each of the front seats. The sensors are identical and the arrow on the sensor faces outward when they are installed.

– Observe all **Warnings** at the beginning of this section.

– Disconnect battery negative (–) cable.

– Remove front seat. See **72 Seats**.

– Remove lower A-pillar trim. See **70 Trim–Interior**.

– Unclip sill panel trim on inside.

– Lift carpeting around sensor.

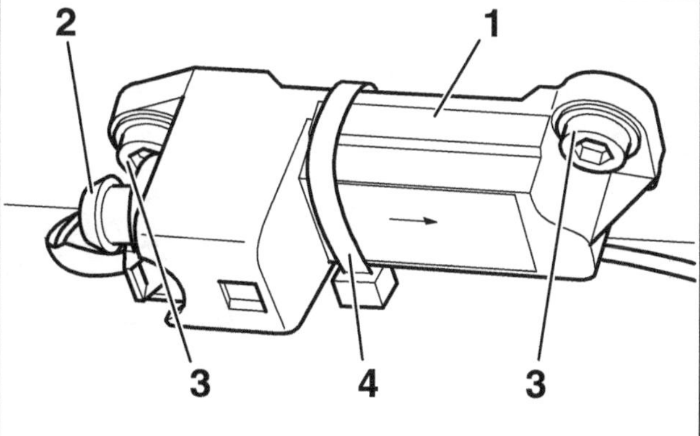

A69-0143

< Cut cable tie (**4**) and unplug harness connector (**2**) from sensor (**1**).

– Remove mounting bolts (**3**) and sensor.

– When installing, replace cable tie.

– Tighten sensor mounting bolts to specification.

Tightening torque	
Airbag sensor to body	6 Nm (53 in-lb)

– After installing airbag crash sensor, switch ignition on and close doors.

> ***WARNING —***
> *• Make sure no one is in vehicle when connecting battery.*

– Connect battery ground (–) strap. See **27 Battery, Alternator, Starter**.

– If the airbag malfunction indicator light signals a malfunction after installing airbag, then the DTC memory must be erased and checked again using a diagnostic scan tool. See **OBD On Board Diagnostics**.

BentleyPublishers
.com
© 2006 BentleyPublishers.com—All Rights Reserved

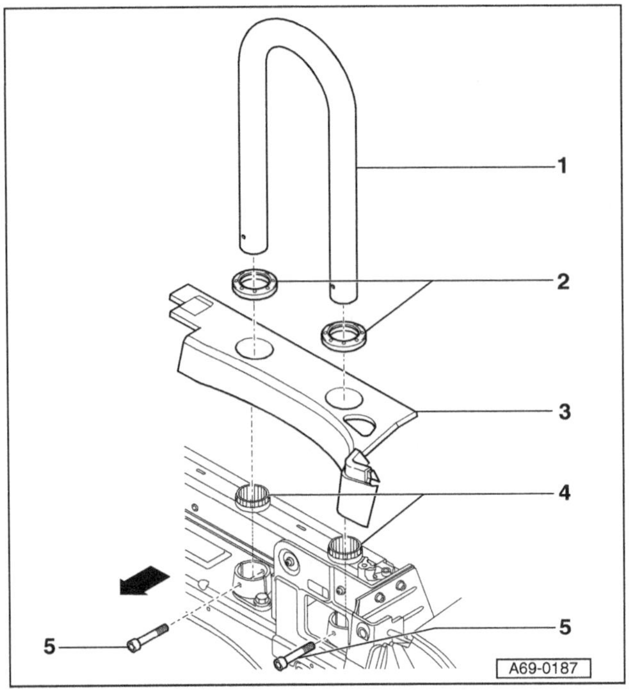

ROLLBAR (ROADSTER)

The convertible driver and passenger are provided with protection in the event of a roll over by a safey cage. Two tubular aluminum rollbars anchored to the vehicle body and projecting beyond the two head restraints, in combination with the ultra-rigid A-pillar / windshield cross member, provide additional crash protection.

Rollbar, removing and installing

< Loosen rollbar mounting collars (**2**).

- Remove rear trim panels. See **70 Trim–Interior**.
- Lift rollbar trim panel (**3**) enough to access rollbar mounting bolts. See **70 Trim–Interior**.
- Remove rollbar mounting bolts (**5**) and rollbar (**1**).

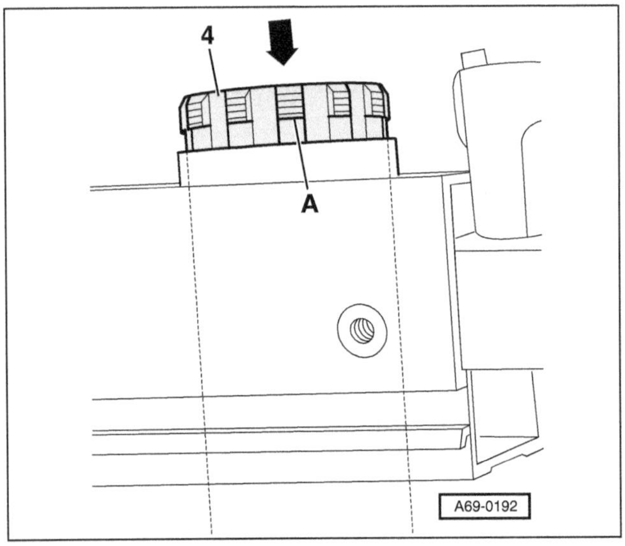

< When installing rollbar, check that rubber guide (**4**) is properly seated in cross member up to edge (**A**).

- Install rollbar using soapy water as a lubricant if necessary. Do not use other lubricants on rubber guides.
- Tighten rollbar mounting bolts to specification.

Tightening torque	
Rollbar to body	20 Nm (15 ft-lb)

- Remaining installation is reverse of removal.

© 2006 BentleyPublishers.com—All Rights Reserved

70 Trim–Interior

BentleyPublishers.com

©2006 BentleyPublishers.com—All Rights Reserved

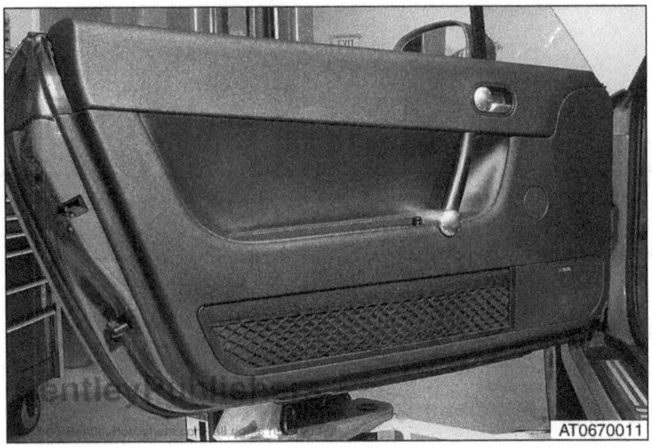

General

This repair group covers interior equipment and trim panels, including center console, door panel and door panel switch removal.

- For information of hood, trunk, and tailgate cables, See **55 Hoods and Lids**.
- For information on doors, power windows, and door window glass, see **57 Doors and Central Locking**.
- For information on radio and speaker removal, see **91 Radio and Communication**.
- For information on interior lights and accessories, see **96 Interior Lights, Switches, Anti-theft**.

Interior Rear View Mirror

Interior rear view mirror without electronics

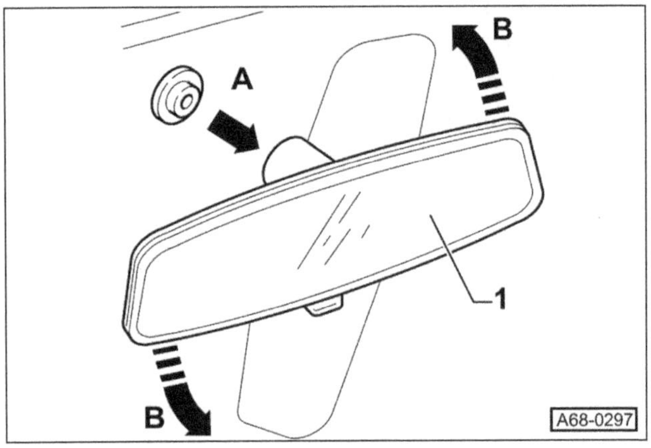

Removing

◀ Firmly grasp mirror assembly (**1**) and rotate counter-clockwise (**arrows B**) approximately 60° to 90° until retaining spring releases.

– Pull mirror (**1**) straight out (**arrow A**) and off of mounting lug.

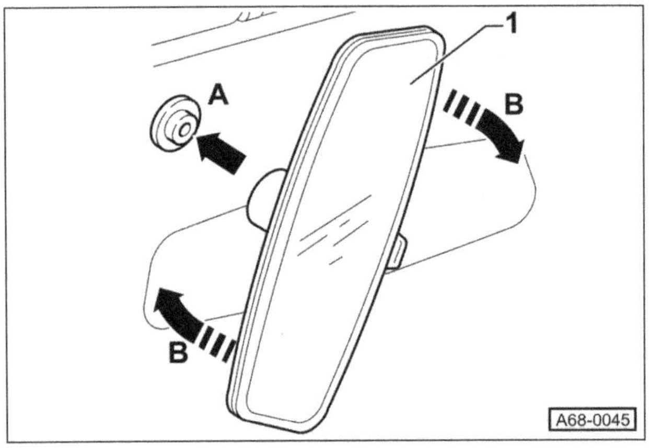

Installing

◀ Fit mirror (**1**) to mounting lug (**arrow A**) on windshield at an angle of approximately 60° to 90° as shown.

– Firmly grasp mirror assembly (**1**) and rotate clockwise (**arrows B**) until retaining spring clicks into place near the horizontal position.

Interior rear view mirror with electronics

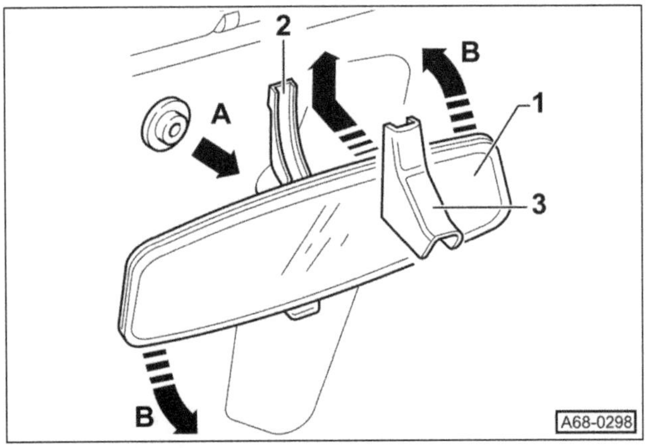

Removing

◀ Unclip cable channel cover (**3**) and remove (**arrow**).

– Remove connector from retainer and disconnect harness connector.

– Firmly grasp mirror assembly (**1**) and rotate counter-clockwise (**arrows B**) approximately 60° to 90° until retaining spring releases.

NOTE —
- *Do not use cable channel (**2**) to turn mirror.*

– Pull mirror (**1**) straight out (**arrow A**) and off of mounting lug.

BentleyPublishers.com
©2006 BentleyPublishers.com—All Rights Reserved

Installing

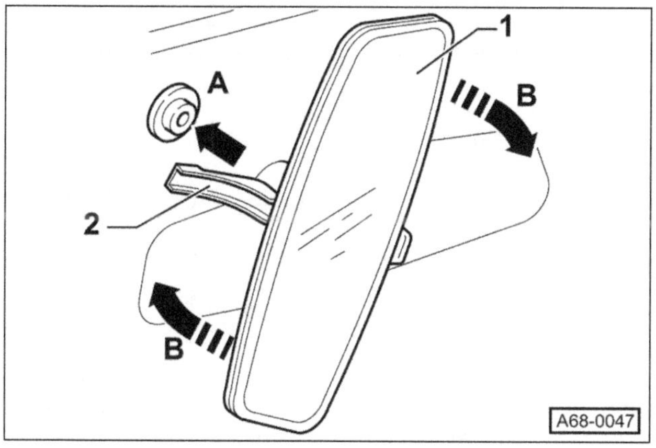

- Fit mirror (**1**) to mounting lug (**arrow A**) on windshield at an angle of approximately 60° to 90° as shown. Firmly grasp mirror assembly (**1**) and rotate clockwise (**arrows B**) until retaining spring clicks into place near the horizontal position.

NOTE —

- *Do not use cable channel (**2**) to turn mirror.*

Mounting lug for interior rear view mirror, installing

– Mounting lug installation requires the following:
  - Glass-metal adhesive kit, Audi part number D 000 703 A1 or equivalent.
  - Commercially available glass scraper with a new, sharp blade.
  - Work area with typical room temperature.
  - Several hours for undisturbed curing time.

– Remove mounting lug from bottom of mirror.

– Remove PUR adhesive from mounting lug with a wire brush.

– Carefully grind off three spacer protrusions from mounting lug bonding surface using 360/400 grit sandpaper on a flat surface. Ensure that sanded surface is clean and free of grease.

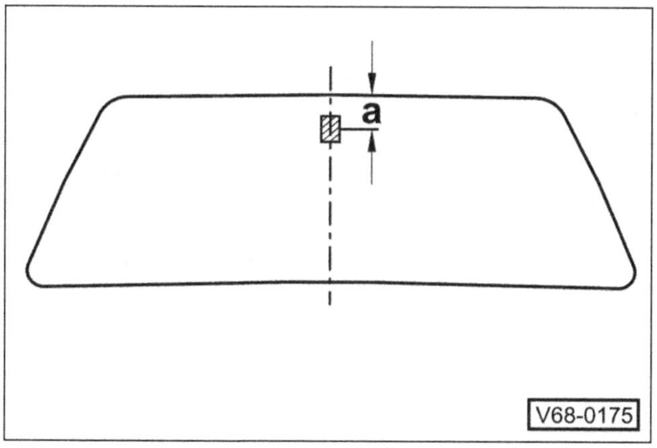

- Note location of mounting lug on windshield.
  - With a crayon or marker from outside, mark center line of windshield glass.
  - Measure from top edge of windshield glass to center point of mounting lug.
  - Dimension **a** = 135 mm (approximately 5 5⁄16 inches).

– Working on the inside of windshield, scrape off any remaining PUR material from windshield.

> ***CAUTION—***
> - *Do not damage ceramic coating on inside of windshield. Scratches will remain permanently visible.*
> - *Observe all manufacturers safety instructions and directions for use of cleaners and adhesives.*
> - *When working with cleaners and adhesives, wear hand protection (gloves).*

– Clean bonding area with adhesive remover D 002 000 10, cleaning solution D 009 401 04, or equivalent.

– Cut nylon mesh fabric from kit exactly to the size of mirror mounting lug.

– Apply adhesive evenly and smoothly to mirror mounting lug.

BentleyPublishers.com
© 2006 BentleyPublishers.com—All Rights Reserved

NOTE—
- *Work quickly but carefully; working time of adhesive is approximately 30 seconds between application and attachment to windshield.*

- Attach nylon mesh fabric onto mirror mounting lug.
- Apply additional adhesive to nylon mesh fabric by dabbing with tube.
- Press mirror mounting lug firmly (but without force) against windshield for approximately 15 seconds.
- Remove excess adhesive with a cloth.
- After approximately 15 minutes at typical room temperature, interior rear view mirror can be installed, but complete curing of adhesive will take several hours depending on temperature. During this curing period, any vibration can affect bond strength.

STORAGE COMPARTMENTS, COVERS AND TRIM

Center console trim surround

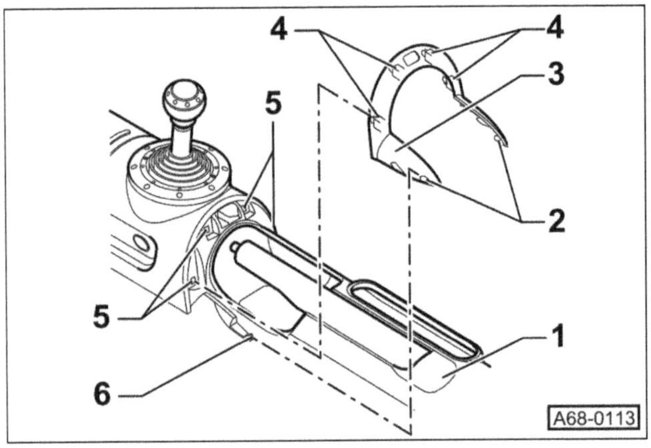

◄ Unsnap clips (**4**) on center console trim surround (**3**) from console (**5**).

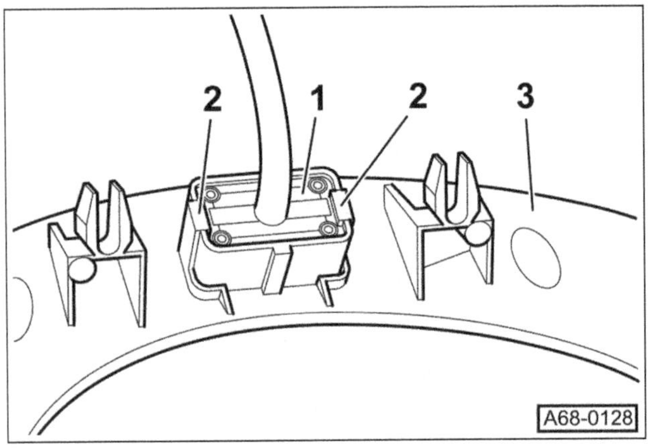

◄ If vehicle is equipped with an **AIRBAG OFF** warning light in center console trim surround, slide trim surround slightly to the rear. With a small screwdriver, pry out retaining tabs (**2**) and push out **AIRBAG OFF** warning light (**1**) out of trim surround (**3**). Slide trim surround (**3**) further to the rear to remove.

- If vehicle is not equipped with a warning light, slide trim surround (**3**) to the rear to remove.

BentleyPublishers.com
©2006 BentleyPublishers.com—All Rights Reserved

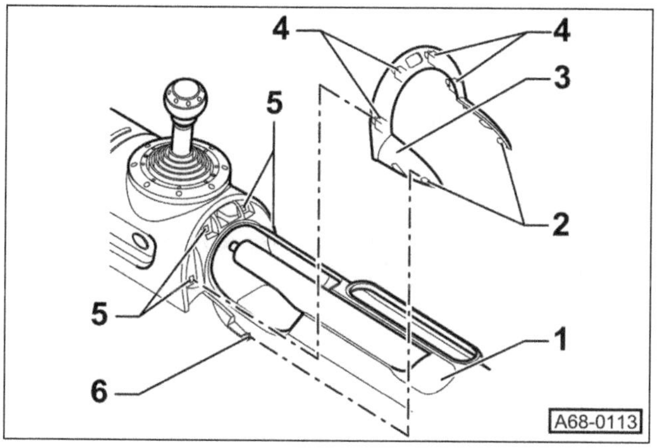

◀ To install, position trim surround on console, snap AIRBAG OFF warning light trim into surround (where equipped).

– Insert tabs (**2**) into mounts (**6**) and slide forward until clips snap into place on center console (**1**).

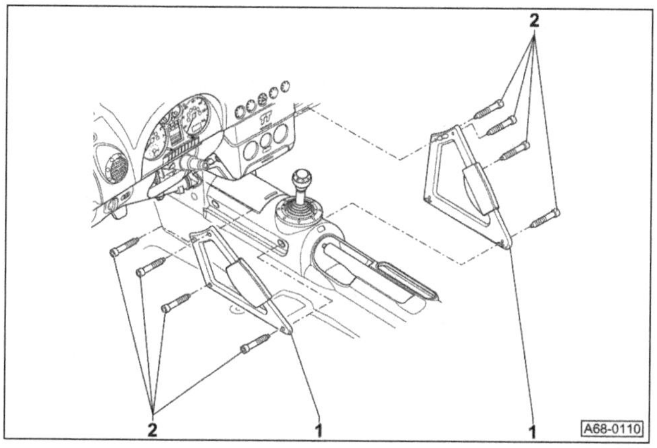

Center console grab handles

◀ Remove securing screws (**2**) from both sides.

- Tighten to 10 Nm (7 ft-lb)

– Pull grab handles (**1**) out from each side.

– Installation is the reverse of removal.

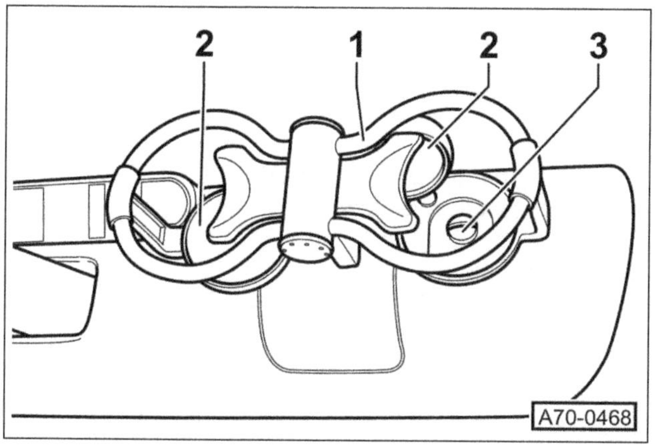

Center console cup holder

◀ Pop open covers both (**2**).

– Release locking both mechanisms (**3**) by turning with a screwdriver.

– Lift off cupholder (**1**).

– Installation is the reverse of removal.

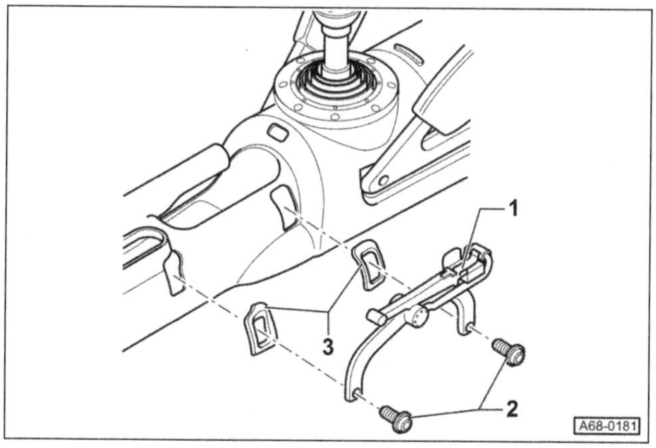

Center console phone brackets

Side phone bracket

◀ Remove both securing screws (**2**).

– Lift off bracket (**1**).

– When installing, ensure proper position of grommets (**3**).

– Install bracket and tighten screws (**2**).

- Tighten to 2.3 Nm (20 in-lb)

BentleyPublishers
.com
©2006 BentleyPublishers.com–All Rights Reserved

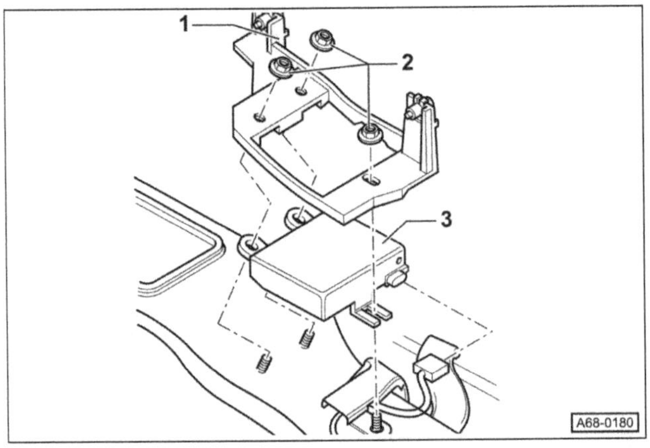

Top phone bracket

◀ Remove rear center console.

– Remove three securing nuts (**2**).

– Lift off bracket (**1**).

– When installing, ensure proper positioning over radar interior monitoring control module 1 (G303), (**3**).

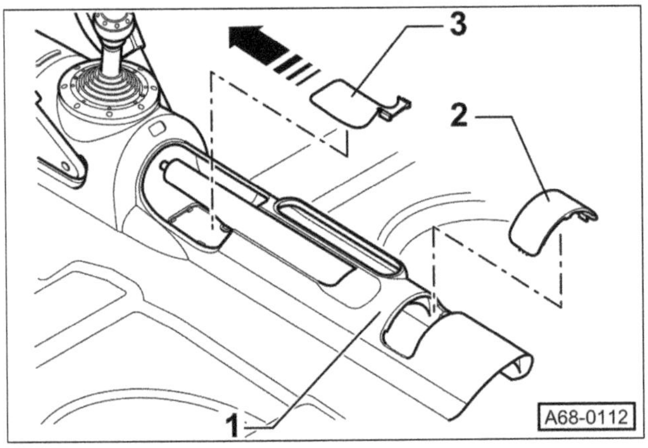

Rear center console covers

◀ Pry up front part of cover (**3**) under parking brake handle.

– Pull cover (**3**) forward and out of center console (**1**).

– On vehicles without phone equipment, pry up rear cover (**2**) and lift off.

– On vehicles with phone equipment, unplug phone handset and pry cover (**2**) up and lift off. Cover with harness connector can be moved aside for access to console screws as needed. Remove harness connectors for phone equipment as needed.

- Tighten screws to 0.8 Nm (7 in-lb).

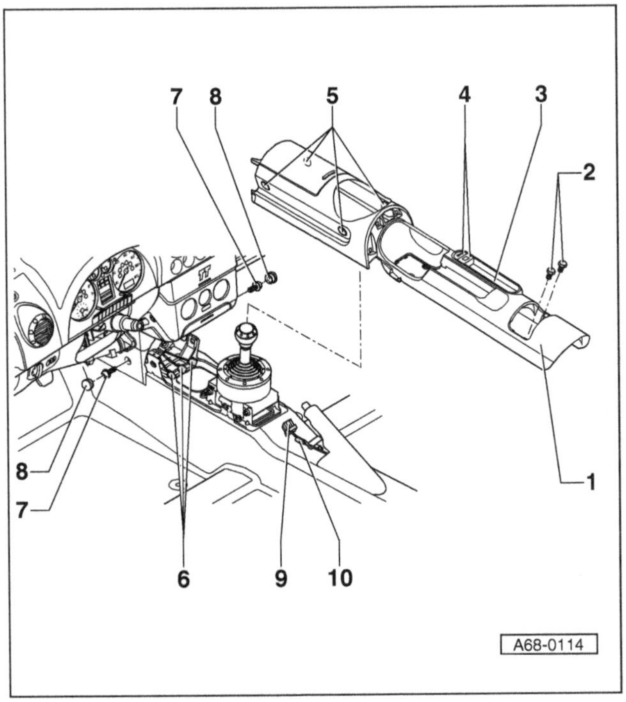

Rear center console

1. Rear section of center console
   - Removing:
     Pull parking brake handle up as far as it will go
     Remove rear center console covers
     Remove screws (item 2)
     Remove grab handles
     Pry out screw caps (item 8) and remove screws (item 7)
     Pull apart footwell vent trim
     Pull apart rear center console at mounts (item 5)
     Lift rear center console over rear seat
     Pull rear center console out from under footwell vent to the rear
     Lift rear center console forward and over parking brake handle
     Disconnect harness connectors (items 6, 9, 10)
   - Installation is the reverse of removal
2. Screws, tighten to 3.5 Nm (31 in-lb)
3. Tray
4. Central locking switch
5. Mounting point for grab handle
6. Harness connectors for switches
7. Screws, tighten to 3.5 Nm (31 in-lb)
8. Screw caps
9. Harness connector for central locking switch (item 4)
10. Harness connector for AIRBAG OFF warning light

BentleyPublishers
.com
© 2006 BentleyPublishers.com—All Rights Reserved

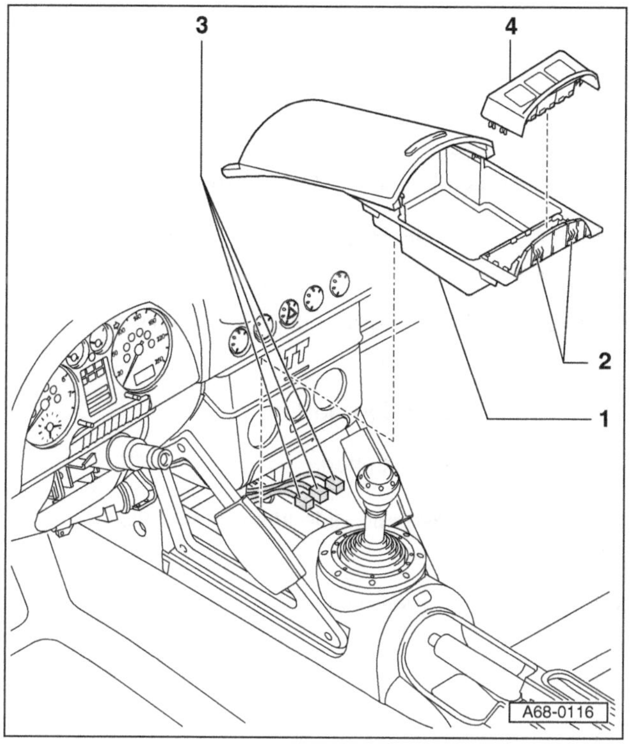

Center console front tray

1. Tray
   - Removing:
     Slightly open lid
     Remove switch console (item 4)
     Disconnect harness connectors (item 3)
     Using a small flat screwdriver, release retaining tabs (item 2)
     Press down on center catch between retainer tabs
   - Install in reverse order of removal
2. Retaining tabs
3. Harness connectors
4. Switch console
   - Unclip at top
   - Disconnect harness connectors

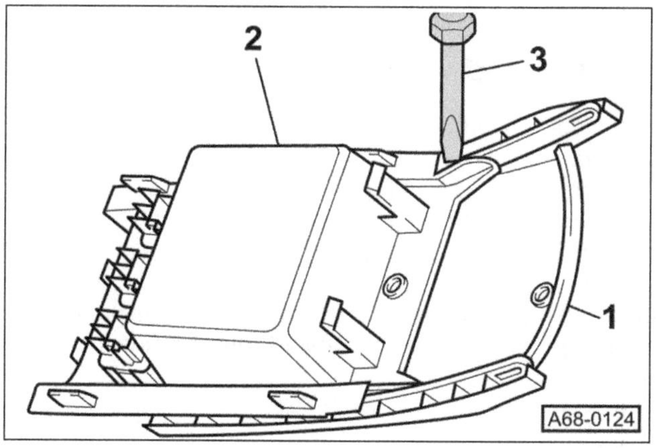

Center console front tray lid

– Remove center console tray (**2**) and turn over.

< Using a screwdriver (**3**), pry lid (**1**) off of tray from the side.

– To install, position lid (**1**) on tray (**2**) and snap into position.

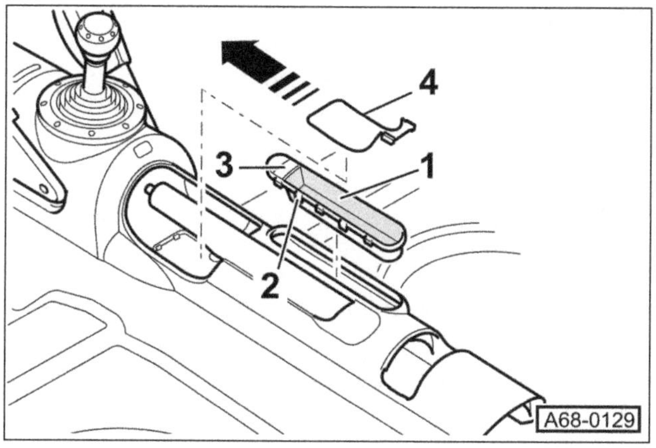

Rear center console rear tray lid

< Unclip cover (**4**).

– Using a screwdriver, pry out tray (**1**) at retaining hooks (**2**).

– Disconnect harness connector for switch (**3**).

BentleyPublishers
.com
© 2006 BentleyPublishers.com—All Rights Reserved

Parking brake handle trim

Handle button, removing and installing

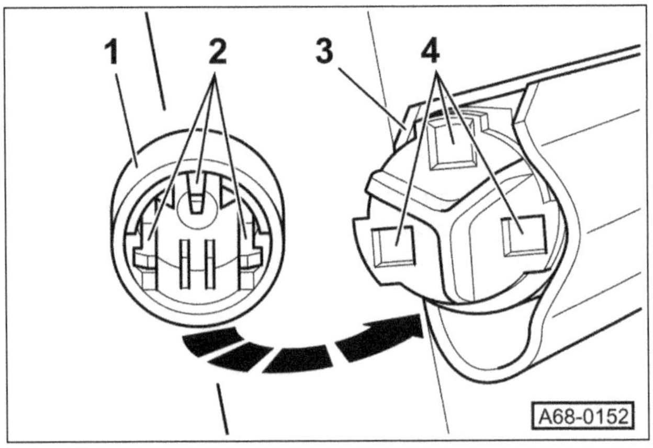

- To remove button, release parking brake (**3**).
- Carefully pivot parking brake button (**1**) off towards the top.
- To install button (**1**), insert into mounts (**4**) with retaining hooks (**2**) positioned as shown.
- Carefully push parking brake handle button fully onto handle (**3**).

Handle trim, removing and installing

- Remove parking brake handle button.

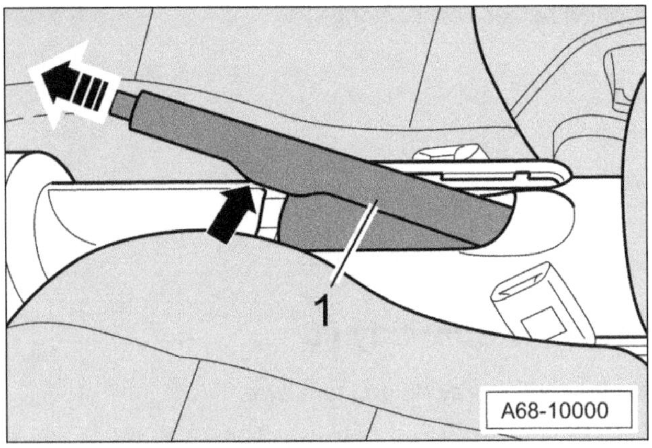

- Remove screw (**solid arrow**) and slide trim (**1**) off of handle (**dashed arrow**).
- Installation is the reverse of removal.

Footwell vent and vent trim

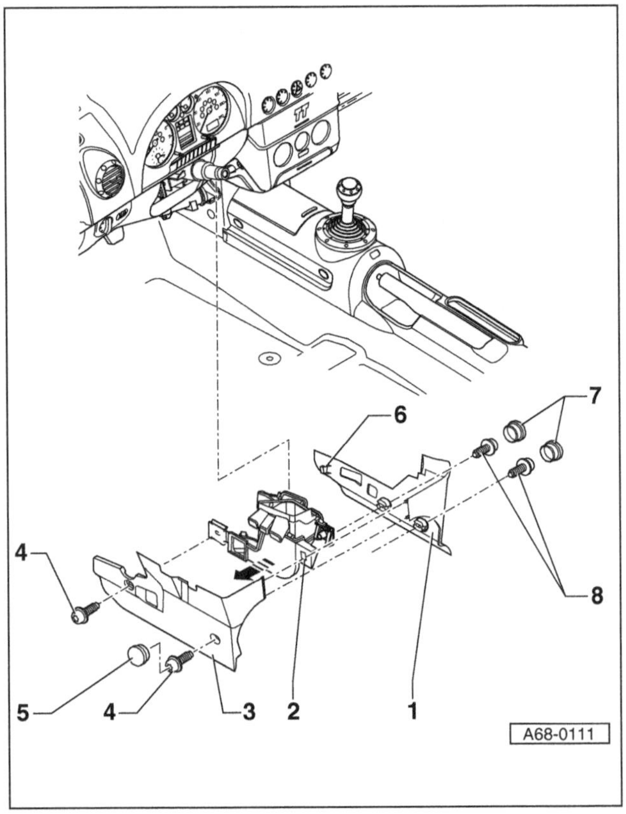

1. Right side footwell trim
   - Removing:
     Pry out screw caps (item 7)
     Remove screws (item 8)
     Pull footwell vent trim (item 1) out and to the rear
   - Installing:
     Insert retaining hook (item 6) into slot in instrument panel
     Connect right side footwell trim (item 1) and left side footwell trim (item 3) in center
2. Footwell vent
   - Removing:
     Remove left and right side footwell trim (items 1, 3)
     Pull tent straight out (arrow)
   - Installing:
     Insert footwell vent and clip into position
     Install trim
3. Left side footwell vent trim
   - Removing:
     Pry out trim cap (item 5)
     Remove screws (item 4)
   - Installing:
     Connect right side footwell trim (item 1) and left side footwell trim (item 3) in center
4. Screws, tighten to 3.5 Nm (31 in-lb)
5. Trim cap
6. Retaining hooks
7. Trim caps
8. Screws, tighten to 3.5 Nm (31 in-lb)

BentleyPublishers.com
© 2006 BentleyPublishers.com—All Rights Reserved

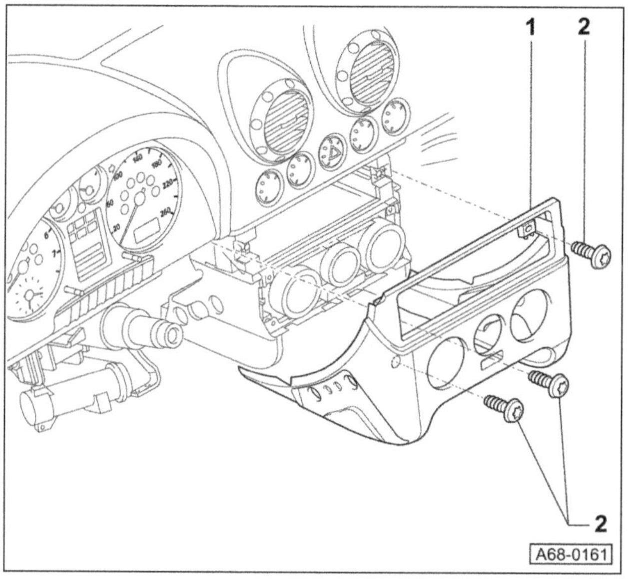

Center console trim

1. Center console trim
   - Removing:
     Remove grab handles
     Remove ashtray
     Remove radio, see **91 Radio and Communication**
     Remove screws (item 2)
     Unclip center console trim
   - Install in reverse order of removal
2. Screws, tighten to 1.5 Nm (13 in-lb)

– Remove storage compartment (**3**) from center console and, depending on vehicle equipment, disconnect wiring harnesses.

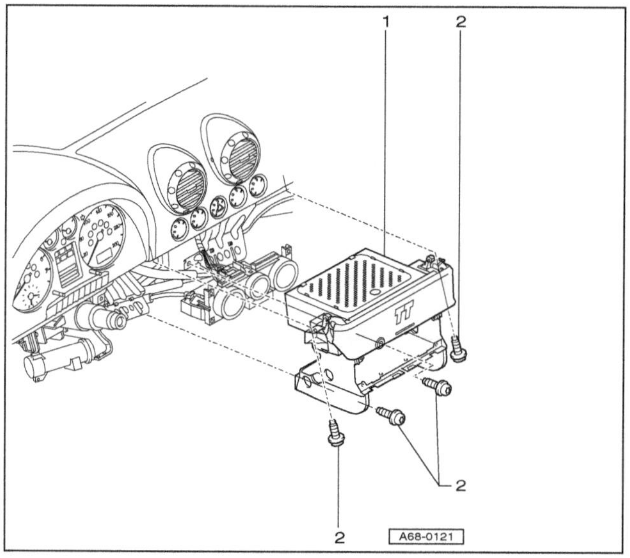

Front center console, removing and installing

1. Front center console
   - Removing:
     Remove glove box
     Remove grab handles
     Remove left side under-dash panel
     Remove front ash tray
     Remove center console trim
     Remove screws (item 2)
   - Install in reverse order of removal
2. Screws, tighten to 1.5 Nm (13 in-lb)

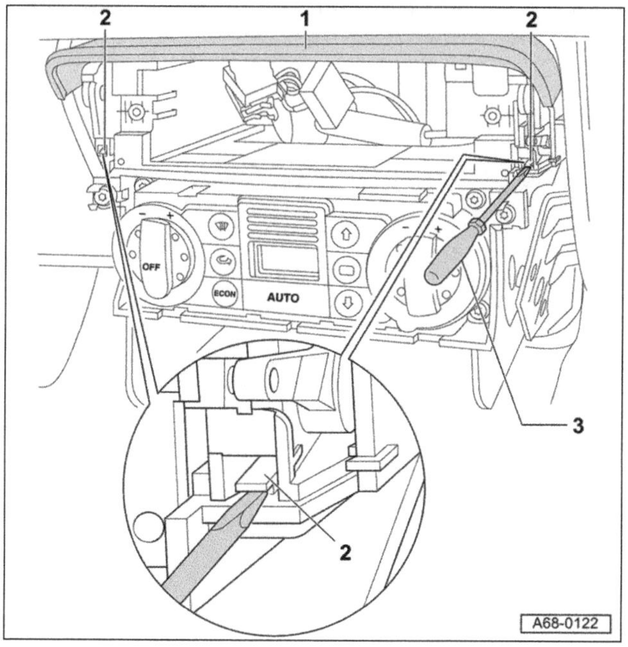

Front center console radio flap

1. Radio flap (in open position)
   - Removing:
     Remove radio, see 91 **Radio and Communications**
     Remove center console trim
     Using a screwdriver (item 3), press up on retainer tabs (item 2)
     Pull radio flap straight out
   - Install in reverse order of removal
2. Retaining tabs (note detail in inset)
3. Screwdriver

BentleyPublishers
.com
© 2006 BentleyPublishers.com—All Rights Reserved

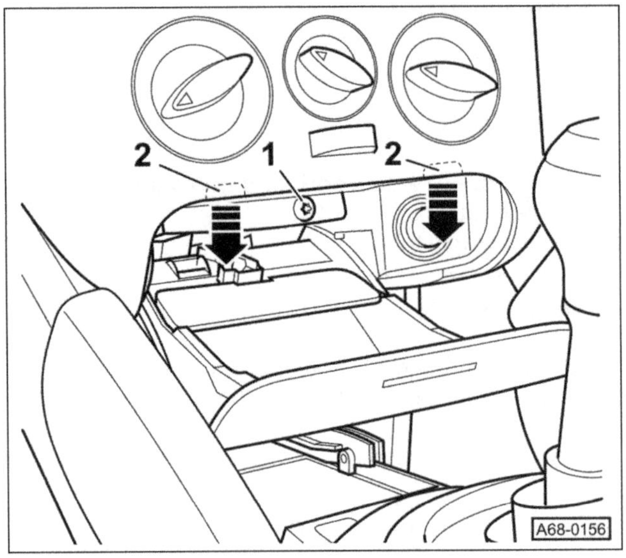

Front ashtray

– Open front ashtray.

◀ Remove screws (**1**).

– Pull ashtray downward (**dashed arrows**) at locating tabs (**2**).

– Disconnect harness connectors.

– When installing, guide locating studs on rear into both mounts.

– Tighten screws (**1**) to 2.0 Nm (18 in-lb)

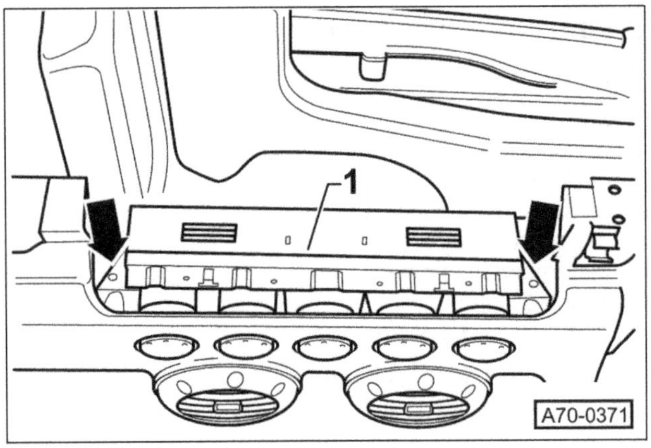

Switch console

◀ Remove mounting screws (**arrows**).

NOTE—
- *View is from underneath, looking up.*

– Remove switch console (**1**) with switches installed.

– Disconnect harness connectors.

– Installation is the reverse of removal.

Glove box

Removing

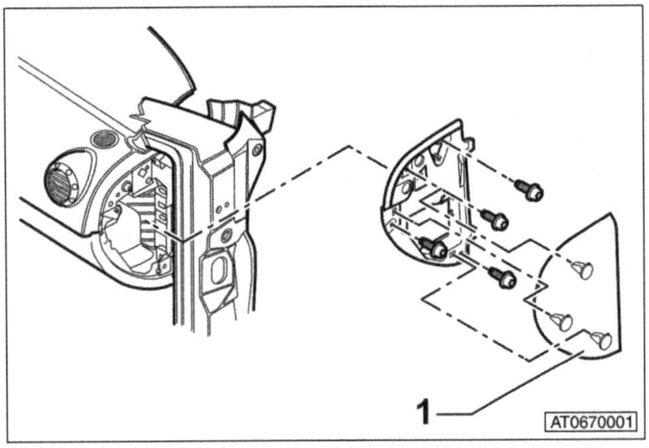

◀ Using a small screwdriver, unclip right side instrument panel end trim (**1**) and remove.

BentleyPublishers.com
©2006 BentleyPublishers.com—All Rights Reserved

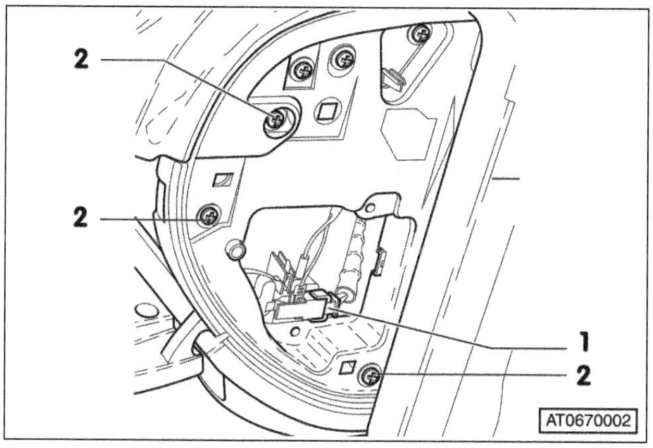

◄ Unplug harness connector for glove box light (**1**).

– Remove three side retaining screws (**2**).

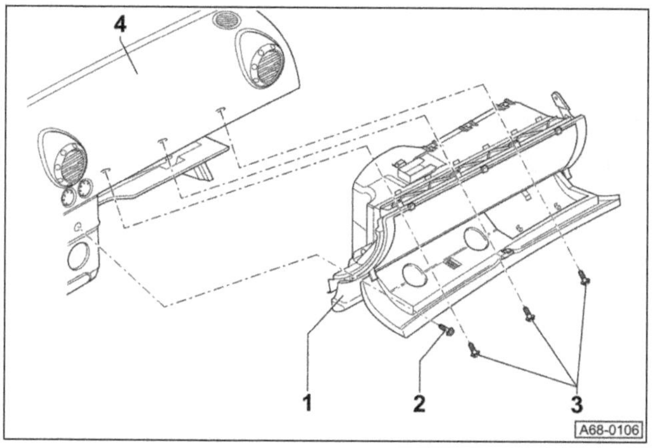

◄ Remove screw (**2**) from glove box (**1**).

– Open glove box door and remove three screws (**3**).

– Pull sides of frame outward from instrument panel (**4**).

– Slide glove box downward.

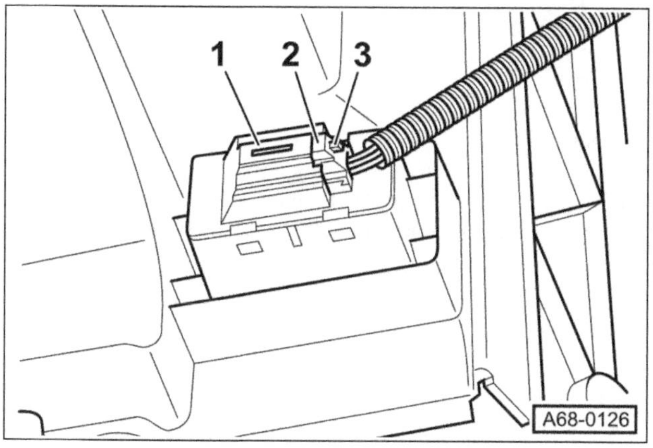

◄ Depress release tab (**3**) and disconnect harness connector (**2**) from front passenger's airbag key switch (**1**).

Installing

– Move glove box into position and connect front passenger's airbag key switch.

– Pull sides of frame outward.

– Install inner side of glove box first.

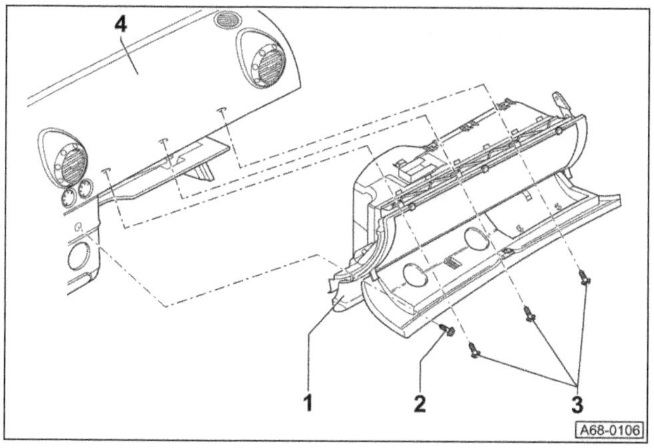

◄ Install screw (**2**), tighten to 3.5 Nm (31 in-lb).

– Install screws (**3**), tighten to 3.5 Nm (31 in-lb).

– Install side retaining screws.

– Connect harness connector for glove box light.

– Snap instrument panel end trim into position.

BentleyPublishers.com
©2006 BentleyPublishers.com—All Rights Reserved

Glove box door and trim

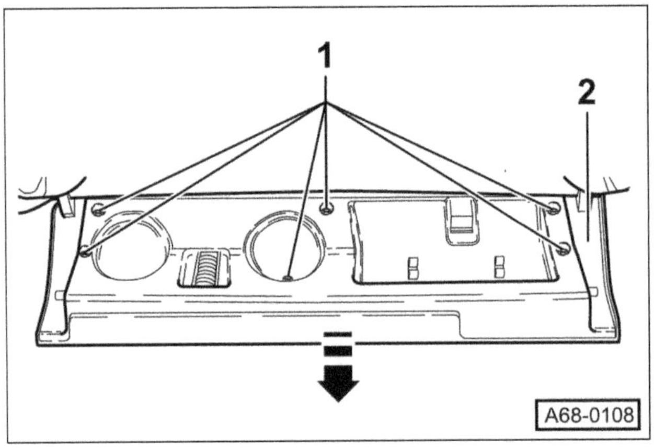

- Remove six retaining screws (**1**).
  - Tighten to 1.0 Nm (9 in-lb)
- Pull outer door (**2**) off (**dashed arrow**).

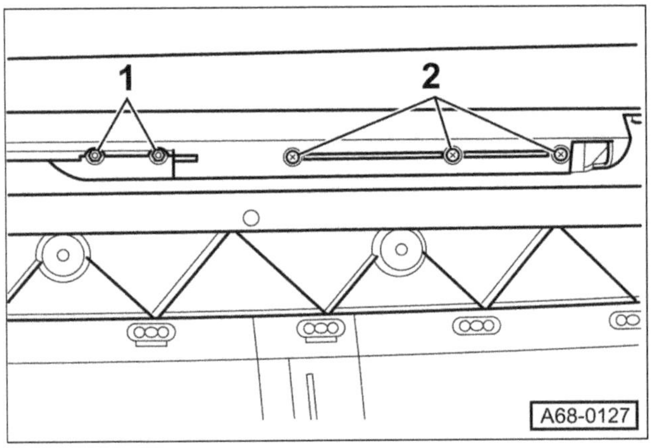

- Remove two screws (**1**).
  - Tighten to 2.0 Nm (18 in-lb)
- Remove three nuts (**2**).
  - Tighten to 1.0 Nm (9 in-lb)
- Detach trim.

Glove box lock

Removing

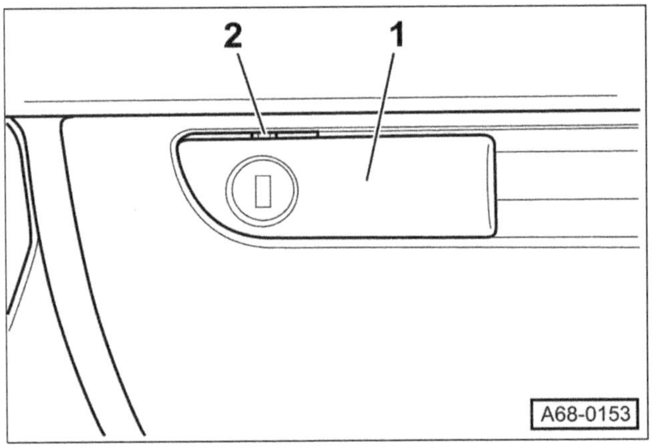

- Insert key into lock cylinder.
- Pry out glove box opening mechanism (**1**) using a small screwdriver. Insert screwdriver into tab (**2**) and carefully twist to remove.

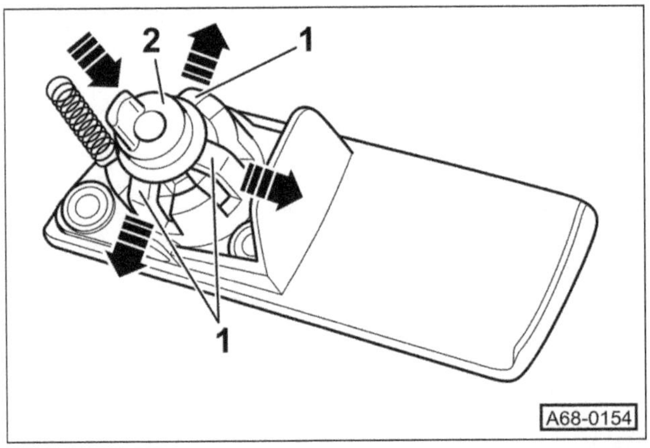

- Remove lock cylinder from opening mechanism. Use a small screwdriver and carefully pry out (**arrows**) on all four plastic retaining tabs (**1**) while at the same time pushing out lock cylinder assembly (**2**).

BentleyPublishers.com
© 2006 BentleyPublishers.com–All Rights Reserved

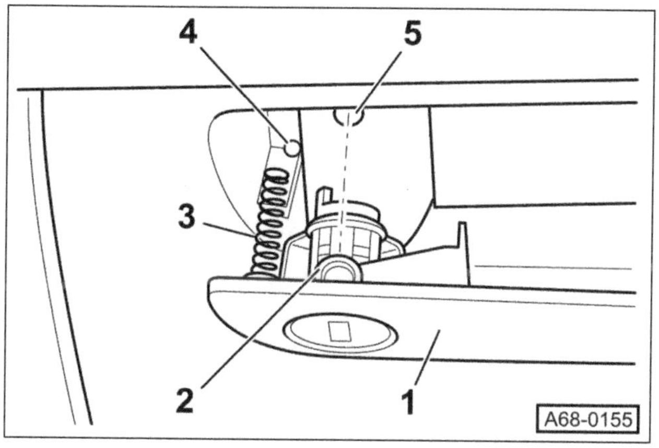

Installing

– With key inserted, push lock cylinder into opening mechanism until all retaining tabs engage.

◄ Position opening mechanism (**1**) so that mounts (**2**) snap into both retaining tabs (**5**) and spring (**3**) aligns and engages in pin (**4**).

– Carefully press glove box opening mechanism (**1**) into glove box lid until both retainer tabs (**5**) completely engage.

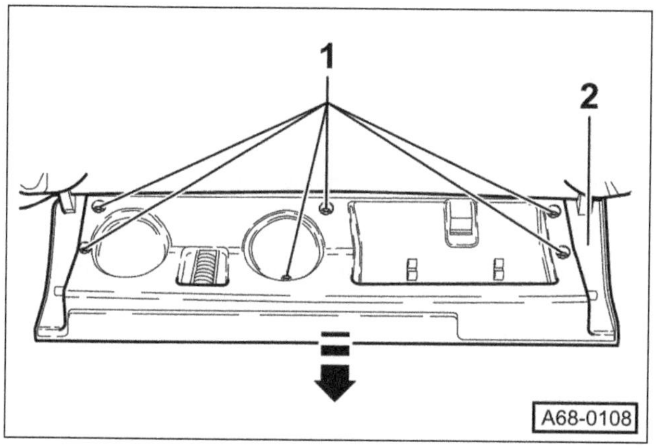

Glove box latch, removing and installing

◄ Remove six retaining screws (**1**).
- Tighten to 1.0 Nm (9 in-lb)

– Pull outer door (**2**) off (**dashed arrow**).

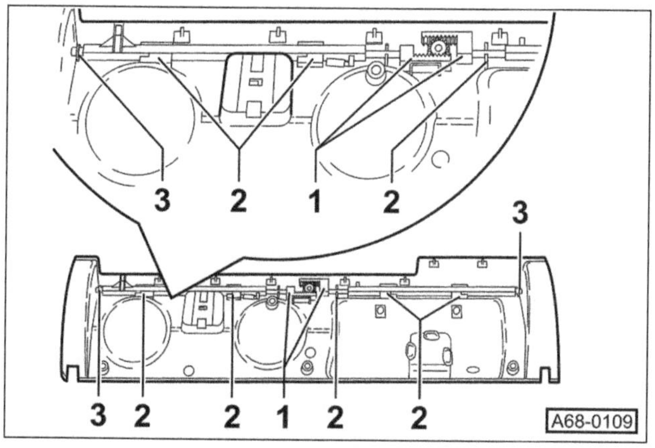

◄ Using a small screwdriver, pry out both linkage rods (**1**) at guides (**2**).

– Pull both linkage rods (**1**) inwards at slots (**3**).

– When installing, insert linkage rods (**1**) onto central drive pinion so that projection at slot (**3**) is approximately 5 mm ($\frac{13}{64}$ in).

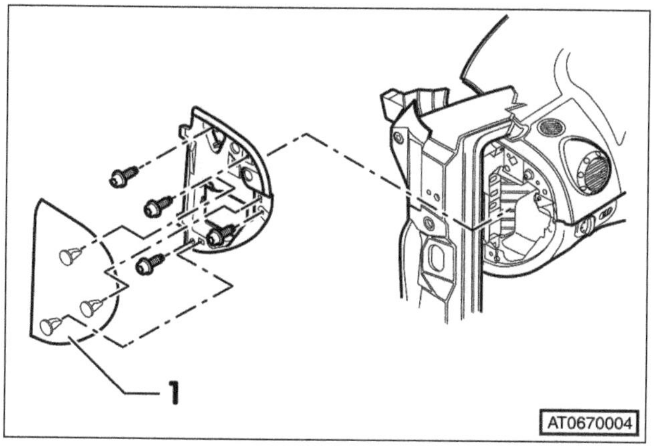

Under-dash panel, left side

Removing

◄ Using a small screwdriver, unclip left side instrument panel end trim panel (**1**) and remove.

– Remove hood latch release cable handle. See **55 Hoods and Lids**.

BentleyPublishers.com
© 2006 BentleyPublishers.com–All Rights Reserved

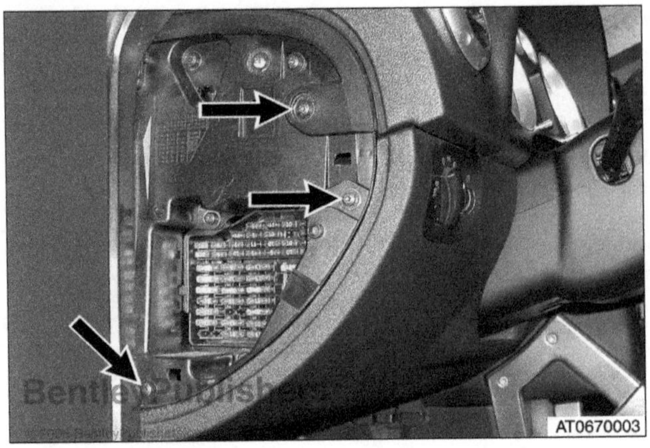

◀ Remove three screws on side of panel (**arrows**).

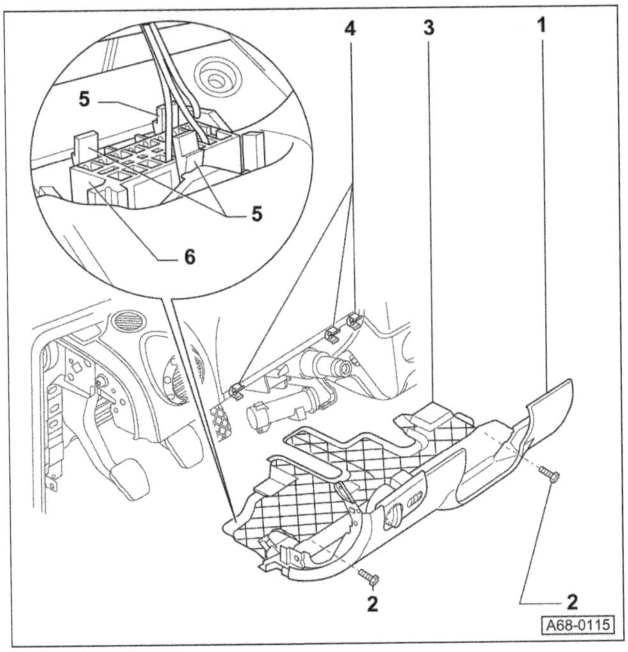

◀ Remove panel screws (**2**).

- Pull under-dash panel (**1**) out of three securing clips (**4**).
- Pull sides of frame outward.
- Carefully position under-dash panel to allow access to electrical components.
- Release three retaining tabs (**5**) and remove data link connector (**6**).

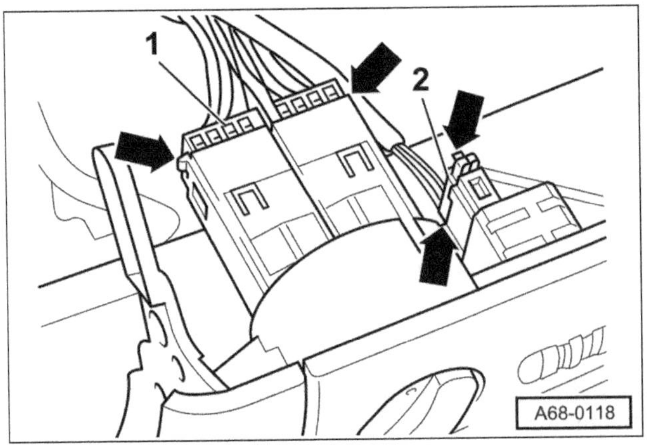

◀ Depress retaining tabs (**arrows**) and disconnect harness connectors for headlight switch (**1**) and dimmer (**2**).

- Remove under-dash panel.

BentleyPublishers
.com
©2006 BentleyPublishers.com—All Rights Reserved

Under dash panel, left side, installing

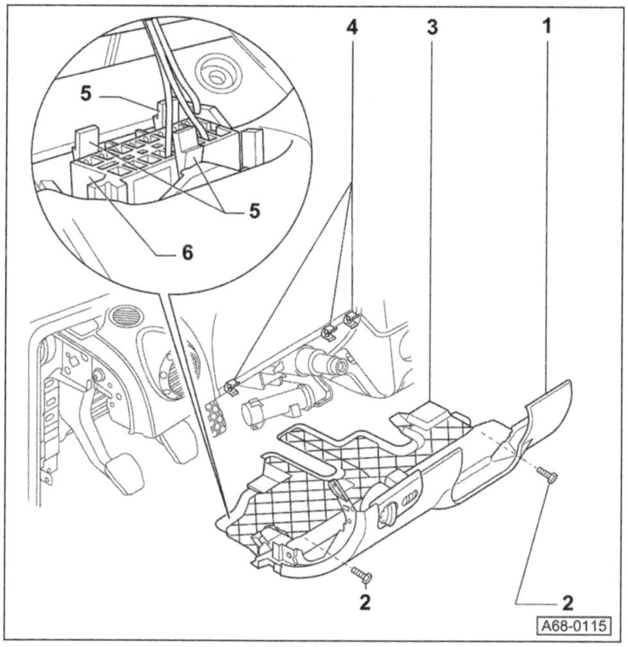

- Check securing clips (**4**) to confirm installation and position.
- Install harness connectors for headlight switch and dimmer.
- Install data link connector (**6**).
- Guide holder (**3**) into mount at pedal cluster and position under-dash panel (**1**) into securing clips (**4**).
- Install panel screws (**2**).
- Install hood latch release cable handle.
- Install three screws on side of panel.
- Install left side instrument panel end trim panel.

Instrument panel center defroster vent trim

Defroster vent will be destroyed during removal process. Removal is required for access to sunlight photo sensor (G107) and phone speaker (R39).

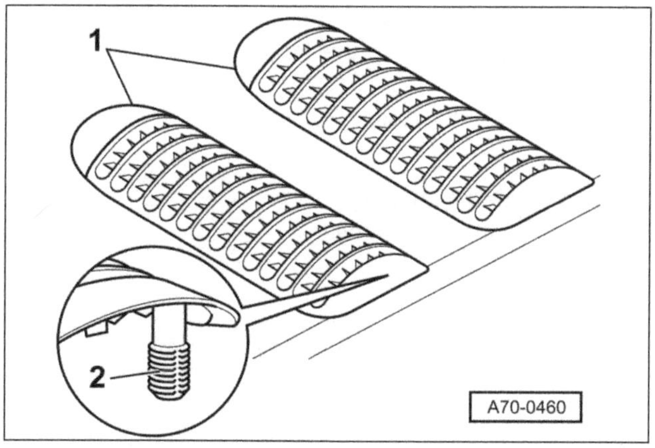

- Pull defroster vents (**1**) upwards, starting at front.
  - Mounting lugs (**2**) will break off.

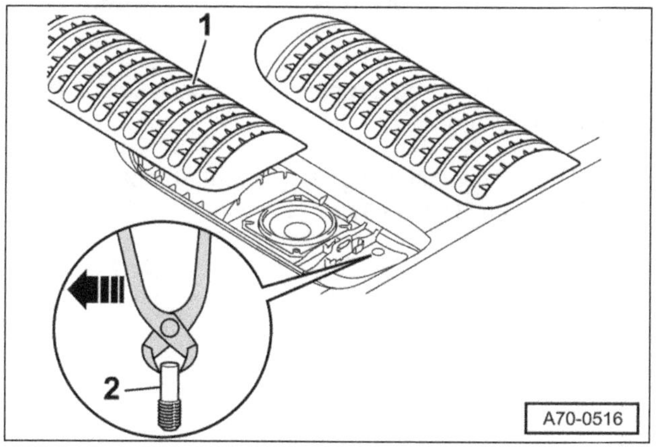

- Using a pair of nippers or equivalent, pull broken pieces (**2**) out of instrument panel using a rocking motion (**arrow**).

BentleyPublishers
.com
© 2006 BentleyPublishers.com—All Rights Reserved

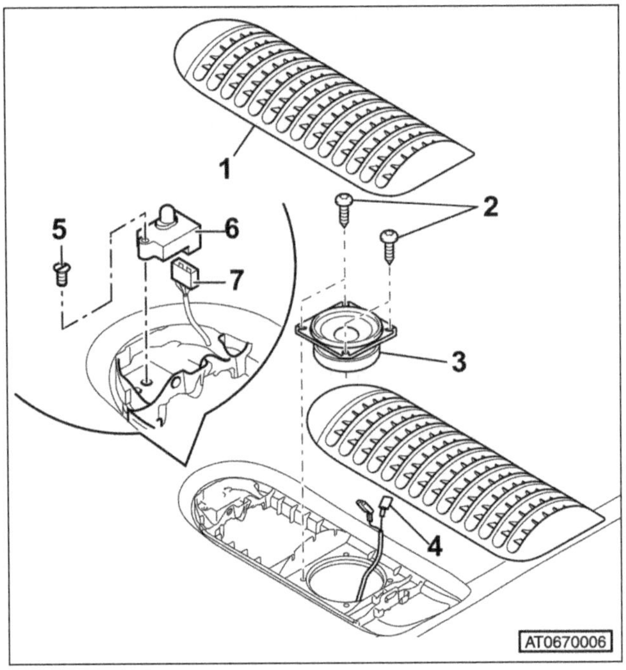

After removing vent (**1**), telephone speaker (**3**) and sunlight photo sensor (**6**) are accessible.

– **Telephone speaker (3)** (through model year 2001 only):
  - Remove screws (**2**) and unplug wiring (**4**).
  - Tighten screws to 0.8 Nm (7 in-lb)

– **Sunlight photo sensor** (**6**):
  - Remove screw (**5**) and unplug harness connector (**7**).
  - Tighten screw to 0.8 Nm (7 in-lb)

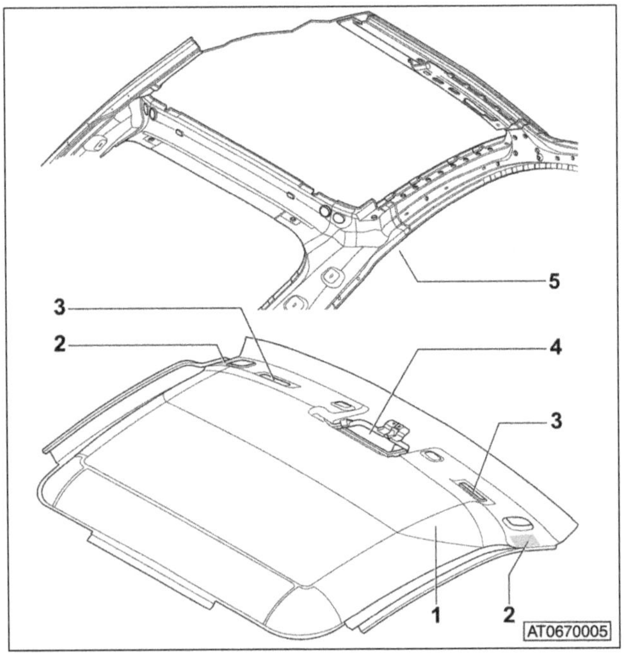

HEADLINER AND ROOF TRIM

Headliner overview

1. Molded headliner panel
   - Remove through tailgate opening
   - Remove rear headliner trim panel for access to roof-mounted antenna
2. Hook and loop (Velcro) strips
3. Make-up mirror lights
4. Light panel in molded headliner
   - Covers cut-out in molded headliner
5. Roof structure
   - View from inside

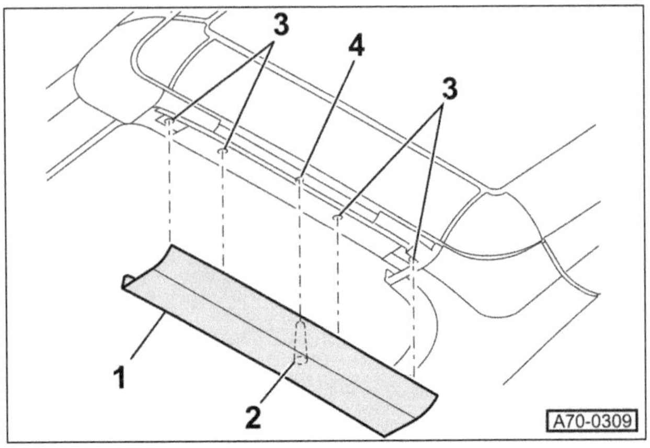

Rear headliner trim panel (coupe), removing and installing

Detach rear headliner trim panel (**1**) clips from four openings (**3**) in roof (**4**) and pull straight out.

– When installing, position locating lug (**2**) into roof (**4**) opening and snap rear headliner trim (**1**) into place.

BentleyPublishers.com

© 2006 BentleyPublishers.com—All Rights Reserved

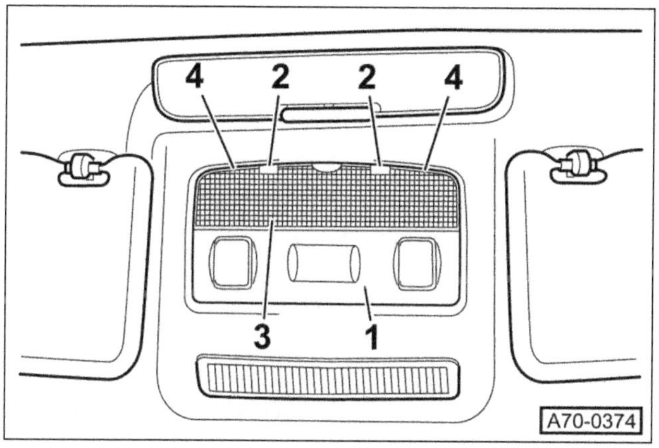

Light panel in headliner (coupe)

- Carefully pry out panel for interior light (**3**) at indentations (**2**) and pull down and out.
- Carefully pry out light panel for molded headliner (**1**) at indentations (**4**) and pull out.
- Disconnect harness connector.
- When installing, clip into position.

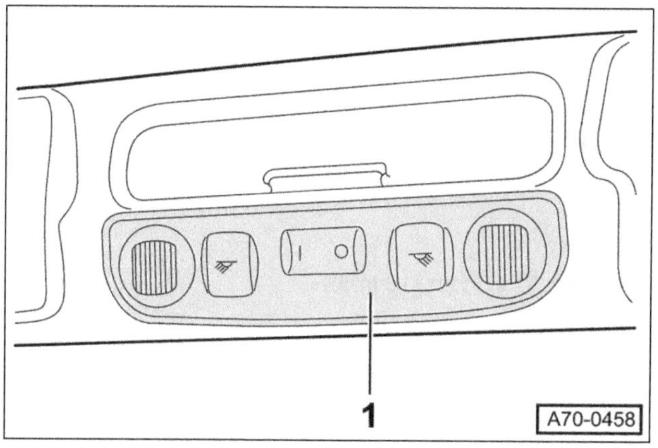

Light panel in upper roof frame (roadster)

- Release interior light (**1**) on front side.
- Disconnect harness connector.
- When installing, snap into position.

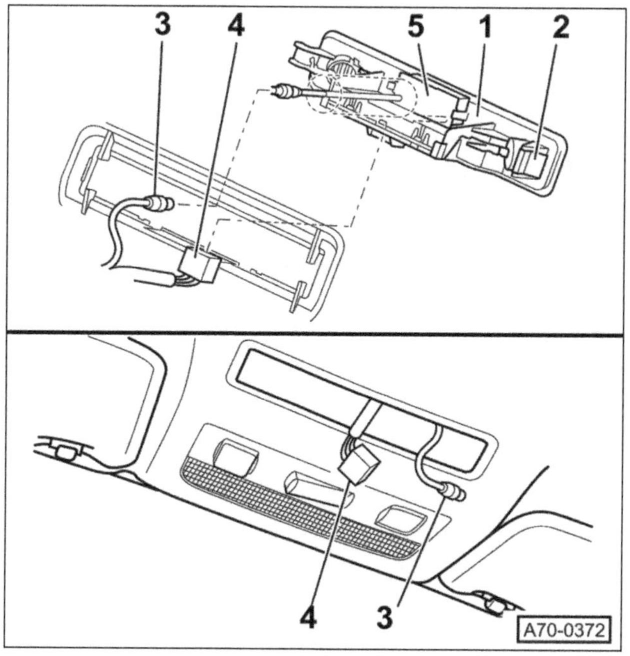

Sensor panel in headliner

- Carefully unclip sensor panel (**1**) at left side (**2**) retaining clip.
- Disconnect harness connector (**4**) for anti-theft alarm sensor.
- Disconnect harness connector (**3**) for hands-free microphone (**5**).

BentleyPublishers.com
© 2006 BentleyPublishers.com—All Rights Reserved

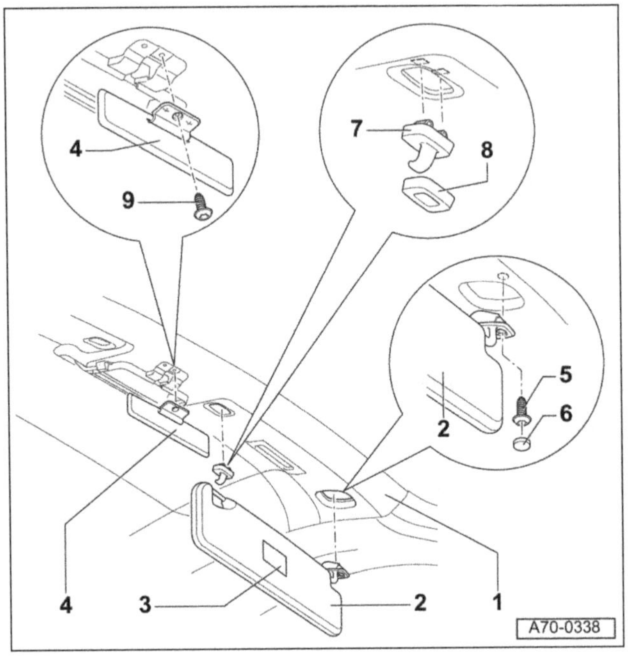

Sun visors (coupe)

1. Molded headliner
2. Sun visor
   - Removing:
     Unclip sun visor from holder (item 7)
     Remove cap (item 6)
     Remove securing screw (item 5)
3. Sticker
   - Usage varies according to market
   - Can be on left and/or right side
   - Replacement visors may come with sticker installed; if not, install new sticker to remain in compliance with regulations
   - If replacing, install in center of visor
4. Center sun visor
5. Screw, tighten to 1.5 Nm (13 in-lb)
6. Cap
   - Carefully pry off
7. Sun visor holder
   - Push firmly to the side and pivot out
8. Cap
   - Carefully pry off
9. Screw, tighten to 1.5 Nm (13 in-lb)

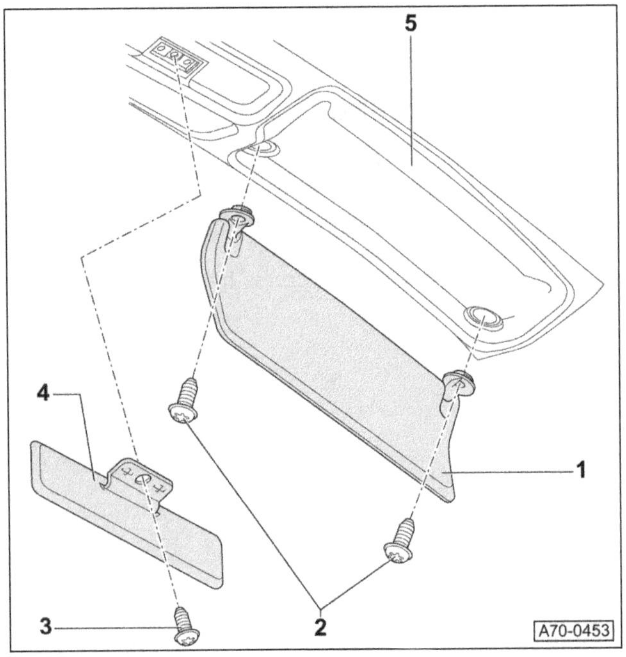

Sun visors (roadster)

1. Sun visor
   - Removing:
     Remove light panel in headliner
     Fold sun visor into open position
     Remove screws (item 2)
     Pull visor down and disconnect harness connector
   - Installation is the reverse of removal.
2. Screws, tighten to 1.5 Nm (13 in-lb)
3. Screw, tighten to 1.5 Nm (13 in-lb)
4. Center sun visor
   - Removing:
     Fold center sun visor into open position
     Remove screws (item 3)
5. Upper roof frame

BentleyPublishers
.com
©2006 BentleyPublishers.com—All Rights Reserved

Interior Trim

Lower A-pillar trim

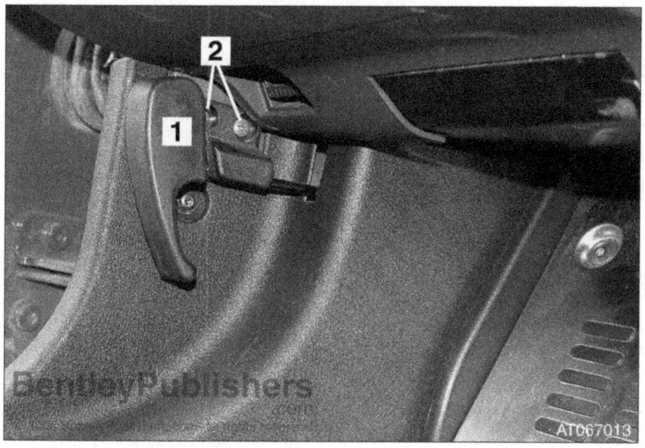

◄ Remove screws (**2**) and pull hood release handle (**1**) aside.
- Screws (**2**), tighten to 4 Nm (35 in-lb)

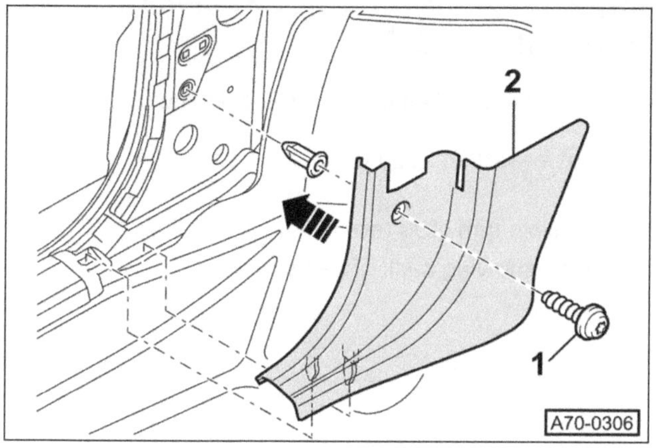

◄ Remove screw (**1**) and unclip (**arrow**) trim panel (**2**).
- Screw (**1**), tighten to 1.5 Nm (13 in-lb)

– Installation is the reverse of removal.

Foot rest and support

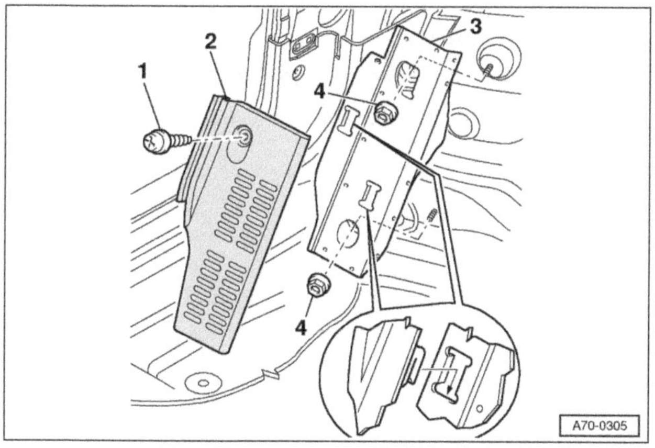

– Remove lower A-pillar trim.

◄ Remove screw (**1**).
- Tighten to 1.5 Nm (13 in-lb)

– Slide foot rest (**2**) up and out of slot.

– Fold back floor covering (carpet) around support.

– Remove both nuts (**4**).
- Tighten to 4 Nm (35 in-lb)

– Pull support bracket (**3**) out.

– Installation is the reverse of removal.

Rear center panel (roadster)

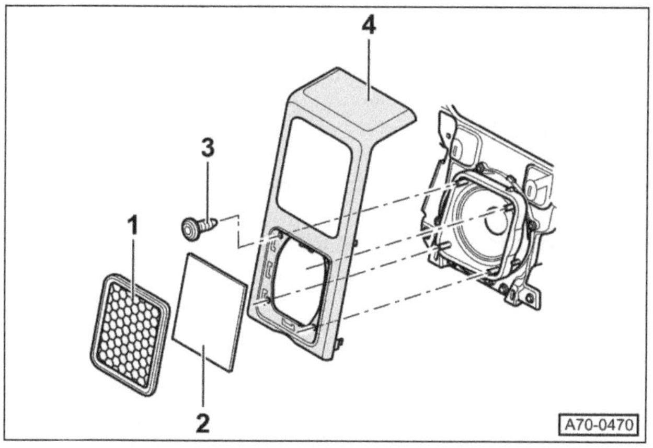

– Move both front seats forward as far as possible and fold backrests.

◄ Determine if trim panel (**4**) has center speaker installed. If no speaker is present, a blank panel (**2**) will be installed.

BentleyPublishers
.com
©2006 BentleyPublishers.com–All Rights Reserved

Vehicles with center speaker

– Use screwdriver to carefully pry out grille for center speaker.

◄ Remove four lower mounting screws (**arrows**).

Continued for all versions

– Use large screwdriver to pry center trim panel away from body.
  - There are three mounting clips on each side of center trim panel.
  - It may be helpful to reach inside center trim panel upper storage compartment and pull trim panel forward.

– Install in reverse order of removal.

Rear side panels (roadster)

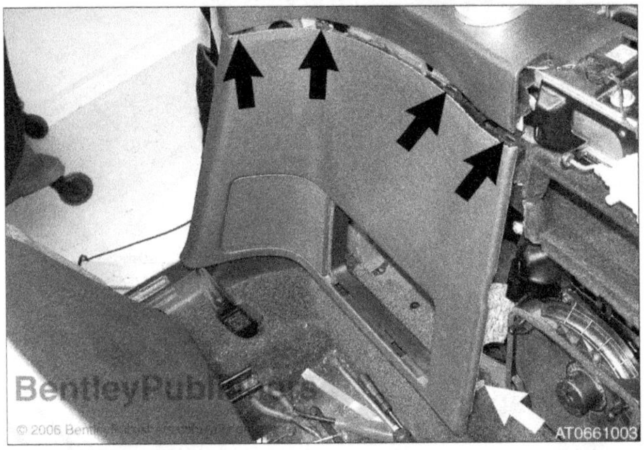

– Remove rear center trim panel as described previously.

◄ Pull rear side panel forward at bottom and then downward from rollbar trim.
  - Note location of visible trim clips (**arrows**) in illustration.
  - Also see illustration showing back of panel below.

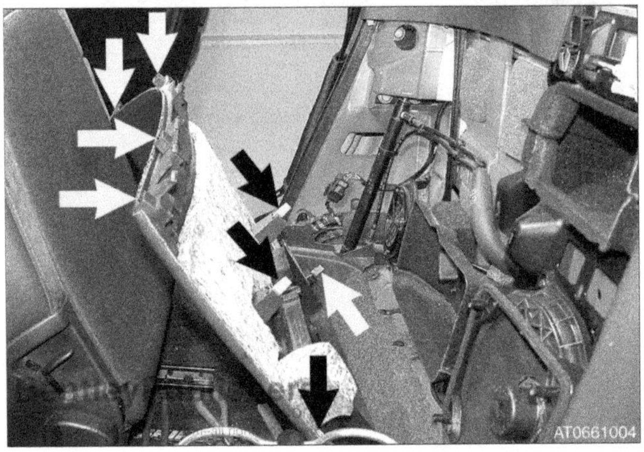

◄ Install in reverse order of removal.
  - Note location of retaining clips (**arrows**) on back of panel.
  - Make sure that transition to door seal and B-pillar trim is smooth.

Sill panel trim

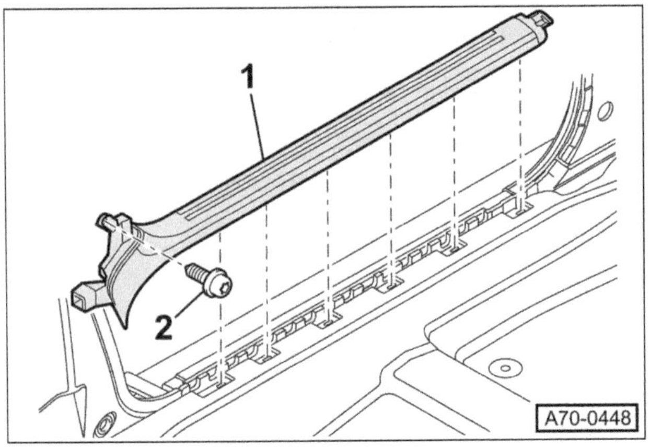

– Remove lower A-pillar trim.

– Remove rear side trim panel.

◄ Remove screw (**2**), as applicable (roadster models).

– Unclip six retaining clips and pivot sill panel (**1**) out and off.

– Installation is the reverse of removal.

BentleyPublishers.com
©2006 BentleyPublishers.com—All Rights Reserved

Door Trim

Door trim overview

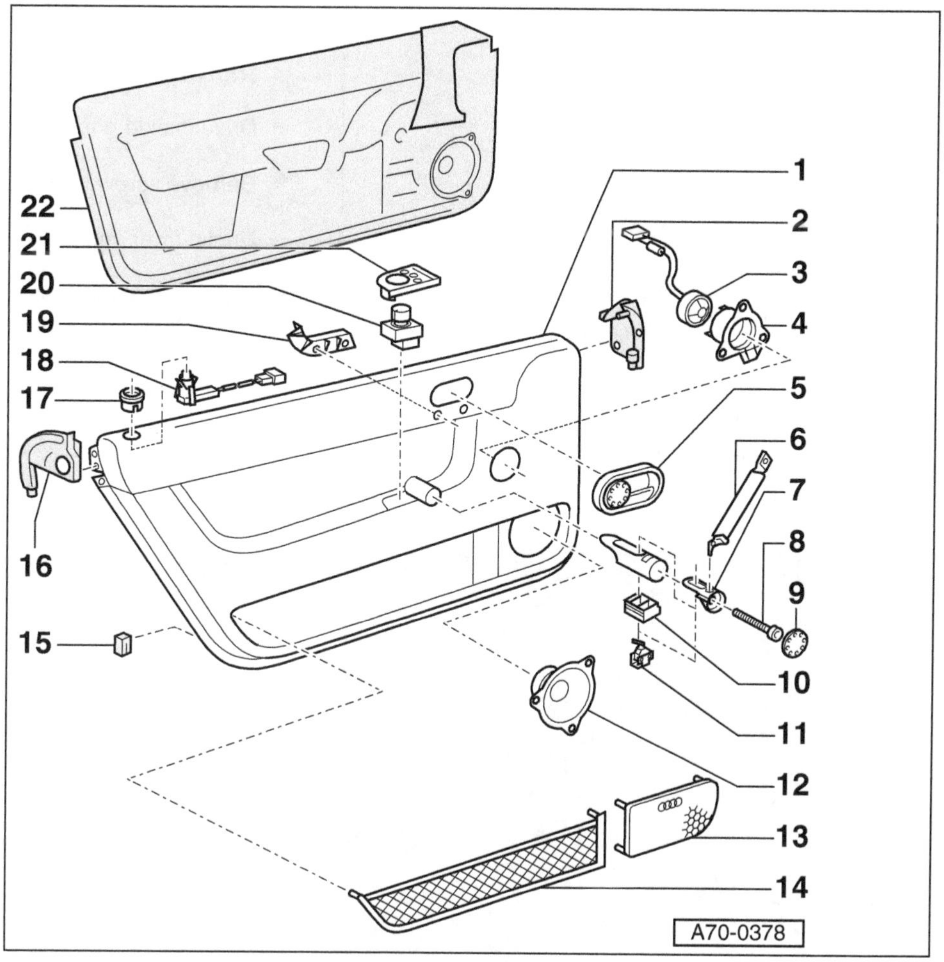

1. **Door panel**
   - See **Door panel, removing and installing**
   - See **Door panel wire routing**
2. **Front connecting piece**
   - Seal
3. **Treble speaker**
   - Removing and installing, see **91 Radio and Communication**
4. **Speaker mounting**
5. **Inside door release lever**
   - See **Inside door release lever**
6. **Inside door grab handle**
   - See **Grab handle and power window switches**
7. **Door switch trim**
   - See **Door handle and switch trim**
8. **Screw**
9. **Screw trim cap**
   - See **Screw trim cap, removing and installing**
10. **Power window switch trim**
11. **Power window switch unit**
    - See **Grab handle and power window switches**
12. **Bass speaker**
    - Removing and installing, see **91 Radio and Communication**
13. **Bass speaker trim**
    - See **Bass speaker trim**
14. **Door pocket**
    - See **Door pocket**
15. **Clip locator**
16. **Rear connecting piece**
    - Seal
17. **LED indicator light trim and mount**
18. **LED indicator light**
    - For alarm system
    - See **LED indicator, replacing**
19. **Inside door release mounting**
20. **Mirror adjustment switch**
    - See **Power mirror switch**
21. **Mirror adjustment switch trim**
22. **Insulation**
    - Form-fitted and attached to inside to door panel
    - See **Insulation, removing and installing**

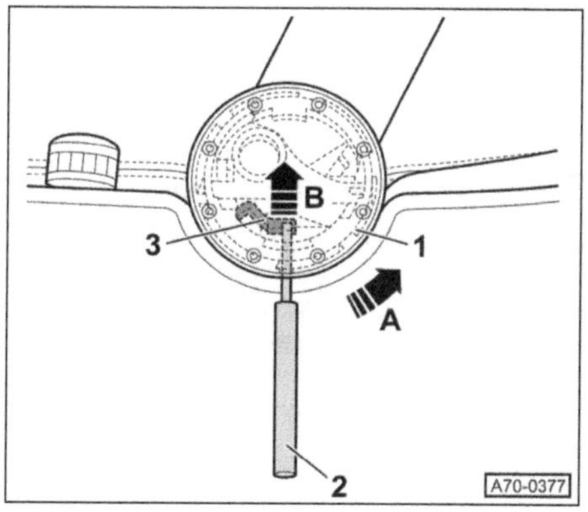

Screw trim cap, removing and installing

- Insert a small drift (**2**) or pin (2 mm or 5/64 in.) through hole in bottom section of trim cap (**1**).
- Use drift to push tab (**3**) slightly upwards while at the same time turning trim cap (**1**) counter-clockwise (**arrow**).
- Trim cap (**1**) should pull off.
- Installation is the reverse of removal.

BentleyPublishers.com
© 2006 BentleyPublishers.com—All Rights Reserved

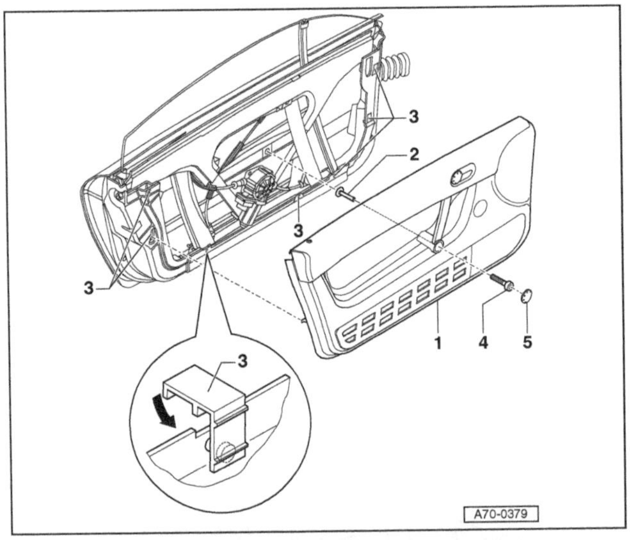

Door panel, removing and installing

Removing

- Driver's side door panel (**1**) shown, passenger's side is similar.
- ◀ Remove screw trim cap (**5**) and screw (**4**).
- Unclip front and rear connecting pieces.
- During removal, note position of clip locators (**3**) and spacer sleeve (**2**) for door panel re-installation.

◀ Lift door panel upwards approximately 20 cm (4 inches) and off of securing lugs (**arrows**) and pull out slightly to allow access to inside door release lever and cable.

NOTE—

- *Number and type of securing lug varies by model year and body type.*

◀ Pull release cable out of guides (**B**).

- Unhook end of release cable (**A**) from eyelet on handle.

NOTE—

- *Hook end of release cable (**arrow**) faces up.*

◀ Lift door panel up and away from door frame a sufficient distance to allow access to harness connectors. Disconnect power window switches, mirror switch, speaker(s), and LED warning light as applicable and ensure that harness connectors are clear of door panel (**circle**).

- Lift door panel up and off vehicle.

BentleyPublishers
.com
©2006 BentleyPublishers.com—All Rights Reserved

Installing

- Installation is the reverse of removal noting the following points.
- Ensure that all switches and speaker(s) are properly installed.
- Ensure that insulation is properly positioned on door panel.
- Connect all harness connectors.
- Ensure clip locators and spacer sleeves are installed.

Hook end of release cable (**A**) onto handle eyelet and snap cable into guide (**B**).

NOTE—

• Install hooked end of cable upwards.

Position door panel against door frame and slightly high so that retaining lugs (**arrows**) can engage slots in door frame. Retaining lugs are along bottom and both sides.

- Push door panel down and into position.

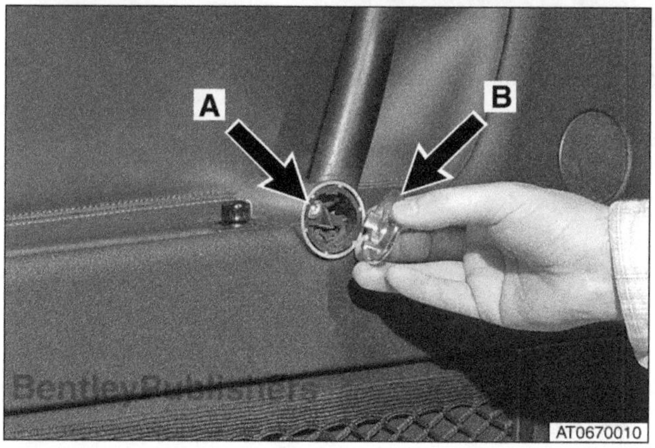

Install screw (**A**) and tighten to 7.5 Nm (66 in-lb)

- Snap trim cap (**B**) into place.
- Install front and rear connecting pieces.
- Check for proper operation of door and electrical components.

Door panel wire routing

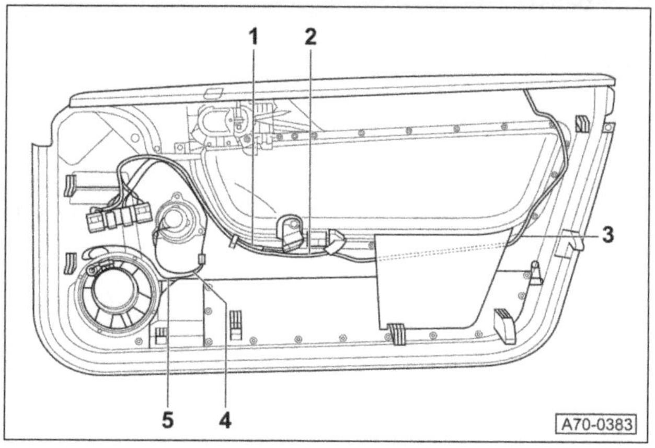

When working with door panel, route wiring as shown to avoid interference with window mechanism and to ensure sufficient length.

1. Power window switch wiring
2. Power mirror switch wiring
3. LED indicator light wiring (driver's side only)
4. Treble speaker wiring
5. Bass speaker wiring

BentleyPublishers
.com
©2006 BentleyPublishers.com—All Rights Reserved

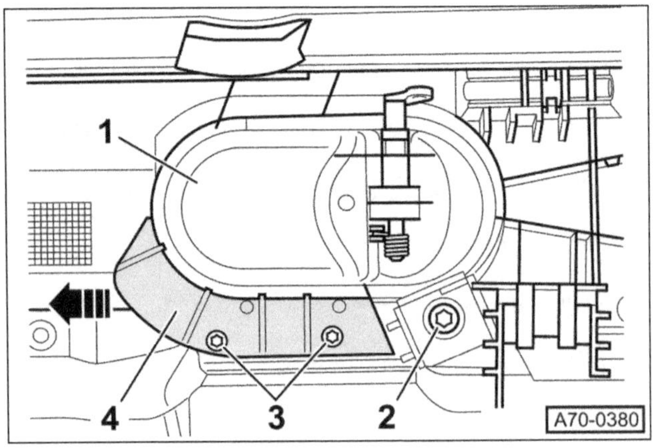

Inside door release lever

– Remove door panel.

– Detach insulation from door panel.

< Remove screws (**3**). Screw (**2**) remains in place unless removing grab handle and power window switches.

– Pull out retainer (**4**) in direction of **arrow**.

– Press out bottom of door handle (**1**).

– Installation is the reverse of removal.

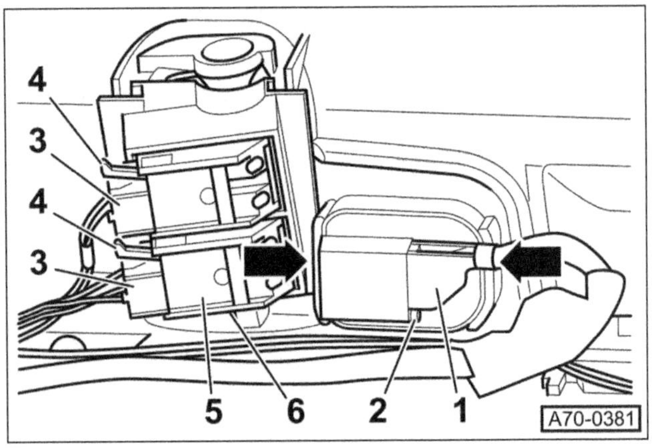

Grab handle and power window switches

– Driver's side is shown, passenger's side is similar.

– Remove inside door release lever including grab handle screw.

< Press retainer hooks (**4**) and disconnect both wiring harness connectors (**3**).

– Using a small screwdriver, carefully pry out both power window switches (**5**) at all four holding points (**6**) and remove.

- Retainer and connector (**1** and **2**) remain in place unless removing power mirror switch.

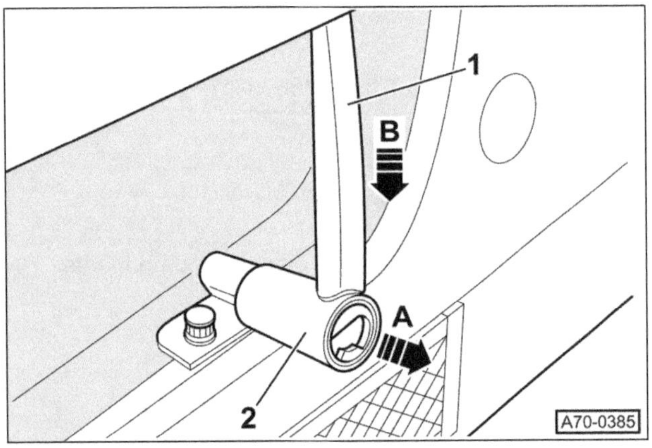

< Slide out (**arrow A**) switch trim (**2**) with grab handle (**1**).

– Pull grab handle (**1**) down (**arrow B**) and off.

– Installation is the reverse of removal.

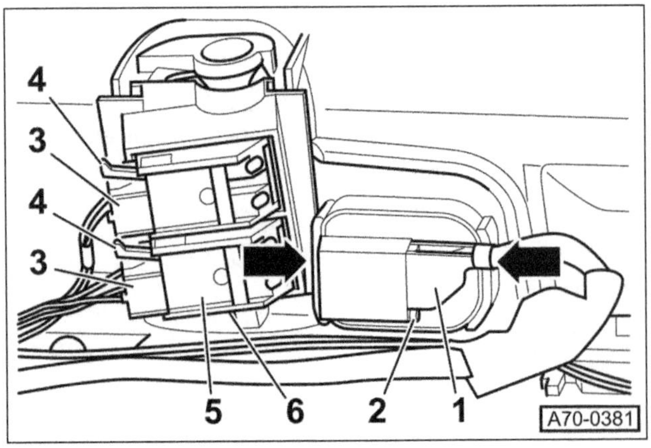

Power mirror switch

– Remove door panel.

– Detach insulation from door panel.

– Remove inside door release lever.

< Disconnect harness connector (**1**) by pressing retainer hook (**2**).

- Additional items (**3, 4, 5, 6**) are for power window switches and do not need to be disturbed for this operation.

– Depress spring clips (**arrows**) and remove trim plate.

BentleyPublishers.com

© 2006 BentleyPublishers.com—All Rights Reserved

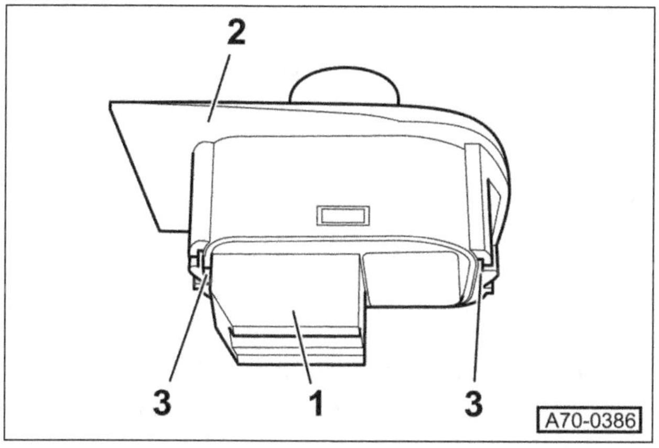

◀ Using a screwdriver, carefully pry trim plate (**2**) away from switch (**1**) at front and back (**3**).

– Installation is the reverse of removal.

Bass speaker trim

If bass speaker trim will be replaced, clips will be required to secure to door panel. If trim is to be re-used, either screws or a soldering iron with appropriate tips will be needed.

– Remove door panel.

– Detach insulation from door panel.

– Remove bass speaker.

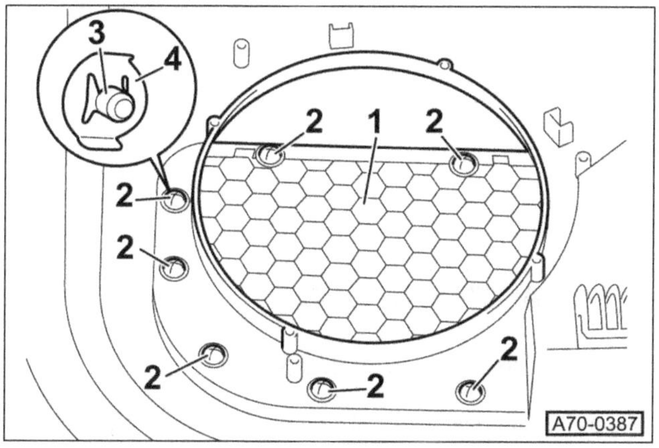

◀ Working from the rear of the speaker trim, proceed as follows:

- **Replace trim** - drill out all seven plastic welds (**2**) and push trim (**1**) out from the rear.
  Instal speaker trim from front (**inset**) and attach pins (**3**) with new clips (**4**).
- **Re-using original** - cut off all seven plastic pins (**3**) down to level of door panel from the rear.
  Secure original speaker trim by using screws or by carefully welding plastic pins with a soldering iron.

– Install bass speaker.

– Install insulation onto door panel.

– Install door panel.

LED indicator, replacing

– Remove door panel.

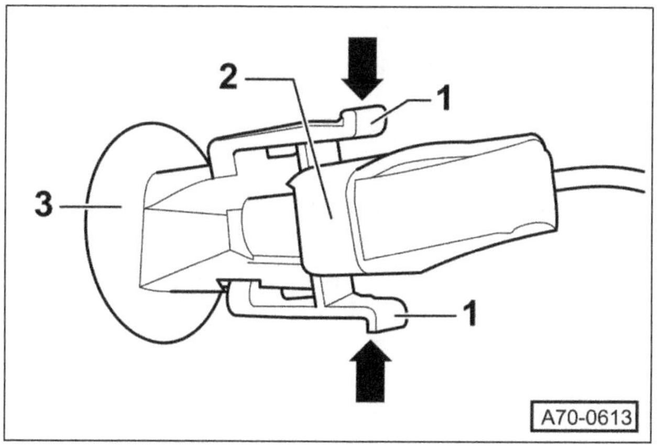

◀ Press retaining tabs (**1**) together (arrows) and pull LED (**2**) out of holder (**3**).

– Cut original LED wiring and splice in new LED and wiring using appropriate components from wiring kit VAS 1978, or equivalent, taking care to match wiring color codes.

– Check for proper operation before completing re-assembly.

– Installation is the reverse of removal.

NOTE—

- *Use care to avoid loosening or damage to the felt pad on rear of door panel.*

BentleyPublishers.com
© 2006 BentleyPublishers.com—All Rights Reserved

Door pocket

If door pocket will be replaced, clips, screws, and washers will be required to secure to door panel. If pocket is to be re-used, a soldering iron with appropriate tips will be needed.

– Remove door panel.

– Detach insulation from door panel.

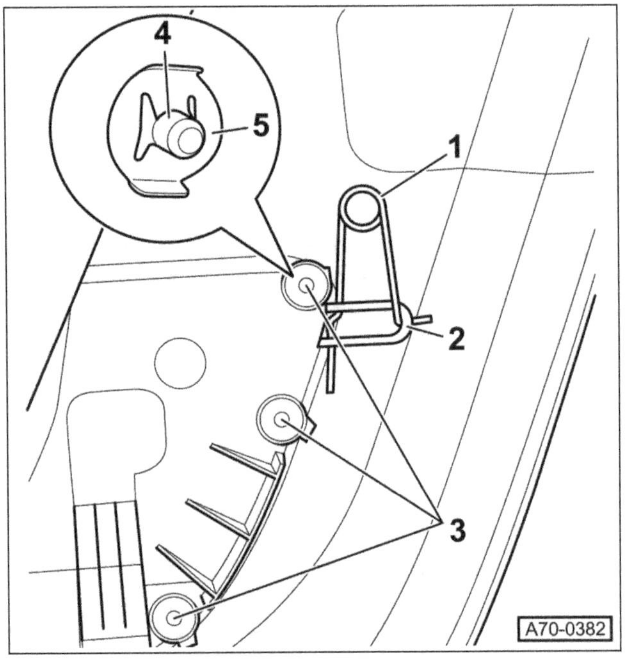

◄ Unclip spring (**1**).

– Working from the rear of the door panel, proceed as follows:

- **Replace pocket** - drill out all eleven plastic welds (**3**)
  Pull link (**2**) forward and out
  Push pocket out from the rear.
  Push retaining pins (**4**) out of new pocket
  Install new pocket from the front and secure with new clips (**5**)
  Connect link (**2**)
  Install pins (**4**) and push in until flush with door panel (**inset**)
  Lock in place with screws and washers
- **Re-using original** - drill out all eleven plastic welds (**3**)
  Pull link (**2**) forward and out
  Push pocket out from the rear.
  Install original pocket
  Connect link (**2**)
  Secure original trim by carefully welding plastic pins with a soldering iron.

– Connect spring (**1**).

– Install insulation onto door panel.

– Install door panel.

Rear Compartment Trim

Rear cross panel trim (coupe)

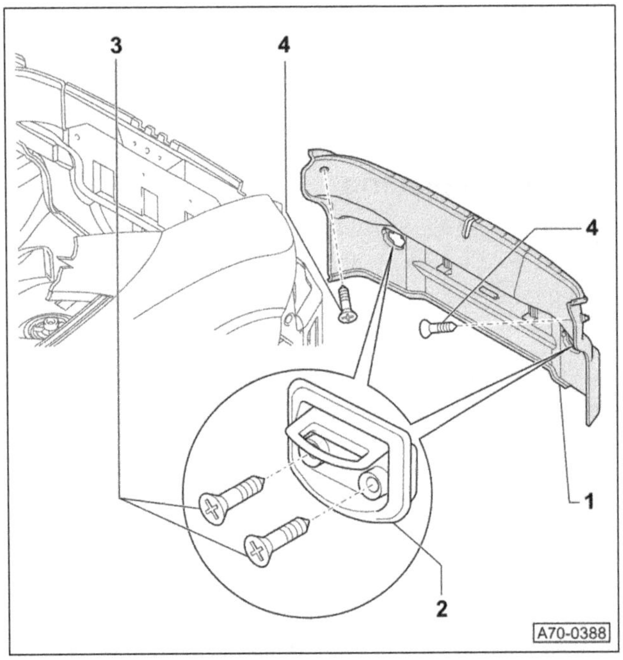

◄ Rear cross panel trim (**4**) is installed over rear metal cross panel and finishes rear compartment.

1. Rear cross panel trim
   - Removing:
     Remove screws (**3**) from tie down loops (**2**) on each side
     Remove both securing screws (**4**)
     Unclip rear cross panel trim (**1**) and lift upward to remove
   - Installation is the reverse of removal
     Upon completion of installation, ensure that seal is installed correctly
2. Tie down loops
3. Screws
   - Screws, tighten to 4.5 Nm (40 in-lb)
4. Screws
   - Screws, tighten to 2.0 Nm (18 in-lb)

BentleyPublishers.com
©2006 BentleyPublishers.com—All Rights Reserved

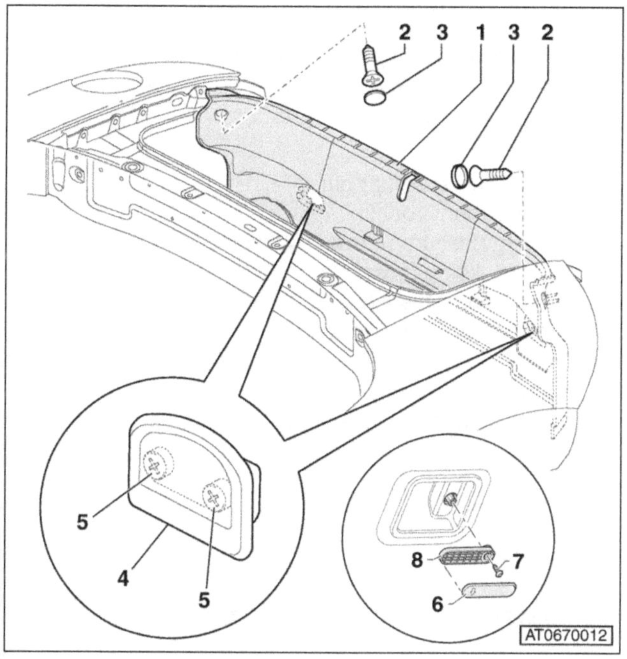

Rear cross panel trim (roadster)

◄ Rear cross panel trim is installed over rear metal cross panel and finishes rear compartment.

1. Rear cross panel trim
   - Removing:
     Carefully pry off both trim caps (**3**)
     Remove both securing screws (**2**)
     Open covers and remove screws (**5**) from tie down loops (**4**) on each side
     Where equipped (**inset**), remove emergency release handle cover (**6**), screw (**7**), and pull off handle (**8**).
     Unclip rear cross panel trim (**1**) and lift upward to remove
   - Installation is the reverse of removal
     Upon completion of installation, ensure that seal is installed correctly
2. Screws
   - Screws, tighten to 2.0 Nm (18 in-lb)
3. Trim caps
4. Tie down loop
5. Screws
   - Screws, tighten to 4.0 Nm (35 in-lb)
6. Cover
7. Screw
8. Emergency release handle

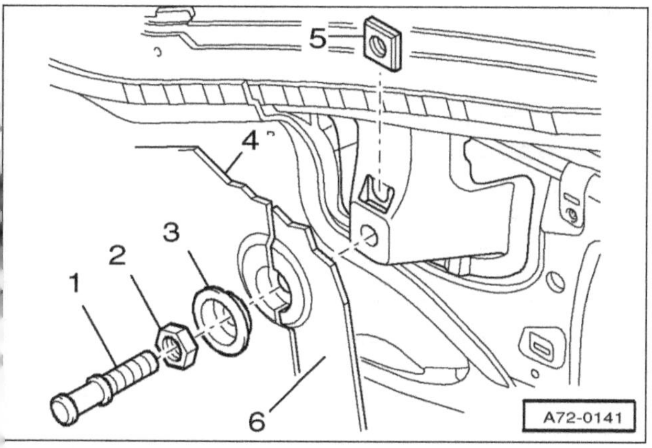

Rear seat backrest striker pin (coupe),

◄ Note exposed length of striker pin (**1**) above locking nut (**2**) before disassembly.

– Loosen locking nut (**2**) and remove striker pin (**1**).

– When installing:
  - Ensure proper position of threaded nut plate (**5**) in body bracket.
  - Ensure proper side panel trim and bezel placement (**3**, **4**, **6**).

– Adjust length of striker pin so that seat backrest latch is fully engaged.

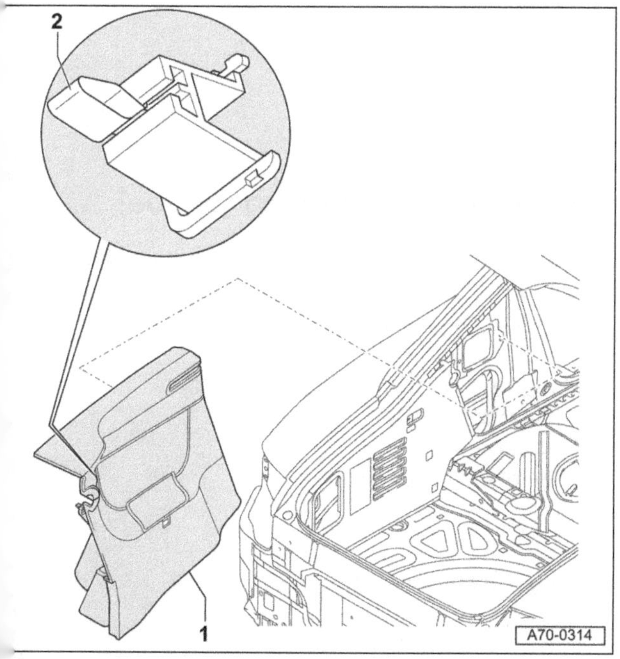

Rear side trim (coupe)

◄ Rear side panel trim covers inner structure and finishes back seat area. Left side shown, right side is similar.

1. Rear side trim
   - Removing:
     Remove rear seat cushion and backrest
     Remove outer seatbelt anchorage point
     Remove rear seat backrest striker pin
     Starting from top rear, unclip eight rear side trim clips (**2**)
   - Installation is the reverse of removal
2. Clips

BentleyPublishers.com
©2006 BentleyPublishers.com—All Rights Reserved

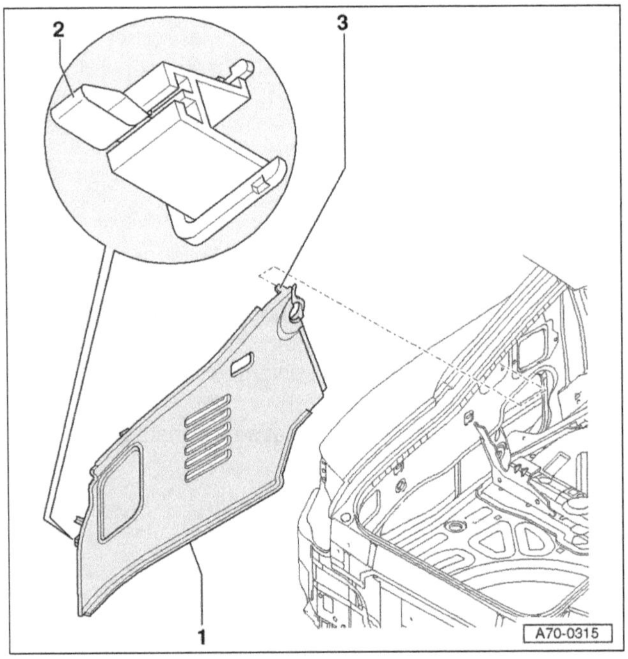

Luggage compartment side trim panel (coupe)

Rear luggage compartment side trim panel covers inner fenders and finishes rear compartment. Left side shown, right side is similar.

1. Luggage compartment side trim
   - Removing:
     Remove rear seat backrest striker pin
     Remove rear compartment floor lining
     Remove left side luggage compartment light by carefully prying out from the rear and sliding forward
     Starting from top rear, unclip six rear side trim clips (**2**)
     Pull luggage compartment side trim panel (**1**) out of retaining tab (**3**) and out
   - Installation is the reverse of removal
     Upon completion of installation, ensure that seal is installed correctly
2. Clips
3. Retainer tab

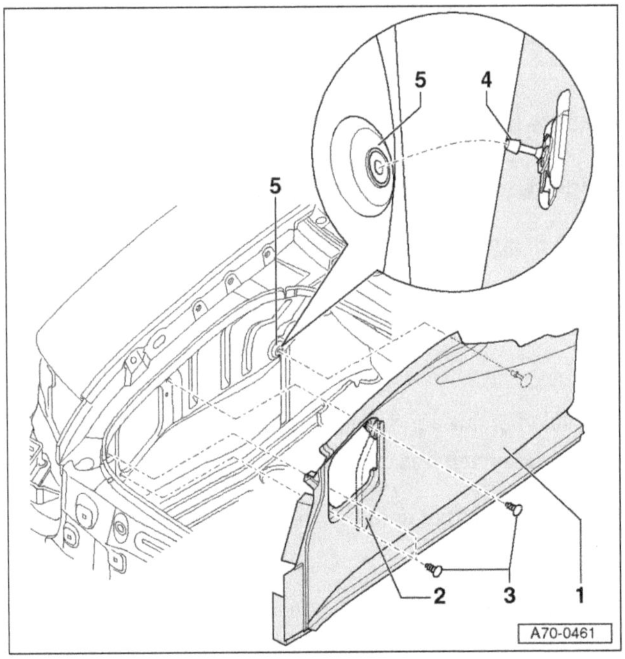

Luggage compartment side trim (roadster)

Rear side panel trim covers inner fenders and finishes rear compartment. Left side shown, right side is similar.

1. Rear side trim
   - Removing:
     Remove rear compartment floor lining
     Remove rear cross panel trim
     Open cover (**2**)
     Remove three clips (**3**)
     Pull panel slightly inward and remove inside clip (**4**) from retainer (**5**)
     Pull rear side trim panel (**1**) up and out
   - Installation is the reverse of removal
2. Cover (door)
3. Clips
4. Clip
5. Mount

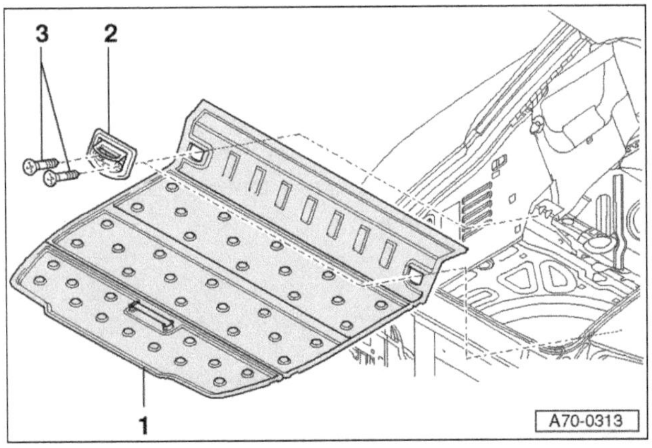

Luggage compartment floor lining (coupe)

Luggage compartment floor lining finishes rear compartment and acts as a sound absorber.

1. Luggage compartment floor lining
   - Removing:
     Remove four bolts (**3**) from both tie down loops (**2**)
     Fold down rear seat backrests
     Pull forward to remove
   - Installation is the reverse of removal
2. Tie down loops
3. Screws
   - Screws, tighten to 4.5 Nm (40 in-lb)

BentleyPublishers.com
© 2006 BentleyPublishers.com—All Rights Reserved

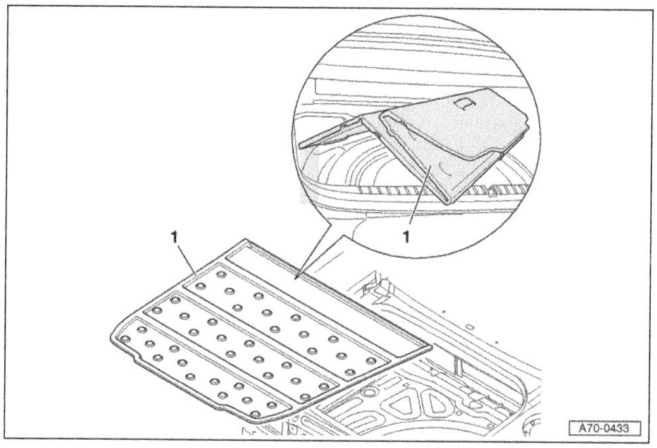

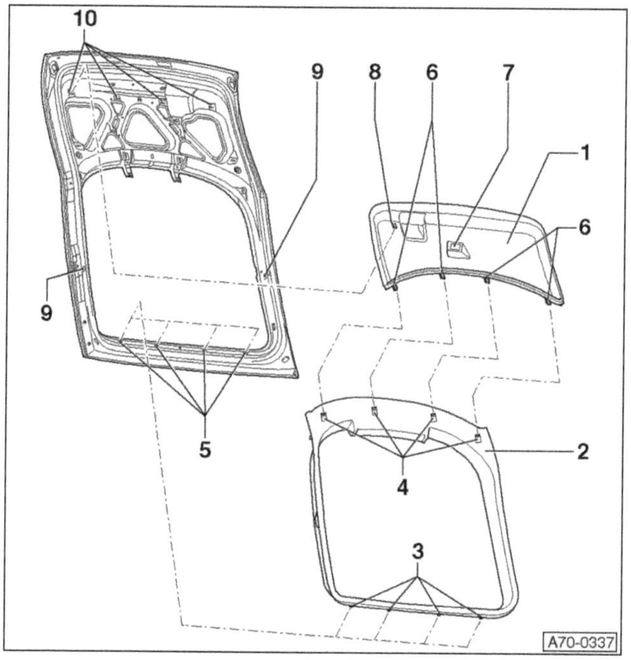

Luggage compartment floor lining (roadster)

◀ Luggage compartment floor lining finishes rear compartment and acts as a sound absorber.

1. Luggage compartment floor lining
   - Removing:
     Disengage lining at front Velcro fastening
     Fold lining as shown (**inset**)
     Rotate panel and remove to the rear
   - Installation is the reverse of removal

Tailgate (hatch lid) trim, overview

◀ Upper and lower sections of tailgate trim panel can only be removed together. Separate both sections after removal.

1. Lower lid trim panel
2. Upper lid trim panel
3. Clip
   - Qty: 8
   - Pull lower trim slightly away from tailgate and use screwdriver to release
4. Mounting points, upper lid
   - Qty.4
5. Mounting points, tailgate
   - Qty: 8
6. Retaining tabs
   - Qty: 4
7. Light
8. Clips
   - Qty: 9
   - Pull lower trim slightly away from tailgate and use screwdriver to release
9. Fasteners
   - Qty: 2
10. Mounting points
    - Qty: 1

© 2006 BentleyPublishers.com—All Rights Reserved

72 Seats

GENERAL

This repair group covers removing and installing the front and rear seats and related components. Both driver and passenger seats are equipped with airbags mounted in the seat backrest frames.

- For airbag precautions and handling procedures, see **69 Seatbelts, Airbags**.

> ***WARNING—***
> - *Do not install aftermarket upholstery on any vehicle equipped with seat mounted airbags. Installation of non-factory upholstery including "beads" and "sheepskins" may cause seat mounted airbags to deploy when they are not supposed to; fail to deploy when they should; or deploy in some manner other than designed. Serious injury or death could result.*

FRONT SEATS

Front seat rail trim covers, removing and installing

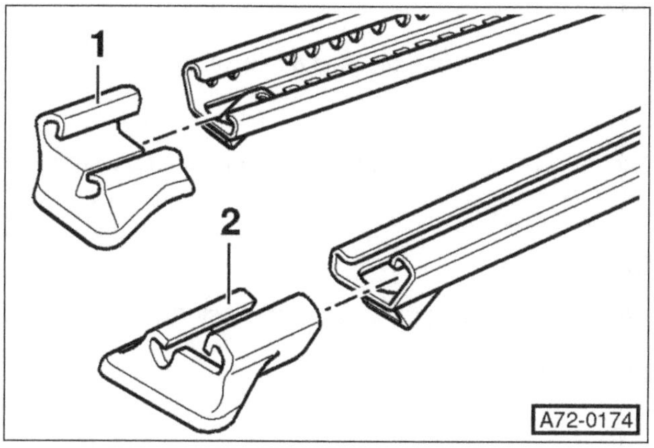

Remove trim covers (**1**, **2**) from front of seat rails (tracks) for access to bolts securing seat to floor. Covers are clipped onto rails; pull forward to remove. Left side shown, right side is similar.

- Inboard trim cover (**1**)
- Outboard trim cover (**2**)

BentleyPublishers.com

© 2006 BentleyPublishers.com—All Rights Reserved

Front seat, removing and installing

NOTE—

- *Left seat shown, right seat is similar.*
- *Several different seat types are installed including manually adjustable, electrically adjustable, and sport type seats. Removal and installation procedures vary slightly, but all versions are similar.*

> *WARNING—*
> - *Observe all airbag warnings. See* **69 Seatbelts, Airbags**.
> - *Before disconnecting the battery, read the warnings and cautions in* **27 Battery, Alternator, Starter**.

Removing

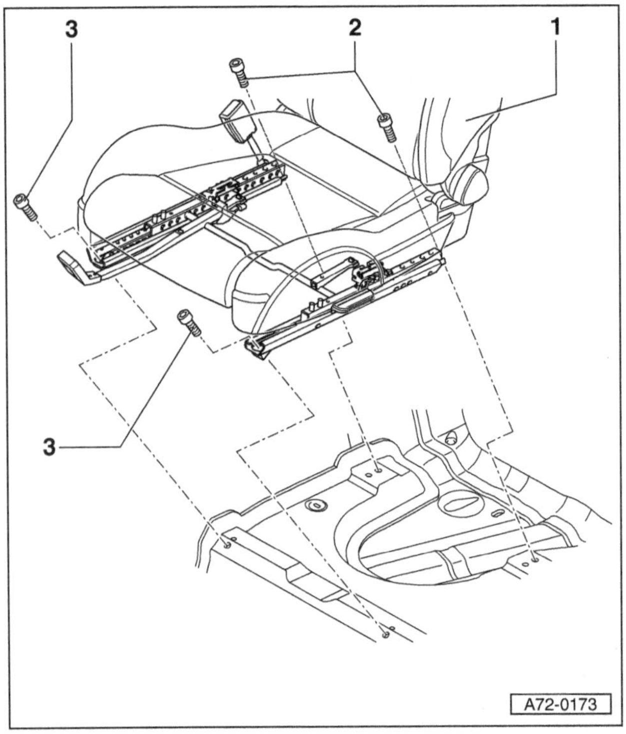

- Adjust seat (**1**) as far forward as possible.
- Remove both rear mounting bolts (**2**).
- Adjust seat as far rearward as possible.
- Remove front seat rail trim covers.
- Remove both front mounting bolts (**3**).
- Disconnect battery negative (–) terminal. See **27 Battery, Alternator, Starter**.
- Tilt seat back and disconnect side airbag harness connectors. See **Side airbag connector, disconnecting**.

> *WARNING—*
> - *Electrostatic charges may result in involuntary deployment of the airbag. Before disconnecting harness connector, discharge static electricity by briefly grasping the chassis or door striker.*
> - *Failure to install VAS 5094 could result in deployment of the side airbag(s).*

- Install adapter VAS 5094 to airbag harness connector on seat and leave in place whenever seat is out of vehicle. If both seats are removed from vehicle, use two adapters. See **Connecting airbag adapter VAS 5094**
- Depending on vehicle equipment, disconnect remaining seat wiring harnesses from appropriate connector station.
- Remove seat from vehicle and place in a safe location.

NOTE—

- *Depending on seat version, removal is easier if seat backrest is folded.*

Installing

- Battery negative (–) terminal remains disconnected.
- Install seat into vehicle.

BentleyPublishers.com
© 2006 BentleyPublishers.com—All Rights Reserved

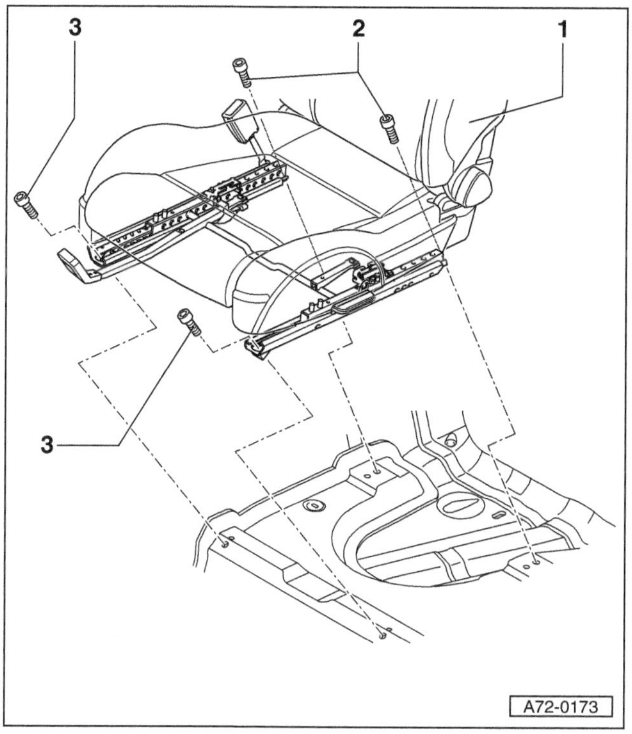

- Depending on vehicle equipment, connect seat wiring harnesses at appropriate connector station.
- Remove adapter VAS 5094 from airbag harness connector on seat and connect seat airbag harness connector to vehicle harness connector.

◀ Install both front mounting bolts (**3**).

Tightening torque	
Front seat to body	40 Nm (30 ft-lb)

- Install front seat rail trim covers.
- With battery disconnected, switch ignition on.
- With ignition on and vehicle unoccupied, reconnect battery negative (–) terminal. See **27 Battery, Alternator, Starter**.

> ***WARNING —***
> *• Make sure no one is in the vehicle*

- Adjust seat (**1**) as far forward as possible.
- Install both rear mounting bolts (**2**) .

Tightening torque	
Front seat to body	40 Nm (30 ft-lb)

- Check operation and function of electrical components that may be affected by disconnecting and reconnecting battery. See **27 Battery, Alternator, Starter**.

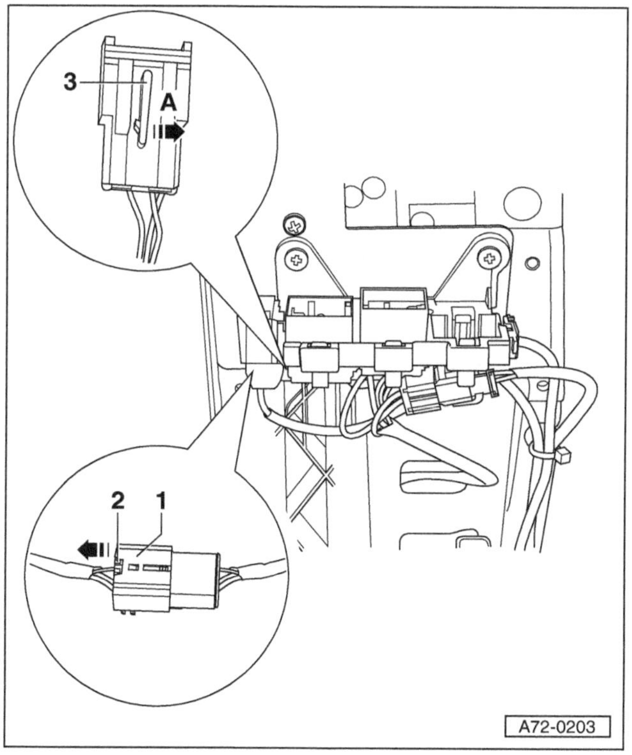

Airbag connector, disconnecting

- Remove seat.

> ***WARNING —***
> *• Electrostatic charges may result in involuntary deployment of the airbag. Before disconnecting harness connector, discharge static electricity by briefly grasping the chassis or door striker.*
> *• Failure to install VAS 5094 could result in deployment of the side airbag(s).*

◀ Working from under seat, pull retainer tab (**3**) on vehicle harness connector with one finger in direction of arrow (**A**). At the same time, pull seat harness connector out to the rear with the other hand (**upper inset**).

- Depress retainer tab (**2**) and pull harness connector (**1**) apart (**arrow, lower inset**).

BentleyPublishers.com
© 2006 BentleyPublishers.com—All Rights Reserved

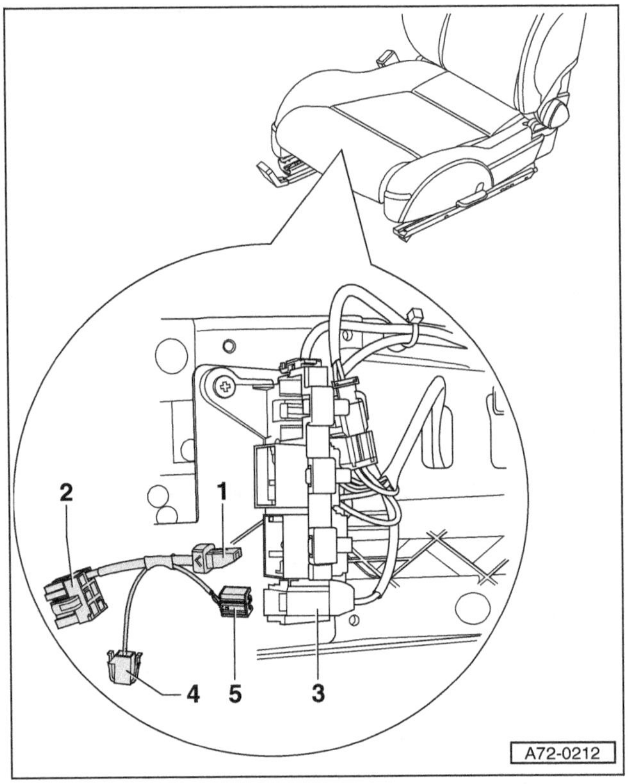

Connecting airbag adapter VAS 5094

- Before removing seat from vehicle, install adapter VAS 5094.
- < Connect igniter wire (**3**) from airbag electronic control module to adapter connector (**5**).
- Disconnect harness connector on vehicles with heated seat.
- Adapter VAS 5094 must remain on airbag harness connector on seat and be in place whenever seat is out of vehicle. If both seats are removed from vehicle, use two adapters
- Unused harness connectors (**1**, **2**, **4**) on adapter VAS 5094 are used for other vehicles and applications.

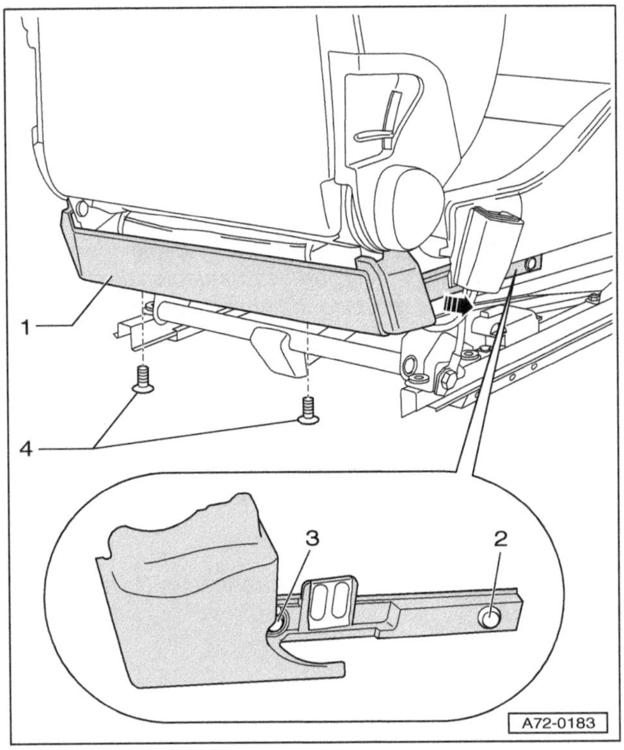

Spring trim cover, removing and installing

- Remove seat.
- < Remove both screws (**4**).
- Using a small screwdriver, carefully pry out clips (**2** and **3**).
- Move seat belt buckle aside (**arrow**) and pull off spring trim cover (**1**) out towards the rear.
- Installation is the reverse of removal.

Tightening torque	
Spring trim cover to frame (**4**)	2.5 Nm (22 in-lb)

BentleyPublishers.com
© 2006 BentleyPublishers.com—All Rights Reserved

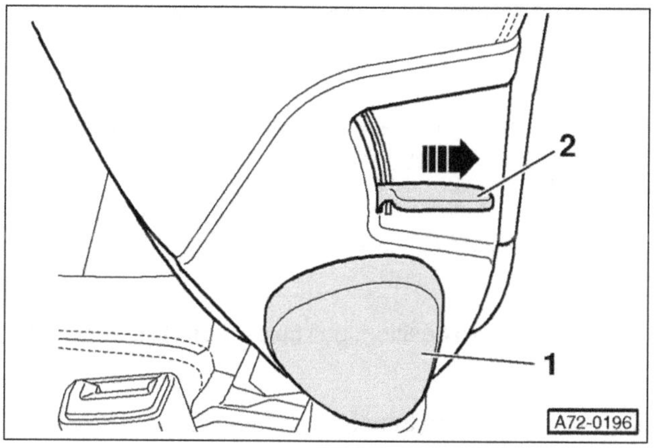

Seat backrest knobs, removing and installing

◀ To remove seat backrest adjuster knob (**1**), pull straight out and off.

– To remove backrest release (**2**), pull straight back (**arrow**) and off of metal mounting tab.

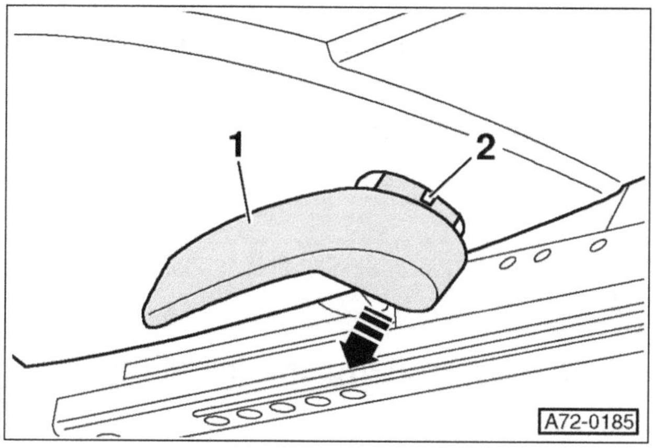

Seat height adjuster knob, removing and installing

◀ Using a screwdriver, depress retainer tab (**2**) while at the same time pulling handle (**1**) off (**arrow**).

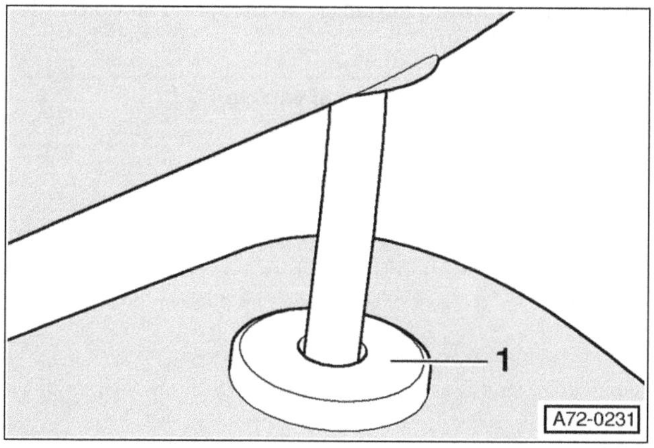

Headrests, removing and installing

Depending on seat version, headrest can be removed.

◀ Carefully pry off left and right covers (**1**) on backrest.

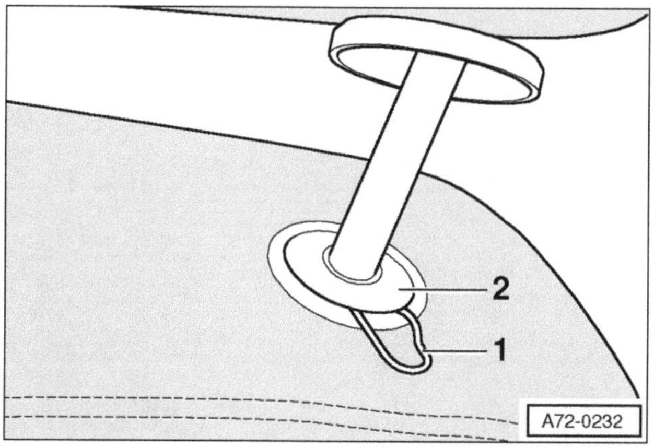

◀ Using a small screwdriver, carefully pull retaining clips (**1**) out of headrest guide tubes (**2**) on left and right side of each seat.

– Slide headrest up and out of guide tubes in backrest.

– When installing, bow in retaining clips (**1**) must face towards outside of seat.

BentleyPublishers
.com
© 2006 BentleyPublishers.com—All Rights Reserved

Headrest guide tube, removing and installing

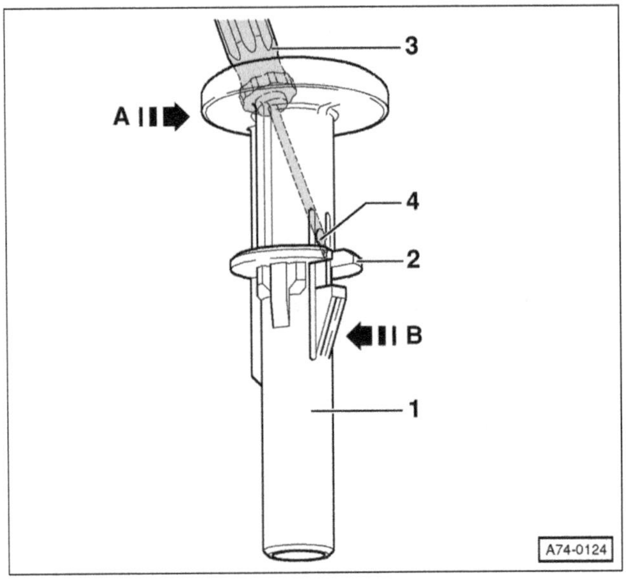

– Remove headrest.

◄ Insert a small screwdriver with a maximum width of 2 mm (**3**) or an upholstery needle, into opening (**4**) opposite retaining clip and engage locking tab (**2**).

– Pry out (**arrow A**) locking tab (**2**) far enough to release it from seat frame (**arrow B**).

– While holding tab in released position, pull guide tube (**1**) straight up and out.

– When installing, note that guide tube (**1**) is keyed to seat frame and only fits properly in one position.

– When in proper position, push guide tube down until it is felt to clip into position.

Rear Seats

Rear seat bottom, removing and installing

Removing

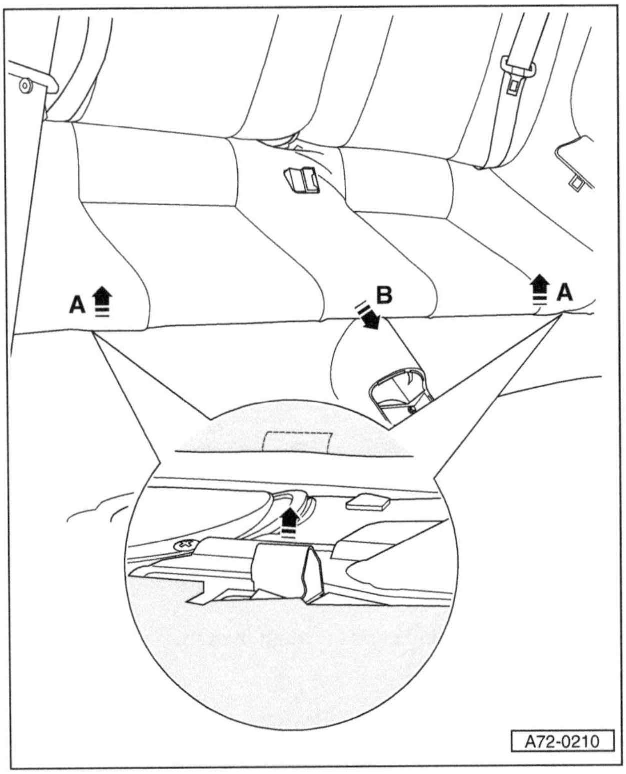

◄ Grip rear seat bottom at forward edges (**A arrows**) and lift up off of mounting points (**inset arrow**).

– Pull seat to the front (**B arrow**) and out of vehicle.

BentleyPublishers
.com
©2006 BentleyPublishers.com—All Rights Reserved

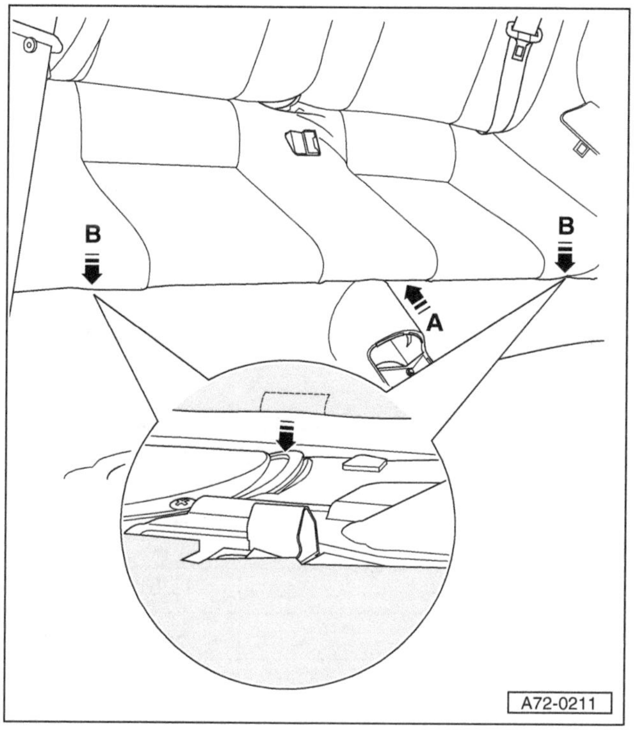

Installing

◄ Slide seat bottom into position at base of backrest as far back as possible (**A arrow**).

– Carefully guide seat down onto plastic clips (**inset arrow**) and press into place (**B arrows**).

NOTE—

- *When installing, take care to avoid damage to plastic clips.*

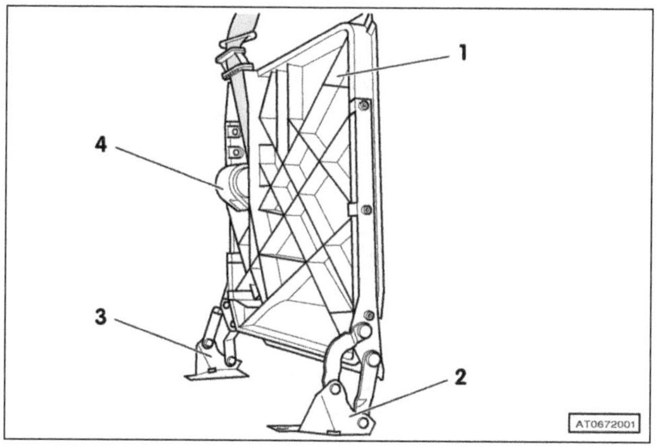

Rear seat backrest, removing and installing

◄ Rear seat backrest (**1**) is secured to the body by means of pivots that allow the seat to be folded. Removal of the outer (**3**) and inner (**2**) pivots or mounts will permit removal of the rear seat backrest. Seat belt reel (**4**) is contained inside of rear seat.

- Upholstery and cushions are shown removed in illustration.

BentleyPublishers
.com
© 2006 BentleyPublishers.com—All Rights Reserved

Center mount, removing and installing

- Remove rear seat bottom.

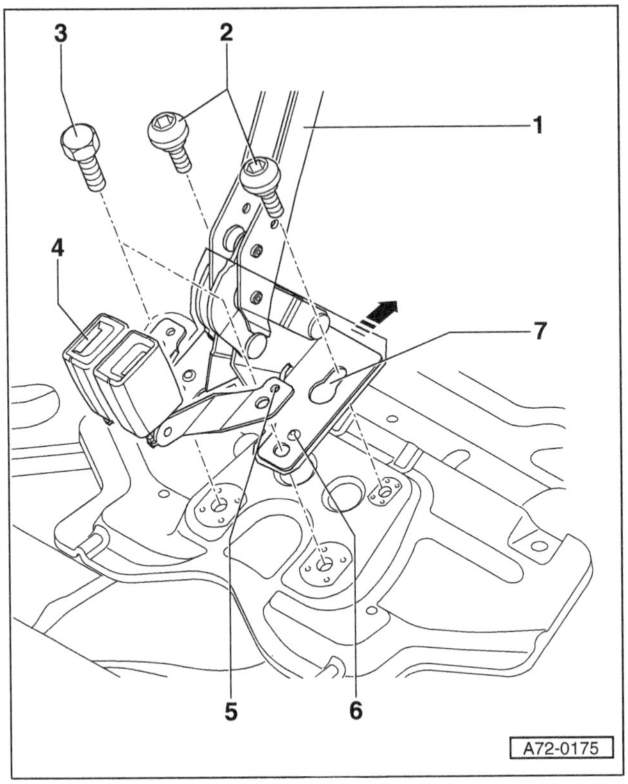

- ◄ Remove both bolts (**3**).
- Remove both seat belt buckles (**4**).
- Slide center mount (**1**) up at an angle (**arrow**) so that bolts (**2**) clear keyhole slots (**7**). Pull center mount out.
- Release seat backrest from latch and remove.
- When installing, position center mount (**1**) onto keyhole slots (**7**) and slide into place under bolts (**2**).
- Install seat belt buckles (**4**) and bolts (**3**). Before tightening bolts (**3**), be sure that locating dimple in center mounting (**6**) aligns with hole (**5**) in buckle bracket.

Tightening torque	
Center mount to body (**2**)	19 Nm (14 ft-lb)
Center mount to body (**3**)	55 Nm (41 ft-lb)

Side mounts, removing and installing

Rear seat side mounts can be removed and installed individually. This procedure covers both sides so that entire rear seat can be removed.

- Remove rear seat bottom.

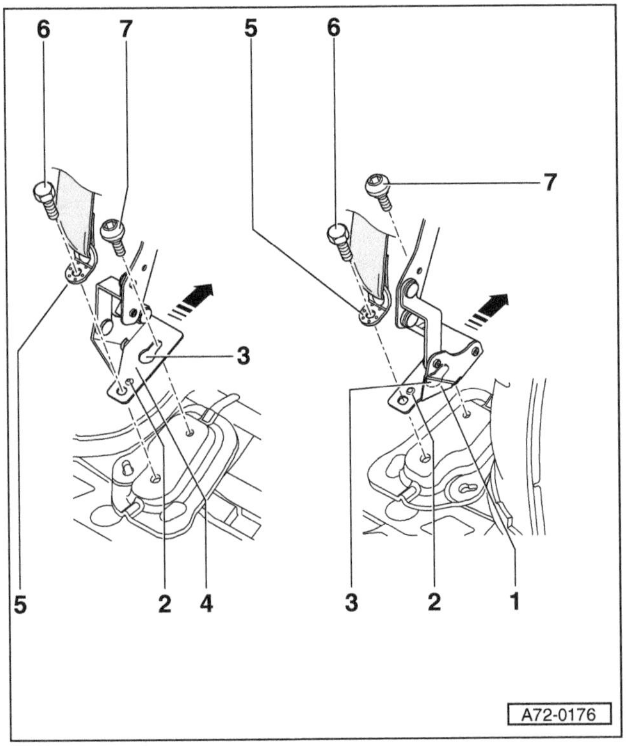

- ◄ Remove seat belt mounting bolts (**6**).
- Remove seat belts (**5**).
- Slide side mounts (**1**, **4**) up at an angle (**arrows**) so that bolts (**7**) clear keyhole slots (**3**). Remove mounts.
- Release seat backrest from latch and remove.
- When installing, position side mounts (**1**, **4**) onto keyhole slots (**3**) and slide into place under bolts (**7**).
- Install seat belts and mounting bolts. Before tightening bolts (**6**), be sure locating dimple on mount (**2**) aligns correctly with seat belt (**5**).

Tightening torque	
Side mount to body (**7**)	19 Nm (14 ft-lb)
Side mount to body (**6**)	55 Nm (41 ft-lb)

BentleyPublishers.com
© 2006 BentleyPublishers.com—All Rights Reserved

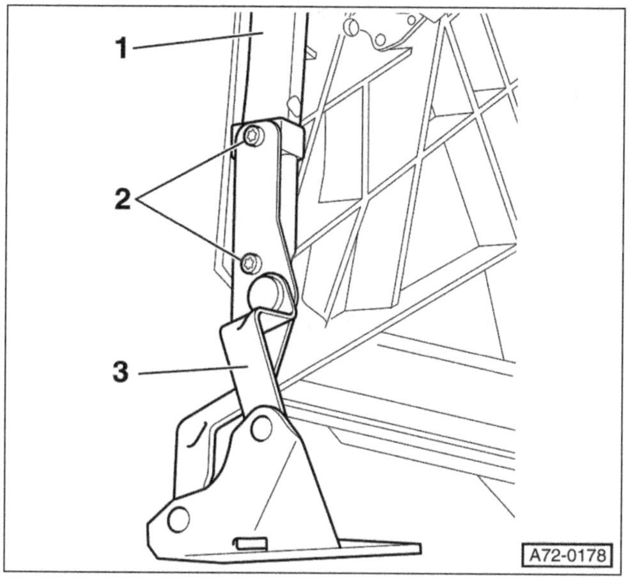

◀ Remove rear seat backrest upholstery and padding in area of hinge (**1**) without removing seat belt trim.

– Remove bolts (**2**) holding hinge (**3**) to seat frame (**1**).

– Repeat for other side as required. Inner and outer hinges, left and right side, are similar and differ primarily in bolt quantity.

– Installation is the reverse of removal.

Tightening torque	
Side mount (hinge) to seat frame (**2**)	7.5 Nm (66 in-lb)

Rear seat latch trim, removing and installing

NOTE—
- *Right side shown, left side similar.*

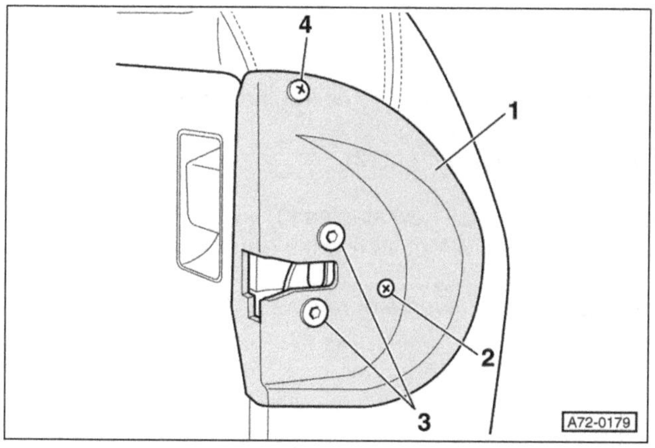

◀ Remove screws (**3** and **4**).

NOTE—
- *Do not remove screw (**2**) for reinforcement plate unless rear seat latch trim is removed. Plate can fall into backrest.*

– Pull off rear seat latch trim (**1**).

Tightening torque	
Latch trim to seat frame (**4**)	2 Nm (18 in-lb)
Latch trim to seat frame (**3**)	10 Nm (41 ft-lb)
Latch trim to reinforcement plate (**2**)	2 Nm (18 in-lb)

– Installation is the reverse of removal.

BentleyPublishers.com
© 2006 BentleyPublishers.com—All Rights Reserved

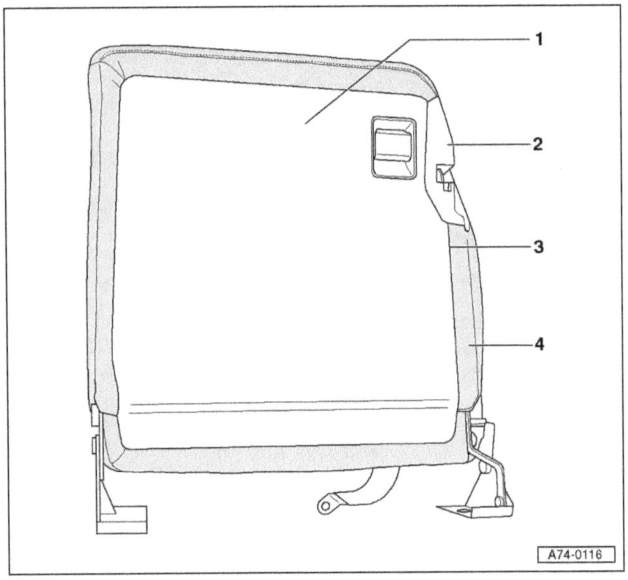

Rear seat backrest upholstery and padding, removing and installing

Right side shown, left side is similar.

– Remove rear seat bottom.

– Remove rear seat backrest.

– Remove latch trim.

◄ Starting at latch trim area (**2**), carefully pry fastening strip (**3**) for upholstery (**4**) away from back side cover (**1**) of backrest.

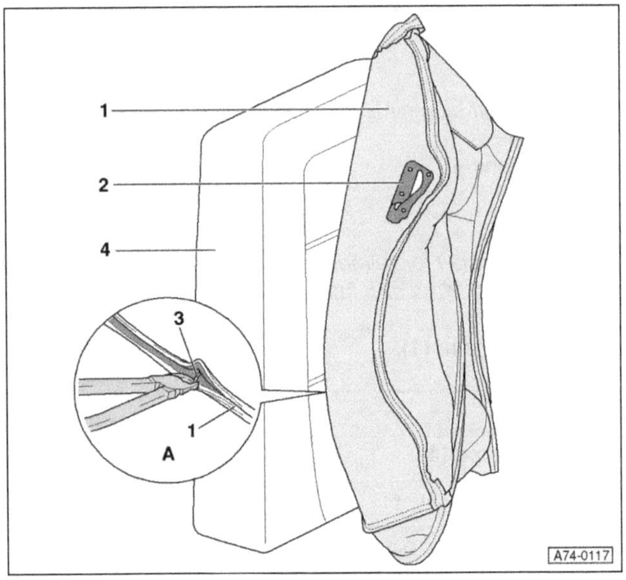

◄ When fastening strip is removed, roll or peel backrest upholstery (**1**) forward and away from cushion (**4**) so that interior staples can be accessed (**inset**).

– Cut staples (**3**) from the front and remove upholstery.

NOTE—

• Upholstery staples are commonly called “hogrings”.

– Pull seat belt out through cut-out (**2**).

– When reassembling, install new staples (**3**) with upholstery clip pliers VAG 1634, or equivalent, in reverse order of removal.

– Balance of installation is reverse of removal.

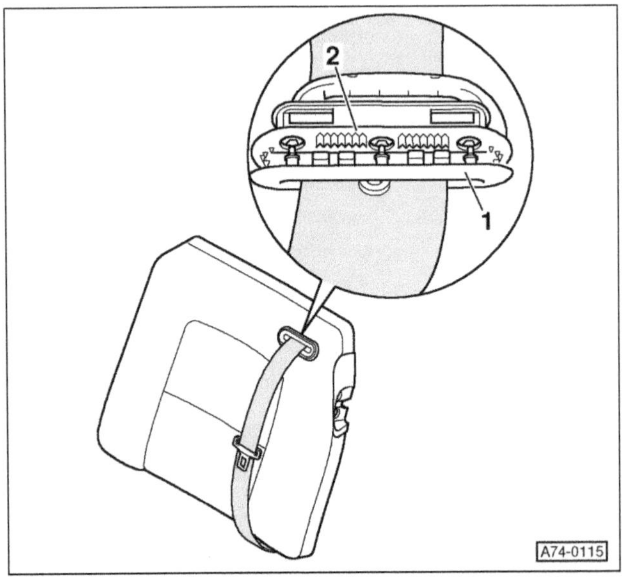

Seat belt trim, removing and installing

Seat belt trim is integral with seat belt and not a separate part. If seat belt trim is damaged, entire seat belt will need to be replaces.

NOTE—

• *Inset view shows upholstery removed for clarification.*

◄ Using a small screwdriver, carefully pry top section of seat belt trim (**1**) away from lower section (**2**).

BentleyPublishers
.com
© 2006 BentleyPublishers.com—All Rights Reserved

87 Heating and Air Conditioning

General

This repair group covers heating and air conditioning (A/C) system controls, components, and refrigerant system.

- For servicing information on belts and dust and pollen filter, see **03 Maintenance**.
- For information on front center console and interior components, see **70 Trim-Interior**.
- For information on radio removal, see **91 Radio and Communication**.

Heating and air conditioning functions in vehicles covered by this manual are combined since air conditioning is standard equipment. The fully automatic climate control system regulates temperature, fan speed, and air distribution once temperature has been set by the driver. All functions have manual overrides.

NOTE—

- *Audi identifies electrical components by a letter and/or a number in electrical schematics. See* **EWD Electrical Wiring Diagrams**. *These electrical identifiers are listed in parentheses as an aid to electrical troubleshooting.*

AT0687001

The control head (E87) is located below the radio head unit and combines an electronic control unit with all related switches.

87

BentleyPublishers.com

© 2006 BentleyPublishers.com—All Rights Reserved

AUTOMATIC CLIMATE CONTROL

Engine compartment and rear body component overview (part 1)

1. **Flow through ventilation air outlets**
   - For proper ventilation of passenger compartment.
   - Ensure free movement and proper sealing of flaps.
   - See **Flow trough vents, checking**.
2. **Coolant fan control module (J293)**
   - Controls operation of radiator fans and A/C compressor clutch.
3. **Outside air temperature sensor (G17)**
   - Remove bumper cover for access.
4. **Condenser**
   - See **Condenser, removing and installing**.
5. **A/C information label**
   - With refrigerant charge information.
   - Can be attached to radiator support, inside under hood, or on driver's side inner fender. See **A/C information label**.
6. **A/C compressor**
   - Type and mounting depends on vehicle equipment level.
   - See **A/C Compressor**.
7. **A/C compressor clutch (N25)**
8. **High pressure service port**
   - Always reinstall cap after service.
   - May require suspension cross brace removal for access depending on engine type.
   - See **Refrigerant circuit, overview**.
9. **High pressure switch (F129) or pressure sensor (G65)**
   - High pressure switch (F129) installed through model year 2000.
   - See **High pressure switch, removing and installing**
   - Pressure sensor (G65) installed from model year 2001.
   - See **High pressure sensor, removing and installing**

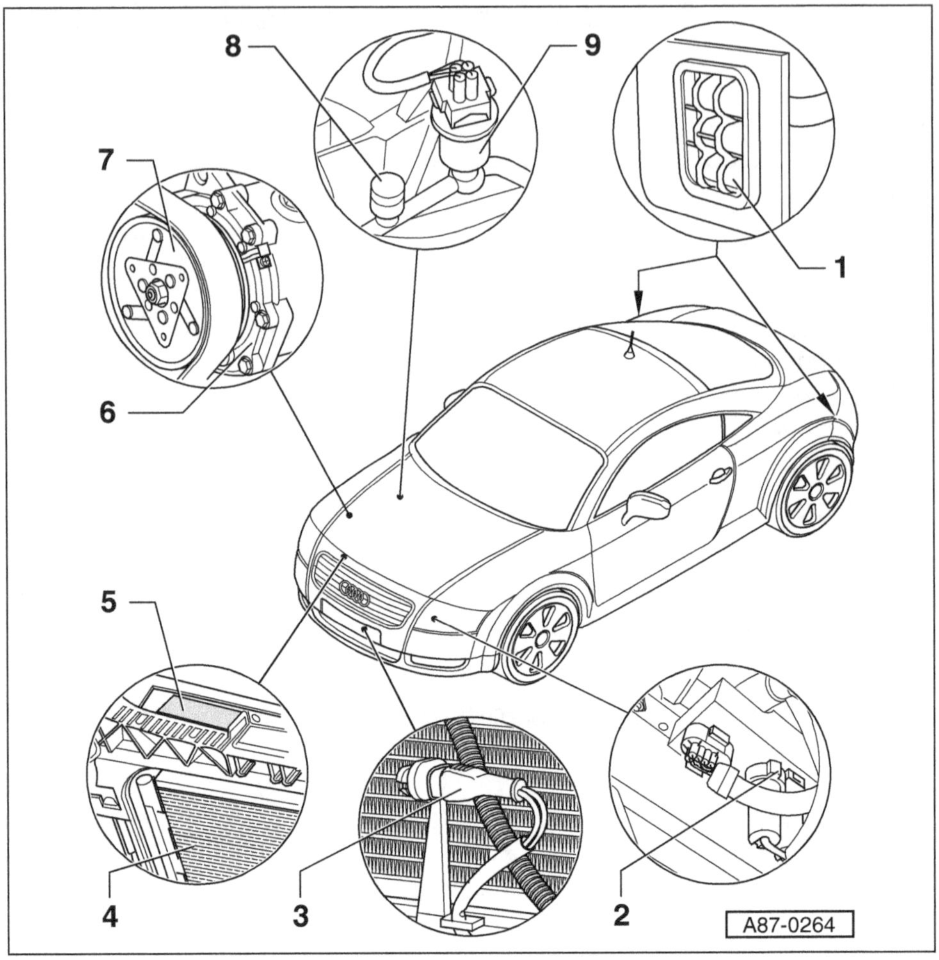

A/C information label

- A/C information sticker (**arrow**) on later production is located on driver's side inner fender.

BentleyPublishers.com

© 2006 BentleyPublishers.com—All Rights Reserved

Engine compartment and rear body component overview (part 2)

10. **Expansion valve**
    - With insulation.
    - See **Expansion valve, removing and installing**.
11. **Evaporator housing condensation drain**
    - Two different types.
    - See **Evaporator drain, removing and installing**.
12. **Receiver drier**
    - See **Receiver drier, removing and installing**.
13. **Damper**
    - Integral with low pressure refrigerant line.
14. **Low pressure service port**
    - Always reinstall cap after service.
    - May require suspension cross brace removal for access depending on engine type.
    - See **Refrigerant circuit, overview**.
15. **Air flow flap**
    - Operation by flap motor V71 with position sensor G113.
16. **Intake duct air temperature sensor (G89)**
17. **Dust and pollen filter**
    - See **03 Maintenance**

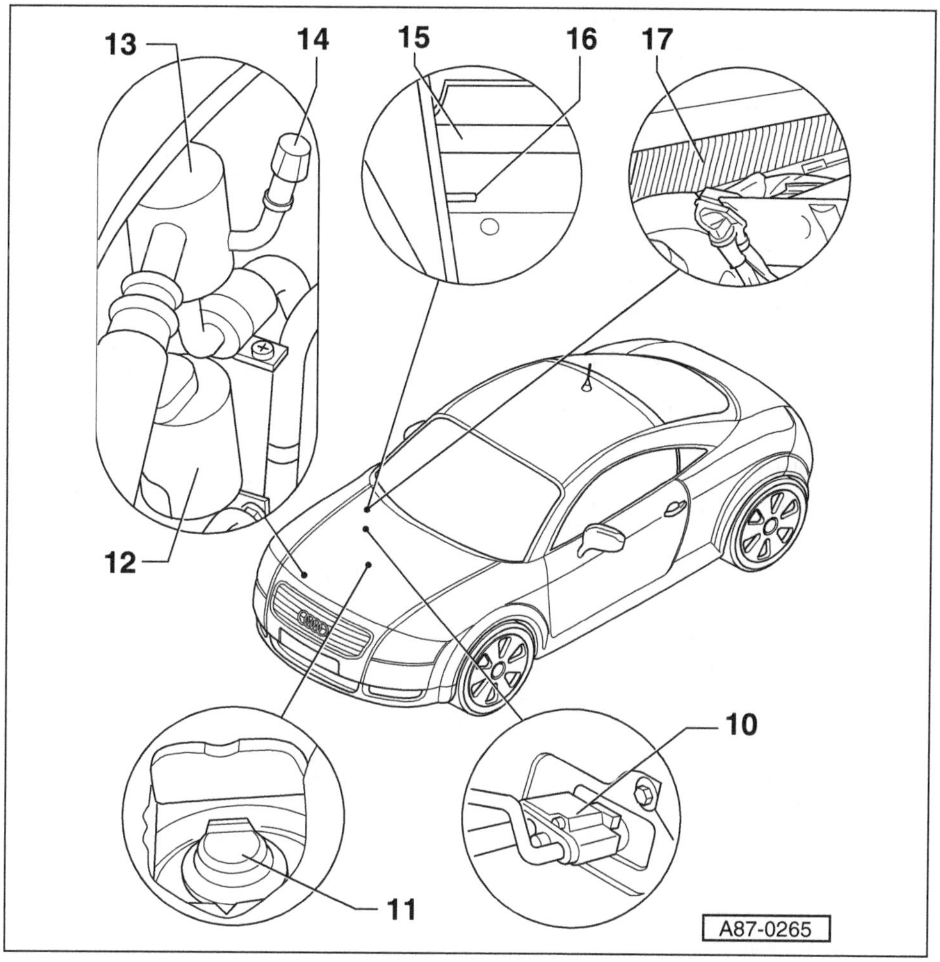

BentleyPublishers
.com
©2006 BentleyPublishers.com—All Rights Reserved

Passenger compartment component overview (part 1)

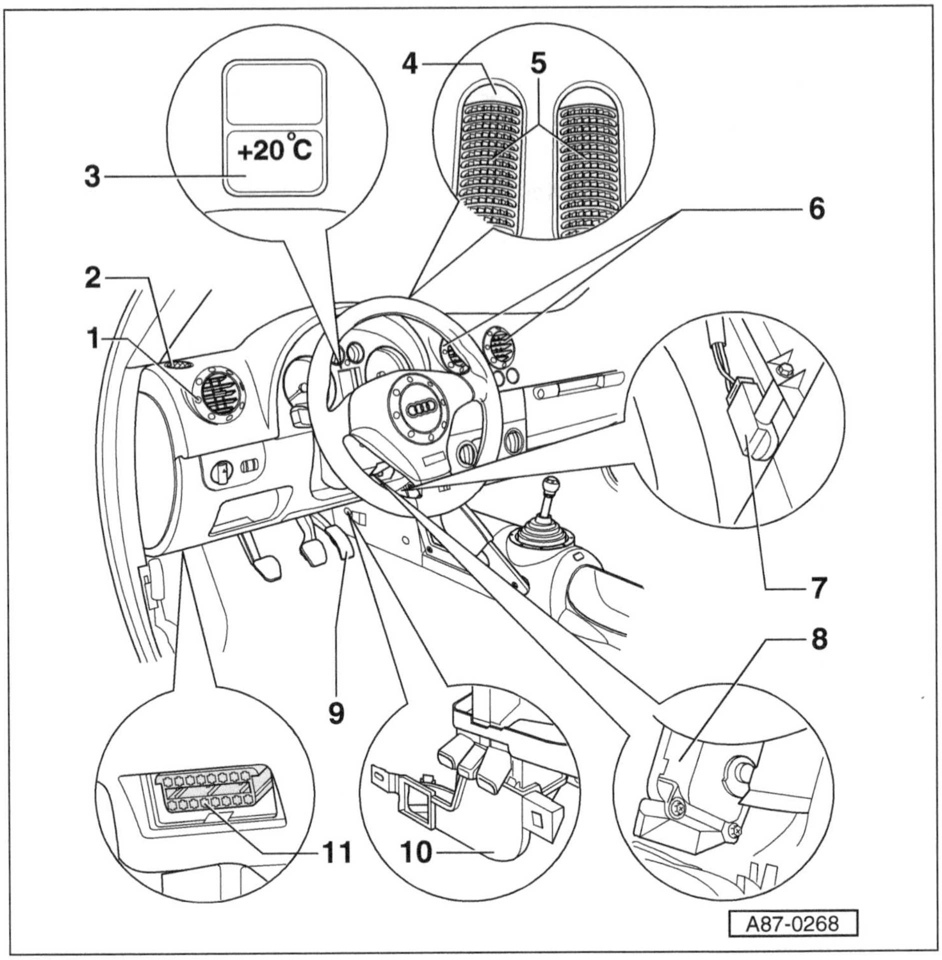

1. **Instrument panel vent, left**
   - Damaged vents can be removed by pulling straight out evenly with hooks. Vents will be destroyed in this procedure.
   - Un-damaged vents that will be reinstalled are removed from the rear. See **Vent removal (reusing)**.
2. **Side window defroster vent, left**
3. **Outside air temperature display (G106)**
   - Integral with instrument cluster
   - Can display temperature in °C or °F depending on control head settings. See **A/C control head, overview**.
4. **Sunlight photo sensor (G107)**
   - Access requires instrument panel center defroster vent trim removal. See **70 Trim-Interior**.
5. **Windshield instrument panel center defroster vent trim**
   - Removing and installing, see **70 Trim-Interior**.
6. **Instrument panel vent, center**
   - Damaged vents can be removed by pulling straight out evenly with hooks. Vents will be destroyed in this procedure.
   - Un-damaged vents that will be reinstalled are removed from the rear. See **Vent removal (reusing)**.
7. **Floor outlet temperature sender (G192)**
8. **Footwell/defroster flap motor (V85) with position sensor (G114)**
   - Actuating arm: yellow
9. **Kickdown switch**
   - Integral with throttle position sensor.
   - Switches A/C compressor clutch off during acceleration via ECM or TCM, depending on equipment level.
10. **Footwell vent**
11. **Data link connector (DLC)**
    - Scan tool connection.
    - See **OBD On Board Diagnosis**.

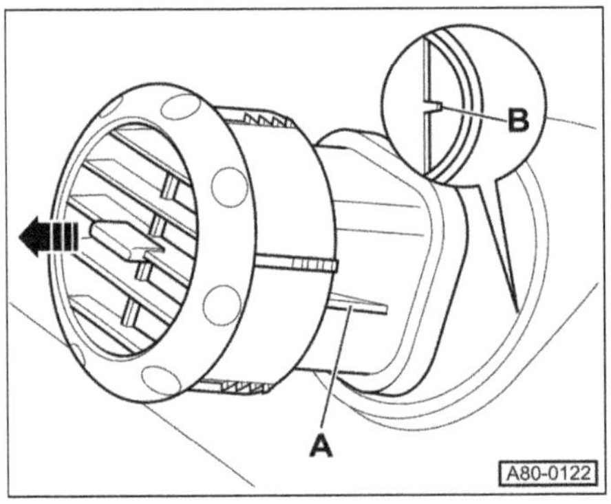

Vent removal (reusing)
- Access back side of instrument panel by removal of appropriate components.
- Reach behind from below instrument panel and firmly push vent straight out (**arrow**).
- When installing, ensure lug (**A**) aligns with slot (**B**) in instrument panel.

BentleyPublishers.com

© 2006 BentleyPublishers.com—All Rights Reserved

Passenger compartment component overview (part 2)

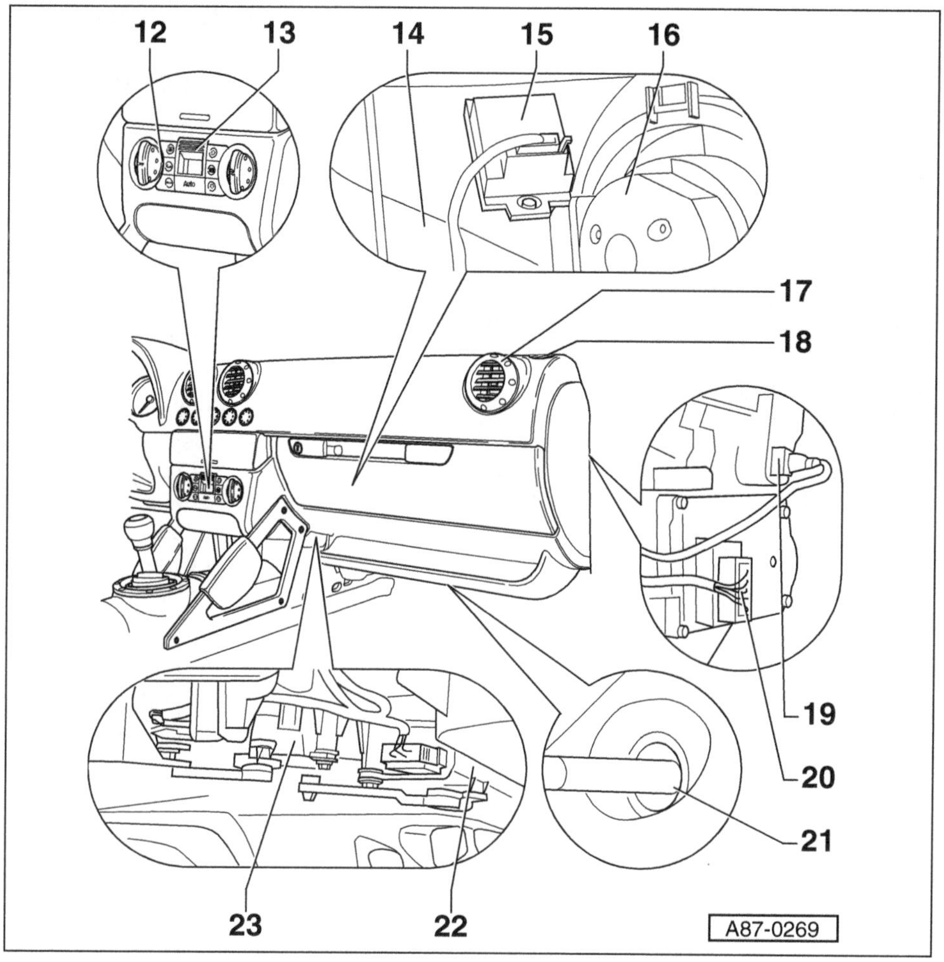

12. **A/C control head (E87)**
    - See **A/C control head**
13. **Interior temperature sensor fan (V42) and instrument panel interior temperature sensor (G56)**
    - Integral with A/C control head.
    - Do not block grill opening.
    - Dust/dirt can obstruct flow.
14. **Heating and A/C unit**
    - With heater core and A/C evaporator.
    - See **Heating and A/C components, overview**.
15. **Fresh air blower control module (J126)**
    - See **Blower motor control module, removing and installing**.
16. **Fresh air blower fan (V2)**
    - See **Blower motor, removing and installing**.
17. **Instrument panel vent, right**
    - Damaged vents can be removed by pulling straight out evenly with hooks. Vents will be destroyed in this procedure.
    - Un-damaged vents that will be reinstalled are removed from the rear. See **Vent removal (reusing)**.
18. **Side window defroster vent, right**
19. **Intake duct air temperature sensor (G89)**
20. **Air flow flap motor (V71) with position sensor (G113)**
21. **Evaporator housing drain valve**
    - Two different types.
    - See **Evaporator drain, removing and installing**.
22. **Temperature regulator flap motor (V68) with position sensor (G92)**
    - Actuating arm color: black or pink
23. **Central flap motor (V70) with position sensor (G112)**
    - Actuating arm: green

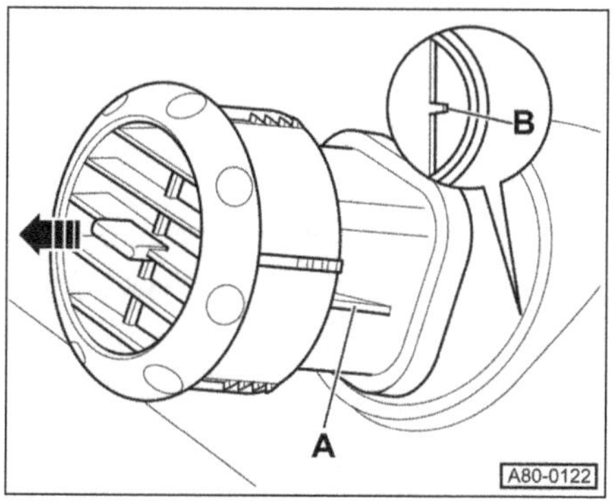

Vent removal (reusing)

- Access back side of instrument panel by removal of appropriate components.
- Reach behind from below instrument panel and firmly push vent straight out (**arrow**).
- When installing, ensure lug (**A**) aligns with slot (**B**) in instrument panel.

BentleyPublishers
.com
© 2006 BentleyPublishers.com—All Rights Reserved

A/C Control Head

A/C control head, overview

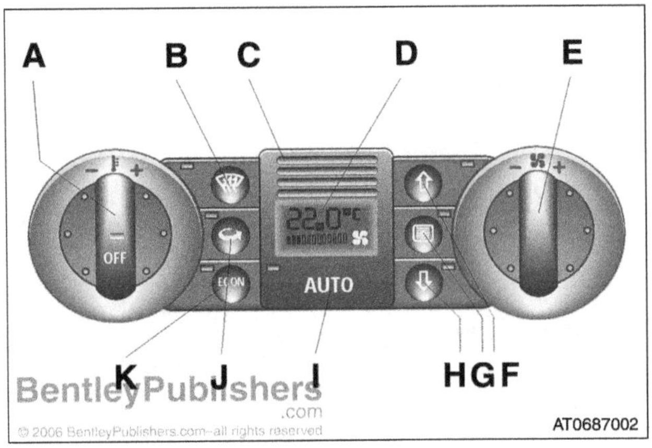

The automatic climate control head panel includes the following switches and controls:

A. System control
- On/Off push/pull switch.
- Temperature control rotary switch (**+** warmer, **-** cooler).

B. Defroster button
- Directs air flow to windshield.

C. Inlet vent for interior temperature sensor fan (V42) and instrument panel interior temperature sensor (G56).

D. Display panel for temperature and blower speed
- Change display between °C and °F by pushing and holding air recirculation button (**I**) while at the same time turning system control knob (**A**) to **+**.
- Changing temperature display at control head changes display in instrument cluster.

E. Blower speed
- Blower motor control rotary switch (**+** faster, **-** slower).

F. Air distribution button: windows

G. Air distribution button: panel outlets

H. Air distribution button: footwell

I. Auto button
- Enables fully automatic operation.

J. Air recirculation button
- When switched, blocks outside air flow to passenger compartment.

K. ECON button
- When switched, A/C compressor is switched off, but operation is otherwise automatic.

A/C control head, removing and installing

Before removing the A/C control head, note the following:

- Always check Diagnostic Trouble Code (DTC) memory before removing.
- After removing, always perform basic setting and check DTC memory.
- If replacing, code A/C control head.
- A/C control head display and control buttons are illuminated by LEDs and cannot be replaced separately.

– Remove radio, see **91 Radio and Communication**.

– Remove center console grab handles, front ashtray, and center console trim. See **70 Trim-Interior**.

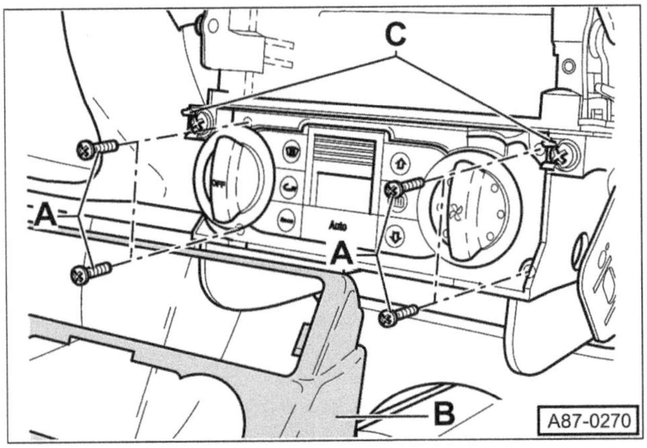

With ashtray and trim (**B**) removed, remove four securing screws (**A**).

– Pull control head out far enough to allow access to harness connectors.

BentleyPublishers
.com
© 2006 BentleyPublishers.com—All Rights Reserved

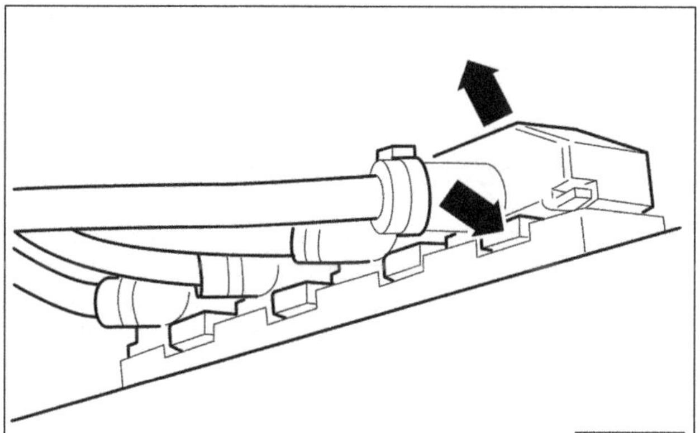
A87-0038

< Push release tabs (**lower arrow**) and pull harness connector straight out (**upper arrow**). Repeat for each harness connector.

– Installation is the reverse of removal.

NOTE —
- *When reinstalling trim (**B**), note proper positioning of spring clips (**C**).*

– Return system to basic settings.

Climate control, resetting basic settings

When the A/C control head has been disconnected from the vehicle electrical system, or when power has been interrupted, stored flap motor limit values will be lost. This can cause erratic operation and/or irregular temperature control. Motor limit values can be restored by allowing the system to adapt using a scan tool to set the basic settings function.

NOTE —
- *If scan tool or scan tool program can save data in memory or by printer, it is advisable to switch this function on before proceeding.*

– Connect scan tool or scan tool computer program.

8N0 820 043 A TT Klimavollautomat	D00
Code 00140	WSC01A00

AT0687101

< Access automatic climate control using address **08 - A/C & Heating Electronics**.

– Advance program past ECU identification to function menu.

– From function menu, select Check DTC memory (function 02) to check for stored codes.

– Evaluate, repair, and erase DTCs as required.

– Advance program past Check DTC Memory to function menu.

– From function menu, select Basic Setting (function 04).

BASIC SETTING	
Input display group number	001

AT0687102

< From Basic setting function, select display group 001 and confirm.
- Flap motors are operated in sequence with end position values from each position sensor stored as a value in A/C control head E87.

BASIC SETTING 1			
XXX	XXX	XXX	XXX

AT0687103

< Movement of each flap motor can be followed on the scan tool display by observing changing values received from each position sensor in the correspond display field from left to right:
- Temperature regulator (V68) with position sensor (G92)
- Central (V70) with position sensor (G112)
- Footwell/defroster (V85) with position sensor (G114)
- Air flow (V71) with position sensor (G113)

BASIC SETTING 1			
0	0	0	0

AT0687104

< When flap motors have stopped, values will be stored and display will switch to zeros for each field. Basic settings are complete.

– Advance program past ECU basic settings to function menu.

– From function menu, select Check DTC memory (function 02) to check for any stored codes.

BentleyPublishers
.com
© 2006 BentleyPublishers.com—All Rights Reserved

– Evaluate, repair, and erase DTCs as required.

– When complete, select End Output from function menu and disconnect scan tool or scan tool computer program.

Climate control, coding control head

If a new control head is installed into the vehicle, it will require coding to match vehicle equipment and to function properly. A control head from a different vehicle may also require coding. If a newly installed A/C control head display flashes for about 2 minutes when the ignition is switched on, it will require coding and basic setting.

NOTE—
- *Scan tool or scan tool computer program may require dealership or user coding before enabling the code control module function (07). Consult scan tool, scan tool computer program documentation, or manufacturer for appropriate procedures.*
- *If scan tool or scan tool program can save data in memory or by printer, it is advisable to switch this function on before proceeding.*

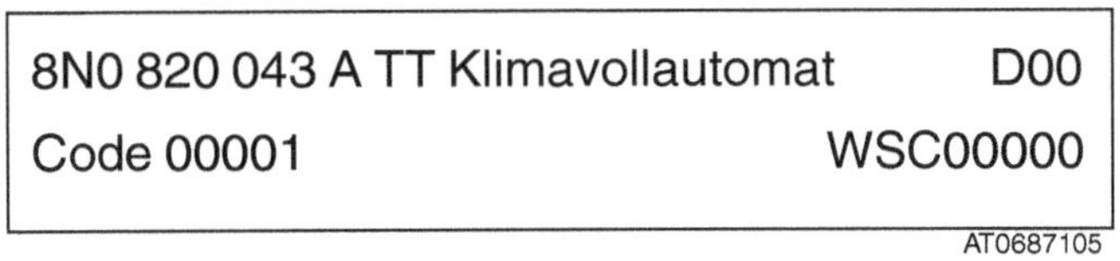

AT0687105

– Connect scan tool or scan tool computer program.

◄ Access automatic climate control using address **08 - A/C & Heating Electronics**.

– Advance program past ECU identification to function menu.

– From function menu, select Code Control Module (function 07).

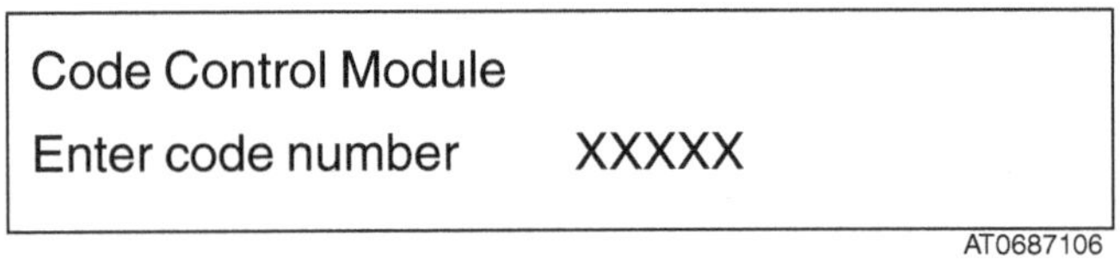

AT0687106

◄ From Code Control Module function, enter appropriate code.
- For North America, only two codes are valid.

Equipment	Code
4-cylinder gasoline engine	00140
6-cylinder gasoline engine	00160
Invalid*	00001
* Any invalid codes entered in the A/C control head are ignored and coding 00001 is stored as a default.	

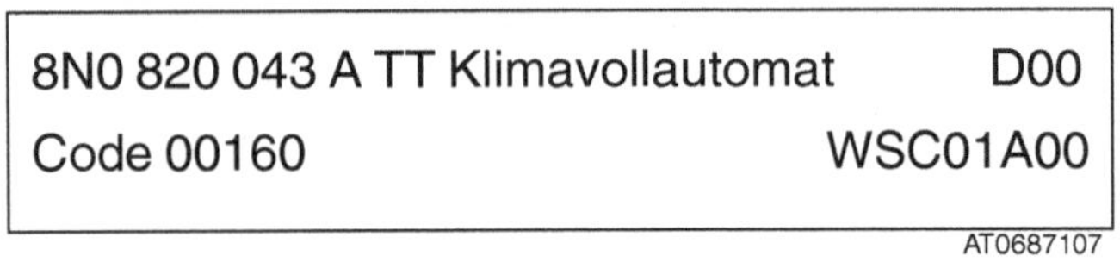

AT0687107

◄ After a short delay, the control module identification, coding and dealership/user number will be displayed.
- Klimavollautomat = fully automatic air conditioner.

– Advance program to function menu.

– From function menu, select Basic Setting (function 04) and initiate basic settings.

– When basic settings are complete, advance to function menu.

– From function menu, select Check DTC memory (function 02) to check for stored codes that may have been stored.

– When complete, select End Output from function menu and disconnect scan tool or scan tool computer program. Evaluate, repair, and erase DTCs as required.

BentleyPublishers.com
© 2006 BentleyPublishers.com—All Rights Reserved

Heating and A/C Unit Components

Heating and A/C components, overview (part 1)

1. **Heater core**
2. **Seal**
   - Always replace.
   - Seals heater core from engine compartment.
3. **Mounting flange**
4. **Screw**
5. **Evaporator housing seal**
   - Always replace.
   - Seals evaporator refrigerant lines from engine compartment.
6. **Defroster duct**
7. **Tensioning strap**
   - Holds assembly together.
   - Installed into channels. See **Tensioning strap, installed position**.
8. **Wiring harness**
   - Ensure proper clearance to moving parts of system. See **Wiring harness, installed position**.
9. **Foam panel**
   - Secured with tensioning strap (item **7**) and screw clips (item **10**).
10. **Screw clips**
11. **Footwell duct**
12. **Heating and A/C unit**

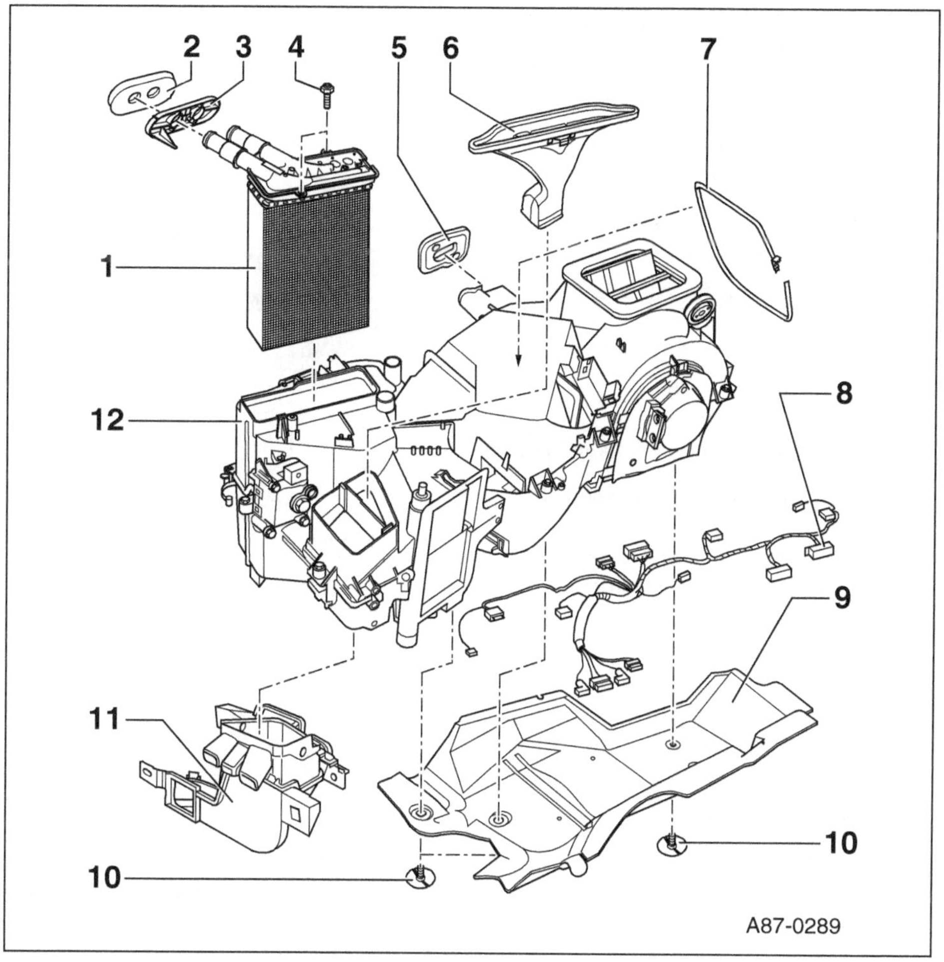

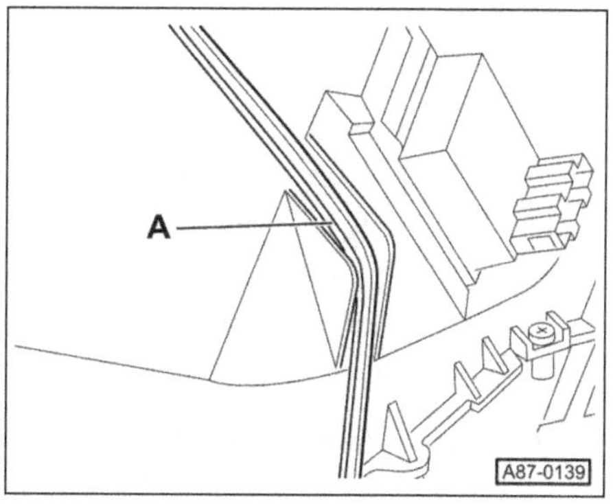

Tensioning strap, installed position

- Ensure correct positioning and fit into channels (**A**).

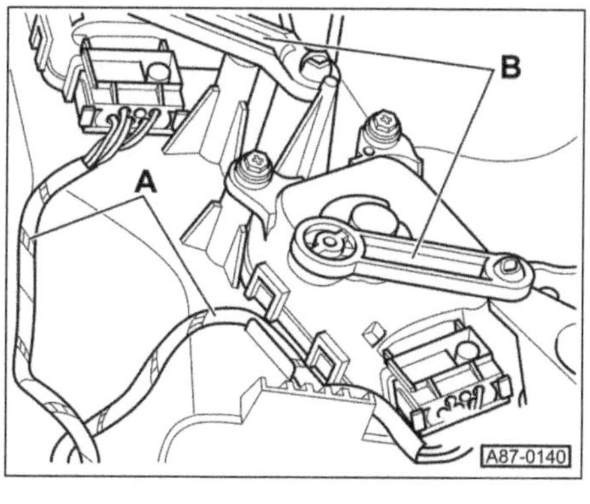

Wiring harness, installed position

- Position harness (**A**) into channels.
- Ensure positioning to avoid contact with positioning motor arms (**B**).
- Secure with tie wraps as needed.

BentleyPublishers.com
© 2006 BentleyPublishers.com—All Rights Reserved

Heating and A/C components, overview (part 2)

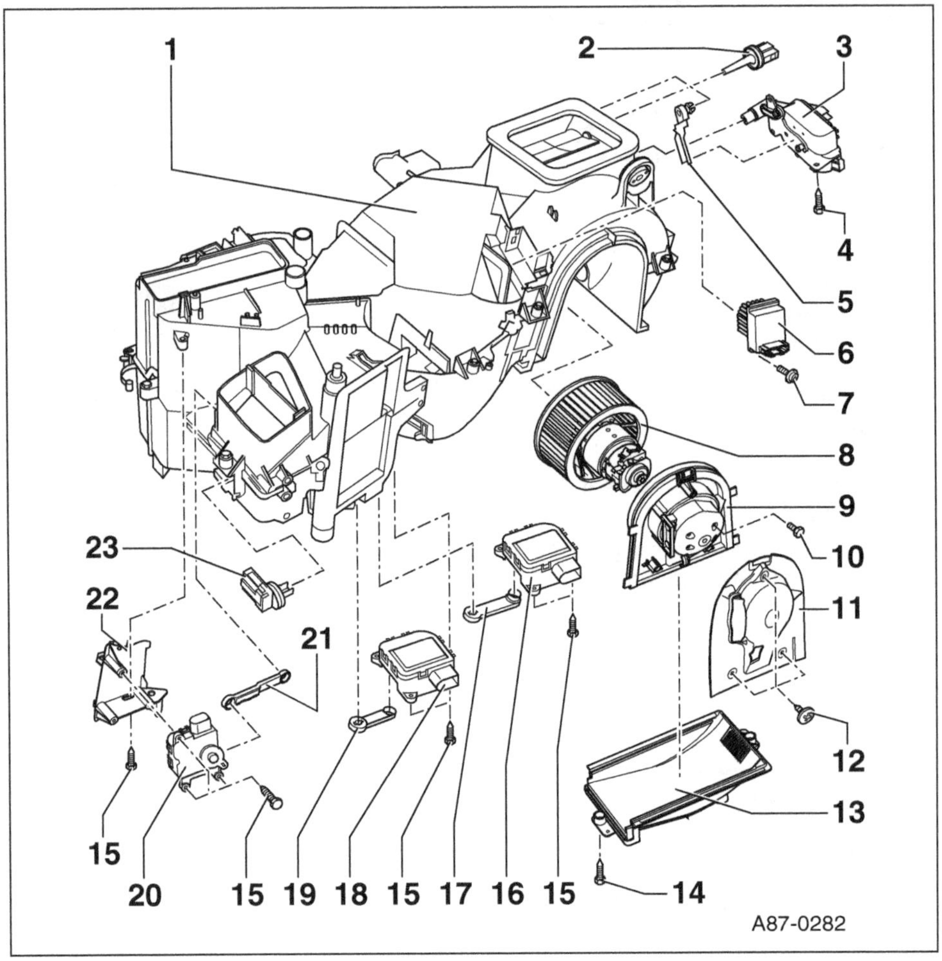

1. **Heating and A/C unit**
2. **Intake duct air temperature sensor (G89)**
3. **Air flow flap motor (V71)**
   - With potentiometer (G113).
   - Operates air flow and recirculation flap.
4. **Screw**
5. **Actuating arm**
   - For air flow and recirculation flap.
6. **Blower motor control module (J126)**
   - See **Blower motor control module, removing and installing**.
7. **Screw**
8. **Blower motor (V2)**
   - See **Blower motor, removing and installing**.
9. **Blower motor carrier**
   - See **Blower motor, removing and installing**.
10. **Screw**
11. **Insulation pad**
    - Installed intermittently.
12. **Screw**
13. **Bottom cover**
14. **Screw**
15. **Screw**
16. **Temperature regulator flap motor (V68)**
    - With potentiometer (G92).
    - Operates flap regulating warm air and cool air flow.
    - Actuating arm color: black or pink.
17. **Actuating arm**
    - For temperature regulation flap.
18. **Central flap motor (V70)**
    - With potentiometer (G112).
    - Actuating arm color: green.
19. **Actuating arm**
    - For central flap.
    - Controls air distribution.
20. **Footwell/defroster flap motor (V85)**
    - With potentiometer (G114)
    - Actuating arm color: yellow.
21. **Actuating arm**
    - For footwell/defroster flap
22. **Mounting bracket**
    - For footwell/defroster flap motor.
23. **Floor outlet temperature sender (G192)**

BentleyPublishers.com
© 2006 BentleyPublishers.com—All Rights Reserved

Heating and A/C components, overview (part 3)

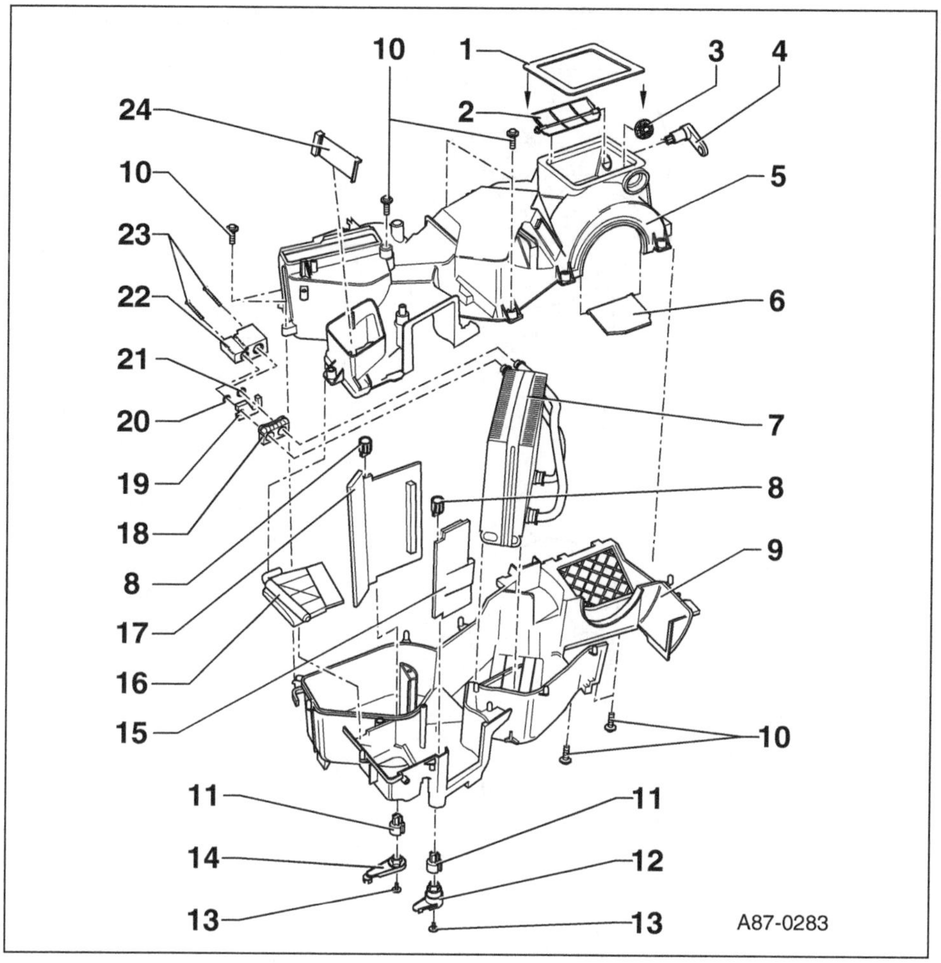

1. **Seal**
   - Seals intake duct from air intake plenum and engine compartment.
   - Check and replace if damaged.
2. **Air flow flap**
   - Lightly lubricate hinge pivots with Audi grease G 052 745 A3 or equivalent.
3. **Grommet**
4. **Actuating arm**
5. **Air distribution housing, upper section**
6. **Recirculation flap**
   - Lightly lubricate hinge pivots with Audi grease G 052 745 A3 or equivalent.
7. **Evaporator**
   - The removed original evaporator contains refrigerant oil. The equivalent amount of oil must be returned to the refrigerant system when replacing evaporator.
8. **Pivot bushing**
   - Lightly lubricate hinge pivots with Audi grease G 052 745 A3 or equivalent.
9. **Air distribution housing, lower section**
10. **Screw**
11. **Pivot bushing**
    - Lightly lubricate hinge pivots with Audi grease G 052 745 A3 or equivalent.
12. **Central flap actuating arm**
13. **Screw**
14. **Temperature regulation actuating arm**
15. **Central flap**
    - Note installed position; shorter section faces up.
16. **Footwell/defroster flap**
    - Lightly lubricate hinge pivots with Audi grease G 052 745 A3 or equivalent.
17. **Temperature regulation flap**
18. **Seal**
19. **Threaded plate**
    - See **Expansion valve, removing and installing**.
20. **O-rings**
    - Always replace.
    - See **O-rings, overview**.
21. **O-rings**
    - Always replace.
    - See **O-rings, overview**.
22. **Expansion valve**
    - See **Expansion valve, removing and installing**.
23. **Bolt**
    - Tighten to 8 Nm (71 in-lb).
24. **Air deflector plate**

BentleyPublishers.com
© 2006 BentleyPublishers.com—All Rights Reserved

Evaporator drain, removing and installing

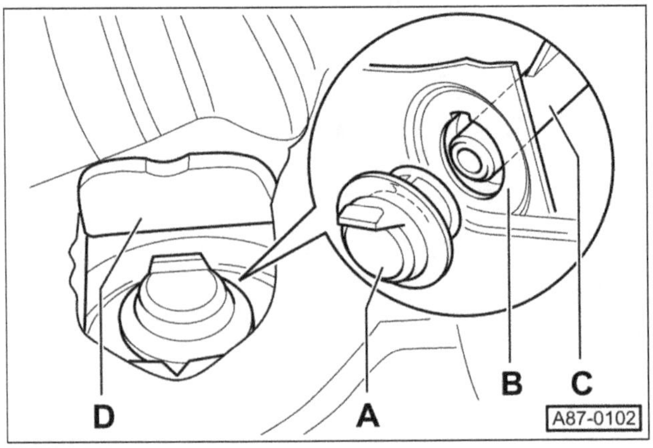

◄ Ensure that flap on valve (**A**) is not damaged, clogged, or obstructed and is able to close freely.

- Flap opening must face downwards.

– To remove, fold back pre-punched cut-out (**D**) in silver bulkhead insulation mat and pull valve (**A**) straight out and down away from bulkhead (**B**) and off of condensate tube (**C**).

– Installation of the evaporator drain valve is the reverse of removal noting the following:

- Valve (**A**) must not touch pre-punched cut-out in insulation mat and must be free of undercoating and rustproofing materials.
- Ensure correct positioning and seating onto condensate tube (**C**) and into bulkhead (**B**). Bulkhead opening is cut to allow small positioning key to fit.

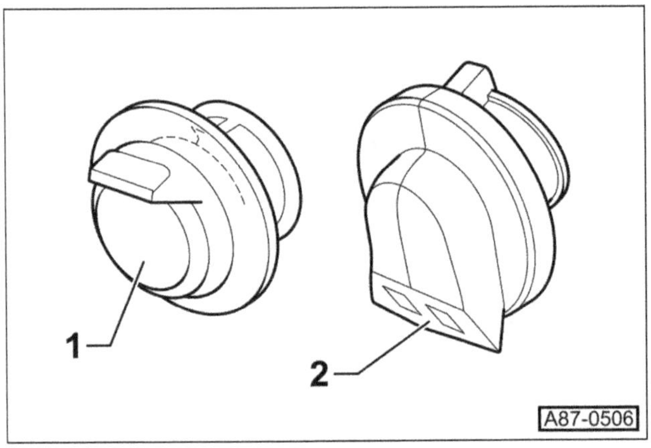

◄ Two versions of evaporator drain valve are used:

- Version **1** installed through model year 2000.
- Version **2** installed from model year 2001 (with two lower openings).

Flow through vents, checking

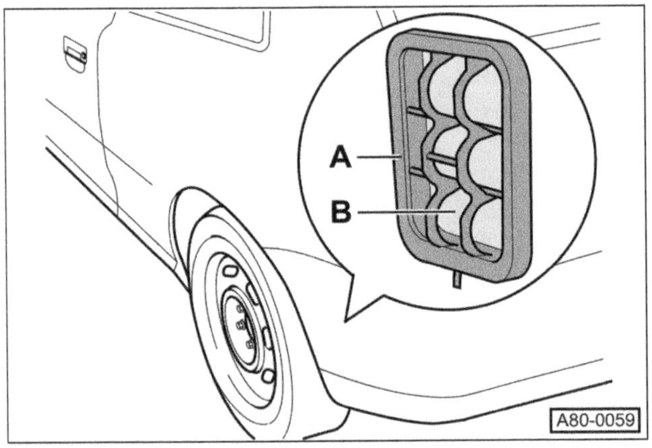

◄ Flow through ventilation flaps are located behind the rear bumper on each side. Left side shown, right side is similar.

– Sealing lips (**B**) on air extraction vent frame (**A**) must be able to move and close freely.

– To ensure proper air extraction, check ventilation slits in luggage compartment side trim panels.

Blower motor control module, removing and installing

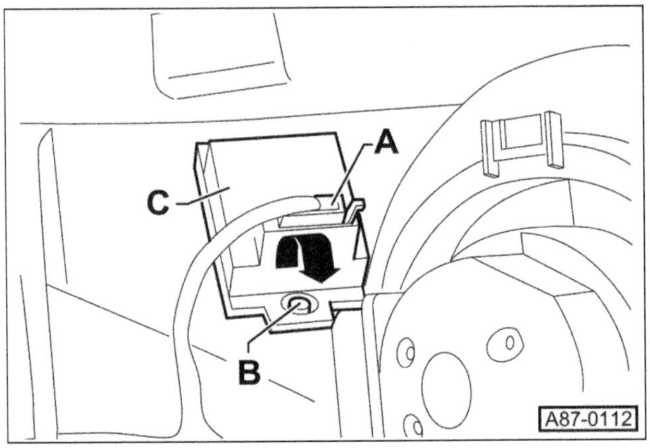

– Remove glove box, see **70 Trim-Interior**.

◄ Disconnect harness connector (**A**) from blower motor control module (J126).

– Remove securing screw (**B**).

– Remove blower motor control module (**C**) in direction of **arrow**.

– Installation is the reverse of removal.

BentleyPublishers.com
©2006 BentleyPublishers.com—All Rights Reserved

Blower motor, removing and installing

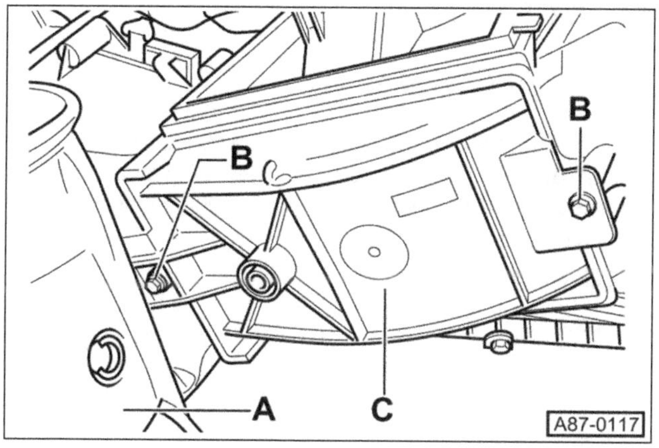

– Remove glove box, see **70 Trim-Interior**.

– Remove screw clip retainer for foam covering.

◄ Fold foam covering (**A**) away and off to the side.

– Remove screws (**B**) and cover (**C**).

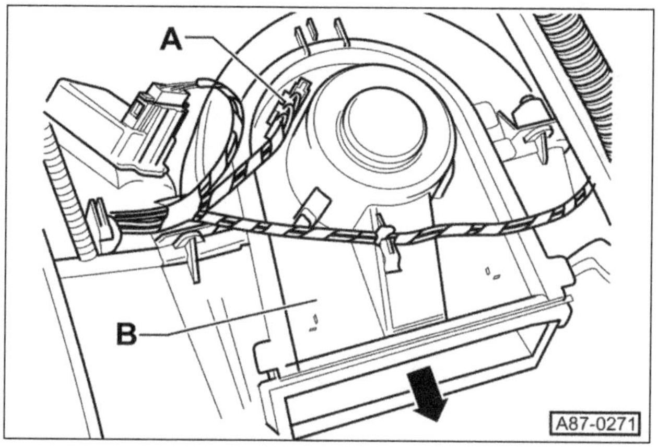

◄ Disconnect harness connector (**A**) from blower motor.

– Pull blower motor carrier (**B**) downwards and out of housing (**arrow**).

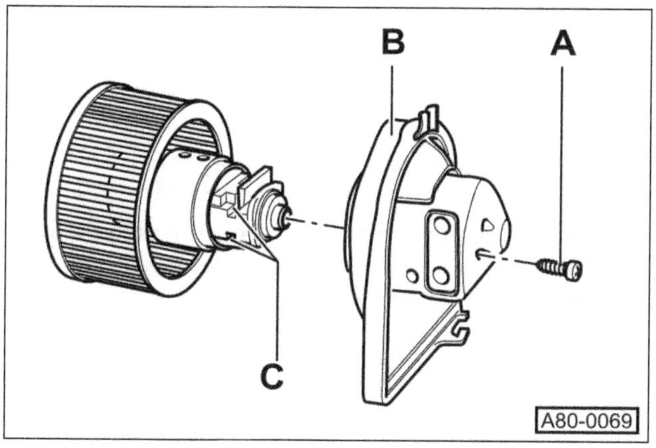

◄ Remove blower fan, early version.
- Remove screw (**A**).
- Slide blower fan (**C**) out of carrier (**B**).
- Carrier (**B**) may be equipped with sound insulation pad.
- If replacement is needed, only late version is supplied.

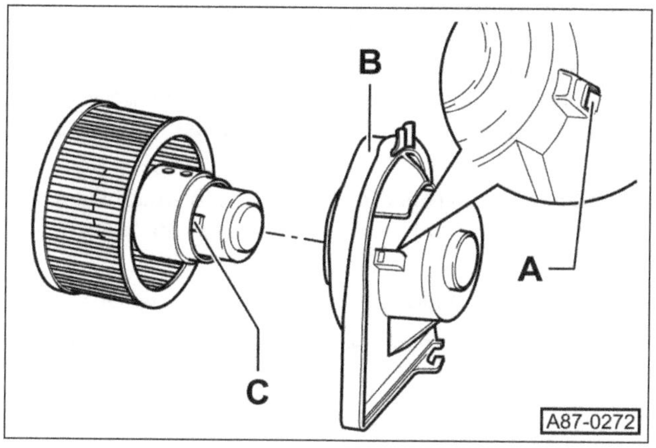

◄ Remove blower fan, late version.
- Depress retainer clips (**A**).
- Slide blower fan (**C**) out of carrier (**B**).

– Install in reverse order of removal noting the following:
- When re-installing blower fan to carrier, ensure mounting lugs (**C**) are aligned in carrier (**B**) and that blower fan is properly seated.
- Ensure correct seating of blower fan to housing.

BentleyPublishers.com
© 2006 BentleyPublishers.com—All Rights Reserved

Refrigerant System

Refrigerant circuit, overview

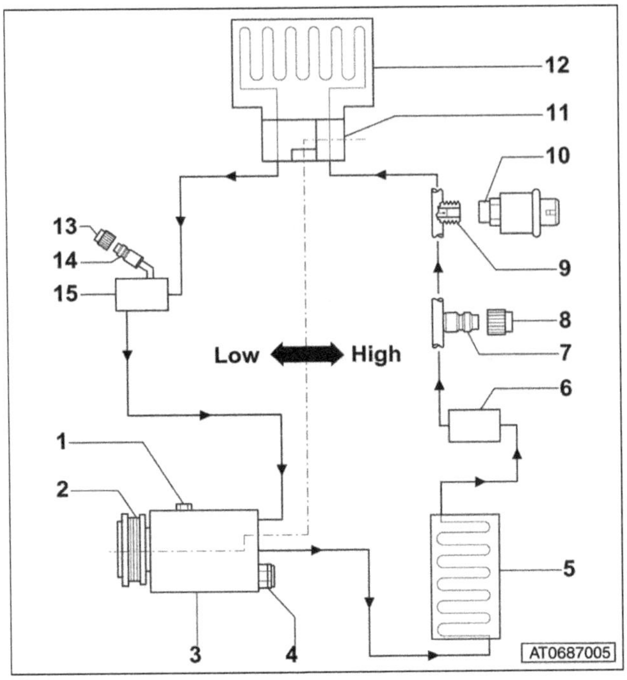

Principle R134a refrigerant components. Small **arrows** indicate direction of refrigerant flow. Large arrow denotes separation between high pressure side of system and low pressure side.

1. Compressor oil fill/drain plug
2. A/C compressor clutch (N25)
3. A/C compressor
4. High pressure relief valve
5. Condenser
6. Receiver drier
7. High pressure service port
8. Cap with seal
9. Pressure switch/sensor port with valve
10. High pressure switch (F129) or pressure sensor (G65)
11. Expansion valve
12. Evaporator
13. Cap with seal
14. Low pressure service port
15. Damper

Refrigerant and refrigerant oil

The air conditioning systems in vehicles covered by this manual use industry-standard refrigerant R134a. Refrigerant R-134a has the following characteristics:

- Colorless and is invisible as a gas.
- Escaped R-134a gases are heavier than air and will gather first in low places, such as under the car.
- R-134a refrigerant gas displaces oxygen and may cause suffocation in low areas or areas of poor air circulation.
- Refrigerant R-134a will deteriorate some plastics. Therefore, when making system repairs, use only genuine VW replacement parts or parts which are specified for use with R-134a refrigerant.
- R-134a in an enclosed container will have a specific temperature/pressure relationship.
- Refrigerant R-134a is not poisonous in any state (liquid or gas) and is safe when used properly.
- Refrigerant R-134a, in its pure state, is chemically stable and will not attack iron, copper, brass or aluminum. However, the mixture of R-134a and PAG oil may deteriorate certain metals (copper). Therefore, when making system repairs, use only genuine VW replacement parts or parts which are specified for use with R-134a refrigerant.
- Liquid R-134a refrigerant will absorb only very minute quantities of moisture. However, R-134a vapor can absorb large amounts of moisture.
- R-134a refrigerant is not flammable.
- DO NOT exceed maximum rated capacity of refrigerant stored in containers.

BentleyPublishers.com
©2006 BentleyPublishers.com—All Rights Reserved

Service to the refrigerant system requires specific training and specialized equipment. In some areas, government-issued licenses are required to purchase and handle R134a refrigerant. Because of this, repairs to the refrigerant system should generally be referred to professionals with the appropriate training and equipment.

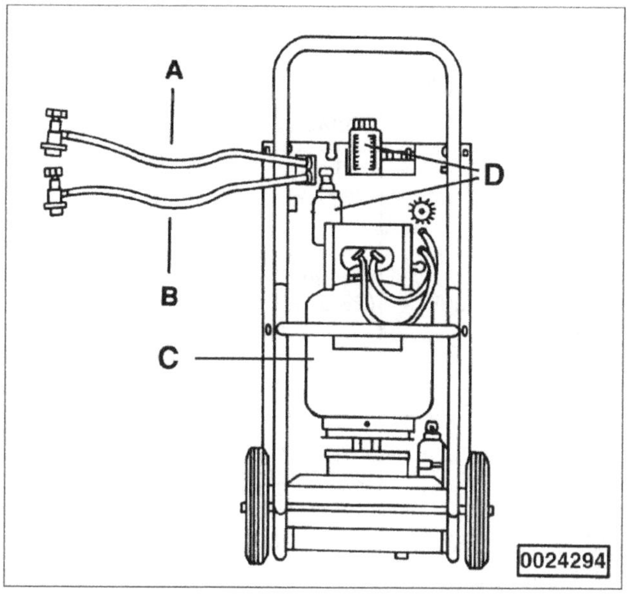

◄ Rear of typical R134a refrigerant recovery and recharging equipment. High pressure hose (**A**) is coded red and low pressure hose (**B**) is coded blue. Refrigerant tank (**C**) sits on a scale for accurate measurement of amount recovered and amount added. This equipment also has provision for adding PAG oil (**D**).

WARNING —

- *Work in a well ventilated area. Refrigerant gases are heavier than air, displace oxygen and may cause suffocation in areas of poor air circulation, for example under the vehicle. Switch on exhaust/ventilation systems when working on refrigerant systems.*
- *Avoid breathing refrigerant vapors. Exposure may irritate eyes, nose and throat.*
- *Pressurized R134a refrigerant in the presence of oxygen may form a combustible mixture. Never introduce compressed air into any R-134a container (full or empty), capped off A/C component, or piece of service equipment.*
- *Always comply with all applicable laws and regulations applicable to the handling of refrigerant R134a.*
- *R-134a refrigerant system oil (PAG oil) absorbs moisture very rapidly. Moisture combines with the refrigerant to form acids which will damage the system. Use only the specified oil from a sealed container and ALWAYS reseal oil container immediately after use. DO NOT use oil if it has become contaminated with moisture.*
- *Immediately plug open connections on A/C components to prevent dirt and especially moisture contamination. Likewise, DO NOT remove new components from packaging until ready to install. Immediately tighten component connections after installation.*
- *Always replace O-rings, DO NOT reuse. Use only correct size and type of O-rings specified for use with R-134a refrigerant. Lubricate O-ring with refrigerant oil before installing.*
- *Always replace damaged and/or leaking A/C system components. Do not attempt repair by soldering or welding.*
- *Always reinstall caps over A/C service valves.*
- *Work area must be extremely clean when working on A/C system components.*

◄ Small containers of R134a are commonly available, however, most professional equipment uses larger containers where refrigerant is charged into the system by weight.

BentleyPublishers.com
©2006 BentleyPublishers.com—All Rights Reserved

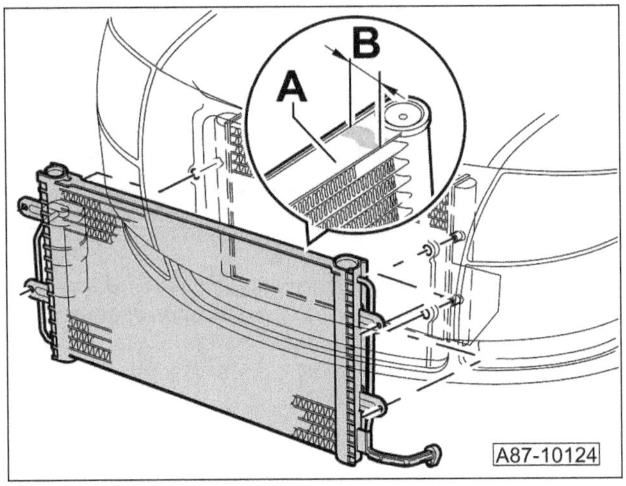

Consult underhood label to identify refrigerant capacity. If label is missing or if there is any doubt about capacity, such as if condenser has been replaced, or an incorrect label is installed, measure width of upper flat pipe and add refrigerant accordingly.

⊲ Measure upper flat pipe width (**B**) of condenser (**A**) if in doubt about refrigerant charge level.

Table a. R134a refrigerant capacity

Application	Capacity
20 mm wide condenser: Through production 10/03, and from 06/04 through 08/04.	750 grams ± 50 grams (26.5 oz ± 1.8 oz)
16 mm wide condenser: From production 11/03 through 05/04 and from 09/04.	650 grams ± 20 grams (23 oz ± 0.7 oz)

Table b. Refrigerant oil

Refrigerant oil	Capacity	Distribution	Audi part number
Synthetic PAG	135 cc ±15 cc (4.6 oz ± 0.5 oz)	n/a*	G 052 154 A2

* No distribution figures are listed for this vehicle, but similar vehicles typically use the following: 50% compressor, 10% condenser, 10% suction hose, 20% evaporator, 10% receiver drier.

High pressure switch (through model year 2000), removing and installing

The high pressure switch (F129) is located in the liquid refrigerant line near the expansion valve. It can be removed without discharging the refrigerant due to the placement of a Schrader valve in the pressure switch port.

This switch, operating through the fan control module (J293), controls the compressor and radiator fans as follows:

- Compressor off if high side pressure exceeds approximately 32 Bar (464 psi). Switches back on when pressure drops below 24 Bar (348 psi).
- Compressor off if high side pressure drops below approximately 1.2 Bar (17 psi). Switches back on when pressure rises to 2.4 Bar (35 psi).
- Radiator fan to speed 2 at approximately 16 Bar (232 psi). Switches back to speed 1 when pressure falls to 12.5 Bar (181 psi).

– Working from the engine compartment, remove passenger's side sound absorber panel. See **03 Maintenance**.

⊲ Locate high pressure switch (**arrow**) and disconnect 4-pin harness connector.

– Unscrew high pressure switch.

BentleyPublishers.com
© 2006 BentleyPublishers.com—All Rights Reserved

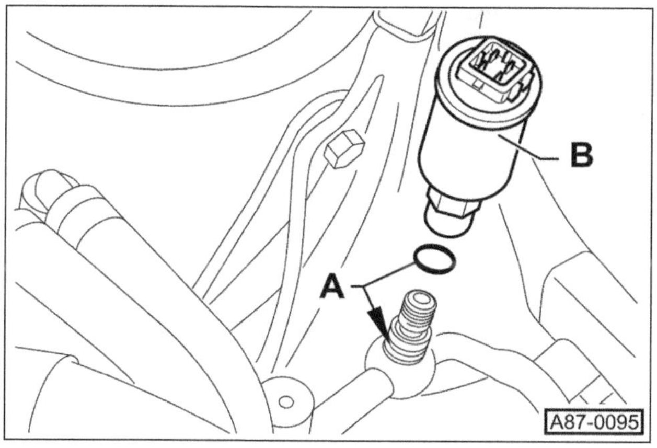

When replacing switch (**B**), always replace O-ring seal on switch port (**A**).

- Tighten to 5 Nm (44 in-lb).

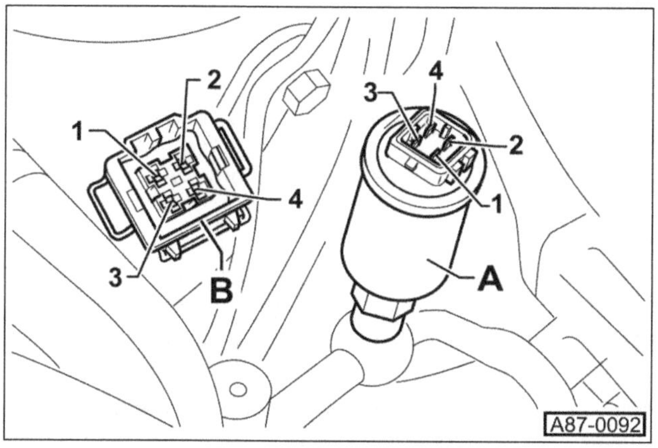

Switch function can be tested by jumpering appropriate pins of harness connector (**B**) with A/C system switched on.

- Jumpering terminals **3** and **4** should run radiator fan at speed 2.
- Jumpering terminals **1** and **2** should engage compressor clutch.

NOTE—

- *If jumpering harness connectors causes system to function in a seemingly normal manner, be sure to check refrigerant charge levels with pressure gauges before replacing switch.*

Pressure sensor (from model year 2001), removing and installing

The pressure sensor (G65) is located in the liquid refrigerant line near the expansion valve at the same location as the high pressure switch (F129) for earlier vehicles. The pressure sensor can be removed without discharging the refrigerant due to the placement of a Schrader valve in the pressure switch port. Switch function cannot be tested by jumpering pins of the harness connector. Testing requires use of a scan tool or scan tool computer program and is outside the scope of this manual.

– Working from the engine compartment, remove passenger's side sound absorber panel. See **03 Maintenance**.

– Locate pressure sensor and disconnect 2-pin harness connector.

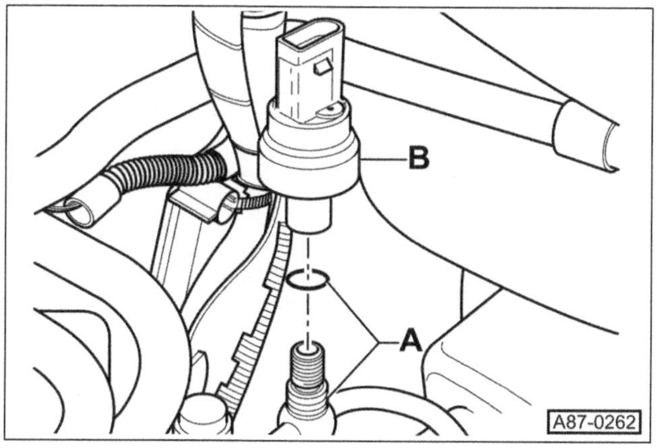

When replacing switch (**B**), always replace O-ring seal on switch port (**A**).

- Tighten to 5 Nm (44 in-lb).

BentleyPublishers.com
©2006 BentleyPublishers.com—All Rights Reserved

Receiver drier, removing and installing

The receiver drier cannot be removed without discharging the refrigerant.

– Discharge refrigerant circuit using approved procedures and equipment according to equipment manufacturer's instructions.

NOTE—

* *Depending on engine version, better access may be obtained by removing support brace between left and right long-member, and/or removing connecting pipe between intercoolers.*

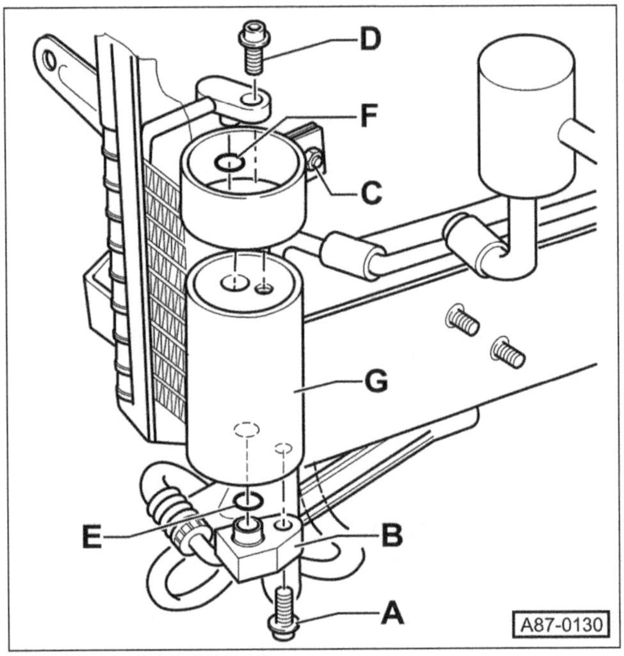

◄ Remove securing bolt (**A**).

– Detach refrigerant pipe connection (**B**).

– Remove bracket clamp nut (**C**).

– Remove refrigerant pipe connection securing bolt (**D**).

– Pull receiver drier (**G**) downwards and out of bracket.

NOTE—

* *Keep receiver drier sealed for as long as possible. Only remove caps immediately before installing. Receiver drier contains desiccant which quickly becomes saturated and rendered useless if the caps are left off, especially in humid conditions.*
* *The removed original receiver drier contains refrigerant oil. The equivalent amount of oil must be returned to the refrigerant system when replacing receiver drier.*
* *It may be sufficient to adjust the quantity of refrigerant oil in the receiver drier. For example, if only slightly damaged in an accident (for example, no refrigerant loss or no moisture or dirt in system), the A/C can be serviced by comparing the weights of the old and new receivers, and then filling the new receiver drier with refrigerant oil until it weighs the same as the damaged one.*

– Installation is the reverse of removal noting the following:
* Always replace O-rings (**E**, **F**).
* Tighten refrigerant line securing bolts (**A**, **D**) to 15 Nm (11 ft-lb).

– Evacuate and recharge system.

Condenser, removing and installing

The condenser cannot be removed without discharging the refrigerant.

– Discharge refrigerant circuit using approved procedures and equipment according to equipment manufacturer's instructions.

– Remove front bumper assembly as a unit, See **63 Bumpers**.

NOTE—

* *Depending on engine version, it is usually necessary to partially loosen and/or move the front end lock carrier forward for access. See* **50 Body-Front**.

BentleyPublishers.com
©2006 BentleyPublishers.com—All Rights Reserved

– Mark installed position of lock carrier and front end components at front fenders and long members with a felt-tipped marker.

– One at a time, remove bolts securing lock carrier to front long members and replace with bolts approximately 100 mm (4 in.) long and washers.

– Pull lock carrier out as far as possible with lines and hoses attached.

– Remove receiver drier. This removes refrigerant lines from passenger's side of condenser.

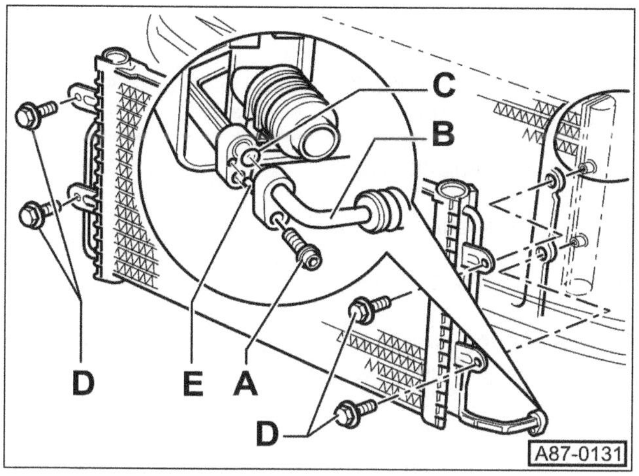

◄ Remove securing bolt (**A**) from refrigerant line (**B**) from driver's side of condenser (**inset**).

– Remove four screws securing condenser to carrier (**D**).

– Remove radiator fans with mountings. See **19-Engine Cooling System**.

– Release radiator from lock carrier and tilt radiator backwards.

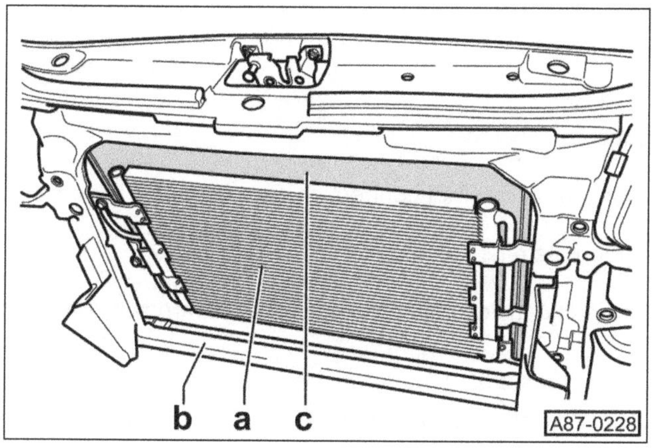

◄ Remove condenser (**a**) from below between lock carrier (**b**) and radiator (**c**).

NOTE —

- *The removed original condenser contains refrigerant oil. The equivalent amount of oil must be returned to the refrigerant system when replacing condenser.*

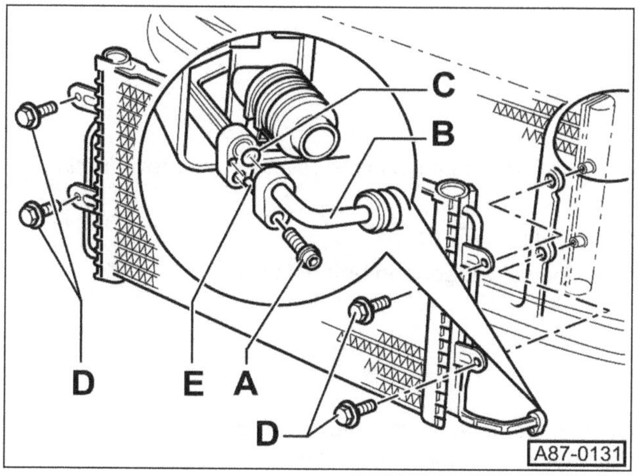

◄ Installation is the reverse of removal noting the following:

- Always replace O-rings (**C**).
- When installing refrigerant line (**B**), ensure that locating pin (**E**) is properly positioned.
- Tighten refrigerant line securing bolt (**A**) to 15 Nm (11 ft-lb).
- Tighten condenser securing bolts (**D**) to 8 Nm (71 in-lb).

– Install lock carrier and front end components in reverse order of removal.

– Evacuate and recharge system.

BentleyPublishers.com
© 2006 BentleyPublishers.com—All Rights Reserved

Expansion valve, removing and installing

– Discharge refrigerant circuit using approved procedures and equipment according to equipment manufacturer's instructions.

NOTE—

- *Depending on engine version, improved access may be obtained by removing front suspension strut tower cross brace.*

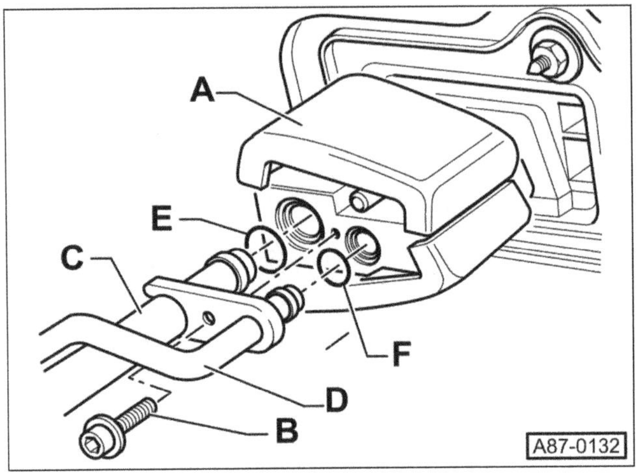

◄ Unclip expansion valve insulation (**A**).

– Remove refrigerant line securing screw (**B**).

– Remove refrigerant lines (**C**, **D**).

– Seal off open ends of refrigerant lines and expansion valve ports to prevent ingress of moisture (humidity) and dirt.

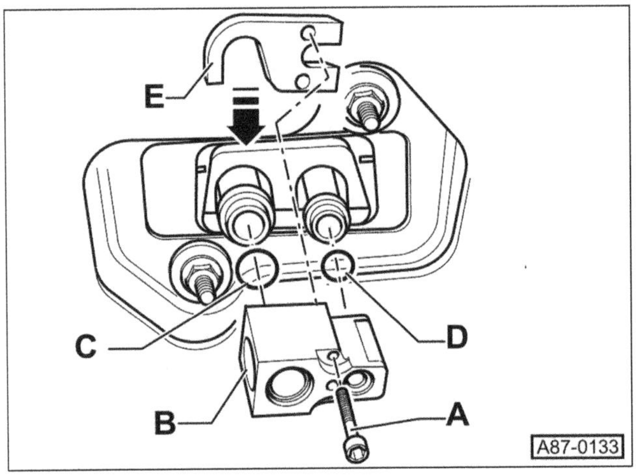

◄ Remove securing screw (**A**).

– Slide expansion valve (**B**) off of connection on evaporator (**arrow**).

– Remove threaded plate (**E**).

– Installation of the expansion valve is the reverse of removal noting the following:

- Always replace O-rings (**C**, **D**).
- Tighten expansion valve securing bolt (**A**) to 8 Nm (71 in-lb).

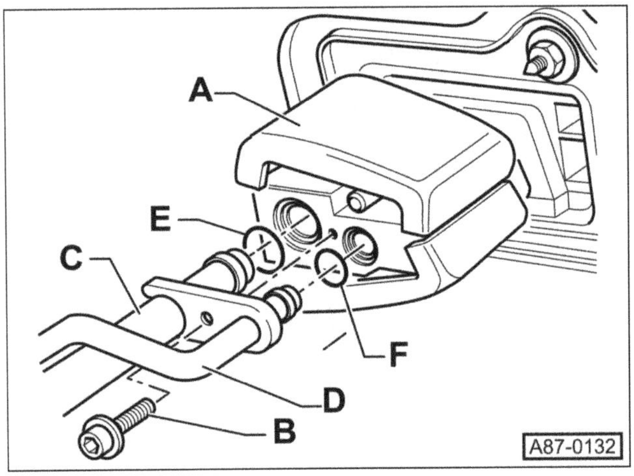

◄ Installation of the refrigerant lines is the reverse of removal noting the following:

- Always replace O-rings (**E**, **F**).
- Tighten expansion valve securing bolt (**B**) to 8 Nm (71 in-lb).

– Install insulation (**A**).

NOTE—

- *When installing expansion valve, ensure the expansion valve insulation is properly positioned and is not damaged.*
- *Due to the thermo-dynamic design of the expansion valve, the operating parameters of the expansion valve are adversely affected when the insulation is missing or improperly installed. A decrease in cooling efficiency will result.*

– Evacuate and recharge system.

BentleyPublishers
.com
© 2006 BentleyPublishers.com—All Rights Reserved

A/C Compressor

Compressor mounting bracket (1.8L engine)

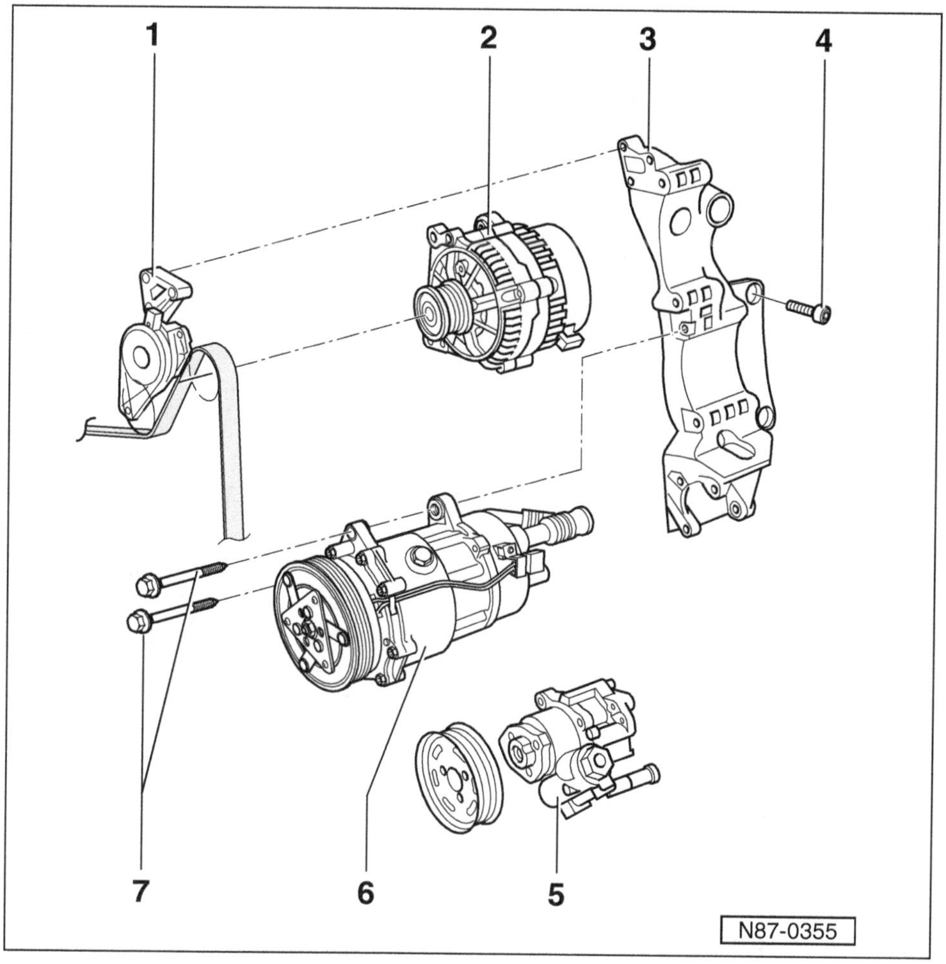

1. **Tensioning roller, ribbed V-belt**
   - See **Ribbed belt routing**.
2. **Generator (GEN) and pulley**
3. **Bracket**
   - Compressor bracket and related components can be removed and installed without opening the refrigerant circuit.
   - Mounting for generator, A/C compressor, and power steering pump.
4. **Bolts**
   - Tighten to 50 Nm (37 ft-lb).
5. **Power steering pump and pulley**
6. **A/C compressor**
   - Can be removed without discharging refrigerant.
   - Secure compressor to body, see **Compressor, securing to body**.
   - See **Compressor overview, Sanden**.
   - See **Compressor overview, Zexel**.
   - See **Compressor refrigerant lines**.
7. **Bolts**
   - Tighten to 45 Nm (33 ft-lb

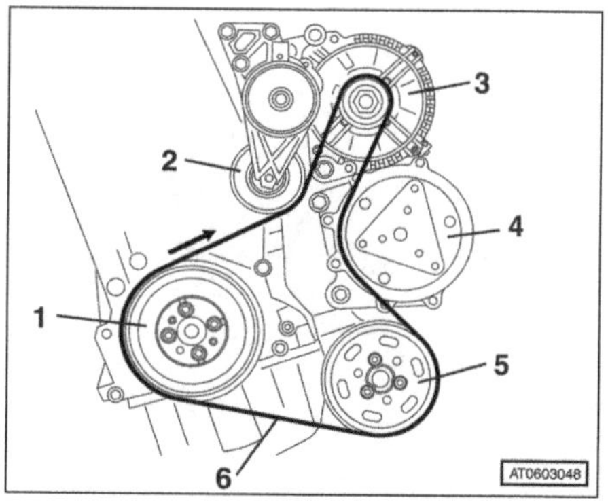

Ribbed belt routing

1. Belt pulley, crankshaft
2. Tensioning roller
3. Belt pulley, generator (alternator)
4. Belt pulley, A/C compressor
5. Belt pulley, power steering pump
6. Ribbed belt
   - **Arrow** indicates direction of travel.

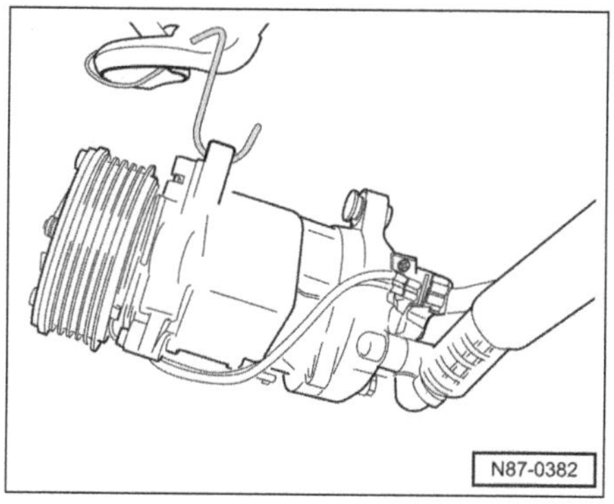

Compressor, securing to body

- Secure to body without discharging refrigerant.
- Do not hang by refrigerant hoses.
- Secure with hook made from welding rod or wire.

BentleyPublishers.com
© 2006 BentleyPublishers.com—All Rights Reserved

Compressor mounting bracket (3.2L engine)

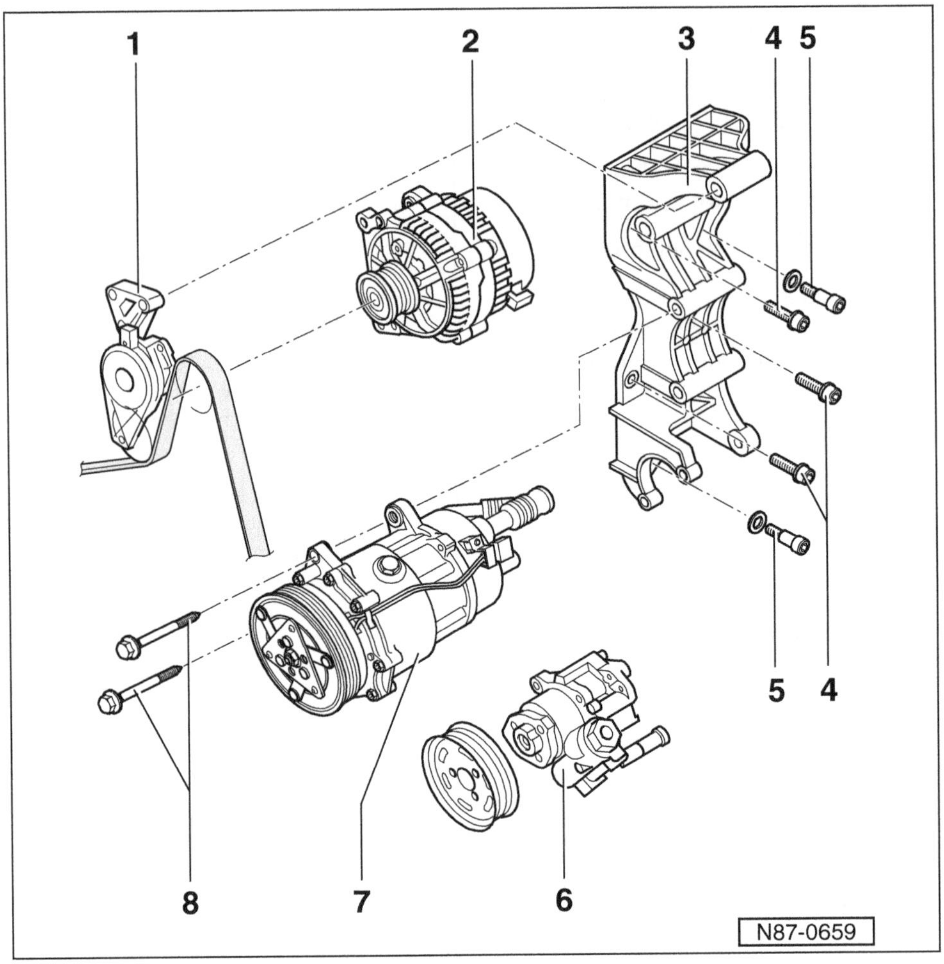

1. **Tensioning roller, ribbed V-belt**
   - See **Ribbed belt routing**.
2. **Generator (GEN) and pulley**
3. **Bracket**
   - Compressor bracket and related components can be removed and installed without opening the refrigerant circuit.
   - Mounting for generator, A/C compressor, and power steering pump.
4. **Bolts**
   - Tighten to 23 Nm (17 ft-lb).
5. **Bolts**
   - Tighten to 23 Nm (17 ft-lb).
   - Note washers where applicable.
6. **Power steering pump and pulley**
7. **A/C compressor**
   - Can be removed without discharging refrigerant.
   - Secure compressor to body, see **Compressor, securing to body**.
   - See **Compressor overview, Sanden**.
   - See **Compressor overview, Zexel**.
   - See **Compressor refrigerant lines**.
8. **Bolts**
   - Tighten to 45 Nm (33 ft-lb)

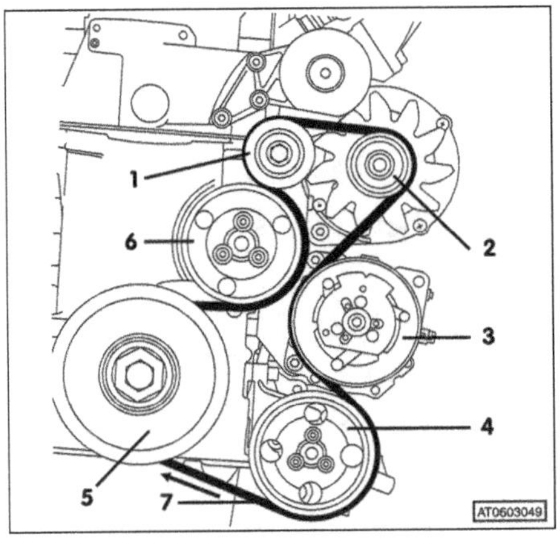

◀ **Ribbed belt routing**

1. Tensioning roller
2. Belt pulley, generator (alternator)
3. Belt pulley, A/C compressor
4. Belt pulley, power steering pump
5. Belt pulley, crankshaft
6. Belt pulley, coolant pump
7. Ribbed belt
   - **Arrow** indicates direction of travel.

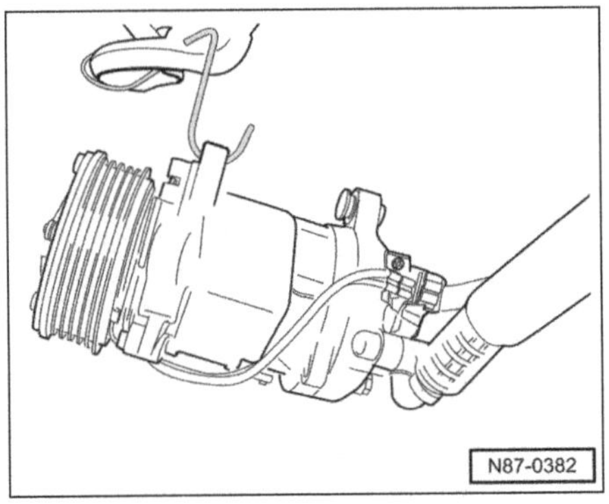

◀ **Compressor, securing to body**

- Secure to body without discharging refrigerant.
- Do not hang by refrigerant hoses.
- Secure with hook made from welding rod or wire.

BentleyPublishers.com

© 2006 BentleyPublishers.com—All Rights Reserved

Compressor overview, Sanden

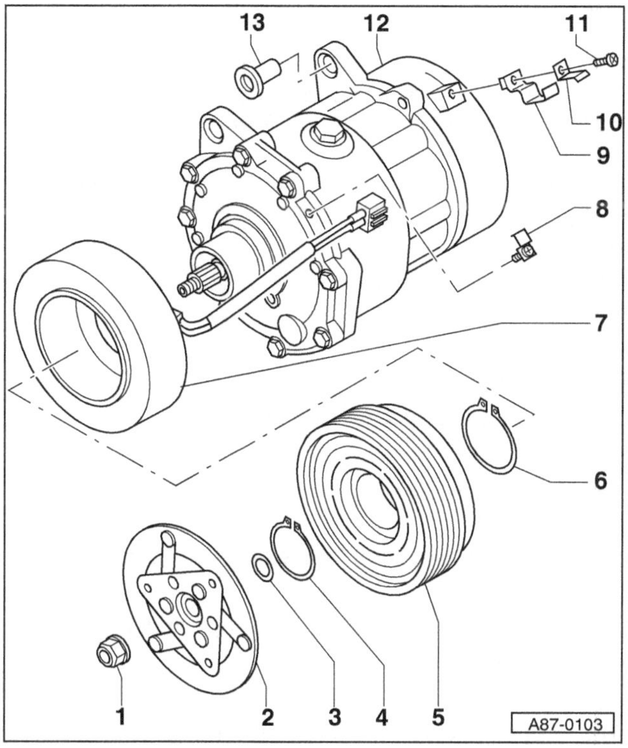

1. Nut, self-locking
   - Always replace.
   - Tighten to 15 Nm (11 ft-lb).
2. Clutch plate
3. Shims
   - For adjusting gap between clutch plate and belt pulley.
   - Adjust installed gap to 0.4 - 0.8 mm (0.016 - 0.032 in).
4. Circlip
   - Always replace.
   - Install flat side towards compressor.
   - Ensure proper seating in groove.
5. Ribbed belt pulley
   - Clean contact surfaces before installing.
6. Circlip
   - Always replace.
   - Install flat side towards compressor.
   - Ensure proper seating in groove.
7. Magnetic pull-in coil
   - With thermal fuse to interrupt operation if coil becomes too hot such as if compressor does not move freely.
   - With protection diode (J201) to absorb surges due to switching.
8. Harness retaining clip with screw
9. Retaining clip
10. Retaining clip
11. Screw
12. Compressor
13. Threaded bushing

Compressor overview, Zexel

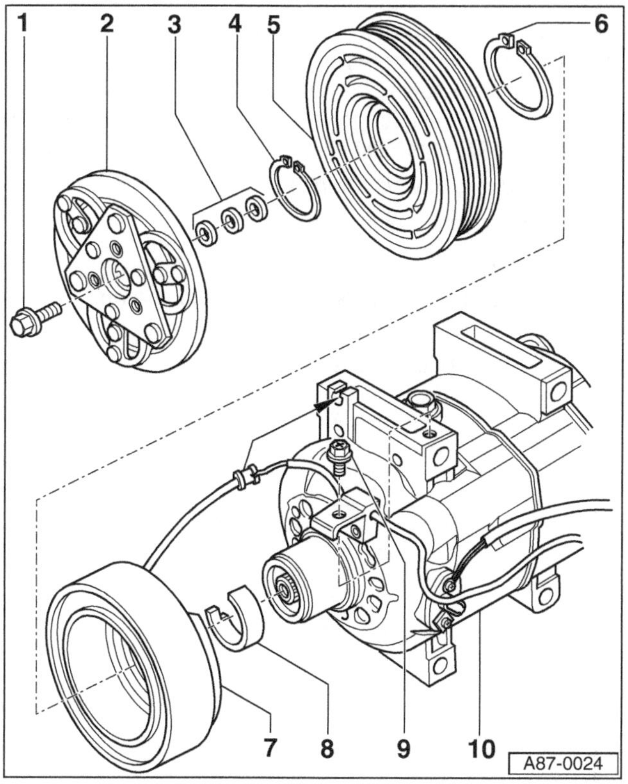

1. Bolt
   - Tighten to 15 Nm (11 ft-lb).
2. Clutch plate
3. Shims
   - For adjusting gap between clutch plate and belt pulley.
   - Adjust installed gap to 0.4 - 0.8 mm (0.016 - 0.032 in).
4. Circlip
   - Always replace.
   - Install flat side towards compressor.
   - Ensure proper seating in groove.
5. Ribbed belt pulley
   - Clean contact surfaces before installing.
6. Circlip
   - Always replace.
   - Install flat side towards compressor.
   - Ensure proper seating in groove.
7. Magnetic pull-in coil
   - Alignment pin fits into hole on compressor.
8. Felt ring
   - Always replace.
9. Screw
10. Compressor

BentleyPublishers
.com
© 2006 BentleyPublishers.com—All Rights Reserved

Compressor refrigerant lines

NOTE—
- *Do not remove refrigerant lines on compressor unless refrigerant has been discharged using approved procedures and equipment according to equipment manufacturer's instructions.*

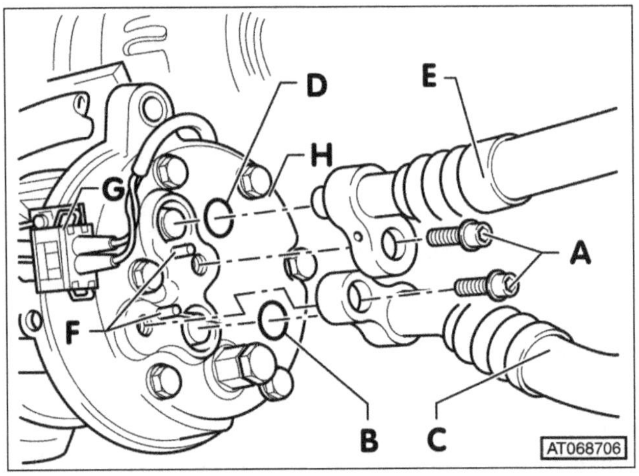

Line connection end of compressor showing refrigerant and electrical connections.

A. Screws
- Tighten to 20 Nm (15 ft-lb).

B. O-ring
- Always replace.
- 10.8 mm x 1.8 mm.

C. Refrigerant line (high pressure) to condenser

D. O-ring
- Always replace.
- 14.3 mm x 2.4 mm

E. Refrigerant line (low pressure) from expansion valve

F. Locating pins
- Ensure correct positioning before installing refrigerant lines.

G. Harness connector

H. Compressor
- Seal off all openings on compressor and lines to avoid ingress of dirt and humidity.

O-rings, overview

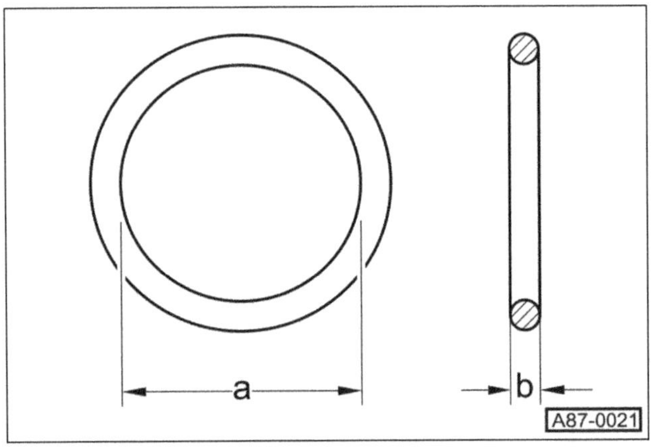

O-rings are used to seal the connections between various components of the A/C system.

Different types of O-rings are available, but not all are suitable for use in R134a systems. To ensure proper sealing, use only o-rings that are compatible with R134a refrigerant and PAG refrigerant oils. Use only O-rings with a known source. Color can be a determining factor with red, green, and violet generally being suitable. In some circumstances, black O-rings may also be used in production.

- Use O-rings only once, always replace.
- Handle in a clean working environment.
- Ensure proper fit into/onto component.
- Use the correct size for the application; dimensions **a** and **b**. Avoid using O-rings that must be stretched to fit.
- Lubricate with fresh PAG oil before installing.

BentleyPublishers.com
© 2006 BentleyPublishers.com—All Rights Reserved

9 Electrical System–General

GENERAL

This repair group presents a brief description of the principal parts of the electrical system, including the various bus systems. Also covered here are basic electrical system troubleshooting tips.

For additional information on electrical systems and components, refer to the following repair groups:

- **27 Battery, Alternator, Starter**
- **57 Doors and Central Locking**
- **64 Door Windows**
- **96 Interior Lights, Switches, Anti-theft**
- **97 Fuses, Relays, Component Locations**
- **EWD Electrical Wiring Diagrams**

TT electrical system

TT models are equipped with a 12-volt, direct current (DC), negative-ground electrical system. The voltage regulator maintains the voltage in the system at approximately 12 volts. All circuits are grounded by direct or indirect connection to the negative (–) terminal of the battery.

All electrical circuits except those required for starting and operating the engine are protected by fuses. Fuses for individual components are housed in the fuse panel at the left end of the dashboard. There are also a number of safety fuses located in the main fuse panel in the front left side of the engine compartment (on top of battery for 4-cylinder vehicles). For a full listing of fuses and fuse locations, see **97 Fuses, Relays, Component Locations**.

Modern Audi vehicles are electrically complex. Many vehicle systems and subsystems are interconnected or integrated. In addition, the requirements of second generation On-Board Diagnostics (OBD II) are such that there are many more circuits and wires in the vehicle than ever before. The components must exchange large volumes of data with one another in order to perform their various functions.

BentleyPublishers.com

© 2006 BentleyPublishers.com—All Rights Reserved

The TT makes extensive use of control module networking using electrical buses. Signals are shared digitally among several control modules on a bus, eliminating the need for separate connections for each pair of control modules. The use of bus communication for controls and accessories reduces wiring complexity and improves system response time. For more information on electrical components that are part of the bus system, see **97 Fuses, Relays, Component Locations**.

Warnings and Cautions

Please read the following before doing any work on vehicle electrical system.

WARNING —

- *Airbags and pyrotechnic seat belt tensioners utilize explosive devices. Handle with extreme care. Refer to the warnings and cautions in* **69 Seatbelts, Airbags**.
- *The ignition system of the car operates at lethal voltages. If you have a weak heart or wear a pacemaker, do not expose yourself to the ignition system electric currents. Take extra precautions when working on the ignition system or when servicing the engine while it is running or the key is ON. See* **28 Ignition System** *for additional ignition system warnings and cautions.*
- *Keep hands, clothing and other objects clear of the electric radiator cooling fan when working on a warm engine. The fan may start at any time, even when the ignition is switched OFF.*

CAUTION—

- *Do not disconnect the battery with the engine running.*
- *Switch the ignition OFF and remove the negative (–) battery cable before removing any electrical components. Connect and disconnect electrical connectors and ignition test equipment leads only while the ignition is switched OFF.*
- *Relay and fuse positions are subject to change and may vary from car to car. If questions arise, an authorized Audi dealer is the best source for the most accurate and up-to-date information.*
- *Use a digital multimeter for electrical tests. Switch the multimeter to the appropriate function and range before making test connections.*
- *Many control modules are static sensitive. Static discharge damages them permanently. Handle the modules using proper static prevention equipment and techniques.*
- *To avoid damaging harness connectors or relay panel sockets, use jumper wires with flat-blade connectors that are the same size as the connector or relay terminals.*
- *Do not try to start the engine of a car which has been heated above 176°F (80°C) (for example, in a paint drying booth). Allow it to cool to normal temperature.*
- *Disconnect the battery before doing any electric welding on the car.*

BentleyPublishers
.com
© 2006 BentleyPublishers.com—All Rights Reserved

CAUTION—
- *Do not wash the engine while it is running, or any time the ignition is ON.*
- *Choose test equipment carefully. Use a digital multimeter with at least 10 MΩ input impedance, or an LED test light. An analog meter (swing-needle) or a test light with a normal incandescent bulb may draw enough current to damage sensitive electronic components.*
- *Do not use an ohmmeter to measure resistance on solid state components such as control modules.*
- *Disconnect the battery before making resistance (ohm) measurements on a circuit.*

ELECTRICAL TROUBLESHOOTING

Four things are required for current to flow in any electrical circuit:
- Voltage source
- Wires or connections to transport voltage
- Load or device that uses electricity
- Connection to ground

Most problems can be found using a digital multimeter (volt / ohm / ammeter) to check the following:
- Voltage supply
- Breaks in the wiring (infinite resistance / no continuity)
- A path to ground that completes the circuit

Electric current is logical in its flow, always moving from the voltage source toward ground. Electrical faults can usually be located through a process of elimination. When troubleshooting a complex circuit, separate the circuit into smaller parts. General tests outlined below may be helpful in finding electrical problems. The information is most helpful when used with wiring diagrams. See **EWD Electrical Wiring Diagrams**.

Be sure to analyze the problem. Use wiring diagrams to determine the most likely cause. Get an understanding of how the circuit works by following the circuit from ground back to the power source.

When making test connections at connectors and components, use care to avoid spreading or damaging the connectors or terminals. Some tests may require jumper wires to bypass components or connections in the wiring harness. When connecting jumper wires, use blade connectors at the wire ends that match the size of the terminal being tested. The small internal contacts are easily spread apart, and this can cause intermittent or faulty connections that can lead to more problems.

BentleyPublishers
.com
© 2006 BentleyPublishers.com—All Rights Reserved

Voltage and voltage drop

Wires, connectors, and switches that carry current are designed with very low resistance so that current flows with a minimum loss of voltage. A voltage drop is caused by higher than normal resistance in a circuit. This additional resistance actually decreases or stops the flow of current. Excessive voltage drop can be noticed by problems ranging from dim headlights to sluggish wipers. Some common sources of voltage drops are corroded or dirty switches, dirty or corroded connections or contacts, and loose or corroded ground wires and ground connections.

A voltage drop test is a good test to perform if current is flowing through the circuit but the circuit is not operating correctly. A voltage drop test helps pinpoint a corroded ground strap or a faulty switch. Normally, there should be less than 1 volt drop across most wires or closed switches. A voltage drop across a connector or short cable should not exceed 0.5 volt.

A voltage drop test is generally more accurate than a simple resistance check because the resistances involved are often too small to measure with most ohmmeters. For example, a resistance as small as 0.02 Ω would results in a 3 volt drop in a typical 150 amp starter circuit. (150 amps x 0.02 Ω = 3 volts).

Keep in mind that voltage with ignition key ON and voltage with engine running are not the same. With ignition ON and engine OFF (battery voltage), voltage should be approximately 12.6 volts. With engine running (charging voltage), voltage should be approximately 14.0 volts. Measure voltage at battery with ignition ON and then with engine running to get exact measurements.

Voltage, measuring

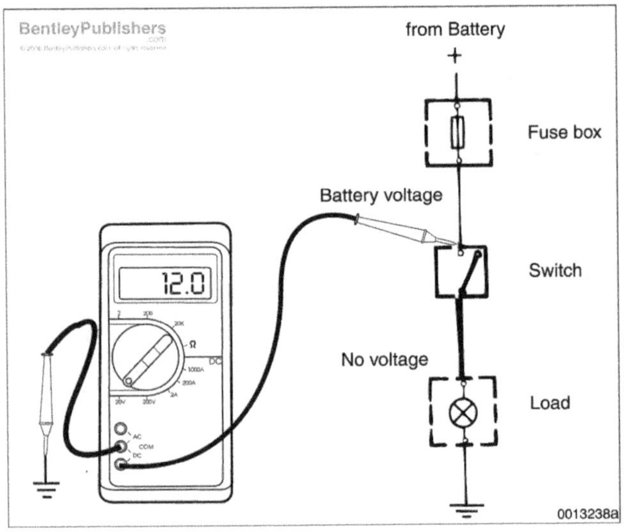

- Connect digital multimeter negative lead to a reliable ground point on car.

NOTE—
- *The negative (–) battery terminal is always a good ground point.*

– Connect digital multimeter positive lead to point in circuit you wish to measure.

– Voltage reading should not deviate more than 1 volt from voltage at battery. If voltage drop is more than this, check for a corroded connector or loose ground wire.

© 2006 BentleyPublishers.com—All Rights Reserved

Voltage drop, testing

Check voltage drop only when there is a load on the circuit, such as when operating the starter motor or turning on the headlights. Use a digital multimeter to ensure accurate readings.

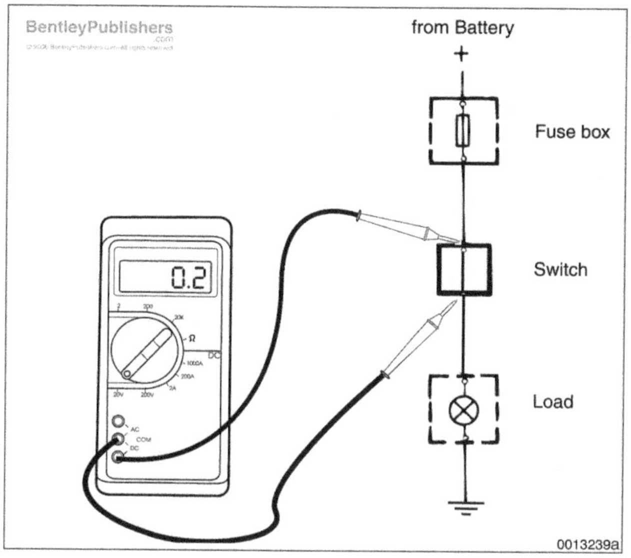

- Connect digital multimeter positive lead to positive (+) battery terminal or a positive power supply close to battery source.

- Connect digital multimeter negative lead to other end of cable or switch being tested.

- With power on and circuit working, meter shows voltage drop (difference between two points). This value should not exceed 1 volt.

- Maximum voltage drop in an automotive circuit, as recommended by the Society of Automotive Engineers (SAE), is as follows:
  - 0 volt for small wire connections
  - 0.1 volt for high current connections
  - 0.2 volt for high current cables
  - 0.3 volt for switch or solenoid contacts

- On longer wires or cables, the drop may be slightly higher. In any case, a voltage drop of more than 1.0 volt usually indicates a problem.

Continuity, checking

Use continuity test to check a circuit or switch. Because most automotive circuits are designed to have little or no resistance, a circuit or part of a circuit can be easily checked for faults using an ohmmeter. An open circuit or a circuit with high resistance does not allow current to flow. A circuit with little or no resistance allows current to flow easily.

When checking continuity, turn ignition OFF. On circuits that are powered at all times, disconnect battery. Using the appropriate wiring diagram, test circuit for faulty connections, wires, switches, relays and engine sensors by checking for continuity.

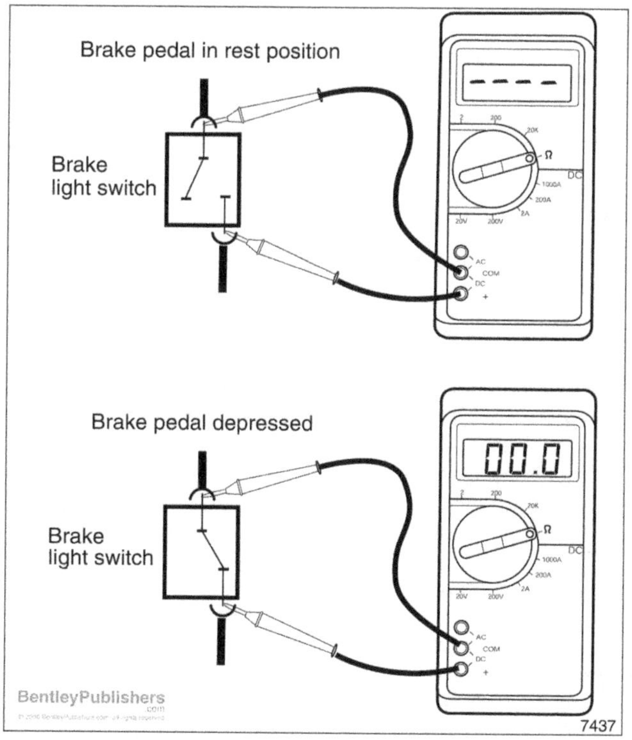

Example: Test brake light switch for continuity:
- With brake pedal in rest position (switch open) there is no continuity (infinite Ω).
- With pedal depressed (switch closed) there is continuity (0 Ω).

Short circuits

Short circuits are exactly what the name implies. The circuit takes a shorter path than it was designed to take. The most common short that causes problems is a short to ground where the insulation on a positive (+) wire wears away and the metal wire is exposed. When the wire rubs against a metal part of the car or other ground source, the circuit is shorted to ground. If the exposed wire is live (positive battery voltage), a fuse blows and the circuit may be damaged.

BentleyPublishers
.com
© 2006 BentleyPublishers.com—All Rights Reserved

Short circuits are often difficult to locate and may vary in nature. Short circuits can be found using a logical approach based on knowledge of the current path.

Use a digital multimeter to locate short circuits.

> ***CAUTION—***
> - *In circuits protected with high rating fuses (25 amp and greater), wires or circuit components may be damaged before the fuse blows. Check for wiring damage before replacing fuses of this rating.*
> - *When replacing blown fuses, confirm correct fuse rating. See* **97 Fuses, Relays, Component Locations**.

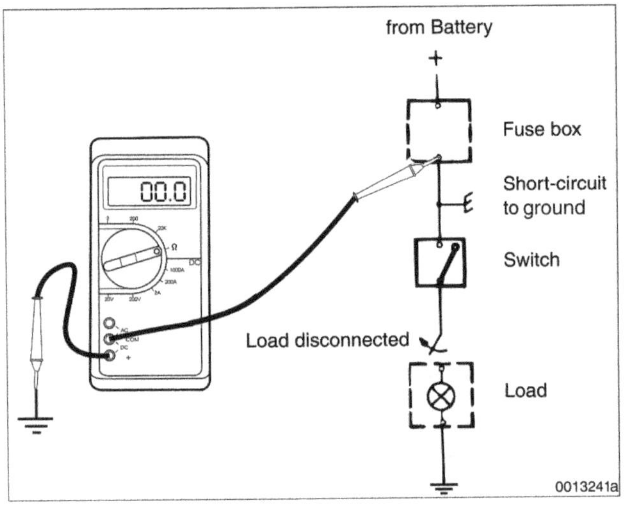

Short circuit, testing with ohmmeter

- Remove blown fuse from circuit and disconnect cables from battery. Disconnect harness connector from circuit load or consumer.
- Using an ohmmeter, connect one test lead to load side of fuse terminal (terminal leading to circuit) and other test lead to ground.
- If there is continuity to ground, there is a short to ground.
- If there is no continuity, work from wire harness nearest to fuse / relay panel and move or wiggle wires while observing meter. Continue to move down harness until meter displays a reading. This is the location of the short to ground.
- Visually inspect wire harness at this point for any faults. If no faults are visible, carefully slice open harness cover or wire insulation for further inspection. Repair any faults found.

Short circuit, testing with voltmeter

- Remove blown fuse from circuit. Disconnect harness connector from circuit load or consumer.

NOTE—
- *Most fuses power more than one consumer. Be sure all consumers are disconnected when checking for a short circuit.*

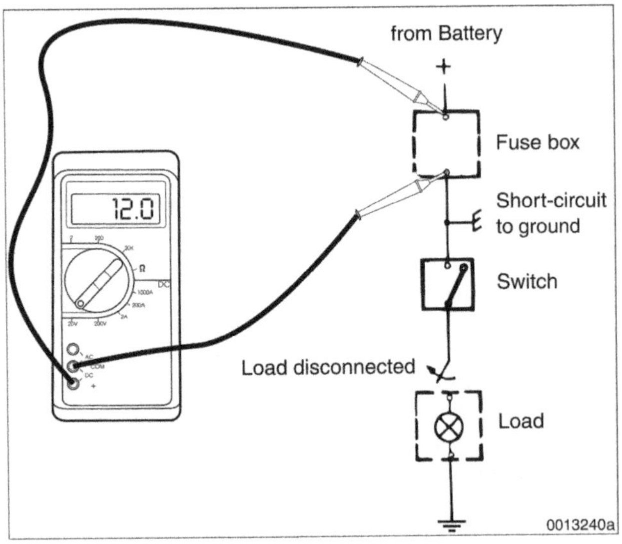

- Using a digital multimeter, connect test leads across fuse terminals. Make sure power is present in circuit. If necessary switch ignition ON.
- If voltage is present at voltmeter, there is a short to ground.
- If voltage is not present, work from wire harness nearest to fuse / relay panel and move or wiggle wires while observing meter. Continue to move down harness until meter displays a reading. This is the location of the short to ground.

  Visually inspect wire harness at this point for any faults. If no faults are visible, carefully slice open harness cover or wire insulation for further inspection. Repair any faults found.

BentleyPublishers.com
©2006 BentleyPublishers.com—All Rights Reserved

90 Instruments

GENERAL

This repair group covers the instrument cluster, harness connector terminal identification on the back of the instrument cluster and warning lights.

Instrument cluster

The instrument cluster contains an electronic speedometer, tachometer, liquid crystal (LCD) displays for odometer, trip odometer, digital clock with date, as well as analog coolant temperature and fuel level gauges. Control and warning lamps are situated within the speedometer and tachometer faces.

A driver's information system LCD is located between the tachometer and speedometer. Displayed information includes an auto check system, outside air temperature, radio frequency and selectable trip computer functions. The auto check system monitors the brake system, coolant temperature, fuel level, engine oil pressure, and displays the ambient temperature as a default. The optional navigation system display is also included in the driver's information system (as applicable).

Warning and indicator lights in the speedometer and tachometer faces are LEDs which cannot be serviced separately. An LED malfunction requires instrument cluster replacement.

The instrument cluster also supplements various warning light displays with acoustic warnings.

On Board Diagnostics (OBD)

OBD program functions on vehicles from 09/2000 production (2001 and later models) must be performed using the VAS 5051/5052 vehicle diagnostic testing and information system in the "Guided Fault Finding" mode. Only the VAS 5051/5052 is capable of processing all display text/data, as well as performing coding, adaptation and related OBD program functions on these vehicles. See **OBD On Board Diagnostics** for more information.

BentleyPublishers.com
© 2006 BentleyPublishers.com—All Rights Reserved

Instrument cluster layout

NOTE—
- *Most indicator lights will illuminate briefly when vehicle is started and go out if there are no malfunctions.*

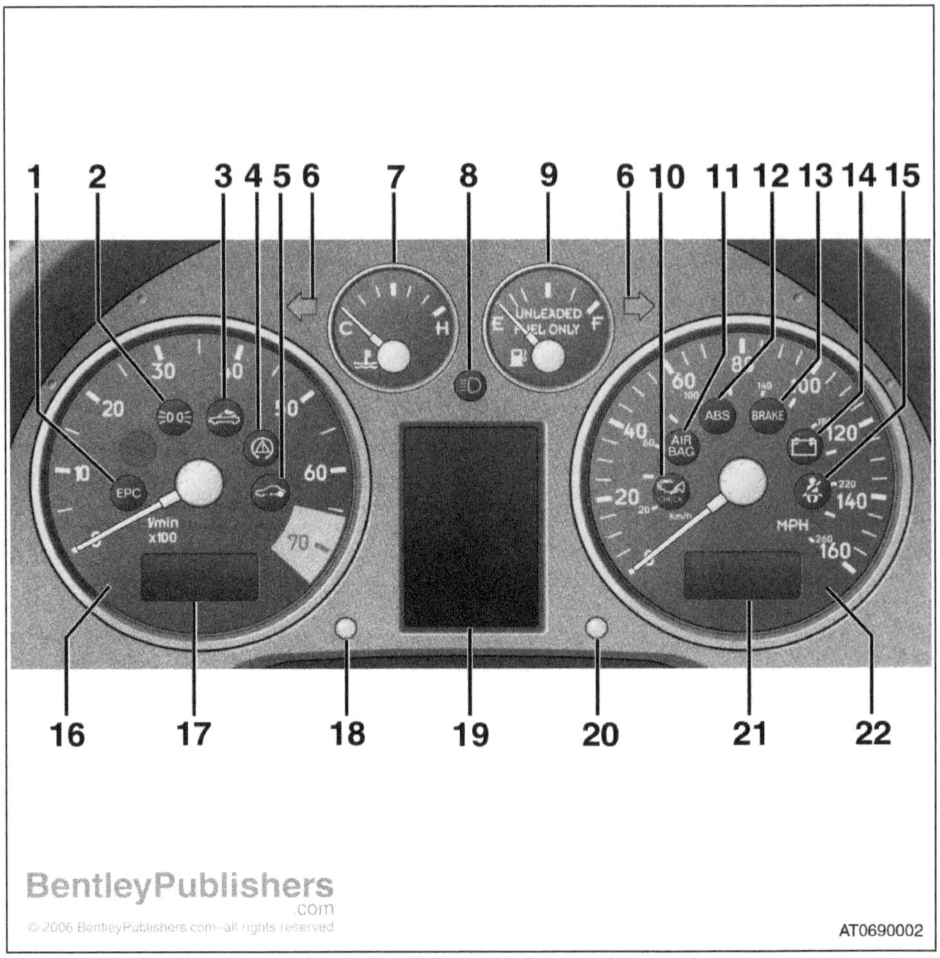

1. **Electronic power control (EPC) indicator**
   - Will illuminate if malfunction detected and EPC will switch to failure-mode program
2. **Side marker/headlight indicator**
3. **Folding top indicator**
   - If light stays on, the folding top latch may not be properly latched
   - If light flashes when operating open/close switch for top, there is a fault in the system
   - Do not operate vehicle without securely latching folding top
4. **Electronic stabilization program (ESP) indicator**
   - Will flash while vehicle is being driven with ESP in closed-loop control
   - Will stay on if ESP system has a malfunction or is off
   - Will also illuminate if ABS system has a malfunction
5. **Electronic immobilizer indicator**
   - Scans key for immobilizer code
   - If a non-coded key is used, light will flash continuously and vehicle will not start
6. **Turn signal indicators**
   - Both indicators will flash when emergency (4-way) flasher is on
7. **Coolant temperature gauge**
8. **High beam indicator**
9. **Fuel gauge**
10. **Malfunction indicator lamp (MIL)**
    - Will illuminate if Diagnostic Trouble Code (DTC) is present. See **OBD On Board Diagnostics**.
11. **Airbag warning indicator**
    - Malfunction indicated by:
      Light not illuminated at start-up
      Light illuminates while driving
      Light is on continuously
    - If malfunction is indicated, have airbag/seatbelt system inspected by authorized Audi dealer
12. **Antilock brake (ABS) indicator**
    - If light does not go out or illuminates while driving, ABS is not working properly. Vehicle can still be safely driven and stopped but ABS protection will not be available.
    - See **45 Antilock Brakes (ABS)** for more information
13. **Brake indicator**
    - Will illuminate if:
      Brake fluid low
      Parking brake on
      Brake system malfunction
14. **Generator/charging system indicator**
15. **Seat belt indicator**
16. **Speedometer**
17. **Digital clock and date display**
18. **Set/check knob**
    - Setting clock
    - Auto-check system
    - Odometer lighting
    - Service interval reset
19. **Driver information system display**
    - Auto-check system
    - Outside air temperature
    - Selector lever position (automatic or DSG transmission)
    - Trip computer display
    - Radio frequency display
    - Service interval
20. **Reset knob for trip odometer**
21. **Odometer/trip odometer display**
22. **Tachometer**

NOTE—
- *The warning lights and information display windows may vary depending on vehicle options and trim level.*

BentleyPublishers.com
© 2006 BentleyPublishers.com—All Rights Reserved

Electrical system safety precautions

Read the following and the additional warnings and cautions in **97 Fuses, Relays, Component Locations** before doing any work on your electrical system.

> ***CAUTION—***
> - *Prior to disconnecting the battery, read the battery disconnection cautions in* **27 Battery, Alternator, Starter**.
> - *Switch the ignition OFF and remove the negative (–) battery cable before removing any electrical components.*
> - *Connect and disconnect electrical connectors and ignition test equipment leads only while the ignition is OFF.*
> - *Only use a digital multimeter for electrical tests.*
> - *To avoid damaging plastic interior trim, use a plastic prying tool or a screwdriver with the tip wrapped with masking tape.*
> - *Some dashboard knobs and buttons are designed to break away in a controlled manner in order to protect occupants. When working on interior components, avoid damage by not pressing on control buttons or displays.*

Immobilizer

TT models produced from 09/2000 (2001 and later models) are equipped with Immobilizer Generation III.

Generation III has the same functionality as the previous immobilizer version. However, in addition to the normal data exchange between the immobilizer control module and ignition key, Generation III includes data exchange between the immobilizer control module and Engine Control Module (ECM).

Data exchange between the Immobilizer control module and ECM takes place over the vehicle CAN-Bus network. For Immobilizer Generation III, the ECM is actively incorporated into the evaluation and monitoring process. The immobilizer control module is integrated into the instrument cluster.

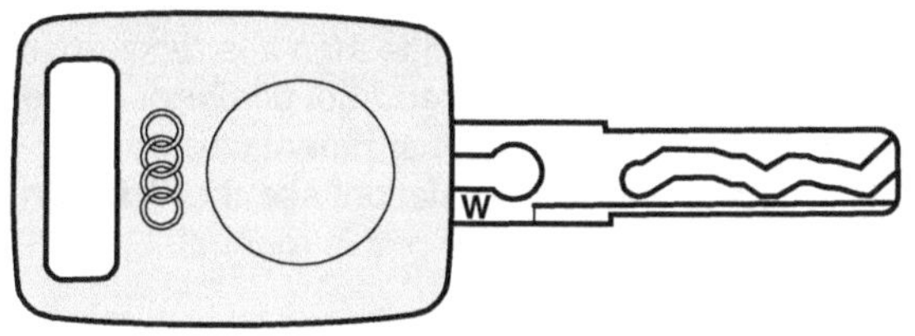

A96-0280

For Immobilizer III, the keys are marked with an inscribed "**W**". The code consists of a fixed code and a variable code portion.

INSTRUMENT CLUSTER SERVICE

Service reminder display

Most vehicles covered by this manual are equipped with a service reminder display (also called a service interval display) that will come on starting at approximately 1200 miles (2000 kilometers) before a scheduled service is required when the engine is started. The display will read "SERVICE IN 1,200 MLS" as the interval approaches. The 1,200 MLS will be reduced at each successive engine start until the service is due. At that point, the display will simply read "SERVICE". Kilometers are substituted where applicable. The display will return to normal operation a few seconds after the engine starts.

BentleyPublishers.com
© 2006 BentleyPublishers.com—All Rights Reserved

The display should be reset after the appropriate service has been completed.

- Switch ignition off.

B8N-316

- Press and hold trip odometer reset button (**2**) while switching ignition on; "SERVICE" will appear in the display.

- Press the digital clock set button (**1**) until the "SERVICE" text is cleared from the display. The next service event should now be showing.

NOTE—
- *Do not reset the display between intervals. This will result in inaccurate displays.*
- *The display information remains in memory even if the battery is disconnected.*
- *Malfunctions in the instrument cluster will result in a continuous display of "dEF" and should be diagnosed as soon as possible.*

Instrument cluster, replacing

If the instrument cluster is to be replaced, it will be necessary to use a VAG 1551/1552 scan tool, 5051/5052 vehicle diagnostic testing and information system, or equivalent aftermarket diagnostic scan tool or computer program. The scan tool is used to code the instrument cluster and diagnostic interface for the data bus as well as adapting the odometer display, service interval data, vehicle immobilizer and reactivate the radio anti-theft system. If you do not have a proper scan tool and/or experience then it is recommended that the vehicle be taken to an authorized Audi dealer or properly equipped independent repair shop for instrument cluster replacement. For more scan tool information, see the **OBD On Board Diagnostics** section at the rear of this manual.

Malfunction recognition and display

The instrument cluster has On-Board Diagnostic (OBD) capability, which is an aid to troubleshooting.

If the instrument cluster control module detects a malfunction with the odometer or speedometer that is non-repairable, then "dEF" will appear in the trip recorder display. Replacement of the instrument cluster will be necessary.

If the instrument cluster control module detects a malfunction and nothing is displayed on the trip recorder display, carry out the following work before removing the instrument cluster:

- Check DTC memory with VAG 1551/5051 or equivalent scan tool .

- Read service interval display values and odometer reading with VAG 1551/5051 or equivalent scan tool and note the values.

If the malfunction indicates that the instrument cluster requires replacement, the recorded values will need to be entered in the new instrument cluster.

BentleyPublishers.com
© 2006 BentleyPublishers.com—All Rights Reserved

Instrument cluster, removing and installing

> ***WARNING —***
> - *Special safety precautions apply to vehicles equipped with airbags.*
> - *Refer to airbag **CAUTIONS** and **WARNINGS** in **69 Seatbelts, Airbags**.*

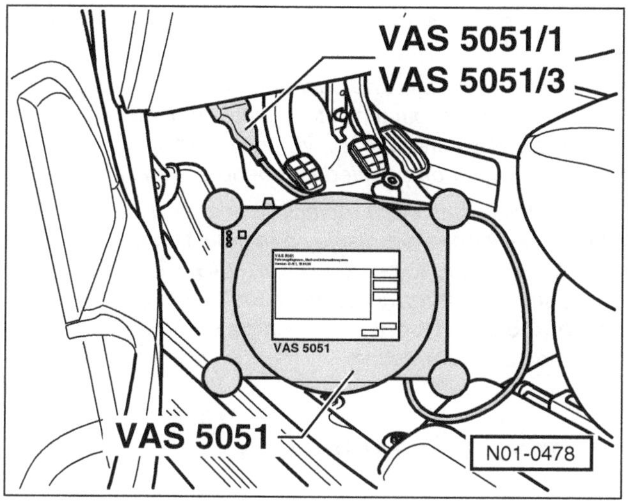

Replacement of instrument cluster **from 2001 m.y.** must be accomplished by properly coding it with the VAS 5051/5052 vehicle diagnostic testing and information system.

> ***WARNING —***
> - *Disconnect battery ground (GND) strap from battery negative (–) terminal.*
> - *Be sure to have the anti-theft radio code on hand before disconnecting the battery.*
> - *Disconnecting the negative (–) battery cable may erase fault codes and basic settings in the engine management and automatic transmission control modules. Some driveability problems may be noticed until the system re-adapts to operating conditions. OBD II readiness codes, which may be required for emissions testing, may also be erased. Convenience electronics (alarm system, interior light control, power locks, mirrors, and windows) may need to be re-set using the VAG 5051/5052 vehicle diagnostic testing and information system.*
> - *See **Malfunction recognition and display** given earlier before removing instrument cluster.*

NOTE —
- *If replacing the instrument cluster, be sure to note the mileage indicated in the old cluster before removal.*
- *For ease of illustration the steering wheel is not shown in the following illustrations. It is not necessary to remove the steering wheel to remove the instrument cluster.*

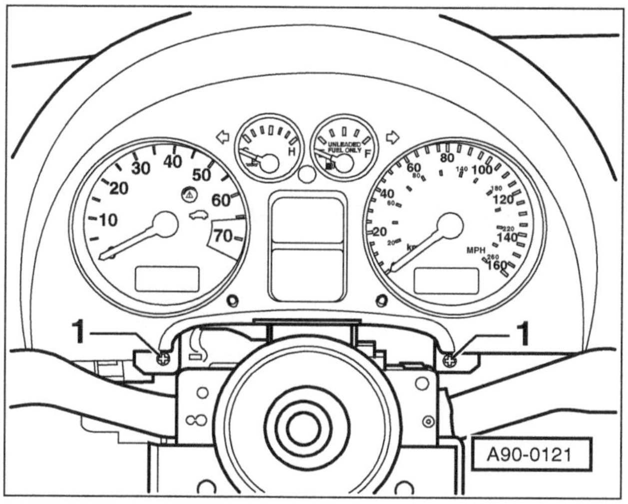

– Release steering wheel position lock, pull steering wheel out completely and lock again in lowest position.

– Remove left side under-dash panel. See **70 Trim-Interior**.

– Remove instrument cluster mounting screws (**arrows**).

– Pull out cluster from bottom to disengage upper mounting clips.
  - Pull out cluster far enough to access harness connectors on rear.

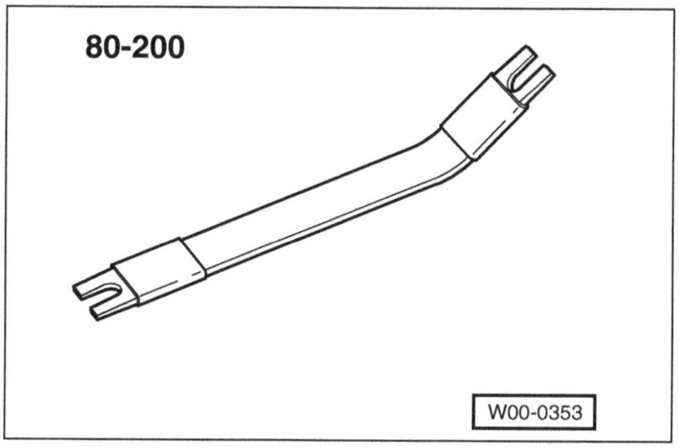

– Disconnect harness connectors at rear of instrument cluster.
  - Use pry lever 80-200, or equivalent, to release harness connector retaining levers.

– Remove cable ties from back of instrument cluster.

BentleyPublishers.com
© 2006 BentleyPublishers.com—All Rights Reserved

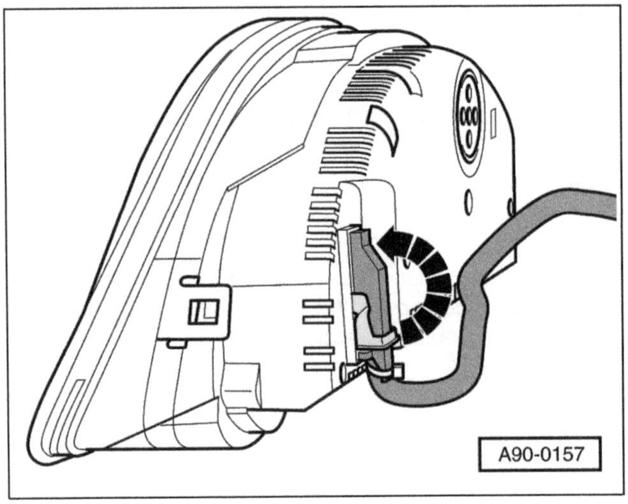

◄ Release harness connector retaining levers in direction of **arrow**.

– Remove instrument cluster.

– Installation is reverse of removal. After completing installation check function of instrument cluster.

NOTE—
- *Due to the large number of instrument cluster variations it always advisable to order a replacement cluster using the part number on the back of the cluster.*
- *Replacement clusters can have the odometer mileage adjusted to match the mileage of the original cluster. Because of the specialized nature of these repairs, servicing should be referred to an authorized Audi dealer or qualified independent repair shop.*
- *If the instrument cluster has been replaced because it was faulty, and a functional check does not display a malfunction the instrument cluster will need to be coded with VAG 1551/1552, 5051/5052, or equivalent aftermarket scan tool or computer program. See the* **OBD On Board Diagnostics** *section at the back of this manual for more information.*

Harness connector identification

NOTE—
- *The instrument cluster must not be disassembled.*
- *For troubleshooting specific systems and circuits see* **EWD Electrical Wiring Diagrams**.

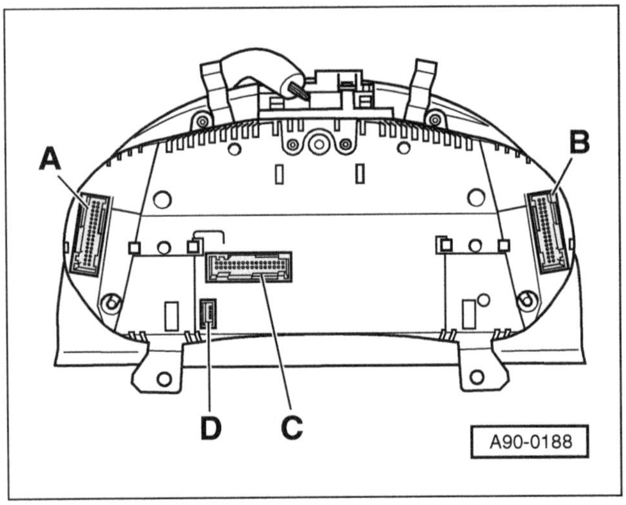

◄ Connectors located on the rear of the instrument cluster:
- **A** - 32-pin connector, green
- **B** - 32-pin connector, blue
- **C** - 32-pin connector, grey
- **D** - 4-pin connector, black (where applicable)

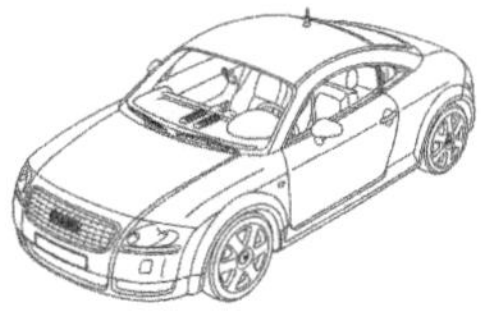

BentleyPublishers.com
© 2006 BentleyPublishers.com—All Rights Reserved

91 Radio and Communication

GENERAL

This repair group covers Audi factory installed sound systems as well as the optional navigation and cellular telephone systems.

After-market sound systems are beyond the scope of this manual. If you plan to install an aftermarket system, see **Cautions** in this repair group.

See **EWD Electrical Wiring Diagrams** for electrical circuit troubleshooting.

Concert sound system

Audi TT models are equipped with a variety of sound systems and features.

Sound systems (radios) offered are:

- Concert (basic) radio, with or without 6-CD changer.
- Bose system, with or without 6-CD changer.
- Navigation system integrated with radio and 6-CD changer.
- Bose system and satellite radio, with or without CD player.

AT0691001

◄ **Early model** radios can be identified by the built in cassette player. (2001 shown)

91

BentleyPublishers.com
© 2006 BentleyPublishers.com—All Rights Reserved

AT0691002

Later model radios have a single CD player integrated into radio. (2006 shown)

Radio anti-theft is built into Audi radios and the navigation system.

Interference suppression is built into standard and optional electrical consumers in the vehicle. Capacitors, coils and diodes for suppression are installed directly in the components.

On Board Diagnostics (OBD). The sound system and the navigation system (if equipped) self-monitors and stores diagnostic trouble codes (DTCs) in memory. Troubleshoot using Audi VAG 1551/1552, VAS 5051/5052 diagnostic scan tool or equivalent. See **OBD On Board Diagnostics**.

Cautions

Read the following cautions as well as those found in **97 Fuses, Relays, Component Locations** before doing any work on your electrical system.

CAUTION—

- *The radio is wired to the vehicle alarm system. If the alarm system is armed, removing the radio activates the alarm even if the proper tools are used. Do not attempt to remove the radio without disarming the alarm.*
- *The factory installed connectors are designed for genuine Audi radios. If installing a different radio, remember that the radio may not fit properly into the space provided, the electrical connections may not be compatible and different terminals may be needed.*
- *Prior to disconnecting the battery, read the battery disconnection cautions in* **27 Battery, Alternator, Starter**.
- *Switch the ignition OFF and remove the negative (–) battery cable before removing any electrical components.*
- *Connect and disconnect electrical connectors and ignition test equipment leads only while the ignition is OFF.*
- *Only use a digital multimeter for electrical tests.*
- *To avoid damaging plastic interior trim, use a plastic prying tool or a screwdriver with the tip wrapped with masking tape.*
- *Some dashboard knobs and buttons are designed to break away in a controlled manner in order to protect occupants. When working on interior components, avoid damage by not pressing on control buttons or displays.*

BentleyPublishers.com
© 2006 BentleyPublishers.com—All Rights Reserved

Sound System Components

NOTE—
- *Not all features are found in every vehicle.*
- *Concert with Bose system shown, with Bose-only components noted. Non-Bose system is similar.*
- *Audi identifies electrical components by a letter and/or a number in the electrical schematics. See* **EWD Electrical Wiring Diagrams**. *These electrical identifiers are listed in parentheses as an aid to electrical troubleshooting.*

Sound system overview (coupe)

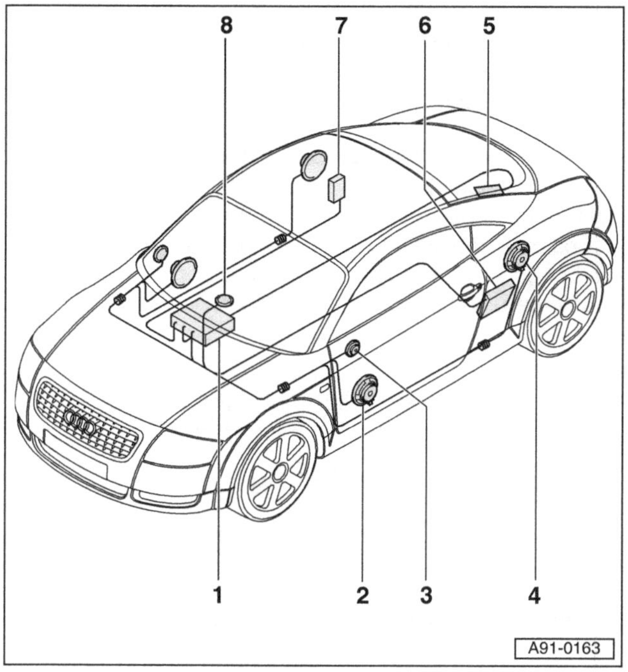

1. Radio
2. Mid-range/bass speaker
   - At bottom of front door trim
3. Treble speaker
   - At top of front door trim
4. Bass speaker
   - In left and right rear side trim
5. Antenna amplifier
   - Below left rear lid trim
6. CD changer
   - Below left rear side trim
7. Bose amplifier
   - Bose system only
   - Below right rear side trim
8. Center speaker
   - Bose system only
   - At front center of dashboard, in left-hand defroster vent

Sound system overview (roadster)

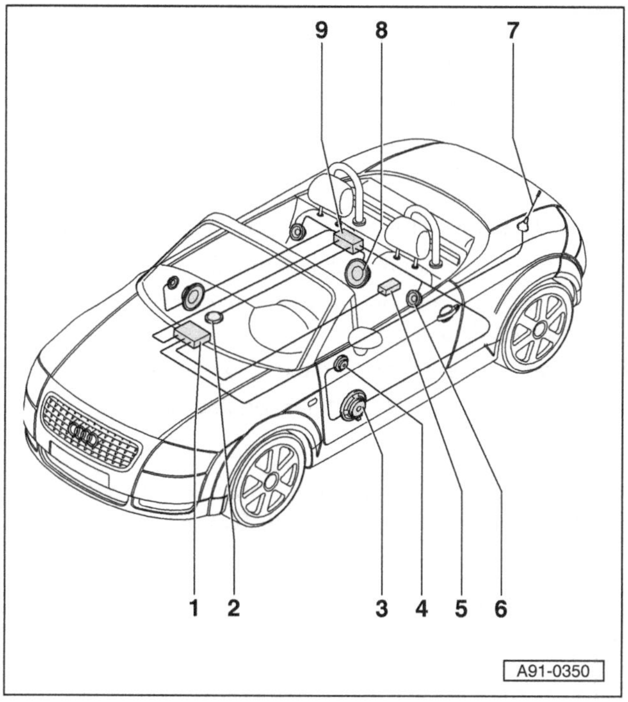

1. Radio
2. Center speaker
   - Bose system only
   - At front center of dashboard, in left-hand defroster vent
3. Mid-range/bass speaker
   - At bottom of front door trim
4. Treble speaker
   - At top of front door trim
5. CD changer
   - Behind driver's seat in rear trim panel
6. Bass speaker
   - In left and right rear side trim
7. Antenna
   - With amplifier
8. Center rear speaker, subwoofer
   - In center rear trim panel
9. Bose amplifier
   - Bose system only
   - Behind storage box in trunk

BentleyPublishers
.com
© 2006 BentleyPublishers.com—All Rights Reserved

Radio anti-theft system

The radio and navigation system (if equipped) are protected with an electronic anti-theft system. Identification data for the radio is stored in the instrument cluster. When the radio is removed (or the power supply is disconnected) and then reinstalled in the same vehicle, it is not necessary to input the anti-theft code.

After initial activation of the electronic anti-theft system, a numeric code is stored in the radio and the instrument cluster. If power is interrupted and then restored (e.g. after removing and installing the unit or disconnecting the battery), data exchange occurs between the radio and the cluster, comparing the numeric code of the unit to the numeric code stored in the instrument cluster. If the numeric code is identical, the instrument cluster recognizes that the unit belongs to the vehicle and it is then ready for operation.

If the radio or navigation unit is replaced, enter the new anti-theft code. If the anti-theft code is entered incorrectly, it can be corrected immediately for one subsequent attempt. If the code is entered incorrectly two times, the system is locked for an hour before it accepts another code. The radio code number and serial number are on the radio card supplied with the vehicle.

> ***CAUTION—***
> *• Keep the radio card in a safe place, not in the vehicle.*

Radio anti-theft code, entering

If the radio has been electronically locked the word "SAFE" will appear in the radio display when the radio is turned on.

– Obtain radio anti-theft card with anti-theft code.

– Turn on radio.

Early models with cassette player in radio

AT0691003

◄ Press the FM2 and RDS buttons (**arrows**) and hold them down until "1000" appears in radio display.

- Do not continue holding buttons after "1000" appears in display
- Do not push buttons again after "1000" appears in display. Doing this would reset the anti-theft code to 1000.

– Use station preset buttons (1 through 4) to enter anti-theft code.

- Use preset button 1 to enter first digit of anti-theft code.
- To enter code number, press preset button repeatedly until desired number appears on display.
- Use preset button 2 to enter second digit of anti-theft code.
- Repeat for remaining digits of anti-theft code using preset buttons 3 and 4.

– Once the correct anti-theft code has been entered, press the FM2 and RDS buttons and hold them down until "SAFE" appears in display.

– Soon afterwards, a radio frequency will be displayed. The radio can now be used.

BentleyPublishers
.com
©2006 BentleyPublishers.com—All Rights Reserved

Later models with single CD player in radio

Press the SAT and SCAN buttons (**arrows**) and hold them down until "1000" appears in radio display.

- Do not continue holding buttons after "1000" appears in display
- Do not push buttons again after "1000" appears in display. Doing this would reset the anti-theft code to 1000.

– Use station preset buttons (**1** through **4**) to enter anti-theft code.

- Use preset button **1** to enter first digit of anti-theft code.
- To enter code number, press preset button repeatedly until desired number appears on display.
- Use preset button **2** to enter second digit of anti-theft code.
- Repeat for remaining digits of anti-theft code using preset buttons **3** and **4**.

– Once the correct anti-theft code has been entered, press the SAT and SCAN buttons and hold them down until "SAFE" appears in display.

– Soon afterwards, a radio frequency will be displayed. The radio can now be used.

Radio, removing and installing

The part number for the replacement radio is printed on a sticker on the unit housing.

– Remove CD from single CD player, if applicable, i.e. later model cars with single CD player integrated into radio.

– Switch electrical consumers OFF. Switch ignition OFF and remove ignition key.

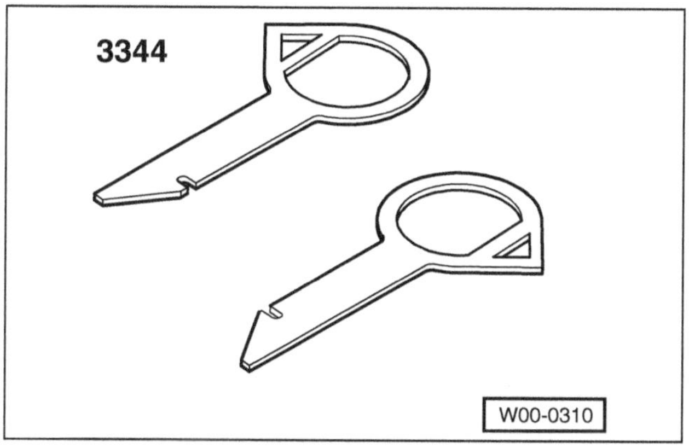

Radio removal tool 3344 is needed to remove the radio.

Insert radio removal tool 3344 into front of radio as shown at **arrows**.

- Left side tool: position so that "Top L" is facing up
- Right side tool: position so that "Top R" is facing up

– Pull radio unit back from opening until connections on rear of unit are accessible. Disengage electrical connections and antenna cable.

– To install, remove radio removal tools from front of radio.

– Plug in harness connectors and antenna.

– Push radio into dash panel until securing tabs lock radio firmly in place.

– If necessary, code new unit to vehicle using Audi diagnostic scan tool or equivalent.

BentleyPublishers.com
©2006 BentleyPublishers.com—All Rights Reserved

CD changer, removing and installing

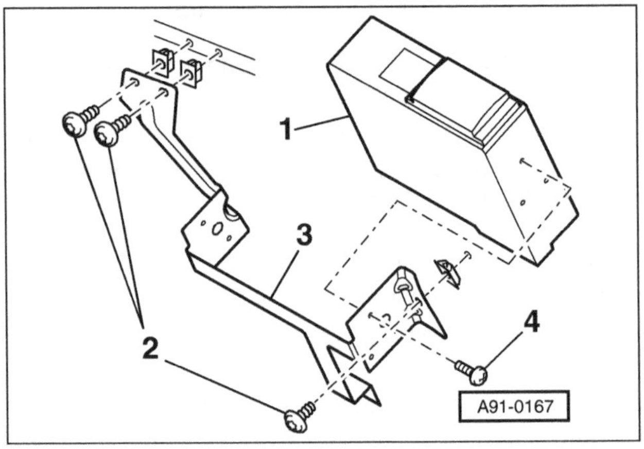

Coupe

- Remove left rear side trim panel. See **70 Trim–Interior**.
- Unplug electrical harness connector from CD changer.
- Remove mounting screws (**2**) from mounting frame (**3**).
- Remove mounting screw (**4**) and CD changer (**1**).
- Install in reverse order of removal.

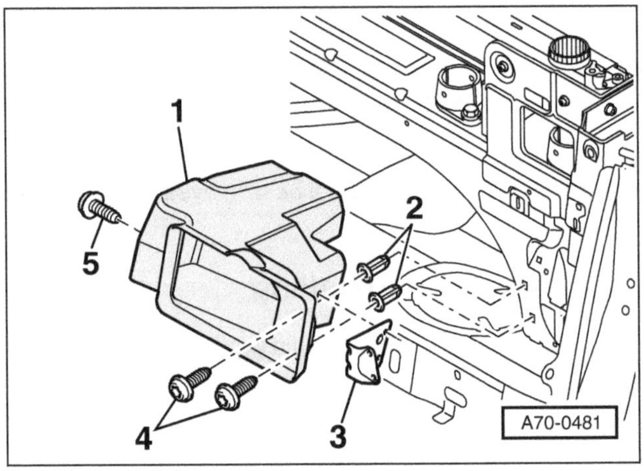

Roadster

- Remove left rear side trim panel.
- Remove center and left rear trim panels. See **70 Trim–Interior**.
- Remove center speaker. See **Speakers** in this repair group.
- Working behind center speaker, remove three mounting bolts (**5**).
- Remove two mounting bolts (**4**).
- Pull storage compartment (**1**) with CD changer out far enough to disconnect harness connectors on back.
- Unplug electrical harness connector from CD changer.

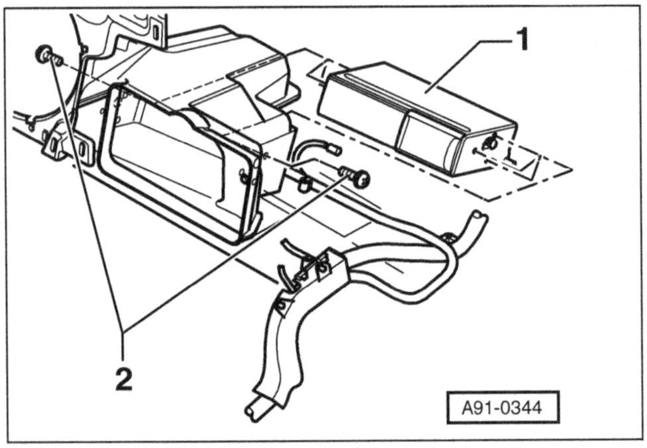

- Remove CD changer mounting screws (**2**).
  - There are three mounting screws on each side of the CD changer.
- Install in reverse order of removal.

CD changer, error messages

Early models with cassette player in radio

Various error messages for the CD changer can appear in the radio display window:

- **CD1 ERR1** = CD dirty or improperly installed
- **CD1 ERR2** = CD has scratches
- **CD1 ERR3** = Mechanical malfunction in CD changer

- Check CDs and CD magazine for damage and correct installation position.

BentleyPublishers
.com
©2006 BentleyPublishers.com—All Rights Reserved

– It is also possible to check the wiring connection between the radio and CD changer using the radio display.

◀ Press and hold MODE button (**arrow**) while turning radio on.

– Release MODE button and check radio display.
- "**CONNECT**" and "**CD**" = wiring connection OK
- "**NO CDC**" = wiring connection not OK

– If wiring connection is not OK, check harness connector on back of CD changer. If still not OK, check wiring using wiring diagrams. See **EWD Electrical Wiring Diagrams**.

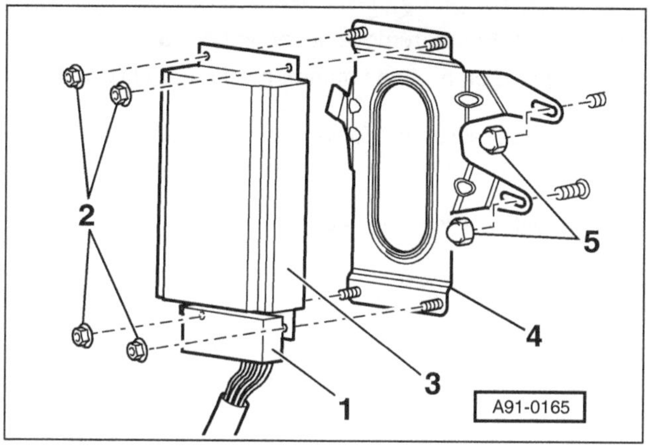

Bose amplifier, removing and installing

Coupe

– Remove right rear trim panel. See **70 Trim–Interior**.

◀ Unplug harness connector (**1**) from amplifier (**3**).

– Remove mounting nuts (**2**) and amplifier.

– To remove mounting bracket (**4**), unscrew mounting nuts (**5**).

– Install in reverse order of removal.

Tightening torque	
Amplifier to bracket	5 Nm (44 in-lb)

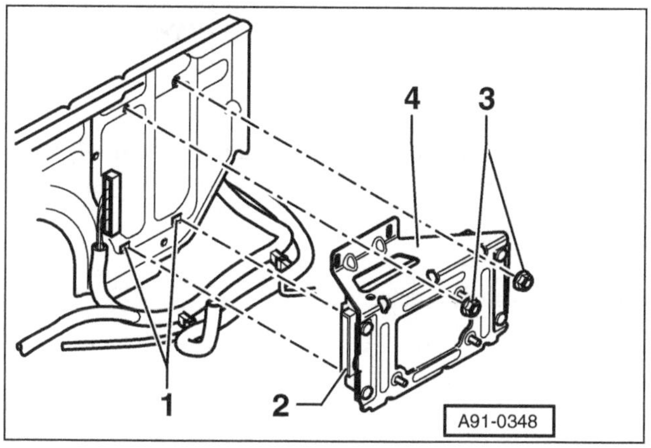

Roadster

– Remove carpet and storage compartment from trunk. See **70 Trim–Interior**.

◀ Unplug harness connector from amplifier (**2**).

– Remove mounting nuts (**3**) and swing mounting bracket (**4**) downward to remove.

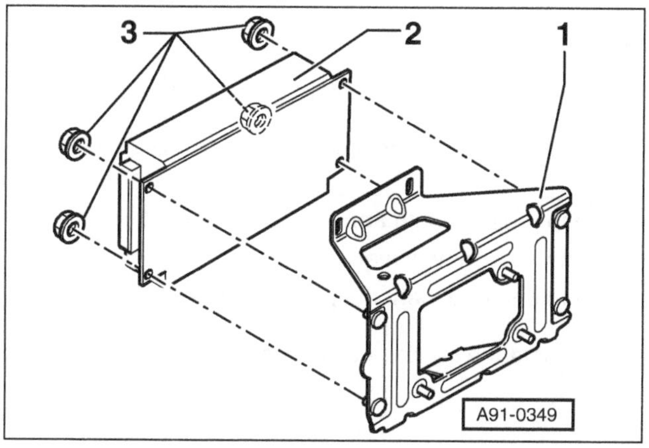

◀ Unbolt mounting nuts (**3**) and remove amplifier (**2**) from bracket (**1**).

– Install in reverse order of removal.

Tightening torque	
Amplifier bracket to body	5 Nm (44 in-lb)
Amplifier to bracket	3 Nm (27 in-lb)

BentleyPublishers .com
© 2006 BentleyPublishers.com–All Rights Reserved

Antenna

On coupe models the antenna is incorporated into the rear window. Models with satellite radio have a "shark fin" style antenna mounted on the roof. Removal of the headliner is necessary to remove the shark fin antenna. See **70 Trim–Interior**.

◀ Roadster models have the antenna mounted on the left rear fender.

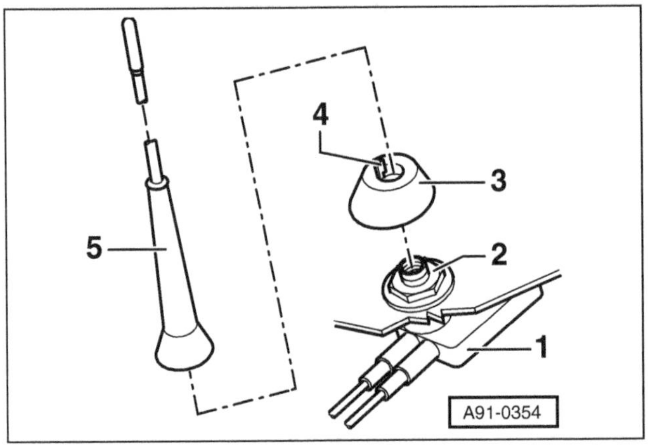

Antenna, removing and installing (roadster)

– Open trunk lid and remove left side trim panel. See **70 Trim–Interior**.

◀ Unscrew antenna mast (**5**) and remove covering cap (**3**).
- Covering cap is clipped on and has an anti-twist mechanism (**4**).

– Remove union nut with seal (**2**) and lower antenna base (**1**) downward.

– Disconnect antenna cable and remove base.

– Install in reverse order of removal.

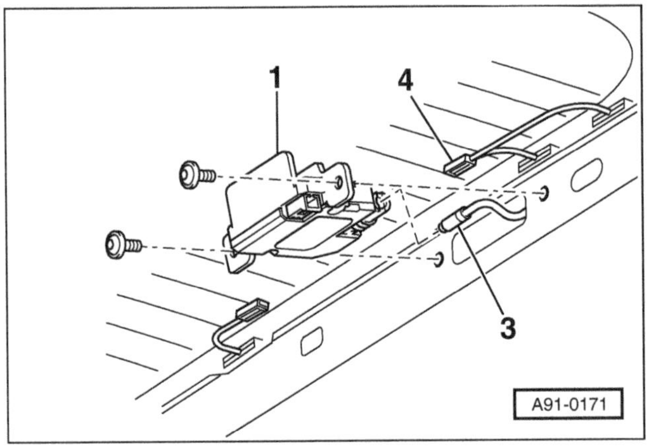

Antenna amplifier, removing and installing (coupe)

– Remove upper and lower tailgate trim panels. See **70 Trim–Interior**.

◀ Unscrew mounting bolts for antenna amplifier (**1**).

– Unscrew antenna wire connection (**3**) and unplug connector (**4**) for rear window from amplifier.

– Remove antenna amplifier.

– Install in reverse order of removal.

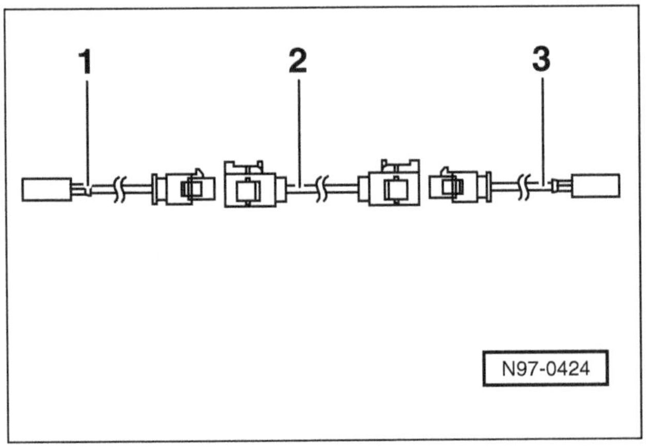

Antenna cable, repairing

The antenna cable can be repaired without complete replacement.

◀ The antenna cable system consists of the following individual components:

1. Adapter cable to radio - length approx. 30 cm (12 in.)
2. Connector cable - available in various lengths
3. Adapter cable to antenna - length approx. 30 cm (12 in.)

– Replace individual antenna sections as necessary.

BentleyPublishers.com
© 2006 BentleyPublishers.com—All Rights Reserved

SPEAKERS, REMOVING AND INSTALLING

Two speakers are located in each front door (midrange/bass and tweeter) as well as two midrange/bass speakers in the rear. Roadsters equipped with the Bose system also have a center mounted rear bass (subwoofer) speaker.

– Before proceeding, switch electrical consumers OFF. Switch ignition OFF and remove ignition key.

– Installation is reverse of removal, unless noted otherwise.

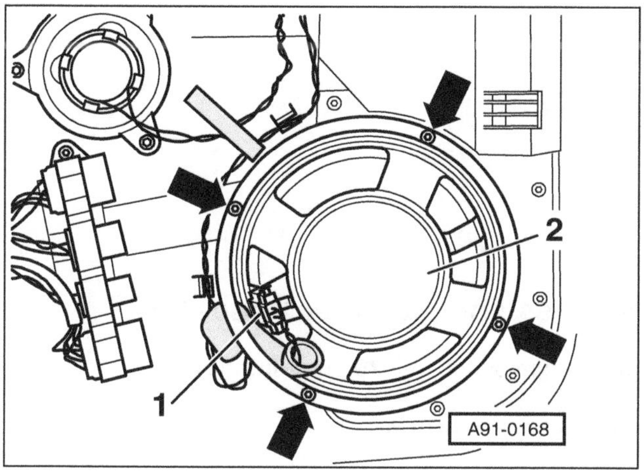

Mid-range/bass speaker (in door)

– Remove inner door panel. See **70 Trim–Interior**.

– Carefully detach insulation material near speaker from back of door trim.

◄ Unplug harness connector (**1**) from speaker (**2**).

– Remove mounting screws (**arrows**) and detach speaker from door panel to remove.

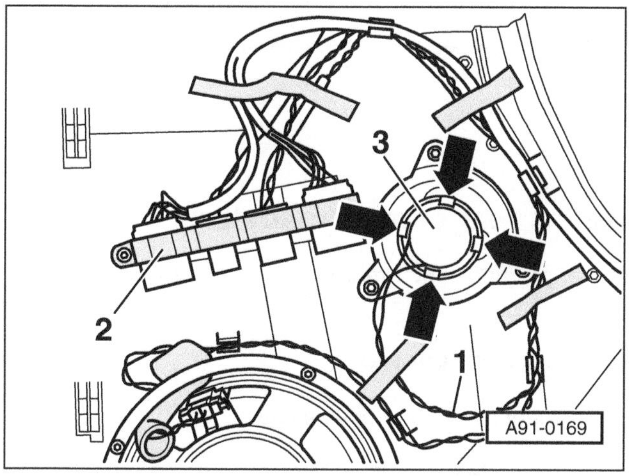

Treble speaker (in door)

– Remove inner door panel. See **70 Trim–Interior**.

– Carefully detach insulation material near speaker from back of door trim.

◄ Follow soldered speaker wire (**1**) from speaker (**3**) back to connector station (**2**). Unplug electrical connector for treble speaker.

– Press retaining clips (**arrows**) away from speaker and remove speaker from door panel.

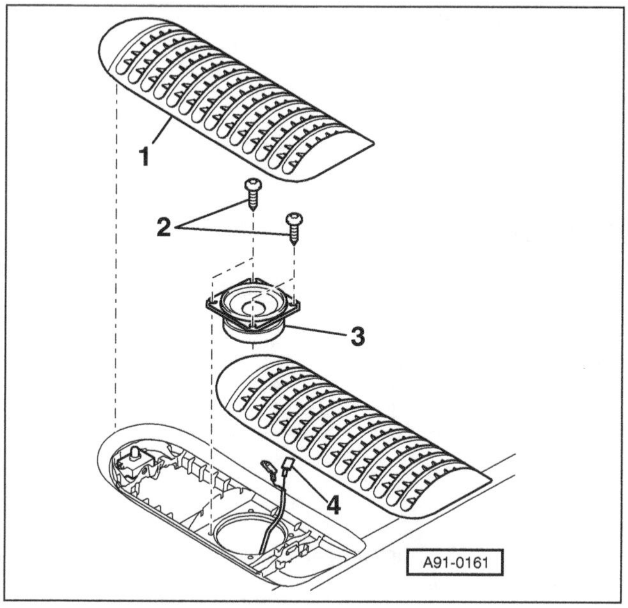

Center dash speaker (through 2001)

A center mounted dash speaker was used for naviagtion and hands-free telephone operation for vehicles with the Bose sound system through model year 2001. Other models incorporate the navigation and hands free speaker operation into the door-mounted speakers.

◄ Pull defroster vent (**1**) upward starting at front.

- Defroster vent may need to be replaced after removal.

– Unscrew mounting screws (**2**) and lift out speaker (**3**).

– Disconnect harness connector (**4**) and remove speaker.

BentleyPublishers.com
© 2006 BentleyPublishers.com—All Rights Reserved

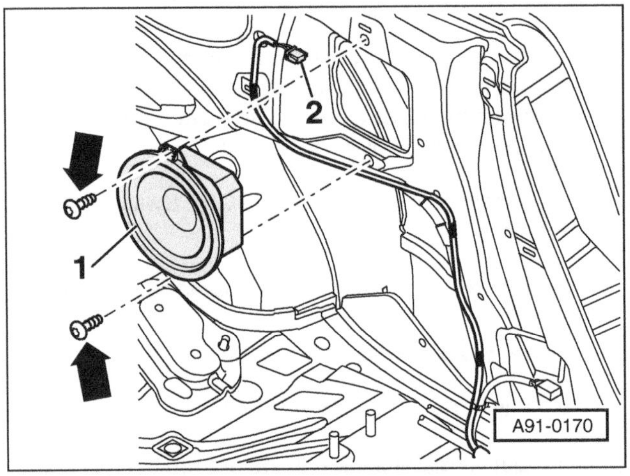

Mid-range/bass speaker, rear (coupe)

- Remove rear side trim panel. See **70 Trim–Interior**.
- Remove mounting screws (**arrows**).
- Unplug harness connector (**2**) and remove speaker (**1**).

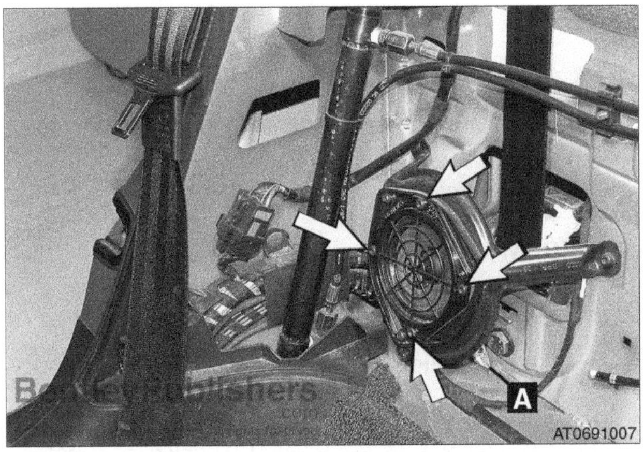

Mid-range/bass speaker, rear (roadster)

- Remove rear side trim panel. See **70 Trim–Interior**.
- Remove mounting screws (**arrows**).
- Unplug harness connector (**A**) and remove speaker.
- When installing position speaker with electrical harness connector at bottom.

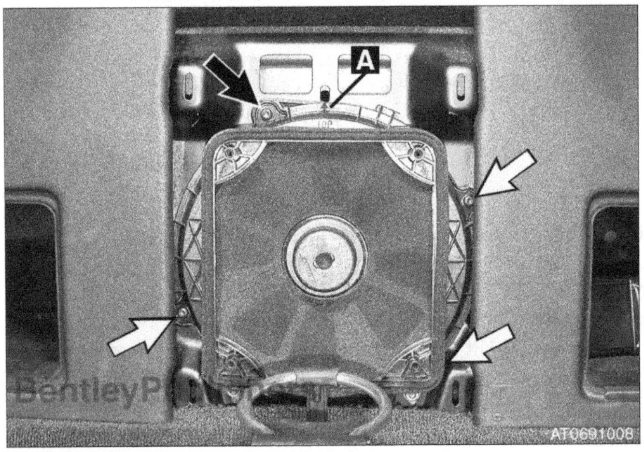

Bass speaker, rear center (roadster)

- Remove speaker grill and center rear trim panel. See **70 Trim–Interior**.
- Remove mounting screws (**arrows**).
- Unplug harness connector and remove speaker.
- When installing, note orientation of speaker.
  - Top arrow (**A**) must be positioned as shown.

NAVIGATION SYSTEM

The Navigation system BNS 4.0 (Nav 4) consists of the following components:

- Multi-function switch (operating unit)
- Instrument cluster (display unit)
- Control module for navigation with CD drive (central unit and navigation unit)
- GPS antenna

All displays appear on the instrument cluster display window which is connected via the CAN-bus.

BentleyPublishers.com
© 2006 BentleyPublishers.com—All Rights Reserved

NOTE —

- *Audi identifies electrical components by a letter and/or a number in the electrical schematics. See* **EWD Electrical Wiring Diagrams**. *These electrical identifiers are listed in parentheses as an aid to electrical troubleshooting.*

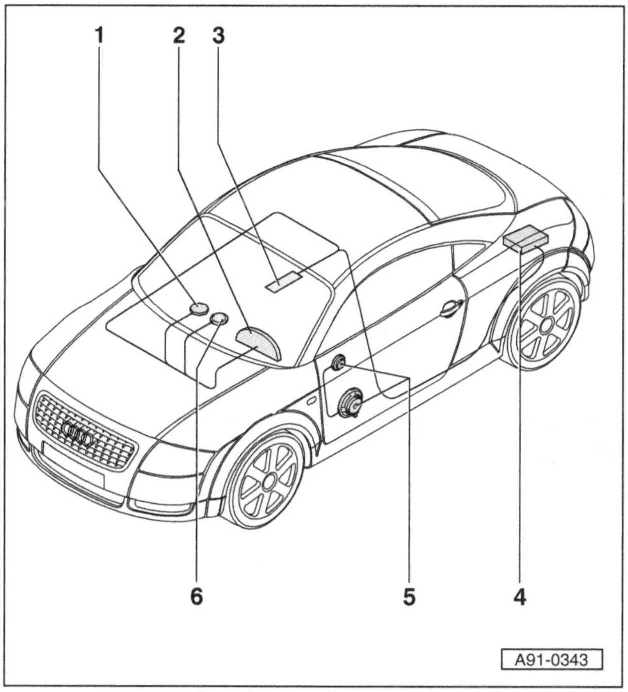

Navigation system BNS 4.0 components

1. Antenna for navigation system (R50)
   - Located under right-side defroster vent at top center of dashboard
2. Display in instrument cluster
3. Function selector switch II (E272)
   - In center console
4. Control module for navigation (J401)
   - With CD mechanism
   - In luggage compartment, left side (coupe) Behind passenger seat (roadster)
5. Speakers in door (part of radio system)
6. Center dash speaker (R39)
   - Under left-side defroster vent
   - Through 2001 m.y.
   - Only for vehicles with Bose sound system

Function selector switch, removing and installing

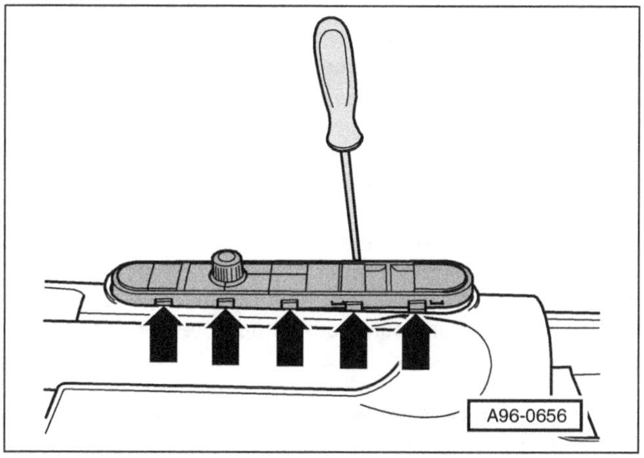

Use flat tipped screwdriver to carefully release retainer tabs (**arrows**) for rear center console switch assembly.

– Disconnect harness connectors and remove switch assembly.

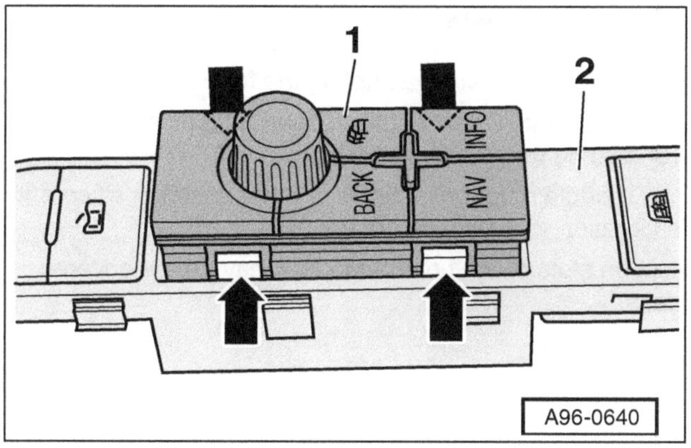

Use small knife or feeler gauge to release retaining tabs (**arrows**).

– Pull switch (**1**) upward and out of switch assembly (**2**).

– Install in reverse order of removal. Be sure to press switch and switch assembly into mount until fasteners engage.

BentleyPublishers
.com
© 2006 BentleyPublishers.com—All Rights Reserved

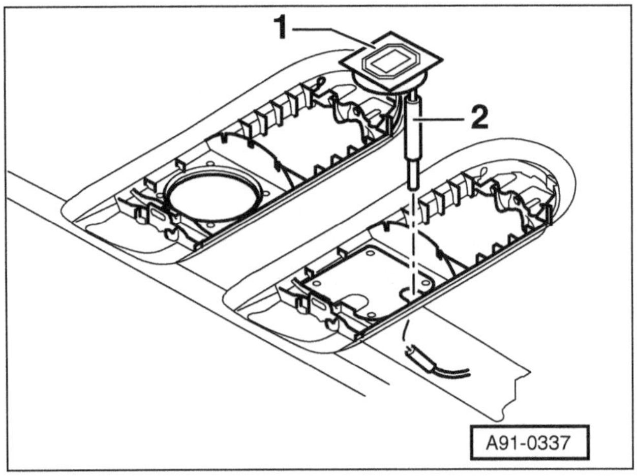

Antenna, removing and installing

– Pull right-side defroster vent upward starting at front.
  - Defroster vent may need to be replaced after removal.

◄ Lift navigation antenna (**1**) slightly upward.
  - Antenna has a magnetic mount

– Slide harness connector foam covering (**2**) out of way and disconnect.

– Remove antenna upward.

– Install in reverse order of removal.

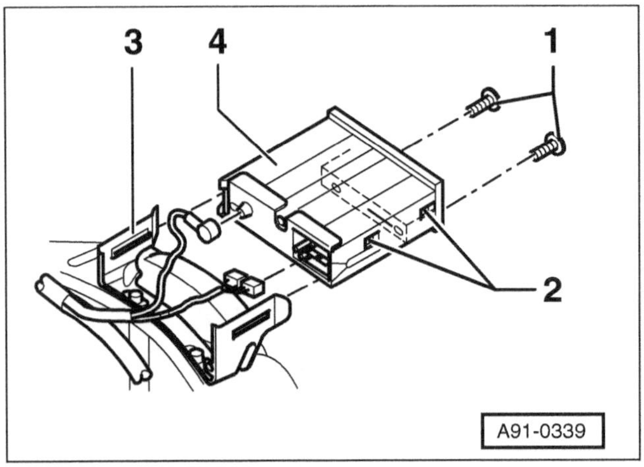

Control module with CD-mechanism, removing and installing

Coupe

– Remove left side trunk trim panel. See **70 Trim–Interior**.

◄ Remove mounting bolts (**1**) and control module with installation frame (**4**).

– Disconnect electrical harness connectors and antenna wire (GPS) at rear of control module.

– Unscrew mounting screws (**2**) at left and right sides and remove control module from installation frame.

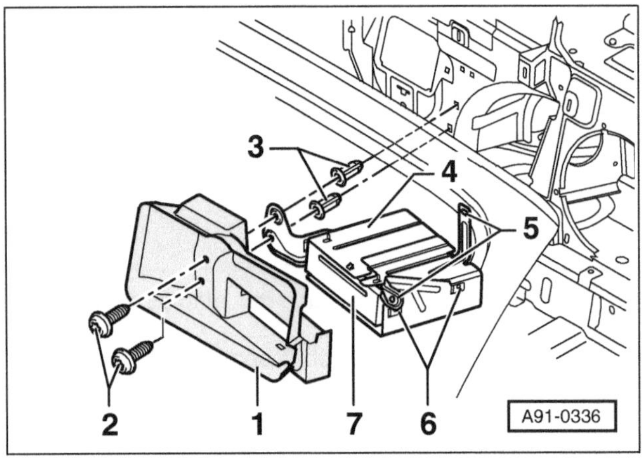

Roadster

– Open right side storage compartment behind passenger seat.

◄ Remove mounting bolts (**2**) from installation frame (**4**).

– Pull control module and mounting frame out far enough to disconnect electrical harness connectors and antenna wire (GPS) from back.

– Unscrew mounting screws (**6**) at left and right sides and remove control module (**7**) from installation frame.

Continued for all models

– Install in reverse order of removal, noting the following:
  - Use only original bolts when installing control module. Otherwise, the control module may be damaged.
  - The control module must be installed in a horizontal position and free from vibration to avoid system trouble.
  - The ventilation slots of the control module may not be covered once installed.

BentleyPublishers
.com
© 2006 BentleyPublishers.com—All Rights Reserved

TELEPHONE SYSTEM

The optional mobile telephone preparation system uses an adapter to mount various Motorola V60 cellphone models for hands-free operation.

Cellphone adapter part number	Motorola phone models
8NO 862 407A	V60, V60I, V60C
8E1 862 407B	V60S, V60X

NOTE —
- *Audi part numbers are given for reference only! Always consult your Audi parts department for the latest parts information.*

System overview (coupe)

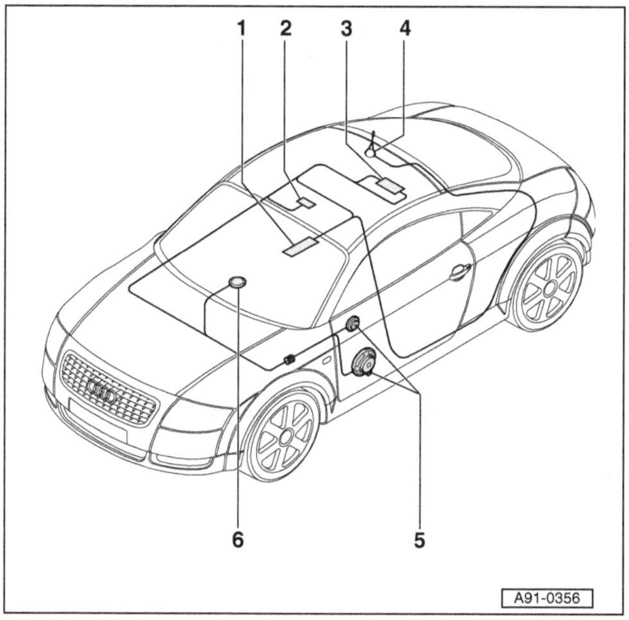

1. Cellular phone mounting bracket with adapter
   - On right side of center console
   - Removal and installation, see **70 Trim–Interior**
2. Hands-free microphone
   - Under front dome light
3. Control module
   - Below rear seat bottom
4. Telephone antenna
   - On roof
   - Remove rear headliner trim panel and part of headliner to access. See **70 Trim–Interior**
   - Different style antennas for 2000-2001 and 2002-on
5. Mid-range/treble speaker and bass speaker
   - In front door trim panel, part of sound system
6. Hands-free speaker (through 2001)
   - Only for 2000-2001 vehicles with Bose sound system
   - Below left defroster vent on dash
   - See **Center dash speaker (through 2001)**

System overview (roadster)

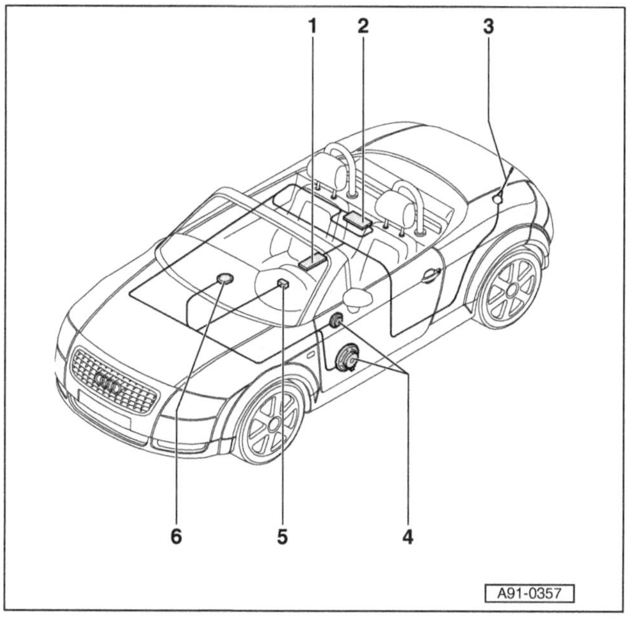

1. Cellular phone mounting bracket with adapter
   - On right side of center console
   - Removal and installation, see **70 Trim–Interior**
2. Control module
   - Behind rear center speaker
3. Telephone antenna
   - Incorporated with radio antenna mounted on left rear fender. See **Antenna** earlier in this repair group.
4. Mid-range/treble speaker and bass speaker
   - In front door trim panel, part of sound system
5. Hands-free microphone
   - In instrument cluster
6. Hands-free speaker (through 2001)
   - Only for 2000-2001 vehicles with Bose sound system
   - Below left defroster vent on dash
   - See **Center dash speaker (through 2001)**

BentleyPublishers.com
©2006 BentleyPublishers.com—All Rights Reserved

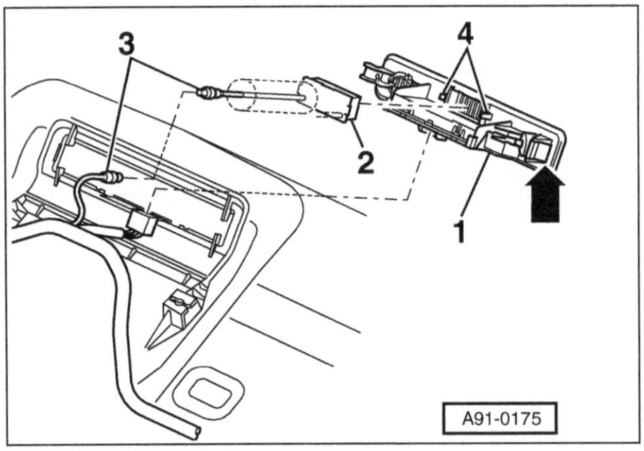

Hands-free microphone, removing and installing

Coupe models

- Remove front dome light assembly (**1**). See **96 Interior Lights, Switches, Anti-theft**.
- Press two retaining tabs (**4**) to side and detach microphone (**2**) from dome light.
- Disconnect harness connector (**3**) and remove micropone from headliner cut-out.
- Install in reverse order of removal.

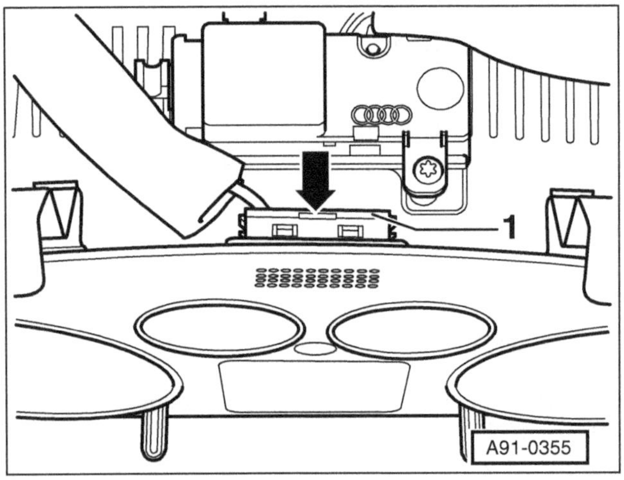

Roadster models

- Remove instrument cluster. See **90 Instruments**.
- Use small screwdriver at arrow to carefully pry hands-free microphone (**1**) out of mounting bracket.
- Install in reverse order of removal.

Control module, removing and installing

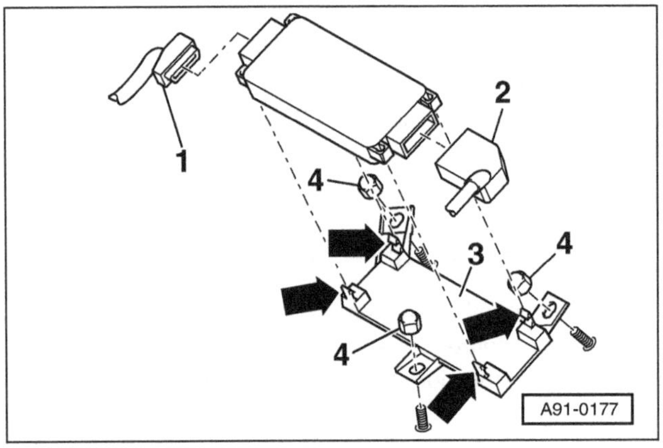

Coupe models

- Remove rear seat bottom. See **72 Seats**.
- Disconnect harness connectors (**1** and **2**) from control module.
- Press four retainers (**arrows**) to the side and remove control module from mounting bracket (**3**).
- Install in reverse order of removal.

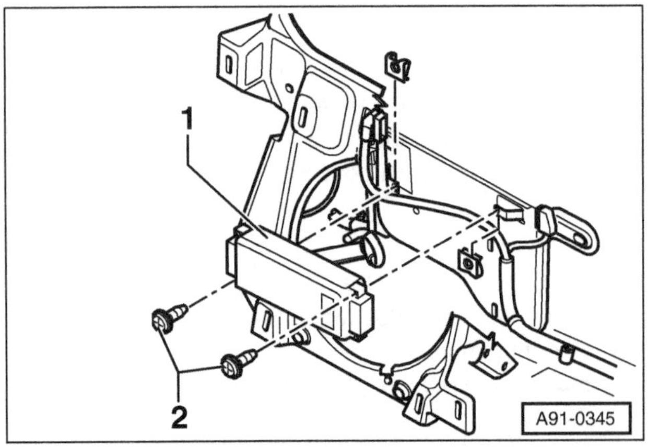

Roadster models

- Remove center rear speaker. See **Bass speaker, rear center (roadster)** earlier in this repair group.
- Disconnect harness connectors (**1**) from control module.
- Remove mounting bolts (**2**) and remove control module.
- Install in reverse order of removal.

BentleyPublishers .com
©2006 BentleyPublishers.com—All Rights Reserved

92 Wipers and Washers

GENERAL

This repair group covers wiper blade service and wiper assembly repairs. Also covered are repairs to the windshield and headlight washer assemblies.

For additional information see the following:

- **03 Maintenance** for wiper blade replacement and windshield spray jet adjustment.
- **48 Steering** for steering column stalk switch assembly replacement.
- **97 Fuses, Relays, Component Locations** for the location of wiper and washer system fuses and control modules.
- **EWD Electrical Wiring Diagrams**

Wipers and washers operation

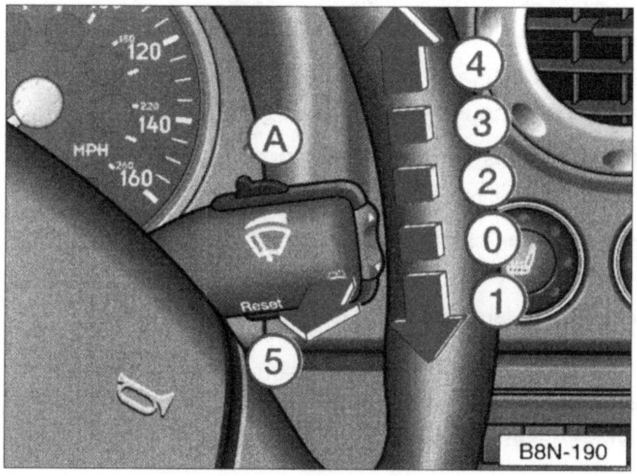

The windshield wipers and washers are operated by the wiper stalk switch on the steering column.

- **0** Rest position
- **1** One-tap wipe
- **2** Intermittent wipe
- **A** Intermittent wipe interval speed
- **3** Slow wipe
- **4** Fast wipe
- **5** Windshield washer (pull)

The following wiper and washer system features are available as options:

- **Heated windshield washer nozzles**. Washer nozzles heated electrically to prevent icing.
- **Headlight washers**. Cleans headlights with high pressure water jets. If the headlights are on, they are washed whenever the wiper/washer stalk is pulled forward (wash). The headlight washer jets automatically come out (under water pressure) when the washers are operated.

BentleyPublishers.com
© 2006 BentleyPublishers.com—All Rights Reserved

Cautions

> ***CAUTION—***
> - *Switch the ignition OFF and disconnect the negative (–) battery cable before removing any electrical components.*
> - *Prior to disconnecting the battery, read the battery disconnection cautions in* **27 Battery, Alternator, Starter**.
> - *Only use a digital multimeter for electrical tests.*
> - *Before using the wipers for the first time in cold weather, make sure the wiper blades are not frozen to the windshield.*
> - *Do not use the windshield washers in freezing weather without first warming the windshield with the defroster. Otherwise the washer solution may freeze on windshield and obscure vision.*
> - *Worn or dirty wiper blades reduce visibility, making driving hazardous. Clean blades regularly to remove road film and car wash wax build-up. Use an alcohol based cleaning solution and a lint free cloth to wipe along the rubber blade.*
> - *To prevent scratching the glass, avoid running the wiper blades over a dry windshield.*
> - *To avoid damaging paint or plastic trim, use a plastic prying tool or a screwdriver with the tip wrapped with masking tape.*

WIPER SYSTEM

Wiper blade replacement is covered **03 Maintenance**.

Wiper blade cleaning problems

Common problems with the windshield wipers include streaking or sheeting, water drops after wiping, and blade chatter.

- **Streaking** is usually caused when wiper blades are coated with road film or car wash wax.

– Clean blades using soapy water. If cleaning does not cure problem, replace blades.

- **Drops** that remain behind after wiping are usually caused by oil, road film, or diesel exhaust residue on the glass.

– Use an alcohol or ammonia solution or other non-abrasive cleaner to clean windshield.

- **Chatter** may be caused by dirty or worn blades, or by wiper arms that are out of alignment.

– Clean blades and windshield as described above.

– Bend wiper arm so that there is even pressure along blade, and so that blade is perpendicular to windshield at rest. See **Wiper blade contact angle, checking** in this repair group.

– If problems persist, replace blades and wiper arms.

BentleyPublishers.com

© 2006 BentleyPublishers.com—All Rights Reserved

Wiper arms, removing and installing

- Switch off ignition and remove ignition key.
- Open hood.
- Note position of wiper blades on windshield.
  - Use chalk or crayon to mark position.
  - This may help with installation of wiper arms.
- Remove plastic trim plug from base of wiper arms.

◄ Remove wiper arm mounting nut: right (**A**) or left (**B**).

- Remove wiper arm by pivoting off windshield and gently rocking arm base back and forth.

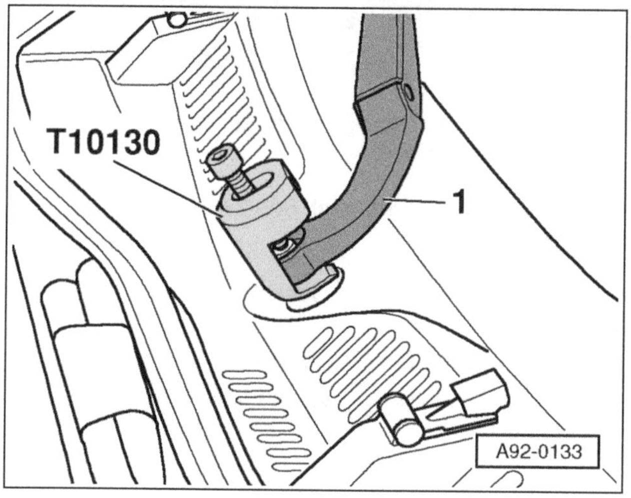

◄ Alternatively, use a small puller, Audi special tool T10130, to remove wiper arm.

- To use puller T10130, leave wiper arm mounting nut threaded on end of wiper shaft.
  - The nut serves as a centering guide for puller spindle.
- Turn spindle screw clockwise until wiper arm releases from shaft.
- Remove nut and wiper arm.
- Installation is reverse of removal.
  - Tighten wiper arm mounting nut.

Tightening torque	
Wiper arm to wiper shaft	16 Nm (12 ft-lb)

- Switch ignition on and operate wipers. Check wiper arm park position.

Wiper arm park position, checking

- Switch windshield wipers on and off, allowing wiper arms to return to park position at base of windshield.
- Switch off ignition.

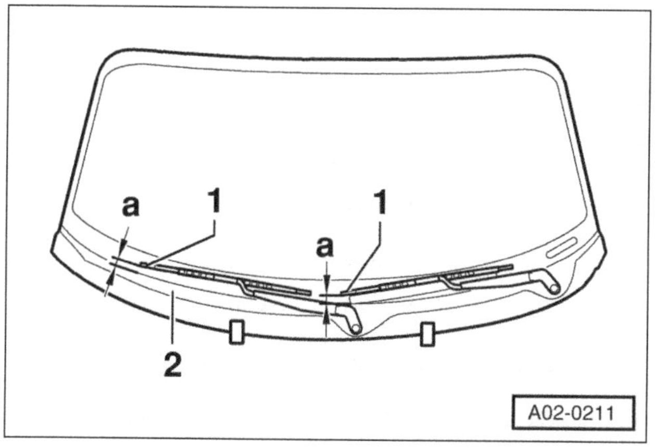

◄ **Early version** (open frame):

- Check distances (**a**) between ends of wiper blades (**1**) and plastic plenum chamber cover (**2**) at bottom of windshield.
  Distance **a** = 15 - 25 mm (9/16 - 1 in)
- If distance **a** is not correct, pull wiper arm off of wiper pivot and reposition arm on pivot. Wiper arm puller tool T10130 is available to facilitate removal of arm.

BentleyPublishers
.com
©2006 BentleyPublishers.com—All Rights Reserved

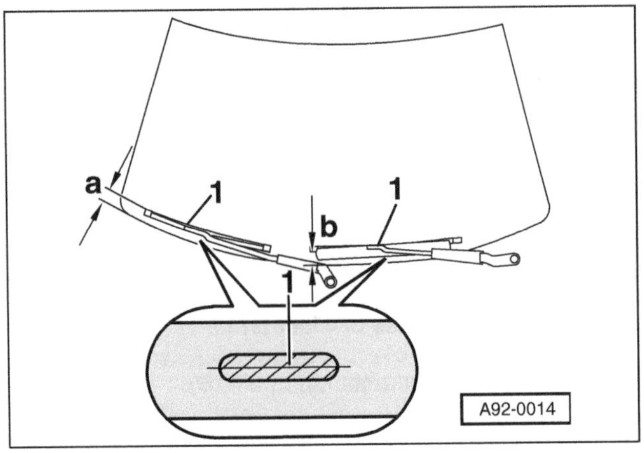

◄ **Later version** (aero):

- Ensure that wiper blades rest against marking (**1**) on windshield.
- Check distances (**a**, **b**) between ends of wiper blades and plastic plenum chamber cover at bottom of windshield.
  Distance **a** = 25 mm ± 5 mm (1 in ± $\frac{3}{16}$)
  Distance **b** = 35 mm ± 5 mm (1 $\frac{3}{8}$ in ± $\frac{3}{16}$)
- If distance **a** or **b** is not correct, pull wiper arm off of wiper pivot and reposition arm on pivot. Wiper arm puller tool T10130 is available to facilitate removal of arm.

– Operate wipers and recheck distances.

Tightening torque	
Wiper arm to wiper shaft	16 Nm (12 ft-lb)

Wiper assembly, removing and installing

– Switch off ignition and remove ignition key.

– Open hood and remove wiper arms. See **Wiper arms, removing and installing** in this repair group.

– Disconnect battery negative (–) terminal. See **27 Battery, Alternator, Starter**.

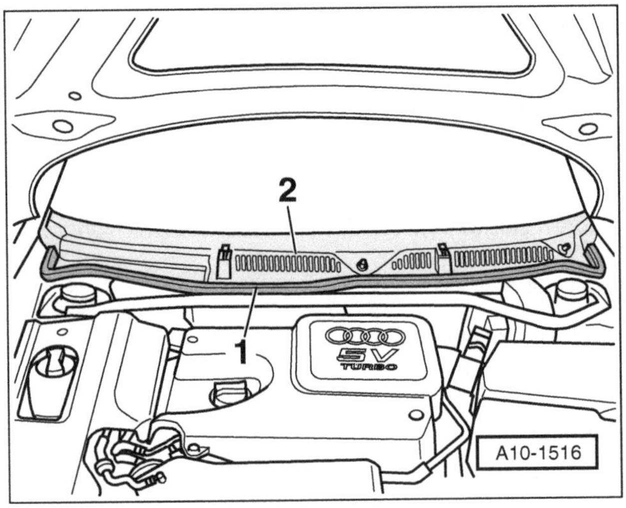

◄ Remove rubber seal (**1**) and plenum cover (**2**) at rear of engine compartment.

- Pry plenum cover at edge of windshield and lift up.
- Detach washer jet connections from bottom of plenum cover.

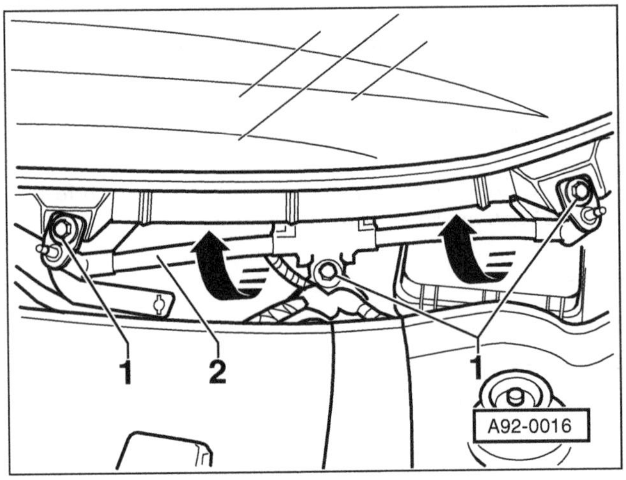

◄ Remove wiper assembly mounting bolts (**1**).

– Disconnect electrical harness connector from wiper motor.

– Carefully tilt front of wiper frame (**2**) upward (**arrows**) and then remove from plenum to left.

– After repairs to wiper assembly or wiper motor:

- Reinstall assembly in reverse order of removal.
- Replace rubber mounting insulators if worn or damaged.
- Reattach electrical connector.

Tightening torque	
Wiper assembly to body	5 Nm (44 in-lb)

– Attach washer jet connection to rear of plenum cover.

– Install plenum cover, inserting into base of windshield.

- Lubricate with soapy water for ease of insertion.

– Install plenum rubber seal, making sure it is seated correctly.

– Check wiper park position. See **Wiper arm park position, checking** in this repair group.

BentleyPublishers.com
©2006 BentleyPublishers.com—All Rights Reserved

Wiper motor, replacing

– Remove wiper arms. See **Wiper arms, removing and installing** in this repair group.

– Remove wiper assembly. See **Wiper assembly, removing and installing** in this repair group.

– Place wiper assembly on workbench.

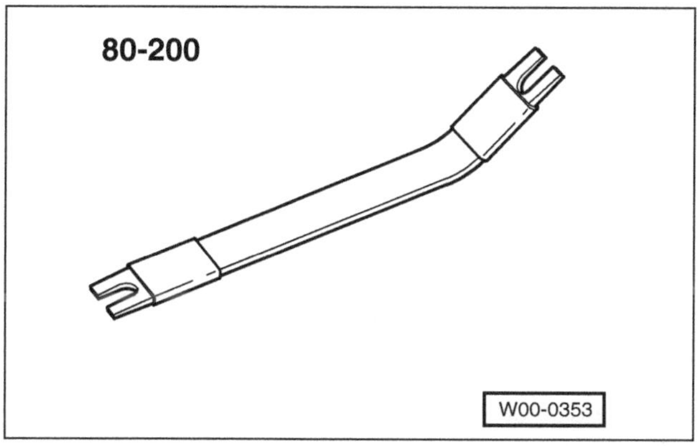

◄ Use VW special tool 80-200 or equivalent to pry wiper linkage rods from ball joints.

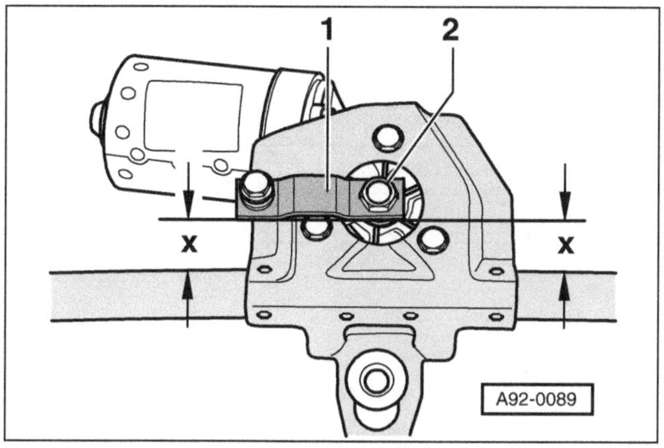

◄ Working at wiper motor:
- Remove motor crank mounting nut (**2**).
- Remove crank (**1**).

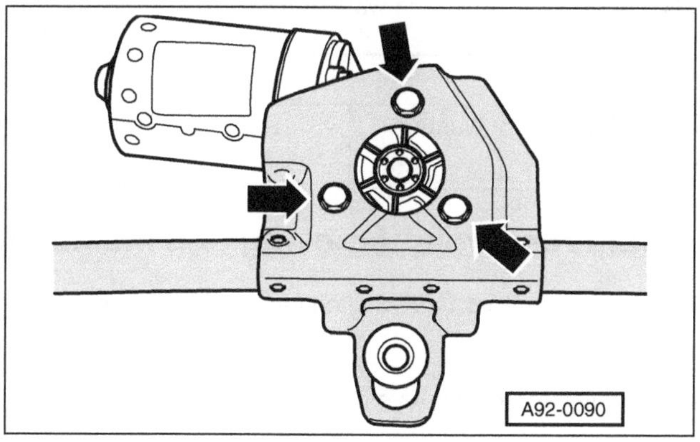

◄ Remove wiper motor mounting bolts (**arrows**) and motor from mounting bracket.

– Install wiper motor and wiper assembly in reverse order of removal.

– Before installing wiper arms, reconnect battery and operate wiper motor.
- Make sure wiper motor is in park position.

– Switch off ignition and remove ignition key.

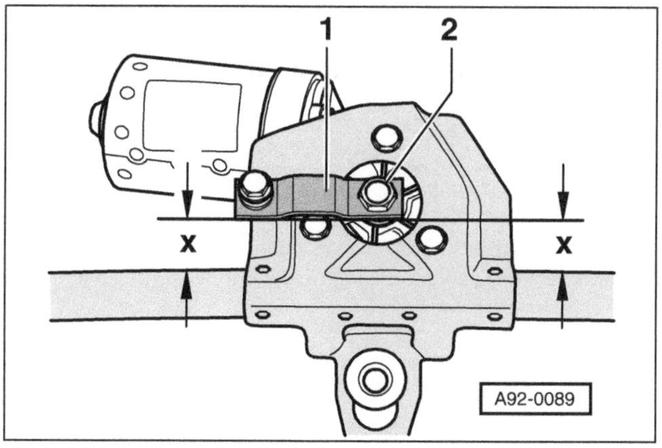

◄ With motor in park position, check that wiper crank arm is parallel to mounting bracket.
- Dimensions (**x**) must be equal.

– Tighten wiper motor and assembly mounting bolts to specification.

Tightening torque	
Wiper motor to mounting bracket	9 Nm (80 in-lb)
Crank to wiper motor	18 Nm (13 ft-lb)
Wiper assembly to body	5 Nm (44 in-lb)

– Remainder of installation is reverse of removal. See **Wiper assembly, removing and installing** in this repair group.

BentleyPublishers.com
©2006 BentleyPublishers.com—All Rights Reserved

Wiper blade contact angle, checking

If the wiper blades are skipping across windshield, it may be necessary to adjust the contact angle of the wiper arms before installing new wiper blades.

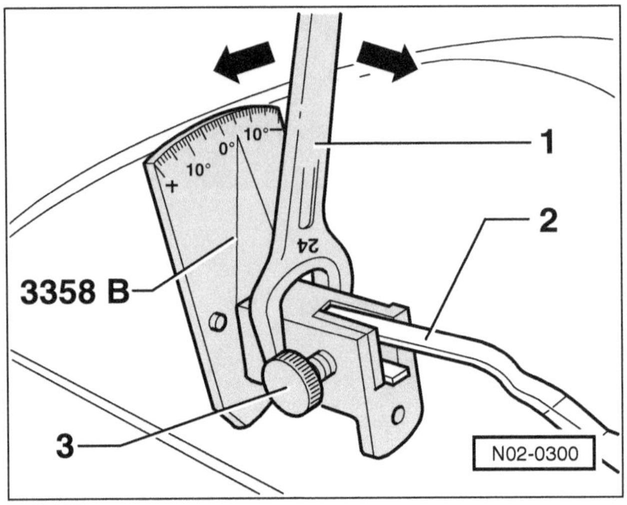

- Move wiper arms to park position (off).
- Remove wiper blade. See **03 Maintenance**.
- Put wiper arm adjusting tool (**3358 B**) onto end of wiper arm (**2**). Secure tool with locking screw (**3**).
- Compare angle indicated with specification.

Wiper blade contact angle specification	
Driver's side	– 4.0°
Passenger side	– 4.5°
Tolerance	– 1.5°

- Use 24mm wrench to adjust wiper arm to specification as necessary.

Washer System

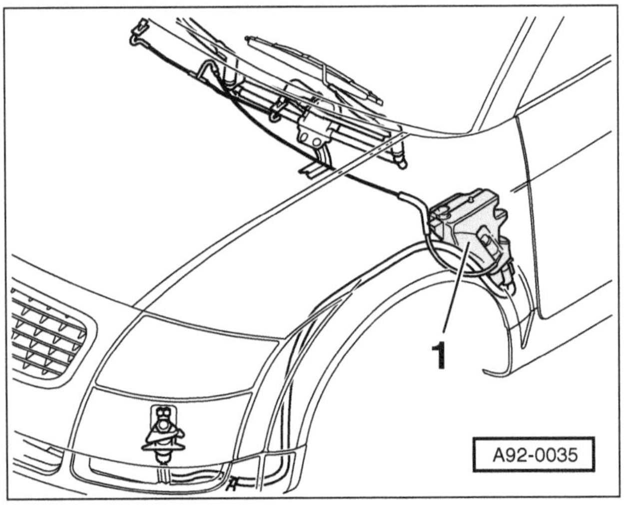

The washer fluid reservoir (**1**) supplies washer fluid to both the windshield washers and headlight washer system.

Washer fluid reservoir	
Windshield and headlight washer fluid capacity	approx. 4.3 liters (4 qt.)

Washer fluid reservoir, removing and installing

- Switch electrical consumers OFF. Switch ignition OFF and remove ignition key.
- Lift washer fluid fill cap and remove circlip from filler neck.
- Remove left front wheel and wheel housing liner. See **03 Maintenance** and **66 Body–Exterior Equipment**.

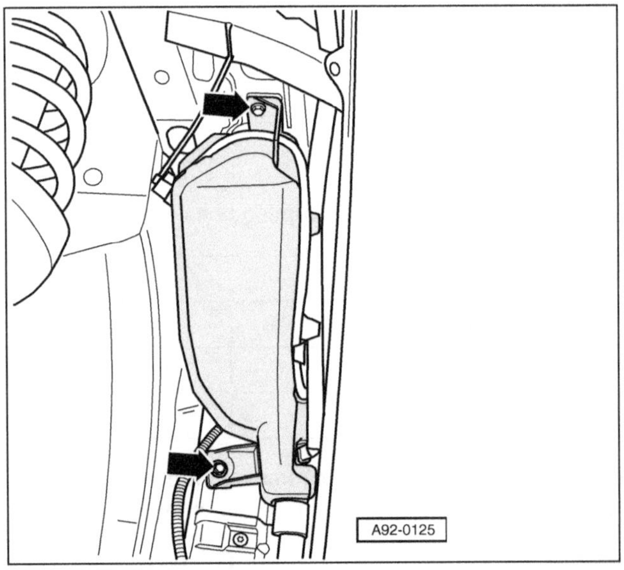

- Remove reservoir mounting screws (**arrows**) and lower reservoir slightly.

BentleyPublishers.com
©2006 BentleyPublishers.com—All Rights Reserved

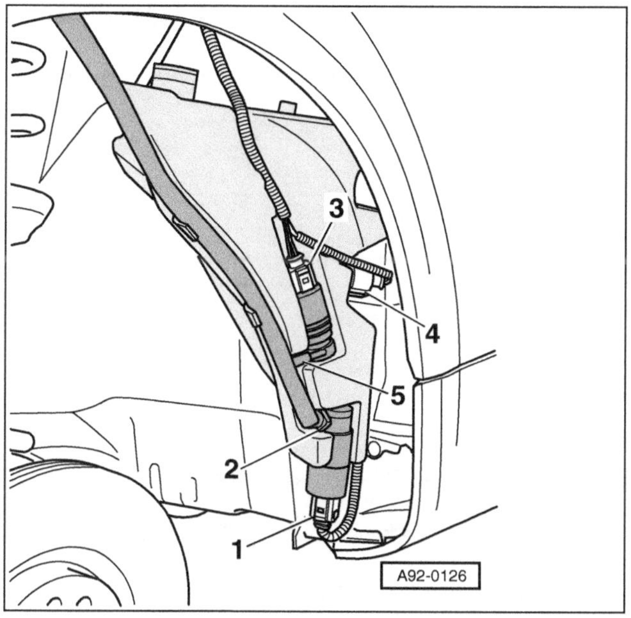

◄ Disconnect harness connectors (**1**, **3** and **4**).

– Depress retainer and remove washer hose coupling (**2**) from headlight washer pump.

– Remove washer hose (**5**) from windshield washer pump.

CAUTION—
• Be prepared to catch dripping washer fluid in a bucket.

– Remove washer hoses from retaining clips.

– Check sealing grommets for wear and damage. Replace as necessary.

– Install in reverse order of removal. Make sure that washer hoses engage securely when installing.

Tightening torque	
Washer fluid reservoir to body	7 Nm (62 in-lb)

Windshield spray jets, removing and installing

– Switch electrical consumers OFF. Switch ignition OFF and remove ignition key.

– Disconnect electrical connectors for spray jet heaters.

– Remove wiper arms and plenum cover. See **Wiper arms, removing and installing** in this repair group.

– Disconnect washer hoses from plenum cover.

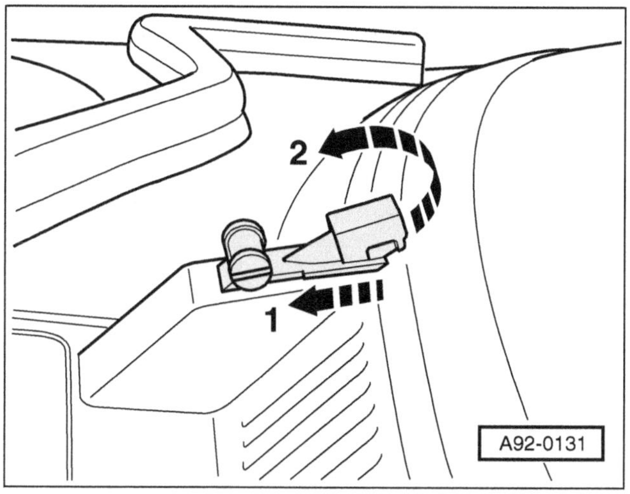

◄ Push spray jet in direction of **arrow 1**.

– Remove spray jet from plenum cover in direction of **arrow 2**.

– Disconnect washer hose from spray jet.

– Install in reverse order of removal, noting the following:
  - Make sure that washer hoses engage securely when installing.
  - Replace any cable ties that were removed.

– Adjust spray jet. See **03 Maintenance**.

BentleyPublishers.com
©2006 BentleyPublishers.com—All Rights Reserved

Headlight spray jet, removing and installing

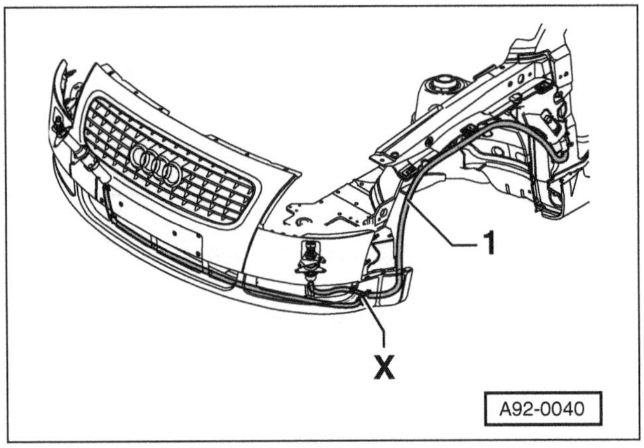

The two headlight washer spray jets are supplied from the washer fluid reservoir via hose (**1**).

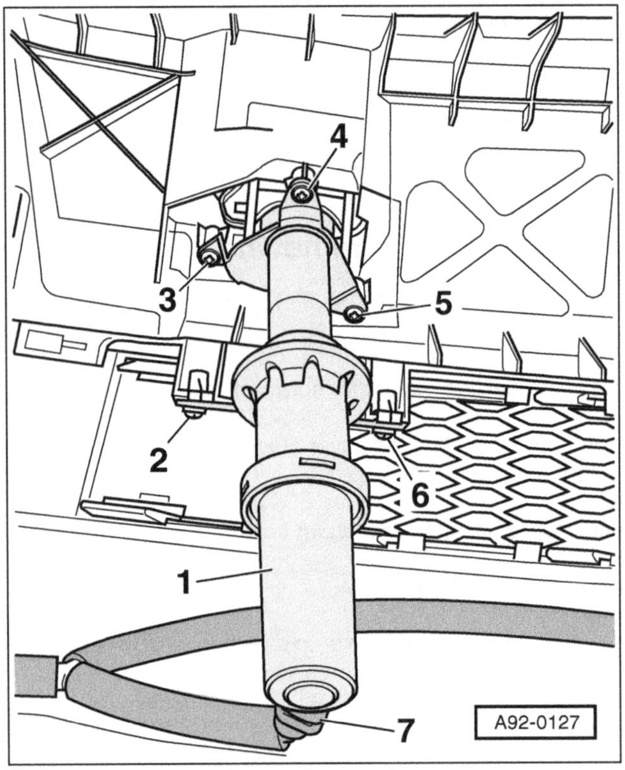

– Remove front bumper cover. See **63 Bumpers**.

Depress retainer and remove hose coupling (**7**) from spray jet (**1**).

– Remove mounting screws (**3**, **4** and **5**).

– Remove mount bracket screws (**2** and **6**).

– Remove spray jet.

– Install in reverse order of removal.

Tightening torque	
Headlight spray jet to bumper	2.5 Nm (23 in-lb)

NOTE —

- *Headlight spray jets are pre-set and do not require adjustment after installation.*

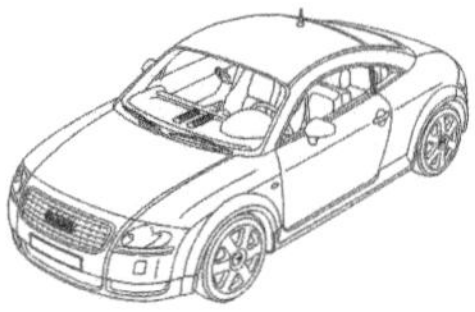

© 2006 BentleyPublishers.com—All Rights Reserved

94 Exterior Lights

TABLES

GENERAL

This repair group covers exterior lighting repairs and bulb replacement. Also see:

- **03 Maintenance** for headlight and foglight adjustment.
- **40 Front Suspension** for vehicle level sensor (Xenon headlights)
- **42 Rear Suspension** for vehicle level sensor (Xenon headlights)
- **45 Antilock Brakes (ABS)** for brake light switch
- **96 Interior Lights, Switches, Anti-theft** for light, steering column and emergency flasher switches
- **97 Fuses, Relays, Component Locations**
- **EWD Electrical Wiring Diagrams**

Headlights

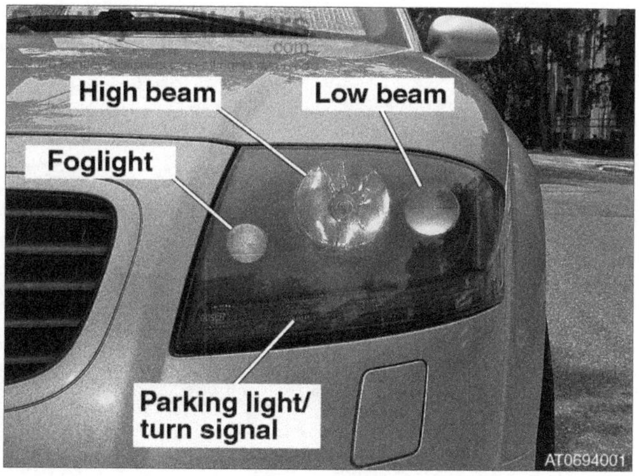

The headlights, front turn signals and foglights are combined in one unit on each side. The headlight lens is made of polycarbonate and care should be taken when cleaning not to scratch the lens.

94

BentleyPublishers.com
©2006 BentleyPublishers.com—All Rights Reserved

Automatic headlight beam adjustment

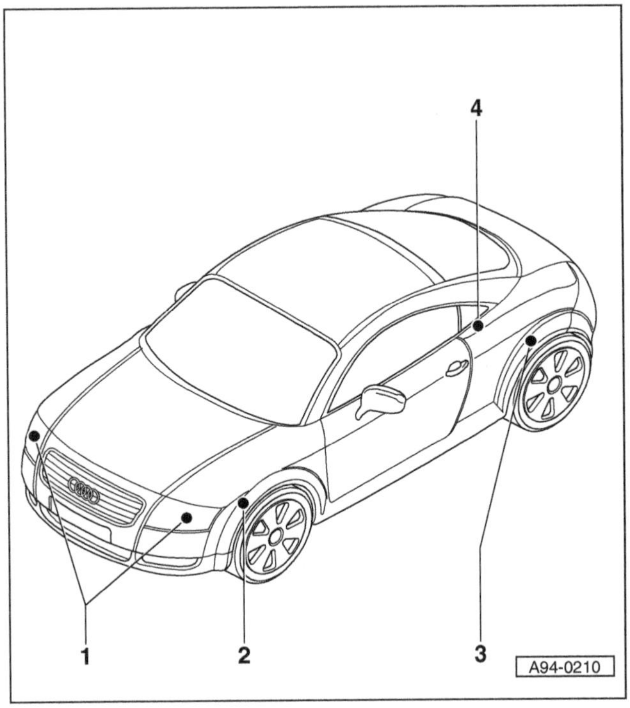

TT models with the optional High-Intensity Discharge (HID), also known as Xenon, headlights are equipped with the automatic headlight adjustment system.

NOTE —
- *Audi identifies electrical components by a letter and/or a number in the electrical schematics. See* **EWD Electrical Wiring Diagrams**. *These electrical identifiers are listed in parentheses as an aid to electrical troubleshooting.*

1. Xenon (HID) headlights with beam adjusting motors (V48 and V49)
   - Only low beam bulbs are Xenon
   - High beam bulbs are Halogen
2. Front suspension level sensor (G78)
   - Located on left front control arm
3. Rear suspension level sensor (G76)
   - Front wheel drive: located at rear axle
   - All wheel drive: located at left rear crossmember
4. Headlight range control module (J431)
   - Coupe: located below left side of rear seat bottom
   - Roadster: located behind right rear trim panel, near access plate for fuel pump

NOTE —
- *Xenon (HID) components are often marked with a yellow high voltage sticker.*
- *Adjustment of Xenon (HID) headlights should be done using the VAS 5051/5052 diagnostic testing and information system and VAS 5107 optical headlight aimer. Use of these devices is beyond the scope of this manual.*
- *See* **03 Maintenance** *for more information on headlight adjustment.*

Taillights

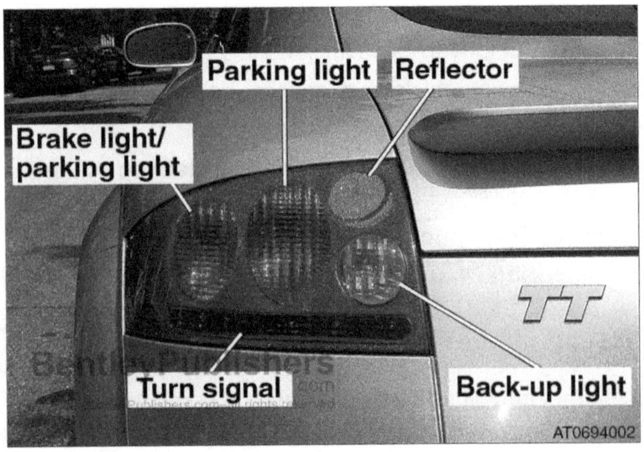

The taillight housing has a brake light/parking light, taillight only, reflector, back-up light and turn signal.

BentleyPublishers
.com
© 2006 BentleyPublishers.com—All Rights Reserved

Exterior bulb applications

Exterior bulb applications for are listed in **Table a**.

Table a. Exterior bulb applications

Function	Type	Wattage
Back-up light	P21W	21
Brake lights (with parking lights)	P21/5W	21/5
Foglight • Front	 H3	 55
Headlight • High beam • Low beam • Xenon (low beam)	 H7 H1 D2S	 55 55 35
License plate light	L	5
Parking light • Front • Rear (with brake light)	 H6 P21/5W	 6 21/5
Side marker • Rear	 W	 5
Turn signal • Front • Rear • Side	 H21 P21/5W WY5W (amber)	 21 21/5 5

Warnings and Cautions

WARNING *—*

- *Xenon (high intensity discharge or HID) bulbs operate at high voltages. When working on xenon headlight components (control module, ignition unit, gas discharge bulb, drive motor), be sure to switch lights OFF.*
- *Some light bulbs, especially headlight bulbs, can become hot enough to cause injury. Allow bulbs to cool before working on the lighting system.*

CAUTION *—*

- *Switch the ignition OFF and disconnect the negative (–) battery cable before removing any electrical components.*
- *Prior to disconnecting the battery, read the battery disconnection precautions in* **27 Battery, Alternator, Starter**.
- *Only use a digital multimeter for electrical tests.*
- *Do not touch the glass of bulbs with bare skin. If necessary wipe bulb using a clean cloth dampened with rubbing alcohol.*
- *Use only original equipment replacement bulbs. Non-original bulbs may cause false failure readings on the dashboard display.*
- *Use only specified bulbs. A bulb with higher wattage may cause damage to bulb housing.*
- *To avoid marring car paint or trim, work with a plastic prying tool.*

BentleyPublishers.com
©2006 BentleyPublishers.com—All Rights Reserved

Headlights

Headlight assembly

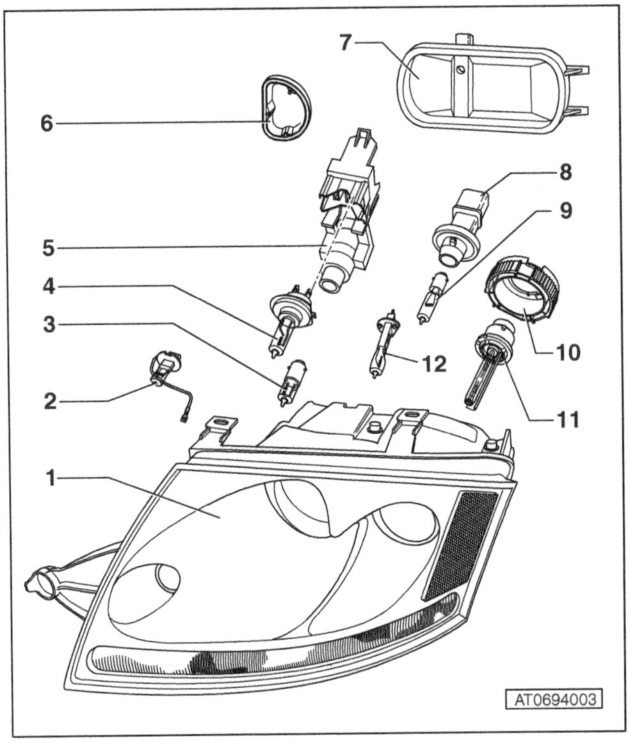

1. Headlight housing
   - Lens bonded to housing, cannot be replaced separately
   - From approx 4/00, Xenon (HID) lamp control modules (J343 and J344) are mounted directly to housing
2. Fog light bulb (L22 and L23)
3. Parking/side marker bulb (M1 and M3)
   - Housing not shown
4. High beam bulb (M30 and M32)
5. High beam and parking/side bulb holder
6. Cover (small)
   - Provides access to fog light bulb
7. Cover (large)
   - Provides access to high and low beam bulbs, including HID (Xenon)
8. Turn signal bulb holder
9. Front turn signal bulb (M5 and M7)
10. Securing ring (bayonet connection)
    - For Xenon (HID) only
11. Xenon (HID) low beam bulb (L13 and L14)
12. Halogen low beam bulb (M29 and M31)
    - For non-Xenon only

Headlight housing, removing and installing

– Switch off all electrical consumers and remove ignition key.

– Make sure light switch is off.

> ***WARNING —***
> - *Disconnect negative (–) battery cable and cover battery terminal to keep cable from accidentally contacting terminal.*

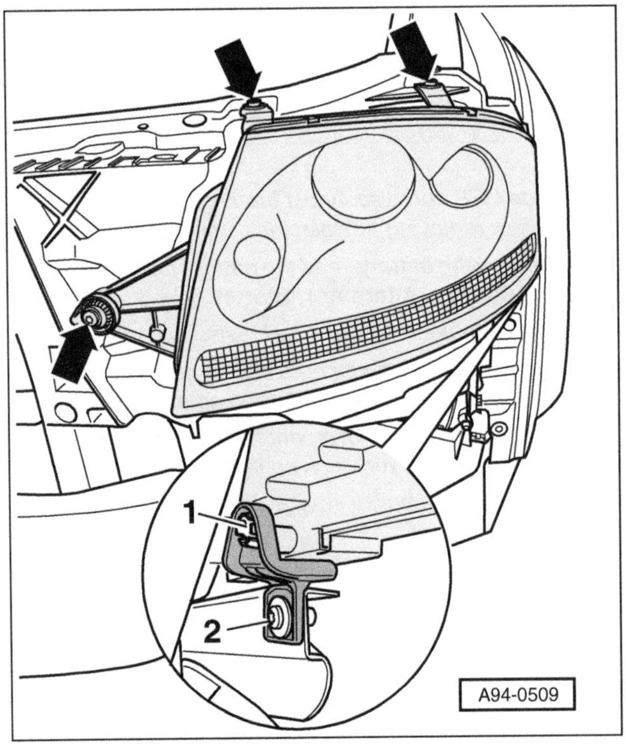

– Remove front bumper cover. See **63 Bumpers**.

◄ Working at headlight assembly, remove 3 mounting screws (**arrows**).

– Slide headlight housing toward radiator and disengage mounting pin (**1**) from retainer (**2**).

– Disconnect electrical harness connectors.

– Remove headlight housing.

BentleyPublishers.com
© 2006 BentleyPublishers.com—All Rights Reserved

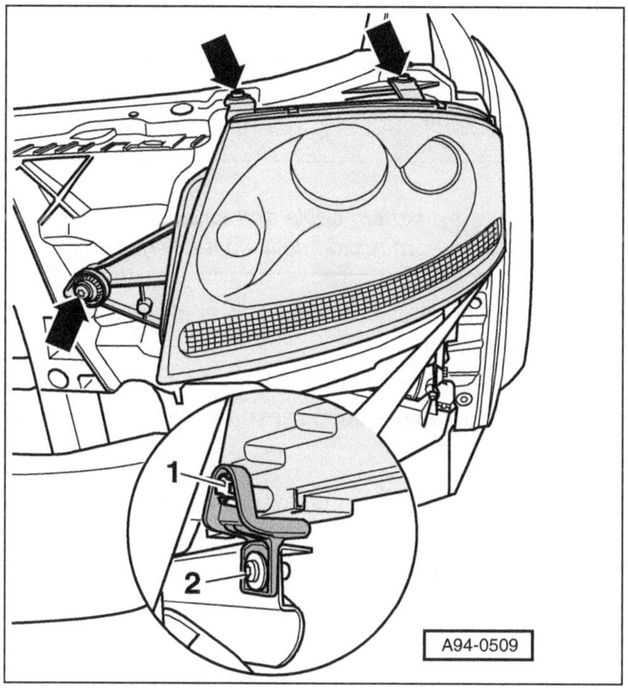

– Begin installation by connecting electrical harness connectors.

◄ Engage mounting pin (**1**) in retainer (**2**).

– Install mounting screws (**arrows**) and hand tighten only.

– Position front bumper cover in place and check position of headlight housing relative to front bumper cover and grill assembly.

– If necessary, adjust position of headlight housing.

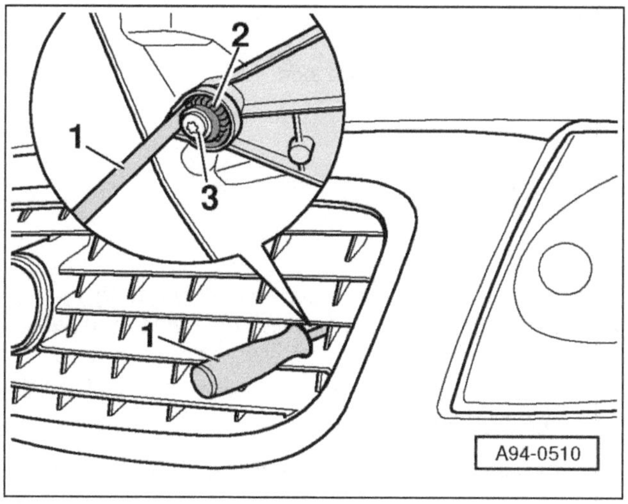

◄ To adjust position of headlight housing, loosen screw (**3**) and turn adjuster (**2**) until headlight housing is aligned properly with bumper cover and grill assembly.

– Remove bumper cover and tighten headlight housing mounting screws to specification.

Tightening torque	
Headlight to lock carrier	6.5 Nm (57 in-lb)

– Install bumper cover.

– Adjust headlight aim. See **03 Maintenance**.

Headlight housing, repairing

A repair kit is available for broken headlight mounting tabs.

Headlight tab repair kit	
Left or right headlight	Part no. 8N0 998 121

NOTE —
- *Part numbers given were correct at publication time but may change over time. Consult an authorized Audi dealer parts department for current part information.*

– Remove headlight housing with broken mounting tab. See **Headlight housing, removing and installing** in this repair group. Remove pieces of broken mount from headlight and body.

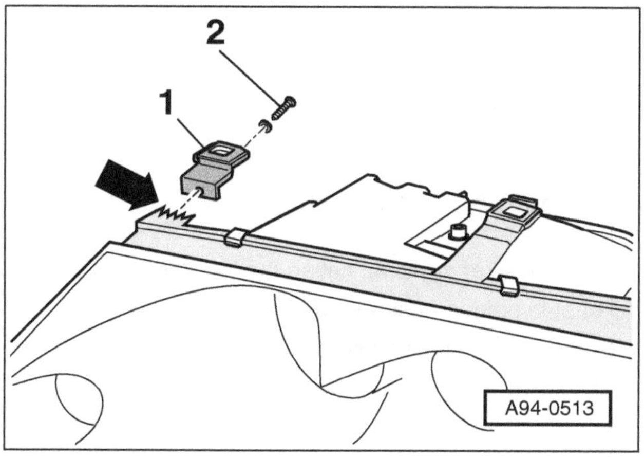

◄ Install new mounting tab (**1**) with screw (**2**) and washer.

– Reinstall headlight housing.

BentleyPublishers.com
© 2006 BentleyPublishers.com—All Rights Reserved

Headlight bulbs, removing and installing

– Switch off all electrical consumers and remove ignition key.

– Make sure light switch is off.

> ***WARNING —***
> *• Disconnect negative (–) battery cable and cover battery terminal to keep cable from accidentally contacting terminal.*

– Remove sound absorber panel(s) to access rear of headlight housing. See **03 Maintenance**.

- For left side headlight, remove sound absorber panel over battery.
- For right side headlight, remove sound absorber panel over coolant reservoir.

– Remove large bulb cover from back of headlight housing.

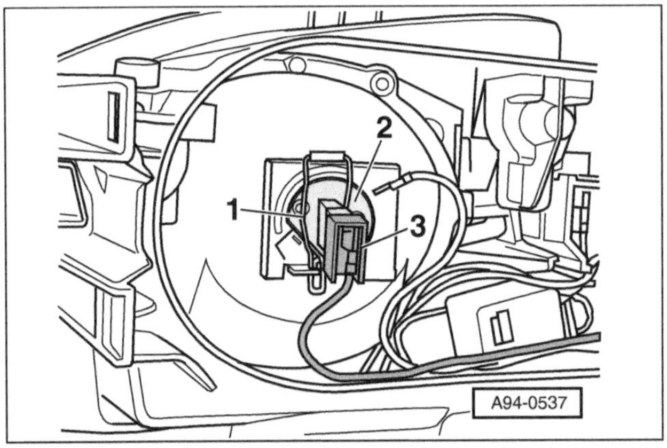

Low beam bulb (Halogen)

◄ Detach harness connector (**3**).

– Remove wire clip (**1**) and take bulb (**2**) out of housing.

> ***CAUTION—***
> *• Do not touch bulb glass with bare skin.*

Low beam bulb (Xenon)

> ***WARNING —***
> *• Disconnect negative (–) battery cable and cover battery terminal to keep cable from accidentally contacting terminal.*

– After disconnecting battery, briefly turn headlights on and off in order to discharge any residual voltage.

◄ Detach harness connector (**1**) by turning counter clockwise (**arrow**).

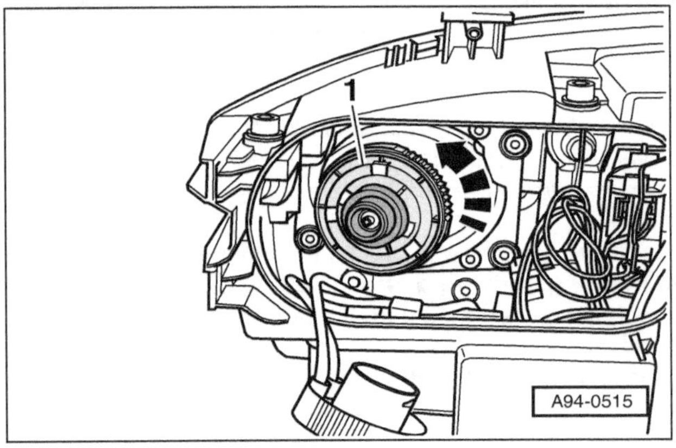

◄ Detach retaining ring (**1**) from bulb by turning counter clockwise (**arrow**).

– Remove bulb from headlight housing.

> ***CAUTION—***
> *• Do not touch bulb glass with bare skin.*

– Install in reverse order of removal, noting the following:

- Lug at top of bulb holder must engage in recess of bulb.
- Bulb retaining ring can only be installed in one position.

BentleyPublishers
.com
© 2006 BentleyPublishers.com—All Rights Reserved

High beam bulb

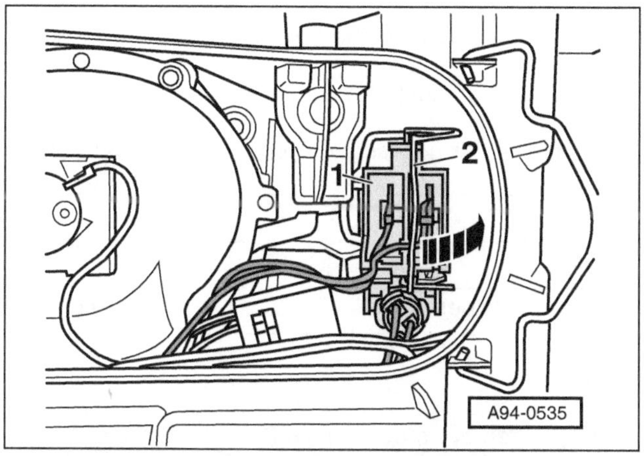

– Detach harness connector.

◄ Pivot wire clip (**2**) in direction of **arrow**.

– Remove bulb holder (**1**) from housing.

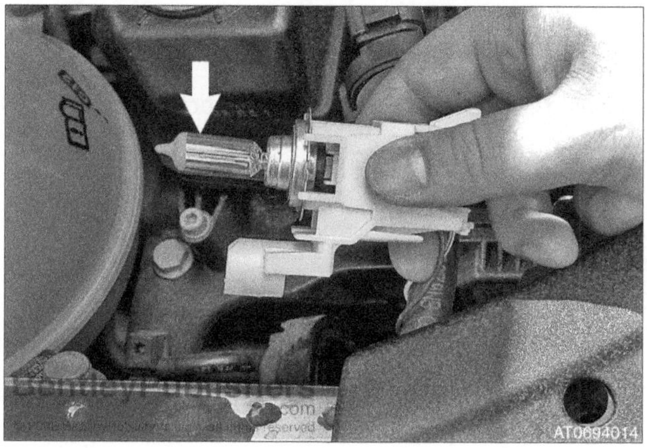

◄ Remove high beam bulb (**arrow**) from holder.

> ***CAUTION—***
> *• Do not touch bulb glass with bare skin.*

– Install in reverse order of removal.

Xenon (HID) lamp control modules, removing and installing

– Switch off all electrical consumers and remove ignition key.

– Make sure light switch is off.

> ***WARNING —***
> *• Disconnect negative (–) battery cable and cover battery terminal to keep cable from accidentally contacting terminal.*

– After disconnecting battery, briefly turn headlights on and off in order to discharge any residual voltage.

– Remove sound absorber panel(s) to access rear of headlight housing. See **03 Maintenance**.
- For left side headlight, remove sound absorber panel over battery.
- For right side headlight, remove sound absorber panel over coolant reservoir.

Through 03/2000

– Remove headlight housing. See **Headlight housing, removing and installing** earlier in this repair group.

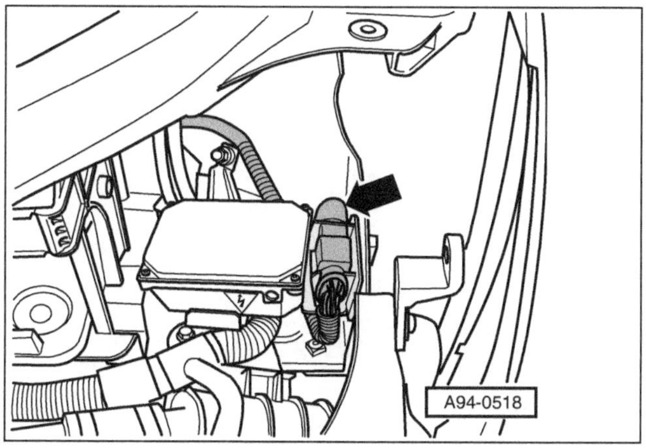

◄ Disconnect and remove harness connector (**arrow**) from retainer.

BentleyPublishers.com
© 2006 BentleyPublishers.com—All Rights Reserved

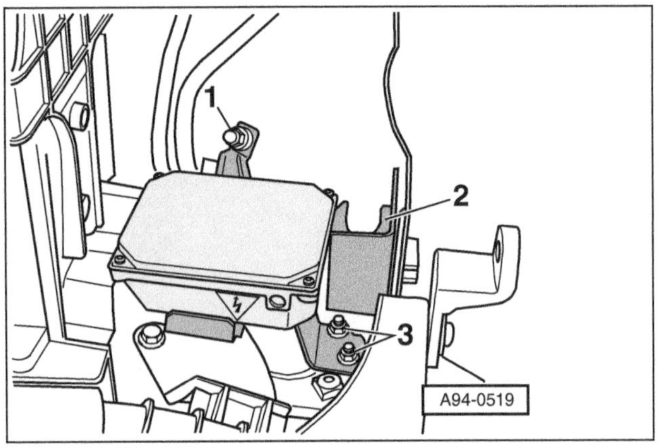

Swivel retainer (**2**) to rear and remove mounting nuts (**3**).

- Remove control module from engine compartment.
- Install in reverse order of removal.
- Replace any cable ties as necessary. Tighten bolts to specification.

Tightening torques	
Retainer to body	6 Nm (53 in-lb)
Control module to retainer	1.2 Nm (9 in-lb)

From 04/2000

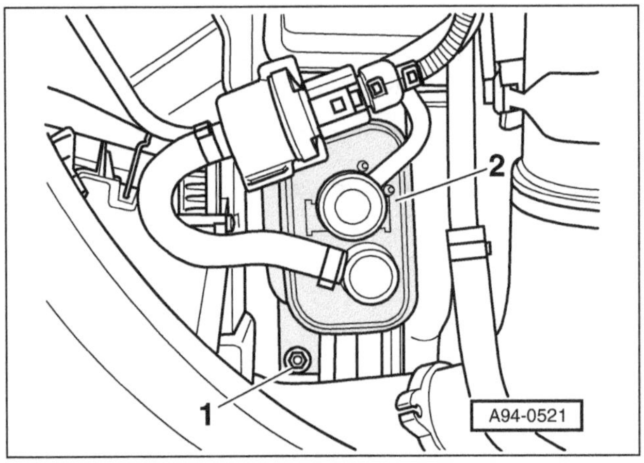

For right headlight, remove mounting nut (**1**) at activated charcoal filter (**2**).

- Unclip breather hoses from retainers.
- Position activated charcoal filter out of way with connecting pipes still attached.
- For left headlight, remove battery. See **27 Battery, Alternator, Starter**.

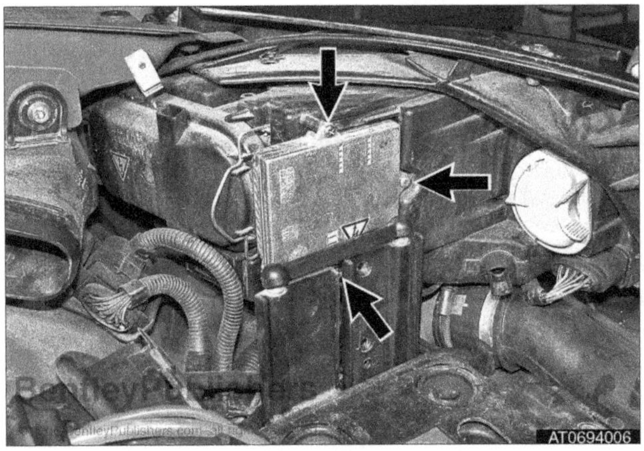

Remove control module mounting screws (**arrows**) and remove module from headlight housing.

- Install in reverse order of removal, noting the following:
  - Check condition of gasket between control module and headlight housing.
- Tighten bolts to specification.

Tightening torques	
Control module to headlight	12 Nm (9 ft-lb)
Charcoal filter to body	10 Nm (7 ft-lb)

BentleyPublishers
.com
©2006 BentleyPublishers.com—All Rights Reserved

FRONT TURN SIGNALS AND PARKING LIGHTS

Front turn signal and parking light bulbs, removing and installing

– Switch off all electrical consumers and remove ignition key.

– Make sure light switch is off.

> ***WARNING —***
> *• Disconnect negative (–) battery cable and cover battery terminal to keep cable from accidentally contacting terminal.*

– Remove sound absorber panel(s) to access rear of headlight housing. See **03 Maintenance**.

- For left side headlight, remove sound absorber panel over battery.
- For right side headlight, remove sound absorber panel over coolant reservoir.

Parking light

◄ Working at inboard corner of headlight assembly, rotate parking light bulb socket (**arrow**) ¼ turn clockwise to remove.

- Battery removed for clarity in illustration at left.

– Push down and rotate bulb ¼ turn to remove from socket. Replace with new bulb and reinstall socket.

> ***CAUTION—***
> *• Do not touch bulb glass with bare skin.*

Turn signal

– For driver's side bulb, remove air filter assembly and/or battery to access turn signal bulb.

◄ Reach under headlight assembly and rotate turn signal bulb socket (**arrow**) ¼ turn clockwise to remove. Disconnect harness connector.

– Push down and rotate bulb ¼ turn to remove from socket. Replace with new bulb and reinstall harness connector and socket.

> ***CAUTION—***
> *• Do not touch bulb glass with bare skin.*

Side turn signal bulb, removing and installing

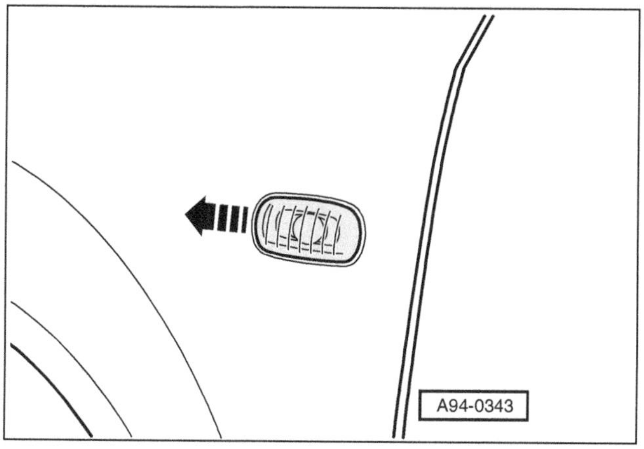

◄ Carefully push side turn signal forward (**arrow**) against spring clip and remove from front fender.

– Detach electrical harness connector. Remove bulb and replace.

> ***CAUTION—***
> *• Do not touch bulb glass with bare skin.*

– Installation is reverse of removal.

BentleyPublishers.com
© 2006 BentleyPublishers.com—All Rights Reserved

Foglights

Audi TT models covered by this manual are not equipped with a rear foglight.

Front foglight bulb, removing and installing

– Switch off all electrical consumers and remove ignition key.

– Make sure light switch is off.

> ***WARNING —***
> • *Disconnect negative (–) battery cable and cover battery terminal to keep cable from accidentally contacting terminal.*

– Remove sound absorber panel(s) to access rear of headlight housing. See **03 Maintenance**.

- For left side headlight, remove sound absorber panel over battery.
- For right side headlight, remove sound absorber panel over coolant reservoir.

◄ Remove small cover (**arrow**) from back of headlight housing by turning counter clockwise.

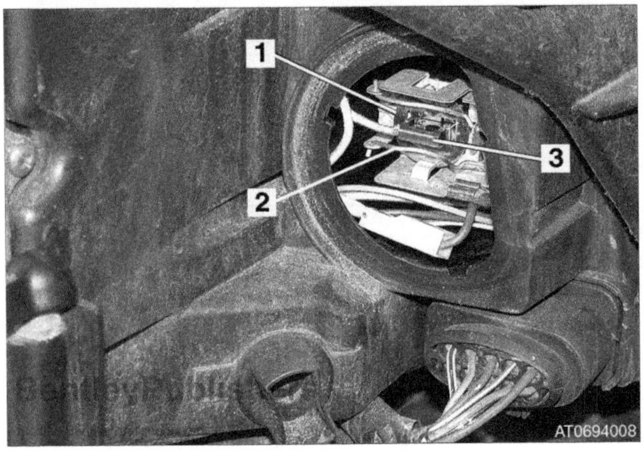

◄ Disengage spring clip (**2**) and remove bulb (**1**) from housing.

– Disconnect electrical harness connector (**3**).

– Install in reverse order of removal. Adjust foglight beam if necessary. See **03 Maintenance** for adjustment.

BentleyPublishers.com
© 2006 BentleyPublishers.com—All Rights Reserved

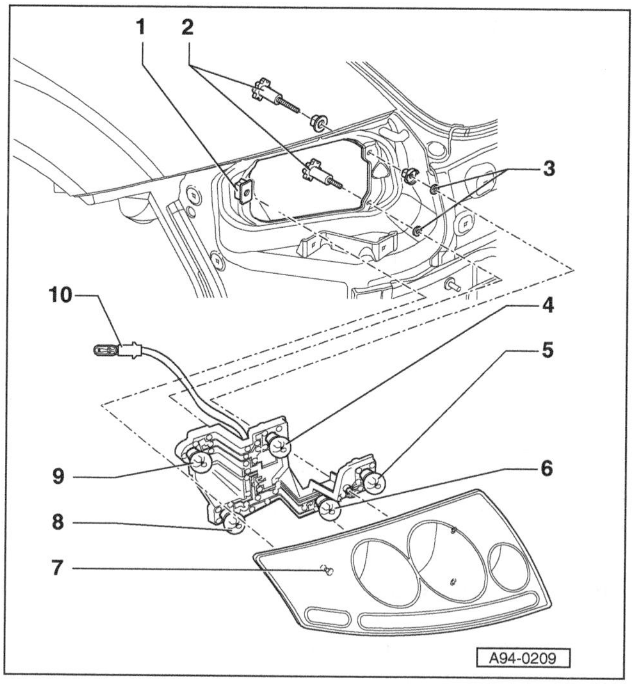

TAILLIGHTS

Taillight assembly

1. Securing clips
2. Mounting knobs
3. Plastic washers
4. Brake light/taillight bulb
5. Back-up bulb
6. Turn signal bulb
7. Retaining pin on housing
   - Engage in plastic clip (**1**)
   - Can be adjusted to align taillight housing with body contours
8. Turn signal bulb
9. Brake light/taillight bulb
10. Side marker bulb

Taillight housing, removing and installing

– Switch off all electrical consumers and remove ignition key.

– Make sure light switch is off.

> ***WARNING —***
> - *Disconnect negative (–) battery cable and cover battery terminal to keep cable from accidentally contacting terminal.*

– Open small lid in luggage area (trunk) side trim panel for access to taillight.

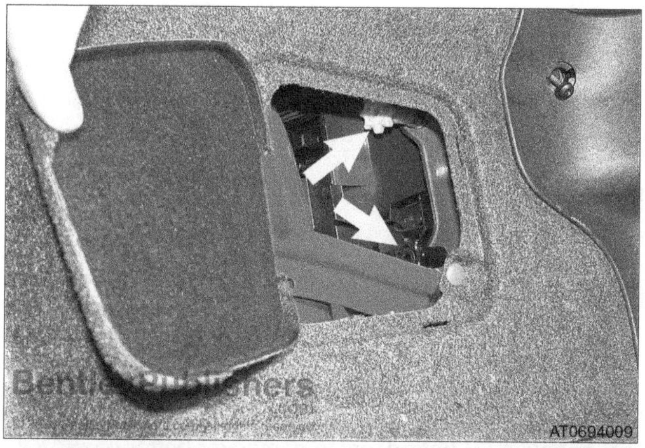

◄ Remove mounting knobs (**arrows**) and disconnect harness connector from back of taillight housing.

– Swing housing out away from body, disengage retaining pin at outer part of housing, and remove housing.

– Install in reverse order of removal, noting the following:
- Ensure tight fit between body and taillight housing.
- Align taillight if necessary. See **Taillight housing, adjusting**.

Taillight bulbs, replacing

– Remove taillight housing. See **Taillight housing, removing and installing** in this repair group.

– Place taillight housing on clean surface.

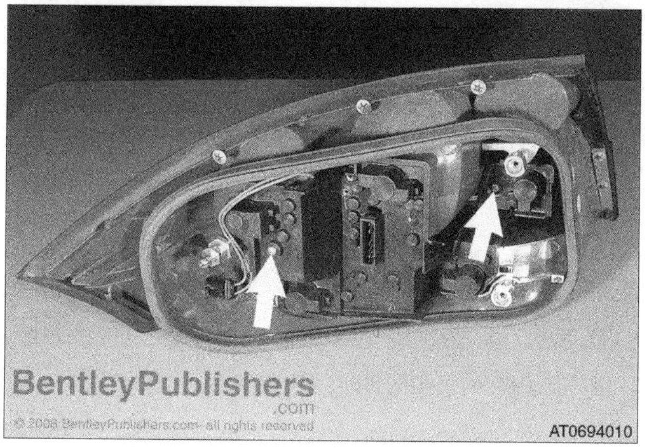

◄ Remove mounting screws (**arrows**) and bulb holder assembly.

– Replace defective bulbs as necessary.

BentleyPublishers
.com
© 2006 BentleyPublishers.com—All Rights Reserved

Taillight housing, adjusting

Front-to-rear adjustment

– Remove taillight housing. See **Taillight housing, removing and installing** in this repair group.

< Gradually alter length of adjusters (**arrows**).

– Install taillight housing and check alignment.
  - Housing must be flush with body contour.

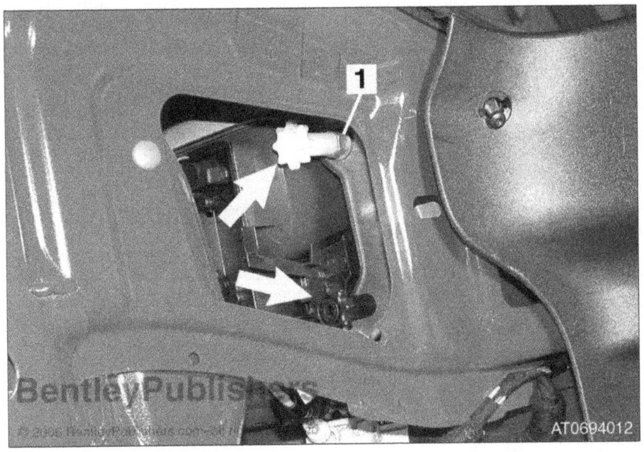

Height adjustment

– Open small lid in luggage area (trunk) side trim panel for access to taillight.

< Loosen mounting knobs (**arrows**).

– Loosen adjustment nut (**1**) approx. 1 turn and move taillight housing as necessary to obtain equal gaps with rear fender and bumper cover.

– Tighten adjustment nut and mounting knobs.

CENTER BRAKE LIGHT

If the center brake light has one or more malfunctioning Light Emitting Diode (LED) then the entire assembly must be replaced.

Center brake light, removing and installing (coupe)

– Switch off all electrical consumers and remove ignition key.

– Make sure light switch is off.

> ***WARNING —***
> - *Disconnect negative (–) battery cable and cover battery terminal to keep cable from accidentally contacting terminal.*

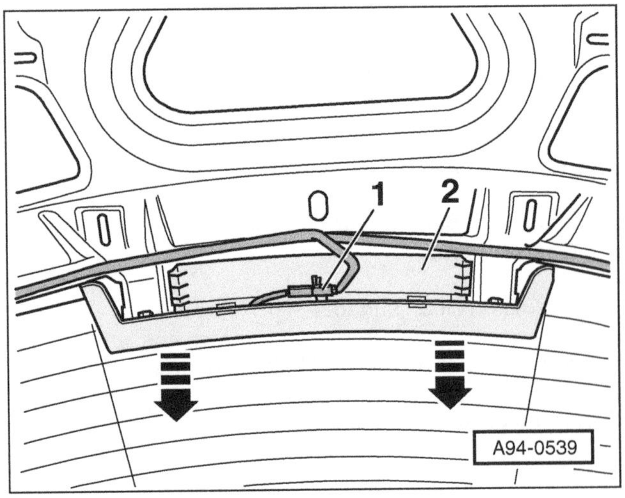

– Open rear lid and remove trim panel. See **70 Trim–Interior**.

< Slide center brake light assembly (**2**) forward and remove from rear window. If necessary, gently pry the assembly out at left and right sides.

– Remove electrical wire from retaining clip and disconnect from center brake light.

– Install in reverse order of removal.
  - Be sure to confirm operation of center brake light.

BentleyPublishers
.com
©2006 BentleyPublishers.com—All Rights Reserved

Center brake light, removing and installing (roadster)

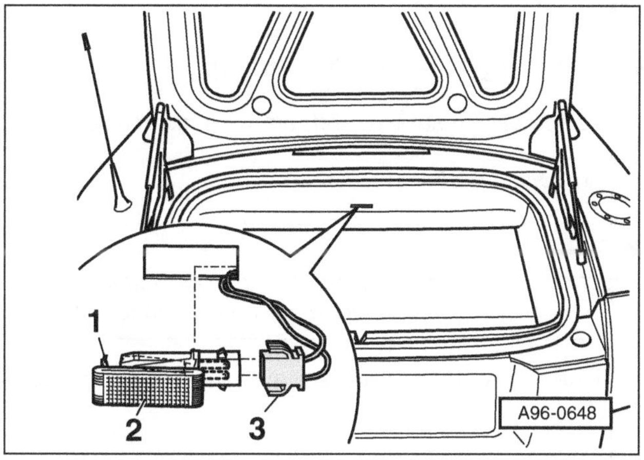

- Switch off all electrical consumers and remove ignition key.
- Make sure light switch is off.

> ***WARNING —***
> *• Disconnect negative (–) battery cable and cover battery terminal to keep cable from accidentally contacting terminal.*

- Open rear lid.
- ◀ Depress locating tabs (**1**) using screwdriver and remove luggage compartment light (**2**).
- Pull luggage compartment light out from trim opening and disconnect electrical harness connector.
- Close rear lid.

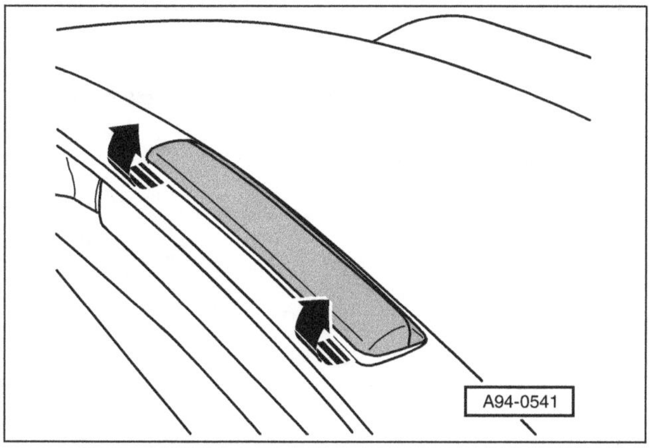

- ◀ Carefully unclip cover for center brake light from the front (leading) edge towards the rear of the vehicle as shown (**arrows**).

> ***CAUTION—***
> *• Do not unclip center brake light cover from rear edge first. Damage to cover and/or light assembly will occur.*

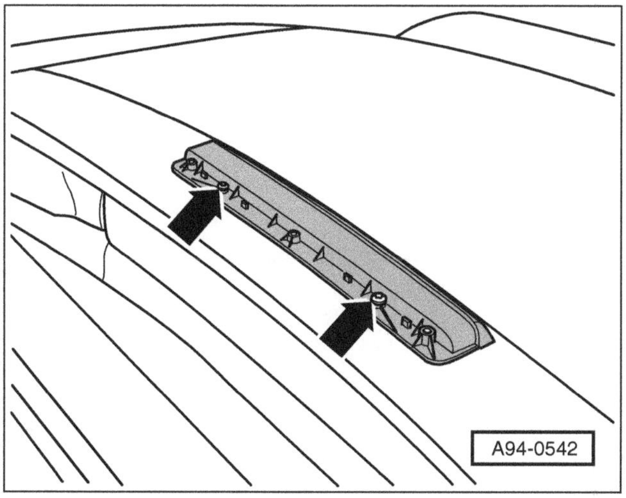

- ◀ Remove mounting screws (**arrows**).
- Remove brake light assembly upwards while routing wiring harness through body opening.
- If defective, replace entire socket. There are no separate bulbs available.
- Install in reverse order of removal, noting the following:
  - Route wiring harness through body opening into luggage compartment using a guide wire if necessary.
  - Reconnect harness at luggage compartment light.
  - Confirm operation of center brake light.
  - Tighten mounting screws to specification.

Tightening torque	
Center brake light to body	2.5 Nm (23 in-lb)

BentleyPublishers
.com
© 2006 BentleyPublishers.com—All Rights Reserved

LICENSE PLATE LIGHT

License plate light bulb, replacing

– Switch off all electrical consumers and remove ignition key.

– Make sure light switch is off.

> ***WARNING —***
> - *Disconnect negative (–) battery cable and cover battery terminal to keep cable from accidentally contacting terminal.*

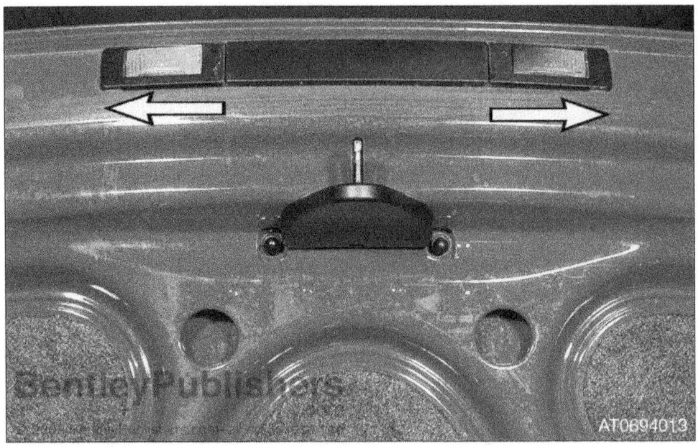

Open trunk lid. Press and slide license plate light outward (**arrows**) and remove.

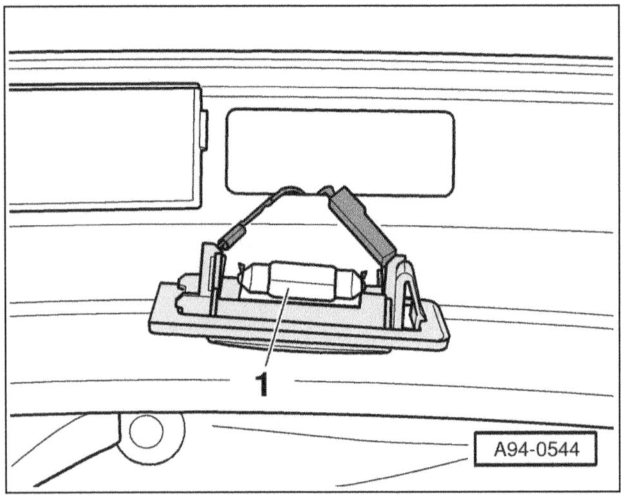

Remove bulb (**1**) from clip holders and replace with new one.

– Install in reverse order of removal.

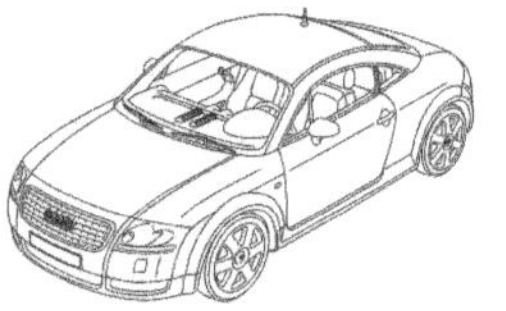

BentleyPublishers
.com
© 2006 BentleyPublishers.com—All Rights Reserved

96 Interior Lights, Switches, Anti-theft

TABLE

GENERAL

This repair group covers interior light repairs, switch replacement, horns and the anti-theft system.

For additional information, see:

- **97 Fuses, Relays, Component Locations**
- **EWD Electrical Wiring Diagrams**

BentleyPublishers.com

© 2006 BentleyPublishers.com—All Rights Reserved

Warnings and cautions

> ***WARNING —***
> - *Some light bulbs can become hot enough to cause injury. Allow bulbs to cool before working on the lighting system.*

> ***CAUTION—***
> - *Switch the ignition OFF and disconnect the negative (–) battery cable before removing any electrical components.*
> - *Prior to disconnecting the battery, read the battery disconnection precautions in* **27 Battery, Alternator, Starter**.
> - *Only use a digital multimeter for electrical tests.*
> - *To avoid marring paint or trim, work with a plastic prying tool or wrap screwdriver tip with tape.*
> - *Do not touch bulb glass with bare skin. Dirt and skin oils may cause the bulb to fail prematurely. If necessary wipe bulb using clean cloth dampened with rubbing alcohol.*
> - *Use only original equipment replacement bulbs. Non-original bulbs may cause false failure readings on the dashboard display.*
> - *Use only specified bulbs. A bulb with higher wattage may cause damage to bulb housing.*

INTERIOR LIGHTS

Interior bulb applications

For convenience, interior bulb applications are listed in **Table a**.

Table a. interior bulb applications

Function	Type	Wattage
Dome light (coupe)	K	10
Glove compartment	W10	3
Make-up mirror (coupe)	L	5
Luggage compartment (coupe)	W10	3
Luggage compartment (roadster)	W	5
Reading lights (coupe and roadster)		5
Rear hatch (coupe)	W	5

Front interior lights

Front dome light/reading light bulbs, replacing (coupe)

– Press interior light switch to position “O” (off).

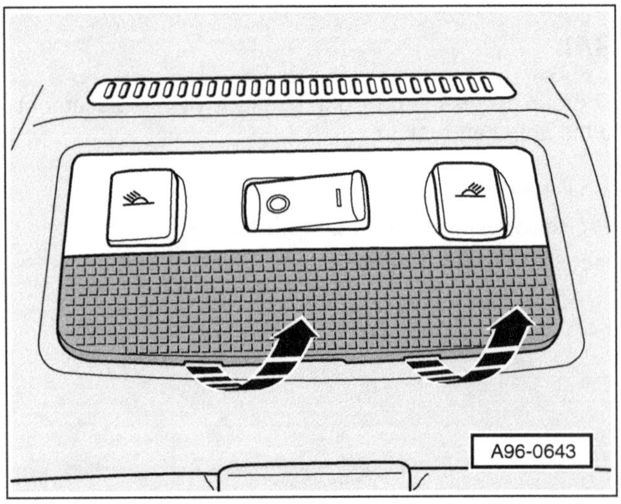

Use plastic prying tool to carefully pry off dome light lens in direction of **arrows**.

BentleyPublishers.com

©2006 BentleyPublishers.com—All Rights Reserved

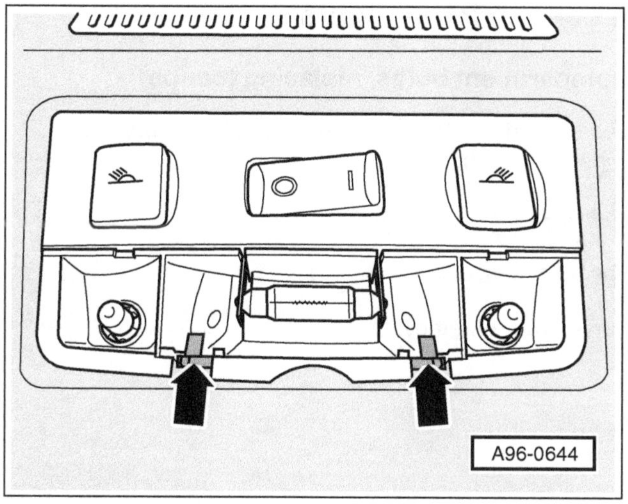

◄ Replace bulbs as necessary.
- Dome light bulb (center) pulls out of spring clamps
- Reading light bulbs (left and right) twist out of sockets

– To remove bulb housing, release retaining tabs (**arrows**) and remove bulb assembly. Disconnect harness connector.

– Install in reverse order of removal.

Front reading light bulbs, replacing (roadster)

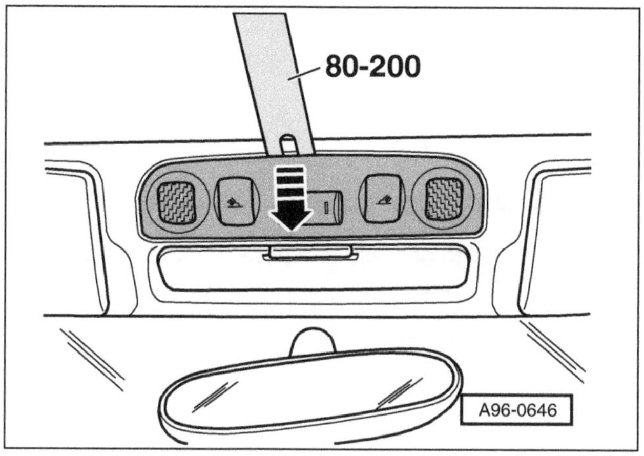

NOTE—
- *In roadster models the reading lights also function as dome lights.*

◄ Use pry lever 80-200, or equivalent, to carefully pry rear edge of interior light assembly from windshield frame in direction of **arrow**.

> ***CAUTION—***
> - *Be careful not to damage retaining tabs when removing light housing.*

– Disconnect harness connector.

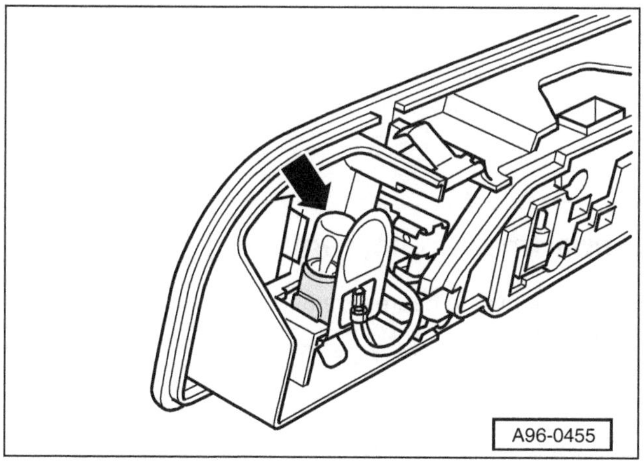

◄ Remove reading light bulb (**arrow**) from back of light assembly and replace as necessary.
- Twist bulb to remove.

– Reattach harness connector.

– When installing assembly into windshield frame, insert connector side first. Press assembly into frame until fully engaged.

BentleyPublishers.com
©2006 BentleyPublishers.com—All Rights Reserved

Luggage compartment lights

Luggage compartment bulbs, replacing (coupe)

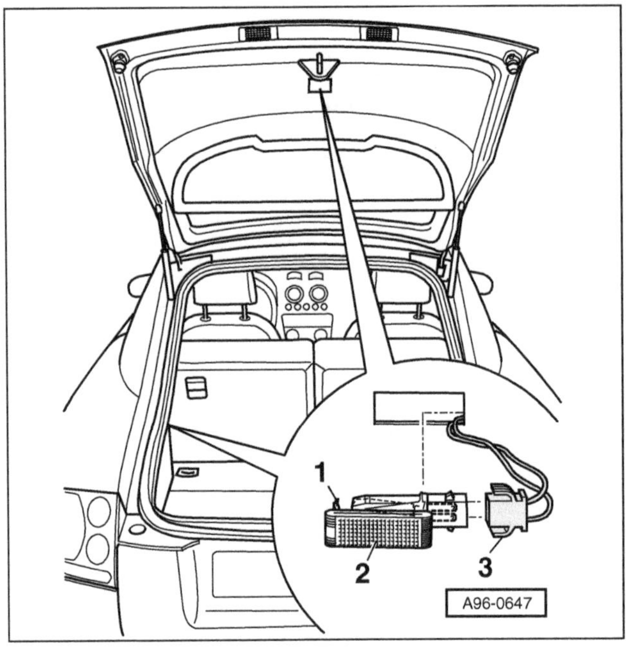

- Use flat tipped screwdriver to depress retainer (**1**) and pry out luggage compartment light (**2**).
- Disconnect harness connector (**3**).
- Replace bulbs as necessary.
- Install in reverse order of removal.

Luggage compartment bulb, replacing (roadster)

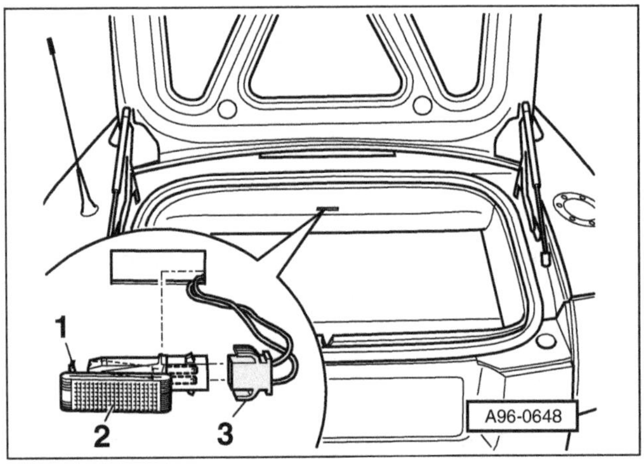

- Use flat tipped screwdriver to depress retainer (**1**) and pry out luggage compartment light (**2**).
- Disconnect harness connector (**3**).
- Replace bulb.
- Install in reverse order of removal.

Glove compartment bulb, replacing

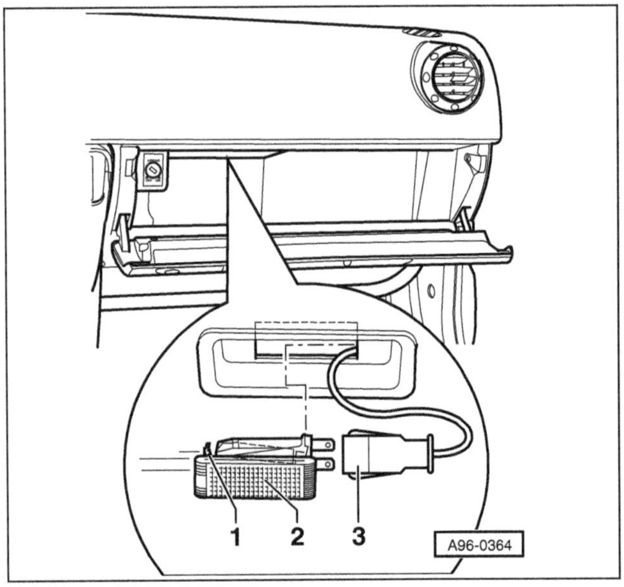

- Use flat tipped screwdriver to depress retainer (**1**) and pry out glove compartment light (**2**).
- Disconnect harness connector (**3**).
- Remove bulb cover and replace bulb.
- Install in reverse order of removal.

BentleyPublishers.com
© 2006 BentleyPublishers.com—All Rights Reserved

Make-up mirror bulbs, replacing (coupe)

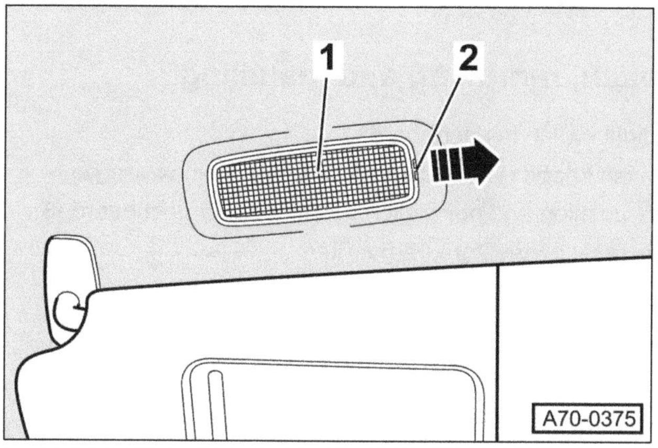

◀ Use screwdriver to carefully pry out make-up mirror light (**1**) at indentation (**2**).

– Slide make-up mirror light to side (**arrow**) and remove.

– Disconnect harness connector.

– Remove bulb cover and replace bulb.

– Install in reverse order of removal.

BACK-UP LIGHTS

Back-up light bulb replacement is covered in **94 Exterior Lights**.

Back-up light switch, manual transmission

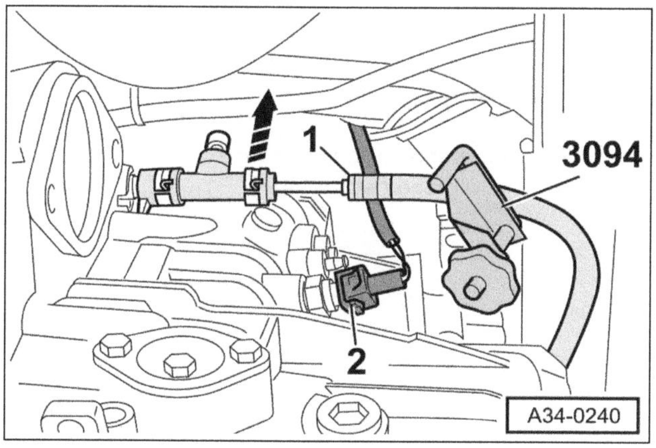

◀ Back-up light switch (**2**) for 5- or 6-speed manual transmission.

Back-up light switch function (DSG and automatic transmission)

The DSG (Direct Shift Gearbox) transmission uses the selector mechanism (E313) and the Mechatronic unit (J743) to operate the back-up light relay which enables the back-up lights to function.

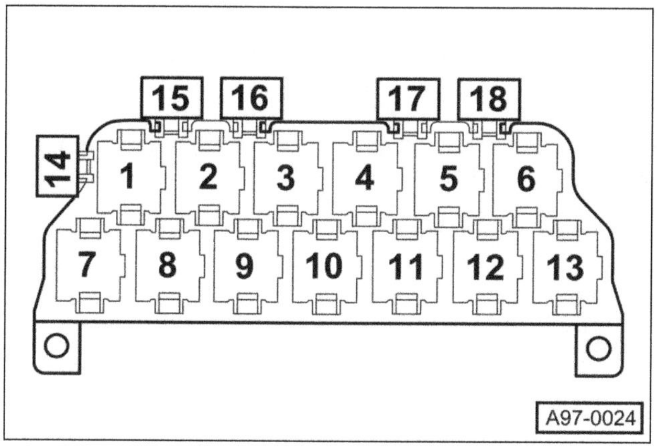

◀ The 13-fold relay panel is located above the micro central electric panel under the driver's side of the padded dash.

- Position 13 - Backup light relay (J219), DSG only

The automatic transmission uses the transmission control module (J217) and the transmission range switch (F125) to enable the back-up lights to function. No external back-up light relay is used.

BentleyPublishers
.com
© 2006 BentleyPublishers.com—All Rights Reserved

Dashboard and Center Console Switches

Light switch, removing and installing

Working at light switch in dashboard:

- Push in switch knob (**1**) and at same time turn it clockwise (**2**).
- Hold in this position and pull switch housing out of dashboard (**3**).
- Detach electrical connectors from switch.

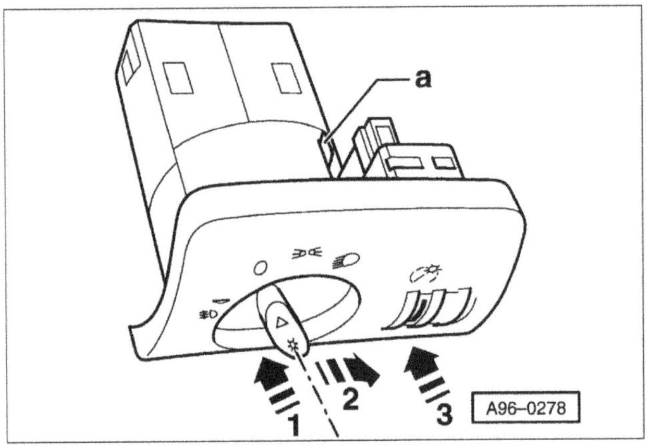

– To reinstall, reattach electrical connectors.

Push switch knob in (**arrow 1**) and at same time turn it clockwise (**arrow 2**) to engage retaining catch (**a**).

– With switch in this position, push switch housing into recess in dashboard until it clicks into place (**arrow 3**).

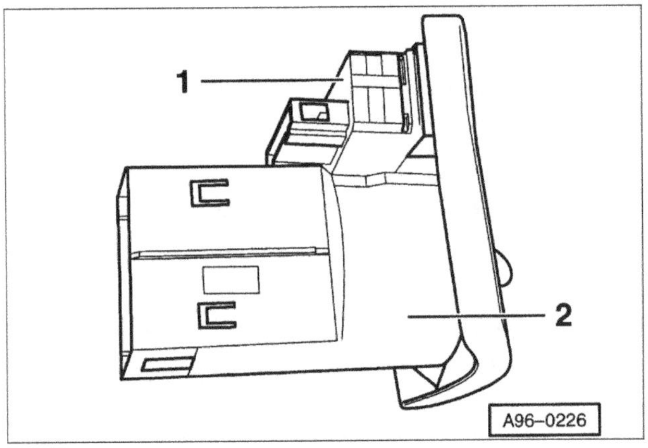

Instrument light dimmer switch, removing and installing

The instrument light dimmer is part of the light switch, but can be replaced separately.

– Remove light switch. See **Light switch, removing and installing** in this repair group.

Use thin flat screwdriver to pry off dimmer switch (**1**) and separate from light switch housing (**2**).

– Installation is reverse of removal

BentleyPublishers.com
© 2006 BentleyPublishers.com—All Rights Reserved

Center dashboard switches, removing and installing

– Remove radio. See **91 Radio and Communication**.

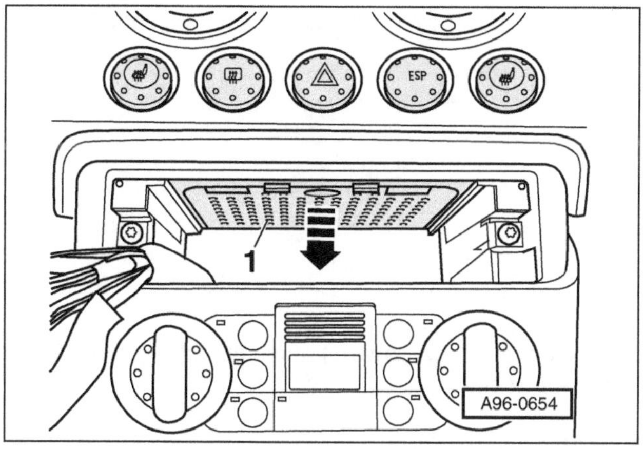

◄ Pull down radio cover plate (**1**) in direction of **arrow** and remove.

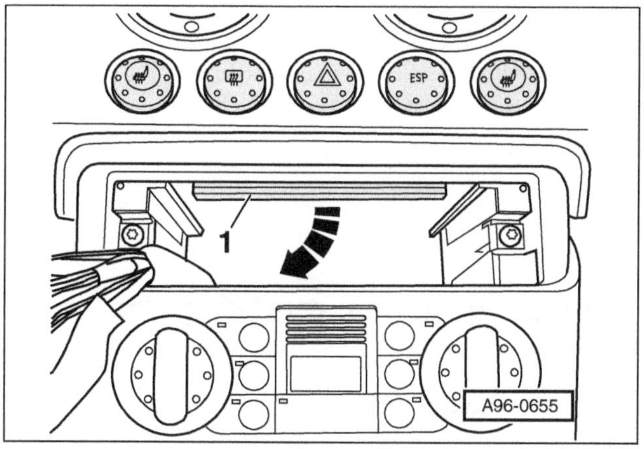

◄ Reach upwards through radio slot and release switch console by pulling retainer clip (**1**) downward in direction of **arrow**.

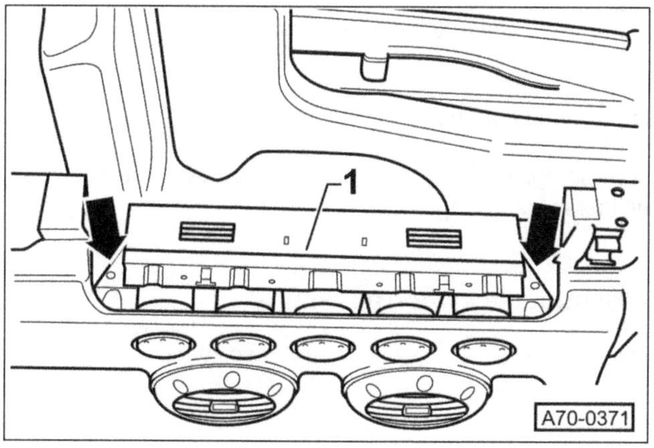

◄ Remove switch console mounting screws (**arrows**).

NOTE —

- *View is from underneath, looking up.*
- *The emergency flasher and turn signal relay are incorporated into the emergency flasher switch.*

– Remove switch console (**1**) with switches installed.

– Disconnect harness connector(s) and remove switch(es) as necessary.

– Installation is the reverse of removal.

BentleyPublishers
.com
© 2006 BentleyPublishers.com—All Rights Reserved

Glove compartment light switch, removing and installing

– Open glove box.

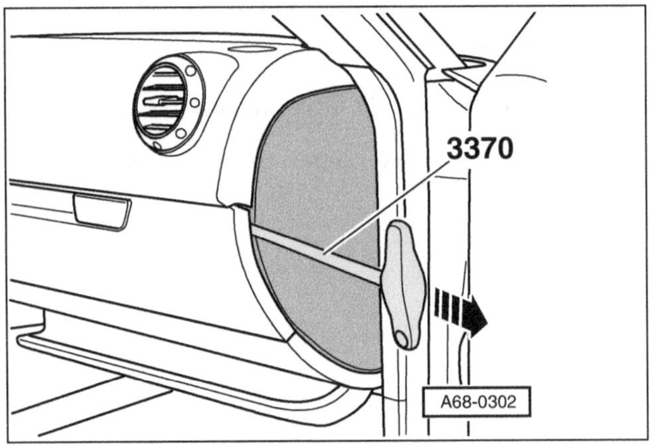

◄ Use hook 3370 to remove right side instrument panel end trim.

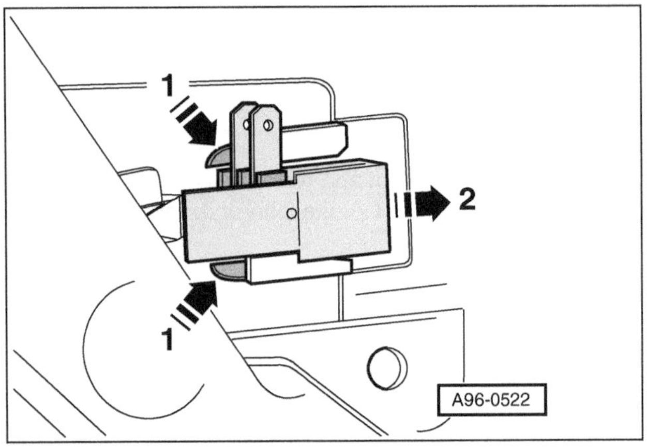

◄ Depress both release tabs (**arrows 1**) and pull switch (**arrow 2**) out of guide in glove box side.

– Disconnect harness connector.

– Install in reverse order of removal.

Center console switches, removing and installing

Front center console switches

– Slide back cover over switch console.

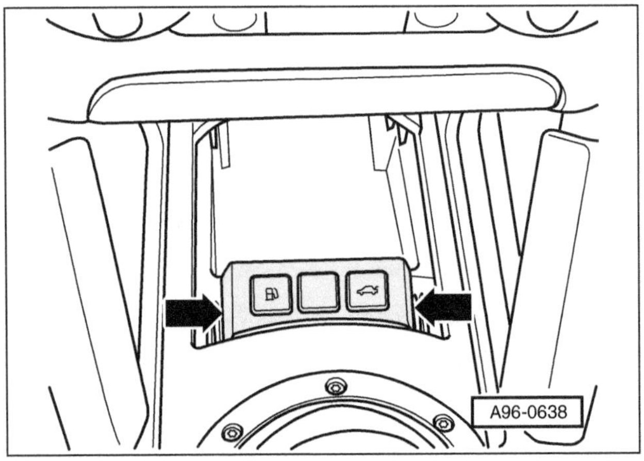

◄ Use flat tipped screwdriver to pry (**arrows**) switches out of center console.

– Disconnect harness connectors.

– Install in reverse order of removal. Be sure to press switches into mount until fasteners engage.

Rear center console switch assembly

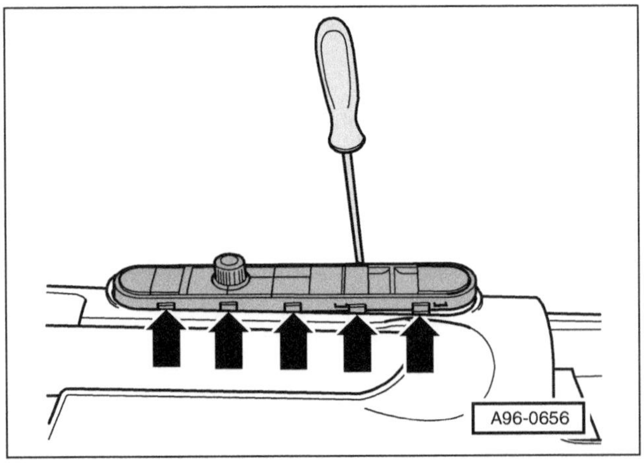

◄ Use flat tipped screwdriver to carefully release retainer tabs (**arrows**) at switch mount.

– Disconnect harness connectors.

– Install in reverse order of removal. Be sure to press switch assembly into mount until fasteners engage.

BentleyPublishers
.com
©2006 BentleyPublishers.com—All Rights Reserved

NOTE —

- *Audi identifies electrical components by a letter and/or a number in the electrical schematics. See* **EWD Electrical Wiring Diagrams**. *These electrical identifiers are listed in parentheses as an aid to electrical troubleshooting.*

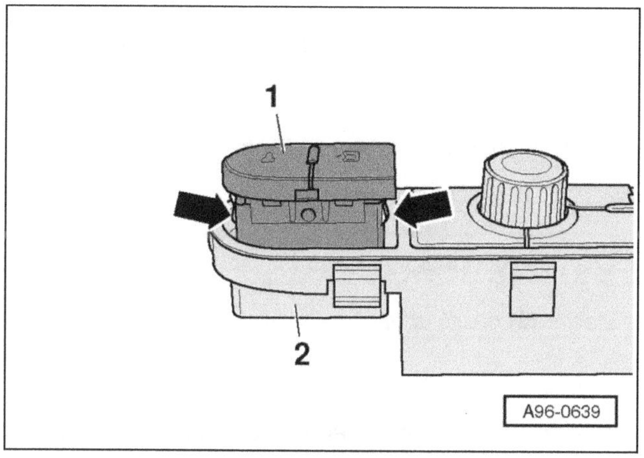

Driver's interior locking switch (E150)

– Remove rear center console switch assembly. See **Center console switches, removing and installing**.

◄ Use small knife or feeler gauge to release retaining tabs (**arrows**).

– Pull switch (**1**) upward and out of switch mount (**2**).

– Install in reverse order of removal. Be sure to press switches into mount until fasteners engage.

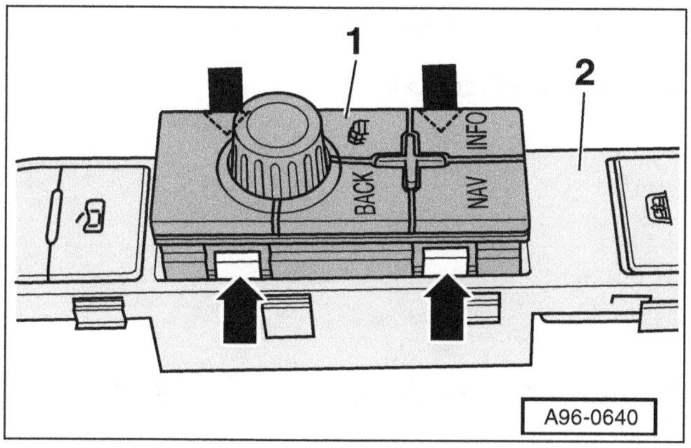

Navigation function selector switch (E272)

– Remove rear center console switch assembly. See **Center console switches, removing and installing**.

◄ Use small knife or feeler gauge to release retaining tabs (**arrows**).

– Pull switch (**1**) upward and out of switch mount (**2**).

– Install in reverse order of removal. Be sure to press switches into mount until fasteners engage.

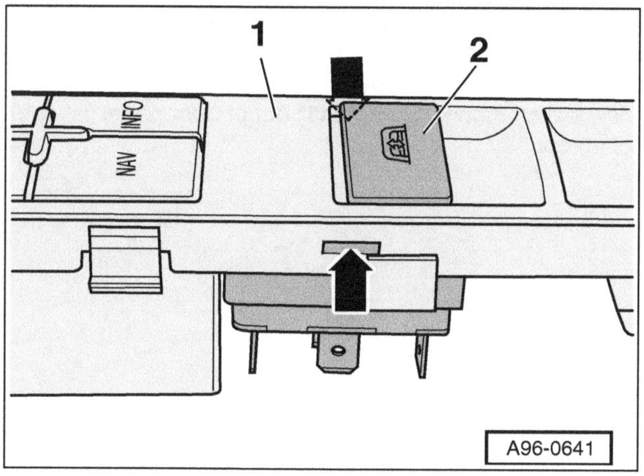

Wind deflector switch (E278)

– Remove rear center console switch assembly. See **Center console switches, removing and installing**.

◄ Use small knife or feeler gauge to release retaining tabs (**arrows**).

– Pull switch (**2**) downward out of switch mount (**1**).

– Install in reverse order of removal. Be sure to press switches into mount until fasteners engage.

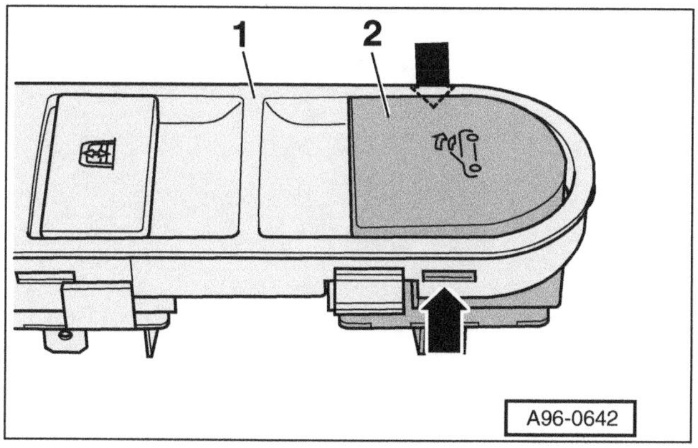

Convertible top switch (E137)

– Remove rear center console switch assembly. See **Center console switches, removing and installing**.

◄ Use small knife or screwdriver to release retaining tabs (**arrows**).

– Pull switch (**2**) downward out of switch mount (**1**).

– Install in reverse order of removal. Be sure to press switches into mount until fasteners engage.

BentleyPublishers
.com
© 2006 BentleyPublishers.com—All Rights Reserved

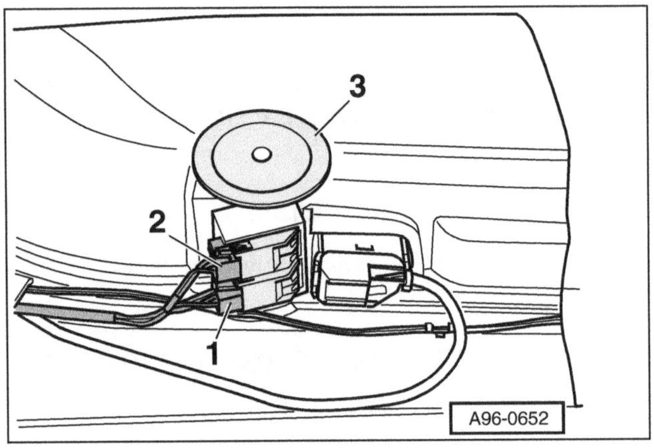

Door Switches

Window switch, removing and installing

NOTE—
- *Driver's side door shown. Passenger side is similar.*

– Remove door panel. See **70 Trim–Interior**.

◄ Disconnect harness connectors (**1** and **2**).

– Pull guide sleeve (**3**) out of door trim.

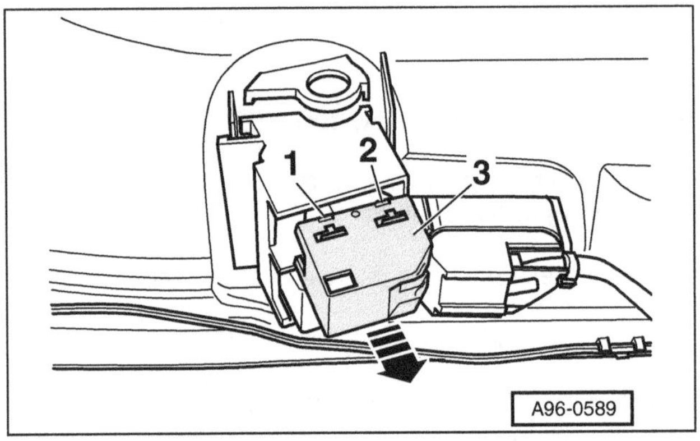

◄ Use small screwdriver to carefully release fasteners (**1** and **2**).

– Remove switch (**3**) from door trim in direction of **arrow**.

– Install in reverse order of removal.

Mirror adjustment switch, removing and installing

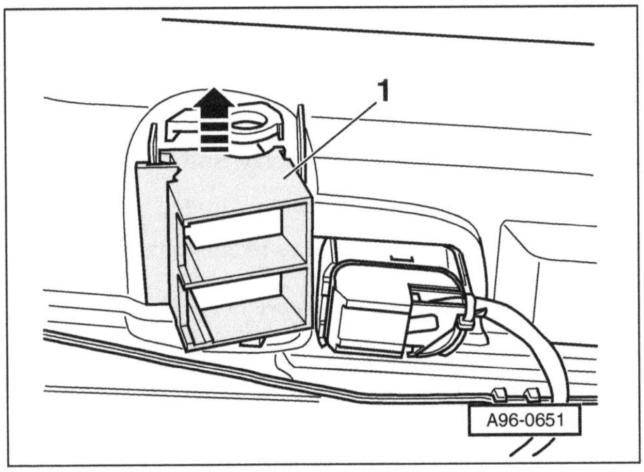

– Remove window switch. See **Window switch, removing and installing**.

◄ Slide window switch mounting frame (**1**) out of door panel in direction of **arrow**.

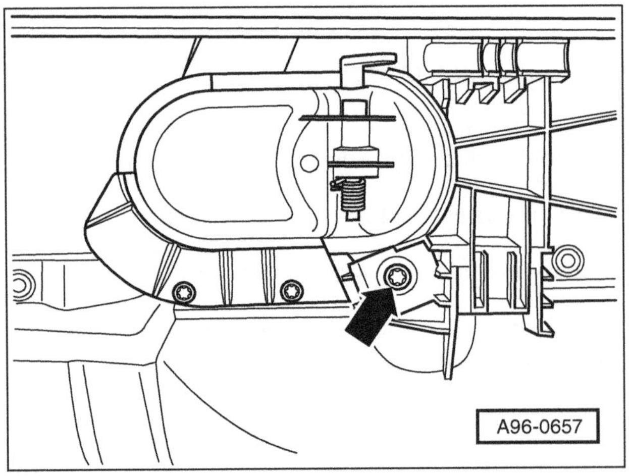

◄ Remove mounting screw (**arrow**).

BentleyPublishers
.com
©2006 BentleyPublishers.com—All Rights Reserved

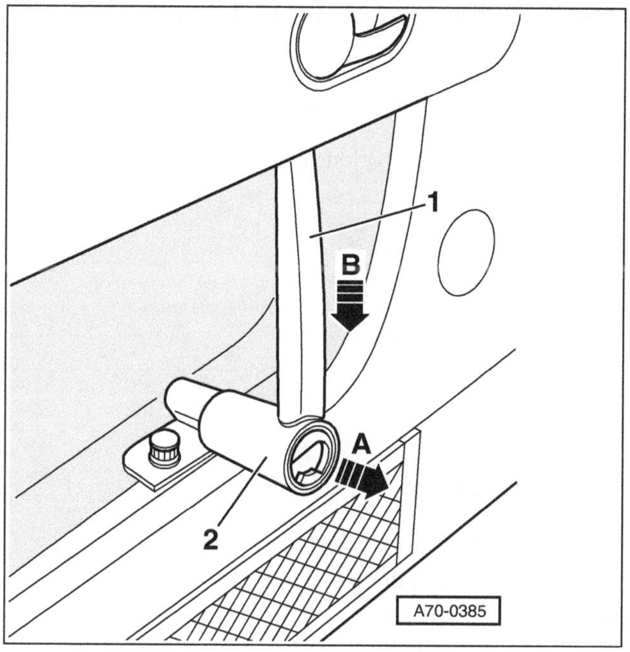

- Slide switch trim (**2**) off door trim panel in direction of **arrow A**.
- Pull grab handle (**1**) out of door panel in direction of **arrow B**.

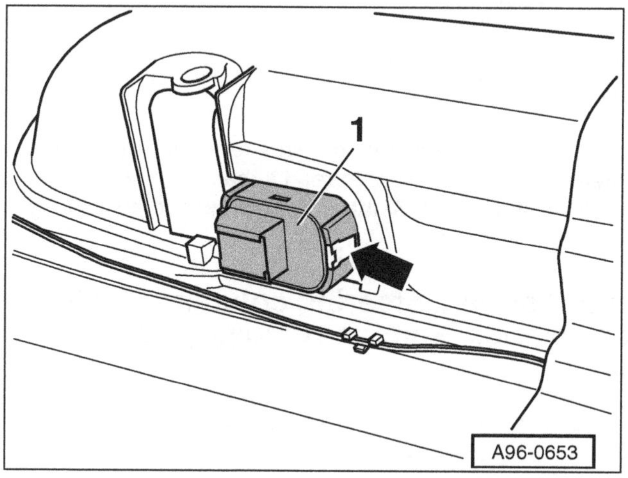

- Carefully release clip (**arrow**) and pull mirror adjustment switch (**1**) upwards from door panel.

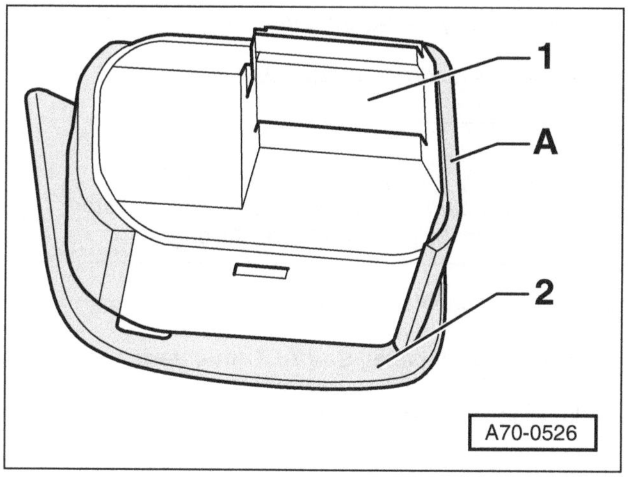

- Separate mirror adjustment switch (**1**) from trim (**2**) by releasing fastener (**A**).
- Install in reverse order of removal, noting the following:
  - Press mirror adjustment switch into trim until fasteners engage.
  - Press switch assembly into mount until fasteners engage.

BentleyPublishers
.com
©2006 BentleyPublishers.com—All Rights Reserved

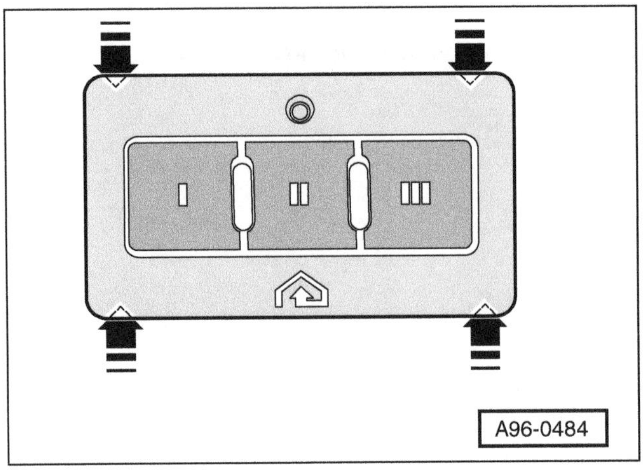

Garage door opener control head, removing and installing

- Use small screwdriver to carefully release fasteners (**arrows**).
- Remove control head from windshield frame and disconnect harness connector.
- Install in reverse order of removal.

Steering Column Switches

The steering column contains the following switches:

- Ignition key cylinder and ignition electrical switch
- Steering column stalk switches:
  - -Turn signal and headlight flasher switch
  - -Cruise control switch
  - -Wiper and washer switch
  - -MFI controller
- Steering angle sensor

Steering column switch replacement requires removal of the driver airbag and steering wheel. Be sure to read airbag warnings and cautions in **48 Steering** and **69 Seatbelts, Airbags** prior to working on steering column switches.

Airbag spiral spring, steering column switches and steering angle sensor, removing and installing

- Pull out adjustable steering column as far as possible. Center steering wheel in straight ahead position.
- Disconnect negative (–) battery cable and cover battery terminal to keep cable from accidentally contacting terminal.
- Remove driver's airbag. See **69 Seatbelts, Airbags**.

> ***WARNING —***
> - *The airbag is inflated by an explosive device. Handled improperly or without adequate safeguards, the airbag system can be very dangerous. See* **69 Seatbelts, Airbags**.

- Remove steering wheel. See **48 Steering**.
- Remove left side under-dash panel. See **70 Trim–Interior.**

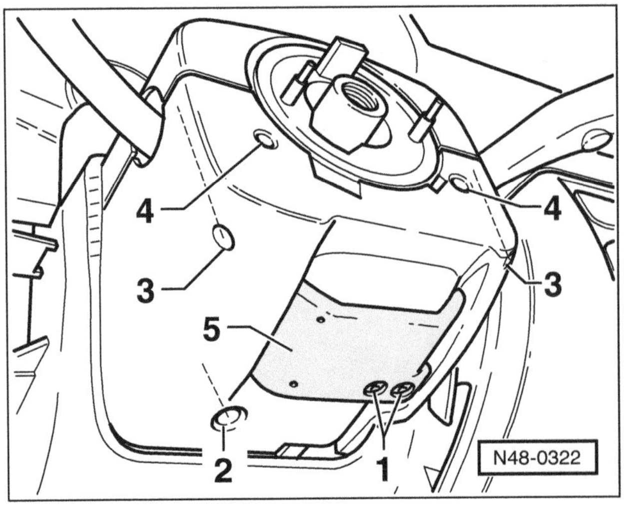

- Remove mounting screws (**1**) and steering column adjustment lever (**5**).
- Remove mounting screws (**2**, **3** and **4**) and upper and lower steering column trim.

BentleyPublishers.com
© 2006 BentleyPublishers.com—All Rights Reserved

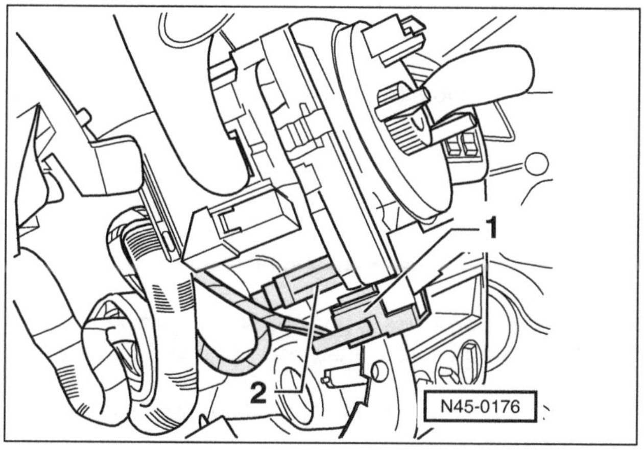

◄ Disconnect harness connectors (**1** and **2**).

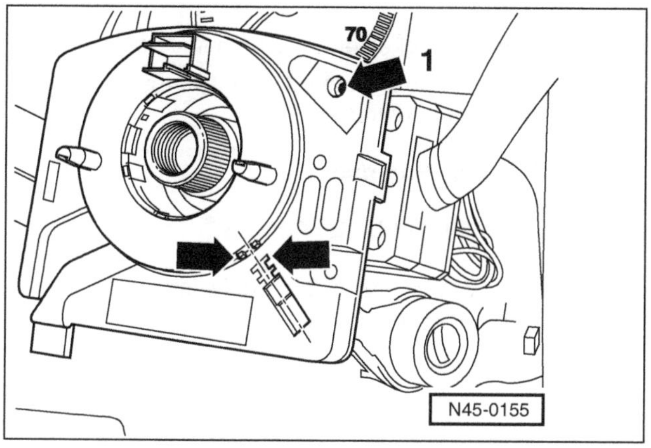

◄ Confirm straight ahead position of steering column and front wheels.

- Yellow dot must be visible in hole (**arrow 1**) and two marks (**arrows**) must be aligned as illustrated.

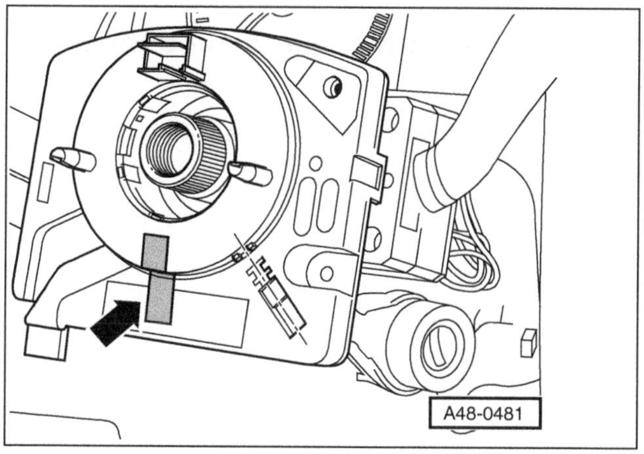

◄ Secure airbag spiral spring in center position with adhesive tape (**arrow**).

◄ Carefully release upper fasteners (**arrows**).

– Carefully release lower fasteners and detach steering angle sensor (with airbag spiral spring still attached) from steering column switch assembly.

BentleyPublishers
.com
©2006 BentleyPublishers.com—All Rights Reserved

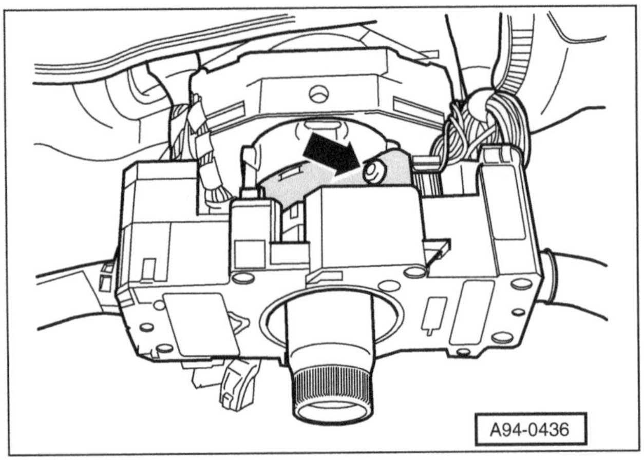

Loosen clamp bolt (**arrow**) enough to remove steering column switches.

– Disconnect harness connectors from steering column switches.

Installing

Install in reverse order of removal, noting the following:

– Slide housing of spiral spring with slip ring and steering angle sensor onto steering column switch until all fasteners are heard to engage.

– Remove adhesive fastening strip or transportation seal if installing new component.

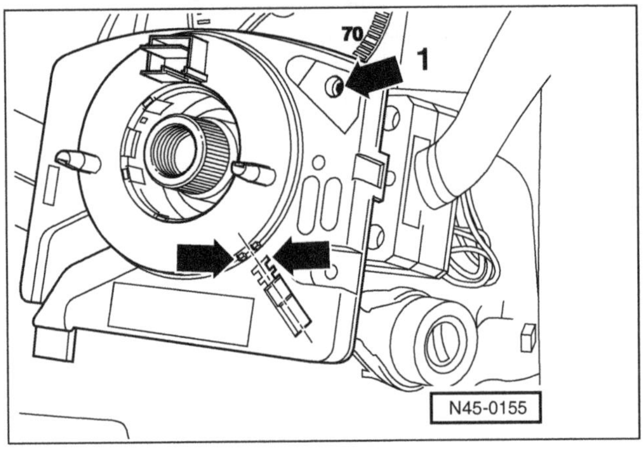

Confirm straight ahead position of steering column and front wheels.

- Yellow dot must be visible in hole (**arrow 1**) and two marks (**arrows**) must be aligned as illustrated.

– Install steering wheel. See **48 Steering**.

– Install driver's airbag. See **69 Seatbelts, Airbags**.

– Use diagnostic scan tool (VAS 5051/5052 or equivalent) to perform basic setting for steering angle sensor.

– Test drive vehicle.

Steering column switches, harness connector identification

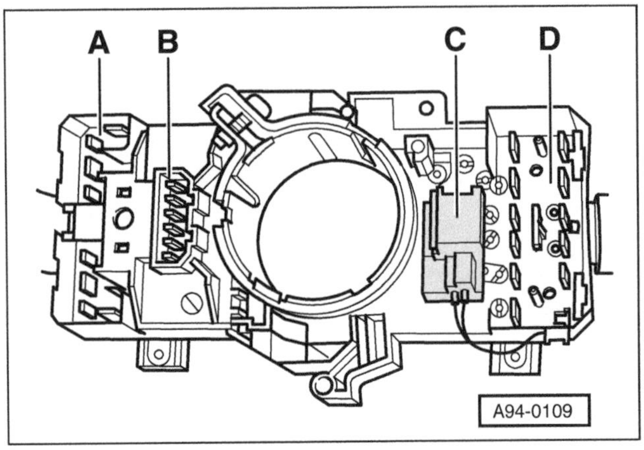

Multi-pin connectors at steering column switches.

- **A** - Windshield wiper switch
- **B** - On board computer and intermittent wiper
- **C** - Cruise control
- **D** - Turn signal switch

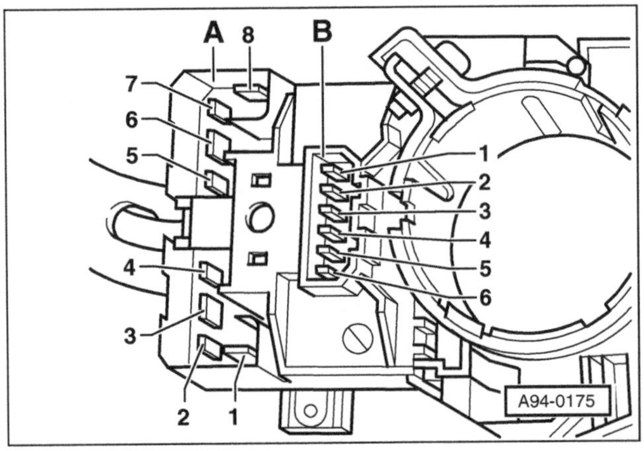

Windshield wiper switch (**A**) terminal assignments.

1. Terminal 53
2. Open
3. Open
4. Terminal 53c
5. Open
6. Terminal 53b
7. Intermittent wipe
8. Terminal 53a

BentleyPublishers.com
© 2006 BentleyPublishers.com—All Rights Reserved

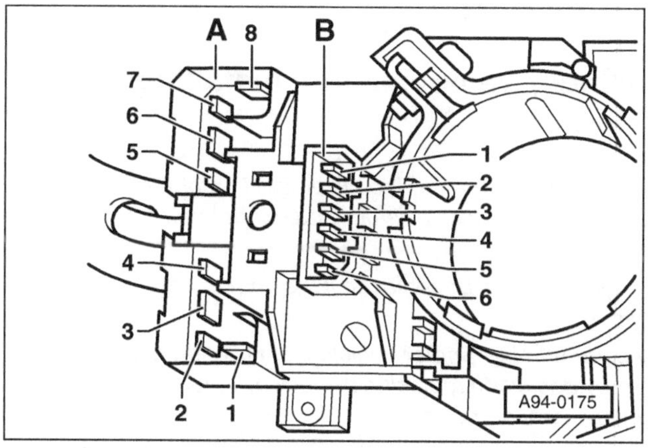

On Board computer switch (**B**) terminal assignments.

1. On Board computer rocker switch, right
2. On Board computer rocker switch, left
3. Terminal 31
4. On Board computer, reset
5. Intermittent wiper relay
6. Terminal 61

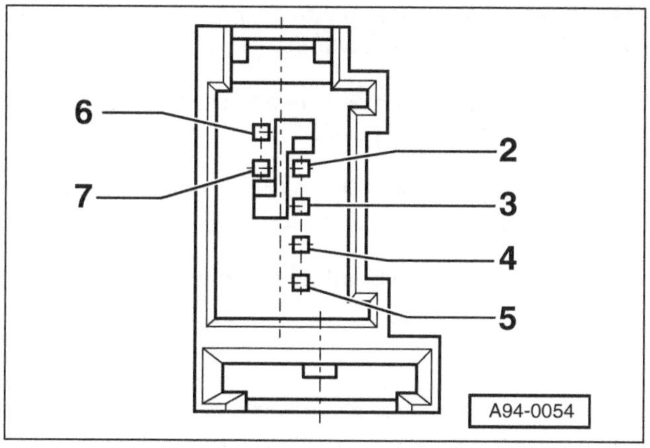

Cruise control system (**C**) terminal assignments.

1. not applicable
2. ECM pin 75 (resume / increase speed)
3. ECM pin 57 (set / reduce speed)
4. ECM pin 38 (on / off + erase; main switch off completely)
5. ECM pin 76 (temporarily off without erasing; main switch not engaged in “off” position)
6. Terminal 15
7. ECM pin 38 (on / off + erase; main switch off completely)

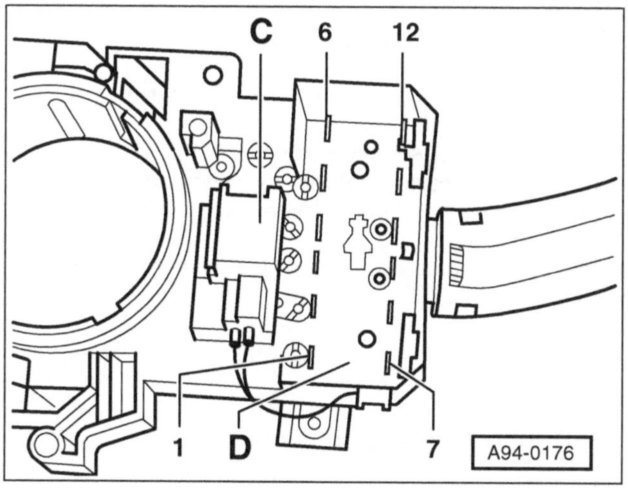

Turn signal switch (**D**) terminal assignments.

1. Terminal 30 (battery +)
2. Terminal 30 (battery +)
3. Terminal L (left turn signals)
4. Terminal P (parking lights)
5. Terminal 49a (turn signals)
6. Open
7. Terminal 56 (headlights)
8. Terminal 56b (low beam headlights)
9. Terminal PL (left parking lights)
10. Terminal RR (right parking lights)
11. Terminal R (right turn signals)
12. Terminal 56a (high beam headlights)

BentleyPublishers.com
© 2006 BentleyPublishers.com—All Rights Reserved

IGNITION SWITCH

Lock cylinder, removing and installing

NOTE—
- *On vehicles with anti-theft immobilizer, the immobilizer induction coil is attached to the lock cylinder and cannot be replaced separately.*
- *The lock cylinder must be replaced if the immobilizer induction coil malfunctions.*
- *The key code number and/or vehicle identification number (VIN) must be provided when replacing the lock cylinder in order to retain a one key system for the vehicle.*
- *The spare key is required for lock cylinder removal. The removal procedures can only be performed with an ignition key with a flat grip (i.e. no light and no remote control).*

– Pull out adjustable steering column as far as possible. Center steering wheel in straight ahead position.

– Disconnect negative (–) battery cable and cover battery terminal to keep cable from accidentally contacting terminal.

– Remove driver's airbag. See **69 Seatbelts, Airbags**.

> ***WARNING—***
> - *The airbag is inflated by an explosive device. Handled improperly or without adequate safeguards, the airbag system can be very dangerous. See* ***69 Seatbelts, Airbags****.*

– Remove steering wheel. See **48 Steering**.

– Remove left side under-dash panel. See **70 Trim–Interior.**

– Remove steering angle sensor and steering column switches. See **Airbag spiral spring, steering column switches and steering angle sensor, removing and installing** earlier in this repair group.

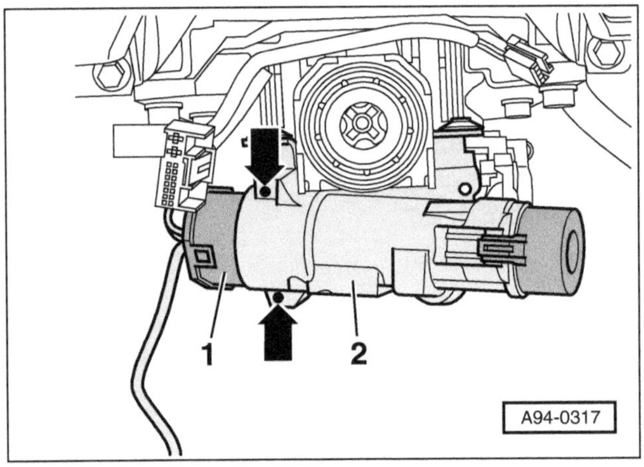

◄ Disconnect electrical harness connector (**1**) from immobilizer induction coil at back of ignition switch.

Vehicles with automatic transmission

– Shift selector lever to "P".

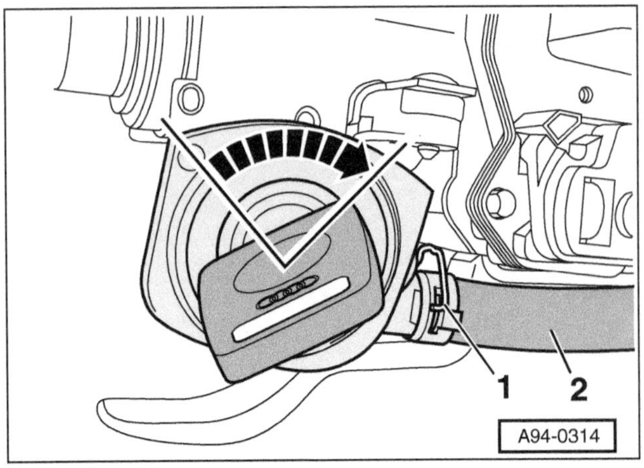

◄ Turn ignition key to on position (**arrow**).

– Lift ignition interlock lever (**1**) slightly and pull cable (**2**) out of ignition lock housing.

BentleyPublishers.com

© 2006 BentleyPublishers.com—All Rights Reserved

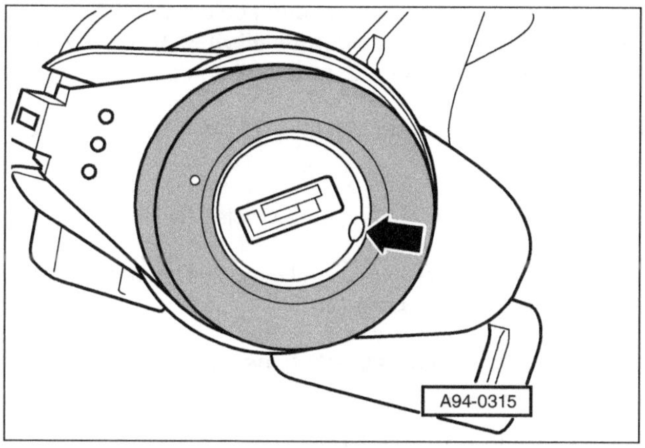

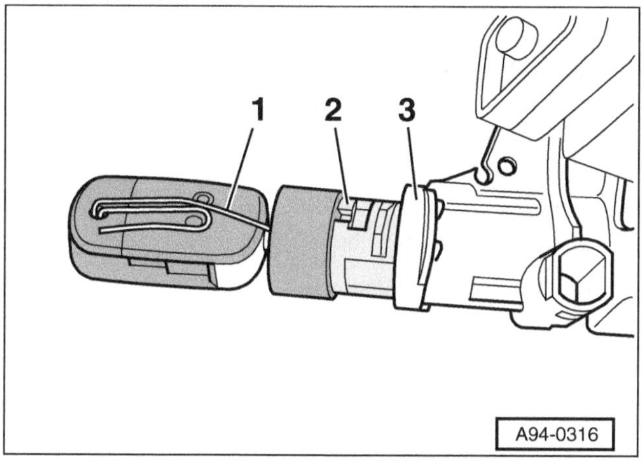

Continued for all vehicles

NOTE —

- *The spare key is required for lock cylinder removal. The following procedures can only be performed with an ignition key with a flat grip (i.e. no light and no remote control).*

◄ Use spare key to turn ignition to on position.

- Recess in ignition trim (**arrow**) should align with hole in ignition lock.
- Illustration at left shows correct lock position without ignition key for clarity.

◄ Using spare key or valet key:

- Insert steel wire (**1**) approx. 1.5 mm (0.06 in) in diameter as far as it will go into hole next to ignition key. A large paper clip is suitable.
- Pull ignition cylinder (**2**) out of steering lock housing (**3**).

CAUTION—

- *Do not lock steering without lock cylinder installed. Steering lock will require replacement if this is done.*

– When installing:

- Insert spare key or valet key into ignition cylinder and turn to on position.
- Insert steel wire into hole next to ignition key. Push wire in as far as it will go.
- Push lock cylinder together with ignition key all the way into steering lock housing.
- Pull wire out and push in lock cylinder firmly until catch engages audibly.
- Install electrical harness connector on ignition key reader coil at back of ignition switch.
- Reinstall steering column switches, steering wheel and airbag.

Ignition electrical switch, removing and installing

– Pull out adjustable steering column as far as possible. Center steering wheel in straight ahead position.

– Disconnect negative (–) battery cable and cover battery terminal to keep cable from accidentally contacting terminal.

– Remove driver's airbag. See **69 Seatbelts, Airbags**.

WARNING —

- *The airbag is inflated by an explosive device. Handled improperly or without adequate safeguards, the airbag system can be very dangerous. See* ***69 Seatbelts, Airbags****.*

– Remove steering wheel. See **48 Steering**.

– Remove left side under-dash panel. See **70 Trim–Interior.**

BentleyPublishers.com
© 2006 BentleyPublishers.com—All Rights Reserved

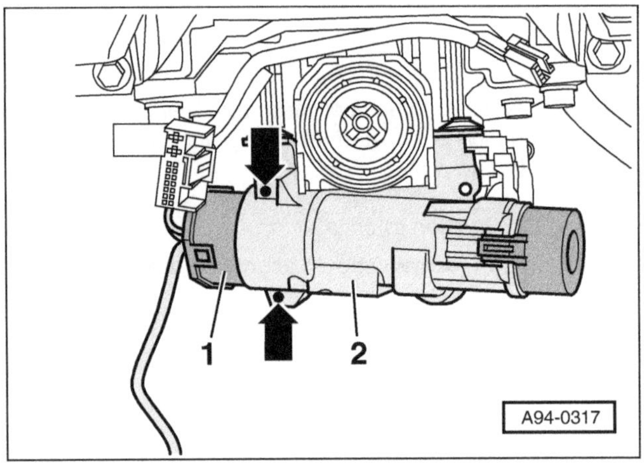

– Remove steering angle sensor and steering column switches. See **Airbag spiral spring, steering column switches and steering angle sensor, removing and installing** earlier in this repair group.

< Disconnect electrical harness connector (**1**) from ignition electrical switch.

– Remove locking fluid from mounting screw holes (**arrows**).

– Loosen mounting screws (**arrows**) slightly and pull ignition electrical switch out of steering lock housing (**2**).

– Install in reverse order of removal, noting the following:
  - Ignition switch must be in the on position when installing.
  - Apply locking fluid to mounting screws.

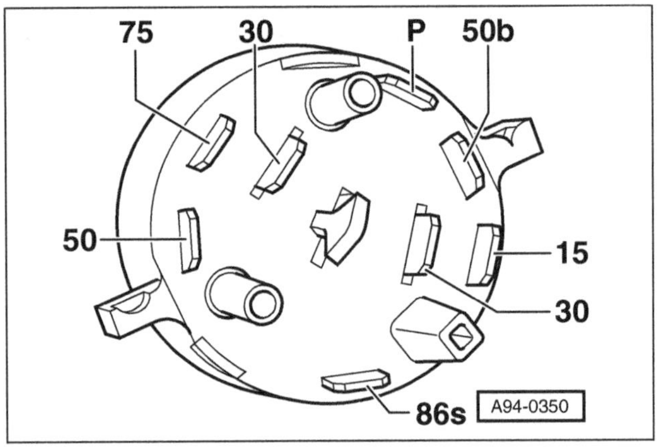

Ignition electrical switch pin assignments

< Terminal assignment for ignition electrical switch.
- **15** - Ignition ON
- **30** - Battery power
- **50** - Power to starter solenoid
- **50b** - Power to starter solenoid
- **75** - Power to foglights, load reduction relay
- **86s** - Power to accessories
- **P** - Park position

Ignition lock housing, removing and installing

– Pull out adjustable steering column as far as possible. Center steering wheel in straight ahead position.

– Disconnect negative (–) battery cable and cover battery terminal to keep cable from accidentally contacting terminal.

– Remove driver's airbag. See **69 Seatbelts, Airbags**.

– Remove steering wheel. See **48 Steering**.

– Remove left side under-dash panel. See **70 Trim–Interior.**

– Remove steering angle sensor and steering column switches. See **Airbag spiral spring, steering column switches and steering angle sensor, removing and installing** earlier in this repair group.

– Remove inner trim covers from steering column, as applicable to access ignition key cylinder.

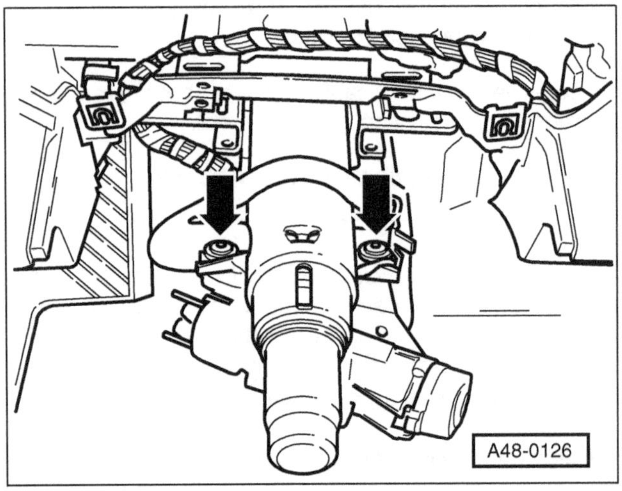

< Drill out or, if necessary, chisel out shear bolts (**arrows**) and remove ignition lock housing from steering column.

– Install in reverse order of removal, noting the following:
  - Use new shear bolts to install lock housing.
  - Tighten shear bolts until heads shear off.

BentleyPublishers
.com
©2006 BentleyPublishers.com—All Rights Reserved

Horns

Horns, removing and installing

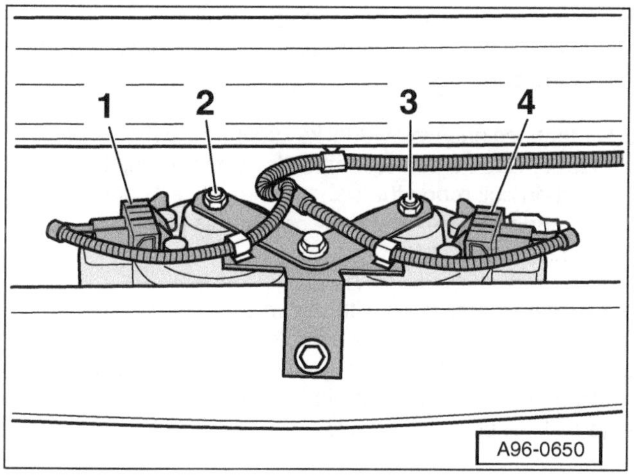

- Disconnect negative (–) battery cable and cover battery terminal to keep cable from accidentally contacting terminal.
- Remove front bumper cover. See **63 Bumpers**.
- ◄ Disconnect electrical harness connectors (**1** and **4**).
- Remove mounting nuts (**2** and **3**) and remove horns from bracket.
- Install in reverse order of removal.

Tightening torque	
Horns to mounting bracket	11 Nm (7 ft-lb)

Anti-theft System

NOTE —
- *For central locking system actuators see* **57 Doors and Central Locking**.
- *The anti-theft alarm horn (H8) is located in the plenum chamber at the base of the windshield and is accessible through the engine compartment.*

Electronic immobilizer

TT models produced from 09/2000 (2001 and later models) are equipped with Immobilizer Generation III. An anti-theft alarm is also available as an option.

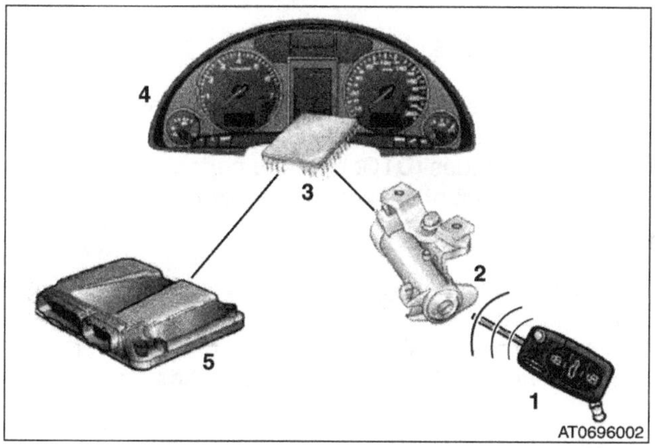

◄ Electronic immobilizer components are as follows:

1. Ignition key
2. Ignition lock cylinder with immobilizer induction coil
3. Immobilizer control module integrated with **4**
4. Instrument cluster
5. Engine control module (ECM)

Electronic immobilization functions as follows:

- With vehicle key inserted into ignition switch, signal from ignition cylinder induction coil energizes key transponder.
- Key transponder sends unique signal to induction coil.
- Key signal is relayed to immobilizer control module in instrument cluster.
- If control module recognizes key as valid, a signal is relayed to engine control module (ECM) via CAN-bus network and vehicle can be started.
- If key is not recognized as valid, starter and engine management system are blocked: vehicle cannot be started.

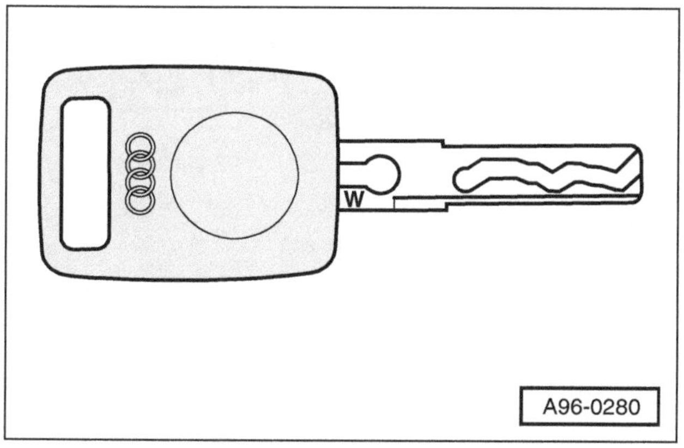

◄ For Immobilizer III, the keys are marked with an inscribed "**W**". The code consists of a fixed code and a variable code portion.

BentleyPublishers.com
© 2006 BentleyPublishers.com—All Rights Reserved

Every immobilizer contains a different computational rule for the variable code that remains the same for its entire service life. During adaptation of the vehicle key, the immobilizer writes this computational rule into the transponder and simultaneously learns the fixed code of the current key.

The fixed code identifies each individual key so that a lost key can be blocked from usage. Each time the ignition is switched on, the immobilizer induction coil reads the transponder's fixed code and then the variable code and determines whether this key is authorized for vehicle start.

The key transponder is integrated with the vehicle key. In case of transponder malfunction, replace the complete key. Once a key is replaced, adapt all the vehicle keys to the vehicle using the VAS 5051/5052 scan tool or equivalent. If only one key is adapted, the remaining keys will not start the vehicle.

The immobilizer warning light illuminates briefly (max. 3 seconds) and then goes out when an authorized vehicle key is used. If an unauthorized vehicle key is used or when a system malfunction occurs, the warning light blinks constantly during ignition on.

The immobilizer control module is integrated with the instrument cluster control module. In case of anti-theft immobilizer control module failure, replace the entire instrument cluster. See **90 Instruments**.

The immobilizer induction coil is integrated with the ignition key cylinder and cannot be replaced separately. In case of induction coil malfunction, replace the ignition cylinder. See **Lock cylinder, removing and installing** in this repair group.

On-board diagnostics

If the On Board Diagnostics (OBD II) system recognizes malfunctions in the immobilizer, anti-theft or central locking systems, it sets diagnostic trouble codes (DTCs). Before performing any troubleshooting or inspection of the anti-theft system, always begin by checking for DTCs.

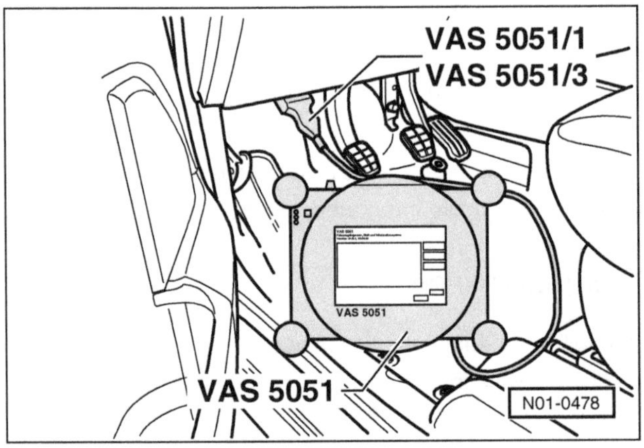

◀ DTCs can be accessed using a diagnostic scan tool such as the VAS 5051/5052. The scan tool is plugged into the Data Link Connector (DLC) under the left side of the dashboard. For more scan tool information see **OBD On Board Diagnostics**.

BentleyPublishers.com
© 2006 BentleyPublishers.com—All Rights Reserved

97 Fuses, Relays, Component Locations

TABLES

GENERAL

This repair group covers fuses, relays, ground locations and electrical component locations, primarily via photos or illustrations.

The complex nature of the TT electrical systems requires a large number of fuses and electrical components. Locating the correct component in this array of equipment is an important first step in electrical diagnosis. Electrical equipment and accessories installed vary depending on model and model year. Always confirm that the proper electrical component has been correctly identified.

Almost all electrical circuits are protected by fuses. For a full listing of fuses and fuse locations, see **Fuses** in this repair group.

Investigating and correcting ground problems often clears mysterious and difficult to trace electrical problems. For ground information, see **Grounds** in this repair group.

Also see:

- **27 Battery, Alternator, Starter,**
- **9 Electrical System–General**
- **EWD Electrical Wiring Diagrams**

BentleyPublishers
.com
©2006 BentleyPublishers.com—All Rights Reserved

Warnings and Cautions

Please read the following before doing any work on your electrical system.

WARNING —

- *Airbags utilize explosive devices. Handle with extreme care. Refer to the warnings and cautions in* **69 Seatbelts, Airbags**.
- *The ignition system of the car operates at lethal voltages. If you have a weak heart or wear a pacemaker, do not expose yourself to the ignition system electric currents. Take extra precautions when working on the ignition system or when servicing the engine while it is running or the key is ON. See* **28 Ignition System** *for additional ignition system warnings and cautions.*

CAUTION—

- *Do not disconnect the battery with the engine running.*
- *Switch the ignition OFF and remove the negative (–) battery cable before removing any electrical components. Connect and disconnect electrical connectors and ignition test equipment leads only while the ignition is switched OFF.*
- *Relay and fuse positions are subject to change and may vary from car to car. If questions arise, an authorized Audi dealer is the best source for the most accurate and up-to-date information.*
- *Use a digital multimeter for electrical tests. Switch the multimeter to the appropriate function and range before making test connections.*
- *Many control modules are static sensitive. Static discharge damages them permanently. Handle the modules using proper static prevention equipment and techniques.*
- *To avoid damaging harness connectors or relay panel sockets, use jumper wires with flat-blade connectors that are the same size as the connector or relay terminals.*
- *Do not try to start the engine of a car which has been heated above 176°F (80°C) (for example, in a paint drying booth). Allow it to cool to normal temperature.*
- *Disconnect the battery before doing any electric welding on the vehicle.*
- *Do not wash the engine while it is running, or any time the ignition is ON.*
- *Choose test equipment carefully. Use a digital multimeter with at least 10 MΩ input impedance, or an LED test light. An analog meter (swing-needle) or a test light with a normal incandescent bulb may draw enough current to damage sensitive electronic components.*
- *Do not use an ohmmeter to measure resistance on solid state components such as control modules.*
- *Disconnect the battery before making resistance (ohm) measurements on a circuit.*

BentleyPublishers.com

©2006 BentleyPublishers.com—All Rights Reserved

Electrical Component Locations

Table a is a cross-referenced listing of electrical components for cars covered by this manual.

Audi identifies electrical components by a letter and/or a number in electrical schematics. These codes, listed in column 2, are also used in **EWD Electrical Wiring Diagrams**.

Figure numbers in column 4 refer to photographs and illustrations immediately following the table.

NOTE—

- *Every component is not installed in every car.*
- *Due to changes in production, component locations may vary from what is illustrated. Consult your Audi dealer for the latest information.*
- *For information about the bus systems and associated control modules, see* **9 Electrical System–General**.

Table a. Electrical component locations

Component	Code	Location	Fig. #	Pg. #
A/C and heater air flow flap motor	V71	On fresh air housing, to right of blower motor, integral with G113		
A/C back pressure flap motor position sensor	G113	On fresh air housing, to right of fresh air blower, integral with V71		
A/C fresh air blower control module	J126	In fresh air housing, to left of blower motor		
A/C fresh air blower motor	V2	Behind glove box		
A/C fresh air blower motor control module (automatic climate control)	J126	In fresh air housing, to left of blower motor		
A/C fresh air blower series resistance with fuse (manual climate control)	N24	In fresh air housing, to left of blower motor		
A/C footwell/defroster flap motor	V85	On heater box, driver's side, forward of footwell outlet duct		
A/C central air flap motor	V70	On bottom of heater box, center		
A/C footwell/defrost flap motor position sensor	G114	On heater box, driver's side, forward of footwell outlet duct, integral with V85		
A/C and heater sender for outlet temperature floor	G192	On left side of heater box, on footwell outlet duct		
A/C temperature regulator position sensor	G92	On bottom of heater box, center, above central tunnel		
A/C and heater temperature regulator flap motor	V68	Integral with G92		
A/C compressor		Lower right front engine compartment		
A/C high pressure sensor A/C high pressure switch	G65 F129	In engine compartment, passenger side, below plenum		
ABS		*see* ESP		
ABS control module (with EDL)	J104	In engine compartment, rear driver's side Integral with V64	**Fig. 3**	97-11
ABS hydraulic pump	V64	In engine compartment, rear driver's side Integral with J104	**Fig. 3**	97-11
ABS inlet/outlet valves	N99 N100 N101 N102 N133 N134 N135 N136	Engine compartment, rear driver's side	**Fig. 3**	97-11

BentleyPublishers.com

©2006 BentleyPublishers.com—All Rights Reserved

Electrical Component Locations

Table a. Electrical component locations

Component	Code	Location	Fig. #	Pg. #
ABS wheel speed sensor left - front	G47	In left front wheel bearing housing		
ABS wheel speed sensor left - rear	G46	In left rear wheel bearing housing		
ABS wheel speed sensor right - front	G45	In right front wheel bearing housing	**Fig. 4**	97-12
ABS wheel speed sensor right - rear	G44	In right rear wheel bearing housing	**Fig. 5**	97-12
Accelerator pedal module	G79 G185	On accelerator pedal assembly. Includes throttle position sensor and pedal position sensor	**Fig. 12**	97-14
Adjusting element for central locking, driver door	F243	In driver's door latch	**Fig. 11**	97-13
Adjusting element for central locking, passenger door	F244	In passenger door latch		
After-run coolant pump	V51	On radiator bracket, near cooling fan, engine compartment front (225hp 1.8L engine only)	**Fig. 13**	97-14
Airbag connector, left front		Under steering column trim covers		
Airbag connector, right front		Underneath right dashboard, above glove compartment		
Airbag control module	J234	On central tunnel, forward of center console, near bulkhead	**Fig. 6**	97-12
Airbag crash sensor, driver's airbag	G179	Under driver's seat, beneath carpet		
Airbag crash sensor, passenger side airbag	G180	Under passenger seat, beneath carpet		
Airbag spiral spring	F138	Steering column, beneath steering wheel		
Airbag, driver		Steering wheel	**Fig. 17**	97-15
Airbag, passenger		Right side dashboard	**Fig. 17**	97-15
Airbag, side		Outer seat bolster		
Alarm system		*see* Immobilizer		
Alarm system horn	H8	In engine compartment, under plenum, center		
All-wheel drive control module		*see* Haldex all wheel drive control module		
Alternator (generator)	C	Right front, on engine	**Fig. 1**	97-11
Amplifier	R12	Coupe: to right of rear seat, beneath trim panel Roadster: below convertible top storage shelf, right side		
Angle sensor 1 for throttle drive	G187	1.8L: at intake manifold, integral with throttle valve control module (J338) 3.2L: at intake manifold, integral with throttle valve control module (J338)		
Angle sensor 2 for throttle drive	G188	1.8L: at intake manifold, integral with throttle valve control module (J338) 3.2L: at intake manifold, integral with throttle valve control module (J338)		
Antenna splitter (telephone system)	R87	In left rear fender, integral with antenna		
Antenna filter		In antenna wiring harness		
Antenna for navigation system	R50	Under right side defroster vent on top of dashboard		
Anti-theft control module		*see* Immobilizer		
ASR switch	E256	On dash board, center *see* ESP		
Automatic climate control		A/C and heater		
Automatic transmission control module	J217	In the plenum chamber, center, near engine control module		
Automatic transmission back-up light switch,		*see* Automatic transmission multifunction switch		
Automatic transmission gear position switch		*see* Automatic transmission multifunction switch		
Automatic transmission multifunction switch	F125	Top of transmission		
Automatic transmission selector lever switch		At shifter		

BentleyPublishers.com

© 2006 BentleyPublishers.com—All Rights Reserved

Table a. Electrical component locations				
Component	**Code**	**Location**	**Fig. #**	**Pg. #**
Battery	A	1.8L: engine compartment, front left 3.2L: luggage compartment, below floor, left rear	**Fig. 1** **Fig. 7**	97-11 97-12
Blower motor		*see* A/C and heater blower motor		
Blower motor resistor		*see* A/C and heater blower motor resistor		
Brake booster vacuum sensor 1 Brake booster vacuum sensor 2	G201 G214	At brake booster, left rear engine compartment At brake booster, left rear engine compartment	**Fig. 14**	97-14
Brake fluid level warning switch	F34	On brake fluid reservoir, left side of engine compartment	**Fig. 14**	97-14
Brake light switch	F	Above brake pedal		
Brake fluid pressure sensor		*see* ESP brake fluid pressure sensor		
Brake pressure release solenoid	F84	In power brake booster, engine compartment left Not equipped on cars with ASB/ESP Mk 60	**Fig. 14**	97-14
Brake pad sensor	G34 G35 G36 G37	In brake pads		
Brake pedal switch	F47	Above brake pedal		
Camshaft adjustment valve 1 (exhaust) (3.2L)	N318	Cylinder head, left side	**Fig. 15**	97-14
Camshaft position sensor 1 (1.8L) Camshaft position sensor 2 (1.8L)	G40 G163	Cylinder head, front, beneath timing belt cover Cylinder head, front, beneath timing belt cover	**Fig. 16**	97-15
Camshaft position sensor 1 (3.2L) Camshaft position sensor 2 (3.2L)	G40 G163	Engine compartment, left, near oil filler Engine compartment, left, near intake air hose	**Fig. 46**	97-22
CAN-bus		*see* **Can-bus system components** in this repair group		
CD changer compartment locking motor	V176	Behind driver's seat, behind trim panel, above CD changer		
CD changer	R41	To left of rear seat, behind trim panel (coupe) Behind driver's seat, behind storage compartment cover (roadster)		
Central flap motor position sensor	G112	ON bottom of heater box, center, integral with V70		
Central locking control module	J429	Coupe: beside left rear seat, behind trim panel Roadster: below convertible top storage shelf, right side	**Fig. 8** **Fig. 9**	97-13 97-13
Central locking motor driver's door	V161	In driver's door latch	**Fig. 11**	97-13
Central locking motor passenger door	V162	In passenger door latch		
Central locking trunk lid switch	F218	In trunk, near trunk latch		
Central locking system activator/switch right - front	F133	In passenger door, on door lock cylinder		
Central locking left - front system activator/switch	F131	In driver's door, on door lock cylinder	**Fig. 11**	97-13
Clutch Pedal Position (CPP) switch	F194	Above clutch pedal		
Clutch vacuum vent valve switch	F36	Above clutch pedal		
Combustion air blower capacitor	C11	In right rear fender, integral with V155		
Connector station A-pillar left and right side		In front footwell right, below front foot well trim In front footwell left, below front foot well trim		
Contact switch for left lock cylinder	F241	In driver's door		
Contact switch for right lock cylinder	F242	In passenger door		
Convertible top cover detector switch, left	F254	To left of driver's seat and bulkhead, in top left bulkhead trim piece.		
Convertible top cover detector switch 2, right	F267	To right of passenger's seat and bulkhead, in top right bulkhead trim piece		
Convertible top front lock switch	F172	On top of windshield, left	**Fig. 18**	97-15

BentleyPublishers
.com
©2006 BentleyPublishers.com—All Rights Reserved

Electrical Component Locations

Table a. Electrical component locations				
Component	**Code**	**Location**	**Fig. #**	**Pg. #**
Convertible top hydraulic pump	V118	Behind driver's seat and bulkhead, below convertible top storage compartment		
Convertible top hydraulic pump relay	J321	Behind driver's seat and bulkhead, below convertible top storage compartment		
Convertible top latch open switch	F205	On top of windshield, right		
Convertible top open position switch	F171	To left of driver's seat and bulkhead, on convertible top frame		
Coolant Fan Control (FC) control module	F293	In engine compartment, left front, next to radiator		
Coolant Fan Control (FC) thermal switch	F54 F18	In radiator, lower left, integral with F18 In radiator, lower left, integral with F54		
Cooling fan		*see* Engine cooling fan		
Crankshaft sensor		*see* Engine speed RPM sensor		
Crash sensor		*see* Airbag crash sensor		
Cruise control stalk switch	E45	On steering column, left	**Fig. 19**	97-15
Data link connector (DLC)	T16	Under left side dashboard		
Diagnostic connector		*see* Data link connector (DLC)		
Differential lock valve 1	N125	In engine compartment, rear, driver's side		
Differential lock valve 2	N126	In engine compartment, rear, driver's side		
Distributor Ignition (DI) capacitor	C6	On cylinder head cover	**Fig. 23**	97-16
Door contact switch drivers side	F2	In driver's door	**Fig. 11**	97-13
Door contact switch passenger's side	F3	In passenger's door		
Direct shift gearbox		*see* DSG		
Door courtesy light		Base of front door		
Door lock		Outer edge of door		
Driver's door central locking system motor	V56	In driver's door latch		
DSG Mechatronics	J743	In direct shift gearbox (2004-on)		
ECM		*see* Engine control module (ECM)		
Electronic power control (EPC)		*see* Accelerator pedal module		
Emergency flasher switch	E3	Center dash		
Engine control module (ECM)	J220	In plenum, center		
Engine control module (ECM) power supply relay	J271	In engine compartment, driver's side, to left of brake fluid reservoir, under cover	**Fig. 20**	97-16
Engine coolant level (ECL) warning switch	F66	In coolant expansion tank		
Engine coolant temperature (ECT) sensor	G2 G62	Coolant flange, left of cylinder head, integral w/G62 Coolant flange, left of cylinder head, integral w/G2	**Fig. 21**	97-16
Engine coolant temperature (ECT) warning switch	F14	In coolant flange, on cylinder head driver's side	**Fig. 22**	97-16
Engine cooling fan Engine cooling fan 2	V7 V177	In front of engine at radiator, larger fan In front of engine at radiator, smaller fan		
Engine cooling fan control module	J293	Engine compartment, left front, next to radiator		
Engine speed (RPM) sensor	G28	On front of engine block, lower left, near transmission flange		
EPC		*see* Electronic power control		
ESP		*see also* ABS		
ESP sensor unit (2002-on)	G419	Under dash, on bracket to right of steering column, integral with G200, G202, G251	**Fig. 42**	97-21
ESP switch	E256	On dash, center		
EVAP canister purge regulator valve (1.8L)	N80	In engine compartment, near power steering reservoir	**Fig. 1**	97-11
EVAP canister purge regulator valve (3.2L)	N80	In engine compartment, near power steering reservoir	**Fig. 2**	97-11
EVAP canister purge solenoid valve	N115	In EVAP canister, above muffler, rear	**Fig. 24**	97-17

BentleyPublishers.com
© 2006 BentleyPublishers.com—All Rights Reserved

Table a. Electrical component locations

Component	Code	Location	Fig. #	Pg. #
Flap motor		*see* A/C		
Flasher		*see* Emergency flasher		
FM frequency filter		Antenna filter		
Fuel injectors cylinders 1 - 4 (1.8L)	N30 - N33	Underneath intake manifold, front of engine	**Fig. 1**	97-11
Fuel injectors cylinders 1 - 6 (3.2L)	N30 - N33 N83 N84	Top of intake manifold, front of engine	**Fig. 1**	97-11
Fuel level sender	G	In fuel tank, accessible under rear seat, passenger side	**Fig. 25**	97-17
Fuel pump	G6	In fuel tank, accessible under rear seat, passenger side	**Fig. 25**	97-17
Fuel pump relay	J17	Micro central electric relay panel, under dash, left	**Fig. 51**	97-27
Fuel tank filler neck ground terminal		Underneath right wheel housing liner		
Fuel tank leak detection pump	V144	In rear of right-front wheel housing	**Fig. 26**	97-17
Fuel tank lid unlock motor	V155	In trunk, right rear, behind wheel housing	**Fig. 10**	97-13
Fuses		*see* **Fuses** in this repair group		
Garage door opener control module (2002-on)	J530	Behind front bumper cover, passenger side	**Fig. 28**	97-18
Generator		*see* Alternator		
GPS		*see* Navigation		
Grounds		*see* **Grounds** in this repair group		
Haldex all-wheel drive control module	J492	On haldex coupling, mounted on rear final drive	**Fig. 45**	97-22
Haldex coupling pump	V181	On haldex coupling below haldex all wheel drive control module, mounted on rear final drive	**Fig. 45**	97-22
Headlight beam adjustment control module (HID / Xenon)	J431	Coupe: under rear seat, left side Roadster: behind right rear trim panel, near access plate for fuel pump		
Headlight beam adjustment motor (HID / Xenon)	V48 V49	Rear of left headlight assembly Rear of right headlight assembly	**Fig. 29**	97-18
Headlight control module (HID / Xenon)	J343 J344	Below left headlight assembly Below right headlight assembly	**Fig. 29**	97-18
Headlight level control system sensor, left - front	G78	Mounted above front axle, left side		
Headlight level control system sensor, left - rear	G76	Front wheel drive: above rear axle beam, left side Quattro: Left rear wheel housing, attached to lower control arm		
Headlight washer pump	V11	Left front wheel housing, in washer fluid reservoir		
Heated oxygen sensor		*see* Oxygen sensor		
Heater for lambda-probe		*see* Oxygen sensor heater		
HID		*see* Headlight		
High intensity discharge headlight		*see* Headlight		
High pressure sensor		*see* A/C high pressure sensor		
Hood alarm switch	F120	On hood latch mechanism		
Horn button		Steering wheel	**Fig. 17**	97-15
Horn relay	J4	Micro central electric relay panel, under dash, left	**Fig. 51**	97-27
Horn, alarm		*see* Alarm system horn		
Horns		Behind front bumper cover		
Hydraulic temperature sensor	G271	In Haldex coupling, mounted to rear final drive, driver's side		

BentleyPublishers
.com
©2006 BentleyPublishers.com—All Rights Reserved

Electrical Component Locations

Table a. Electrical component locations				
Component	**Code**	**Location**	**Fig. #**	**Pg. #**
Ignitor for driver's side airbag	N199	On left-side of front driver's seat backrest, beneath seat cover		
Ignitor for passenger's side airbag	N200	On right-side of front passenger seat backrest, beneath seat cover		
Ignition coil	N	On cylinder head, above # 1 cylinder (1.8L)	**Fig. 1**	97-11
Ignition coil 1 with power output stage	N70	On cylinder head, above # 1 cylinder (3.2L)	**Fig. 2**	97-11
Ignition coil 2	N128	On cylinder head, above # 2 cylinder (1.8L)	**Fig. 1**	97-11
Ignition coil 2 with power output stage	N127	On cylinder head, above # 2 cylinder (3.2L)	**Fig. 2**	97-11
Ignition coil 3	N158	On cylinder head, above # 3 cylinder (1.8L)	**Fig. 1**	97-11
Ignition coil 3 with power output stage	N291	On cylinder head, above # 3 cylinder (3.2L)	**Fig. 2**	97-11
Ignition coil 4	N163	On cylinder head, above # 4 cylinder (1.8L)	**Fig. 1**	97-11
Ignition coil 4 with power output stage	N292	On cylinder head, above # 4 cylinder (3.2L)	**Fig. 2**	97-11
Ignition coil 5 with power output stage	N323	On cylinder head, above # 5 cylinder (3.2L)	**Fig. 2**	97-11
Ignition coil 6 with power output stage	N324	On cylinder head, above # 6 cylinder (3.2L)	**Fig. 2**	97-11
Immobilizer control module		Integral with instrument cluster		
Immobilizer induction coil	D2	Around ignition lock cylinder		
Induction coil		*see* Immobilizer induction coil		
Instrument cluster control module	J285	Instrument cluster		
Intake air temperature sensor (IAT) (1.8L)	G42	Intake manifold, near throttle valve housing		
Intake manifold change-over valve (3.2L)	N156	Under intake manifold at front	**Fig. 33**	97-19
Kickdown switch	F8	Accelerator pedal module		
Knock sensor 1 Knock sensor 2	G61 G66	On front of engine block, passenger side On front of engine block, driver's side, behind oil filter housing	**Fig. 34**	97-19
Leak detection pump		*see* Fuel tank leak detection pump		
Light switch	E1	Left of steering wheel on dashboard	**Fig. 19**	97-15
Lighter		Below center dashboard, in ashtray		
Load reduction relay	J59	Micro central electric relay panel, under dash, left	**Fig. 51**	97-27
Longitudinal acceleration sensor	G251	2002 - 2006: under dash, on bracket to right of steering column, integral with G419, G200, G202 2000 - 2001: next to glove box, behind instrument panel, near right side dash end panel	**Fig. 42**	97-21
Magnetic coil for brake pressure (in brake booster)	N247	In power brake booster, engine compartment, left	**Fig. 14**	97-14
Manifold absolute pressure (MAP) sensor	G71	Below intake air hose	**Fig. 1**	97-11
Manual transmission back-up light switch	F4	Top front of transmission case	**Fig. 35**	97-19
MAP sensor		*see* Manifold absolute pressure (MAP) sensor		
Mass air flow sensor	G70	In left rear of engine compartment		
Mechatronics		*see* DSG Mechatronics		
Motronic control module		*see* Engine control module (ECM)		
Motronic main relay		*see* Engine control module power supply relay		
Navigation system with CD drive control module	J401	Coupe: rear luggage compartment, left side Roadster: behind left rear trim panel	**Fig. 36**	97-20
Navigation function selector switch II	E272	In center console	**Fig. 27**	97-17
Neutral safety switch		*see* Automatic transmission gear position switch		
OBD II plug		*see* Data link connector (DLC)		
Oil pressure switch	F1	In front of engine, left side, on oil cooler flange		
Oil thermal sensor	G266	On bottom of oil pan	**Fig. 37**	97-20
Outside air temperature sensor	G17	Left front, behind front bumper cover		

BentleyPublishers
.com
©2006 BentleyPublishers.com—All Rights Reserved

Table a. Electrical component locations

Component	Code	Location	Fig. #	Pg. #
Oxygen sensor, pre-catalyst, bank 1 sensor 1	G39	1.8L: exhaust pipe, in front of catalytic converter 3.2L: left front exhaust manifold	**Fig. 30**	97-18
Oxygen sensor, pre-catalyst, bank 2 sensor 1	G108	3.2L: Right front exhaust manifold		
Oxygen sensor, post-catalyst, bank 1 sensor 1	G130	1.8L: exhaust pipe, behind catalytic converter 3.2L: at rear of catalytic converter		
Oxygen sensor heater, pre-catalyst, bank 1 sensor 1	Z19	Integral with oxygen sensor G39		
Oxygen sensor heater, pre-catalyst, bank 2 sensor 1	Z28	3.2L: integral with oxygen sensor G108		
Oxygen sensor heater, post-catalyst, bank 1 sensor 2	Z29	Integral with oxygen sensor G130		
Parking brake warning light switch	F9	On central tunnel, below parking brake lever		
Passenger's door central locking system motor	V57	In passenger door latch		
Power convertible top valve 1 (roadster)	N272	Behind left rear trim panel, on top of hydraulic pump		
Power window control module left - front	J295	In driver's door, center	**Fig. 31**	97-18
Power window control module right - front	J296	In passenger door, center		
Power steering pressure switch	F88	On power steering pump		
Pressure sensor		*see* ESP brake fluid pressure sensor		
Pressure switch		*see* A/C high pressure sensor		
Radar interior monitoring control module 1	G303	In center console under recess for hand brakes lever (roadster)	**Fig. 38**	97-20
Radio	R	Center dashboard	**Fig. 17**	97-15
Rear window defogger with window antenna (coupe)	Z24	In rear lid, integral with rear window glass		
Rear window defogger relay	J9	13-fold relay panel, under dash, left	**Fig. 52**	97-27
Recirculating valve for turbocharger (1.8L)	N249	225 hp: on cylinder head cover, engine compartment, left 180 hp: in front of engine, on bracket, below intake manifold	**Fig. 1**	97-11
Relays		*see* **Relays** in this repair group		
Remote unlock motor rear lid	V151	Rear lid latch		
Rotation rate sensor	G202	2002-2006: under dash, on bracket to right of steering column, integral with G419, G200, G251 2000-2001: on bracket to right of steering column	**Fig. 42** **Fig. 39**	97-21 97-20
Satellite radio	R146	Center dashboard (2005-on)		
Seatbelt switch, left	E24	On right side of driver's seat cushion		
Seatbelt switch, right	E25	On left side of passenger's seat cushion		
Seatbelt tensioner igniter, left	N153	In seat belt retractor, driver's side		
Seatbelt tensioner igniter, right	N154	In seat belt retractor, passenger side		
Secondary air injection pump relay	J299	Engine compartment, driver's side, behind brake fluid reservoir, under cover	**Fig. 20**	97-16
Secondary air injection pump	V101	On front of engine, left side		
Secondary air injection solenoid valve	N112	1.8L: under intake manifold at front of engine 3.2L: under intake manifold at front of engine	**Fig. 40**	97-21
Speedometer vehicle speed sensor	G22	In transmission housing, at passenger side output flange		
Shiftlock solenoid	N110	In the shift mechanism near the selector lever		
Starter	B	Top of transmission		
Storage compartment central locking motor	V175	Between roll bar hoops at top of bulkhead, behind center trim		
Steering angle sensor	G85	Steering column	**Fig. 41**	97-21
Stop light switch		*see* Brake light switch		
Sunlight photo sensor	G107	On instrumental panel, center, under driver's side defroster grille		

BentleyPublishers.com
© 2006 BentleyPublishers.com—All Rights Reserved

Electrical Component Locations

Table a. Electrical component locations				
Component	**Code**	**Location**	**Fig. #**	**Pg. #**
Telephone control module	J412	Coupe: below rear seat bottom Roadster: behind center rear speaker		
Telephone hands-free microphone	R38	Coupe: in front dome light housing Roadster: in instrument cluster		
Telephone transceiver	R36	In left rear quarter panel, near taillight		
Temperature regulator flap motor		*see* A/C and heater temperature regulator flap motor		
Throttle position sensor	G79	On bulkhead, above accelerator pedal		
Throttle valve control module	J338	1.8L: at intake manifold on throttle valve housing 3.2L: at intake manifold on throttle valve housing	**Fig. 1** **Fig. 2**	97-11 97-11
Throttle drive (power accelerator actuation)	G186	1.8L: at intake manifold on throttle valve housing 3.2L: at intake manifold on throttle valve housing	**Fig. 1** **Fig. 2**	97-11 97-11
Tiptronic gear selector		*see* Automatic transmission multifunction switch		
Tiptronic downshift button (on steering wheel)	E439	On steering wheel, left side		
Tiptronic upshift button (on steering wheel)	E438	On steering wheel, right side		
Tiptronic switch	F189	Integral with gear selector assembly		
Traction control		*see* ABS / ESP		
Transfer fuel pump		*see* Fuel pump		
Transverse acceleration sensor	G200	2002-2006: under dash, on bracket to right of steering column, integral with G419, G202, G251 2000-2001: on bracket to right of steering column	**Fig. 42**	97-19
Transmission control module		*see* Automatic transmission control module		
Turn signal / high beam stalk switch	E2 E4	On steering column, left side		
Turn signal flasher		Integrated in emergency flasher switch, center dashboard		
Ultra-sound sensor for anti-theft warning	G209	In front dome light housing		
Valve 1 for camshaft adjustment (intake)	N205	1.8L: under timing belt cover 3.2L: on cylinder head, left end	**Fig. 16** **Fig. 15**	97-15 97-14
Vehicle speed sensor (VSS)	G68	On top of transmission near transmission cooler		
Voltage regulator	C1	On back side of alternator		
Wastegate bypass regulator valve	N75	On air intake hose, to left of brake fluid reservoir	**Fig. 1**	97-11
Wheel speed sensor		*see* ABS wheel speed sensor		
Wind deflector control module	J531	At bottom of wind deflector frame, integral with V186		
Wind deflector motor	V186	At bottom of wind deflector frame jack screw, left side		
Window motor, left	V14	In driver's door		
Window motor, right	V15	In passenger door		
Window regulator switch, in passenger door	E107	In passenger's door, near pull handle	**Fig. 43**	97-21
Windshield washer fluid level warning switch	F77	On windshield washer reservoir, in left front wheel housing	**Fig. 44**	97-22
Windshield washer pump	V5	Left front wheel housing, in washer fluid reservoir		
Windshield washer nozzle, heated	Z20 Z21	On plenum panel at base of windshield		
Windshield wiper motor		In plenum at base of windshield		
Wiper / washer intermittent relay	J31	Micro central electric relay panel, under dash, left	**Fig. 51**	97-27
Wiper / washer stalk switch		On steering column, right		
Xenon headlight		*see* Headlight		
Yaw sensor		*see* ESP sensor unit		

BentleyPublishers
.com
©2006 BentleyPublishers.com—All Rights Reserved

Component figures

Fig. 1 1.8L 225hp (code: AMU and BEA) engine compartment

Engine cover removed.

1. Ignition coils # 1 - 4 (N, N128, N158, N163)
2. Recirculating valve for turbocharger (N249)
3. Manifold absolute pressure (MAP) sensor (G71)
4. Wastegate bypass regulator valve (N75)
5. Fuel injectors (N30 - N33)
6. Battery (A)
7. EVAP canister purge regulator valve (1.8L) (N80)
8. Alternator (generator) (C)
9. Angle sensor 1 for throttle drive (G188)
   Throttle valve control module (J338)
   Throttle drive, power accelerator actuation (G186)

Fig. 2 3.2L (code: BHE) engine compartment

Engine cover removed.

1. EVAP canister purge regulator valve (3.2L) (N80)
2. Angle sensor 1 for throttle drive (G188)
   Throttle valve control module (J338)
   Throttle drive, power accelerator actuation (G186)
3. Manifold absolute pressure (MAP) sensor (G71)
4. Ignition coils 1 - 4 (N, N128, N158, N163)
5. Fuel injectors (N30 - N33, N83, N84)

Fig. 3 Engine compartment, rear

Air box removed.

1. ABS inlet / outlet valves (N99 - N102, N133 - N136)
2. ABS hydraulic pump (V64)
3. ABS control module, with EDL (J104)

BentleyPublishers
.com
©2006 BentleyPublishers.com—All Rights Reserved

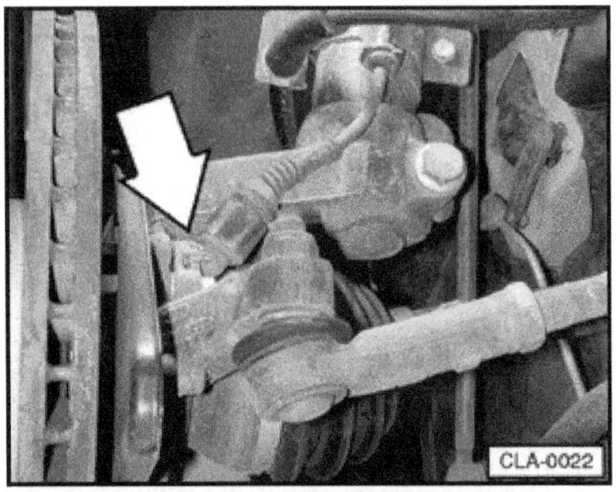

Fig. 4 Right front wheel hub

Wheel removed.

- **Arrow**: ABS wheel speed sensor, right front (G45)

Fig. 5 Right rear suspension

Under side of vehicle.

1. ABS wheel speed sensor, right rear (G44)

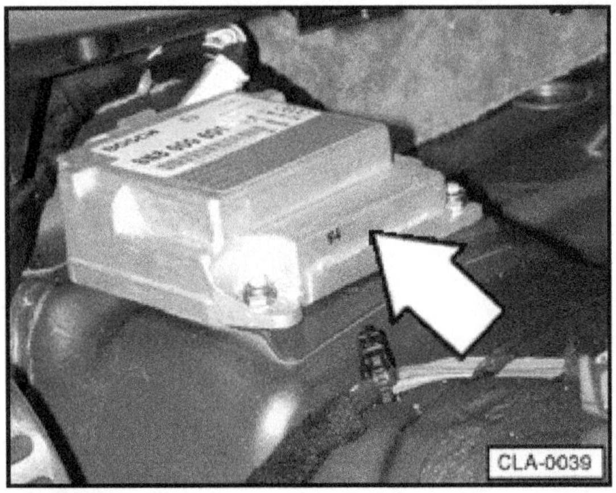

Fig. 6 Behind center console

Center console removed.

- **Arrow**: Airbag control module (J234)

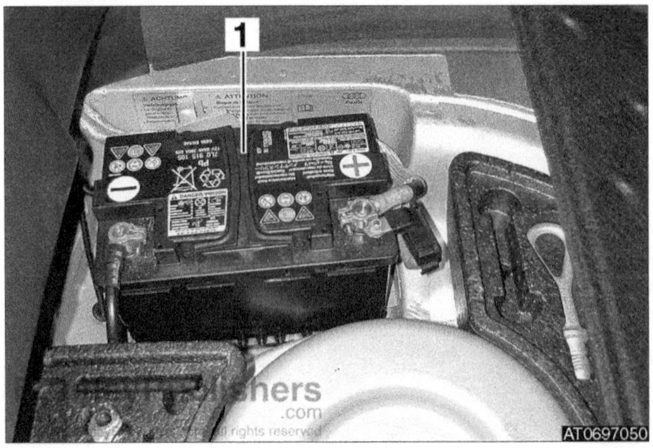

Fig. 7 Rear luggage compartment (3.2L)

Carpet removed from luggage compartment.

1. Battery (A), vehicles with 3.2L engine

BentleyPublishers
.com
©2006 BentleyPublishers.com—All Rights Reserved

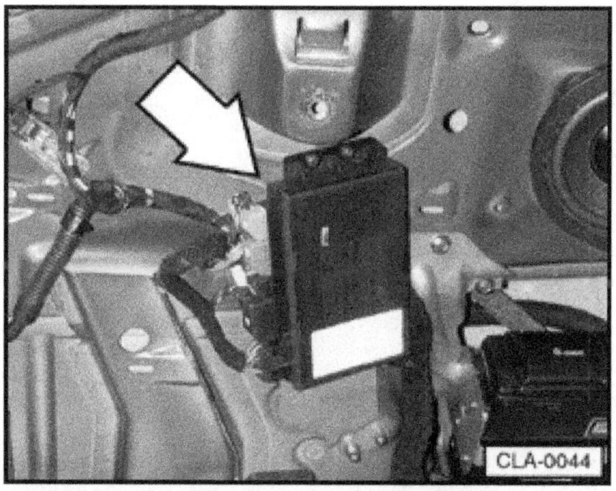

Fig. 8 Driver's side rear interior

Rear seat removed.

- **Arrow**: Control module for central locking (coupe) (J429)

Fig. 9 Convertible top storage shelf

Passenger side access hole in convertible top storage shelf.

1. Control module for central locking (roadster) (J429)

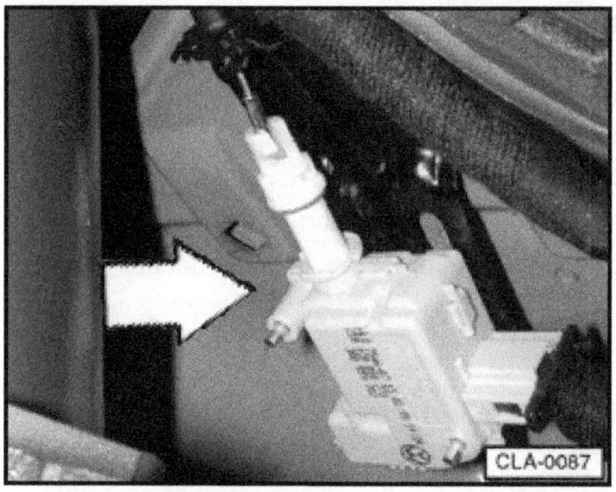

Fig. 10 Left rear wheel housing

Wheel housing liner removed.

- **Arrow**: Fuel tank lid unlock motor (V155)

Fig. 11 Central locking

Door panel removed.

1. Central locking system activator switch (F131)
   Contact switch for left lock cylinder (F241)
   Door contact switch (F2)
2. Adjusting element for central locking (F243)
   Central locking motor (V161)

BentleyPublishers
.com
© 2006 BentleyPublishers.com—All Rights Reserved

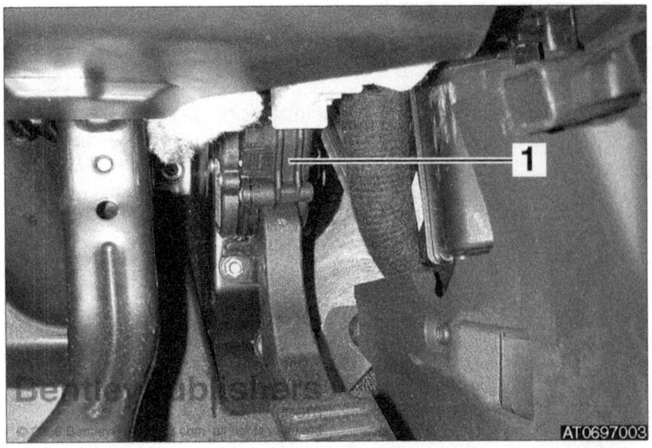

Fig. 12 Driver's footwell

Accelerator pedal.

1. Accelerator pedal module (G79, G185)

Fig. 13 Front of engine

Lock carrier trim removed.

1. After-run coolant pump (V51)

Fig. 14 Brake master cylinder

Brake booster and master cylinder removed from car.

1. Brake fluid level warning switch (F34)
2. Sender 1 for brake booster (G201)
3. Brake pressure release solenoid (F84)
4. Magnetic coil for brake pressure, in brake booster (N427)

Fig. 15 Camshaft adjustment (3.2L)

On left end of 3.2L cylinder head.

1. Valve 1 for camshaft adjustment (N205) (intake cam)
2. Camshaft adjustment valve 1 (N318) (exhaust cam)

BentleyPublishers.com
©2006 BentleyPublishers.com—All Rights Reserved

Fig. 16 Camshaft adjustment (1.8L)

Upper timing chain cover and upper engine cover removed.

1. Camshaft position sensor (G40, G163)
   Valve 1 for camshaft adjustment (N205) (intake cam)

Fig. 17 Dash

Interior of car.

1. A/C and heating controls
2. Passenger's air bag (in dash)
3. Horn button
4. Radio (R)
5. Driver's air bag (in steering wheel)

Fig. 18 Top of windshield molding (roadster)

Convertible top down.

1. Convertible top front lock switch (F172)

Fig. 19 Steering column

Behind steering wheel.

1. Light switch (E1)
2. Cruise control stalk switch (E45)

BentleyPublishers
.com
©2006 BentleyPublishers.com—All Rights Reserved

Fig. 20 Engine compartment relay panel

◄ Relay panel cover removed.

1. Secondary air injection pump relay (J299)
2. Engine control module (ECM) power supply relay (J271)

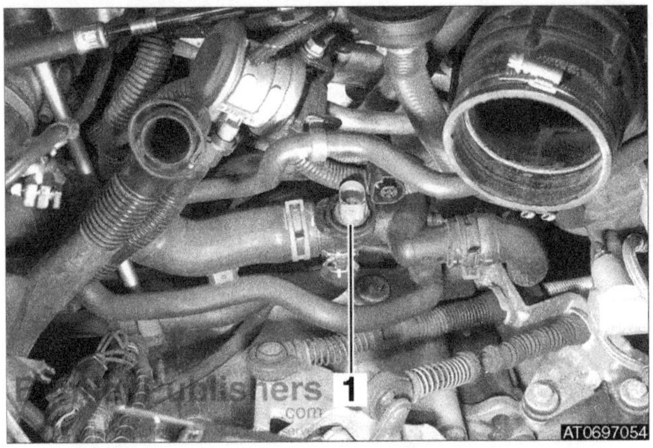

Fig. 21 Engine compartment, left

◄ Air filter assembly removed.

1. Engine coolant temperature (ECT) sensor (G62)

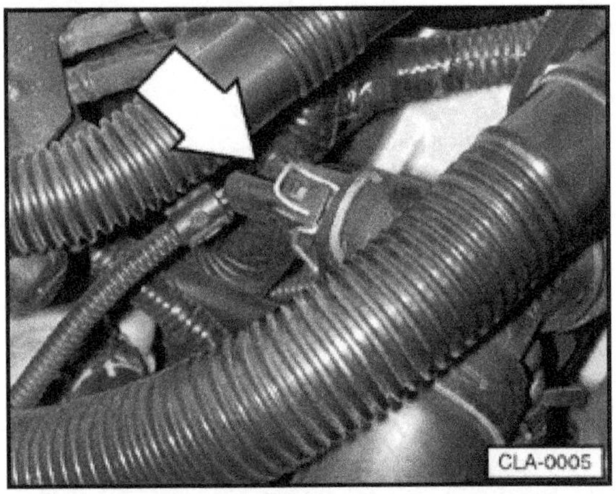

Fig. 22 Engine compartment, left

◄ To the left of cylinder head.

- **Arrow**: Engine coolant temperature (ECT) warning switch (F14)

Fig. 23 Top of engine (1.8L)

◄ Engine cover removed.

1. Distributor Ignition (DI) capacitor (C6)

BentleyPublishers
.com
© 2006 BentleyPublishers.com—All Rights Reserved

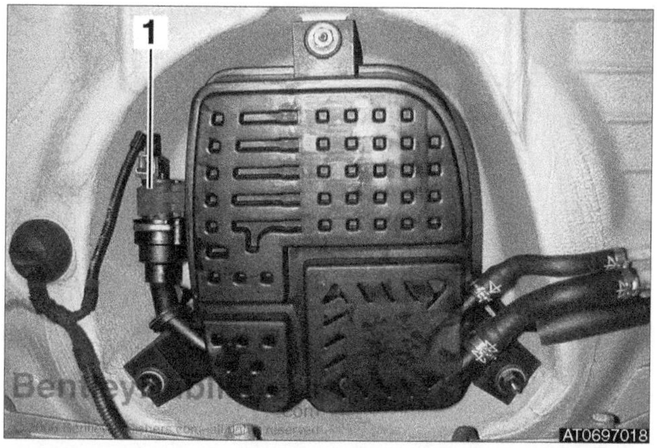

Fig. 24 Under car, rear

◄ Above muffler on the underside of the car.

1. EVAP canister purge solenoid valve (N115)

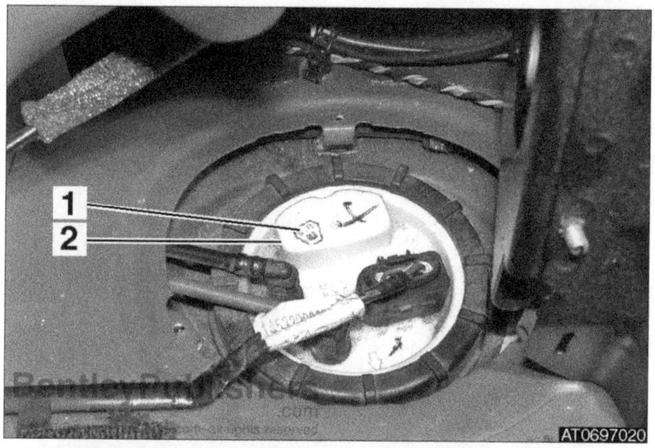

Fig. 25 Fuel tank

◄ Under rear seat, passenger side (coupe).
Under left rear trim panel (roadster).

1. Fuel level sender (G)
2. Fuel pump (G6)

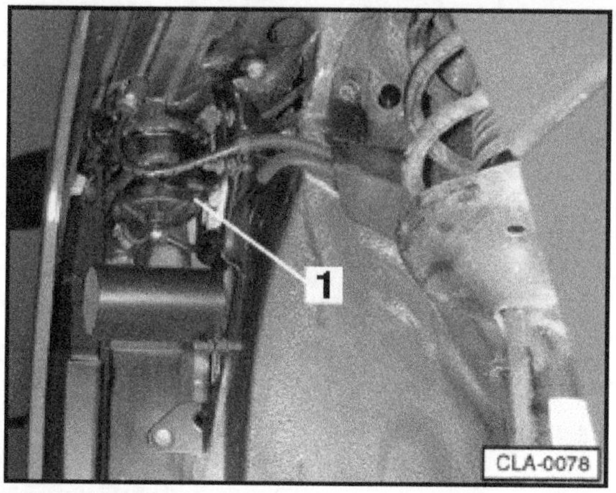

Fig. 26 Right front wheel housing

◄ Wheel housing liner removed.

1. Fuel tank leak detection pump (V144)

Fig. 27 Center console

◄ Between front seats.

1. Navigation function selector switch II (E272)

BentleyPublishers.com
© 2006 BentleyPublishers.com—All Rights Reserved

Fig. 28 Bumper

Front bumper cover removed, bellow right headlight.

1. Garage door opener control module (J530) (2002 - on)

Fig. 29 Headlights

Battery and air filter assembly removed.

1. Headlight aim adjuster motor (HID / Xenon) (V48)
2. Headlight control module (HID Xenon) (J343)

Fig. 30 Exhaust system (1.8L)

Underside of car.

1. Heated oxygen sensor (G39)

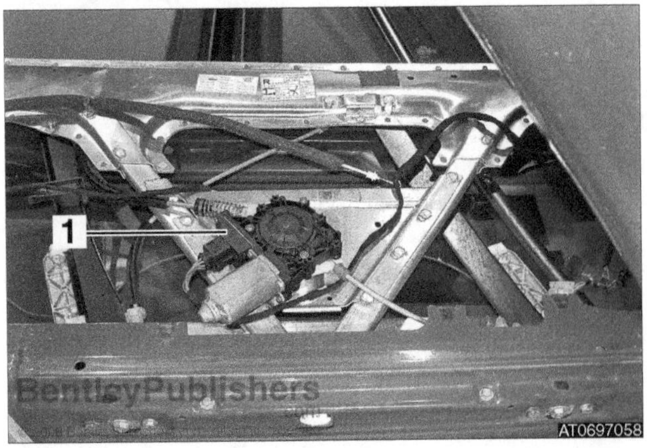

Fig. 31 Driver's door

Driver's door panel removed.

1. Power window control module left - front (J295)

BentleyPublishers.com
©2006 BentleyPublishers.com—All Rights Reserved

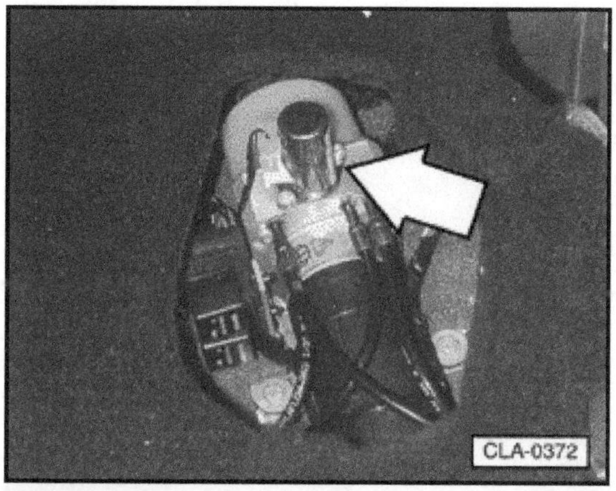

Fig. 32 Convertible top storage shelf (roadster)

Behind driver's seat. Left side access hole cover removed.

1. Power convertible top valve 1 (N272)

Fig. 33 Intake manifold (3.2L)

Right side of intake manifold (3.2L).

1. Intake manifold change - over value (N156)

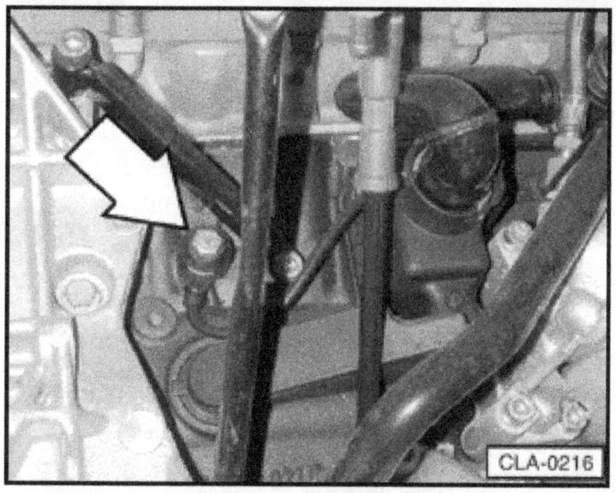

Fig. 34 Front of engine block

Engine block, passenger side.

- **Arrow**: Knock sensor 1 (G61)

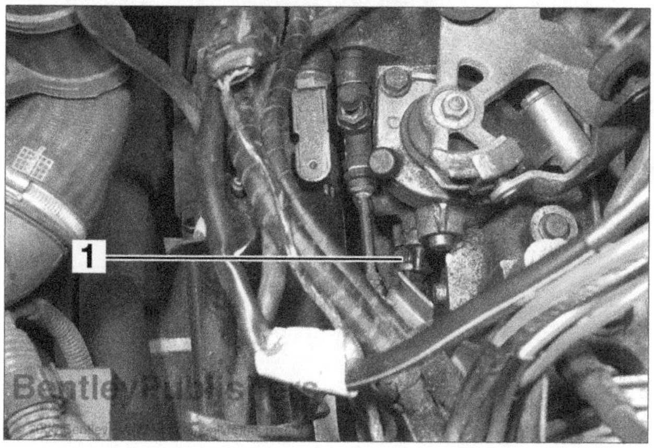

Fig. 35 Manual transmission (1.8L)

Battery removed.

1. Manual transmission back-up light switch (F4)

BentleyPublishers
.com
©2006 BentleyPublishers.com—All Rights Reserved

Electrical Component Locations

Fig. 36 Luggage compartment (coupe)

◀ Left trim panel removed.

1. Navigation system with CD drive control module (J401)

Fig. 37 Bottom of engine

◀ Lower sound absorber panel removed.

1. Oil thermal sensor (G266)

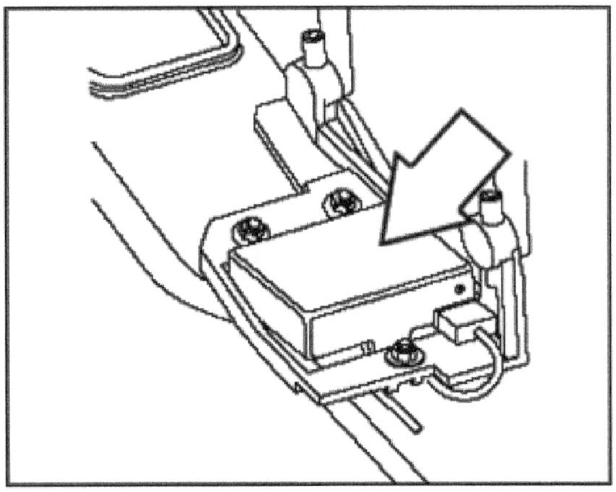

Fig. 38 Interior

◀ In center console under recess for hand brake lever.

- **Arrow**: Radar interior monitoring control module 1 (G303)

Fig. 39 Under dash

◀ On bracket to right of steering wheel. Under dash trim panel removed.

1. Rotation rate sensor (G202)

BentleyPublishers
.com
© 2006 BentleyPublishers.com—All Rights Reserved

Fig. 40 Front of intake manifold (1.8L)

Engine cover removed. Mounting bracket for solenoid valves unbolted and turned over.

1. Secondary air injection solenoid valve (N112)

Fig. 41 Steering column

Steering wheel removed.

1. Steering angle sensor (G85)

Fig. 42 Under dash, pedal cluster

Under dash trim panel removed.

1. ESP sensor unit (G419)
   Transverse acceleration sensor (G200)
   Sender for rotation rate (G202)
   Longitudinal acceleration sensor (G251) (2002 on)

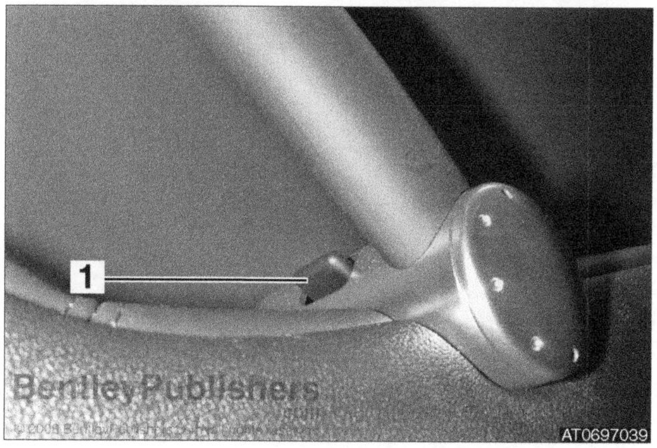

Fig. 43 Passenger door

Underside of passenger door handle interior.

1. Window regulator switch, in passenger door (E107)

BentleyPublishers.com
©2006 BentleyPublishers.com—All Rights Reserved

Fig. 44 Left front wheel well

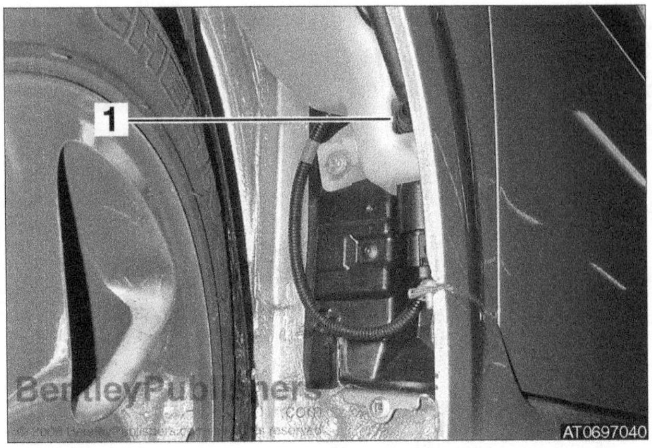

◄ Wheel housing liner removed.

1. Windshield washer fluid level warning switch (F77)

Fig. 45 Rear underside of vehicle (quattro)

◄ Center of vehicle, rear final drive.

1. Haldex all-wheel drive control module (J492)
2. Haldex coupling pump (V181)

Fig. 46 Engine compartment, left (3.2L)

◄ Left side of engine compartment.

- **G40**: Camshaft position sensor 1 (intake)
- **G163**: Camshaft position sensor 2 (exhaust)

BentleyPublishers.com

© 2006 BentleyPublishers.com—All Rights Reserved

Can-bus system components

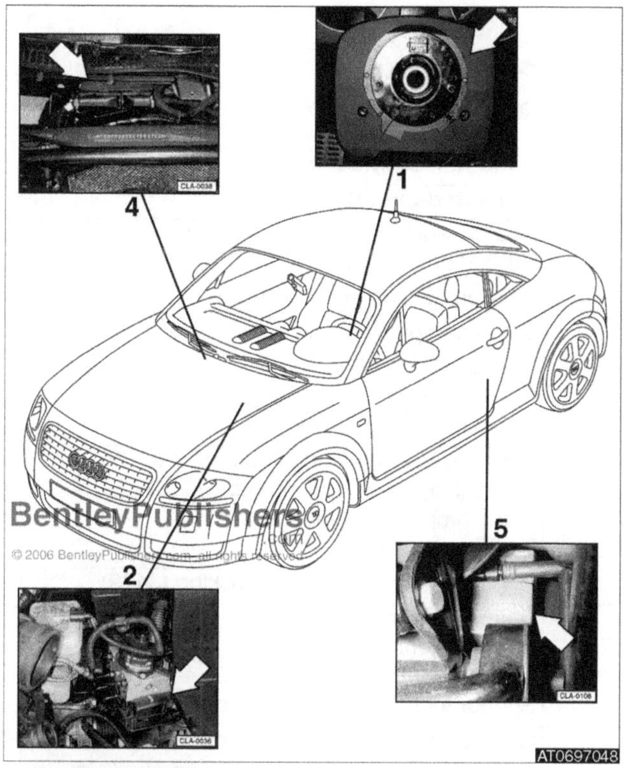

Fig. 47 CAN-bus components (2000 - 2006)

Components connected via drivetrain CAN-bus:

1. Steering angle sensor (G85)
2. ABS control module (with EDL) (J104)
3. Instrument cluster combination processor (J218): *not shown*
4. Motronic engine control module (J220)
5. All wheel drive control module (J492)

NOTE—

- *2000 - 2002 models have components 1 - 5 connected to CAN-bus.*

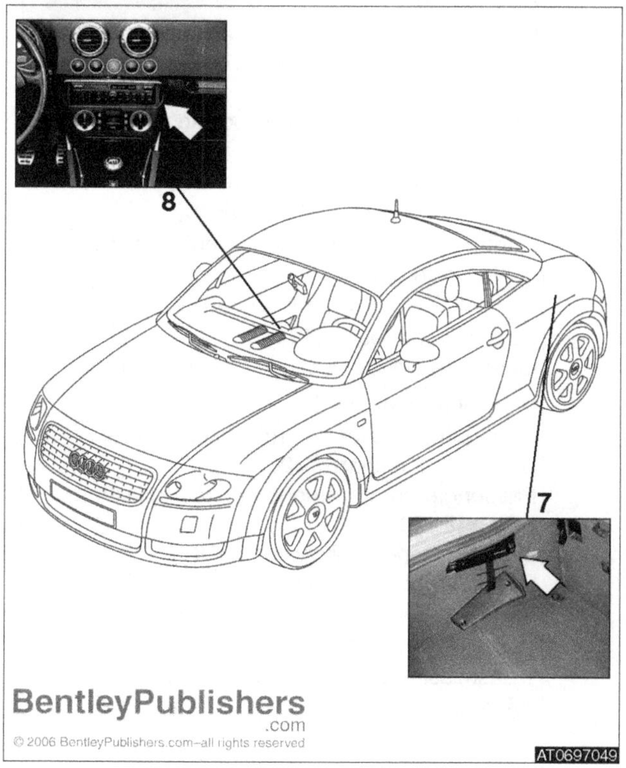

Fig. 48 Additional CAN-bus components (2002 - 2006)

Components connected via drivetrain CAN-bus:

6. Control module with indicator unit in instrument panel insert (J285): *not shown*
7. Control unit for navigation with CD drive (J401)
8. Radio (R)

NOTE—

- *2002 - 2003 have components 1 - 5 shown above, in addition to components 6 - 8, connected to CAN-bus.*

Additional CAN-bus components (2004 - 2006)

9. Selector lever (E313): *not shown*
10. Tiptronic switch (F189): *not shown*
11. Transmission control module (TCM) (J217): *not shown*
12. Direct shift gear box (DSG) mechatronic (J743): *not shown*

NOTE—

- *2004 - 2006 models have components 1 - 8, shown above, in addition to components 9 - 12, connected to CAN-bus.*

Drivetrain CAN-bus specifications	
Data transfer rate	500 kbits / second
Wire colors • CAN high • CAN low	 orange / black orange / brown

BentleyPublishers
.com
©2006 BentleyPublishers.com—All Rights Reserved

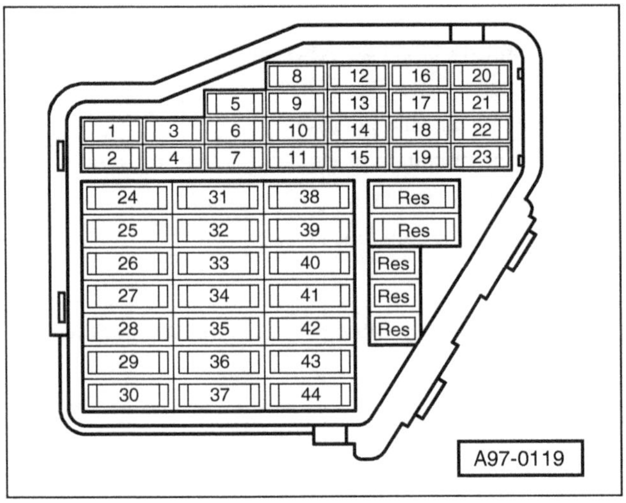

FUSES

Dashboard fuse panel

◄ Fuse panel located at driver's side end of dashboard.

Table b. Dashboard fuse identification

Fuse	Amps	Protected circuit
1	10	Heated washer nozzles (Z20, Z21) Heated outside mirrors (Z4, Z5)
2	10	Emergency flasher switch (E3)
3	5	Glove compartment light (W6) Fog light relay (J5) A/C control head (E87)
4	5	License plate light (X4, X5)
5	7.5	Fresh air / recirculating flap switch (E159) Coolant circulation pump relay (J151) Front lamp control module (J123) Brake pedal switch (F47)
6	5	Control module for central locking (J429)
7	10	Data link connector (T16) Back-up light switch (F4) Emergency flasher switch (E3) Speedometer vehicle speed sensor (G22)
8	5	Operating electronics, telephone module (J412)
9	5	ABS control module (J104)
10	10	Engine control module (ECM) (J220) ECM power supply relay (J271) 86s-Contact (2002-on) Control module for central locking (2002-on) (J429) Control module with indicator unit in instrument panel insert (2002-on) (J285)
11	5	Control module with indicator unit in instrument panel insert (J285) Oil level thermal sensor (G266)
12	7.5	Data link connector (T16) Operating electronics, telephone module (J412)
13	10	Brake light switch (F)
14	10	Control module for central locking (J429) Door warning lights (M27, M28) Passenger compartment monitoring system switch (E183) Driver's interior lock switch (E150) Switch for remote unlock, rear lid (E188)
15	5	Control module with indicator unit in instrument panel insert (J285) Transmission control module (J217)
16	10	Coolant fan control, control module (J293)
17	15	Daytime running lights Fog light relay (J5)
18	10	Right high beam headlight (M32) Fog light relay (J5) Control module with indicator unit in instrument panel insert (J285)
19	10	Left headlight, high beam (M30)

BentleyPublishers.com

© 2006 BentleyPublishers.com—All Rights Reserved

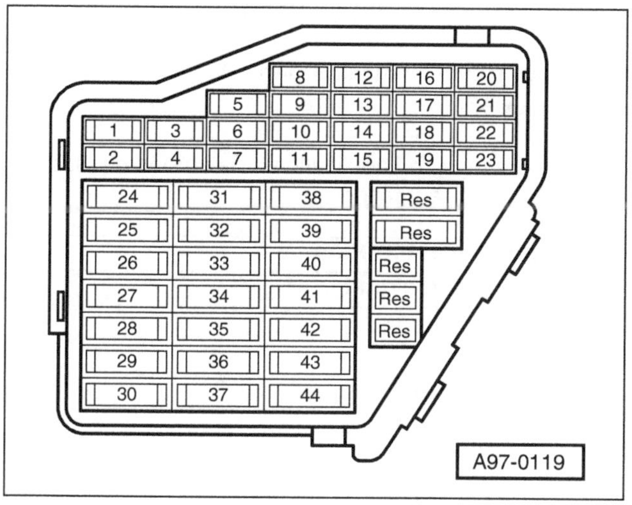

Table b. Dashboard fuse identification		
Fuse	**Amps**	**Protected circuit**
20	15	Right headlight beam adjusting motor (V49) Headlight adjuster (E102) Left headlight beam adjusting motor (V48) Front lamp control module (J123)
21	15	Front lamp control module (J123)
22	5	Front lamp control module (J123) Right parking light (M3) Control module with indicator unit in instrument panel insert (J285)
23	5	Front lamp control module (J123) Left parking light (M1) Control module with indicator unit in instrument panel insert (J285)
24	20	Wiper / washer intermittent relay (J31) Windshield wiper / washer switch (E22)
25	25	A/C control head (E87) Rear window defogger relay (J9) Fresh air blower (V2) Control module for fresh air blower (J126)
26	25	Rear window defogger (E15)
27	30	Wind deflector module (convertible) (J531) Wind deflector switch (convertible) (E278)
28	20	Fuel pump (FP) (G6)
29	15	Motronic engine control module (J220)
30		Not used
31	5	All-wheel drive control module (Haldex) (J429) Transmission control module (J217)
32	10	Fuel injectors (N...)
33	20 (1.8L) 30 (3.2L)	Wiper/washer intermittent relay (J31)
34	10	Wastegate bypass regulator valve (N75) Valve 1 for camshaft adjustment (N205) Recirculation valve for turbocharger (N249)
35		Not used
36	15	Fog light switch (E23)
37	10 20	86s-Contact Control module for central locking (J429) Control module with indicator unit in instrument panel insert (J285) Engine control module (ECM)(2002-on) (J220) ECM power supply relay (2002-on) (J271)
38	15	Control module for central locking (J429) Alarm horn (H8)
39	15	Emergency flasher switch (E3)
40	20	Horn relay (J4)
41	15	Cigarette lighter (U1)
42	20	Radio (R), Amplifier (R12),Satellite radio (R146)
43	10 (1.8L) 15 (3.2L)	EVAP canister purge regulator valve (N80) Secondary air injection solenoid valve (N112) Secondary air injection pump relay (J299) Mass air flow sensor (G70) Exhaust gas temperature sensor (G235)
44	15	Driver's heated seat adjuster (E94) Front passenger's heated seat adjuster (E95)

BentleyPublishers.com
© 2006 BentleyPublishers.com—All Rights Reserved

Fuse panels

Fig. 49 Main fuse panel on top of battery box

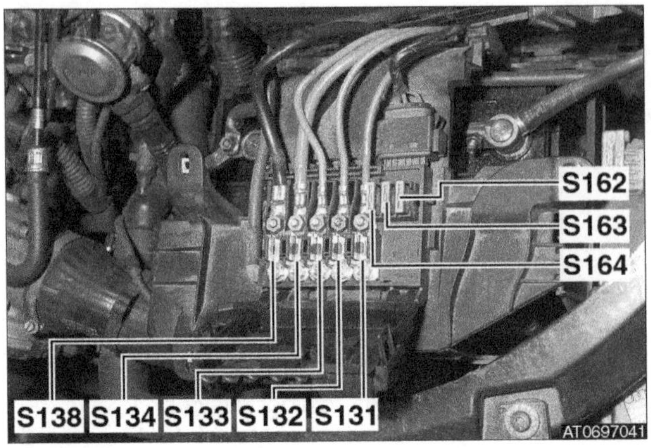

Front left side of engine compartment, shown with cover removed.

- 1.8L shown, 3.2L is similar.

See **Table c** for fuse ratings and applications. Be sure to replace a bad fuse with one of equivalent rating.

Table c. Battery fuse identification		
Fuse	**Amps**	**Protected circuit**
S131		Safety fuse 1
S132		Safety fuse 2
S133		Safety fuse 3
S134		Safety fuse 4
S138		Safety fuse 5
S162	30	Fuse 1 in fuse bracket / battery
S163	30	Fuse 2 in fuse bracket / battery
S164	30	Fuse 3 in fuse bracket / battery

Fig. 50 Dashboard fuse panel

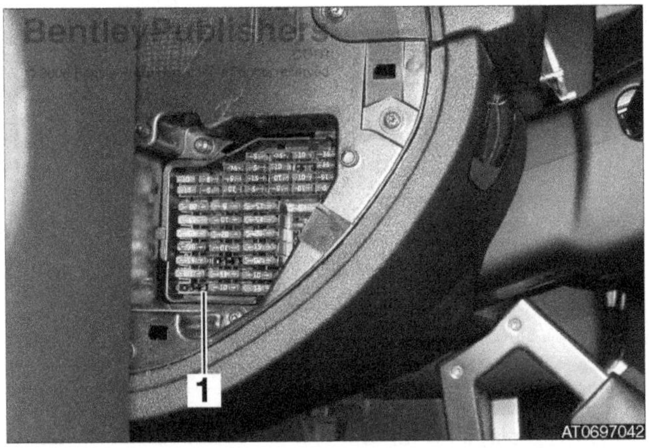

Left end of dashboard, under cover

1. Dash fuse panel

See **Table b** for fuse ratings and applications. Be sure to replace a bad fuse with one of equivalent rating.

RELAYS

Electrical relays are located in various relay panels in the vehicle. See **Table d** for relay panel location information.

Table d. Relay panels		
Year: Model	**Relay panel**	**Location**
All models	Micro central electric relay panel	Under dashboard, left side
All models	13-fold relay panel	Under dashboard, above micro central electric panel
All models	Engine compartment fuse/relay box (w-box)	Left rear engine compartment, on plenum bulkhead

CAUTION—

- *Relay positions are subject to change and may vary from car to car. If questions arise, consult an authorized Audi dealer for the most up-to-date information.*

BentleyPublishers.com
©2006 BentleyPublishers.com—All Rights Reserved

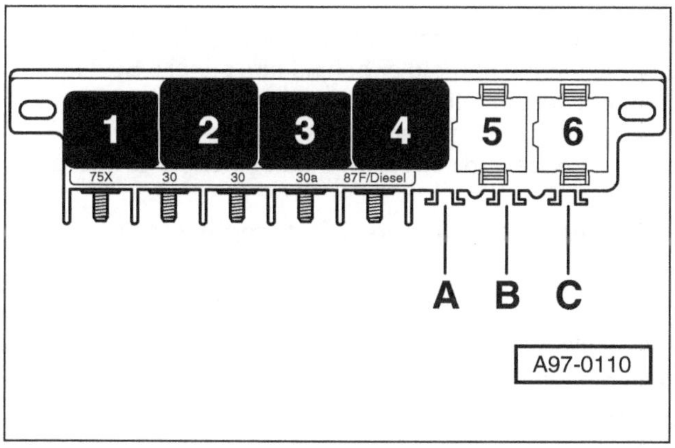

Fig. 51 Micro central electric relay panel

Under dashboard, left side.

Relay	Circuit
1	Horn relay (J4)
2	Load reduction relay (J59)
3	not used
4	Fuel pump (FP) relay (J17)
5	Wiper/washer intermittent relay
6	Wiper/washer intermittent relay
A	Convertible top operation fuse (S67)
B	Hydraulic pump relay fuse (S279)
C	Power window circuit breaker (S43)

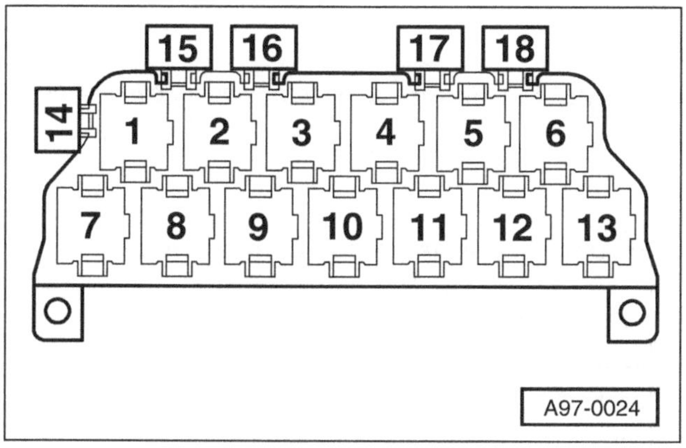

Fig. 52 13-fold relay panel

Under dashboard, left side, above micro central electric relay panel

Relay	Circuit
1	not used
2	Lamp control module (J123)
3	Lamp control module (J123)
4	Fog light relay (J5)
5	Starter interlock relay (J207)
6	not used
7	Brake light disable relay (J9) for ESP not used from 2003-on
8	not used
9	not used
10	Rear window defogger relay (J9)
11	Starter interlock relay (J207)
12	Wind deflector control module (J531) (roadster)
13	Coolant circulation pump relay (J151) Back-up light relay (J219)
14-18	not used

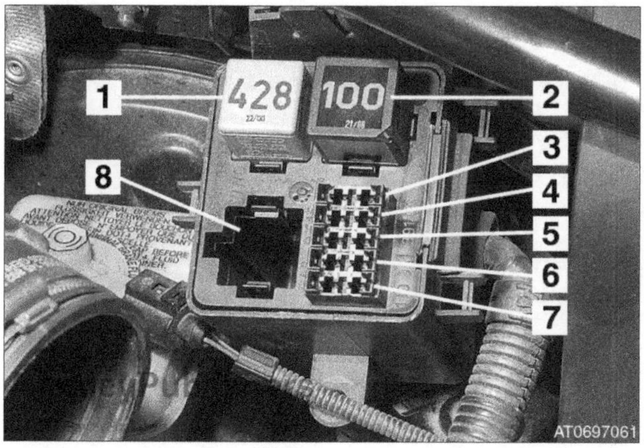

Fig. 53 Engine compartment fuse/relay box (w-box)

In rear left engine compartment

Relay/fuse	Circuit
1	Secondary air injection (AIR) pump relay (J299)
2	Motronic engine control module (ECM) power supply relay (J271)
3	Hydraulic pump relay fuse (S279)
4	Fuse for secondary air pump (S130)
5	not used
6	not used
7	Fuse for automatic transmission control module (S113)
8	not used

NOTE—

- *Fuse positions #3 through #7 are not used on some models.*

©2006 BentleyPublishers.com—All Rights Reserved

GROUNDS

Grounds are distributed throughout the vehicle body. Many are found under the interior carpets or behind trim panels. Several components grounds are often ganged. Ground positions vary among models. Lugs and connectors attached to ground are susceptible to damage and corrosion. Clean or renew as necessary.

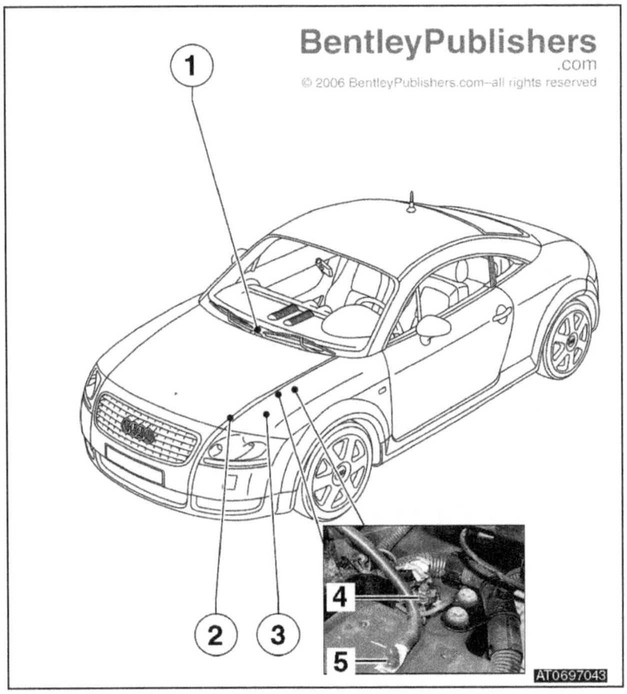

Fig. 54 Ground locations engine compartment

Wiring diagram current tracks indicate ground numbers.

1. Ground connection (in center plenum chamber)
2. Ground strap, battery to body
3. Ground strap, transmission to body
4. Ground connection, in left side of engine compartment
5. Ground connection, in left side of engine compartment

NOTE —

- *Remove battery (1.8L only) to access ground connections 4 and 5.*

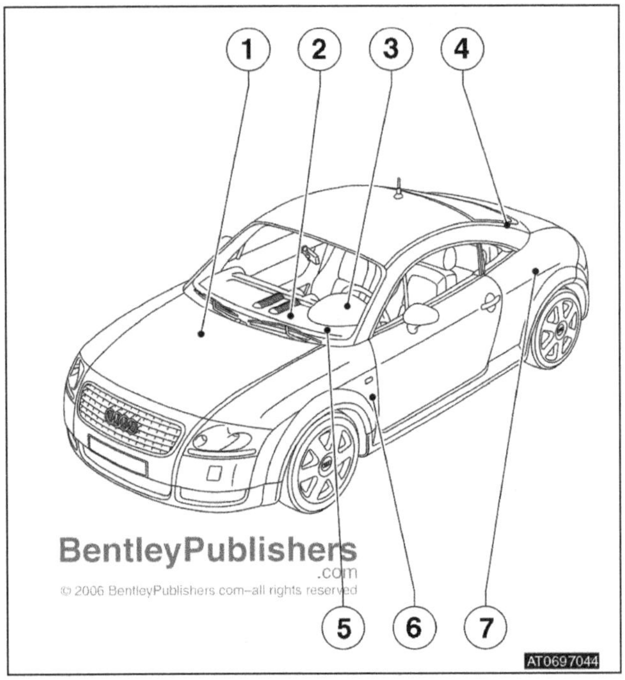

Fig. 55 Ground locations engine compartment

Wiring diagram current tracks indicate ground numbers.

1. Ground connection, on left A - pillar lower part
2. Ground connection, 1, beside fuse / relay panel
3. Ground connection, beside steering column
4. Ground connection, in rear lid, left
5. Ground connection behind instrument panel, center
6. Ground connection, on left A-pillar, lower part
7. Ground connection, in luggage compartment, left

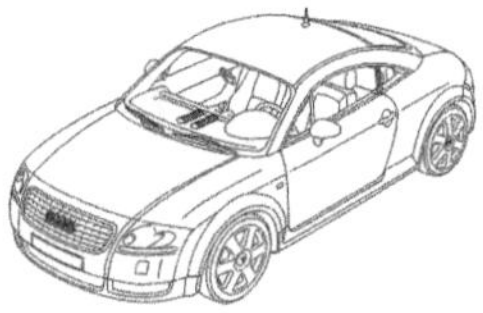

BentleyPublishers
.com
©2006 BentleyPublishers.com—All Rights Reserved

EWD Electrical Wiring Diagrams

© 2006 BentleyPublishers.com—All Rights Reserved

General

This section contains electrical wiring diagrams for vehicles covered by this manual. These diagrams are provided by Audi of America, Inc. The publisher cannot vouch for the accuracy of these diagrams.

For a comprehensive list of fuses, relays and electrical component locations, see **97 Fuses, Relays, Component Locations**.

Warnings and Cautions

WARNING —

- *Airbags and pyrotechnic seat belt tensioners utilize explosive devices. Handle with extreme care. Refer to the warnings and cautions in* **69 Seat Belts, Airbags**.
- *The ignition system of the car operates at lethal voltages. If you have a weak heart or wear a pacemaker, do not expose yourself to the ignition system electric currents. Take extra precautions when working on the ignition system or when servicing the engine while it is running or the key is ON. See* **28 Ignition System** *for additional ignition system warnings and cautions.*
- *Keep hands, clothing and other objects clear of the electric radiator cooling fan when working on a warm engine. The fan may start at any time, even when the ignition is switched OFF.*

CAUTION—

- *Do not disconnect the battery with the engine running.*
- *Switch the ignition OFF and remove the negative (-) battery cable before removing any electrical components. Connect and disconnect electrical connectors and ignition test equipment leads only while the ignition is switched OFF.*
- *Relay and fuse positions are subject to change and may vary from car to car. If questions arise, an authorized Audi dealer is the best source for the most accurate and up-to-date information.*
- *Use a digital multimeter for electrical tests. Switch multimeter to appropriate function and range before making test connections.*
- *Many control modules are static sensitive. Static discharge damages them permanently. Handle the modules using proper static prevention equipment and techniques.*
- *To avoid damaging harness connectors or relay panel sockets, use jumper wires with flat-blade connectors that are the same size as the connector or relay terminals.*
- *Do not try to start the engine of a car which has been heated above 176°F (80°C) (for example, in a paint drying booth). Allow it to cool to normal temperature.*
- *Disconnect battery before doing any electric welding on the car.*
- *Do not wash engine while it is running, or any time ignition is ON.*
- *Choose test equipment carefully. Use a digital multimeter with at least 10 MΩ input impedance, or an LED test light. An analog meter (swing-needle) or a test light with a normal incandescent bulb may draw enough current to damage sensitive electronic components.*
- *Do not use an ohmmeter to measure resistance on solid state components such as control modules.*
- *Disconnect the battery before making resistance (ohm) measurements on a circuit.*

© 2006 BentleyPublishers.com—All Rights Reserved

How to read wiring diagrams

The diagram below shows how a typical wiring diagram page is laid out.

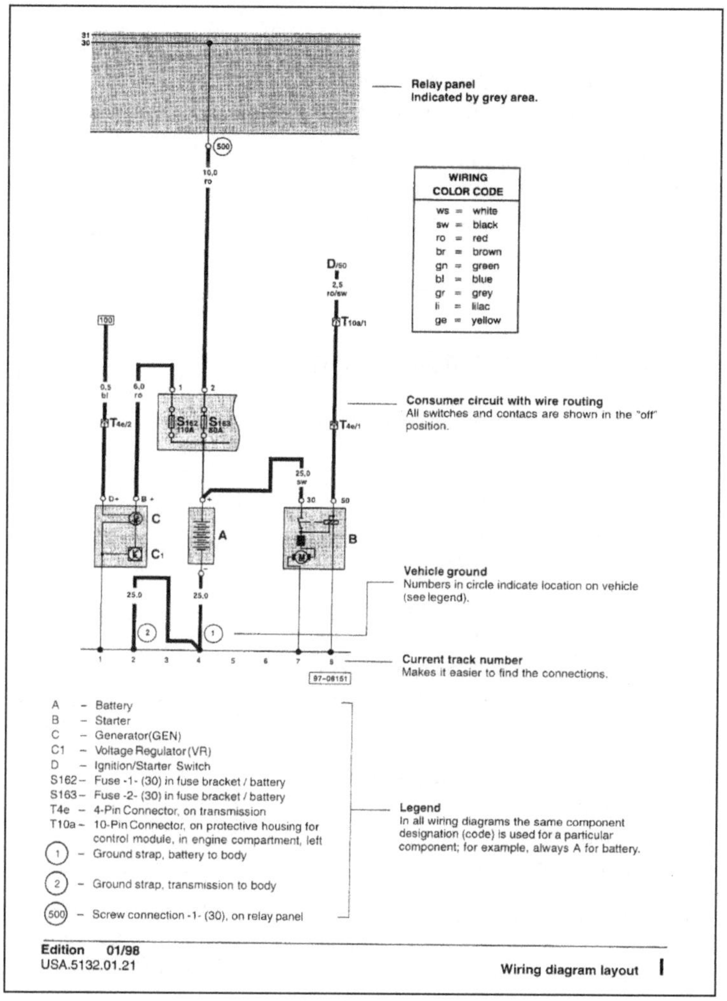

Terminal (circuit) identification

Several circuits in the vehicle's electrical system are identified with a number or letter designation. These circuits are universally identified in many wiring diagrams and may also be used to identify switch connector terminals.

Terminal	Identification
1	Ignition coil/distributor low voltage (i.e. RPM signal)
15	Switched battery positive (B+) from ignition switch
30	Battery positive (B+), hot at all times
31	Ground (–)
31b	Switched ground (–)
50	Starter control; switched B+ from ignition switch
58	Switched parking light, taillight, B+ from light switch
S (SU)	Key in ignition, switched B+ from ignition switch
X	Switched B+ from load reduction relay

© 2006 BentleyPublishers.com—All Rights Reserved

Wiring diagram details

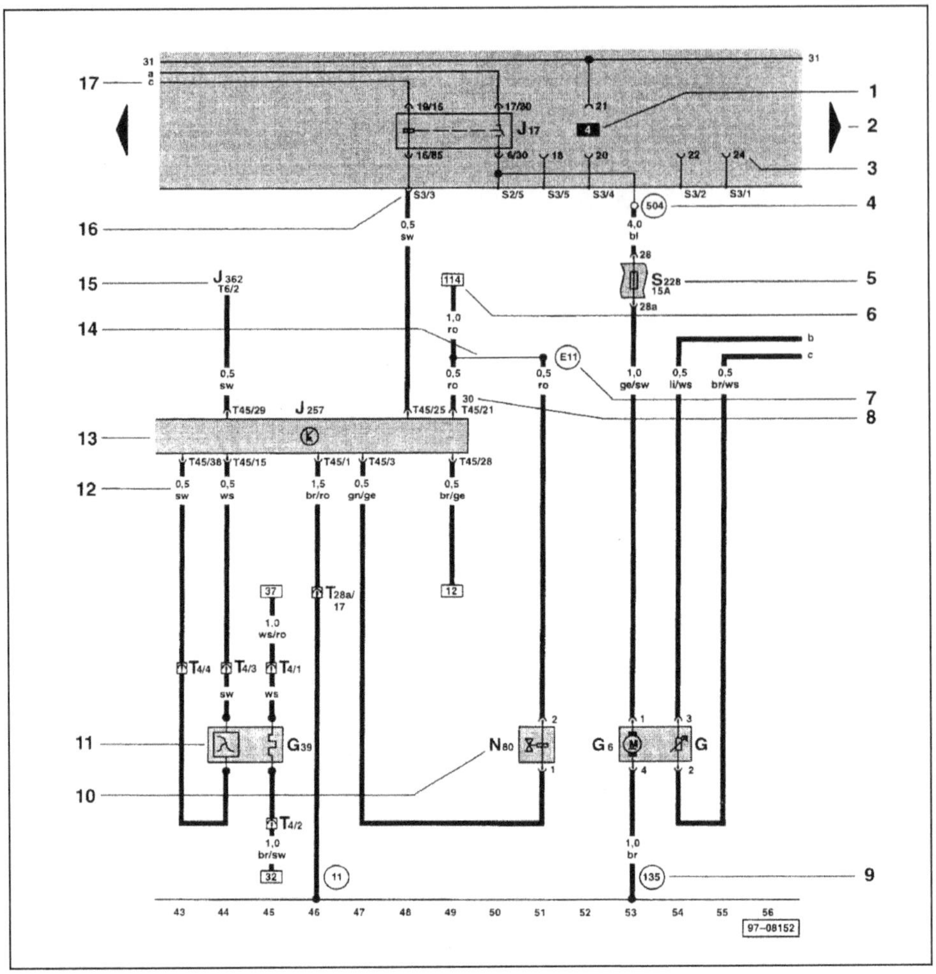

1. **Relay location number**
   - Indicates location on relay panel.
2. **Arrow**
   - Indicates wiring circuit is continued on the previous and/or next page.
3. **Connection designation - relay control module on relay panel**
   - Shows individual terminals in a multi-point connector.
   - For example: terminal 24 indicated.
4. **Diagram of threaded pin on relay panel**
   - White circle shows detachable connection.
5. **Fuse designation**
   - For example: S228 = fuse number 228, 15 amps, in fuse holder.
6. **Reference of wire continuation (current track number)**
   - Number in frame indicates current track where wire is continued on another wiring diagram page.
7. **Wire connection designation in wiring harness**
   - Location of wire connections are indicated in legend.
8. **Terminal designation**
   - Designation which appears on actual component and/or terminal number of a multi-point connector.
9. **Ground connection designation in wire harness**
   - Locations of ground connections are indicated in legend.
10. **Component designation**
    - Use legend to identify component code.
11. **Component symbols**
    - See **Symbols used in wiring diagrams** in this repair group.
12. **Wire cross-section size (in mm^2) and wire colors**
    - Abbreviations are explained in color chart beside wiring diagram.
13. **Component symbol with open drawing side**
    - Indicated component is continued on another wiring diagram page.
14. **Internal connections (thin lines)**
    - These connections are not wires.
    - Internal connections are current carrying and are shown to allow tracing of current flow inside components or wiring harnesses.
15. **Reference of continuation of wire to component**
    - For example: Anti-theft immobilizer (J362) control module on 6-pin connector, terminal 2.
16. **Relay panel connectors**
    - Shows wiring of multi-point or single connectors on relay panel.
    - For example: S3/3 = multi-point connector S3, terminal 3.
17. **Reference of internal connection continuation**
    - Letters indicate where connection continues on previous and/or next page.

BentleyPublishers.com
© 2006 BentleyPublishers.com—All Rights Reserved

Symbols used in wiring diagrams

Various symbols identified below are used in the wiring diagrams.

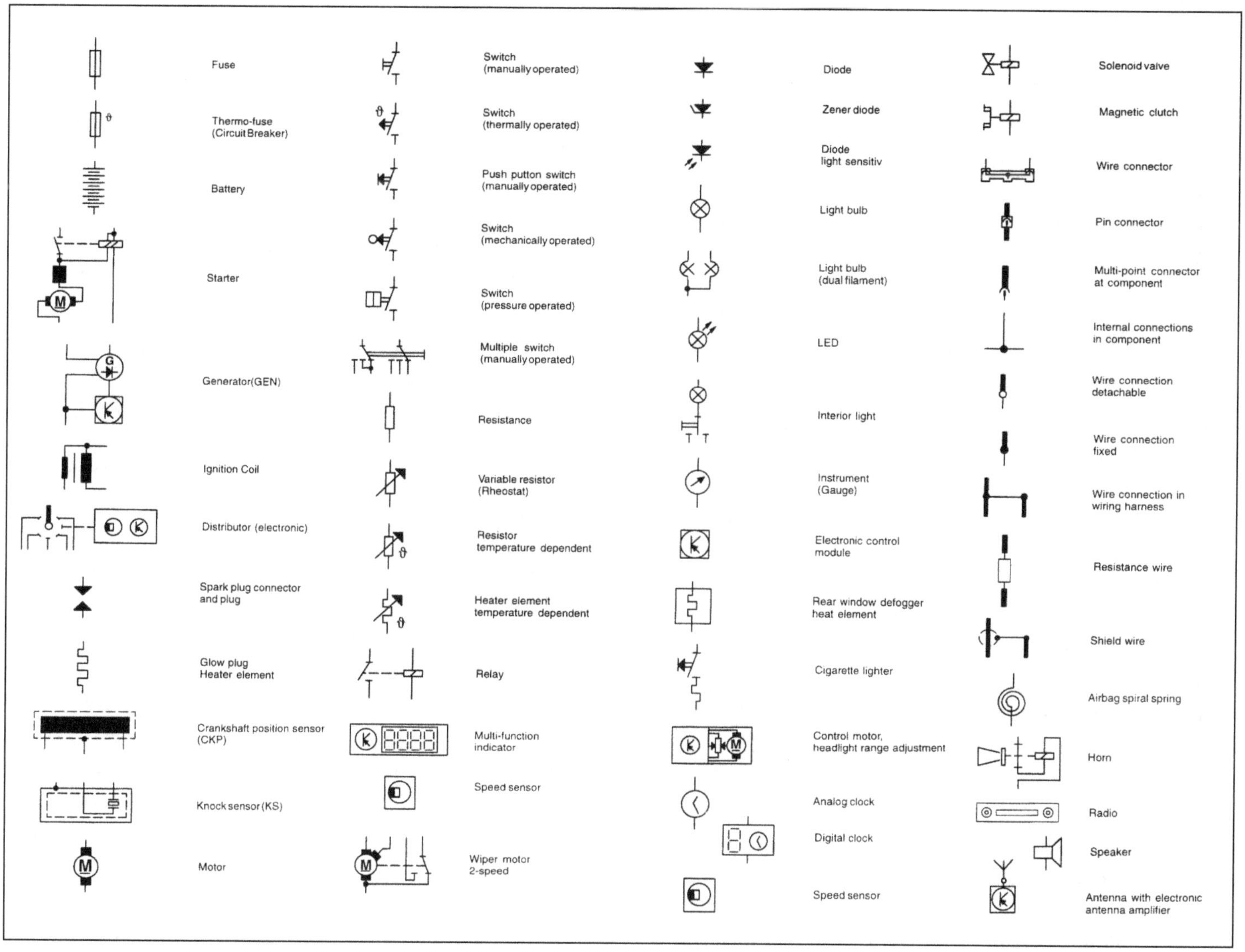

© 2006 BentleyPublishers.com—All Rights Reserved

Electrical Wiring Diagrams and Detailed Indexes

© 2006 BentleyPublishers.com—All Rights Reserved

© 2006 BentleyPublishers.com—All Rights Reserved

Electrical Wiring Diagrams and Detailed Indexes

© 2006 BentleyPublishers.com—All Rights Reserved

© 2006 BentleyPublishers.com—All Rights Reserved

Electrical Wiring Diagrams and Detailed Indexes

©2006 BentleyPublishers.com—All Rights Reserved

©2006 BentleyPublishers.com—All Rights Reserved

Electrical Wiring Diagrams and Detailed Indexes

© 2006 BentleyPublishers.com—All Rights Reserved

© 2006 BentleyPublishers.com—All Rights Reserved

Electrical Wiring Diagrams and Detailed Indexes

© 2006 BentleyPublishers.com—All Rights Reserved

Standard Equipment

2000 m. y.

Fuse Panel

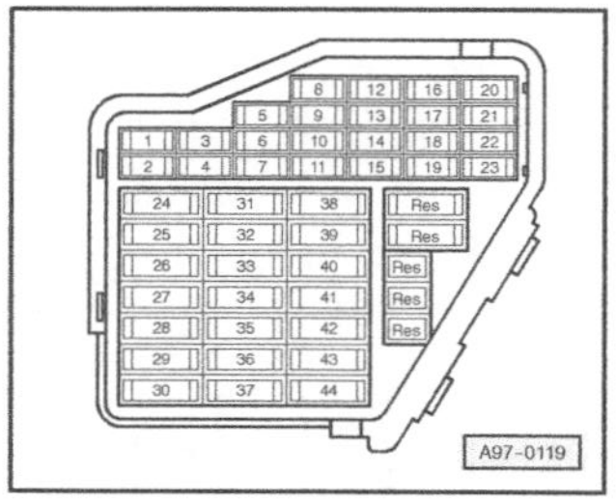

Fuse Colors:

30 A - Green
25 A - White
20 A - Yellow
15 A - Blue
10 A - Red
7,5 A - Brown
5 A - Beige

Starting with fuse position 23, fuses in the fuse holder are identified with 223 in the wiring diagram.

Micro Central Electric Panel

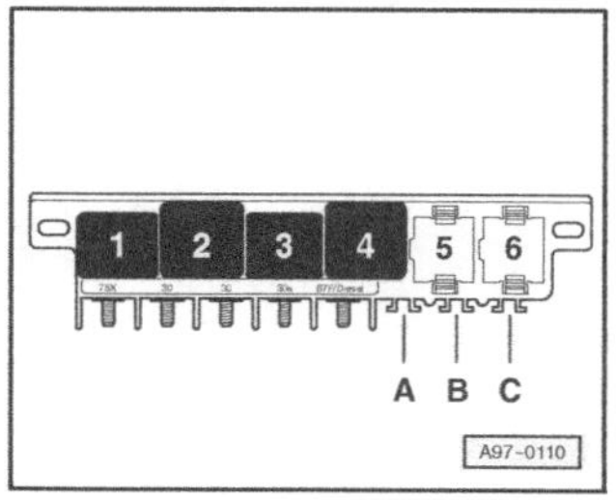

Relay Location:

- **1** -Dual Horn Relay, J4
- **2** -Load Reduction Relay, J59
- **4** -Fuel Pump (FP) Relay, J17
- **5** -Wiper/Washer Intermittent Relay, J31
- **6** -Wiper/Washer Intermittent Relay, J31

13 - Fold Relay Panel

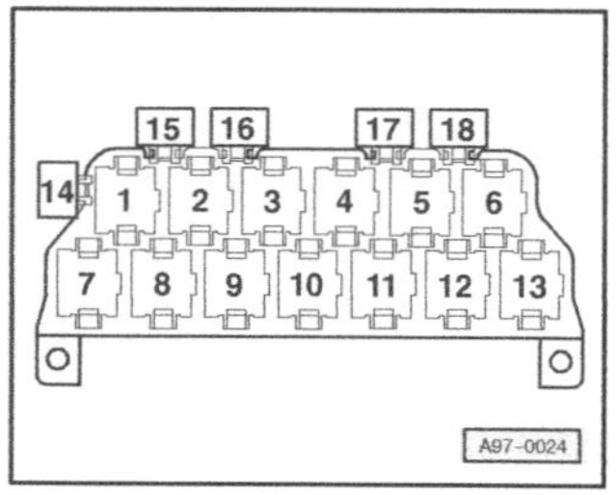

Relay Location:

- **4** -Fog Light Relay, J5
- **5** -Starting Interlock Relay, J207
- **7** -Brake light disable relay (ESP), J508
- **10** -Rear Window Defogger Relay, J9

Main Fuse Box/Battery

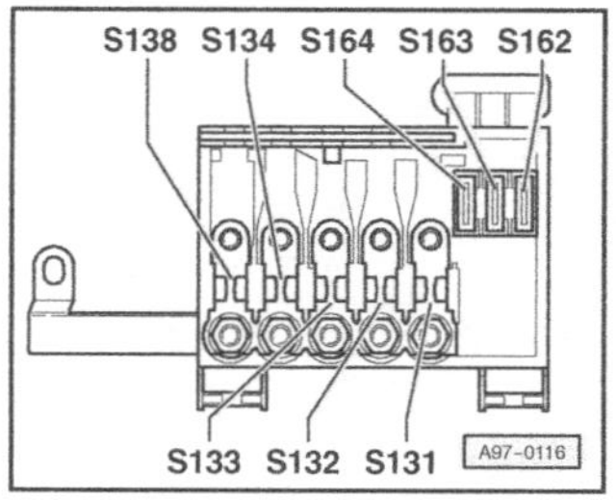

Fuses Location:

S131 - Safety fuse 1
S132 - Safety fuse 2
S133 - Safety fuse 3
S134 - Safety fuse 4
S138 - Safety fuse 5
S162 - Fuse 1 in fuse bracket/battery
S163 - Fuse 2 in fuse bracket/battery
S164 - Fuse 3 in fuse bracket/battery

©2006 BentleyPublishers.com—All Rights Reserved

Audi TT — Wiring diagram — No. 1/3

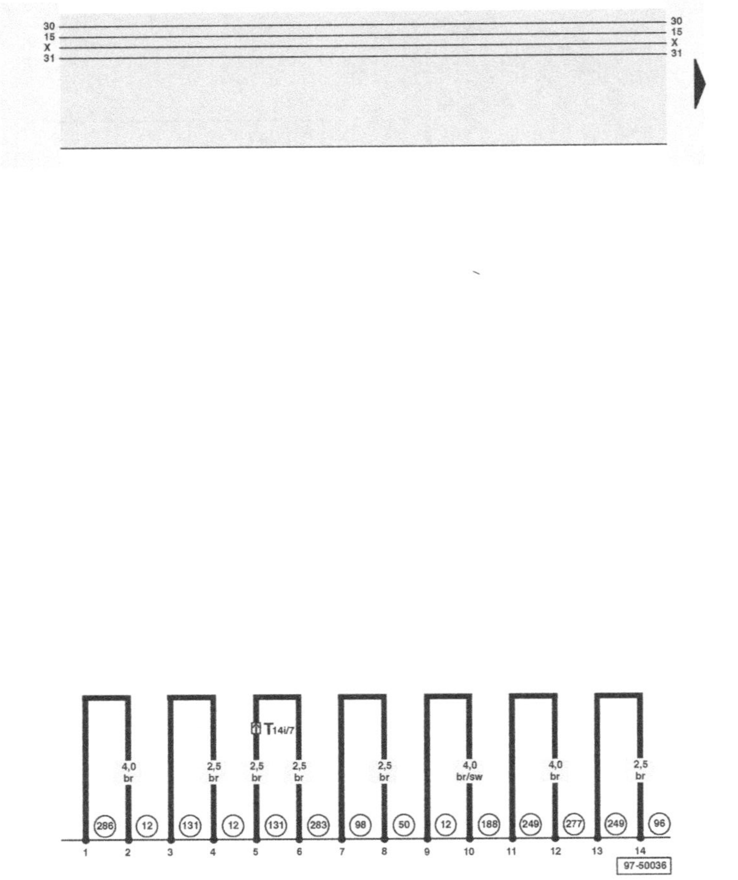

ws = white
sw = black
ro = red
br = brown
gn = green
bl = blue
gr = grey
li = violet
ge = yellow

- T14i - Connector 14 pin, in engine compartment, left
- (12) - Ground connection, in engine compartment, left
- (50) - Ground connection, in luggage compartment, left
- (96) - Ground connection -1-, in heated seats wiring harness
- (98) - Ground connection, in rear lid wiring harness
- (131) - Ground connection -2-, in engine compartment wiring harness
- (188) - Ground connection -3-, in A/C wiring harness
- (249) - Ground connection -2-, in wiring harness interior
- (277) - Ground connector -3-, in wiring harness interior
- (283) - Ground connection 2, in wiring harness engine pre-wiring
- (286) - Ground connection -8- in wiring harness interior

No. 1/4 — Wiring diagram — Audi TT

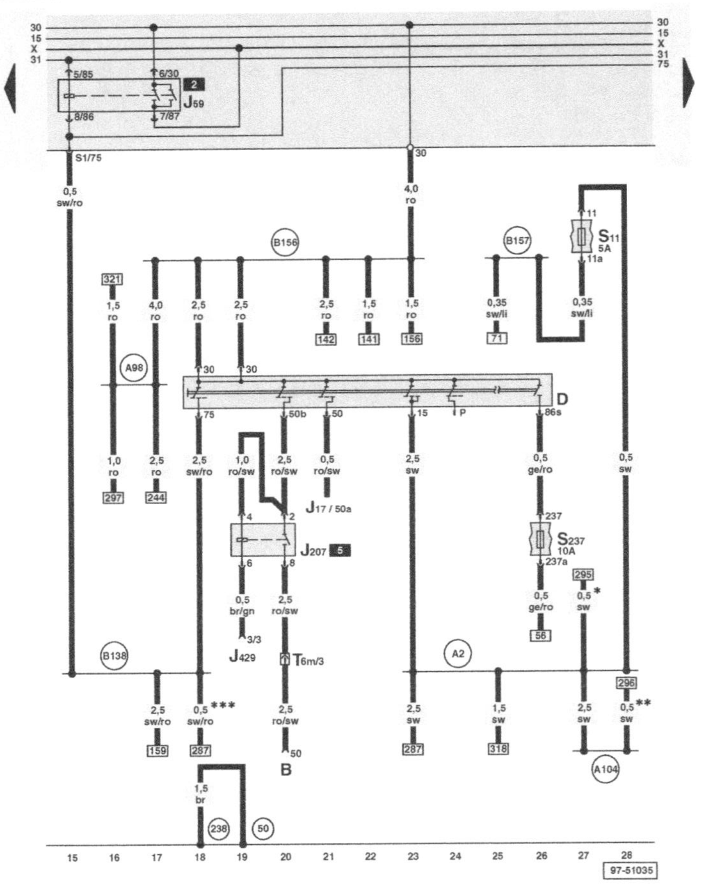

ws = white
sw = black
ro = red
br = brown
gn = green
bl = blue
gr = grey
li = violet
ge = yellow

- B - Starter
- D - Ignition/Starter Switch
- J17 - Fuel Pump (FP) Relay
- J59 - Load Reduction Relay
- J207 - Starting Interlock Relay
- J429 - Control module for central locking
- S11 - Fuse
- S237 - Fuse in fuse holder
- T6m - Connector 6 pin, brown, in E-box plenum chamber

- (50) - Ground connection, in luggage compartment, left
- (238) - Ground connection -1-, in wiring harness interior
- (A2) - Plus connection (15), in instrument panel wiring harness
- (A98) - Plus connector -4- (30), in instrument panel wiring harness
- (A104) - Plus connector -2- (15), in instrument panel wiring harness
- (B138) - Plus connector (X), in wiring harness, interior
- (B156) - Plus connector (30a), in wiring harness interior
- (B157) - Plus connector -3- (15a), in wiring harness interior

- * - Vehicles with ABS
- ** - Vehicles with ABS and ESP
- *** - Roadster

© 2006 BentleyPublishers.com—All Rights Reserved

Audi TT — Wiring diagram — No. 1/5

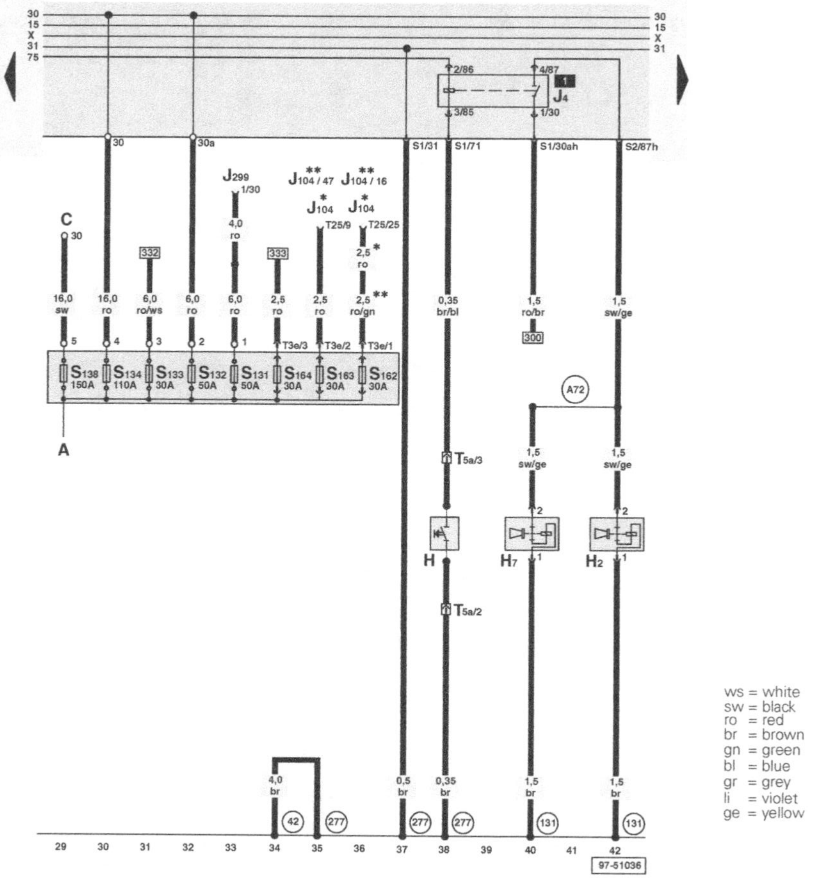

ws = white
sw = black
ro = red
br = brown
gn = green
bl = blue
gr = grey
li = violet
ge = yellow

A - Battery
C - Generator (GEN)
H - Horn Button
H2 - High Tone Horn
H7 - Low Tone Horn
J4 - Dual Horn Relay
J104 - ABS Control Module (w/EDL)
J299 - Secondary Air Injection (AIR) Pump Relay
S131 - Safety fuse 1
S132 - Safety fuse 2
S133 - Safety fuse 3
S134 - Safety fuse 4
S138 - Safety fuse 5
S162 - Fuse -1- in fuse bracket/battery
S163 - Fuse -2- in fuse bracket/battery
S164 - Fuse -3- in fuse bracket/battery
T3e - Connector 3 pin, black, on Main Fuse Box/Battery
T5a - Connector 5 pin, yellow, in steering wheel
T25 - Connector 25 pin, on ABS Control Module (w/EDL)

(42) - Ground connection, on steering column
(131) - Ground connection -2-, in engine compartment wiring harness
(277) - Ground connector -3-, in wiring harness interior
(A72) - Connector (71), in instrument panel wiring harness

* - Vehicles with ABS
** - Vehicles with ABS and ESP

No. 1/6 — Wiring diagram — Audi TT

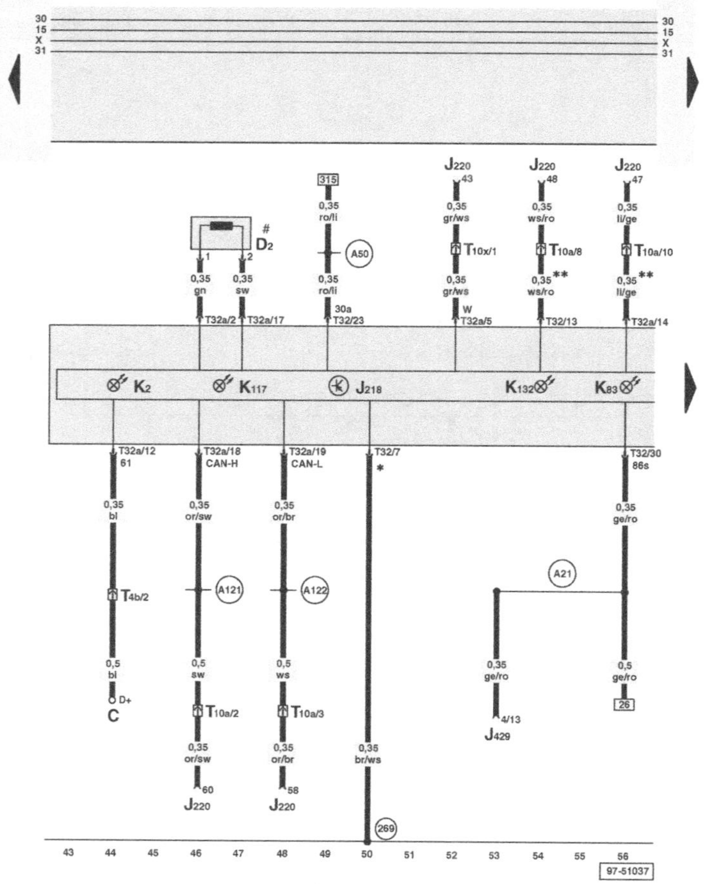

ws = white
sw = black
ro = red
br = brown
gn = green
bl = blue
gr = grey
li = violet
ge = yellow
or = orange

C - Generator (GEN)
D2 - Induction coil of antitheft immobilizer
J218 - Instrument Cluster Combination Processor
J220 - Motronic Engine Control Module (ECM)
J429 - Control module for central locking
K2 - Generator (GEN) Warning Light
K83 - Malfunction Indicator Lamp (MIL)
K117 - Warning light for anti-theft immobilizer
K132 - Electronic Power Control (EPC) Warning Lamp
T4b - Connector 4 pin, in engine compartment, left
T10a - Connector 10 pin, white, in E-box plenum chamber
T10x - Connector 10 pin, orange, in E-box plenum chamber
T32 - Connector 32 pin, blue, on instrument cluster
T32a - Connector 32 pin, green, on instrument cluster

(269) - Ground connector (sensor ground) -1-, in instrument panel wiring harness
(A21) - Wire connection (86s), in instrument panel wiring harness
(A50) - Plus connection (30a), in instrument panel wiring harness
(A121) - Connection (high bus) in instrument panel wiring harness
(A122) - Connection (low bus) in instrument panel wiring harness

* - Sensor ground output
** - only instrument cluster without CAN-Bus
# - vehicles with Anti-theft Immobilizer

©2006 BentleyPublishers.com—All Rights Reserved

Audi TT Wiring diagram **No. 1/7**

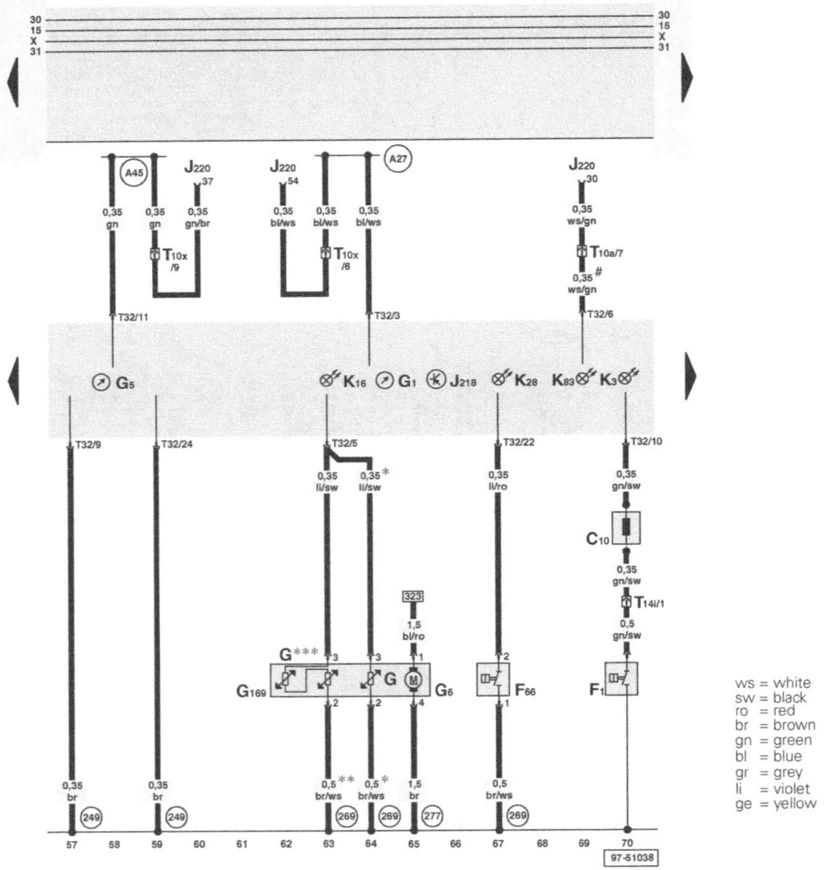

ws = white
sw = black
ro = red
br = brown
gn = green
bl = blue
gr = grey
li = violet
ge = yellow

C10 - Suppressor Filter
F1 - Oil Pressure Switch
F66 - Engine Coolant Level (ECL) Warning Switch
G - Fuel Level Sensor
G1 - Fuel Gauge
G5 - Tachometer
G6 - Fuel Pump (FP)
G169 - Fuel Level Sensor 2
J218 - Instrument Cluster Combination Processor
J220 - Motronic Engine Control Module (ECM)
K3 - Oil Pressure Warning Light
K16 - Fuel Reserve Warning Light
K28 - Engine Coolant Level/Temperature (ECL/ECT) Warning Light
K83 - Malfunction Indicator Lamp (MIL)
T10a - Connector 10 pin, white, in E-box plenum chamber
T10x - Connector 10 pin, orange, in E-box plenum chamber
T14i - Connector 14 pin, in engine compartment, left
T32 - Connector 32 pin, blue, on instrument cluster

(249) - Ground connection -2-, in wiring harness interior
(269) - Ground connector (sensor ground) -1-, in instrument panel wiring harness
(277) - Ground connector -3-, in wiring harness interior
(A27) - Wire connection (vehicle speed signal), in instrument panel wiring harness
(A45) - Wire connection (RPM-signal), in instrument panel wiring harness

* - Vehicles with front wheel drive
** - Vehicles with all wheel drive
*** - Wiring inside fuel tank
# - only instrument cluster without CAN-Bus

Edition 12/02
W42.USA.5702.12.21

Instrument Cluster, Oil Pressure Switch, Engine Coolant Level Warning Switch, Fuel Pump

No. 1/8 Wiring diagram **Audi TT**

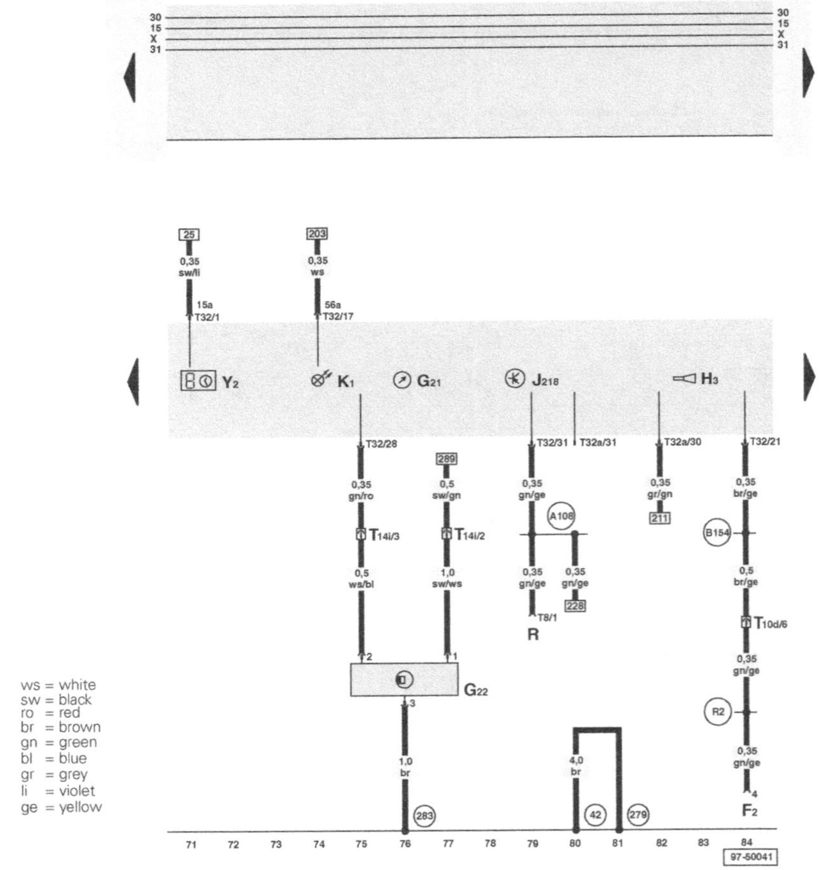

ws = white
sw = black
ro = red
br = brown
gn = green
bl = blue
gr = grey
li = violet
ge = yellow

F2 - Door Contact Switch - Driver's side
G21 - Speedometer
G22 - Speedometer Vehicle Speed Sensor (VSS)
H3 - Warning Buzzer
J218 - Instrument Cluster Combination Processor
K1 - Headlight High Beam Indicator Light
R - Radio
T8 - Connector 8 pin, black, connector radio III
T10d - Connector 10 pin, brown, connector station A pillar, left
T14i - Connector 14 pin, in engine compartment, left
T32 - Connector 32 pin, blue, on instrument cluster
T32a - Connector 32 pin, green, on instrument cluster
Y2 - Digital Clock

(42) - Ground connection, on steering column
(279) - Ground connector -5-, in wiring harness interior
(283) - Ground connection 2, in wiring harness engine pre-wiring
(A108) - Connector (speed signal), in instrument panel wiring harness
(B154) - Connector -1- (TK), in wiring harness interior
(R2) - Wire connection (door contact switch, left), in power window/central locking system and door contact switch wiring harness

Instrument Cluster, Speedometer Vehicle Speed Sensor

Edition 12/02
W42.USA.5702.12.21

© 2006 BentleyPublishers.com—All Rights Reserved

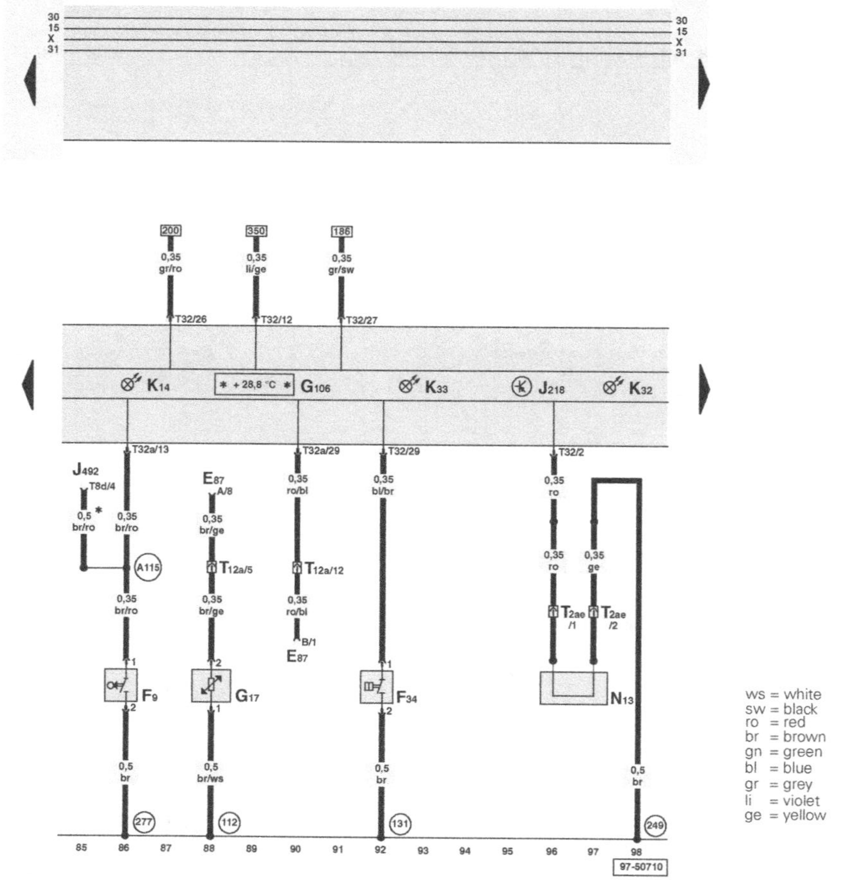

ws = white
sw = black
ro = red
br = brown
gn = green
bl = blue
gr = grey
li = violet
ge = yellow

E87 - A/C Control Head
F9 - Parking Brake Warning Light Switch
F34 - Brake Fluid Level Warning Switch
G17 - Outside Air Temperature Sensor
G106 - Outside Air Temperature Display
J218 - Instrument Cluster Combination Processor
J492 - Control module for all wheel drive
K14 - Parking Brake Indicator Light
K32 - Brake Pad Wear Indicator Light
K33 - Brake Fluid Level Warning Light
N13 - Left Brake Pad Wear Indicator (Shear Element)
T2ae - Connector 2 pin, on firewall, left
T8d - Connector 8 pin, on Control module for all wheel drive
T12a - Connector 12 pin, behind instrument panel, right
T32 - Connector 32 pin, blue, on instrument cluster
T32a - Connector 32 pin, green, on instrument cluster

(112) - Ground connection -2-, in A/C wiring harness
(131) - Ground connection -2-, in engine compartment wiring harness
(249) - Ground connection -2-, in wiring harness interior
(277) - Ground connector -3-, in wiring harness interior
(A115) - Connection (parking brake control) in instrument panel wiring harness
* - Vehicles with all-wheel drive

Edition 07.00
W42.USA.5702.04.21

Instrument Cluster, Parking Brake Warning Light Switch, Brake Fluid Level Warning Switch, Outside Air Temperature Sensor

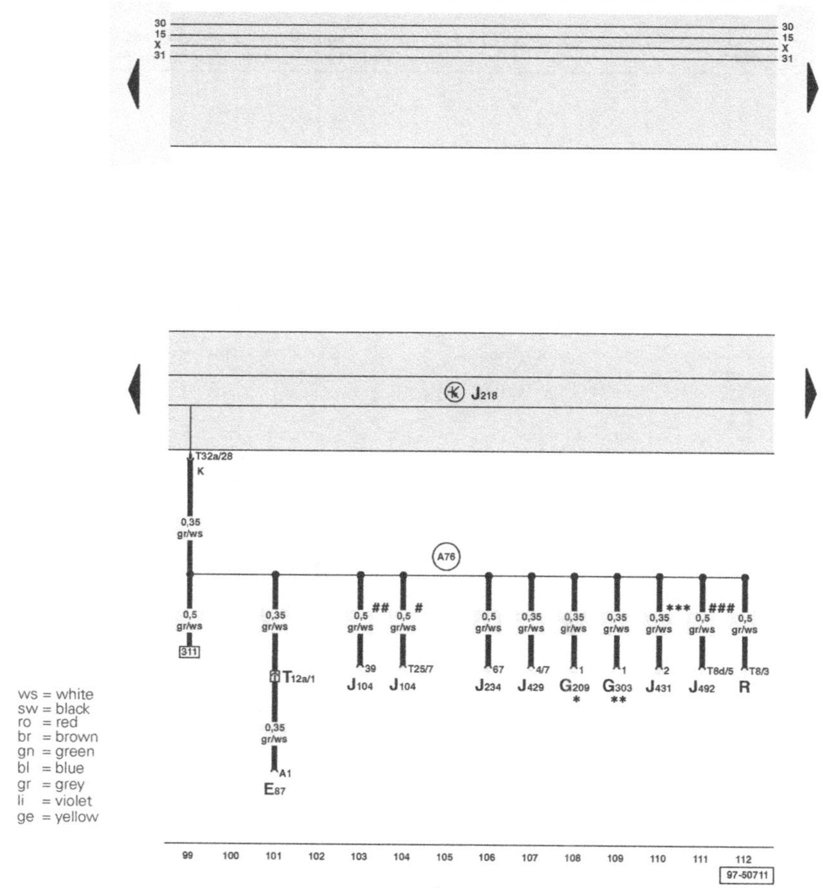

ws = white
sw = black
ro = red
br = brown
gn = green
bl = blue
gr = grey
li = violet
ge = yellow

E87 - A/C Control Head
G209 - Ultra-sound sensor for anti-theft warning
G303 - Radar Interior Monitoring Control Module 1
J104 - ABS Control Module (w/EDL)
J218 - Instrument Cluster Combination Processor
J234 - Airbag Control Module
J429 - Control module for central locking
J431 - Control module for Headlight Beam Adjusting
J492 - Control module for all wheel drive
R - Radio
T8 - Connector 8 pin, black, connector radio III
T8d - Connector 8 pin, on Control module for all wheel drive
T12a - Connector 12 pin, behind instrument panel, right
T25 - Connector 25 pin, on ABS Control Module (w/EDL)
T32a - Connector 32 pin, green, on instrument cluster

(A76) - Connector (K-diagnosis wire), in instrument panel wiring harness

* - Coupé
** - Roadster
*** - Vehicles with Automatic Headlight Beam Adjusting
# - Vehicles with ABS
- Vehicles with ABS and ESP
- Vehicles with all-wheel drive

Instrument Cluster, Connector (K-diagnosis wire) for Data Link Connector (DLC)

Edition 07.00
W42.USA.5702.04.21

© 2006 BentleyPublishers.com—All Rights Reserved

Audi TT Wiring diagram **No. 1/11**

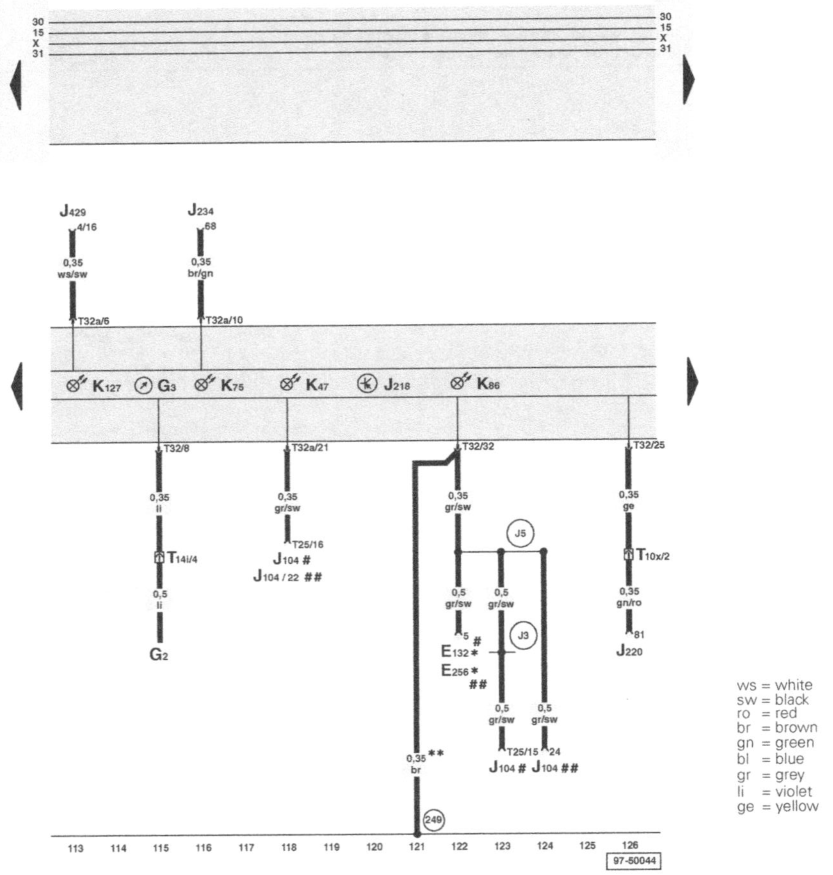

ws = white
sw = black
ro = red
br = brown
gn = green
bl = blue
gr = grey
li = violet
ge = yellow

- E132 - Anti-Slip Control Switch
- E256 - ASR/ESP Button
- G2 - Engine Coolant Temperature (ECT) Sensor
- G3 - Engine Coolant Temperature (ECT) Gauge
- J104 - ABS Control Module (w/EDL)
- J218 - Instrument Cluster Combination Processor
- J220 - Motronic Engine Control Module (ECM)
- J234 - Airbag Control Module
- J429 - Control module for central locking
- K47 - ABS Warning Light
- K75 - Airbag Malfunction Indicator Lamp (MIL)
- K86 - Traction Control Indicator Light
- K127 - Indicator light for open trunk
- T10x - Connector 10 pin, orange, in E-box plenum chamber
- T14i - Connector 14 pin, in engine compartment, left
- T25 - Connector 25 pin, on ABS Control Module (w/EDL)
- T32 - Connector 32 pin, blue, on instrument cluster
- T32a - Connector 32 pin, green, on instrument cluster

- (249) - Ground connection -2-, in wiring harness interior
- (J3) - Wire connection -1-, in ABS wiring harness
- (J5) - Wire connection -2-, in ABS wiring harness, near anti-slip control switch

- * - Vehicles with Traction Control
- ** - Vehicles with all-wheel drive
- # - Vehicles with ABS
- ## - Vehicles with ABS and ESP

Edition 07.00
W42.USA.5702.04.21

Instrument Cluster, Engine Coolant Temperature Gauge

No. 1/12 Wiring diagram **Audi TT**

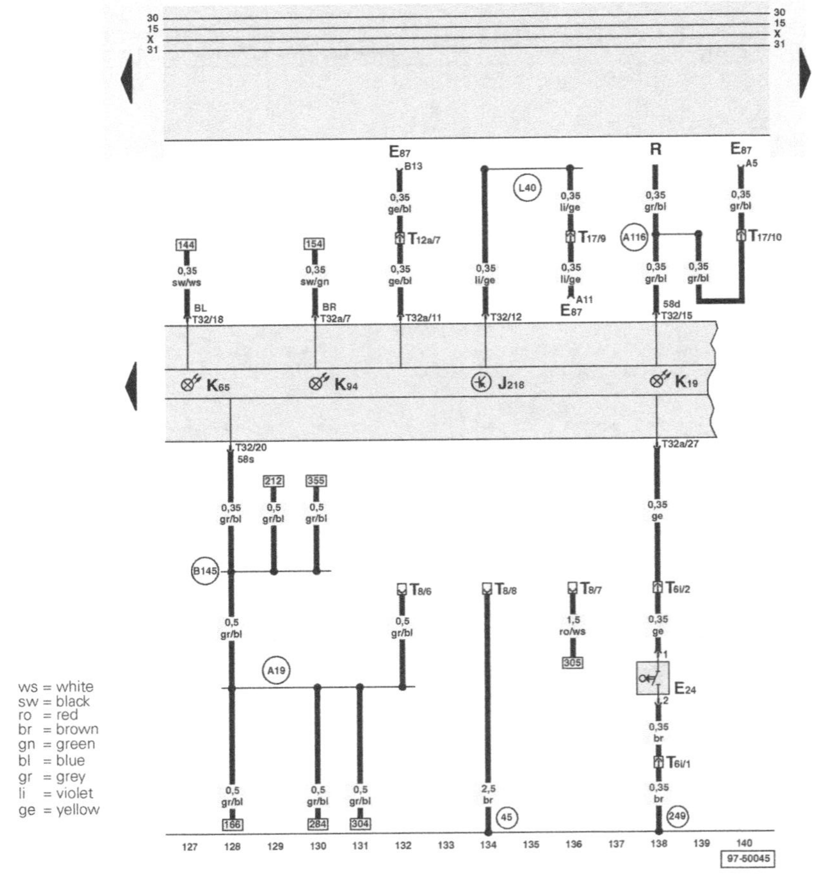

ws = white
sw = black
ro = red
br = brown
gn = green
bl = blue
gr = grey
li = violet
ge = yellow

- E24 - Left Seat Belt Switch
- E87 - A/C Control Head
- J218 - Instrument Cluster Combination Processor
- K19 - Seat Belt Warning Light
- K65 - Left Turn Signal Indicator Light
- K94 - Right Turn Signal Indicator Light
- R - Radio
- T6i - Connector 6 pin, below driver's seat
- T8 - Connector 8 pin, black, connector radio III
- T12a - Connector 12 pin, behind instrument panel, right
- T17 - Connector 17 pin, behind instrument panel
- T32 - Connector 32 pin, blue, on instrument cluster
- T32a - Connector 32 pin, green, on instrument cluster

- (45) - Ground connection, behind instrument panel, center
- (249) - Ground connection -2-, in wiring harness interior
- (A19) - Wire connection (58s), in instrument panel wiring harness
- (A116) - Connection -2- (58d) in instrument panel wiring harness
- (B145) - Plus connector (58s), in wiring harness interior
- (L40) - Wire connection -2- (A/C Pressure Switch), in Climatronic wiring harness

Instrument Cluster, Turn Signal Indicator Lights, Connector Radio, Seat Belt Warning Light

Edition 07.00
W42.USA.5702.04.21

© 2006 BentleyPublishers.com—All Rights Reserved

Audi TT **Wiring diagram** **No. 1/13**

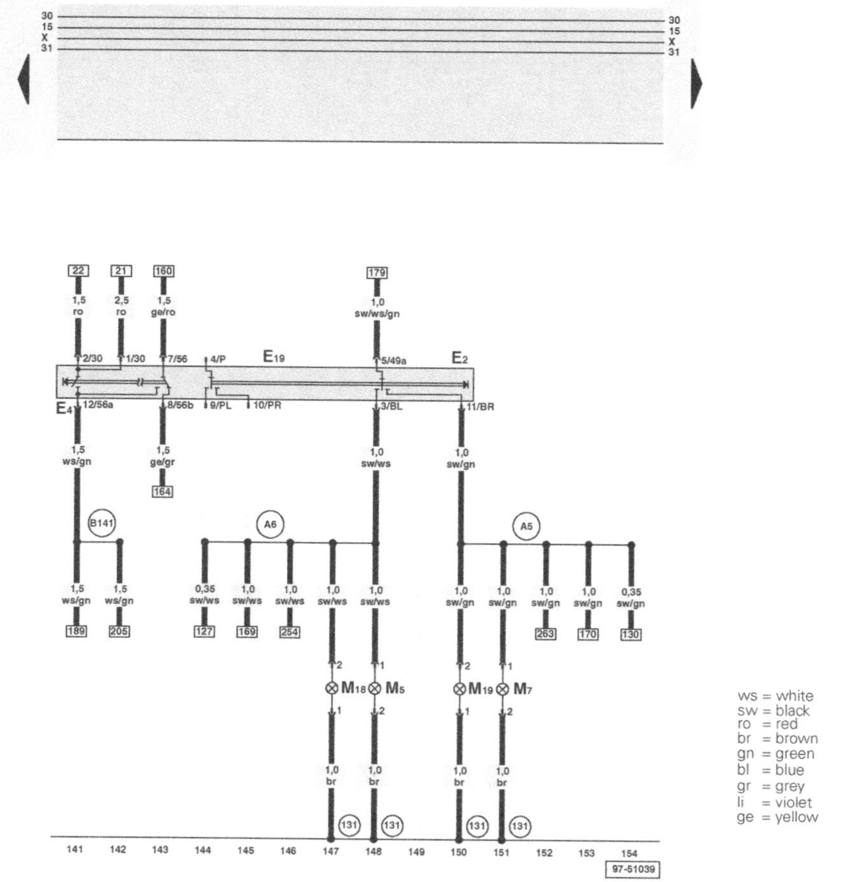

ws = white
sw = black
ro = red
br = brown
gn = green
bl = blue
gr = grey
li = violet
ge = yellow

- E2 - Turn Signal Switch
- E4 - Headlight Dimmer/Flasher Switch
- E19 - Park Light Switch
- M5 - Left Front Turn Signal Light
- M7 - Right Front Turn Signal Light
- M18 - Left, Side Turn Signal Light
- M19 - Right, Side Turn Signal Light

- (131) - Ground connection -2-, in engine compartment wiring harness
- (A5) - Plus connection (right turn signal), in instrument panel wiring harness
- (A6) - Plus connection (left turn signal), in instrument panel wiring harness
- (B141) - Plus connector -1- (56a), in wiring harness interior

Turn Signal Switch, Headlight Dimmer/Flasher Switch, Park Light Switch, Front Turn Signal Light

No. 1/14 **Wiring diagram** **Audi TT**

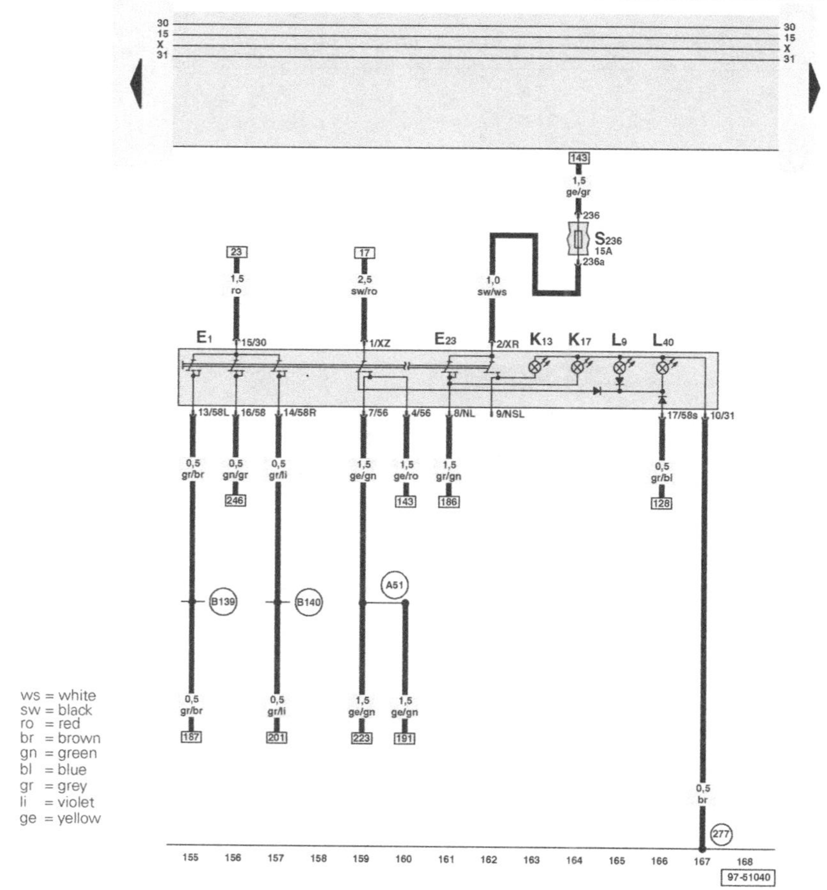

ws = white
sw = black
ro = red
br = brown
gn = green
bl = blue
gr = grey
li = violet
ge = yellow

- E1 - Light Switch
- E23 - Fog Light Switch
- K13 - Rear Fog Light Indicator Light
- K17 - Fog Light Indicator Light
- L9 - Headlight Switch Light
- L40 - Front And Rear Fog Light Switch Light
- S236 - Fuse in fuse holder

- (277) - Ground connector -3-, in wiring harness interior
- (A51) - Wire connection (56), in instrument panel wiring harness
- (B139) - Connector (PL), in wiring harness, interior
- (B140) - Connector (PR), in wiring harness, interior

Light Switch, Fog Light Switch

© 2006 BentleyPublishers.com—All Rights Reserved

Audi TT Wiring diagram **No. 1/15**

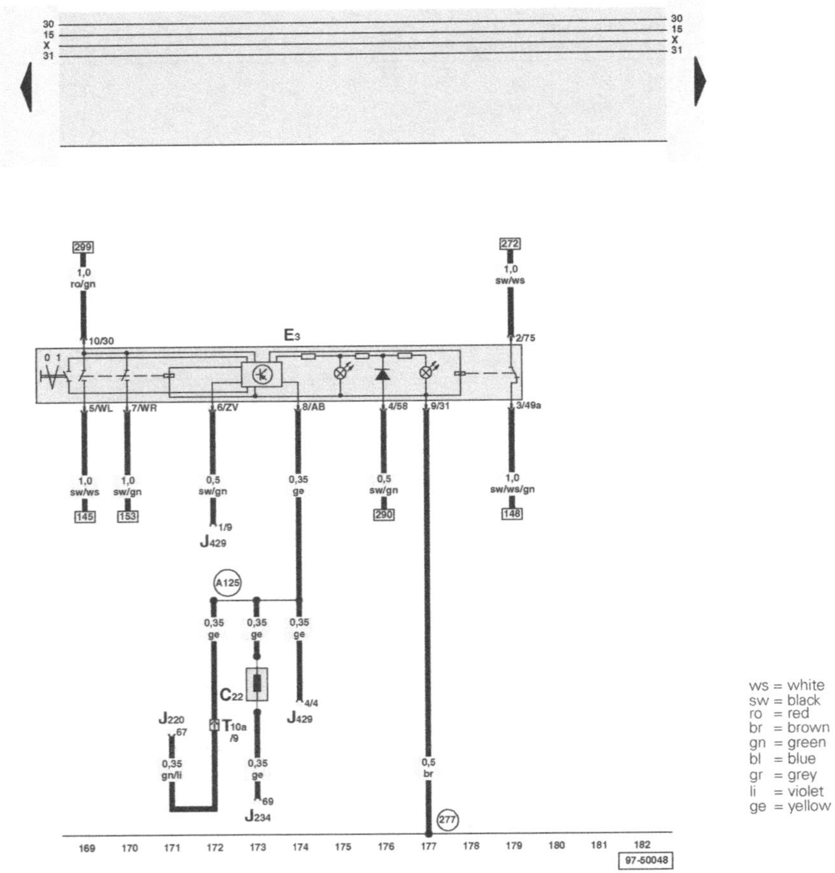

ws = white
sw = black
ro = red
br = brown
gn = green
bl = blue
gr = grey
li = violet
ge = yellow

C22 - Airbag Suppresser
E3 - Emergency Flasher Switch
J220 - Motronic Engine Control Module (ECM)
J234 - Airbag Control Module
J429 - Control module for central locking
T10a - Connector 10 pin, white, in E-box plenum chamber

(277) - Ground connector -3-, in wiring harness interior
(A125) - Connection (crash signal) in instrument panel wiring harness

Edition 12/02
W42.USA.5702.12.21
Emergency Flasher Switch

No. 1/16 Wiring diagram **Audi TT**

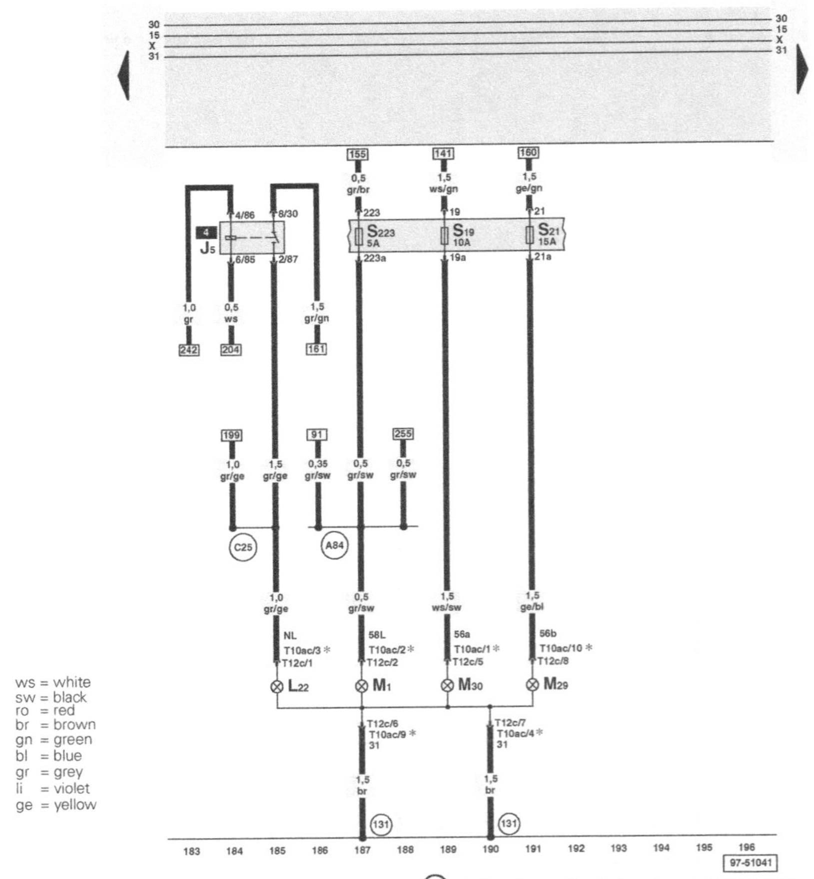

ws = white
sw = black
ro = red
br = brown
gn = green
bl = blue
gr = grey
li = violet
ge = yellow

J5 - Fog Light Relay
L22 - Left Front Fog Light
M1 - Left Parking Light
M29 - Left Low Beam Headlight
M30 - Left High Beam Headlight
S19 - Fuse
S21 - Fuse
S223 - Fuse in fuse holder
T10ac - Connector 10 pin, on headlight, left
T12c - Connector 12 pin, on headlight, left

(131) - Ground connection -2-, in engine compartment wiring harness
(A84) - Connector (58L), in instrument panel wiring harness
(C25) - Wire connection (fog lights), in front light wiring harness
* - from VIN Y1 010 000

Fog Light Relay, Left Headlight, Fuses
Edition 12/02
W42.USA.5702.12.21

© 2006 BentleyPublishers.com—All Rights Reserved

Audi TT Wiring diagram **No. 1/17**

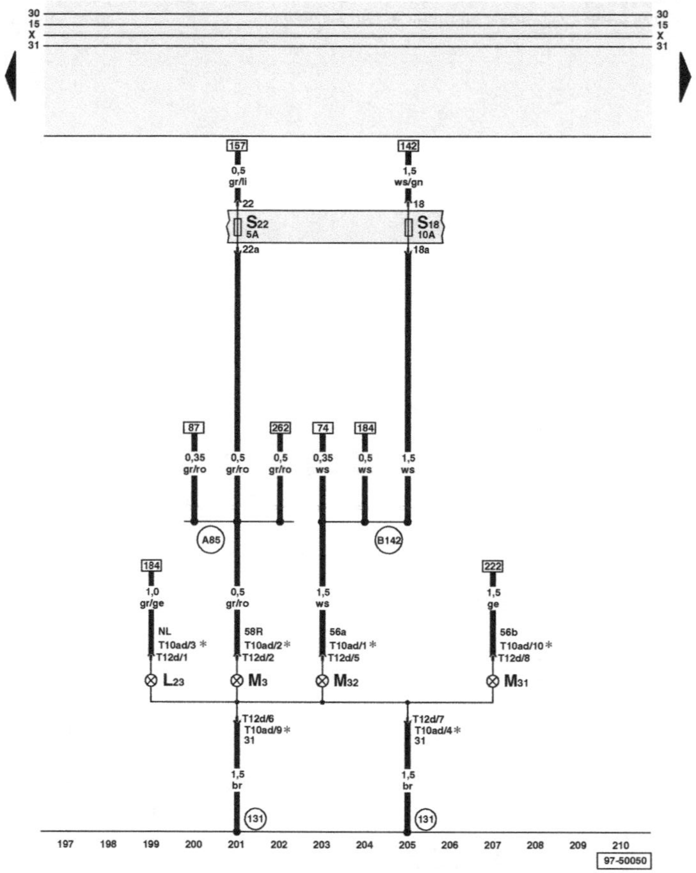

ws = white
sw = black
ro = red
br = brown
gn = green
bl = blue
gr = grey
li = violet
ge = yellow

- L23 - Right Front Fog Light
- M3 - Right Parking Light
- M31 - Right Low Beam Headlight
- M32 - Right High Beam Headlight
- S18 - Fuse
- S22 - Fuse
- T10ad - Connector 10 pin, on headlight, right
- T12d - Connector 12 pin, on headlight, right

- (131) - Ground connection -2-, in engine compartment wiring harness
- (A85) - Connector (58R), in instrument panel wiring harness
- (B142) - Plus connector -2- (56a), in wiring harness interior
- * - from VIN Y1 010 000

Edition 12/02
W42.USA.5702.12.21

Right Headlight, Fuses

No. 1/18 Wiring diagram **Audi TT**

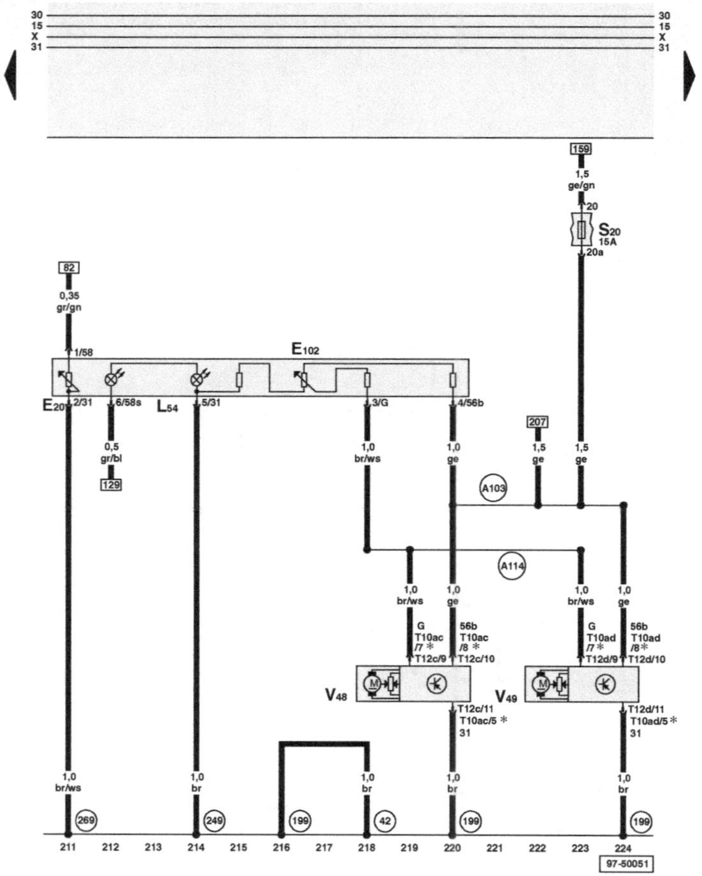

ws = white
sw = black
ro = red
br = brown
gn = green
bl = blue
gr = grey
li = violet
ge = yellow

- E20 - Instrument Panel Light Dimmer Switch
- E102 - Headlight Adjuster
- L54 - Headlight Adjuster Control Light
- S20 - Fuse
- T10ac - Connector 10 pin, on headlight, left
- T10ad - Connector 10 pin, on headlight, right
- T12c - Connector 12 pin, on headlight, left
- T12d - Connector 12 pin, on headlight, right
- V48 - Left Headlight Beam Adjusting Motor
- V49 - Right Headlight Beam Adjusting Motor

- (42) - Ground connection, on steering column
- (199) - Ground connection -3-, in instrument panel wiring harness
- (249) - Ground connection -2-, in wiring harness interior
- (269) - Ground connector (sensor ground) -1-, in instrument panel wiring harness
- (A103) - Connector -2- (56), in instrument panel wiring harness
- (A114) - Connection (Headlight adjusting), in instrument panel wiring harness
- * - from VIN Y1 010 000

Headlight Adjuster, Headlight Beam Adjusting Motors

Edition 12/02
W42.USA.5702.12.21

© 2006 BentleyPublishers.com—All Rights Reserved

Audi TT — Wiring diagram — No. 1/19

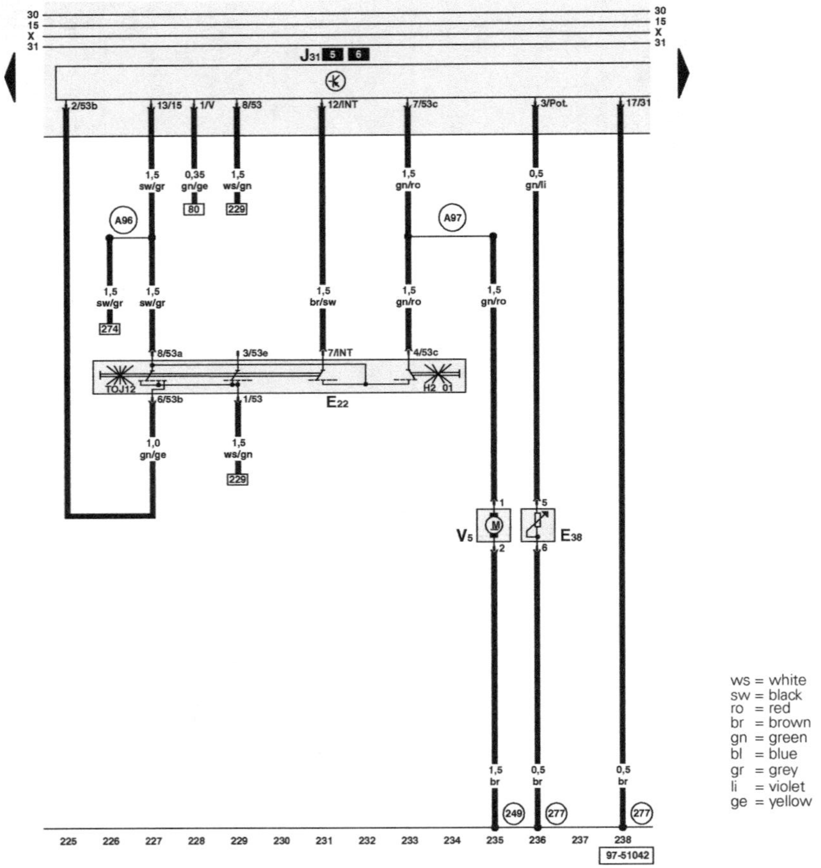

ws = white
sw = black
ro = red
br = brown
gn = green
bl = blue
gr = grey
li = violet
ge = yellow

- E22 - Windshield Wiper/Washer Switch
- E38 - Windshield Wiper Intermittent Regulator
- J31 - Wiper/Washer Intermittent Relay
- V5 - Windshield Washer Pump

- (249) - Ground connection -2-, in wiring harness interior
- (277) - Ground connector -3-, in wiring harness interior
- (A96) - Connector (53a), in instrument panel wiring harness
- (A97) - Connector (53c), in instrument panel wiring harness

Wiper/Washer Intermittent Relay, Windshield Wiper/Washer Switch, Windshield Washer Pump

No. 1/20 — Wiring diagram — Audi TT

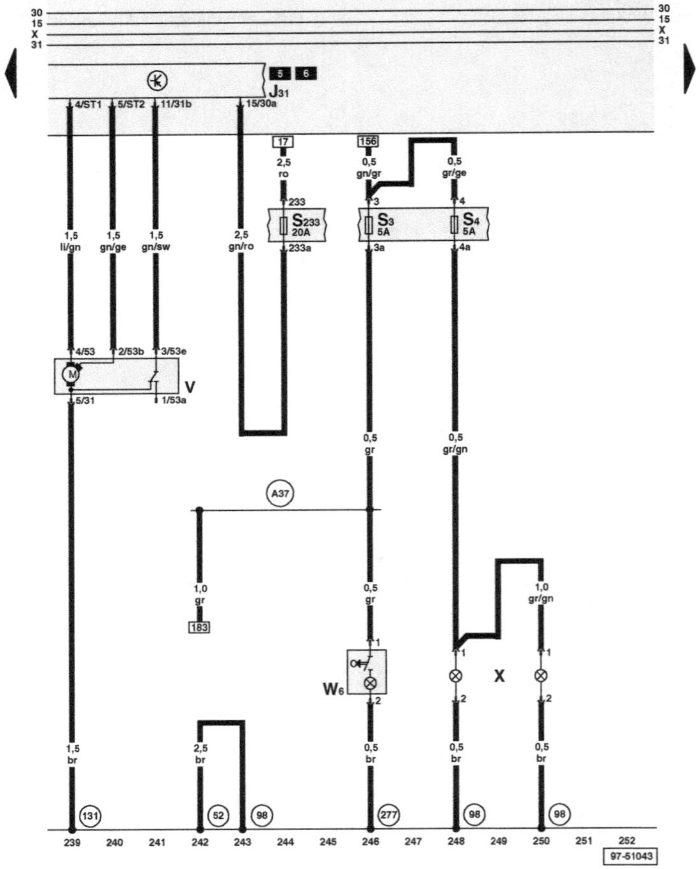

ws = white
sw = black
ro = red
br = brown
gn = green
bl = blue
gr = grey
li = violet
ge = yellow

- J31 - Wiper/Washer Intermittent Relay
- S3 - Fuse
- S4 - Fuse
- S233 - Fuse in fuse holder
- V - Windshield Wiper Motor
- W6 - Glove Compartment Light
- X - License Plate Light

- (52) - Ground connection, in rear lid, left
- (98) - Ground connection, in rear lid wiring harness
- (131) - Ground connection -2-, in engine compartment wiring harness
- (277) - Ground connector -3-, in wiring harness interior
- (A37) - Wire connection (58a), in instrument panel wiring harness

Wiper/Washer Intermittent Relay, Windshield Wiper Motor, Glove Compartment Light, License Plate Light, Fuses

© 2006 BentleyPublishers.com—All Rights Reserved

Audi TT — Wiring diagram — No. 1/21

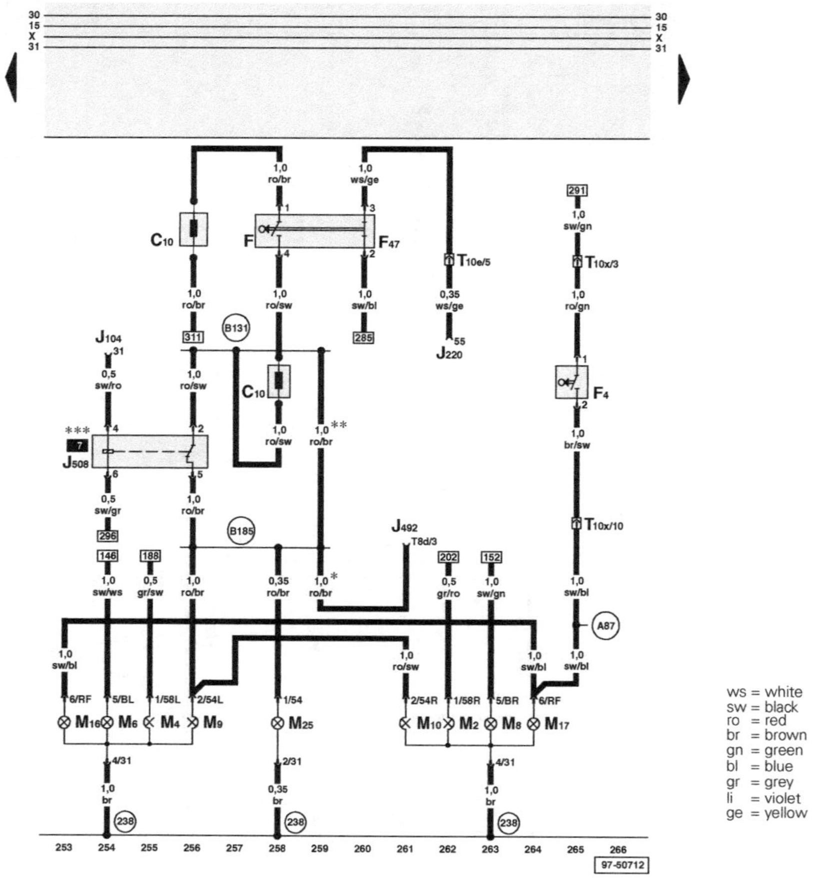

C10 - Suppressor Filter
F - Brake Light Switch
F4 - Back-Up Light Switch
F47 - Vacuum Vent Valve, Brake
J104 - ABS Control Module (w/EDL)
J220 - Motronic Engine Control Module (ECM)
J492 - Control module for all wheel drive
J508 - Brake light disable relay (ESP)
M2 - Right Tail Light
M4 - Left Tail Light
M6 - Left Rear Turn Signal Light
M8 - Right Rear Turn Signal Light
M9 - Left Brake Light
M10 - Right Brake Light
M16 - Left Back-Up Light
M17 - Right Back-Up Light
M25 - High-mount Brake Light
T8d - Connector 8 pin, on Control module for all wheel drive
T10e - Connector 10 pin, black, in E-box plenum chamber
T10x - Connector 10 pin, orange, in E-box plenum chamber

(238) - Ground connection -1-, in wiring harness interior
(A87) - Connector (RF), in instrument panel wiring harness
(B131) - Connector (54), in wiring harness interior
(B185) - Connection 2 (54) in passenger compartment wiring harness

* - Vehicles with all-wheel drive
** - Vehicles with ABS
*** - Vehicles with ABS and ESP

Edition 12/02
W42.USA.5702.12.21

Tail Light, Brake Light, Rear Fog Light

No. 1/22 — Wiring diagram — Audi TT

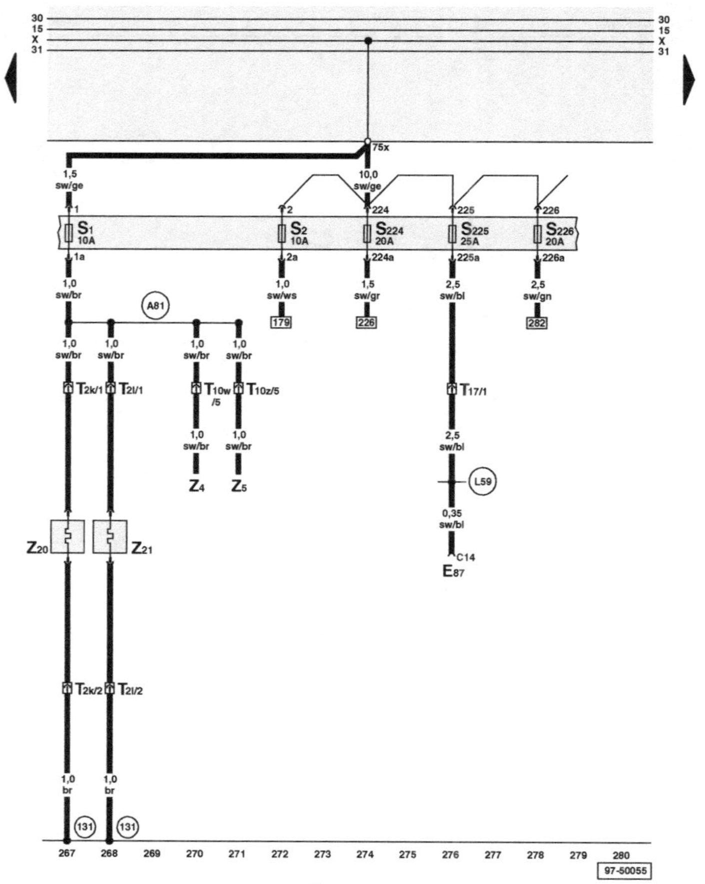

E87 - A/C Control Head
S1 - Fuse
S2 - Fuse
S224 - Fuse in fuse holder
S225 - Fuse in fuse holder
S226 - Fuse in fuse holder
T2k - Connector 2 pin, black, on engine hood
T2l - Connector 2 pin, black, on engine hood
T10w - Connector 10 pin, blue, connector station A pillar, left
T10z - Connector 10 pin, blue, connector station A pillar, right
T17 - Connector 17 pin, behind instrument panel
Z4 - Heated outside mirror, driver side
Z5 - Heated outside mirror, passenger side
Z20 - Left Washer Nozzle Heater
Z21 - Right Washer Nozzle Heater

(131) - Ground connection -2-, in engine compartment wiring harness
(A81) - Connector -2- (X), in instrument panel wiring harness
(L59) - Plus connection (30), in wiring harness heater

Washer Nozzle Heater, Fuses

Edition 12/02
W42.USA.5702.12.21

© 2006 BentleyPublishers.com—All Rights Reserved

Audi TT — Wiring diagram — No. 1/23

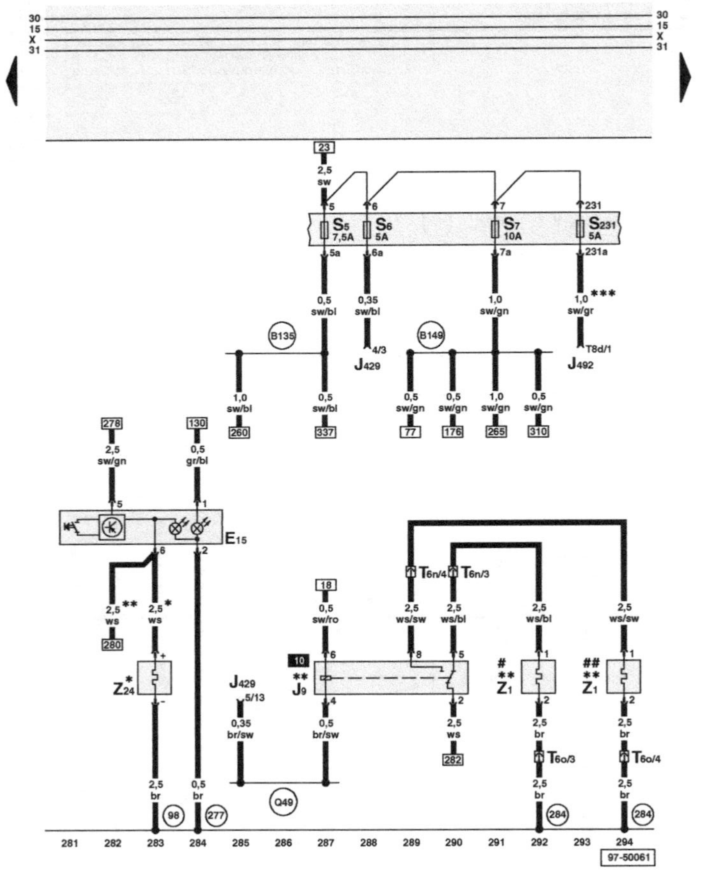

ws = white
sw = black
ro = red
br = brown
gn = green
bl = blue
gr = grey
li = violet
ge = yellow

E15 - Rear Window Defogger Switch
J9 - Rear Window Defogger Relay
J429 - Control module for central locking
J492 - Control module for all wheel drive
S5 - Fuse
S6 - Fuse
S7 - Fuse
S231 - Fuse in fuse holder
T6n - Connector 6 pin, Hardtop left
T6o - Connector 6 pin, Hardtop right
T8d - Connector 8 pin, on Control module for all wheel drive
Z1 - Heated rear window
Z24 - Rear Window Defogger With Window Antenna

(98) - Ground connection, in rear lid wiring harness
(277) - Ground connector -3-, in wiring harness interior
(284) - Ground connection -6- in wiring harness interior
(B135) - Connector (15a), in wiring harness interior
(B149) - Plus connector -2- (15a), in wiring harness interior
(Q49) - Connection (closed), in wiring harness interior

* - Coupé
** - Roadster
*** - Vehicles with all-wheel drive
# - Softtop
- Hardtop

No. 1/24 — Wiring diagram — Audi TT

ws = white
sw = black
ro = red
br = brown
gn = green
bl = blue
gr = grey
li = violet
ge = yellow

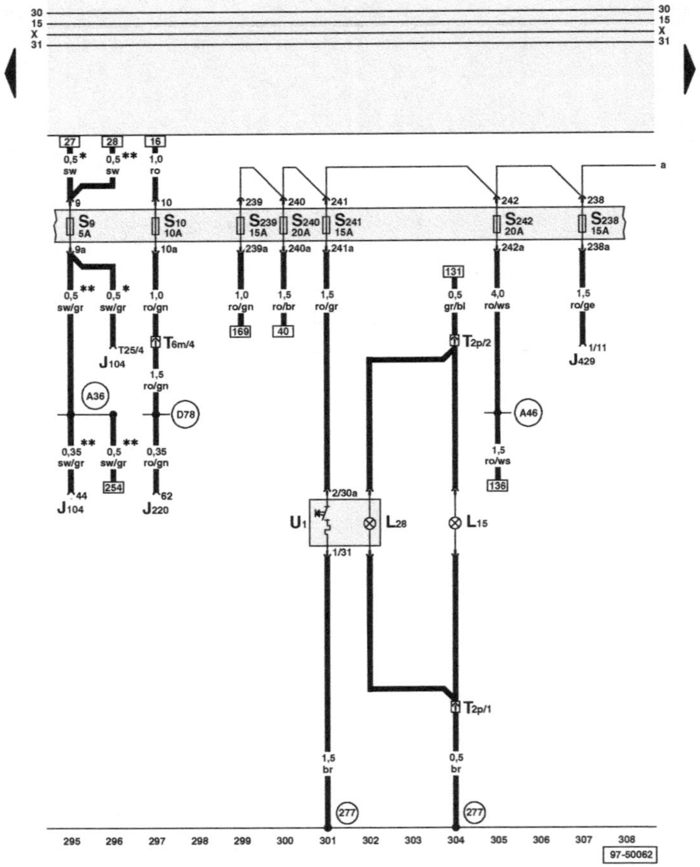

J104 - ABS Control Module (w/EDL)
J220 - Motronic Engine Control Module (ECM)
J429 - Control module for central locking
L15 - Ashtray Light
L28 - Cigarette Lighter Light
S9 - Fuse
S10 - Fuse
S238 - Fuse in fuse holder
S239 - Fuse in fuse holder
S240 - Fuse in fuse holder
S241 - Fuse in fuse holder
S242 - Fuse in fuse holder
T2p - Connector 2 pin, behind console
T6m - Connector 6 pin, brown, in E-box plenum chamber
T25 - Connector 25 pin, on ABS Control Module (w/EDL)
U1 - Cigarette Lighter

(277) - Ground connector -3-, in wiring harness interior
(A36) - Wire connection (75a), in instrument panel wiring harness
(A46) - Plus connection (30-from radio), in instrument panel wiring harness
(D78) - Plus connection -1- (30a), in engine compartment wiring harness

* - Vehicles with ABS
** - Vehicles with ABS and ESP

© 2006 BentleyPublishers.com—All Rights Reserved

Audi TT — Wiring diagram — No. 1/25

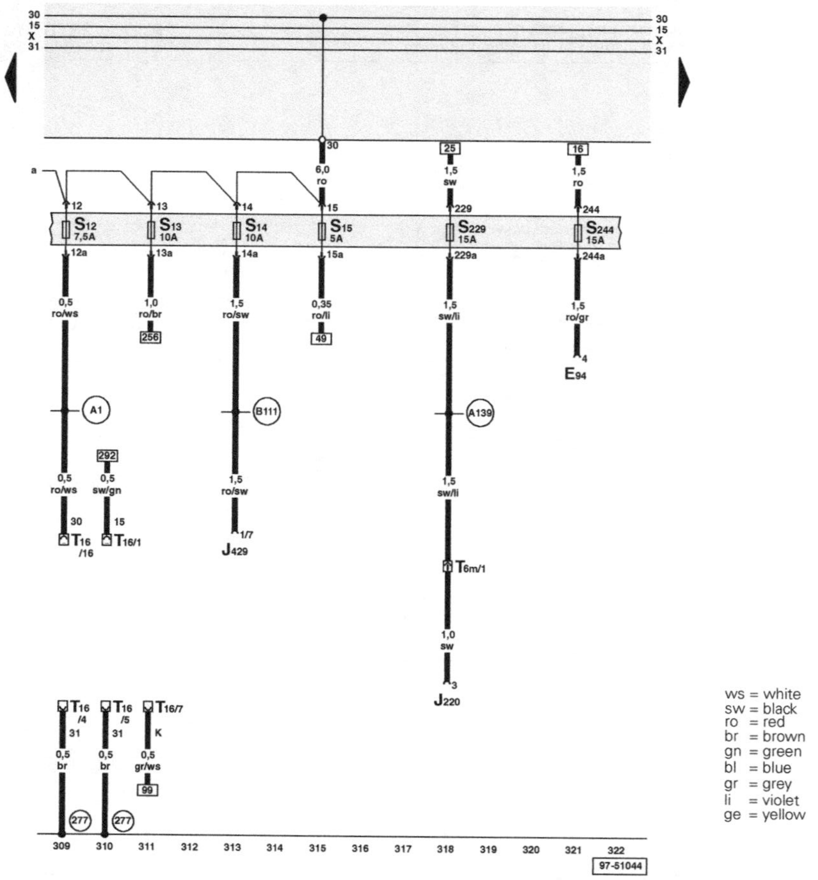

ws = white
sw = black
ro = red
br = brown
gn = green
bl = blue
gr = grey
li = violet
ge = yellow

- E94 - Adjuster for heated driver's seat
- J220 - Motronic Engine Control Module (ECM)
- J429 - Control module for central locking
- S12 - Fuse
- S13 - Fuse
- S14 - Fuse
- S15 - Fuse
- S229 - Fuse in fuse holder
- S244 - Fuse in fuse holder
- T6m - Connector 6 pin, brown, in E-box plenum chamber
- T16 - Connector 16 pin, black, Data Link Connector (DLC), below storage compartment, driver's side

- (277) - Ground connector -3-, in wiring harness interior
- (A1) - Plus connection (30a), in instrument panel wiring harness
- (A139) - Connection 3 (15), in instrument panel wiring harness
- (B111) - Connector (30a), in wiring harness interior

Edition 07.00
W42.USA.5702.04.21

Data Link Connector, Fuses

No. 1/26 — Wiring diagram — Audi TT

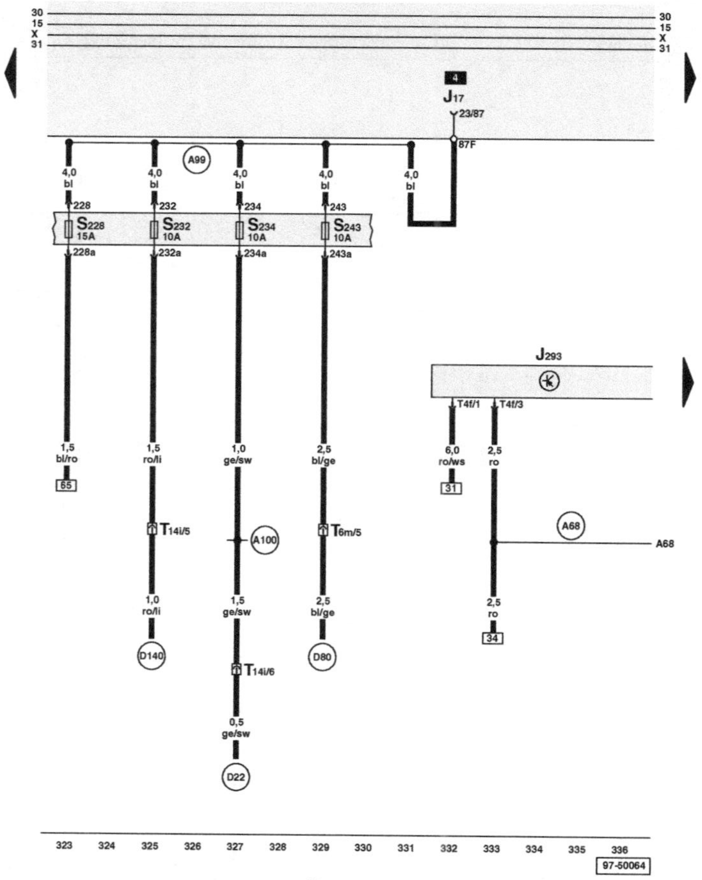

ws = white
sw = black
ro = red
br = brown
gn = green
bl = blue
gr = grey
li = violet
ge = yellow

- J17 - Fuel Pump (FP) Relay
- J293 - Coolant FC (Fan Control) Control Module, in engine compartment, left front
- S228 - Fuse in fuse holder
- S232 - Fuse in fuse holder
- S234 - Fuse in fuse holder
- S243 - Fuse in fuse holder
- T4f - Connector 4 pin, connector B on J293
- T6m - Connector 6 pin, brown, in E-box plenum chamber
- T14i - Connector 14 pin, in engine compartment, left

- (A68) - Connector (30, A/C), in instrument panel wiring harness
- (A99) - Connector -1- (87), in instrument panel wiring harness
- (A100) - Connector -2- (87), in instrument panel wiring harness
- (D22) - Connector (over fuse 234), in wiring harness front, right
- (D80) - Plus connection (87a- for EVAP system solenoid), in engine compartment wiring harness
- (D140) - Connector (injectors), in wiring harness, engine pre-wiring

Coolant (Fan Control) Control Module, Fuses

Edition 07.00
W42.USA.5702.04.21

© 2006 BentleyPublishers.com—All Rights Reserved

Audi TT Wiring diagram **No. 1/27**

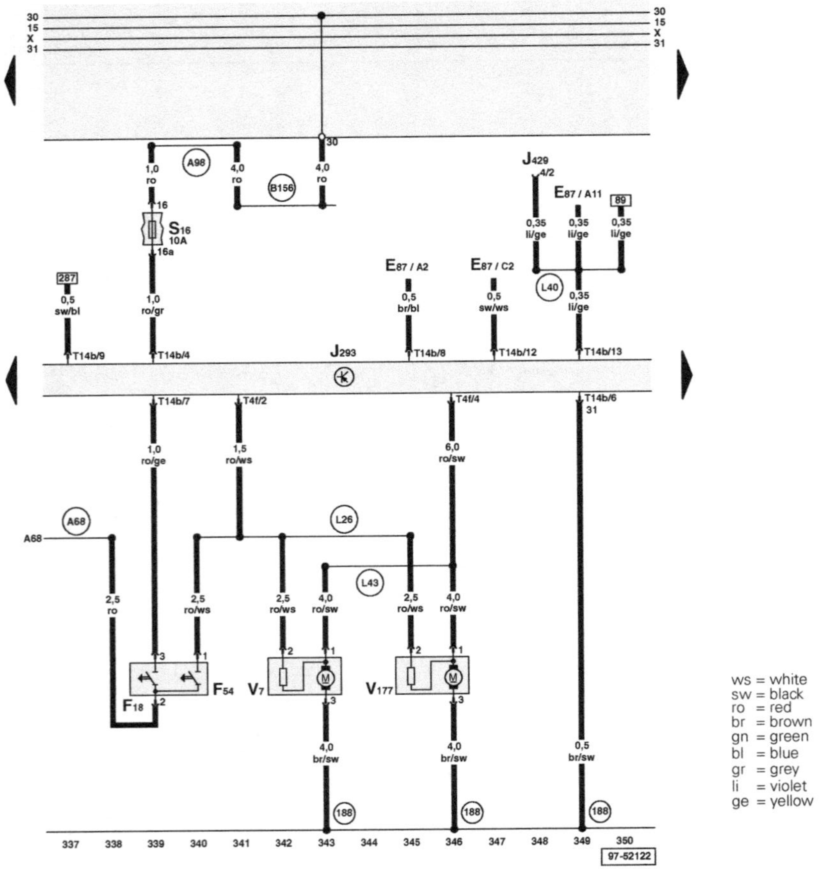

ws = white
sw = black
ro = red
br = brown
gn = green
bl = blue
gr = grey
li = violet
ge = yellow

E87 - A/C Control Head
F18 - Coolant Fan Control (FC) Thermal Switch
F54 - Coolant Fan Control (FC) Thermal Switch
J293 - Coolant FC (Fan Control) Control Module, in engine compartment, left front
J429 - Control module for central locking
S16 - Fuse
T4f - Connector 4 pin, connector B on J293
T14b - Connector 14 pin, connector A on J293
V7 - Coolant Fan
V177 - Coolant Fan -2-

(188) - Ground connection -3-, in A/C wiring harness
(A68) - Connector (30, A/C), in instrument panel wiring harness
(A98) - Plus connector -4- (30), in instrument panel wiring harness
(B156) - Plus connector -4- (30), in wiring harness interior
(L26) - Wire connection (coolant fan), in A/C-climatronic/engine compartment wiring harness
(L40) - Wire connection -2- (A/C Pressure Switch), in Climatronic wiring harness
(L43) - Wire connection (coolant fan), in A/C system-coolant fan wiring harness

Coolant (Fan Control) Control Module, Coolant Fan Control Thermal Switch, Coolant Fan

No. 1/28 Wiring diagram **Audi TT**

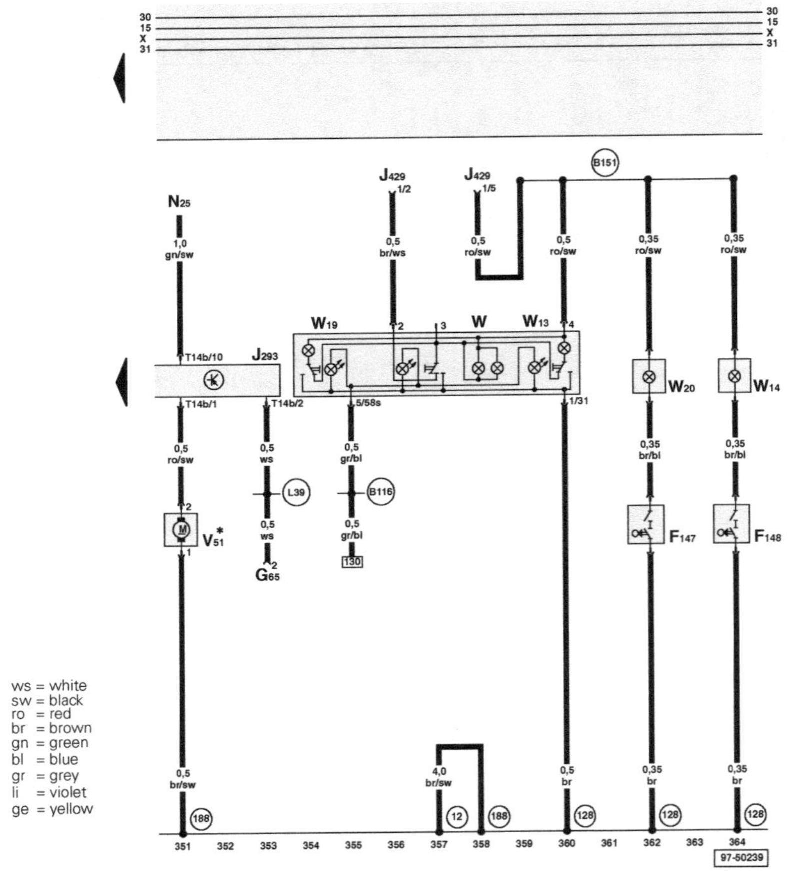

ws = white
sw = black
ro = red
br = brown
gn = green
bl = blue
gr = grey
li = violet
ge = yellow

F147 - Left Make-Up Mirror Light Switch
F148 - Right Make-Up Mirror Light Switch
G65 - High Pressure Sensor
J293 - Coolant FC (Fan Control) Control Module, in engine compartment, left front
J429 - Control module for central locking
N25 - A/C Clutch
T14b - Connector 14 pin, connector A on J293
V51 - After-Run Coolant Pump
W - Front Interior Light
W13 - Right Front Map/Reading Light
W14 - Right Make-up Mirror Light
W19 - Left Front Reading Light
W20 - Left Make-up Mirror Light

(12) - Ground connection, in engine compartment, left
(128) - Ground connection -1-, in interior light wiring harness
(188) - Ground connection -3-, in A/C wiring harness
(B116) - Connector (58s), in wiring harness interior
(B151) - Plus connector -3- (30a), in wiring harness interior
(L39) - Wire connection -1- (A/C Pressure Switch), in Climatronic wiring harness
* - Vehicles with 165 kW engine, code AMU

Interior Light, Make-Up Mirror Light, After-Run Coolant Pump

© 2006 BentleyPublishers.com—All Rights Reserved

Audi TT Wiring Diagram **No. 2/1**

1.8 L - Engine Motronic Multiport Fuel Injection (MFI)/ 132 kW, code ATC/AWP & 165 kW, code AMU

2000 m.y.

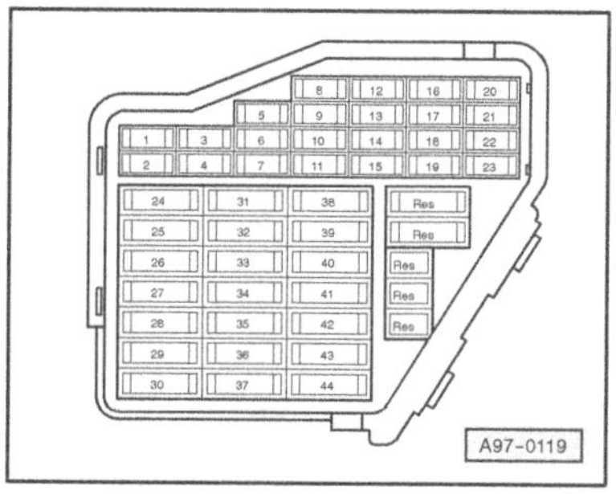

Fuse Panel

Fuse Colors:
- 30 A - Green
- 25 A - White
- 20 A - Yellow
- 15 A - Blue
- 10 A - Red
- 7,5 A - Brown
- 5 A - Beige
- 1 A - Black

Starting with fuse position 23, fuses in the fuse holder are identified with 223 in the wiring diagram.

Audi TT Wiring Diagram **No. 2/2**

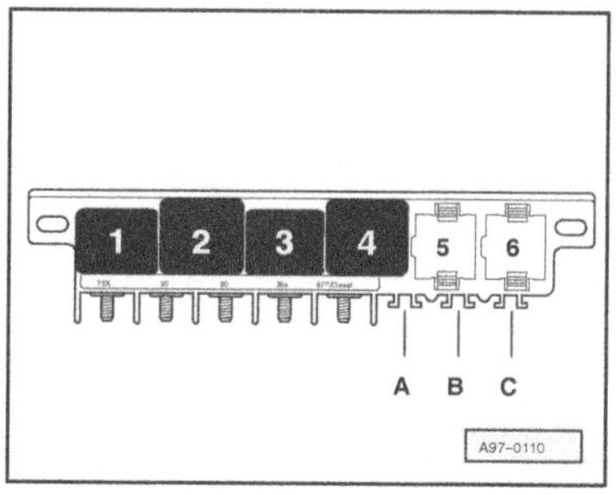

Micro Central Electric Panel

Relay Location:
- 4 - Fuel Pump (FP) Relay, J17

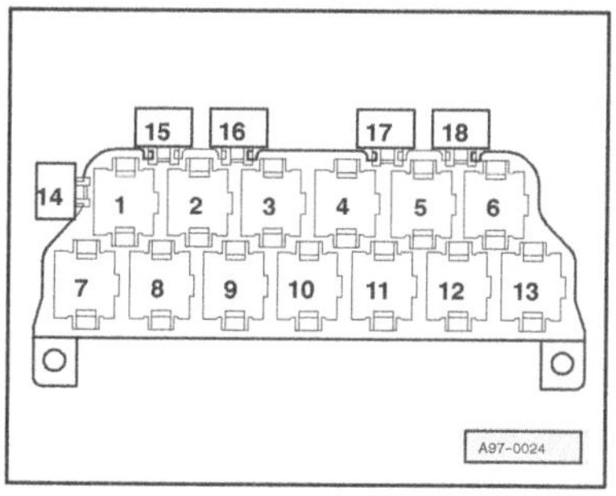

13 - Fold Relay Panel

Relay Location:
- 7 - Brake Light Disable Relay (ESP), J508

© 2006 BentleyPublishers.com—All Rights Reserved

Audi TT — Wiring Diagram — No. 2/3

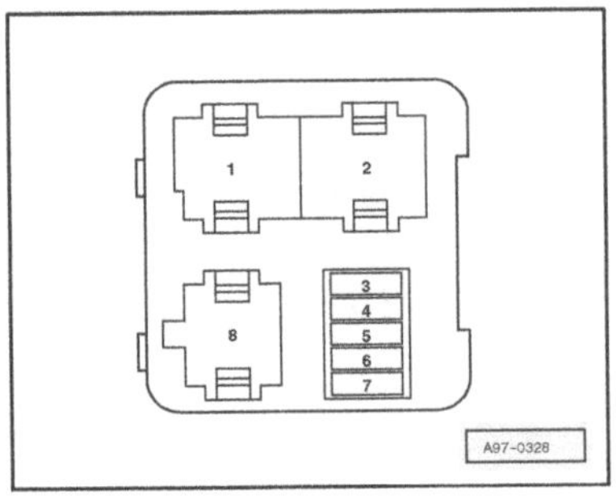

W-Box in engine compartment

Relay Location:

1 - Secondary Air Injection (AIR) Pump Relay, J299

2 - Motronic Engine Control Module (ECM) Power Supply Relay, J271

4 - Fuse for Secondary Air Pump, S130

Audi TT — Wiring Diagram — No. 2/4

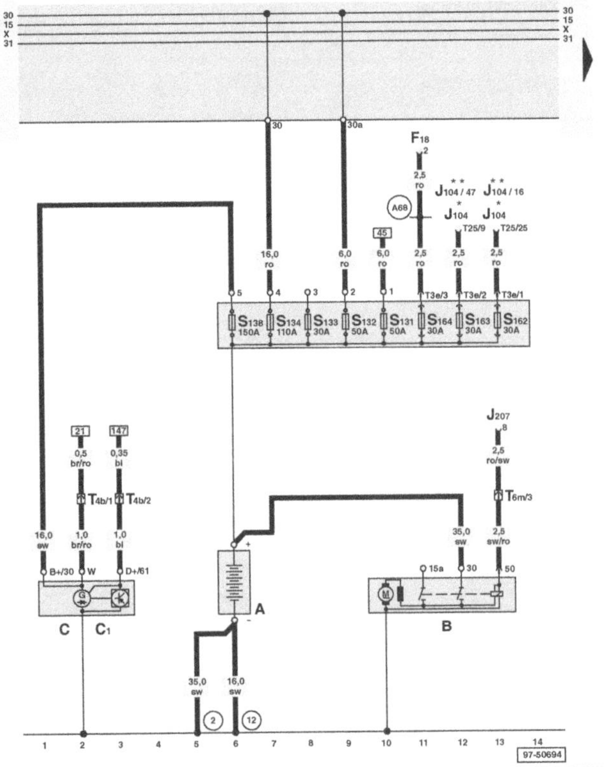

Battery, Starter, Generator, Main Fuse Box/Battery, Starting Interlock Relay

A	- Battery
B	- Starter
C	- Generator (GEN)
C1	- Voltage Regulator (VR)
F18	- Coolant Fan Control (FC) Thermal Switch
J104	- ABS Control Module (w/EDL)
J207	- Starting Interlock Relay
S131	- Safety fuse 1
S132	- Safety fuse 2
S133	- Safety fuse 3
S134	- Safety fuse 4
S138	- Safety fuse 5
S162	- Fuse -1- in fuse bracket/battery
S163	- Fuse -2- in fuse bracket/battery
S164	- Fuse -3- in fuse bracket/battery
T3e	- Connector 3 pin, black, on Main Fuse Box/Battery
T4b	- Connector 4 pin, in engine compartment, left
T6m	- Connector 6 pin, brown, in E-box plenum chamber
T25	- Connector 25 pin, on ABS Control Module (w/EDL)
(2)	- Ground strap, transmission to body
(12)	- Ground connection, in engine compartment, left
(A68)	- Connector (C15, A/C), in instrument panel wiring harness
*	- Vehicles with ABS
**	- Vehicles with ABS and ESP

ws = white
sw = black
ro = red
br = brown
gn = green
bl = blue
gr = grey
li = violet
ge = yellow
or = orange

© 2006 BentleyPublishers.com—All Rights Reserved

Engine Management, 1.8L, engine code: ATC, AWP, AMU (2000-2001)

Audi TT **Wiring Diagram** **No. 2/5**

Motronic Engine Control Module, Fuel Pump Relay, Fuel Injectors, Brake Light Switch

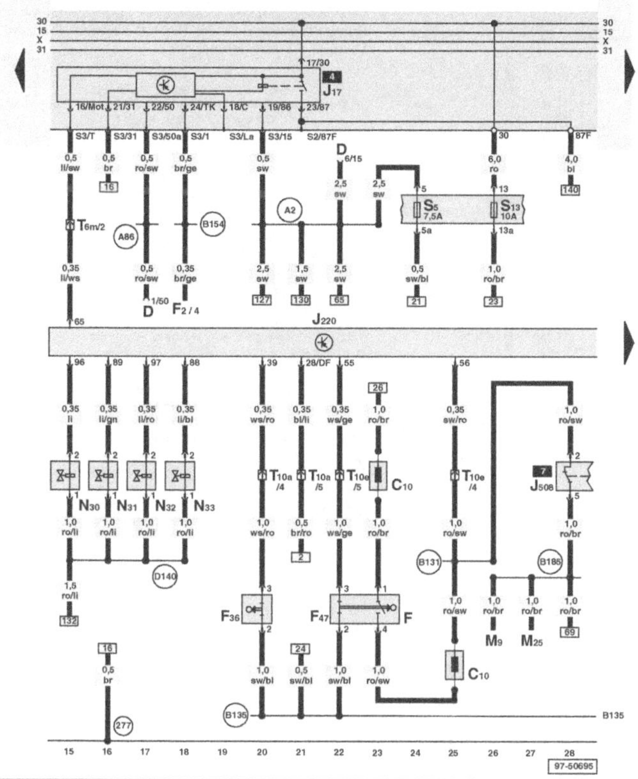

- C10 - Suppressor Filter
- D - Ignition/Starter Switch
- F - Brake Light Switch
- F2 - Door Contact Switch - Driver's side
- F36 - Clutch Vacuum Vent Valve Switch
- F47 - Brake Vacuum Vent Valve Switch for Cruise Control
- J17 - Fuel Pump (FP) Relay
- J220 - Motronic Engine Control Module (ECM)
- J508 - Brake light disable relay (ESP)
- M9 - Left Brake Light
- M25 - High-mount Brake Light
- N30 - Cylinder 1 Fuel Injector
- N31 - Cylinder 2 Fuel Injector
- N32 - Cylinder 3 Fuel Injector
- N33 - Cylinder 4 Fuel Injector
- S5 - Fuse
- S13 - Fuse
- T6m - Connector 6 pin, brown, in E-box plenum chamber
- T10a - Connector 10 pin, white, in E-box plenum chamber
- T10e - Connector 10 pin, black, in E-box plenum chamber
- (277) - Ground connector -3-, in wiring harness interior
- (A2) - Plus connection (15), in instrument panel wiring harness
- (A86) - Connection (50a), in instrument panel wiring harness
- (B131) - Connector (54), in wiring harness interior
- (B135) - Connector (15a), in wiring harness interior
- (B154) - Connector -1- (TK), in wiring harness interior
- (B185) - Connection 2 (54), in passenger compartment wiring harness
- (D140) - Connector (injectors), in wiring harness, engine pre-wiring

ws = white
sw = black
ro = red
br = brown
gn = green
bl = blue
gr = grey
li = violet
ge = yellow
or = orange

W42.USA.5702.13.21 **Edition 06/03**

Audi TT **Wiring Diagram** **No. 2/6**

Motronic Engine Control Module, Power Steering Pressure Switch, Cruise Control Switch, Throttle Position Sensor

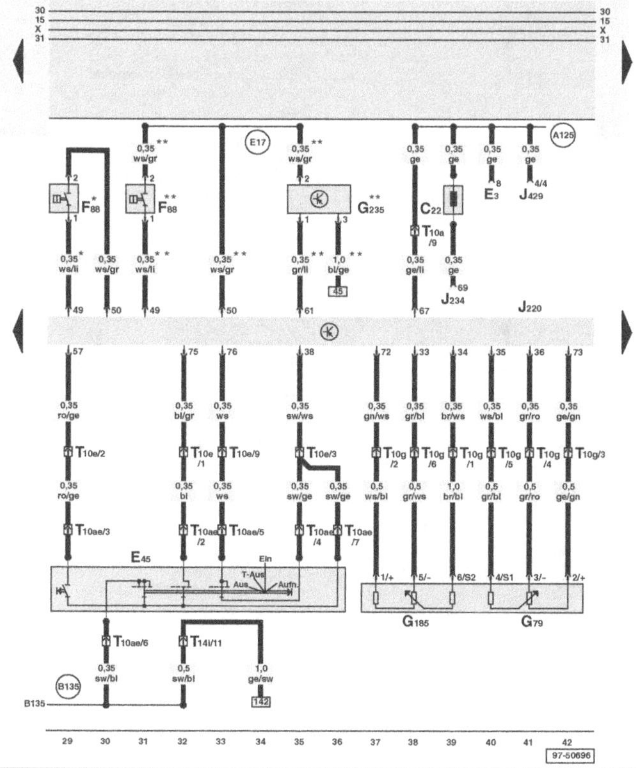

- C22 - Airbag Suppresser
- E3 - Emergency Flasher Switch
- E45 - Cruise Control Switch
- F88 - Power Steering Pressure (PSP) Switch
- G79 - Throttle Position (TP) Sensor
- G185 - Sensor -2- for accelerator pedal position
- G235 - Sensor -1- for exhaust temperature
- J220 - Motronic Engine Control Module (ECM)
- J234 - Airbag Control Module
- J429 - Control module for central locking
- T10a - Connector 10 pin, white, in E-box plenum chamber
- T10e - Connector 10 pin, black, in E-box plenum chamber
- T10g - Connector 10 pin, blue, in E-box plenum chamber
- T10ae - Connector 10 pin, on Cruise Control Switch
- T14i - Connector 14 pin, in engine compartment, left
- (A125) - Connection (crash signal) in instrument panel wiring harness
- (B135) - Connector (15a), in wiring harness interior
- (E17) - Wire connection -2-, in Motronic Multiport Fuel Injection (MFI) wiring harness
- * - only engine code ATC,AWP
- ** - only engine code AMU

ws = white
sw = black
ro = red
br = brown
gn = green
bl = blue
gr = grey
li = violet
ge = yellow
or = orange

W42.USA.5702.13.21 **Edition 06/03**

© 2006 BentleyPublishers.com—All Rights Reserved

Engine Management, 1.8L, engine code: ATC, AWP, AMU (2000-2001)

Audi TT Wiring Diagram **No. 2/7**

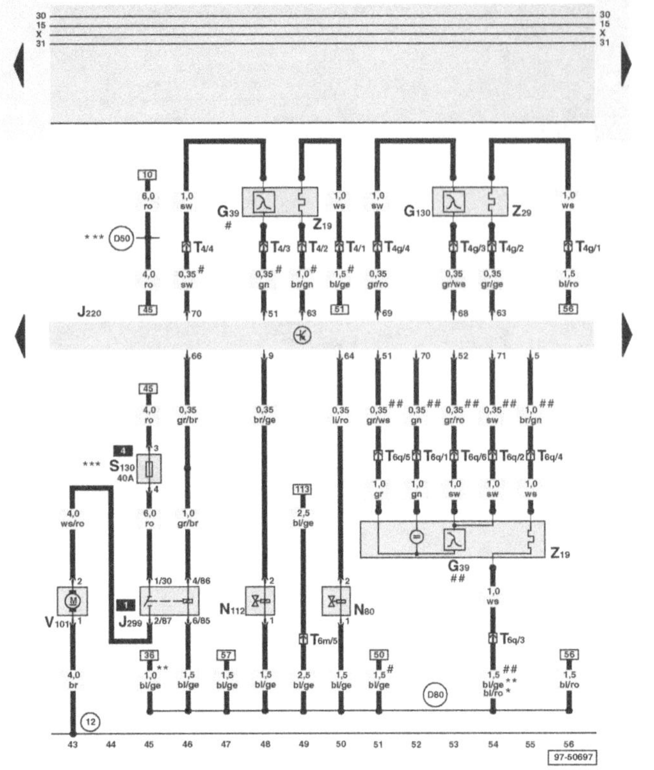

ws = white
sw = black
ro = red
br = brown
gn = green
bl = blue
gr = grey
li = violet
ge = yellow
or = orange

Motronic Engine Control Module, Secondary Air Injection Pump Relay, Evaporative Emission Canister Purge Regulator Valve, Heated Oxygen Sensors, Oxygen Sensor Heater

- G39 - Heated Oxygen Sensor (HO2S)
- G130 - Heated Oxygen Sensor (O2S) Behind Three Way Catalytic Converter (TWC)
- J220 - Motronic Engine Control Module (ECM)
- J299 - Secondary Air Injection (AIR) Pump Relay
- N80 - Evaporative Emission (EVAP) Canister Purge Regulator Valve
- N112 - Secondary Air Injection (AIR) Solenoid Valve
- S130 - Fuse for secondary air pump
- T4 - Connector 4 pin, in engine compartment, right
- T4g - Connector 4 pin, in engine compartment, right, for Oxygen Sensor behind Three Way Catalytic Converter
- T6m - Connector 6 pin, brown, in E-box plenum chamber
- T6q - Connector 6 pin, in engine compartment right
- V101 - Secondary Air Injection (AIR) Pump Motor
- Z19 - Oxygen Sensor (O2S) Heater
- Z29 - Oxygen Sensor (O2S) Heater 1 (behind Three Way Catalytic Converter (TWC))

- (12) - Ground connection, in engine compartment, left
- (D50) - Plus connection (30), in engine compartment wiring harness
- (D80) - Plus connection (87a- for EVAP system solenoid), in engine compartment wiring harness

- * - only engine code ATC, AWP
- ** - only engine code AMU
- *** - running production change from 10.2000
- # - Heated Oxygen Sensor with 4-pin connector
- ## - Heated Oxygen Sensor with 6-pin connector

W42.USA.5702.13.21 Edition 06/03

Audi TT Wiring Diagram **No. 2/8**

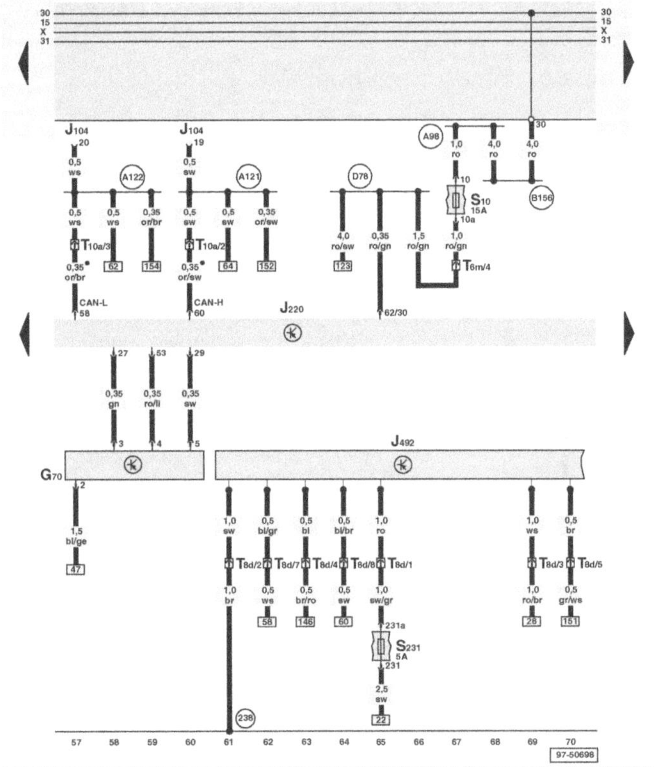

ws = white
sw = black
ro = red
br = brown
gn = green
bl = blue
gr = grey
li = violet
ge = yellow
or = orange

Motronic Engine Control Module, All-wheel Drive Control Module, Mass Air Flow (MAF) Sensor

- G70 - Mass Air Flow (MAF) Sensor
- J104 - ABS Control Module (w/EDL)
- J220 - Motronic Engine Control Module (ECM)
- J492 - All-wheel drive control module
- S10 - Fuse
- S231 - Fuse 31 in fuse holder
- T6m - Connector 6 pin, brown, in E-box plenum chamber
- T8d - Connector 8 pin, on Control module for all wheel drive
- T10a - Connector 10 pin, white, in E-box plenum chamber

- (238) - Ground connection -1-, in wiring harness interior
- (A98) - Plus connector -4- (30), in instrument panel wiring harness
- (A121) - Connection (high bus) in instrument panel wiring harness
- (A122) - Connection (low bus) in instrument panel wiring harness
- (B156) - Plus connector (30a), in wiring harness interior
- (D78) - Plus connection -1- (30a), in engine compartment wiring harness

- • - CAN-Bus (Data-Bus)

W42.USA.5702.13.21 Edition 06/03

© 2006 BentleyPublishers.com—All Rights Reserved

Engine Management, 1.8L, engine code: ATC, AWP, AMU (2000-2001)

Audi TT

Wiring Diagram

No. 2/9

Motronic Engine Control Module, Engine Speed Sensor, Charge Air Pressure Sensor, Camshaft Position Sensor 2, Knock Sensor 1

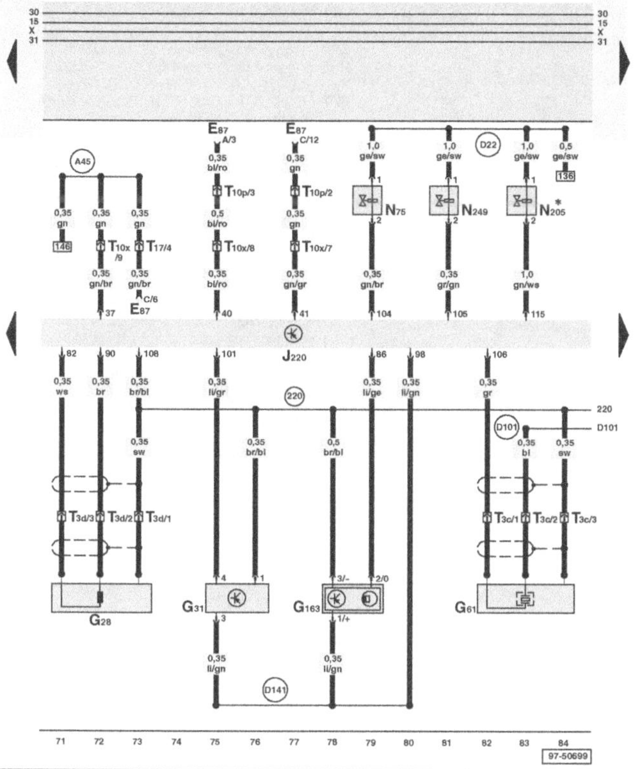

E87	- A/C Control Head
G28	- Engine Speed (RPM) Sensor
G31	- Charge Air Pressure Sensor
G61	- Knock Sensor (KS) 1
G163	- Camshaft Position (CMP) Sensor 2
J220	- Motronic Engine Control Module (ECM)
N75	- Wastegate Bypass Regulator Valve
N205	- Valve -1- for camshaft adjustment
N249	- Recirculating valve for turbocharger
T3c	- Connector 3 pin, brown, in engine compartment
T3d	- Connector 3 pin, grey, in engine compartment
T10p	- Connector 10 pin, behind instrument panel
T10x	- Connector 10 pin, orange, in E-box plenum chamber
T17	- Connector 17 pin, behind instrument panel
(220)	- Ground connection (sensor ground), in engine compartment wiring harness
(A45)	- Wire connection (RPM-signal), in instrument panel wiring harness
(D22)	- Connector (over fuse 234), in wiring harness front, right
(D101)	- Wire connection -1-, in engine compartment wiring harness
(D141)	- Connector (5V), in wiring harness, engine pre-wiring
*	- only engine code ATC, AWP

ws = white
sw = black
ro = red
br = brown
gn = green
bl = blue
gr = grey
li = violet
ge = yellow
or = orange

W42.USA.5702.13.21

Edition 06/03

Audi TT

Wiring Diagram

No. 2/10

Motronic Engine Control Module, Knock Sensor 2, Intake Air Temperature Sensor, Throttle Valve Control Module

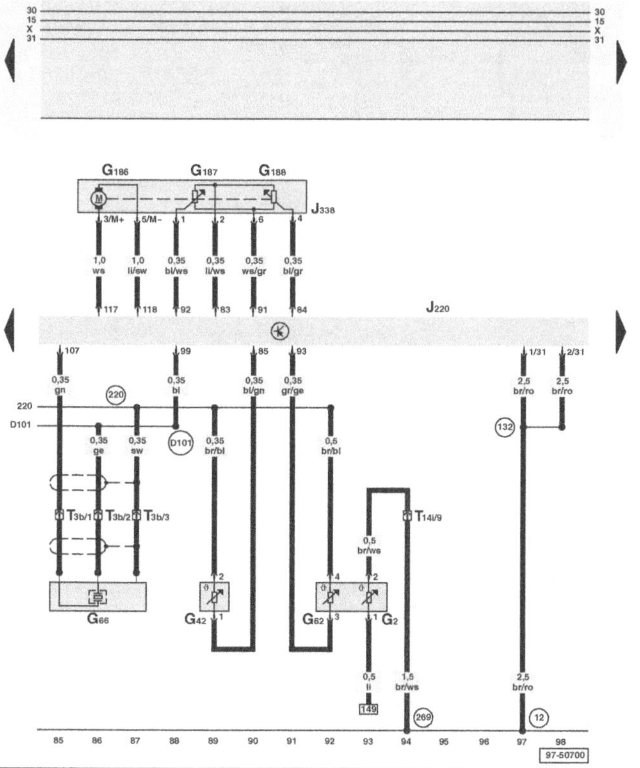

G2	- Engine Coolant Temperature (ECT) Sensor
G42	- Intake Air Temperature (IAT) Sensor
G62	- Engine Coolant Temperature (ECT) Sensor
G66	- Knock Sensor (KS) 2
G186	- Throttle drive (power accelerator actuation)
G187	- Angel sensor -1- for throttle drive (power accelerator actuation)
G188	- Angel sensor -2- for throttle drive (power accelerator actuation)
J220	- Motronic Engine Control Module (ECM)
J338	- Throttle Valve Control Module
T3b	- Connector 3 pin, black, in engine compartment
T14i	- Connector 14 pin, in engine compartment, left
(12)	- Ground connection, in engine compartment, left
(132)	- Ground connection -3-, in engine compartment wiring harness
(220)	- Ground connection (sensor ground), in engine compartment wiring harness
(269)	- Ground connector (sensor ground) -1-, in instrument panel wiring harness
(D101)	- Wire connection -1-, in engine compartment wiring harness

ws = white
sw = black
ro = red
br = brown
gn = green
bl = blue
gr = grey
li = violet
ge = yellow
or = orange

W42.USA.5702.13.21

Edition 06/03

©2006 BentleyPublishers.com—All Rights Reserved

Engine Management, 1.8L, engine code: ATC, AWP, AMU (2000-2001)

Audi TT — Wiring Diagram — No. 2/11

Motronic Engine Control Module, Ignition Coils with Power Output Stage

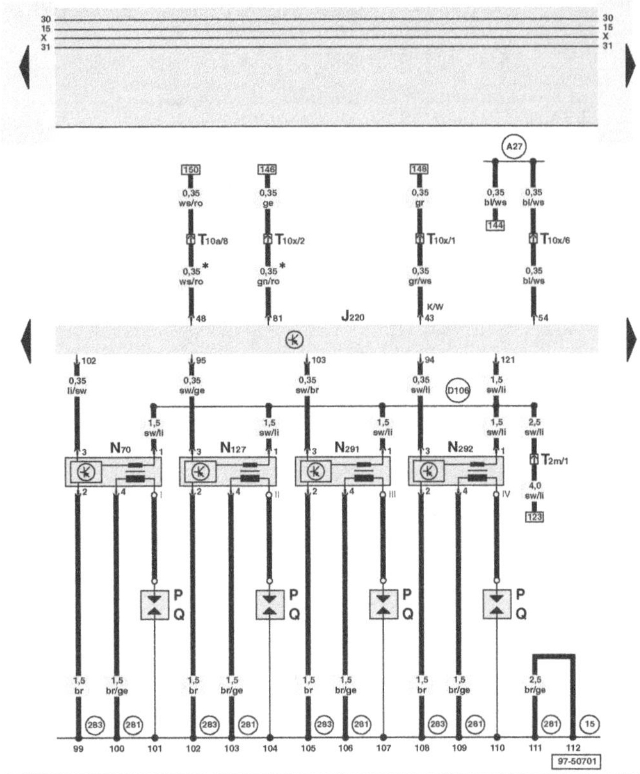

- J220 - Motronic Engine Control Module (ECM)
- N70 - Ignition Coil 1 with Power Output Stage
- N127 - Ignition Coil 2 with Power Output Stage
- N291 - Ignition Coil 3 with Power Output Stage
- N292 - Ignition Coil 4 with Power Output Stage
- P - Spark Plug Connectors
- Q - Spark Plugs
- T2m - Connector 2 pin, in engine compartment, left
- T10a - Connector 10 pin, white, in E-box plenum chamber
- T10x - Connector 10 pin, orange, in E-box plenum chamber
- (15) - Ground connection, on cylinder head
- (281) - Ground connector -1-, in wiring harness engine pre-wiring
- (283) - Ground connection 2, in wiring harness engine pre-wiring
- (A27) - Wire connection (vehicle speed signal), in instrument panel wiring harness
- (D106) - Wire connection -4-, in engine compartment wiring harness
- * - only instrument cluster without CAN-Bus

ws = white
sw = black
ro = red
br = brown
gn = green
bl = blue
gr = grey
li = violet
ge = yellow
or = orange

W42.USA.5702.13.21 Edition 06/03

Audi TT — Wiring Diagram — No. 2/12

Motronic Engine Control Module, Leak Detection Pump, Motronic Engine Control Module Power Supply Relay, Evaporative Emission Canister Purge Solenoid Valve

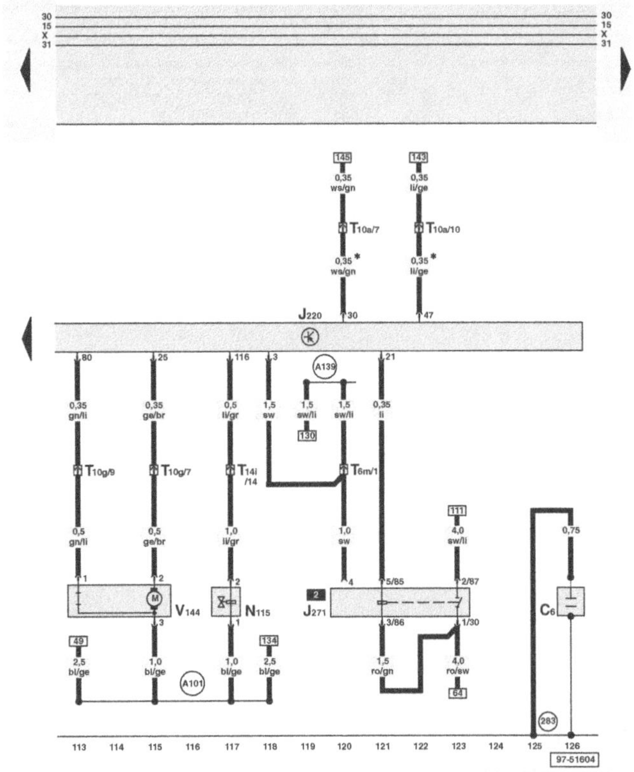

- C6 - Distributor Ignition (DI) Capacitor
- J220 - Motronic Engine Control Module (ECM)
- J271 - Motronic Engine Control Module (ECM) Power Supply Relay
- N115 - Evaporative Emission (EVAP) Canister Purge Solenoid Valve
- T6m - Connector 6 pin, brown, in E-box plenum chamber
- T10a - Connector 10 pin, white, in E-box plenum chamber
- T10g - Connector 10 pin, blue, in E-box plenum chamber
- T14i - Connector 14 pin, in engine compartment, left
- V144 - Leak Detection Pump (LDP)
- (283) - Ground connection 2, in wiring harness engine pre-wiring
- (A101) - Connector -3- (87), in instrument panel wiring harness
- (A139) - Connection 3 (15) (in instrument panel wiring harness)
- * - only instrument cluster without CAN-Bus

ws = white
sw = black
ro = red
br = brown
gn = green
bl = blue
gr = grey
li = violet
ge = yellow
or = orange

W42.USA.5702.13.21 Edition 06/03

© 2006 BentleyPublishers.com—All Rights Reserved

Engine Management, 1.8L, engine code: ATC, AWP, AMU (2000-2001)

Audi TT

Wiring Diagram **No. 2/13**

Fuses, Speedometer Vehicle Speed Sensor, Fuel Level Sensor, Fuel Pump

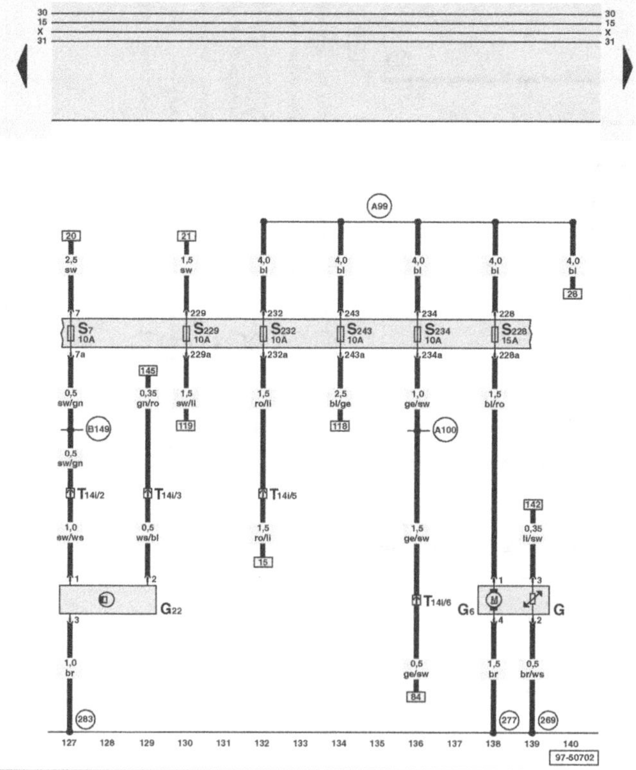

- G - Sender for fuel gauge
- G6 - Fuel Pump (FP)
- G22 - Speedometer Vehicle Speed Sensor (VSS)
- S7 - Fuse
- S228 - Fuse 28 in fuse holder
- S229 - Fuse 29 in fuse holder
- S232 - Fuse 32 in fuse holder
- S234 - Fuse 34 in fuse holder
- S243 - Fuse 43 in fuse holder
- T14i - Connector 14 pin, in engine compartment, left

- (269) - Ground connector (sensor ground) -1-, in instrument panel wiring harness
- (277) - Ground connector -3-, in wiring harness interior
- (283) - Ground connection 2, in wiring harness engine pre-wiring
- (A99) - Connector -1- (87), in instrument panel wiring harness
- (A100) - Connector -2- (87), in instrument panel wiring harness
- (B149) - Plus connector -2- (15a), in wiring harness interior

ws = white
sw = black
ro = red
br = brown
gn = green
bl = blue
gr = grey
li = violet
ge = yellow
or = orange

W42.USA.5702.13.21 Edition 06/03

Audi TT

Wiring Diagram **No. 2/14**

Instrument Cluster, Parking Brake Warning Light Switch, Oil Pressure Switch

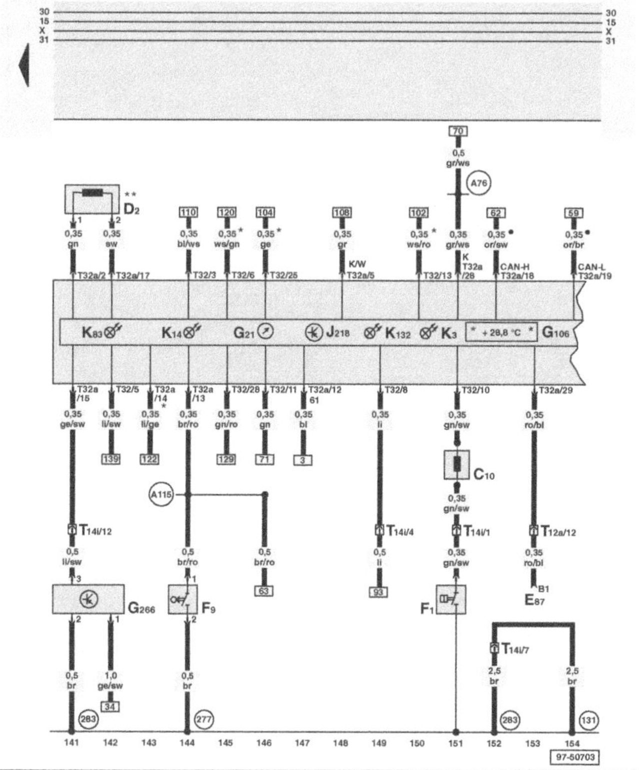

- C10 - Suppressor Filter
- D2 - Induction coil of antitheft immobilizer
- E87 - A/C Control Head
- F1 - Oil Pressure Switch
- F9 - Parking Brake Warning Light Switch
- G21 - Speedometer
- G106 - Outside Air Temperature Display
- G266 - Oil Level Thermal Sensor (SRI)
- J218 - Instrument Cluster Combination Processor
- K3 - Oil Pressure Warning Light
- K14 - Parking Brake Indicator Light
- K83 - Malfunction Indicator Lamp (MIL)
- K132 - Electronic Power Control (EPC) Warning Lamp
- T12a - Connector 12 pin, behind instrument panel, right
- T14i - Connector 14 pin, in engine compartment, left
- T32 - Connector 32 pin, blue, on instrument cluster
- T32a - Connector 32 pin, green, on instrument cluster

- (131) - Ground connection -2-, in engine compartment wiring harness
- (277) - Ground connector -3-, in wiring harness interior
- (283) - Ground connection 2, in wiring harness engine pre-wiring
- (A76) - Connector (K-diagnosis wire), in instrument panel wiring harness
- (A115) - Connection (parking brake control) in instrument panel wiring harness

- • - CAN-Bus (Data-Bus)
- * - only instrument cluster without CAN-Bus
- ** - with Anti-theft immobilizer III

ws = white
sw = black
ro = red
br = brown
gn = green
bl = blue
gr = grey
li = violet
ge = yellow
or = orange

W42.USA.5702.13.21 Edition 06/03

© 2006 BentleyPublishers.com—All Rights Reserved

Power Heated Mirrors

2000 m. y.

Fuse Panel

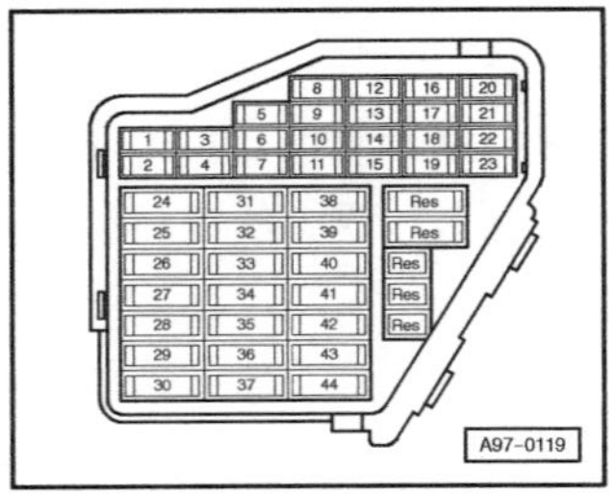

Fuse Colors:

30 A - Green
25 A - White
20 A - Yellow
15 A - Blue
10 A - Red
7,5 A - Brown
5 A - Beige

Starting with fuse position 23, fuses in the fuse holder are identified with 223 in the wiring diagram.

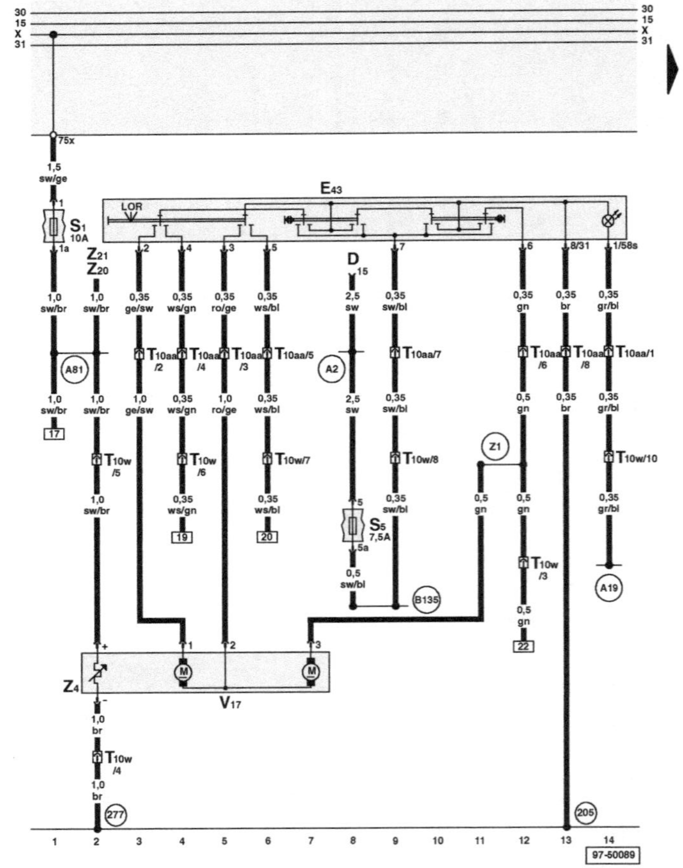

ws = white
sw = black
ro = red
br = brown
gn = green
bl = blue
gr = grey
li = violet
ge = yellow

D - Ignition/Starter Switch
E43 - Mirror Adjustment Switch
S1 - Fuse
S5 - Fuse
T10w - Connector 10 pin, blue, connector station A pillar, left
T10aa - Connector 10 pin, blue, in left door
V17 - Driver's Side Mirror Adjustment Motor
Z4 - Heated outside mirror, driver side
Z20 - Left Washer Nozzle Heater
Z21 - Right Washer Nozzle Heater

(205) - Ground connection, in driver's door wiring harness
(277) - Ground connector -3-, in wiring harness interior
(A2) - Plus connection (15), in instrument panel wiring harness
(A19) - Wire connection (58s), in instrument panel wiring harness
(A81) - Connector -2- (X), in instrument panel wiring harness
(B135) - Connector (15a), in wiring harness interior
(Z1) - Wire connection -1-, in power heated mirror wiring harness

© 2006 BentleyPublishers.com—All Rights Reserved

Audi TT — Wiring diagram — **No. 3/3**

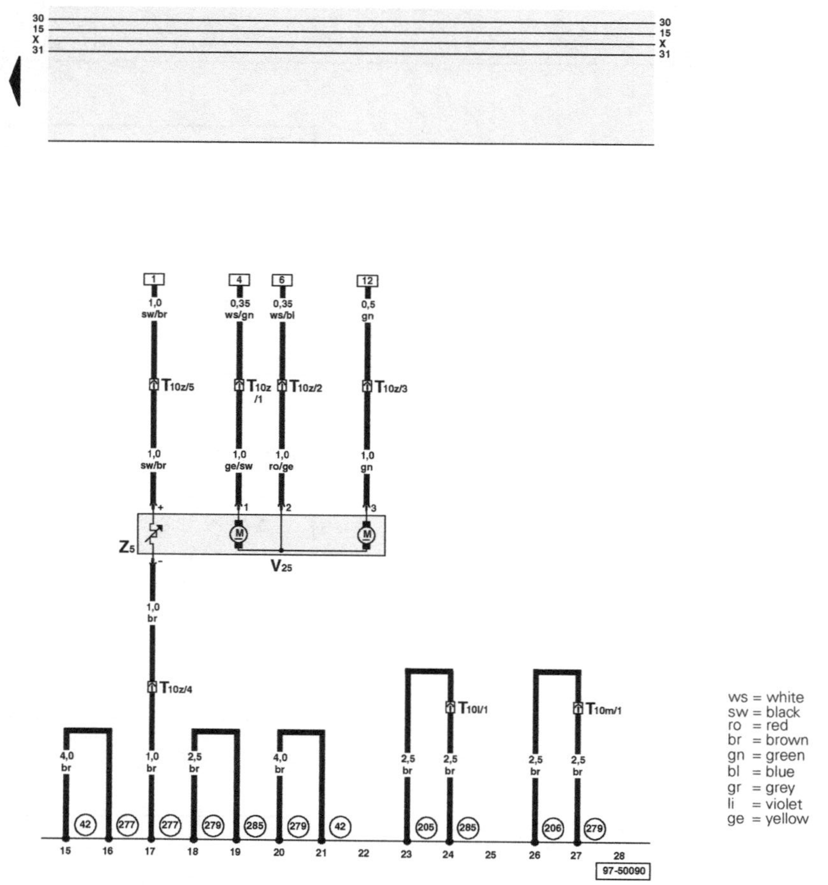

ws = white
sw = black
ro = red
br = brown
gn = green
bl = blue
gr = grey
li = violet
ge = yellow

T10l - Connector 10 pin, red, connector station A pillar, left
T10m - Connector 10 pin, red, connector station A pillar, right
T10z - Connector 10 pin, blue, connector station A pillar, right
V25 - Passenger's Side Mirror Adjustment Motor
Z5 - Heated outside mirror, passenger side

(42) - Ground connection, on steering column
(205) - Ground connection, in driver's door wiring harness
(206) - Ground connection, in passenger's door wiring harness
(277) - Ground connector -3-, in wiring harness interior
(279) - Ground connector -5-, in wiring harness interior
(285) - Ground connection -7- in wiring harness interior

Edition 01.00
W42.USA.5702.03.21

Power Heated Mirrors, Passenger's Side

THIS SPACE INTENTIONALLY LEFT BLANK

© 2006 BentleyPublishers.com—All Rights Reserved

Central Locking System / Anti Theft System / Ultra-Sound Interior Monitor (Coupé)

2000 m. y.

Fuse Panel

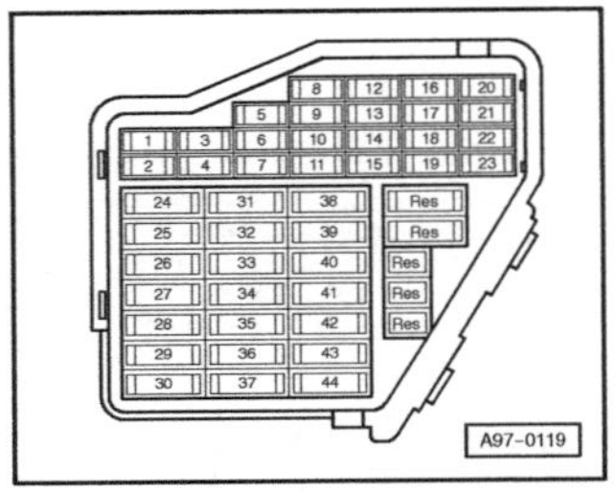

Fuse Colors:

30 A - Green
25 A - White
20 A - Yellow
15 A - Blue
10 A - Red
7,5 A - Brown
5 A - Beige

Starting with fuse position 23, fuses in the fuse holder are identified with 223 in the wiring diagram.

Micro Central Electric Panel

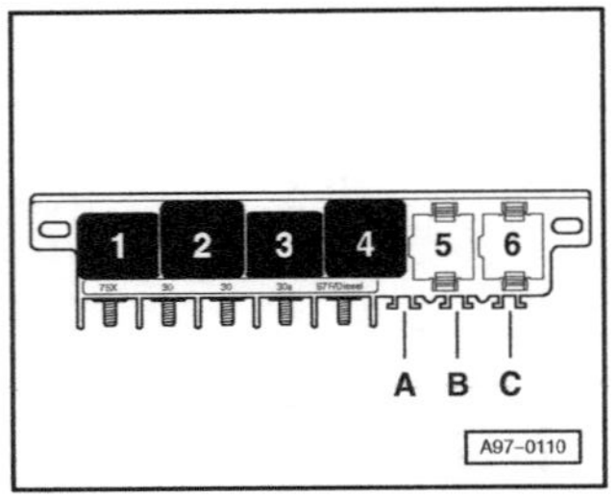

13 - Fold Relay Panel

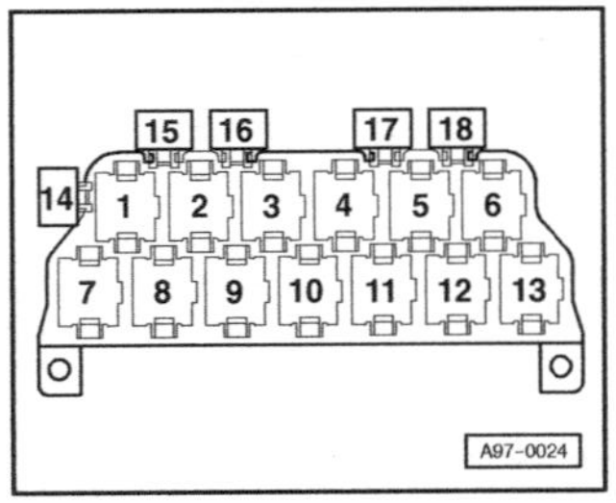

Relay Location:

5 -Starting Interlock Relay, J207

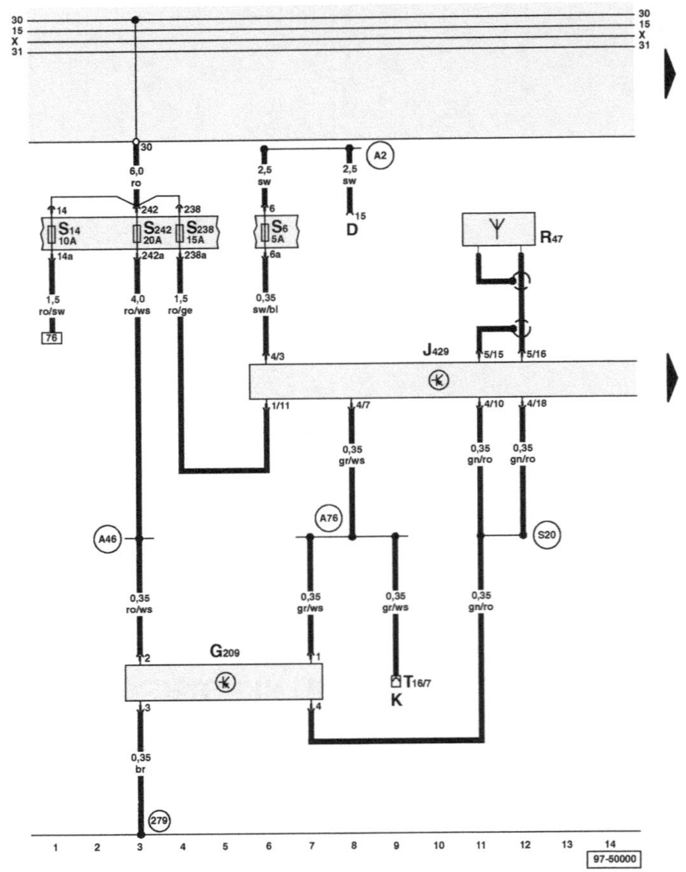

ws = white
sw = black
ro = red
br = brown
gn = green
bl = blue
gr = grey
li = violet
ge = yellow

D - Ignition/Starter Switch
G209 - Ultra-sound sensor for anti-theft warning
J429 - Control module for central locking
K - Data Link Connector (DLC) Rapid Data Transfer
R47 - Antenna wire for central locking and anti-theft warning system
S6 - Fuse
S14 - Fuse
S238 - Fuse 38 in fuse holder
S242 - Fuse 42 in fuse holder
T16 - Connector 16 pin, black, Data Link Connector (DLC), under steering column, left

(279) - Ground connector -5-, in wiring harness interior
(A2) - Plus connection (15), in instrument panel wiring harness
(A46) - Plus connection (30a), in instrument panel wiring harness
(A76) - Connector (K-diagnosis wire), in instrument panel wiring harness
(S20) - Connector -4- in wiring harness central locking

Control Module for Central Locking, Fuses, Ultra-Sound Sensor for Anti-theft Warning

© 2006 BentleyPublishers.com—All Rights Reserved

Central locking, Anti-theft, coupe (2000-2001)

Audi TT Wiring diagram **No. 4/3**

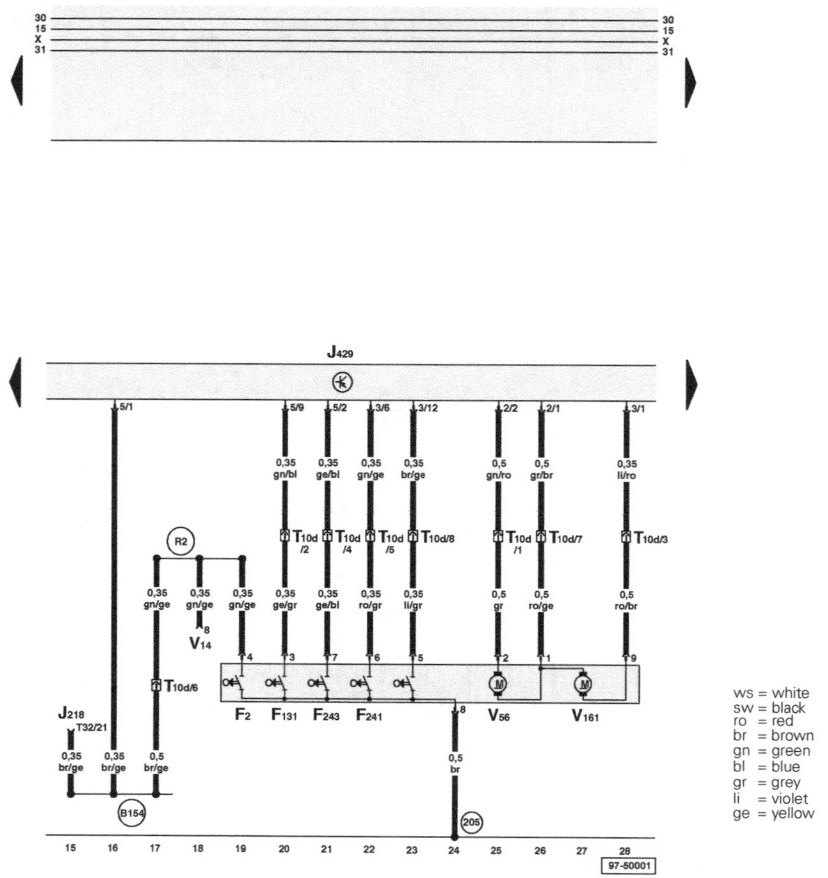

ws = white
sw = black
ro = red
br = brown
gn = green
bl = blue
gr = grey
li = violet
ge = yellow

F2 - Door Contact Switch - Driver's side
F131 - Left Front Central Locking System Activator/Switch
F241 - Contact switch for lock cylinder, driver side
F243 - Adjusting element for central locking (safe), driver door
J218 - Instrument Cluster Combination Processor
J429 - Control module for central locking
T10d - Connector 10 pin, brown, connector station A pillar, left
T32 - Connector 32 pin, blue, on instrument cluster
V14 - Left Window Motor
V56 - Driver's Door Central Locking System Motor
V161 - Motor for central locking (safe), driver door

(205) - Ground connection, in driver's door wiring harness

(B154) - Connector -1- (TK), in wiring harness interior

(R2) - Wire connection (door contact switch, left), in power window/central locking system and door contact switch wiring harness

Edition 01.00
W42.USA.5702.03.21

Control Module for Central Locking,
Lock Unit for Central Locking Driver's Side

No. 4/4 Wiring diagram **Audi TT**

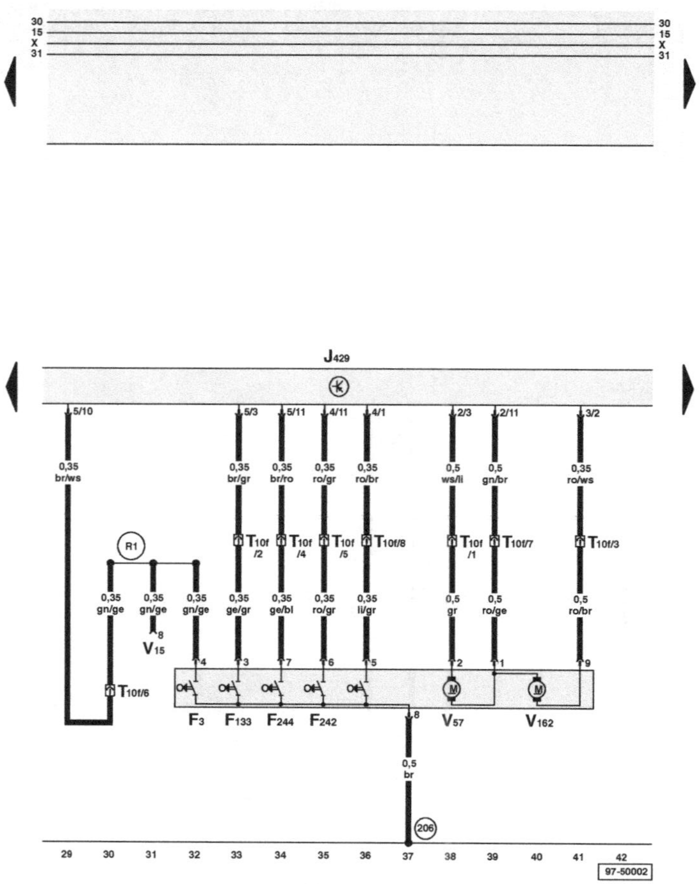

ws = white
sw = black
ro = red
br = brown
gn = green
bl = blue
gr = grey
li = violet
ge = yellow

F3 - Door Contact Switch - Passenger side
F133 - Right Front Central Locking System Activator/Switch
F242 - Contact switch for lock cylinder, passenger side
F244 - Adjusting element for central locking (safe), passenger door
J429 - Control module for central locking
T10f - Connector 10 pin, brown, connector station A pillar, right
V15 - Right Window Motor
V57 - Passenger's Door Central Locking System Motor
V162 - Motor for central locking (safe), passenger door

(206) - Ground connection, in passenger's door wiring harness

(R1) - Wire connection (door contact switch, right), in power window/central locking system and door contact switch wiring harness

Control Module for Central Locking,
Lock Unit for Central Locking Passenger's Side

Edition 01.00
W42.USA.5702.03.21

©2006 BentleyPublishers.com—All Rights Reserved

Central locking, Anti-theft, coupe (2000-2001)

Audi TT — Wiring diagram — No. 4/5

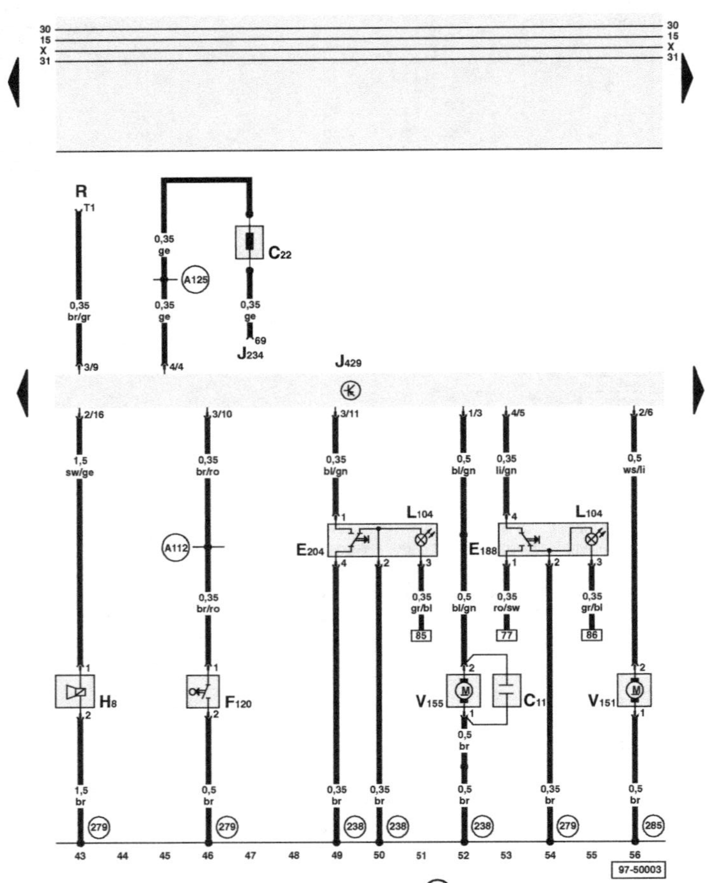

ws = white
sw = black
ro = red
br = brown
gn = green
bl = blue
gr = grey
li = violet
ge = yellow

C11 - Condensor
C22 - Airbag Suppresser
E188 - Switch for remote unlock, rear lid
E204 - Switch for remote/fuel tank door
F120 - Hood Alarm Switch
H8 - Alarm Horn
J234 - Airbag Control Module
J429 - Control module for central locking
L104 - Illumination for switch remote unlock
R - Radio
T1 - Connector single, brown, on radio
V151 - Motor for remote unlock rear lid
V155 - Motor for fuel tank lid unlock

(238) - Ground connection -1-, in wiring harness interior

(279) - Ground connector -5-, in wiring harness interior

(285) - Ground connection -7- in wiring harness interior

(A112) - Connection (Hood alarm switch), in instrument panel wiring harness

(A125) - Connection (crash signal) in instrument panel wiring harness

Control Module for Central Locking, Switch for Remote Unlock rear lid, Switch for Remote/Fuel Tank Door

No. 4/6 — Wiring diagram — Audi TT

ws = white
sw = black
ro = red
br = brown
gn = green
bl = blue
gr = grey
li = violet
ge = yellow

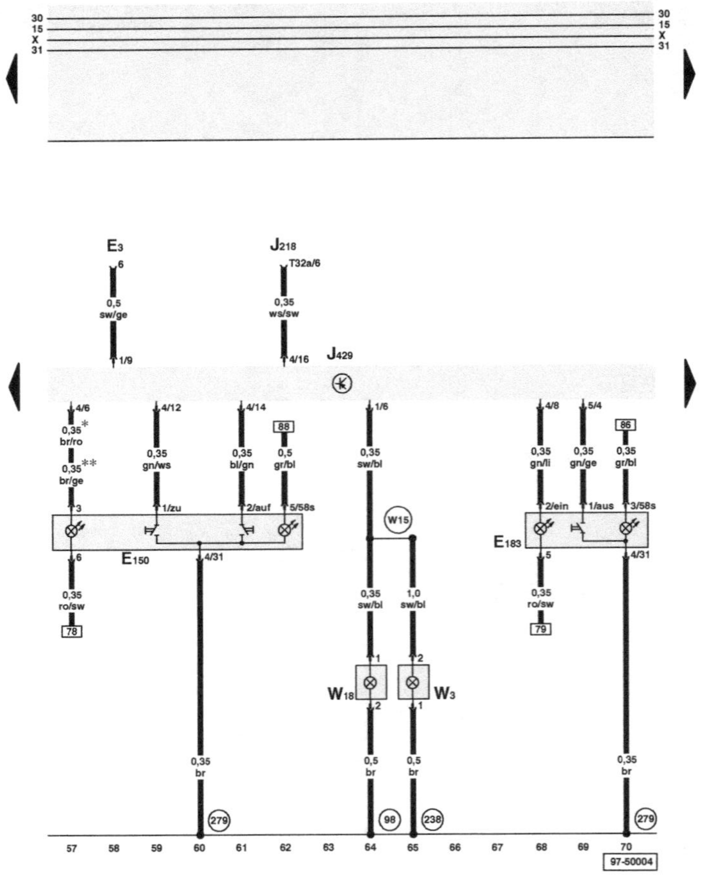

E3 - Emergency Flasher Switch
E150 - Switch for interior lock, driver side
E183 - Switch for passenger compartment monitoring
J218 - Instrument Cluster Combination Processor
J429 - Control module for central locking
T32a - Connector 32 pin, green, on instrument cluster
W3 - Luggage Compartment Light
W18 - Left Luggage Compartment Light in rear lid

(98) - Ground connection, in rear lid wiring harness

(238) - Ground connection -1-, in wiring harness interior

(279) - Ground connector -5-, in wiring harness interior

(W15) - Connection -1- (luggage compartment light), in rear wiring harness

* - 2000 m.y.
** - 2001 m.y.

Control Module for Central Locking, Switch for Interior Lock Driver Side, Switch for Passenger Compartment Monitoring, Luggage Compartment Light

© 2006 BentleyPublishers.com—All Rights Reserved

Audi TT Wiring diagram **No. 4/7**

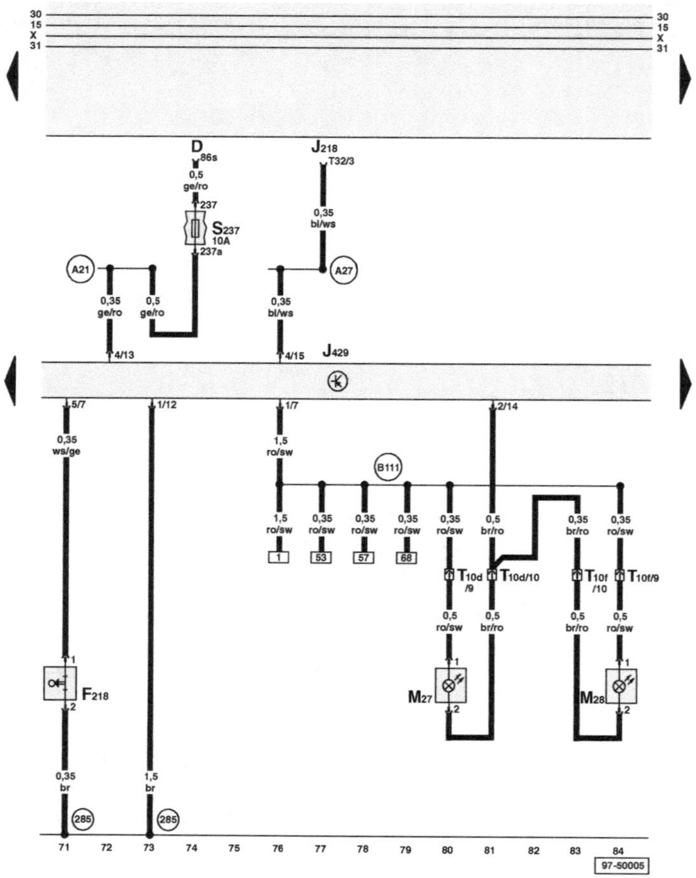

ws = white
sw = black
ro = red
br = brown
gn = green
bl = blue
gr = grey
li = violet
ge = yellow

D - Ignition/Starter Switch
F218 - Switch for central locking, trunk lid
J218 - Instrument Cluster Combination Processor
J429 - Control module for central locking
M27 - Left Door Warning Light
M28 - Right Door Warning Light
S237 - Fuse 37 in fuse holder
T10d - Connector 10 pin, brown, connector station A pillar, left
T10f - Connector 10 pin, brown, connector station A pillar, right
T32 - Connector 32 pin, blue, on instrument cluster

(285) - Ground connection -7- in wiring harness interior
(A21) - Wire connection (86s), in instrument panel wiring harness
(A27) - Wire connection (vehicle speed signal), in instrument panel wiring harness
(B111) - Connector (30a), in wiring harness interior

Edition 07.00
W42.USA.5702.04.21

Control Module for Central Locking, Door Warning Lights

No. 4/8 Wiring diagram **Audi TT**

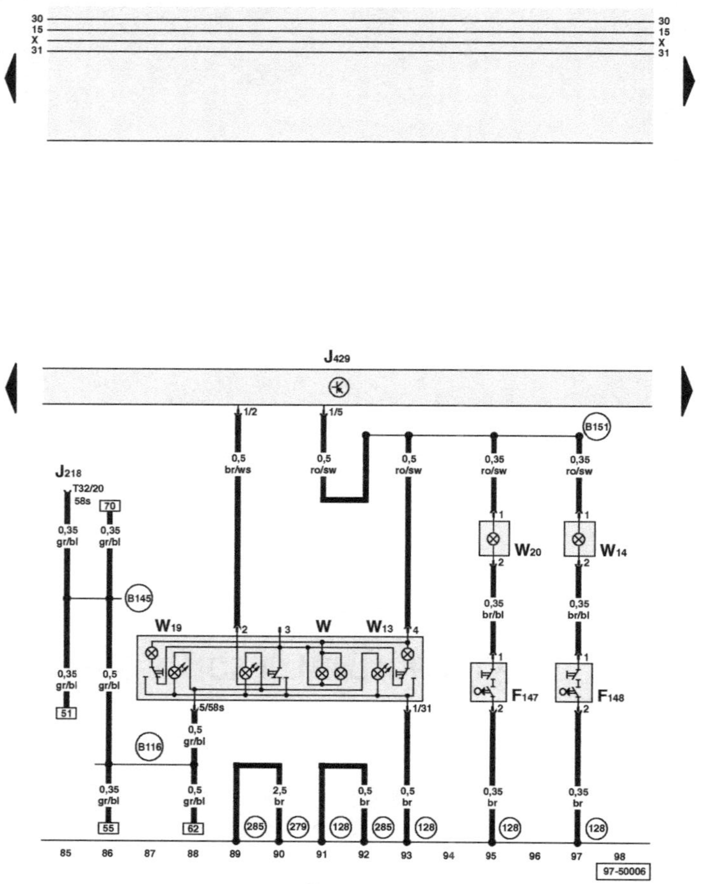

ws = white
sw = black
ro = red
br = brown
gn = green
bl = blue
gr = grey
li = violet
ge = yellow

F147 - Left Make-Up Mirror Light Switch
F148 - Right Make-Up Mirror Light Switch
J218 - Instrument Cluster Combination Processor
J429 - Control module for central locking
T32 - Connector 32 pin, blue, on instrument cluster
W - Front Interior Light
W13 - Right Front Map/Reading Light
W14 - Right Make-up Mirror Light
W19 - Left Front Reading Light
W20 - Left Make-up Mirror Light

(128) - Ground connection -1-, in interior light wiring harness
(279) - Ground connector -5-, in wiring harness interior
(285) - Ground connection -7- in wiring harness interior
(B116) - Connector (58s), in wiring harness interior
(B145) - Plus connector (58s), in wiring harness interior
(B151) - Plus connector -3- (30a), in wiring harness interior

Control Module for Central Locking, Front Interior Light, Reading Lights, Make-Up Mirror Lights

Edition 07.00
W42.USA.5702.04.21

©2006 BentleyPublishers.com—All Rights Reserved

Audi TT Wiring diagram **No. 4/9**

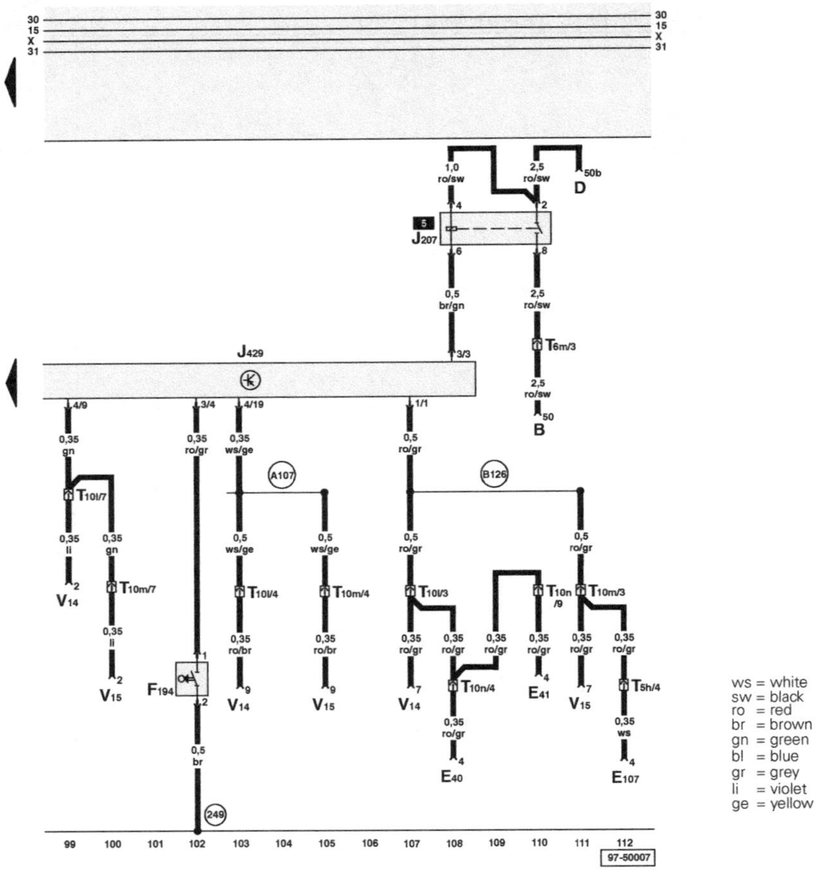

ws = white
sw = black
ro = red
br = brown
gn = green
bl = blue
gr = grey
li = violet
ge = yellow

B - Starter
D - Ignition/Starter Switch
E40 - Left Front Window Switch
E41 - Right Front Window Switch
E107 - Switch for window regulator, in passenger door
F194 - Clutch Pedal Position (CPP) Switch
J207 - Starting Interlock Relay
J429 - Control module for central locking
T5h - Connector 5 pin, in right door
T6m - Connector 6 pin, brown, in E-box plenum chamber
T10l - Connector 10 pin, red, connector station A pillar, left
T10m - Connector 10 pin, red, connector station A pillar, right
T10n - Connector 10 pin, in left door
V14 - Left Window Motor
V15 - Right Window Motor

(249) - Ground connection -2-, in wiring harness interior
(A107) - Connector (comfort close), in instrument panel wiring harness
(B126) - Connector (87, driver's door) , in wiring harness interior

Edition 07.00
W42.USA.5702.04.21

Control Module for Central Locking, Starting Interlock Relay

THIS SPACE INTENTIONALLY LEFT BLANK

© 2006 BentleyPublishers.com—All Rights Reserved

Power Windows

2000 m. y.

Fuse Panel

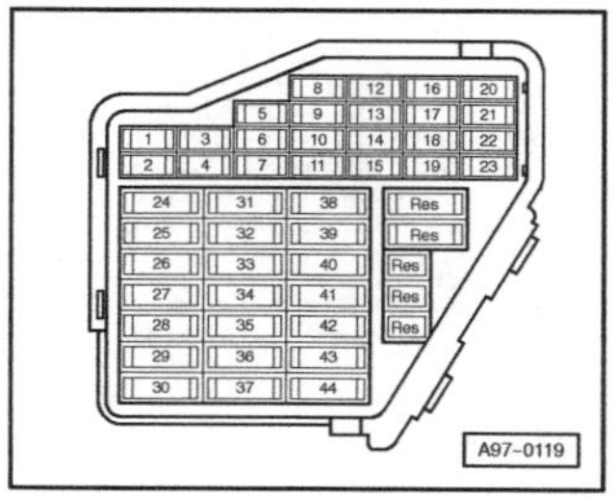

Fuse Colors:

30 A - Green
25 A - White
20 A - Yellow
15 A - Blue
10 A - Red
7,5 A - Brown
5 A - Beige

Starting with fuse position 23, fuses in the fuse holder are identified with 223 in the wiring diagram.

Micro Central Electric Panel

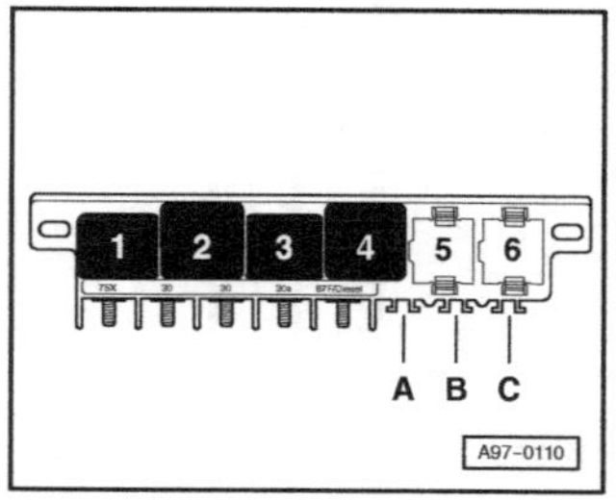

Relay Location:

C -Power Window Circuit Breaker Rear, S43

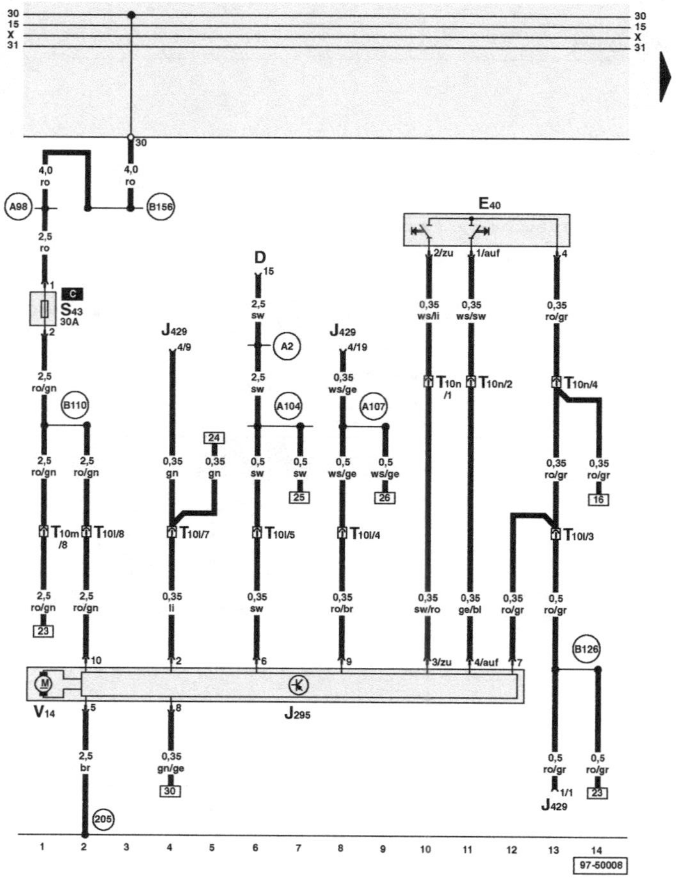

ws = white
sw = black
ro = red
br = brown
gn = green
bl = blue
gr = grey
li = violet
ge = yellow

D - Ignition/Starter Switch
E40 - Left Front Window Switch
J295 - Left Front Power Window Control Module
J429 - Control module for central locking
S43 - Power Window Circuit Breaker Rear
T10l - Connector 10 pin, red, connector station A pillar, left
T10m - Connector 10 pin, red, connector station A pillar, right
T10n - Connector 10 pin, in left door
V14 - Left Window Motor

(205) - Ground connection, in driver's door wiring harness
(A2) - Plus connection (15), in instrument panel wiring harness
(A98) - Plus connector -4- (30), in instrument panel wiring harness
(A104) - Plus connector -2- (15), in instrument panel wiring harness
(A107) - Connector (comfort close), in instrument panel wiring harness
(B110) - Connection (30, window regulator), in wiring harness interior
(B126) - Connector (87, driver's door), in wiring harness interior
(B156) - Plus connector (30a), in wiring harness interior

© 2006 BentleyPublishers.com—All Rights Reserved

Audi TT — Wiring diagram — No. 5/3

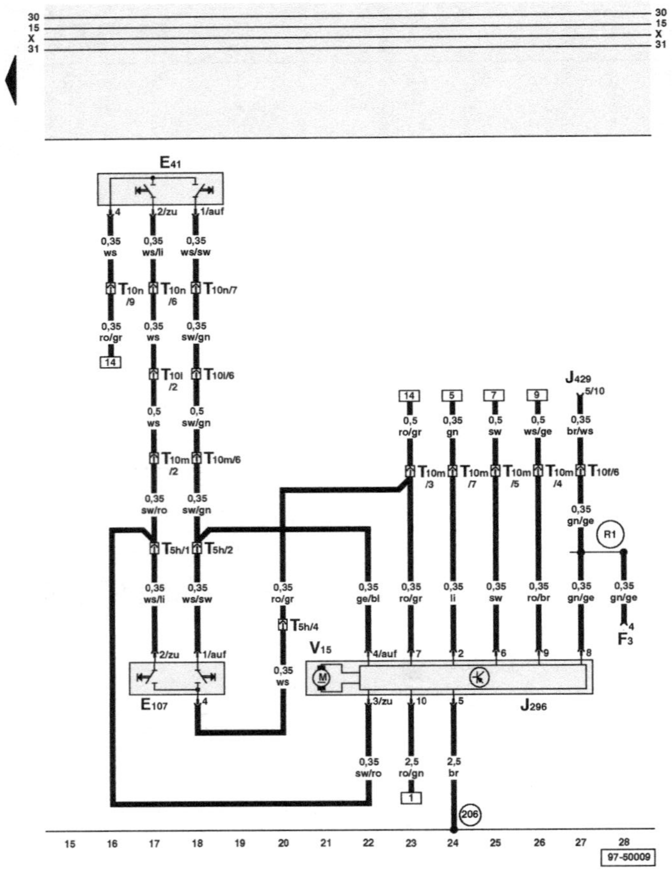

ws = white
sw = black
ro = red
br = brown
gn = green
bl = blue
gr = grey
li = violet
ge = yellow

E41 - Right Front Window Switch
E107 - Switch for window regulator, in passenger door
F3 - Door Contact Switch - Passenger side
J296 - Right Front Power Window Control Module
J429 - Control module for central locking
T5h - Connector 5 pin, in right door
T10f - Connector 10 pin, brown, connector station A pillar, right
T10l - Connector 10 pin, red, connector station A pillar, left
T10m - Connector 10 pin, red, connector station A pillar, right
T10n - Connector 10 pin, in left door
V15 - Right Window Motor

(206) - Ground connection, in passenger's door wiring harness

(R1) - Wire connection (door contact switch, right), in power window/central locking system and door contact switchwiring harness

Edition 07.00
W42.USA.5702.04.21

Power Windows, Right Front Power Window Control Module, Right Front Window Switch

No. 5/4 — Wiring diagram — Audi TT

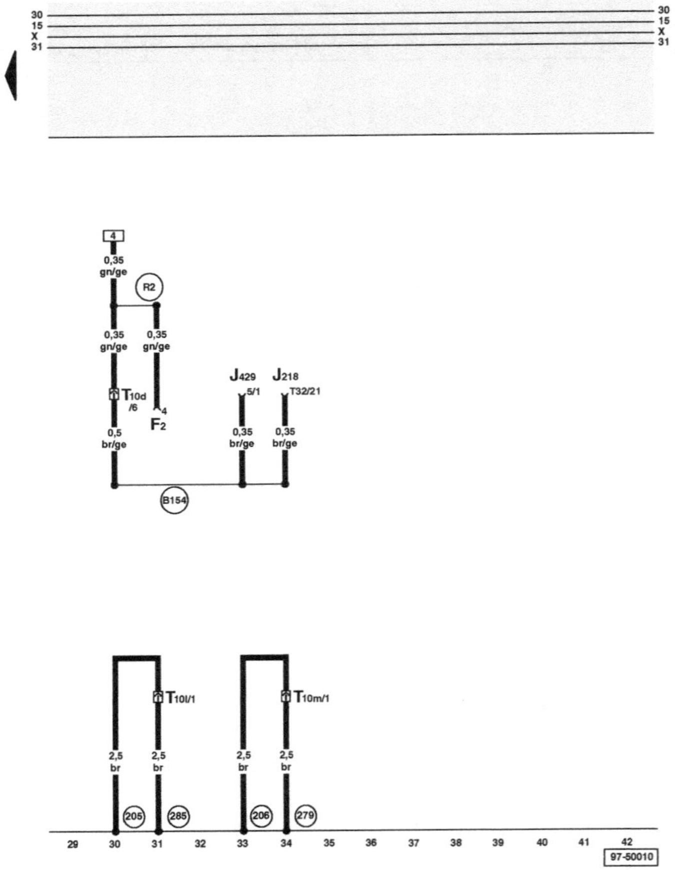

ws = white
sw = black
ro = red
br = brown
gn = green
bl = blue
gr = grey
li = violet
ge = yellow

F2 - Door Contact Switch - Driver's side
J218 - Instrument Cluster Combination Processor
J429 - Control module for central locking
T10d - Connector 10 pin, brown, connector station A pillar, left
T10l - Connector 10 pin, red, connector station A pillar, left
T10m - Connector 10 pin, red, connector station A pillar, right
T32 - Connector 32 pin, blue, on instrument cluster

(205) - Ground connection, in driver's door wiring harness

(206) - Ground connection, in passenger's door wiring harness

(279) - Ground connector -5-, in wiring harness interior

(285) - Ground connection -7- in wiring harness interior

(B154) - Connector -1- (TK), in wiring harness interior

(R2) - Wire connection (door contact switch, left), in power window/central locking system and door contact switch wiring harness

Power Windows

Edition 07.00
W42.USA.5702.04.21

© 2006 BentleyPublishers.com—All Rights Reserved

Anti-Lock Brake System (ABS) with Electronic Differential Lock (EDL) and Anti-Slip Regulation (ASR - ASC)

2000 m. y.

Fuse Panel

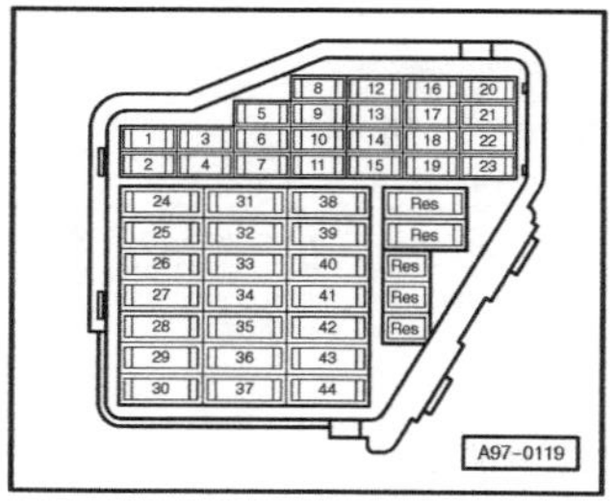

Fuse Colors:

30 A - Green
25 A - White
20 A - Yellow
15 A - Blue
10 A - Red
7,5 A - Brown
5 A - Beige

Starting with fuse position 23, fuses in the fuse holder are identified with 223 in the wiring diagram.

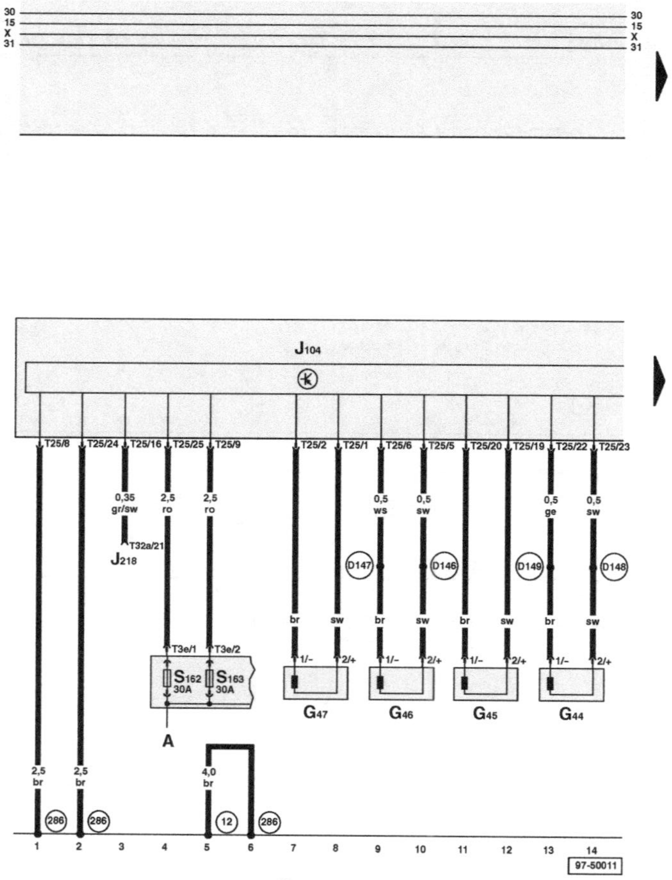

ws = white
sw = black
ro = red
br = brown
gn = green
bl = blue
gr = grey
li = violet
ge = yellow

A - Battery
G44 - Right Rear ABS Wheel Speed Sensor
G45 - Right Front ABS Wheel Speed Sensor
G46 - Left Rear ABS Wheel Speed Sensor
G47 - Left Front ABS Wheel Speed Sensor
J104 - ABS Control Module (w/EDL)
J218 - Instrument Cluster Combination Processor
S162 - Fuse -1- in fuse bracket/battery
S163 - Fuse -2- in fuse bracket/battery
T3e - Connector 3 pin, black, on Main Fuse Box/Battery
T25 - Connector 25 pin, on ABS Control Module (w/EDL)
T32a - Connector 32 pin, green, on instrument cluster

(12) - Ground connection, in engine compartment, left
(286) - Ground connection -8- in wiring harness interior
(D146) - Connector (speed sensor rear, left +), in wiring harness engine compartment
(D147) - Connector (speed sensor rear, left -), in wiring harness engine compartment
(D148) - Connector (speed sensor rear, right +), in wiring harness engine compartment
(D149) - Connector (speed sensor rear, right -), in wiring harness engine compartment

ABS Control Module, Fuses, Wheel Speed Sensors

©2006 BentleyPublishers.com—All Rights Reserved

Anti-Lock Brake System, ABS/EDL/ASR (2000)

Audi TT — Wiring diagram — **No. 6/3**

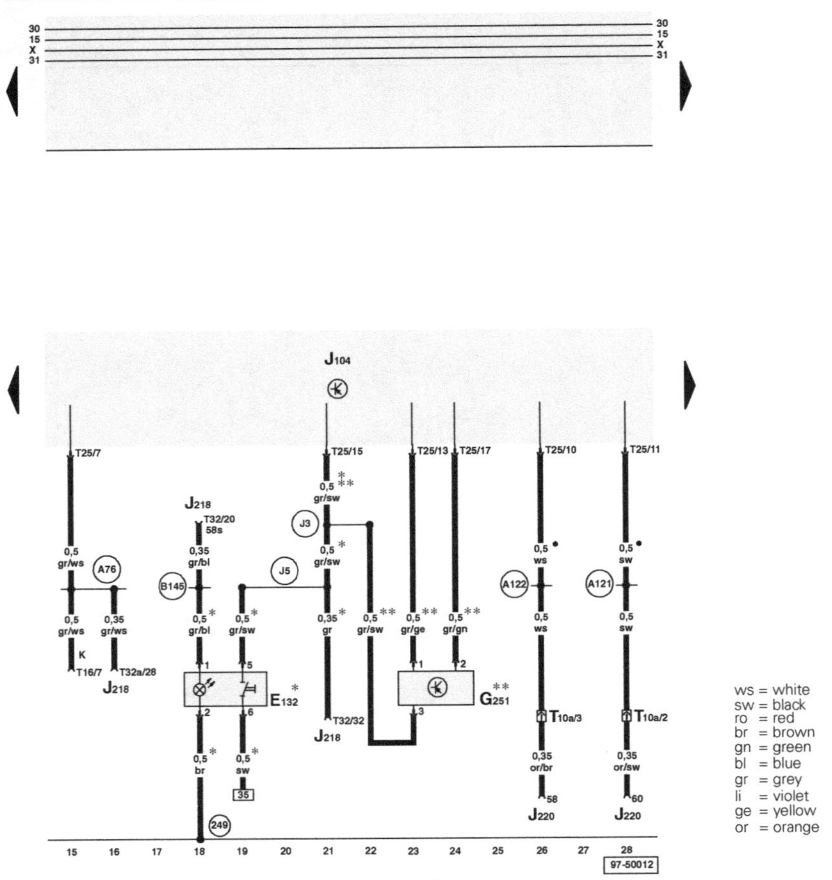

ws = white
sw = black
ro = red
br = brown
gn = green
bl = blue
gr = grey
li = violet
ge = yellow
or = orange

E132 - Anti-Slip Control Switch
G251 - Longitudinal acceleration sensor
J104 - ABS Control Module (w/EDL)
J218 - Instrument Cluster Combination Processor
J220 - Motronic Engine Control Module (ECM)
T16 - Connector 16 pin, black, Data Link Connector (DLC), below steering column, left
T25 - Connector 25 pin, on ABS Control Module (w/EDL)
T32 - Connector 32 pin, blue, on instrument cluster
T32a - Connector 32 pin, green, on instrument cluster

(249) - Ground connection -2-, in wiring harness interior

(A76) - Connector (K-diagnosis wire), in instrument panel wiring harness

(A121) - Connection (high bus) in instrument panel wiring harness

(A122) - Connection (low bus) in instrument panel wiring harness

(B145) - Plus connector (58s), in wiring harness interior

(J3) - Wire connection -1-, in ABS wiring harness

(J5) - Wire connection -2-, in ABS wiring harness, near anti-slip control switch

- CAN-Bus (Data-Bus)
* - Vehicles with Anti-Slip Regulation
** - Vehicles with all-wheel drive

Edition 12/02
W42.USA.5702.12.21

ABS Control Module, Anti-Slip Control Switch

No. 6/4 — Wiring diagram — **Audi TT**

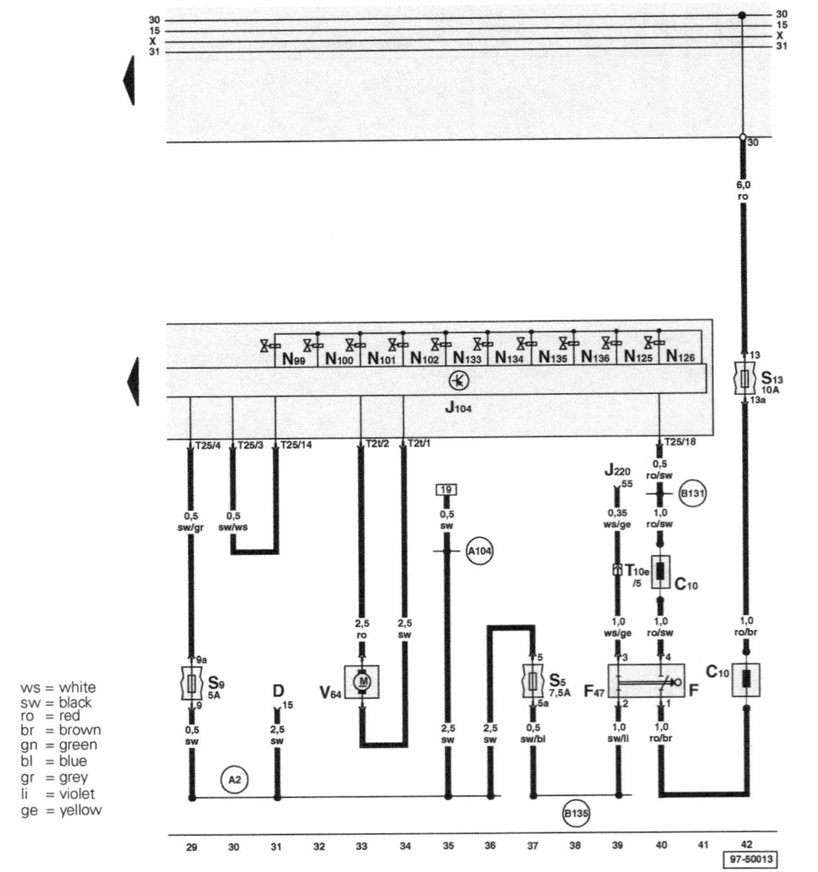

ws = white
sw = black
ro = red
br = brown
gn = green
bl = blue
gr = grey
li = violet
ge = yellow

C10 - Suppressor Filter
D - Ignition/Starter Switch
F - Brake Light Switch
F47 - Vacuum Vent Valve, Brake
J104 - ABS Control Module (w/EDL)
J220 - Motronic Engine Control Module (ECM)
N99 - Right Front ABS Inlet Valve
N100 - Right Front ABS Outlet Valve
N101 - Left Front ABS Inlet Valve
N102 - Left Front ABS Outlet Valve
N125 - Differential Lock Valve 1
N126 - Differential Lock Valve 2
N133 - Right Rear ABS Inlet Valve
N134 - Left Rear ABS Inlet Valve
N135 - Right Rear ABS Outlet Valve
N136 - Left Rear ABS Outlet Valve
S5 - Fuse
S9 - Fuse
S13 - Fuse

T2t - Connector 2 pin, on Airbag Control Module
T10e - Connector 10 pin, black, in E-box plenum chamber
T25 - Connector 25 pin, on ABS Control Module (w/EDL)
V64 - ABS Hydraulic Pump

(A2) - Plus connection (15), in instrument panel wiring harness

(A104) - Plus connector -2- (15), in instrument panel wiring harness

(B131) - Connector (54), in wiring harness interior

(B135) - Connector (15a), in wiring harness interior

ABS Control Module, Brake Light Switch, ABS Hydraulic Pump

Edition 12/02
W42.USA.5702.12.21

© 2006 BentleyPublishers.com—All Rights Reserved

Airbag

2000 m. y.

Fuse Panel

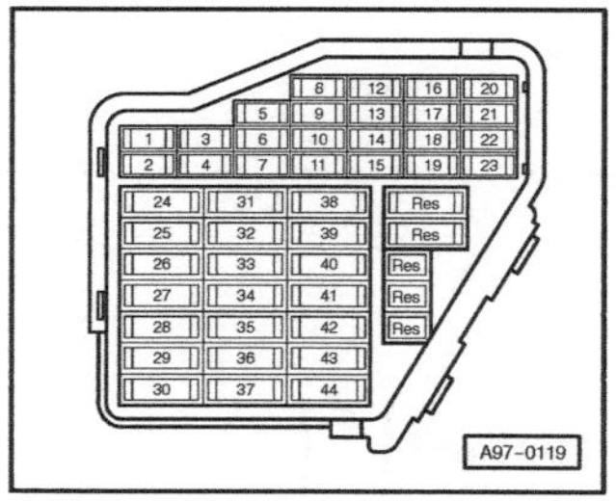

Fuse Colors:

30	A	-	Green
25	A	-	White
20	A	-	Yellow
15	A	-	Blue
10	A	-	Red
7,5	A	-	Brown
5	A	-	Beige

Starting with fuse position 23, fuses in the fuse holder are identified with 223 in the wiring diagram.

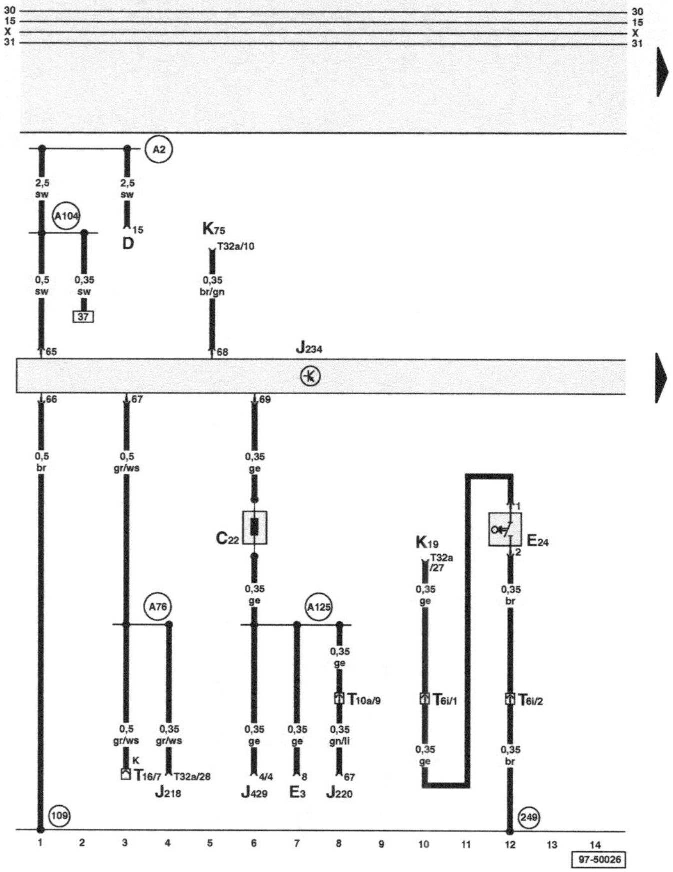

ws = white
sw = black
ro = red
br = brown
gn = green
bl = blue
gr = grey
li = violet
ge = yellow

C22	-	Airbag Suppresser
D	-	Ignition/Starter Switch
E3	-	Emergency Flasher Switch
E24	-	Left Seat Belt Switch
J218	-	Instrument Cluster Combination Processor
J220	-	Motronic Engine Control Module (ECM)
J234	-	Airbag Control Module
J429	-	Control module for central locking
K19	-	Seat Belt Warning Light
K75	-	Airbag Malfunction Indicator Lamp (MIL)
T6i	-	Connector 6 pin, below driver's seat
T10a	-	Connector 10 pin, white, in E-box plenum chamber
T16	-	Connector 16 pin, black, Data Link Connector (DLC), below storage compartment, driver's side
T32a	-	Connector 32 pin, green, on instrument cluster

(109)	-	Ground connection, in airbag wiring harness
(249)	-	Ground connection -2-, in wiring harness interior
(A2)	-	Plus connection (15), in instrument panel wiring harness
(A76)	-	Connector (K-diagnosis wire), in instrument panel wiring harness
(A104)	-	Plus connector -2- (15), in instrument panel wiring harness
(A125)	-	Connection (crash signal) in instrument panel wiring harness

©2006 BentleyPublishers.com—All Rights Reserved

Airbags (2000-2001)

Audi TT — Wiring diagram — No. 7/3

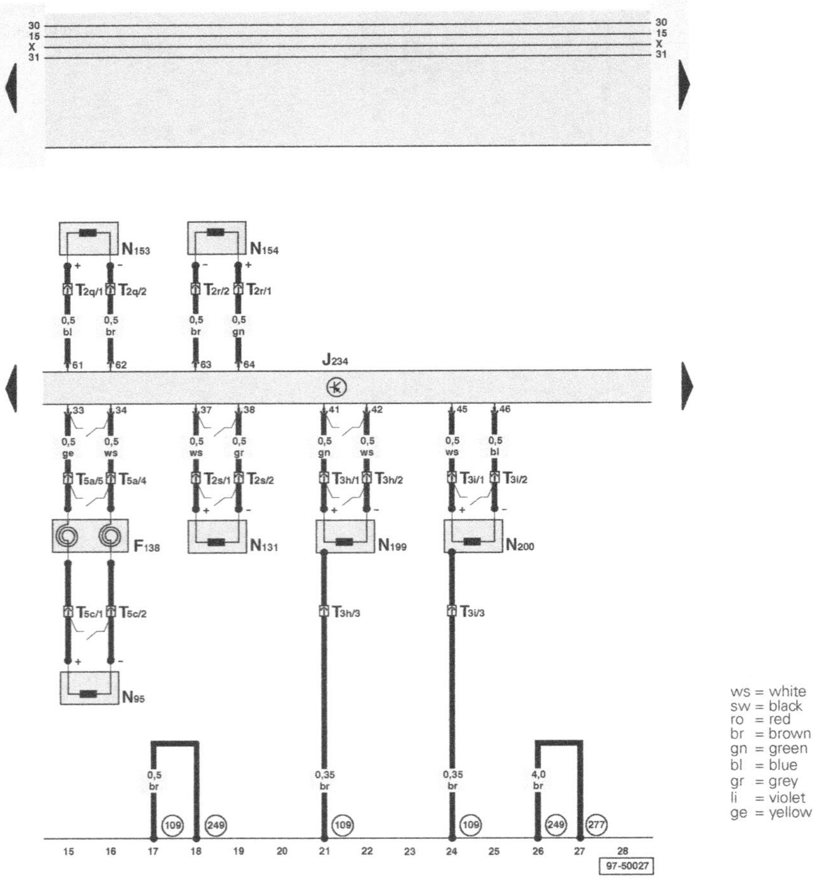

ws = white
sw = black
ro = red
br = brown
gn = green
bl = blue
gr = grey
li = violet
ge = yellow

F138 - Airbag Spiral Spring/Return Spring With Slip Ring
J234 - Airbag Control Module
N95 - Driver's Side Airbag Igniter
N131 - Passenger's Side Airbag Igniter 1
N153 - Left Seat Belt Tensioner Igniter
N154 - Right Seat Belt Tensioner Igniter
N199 - Igniter for side airbag, driver's side
N200 - Igniter for side airbag, passenger side
T2q - Connector 2 pin, on left B-pillar, lower part
T2r - Connector 2 pin, on right B-pillar, lower part
T2s - Connector 2 pin, behind instrument panel, right
T3h - Connector 3 pin, below dirver's seat (for side airbag)
T3i - Connector 3 pin, below passenger's seat (for side airbag)
T5a - Connector 5 pin, yellow, in steering wheel
T5c - Connector 5 pin, yellow, in steering wheel

(109) - Ground connection, in airbag wiring harness

(249) - Ground connection -2-, in wiring harness interior

(277) - Ground connector -3-, in wiring harness interior

Bridge closes when disconnecting connector (safety feature)

No. 7/4 — Wiring diagram — Audi TT

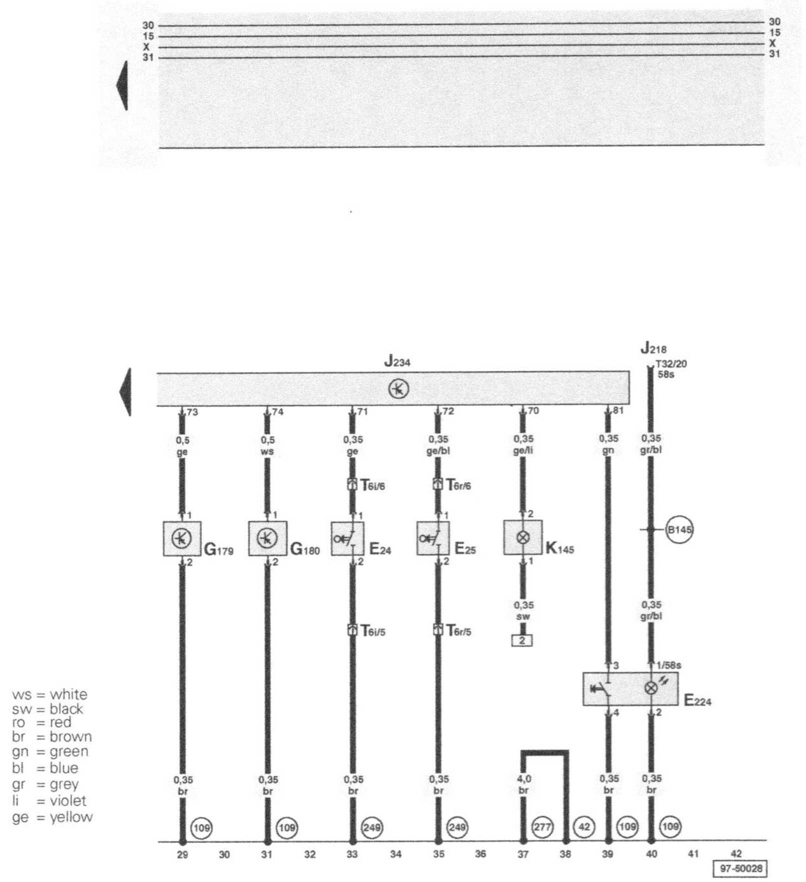

ws = white
sw = black
ro = red
br = brown
gn = green
bl = blue
gr = grey
li = violet
ge = yellow

E24 - Left Seat Belt Switch
E25 - Right Seat Belt Switch
E224 - Key switch for switching off passenger side airbag
G179 - Crash sensor for side airbag, driver's side
G180 - Crash sensor for side airbag, passenger side
J218 - Instrument Cluster Combination Processor
J234 - Airbag Control Module
K145 - Warning light for airbag off, passenger side
T6i - Connector 6 pin, below driver's seat
T6r - Connector 6 pin, below passenger's seat
T32 - Connector 32 pin, blue, on instrument cluster

(42) - Ground connection, on steering column

(109) - Ground connection, in airbag wiring harness

(249) - Ground connection -2-, in wiring harness interior

(277) - Ground connector -3-, in wiring harness interior

(B145) - Plus connector (58s), in wiring harness interior

© 2006 BentleyPublishers.com—All Rights Reserved

Heated Seats

2000 m. y.

Fuse Panel

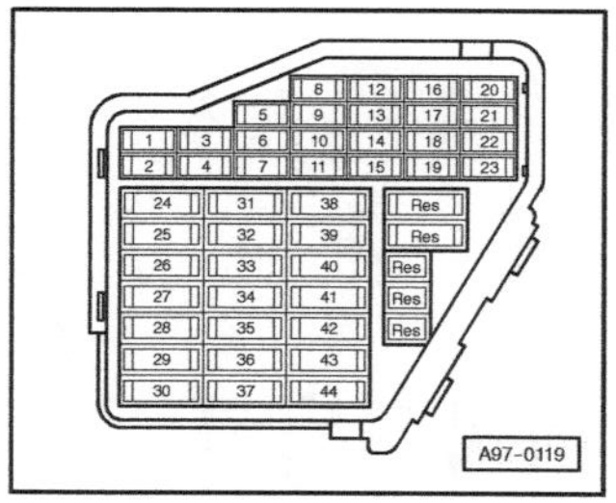

Fuse Colors:

30 A - Green
25 A - White
20 A - Yellow
15 A - Blue
10 A - Red
7,5 A - Brown
5 A - Beige

Starting with fuse position 23, fuses in the fuse holder are identified with 223 in the wiring diagram.

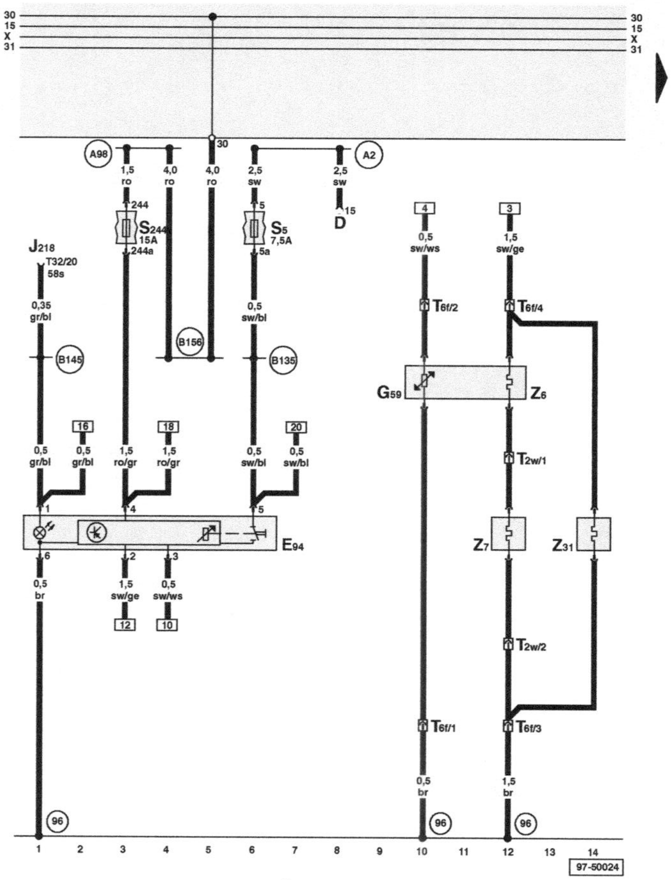

ws = white
sw = black
ro = red
br = brown
gn = green
bl = blue
gr = grey
li = violet
ge = yellow

D - Ignition/Starter Switch
E94 - Adjuster for heated driver's seat
G59 - Driver's Heated Seat Temperature Sensor
J218 - Instrument Cluster Combination Processor
S5 - Fuse
S244 - Fuse in fuse holder
T2w - Connector 2 pin, in driver's seat
T6f - Connector 6 pin, below driver's seat
T32 - Connector 32 pin, blue, on instrument cluster
Z6 - Driver's Seat Heating Element
Z7 - Driver's Backrest Heating Element
Z31 - Driver's Seat Side Bolster Heater

(96) - Ground connection -1-, in heated seats wiring harness
(A2) - Plus connection (15), in instrument panel wiring harness
(A98) - Plus connector -4- (30), in instrument panel wiring harness
(B135) - Connector (15a), in wiring harness interior
(B145) - Plus connector (58s), in wiring harness interior
(B156) - Plus connector (30a), in wiring harness interior

©2006 BentleyPublishers.com—All Rights Reserved

Audi TT — Wiring diagram — No. 8/3

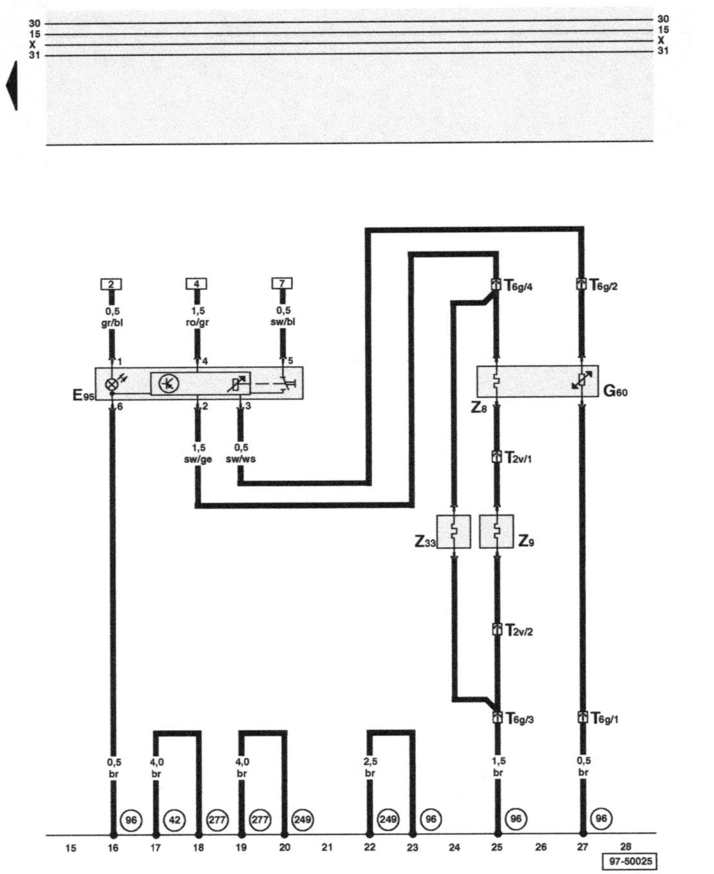

ws = white
sw = black
ro = red
br = brown
gn = green
bl = blue
gr = grey
li = violet
ge = yellow

E95 - Adjuster for heated passenger's seat
G60 - Passenger's Heated Seat Temperature Sensor
T2v - Connector 2 pin, in passenger's seat
T6g - Connector 6 pin, below passenger's seat
Z8 - Passenger's Seat Heating Element
Z9 - Passenger's Backrest Heating Element
Z33 - Passenger's Seat Side Bolster Heater

(42) - Ground connection, on steering column
(96) - Ground connection -1-, in heated seats wiring harness
(249) - Ground connection -2-, in wiring harness interior
(277) - Ground connector -3-, in wiring harness interior

THIS SPACE INTENTIONALLY LEFT BLANK

© 2006 BentleyPublishers.com—All Rights Reserved

Headlight with High Intensity Gas discharge Lamps and automatic Headlight Beam Adjusting

2000 m. y.O

Fuse Panel

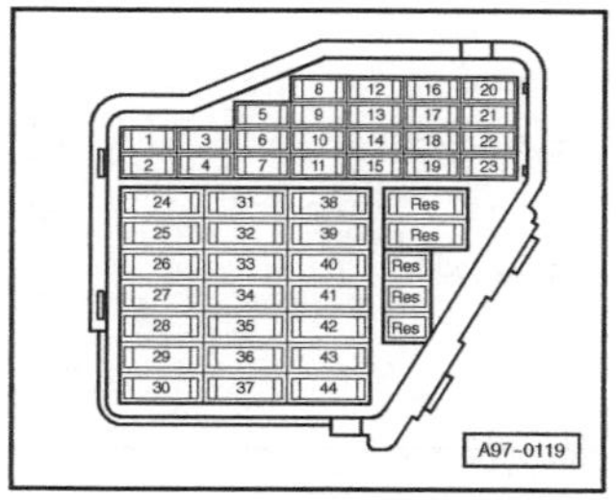

Fuse Colors:

30 A - Green
25 A - White
20 A - Yellow
15 A - Blue
10 A - Red
7,5 A - Brown
5 A - Beige

Starting with fuse position 23, fuses in the fuse holder are identified with 223 in the wiring diagram.

Micro Central Electric Panel

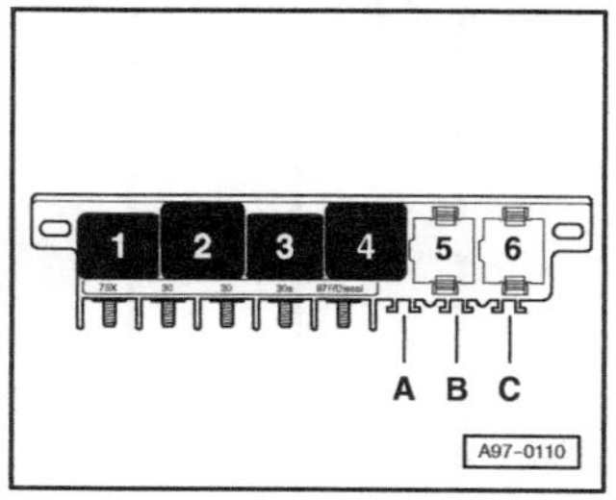

Relay Location:

2 -Load Reduction Relay, J59

13 - Fold Relay Panel

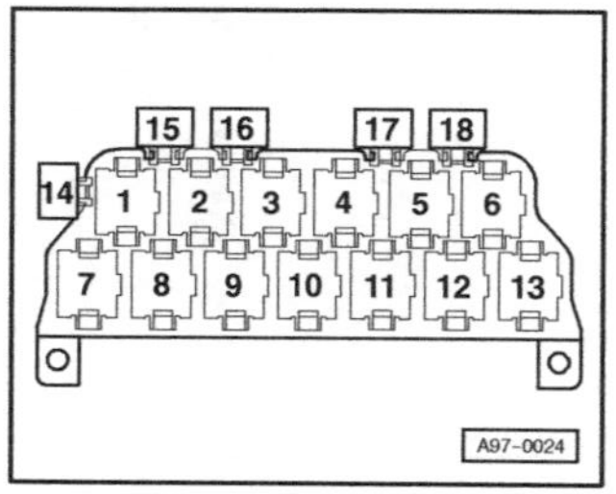

Relay Location:

2 -Lamp Control Module, J123

3 -Lamp Control Module, J123

4 -Fog Light Relay, J5

5 -Starting Interlock Relay, J207

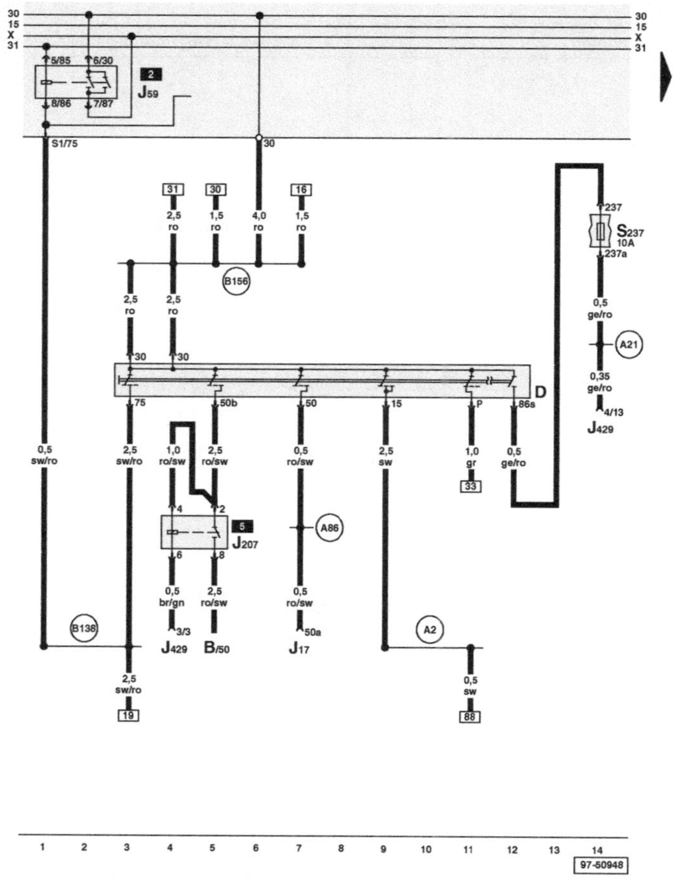

ws = white
sw = black
ro = red
br = brown
gn = green
bl = blue
gr = grey
li = violet
ge = yellow

B - Starter
D - Ignition/Starter Switch
J17 - Fuel Pump (FP) Relay
J59 - Load Reduction Relay
J207 - Starting Interlock Relay
J429 - Control module for central locking
S237 - Fuse in fuse holder

A2 - Plus connection (15), in instrument panel wiring harness
A21 - Wire connection (86s), in instrument panel wiring harness
A86 - Connection (50a), in instrument panel wiring harness
B138 - Plus connector (X), in wiring harness, interior
B156 - Plus connector (30a), in wiring harness interior

Ignition/Starter Switch, Load Reduction Relay, Starting Interlock Relay

© 2006 BentleyPublishers.com—All Rights Reserved

Headlights, Xenon (HID) (2000-2001)

Audi TT — Wiring diagram — No. 9/3

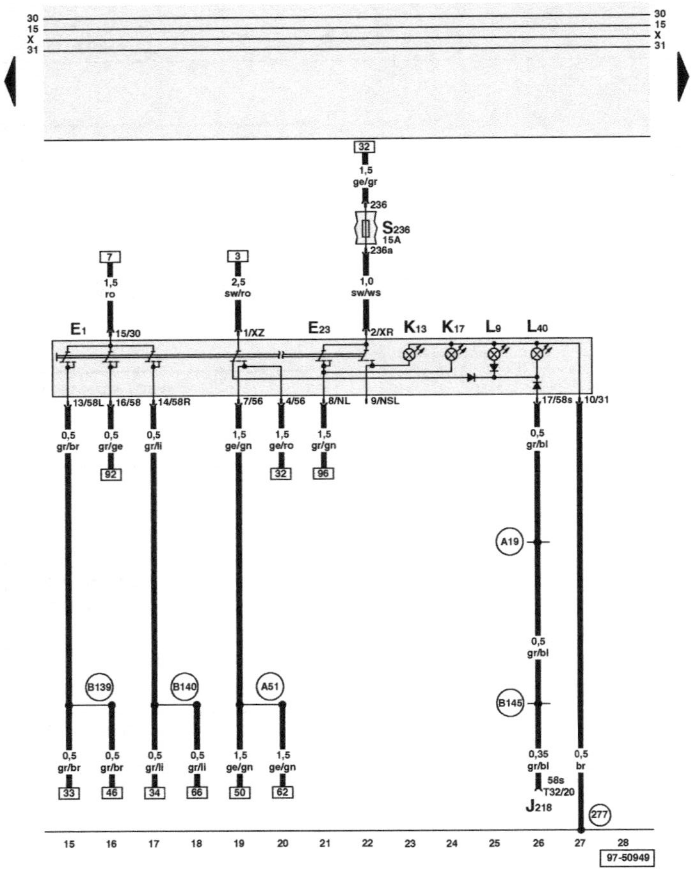

ws = white
sw = black
ro = red
br = brown
gn = green
bl = blue
gr = grey
li = violet
ge = yellow

E1 - Light Switch
E23 - Fog Light Switch
J218 - Instrument Cluster Combination Processor
K13 - Rear Fog Light Indicator Light
K17 - Fog Light Indicator Light
L9 - Headlight Switch Light
L40 - Front And Rear Fog Light Switch Light
S236 - Fuse in fuse holder
T32 - Connector 32 pin, blue, on instrument cluster

(277) - Ground connector -3-, in wiring harness interior
(A19) - Wire connection (58s), in instrument panel wiring harness
(A51) - Wire connection (56), in instrument panel wiring harness
(B139) - Connector (PL), in wiring harness, interior
(B140) - Connector (PR), in wiring harness, interior
(B145) - Plus connector (58s), in wiring harness interior

Edition 01.00
W42.USA.5702.03.21

Light Switch

No. 9/4 — Wiring diagram — Audi TT

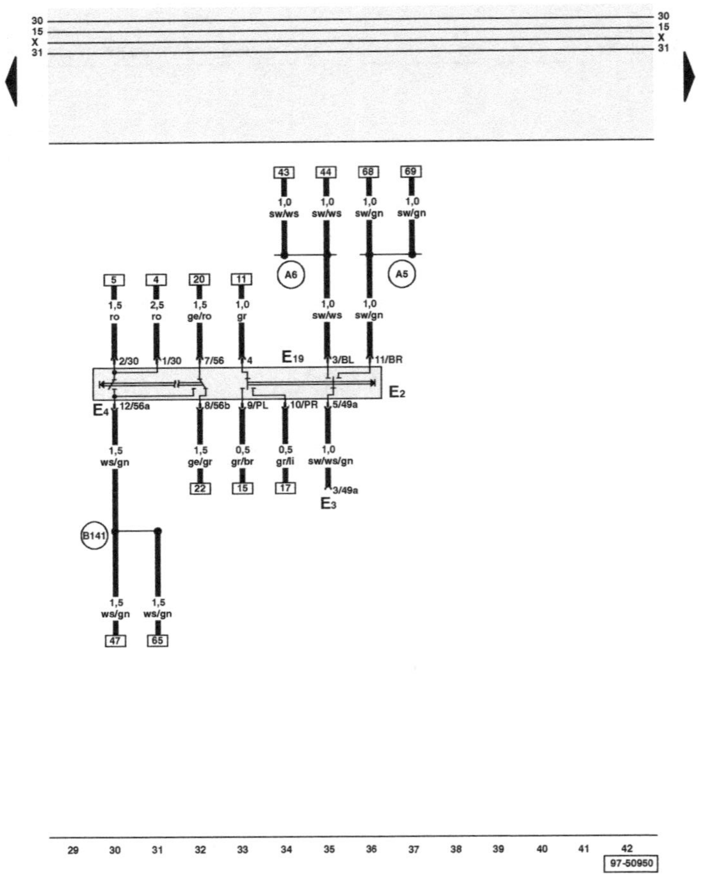

ws = white
sw = black
ro = red
br = brown
gn = green
bl = blue
gr = grey
li = violet
ge = yellow

E2 - Turn Signal Switch
E3 - Emergency Flasher Switch
E4 - Headlight Dimmer/Flasher Switch
E19 - Park Light Switch

(A5) - Plus connection (right turn signal), in instrument panel wiring harness
(A6) - Plus connection (left turn signal), in instrument panel wiring harness
(B141) - Plus connector -1- (56a), in wiring harness interior

Turn Signal Switch, Headlight Dimmer/Flasher Switch

Edition 01.00
W42.USA.5702.03.21

© 2006 BentleyPublishers.com—All Rights Reserved

Audi TT — Wiring diagram — No. 9/5

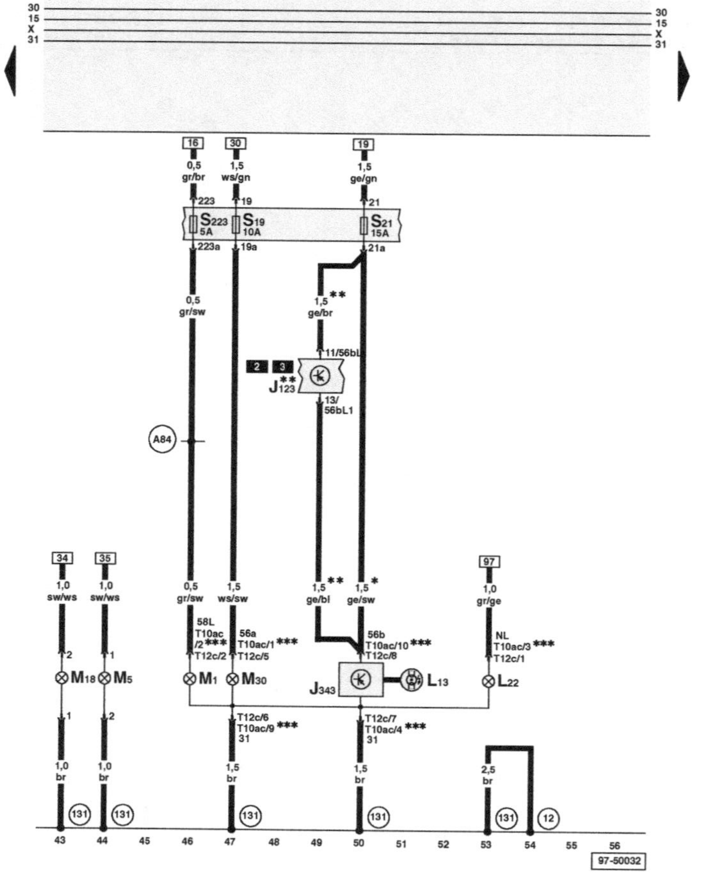

ws = white
sw = black
ro = red
br = brown
gn = green
bl = blue
gr = grey
li = violet
ge = yellow

J123 - Lamp Control Module
J343 - Control Module for High-intensity gas discharge lamp
L13 - High intensity gas discharge lamp left
L22 - Left Front Fog Light
M1 - Left Parking Light
M5 - Left Front Turn Signal Light
M18 - Left, Side Turn Signal Light
M30 - Left High Beam Headlight
S19 - Fuse
S21 - Fuse
S223 - Fuse in fuse holder
T10ac - Connector 10 pin, on headlight, left
T12c - Connector 12 pin, on headlight, left

(12) - Ground connection, in engine compartment, left

(131) - Ground connection -2-, in engine compartment wiring harness

(A84) - Connector (58L), in instrument panel wiring harness

* - Vehicles without Driver's Information System
** - Vehicles with Driver's Information System
*** - from VIN Y1 010 000

Edition 01.00
W42.USA.5702.03.21

Fuses, High Intensity Gas Discharge Lamp left, Control Module for High-Intensity Gas Discharge Lamp

No. 9/6 — Wiring diagram — Audi TT

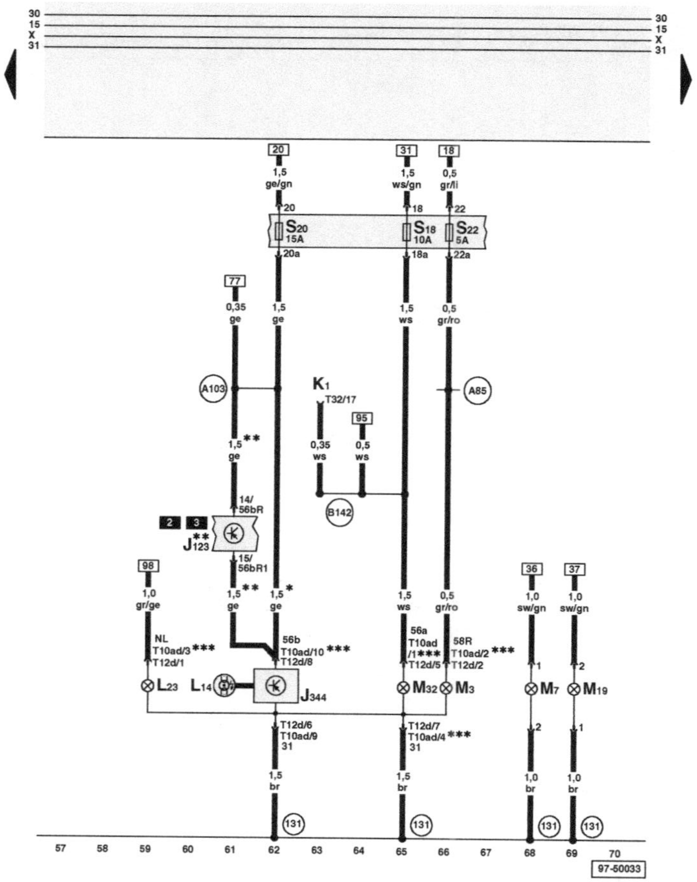

ws = white
sw = black
ro = red
br = brown
gn = green
bl = blue
gr = grey
li = violet
ge = yellow

J123 - Lamp Control Module
J344 - Control Module for High-intensity gas discharge lamp
K1 - Headlight High Beam Indicator Light
L14 - High intensity gas discharge lamp right
L23 - Right Front Fog Light
M3 - Right Parking Light
M7 - Right Front Turn Signal Light
M19 - Right, Side Turn Signal Light
M32 - Right High Beam Headlight
S18 - Fuse
S20 - Fuse
S22 - Fuse
T10ad - Connector 10 pin, on headlight, right
T12d - Connector 12 pin, on headlight, right
T32 - Connector 32 pin, blue, on instrument cluster

(131) - Ground connection -2-, in engine compartment wiring harness

(A85) - Connector (58R), in instrument panel wiring harness

(A103) - Connector -2- (56), in instrument panel wiring harness

(B142) - Plus connector -2- (56a), in wiring harness interior

* - Vehicles without Driver's Information System
** - Vehicles with Driver's Information System
*** - from VIN Y1 010 000

Fuses, High Intensity Gas Discharge Lamp right, Control Module for High-Intensity Gas Discharge Lamp

Edition 01.00
W42.USA.5702.03.21

© 2006 BentleyPublishers.com—All Rights Reserved

Headlights, Xenon (HID) (2000-2001)

Audi TT — Wiring diagram — **No. 9/7**

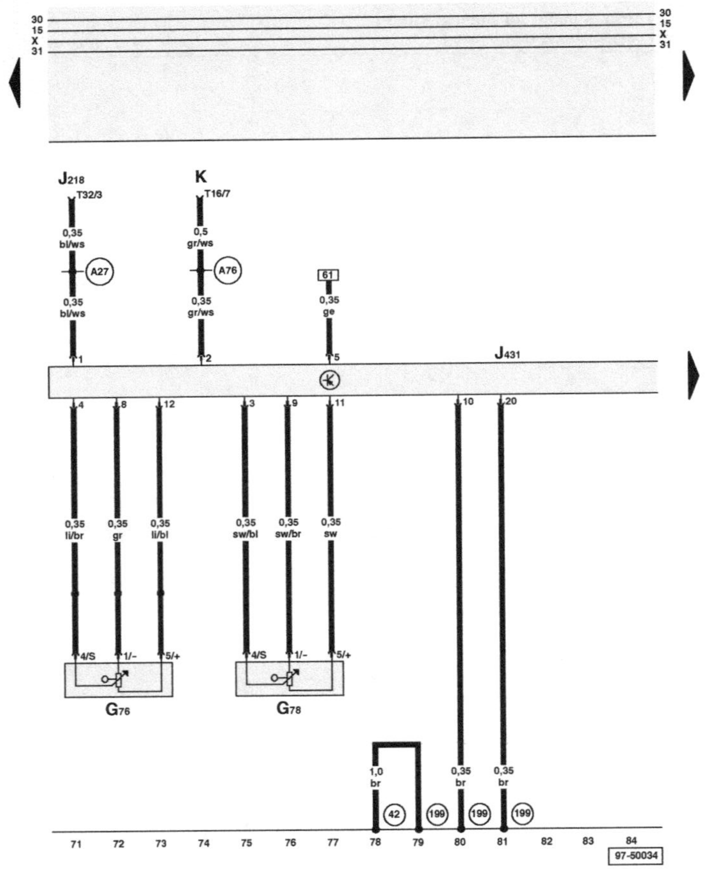

ws = white
sw = black
ro = red
br = brown
gn = green
bl = blue
gr = grey
li = violet
ge = yellow

G76 - Left Rear Level Control System Sensor
G78 - Left Front Level Control System Sensor
J218 - Instrument Cluster Combination Processor
J431 - Control module for Headlight Beam Adjusting
K - Data Link Connector (DLC) Rapid Data Transfer
T16 - Connector 16 pin, black, Data Link Connector (DLC), under steering column, left
T32 - Connector 32 pin, blue, on instrument cluster

(42) - Ground connection, on steering column
(199) - Ground connection -3-, in instrument panel wiring harness
(A27) - Wire connection (vehicle speed signal), in instrument panel wiring harness
(A76) - Connector (K-diagnosis wire), in instrument panel wiring harness

Edition 01.00
W42.USA.5702.03.21

Control Module for Headlight Beam Adjusting, Level Control System Sensors

No. 9/8 — Wiring diagram — **Audi TT**

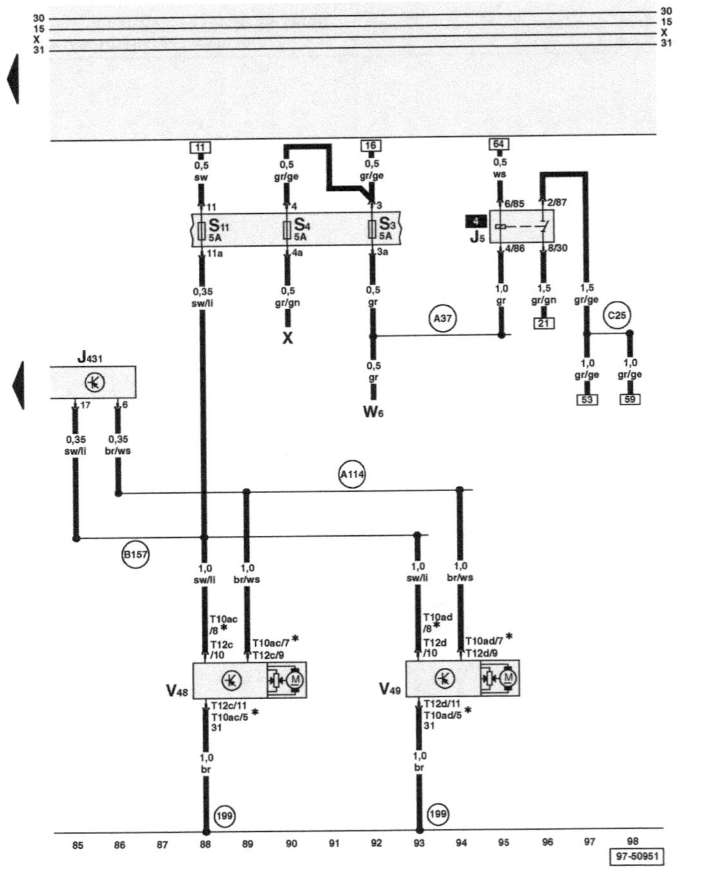

ws = white
sw = black
ro = red
br = brown
gn = green
bl = blue
gr = grey
li = violet
ge = yellow

J5 - Fog Light Relay
J431 - Control module for Headlight Beam Adjusting
S3 - Fuse
S4 - Fuse
S11 - Fuse
T10ac - Connector 10 pin, on headlight, left
T10ad - Connector 10 pin, on headlight, right
T12c - Connector 12 pin, on headlight, left
T12d - Connector 12 pin, on headlight, right
V48 - Left Headlight Beam Adjusting Motor
V49 - Right Headlight Beam Adjusting Motor
W6 - Glove Compartment Light
X - License Plate Light

(199) - Ground connection -3-, in instrument panel wiring harness
(A37) - Wire connection (58a), in instrument panel wiring harness
(A114) - Connection (Headlight adjusting), in instrument panel wiring harness
(B157) - Plus connector -3- (15a), in wiring harness interior
(C25) - Wire connection (fog lights), in front light wiring harness
* - from VIN Y1 010 000

Headlight Beam Adjusting Motors, Fog Light Relay

Edition 01.00
W42.USA.5702.03.21

© 2006 BentleyPublishers.com—All Rights Reserved

Daytime Running Lights with Fog Light (Canada)

2000 m. y.

Fuse Panel

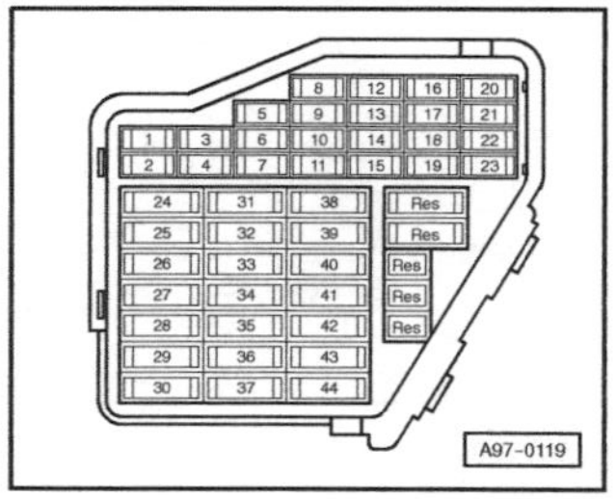

Fuse Colors:

30 A - Green
25 A - White
20 A - Yellow
15 A - Blue
10 A - Red
7,5 A - Brown
5 A - Beige

Starting with fuse position 23, fuses in the fuse holder are identified with 223 in the wiring diagram.

Micro Central Electric Panel

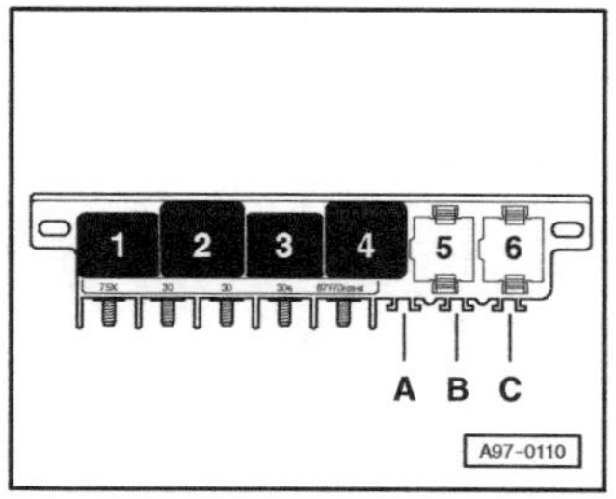

Relay Location:

2 -Load Reduction Relay, J59

13 - Fold Relay Panel

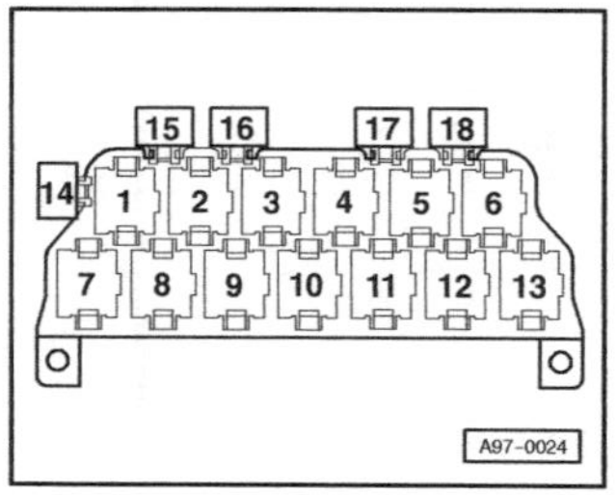

Relay Location:

4 -Fog Light Relay, J5

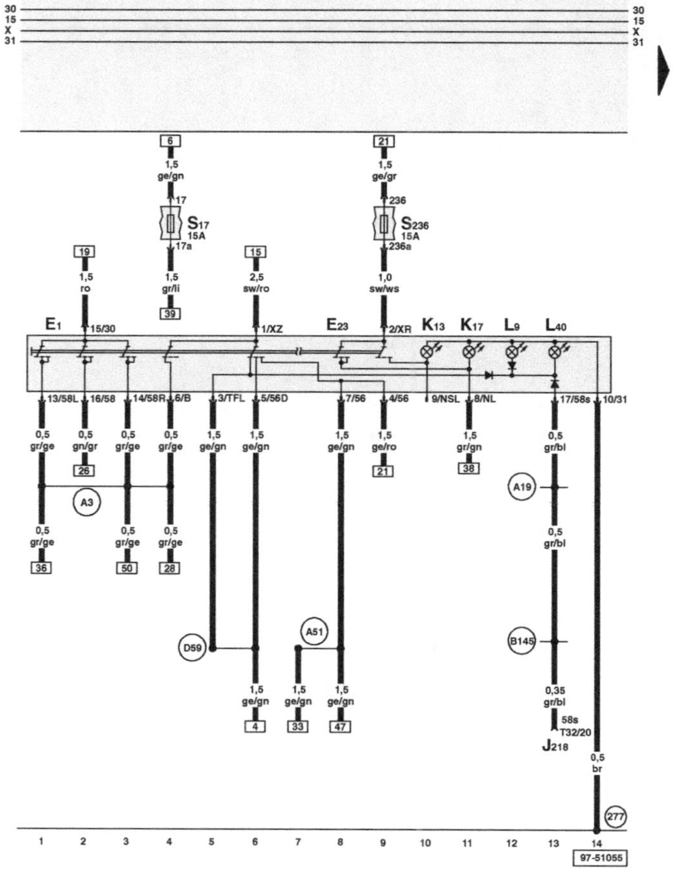

ws = white
sw = black
ro = red
br = brown
gn = green
bl = blue
gr = grey
li = violet
ge = yellow

E1 - Light Switch
E23 - Fog Light Switch
J218 - Instrument Cluster Combination Processor
K13 - Rear Fog Light Indicator Light
K17 - Fog Light Indicator Light
L9 - Headlight Switch Light
L40 - Front And Rear Fog Light Switch Light
S17 - Fuse
S236 - Fuse in fuse holder
T32 - Connector 32 pin, blue, on instrument cluster

(277) - Ground connector -3-, in wiring harness interior
(A3) - Plus connection (58), in instrument panel wiring harness
(A19) - Wire connection (58s), in instrument panel wiring harness
(A51) - Wire connection (56), in instrument panel wiring harness
(B145) - Plus connector (58s), in wiring harness interior
(D59) - Wire connection (Fog light), in engine compartment wiring harness

© 2006 BentleyPublishers.com—All Rights Reserved

Daytime Running Lights and Rear Fog Light, Canada (2000-2001)

Audi TT — Wiring diagram — No. 10/3

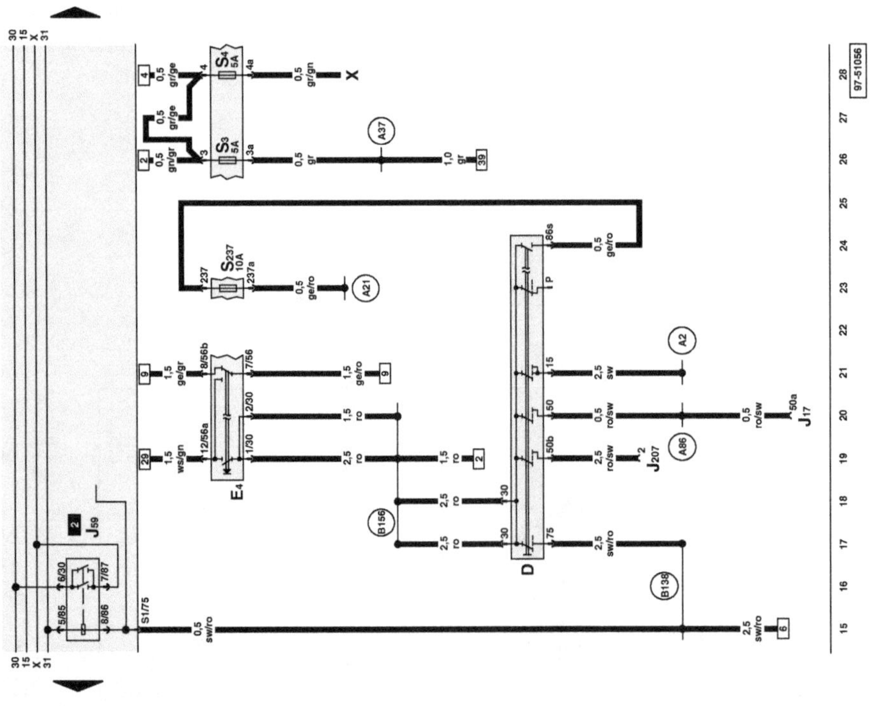

ws = white
sw = black
ro = red
br = brown
gn = green
bl = blue
gr = grey
li = violet
ge = yellow

- D - Ignition/Starter Switch
- E4 - Headlight Dimmer/Flasher Switch
- J17 - Fuel Pump (FP) Relay
- J59 - Load Reduction Relay
- J207 - Starting Interlock Relay
- S3 - Fuse
- S4 - Fuse
- S237 - Fuse in fuse holder
- X - License Plate Light

- (A2) - Plus connection (15), in instrument panel wiring harness
- (A21) - Wire connection (86s), in instrument panel wiring harness
- (A37) - Wire connection (58a), in instrument panel wiring harness
- (A86) - Connection (50a), in instrument panel wiring harness
- (B138) - Plus connector (X), in wiring harness, interior
- (B156) - Plus connector (30a), in wiring harness interior

Ignition/Starter Switch, Load Reduction Relay, Headlight Dimmer/Flasher Switch

Edition 01.00
W42.USA.5702.03.21

No. 10/4 — Wiring diagram — Audi TT

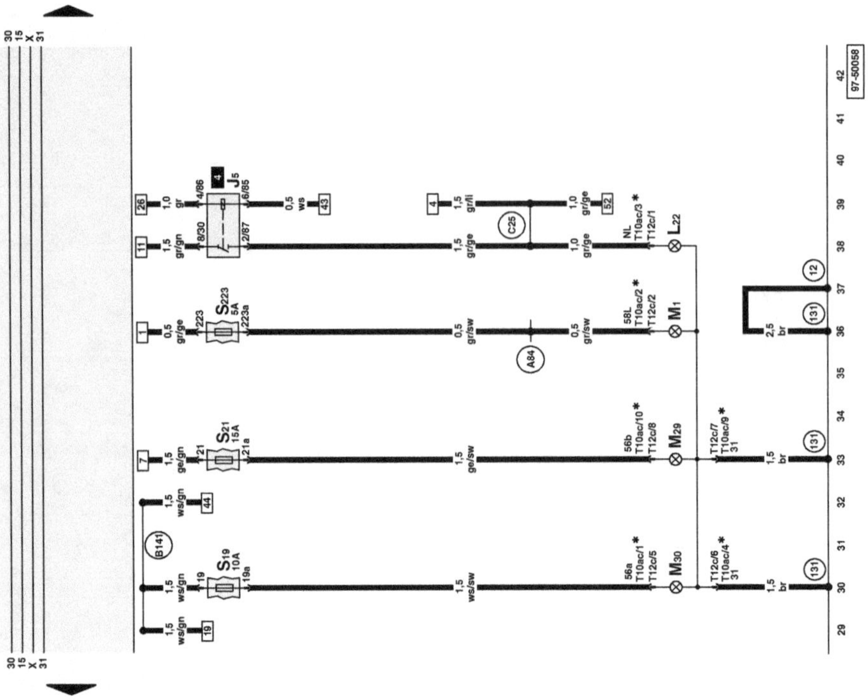

ws = white
sw = black
ro = red
br = brown
gn = green
bl = blue
gr = grey
li = violet
ge = yellow

- J5 - Fog Light Relay
- L22 - Left Front Fog Light
- M1 - Left Parking Light
- M29 - Left Low Beam Headlight
- M30 - Left High Beam Headlight
- S19 - Fuse
- S21 - Fuse
- S223 - Fuse in fuse holder
- T10ac - Connector 10 pin, on headlight, left
- T12c - Connector 12 pin, on headlight, left

- (12) - Ground connection, in engine compartment, left
- (131) - Ground connection -2-, in engine compartment wiring harness
- (A84) - Connector (58L), in instrument panel wiring harness
- (B141) - Plus connector -1- (56a), in wiring harness interior
- (C25) - Wire connection (fog lights), in front light wiring harness
- * - from VIN Y1 010 000

Left Headlight

Edition 01.00
W42.USA.5702.03.21

© 2006 BentleyPublishers.com—All Rights Reserved

Audi TT **Wiring diagram** **No. 10/5**

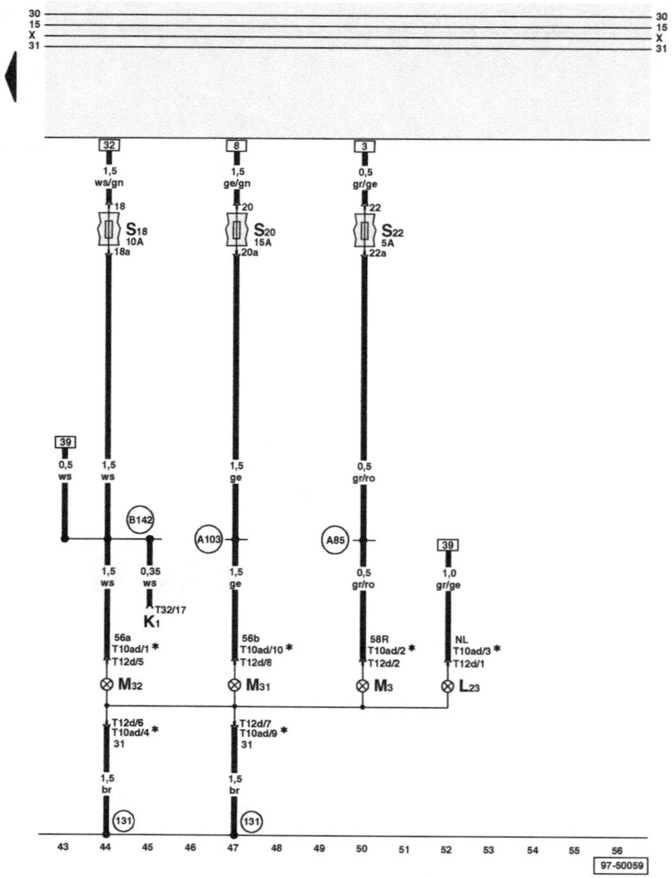

ws = white
sw = black
ro = red
br = brown
gn = green
bl = blue
gr = grey
li = violet
ge = yellow

K1 - Headlight High Beam Indicator Light
L23 - Right Front Fog Light
M3 - Right Parking Light
M31 - Right Low Beam Headlight
M32 - Right High Beam Headlight
S18 - Fuse
S20 - Fuse
S22 - Fuse
T10ad - Connector 10 pin, on headlight, right
T12d - Connector 12 pin, on headlight, right
T32 - Connector 32 pin, blue, on instrument cluster

(131) - Ground connection -2-, in engine compartment wiring harness
(A85) - Connector (58R), in instrument panel wiring harness
(A103) - Connector -2- (56), in instrument panel wiring harness
(B142) - Plus connector -2- (56a), in wiring harness interior

* - from VIN Y1 010 000

Edition 01.00
W42.USA.5702.03.21

Right Headlight

THIS SPACE INTENTIONALLY LEFT BLANK

© 2006 BentleyPublishers.com—All Rights Reserved

Headlight Washer

2000 m. y.

Fuse Panel

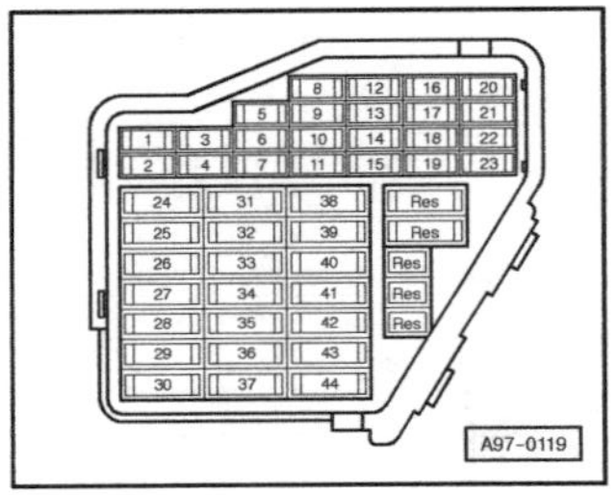

Fuse Colors:

30 A - Green
25 A - White
20 A - Yellow
15 A - Blue
10 A - Red
7,5 A - Brown
5 A - Beige

Starting with fuse position 23, fuses in the fuse holder are identified with 223 in the wiring diagram.

Micro Central Electric Panel

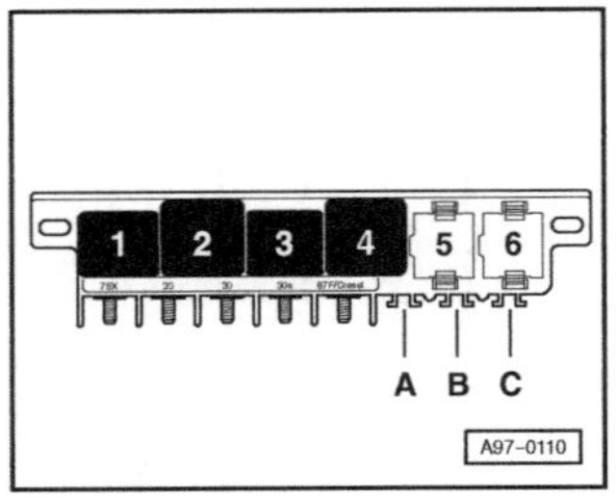

Relay Location:

5 -Wiper/Washer Intermittent Relay, J31

6 -Wiper/Washer Intermittent Relay, J31

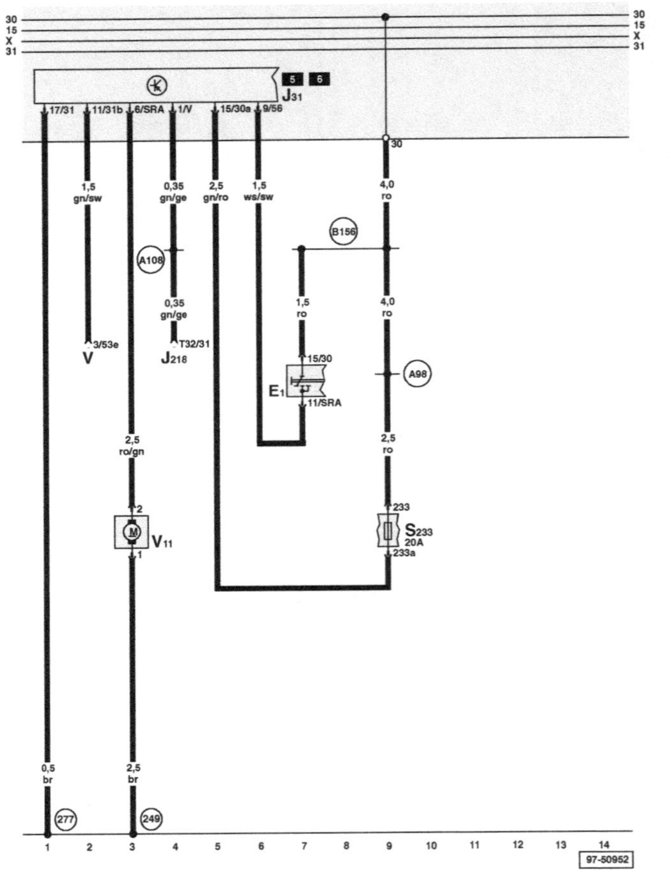

ws = white
sw = black
ro = red
br = brown
gn = green
bl = blue
gr = grey
li = violet
ge = yellow

E1 - Light Switch
J31 - Wiper/Washer Intermittent Relay
J218 - Instrument Cluster Combination Processor
S233 - Fuse in fuse holder
T32 - Connector 32 pin, blue, on instrument cluster
V - Windshield Wiper Motor
V11 - Headlight Washer Pump

(249) - Ground connection -2-, in wiring harness interior
(277) - Ground connector -3-, in wiring harness interior
(A98) - Plus connector -4- (30), in instrument panel wiring harness
(A108) - Connector (speed signal), in instrument panel wiring harness
(B156) - Plus connector (30a), in wiring harness interior

Headlight Washer Pump

© 2006 BentleyPublishers.com—All Rights Reserved

Central Locking System / Anti Theft System / Radar Interior Monitor (Roadster)

2000 m. y.

Fuse Panel

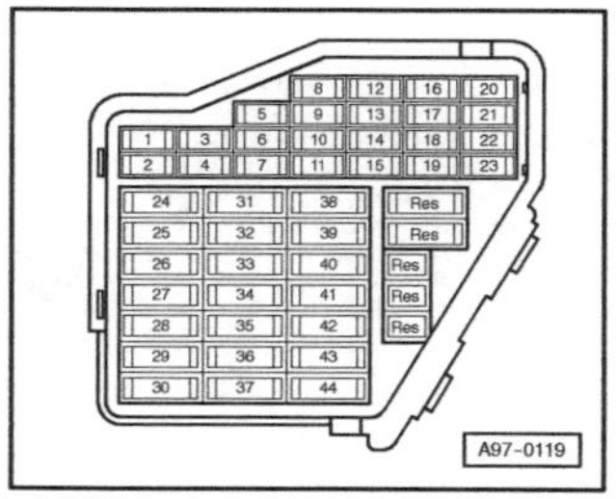

Fuse Colors:

30 A - Green
25 A - White
20 A - Yellow
15 A - Blue
10 A - Red
7,5 A - Brown
5 A - Beige

Starting with fuse position 23, fuses in the fuse holder are identified with 223 in the wiring diagram.

Micro Central Electric Panel

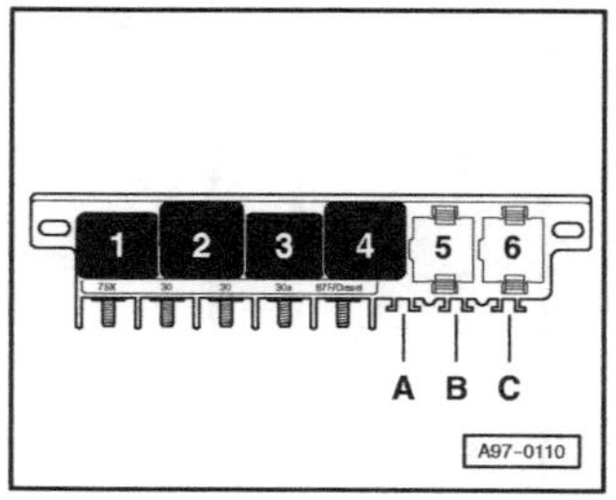

Fuse Location:

A -Convertible Top Fuse, S67

13 - Fold Relay Panel

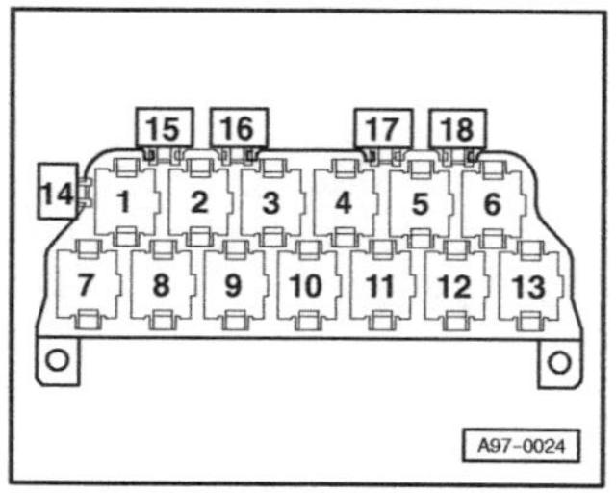

Relay Location:

5 -Starting Interlock Relay, J207

10 -Rear Window Defogger Relay, J9

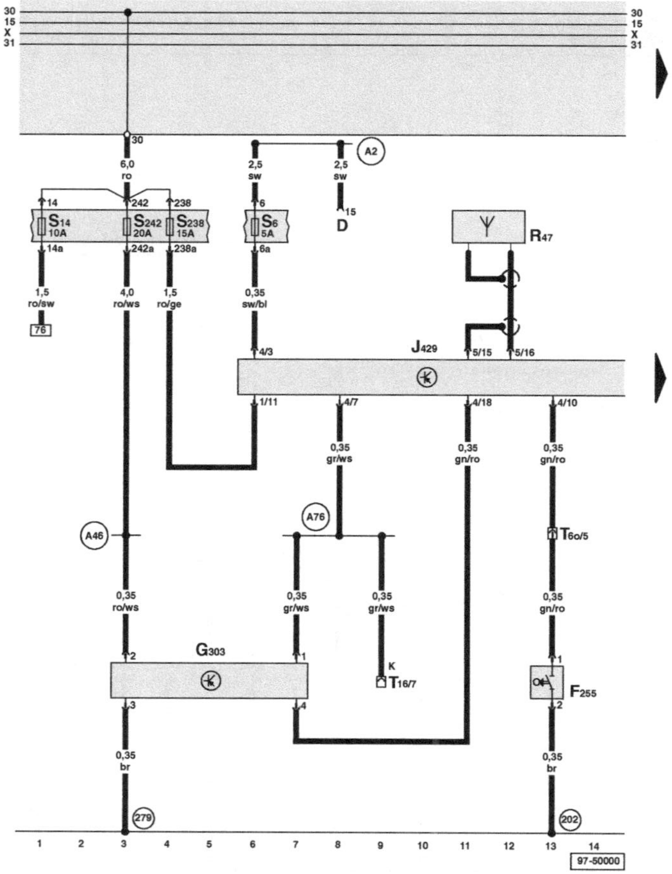

ws = white
sw = black
ro = red
br = brown
gn = green
bl = blue
gr = grey
li = violet
ge = yellow

D - Ignition/Starter Switch
F255 - Hardtop Detector Switch
G303 - Radar Interior Monitoring Control Module 1
J429 - Control module for central locking
R47 - Antenna wire for central locking and anti-theft warning system
S6 - Fuse
S14 - Fuse
S238 - Fuse 38 in fuse holder
S242 - Fuse 42 in fuse holder
T6o - Connector 6 pin, Hardtop right
T16 - Connector 16 pin, black, Data Link Connector (DLC), below storage compartment, driver s side

(202) - Ground connection, in convertible top wiring harness
(279) - Ground connector -5-, in wiring harness interior
(A2) - Plus connection (15), in instrument panel wiring harness
(A46) - Plus connection (30a), in instrument panel wiring harness
(A76) - Connector (K-diagnosis wire), in instrument panel wiring harness

© 2006 BentleyPublishers.com—All Rights Reserved

Central locking, Anti-theft, roadster (2000-2001)

Audi TT Wiring diagram **No. 12/3**

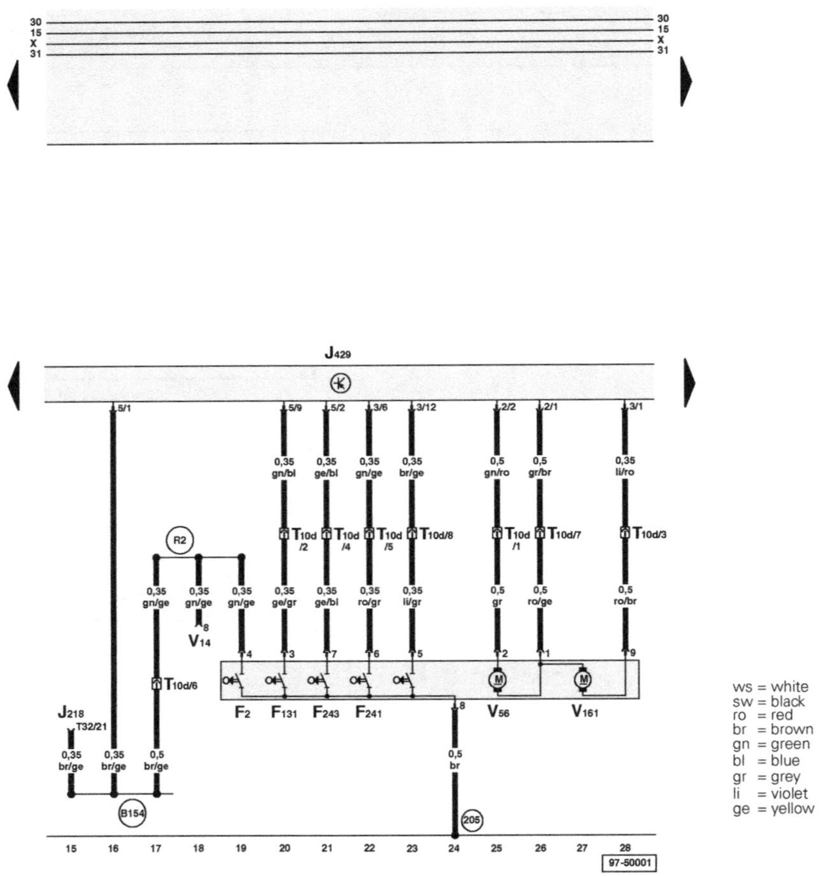

ws = white
sw = black
ro = red
br = brown
gn = green
bl = blue
gr = grey
li = violet
ge = yellow

F2 - Door Contact Switch - Driver's side
F131 - Left Front Central Locking System Activator/Switch
F241 - Contact switch for lock cylinder, driver side
F243 - Adjusting element for central locking (safe), driver door
J218 - Instrument Cluster Combination Processor
J429 - Control module for central locking
T10d - Connector 10 pin, brown, connector station A pillar, left
T32 - Connector 32 pin, blue, on instrument cluster
V14 - Left Window Motor
V56 - Driver's Door Central Locking System Motor
V161 - Motor for central locking (safe), driver door

(205) - Ground connection, in driver's door wiring harness
(B154) - Connector -1- (TK), in wiring harness interior
(R2) - Wire connection (door contact switch, left), in power window/central locking system and door contact switch wiring harness

Edition 01.00
W42.USA.5702.03.21

Control Module for Central Locking, Lock Unit for Central Locking Driver's Side

No. 12/4 Wiring diagram **Audi TT**

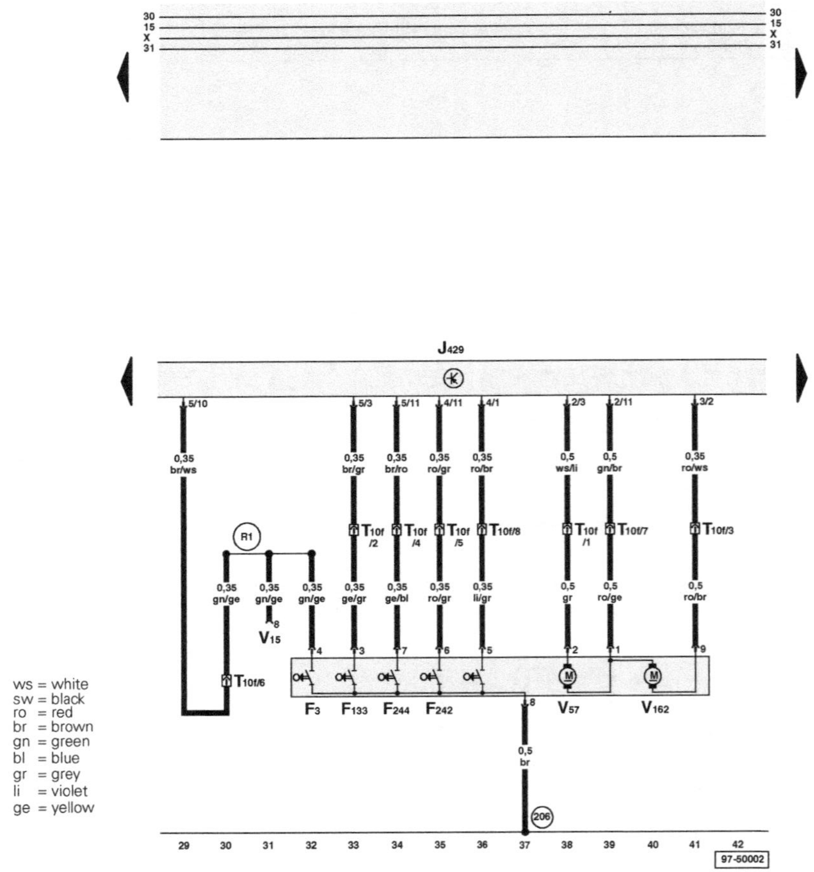

ws = white
sw = black
ro = red
br = brown
gn = green
bl = blue
gr = grey
li = violet
ge = yellow

F3 - Door Contact Switch - Passenger side
F133 - Right Front Central Locking System Activator/Switch
F242 - Contact switch for lock cylinder, passenger side
F244 - Adjusting element for central locking (safe), passenger door
J429 - Control module for central locking
T10f - Connector 10 pin, brown, connector station A pillar, right
V15 - Right Window Motor
V57 - Passenger's Door Central Locking System Motor
V162 - Motor for central locking (safe), passenger door

(206) - Ground connection, in passenger's door wiring harness
(R1) - Wire connection (door contact switch, right), in power window/central locking system and door contact switch wiring harness

Control Module for Central Locking, Lock Unit for Central Locking Passenger's Side

Edition 01.00
W42.USA.5702.03.21

© 2006 BentleyPublishers.com—All Rights Reserved

Audi TT — Wiring diagram — No. 12/5

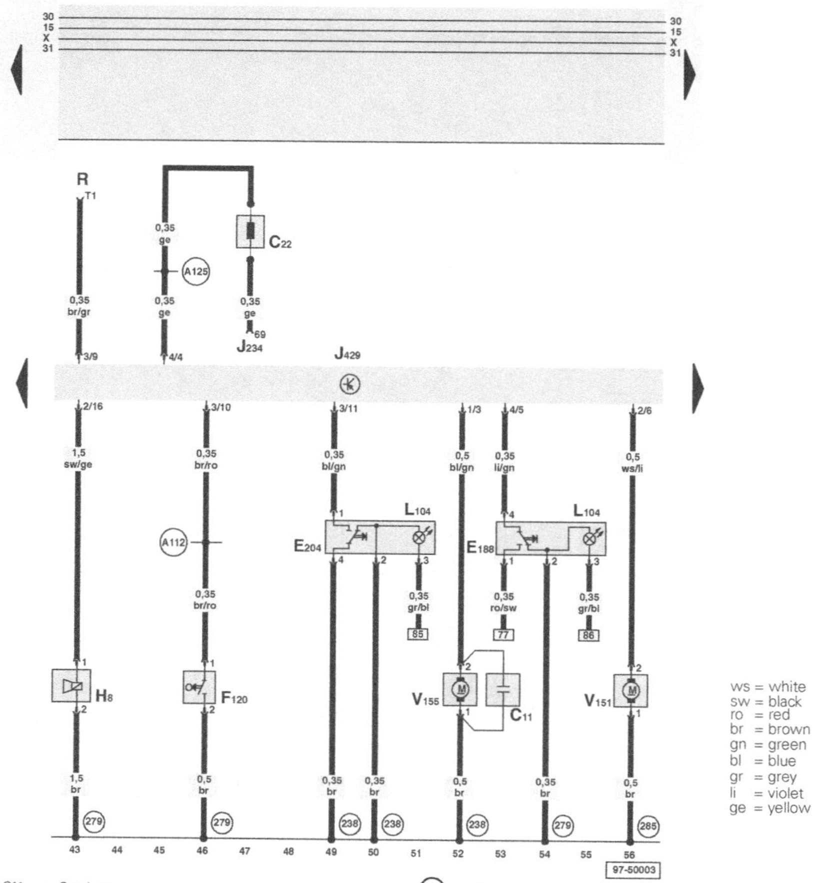

ws = white
sw = black
ro = red
br = brown
gn = green
bl = blue
gr = grey
li = violet
ge = yellow

- C11 - Condensor
- C22 - Airbag Suppresser
- E188 - Switch for remote unlock, rear lid
- E204 - Switch for remote/fuel tank door
- F120 - Hood Alarm Switch
- H8 - Alarm Horn
- J234 - Airbag Control Module
- J429 - Control module for central locking
- L104 - Illumination for switch remote unlock
- R - Radio
- T1 - Connector single, brown, on radio
- V151 - Motor for remote unlock rear lid
- V155 - Motor for fuel tank lid unlock

- (238) - Ground connection -1-, in wiring harness interior
- (279) - Ground connector -5-, in wiring harness interior
- (285) - Ground connection -7- in wiring harness interior
- (A112) - Connection (Hood alarm switch), in instrument panel wiring harness
- (A125) - Connection (crash signal) in instrument panel wiring harness

Edition 12/02
W42.USA.5702.12.21

Control Module for Central Locking, Switch for Remote Unlock rear lid, Switch for Remote/Fuel Tank Door

No. 12/6 — Wiring diagram — Audi TT

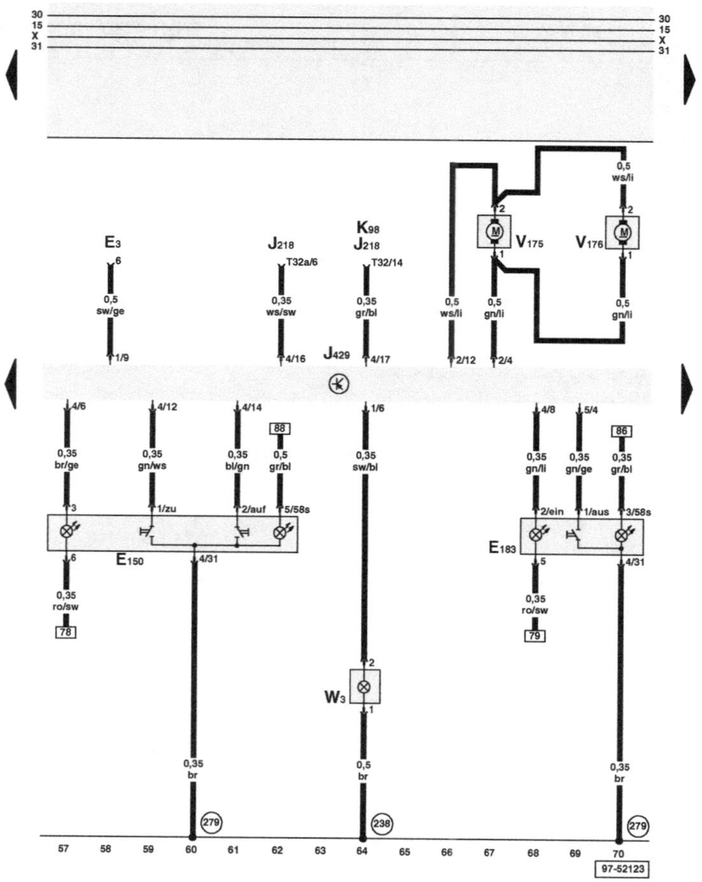

ws = white
sw = black
ro = red
br = brown
gn = green
bl = blue
gr = grey
li = violet
ge = yellow

- E3 - Emergency Flasher Switch
- E150 - Switch for interior lock, driver side
- E183 - Switch for passenger compartment monitoring
- J218 - Instrument Cluster Combination Processor
- J429 - Control module for central locking
- K98 - Convertible Top Locked Indicator Light
- T32 - Connector 32 pin, blue, on instrument cluster
- T32a - Connector 32 pin, green, on instrument cluster
- V175 - Storage Compartment Central Locking Motor
- V176 - CD Changer Compartment Central Locking Motor
- W3 - Luggage Compartment Light

- (238) - Ground connection -1-, in wiring harness interior
- (279) - Ground connector -5-, in wiring harness interior

Control Module for Central Locking, Switch for Interior Lock Driver Side, Switch for Passenger Compartment Monitoring, Luggage Compartment Light

Edition 12/02
W42.USA.5702.12.21

©2006 BentleyPublishers.com—All Rights Reserved

Central locking, Anti-theft, roadster (2000-2001)

Audi TT | **Wiring diagram** | **No. 12/7**

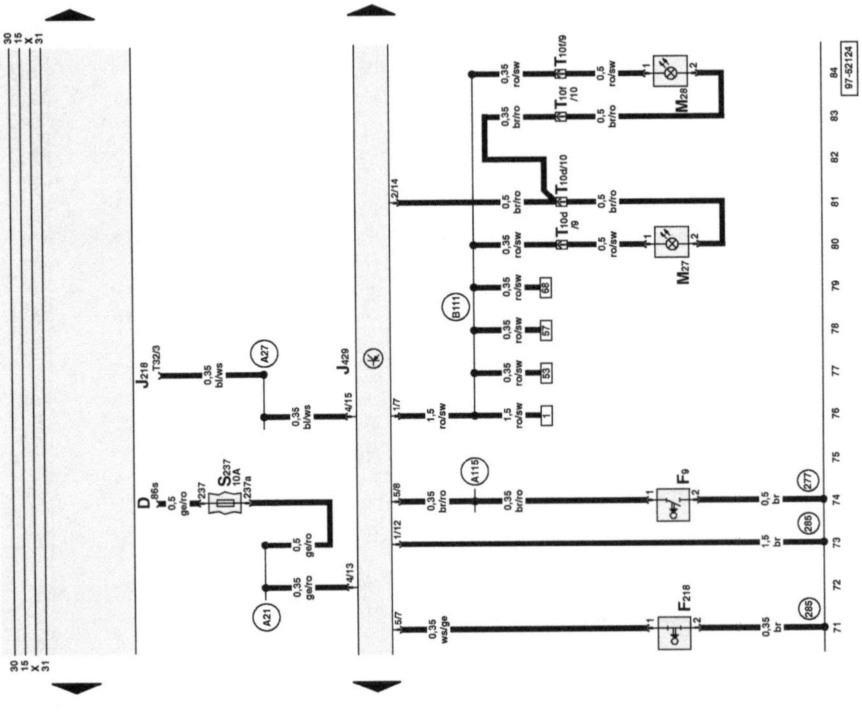

ws = white
sw = black
ro = red
br = brown
gn = green
bl = blue
gr = grey
li = violet
ge = yellow

- D - Ignition/Starter Switch
- F9 - Parking Brake Warning Light Switch
- F218 - Switch for central locking, trunk lid
- J218 - Instrument Cluster Combination Processor
- J429 - Control module for central locking
- M27 - Left Door Warning Light
- M28 - Right Door Warning Light
- S237 - Fuse 37 in fuse holder
- T10d - Connector 10 pin, brown, connector station A pillar, left
- T10f - Connector 10 pin, brown, connector station A pillar, right
- T32 - Connector 32 pin, blue, on instrument cluster

- (277) - Ground connector -3-, in wiring harness interior
- (285) - Ground connection -7- in wiring harness interior
- (A21) - Wire connection (86s), in instrument panel wiring harness
- (A27) - Wire connection (vehicle speed signal), in instrument panel wiring harness
- (A115) - Connection (parking brake control) in instrument panel wiring harness
- (B111) - Connector (30a), in wiring harness interior

Edition 07.00
W42.USA.5702.04.21

Control Module for Central Locking, Door Warning Lights

Audi TT | **Wiring diagram** | **No. 12/8**

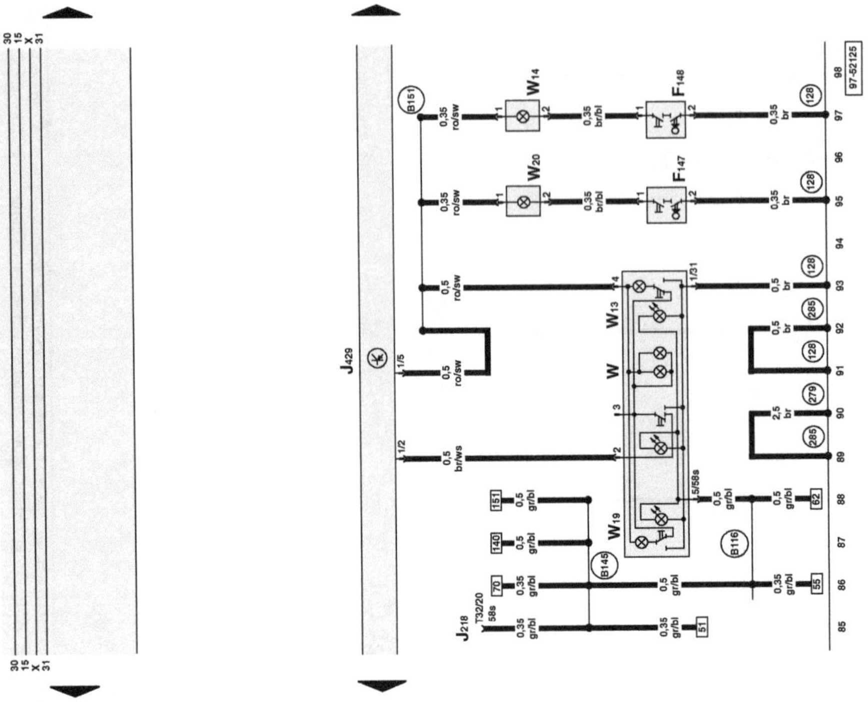

ws = white
sw = black
ro = red
br = brown
gn = green
bl = blue
gr = grey
li = violet
ge = yellow

- F147 - Left Make-Up Mirror Light Switch
- F148 - Right Make-Up Mirror Light Switch
- J218 - Instrument Cluster Combination Processor
- J429 - Control module for central locking
- T32 - Connector 32 pin, blue, on instrument cluster
- W - Front Interior Light
- W13 - Right Front Map/Reading Light
- W14 - Right Make-up Mirror Light
- W19 - Left Front Reading Light
- W20 - Left Make-up Mirror Light

- (128) - Ground connection -1-, in interior light wiring harness
- (279) - Ground connection -5-, in wiring harness interior
- (285) - Ground connection -7- in wiring harness interior
- (B116) - Connector (58s), in wiring harness interior
- (B145) - Plus connector (58s), in wiring harness interior
- (B151) - Plus connector -3- (30a), in wiring harness interior

Control Module for Central Locking, Front Interior Light, Reading Lights, Make-Up Mirror Lights

Edition 07.00
W42.USA.5702.04.21

© 2006 BentleyPublishers.com—All Rights Reserved

Audi TT   Wiring diagram   No. 12/9

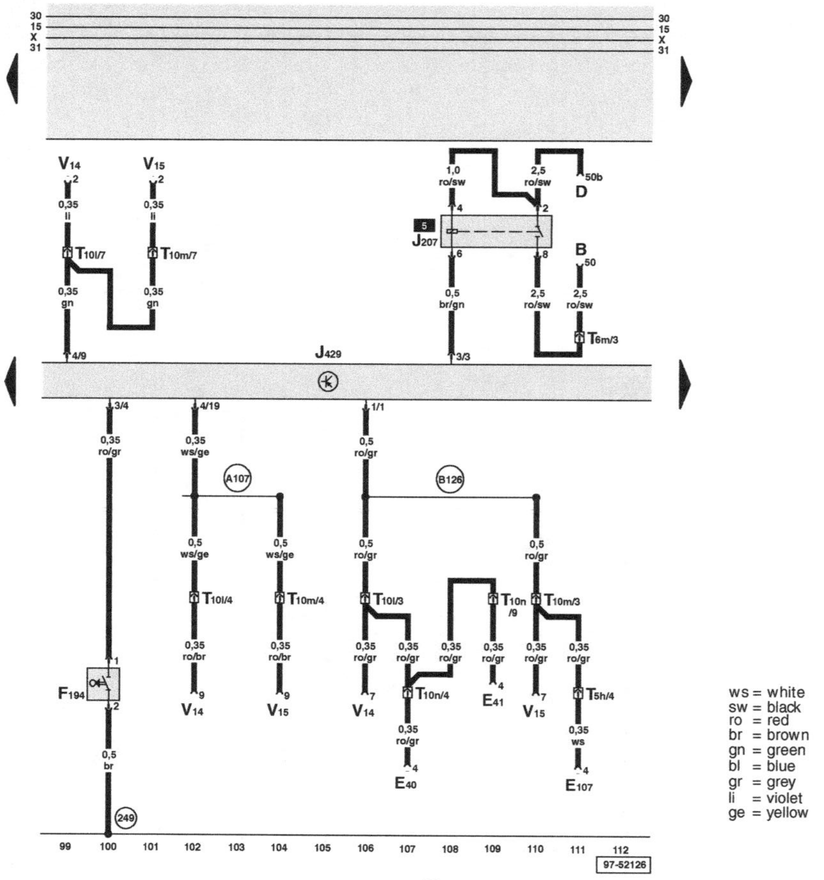

ws = white
sw = black
ro = red
br = brown
gn = green
bl = blue
gr = grey
li = violet
ge = yellow

- B - Starter
- D - Ignition/Starter Switch
- E40 - Left Front Window Switch
- E41 - Right Front Window Switch
- E107 - Switch for window regulator, in passenger door
- F194 - Clutch Pedal Position (CPP) Switch
- J207 - Starting Interlock Relay
- J429 - Control module for central locking
- T5h - Connector 5 pin, in right door
- T6m - Connector 6 pin, brown, in E-box plenum chamber
- T10l - Connector 10 pin, red, connector station A pillar, left
- T10m - Connector 10 pin, red, connector station A pillar, right
- T10n - Connector 10 pin, in left door
- V14 - Left Window Motor
- V15 - Right Window Motor

- (249) - Ground connection -2-, in wiring harness interior
- (A107) - Connector (comfort close), in instrument panel wiring harness
- (B126) - Connector (87, driver s door) , in wiring harness interior

Edition 05/01
W42.USA.5702.07.21

Control Module for Central Locking, Starting Interlock Relay

No. 12/10   Wiring diagram   Audi TT

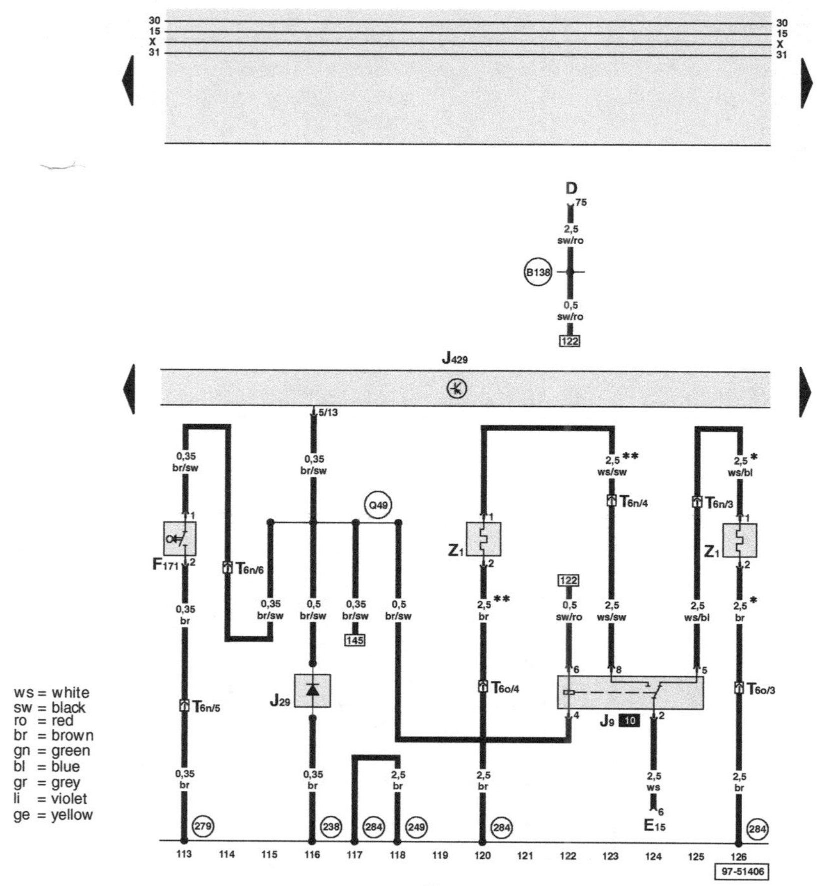

ws = white
sw = black
ro = red
br = brown
gn = green
bl = blue
gr = grey
li = violet
ge = yellow

- D - Ignition/Starter Switch
- E15 - Rear Window Defogger Switch
- F171 - Convertible Top Open Position Switch
- J9 - Rear Window Defogger Relay
- J29 - Blocking Diode
- J429 - Control module for central locking
- T6n - Connector 6 pin, Hardtop left
- T6o - Connector 6 pin, Hardtop right
- Z1 - Heated rear window

- (238) - Ground connection -1-, in wiring harness interior
- (249) - Ground connection -2-, in wiring harness interior
- (279) - Ground connector -5-, in wiring harness interior
- (284) - Ground connection -6- in wiring harness interior
- (B138) - Plus connector (X), in wiring harness, interior
- (Q49) - Connection (closed), in wiring harness interior

- * - Softtop
- ** - Hardtop (where applicable)

Top, Heated Rear Window

Edition 05/01
W42.USA.5702.07.21

©2006 BentleyPublishers.com—All Rights Reserved

Central locking, Anti-theft, roadster (2000-2001)

Audi TT — No. 12/11 — Wiring diagram

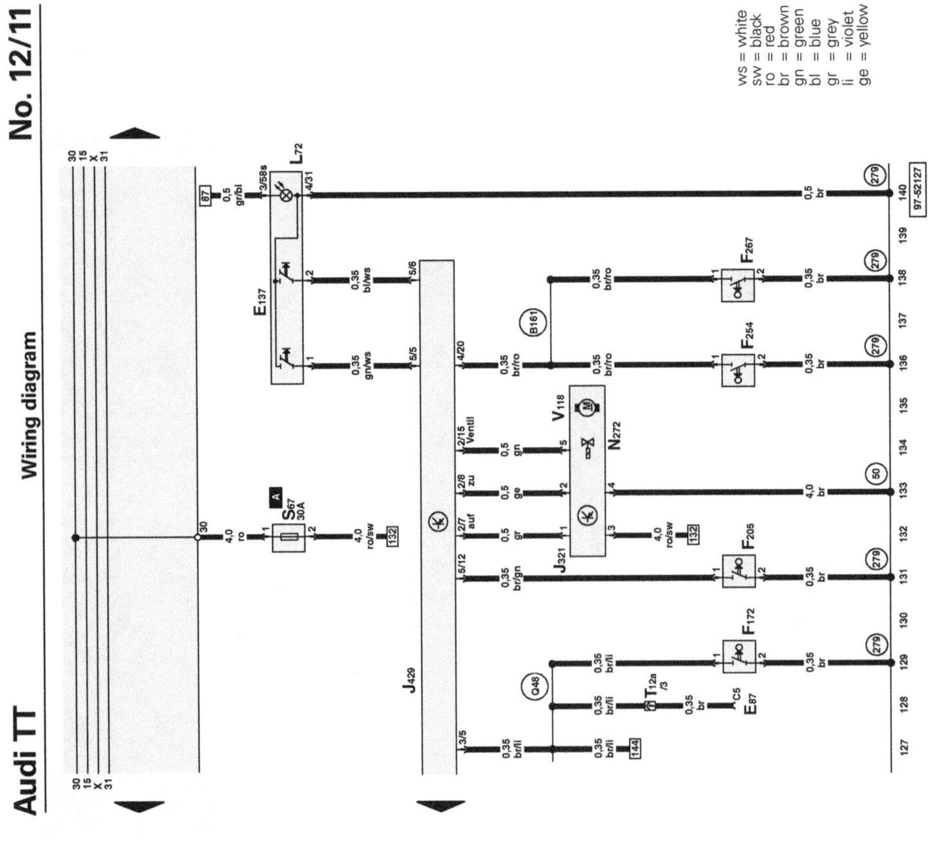

ws = white
sw = black
ro = red
br = brown
gn = green
bl = blue
gr = grey
li = violet
ge = yellow

- E87 - A/C Control Head
- E137 - Convertible Top Operation Switch
- F172 - Convertible Top Front Lock Switch
- F205 - Convertible Top Latch Open Position Switch
- F254 - Convertible Top Cover Detector Switch
- F267 - Convertible Top Cover Detector Switch 2
- J321 - Convertible Top Hydraulic Pump Relay
- J429 - Control module for central locking
- L72 - Convertible Top Switch Light
- N272 - Power Convertible Top Valve -1-
- S67 - Convertible Top Fuse
- T12a - Connector 12 pin, behind instrument panel, right
- V118 - Convertible Top Hydraulic Pump

- (50) - Ground connection, in luggage compartment, left
- (279) - Ground connector -5-, in wiring harness interior
- (B161) - Connector (75a), in wiring harness interior
- (Q48) - Connection (open), in wiring harness interior

Top

Edition 01.00
W42.USA.5702.03.21

Audi TT — No. 12/12 — Wiring diagram

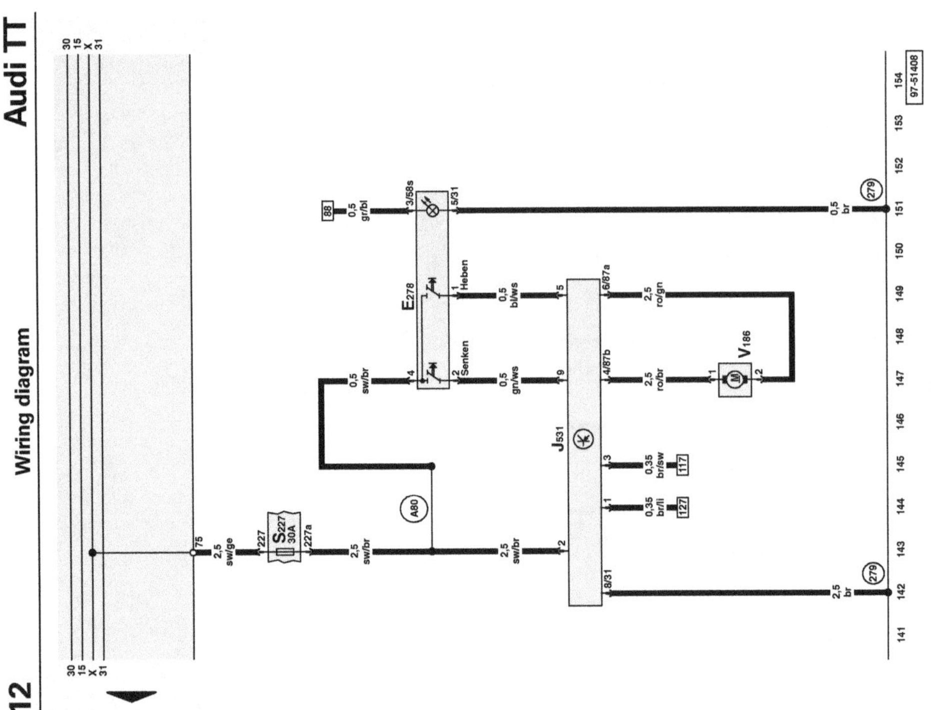

ws = white
sw = black
ro = red
br = brown
gn = green
bl = blue
gr = grey
li = violet
ge = yellow

- E278 - Wind Deflector Switch
- J531 - Wind Deflector Control Module
- S227 - Fuse 27 in fuse holder
- V186 - Wind Deflector Motor

- (279) - Ground connector -5-, in wiring harness interior
- (A80) - Connector -1- (X), in instrument panel wiring harness

Wind Deflector

Edition 01.00
W42.USA.5702.03.21

© 2006 BentleyPublishers.com—All Rights Reserved

Driver's Information System

2000 m. y.

Fuse Panel

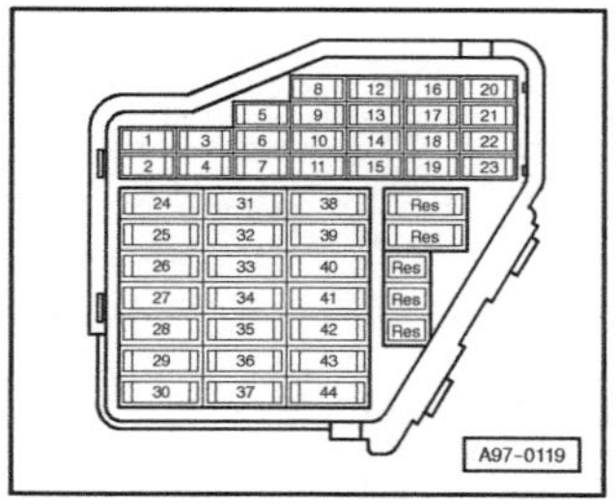

Fuse Colors:

30 A - Green
25 A - White
20 A - Yellow
15 A - Blue
10 A - Red
7,5 A - Brown
5 A - Beige

Starting with fuse position 23, fuses in the fuse holder are identified with 223 in the wiring diagram.

13 - Fold Relay Panel

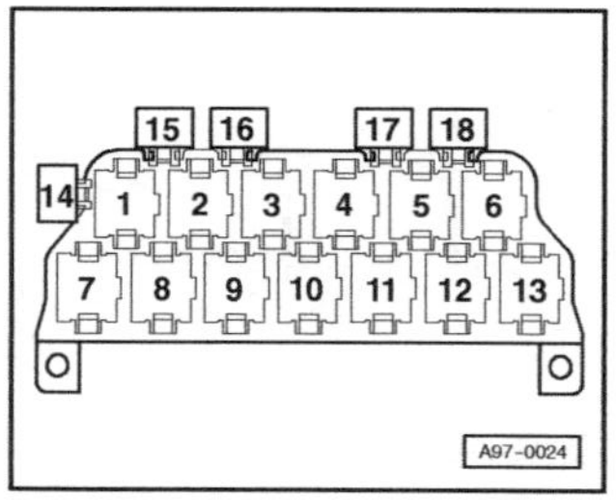

Relay Location:

- **2** -Lamp Control Module, J123
- **3** -Lamp Control Module, J123
- **7** -Brake light disable relay (ESP), J508

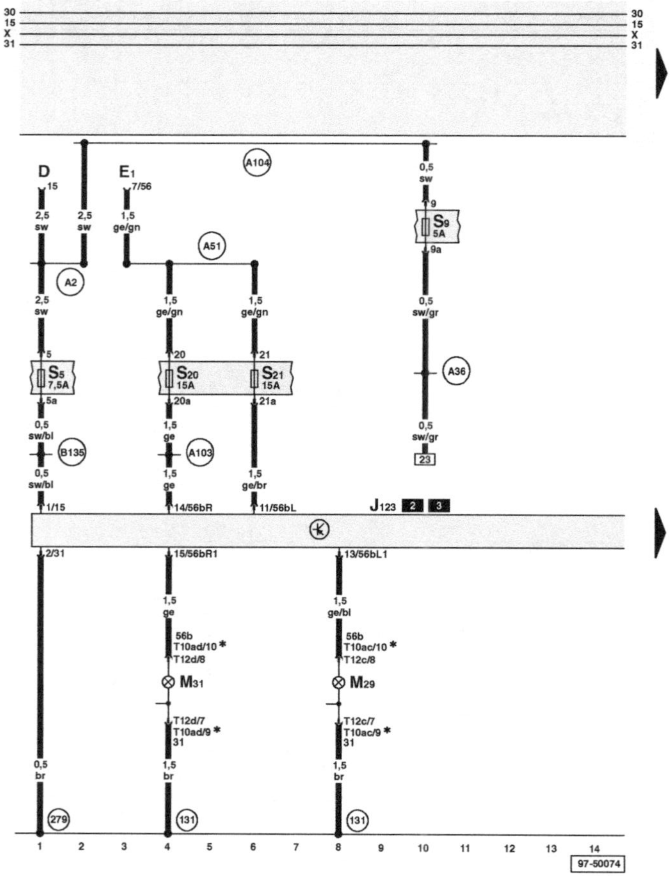

ws = white
sw = black
ro = red
br = brown
gn = green
bl = blue
gr = grey
li = violet
ge = yellow

D - Ignition/Starter Switch
E1 - Light Switch
J123 - Lamp Control Module
M29 - Left Low Beam Headlight
M31 - Right Low Beam Headlight
S5 - Fuse
S9 - Fuse
S20 - Fuse
S21 - Fuse
T10ac - Connector 10 pin, on headlight, left
T10ad - Connector 10 pin, on headlight, right
T12c - Connector 12 pin, on headlight, left
T12d - Connector 12 pin, on headlight, right

(131) - Ground connection -2-, in engine compartment wiring harness
(279) - Ground connector -5-, in wiring harness interior
(A2) - Plus connection (15), in instrument panel wiring harness
(A36) - Wire connection (75a), in instrument panel wiring harness
(A51) - Wire connection (56), in instrument panel wiring harness
(A103) - Connector -2- (56), in instrument panel wiring harness
(A104) - Plus connector -2- (15), in instrument panel wiring harness
(B135) - Connector (15a), in wiring harness interior

* - from VIN Y1 010 000

Lamp Control Module, Low Beam Headlight

© 2006 BentleyPublishers.com—All Rights Reserved

Audi TT Wiring diagram **No. 13/3**

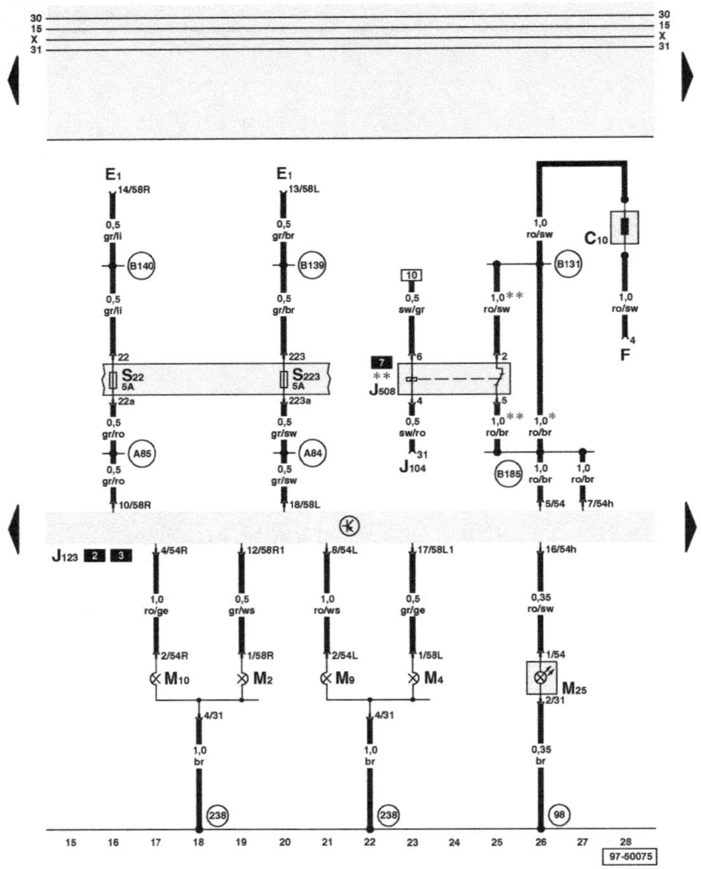

ws = white
sw = black
ro = red
br = brown
gn = green
bl = blue
gr = grey
li = violet
ge = yellow

C10 - Suppressor Filter
E1 - Light Switch
F - Brake Light Switch
J104 - ABS Control Module (w/EDL)
J123 - Lamp Control Module
J508 - Brake light disable relay (ESP)
M2 - Right Tail Light
M4 - Left Tail Light
M9 - Left Brake Light
M10 - Right Brake Light
M25 - High-mount Brake Light
S22 - Fuse
S223 - Fuse in fuse holder

(98) - Ground connection, in rear lid wiring harness
(238) - Ground connection -1-, in wiring harness interior
(A84) - Connector (58L), in instrument panel wiring harness
(A85) - Connector (58R), in instrument panel wiring harness
(B131) - Connector (54), in wiring harness interior
(B139) - Connector (PL), in wiring harness, interior
(B140) - Connector (PR), in wiring harness, interior
(B185) - Connection 2 (54), in passenger compartment wiring harness

* - Vehicles with ABS
** - Vehicles with ABS and ESP

Edition 12/02
W42.USA.5702.12.21

Lamp Control Module, Tail Light, Brake Light

No. 13/4 Wiring diagram **Audi TT**

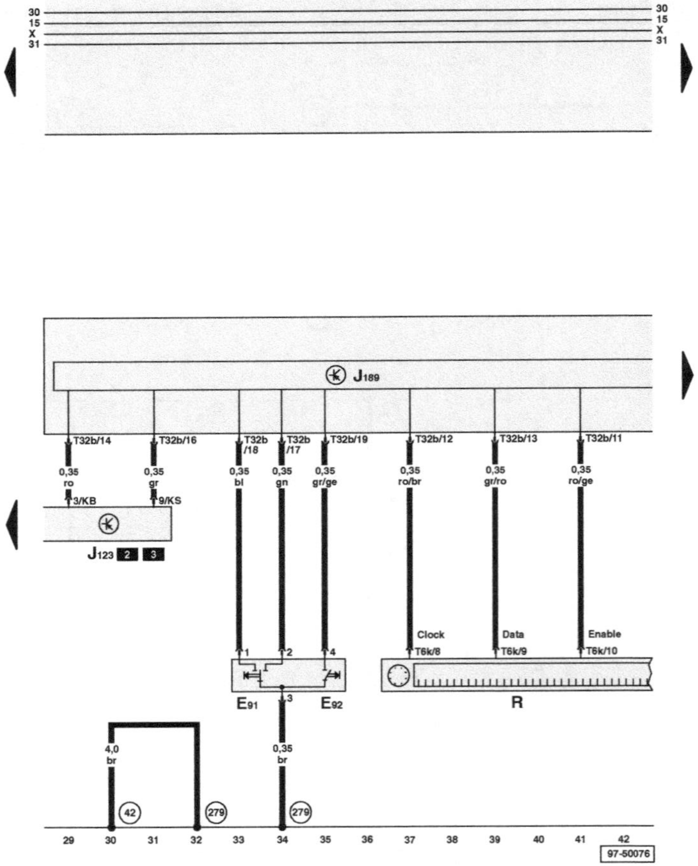

ws = white
sw = black
ro = red
br = brown
gn = green
bl = blue
gr = grey
li = violet
ge = yellow

E91 - Board Computer Function Selector Switch
E92 - Board Computer Reset Button
J123 - Lamp Control Module
J189 - Auto Check System
R - Radio
T6k - Connector 6 pin, green, connector radio I-2
T32b - Connector 32 pin, gray, on instrument cluster

(42) - Ground connection, on steering column
(279) - Ground connector -5-, in wiring harness interior

Auto Check System, Board Computer Function Selector Switch, Board Computer Reset Button

Edition 12/02
W42.USA.5702.12.21

© 2006 BentleyPublishers.com—All Rights Reserved

Audi TT Wiring diagram **No. 13/5**

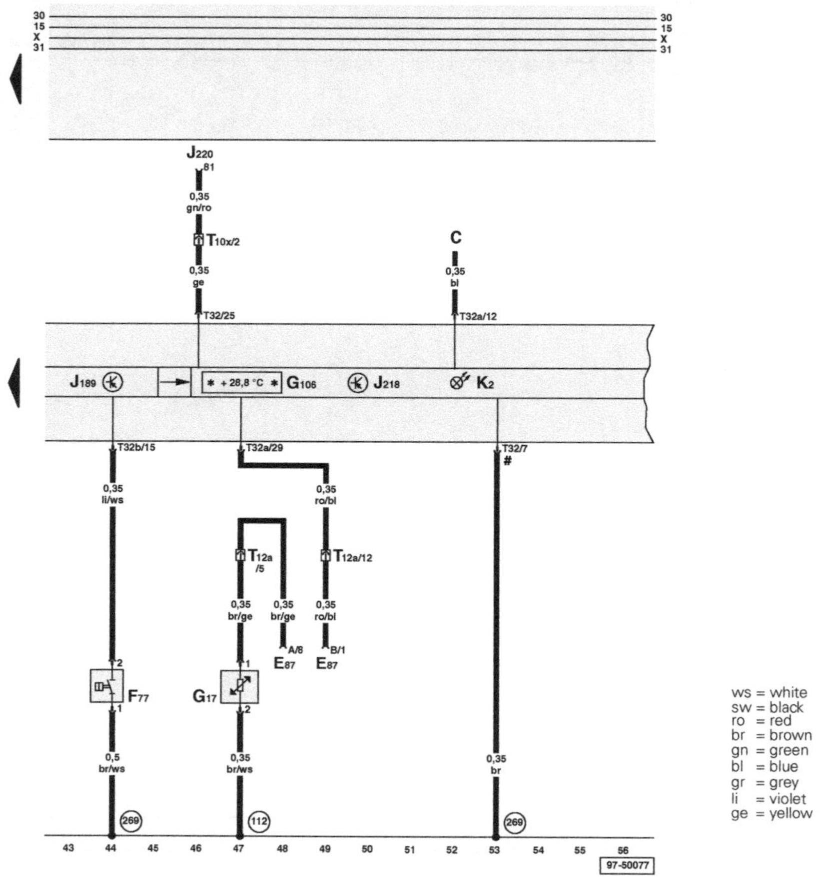

ws = white
sw = black
ro = red
br = brown
gn = green
bl = blue
gr = grey
li = violet
ge = yellow

C - Generator (GEN)
E87 - A/C Control Head
F77 - Windshield Washer Fluid Level Warning Switch
G17 - Outside Air Temperature Sensor
G106 - Outside Air Temperature Display
J189 - Auto Check System
J218 - Instrument Cluster Combination Processor
J220 - Motronic Engine Control Module (ECM)
K2 - Generator (GEN) Warning Light
T10x - Connector 10 pin, orange, in E-box plenum chamber
T12a - Connector 12 pin, behind instrument panel, right
T32 - Connector 32 pin, blue, on instrument cluster
T32a - Connector 32 pin, green, on instrument cluster
T32b - Connector 32 pin, gray, on instrument cluster

(112) - Ground connection -2-, in A/C wiring harness

(269) - Ground connector (sensor ground) -1-, in instrument panel wiring harness

- Sensor ground output

Edition 01.00
W42.USA.5702.03.21

Instrument Cluster, Windshield Washer Fluid Level Warning Switch, Outside Air Temperature Sensor

THIS SPACE INTENTIONALLY LEFT BLANK

© 2006 BentleyPublishers.com—All Rights Reserved

Navigation System

2000 m. y.

Fuse Panel

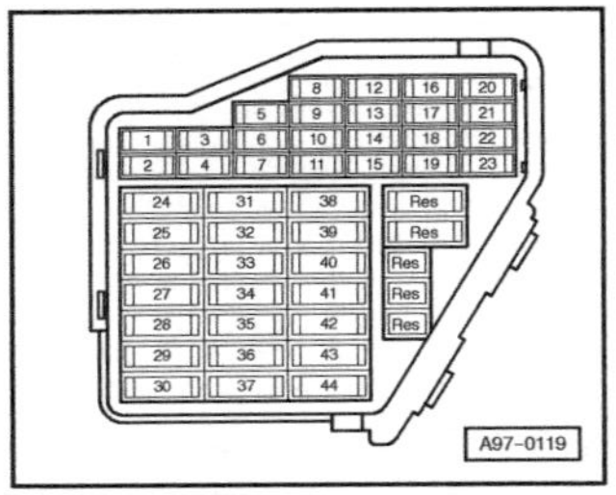

Fuse Colors:

30	A	-	Green
25	A	-	White
20	A	-	Yellow
15	A	-	Blue
10	A	-	Red
7,5	A	-	Brown
5	A	-	Beige

Starting with fuse position 23, fuses in the fuse holder are identified with 223 in the wiring diagram.

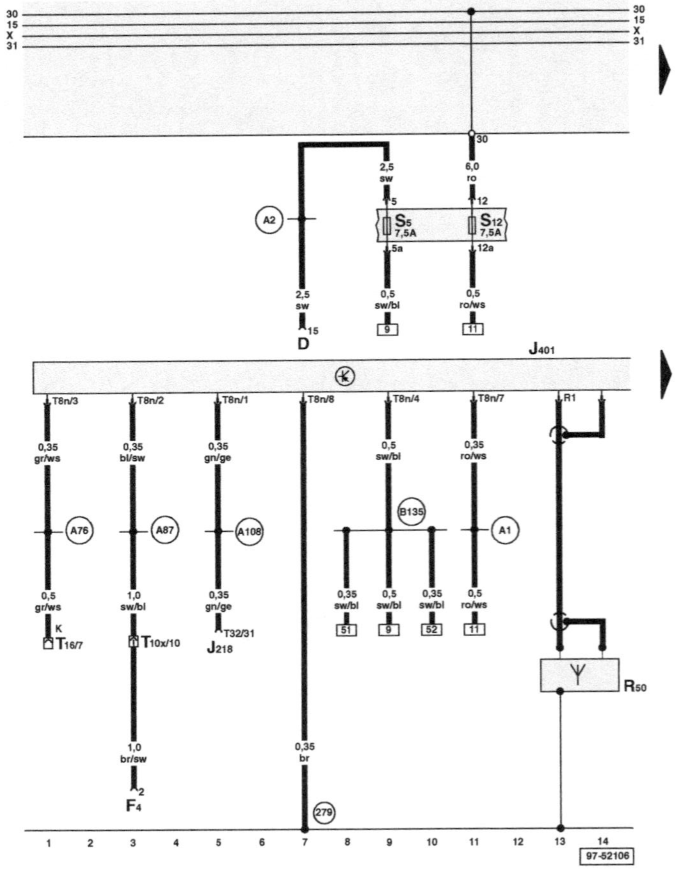

ws = white
sw = black
ro = red
br = brown
gn = green
bl = blue
gr = grey
li = violet
ge = yellow

D - Ignition/Starter Switch
F4 - Back-Up Light Switch
J218 - Instrument Cluster Combination Processor
J401 - Control unit for navigation with CD drive
R1 - Antenna Connection
R50 - Antenna for Navigation System (GPS)
S5 - Fuse
S12 - Fuse
T8n - Connector 8 pin, black, connector A on Control Module for Navigation
T10x - Connector 10 pin, orange, in E-box plenum chamber
T16 - Connector 16 pin, black, Data Link Connector (DLC), below storage compartment, driver's side
T32 - Connector 32 pin, blue, on instrument cluster

(279) - Ground connector -5-, in wiring harness interior
(A1) - Plus connection (30a), in instrument panel wiring harness
(A2) - Plus connection (15), in instrument panel wiring harness
(A76) - Connector (K-diagnosis wire), in instrument panel wiring harness
(A87) - Connector (RF), in instrument panel wiring harness
(A108) - Connector (speed signal), in instrument panel wiring harness
(B135) - Connector (15a), in wiring harness interior

Control Unit for Navigation with CD Drive, Antenna for Navigation System, Fuses

© 2006 BentleyPublishers.com—All Rights Reserved

Audi TT — Wiring diagram — No. 14/3

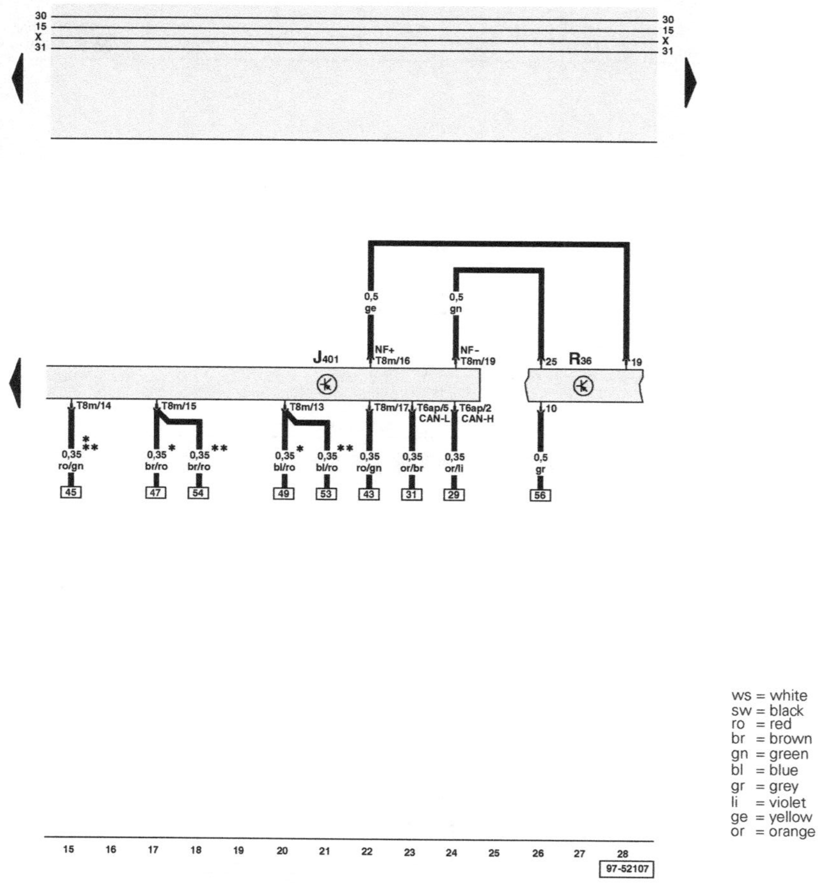

ws = white
sw = black
ro = red
br = brown
gn = green
bl = blue
gr = grey
li = violet
ge = yellow
or = orange

J401 - Control unit for navigation with CD drive
R36 - Telephone Transceiver
T6ap - Connector 6 pin, Connector C I on Control Module for Navigation
T8m - Connector 8 pin, C III on Control Module for Navigation

\- CAN-Bus (Data-Bus)
* - Vehicles with Amplified Speaker
** - Vehicles with Bose Sound System

Edition 01.00
W42.USA.5702.03.21

Control Unit for Navigation with CD Drive, Telephone Transceiver

No. 14/4 — Wiring diagram — Audi TT

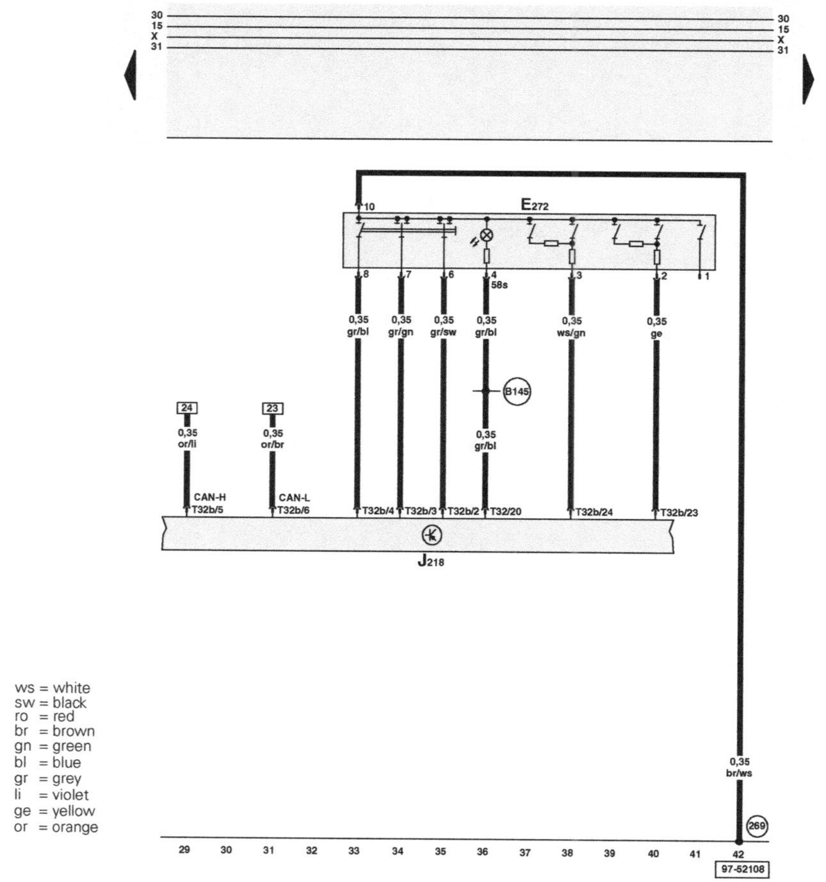

ws = white
sw = black
ro = red
br = brown
gn = green
bl = blue
gr = grey
li = violet
ge = yellow
or = orange

E272 - Function Selector Switch II
J218 - Instrument Cluster Combination Processor
T32 - Connector 32 pin, blue, on instrument cluster
T32b - Connector 32 pin, gray, on instrument cluster

(269) - Ground connector (sensor ground) -1-, in instrument panel wiring harness
(B145) - Plus connector (58s), in wiring harness interior

\- CAN-Bus (Data-Bus)

Instrument Cluster Combination Processor, Function Selector Switch II

Edition 01.00
W42.USA.5702.03.21

© 2006 BentleyPublishers.com—All Rights Reserved

Navigation System (2000-2001)

Audi TT **Wiring diagram** **No. 14/5**

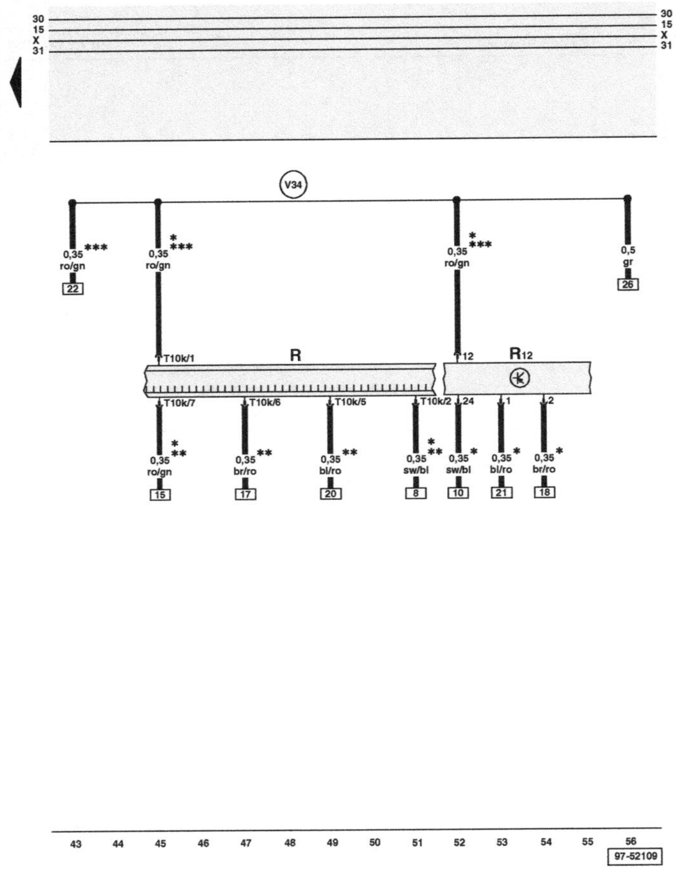

ws = white
sw = black
ro = red
br = brown
gn = green
bl = blue
gr = grey
li = violet
ge = yellow

R - Radio
R12 - Amplifier
T10k - Connector 10 pin, red, on radio

V34 - Connector (mute button), in wiring harness telephone

* - Vehicles with Bose Sound System
** - Vehicles with Amplified Speaker
*** - Vehicles with Telephone

Edition 01.00
W42.USA.5702.03.21

Radio, Amplifier

THIS SPACE INTENTIONALLY
LEFT BLANK

© 2006 BentleyPublishers.com—All Rights Reserved

Radio-Stereo Concert with and without CD-Changer (Coupé)

2000 m. y.

Fuse Panel

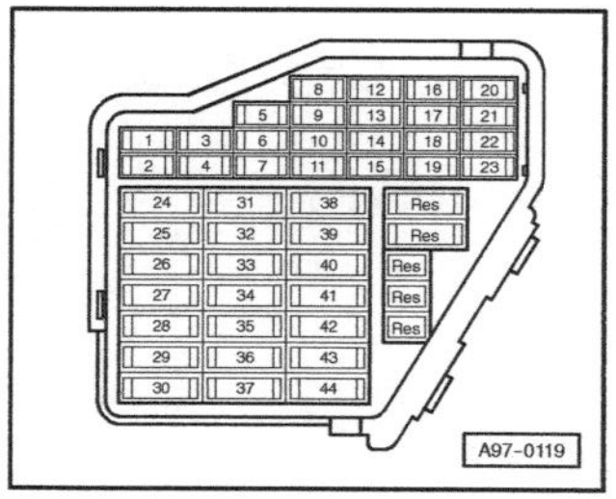

Fuse Colors:

30 A - Green
25 A - White
20 A - Yellow
15 A - Blue
10 A - Red
7,5 A - Brown
5 A - Beige

Starting with fuse position 23, fuses in the fuse holder are identified with 223 in the wiring diagram.

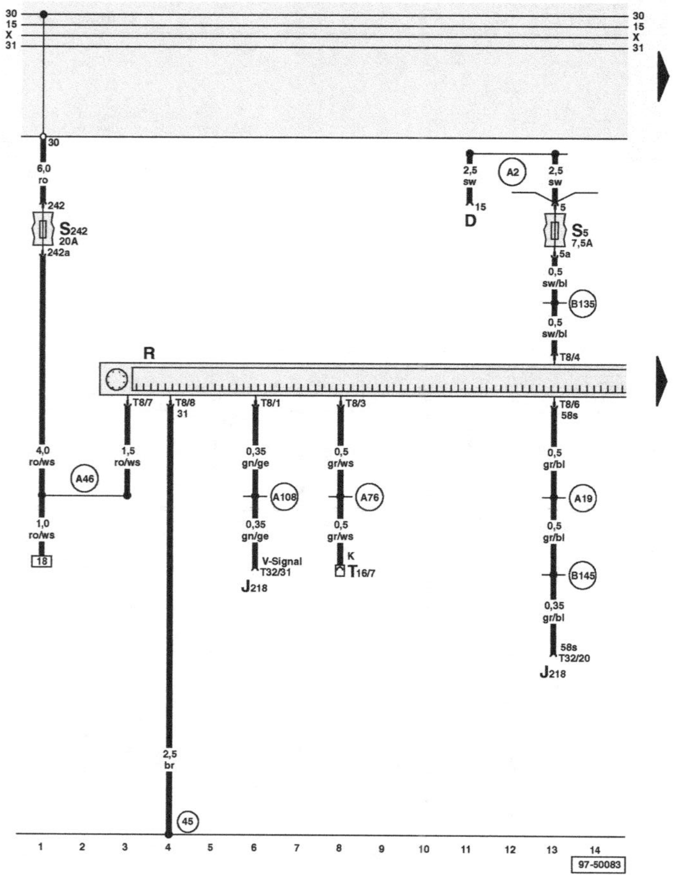

ws = white
sw = black
ro = red
br = brown
gn = green
bl = blue
gr = grey
li = violet
ge = yellow

D - Ignition/Starter Switch
J218 - Instrument Cluster Combination Processor
R - Radio
S5 - Fuse
S242 - Fuse in fuse holder
T8 - Connector 8 pin, black, connector radio III
T16 - Connector 16 pin, black, Data Link Connector (DLC), below storage compartment, driver's side
T32 - Connector 32 pin, blue, on instrument cluster

(45) - Ground connection, behind instrument panel, center
(A2) - Plus connection (15), in instrument panel wiring harness
(A19) - Wire connection (58s), in instrument panel wiring harness
(A46) - Plus connection (30-from radio), in instrument panel wiring harness
(A76) - Connector (K-diagnosis wire), in instrument panel wiring harness
(A108) - Connector (speed signal), in instrument panel wiring harness
(B135) - Connector (15a), in wiring harness interior
(B145) - Plus connector (58s), in wiring harness interior

© 2006 BentleyPublishers.com—All Rights Reserved

Sound System, Concert, coupe (2000-2001)

Audi TT — Wiring diagram — No. 15/3

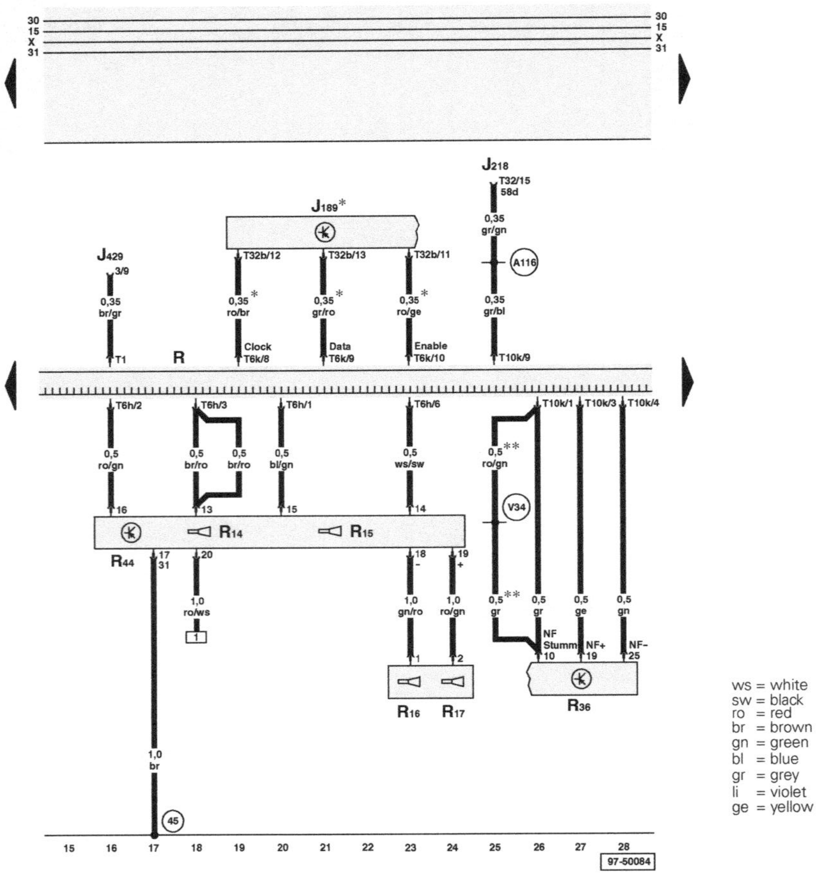

ws = white
sw = black
ro = red
br = brown
gn = green
bl = blue
gr = grey
li = violet
ge = yellow

J189 - Auto Check System
J218 - Instrument Cluster Combination Processor
J429 - Control module for central locking
R - Radio
R14 - Left Rear Tweeter
R15 - Left Rear Woofer
R16 - Right Rear Tweeter
R17 - Right Rear Woofer
R36 - Telephone Transceiver
R44 - Left Rear Woofer/Amplifier
T1 - Connector single, brown, on radio
T6h - Connector 6 pin, yellow, connector radio I-1
T6k - Connector 6 pin, green, connector radio I-2
T10k - Connector 10 pin, red, on radio
T32 - Connector 32 pin, blue, on instrument cluster
T32b - Connector 32 pin, gray, on instrument cluster

(45) - Ground connection, behind instrument panel, center
(A116) - Connection -2- (58d) in instrument panel wiring harness
(V34) - Connector (mute button), in wiring harness telephone

* - Vehicles with Auto Check System
** - Vehicles with navigation system

Edition 12/02
W42.USA.5702.12.21

Radio, Left Rear Woofer/Amplifier

No. 15/4 — Wiring diagram — Audi TT

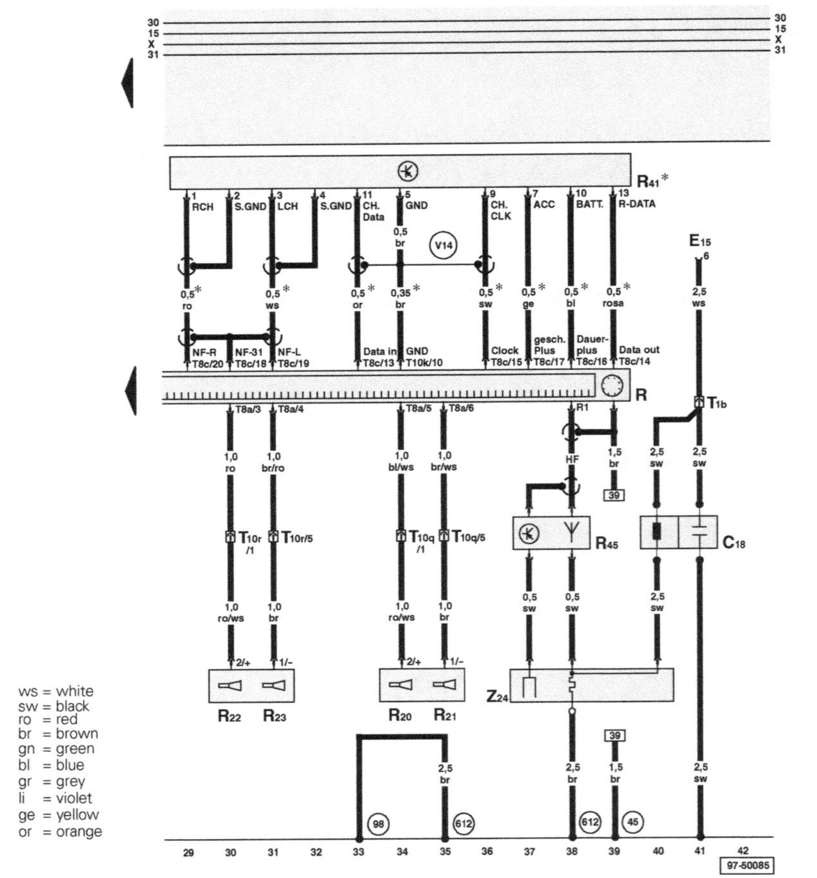

ws = white
sw = black
ro = red
br = brown
gn = green
bl = blue
gr = grey
li = violet
ge = yellow
or = orange

C18 - Windshield Antenna Suppression Filter
E15 - Rear Window Defogger Switch
R - Radio
R1 - Antenna Connection
R20 - Left Front Tweeter
R21 - Left Front Woofer
R22 - Right Front Tweeter
R23 - Right Front Woofer
R41 - CD Changer Unit
R45 - Telephone/Radio Antenna Amplifier
T1b - Connector single, black, in rear lid left
T8a - Connector 8 pin, brown, connector radio II
T8c - Connector 8 pin, blue, connector radio I-3
T10k - Connector 10 pin, red, on radio
T10q - Connector 10 pin, green, connector station A pillar, left
T10r - Connector 10 pin, green, connector station A pillar, right
Z24 - Rear Window Defogger With Window Antenna

(45) - Ground connection, behind instrument panel, center
(98) - Ground connection, in rear lid wiring harness
(612) - Ground Point in rear lid, center
(V14) - Wire connection (shielding), in CD Changer wiring harness

* - Vehicles with CD-Changer

Radio, CD Changer Unit, Antenna

Edition 12/02
W42.USA.5702.12.21

© 2006 BentleyPublishers.com—All Rights Reserved

Radio Bose with and without CD-Changer

2000 m. y.

Fuse Panel

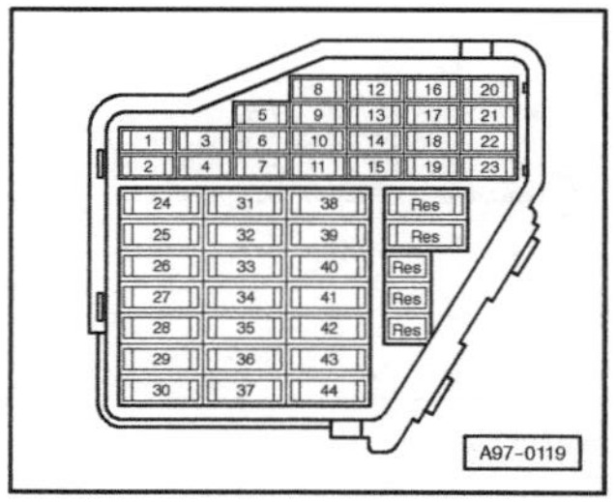

Fuse Colors:

30 A - Green
25 A - White
20 A - Yellow
15 A - Blue
10 A - Red
7,5 A - Brown
5 A - Beige

Starting with fuse position 23, fuses in the fuse holder are identified with 223 in the wiring diagram.

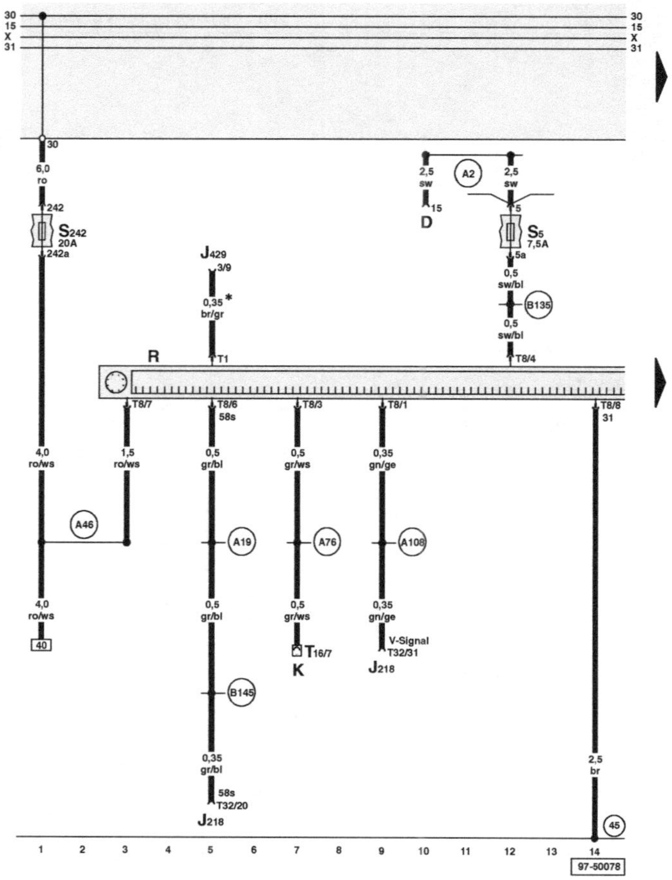

ws = white
sw = black
ro = red
br = brown
gn = green
bl = blue
gr = grey
li = violet
ge = yellow

D - Ignition/Starter Switch
J218 - Instrument Cluster Combination Processor
J429 - Control module for central locking
K - Data Link Connector (DLC) Rapid Data Transfer
R - Radio
S5 - Fuse
S242 - Fuse in fuse holder
T1 - Connector single, brown, on radio
T8 - Connector 8 pin, black, connector radio III
T16 - Connector 16 pin, black, Data Link Connector (DLC), under steering column, left
T32 - Connector 32 pin, blue, on instrument cluster

(45) - Ground connection, behind instrument panel, center
(A2) - Plus connection (15), in instrument panel wiring harness
(A19) - Wire connection (58s), in instrument panel wiring harness
(A46) - Plus connection (30-from radio), in instrument panel wiring harness
(A76) - Connector (K-diagnosis wire), in instrument panel wiring harness
(A108) - Connector (speed signal), in instrument panel wiring harness
(B135) - Connector (15a), in wiring harness interior
(B145) - Plus connector (58s), in wiring harness interior
* - Vehicles with Anti theft Alarm System

©2006 BentleyPublishers.com—All Rights Reserved

Sound System, Bose, coupe (2000-2001)

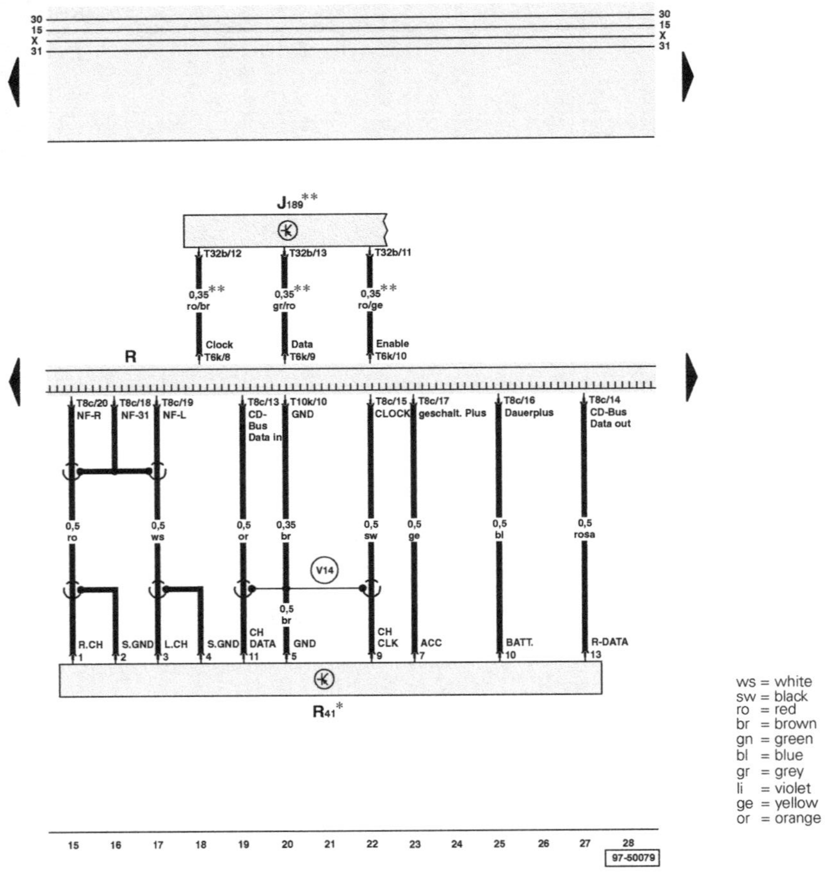

ws = white
sw = black
ro = red
br = brown
gn = green
bl = blue
gr = grey
li = violet
ge = yellow
or = orange

J189 - Auto Check System
R - Radio
R41 - CD Changer Unit
T6k - Connector 6 pin, green, connector radio I-2
T8c - Connector 8 pin, blue, connector radio I-3
T10k - Connector 10 pin, red, on radio
T32b - Connector 32 pin, gray, on instrument cluster

(V14) - Wire connection (shielding), in CD Changer wiring harness

* - Vehicles with CD-Changer
** - Vehicles with Auto Check System

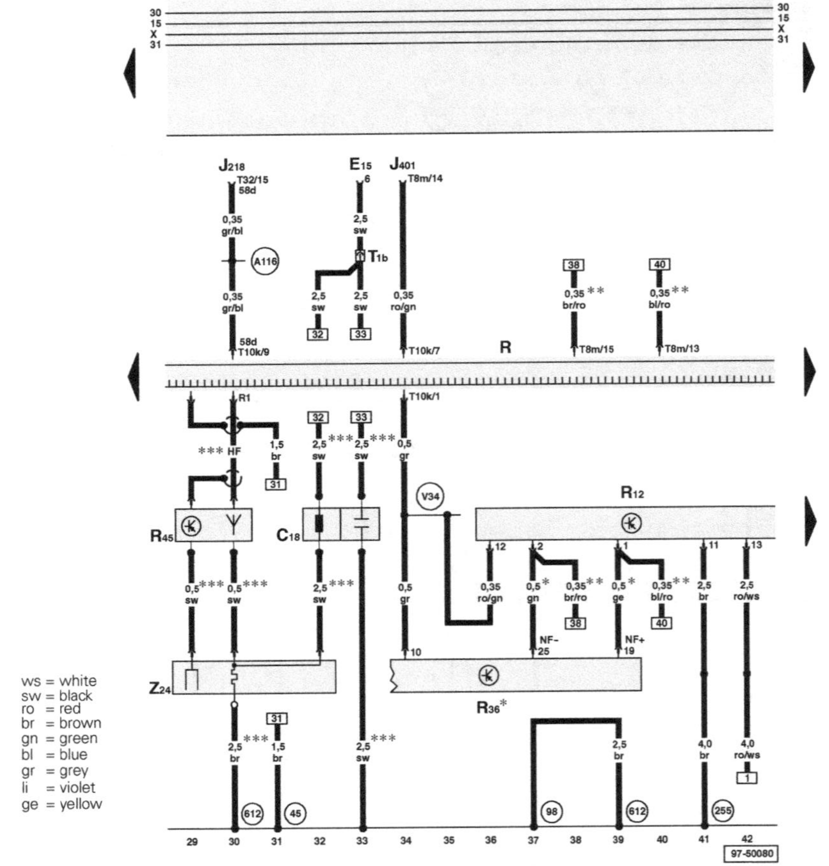

ws = white
sw = black
ro = red
br = brown
gn = green
bl = blue
gr = grey
li = violet
ge = yellow

C18 - Windshield Antenna Suppression Filter
E15 - Rear Window Defogger Switch
J218 - Instrument Cluster Combination Processor
J401 - Control unit for navigation with CD drive
R - Radio
R1 - Antenna Connection
R12 - Amplifier
R36 - Telephone Transceiver
R45 - Telephone/Radio Antenna Amplifier
T1b - Connector single, black, in rear lid left
T8m - Connector 8 pin, C III on Control Module for Navigation
T10k - Connector 10 pin, red, on radio
T32 - Connector 32 pin, blue, on instrument cluster
Z24 - Rear Window Defogger With Window Antenna

(45) - Ground connection, behind instrument panel, center
(98) - Ground connection, in rear lid wiring harness
(255) - Ground connection -1-, in wiring harness Radio
(612) - Ground Point in rear lid, center
(A116) - Connection -2- (58d) in instrument panel wiring harness
(V34) - Connector (mute button), in wiring harness telephone

* - Vehicles without navigation system
** - Vehicles with navigation system
*** - only Coupé

© 2006 BentleyPublishers.com—All Rights Reserved

Audi TT — Wiring diagram — No. 16/5

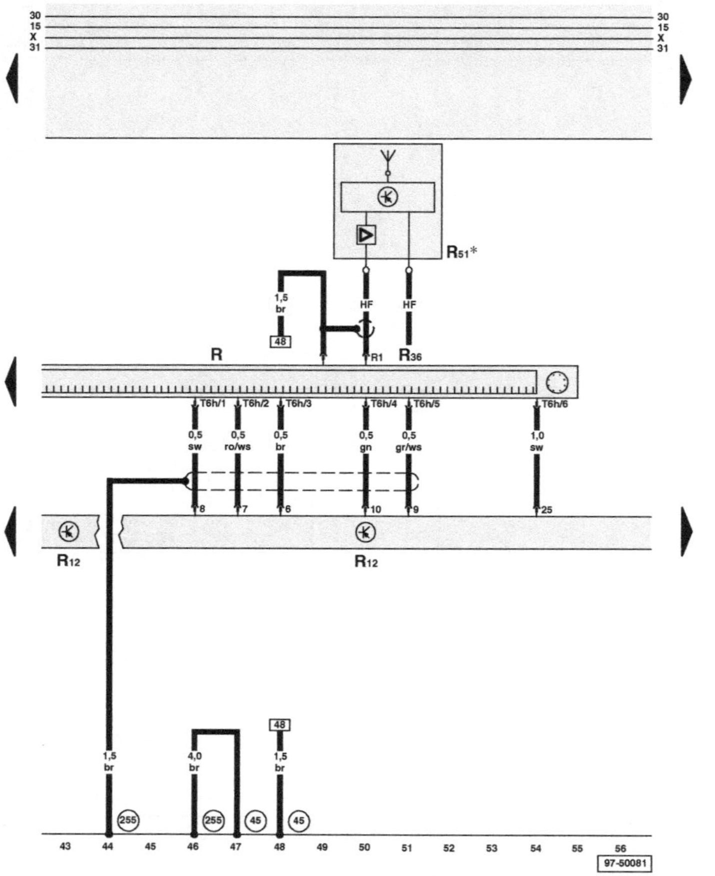

ws = white
sw = black
ro = red
br = brown
gn = green
bl = blue
gr = grey
li = violet
ge = yellow

R - Radio
R1 - Antenna Connection
R12 - Amplifier
R36 - Telephone Transceiver
R51 - Antenna for radio/telephone/park heating
T6h - Connector 6 pin, yellow, connector radio I-1

(45) - Ground connection, behind instrument panel, center

(255) - Ground connection -1-, in wiring harness Radio

* - only Roadster

Edition 12/02
W42.USA.5702.12.21

Radio, Amplifier

No. 16/6 — Wiring diagram — Audi TT

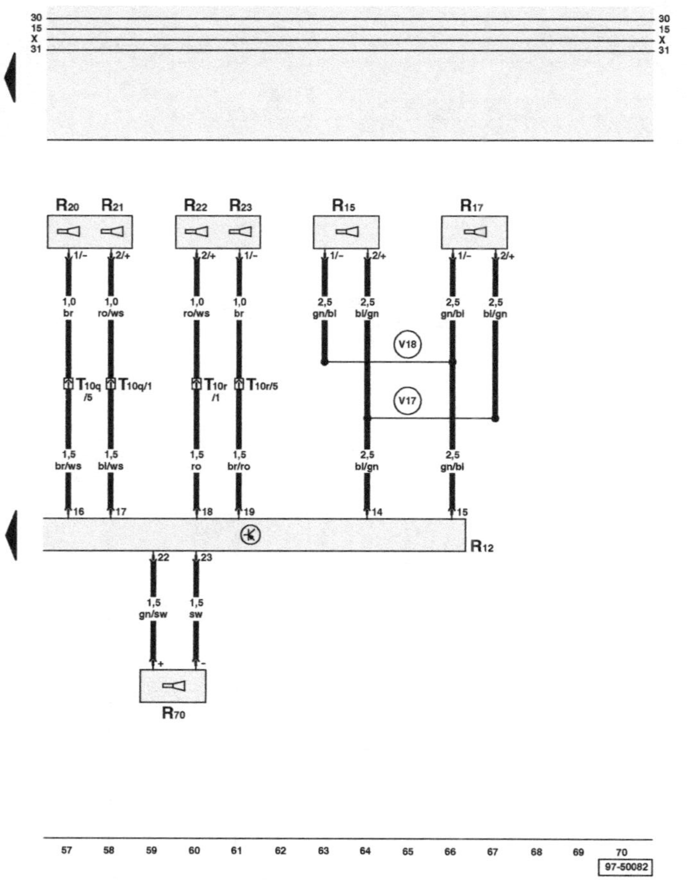

ws = white
sw = black
ro = red
br = brown
gn = green
bl = blue
gr = grey
li = violet
ge = yellow

R12 - Amplifier
R15 - Left Rear Woofer
R17 - Right Rear Woofer
R20 - Left Front Tweeter
R21 - Left Front Woofer
R22 - Right Front Tweeter
R23 - Right Front Woofer
R70 - Inside loudspeaker, center
T10q - Connector 10 pin, green, connector station A pillar, left
T10r - Connector 10 pin, green, connector station A pillar, right

(V17) - Connector (plus, speaker rear), in wiring harness speaker

(V18) - Connector (minus, speaker rear), in wiring harness speaker

Amplifier, Loudspeaker

Edition 12/02
W42.USA.5702.12.21

© 2006 BentleyPublishers.com—All Rights Reserved

Automatic Climate Control

2000 m. y.

Fuse Panel

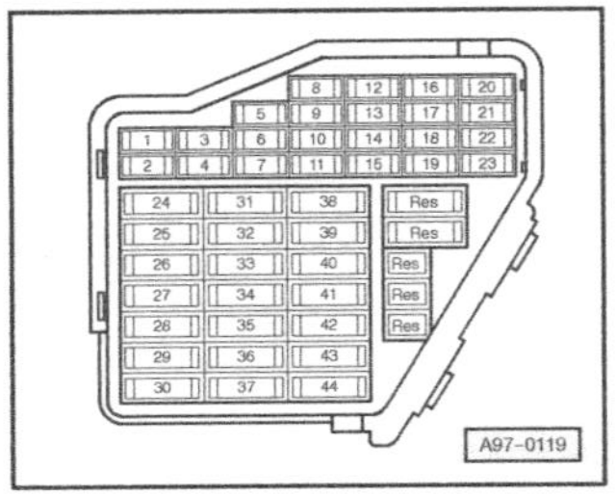

Fuse Colors:

30 A - Green
25 A - White
20 A - Yellow
15 A - Blue
10 A - Red
7,5 A - Brown
5 A - Beige

Starting with fuse position 23, fuses in the fuse holder are identified with 223 in the wiring diagram.

Main Fuse Box/Battery

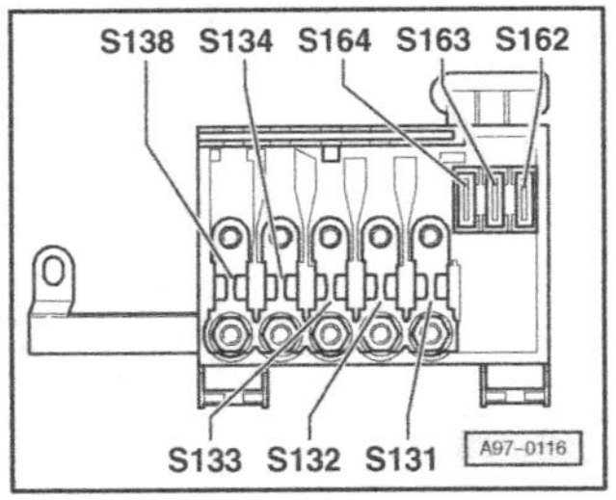

Fuses Location:

S131 - Safety fuse 1
S132 - Safety fuse 2
S133 - Safety fuse 3
S134 - Safety fuse 4
S138 - Safety fuse 5
S162 - Fuse 1 in fuse bracket/battery
S163 - Fuse 2 in fuse bracket/battery
S164 - Fuse 3 in fuse bracket/battery

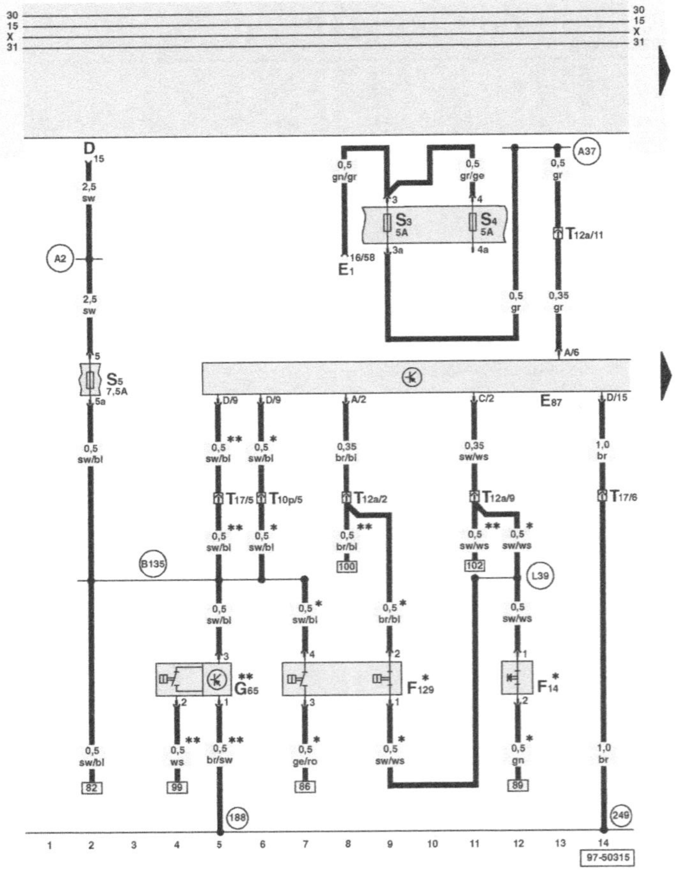

ws = white
sw = black
ro = red
br = brown
gn = green
bl = blue
gr = grey
li = violet
ge = yellow

D - Ignition/Starter Switch
E1 - Light Switch
E87 - A/C Control Head
F14 - Engine Coolant Temperature (ECT) Warning Switch
F129 - A/C Pressure Switch
G65 - High Pressure Sensor
S3 - Fuse
S4 - Fuse
S5 - Fuse
T10p - Connector 10 pin, behind instrument panel
T12a - Connector 12 pin, behind instrument panel, right
T17 - Connector 17 pin, behind instrument panel

(85) - Ground connection -1-, in engine compartment wiring harness
(188) - Ground connection -3-, in A/C wiring harness
(249) - Ground connection -2-, in wiring harness interior
(A2) - Plus connection (15), in instrument panel wiring harness
(A37) - Wire connection (58a), in instrument panel wiring harness
(B135) - Connector (15a), in wiring harness interior
(L39) - Wire connection -1- (A/C Pressure Switch), in Climatronic wiring harness

* - Automatic Climate Control with A/C Pressure switch
** - Automatic Climate Control with High pressure sensor, from 12.12.1999 or VIN 8N ZY1 055070

A/C Control Head, High Pressure Sensor

© 2006 BentleyPublishers.com—All Rights Reserved

Audi TT — Wiring diagram — No. 17/3

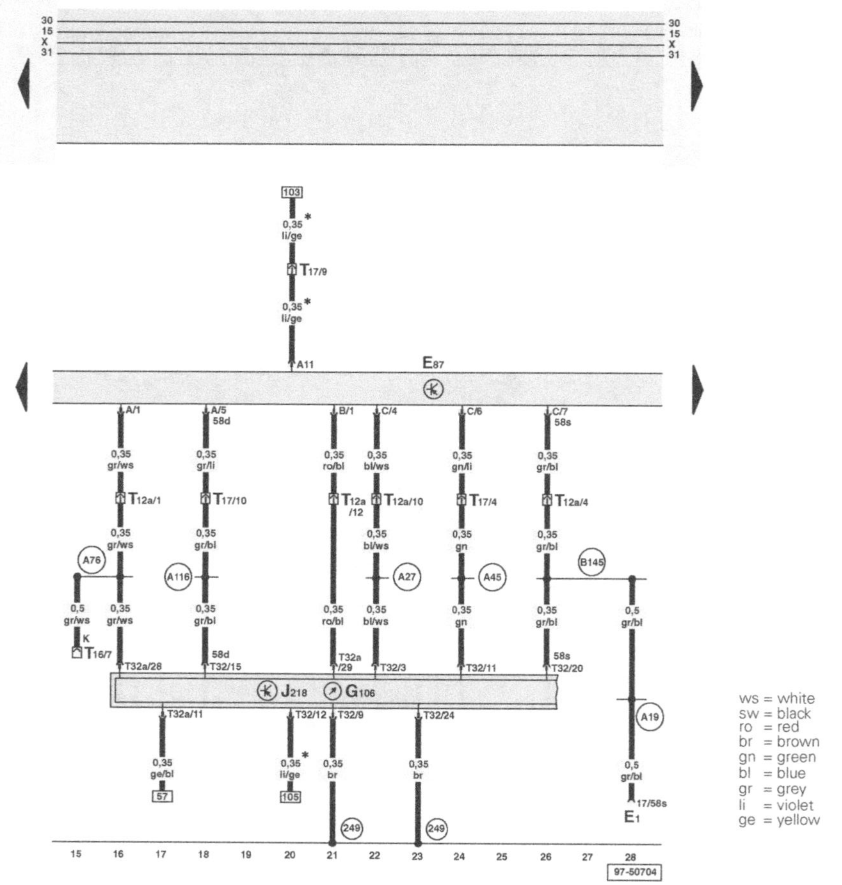

ws = white
sw = black
ro = red
br = brown
gn = green
bl = blue
gr = grey
li = violet
ge = yellow

- E1 - Light Switch
- E87 - A/C Control Head
- G106 - Outside Air Temperature Display
- J218 - Instrument Cluster Combination Processor
- T12a - Connector 12 pin, behind instrument panel, right
- T16 - Connector 16 pin, black, Data Link Connector (DLC), below storage compartment, driver's side
- T17 - Connector 17 pin, behind instrument panel
- T32 - Connector 32 pin, blue, on instrument cluster
- T32a - Connector 32 pin, green, on instrument cluster

- (249) - Ground connection -2-, in wiring harness interior
- (A19) - Wire connection (58s), in instrument panel wiring harness
- (A27) - Wire connection (vehicle speed signal), in instrument panel wiring harness
- (A45) - Wire connection (RPM-signal), in instrument panel wiring harness
- (A76) - Connector (K-diagnosis wire), in instrument panel wiring harness
- (A116) - Connection -2- (58d) in instrument panel wiring harness

* - Automatic climate control with High pressure sensor, from 12.12.1999 or VIN 8N ZY1 055070

No. 17/4 — Wiring diagram — Audi TT

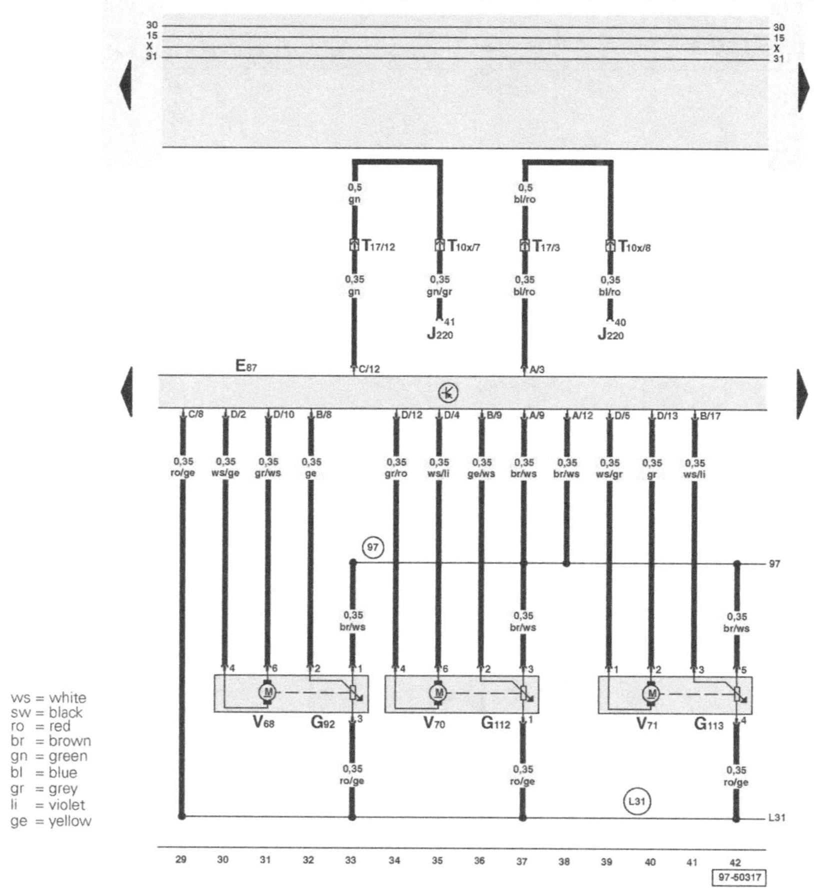

ws = white
sw = black
ro = red
br = brown
gn = green
bl = blue
gr = grey
li = violet
ge = yellow

- E87 - A/C Control Head
- G92 - Temperature Regulator Flap Motor Position Sensor
- G112 - Central Flap Motor Position Sensor
- G113 - Back Pressure Flap Motor Position Sensor
- J220 - Motronic Engine Control Module (ECM)
- T10x - Connector 10 pin, orange, in E-box plenum chamber
- T17 - Connector 17 pin, behind instrument panel
- V68 - Temperature Regulator Flap Motor
- V70 - Central Air Flap Motor
- V71 - Air Flow Flap Motor

- (97) - Ground connection -1-, in A/C wiring harness
- (L31) - Wire connection (5 Volts), in A/C wiring harness

©2006 BentleyPublishers.com—All Rights Reserved

Audi TT Wiring diagram **No. 17/5**

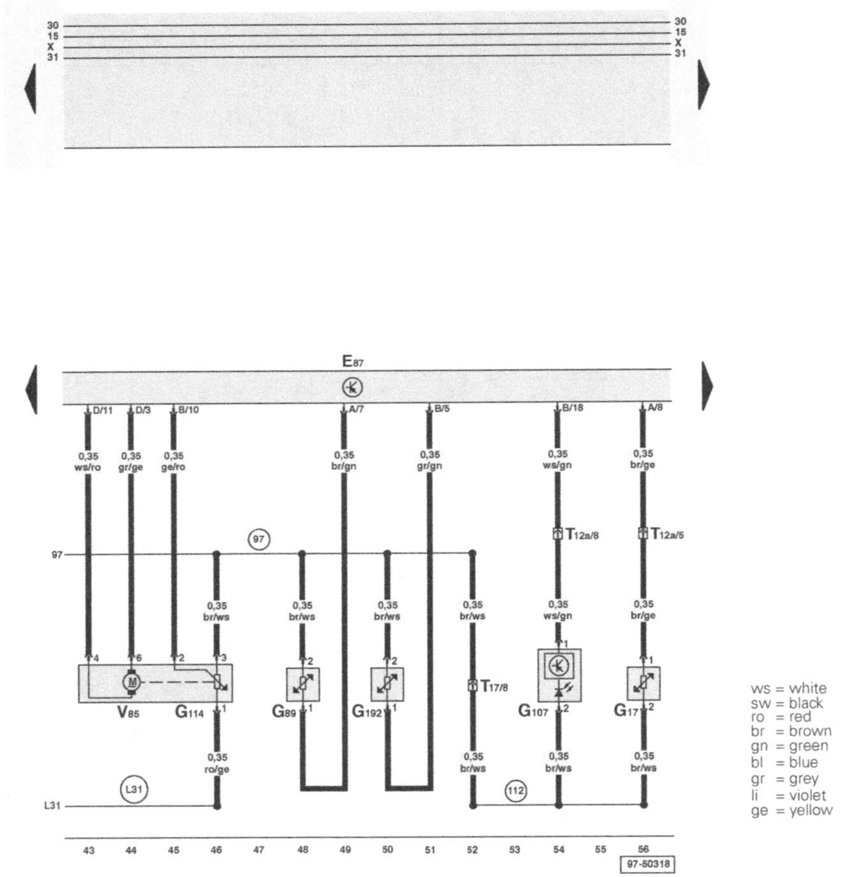

ws = white
sw = black
ro = red
br = brown
gn = green
bl = blue
gr = grey
li = violet
ge = yellow

E87 - A/C Control Head
G17 - Outside Air Temperature Sensor
G89 - Fresh Air Intake Duct Temperature Sensor
G107 - Sunlight Photo Sensor (3241)
G114 - Footwell/Defrost Flap Motor Position Sensor
G192 - Sender for outlet temperature, floor outlet
T12a - Connector 12 pin, behind instrument panel, right
T17 - Connector 17 pin, behind instrument panel
V85 - Footwell/Defroster Flap Motor

(97) - Ground connection -1-, in A/C wiring harness

(112) - Ground connection -2-, in A/C wiring harness

(L31) - Wire connection (5 Volts), in A/C wiring harness

No. 17/6 Wiring diagram **Audi TT**

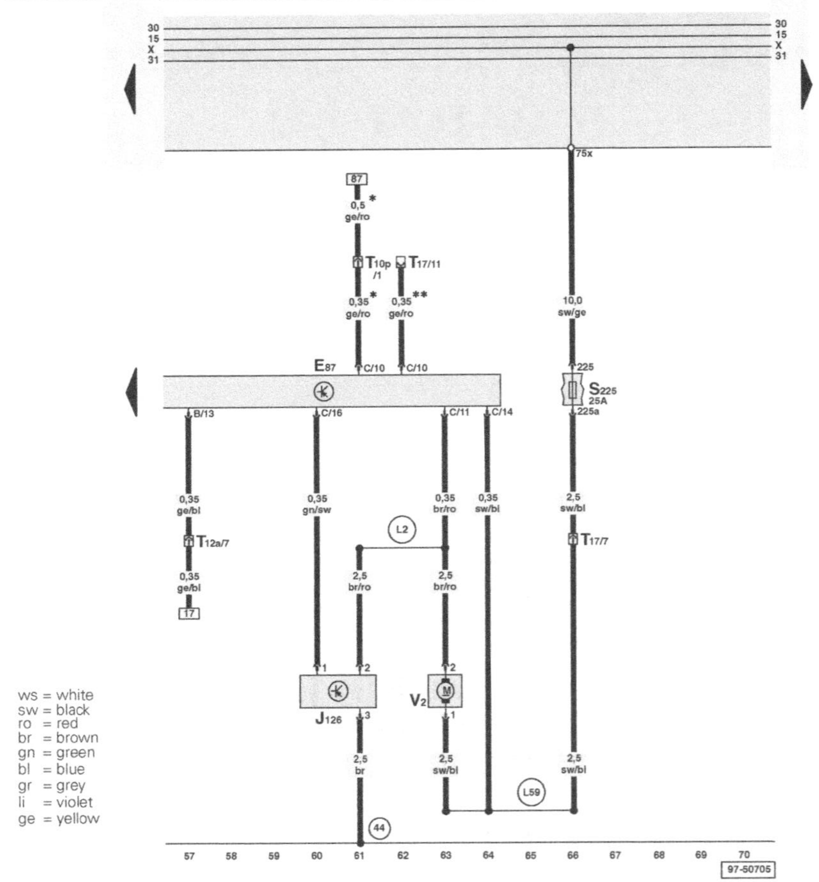

ws = white
sw = black
ro = red
br = brown
gn = green
bl = blue
gr = grey
li = violet
ge = yellow

E87 - A/C Control Head
J126 - Control module for fresh air blower
S225 - Fuse in fuse holder
T10p - Connector 10 pin, behind instrument panel
T12a - Connector 12 pin, behind instrument panel, right
T17 - Connector 17 pin, behind instrument panel
V2 - Fresh Air Blower

(44) - Ground connection, on left A-pillar, lower part

(L2) - Wire connection, in A/C wiring harness

(L59) - Plus connection (30), in wiring harness heater

* - Automatic climate control with A/C Pressure switch
** - Automatic climate control with High pressure sensor, from 12.12.1999 or VIN 8N ZY1 055070

© 2006 BentleyPublishers.com—All Rights Reserved

Audi TT Wiring diagram **No. 17/7**

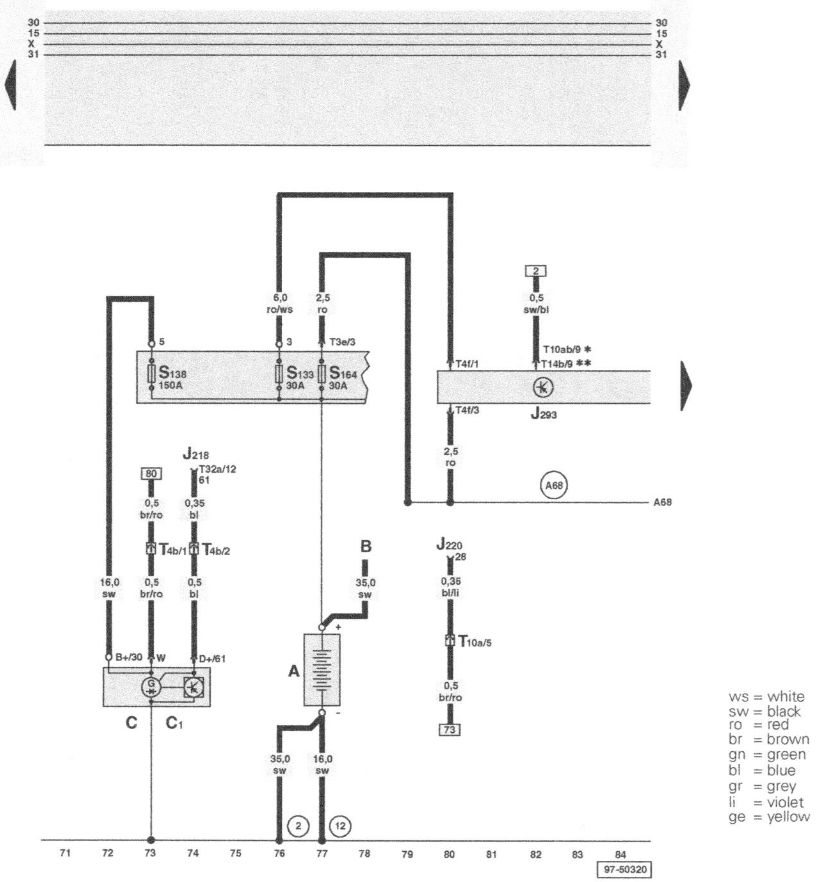

ws = white
sw = black
ro = red
br = brown
gn = green
bl = blue
gr = grey
li = violet
ge = yellow

A - Battery
B - Starter
C - Generator (GEN)
C1 - Voltage Regulator (VR)
J218 - Instrument Cluster Combination Processor
J220 - Motronic Engine Control Module (ECM)
J293 - Coolant FC (Fan Control) Control Module, in engine compartment, left front
S133 - Safety fuse 3
S138 - Safety fuse 5
S164 - Fuse -3- in fuse bracket/battery
T3e - Connector 3 pin, black, on Main Fuse Box/Battery
T4b - Connector 4 pin, in engine compartment, left
T4f - Connector 4 pin, connector B on J293
T10a - Connector 10 pin, white, in E-box plenum chamber
T10ab - Connector 10 pin, connector A on J293
T14b - Connector 14 pin, connector A on J293
T32a - Connector 32 pin, green, on instrument cluster

(2) - Ground strap, transmission to body
(12) - Ground connection, in engine compartment, left
(A68) - Connector (30, A/C), in instrument panel wiring harness

* - Automatic climate control with A/C pressure switch
** - Automatic climate control with high pressure sensor, from 12.12.1999 or VIN 8N ZY1 055070

Edition 11.00
W42.USA.5702.05.21

Coolant FC Control Module in Engine Compartment left front, Fuses

No. 17/8 Wiring diagram **Audi TT**

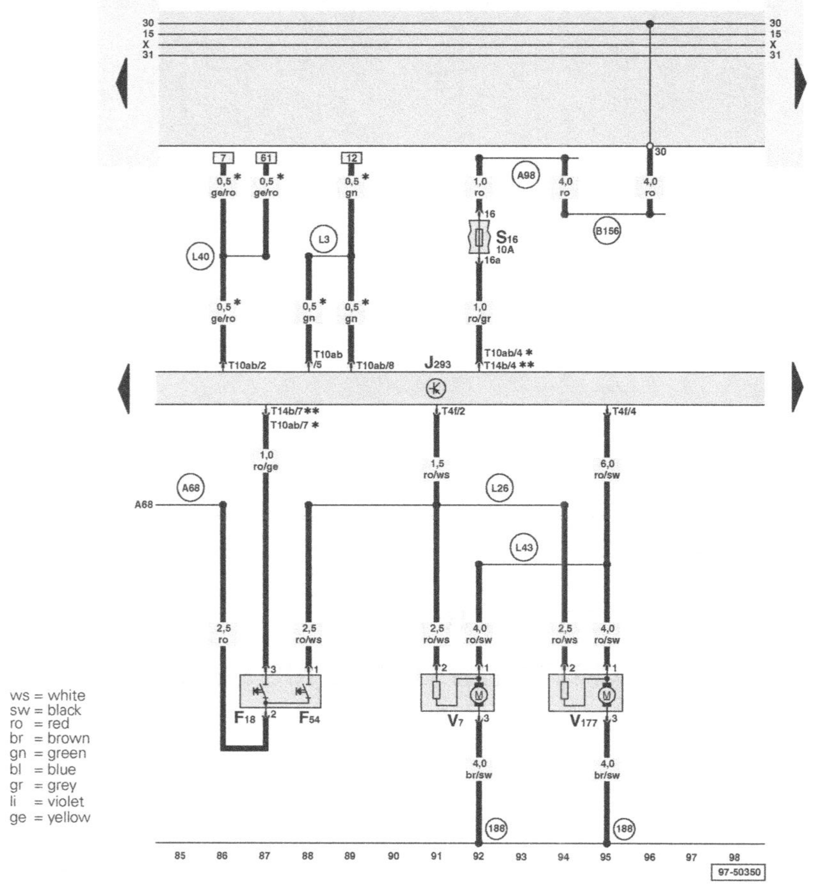

ws = white
sw = black
ro = red
br = brown
gn = green
bl = blue
gr = grey
li = violet
ge = yellow

F18 - Coolant Fan Control (FC) Thermal Switch
F54 - Coolant Fan Control (FC) Thermal Switch
J293 - Coolant FC (Fan Control) Control Module, in engine compartment, left front
S16 - Fuse
T4f - Connector 4 pin, connector B on J293
T10ab - Connector 10 pin, connector A on J293
T14b - Connector 14 pin, connector A on J293
V7 - Coolant Fan
V177 - Coolant Fan -2-

(188) - Ground connection -3-, in A/C wiring harness
(A68) - Connector (30, A/C), in instrument panel wiring harness

(A98) - Plus connector -4- (30), in instrument panel wiring harness
(B156) - Plus connector -4- (30), in wiring harness interior
(L3) - Wire connection (temperature switch), in A/C wiring harness
(L26) - Wire connection (coolant fan), in A/C-climatronic/engine compartment wiring harness
(L40) - Wire connection -2- (A/C Pressure Switch), in Climatronic wiring harness
(L43) - Wire connection (coolant fan), in A/C system-coolant fan wiring harness

* - Automatic Climate Control with A/C Pressure switch
** - Automatic Climate Control with High pressure sensor, from 12.12.1999 or VIN 8N ZY1 055070

Coolant FC Control Module in Engine Compartment left front, Coolant Fan Control Thermal Switch, Coolant Fan

Edition 11.00
W42.USA.5702.05.21

© 2006 BentleyPublishers.com—All Rights Reserved

Audi TT Wiring diagram **No. 17/9**

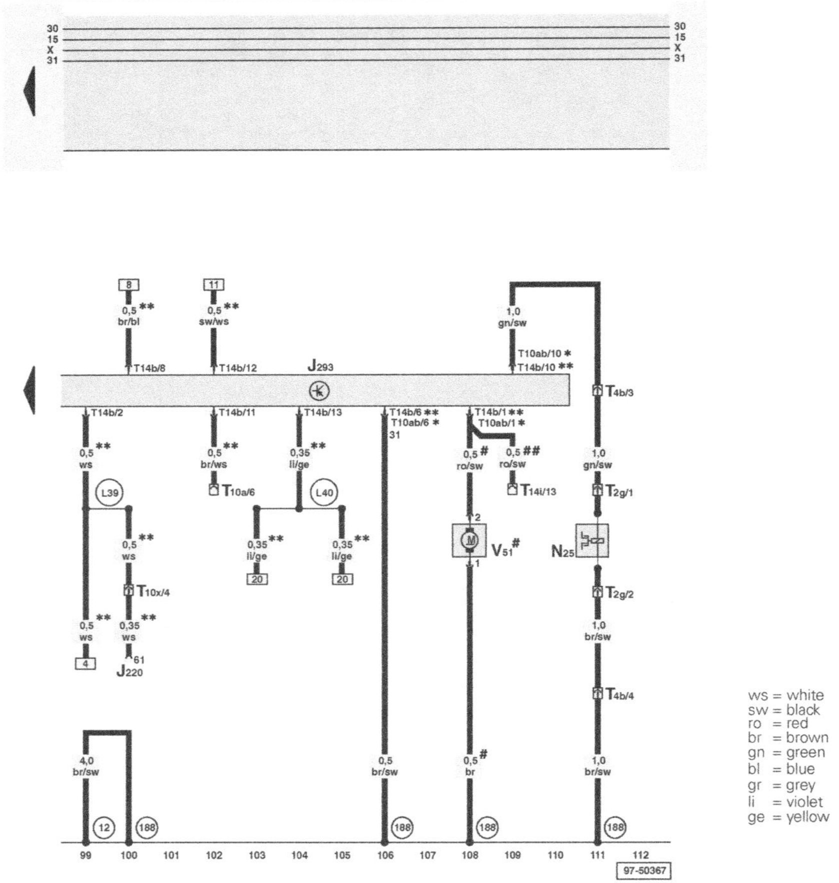

ws = white
sw = black
ro = red
br = brown
gn = green
bl = blue
gr = grey
li = violet
ge = yellow

J220 - Motronic Engine Control Module (ECM)
J293 - Coolant FC (Fan Control) Control Module, in engine compartment, left front
N25 - A/C Clutch
T2g - Connector 2 pin, near compressor
T4b - Connector 4 pin, in engine compartment, left
T10a - Connector 10 pin, white, in E-box plenum chamber
T10p - Connector 10 pin, behind instrument panel
T10x - Connector 10 pin, orange, in E-box plenum chamber
T10ab - Connector 10 pin, connector A on J293
T14b - Connector 14 pin, connector A on J293
T14i - Connector 14 pin, in engine compartment, left
V51 - After-Run Coolant Pump

(12) - Ground connection, in engine compartment, left
(188) - Ground connection -3-, in A/C wiring harness
(L39) - Wire connection -1- (A/C Pressure Switch), in Climatronic wiring harness
(L40) - Wire connection -2- (A/C Pressure Switch), in Climatronic wiring harness

* - Automatic Climate Control with A/C Pressure switch
** - Automatic Climate Control with High pressure sensor, from 12.12.1999 or VIN 8N ZY1 055070
\# - Vehicles with 165 kW engine, code AMU
\#\# - Vehicles with 132 kW engine, code ATC

Edition 11.00
W42.USA.5702.05.21

Coolant FC Control Module in Engine Compartment left front, A/C Clutch, After-Run Coolant Pump

THIS SPACE INTENTIONALLY

LEFT BLANK

© 2006 BentleyPublishers.com—All Rights Reserved

Haldex Coupling (02D) for All-Wheel Drive

2000 m. y.

Fuse Panel

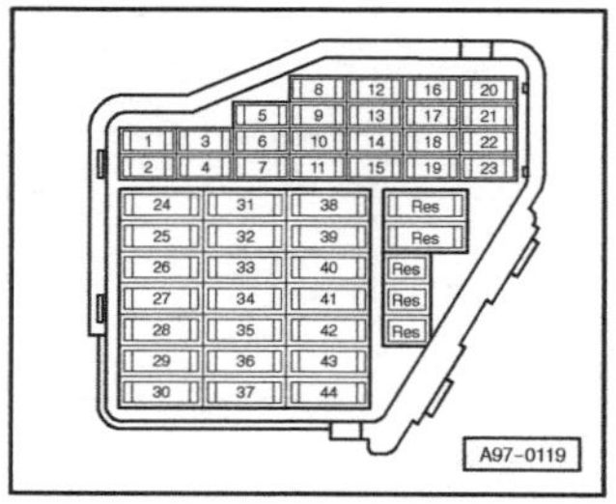

Fuse Colors:

30 A - Green
25 A - White
20 A - Yellow
15 A - Blue
10 A - Red
7,5 A - Brown
5 A - Beige

Starting with fuse position 23, fuses in the fuse holder are identified with 223 in the wiring diagram.

Mirco Central Electric Panel

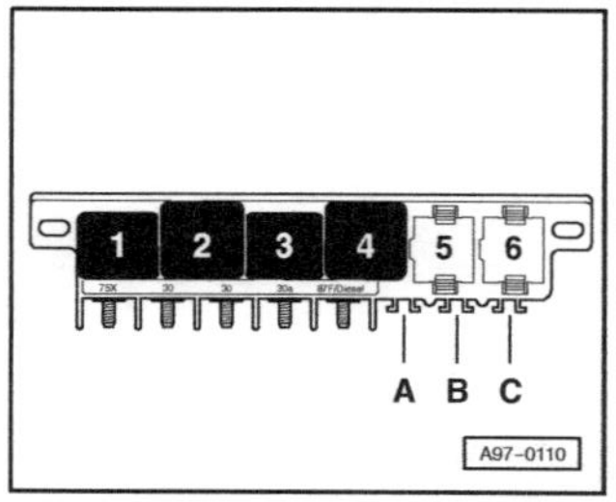

13 - Fold Relay Panel

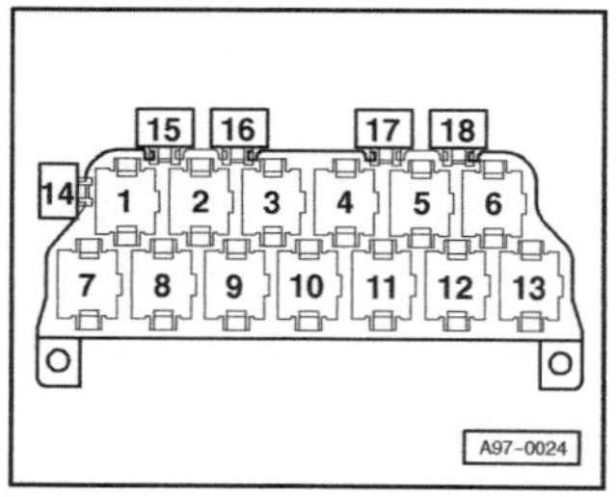

Relay Location:

7 -Brake light disable relay (ESP), J508

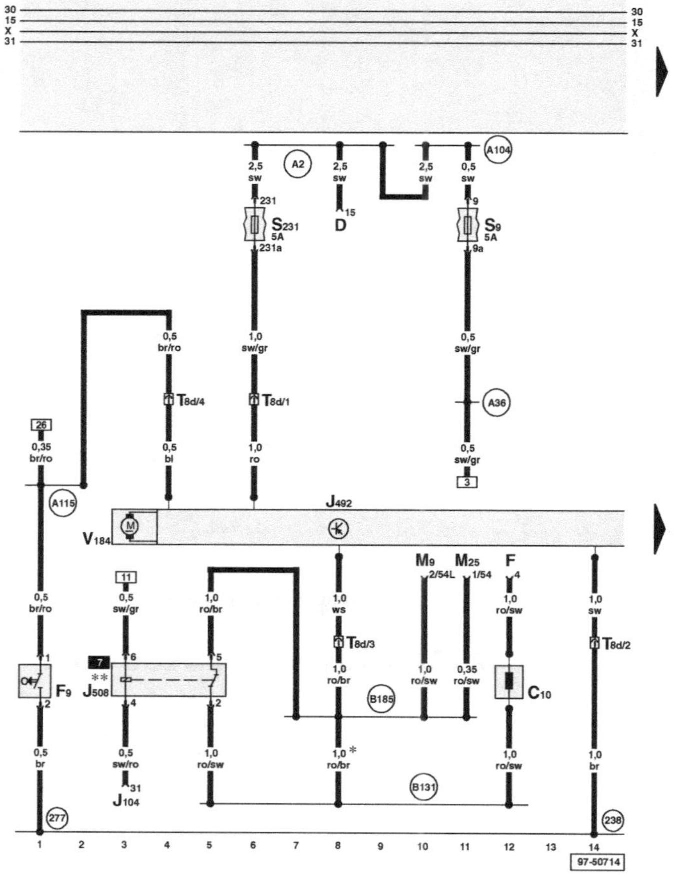

ws = white
sw = black
ro = red
br = brown
gn = green
bl = blue
gr = grey
li = violet
ge = yellow

C10 - Suppressor Filter
D - Ignition/Starter Switch
F - Brake Light Switch
F9 - Parking Brake Warning Light Switch
J104 - ABS Control Module (w/EDL)
J492 - Control module for all wheel drive
J508 - Brake light disable relay (ESP)
M9 - Left Brake Light
M25 - High-mount Brake Light
S9 - Fuse
S231 - Fuse 31 in fuse holder
T8d - Connector 8 pin, on Control module for all wheel drive
V184 - Positioning motor for Oil pressure

(238) - Ground connection -1-, in wiring harness interior
(277) - Ground connector -3-, in wiring harness interior

(A2) - Plus connection (15), in instrument panel wiring harness
(A36) - Wire connection (75a), in instrument panel wiring harness
(A104) - Plus connector -2- (15), in instrument panel wiring harness
(A115) - Connection (parking brake control) in instrument panel wiring harness
(B131) - Connector (54), in wiring harness interior
(B185) - Connection 2 (54) in passenger compartment wiring harness

* - Vehicles with ABS
** - Vehicles with ABS and ESP

© 2006 BentleyPublishers.com—All Rights Reserved

Audi TT — Wiring diagram — No. 18/3

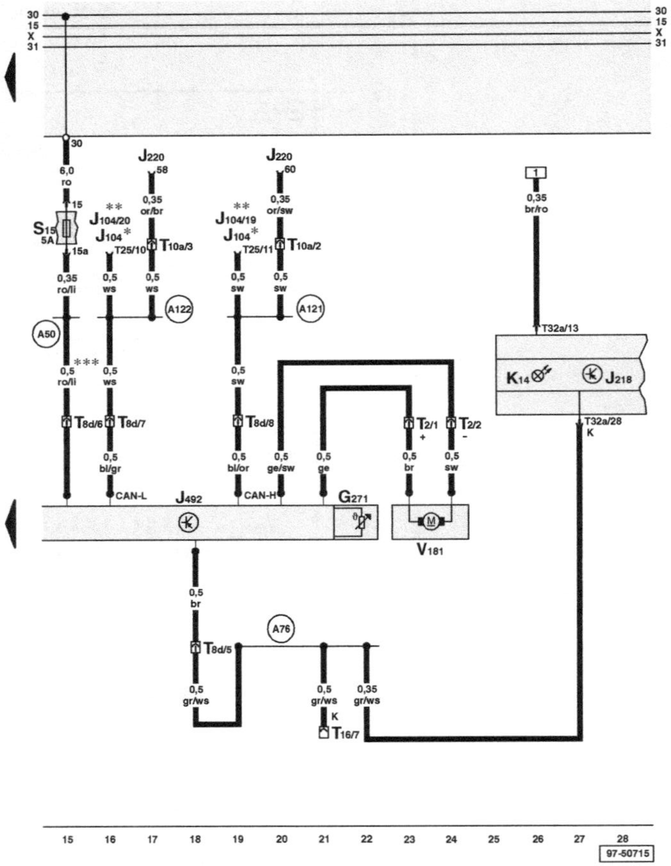

ws = white
sw = black
ro = red
br = brown
gn = green
bl = blue
gr = grey
li = violet
ge = yellow
or = orange

G271 - Hydraulic Temperature Sensor
J104 - ABS Control Module (w/EDL)
J218 - Instrument Cluster Combination Processor
J220 - Motronic Engine Control Module (ECM)
J492 - Control module for all wheel drive
K14 - Parking Brake Indicator Light
S15 - Fuse
T2 - Connector 2 pin, on all-wheel drive
T8d - Connector 8 pin, on Control module for all wheel drive
T10a - Connector 10 pin, white, in E-box plenum chamber
T16 - Connector 16 pin, black, Data Link Connector (DLC), below storage compartment, driver's side
T25 - Connector 25 pin, on ABS Control Module (w/EDL)
T32a - Connector 32 pin, green, on instrument cluster
V181 - Pump for Haldex Coupling

(A50) - Plus connection (30a), in instrument panel wiring harness
(A76) - Connector (K-diagnosis wire), in instrument panel wiring harness
(A121) - Connection (high bus) in instrument panel wiring harness
(A122) - Connection (low bus) in instrument panel wiring harness

- CAN-Bus (Data-Bus)
* - Vehicles with ABS
** - Vehicles with ABS and ESP
*** - deleted from 2001 m.y.

Edition 12/02
W42.USA.5702.12.21

Control Module for All Wheel Drive, Pump for Haldex Coupling

THIS SPACE INTENTIONALLY LEFT BLANK

© 2006 BentleyPublishers.com—All Rights Reserved

Telephone

2000 m. y.

Fuse Panel

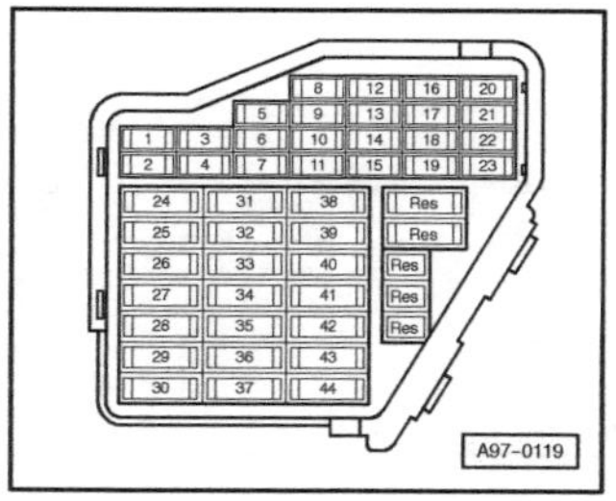

Fuse Colors:

30 A - Green
25 A - White
20 A - Yellow
15 A - Blue
10 A - Red
7,5 A - Brown
5 A - Beige

Starting with fuse position 23, fuses in the fuse holder are identified with 223 in the wiring diagram.

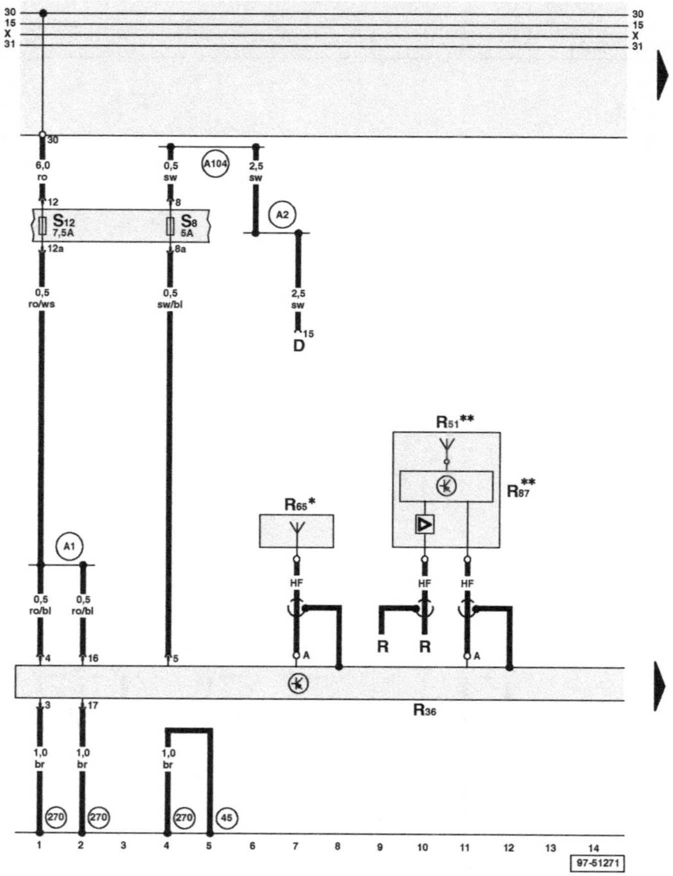

ws = white
sw = black
ro = red
br = brown
gn = green
bl = blue
gr = grey
li = violet
ge = yellow

D - Ignition/Starter Switch
R36 - Telephone Transceiver
R51 - Antenna for radio/telephone
R65 - Antenna for telephone
R87 - Antenna Splitter
S8 - Fuse
S12 - Fuse

(45) - Ground connection, behind instrument panel, center
(270) - Ground connector, in wiring harness telephone
(A1) - Plus connection (30a), in instrument panel wiring harness
(A2) - Plus connection (15), in instrument panel wiring harness
(A104) - Plus connector -2- (15), in instrument panel wiring harness

* - Coupé, roof antenna
** - Roadster, rod antenna

Telephone Transceiver, Antenna for Radio/Telephone

©2006 BentleyPublishers.com—All Rights Reserved

Audi TT — Wiring diagram — No. 19/3

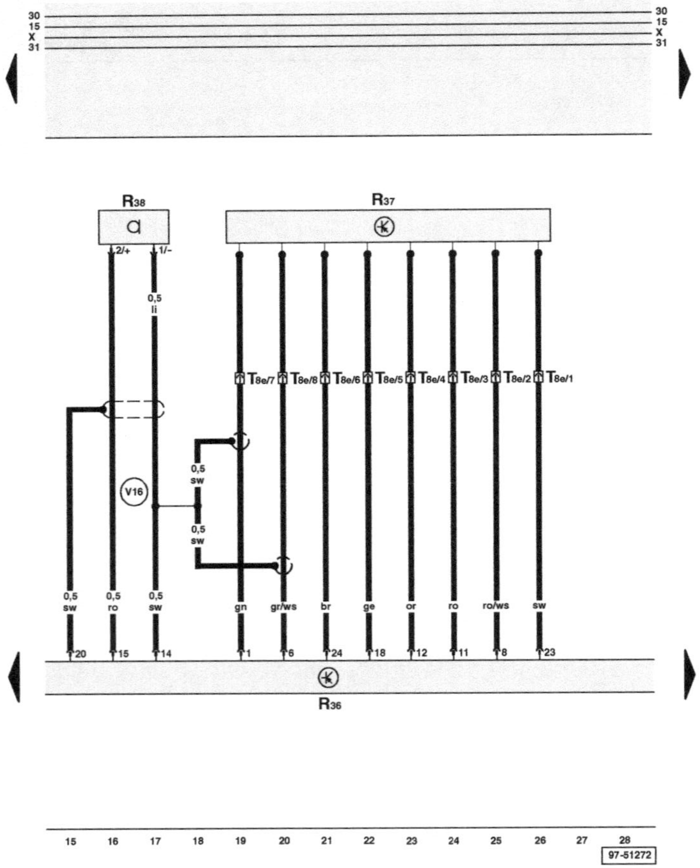

ws = white
sw = black
ro = red
br = brown
gn = green
bl = blue
gr = grey
li = violet
ge = yellow
or = orange

R36 - Telephone Transceiver
R37 - Telephone Handset
R38 - Telephone Microphone
T8e - Connector 8 pin, on Telephone Handset

(V16) - Wire connection (shielding), in telephone wiring harness

Edition 07.00
W42.USA.5702.04.21

Telephone Transceiver, Telephone Handset, Telephone Microphone

No. 19/4 — Wiring diagram — Audi TT

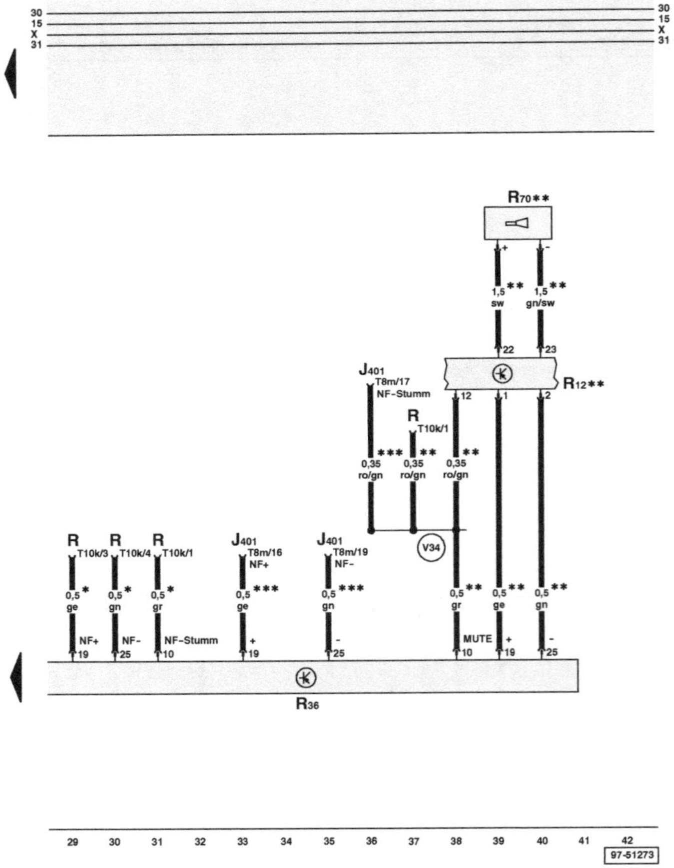

ws = white
sw = black
ro = red
br = brown
gn = green
bl = blue
gr = grey
li = violet
ge = yellow

J401 - Control unit for navigation with CD drive
R - Radio
R12 - Amplifier
R36 - Telephone Transceiver
R70 - Inside loudspeaker, center
T8m - Connector 8 pin, C III on Control Module for Navigation
T10k - Connector 10 pin, red, on radio

(V34) - Connector (mute button), in wiring harness telephone

* - Vehicles with Radio Concert
** - Vehicles with Radio Bose
*** - Vehicles with navigation

Telephone Transceiver, Inside Loudspeaker, Center

Edition 07.00
W42.USA.5702.04.21

© 2006 BentleyPublishers.com—All Rights Reserved

Anti-Lock Brake System (ABS) MK20 with Electronic Differential Lock (EDL), Anti-Slip Regulation (ASR - ASC) and Electronic Stabilization Program (ESP)

2000 m. y.

Fuse Panel

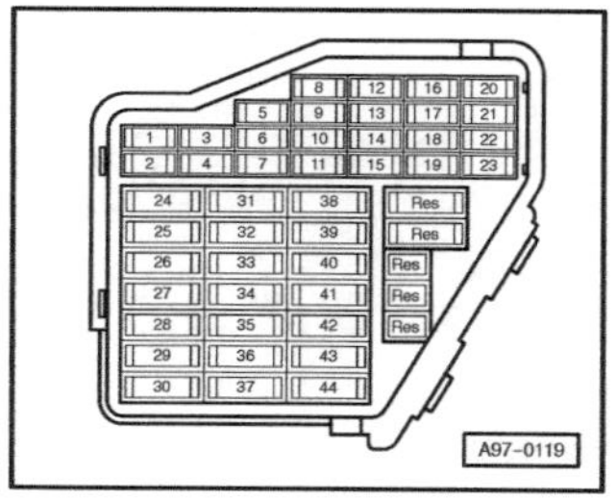

Fuse Colors:

30 A - Green
25 A - White
20 A - Yellow
15 A - Blue
10 A - Red
7,5 A - Brown
5 A - Beige

Starting with fuse position 23, fuses in the fuse holder are identified with 223 in the wiring diagram.

13 - Fold Relay Panel

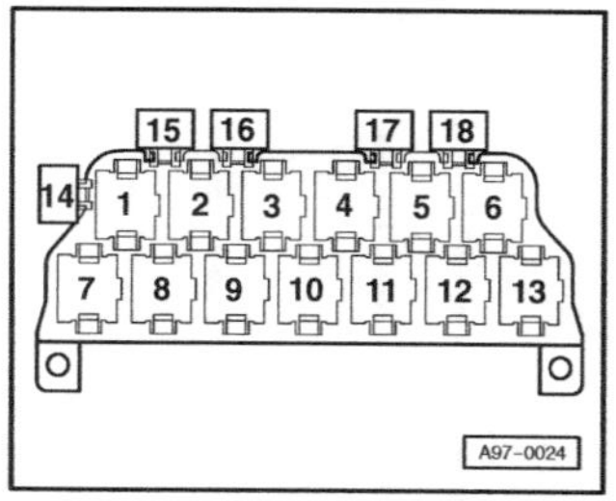

Relay Location:

7 -Brake Light Disable Relay (ESP), J508

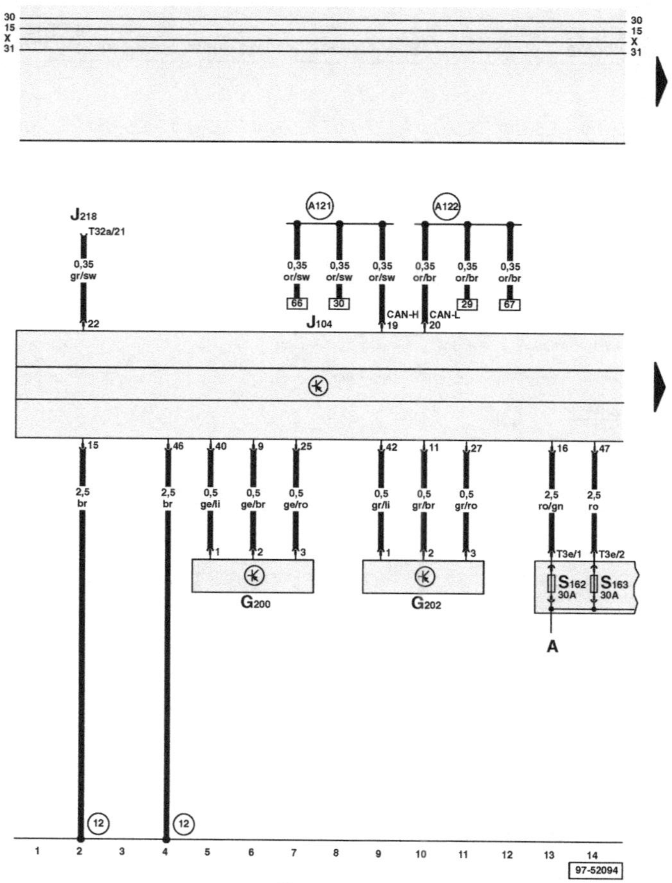

ws = white
sw = black
ro = red
br = brown
gn = green
bl = blue
gr = grey
li = violet
ge = yellow
or = orange

A - Battery
G200 - Sensor for transverse acceleration
G202 - Sender for rotation rate
J104 - ABS Control Module (w/EDL)
J218 - Instrument Cluster Combination Processor
S162 - Fuse -1- in fuse bracket/battery
S163 - Fuse -2- in fuse bracket/battery
T3e - Connector 3 pin, black, on Main Fuse Box/Battery
T32a - Connector 32 pin, green, on instrument cluster

(12) - Ground connection, in engine compartment, left
(A121) - Connection (high bus) in instrument panel wiring harness
(A122) - Connection (low bus) in instrument panel wiring harness
- CAN-Bus (Data-Bus)

ABS Control Module, Sensor for Transverse Acceleration, Sender for Rotation Rate

© 2006 BentleyPublishers.com—All Rights Reserved

Audi TT — Wiring diagram — No. 20/3

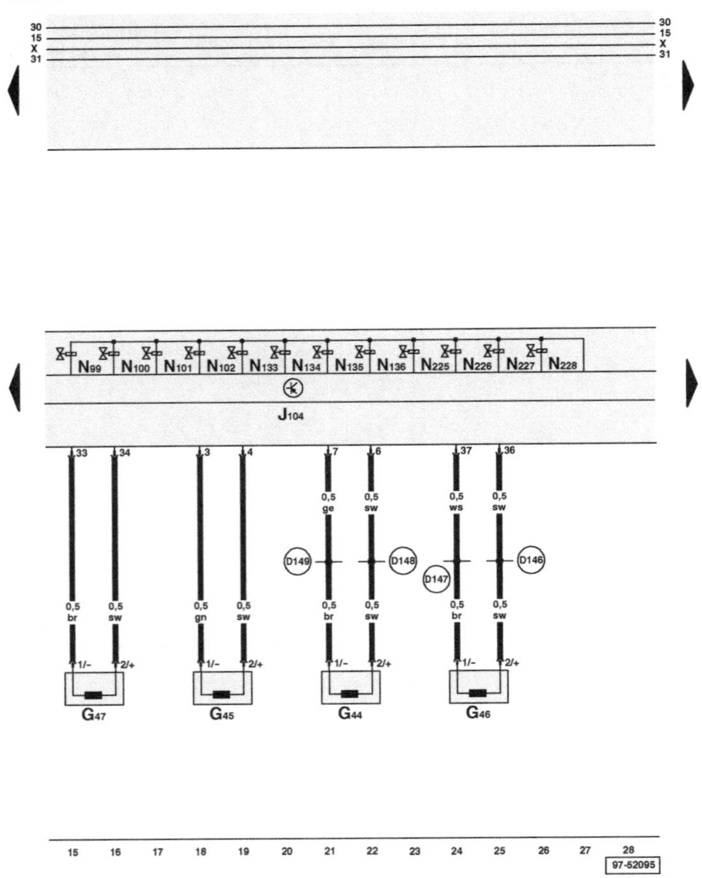

ws = white
sw = black
ro = red
br = brown
gn = green
bl = blue
gr = grey
li = violet
ge = yellow

- G44 - Right Rear ABS Wheel Speed Sensor
- G45 - Right Front ABS Wheel Speed Sensor
- G46 - Left Rear ABS Wheel Speed Sensor
- G47 - Left Front ABS Wheel Speed Sensor
- J104 - ABS Control Module (w/EDL)
- N99 - Right Front ABS Inlet Valve
- N100 - Right Front ABS Outlet Valve
- N101 - Left Front ABS Inlet Valve
- N102 - Left Front ABS Outlet Valve
- N133 - Right Rear ABS Inlet Valve
- N134 - Left Rear ABS Inlet Valve
- N135 - Right Rear ABS Outlet Valve
- N136 - Left Rear ABS Outlet Valve
- N225 - Pilot valve -1- traction control
- N226 - Pilot valve -2- traction control
- N227 - High pressure switch valve -1- traction control
- N228 - High pressure switch valve -2- traction control

- (D146) - Connector (speed sensor rear, left +), in wiring harness engine compartment
- (D147) - Connector (speed sensor rear, left -), in wiring harness engine compartment
- (D148) - Connector (speed sensor rear, right +), in wiring harness engine compartment
- (D149) - Connector (speed sensor rear, right -), in wiring harness engine compartment

Edition 12/02
W42.USA.5702.12.21

ABS Control Module, Wheel Speed Sensors, Valves for ABS

No. 20/4 — Wiring diagram — Audi TT

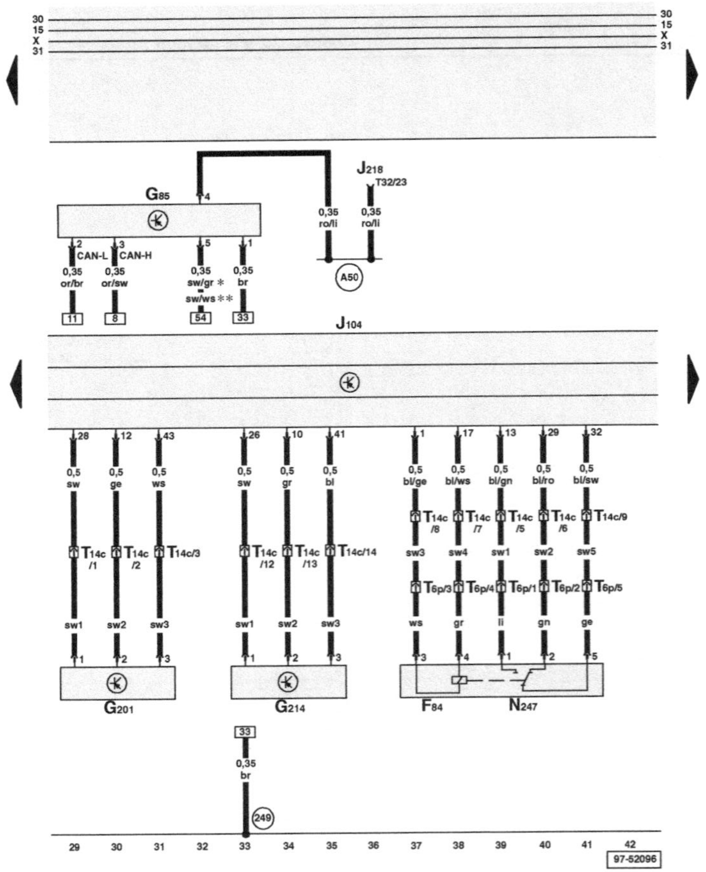

ws = white
sw = black
ro = red
br = brown
gn = green
bl = blue
gr = grey
li = violet
ge = yellow

- F84 - Brake Pressure Release Solenoid
- G85 - Steering Angle Sensor
- G201 - Sender 1 for brake booster
- G214 - Sensor -2- for brake pressure
- J104 - ABS Control Module (w/EDL)
- J218 - Instrument Cluster Combination Processor
- N247 - Magnetic coil for brake pressure, in brake booster
- T6p - Connector 6 pin, black, on brake booster
- T14c - Connector 14 pin, connector in wiring harness ESP
- T32 - Connector 32 pin, blue, on instrument cluster

- (249) - Ground connection -2-, in wiring harness interior
- (A50) - Plus connection (30a), in instrument panel wiring harness

- * - 2000 m.y.
- ** - 2001 m.y.

Brake Pressure Release Solenoid, ABS Control Module, Steering Angle Sensor, Sensors for Brake Pressure

Edition 12/02
W42.USA.5702.12.21

© 2006 BentleyPublishers.com—All Rights Reserved

Audi TT — Wiring diagram — No. 20/5

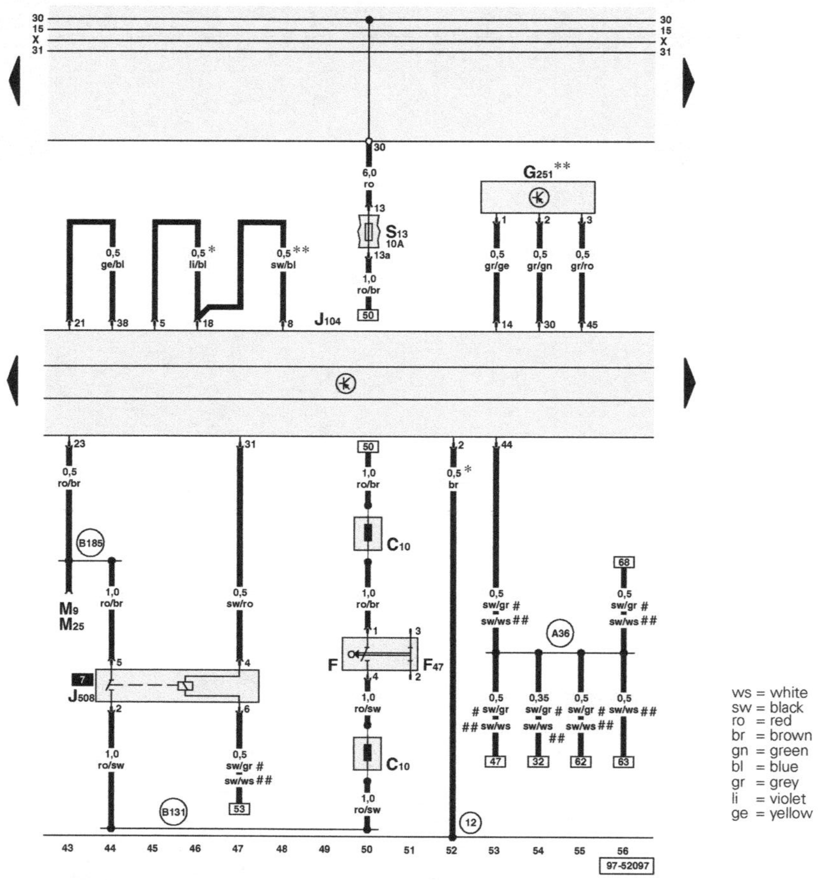

ws = white
sw = black
ro = red
br = brown
gn = green
bl = blue
gr = grey
li = violet
ge = yellow

C10 - Suppressor Filter
F - Brake Light Switch
F47 - Vacuum Vent Valve, Brake
G251 - Longitudinal Acceleration Sensor
J104 - ABS Control Module (w/EDL)
J508 - Brake Light Disable Relay (ESP)
M9 - Left Brake Light
M25 - High-mount Brake Light
S13 - Fuse

(12) - Ground connection, in engine compartment, left
(A36) - Wire connection (75a), in instrument panel wiring harness
(B131) - Connector (54), in wiring harness interior
(B185) - Connection 2 (54) in passenger compartment wiring harness

* - Vehicles with front-wheel drive
** - Vehicles with all-wheel drive
# - 2000 m.y.
- 2001 m.y.

No. 20/6 — Wiring diagram — Audi TT

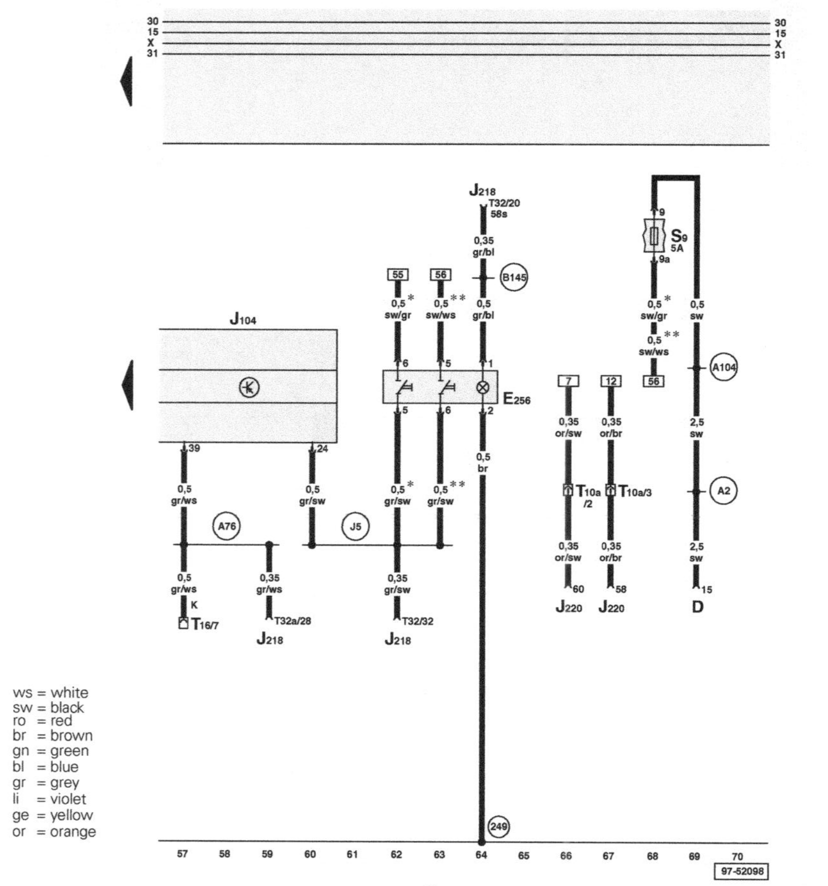

ws = white
sw = black
ro = red
br = brown
gn = green
bl = blue
gr = grey
li = violet
ge = yellow
or = orange

E256 - ASR/ESP Button
J104 - ABS Control Module (w/EDL)
J218 - Instrument Cluster Combination Processor
J220 - Motronic Engine Control Module (ECM)
S9 - Fuse
T10a - Connector 10 pin, white, in E-box plenum chamber
T16 - Connector 16 pin, black, Data Link Connector (DLC), below storage compartment, driver's side
T32 - Connector 32 pin, blue, on instrument cluster
T32a - Connector 32 pin, green, on instrument cluster

(249) - Ground connection -2-, in wiring harness interior
(A2) - Plus connection (15), in instrument panel wiring harness
(A76) - Connector (K-diagnosis wire), in instrument panel wiring harness
(A104) - Plus connector -2- (15), in instrument panel wiring harness
(B145) - Plus connector (58s), in wiring harness interior
(J5) - Wire connection -2-, in ABS wiring harness, near anti-slip control switch

* - 2000 m.y.
** - 2001 m.y.

©2006 BentleyPublishers.com—All Rights Reserved

Radio Stereo Concert with and without CD-Changer (Roadster)

2000 m. y.

Fuse Panel

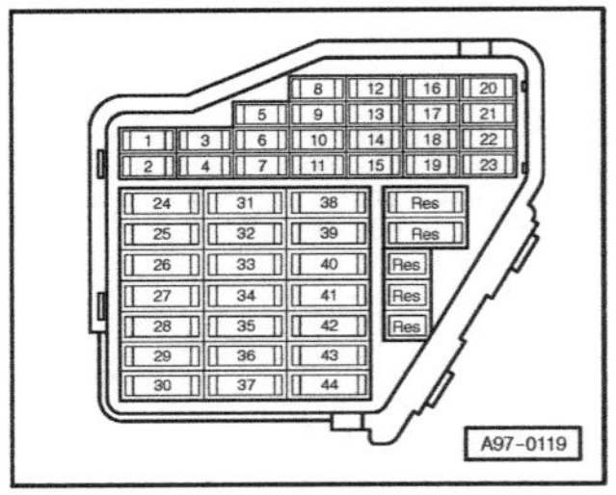

Fuse Colors:

30 A - Green
25 A - White
20 A - Yellow
15 A - Blue
10 A - Red
7,5 A - Brown
5 A - Beige

Starting with fuse position 23, fuses in the fuse holder are identified with 223 in the wiring diagram.

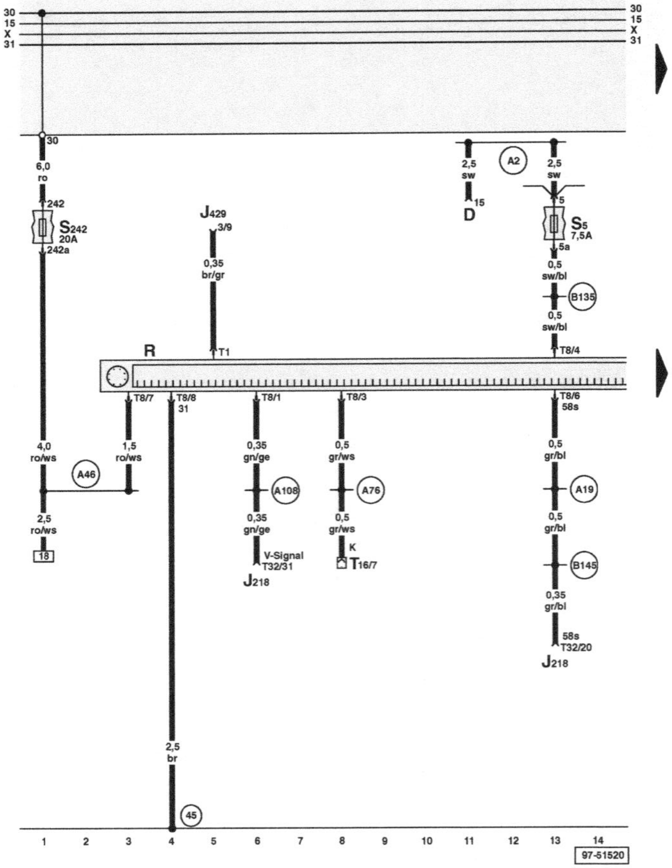

ws = white
sw = black
ro = red
br = brown
gn = green
bl = blue
gr = grey
li = violet
ge = yellow

D - Ignition/Starter Switch
J218 - Instrument Cluster Combination Processor
J429 - Control module for central locking
R - Radio
S5 - Fuse
S242 - Fuse 42 in fuse holder
T1 - Connector single, brown, on radio
T8 - Connector 8 pin, black, connector radio III
T16 - Connector 16 pin, black, Data Link Connector (DLC), below storage compartment, driver's side
T32 - Connector 32 pin, blue, on instrument cluster

(45) - Ground connection, behind instrument panel, center
(A2) - Plus connection (15), in instrument panel wiring harness
(A19) - Wire connection (58s), in instrument panel wiring harness
(A46) - Plus connection (30-from radio), in instrument panel wiring harness
(A76) - Connector (K-diagnosis wire), in instrument panel wiring harness
(A108) - Connector (speed signal), in instrument panel wiring harness
(B135) - Connector (15a), in wiring harness interior
(B145) - Plus connector (58s), in wiring harness interior

© 2006 BentleyPublishers.com—All Rights Reserved

Audi TT Wiring diagram **No. 21/3**

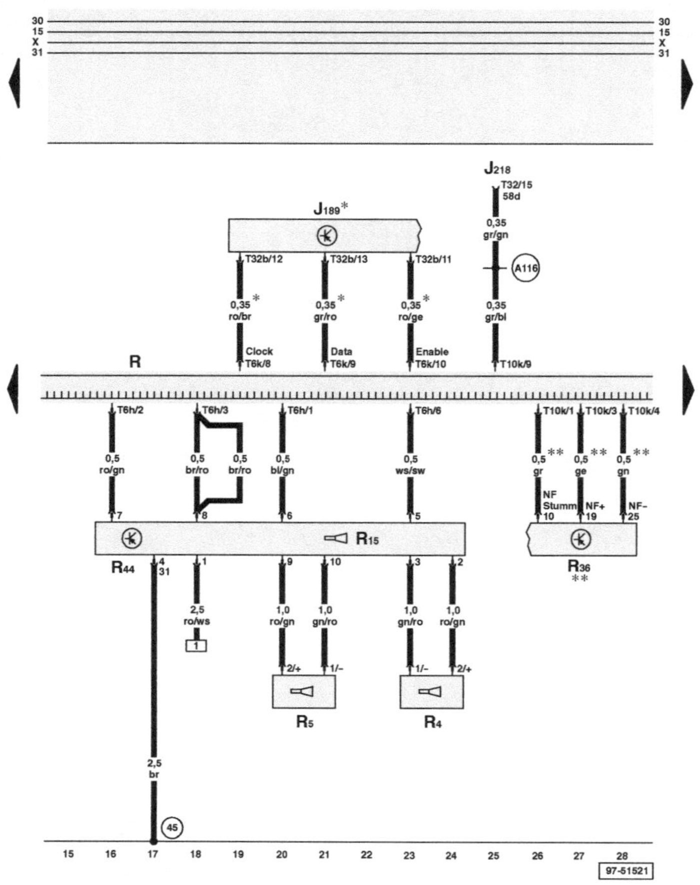

ws = white
sw = black
ro = red
br = brown
gn = green
bl = blue
gr = grey
li = violet
ge = yellow

J189 - Auto Check System
J218 - Instrument Cluster Combination Processor
R - Radio
R4 - Left Rear Speaker
R5 - Right Rear Speaker
R15 - Center Rear Woofer
R36 - Telephone Transceiver
R44 - Center Rear Woofer/Amplifier
T6h - Connector 6 pin, yellow, connector radio I-1
T6k - Connector 6 pin, green, connector radio I-2
T10k - Connector 10 pin, red, on radio
T32 - Connector 32 pin, blue, on instrument cluster
T32b - Connector 32 pin, gray, on instrument cluster

(45) - Ground connection, behind instrument panel, center

(A116) - Connection -2- (58d) in instrument panel wiring harness

* - Vehicles with Auto Check System
** - Vehicles with telephone

Edition 12/02
W42.USA.5702.12.21

Radio Stereo Concert

No. 21/4 Wiring diagram **Audi TT**

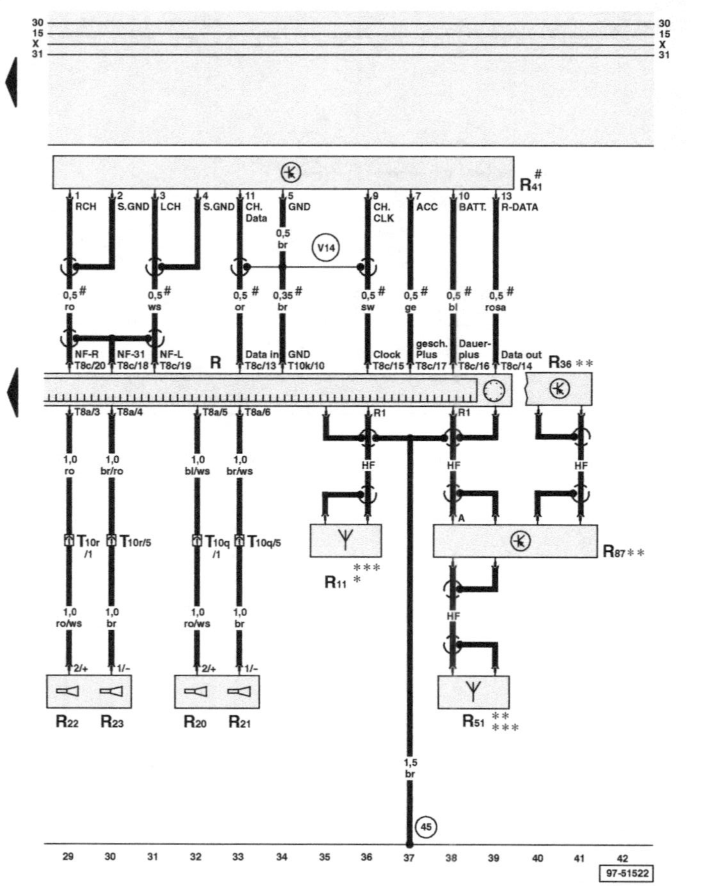

ws = white
sw = black
ro = red
br = brown
gn = green
bl = blue
gr = grey
li = violet
ge = yellow
or = orange

R - Radio
R1 - Antenna Connection
R11 - Antenna
R20 - Left Front Tweeter
R21 - Left Front Woofer
R22 - Right Front Tweeter
R23 - Right Front Woofer
R36 - Telephone Transceiver
R41 - CD Changer Unit
R51 - Antenna for radio/telephone/park heating
R87 - Antenna Splitter
T8a - Connector 8 pin, brown, connector radio II
T10q - Connector 10 pin, green, connector station A pillar, left
T10r - Connector 10 pin, green, connector station A pillar, right

(45) - Ground connection, behind instrument panel, center

(V14) - Wire connection (shielding), in CD Changer wiring harness

* - Vehicles without telephone
** - Vehicles with telephone
*** - Rod antenna
\# - Vehicles with CD Changer Unit

Radio Stereo Concert, CD Changer Unit, Antenna

Edition 12/02
W42.USA.5702.12.21

© 2006 BentleyPublishers.com—All Rights Reserved

Radio Bose with and without CD-Changer (Roadster)

2000 m. y.

Fuse Panel

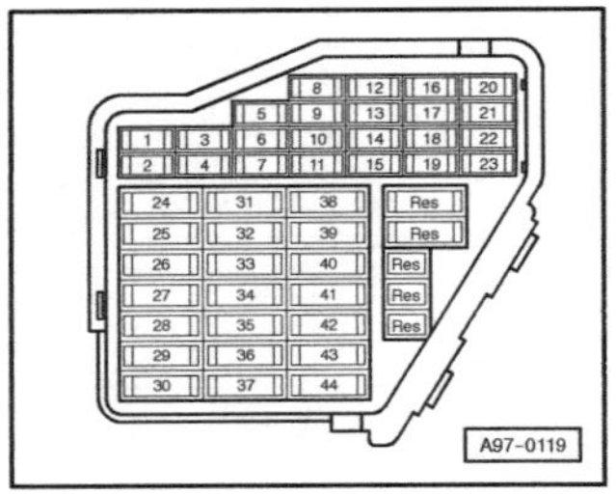

Fuse Colors:

30 A	-	Green
25 A	-	White
20 A	-	Yellow
15 A	-	Blue
10 A	-	Red
7,5 A	-	Brown
5 A	-	Beige

Starting with fuse position 23, fuses in the fuse holder are identified with 223 in the wiring diagram.

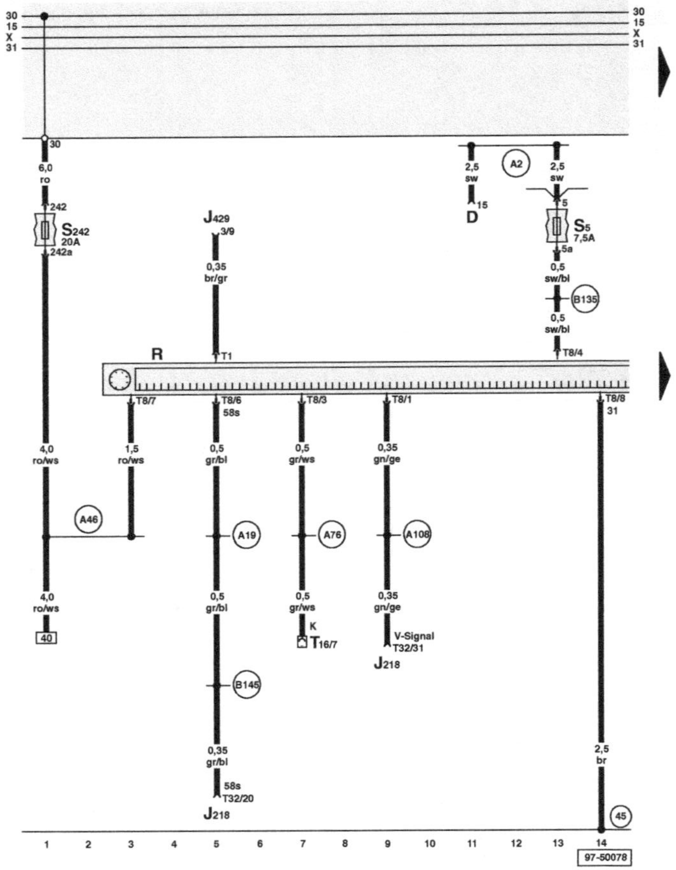

ws = white
sw = black
ro = red
br = brown
gn = green
bl = blue
gr = grey
li = violet
ge = yellow

D - Ignition/Starter Switch
J218 - Instrument Cluster Combination Processor
J429 - Control module for central locking
R - Radio
S5 - Fuse
S242 - Fuse 42 in fuse holder
T1 - Connector single, brown, on radio
T8 - Connector 8 pin, black, connector radio III
T16 - Connector 16 pin, black, Data Link Connector (DLC), below storage compartment, driver's side
T32 - Connector 32 pin, blue, on instrument cluster

(45) - Ground connection, behind instrument panel, center
(A2) - Plus connection (15), in instrument panel wiring harness
(A19) - Wire connection (58s), in instrument panel wiring harness
(A46) - Plus connection (30-from radio), in instrument panel wiring harness
(A76) - Connector (K-diagnosis wire), in instrument panel wiring harness
(A108) - Connector (speed signal), in instrument panel wiring harness
(B135) - Connector (15a), in wiring harness interior
(B145) - Plus connector (58s), in wiring harness interior

© 2006 BentleyPublishers.com—All Rights Reserved

Sound System, Bose, roadster (2000-2001)

Audi TT — Wiring diagram — **No. 22/3**

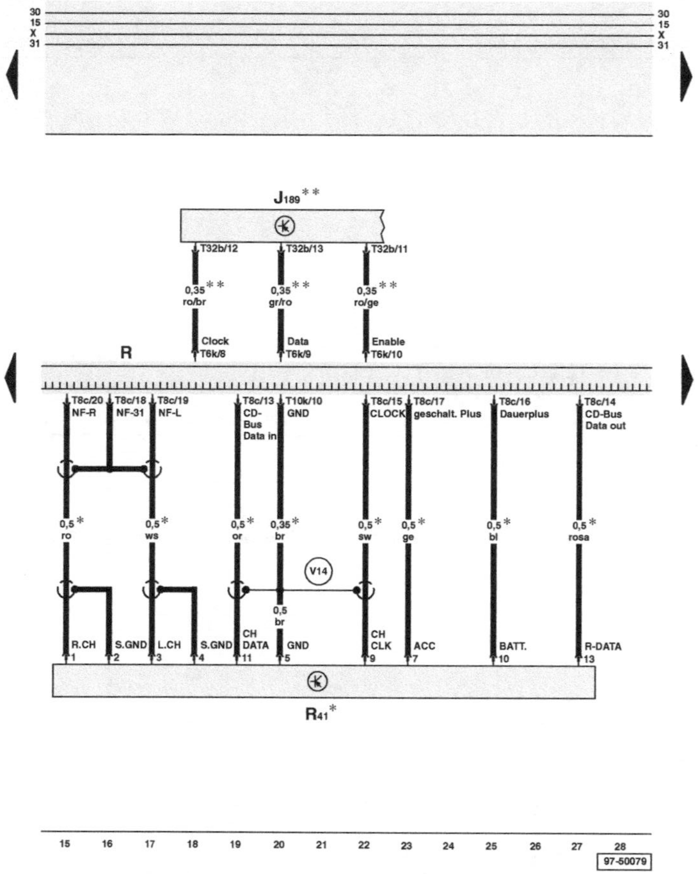

ws = white
sw = black
ro = red
br = brown
gn = green
bl = blue
gr = grey
li = violet
ge = yellow
or = orange

J189 - Auto Check System
R - Radio
R41 - CD Changer Unit
T6k - Connector 6 pin, green, connector radio I-2
T8c - Connector 8 pin, blue, connector radio I-3
T10k - Connector 10 pin, red, on radio
T32b - Connector 32 pin, gray, on instrument cluster

(V14) - Wire connection (shielding), in CD Changer wiring harness

* - Vehicles with CD-Changer
** - Vehicles with Auto Check System

No. 22/4 — Wiring diagram — **Audi TT**

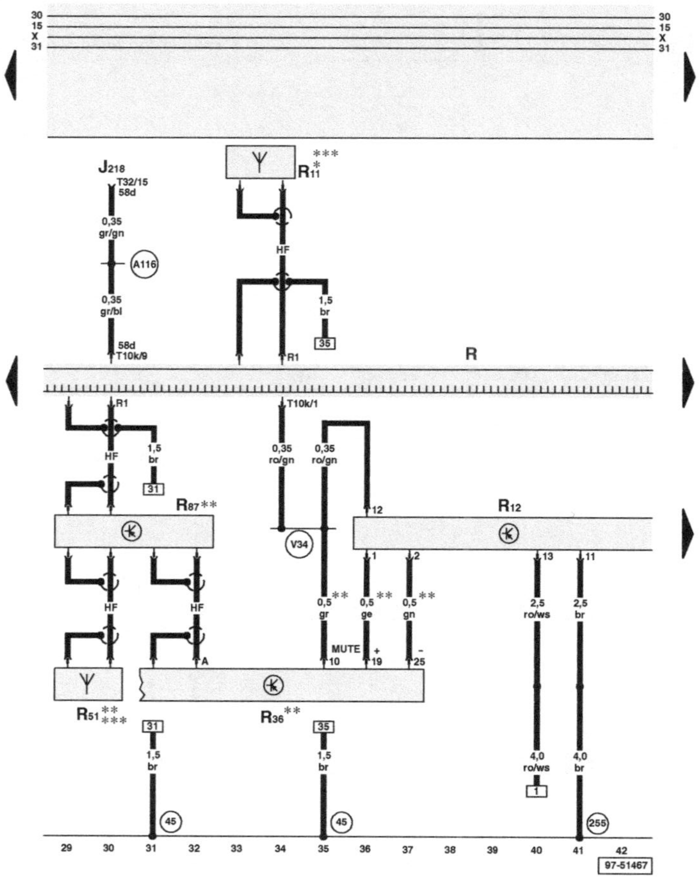

ws = white
sw = black
ro = red
br = brown
gn = green
bl = blue
gr = grey
li = violet
ge = yellow

J218 - Instrument Cluster Combination Processor
R - Radio
R1 - Antenna Connection
R11 - Antenna
R12 - Amplifier
R36 - Telephone Transceiver
R51 - Antenna for radio/telephone/park heating
R87 - Antenna Splitter
T10k - Connector 10 pin, red, on radio
T32 - Connector 32 pin, blue, on instrument cluster

(45) - Ground connection, behind instrument panel, center
(98) - Ground connection, in rear lid wiring harness
(255) - Ground connection -1-, in wiring harness Radio
(A116) - Connection -2- (58d) in instrument panel wiring harness
(V34) - Connector (mute button), in wiring harness telephone

* - Vehicles without Telephone
** - Vehicles with Telephone
*** - Rod antenna

© 2006 BentleyPublishers.com—All Rights Reserved

Sound System, Bose, roadster (2000-2001)

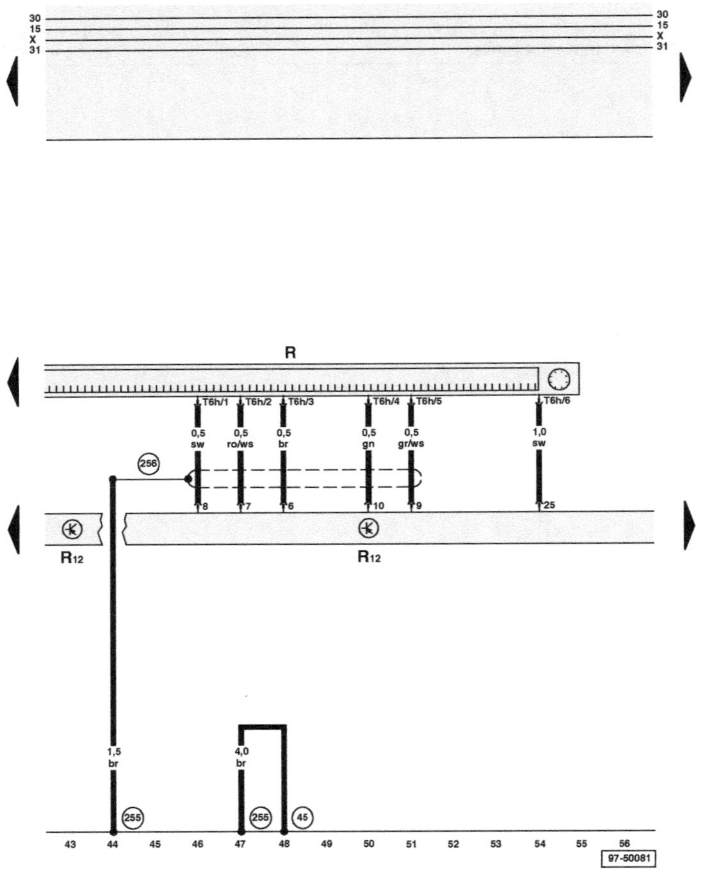

ws = white
sw = black
ro = red
br = brown
gn = green
bl = blue
gr = grey
li = violet
ge = yellow

R - Radio
R12 - Amplifier
T6h - Connector 6 pin, yellow, connector radio I-1

(45) - Ground connection, behind instrument panel, center
(255) - Ground connection -1-, in wiring harness Radio
(256) - Ground connection -2-, in wiring harness Radio

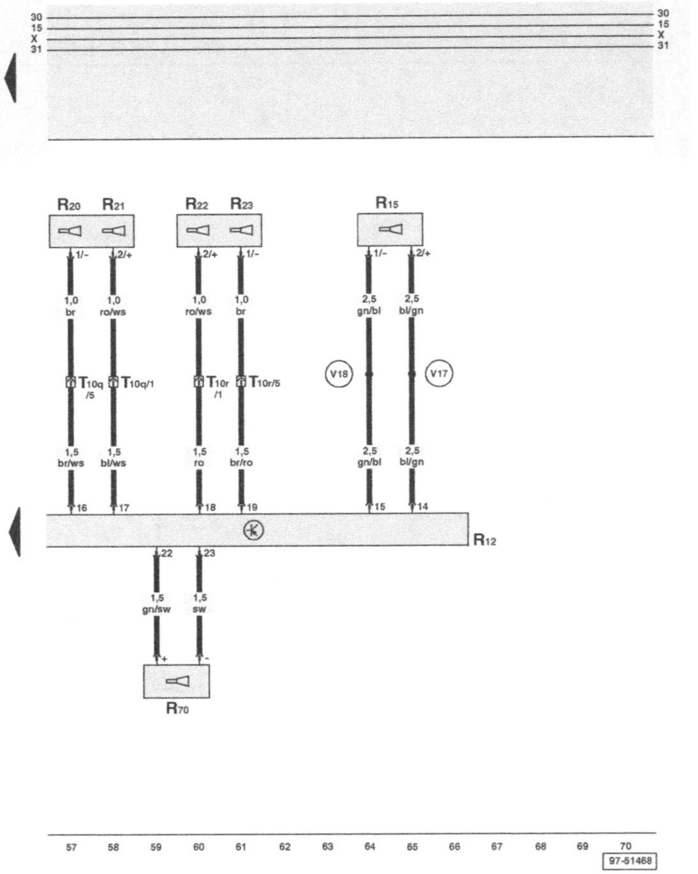

ws = white
sw = black
ro = red
br = brown
gn = green
bl = blue
gr = grey
li = violet
ge = yellow

R12 - Amplifier
R15 - Left Rear Woofer
R20 - Left Front Tweeter
R21 - Left Front Woofer
R22 - Right Front Tweeter
R23 - Right Front Woofer
R70 - Inside loudspeaker
T10q - Connector 10 pin, green, connector station A pillar, left
T10r - Connector 10 pin, green, connector station A pillar, right

(V17) - Connector (plus, speaker rear), in wiring harness speaker
(V18) - Connector (minus, speaker rear), in wiring harness speaker

© 2006 BentleyPublishers.com—All Rights Reserved

CAN-Bus Wiring

2000 m. y.

Fuse Panel

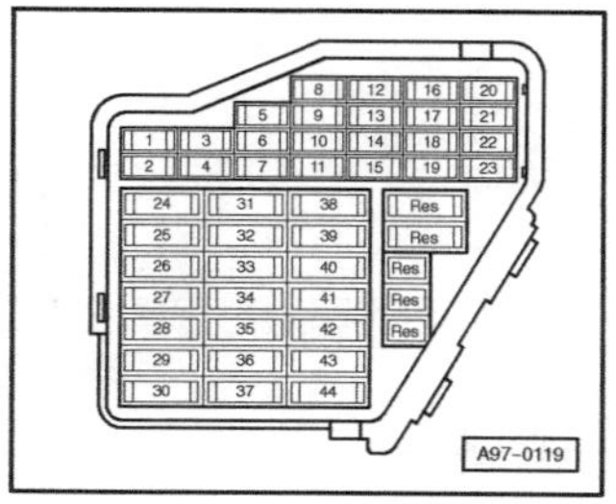

Fuse Colors:

30 A - Green
25 A - White
20 A - Yellow
15 A - Blue
10 A - Red
7,5 A - Brown
5 A - Beige

Starting with fuse position 23, fuses in the fuse holder are identified with 223 in the wiring diagram.

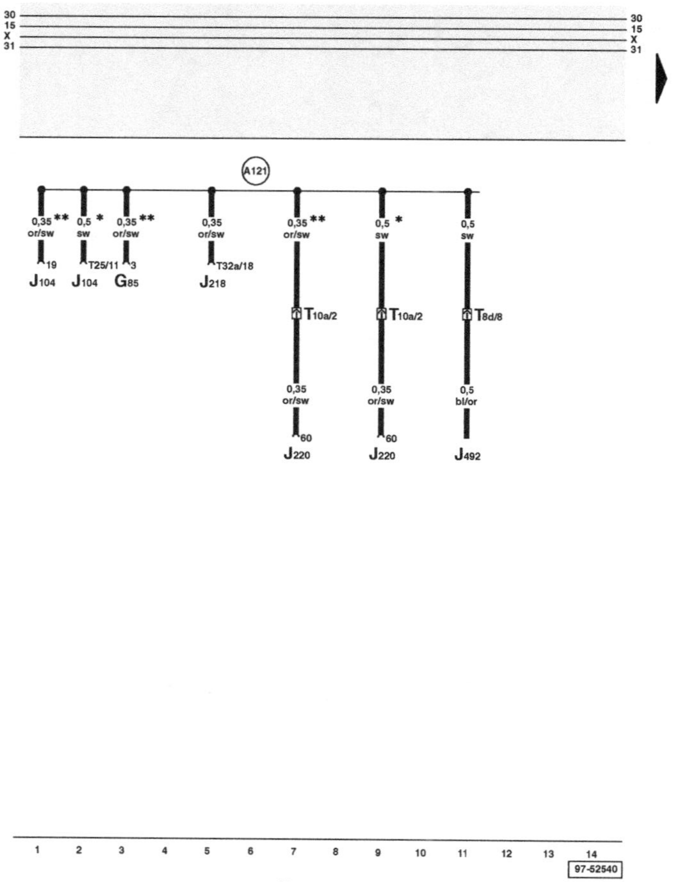

ws = white
sw = black
ro = red
br = brown
gn = green
bl = blue
gr = grey
li = violet
ge = yellow
or = orange

G85 - Steering Angle Sensor
J104 - ABS Control Module (w/EDL)
J218 - Instrument Cluster Combination Processor
J220 - Motronic Engine Control Module (ECM)
J492 - All-wheel drive control module
T8d - Connector 8 pin, on Control module for all wheel drive
T10a - Connector 10 pin, white, in E-box plenum chamber
T25 - Connector 25 pin, on ABS Control Module (w/EDL)
T32a - Connector 32 pin, green, on instrument cluster

(A121) - Connection (high bus) in instrument panel wiring harness

* - Vehicles with ABS
** - Vehicles with ABS and ESP

©2006 BentleyPublishers.com—All Rights Reserved

Audi TT **Wiring diagram** **No. 23/3**

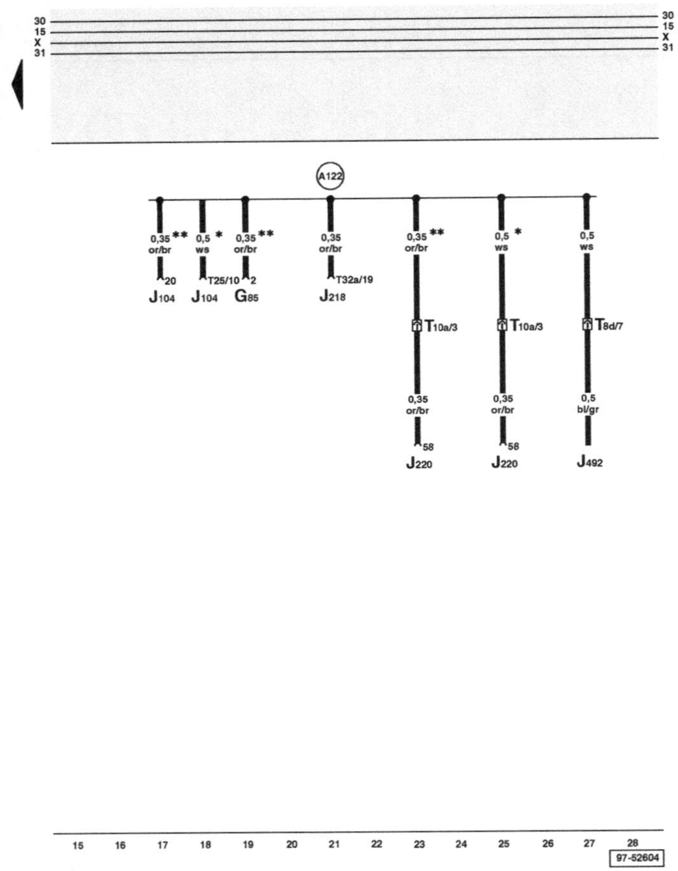

ws = white
sw = black
ro = red
br = brown
gn = green
bl = blue
gr = grey
li = violet
ge = yellow
or = orange

G85 - Steering Angle Sensor
J104 - ABS Control Module (w/EDL)
J218 - Instrument Cluster Combination Processor
J220 - Motronic Engine Control Module (ECM)
J492 - All-wheel drive control module
T8d - Connector 8 pin, on Control module for all wheel drive
T10a - Connector 10 pin, white, in E-box plenum chamber
T25 - Connector 25 pin, on ABS Control Module (w/EDL)
T32a - Connector 32 pin, green, on instrument cluster

(A122) - Connection (low bus) in instrument panel wiring harness

* - Vehicles with ABS
** - Vehicles with ABS and ESP

Edition 07.00
W42.USA.5702.04.21

CAN-Bus Wiring, Connection Low-Bus

THIS SPACE INTENTIONALLY LEFT BLANK

© 2006 BentleyPublishers.com—All Rights Reserved

Anti-Lock Brake System (ABS) Mk 20 with Electronic Differential Lock (EDL), Anti-Slip Regulation (ASR - ASC) and Electronic Stabilization Program (ESP), Retrofit

2000 m. y.

Fuse Panel

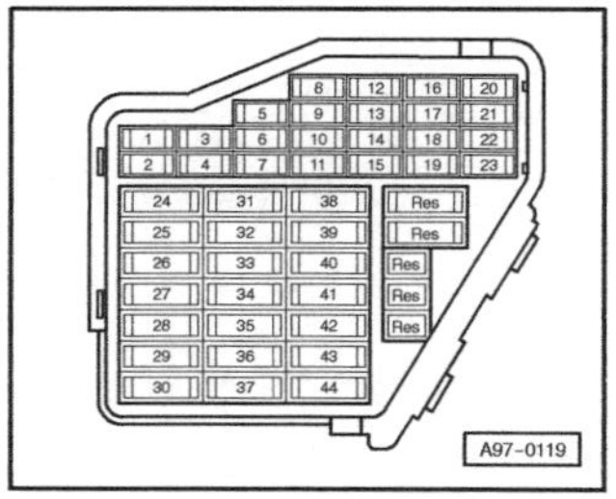

Fuse Colors:

30 A - Green
25 A - White
20 A - Yellow
15 A - Blue
10 A - Red
7,5 A - Brown
5 A - Beige

Starting with fuse position 23, fuses in the fuse holder are identified with 223 in the wiring diagram.

13 - Fold Relay Panel

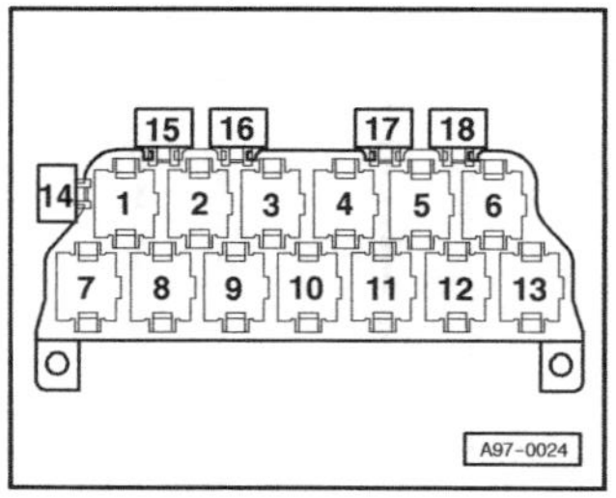

Relay Location:

7 -Brake Light Disable Relay (ESP), J508

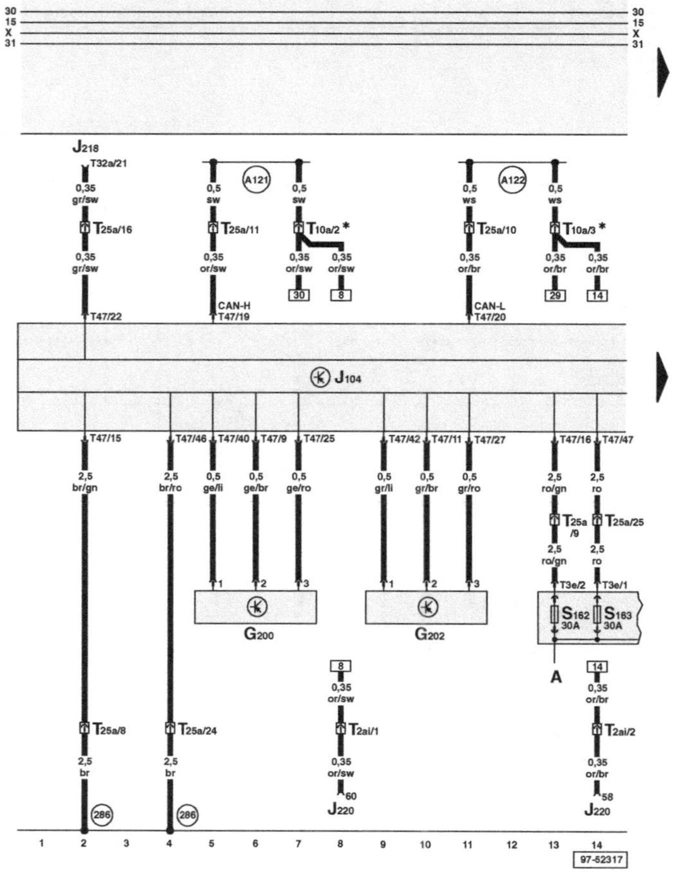

ws = white
sw = black
ro = red
br = brown
gn = green
bl = blue
gr = grey
li = violet
ge = yellow
or = orange

A - Battery
G200 - Sensor for transverse acceleration
G202 - Sender for rotation rate
J104 - ABS Control Module (w/EDL)
J218 - Instrument Cluster Combination Processor
J220 - Motronic Engine Control Module (ECM)
S162 - Fuse -1- in fuse bracket/battery
S163 - Fuse -2- in fuse bracket/battery
T2ai - Connector 2 pin, black, in E-Box plenum chamber
T3e - Connector 3 pin, black, on Main Fuse Box/Battery
T10a - Connector 10 pin, white, in E-box plenum chamber
T25a - Connector 25 pin, next to ABS Control Module (w/EDL)
T32a - Connector 32 pin, green, on instrument cluster
T47 - 47-Pin Connector, on ABS Control Module (w/EDL)

(286) - Ground connection -8- in wiring harness interior
(A121) - Connection (high bus) in instrument panel wiring harness
(A122) - Connection (low bus) in instrument panel wiring harness
- CAN-Bus (Data-Bus)
* - m.y. 1999 Connector T10a, green
m.y. 2000 Connector T10a, white

© 2006 BentleyPublishers.com—All Rights Reserved

Anti-Lock Brake System, ABS/EDL/ASR/ESP, Mk. 20, retrofit (2000-2001)

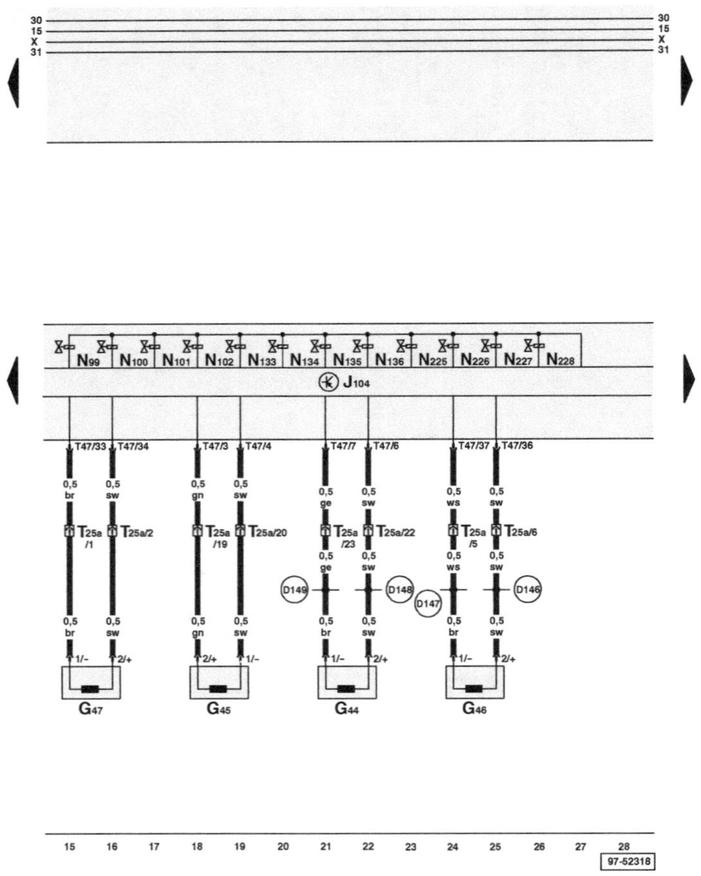

ws = white
sw = black
ro = red
br = brown
gn = green
bl = blue
gr = grey
li = violet
ge = yellow

G44 - Right Rear ABS Wheel Speed Sensor
G45 - Right Front ABS Wheel Speed Sensor
G46 - Left Rear ABS Wheel Speed Sensor
G47 - Left Front ABS Wheel Speed Sensor
J104 - ABS Control Module (w/EDL)
N99 - Right Front ABS Inlet Valve
N100 - Right Front ABS Outlet Valve
N101 - Left Front ABS Inlet Valve
N102 - Left Front ABS Outlet Valve
N133 - Right Rear ABS Inlet Valve
N134 - Left Rear ABS Inlet Valve
N135 - Right Rear ABS Outlet Valve
N136 - Left Rear ABS Outlet Valve
N225 - Pilot valve -1- traction control (01418)
N226 - Pilot valve -2- traction control (01419)
N227 - High pressure switch valve -1- traction control (01420)
N228 - High pressure switch valve -2- traction control (01421)
T25a - Connector 25 pin, next to ABS Control Module (w/EDL)
T47 - 47-Pin Connector, on ABS Control Module (w/EDL)

(D146) - Connector (speed sensor rear, left +), in wiring harness engine compartment
(D147) - Connector (speed sensor rear, left -), in wiring harness engine compartment
(D148) - Connector (speed sensor rear, right +), in wiring harness engine compartment
(D149) - Connector (speed sensor rear, right -), in wiring harness engine compartment

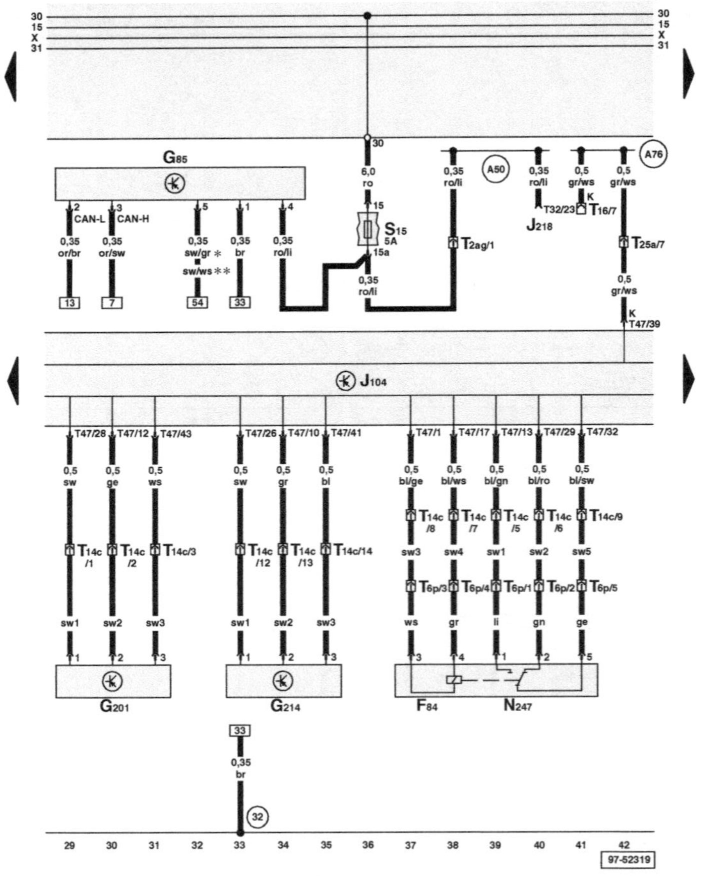

ws = white
sw = black
ro = red
br = brown
gn = green
bl = blue
gr = grey
li = violet
ge = yellow
or = orange

F84 - Brake Pressure Release Solenoid
G85 - Steering Angle Sensor
G201 - Sender 1 for brake booster
G214 - Sensor -2- for brake pressure
J104 - ABS Control Module (w/EDL)
J218 - Instrument Cluster Combination Processor
N247 - Magnetic coil for brake pressure, in brake booster
S15 - Fuse
T2ag - Connector 2 pin, black, near fuse panel
T6p - Connector 6 pin, black, on brake booster
T14c - Connector 14 pin, connector in wiring harness ESP
T16 - Connector 16 pin, black, Data Link Connector (DLC), below storage compartment, driver's side
T25a - Connector 25 pin, next to ABS Control Module (w/EDL)
T32 - Connector 32 pin, blue, on instrument cluster
T47 - 47-Pin Connector, on ABS Control Module (w/EDL)

(32) - Ground connection, behind instrument panel, left
(A50) - Plus connection (30a), in instrument panel wiring harness
(A76) - Connector (K-diagnosis wire), in instrument panel wiring harness
* - 2000 m.y.
** - 2001 m.y.

© 2006 BentleyPublishers.com—All Rights Reserved

Anti-Lock Brake System, ABS/EDL/ASR/ESP, Mk. 20, retrofit (2000-2001)

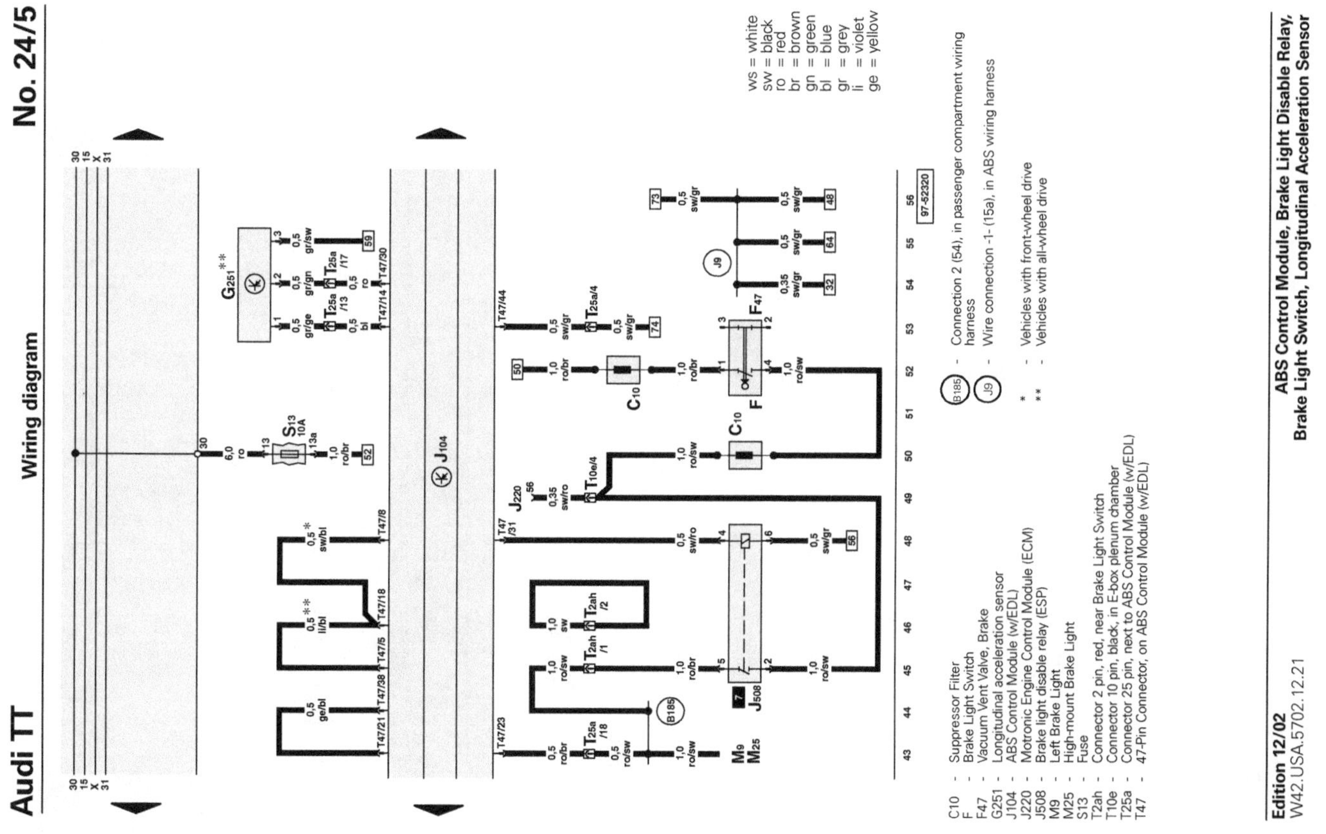

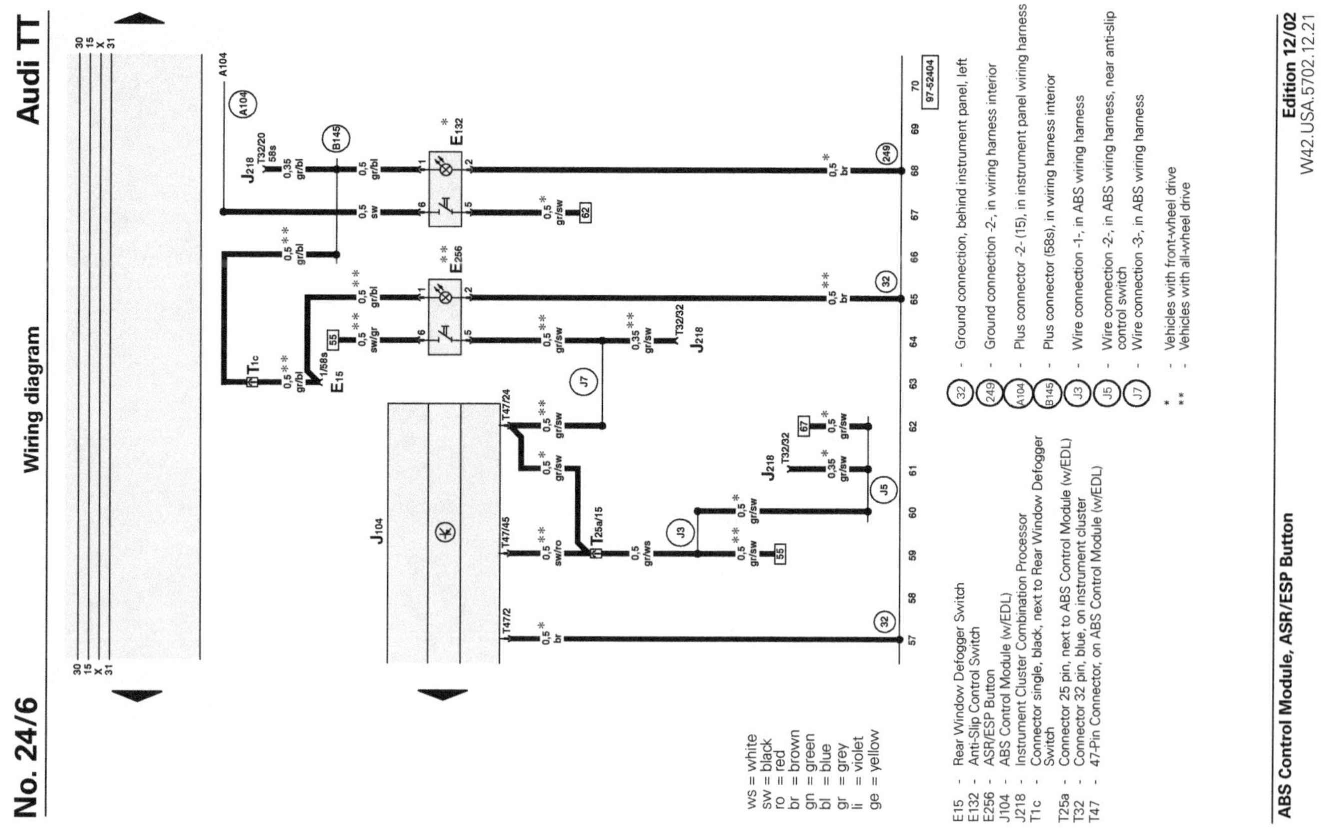

© 2006 BentleyPublishers.com—All Rights Reserved

Audi TT Wiring diagram **No. 24/7**

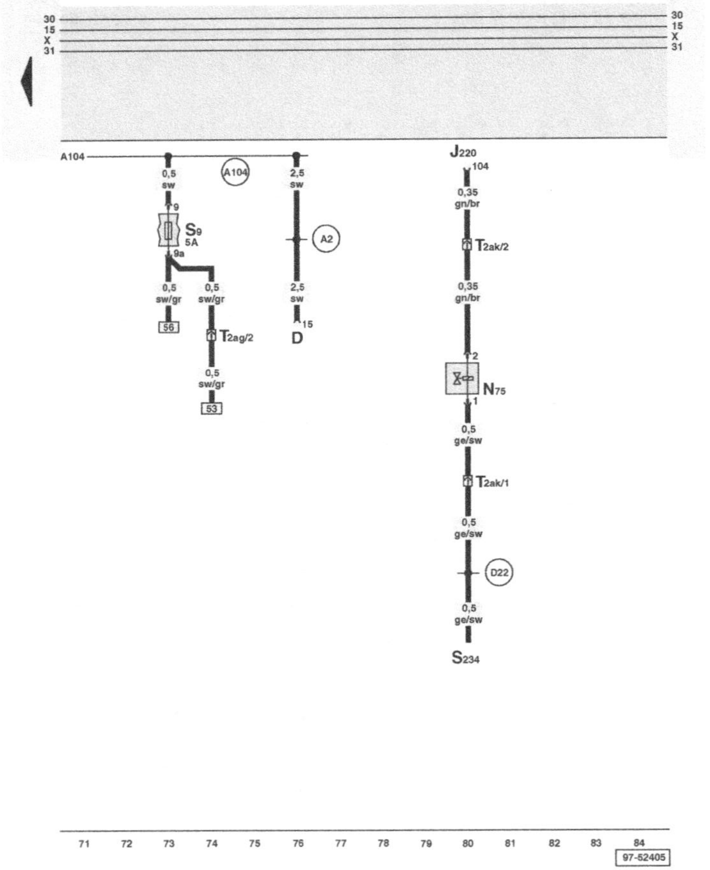

ws = white
sw = black
ro = red
br = brown
gn = green
bl = blue
gr = grey
li = violet
ge = yellow

D - Ignition/Starter Switch
J220 - Motronic Engine Control Module (ECM)
N75 - Wastegate Bypass Regulator Valve
S9 - Fuse
S234 - Fuse 34 in fuse holder
T2ag - Connector 2 pin, black, near fuse panel
T2ak - Connector 2 pin, on wire extension for Wastegate Bypass Regulator Valve

(A2) - Plus connection (15), in instrument panel wiring harness
(A104) - Plus connector -2- (15), in instrument panel wiring harness
(D22) - Connector (over fuse 234), in wiring harness front, right

Edition 11.00
W42.USA.5702.05.21

Wastegate Bypass Regulator Valve

THIS SPACE INTENTIONALLY LEFT BLANK

© 2006 BentleyPublishers.com—All Rights Reserved

Standard Equipment

2002 m. y.

Fuse Panel

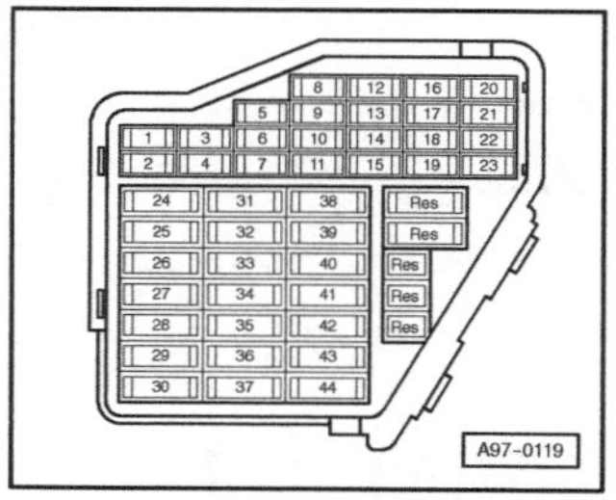

Fuse Colors:

30 A - Green
25 A - White
20 A - Yellow
15 A - Blue
10 A - Red
7,5 A - Brown
5 A - Beige

Starting with fuse position 23, fuses in the fuse holder are identified with 223 in the wiring diagram.

Micro Central Electric Panel

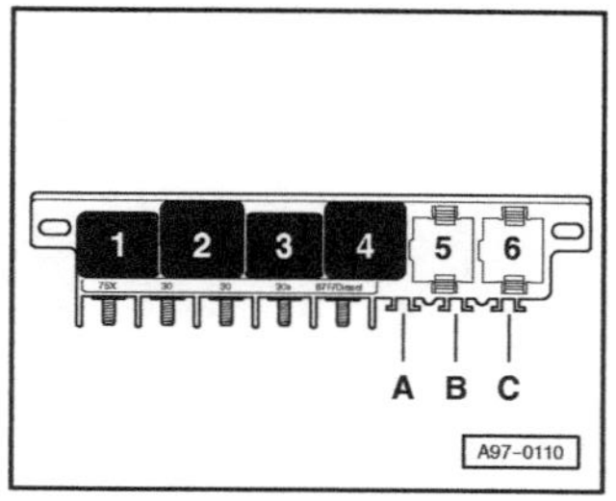

Relay Location:

1 -Dual Horn Relay, J4
2 -Load Reduction Relay, J59
4 -Fuel Pump (FP) Relay, J17
5 -Wiper/Washer Intermittent Relay, J31
6 -Wiper/Washer Intermittent Relay, J31

13 - Fold Relay Panel

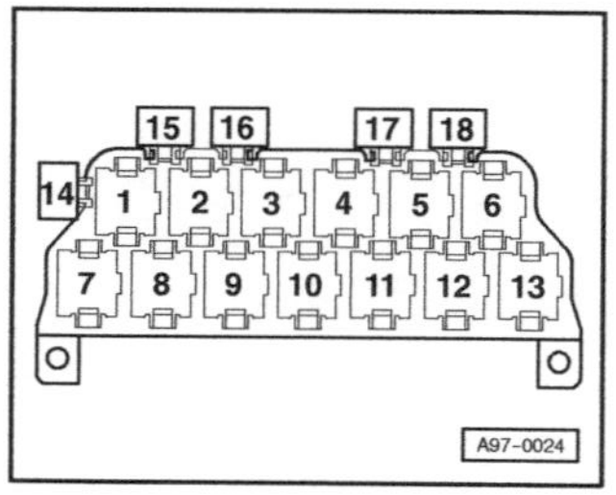

Relay Location:

4 -Fog Light Relay, J5
5 -Starting Interlock Relay, J207
10 -Rear Window Defogger Relay, J9

Main Fuse Box/ Battery

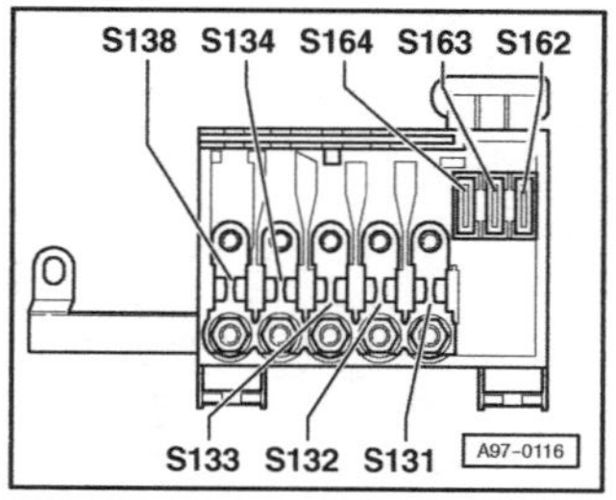

Fuses Location:

S131 - Safety fuse 1
S132 - Safety fuse 2
S133 - Safety fuse 3
S134 - Safety fuse 4
S138 - Safety fuse 5
S162 - Fuse 1 in fuse bracket/battery
S163 - Fuse 2 in fuse bracket/battery
S164 - Fuse 3 in fuse bracket/battery

© 2006 BentleyPublishers.com—All Rights Reserved

Audi TT — Wiring diagram — No. 25/3

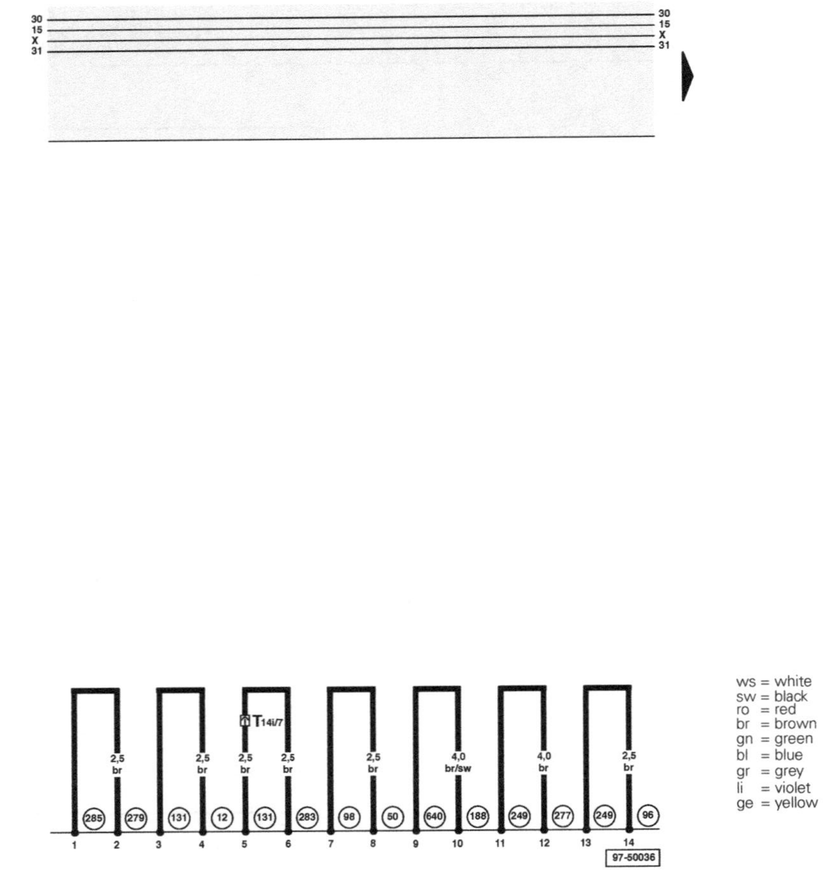

ws = white
sw = black
ro = red
br = brown
gn = green
bl = blue
gr = grey
li = violet
ge = yellow

- T14i - Connector 14 pin, in engine compartment, left
- (12) - Ground connection, in engine compartment, left
- (50) - Ground connection, in luggage compartment, left
- (96) - Ground connection -1-, in heated seats wiring harness
- (98) - Ground connection, in rear lid wiring harness
- (131) - Ground connection -2-, in engine compartment wiring harness
- (188) - Ground connection -3-, in A/C wiring harness
- (249) - Ground connection -2-, in wiring harness interior
- (277) - Ground connector -3-, in wiring harness interior
- (279) - Ground connector -5-, in wiring harness interior
- (283) - Ground connection 2, in wiring harness engine pre-wiring
- (285) - Ground connection -7- in wiring harness interior
- (640) - Ground connection -2-, in engine compartment, left

Edition 10/02 W42.USA.5702.11.21 **Ground Connections**

No. 25/4 — Wiring diagram — Audi TT

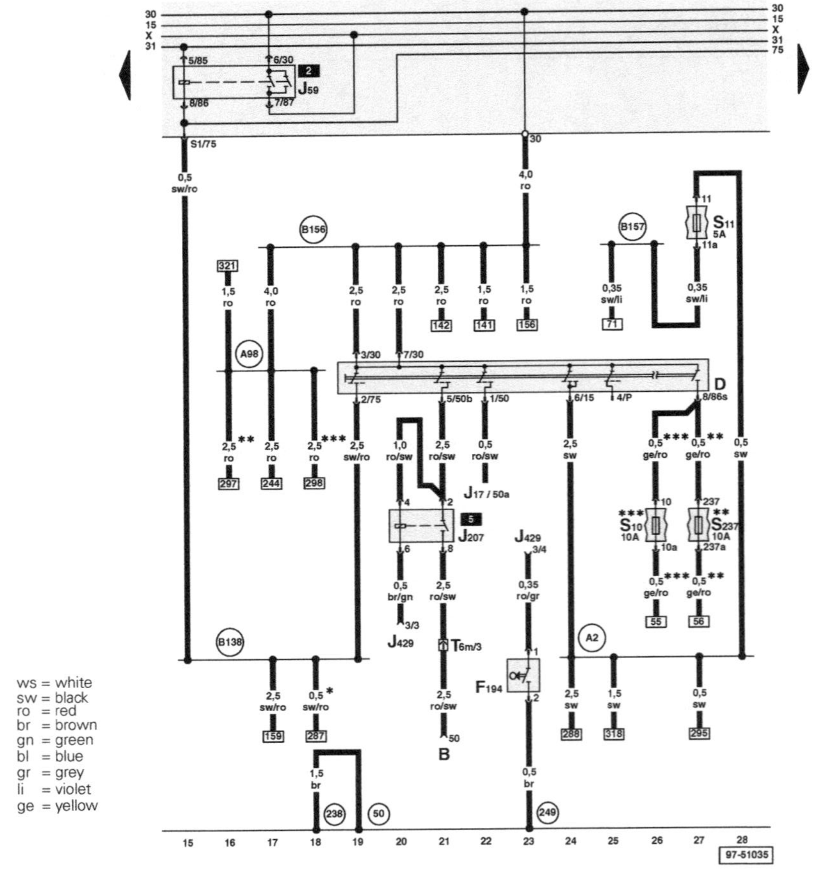

ws = white
sw = black
ro = red
br = brown
gn = green
bl = blue
gr = grey
li = violet
ge = yellow

- B - Starter
- D - Ignition/Starter Switch
- F194 - Clutch Pedal Position (CPP) Switch
- J17 - Fuel Pump (FP) Relay
- J59 - Load Reduction Relay
- J207 - Starting Interlock Relay
- J429 - Control module for central locking
- S10 - Fuse
- S11 - Fuse
- S237 - Fuse in fuse holder
- T6m - Connector 6 pin, brown, in E-box plenum chamber
- (50) - Ground connection, in luggage compartment, left
- (238) - Ground connection -1-, in wiring harness interior
- (249) - Ground connection -2-, in wiring harness interior
- (A2) - Plus connection (15), in instrument panel wiring harness
- (A98) - Plus connector -4- (30), in instrument panel wiring harness
- (B138) - Plus connector (X), in wiring harness, interior
- (B156) - Plus connector (30a), in wiring harness interior
- (B157) - Plus connector -3- (15a), in wiring harness interior

- * - Roadster
- ** - through 10.2001
- *** - from 11.2001

Load Reduction Relay, Ignition/Starter Switch, Starting Interlock Relay **Edition 10/02** W42.USA.5702.11.21

© 2006 BentleyPublishers.com—All Rights Reserved

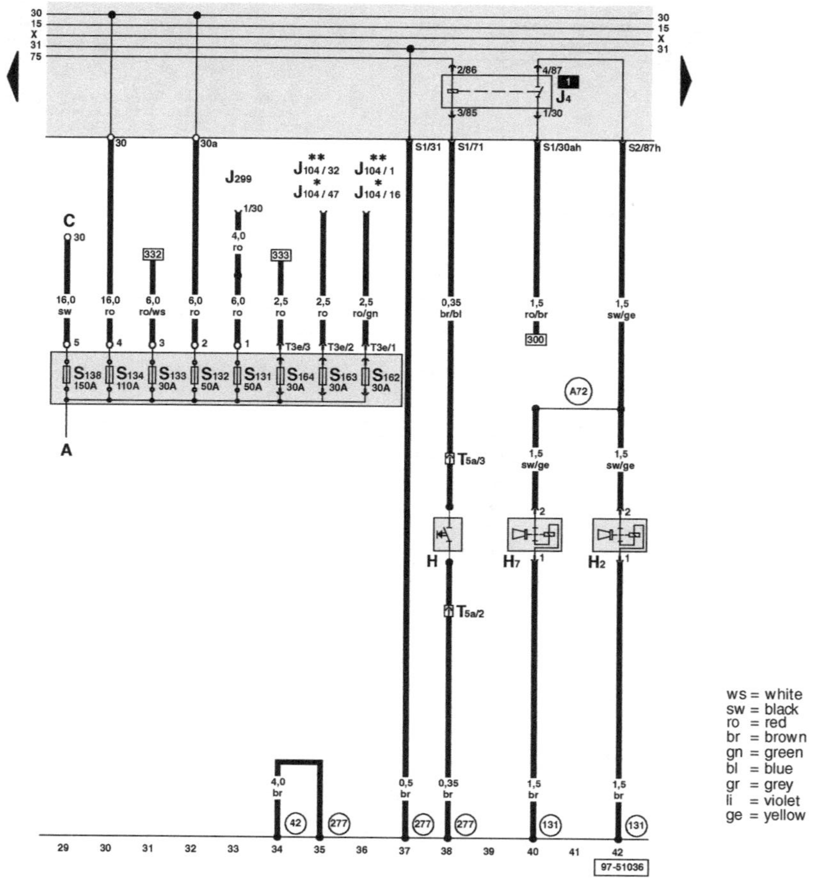

ws = white
sw = black
ro = red
br = brown
gn = green
bl = blue
gr = grey
li = violet
ge = yellow

A - Battery
C - Generator (GEN)
H - Signal horn activation
H2 - High Tone Horn
H7 - Low Tone Horn
J4 - Dual Horn Relay
J104 - ABS Control Module (w/EDL)
J299 - Secondary Air Injection (AIR) Pump Relay
S131 - Safety fuse 1
S132 - Safety fuse 2
S133 - Safety fuse 3
S134 - Safety fuse 4
S138 - Safety fuse 5
S162 - Fuse -1- in fuse bracket/battery
S163 - Fuse -2- in fuse bracket/battery
S164 - Fuse -3- in fuse bracket/battery
T3e - Connector 3 pin, black, on Main Fuse Box/Battery
T5a - Connector 5 pin, yellow, in steering wheel

(42) - Ground connection, on steering column

(131) - Ground connection -2-, in engine compartment wiring harness

(277) - Ground connector -3-, in wiring harness interior

(A72) - Connector (71), in instrument panel wiring harness

* - ABS MK20 with EDL/ASR/ESP
** - ABS MK60 with EDL/ASR/ESP

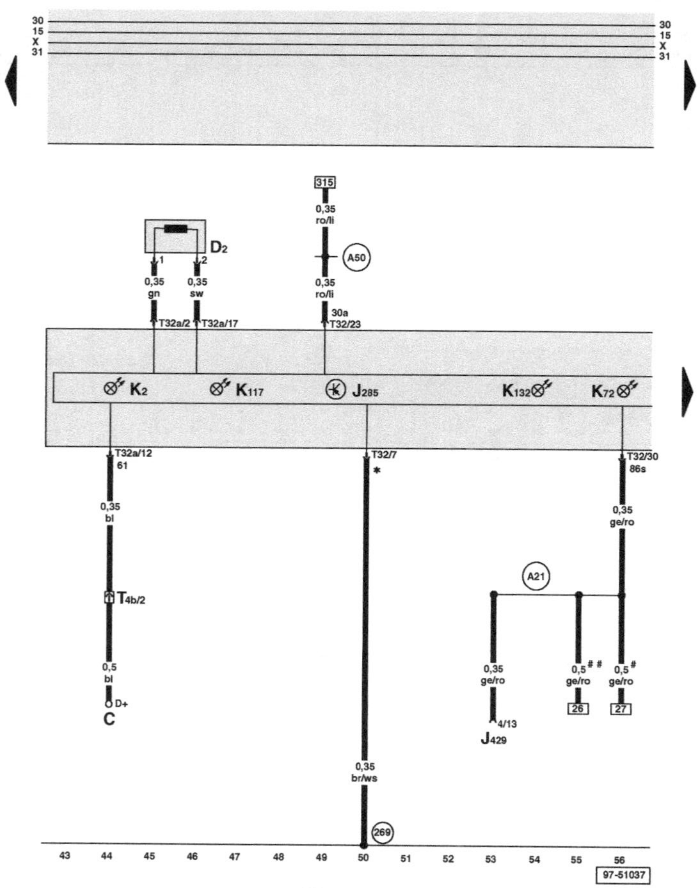

ws = white
sw = black
ro = red
br = brown
gn = green
bl = blue
gr = grey
li = violet
ge = yellow

C - Generator (GEN)
D2 - Induction coil of antitheft immobilizer
J285 - Control Module with indicator unit in instrument panel insert
J429 - Control module for central locking
K2 - Generator (GEN) Warning Light
K72 - Warning light for catalyst monitoring
K117 - Warning light for anti-theft immobilizer sensor
K132 - Electronic Power Control (EPC) Warning Lamp
T4b - Connector 4 pin, in engine compartment, left
T32 - Connector 32 pin, blue, on instrument cluster
T32a - Connector 32 pin, green, on instrument cluster

(269) - Ground connector (sensor ground) -1-, in instrument panel wiring harness

(A21) - Wire connection (86s), in instrument panel wiring harness

(A50) - Plus connection (30a), in instrument panel wiring harness

* - Sensor ground output
\# - to 11.2001
\#\# - from 11.2001

©2006 BentleyPublishers.com—All Rights Reserved

Audi TT **Wiring diagram** **No. 25/7**

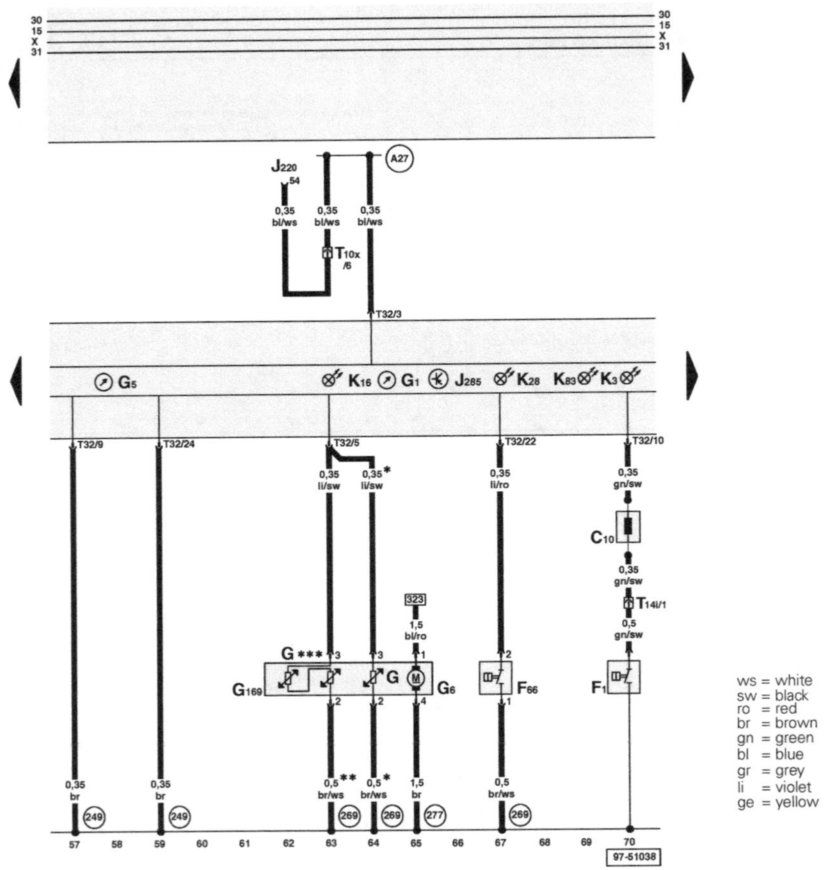

ws = white
sw = black
ro = red
br = brown
gn = green
bl = blue
gr = grey
li = violet
ge = yellow

C10 - Suppressor Filter
F1 - Oil Pressure Switch
F66 - Engine Coolant Level (ECL) Warning Switch
G - Sender for fuel gauge
G1 - Fuel Gauge
G5 - Tachometer
G6 - Fuel Pump (FP)
G169 - Fuel Level Sensor 2
J220 - Motronic Engine Control Module (ECM)
J285 - Control Module with indicator unit in instrument panel insert
K3 - Oil Pressure Warning Light
K16 - Fuel Reserve Warning Light
K28 - Engine Coolant Level/Temperature (ECL/ECT) Warning Light
K83 - Malfunction Indicator Lamp (MIL)
T10x - Connector 10 pin, orange, in E-box plenum chamber
T14i - Connector 14 pin, in engine compartment, left
T32 - Connector 32 pin, blue, on instrument cluster

(249) - Ground connection -2-, in wiring harness interior
(269) - Ground connector (sensor ground) -1-, in instrument panel wiring harness
(277) - Ground connector -3-, in wiring harness interior
(A27) - Wire connection (vehicle speed signal), in instrument panel wiring harness

* - Vehicles with front wheel drive
** - Vehicles with all wheel drive
*** - Wiring inside fuel tank

Edition 10/02
W42.USA.5702.11.21

Instrument Cluster, Oil Pressure Switch, Engine Coolant Level Warning Switch, Fuel Pump

No. 25/8 **Wiring diagram** **Audi TT**

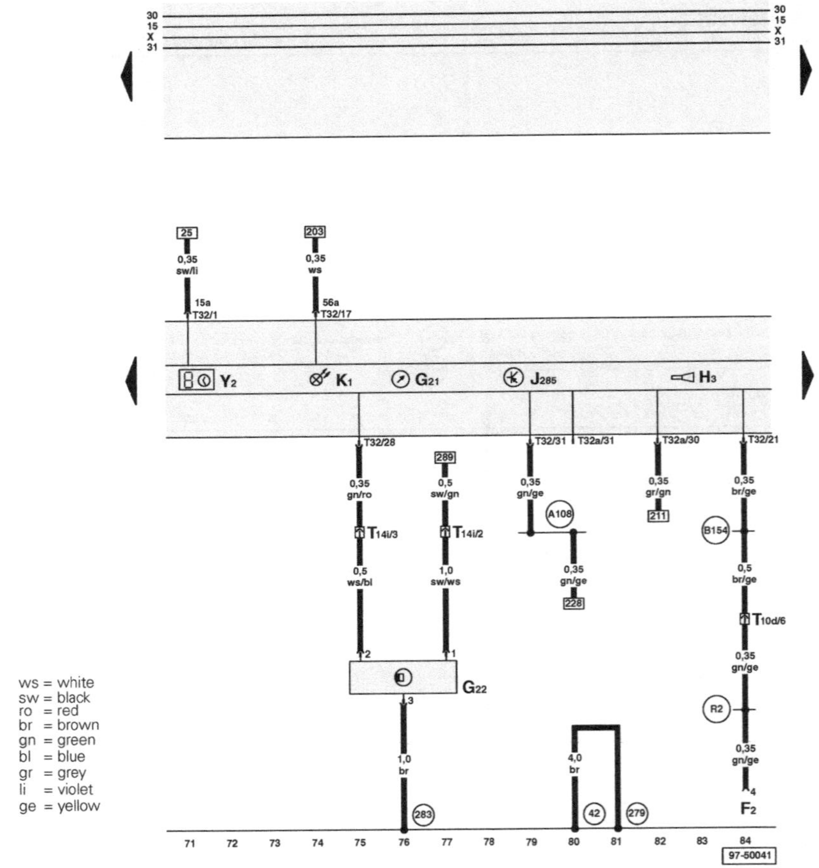

ws = white
sw = black
ro = red
br = brown
gn = green
bl = blue
gr = grey
li = violet
ge = yellow

C10 - Suppressor Filter
F2 - Door Contact Switch - Driver's side
G21 - Speedometer
G22 - Speedometer Vehicle Speed Sensor (VSS)
H3 - Warning Buzzer
J285 - Control Module with indicator unit in instrument panel insert
K1 - Headlight High Beam Indicator Light
T10d - Connector 10 pin, brown, connector station A pillar, left
T14i - Connector 14 pin, in engine compartment, left
T32 - Connector 32 pin, blue, on instrument cluster
T32a - Connector 32 pin, green, on instrument cluster
Y2 - Digital Clock

(42) - Ground connection, on steering column
(279) - Ground connector -5-, in wiring harness interior
(283) - Ground connection 2, in wiring harness engine pre-wiring
(A108) - Connector (speed signal), in instrument panel wiring harness
(B154) - Connector -1- (TK), in wiring harness interior
(R2) - Wire connection (door contact switch, left), in power window/central locking system and door contact switch wiring harness

Instrument Cluster, Speedometer Vehicle Speed Sensor

Edition 10/02
W42.USA.5702.11.21

© 2006 BentleyPublishers.com—All Rights Reserved

Audi TT — Wiring diagram — No. 25/9

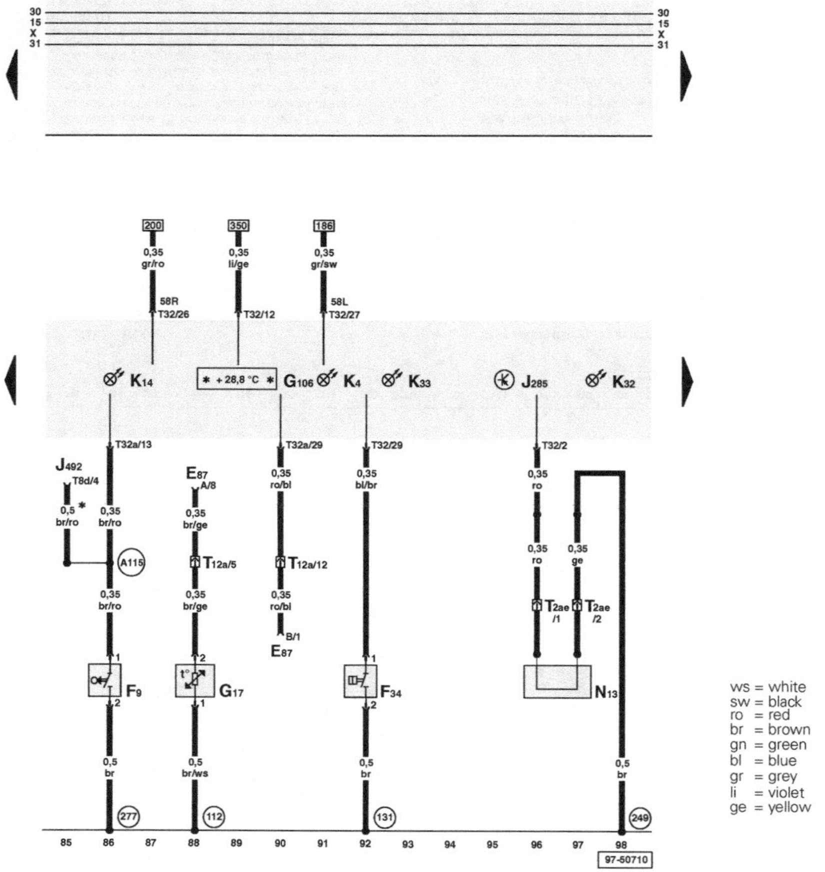

ws = white
sw = black
ro = red
br = brown
gn = green
bl = blue
gr = grey
li = violet
ge = yellow

E87 - A/C Control Head
F9 - Parking Brake Warning Light Switch
F34 - Brake Fluid Level Warning Switch
G17 - Outside Air Temperature Sensor
G106 - Outside Air Temperature Display
J285 - Control Module with indicator unit in instrument panel insert
J492 - Control module for all wheel drive
K4 - Park Light Indicator Light
K14 - Parking Brake Indicator Light
K32 - Brake Pad Wear Indicator Light
K33 - Brake Fluid Level Warning Light
N13 - Left Brake Pad Wear Indicator (Shear Element)
T2ae - Connector 2 pin, on firewall, left
T8d - Connector 8 pin, on Control module for all wheel drive
T12a - Connector 12 pin, behind instrument panel, right
T32 - Connector 32 pin, blue, on instrument cluster
T32a - Connector 32 pin, green, on instrument cluster

(112) - Ground connection -2-, in A/C wiring harness
(131) - Ground connection -2-, in engine compartment wiring harness
(249) - Ground connection -2-, in wiring harness interior
(277) - Ground connector -3-, in wiring harness interior
(A115) - Connection (parking brake control) in instrument panel wiring harness
* - Vehicles with all-wheel drive

Edition 10/01
W42.USA.5702.09.21

Instrument Cluster, Parking Brake Warning Light Switch, Brake Fluid Level Warning Switch, Outside Air Temperature Sensor

No. 25/10 — Wiring diagram — Audi TT

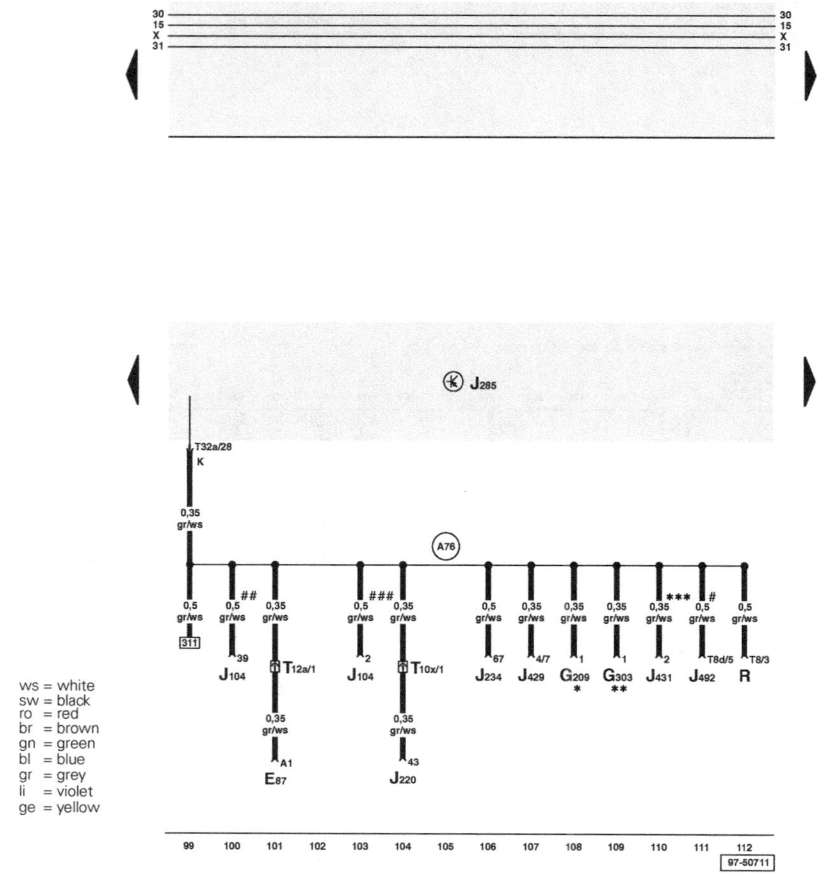

ws = white
sw = black
ro = red
br = brown
gn = green
bl = blue
gr = grey
li = violet
ge = yellow

E87 - A/C Control Head
G209 - Ultra-sound sensor for anti-theft warning
G303 - Radar Interior Monitoring Control Module 1
J104 - ABS Control Module (w/EDL)
J220 - Motronic Engine Control Module (ECM)
J234 - Airbag Control Module
J285 - Control Module with indicator unit in instrument panel insert
J429 - Control module for central locking
J431 - Control module for Headlight Beam Adjusting
J492 - Control module for all wheel drive
R - Radio
T8 - Connector 8 pin, black, connector radio III
T8d - Connector 8 pin, on Control module for all wheel drive
T10x - Connector 10 pin, orange, in E-box plenum chamber
T12a - Connector 12 pin, behind instrument panel, right
T32a - Connector 32 pin, green, on instrument cluster

(A76) - Connector (K-diagnosis wire), in instrument panel wiring harness

* - Coupé
** - Roadster
*** - Vehicles with Automatic Headlight Beam Adjusting
\# - Vehicles with all-wheel drive
\#\# - ABS MK20 with EDL/ASR/ESP
\#\#\# - ABS MK60 with EDL/ASR/ESP

Instrument Cluster, Connector (K-diagnosis wire) for Data Link Connector (DLC)

Edition 10/01
W42.USA.5702.09.21

© 2006 BentleyPublishers.com—All Rights Reserved

Audi TT — Wiring diagram — No. 25/11

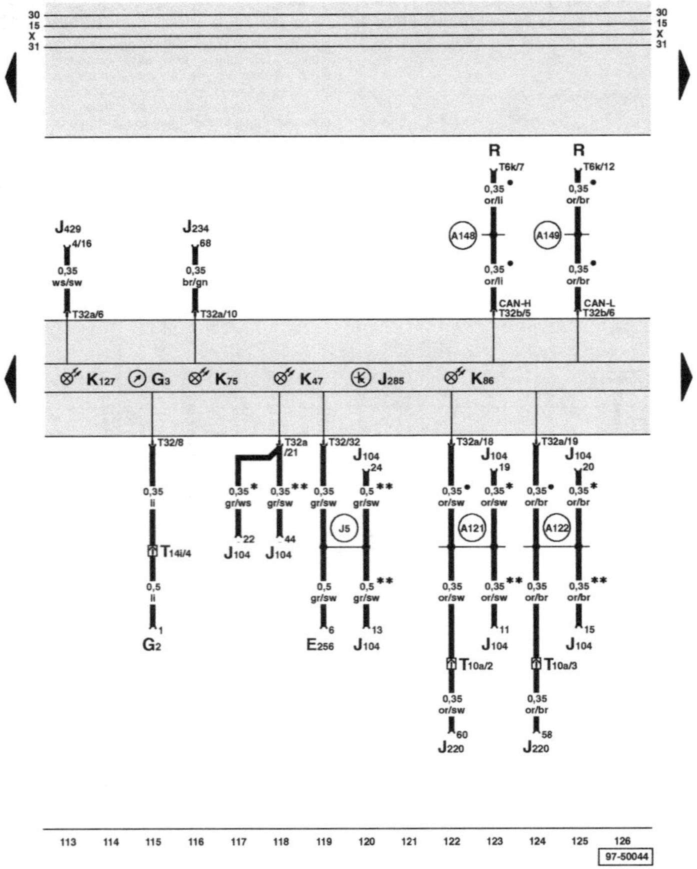

ws = white
sw = black
ro = red
br = brown
gn = green
bl = blue
gr = grey
li = violet
ge = yellow
or = orange

E256 - ASR/ESP Button
G2 - Engine Coolant Temperature (ECT) Sensor
G3 - Engine Coolant Temperature (ECT) Gauge
J104 - ABS Control Module (w/EDL)
J220 - Motronic Engine Control Module (ECM)
J285 - Control Module with indicator unit in instrument panel insert
J234 - Airbag Control Module
J429 - Control module for central locking
K47 - ABS Warning Light
K75 - Airbag Malfunction Indicator Lamp (MIL)
K86 - Traction Control Indicator Light
K127 - Indicator light for open trunk
R - Radio
T6k - Connector 6 pin, green, connector radio I-2
T10a - Connector 10 pin, white, in E-box plenum chamber
T14i - Connector 14 pin, in engine compartment, left
T32 - Connector 32 pin, blue, on instrument cluster
T32a - Connector 32 pin, green, on instrument cluster
T32b - Connector 32 pin, gray, on instrument cluster

(A121) - Connection (high bus) in instrument panel wiring harness
(A122) - Connection (low bus) in instrument panel wiring harness
(A148) - Connection (high bus, navigation), in instrument panel wiring harness
(A149) - Connection (low bus, navigation), in instrument panel wiring harness
(J5) - Wire connection -2-, in ABS wiring harness, near anti-slip control switch

• - CAN-Bus (Data-Bus)
* - ABS MK20 with EDL/ASR/ESP
** - ABS MK60 with EDL/ASR/ESP

No. 25/12 — Wiring diagram — Audi TT

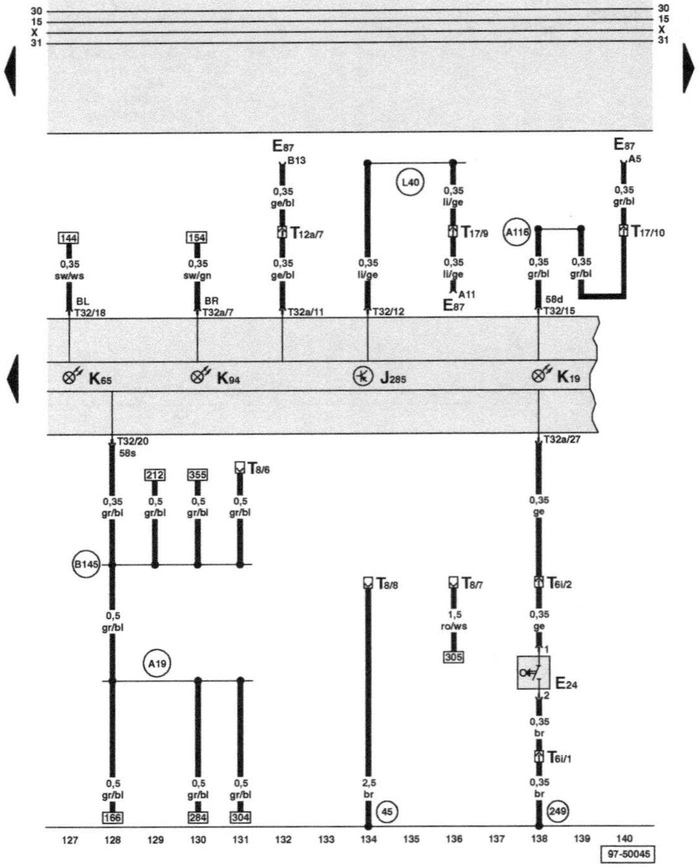

ws = white
sw = black
ro = red
br = brown
gn = green
bl = blue
gr = grey
li = violet
ge = yellow

E24 - Left Seat Belt Switch
E87 - A/C Control Head
J285 - Control Module with indicator unit in instrument panel insert
K19 - Seat Belt Warning Light
K65 - Left Turn Signal Indicator Light
K94 - Right Turn Signal Indicator Light
T6i - Connector 6 pin, below driver s seat
T8 - Connector 8 pin, black, connector radio III
T12a - Connector 12 pin, behind instrument panel, right
T17 - Connector 17 pin, behind instrument panel
T32 - Connector 32 pin, blue, on instrument cluster
T32a - Connector 32 pin, green, on instrument cluster

(45) - Ground connection, behind instrument panel, center
(249) - Ground connection -2-, in wiring harness interior
(A19) - Wire connection (58s), in instrument panel wiring harness
(A116) - Connection -2- (58d) in instrument panel wiring harness
(B145) - Plus connector (58s), in wiring harness interior
(L40) - Wire connection -2- (A/C Pressure Switch), in Climatronic wiring harness

© 2006 BentleyPublishers.com—All Rights Reserved

Audi TT — Wiring diagram — No. 25/13

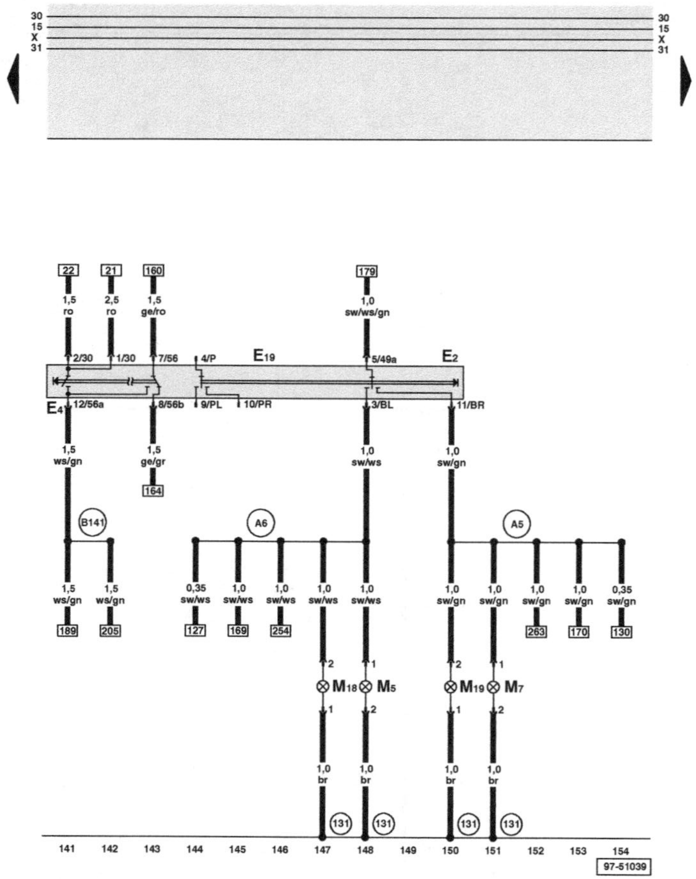

ws = white
sw = black
ro = red
br = brown
gn = green
bl = blue
gr = grey
li = violet
ge = yellow

E2 - Turn Signal Switch
E4 - Headlight Dimmer/Flasher Switch
E19 - Park Light Switch
M5 - Left Front Turn Signal Light
M7 - Right Front Turn Signal Light
M18 - Left, Side Turn Signal Light
M19 - Right, Side Turn Signal Light

(131) - Ground connection -2-, in engine compartment wiring harness
(A5) - Plus connection (right turn signal), in instrument panel wiring harness
(A6) - Plus connection (left turn signal), in instrument panel wiring harness
(B141) - Plus connector -1- (56a), in wiring harness interior

Turn Signal Switch, Headlight Dimmer/Flasher Switch, Park Light Switch, Front Turn Signal Light

No. 25/14 — Wiring diagram — Audi TT

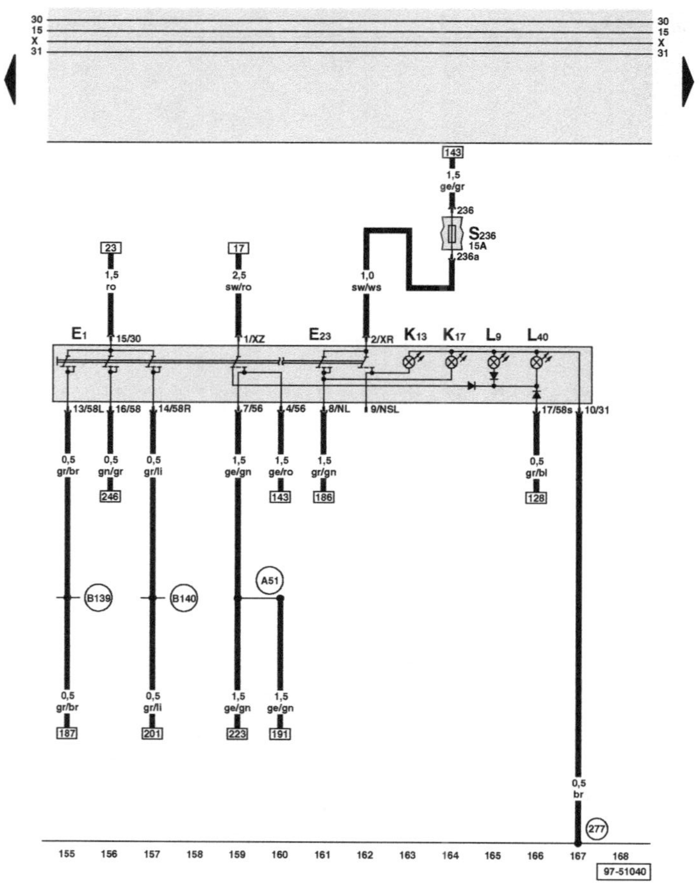

ws = white
sw = black
ro = red
br = brown
gn = green
bl = blue
gr = grey
li = violet
ge = yellow

E1 - Light Switch
E23 - Fog Light Switch
K13 - Rear Fog Light Indicator Light
K17 - Fog Light Indicator Light
L9 - Headlight Switch Light
L40 - Front And Rear Fog Light Switch Light
S236 - Fuse in fuse holder

(277) - Ground connector -3-, in wiring harness interior
(A51) - Wire connection (56), in instrument panel wiring harness
(B139) - Connector (PL), in wiring harness, interior
(B140) - Connector (PR), in wiring harness, interior

Light Switch, Fog Light Switch

© 2006 BentleyPublishers.com—All Rights Reserved

Audi TT Wiring diagram **No. 25/15**

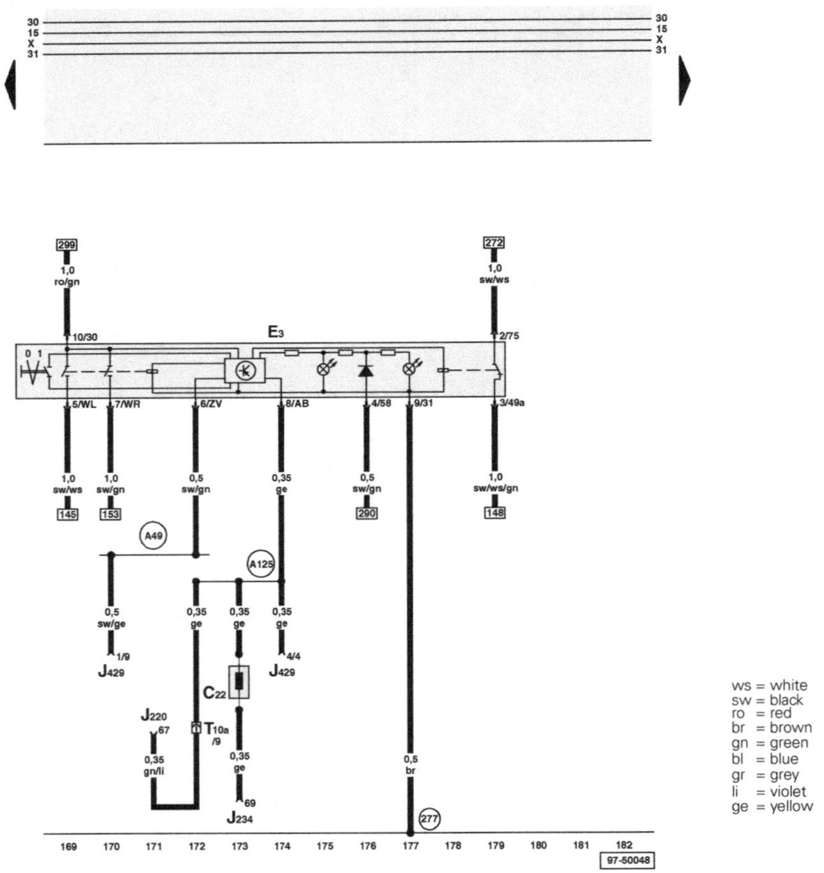

ws = white
sw = black
ro = red
br = brown
gn = green
bl = blue
gr = grey
li = violet
ge = yellow

C22 - Airbag Suppresser
E3 - Emergency Flasher Switch
J220 - Motronic Engine Control Module (ECM)
J234 - Airbag Control Module
J429 - Control module for central locking
T10a - Connector 10 pin, white, in E-box plenum chamber

(277) - Ground connector -3-, in wiring harness interior

(A49) - Wire connection -1-, in instrument panel wiring harness

(A125) - Connection (crash signal) in instrument panel wiring harness

No. 25/16 Wiring diagram **Audi TT**

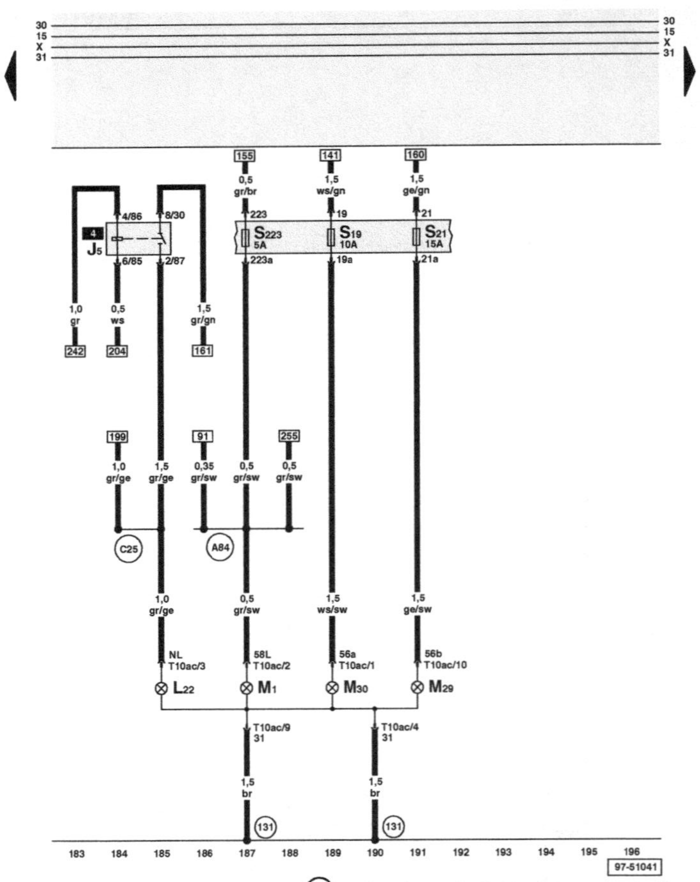

ws = white
sw = black
ro = red
br = brown
gn = green
bl = blue
gr = grey
li = violet
ge = yellow

J5 - Fog Light Relay
L22 - Left Front Fog Light
M1 - Left Parking Light
M29 - Left Low Beam Headlight
M30 - Left High Beam Headlight
S19 - Fuse
S21 - Fuse
S223 - Fuse in fuse holder
T10ac - Connector 10 pin, on headlight, left

(131) - Ground connection -2-, in engine compartment wiring harness

(A84) - Connector (58L), in instrument panel wiring harness

(C25) - Wire connection (fog lights), in front light wiring harness

© 2006 BentleyPublishers.com—All Rights Reserved

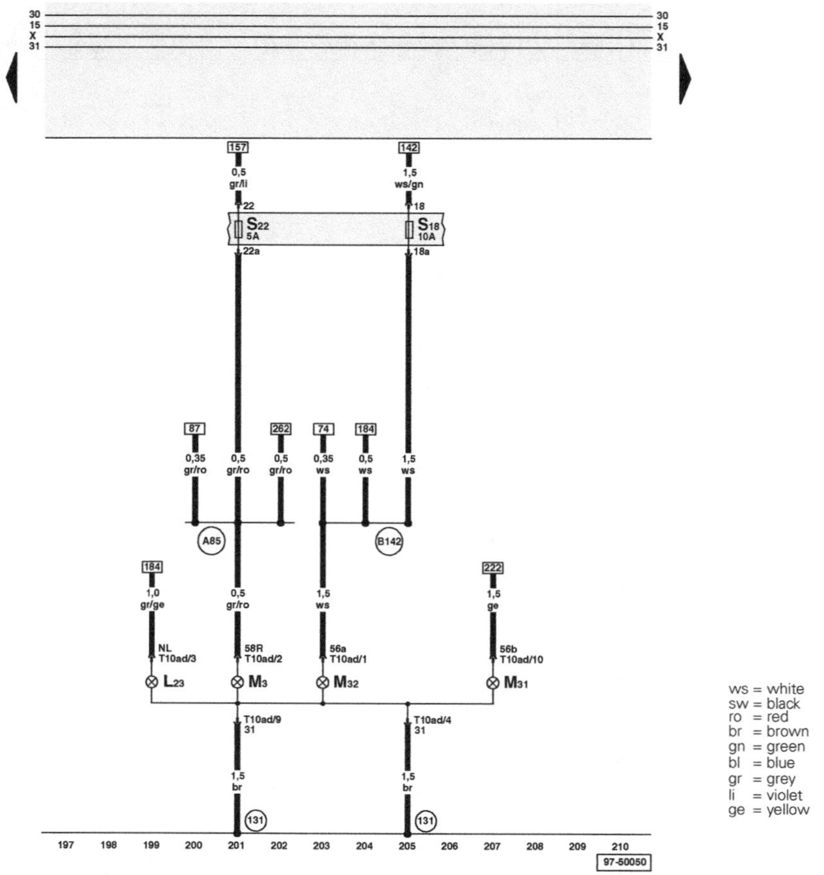

ws = white
sw = black
ro = red
br = brown
gn = green
bl = blue
gr = grey
li = violet
ge = yellow

L23 - Right Front Fog Light
M3 - Right Parking Light
M31 - Right Low Beam Headlight
M32 - Right High Beam Headlight
S18 - Fuse
S22 - Fuse
T10ad - Connector 10 pin, on headlight, right

(131) - Ground connection -2-, in engine compartment wiring harness
(A85) - Connector (58R), in instrument panel wiring harness
(B142) - Plus connector -2- (56a), in wiring harness interior

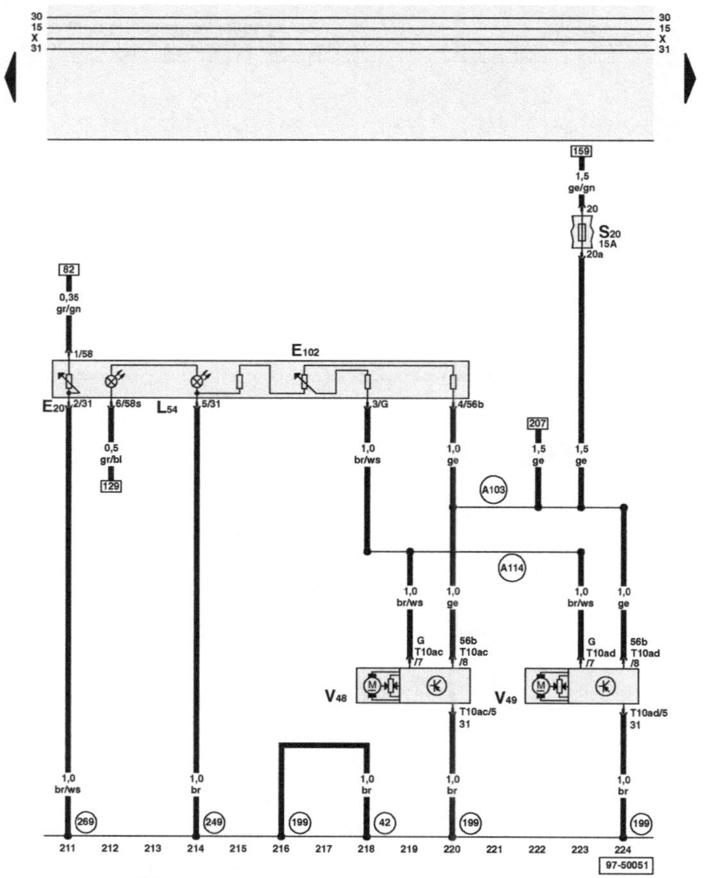

ws = white
sw = black
ro = red
br = brown
gn = green
bl = blue
gr = grey
li = violet
ge = yellow

E20 - Instrument Panel Light Dimmer Switch
E102 - Headlight Adjuster
L54 - Headlight Adjuster Control Light
S20 - Fuse
T10ac - Connector 10 pin, on headlight, left
T10ad - Connector 10 pin, on headlight, right
V48 - Left Headlight Beam Adjusting Motor
V49 - Right Headlight Beam Adjusting Motor

(42) - Ground connection, on steering column
(199) - Ground connection -3-, in instrument panel wiring harness
(249) - Ground connection -2-, in wiring harness interior
(269) - Ground connector (sensor ground) -1-, in instrument panel wiring harness
(A103) - Connector -2- (56), in instrument panel wiring harness
(A114) - Connection (Headlight adjusting), in instrument panel wiring harness

© 2006 BentleyPublishers.com—All Rights Reserved

Audi TT — Wiring diagram — No. 25/19

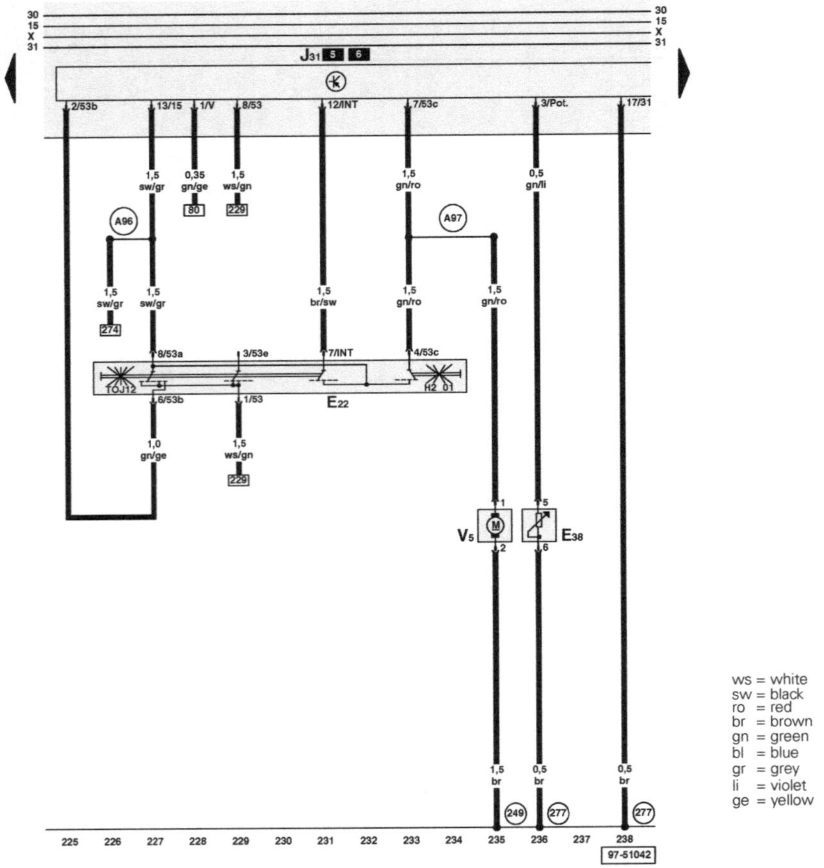

ws = white
sw = black
ro = red
br = brown
gn = green
bl = blue
gr = grey
li = violet
ge = yellow

E22 - Windshield Wiper/Washer Switch
E38 - Windshield Wiper Intermittent Regulator
J31 - Wiper/Washer Intermittent Relay
V5 - Windshield Washer Pump

(249) - Ground connection -2-, in wiring harness interior
(277) - Ground connector -3-, in wiring harness interior
(A96) - Connector (53a), in instrument panel wiring harness
(A97) - Connector (53c), in instrument panel wiring harness

No. 25/20 — Wiring diagram — Audi TT

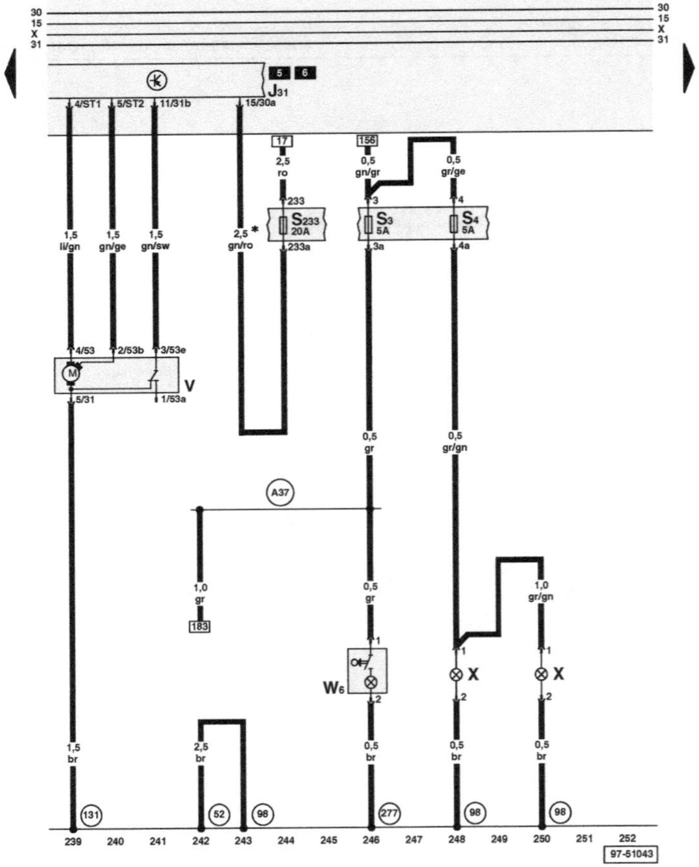

ws = white
sw = black
ro = red
br = brown
gn = green
bl = blue
gr = grey
li = violet
ge = yellow

J31 - Wiper/Washer Intermittent Relay
S3 - Fuse
S4 - Fuse
S233 - Fuse in fuse holder
V - Windshield Wiper Motor
W6 - Glove Compartment Light
X - License Plate Light

(52) - Ground connection, in rear lid, left
(98) - Ground connection, in rear lid wiring harness
(131) - Ground connection -2-, in engine compartment wiring harness
(277) - Ground connector -3-, in wiring harness interior
(A37) - Wire connection (58a), in instrument panel wiring harness
* - Vehicles with Headlight Washer

© 2006 BentleyPublishers.com—All Rights Reserved

Audi TT — Wiring diagram — **No. 25/21**

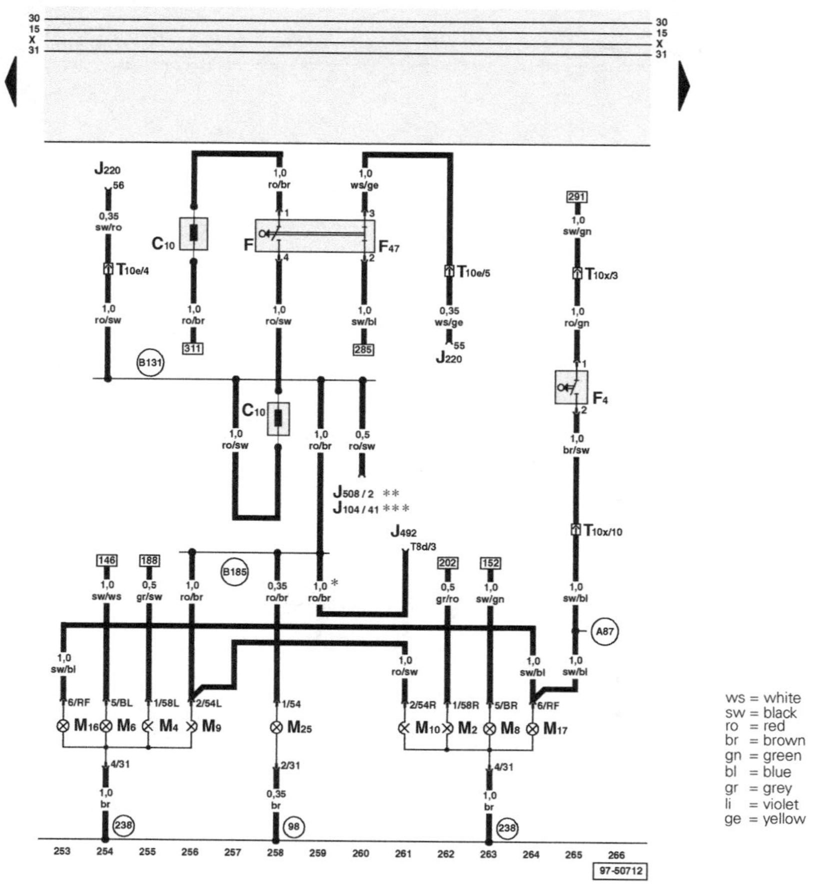

ws = white
sw = black
ro = red
br = brown
gn = green
bl = blue
gr = grey
li = violet
ge = yellow

- C10 - Suppressor Filter
- F - Brake Light Switch
- F4 - Back-Up Light Switch
- F47 - Brake Pedal Switch (Cruise Control/Diesel Direct Fuel Injection)
- J104 - ABS Control Module (w/EDL)
- J220 - Motronic Engine Control Module (ECM)
- J492 - Control module for all wheel drive
- J508 - Brake light disable relay (ESP)
- M2 - Right Tail Light
- M4 - Left Tail Light
- M6 - Left Rear Turn Signal Light
- M8 - Right Rear Turn Signal Light
- M9 - Left Brake Light
- M10 - Right Brake Light
- M16 - Left Back-Up Light
- M17 - Right Back-Up Light
- M25 - High-mount Brake Light
- T8d - Connector 8 pin, on Control module for all wheel drive
- T10e - Connector 10 pin, black, in E-box plenum chamber
- T10x - Connector 10 pin, orange, in E-box plenum chamber

- (98) - Ground connection, in rear lid wiring harness
- (238) - Ground connection -1-, in wiring harness interior
- (A87) - Connector (RF), in instrument panel wiring harness
- (B131) - Connector (54), in wiring harness interior
- (B185) - Connection 2 (54) in passenger compartment wiring harness

- * - Vehicles with all-wheel drive
- ** - ABS MK20 with EDL/ASR/ESP
- *** - ABS MK60 with EDL/ASR/ESP

Edition 12/02
W42.USA.5702.12.21

Tail Light, Brake Light, Rear Fog Light

No. 25/22 — Wiring diagram — **Audi TT**

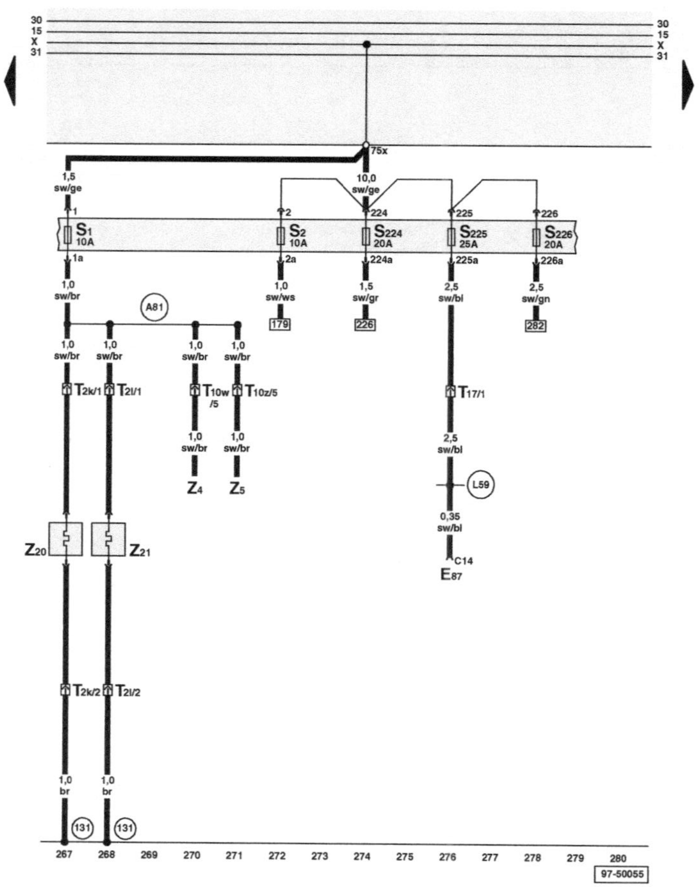

ws = white
sw = black
ro = red
br = brown
gn = green
bl = blue
gr = grey
li = violet
ge = yellow

- E87 - A/C Control Head
- S1 - Fuse
- S2 - Fuse
- S224 - Fuse in fuse holder
- S225 - Fuse in fuse holder
- S226 - Fuse in fuse holder
- T2k - Connector 2 pin, black, on engine hood
- T2l - Connector 2 pin, black, on engine hood
- T10w - Connector 10 pin, blue, connector station A pillar, left
- T10z - Connector 10 pin, blue, connector station A pillar, right
- T17 - Connector 17 pin, behind instrument panel
- Z4 - Heated outside mirror, driver side
- Z5 - Heated outside mirror, passenger side
- Z20 - Left Washer Nozzle Heater
- Z21 - Right Washer Nozzle Heater

- (131) - Ground connection -2-, in engine compartment wiring harness
- (A81) - Connector -2- (X), in instrument panel wiring harness
- (L59) - Plus connection (30), in wiring harness heater

Washer Nozzle Heater, Fuses

Edition 12/02
W42.USA.5702.12.21

© 2006 BentleyPublishers.com—All Rights Reserved

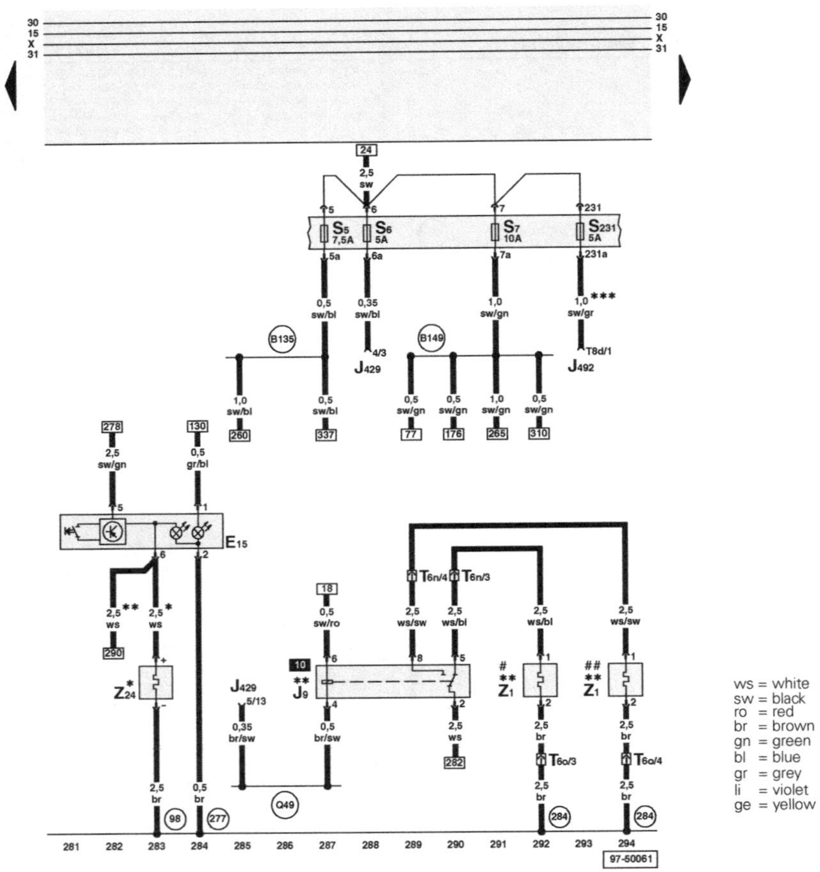

ws = white
sw = black
ro = red
br = brown
gn = green
bl = blue
gr = grey
li = violet
ge = yellow

E15 - Rear Window Defogger Switch
J9 - Rear Window Defogger Relay
J429 - Control module for central locking
J492 - Control module for all wheel drive
S5 - Fuse
S6 - Fuse
S7 - Fuse
S231 - Fuse in fuse holder
T6n - Connector 6 pin, Hardtop left
T6o - Connector 6 pin, Hardtop right
T8d - Connector 8 pin, on Control module for all wheel drive
Z1 - Heated rear window
Z24 - Rear Window Defogger With Window Antenna

(98) - Ground connection, in rear lid wiring harness
(277) - Ground connector -3-, in wiring harness interior
(284) - Ground connection -6- in wiring harness interior
(B135) - Connector (15a), in wiring harness interior
(B149) - Plus connector -2- (15a), in wiring harness interior
(Q49) - Connection (closed), in wiring harness interior

* - Coupé
** - Roadster
*** - Vehicles with all-wheel drive
# - Softtop
- Hardtop

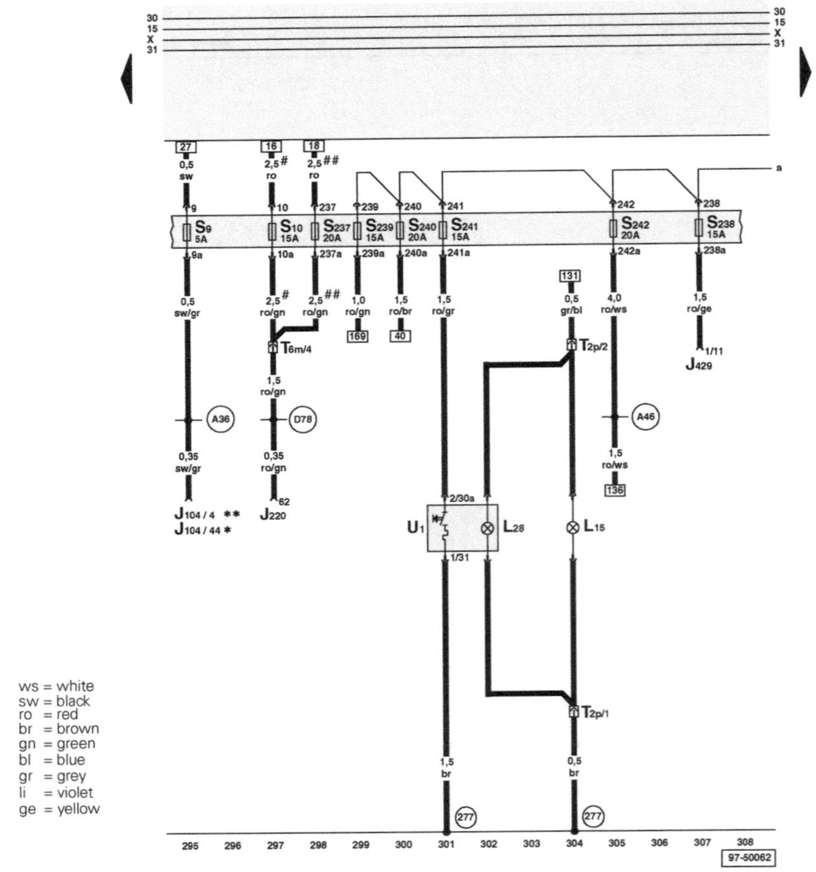

ws = white
sw = black
ro = red
br = brown
gn = green
bl = blue
gr = grey
li = violet
ge = yellow

J104 - ABS Control Module (w/EDL)
J220 - Motronic Engine Control Module (ECM)
J429 - Control module for central locking
L15 - Ashtray Light
L28 - Cigarette Lighter Light
S9 - Fuse
S10 - Fuse
S237 - Fuse 37 in fuse holder
S238 - Fuse in fuse holder
S239 - Fuse in fuse holder
S240 - Fuse in fuse holder
S241 - Fuse in fuse holder
S242 - Fuse in fuse holder
T2p - Connector 2 pin, behind console
T6m - Connector 6 pin, brown, in E-box plenum chamber
U1 - Cigarette Lighter

(277) - Ground connector -3-, in wiring harness interior
(A36) - Wire connection (75a), in instrument panel wiring harness
(A46) - Plus connection (30-from radio), in instrument panel wiring harness
(D78) - Plus connection -1- (30a), in engine compartment wiring harness

* - ABS MK20 with EDL/ASR/ESP
** - ABS MK60 with EDL/ASR/ESP
# - through 10.2001
- from 11.2001

© 2006 BentleyPublishers.com—All Rights Reserved

Audi TT — Wiring diagram — No. 25/25

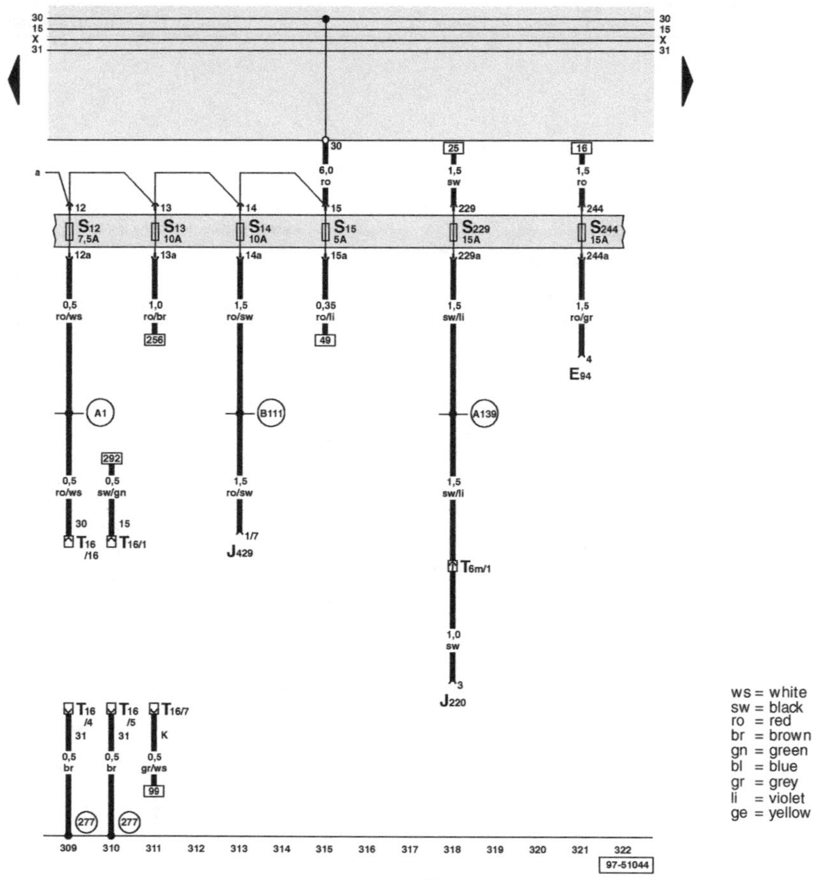

ws = white
sw = black
ro = red
br = brown
gn = green
bl = blue
gr = grey
li = violet
ge = yellow

E94 - Adjuster for heated driver s seat
J220 - Motronic Engine Control Module (ECM)
J429 - Control module for central locking
S12 - Fuse
S13 - Fuse
S14 - Fuse
S15 - Fuse
S229 - Fuse in fuse holder
S244 - Fuse in fuse holder
T6m - Connector 6 pin, brown, in E-box plenum chamber
T16 - Connector 16 pin, black, Data Link Connector (DLC), below storage compartment, driver s side

(277) - Ground connector -3-, in wiring harness interior
(A1) - Plus connection (30a), in instrument panel wiring harness
(A139) - Connection 3 (15), in instrument panel wiring harness
(B111) - Connector (30a), in wiring harness interior

Edition 05/01
W42.USA.5702.07.21

Data Link Connector (DLC), Fuses

No. 25/26 — Wiring diagram — Audi TT

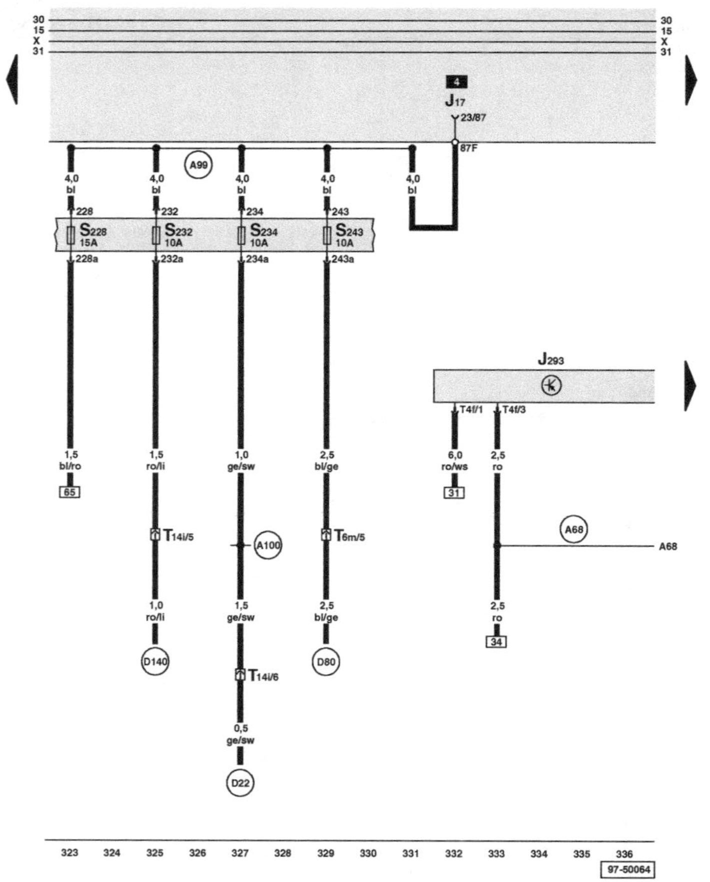

ws = white
sw = black
ro = red
br = brown
gn = green
bl = blue
gr = grey
li = violet
ge = yellow

J17 - Fuel Pump (FP) Relay
J293 - Coolant FC (Fan Control) Control Module, in engine compartment, left front
S228 - Fuse in fuse holder
S232 - Fuse in fuse holder
S234 - Fuse in fuse holder
S243 - Fuse in fuse holder
T4f - Connector 4 pin, connector B on J293
T6m - Connector 6 pin, brown, in E-box plenum chamber
T14i - Connector 14 pin, in engine compartment, left

(A68) - Connector (30, A/C), in instrument panel wiring harness
(A99) - Connector -1- (87), in instrument panel wiring harness
(A100) - Connector -2- (87), in instrument panel wiring harness
(D22) - Connector (over fuse 234), in wiring harness front, right
(D80) - Plus connection (87a- for EVAP system solenoid), in engine compartment wiring harness
(D140) - Connector (injectors), in wiring harness, engine pre-wiring

Coolant (Fan Control) Control Module, Fuses

Edition 05/01
W42.USA.5702.07.21

© 2006 BentleyPublishers.com—All Rights Reserved

Audi TT — Wiring diagram — No. 25/27

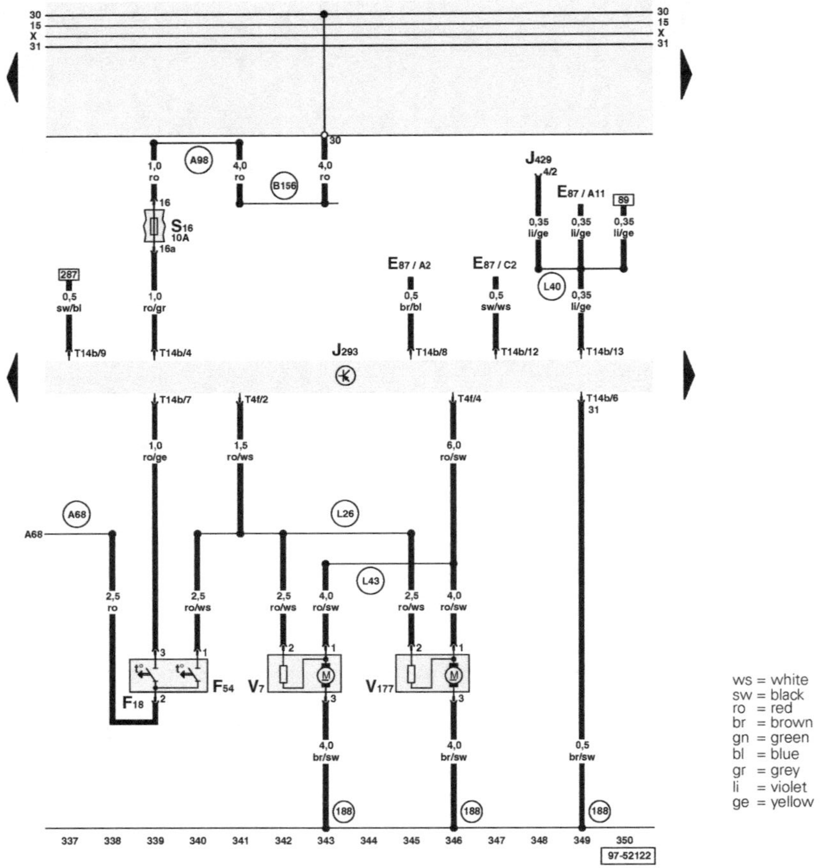

ws = white
sw = black
ro = red
br = brown
gn = green
bl = blue
gr = grey
li = violet
ge = yellow

E87 - A/C Control Head
F18 - Coolant Fan Control (FC) Thermal Switch
F54 - Coolant Fan Control (FC) Thermal Switch
J293 - Coolant FC (Fan Control) Control Module, in engine compartment, left front
J429 - Control module for central locking
S16 - Fuse
T4f - Connector 4 pin, connector B on J293
T14b - Connector 14 pin, connector A on J293
V7 - Coolant Fan
V177 - Coolant Fan -2-

(188) - Ground connection -3-, in A/C wiring harness
(A68) - Connector (30, A/C), in instrument panel wiring harness
(A98) - Plus connector -4- (30), in instrument panel wiring harness
(B156) - Plus connector -4- (30), in wiring harness interior
(L26) - Wire connection (coolant fan), in A/C-climatronic/engine compartment wiring harness
(L40) - Wire connection -2- (A/C Pressure Switch), in Climatronic wiring harness
(L43) - Wire connection (coolant fan), in A/C system-coolant fan wiring harness

Edition 07/01
W42.USA.5702.08.21

Coolant (Fan Control) Control Module, Coolant Fan Control Thermal Switch, Coolant Fan

No. 25/28 — Wiring diagram — Audi TT

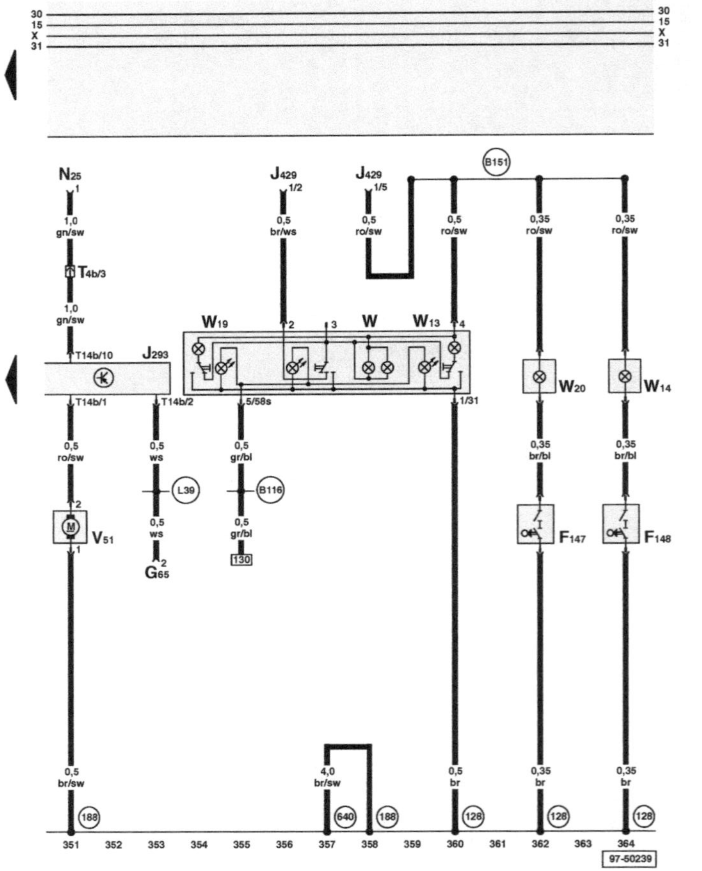

ws = white
sw = black
ro = red
br = brown
gn = green
bl = blue
gr = grey
li = violet
ge = yellow

F147 - Left Make-Up Mirror Light Switch
F148 - Right Make-Up Mirror Light Switch
G65 - High Pressure Sensor
J293 - Coolant FC (Fan Control) Control Module, in engine compartment, left front
J429 - Control module for central locking
N25 - A/C Clutch
T4b - Connector 4 pin, in engine compartment, left
T14b - Connector 14 pin, connector A on J293
V51 - After-Run Coolant Pump
W - Front Interior Light
W13 - Passenger's Map/Reading Light
W14 - Right Make-up Mirror Light
W19 - Left Front Reading Light
W20 - Left Make-up Mirror Light

(128) - Ground connection -1-, in interior light wiring harness
(188) - Ground connection -3-, in A/C wiring harness
(640) - Ground connection -2-, in engine compartment, left
(B116) - Connector (58s), in wiring harness interior
(B151) - Plus connector -3- (30a), in wiring harness interior
(L39) - Wire connection -1- (A/C Pressure Switch), in Climatronic wiring harness

Interior Light, Make-Up Mirror Light, After-Run Coolant Pump

Edition 07/01
W42.USA.5702.08.21

© 2006 BentleyPublishers.com—All Rights Reserved

Audi TT Wiring Diagram **No. 26/1**

1.8 L - Engine Motronic Multiport Fuel Injection (MFI)/ 132 kW, code ATC/AWP & 165 kW, code AMU

2002 m.y.

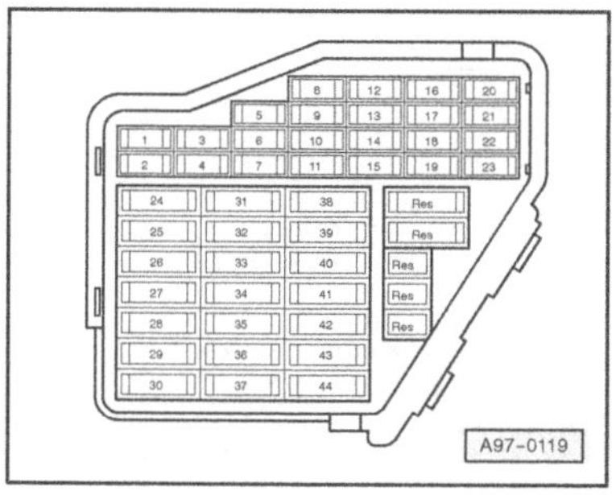

Fuse Panel

Fuse Colors:

30 A - Green
25 A - White
20 A - Yellow
15 A - Blue
10 A - Red
7,5 A - Brown
5 A - Beige
1 A - Black

Starting with fuse position 23, fuses in the fuse holder are identified with 223 in the wiring diagram.

Audi TT Wiring Diagram **No. 26/2**

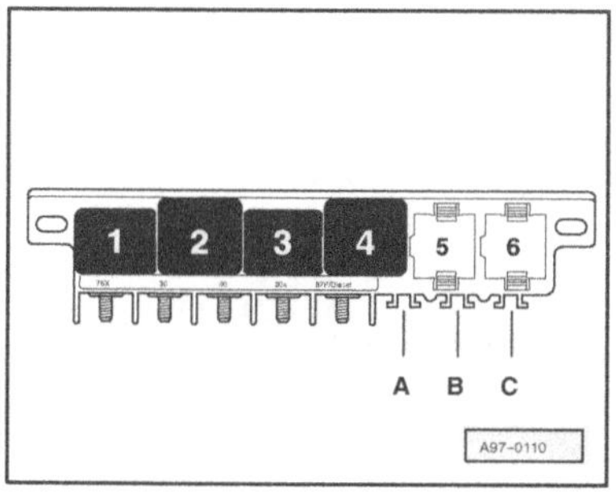

Micro Central Electric Panel

Relay Location:

4 - Fuel Pump (FP) Relay, J17

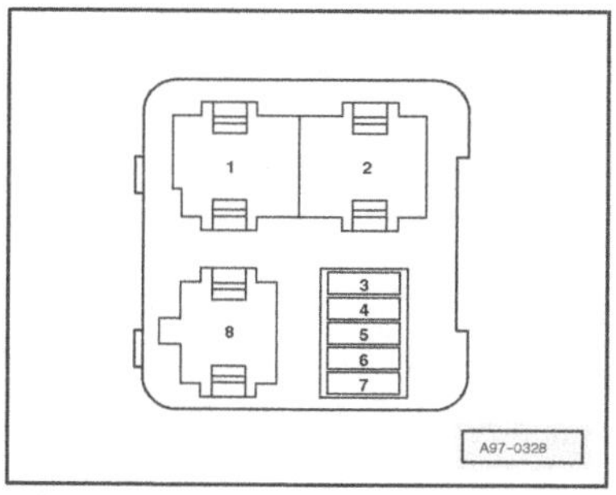

W-Box in engine compartment

Relay Location:

1 - Secondary Air Injection (AIR) Pump Relay, J299
2 - Motronic Engine Control Module (ECM) Power Supply Relay, J271
4 - Fuse for secondary air pump, S130

©2006 BentleyPublishers.com—All Rights Reserved

Engine Management, 1.8L, engine code: ATC, AWP, AMU (2002)

Audi TT — Wiring Diagram — No. 26/3

Battery, Starter, Generator, Main Fuse Box/Battery, Starting Interlock Relay

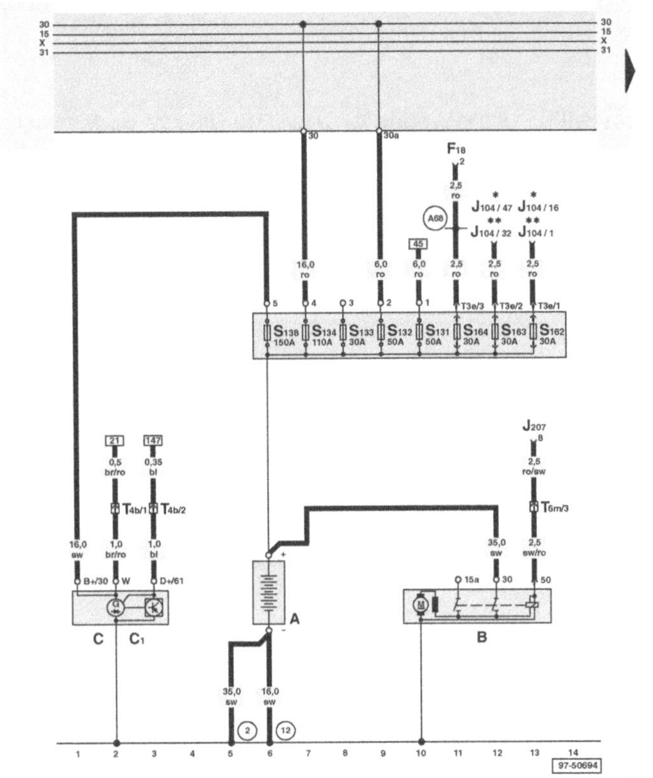

- A - Battery
- B - Starter
- C - Generator (GEN)
- C1 - Voltage Regulator (VR)
- F18 - Coolant Fan Control (FC) Thermal Switch
- J104 - ABS Control Module (w/EDL)
- J207 - Starting Interlock Relay
- S131 - Safety fuse 1
- S132 - Safety fuse 2
- S133 - Safety fuse 3
- S134 - Safety fuse 4
- S138 - Safety fuse 5
- S162 - Fuse -1- in fuse bracket/battery
- S163 - Fuse -2- in fuse bracket/battery
- S164 - Fuse -3- in fuse bracket/battery
- T3e - Connector 3 pin, black, on Main Fuse Box/Battery
- T4b - Connector 4 pin, in engine compartment, left
- T6m - Connector 6 pin, brown, in E-box plenum chamber
- (2) - Ground strap, transmission to body
- (12) - Ground connection, in engine compartment, left
- (A68) - Connector (C15, A/C), in instrument panel wiring harness
- * - ABS MK20 with EDL/ASR/ESP
- ** - ABS MK60 with EDL/ASR/ESP

ws = white
sw = black
ro = red
br = brown
gn = green
bl = blue
gr = grey
li = violet
ge = yellow
or = orange

W42.USA.5702.13.21 — Edition 06/03

Audi TT — Wiring Diagram — No. 26/4

Motronic Engine Control Module, Fuel Pump Relay, Fuel Injectors, Brake Light Switch

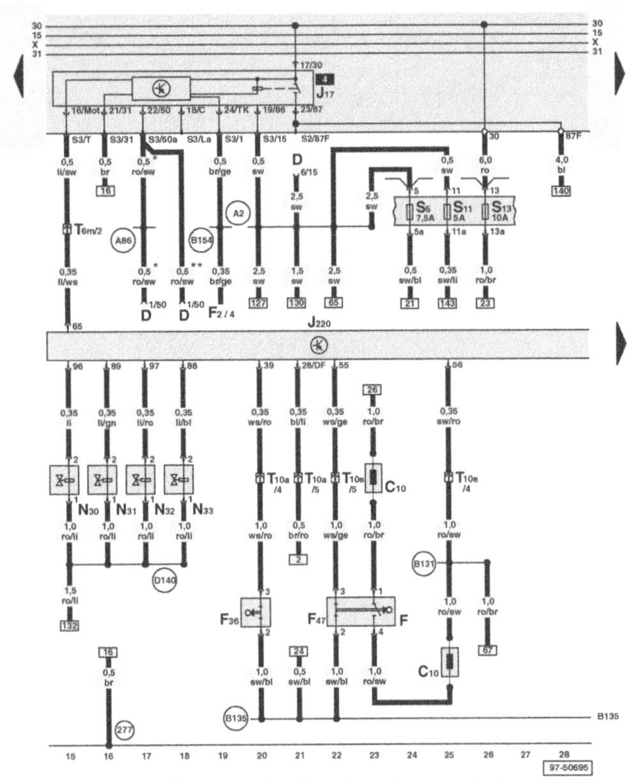

- C10 - Suppressor Filter
- D - Ignition/Starter Switch
- F - Brake Light Switch
- F2 - Door Contact Switch - Driver's side
- F36 - Clutch Vacuum Vent Valve Switch
- F47 - Brake Vacuum Vent Valve Switch for Cruise Control
- J17 - Fuel Pump (FP) Relay
- J220 - Motronic Engine Control Module (ECM)
- N30 - Cylinder 1 Fuel Injector
- N31 - Cylinder 2 Fuel Injector
- N32 - Cylinder 3 Fuel Injector
- N33 - Cylinder 4 Fuel Injector
- S5 - Fuse
- S11 - Fuse
- S13 - Fuse
- T6m - Connector 6 pin, brown, in E-box plenum chamber
- T10a - Connector 10 pin, white, in E-box plenum chamber
- T10e - Connector 10 pin, black, in E-box plenum chamber
- (277) - Ground connector -3-, in wiring harness interior
- (A2) - Plus connection (15), in instrument panel wiring harness
- (A86) - Connection (50a), in instrument panel wiring harness
- (B131) - Connector (54), in wiring harness interior
- (B135) - Connector (15a), in wiring harness interior
- (B154) - Connector -1- (TK), in wiring harness interior
- (D140) - Connector (injectors), in wiring harness, engine pre-wiring
- * - through 10.2001
- ** - from 11.2001

ws = white
sw = black
ro = red
br = brown
gn = green
bl = blue
gr = grey
li = violet
ge = yellow
or = orange

W42.USA.5702.13.21 — Edition 06/03

© 2006 BentleyPublishers.com—All Rights Reserved

Engine Management, 1.8L, engine code: ATC, AWP, AMU (2002)

Audi TT **Wiring Diagram** **No. 26/5**

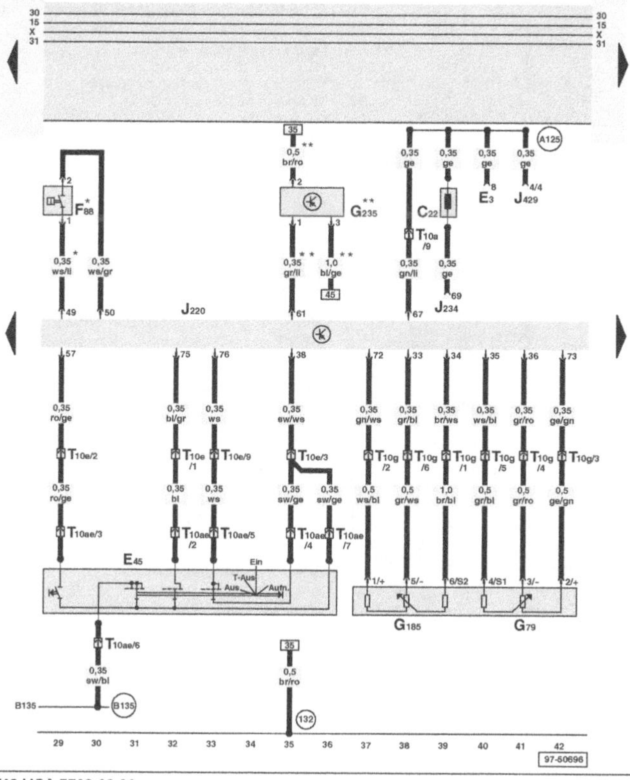

Motronic Engine Control Module, Power Steering Pressure Switch, Cruise Control Switch, Throttle Position Sensor

- C22 - Airbag Suppresser
- E3 - Emergency Flasher Switch
- E45 - Cruise Control Switch
- F88 - Power Steering Pressure (PSP) Switch
- G79 - Throttle Position (TP) Sensor
- G185 - Sensor -2- for accelerator pedal position
- G235 - Sensor -1- for exhaust temperature
- J220 - Motronic Engine Control Module (ECM)
- J234 - Airbag Control Module
- J429 - Control module for central locking
- T10a - Connector 10 pin, white, in E-box plenum chamber
- T10e - Connector 10 pin, black, in E-box plenum chamber
- T10g - Connector 10 pin, blue, in E-box plenum chamber
- T10ae - Connector 10 pin, on Cruise Control Switch

- (132) - Ground connection -3-, in engine compartment wiring harness
- (A125) - Connection (crash signal) in instrument panel wiring harness
- (B135) - Connector (15a), in wiring harness interior

- * - only engine code ATC, AWP
- ** - only engine code AMU

ws = white
sw = black
ro = red
br = brown
gn = green
bl = blue
gr = grey
li = violet
ge = yellow
or = orange

W42.USA.5702.13.21 **Edition 06/03**

Audi TT **Wiring Diagram** **No. 26/6**

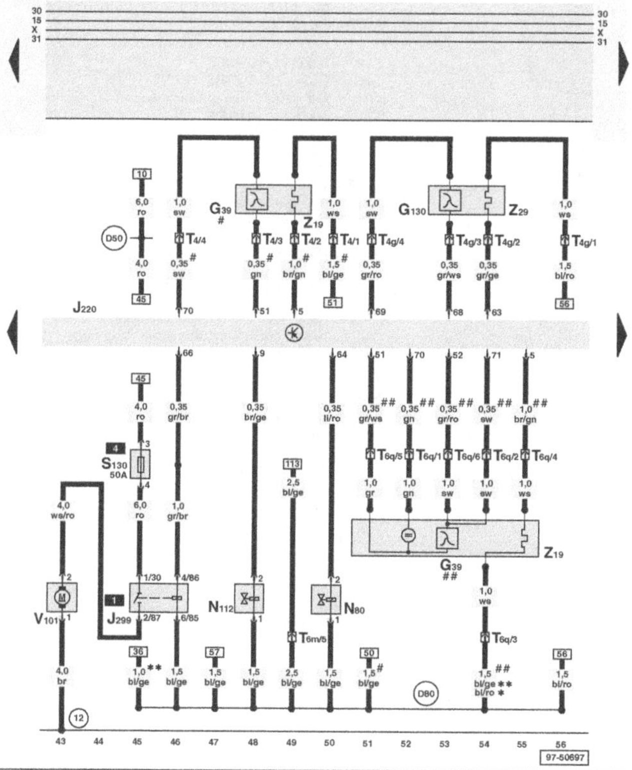

Motronic Engine Control Module, Secondary Air Injection Pump, Heated Oxygen Sensors

- G39 - Heated Oxygen Sensor (HO2S)
- G130 - Heated Oxygen Sensor (O2S) Behind Three Way Catalytic Converter (TWC)
- J220 - Motronic Engine Control Module (ECM)
- J299 - Secondary Air Injection (AIR) Pump Relay
- N80 - Evaporative Emission (EVAP) Canister Purge Regulator Valve
- N112 - Secondary Air Injection (AIR) Solenoid Valve
- S130 - Fuse for secondary air pump
- T4 - Connector 4 pin, in engine compartment, right
- T4g - Connector 4 pin, in engine compartment, right, for Oxygen Sensor behind Three Way Catalytic Converter
- T6m - Connector 6 pin, brown, in E-box plenum chamber
- T6q - Connector 6 pin, in engine compartment right
- V101 - Secondary Air Injection (AIR) Pump Motor
- Z19 - Oxygen Sensor (O2S) Heater
- Z29 - Oxygen Sensor (O2S) Heater 1 (behind Three Way Catalytic Converter (TWC))

- (12) - Ground connection, in engine compartment, left
- (D50) - Plus connection (30), in engine compartment wiring harness
- (D80) - Plus connection (87a- for EVAP system solenoid), in engine compartment wiring harness

- * - only engine code ATC, AWP
- ** - only engine code AMU
- # - Heated Oxygen Sensor with 4-pin connector
- ## - Heated Oxygen Sensor with 6-pin connector

ws = white
sw = black
ro = red
br = brown
gn = green
bl = blue
gr = grey
li = violet
ge = yellow
or = orange

W42.USA.5702.13.21 **Edition 06/03**

©2006 BentleyPublishers.com—All Rights Reserved

Engine Management, 1.8L, engine code: ATC, AWP, AMU (2002)

Audi TT — Wiring Diagram — No. 26/7

Motronic Engine Control Module, All-wheel Drive Control Module, Mass Air Flow (MAF) Sensor

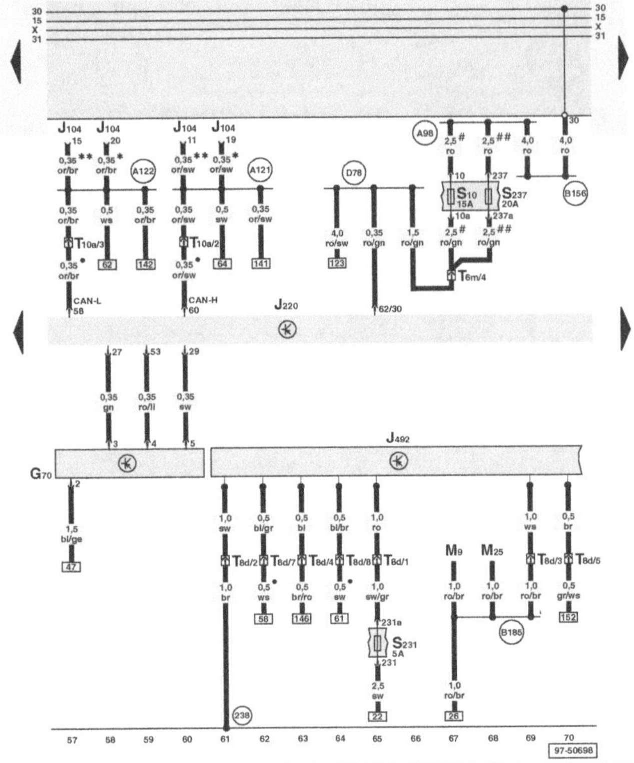

- G70 - Mass Air Flow (MAF) Sensor
- J104 - ABS Control Module (w/EDL)
- J220 - Motronic Engine Control Module (ECM)
- J492 - All-wheel drive control module
- M9 - Left Brake Light
- M25 - High-mount Brake Light
- S10 - Fuse
- S231 - Fuse 31 in fuse holder
- S237 - Fuse 37 in fuse holder
- T6m - Connector 6 pin, brown, in E-box plenum chamber
- T8d - Connector 8 pin, on Control module for all wheel drive
- T10a - Connector 10 pin, white, in E-box plenum chamber

- (238) - Ground connection -1-, in wiring harness interior
- (A98) - Plus connector -4- (30), in instrument panel wiring harness
- (A121) - Connection (high bus) in instrument panel wiring harness
- (A122) - Connection (low bus) in instrument panel wiring harness
- (B156) - Plus connector (30a), in wiring harness interior
- (B185) - Connection 2 (54), in passenger compartment wiring harness
- (D78) - Plus connection -1- (30a), in engine compartment wiring harness

- • - CAN-Bus (Data-Bus)
- * - ABS MK20 with EDL/ASR/ESP
- ** - ABS MK60 with EDL/ASR/ESP
- # - through 10.2001
- ## - from 11.2001

ws = white
sw = black
ro = red
br = brown
gn = green
bl = blue
gr = grey
li = violet
ge = yellow
or = orange

W42.USA.5702.13.21 Edition 06/03

Audi TT — Wiring Diagram — No. 26/8

Motronic Engine Control Module, Engine Speed Sensor, Charge Air Pressure Sensor, Camshaft Position Sensor 2

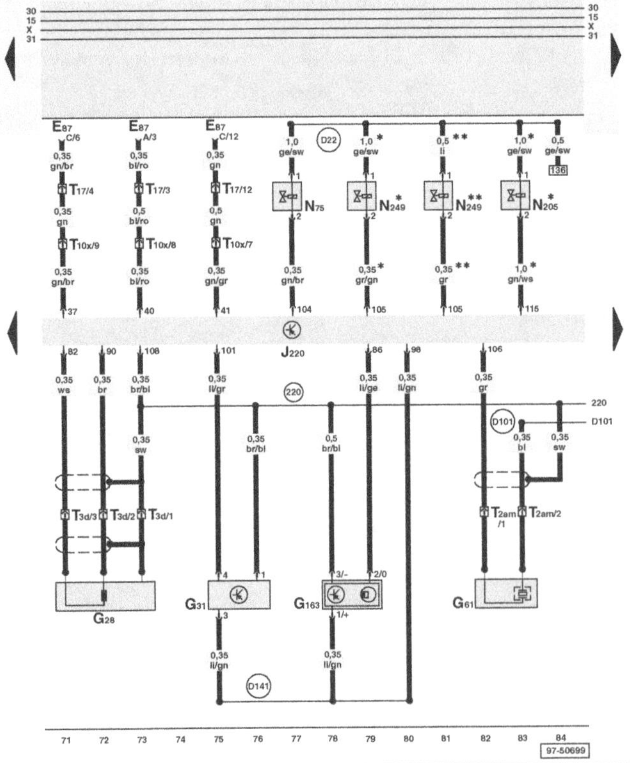

- E87 - A/C Control Head
- G28 - Engine Speed (RPM) Sensor
- G31 - Charge Air Pressure Sensor
- G61 - Knock Sensor (KS) 1
- G163 - Camshaft Position (CMP) Sensor 2
- J220 - Motronic Engine Control Module (ECM)
- N75 - Wastegate Bypass Regulator Valve
- N205 - Valve -1- for camshaft adjustment
- N249 - Recirculating valve for turbocharger
- T2am - Connector 2 pin, black, in engine compartment
- T3d - Connector 3 pin, grey, in engine compartment
- T10x - Connector 10 pin, orange, in E-box plenum chamber
- T17 - Connector 17 pin, behind instrument panel

- (220) - Ground connection (sensor ground), in engine compartment wiring harness
- (D22) - Connector (over fuse 234), in wiring harness front, right
- (D101) - Wire connection -1-, in engine compartment wiring harness
- (D141) - Connector (5V), in wiring harness, engine pre-wiring

- * - engine code ATC, AWP
- ** - engine code AMU

ws = white
sw = black
ro = red
br = brown
gn = green
bl = blue
gr = grey
li = violet
ge = yellow
or = orange

W42.USA.5702.13.21 Edition 06/03

© 2006 BentleyPublishers.com—All Rights Reserved

Audi TT — Wiring Diagram — No. 26/9

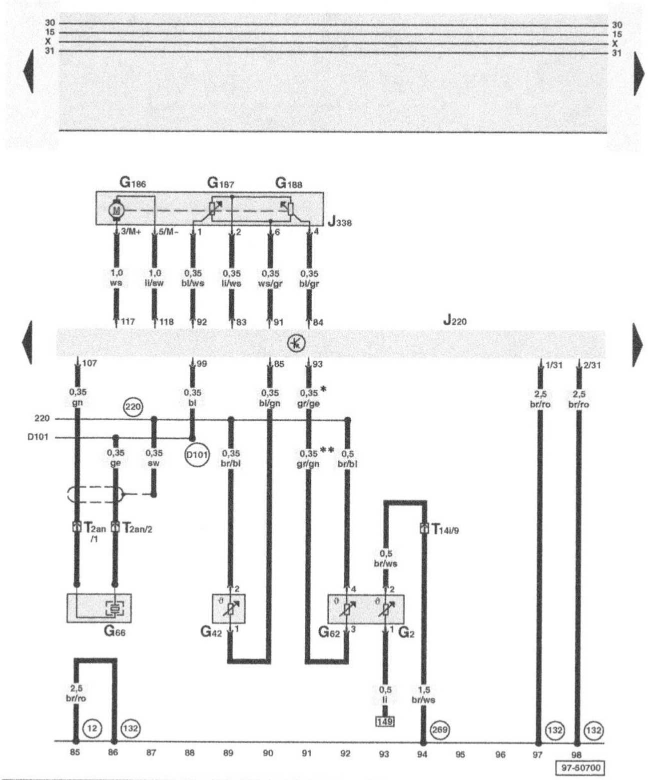

Motronic Engine Control Module, Knock Sensor 2, Intake Air Temperature Sensor, Throttle Valve Control Module

- G2 - Engine Coolant Temperature (ECT) Sensor
- G42 - Intake Air Temperature (IAT) Sensor
- G62 - Engine Coolant Temperature (ECT) Sensor
- G66 - Knock Sensor (KS) 2
- G186 - Throttle drive (power accelerator actuation)
- G187 - Angel sensor -1- for throttle drive (power accelerator actuation)
- G188 - Angel sensor -2- for throttle drive (power accelerator actuation)
- J220 - Motronic Engine Control Module (ECM)
- J338 - Throttle Valve Control Module
- T2an - Connector 2 pin, black, in engine compartment
- T14i - Connector 14 pin, in engine compartment, left
- (12) - Ground connection, in engine compartment, left
- (132) - Ground connection -3-, in engine compartment wiring harness
- (220) - Ground connection (sensor ground), in engine compartment wiring harness
- (269) - Ground connector (sensor ground) -1-, in instrument panel wiring harness
- (D101) - Wire connection -1-, in engine compartment wiring harness

* - engine code ATC, AWP
** - engine code AMU

ws = white
sw = black
ro = red
br = brown
gn = green
bl = blue
gr = grey
li = violet
ge = yellow
or = orange

W42.USA.5702.13.21 — Edition 06/03

Audi TT — Wiring Diagram — No. 26/10

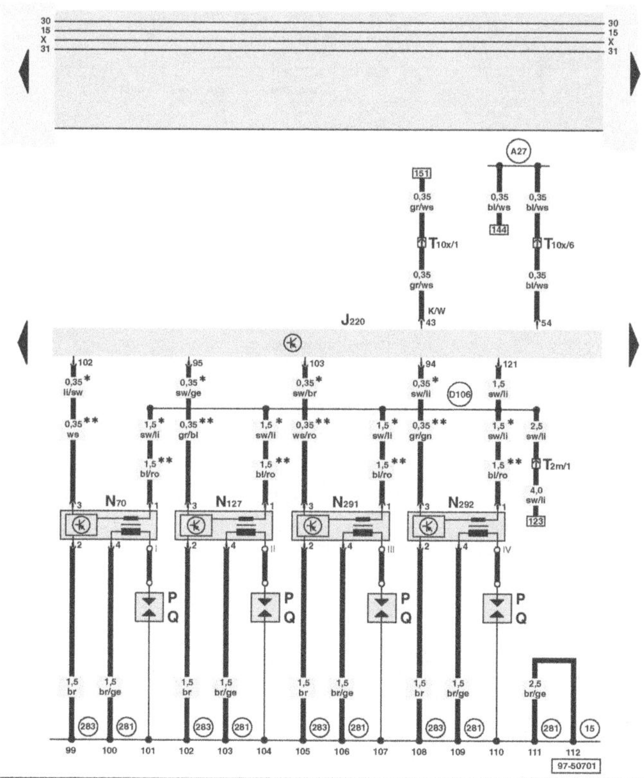

Motronic Engine Control Module, Ignition Coils

- J220 - Motronic Engine Control Module (ECM)
- N70 - Ignition Coil 1 with Power Output Stage
- N127 - Ignition Coil 2 with Power Output Stage
- N291 - Ignition Coil 3 with Power Output Stage
- N292 - Ignition Coil 4 with Power Output Stage
- P - Spark Plug Connectors
- Q - Spark Plugs
- T2m - Connector 2 pin, in engine compartment, left
- T10x - Connector 10 pin, orange, in E-box plenum chamber
- (15) - Ground connection, on cylinder head
- (281) - Ground connector -1-, in wiring harness engine pre-wiring
- (283) - Ground connection 2, in wiring harness engine pre-wiring
- (A27) - Wire connection (vehicle speed signal), in instrument panel wiring harness
- (D106) - Wire connection -4-, in engine compartment wiring harness

* - engine code ATC, AWP
** - engine code AMU

ws = white
sw = black
ro = red
br = brown
gn = green
bl = blue
gr = grey
li = violet
ge = yellow
or = orange

W42.USA.5702.13.21 — Edition 06/03

© 2006 BentleyPublishers.com—All Rights Reserved

Engine Management, 1.8L, engine code: ATC, AWP, AMU (2002)

Audi TT — Wiring Diagram — No. 26/11

Motronic Engine Control Module, Motronic Engine Control Module Power Supply Relay, Leak Detection Pump, Evaporative Emission (EVAP) Canister Purge Solenoid Valve

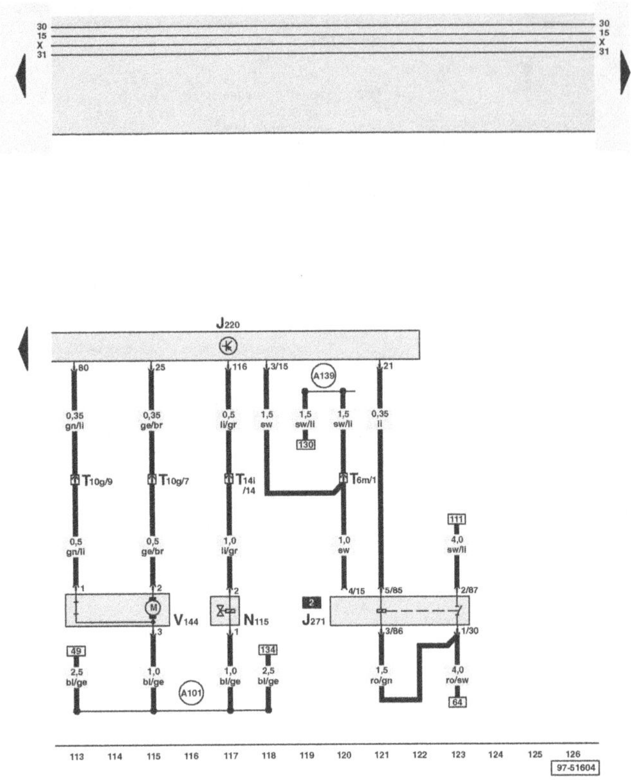

- J220 - Motronic Engine Control Module (ECM)
- J271 - Motronic Engine Control Module (ECM) Power Supply Relay
- N115 - Evaporative Emission (EVAP) Canister Purge Solenoid Valve
- T6m - Connector 6 pin, brown, in E-box plenum chamber
- T10g - Connector 10 pin, blue, in E-box plenum chamber
- T14i - Connector 14 pin, in engine compartment, left
- V144 - Leak Detection Pump (LDP)
- (A101) - Connector -3- (87), in instrument panel wiring harness
- (A139) - Connection 3 (15) (in instrument panel wiring harness)

ws = white
sw = black
ro = red
br = brown
gn = green
bl = blue
gr = grey
li = violet
ge = yellow
or = orange

W42.USA.5702.13.21 — Edition 06/03

Audi TT — Wiring Diagram — No. 26/12

Fuses, Speedometer Vehicle Speed Sensor (VSS), Sender for Fuel Gauge, Fuel Pump (FP)

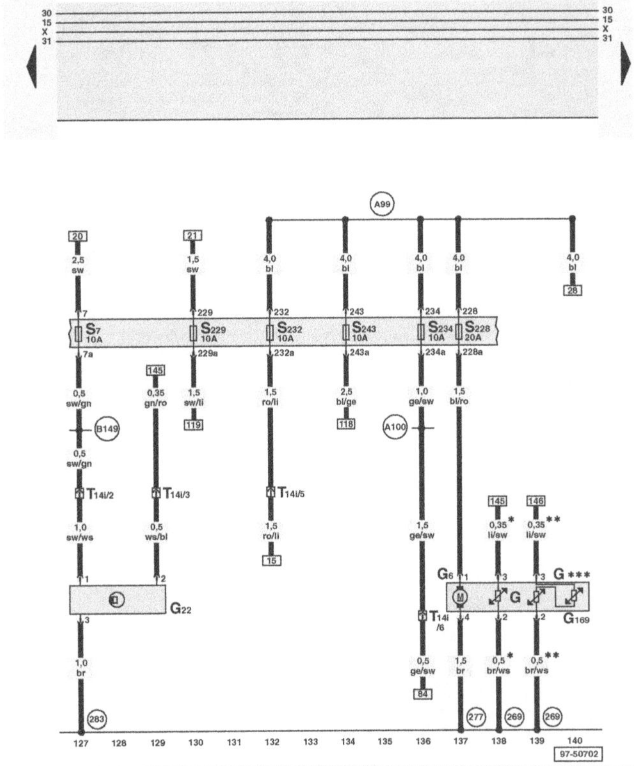

- G - Sender for fuel gauge
- G6 - Fuel Pump (FP)
- G22 - Speedometer Vehicle Speed Sensor (VSS)
- G169 - Fuel Level Sensor 2
- S7 - Fuse
- S228 - Fuse 28 in fuse holder
- S229 - Fuse 29 in fuse holder
- S232 - Fuse 32 in fuse holder
- S234 - Fuse 34 in fuse holder
- S243 - Fuse 43 in fuse holder
- T14i - Connector 14 pin, in engine compartment, left
- (269) - Ground connector (sensor ground) -1-, in instrument panel wiring harness
- (277) - Ground connector -3-, in wiring harness interior
- (283) - Ground connection 2, in wiring harness engine pre-wiring
- (A99) - Connector -1- (87), in instrument panel wiring harness
- (A100) - Connector -2- (87), in instrument panel wiring harness
- (B149) - Plus connector -2- (15a), in wiring harness interior

- * - Vehicles with front wheel drive
- ** - Vehicles with all wheel drive
- *** - Wiring inside fuel tank

ws = white
sw = black
ro = red
br = brown
gn = green
bl = blue
gr = grey
li = violet
ge = yellow
or = orange

W42.USA.5702.13.21 — Edition 06/03

© 2006 BentleyPublishers.com—All Rights Reserved

Audi TT **Wiring Diagram** **No. 26/13**

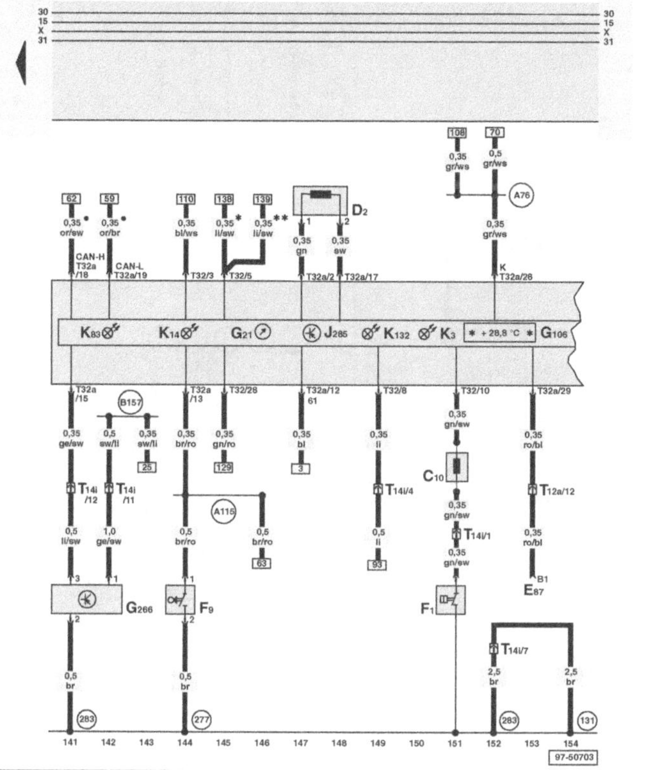

ws = white
sw = black
ro = red
br = brown
gn = green
bl = blue
gr = grey
li = violet
ge = yellow
or = orange

Instrument Cluster, Parking Brake Warning Light Switch, Oil Pressure Switch, Induction Coil of Anti-theft Immobilizer

- C10 - Suppressor Filter
- D2 - Induction coil of antitheft immobilizer
- E87 - A/C Control Head
- F1 - Oil Pressure Switch
- F9 - Parking Brake Warning Light Switch
- G21 - Speedometer
- G106 - Outside Air Temperature Display
- G266 - Oil Level Thermal Sensor (SRI)
- J285 - Control Module with indicator unit in instrument panel insert
- K3 - Oil Pressure Warning Light
- K14 - Parking Brake Indicator Light
- K83 - Malfunction Indicator Lamp (MIL)
- K132 - Electronic Power Control (EPC) Warning Lamp
- T12a - Connector 12 pin, behind instrument panel, right
- T14i - Connector 14 pin, in engine compartment, left
- T32 - Connector 32 pin, blue, on instrument cluster
- T32a - Connector 32 pin, green, on instrument cluster

- (131) - Ground connection -2-, in engine compartment wiring harness
- (277) - Ground connector -3-, in wiring harness interior
- (283) - Ground connection 2, in wiring harness engine pre-wiring
- (A76) - Connector (K-diagnosis wire), in instrument panel wiring harness
- (A115) - Connection (parking brake control) in instrument panel wiring harness
- (B157) - Plus connector -3- (15a), in wiring harness interior

- • - CAN-Bus (Data-Bus)
- * - Vehicles with front wheel drive
- ** - Vehicles with all wheel drive

W42.USA.5702.13.21 Edition 06/03

THIS SPACE INTENTIONALLY

LEFT BLANK

© 2006 BentleyPublishers.com—All Rights Reserved

Central Locking System / Anti Theft System / Ultra-Sound Interior Monitor (Coupé)

2002 m. y.

Fuse Panel

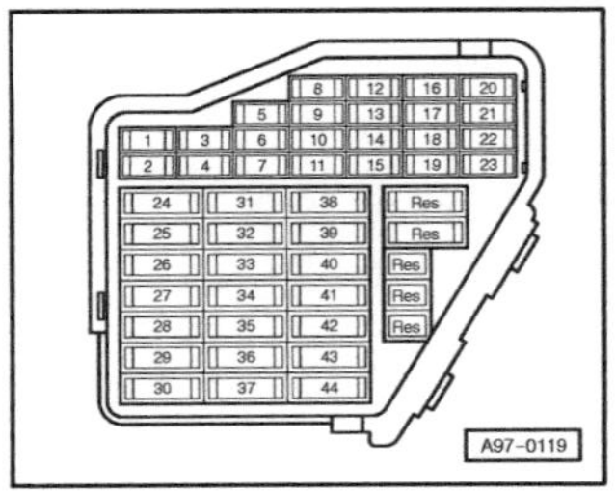

Fuse Colors:

30 A - Green
25 A - White
20 A - Yellow
15 A - Blue
10 A - Red
7,5 A - Brown
5 A - Beige

Starting with fuse position 23, fuses in the fuse holder are identified with 223 in the wiring diagram.

Micro Central Electric Panel

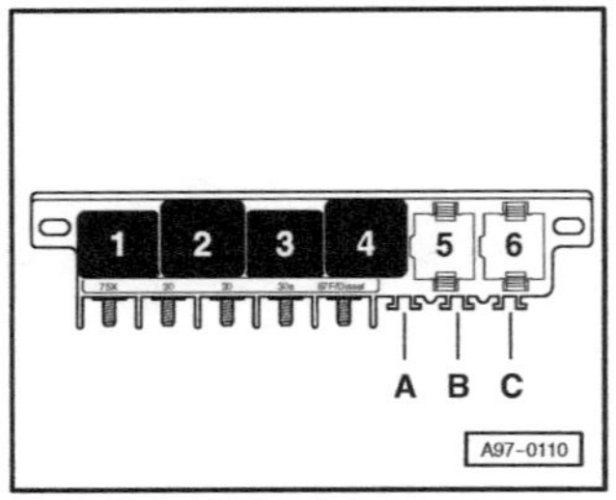

13 - Fold Relay Panel

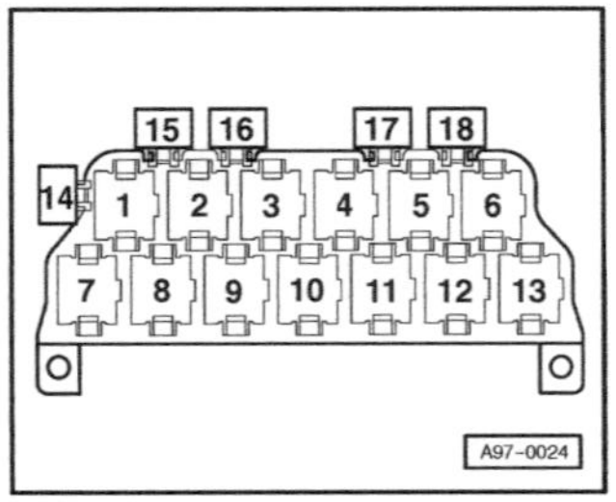

Relay Location:

5 -Starting Interlock Relay, J207

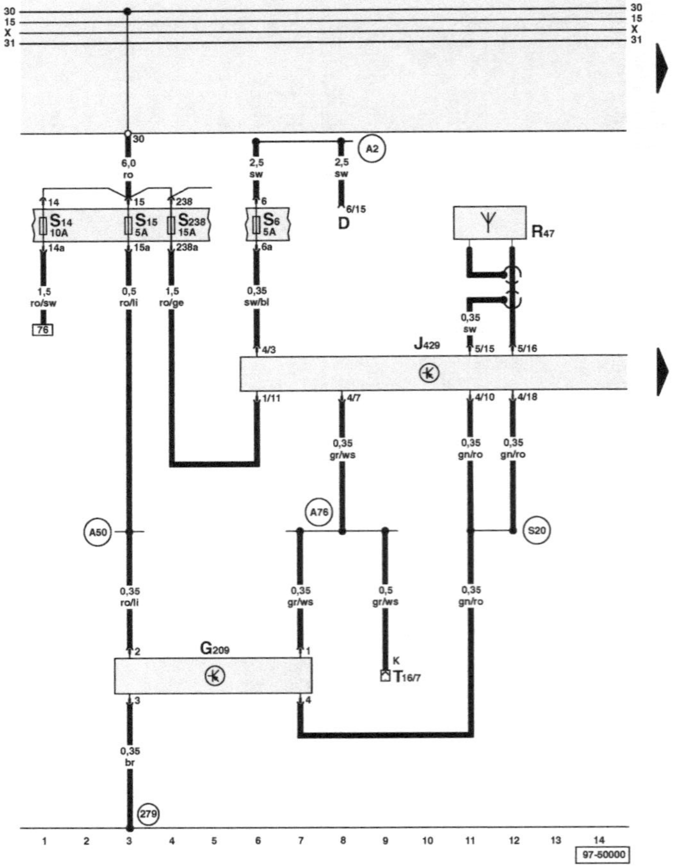

ws = white
sw = black
ro = red
br = brown
gn = green
bl = blue
gr = grey
li = violet
ge = yellow

D - Ignition/Starter Switch
G209 - Ultra-sound sensor for anti-theft warning
J429 - Control module for central locking
R47 - Antenna wire for central locking and anti-theft warning system
S6 - Fuse
S14 - Fuse
S15 - Fuse
S238 - Fuse 38 in fuse holder
T16 - Connector 16 pin, black, Data Link Connector (DLC), below storage compartment, driver's side

(279) - Ground connector -5-, in wiring harness interior
(A2) - Plus connection (15), in instrument panel wiring harness
(A50) - Plus connection (30a), in instrument panel wiring harness
(A76) - Connector (K-diagnosis wire), in instrument panel wiring harness
(S20) - Connector -4- in wiring harness central locking

© 2006 BentleyPublishers.com—All Rights Reserved

Audi TT — Wiring diagram — No. 27/3

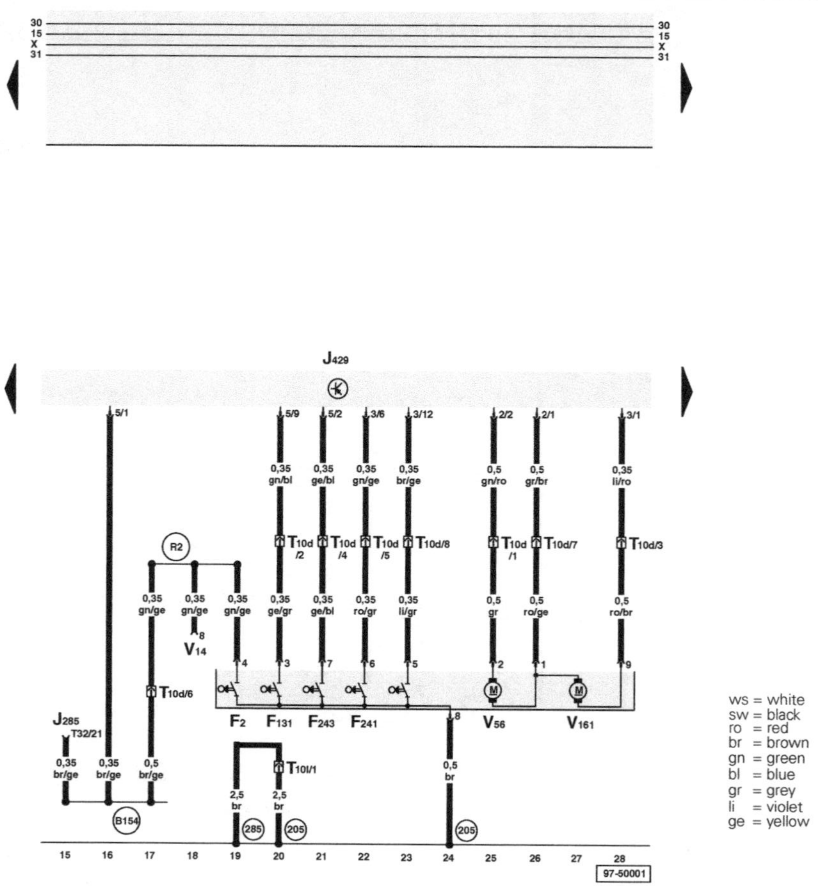

ws = white
sw = black
ro = red
br = brown
gn = green
bl = blue
gr = grey
li = violet
ge = yellow

F2 - Door Contact Switch - Driver's side
F131 - Left Front Central Locking System Activator/Switch
F241 - Contact switch for lock cylinder, driver side
F243 - Adjusting element for central locking (safe), driver door
J285 - Control Module with indicator unit in instrument panel insert
J429 - Control module for central locking
T10d - Connector 10 pin, brown, connector station A pillar, left
T10l - Connector 10 pin, red, connector station A pillar, left
T32 - Connector 32 pin, blue, on instrument cluster
V14 - Left Window Motor
V56 - Driver's Door Central Locking System Motor
V161 - Driver's Door Central Locking (Safe) Motor

(205) - Ground connection, in driver's door wiring harness
(285) - Ground connection -7- in wiring harness interior
(B154) - Connector -1- (TK), in wiring harness interior
(R2) - Wire connection (door contact switch, left), in power window/central locking system and door contact switch wiring harness

Edition 10/01
W42.USA.5702.09.21

Control Module for Central Locking, Lock Unit for Central Locking Driver's Side

No. 27/4 — Wiring diagram — Audi TT

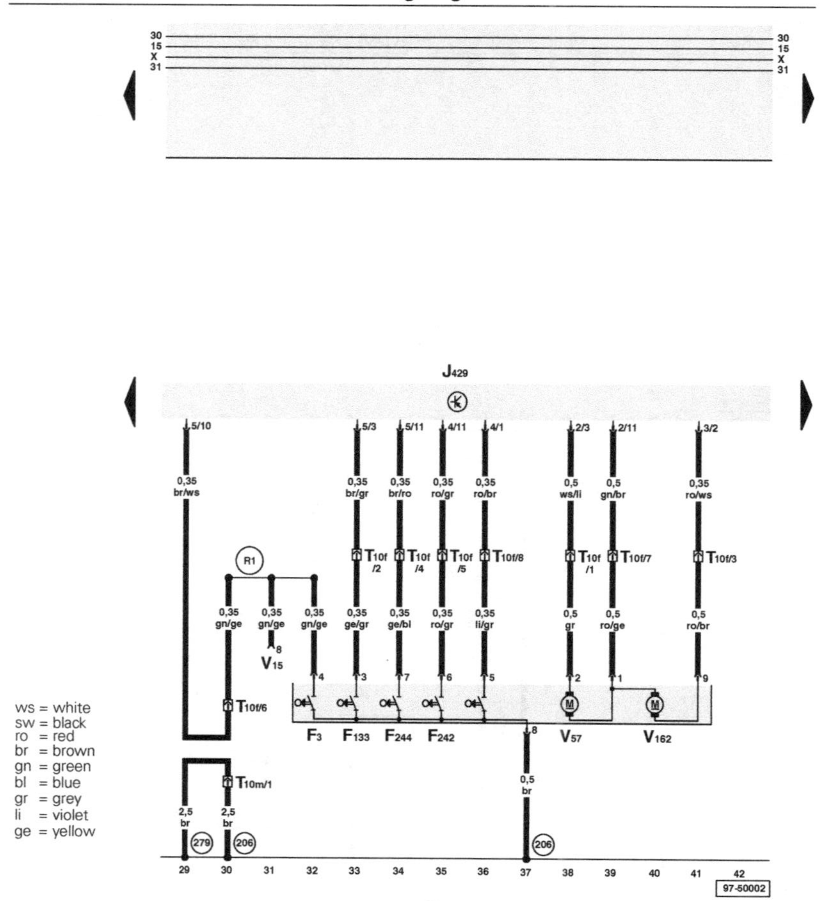

ws = white
sw = black
ro = red
br = brown
gn = green
bl = blue
gr = grey
li = violet
ge = yellow

F3 - Door Contact Switch - Passenger side
F133 - Right Front Central Locking System Activator/Switch
F242 - Contact switch for lock cylinder, passenger side
F244 - Adjusting element for central locking (safe), passenger door
J429 - Control module for central locking
T10f - Connector 10 pin, brown, connector station A pillar, right
T10m - Connector 10 pin, red, connector station A pillar, right
V15 - Right Window Motor
V57 - Passenger's Door Central Locking System Motor
V162 - Front Passenger's Central Locking (Safe) Motor

(206) - Ground connection, in passenger's door wiring harness
(279) - Ground connector -5-, in wiring harness interior
(R1) - Wire connection (door contact switch, right), in power window/central locking system and door contact switch wiring harness

Control Module for Central Locking, Lock Unit for Central Locking Passenger's Side

Edition 10/01
W42.USA.5702.09.21

© 2006 BentleyPublishers.com—All Rights Reserved

Audi TT — Wiring diagram — No. 27/5

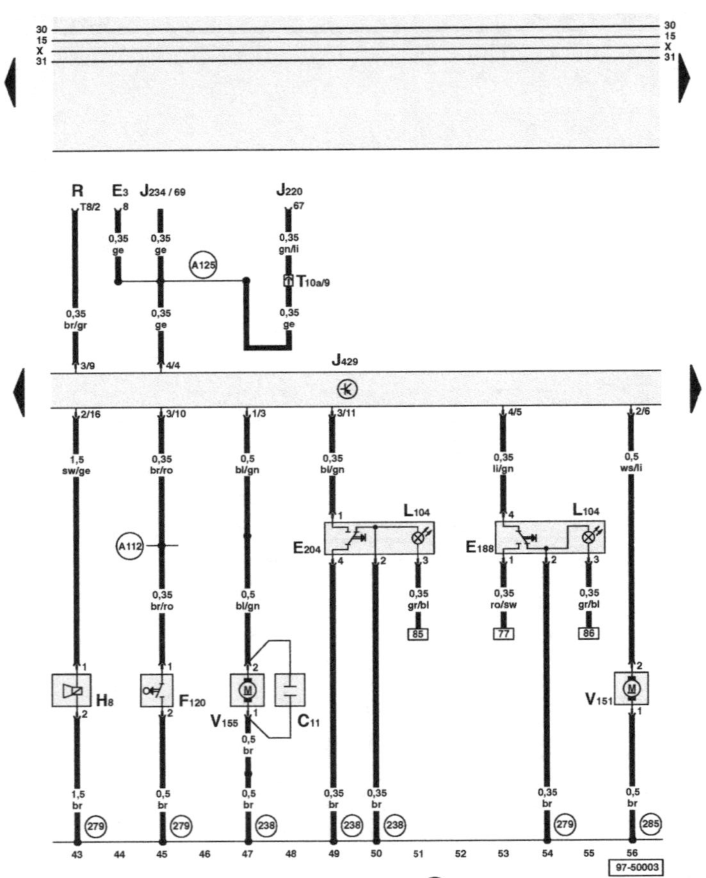

ws = white
sw = black
ro = red
br = brown
gn = green
bl = blue
gr = grey
li = violet
ge = yellow

C11 - Condensor
E3 - Emergency Flasher Switch
E188 - Switch for remote unlock, rear lid
E204 - Switch for remote/fuel tank door
F120 - Hood Alarm Switch
H8 - Alarm Horn
J220 - Motronic Engine Control Module (ECM)
J234 - Airbag Control Module
J429 - Control module for central locking
L104 - Illumination for switch remote unlock
R - Radio
T8 - Connector 8 pin, black, connector radio III
T10a - Connector 10 pin, white, in E-box plenum chamber
V151 - Motor for remote unlock rear lid
V155 - Motor for fuel tank lid unlock

(238) - Ground connection -1-, in wiring harness interior
(279) - Ground connector -5-, in wiring harness interior
(285) - Ground connection -7- in wiring harness interior
(A112) - Connection (Hood alarm switch), in instrument panel wiring harness
(A125) - Connection (crash signal) in instrument panel wiring harness

Edition 10/02
W42.USA.5702.11.21

Control Module for Central Locking, Switch for Remote Unlock rear lid, Switch for Remote/Fuel Tank Door

No. 27/6 — Wiring diagram — Audi TT

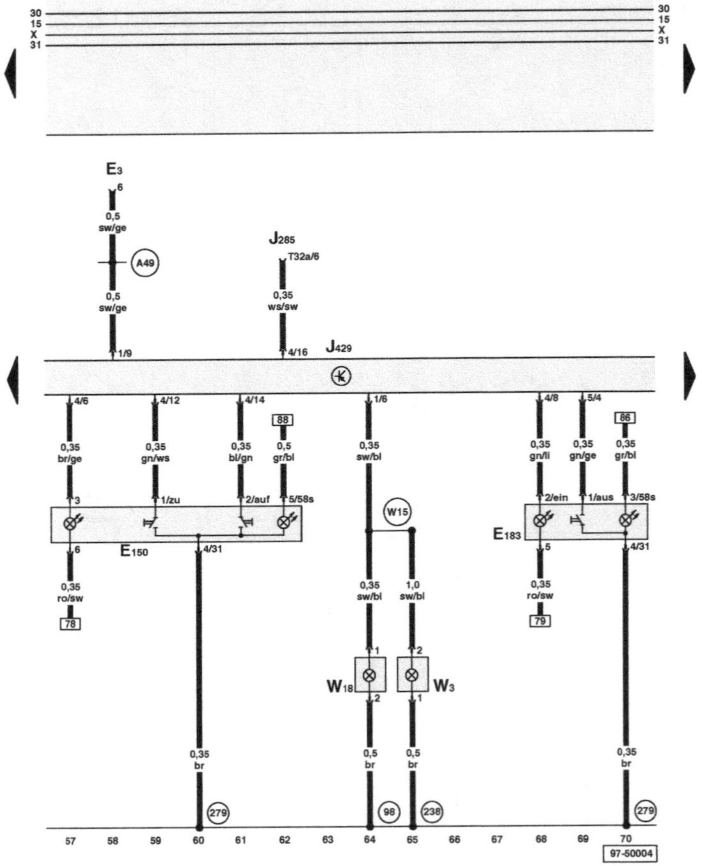

ws = white
sw = black
ro = red
br = brown
gn = green
bl = blue
gr = grey
li = violet
ge = yellow

E3 - Emergency Flasher Switch
E150 - Switch for interior lock, driver side
E183 - Switch for passenger compartment monitoring
J285 - Control Module with indicator unit in instrument panel insert
J429 - Control module for central locking
T32a - Connector 32 pin, green, on instrument cluster
W3 - Luggage Compartment Light
W18 - Left Luggage Compartment Light in rear lid

(98) - Ground connection, in rear lid wiring harness
(238) - Ground connection -1-, in wiring harness interior
(279) - Ground connector -5-, in wiring harness interior
(A49) - Wire connection -1-, in instrument panel wiring harness
(W15) - Connection -1- (luggage compartment light), in rear wiring harness

Control Module for Central Locking, Switch for Interior Lock Driver Side, Switch for Passenger Compartment Monitoring, Luggage Compartment Light

Edition 10/02
W42.USA.5702.11.21

©2006 BentleyPublishers.com—All Rights Reserved

Audi TT — Wiring diagram — No. 27/7

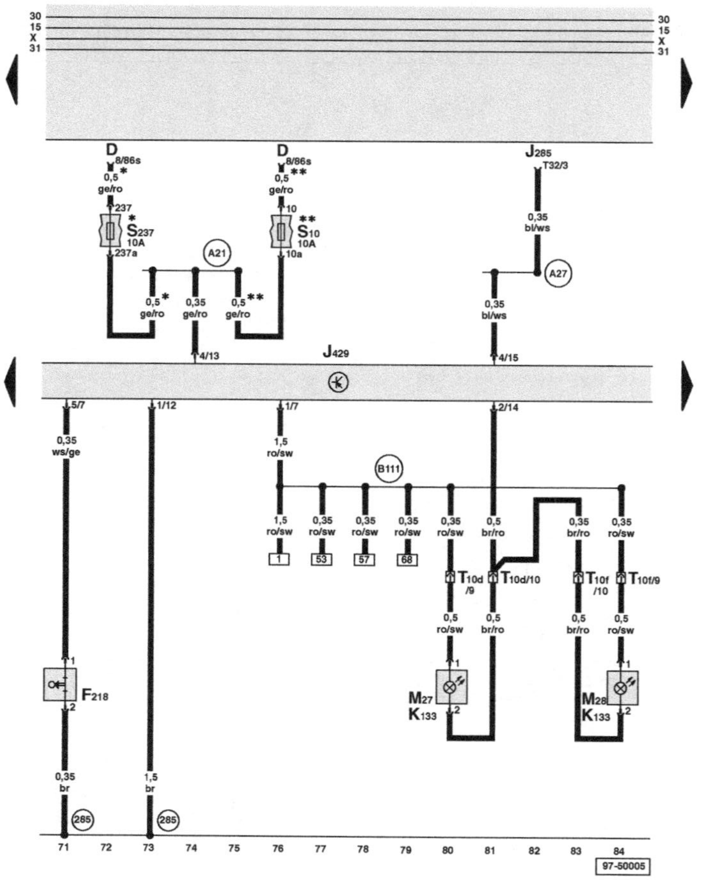

ws = white
sw = black
ro = red
br = brown
gn = green
bl = blue
gr = grey
li = violet
ge = yellow

- D - Ignition/Starter Switch
- F218 - Switch for central locking, trunk lid
- J285 - Control Module with indicator unit in instrument panel insert
- J429 - Control module for central locking
- K133 - Warning light for central locking -SAFE-
- M27 - Left Door Warning Light
- M28 - Right Door Warning Light
- S10 - Fuse
- S237 - Fuse 37 in fuse holder
- T10d - Connector 10 pin, brown, connector station A pillar, left
- T10f - Connector 10 pin, brown, connector station A pillar, right
- T32 - Connector 32 pin, blue, on instrument cluster

- (285) - Ground connection -7- in wiring harness interior
- (A21) - Wire connection (86s), in instrument panel wiring harness
- (A27) - Wire connection (vehicle speed signal), in instrument panel wiring harness
- (B111) - Connector (30a), in wiring harness interior
- * - to 11.2001
- ** - from 11.2001

Edition 11/01
W42.USA.5702.10.21

Control Module for Central Locking, Door Warning Lights

No. 27/8 — Wiring diagram — Audi TT

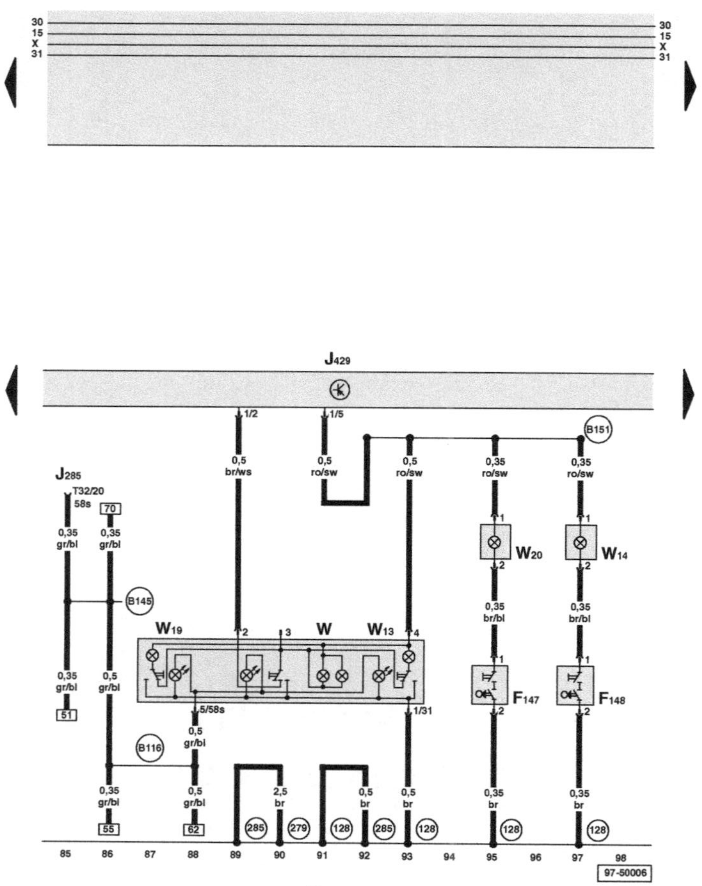

ws = white
sw = black
ro = red
br = brown
gn = green
bl = blue
gr = grey
li = violet
ge = yellow

- F147 - Left Make-Up Mirror Light Switch
- F148 - Right Make-Up Mirror Light Switch
- J285 - Control Module with indicator unit in instrument panel insert
- J429 - Control module for central locking
- T32 - Connector 32 pin, blue, on instrument cluster
- W - Front Interior Light
- W13 - Passenger s Map/Reading Light
- W14 - Right Make-up Mirror Light
- W19 - Left Front Reading Light
- W20 - Left Make-up Mirror Light

- (128) - Ground connection -1-, in interior light wiring harness
- (279) - Ground connector -5-, in wiring harness interior
- (285) - Ground connection -7- in wiring harness interior
- (B116) - Connector (58s), in wiring harness interior
- (B145) - Plus connector (58s), in wiring harness interior
- (B151) - Plus connector -3- (30a), in wiring harness interior

Control Module for Central Locking, Front Interior Light, Reading Lights, Make-Up Mirror Lights

Edition 11/01
W42.USA.5702.10.21

© 2006 BentleyPublishers.com—All Rights Reserved

Central locking, Anti-theft, coupe (2002-2003)

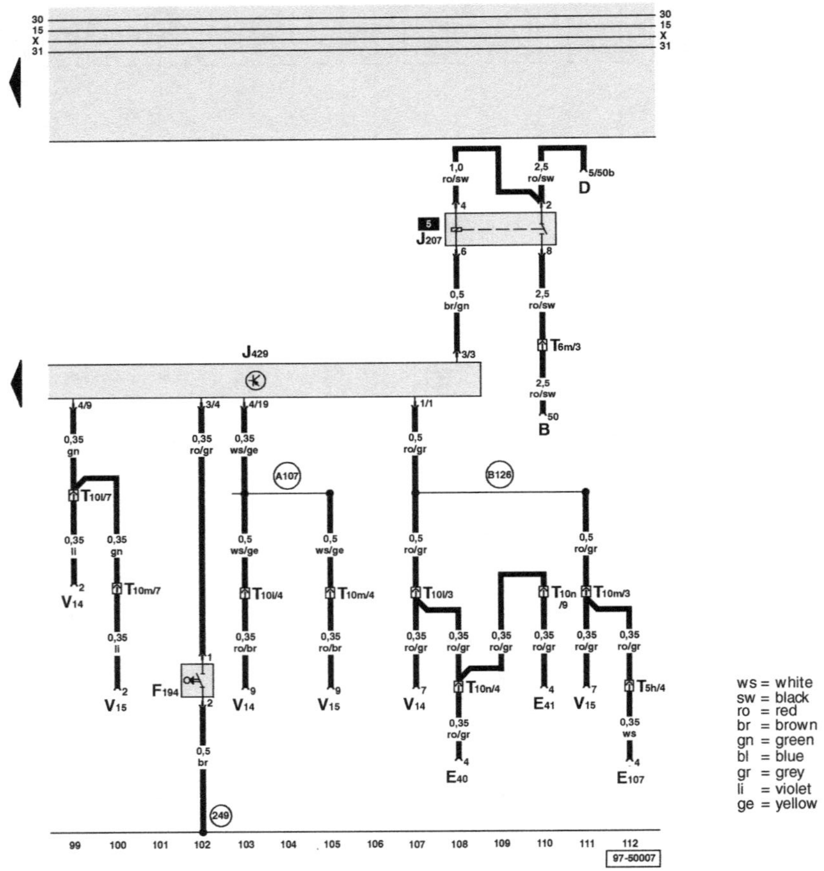

ws = white
sw = black
ro = red
br = brown
gn = green
bl = blue
gr = grey
li = violet
ge = yellow

- B - Starter
- D - Ignition/Starter Switch
- E40 - Left Front Window Switch
- E41 - Right Front Window Switch
- E107 - Switch for window regulator, in passenger door
- F194 - Clutch Pedal Position (CPP) Switch
- J207 - Starting Interlock Relay
- J429 - Control module for central locking
- T5h - Connector 5 pin, in right door
- T6m - Connector 6 pin, brown, in E-box plenum chamber
- T10l - Connector 10 pin, red, connector station A pillar, left
- T10m - Connector 10 pin, red, connector station A pillar, right
- T10n - Connector 10 pin, in left door
- V14 - Left Window Motor
- V15 - Right Window Motor

- (249) - Ground connection -2-, in wiring harness interior
- (A107) - Connector (comfort close), in instrument panel wiring harness
- (B126) - Connector (87, driver s door) , in wiring harness interior

Edition 05/01
W42.USA.5702.07.21

Control Module for Central Locking, Starting Interlock Relay

THIS SPACE INTENTIONALLY LEFT BLANK

© 2006 BentleyPublishers.com—All Rights Reserved

Headlight with High Intensity Gas Discharge (HID) Lamps and automatic Headlight Beam Adjusting

2002 m. y.

Fuse Panel

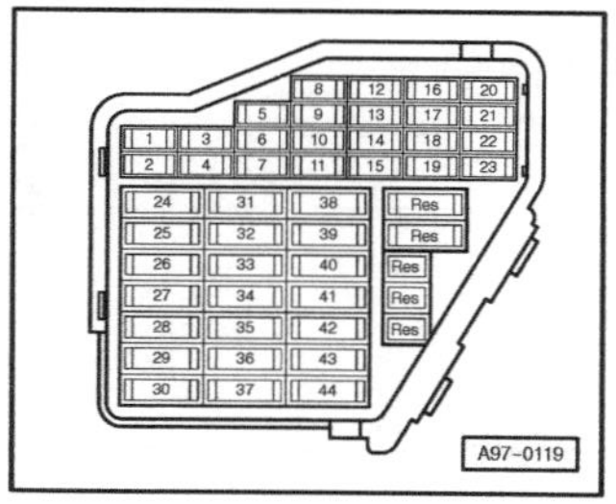

Fuse Colors:

30 A - Green
25 A - White
20 A - Yellow
15 A - Blue
10 A - Red
7,5 A - Brown
5 A - Beige

Starting with fuse position 23, fuses in the fuse holder are identified with 223 in the wiring diagram.

Micro Central Electric Panel

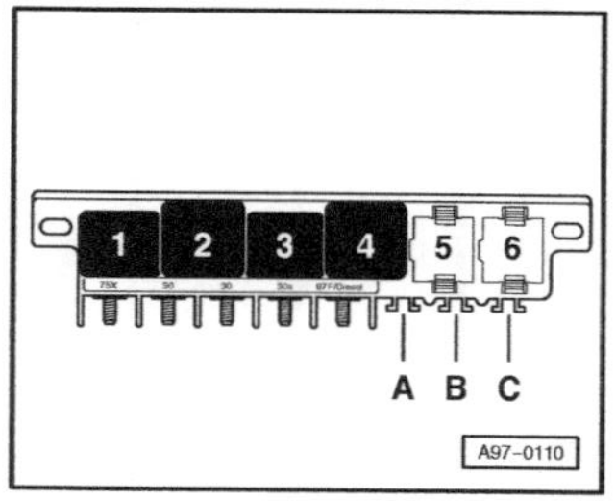

Relay Location:

2 -Load Reduction Relay, J59

13 - Fold Relay Panel

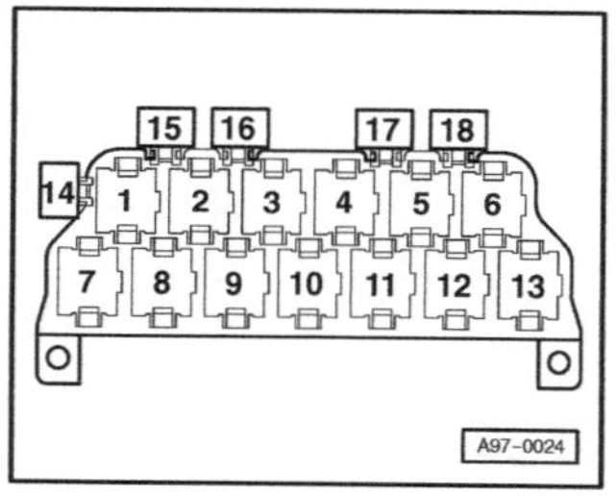

Relay Location:

2 -Lamp Control Module, J123

3 -Lamp Control Module, J123

4 -Fog Light Relay, J5

5 -Starting Interlock Relay, J207

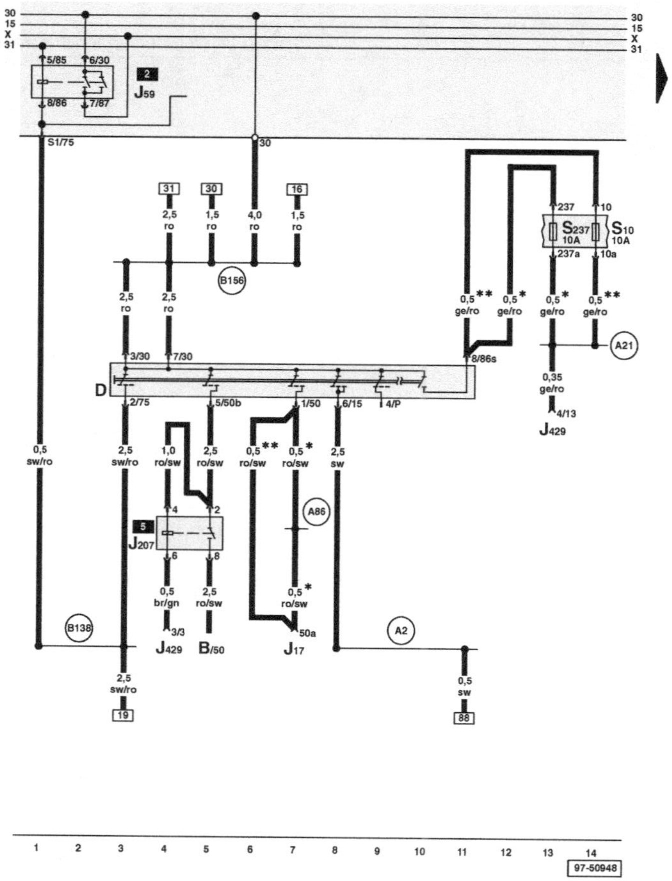

ws = white
sw = black
ro = red
br = brown
gn = green
bl = blue
gr = grey
li = violet
ge = yellow

B - Starter
D - Ignition/Starter Switch
J17 - Fuel Pump (FP) Relay
J59 - Load Reduction Relay
J207 - Starting Interlock Relay
J429 - Control module for central locking
S10 - Fuse
S237 - Fuse 37 in fuse holder

A2 - Plus connection (15), in instrument panel wiring harness
A21 - Wire connection (86s), in instrument panel wiring harness
A86 - Connection (50a), in instrument panel wiring harness
B138 - Plus connector (X), in wiring harness, interior
B156 - Plus connector (30a), in wiring harness interior

* - to 11.2001
** - from 11.2001

Ignition/Starter Switch, Load Reduction Relay, Starting Interlock Relay

© 2006 BentleyPublishers.com—All Rights Reserved

Headlights, Xenon (HID) (2002-2003)

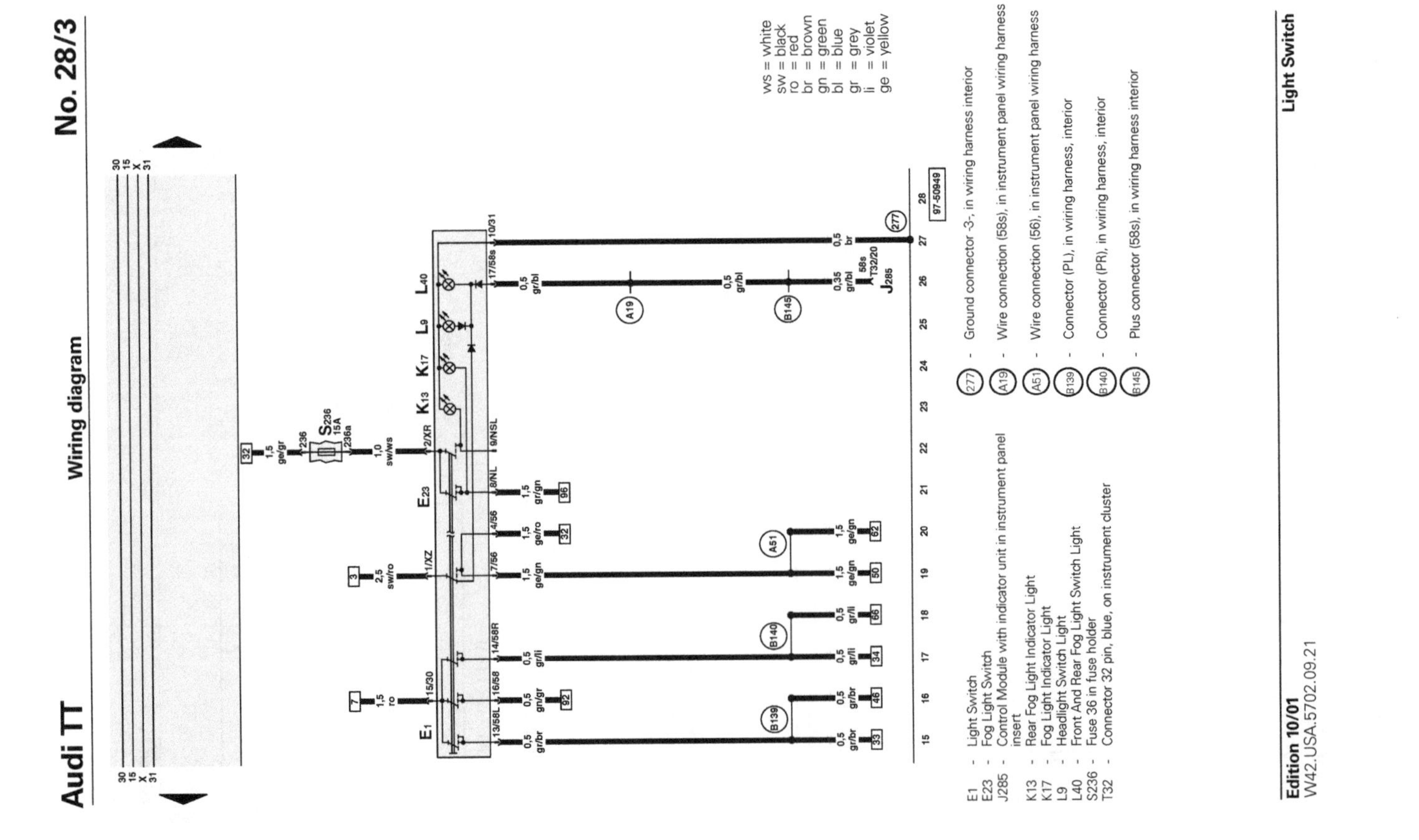

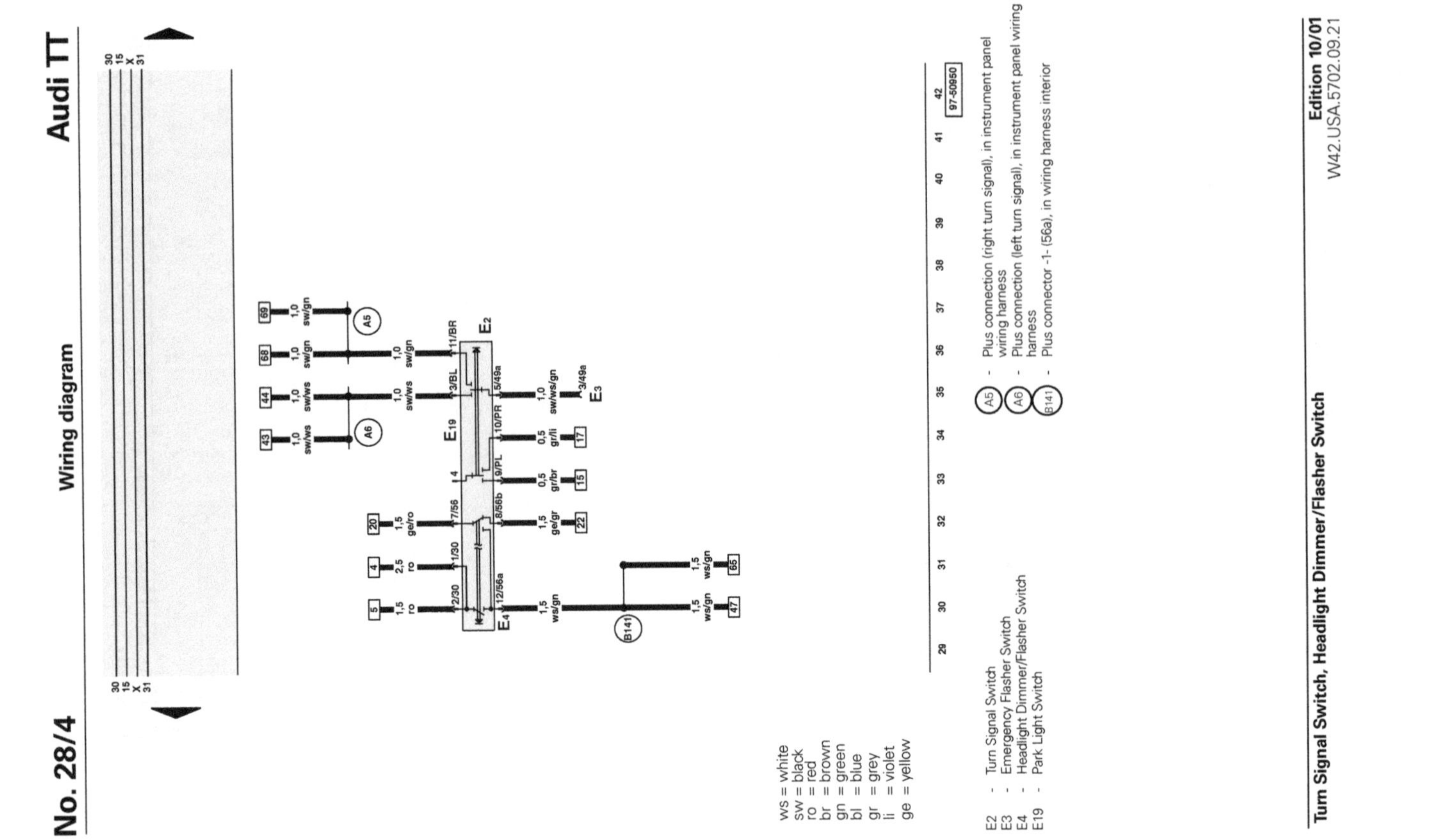

© 2006 BentleyPublishers.com—All Rights Reserved

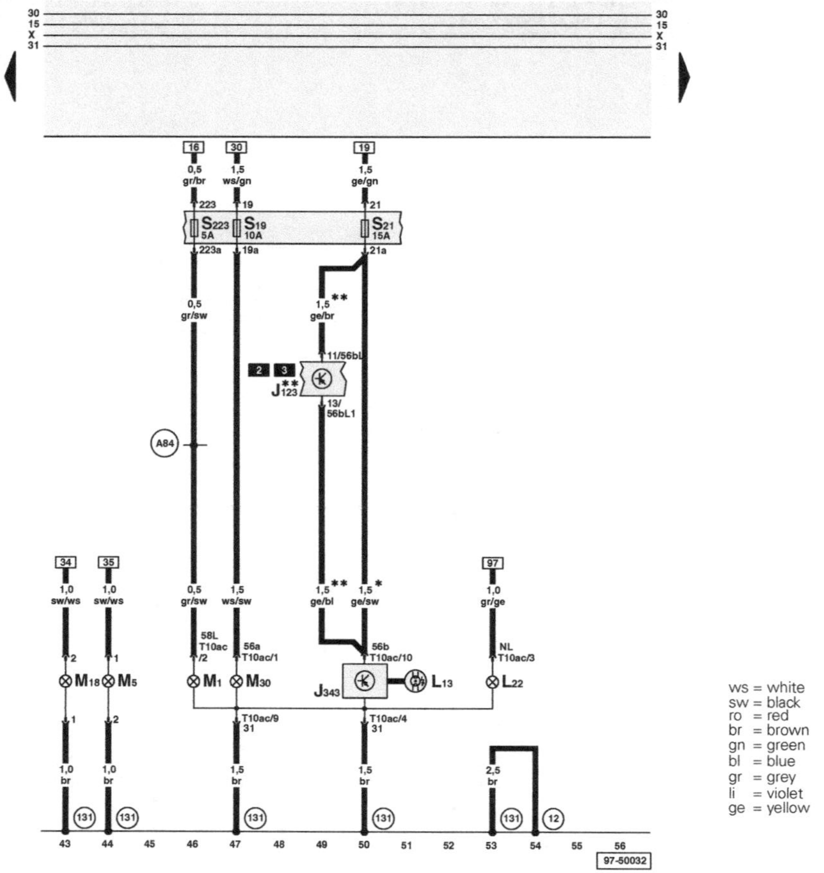

ws = white
sw = black
ro = red
br = brown
gn = green
bl = blue
gr = grey
li = violet
ge = yellow

J123 - Lamp Control Module
J343 - Control Module for High-intensity gas discharge lamp
L13 - High intensity gas discharge lamp left
L22 - Left Front Fog Light
M1 - Left Parking Light
M5 - Left Front Turn Signal Light
M18 - Left, Side Turn Signal Light
M30 - Left High Beam Headlight
S19 - Fuse
S21 - Fuse
S223 - Fuse 23 in fuse holder
T10ac - Connector 10 pin, on headlight, left

(12) - Ground connection, in engine compartment, left

(131) - Ground connection -2-, in engine compartment wiring harness

(A84) - Connector (58L), in instrument panel wiring harness

* - Vehicles without Driver's Information System
** - Vehicles with Driver's Information System

Fuses, High Intensity Gas Discharge Lamp left, Control Module for High-Intensity Gas Discharge Lamp

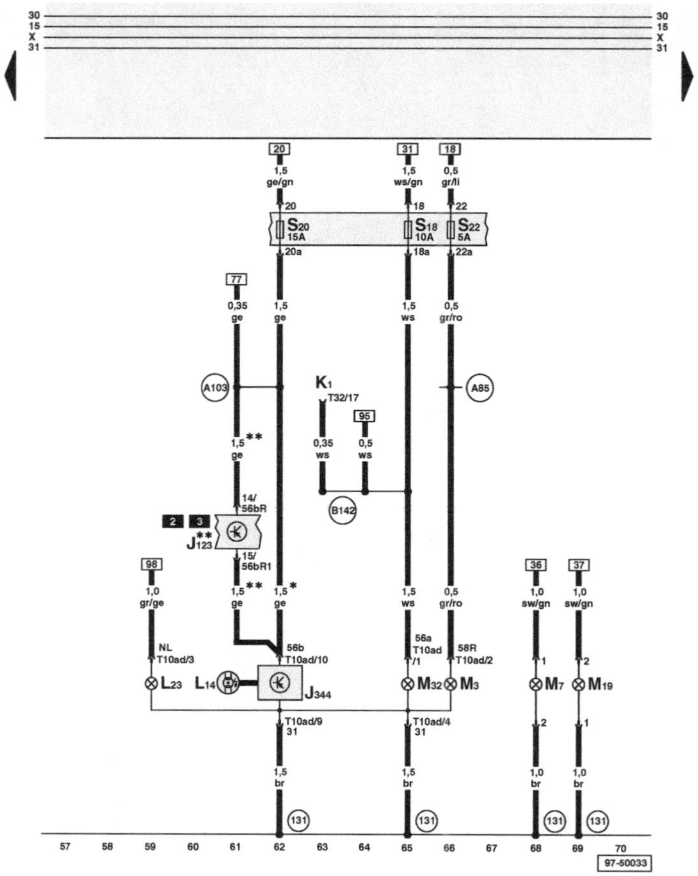

ws = white
sw = black
ro = red
br = brown
gn = green
bl = blue
gr = grey
li = violet
ge = yellow

J123 - Lamp Control Module
J344 - Control Module for High-intensity gas discharge lamp
K1 - Headlight High Beam Indicator Light
L14 - High intensity gas discharge lamp right
L23 - Right Front Fog Light
M3 - Right Parking Light
M7 - Right Front Turn Signal Light
M19 - Right, Side Turn Signal Light
M32 - Right High Beam Headlight
S18 - Fuse
S20 - Fuse
S22 - Fuse
T10ad - Connector 10 pin, on headlight, right
T32 - Connector 32 pin, blue, on instrument cluster

(131) - Ground connection -2-, in engine compartment wiring harness

(A85) - Connector (58R), in instrument panel wiring harness

(A103) - Connector -2- (56), in instrument panel wiring harness

(B142) - Plus connector -2- (56a), in wiring harness interior

* - Vehicles without Driver's Information System
** - Vehicles with Driver's Information System

Fuses, High Intensity Gas Discharge Lamp right, Control Module for High-Intensity Gas Discharge Lamp

© 2006 BentleyPublishers.com—All Rights Reserved

Headlights, Xenon (HID) (2002-2003)

Audi TT Wiring diagram **No. 28/7**

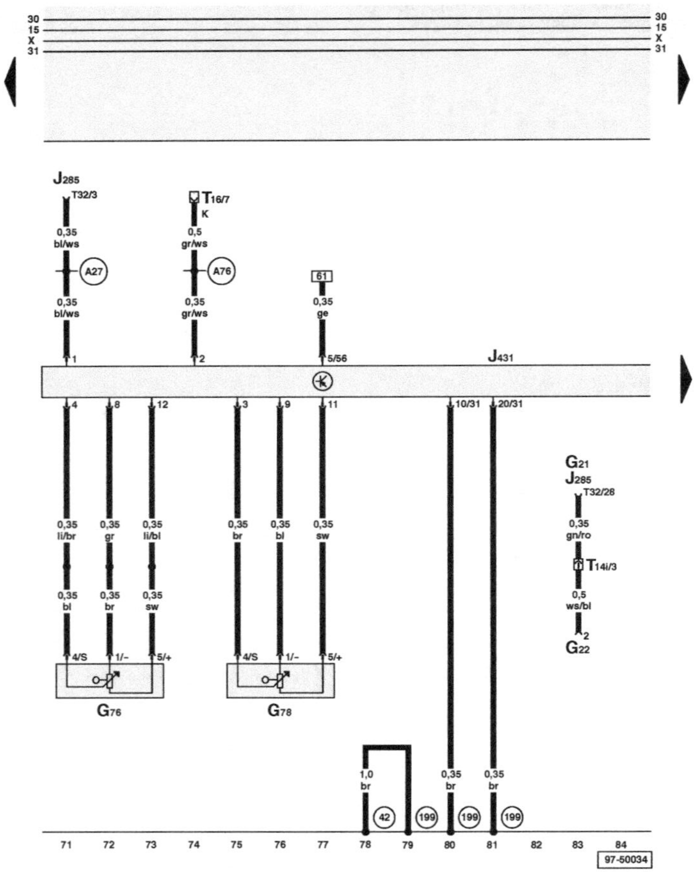

ws = white
sw = black
ro = red
br = brown
gn = green
bl = blue
gr = grey
li = violet
ge = yellow

G21 - Speedometer
G22 - Speedometer Vehicle Speed Sensor (VSS)
G76 - Left Rear Level Control System Sensor
G78 - Left Front Level Control System Sensor
J285 - Control Module with indicator unit in instrument panel insert
J431 - Control module for Headlight Beam Adjusting
T14i - Connector 14 pin, in engine compartment, left
T16 - Connector 16 pin, black, Data Link Connector (DLC), below storage compartment, driver's side
T32 - Connector 32 pin, blue, on instrument cluster

(42) - Ground connection, on steering column
(199) - Ground connection -3-, in instrument panel wiring harness
(A27) - Wire connection (vehicle speed signal), in instrument panel wiring harness
(A76) - Connector (K-diagnosis wire), in instrument panel wiring harness

Control Module for Headlight Beam Adjusting, Level Control System Sensors

No. 28/8 Wiring diagram **Audi TT**

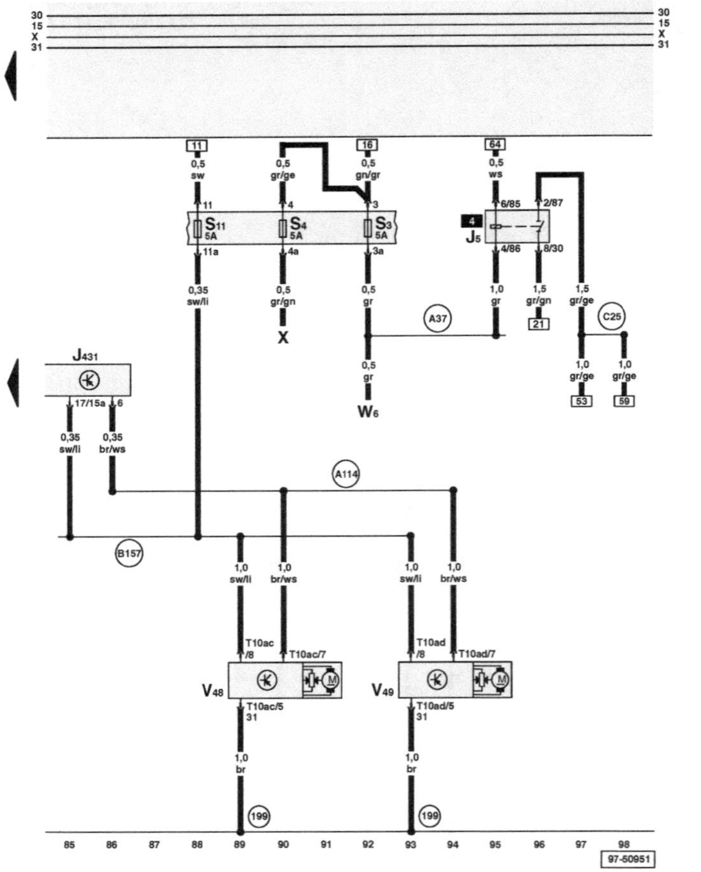

ws = white
sw = black
ro = red
br = brown
gn = green
bl = blue
gr = grey
li = violet
ge = yellow

J5 - Fog Light Relay
J431 - Control module for Headlight Beam Adjusting
S3 - Fuse
S4 - Fuse
S11 - Fuse
T10ac - Connector 10 pin, on headlight, left
T10ad - Connector 10 pin, on headlight, right
V48 - Left Headlight Beam Adjusting Motor
V49 - Right Headlight Beam Adjusting Motor
W6 - Glove Compartment Light
X - License Plate Light

(199) - Ground connection -3-, in instrument panel wiring harness
(A37) - Wire connection (58a), in instrument panel wiring harness
(A114) - Connection (Headlight adjusting), in instrument panel wiring harness
(B157) - Plus connector -3- (15a), in wiring harness interior
(C25) - Wire connection (fog lights), in front light wiring harness

Headlight Beam Adjusting Motors, Fog Light Relay

© 2006 BentleyPublishers.com—All Rights Reserved

Daytime Running Lights with Fog Light (Canada)

2002 m. y.

Fuse Panel

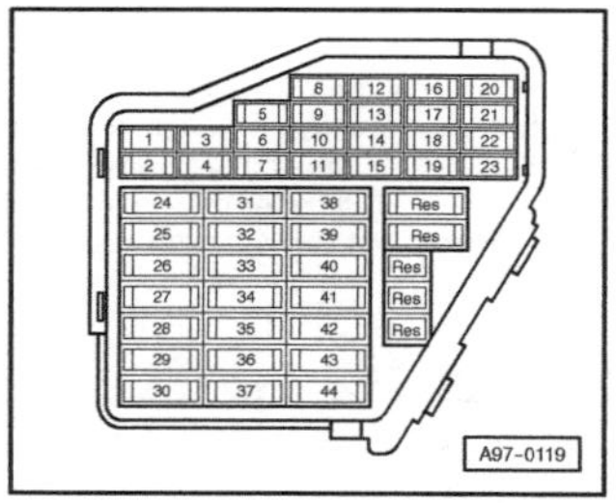

Fuse Colors:

30 A - Green
25 A - White
20 A - Yellow
15 A - Blue
10 A - Red
7,5 A - Brown
5 A - Beige

Starting with fuse position 23, fuses in the fuse holder are identified with 223 in the wiring diagram.

Micro Central Electric Panel

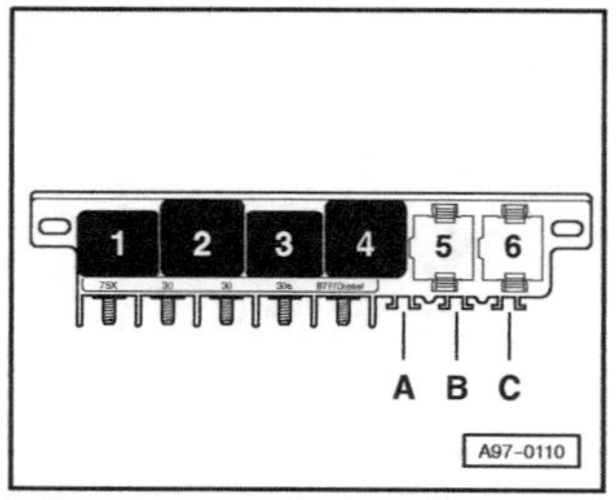

Relay Location:

2 -Load Reduction Relay, J59

13 - Fold Relay Panel

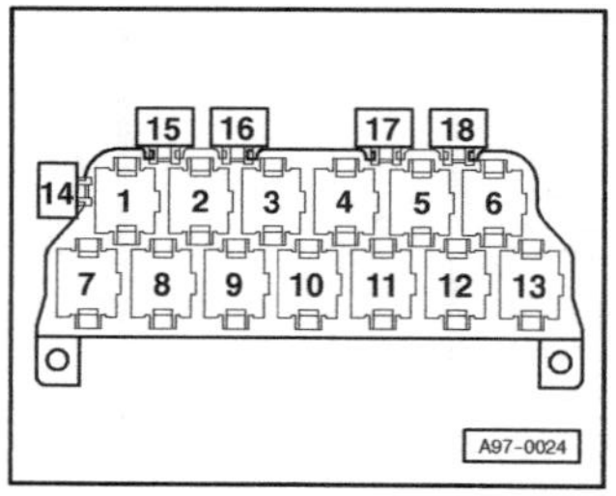

Relay Location:

4 -Fog Light Relay, J5

Edition 05/01
W42.USA.5702.07.21

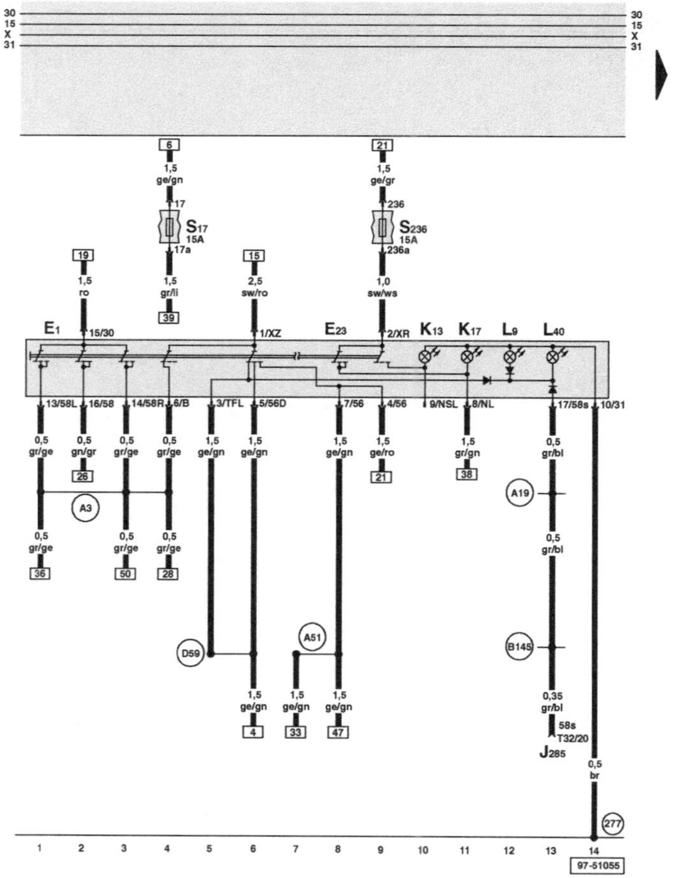

ws = white
sw = black
ro = red
br = brown
gn = green
bl = blue
gr = grey
li = violet
ge = yellow

E1 - Light Switch
E23 - Fog Light Switch
J285 - Control Module With Display Unit In Instrument Cluster
K13 - Rear Fog Light Indicator Light
K17 - Fog Light Indicator Light
L9 - Headlight Switch Light
L40 - Front And Rear Fog Light Switch Light
S17 - Fuse
S236 - Fuse in fuse holder
T32 - Connector 32 pin, blue, on instrument cluster

(277) - Ground connector -3-, in wiring harness interior
(A3) - Plus connection (58), in instrument panel wiring harness
(A19) - Wire connection (58s), in instrument panel wiring harness
(A51) - Wire connection (56), in instrument panel wiring harness
(B145) - Plus connector (58s), in wiring harness interior
(D59) - Wire connection (Fog light), in engine compartment wiring harness

Light Switch, Fog Light Switch

Edition 05/01
W42.USA.5702.07.21

© 2006 BentleyPublishers.com—All Rights Reserved

Daytime Running Lights and Rear Fog Light, Canada (2002)

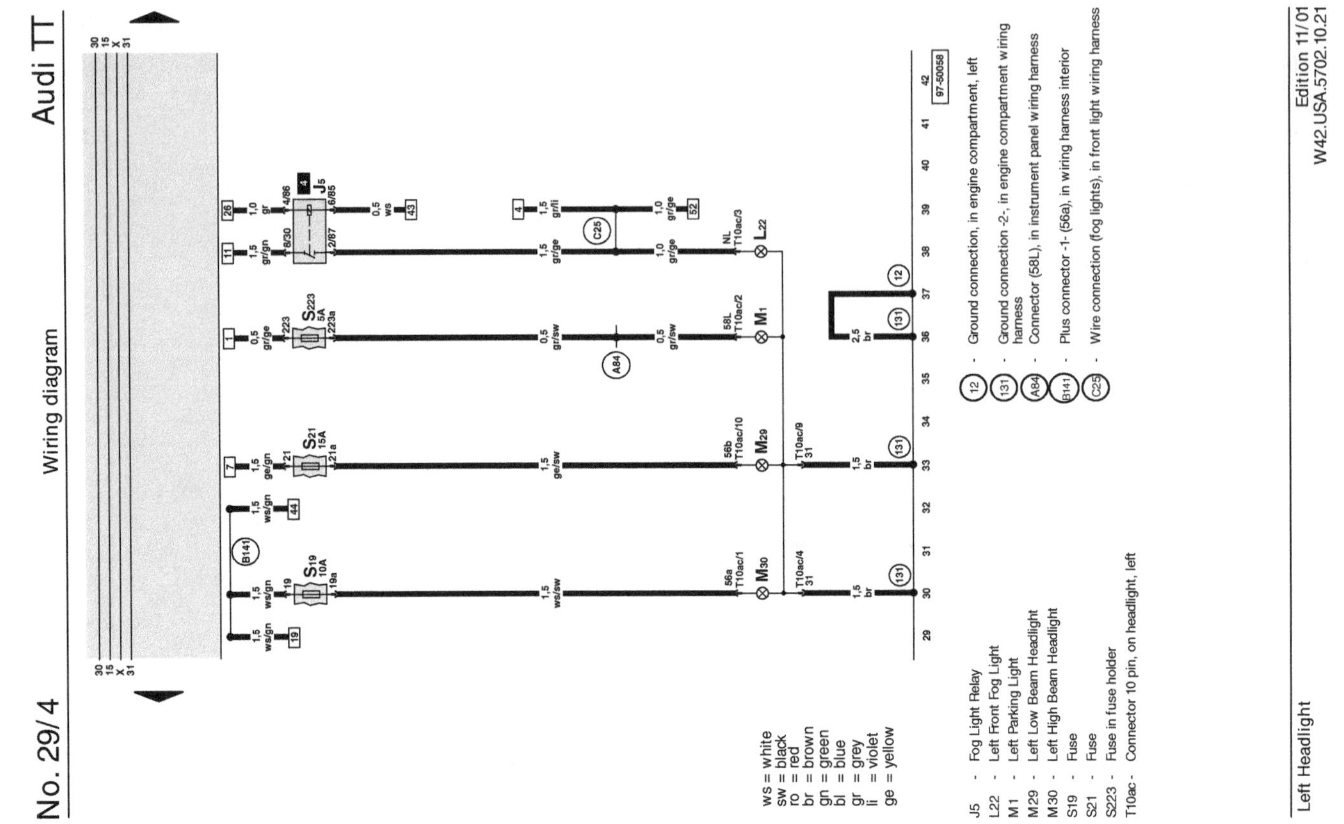

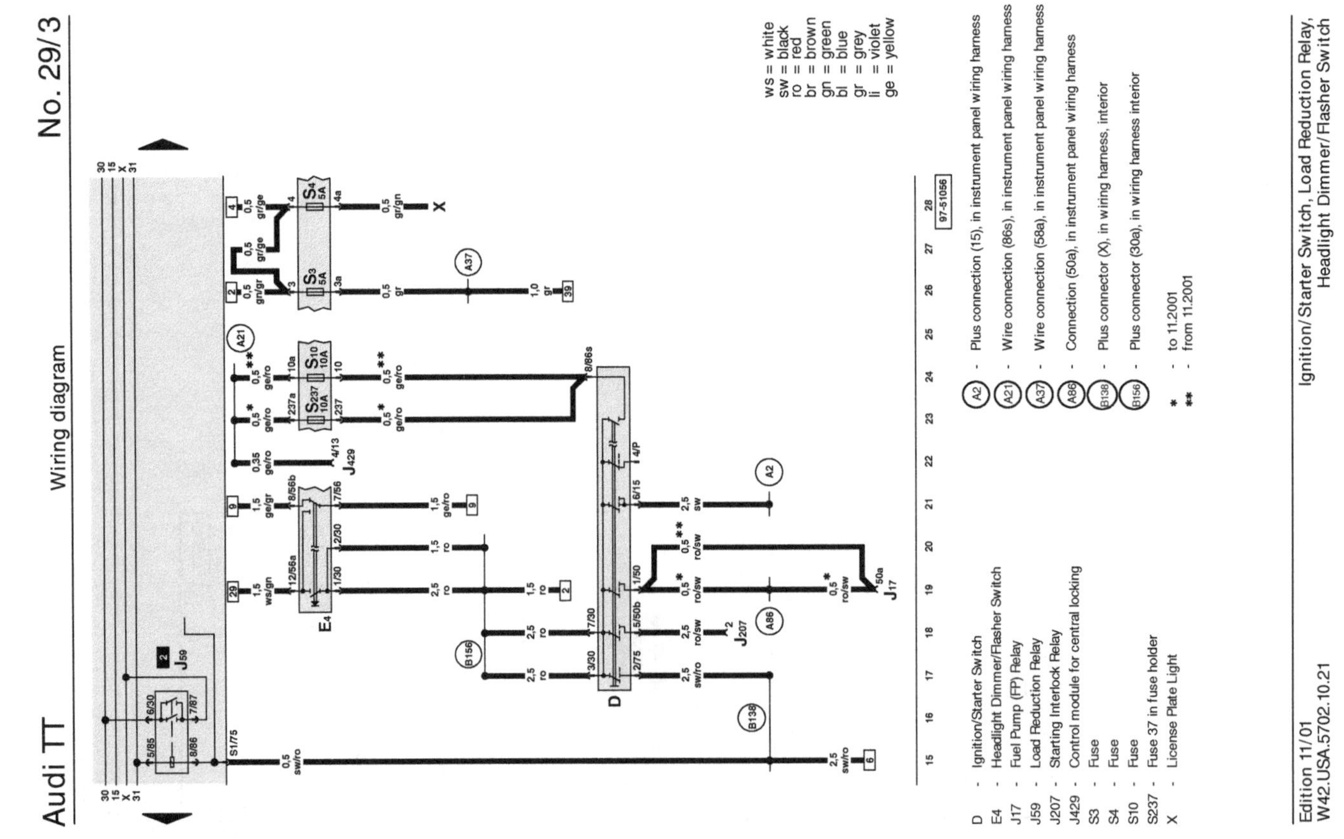

© 2006 BentleyPublishers.com—All Rights Reserved

Audi TT — Wiring diagram — No. 29/5

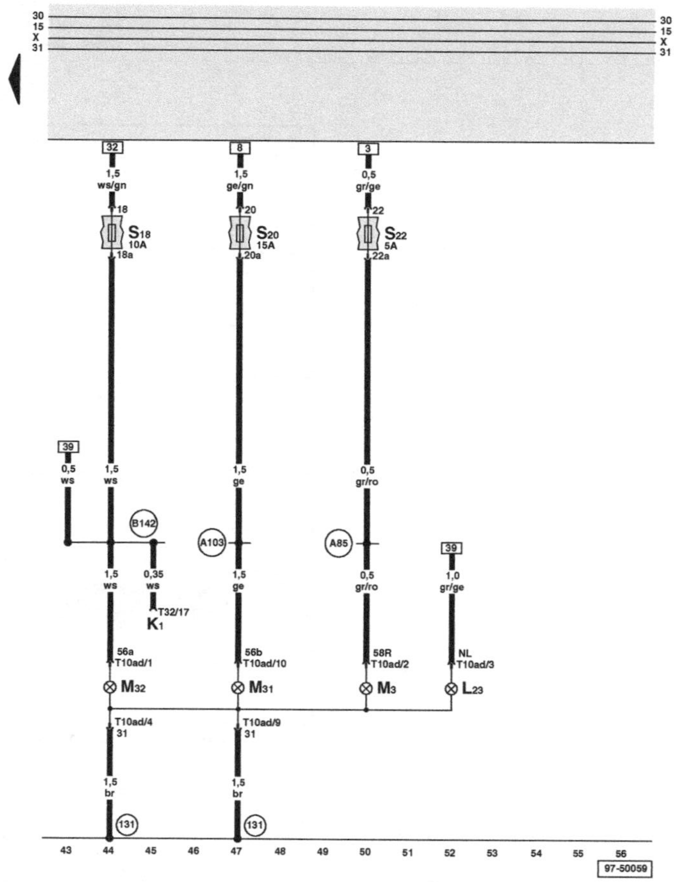

ws = white
sw = black
ro = red
br = brown
gn = green
bl = blue
gr = grey
li = violet
ge = yellow

K1 - Headlight High Beam Indicator Light
L23 - Right Front Fog Light
M3 - Right Parking Light
M31 - Right Low Beam Headlight
M32 - Right High Beam Headlight
S18 - Fuse
S20 - Fuse
S22 - Fuse
T10ad - Connector 10 pin, on headlight, right
T32 - Connector 32 pin, blue, on instrument cluster

(131) - Ground connection -2-, in engine compartment wiring harness
(A85) - Connector (58R), in instrument panel wiring harness
(A103) - Connector -2- (56), in instrument panel wiring harness
(B142) - Plus connector -2- (56a), in wiring harness interior

Edition 05/01
W42.USA.5702.07.21

Right Headlight

THIS SPACE INTENTIONALLY LEFT BLANK

© 2006 BentleyPublishers.com—All Rights Reserved

Central Locking System / Anti Theft System / Radar Interior Monitor (Roadster)

2002 m. y.

Fuse Panel

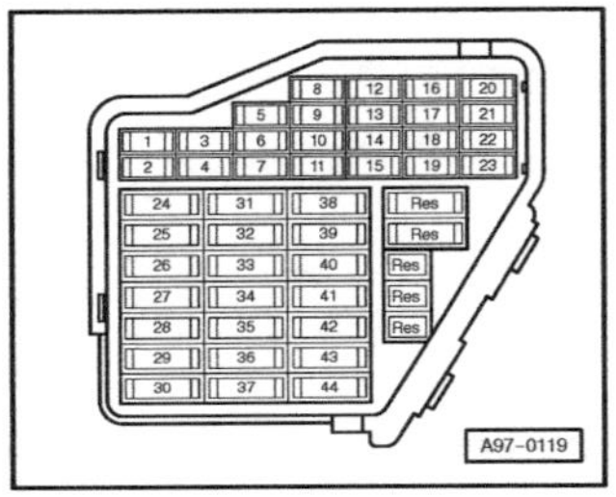

Fuse Colors:

30 A - Green
25 A - White
20 A - Yellow
15 A - Blue
10 A - Red
7,5 A - Brown
5 A - Beige

Starting with fuse position 23, fuses in the fuse holder are identified with 223 in the wiring diagram.

Micro Central Electric Panel

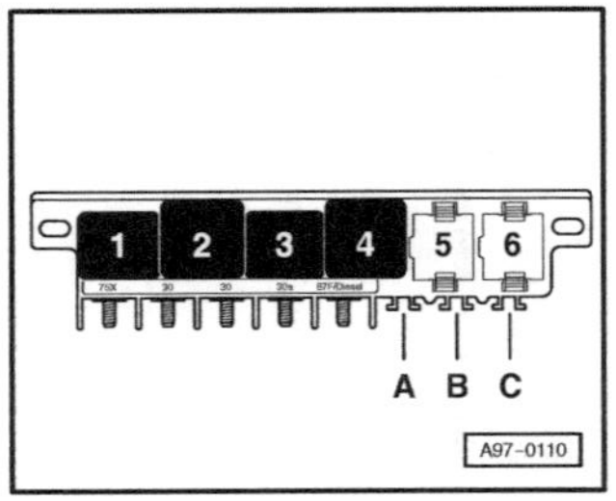

Fuse Location:

A -Convertible Top Fuse, S67

13 - Fold Relay Panel

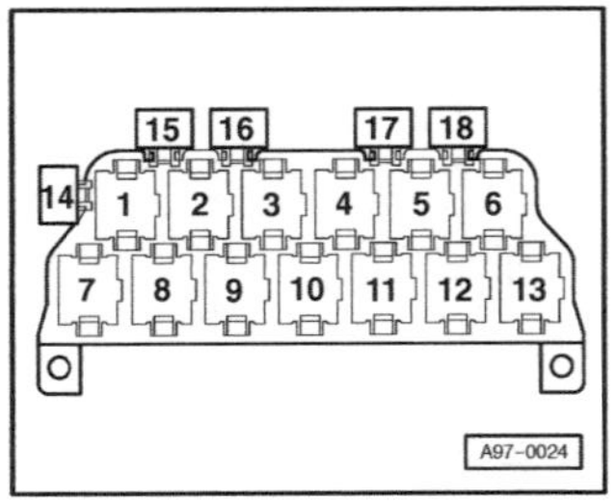

Relay Location:

5 -Starting Interlock Relay, J207

10 -Rear Window Defogger Relay, J9

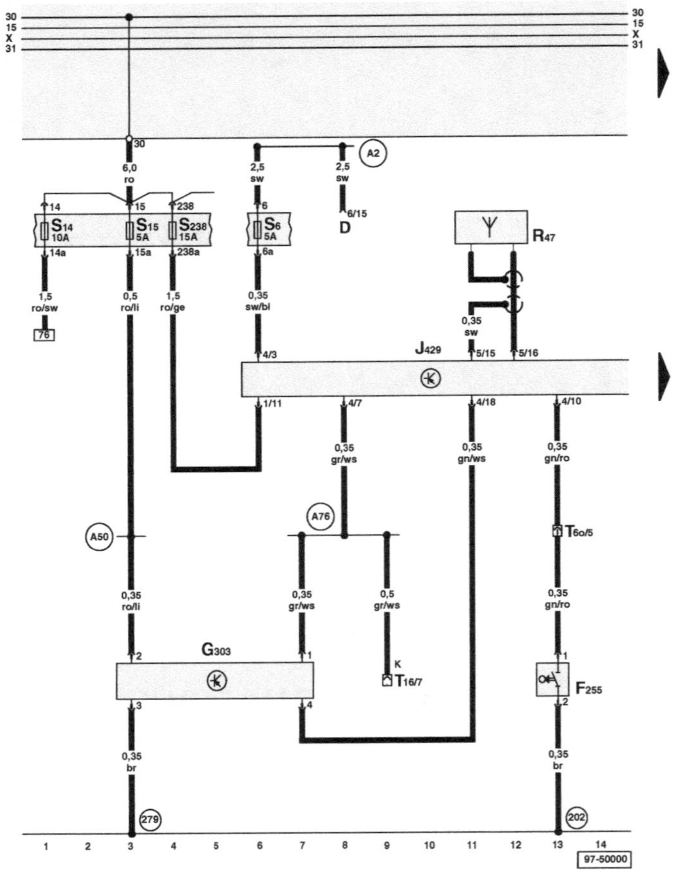

ws = white
sw = black
ro = red
br = brown
gn = green
bl = blue
gr = grey
li = violet
ge = yellow

D - Ignition/Starter Switch
F255 - Hardtop Detector Switch
G303 - Radar Interior Monitoring Control Module 1
J429 - Control module for central locking
R47 - Antenna wire for central locking and anti-theft warning system
S6 - Fuse
S14 - Fuse
S15 - Fuse
S238 - Fuse 38 in fuse holder
T6o - Connector 6 pin, Hardtop right
T16 - Connector 16 pin, black, Data Link Connector (DLC), below storage compartment, driver's side

(202) - Ground connection, in convertible top wiring harness
(279) - Ground connector -5-, in wiring harness interior
(A2) - Plus connection (15), in instrument panel wiring harness
(A50) - Plus connection (30a), in instrument panel wiring harness
(A76) - Connector (K-diagnosis wire), in instrument panel wiring harness

© 2006 BentleyPublishers.com—All Rights Reserved

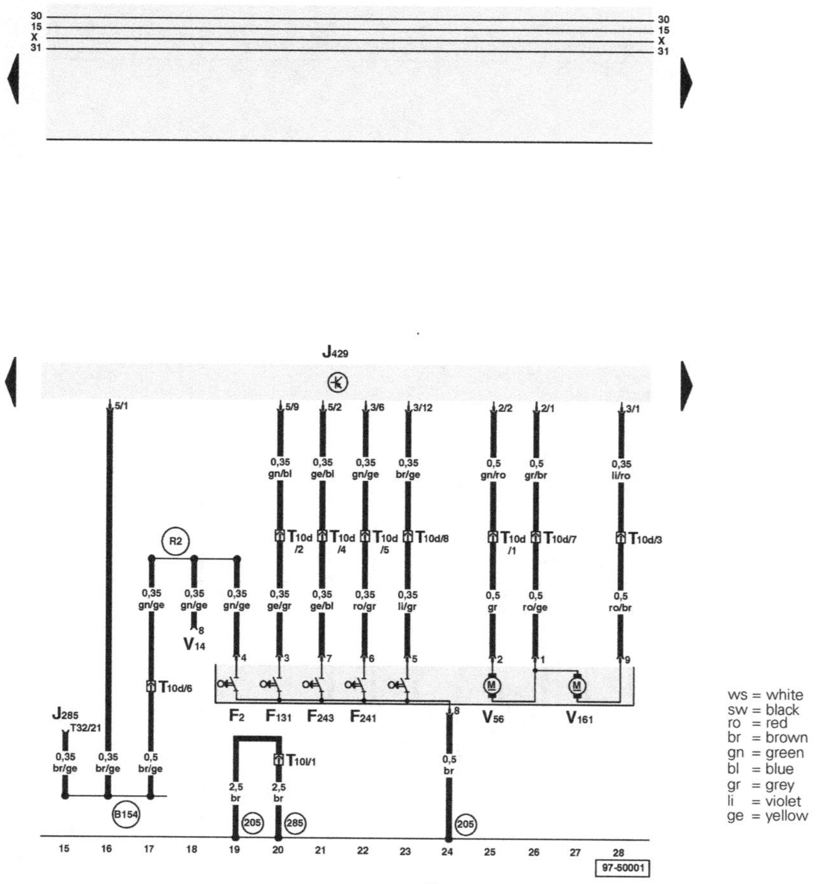

ws = white
sw = black
ro = red
br = brown
gn = green
bl = blue
gr = grey
li = violet
ge = yellow

- F2 - Door Contact Switch - Driver's side
- F131 - Left Front Central Locking System Activator/Switch
- F241 - Contact switch for lock cylinder, driver side
- F243 - Adjusting element for central locking (safe), driver door
- J285 - Control Module with indicator unit in instrument panel insert
- J429 - Control module for central locking
- T10d - Connector 10 pin, brown, connector station A pillar, left
- T10l - Connector 10 pin, red, connector station A pillar, left
- T32 - Connector 32 pin, blue, on instrument cluster
- V14 - Left Window Motor
- V56 - Driver's Door Central Locking System Motor
- V161 - Driver's Door Central Locking (Safe) Motor

- (205) - Ground connection, in driver's door wiring harness
- (285) - Ground connection -7- in wiring harness interior
- (B154) - Connector -1- (TK), in wiring harness interior
- (R2) - Wire connection (door contact switch, left), in power window/central locking system and door contact switch wiring harness

Control Module for Central Locking, Lock Unit for Central Locking Driver's Side

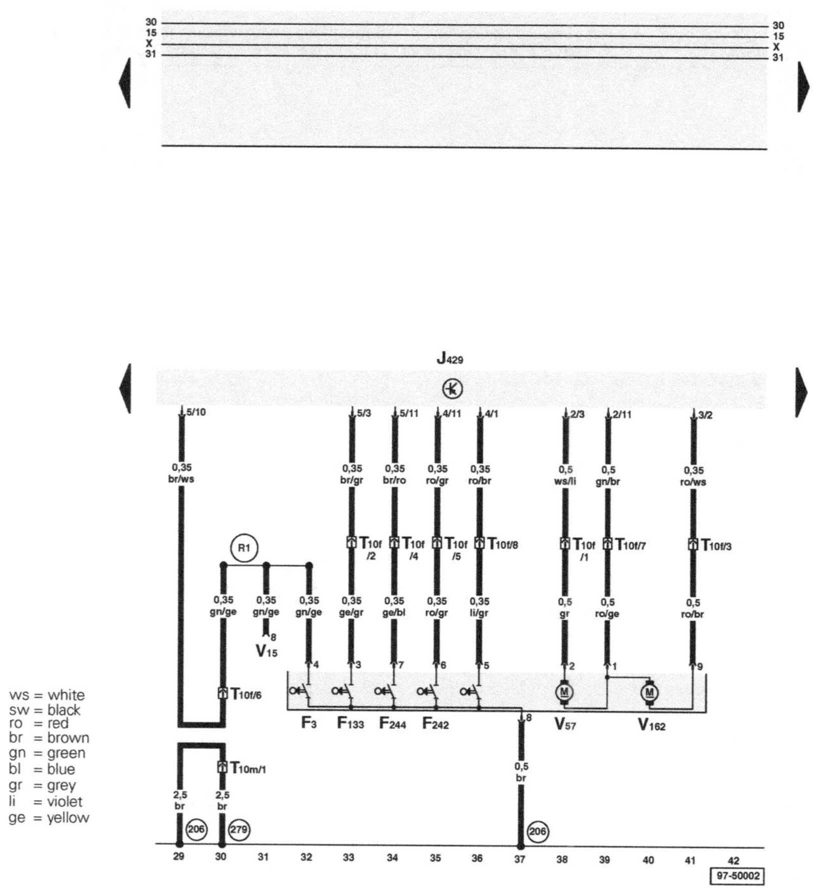

ws = white
sw = black
ro = red
br = brown
gn = green
bl = blue
gr = grey
li = violet
ge = yellow

- F3 - Door Contact Switch - Passenger side
- F133 - Right Front Central Locking System Activator/Switch
- F242 - Contact switch for lock cylinder, passenger side
- F244 - Adjusting element for central locking (safe), passenger door
- J429 - Control module for central locking
- T10f - Connector 10 pin, brown, connector station A pillar, right
- T10m - Connector 10 pin, red, connector station A pillar, right
- V15 - Right Window Motor
- V57 - Passenger's Door Central Locking System Motor
- V162 - Front Passenger's Central Locking (Safe) Motor

- (206) - Ground connection, in passenger's door wiring harness
- (279) - Ground connector -5-, in wiring harness interior
- (R1) - Wire connection (door contact switch, right), in power window/central locking system and door contact switch wiring harness

Control Module for Central Locking, Lock Unit for Central Locking Passenger's Side

© 2006 BentleyPublishers.com—All Rights Reserved

Audi TT | **Wiring diagram** | **No. 30/5**

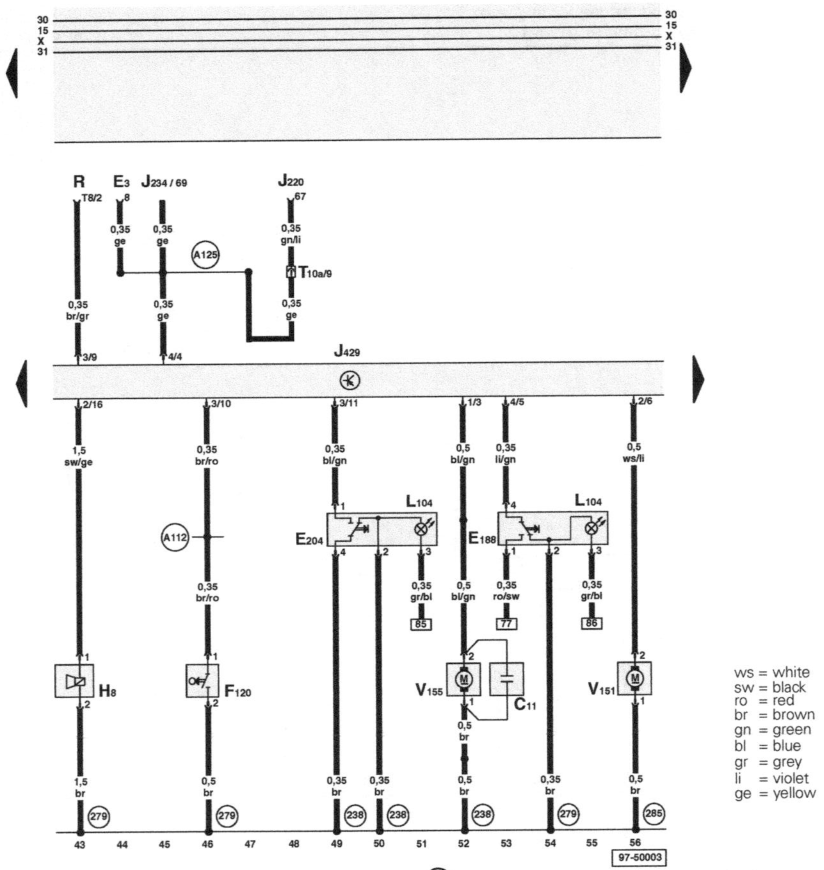

ws = white
sw = black
ro = red
br = brown
gn = green
bl = blue
gr = grey
li = violet
ge = yellow

- C11 - Condensor
- E3 - Emergency Flasher Switch
- E188 - Switch for remote unlock, rear lid
- E204 - Switch for remote/fuel tank door
- F120 - Hood Alarm Switch
- H8 - Alarm Horn
- J220 - Motronic Engine Control Module (ECM)
- J234 - Airbag Control Module
- J429 - Control module for central locking
- L104 - Illumination for switch remote unlock
- R - Radio
- T8 - Connector 8 pin, black, connector radio III
- T10a - Connector 10 pin, white, in E-box plenum chamber
- V151 - Motor for remote unlock rear lid
- V155 - Motor for fuel tank lid unlock

- (238) - Ground connection -1-, in wiring harness interior
- (279) - Ground connector -5-, in wiring harness interior
- (285) - Ground connection -7- in wiring harness interior
- (A112) - Connection (Hood alarm switch), in instrument panel wiring harness
- (A125) - Connection (crash signal) in instrument panel wiring harness

Edition 10/02
W42.USA.5702.11.21

Control Module for Central Locking, Switch for Remote Unlock rear lid, Switch for Remote/Fuel Tank Door

No. 30/6 | **Wiring diagram** | **Audi TT**

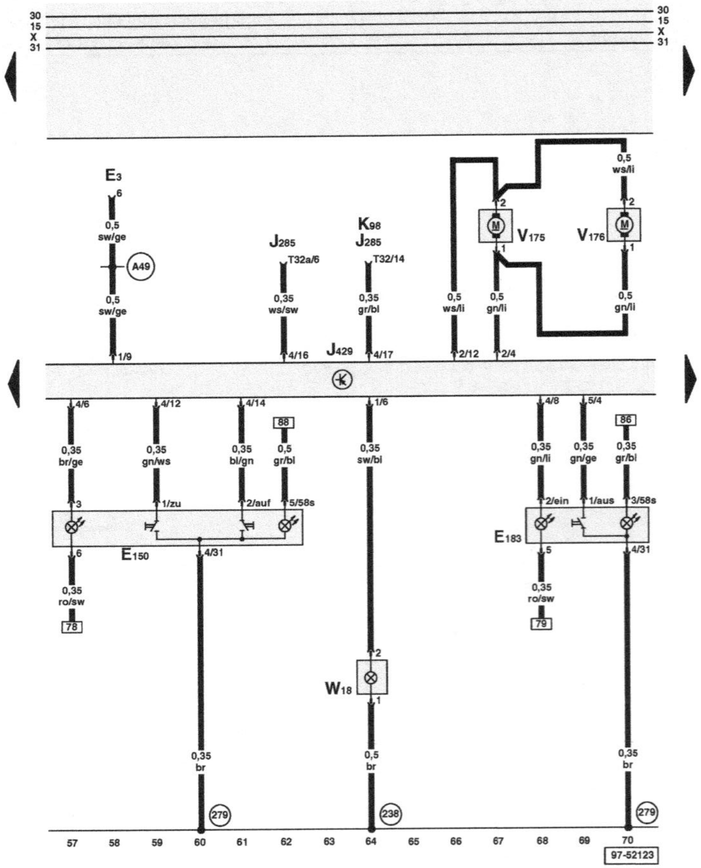

ws = white
sw = black
ro = red
br = brown
gn = green
bl = blue
gr = grey
li = violet
ge = yellow

- E3 - Emergency Flasher Switch
- E150 - Switch for interior lock, driver side
- E183 - Switch for passenger compartment monitoring
- J285 - Control Module with indicator unit in instrument panel insert
- J429 - Control module for central locking
- K98 - Convertible Top Locked Indicator Light
- T32 - Connector 32 pin, blue, on instrument cluster
- T32a - Connector 32 pin, green, on instrument cluster
- V175 - Storage Compartment Central Locking Motor
- V176 - CD Changer Compartment Central Locking Motor
- W18 - Left Luggage Compartment Light in rear lid

- (238) - Ground connection -1-, in wiring harness interior
- (279) - Ground connector -5-, in wiring harness interior
- (A49) - Wire connection -1-, in instrument panel wiring harness

Control Module for Central Locking, Switch for Interior Lock Driver Side, Switch for Passenger Compartment Monitoring, Luggage Compartment Light

Edition 10/02
W42.USA.5702.11.21

© 2006 BentleyPublishers.com—All Rights Reserved

Audi TT — Wiring diagram — No. 30/7

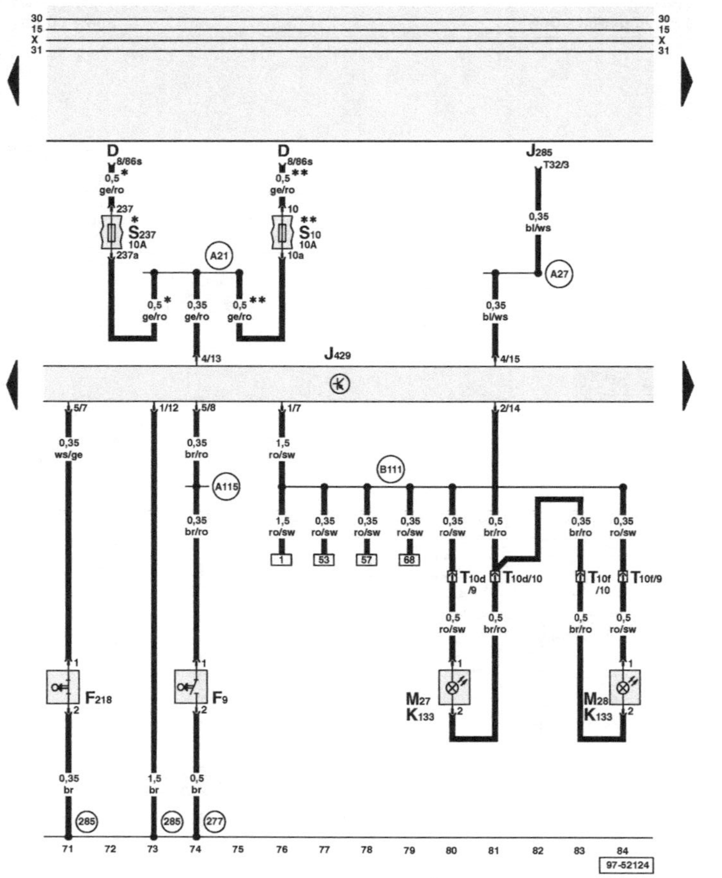

ws = white
sw = black
ro = red
br = brown
gn = green
bl = blue
gr = grey
li = violet
ge = yellow

D - Ignition/Starter Switch
F9 - Parking Brake Warning Light Switch
F218 - Switch for central locking, trunk lid
J285 - Control Module with indicator unit in instrument panel insert
J429 - Control module for central locking
K133 - Warning light for central locking -SAFE-
M27 - Left Door Warning Light
M28 - Right Door Warning Light
S10 - Fuse
S237 - Fuse 37 in fuse holder
T10d - Connector 10 pin, brown, connector station A pillar, left
T10f - Connector 10 pin, brown, connector station A pillar, right
T32 - Connector 32 pin, blue, on instrument cluster

(277) - Ground connector -3-, in wiring harness interior
(285) - Ground connection -7- in wiring harness interior
(A21) - Wire connection (86s), in instrument panel wiring harness
(A27) - Wire connection (vehicle speed signal), in instrument panel wiring harness
(A115) - Connection (parking brake control) in instrument panel wiring harness
(B111) - Connector (30a), in wiring harness interior

* - through 10.2001
** - from 11.2001

Edition 10/02
W42.USA.5702.11.21

Control Module for Central Locking, Door Warning Lights

No. 30/8 — Wiring diagram — Audi TT

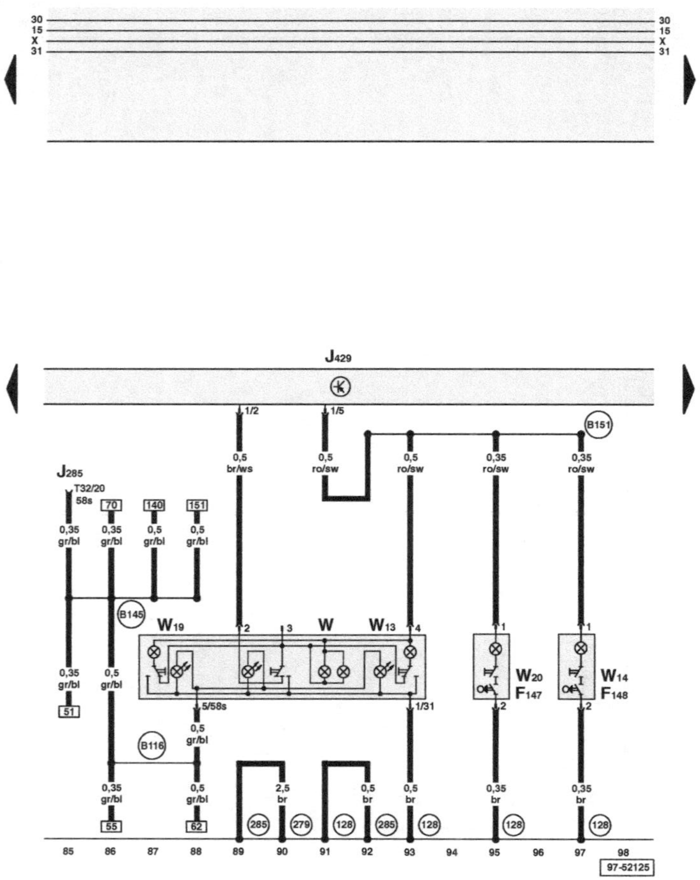

ws = white
sw = black
ro = red
br = brown
gn = green
bl = blue
gr = grey
li = violet
ge = yellow

F147 - Left Make-Up Mirror Light Switch
F148 - Right Make-Up Mirror Light Switch
J285 - Control Module with indicator unit in instrument panel insert
J429 - Control module for central locking
T32 - Connector 32 pin, blue, on instrument cluster
W - Front Interior Light
W13 - Passenger's Map/Reading Light
W14 - Right Make-up Mirror Light
W19 - Left Front Reading Light
W20 - Left Make-up Mirror Light

(128) - Ground connection -1-, in interior light wiring harness
(279) - Ground connector -5-, in wiring harness interior
(285) - Ground connection -7- in wiring harness interior
(B116) - Connector (58s), in wiring harness interior
(B145) - Plus connector (58s), in wiring harness interior
(B151) - Plus connector -3- (30a), in wiring harness interior

Control Module for Central Locking, Front Interior Light, Reading Lights, Make-Up Mirror Lights

Edition 10/02
W42.USA.5702.11.21

© 2006 BentleyPublishers.com—All Rights Reserved

Audi TT — Wiring diagram — No. 30/9

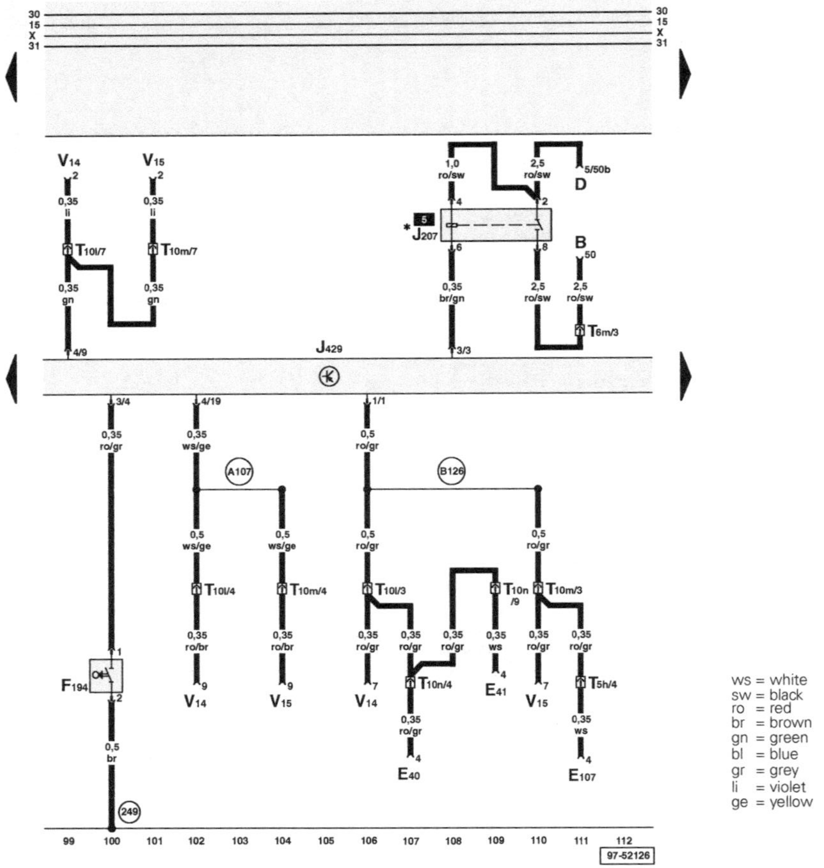

ws = white
sw = black
ro = red
br = brown
gn = green
bl = blue
gr = grey
li = violet
ge = yellow

B - Starter
D - Ignition/Starter Switch
E40 - Left Front Window Switch
E41 - Right Front Window Switch
E107 - Switch for window regulator, in passenger door
F194 - Clutch Pedal Position (CPP) Switch
J207 - Starting Interlock Relay
J429 - Control module for central locking
T5h - Connector 5 pin, in right door
T6m - Connector 6 pin, brown, in E-box plenum chamber
T10l - Connector 10 pin, red, connector station A pillar, left
T10m - Connector 10 pin, red, connector station A pillar, right
T10n - Connector 10 pin, in left door
V14 - Left Window Motor
V15 - Right Window Motor

(249) - Ground connection -2-, in wiring harness interior
(A107) - Connector (comfort close), in instrument panel wiring harness
(B126) - Connector (87, driver's door) , in wiring harness interior
* - Vehicles with Manual Transmission

No. 30/10 — Wiring diagram — Audi TT

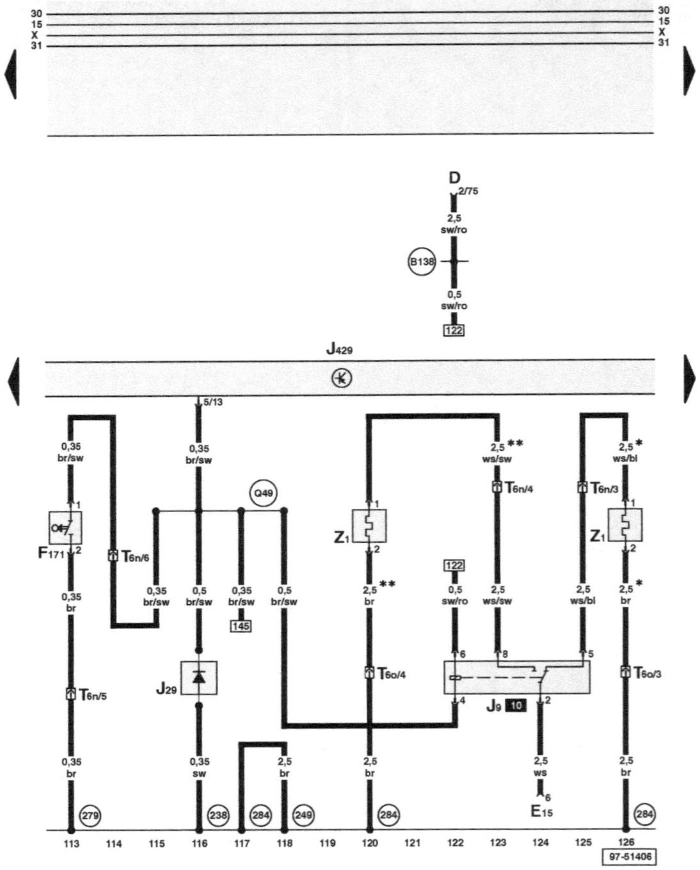

ws = white
sw = black
ro = red
br = brown
gn = green
bl = blue
gr = grey
li = violet
ge = yellow

D - Ignition/Starter Switch
E15 - Rear Window Defogger Switch
F171 - Convertible Top Open Position Switch
J9 - Rear Window Defogger Relay
J29 - Blocking Diode
J429 - Control module for central locking
T6n - Connector 6 pin, Hardtop left
T6o - Connector 6 pin, Hardtop right
Z1 - Heated rear window

(238) - Ground connection -1-, in wiring harness interior
(249) - Ground connection -2-, in wiring harness interior
(279) - Ground connector -5-, in wiring harness interior
(284) - Ground connection -6- in wiring harness interior
(B138) - Plus connector (X), in wiring harness, interior
(Q49) - Connection (closed), in wiring harness interior

* - Softtop
** - Hardtop (where applicable)

© 2006 BentleyPublishers.com—All Rights Reserved

Audi TT — Wiring diagram — No. 30/11

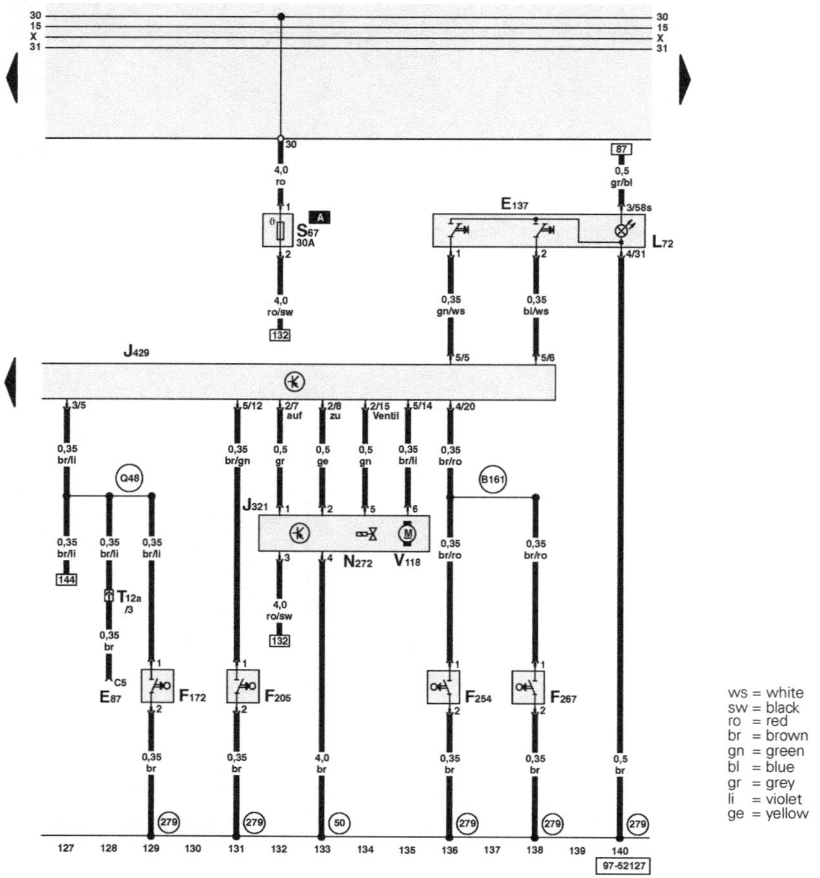

ws = white
sw = black
ro = red
br = brown
gn = green
bl = blue
gr = grey
li = violet
ge = yellow

E87 - A/C Control Head
E137 - Convertible Top Operation Switch
F172 - Convertible Top Front Lock Switch
F205 - Convertible Top Latch Open Position Switch
F254 - Convertible Top Cover Detector Switch
F267 - Convertible Top Cover Detector Switch 2
J321 - Convertible Top Hydraulic Pump Relay
J429 - Control module for central locking
L72 - Convertible Top Switch Light
N272 - Power Convertible Top Valve -1-
S67 - Convertible Top Fuse
T12a - Connector 12 pin, behind instrument panel, right
V118 - Convertible Top Hydraulic Pump

(50) - Ground connection, in luggage compartment, left
(279) - Ground connector -5-, in wiring harness interior
(B161) - Connector (75a), in wiring harness interior
(Q48) - Connection (open), in wiring harness interior

Edition 10/02
W42.USA.5702.11.21

Top

Audi TT — Wiring diagram — No. 30/12

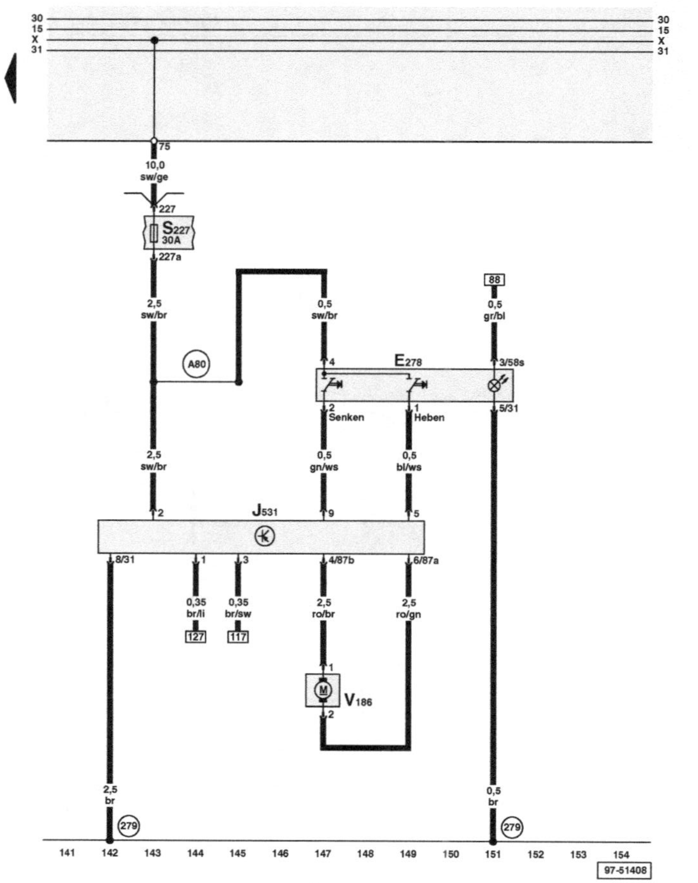

ws = white
sw = black
ro = red
br = brown
gn = green
bl = blue
gr = grey
li = violet
ge = yellow

E278 - Wind Deflector Switch
J531 - Wind Deflector Control Module
S227 - Fuse 27 in fuse holder
V186 - Wind Deflector Motor

(279) - Ground connector -5-, in wiring harness interior
(A80) - Connector -1- (X), in instrument panel wiring harness

Wind Deflector

Edition 10/02
W42.USA.5702.11.21

© 2006 BentleyPublishers.com—All Rights Reserved

Driver s Information System

2002 m. y.

Fuse Panel

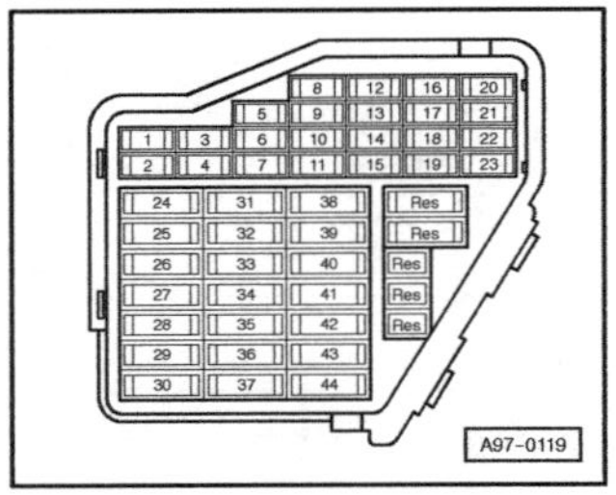

Fuse Colors:

30 A - Green
25 A - White
20 A - Yellow
15 A - Blue
10 A - Red
7,5 A - Brown
5 A - Beige

Starting with fuse position 23, fuses in the fuse holder are identified with 223 in the wiring diagram.

13 - Fold Relay Panel

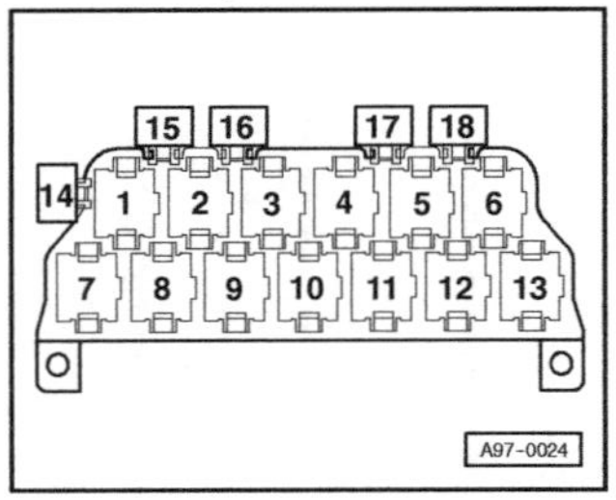

Relay Location:

2 -Lamp Control Module, J123

3 -Lamp Control Module, J123

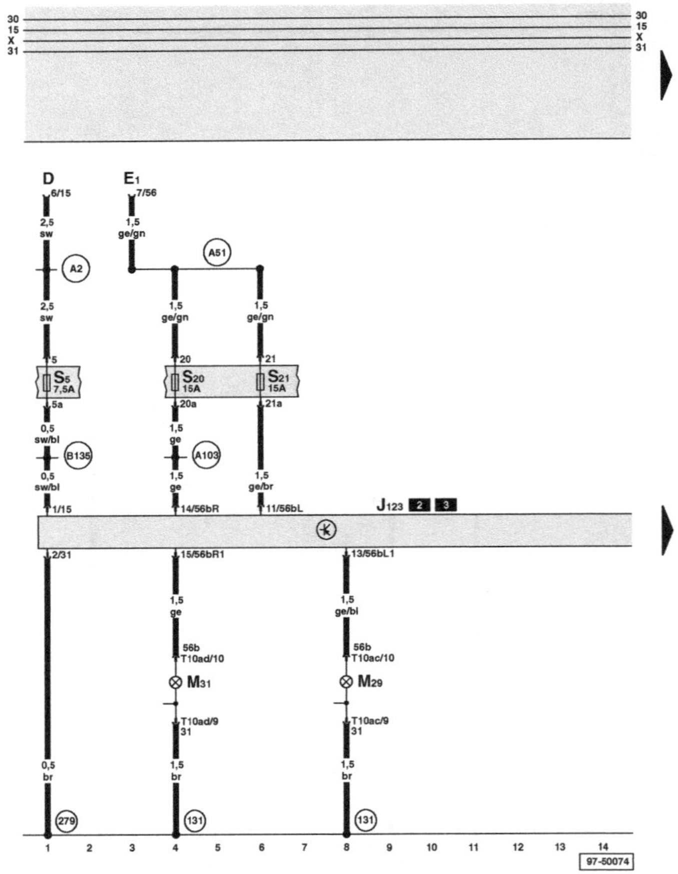

ws = white
sw = black
ro = red
br = brown
gn = green
bl = blue
gr = grey
li = violet
ge = yellow

D - Ignition/Starter Switch
E1 - Light Switch
J123 - Lamp Control Module
M29 - Left Low Beam Headlight
M31 - Right Low Beam Headlight
S5 - Fuse
S20 - Fuse
S21 - Fuse
T10ac - Connector 10 pin, on headlight, left
T10ad - Connector 10 pin, on headlight, right

(131) - Ground connection -2-, in engine compartment wiring harness
(279) - Ground connector -5-, in wiring harness interior
(A2) - Plus connection (15), in instrument panel wiring harness
(A51) - Wire connection (56), in instrument panel wiring harness
(A103) - Connector -2- (56), in instrument panel wiring harness
(B135) - Connector (15a), in wiring harness interior

Lamp Control Module, Low Beam Headlight

© 2006 BentleyPublishers.com—All Rights Reserved

Audi TT — Wiring diagram — No. 31/3

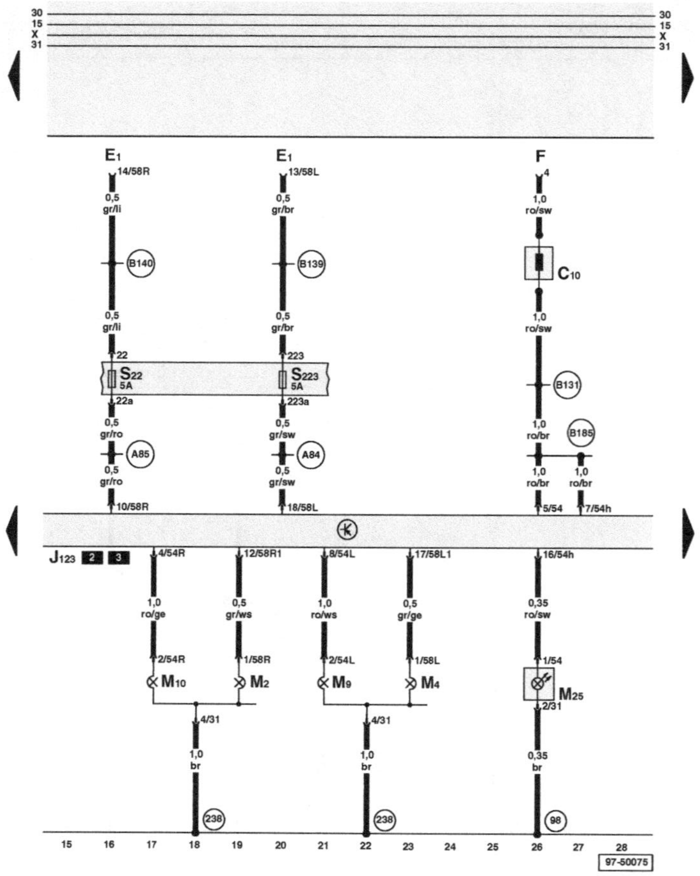

ws = white
sw = black
ro = red
br = brown
gn = green
bl = blue
gr = grey
li = violet
ge = yellow

C10 - Suppressor Filter
E1 - Light Switch
F - Brake Light Switch
J123 - Lamp Control Module
M2 - Right Tail Light
M4 - Left Tail Light
M9 - Left Brake Light
M10 - Right Brake Light
M25 - High-mount Brake Light
S22 - Fuse
S223 - Fuse 23 in fuse holder

(98) - Ground connection, in rear lid wiring harness
(238) - Ground connection -1-, in wiring harness interior
(A84) - Connector (58L), in instrument panel wiring harness
(A85) - Connector (58R), in instrument panel wiring harness
(B131) - Connector (54), in wiring harness interior
(B139) - Connector (PL), in wiring harness, interior
(B140) - Connector (PR), in wiring harness, interior
(B185) - Connection 2 (54), in passenger compartment wiring harness

No. 31/4 — Wiring diagram — Audi TT

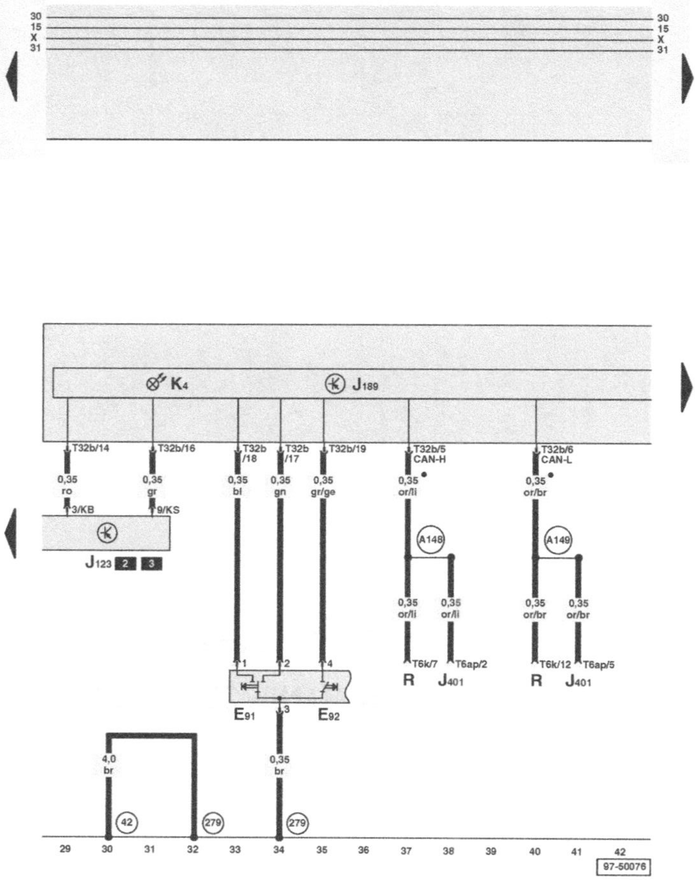

ws = white
sw = black
ro = red
br = brown
gn = green
bl = blue
gr = grey
li = violet
ge = yellow
or = orange

E91 - Board Computer Function Selector Switch
E92 - Board Computer Reset Button
J123 - Lamp Control Module
J189 - Auto Check System
J401 - Control unit for navigation with CD drive
K4 - Park Light Indicator Light
R - Radio
T6k - Connector 6 pin, green, connector radio I-2
T6ap - Connector 6 pin, Connector C I on Control Module for Navigation
T32b - Connector 32 pin, gray, on instrument cluster

(42) - Ground connection, on steering column
(279) - Ground connector -5-, in wiring harness interior
(A148) - Connection (high bus, navigation), in instrument panel wiring harness
(A149) - Connection (low bus, navigation), in instrument panel wiring harness
• - CAN-Bus (Data-Bus)

© 2006 BentleyPublishers.com—All Rights Reserved

THIS SPACE INTENTIONALLY LEFT BLANK

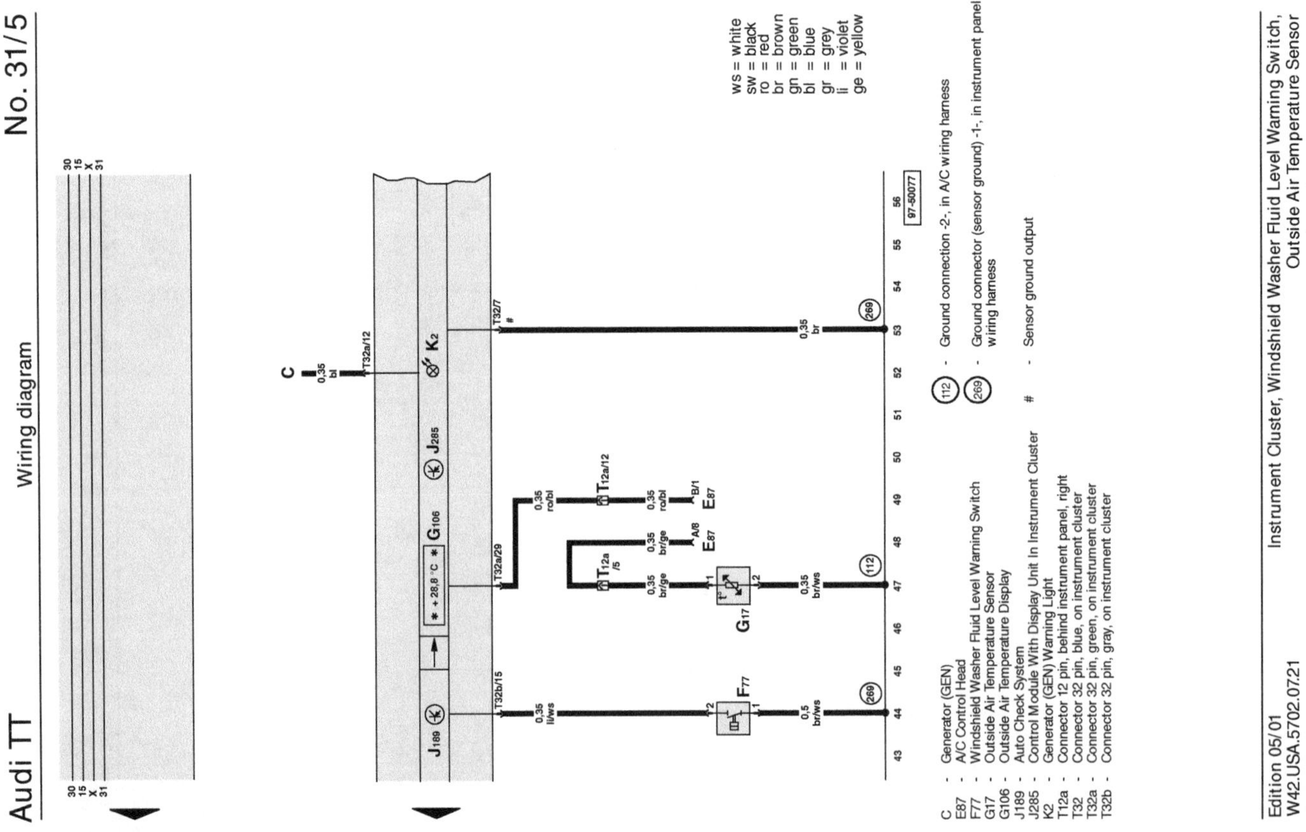

© 2006 BentleyPublishers.com—All Rights Reserved

Navigation System

2002 m. y.

Fuse Panel

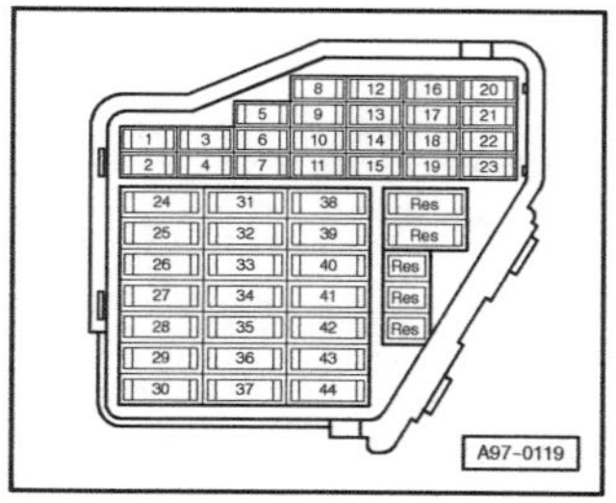

Fuse Colors:

30 A - Green
25 A - White
20 A - Yellow
15 A - Blue
10 A - Red
7,5 A - Brown
5 A - Beige

Starting with fuse position 23, fuses in the fuse holder are identified with 223 in the wiring diagram.

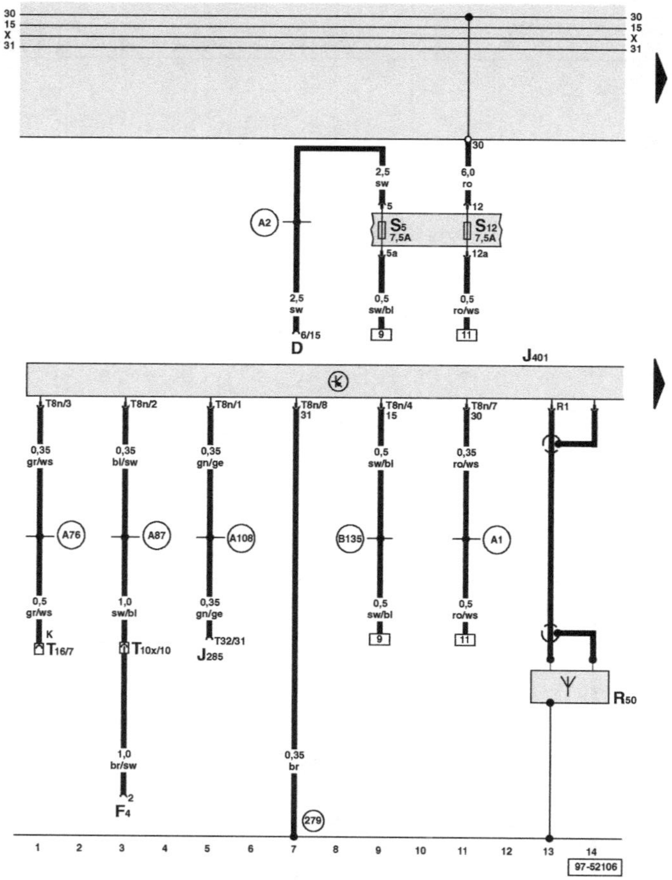

ws = white
sw = black
ro = red
br = brown
gn = green
bl = blue
gr = grey
li = violet
ge = yellow

D - Ignition/Starter Switch
F4 - Back-Up Light Switch
J285 - Control Module With Display Unit In Instrument Cluster
J401 - Control unit for navigation with CD drive
R1 - Antenna Connection
R50 - Antenna for Navigation System (GPS)
S5 - Fuse
S12 - Fuse
T8n - Connector 8 pin, black, connector A on Control Module for Navigation
T10x - Connector 10 pin, orange, in E-box plenum chamber
T16 - Connector 16 pin, black, Data Link Connector (DLC), below storage compartment, driver s side
T32 - Connector 32 pin, blue, on instrument cluster

(279) - Ground connector -5-, in wiring harness interior
(A1) - Plus connection (30a), in instrument panel wiring harness
(A2) - Plus connection (15), in instrument panel wiring harness
(A76) - Connector (K-diagnosis wire), in instrument panel wiring harness
(A87) - Connector (RF), in instrument panel wiring harness
(A108) - Connector (speed signal), in instrument panel wiring harness
(B135) - Connector (15a), in wiring harness interior

©2006 BentleyPublishers.com—All Rights Reserved

Navigation System (2002)

Audi TT — Wiring diagram — No. 32/3

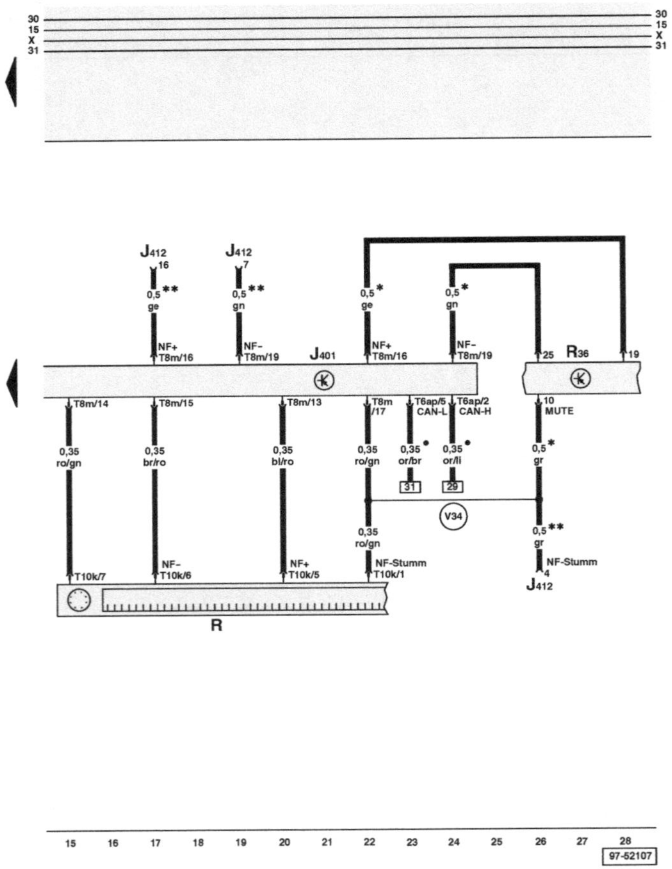

ws = white
sw = black
ro = red
br = brown
gn = green
bl = blue
gr = grey
li = violet
ge = yellow
or = orange

J401 - Control unit for navigation with CD drive
J412 - Operating electronics and Telephone Control Module
R - Radio
R36 - Telephone Transceiver
T6ap - Connector 6 pin, Connector C I on Control Module for Navigation
T8m - Connector 8 pin, C III on Control Module for Navigation
T10k - Connector 10 pin, red, on radio

(V34) - Connector (mute button), in wiring harness telephone

• - CAN-Bus (Data-Bus)
* - Vehicles with Telephone
** - Vehicles with Cellular Telephone Preparation

Edition 10/01
W42.USA.5702.09.21

Control Unit for Navigation with CD Drive, Radio ,Telephone Transceiver

No. 32/4 — Wiring diagram — Audi TT

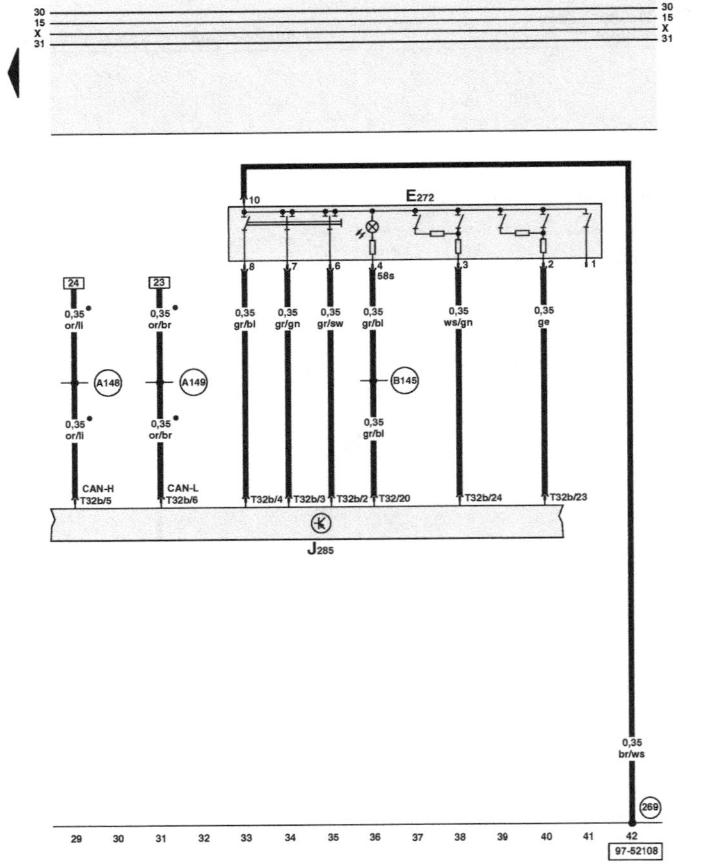

ws = white
sw = black
ro = red
br = brown
gn = green
bl = blue
gr = grey
li = violet
ge = yellow
or = orange

E272 - Function Selector Switch II, in console
J285 - Control Module with indicator unit in instrument panel insert
T32 - Connector 32 pin, blue, on instrument cluster
T32b - Connector 32 pin, gray, on instrument cluster

(269) - Ground connector (sensor ground) -1-, in instrument panel wiring harness
(A148) - Connection (high bus, navigation), in instrument panel wiring harness
(A149) - Connection (low bus, navigation), in instrument panel wiring harness
(B145) - Plus connector (58s), in wiring harness interior

• - CAN-Bus (Data-Bus)

Control Module with Indicator Unit in Instrument Panel Insert, Function Selector Switch II

Edition 10/01
W42.USA.5702.09.21

© 2006 BentleyPublishers.com—All Rights Reserved

Audi TT Wiring Diagram **No. 33/1**

Radio-Stereo Concert with and without CD-Changer (Coupé)

2002 m.y.

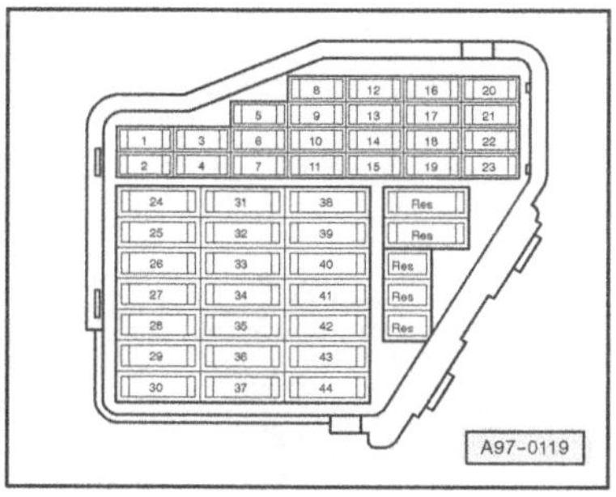

Fuse Panel

Fuse Colors:

30 A - Green
25 A - White
20 A - Yellow
15 A - Blue
10 A - Red
7,5 A - Brown
5 A - Beige
1 A - Black

Starting with fuse position 23, fuses in the fuse holder are identified with 223 in the wiring diagram.

W42.USA.5702.13.21 **Edition 06/03**

Audi TT Wiring Diagram **No. 33/2**

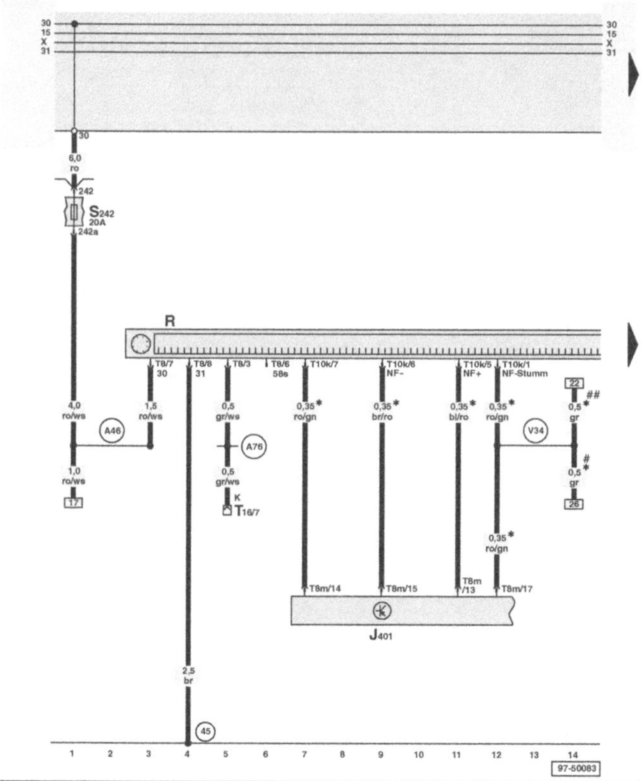

Radio

J401 - Control unit for navigation with CD-mechanism
R - Radio
S242 - Fuse 42 in fuse holder
T6k - Connector 6 pin, green, connector radio I-2
T8 - Connector 8 pin, black, connector radio III
T8m - Connector 8 pin, C III on Control Module for Navigation
T10k - Connector 10 pin, red, on radio
T16 - Connector 16 pin, black, Data Link Connector (DLC), below storage compartment, driver's side

(45) - Ground connection, behind instrument panel, center
(A46) - Plus connection (30-from radio), in instrument panel wiring harness
(A76) - Connector (K-diagnosis wire), in instrument panel wiring harness
(V34) - Connector (mute button), in wiring harness telephone

* - Vehicles with Navigation System
# - Vehicles with Telephone
- Vehicles with Cellular Telephone Preparation

ws = white
sw = black
ro = red
br = brown
gn = green
bl = blue
gr = grey
li = violet
ge = yellow
or = orange

W42.USA.5702.13.21 **Edition 06/03**

©2006 BentleyPublishers.com—All Rights Reserved

Sound System, Concert, coupe (2002-2003)

Audi TT — Wiring Diagram — No. 33/3

Radio, Left Rear Woofer/Amplifier

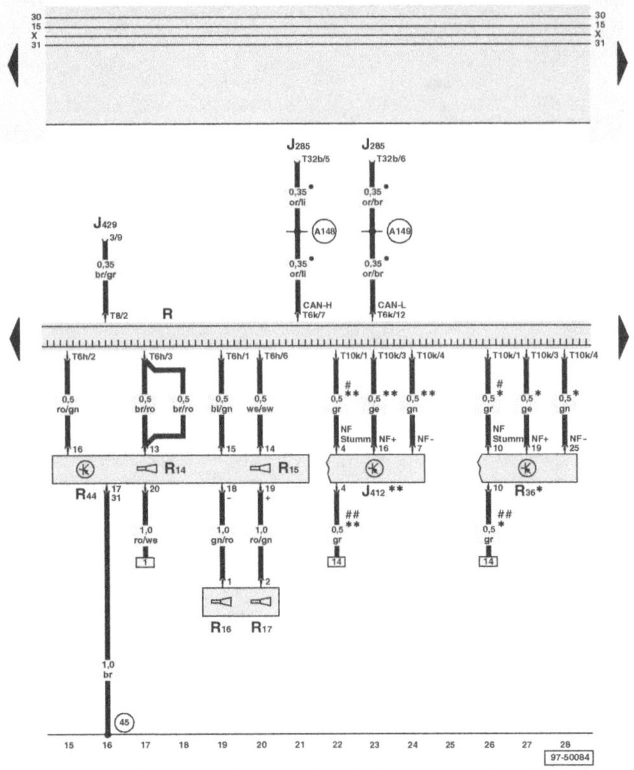

- J285 - Control Module with indicator unit in instrument panel insert
- J412 - Operating electronics and Telephone Control Module
- J429 - Control module for central locking
- R - Radio
- R14 - Treble speaker, left rear
- R15 - Bass speaker, left rear
- R16 - Treble speaker, right rear
- R17 - Bass speaker, right rear
- R36 - Telephone Transceiver
- R44 - Left Rear Woofer/Amplifier
- T6h - Connector 6 pin, yellow, connector radio I-1
- T6k - Connector 6 pin, green, connector radio I-2
- T8 - Connector 8 pin, black, connector radio III
- T10k - Connector 10 pin, red, on radio
- T32b - Connector 32 pin, grey, on instrument cluster
- (45) - Ground connection, behind instrument panel, center
- (A148) - Connection (high bus, navigation), in instrument panel wiring harness
- (A149) - Connection (low bus, navigation), in instrument panel wiring harness
- • - CAN-Bus (Data-Bus)
- * - Vehicles with Telephone
- ** - Vehicles with Cellular Telephone Preparation
- # - Vehicles without Navigation System
- ## - Vehicles with Navigation System

ws = white
sw = black
ro = red
br = brown
gn = green
bl = blue
gr = grey
li = violet
ge = yellow
or = orange

W42.USA.5702.13.21 — Edition 06/03

Audi TT — Wiring Diagram — No. 33/4

Radio, CD Changer Unit, Antenna

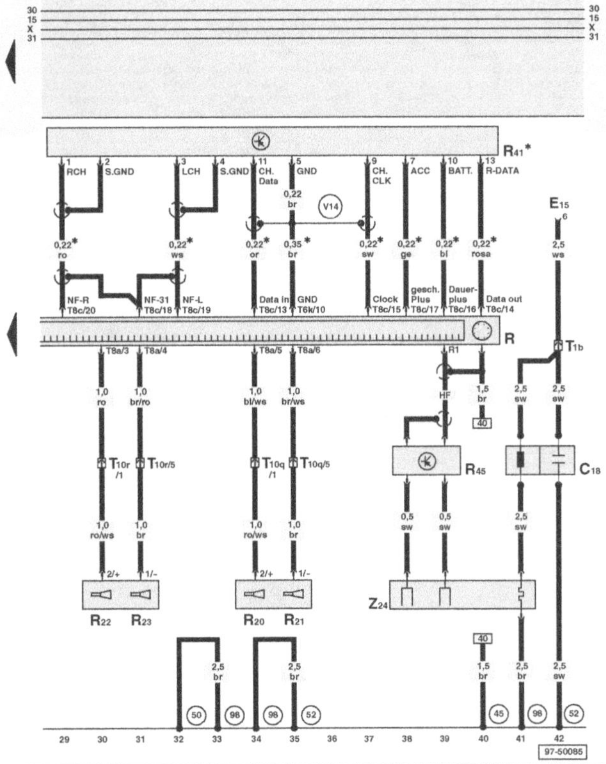

- C18 - Windshield Antenna Suppression Filter
- E15 - Rear Window Defogger Switch
- R - Radio
- R1 - Antenna Connection
- R20 - Treble speaker, left front
- R21 - Bass speaker, left front
- R22 - Treble speaker, right front
- R23 - Bass speaker, right front
- R41 - CD Changer Unit
- R45 - Telephone/Radio Antenna Amplifier
- T1b - Connector single, black, in rear lid left
- T6k - Connector 6 pin, green, connector radio I-2
- T8a - Connector 8 pin, brown, connector radio II
- T8c - Connector 8 pin, blue, connector radio I-3
- T10q - Connector 10 pin, green, connector station A pillar, left
- T10r - Connector 10 pin, green, connector station A pillar, right
- Z24 - Rear Window Defogger With Window Antenna
- (45) - Ground connection, behind instrument panel, center
- (50) - Ground connection, in luggage compartment, left
- (52) - Ground connection, in rear lid, left
- (98) - Ground connection, in rear lid wiring harness
- (V14) - Wire connection (shielding), in CD-changer wiring harness
- * - Vehicles with CD-Changer

ws = white
sw = black
ro = red
br = brown
gn = green
bl = blue
gr = grey
li = violet
ge = yellow
or = orange

W42.USA.5702.13.21 — Edition 06/03

© 2006 BentleyPublishers.com—All Rights Reserved

Audi TT Wiring Diagram **No. 34/1**

Radio Bose with and without CD-Changer (Coupé)

2002 m.y.

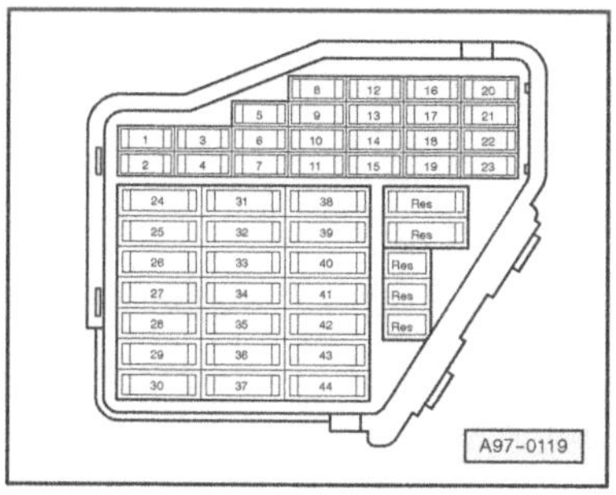

Fuse Panel

Fuse Colors:
30 A - Green
25 A - White
20 A - Yellow
15 A - Blue
10 A - Red
7,5 A - Brown
5 A - Beige
1 A - Black

Starting with fuse position 23, fuses in the fuse holder are identified with 223 in the wiring diagram.

W42.USA.5702.13.21 **Edition 06/03**

Audi TT Wiring Diagram **No. 34/2**

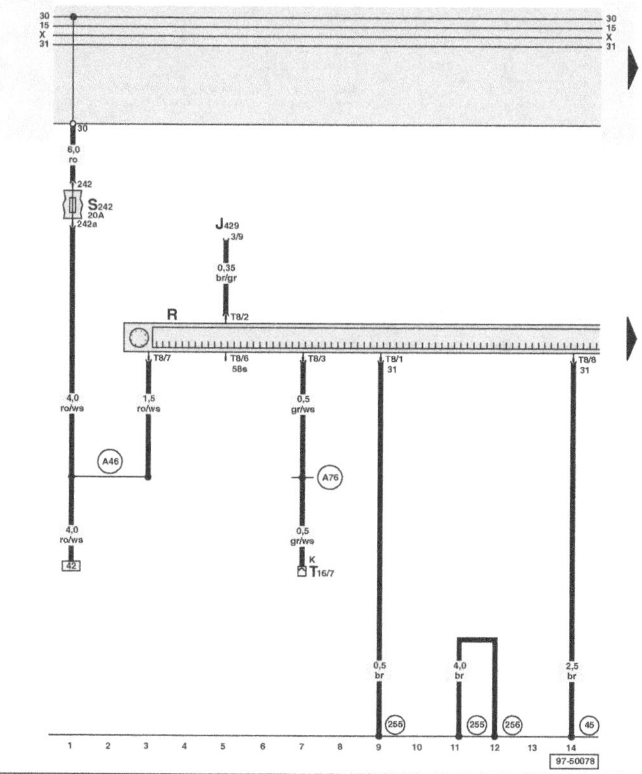

Radio

J429 - Control module for central locking
R - Radio
S242 - Fuse 42 in fuse holder
T8 - Connector 8 pin, black, connector radio III
T16 - Connector 16 pin, black, Data Link Connector (DLC), below storage compartment, driver's side

(45) - Ground connection, behind instrument panel, center
(255) - Ground connection, -1-, in wiring harness Radio
(256) - Ground connection, -2-, in wiring harness Radio
(A46) - Plus connection (30-from radio), in instrument panel wiring harness
(A76) - Connector (K-diagnosis wire), in instrument panel wiring harness

ws = white
sw = black
ro = red
br = brown
gn = green
bl = blue
gr = grey
li = violet
ge = yellow
or = orange

W42.USA.5702.13.21 **Edition 06/03**

©2006 BentleyPublishers.com—All Rights Reserved

Sound System, Bose, coupe (2002)

Audi TT — Wiring Diagram — No. 34/3

Radio, CD Changer Unit

J285 - Control Module with indicator unit in instrument panel insert
R - Radio
R41 - CD Changer Unit
T6k - Connector 6 pin, green, connector radio I-2
T8c - Connector 8 pin, blue, connector radio I-3
T32b - Connector 32 pin, grey, on instrument cluster

(A148) - Connection (high bus, navigation), in instrument panel wiring harness
(A149) - Connection (low bus, navigation), in instrument panel wiring harness
(V14) - Wire connection (shielding), in CD-changer wiring harness

• - CAN-Bus (Data-Bus)
* - Vehicles with CD-Changer

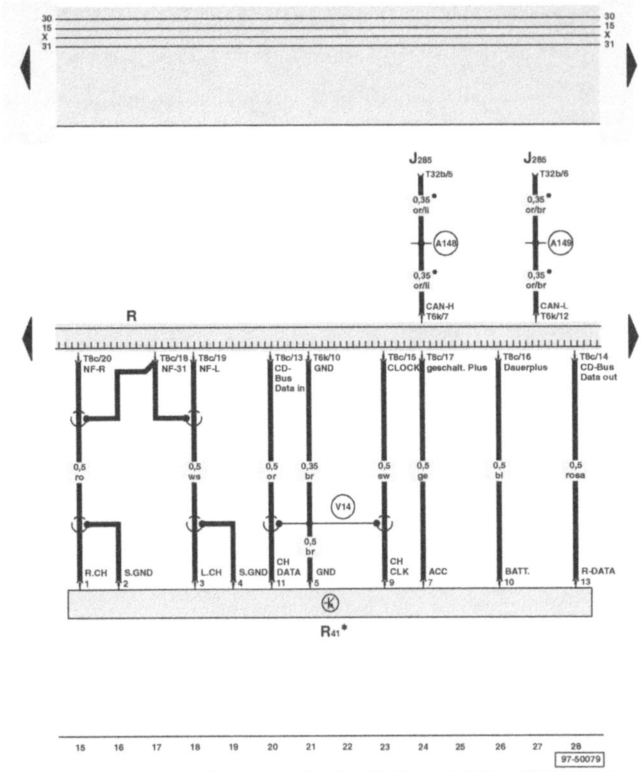

ws = white
sw = black
ro = red
br = brown
gn = green
bl = blue
gr = grey
li = violet
ge = yellow
or = orange

W42.USA.5702.13.21 Edition 06/03

Audi TT — Wiring Diagram — No. 34/4

Radio, Rear Window Defogger with Window Antenna

C18 - Windshield Antenna Suppression Filter
E15 - Rear Window Defogger Switch
J401 - Control unit for navigation with CD-mechanism
J412 - Operating electronics and Telephone Control Module
R - Radio
R1 - Antenna Connection
R12 - Amplifier
R36 - Telephone Transceiver
R45 - Telephone/Radio Antenna Amplifier
T1b - Connector single, black, in rear lid left
T8m - Connector 8 pin, C III on Control Module for Navigation
T10k - Connector 10 pin, red, on radio
Z24 - Rear Window Defogger With Window Antenna

(45) - Ground connection, behind instrument panel, center
(50) - Ground connection, in luggage compartment, left
(52) - Ground connection, in rear lid, left
(98) - Ground connection, in rear lid wiring harness
(256) - Ground connection, -2-, in wiring harness Radio
(V34) - Connector (mute button), in wiring harness telephone

* - Vehicles without navigation system
** - Vehicles with navigation system
*** - Vehicles with telephone
# - Vehicles with Cellular Telephone Preparation

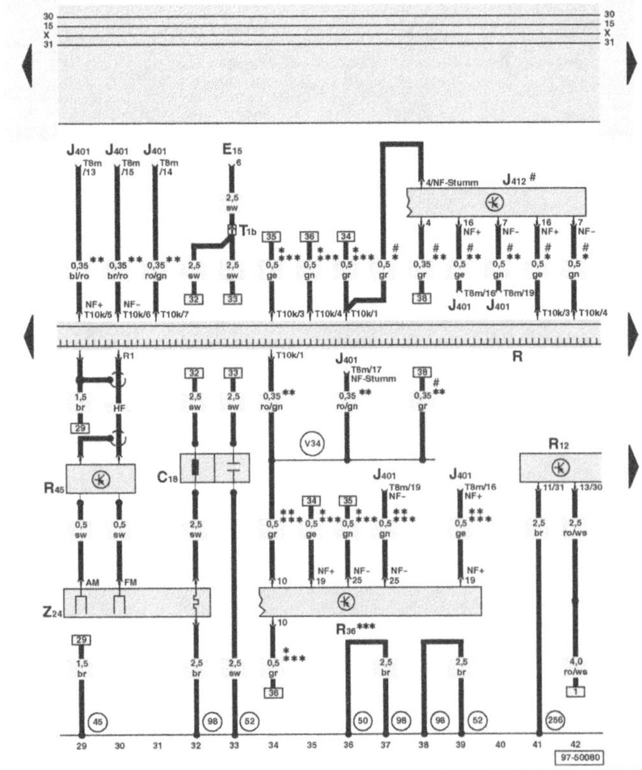

ws = white
sw = black
ro = red
br = brown
gn = green
bl = blue
gr = grey
li = violet
ge = yellow
or = orange

W42.USA.5702.13.21 Edition 06/03

© 2006 BentleyPublishers.com—All Rights Reserved

Audi TT

Wiring Diagram **No. 34/5**

Radio, Amplifier

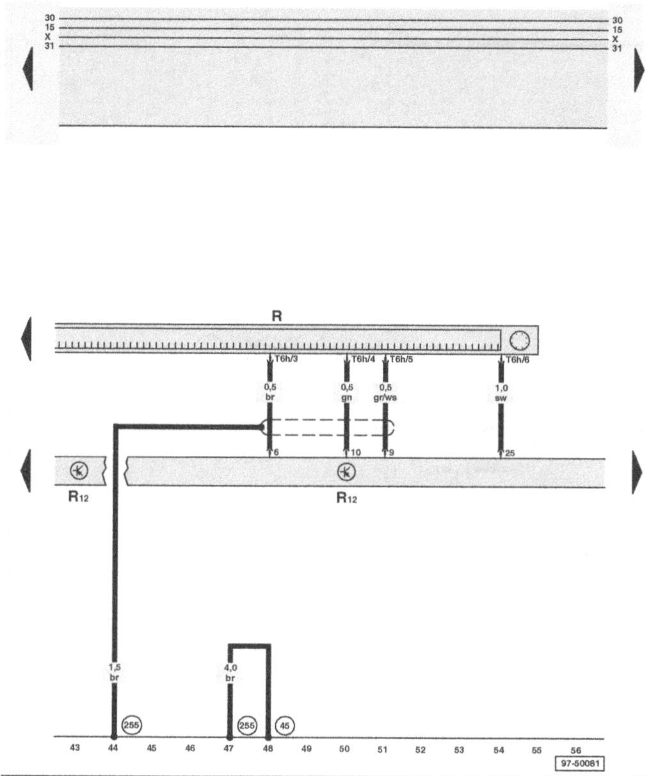

R - Radio
R12 - Amplifier
T6h - Connector 6 pin, yellow, connector radio I-1

(45) - Ground connection, behind instrument panel, center
(255) - Ground connection, -1-, in wiring harness Radio

ws = white
sw = black
ro = red
br = brown
gn = green
bl = blue
gr = grey
li = violet
ge = yellow
or = orange

W42.USA.5702.13.21 **Edition 06/03**

Audi TT

Wiring Diagram **No. 34/6**

Amplifier, Loudspeaker

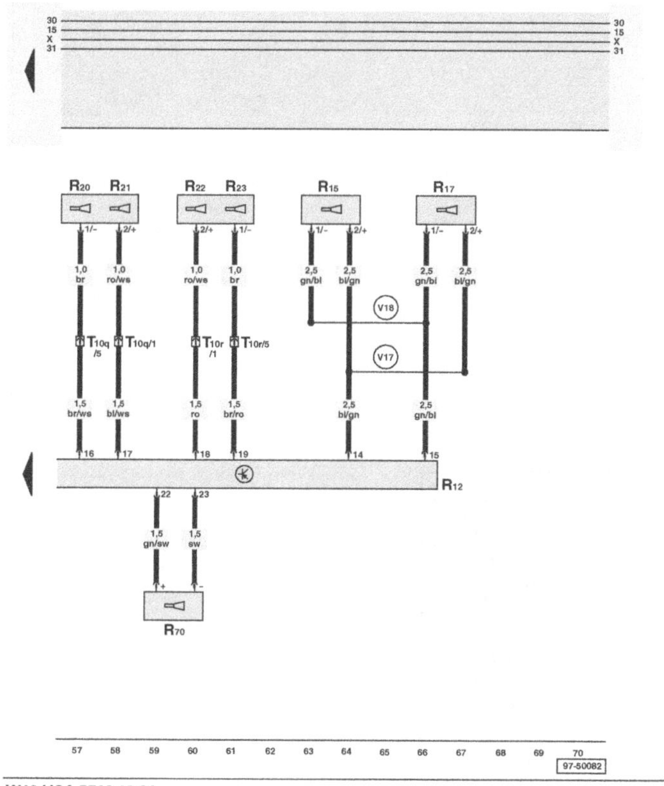

R12 - Amplifier
R15 - Bass speaker, left rear
R17 - Bass speaker, right rear
R20 - Treble speaker, left front
R21 - Bass speaker, left front
R22 - Treble speaker, right front
R23 - Bass speaker, right front
R70 - Inside loudspeaker
T10q - Connector 10 pin, green, connector station A pillar, left
T10r - Connector 10 pin, green, connector station A pillar, right

(V17) - Connector (plus, speaker rear left) in wiring harness speaker
(V18) - Connector (minus, speaker rear left) in wiring harness speaker

ws = white
sw = black
ro = red
br = brown
gn = green
bl = blue
gr = grey
li = violet
ge = yellow
or = orange

W42.USA.5702.13.21 **Edition 06/03**

© 2006 BentleyPublishers.com—All Rights Reserved

Automatic Climate Control

2002 m. y.

Fuse Panel

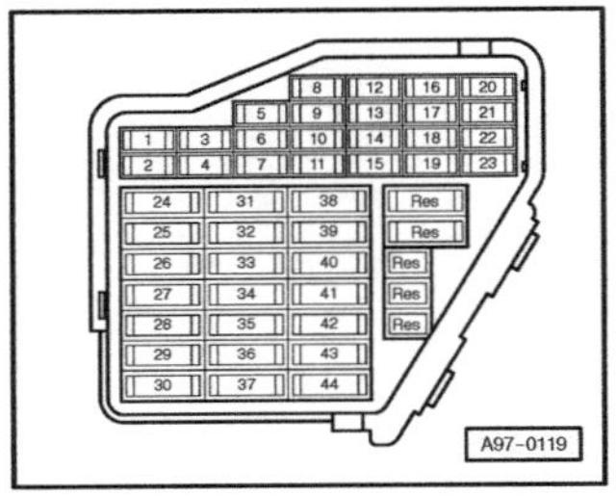

Fuse Colors:

30 A - Green
25 A - White
20 A - Yellow
15 A - Blue
10 A - Red
7,5 A - Brown
5 A - Beige

Starting with fuse position 23, fuses in the fuse holder are identified with 223 in the wiring diagram.

Main Fuse Box/Battery

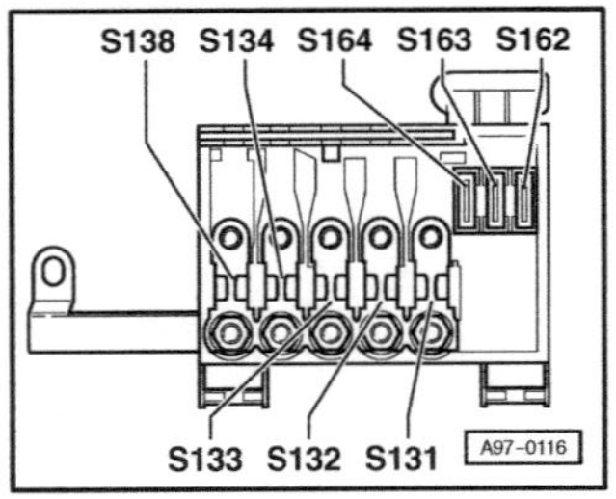

Fuses Location:

S131 - Safety fuse 1
S132 - Safety fuse 2
S133 - Safety fuse 3
S134 - Safety fuse 4
S138 - Safety fuse 5
S162 - Fuse 1 in fuse bracket/battery
S163 - Fuse 2 in fuse bracket/battery
S164 - Fuse 3 in fuse bracket/battery

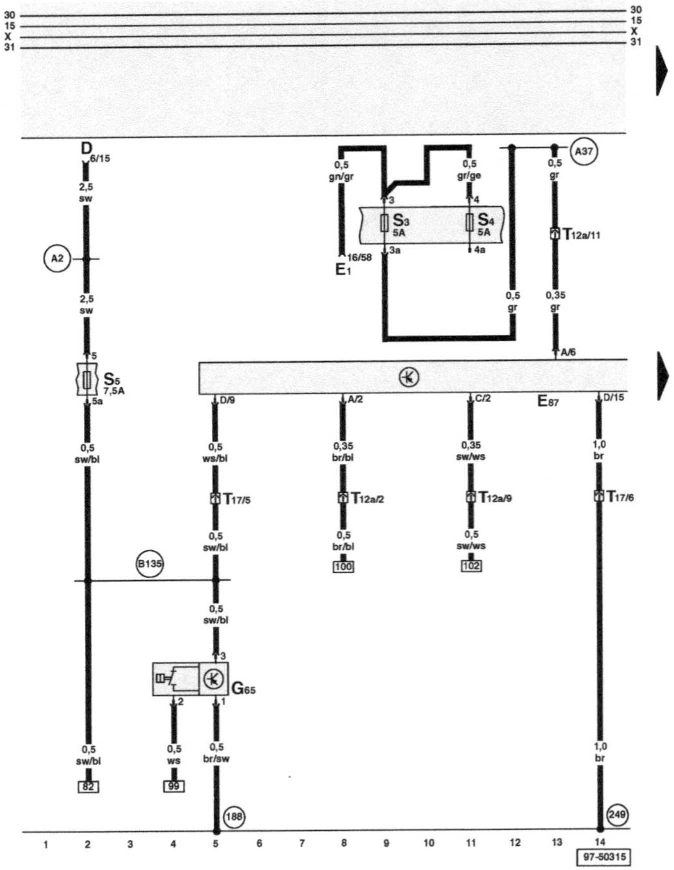

ws = white
sw = black
ro = red
br = brown
gn = green
bl = blue
gr = grey
li = violet
ge = yellow

D - Ignition/Starter Switch
E1 - Light Switch
E87 - A/C Control Head
G65 - High Pressure Sensor
S3 - Fuse
S4 - Fuse
S5 - Fuse
T12a - Connector 12 pin, behind instrument panel, right
T17 - Connector 17 pin, behind instrument panel

(188) - Ground connection -3-, in A/C wiring harness
(249) - Ground connection -2-, in wiring harness interior
(A2) - Plus connection (15), in instrument panel wiring harness
(A37) - Wire connection (58a), in instrument panel wiring harness
(B135) - Connector (15a), in wiring harness interior

© 2006 BentleyPublishers.com—All Rights Reserved

Audi TT — Wiring diagram — No. 35/3

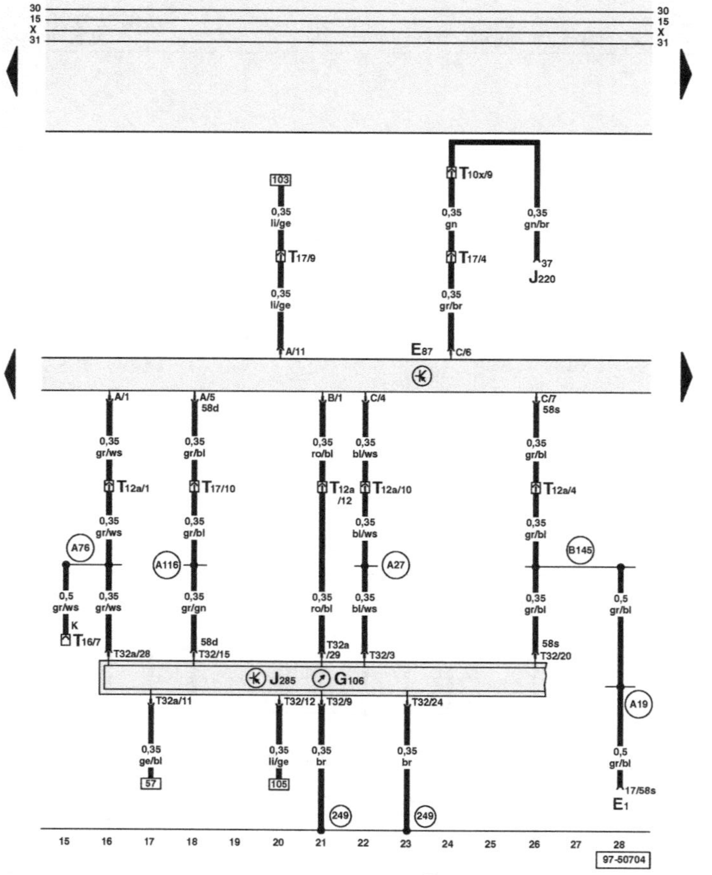

ws = white
sw = black
ro = red
br = brown
gn = green
bl = blue
gr = grey
li = violet
ge = yellow

- E1 - Light Switch
- E87 - A/C Control Head
- G106 - Outside Air Temperature Display
- J220 - Motronic Engine Control Module (ECM)
- J285 - Control Module with indicator unit in instrument panel insert
- T10x - Connector 10 pin, orange, in E-box plenum chamber
- T12a - Connector 12 pin, behind instrument panel, right
- T16 - Connector 16 pin, black, Data Link Connector (DLC), below storage compartment, driver's side
- T17 - Connector 17 pin, behind instrument panel
- T32 - Connector 32 pin, blue, on instrument cluster
- T32a - Connector 32 pin, green, on instrument cluster

- (249) - Ground connection -2-, in wiring harness interior
- (A19) - Wire connection (58s), in instrument panel wiring harness
- (A27) - Wire connection (vehicle speed signal), in instrument panel wiring harness
- (A76) - Connector (K-diagnosis wire), in instrument panel wiring harness
- (A116) - Connection -2- (58d) in instrument panel wiring harness
- (B145) - Plus connector (58s), in wiring harness interior

Edition 10/02
W42.USA.5702.11.21

A/C Control Head, Control Module with Indicator Unit in Instrument Panel Insert

No. 35/4 — Wiring diagram — Audi TT

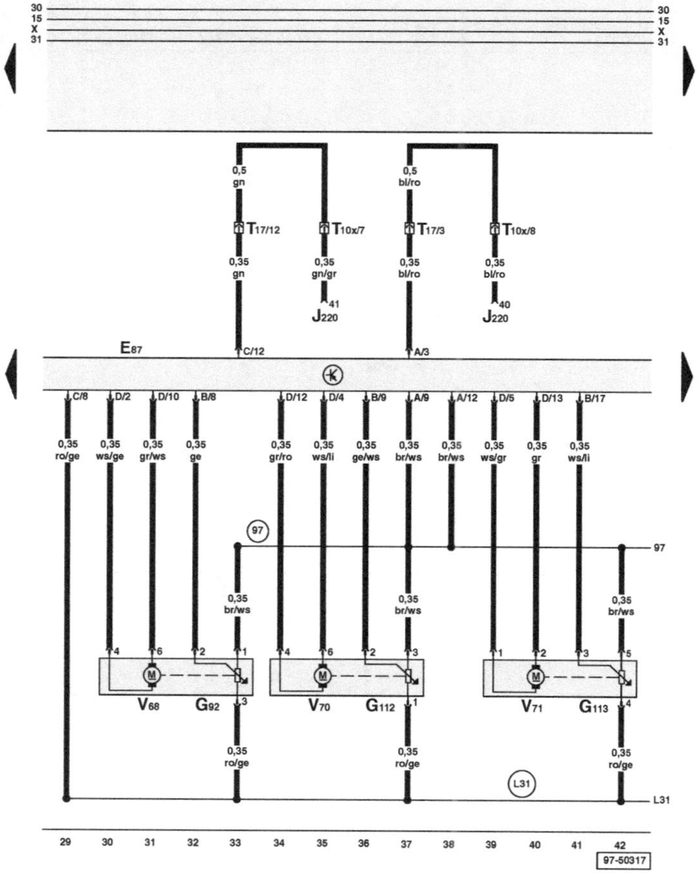

ws = white
sw = black
ro = red
br = brown
gn = green
bl = blue
gr = grey
li = violet
ge = yellow

- E87 - A/C Control Head
- G92 - Temperature Regulator Flap Motor Position Sensor
- G112 - Central Flap Motor Position Sensor
- G113 - Back Pressure Flap Motor Position Sensor
- J220 - Motronic Engine Control Module (ECM)
- T10x - Connector 10 pin, orange, in E-box plenum chamber
- T17 - Connector 17 pin, behind instrument panel
- V68 - Temperature Regulator Flap Motor
- V70 - Central Air Flap Motor
- V71 - Air Flow Flap Motor

- (97) - Ground connection -1-, in A/C wiring harness
- (L31) - Wire connection (5 Volts), in A/C wiring harness

A/C Control Head, Flap Motors

Edition 10/02
W42.USA.5702.11.21

©2006 BentleyPublishers.com—All Rights Reserved

Audi TT — Wiring diagram — No. 35/5

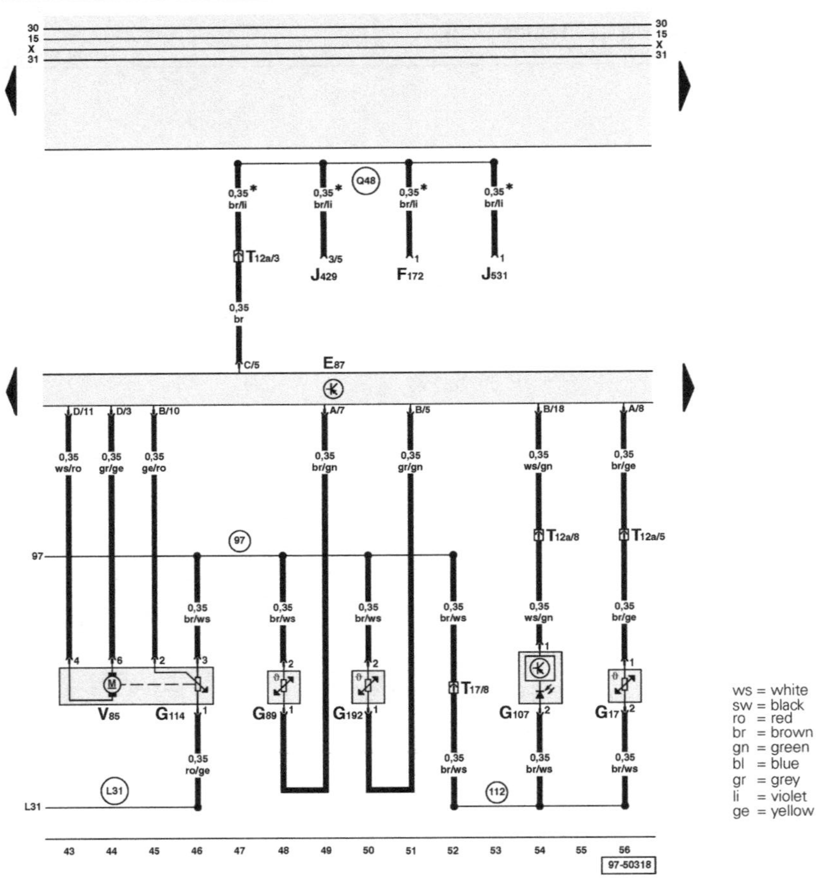

ws = white
sw = black
ro = red
br = brown
gn = green
bl = blue
gr = grey
li = violet
ge = yellow

E87 - A/C Control Head
F172 - Convertible Top Front Lock Switch
G17 - Outside Air Temperature Sensor
G89 - Fresh Air Intake Duct Temperature Sensor
G107 - Sunlight Photo Sensor (3241)
G114 - Footwell/Defrost Flap Motor Position Sensor
G192 - Sender for outlet temperature, floor outlet
J429 - Control module for central locking
J531 - Wind Deflector Control Module (Cabrio)
T12a - Connector 12 pin, behind instrument panel, right
T17 - Connector 17 pin, behind instrument panel
V85 - Footwell/Defroster Flap Motor

(97) - Ground connection -1-, in A/C wiring harness

(112) - Ground connection -2-, in A/C wiring harness

(L31) - Wire connection (5 Volts), in A/C wiring harness

(Q48) - Connection (open), in wiring harness interior

* - only Roadster

A/C Control Head, Flap Motor, Outside Air Temperature Sensor, Sunlight Photo Sensor

No. 35/6 — Wiring diagram — Audi TT

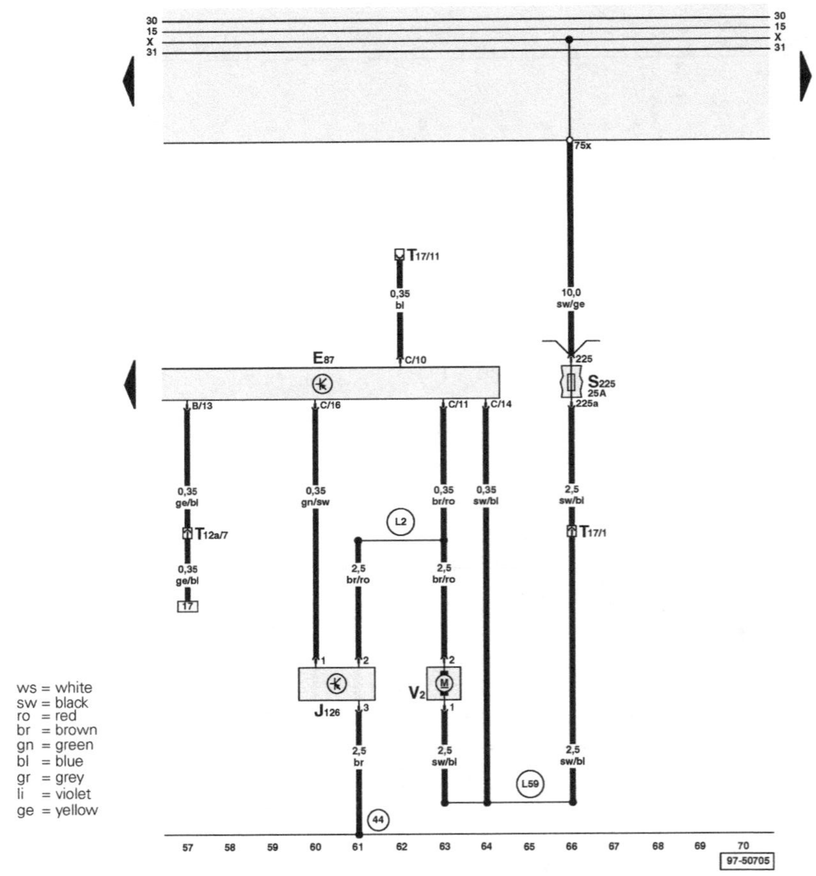

ws = white
sw = black
ro = red
br = brown
gn = green
bl = blue
gr = grey
li = violet
ge = yellow

E87 - A/C Control Head
J126 - Control module for fresh air blower
S225 - Fuse in fuse holder
T12a - Connector 12 pin, behind instrument panel, right
T17 - Connector 17 pin, behind instrument panel
V2 - Fresh Air Blower

(44) - Ground connection, on left A-pillar, lower part

(L2) - Wire connection, in A/C wiring harness

(L59) - Plus connection (30), in wiring harness heater

A/C Control Head, Control Module for Fresh Air Blower, Fresh Air Blower

© 2006 BentleyPublishers.com—All Rights Reserved

Audi TT **Wiring diagram** **No. 35/7**

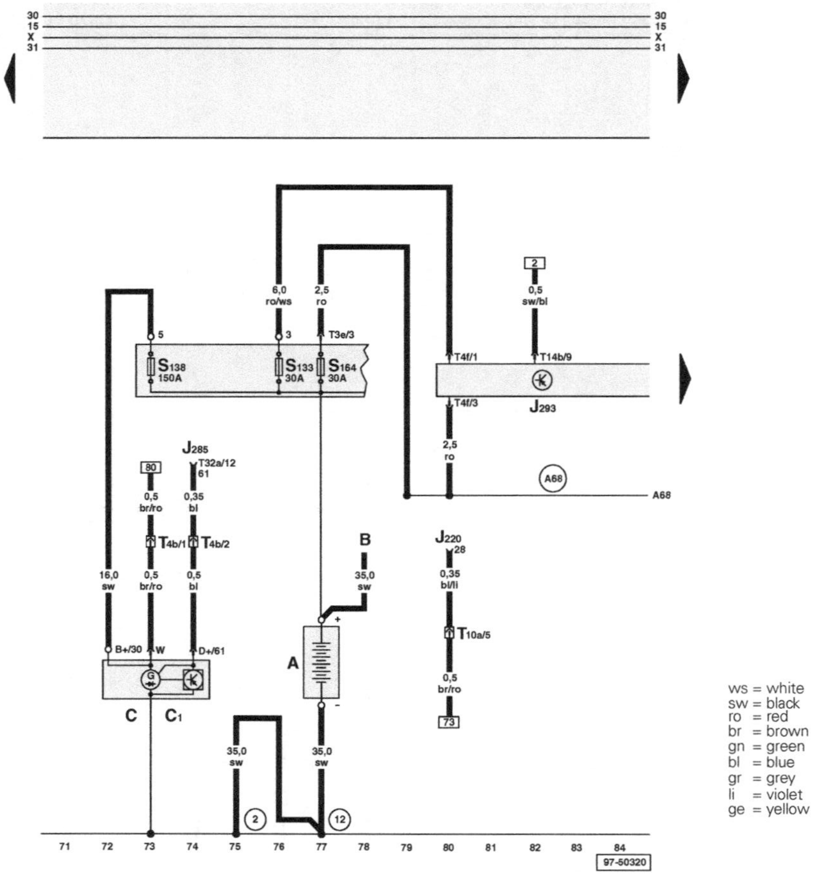

ws = white
sw = black
ro = red
br = brown
gn = green
bl = blue
gr = grey
li = violet
ge = yellow

A - Battery
B - Starter
C - Generator (GEN)
C1 - Voltage Regulator (VR)
J220 - Motronic Engine Control Module (ECM)
J285 - Control Module with indicator unit in instrument panel insert
J293 - Coolant FC (Fan Control) Control Module, in engine compartment, left front
S133 - Safety fuse 3
S138 - Safety fuse 5
S164 - Fuse -3- in fuse bracket/battery
T3e - Connector 3 pin, black, on Main Fuse Box/Battery
T4b - Connector 4 pin, in engine compartment, left
T4f - Connector 4 pin, connector B on J293
T10a - Connector 10 pin, white, in E-box plenum chamber
T14b - Connector 14 pin, connector A on J293
T32a - Connector 32 pin, green, on instrument cluster

(2) - Ground strap, transmission to body
(12) - Ground connection, in engine compartment, left
(A68) - Connector (30, A/C), in instrument panel wiring harness

Edition 10/02
W42.USA.5702.11.21

Coolant FC Control Module in Engine Compartment left front, Fuses

No. 35/8 **Wiring diagram** **Audi TT**

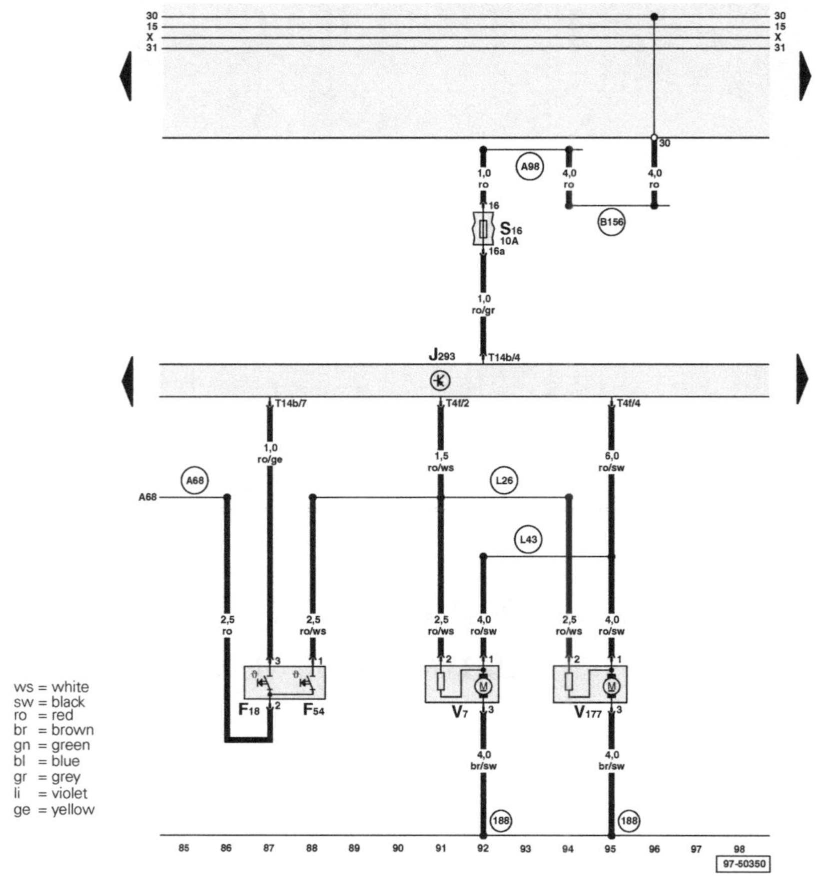

ws = white
sw = black
ro = red
br = brown
gn = green
bl = blue
gr = grey
li = violet
ge = yellow

F18 - Coolant Fan Control (FC) Thermal Switch
F54 - Coolant Fan Control (FC) Thermal Switch
J293 - Coolant FC (Fan Control) Control Module, in engine compartment, left front
S16 - Fuse
T4f - Connector 4 pin, connector B on J293
T14b - Connector 14 pin, connector A on J293
V7 - Coolant Fan
V177 - Coolant Fan -2-

(188) - Ground connection -3-, in A/C wiring harness
(A68) - Connector (30, A/C), in instrument panel wiring harness
(A98) - Plus connector -4- (30), in instrument panel wiring harness
(B156) - Plus connector -4- (30), in wiring harness interior
(L26) - Wire connection (coolant fan), in A/C-climatronic/engine compartment wiring harness
(L43) - Wire connection (coolant fan), in A/C system-coolant fan wiring harness

Coolant FC Control Module in Engine Compartment left front, Coolant Fan Control Thermal Switch, Coolant Fan

Edition 10/02
W42.USA.5702.11.21

© 2006 BentleyPublishers.com—All Rights Reserved

Audi TT — **Wiring diagram** — **No. 35/9**

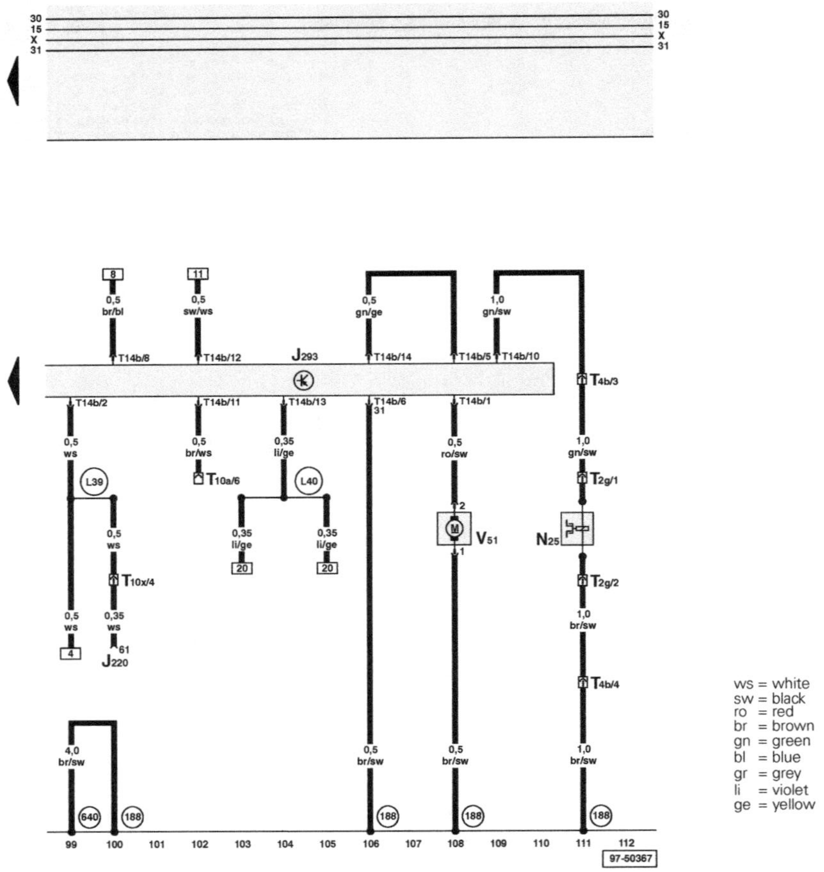

ws = white
sw = black
ro = red
br = brown
gn = green
bl = blue
gr = grey
li = violet
ge = yellow

- J220 - Motronic Engine Control Module (ECM)
- J293 - Coolant FC (Fan Control) Control Module, in engine compartment, left front
- N25 - A/C Clutch
- T2g - Connector 2 pin, near compressor
- T4b - Connector 4 pin, in engine compartment, left
- T10a - Connector 10 pin, white, in E-box plenum chamber
- T10x - Connector 10 pin, orange, in E-box plenum chamber
- T14b - Connector 14 pin, connector A on J293
- V51 - After-Run Coolant Pump

- (188) - Ground connection -3-, in A/C wiring harness
- (640) - Ground connection -2-, in engine compartment, left
- (L39) - Wire connection -1- (A/C Pressure Switch), in Climatronic wiring harness
- (L40) - Wire connection -2- (A/C Pressure Switch), in Climatronic wiring harness

Edition 07/01
W42.USA.5702.08.21

Coolant FC Control Module in Engine Compartment left front, A/C Clutch, After-Run Coolant Pump

THIS SPACE INTENTIONALLY LEFT BLANK

© 2006 BentleyPublishers.com—All Rights Reserved

Haldex Coupling (02D) for All-Wheel Drive

2002 m. y.

Fuse Panel

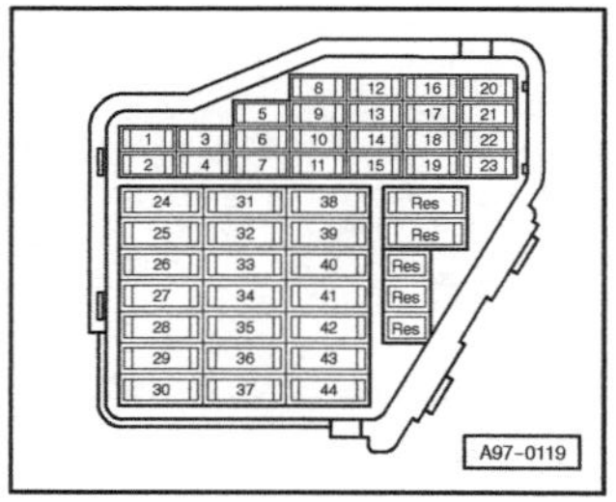

Fuse Colors:

30 A - Green
25 A - White
20 A - Yellow
15 A - Blue
10 A - Red
7,5 A - Brown
5 A - Beige

Starting with fuse position 23, fuses in the fuse holder are identified with 223 in the wiring diagram.

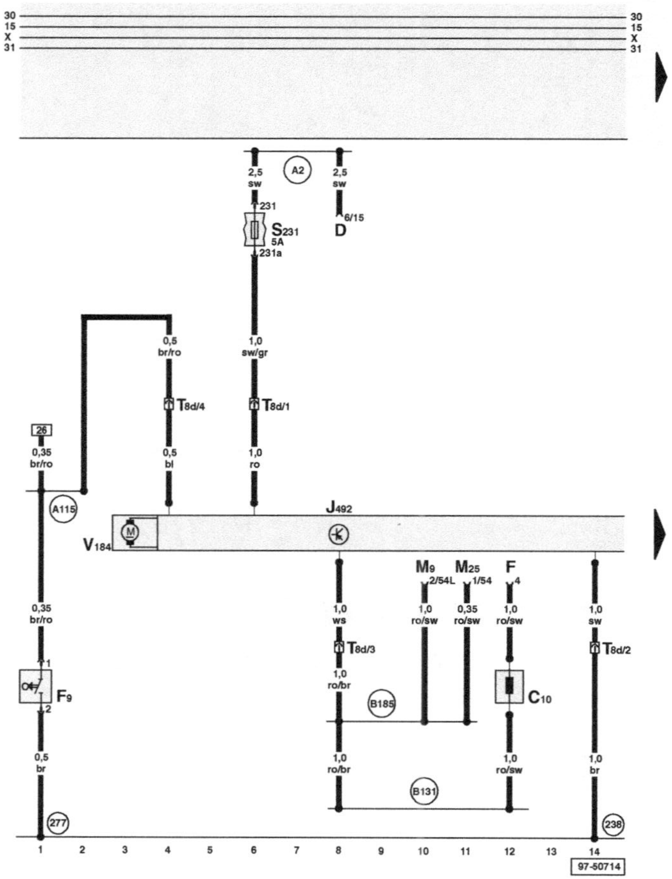

ws = white
sw = black
ro = red
br = brown
gn = green
bl = blue
gr = grey
li = violet
ge = yellow

C10 - Suppressor Filter
D - Ignition/Starter Switch
F - Brake Light Switch
F9 - Parking Brake Warning Light Switch
J492 - Control module for all wheel drive
M9 - Left Brake Light
M25 - High-mount Brake Light
S231 - Fuse 31 in fuse holder
T8d - Connector 8 pin, on Control module for all wheel drive
V184 - Positioning motor for Oil pressure

(238) - Ground connection -1-, in wiring harness interior
(277) - Ground connector -3-, in wiring harness interior
(A2) - Plus connection (15), in instrument panel wiring harness
(A115) - Connection (parking brake control) in instrument panel wiring harness
(B131) - Connector (54), in wiring harness interior
(B185) - Connection 2 (54) in passenger compartment wiring harness

Control Module for All Wheel Drive, Positioning Motor for Oil Pressure

© 2006 BentleyPublishers.com—All Rights Reserved

Audi TT **Wiring diagram** **No. 36/3**

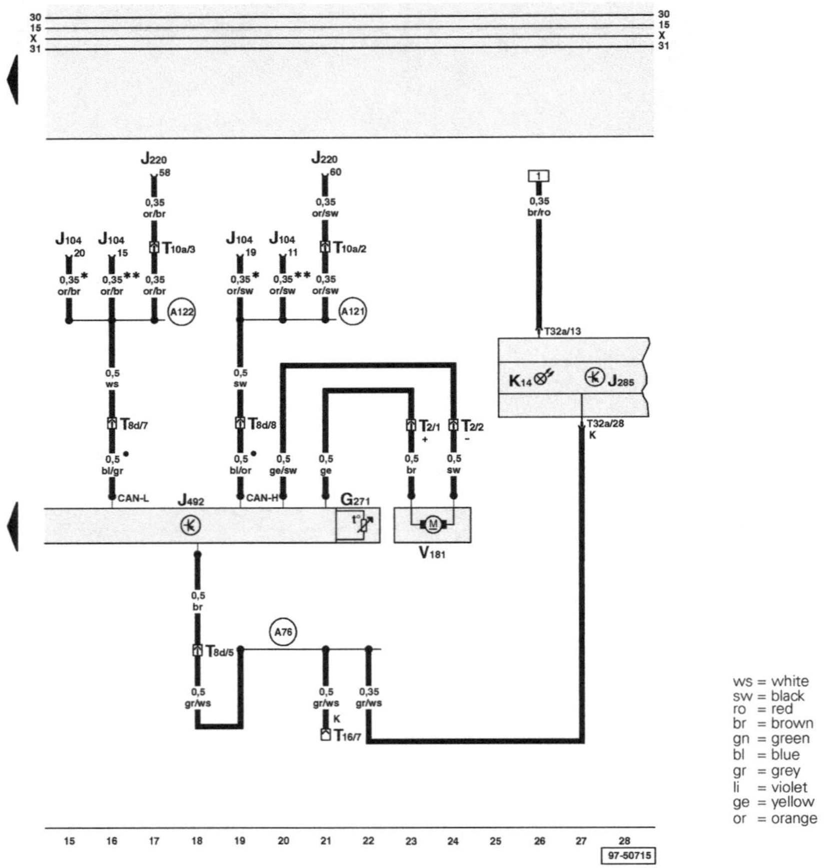

ws = white
sw = black
ro = red
br = brown
gn = green
bl = blue
gr = grey
li = violet
ge = yellow
or = orange

G271 - Hydraulic Temperature Sensor
J104 - ABS Control Module (w/EDL)
J220 - Motronic Engine Control Module (ECM)
J285 - Control Module with indicator unit in instrument panel insert
J492 - Control module for all wheel drive
K14 - Parking Brake Indicator Light
T2 - Connector 2 pin, on all-wheel drive
T8d - Connector 8 pin, on Control module for all wheel drive
T10a - Connector 10 pin, white, in E-box plenum chamber
T16 - Connector 16 pin, black, Data Link Connector (DLC), below storage compartment, driver's side
T25 - Connector 25 pin, on ABS Control Module (w/EDL)
T32a - Connector 32 pin, green, on instrument cluster
V181 - Pump for Haldex Coupling

(A76) - Connector (K-diagnosis wire), in instrument panel wiring harness
(A121) - Connection (high bus) in instrument panel wiring harness
(A122) - Connection (low bus) in instrument panel wiring harness

• - CAN-Bus (Data-Bus)
* - ABS MK20 with EDL/ASR/ESP
** - ABS MK 60 wit EDL/ASR/ESP

Edition 10/01
W42.USA.5702.09.21

Control Module for All Wheel Drive, Pump for Haldex Coupling

THIS SPACE INTENTIONALLY LEFT BLANK

© 2006 BentleyPublishers.com—All Rights Reserved

Telephone

2002 m. y.

Fuse Panel

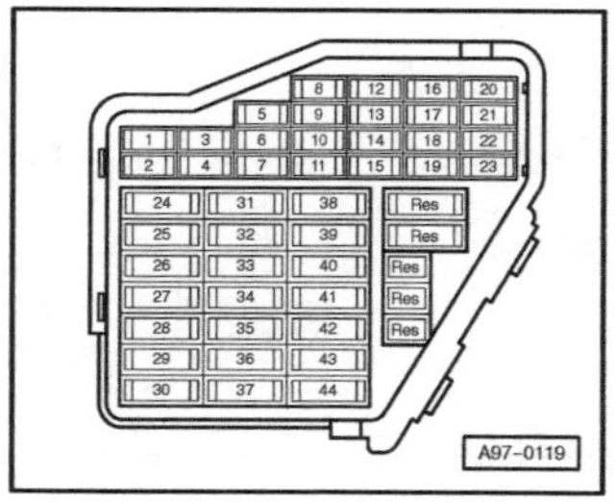

Fuse Colors:

30 A - Green
25 A - White
20 A - Yellow
15 A - Blue
10 A - Red
7,5 A - Brown
5 A - Beige

Starting with fuse position 23, fuses in the fuse holder are identified with 223 in the wiring diagram.

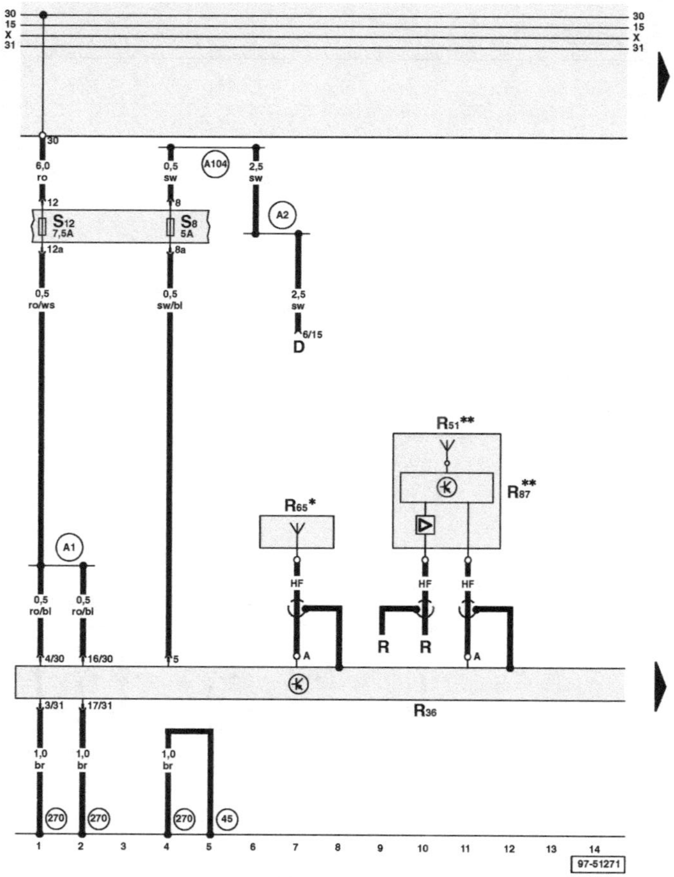

ws = white
sw = black
ro = red
br = brown
gn = green
bl = blue
gr = grey
li = violet
ge = yellow

D - Ignition/Starter Switch
R - Radio
R36 - Telephone Transceiver
R51 - Antenna for radio/telephone
R65 - Antenna for telephone
R87 - Antenna Splitter
S8 - Fuse
S12 - Fuse

(45) - Ground connection, behind instrument panel, center
(270) - Ground connector, in wiring harness telephone
(A1) - Plus connection (30a), in instrument panel wiring harness
(A2) - Plus connection (15), in instrument panel wiring harness
(A104) - Plus connector -2- (15), in instrument panel wiring harness

** - Coupé, roof antenna*
*** - Roadster, rod antenna*

Telephone Transceiver, Antenna for Radio/Telephone

©2006 BentleyPublishers.com—All Rights Reserved

Audi TT — Wiring diagram — No. 37/3

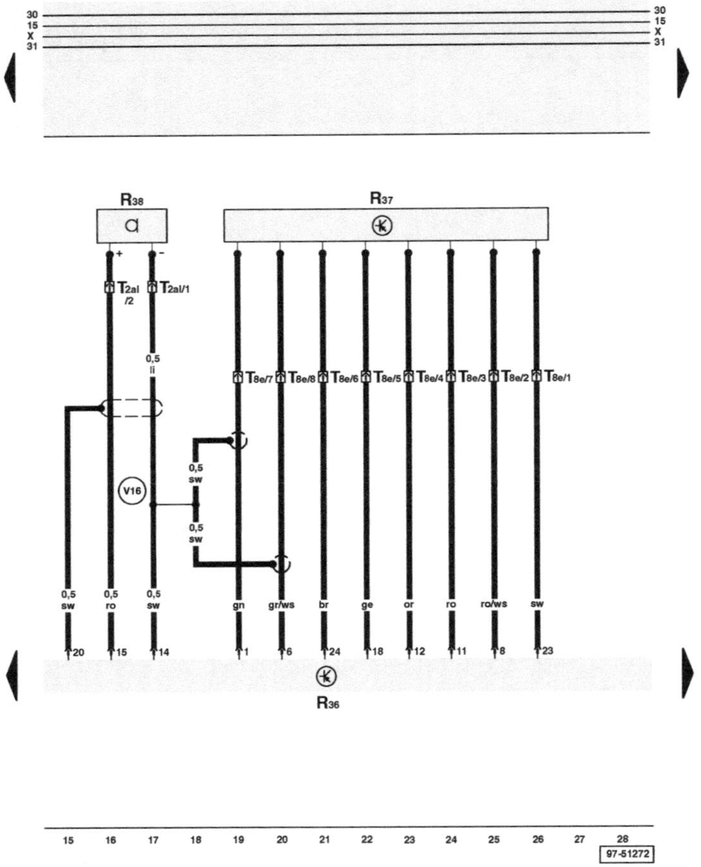

ws = white
sw = black
ro = red
br = brown
gn = green
bl = blue
gr = grey
li = violet
ge = yellow
or = orange

R36 - Telephone Transceiver
R37 - Telephone Handset
R38 - Telephone Microphone
T2al - Connector 2 pin, behind instrument cluster
T8e - Connector 8 pin, on Telephone Handset

(V16) - Wire connection (shielding), in telephone wiring harness

Edition 10/01
W42.USA.5702.09.21

Telephone Transceiver, Telephone Handset, Telephone Microphone

No. 37/4 — Wiring diagram — Audi TT

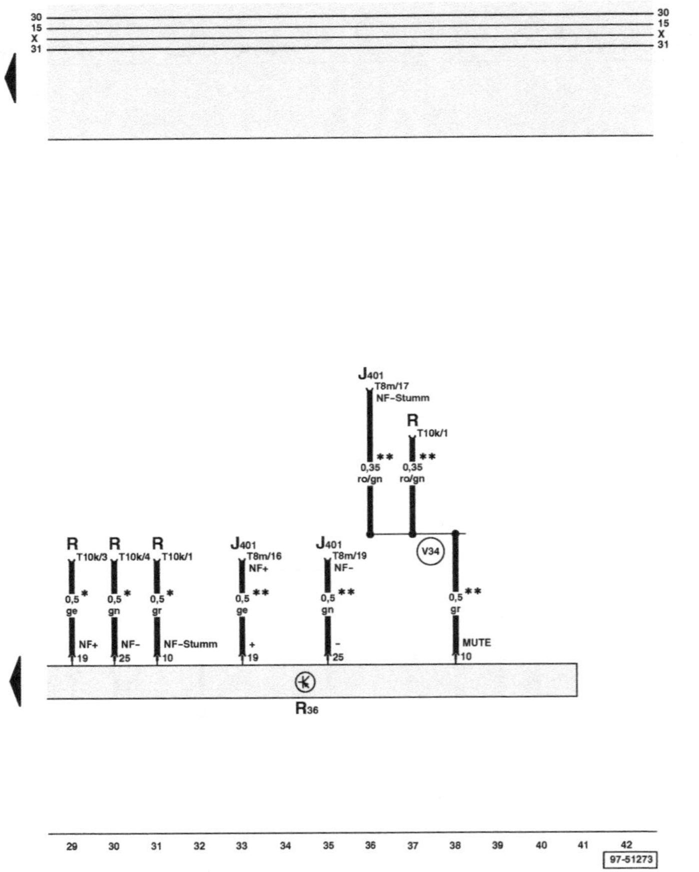

ws = white
sw = black
ro = red
br = brown
gn = green
bl = blue
gr = grey
li = violet
ge = yellow

J401 - Control unit for navigation with CD drive
R - Radio
R36 - Telephone Transceiver
T8m - Connector 8 pin, C III on Control Module for Navigation
T10k - Connector 10 pin, red, on radio

(V34) - Connector (mute button), in wiring harness telephone

* - Vehicles with Radio Concert
** - Vehicles with navigation

Telephone Transceiver

Edition 10/01
W42.USA.5702.09.21

© 2006 BentleyPublishers.com—All Rights Reserved

Anti-Lock Brake System (ABS) MK60 with Electronic Differential Lock (EDL), Anti-Slip Regulation (ASR - ASC) and Electronic Stabilization Program (ESP)

2002 m. y.

Fuse Panel

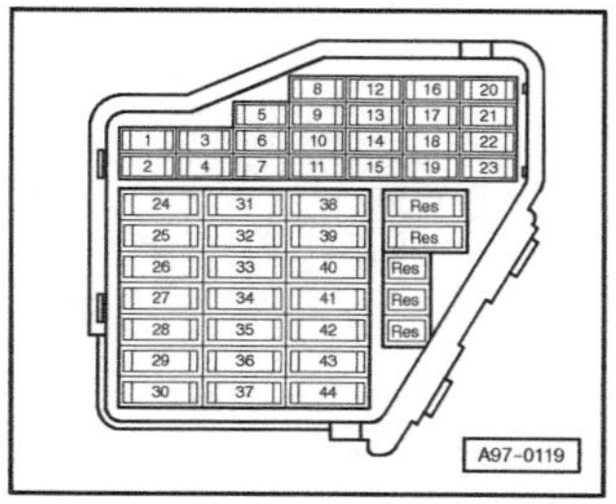

Fuse Colors:

30	A	-	Green
25	A	-	White
20	A	-	Yellow
15	A	-	Blue
10	A	-	Red
7,5	A	-	Brown
5	A	-	Beige

Starting with fuse position 23, fuses in the fuse holder are identified with 223 in the wiring diagram.

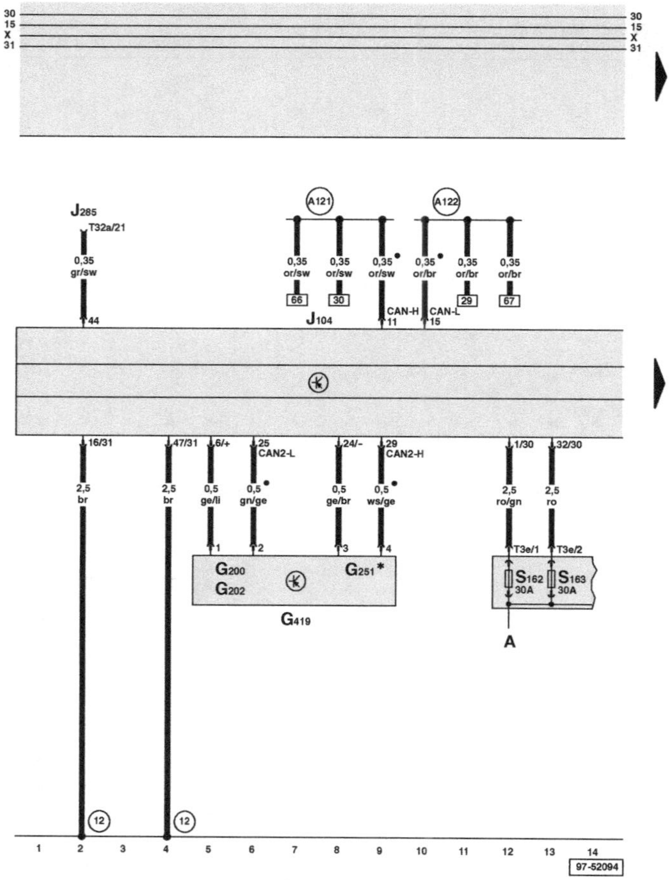

ws = white
sw = black
ro = red
br = brown
gn = green
bl = blue
gr = grey
li = violet
ge = yellow
or = orange

A - Battery
G200 - Sensor for transverse acceleration
G202 - Sender for rotation rate
G251 - Longitudinal acceleration sensor
G419 - ESP - Sensor Unit
J104 - ABS Control Module (w/EDL)
J285 - Control Module with indicator unit in instrument panel insert
S162 - Fuse -1- in fuse bracket/battery
S163 - Fuse -2- in fuse bracket/battery
T3e - Connector 3 pin, black, on Main Fuse Box/Battery
T32a - Connector 32 pin, green, on instrument cluster

(12) - Ground connection, in engine compartment, left
(A121) - Connection (high bus) in instrument panel wiring harness
(A122) - Connection (low bus) in instrument panel wiring harness

• - CAN-Bus (Data-Bus)
* - only Quattro

© 2006 BentleyPublishers.com—All Rights Reserved

Audi TT | Wiring diagram | No. 38/3

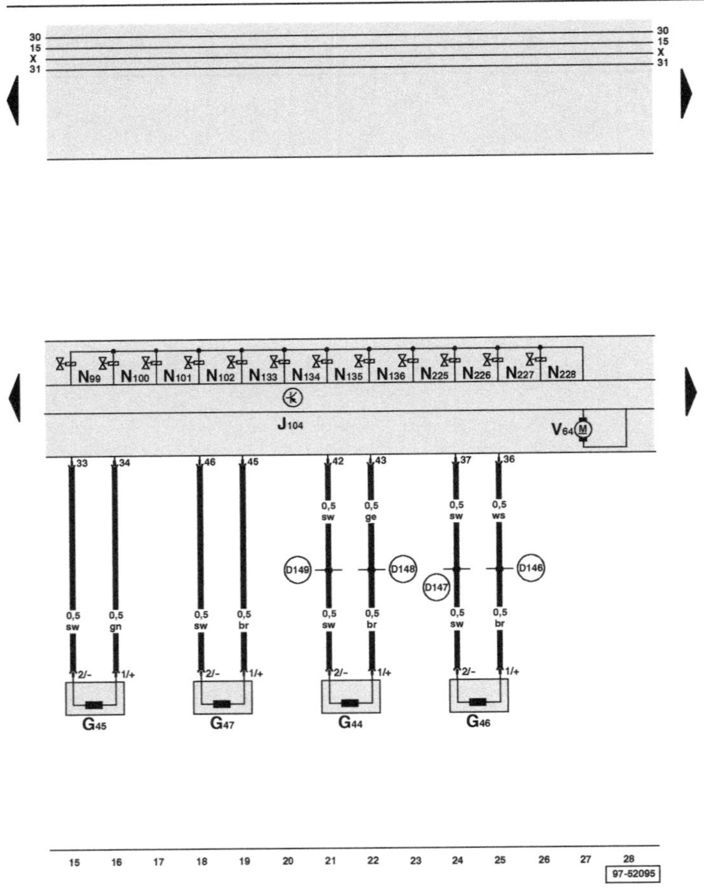

ws = white
sw = black
ro = red
br = brown
gn = green
bl = blue
gr = grey
li = violet
ge = yellow

- G44 - Right Rear ABS Wheel Speed Sensor
- G45 - Right Front ABS Wheel Speed Sensor
- G46 - Left Rear ABS Wheel Speed Sensor
- G47 - Left Front ABS Wheel Speed Sensor
- J104 - ABS Control Module (w/EDL)
- N99 - Right Front ABS Inlet Valve
- N100 - Right Front ABS Outlet Valve
- N101 - Left Front ABS Inlet Valve
- N102 - Left Front ABS Outlet Valve
- N133 - Right Rear ABS Inlet Valve
- N134 - Left Rear ABS Inlet Valve
- N135 - Right Rear ABS Outlet Valve
- N136 - Left Rear ABS Outlet Valve
- N225 - Pilot valve -1- traction control (01418)
- N226 - Pilot valve -2- traction control (01419)
- N227 - High pressure switch valve -1- traction control (01420)
- N228 - High pressure switch valve -2- traction control (01421)
- V64 - ABS Hydraulic Pump

- (D146) - Connector (speed sensor rear, left +), in wiring harness engine compartment
- (D147) - Connector (speed sensor rear, left -), in wiring harness engine compartment
- (D148) - Connector (speed sensor rear, right +), in wiring harness engine compartment
- (D149) - Connector (speed sensor rear, right -), in wiring harness engine compartment

Edition 11/01
W42.USA.5702.10.21

ABS Control Module, Wheel Speed Sensors, Valves for ABS, ABS Hydraulic Pump

No. 38/4 | Wiring diagram | Audi TT

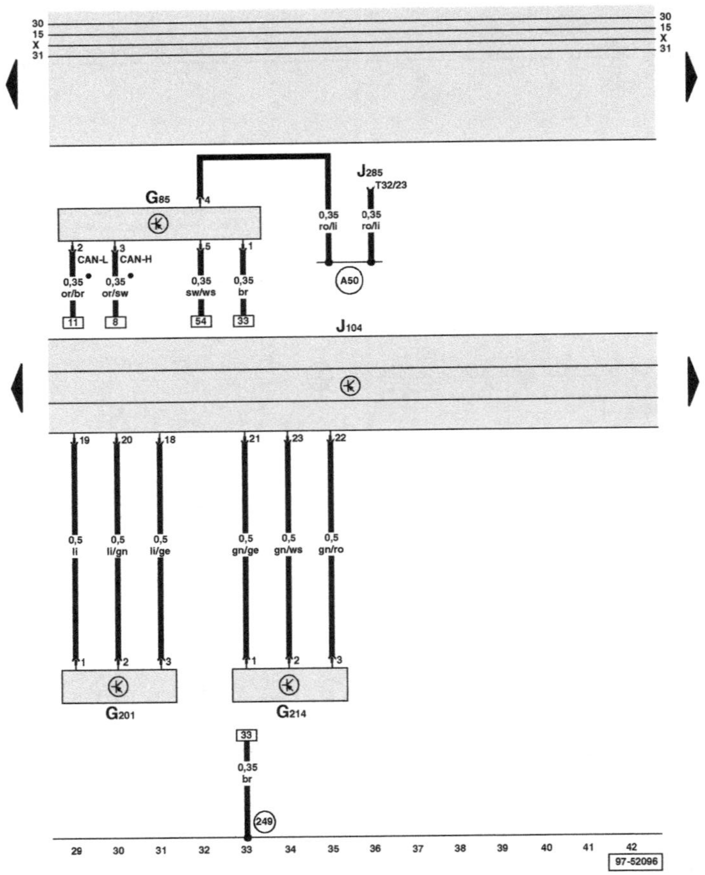

ws = white
sw = black
ro = red
br = brown
gn = green
bl = blue
gr = grey
li = violet
ge = yellow

- G85 - Steering Angle Sensor
- G201 - Sender 1 for brake booster
- G214 - Sensor -2- for brake pressure
- J104 - ABS Control Module (w/EDL)
- J285 - Control Module with indicator unit in instrument panel insert
- T32 - Connector 32 pin, blue, on instrument cluster

- (249) - Ground connection -2-, in wiring harness interior
- (A50) - Plus connection (30a), in instrument panel wiring harness
- • - CAN-Bus (Data-Bus)

ABS Control Module, Steering Angle Sensor, Sensors for Brake Pressure

Edition 11/01
W42.USA.5702.10.21

© 2006 BentleyPublishers.com—All Rights Reserved

Audi TT **Wiring diagram** **No. 38/5**

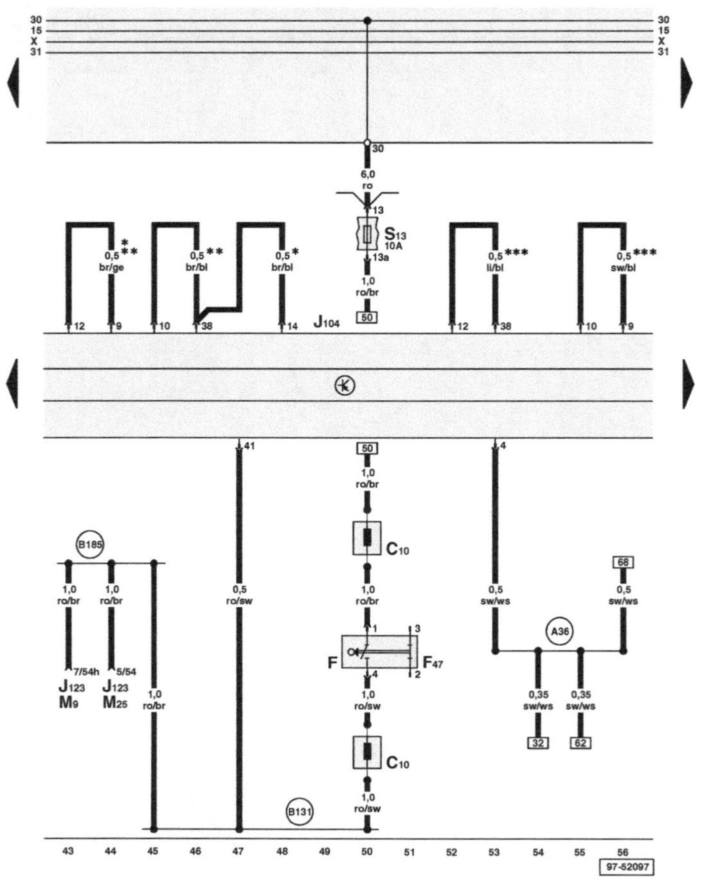

ws = white
sw = black
ro = red
br = brown
gn = green
bl = blue
gr = grey
li = violet
ge = yellow

C10 - Suppressor Filter
F - Brake Light Switch
F47 - Brake Pedal Switch (Cruise Control/Diesel Direct Fuel Injection)
J104 - ABS Control Module (w/EDL)
J123 - Lamp Control Module
M9 - Left Brake Light
M25 - High-mount Brake Light
S13 - Fuse

(A36) - Wire connection (75a), in instrument panel wiring harness
(B131) - Connector (54), in wiring harness interior
(B185) - Connection 2 (54) in passenger compartment wiring harness

* - Vehicles with front-wheel drive
** - Vehicles with all-wheel drive and 132 kW engine, code ATC, AWP
*** - Vehicles with all-wheel drive and 165 kW engine, code AMU

Edition 10/02
W42.USA.5702.11.21

ABS Control Module, Brake Light Switch

No. 38/6 **Wiring diagram** **Audi TT**

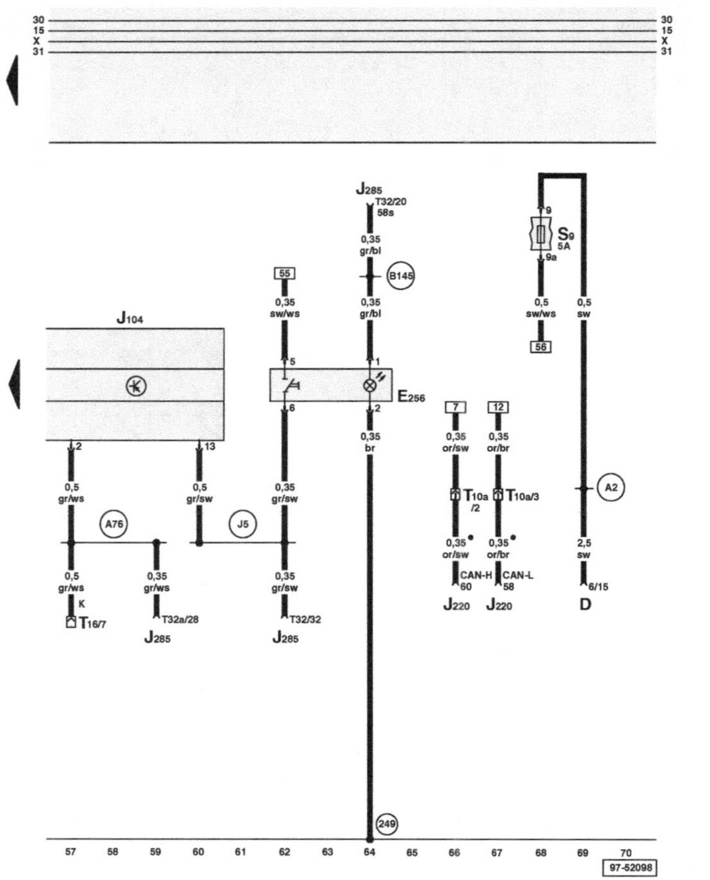

ws = white
sw = black
ro = red
br = brown
gn = green
bl = blue
gr = grey
li = violet
ge = yellow
or = orange

E256 - ASR/ESP Button
J104 - ABS Control Module (w/EDL)
J285 - Control Module with indicator unit in instrument panel insert
J220 - Motronic Engine Control Module (ECM)
S9 - Fuse
T10a - Connector 10 pin, white, in E-box plenum chamber
T16 - Connector 16 pin, black, Data Link Connector (DLC), below storage compartment, driver's side
T32 - Connector 32 pin, blue, on instrument cluster
T32a - Connector 32 pin, green, on instrument cluster

(249) - Ground connection -2-, in wiring harness interior
(A2) - Plus connection (15), in instrument panel wiring harness
(A76) - Connector (K-diagnosis wire), in instrument panel wiring harness
(B145) - Plus connector (58s), in wiring harness interior
(J5) - Wire connection -2-, in ABS wiring harness, near anti-slip control switch
• - CAN-Bus (Data-Bus)

ABS Control Module, ASR/ESP Button

Edition 10/02
W42.USA.5702.11.21

© 2006 BentleyPublishers.com—All Rights Reserved

Radio Stereo Concert with and without CD-Changer (Roadster)

2002 m. y.

Fuse Panel

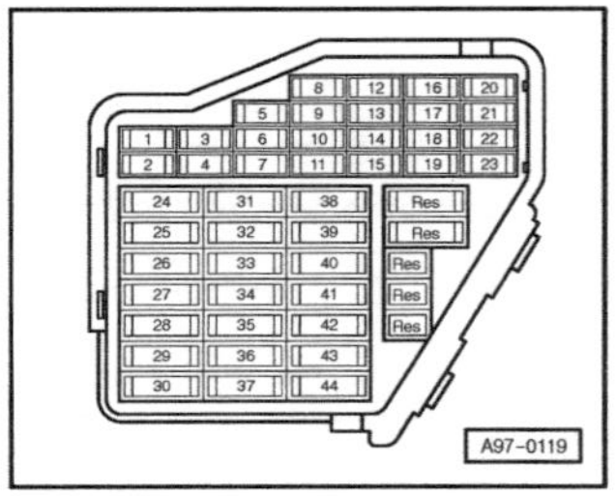

Fuse Colors:

30 A - Green
25 A - White
20 A - Yellow
15 A - Blue
10 A - Red
7,5 A - Brown
5 A - Beige

Starting with fuse position 23, fuses in the fuse holder are identified with 223 in the wiring diagram.

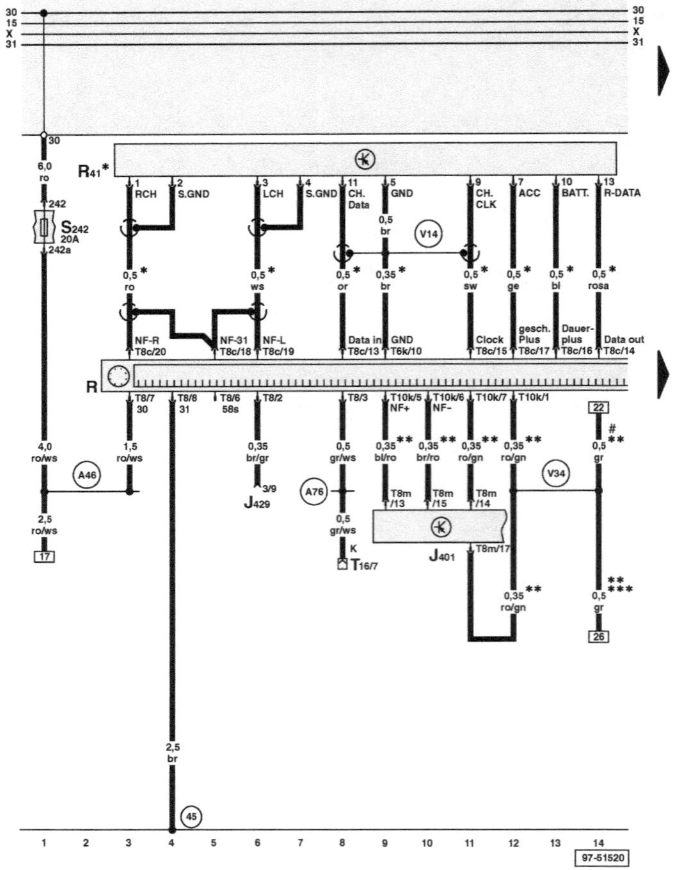

ws = white
sw = black
ro = red
br = brown
gn = green
bl = blue
gr = grey
li = violet
ge = yellow

J401 - Control unit for navigation with CD drive
J429 - Control module for central locking
R - Radio
R41 - CD Changer Unit
S242 - Fuse 42 in fuse holder
T6k - Connector 6 pin, green, connector radio I-2
T8 - Connector 8 pin, black, connector radio III
T8c - Connector 8 pin, blue, connector radio I-3
T8m - Connector 8 pin, C III on Control Module for Navigation
T10k - Connector 10 pin, red, on radio
T16 - Connector 16 pin, black, Data Link Connector (DLC), below storage compartment, driver's side

(45) - Ground connection, behind instrument panel, center
(A46) - Plus connection (30-from radio), in instrument panel wiring harness
(A76) - Connector (K-diagnosis wire), in instrument panel wiring harness
(V14) - Wire connection (shielding), in CD Changer wiring harness
(V34) - Connector (mute button), in wiring harness telephone

* - Vehicles with CD Changer
** - Vehicles with Navigation System
*** - Vehicles with Telephone
# - Vehicles with Cellular Telephone Preparation

© 2006 BentleyPublishers.com—All Rights Reserved

Sound System, Concert, roadster (2002-2003)

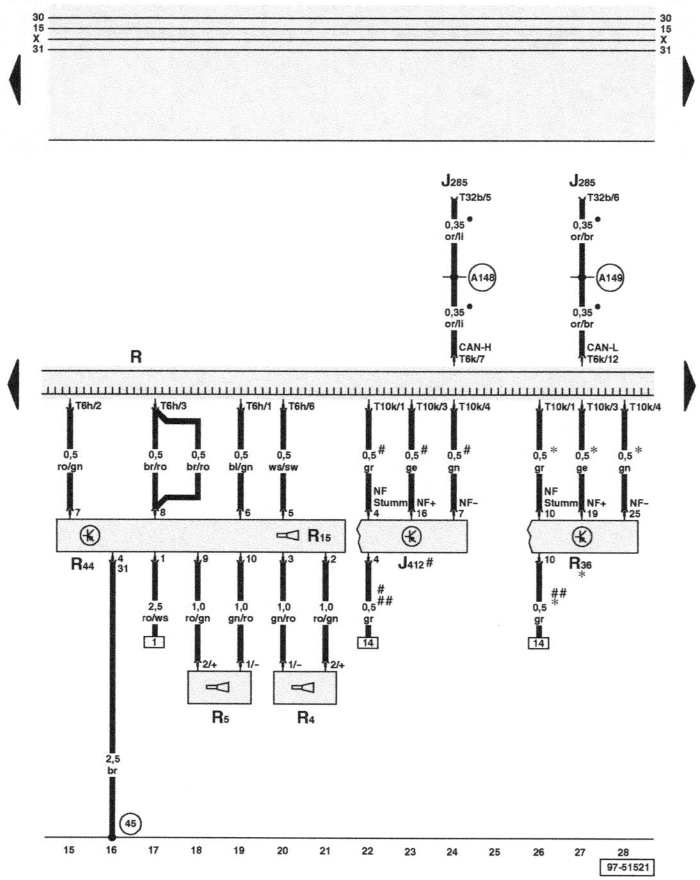

ws = white
sw = black
ro = red
br = brown
gn = green
bl = blue
gr = grey
li = violet
ge = yellow
or = orange

J285 - Control Module with indicator unit in instrument panel insert
J412 - Operating electronics and Telephone Control Module
R - Radio
R4 - Left Rear Speaker
R5 - Right Rear Speaker
R15 - Center Rear Woofer
R36 - Telephone Transceiver
R44 - Center Rear Woofer/Amplifier
T6h - Connector 6 pin, yellow, connector radio I-1
T6k - Connector 6 pin, green, connector radio I-2
T10k - Connector 10 pin, red, on radio
T32b - Connector 32 pin, gray, on instrument cluster

(45) - Ground connection, behind instrument panel, center
(A148) - Connection (high bus, navigation), in instrument panel wiring harness
(A149) - Connection (low bus, navigation), in instrument panel wiring harness

● - CAN-Bus (Data-Bus)
* - Vehicles with telephone
# - Vehicles with Cellular Telephone Preparation
- Vehicles with Navigation System

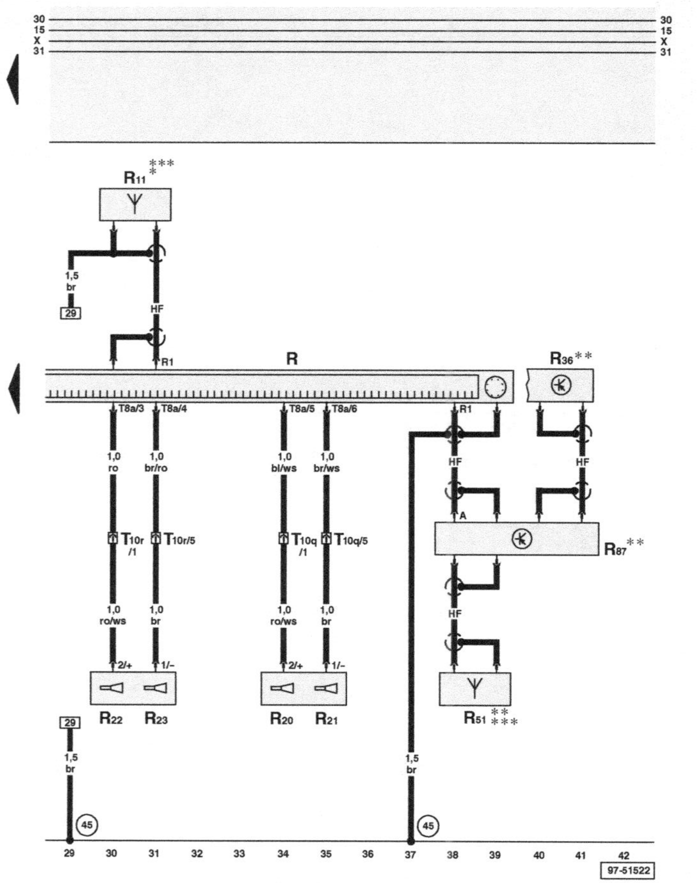

ws = white
sw = black
ro = red
br = brown
gn = green
bl = blue
gr = grey
li = violet
ge = yellow

R - Radio
R1 - Antenna Connection
R11 - Antenna
R20 - Treble speaker, left front
R21 - Bass speaker, left front
R22 - Treble speaker, right front
R23 - Bass speaker, right front
R36 - Telephone Transceiver
R51 - Antenna for radio/telephone
R87 - Antenna Splitter
T8a - Connector 8 pin, brown, connector radio II
T10q - Connector 10 pin, green, connector station A pillar, left
T10r - Connector 10 pin, green, connector station A pillar, right

(45) - Ground connection, behind instrument panel, center

* - Vehicles without telephone
** - Vehicles with telephone
*** - Rod antenna

© 2006 BentleyPublishers.com—All Rights Reserved

Radio Bose with and without CD-Changer (Roadster)

2002 m. y.

Fuse Panel

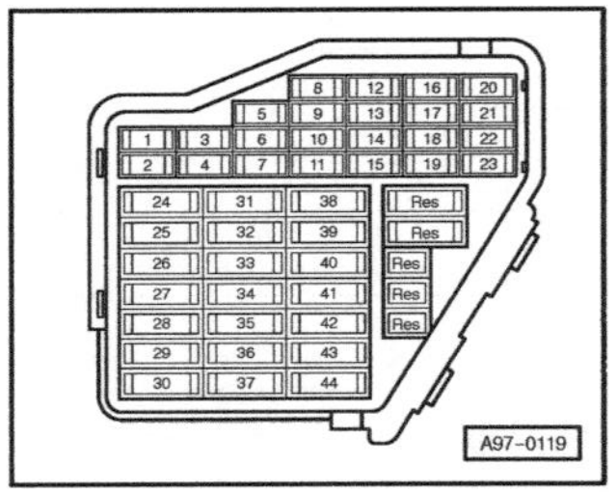

Fuse Colors:

30 A - Green
25 A - White
20 A - Yellow
15 A - Blue
10 A - Red
7,5 A - Brown
5 A - Beige

Starting with fuse position 23, fuses in the fuse holder are identified with 223 in the wiring diagram.

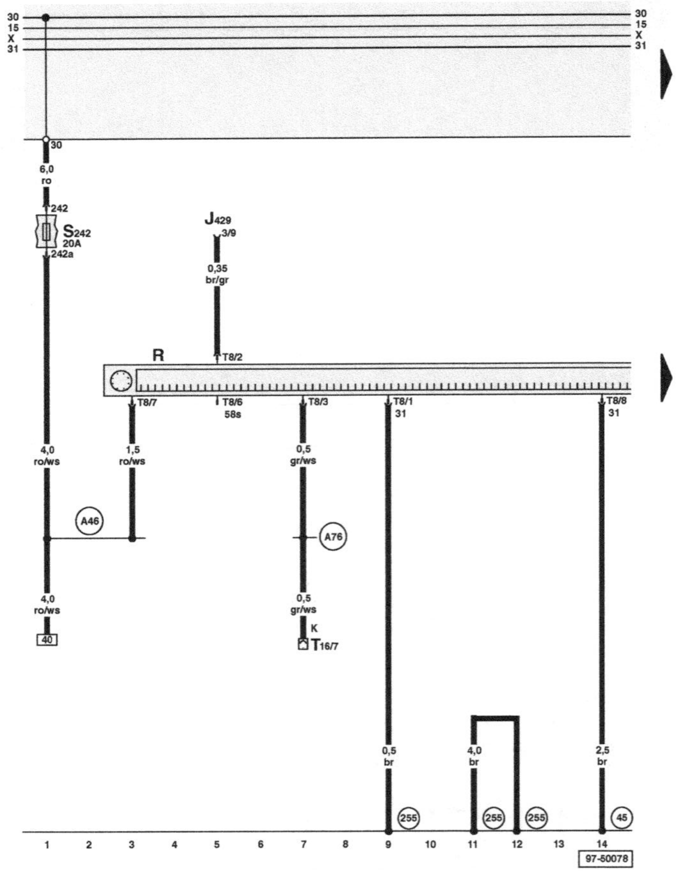

ws = white
sw = black
ro = red
br = brown
gn = green
bl = blue
gr = grey
li = violet
ge = yellow

J429 - Control module for central locking
R - Radio
S242 - Fuse 42 in fuse holder
T8 - Connector 8 pin, black, connector radio III
T16 - Connector 16 pin, black, Data Link Connector (DLC), below storage compartment, driver's side

(45) - Ground connection, behind instrument panel, center
(255) - Ground connection -1-, in wiring harness Radio
(256) - Ground connection -2-, in wiring harness Radio
(A46) - Plus connection (30-from radio), in instrument panel wiring harness
(A76) - Connector (K-diagnosis wire), in instrument panel wiring harness

© 2006 BentleyPublishers.com—All Rights Reserved

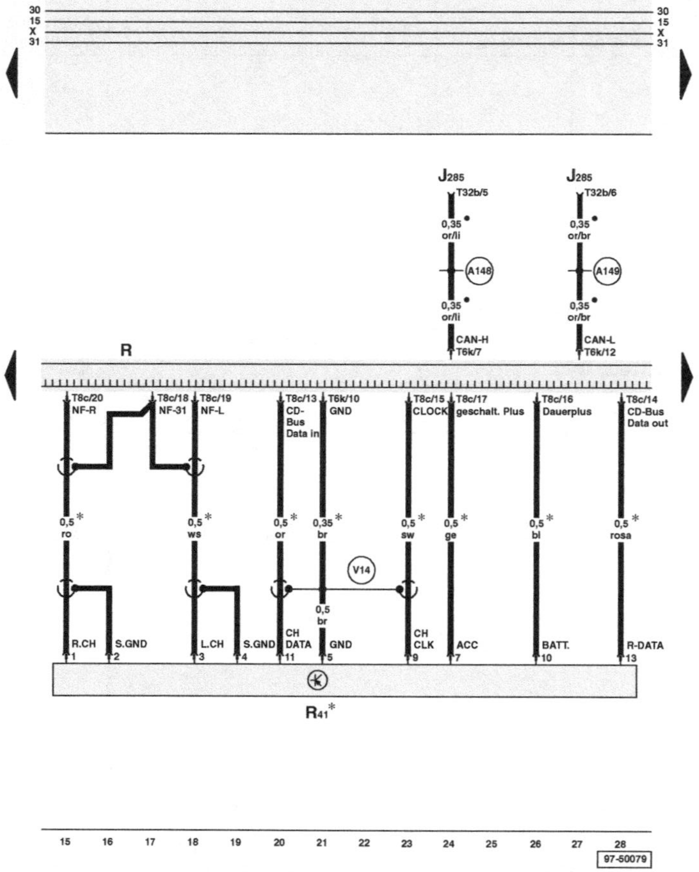

ws = white
sw = black
ro = red
br = brown
gn = green
bl = blue
gr = grey
li = violet
ge = yellow
or = orange

J285 - Control Module with indicator unit in instrument panel insert
R - Radio
R41 - CD Changer Unit
T6k - Connector 6 pin, green, connector radio I-2
T8c - Connector 8 pin, blue, connector radio I-3
T32b - Connector 32 pin, gray, on instrument cluster

(A148) - Connection (high bus, navigation), in instrument panel wiring harness
(A149) - Connection (low bus, navigation), in instrument panel wiring harness
(V14) - Wire connection (shielding), in CD Changer wiring harness

• - CAN-Bus (Data-Bus)
* - Vehicles with CD-Changer

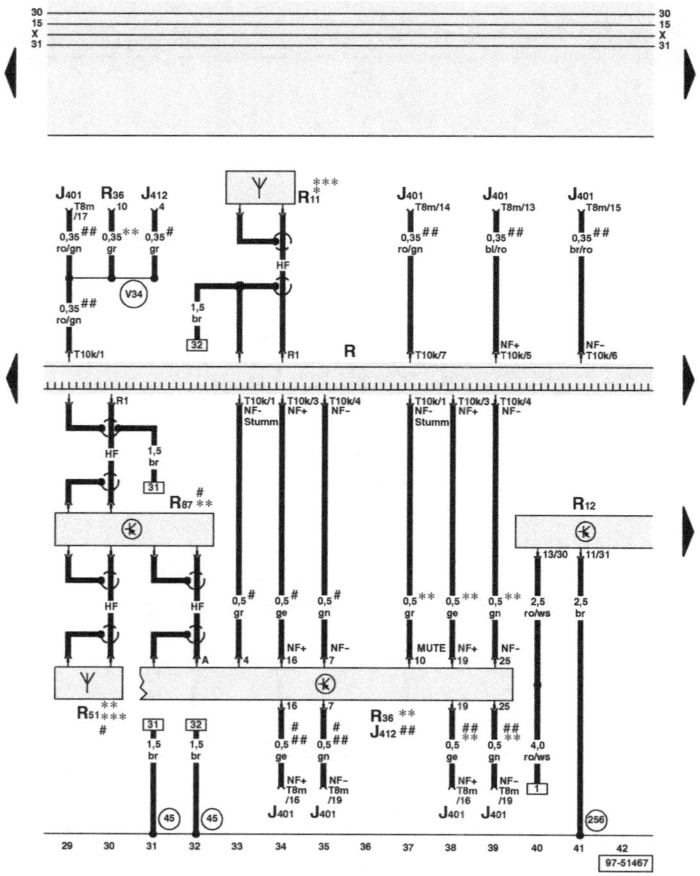

ws = white
sw = black
ro = red
br = brown
gn = green
bl = blue
gr = grey
li = violet
ge = yellow

J401 - Control unit for navigation with CD drive
J412 - Operating electronics and Telephone Control Module
R - Radio
R1 - Antenna Connection
R11 - Antenna
R12 - Amplifier
R36 - Telephone Transceiver
R51 - Antenna for radio/telephone
R87 - Antenna Splitter
T8m - Connector 8 pin, C III on Control Module for Navigation
T10k - Connector 10 pin, red, on radio

(45) - Ground connection, behind instrument panel, center
(256) - Ground connection -2-, in wiring harness Radio

* - Vehicles without Telephone
** - Vehicles with Telephone
*** - Rod antenna
# - Vehicles with Cellular Telephone Preparation
- Vehicles with Navigation System

© 2006 BentleyPublishers.com—All Rights Reserved

Sound System, Bose, roadster (2002)

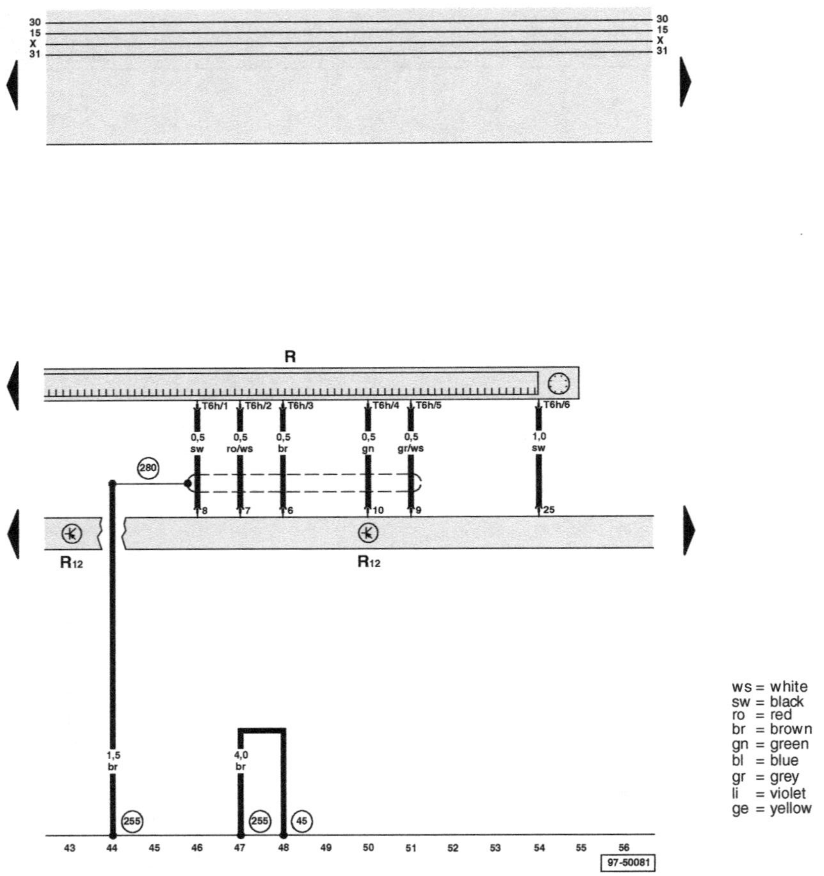

ws = white
sw = black
ro = red
br = brown
gn = green
bl = blue
gr = grey
li = violet
ge = yellow

R - Radio
R12 - Amplifier
T6h - Connector 6 pin, yellow, connector radio I-1

(45) - Ground connection, behind instrument panel, center
(255) - Ground connection -1-, in wiring harness Radio
(280) - Ground connection, in communication wiring harness

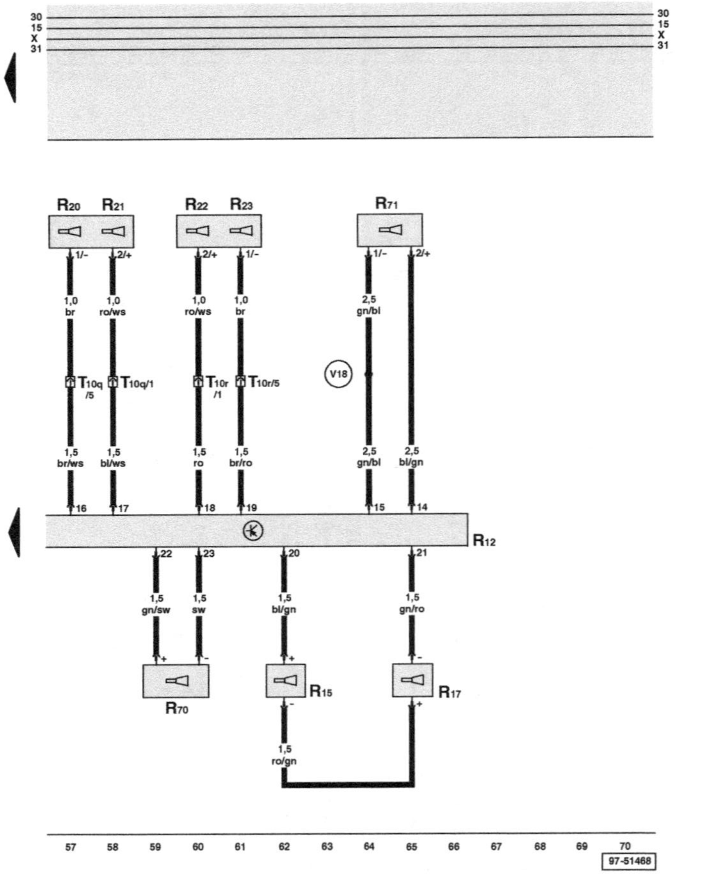

ws = white
sw = black
ro = red
br = brown
gn = green
bl = blue
gr = grey
li = violet
ge = yellow

R12 - Amplifier
R15 - Left Rear Woofer
R17 - Right Rear Woofer
R20 - Left Front Tweeter
R21 - Left Front Woofer
R22 - Right Front Tweeter
R23 - Right Front Woofer
R70 - Loudspeaker front, center
R71 - Loudspeaker rear, center
T10q - Connector 10 pin, green, connector station A pillar, left
T10r - Connector 10 pin, green, connector station A pillar, right

(V18) - Connector (minus, speaker rear), in wiring harness speaker

© 2006 BentleyPublishers.com—All Rights Reserved

CAN-Bus Wiring

2002 m. y.

Fuse Panel

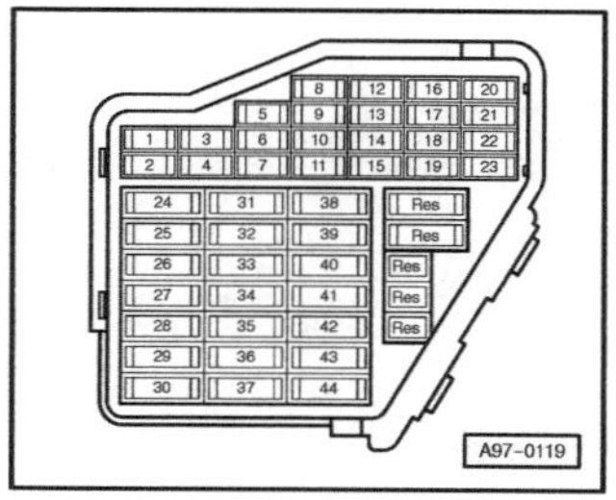

Fuse Colors:

30 A - Green
25 A - White
20 A - Yellow
15 A - Blue
10 A - Red
7,5 A - Brown
5 A - Beige

Starting with fuse position 23, fuses in the fuse holder are identified with 223 in the wiring diagram.

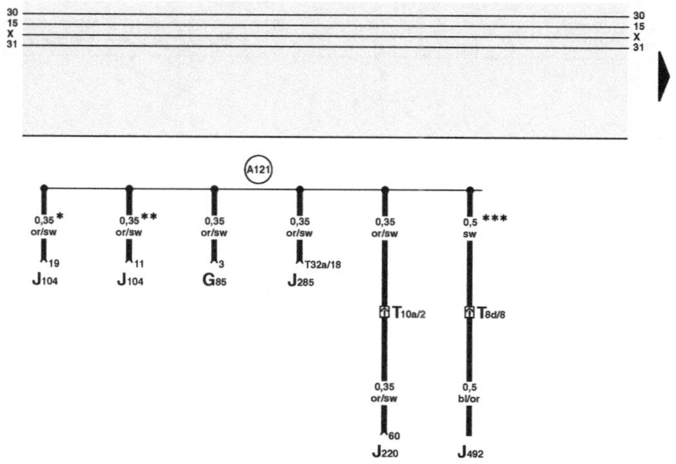

ws = white
sw = black
ro = red
br = brown
gn = green
bl = blue
gr = grey
li = violet
ge = yellow
or = orange

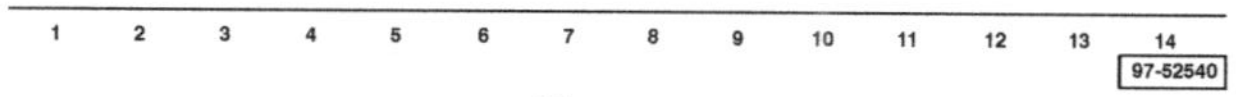

G85 - Steering Angle Sensor
J104 - ABS Control Module (w/EDL)
J220 - Motronic Engine Control Module (ECM)
J285 - Control Module With Display Unit In Instrument Cluster
J492 - All-wheel drive control module
T8d - Connector 8 pin, on Control module for all wheel drive
T10a - Connector 10 pin, white, in E-box plenum chamber
T32a - Connector 32 pin, green, on instrument cluster

(A121) - Connection (high bus) in instrument panel wiring harness

* - Anti-lock Brake System (ABS) MK20 with EDL/ASR/ESP
** - Anti-lock Brake System (ABS) MK60 with EDL/ASR/ESP
*** - Vehicles with all-wheel drive

© 2006 BentleyPublishers.com—All Rights Reserved

Audi TT Wiring diagram **No. 41/3**

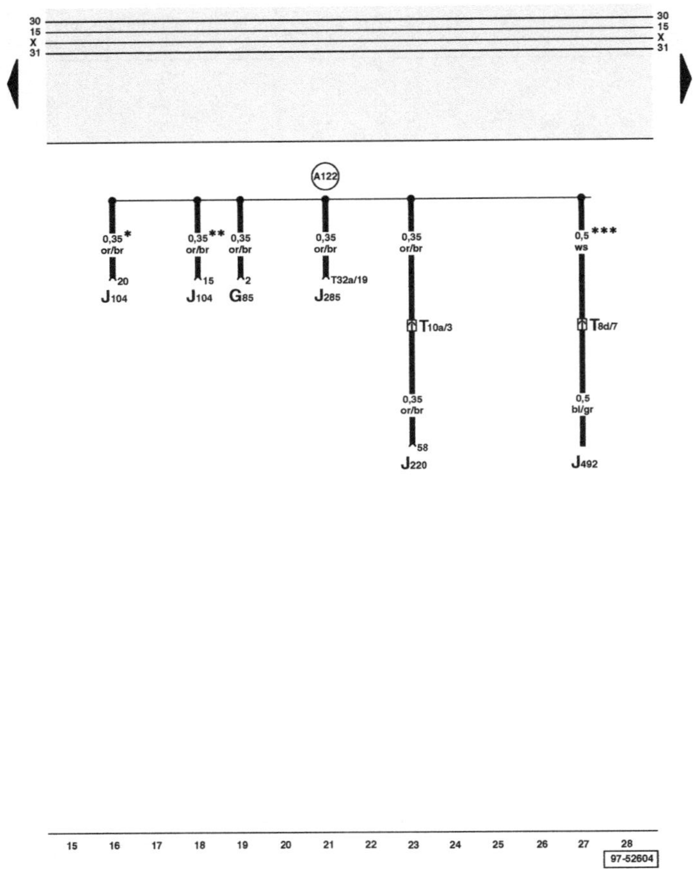

ws = white
sw = black
ro = red
br = brown
gn = green
bl = blue
gr = grey
li = violet
ge = yellow
or = orange

- G85 - Steering Angle Sensor
- J104 - ABS Control Module (w/EDL)
- J220 - Motronic Engine Control Module (ECM)
- J285 - Control Module with indicator unit in instrument panel insert
- J492 - All-wheel drive control module
- T8d - Connector 8 pin, on Control module for all wheel drive
- T10a - Connector 10 pin, white, in E-box plenum chamber
- T32a - Connector 32 pin, green, on instrument cluster

- (A122) - Connection (low bus) in instrument panel wiring harness

- * - ABS MK20 with EDL/ASR/ESP
- ** - ABS MK60 with EDL/ASR/ESP
- *** - Vehicles with all-wheel drive

Edition 11/01 W42.USA.5702.10.21 **CAN-Bus Wiring, Connection Low-Bus**

No. 41/4 Wiring diagram **Audi TT**

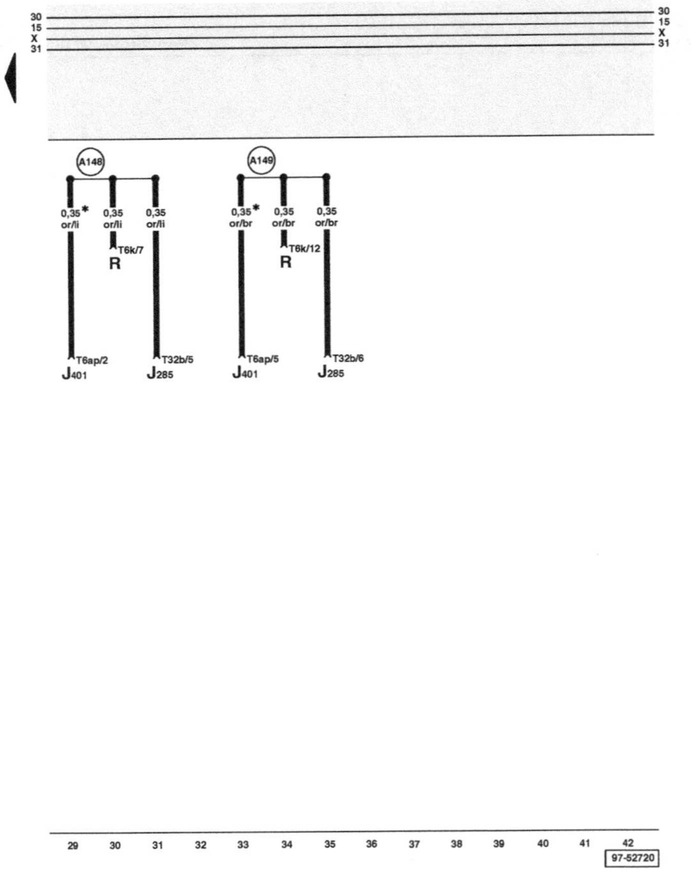

ws = white
sw = black
ro = red
br = brown
gn = green
bl = blue
gr = grey
li = violet
ge = yellow
or = orange

- J285 - Control Module with indicator unit in instrument panel insert
- J401 - Control unit for navigation with CD drive
- R - Radio
- T6k - Connector 6 pin, green, connector radio I-2
- T6ap - Connector 6 pin, Connector C I on Control Module for Navigation
- T32b - Connector 32 pin, gray, on instrument cluster

- (A148) - Connection (high bus, navigation), in instrument panel wiring harness
- (A149) - Connection (low bus, navigation), in instrument panel wiring harness

- * - Vehicles with navigation system

CAN-Bus Wiring, Connection Comfort System, Connection Navigation System (High-Bus, Low-Bus) **Edition 11/01** W42.USA.5702.10.21

© 2006 BentleyPublishers.com—All Rights Reserved

Cellular Telephone - Preparation

2002 m. y.

Fuse Panel

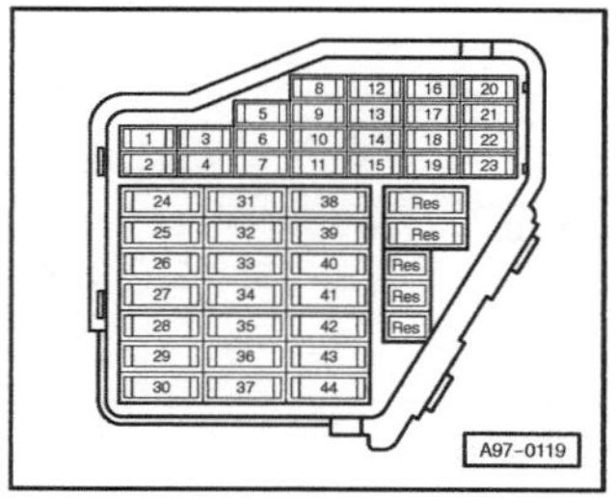

Fuse Colors:

30 A - Green
25 A - White
20 A - Yellow
15 A - Blue
10 A - Red
7,5 A - Brown
5 A - Beige

Starting with fuse position 23, fuses in the fuse holder are identified with 223 in the wiring diagram.

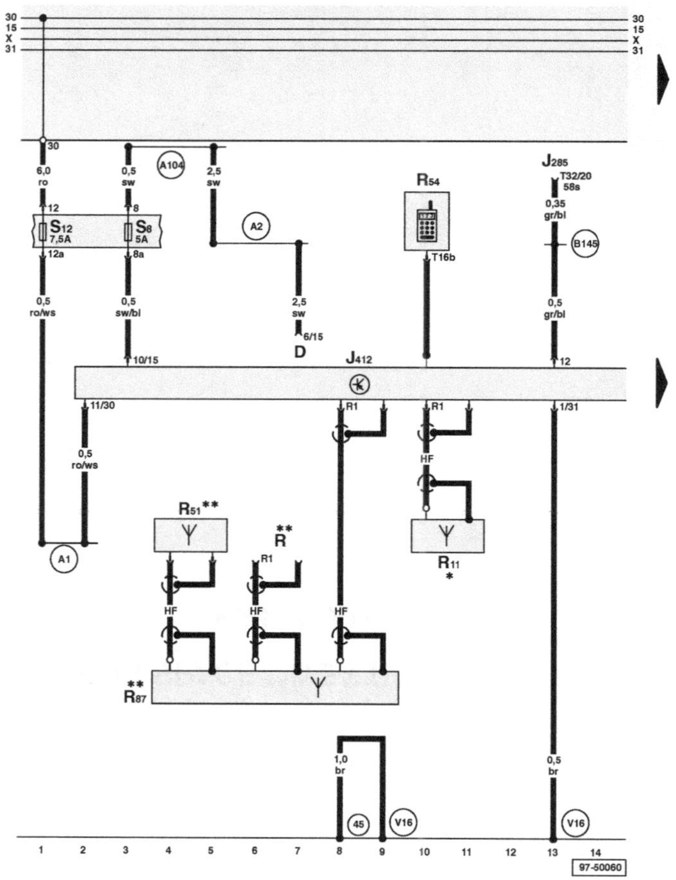

ws = white
sw = black
ro = red
br = brown
gn = green
bl = blue
gr = grey
li = violet
ge = yellow

D - Ignition/Starter Switch
J285 - Control Module with indicator unit in instrument panel insert
J412 - Operating electronics and Telephone Control Module
R - Radio
R1 - Antenna Connection
R11 - Antenna
R51 - Antenna for radio/telephone/park heating
R54 - Cellular telephone
R87 - Antenna Splitter
S8 - Fuse
S12 - Fuse
T16b - Connector 16 pin, on Cellular Phone
T32 - Connector 32 pin, blue, on instrument cluster

(45) - Ground connection, behind instrument panel, center
(A1) - Plus connection (30a), in instrument panel wiring harness
(A2) - Plus connection (15), in instrument panel wiring harness
(A104) - Plus connector -2- (15), in instrument panel wiring harness
(B145) - Plus connector (58s), in wiring harness interior
(V16) - Wire connection (shielding), in telephone wiring harness

* - Coupé, roof antenna
** - Roadster, rod antenna

©2006 BentleyPublishers.com—All Rights Reserved

Audi TT **Wiring diagram** **No. 42/3**

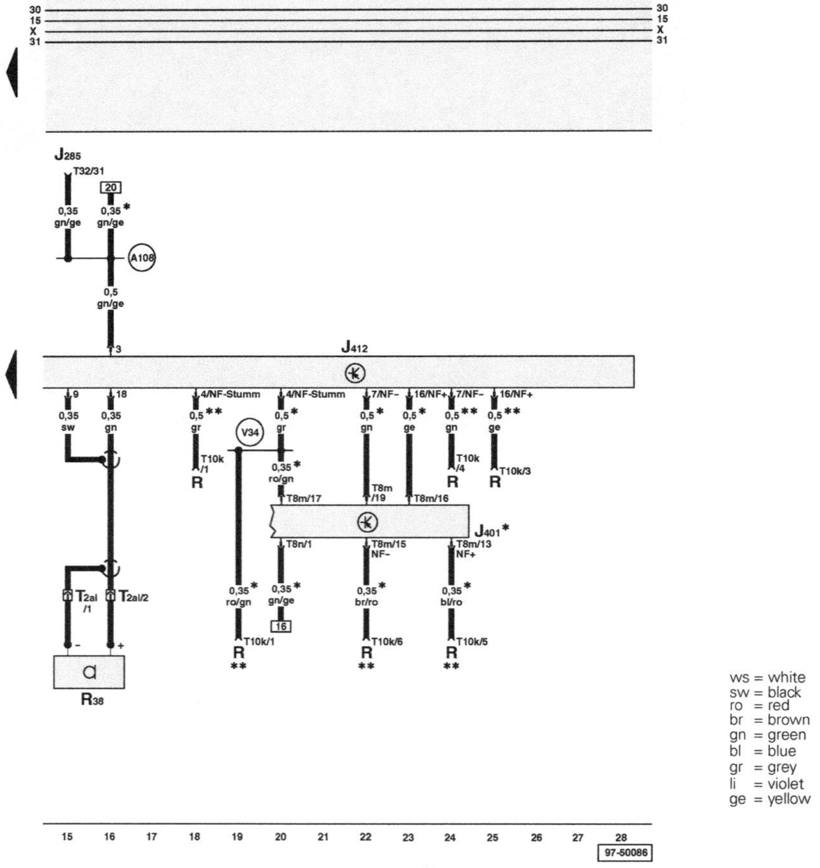

ws = white
sw = black
ro = red
br = brown
gn = green
bl = blue
gr = grey
li = violet
ge = yellow

J285 - Control Module with indicator unit in instrument panel insert
J401 - Control unit for navigation with CD drive
J412 - Operating electronics and Telephone Control Module
R - Radio
R38 - Telephone Microphone
T2al - Connector 2 pin, behind instrument cluster
T8m - Connector 8 pin, C III on Control Module for Navigation
T8n - Connector 8 pin, black, connector A on Control Module for Navigation
T10k - Connector 10 pin, red, on radio
T32 - Connector 32 pin, blue, on instrument cluster

(A108) - Connector (speed signal), in instrument panel wiring harness

(V34) - Connector (mute button), in wiring harness telephone

* - Vehicles with Radio and Navigation
** - Vehicles with Radio

THIS SPACE INTENTIONALLY LEFT BLANK

© 2006 BentleyPublishers.com—All Rights Reserved

Garage Door Opener - HomeLink

2002 m. y.

Fuse Panel

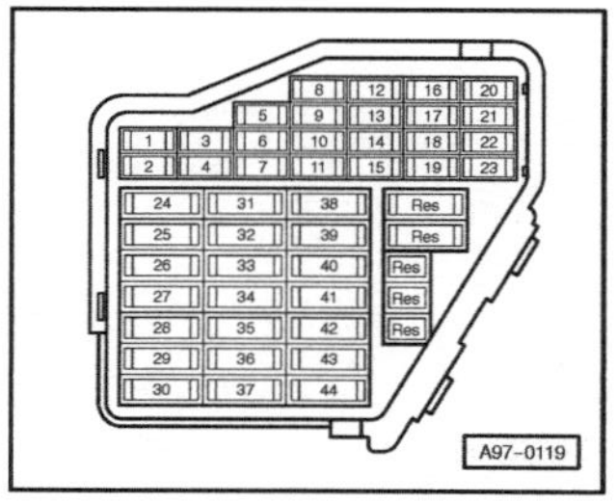

Fuse Colors:

30 A	-	Green
25 A	-	White
20 A	-	Yellow
15 A	-	Blue
10 A	-	Red
7,5 A	-	Brown
5 A	-	Beige

Starting with fuse position 23, fuses in the fuse holder are identified with 223 in the wiring diagram.

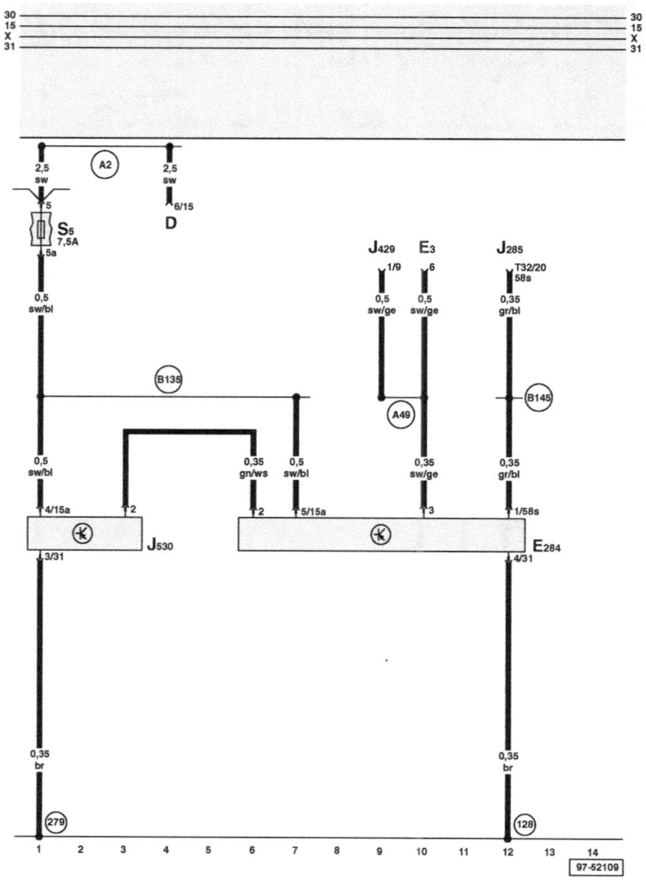

ws = white
sw = black
ro = red
br = brown
gn = green
bl = blue
gr = grey
li = violet
ge = yellow

D - Ignition/Starter Switch
E3 - Emergency Flasher Switch
E284 - Garage Door Opener Control Head
J285 - Control Module With Display Unit In Instrument Cluster
J429 - Control module for central locking
J530 - Garage Door Opener Control Module
S5 - Fuse
T32 - Connector 32 pin, blue, on instrument cluster

(128) - Ground connection -1-, in interior light wiring harness
(279) - Ground connector -5-, in wiring harness interior
(A2) - Plus connection (15), in instrument panel wiring harness
(A49) - Wire connection -1-, in instrument panel wiring harness
(B135) - Connector (15a), in wiring harness interior
(B145) - Plus connector (58s), in wiring harness interior

© 2006 BentleyPublishers.com—All Rights Reserved

Airbag

2002 m. y.

Fuse Panel

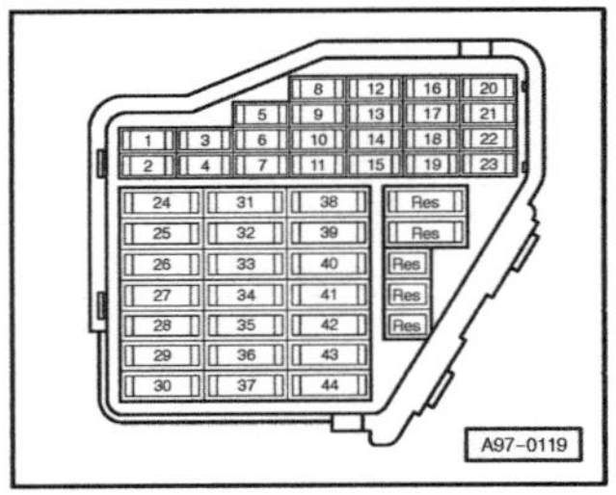

Fuse Colors:

30 A - Green
25 A - White
20 A - Yellow
15 A - Blue
10 A - Red
7,5 A - Brown
5 A - Beige

Starting with fuse position 23, fuses in the fuse holder are identified with 223 in the wiring diagram.

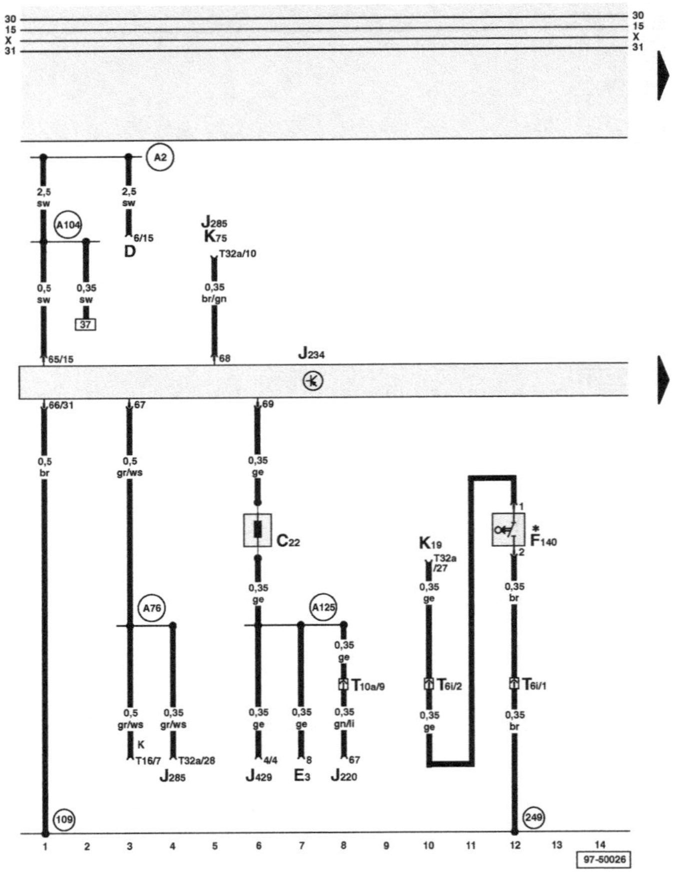

ws = white
sw = black
ro = red
br = brown
gn = green
bl = blue
gr = grey
li = violet
ge = yellow

C22 - Airbag Suppresser
D - Ignition/Starter Switch
E3 - Emergency Flasher Switch
F140 - Left Front Seatbelt Microswitch
J220 - Motronic Engine Control Module (ECM)
J234 - Airbag Control Module
J285 - Control Module with indicator unit in instrument panel insert
J429 - Control module for central locking
K19 - Seat Belt Warning Light
K75 - Airbag Malfunction Indicator Lamp (MIL)
T6i - Connector 6 pin, below driver's seat
T10a - Connector 10 pin, white, in E-box plenum chamber
T16 - Connector 16 pin, black, Data Link Connector (DLC), below storage compartment, driver's side
T32a - Connector 32 pin, green, on instrument cluster

(109) - Ground connection, in airbag wiring harness
(249) - Ground connection -2-, in wiring harness interior
(A2) - Plus connection (15), in instrument panel wiring harness
(A76) - Connector (K-diagnosis wire), in instrument panel wiring harness
(A104) - Plus connector -2- (15), in instrument panel wiring harness
(A125) - Connection (crash signal) in instrument panel wiring harness

* - Vehicles with seat belt warning lamp

Airbag Control Module

© 2006 BentleyPublishers.com—All Rights Reserved

Audi TT — Wiring diagram — No. 44/3

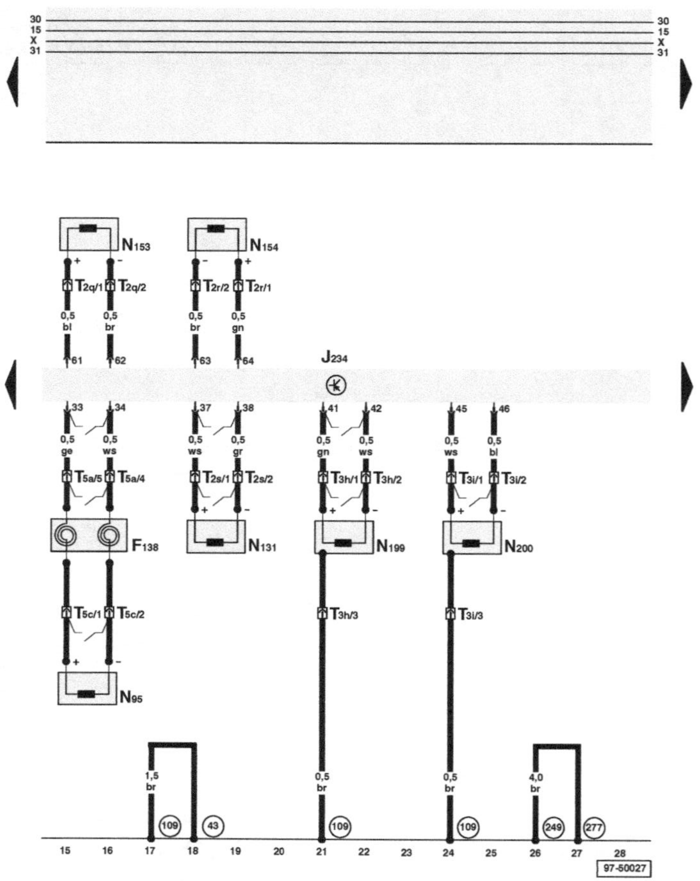

ws = white
sw = black
ro = red
br = brown
gn = green
bl = blue
gr = grey
li = violet
ge = yellow

F138 - Airbag Spiral Spring/Return Spring With Slip Ring
J234 - Airbag Control Module
N95 - Driver's Side Airbag Igniter
N131 - Passenger's Side Airbag Igniter 1
N153 - Driver's Seat Belt Tensioner Igniter 1
N154 - Passenger's Seat Belt Tensioner Igniter 1
N199 - Igniter for side airbag, driver's side
N200 - Igniter for side airbag, passenger side
T2q - Connector 2 pin, on left B-pillar, lower part
T2r - Connector 2 pin, on right B-pillar, lower part
T2s - Connector 2 pin, behind instrument panel, right
T3h - Connector 3 pin, below dirver's seat (for side airbag)
T3i - Connector 3 pin, below passenger's seat (for side airbag)
T5a - Connector 5 pin, yellow, in steering wheel
T5c - Connector 5 pin, yellow, in steering wheel

(43) - Ground connection, on right A-pillar, lower part
(109) - Ground connection, in airbag wiring harness
(249) - Ground connection -2-, in wiring harness interior
(277) - Ground connector -3-, in wiring harness interior
/\ - Bridge closes when disconnecting connector (safety feature)

Edition 10/01
W42.USA.5702.09.21

Airbag Control Module, Airbag, Side Airbag, Belt Tensioner Igniter

Audi TT — Wiring diagram — No. 44/4

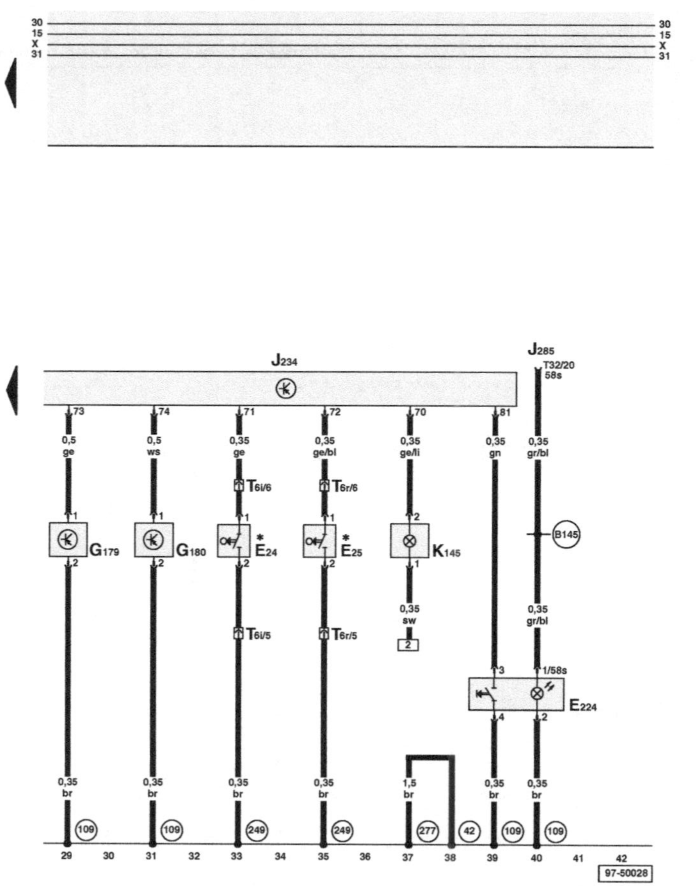

ws = white
sw = black
ro = red
br = brown
gn = green
bl = blue
gr = grey
li = violet
ge = yellow

E24 - Left Seat Belt Switch
E25 - Right Seat Belt Switch
E224 - Key switch for switching off passenger side airbag
G179 - Crash sensor for side airbag, driver's side
G180 - Crash sensor for side airbag, passenger side
J234 - Airbag Control Module
J285 - Control Module with indicator unit in instrument panel insert
K145 - Warning light for airbag off, passenger side
T6i - Connector 6 pin, below driver's seat
T6r - Connector 6 pin, below passenger's seat
T32 - Connector 32 pin, blue, on instrument cluster

(42) - Ground connection, on steering column
(109) - Ground connection, in airbag wiring harness
(249) - Ground connection -2-, in wiring harness interior
(277) - Ground connector -3-, in wiring harness interior
(B145) - Plus connector (58s), in wiring harness interior
* - Where applicable

Crash Sensors for Side Airbags, Key Switch for Switching off Passenger Side Airbag, Seat Belt Switch

Edition 10/01
W42.USA.5702.09.21

©2006 BentleyPublishers.com—All Rights Reserved

Anti-Lock Brake System (ABS) MK20 with Electronic Differential Lock (EDL), Anti-Slip Regulation (ASR - ASC) and Electronic Stabilization Program (ESP)

2002 m. y.

Fuse Panel

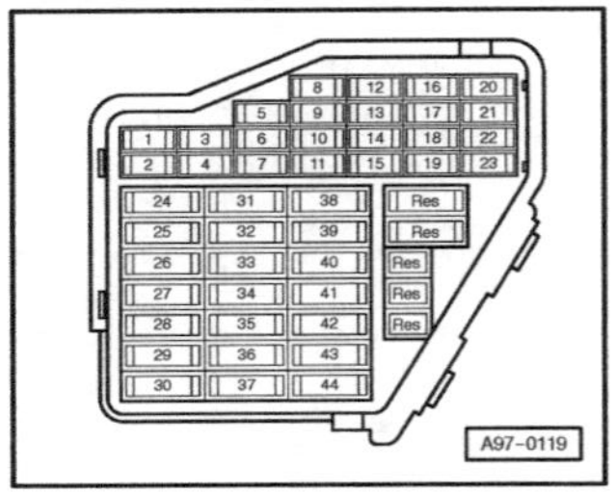

Fuse Colors:

30 A - Green
25 A - White
20 A - Yellow
15 A - Blue
10 A - Red
7,5 A - Brown
5 A - Beige

Starting with fuse position 23, fuses in the fuse holder are identified with 223 in the wiring diagram.

13 - Fold Relay Panel

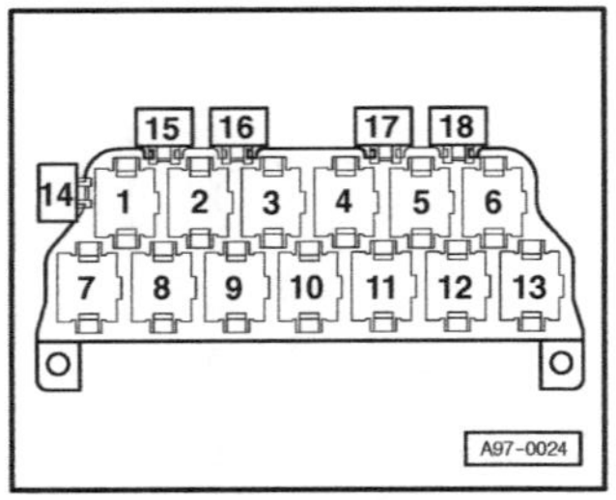

Relay Location:

7 -Brake Light Disable Relay (ESP), J508

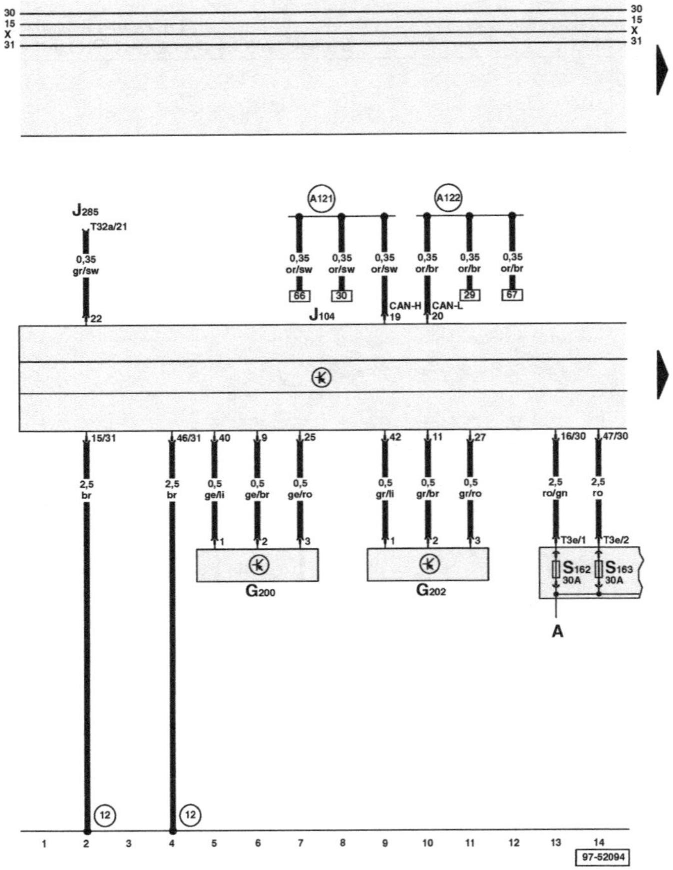

ws = white
sw = black
ro = red
br = brown
gn = green
bl = blue
gr = grey
li = violet
ge = yellow
or = orange

A - Battery
G200 - Sensor for transverse acceleration
G202 - Sender for rotation rate
J104 - ABS Control Module (w/EDL)
J285 - Control Module With Display Unit In Instrument Cluster
S162 - Fuse -1- in fuse bracket/battery
S163 - Fuse -2- in fuse bracket/battery
T3e - Connector 3 pin, black, on Main Fuse Box/Battery
T32a - Connector 32 pin, green, on instrument cluster

(12) - Ground connection, in engine compartment, left
(A121) - Connection (high bus) in instrument panel wiring harness
(A122) - Connection (low bus) in instrument panel wiring harness
- CAN-Bus (Data-Bus)

© 2006 BentleyPublishers.com—All Rights Reserved

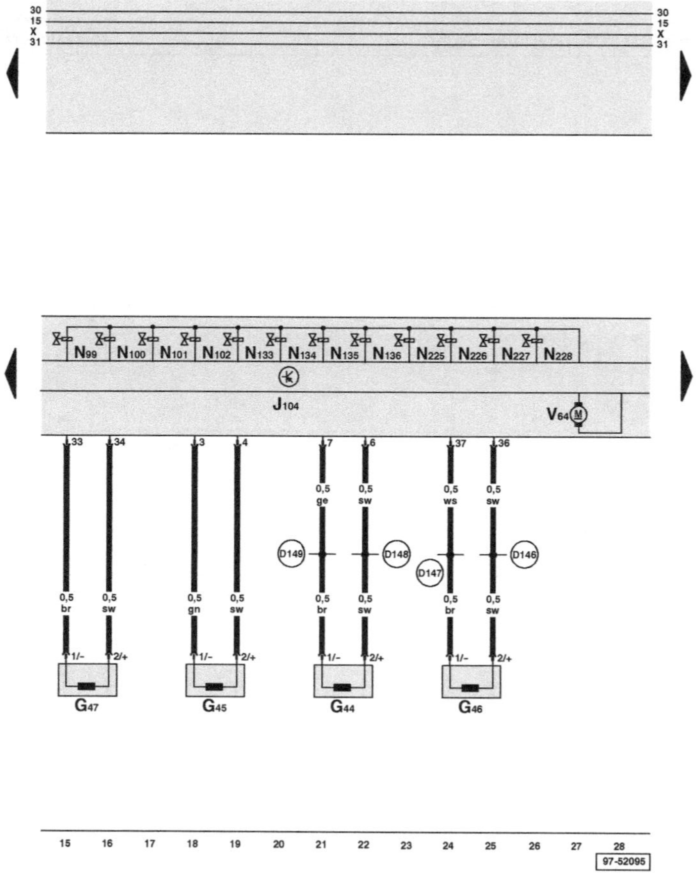

ws = white
sw = black
ro = red
br = brown
gn = green
bl = blue
gr = grey
li = violet
ge = yellow

G44 - Right Rear ABS Wheel Speed Sensor
G45 - Right Front ABS Wheel Speed Sensor
G46 - Left Rear ABS Wheel Speed Sensor
G47 - Left Front ABS Wheel Speed Sensor
J104 - ABS Control Module (w/EDL)
N99 - Right Front ABS Inlet Valve
N100 - Right Front ABS Outlet Valve
N101 - Left Front ABS Inlet Valve
N102 - Left Front ABS Outlet Valve
N133 - Right Rear ABS Inlet Valve
N134 - Left Rear ABS Inlet Valve
N135 - Right Rear ABS Outlet Valve
N136 - Left Rear ABS Outlet Valve
N225 - Pilot valve -1- traction control (01418)
N226 - Pilot valve -2- traction control (01419)
N227 - High pressure switch valve -1- traction control (01420)
N228 - High pressure switch valve -2- traction control (01421)
V64 - ABS Hydraulic Pump

(D146) - Connector (speed sensor rear, left +), in wiring harness engine compartment
(D147) - Connector (speed sensor rear, left -), in wiring harness engine compartment
(D148) - Connector (speed sensor rear, right +), in wiring harness engine compartment
(D149) - Connector (speed sensor rear, right -), in wiring harness engine compartment

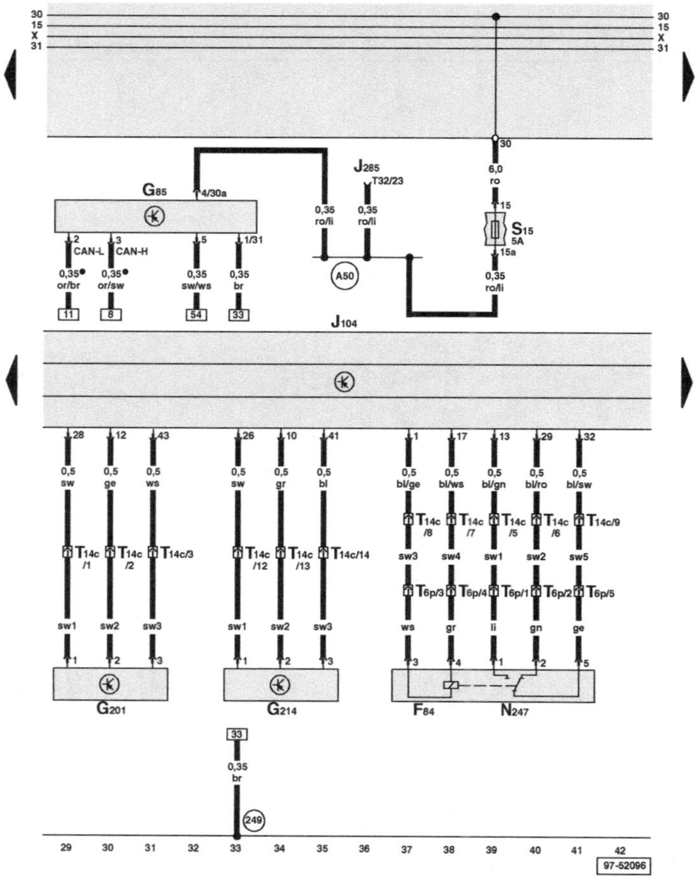

ws = white
sw = black
ro = red
br = brown
gn = green
bl = blue
gr = grey
li = violet
ge = yellow
or = orange

F84 - Brake Pressure Release Solenoid
G85 - Steering Angle Sensor
G201 - Sender 1 for brake booster
G214 - Sensor -2- for brake pressure
J104 - ABS Control Module (w/EDL)
J285 - Control Module with indicator unit in instrument panel insert
N247 - Magnetic coil for brake pressure, in brake booster
S15 - Fuse
T6p - Connector 6 pin, black, on brake booster
T14c - Connector 14 pin, connector in wiring harness ESP
T32 - Connector 32 pin, blue, on instrument cluster

(249) - Ground connection -2-, in wiring harness interior
(A50) - Plus connection (30a), in instrument panel wiring harness
• - CAN-Bus (Data-Bus)

©2006 BentleyPublishers.com—All Rights Reserved

Anti-Lock Brake System, ABS/EDL/ASR/ESP Mk. 20 (2002)

Audi TT Wiring diagram **No. 45/5**

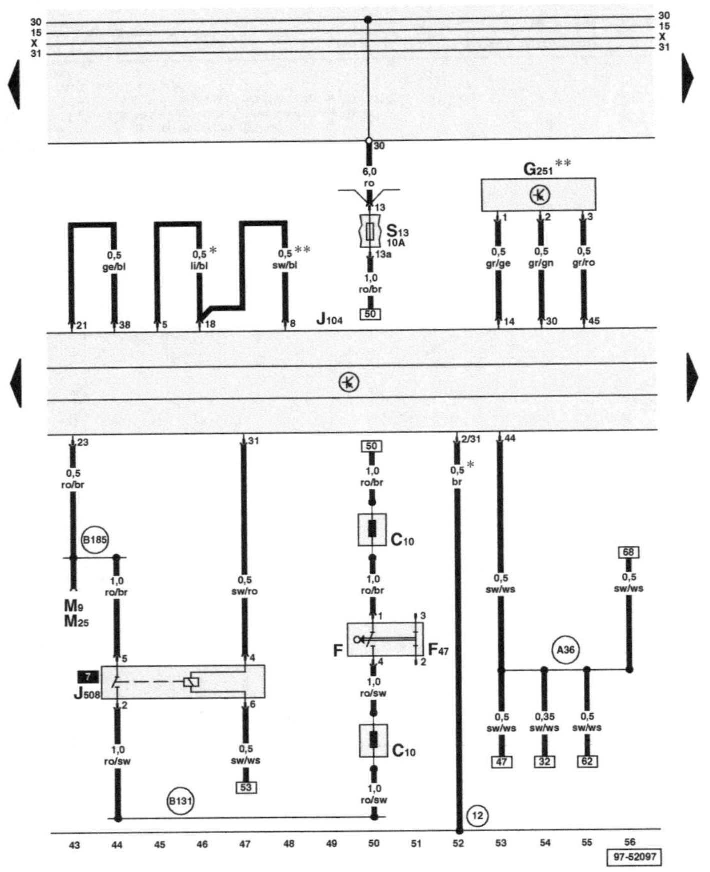

ws = white
sw = black
ro = red
br = brown
gn = green
bl = blue
gr = grey
li = violet
ge = yellow

C10 - Suppressor Filter
F - Brake Light Switch
F47 - Brake Pedal Switch (Cruise Control/Diesel Direct Fuel Injection)
G251 - Longitudinal Acceleration Sensor
J104 - ABS Control Module (w/EDL)
J508 - Brake Light Disable Relay (ESP)
M9 - Left Brake Light
M25 - High-mount Brake Light
S13 - Fuse

(12) - Ground connection, in engine compartment, left
(A36) - Wire connection (75a), in instrument panel wiring harness
(B131) - Connector (54), in wiring harness interior
(B185) - Connection 2 (54) in passenger compartment wiring harness

* - Vehicles with front-wheel drive
** - Vehicles with all-wheel drive

Edition 12/02
W42.USA.5702.12.21

ABS Control Module, Brake Light Disable Relay, Brake Light Switch

No. 45/6 Wiring diagram **Audi TT**

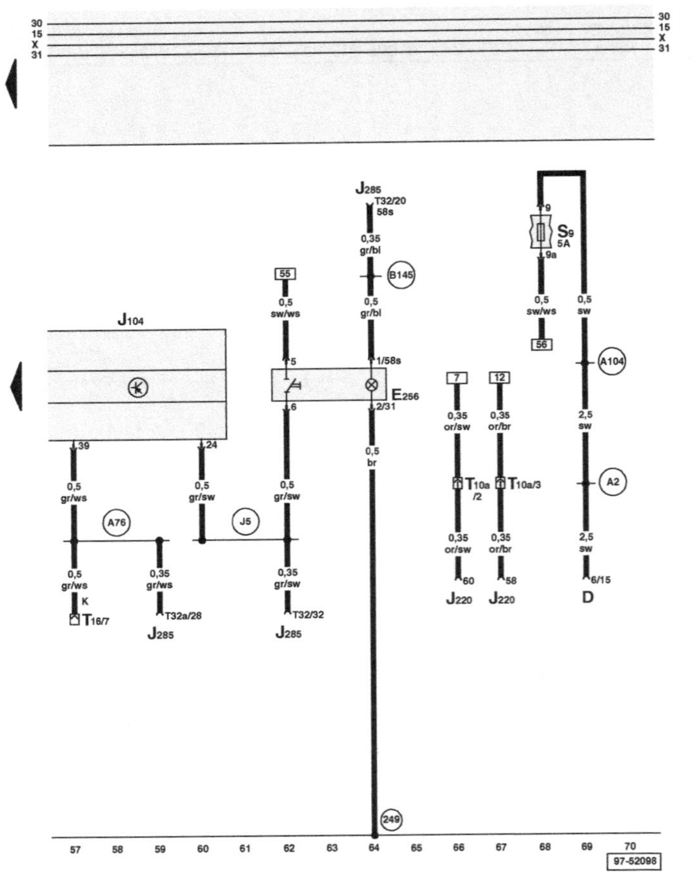

ws = white
sw = black
ro = red
br = brown
gn = green
bl = blue
gr = grey
li = violet
ge = yellow
or = orange

E256 - ASR/ESP Button
J104 - ABS Control Module (w/EDL)
J220 - Motronic Engine Control Module (ECM)
J285 - Control Module with indicator unit in instrument panel insert
S9 - Fuse
T10a - Connector 10 pin, white, in E-box plenum chamber
T16 - Connector 16 pin, black, Data Link Connector (DLC), below storage compartment, driver's side
T32 - Connector 32 pin, blue, on instrument cluster
T32a - Connector 32 pin, green, on instrument cluster

(249) - Ground connection -2-, in wiring harness interior
(A2) - Plus connection (15), in instrument panel wiring harness
(A76) - Connector (K-diagnosis wire), in instrument panel wiring harness
(A104) - Plus connector -2- (15), in instrument panel wiring harness
(B145) - Plus connector (58s), in wiring harness interior
(J5) - Wire connection -2-, in ABS wiring harness, near anti-slip control switch

ABS Control Module, ASR/ESP Button

Edition 12/02
W42.USA.5702.12.21

© 2006 BentleyPublishers.com—All Rights Reserved

Audi TT **Wiring Diagram** **No. 1/1**

Standard Equipment

2003 m.y.

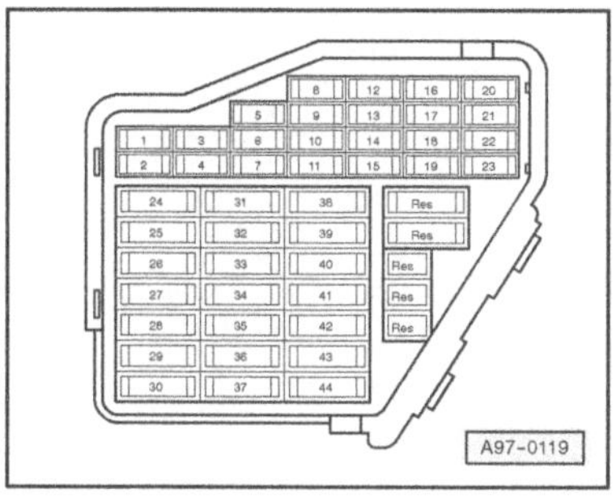

Fuse Panel

Fuse Colors:

30 A - Green
25 A - White
20 A - Yellow
15 A - Blue
10 A - Red
7,5 A - Brown
5 A - Beige
1 A - Black

Starting with fuse position 23, fuses in the fuse holder are identified with 223 in the wiring diagram.

Audi TT **Wiring Diagram** **No. 1/2**

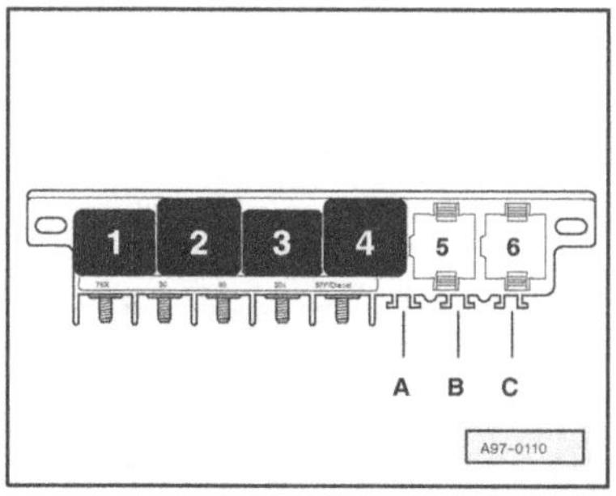

Micro Central Electric Panel

Relay Location:

1 - Dual Horn Relay, J4
2 - Load Reduction Relay, J59
4 - Fuel Pump (FP) Relay, J17
5 - Wiper/Washer Intermittent Relay, J31
6 - Wiper/Washer Intermittent Relay, J31

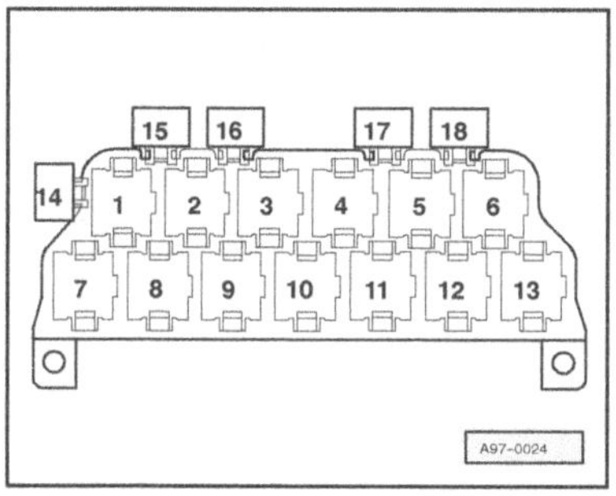

13 - Fold Relay Panel

Relay Location:

2 - Lamp Control Module, J123
3 - Lamp Control Module, J123
4 - Fog Light Relay, J5
5 - Starting Interlock Relay, J207*
10 - Rear Window Defogger Relay, J9
11 - Starting Interlock Relay, J207**
13 - Back-up Light Relay, J219***

* - Vehicles with Manual Transmission
** - Vehicles with Automatic Transmission 09G
*** - Vehicles with Direct Shift Gearbox (DSG) 02E

© 2006 BentleyPublishers.com—All Rights Reserved

Audi TT — Wiring Diagram — No. 1/3

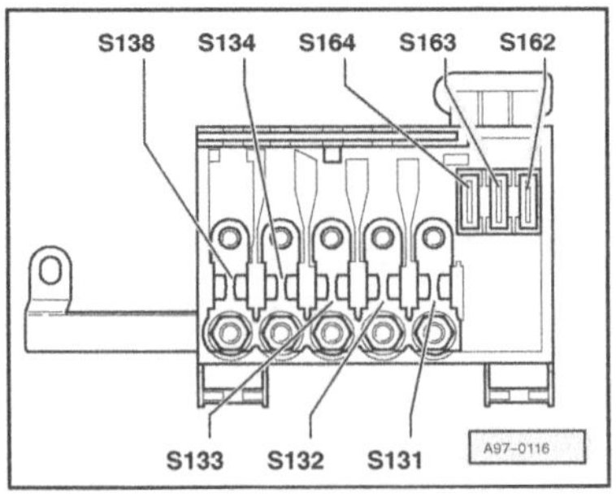

Main Fuse Box/Battery

Fuses Location:

- S131 - Safety fuse 1
- S132 - Safety fuse 2
- S133 - Safety fuse 3
- S134 - Safety fuse 4
- S138 - Safety fuse 5
- S162 - Fuse 1 in fuse bracket/battery
- S163 - Fuse 2 in fuse bracket/battery
- S164 - Fuse 3 in fuse bracket/battery

USA.5703.02.21 Edition 04/03

Audi TT — Wiring Diagram — No. 1/4

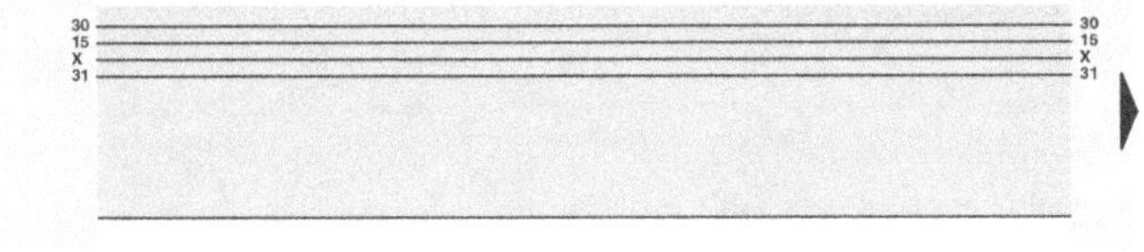

Ground Connections

- T14i - Connector 14 pin, in engine compartment, left
- (12) - Ground connection, in engine compartment, left
- (50) - Ground connection, in luggage compartment, left
- (96) - Ground connection -1-, in heated seats wiring harness
- (98) - Ground connection, in rear lid wiring harness
- (131) - Ground connection -2-, in engine compartment wiring harness
- (188) - Ground connection -3-, in A/C wiring harness
- (249) - Ground connection -2-, in wiring harness interior
- (277) - Ground connector -3-, in wiring harness interior
- (279) - Ground connector -5-, in wiring harness interior
- (283) - Ground connection 2, in wiring harness engine pre-wiring
- (285) - Ground connection -7- in wiring harness interior
- (640) - Ground Connection (in engine compartment, left)

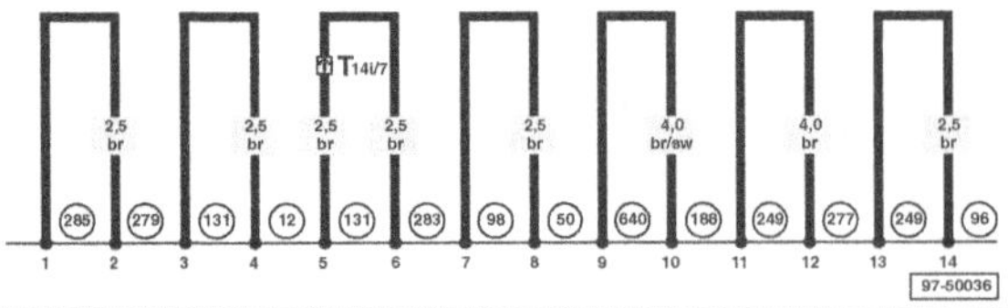

ws = white
sw = black
ro = red
br = brown
gn = green
bl = blue
gr = grey
li = violet
ge = yellow
or = orange

USA.5703.02.21 Edition 04/03

© 2006 BentleyPublishers.com—All Rights Reserved

Audi TT — Wiring Diagram — No. 1/5

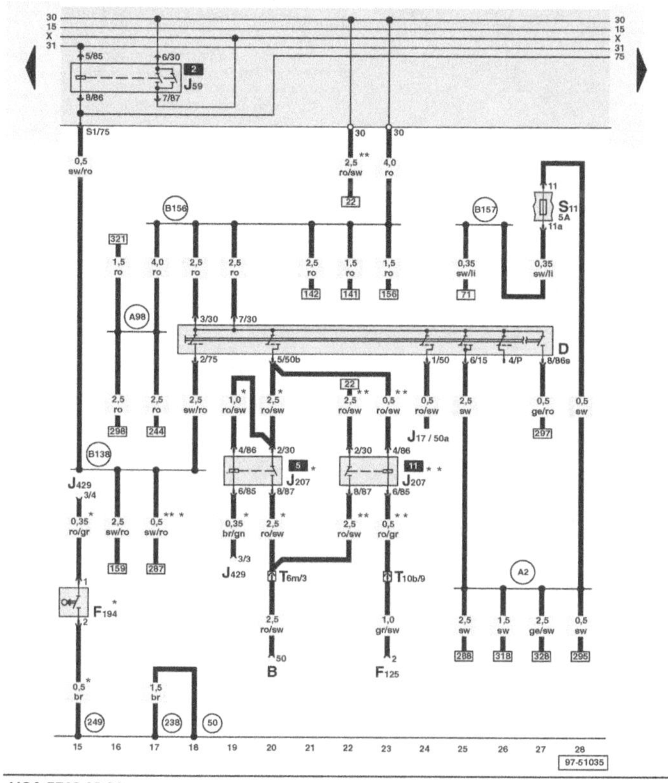

Load Reduction Relay, Ignition/Starter Switch, Starting Interlock Relay

- B - Starter
- D - Ignition/Starter Switch
- F125 - Multi-Function Transmission Range (TR) Switch
- F194 - Clutch Pedal Position (CPP) Switch
- J17 - Fuel Pump (FP) Relay
- J59 - Load Reduction Relay
- J207 - Starting Interlock Relay
- J429 - Control module for central locking
- S11 - Fuse
- T6m - Connector 6 pin, brown, in E-box plenum chamber
- T10b - Connector 10 pin, grey, in E-box plenum chamber

- (50) - Ground connection, in luggage compartment, left
- (238) - Ground connection -1-, in wiring harness interior
- (249) - Ground connection -2-, in wiring harness interior
- (A2) - Plus connection (15), in instrument panel wiring harness
- (A98) - Plus connector -4- (30), in instrument panel wiring harness
- (B138) - Plus connector (X), in wiring harness, interior
- (B156) - Plus connector (30a), in wiring harness interior
- (B157) - Plus connector -3- (15a), in wiring harness interior

- * - Vehicles with Manual Transmission
- ** - Vehicles with Automatic Transmission 09G
- *** - Roadster

ws = white
sw = black
ro = red
br = brown
gn = green
bl = blue
gr = grey
li = violet
ge = yellow
or = orange

USA.5703.03.21 Edition 10/03

Audi TT — Wiring Diagram — No. 1/6

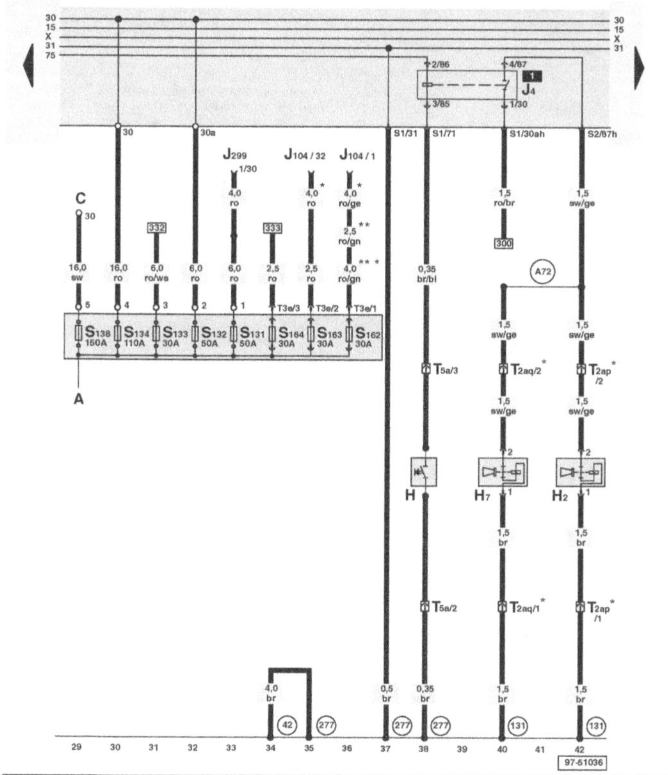

Main Fuse Box/Battery, Signal Horn

- A - Battery
- C - Generator (GEN)
- H - Signal Horn Activation
- H2 - High Tone Horn
- H7 - Low Tone Horn
- J4 - Dual Horn Relay
- J104 - ABS Control Module (w/EDL)
- J299 - Secondary Air Injection (AIR) Pump Relay
- S131 - Safety fuse 1
- S132 - Safety fuse 2
- S133 - Safety fuse 3
- S134 - Safety fuse 4
- S138 - Safety fuse 5
- S162 - Fuse -1- in fuse bracket/battery
- S163 - Fuse -2- in fuse bracket/battery
- S164 - Fuse -3- in fuse bracket/battery
- T2ap - Connector 2 pin, in engine compartment left, for High Tone Horn
- T2aq - Connector 2 pin, in engine compartment left, for Low Tone Horn
- T3e - Connector 3 pin, black, on Main Fuse Box/Battery
- T5a - Connector 5 pin, yellow, in steering wheel

- (42) - Ground connection, on steering column
- (131) - Ground connection -2-, in engine compartment wiring harness
- (277) - Ground connector -3-, in wiring harness interior
- (A72) - Connector (71), in instrument panel wiring harness

- * - only 3.2L engine, code BHE
- ** - through 2003 m.y.
- *** - from 2004 m.y.

ws = white
sw = black
ro = red
br = brown
gn = green
bl = blue
gr = grey
li = violet
ge = yellow
or = orange

USA.5703.03.21 Edition 10/03

© 2006 BentleyPublishers.com—All Rights Reserved

Audi TT

Wiring Diagram

No. 1/7

Instrument Cluster, Warning Lights

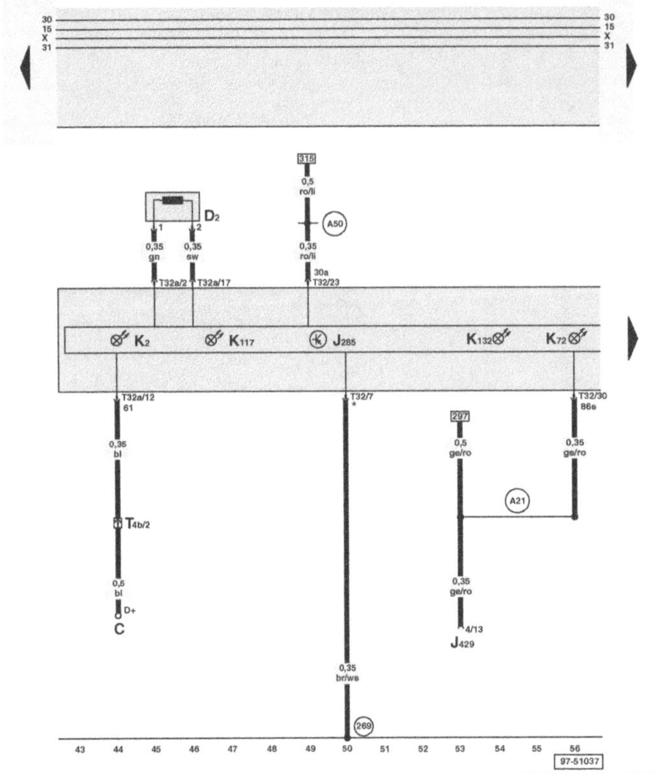

C - Generator (GEN)
D2 - Induction coil of antitheft immobilizer
J285 - Control Module with indicator unit in instrument panel insert
J429 - Control module for central locking
K2 - Generator (GEN) Warning Light
K72 - Warning light for catalyst monitoring
K117 - Warning light for anti-theft immobilizer sensor
K132 - Electronic Power Control (EPC) Warning Lamp
T4b - Connector 4 pin, in engine compartment, left
T32 - Connector 32 pin, blue, on instrument cluster
T32a - Connector 32 pin, green, on instrument cluster

(269) - Ground connector (sensor ground) -1-, in instrument panel wiring harness
(A21) - Wire connection (86s), in instrument panel wiring harness
(A50) - Plus connection (30as), in instrument panel wiring harness

* - Sensor ground output

ws = white
sw = black
ro = red
br = brown
gn = green
bl = blue
gr = grey
li = violet
ge = yellow
or = orange

USA.5703.02.21

Edition 04/03

Audi TT

Wiring Diagram

No. 1/8

Instrument Cluster, Oil Pressure Switch, Engine Coolant Level Warning Switch, Fuel Pump

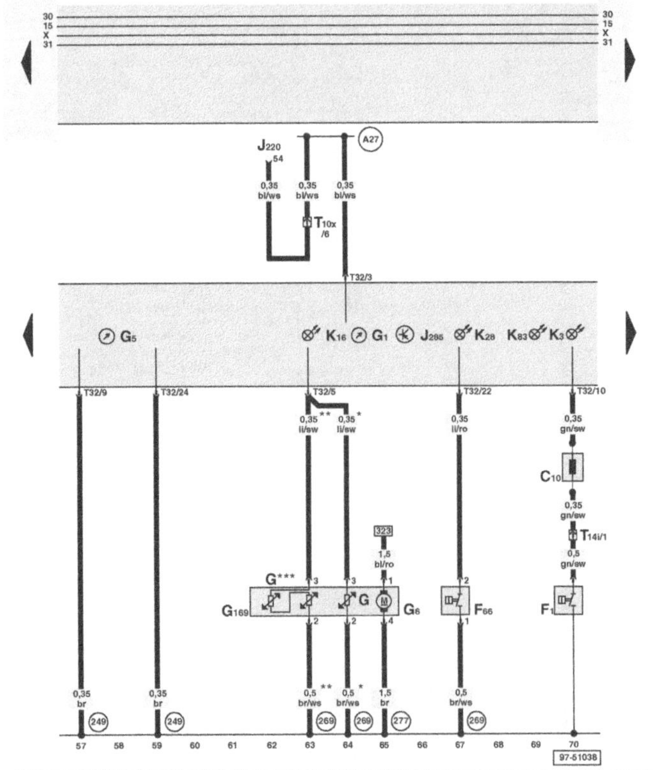

C10 - Suppressor Filter
F1 - Oil Pressure Switch
F66 - Engine Coolant Level (ECL) Warning Switch
G - Sender for fuel gauge
G1 - Fuel Gauge
G5 - Tachometer
G6 - Fuel Pump (FP)
G169 - Fuel Level Sensor 2
J220 - Motronic Engine Control Module (ECM)
J285 - Control Module with indicator unit in instrument panel insert
K3 - Oil Pressure Warning Light
K16 - Fuel Reserve Warning Light
K28 - Engine Coolant Level/Temperature (ECL/ECT) Warning Light
K83 - Malfunction Indicator Lamp (MIL)
T10x - Connector 10 pin, orange, in E-box plenum chamber
T14i - Connector 14 pin, in engine compartment, left
T32 - Connector 32 pin, blue, on instrument cluster

(249) - Ground connection -2-, in wiring harness interior
(269) - Ground connector (sensor ground) -1-, in instrument panel wiring harness
(277) - Ground connector -3-, in wiring harness interior
(A27) - Wire connection (vehicle speed signal), in instrument panel wiring harness

* - Vehicles with front wheel drive
** - Vehicles with quattro
*** - Wiring inside fuel tank

ws = white
sw = black
ro = red
br = brown
gn = green
bl = blue
gr = grey
li = violet
ge = yellow
or = orange

USA.5703.02.21

Edition 04/03

© 2006 BentleyPublishers.com—All Rights Reserved

Audi TT — Wiring Diagram — No. 1/9

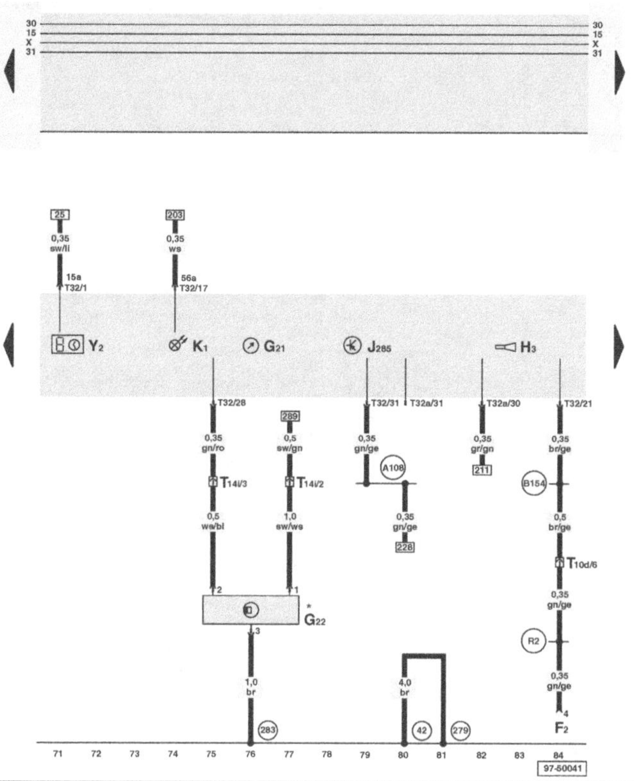

ws = white
sw = black
ro = red
br = brown
gn = green
bl = blue
gr = grey
li = violet
ge = yellow
or = orange

Instrument Cluster, Speedometer Vehicle Speed Sensor

- F2 - Door Contact Switch - Driver's side
- G21 - Speedometer
- G22 - Speedometer Vehicle Speed Sensor (VSS)
- H3 - Warning Buzzer
- J285 - Control Module with indicator unit in instrument panel insert
- K1 - Headlight High Beam Indicator Light
- T10d - Connector 10 pin, brown, connector station A pillar, left
- T14i - Connector 14 pin, in engine compartment, left
- T32 - Connector 32 pin, blue, on instrument cluster
- T32a - Connector 32 pin, green, on instrument cluster
- Y2 - Digital Clock

- (42) - Ground connection, on steering column
- (279) - Ground connector -5-, in wiring harness interior
- (283) - Ground connection 2, in wiring harness engine pre-wiring
- (A108) - Connector (speed signal), in instrument panel wiring harness
- (B154) - Connector -1- (TK), in wiring harness interior
- (R2) - Wire connection (door contact switch, left), in power window/central locking system and door contact switch wiring harness

* - Vehicles with manual transmission

USA.5703.03.21 — Edition 10/03

Audi TT — Wiring Diagram — No. 1/10

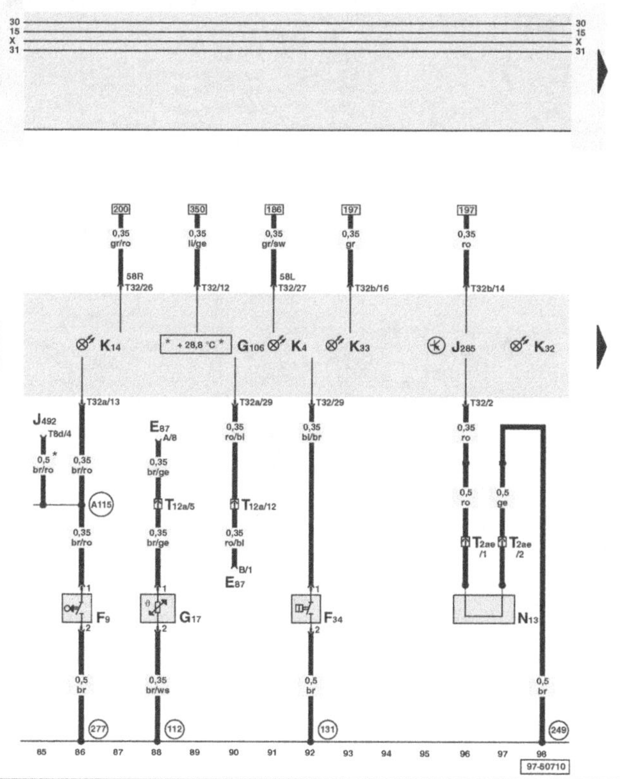

ws = white
sw = black
ro = red
br = brown
gn = green
bl = blue
gr = grey
li = violet
ge = yellow
or = orange

Instrument Cluster, Parking Brake Warning Light Switch, Brake Fluid Level Warning Switch, Outside Air Temperature Sensor

- E87 - A/C Control Head
- F9 - Parking Brake Warning Light Switch
- F34 - Brake Fluid Level Warning Switch
- G17 - Outside Air Temperature Sensor
- G106 - Outside Air Temperature Display
- J285 - Control Module with indicator unit in instrument panel insert
- J492 - All-wheel drive control module
- K4 - Park Light Indicator Light
- K14 - Parking Brake Indicator Light
- K32 - Brake Pad Wear Indicator Light
- K33 - Brake Fluid Level Warning Light
- N13 - Left Brake Pad Wear Indicator (Shear Element)
- T2ae - Connector 2 pin, on firewall, left
- T8d - Connector 8 pin, on Control module for all wheel drive
- T12a - Connector 12 pin, behind instrument panel, right
- T32 - Connector 32 pin, blue, on instrument cluster
- T32a - Connector 32 pin, green, on instrument cluster
- T32b - Connector 32 pin, grey, on instrument cluster

- (112) - Ground connection -2-, in A/C wiring harness
- (131) - Ground connection -2-, in engine compartment wiring harness
- (249) - Ground connection -2-, in wiring harness interior
- (277) - Ground connector -3-, in wiring harness interior
- (A115) - Connection (parking brake control) in instrument panel wiring harness

* - Vehicles with quattro

USA.5703.03.21 — Edition 10/03

© 2006 BentleyPublishers.com—All Rights Reserved

Audi TT — Wiring Diagram — No. 1/11

Instrument Cluster, Connector (K-diagnosis wire) for Data Link Connector (DLC)

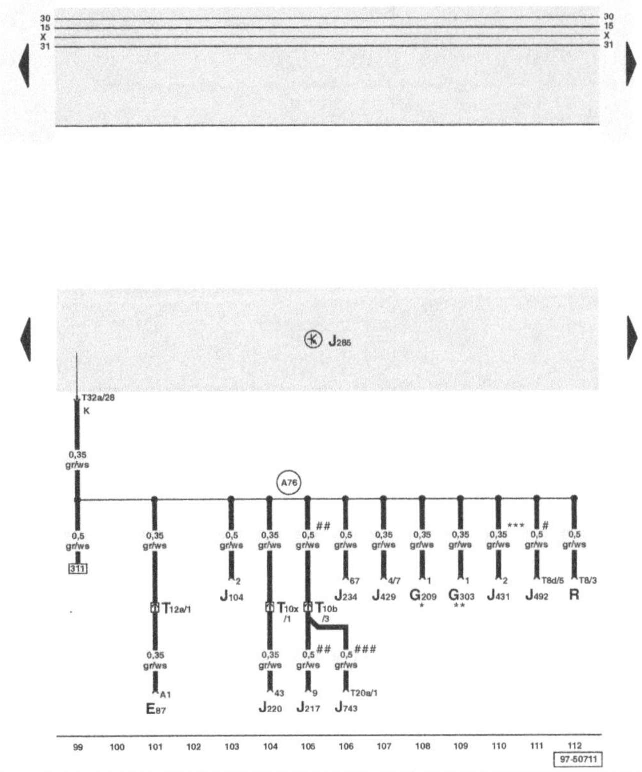

- E87 - A/C Control Head
- G209 - Ultra-sound sensor for anti-theft warning
- G303 - Radar Interior Monitoring Control Module 1
- J104 - ABS Control Module (w/EDL)
- J217 - Transmission Control Module (TCM)
- J220 - Motronic Engine Control Module (ECM)
- J234 - Airbag Control Module
- J285 - Control Module with indicator unit in instrument panel insert
- J429 - Control module for central locking
- J431 - Control module for Headlight Beam Adjusting
- J492 - All-wheel drive control module
- J743 - Direct Shift Gearbox (DSG) Mechatronics
- R - Radio
- T8 - Connector 8 pin, black, connector radio III
- T8d - Connector 8 pin, on Control module for all wheel drive
- T10b - Connector 10 pin, grey, in E-box plenum chamber
- T10x - Connector 10 pin, orange, in E-box plenum chamber
- T12a - Connector 12 pin, behind instrument panel, right
- T20a - Connector 20 pin, on Direct Shift Gearbox (DSG) Mechatronics
- T32a - Connector 32 pin, green, on instrument cluster
- (A76) - Connector (K-diagnosis wire), in instrument panel wiring harness
- * - Coupé
- ** - Roadster
- *** - Vehicles with Automatic Headlight Beam Adjusting
- # - Vehicles with quattro
- ## - Vehicles with Automatic Transmission 09G
- ### - Vehicles with Direct Shift Gearbox (DSG) 02E

ws = white
sw = black
ro = red
br = brown
gn = green
bl = blue
gr = grey
li = violet
ge = yellow
or = orange

Audi TT — Wiring Diagram — No. 1/12

Instrument Cluster, Engine Coolant Temperature Gauge

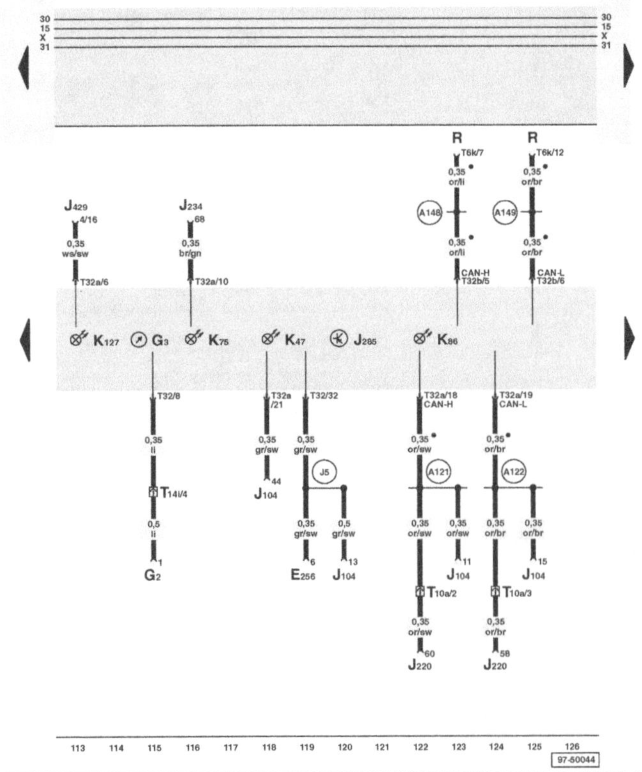

- E256 - ASR/ESP button
- G2 - Engine Coolant Temperature (ECT) Sensor
- G3 - Engine Coolant Temperature (ECT) Gauge
- J104 - ABS Control Module (w/EDL)
- J220 - Motronic Engine Control Module (ECM)
- J234 - Airbag Control Module
- J285 - Control Module with indicator unit in instrument panel insert
- J429 - Control module for central locking
- K47 - ABS Warning Light
- K75 - Airbag Malfunction Indicator Lamp (MIL)
- K86 - Traction Control Indicator Light
- K127 - Indicator light for open trunk
- R - Radio
- T6k - Connector 6 pin, green, connector radio I-2
- T10a - Connector 10 pin, white, in E-box plenum chamber
- T14i - Connector 14 pin, in engine compartment, left
- T32 - Connector 32 pin, blue, on instrument cluster
- T32a - Connector 32 pin, green, on instrument cluster
- T32b - Connector 32 pin, grey, on instrument cluster
- (A121) - Connection (high bus) in instrument panel wiring harness
- (A122) - Connection (low bus) in instrument panel wiring harness
- (A148) - Connection (high bus, navigation), in instrument panel wiring harness
- (A149) - Connection (low bus, navigation), in instrument panel wiring harness
- (J5) - Wire connection -2-, in ABS wiring harness
- • - CAN-Bus (Data-Bus)

ws = white
sw = black
ro = red
br = brown
gn = green
bl = blue
gr = grey
li = violet
ge = yellow
or = orange

© 2006 BentleyPublishers.com—All Rights Reserved

Audi TT — Wiring Diagram — No. 1/13

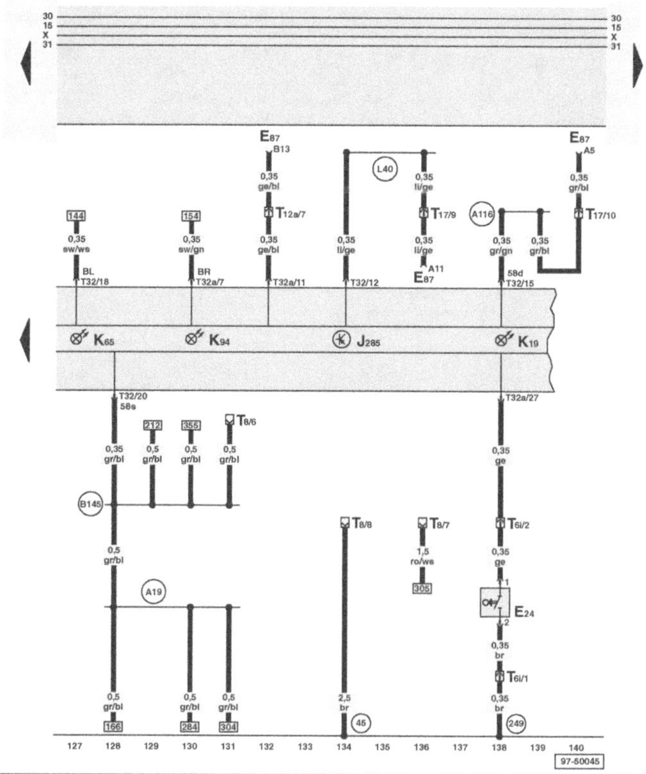

Instrument Cluster, Turn Signal Indicator Lights, Connector Radio, Seat Belt Warning Light

- E24 - Left Seat Belt Switch
- E87 - A/C Control Head
- J285 - Control Module with indicator unit in instrument panel insert
- K19 - Seat Belt Warning Light
- K65 - Left Turn Signal Indicator Light
- K94 - Right Turn Signal Indicator Light
- T6i - Connector 6 pin, below driver's seat
- T8 - Connector 8 pin, black, connector radio III
- T12a - Connector 12 pin, behind instrument panel, right
- T17 - Connector 17 pin, behind instrument panel
- T32 - Connector 32 pin, blue, on instrument cluster
- T32a - Connector 32 pin, green, on instrument cluster

- (45) - Ground connection, behind instrument panel, center
- (249) - Ground connection -2-, in wiring harness interior
- (A19) - Wire connection (58s), in instrument panel wiring harness
- (A116) - Connection -2- (58d) in instrument panel wiring harness
- (B145) - Plus connector (58b), in wiring harness interior
- (L40) - Wire connection -2- (A/C Pressure Switch), in Climatronic wiring harness

ws = white
sw = black
ro = red
br = brown
gn = green
bl = blue
gr = grey
li = violet
ge = yellow
or = orange

USA.5703.02.21 — Edition 04/03

Audi TT — Wiring Diagram — No. 1/14

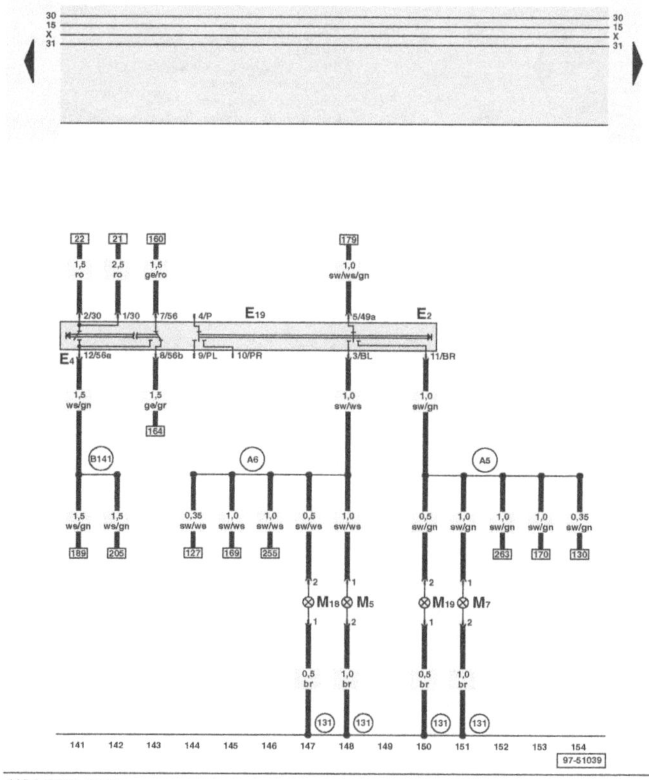

Turn Signal Switch, Headlight Dimmer/Flasher Switch, Park Light Switch, Front Turn Signal Light

- E2 - Turn Signal Switch
- E4 - Headlight Dimmer/Flasher Switch
- E19 - Park Light Switch
- M5 - Left Front Turn Signal Light
- M7 - Right Front Turn Signal Light
- M18 - Left, Side Turn Signal Light
- M19 - Right, Side Turn Signal Light

- (131) - Ground connection -2-, in engine compartment wiring harness
- (A5) - Plus connection (right turn signal), in instrument panel wiring harness
- (A6) - Plus connection (left turn signal), in instrument panel wiring harness
- (B141) - Plus connector -1- (56a), in wiring harness interior

ws = white
sw = black
ro = red
br = brown
gn = green
bl = blue
gr = grey
li = violet
ge = yellow
or = orange

USA.5703.02.21 — Edition 04/03

©2006 BentleyPublishers.com—All Rights Reserved

Audi TT

Wiring Diagram

No. 1/15

Light Switch, Fog Light Switch

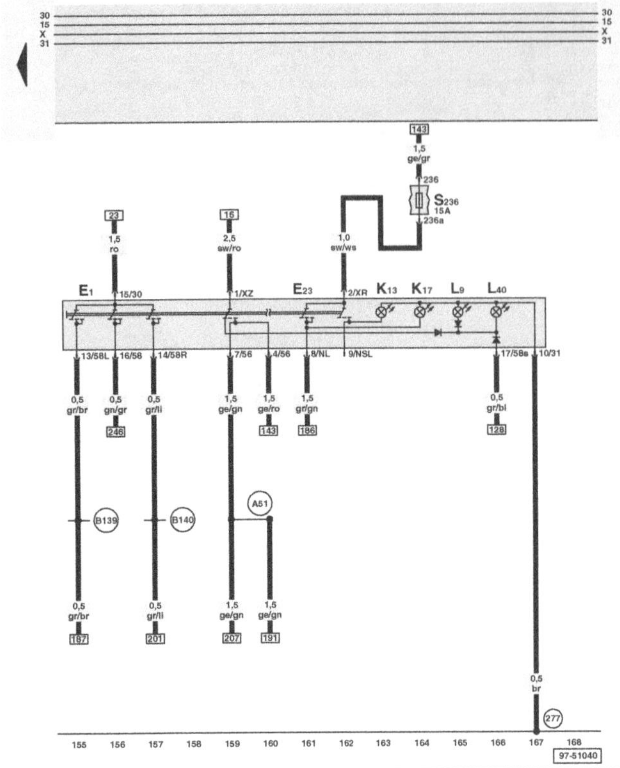

- E1 - Light Switch
- E23 - Fog Light Switch
- K13 - Rear Fog Light Indicator Light
- K17 - Fog Light Indicator Light
- L9 - Headlight Switch Light
- L40 - Front And Rear Fog Light Switch Light
- S236 - Fuse 36 in fuse holder
- (277) - Ground connector -3-, in wiring harness interior
- (A51) - Wire connection (56), in instrument panel wiring harness
- (B139) - Connector (PL), in wiring harness, interior
- (B140) - Connector (PR), in wiring harness, interior

ws = white
sw = black
ro = red
br = brown
gn = green
bl = blue
gr = grey
li = violet
ge = yellow
or = orange

USA.5703.02.21

Edition 04/03

Audi TT

Wiring Diagram

No. 1/16

Emergency Flasher Switch

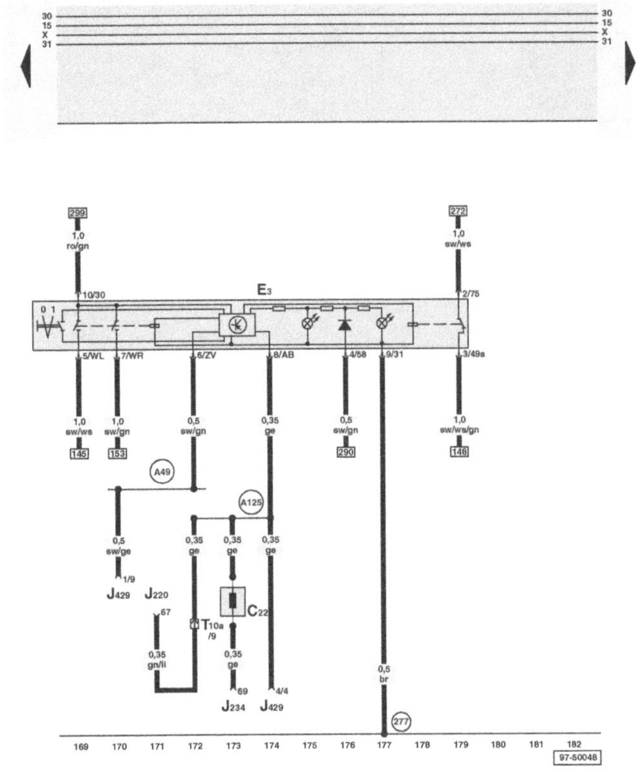

- C22 - Airbag Suppresser
- E3 - Emergency Flasher Switch
- J220 - Motronic Engine Control Module (ECM)
- J234 - Airbag Control Module
- J429 - Control module for central locking
- T10a - Connector 10 pin, white, in E-box plenum chamber
- (277) - Ground connector -3-, in wiring harness interior
- (A49) - Wire connection -1-, in instrument panel wiring harness
- (A125) - Connection (crash signal) in instrument panel wiring harness

ws = white
sw = black
ro = red
br = brown
gn = green
bl = blue
gr = grey
li = violet
ge = yellow
or = orange

USA.5703.02.21

Edition 04/03

© 2006 BentleyPublishers.com—All Rights Reserved

Audi TT

Wiring Diagram

No. 1/17

Fog Light Relay, Left Headlight, Fuses, Lamp Control Module

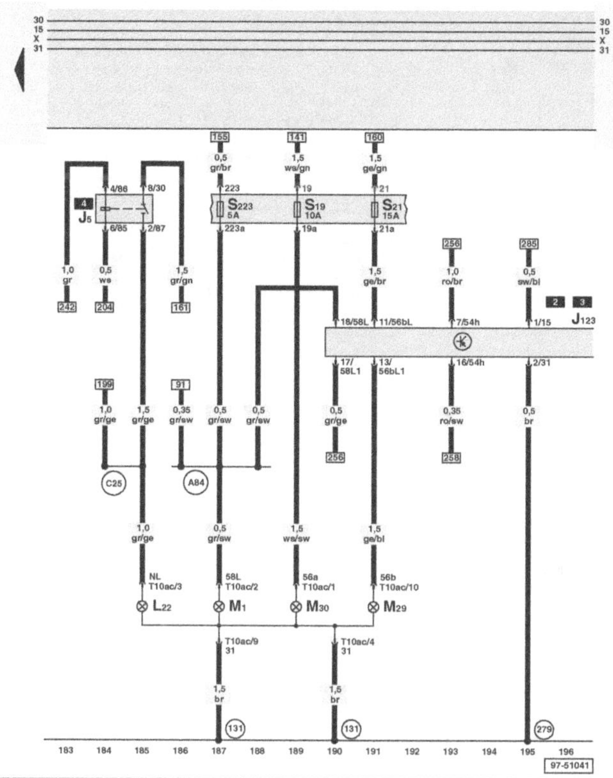

J5 - Fog Light Relay
J123 - Lamp Control Module
L22 - Left Front Fog Light
M1 - Left Parking Light
M29 - Left Low Beam Headlight
M30 - Left High Beam Headlight
S19 - Fuse
S21 - Fuse
S223 - Fuse 23 in fuse holder
T10ac - Connector 10 pin, on headlight, left

(131) - Ground connection -2-, in engine compartment wiring harness
(279) - Ground connector -5-, in wiring harness interior
(A84) - Connector (58L), in instrument panel wiring harness
(C25) - Wire connection (fog lights), in front light wiring harness

ws = white
sw = black
ro = red
br = brown
gn = green
bl = blue
gr = grey
li = violet
ge = yellow
or = orange

USA.5703.02.21

Edition 04/03

Audi TT

Wiring Diagram

No. 1/18

Right Headlight, Fuses, Lamp Control Module

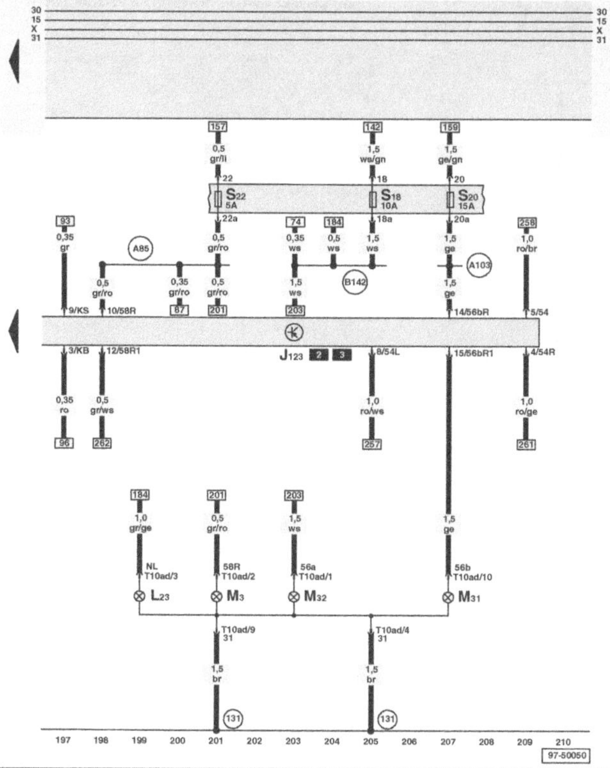

J123 - Lamp Control Module
L23 - Right Front Fog Light
M3 - Right Parking Light
M31 - Right Low Beam Headlight
M32 - Right High Beam Headlight
S18 - Fuse
S20 - Fuse
S22 - Fuse
T10ad - Connector 10 pin, on headlight, right

(131) - Ground connection -2-, in engine compartment wiring harness
(A85) - Connector (58R), in instrument panel wiring harness
(A103) - Connector -2- (56), in instrument panel wiring harness
(B142) - Plus connector -2- (56a), in wiring harness interior

ws = white
sw = black
ro = red
br = brown
gn = green
bl = blue
gr = grey
li = violet
ge = yellow
or = orange

USA.5703.02.21

Edition 04/03

©2006 BentleyPublishers.com—All Rights Reserved

Audi TT — Wiring Diagram — No. 1/19

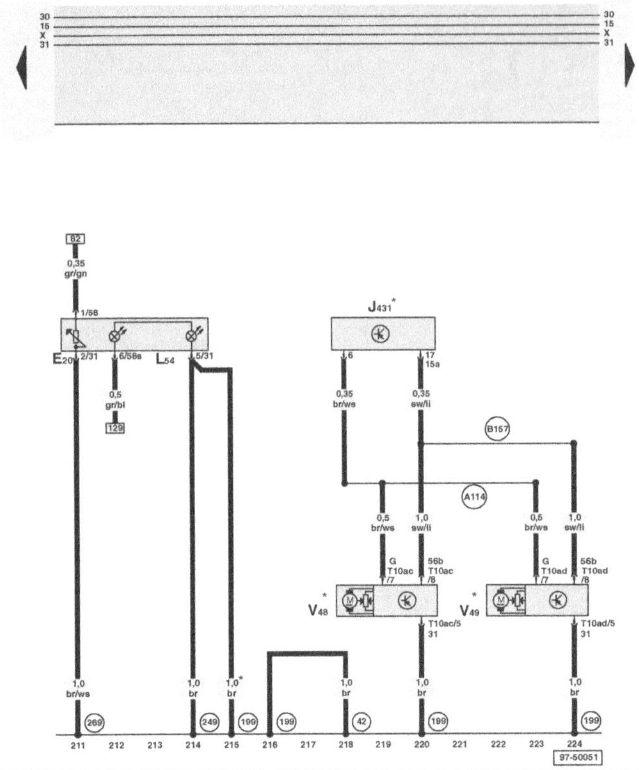

Headlight Adjuster, Headlight Beam Adjusting Motors

- E20 - Instrument Panel Light Dimmer Switch
- J431 - Control module for Headlight Beam Adjusting
- L54 - Headlight Adjuster Control Light
- T10ac - Connector 10 pin, on headlight, left
- T10ad - Connector 10 pin, on headlight, right
- V48 - Left Headlight Beam Adjusting Motor
- V49 - Right Headlight Beam Adjusting Motor

- (42) - Ground connection, on steering column
- (199) - Ground connection -3-, in instrument panel wiring harness
- (249) - Ground connection -2-, in wiring harness interior
- (269) - Ground connector (sensor ground) -1-, in instrument panel wiring harness
- (A114) - Connection (headlight beam adjusting) in instrument panel wiring harness
- (B157) - Plus connector -3- (15a), in wiring harness interior

- * - Where applicable

ws = white
sw = black
ro = red
br = brown
gn = green
bl = blue
gr = grey
li = violet
ge = yellow
or = orange

USA.5703.02.21 — Edition 04/03

Audi TT — Wiring Diagram — No. 1/20

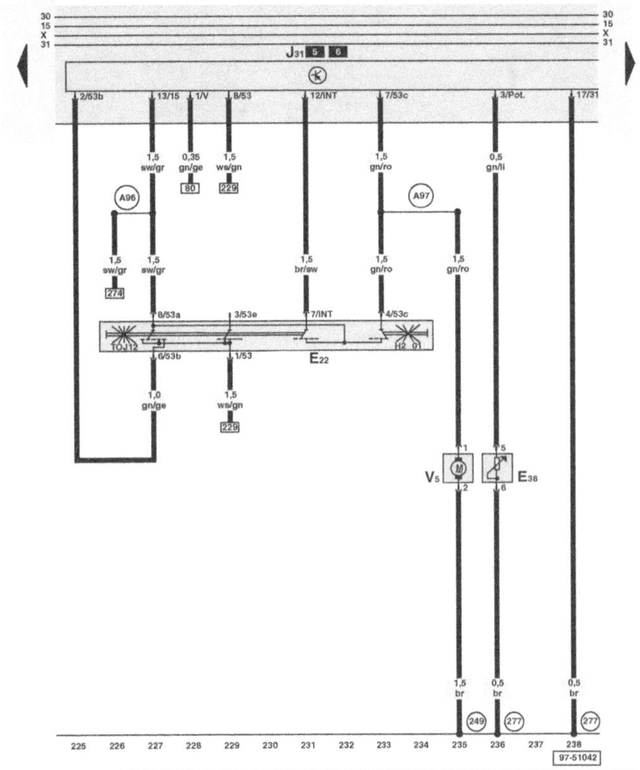

Wiper/Washer Intermittent Relay, Windshield Wiper/Washer Switch, Windshield Washer Pump

- E22 - Windshield Wiper/Washer Switch
- E38 - Windshield Wiper Intermittent Regulator
- J31 - Wiper/Washer Intermittent Relay
- V5 - Windshield Washer Pump

- (249) - Ground connection -2-, in wiring harness interior
- (277) - Ground connector -3-, in wiring harness interior

- (A96) - Connector (53a), in instrument panel wiring harness
- (A97) - Connector (53), in instrument panel wiring harness

ws = white
sw = black
ro = red
br = brown
gn = green
bl = blue
gr = grey
li = violet
ge = yellow
or = orange

USA.5703.02.21 — Edition 04/03

© 2006 BentleyPublishers.com—All Rights Reserved

Audi TT

Wiring Diagram

No. 1/21

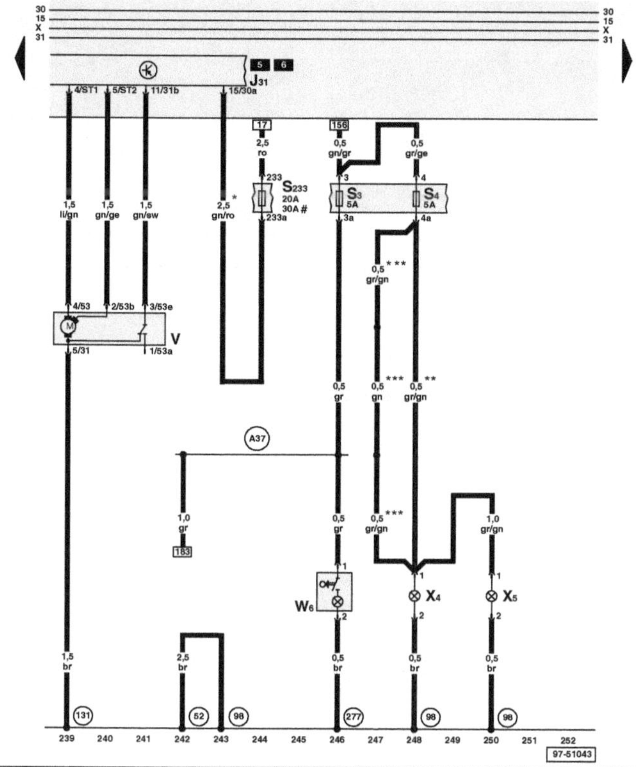

Wiper/Washer Intermittent Relay, Windshield Wiper Motor, Glove Compartment Light, License Plate Light, Fuses

J31	- Wiper/Washer Intermittent Relay
S3	- Fuse
S4	- Fuse
S233	- Fuse 33 in fuse holder
V	- Windshield Wiper Motor
W6	- Glove Compartment Light
X4	- Left License Plate Light
X5	- Right License Plate Light
(52)	- Ground connection, in rear lid, left
(98)	- Ground connection, in rear lid wiring harness
(131)	- Ground connection -2-, in engine compartment wiring harness
(277)	- Ground connector -3-, in wiring harness interior
(A37)	- Wire connection (58a), in instrument panel wiring harness
*	- Vehicles with Headlight Washer
**	- Coupé
***	- Roadster
#	- from 2005 m.y.

ws = white
sw = black
ro = red
br = brown
gn = green
bl = blue
gr = grey
li = violet
ge = yellow
or = orange

USA.5703.04.21

Edition 06/04

Audi TT

Wiring Diagram

No. 1/22

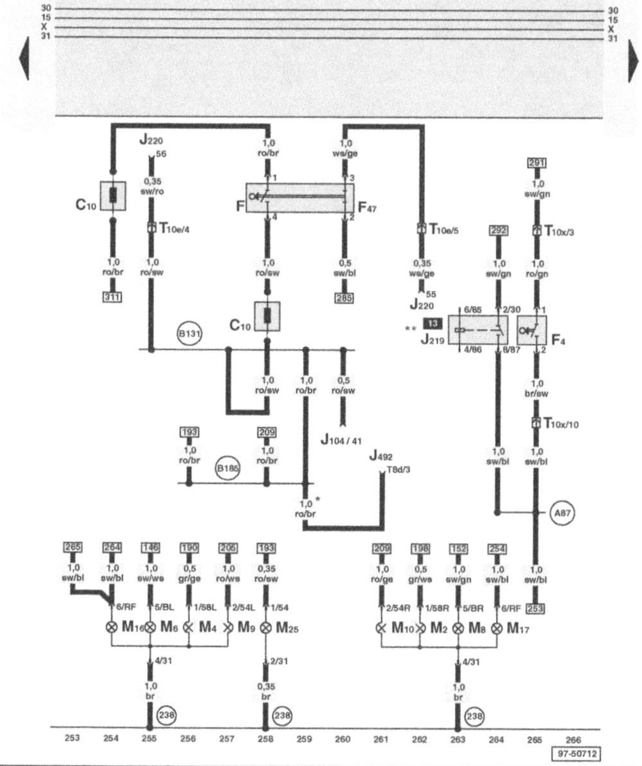

Tail Light, Brake Light, Rear Fog Light

C10	- Suppressor Filter
F	- Brake Light Switch
F4	- Back-Up Light Switch
F47	- Brake Vacuum Vent Valve Switch for Cruise Control/Diesel Fuel Injection
J104	- ABS Control Module (w/EDL)
J219	- Back-Up Light Relay
J220	- Motronic Engine Control Module (ECM)
J492	- All-wheel drive control module
M2	- Right Tail Light
M4	- Left Tail Light
M6	- Left Rear Turn Signal Light
M8	- Right Rear Turn Signal Light
M9	- Left Brake Light
M10	- Right Brake Light
M16	- Left Back-Up Light
M17	- Right Back-Up Light
M25	- High-mount Brake Light
T8d	- Connector 8 pin, on Control module for all wheel drive
T10e	- Connector 10 pin, black, in E-box plenum chamber
T10x	- Connector 10 pin, orange, in E-box plenum chamber
(238)	- Ground connection -1-, in wiring harness interior
(A87)	- Connector (RF), in instrument panel wiring harness
(B131)	- Connector (54), in wiring harness interior
(B185)	- Connection 2 (54), in passenger compartment wiring harness
*	- Vehicles with quattro
**	- Vehicles with Direct Shift Gearbox (DSG) 02E

ws = white
sw = black
ro = red
br = brown
gn = green
bl = blue
gr = grey
li = violet
ge = yellow
or = orange

USA.5703.02.21

Edition 04/03

©2006 BentleyPublishers.com—All Rights Reserved

Standard Equipment (2003-2006)

Audi TT

Wiring Diagram **No. 1/23**

Washer Nozzle Heater, Fuses

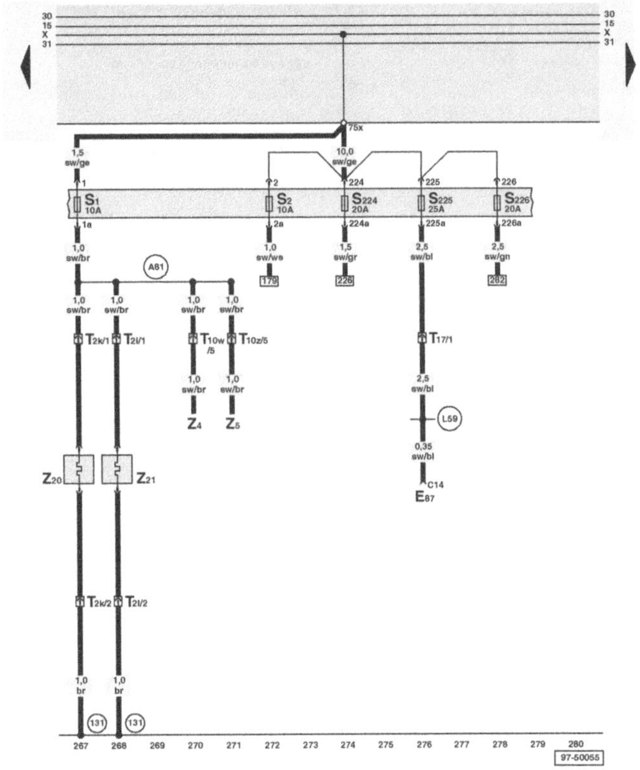

E87 - A/C Control Head
S1 - Fuse
S2 - Fuse
S224 - Fuse 24 in fuse holder
S225 - Fuse 25 in fuse holder
S226 - Fuse 26 in fuse holder
T2k - Connector 2 pin, black, on engine hood
T2l - Connector 2 pin, black, on engine hood
T10w - Connector 10 pin, blue, connector station A pillar, left
T10z - Connector 10 pin, blue, connector station A pillar, right
T17 - Connector 17 pin, behind instrument panel
Z4 - Heated outside mirror, driver side
Z5 - Heated outside mirror, passenger side
Z20 - Left Washer Nozzle Heater
Z21 - Right Washer Nozzle Heater

(131) - Ground connection -2-, in engine compartment wiring harness
(A81) - Connector -2- (X), in instrument panel wiring harness
(L59) - Plus connection (30), in wiring harness heater

ws = white
sw = black
ro = red
br = brown
gn = green
bl = blue
gr = grey
li = violet
ge = yellow
or = orange

USA.5703.02.21 **Edition 04/03**

Audi TT

Wiring Diagram **No. 1/24**

Heated Rear Window, Fuses

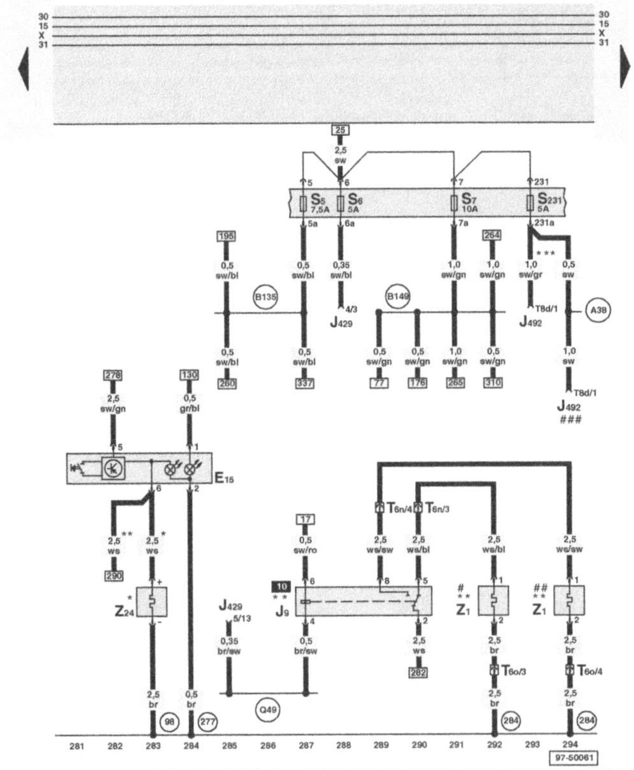

E15 - Rear Window Defogger Switch
J9 - Rear Window Defogger Relay
J429 - Control module for central locking
J492 - All-wheel drive control module
S5 - Fuse
S6 - Fuse
S7 - Fuse
S231 - Fuse 31 in fuse holder
T6n - Connector 6 pin, Hardtop left
T6o - Connector 6 pin, Hardtop right
T8d - Connector 8 pin, on Control module for all wheel drive
Z1 - Heated rear window
Z24 - Rear Window Defogger With Window Antenna

(98) - Ground connection, in rear lid wiring harness
(277) - Ground connector -3-, in wiring harness interior
(284) - Ground connection -6- in wiring harness interior
(A38) - Plus connection -2- (15a), in instrument panel wiring harness
(B135) - Connector (15a), in wiring harness interior
(B149) - Plus connector -2- (15a), in wiring harness interior
(Q49) - Connection (closed), in wiring harness interior

* - Coupé
** - Roadster
*** - Vehicles with quattro
# - Softtop
- Hardtop (where applicable)
- Vehicles with S-tronic Transmission 02E

ws = white
sw = black
ro = red
br = brown
gn = green
bl = blue
gr = grey
li = violet
ge = yellow
or = orange

USA.5703.02.21 **Edition 04/03**

©2006 BentleyPublishers.com—All Rights Reserved

Audi TT

Wiring Diagram **No. 1/25**

Cigarette Lighter, Fuses

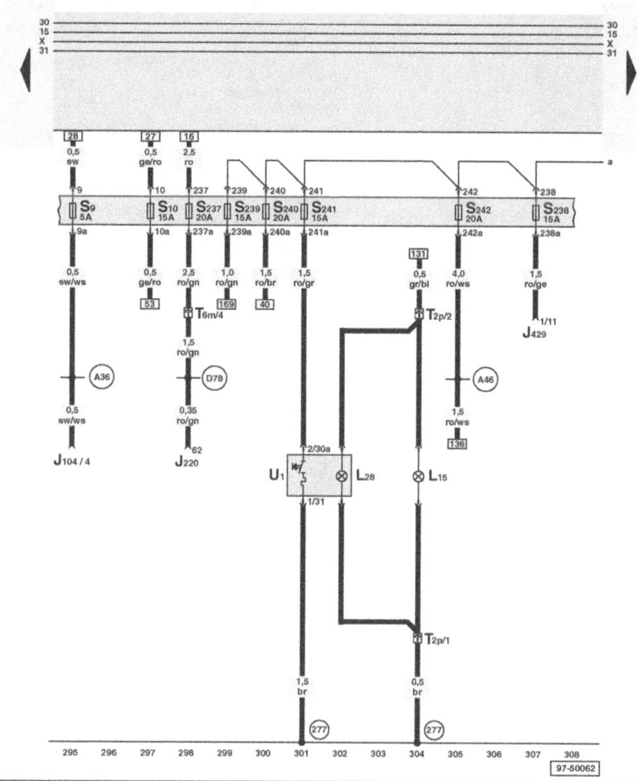

J104	- ABS Control Module (w/EDL)
J220	- Motronic Engine Control Module (ECM)
J429	- Control module for central locking
L15	- Ashtray Light
L28	- Cigarette Lighter Light
S9	- Fuse
S10	- Fuse
S237	- Fuse 37 in fuse holder
S238	- Fuse 38 in fuse holder
S239	- Fuse 39 in fuse holder
S240	- Fuse 40 in fuse holder
S241	- Fuse 41 in fuse holder
S242	- Fuse 42 in fuse holder
T2p	- Connector 2 pin, behind console
T6m	- Connector 6 pin, brown, in E-box plenum chamber
U1	- Cigarette Lighter
(277)	- Ground connector -3-, in wiring harness interior
(A36)	- Wire connection (75a), in instrument panel wiring harness
(A46)	- Plus connection (30-from radio), in instrument panel wiring harness
(D78)	- Plus connection -1- (30a), in engine compartment wiring harness

ws = white
sw = black
ro = red
br = brown
gn = green
bl = blue
gr = grey
li = violet
ge = yellow
or = orange

USA.5703.02.21 Edition 04/03

Audi TT

Wiring Diagram **No. 1/26**

Data Link Connector (DLC), Fuses

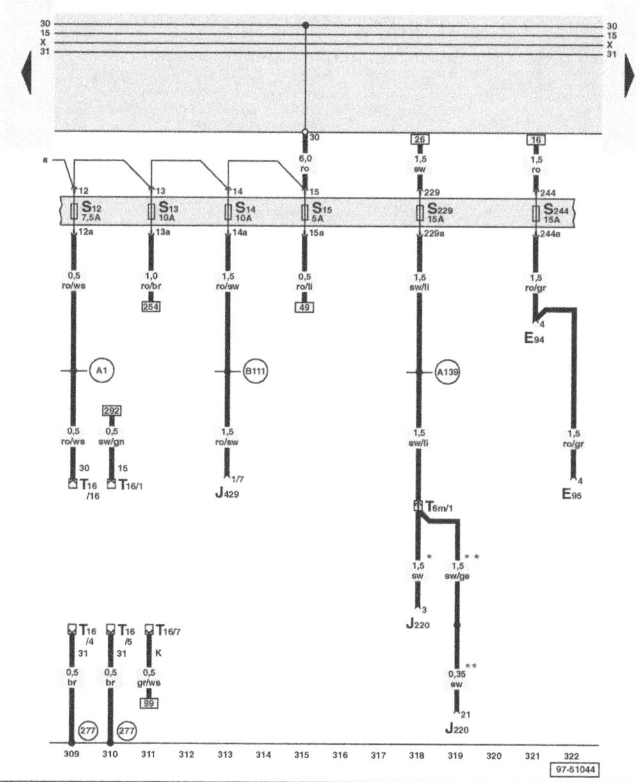

E94	- Driver's Heated Seat Adjuster
E95	- Passenger's Heated Seat Adjuster
J220	- Motronic Engine Control Module (ECM)
J429	- Control module for central locking
S12	- Fuse
S13	- Fuse
S14	- Fuse
S15	- Fuse
S229	- Fuse 29 in fuse holder
S244	- Fuse 44 in fuse holder
T6m	- Connector 6 pin, brown, in E-box plenum chamber
T16	- Connector 16 pin, black, Data Link Connector (DLC), below storage compartment, driver's side
(277)	- Ground connector -3-, in wiring harness interior
(A1)	- Plus connection (30a), in instrument panel wiring harness
(A139)	- Connection 3 (15) (in instrument panel wiring harness)
(B111)	- Connector (30a), in wiring harness interior
*	- only 1.8L engines
**	- only 3.2L engine

ws = white
sw = black
ro = red
br = brown
gn = green
bl = blue
gr = grey
li = violet
ge = yellow
or = orange

USA.5703.02.21 Edition 04/03

©2006 BentleyPublishers.com—All Rights Reserved

Audi TT — Wiring Diagram — No. 1/27

Coolant (Fan Control) Control Module, Fuses

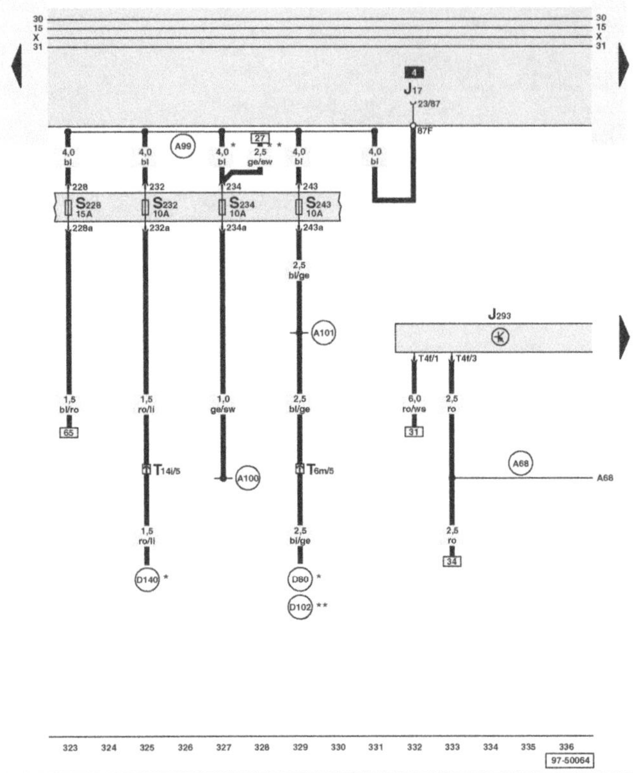

J17 - Fuel Pump (FP) Relay
J293 - Coolant FC (Fan Control) Control Module, in engine compartment, left front
S228 - Fuse 28 in fuse holder
S232 - Fuse 32 in fuse holder
S234 - Fuse 34 in fuse holder
S243 - Fuse 43 in fuse holder
T4f - Connector 4 pin, connector B on J293
T6m - Connector 6 pin, brown, in E-box plenum chamber
T14i - Connector 14 pin, in engine compartment, left

(A68) - Connector (C15, A/C), in instrument panel wiring harness
(A99) - Connector -1- (87), in instrument panel wiring harness
(A100) - Connector -2- (87), in instrument panel wiring harness
(A101) - Connector -3- (87), in instrument panel wiring harness
(D80) - Plus connection (87a- for EVAP system solenoid), in engine compartment wiring harness
(D102) - Wire connection -2-, in engine compartment wiring harness
(D140) - Connector (injectors), in wiring harness, engine pre-wiring

* - only 1.8L engines
** - only 3.2L engine

ws = white
sw = black
ro = red
br = brown
gn = green
bl = blue
gr = grey
li = violet
ge = yellow
or = orange

USA.5703.02.21 — Edition 04/03

Audi TT — Wiring Diagram — No. 1/28

Coolant (Fan Control) Control Module, Coolant Fan Control Thermal Switch, Coolant Fan

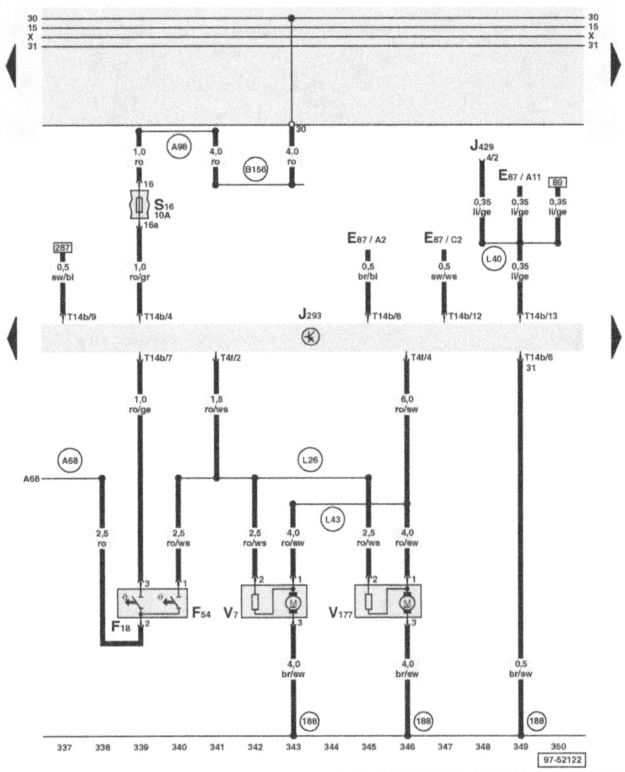

E87 - A/C Control Head
F18 - Coolant Fan Control (FC) Thermal Switch
F54 - Coolant Fan Control (FC) Thermal Switch
J293 - Coolant FC (Fan Control) Control Module, in engine compartment, left front
J429 - Control module for central locking
S16 - Fuse
T4f - Connector 4 pin, connector B on J293
T14b - Connector 14 pin, connector A on J293
V7 - Coolant Fan
V177 - Coolant fan -2-

(188) - Ground connection -3-, in A/C wiring harness
(A68) - Connector (C15, A/C), in instrument panel wiring harness
(A98) - Plus connector -4- (30), in instrument panel wiring harness
(B156) - Plus connector (30a), in wiring harness interior
(L26) - Wire connection (coolant fan), in A/C-climatronic/engine compartment wiring harness
(L40) - Wire connection -2- (A/C Pressure Switch), in Climatronic wiring harness
(L43) - Wire connection (coolant fan), in A/C system-coolant fan wiring harness

ws = white
sw = black
ro = red
br = brown
gn = green
bl = blue
gr = grey
li = violet
ge = yellow
or = orange

USA.5703.02.21 — Edition 04/03

© 2006 BentleyPublishers.com—All Rights Reserved

Audi TT **Wiring Diagram** **No. 1/29**

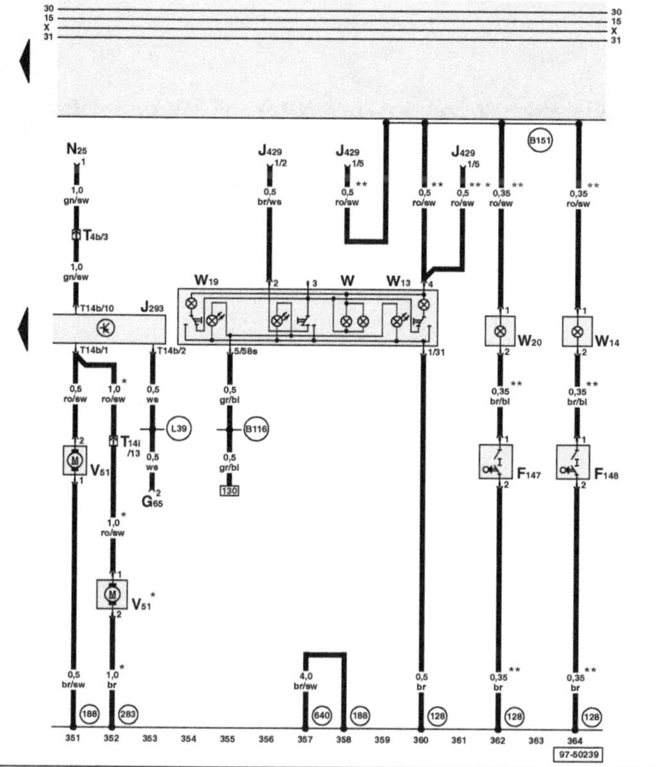

ws = white
sw = black
ro = red
br = brown
gn = green
bl = blue
gr = grey
li = violet
ge = yellow
or = orange

Interior Light, Make-Up Mirror Light, After-Run Coolant Pump

- F147 - Left Make-Up Mirror Light Switch
- F148 - Right Make-Up Mirror Light Switch
- G65 - High Pressure Sensor
- J293 - Coolant FC (Fan Control) Control Module, in engine compartment, left front
- J429 - Control module for central locking
- N25 - A/C Clutch
- T4b - Connector 4 pin, in engine compartment, left
- T14b - Connector 14 pin, connector A on J293
- T14i - Connector 14 pin, in engine compartment, left
- V51 - After-Run Coolant Pump
- W - Front Interior Light
- W13 - Passenger's Map/Reading Light
- W14 - Right Make-up Mirror Light
- W19 - Left Front Reading Light
- W20 - Left Make-up Mirror Light

- (128) - Ground connection -1-, in interior light wiring harness
- (188) - Ground connection -3-, in A/C wiring harness
- (283) - Ground connection 2, in wiring harness engine pre-wiring
- (640) - Ground Connection (in engine compartment, left)
- (B116) - Connector (58s), in wiring harness interior
- (B151) - Plus connector -3- (30a), in wiring harness interior
- (L39) - Wire connection -1- (A/C Pressure Switch), in Climatronic wiring harness

- * - only 3.2L engine
- ** - not applicable to Roadster from 2005 m.y.
- *** - only Roadster from 2005 m.y.

USA.5703.04.21 Edition 06/04

THIS SPACE INTENTIONALLY

LEFT BLANK

©2006 BentleyPublishers.com—All Rights Reserved

1.8L - Engine Motronic Multiport Fuel Injection (MFI)/ 132 kW, codes ATC/AWP & 165 kW, code BEA

2003 m.y.

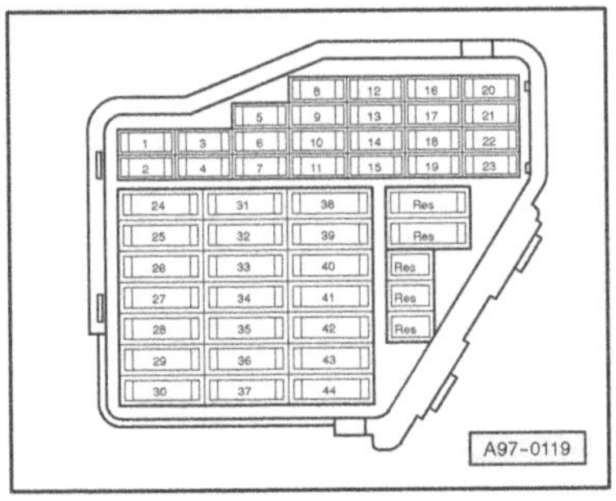

Fuse Panel

Fuse Colors:

30 A - Green
25 A - White
20 A - Yellow
15 A - Blue
10 A - Red
7,5 A - Brown
5 A - Beige
1 A - Black

Starting with fuse position 23, fuses in the fuse holder are identified with 223 in the wiring diagram.

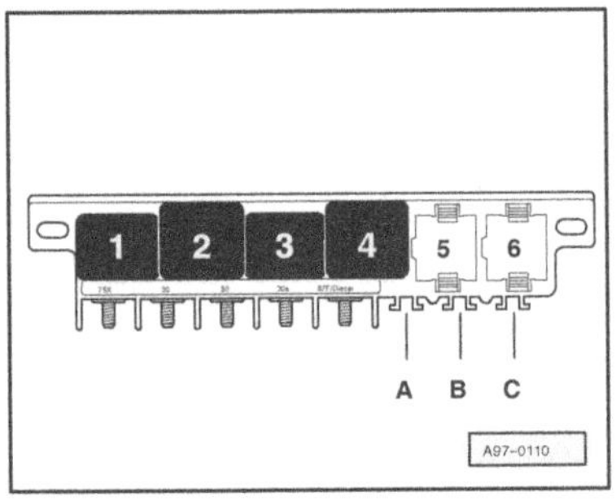

Micro Central Electric Panel

Relay Location:

4 - Fuel Pump (FP) Relay, J17
B - Hydraulic Pump Relay Fuse, S279

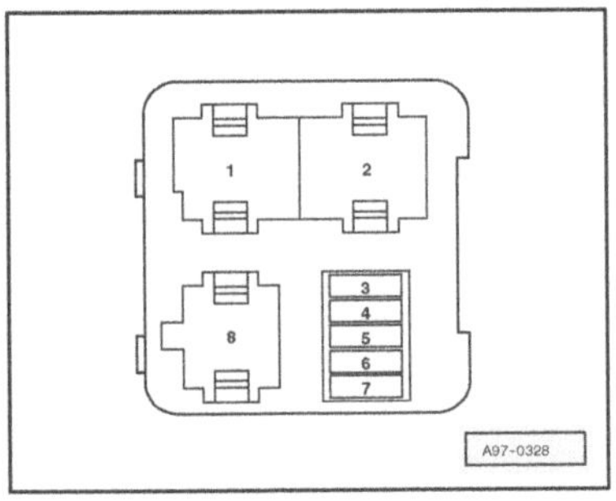

W-Box in engine compartment

Relay Location:

1 - Secondary Air Injection (AIR) Pump Relay, J299
2 - Motronic Engine Control Module (ECM) Power Supply Relay, J271

© 2006 BentleyPublishers.com—All Rights Reserved

Audi TT **Wiring Diagram** **No. 2/3**

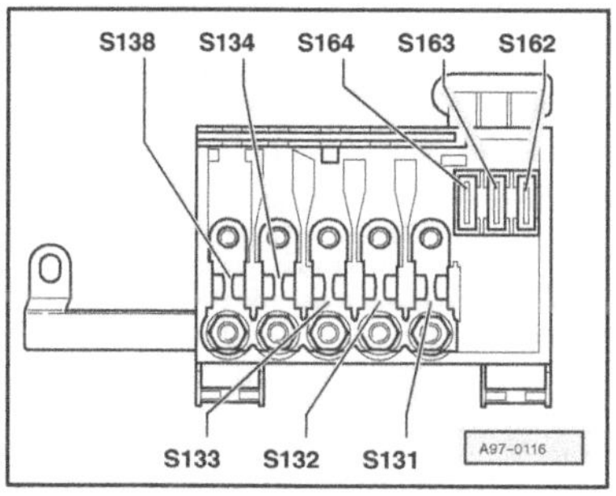

Main Fuse Box/Battery

Fuses Location:

S131 - Safety fuse 1
S132 - Safety fuse 2
S133 - Safety fuse 3
S134 - Safety fuse 4
S138 - Safety fuse 5
S162 - Fuse 1 in fuse bracket/battery
S163 - Fuse 2 in fuse bracket/battery
S164 - Fuse 3 in fuse bracket/battery

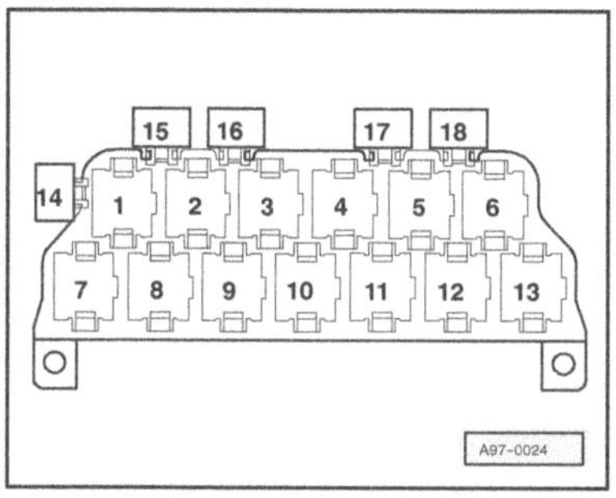

13 - Fold Relay Panel

Relay Location:

5 - Starting Interlock Relay, J207*
11 - Starting Interlock Relay, J207**

* - Vehicles with Manual Transmission
** - Vehicles with Automatic Transmission 09G

Audi TT **Wiring Diagram** **No. 2/4**

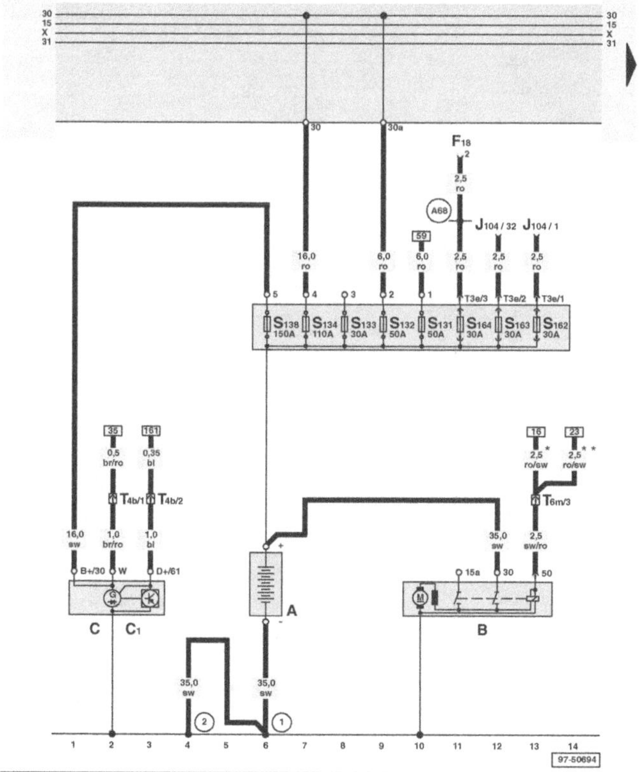

Battery, Starter, Generator, Main Fuse Box/Battery

A - Battery
B - Starter
C - Generator (GEN)
C1 - Voltage Regulator (VR)
F18 - Coolant Fan Control (FC) Thermal Switch
J104 - ABS Control Module (w/EDL)
S131 - Safety fuse 1
S132 - Safety fuse 2
S133 - Safety fuse 3
S134 - Safety fuse 4
S138 - Safety fuse 5
S162 - Fuse -1- in fuse bracket/battery
S163 - Fuse -2- in fuse bracket/battery
S164 - Fuse -3- in fuse bracket/battery
T3e - Connector 3 pin, black, on Main Fuse Box/Battery
T4b - Connector 4 pin, in engine compartment, left
T6m - Connector 6 pin, brown, in E-box plenum chamber

(1) - Ground strap, battery to body

(2) - Ground strap, transmission to body

(A68) - Connector (C15, A/C), in instrument panel wiring harness

* - Vehicles with Manual Transmission
** - Vehicles with Automatic Transmission 09G

ws = white
sw = black
ro = red
br = brown
gn = green
bl = blue
gr = grey
li = violet
ge = yellow
or = orange

©2006 BentleyPublishers.com—All Rights Reserved

Engine Management, 1.8L, engine code: ATC, AWP, BEA, (2003-2006)

Audi TT — Wiring Diagram — No. 2/5

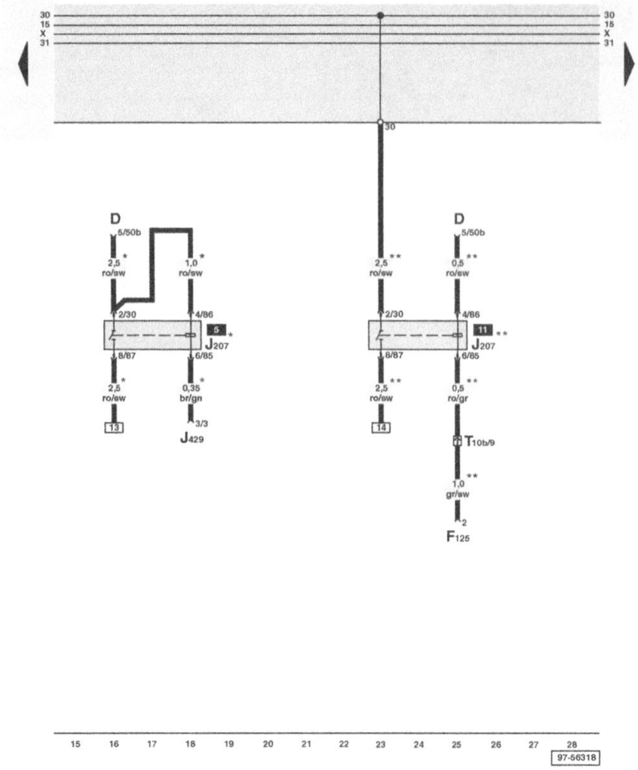

Starting Interlock Relay

D - Ignition/Starter Switch
F125 - Multi-Function Transmission Range (TR) Switch
J207 - Starting Interlock Relay
J429 - Control module for central locking
T10b - Connector 10 pin, grey, in E-box plenum chamber

* - Vehicles with Manual Transmission
** - Vehicles with Automatic Transmission 09G

ws = white
sw = black
ro = red
br = brown
gn = green
bl = blue
gr = grey
li = violet
ge = yellow
or = orange

USA.5703.02.21 Edition 04/03

Audi TT — Wiring Diagram — No. 2/6

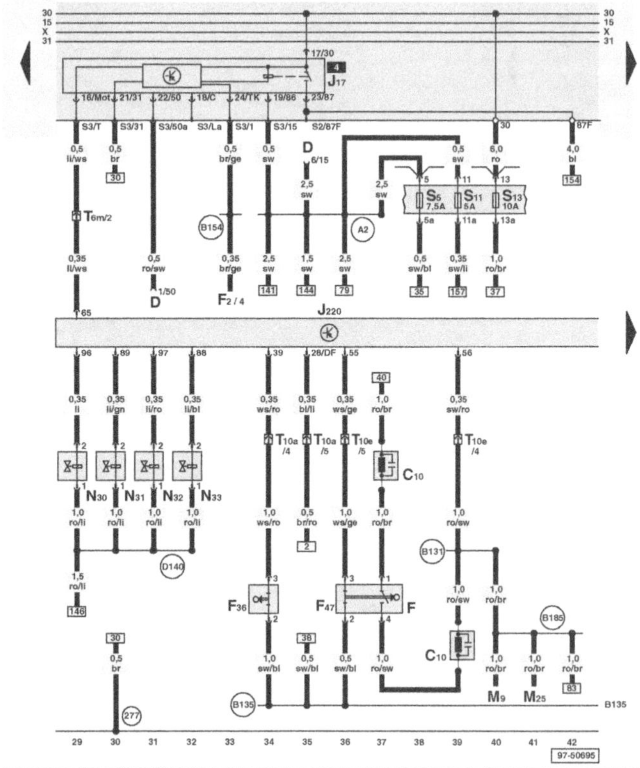

Motronic Engine Control Module, Fuel Pump Relay, Fuel Injectors, Brake Light Switch

C10 - Suppressor Filter
D - Ignition/Starter Switch
F - Brake Light Switch
F2 - Door Contact Switch - Driver's side
F36 - Clutch Vacuum Vent Valve Switch
F47 - Brake Vacuum Vent Valve Switch for Cruise Control/Diesel Fuel Injection
J17 - Fuel Pump (FP) Relay
J220 - Motronic Engine Control Module (ECM)
M9 - Left Brake Light
M25 - High-mount Brake Light
N30 - Cylinder 1 Fuel Injector
N31 - Cylinder 2 Fuel Injector
N32 - Cylinder 3 Fuel Injector
N33 - Cylinder 4 Fuel Injector
S5 - Fuse
S11 - Fuse
S13 - Fuse
T6m - Connector 6 pin, brown, in E-box plenum chamber
T10a - Connector 10 pin, white, in E-box plenum chamber
T10e - Connector 10 pin, black, in E-box plenum chamber

(277) - Ground connector -3-, in wiring harness interior
(A2) - Plus connection (15), in instrument panel wiring harness
(B131) - Connector (54), in wiring harness interior
(B135) - Connector (15a), in wiring harness interior
(B154) - Connector -1- (TK), in wiring harness interior
(B185) - Connection 2 (54), in passenger compartment wiring harness
(D140) - Connector (injectors), in wiring harness, engine pre-wiring

ws = white
sw = black
ro = red
br = brown
gn = green
bl = blue
gr = grey
li = violet
ge = yellow
or = orange

USA.5703.02.21 Edition 04/03

© 2006 BentleyPublishers.com—All Rights Reserved

Audi TT — Wiring Diagram — No. 2/7

Motronic Engine Control Module, Power Steering Pressure Switch, Cruise Control Switch, Throttle Position Sensor

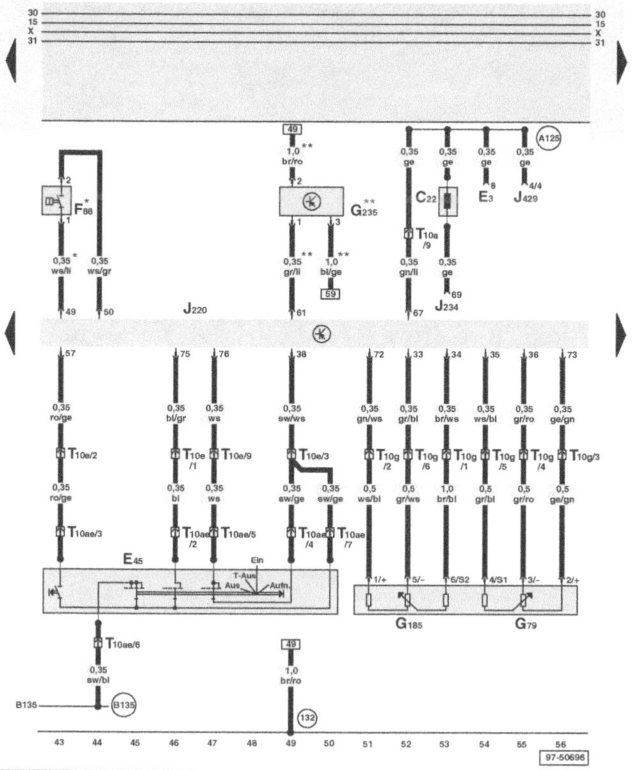

- C22 - Airbag Suppresser
- E3 - Emergency Flasher Switch
- E45 - Cruise Control Switch
- F88 - Power Steering Pressure (PSP) Switch
- G79 - Throttle Position (TP) Sensor
- G185 - Sensor -2- for accelerator pedal position
- G235 - Sensor -1- for exhaust temperature
- J220 - Motronic Engine Control Module (ECM)
- J234 - Airbag Control Module
- J429 - Control module for central locking
- T10a - Connector 10 pin, white, in E-box plenum chamber
- T10e - Connector 10 pin, black, in E-box plenum chamber
- T10g - Connector 10 pin, blue, in E-box plenum chamber
- T10ae - Connector 10 pin, on Cruise Control Switch
- (132) - Ground connection -3-, in engine compartment wiring harness
- (A125) - Connection (crash signal) in instrument panel wiring harness
- (B135) - Connector (15a), in wiring harness interior

* - only 1.8L engine code ATC, AWP
** - only 1.8L engine code BEA

ws = white
sw = black
ro = red
br = brown
gn = green
bl = blue
gr = grey
li = violet
ge = yellow
or = orange

USA.5703.02.21 — Edition 04/03

Audi TT — Wiring Diagram — No. 2/8

Motronic Engine Control Module, Secondary Air Injection Pump, Heated Oxygen Sensors

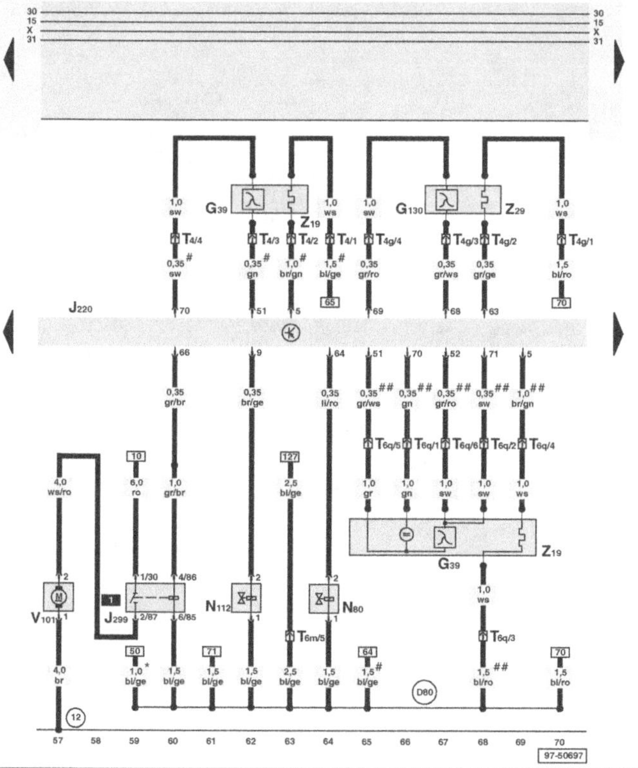

- G39 - Heated Oxygen Sensor (HO2S)
- G130 - Heated Oxygen Sensor (O2S) Behind Three Way Catalytic Converter (TWC)
- J220 - Motronic Engine Control Module (ECM)
- J299 - Secondary Air Injection (AIR) Pump Relay
- N80 - Evaporative Emission (EVAP) Canister Purge Regulator Valve
- N112 - Secondary Air Injection (AIR) Solenoid Valve
- T4 - Connector 4 pin, in engine compartment, right
- T4g - Connector 4 pin, in engine compartment, right, for Oxygen Sensor behind Three Way Catalytic Converter
- T6m - Connector 6 pin, brown, in E-box plenum chamber
- T6q - Connector 6 pin, in engine compartment right
- V101 - Secondary Air Injection (AIR) Pump Motor
- Z19 - Oxygen Sensor (O2S) Heater
- Z29 - Oxygen Sensor (O2S) Heater 1
- (12) - Ground connection, in engine compartment, left
- (D80) - Plus connection (87a- for EVAP system solenoid), in engine compartment wiring harness

* - only engine code BEA
# - Heated Oxygen Sensor with 4-pin connector
- Heated Oxygen Sensor with 6-pin connector

ws = white
sw = black
ro = red
br = brown
gn = green
bl = blue
gr = grey
li = violet
ge = yellow
or = orange

USA.5703.02.21 — Edition 04/03

©2006 BentleyPublishers.com—All Rights Reserved

Engine Management, 1.8L, engine code: ATC, AWP, BEA, (2003-2006)

Audi TT — Wiring Diagram — No. 2/9

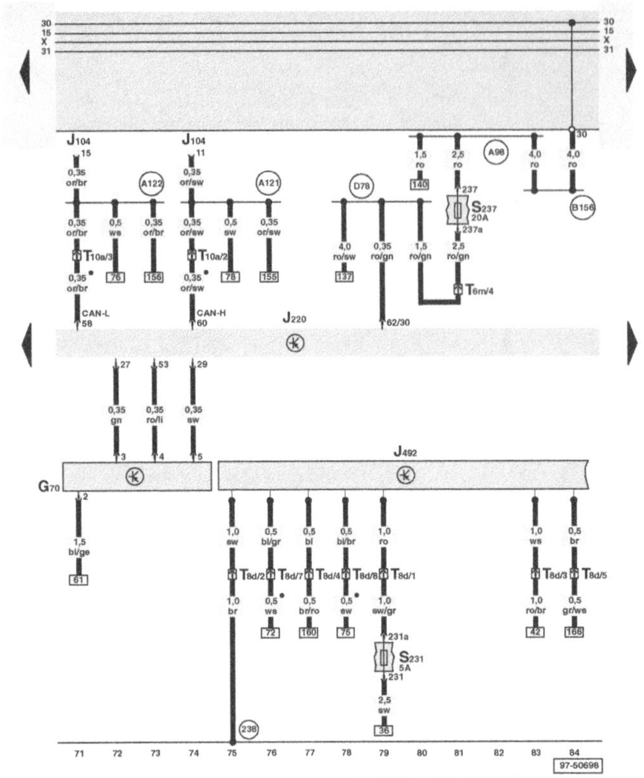

Motronic Engine Control Module, All-wheel Drive Control Module, Mass Air Flow (MAF) Sensor

G70 - Mass Air Flow (MAF) Sensor
J104 - ABS Control Module (w/EDL)
J220 - Motronic Engine Control Module (ECM)
J492 - All-wheel drive control module
S231 - Fuse 31 in fuse holder
S237 - Fuse 37 in fuse holder
T6m - Connector 6 pin, brown, in E-box plenum chamber
T8d - Connector 8 pin, on Control module for all wheel drive
T10a - Connector 10 pin, white, in E-box plenum chamber

238 - Ground connection -1-, in wiring harness interior
A98 - Plus connector -4- (30), in instrument panel wiring harness
A121 - Connection (high bus) in instrument panel wiring harness
A122 - Connection (low bus) in instrument panel wiring harness
B156 - Plus connector (30a), in wiring harness interior
D78 - Plus connection -1- (30a), in engine compartment wiring harness

• - CAN-Bus (Data-Bus)

ws = white
sw = black
ro = red
br = brown
gn = green
bl = blue
gr = grey
li = violet
ge = yellow
or = orange

97-50698

USA.5703.02.21 — Edition 04/03

Audi TT — Wiring Diagram — No. 2/10

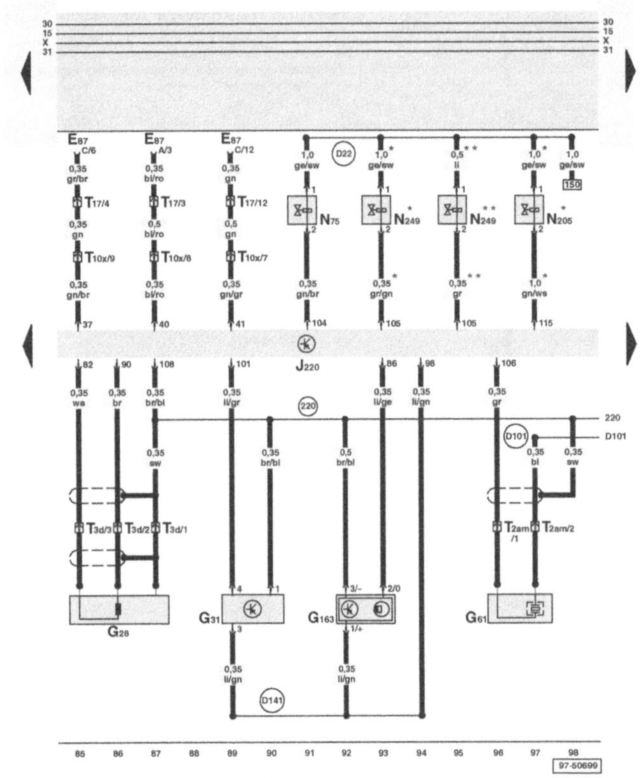

Motronic Engine Control Module, Engine Speed Sensor, Charge Air Pressure Sensor, Camshaft Position Sensor 2

E87 - A/C Control Head
G28 - Engine Speed (RPM) Sensor
G31 - Charge Air Pressure Sensor
G61 - Knock Sensor (KS) 1
G163 - Camshaft Position (CMP) Sensor 2
J220 - Motronic Engine Control Module (ECM)
N75 - Wastegate Bypass Regulator Valve
N205 - Valve -1- for camshaft adjustment
N249 - Recirculating valve for turbocharger
T2am - Connector 2 pin, black, in engine compartment
T3d - Connector 3 pin, grey, in engine compartment
T10x - Connector 10 pin, orange, in E-box plenum chamber
T17 - Connector 17 pin, behind instrument panel

220 - Ground connection (sensor ground), in engine compartment wiring harness
D22 - Connector (over fuse 234), in wiring harness front, right
D101 - Wire connection -1-, in engine compartment wiring harness
D141 - Connector (5V), in wiring harness, engine pre-wiring

* - only 1.8L engine code ATC, AWP
** - only 1.8L engine code BEA

ws = white
sw = black
ro = red
br = brown
gn = green
bl = blue
gr = grey
li = violet
ge = yellow
or = orange

97-50699

USA.5703.02.21 — Edition 04/03

© 2006 BentleyPublishers.com—All Rights Reserved

Audi TT — Wiring Diagram — No. 2/11

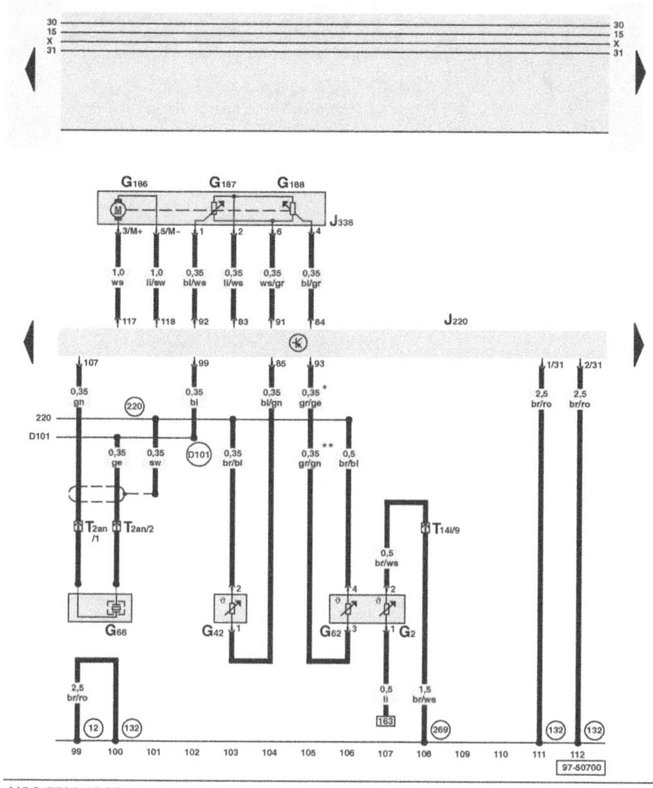

Motronic Engine Control Module, Knock Sensor 2, Intake Air Temperature Sensor, Throttle Valve Control Module

- G2 - Engine Coolant Temperature (ECT) Sensor
- G42 - Intake Air Temperature (IAT) Sensor
- G62 - Engine Coolant Temperature (ECT) Sensor
- G66 - Knock Sensor (KS) 2
- G186 - Throttle drive (power accelerator actuation)
- G187 - Angel sensor -1- for throttle drive (power accelerator actuation)
- G188 - Angel sensor -2- for throttle drive (power accelerator actuation)
- J220 - Motronic Engine Control Module (ECM)
- J338 - Throttle Valve Control Module
- T2an - Connector 2 pin, black, in engine compartment
- T14i - Connector 14 pin, in engine compartment, left

- (12) - Ground connection, in engine compartment, left
- (132) - Ground connection -3-, in engine compartment wiring harness
- (220) - Ground connection (sensor ground), in engine compartment wiring harness
- (269) - Ground connector (sensor ground) -1-, in instrument panel wiring harness
- (D101) - Wire connection -1-, in engine compartment wiring harness

- * - only 1.8L engine code ATC, AWP
- ** - only 1.8L engine code BEA

ws = white
sw = black
ro = red
br = brown
gn = green
bl = blue
gr = grey
li = violet
ge = yellow
or = orange

USA.5703.02.21 Edition 04/03

Audi TT — Wiring Diagram — No. 2/12

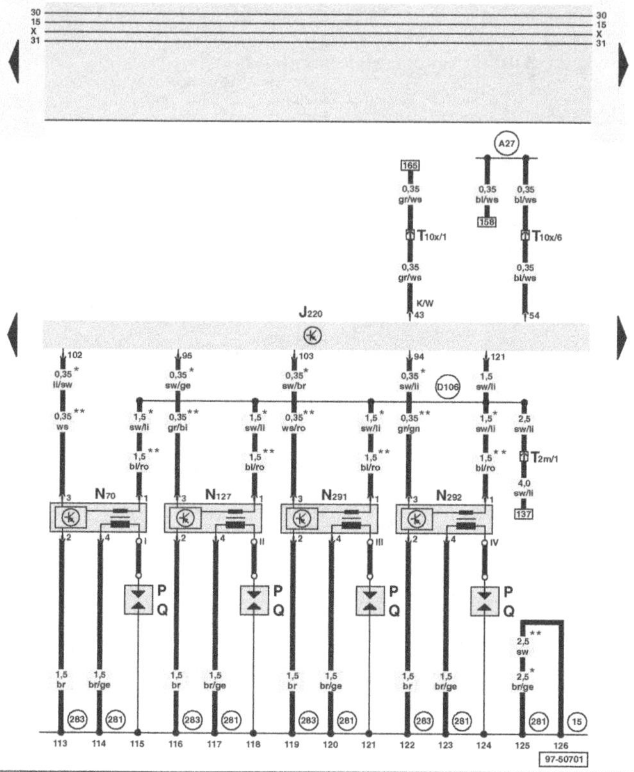

Motronic Engine Control Module, Ignition Coils

- J220 - Motronic Engine Control Module (ECM)
- N70 - Ignition Coil 1 with Power Output Stage
- N127 - Ignition Coil 2 with Power Output Stage
- N291 - Ignition Coil 3 with Power Output Stage
- N292 - Ignition Coil 4 with Power Output Stage
- P - Spark Plug Connectors
- Q - Spark Plugs
- T2m - Connector 2 pin, in engine compartment, left
- T10x - Connector 10 pin, orange, in E-box plenum chamber

- (15) - Ground connection, on cylinder head
- (281) - Ground connector -1-, in wiring harness engine pre-wiring
- (283) - Ground connection 2, in wiring harness engine pre-wiring
- (A27) - Wire connection (vehicle speed signal), in instrument panel wiring harness
- (D106) - Wire connection -4-, in engine compartment wiring harness

- * - only 1.8L engines code ATC, AWP
- ** - only 1.8L engine code BEA

ws = white
sw = black
ro = red
br = brown
gn = green
bl = blue
gr = grey
li = violet
ge = yellow
or = orange

USA.5703.02.21 Edition 04/03

© 2006 BentleyPublishers.com—All Rights Reserved

Engine Management, 1.8L, engine code: ATC, AWP, BEA, (2003-2006)

Audi TT — Wiring Diagram — No. 2/13

Motronic Engine Control Module, Leak Detection Pump, Motronic Engine Control Module Power Supply Relay, Brake Booster Relay, Evaporative Emission Canister Purge Solenoid Valve

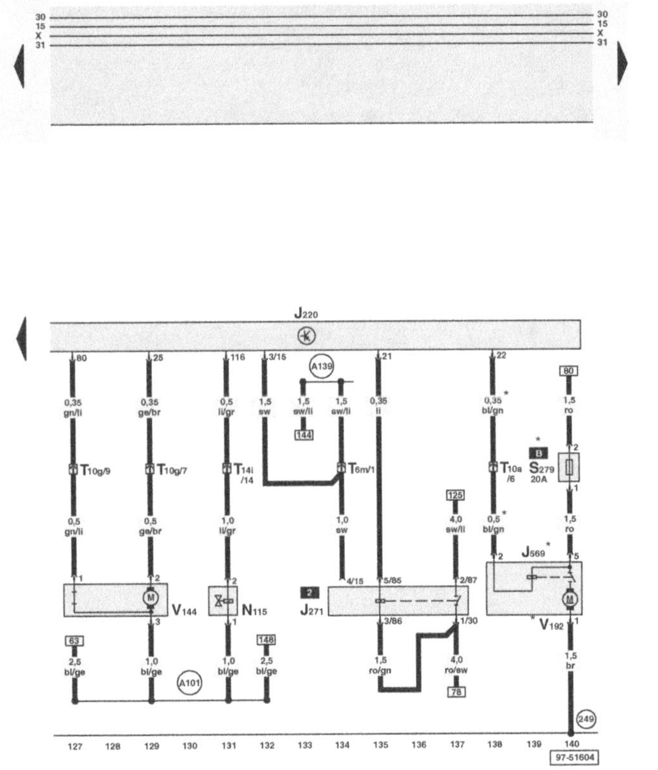

ws = white
sw = black
ro = red
br = brown
gn = green
bl = blue
gr = grey
li = violet
ge = yellow
or = orange

- J220 - Motronic Engine Control Module (ECM)
- J271 - Motronic Engine Control Module (ECM) Power Supply Relay
- J569 - Brake Booster Relay
- N115 - Evaporative Emission (EVAP) Canister Purge Solenoid Valve
- S279 - Hydraulic Pump Relay Fuse
- T6m - Connector 6 pin, brown, in E-box plenum chamber
- T10g - Connector 10 pin, blue, in E-box plenum chamber
- T14i - Connector 14 pin, in engine compartment, left
- V144 - Leak Detection Pump (LDP)
- V192 - Brake System Vacuum Pump
- (249) - Ground connection -2-, in wiring harness interior
- (A101) - Connector -3- (87), in instrument panel wiring harness
- (A139) - Connection 3 (15) (in instrument panel wiring harness)
- * - Vehicles with Automatic Transmission 09G

USA.5703.02.21 — Edition 04/03

Audi TT — Wiring Diagram — No. 2/14

Fuses, Speedometer Vehicle Speed Sensor, Sender for Fuel Gauge, Fuel Pump

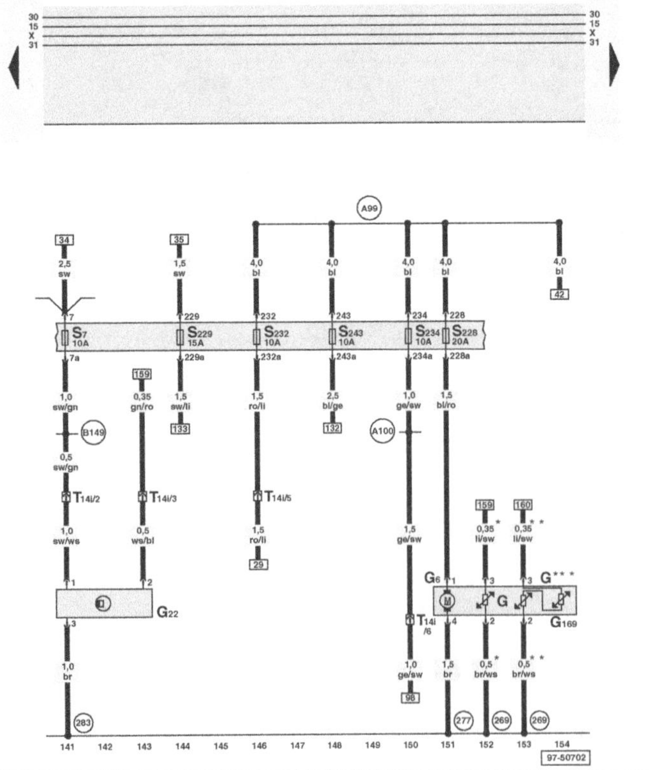

ws = white
sw = black
ro = red
br = brown
gn = green
bl = blue
gr = grey
li = violet
ge = yellow
or = orange

- G - Sender for fuel gauge
- G6 - Fuel Pump (FP)
- G22 - Speedometer Vehicle Speed Sensor (VSS)
- G169 - Fuel Level Sensor 2
- S7 - Fuse
- S228 - Fuse 28 in fuse holder
- S229 - Fuse 29 in fuse holder
- S232 - Fuse 32 in fuse holder
- S234 - Fuse 34 in fuse holder
- S243 - Fuse 43 in fuse holder
- T14i - Connector 14 pin, in engine compartment, left
- (269) - Ground connector (sensor ground) -1-, in instrument panel wiring harness
- (277) - Ground connector -3-, in wiring harness interior
- (283) - Ground connection 2, in wiring harness engine pre-wiring
- (A99) - Connector -1- (87), in instrument panel wiring harness
- (A100) - Connector -2- (87), in instrument panel wiring harness
- (B149) - Plus connector -2- (15a), in wiring harness interior
- * - Vehicles with front wheel drive
- ** - Vehicles with quattro
- *** - Wiring inside fuel tank

USA.5703.02.21 — Edition 04/03

© 2006 BentleyPublishers.com—All Rights Reserved

Engine Management, 1.8L, engine code: ATC, AWP, BEA, (2003-2006)

Audi TT **Wiring Diagram** **No. 2/15**

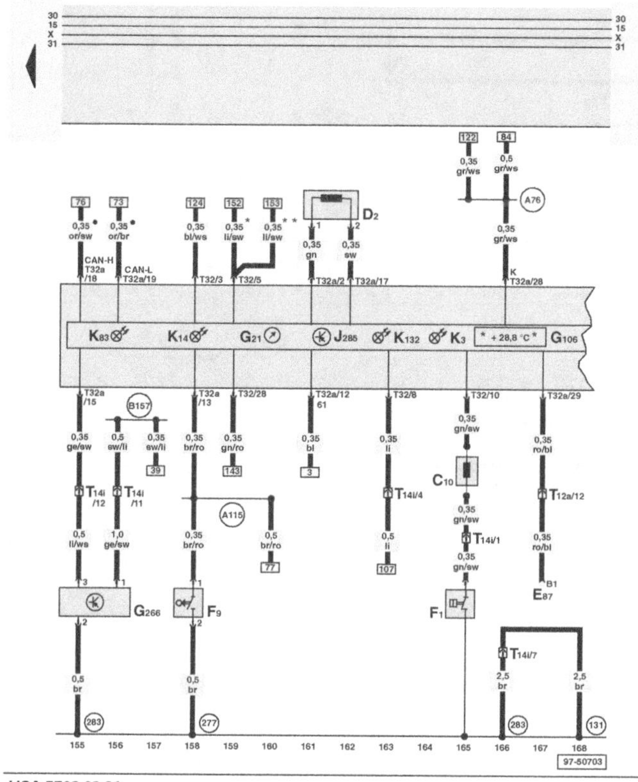

Instrument Cluster, Parking Brake Warning Light Switch, Oil Pressure Switch, Induction Coil of Anti-theft Immobilizer

C10 - Suppressor Filter
D2 - Induction coil of antitheft immobilizer
E87 - A/C Control Head
F1 - Oil Pressure Switch
F9 - Parking Brake Warning Light Switch
G21 - Speedometer
G106 - Outside Air Temperature Display
G266 - Oil Level Thermal Sensor (SRI)
J285 - Control Module with indicator unit in instrument panel insert
K3 - Oil Pressure Warning Light
K14 - Parking Brake Indicator Light
K83 - Malfunction Indicator Lamp (MIL)
K132 - Electronic Power Control (EPC) Warning Lamp
T12a - Connector 12 pin, behind instrument panel, right
T14i - Connector 14 pin, in engine compartment, left
T32 - Connector 32 pin, blue, on instrument cluster
T32a - Connector 32 pin, green, on instrument cluster

(131) - Ground connection -2-, in engine compartment wiring harness
(277) - Ground connector -3-, in wiring harness interior
(283) - Ground connection 2, in wiring harness engine pre-wiring
(A76) - Connector (K-diagnosis wire), in instrument panel wiring harness
(A115) - Connection (parking brake control) in instrument panel wiring harness
(B157) - Plus connector -3- (15a), in wiring harness interior

• - CAN-Bus (Data-Bus)
* - Vehicles with front wheel drive
** - Vehicles with quattro

ws = white
sw = black
ro = red
br = brown
gn = green
bl = blue
gr = grey
li = violet
ge = yellow
or = orange

USA.5703.02.21 **Edition 04/03**

THIS SPACE INTENTIONALLY LEFT BLANK

©2006 BentleyPublishers.com—All Rights Reserved

Automatic Transmission 09G (6-speed) with Tiptronic

2003 m. y.

Fuse Panel

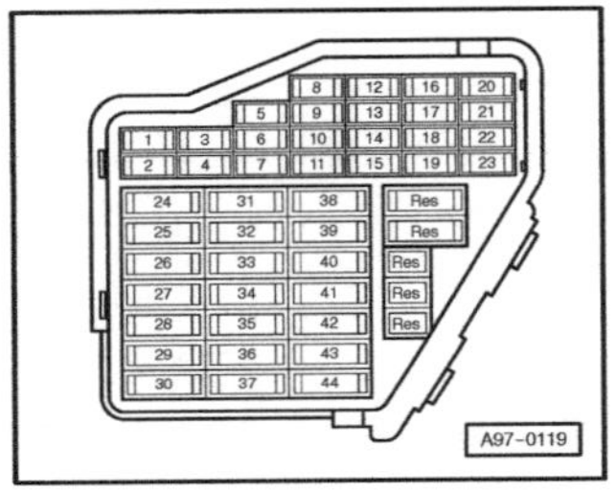

Fuse Colors:

30 A - Green
25 A - White
20 A - Yellow
15 A - Blue
10 A - Red
7,5 A - Brown
5 A - Beige

Starting with fuse position 23, fuses in the fuse holder are identified with 223 in the wiring diagram.

Micro Central Electric Panel

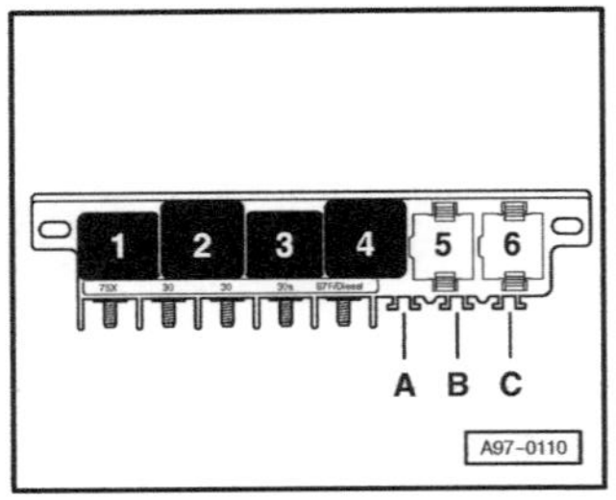

Relay Location:

B -Hydraulic Pump Relay Fuse, S279

13 - Fold Relay Panel

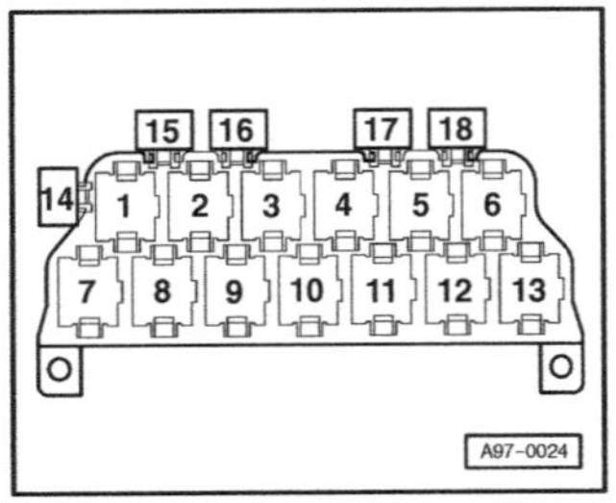

Relay Location:

11 -Starting Interlock Relay, J207

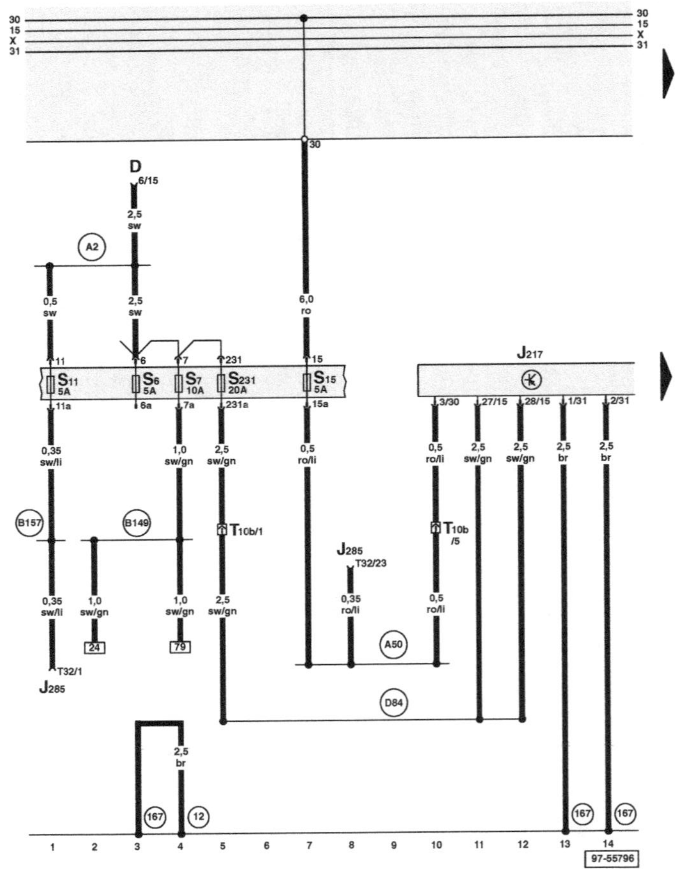

ws = white
sw = black
ro = red
br = brown
gn = green
bl = blue
gr = grey
li = violet
ge = yellow

D - Ignition/Starter Switch
J217 - Transmission Control Module (TCM)
J285 - Control module with indicator unit in instrument panel insert
S6 - Fuse
S7 - Fuse
S11 - Fuse
S15 - Fuse
S231 - Fuse 31 in fuse holder
T10b - Connector 10 pin, gray, in E-box plenum chamber
T32 - Connector 32 pin, blue, on instrument cluster

(12) - Ground connection, in engine compartment, left
(167) - Ground connection -4-, in engine compartment wiring harness
(A2) - Plus connection (15), in instrument panel wiring harness
(A50) - Plus connection (30a), in instrument panel wiring harness
(B149) - Plus connector -2- (15a), in wiring harness interior
(B157) - Plus connector -3- (15a), in wiring harness interior
(D84) - Wire connection (Auto Trans.), in engine compartment wiring harness

© 2006 BentleyPublishers.com—All Rights Reserved

Automatic Transmission, 6-speed 09G with Tiptronic (2003-2006)

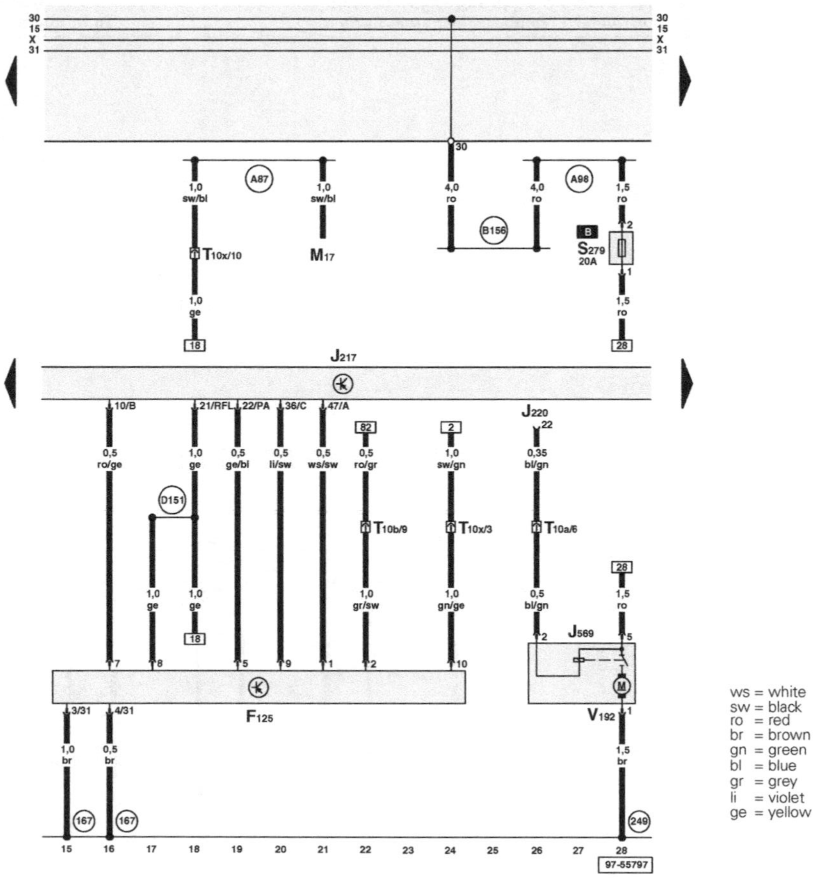

ws = white
sw = black
ro = red
br = brown
gn = green
bl = blue
gr = grey
li = violet
ge = yellow

F125 - Multi-Function Transmission Range (TR) Switch
J217 - Transmission Control Module (TCM)
J220 - Motronic Engine Control Module (ECM)
J569 - Brake Booster relay
M17 - Right Back-Up Light
S279 - Hydraulic Pump Relay Fuse
T10a - Connector 10 pin, white, in E-box plenum chamber
T10b - Connector 10 pin, gray, in E-box plenum chamber
T10x - Connector 10 pin, orange, in E-box plenum chamber
V192 - Brake System Vacuum Pump

(167) - Ground connection -4-, in engine compartment wiring harness
(249) - Ground connection -2-, in wiring harness interior
(A87) - Connector (RF), in instrument panel wiring harness
(A98) - Plus connector -4- (30), in instrument panel wiring harness
(B156) - Plus connector -4- (30), in wiring harness interior
(D151) - Connection (RF), in engine compartment wiring harness

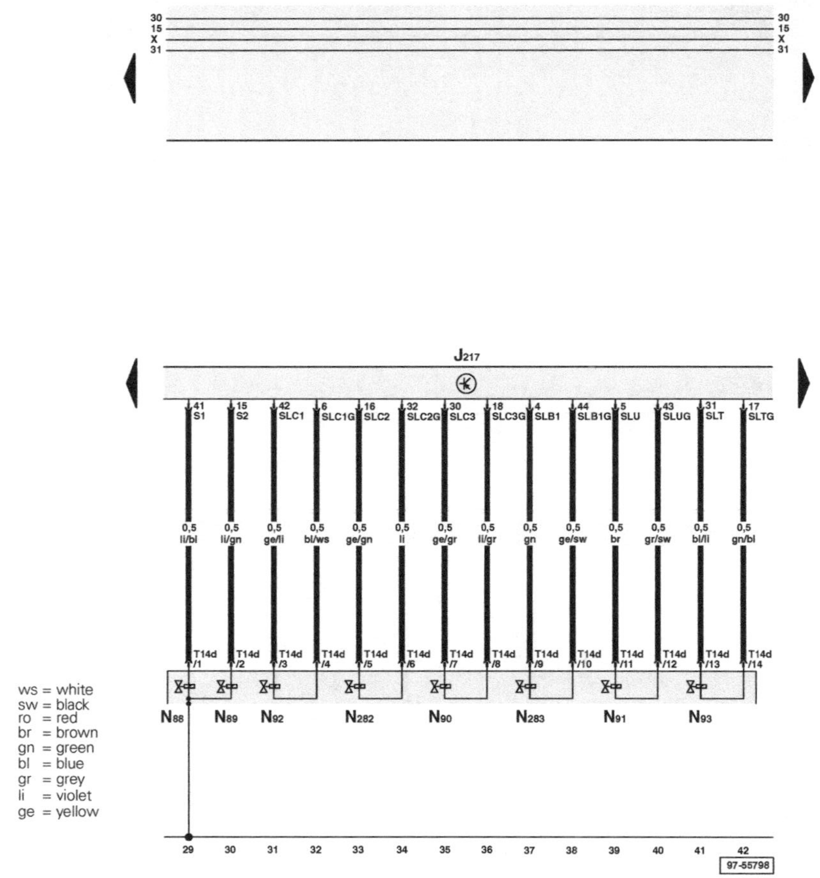

ws = white
sw = black
ro = red
br = brown
gn = green
bl = blue
gr = grey
li = violet
ge = yellow

J217 - Transmission Control Module (TCM)
N88 - Solenoid Valve 1
N89 - Solenoid Valve 2
N90 - Solenoid Valve 3
N91 - Solenoid Valve 4
N92 - Solenoid Valve 5
N93 - Solenoid Valve 6
N282 - Solenoid Valve 9
N283 - Solenoid Valve 10
T14d - Connector 14 pin, on transmission

© 2006 BentleyPublishers.com—All Rights Reserved

Audi TT — Wiring diagram — No. 3/5

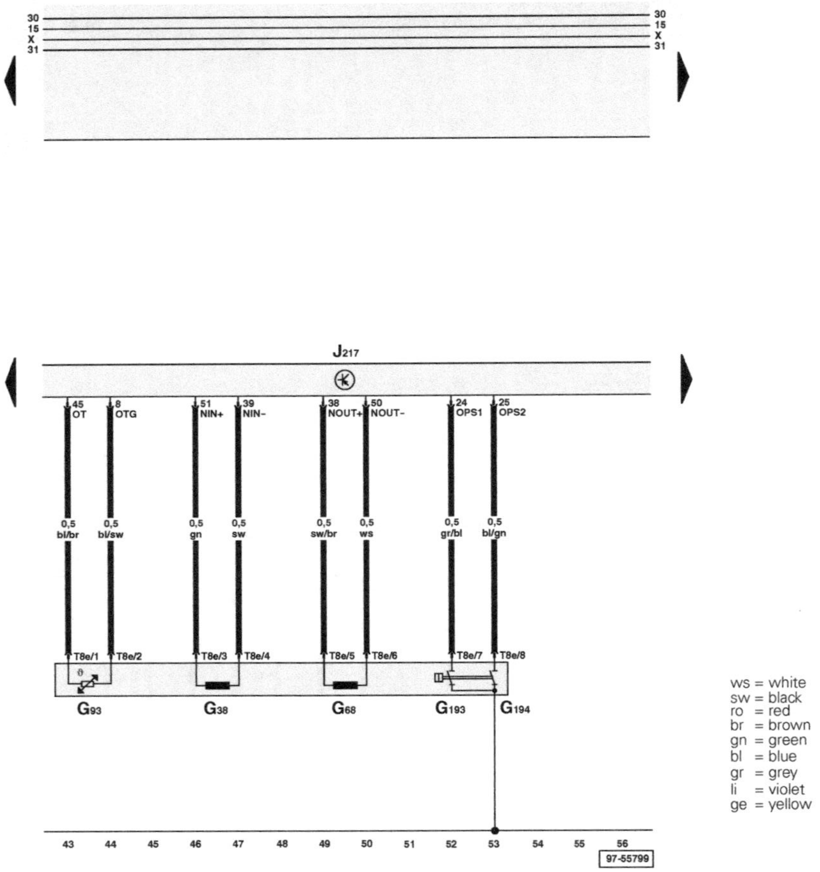

ws = white
sw = black
ro = red
br = brown
gn = green
bl = blue
gr = grey
li = violet
ge = yellow

G38 - Transmission Vehicle Speed Sensor (VSS)
G68 - Vehicle Speed Sensor (VSS)
G93 - Transmission Fluid Temperature Sensor
G193 - Sender -1- for hydraulic pressure, automatic transmission
G194 - Sender -2- for hydraulic pressure, automatic transmission
J217 - Transmission Control Module (TCM)
T8e - Connector 8 pin, on transmission

No. 3/6 — Wiring diagram — Audi TT

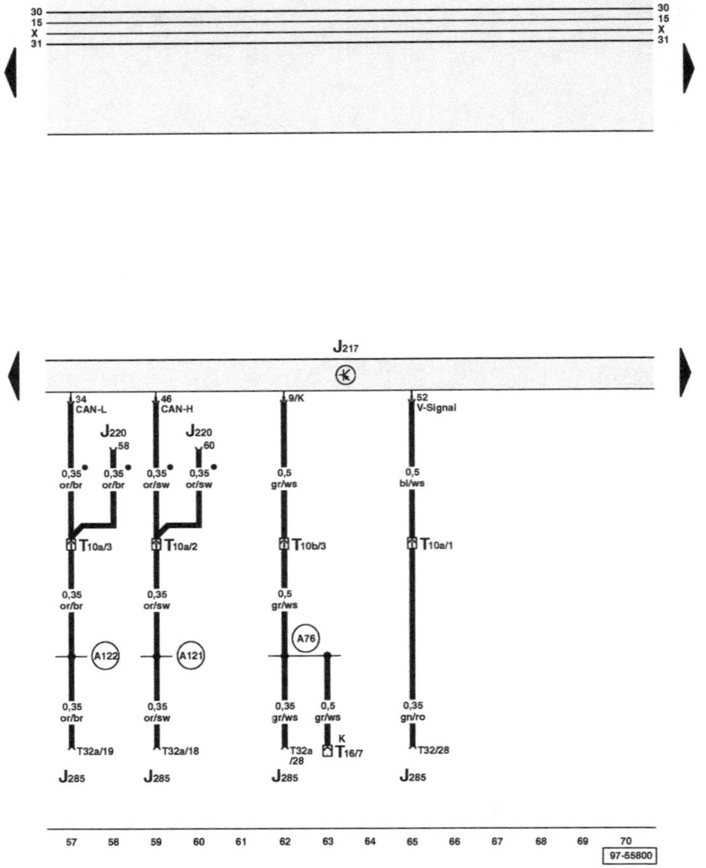

ws = white
sw = black
ro = red
br = brown
gn = green
bl = blue
gr = grey
li = violet
ge = yellow
or = orange

J217 - Transmission Control Module (TCM)
J220 - Motronic Engine Control Module (ECM)
J285 - Control module with indicator unit in instrument panel insert
T10a - Connector 10 pin, white, in E-box plenum chamber
T10b - Connector 10 pin, grey, in E-box plenum chamber
T16 - Connector 16 pin, black, Data Link Connector (DLC), below storage compartment, driver's side
T32 - Connector 32 pin, blue, on instrument cluster
T32a - Connector 32 pin, green, on instrument cluster

A76 - Connector (K-diagnosis wire), in instrument panel wiring harness
A121 - Connection (high bus) in instrument panel wiring harness
A122 - Connection (low bus) in instrument panel wiring harness
- CAN-Bus (Data-Bus)

© 2006 BentleyPublishers.com—All Rights Reserved

Audi TT — Wiring diagram — **No. 3/7**

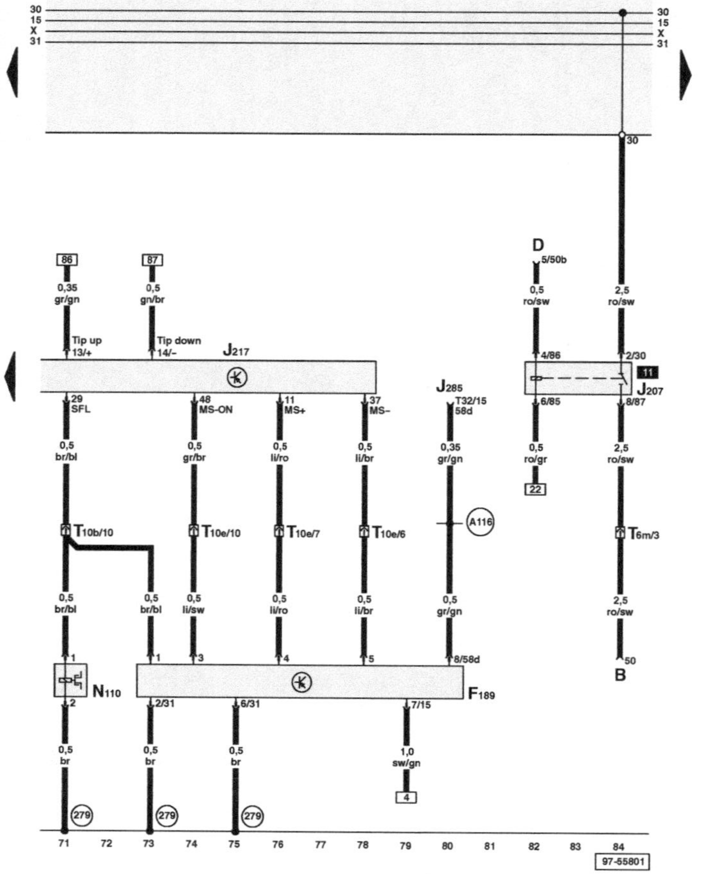

ws = white
sw = black
ro = red
br = brown
gn = green
bl = blue
gr = grey
li = violet
ge = yellow

- B - Starter
- D - Ignition/Starter Switch
- F189 - Tiptronic Switch
- J207 - Starting Interlock Relay
- J217 - Transmission Control Module (TCM)
- J285 - Control module with indicator unit in instrument panel insert
- N110 - Shift Lock Solenoid
- T6m - Connector 6 pin, brown, in E-box plenum chamber
- T10b - Connector 10 pin, grey, in E-box plenum chamber
- T10e - Connector 10 pin, black, in E-box plenum chamber
- T32 - Connector 32 pin, blue, on instrument cluster

- (279) - Ground connector -5-, in wiring harness interior
- (A116) - Connection -2- (58d) in instrument panel wiring harness

Edition 10/02
USA.5703.01.21

Transmission Control Module, Starting Interlock Relay, Shift Lock Solenoid, Tiptronic Switch

No. 3/8 — Wiring diagram — **Audi TT**

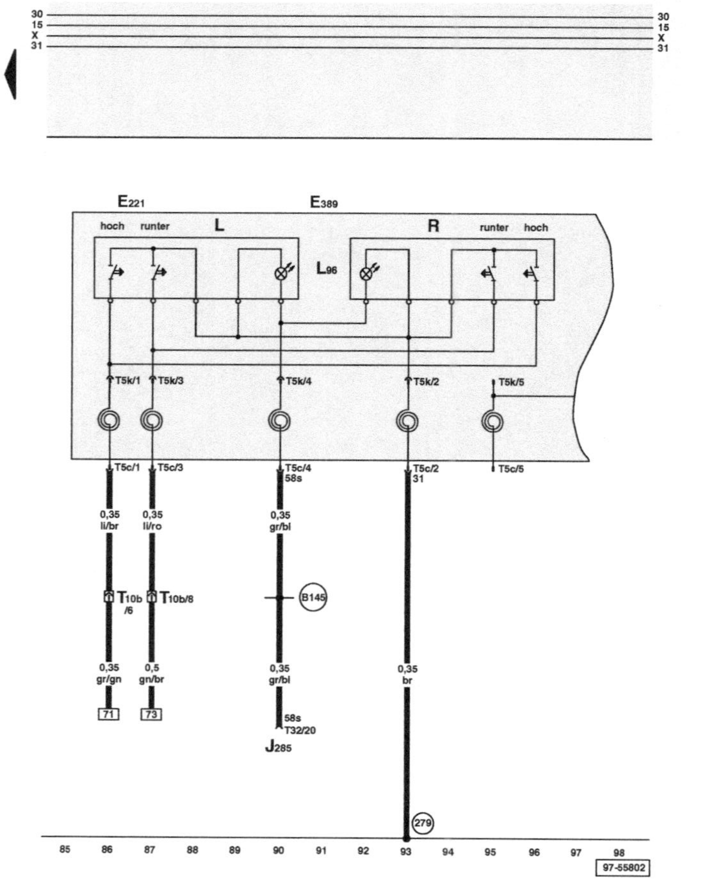

ws = white
sw = black
ro = red
br = brown
gn = green
bl = blue
gr = grey
li = violet
ge = yellow

- E221 - Control unit in steering wheel
- E389 - Steering Wheel Tiptronic Switch
- J285 - Control module with indicator unit in instrument panel insert
- L96 - Light for +/- indicator (automatic transmission)
- T5c - Connector 5 pin, yellow, in steering wheel
- T5k - Connector 5 pin, yellow, on steering column
- T10b - Connector 10 pin, grey, in E-box plenum chamber
- T32 - Connector 32 pin, blue, on instrument cluster

- (279) - Ground connector -5-, in wiring harness interior
- (B145) - Plus connector (58s), in wiring harness interior

Control Unit in Steering Wheel, Steering Wheel Tiptronic Switch

Edition 10/02
USA.5703.01.21

© 2006 BentleyPublishers.com—All Rights Reserved

Audi TT **Wiring Diagram** **No. 4/1**

CAN-Bus Wiring

2003 m.y.

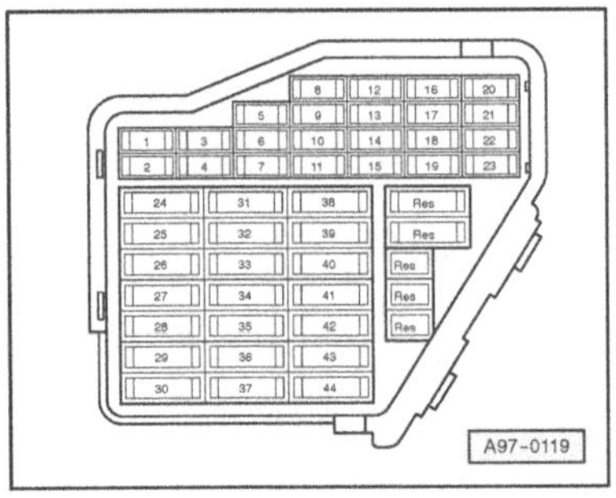

Fuse Panel

Fuse Colors:
- 30 A - Green
- 25 A - White
- 20 A - Yellow
- 15 A - Blue
- 10 A - Red
- 7,5 A - Brown
- 5 A - Beige
- 1 A - Black

Starting with fuse position 23, fuses in the fuse holder are identified with 223 in the wiring diagram.

USA.5703.03.21 Edition 10/03

Audi TT **Wiring Diagram** **No. 4/2**

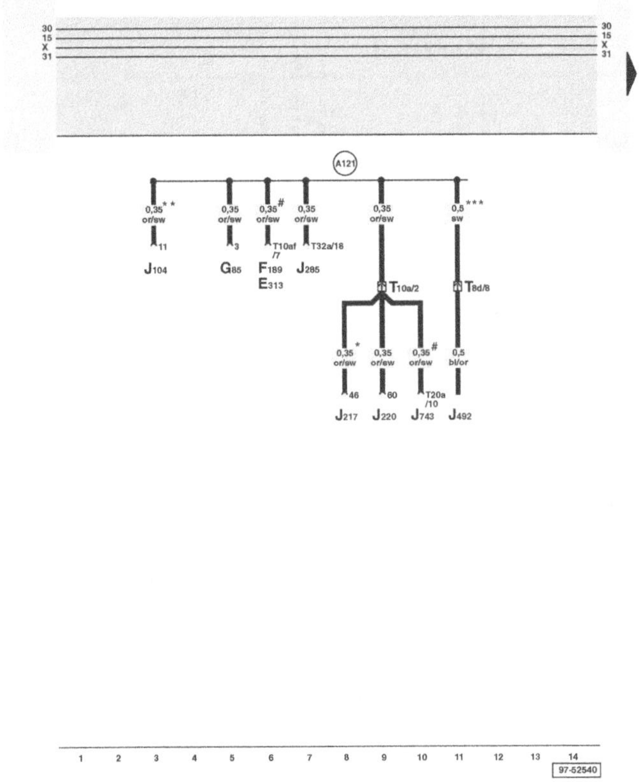

CAN-Bus Wiring, Connection High-Bus

- E313 - Selector Lever
- F189 - Tiptronic Switch
- G85 - Steering Angle Sensor
- J104 - ABS Control Module (w/EDL)
- J217 - Transmission Control Module (TCM)
- J220 - Motronic Engine Control Module (ECM)
- J285 - Control Module with indicator unit in instrument panel insert
- J492 - All-wheel drive control module
- J743 - Direct Shift Gearbox (DSG) Mechatronic
- T8d - Connector 8 pin, on Control module for all wheel drive
- T10a - Connector 10 pin, white, in E-box plenum chamber
- T10af - Connector 10 pin, on Selector Lever
- T20a - Connector 20 pin, on Direct Shift Gearbox (DSG) Mechatronic
- T32a - Connector 32 pin, green, on instrument cluster
- (A121) - Connection (high bus) in instrument panel wiring harness

- * - Vehicles with Automatic Transmission 09G
- ** - ABS MK60 with EDL/ASR/ESP
- *** - Vehicles with quattro
- # - Vehicles with Direct Shift Gearbox (DSG) 02E

ws = white
sw = black
ro = red
br = brown
gn = green
bl = blue
gr = grey
li = violet
ge = yellow
or = orange

USA.5703.03.21 Edition 10/03

© 2006 BentleyPublishers.com—All Rights Reserved

Audi TT

Wiring Diagram **No. 4/3**

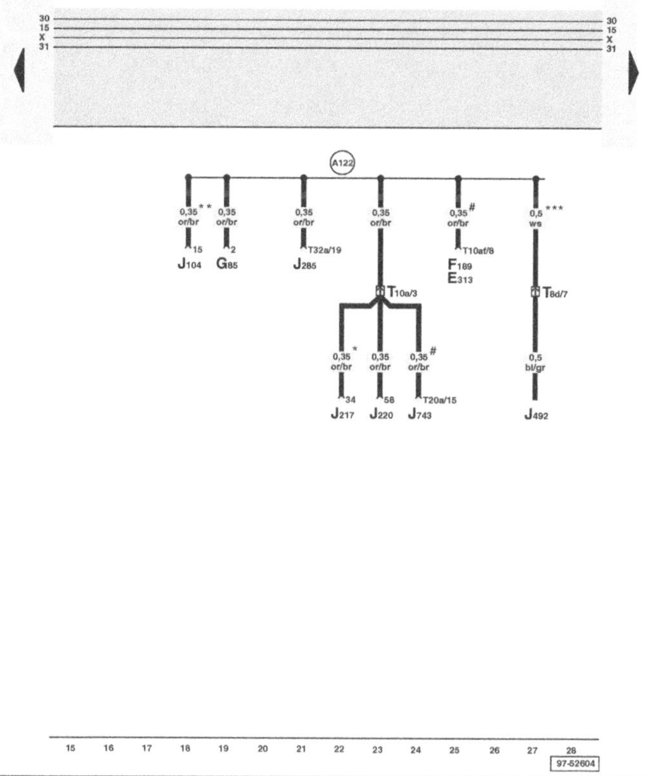

CAN-Bus Wiring, Connection Low-Bus

- E313 - Selector Lever
- F189 - Tiptronic Switch
- G85 - Steering Angle Sensor
- J104 - ABS Control Module (w/EDL)
- J217 - Transmission Control Module (TCM)
- J220 - Motronic Engine Control Module (ECM)
- J285 - Control Module with indicator unit in instrument panel insert
- J492 - All-wheel drive control module
- J743 - Direct Shift Gearbox (DSG) Mechatronic
- T8d - Connector 8 pin, on Control module for all wheel drive
- T10a - Connector 10 pin, white, in E-box plenum chamber
- T10af - Connector 10 pin, on Selector Lever
- T20a - Connector 20 pin, on Direct Shift Gearbox (DSG) Mechatronic
- T32a - Connector 32 pin, green, on instrument cluster
- (A122) - Connection (low bus) in instrument panel wiring harness

* - Vehicles with Automatic Transmission 09G
** - ABS MK60 with EDL/ASR/ESP
*** - Vehicles with quattro
# - Vehicles with Direct Shift Gearbox (DSG) 02E

ws = white
sw = black
ro = red
br = brown
gn = green
bl = blue
gr = grey
li = violet
ge = yellow
or = orange

USA.5703.03.21 **Edition 10/03**

Audi TT

Wiring Diagram **No. 4/4**

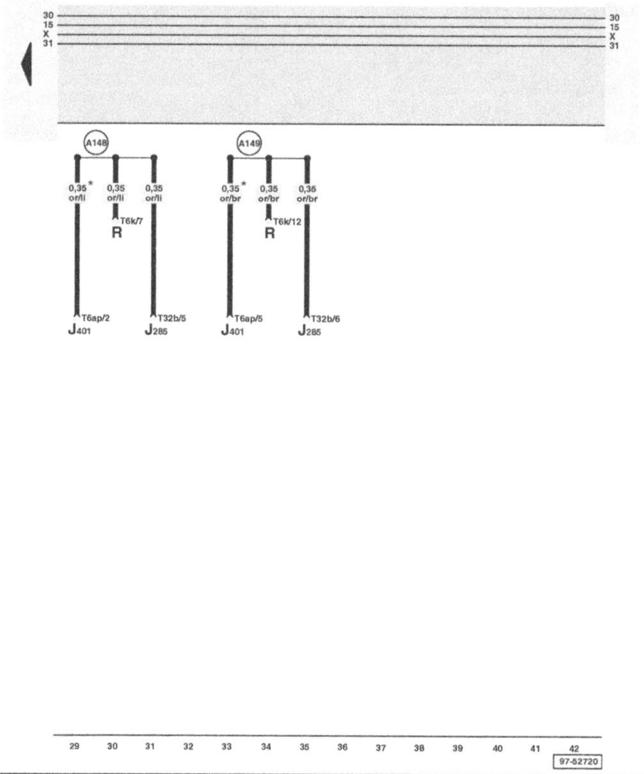

CAN-Bus Wiring, Connection Comfort System, Connection Navigation System (High-Bus, Low-Bus)

- J285 - Control Module with indicator unit in instrument panel insert
- J401 - Control unit for navigation with CD-mechanism
- R - Radio
- T6k - Connector 6 pin, green, connector radio I-2
- T6ap - Connector 6 pin, Connector C I on Control Module for Navigation
- T32b - Connector 32 pin, grey, on instrument cluster
- (A148) - Connection (high bus, navigation), in instrument panel wiring harness
- (A149) - Connection (low bus, navigation), in instrument panel wiring harness

* - Vehicles with Navigation System

ws = white
sw = black
ro = red
br = brown
gn = green
bl = blue
gr = grey
li = violet
ge = yellow
or = orange

USA.5703.03.21 **Edition 10/03**

© 2006 BentleyPublishers.com—All Rights Reserved

Audi TT Wiring Diagram **No. 5/1**

Steering Wheel with Tiptronic Switch

2003 m.y.

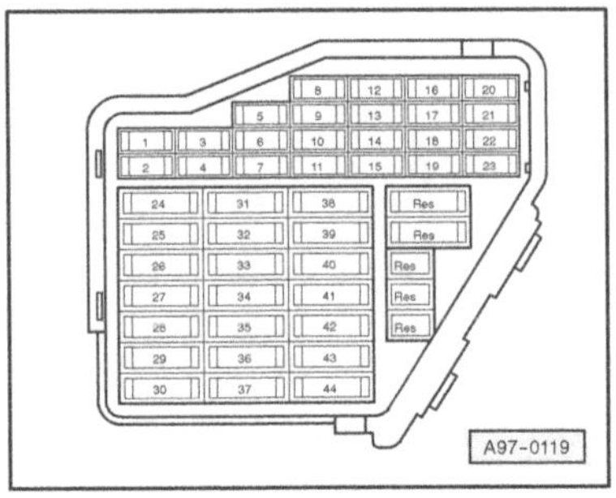

Fuse Panel

Fuse Colors:

30 A - Green
25 A - White
20 A - Yellow
15 A - Blue
10 A - Red
7,5 A - Brown
5 A - Beige
1 A - Black

Starting with fuse position 23, fuses in the fuse holder are identified with 223 in the wiring diagram.

Audi TT Wiring Diagram **No. 5/2**

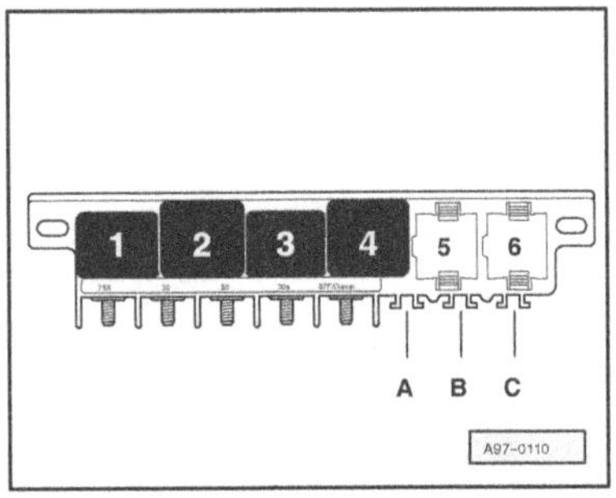

Micro Central Electric Panel

Relay Location:

1 - Dual Horn Relay, J4

© 2006 BentleyPublishers.com—All Rights Reserved

Audi TT

Wiring Diagram **No. 5/3**

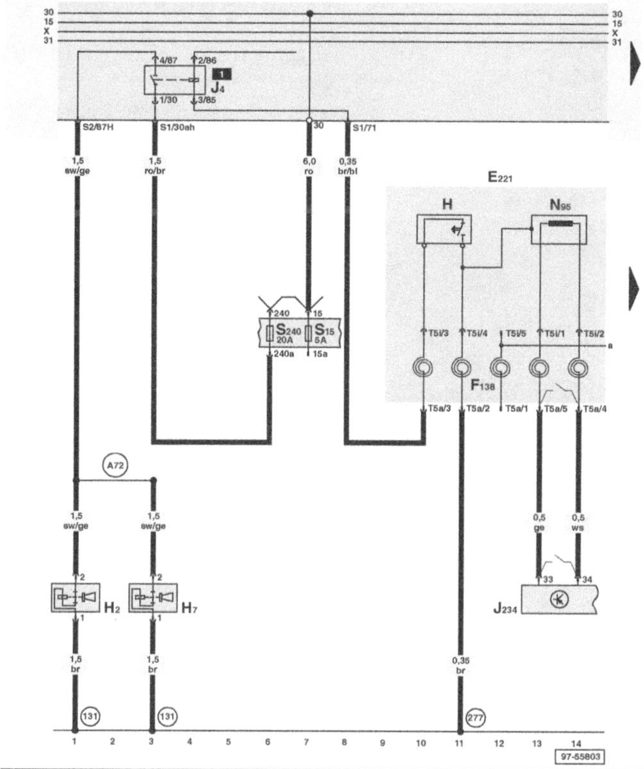

Dual Horn Relay, High Tone Horn, Low Tone Horn, Control Unit in Steering Wheel

- E221 - Control unit in steering wheel
- F138 - Airbag Spiral Spring/Return Spring With Slip Ring
- H - Signal Horn Activation
- H2 - High Tone Horn
- H7 - Low Tone Horn
- J4 - Dual Horn Relay
- J234 - Airbag Control Module
- N95 - Driver's Side Airbag Igniter
- S15 - Fuse
- S240 - Fuse 40 in fuse holder
- T5a - Connector 5 pin, yellow, in steering wheel
- T5i - Connector 5 pin, yellow, on steering column
- (131) - Ground connection -2-, in engine compartment wiring harness
- (277) - Ground connector -3-, in wiring harness interior
- (A72) - Connector (71), in instrument panel wiring harness
- Bridge closes when disconnecting connector (safety feature)

ws = white
sw = black
ro = red
br = brown
gn = green
bl = blue
gr = grey
li = violet
ge = yellow
or = orange

USA.5703.02.21 Edition 04/03

Audi TT

Wiring Diagram **No. 5/4**

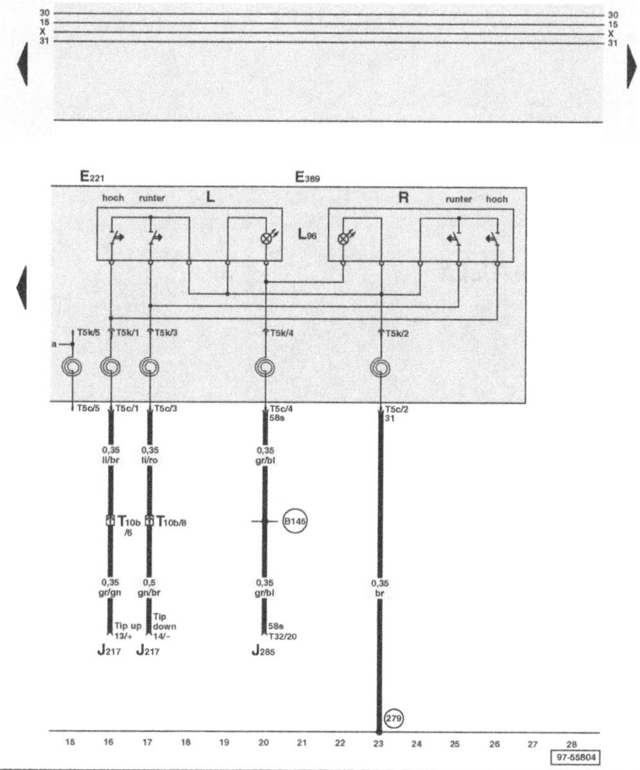

Control Unit in Steering Wheel, Steering Wheel Tiptronic Switch for Vehicles with Automatic Transmission 09G

- E221 - Control unit in steering wheel
- E389 - Steering Wheel Tiptronic Switch
- J217 - Transmission Control Module (TCM)
- J285 - Control Module with indicator unit in instrument panel insert
- L96 - Light for +/- indicator (automatic transmission)
- T5c - Connector 5 pin, yellow, in steering wheel
- T5k - Connector 5 pin, yellow, on steering column
- T10b - Connector 10 pin, grey, in E-box plenum chamber
- T32 - Connector 32 pin, blue, on instrument cluster
- (279) - Ground connector -5-, in wiring harness interior
- (B145) - Plus connector (58b), in wiring harness interior

ws = white
sw = black
ro = red
br = brown
gn = green
bl = blue
gr = grey
li = violet
ge = yellow
or = orange

USA.5703.02.21 Edition 04/03

©2006 BentleyPublishers.com—All Rights Reserved

Steering Wheel with Tiptronic Switch (2003-2006)

Audi TT **Wiring Diagram** **No. 5/5**

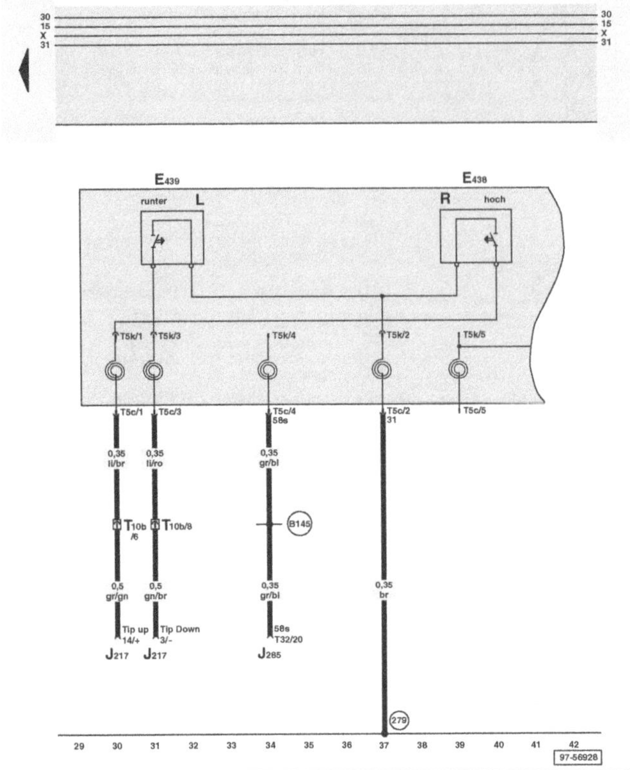

Tiptronic Upshift and Downshift Button in Steering Wheel for Vehicles with Direct Manual Transmission 02E

- E438 - Tiptronic Upshift Button (in steering wheel)
- E439 - Tiptronic Downshift Buttons (on steering wheel)
- J217 - Transmission Control Module (TCM)
- J285 - Control Module with indicator unit in instrument panel insert
- T5c - Connector 5 pin, yellow, in steering wheel
- T5k - Connector 5 pin, yellow, on steering column
- T10b - Connector 10 pin, grey, in E-box plenum chamber
- T32 - Connector 32 pin, blue, on instrument cluster
- (279) - Ground connector -5-, in wiring harness interior
- (B145) - Plus connector (58b), in wiring harness interior

ws = white
sw = black
ro = red
br = brown
gn = green
bl = blue
gr = grey
li = violet
ge = yellow
or = orange

USA.5703.02.21 Edition 04/03

THIS SPACE INTENTIONALLY

LEFT BLANK

© 2006 BentleyPublishers.com—All Rights Reserved

Navigation System

2003 m. y.

Fuse Panel

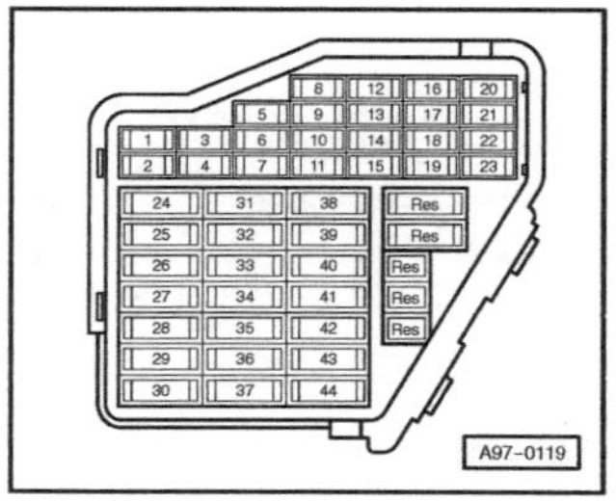

Fuse Colors:

30 A - Green
25 A - White
20 A - Yellow
15 A - Blue
10 A - Red
7,5 A - Brown
5 A - Beige

Starting with fuse position 23, fuses in the fuse holder are identified with 223 in the wiring diagram.

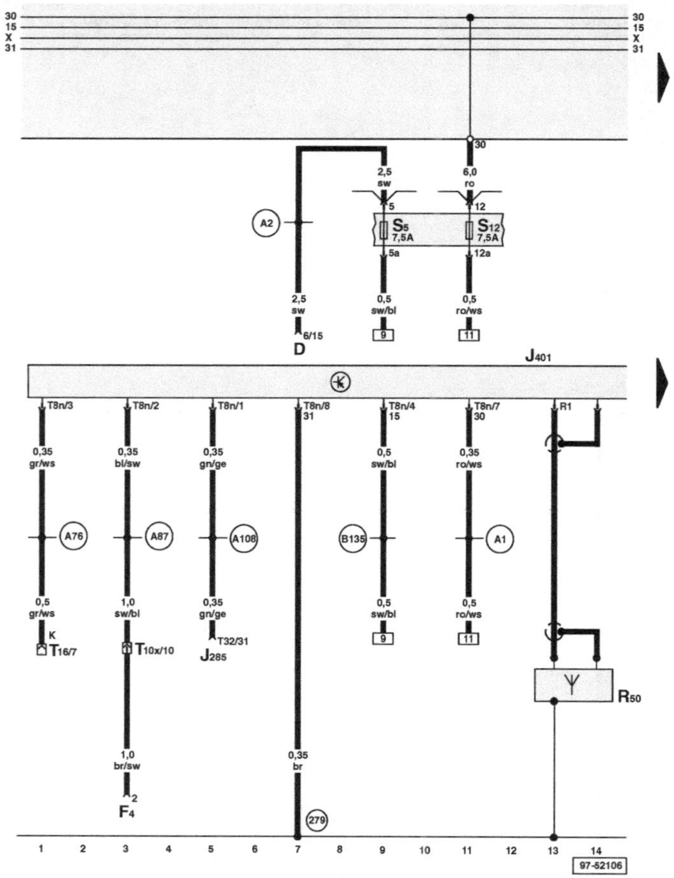

ws = white
sw = black
ro = red
br = brown
gn = green
bl = blue
gr = grey
li = violet
ge = yellow

D - Ignition/Starter Switch
F4 - Back-Up Light Switch
J285 - Control Module with indicator unit in instrument panel insert
J401 - Control unit for navigation with CD drive
R1 - Antenna Connection
R50 - Antenna for Navigation System (GPS)
S5 - Fuse
S12 - Fuse
T8n - Connector 8 pin, black, connector A on Control Module for Navigation
T10x - Connector 10 pin, orange, in E-box plenum chamber
T16 - Connector 16 pin, black, Data Link Connector (DLC), below storage compartment, driver's side
T32 - Connector 32 pin, blue, on instrument cluster

(279) - Ground connector -5-, in wiring harness interior
(A1) - Plus connection (30a), in instrument panel wiring harness
(A2) - Plus connection (15), in instrument panel wiring harness
(A76) - Connector (K-diagnosis wire), in instrument panel wiring harness
(A87) - Connector (RF), in instrument panel wiring harness
(A108) - Connector (speed signal), in instrument panel wiring harness
(B135) - Connector (15a), in wiring harness interior

Control Unit for Navigation with CD Drive, Antenna for Navigation System, Fuses

© 2006 BentleyPublishers.com—All Rights Reserved

Audi TT — Wiring diagram — No. 6/3

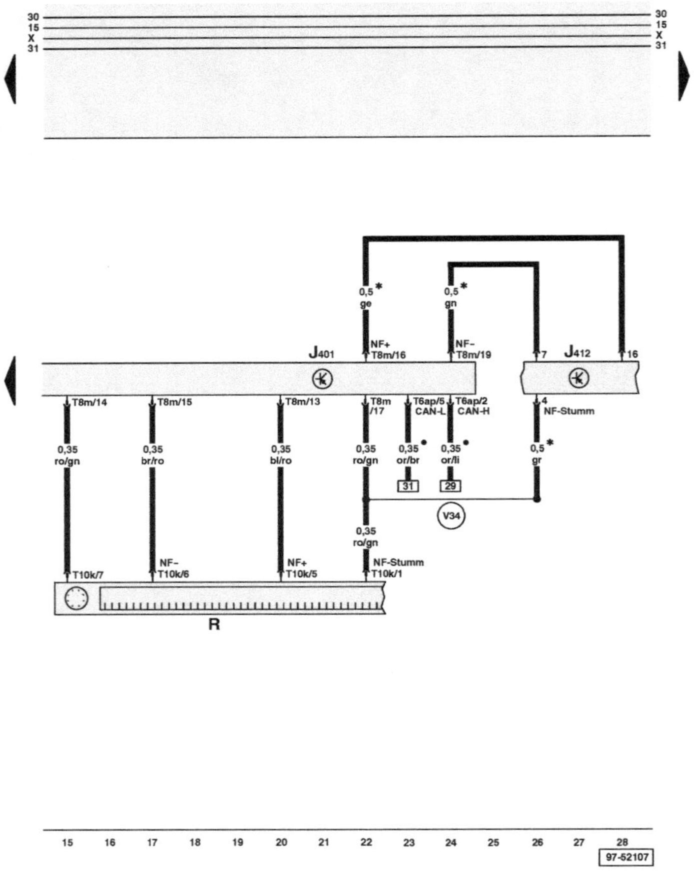

ws = white
sw = black
ro = red
br = brown
gn = green
bl = blue
gr = grey
li = violet
ge = yellow
or = orange

J401 - Control unit for navigation with CD drive
J412 - Operating electronics and Telephone Control Module
R - Radio
T6ap - Connector 6 pin, Connector C I on Control Module for Navigation
T8m - Connector 8 pin, C III on Control Module for Navigation
T10k - Connector 10 pin, red, on radio

(V34) - Connector (mute button), in wiring harness telephone

• - CAN-Bus (Data-Bus)
* - Vehicles with Cellular Telephone Preparation

Edition 10/02
USA.5703.01.21

Control Unit for Navigation with CD Drive, Radio, Operating Electronics and Telephone Control Module

No. 6/4 — Wiring diagram — Audi TT

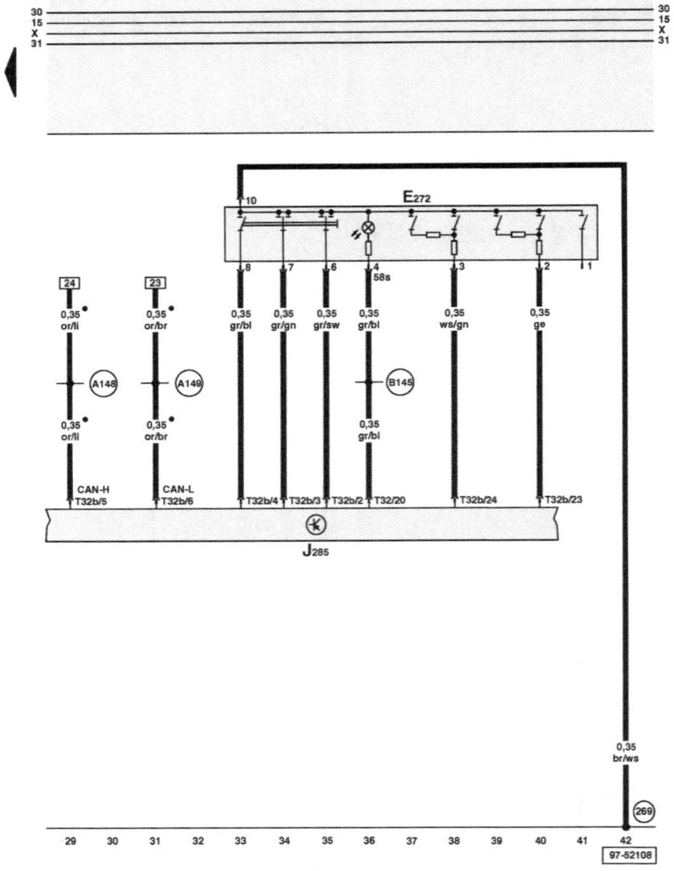

ws = white
sw = black
ro = red
br = brown
gn = green
bl = blue
gr = grey
li = violet
ge = yellow
or = orange

E272 - Function Selector Switch II, in console
J285 - Control Module with indicator unit in instrument panel insert
T32 - Connector 32 pin, blue, on instrument cluster
T32b - Connector 32 pin, gray, on instrument cluster

(269) - Ground connector (sensor ground) -1-, in instrument panel wiring harness
(A148) - Connection (high bus, navigation), in instrument panel wiring harness
(A149) - Connection (low bus, navigation), in instrument panel wiring harness
(B145) - Plus connector (58s), in wiring harness interior

• - CAN-Bus (Data-Bus)

Control Module with Indicator Unit in Instrument Panel Insert, Function Selector Switch II

Edition 10/02
USA.5703.01.21

© 2006 BentleyPublishers.com—All Rights Reserved

Audi TT | **Wiring Diagram** | **No. 7/1**

Radio Bose with and without CD-Changer (Coupé)

2003 m.y.

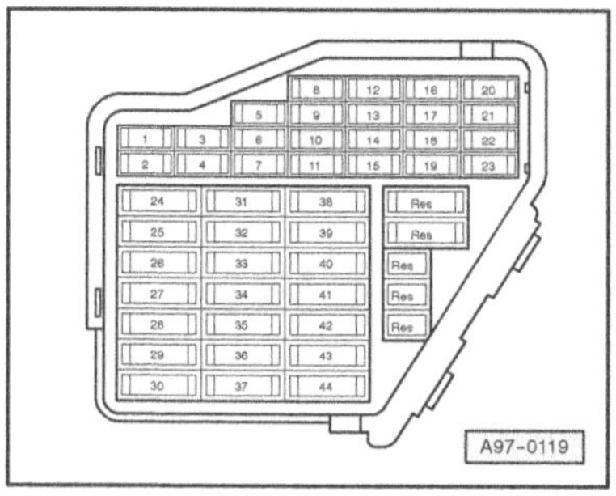

Fuse Panel

Fuse Colors:
- 30 A - Green
- 25 A - White
- 20 A - Yellow
- 15 A - Blue
- 10 A - Red
- 7,5 A - Brown
- 5 A - Beige
- 1 A - Black

Starting with fuse position 23, fuses in the fuse holder are identified with 223 in the wiring diagram.

Audi TT | **Wiring Diagram** | **No. 7/2**

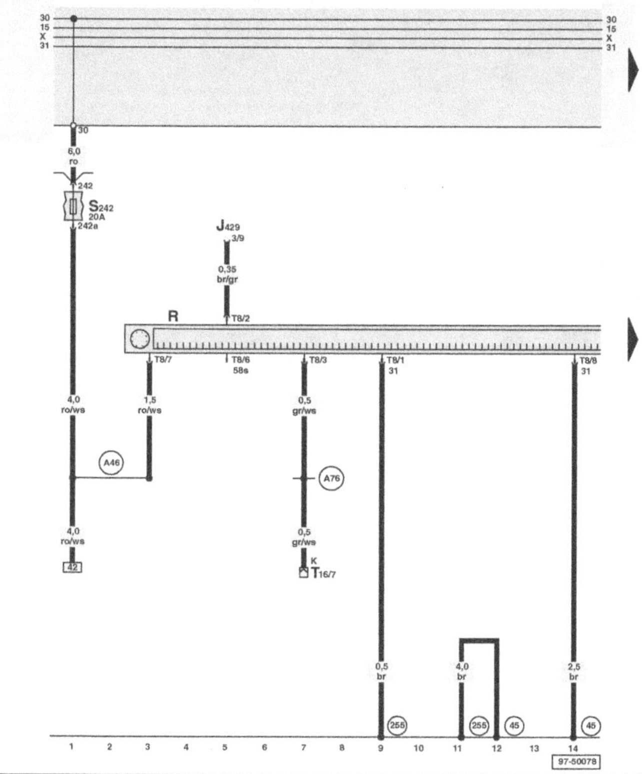

Radio

- J429 - Control module for central locking
- R - Radio
- S242 - Fuse 42 in fuse holder
- T8 - Connector 8 pin, black, connector radio III
- T16 - Connector 16 pin, black, Data Link Connector (DLC), below storage compartment, driver's side

- (45) - Ground connection, behind instrument panel, center
- (255) - Ground connection, -1-, in wiring harness Radio
- (A46) - Plus connection (30-from radio), in instrument panel wiring harness
- (A76) - Connector (K-diagnosis wire), in instrument panel wiring harness

ws = white
sw = black
ro = red
br = brown
gn = green
bl = blue
gr = grey
li = violet
ge = yellow
or = orange

©2006 BentleyPublishers.com—All Rights Reserved

Sound System, Bose, coupe (2003-2006)

Audi TT — Wiring Diagram — No. 7/3

Radio, CD Changer Unit

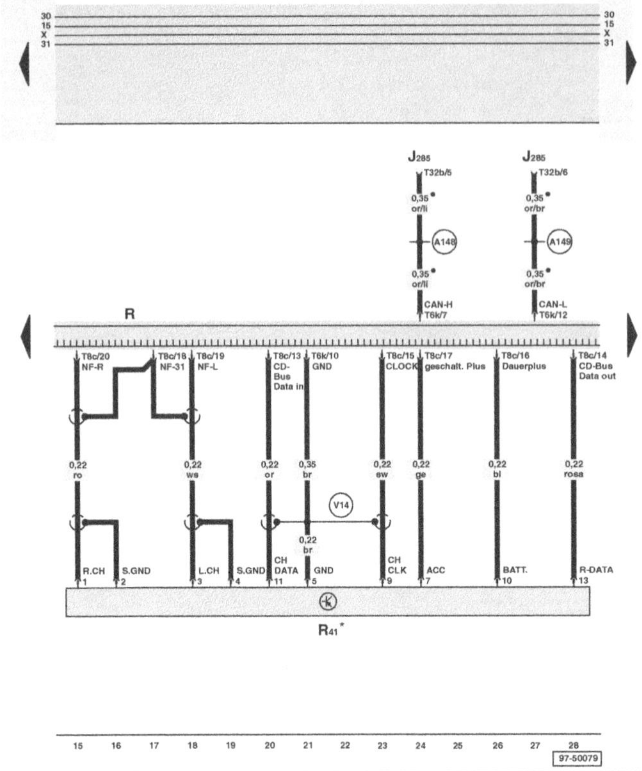

- J285 - Control Module with indicator unit in instrument panel insert
- R - Radio
- R41 - CD Changer Unit
- T6k - Connector 6 pin, green, connector radio I-2
- T8c - Connector 8 pin, blue, connector radio I-3
- T32b - Connector 32 pin, grey, on instrument cluster
- (A148) - Connection (high bus, navigation), in instrument panel wiring harness
- (A149) - Connection (low bus, navigation), in instrument panel wiring harness
- (V14) - Wire connection (shielding), in CD-changer wiring harness
- • - CAN-Bus (Data-Bus)
- * - Vehicles with CD-Changer

ws = white
sw = black
ro = red
br = brown
gn = green
bl = blue
gr = grey
li = violet
ge = yellow
or = orange

USA.5703.02.21 — Edition 04/03

Audi TT — Wiring Diagram — No. 7/4

Radio, Rear Window Defogger with Window Antenna

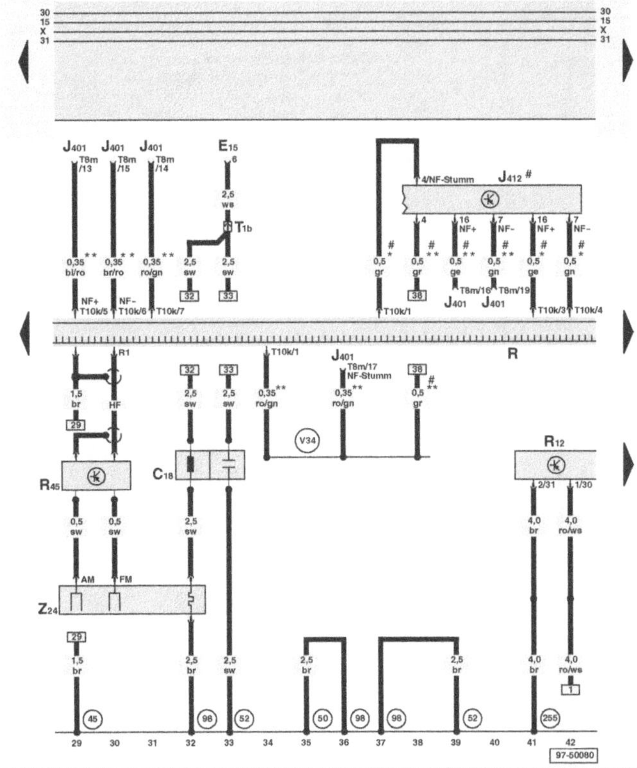

- C18 - Windshield Antenna Suppression Filter
- E15 - Rear Window Defogger Switch
- J401 - Control unit for navigation with CD-mechanism
- J412 - Operating electronics and Telephone Control Module
- R - Radio
- R1 - Antenna Connection
- R12 - Amplifier
- R45 - Telephone/Radio Antenna Amplifier
- T1b - Connector single, black, in rear lid left
- T8m - Connector 8 pin, C III on Control Module for Navigation
- T10k - Connector 10 pin, red, on radio
- Z24 - Rear Window Defogger With Window Antenna
- (45) - Ground connection, behind instrument panel, center
- (50) - Ground connection, in luggage compartment, left
- (52) - Ground connection, in rear lid, left
- (98) - Ground connection, in rear lid wiring harness
- (255) - Ground connection, -1-, in wiring harness Radio
- (V34) - Connector (mute button), in wiring harness telephone
- * - Vehicles without Navigation System
- ** - Vehicles with Navigation System
- # - Vehicles with Cellular Telephone Preparation

ws = white
sw = black
ro = red
br = brown
gn = green
bl = blue
gr = grey
li = violet
ge = yellow
or = orange

USA.5703.02.21 — Edition 04/03

© 2006 BentleyPublishers.com—All Rights Reserved

Audi TT — Wiring Diagram — No. 7/5

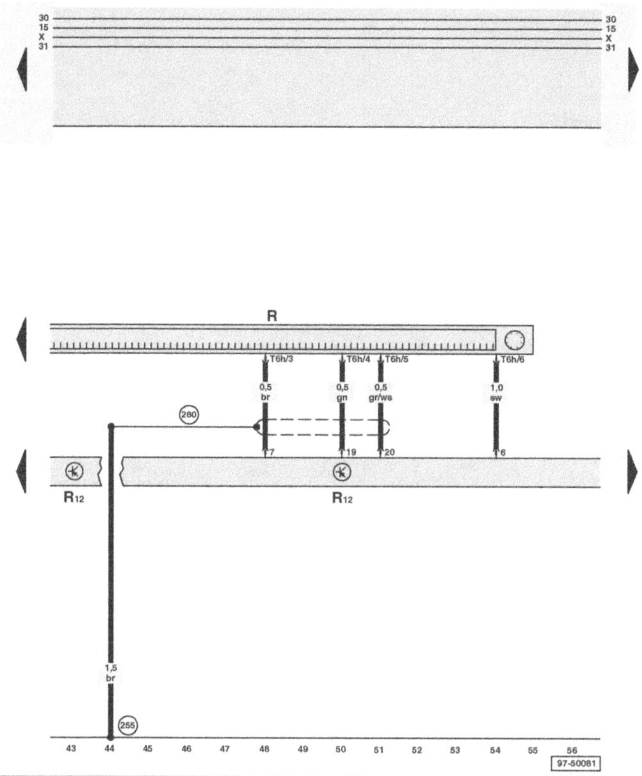

Radio, Amplifier

R - Radio
R12 - Amplifier
T6h - Connector 6 pin, yellow, connector radio I-1
(255) - Ground connection, -1-, in wiring harness Radio
(280) - Ground connector -2-, in wiring harness engine compartment

ws = white
sw = black
ro = red
br = brown
gn = green
bl = blue
gr = grey
li = violet
ge = yellow
or = orange

USA.5703.02.21 Edition 04/03

Audi TT — Wiring Diagram — No. 7/6

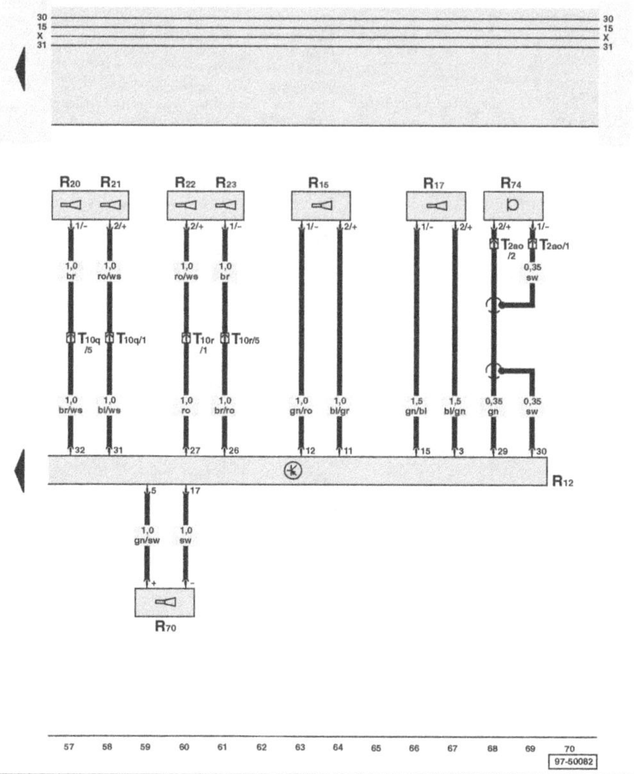

Amplifier, Loudspeaker

R12 - Amplifier
R15 - Bass speaker, left rear
R17 - Bass speaker, right rear
R20 - Treble speaker, left front
R21 - Bass speaker, left front
R22 - Treble speaker, right front
R23 - Bass speaker, right front
R70 - Inside loudspeaker, center
R74 - Inside microphone
T2ao - Connector 2 pin, near right rear ashtray
T10q - Connector 10 pin, green, connector station A pillar, left
T10r - Connector 10 pin, green, connector station A pillar, right

ws = white
sw = black
ro = red
br = brown
gn = green
bl = blue
gr = grey
li = violet
ge = yellow
or = orange

USA.5703.02.21 Edition 04/03

© 2006 BentleyPublishers.com—All Rights Reserved

Sound System, Bose, roadster (2003-2006)

Audi TT Wiring Diagram **No. 8/1**

Radio Bose with and without CD-Changer (Roadster)

2003 m.y.

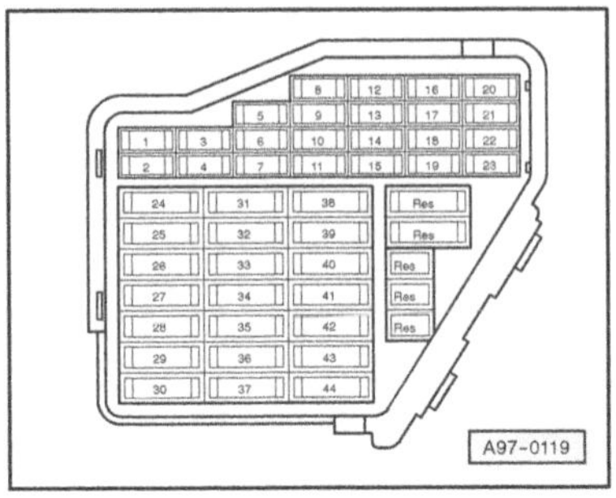

Fuse Panel

Fuse Colors:
- 30 A - Green
- 25 A - White
- 20 A - Yellow
- 15 A - Blue
- 10 A - Red
- 7,5 A - Brown
- 5 A - Beige
- 1 A - Black

Starting with fuse position 23, fuses in the fuse holder are identified with 223 in the wiring diagram.

Audi TT Wiring Diagram **No. 8/2**

Radio Bose with and without CD-Changer

- J429 - Control module for central locking
- R - Radio
- S242 - Fuse 42 in fuse holder
- T8 - Connector 8 pin, black, connector radio III
- T16 - Connector 16 pin, black, Data Link Connector (DLC), below storage compartment, driver's side

- (45) - Ground connection, behind instrument panel, center
- (255) - Ground connection, -1-, in wiring harness Radio
- (A46) - Plus connection (30-from radio), in instrument panel wiring harness
- (A76) - Connector (K-diagnosis wire), in instrument panel wiring harness

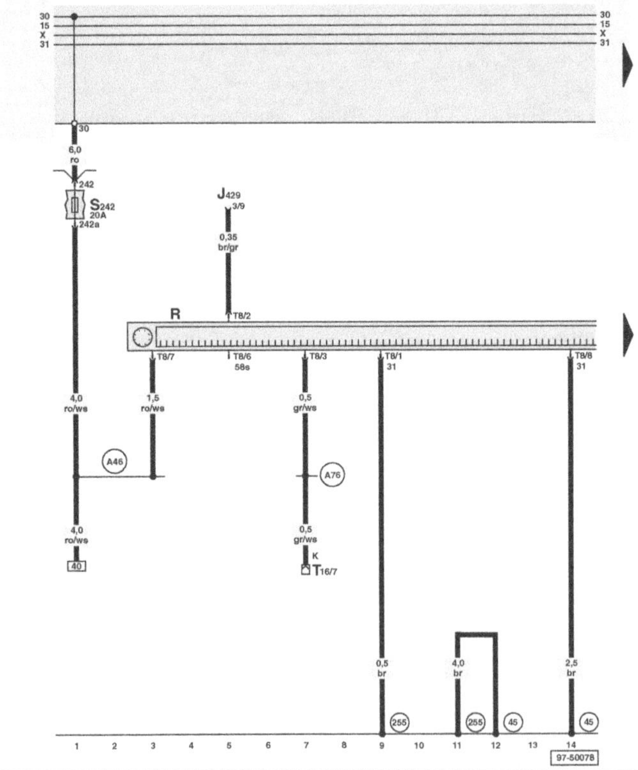

ws = white
sw = black
ro = red
br = brown
gn = green
bl = blue
gr = grey
li = violet
ge = yellow
or = orange

© 2006 BentleyPublishers.com—All Rights Reserved

Audi TT

Wiring Diagram

No. 8/3

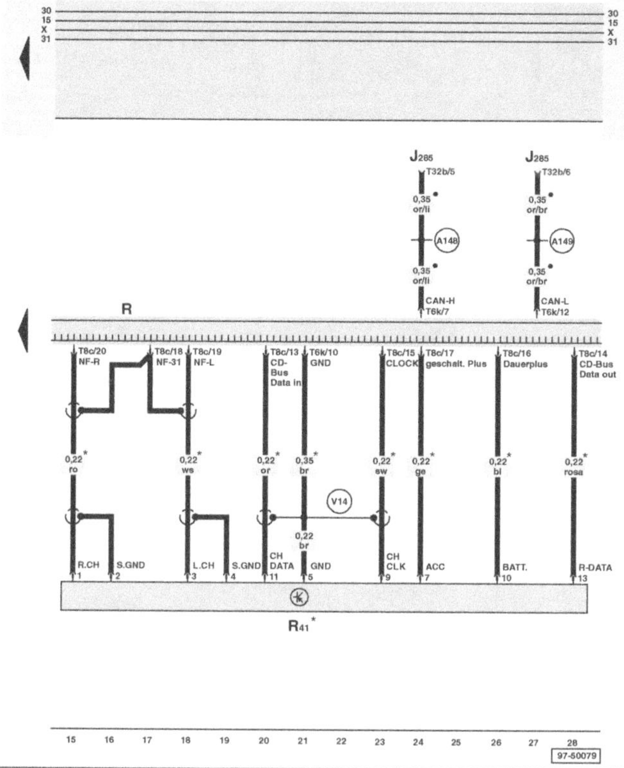

Radio Bose with and without CD-Changer

- J285 - Control Module with indicator unit in instrument panel insert
- R - Radio
- R41 - CD Changer Unit
- T6k - Connector 6 pin, green, connector radio I-2
- T8c - Connector 8 pin, blue, connector radio I-3
- T32b - Connector 32 pin, grey, on instrument cluster

- (A148) - Connection (high bus, navigation), in instrument panel wiring harness
- (A149) - Connection (low bus, navigation), in instrument panel wiring harness
- (V14) - Wire connection (shielding), in CD-changer wiring harness

- • - CAN-Bus (Data-Bus)
- * - Vehicles with CD-Changer

ws = white
sw = black
ro = red
br = brown
gn = green
bl = blue
gr = grey
li = violet
ge = yellow
or = orange

USA.5703.02.21

Edition 04/03

Audi TT

Wiring Diagram

No. 8/4

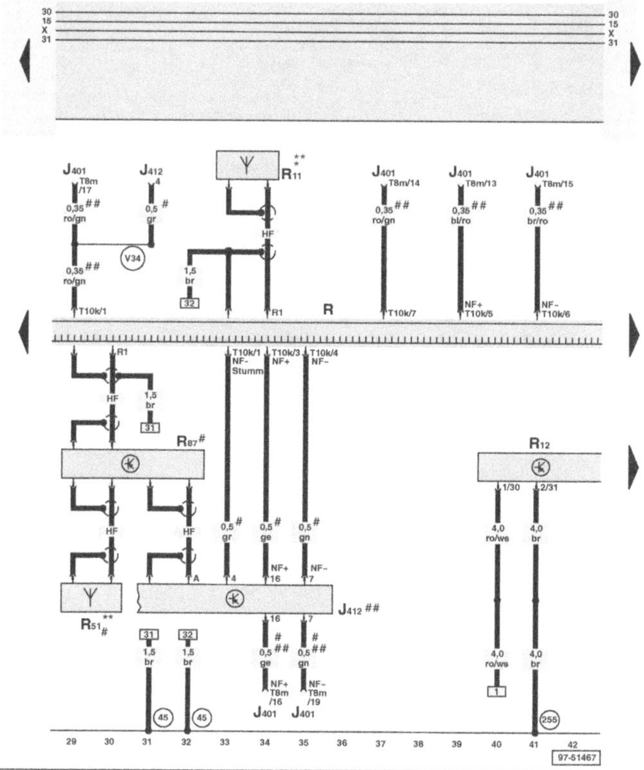

Radio Bose with and without CD-Changer

- J401 - Control unit for navigation with CD-mechanism
- J412 - Operating electronics and Telephone Control Module
- R - Radio
- R1 - Antenna Connection
- R11 - Antenna
- R12 - Amplifier
- R51 - Radio/Telephone Antenna
- R87 - Antenna Splitter
- T8m - Connector 8 pin, C III on Control Module for Navigation
- T10k - Connector 10 pin, red, on radio

- (45) - Ground connection, behind instrument panel, center
- (255) - Ground connection, -1-, in wiring harness Radio
- (V34) - Connector (mute button), in wiring harness telephone

- * - Vehicles without Cellular Telephone Preparation
- ** - Rod Antenna
- # - Vehicles with Cellular Telephone Preparation
- ## - Vehicles with Navigation System

ws = white
sw = black
ro = red
br = brown
gn = green
bl = blue
gr = grey
li = violet
ge = yellow
or = orange

USA.5703.02.21

Edition 04/03

© 2006 BentleyPublishers.com—All Rights Reserved

Sound System, Bose, roadster (2003-2006)

Audi TT

Wiring Diagram

No. 8/5

Radio Bose with and without CD-Changer

F172 - Convertible Top Front Lock Switch
J429 - Control module for central locking
R - Radio
R12 - Amplifier
T6h - Connector 6 pin, yellow, connector radio I-1

(255) - Ground connection, -1-, in wiring harness Radio
(279) - Ground connector -5-, in wiring harness interior
(280) - Ground connector -2-, in wiring harness engine compartment
(Q48) - Connection (open), in wiring harness interior

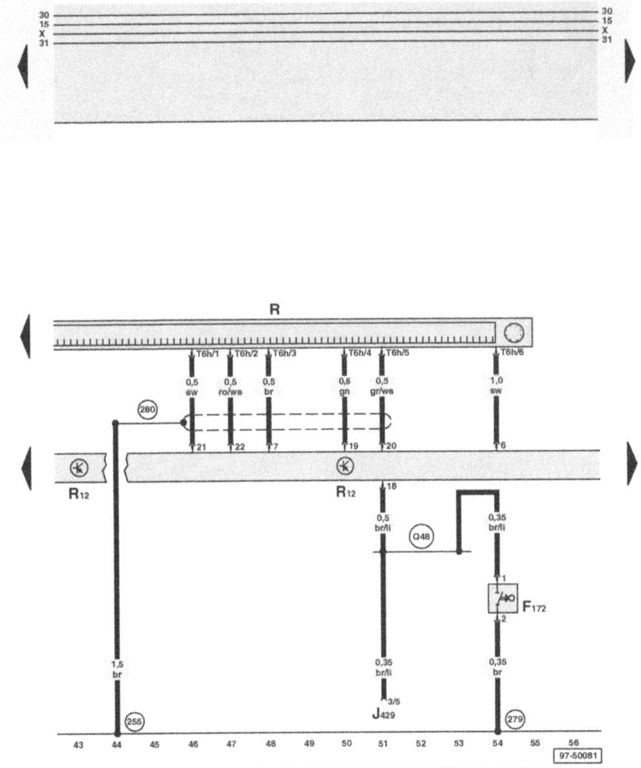

ws = white
sw = black
ro = red
br = brown
gn = green
bl = blue
gr = grey
li = violet
ge = yellow
or = orange

USA.5703.02.21 — Edition 04/03

Audi TT

Wiring Diagram

No. 8/6

Radio Bose with and without CD-Changer

R12 - Amplifier
R15 - Bass speaker, left rear
R17 - Bass speaker, right rear
R20 - Treble speaker, left front
R21 - Bass speaker, left front
R22 - Treble speaker, right front
R23 - Bass speaker, right front
R70 - Loudspeaker front, center
R71 - Loudspeaker rear, center
R74 - Inside microphone
T2al - Connector 2 pin, behind instrument cluster
T10q - Connector 10 pin, green, connector station A pillar, left
T10r - Connector 10 pin, green, connector station A pillar, right

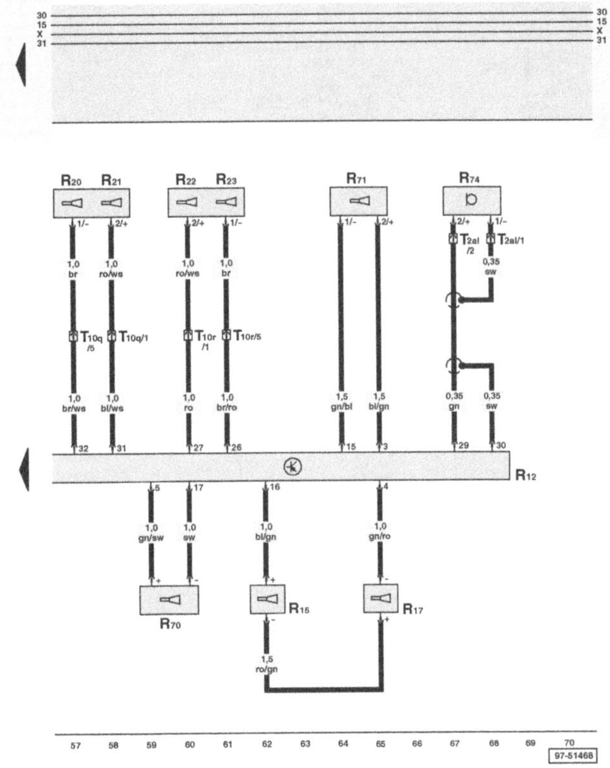

ws = white
sw = black
ro = red
br = brown
gn = green
bl = blue
gr = grey
li = violet
ge = yellow
or = orange

USA.5703.02.21 — Edition 04/03

© 2006 BentleyPublishers.com—All Rights Reserved

Daytime Running Lights with Fog Light (Canada)

2003 m. y.

Fuse Panel

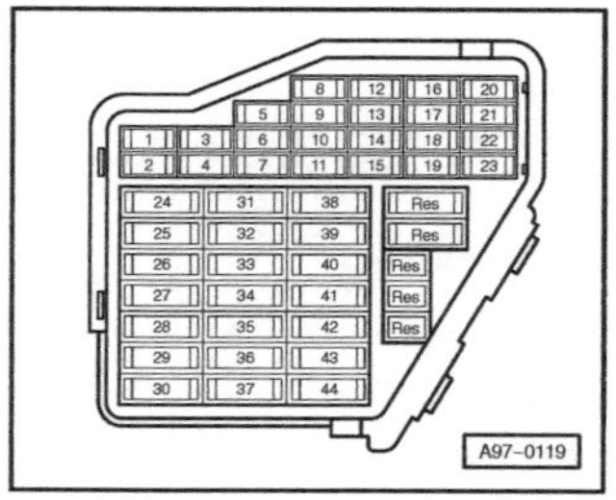

Fuse Colors:

30 A - Green
25 A - White
20 A - Yellow
15 A - Blue
10 A - Red
7,5 A - Brown
5 A - Beige

Starting with fuse position 23, fuses in the fuse holder are identified with 223 in the wiring diagram.

Micro Central Electric Panel

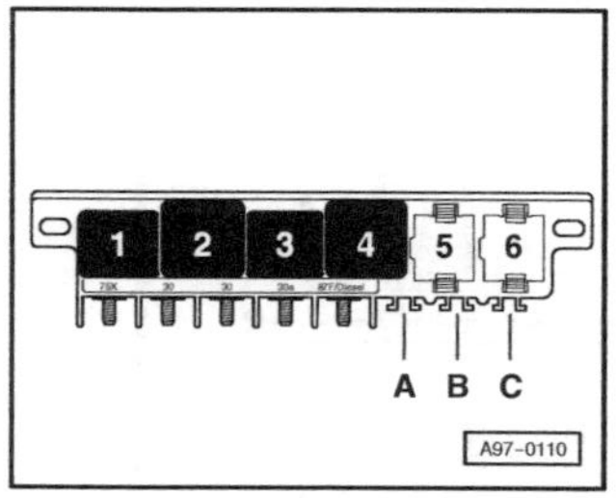

Relay Location:

2 -Load Reduction Relay, J59

13 - Fold Relay Panel

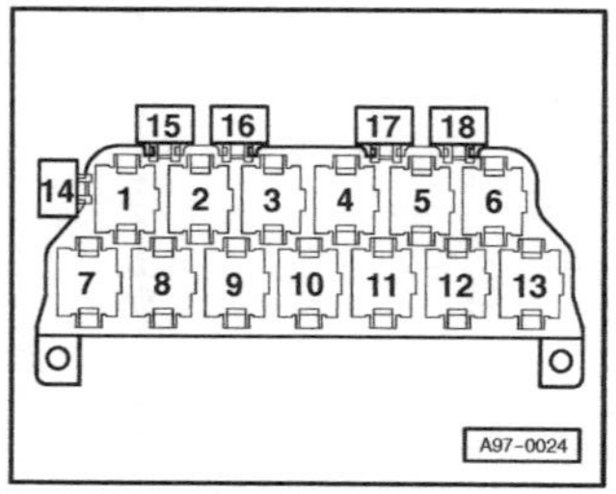

Relay Location:

2 -Lamp Control Module, J123

3 -Lamp Control Module, J123

4 -Fog Light Relay, J5

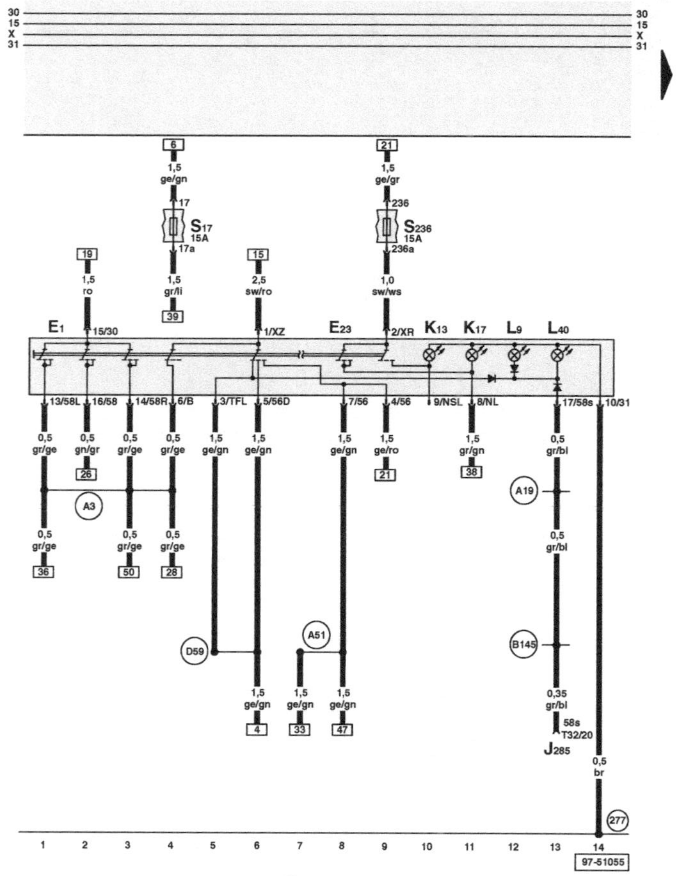

ws = white
sw = black
ro = red
br = brown
gn = green
bl = blue
gr = grey
li = violet
ge = yellow

E1 - Light Switch
E23 - Fog Light Switch
J285 - Control Module with indicator unit in instrument panel insert
K13 - Rear Fog Light Indicator Light
K17 - Fog Light Indicator Light
L9 - Headlight Switch Light
L40 - Front And Rear Fog Light Switch Light
S17 - Fuse
S236 - Fuse in fuse holder
T32 - Connector 32 pin, blue, on instrument cluster

(277) - Ground connector -3-, in wiring harness interior
(A3) - Plus connection (58), in instrument panel wiring harness
(A19) - Wire connection (58s), in instrument panel wiring harness
(A51) - Wire connection (56), in instrument panel wiring harness
(B145) - Plus connector (58s), in wiring harness interior
(D59) - Wire connection (Fog light), in engine compartment wiring harness

Light Switch, Fog Light Switch

©2006 BentleyPublishers.com—All Rights Reserved

Daytime Running Lights and Rear Fog Light, Canada (2003-2006)

Audi TT — Wiring diagram — No. 9/3

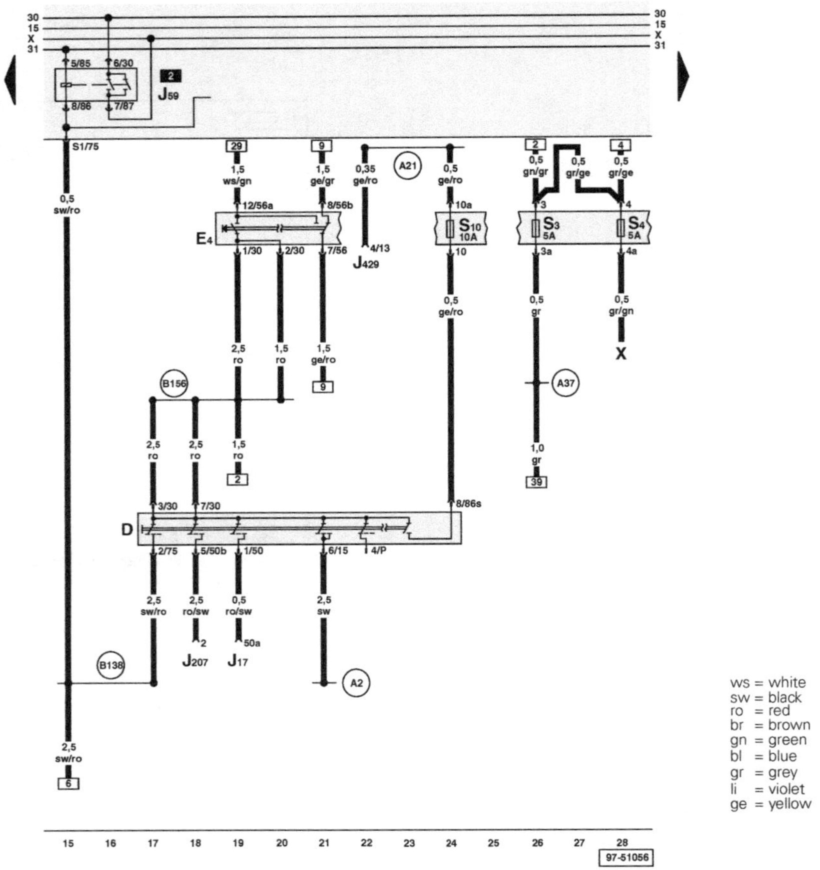

ws = white
sw = black
ro = red
br = brown
gn = green
bl = blue
gr = grey
li = violet
ge = yellow

- D - Ignition/Starter Switch
- E4 - Headlight Dimmer/Flasher Switch
- J17 - Fuel Pump (FP) Relay
- J59 - Load Reduction Relay
- J207 - Starting Interlock Relay
- J429 - Control module for central locking
- S3 - Fuse
- S4 - Fuse
- S10 - Fuse
- X - License Plate Light

- A2 - Plus connection (15), in instrument panel wiring harness
- A21 - Wire connection (86s), in instrument panel wiring harness
- A37 - Wire connection (58a), in instrument panel wiring harness
- B138 - Plus connector (X), in wiring harness, interior
- B156 - Plus connector (30a), in wiring harness interior

Edition 10/02
USA.5703.01.21

Ignition/Starter Switch, Load Reduction Relay, Headlight Dimmer/Flasher Switch

No. 9/4 — Wiring diagram — Audi TT

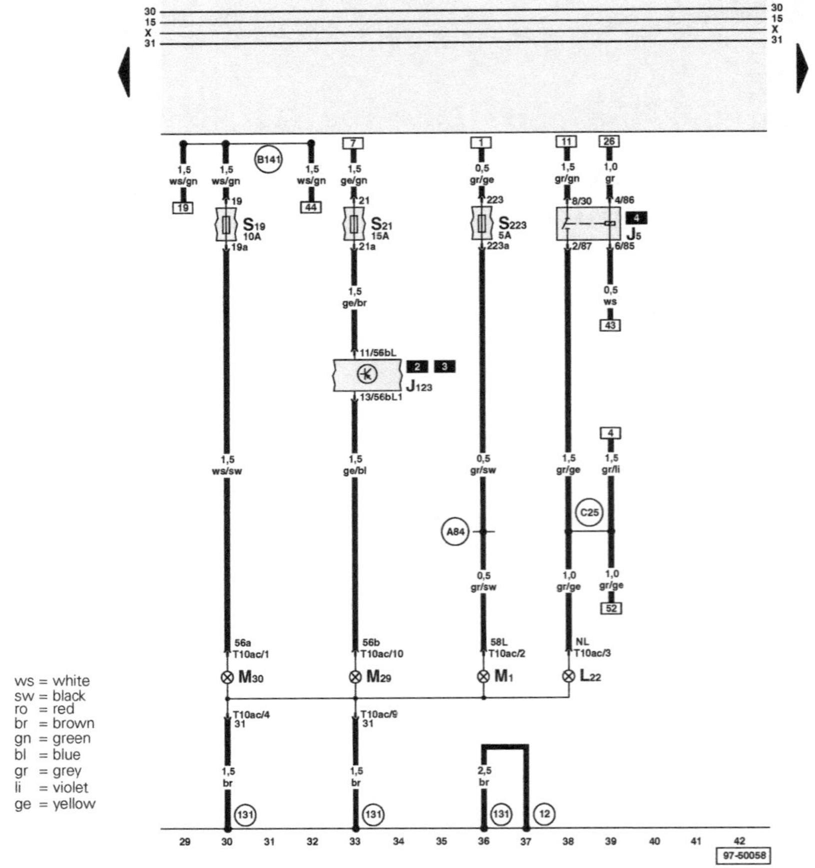

ws = white
sw = black
ro = red
br = brown
gn = green
bl = blue
gr = grey
li = violet
ge = yellow

- J5 - Fog Light Relay
- J123 - Lamp Control Module
- L22 - Left Front Fog Light
- M1 - Left Parking Light
- M29 - Left Low Beam Headlight
- M30 - Left High Beam Headlight
- S19 - Fuse
- S21 - Fuse
- S223 - Fuse in fuse holder
- T10ac - Connector 10 pin, on headlight, left

- 12 - Ground connection, in engine compartment, left
- 131 - Ground connection -2-, in engine compartment wiring harness
- A84 - Connector (58L), in instrument panel wiring harness
- B141 - Plus connector -1- (56a), in wiring harness interior
- C25 - Wire connection (fog lights), in front light wiring harness

Left Headlight

Edition 10/02
USA.5703.01.21

© 2006 BentleyPublishers.com—All Rights Reserved

Audi TT Wiring diagram **No. 9/5**

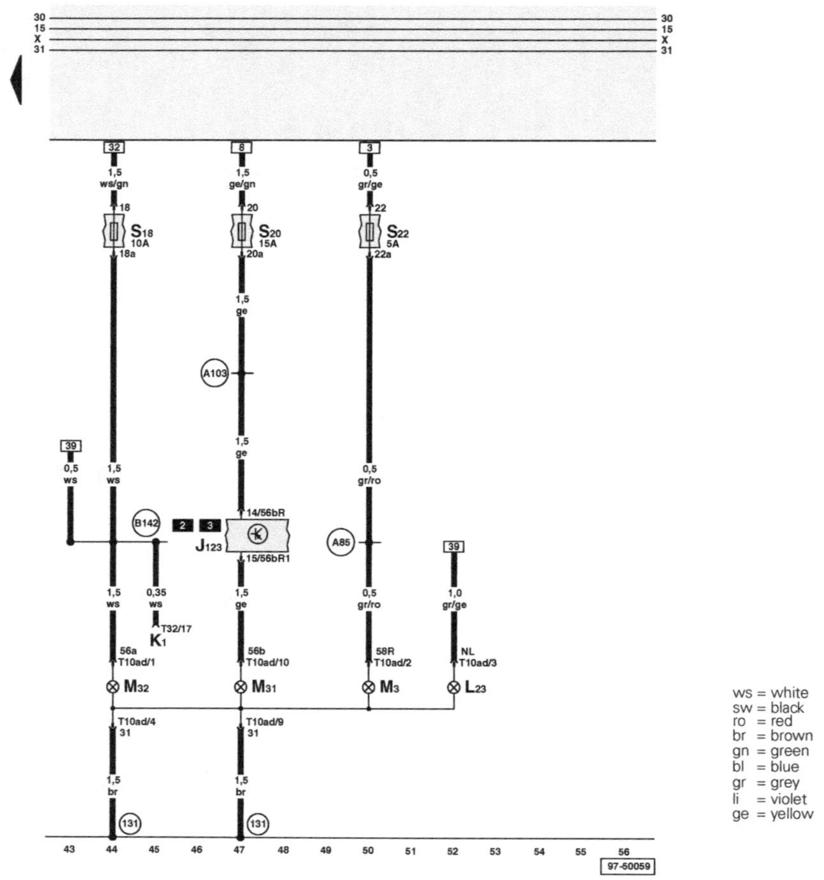

ws = white
sw = black
ro = red
br = brown
gn = green
bl = blue
gr = grey
li = violet
ge = yellow

J123 - Lamp Control Module
K1 - Headlight High Beam Indicator Light
L23 - Right Front Fog Light
M3 - Right Parking Light
M31 - Right Low Beam Headlight
M32 - Right High Beam Headlight
S18 - Fuse
S20 - Fuse
S22 - Fuse
T10ad - Connector 10 pin, on headlight, right
T32 - Connector 32 pin, blue, on instrument cluster

(131) - Ground connection -2-, in engine compartment wiring harness

(A85) - Connector (58R), in instrument panel wiring harness

(A103) - Connector -2- (56), in instrument panel wiring harness

(B142) - Plus connector -2- (56a), in wiring harness interior

Edition 10/02
USA.5703.01.21

Right Headlight

THIS SPACE INTENTIONALLY LEFT BLANK

© 2006 BentleyPublishers.com—All Rights Reserved

Audi TT **Wiring Diagram** **No. 10/1**

Airbag

2004 m.y.

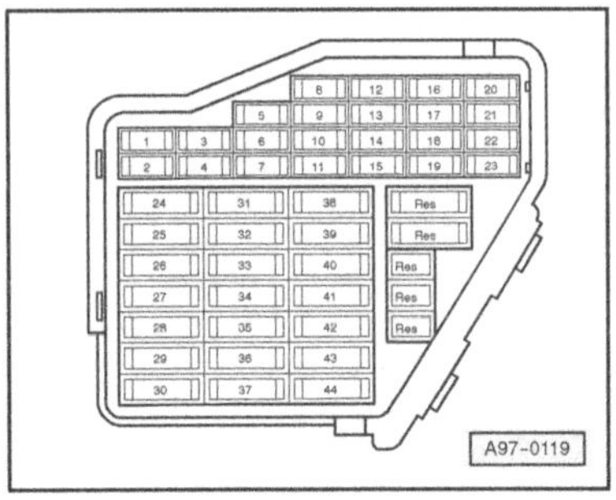

Fuse Panel

Fuse Colors:
- 30 A - Green
- 25 A - White
- 20 A - Yellow
- 15 A - Blue
- 10 A - Red
- 7,5 A - Brown
- 5 A - Beige
- 1 A - Black

Starting with fuse position 23, fuses in the fuse holder are identified with 223 in the wiring diagram.

Audi TT **Wiring Diagram** **No. 10/2**

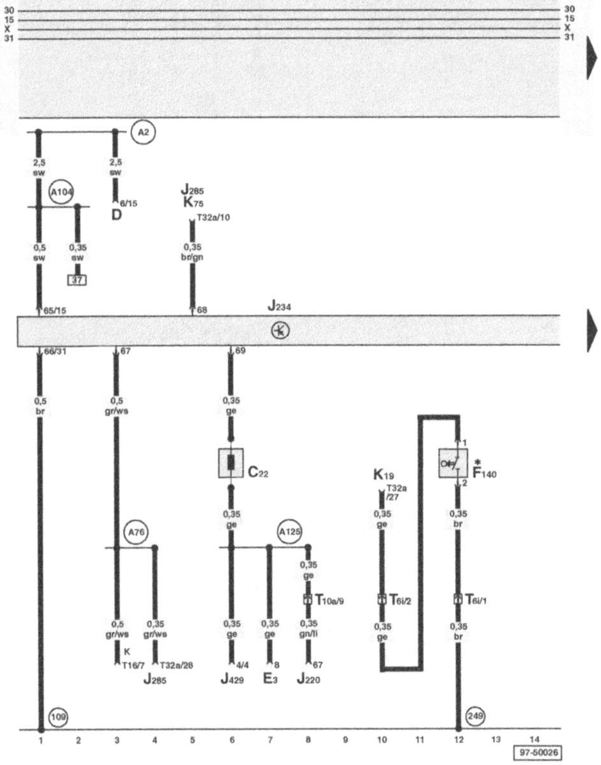

Airbag Control Module

- C22 - Airbag Suppresser
- D - Ignition/Starter Switch
- E3 - Emergency Flasher Switch
- F140 - Left Front Seatbelt Microswitch
- J220 - Motronic Engine Control Module (ECM)
- J234 - Airbag Control Module
- J285 - Control Module with indicator unit in instrument panel insert
- J429 - Control module for central locking
- K19 - Seat Belt Warning Light
- K75 - Airbag Malfunction Indicator Lamp (MIL)
- T6i - Connector 6 pin, below driver's seat
- T10a - Connector 10 pin, white, in E-box plenum chamber
- T16 - Connector 16 pin, black, Data Link Connector (DLC), below storage compartment, driver's side
- T32a - Connector 32 pin, green, on instrument cluster
- (109) - Ground connection, in airbag wiring harness
- (249) - Ground connection -2-, in wiring harness interior
- (A2) - Plus connection (15), in instrument panel wiring harness
- (A76) - Connector (K-diagnosis wire), in instrument panel wiring harness
- (A104) - Plus connector -2- (15), in instrument panel wiring harness
- (A125) - Connection (crash signal) in instrument panel wiring harness
- * - Vehicles with seat belt warning lamp

ws = white
sw = black
ro = red
br = brown
gn = green
bl = blue
gr = grey
li = violet
ge = yellow
or = orange

©2006 BentleyPublishers.com—All Rights Reserved

Audi TT

Wiring Diagram

No. 10/3

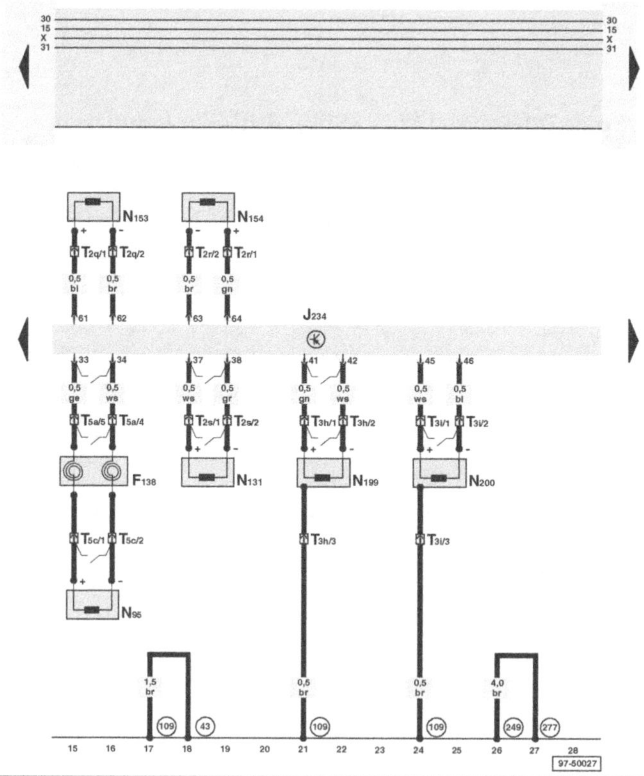

Airbag Control Module, Side Airbag Igniter, Seat Belt Tensioner Igniter

F138 - Airbag Spiral Spring/Return Spring With Slip Ring
J234 - Airbag Control Module
N95 - Driver's Side Airbag Igniter
N131 - Passenger's Side Airbag Igniter 1
N153 - Driver's seat Belt Tensioner Igniter 1
N154 - Passenger's Seat Belt Tensioner Igniter 1
N199 - Igniter for side airbag, driver's side
N200 - Igniter for side airbag, passenger's side
T2q - Connector 2 pin, on left B-pillar, lower part
T2r - Connector 2 pin, on right B-pillar, lower part
T2s - Connector 2 pin, behind instrument panel, right
T3h - Connector 3 pin, below driver's seat (for side airbag)
T3i - Connector 3 pin, below passenger's seat (for side airbag)
T5a - Connector 5 pin, yellow, in steering wheel
T5c - Connector 5 pin, yellow, in steering wheel

(43) - Ground connection, on right A-pillar, lower part
(109) - Ground connection, in airbag wiring harness
(249) - Ground connection -2-, in wiring harness interior
(277) - Ground connector -3-, in wiring harness interior
- Bridge closes when disconnecting connector (safety feature)

ws = white
sw = black
ro = red
br = brown
gn = green
bl = blue
gr = grey
li = violet
ge = yellow
or = orange

USA.5703.02.21 Edition 04/03

Audi TT

Wiring Diagram

No. 10/4

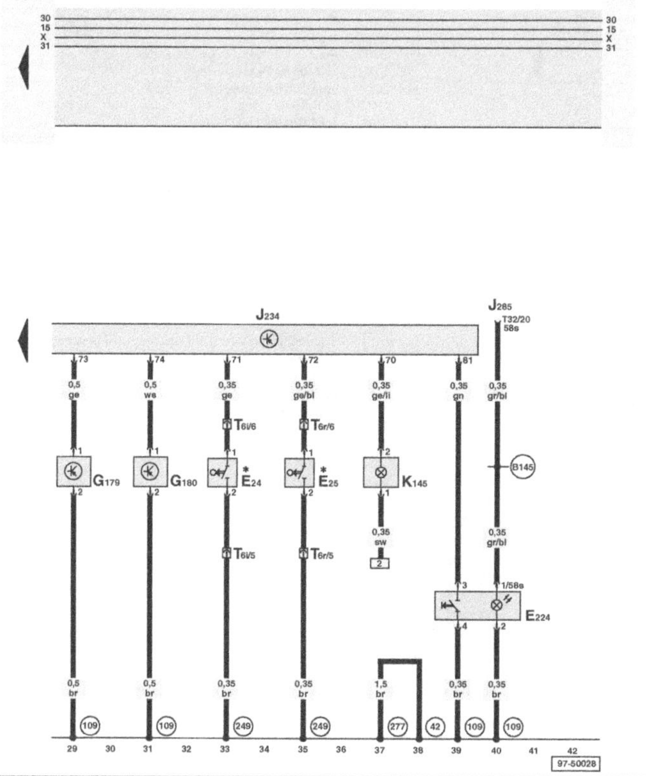

Airbag Control Module, Crash Sensors for Side Airbags, Seat Belt Switch, Key Switch for Switching Off Passenger Side Airbag

E24 - Left Seat Belt Switch
E25 - Right Seat Belt Switch
E224 - Key switch for switching off passenger side airbag
G179 - Crash sensor for side airbag, driver's side
G180 - Crash sensor for side airbag, passenger's side
J234 - Airbag Control Module
J285 - Control Module with indicator unit in instrument panel insert
K145 - Warning light for airbag off, passenger side
T6i - Connector 6 pin, below driver's seat
T6r - Connector 6 pin, below passenger's seat
T32 - Connector 32 pin, blue, on instrument cluster

(42) - Ground connection, on steering column
(109) - Ground connection, in airbag wiring harness
(249) - Ground connection -2-, in wiring harness interior
(277) - Ground connector -3-, in wiring harness interior
(B145) - Plus connector (58b), in wiring harness interior
* - Vehicles with seat belt warning

ws = white
sw = black
ro = red
br = brown
gn = green
bl = blue
gr = grey
li = violet
ge = yellow
or = orange

USA.5703.02.21 Edition 04/03

© 2006 BentleyPublishers.com—All Rights Reserved

Audi TT | **Wiring Diagram** | **No. 11/1**

Anti-Lock Brake System (ABS) MK60 with Electronic Differential Lock (EDL), Anti-Slip Regulation (ASR - ASC) and Electronic Stabilization Program (ESP)

2004 m.y.

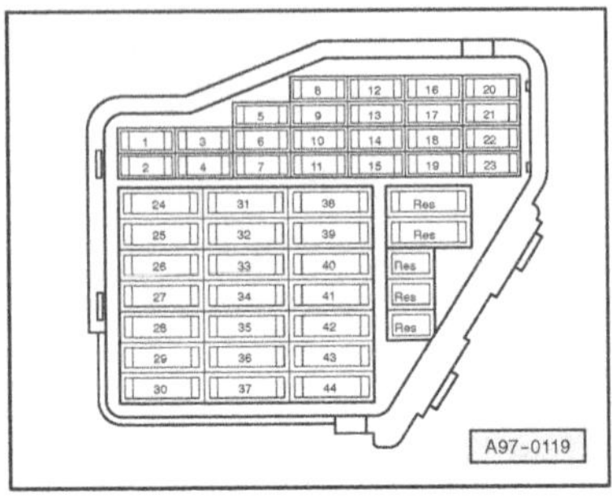

Fuse Panel

Fuse Colors:
- 30 A - Green
- 25 A - White
- 20 A - Yellow
- 15 A - Blue
- 10 A - Red
- 7,5 A - Brown
- 5 A - Beige
- 1 A - Black

Starting with fuse position 23, fuses in the fuse holder are identified with 223 in the wiring diagram.

Audi TT | **Wiring Diagram** | **No. 11/2**

ABS Control Module, ESP - Sensor Unit

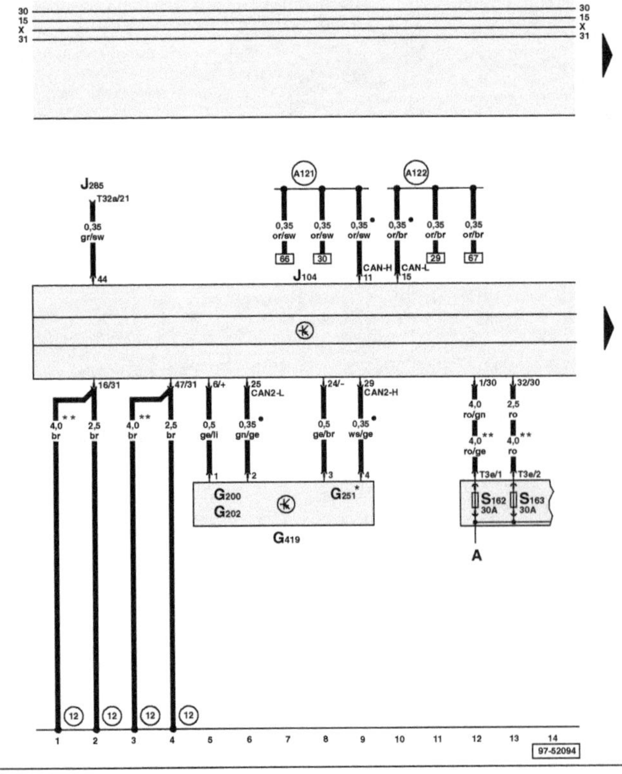

- A - Battery
- G200 - Sensor for transverse acceleration
- G202 - Sender for Rotation Rate
- G251 - Longitudinal acceleration sensor
- G419 - ESP-Sensor Unit
- J104 - ABS Control Module (w/EDL)
- J285 - Control Module with indicator unit in instrument panel insert
- S162 - Fuse -1- in fuse bracket/battery
- S163 - Fuse -2- in fuse bracket/battery
- T3e - Connector 3 pin, black, on Main Fuse Box/Battery
- T32a - Connector 32 pin, green, on instrument cluster
- (12) - Ground connection, in engine compartment, left
- (A121) - Connection (high bus) in instrument panel wiring harness
- (A122) - Connection (low bus) in instrument panel wiring harness
- • - CAN-Bus (Data-Bus)
- * - only quattro
- ** - only 3.2L engine, code BHE, BPF

ws = white
sw = black
ro = red
br = brown
gn = green
bl = blue
gr = grey
li = violet
ge = yellow
or = orange

© 2006 BentleyPublishers.com—All Rights Reserved

Audi TT

Wiring Diagram

No. 11/3

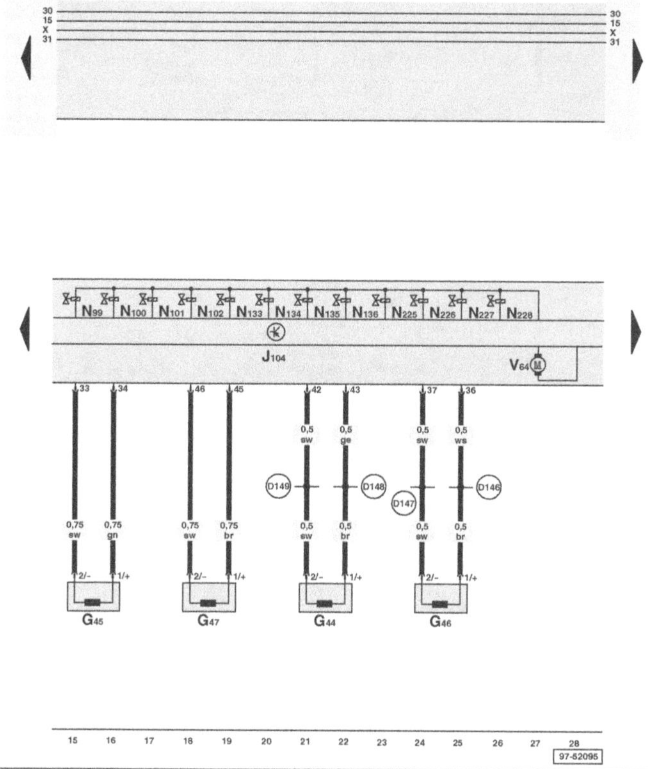

ABS Control Module, Wheel Speed Sensors, Valves

- G44 - Right Rear ABS Wheel Speed Sensor
- G45 - Right Front ABS Wheel Speed Sensor
- G46 - Left Rear ABS Wheel Speed Sensor
- G47 - Left Front ABS Wheel Speed Sensor
- J104 - ABS Control Module (w/EDL)
- N99 - Right Front ABS Inlet Valve
- N100 - Right Front ABS Outlet Valve
- N101 - Left Front ABS Inlet Valve
- N102 - Left Front ABS Outlet Valve
- N133 - Right Rear ABS Inlet Valve
- N134 - Left Rear ABS Inlet Valve
- N135 - Right Rear ABS Outlet Valve
- N136 - Left Rear ABS Outlet Valve
- N225 - Pilot valve -1- traction control (01418)
- N226 - Pilot valve -2- traction control (01419)
- N227 - High pressure switch valve -1- traction control
- N228 - High pressure switch valve -2- traction control
- V64 - ABS Hydraulic Pump

- (D146) - Connector (speed sensor rear, left +), in wiring harness engine compartment
- (D147) - Connector (speed sensor rear, left -), in wiring harness engine compartment
- (D148) - Connector (speed sensor rear, right +), in wiring harness engine compartment
- (D149) - Connector (speed sensor rear, right -), in wiring harness engine compartment

ws = white
sw = black
ro = red
br = brown
gn = green
bl = blue
gr = grey
li = violet
ge = yellow
or = orange

USA.5703.02.21 Edition 04/03

Audi TT

Wiring Diagram

No. 11/4

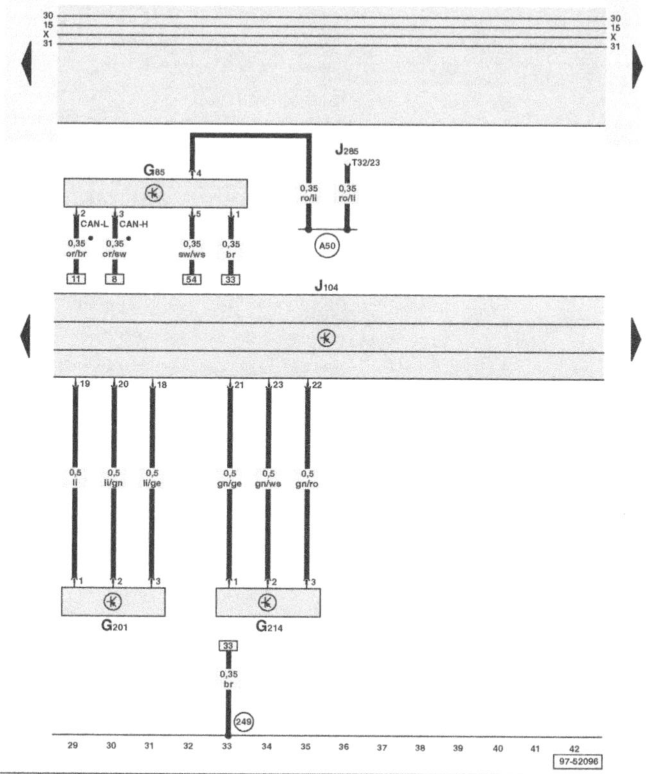

ABS Control Module, Steering Angle Sensor, Sensor for Brake Pressure, Sender for Brake Booster

- G85 - Steering Angle Sensor
- G201 - Sender 1 for brake booster
- G214 - Sensor -2- for brake pressure
- J104 - ABS Control Module (w/EDL)
- J285 - Control Module with indicator unit in instrument panel insert
- T32 - Connector 32 pin, blue, on instrument cluster

- (249) - Ground connection -2-, in wiring harness interior
- (A50) - Plus connection (30as), in instrument panel wiring harness

- • - CAN-Bus (Data-Bus)

ws = white
sw = black
ro = red
br = brown
gn = green
bl = blue
gr = grey
li = violet
ge = yellow
or = orange

USA.5703.02.21 Edition 04/03

©2006 BentleyPublishers.com—All Rights Reserved

Audi TT

Wiring Diagram

No. 11/5

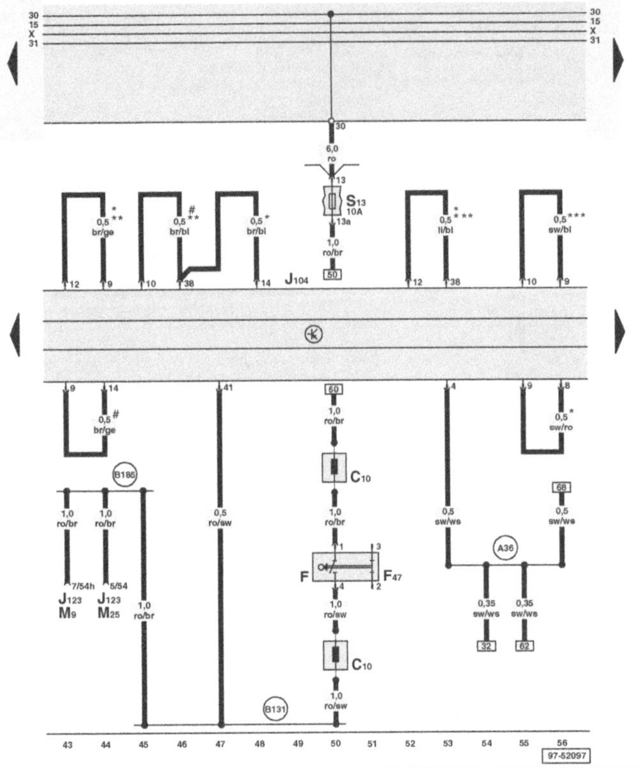

ws = white
sw = black
ro = red
br = brown
gn = green
bl = blue
gr = grey
li = violet
ge = yellow
or = orange

ABS Control Module, Brake Light Switch

- C10 - Suppressor Filter
- F - Brake Light Switch
- F47 - Brake Vacuum Vent Valve Switch for Cruise Control/Diesel Fuel Injection
- J104 - ABS Control Module (w/EDL)
- J123 - Lamp Control Module
- M9 - Left Brake Light
- M25 - High-mount Brake Light
- S13 - Fuse

- (A36) - Wire connection (75a), in instrument panel wiring harness
- (B131) - Connector (54), in wiring harness interior
- (B185) - Connection 2 (54), in passenger compartment wiring harness

- * - Vehicles with front wheel drive
- ** - Vehicles with quattro and 1.8L engine, code ATC, AWP
- *** - Vehicles with quattro and 1.8L engine, code AMU
- # - Vehicles with quattro 3.2L engine, code BHE

USA.5703.02.21 — Edition 04/03

Audi TT

Wiring Diagram

No. 11/6

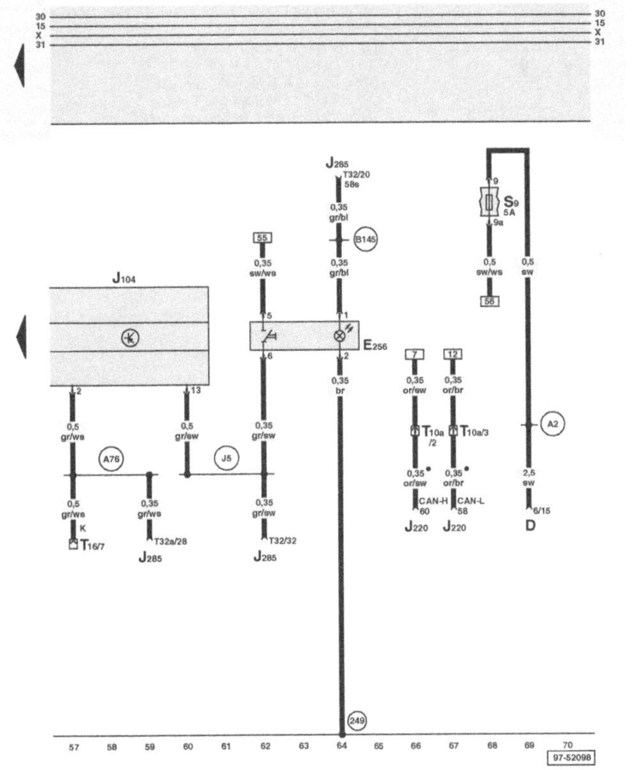

ws = white
sw = black
ro = red
br = brown
gn = green
bl = blue
gr = grey
li = violet
ge = yellow
or = orange

ABS Control Module, ASR/ESP Button

- E256 - ASR/ESP button
- J104 - ABS Control Module (w/EDL)
- J220 - Motronic Engine Control Module (ECM)
- J285 - Control Module with indicator unit in instrument panel insert
- S9 - Fuse
- T10a - Connector 10 pin, white, in E-box plenum chamber
- T16 - Connector 16 pin, black, Data Link Connector (DLC), below storage compartment, driver's side
- T32 - Connector 32 pin, blue, on instrument cluster
- T32a - Connector 32 pin, green, on instrument cluster

- (249) - Ground connection -2-, in wiring harness interior
- (A2) - Plus connection (15), in instrument panel wiring harness
- (A76) - Connector (K-diagnosis wire), in instrument panel wiring harness
- (B145) - Plus connector (58b), in wiring harness interior
- (J5) - Wire connection -2-, in ABS wiring harness

- • - CAN-Bus (Data-Bus)

USA.5703.02.21 — Edition 04/03

© 2006 BentleyPublishers.com—All Rights Reserved

Audi TT Wiring Diagram **No. 12/1**

Automatic Climate Control

2004 m.y.

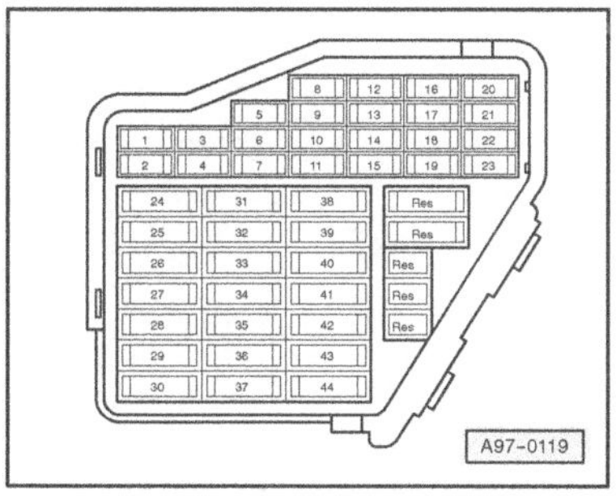

Fuse Panel

Fuse Colors:

30 A - Green
25 A - White
20 A - Yellow
15 A - Blue
10 A - Red
7,5 A - Brown
5 A - Beige
1 A - Black

Starting with fuse position 23, fuses in the fuse holder are identified with 223 in the wiring diagram.

Audi TT Wiring Diagram **No. 12/2**

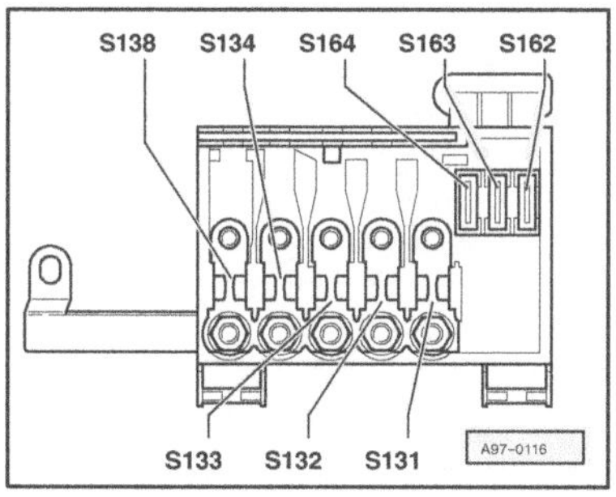

Main Fuse Box/Battery

Fuses Location:

S131 - Safety fuse 1
S132 - Safety fuse 2
S133 - Safety fuse 3
S134 - Safety fuse 4
S138 - Safety fuse 5
S162 - Fuse 1 in fuse bracket/battery
S163 - Fuse 2 in fuse bracket/battery
S164 - Fuse 3 in fuse bracket/battery

© 2006 BentleyPublishers.com—All Rights Reserved

Audi TT — Wiring Diagram — No. 12/3

A/C Control Head, High Pressure Sensor

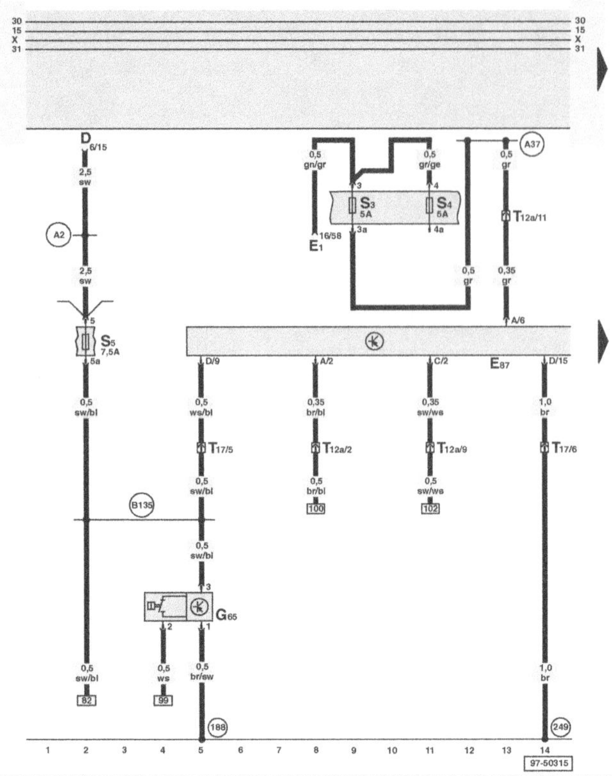

- D - Ignition/Starter Switch
- E1 - Light Switch
- E87 - A/C Control Head
- G65 - High Pressure Sensor
- S3 - Fuse
- S4 - Fuse
- S5 - Fuse
- T12a - Connector 12 pin, behind instrument panel, right
- T17 - Connector 17 pin, behind instrument panel
- (188) - Ground connection -3-, in A/C wiring harness
- (249) - Ground connection -2-, in wiring harness interior
- (A2) - Plus connection (15), in instrument panel wiring harness
- (A37) - Wire connection (58a), in instrument panel wiring harness
- (B135) - Connector (15a), in wiring harness interior

ws = white
sw = black
ro = red
br = brown
gn = green
bl = blue
gr = grey
li = violet
ge = yellow
or = orange

USA.5703.02.21 — Edition 04/03

Audi TT — Wiring Diagram — No. 12/4

A/C Control Head, Control Module with Indicator Unit in Instrument Panel Insert

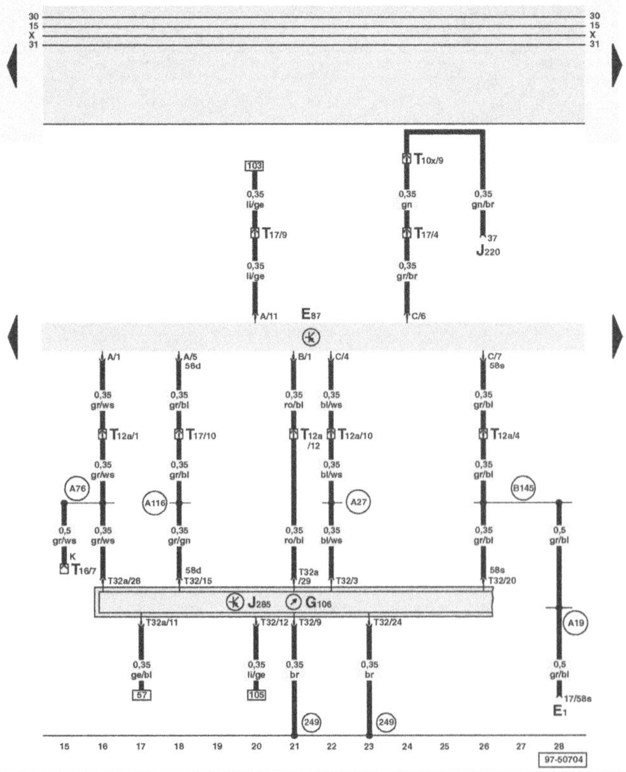

- E1 - Light Switch
- E87 - A/C Control Head
- G106 - Outside Air Temperature Display
- J220 - Motronic Engine Control Module (ECM)
- J285 - Control Module with indicator unit in instrument panel insert
- T10x - Connector 10 pin, orange, in E-box plenum chamber
- T12a - Connector 12 pin, behind instrument panel, right
- T16 - Connector 16 pin, black, Data Link Connector (DLC), below storage compartment, driver's side
- T17 - Connector 17 pin, behind instrument panel
- T32 - Connector 32 pin, blue, on instrument cluster
- T32a - Connector 32 pin, green, on instrument cluster
- (249) - Ground connection -2-, in wiring harness interior
- (A19) - Wire connection (58s), in instrument panel wiring harness
- (A27) - Wire connection (vehicle speed signal), in instrument panel wiring harness
- (A76) - Connector (K-diagnosis wire), in instrument panel wiring harness
- (A116) - Connection -2- (58d) in instrument panel wiring harness
- (B145) - Plus connector (58b), in wiring harness interior

ws = white
sw = black
ro = red
br = brown
gn = green
bl = blue
gr = grey
li = violet
ge = yellow
or = orange

USA.5703.02.21 — Edition 04/03

© 2006 BentleyPublishers.com—All Rights Reserved

Audi TT — Wiring Diagram — No. 12/5

A/C Control Head, Flap Motors

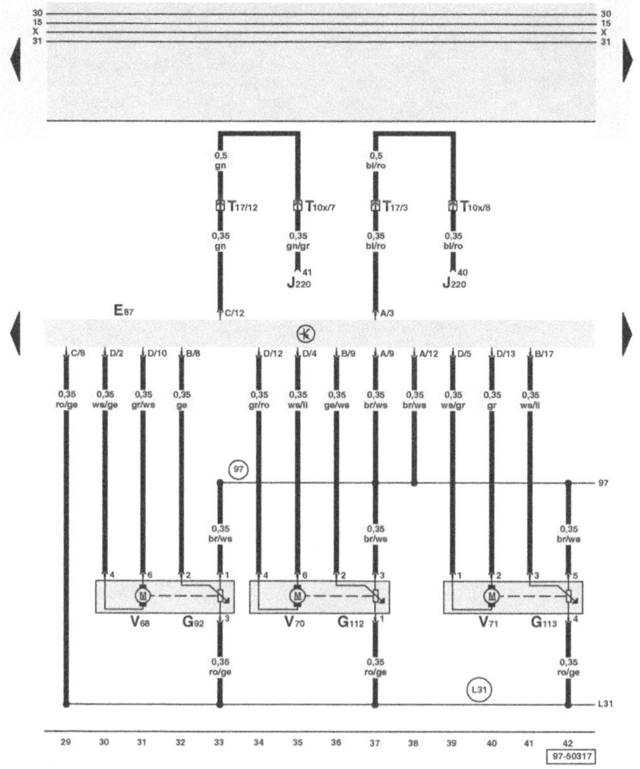

- E87 - A/C Control Head
- G92 - Temperature Regulator Flap Motor Position Sensor
- G112 - Central Flap Motor Position Sensor
- G113 - Back Pressure Flap Motor Position Sensor
- J220 - Motronic Engine Control Module (ECM)
- T10x - Connector 10 pin, orange, in E-box plenum chamber
- T17 - Connector 17 pin, behind instrument panel
- V68 - Temperature Regulator Flap Motor
- V70 - Central Air Flap Motor
- V71 - Air Flow Flap Motor
- (97) - Ground connection -1-, in A/C wiring harness
- (L31) - Wire connection (5 Volts), in A/C wiring harness

ws = white
sw = black
ro = red
br = brown
gn = green
bl = blue
gr = grey
li = violet
ge = yellow
or = orange

Audi TT — Wiring Diagram — No. 12/6

A/C Control Head, Flap Motor, Outside Air Temperature Sensor, Sunlight Photo Sensor

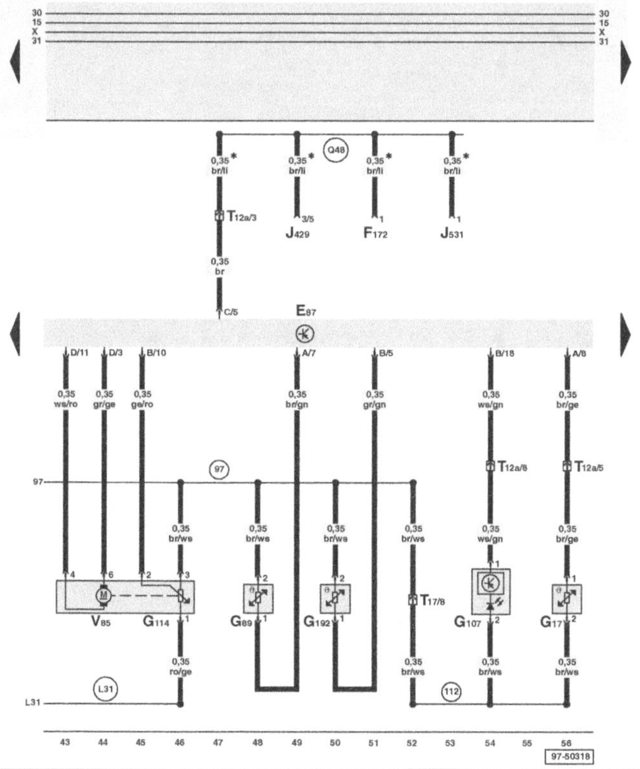

- E87 - A/C Control Head
- F172 - Convertible Top Front Lock Switch
- G17 - Outside Air Temperature Sensor
- G89 - Fresh Air Intake Duct Temperature Sensor
- G107 - Sunlight Photo Sensor (3241)
- G114 - Footwell/Defrost Flap Motor Position Sensor
- G192 - Sensor for outlet temperature, center
- J429 - Control module for central locking
- J531 - Wind Deflector Control Module (Cabrio)
- T12a - Connector 12 pin, behind instrument panel, right
- T17 - Connector 17 pin, behind instrument panel
- V85 - Footwell/Defroster Flap Motor
- (97) - Ground connection -1-, in A/C wiring harness
- (112) - Ground connection -2-, in A/C wiring harness
- (L31) - Wire connection (5 Volts), in A/C wiring harness
- (Q48) - Connection (open), in wiring harness interior
- * - only Roadster

ws = white
sw = black
ro = red
br = brown
gn = green
bl = blue
gr = grey
li = violet
ge = yellow
or = orange

© 2006 BentleyPublishers.com—All Rights Reserved

Automatic Climate Control (2004-2006)

Audi TT — Wiring Diagram — No. 12/7

A/C Control Head, Control Module for Fresh Air Blower, Fresh Air Blower

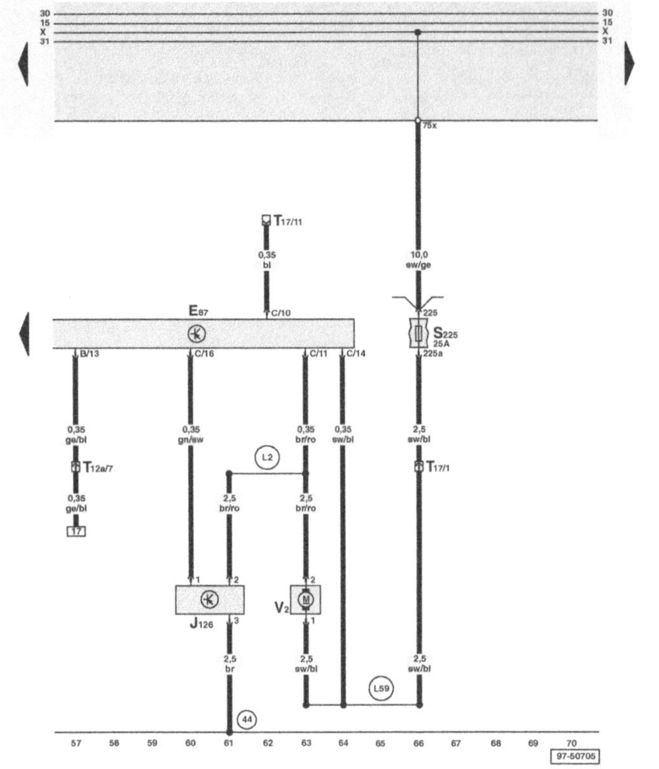

- E87 - A/C Control Head
- J126 - Control module for fresh air blower
- S225 - Fuse 25 in fuse holder
- T12a - Connector 12 pin, behind instrument panel, right
- T17 - Connector 17 pin, behind instrument panel
- V2 - Fresh Air Blower
- (44) - Ground connection, on left A-pillar, lower part
- (L2) - Wire connection, in A/C wiring harness
- (L59) - Plus connection (30), in wiring harness heater

ws = white
sw = black
ro = red
br = brown
gn = green
bl = blue
gr = grey
li = violet
ge = yellow
or = orange

USA.5703.02.21 — Edition 04/03

Audi TT — Wiring Diagram — No. 12/8

Coolant FC (Fan Control) Control Module, Fuses

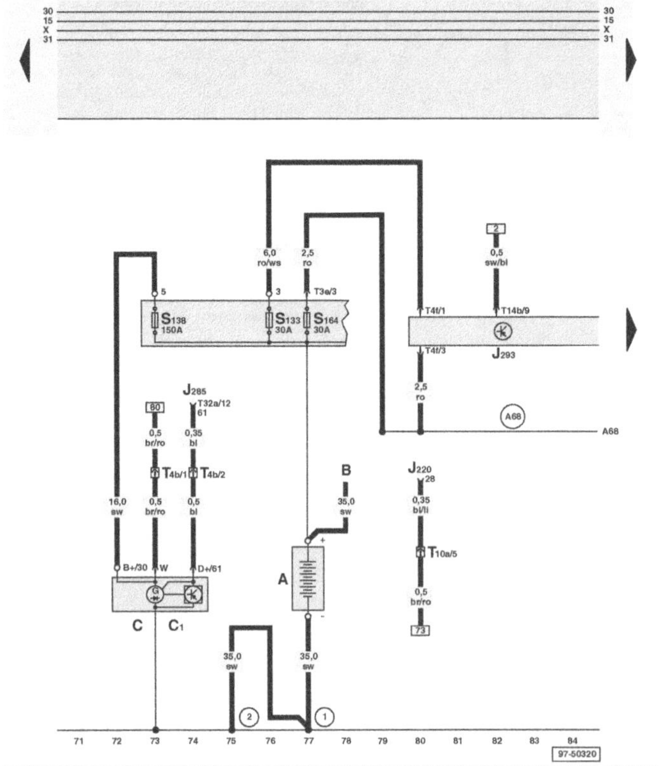

- A - Battery
- B - Starter
- C - Generator (GEN)
- C1 - Voltage Regulator (VR)
- J220 - Motronic Engine Control Module (ECM)
- J285 - Control Module with indicator unit in instrument panel insert
- J293 - Coolant FC (Fan Control) Control Module, in engine compartment, left front
- S133 - Safety fuse 3
- S138 - Safety fuse 5
- S164 - Fuse -3- in fuse bracket/battery
- T3e - Connector 3 pin, black, on Main Fuse Box/Battery
- T4b - Connector 4 pin, in engine compartment, left
- T4f - Connector 4 pin, connector B on J293
- T10a - Connector 10 pin, white, in E-box plenum chamber
- T14b - Connector 14 pin, connector A on J293
- T32a - Connector 32 pin, green, on instrument cluster
- (1) - Ground strap, battery to body
- (2) - Ground strap, transmission to body
- (A68) - Connector (C15, A/C), in instrument panel wiring harness

ws = white
sw = black
ro = red
br = brown
gn = green
bl = blue
gr = grey
li = violet
ge = yellow
or = orange

USA.5703.02.21 — Edition 04/03

© 2006 BentleyPublishers.com—All Rights Reserved

Audi TT — Wiring Diagram — No. 12/9

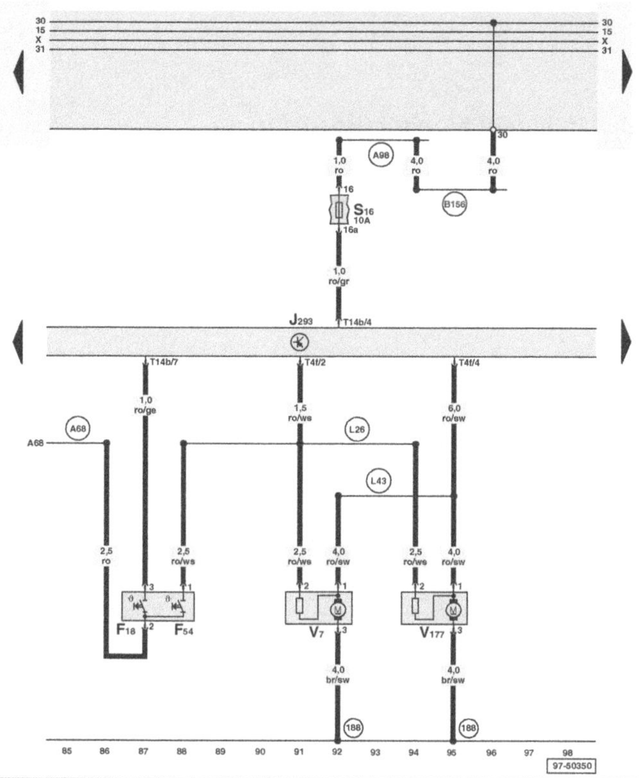

Coolant FC (Fan Control) Control Module, Coolant Fan Control Thermal Switch, Coolant Fan

- F18 - Coolant Fan Control (FC) Thermal Switch
- F54 - Coolant Fan Control (FC) Thermal Switch
- J293 - Coolant FC (Fan Control) Control Module, in engine compartment, left front
- S16 - Fuse
- T4f - Connector 4 pin, connector B on J293
- T14b - Connector 14 pin, connector A on J293
- V7 - Coolant Fan
- V177 - Coolant fan -2-

- (188) - Ground connection -3-, in A/C wiring harness
- (A68) - Connector (C15, A/C), in instrument panel wiring harness
- (A98) - Plus connector -4- (30), in instrument panel wiring harness
- (B156) - Plus connector (30a), in wiring harness interior
- (L26) - Wire connection (coolant fan), in A/C-climatronic/engine compartment wiring harness
- (L43) - Wire connection (coolant fan), in A/C system-coolant fan wiring harness

ws = white
sw = black
ro = red
br = brown
gn = green
bl = blue
gr = grey
li = violet
ge = yellow
or = orange

USA.5703.02.21 — Edition 04/03

Audi TT — Wiring Diagram — No. 12/10

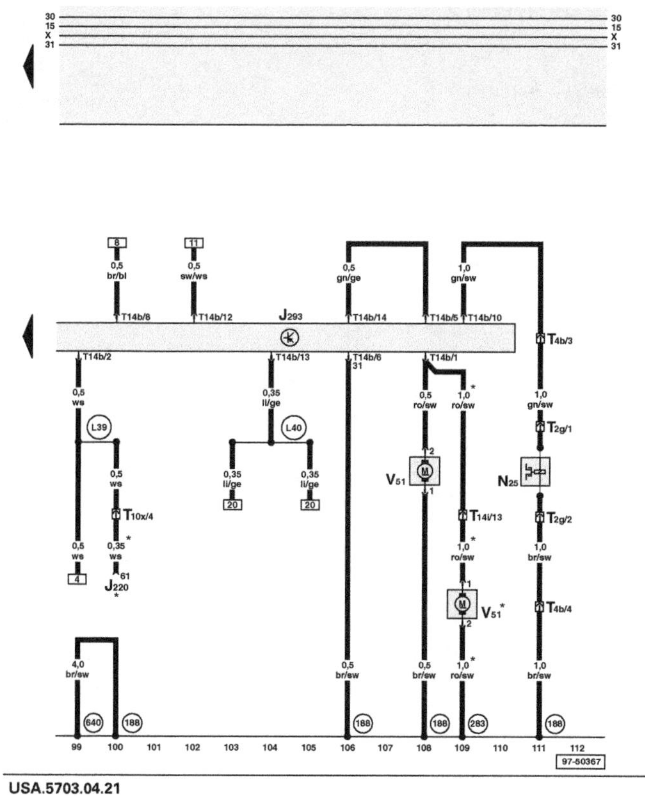

Coolant FC (Fan Control) Control Module, A/C Clutch, After-Run Coolant Pump

- J220 - Motronic Engine Control Module (ECM)
- J293 - Coolant FC (Fan Control) Control Module, in engine compartment, left front
- N25 - A/C Clutch
- T2g - Connector 2 pin, near compressor
- T4b - Connector 4 pin, in engine compartment, left
- T10x - Connector 10 pin, orange, in E-box plenum chamber
- T14b - Connector 14 pin, connector A on J293
- T14i - Connector 14 pin, in engine compartment, left
- V51 - After-Run Coolant Pump

- (188) - Ground connection -3-, in A/C wiring harness
- (283) - Ground connection 2, in wiring harness engine pre-wiring
- (640) - Ground Connection (in engine compartment, left)
- (L39) - Wire connection -1- (A/C Pressure Switch), in Climatronic wiring harness
- (L40) - Wire connection -2- (A/C Pressure Switch), in Climatronic wiring harness

* - only 3.2L engine, code BHE, BPF

ws = white
sw = black
ro = red
br = brown
gn = green
bl = blue
gr = grey
li = violet
ge = yellow
or = orange

USA.5703.04.21 — Edition 06/04

© 2006 BentleyPublishers.com—All Rights Reserved

Audi TT **Wiring Diagram** **No. 13/1**

Central Locking System / Anti Theft System / Radar Interior Monitor (Roadster)

2004 m.y.

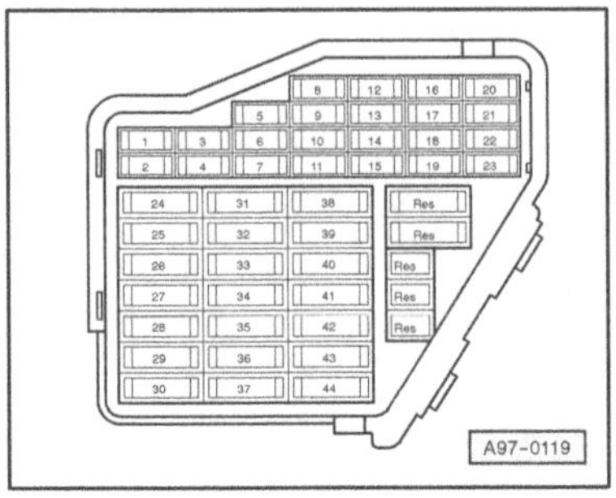

Fuse Panel

Fuse Colors:

30 A - Green
25 A - White
20 A - Yellow
15 A - Blue
10 A - Red
7,5 A - Brown
5 A - Beige
1 A - Black

Starting with fuse position 23, fuses in the fuse holder are identified with 223 in the wiring diagram.

Audi TT **Wiring Diagram** **No. 13/2**

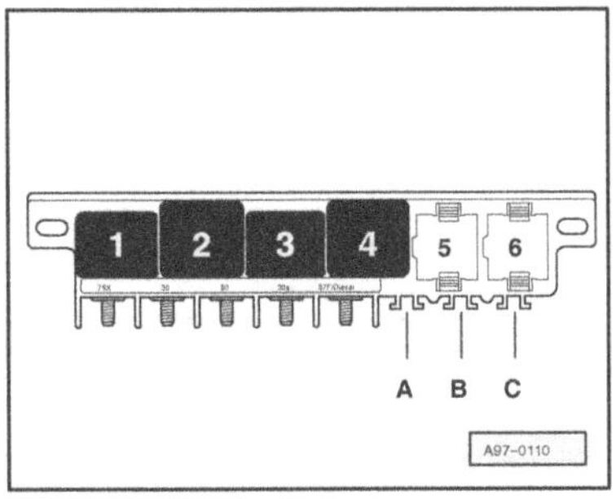

Micro Central Electric Panel

Fuse Location:

A - Convertible Top Fuse, S67

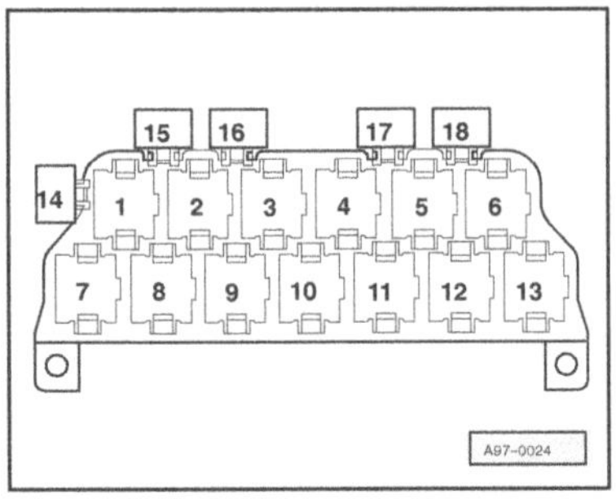

13 - Fold Relay Panel

Relay Location:

5 - Starting Interlock Relay, J207
10 - Rear Window Defogger Relay, J9

© 2006 BentleyPublishers.com—All Rights Reserved

Audi TT

Wiring Diagram **No. 13/3**

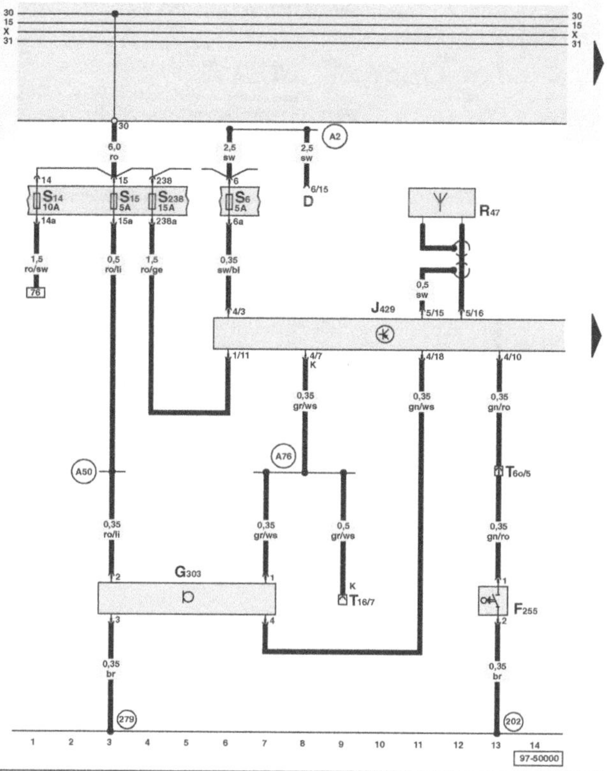

Control Module for Central Locking, Fuses, Radar Interior Monitoring Control Module for Anti-theft Warning

- D - Ignition/Starter Switch
- F255 - Hardtop detector switch
- G303 - Radar Interior Monitoring Control Module 1
- J429 - Control module for central locking
- R47 - Window antenna for central locking and anti-theft warning system
- S6 - Fuse
- S14 - Fuse
- S15 - Fuse
- S238 - Fuse 38 in fuse holder
- T6o - Connector 6 pin, Hardtop right
- T16 - Connector 16 pin, black, Data Link Connector (DLC), below storage compartment, driver's side

- (202) - Ground connection, in convertible top wiring harness
- (279) - Ground connector -5-, in wiring harness interior
- (A2) - Plus connection (15), in instrument panel wiring harness
- (A50) - Plus connection (30as), in instrument panel wiring harness
- (A76) - Connector (K-diagnosis wire), in instrument panel wiring harness

ws = white
sw = black
ro = red
br = brown
gn = green
bl = blue
gr = grey
li = violet
ge = yellow
or = orange

USA.5703.02.21 Edition 04/03

Audi TT

Wiring Diagram **No. 13/4**

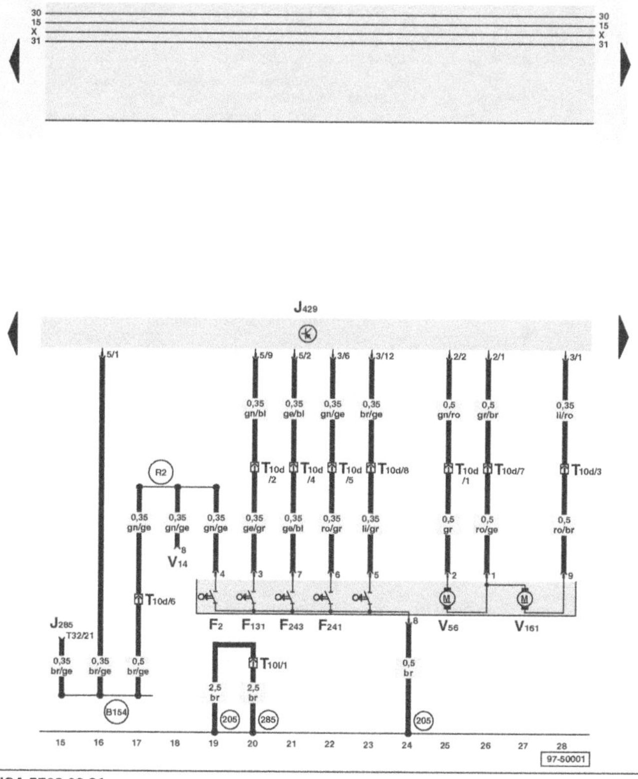

Control Module for Central Locking, Lock Unit for Central Locking Driver's Side

- F2 - Door Contact Switch - Driver's side
- F131 - Left Front Central Locking System Activator/Switch
- F241 - Contact switch for lock cylinder, driver side
- F243 - Adjusting element for central locking (safe), driver door
- J285 - Control Module with indicator unit in instrument panel insert
- J429 - Control module for central locking
- T10d - Connector 10 pin, brown, connector station A pillar, left
- T10l - Connector 10 pin, red, connector station A pillar, left
- T32 - Connector 32 pin, blue, on instrument cluster
- V14 - Left Window Motor
- V56 - Driver's Door Central Locking System Motor
- V161 - Motor for central locking (safe), driver door

- (205) - Ground connection, in driver's door wiring harness
- (285) - Ground connection -7- in wiring harness interior
- (B154) - Connector -1- (TK), in wiring harness interior
- (R2) - Wire connection (door contact switch, left), in power window/central locking system and door contact switch wiring harness

ws = white
sw = black
ro = red
br = brown
gn = green
bl = blue
gr = grey
li = violet
ge = yellow
or = orange

USA.5703.02.21 Edition 04/03

© 2006 BentleyPublishers.com—All Rights Reserved

Audi TT

Wiring Diagram

No. 13/5

Control Module for Central Locking, Lock Unit for Central Locking Passenger's Side

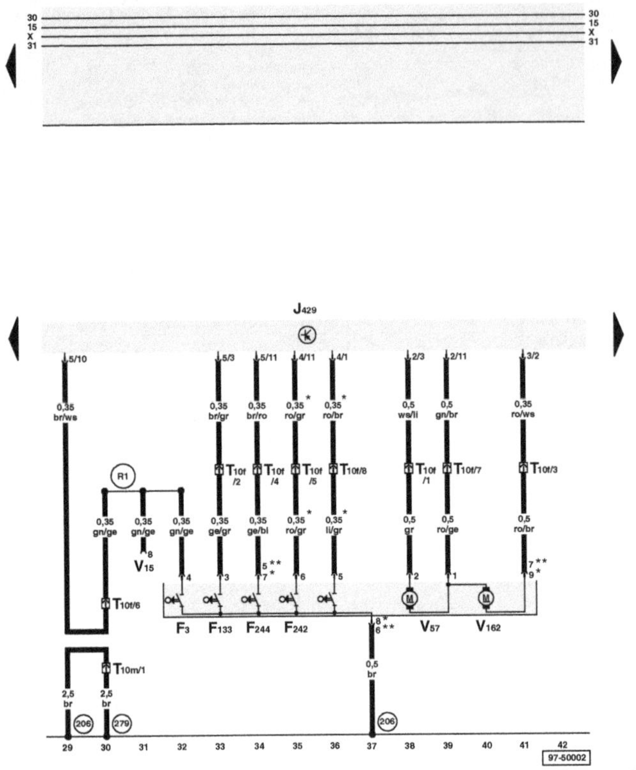

- F3 - Door Contact Switch - Passenger's side
- F133 - Right Front Central Locking System Activator/Switch
- F242 - Contact switch for lock cylinder, passenger side
- F244 - Adjusting element for central locking (safe), passenger door
- J429 - Control module for central locking
- T10f - Connector 10 pin, brown, connector station A pillar, right
- T10m - Connector 10 pin, red, connector station A pillar, right
- V15 - Right Window Motor
- V57 - Passenger's Door Central Locking System Motor
- V162 - Motor for central locking (safe), passenger door
- (206) - Ground connection, in passenger's door wiring harness
- (279) - Ground connector -5-, in wiring harness interior
- (R1) - Wire connection (door contact switch, right), in power window/central locking system and door contact switch wiring harness
- * - to 2004 m.y.
- ** - from 2005 m.y.

ws = white
sw = black
ro = red
br = brown
gn = green
bl = blue
gr = grey
li = violet
ge = yellow
or = orange

USA.5703.04.21

Edition 06/04

Audi TT

Wiring Diagram

No. 13/6

Control Module for Central Locking, Switch for Remote Unlock rear lid, Switch for Remote/Fuel Tank Door

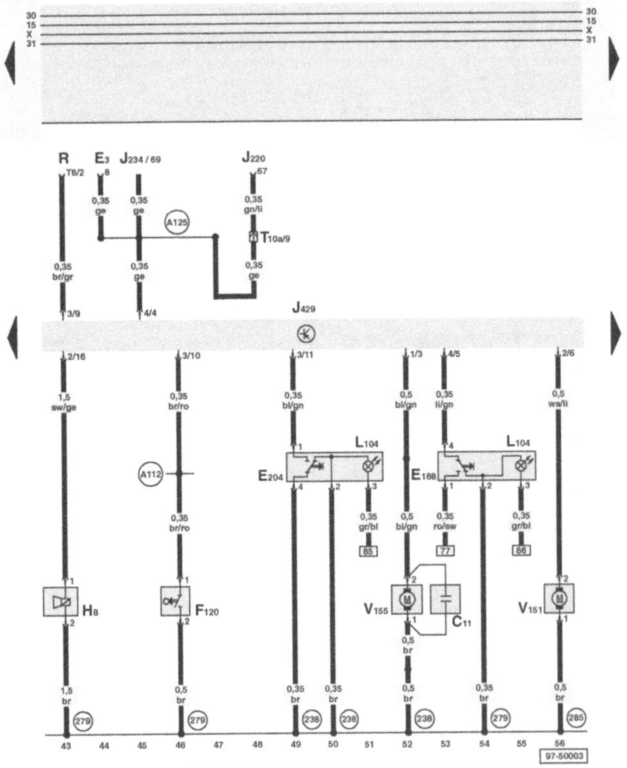

- C11 - Capacitor
- E3 - Emergency Flasher Switch
- E188 - Switch for remote unlock, rear lid
- E204 - Switch for remote/fuel tank door
- F120 - Hood Alarm Switch
- H8 - Alarm Horn
- J220 - Motronic Engine Control Module (ECM)
- J234 - Airbag Control Module
- J429 - Control module for central locking
- L104 - Illumination for switch remote unlock
- R - Radio
- T8 - Connector 8 pin, black, connector radio III
- T10a - Connector 10 pin, white, in E-box plenum chamber
- V151 - Motor for remote unlock rear lid
- V155 - Motor for fuel tank lid unlock
- (238) - Ground connection -1-, in wiring harness interior
- (279) - Ground connector -5-, in wiring harness interior
- (285) - Ground connection -7- in wiring harness interior
- (A112) - Connection (driver's cab lock), in instrument panel wiring harness
- (A125) - Connection (crash signal) in instrument panel wiring harness

ws = white
sw = black
ro = red
br = brown
gn = green
bl = blue
gr = grey
li = violet
ge = yellow
or = orange

USA.5703.02.21

Edition 04/03

© 2006 BentleyPublishers.com—All Rights Reserved

Audi TT

Wiring Diagram

No. 13/7

Control Module for Central Locking, Switch for Interior Lock Driver Side, Switch for Passenger Compartment Monitoring, Luggage Compartment Light

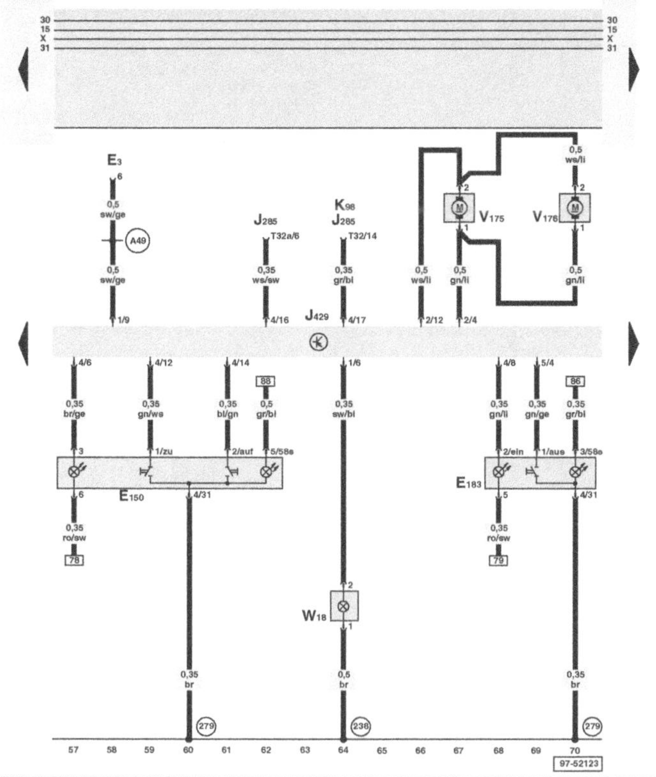

- E3 - Emergency Flasher Switch
- E150 - Switch for interior lock, driver side
- E183 - Switch for passenger compartment monitoring
- J285 - Control Module with indicator unit in instrument panel insert
- J429 - Control module for central locking
- K98 - Convertible Top Locked Indicator Light
- T32 - Connector 32 pin, blue, on instrument cluster
- T32a - Connector 32 pin, green, on instrument cluster
- V175 - Storage compartment central locking motor
- V176 - CD changer compartment central locking motor
- W18 - Luggage Compartment light, left
- (238) - Ground connection -1-, in wiring harness interior
- (279) - Ground connector -5-, in wiring harness interior
- (A49) - Wire connection -1-, in instrument panel wiring harness

ws = white
sw = black
ro = red
br = brown
gn = green
bl = blue
gr = grey
li = violet
ge = yellow
or = orange

USA.5703.02.21 **Edition 04/03**

Audi TT

Wiring Diagram

No. 13/8

Control Module for Central Locking, Door Warning Lights

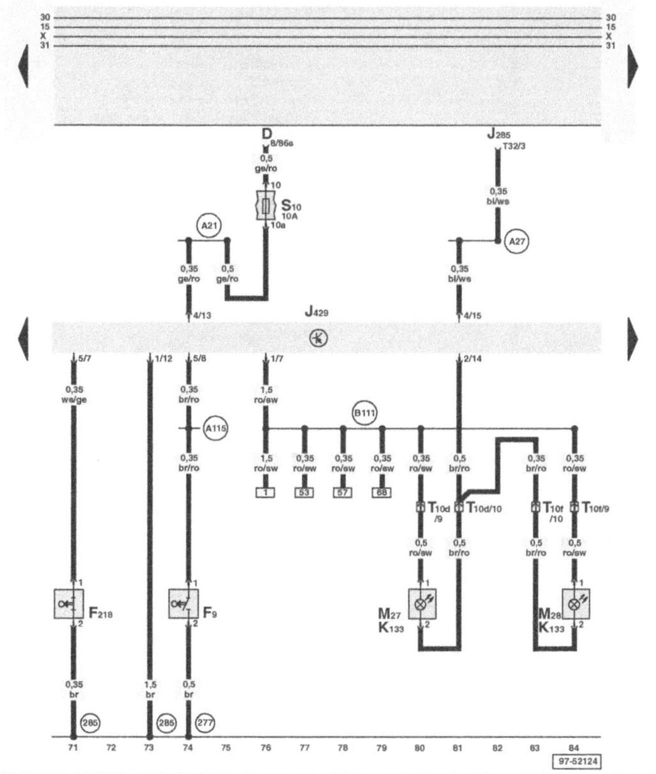

- D - Ignition/Starter Switch
- F9 - Parking Brake Warning Light Switch
- F218 - Switch for central locking, trunk lid
- J285 - Control Module with indicator unit in instrument panel insert
- J429 - Control module for central locking
- K133 - Warning light for central locking -SAFE-
- M27 - Left Door Warning Light
- M28 - Right Door Warning Light
- S10 - Fuse
- T10d - Connector 10 pin, brown, connector station A pillar, left
- T10f - Connector 10 pin, brown, connector station A pillar, right
- T32 - Connector 32 pin, blue, on instrument cluster
- (277) - Ground connector -3-, in wiring harness interior
- (285) - Ground connection -7- in wiring harness interior
- (A21) - Wire connection (86s), in instrument panel wiring harness
- (A27) - Wire connection (vehicle speed signal), in instrument panel wiring harness
- (A115) - Connection (parking brake control) in instrument panel wiring harness
- (B111) - Connector (30a), in wiring harness interior

ws = white
sw = black
ro = red
br = brown
gn = green
bl = blue
gr = grey
li = violet
ge = yellow
or = orange

USA.5703.02.21 **Edition 04/03**

© 2006 BentleyPublishers.com—All Rights Reserved

Central locking, Anti-theft, roadster (2004-2006)

Audi TT — Wiring Diagram — No. 13/9

Control Module for Central Locking, Front Interior Light, Reading Lights, Make-Up Mirror Lights

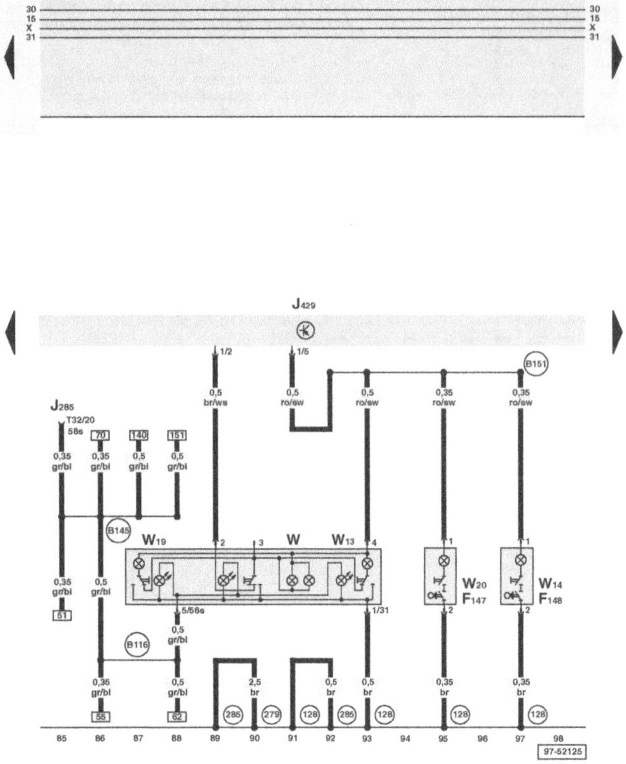

- F147 - Left Make-Up Mirror Light Switch
- F148 - Right Make-Up Mirror Light Switch
- J285 - Control Module with indicator unit in instrument panel insert
- J429 - Control module for central locking
- T32 - Connector 32 pin, blue, on instrument cluster
- W - Front Interior Light
- W13 - Passenger's Map/Reading Light
- W14 - Right Make-up Mirror Light
- W19 - Left Front Reading Light
- W20 - Left Make-up Mirror Light

- (128) - Ground connection -1-, in interior light wiring harness
- (279) - Ground connector -5-, in wiring harness interior
- (285) - Ground connection -7- in wiring harness interior
- (B116) - Connector (58s), in wiring harness interior
- (B145) - Plus connector (58b), in wiring harness interior
- (B151) - Plus connector -3- (30a), in wiring harness interior

ws = white
sw = black
ro = red
br = brown
gn = green
bl = blue
gr = grey
li = violet
ge = yellow
or = orange

97-52125

USA.5703.02.21 — Edition 04/03

Audi TT — Wiring Diagram — No. 13/10

Control Module for Central Locking, Starting Interlock Relay

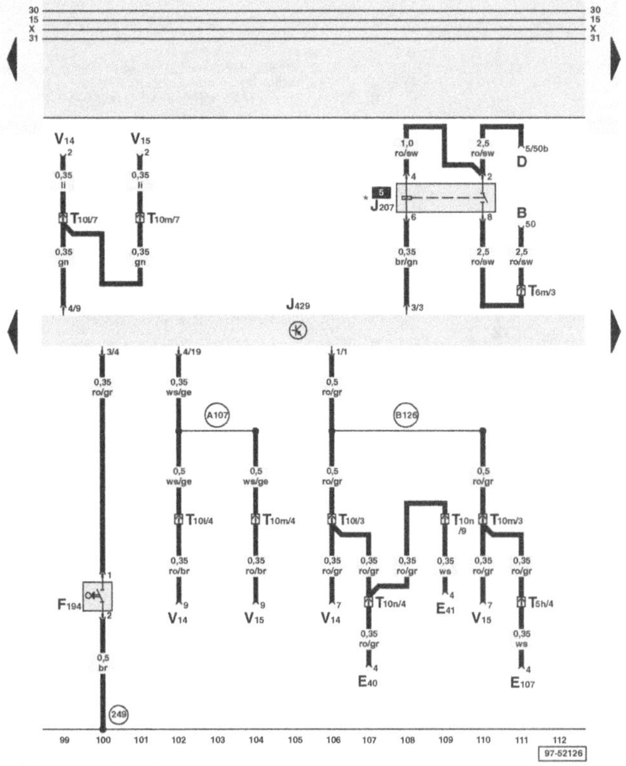

- B - Starter
- D - Ignition/Starter Switch
- E40 - Left Front Window Switch
- E41 - Right Front Window Switch
- E107 - Switch for window regulator, in passenger door
- F194 - Clutch Pedal Position (CPP) Switch
- J207 - Starting Interlock Relay
- J429 - Control module for central locking
- T5h - Connector 5 pin, in right door
- T6m - Connector 6 pin, brown, in E-box plenum chamber
- T10l - Connector 10 pin, red, connector station A pillar, left
- T10m - Connector 10 pin, red, connector station A pillar, right
- T10n - Connector 10 pin, in left door
- V14 - Left Window Motor
- V15 - Right Window Motor

- (249) - Ground connection -2-, in wiring harness interior
- (A107) - Connector (comfort close), in instrument panel wiring harness
- (B126) - Connector (87, driver's door), in wiring harness interior

* - Vehicles with Manual Transmission

ws = white
sw = black
ro = red
br = brown
gn = green
bl = blue
gr = grey
li = violet
ge = yellow
or = orange

97-52126

USA.5703.02.21 — Edition 04/03

© 2006 BentleyPublishers.com—All Rights Reserved

Audi TT

Wiring Diagram

No. 13/11

Top, Heated Rear Window

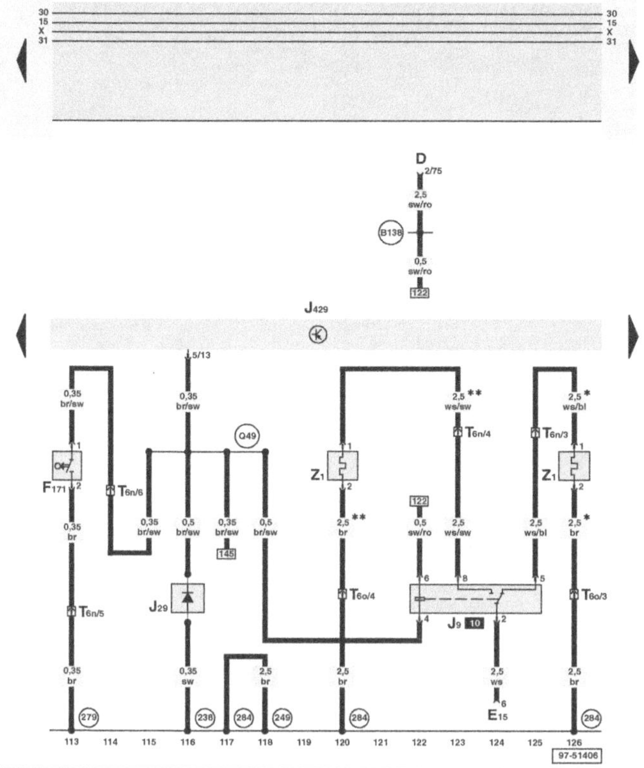

ws = white
sw = black
ro = red
br = brown
gn = green
bl = blue
gr = grey
li = violet
ge = yellow
or = orange

D - Ignition/Starter Switch
E15 - Rear Window Defogger Switch
F171 - Convertible Top Open Position Switch
J9 - Rear Window Defogger Relay
J29 - Blocking Diode
J429 - Control module for central locking
T6n - Connector 6 pin, Hardtop left
T6o - Connector 6 pin, Hardtop right
Z1 - Heated rear window

(238) - Ground connection -1-, in wiring harness interior
(249) - Ground connection -2-, in wiring harness interior
(279) - Ground connector -5-, in wiring harness interior
(284) - Ground connection -6- in wiring harness interior
(B138) - Plus connector (X), in wiring harness, interior
(Q49) - Connection (closed), in wiring harness interior

* - Softtop
** - Hardtop (where applicable)

Audi TT

Wiring Diagram

No. 13/12

Top

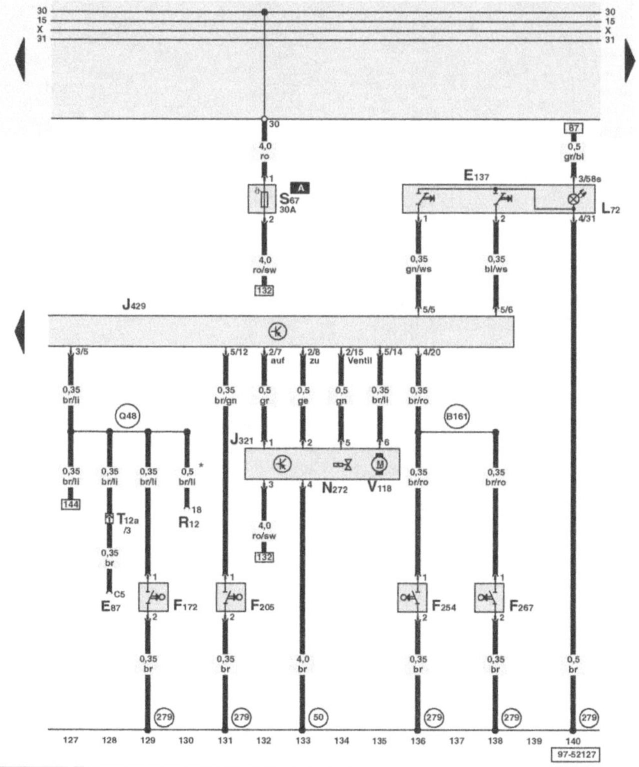

ws = white
sw = black
ro = red
br = brown
gn = green
bl = blue
gr = grey
li = violet
ge = yellow
or = orange

E87 - A/C Control Head
E137 - Convertible Top Operation Switch
F172 - Convertible Top Front Lock Switch
F205 - Convertible Top Latch Open Position Switch
F254 - Convertible top cover detector switch
F267 - Convertible Top Cover Detector Switch 2
J321 - Convertible Top Hydraulic Pump Relay
J429 - Control module for central locking
L72 - Convertible Top Switch Light
N272 - Power convertible top valve -1-
R12 - Amplifier
S67 - Convertible Top Fuse
T12a - Connector 12 pin, behind instrument panel, right
V118 - Convertible Top Hydraulic Pump

(50) - Ground connection, in luggage compartment, left
(279) - Ground connector -5-, in wiring harness interior
(B161) - Connector (75a), in wiring harness interior
(Q48) - Connection (open), in wiring harness interior

* - Vehicles with Radio Bose

© 2006 BentleyPublishers.com—All Rights Reserved

Central locking, Anti-theft, roadster (2004-2006)

Audi TT — Wiring Diagram — **No. 13/13**

Wind Deflector

E278 - Wind Deflector Switch (Cabrio)
J531 - Wind Deflector Control Module (Cabrio)
S227 - Fuse 27 in fuse holder
V186 - Wind Deflector Motor (Cabrio)

(279) - Ground connector -5-, in wiring harness interior

(A80) - Connector -1- (X), in instrument panel wiring harness

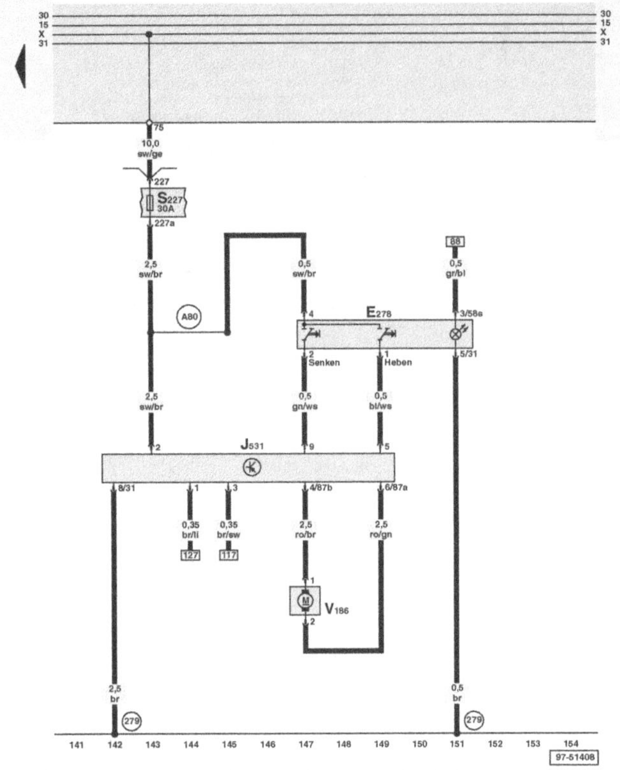

ws = white
sw = black
ro = red
br = brown
gn = green
bl = blue
gr = grey
li = violet
ge = yellow
or = orange

USA.5703.02.21 — Edition 04/03

THIS SPACE INTENTIONALLY

LEFT BLANK

© 2006 BentleyPublishers.com—All Rights Reserved

Audi TT Wiring Diagram **No. 14/1**

Central Locking System / Anti Theft System / Ultra-Sound Interior Monitor (Coupé)

2004 m.y.

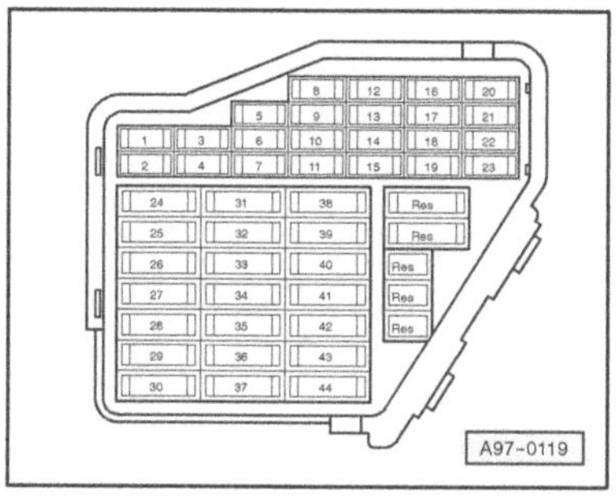

Fuse Panel

Fuse Colors:

30 A - Green
25 A - White
20 A - Yellow
15 A - Blue
10 A - Red
7,5 A - Brown
5 A - Beige
1 A - Black

Starting with fuse position 23, fuses in the fuse holder are identified with 223 in the wiring diagram.

Audi TT Wiring Diagram **No. 14/2**

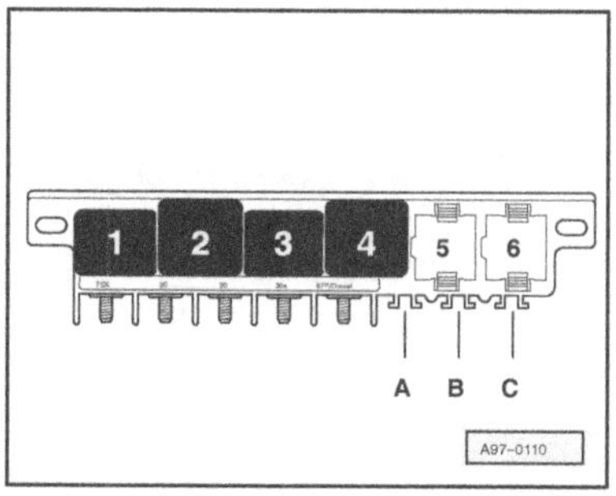

Micro Central Electric Panel

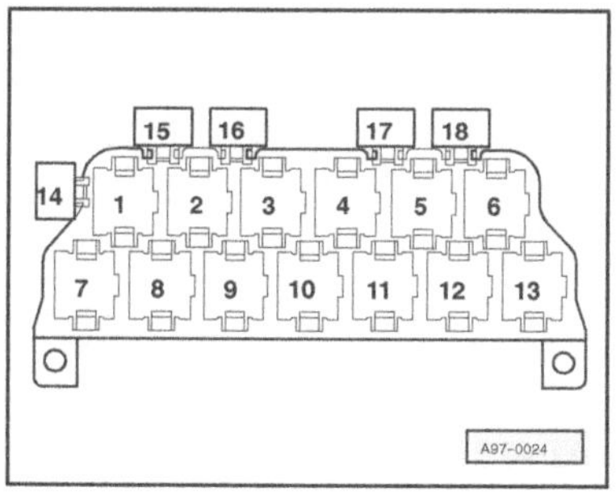

13 - Fold Relay Panel

Relay Location:

5 - Starting Interlock Relay, J207

©2006 BentleyPublishers.com—All Rights Reserved

Central locking, Anti-theft, coupe (2004-2006)

Audi TT — Wiring Diagram — No. 14/3

Control Module for Central Locking, Fuses, Ultra-Sound Sensor for Anti-theft Warning

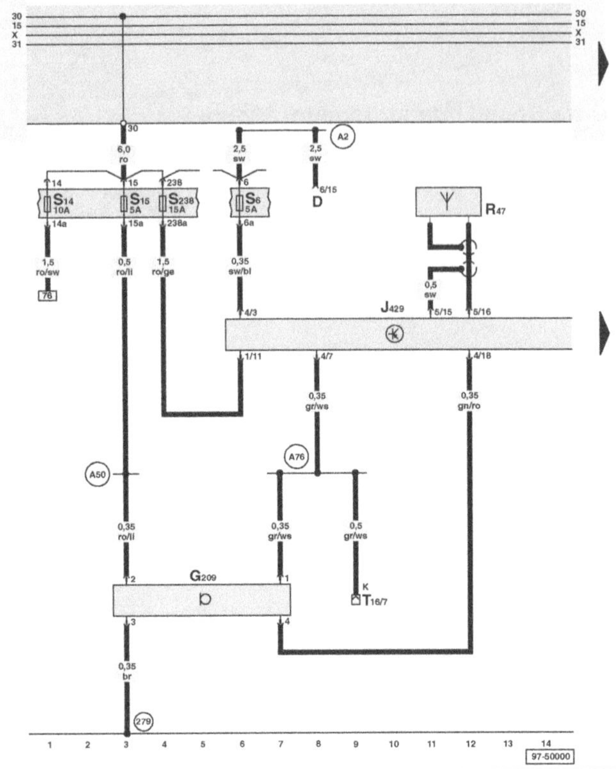

- D - Ignition/Starter Switch
- G209 - Ultra-sound sensor for anti-theft warning
- J429 - Control module for central locking
- R47 - Window antenna for central locking and anti-theft warning system
- S6 - Fuse
- S14 - Fuse
- S15 - Fuse
- S238 - Fuse 38 in fuse holder
- T16 - Connector 16 pin, black, Data Link Connector (DLC), below storage compartment, driver's side

- (279) - Ground connector -5-, in wiring harness interior
- (A2) - Plus connection (15), in instrument panel wiring harness
- (A50) - Plus connection (30as), in instrument panel wiring harness
- (A76) - Connector (K-diagnosis wire), in instrument panel wiring harness

ws = white
sw = black
ro = red
br = brown
gn = green
bl = blue
gr = grey
li = violet
ge = yellow
or = orange

USA.5703.02.21 — Edition 04/03

Audi TT — Wiring Diagram — No. 14/4

Control Module for Central Locking, Lock Unit for Central Locking Driver's Side

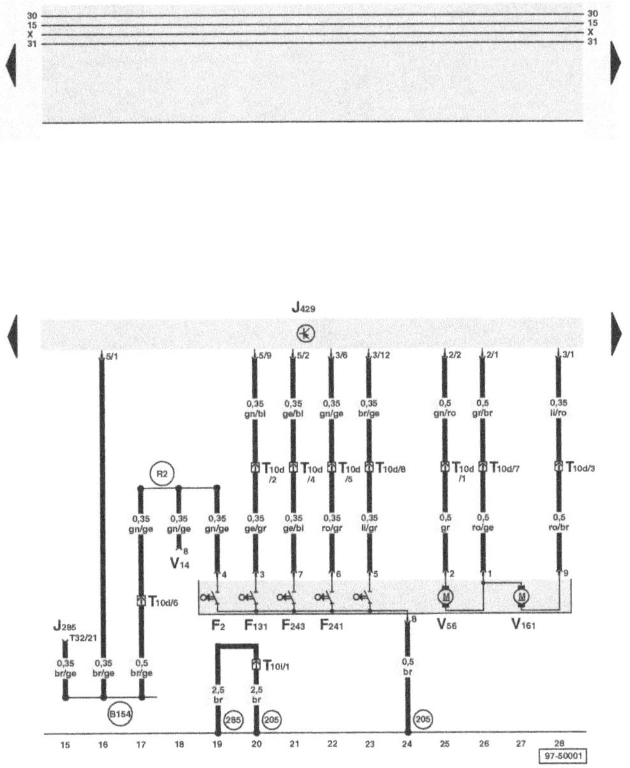

- F2 - Door Contact Switch - Driver's side
- F131 - Left Front Central Locking System Activator/Switch
- F241 - Contact switch for lock cylinder, driver side
- F243 - Adjusting element for central locking (safe), driver door
- J285 - Control Module with indicator unit in instrument panel insert
- J429 - Control module for central locking
- T10d - Connector 10 pin, brown, connector station A pillar, left
- T10l - Connector 10 pin, red, connector station A pillar, left
- T32 - Connector 32 pin, blue, on instrument cluster
- V14 - Left Window Motor
- V56 - Driver's Door Central Locking System Motor
- V161 - Motor for central locking (safe), driver door

- (205) - Ground connection, in driver's door wiring harness
- (285) - Ground connection -7- in wiring harness interior
- (B154) - Connector -1- (TK), in wiring harness interior
- (R2) - Wire connection (door contact switch, left), in power window/central locking system and door contact switch wiring harness

ws = white
sw = black
ro = red
br = brown
gn = green
bl = blue
gr = grey
li = violet
ge = yellow
or = orange

USA.5703.02.21 — Edition 04/03

© 2006 BentleyPublishers.com—All Rights Reserved

Audi TT — Wiring Diagram — No. 14/5

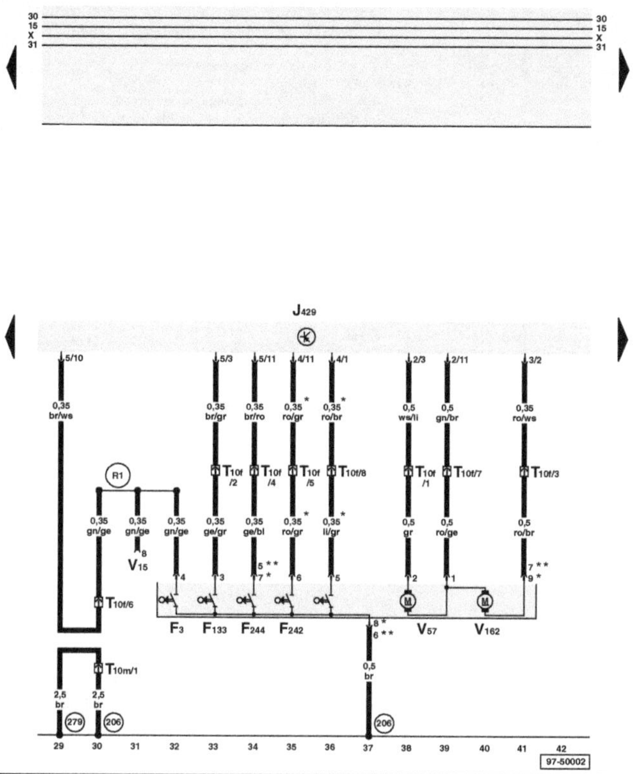

Control Module for Central Locking, Lock Unit for Central Locking Passenger's Side

- F3 - Door Contact Switch - Passenger's side
- F133 - Right Front Central Locking System Activator/Switch
- F242 - Contact switch for lock cylinder, passenger side
- F244 - Adjusting element for central locking (safe), passenger door
- J429 - Control module for central locking
- T10f - Connector 10 pin, brown, connector station A pillar, right
- T10m - Connector 10 pin, red, connector station A pillar, right
- V15 - Right Window Motor
- V57 - Passenger's Door Central Locking System Motor
- V162 - Motor for central locking (safe), passenger door
- (206) - Ground connection, in passenger's door wiring harness
- (279) - Ground connector -5-, in wiring harness interior
- (R1) - Wire connection (door contact switch, right), in power window/central locking system and door contact switch wiring harness
- * - to 2004 m.y.
- ** - from 2005 m.y.

ws = white
sw = black
ro = red
br = brown
gn = green
bl = blue
gr = grey
li = violet
ge = yellow
or = orange

USA.5703.04.21 — Edition 06/04

Audi TT — Wiring Diagram — No. 14/6

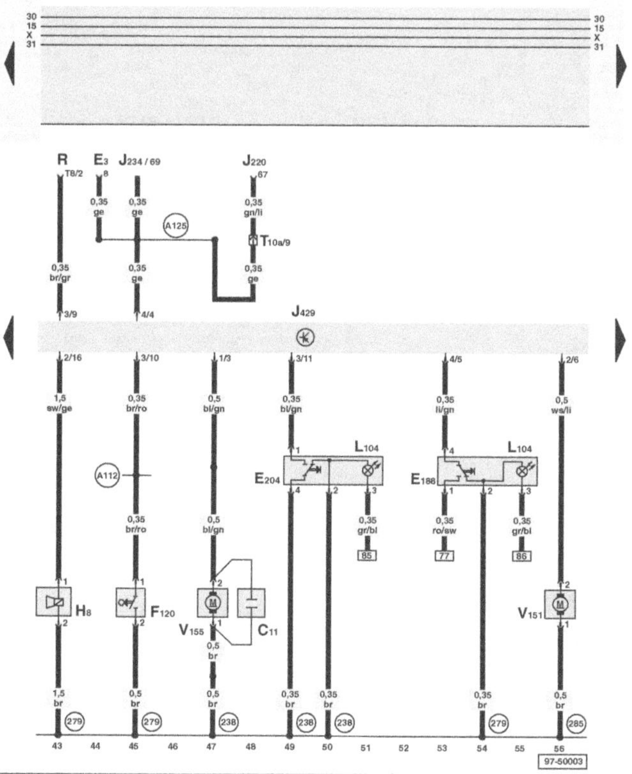

Control Module for Central Locking, Switch for Remote Unlock rear lid, Switch for Remote/Fuel Tank Door

- C11 - Capacitor
- E3 - Emergency Flasher Switch
- E188 - Switch for remote unlock, rear lid
- E204 - Switch for remote/fuel tank door
- F120 - Hood Alarm Switch
- H8 - Alarm Horn
- J220 - Motronic Engine Control Module (ECM)
- J234 - Airbag Control Module
- J429 - Control module for central locking
- L104 - Illumination for switch remote unlock
- R - Radio
- T8 - Connector 8 pin, black, connector radio III
- T10a - Connector 10 pin, white, in E-box plenum chamber
- V151 - Motor for remote unlock rear lid
- V155 - Motor for fuel tank lid unlock
- (238) - Ground connection -1-, in wiring harness interior
- (279) - Ground connector -5-, in wiring harness interior
- (285) - Ground connection -7- in wiring harness interior
- (A112) - Connection (driver's cab lock), in instrument panel wiring harness
- (A125) - Connection (crash signal) in instrument panel wiring harness

ws = white
sw = black
ro = red
br = brown
gn = green
bl = blue
gr = grey
li = violet
ge = yellow
or = orange

USA.5703.02.21 — Edition 04/03

© 2006 BentleyPublishers.com—All Rights Reserved

Central locking, Anti-theft, coupe (2004-2006)

Audi TT — Wiring Diagram — No. 14/7

Control Module for Central Locking, Switch for Interior Lock Driver Side, Switch for Passenger Compartment Monitoring, Luggage Compartment Light

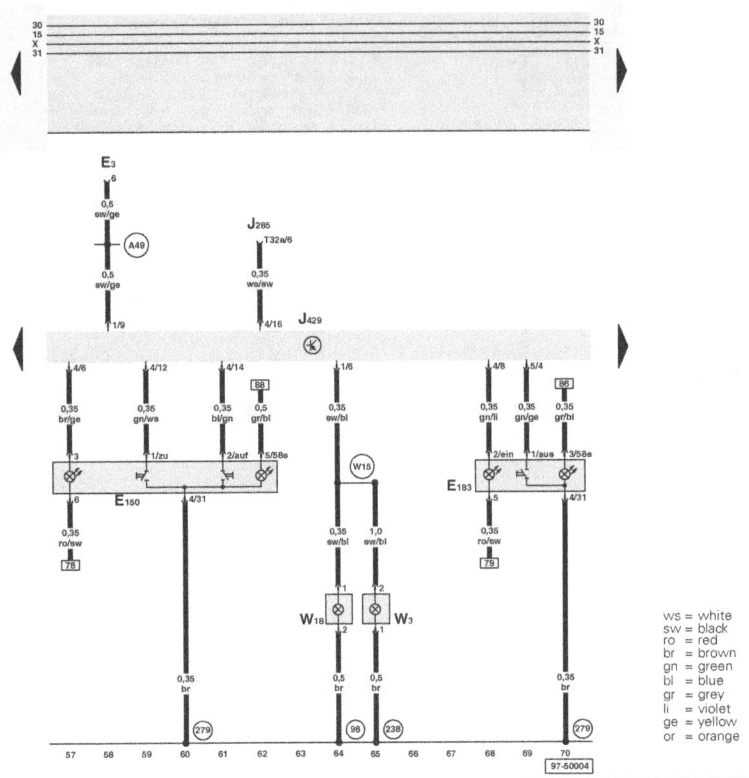

- E3 - Emergency Flasher Switch
- E150 - Switch for interior lock, driver side
- E183 - Switch for passenger compartment monitoring
- J285 - Control Module with indicator unit in instrument panel insert
- J429 - Control module for central locking
- T32a - Connector 32 pin, green, on instrument cluster
- W3 - Luggage Compartment Light
- W18 - Luggage Compartment light, left
- (98) - Ground connection, in rear lid wiring harness
- (238) - Ground connection -1-, in wiring harness interior
- (279) - Ground connector -5-, in wiring harness interior
- (A49) - Wire connection -1-, in instrument panel wiring harness
- (W15) - Connection -1- (luggage compartment light), in rear wiring harness

Audi TT — Wiring Diagram — No. 14/8

Control Module for Central Locking, Door Warning Lights

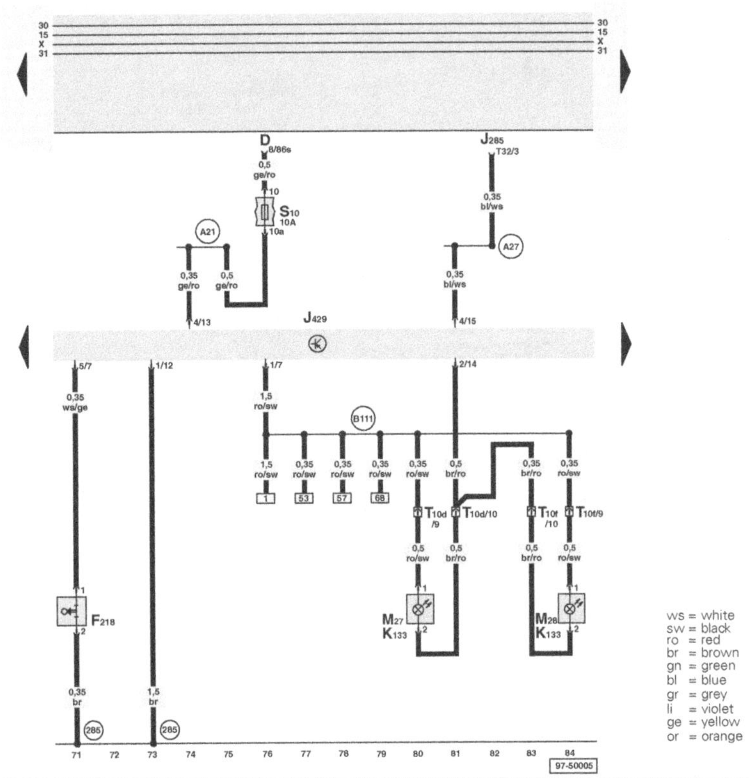

- D - Ignition/Starter Switch
- F218 - Switch for central locking, trunk lid
- J285 - Control Module with indicator unit in instrument panel insert
- J429 - Control module for central locking
- K133 - Warning light for central locking -SAFE-
- M27 - Left Door Warning Light
- M28 - Right Door Warning Light
- S10 - Fuse
- T10d - Connector 10 pin, brown, connector station A pillar, left
- T10f - Connector 10 pin, brown, connector station A pillar, right
- T32 - Connector 32 pin, blue, on instrument cluster
- (285) - Ground connection -7- in wiring harness interior
- (A21) - Wire connection (86s), in instrument panel wiring harness
- (A27) - Wire connection (vehicle speed signal), in instrument panel wiring harness
- (B111) - Connector (30a), in wiring harness interior

© 2006 BentleyPublishers.com—All Rights Reserved

Audi TT

Wiring Diagram

No. 14/9

Control Module for Central Locking, Front Interior Light, Reading Lights, Make-Up Mirror Lights

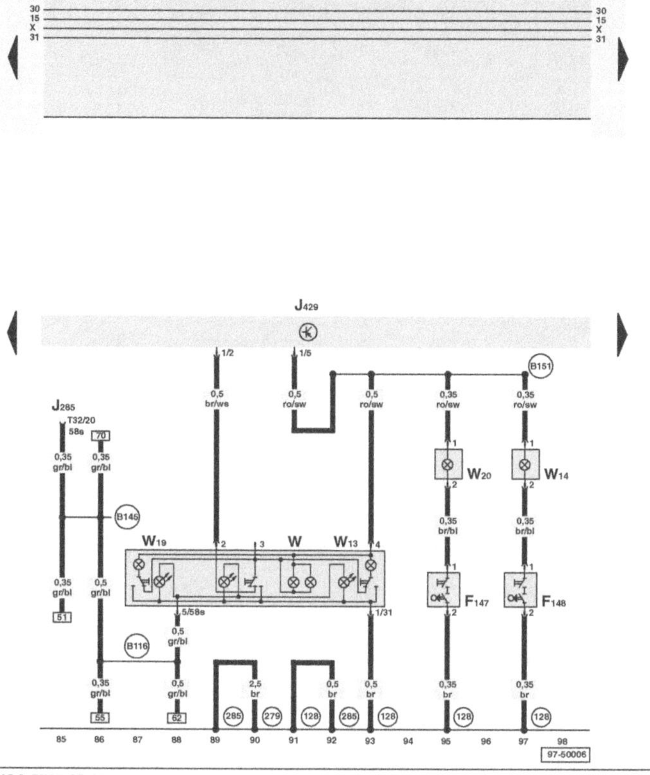

- F147 - Left Make-Up Mirror Light Switch
- F148 - Right Make-Up Mirror Light Switch
- J285 - Control Module with indicator unit in instrument panel insert
- J429 - Control module for central locking
- T32 - Connector 32 pin, blue, on instrument cluster
- W - Front Interior Light
- W13 - Passenger's Map/Reading Light
- W14 - Right Make-up Mirror Light
- W19 - Left Front Reading Light
- W20 - Left Make-up Mirror Light

- (128) - Ground connection -1-, in interior light wiring harness
- (279) - Ground connector -5-, in wiring harness interior
- (285) - Ground connection -7- in wiring harness interior
- (B116) - Connector (58s), in wiring harness interior
- (B145) - Plus connector (58b), in wiring harness interior
- (B151) - Plus connector -3- (30a), in wiring harness interior

ws = white
sw = black
ro = red
br = brown
gn = green
bl = blue
gr = grey
li = violet
ge = yellow
or = orange

USA.5703.02.21 Edition 04/03

Audi TT

Wiring Diagram

No. 14/10

Control Module for Central Locking, Starting Interlock Relay

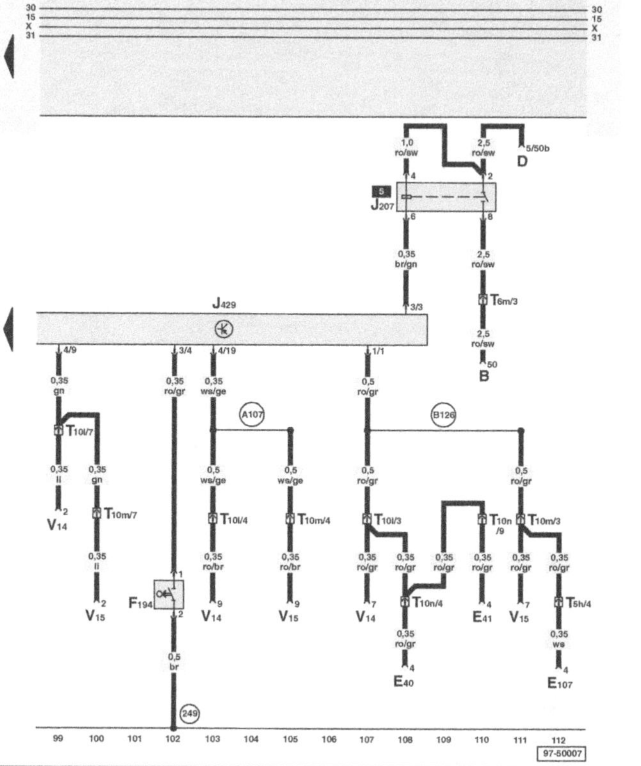

- B - Starter
- D - Ignition/Starter Switch
- E40 - Left Front Window Switch
- E41 - Right Front Window Switch
- E107 - Switch for window regulator, in passenger door
- F194 - Clutch Pedal Position (CPP) Switch
- J207 - Starting Interlock Relay
- J429 - Control module for central locking
- T5h - Connector 5 pin, in right door
- T6m - Connector 6 pin, brown, in E-box plenum chamber
- T10l - Connector 10 pin, red, connector station A pillar, left
- T10m - Connector 10 pin, red, connector station A pillar, right
- T10n - Connector 10 pin, in left door
- V14 - Left Window Motor
- V15 - Right Window Motor

- (249) - Ground connection -2-, in wiring harness interior
- (A107) - Connector (comfort close), in instrument panel wiring harness
- (B126) - Connector (87, driver's door), in wiring harness interior

ws = white
sw = black
ro = red
br = brown
gn = green
bl = blue
gr = grey
li = violet
ge = yellow
or = orange

USA.5703.02.21 Edition 04/03

© 2006 BentleyPublishers.com—All Rights Reserved

Audi TT Wiring Diagram **No. 15/1**

Haldex Coupling (02D) for All-Wheel Drive

2004 m.y.

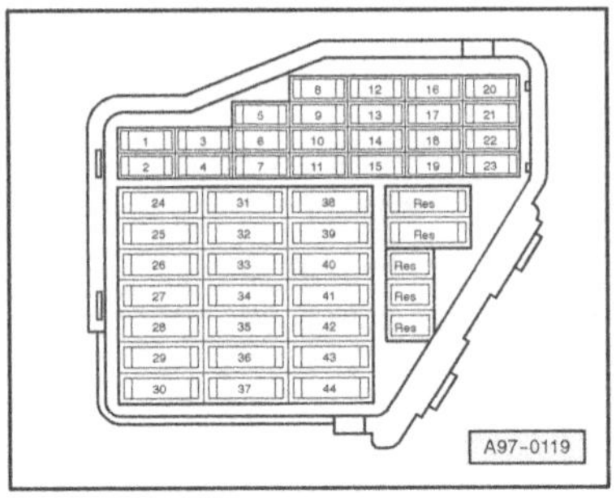

Fuse Panel

Fuse Colors:
- 30 A - Green
- 25 A - White
- 20 A - Yellow
- 15 A - Blue
- 10 A - Red
- 7,5 A - Brown
- 5 A - Beige
- 1 A - Black

Starting with fuse position 23, fuses in the fuse holder are identified with 223 in the wiring diagram.

USA.5703.02.21 Edition 04/03

Audi TT Wiring Diagram **No. 15/2**

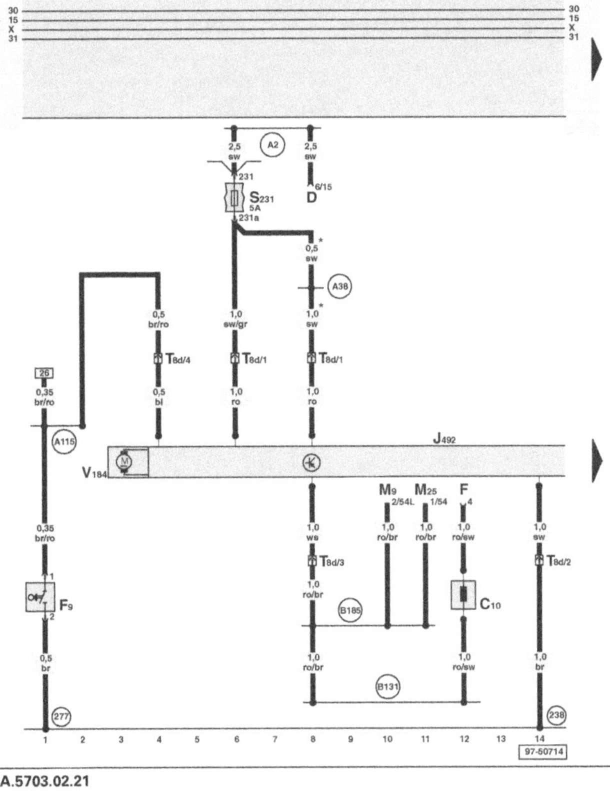

All-Wheel Drive Control Module, Oil Pressure Actuator

- C10 - Suppressor Filter
- D - Ignition/Starter Switch
- F - Brake Light Switch
- F9 - Parking Brake Warning Light Switch
- J492 - All-wheel drive control module
- M9 - Left Brake Light
- M25 - High-mount Brake Light
- S231 - Fuse 31 in fuse holder
- T8d - Connector 8 pin, on Control module for all wheel drive
- V184 - Oil pressure actuator

- (238) - Ground connection -1-, in wiring harness interior
- (277) - Ground connector -3-, in wiring harness interior
- (A2) - Plus connection (15), in instrument panel wiring harness
- (A38) - Plus connection -2- (15a), in instrument panel wiring harness
- (A115) - Connection (parking brake control) in instrument panel wiring harness
- (B131) - Connector (54), in wiring harness interior
- (B185) - Connection 2 (54), in passenger compartment wiring harness

* - only 3.2L engine, code BHE

ws = white
sw = black
ro = red
br = brown
gn = green
bl = blue
gr = grey
li = violet
ge = yellow
or = orange

USA.5703.02.21 Edition 04/03

© 2006 BentleyPublishers.com—All Rights Reserved

Audi TT **Wiring Diagram** **No. 15/3**

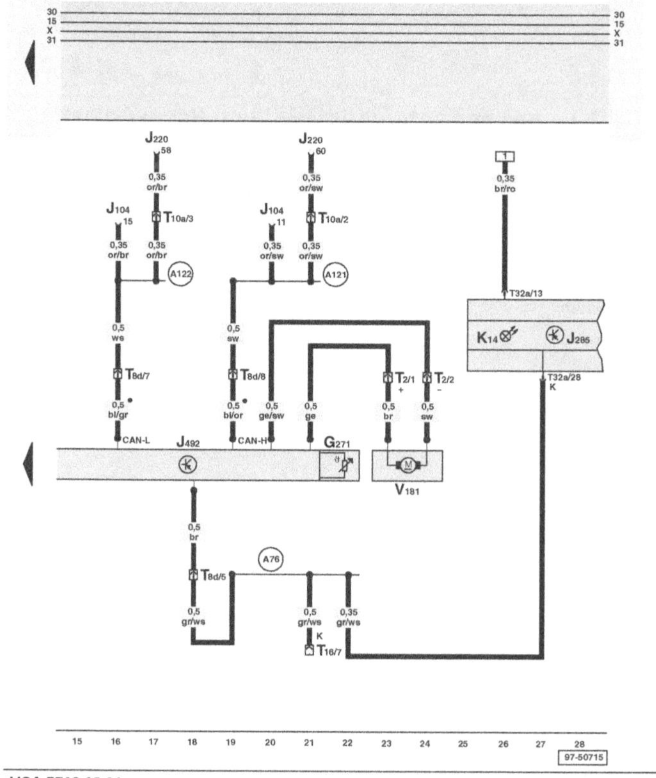

All-Wheel Drive Control Module, Haldex Clutch Pump

- G271 - Hydraulic temperature sensor
- J104 - ABS Control Module (w/EDL)
- J220 - Motronic Engine Control Module (ECM)
- J285 - Control Module with indicator unit in instrument panel insert
- J492 - All-wheel drive control module
- K14 - Parking Brake Indicator Light
- T2 - Connector 2 pin, on all-wheel drive
- T8d - Connector 8 pin, on Control module for all wheel drive
- T10a - Connector 10 pin, white, in E-box plenum chamber
- T16 - Connector 16 pin, black, Data Link Connector (DLC), below storage compartment, driver's side
- T32a - Connector 32 pin, green, on instrument cluster
- V181 - Haldex clutch pump
- (A76) - Connector (K-diagnosis wire), in instrument panel wiring harness
- (A121) - Connection (high bus) in instrument panel wiring harness
- (A122) - Connection (low bus) in instrument panel wiring harness
- • - CAN-Bus (Data-Bus)

ws = white
sw = black
ro = red
br = brown
gn = green
bl = blue
gr = grey
li = violet
ge = yellow
or = orange

USA.5703.02.21 Edition 04/03

THIS SPACE INTENTIONALLY LEFT BLANK

©2006 BentleyPublishers.com—All Rights Reserved

Audi TT Wiring Diagram **No. 16/1**

Headlight with High Intensity Gas Discharge (HID) Lamps and automatic Headlight Beam Adjusting

2004 m.y.

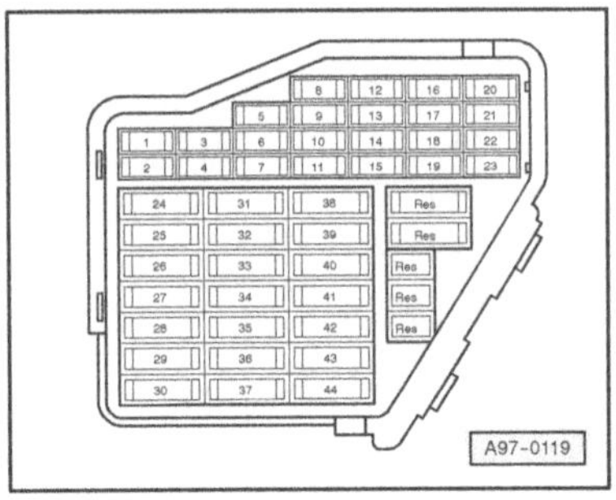

Fuse Panel

Fuse Colors:
- 30 A - Green
- 25 A - White
- 20 A - Yellow
- 15 A - Blue
- 10 A - Red
- 7,5 A - Brown
- 5 A - Beige
- 1 A - Black

Starting with fuse position 23, fuses in the fuse holder are identified with 223 in the wiring diagram.

Audi TT Wiring Diagram **No. 16/2**

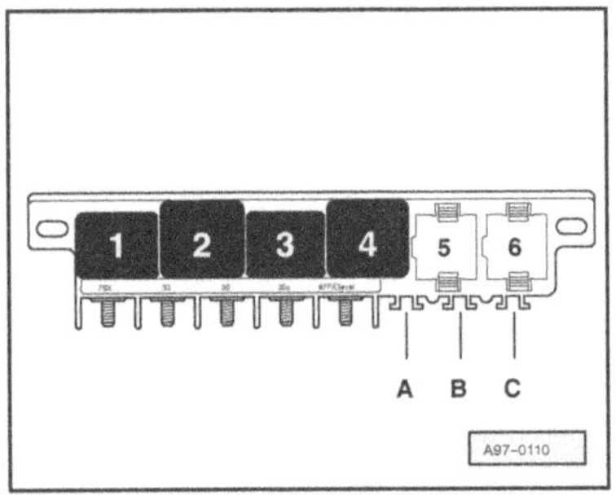

Micro Central Electric Panel

Relay Location:
- 2 - Load Reduction Relay, J59
- 4 - Fuel Pump (FP) Relay, J17

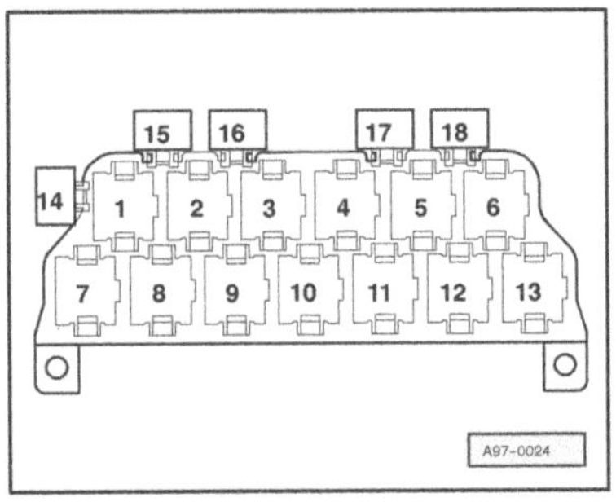

13 - Fold Relay Panel

Relay Location:
- 2 - Lamp Control Module, J123
- 3 - Lamp Control Module, J123
- 4 - Fog Light Relay, J5
- 5 - Starting Interlock Relay, J207

© 2006 BentleyPublishers.com—All Rights Reserved

Audi TT — Wiring Diagram — No. 16/3

Ignition/Starter Switch, Load Reduction Relay, Starting Interlock Relay

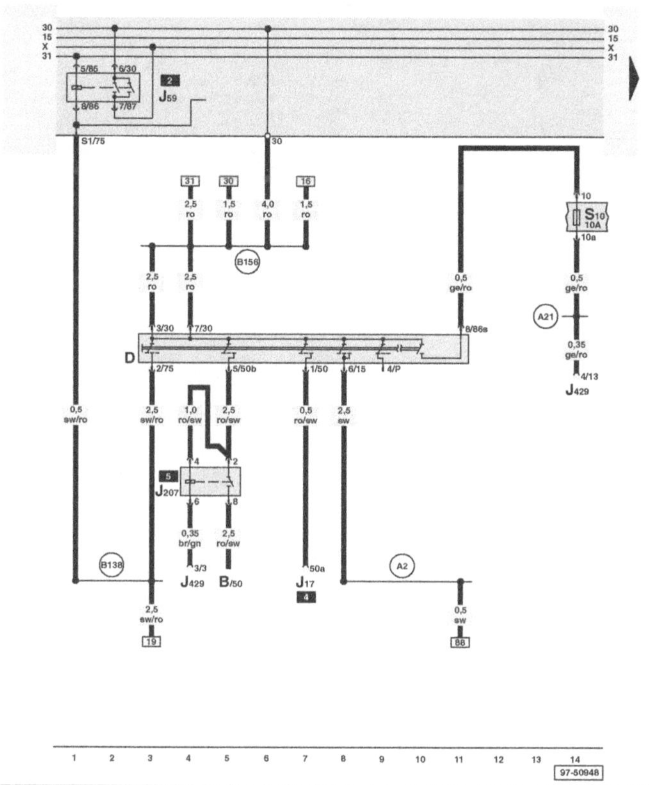

- B - Starter
- D - Ignition/Starter Switch
- J17 - Fuel Pump (FP) Relay
- J59 - Load Reduction Relay
- J207 - Starting Interlock Relay
- J429 - Control module for central locking
- S10 - Fuse

- (A2) - Plus connection (15), in instrument panel wiring harness
- (A21) - Wire connection (86s), in instrument panel wiring harness
- (B138) - Plus connector (X), in wiring harness, interior
- (B156) - Plus connector (30a), in wiring harness interior

ws = white
sw = black
ro = red
br = brown
gn = green
bl = blue
gr = grey
li = violet
ge = yellow
or = orange

USA.5703.02.21 — Edition 04/03

Audi TT — Wiring Diagram — No. 16/4

Light Switch

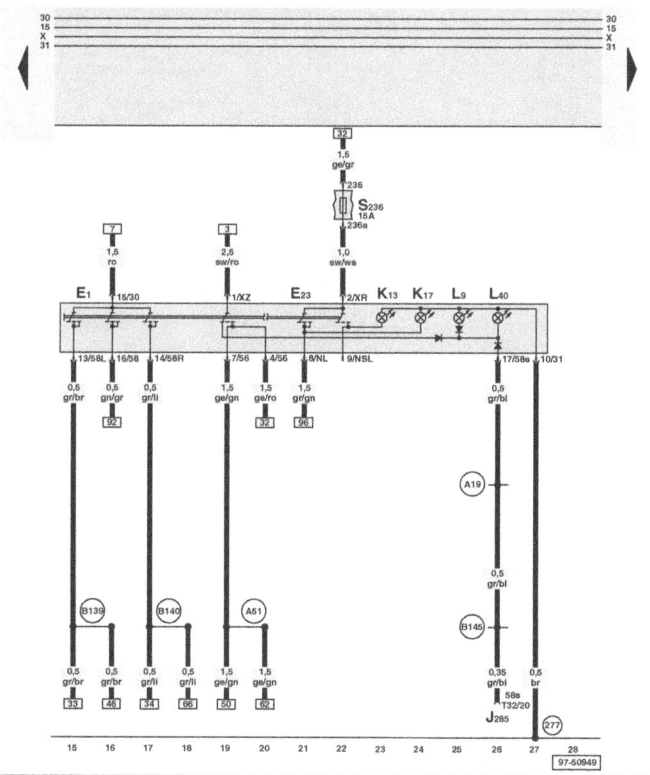

- E1 - Light Switch
- E23 - Fog Light Switch
- J285 - Control Module with indicator unit in instrument panel insert
- K13 - Rear Fog Light Indicator Light
- K17 - Fog Light Indicator Light
- L9 - Headlight Switch Light
- L40 - Front And Rear Fog Light Switch Light
- S236 - Fuse 36 in fuse holder
- T32 - Connector 32 pin, blue, on instrument cluster

- (277) - Ground connector -3-, in wiring harness interior
- (A19) - Wire connection (58s), in instrument panel wiring harness
- (A51) - Wire connection (56), in instrument panel wiring harness
- (B139) - Connector (PL), in wiring harness, interior
- (B140) - Connector (PR), in wiring harness, interior
- (B145) - Plus connector (58b), in wiring harness interior

ws = white
sw = black
ro = red
br = brown
gn = green
bl = blue
gr = grey
li = violet
ge = yellow
or = orange

USA.5703.02.21 — Edition 04/03

©2006 BentleyPublishers.com—All Rights Reserved

Headlights, Xenon (HID) (2004-2006)

Audi TT — Wiring Diagram — **No. 16/5**

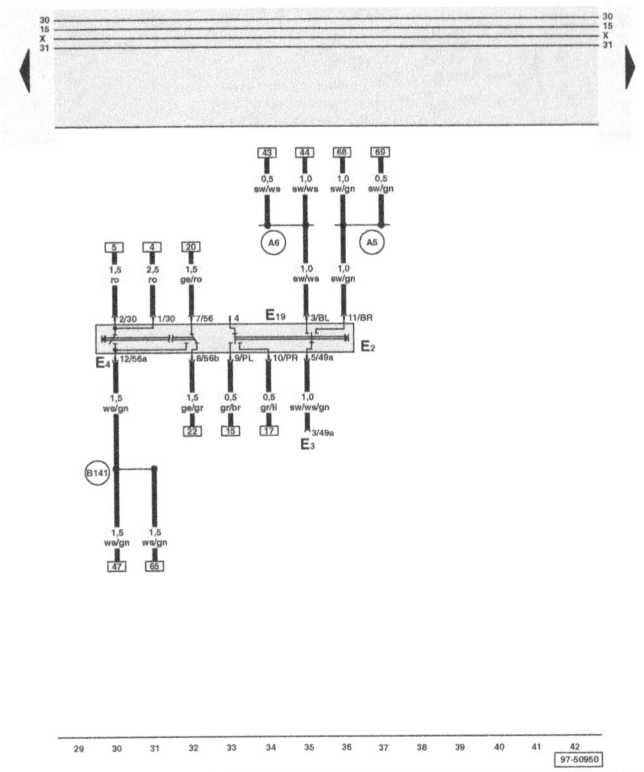

Turn Signal Switch, Headlight Dimmer/Flasher Switch

E2 - Turn Signal Switch
E3 - Emergency Flasher Switch
E4 - Headlight Dimmer/Flasher Switch
E19 - Park Light Switch

(A5) - Plus connection (right turn signal), in instrument panel wiring harness
(A6) - Plus connection (left turn signal), in instrument panel wiring harness
(B141) - Plus connector -1- (56a), in wiring harness interior

ws = white
sw = black
ro = red
br = brown
gn = green
bl = blue
gr = grey
li = violet
ge = yellow
or = orange

USA.5703.02.21 — Edition 04/03

Audi TT — Wiring Diagram — **No. 16/6**

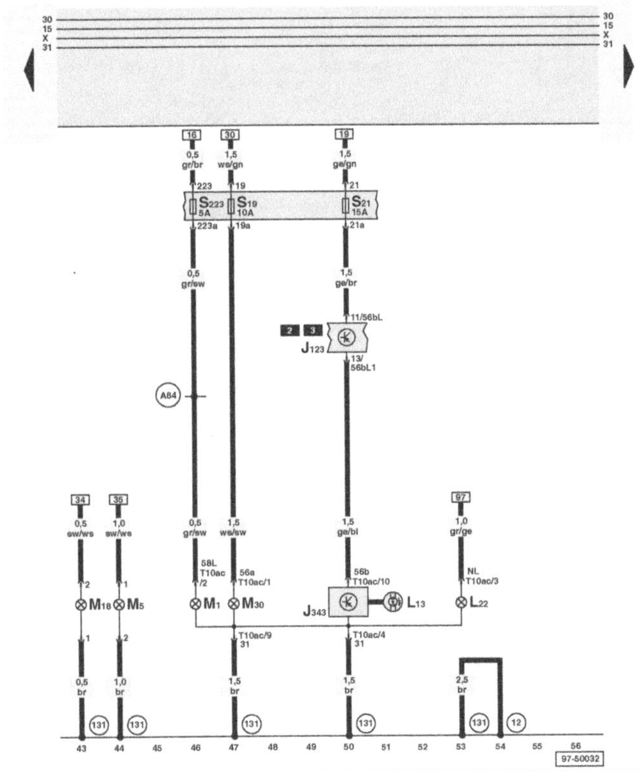

Fuses, High Intensity Gas Discharge Lamp left, Left High-intensity Gas Discharge Lamp Control Module

J123 - Lamp Control Module
J343 - Left High-intensity Gas Discharge Lamp Control Module
L13 - High intensity gas discharge lamp left
L22 - Left Front Fog Light
M1 - Left Parking Light
M5 - Left Front Turn Signal Light
M18 - Left, Side Turn Signal Light
M30 - Left High Beam Headlight
S19 - Fuse
S21 - Fuse
S223 - Fuse 23 in fuse holder
T10ac - Connector 10 pin, on headlight, left

(12) - Ground connection, in engine compartment, left
(131) - Ground connection -2-, in engine compartment wiring harness
(A84) - Connector (58L), in instrument panel wiring harness

ws = white
sw = black
ro = red
br = brown
gn = green
bl = blue
gr = grey
li = violet
ge = yellow
or = orange

USA.5703.02.21 — Edition 04/03

© 2006 BentleyPublishers.com—All Rights Reserved

Audi TT

Wiring Diagram

No. 16/7

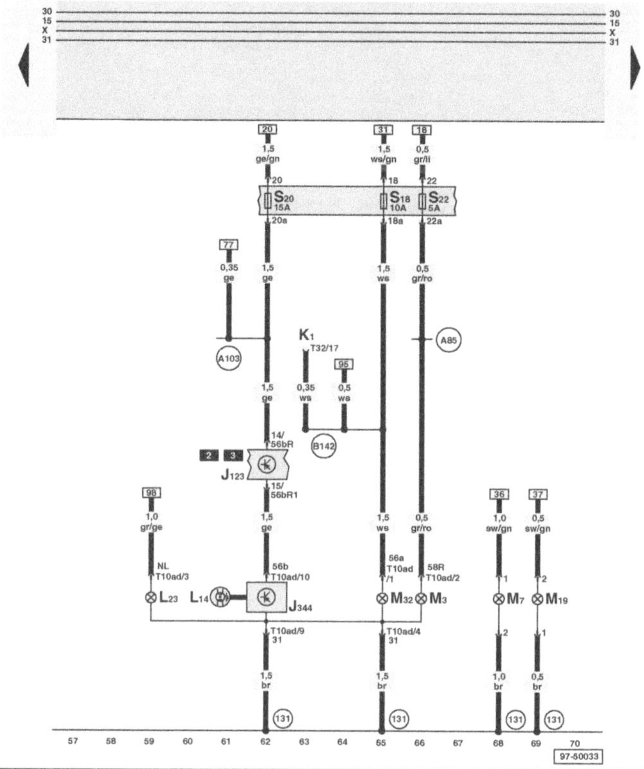

Fuses, High Intensity Gas Discharge Lamp Right, Right High-intensity Gas Discharge Lamp Control Module

- J123 - Lamp Control Module
- J344 - Right High-intensity Gas Discharge Lamp Control Module
- K1 - Headlight High Beam Indicator Light
- L14 - High intensity gas discharge lamp right
- L23 - Right Front Fog Light
- M3 - Right Parking Light
- M7 - Right Front Turn Signal Light
- M19 - Right, Side Turn Signal Light
- M32 - Right High Beam Headlight
- S18 - Fuse
- S20 - Fuse
- S22 - Fuse
- T10ad - Connector 10 pin, on headlight, right
- T32 - Connector 32 pin, blue, on instrument cluster

- (131) - Ground connection -2-, in engine compartment wiring harness
- (A85) - Connector (58R), in instrument panel wiring harness
- (A103) - Connector -2- (56), in instrument panel wiring harness
- (B142) - Plus connector -2- (56a), in wiring harness interior

ws = white
sw = black
ro = red
br = brown
gn = green
bl = blue
gr = grey
li = violet
ge = yellow
or = orange

USA.5703.02.21

Edition 04/03

Audi TT

Wiring Diagram

No. 16/8

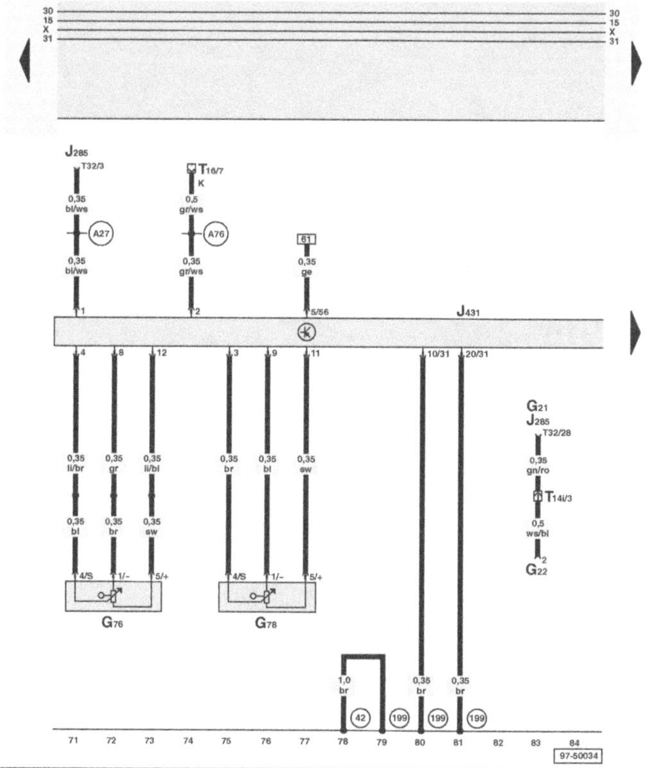

Control Module for Headlight Beam Adjusting, Level Control System Sensors

- G21 - Speedometer
- G22 - Speedometer Vehicle Speed Sensor (VSS)
- G76 - Left Rear Level Control System Sensor
- G78 - Left Front Level Control System Sensor
- J285 - Control Module with indicator unit in instrument panel insert
- J431 - Control module for Headlight Beam Adjusting
- T14i - Connector 14 pin, in engine compartment, left
- T16 - Connector 16 pin, black, Data Link Connector (DLC), below storage compartment, driver's side
- T32 - Connector 32 pin, blue, on instrument cluster

- (42) - Ground connection, on steering column
- (199) - Ground connection -3-, in instrument panel wiring harness
- (A27) - Wire connection (vehicle speed signal), in instrument panel wiring harness
- (A76) - Connector (K-diagnosis wire), in instrument panel wiring harness

ws = white
sw = black
ro = red
br = brown
gn = green
bl = blue
gr = grey
li = violet
ge = yellow
or = orange

USA.5703.02.21

Edition 04/03

© 2006 BentleyPublishers.com—All Rights Reserved

Headlights, Xenon (HID) (2004-2006)

Audi TT — Wiring Diagram — **No. 16/9**

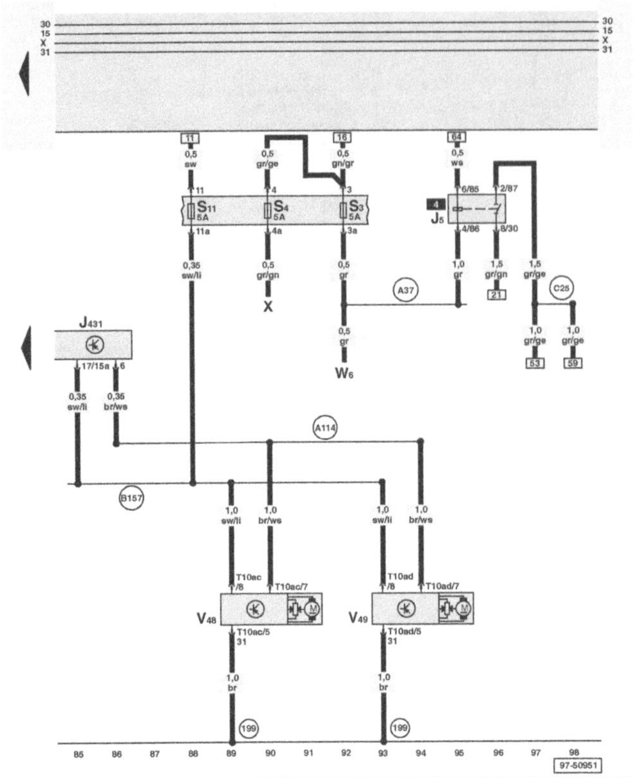

Headlight Beam Adjusting Motors, Fog Light Relay

J5 - Fog Light Relay
J431 - Control module for Headlight Beam Adjusting
S3 - Fuse
S4 - Fuse
S11 - Fuse
T10ac - Connector 10 pin, on headlight, left
T10ad - Connector 10 pin, on headlight, right
V48 - Left Headlight Beam Adjusting Motor
V49 - Right Headlight Beam Adjusting Motor
W6 - Glove Compartment Light
X - License Plate Light

(199) - Ground connection -3-, in instrument panel wiring harness
(A37) - Wire connection (58a), in instrument panel wiring harness
(A114) - Connection (headlight beam adjusting) in instrument panel wiring harness
(B157) - Plus connector -3- (15a), in wiring harness interior
(C25) - Wire connection (fog lights), in front light wiring harness

ws = white
sw = black
ro = red
br = brown
gn = green
bl = blue
gr = grey
li = violet
ge = yellow
or = orange

USA.5703.02.21 — Edition 04/03

THIS SPACE INTENTIONALLY LEFT BLANK

©2006 BentleyPublishers.com—All Rights Reserved

Audi TT Wiring Diagram **No. 17/1**

Radio Stereo Concert with and without CD-Changer (Roadster)

2004 m.y.

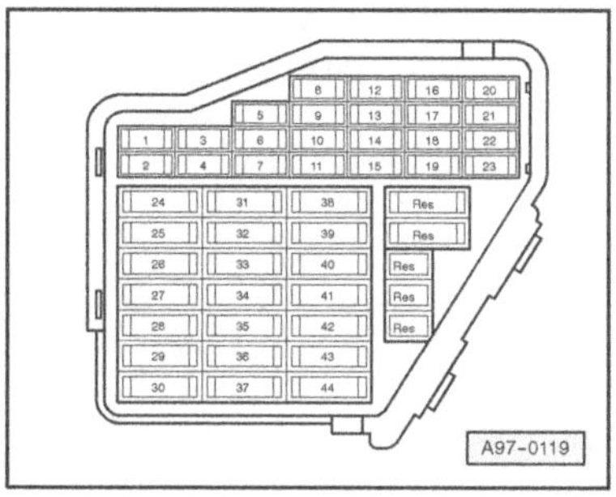

Fuse Panel

Fuse Colors:
- 30 A - Green
- 25 A - White
- 20 A - Yellow
- 15 A - Blue
- 10 A - Red
- 7,5 A - Brown
- 5 A - Beige
- 1 A - Black

Starting with fuse position 23, fuses in the fuse holder are identified with 223 in the wiring diagram.

USA.5703.02.21 **Edition 04/03**

Audi TT Wiring Diagram **No. 17/2**

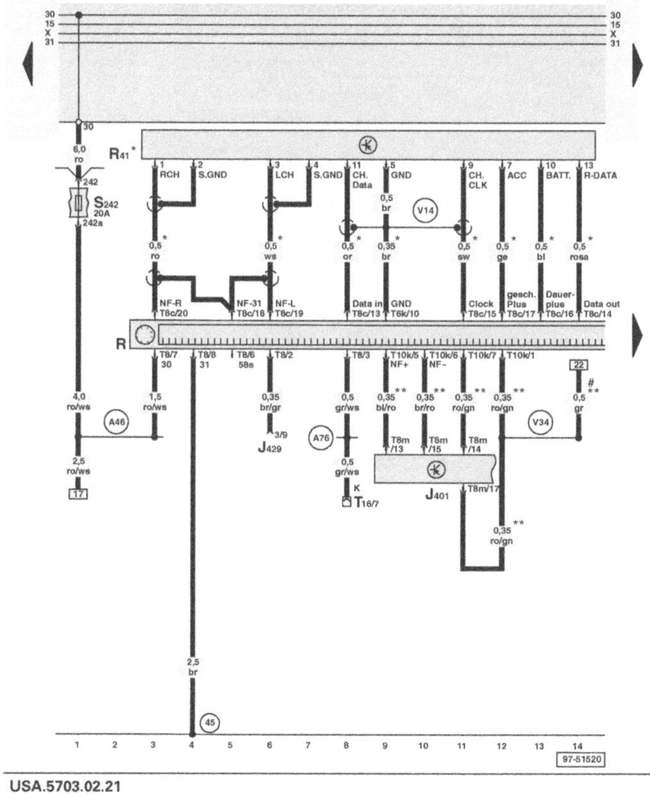

Radio, CD Changer Unit

- J401 - Control unit for navigation with CD-mechanism
- J429 - Control module for central locking
- R - Radio
- R41 - CD Changer Unit
- S242 - Fuse 42 in fuse holder
- T6k - Connector 6 pin, green, connector radio I-2
- T8 - Connector 8 pin, black, connector radio III
- T8c - Connector 8 pin, blue, connector radio I-3
- T8m - Connector 8 pin, C III on Control Module for Navigation
- T10k - Connector 10 pin, red, on radio
- T16 - Connector 16 pin, black, Data Link Connector (DLC), below storage compartment, driver's side

- (45) - Ground connection, behind instrument panel, center
- (A46) - Plus connection (30-from radio), in instrument panel wiring harness
- (A76) - Connector (K-diagnosis wire), in instrument panel wiring harness
- (V14) - Wire connection (shielding), in CD-changer wiring harness
- (V34) - Connector (mute button), in wiring harness telephone

- * - Vehicles with CD Changer
- ** - Vehicles with Navigation System
- # - Vehicles with Cellular Telephone Preparation

ws = white
sw = black
ro = red
br = brown
gn = green
bl = blue
gr = grey
li = violet
ge = yellow
or = orange

USA.5703.02.21 **Edition 04/03**

© 2006 BentleyPublishers.com—All Rights Reserved

Sound System, Concert, roadster (2004-2006)

Audi TT Wiring Diagram **No. 17/3**

Radio, Center Rear Woofer/Amplifier

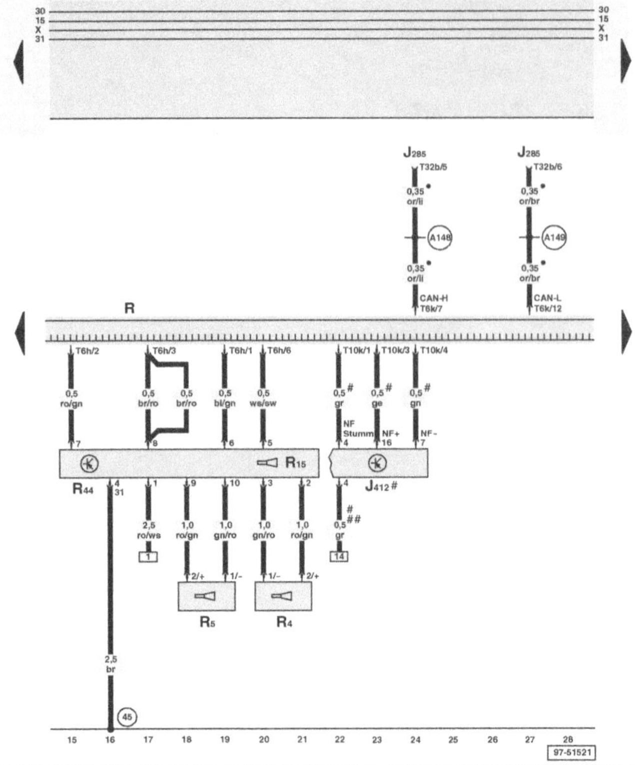

J285 - Control Module with indicator unit in instrument panel insert
J412 - Operating electronics and Telephone Control Module
R - Radio
R4 - Left Rear Speaker
R5 - Right Rear Speaker
R15 - Bass speaker, left rear
R44 - Center Rear Woofer/Amplifier
T6h - Connector 6 pin, yellow, connector radio I-1
T6k - Connector 6 pin, green, connector radio I-2
T10k - Connector 10 pin, red, on radio
T32b - Connector 32 pin, grey, on instrument cluster

(45) - Ground connection, behind instrument panel, center
(A148) - Connection (high bus, navigation), in instrument panel wiring harness
(A149) - Connection (low bus, navigation), in instrument panel wiring harness

• - CAN-Bus (Data-Bus)
- Vehicles with Cellular Telephone Preparation
- Vehicles with Navigation System

ws = white
sw = black
ro = red
br = brown
gn = green
bl = blue
gr = grey
li = violet
ge = yellow
or = orange

USA.5703.02.21 Edition 04/03

Audi TT Wiring Diagram **No. 17/4**

Radio, Antenna, Bass Speaker Front, Treble Speaker Front, Antenna Splitter, Antenna

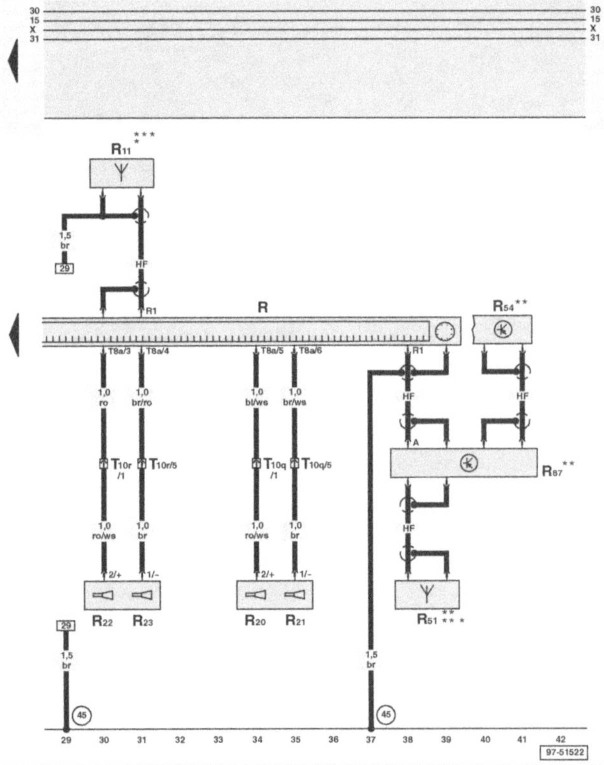

R - Radio
R1 - Antenna Connection
R11 - Antenna
R20 - Treble speaker, left front
R21 - Bass speaker, left front
R22 - Treble speaker, right front
R23 - Bass speaker, right front
R51 - Radio/Telephone Antenna
R54 - Cellular telephone
R87 - Antenna Splitter
T8a - Connector 8 pin, brown, connector radio II
T10q - Connector 10 pin, green, connector station A pillar, left
T10r - Connector 10 pin, green, connector station A pillar, right

(45) - Ground connection, behind instrument panel, center

* - Vehicles without Cellular Telephone Preparation
** - Vehicles with Cellular Telephone Preparation
*** - Rod Antenna

ws = white
sw = black
ro = red
br = brown
gn = green
bl = blue
gr = grey
li = violet
ge = yellow
or = orange

USA.5703.02.21 Edition 04/03

© 2006 BentleyPublishers.com—All Rights Reserved

Audi TT Wiring Diagram **No. 18/1**

Radio-Stereo Concert with and without CD-Changer (Coupé)

2004 m.y.

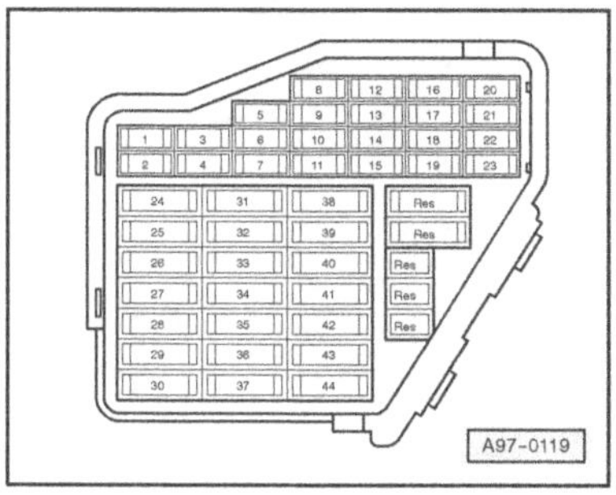

Fuse Panel

Fuse Colors:
- 30 A - Green
- 25 A - White
- 20 A - Yellow
- 15 A - Blue
- 10 A - Red
- 7,5 A - Brown
- 5 A - Beige
- 1 A - Black

Starting with fuse position 23, fuses in the fuse holder are identified with 223 in the wiring diagram.

USA.5703.02.21 **Edition 04/03**

Audi TT Wiring Diagram **No. 18/2**

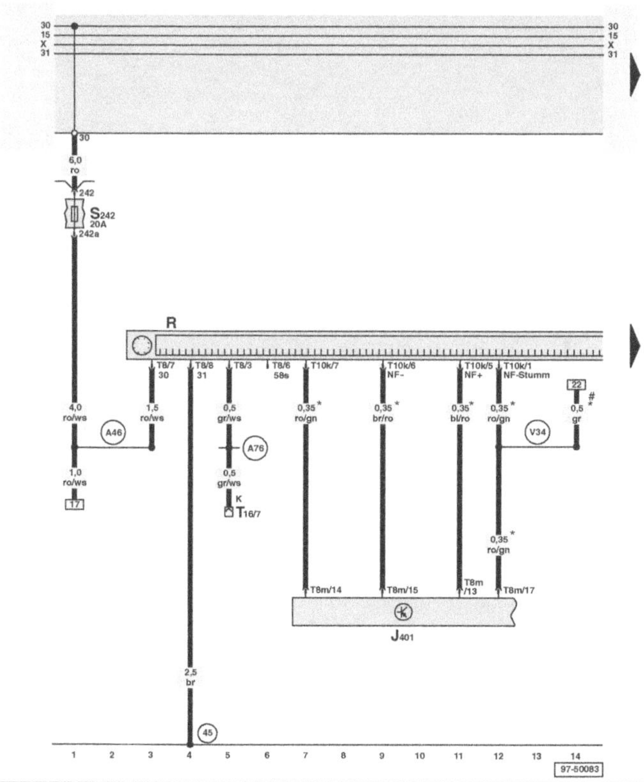

Radio

- J401 - Control unit for navigation with CD-mechanism
- R - Radio
- S242 - Fuse 42 in fuse holder
- T8 - Connector 8 pin, black, connector radio III
- T8m - Connector 8 pin, C III on Control Module for Navigation
- T10k - Connector 10 pin, red, on radio
- T16 - Connector 16 pin, black, Data Link Connector (DLC), below storage compartment, driver's side
- (45) - Ground connection, behind instrument panel, center
- (A46) - Plus connection (30-from radio), in instrument panel wiring harness
- (A76) - Connector (K-diagnosis wire), in instrument panel wiring harness
- (V34) - Connector (mute button), in wiring harness telephone

- * - Vehicles with Navigation System
- # - Vehicles with Cellular Telephone Preparation

ws = white
sw = black
ro = red
br = brown
gn = green
bl = blue
gr = grey
li = violet
ge = yellow
or = orange

USA.5703.02.21 **Edition 04/03**

© 2006 BentleyPublishers.com—All Rights Reserved

Sound System, Concert, coupe (2004-2006)

Audi TT — Wiring Diagram — No. 18/3

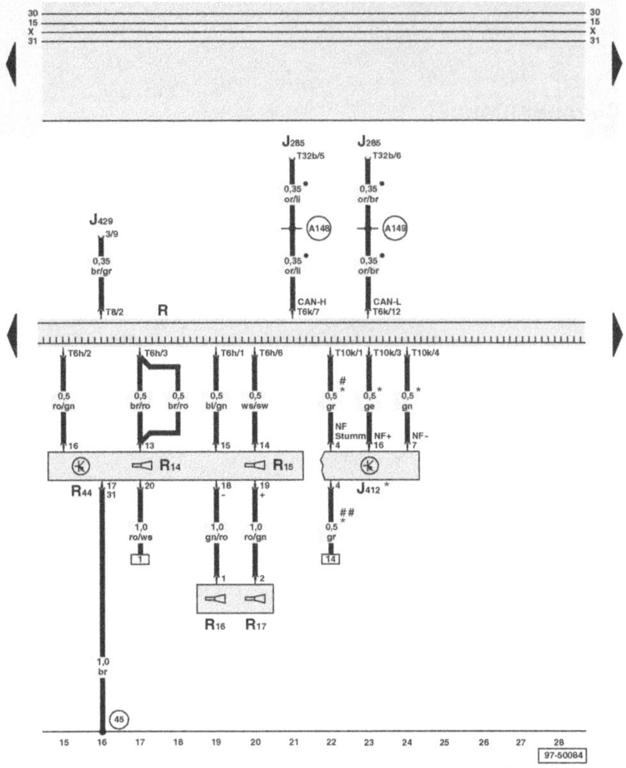

ws = white
sw = black
ro = red
br = brown
gn = green
bl = blue
gr = grey
li = violet
ge = yellow
or = orange

Radio, Left Rear Woofer/Amplifier

- J285 - Control Module with indicator unit in instrument panel insert
- J412 - Operating electronics and Telephone Control Module
- J429 - Control module for central locking
- R - Radio
- R14 - Treble speaker, left rear
- R15 - Bass speaker, left rear
- R16 - Treble speaker, right rear
- R17 - Bass speaker, right rear
- R44 - Left Rear Woofer/Amplifier
- T6h - Connector 6 pin, yellow, connector radio I-1
- T6k - Connector 6 pin, green, connector radio I-2
- T8 - Connector 8 pin, black, connector radio III
- T10k - Connector 10 pin, red, on radio
- T32b - Connector 32 pin, grey, on instrument cluster

- (45) - Ground connection, behind instrument panel, center
- (A148) - Connection (high bus, navigation), in instrument panel wiring harness
- (A149) - Connection (low bus, navigation), in instrument panel wiring harness

- • - CAN-Bus (Data-Bus)
- * - Vehicles with Cellular Telephone Preparation
- # - Vehicles without Navigation System
- ## - Vehicles with Navigation System

USA.5703.02.21 — Edition 04/03

Audi TT — Wiring Diagram — No. 18/4

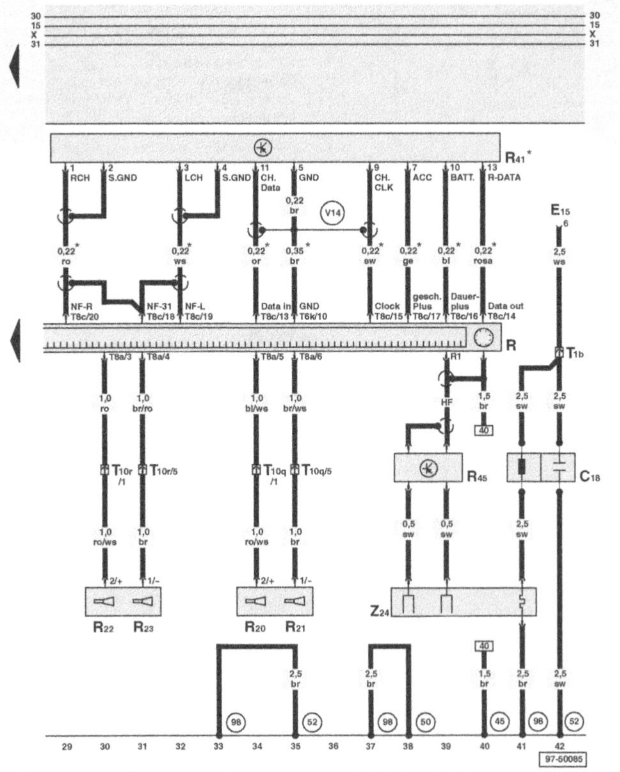

ws = white
sw = black
ro = red
br = brown
gn = green
bl = blue
gr = grey
li = violet
ge = yellow
or = orange

Radio, CD Changer Unit, Antenna

- C18 - Windshield Antenna Suppression Filter
- E15 - Rear Window Defogger Switch
- R - Radio
- R1 - Antenna Connection
- R20 - Treble speaker, left front
- R21 - Bass speaker, left front
- R22 - Treble speaker, right front
- R23 - Bass speaker, right front
- R41 - CD Changer Unit
- R45 - Telephone/Radio Antenna Amplifier
- T1b - Connector single, black, in rear lid left
- T6k - Connector 6 pin, green, connector radio I-2
- T8a - Connector 8 pin, brown, connector radio II
- T8c - Connector 8 pin, blue, connector radio I-3
- T10q - Connector 10 pin, green, connector station A pillar, left
- T10r - Connector 10 pin, green, connector station A pillar, right
- Z24 - Rear Window Defogger With Window Antenna

- (45) - Ground connection, behind instrument panel, center
- (50) - Ground connection, in luggage compartment, left
- (52) - Ground connection, in rear lid, left
- (98) - Ground connection, in rear lid wiring harness
- (V14) - Wire connection (shielding), in CD-changer wiring harness

- * - Vehicles with CD-Changer

USA.5703.02.21 — Edition 04/03

©2006 BentleyPublishers.com—All Rights Reserved

Audi TT Wiring Diagram **No. 19/1**

3.2L - Engine - Motronic Multiport Fuel Injection (MFI) / 184kW , codes BHE, BPF

2004 m.y.

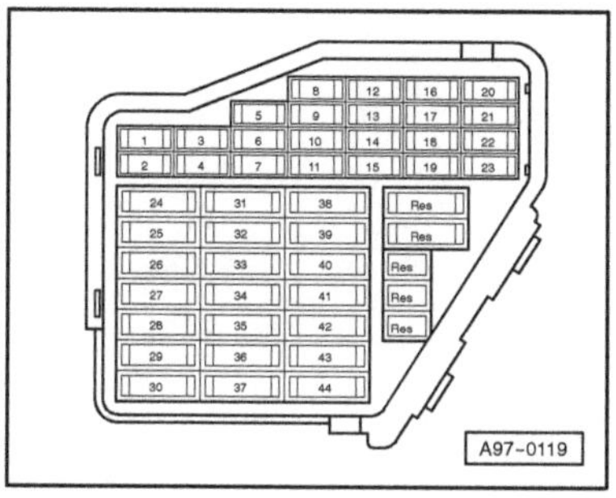

Fuse Panel

Fuse Colors:

30 A - Green
25 A - White
20 A - Yellow
15 A - Blue
10 A - Red
7,5 A - Brown
5 A - Beige
1 A - Black

Starting with fuse position 23, fuses in the fuse holder are identified with 223 in the wiring diagram.

USA.5703.04.21 Edition 06/04

Audi TT Wiring Diagram **No. 19/2**

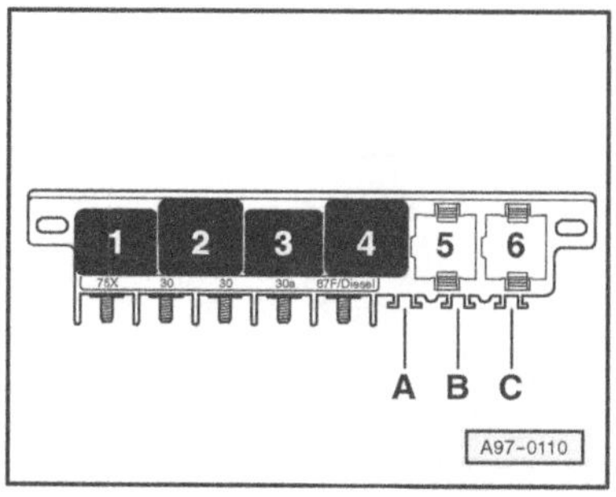

Micro Central Electric Panel

Relay Location:

4 - Fuel Pump (FP) Relay, J17
B - Hydraulic Pump Relay Fuse, S279

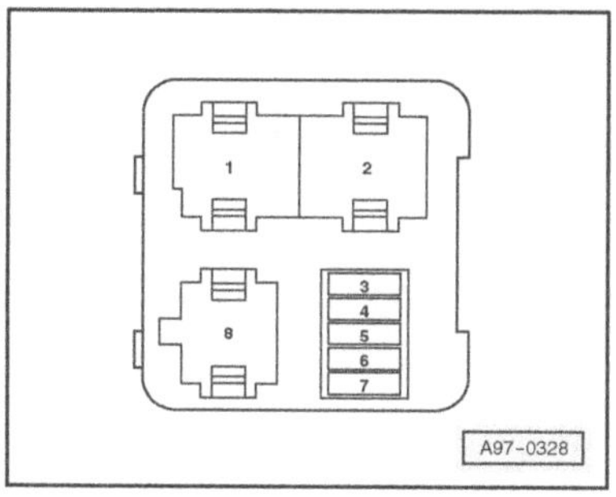

W-Box in engine compartment

Relay Location:

1 - Secondary Air Injection (AIR) Pump Relay, J299
2 - Motronic Engine Control Module (ECM) Power Supply Relay, J271

USA.5703.02.21 Edition 04/03

© 2006 BentleyPublishers.com—All Rights Reserved

Engine Management, 3.2L, engine code: BHE, BPF (2004-2006)

Audi TT **Wiring Diagram** **No. 19/3**

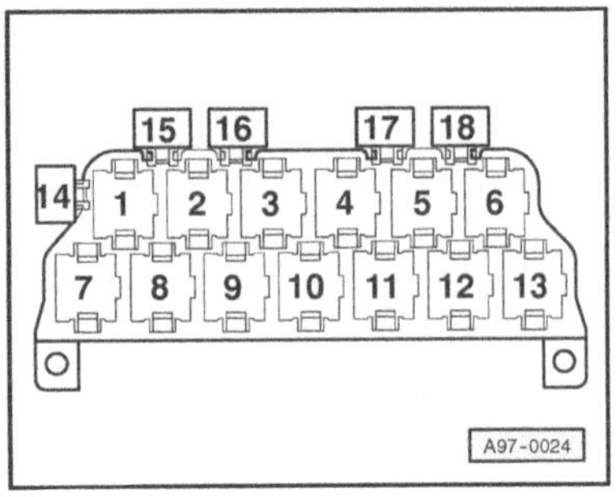

13 - Fold Relay Panel

Relay Location:

11 - Starting Interlock Relay, J207

USA.5703.03.21 Edition 10/03

Audi TT **Wiring Diagram** **No. 19/4**

Battery, Starter, Generator, Main Fuse Box/Battery, Starting Interlock Relay

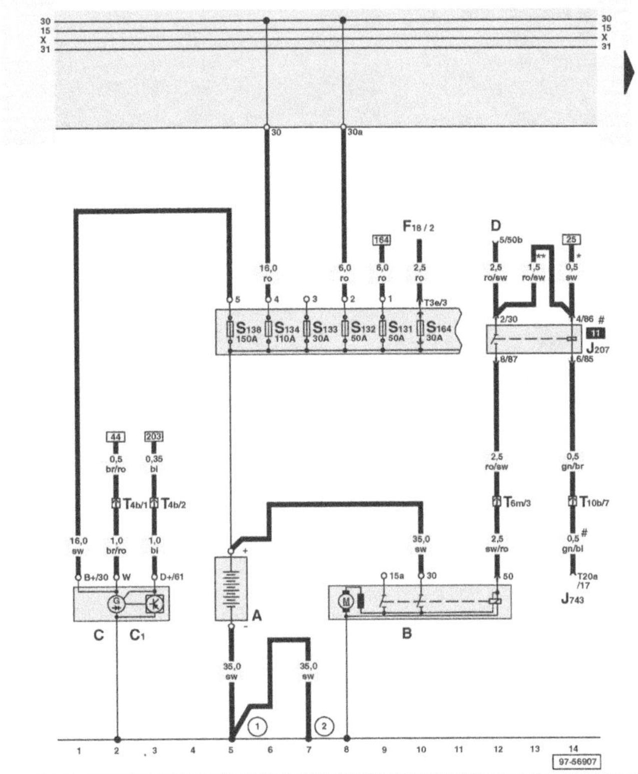

A - Battery
B - Starter
C - Generator (GEN)
C1 - Voltage Regulator (VR)
D - Ignition/Starter Switch
F18 - Coolant Fan Control (FC) Thermal Switch
J207 - Starting Interlock Relay
J743 - Direct Shift Gearbox (DSG) Mechatronic
S131 - Safety fuse 1
S132 - Safety fuse 2
S133 - Safety fuse 3
S134 - Safety fuse 4
S138 - Safety fuse 5
S164 - Fuse -3- in fuse bracket/battery
T3e - Connector 3 pin, black, on Main Fuse Box/Battery
T4b - Connector 4 pin, in engine compartment, left
T6m - Connector 6 pin, brown, in E-box plenum chamber
T10b - Connector 10 pin, grey, in E-box plenum chamber
T20a - Connector 20 pin, on Direct Shift Gearbox (DSG) Mechatronic

(1) - Ground strap, battery to body
(2) - Ground strap, transmission to body

* - through 2003 m.y.
** - from 2004 m.y.
# - Vehicles with Direct Shift Gearbox (DSG) 02E

ws = white
sw = black
ro = red
br = brown
gn = green
bl = blue
gr = grey
li = violet
ge = yellow
or = orange

USA.5703.03.21 Edition 10/03

© 2006 BentleyPublishers.com—All Rights Reserved

Audi TT

Wiring Diagram

No. 19/5

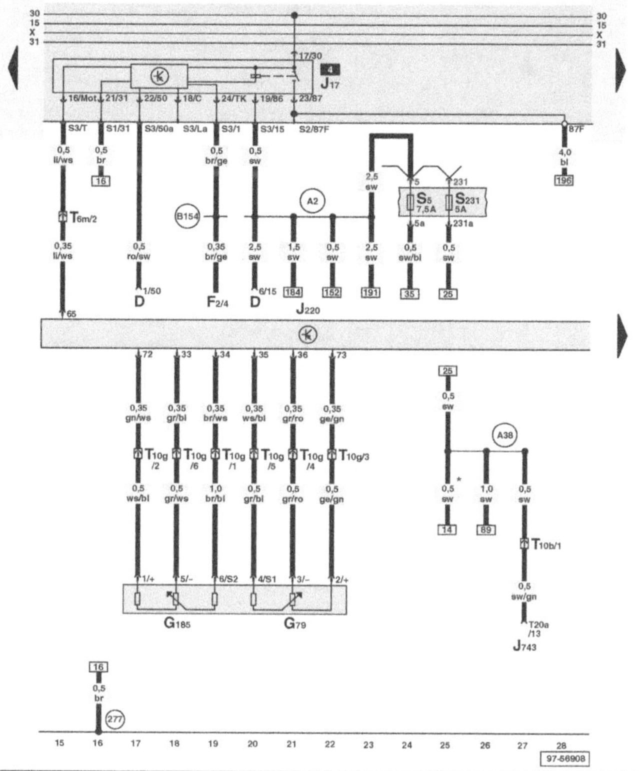

Motronic Engine Control Module, Fuel Pump (FP) Relay, Throttle Position (TP) Sensor

- D - Ignition/Starter Switch
- F2 - Door Contact Switch - Driver's side
- G79 - Throttle Position (TP) Sensor
- G185 - Sensor -2- for accelerator pedal position
- J17 - Fuel Pump (FP) Relay
- J220 - Motronic Engine Control Module (ECM)
- J743 - Direct Shift Gearbox (DSG) Mechatronic
- S5 - Fuse
- S231 - Fuse 31 in fuse holder
- T6m - Connector 6 pin, brown, in E-box plenum chamber
- T10b - Connector 10 pin, grey, in E-box plenum chamber
- T10g - Connector 10 pin, blue, in E-box plenum chamber
- T20a - Connector 20 pin, on Direct Shift Gearbox (DSG) Mechatronic
- (277) - Ground connector -3-, in wiring harness interior
- (A2) - Plus connection (15), in instrument panel wiring harness
- (A38) - Plus connection -2- (15a), in instrument panel wiring harness
- (B154) - Connector -1- (TK), in wiring harness interior
- * - through 2003 m.y.

ws = white
sw = black
ro = red
br = brown
gn = green
bl = blue
gr = grey
li = violet
ge = yellow
or = orange

USA.5703.03.21

Edition 10/03

Audi TT

Wiring Diagram

No. 19/6

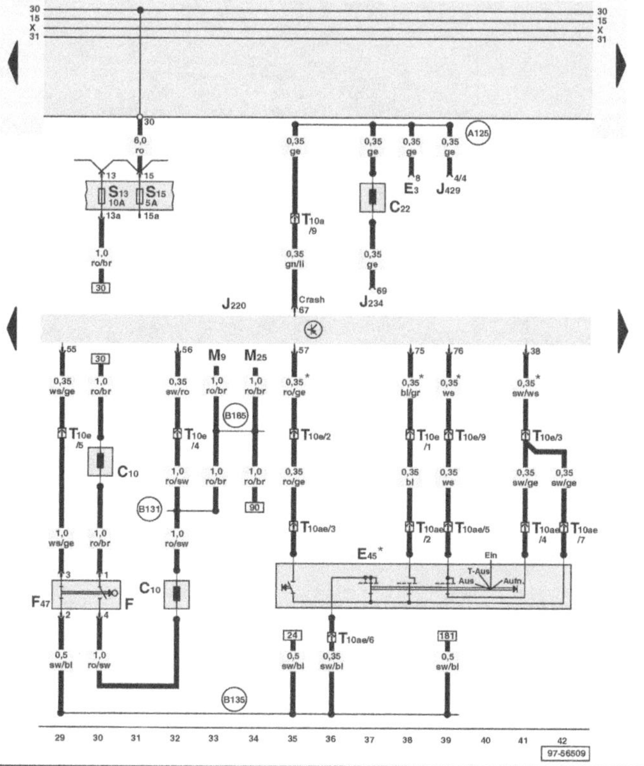

Motronic Engine Control Module, Cruise Control Switch, Brake Light Switch

- C10 - Suppressor Filter
- C22 - Airbag Suppresser
- E3 - Emergency Flasher Switch
- E45 - Cruise Control Switch
- F - Brake Light Switch
- F47 - Brake Vacuum Vent Valve Switch for Cruise Control
- J220 - Motronic Engine Control Module (ECM)
- J234 - Airbag Control Module
- J429 - Control module for central locking
- M9 - Left Brake Light
- M25 - High-mount Brake Light
- S13 - Fuse
- S15 - Fuse
- T10a - Connector 10 pin, white, in E-box plenum chamber
- T10e - Connector 10 pin, black, in E-box plenum chamber
- T10ae - Connector 10 pin, on Cruise Control Switch
- (A125) - Connection (crash signal) in instrument panel wiring harness
- (B131) - Connector (54), in wiring harness interior
- (B135) - Connector (15a), in wiring harness interior
- (B185) - Connection 2 (54), in passenger compartment wiring harness
- * - Vehicles with Cruise Control

ws = white
sw = black
ro = red
br = brown
gn = green
bl = blue
gr = grey
li = violet
ge = yellow
or = orange

USA.5703.03.21

Edition 10/03

©2006 BentleyPublishers.com—All Rights Reserved

Engine Management, 3.2L, engine code: BHE, BPF (2004-2006)

Audi TT — Wiring Diagram — No. 19/7

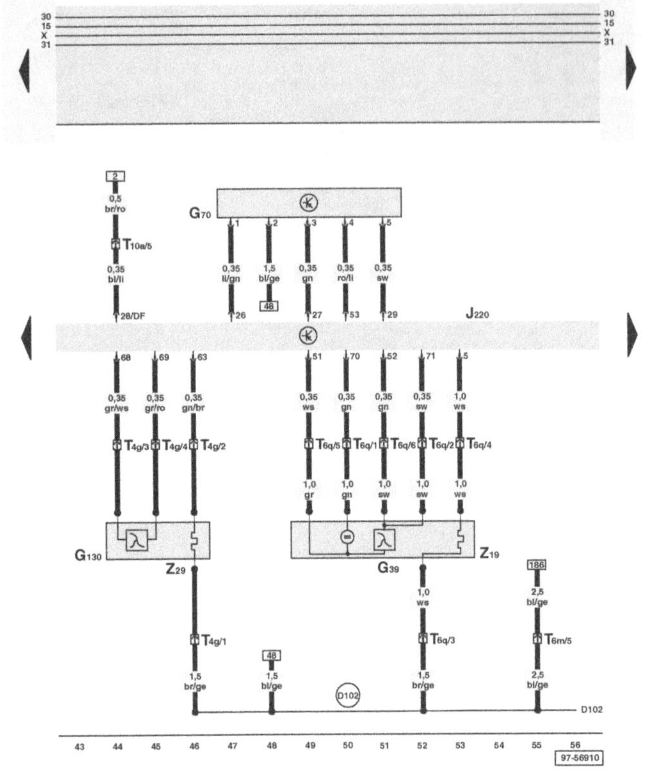

ws = white
sw = black
ro = red
br = brown
gn = green
bl = blue
gr = grey
li = violet
ge = yellow
or = orange

Motronic Engine Control Module, Mass Air Flow (MAF) Sensor, Heated Oxygen Sensor

- G39 - Heated Oxygen Sensor (HO2S)
- G70 - Mass Air Flow (MAF) Sensor
- G130 - Heated Oxygen Sensor (O2S) Behind Three Way Catalytic Converter (TWC)
- J220 - Motronic Engine Control Module (ECM)
- T4g - Connector 4 pin, in engine compartment, right, for Oxygen Sensor behind Three Way Catalytic Converter
- T6m - Connector 6 pin, brown, in E-box plenum chamber
- T6q - Connector 6 pin, in engine compartment right
- T10a - Connector 10 pin, white, in E-box plenum chamber
- Z19 - Oxygen Sensor (O2S) Heater
- Z29 - Oxygen Sensor (O2S) Heater 1
- (D102) - Wire connection -2-, in engine compartment wiring harness

USA.5703.02.21 Edition 04/03

Audi TT — Wiring Diagram — No. 19/8

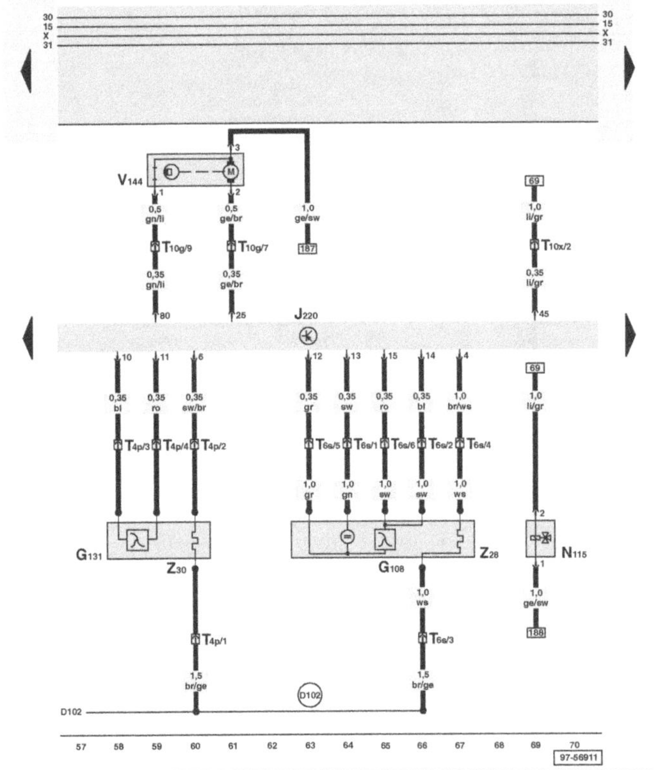

ws = white
sw = black
ro = red
br = brown
gn = green
bl = blue
gr = grey
li = violet
ge = yellow
or = orange

Motronic Engine Control Module, Heated Oxygen Sensor 2, Leak Detection Pump

- G108 - Heated Oxygen Sensor (HO2S) 2
- G131 - Heated Oxygen Sensor (O2S) 2 Behind Three Way Catalytic Converter (TWC)
- J220 - Motronic Engine Control Module (ECM)
- N115 - Evaporative Emission (EVAP) Canister Purge Solenoid Valve
- T4p - Connector 4 pin, in engine compartment right, for Heated Oxygen Sensor 2 Behind Three Way Catalytic Converter
- T6s - Connector 6 pin, in engine compartment right, for Heated Oxygen Sensor 2
- T10g - Connector 10 pin, blue, in E-box plenum chamber
- T10x - Connector 10 pin, orange, in E-box plenum chamber
- V144 - Leak Detection Pump (LDP)
- Z28 - Oxygen Sensor (O2S) 2 Heater
- Z30 - Oxygen Sensor (O2S) Heater 2
- (D102) - Wire connection -2-, in engine compartment wiring harness

USA.5703.02.21 Edition 04/03

© 2006 BentleyPublishers.com—All Rights Reserved

Audi TT — Wiring Diagram — No. 19/9

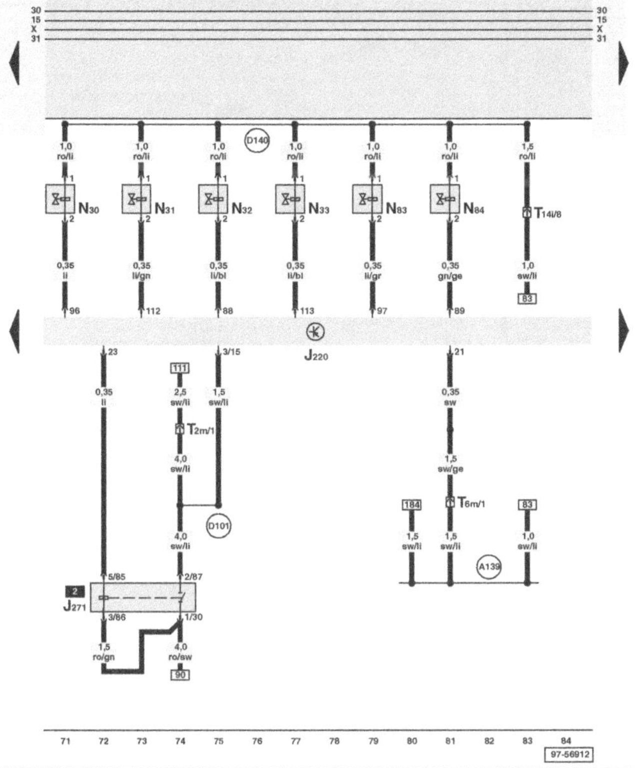

Motronic Engine Control Module, Motronic Engine Control Module Power Supply Relay, Cylinder 1 - 6 Fuel Injectors

- J220 - Motronic Engine Control Module (ECM)
- J271 - Motronic Engine Control Module (ECM) Power Supply Relay
- N30 - Cylinder 1 Fuel Injector
- N31 - Cylinder 2 Fuel Injector
- N32 - Cylinder 3 Fuel Injector
- N33 - Cylinder 4 Fuel Injector
- N83 - Cylinder 5 Fuel Injector
- N84 - Cylinder 6 Fuel Injector
- T2m - Connector 2 pin, in engine compartment, left
- T6m - Connector 6 pin, brown, in E-box plenum chamber
- T14i - Connector 14 pin, in engine compartment, left
- (A139) - Connection 3 (15) (in instrument panel wiring harness)
- (D101) - Wire connection -1-, in engine compartment wiring harness
- (D140) - Connector (injectors), in wiring harness, engine pre-wiring

ws = white
sw = black
ro = red
br = brown
gn = green
bl = blue
gr = grey
li = violet
ge = yellow
or = orange

USA.5703.02.21 Edition 04/03

Audi TT — Wiring Diagram — No. 19/10

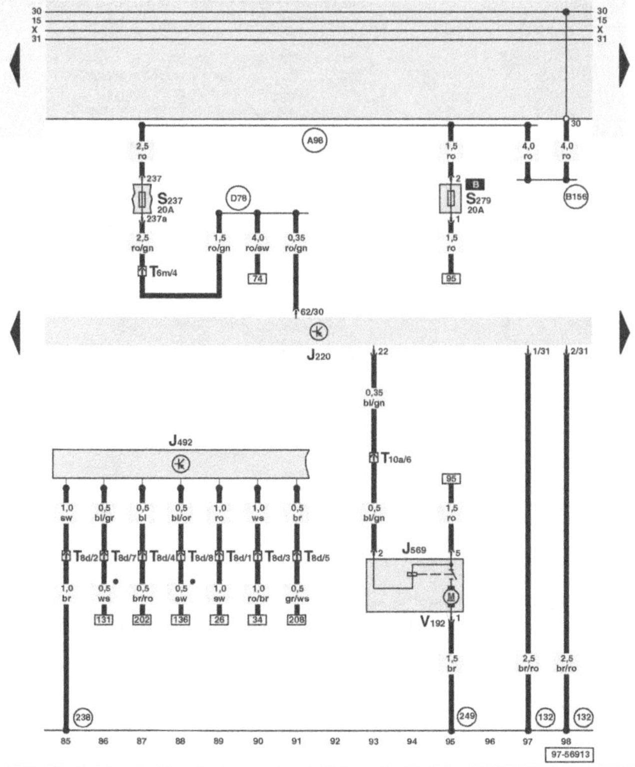

Motronic Engine Control Module, Brake Booster Relay, Brake System Vacuum Pump, All-Wheel Drive Control Module, Hydraulic Pump Relay Fuse

- J220 - Motronic Engine Control Module (ECM)
- J492 - All-wheel drive control module
- J569 - Brake Booster relay
- S237 - Fuse 37 in fuse holder
- S279 - Hydraulic Pump Relay Fuse
- T6m - Connector 6 pin, brown, in E-box plenum chamber
- T8d - Connector 8 pin, on Control module for all wheel drive
- T10a - Connector 10 pin, white, in E-box plenum chamber
- V192 - Brake System Vacuum Pump
- (132) - Ground connection -3-, in engine compartment wiring harness
- (238) - Ground connection -1-, in wiring harness interior
- (249) - Ground connection -2-, in wiring harness interior
- (A98) - Plus connector -4- (30), in instrument panel wiring harness
- (B156) - Plus connector (30a), in wiring harness interior
- (D78) - Plus connection -1- (30a), in engine compartment wiring harness
- • - CAN-Bus (Data-Bus)

ws = white
sw = black
ro = red
br = brown
gn = green
bl = blue
gr = grey
li = violet
ge = yellow
or = orange

USA.5703.02.21 Edition 04/03

© 2006 BentleyPublishers.com—All Rights Reserved

Engine Management, 3.2L, engine code: BHE, BPF (2004-2006)

Audi TT **Wiring Diagram** **No. 19/11**

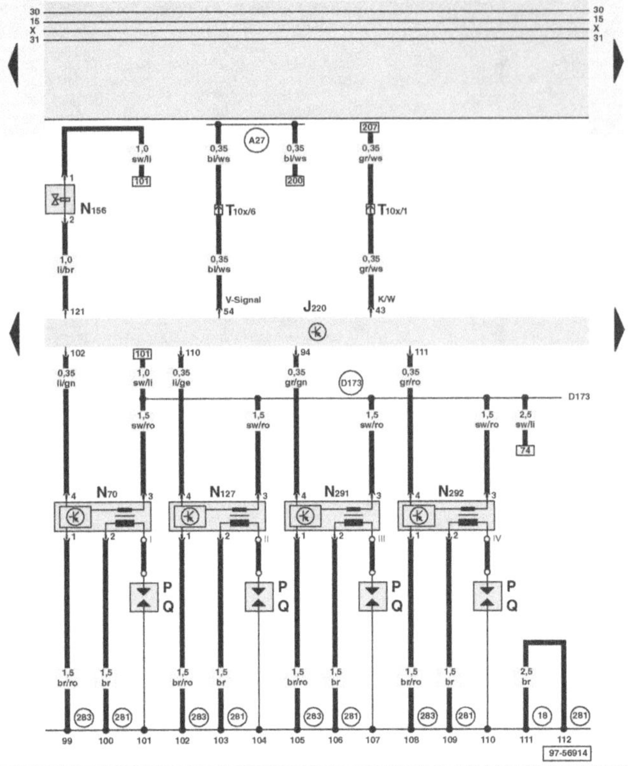

Motronic Engine Control Module, Ignition Coil 1 - 4 with Power Output Stage, Intake Change-Over Valve

J220 - Motronic Engine Control Module (ECM)
N70 - Ignition Coil 1 with Power Output Stage
N127 - Ignition Coil 2 with Power Output Stage
N156 - Intake Change-Over Valve
N291 - Ignition Coil 3 with Power Output Stage
N292 - Ignition Coil 4 with Power Output Stage
T10x - Connector 10 pin, orange, in E-box plenum chamber

(18) - Ground connection, on engine block

(281) - Ground connector -1-, in wiring harness engine pre-wiring
(283) - Ground connection 2, in wiring harness engine pre-wiring
(A27) - Wire connection (vehicle speed signal), in instrument panel wiring harness
(D173) - Connection (87) (in engine compartment wiring harness)

ws = white
sw = black
ro = red
br = brown
gn = green
bl = blue
gr = grey
li = violet
ge = yellow
or = orange

USA.5703.02.21 Edition 04/03

Audi TT **Wiring Diagram** **No. 19/12**

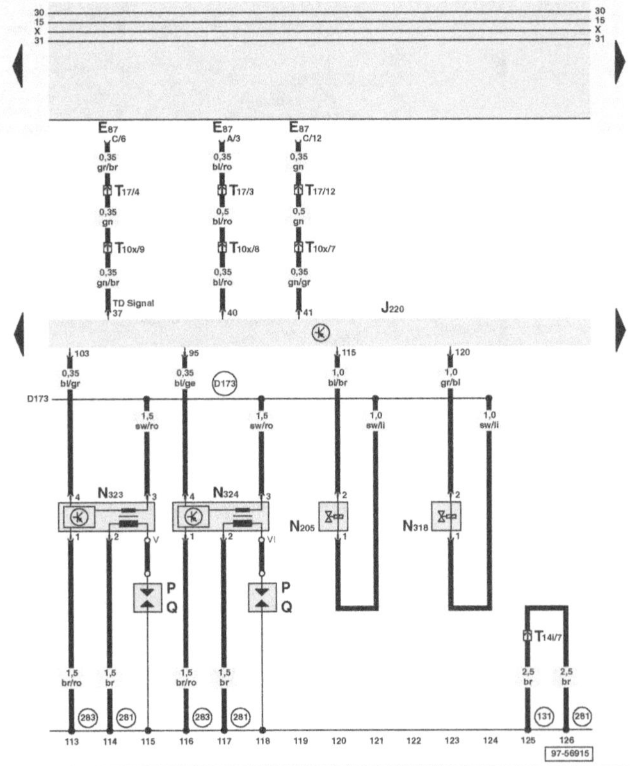

Motronic Engine Control Module, Ignition Coil 5 and 6 with Power Output Stage, Valve for Camshaft Adjustment

E87 - A/C Control Head
J220 - Motronic Engine Control Module (ECM)
N205 - Valve -1- for camshaft adjustment
N318 - Camshaft Adjustment Valve 1 (exhaust)
N323 - Ignition Coil 5 with Power Output Stage
N324 - Ignition Coil 6 with Power Output Stage
P - Spark Plug Connectors
Q - Spark Plugs
T10x - Connector 10 pin, orange, in E-box plenum chamber
T14i - Connector 14 pin, in engine compartment, left
T17 - Connector 17 pin, behind instrument panel

(131) - Ground connection -2-, in engine compartment wiring harness
(281) - Ground connector -1-, in wiring harness engine pre-wiring
(283) - Ground connection 2, in wiring harness engine pre-wiring
(D173) - Connection (87) (in engine compartment wiring harness)

ws = white
sw = black
ro = red
br = brown
gn = green
bl = blue
gr = grey
li = violet
ge = yellow
or = orange

USA.5703.02.21 Edition 04/03

© 2006 BentleyPublishers.com—All Rights Reserved

Audi TT — Wiring Diagram — No. 19/13

Motronic Engine Control Module, Engine Speed Sensor, Camshaft Position Sensor, Knock Sensor 1

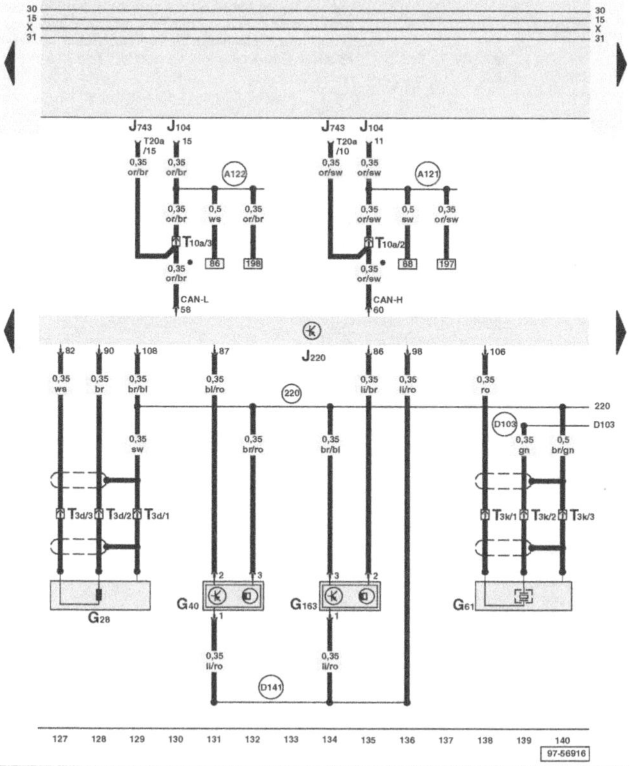

- G28 - Engine Speed (RPM) Sensor
- G40 - Camshaft Position (CMP) Sensor
- G61 - Knock Sensor (KS) 1
- G163 - Camshaft Position (CMP) Sensor 2
- J104 - ABS Control Module (w/EDL)
- J220 - Motronic Engine Control Module (ECM)
- J743 - Direct Shift Gearbox (DSG) Mechatronic
- T3d - Connector 3 pin, grey, in engine compartment
- T3k - Connector 3 pin, black, in engine compartment
- T10a - Connector 10 pin, white, in E-box plenum chamber
- T20a - Connector 20 pin, on Mechatronics for Direct Shift Gearbox (DSG)

- (220) - Ground connection (sensor ground), in engine compartment wiring harness
- (A121) - Connection (high bus) in instrument panel wiring harness
- (A122) - Connection (low bus) in instrument panel wiring harness
- (D103) - Wire connection -3-, in engine compartment wiring harness
- (D141) - Connector (5V), in wiring harness, engine pre-wiring

ws = white
sw = black
ro = red
br = brown
gn = green
bl = blue
gr = grey
li = violet
ge = yellow
or = orange

USA.5703.03.21 — Edition 10/03

Audi TT — Wiring Diagram — No. 19/14

Motronic Engine Control Module, Throttle Valve Control Module, Knock Sensor 2, Engine Coolant Temperature Sensor, Oil Level Thermal Sensor

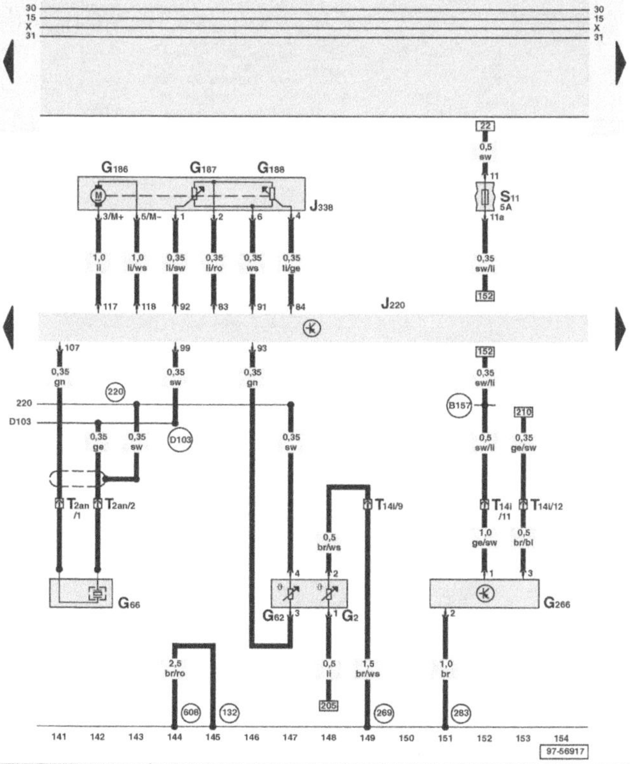

- G2 - Engine Coolant Temperature (ECT) Sensor
- G62 - Engine Coolant Temperature (ECT) Sensor
- G66 - Knock Sensor (KS) 2
- G186 - Throttle drive (power accelerator actuation)
- G187 - Angle sensor -1- for throttle drive (power accelerator actuation)
- G188 - Angle sensor -2- for throttle drive (power accelerator actuation)
- G266 - Oil Level Thermal Sensor (SRI)
- J220 - Motronic Engine Control Module (ECM)
- J338 - Throttle Valve Control Module
- S11 - Fuse
- T2an - Connector 2 pin, black, in engine compartment
- T14i - Connector 14 pin, in engine compartment, left

- (132) - Ground connection -3-, in engine compartment wiring harness
- (220) - Ground connection (sensor ground), in engine compartment wiring harness
- (269) - Ground connector (sensor ground) -1-, in instrument panel wiring harness
- (283) - Ground connection 2, in wiring harness engine pre-wiring
- (608) - Ground Connection (in center plenum chamber)
- (B157) - Plus connector -3- (15a), in wiring harness interior
- (D103) - Wire connection -3-, in engine compartment wiring harness

ws = white
sw = black
ro = red
br = brown
gn = green
bl = blue
gr = grey
li = violet
ge = yellow
or = orange

USA.5703.03.21 — Edition 10/03

© 2006 BentleyPublishers.com—All Rights Reserved

Engine Management, 3.2L, engine code: BHE, BPF (2004-2006)

Audi TT — Wiring Diagram — No. 19/15

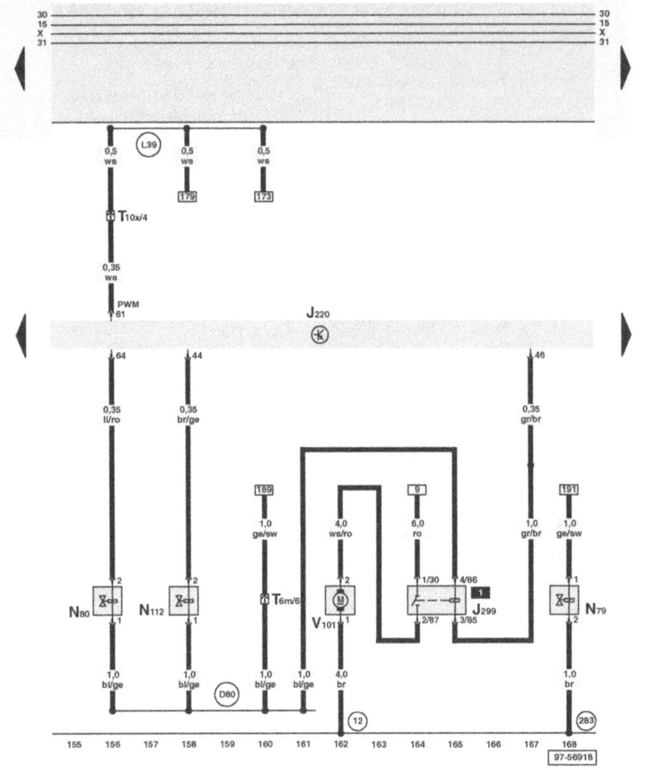

Motronic Engine Control Module, Secondary Air Injection Pump Relay, Secondary Air Injection Pump Motor, Evaporative Emission Canister Purge Regulator Valve, Secondary Air Injection Solenoid Valve, Positive Crankcase Ventilation Heating Element

- J220 - Motronic Engine Control Module (ECM)
- J299 - Secondary Air Injection (AIR) Pump Relay
- N79 - Positive Crankcase Ventilation (PCV) Heating Element
- N80 - Evaporative Emission (EVAP) Canister Purge Regulator Valve
- N112 - Secondary Air Injection (AIR) Solenoid Valve
- T6m - Connector 6 pin, brown, in E-box plenum chamber
- T10x - Connector 10 pin, orange, in E-box plenum chamber
- V101 - Secondary Air Injection (AIR) Pump Motor
- (12) - Ground connection, in engine compartment, left
- (283) - Ground connection 2, in wiring harness engine pre-wiring
- (D80) - Plus connection (87a- for EVAP system solenoid), in engine compartment wiring harness
- (L39) - Wire connection -1- (A/C Pressure Switch), in Climatronic wiring harness

ws = white
sw = black
ro = red
br = brown
gn = green
bl = blue
gr = grey
li = violet
ge = yellow
or = orange

USA.5703.02.21 — Edition 04/03

Audi TT — Wiring Diagram — No. 19/16

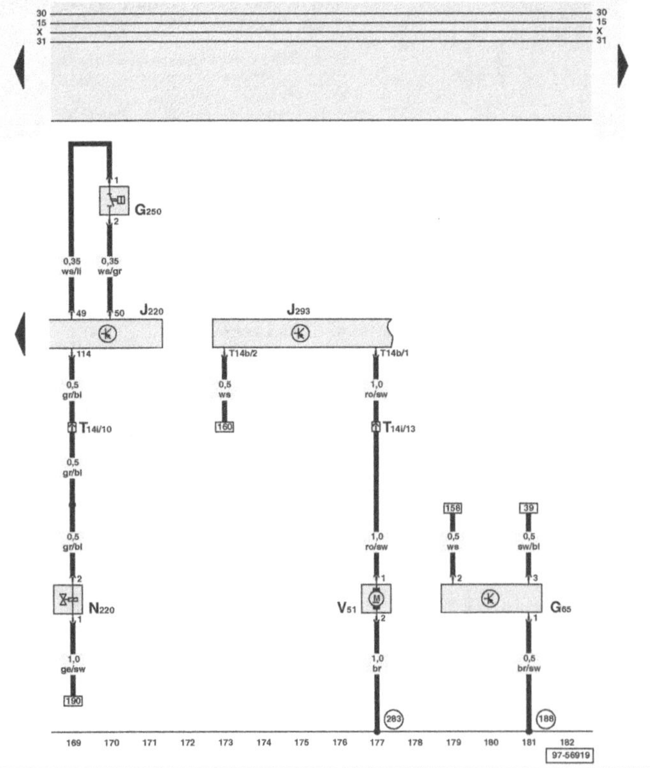

Motronic Engine Control Module, Valve for Exhaust Flap, Coolant FC (Fan Control) Control Module, After-Run Coolant Pump, High Pressure Sensor, Power Steering Sensor

- G65 - High Pressure Sensor
- G250 - Power steering sensor
- J220 - Motronic Engine Control Module (ECM)
- J293 - Coolant FC (Fan Control) Control Module, in engine compartment, left front
- N220 - Valve for exhaust flap
- T14b - Connector 14 pin, connector A on J293
- T14i - Connector 14 pin, in engine compartment, left
- V51 - After-Run Coolant Pump
- (188) - Ground connection -3-, in A/C wiring harness
- (283) - Ground connection 2, in wiring harness engine pre-wiring

ws = white
sw = black
ro = red
br = brown
gn = green
bl = blue
gr = grey
li = violet
ge = yellow
or = orange

USA.5703.02.21 — Edition 04/03

© 2006 BentleyPublishers.com—All Rights Reserved

Audi TT

Wiring Diagram

No. 19/17

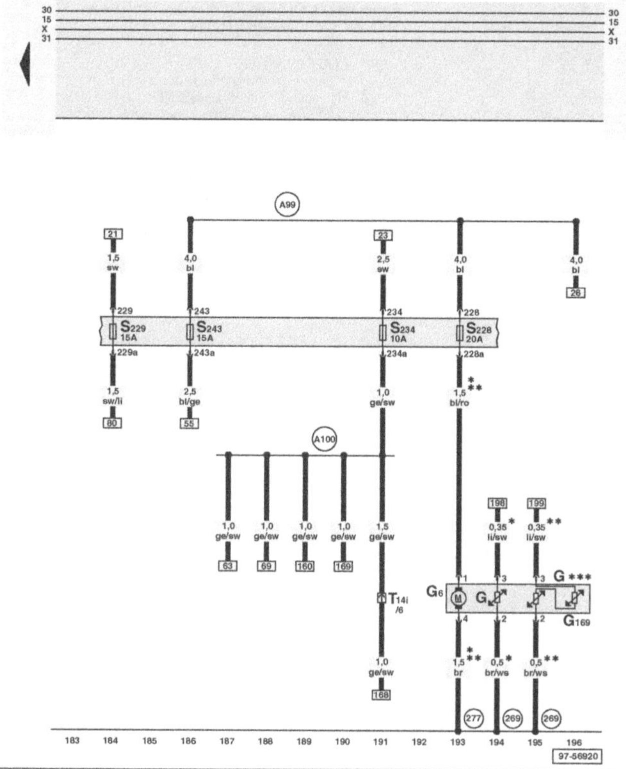

Fuses, Sender for Fuel Gauge, Fuel Pump

- G - Sender for fuel gauge
- G6 - Fuel Pump (FP)
- G169 - Fuel Level Sensor 2
- S228 - Fuse 28 in fuse holder
- S229 - Fuse 29 in fuse holder
- S234 - Fuse 34 in fuse holder
- S243 - Fuse 43 in fuse holder
- T14i - Connector 14 pin, in engine compartment, left

- (269) - Ground connector (sensor ground) -1-, in instrument panel wiring harness
- (277) - Ground connector -3-, in wiring harness interior
- (A99) - Connector -1- (87), in instrument panel wiring harness
- (A100) - Connector -2- (87), in instrument panel wiring harness

- * - Vehicles with Front Wheel Drive
- ** - Vehicles with quattro
- *** - Wiring inside Fuel Tank

ws = white
sw = black
ro = red
br = brown
gn = green
bl = blue
gr = grey
li = violet
ge = yellow
or = orange

USA.5703.02.21 **Edition 04/03**

Audi TT

Wiring Diagram

No. 19/18

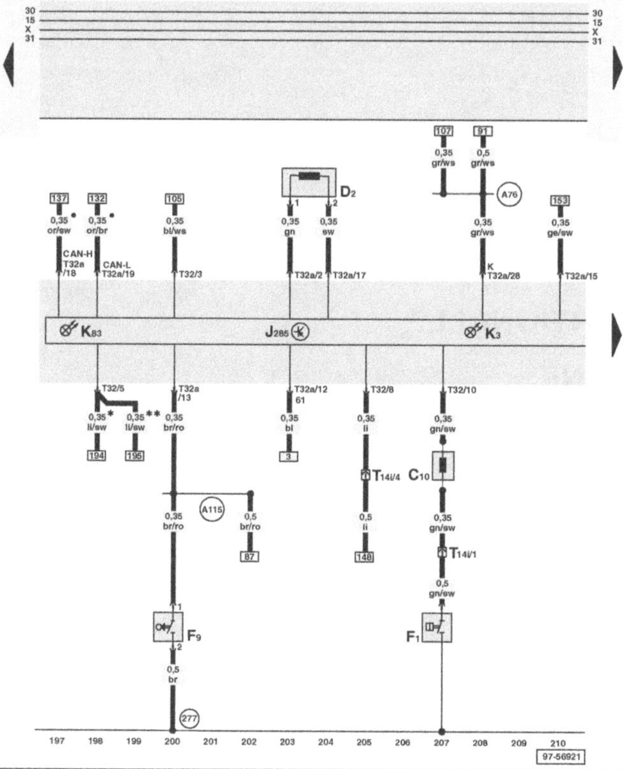

Control Module with Indicator Unit in Instrument Panel Insert, Parking Brake Warning Light Switch, Oil Pressure Switch, Induction Coil of Antitheft Immobilizer

- C10 - Suppressor Filter
- D2 - Induction coil of antitheft immobilizer
- F1 - Oil Pressure Switch
- F9 - Parking Brake Warning Light Switch
- J285 - Control Module with indicator unit in instrument panel insert
- K3 - Oil Pressure Warning Light
- K83 - Malfunction Indicator Lamp (MIL)
- T14i - Connector 14 pin, in engine compartment, left
- T32 - Connector 32 pin, blue, on instrument cluster
- T32a - Connector 32 pin, green, on instrument cluster

- (277) - Ground connector -3-, in wiring harness interior
- (A76) - Connector (K-diagnosis wire), in instrument panel wiring harness
- (A115) - Connection (parking brake control) in instrument panel wiring harness

- • - CAN-Bus (Data-Bus)
- * - Vehicles with Front Wheel Drive
- ** - Vehicles with quattro

ws = white
sw = black
ro = red
br = brown
gn = green
bl = blue
gr = grey
li = violet
ge = yellow
or = orange

USA.5703.02.21 **Edition 04/03**

© 2006 BentleyPublishers.com—All Rights Reserved

Audi TT Wiring Diagram **No. 19/19**

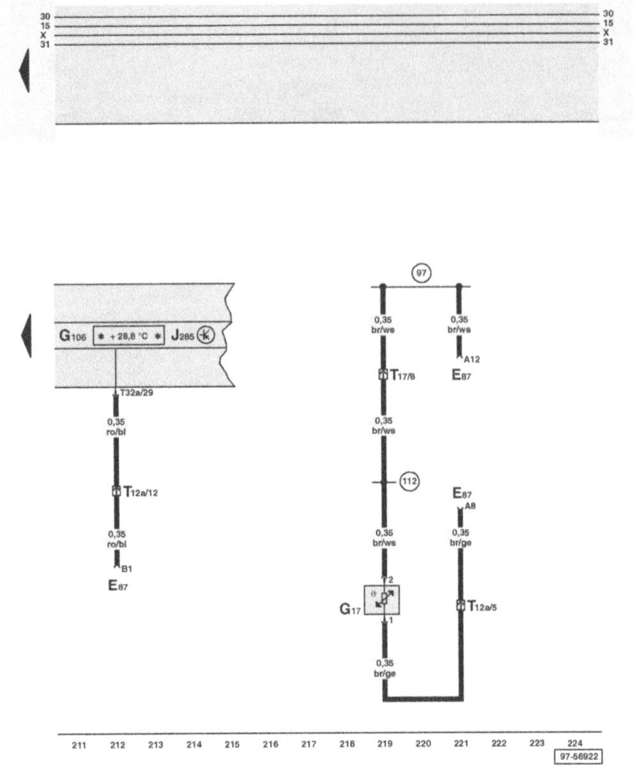

Control Module with Indicator Unit in Instrument Panel Insert, Outside Air Temperature Sensor, Outside Air Temperature Display

- E87 - A/C Control Head
- G17 - Outside Air Temperature Sensor
- G106 - Outside Air Temperature Display
- J285 - Control Module with indicator unit in instrument panel insert
- T12a - Connector 12 pin, behind instrument panel, right
- T17 - Connector 17 pin, behind instrument panel
- T32a - Connector 32 pin, green, on instrument cluster
- (97) - Ground connection -1-, in A/C wiring harness
- (112) - Ground connection -2-, in A/C wiring harness

ws = white
sw = black
ro = red
br = brown
gn = green
bl = blue
gr = grey
li = violet
ge = yellow
or = orange

USA.5703.02.21 Edition 04/03

THIS SPACE INTENTIONALLY LEFT BLANK

© 2006 BentleyPublishers.com—All Rights Reserved

Direct Shift Gearbox (DSG) 02E

2004 m.y.

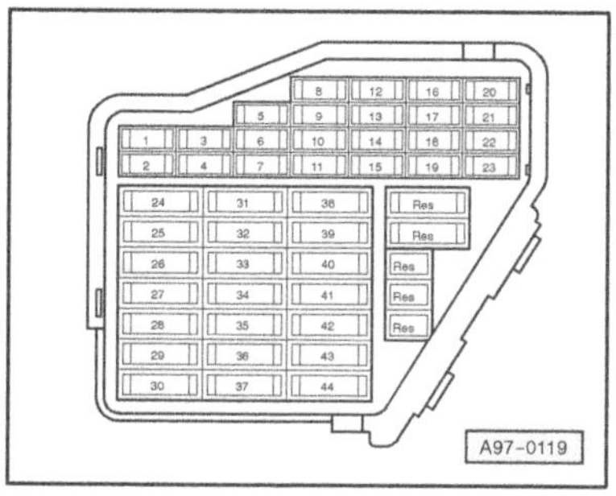

Fuse Panel

Fuse Colors:
- 30 A - Green
- 25 A - White
- 20 A - Yellow
- 15 A - Blue
- 10 A - Red
- 7,5 A - Brown
- 5 A - Beige
- 1 A - Black

Starting with fuse position 23, fuses in the fuse holder are identified with 223 in the wiring diagram.

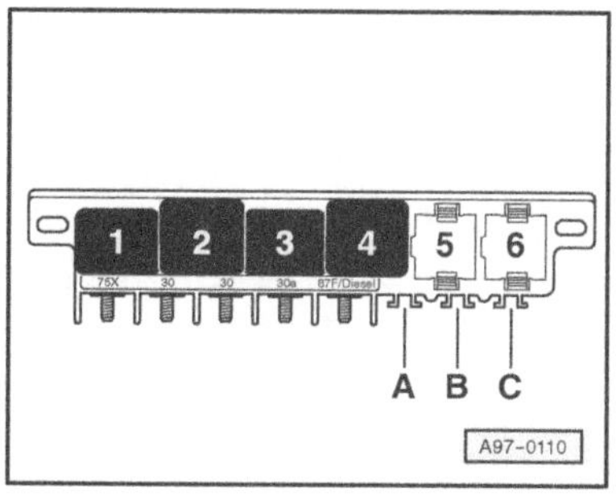

Micro Central Electric Panel

Relay Location:
- B - Hydraulic Pump Relay Fuse, S279

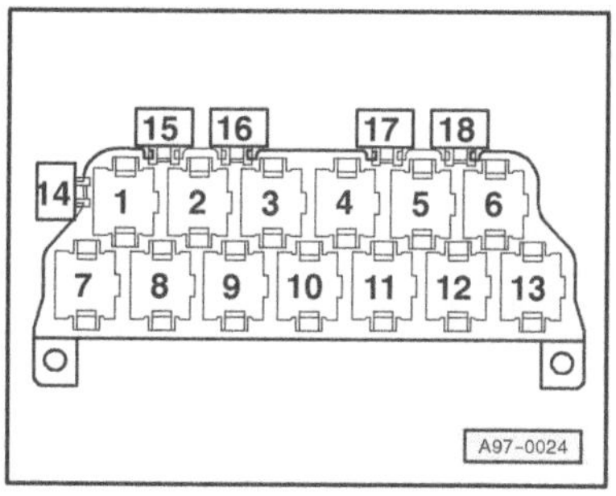

13 - Fold Relay Panel

Relay Location:
- 11 - Starting Interlock Relay, J207
- 13 - Back-Up Light Relay, J219

© 2006 BentleyPublishers.com—All Rights Reserved

Audi TT

Wiring Diagram

No. 20/3

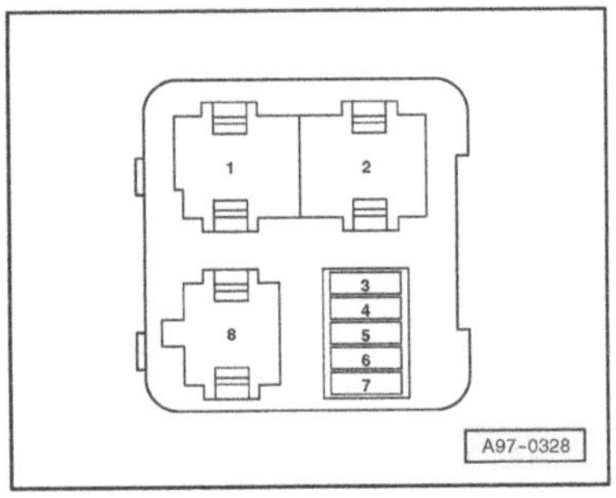

W-Box in engine compartment

Fuses Location:

7 - Fuse for Control Module - Automatic Transmission, S113

Audi TT

Wiring Diagram

No. 20/4

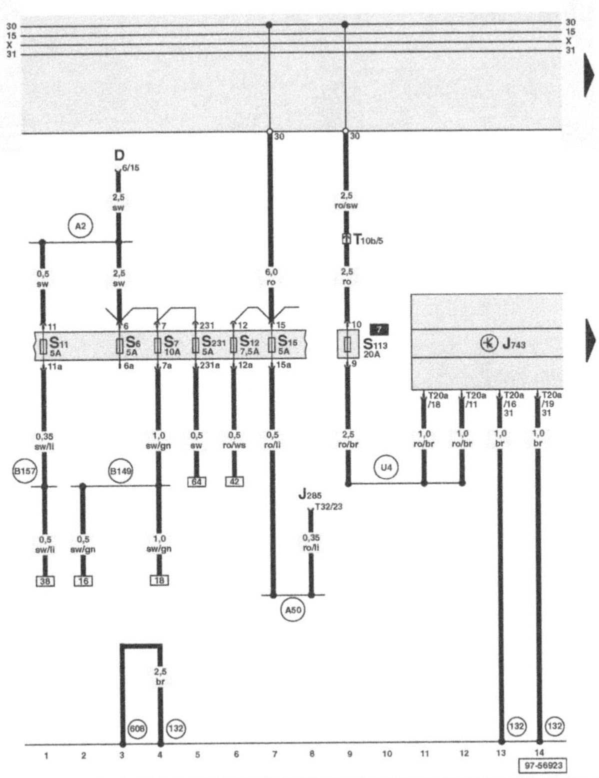

Direct Shift Gearbox (DSG) Mechatronic, Fuse for Control Module - Automatic Transmission, Fuses

- D - Ignition/Starter Switch
- J285 - Control Module with indicator unit in instrument panel insert
- J743 - Direct Shift Gearbox (DSG) Mechatronic
- S6 - Fuse
- S7 - Fuse
- S11 - Fuse
- S12 - Fuse
- S15 - Fuse
- S113 - Fuse for control module - Automatic Transmission
- S231 - Fuse 31 in fuse holder
- T10b - Connector 10 pin, grey, in E-box plenum chamber
- T20a - Connector 20 pin, on Mechatronics for Direct Shift Gearbox (DSG)
- T32 - Connector 32 pin, blue, on instrument cluster
- (132) - Ground connection -3-, in engine compartment wiring harness
- (608) - Ground Connection (in center plenum chamber)
- (A2) - Plus connection (15), in instrument panel wiring harness
- (A50) - Plus connection (30as), in instrument panel wiring harness
- (B149) - Plus connector -2- (15a), in wiring harness interior
- (B157) - Plus connector -3- (15a), in wiring harness interior
- (U4) - Wire connection (30), in automatic transmission wiring harness

ws = white
sw = black
ro = red
br = brown
gn = green
bl = blue
gr = grey
li = violet
ge = yellow
or = orange

© 2006 BentleyPublishers.com—All Rights Reserved

Audi TT

Wiring Diagram

No. 20/5

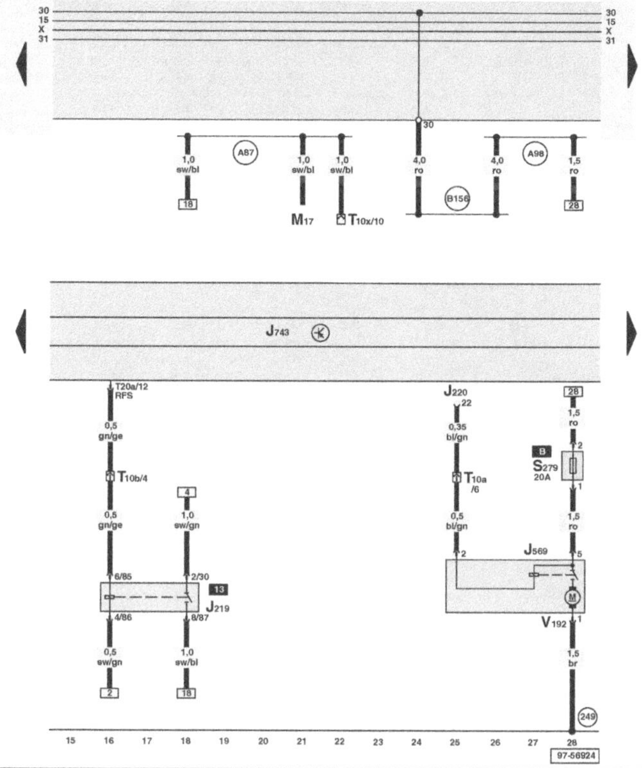

Direct Shift Gearbox (DSG) Mechatronic, Back-Up Light Relay, Brake System Vacuum Pump, Brake Booster Relay

- J219 - Back-Up Light Relay
- J220 - Motronic Engine Control Module (ECM)
- J569 - Brake Booster relay
- J743 - Direct Shift Gearbox (DSG) Mechatronic
- M17 - Right Back-Up Light
- S279 - Hydraulic Pump Relay Fuse
- T10a - Connector 10 pin, white, in E-box plenum chamber
- T10b - Connector 10 pin, grey, in E-box plenum chamber
- T10x - Connector 10 pin, orange, in E-box plenum chamber
- T20a - Connector 20 pin, on Mechatronics for Direct Shift Gearbox (DSG)
- V192 - Brake System Vacuum Pump
- (249) - Ground connection -2-, in wiring harness interior
- (A87) - Connector (RF), in instrument panel wiring harness
- (A98) - Plus connector -4- (30), in instrument panel wiring harness
- (B156) - Plus connector (30a), in wiring harness interior

ws = white
sw = black
ro = red
br = brown
gn = green
bl = blue
gr = grey
li = violet
ge = yellow
or = orange

USA.5703.03.21 | Edition 10/03

Audi TT

Wiring Diagram

No. 20/6

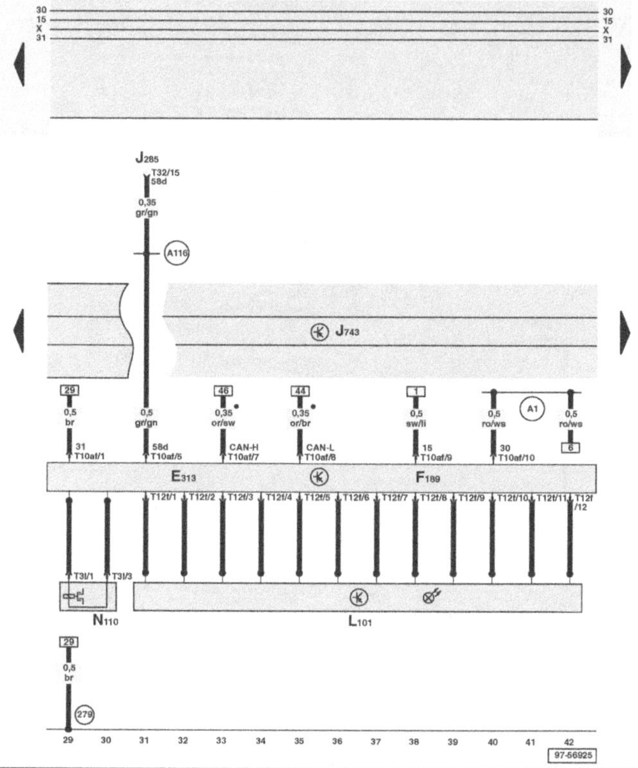

Direct Shift Gearbox (DSG) Mechatronic, Shift Lock Solenoid, Selector Lever, Tiptronic Switch, Illumination for Selector Lever Scale

- E313 - Selector Lever
- F189 - Tiptronic Switch
- J285 - Control Module with indicator unit in instrument panel insert
- J743 - Direct Shift Gearbox (DSG) Mechatronic
- L101 - Illumination for selector lever scale
- N110 - Shift Lock Solenoid
- T3l - Connector 3 pin, on Shift Lock Solenoid
- T10af - Connector 10 pin, on Selector Lever
- T12f - Connector 12 pin, on Selector Lever
- (279) - Ground connector -5-, in wiring harness interior
- (A1) - Plus connection (30a), in instrument panel wiring harness
- (A116) - Connection -2- (58d) in instrument panel wiring harness
- • - CAN-Bus (Data-Bus)

ws = white
sw = black
ro = red
br = brown
gn = green
bl = blue
gr = grey
li = violet
ge = yellow
or = orange

USA.5703.03.21 | Edition 10/03

© 2006 BentleyPublishers.com—All Rights Reserved

DSG Transmission, code: 02E (2004-2006)

Audi TT — Wiring Diagram — No. 20/7

Direct Shift Gearbox (DSG) Mechatronic, CAN-Bus Wiring

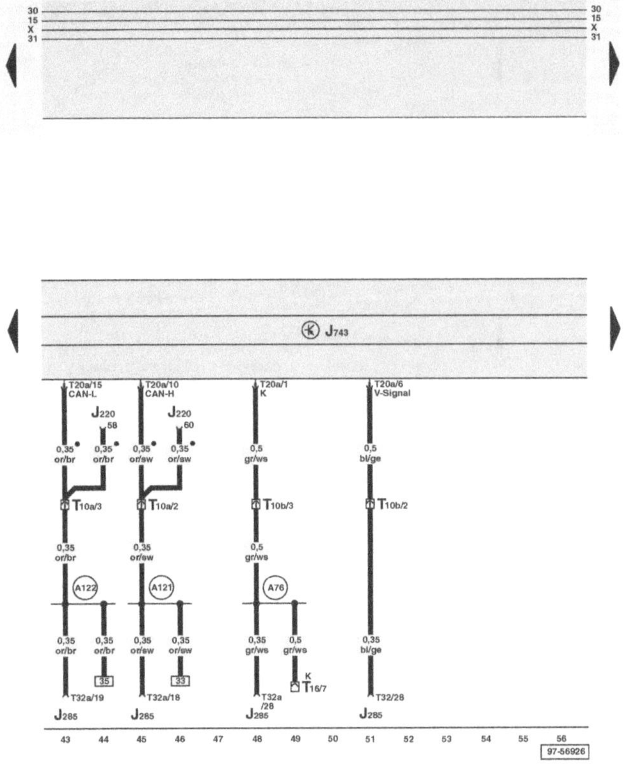

J220	- Motronic Engine Control Module (ECM)
J285	- Control Module with indicator unit in instrument panel insert
J743	- Direct Shift Gearbox (DSG) Mechatronic
T10a	- Connector 10 pin, white, in E-box plenum chamber
T10b	- Connector 10 pin, grey, in E-box plenum chamber
T16	- Connector 16 pin, black, Data Link Connector (DLC), below storage compartment, driver's side
T20a	- Connector 20 pin, on Mechatronics for Direct Shift Gearbox (DSG)
T32	- Connector 32 pin, blue, on instrument cluster
T32a	- Connector 32 pin, green, on instrument cluster
(A76)	- Connector (K-diagnosis wire), in instrument panel wiring harness
(A121)	- Connection (high bus) in instrument panel wiring harness
(A122)	- Connection (low bus) in instrument panel wiring harness
•	- CAN-Bus (Data-Bus)

ws = white
sw = black
ro = red
br = brown
gn = green
bl = blue
gr = grey
li = violet
ge = yellow
or = orange

USA.5703.03.21 — Edition 10/03

Audi TT — Wiring Diagram — No. 20/8

Direct Shift Gearbox (DSG) Mechatronic, Starting Interlock Relay

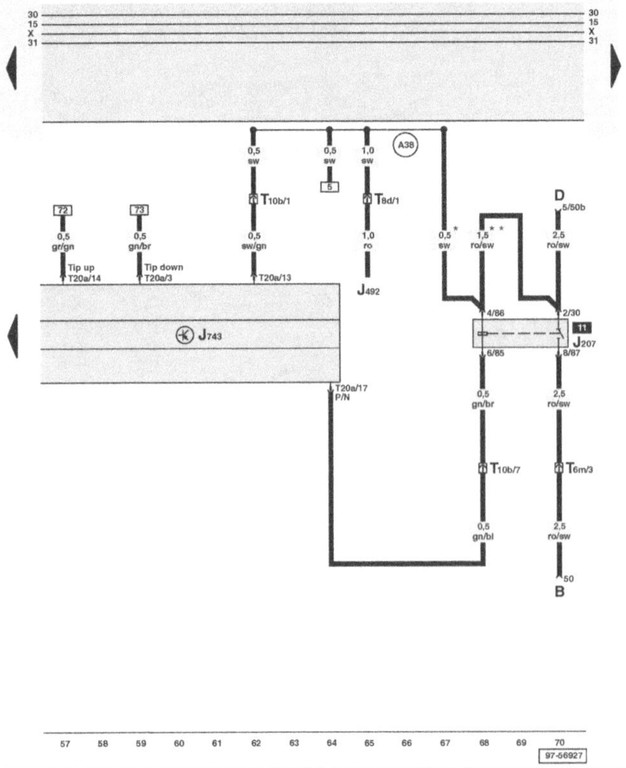

B	- Starter
D	- Ignition/Starter Switch
J207	- Starting Interlock Relay
J217	- Transmission Control Module (TCM)
J492	- All-wheel drive control module
J743	- Direct Shift Gearbox (DSG) Mechatronic
T6m	- Connector 6 pin, brown, in E-box plenum chamber
T8d	- Connector 8 pin, on Control module for all wheel drive
T10b	- Connector 10 pin, grey, in E-box plenum chamber
T20a	- Connector 20 pin, on Mechatronics for Direct Shift Gearbox (DSG)
(A38)	- Plus connection -2- (15a), in instrument panel wiring harness
*	- through 2003 m.y.
**	- from 2004 m.y.

ws = white
sw = black
ro = red
br = brown
gn = green
bl = blue
gr = grey
li = violet
ge = yellow
or = orange

USA.5703.03.21 — Edition 10/03

© 2006 BentleyPublishers.com—All Rights Reserved

Audi TT **Wiring Diagram** **No. 20/9**

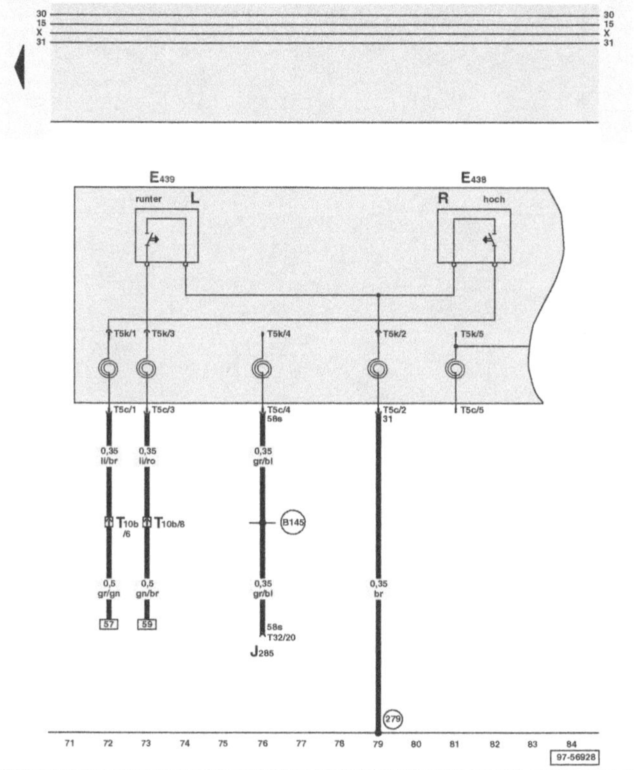

Tiptronic Upshift and Downshift Button in Steering Wheel

E438 - Tiptronic Upshift Button (in steering wheel)
E439 - Tiptronic Downshift Buttons (on steering wheel)
J285 - Control Module with indicator unit in instrument panel insert
T5c - Connector 5 pin, yellow, in steering wheel
T5k - Connector 5 pin, yellow, on steering column
T10b - Connector 10 pin, grey, in E-box plenum chamber
T32 - Connector 32 pin, blue, on instrument cluster
(279) - Ground connector -5-, in wiring harness interior
(B145) - Plus connector (58b), in wiring harness interior

ws = white
sw = black
ro = red
br = brown
gn = green
bl = blue
gr = grey
li = violet
ge = yellow
or = orange

USA.5703.02.21 Edition 04/03

THIS SPACE INTENTIONALLY LEFT BLANK

© 2006 BentleyPublishers.com—All Rights Reserved

Audi TT Wiring Diagram **No. 21/1**

Cellular Telephone - Preparation

2003 m.y.

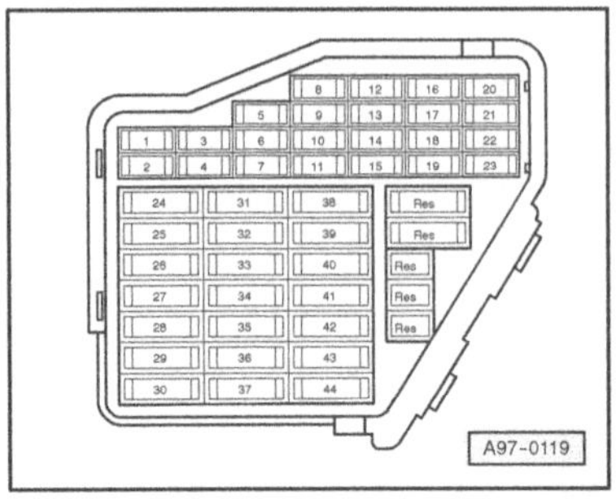

Fuse Panel

Fuse Colors:

30 A - Green
25 A - White
20 A - Yellow
15 A - Blue
10 A - Red
7,5 A - Brown
5 A - Beige
1 A - Black

Starting with fuse position 23, fuses in the fuse holder are identified with 223 in the wiring diagram.

Audi TT Wiring Diagram **No. 21/2**

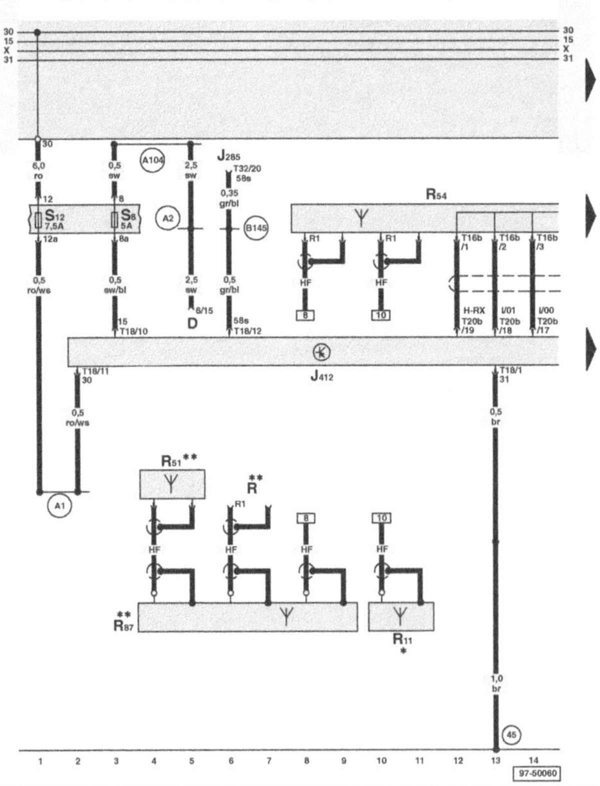

ws = white
sw = black
ro = red
br = brown
gn = green
bl = blue
gr = grey
li = violet
ge = yellow
or = orange

Cellular Telephone - Preparation

D - Ignition/Starter Switch
J285 - Control Module with indicator unit in instrument panel insert
J412 - Operating electronics and Telephone Control Module
R - Radio
R1 - Antenna Connection
R11 - Antenna
R51 - Radio/Telephone Antenna
R54 - Cellular telephone
R87 - Antenna Splitter
S8 - Fuse
S12 - Fuse
T16b - Connector 16 pin, on Cellular Telephone
T18 - Connector 18 pin, on Operating Electronics and Telephone Control Module
T20b - Connector 20 pin, on Operating Electronics and Telephone Control Module
T32 - Connector 32 pin, blue, on instrument cluster

(45) - Ground connection, behind instrument panel, center
(A1) - Plus connection (30a), in instrument panel wiring harness
(A2) - Plus connection (15), in instrument panel wiring harness
(A104) - Plus connector -2- (15), in instrument panel wiring harness
(B145) - Plus connector (58b), in wiring harness interior

* - Coupé, roof antenna
** - Roadster, rod antenna

© 2006 BentleyPublishers.com—All Rights Reserved

Audi TT **Wiring Diagram** **No. 21/3**

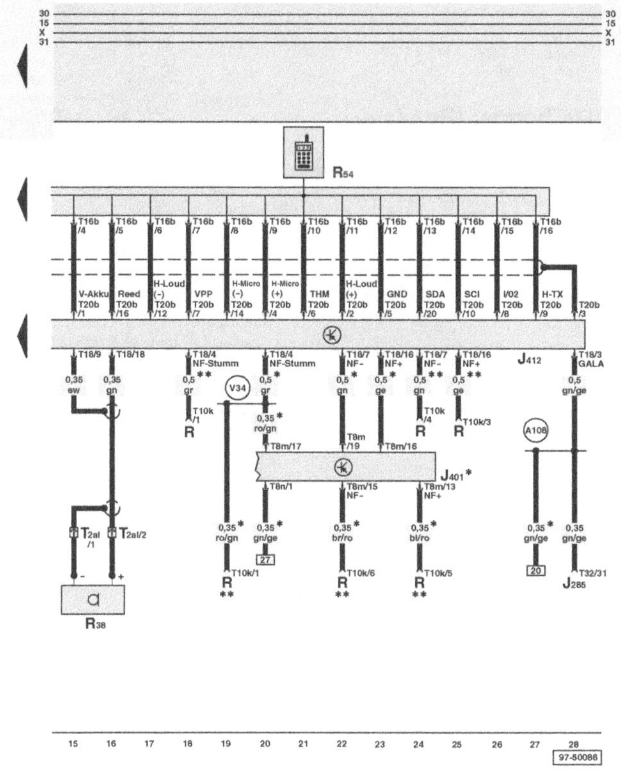

Cellular Telephone - Preparation

J285 - Control Module with indicator unit in instrument panel insert
J401 - Control unit for navigation with CD-mechanism
J412 - Operating electronics and Telephone Control Module
R - Radio
R38 - Telephone Microphone
R54 - Cellular telephone
T2al - Connector 2 pin, behind instrument cluster
T8m - Connector 8 pin, C III on Control Module for Navigation
T8n - Connector 8 pin, black, connector A on Control Module for Navigation
T10k - Connector 10 pin, red, on radio
T16b - Connector 16 pin, on Cellular Telephone
T18 - Connector 18 pin, on Operating Electronics and Telephone Control Module
T20b - Connector 20 pin, on Operating Electronics and Telephone Control Module
T32 - Connector 32 pin, blue, on instrument cluster

(A108) - Connector (speed signal), in instrument panel wiring harness
(V34) - Connector (mute button), in wiring harness telephone

* - Vehicles with Radio and Navigation
** - Vehicles with Radio

ws = white
sw = black
ro = red
br = brown
gn = green
bl = blue
gr = grey
li = violet
ge = yellow
or = orange

USA.5703.02.21 Edition 04/03

THIS SPACE INTENTIONALLY

LEFT BLANK

©2006 BentleyPublishers.com—All Rights Reserved

Sound System, Bose satellite, roadster (2005-2006)

Audi TT Wiring Diagram **No. 22/1**

Radio, Satellite Radio - Bose with and without CD-Changer (Roadster)

2005 m.y.

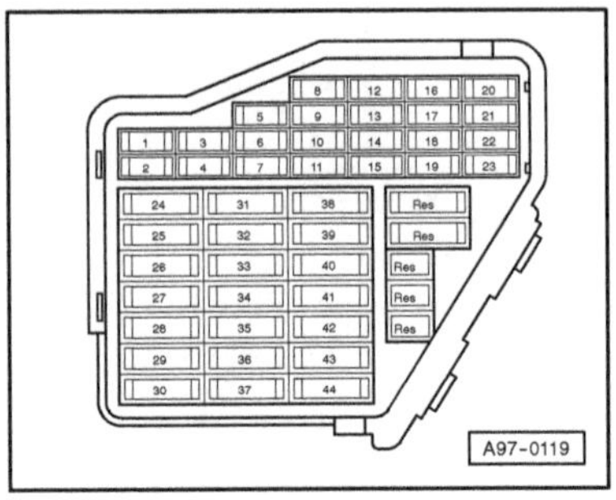

Fuse Panel

Fuse Colors:
- 30 A - Green
- 25 A - White
- 20 A - Yellow
- 15 A - Blue
- 10 A - Red
- 7,5 A - Brown
- 5 A - Beige
- 1 A - Black

Starting with fuse position 23, fuses in the fuse holder are identified with 223 in the wiring diagram.

USA.5703.04.21 Edition 06.04

Audi TT Wiring Diagram **No. 22/2**

Radio, Fuses

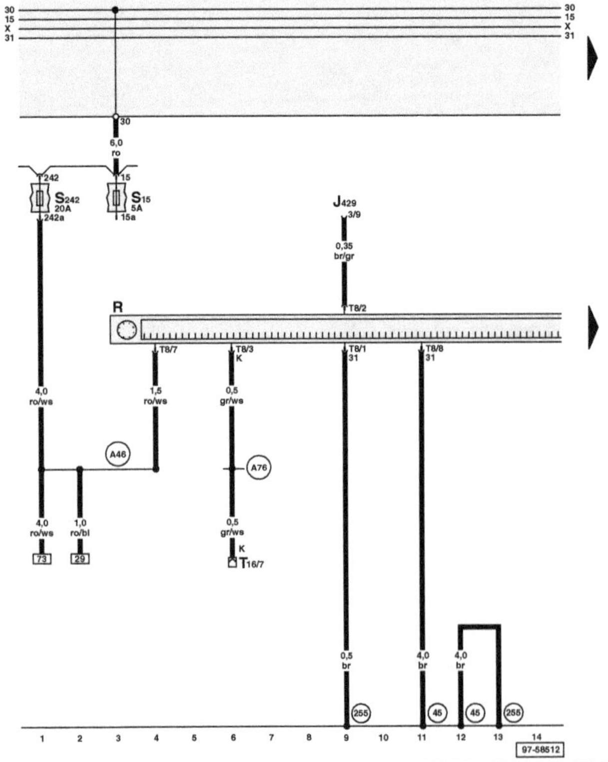

- J429 - Control module for central locking
- R - Radio
- S15 - Fuse
- S242 - Fuse 42 in fuse holder
- T8 - Connector 8 pin, black, connector radio III
- T16 - Connector 16 pin, black, Data Link Connector (DLC), below storage compartment, driver's side
- (45) - Ground connection, behind instrument panel, center
- (255) - Ground connection, -1-, in wiring harness Radio
- (A46) - Plus connection (30-from radio), in instrument panel wiring harness
- (A76) - Connector (K-diagnosis wire), in instrument panel wiring harness

ws = white
sw = black
ro = red
br = brown
gn = green
bl = blue
gr = grey
li = violet
ge = yellow
or = orange

USA.5703.04.21 Edition 06.04

© 2006 BentleyPublishers.com—All Rights Reserved

Audi TT — Wiring Diagram — No. 22/3

Radio, CD Changer Unit

- R - Radio
- R41 - CD Changer Unit
- T6k - Connector 6 pin, green, connector radio I-2
- T8c - Connector 8 pin, blue, connector radio I-3
- V14 - Wire connection (shielding), in CD-changer wiring harness
- * - Vehicles with CD Changer Unit

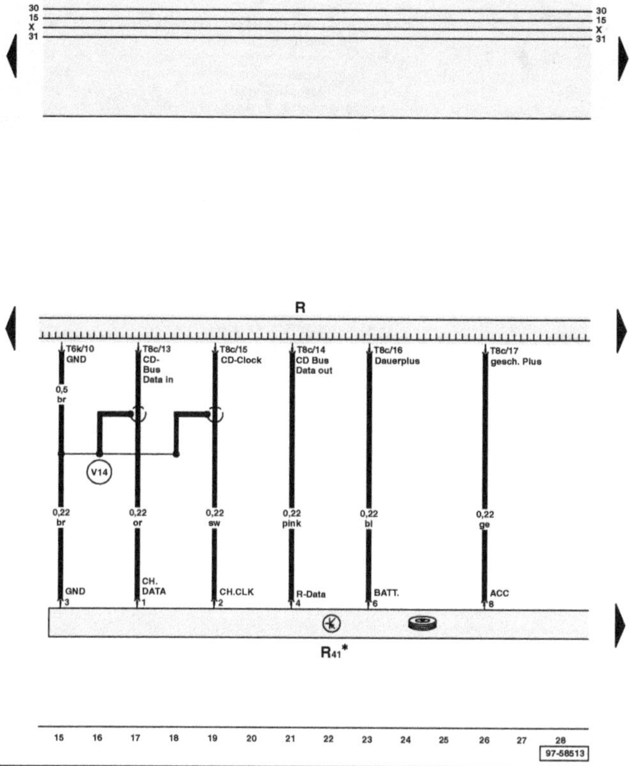

ws = white
sw = black
ro = red
br = brown
gn = green
bl = blue
gr = grey
li = violet
ge = yellow
or = orange

USA.5703.04.21 Edition 06.04

Audi TT — Wiring Diagram — No. 22/4

Radio, Satellite Radio, CD Changer Unit

- R - Radio
- R41 - CD Changer Unit
- R146 - Satellite Radio
- T8c - Connector 8 pin, blue, connector radio I-3
- T8g - Connector 8 pin, Connector 2, on Satellite Radio
- T8h - Connector 8 pin, Connector 1, on Satellite Radio
- 45 - Ground connection, behind instrument panel, center
- 50 - Ground connection, in luggage compartment, left
- V45 - Connection 2 (shielding) (in CD changer wiring harness)
- * - Vehicles without Satellite Radio to bridge with counter plug
- ** - Vehicles with CD Changer Unit

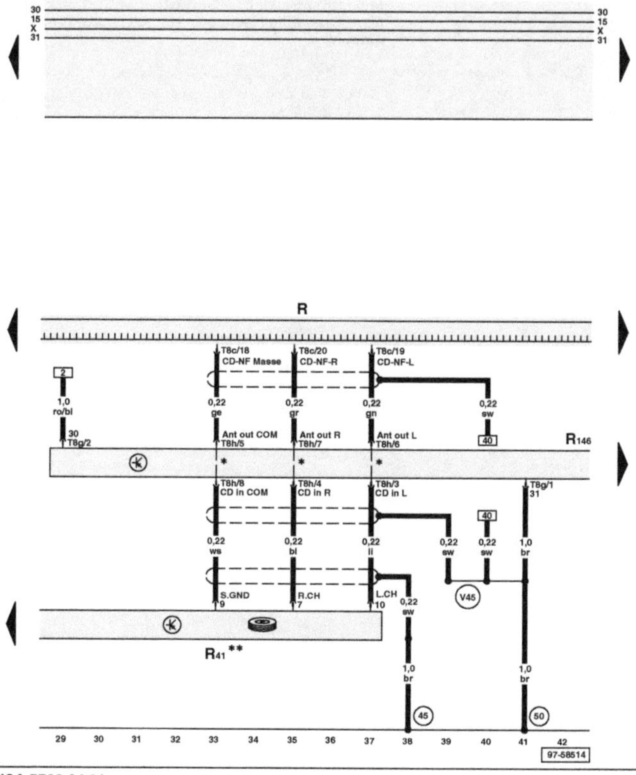

ws = white
sw = black
ro = red
br = brown
gn = green
bl = blue
gr = grey
li = violet
ge = yellow
or = orange

USA.5703.04.21 Edition 06.04

© 2006 BentleyPublishers.com—All Rights Reserved

Sound System, Bose satellite, roadster (2005-2006)

Audi TT — Wiring Diagram — No. 22/5

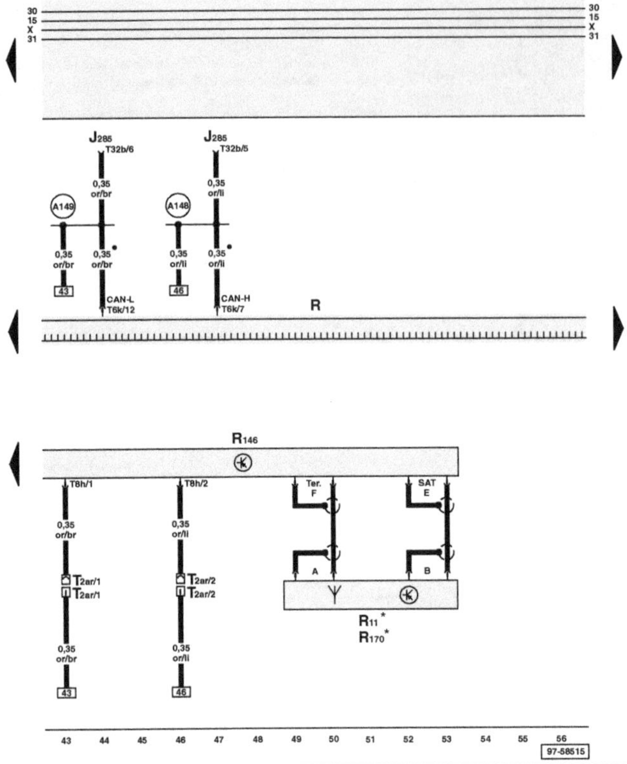

Radio, Satellite Radio, Antenna

- J285 - Control Module with indicator unit in instrument panel insert
- R - Radio
- R11 - Antenna
- R146 - Satellite Radio
- R170 - Satellite Antenna
- T2ar - Connector 2 pin, CAN-Bus Disconnection plug, A-pillar left
- T6k - Connector 6 pin, green, connector radio I-2
- T8h - Connector 8 pin, Connector 1, on Satellite Radio
- T32b - Connector 32 pin, grey, on instrument cluster
- (A148) - Connection (high bus, navigation), in instrument panel wiring harness
- (A149) - Connection (low bus, navigation), in instrument panel wiring harness
- • - CAN-Bus (Data-Bus)
- * - on rear lid

ws = white
sw = black
ro = red
br = brown
gn = green
bl = blue
gr = grey
li = violet
ge = yellow
or = orange

USA.5703.04.21 — Edition 06.04

Audi TT — Wiring Diagram — No. 22/6

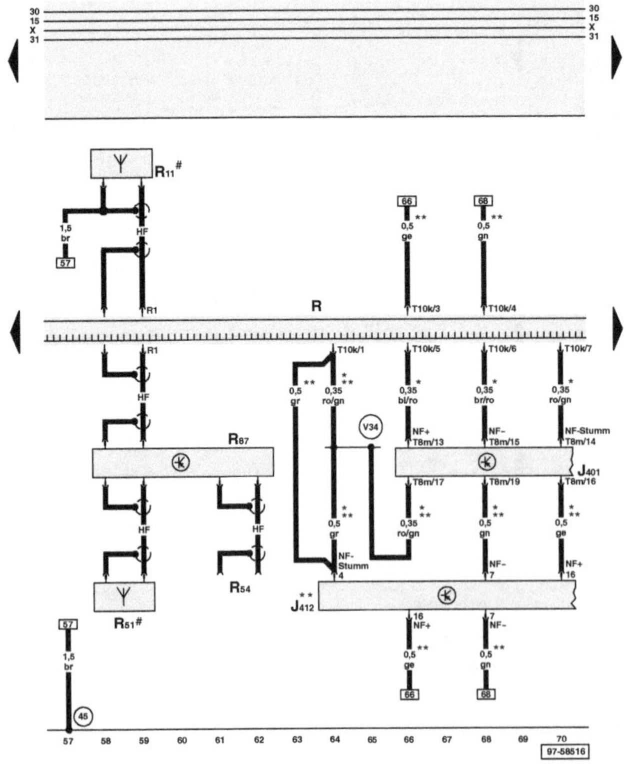

Radio, Antenna Splitter, Antenna, Cellular Telephone, Control Unit for Navigation with CD-mechanism, Operating Electronics and Telephone Control Module

- J401 - Control unit for navigation with CD-mechanism
- J412 - Operating electronics and Telephone Control Module
- R - Radio
- R11 - Antenna
- R51 - Radio/Telephone Antenna
- R54 - Cellular telephone
- R87 - Antenna Splitter
- T8m - Connector 8 pin, C III on Control Module for Navigation
- T10k - Connector 10 pin, red, on radio
- (45) - Ground connection, behind instrument panel, center
- (V34) - Connector (mute button), in wiring harness telephone
- * - Vehicles with Navigation
- ** - Vehicles with Telephone and without Navigation
- # - Vehicles with Rod Antenna

ws = white
sw = black
ro = red
br = brown
gn = green
bl = blue
gr = grey
li = violet
ge = yellow
or = orange

USA.5703.04.21 — Edition 06.04

© 2006 BentleyPublishers.com—All Rights Reserved

Audi TT — Wiring Diagram — No. 22/7

Radio, Amplifier, Convertible Top Front Lock Switch

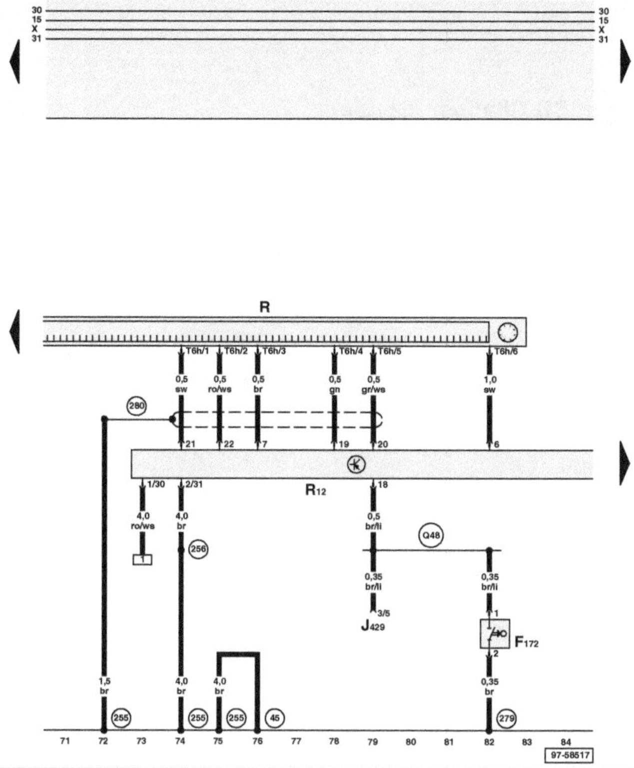

- F172 - Convertible Top Front Lock Switch
- J429 - Control module for central locking
- R - Radio
- R12 - Amplifier
- T6h - Connector 6 pin, yellow, connector radio I-1
- (45) - Ground connection, behind instrument panel, center
- (255) - Ground connection, -1-, in wiring harness Radio
- (256) - Ground connection, -2-, in wiring harness Radio
- (279) - Ground connector -5-, in wiring harness interior
- (280) - Ground connector -2-, in wiring harness engine compartment
- (Q48) - Connection (open), in wiring harness interior

ws = white
sw = black
ro = red
br = brown
gn = green
bl = blue
gr = grey
li = violet
ge = yellow
or = orange

USA.5703.04.21 Edition 06.04

Audi TT — Wiring Diagram — No. 22/8

Amplifier, Treble Speaker, Bass Speaker, Inside Loudspeaker, Inside Microphone

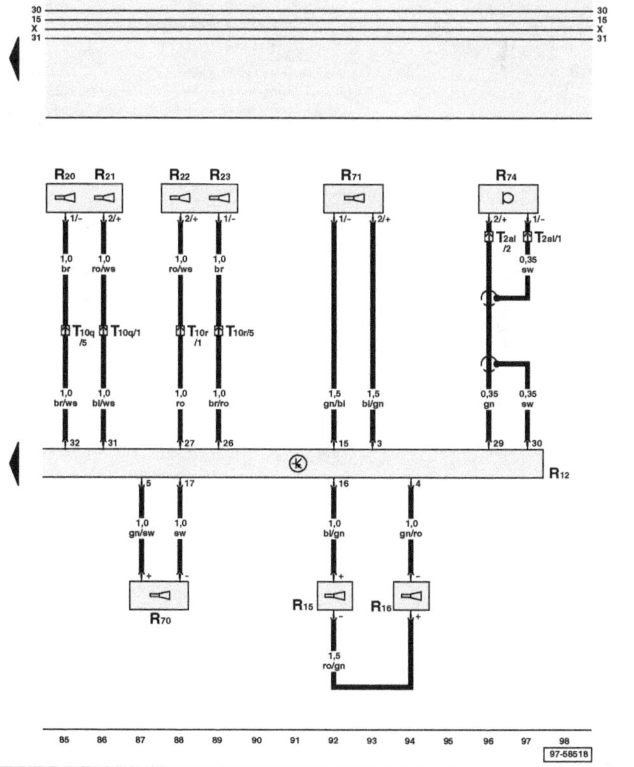

- R12 - Amplifier
- R15 - Left Rear Bass Speaker
- R16 - Right Rear Treble Speaker
- R20 - Left Front Treble Speaker
- R21 - Left Front Bass Speaker
- R22 - Right Front Treble Speaker
- R23 - Right Front Bass Speaker
- R70 - Inside Loudspeaker
- R71 - Inside Loudspeaker 2
- R74 - Inside Microphone
- T2al - Connector 2 pin, behind instrument cluster
- T10q - Connector 10 pin, green, connector station A pillar, left
- T10r - Connector 10 pin, green, connector station A pillar, right

ws = white
sw = black
ro = red
br = brown
gn = green
bl = blue
gr = grey
li = violet
ge = yellow
or = orange

USA.5703.04.21 Edition 06.04

© 2006 BentleyPublishers.com—All Rights Reserved

Audi TT | **Wiring Diagram** | **No. 23/1**

Radio, Satellite Radio - Bose , with and without CD-Changer (Coupé)

2005 m.y.

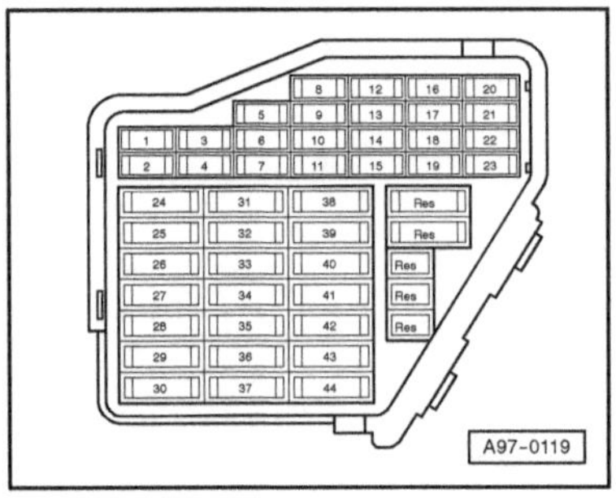

Fuse Panel

Fuse Colors:
30 A - Green
25 A - White
20 A - Yellow
15 A - Blue
10 A - Red
7,5 A - Brown
5 A - Beige
1 A - Black

Starting with fuse position 23, fuses in the fuse holder are identified with 223 in the wiring diagram.

USA.5703.04.21 | **Edition 06.04**

Audi TT | **Wiring Diagram** | **No. 23/2**

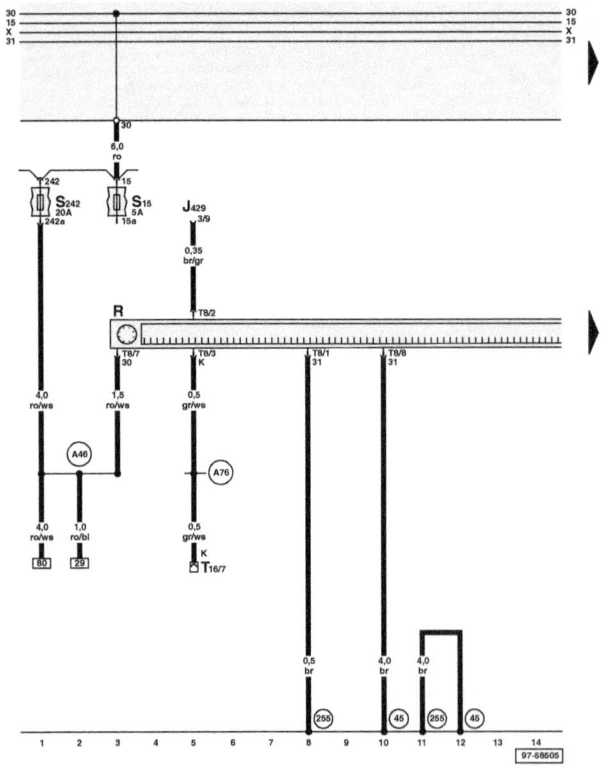

Radio, Fuses

J429 - Control module for central locking
R - Radio
S15 - Fuse
S242 - Fuse 42 in fuse holder
T8 - Connector 8 pin, black, connector radio III
T16 - Connector 16 pin, black, Data Link Connector (DLC), below storage compartment, driver's side

(45) - Ground connection, behind instrument panel, center
(255) - Ground connection, -1-, in wiring harness Radio
(A46) - Plus connection (30-from radio), in instrument panel wiring harness
(A76) - Connector (K-diagnosis wire), in instrument panel wiring harness

ws = white
sw = black
ro = red
br = brown
gn = green
bl = blue
gr = grey
li = violet
ge = yellow
or = orange

USA.5703.04.21 | **Edition 06.04**

© 2006 BentleyPublishers.com—All Rights Reserved

Audi TT

Wiring Diagram

No. 23/3

Radio, CD Changer Unit

R - Radio
R41 - CD Changer Unit
T6k - Connector 6 pin, green, connector radio I-2
T8c - Connector 8 pin, blue, connector radio I-3

(V14) - Wire connection (shielding), in CD-changer wiring harness

* - Vehicles with CD Changer Unit

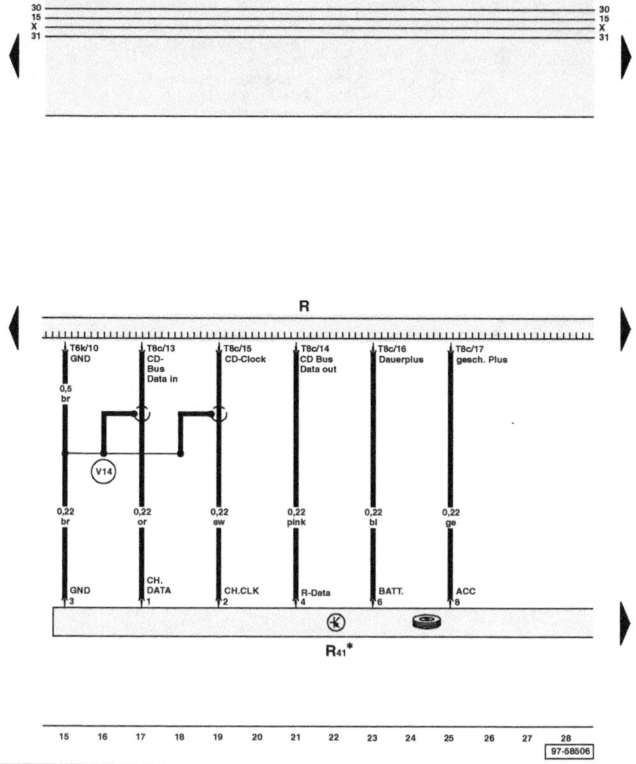

ws = white
sw = black
ro = red
br = brown
gn = green
bl = blue
gr = grey
li = violet
ge = yellow
or = orange

Audi TT

Wiring Diagram

No. 23/4

Radio, Satellite Radio, CD Changer Unit

R - Radio
R41 - CD Changer Unit
R146 - Satellite Radio
T8c - Connector 8 pin, blue, connector radio I-3
T8g - Connector 8 pin, Connector 2, on Satellite Radio
T8h - Connector 8 pin, Connector 1, on Satellite Radio

(45) - Ground connection, behind instrument panel, center
(50) - Ground connection, in luggage compartment, left
(V45) - Connection 2 (shielding) (in CD changer wiring harness)

* - Vehicles without Satellite Radio to bridge with counter plug
** - Vehicles with CD Changer Unit

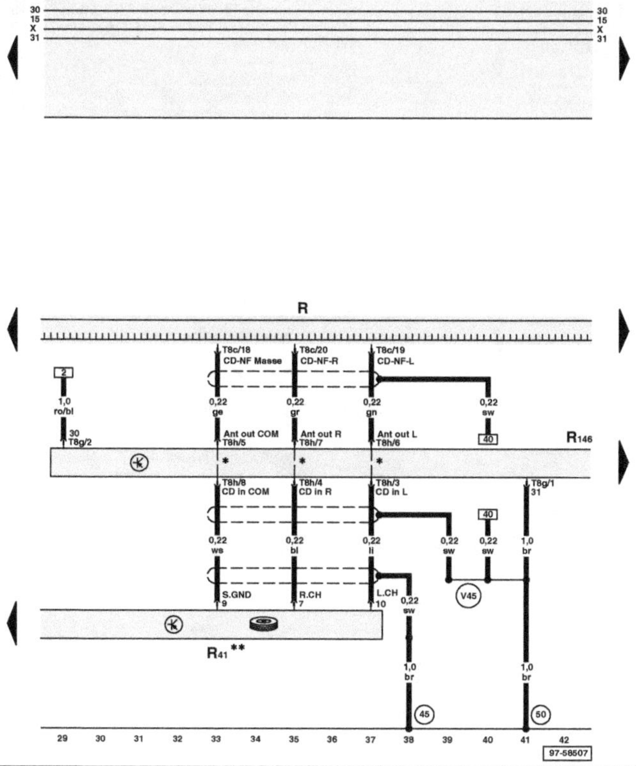

ws = white
sw = black
ro = red
br = brown
gn = green
bl = blue
gr = grey
li = violet
ge = yellow
or = orange

© 2006 BentleyPublishers.com—All Rights Reserved

Audi TT — Wiring Diagram — No. 23/5

Radio, Satellite Radio, Radio Antenna

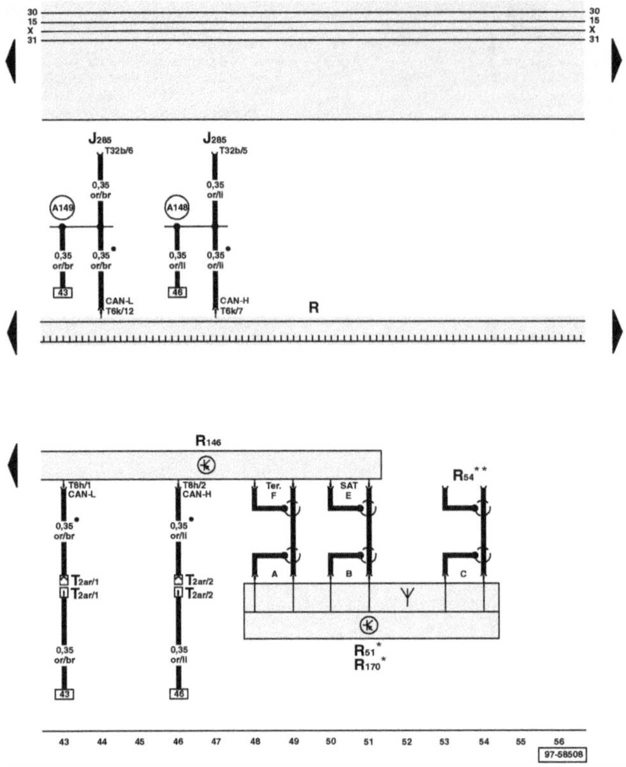

J285 - Control Module with indicator unit in instrument panel insert
R - Radio
R51 - Radio/Telephone Antenna
R54 - Cellular Telephone
R146 - Satellite Radio
R170 - Satellite Antenna
T2ar - Connector 2 pin, CAN-Bus Disconnection plug, A-pillar left
T6k - Connector 6 pin, green, connector radio I-2
T8h - Connector 8 pin, Connector 1, on Satellite Radio
T32b - Connector 32 pin, grey, on instrument cluster

A148 - Connection (high bus, navigation), in instrument panel wiring harness
A149 - Connection (low bus, navigation), in instrument panel wiring harness

• - CAN-Bus (Data-Bus)
* - Vehicles with Roof Antenna
** - Vehicles with Cellular Telephone

ws = white
sw = black
ro = red
br = brown
gn = green
bl = blue
gr = grey
li = violet
ge = yellow
or = orange

USA.5703.04.21 Edition 06.04

Audi TT — Wiring Diagram — No. 23/6

Radio, Control Unit for Navigation with CD-Mechanism, Operating Electronics and Telephone Control Module, Telephone/Radio Antenna Amplifier, Rear Window Defogger

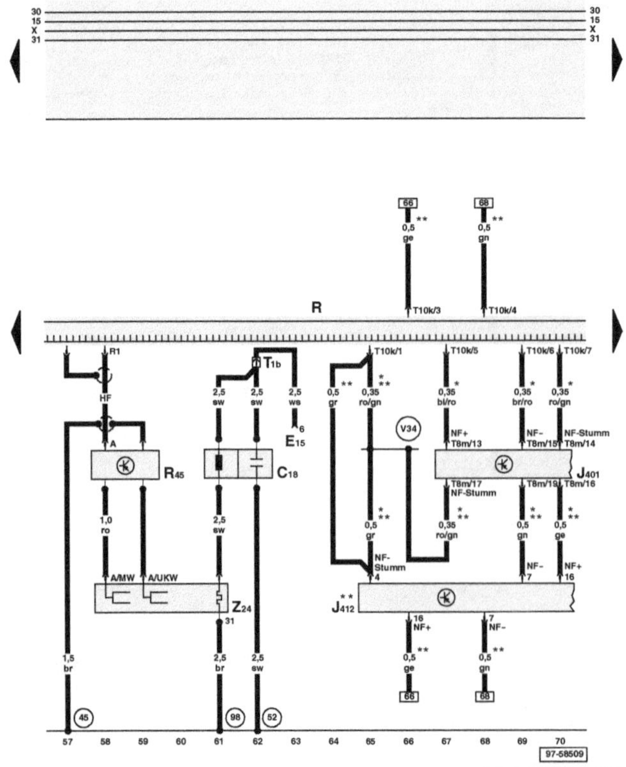

C18 - Windshield Antenna Suppression Filter
E15 - Rear Window Defogger Switch
J401 - Control unit for navigation with CD-mechanism
J412 - Operating electronics and Telephone Control Module
R - Radio
R45 - Telephone/Radio Antenna Amplifier
T1b - Connector single, black, in rear lid left
T8m - Connector 8 pin, C III on Control Module for Navigation
T10k - Connector 10 pin, red, on radio
Z24 - Rear Window Defogger With Window Antenna

45 - Ground connection, behind instrument panel, center
52 - Ground connection, in rear lid, left
98 - Ground connection, in rear lid wiring harness
V34 - Connector (mute button), in wiring harness telephone

* - Vehicles with Navigation
** - Vehicles with Cellular Telephone

ws = white
sw = black
ro = red
br = brown
gn = green
bl = blue
gr = grey
li = violet
ge = yellow
or = orange

USA.5703.04.21 Edition 06.04

© 2006 BentleyPublishers.com—All Rights Reserved

Audi TT

Wiring Diagram

No. 23/7

Radio, Amplifier

- R - Radio
- R12 - Amplifier
- T6h - Connector 6 pin, yellow, connector radio I-1
- (45) - Ground connection, behind instrument panel, center
- (255) - Ground connection, -1-, in wiring harness Radio
- (280) - Ground connector -2-, in wiring harness engine compartment

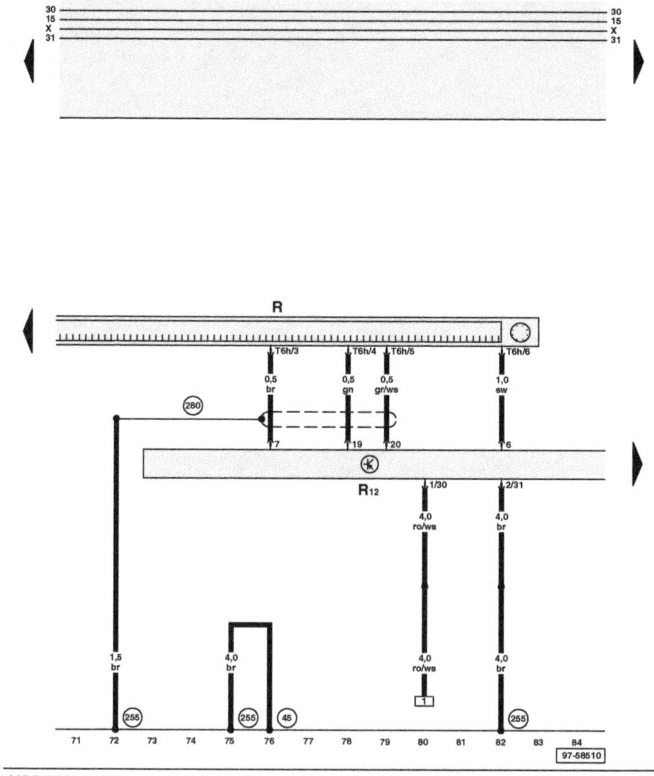

ws = white
sw = black
ro = red
br = brown
gn = green
bl = blue
gr = grey
li = violet
ge = yellow
or = orange

USA.5703.04.21

Edition 06.04

Audi TT

Wiring Diagram

No. 23/8

Amplifier, Bass Speaker, Treble Speaker, Inside Loudspeaker, Inside Microphone

- R12 - Amplifier
- R15 - Left Rear Bass Speaker
- R17 - Right Rear Bass Speaker
- R20 - Left Front Treble Speaker
- R21 - Left Front Bass Speaker
- R22 - Right Front Treble Speaker
- R23 - Right Front Bass Speaker
- R70 - Inside Loudspeaker
- R74 - Inside Microphone
- T2al - Connector 2 pin, behind instrument cluster
- T10q - Connector 10 pin, green, connector station A pillar, left
- T10r - Connector 10 pin, green, connector station A pillar, right

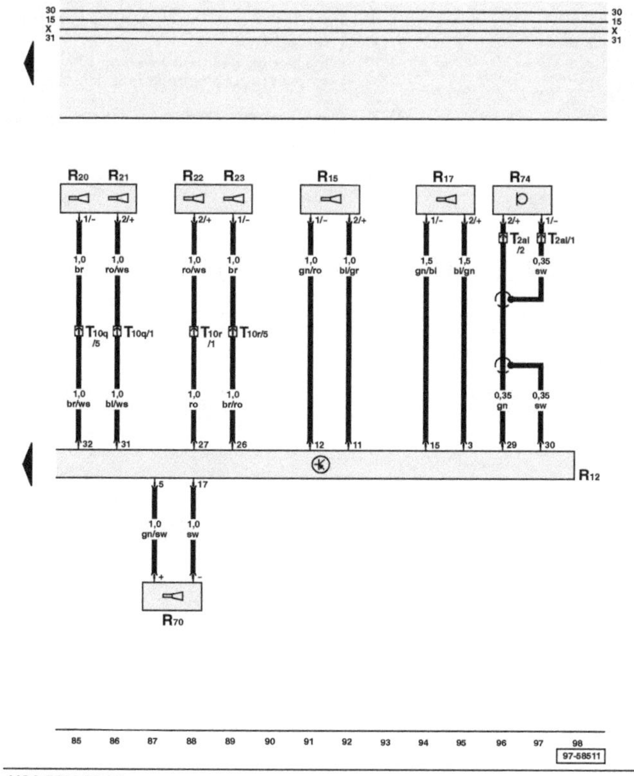

ws = white
sw = black
ro = red
br = brown
gn = green
bl = blue
gr = grey
li = violet
ge = yellow
or = orange

USA.5703.04.21

Edition 06.04

© 2006 BentleyPublishers.com—All Rights Reserved

Audi TT **Wiring Diagram** **No. 24/1**

CD Changer (Coupé and Roadster)

2005 m.y.

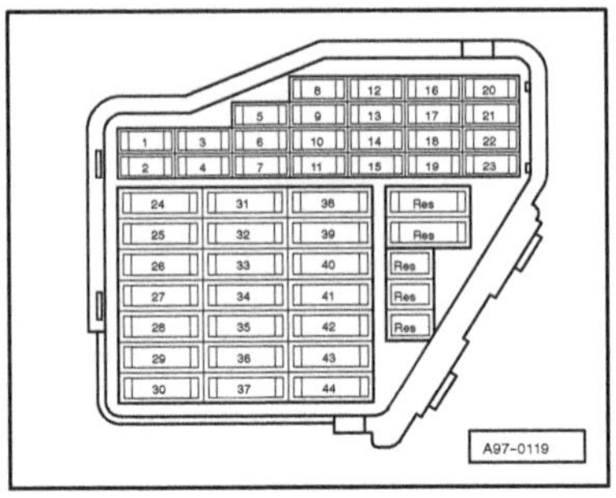

Fuse Panel

Fuse Colors:

30 A - Green
25 A - White
20 A - Yellow
15 A - Blue
10 A - Red
7,5 A - Brown
5 A - Beige
1 A - Black

Starting with fuse position 23, fuses in the fuse holder are identified with 223 in the wiring diagram.

USA.5703.04.21 **Edition 06.04**

Audi TT **Wiring Diagram** **No. 24/2**

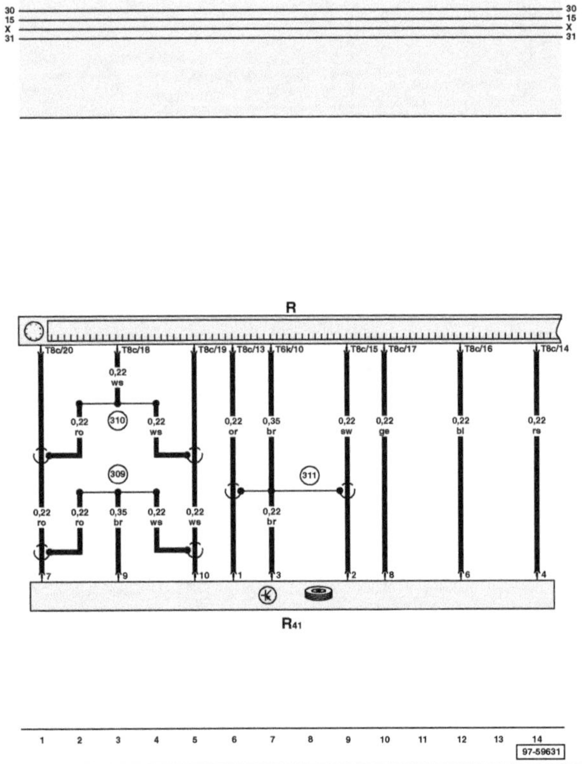

Radio, CD Changer Unit

R - Radio
R41 - CD Changer Unit
T6k - Connector 6 pin, green, connector radio I-2
T8c - Connector 8 pin, blue, connector radio I-3

(309) - Ground Connection (shielding) (in CD changer wiring harness)
(310) - Ground Connection 2 (shielding) (in CD changer wiring harness)
(311) - Ground Connection 3 (shielding) (in CD changer wiring harness)

ws = white
sw = black
ro = red
br = brown
gn = green
bl = blue
gr = grey
li = violet
ge = yellow
or = orange
rs = pink

USA.5703.04.21 **Edition 06.04**

© 2006 BentleyPublishers.com—All Rights Reserved

OBD On Board Diagnostics

General

This section gives a brief overview of On Board Diagnostics and a list of Diagnostic Trouble Codes (DTCs) applicable to the powertrain (P-codes) and related network communications (U-codes).

While most DTCs can be accessed with a Generic Scan Tool (GST), full and complete vehicle system information can only be obtained using a Audi supplied diagnostic scan tool computer such as the VAS 5051/5052 and derivatives, or an equivalent aftermarket scan tool or computer based scan tool program. The function for advanced diagnosis with these tools is called guided fault finding. A listing of scan tool and special tool suppliers is at the end of this section.

On Board Diagnostics

All vehicles covered by this manual are equipped with On Board Diagnostics (OBD). All system access is by way of the Data Link Connector (DLC) located under the left lower side of the dashboard. The DLC uses the industry standard SAE 16-pin connector.

Data link connector (**arrow**) is located under left lower side of dashboard. Connector is recessed into panel as shown and does not use a cover panel.

On Board Diagnostics include the United States government mandated second generation vehicle emissions monitoring system (OBD II) and Audi factory system diagnostics. OBD II codes are identified as P-codes and U-codes. Audi factory codes are identified as VAG codes. Most OBD II codes also have a corresponding VAG code. Both codes are listed in the DTC tables in this section as an aid to identification.

The vehicles covered by this manual are equipped with diagnostics that are far more extensive than the federal requirements. Communication is by way of twisted pair wiring called a "bus". Controller Area Network (CAN) communication over the twisted pair wiring is called CAN-bus. There are different CAN-bus lines covering most of the electronically controlled systems in the vehicle.

On Board Diagnostics

Terminal assignment for 16-pin DLC (T16). Terminals not listed are not currently in use.

1 - Terminal 15 (power with ignition on and in crank position)
4 - Terminal 31 (ground)
5 - Terminal 31 (ground)
7 - Diagnosis K-wire
16 - Terminal 30 (battery positive power)

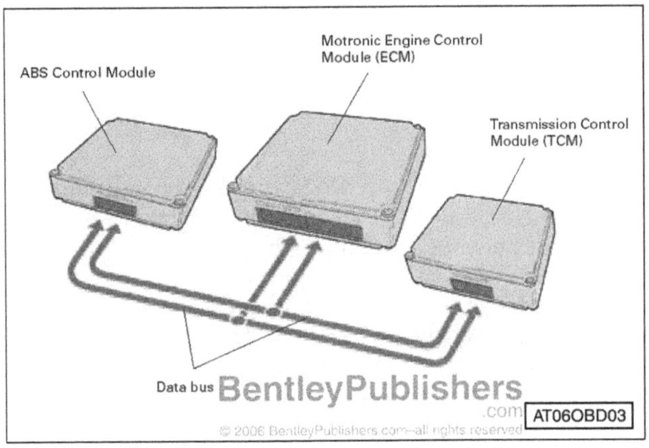

CAN-bus interface links ABS, engine, and transmission control modules. A similar CAN-bus connects control units of the convenience system.

Access to systems other than OBD II may require an advanced scan tool (VAS 5051/5052) and is outside the scope of this manual.

OBD II

On Board Diagnostics II monitors operation and function of engine management, and automatic transmission. Other emissions-related systems or parts of them, are monitored to insure compliance with specified emission levels.

Vehicle emission levels are continually monitored by the OBD II system and problems are recognized and recorded. A Malfunction Indicator Light (MIL) in the instrument cluster alerts the driver to the problem and the need to have the system checked. When a DTC is stored, or a potential problem exists, any of several indicator lights in the instrument cluster could be illuminated to notify the driver.

It is important to note that a diagnostic trouble code does not necessarily mean that an electronic component has failed. Even running low on gasoline can set a DTC as the system tries to cope with the ever-lessening fuel volume and pressure.

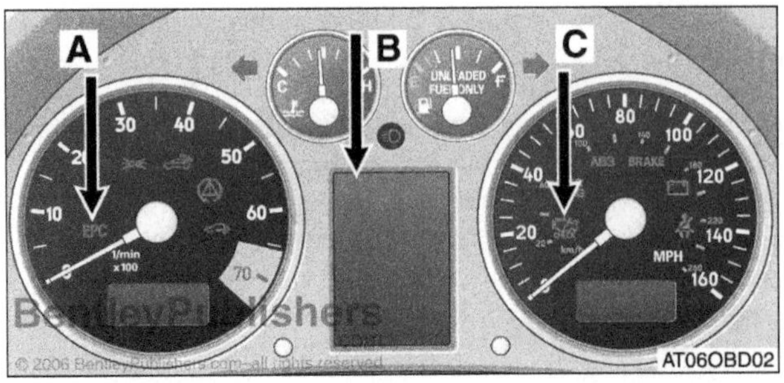

Instrument cluster warning lights:

A. EPC - Electronic Power Control
Illuminated for drive-by-wire malfunctions

B. Multi-function indicator
Illuminated with text "Emissions Workshop" and warning buzzer in conjunction with MIL operation

C. MIL - Malfunction Indicator Light
Flashes or steady illumination for emission malfunction

Diagnostic trouble code structure

When a system operates outside assigned operating parameters, a code is stored denoting that failure mode. This code is known as a diagnostic trouble code or DTC.

Each DTC is assigned two codes. The first code is a numerical code assigned by Audi (VAG code). The second code is referred to as a P-code or U-code and follows a structure required by law and defined by the Society of Automotive Engineers (SAE). Diagnostic Trouble Codes (DTCs) follow a standard format. This standard uses a letter to designate the system and four numbers or letters to further identify and detail the malfunction as listed below.

First digit structure is as follows:

- Pxxxx for powertrain
- Uxxxx for vehicle network communications
- Bxxxx for body
- Cxxxx for chassis

Second digit structure is:

- P0xxx Government required codes
- P1xxx Manufacturer codes for additional emission system function; not required but reported to government
- P2xxx Fuel and air mixture. Additional emissions regulations
- P3xxx Fuel and air mixture. Additional emissions regulations

Third digit structure is:

- Px1xx measurement of air and fuel
- Px2xx measurement of air and fuel
- Px3xx ignition system
- Px4xx additional emission control
- Px5xx speed and idle regulation
- Px6xx computer and output signals
- Px7xx transmission
- Px8xx transmission
- Px9xx control modules, input and output signals

The fourth and fifth digits denote the individual components and systems. In some cases, the fifth digit can be a letter. Currently A through D are used.

The digit structure shown above has "P" as the first digit, however, letters "B", "C", and "U" could be used where applicable. Only "P" and "U" DTCs are currently used in vehicles covered by this manual.

BentleyPublishers.com
© 2006 BentleyPublishers.com All rights reserved

VAS 5052

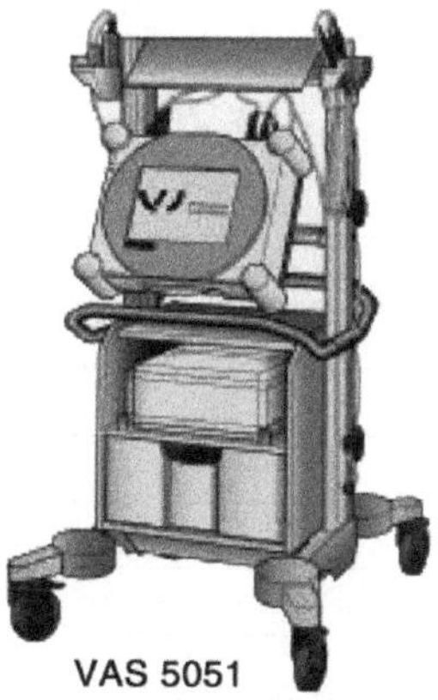

VAS 5051

VAG 1552

VAG1551

0024463

◄ Sampling of Audi supplied scan tools. VAS 5051 has been supplemented by a "B" version. Also note that VAG 1551 and VAG 1552 may not have the ability to access all vehicle systems and may not display the latest DTC text.

Diagnostic Trouble Codes

NOTE—
- *The following table contains a list of available scan tool codes for Volkswagen and Audi vehicles available at the time of publication. Some of the codes do not apply to vehicles covered by this manual.*

Diagnostic Trouble Codes (Powertrain)

V.A.G. code	SAE P-code	Description
16394	P0010	Intake "A" Camshaft Position Actuator Circuit Bank 1 Malfunction
16395	P0011	Intake "A" Camshaft Position, Timing Over Advanced or System Performance Bank 1
16396	P0012	Intake "A" Camshaft Position, Timing Over Retarded Bank 1
16397	P0013	Intake "B" Camshaft Position Actuator Circuit Bank 1 Malfunction
16398	P0014	Intake "B" Camshaft Position, Timing Over Advanced or System Performance Bank 1
16399	P0015	Intake "B" Camshaft Position, Timing Over Retarded Bank 1
16100	P0016	Camshaft Position, Camshaft Position Correlation B1 Sensor "A"
16402	P0018	Camshaft Position, Camshaft Position Correlation B1 Sensor "B"
16406	P0020	-A- Camshaft Pos. Actuator Circ. Bank 2 Malfunction
16414	P0030	HO2S Heater Control Circuit, B1S1
16415	P0031	HO2S Heater Control Circuit Low, B1S1
16416	P0032	HO2S Heater Control Circuit High, B1S1
16420	P0036	HO2S Heater Control Circuit B1S2
16421	P0037	HO2S Heater Control Circuit Low B1S2
16422	P0038	HO2S Heater Control Circuit High B1S2
16424	P0040	O2 Sensor Signals Swapped B1S1/B2S1
16425	P0041	O2 Sensor Signals Swapped B2S1/B2S2
16426	P0042	HO2S Heater Control Circuit B1S3
16427	P0043	HO2S Heater Control Circuit Low B1S3
16428	P0044	HO2S Heater Control Circuit High B1S3
16434	P0050	HO2S Heater Control Circuit B2S1
16435	P0051	HO2S Heater Control Circuit Low B2S1
16436	P0052	HO2S Heater Control Circuit High B2S1
16440	P0056	HO2S Heater Control Circuit B2S2
16441	P0057	HO2S Heater Control Circuit Low B2S2
16442	P0058	HO2S Heater Control Circuit High B2S2
16449	P0065	Air Assisted Injector Control Range/Performance
16450	P0066	Air Assisted Injector Control Low Input/Short to ground
16451	P0067	Air Assisted Injector Control Input/Short to B+
16452	P0068	MAP/MAF Throttle Position Correlation
16453	P0069	MAP/Barometric Pressure Correlation
16484	P0100	Mass or Volume Air Flow Circuit
16485	P0101	Mass or Volume Air Flow Circ Range/Performance
16486	P0102	Mass or Volume Air Flow Circ Low Input
16487	P0103	Mass or Volume Air Flow Circ High Input
16489	P0105	Manifold Abs.Pressure or Bar.Pressure Voltage supply
16490	P0106	Manifold Abs.Pressure or Bar.Pressure Range/Performance
16491	P0107	Manifold Abs.Pressure or Bar.Pressure Low Input

Diagnostic Trouble Codes (Powertrain)

V.A.G. code	SAE P-code	Description
16492	P0108	Manifold Abs.Pressure or Bar.Pressure High Input
16495	P0111	Intake Air Temperature Sensor Circuit Range/Performance
16496	P0112	Intake Air Temp.Circ Low Input
16497	P0113	Intake Air Temp.Circ High Input
16500	P0116	Engine Coolant Temp.Circ Range/Performance
16501	P0117	Engine Coolant Temp.Circ Low Input
16502	P0118	Engine Coolant Temp.Circ High Input
16504	P0120	Throttle/Pedal Pos.Sensor A Circ Malfunction
16505	P0121	Throttle/Pedal Pos.Sensor A Circ Range/Performance
16506	P0122	Throttle/Pedal Pos.Sensor A Circ Low Input
16507	P0123	Throttle/Pedal Pos.Sensor A Circ High Input
16509	P0125	Insufficient Coolant Temp.for Closed Loop Fuel Control
16512	P0128	Coolant Thermostat/Valve Temperature below control range
16514	P0130	O2 Sensor Circ., B1S1 Malfunction
16515	P0131	O2 Sensor Circ., B1S1 Low Voltage
16516	P0132	O2 Sensor Circ., B1S1 High Voltage
16517	P0133	O2 Sensor Circ., B1S1 Slow Response
16518	P0134	O2 Sensor Circ., B1S1 No Activity Detected
16519	P0135	O2 Sensor Heater Circ.,Bank 1-Sensor 1 Malfunction
16520	P0136	O2 Sensor Circ., B1S2 Malfunction
16521	P0137	O2 Sensor Circ., B1S2 Low Voltage
16522	P0138	O2 Sensor Circ., B1S2 High Voltage
16523	P0139	O2 Sensor Circ., B1S2 Slow Response
16524	P0140	O2 Sensor Circ., B1S2 No Activity Detected
16525	P0141	O2 Sensor Heater Circ., B1S2 Malfunction
16526	P0142	O2 Sensor Circ., B1S3
16427	P0143	O2 Sensor Circ., Low Voltage B1S3
16428	P0144	O2 Sensor Circ., High Voltage B1S3
16429	P0145	O2 Sensor Circ., Slow Response B1S3
16530	P0146	O2 Sensor Circ., No Activity Detected B1S3
16531	P0147	O2 Sensor Heater Circ. B1S3
16534	P0150	O2 Sensor Circ.,Bank2-Sensor1 Malfunction
16535	P0151	O2 Sensor Circ.,Bank2-Sensor1 Low Voltage
16536	P0152	O2 Sensor Circ.,Bank2-Sensor1 High Voltage
16537	P0153	O2 Sensor Circ.,Bank2-Sensor1 Slow Response
16538	P0154	O2 Sensor Circ.,Bank2-Sensor1 No Activity Detected
16539	P0155	O2 Sensor Heater Circ.,Bank2-Sensor1 Malfunction
16540	P0156	O2 Sensor Circ.,Bank2-Sensor2 Malfunction
16541	P0157	O2 Sensor Circ.,Bank2-Sensor2 Low Voltage
16542	P0158	O2 Sensor Circ.,Bank2-Sensor2 High Voltage
16543	P0159	O2 Sensor Circ.,Bank2-Sensor2 Slow Response
16544	P0160	O2 Sensor Circ.,Bank2-Sensor2 No Activity Detected
16545	P0161	O2 Sensor Heater Circ.,Bank2-Sensor2 Malfunction
16554	P0170	Fuel Trim,Bank1 Malfunction

Diagnostic Trouble Codes (Powertrain)

V.A.G. code	SAE P-code	Description
16555	P0171	Fuel Trim,Bank1 System too Lean
16556	P0172	Fuel Trim,Bank1 System too Rich
16557	P0173	Fuel Trim,Bank2 Malfunction
16558	P0174	Fuel Trim,Bank2 System too Lean
16559	P0175	Fuel Trim,Bank2 System too Rich
16565	P0181	Fuel temperature sender A Circuit Range/Performance
16566	P0182	Fuel temperature sender A Circuit Low
16567	P0183	Fuel temperature sender A Circuit High
16576	P0192	Fuel Rail Pressure Sensor Circuit Low
16581	P0197	Engine Oil Temperature Circuit Low Input
16582	P0198	Engine Oil Temperature Circuit High Input
16584	P0200	Injector Circuit Open
16585	P0201	Cyl.1, Injector Circuit Fault in electrical circuit
16586	P0202	Cyl.2, Injector Circuit Fault in electrical circuit
16587	P0203	Cyl.3, Injector Circuit Fault in electrical circuit
16588	P0204	Cyl.4, Injector Circuit Fault in electrical circuit
16589	P0205	Cyl.5 Injector Circuit Fault in electrical circuit
16590	P0206	Cyl.6 Injector Circuit Fault in electrical circuit
16591	P0207	Cyl.7 Injector Circuit Fault in electrical circuit
16592	P0208	Cyl.8 Injector Circuit Fault in electrical circuit
16599	P0215	Engine Shut-Off Solenoid Malfunction
16600	P0216	Injector/Injection Timing Control Malfunction
16603	P0219	Engine Overspeed Condition
16605	P0221	Throttle Pos. Sensor -B- Circuit Range/Performance
16606	P0222	Throttle Pos. Sensor -B- Circuit Low Input
16607	P0223	Throttle Pos. Sensor -B- Circuit High Input
16609	P0225	Throttle Pos. Sensor -C- Circuit Voltage supply
16610	P0226	Throttle Pos. Sensor -C- Circuit Range/Performance
16611	P0227	Throttle Pos. Sensor -C- Circuit Low Input
16612	P0228	Throttle Pos. Sensor -C- Circuit High Input
16614	P0230	Fuel Pump Primary Circuit Fault in electrical circuit
16615	P0231	Fuel Pump Secondary Circuit Low
16616	P0232	Fuel Pump Secondary Circuit High
16618	P0234	Turbocharger Overboost Condition Control limit exceeded
16619	P0235	Turbocharger Boost Sensor (A) Circ Control limit not reached
16620	P0236	Turbocharger Boost Sensor (A) Circ Range/Performance
16621	P0237	Turbocharger Boost Sensor (A) Circ Low Input
16622	P0238	Turbocharger Boost Sensor (A) Circ High Input
16627	P0243	Turbocharger Wastegate Solenoid (A) Open/Short Circuit to Ground
16629	P0245	Turbocharger Wastegate Solenoid (A) Low Input/Short to ground
16630	P0246	Turbocharger Wastegate Solenoid (A) High Input/Short to B+
16645	P0261	Cyl.1 Injector Circuit Low Input/Short to ground
16646	P0262	Cyl.1 Injector Circuit High Input/Short to B+
16647	P0263	Cyl. 1 Contribution/Balance

Diagnostic Trouble Codes (Powertrain)

V.A.G. code	SAE P-code	Description
16648	P0264	Cyl.2 Injector Circuit Low Input/Short to ground
16649	P0265	Cyl.2 Injector Circuit High Input/Short to B+
16650	P0266	Cyl. 2 Contribution/Balance
16651	P0267	Cyl.3 Injector Circuit Low Input/Short to ground
16652	P0268	Cyl.3 Injector Circuit High Input/Short to B+
16653	P0269	Cyl. 3 Contribution/Balance
16654	P0270	Cyl.4 Injector Circuit Low Input/Short to ground
16655	P0271	Cyl.4 Injector Circuit High Input/Short to B+
16656	P0272	Cyl.4 Contribution/Balance
16657	P0273	Cyl.5 Injector Circuit Low Input/Short to ground
16658	P0274	Cyl.5 Injector Circuit High Input/Short to B+
16660	P0276	Cyl.6 Injector Circuit Low Input/Short to ground
16661	P0277	Cyl.6 Injector Circuit High Input/Short to B+
16683	P0299	Turbo/Supercharger underboost
16684	P0300	Random/Multiple Cylinder Misfire Detected
16685	P0301	Cyl.1 Misfire Detected
16686	P0302	Cyl.2 Misfire Detected
16687	P0303	Cyl.3 Misfire Detected
16688	P0304	Cyl.4 Misfire Detected
16689	P0305	Cyl.5 Misfire Detected
16690	P0306	Cyl.6 Misfire Detected
16697	P0313	Misfire Detected Low Fuel Level
16698	P0314	Single Cylinder Misfire
16702	P0318	Rough Road Sensor "A" Signal Circuit
16705	P0321	Ign./Distributor Eng.Speed Inp.Circ Range/Performance
16706	P0322	Ign./Distributor Eng.Speed Inp.Circ No Signal
16708	P0324	Knock Control System Error
16709	P0325	Knock Sensor 1 Circuit Electrical Fault in Circuit
16710	P0326	Knock Sensor 1 Circuit Range/Performance
16711	P0327	Knock Sensor 1 Circ Low Input
16712	P0328	Knock Sensor 1 Circ High Input
16716	P0332	Knock Sensor 2 Circ Low Input
16717	P0333	Knock Sensor 2 Circ High Input
16719	P0335	Crankshaft Pos. Sensor (A) Circ Malfunction
16720	P0336	Crankshaft Pos. Sensor (A) Circ Range/Performance/Missing tooth
16721	P0337	Crankshaft Pos.Sensor (A) Circ Low Input
16724	P0340	Camshaft Pos. Sensor (A) Circ Incorrect allocation
16725	P0341	Camshaft Pos.Sensor Circ Range/Performance
16726	P0342	Camshaft Pos.Sensor Circ Low Input
16727	P0343	Camshaft Pos.Sensor Circ High Input
16729	P0345	Camshaft Position Sensor "A" Circuit Bank 2
16730	P0346	Camshaft Position Sensor "A" Circuit Range/Performance Bank 2
16731	P0347	Camshaft Position Sensor "A" Circuit Low Bank 2
16732	P0348	Camshaft Position Sensor "A" Circuit High Bank 2

Diagnostic Trouble Codes (Powertrain)

V.A.G. code	SAE P-code	Description
16735	P0351	Ignition Coil (A) Cyl.1 Prim./Sec. Circ Malfunction
16736	P0352	Ignition Coil (B) Cyl.2 Prim./Sec. Circ Malfunction
16737	P0353	Ignition Coil (C) Cyl.3 Prim./Sec. Circ Malfunction
16738	P0354	Ignition Coil (D) Cyl.4 Prim./Sec. Circ Malfunction
16739	P0355	Ignition Coil (E) Cyl.5 Prim./Sec. Circ Malfunction
16740	P0356	Ignition Coil (F) Cyl.6 Prim./Sec. Circ Malfunction
16764	P0380	Glow Plug/Heater Circuit (A) Electrical Fault in Circuit
16765	P0381	Glow Plug/Heater indicator Circuit
16784	P0400	Exhaust Gas Recirc.Flow Malfunction
16785	P0401	Exhaust Gas Recirc.Flow Insufficient Detected
16786	P0402	Exhaust Gas Recirc.Flow Excessive Detected
16787	P0403	Exhaust Gas Recirc. Contr. Circ Malfunction
16788	P0404	Exhaust Gas Recirc. Contr. Circ Range/Performance
16789	P0405	Exhaust Gas Recirc. Sensor (A) Circ Low Input
16790	P0406	Exhaust Gas Recirc. Sensor (A) Circ High Input
16791	P0407	Exhaust Gas Recirc. Sensor (B) Circ Low Input
16792	P0408	Exhaust Gas Recirc. Sensor (B) Circ High Input
16794	P0410	Sec.Air Inj.Sys Malfunction
16795	P0411	Sec.Air Inj.Sys. Incorrect Flow Detected
16796	P0412	Sec.Air Inj.Sys.Switching Valve A Circ Malfunction
16797	P0413	Secondary Air Injection System Switching Valve “A” Circuit Open
16798	P0414	Secondary Air Injection System Switching Valve “A” Circuit Shorted
16802	P0418	Sec. Air Inj. Sys. Relay (A) Contr. Circ Malfunction
16804	P0420	Catalyst System,Bank1 Efficiency Below Threshold
16806	P0422	Main Catalyst,Bank1 Below Threshold
16811	P0427	Catalyst Temperature Sensor, Bank 1 Low Input/Short to ground
16812	P0428	Catalyst Temperature Sensor, Bank 1 High Input/Open/Short Circuit to B+
16814	P0430	Catalyst System Efficiency Below Threshold Bank 2
16816	P0432	Main Catalyst,Bank2 Efficiency Below Threshold
16820	P0436	Catalyst Temperature Sensor, Bank 2 Range/Performance
16821	P0437	Catalyst Temperature Sensor, Bank 2 Low Input/Short to ground
16822	P0438	Catalyst Temperature Sensor, Bank 2 High Input/Open/Short Circuit to B+
16824	P0440	EVAP Emission Contr.Sys. Malfunction
16825	P0441	EVAP Emission Contr.Sys.Incorrect Purge Flow
16826	P0442	EVAP Emission Contr.Sys.(Small Leak) Leak Detected
16827	P0443	EVAP Emiss. Contr. Sys. Purge Valve Circ Electrical Fault in Circuit
16828	P0444	Evaporative Emission System Purge Control Valve Circuit Open
16829	P0445	Evaporative Emission System Purge Control Valve Circuit Shorted
16833	P0449	Evaporative Emission System Vent Valve/Solenoid Circuit
16836	P0452	EVAP Emission Contr.Sys.Press.Sensor Low Input
16837	P0453	EVAP Emission Contr.Sys.Press.Sensor High Input
16839	P0455	EVAP Emission Contr.Sys. Gross Leak Detected
16840	P0456	EVAP Emission Contr.Sys.Very Small Leak Detected
16842	P0458	Evaporative Emission System Purge Regulator Valve (N80) Short to Ground

Diagnostic Trouble Codes (Powertrain)

V.A.G. code	SAE P-code	Description
16843	P0459	Evaporative Emission System Purge Regulator Valve (N80) Short to Positive
16845	P0461	Fuel Level Sensor Circ Range/Performance
16846	P0462	Fuel Level Sensor Circuit Low Input
16847	P0463	Fuel Level Sensor Circuit High Input
16864	P0480	Fan 1 Control Circuit
16873	P0489	Exhaust Gas Recirculation Control Circuit Low
16874	P0490	Exhaust Gas Recirculation Control Circuit High
16875	P0491	Secondary Air Injection System Insufficient Flow Bank 1
16876	P0492	Secondary Air Injection System Insufficient Flow Bank 2
16882	P0498	Evaporative Emission System Vent Valve Control Circuit Low
16883	P0499	Evaporative Emission System Vent Valve Control Circuit High
16885	P0501	Vehicle Speed Sensor Range/Performance
16887	P0503	Vehicle Speed Sensor Intermittent/Erratic/High Input
16889	P0505	Idle Control System Malfunction
16890	P0506	Idle Control System RPM Lower than Expected
16891	P0507	Idle Control System Higher than Expected
16894	P0510	Closed Throttle Pos.Switch Malfunction
16897	P0513	Incorrect Immobilizer Key
16915	P0531	A/C Refrigerant Pressure Sensor Circuit Range/Performance
16916	P0532	A/C Refrigerant Pressure Sensor Circuit Low Input
16917	P0533	A/C Refrigerant Pressure Sensor Circuit High Input
16928	P0544	Exhaust Gas Temperature Sensor Circuit B1S1
16929	P0545	Exhaust Gas Temperature Sensor Circuit Low B1S1
16930	P0546	Exhaust Gas Temperature Sensor Circuit High B1S1
16935	P0551	Power Steering Pressure Sensor Circuit Range/Performance
16944	P0560	System Voltage Malfunction
16946	P0562	System Voltage Low Voltage
16947	P0563	System Voltage High Voltage
16952	P0568	Cruise Control Set Signal Incorrect Signal
16955	P0571	Cruise/Brake Switch (A) Circ Malfunction
16984	P0600	Serial Comm. Link (Data Bus) Message Missing
16985	P0601	Internal Contr.Module Memory Check Sum Error
16986	P0602	Control Module Programming Error/Malfunction
16987	P0603	Internal Contr.Module (KAM) Error
16988	P0604	Internal Contr.Module Random Access Memory (RAM) Error
16989	P0605	Internal Contr.Module ROM Test Error
16990	P0606	ECM/PCM Processor
16991	P0607	Control Module performance
16992	P0608	Control Module Voltage Supply "A" Sensors/Actuators
16993	P0609	Control Module Voltage Supply "B" Sensors/Actuators
16994	P0610	Control Module Vehicle Options Error
16997	P0613	TCM Processor Malfunction
16998	P0614	ECM/TCM Incompatible (mismatch)
17011	P0627	Fuel Pump (FP) Relay Electrical Malfunction/Open Circuit

Diagnostic Trouble Codes (Powertrain)

V.A.G. code	SAE P-code	Description
17012	P0628	Fuel Pump (FP) Relay Short Circuit to Ground
17084	P0634	PCM/ECM/TCM Internal Temperature Too High
17022	P0638	Throttle Actuator Control Range/Performance B1
17023	P0639	Throttle Actuator Control Range/Performance B2
17025	P0641	Sensor Reference Voltage “A” Circuit Open
17026	P0642	Sensor Reference Voltage “A” Circuit Low
17027	P0643	Sensor Reference Voltage “A” Circuit High
17029	P0645	A/C Clutch Relay Control Circuit
17034	P0650	MIL Control Circuit Electrical Fault in Circuit
17035	P0651	Sensor Reference Voltage “B” Circuit Open
17036	P0652	Sensor Reference Voltage “B” Circuit Low
17037	P0653	Sensor Reference Voltage “B” Circuit High
17038	P0654	Engine RPM Output Circuit Electrical Fault in Circuit
17040	P0656	Fuel Level Output Circuit Electrical Fault in Circuit
17044	P0660	Intake Manifold Tuning Valve Control Circuit Open Bank 1
17045	P0661	Intake Manifold Tuning Valve Control Circuit Low Bank 1
17046	P0662	Intake Manifold Tuning Valve Control Circuit High Bank 1
17052	P0668	PCM/ECM/TCM Internal Temperature Sensor Circuit Voltage Too Low
17053	P0669	PCM/ECM/TCM Internal Temperature Sensor Circuit Voltage Too High
17054	P0670	Glow Plug Module Control Circuit
17055	P0671	Cylinder 1 Glow Plug Circuit
17056	P0672	Cylinder 2 Glow Plug Circuit
17057	P0673	Cylinder 3 Glow Plug Circuit
17058	P0674	Cylinder 4 Glow Plug Circuit
17068	P0684	Glow Plug Module to PCM Communication Circuit Range/Performance
17069	P0685	ECM/PCM Power Relay Control Circuit Open
17070	P0686	ECM/PCM Power Relay Control Low
17071	P0687	ECM/PCM Power Relay Control High
17072	P0688	ECM/PCM Power Relay Sense Circuit Open
17075	P0691	Fan Control Circuit Low
17076	P0692	Fan Control Circuit High
17081	P0697	Sensor Reference Voltage “C” Circuit Open
17082	P0698	Sensor Reference Voltage “C” Circuit Low
17083	P0699	Sensor Reference Voltage “C” Circuit High
17084	P0700	Transm.Contr.System Malfunction
17085	P0701	Transmission Control System Range/Performance
17086	P0702	Transm.Contr.System Electrical
17087	P0703	Torque Converter/Brake Switch B Circ Malfunction
17088	P0704	Clutch Switch Input Circuit Malfunction
17089	P0705	Transm.Range Sensor Circ.(PRNDL Inop.) Malfunction
17090	P0706	Transm.Range Sensor Circ Range/Performance
17091	P0707	Transm.Range Sensor Circ Low Input
17092	P0708	Transm.Range Sensor Circ High Input
17094	P0710	Transm.Fluid Temp.Sensor Circ. Malfunction

Diagnostic Trouble Codes (Powertrain)

V.A.G. code	SAE P-code	Description
17095	P0711	Transm.Fluid Temp.Sensor Circ. Range/Performance
17096	P0712	Transm.Fluid Temp.Sensor Circ. Low Input
17097	P0713	Transm.Fluid Temp.Sensor Circ. High Input
17098	P0714	Transm.Fluid Temp.Sensor Circ. Intermittent
17099	P0715	Input Turbine/Speed Sensor Circ. Malfunction
17100	P0716	Input Turbine/Speed Sensor Circ. Range/Performance
17101	P0717	Input Turbine/Speed Sensor Circ. No Signal
17104	P0720	Output Speed Sensor Circ Malfunction
17105	P0721	Output Speed Sensor Circ Range/Performance
17106	P0722	Output Speed Sensor Circ No Signal
17109	P0725	Engine Speed Inp.Circ. Malfunction
17110	P0726	Engine Speed Inp.Circ. Range/Performance
17111	P0727	Engine Speed Inp.Circ. No Signal
17113	P0729	Gear 6 Incorrect Ratio
17114	P0730	Gear Incorrect Ratio
17115	P0731	Gear 1 Incorrect Ratio
17116	P0732	Gear 2 Incorrect Ratio
17117	P0733	Gear 3 Incorrect Ratio
17118	P0734	Gear 4 Incorrect Ratio
17119	P0735	Gear 5 Incorrect Ratio
17124	P0740	Torque Converter Clutch Circuit Malfunction
17125	P0741	Torque Converter Clutch Circuit Performance or Stuck Off
17127	P0743	Torque Converter Clutch Circuit Electrical
17130	P0746	Pressure Control Solenoid "A" Performance or Stuck Off
17131	P0747	Pressure Control Solenoid "A" Stuck On
17132	P0748	Pressure Control Solenoid "A" Electrical
17133	P0749	Pressure Control Solenoid "A" Intermittent
17134	P0750	Shift Solenoid A malfunction
17135	P0751	Shift Solenoid A Performance or Stuck Off
17136	P0752	Shift Solenoid A Stuck On
17137	P0753	Shift Solenoid A Electrical
17140	P0756	Shift Solenoid B Performance or Stuck Off
17141	P0757	Shift Solenoid B Stuck On
17142	P0758	Shift Solenoid B Electrical
17145	P0761	Shift Solenoid C Performance or Stuck Off
17146	P0762	Shift Solenoid C Stuck On
17147	P0763	Shift Solenoid C Electrical
17150	P0766	Shift Solenoid D Performance or Stuck Off
17152	P0768	Shift Solenoid D Electrical
17155	P0771	Shift Solenoid E Performance or Stuck Off
17157	P0773	Shift Solenoid E Electrical
17160	P0776	Pressure Control Solenoid B Performance or Stuck Off
17161	P0777	Pressure Control Solenoid B Stuck On
17162	P0778	Pressure Control Solenoid B Electrical

Diagnostic Trouble Codes (Powertrain)

V.A.G. code	SAE P-code	Description
17164	P0780	Shift monitoring
17165	P0781	1-2 Shift
17166	P0782	2-3 Shift
17167	P0783	3-4 Shift
17168	P0784	4-5 Shift
17169	P0785	Shift/Timing Solenoid “A” Electrical
17174	P0790	Normal/Performance Switch Circ Malfunction
17175	P0791	Intermediate Shaft Speed Sensor “A” Circuit
17180	P0796	Pressure Control Solenoid “C” Performance or Stuck Off
17181	P0797	Pressure Control Solenoid “C” Stuck On
17182	P0798	Pressure Control Solenoid “C” Electrical
17183	P0799	Pressure Control Solenoid “C” Intermittent
17195	P0811	Excessive Clutch Slippage
17204	P0820	Gear Lever X-Y Position Sensor Circuit
17213	P0829	5-6 Shift
17224	P0840	Transmission Fluid Pressure Sensor/Switch “A” Circuit Electrical Malfunction
17225	P0841	Transmission Fluid Pressure Sensor/Switch “A” Circuit Range/Performance
17226	P0842	Transmission Fluid Pressure Sensor/Switch "A" Circuit Low
17227	P0843	Transmission Fluid Pressure Sensor/Switch "A" Circuit High
17228	P0844	Transmission Fluid Pressure Sensor/Switch "A" Circuit Intermittent
17229	P0845	Transmission Fluid Pressure Sensor/Switch “B” Circuit Electrical Malfunction
17230	P0846	Transmission Fluid Pressure Sensor/Switch “B” Circuit Range/Performance
17247	P0863	TCM Communication Circuit
17248	P0864	TCM Communication Circuit Range/Performance
17249	P0865	TCM Communication Circuit Low
17273	P0889	TCM Power Relay Sense Circuit Range/Performance
17274	P0890	TCM Power Relay Sense Circuit Low
17275	P0891	TCM Power Relay Sense Circuit High
17276	P0892	TCM Power Relay Sense Circuit Intermittent
17298	P0914	Gear Shift Position Circuit Electrical Malfunction
17303	P0919	Gear Shift Position Control Error
17313	P0929	Gear Shift Lock Solenoid Control Circuit Range/Performance
17433	P1025	Fuel Pressure Regulator Valve (N276) Mechanical malfunction
17434	P1026	Activation intake manifold flap for airstream regulation, Short Circuit to B+
17435	P1027	Activation intake manifold flap for airstream regulation, Short Circuit to Ground
17436	P1028	Activation intake manifold flap for airstream regulation, Open Circuit
17501	P1093	Bank 1, fuel measuring system 2 Malfunction
17506	P1098	Intake Manifold Tuning (IMT) Valve Position Sensor (G336) Open circuit/short circuit to B+
17509	P1101	O2 Sensor Circ.,Bank1-Sensor1Voltage too Low/Air Leak
17510	P1102	O2 Sensor Heating Circ.,Bank1-Sensor1 Short to B+
17511	P1103	O2 Sensor Heating Circ.,Bank1-Sensor1 Output too Low
17512	P1104	Bank1-Sensor2 Voltage too Low/Air Leak
17513	P1105	O2 Sensor Heating Circ.,Bank1-Sensor 2 High
17514	P1106	O2 Sensor Circ.,Bank2-Sensor1 Voltage too Low/Air Leak

Diagnostic Trouble Codes (Powertrain)

V.A.G. code	SAE P-code	Description
17515	P1107	O2 Sensor Heating Circ.,Bank2-Sensor1 Short to B+
17516	P1108	O2 Sensor Heating Circ.,Bank2-Sensor1 Output too Low
17517	P1109	O2 Sensor Circ.,Bank2-Sensor2 Voltage too Low/Air Leak
17518	P1110	O2 Sensor Heating Circ.,Bank2-Sensor2 Short to B+
17519	P1111	O2 Control (Bank 1) System too lean
17520	P1112	O2 Control (Bank 1) System too rich
17521	P1113	Bank1-Sensor1 Internal Resistance too High
17522	P1114	Bank1-Sensor2 Internal Resistant too High
17523	P1115	O2 Sensor Heater Circ.,Bank1-Sensor1 Short to Ground
17524	P1116	O2 Sensor Heater Circ.,Bank1-Sensor1 Open
17525	P1117	O2 Sensor Heater Circ.,Bank1-Sensor2 Short to Ground
17526	P1118	O2 Sensor Heater Circ.,Bank1-Sensor2 Open
17527	P1119	O2 Sensor Heater Circ.,Bank2-Sensor1 Short to Ground
17528	P1120	O2 Sensor Heater Circ.,Bank2-Sensor1 Open
17529	P1121	O2 Sensor Heater Circ.,Bank2-Sensor2 Short to Ground
17530	P1122	O2 Sensor Heater Circ.,Bank2-Sensor2 Open
17531	P1123	Long Term Fuel Trim Add.Air.,Bank1 System too Rich
17532	P1124	Long Term Fuel Trim Add.Air.,Bank1 System too Lean
17533	P1125	Long Term Fuel Trim Add.Air.,Bank2 System too Rich
17534	P1126	Long Term Fuel Trim Add.Air.,Bank2 System too Lean
17535	P1127	Long Term Fuel Trim mult.,Bank1 System too Rich
17536	P1128	Long Term Fuel Trim mult.,Bank1 System too Lean
17537	P1129	Long Term Fuel Trim mult.,Bank2 System too Rich
17538	P1130	Long Term Fuel Trim mult.,Bank2 System too Lean
17539	P1131	Bank2-Sensor1 Internal Resistance too High
17540	P1132	O2 Sensor Heating Circ.,Bank1+2-Sensor1 Short to B+
17541	P1133	O2 Sensor Heating Circ.,Bank1+2-Sensor1 Electrical Malfunction
17542	P1134	O2 Sensor Heating Circ.,Bank1+2-Sensor2 Short to B+
17543	P1135	O2 Sensor Heating Circ.,Bank1+2-Sensor2 Electrical Malfunction
17544	P1136	Long Term Fuel Trim Add.Fuel,Bank1 System too Lean
17545	P1137	Long Term Fuel Trim Add.Fuel,Bank1 System too Rich
17546	P1138	Long Term Fuel Trim Add.Fuel,Bank2 System too Lean
17547	P1139	Long Term Fuel Trim Add.Fuel,Bank2 System too Rich
17548	P1140	Bank2-Sensor2 Internal Resistance too High
17549	P1141	Load Calculation Cross Check Range/Performance
17550	P1142	Load Calculation Cross Check Lower Limit Exceeded
17551	P1143	Load Calculation Cross Check Upper Limit Exceeded
17552	P1144	Mass or Volume Air Flow Circ Open/Short to Ground
17553	P1145	Mass or Volume Air Flow Circ Short to B+
17554	P1146	Mass or Volume Air Flow Circ Supply Malfunction
17555	P1147	O2 Control (Bank 2) System too lean
17556	P1148	O2 Control (Bank 2) System too rich
17557	P1149	O2 Control (Bank 1) Out of range
17558	P1150	O2 Control (Bank 2) Out of range

Diagnostic Trouble Codes (Powertrain)

V.A.G. code	SAE P-code	Description
17559	P1151	Bank1, Long Term Fuel Trim, Range 1 Leanness Lower Limit Exceeded
17560	P1152	Bank1, Long Term Fuel Trim, Range 2 Leanness Lower Limit Exceeded
17562	P1154	Manifold Switch Over Malfunction
17563	P1155	Manifold Abs.Pressure Sensor Circ. Short to B+
17564	P1156	Manifold Abs.Pressure Sensor Circ. Open/Short to Ground
17565	P1157	Manifold Abs.Pressure Sensor Circ. Power Supply Malfunction
17566	P1158	Manifold Abs.Pressure Sensor Circ. Range/Performance
17568	P1160	Manifold Temp.Sensor Circ. Short to Ground
17569	P1161	Manifold Temp.Sensor Circ. Open/Short to B+
17570	P1162	Fuel Temp.Sensor Circ. Short to Ground
17571	P1163	Fuel Temp.Sensor Circ. Open/Short to B+
17572	P1164	Fuel Temperature Sensor Range/Performance/Incorrect Signal
17573	P1165	Bank1, Long Term Fuel Trim, Range 1 Rich Limit Exceeded
17574	P1166	Bank1, Long Term Fuel Trim, Range 2 Rich Limit Exceeded
17579	P1171	Throttle Actuation Potentiometer Sign.2 Range/Performance
17580	P1172	Throttle Actuation Potentiometer Sign.2 Signal too Low
17581	P1173	Throttle Actuation Potentiometer Sign.2 Signal too High
17582	P1174	Fuel Trim, Bank 1 Different injection times
17584	P1176	O2 Correction Behind Catalyst,B1 Limit Attained
17585	P1177	O2 Correction Behind Catalyst,B2 Limit Attained
17586	P1178	Linear 02 Sensor / Pump Current Open Circuit
17587	P1179	Linear 02 Sensor / Pump Current Short to ground
17588	P1180	Linear 02 Sensor / Pump Current Short to B+
17589	P1181	Linear 02 Sensor / Reference Voltage Open Circuit
17590	P1182	Linear 02 Sensor / Reference Voltage Short to ground
17591	P1183	Linear 02 Sensor / Reference Voltage Short to B+
17592	P1184	Linear 02 Sensor / Common Ground Wire Open Circuit
17593	P1185	Linear 02 Sensor / Common Ground Wire Short to ground
17594	P1186	Linear 02 Sensor / Common Ground Wire Short to B+
17595	P1187	Linear 02 Sensor / Compens. Resistor Open Circuit
17596	P1188	Linear 02 Sensor / Compens. Resistor Short to ground
17597	P1189	Linear 02 Sensor / Compens. Resistor Short to B+
17598	P1190	Linear 02 Sensor / Reference Voltage Incorrect Signal
17604	P1196	O2 Sensor Heater Circ.,Bank1-Sensor1 Electrical Malfunction
17605	P1197	O2 Sensor Heater Circ.,Bank2-Sensor1 Electrical Malfunction
17606	P1198	O2 Sensor Heater Circ.,Bank1-Sensor2 Electrical Malfunction
17607	P1199	O2 Sensor Heater Circ.,Bank2-Sensor2 Electrical Malfunction
17609	P1201	Cyl.1-Fuel Inj.Circ. Electrical Malfunction
17610	P1202	Cyl.2-Fuel Inj.Circ. Electrical Malfunction
17611	P1203	Cyl.3-Fuel Inj.Circ. Electrical Malfunction
17612	P1204	Cyl.4-Fuel Inj.Circ. Electrical Malfunction
17613	P1205	Cyl.5-Fuel Inj.Circ. Electrical Malfunction
17614	P1206	Cyl.6-Fuel Inj.Circ. Electrical Malfunction
17617	P1209	Intake valves for cylinder shut-off Short circuit to ground

Diagnostic Trouble Codes (Powertrain)

V.A.G. code	SAE P-code	Description
17618	P1210	Intake valves for cylinder shut-off Short to B+
17619	P1211	Intake valves for cylinder shut-off Open circuit
17621	P1213	Cyl.1-Fuel Inj.Circ. Short to B+
17622	P1214	Cyl.2-Fuel Inj.Circ. Short to B+
17623	P1215	Cyl.3-Fuel Inj.Circ. Short to B+
17624	P1216	Cyl.4-Fuel Inj.Circ. Short to B+
17625	P1217	Cyl.5-Fuel Inj.Circ. Short to B+
17626	P1218	Cyl.6-Fuel Inj.Circ. Short to B+
17629	P1221	Cylinder shut-off exhaust valves Short circuit to ground
17630	P1222	Cylinder shut-off exhaust valves Short to B+
17631	P1223	Cylinder shut-off exhaust valves Open circuit
17633	P1225	Cyl.1-Fuel Inj.Circ. Short to Ground
17634	P1226	Cyl.2-Fuel Inj.Circ. Short to Ground
17635	P1227	Cyl.3-Fuel Inj.Circ. Short to Ground
17636	P1228	Cyl.4-Fuel Inj.Circ. Short to Ground
17637	P1229	Cyl.5-Fuel Inj.Circ. Short to Ground
17638	P1230	Cyl.6-Fuel Inj.Circ. Short to Ground
17645	P1237	Cyl.1-Fuel Inj.Circ. Open Circ.
17646	P1238	Cyl.2-Fuel Inj.Circ. Open Circ.
17647	P1239	Cyl.3-Fuel Inj.Circ. Open Circ.
17648	P1240	Cyl.4-Fuel Inj.Circ. Open Circ.
17649	P1241	Cyl.5-Fuel Inj.Circ. Open Circ.
17650	P1242	Cyl.6-Fuel Inj.Circ. Open Circ.
17656	P1248	Injection Start Control Deviation
17657	P1249	Fuel consumption signal Electrical Fault in Circuit
17658	P1250	Fuel Level Too Low
17659	P1251	Start of Injection Solenoid Circ Short to B+
17660	P1252	Start of Injection Solenoid Circ Open/Short to Ground
17661	P1253	Fuel consumption signal Short to ground
17662	P1254	Fuel consumption signal Short to B+
17663	P1255	Engine Coolant Temp.Circ Short to Ground
17664	P1256	Engine Coolant Temp.Circ Open/Short to B+
17665	P1257	Engine Coolant System Valve Open
17666	P1258	Engine Coolant System Valve Short to B+
17667	P1259	Engine Coolant System Valve Short to Ground
17688	P1280	Fuel Inj.Air Contr.Valve Circ. Flow too Low
17691	P1283	Fuel Inj.Air Contr.Valve Circ. Electrical Malfunction
17692	P1284	Fuel Inj.Air Contr.Valve Circ. Open
17693	P1285	Fuel Inj.Air Contr.Valve Circ. Short to Ground
17694	P1286	Fuel Inj.Air Contr.Valve Circ. Short to B+
17695	P1287	Turbocharger bypass valve open
17696	P1288	Turbocharger bypass valve short to B+
17697	P1289	Turbocharger bypass valve short to ground
17703	P1295	Bypass Turbocharger Throughput Faulty

Diagnostic Trouble Codes (Powertrain)

V.A.G. code	SAE P-code	Description
17704	P1296	Cooling system malfunction
17705	P1297	Connection turbocharger - throttle valve pressure hose
17708	P1300	Misfire detected Reason: Fuel level too low
17721	P1319	Knock Sensor 1 Circ. Short to Ground
17728	P1320	Knock Sensor 2 Circ. Short to Ground
17729	P1321	Knock Sensor 3 Circ. Low Input
17730	P1322	Knock Sensor 3 Circ. High Input
17731	P1323	Knock Sensor 4 Circ. Low Input
17732	P1324	Knock Sensor 4 Circ. High Input
17733	P1325	Cyl.1-Knock Contr. Limit Attained
17734	P1326	Cyl.2-Knock Contr. Limit Attained
17735	P1327	Cyl.3-Knock Contr. Limit Attained
17736	P1328	Cyl.4-Knock Contr. Limit Attained
17737	P1329	Cyl.5-Knock Contr. Limit Attained
17738	P1330	Cyl.6-Knock Contr. Limit Attained
17743	P1335	Engine Torque Monitoring 2 Control Limit Exceeded
17744	P1336	Engine Torque Monitoring Adaptation at limit
17745	P1337	Camshaft Pos.Sensor,Bank1 Short to Ground
17746	P1338	Camshaft Pos.Sensor,Bank1 Open Circ./Short to B+
17747	P1339	Crankshaft Pos./Engine Speed Sensor Cross Connected
17748	P1340	Crankshaft-/Camshaft Pos.Sens.Signals Out of Sequence
17749	P1341	Ignition Coil Power Output Stage 1 Short to Ground
17750	P1342	Ignition Coil Power Output Stage 1 Short to B+
17751	P1343	Ignition Coil Power Output Stage 2 Short to Ground
17752	P1344	Ignition Coil Power Output Stage 2 Short to B+
17753	P1345	Ignition Coil Power Output Stage 3 Short to Ground
17754	P1346	Ignition Coil Power Output Stage 3 Short to B+
17755	P1347	Bank2,Crankshaft-/Camshaft os.Sens.Sign. Out of Sequence
17756	P1348	Ignition Coil Power Output Stage 1 Open Circuit
17757	P1349	Ignition Coil Power Output Stage 2 Open Circuit
17758	P1350	Ignition Coil Power Output Stage 3 Open Circuit
17762	P1354	Modulation Piston Displ.Sensor Circ. Malfunction
17763	P1355	Cyl. 1, ignition circuit Open Circuit
17764	P1356	Cyl. 1, ignition circuit Short to B+
17765	P1357	Cyl. 1, ignition circuit Short to ground
17766	P1358	Cyl. 2, ignition circuit Open Circuit
17767	P1359	Cyl. 2, ignition circuit Short Circuit to B+
17768	P1360	Cyl. 2, ignition circuit Short Circuit to Ground
17769	P1361	Cyl. 3, ignition circuit Open Circuit
17770	P1362	Cyl. 3, ignition circuit Short Circuit to B+
17771	P1363	Cyl. 3, ignition circuit Short Circuit to ground
17772	P1364	Cyl. 4 ignition circuit Open Circuit
17773	P1365	Cyl. 4 ignition circuit Short circuit to B+
17774	P1366	Cyl. 4 ignition circuit Short circuit to ground

Diagnostic Trouble Codes (Powertrain)

V.A.G. code	SAE P-code	Description
17775	P1367	Cyl. 5, ignition circuit Open Circuit
17776	P1368	Cyl. 5, ignition circuit Short Circuit to B+
17777	P1369	Cyl. 5, ignition circuit short to ground
17778	P1370	Cyl. 6, ignition circuit Open Circuit
17779	P1371	Cyl. 6, ignition circuit Short Circuit to B+
17780	P1372	Cyl. 6, ignition circuit short to ground
17793	P1385	Internal Control Module Memory Check Sum Error
17794	P1386	Internal Control Module Knock Control Circ.Error
17795	P1387	Internal Contr. Module altitude sensor error
17796	P1388	Internal Contr. Module drive by wire error
17799	P1391	Camshaft Pos.Sensor,Bank2 Short to Ground
17800	P1392	Camshaft Pos.Sensor,Bank2 Open Circ./Short to B+
17801	P1393	Ignition Coil Power Output Stage 1 Electrical Malfunction
17802	P1394	Ignition Coil Power Output Stage 2 Electrical Malfunction
17803	P1395	Ignition Coil Power Output Stage 3 Electrical Malfunction
17804	P1396	Engine Speed Sensor Missing Tooth
17805	P1397	Engine speed wheel Adaptation limit reached
17806	P1398	Engine RPM signal, TD Short to ground
17807	P1399	Engine RPM signal, TD Short Circuit to B+
17808	P1400	EGR Valve Circ Electrical Malfunction
17809	P1401	EGR Valve Circ Short to Ground
17810	P1402	EGR Valve Circ Short to B+
17811	P1403	EGR Flow Deviation
17812	P1404	EGR Flow Basic Setting not carried out
17814	P1406	EGR Temp.Sensor Range/Performance
17815	P1407	EGR Temp.Sensor Signal too Low
17816	P1408	EGR Temp.Sensor Signal too High
17817	P1409	Tank Ventilation Valve Circ. Electrical Malfunction
17818	P1410	Tank Ventilation Valve Circ. Short to B+
17819	P1411	Sec.Air Inj.Sys.,Bank2 Flow too Flow
17820	P1412	EGR Different.Pressure Sensor Signal too Low
17821	P1413	EGR Different.Pressure Sensor Signal too High
17822	P1414	Sec.Air Inj.Sys.,Bank2 Leak Detected
17825	P1417	Fuel Level Sensor Circ Signal too Low
17826	P1418	Fuel Level Sensor Circ Signal too High
17828	P1420	Sec.Air Inj.Valve Circ Electrical Malfunction
17829	P1421	Sec.Air Inj.Valve Circ Short to Ground
17830	P1422	Sec.Air Inj.Sys.Contr.Valve Circ Short to B+
17831	P1423	Sec.Air Inj.Sys.,Bank1 Flow too Low
17832	P1424	Sec.Air Inj.Sys.,Bank1 Leak Detected
17833	P1425	Evaporative Emission (EVAP) Canister Purge Regulator Valve 1 Short To Ground
17834	P1426	Evaporative Emission (EVAP) Canister Purge Regulator Valve 1 Open Circuit
17840	P1432	Sec.Air Inj.Valve Open
17841	P1433	Sec.Air Inj.Sys.Pump Relay Circ. open

Diagnostic Trouble Codes (Powertrain)

V.A.G. code	SAE P-code	Description
17842	P1434	Sec.Air Inj.Sys.Pump Relay Circ. Short to B+
17843	P1435	Sec.Air Inj.Sys.Pump Relay Circ. Short to ground
17844	P1436	Sec.Air Inj.Sys.Pump Relay Circ. Electrical Malfunction
17847	P1439	EGR Potentiometer Error in Basic Setting
17848	P1440	EGR Valve Power Stage Open
17849	P1441	EGR Valve Circ Open/Short to Ground
17850	P1442	EGR Valve Position Sensor Signal too high
17851	P1443	EGR Valve Position Sensor Signal too low
17852	P1444	EGR Valve Position Sensor range/performance
17853	P1445	Catalyst Temp.Sensor 2 Circ. Range/Performance
17854	P1446	Catalyst Temp.Circ Short to Ground
17855	P1447	Catalyst Temp.Circ Open/Short to B+
17856	P1448	Catalyst Temp.Sensor 2 Circ. Short to Ground
17857	P1449	Catalyst Temp.Sensor 2 Circ. Open/Short to B+
17858	P1450	Sec.Air Inj.Sys.Circ Short to B+
17859	P1451	Sec.Air Inj.Sys.Circ Short to Ground
17860	P1452	Sec.Air Inj.Sys. Open Circ.
17861	P1453	Exhaust gas temperature sensor 1 open/short to B+
17862	P1454	Exhaust gas temperature sensor short 1 to ground
17863	P1455	Exhaust gas temperature sensor 1 range/performance
17864	P1456	Exhaust gas temperature control bank 1 limit attained
17865	P1457	Exhaust gas temperature sensor 2 open/short to B+
17866	P1458	Exhaust gas temperature sensor 2 short to ground
17867	P1459	Exhaust gas temperature sensor 2 range/performance
17868	P1460	Exhaust gas temperature control bank 2 limit attained
17869	P1461	Exhaust gas temperature control bank 1 Range/Performance
17870	P1462	Exhaust gas temperature control bank 2 Range/Performance
17873	P1465	Additive Pump Short Circuit to B+
17874	P1466	Additive Pump Open/Short to Ground
17875	P1467	EVAP Canister Purge Solenoid Valve Short Circuit to B+
17876	P1468	EVAP Canister Purge Solenoid Valve Short Circuit to Ground
17877	P1469	EVAP Canister Purge Solenoid Valve Open Circuit
17878	P1470	EVAP Emission Contr.LDP Circ Electrical Malfunction
17879	P1471	EVAP Emission Contr.LDP Circ Short to B+
17880	P1472	EVAP Emission Contr.LDP Circ Short to Ground
17881	P1473	EVAP Emission Contr.LDP Circ Open Circ.
17882	P1474	EVAP Canister Purge Solenoid Valve electrical malfunction
17883	P1475	EVAP Emission Contr.LDP Circ Malfunction/Signal Circ.Open
17884	P1476	EVAP Emission Contr.LDP Circ Malfunction/Insufficient Vacuum
17885	P1477	EVAP Emission Contr.LDP Circ Malfunction
17886	P1478	EVAP Emission Contr.LDP Circ Clamped Tube Detected
17898	P1490	Evaporative Emission (EVAP) Canister Purge Regulator Valve 2 Short To Ground
17899	P1491	Evaporative Emission (EVAP) Canister Purge Regulator Valve 2 Open Circuit
17903	P1495	Radiator Exhaust Gas Recirculation Switch-over Valve (N345), Open circuit/short circuit to ground

Diagnostic Trouble Codes (Powertrain)

V.A.G. code	SAE P-code	Description
17904	P1496	Radiator Exhaust Gas Recirculation Switch-over Valve (N345), Short to B+
17908	P1500	Fuel Pump Relay Circ. Electrical Malfunction
17909	P1501	Fuel Pump Relay Circ. Short to Ground
17910	P1502	Fuel Pump Relay Circ. Short to B+
17911	P1503	Load signal from Alternator Term. DF Range/performance/Incorrect Signal
17912	P1504	Intake Air Sys.Bypass Leak Detected
17913	P1505	Closed Throttle Pos. Does Not Close/Open Circ
17914	P1506	Closed Throttle Pos.Switch Does Not Open/Short to Ground
17915	P1507	Idle Sys.Learned Value Lower Limit Attained
17916	P1508	Idle Sys.Learned Value Upper Limit Attained
17917	P1509	Idle Air Control Circ. Electrical Malfunction
17918	P1510	Idle Air Control Circ. Short to B+
17919	P1511	Intake Manifold Changeover Valve circuit electrical malfunction
17920	P1512	Intake Manifold Changeover Valve circuit Short to B+
17921	P1513	Intake Manifold Changeover Valve2 circuit Short to B+
17922	P1514	Intake Manifold Changeover Valve2 circuit Short to ground
17923	P1515	Intake Manifold Changeover Valve circuit Short to Ground
17924	P1516	Intake Manifold Changeover Valve circuit Open
17925	P1517	Main Relay Circ. Electrical Malfunction
17926	P1518	Main Relay Circ. Short to B+
17927	P1519	Intake Camshaft Contr.,Bank1 Malfunction
17928	P1520	Intake Manifold Changeover Valve2 circuit Open
17929	P1521	Intake Manifold Changeover Valve2 circuit electrical malfunction
17930	P1522	Intake Camshaft Contr.,Bank2 Malfunction
17931	P1523	Crash Signal from Airbag Control Unit range/performance
17933	P1525	Intake Camshaft Contr.Circ.,Bank1 Electrical Malfunction
17934	P1526	Intake Camshaft Contr.Circ.,Bank1 Short to B+
17935	P1527	Intake Camshaft Contr.Circ.,Bank1 Short to Ground
17936	P1528	Intake Camshaft Contr.Circ.,Bank1 Open
17937	P1529	Camshaft Control Circuit Short to B+
17938	P1530	Camshaft Control Circuit Short to ground
17939	P1531	Camshaft Control Circuit open
17941	P1533	Intake Camshaft Contr.Circ.,Bank2 Electrical Malfunction
17942	P1534	Intake Camshaft Contr.Circ.,Bank2 Short to B+
17943	P1535	Intake Camshaft Contr.Circ.,Bank2 Short to Ground
17944	P1536	Intake Camshaft Contr.Circ.,Bank2 Open
17945	P1537	Engine Shutoff Solenoid Malfunction
17946	P1538	Engine Shutoff Solenoid Open/Short to Ground
17947	P1539	Clutch Vacuum Vent Valve Switch Incorrect signal
17948	P1540	Vehicle Speed Sensor High Input
17949	P1541	Fuel Pump Relay Circ Open
17950	P1542	Throttle Actuation Potentiometer Range/Performance
17951	P1543	Throttle Actuation Potentiometer Signal too Low
17952	P1544	Throttle Actuation Potentiometer Signal too High

Diagnostic Trouble Codes (Powertrain)

V.A.G. code	SAE P-code	Description
17953	P1545	Throttle Pos.Contr Malfunction
17954	P1546	Boost Pressure Contr.Valve Short to B+
17955	P1547	Boost Pressure Contr.Valve Short to Ground
17956	P1548	Boost Pressure Contr.Valve Open
17957	P1549	Boost Pressure Contr.Valve Short to Ground
17958	P1550	Charge Pressure Deviation
17959	P1551	Barometric Pressure Sensor Circ. Short to B+
17960	P1552	Barometric Pressure Sensor Circ. Open/Short to Ground
17961	P1553	Barometric/manifold pressure signal ratio out of range
17962	P1554	Idle Speed Contr.Throttle Pos. Basic Setting Conditions not met
17963	P1555	Charge Pressure Upper Limit exceeded
17964	P1556	Charge Pressure Contr. Negative Deviation
17965	P1557	Charge Pressure Contr. Positive Deviation
17966	P1558	Throttle Actuator Electrical Malfunction
17967	P1559	Idle Speed Contr.Throttle Pos. Adaptation Malfunction
17968	P1560	Maximum Engine Speed Exceeded
17969	P1561	Quantity Adjuster Deviation
17970	P1562	Quantity Adjuster Upper Limit Attained
17971	P1563	Quantity Adjuster Lower Limit Attained
17972	P1564	Idle Speed Contr.Throttle Pos. Low Voltage During Adaptation
17973	P1565	Idle Speed Control Throttle Position lower limit not attained
17974	P1566	Load signal from A/C compressor range/performance
17975	P1567	Load signal from A/C compressor no signal
17976	P1568	Idle Speed Contr.Throttle Pos. mechanical Malfunction
17977	P1569	Cruise control switch Incorrect signal
17978	P1570	Contr.Module Locked
17979	P1571	Left Eng. Mount Solenoid Valve Short to B+
17980	P1572	Left Eng. Mount Solenoid Valve Short to ground
17981	P1573	Left Eng. Mount Solenoid Valve Open circuit
17982	P1574	Left Eng. Mount Solenoid Valve Electrical fault in circuit
17983	P1575	Right Eng. Mount Solenoid Valve Short to B+
17984	P1576	Right Eng. Mount Solenoid Valve Short to ground
17985	P1577	Right Eng. Mount Solenoid Valve Open circuit
17986	P1578	Right Eng. Mount Solenoid Valve Electrical fault in circuit
17987	P1579	Idle Speed Contr.Throttle Pos. Adaptation not started
17988	P1580	Throttle Actuator B1 Malfunction
17989	P1581	Idle Speed Contr.Throttle Pos. Basic Setting Not Carried Out
17990	P1582	Idle Adaptation at Limit
17991	P1583	Transmission mount valves Short to B+
17992	P1584	Transmission mount valves Short to ground
17993	P1585	Transmission mount valves Open circuit
17994	P1586	Engine mount solenoid valves Short to B+
17995	P1587	Engine mount solenoid valves Short to ground
17996	P1588	Engine mount solenoid valves Open circuit

Diagnostic Trouble Codes (Powertrain)

V.A.G. code	SAE P-code	Description
18000	P1592	Height sensor/charge pressure sensor, Unplausible ratio
18008	P1600	Power Supply (B+) Terminal 15 Low Voltage
18009	P1601	Power Supply Relay (J317), Unplausible signal
18010	P1602	Power Supply (B+) Terminal 30 Low Voltage
18011	P1603	Internal Control Module Malfunction
18012	P1604	Internal Control Module Driver Error
18013	P1605	Rough Road/Acceleration Sensor Electrical Malfunction
18014	P1606	Rough Road Spec Engine Torque ABS-ECU Electrical Malfunction
18015	P1607	Vehicle speed signal Error message from instrument cluster
18016	P1608	Steering angle signal Error message from steering angle sensor
18017	P1609	Crash shut-down activated
18018	P1610	Internal Control Module Memory Check Sum Error
18019	P1611	MIL Call-up Circ./Transm.Contr.Module Short to Ground
18020	P1612	Electronic Control Module Incorrect Coding
18021	P1613	MIL Call-up Circ Open/Short to B+
18022	P1614	MIL Call-up Circ./Transm.Contr.Module Range/Performance
18023	P1615	Engine Oil Temperature Sensor Circuit range/performance
18028	P1620	Engine coolant temperature signal open/short to B+
18029	P1621	Engine coolant temperature signal short to ground
18030	P1622	Engine coolant temperature signal range/performance
18031	P1623	Data Bus Powertrain No Communication
18032	P1624	MIL Request Sign.active
18033	P1625	Data-Bus Powertrain Implausible Message from Transm.Contr.
18034	P1626	Data-Bus Powertrain Missing Message from Transm.Contr.
18035	P1627	Data-Bus Powertrain missing message from fuel injection pump
18036	P1628	Data-Bus Powertrain missing message from steering sensor
18037	P1629	Data-Bus Powertrain missing message from distance control
18038	P1630	Accelera.Pedal Pos.Sensor 1 Signal too Low
18039	P1631	Accelera.Pedal Pos.Sensor 1 Signal too High
18040	P1632	Accelera.Pedal Pos.Sensor 1 Power Supply Malfunction
18041	P1633	Accelera.Pedal Pos.Sensor 2 Signal too Low
18042	P1634	Accelera.Pedal Pos.Sensor 2 Signal too High
18043	P1635	Data Bus Powertrain missing message f.air condition control
18044	P1636	Data Bus Powertrain missing message from Airbag control
18045	P1637	Data Bus Powertrain missing message f.central electr.control
18046	P1638	Data Bus Powertrain missing message from clutch control
18047	P1639	Accelera.Pedal Pos.Sensor 1+2 Range/Performance
18048	P1640	Internal Contr.Module (EEPROM) Error
18049	P1641	Please check DTC Memory of Air Condition ECU
18050	P1642	Please check DTC Memory of Airbag ECU
18051	P1643	Please check DTC Memory of central electric ECU
18052	P1644	Please check DTC Memory of clutch ECU
18053	P1645	Data Bus Powertrain missing message f.all wheel drive contr.
18054	P1646	Check DTC Memory of all wheel drive ECU

Diagnostic Trouble Codes (Powertrain)

V.A.G. code	SAE P-code	Description
18055	P1647	Check coding of ECUs in Data Bus Powertrain
18056	P1648	Data Bus Powertrain Malfunction
18057	P1649	Data Bus Powertrain Missing message from ABS Control Module
18058	P1650	Data Bus Powertrain Missing message fr.instrument panel ECU
18059	P1651	Data Bus Powertrain missing messages
18060	P1652	Please check DTC Memory of transmission ECU
18061	P1653	Please check DTC Memory of ABS Control Module
18062	P1654	Please check DTC Memory of control panel ECU
18063	P1655	Please check DTC Memory of ADR Control Module
18064	P1656	A/C clutch relay circuit short to ground
18065	P1657	A/C clutch relay circuit short to B+
18066	P1658	Data Bus Powertrain Incorrect signal from ADR Control Module
18082	P1674	Data-bus-drivetrain implausible message from instrument panel
18084	P1676	Drive by Wire-MIL Circ. Electrical Malfunction
18085	P1677	Drive by Wire-MIL Circ. Short to B+
18086	P1678	Drive by Wire-MIL Circ. Short to Ground
18087	P1679	Drive by Wire-MIL Circ. Open
18089	P1681	Contr.Unit Programming, Programming not Finished
18090	P1682	Powertrain databus Unplausible message from ABS control module
18091	P1683	Powertrain databus Unplausible message from Airbag Control Modules
18092	P1684	Contr.Unit Programming Communication Error
180093	P1685	Data-bus-drivetrain, implausible message from steering angle sensor
18094	P1686	Contr.Unit Error Programming Error
18095	P1687	Data-bus-driving gear, Implausible message from climate control module
18097	P1689	Data-bus-drivetrain, implausible message from electronic central electric
18098	P1690	Malfunction Indication Light Malfunction
18099	P1691	Malfunction Indication Light Open
18100	P1692	Malfunction Indication Light Short to Ground
18101	P1693	Malfunction Indication Light Short to B+
18102	P1694	Malfunction Indication Light Open/Short to Ground
18104	P1696	Powertrain databus, implausible signal from Steering Column Electronic
18106	P1698	Read fault memory of Steering Column Electronic
18110	P1702	Malfunction cannot activate control module replacement function, since another malfunction with equal priority is present
18112	P1704	Kick Down Switch (F8) Malfunction
18113	P1705	Gear/Ratio Monitoring Adaptation limit reached
18114	P1706	Transmission faulty
18115	P1707	Interference in Mechatronic unit
18119	P1711	Wheel Speed Signal 1 Range/Performance
18124	P1716	Wheel Speed Signal 2 Range/Performance
18129	P1721	Wheel Speed Signal 3 Range/Performance
18131	P1723	Starter Interlock Circ. Open
18132	P1724	Starter Interlock Circ. Short to Ground
18134	P1726	Wheel Speed Signal 4 Range/Performance

Diagnostic Trouble Codes (Powertrain)

V.A.G. code	SAE P-code	Description
18136	P1728	Different Wheel Speed Signals Range/Performance
18137	P1729	Starter Interlock Circ. Short to B+
18141	P1733	Tiptronic Switch Down Circ. Short to Ground
18147	P1739	Tiptronic Switch up Circ. Short to Ground
18148	P1740	Clutch temperature control
18149	P1741	Clutch pressure adaptation at limit
18150	P1742	Clutch torque adaptation at limit
18151	P1743	Clutch slip control signal too high
18152	P1744	Tiptronic Switch Recognition Circ. Short to Ground
18153	P1745	Transm.Contr.Unit Relay Short to B+
18154	P1746	Transm.Contr.Unit Relay Malfunction
18155	P1747	Transm.Contr.Unit Relay Open/Short to Ground
18156	P1748	Transm.Contr.Unit Self-Check
18157	P1749	Transm.Contr.Unit Incorrect Coded
18158	P1750	Power Supply Voltage Low Voltage
18159	P1751	Power Supply Voltage High Voltage
18160	P1752	Power Supply Malfunction
18161	P1753	Tiptronic Switch (F189), Unplausible signal
18162	P1754	Tiptronic Switch (F189), up function, Open circuit/short circuit to B+
18163	P1755	Tiptronic Switch (F189), down function, Open circuit/short circuit to B+
18164	P1756	Tiptronic Switch (F189), recognition function, Open circuit/short circuit to B+
18167	P1759	Transmission fluid temperature, Exceeded multiple times!
18168	P1760	Shift Lock Solenoid (N110) Malfunction
18169	P1761	Shift Lock Solenoid (N110) Short to Ground
18170	P1762	Shift Lock Solenoid (N110) Short to B+
18171	P1763	Shift Lock Solenoid (N110) Open
18172	P1764	Transmission temperature control
18173	P1765	Hydraulic Pressure Sensor 2 adaptation at limit
18174	P1766	Throttle Angle Signal Stuck Off
18175	P1767	Throttle Angle Signal Stuck On
18176	P1768	Hydraulic Pressure Sensor 2 Too High
18177	P1769	Hydraulic Pressure Sensor 2 Too Low
18178	P1770	Load Signal Range/Performance
18179	P1771	Load Signal Stuck Off
18180	P1772	Load Signal Stuck On
18181	P1773	Hydraulic Pressure Sensor 1 Too High
18182	P1774	Hydraulic Pressure Sensor 1 Too Low
18183	P1775	Hydraulic Pressure Sensor 1 adaptation at limit
18184	P1776	Hydraulic Pressure Sensor 1 range/performance
18185	P1777	Hydraulic Pressure Sensor 2 range/performance
18186	P1778	Solenoid EV7 Electrical Malfunction
18188	P1780	Torque retraction malfunctioning
18189	P1781	Engine Torque Reduction Open/Short to Ground
18190	P1782	Engine Torque Reduction Short to B+

Diagnostic Trouble Codes (Powertrain)

V.A.G. code	SAE P-code	Description
18192	P1784	Shift up/down Wire Open/Short to Ground
18193	P1785	Shift up/down Wire Short to B+
18194	P1786	Reversing Light Circ. Open
18195	P1787	Reversing Light Circ. Short to Ground
18196	P1788	Reversing Light Circ. Short to B+
18197	P1789	Idle Speed Intervention Circ. Error Message from Engine Contr.
18198	P1790	Transmission Range Display Circ. Open
18199	P1791	Transmission Range Display Circ. Short to Ground
18200	P1792	Transmission Range Display Circ. Short to B+
18201	P1793	Output Speed Sensor 2 Circ. No Signal
18203	P1795	Vehicle Speed Signal Circ. Open
18204	P1796	Vehicle Speed Signal Circ. Short to Ground
18205	P1797	Vehicle Speed Signal Circ. Short to B+
18206	P1798	Output Speed Sensor 2 Circ. Range/Performance
18207	P1799	Output Speed Sensor 2 Circ. Rpm too High
18221	P1813	Pressure Contr.Solenoid 1 Electrical
18222	P1814	Pressure Contr.Solenoid 1 Open/Short to Ground
18223	P1815	Pressure Contr.Solenoid 1 Short to B+
18226	P1818	Pressure Contr.Solenoid 2 Electrical
18227	P1819	Pressure Contr.Solenoid 2 Open/Short to Ground
18228	P1820	Pressure Contr.Solenoid 2 Short to B+
18231	P1823	Pressure Contr.Solenoid 3 Electrical
18232	P1824	Pressure Contr.Solenoid 3 Open/Short to Ground
18233	P1825	Pressure Contr.Solenoid 3 Short to B+
18236	P1828	Pressure Contr.Solenoid 4 Electrical
18237	P1829	Pressure Contr.Solenoid 4 Open/Short to Ground
18238	P1830	Pressure Contr.Solenoid 4 Short to B+
18241	P1833	Pressure regulating valve 5 for automatic transmission, electrical malfunction in circuit
18242	P1834	Pressure Contr.Solenoid 5 Open/Short to Ground
18243	P1835	Pressure Contr.Solenoid 5 Short to B+
18246	P1838	Pressure regulating valve 6 (N371), electrical malfunction in circuit
18247	P1839	Pressure regulating valve 6 (N371), Open circuit/short circuit to Ground (GND)
18248	P1840	Pressure Regulating Valve 6 (N371), Short circuit to B+
18249	P1841	Engine/Transmission Control Modules Versions do not match
18250	P1842	Please check DTC Memory of instrument panel ECU
18251	P1843	Please check DTC Memory of ADR Control Module
18252	P1844	Please check DTC Memory of central electric control ECU
18253	P1845	Read DTC memory of Steering Column Electronic Systems Control Module (J527)
18255	P1847	Please check DTC Memory of brake system ECU
18256	P1848	Please check DTC Memory of engine ECU
18257	P1849	Please check DTC Memory of transmission ECU
18258	P1850	Data-Bus Powertrain Missing Message from Engine Contr.
18259	P1851	Data-Bus Powertrain Missing Message from Brake Contr.
18260	P1852	Data-Bus Powertrain Implausible Message from Engine Contr.

Diagnostic Trouble Codes (Powertrain)

V.A.G. code	SAE P-code	Description
18261	P1853	Data-Bus Powertrain Implausible Message from Brake Contr.
18262	P1854	Data-Bus Powertrain Hardware Defective
18263	P1855	Data-Bus Powertrain Software version Contr.
18264	P1856	Throttle/Pedal Pos.Sensor A Circ. Error Message from Engine Contr.
18265	P1857	Load Signal Error Message from Engine Contr.
18266	P1858	Engine Speed Input Circ. Error Message from Engine Contr.
18267	P1859	Brake Switch Circ. Error Message from Engine Contr.
18268	P1860	Kick Down Switch Error Message from Engine Contr.
18269	P1861	Throttle Position (TP) sensor Error Message from ECM
18270	P1862	Data Bus Powertrain Missing message from instr. panel ECU
18271	P1863	Data Bus Powertrain Missing Message from St. Angle Sensor
18272	P1864	Data Bus Powertrain Missing message from ADR control module
18273	P1865	Data Bus Powertrain Missing message from central electronics
18274	P1866	Data Bus Powertrain Missing messages
18320	P1912	Brake Booster pressure Sensor Short To Positive
18321	P1913	Brake Booster pressure Sensor Short To Ground
18322	P1914	Brake Booster pressure Sensor Implausible Signal
18332	P1924	Control Module pin coding, Unplausible signal
18440	P2008	Intake Manifold Runner Control Circuit Open Circuit Bank 1
18520	P2088	Intake "A" Camshaft Position Actuator Control Circuit Low Bank 1
18521	P2089	Intake "A" Camshaft Position Actuator Control Circuit High Bank 1
18522	P2090	"B" Camshaft Position Actuator Control Circuit Low Bank 1
18523	P2091	"B" Camshaft Position Actuator Control Circuit High Bank 1
18528	P2096	Post Catalyst Fuel Trim System Too Lean Bank 1
18529	P2097	Post Catalyst Fuel Trim System Too Rich Bank 1
18532	P2100	Throttle Actuator Control Motor Circuit Open (Motor for Intake Flap V157 implausible signal)
18533	P2101	Throttle Actuator Control Motor Circuit Range/Performance
18534	P2102	Throttle Actuator Control Motor Circuit Low (Motor for Intake Flap V157 signal Too Low)
18535	P2103	Throttle Actuator Control Motor Circuit High (Motor for Intake Flap V157 signal Too High)
18538	P2106	Throttle Actuator Control Motor Circuit Forced Limited Power
18540	P2108	Throttle Actuator Control Module Performance (Motor for Intake Flap V157 Malfunction)
18544	P2112	Throttle Actuator Control System, Stuck Closed
18554	P2122	Throttle/Pedal position Sensor/Switch "D" Circuit Low Input
18555	P2123	Throttle/Pedal position Sensor/Switch "D" Circuit High Input
18559	P2127	Throttle/Pedal position Sensor/Switch "E" Circuit Low Input
18560	P2128	Throttle/Pedal position Sensor/Switch "E" Circuit High Input
18565	P2133	Throttle/Pedal position Sensor/Switch "F" Circuit Low Input
18570	P2138	Throttle/Pedal position Sensor/Switch "D"/"F" Circuit Voltage Correlation
18578	P2146	Fuel Injector Group A Supply Voltage Circuit/Open
18581	P2149	Fuel Injector Group B Supply Voltage Circuit/Open
18609	P2177	System Too Lean Off Idle, Bank 1
18610	P2178	System Too Rich Off Idle, Bank 1
18611	P2179	System Too Lean Off Idle, Bank 2
18612	P2180	System Too Rich Off Idle, Bank 3

Diagnostic Trouble Codes (Powertrain)

V.A.G. code	SAE P-code	Description
18613	P2181	Cooling System Performance
18616	P2184	Engine Coolant Temperature (ECT) Sensor (on radiator) Circuit 2, Signal too low
18617	P2185	Engine Coolant Temperature (ECT) Sensor (on radiator) Circuit 2, Signal too high
18619	P2187	System Too Lean At Idle, Bank 1
18620	P2188	System Too Rich At Idle, Bank 1
18621	P2189	System Too Lean At Idle, Bank 2
18622	P2190	System Too Rich At Idle, Bank 2
18623	P2191	System Too Lean At Higher Load, Bank 1
18624	P2192	System Too Rich At Higher Load, Bank 1
18625	P2193	System Too Lean At Higher Load, Bank 2
18626	P2194	System Too Rich At Higher Load, Bank 2
18627	P2195	O2 Sensor Signal Stuck Lean B1S1
18628	P2196	O2 Sensor Signal Stuck Rich B1S1
18663	P2231	O2 Sensor Signal Circuit Shorted to Heater Circuit B1S1
18666	P2234	O2 Sensor Signal Circuit Shorted to Heater Circuit B2S1
18669	P2237	O2 Sensor Positive Current Control Circuit Open B1S1
18670	P2238	O2 Sensor Positive Current Control Circuit Low B1S1
18671	P2239	O2 Sensor Positive Current Control Circuit High B1S1
18672	P2240	O2 Sensor Positive Current Control Circuit Open B2S1
18675	P2243	O2 Sensor Reference Voltage Circuit Open B1S1
18677	P2245	O2 Sensor Reference Voltage Circuit Low B1S1
18678	P2246	O2 Sensor Reference Voltage Circuit High B1S1
18679	P2247	O2 Sensor Reference Voltage Circuit Open B2S1
18683	P2251	O2 Sensor Negative Current Control Circuit Open B1S1
18684	P2252	O2 Sensor Negative Current Control Circuit Low B1S1
18685	P2253	O2 Sensor Negative Current Control Circuit High B1S1
18686	P2254	O2 Sensor Negative Current Control Circuit Open B2S1
18689	P2257	Secondary Air Injection System Control “A” Circuit Low
18690	P2258	Secondary Air Injection System Control “A” Circuit High
18693	P2261	Turbo Boost Pressure Not Detected, Mechanical
18695	P2263	Turbo/Super Charger Boost System Performance
18702	P2270	O2 Sensor Signal Stuck Lean B1S2
18703	P2271	O2 Sensor Signal Stuck Rich B1S2
18704	P2272	O2 Sensor Signal Stuck Lean B2S2
18705	P2273	O2 Sensor Signal Stuck Rich B2S2
18706	P2274	O2 Sensor Signal Stuck Lean B1S3
18707	P2275	O2 Sensor Signal Stuck Rich B1S3
18711	P2279	Intake Air System Leak
18725	P2293	Fuel Pressure Regulator (N276) Mechanical Malfunction
18726	P2294	Fuel Pressure Regulator (N276) Open circuit
18727	P2295	Fuel Pressure Regulator (N276) Short Circuit to Ground
18728	P2296	Fuel Pressure Regulator (N276) Short Circuit to B+
18768	P2336	Cylinder #1 Above Knock Threshold
18769	P2337	Cylinder #2 Above Knock Threshold

Diagnostic Trouble Codes (Powertrain)

V.A.G. code	SAE P-code	Description
18770	P2338	Cylinder #3 Above Knock Threshold
18771	P2339	Cylinder #4 Above Knock Threshold
18772	P2340	Cylinder #5 Above Knock Threshold
18773	P2341	Cylinder #6 Above Knock Threshold
18832	P2400	Evaporative Emission System Leak Detection Pump Control Circuit Open
18833	P2401	Evaporative Emission System Leak Detection Pump Control Circuit Low
18834	P2402	Evaporative Emission System Leak Detection Pump Control Circuit High
18835	P2403	Evaporative Emission System Leak Detection Pump Sense Circuit Open
18836	P2404	Evaporative Emission System Leak Detection Pump Sense Circuit Range/Performance
18845	P2413	Exhaust Gas Recirculation System Performance
18846	P2414	O2 Sensor Exhaust Sample Error B1S1
18854	P2422	Evaporative Emission System Vent Valve Stuck Closed
18857	P2425	Exhaust Gas Recirculation Cooling Valve Control Circuit Open
18858	P2426	Exhaust Gas Recirculation Cooling Valve Control Circuit Low
18859	P2427	Exhaust Gas Recirculation Cooling Valve Control Circuit High
18973	P2541	Low Fuel Pressure Sensor (G410) Open Circuit
18996	P2564	Turbocharger Boost Control Position Sensor Circuit Low
18997	P2565	Turbocharger Boost Control Position Sensor Circuit High
19058	P2626	O2 Sensor Pumping Current Trim Circuit Open B1S1
19061	P2629	O2 Sensor Pumping Current Trim Circuit Open B2S1
19069	P2637	Torque Management Feedback Signal “A”
19140	P2708	Shift Solenoid "F" Stuck On
19143	P2711	Unexpected Mechanical Gear Disengagement
19146	P2714	Pressure Control Solenoid "D" Performance or Stuck Off
19147	P2715	Pressure Control Solenoid "D" Stuck On
19141	P2716	Pressure Control Solenoid “D” Electrical
19142	P2717	Pressure Control Solenoid "D" Intermittent
19155	P2723	Pressure Control Solenoid ”E” Performance or Stuck On
19156	P2724	Pressure Control Solenoid ”E” Stuck On
19157	P2725	Pressure Control Solenoid ”E” Electrical
19158	P2726	Pressure Control Solenoid ”E” Intermittent
19164	P2732	Pressure Control Solenoid ”F” Performance or Stuck On
19165	P2733	Pressure Control Solenoid ”F” Stuck On
19166	P2734	Pressure Control Solenoid ”F” Electrical
19167	P2735	Pressure Control Solenoid ”F” Intermittent
19473	P3017	Electronic Differential Lock Intake Valve for Left Front and Right Rear Short to Positive
19474	P3018	Exhaust Flap Valve 1 Short to Ground
19475	P3019	Exhaust Flap Valve 1 Open Circuit
19484	P3028	Angle sensor 2 (on throttle drive 2 power accelerator actuation) (G298), Unplausible Signal
19487	P3031	Throttle Drive 2 (power accelerator actuation) (G296), Electrical malfunction in circuit
19488	P3032	Throttle Valve Control Module 2 (J544), Malfunction in basic setting
19491	P3035	Throttle Valve Control Module 2 (J544), Mechanical malfunction
19501	P3045	Fuel Pump Electronic Faulty
19503	P3047	Activation Starter Relay 2, Short to Positive

Diagnostic Trouble Codes (Powertrain)

V.A.G. code	SAE P-code	Description
19504	P3048	Activation Starter Relay 2, Short to Negative
19505	P3049	Activation Starter Relay 2, Open Circuit
19506	P3050	Starter Relay 2 Electrical Malfunction in Circuit (Stuck/Inoperative Relay)
19534	P3078	Thrfottle Valve Control Module Airflow at Idle Too Low
19537	P3081	Engine Temperature Too Low
19548	P3092	Internal Control Module Memory Checksum Error
19549	P3093	Internal Control Module Memory Checksum Error
19552	P3096	Internal Control Module Memory Checksum Error
19553	P3097	Internal Control Module Memory Checksum Error
19593	P3137	Intake Manifold Runner Control (IMRC) Basic setting not carried out
19594	P3138	Intake Manifold Runner Control (IMRC) Deviation
19647	P3191	Intake Manifold Runner Control (IMRC) Bank 1 Control, Basic setting aborted with mechanical impact open
19648	P3192	Intake Manifold Runner Control (IMRC) Bank 1 Control, Basic setting aborted with mechanical impact closed
19649	P3193	Intake Manifold Runner Control (IMRC) Bank 1 Control Circuit outside valid range
19667	P3211	Exhaust B1S1 Heater return Connection
19668	P3212	Exhaust B2S1 Heater return Connection
19711	P3255	Bank 1, oxygen sensor before catalytic converter, heating circuit Regulation at upper impact
19712	P3256	Bank 1, oxygen sensor before catalytic converter, heating circuit Regulation at lower impact
19722	P3266	Bank 1, sensor 1 Internal resistance too large

Diagnostic Trouble Codes (Vehicle Network Communications)

V.A.G. code	SAE U-code	Description
28673	U0001	High Speed CAN Communication Bus
28772	U0100	Lost Communication With ECM/PCM "A"
28773	U0101	Lost Communication with Transmission Control Module (TCM)
28775	U0103	Lost Communication With Gear Shift Control Module
28793	U0121	Lost Communication with Anti-Lock Brake System (ABS) Control Module
28798	U0126	Lost Communication with Steering Angle Sensor (G85)
28818	U0146	Lost Communication with Diagnostic Interface
28823	U0151	Lost Communication with Airbag Control Module
28827	U0155	Lost Communication with Instrument Panel Cluster (IPC) Control Module
2836	U0164	Lost Communication with A/C Control Module
28974	U0302	Software Incompatibility with Transmission Control Module (TCM)
28987	U0315	Software Incompatibility with Anti-Lock Brake System (ABS) Control Module
29074	U0402	Invalid Data Received From Transmission Control (TCM) Module
29076	U0404	Invalid Data Received From Gear Shift Control Module
29087	U0415	Invalid Data Received From Anti-Lock Brake System (ABS) Control Module
29705	U1009	Gateway, Range/Performance
65535	n/a	Internal control module memory check sum error

Diagnostic Scan Tool and Special Tool Suppliers

The following suppliers offer diagnostic scan tools or computerized scan tool programs. Some of them also offer special tools and equipment for specialized repair work.

Assenmacher Specialty Tools, Inc.
6440 Odell Place
Boulder, CO 80301
(800) 525-2943
www.asttool.com

Baum Tools Unlimited, Inc.
PO Box 5867
Sarasota, FL 34277
(800) 848-6657
www.baumtools.com

Equipment Solutions
P.O. Box 1450
Kenosha, WI 53141-1450
(800) 892-9650

Metalnerd
5113 Crestview Drive, Suite B
Greensburg, PA, 15601 USA
(412) 601-4270
www.metalnerd.com

Ross-Tech
920 South Broad Street
Lansdale, PA 19446
215-361-8942
www.ross-tech.com

Samstag Sales
115 Main St. N., Suite 216
Carthage, TN 37030
(615) 735-3388
www.samstagsales.com

Shade Tree Software
4186 Culebra Ct.
Boulder, CO 80301
(303) 449-1664
(303) 940-2468
www.shadetreesoftware.com

ZDMAK Tools
P.O. Box 5100
Sarasota, FL 34277
(877) 938-6657
www.zdmak.com

Zelenda Automotive, Inc.
66-02 Austin St.
Forest Hills, NY 11374
(888) 892-8348
www.zelenda.com

VAG 1551/1552 OPERATING OVERVIEW FLOW CHART

MODE MENU

1 - Rapid Data =>
2 - Blink Code
3 - Self Test
4 - Dealership Code
5 - LT - Diagnosis

ADDRESS WORDS MENU (V9)

01 - Engine Electronics =>
11 - Engine Electronics II
41 - Diesel Pump Electronics
51 - Electric Drive
61 - Battery Control
71 - Battery Charger
02 - Transmission Electronics
12 - Clutch Electronics
22 - 4 Wheel Drive Electronics
03 - Brake Electronics
13 - Distance Control
33 - OBD II
14 - Suspension Electronics
24 - Anti - Slip Control
34 - Level Control
44 - Steering Assist
54 - Rear Spoiler
15 - Airbag
25 - Anti-Theft Immobilizer Sensor
35 - Central Locking
45 - Interior Monitor
55 - Headlamp Vert. Aim Control
65 - Tire Pressure Monitoring
75 - Emergency Call Module
16 - Steering Wheel Electronics
26 - Automatic Roof Control
36 - Seat Adjustment Driver's Side
46 - Comfort System (AC CM)
56 - Radio
66 - Seat & Mirror Adjustment
76 - Parking Aid
17 - Instrument Cluster
37 - Navigation System
47 - Sound System
57 - Airbag (Cabriolet only)
67 - Language Control
08 - A/C & Heating Electronics
18 - Auxiliary Heater
09 - Electronic Control Module
19 - Diagnostic Interface for Data Bus
29 - Left Light Control
39 - Right Light Control
49 - Auto Light Switch
00 - Automatic Test Sequence

FUNCTIONS MENU

01 - Check Control Module Versions
02 - Check DTC Memory
03 - Output Diagnostic Test Mode
04 - Basic Setting
    Display Group xxx =>
05 - Erase DTC Memory
06 - End Output
07 - Code Control Module
08 - Read Measuring Value Block
    Display Group xxx =>
09 - Read Individual Measuring Values
    Channel xx
10 - Adaptation
    Channel xx
11 - Log-In Procedure
15 - Readiness Code

DISPLAY GROUP FORMAT

Display Group 000 (only)
xx xx xx xx xx xx xx xx xx xx

Display Group 001 (and higher)
xxxxx xxxxx xxxxx xxxxx

WARNING

Your common sense, good judgement and general alertness are crucial to safe and successful service work. Before attempting any work on your Audi, be sure to read **00 Warnings and Cautions** *and the copyright page at the front of the manual. Review these warnings and cautions each time you prepare to work on your car. Please also read any warnings and cautions that accompany the procedures in the manual.*

A

B

C

WARNING

Your common sense, good judgement and general alertness are crucial to safe and successful service work. Before attempting any work on your Audi, be sure to read **00 Warnings and Cautions** *and the copyright page at the front of the manual. Review these warnings and cautions each time you prepare to work on your car. Please also read any warnings and cautions that accompany the procedures in the manual.*

> **_WARNING_**
>
> *Your common sense, good judgement and general alertness are crucial to safe and successful service work. Before attempting any work on your Audi, be sure to read* **00 Warnings and Cautions** *and the copyright page at the front of the manual. Review these warnings and cautions each time you prepare to work on your car. Please also read any warnings and cautions that accompany the procedures in the manual.*

NOTES

Selected Books and Repair Information From Bentley Publishers

Motorsports

Alex Zanardi: My Sweetest Victory
Alex Zanardi and Gianluca Gasparini
ISBN 978-0-8376-1249-2

The Unfair Advantage
Mark Donohue and Paul van Valkenburgh
ISBN 978-0-8376-0069-7

Equations of Motion - Adventure, Risk and Innovation *William F. Milliken*
ISBN 978-0-8376-1570-7

Engineering

Hack Mechanic Guide to European Automotive Electrical Systems
Rob Siegel ISBN 978-0-8376-1751-0

Bosch Automotive Handbook *Robert Bosch GmbH* ISBN 978-0-8376-1732-9

Bosch Fuel Injection and Engine Management *Charles O. Probst, SAE*
ISBN 978-0-8376-0300-1

Maximum Boost: Designing, Testing, and Installing Turbocharger Systems
Corky Bell ISBN 978-0-8376-0160-1

Supercharged! Design, Testing and Installation of Supercharger Systems
Corky Bell ISBN 978-0-8376-0168-7

Scientific Design of Exhaust and Intake Systems *Phillip H. Smith & John C. Morrison* ISBN 978-0-8376-0309-4

Physics for Gearheads
Randy Beikmann ISBN 978-0-8376-1615-5

Audi

Audi A4 Service Manual: 2002-2008, 1.8L Turbo, 2.0L Turbo, 3.0L, 3.2L
Bentley Publishers
ISBN 978-0-8376-1574-5

Audi A4 Service Manual: 1996-2001, 1.8L Turbo, 2.8L *Bentley Publishers*
ISBN 978-0-8376-1675-9

Audi TT Service Manual: 2000-2006, 1.8L turbo, 3.2 L *Bentley Publishers*
ISBN 978-0-8376-1625-4

Audi A6 (C5 platform) Service Manual: 1998-2004 *Bentley Publishers*
ISBN 978-0-8376-1670-4

BMW

Memoirs of a Hack Mechanic
Rob Siegel ISBN 978-0-8376-1720-6

BMW 3 Series (F30, F31, F34) Service Manual: 2012-2015 *Bentley Publishers*
ISBN 978-0-8376-1752-7

BMW X3 (E83) Service Manual: 2004-2010 *Bentley Publishers*
ISBN 978-0-8376-1731-2

BMW X5 (53) Service Manual: 2000-2006 *Bentley Publishers*
ISBN 978-0-8376-1643-8

BMW 3 Series (E90, E91, E92, E93) Service Manual: 2006-2011
Bentley Publishers ISBN 978-0-8376-1723-7

BMW 3 Series (E46) Service Manual: 1999-2005 *Bentley Publishers*
ISBN 978-0-8376-1657-5

BMW 3 Series (E36) Service Manual: 1992-1998 *Bentley Publishers*
ISBN 978-0-8376-1709-1

BMW Z3 (E36/7) Service Manual: 1996-2002 *Bentley Publishers*
ISBN 978-0-8376-1617-9

BMW 5 Series (E60, E61) Service Manual: 2004-2010 *Bentley Publishers*
ISBN 978-0-8376-1689-6

BMW 5 Series (E39) Service Manual: 1997-2003 *Bentley Publishers*
ISBN 978-0-8376-1672-8

Corvette

Corvette: America's Star-Spangled Sports Car *Karl Ludvigsen*
ISBN 978-0-8376-1659-9

Corvette by the Numbers: The Essential Corvette Parts Reference 1955-1982 *Alan Covin*
ISBN 978-0-8376-0228-2

Zora Arkus-Duntov: The Legend Behind Corvette *Jerry Burton*
ISBN 978-08376-0858-7

Porsche

Porsche 911 (996) Service Manual: 1999-2005 *Bentley Publishers*
ISBN 978-0-8376-1710-7

Porsche 911 (993) Service Manual: 1995-1998 *Bentley Publishers*
ISBN 978-0-8376-1719-0

Porsche Boxster Service Manual: 1997-2004 *Bentley Publishers*
ISBN 978-0-8376-1645-2

Porsche 911 Carrera Service Manual: 1984-1989 *Bentley Publishers*
ISBN 978-0-8376-1696-4

Porsche 911 SC Service Manual: 1978-1983 *Bentley Publishers*
ISBN 978-0-8376-1705-3

Porsche: Excellence Was Expected
Karl Ludvigsen
ISBN 978-0-8376-0235-6

Porsche — Origin of the Species
Karl Ludvigsen
ISBN 978-0-8376-1331-4

Volkswagen

Volkswagen Rabbit, GTI Service Manual: 2006-2009 *Bentley Publishers*
ISBN 978-0-8376-1664-3

Volkswagen Jetta Service Manual: 2005-2010 *Bentley Publishers*
ISBN 978-0-8376-1616-2

Volkswagen Jetta, Golf, GTI Service Manual: 1999-2005 *Bentley Publishers*
ISBN 978-0-8376-1678-0

Volkswagen Jetta, Golf, GTI: 1993-1999, Cabrio: 1995-2002 Service Manual *Bentley Publishers*
ISBN 978-0-8376-1660-5

Volkswagen GTI, Golf, Jetta Service Manual: 1985-1992 *Bentley Publishers*
ISBN 978-0-8376-1637-7

Volkswagen Corrado Repair Manual: 1990-1994 *Bentley Publishers*
ISBN 978-0-8376-1699-5

Volkswagen Passat, Passat Wagon Service Manual: 1998-2005
Bentley Publishers
ISBN 978-0-8376-1669-8

MINI Repair Manuals

MINI Cooper Service Manual: 2007-2013 *Bentley Publishers*
ISBN 978-0-8376-1730-5

MINI Cooper Service Manual: 2002-2006 *Bentley Publishers*
ISBN 978-0-8376-1639-1

Mercedes-Benz

Mercedes-Benz C-Class (W202) Service Manual 1994-2000
Bentley Publishers
ISBN 978-0-8376-1692-6

Mercedes Benz E-Class (W124) Owner's Bible: 1986-1995
Bentley Publishers
ISBN 978-0-8376-0230-1

Mercedes-Benz Technical Companion
Staff of The Star and members of Mercedes-Benz Club of America
ISBN 978-0-8376-1033-7

BentleyPublishers®
Automotive Reference

Bentley Publishers has published service manuals and automobile books since 1950. For more information, please contact Bentley Publishers at 1734 Massachusetts Avenue, Cambridge, MA 02138 USA, or visit our web site at
BentleyPublishers.com